Biology

Concepts and Applications

SEVENTH EDITION

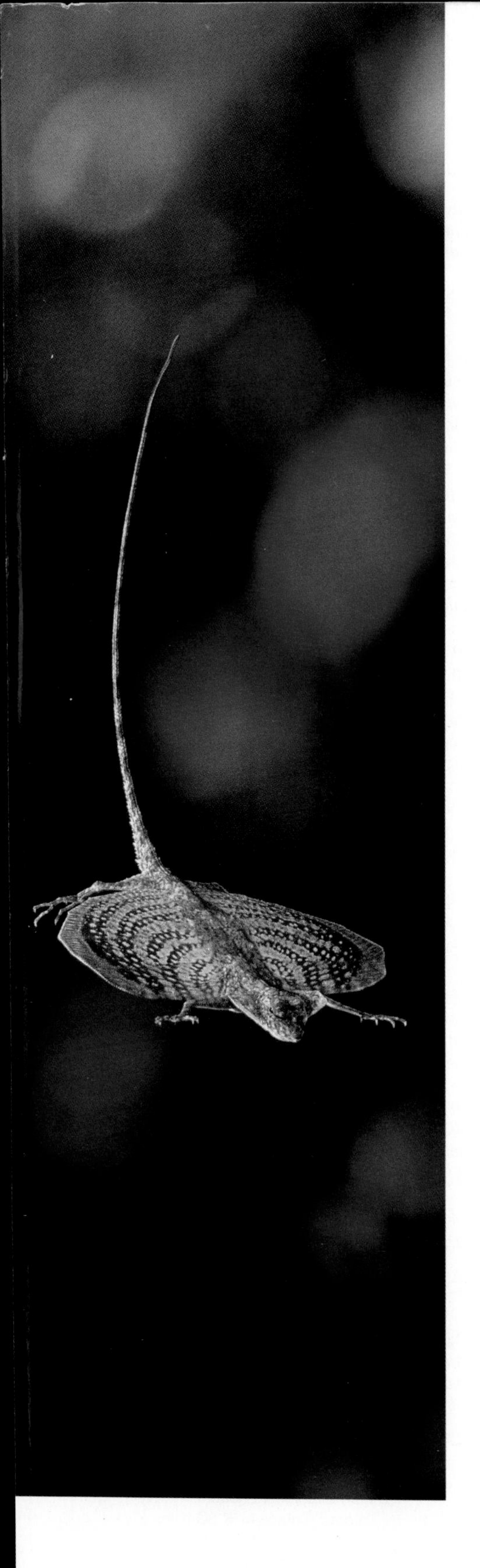

Cecie Starr | Christine A. Evers | Lisa Starr

Biology

Concepts and Applications

SEVENTH EDITION

THOMSON
BROOKS/COLE™

Australia • Brazil • Canada • Mexico • Singapore • Spain • United Kingdom • United States

PUBLISHER Jack Carey and Yolanda Cossio
MANAGING DEVELOPMENT EDITOR Peggy Williams
ASSISTANT EDITOR Jessica Kuhn
EDITORIAL ASSISTANT Rose Barlow
TECHNOLOGY PROJECT MANAGER Kristina Razmara
MARKETING MANAGER Kara Kindstrom
MARKETING COMMUNICATIONS MANAGER Stacy Pratt
PROJECT MANAGER, EDITORIAL PRODUCTION Andy Marinkovich
CREATIVE DIRECTOR Rob Hugel
ART DIRECTOR John Walker
PRINT BUYER Karen Hunt
PERMISSIONS EDITOR Bob Kauser
PRODUCTION SERVICE Grace Davidson & Associates
TEXT DESIGNER Chris Keeney, Yvo Riezebos, John Walker
PHOTO RESEARCHER Myrna Engler Photo Research Inc.
COPY EDITOR Anita Wagner
ILLUSTRATORS Gary Head, ScEYEnce Studios, Lisa Starr
COVER DESIGNER Dare Porter, John Walker
COVER PRINTER Quebecor World/Versailles
COMPOSITOR Lachina Publishing Services
PRINTER Quebecor World/Dubuque

COVER IMAGE *Common flying dragon* (Draco Volans) *gliding by extending ribbed wings; native to Indonesia, India, and Philippine Islands (Stephen Dalton/Minden Pictures).*

Printed in the United States of America
1 2 3 4 5 6 7 11 10 09 08 07

Library of Congress Control Number: 2007932373

Paperback Edition:
ISBN-13: 978-0-495-11981-4
ISBN-10: 0-495-11981-4

Hardcover Edition:
ISBN-13: 978-0-495-11997-5
ISBN-10: 0-495-11997-0

For more information about our products, contact us at:
Thomson Learning Academic Resource Center
1-800-423-0563

For permission to use material from this text or product, submit a request online at http://www.thomsonrights.com.
Any additional questions about permissions can be submitted by e-mail to thomsonrights@thomson.com.

BOOKS IN THE BROOKS/COLE BIOLOGY SERIES

Biology: The Unity and Diversity of Life, Eleventh, Starr/Taggart
Biology: Concepts and Applications, Seventh, Starr/Evers/Starr
Biology: Concepts and Applications Without Physiology, Seventh, Starr/Evers/Starr
Biology Today and Tomorrow, Second, Starr/Evers/Starr
Biology, the Dynamic Science, First, Russell/Wolfe/Hertz/Starr/McMillan
Biology, Eighth, Solomon/Berg/Martin
Human Biology, Seventh, Starr/McMillan
Biology: A Human Emphasis, Seventh, Starr/Evers/Starr
Human Physiology, Fifth, Sherwood
Fundamentals of Physiology, Second, Sherwood
Human Physiology, Fourth, Rhoades/Pflanzer

Laboratory Manual for Biology, Fifth, Perry/Morton/Perry
Laboratory Manual for Human Biology, Morton/Perry/Perry
Photo Atlas for Biology, Perry/Morton
Photo Atlas for Anatomy and Physiology, Morton/Perry
Photo Atlas for Botany, Perry/Morton
Virtual Biology Laboratory, Beneski/Waber
Introduction to Cell and Molecular Biology, Wolfe
Molecular and Cellular Biology, Wolfe
Biotechnology: An Introduction, Second, Barnum

Introduction to Microbiology, Third, Ingraham/Ingraham
Microbiology: An Introduction, Batzing
Genetics: The Continuity of Life, Fairbanks/Anderson
Human Heredity, Seventh, Cummings
Current Perspectives in Genetics, Second, Cummings
Gene Discovery Lab, Benfey

Animal Physiology, Sherwood, Kleindorf, Yarcey
Invertebrate Zoology, Seventh, Ruppert/Fox/Barnes
Mammalogy, Fourth, Vaughan/Ryan/Czaplewski
Biology of Fishes, Third, Bond
Vertebrate Dissection, Ninth, Homberger/Walker

Plant Biology, Second, Rost/Barbour/Stocking/Murphy
Plant Physiology, Fourth, Salisbury/Ross
Introductory Botany, Berg

General Ecology, Second, Krohne
Essentials of Ecology, Fourth, Miller
Terrestrial Ecosystems, Second, Aber/Melillo
Living in the Environment, Fifteenth, Miller
Environmental Science, Twelfth, Miller/Spoolman
Sustaining the Earth, Eighth, Miller
Case Studies in Environmental Science, Second, Underwood
Environmental Ethics, Third, Des Jardins
Watersheds 3—Ten Cases in Environmental Ethics, Third, Newton/Dillingham

Problem-Based Learning Activities for General Biology, Allen/Duch
The Pocket Guide to Critical Thinking, Second, Epstein

Thomson Higher Education
10 Davis Drive
Belmont, CA 94002-3098
USA

Asia (including India)
Thomson Learning
5 Shenton Way
#01-01 UIC Building
Singapore 068808

Australia/New Zealand
Thomson Learning Australia
102 Dodds Street
Southbank, Victoria 3006
Australia

Canada
Thomson Nelson
1120 Birchmount Road
Toronto, Ontario M1K 5G4

UK/Europe/Middle East/Africa
Thomson Learning
High Holborn House
50/51 Bedford Row
London WC1R 4LR
United Kingdom

CONTENTS IN BRIEF

DETAILED CONTENTS

INTRODUCTION

UNIT I PRINCIPLES OF CELLULAR LIFE

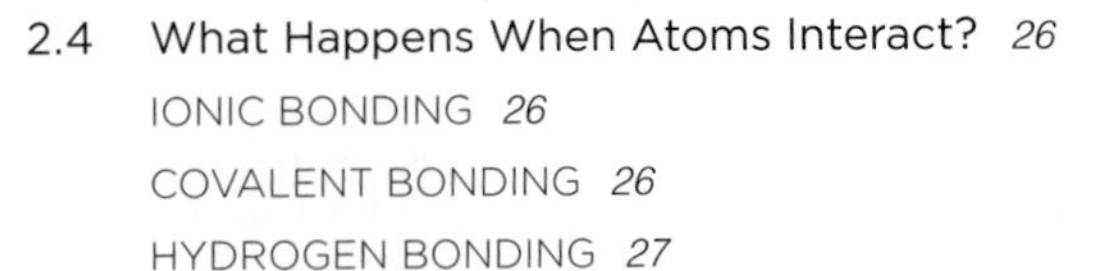

UNIT II PRINCIPLES OF INHERITANCE

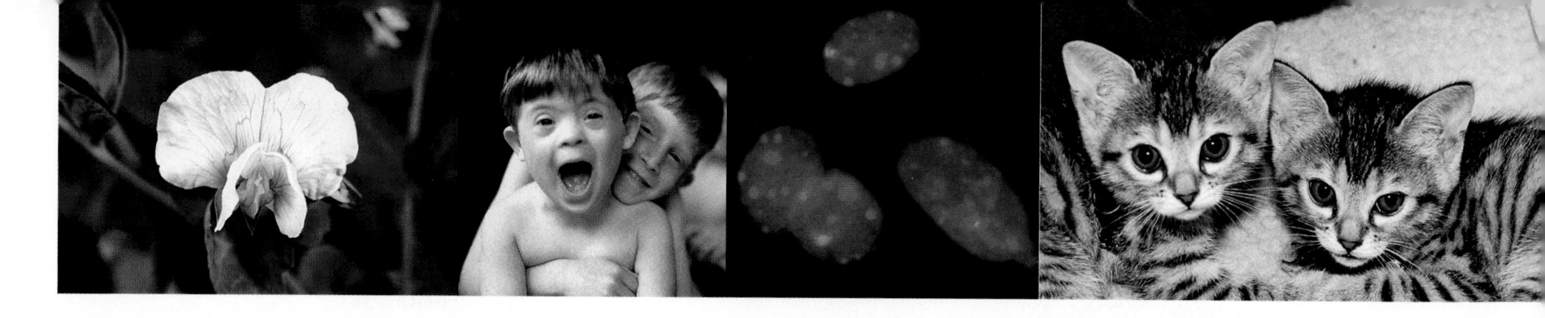

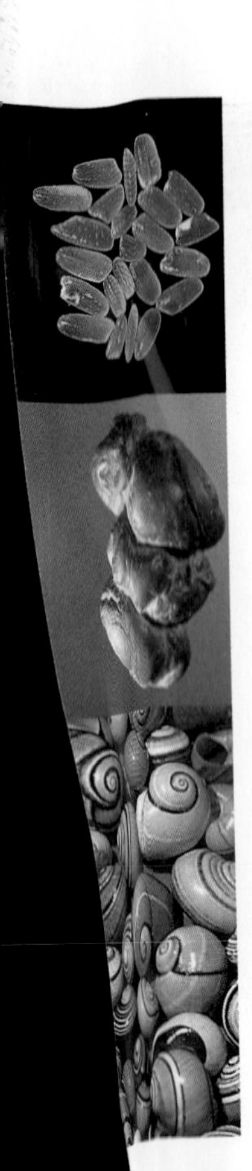

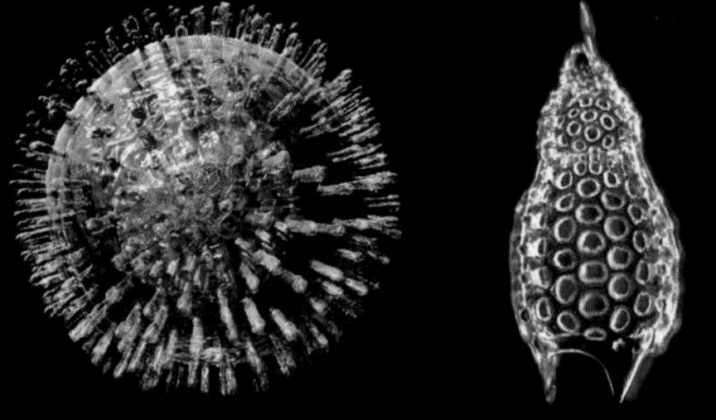

24 Animal Evolution—The Vertebrates

25 Plants and Animals—Common Challenges

UNIT V HOW PLANTS WORK

26 Plant Tissues

27 Plant Nutrition and Transport

28 Plant Reproduction and Development

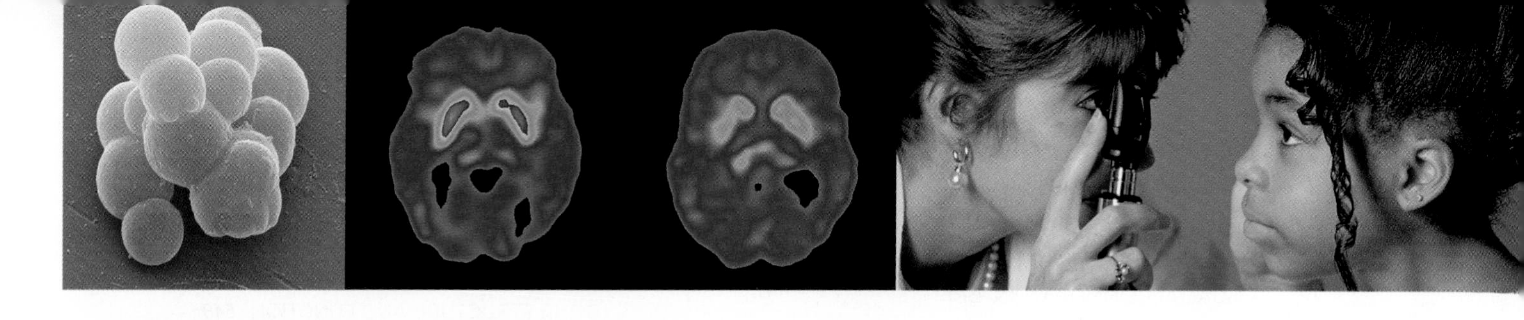

UNIT VI HOW ANIMALS WORK

29 Animal Tissues and Organ Systems

30 Neural Control

31 Sensory Perception

32 Endocrine Control

33 Structural Support and Movement

34 Circulation

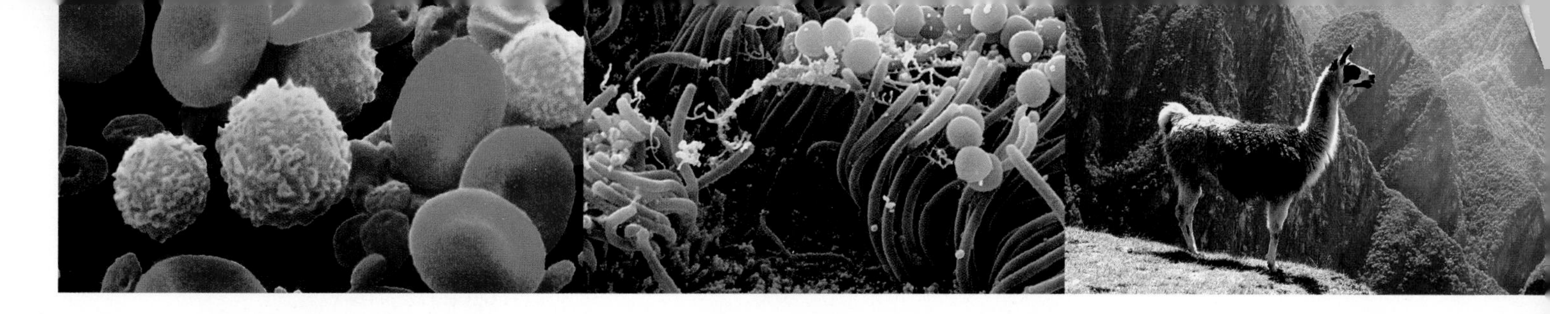

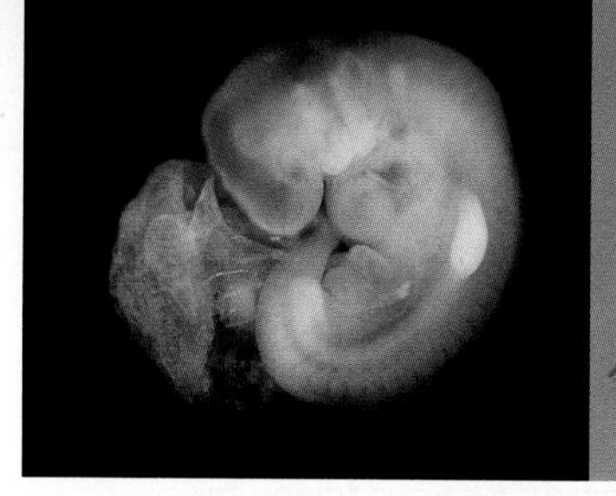

UNIT VII PRINCIPLES OF ECOLOGY

40 Population Ecology

41 Community Structure and Biodiversity

42 Ecosystems

43 The Biosphere

Preface

In preparation for this revision, we invited instructors who teach introductory biology for non-majors students to meet with with us and discuss the goals of their course. Nearly always, their goal was something like this: "To familiarize students with the way that science works and provide them with the tools they need to make well-informed choices as consumers and as voters." This makes sense. Most students who use this book will not be biologists, and many will never take another science course. Yet they certainly need to make decisions that require an understanding of the process of science and of basic biological principles.

We provide these future decision-makers with an accessible introduction to science. Throughout this edition, we emphasize that biology is not a body of facts, but rather an ongoing endeavor carried out by a diverse community of people. We underscore this point by describing current research and providing photos and videos of the scientists who do it. We explain not only what is known, but also how it was discovered, and how our understanding has changed over time. At the same time, we highlight the role of longstanding scientific theories, most notably the theory of evolution, which is a unifying theme in this book.

We revised every page of text to make it as straightforward and clear as possible, keeping in mind that English is a second language for many students. We added new tables to summarize important points and streamlined figures to eliminate unnecessary complexity.

CHANGES FOR THIS EDITION

Links to Key Concepts New to this edition are tools that link concepts within and between chapters. These tools reinforce the concept that each new idea in science rests on a foundation of other ideas.

Every chapter introduction has a section-by-section list of *Key Concepts*, each with a simple title. We repeat the titles at the top of appropriate text pages as ongoing reminders of the chapter's conceptual organization. A brief list of *Links to Earlier Concepts* helps remind students of relevant concepts that they encountered in previous chapters. For instance, students are advised that before reading about neural function, they may wish to scan an earlier chapter section on active transport. Icons are repeated in text page margins.

Media-Integrated Summaries We have always offered a wealth of online media for students. With this edition, we have made it easier for students to determine which online material supports each section. We have integrated information about the relevant animations, tutorials, and videos into the section summaries.

Chapter-Specific Changes Every chapter was extensively revised for clarity; this edition has 350 new photos and almost 170 new or updated figures. A page-by-page guide to new content and figures is available upon request, but we summarize the highlights here.

• *Chapter 1, Invitation to Biology* New essay about discovery of new species. Greatly expanded coverage of critical thinking and the process of science.
• *Chapter 2, Life's Chemical Basis* Chemistry of bonding revised to include electronegativity; new pH art.
• *Chapter 3, Molecules of Life* New art demonstrating protein structural organization; other art reorganized.
• *Chapter 4, Cell Structure and Function* Microscopy section updated; plasma membrane art simplified; new focus section on biofilms; cytoskeleton section reworked.
• *Chapter 5, Ground Rules of Metabolism* Energy and metabolism sections reorganized and rewritten; much new art, including molecular model of active site.
• *Chapter 6, Where It Starts—Photosynthesis* New essay on global warming emphasizes role of photosynthesis in the cycling of atmospheric carbon dioxide.
• *Chapter 7, How Cells Release Chemical Energy* All art showing metabolic pathways revised and simplified.
• *Chapter 8, How Cells Reproduce* Updated micrographs of mitosis; cancer section updated.
• *Chapter 9, Meiosis and Sexual Reproduction* Opener revised to include Red Queen hypothesis; new essay on evolutionary connection between mitosis and meiosis.
• *Chapter 10, Observing Patterns in Inherited Traits* Updated essay on cystic fibrosis; new figures for coat color genetics in dogs, and environmental effects on *Daphnia* phenotype.
• *Chapter 11, Chromosomes and Human Inheritance* Chapter reorganized; expanded discussion and new figure on the evolution of chromosome structure.
• *Chapter 12, DNA Structure and Function* New opener essay on pet cloning; adult cloning section updated.
• *Chapter 13, From DNA to Protein* New, simplified figures for transcription and translation.
• *Chapter 14, Controls Over Genes* Chapter reorganized; eukaryotic gene control section rewritten; updated X chromosome inactivation photos; new lac operon art.
• *Chapter 15, Studying and Manipulating Genomes* Chapter reorganized; gene library and PCR section rewritten; genetic engineering sections updated and expanded.
• *Chapter 16, Evidence of Evolution* Heavily revised; reorganized with Chapter 17 to emphasize evidence-based thinking. Revised opener essay on evidence leading to inference; updated geologic time scale; comparative morphology section rewritten with new figure; comparative embryology photo series added; cladistics section rewritten; new, updated tree of life.
• *Chapter 17, Processes of Evolution* Heavily revised; reorganized with chapter 16 to emphasize evolution as a process. Revised rats/warfarin essay; sections on sexual selection, reproductive isolation, sympatric speciation, and macroevolution rewritten; examples added: directional selection in the peppered moth, reproductive isolation in stalk-eyed flies, genetic drift in flour beetles, mechanical isolation in sage, sympatric speciation in palms, and ring species.
• *Chapter 18, Life's Origin and Early Evolution* Information about origin of agents of metabolism updated. New discussion of ribozymes as evidence for RNA world.

• *Chapter 19, Prokaryotes and Viruses* New art of viral structure. Herpes virus replication added. New section on discovery of viroids and prions.

• *Chapter 20, Protists—The Simplest Eukaryotes* Figure showing different protist life cycles added. New section about amoebozoans. Fungi now in separate chapter.

• *Chapter 21, Plant Evolution* Plant life cycle diagram added. Whisk fern coverage added. More about ferns. New section about quinoa, the most nutritious plant.

• *Chapter 22, Fungi* New chapter devoted to the fungi. Includes information on chytrids and microsporidians, a separate section for each major fungal group.

• *Chapter 23, Animal Evolution—the Invertebrates* Improved coverage of animal origins and of crustacean diversity. New section about invertebrate pests and parasites.

• *Chapter 24, Animal Evolution—the Vertebrates* Updated figure for fish-to-tetrapod limb evolution. Sections on primate and human evolution revised and updated.

• *Chapter 25, Plants and Animals—Common Challenges* More information about plant defensive mechanisms.

• *Chapter 26, Plant Tissues* Primary structure of roots reorganized; new section on tree rings and past climate.

• *Chapter 27, Plant Nutrition and Transport* New essay on effects of ozone on plants.

• *Chapter 28, Plant Reproduction and Development* New section on plant responses to seasonal changes.

• *Chapter 29, Animal Tissues and Organ Systems* Opener about stem cells updated. Improved coverage of embryonic tissues, development of body cavities.

• *Chapter 30, Neural Control* Chapter reorganized to begin with overview of nervous systems. New sections cover neurotransmitters and the role of neuroglia.

• *Chapter 31, Sensory Perception* New art of vestibular apparatus, image formation in eyes, and accommodation.

• *Chapter 32, Endocrine Control* Chapter reorganized into smaller sections focused on specific glands. New graphic for insulin/glucagon effects. More on diabetes.

• *Chapter 33, Structural Support and Movement* Improved coverage of joints, clarified discussion of sliding-filament model.

• *Chapter 34, Circulation* Clearer discussion of Rh factor and risks with pregnancy. New art of cardiac muscle.

• *Chapter 35, Immunity* Heavily reorganized, updated to reflect current paradigms, and rewritten to emphasize integrated actions of the immune system. Opener essay updated to include vaccine development; new, simplified art of adaptive immune responses; AIDS section updated.

• *Chapter 36, Respiration* New section about respiration in extreme habitats (high altitude and deep dives).

• *Chapter 37, Digestion and Human Nutrition* Nutritional information and obesity research sections updated.

• *Chapter 38, The Internal Environment* New figure of fluid distribution in the body. Coverage of nephron anatomy and urine formation completely revised.

• *Chapter 39, Animal Reproduction and Development* The chapter has been shortened by tightening sections about classical embryology and birth defects. New section about female reproductive disorders.

• *Chapter 40, Population Ecology* Exponential and logistic growth clarified. Effect of fishing on Atlantic cod added.

• *Chapter 41, Community Structure and Biodiversity* Whirling disease in trout, salamander competition study added. Updated coverage of succession and stability.

• *Chapter 42, Ecosystems* New figures for food chain and food webs. Updated greenhouse gas coverage.

• *Chapter 43, The Biosphere* New section about soils and desertification. New section about rain forests. More on coral reefs and threats to them. More on ocean life.

• *Chapter 44, Behavioral Ecology* Chapter reorganized and shortened.

Appendix V, Molecular Models New art and text explain why we use different types of molecular models.

Appendix VI, Closer Look at Some Major Metabolic Pathways New art shows details of electron transport chains in thylakoid membranes.

Appendix VIII, Restless Earth—Life's Changing Geologic Stage A new map from NASA summarizes Earth's tectonic and volcanic activity.

Appendix X, A Comparitive View of Mitosis in Plant and Animal Cells A new figure shows the stages of plant and animal mitosis side-by-side for easy comparison.

ACKNOWLEDGMENTS

Thanks to our advisors for their ongoing impact on the book's content. John Jackson, Jean deSaix, David Rintoul, and Michael Plotkin all deserve recognition for their deep commitment to excellence in education. This edition also reflects many influential contributions of the instructors, listed on the following page, who helped shape our thinking. *Impacts/Issues* essays, *Key Concepts*, custom videos—such features are direct responses to their insights from the classroom.

Thomson Learning continues to prove why it is one of the world's foremost publishers; Michelle Julet, thank you again for supporting our ideals and our creativity. Keli Amann and Kristina Razmara created a world-class technology package for both students and instructors. Peggy Williams, with her clarity, humor, intelligence, and patience, has been truly inspiring. Grace Davidson calmly kept us on track and put all of the pieces together, and Andy Marinkovich made sure that production went smoothly. Thanks also to our marketing manager Kara Kindstrom, Paul Forkner in photo research, and Jessica Kuhn and Rose Barlow, our tireless editorial assistants.

It takes a dedicated group of publishing professionals to produce a textbook, yet no listing conveys how this team interacted to create something extraordinary. And thank you, Jack Carey, for being the first to identify the need for features, including student voting, that can further biology education.

CECIE STARR, CHRIS EVERS, AND LISA STARR *July 2007*

CONTRIBUTORS TO THIS EDITION: INFLUENTIAL CLASS TESTS AND REVIEWS

Brenda Alston-Mills
North Carolina State University

Norris Armstrong
University of Georgia

Dave Bachoon
Georgia College & State University

Andrew Baldwin
Mesa Community College

Lisa Lynn Boggs
Southwestern Oklahoma State University

Gail Breen
University of Texas at Dallas

Marguerite "Peggy" Brickman
University of Georgia

David William Bryan
Cincinnati State College

Uriel Buitrago-Suarez
Harper College

Sharon King Bullock
Virginia Commonwealth University

John Capehart
University of Houston - Downtown

Daniel Ceccoli
American InterContinental University

Tom Clark
Indiana University South Bend

Heather Collins
Greenville Technical College

Cynthia Lynn Dassler
Ohio State University

Carole Davis
Kellogg Community College

Lewis E. Deaton
University of Louisiana - Lafayette

Jean Swaim DeSaix
University of North Carolina - Chapel Hill

(Joan) Lee Edwards
Greenville Technical College

Hamid M. Elhag
Clayton State University

Patrick Enderle
East Carolina University

Daniel J. Fairbanks
Brigham Young University

Amy Fenster
Virginia Western Community College

Kathy E. Ferrell
Greenville Technical College

Rosa Gambier
Suffok Community College - Ammerman

Tim D. Gaskin
Cuyahoga Community College - Metropolitano

Stephen J. Gould
Johns Hopkins University

Marcella Hackney
Baton Rouge Community College

Gale R. Haigh
McNeese State University

John Hamilton
Gainsville State

Richard Hanke
Rose State Community College

Chris Haynes
Shelton St. Community College

Kendra M. Hill
South Dakota State University

Juliana Guillory Hinton
McNeese State University

Kelly Hogan
University of North Carolina

Robert Hunter
Trident Technical College

John Ireland
Jackson Community College

Thomas M. Justice
McLennan College

Timothy Owen Koneval
Laredo Community College

Sherry Krayesky
University of Louisiana - Lafayette

Dubear Kroening
University of Wisconsin - Fox Valley

Jerome Krueger
South Dakota State University

Jim Krupa
University of Kentucky

Mary Lynn LaMantia
Golden West College

Kevin T. Lampe
Bucks County Community College

Susanne W. Lindgren
Sacramento State University

Madeline Love
New River Community College

Dr. Kevin C. McGarry
Kaiser College - Melbourne

Jeanne Mitchell
Truman State University

Alice J. Monroe
St. Petersburg College - Clearwater

Brenda Moore
Truman State University

Rajkumar "Raj" Nathaniel
Nicholls State University

Francine Natalie Norflus
Clayton State University

Alexander E. Olvido
Virginia State University

Bob Patterson
North Carolina State University

Shelley Penrod
North Harris College

Mary A. (Molly) Perry
Kaiser College - Corporate

John S. Peters
College of Charleston

Michael Plotkin
Mt. San Jacinto College

Ron Porter
Penn State University

Karen Raines
Colorado State University

Larry A. Reichard
Metropolitan Community College - Maplewood

Jill D. Reid
Virginia Commonwealth University

Robert Reinswold
University of Northern Colorado

David Rintoul
Kansas State University

Darryl Ritter
Okaloosa Walton Junior College

Amy Wolf Rollins
Clayton State University

Robin Searles-Adenegan
Morgan State University

Julie Shepker
Kaiser College - Melbourne

Rainy Shorey
Illinois Central College

Eric Sikorski
University of South Florida

Robert (Bob) Speed
Wallace Junior College

Tony Stancampiano
Oklahoma City Community College

Jon R. Stoltzfus
Michigan State University

Peter Svensson
West Valley Collegee

Jeffrey L. Travis
University at Albany

Nels H. Troelstrup, Jr.
South Dakota State University

Allen Adair Tubbs
Troy University

Will Unsell
University of Central Oklahoma

Rani Vajravelu
University of Central Florida

Jack Waber
West Chester University of Pennsylvania

Kathy Webb
Bucks County Community College

Virginia White
Riverside Community College

Kathleen Lucy Wilsenn
University of Northern Colorado

Penni Jo Wilso
Cleveland State Community College

Michael L. Womack
Macon State College

Mark L. Wygoda
McNeese State University

Lan Xu
South Dakota State University

Poksyn ("Grace") Yoon
Johnson and Wales University

Introduction

Current configurations of the Earth's oceans and land masses—the geologic stage upon which life's drama continues to unfold. This composite satellite image reveals global energy use at night by the human population. Just as biological science does, it invites you to think more deeply about the world of life—and about our impact upon it.

1 INVITATION TO BIOLOGY

IMPACTS, ISSUES Lost Worlds and Other Wonders

In this era of satellites, submarines, and global positioning systems, could there possibly be any more places on Earth that we have not explored? Well, yes. In 2005, for instance, helicopters dropped a team of biologists into a swamp in the middle of a vast and otherwise inaccessible tropical forest in New Guinea. Later, team member Bruce Beehler remarked, "Everywhere we looked, we saw amazing things we had never seen before. I was shouting. This trip was a once-in-a lifetime series of shouting experiences."

The team discovered dozens of animals and plants unknown to science, including a rhododendron with plate-sized flowers. They found animals that are being hunted to extinction in other parts of the world, and a bird that supposedly was extinct.

The expedition fired the imagination of people all over the world. It is not that finding new kinds of organisms is such a rare event. Almost every week, biologists discover many kinds of insects and other small organisms. However, the animals in this particular rain forest—mammals and birds especially—seem too big to have gone unnoticed before. Had people just missed them? Perhaps not. No trails or other human disturbances cut through that part of the forest. The animals had never learned to be afraid of humans, so the biologists could simply walk over and pick them up (Figure 1.1).

Other animals have turned up in the past few years, including lemurs in Madagascar (Figure 1.2), monkeys in India and Tanzania, and whales and giant jellylike animals in the seas. Most came to light during survey trips similar to the New Guinea expedition—when biologists simply were attempting to find out what lives where.

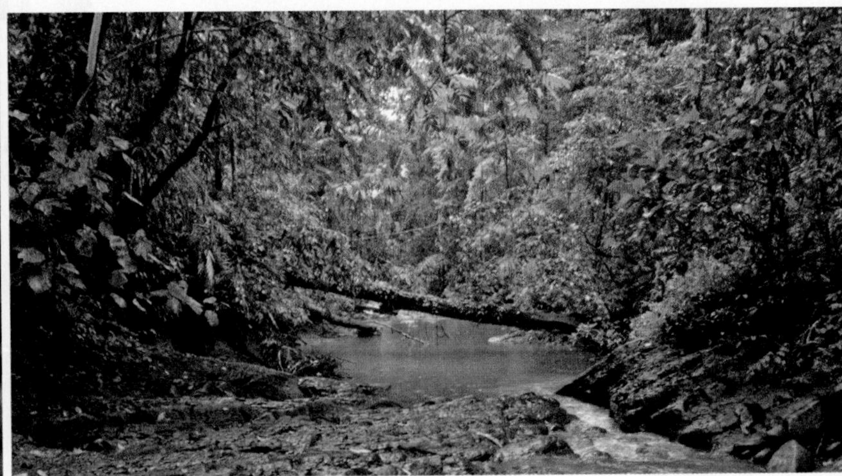

Figure 1.1 Biologist Kris Helgen and a rare golden-mantled tree kangaroo in a tropical rain forest in the Foja Mountains of New Guinea. There, in 2005, explorers discovered dozens of previously unknown species.

How would you vote? The discoverer of a new species usually is the one who gives it a scientific name. In 2005, a Canadian casino bought the right to name a monkey species. Should naming rights be sold? See ThomsonNOW for details, then vote online.

Exploring and making sense of nature is nothing new. We humans and our immediate ancestors have been at it for at least 2 million years. We observe, come up with explanations about what the observations mean, and then test the explanations. Ironically, the more we learn about nature, the more we realize how much we have yet to learn.

You might choose to let others tell you what to think about the world around you. Or you might choose to develop your own understanding of it. Perhaps, like the New Guinea explorers, you are interested in animals and where they live. Maybe you are interested in aspects that affect your health, the food you eat, or your home and family. Whatever your focus may be, the scientific study of life—biology—can deepen your perspective on the world.

Throughout this book, you will find examples of how organisms are constructed, where they live, and what they do. These examples support concepts that, when taken together, convey what "life" is. This chapter gives you an overview of basic concepts. It sets the stage for upcoming descriptions of scientific observations and applications that can help you refine your understanding of life.

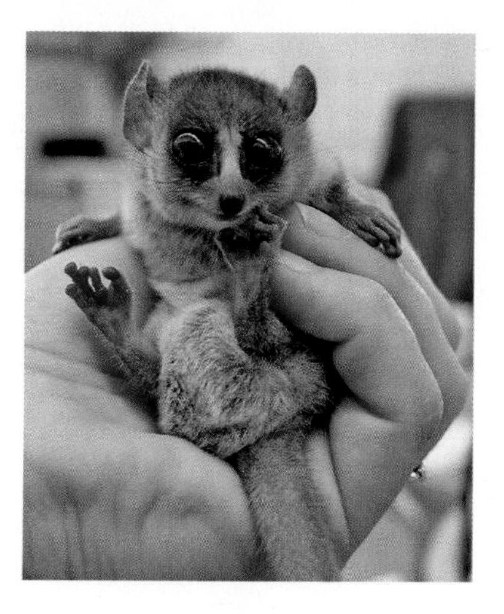

Figure 1.2 Goodman's mouse lemur (*Microcebus lehilahytsara*). Explorers discovered this small mammal in a Madagascar rain forest in 2005.

Key Concepts

LEVELS OF ORGANIZATION

We study the world of life at different levels of organization, which extend from atoms and molecules to the biosphere. The quality known as "life" emerges at the level of cells. **Section 1.1**

LIFE'S UNDERLYING UNITY

The world of life shows unity, because all organisms are alike in key respects. They consist of one or more cells, which stay alive through ongoing inputs of energy and raw materials. They sense and respond to changes in their external and internal environments. Their cells contain DNA, a type of molecule that offspring inherit from parents and that encodes information necessary for growth, survival, and reproduction. **Section 1.2**

LIFE'S DIVERSITY

The world of life also shows great diversity. Many millions of kinds of organisms, or species, have appeared and disappeared over time. Each species is unique in at least one trait—in some aspect of its body form or behavior. **Section 1.3**

EXPLAINING UNITY IN DIVERSITY

Theories of evolution, especially a theory of evolution by natural selection, help explain why life shows both unity and diversity. Evolutionary theories guide research in all fields of biology. **Section 1.4**

HOW WE KNOW

Biologists make systematic observations, predictions, and tests in the laboratory and in the field. They report their results so others may repeat their work and check their reasoning. **Sections 1.5–1.8**

Links to Earlier Concepts

This book parallels nature's levels of organization, from atoms to the biosphere. Learning about the structure and function of atoms and molecules primes you to understand the structure of living cells. Learning about processes that keep a single cell alive can help you understand how large organisms survive, because their many living cells use the same processes. Knowing what it takes for large organisms to survive can help you see why and how they interact with one another and with the environment.

At the start of each chapter, we will be reminding you of such connections. Within chapters, key icons and cross-references will link you to relevant sections in earlier chapters.

1.1 Life's Levels of Organization

Suppose someone asks you to explain how "life" differs from "nonlife." Where would you start? Life's building blocks are as ordinary as the ones you find in rocks and the seas. However, the quality of life emerges as particular building blocks join up and interact in organized units called cells.

MAKING SENSE OF THE WORLD

Most of us intuitively understand what nature means, but could you define it? Nature is everything in the universe *except* what humans have manufactured. It encompasses every substance, event, force, and energy—sunlight, flowers, animals, bacteria, rocks, thunder, waves, and so on. It excludes everything artificial.

Scientists, clerics, farmers, astronauts, and anyone else who is of a mind to do so attempt to make sense of nature. Interpretations differ, for no one can be expert in everything learned so far or have foreknowledge of all that remains hidden. If you are reading this book, you are starting to explore how a subset of scientists, the biologists, think about things, what they found out, and what they are up to now.

A PATTERN IN LIFE'S ORGANIZATION

Biologists look at all aspects of life, past and present. Their focus takes them all the way down to atoms, and all the way up to global relationships among organisms and the environment. Through their work, we glimpse a great pattern of organization in nature.

The pattern starts at the level of atoms. Atoms are fundamental building blocks of all substances, living and nonliving (Figure 1.3*a*).

At the next level of organization are molecules, or units in which atoms are joined together (Figure 1.3*b*). Among the molecules are complex carbohydrates and lipids, proteins, DNA, and RNA. In nature, only living cells now make these "molecules of life."

The pattern crosses the threshold to life when many molecules are organized as cells (Figure 1.3*c*). A cell is the smallest unit of life that can survive and reproduce on its own, given information in DNA, energy inputs, raw materials, and suitable environmental conditions.

An organism is an individual that consists of one or more cells. In larger multicelled organisms, trillions of

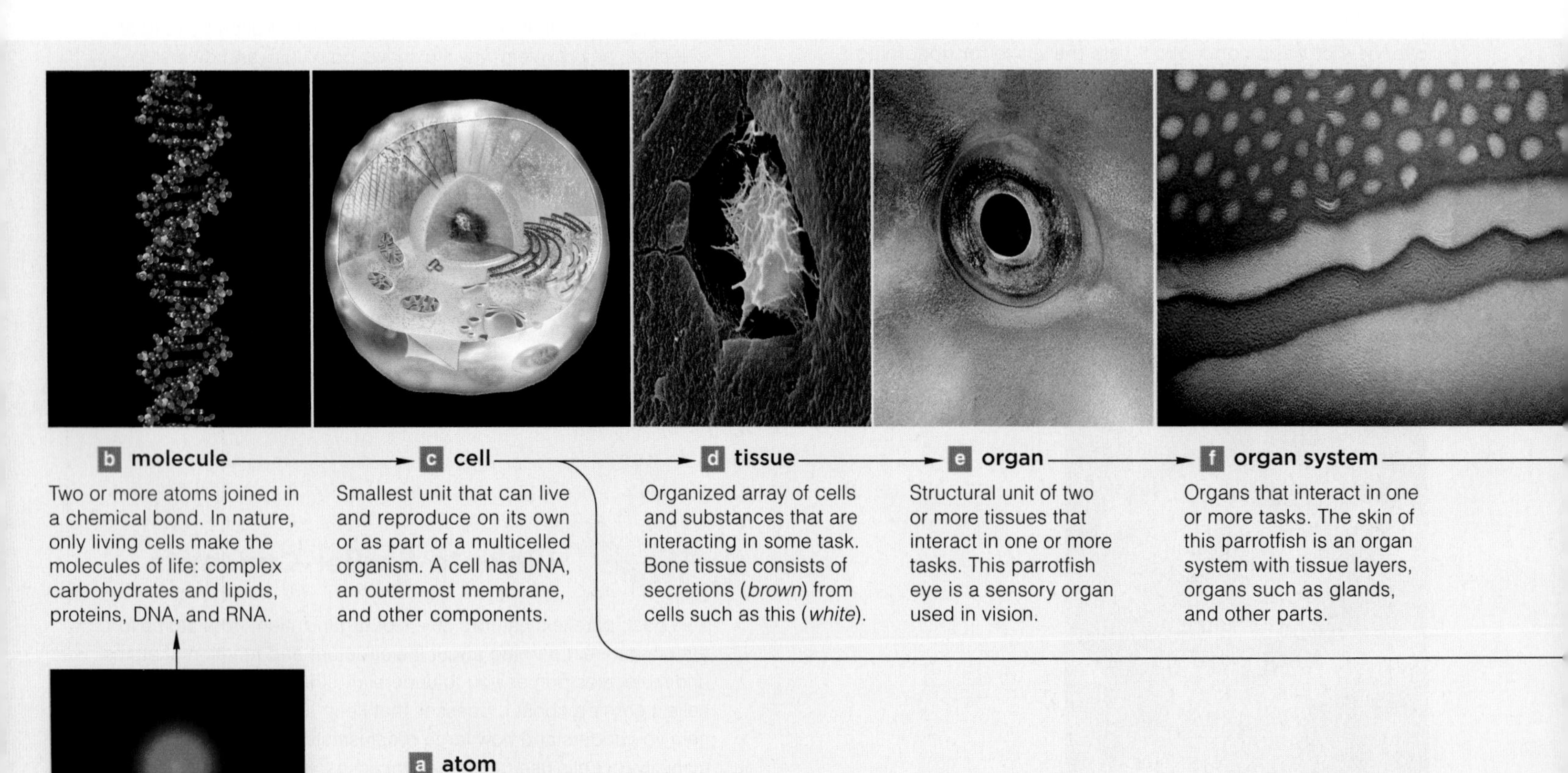

Figure 1.3 Animated! Levels of organization in nature.

cells organize into tissues, organs, and organ systems, all interacting in tasks that keep the whole body alive. Figure 1.3*d*–*g* defines these body parts.

Populations are at a greater level of organization. Each population is a group of individuals of the same kind of organism, or species, in a specified area (Figure 1.3*h*). Examples are all humphead parrotfish living on Shark Reef in the Red Sea or all California poppies in California's Antelope Valley Poppy Reserve.

Communities are at the next level. A community consists of all populations of all species in a specified area. As an example, Figure 1.3*i* shows a sampling of the Shark Reef's species. This underwater community includes many kinds of seaweeds, fishes, corals, sea anemones, shrimps, and other living organisms that make their home in or on the reef. Communities may be large or small, depending on the area defined.

The next level of organization is the ecosystem, or a community interacting with its physical and chemical environment. The biosphere—the most inclusive level —encompasses all regions of Earth's crust, waters, and atmosphere in which organisms live.

Bear in mind, life is more than the sum of its parts. In other words, emergent properties occur at successive levels of life's organization. Emergent properties are characteristics of a system that do not appear in any of its component parts. As one example, molecules are not alive. Considering them separately, no one could predict that a particular quantity and arrangement of molecules will form a living cell. Life—an emergent property—appears first at the level of the cell but not at any lower level of organization in nature.

This book is a journey through the globe-spanning organization of life. Take a moment to study Figure 1.3. You can use it as a road map showing where each part fits into the great scheme of nature.

Nature shows levels of organization, from the simple to the increasingly complex.

The unique properties of life emerge as certain kinds of molecules become organized into cells. Greater levels of organization include multicelled organisms, populations, communities, ecosystems, and the biosphere.

g multicelled organism

Individual made of different types of cells. Cells of most multicelled organisms, such as this Red Sea parrotfish, make up tissues, organs, and organ systems.

h population

Group of single-celled or multicelled individuals of a species in a given area. This is a population of one fish species in the Red Sea.

i community

All populations of all species in a specified area. These populations belong to a coral reef community in a gulf of the Red Sea.

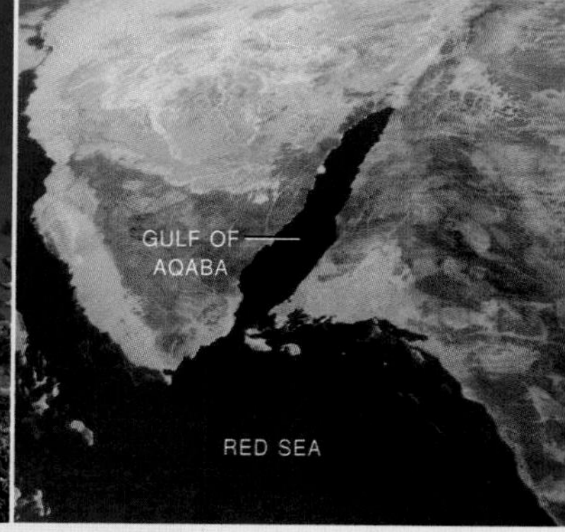

j ecosystem

A community that is interacting with its physical environment through inputs and outputs of energy and materials. Reef ecosystems flourish in warm, clear seawater throughout the Middle East.

k biosphere

All regions of Earth's waters, crust, and atmosphere that hold organisms. In the vast universe, Earth is a rare planet. Life as we know it is impossible without its abundance of free-flowing water.

1.2 Overview of Life's Unity

Never-ending infusions of energy and materials maintain life's complex organization. Without those vital inputs, organisms could not sense and respond to changes that might disrupt their organization. They could not build and maintain DNA and all of the other complex molecules that help them stay alive, grow, and reproduce.

ENERGY AND LIFE'S ORGANIZATION

As you know, giving up eating would be a bad idea, because you would run out of the energy and nutrients that keep your body organized and functioning. Energy is the capacity to do work. A nutrient is a particular type of atom or molecule that has an essential role in growth and survival.

All single-celled and multicelled organisms spend a lot of time getting energy and nutrients, although they get them from different sources. The differences allow us to put organisms into one of two broad categories: producers or consumers.

Producers get energy and simple raw materials from environmental sources and make their own food. Plants are producers. By a process called photosynthesis, they use energy from the sun to make sugars from carbon dioxide and water. Those sugars function as packets of immediately available energy or as building blocks for larger molecules.

Consumers cannot make their own food; they get energy and nutrients indirectly—by eating producers and other organisms. Animals fall within the consumer category. So do decomposers, which feed on wastes or remains of organisms. We find leftovers of their meals in the environment. Producers take up the leftovers as sources of nutrients. Said another way, producers and consumers cycle nutrients among themselves.

Energy, however, is not cycled. It flows through the world of life in one direction—from the environment, through producers, then through consumers. This flow maintains the organization of individual organisms, and also it is the basis of life's organization within the biosphere (Figure 1.4). It is a one-way flow, because with each transfer, some energy escapes as heat. Cells do not use heat to do work. Thus, energy that enters the world of life ultimately leaves it—permanently.

ORGANISMS SENSE AND RESPOND TO CHANGE

Organisms sense and respond to changes in conditions inside and outside the body by way of receptors. Each receptor is a molecule or cellular structure that responds to a specific form of stimulation, such as the energy of sunlight or the mechanical energy of a bite (Figure 1.5).

Stimulated receptors trigger changes in activities of organisms. For example, after you eat, the sugars from your meal become added to the sugars that are already circulating in your blood. Your body responds to this input. Blood and tissue fluids form the body's *internal*

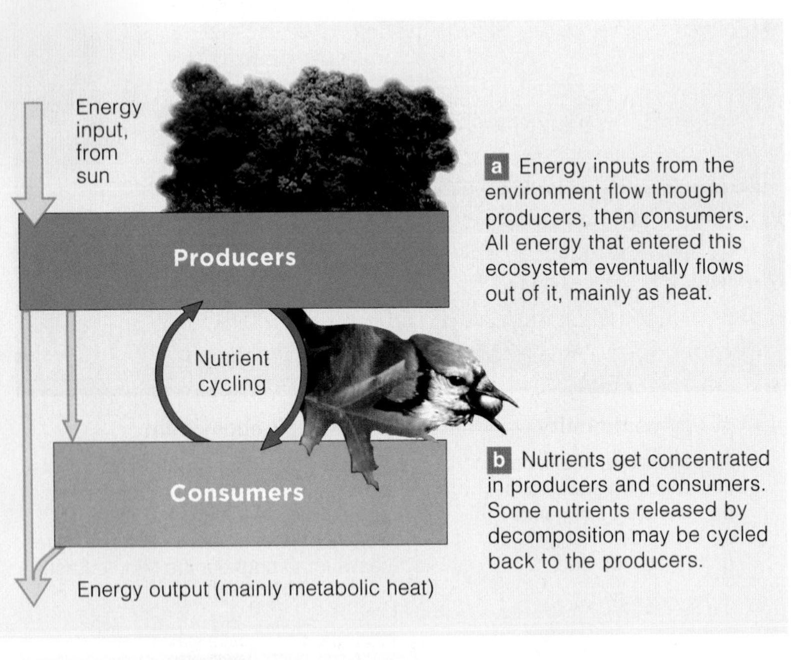

Figure 1.4 Animated! The one-way flow of energy and cycling of materials through an ecosystem.

Figure 1.5 A roaring response to signals from pain receptors, activated by a lion cub flirting with disaster.

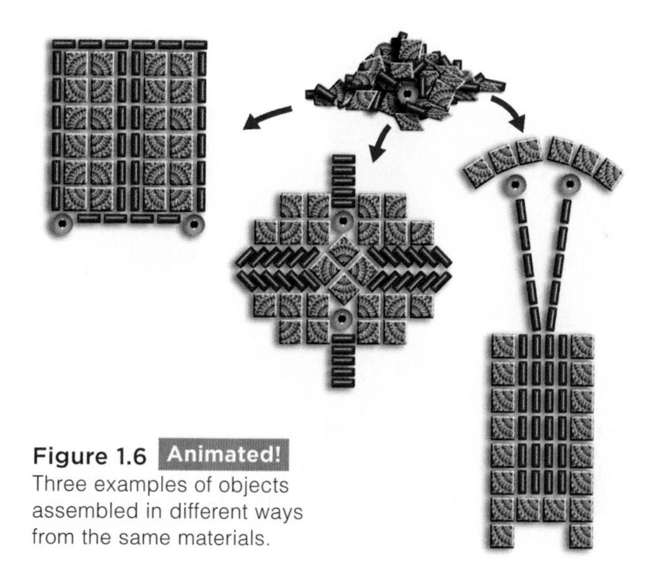

Figure 1.6 Animated! Three examples of objects assembled in different ways from the same materials.

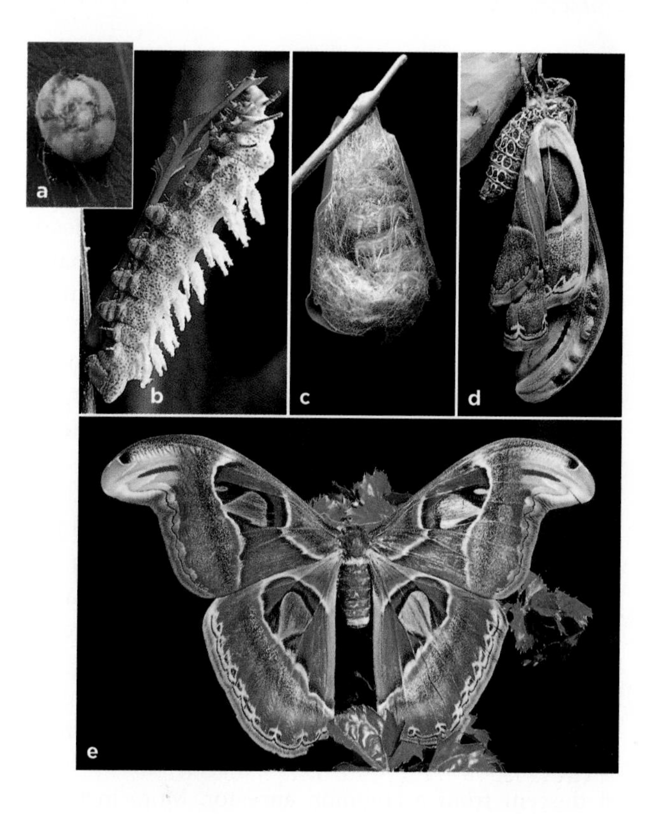

Figure 1.7 Silkworm moth development. Instructions in DNA guide the development of this insect through a series of stages, from a fertilized egg (**a**), to a larval stage called a caterpillar (**b**), to a pupal stage (**c**), to the winged adult form (**d**,**e**).

environment. Unless that environment's composition is kept within a certain range, cells in the body will die. In this case, the added sugars bind to receptors on cells of your pancreas, a large organ. Binding sets in motion a series of events that causes cells throughout the body to take up sugar faster, so the sugar level in your blood returns to normal.

By sensing and adjusting to change, organisms keep conditions in their internal environment within a range that favors cell survival. This process is homeostasis, and it is a defining feature of life.

ORGANISMS GROW AND REPRODUCE

Organisms grow and reproduce based on information in DNA, a nucleic acid. DNA is *the* signature molecule of life. No chunk of granite or quartz has it.

Why is DNA so important? It is the basis of growth, survival, and reproduction. It is also the source of each organism's distinct features, or traits.

DNA contains instructions. Cells use some of those instructions to make proteins, which are long chains of amino acids. There are only 20 kinds of amino acids, but cells string them together in different sequences to make a tremendous variety of proteins. By analogy, a few different kinds of tiles can be organized into many different patterns (Figure 1.6).

Different proteins have structural or functional roles. For instance, certain proteins are enzymes—functional molecules that make cell activities occur much faster than they would on their own. Without enzymes, such activities would not happen fast enough for a cell to survive. There would be no more cells—and no life.

In nature, an organism inherits DNA—the basis of its traits—from parents. Inheritance is the transmission of DNA from parents to offspring. Why do baby storks look like storks and not like pelicans? Because they inherited stork DNA, which differs from pelican DNA.

Reproduction refers to actual mechanisms by which parents transmit DNA to offspring. For all multicelled individuals, DNA has information that guides growth and development—the orderly transformation of the first cell of a new individual into an adult (Figure 1.7).

A one-way flow of energy and a cycling of nutrients through organisms and the environment sustain life's organization.

Organisms maintain homeostasis by sensing and responding to changing conditions. They make adjustments that keep conditions in their internal environment within a range that favors cell survival.

Organisms grow and reproduce based on information in DNA molecules, which they inherit from their parents.

Taken together, these characteristics reinforce a global concept: Unity underlies the world of life.

1.3 If So Much Unity, Why So Many Species?

Superimposed on life's unity is tremendous diversity. Of an estimated 100 billion kinds of organisms that have ever lived on Earth, as many as 100 million are with us today.

How is it possible to organize information about so many species, or kinds of organisms? Each species is assigned a two-part name. The first part of the name specifies the genus (plural, genera), which is a group of species that share a unique set of features. When combined with the second part, the name designates one species. Individuals of a species share one or more traits, and can interbreed successfully if the species is a sexually reproducing one.

For example, *Scarus* is one genus of parrotfish. The name of the humphead parrotfish shown in Figure 1.3*g* is *S. gibbus*. A different species in the same genus, the midnight parrotfish, is *S. coelestinus*. Notice the *S.* as an abbreviation for *Scarus*. You can abbreviate any genus name in a document after you first spell it out.

We organize and retrieve information about species with classification systems. The main systems group species on the basis of observable traits and evidence of descent from a common ancestor. More inclusive groupings above the level of genus include phylum (plural, phyla), kingdom, and domain. Table 1.1 and Figure 1.8 showcase a currently favored system that classifies species into one of three domains: Bacteria, Archaea, and Eukarya. The protists, plants, fungi, and animals make up domain Eukarya.

All bacteria (singular, bacterium) and archaeans are single-celled organisms. All are *prokaryotic*, meaning they do not have a nucleus. In all other organisms, this membrane-enclosed sac holds DNA. Prokaryotes as a group have the most diverse ways of procuring energy and nutrients. They are producers and consumers in nearly all of Earth's environments, including extreme ones such as frozen desert rocks and boiling, sulfur-clogged lakes. They probably resemble the first cells.

Structurally, the protists are the simplest *eukaryotic* organisms, which means their cells contain a nucleus. Different kinds are producers or consumers. Many are single cells that are larger and far more complex than prokaryotes. Some are tree-sized, multicelled seaweeds.

Table 1.1 Comparison of Life's Three Domains	
Bacteria	Single cells, prokaryotic (no nucleus). Most ancient lineage.
Archaea	Single cells, prokaryotic. Evolutionarily closer to eukaryotes.
Eukarya	Eukaryotic cells (with a nucleus). Single-celled and multicelled species categorized as protists, plants, fungi, and animals.

a Bacteria

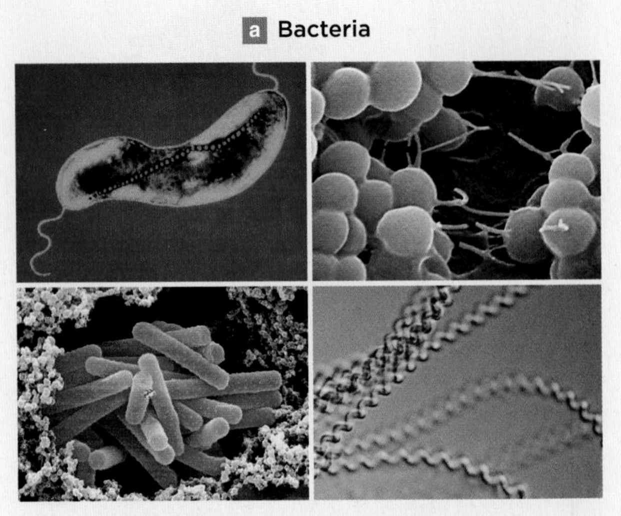

Compared with other species, these single prokaryotic cells tap more diverse sources of energy and nutrients. *Clockwise from upper left*, a bacterium with a tiny compass—a row of iron crystals; bacteria living on human skin; spiral cyanobacteria that are aquatic producers; and *Lactobacillus* cells in yogurt.

b Archaea

These prokaryotes are evolutionarily closer to eukaryotes than to bacteria. *Left*, a colony of methane-producing cells. *Right*, two species from a hydrothermal vent on the seafloor.

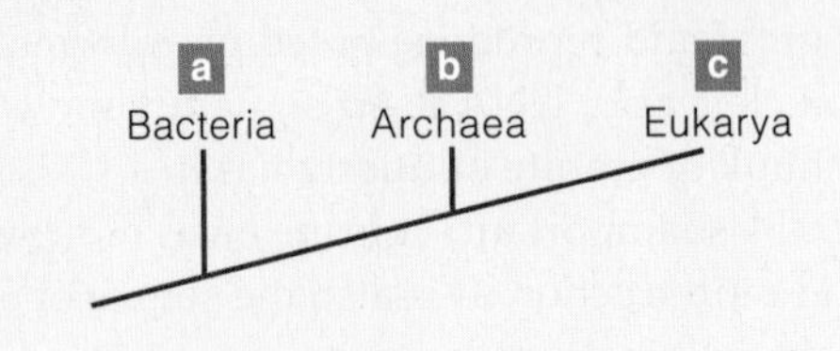

Figure 1.8 **Animated!** Representatives of diversity from the three most inclusive branchings of the tree of life.

Actually, the protists are so diverse that they are being reclassified into a number of separate major lineages.

Cells of fungi, plants, and animals are eukaryotic. Most fungi, such as the types that form mushrooms, are multicelled. Many are decomposers, and all secrete enzymes that digest food outside the body. Their cells then absorb the released nutrients.

Plants are multicelled species. Most of them live on land or in freshwater environments. Nearly all plants

C Eukarya

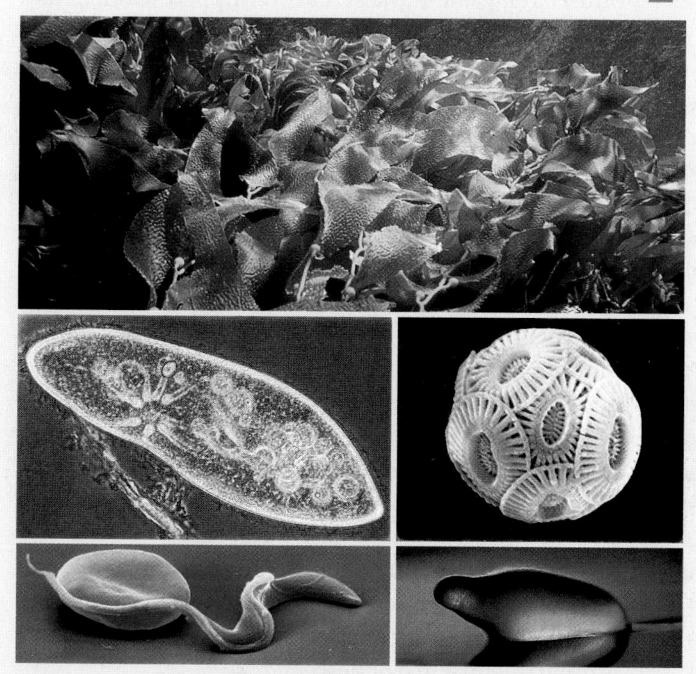

Protists Single-celled and multicelled eukaryotic species that range from the microscopic to giant seaweeds. Many biologists are now viewing the "protists" as many major lineages.

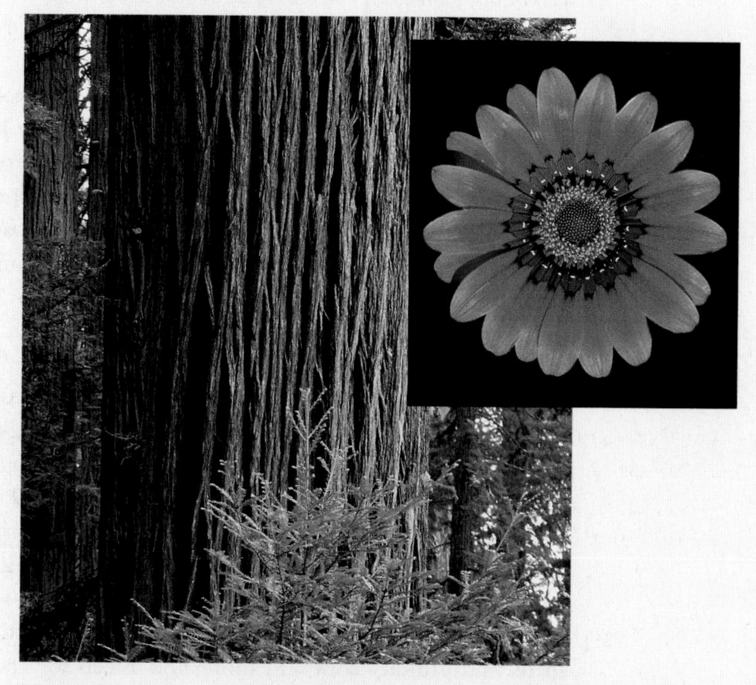

Plants Multicelled eukaryotes. Nearly all are photosynthetic; most have roots, stems, and leaves. Plants are the primary producers for ecosystems on land. Redwoods and flowering plants are examples.

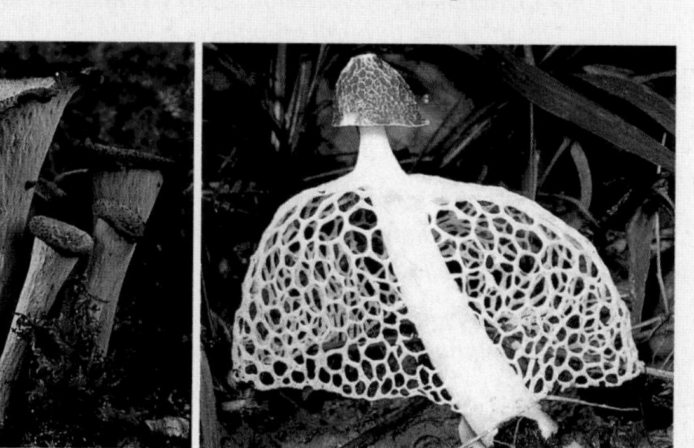

Fungi Single-celled and multicelled eukaryotes. Different kinds are decomposers, parasites, or pathogens. Without decomposers, communities would become buried in their own wastes.

Animals Multicelled eukaryotes that ingest tissues or juices of other organisms. Like this basilisk lizard, they actively move about during at least part of their life.

are photosynthetic: They harness the energy in sunlight to drive the production of sugars from carbon dioxide and water. Besides feeding themselves, plants also are producers that feed much of the biosphere.

The animals are multicelled consumers that ingest tissues or juices of other organisms. Herbivores graze, carnivores eat meat, scavengers eat remains of other organisms, and parasites pilfer nutrients from a host's tissues. Animals grow and develop through a series of stages that lead to the adult form. Most kinds actively move about during at least part of their lives.

Pulling this overview together, are you starting to get a sense of what it means when someone states that life shows unity *and* diversity?

We group species on the basis of shared traits and evidence of descent from a common ancestor. The most inclusive groupings are domains Bacteria, Archaea, and Eukarya.

Although unity underlies the world of life, we also observe great diversity. Organisms differ in their details; they show tremendous variation in traits.

1.4 An Evolutionary View of Diversity

How can organisms be so much alike and still show tremendous diversity? A theory of evolution by way of natural selection is one explanation.

Individuals of a population are alike in certain aspects of their body form, function, and behavior. Rarely are these traits exactly alike; their details differ from one individual to the next. For instance, except for identical twins, all 6.5 billion individuals of the human species (*Homo sapiens*) show variation in height, hair color, and other traits.

Variations in most traits arise through mutations, or changes in DNA. Most mutations have neutral or bad effects, but some cause a trait to change in a way that makes an individual of a population better adapted to its environment than individuals without the mutation. Such traits are *adaptive*. An individual with an adaptive form of a trait is more likely to survive and pass on its DNA to offspring. Charles Darwin, a naturalist, might have expressed it this way:

First, a natural population tends to increase in size, so its individuals compete more and more for food, shelter, and other limited environmental resources.

Second, those individuals differ from one another in the details of shared traits. Most traits are heritable; they can be passed to offspring (by way of DNA).

Third, adaptive forms of traits make their bearers more competitive, and so they tend to become more common over generations. The differential survival and reproduction of individuals in a population that differ in the details of their heritable traits is called natural selection.

rock pigeon

Think of how pigeons differ in feather color, size, and other traits (Figure 1.9*a*). Suppose a pigeon breeder prefers black, curly-tipped feathers. She selects captive birds having the darkest, curliest-tipped feathers and lets only those birds mate. Over time, more and more pigeons in the breeder's captive population will have black, curly-tipped feathers.

Pigeon breeding is a case of *artificial* selection. One form of a trait is favored over others under contrived, manipulated conditions—in an artificial environment. Darwin saw that breeding practices could be an easily understood model for *natural* selection, a favoring of some forms of a given trait over others in nature.

Just as breeders are "selective agents" that promote reproduction of certain pigeons, agents of selection act on the range of variation in the wild. Among them are pigeon-eating peregrine falcons (Figure 1.9*b*). Swifter or better camouflaged pigeons are more likely to avoid falcons and live long enough to reproduce, compared with not-so-swift or too-flashy pigeons.

When different forms of a trait are becoming more or less common over successive generations, evolution is under way. In biology, evolution simply means change is occurring in a line of descent.

Individuals of a population show variation in heritable traits, which arises through mutations in DNA.

Because adaptive forms of traits tend to improve chances for survival and reproduction, they become more common in a population over successive generations.

Differential survival and reproduction among individuals of a population that differ in the details of one or more heritable traits is called natural selection.

In biology, evolution means change in a line of descent. Evolutionary processes and events underlie life's diversity.

Figure 1.9 (**a**) Outcome of artificial selection: a few of the hundreds of varieties of domesticated pigeons descended from captive populations of wild rock pigeons (*Columba livia*). (**b**) Peregrine falcons (*left*) prey on pigeons (*right*) and thus act as agents of natural selection in the wild.

1.5 Critical Thinking and Science

Earlier sections introduced some big concepts. Consider approaching these views of nature—or any others—with a critical attitude: "Why would I accept these views?"

THINKING ABOUT THINKING

Most of us assume that we do our own thinking—but do we, really? You might be surprised to find out just how often we allow others to think for us. For instance, a school's job, which is to impart as much information as possible to students, meshes with a student's job, which is to acquire as much knowledge as possible. In the rapid-fire exchange of information, it is all too easy to forget about the *quality* of what is being exchanged. Accept information without question, and you allow someone else to do your thinking for you.

Critical thinking means judging information before accepting it. "Critical" comes from the Greek *kriticos* (discerning judgment). When you think this way, you move beyond the content of new information. You are looking for underlying assumptions, evaluating the supporting statements, and thinking of alternatives (Table 1.2).

How does the busy student manage this? Be aware of what you intend to learn from new information. Be conscious of bias or underlying agendas in books or lectures. Consider your own biases—what you want to believe—and realize they influence your learning. Question authority figures. Decide whether ideas are based on opinion or evidence. Such practices will help you decide whether to accept or reject the information, or postpone your judgment about it.

Table 1.2 A Guide to Evidence-Based Thinking

Be able to state clearly your view on a subject.
Be aware of the evidence that led you to hold this view.
Ask yourself if there are alternative ways to interpret the evidence.
Think about the kind of information that might make you reconsider your view.
If you decide that nothing can ever persuade you to alter your view, recognize that you are not being objective about this subject.

THE SCOPE AND LIMITS OF SCIENCE

Because each of us is unique, there are as many ways to think about the natural world as there are people. Science, the systematic study of nature, is one way. It helps us be objective about our observations of nature, in part because of its limitations. We limit science to a subset of the world—*only that which is observable.*

Science does not address some questions, such as "Why do I exist?" Most answers to such questions are subjective; they come from within as an integration of the personal experiences and mental connections that shape our consciousness. This is not to say subjective answers have no value. No human society functions for very long unless its individuals share standards for making judgments, even if they are subjective. Moral, aesthetic, and philosophical standards vary from one society to the next, but all help people decide what is important and good. All give meaning to what we do.

Also, science does not address the supernatural, or anything that is "beyond nature." Science does not assume or deny that supernatural phenomena occur, but scientists may still cause controversy when they discover a natural explanation for something that was thought to be unexplainable. Such controversy often arises when a society's moral standards have become interwoven with traditional interpretations of nature.

As one example, centuries ago in Europe, Nikolaus Copernicus studied the planets and decided that Earth circles the sun. Today this seems obvious. Back then, it was heresy. The prevailing belief was that the Creator made Earth—and, by extension, humans—as the fixed center of the universe. Galileo Galilei, another scholar, found evidence for the Copernican model of the solar system and published his findings. He was publicly forced to put Earth back as the center of things.

Exploring a traditional view of the natural world from a scientific perspective might be misinterpreted as questioning morality even though the two are not the same. As a group, scientists are no less moral, less lawful, or less compassionate than anyone else. As you will see next, however, they follow a certain standard: *Explanations must be testable in the natural world in ways that others can repeat.*

Science helps us communicate experiences without bias; it may be as close as we can get to a universal language. We are fairly sure, for example, that laws of gravity apply everywhere in the universe. Intelligent beings on a distant planet would likely understand the concept of gravity. We might well use such concepts to communicate with them—or anyone—anywhere. The point of science, however, is not to communicate with aliens. It is to find common ground here on Earth.

Critical thinking means systematically judging the quality of information as you learn its content and implications.

Science looks for natural explanations of objects and events. It does not address the supernatural.

1.6 How Science Works

Scientists make potentially falsifiable predictions about how the natural world works. They search for evidence that may disprove or lend support to an explanation.

OBSERVATIONS, HYPOTHESES, AND TESTS

Science, again, is the systematic study of nature. To get a sense of how to do science, consider Table 1.3 and this list of practices, which are common in research:

1. Observe some aspect of nature.
2. Frame a question that relates to your observation.
3. Check to see what others have found out about the subject, then propose a hypothesis, a testable answer to your question.
4. Using the hypothesis as a guide, make a prediction: a statement of some condition that should exist if the hypothesis is not wrong. Making predictions is called the if–then process—with "if" being the hypothesis and "then" being the prediction. All predictions are potentially falsifiable, in that tests may disprove them.
5. Devise ways to test the accuracy of your prediction by making systematic observations or by conducting experiments. You may perform your tests on a model, an analogous system, if you are not able to observe or test an object or event directly.
6. Assess the results of your tests. Results that confirm your prediction are evidence—data—in support of the hypothesis. Results that disprove your prediction are evidence that the hypothesis may be flawed.
7. Report all the steps of your work, along with any conclusions you drew, to the scientific community.

Table 1.3 Example of a Scientific Approach to a Question

1. Observation	People get cancer.
2. Question	Why do people get cancer?
3. Hypothesis	Smoking cigarettes causes cancer.
4. Prediction	If smoking causes cancer, then individuals who smoke will get cancer more often than those who do not.
5. Observational test	Conduct a survey of individuals who smoke and individuals who do not smoke. Determine which group has the highest incidence of cancers.
6. Experimental test	Establish identical groups of laboratory rats. Expose one group (the model system) to cigarette smoke and compare the incidence of new cancers (if any) with the incidence in the control group.
7. Report	Report the test results, quantitatively if possible, and the conclusions drawn from them.

You might hear someone refer to these practices as "the scientific method," as if all scientists march to the drumbeat of a fixed procedure. They do not. There are different ways to do research, particularly in biology (Figure 1.10). Some biologists do surveys; they observe without making hypotheses. Others make hypotheses and leave tests to others. Some stumble onto valuable information they are not even looking for. Of course, it is not only a matter of luck. Chance favors a mind that is already prepared, by education and experience, to recognize what the new information might mean.

Regardless of the variation, one thing is constant: Scientists do not accept information simply because someone says it is the truth. They evaluate evidence, biases, and find potential alternatives. Does this sound familiar? It should—it is critical thinking.

ABOUT THE WORD "THEORY"

Most scientists carefully avoid the word "truth" when discussing science. Instead, they prefer to say that data either support or do not support a hypothesis.

Suppose a hypothesis still stands even after years of tests. It is consistent with all evidence gathered to date. It proves useful in helping us make predictions about other phenomena, and its predictive power has been tested many times. When any hypothesis meets these criteria, it becomes a scientific theory.

To give an example, observations for all of recorded history have favored the hypothesis that gravity pulls objects toward Earth. Scientists no longer spend time testing the hypothesis for the simple reason that, after many thousands of years of observation, no one has seen otherwise. This hypothesis is an accepted theory, but it is not an "absolute truth." Why not? An infinite number of tests would be necessary to confirm that it holds under every possible circumstance.

However, a single observation or result that is *not* consistent with a theory opens that theory to revision. If gravity does cause apples to fall down, it would be logical to predict that apples will fall down tomorrow. However, a scientist might well see tomorrow as an opportunity for the prediction to fail. Think about it. If even one apple falls up instead of down tomorrow, the theory of gravity would be re-evaluated. Like every other theory, this one remains open to revision.

A well-tested theory is as close to the "truth" as scientists will venture. Table 1.4 lists a few established theories. One of them, the theory of natural selection, holds after more than a century of testing. We cannot be sure that it will hold under all possible conditions. We can say it has a very high probability of not being

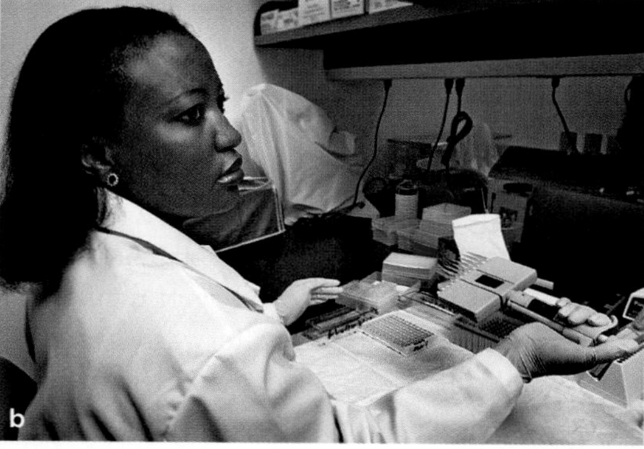
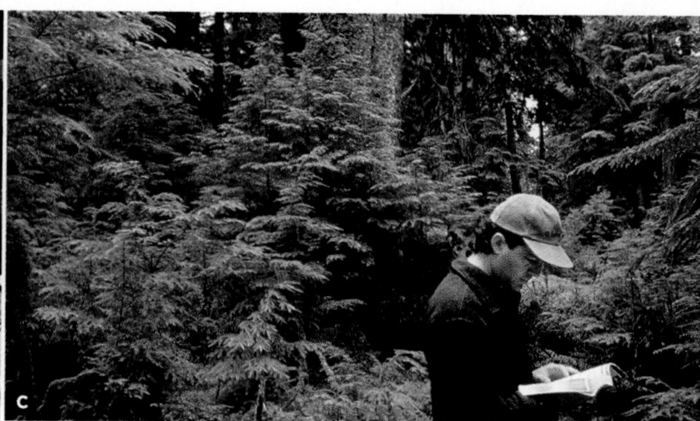

Figure 1.10 Scientists doing research in the laboratory and in the field. (**a**) Analyzing data with computers. (**b**) At the Centers for Disease Control, Mary Ari testing a sample for the presence of dangerous bacteria. (**c**) Making field observations in an old-growth forest.

wrong. In the future, if any evidence turns up that is inconsistent with the theory of natural selection, then biologists will revise it. Such a willingness to modify or discard even an entrenched theory is a strength of science, not a weakness.

You may hear people apply the word "theory" to a speculative idea, as in the phrase "It's just a theory." Speculation is opinion or belief, a personal conviction that is not necessarily supported by testable evidence. A scientific theory is not just an opinion. By definition, it must be supported by many different kinds of tests and have wide-ranging predictive power.

Unlike theories, many beliefs and opinions cannot be tested. Without being able to test something, there is no way to disprove it. Although personal conviction often has tremendous value in our lives, it should not be confused with scientific theory.

Table 1.4 Examples of Scientific Theories

Gravitational theory	Objects attract one another with a force that depends on their mass and how close together they are.
Cell theory	All organisms consist of one or more cells, the cell is the basic unit of life, and all cells arise from existing cells.
Germ theory	Germs cause infectious diseases.
Plate tectonics theory	Earth's crust is like a cracked eggshell, and its huge, fragmented slabs slowly collide and move apart.
Theory of evolution	Change can occur in lines of descent.
Theory of natural selection	Variation in heritable traits influences which individuals of a population reproduce in each generation.

SOME TERMS USED IN EXPERIMENTS

Careful observations are one way to test predictions that flow from a hypothesis. So are experiments. You will find examples of experiments in the next section. For now, just become acquainted with some important terms that researchers commonly use:

1. Experiments are tests designed to support or falsify a prediction.

2. Scientists simplify their observations by designing experiments to test one variable at a time. A variable is some characteristic or an event that differs among individuals or systems and that may change over time. Experimenters measure and manipulate variables.

3. Researchers design experiments to demonstrate the effects of a certain variable on an *experimental* group. Biological systems have so many interacting variables that it is often impossible to separate one from the rest. Instead, researchers test an experimental group side by side with a *control* group, which is identical to the experimental group except for the one variable being tested. The complexity of the two groups is the same, so presumably any differences in the results of the test on the two groups will be due to the variable alone.

Scientific inquiry involves asking questions about some aspect of nature, then formulating hypotheses, making and testing predictions, and reporting the results.

A scientific theory is a concept of cause and effect that is consistent with a large body of evidence, and is used to make useful predictions about other related phenomena.

Because we cannot prove a theory will hold under every possible condition, it is always open to tests and revision. The external world, not internal conviction, is the testing ground for scientific theories.

1.7 The Power of Experimental Tests

Natural processes often are interrelated. Researchers unravel how processes work together by studying one variable at a time. They design experiments to identify the function, cause, or effect of that variable in isolation. Here we summarize two published experiments.

POTATO CHIPS AND GAS

In 1996 the FDA approved Olestra®, a type of synthetic fat replacement made from sugar and vegetable oil, as a food additive. Potato chips were the first Olestra-laced food product on the market in the United States. Controversy soon raged. Some people complained of intestinal cramps after eating the chips and concluded that Olestra caused them. Two years later, researchers at Johns Hopkins University designed an experiment to test the hypothesis that this food additive can cause such a problem. They predicted that *if* Olestra causes cramps, *then* people who eat Olestra are more likely to get cramps than people who do not.

To test the prediction, they used a Chicago theater as the "laboratory." They asked more than 1,100 people between ages thirteen and thirty-eight to watch a movie and eat their fill of potato chips. Each person got an unmarked bag that contained 13 ounces of chips. The individuals who received a bag of Olestra-laced potato chips were the experimental group. Individuals who got a bag of regular chips were the control group.

Afterward, researchers contacted all of the people and tabulated the reports of gastrointestinal cramps. Of 563 people making up the experimental group, 89 (15.8 percent) complained about problems. However, so did 93 of the 529 people (17.6 percent) making up the control group—who had munched on regular chips! This simple experiment disproved the prediction that eating Olestra-laced potato chips at a single sitting can cause gastrointestinal cramps (Figure 1.11).

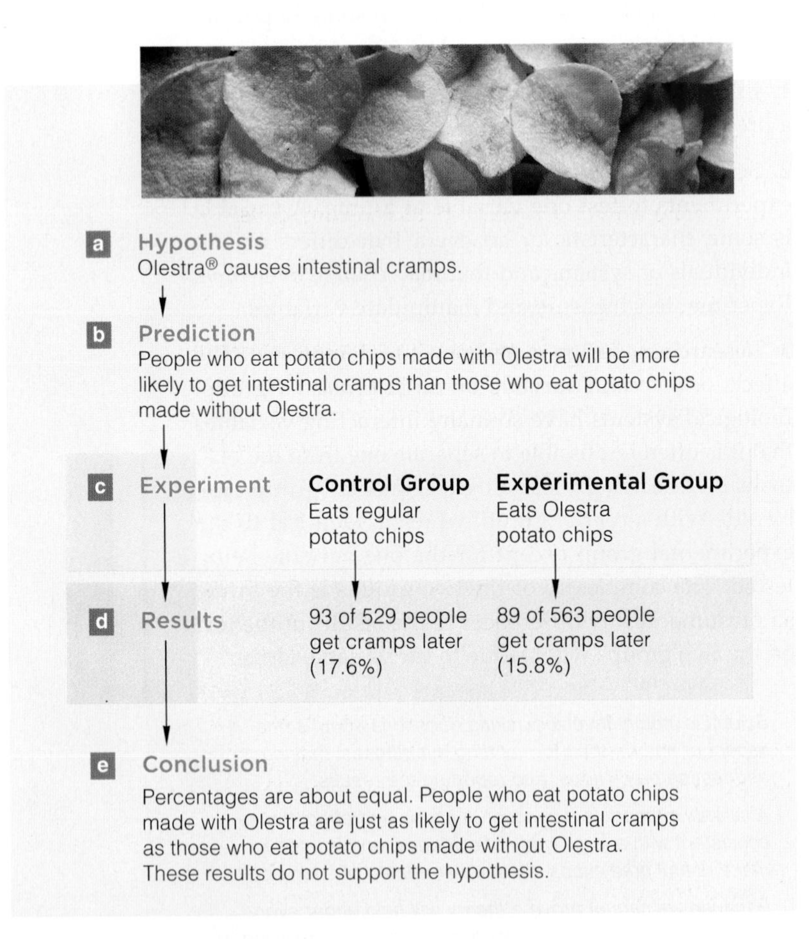

Figure 1.11 **Animated!** The steps in a scientific experiment to determine if Olestra causes cramps. A report of this study was published in the *Journal of the American Medical Association* in January 1998.

BUTTERFLIES AND BIRDS

Consider the peacock butterfly. This winged insect has a long life span, for a butterfly. It hibernates through cold winter months in protected spots. The longer life span gives butterfly-eating birds a bigger window of opportunity to eat individual butterflies. Do the birds act as selective agents for butterfly defenses? Probably.

In 2005, researchers published a report on their tests to identify factors that help peacock butterflies defend themselves against blue tits—small, insect-eating birds that commonly prey on butterflies. Follow the thought process that led to the experimental design.

The researchers made two key observations. First, when a peacock butterfly rests, it folds its ragged-edged wings, so only the dark underside shows (Figure 1.12*a*). Second, when a butterfly sees a predator approaching, it repeatedly flicks its paired forewings and hindwings wide open, then closes them. At the same time, each forewing slides over the hindwing, which produces a hissing sound and a series of clicks.

The researchers asked this question, "Why does the peacock butterfly flick its wings?" After they reviewed earlier studies, they formulated three hypotheses that might explain the wing-flicking behavior:

1. When folded, the butterfly wings resemble a dead leaf. They may camouflage the butterfly—help it hide in the open—from some predators in its forest habitat.

2. Although wing-flicking attracts birds, opening the wings exposes brilliant spots that resemble owl eyes (Figure 1.12*b*). Anything that looks like the eyes of an owl is known to startle small, butterfly-eating birds, so flicking wing spots might scare off predators.

a

b

c

Figure 1.12 Peacock butterfly defenses against predatory birds. (**a**) With wings folded, a resting peacock butterfly looks like a dead leaf. (**b**) When a bird approaches, the butterfly repeatedly flicks its wings open and closed. This defensive behavior exposes brilliant spots. It also produces hissing and clicking sounds.

Researchers tested whether the behavior deters blue tits (**c**). They painted over the spots of some butterflies, cut the sound-making part of the wings on other butterflies, and did both to a third group; then the biologists exposed each butterfly to a hungry bird.

The results, listed in Table 1.5, support the hypotheses that peacock butterfly spots and sounds can deter predatory birds. The study was reported in *Proceedings of the Royal Society (B)* in June 2005.

Table 1.5 Results for Peacock Butterfly Experiment

Wing Spots Painted Out	Wing Sound Silenced	Number of Survivors	Number Eaten	Survival Rate (percent)
No	No	9	0	100
Yes	No	5	5	50
No	Yes	8	0	100
Yes	Yes	2	8	20

3. The hissing and clicking sounds produced when the peacock butterfly rubs the sections of its wings together deter predatory birds.

The researchers decided to test hypotheses 2 and 3. They made the following predictions:

1. *If* the brilliant wing spots of peacock butterflies deter predatory birds, *then* individuals having wings with no spots will be more likely to get eaten by predatory birds than individuals with wing spots.

2. *If* the sounds that peacock butterflies produce deter predatory birds, *then* individuals that cannot make the sounds will be more likely to be eaten by predatory birds than individuals that can make the sounds.

The next step was the experiment. The researchers painted the wing spots of some butterflies black, cut off the sound-making part of the hindwings of others, and did both to a third group. They put each butterfly in a large cage with a hungry blue tit (Figure 1.12*c*) and then watched the pair for thirty minutes.

Table 1.5 lists the results of the experiment. All of the butterflies with unmodified wing spots survived, regardless of whether they made sounds. By contrast, only half of the butterflies that had spots painted out but could make sounds survived. Most butterflies with neither spots nor sound structures were eaten.

The test results confirmed both predictions, so they support the hypotheses. Birds are deterred by peacock butterfly sounds, and even more so by wing spots.

ASKING USEFUL QUESTIONS

Experimenters risk interpreting their results in terms of what they want to find out. That is why they often design experiments to yield *quantitative* results, which are counts or some other data that can be measured or gathered objectively. Such results give other scientists an opportunity to repeat the experiments and check the conclusions drawn from them.

This last point gets us back to the value of thinking critically. Scientists expect one another to put aside bias and test hypotheses in ways that may prove them wrong. If some individual will not do so, others will—because science is a competitive community. It is also cooperative. Scientists share ideas, knowing it is just as useful to expose errors as to applaud insights.

Scientific experiments can simplify the study of a complex natural process by restricting the researcher's focus to a single aspect of that process.

Researchers try to design experiments carefully in order to minimize the potential for bias.

1.8 Sampling Error in Experiments

FOCUS ON SCIENCE

In most cases, experiments cannot be performed on all individuals of a group or in each part of the places where organisms live. Researchers generalize from samplings—which opens the door for mistakes.

Rarely can researchers observe all individuals of a group. For example, remember the explorers you read about in the chapter introduction? They could not sample the entire rain forest, which cloaks more than 2 million acres of New Guinea's Foja Mountains. Doing so would take unrealistic amounts of time and effort. Besides, tromping about even in a small area can damage forest ecosystems.

Given such constraints, researchers tend to experiment on subsets of a population, event, or some other aspect of nature that they select to represent the whole. They test the subsets and use the results to make generalizations about the whole population.

Suppose they design an experiment to identify variables that influence the population growth of golden-mantled tree kangaroos. They might focus only on the population living in one acre of the Foja Mountains. If they identify only 5 golden-mantled tree kangaroos in that specified area, then they might extrapolate that there are 50 in every ten acres, 100 in every twenty acres, and so forth.

However, generalizing from a subset can be risky: The subset may not be representative of the whole. If the only population of golden-mantled tree kangaroos in the forest just happens to be living in the surveyed acre, then the researchers' assumptions about the number of kangaroos in the rest of the forest will be wrong.

Sampling error is a difference between results from a subset and results from the whole. It happens most often when sample sizes are small. Starting with a large sample or repeating the experiment many times helps minimize sampling error (Figure 1.13). To understand why, imagine flipping a coin. There are two possible results: The coin lands heads up, or it lands tails up. You might predict that the coin will land heads up as often as it lands tails up. When you actually flip the coin, though, often it will land heads up, or tails up, several times in a row. If you flip the coin only a few times, the results may differ greatly from your prediction. Flip it many times, and you probably will come closer to having equal numbers of heads and tails.

Sampling error is an important consideration in the design of most, if not all, experiments. The possibility that it occurred should be part of the critical thinking process as you read about experiments. Remember to ask: If the experimenters used a subset of the whole, did they select a large enough sample? Did they repeat the experiment many times? Thinking about these possibilities will help you evaluate the results and conclusions reached.

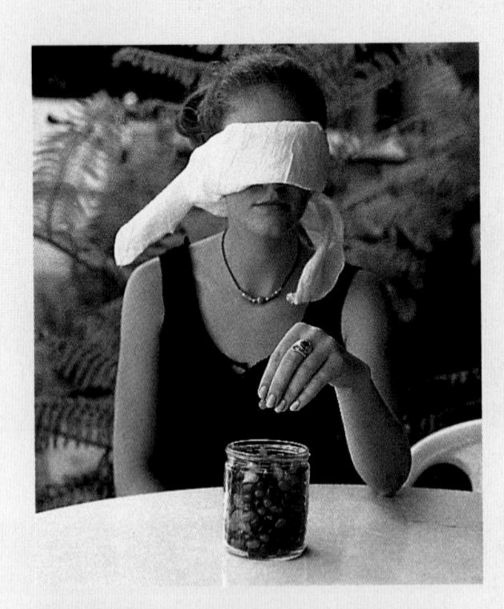

a Natalie, blindfolded, randomly plucks a jelly bean from a jar. There are 120 green and 280 black jelly beans in that jar, so 30 percent of the jelly beans in the jar are green, and 70 percent are black.

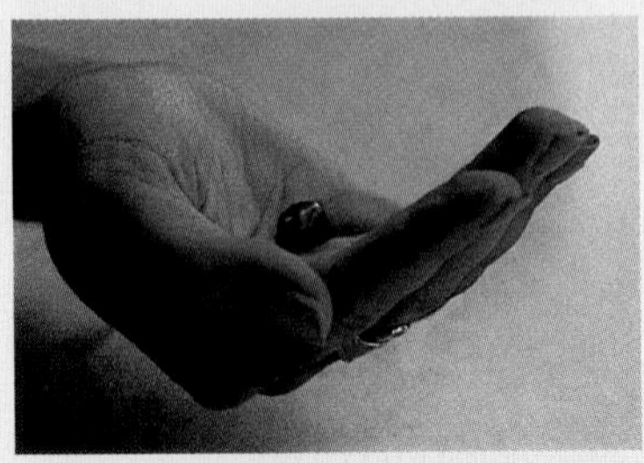

b The jar is hidden from Natalie's view before she removes her blindfold. She sees only one green jelly bean in her hand and assumes that the jar must hold only green jelly beans.

c Still blindfolded, Natalie randomly picks out 50 jelly beans from the jar and ends up with 10 green and 40 black ones.

d The larger sample leads Natalie to assume that one-fifth of the jar's jelly beans are green (20 percent) and four-fifths are black (80 percent). The sample more closely approximates the jar's actual green-to-black ratio of 30 percent to 70 percent. The more times Natalie repeats the sampling, the greater the chance she will come close to knowing the actual ratio.

Figure 1.13 **Animated!** Demonstration of sampling error.

www.thomsonedu.com Thomson NOW!

Summary

Section 1.1 Nature has levels of organization, and unique properties emerge at successively higher levels. Life emerges at the cellular level. All organisms consist of one or more cells. Most multicelled species have cells organized into tissues, organs, and organ systems. A population is a group of all individuals of one species in a specified area; a community consists of all populations in a specified area. An ecosystem is a community interacting with its environment. The biosphere includes all regions of Earth that hold life—land, water, and atmosphere.

- *Explore levels of biological organization with the interaction on ThomsonNOW.*

Section 1.2 The world of life shows underlying unity (Table 1.6). All organisms require inputs of energy and materials, which sustain their organization and activities. Organisms sense change. Their responses keep conditions in the internal environment within ranges that cells can tolerate, a state called homeostasis. Organisms also grow and reproduce, based on information encoded in DNA.

- *Use instructions with the animation on ThomsonNOW to see how different objects are assembled from the same materials. Also view energy flow and materials cycling.*

Section 1.3 The world of life, past and present, shows great diversity. Classification systems organize species in ever more inclusive groups. Each species has a two-part name. The first part is the genus name. When combined with the second part, it designates one particular species. A species is one kind of organism. A current classification system groups all species into three domains: Bacteria, Archaea, and Eukarya. Eukarya includes protists, plants, fungi, and animals.

- *Use the interaction on ThomsonNOW to explore characteristics of the three domains of life.*

Section 1.4 Life's diversity arises as an outcome of mutations. Mutations are changes in molecules of DNA, which offspring inherit from their parents. In natural populations, mutations introduce variation in the details of heritable traits among individuals (Table 1.6).

Some forms of traits are more adaptive than others, so their bearers are more likely to survive and reproduce. Over generations, adaptive forms of traits tend to become more common in a population; less adaptive forms of the same traits become less common or are lost.

Thus, traits that help characterize a population (and a species) can change over the generations; the population can evolve. In biology, evolution means that change is occurring in a line of descent.

For natural populations, the differential survival and reproduction among individuals that vary in the details of one or more heritable traits is called natural selection.

- *Learn more about natural selection and evolution with InfoTrac readings on ThomsonNOW.*
- *Read the InfoTrac article "Will We Keep Evolving?" Ian Tattersall,* Time, *April 2000.*

Section 1.5 Critical thinking is a self-directed act of judging the quality of information as one learns.

Science is one way of looking at the natural world. It helps us minimize bias in our judgments by focusing on only testable ideas about observable aspects of nature.

Section 1.6 Scientific methods differ, but researchers generally observe something in nature, form hypotheses (testable assumptions) about it, then make predictions about what might occur if the hypothesis is not wrong. They test their predictions by observations, experiments, or both. A hypothesis that is not consistent with results of scientific tests (evidence) is modified or discarded.

Each scientific theory is a well-tested hypothesis that explains a broad range of observations and can be used to make useful predictions about other phenomena. Opinion and belief have value in human culture, but neither can be disproved by experiment. Thus, opinion and belief are different from scientific theory.

- *See an annotated scientific paper in Appendix II.*

Section 1.7 Biological systems are usually influenced by many interacting variables. Scientific experiments can simplify observations of nature by focusing on the cause, effect, or function of one variable at a time. Researchers design experiments carefully to minimize potential bias in interpreting the results.

Section 1.8 Small sample size increases the likelihood of sampling error in experiments. In such cases, a subset may be tested that is not representative of the whole.

Table 1.6 Summary of Life's Characteristics

Shared characteristics that reflect life's unity

1. In nature's great pattern of organization, the quality of life emerges at the level of cells. All organisms consist of one or more cells.
2. Organisms make the molecules of life: complex carbohydrates and lipids, proteins, and nucleic acids (DNA and RNA).
3. Ongoing inputs of energy and nutrients sustain the organization, growth, survival, and reproduction of all organisms.
4. Organisms sense and respond to changing conditions in ways that maintain homeostasis; they keep their internal environment within a range that favors cell survival.
5. Organisms grow and reproduce based on heritable information encoded in DNA.
6. The traits that characterize a population of organisms can change over the generations; the population can evolve.

Foundations for life's diversity

1. Mutations (heritable changes in DNA) give rise to variation in details of body form, the functioning of body parts, and behavior.
2. Diversity is the sum total of variations that have accumulated, since the time of life's origin, in different lines of descent. It is an outcome of natural selection and other processes of evolution.

Self-Quiz

Answers in Appendix III

1. The smallest unit of life is the ______ .

2. ______ and ______ are required to maintain levels of biological organization, from cells to populations and communities, even entire ecosystems.

3. ______ is a state in which conditions in the internal environment are being maintained within ranges that individual cells can tolerate.

4. Bacteria, Archaea, and Eukarya are three ______ .

5. DNA ______ .
   a. contains instructions for building proteins
   b. undergoes mutation
   c. is transmitted from parents to offspring
   d. all of the above

6. ______ is the transmission of DNA to offspring.
   a. Reproduction
   b. Development
   c. Homeostasis
   d. Inheritance

7. ______ are the original source of variation in traits.

8. A trait is ______ if it improves an organism's chances to survive and reproduce in its environment.

9. A control group is ______ .
   a. the standard against which experimental groups can be compared
   b. the experiment that gives conclusive results
   c. both a and b

10. Match the terms with the most suitable description.

___ emergent property	a. statement of what a hypothesis leads you to expect to see in nature
___ natural selection	b. testable explanation
___ scientific theory	c. occurs at a higher organizational level in nature, not at levels below it
___ hypothesis	d. time-tested hypothesis that can explain a range of observations
___ prediction	e. differential survival and reproduction among individuals of a population that vary in details of shared traits

■ *Visit ThomsonNOW for additional questions.*

Critical Thinking

1. It is often said that only living things respond to the environment. Yet even a rock shows responsiveness, as when it yields to gravity's force and tumbles down a hill or changes its shape slowly under the repeated batterings of wind, rain, or tides. So how do living things differ from rocks in their responsiveness?

2. Why would you think twice about ordering from a cafe menu that lists only the second part of the species name (not the genus) of its offerings? *Hint:* Look up *Ursus americanus*, *Ceanothus americanus*, *Bufus americanus*, and *Lepus americanus*.

3. Witnesses in a court of law are asked to "swear to tell the truth, the whole truth, and nothing but the truth." Can you think of a less subjective alternative for this oath?

4. Procter & Gamble makes Olestra and financed the study described in Section 1.7. The main researcher, Lawrence Cheskin of Johns Hopkins University, was a consultant to Procter & Gamble during the study. What do you think about scientific information that comes from tests financed by companies with a vested interest in the outcome?

5. Suppose an outcome of some event has been observed to happen with great regularity. Can we predict that the same thing will always happen? Not really, because there is no way to account for all of the possible variables that might affect the outcome. To illustrate this point, Garvin McCain and Erwin Segal offer a parable:

Once there was a highly intelligent turkey. The turkey lived in a pen, attended by a kind, thoughtful master. It had nothing to do but reflect on the world's wonders and regularities. It observed some major regularities. Morning always started out with the sky turning light, followed by the clop, clop, clop of the master's footsteps, which was always followed by the appearance of food. Other things varied—sometimes the morning was warm and sometimes cold—but food always followed footsteps. The sequence of events was so predictable that it eventually became the basis of the turkey's theory about the goodness of the world. One morning, after more than 100 confirmations of the goodness theory, the turkey listened for the clop, clop, clop, heard it, and had its head chopped off.

Any scientific theory is modified or discarded when contradictory evidence becomes available. The absence of absolute certainty has led some people to conclude that "facts are irrelevant—facts change." If that is so, should we just stop doing scientific research? Why or why not?

6. In 2005 a South Korean scientist, Woo-suk Hwang, reported that he made immortal stem cells from eleven human patients. His research was hailed as a breakthrough for people affected by currently incurable degenerative diseases, because such stem cells might be used to repair a person's own damaged tissues. Hwang published his results in a respected scientific journal. In 2006, the journal retracted his paper after other scientists discovered that Hwang and his colleagues had faked their results. Some people think this incident shows that scientists are not telling the truth about the natural world. However, others think that the incident helps confirm the usefulness of a scientific approach, because other scientists quickly discovered and exposed the fraud. What do you think?

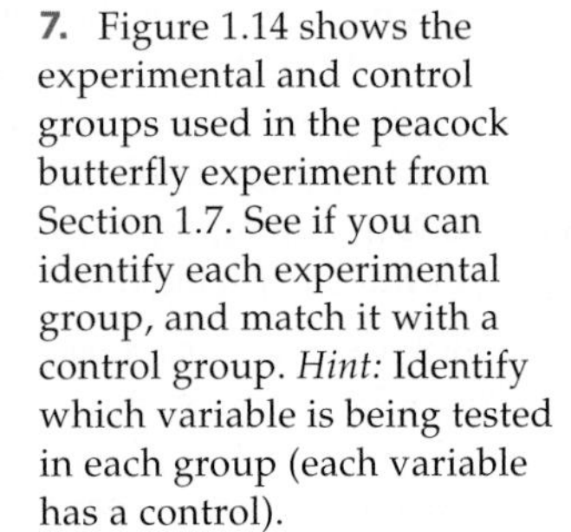
7. Figure 1.14 shows the experimental and control groups used in the peacock butterfly experiment from Section 1.7. See if you can identify each experimental group, and match it with a control group. *Hint:* Identify which variable is being tested in each group (each variable has a control).

a Wing spots painted out

b Wing spots visible; wings silenced

c Wing spots painted out; wings silenced

d Wings painted but spots visible

e Wings cut but not silenced

f Wings painted but spots visible; wings cut but not silenced

Figure 1.14 Experimental peacock butterflies modified with a black marker pen and scissors.

I Principles of Cellular Life

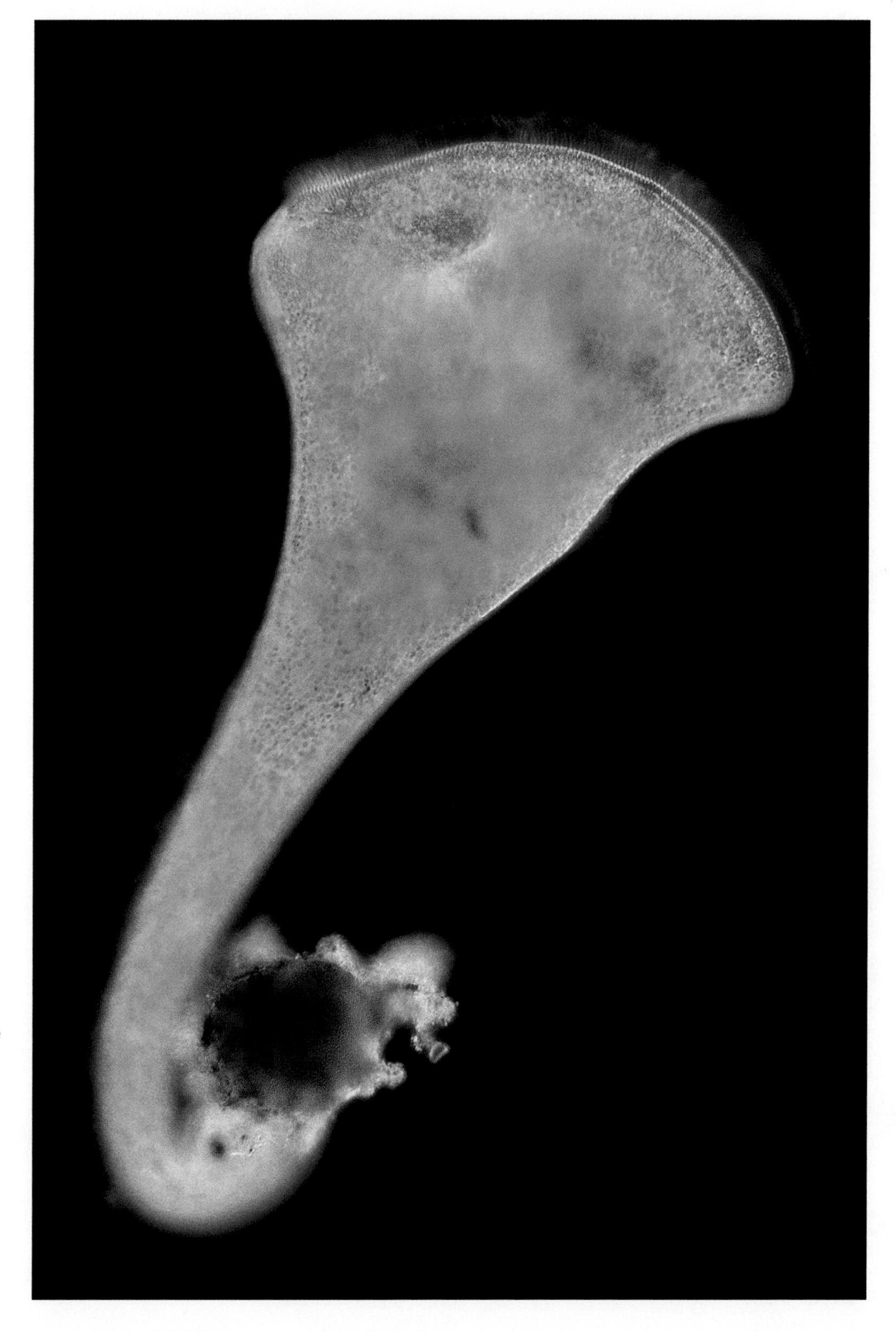

Staying alive means securing energy and raw materials from the environment. Shown here, a living cell of the genus *Stentor*. This protist has hairlike projections around an opening to a cavity in its body, which is about 2 millimeters long. Its "hairs" of fused-together cilia beat the surrounding water. They create a current that wafts food into the cavity.

2 LIFE'S CHEMICAL BASIS

IMPACTS, ISSUES What Are You Worth?

Hollywood thinks actress Julia Roberts is worth $20 million per movie, the Yankees think shortstop Alex Rodriguez is worth $252 million per decade, and the United States thinks the average public school teacher is worth $46,597 a year. How much is one human body really worth? You can buy the entire collection of ingredients that make up an average seventy-kilogram body for $118.63 (Figure 2.1). Of course, all you have to do is watch Julia, Alex, or any teacher for a bit to know that a human body is far more than a list of its ingredients. What makes us worth more than the sum of our parts?

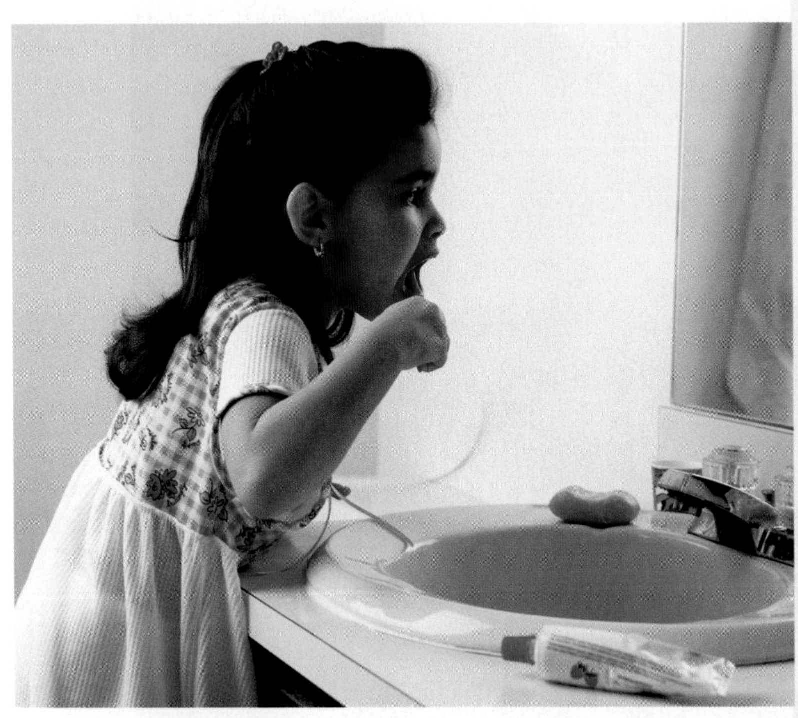

Elements in a Human Body

Element	Number of Atoms ($\times 10^{15}$)	Retail Cost
Hydrogen	41,808,044,129,611	$ 0.028315
Oxygen	16,179,356,725,877	0.021739
Carbon	8,019,515,931,628	6.400000
Nitrogen	773,627,553,592	9.706929
Phosphorus	151,599,284,310	68.198594
Calcium	150,207,096,162	15.500000
Sulfur	26,283,290,713	0.011623
Sodium	26,185,559,925	2.287748
Potassium	21,555,924,426	4.098737
Chlorine	16,301,156,188	1.409496
Magnesium	4,706,027,566	0.444909
Fluorine	823,858,713	7.917263
Iron	452,753,156	0.054600
Silicon	214,345,481	0.370000
Zinc	211,744,915	0.088090
Rubidium	47,896,401	1.087153
Strontium	21,985,848	0.177237
Bromine	19,588,506	0.012858
Boron	10,023,125	0.002172
Copper	6,820,886	0.012961
Lithium	6,071,171	0.024233
Lead	3,486,486	0.003960
Cadmium	2,677,674	0.010136
Titanium	2,515,303	0.010920
Cerium	1,718,576	0.043120
Chromium	1,620,894	0.003402
Nickel	1,538,503	0.031320
Manganese	1,314,936	0.001526
Selenium	1,143,617	0.037949
Tin	1,014,236	0.005387
Iodine	948,745	0.094184
Arsenic	562,455	0.023576
Germanium	414,543	0.130435
Molybdenum	313,738	0.001260
Cobalt	306,449	0.001509
Cesium	271,772	0.000016
Mercury	180,069	0.004718
Silver	111,618	0.013600
Antimony	98,883	0.000243
Niobium	97,195	0.000624
Barium	96,441	0.028776
Gallium	60,439	0.003367
Yttrium	40,627	0.005232
Lanthanum	34,671	0.000566
Tellurium	33,025	0.000722
Scandium	26,782	0.058160
Beryllium	24,047	0.000218
Indium	20,972	0.000600
Thallium	14,727	0.000894
Bismuth	14,403	0.000119
Vanadium	12,999	0.000322
Tantalum	6,654	0.001631
Zirconium	6,599	0.000830
Gold	6,113	0.001975
Samarium	2,002	0.000118
Tungsten	655	0.000007
Thorium	3	0.004948
Uranium	3	0.000103
Total	67,179,218,505,055 $\times 10^{15}$	**$118.63**

See the video! **Figure 2.1** Composition of an average-sized adult human body, by weight and retail cost. Manufacturers commonly add fluoride to toothpaste. Fluoride is a form of fluorine, one of several elements with vital functions—but only in trace amounts. Too much can be toxic.

How would you vote? Fluoride helps prevent tooth decay, but too much wrecks bones and teeth, and causes birth defects. A lot can kill you. Many communities in the United States add fluoride to drinking water. Do you want it in yours? See ThomsonNOW for details, then vote online.

Human		Earth		Seawater	
Hydrogen	62.0%	Hydrogen	3.1%	Hydrogen	66.0%
Oxygen	24.0	Oxygen	60.0	Oxygen	33.0
Carbon	12.0	Carbon	0.3	Carbon	< 0.1
Nitrogen	1.2	Nitrogen	< 0.1	Nitrogen	< 0.1
Phosphorus	0.2	Phosphorus	< 0.1	Phosphorus	< 0.1
Calcium	0.2	Calcium	2.6	Calcium	< 0.1
Sodium	< 0.1	Sodium	< 0.1	Sodium	0.3
Potassium	< 0.1	Potassium	0.8	Potassium	< 0.1
Chlorine	< 0.1	Chlorine	< 0.1	Chlorine	0.3

Figure 2.2 Comparison of the abundance of some elements in a human, Earth's crust, and typical seawater. Each number is the percent of the total number of atoms in each source. For instance, 120 of every 1,000 atoms in a human body are carbon, compared with only 3 carbon atoms in every 1,000 atoms of dirt.

The pure substances listed in Figure 2.1 are known as elements, each of which consists entirely of the same kind of atom. You will find the same elements that make up the human body in, say, dirt or seawater. However, the proportions of those elements differ between living and nonliving things (Figure 2.2). For example, you contain far more carbon. Rocks and seawater have no more than a trace of it.

This chapter will start you thinking about elements that are the building blocks for life in general. Foremost among them are oxygen, carbon, hydrogen, nitrogen, and calcium. Next are phosphorus, potassium, sulfur, sodium, and chlorine. Also necessary are many trace elements, each making up less than 0.01 percent of the body's weight.

The chapter also invites you to look closely at another big difference. Living and nonliving things have the same kinds of atoms joined together as molecules, but those molecules differ in their proportions of elements and in how the atoms of those elements are arranged.

We are only starting to understand the processes by which a collection of elements becomes assembled as a living body. We do know that life's unique organization starts with the properties of atoms that make up certain elements. This is your chemistry. It makes you far more than the sum of your body's ingredients—a handful of lifeless chemicals.

Key Concepts

ATOMS AND ELEMENTS

Atoms are fundamental units of all matter. Protons, electrons, and neutrons are their building blocks. Elements are pure substances, each consisting entirely of atoms that have the same number of protons. Isotopes are atoms of the same element that have different numbers of neutrons. **Sections 2.1, 2.2**

WHY ELECTRONS MATTER

Whether one atom will bond with others depends on the number and arrangement of its electrons. **Section 2.3**

ATOMS BOND

Atoms of many elements interact by acquiring, sharing, and giving up electrons. Ionic, covalent, and hydrogen bonds are the main interactions between atoms in biological molecules. **Section 2.4**

NO WATER, NO LIFE

Life originated in water and is adapted to its properties. It has temperature-stabilizing effects, cohesion, and a capacity to act as a solvent for so many other substances. These properties make life possible on Earth. **Section 2.5**

HYDROGEN IONS RULE

Life is responsive to changes in the amounts of hydrogen ions and other substances dissolved in water. **Section 2.6**

Links to Earlier Concepts

With this chapter, we turn to the first of life's levels of organization, so take a moment to review Section 1.1. It all starts with atoms and energy. Life's organization requires continuous inputs of energy (1.2), and organisms store that energy in bonds between atoms. You will come across a simple example of how the body's built-in mechanisms help return the internal environment to homeostasis when conditions shift beyond ranges that cells can tolerate (1.2). You will also come across examples of how scientists have made major discoveries (1.6).

2.1 Start With Atoms

Know a bit about the structure of atoms, and you have a clue to why the elements that make up living things behave as they do.

CHARACTERISTICS OF ATOMS

LINKS TO SECTIONS 1.1, 1.6

Atoms are particles that are the building blocks of all substances. They are less than one-billionth of a meter wide. Each consists of smaller subatomic particles—protons, electrons, and neutrons. The first two carry a charge, an electrical property that attracts or repels other subatomic particles. Protons (p^+) carry a positive charge, and electrons (e^-) carry a negative charge. The third kind, neutrons, have no charge.

Atoms differ in the number of subatomic particles, but all have a nucleus: a core of one or more protons and (except for hydrogen) neutrons. Electrons occupy defined spaces around the nucleus (Figure 2.3).

Elements are pure substances; each consists only of atoms with the same number of protons. That number defines the element, and we call it the atomic number. For example, a chunk of carbon contains only carbon atoms, all of which have six protons in their nucleus. The atomic number of carbon is 6. All atoms with six protons in their nucleus are carbon atoms, no matter how many electrons or neutrons they have.

All atoms of an element that differ in their number of neutrons are called isotopes. We define isotopes by their mass number, the total number of protons and neutrons in their nucleus. All elements have isotopes, which are identified by a superscript number to the left of an element's symbol. For instance, carbon's most common isotope is ^{12}C (six protons and six neutrons). Another is ^{13}C (six protons, seven neutrons).

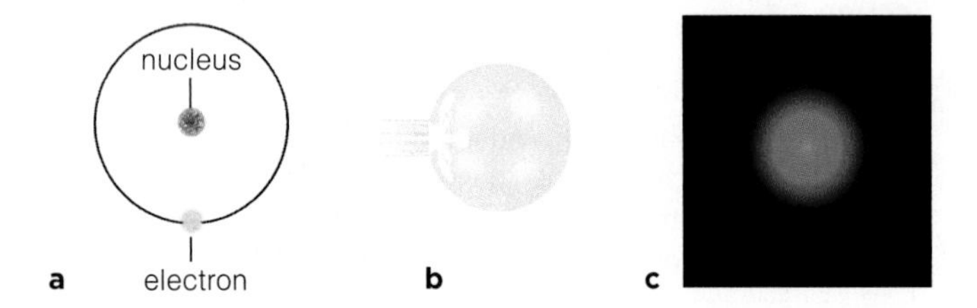

Figure 2.3 Representing atoms. (**a**) Shell models show the number of electrons. (**b**) Ball-and-stick models show sizes of atoms relative to one another. (**c**) A cloud of variable density depicts the volume of space around the nucleus where an electron is most likely to be. Regarding the scale, if electrons were as big as apples, you would be 3.5 times taller than our solar system is wide.

THE PERIODIC TABLE

Knowing about the numbers of electrons, protons, and neutrons helps us predict how elements will behave. Scientists started to classify elements on the basis of chemical behavior before subatomic particles were discovered. In 1869, Dmitry Mendeleev arranged the known elements into a table based on their chemical properties. He had constructed a periodic table of the elements. Until he came up with the table, Mendeleev was known mainly for his extravagant hair; he cut it only once a year.

In the periodic table, the symbol for each element is an abbreviation of its name. Elements are ordered by atomic number (Figure 2.4). Those in each vertical column behave in similar ways. For instance, all of the elements in the far right column of the table are inert gases; they do not interact with other atoms. In nature, such elements occur only as solitary atoms.

We can find the first ninety-four elements in nature. The others are so unstable that we see them only by making one atom at a time, and even then they wink out of existence fast. An atomic nucleus is not altered by heat or other ordinary means, so nuclear physicists are the only people who can make such elements.

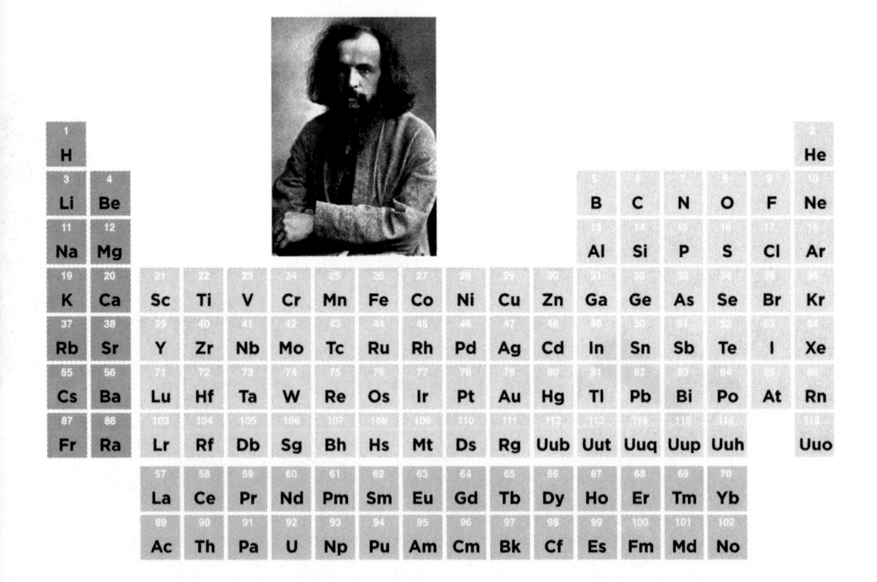

Figure 2.4 Periodic table of the elements shown with Dmitry Mendeleev, who created it. Some symbols for elements are abbreviations for their Latin names. For instance, Pb (lead) is short for *plumbum;* the word "plumbing" is related—ancient Romans made their water pipes with lead. Appendix IV has a more detailed table.

Atoms are the basic building blocks of all substances. Each has a nucleus of one or more protons and (except for hydrogen) neutrons. Each also has one or more electrons that occupy defined spaces around the nucleus.

An element is a pure substance. Each consists of atoms that all have the same number of protons.

2.2 Putting Radioisotopes To Use

FOCUS ON SCIENCE

The number of protons in the atomic nucleus defines the element, and the number of neutrons defines the isotope. We use some of the radioactive isotopes in research and in medical applications.

In 1896, Henri Becquerel made a chance discovery. He left some crystals of a uranium salt in a desk drawer, on top of a metal screen. Under the screen was a photographic plate wrapped tightly in black paper. Becquerel developed the film a few days later and was surprised to see a negative image of the screen. He realized that "invisible radiations" coming from the uranium salts had passed through the paper and exposed the film around the screen.

Becquerel's images were evidence that uranium has radioisotopes, or radioactive isotopes. So do many other elements. The atoms of radioisotopes spontaneously emit subatomic particles or energy when their nucleus breaks down. This process, radioactive decay, can transform one element into another. For example, ^{14}C is a radioisotope of carbon. It decays when one of its neutrons spontaneously splits into a proton and an electron. Its nucleus emits the electron, and so an atom of ^{14}C (with eight neutrons and six protons) becomes an atom of ^{14}N (nitrogen 14, with seven neutrons and seven protons).

Radioactive decay occurs independently of external factors such as temperature, pressure, or whether the atoms are part of molecules. A radioisotope always decays at a constant rate into the same products. For example, after 5,730 years, half of the atoms in any sample of ^{14}C will be ^{14}N atoms. Researchers use this predictability to estimate the age of rocks and fossils by their radioisotope content. We return to this topic in Section 16.5.

Researchers and clinicians also use radioisotopes in living organisms. Remember, isotopes are atoms of the same element. An isotope of, say, carbon generally has the same properties regardless of how many neutrons it has. Organisms use atoms of ^{14}C in metabolic reactions, the same way that they use atoms of ^{12}C. The consistent chemical behavior of isotopes allows researchers to use radioisotopes in tracers.

LINK TO SECTION 1.6

A tracer is any molecule with a detectable substance attached. Typically, a radioactive tracer is a molecule in which radioisotopes have been swapped for one or more atoms. Radioactive tracers are delivered into a biological system such as a cell or a multicelled body. Instruments that can detect radioactivity let researchers follow the tracer as it moves through the system.

For example, Melvin Calvin and his colleagues used a radioactive tracer to identify specific reaction steps of photosynthesis. These researchers let growing plants take up a radioactive gas—carbon dioxide that had been made with ^{14}C. Using instruments that detected the radioactive decay of ^{14}C, they tracked carbon through steps by which plants make simple sugars and starches.

Here is an example of how radioisotopes have uses in medicine. PET (short for *P*ositron-*E*mission *T*omography) helps us "see" cell activity. By this procedure, a tracer, such as a radioactive sugar, is injected into a patient, who is then moved into a PET scanner (Figure 2.5*a*). Cells in different parts of the patient's body take up the tracer at different rates. The scanner detects radioactive decay wherever the tracer is, then translates that radiation into an image on a computer monitor, as in Figure 2.5*d*. Such images can reveal subtle variations and abnormalities in cell activity.

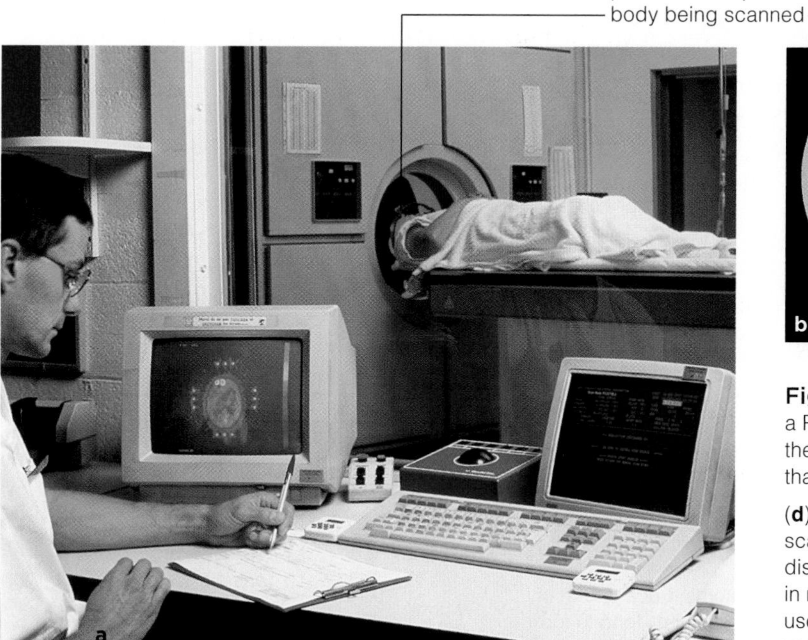

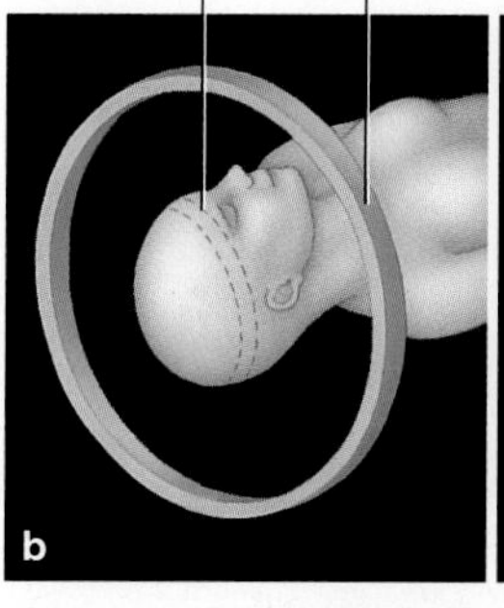

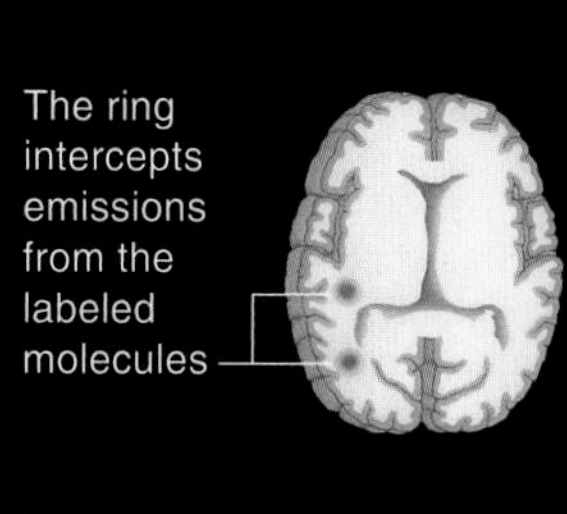

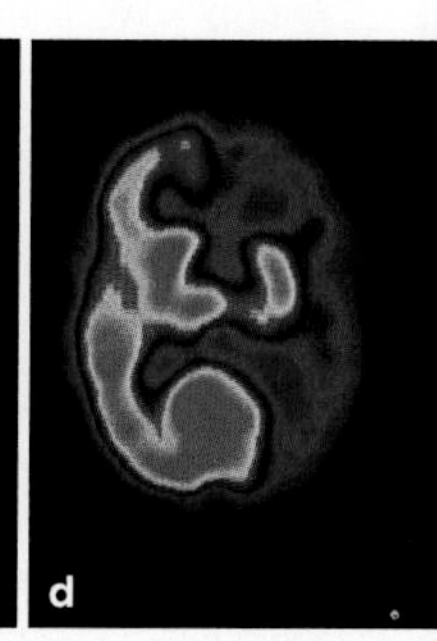

Figure 2.5 Animated! (**a**) Patient whose brain is being examined in a PET scanner. (**b,c**) The patient is injected with a tracer and placed into the scanner so that the body part of interest is surrounded by detectors that can intercept radioactive emissions.

(**d**) Computers analyze the number of emissions from each location in the scanned region. Results are converted into color-coded digital images and displayed on computer screens. Different colors in a scan signify differences in metabolic activity. Cells in the left half of this person's brain absorbed and used labeled molecules at expected rates, but cells in the right half showed very little activity. This particular patient has a neurological disorder.

2.3 Why Electrons Matter

Atoms acquire, share, and donate electrons. The atoms of some elements do so quite easily, and others do not. Why? To come up with a possible explanation, start with the number and arrangement of electrons in atoms.

ELECTRONS AND ENERGY LEVELS

LINK TO SECTION 1.1

In our world, simple physics explains the motion of, say, an apple falling from a tree. Electrons belong to a strange world where everyday physics does not apply. Different forces govern electrons, which somehow get from one place to another without going in between!

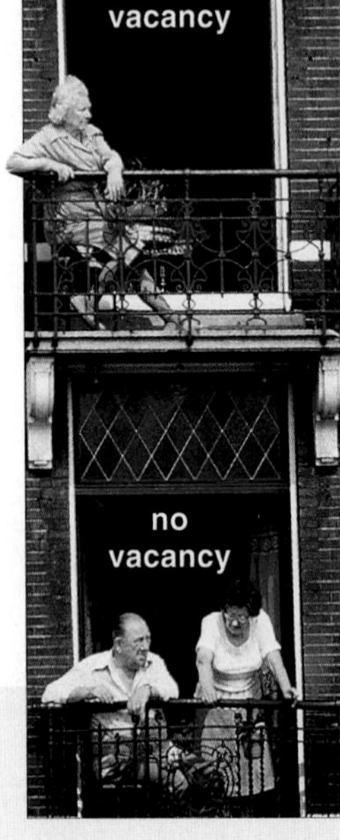

In general, atoms have about the same number of electrons as protons. Most have many electrons. How are all of the electrons arranged around a single nucleus, given that they repel each other? The answer is that they travel in different orbitals, which are defined volumes of space around the atomic nucleus.

Each atom is like a multilevel apartment building, with many rooms available to rent to electrons and a nucleus in the basement. Each "room" is one orbital, and it rents out to no more than two electrons at a time. An orbital that holds only a single electron has a vacancy, and another electron can move in.

Each floor in the apartment building corresponds to one energy level. There is only one room on the first floor: one orbital at the lowest energy level, closest to the nucleus. It fills up first. In hydrogen, the simplest atom, a single electron occupies that room. Helium has two electrons, so it has no vacancies at the lowest energy level. In larger atoms, more electrons rent the second-floor rooms. When the second floor fills, more electrons rent third-floor rooms, and so on. Electrons fill orbitals at successively higher energy levels.

The farther an electron is from the basement (the nucleus), the greater its energy. An electron in a first-floor room cannot move to the second or third floor, let alone the penthouse, unless an input of energy gives it a boost. Suppose an electron absorbs enough energy from, say, sunlight, to get excited about moving up. Move it does. If nothing fills that lower room, though, the electron immediately returns to it and emits extra energy as it does. In later chapters, you will see how some types of cells harvest that released energy.

WHY ATOMS INTERACT

Shells and Electrons We use a shell model to help us check an atom for vacancies (Figure 2.6). With this model, nested "shells" correspond to successive energy

c Third shell This shell corresponds to the third energy level. It has four orbitals with room for eight electrons. Sodium has one electron in the third shell; chlorine has seven. Both have vacancies, so both form chemical bonds. Argon, with no vacancies, does not.

b Second shell This shell, which corresponds to the second energy level, has four orbitals—room for a total of eight electrons. Carbon has six electrons: two in the first shell and four in the second. It has four vacancies. Oxygen has two vacancies. Both carbon and oxygen form chemical bonds. Neon, with no vacancies, does not.

a First shell A single shell corresponds to the first energy level, which has a single orbital that can hold two electrons. Hydrogen has only one electron in this shell and gives it up easily. A helium atom has two electrons (no vacancies), so it does not form bonds.

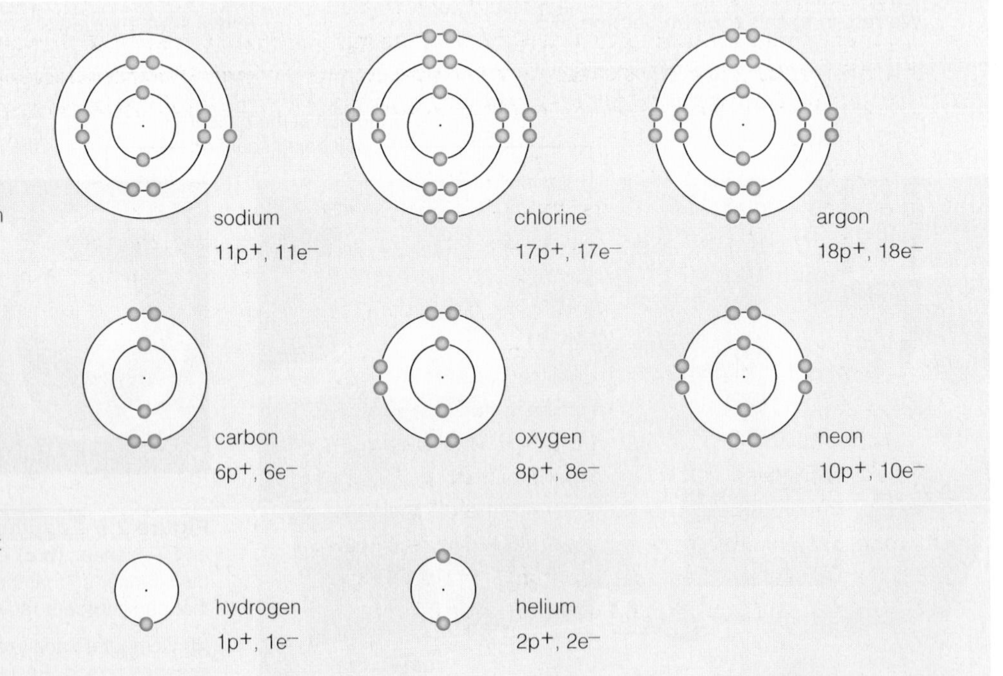

Figure 2.6 **Animated!** Shell models, which help us check for vacancies in atoms. Each circle, or shell, represents all orbitals at one energy level. Atoms with vacancies in the outermost shell tend to form bonds. Bear in mind, atoms are three-dimensional; they do not look anything like these flat diagrams.

levels. Each shell includes all rooms on one floor of the atomic apartment building. We draw an atom's shells by filling them with electrons (represented as dots). Shells are filled from the bottom floor up until there are as many electrons as the atom has protons.

If an atom's outermost shell is full of electrons, it has no vacancies. Atoms of such elements are chemically inactive because they are very stable as single atoms. Helium, neon, and the other inert gases are like this.

If an atom's outermost shell has room for an extra electron, it has a vacancy. Atoms with vacancies tend to interact with other atoms; they give up, acquire, or share electrons until they have no vacancies in their outermost shell. Any atom is in its most stable state when it has no vacancies.

Atoms and Ions An atom is uncharged only when it has as many electrons as protons; the negative charge of an electron cancels the positive charge of a proton. However, an atom can gain or lose electrons, so it no longer has the same number of electrons as protons. When that happens, the atom becomes an ion. An ion is an atom that carries a charge; it acquired a negative charge by pulling an electron away from another atom, or it acquired a positive charge by losing an electron.

Electronegativity is a measure of an atom's ability to pull electrons from other atoms. Whether the pull is strong or weak depends on the atom's size and how many vacancies it has; it is not a measure of charge.

As an example, when a chlorine atom is uncharged, it has 17 protons and 17 electrons. Seven electrons are in its outer (third) shell, which can hold eight (Figure 2.7). It has one vacancy. An uncharged chlorine atom is highly electronegative; it can pull an electron away from another atom and fill its third shell. When that happens, the atom becomes a chloride ion (Cl^-) with 17 protons, 18 electrons, and a net negative charge.

As another example, an uncharged sodium atom has 11 protons and 11 electrons. This atom has one electron in its outer (third) shell, which can hold eight. It has seven vacancies. An uncharged sodium atom is weakly electronegative; it cannot pull seven electrons from other atoms to fill its third shell. Instead, it tends to lose the single electron in its third shell. When that happens, two full shells—and no vacancies—remain. The atom has now become a sodium ion (Na^+), with 11 protons, 10 electrons, and a net positive charge.

From Atoms to Molecules Bonds occur because atoms tend to reach the state in which they have no vacancies. A chemical bond is an attractive force that arises between two atoms when their electrons interact.

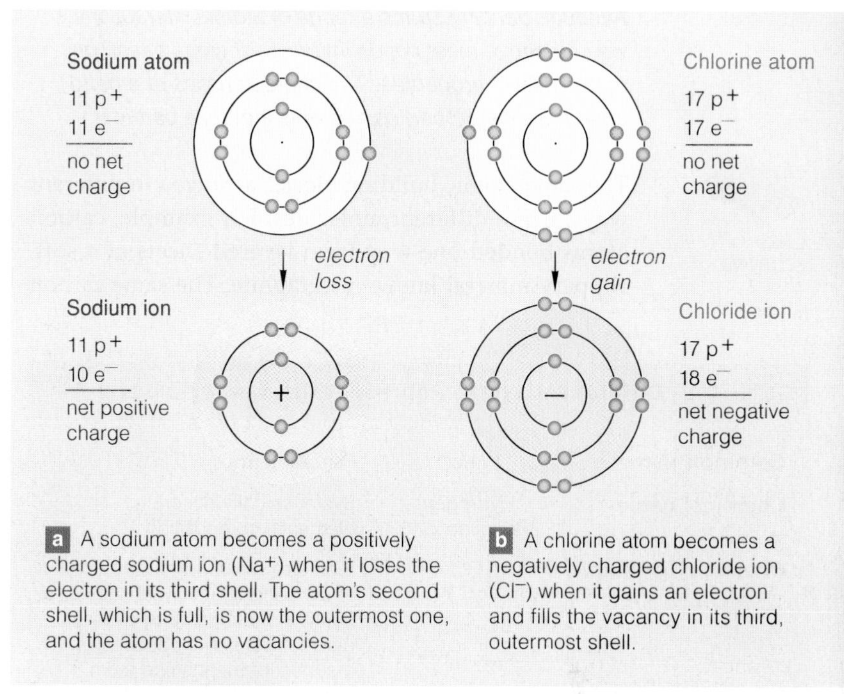

a A sodium atom becomes a positively charged sodium ion (Na^+) when it loses the electron in its third shell. The atom's second shell, which is full, is now the outermost one, and the atom has no vacancies.

b A chlorine atom becomes a negatively charged chloride ion (Cl^-) when it gains an electron and fills the vacancy in its third, outermost shell.

Figure 2.7 Animated! Ion formation.

A molecule forms when two or more atoms of the same or different elements join in chemical bonds. The next section explains the main types of bonds in biological molecules.

Compounds are molecules that consist of two or more different elements in proportions that do not vary. Water is an example. All water molecules have one oxygen atom bonded to two hydrogen atoms. The water in rain clouds, the seas, a Siberian lake, petals, your bathtub, or anywhere else has twice as many hydrogen as oxygen atoms. By contrast, in a mixture, two or more substances intermingle, and their proportions can vary because the substances do not bond with each other. For example, you can make a mixture by swirling sugar into water. The sugar dissolves, but no chemical bonds form.

Always two H for every O

A maximum of two electrons can occupy an orbital around an atom's nucleus. A shell model represents all of the orbitals at the same energy level as one shell. Nested shells correspond to higher energy levels.

Atoms with vacancies in their outermost shell tend to interact with other atoms. Atoms get rid of vacancies in their outermost shell by gaining or losing electrons (thus becoming ions), or by sharing electrons with other atoms.

Chemical bonds connect atoms into molecules. In compounds, atoms of two or more elements bond in unvarying proportions.

2.4 What Happens When Atoms Interact?

Although bonding spans a range of atomic interactions, we categorize most bonds into distinct types based on their different properties. The characteristics of a bond arise from the properties of atoms that take part in it.

LINKS TO SECTIONS 1.1, 1.2

The same atomic building blocks, arranged in different ways, make different molecules. For example, carbon atoms bonded one way form layered sheets of a soft, slippery mineral known as graphite. The same carbon atoms bonded a different way form the rigid crystal lattice of diamond—the hardest mineral. Bond oxygen and hydrogen atoms to carbon and you get sugar.

The bonds themselves vary. Three types of bonds—ionic, covalent, and hydrogen—are most common in biological molecules. Which type forms depends on the vacancy state and electronegativity of atoms that take part in it. Table 2.1 compares some different ways to represent molecules and their bonds.

Table 2.1 Different Ways To Represent the Same Molecule

Common name	Water	Familiar term.
Chemical name	Hydrogen oxide	Systematically describes elemental composition.
Chemical formula	H_2O	Indicates unvarying proportions of elements. Subscripts show number of atoms of an element per molecule. The absence of a subscript means one atom.
Structural formula	H—O—H H/O\H (angled)	Represents each covalent bond as a single line between atoms. The bond angles also may be represented.
Structural model		Shows the positions and relative sizes of atoms.
Shell model		Shows how pairs of electrons are shared in covalent bonds.

IONIC BONDING

A weakly electronegative atom tends to lose one of its electrons, thus becoming a positively charged ion. By contrast, a strongly electronegative atom tends to gain an electron, thus becoming a negatively charged ion. Two atoms with a large difference in electronegativity may stay together in an ionic bond, which is a strong mutual attraction of two oppositely charged ions. Such bonds do not usually form by the direct transfer of an electron from one atom to another; rather, atoms that have already become ions stay close together because of their opposite charges.

Figure 2.8 shows a model for a solid crystal of table salt (sodium chloride, or NaCl). Ionic bonds in such crystals hold sodium and chloride ions in an orderly, cubic arrangement.

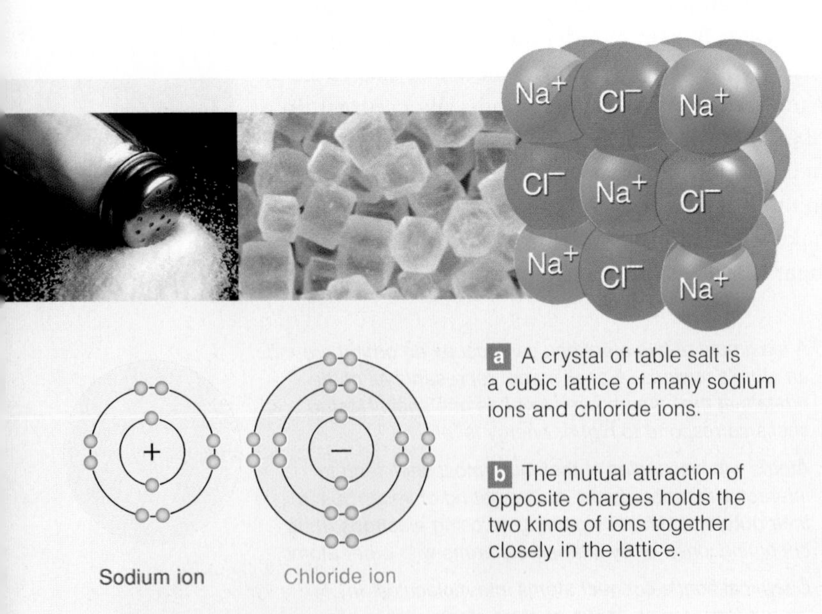

a A crystal of table salt is a cubic lattice of many sodium ions and chloride ions.

b The mutual attraction of opposite charges holds the two kinds of ions together closely in the lattice.

Figure 2.8 **Animated!** Ionic bonds.

COVALENT BONDING

In a covalent bond, two atoms share a pair of electrons. Such bonds typically form between atoms with similar electronegativity and unpaired electrons. By sharing their electrons, each atom's vacancy becomes partially filled (Figure 2.9). Two atoms can share one, two, or three pairs of electrons. Covalent bonds can be much stronger than ionic bonds, but are not always so.

Take a look at the structural formula in Table 2.1. In such formulas, covalent bonds indicate the physical arrangement of atoms in a molecule. A line between two atoms represents a *single* covalent bond, which is a sharing of one pair of electrons between two atoms. A simple example is molecular hydrogen (H_2), with one covalent bond between hydrogen atoms (H—H).

Two lines between atoms represent a *double* covalent bond, in which two pairs of electrons are being shared. Molecular oxygen (O=O) is like this. Three lines are a *triple* covalent bond; the atoms share three pairs of electrons. Molecular nitrogen (N≡N) has this bond.

Covalent bonds can be polar or nonpolar. A *polar* covalent bond forms between two atoms with a small difference in electronegativity. In such bonds, the two atoms do not share electrons equally. The atom that is more electronegative pulls the electrons a bit more toward its "end" of the bond, so that atom bears a slightly negative charge. The atom at the other end of the bond bears a slightly positive charge.

For example, the water molecule shown in Table 2.1 has two polar covalent bonds (H—O—H). The oxygen atom carries a slight negative charge, but each of the hydrogen atoms carries a slight positive charge. Any such separation of charge into distinct positive and negative regions is called polarity. As you will see in the next section, the polarity of the water molecule is very important for the world of life.

In a *nonpolar* covalent bond, two atoms of identical electronegativity share electrons equally. There is no difference in charge between the two ends of the bond. Such bonds occur in the molecular hydrogen, oxygen, and nitrogen mentioned above. These molecules have the chemical formulas H_2, O_2, and N_2. All three are simple gases that you breathe in from the air.

HYDROGEN BONDING

Hydrogen bonds often form between polar regions of two molecules or two regions of the same molecule. A hydrogen bond is an attraction between a hydrogen atom and an electronegative atom, both of which are taking part in separate polar covalent bonds.

Hydrogen bonds are not chemical bonds; they do not make molecules out of atoms. Even so, they fall within a spectrum of interactions that characterize bonds. Like ionic bonds, they form by mutual attraction of opposite charges. The hydrogen atom bears a slight positive charge; the other atom bears a slight negative charge. Hydrogen bonds are much weaker than ionic or covalent bonds, and they form and break easily. Even so, many form in all of the large molecules of life. Collectively, they are strong enough to help hold molecules in three-dimensional shapes (Figure 2.10).

An ionic bond is a strong mutual attraction that keeps ions of opposite charges close together.

Atoms share a pair of electrons in a covalent bond. When the atoms share electrons equally, the bond is nonpolar. When they do not share them equally, the bond is polar: slightly positive at one end and slightly negative at the other.

A hydrogen bond is an attraction between a hydrogen atom and an electronegative atom, both of which are taking part in separate polar covalent bonds. Though individually weak, hydrogen bonds collectively stabilize the structures of large biological molecules.

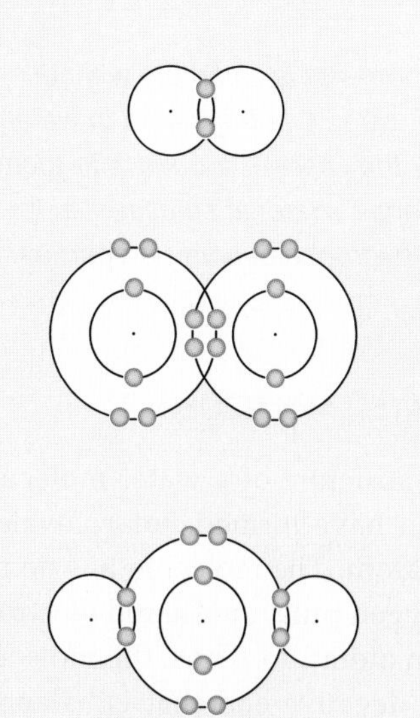

Figure 2.9 Animated! Covalent bonds. Two atoms with unpaired electrons in their outermost shell become more stable by sharing electrons. Two electrons are shared in each covalent bond. When the electrons are shared equally, the covalent bond is nonpolar. When one atom exerts a greater pull on the shared electrons, the covalent bond is polar.

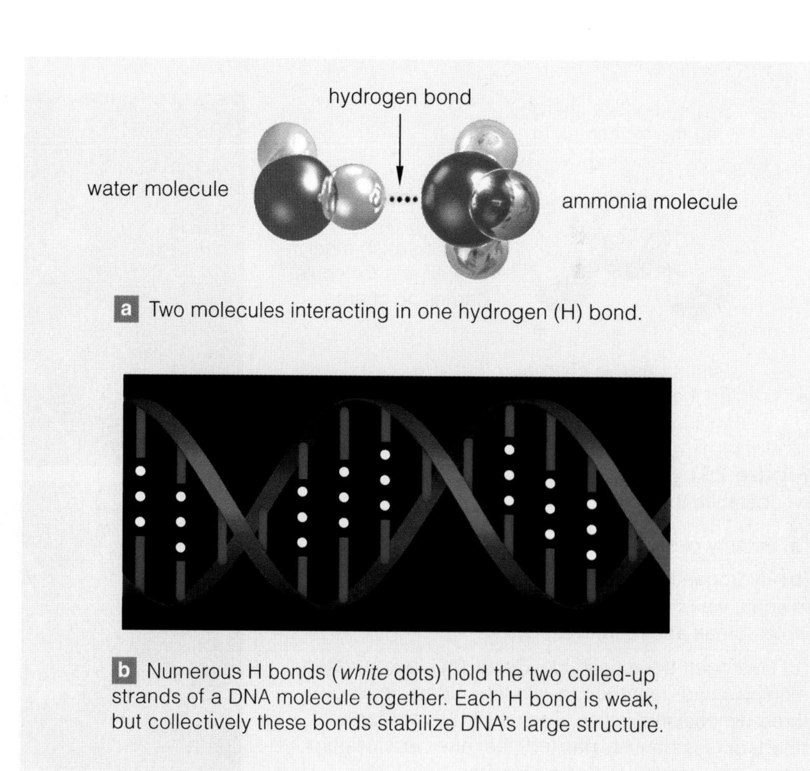

a Two molecules interacting in one hydrogen (H) bond.

b Numerous H bonds (*white* dots) hold the two coiled-up strands of a DNA molecule together. Each H bond is weak, but collectively these bonds stabilize DNA's large structure.

Figure 2.10 Animated! Hydrogen bonds. Hydrogen bonds form at a hydrogen atom taking part in a polar covalent bond. The hydrogen atom's slight positive charge weakly attracts an electronegative atom taking part in a separate polar covalent bond. As shown here, hydrogen (H) bonds can form between molecules and between different parts of the same molecule.

2.5 Water's Life-Giving Properties

No sprint through basic chemistry is complete without a tour of the properties of water. Life originated in water. Organisms still live in it or they carry it around with them inside cells and tissue spaces. Water is so essential for life because of its unique hydrogen bonding properties; no other liquid is like it.

POLARITY OF THE WATER MOLECULE

Figure 2.11*a* shows the structure of a water molecule. Two atoms of hydrogen have formed polar covalent bonds with an oxygen atom. The molecule has no net charge. Even so, the oxygen pulls the shared electrons more than the hydrogen atoms do. Thus, the molecule of water has a slightly negative end that is balanced by its slightly positive end.

The polarity of each water molecule attracts other water molecules. Hydrogen bonds form between them in tremendous numbers (Figure 2.11*b*). Such extensive hydrogen bonding between water molecules imparts unique properties to liquid water.

Water also forms hydrogen bonds with sugars and other polar molecules. That is why polar molecules are hydrophilic (water-loving) substances.

The polarity repels other nonpolar molecules, such as oils, which are called hydrophobic (water-dreading) substances. Shake a bottle filled with water and salad oil, then set it on a table. Soon, new hydrogen bonds replace the ones broken by the shaking. The reunited water molecules push out oil molecules, which cluster as oil droplets or as an oily film at the water's surface.

The same kinds of interactions proceed at the thin, oily membrane between the water inside and outside cells. Membrane organization—and life itself—starts with such hydrophilic and hydrophobic interactions. You will read about membrane structure in Chapter 4.

WATER'S TEMPERATURE-STABILIZING EFFECTS

All molecules vibrate nonstop, and they move faster as they absorb heat. Temperature is a way to measure the energy of molecular motion. In liquid water, extensive hydrogen bonding restricts the jiggling of individual water molecules by absorbing some of their energy. Thus, compared with other liquids, water can absorb more heat before it gets hotter. This property means that water functions as a heat reservoir, so it keeps the temperature of the surrounding air relatively stable.

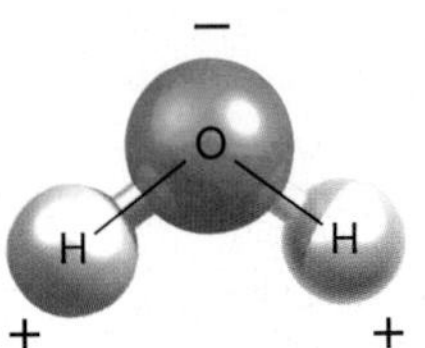

Figure 2.11 **Animated!** Characteristics of water, a substance that is essential for life.

(**a**) Polarity of an individual water molecule.

(**b**) Hydrogen bonding pattern among water molecules in liquid water. Dashed lines signify hydrogen bonds, which break and re-form rapidly.

(**c**) Hydrogen bonding in ice. Below 0°C, every water molecule hydrogen-bonds with four others, in a rigid three-dimensional lattice. The molecules are farther apart, or less densely packed, than they are in liquid water. As a result, ice floats on water.

Thanks to rising levels of methane and other greenhouse gases that are contributing to global warming, the Arctic ice cap is melting. At current rates, it will be gone in fifty years, and so will the polar bears. Already the seal-hunting season is shorter. The bears are becoming thinner, and they are giving birth to fewer cubs.

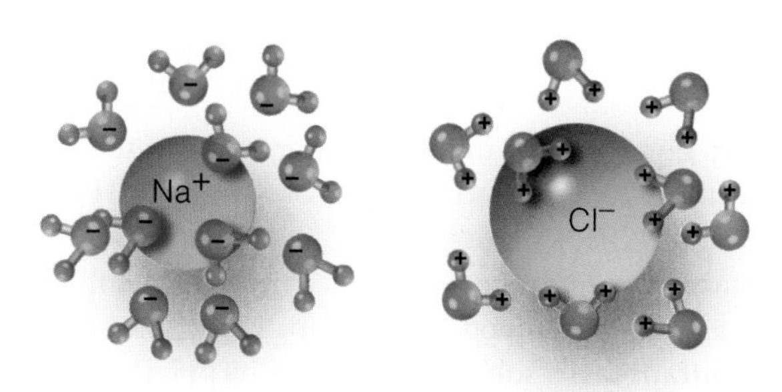

Figure 2.12 **Animated!** Spheres of hydration around ions. Water molecules that surround an ionic solid pull its atoms apart, thereby dissolving them.

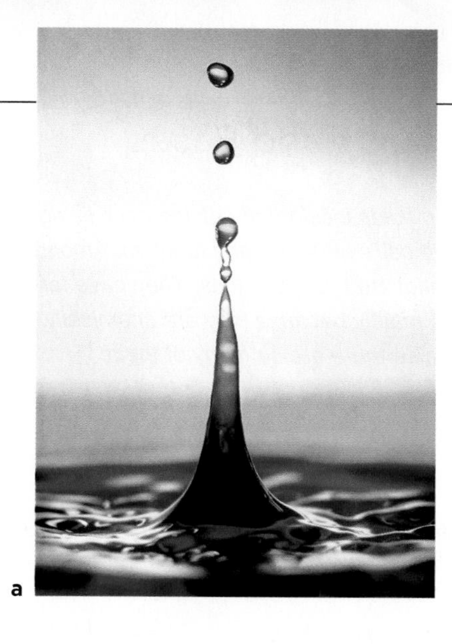

Figure 2.13 Examples of water's cohesion. (**a**) After a pebble hits liquid water and forces molecules from the surface, individual water molecules do not fly every which way. They stay together in droplets, because countless hydrogen bonds continuously pull individual molecules at the surface inward. (**b**) How can water rise to the top of trees? Cohesion, and evaporation from the leaves, pulls water upward.

When the temperature of water is below its boiling point, hydrogen bonds form as fast as they break. As water gets hotter, the increase in molecular motion can keep bonds from forming, so individual molecules at the water's surface escape into the air. By this process, called evaporation, heat energy converts liquid water to a gas. The energy increase overcomes the attraction between water molecules, which break free.

The surface temperature of water decreases during evaporation. Evaporative water loss can help you and some other mammals cool off when you sweat in hot, dry weather. Sweat is about 99 percent water, and it evaporates from skin.

Below 0°C (32°F), molecules of water do not move enough to break hydrogen bonds. They become locked in the latticelike bonding pattern of ice (Figure 2.11*c*). Ice is less dense than water. During winter freezes, ice sheets may form near the surface of ponds, lakes, and streams. The ice "blanket" insulates the liquid water beneath it and helps protect many fishes, frogs, and other aquatic organisms against freezing.

WATER'S SOLVENT PROPERTIES

A solvent is a substance, usually liquid, that dissolves other substances. We refer to dissolved substances as solutes. Water is an excellent solvent; ionic and polar substances easily dissolve in it. In general, a substance is said to be dissolved after solvent molecules cluster around its ions or molecules and keep them dispersed.

A clustering of water molecules around a solute is a sphere of hydration. Such spheres form around any solute in cellular fluids, tree sap, blood, the fluid in your gut, and every other fluid associated with life. Watch it happen after you pour table salt (NaCl) into a cup of water. The crystals of salt will separate into ions of sodium (Na^+) and chloride (Cl^-). Each Na^+ attracts the negative end of some water molecules as Cl^- attracts the positive end of others (Figure 2.12). In general, spheres of hydration that form in such a way keep a great variety of ions dispersed in pond water, soil water, the seas, sap, blood, and other fluids.

WATER'S COHESION

Another life-sustaining property of water is cohesion. Cohesion means molecules resist separating from one another. You see its effect as surface tension when you toss a pebble into a pond (Figure 2.13*a*). Although the water ripples and sprays, its individual molecules do not fly apart. Its hydrogen bonds collectively exert a continuous pull on individual water molecules. This pull is so strong that the molecules stay together rather than spreading out in a thin film like other liquids do.

Cohesion works inside organisms, too. For example, plants absorb nutrient-laden water while they grow. Columns of liquid water rise inside narrow pipelines of vascular tissues, which extend from roots to leaves. Water evaporates from leaves when molecules break free and diffuse into air (Figure 2.13*b*). The cohesive force of hydrogen bonds pulls replacements into the leaf cells, as Section 27.3 explains.

b

Being polar, water molecules hydrogen-bond with one another and with other polar (hydrophilic) substances. They tend to repel nonpolar (hydrophobic) substances.

Extensive hydrogen bonding between water molecules gives liquid water unique properties that make life possible. Water has temperature-stabilizing effects, solvency, and cohesion.

2.6 Acids and Bases

LINK TO SECTION 1.2

Ions dissolved in fluids inside and outside each living cell influence the cell's structure and function. Among the most influential are hydrogen ions. They have far-reaching effects mainly because they are chemically active and because there are so many of them.

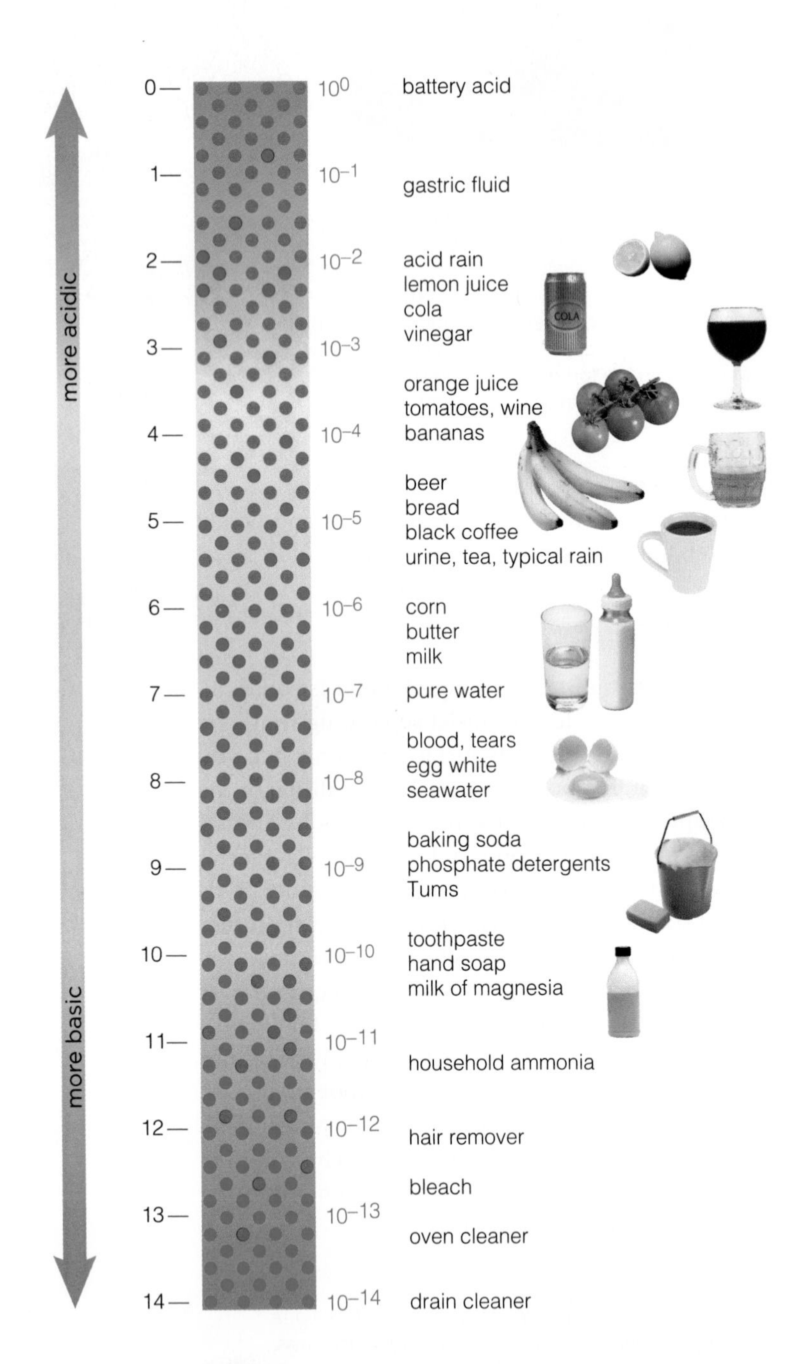

Figure 2.14 **Animated!** A pH scale, which represents the concentration of hydrogen ions. Here, *red* dots signify hydrogen ions (H^+) and *blue* dots signify hydroxyl ions (OH^-). Also shown are approximate pH values for some solutions. This pH scale ranges from 0 (most acidic) to 14 (most basic). A change of one unit on the scale means a tenfold change in H^+ concentration (*blue* numbers).

THE pH SCALE

At any given instant in liquid water, some of the water molecules are separated into ions of hydrogen (H^+) and hydroxide (OH^-). Figure 2.14 shows a pH scale, which is a measure of hydrogen ion concentration in solutions such as seawater, blood, or tree sap. The greater the H^+ concentration of a solution, the lower its pH. Pure water (not rainwater or tap water) always has as many H^+ as OH^- ions. Such a balance is pH 7, the point of neutrality on this pH scale, which ranges from 0 to 14.

A one-unit decrease in pH corresponds to a tenfold increase in H^+ concentration, and a one-unit increase corresponds to a tenfold decrease in H^+ concentration. One way to get a sense of what this means is to taste dissolved baking soda (pH 9), distilled water (pH 7), and lemon juice (pH 2).

Nearly all of life's chemistry occurs near pH 7. Most of your body's internal environment (tissue fluids and blood) is between pH 7.3 and 7.5.

HOW DO ACIDS AND BASES DIFFER?

Substances called acids *donate* hydrogen ions, and bases *accept* hydrogen ions as they dissolve in water. *Acidic* solutions, such as lemon juice, gastric fluid, and coffee, release H^+ in water; their pH is below 7. *Basic* solutions, such as seawater and egg white, contain more OH^- than H^+. Basic, or alkaline, solutions have a pH above 7.

Acids and bases can be weak or strong. Weak acids, such as carbonic acid (H_2CO_3), are stingy H^+ donors. Strong acids readily give up H^+ in water. An example is the hydrochloric acid that dissociates into H^+ and Cl^- inside your stomach. The H^+ makes gastric fluid very acidic. The acidity activates enzymes that digest proteins in your food.

HCl splashing up out of the stomach can provoke *acid indigestion*. Milk of magnesia and other antacids help by releasing OH^- ions, which combine with H^+ to reduce the pH of the stomach's contents.

Exposure to strong acids or bases can cause severe chemical burns. That is why you are supposed to read the labels on containers of ammonia, drain cleaner, and many other common household products. That is why you are not supposed to let a car battery's sulfuric acid drip on your skin.

Also, strong acids or bases that accumulate to high concentrations in ecosystems can kill organisms. For instance, fossil fuel burning and nitrogen-containing fertilizers release strong acids that lower the pH of rainwater (Figure 2.15). Some regions are sensitive to this acid rain. Alterations in the chemical composition

Figure 2.15 Emissions of sulfur dioxide from a coal-burning power plant. Airborne pollutants such as sulfur dioxide dissolve in water vapor and form acidic solutions. They are a component of acid rain. The far-right photograph shows how acid rain can corrode stone sculptures.

of soil and water harm fishes and other organisms in these regions. We return to this topic in Section 43.2.

SALTS AND WATER

A salt is any compound that dissolves easily in water and releases ions *other than* H^+ and OH^-. It commonly forms when an acid interacts with a base. For example:

$$\underset{\substack{\text{HYDROCHLORIC}\\\text{ACID}\\\text{(acid)}}}{HCl} + \underset{\substack{\text{SODIUM}\\\text{HYDROXIDE}\\\text{(base)}}}{NaOH} \rightleftharpoons \underset{\substack{\text{SODIUM}\\\text{CHLORIDE}\\\text{(salt)}}}{NaCl} + \underset{\text{WATER}}{H_2O}$$

NaCl, the salt product of this reaction, dissociates into sodium ions (Na^+) and chloride ions (Cl^-) when it is dissolved in water. Many ions that are released when salts dissolve in fluid are important components of all cellular processes. For example, sodium, potassium, and calcium ions are vital for nerve and muscle cell functions. As another example, potassium ions affect the amount of water a plant loses on hot, dry days.

BUFFERS AGAINST SHIFTS IN pH

Cells must respond quickly to even slight shifts in pH. Why? Enzymes and many other biological molecules can function properly only within a narrow range of pH. A slight deviation from that range halts cellular processes completely.

Most body fluids maintain a consistent pH because they are buffered. A buffer system is a set of chemicals, often a weak acid or base and its salt, that keeps the pH of a solution stable. It works because the chemicals can donate or accept ions that contribute to pH.

For example, when base is added to a fluid, OH^- is released, and the fluid's pH rises. However, if the fluid is buffered, the weak acid partner gives up H^+. The H^+ combines with the OH^-, forming a small amount of water that does not affect pH. So, a buffered fluid's pH stays constant even when base is added.

Carbon dioxide, which forms in many reactions, takes part in an important buffer system. It combines with water in human blood to form carbonic acid and bicarbonate. When the pH of blood starts to rise due to other factors, the carbonic acid neutralizes the excess OH^- by releasing H^+. The two kinds of ions combine and form water:

$$OH^- + \underset{\substack{\text{CARBONIC ACID}\\\text{(acid)}}}{H_2CO_3} \longrightarrow \underset{\substack{\text{BICARBONATE}\\\text{(base)}}}{HCO_3^-} + \underset{\text{WATER}}{H_2O}$$

When the blood becomes more acidic, the bicarbonate absorbs excess H^+ and thus shifts the balance of the buffer system toward carbonic acid:

$$\underset{\substack{\text{BICARBONATE}\\\text{(base)}}}{HCO_3^-} + H^+ \longrightarrow \underset{\substack{\text{CARBONIC ACID}\\\text{(acid)}}}{H_2CO_3}$$

Together, these reactions keep the blood pH between 7.3 and 7.5, but only up to a point. A buffer system can neutralize only so many ions. Even slightly more than that limit causes the pH to swing widely.

A buffer system failure in a biological system can cause big problems. In *acute respiratory acidosis*, carbon dioxide accumulates, and excess carbonic acid forms in blood. The steep decline in blood pH may cause an individual to enter a *coma*, a level of unconsciousness that is dangerous. *Alkalosis*, a potentially lethal rise in blood pH, can also invite coma. Even an increase to 7.8 can result in *tetany*, or prolonged muscle spasm.

Ions dissolved in fluids on the inside and outside of cells have important roles in cell function. When dissolved, acidic substances release hydrogen ions, and basic substances accept them. Salts release ions other than H^+ and OH^-.

pH reflects hydrogen ion concentration in a fluid. Buffer systems help maintain homeostasis by keeping the pH of body fluids within a range that is suitable for life.

www.thomsonedu.com Thomson NOW!

Summary

Section 2.1 Atoms, fundamental building blocks of matter, consist of negatively charged electrons that move around a nucleus of positively charged protons and (except for hydrogen) uncharged neutrons. An element is a pure substance consisting of atoms that have the same number of protons. Isotopes are atoms of an element that differ in the number of neutrons (Table 2.2).

Section 2.2 Radioisotopes are radioactive isotopes. They are not stable, and emit particles and energy as they decay spontaneously into other elements.

■ *Use the animation on ThomsonNOW to learn how radioisotopes are used in making PET scans.*

Table 2.2 Summary of Players in the Chemistry of Life

Atom	Particles that are basic building blocks of all matter; the smallest unit that retains an element's properties.
Element	One of ninety-two naturally occurring pure substances. Each consists entirely of atoms that have the same, characteristic number of protons.
Proton (p^+)	Positively charged particle of an atom's nucleus
Electron (e^-)	Negatively charged particle that can occupy a volume of space (orbital) around an atom's nucleus
Neutron	Uncharged particle of an atom's nucleus
Isotope	One of two or more forms of an element's atoms that differ in the number of neutrons
Radioisotope	Unstable isotope that emits particles and energy when its nucleus disintegrates.
Tracer	Molecule that has a detectable substance (such as a radioisotope) attached. Used with tracking devices to identify the movement or destination of the molecule in a metabolic pathway, the body, or some other system
Ion	Atom that carries a charge after it has gained or lost one or more electrons. A single proton without an electron is a hydrogen ion (H^+)
Molecule	Two or more atoms joined in a chemical bond
Compound	Molecule of two or more different elements in unvarying proportions (for example, water)
Mixture	Intermingling of two or more elements or compounds in proportions that can vary
Solute	Molecule or ion dissolved in some solvent
Hydrophilic substance	Polar substance (or molecular region) that readily dissolves in water
Hydrophobic substance	Nonpolar substance (or molecular region) that resists dissolving in water
Acid	Substance that releases H^+ when dissolved in water
Base	Substance that accepts H^+ when dissolved in water
Salt	Compound that releases ions other than H^+ or OH^- when dissolved in water

Section 2.3 Electrons occupy orbitals (volumes of space) around the nucleus. Up to two electrons occupy each orbital. The shell model represents orbital energy levels as successively larger circles, or shells. We use it to view an atom's electron structure. Atoms with unpaired electrons in their outermost shell tend to interact with other atoms; they donate, accept, or share electrons so that their vacancies go away.

An atom with equal numbers of protons and electrons has no net charge. Atoms that have either gained or lost electrons are ions. Electronegativity is a measure of how strongly an atom attracts electrons from other atoms.

A chemical bond is an attractive force that unites two atoms into a molecule. Compounds are molecules that consist of two or more elements.

■ *Use the animation and interaction on ThomsonNOW to study electron distribution and the shell model.*

Section 2.4 Ionic, covalent, and hydrogen bonding are the most common atomic interactions in biological molecules. An ionic bond is a strong association between a positively charged ion and a negatively charged ion; it arises from the mutual attraction of opposite charges.

In a covalent bond, two atoms share a pair of electrons. The atoms share electrons equally in a nonpolar covalent bond. The sharing is unequal in a polar covalent bond, so there is a slight negative charge at one end of the bond and a slight positive charge at the other. A molecule that has a separation of charge is said to show polarity.

A hydrogen bond forms between a hydrogen atom and an electronegative atom, both of which are taking part in separate polar covalent bonds. Hydrogen bonds are not chemical bonds because they do not make atoms into molecules. Individually, they are weak. Collectively, they stabilize the structures of large molecules.

■ *Use the animation on ThomsonNOW to compare the types of chemical bonds in biological molecules.*

Section 2.5 The polarity of water molecules invites the extensive hydrogen bonding that gives liquid water its unique properties: resistance to temperature changes, internal cohesion, and the capacity to dissolve polar and ionic substances. These properties make life possible.

■ *Use the animation on ThomsonNOW to view the structure of the water molecule and properties of liquid water.*

Section 2.6 A pH scale indicates the hydrogen ion (H^+) concentration of a solution. Typical pH scales range from 0 (most acidic) to 14 (most basic or alkaline). At pH 7, or neutrality, H^+ and OH^- concentrations are equal.

Salts are compounds that dissolve easily in water and release ions other than H^+ and OH^-. Acids release H^+ in water; bases accept H^+. A buffer system is a dynamic chemical partnership between a weak acid or base and its salt. The pH of the solution remains stable because one donates ions and the other accepts them. Buffers help maintain homeostasis. Most biological processes proceed only within a narrow pH range, usually near neutrality.

■ *Use the interaction on ThomsonNOW to investigate the pH of common solutions.*

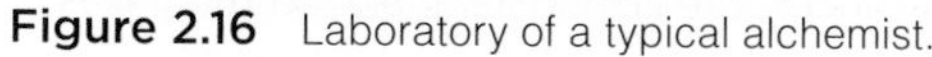

Figure 2.16 Laboratory of a typical alchemist.

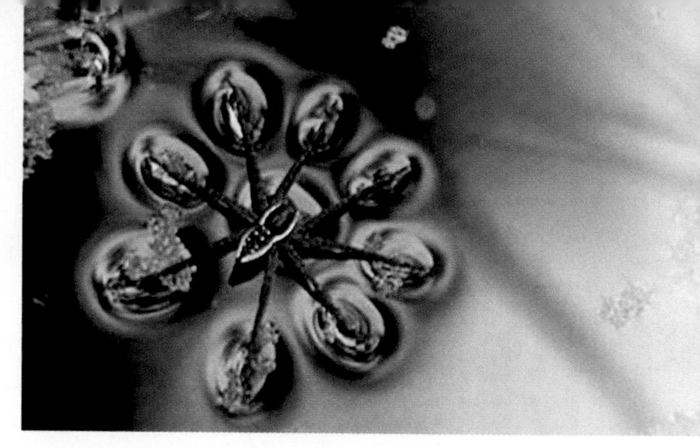

Figure 2.17 Fishing spider, not sinking.

Self-Quiz

Answers in Appendix III

1. Is this statement true or false? All atoms consist of electrons, protons, and neutrons.

2. Electrons carry a ______ charge.
   a. positive b. negative c. zero

3. A(n) ______ is a molecule into which a radioisotope has been incorporated.
   a. ion b. isotope c. element d. tracer

4. An ion is an atom that has ______ an electron.
   a. gained c. hydrogen-bonded
   b. lost d. a or b

5. The mutual attraction of opposite charges holds atoms together as molecules in a(n) ______ bond.
   a. ionic c. polar covalent
   b. hydrogen d. nonpolar covalent

6. Atoms share electrons unequally in a(n) ______ bond.
   a. ionic c. polar covalent
   b. hydrogen d. nonpolar covalent

7. In a hydrogen bond, a covalently bound hydrogen atom weakly attracts an ______ in a different molecule or a different region of the same molecule.
   a. ion b. electronegative atom

8. Liquid water has ______ .
   a. polarity d. resistance to increases in temperature
   b. a profusion of hydrogen bonds e. b through d
   c. cohesion f. all of the above

9. Hydrogen ions (H^+) are ______ .
   a. indicated by a pH scale c. dissolved in blood
   b. unbound protons d. all of the above

10. When dissolved in water, a(n) ______ donates H^+, and a(n) ______ accepts H^+.

11. A(n) ______ is a dynamic chemical partnership between a weak acid or base and its salt.
   a. ionic bond c. buffer system
   b. solute d. solvent

12. Match the terms with their most suitable description.
   ___ trace element a. components of atomic nucleus
   ___ salt b. two atoms sharing electrons
   ___ covalent bond c. any polar molecule that readily dissolves in water
   ___ hydrophilic substance d. releases ions other than H^+ and OH^- when dissolved in water
   ___ protons, neutrons e. makes up less than 0.01 percent of the body's weight

■ *Visit ThomsonNOW for additional questions.*

Critical Thinking

1. Alchemists were medieval scholars and philosophers who were the forerunners of modern-day chemists (Figure 2.16). Many tried repeatedly to transform lead (atomic number 82) into gold (atomic number 79). Explain why they never did succeed.

2. For centuries and perhaps longer, people have waged battles with bacteria, fungi, oxygen, and other agents that can spoil beef, pork, and other meats. That is why meats are often "cured," or salted, dried, smoked, pickled, or treated with chemicals that can delay the attacks. Ever since the mid-1800s, sodium nitrite ($NaNO_2$) has been used in processed meat products such as hot dogs, bologna, sausages, jerky, bacon, and ham. Nitrites help keep the meats from turning an unappetizing gray color. They also prevent growth of *Clostridium botulinum*. If ingested, this bacterium can cause a form of food poisoning called botulism. It makes a toxin that locks muscles in contraction and may lead to death.

In water, sodium nitrite dissociates into sodium ions (Na^+) and nitrite ions (NO_2^-), which are called nitrites. Nitrites are rapidly converted to nitric oxide (NO), the compound that gives nitrites their preservative qualities. Eating preserved meats increases the risk of cancer, but nitrites may not be at fault. It turns out that nitric oxide has several important functions, including blood vessel dilation (for example, inside a penis during an erection), cell to cell signaling, and antimicrobial activities of the immune system. Draw a shell model for nitric oxide and then use it to explain why the molecule is so reactive.

3. Ozone is a chemically active form of oxygen gas. High in Earth's atmosphere, it forms a layer that absorbs about 98 percent of the sun's harmful rays. Oxygen gas consists of two covalently bonded oxygen atoms: O=O. Ozone has three covalently bonded oxygen atoms: O=O—O. It reacts easily with many substances, and gives up an oxygen atom and releases gaseous oxygen (O=O). From what you know about chemistry, why do you suppose ozone is so reactive?

4. David, an inquisitive three-year-old, poked his fingers into warm water in a metal pan on the stove and did not sense anything hot. Then he touched the pan itself and got a nasty burn. Explain why water in a metal pan heats up far more slowly than the pan itself.

5. How can fishing spiders (Figure 2.17) and other insects such as water striders walk on water?

6. Some undiluted acids are more corrosive when diluted with water. That is why lab workers are told to wipe off splashes with a towel before washing. Explain.

3 MOLECULES OF LIFE

IMPACTS, ISSUES Science or Supernatural?

About 2,000 years ago in the mountains of Greece, the oracle of Delphi was famous for her rambling prophecies. The young woman spoke in a trance after inhaling fumes that collected near the floor of her temple. As we now know, her temple was built on earthquake-prone faults. When the faults slipped, hallucinogenic gases escaped from the ground (Figure 3.1). The ancient Greeks thought that their god Apollo spoke to them through the oracle; they believed in the supernatural. Scientists looked for a natural explanation and found carbon compounds behind her words.

Starting with information on the structure and effects of natural substances, the scientists analyzed the gases at the temple. These gases consist only of carbon and hydrogen atoms; they are hydrocarbons. We now know a lot about them. Methane, for instance, was around when Earth formed. It is released when volcanoes erupt, when we burn wood, peat, or fossil fuels, and when termites, people, and cattle pass gas. It collects deep in the sea next to continents. It collects in the atmosphere and is now contributing to global warming.

Big methane deposits may be big trouble. The remains of ancient marine organisms are a source of energy for archaeans that live kilometers under the seafloor. These prokaryotes release methane as they break down organic molecules in the remains. Great quantities of methane bubble up from the ocean floor (Figure 3.1). There, low temperatures and high pressure "freeze" the gas into icy, unstable crystals of methane hydrate.

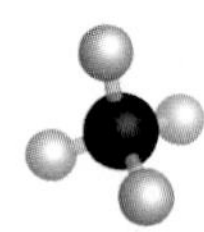

methane—only one carbon and four hydrogen atoms, but a molecule with global impact

See the video! **Figure 3.1** *Left,* ruins of the Temple of Apollo, where hydrocarbon gases escape from the ground. *Right,* microorganisms and bubbles of methane gas almost 230 meters (750 feet) below sea level in the Black Sea. Archaeans far beneath the seafloor produce the methane, which seeps upward into deep ocean water.

How would you vote? Should companies work toward developing the vast undersea methane deposits as an energy source, given that the environmental costs and risks to life are unknown? See ThomsonNOW for details, then vote online.

There may be a thousand billion tons of frozen methane hydrate on the seafloor. It is the world's largest reservoir of natural gas, but no one has figured out how to tap it. If the water temperature rises a few degrees or the pressure falls, the unstable crystals will disintegrate explosively into methane gas and liquid water. Surrounding deposits will vaporize in an irreversible chain reaction.

In the past, immense explosions caused underwater landslides that stretched from one continent to another. About 250 million years ago, at the end of the Permian, one may have caused the greatest of all mass extinctions. All but about 5 percent of life in the seas and 70 percent of life on land abruptly vanished. Scientists, who are not given to exaggeration, call it The Great Dying.

Chemical clues in fossils dating from that time point to a sharp spike in atmospheric carbon dioxide—not just any carbon dioxide, but molecules that living things had assembled. In one gargantuan burp, millions of tons of methane hydrate exploded from the seafloor. Methane-eating bacteria converted most of it to carbon dioxide, which displaced most of the oxygen in the seas and sky.

Too much carbon dioxide, too little oxygen. Before The Great Dying, oxygen made up about 35 percent of Earth's atmosphere. When its concentration plummeted quickly to 12 percent, most animals in the seas and on land suffocated.

Today, as one potential outcome of global warming, great currents that move between the ocean's floor and its surface may shift. Will such a shift disturb the methane hydrates? We already know about vast deposits 12 miles off California's southern coast, 60 miles off the Oregon coast, and 200 miles off the South Carolina coast. Will a new methane burp take us out in the next Great Dying?

No matter where you look, knowledge about lifeless molecules can tell you a lot about life, including your own. It can guide you into the past, present, and future—from ancient myths, to health or disease, to forests, to physical and chemical conditions that affect life everywhere. This chapter is your survey of the molecules which, by their interactions, give rise to the processes of life.

Key Concepts

STRUCTURE DICTATES FUNCTION

We define cells partly by their capacity to build complex carbohydrates and lipids, proteins, and nucleic acids. The main building blocks are simple sugars, fatty acids, amino acids, and nucleotides. All of these organic compounds have a backbone of carbon atoms with functional groups attached. **Section 3.1**

CARBOHYDRATES

Carbohydrates are the most abundant biological molecules. Simple sugars function as transportable forms of energy or as quick energy sources. The complex carbohydrates are structural materials or energy reservoirs. **Section 3.2**

LIPIDS

Complex lipids function as energy reservoirs, structural materials of cell membranes, signaling molecules, and waterproofing or lubricating substances. **Section 3.3**

PROTEINS

Structurally and functionally, proteins are the most diverse molecules of life. They include enzymes, structural materials, signaling molecules, and transporters. **Sections 3.4, 3.5**

NUCLEOTIDES AND NUCLEIC ACIDS

Nucleotides have major metabolic roles and are building blocks of nucleic acids. Two kinds of nucleic acids, DNA and RNA, interact as the cell's system of storing, retrieving, and translating information about building proteins. **Section 3.6**

Links to Earlier Concepts

Having learned about atoms, you are about to enter the next level of organization in nature, as represented by the molecules of life. Keep the big picture in mind by quickly scanning Section 1.1 once again. You will be building on your understanding of how electrons are arranged in atoms (2.3) as well as the nature of covalent bonding and hydrogen bonding (2.4). Here again, you will be considering one of the consequences of mutation in DNA (1.4), this time with sickle-cell anemia as the example.

3.1 Molecules of Life—From Structure to Function

Under present-day conditions in nature, only living cells make complex carbohydrates and lipids, proteins, and nucleic acids. These molecules of life have a carbon backbone, and their structure dictates how they function.

LINKS TO SECTIONS 1.1, 2.3, 2.4, 2.5

The molecules of life are organic compounds, which contain carbon and at least one hydrogen atom. Most of these molecules also have one or more functional groups: certain atoms or clusters of atoms covalently bonded to carbon. Methane and other hydrocarbons are organic compounds that consist only of hydrogen covalently bonded to carbon.

CARBON'S BONDING BEHAVIOR

Living things consist mainly of oxygen, hydrogen, and carbon. Most of their oxygen and hydrogen are in the form of water. Put water aside, and carbon makes up more than half of what is left.

Carbon's importance to life starts with its versatile bonding behavior: A carbon atom can form covalent bonds with as many as four other atoms. Most organic compounds have a backbone of carbon atoms to which functional groups attach. This arrangement gives rise to the diverse shapes of organic compounds.

As the ball-and-stick model on page 34 indicates, methane has four hydrogen atoms covalently bound to one carbon atom (CH_4). Figure 3.2*a* shows a model for glucose, an organic compound with a backbone of six carbon atoms. In cells, the backbone often forms a ring structure (Figure 3.2*b*). A flat structural model shows the atoms in such rings connected by lines that represent bonds (Figure 3.2*c*). Some models identify an atom at a given position in the ring only when it is not a carbon atom (Figure 3.2*d*).

Considering a molecule's structural features gives us insight into how it functions. For instance, viruses infect cells by docking at proteins on the cell surface. Certain viral proteins have ridges, clefts, and charged regions that fit into complementary ridges, clefts, and charged regions of cell surface proteins.

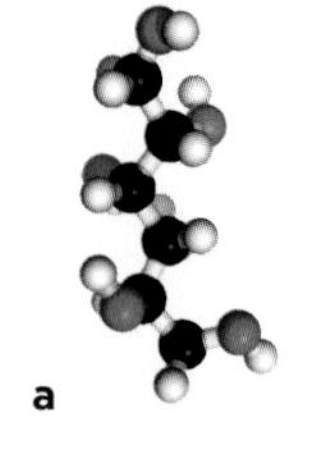
a

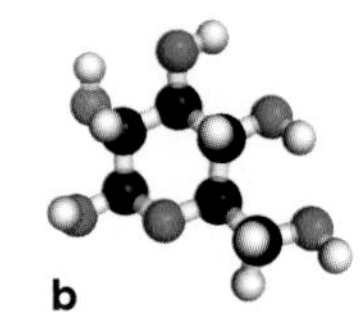
b

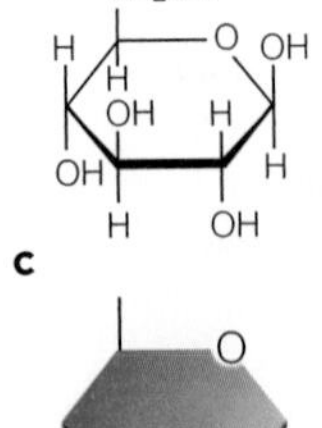

c

d

Figure 3.2 A few models for the same organic compound—glucose (**a**) Ball-and-stick model for the linear form of glucose. Carbon atoms are coded black, hydrogen white, and oxygen red. (**b**) Ring structure, (**c**) flat structural formula, and (**d**) simple icon. Appendix V explains how different models reveal different information about molecules of life.

FUNCTIONAL GROUPS

Figure 3.3 lists functional groups that are common in carbohydrates, lipids, proteins, and nucleic acids. The number, kind, and arrangement of these groups give rise to specific properties, such as polarity and acidity.

For example, sugars are among the alcohols, a class of organic compounds that have one or more *hydroxyl* groups (—OH). These polar groups hydrogen-bond with water, so alcohols—at least small ones—dissolve fast. Larger alcohols do not dissolve as easily, because their long nonpolar hydrocarbon chains repel water. Fatty acids also are like this, which is why lipids that have fatty acid tails do not dissolve easily in water.

Highly reactive *carbonyl* groups (—C=O) are part of fats and carbohydrates. *Carboxyl* groups (—COOH)

Group	Structure	Where found / properties
hydroxyl	—OH	In alcohols (e.g., sugars, amino acids); water soluble
methyl	—CH_3	In fatty acid chains; insoluble in water
carbonyl	—CHO (aldehyde); >CO (ketone)	In sugars, amino acids, nucleotides; water soluble. An aldehyde if at end of a carbon backbone; a ketone if attached to an interior carbon of backbone
carboxyl	—COOH (non-ionized); —COO^- (ionized)	In amino acids, fatty acids, carbohydrates; water soluble. Highly polar; acts as an acid (releases H^+)
amino	—NH_2 (non-ionized); —NH_3^+ (ionized)	In amino acids and certain nucleotide bases; water soluble, acts as a weak base (accepts H^+)
phosphate	—O—P(O^-)(=O)—O^-; icon: P	In nucleotides (e.g., ATP), also in DNA, RNA, many proteins, phospholipids; water soluble, acidic

Figure 3.3 Animated! Common functional groups in biological molecules, with examples of where they occur.

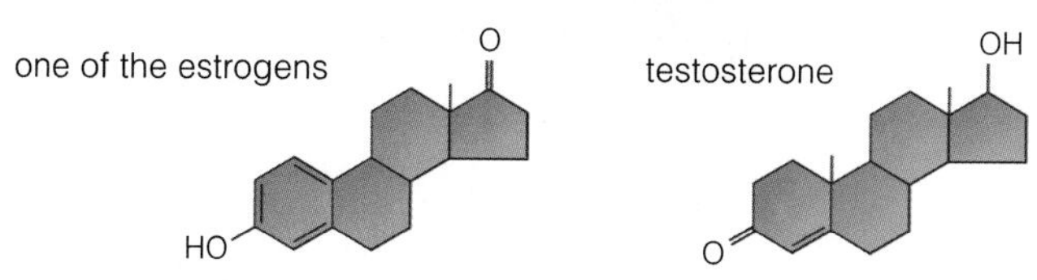

Figure 3.4 Estrogen and testosterone, sex hormones that cause differences in traits between males and females of many species such as wood ducks (*Aix sponsa*). These hormones differ only in the position of two functional groups.

make amino acids and fatty acids acidic. ATP releases chemical energy as it donates a *phosphate* group (PO_4). DNA and RNA backbones also have phosphate groups.

How much can one functional group do? Consider a seemingly minor difference in the functional groups of two structurally similar sex hormones (Figure 3.4). Early on, an embryo of a wood duck, human, or any other vertebrate is neither male nor female. If it starts making the hormone testosterone, a set of tubes and ducts will become male sex organs and male traits will develop. Without testosterone, those ducts and tubes become female sex organs, and hormones called estrogens will guide the development of female traits.

Table 3.1 What Cells Do to Organic Compounds

Class of Reaction	What Happens
Condensation	Two molecules covalently bond into a larger one.
Cleavage	A molecule splits into two smaller ones, as by hydrolysis.
Functional group transfer	One molecule gives up a functional group entirely, and a different molecule immediately accepts it.
Electron transfer	One or more electrons taken from one molecule are donated to another molecule.
Rearrangement	Juggling of internal bonds converts one type of organic compound to another.

WHAT CELLS DO TO ORGANIC COMPOUNDS

Metabolism refers to activities by which cells acquire and use energy as they construct, rearrange, and split organic compounds. These activities help each cell stay alive, grow, and reproduce. They require enzymes, or proteins that make reactions proceed faster than they would on their own. Table 3.1 lists the main metabolic reactions. For now, start thinking about two of them.

With condensation, two molecules covalently bond into a larger one. Water usually forms as a product of condensation when enzymes remove an —OH group from one of the molecules and a hydrogen atom from the other (Figure 3.5*a*). Some large molecules such as starch form by repeated condensation reactions.

One cleavage reaction, hydrolysis, is the reverse of condensation (Figure 3.5*b*). Enzymes break a bond and attach an —OH group to one of the exposed bonding sites and an H atom to the other. The —OH and H are derived from a water molecule. Hydrolysis helps cells break large molecules into smaller ones.

Cells maintain pools of small organic molecules. Some of these molecules are used as sources of energy. Others are used as subunits, or monomers, to build larger molecules that are the structural and functional parts of cells. These larger molecules, or polymers, are chains of three to millions of monomers. When cells break down a polymer, the released monomers may be used for energy, or they may reenter cellular pools.

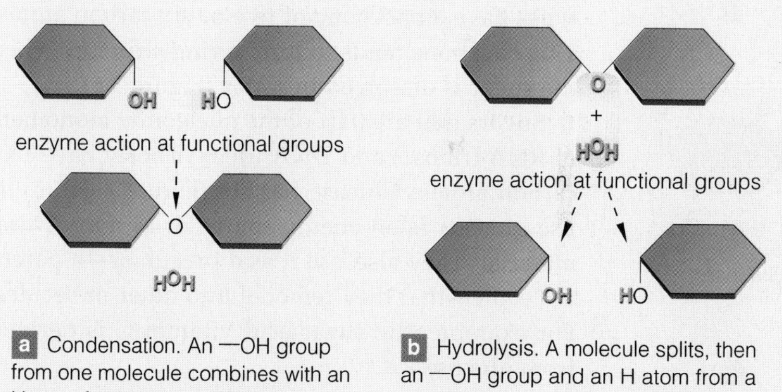

a Condensation. An —OH group from one molecule combines with an H atom from another. Water forms as the two molecules bond covalently.

b Hydrolysis. A molecule splits, then an —OH group and an H atom from a water molecule become attached to sites exposed by the reaction.

Figure 3.5 Animated! Two examples of what happens to the organic molecules in cells. (**a**) Condensation, with two molecules being covalently bonded into a larger one. (**b**) Hydrolysis, a water-requiring cleavage reaction in which a larger molecule is split into two smaller molecules.

Under present-day conditions in nature, only living cells make complex carbohydrates and lipids, proteins, and nucleic acids—the molecules of life. The main building blocks are smaller organic compounds known as simple sugars, fatty acids, amino acids, and nucleotides.

The structure of an organic molecule starts with its carbon backbone and the functional groups attached to it.

3.2 Carbohydrates—The Most Abundant Ones

Carbohydrates, lipids, proteins—you already know about these organic compounds, because they are what you eat. Foods rich in carbohydrates, such as corn, beans, peas, rice, and fruit, dominate most human diets. We consider these molecules first, then fats and other lipids, and then proteins. Nucleic acids, not on anybody's menu, are last.

Carbohydrates are organic compounds that consist of carbon, hydrogen, and oxygen in a 1:2:1 ratio. Sugars and other carbohydrates are the most abundant of all biological molecules on the planet. In cells, different kinds are stored or put to use as structural materials and as sources of instant energy. Three main types of carbohydrates are monosaccharides, oligosaccharides, and polysaccharides.

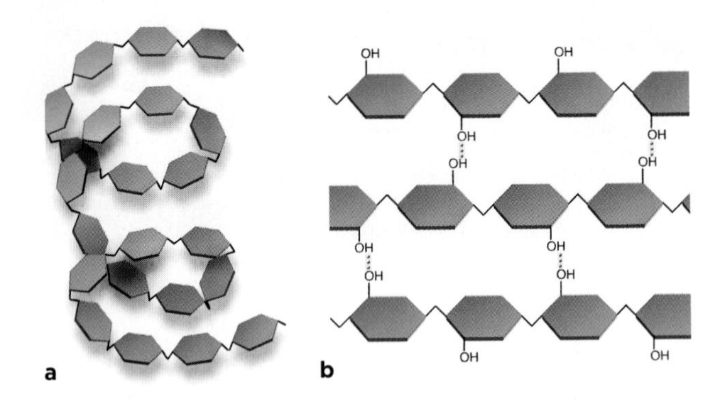

Figure 3.7 Bonding patterns for glucose units in (**a**) starch, and (**b**) cellulose. In amylose, a form of starch, a series of covalently bonded glucose units form a chain that coils. In cellulose, hydrogen bonds form between glucose chains. The pattern stabilizes the chains, which can pack tightly.

SIMPLE SUGARS

*Mono*saccharides (one sugar unit) are the simplest carbohydrates. "Saccharide" is from a Greek word that means sugar. Monosaccharides have at least two hydroxyl groups and one ketone or aldehyde group bonded to a carbon backbone. Most are water soluble and easily transported in fluids. Common types have a backbone of five or six carbon atoms. This backbone tends to form a ring structure when the sugar is dissolved in water (Figure 3.6).

Sugars that are part of the nucleotide monomers of RNA (ribose) and DNA (deoxyribose) have five carbon atoms. Glucose has six (Figure 3.6*a*). Cells use glucose as an energy source or as a structural material. They also use it as a precursor—a parent molecule—that they remodel into other molecules. For example, the sugar acid vitamin C is derived from glucose.

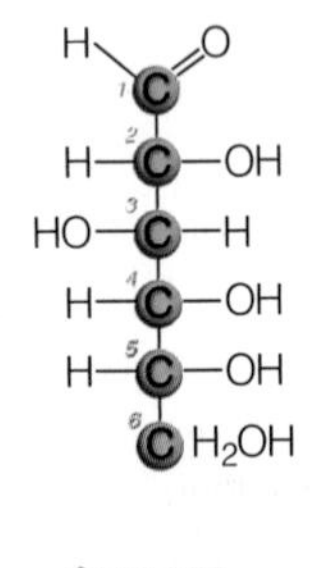

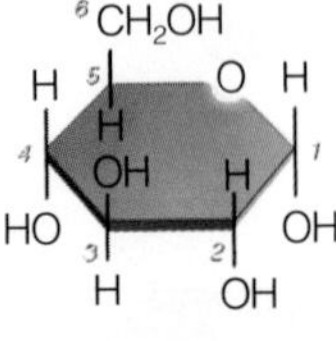

a Glucose

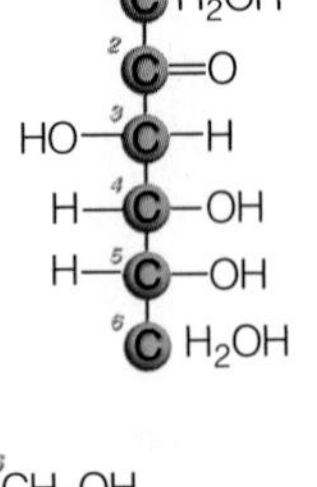

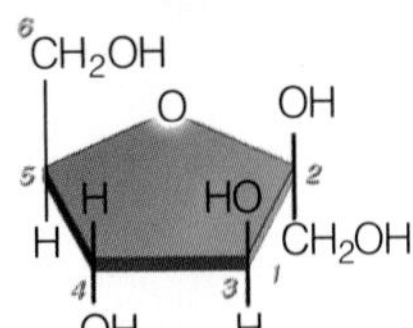

b Fructose

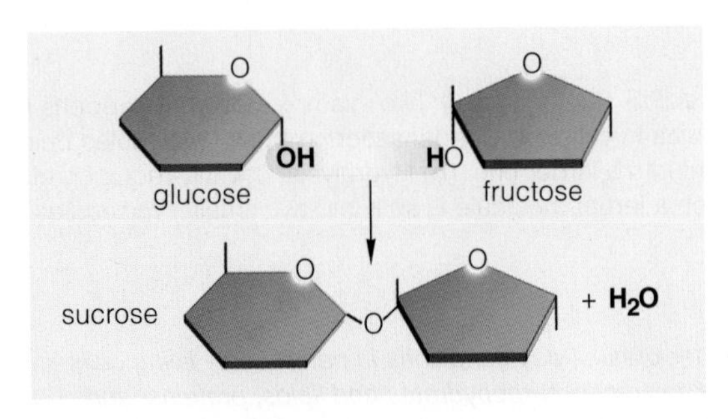

c Formation of a sucrose molecule

Figure 3.6 (**a**,**b**) Straight-chain and ring forms of glucose and fructose. For reference purposes, the carbon atoms of these simple sugars are numbered in sequence, starting at the end closest to the molecule's aldehyde or ketone group. (**c**) Condensation of two monosaccharides into a disaccharide.

SHORT-CHAIN CARBOHYDRATES

Unlike the simple sugars, an *oligo*saccharide is a short chain of covalently bonded sugar monomers. *Oligo*– means a few. As examples, the *di*saccharides have two sugar monomers. The lactose in milk is a disaccharide with a glucose and a galactose unit. Sucrose, the most plentiful sugar in nature, has a glucose and a fructose unit (Figure 3.6). Sucrose extracted from sugarcane or sugar beets is our table sugar. Oligosaccharides with three or more sugar units are often attached to lipids or proteins that have important immune functions.

COMPLEX CARBOHYDRATES

The "complex" carbohydrates, or *poly*saccharides, are straight or branched chains of many sugar monomers —often hundreds or thousands. There may be one type or many types of monomers in a polysaccharide. The most common polysaccharides are cellulose, glycogen, and starch. All consist of glucose monomers, but they differ in their chemical properties. Why? The answer begins with differences in covalent bonding patterns that link their glucose units (Figures 3.7 and 3.8).

For example, the covalent bonding pattern of starch puts each glucose unit at an angle relative to the next unit in the chain, which coils up like a spiral staircase (Figure 3.7*a*). Starch does not dissolve easily in water, so it resists hydrolysis. This stability is one reason why starch is used to store chemical energy in the water-based environment of a cell's interior.

Plant cells store their photosynthetically produced glucose as starch. However, because of its insolubility, starch cannot be transported out of the photosynthetic cells and distributed to other parts of the plant. When sugars are in short supply, hydrolysis enzymes nibble

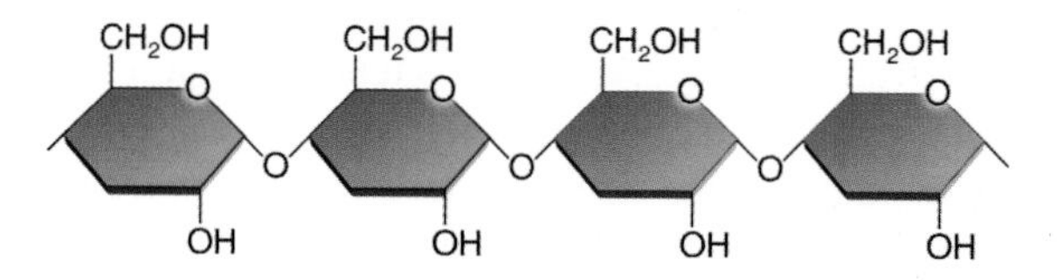

a Structure of amylose, a form of starch. Cells inside tree leaves briefly store excess glucose monomers as starch in their chloroplasts, which are tiny, membrane-bound sacs that specialize in photosynthesis.

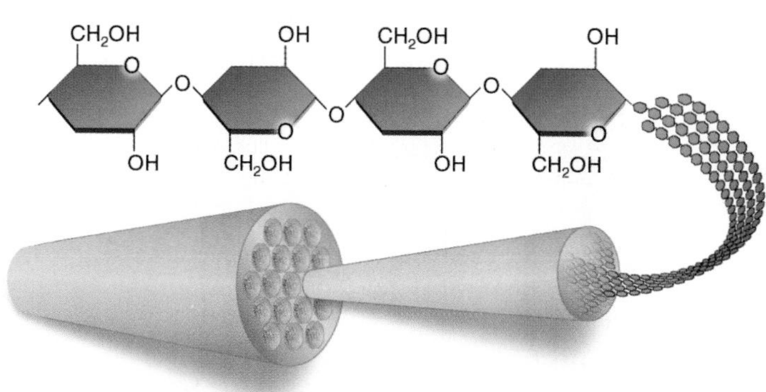

b Structure of cellulose. In cellulose fibers, chains of glucose units stretch side by side and hydrogen-bond at —OH groups. The many hydrogen bonds stabilize the chains in tight bundles that form long fibers. Few organisms produce enzymes that can digest this insoluble material. Cellulose is a structural component of plants and plant products, such as wood and cotton dresses.

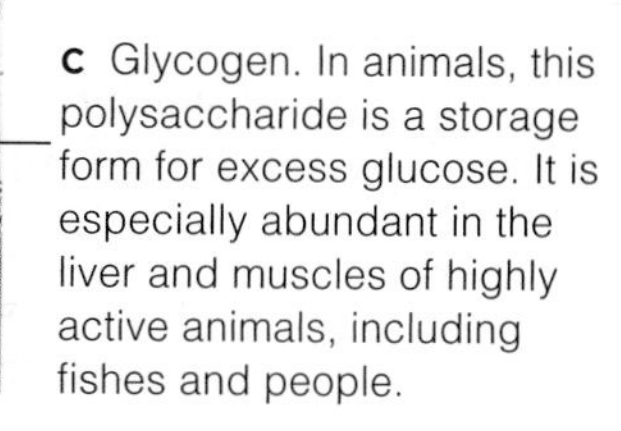

c Glycogen. In animals, this polysaccharide is a storage form for excess glucose. It is especially abundant in the liver and muscles of highly active animals, including fishes and people.

Figure 3.8 Molecular structure of (**a**) starch, (**b**) cellulose, and (**c**) glycogen, and their typical locations in a few organisms. All three carbohydrates consist only of glucose units.

at bonds between the sugar monomers in starch. The monomers are released as glucose molecules, which can be used at once as building blocks for sucrose—which is an easily transported sugar.

In cellulose, glucose chains stretch side by side and hydrogen-bond to one another, as in Figure 3.7*b*. The bonding arrangement stabilizes the chains in a tightly bundled pattern, which can resist hydrolysis by most enzymes. Plant cell walls contain long cellulose fibers (Figure 3.8*b*). Like steel rods in reinforced concrete, the tough, insoluble fibers help tall stems resist winds and other forms of mechanical stress.

In animals, glycogen is the sugar-storage equivalent of starch in plants (Figure 3.8*c*). Muscle and liver cells have large stores of it. When the sugar level in blood falls, liver cells degrade glycogen and release glucose into the blood. Exercise strenuously but briefly, and muscle cells tap glycogen for a burst of energy.

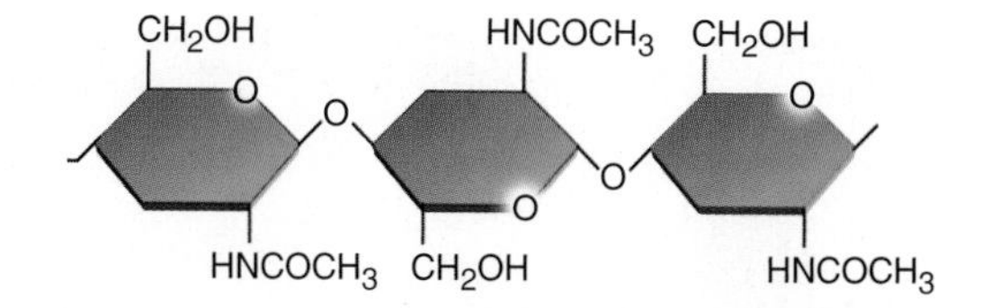

Figure 3.9 Chitin. This polysaccharide strengthens the hard parts of many small invertebrates, such as ticks.

Chitin is a modified polysaccharide that has nitrogen-containing groups attached to its many glucose monomers. Chitin strengthens the hard parts of many animals, including the cuticle of crabs, earthworms, insects, and ticks (Figure 3.9). It also reinforces the cell wall of many fungi.

Carbohydrates include monosaccharides (simple sugars such as glucose), oligosaccharides (such as sucrose), and polysaccharides (such as starch). Cells use carbohydrates as structural materials, easily transported packets of instant energy, and storage forms of energy.

3.3 Greasy, Oily—Must Be Lipids

Lipids function as the body's major energy reservoirs as well as structural materials, as in cell membranes.

Lipids are fatty, oily, or waxy organic compounds that are insoluble in water. Many lipids incorporate fatty acids: simple organic compounds with a carboxyl group that is joined to a backbone of four to thirty-six carbon atoms (Figure 3.10). The omega-3 and omega-6 fatty acids are "essential fatty acids." Your body does not make them, so they must come from food.

FATS

Fats are lipids with one, two, or three fatty acids that dangle like tails from a small alcohol called glycerol. Most *neutral* fats, such as butter and vegetable oils, are triglycerides. Triglycerides have three fatty acid tails linked to the glycerol (Figure 3.11). In vertebrates, they are the most abundant energy source, and the richest. They are concentrated in adipose tissue that insulates and cushions parts of the body, as in penguins. Gram for gram, triglycerides contain more than twice the energy of glycogen, a complex carbohydrate.

In *saturated* fats, the fatty acid backbones have only single covalent bonds. Animal fats tend to remain solid at room temperature because their saturated fatty acid tails pack tightly. The fatty acid tails of *unsaturated* fats have one or more double covalent bonds. Such rigid bonds make kinks that prevent unsaturated fats from packing tightly. Most vegetable oils are unsaturated; they tend to remain liquid at room temperature.

Some unsaturated fats are bad for you. A double bond in *cis* fatty acids keeps them kinked, but in *trans* fatty acids, a double bond keeps them straight (Figure

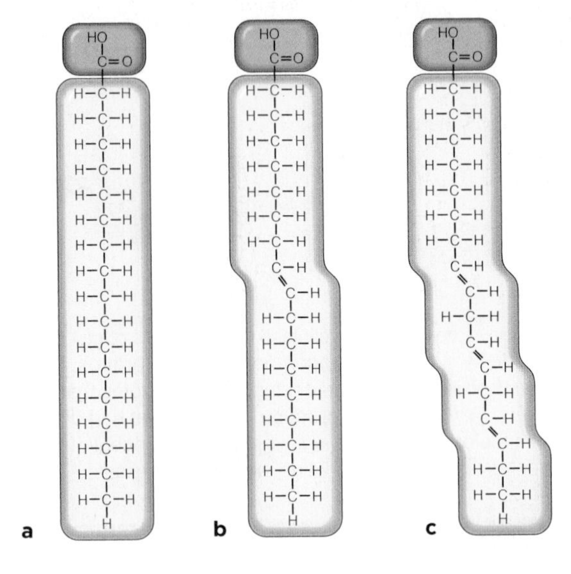

Figure 3.10 Structural formulas for three fatty acids. (**a**) Stearic acid. This backbone is fully saturated with hydrogen atoms. (**b**) Oleic acid, with a double bond in its backbone, is an unsaturated fatty acid. (**c**) Linolenic acid, also unsaturated, has three double bonds.

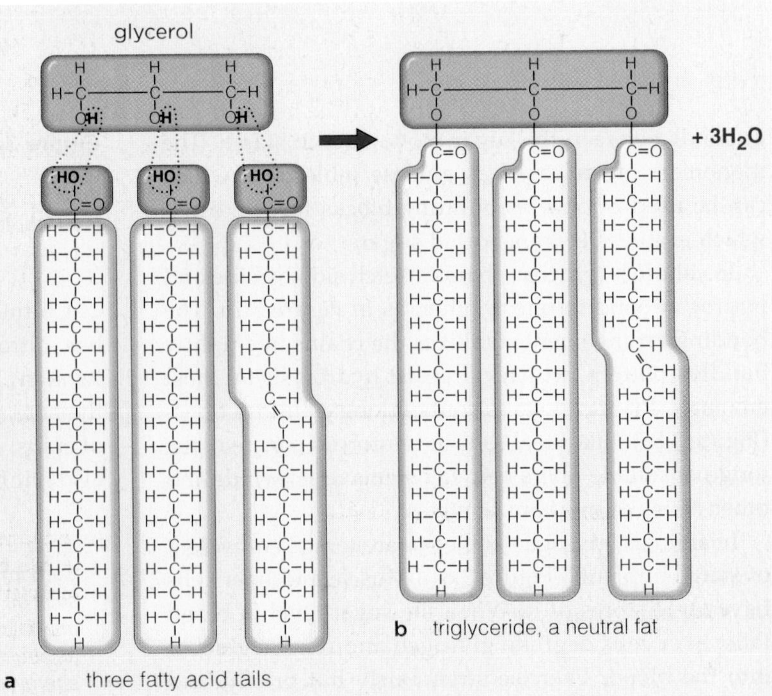

Figure 3.11 Animated! Triglyceride formation by the condensation of three fatty acids with one glycerol molecule. The photograph shows triglyceride-insulated emperor penguins during an Antarctic blizzard.

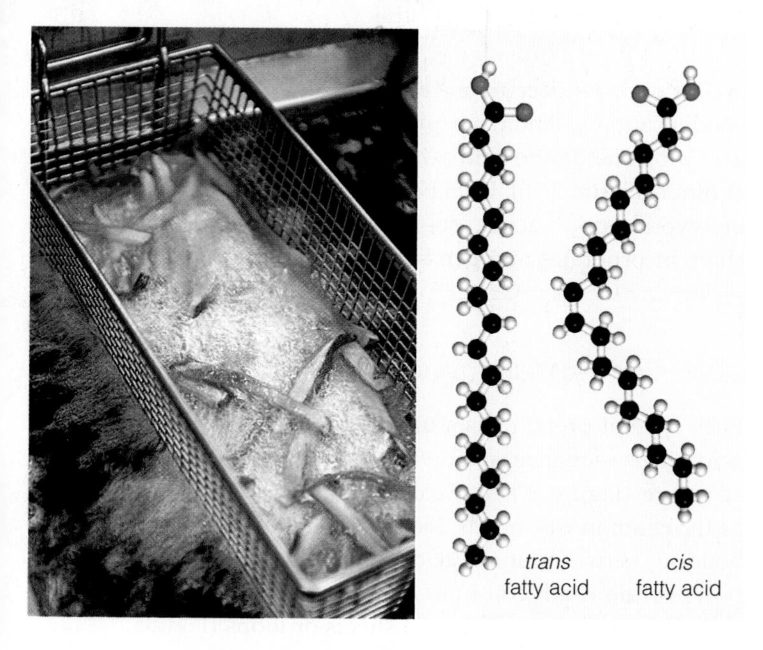

Figure 3.12 Maybe rethink the french fries? Ball-and-stick models for a *trans* fatty acid and a *cis* fatty acid.

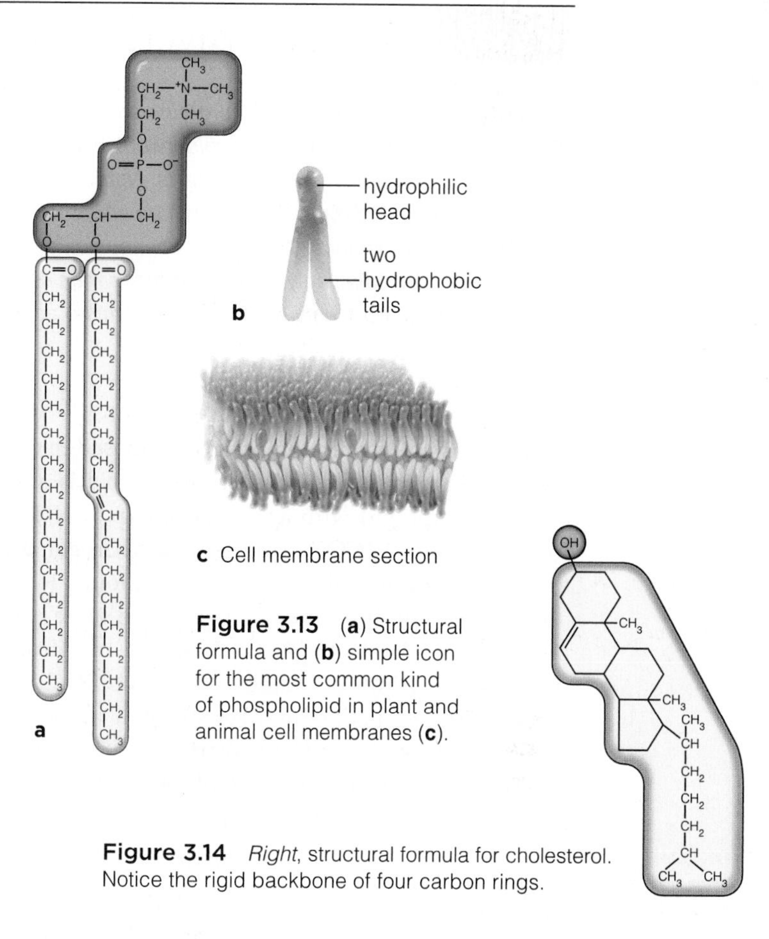

Figure 3.13 (**a**) Structural formula and (**b**) simple icon for the most common kind of phospholipid in plant and animal cell membranes (**c**).

Figure 3.14 *Right*, structural formula for cholesterol. Notice the rigid backbone of four carbon rings.

3.12). Some *trans* fatty acids occur naturally in beef, but most form by manufacturing processes that solidify vegetable oils for margarines and shortenings that are widely used in prepared foods. French fries and other fast-food products have an abundance of them. A diet high in *trans* fatty acids increases risk of heart attack.

PHOSPHOLIPIDS

Phospholipids have a polar head with a phosphate in it, and two nonpolar fatty acid tails. They are the most abundant lipids in cell membranes, which have two phospholipid layers (Figure 3.13*a–c*). The heads of one layer are dissolved in the cell's fluid interior, and the heads of the other layer are dissolved in the cell's fluid surroundings. Sandwiched between the two are all of the hydrophobic tails. You will read about membrane structure and function in Chapters 4 and 5.

WAXES

All waxes are firm, water-repellent lipids with long, tightly packed fatty acid tails bonded to long-chain alcohols or carbon rings. Waxes in the plant cuticle that covers exposed surfaces help restrict water loss

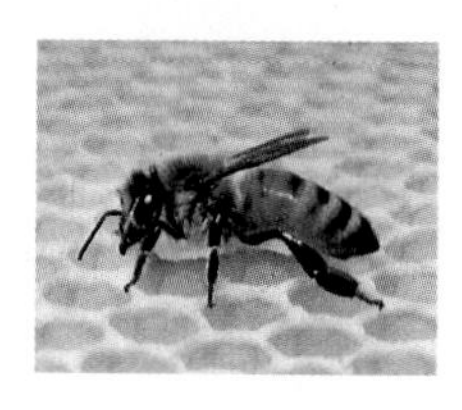

and keep out some parasites and other pests. Other waxes protect, lubricate, and soften the skin and hair. Waxes, together with fats and fatty acids, make feathers waterproof. Bees store honey and raise new generations of bees in honeycomb, which they make from beeswax.

CHOLESTEROL AND OTHER STEROLS

Sterols are lipids with a rigid backbone of four carbon rings and no fatty acid tails. They differ in the type, number, and position of their functional groups. All eukaryotic cell membranes contain sterols. In animal tissues, cholesterol is the most common type (Figure 3.14). It gets remodeled into many molecules, such as bile salts (which help digest fats) and vitamin D (required to keep teeth and bones strong). The steroid hormones also are derived from cholesterol. Estrogens and testosterone, hormones that govern reproduction and secondary sexual traits, are examples (Figure 3.5).

Being largely hydrocarbon, lipids can dissolve in other nonpolar substances, but they resist dissolving in water.

Triglycerides, or neutral fats, are the major reservoirs of energy in the vertebrate body.

Phospholipids are the main component of cell membranes.

Waxes are firm yet pliable components of water-repelling and lubricating substances.

Sterols are membrane components and precursors of steroid hormones and other important molecules.

3.4 Proteins—Diversity in Structure and Function

Of all large biological molecules, proteins are the most diverse. Structural types make up spiderwebs and feathers, hooves, hair, and many other body parts. Nutritious types abound in seeds and eggs. Most enzymes are proteins. Proteins move substances, help cells communicate, and defend the body. Amazingly, cells build thousands of different proteins from only twenty kinds of amino acids.

PROTEINS AND AMINO ACIDS

LINK TO SECTION 2.4

A protein is an organic compound composed of one or more chains of amino acids. An amino acid is a small organic compound with an amino group ($-NH_3^+$), a carboxyl group ($-COO^-$, the acid), a hydrogen atom, and one or more atoms called an R group. In most amino acids, all three are attached to the same carbon atom (Figure 3.15). Appendix V shows the structure of the twenty kinds of biological amino acids.

Protein construction involves stringing amino acids together. For each type of protein, DNA instructions specify the order in which any of the twenty kinds of amino acids will occur, one after the other. Through a condensation reaction, a peptide bond joins the amino group of one amino acid with the carboxyl group of another (Figure 3.16). Each polypeptide chain consists of several amino acids. The carbon backbone of the chain incorporates nitrogen atoms in a regular pattern: —N—C—C—N—C—C—.

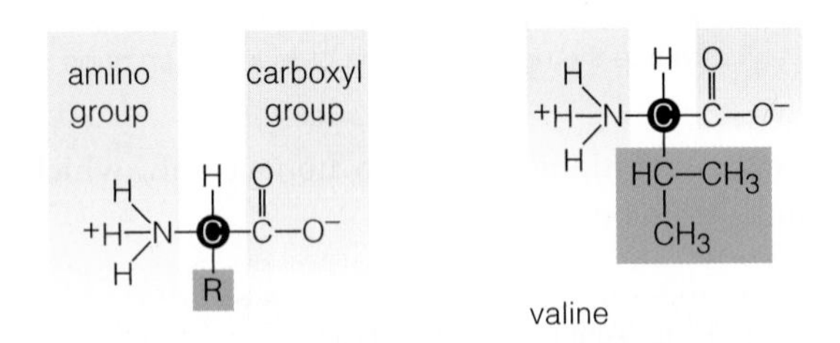

Figure 3.15 Generalized structural formula for amino acids, together with one example. *Green* boxes highlight R groups. Appendix V shows ball-and-stick models for all twenty of the common amino acids.

LEVELS OF PROTEIN STRUCTURE

Each type of protein has a unique sequence of amino acids. This sequence is known as the protein's *primary* structure (Figure 3.17*a*). *Secondary* structure emerges as the chain twists, bends, loops, and folds. Hydrogen bonding between amino acids makes stretches of the polypeptide chain coil into a helix, a bit like a spiral staircase, or makes them form sheets or loops (Figure 3.17*b*). Bear in mind, the primary structure for each type of protein is unique, but similar patterns of coils, sheets, and loops occur in most proteins.

Much as an overly twisted rubber band coils back on itself, the coils, sheets, and loops of a protein fold up even more, into compact domains. A "domain" is a part of a protein that is organized as a structurally stable unit. Such units are a protein's *tertiary* structure, its third level of organization. Tertiary structure makes a protein a working molecule. For instance, the barrel-shaped domains of some proteins function as tunnels through membranes (Figure 3.17*c*).

Many proteins have a fourth level of organization, or *quaternary* structure: They consist of two or more polypeptide chains bound together or associating with one another (Figure 3.17*d*). Most enzymes and many

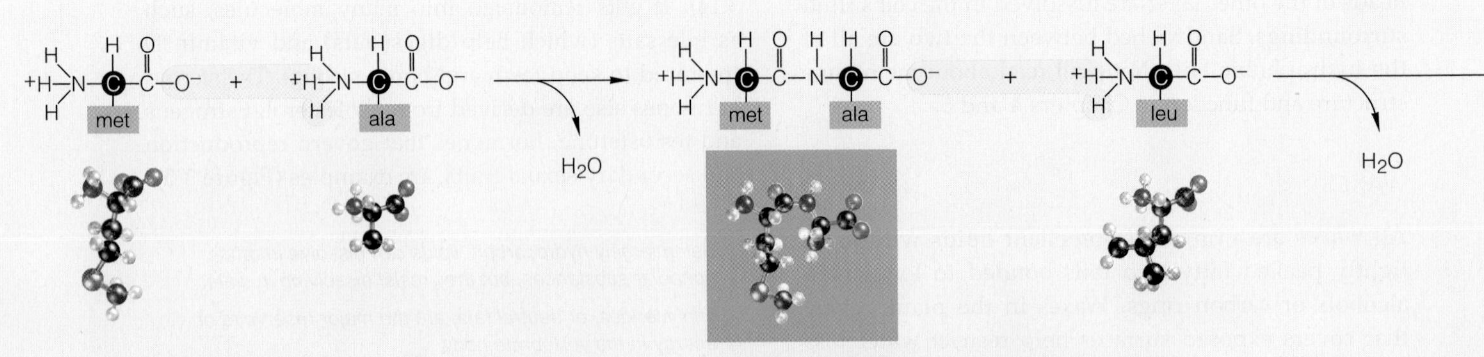

a DNA encodes the order of amino acids in a new polypeptide chain. Methionine (met) normally is the first amino acid.

b In a condensation reaction, a peptide bond forms between the methionine and the next amino acid, alanine (ala). Leucine (leu) will be next. Think about the polarity, charge, and other properties of atoms that are becoming new neighbors in the growing chain.

Figure 3.16 Animated! Example of peptide bond formation during protein synthesis in eukaryotic cells. Chapter 10 offers a closer look at protein synthesis.

other proteins are globular, with several polypeptide chains folded into shapes that are roughly spherical. Hemoglobin, described shortly, is an example.

Protein structure does not stop here. Enzymes often attach short, linear, or branched oligosaccharides to a new polypeptide chain, making a *glyco*protein. Many glycoproteins occur at the cell surface. Other enzymes attach lipids to proteins; *lipo*proteins are the result.

Some proteins aggregate by many thousands into much larger structures, with their polypeptide chains organized into strands or sheets. Some of these *fibrous* proteins contribute to the structure and organization of cells and tissues. The keratin in your fingernails is an example. Other fibrous proteins, such as the actin and myosin filaments in muscle cells, are part of the mechanisms that help cells and cell parts move.

Peptide bonds are covalent bonds that join a sequence of amino acids into a polypeptide chain. The linear sequence of amino acids in a chain is a protein's primary structure.

Local regions of a polypeptide chain become twisted and folded into helical coils, sheetlike arrays, and loops. These arrangements are the protein's secondary structure.

A polypeptide chain or parts of it become organized as structurally stable, compact, functional domains. Such domains are a protein's tertiary structure.

Many proteins have quaternary structure; they consist of two or more polypeptide chains.

a Protein primary structure: Amino acids bonded in a polypeptide chain.

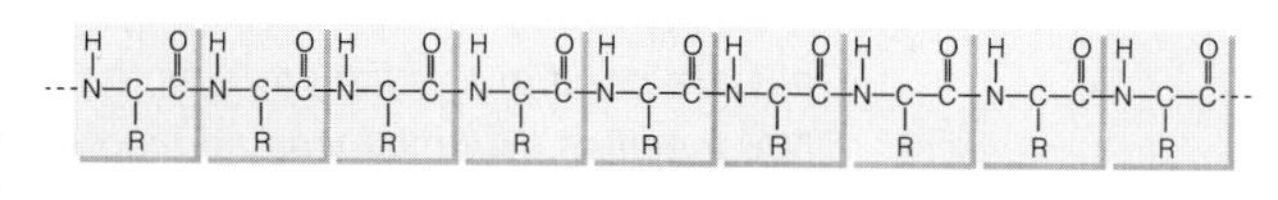

b Protein secondary structure: A coiled (helical) or sheetlike array, held in place by hydrogen bonds (*dotted lines*) between different parts of the polypeptide chain.

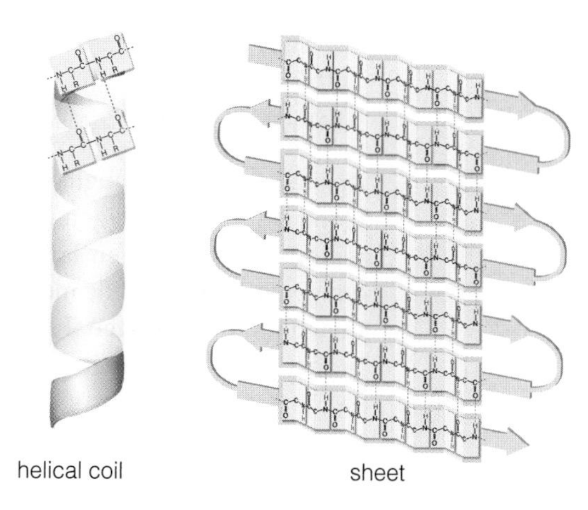

c Protein tertiary structure: A chain's coiled parts, sheetlike arrays, or both have folded and twisted into stable, functional domains, including clusters, pockets, and barrels.

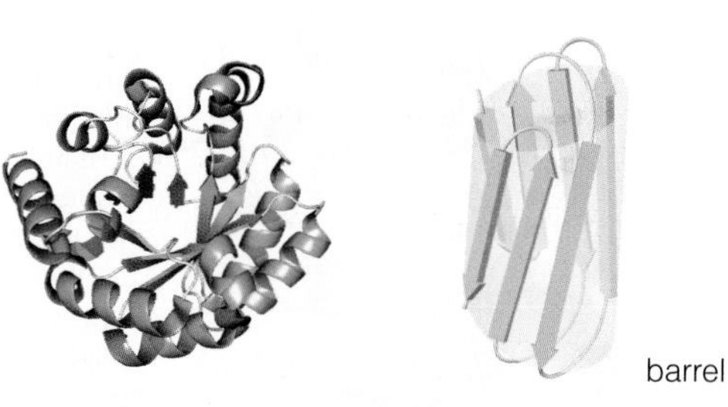

d Protein quaternary structure: Many weak interactions hold two or more polypeptide chains together as a single molecule.

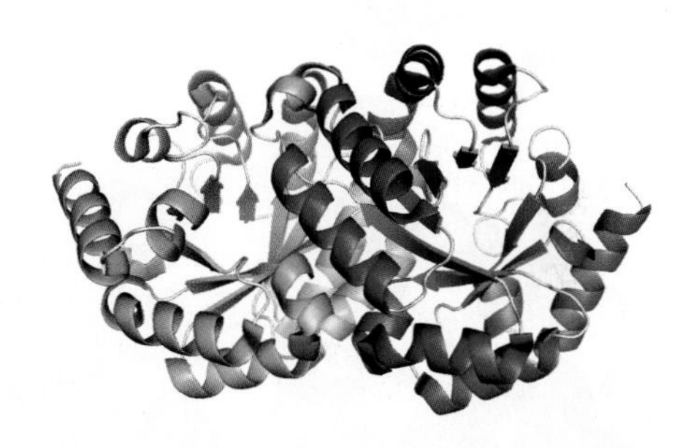

Figure 3.17 Four levels of a protein's structural organization.

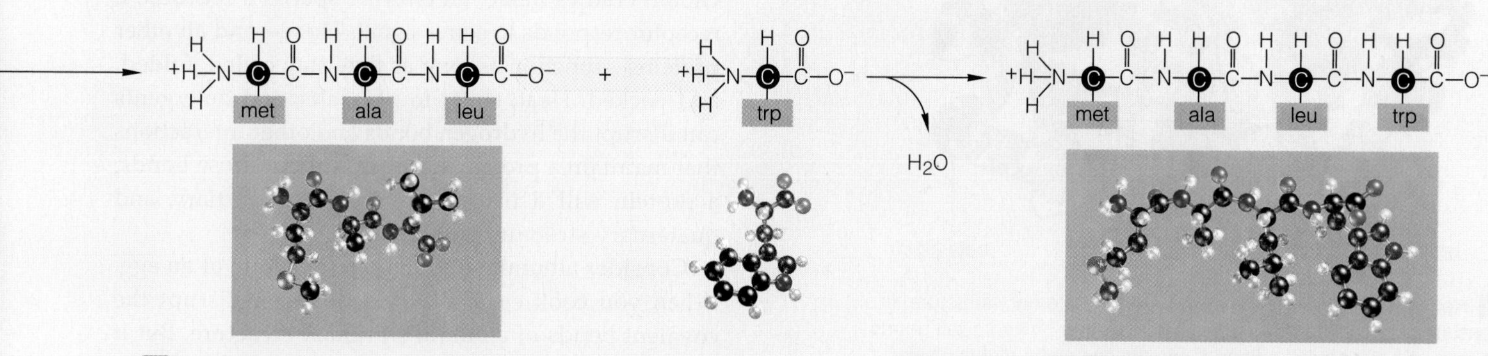

c A peptide bond forms between the alanine and leucine. Tryptophan (trp) will be next. The chain is starting to twist and fold as atoms swivel around some bonds and weakly attract or repel their neighbors.

d The sequence of amino acid subunits in this newly forming peptide chain is now met–ala–leu–trp. The process may continue until there are hundreds or thousands of amino acids in the chain.

3.5 Why Is Protein Structure So Important?

Cells are good at making proteins that are just what their DNA specifies. But mistakes and mutations happen, and they may alter a protein's primary structure. Changes in amino acid sequence may have drastic consequences.

JUST ONE WRONG AMINO ACID . . .

LINK TO SECTIONS 1.2, 2.6

As blood moves through lungs, hemoglobin inside red blood cells binds oxygen gas, then gives it up in body regions where oxygen levels are low. After releasing oxygen, red blood cells move back to the lungs, where they bind more oxygen. Hemoglobin's oxygen-binding properties depend on its structure.

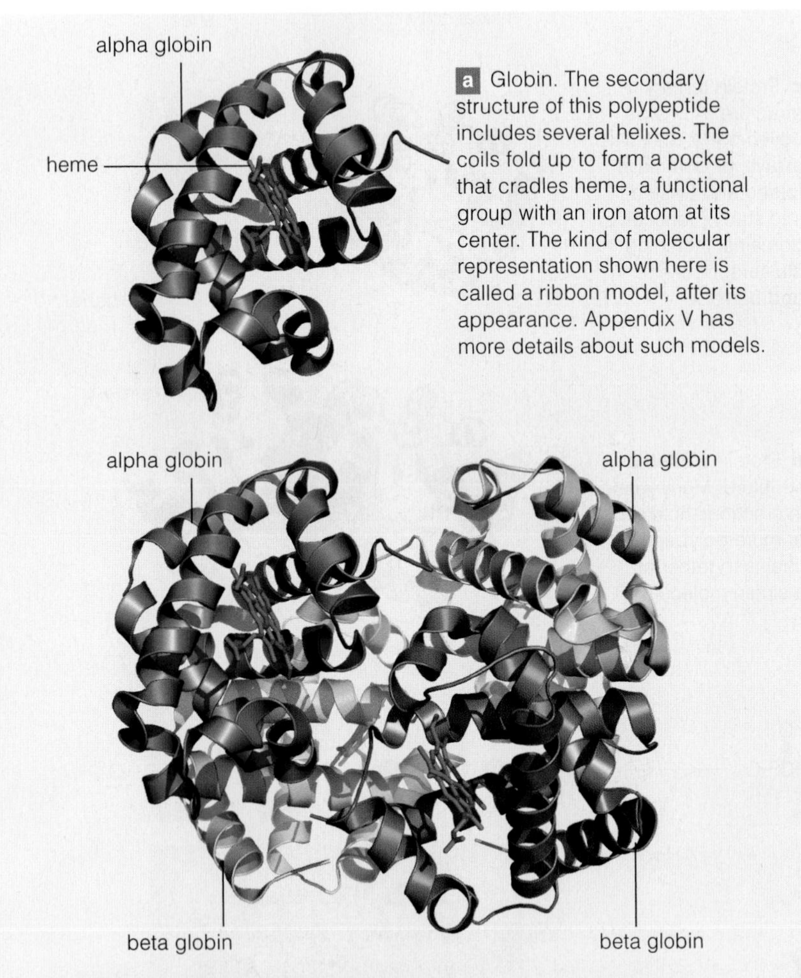

a Globin. The secondary structure of this polypeptide includes several helixes. The coils fold up to form a pocket that cradles heme, a functional group with an iron atom at its center. The kind of molecular representation shown here is called a ribbon model, after its appearance. Appendix V has more details about such models.

b Hemoglobin is one of the proteins with quaternary structure. It consists of four globin molecules held together by hydrogen bonds. To help you distinguish among them, the two alpha globin chains are shown here in *green*, and the two beta globins are in *brown*.

Figure 3.18 Animated! (**a**) Globin, a coiled polypeptide chain. The chain cradles heme, a functional group that contains an iron atom. (**b**) Hemoglobin, an oxygen-transport protein in red blood cells.

Each of the four globin chains in the hemoglobin protein forms a pocket that holds an iron-containing heme group (Figure 3.18). During its life span, each red blood cell transports billions of oxygen molecules bound to the hemes in hemoglobin molecules.

Globin comes in two slightly different forms, alpha and beta. In adult humans, two of each form make up each hemoglobin molecule. Glutamate is normally the sixth amino acid in the beta globin chain, but a DNA mutation sometimes puts a different amino acid—valine—in the chain's sixth position (Figure 3.19*a*,*b*). Unlike glutamate, which carries an overall negative charge, valine has no net charge. As a result of that one substitution, a tiny patch of the protein changes from polar to nonpolar—which in turn causes globin's behavior to change slightly. Hemoglobin that has this mutation is called HbS. Under some conditions, HbS molecules form large, stable, rod-shaped clumps. Red blood cells containing these clumps become distorted into a sickled shape (Figure 3.19*c*). Sickled cells clog tiny blood vessels and disrupt blood circulation.

Every human inherits two genes for beta globin, one from each of the two parents. (Genes are units of DNA that encode proteins.) Cells use both genes to make beta globin. If one gene is normal and the other has the valine mutation, a person can make enough normal hemoglobin to lead a relatively normal life. However, someone who has two mutant genes can only make the mutant HbS hemoglobin. The outcome is sickle-cell anemia, a severe genetic disorder. Figure 3.19*d* lists far-reaching effects of sickle-cell anemia.

PROTEINS UNDONE—DENATURATION

The shape of a protein defines its biological activity: Globin cradles heme, an enzyme speeds a reaction, a receptor responds to some signal. These—and all other proteins—function as long as they stay coiled, folded, and packed. Heat, shifts in pH, salts, and detergents can disrupt the hydrogen bonds and other interactions that maintain a protein's shape. Without those bonds, a protein will denature—its secondary, tertiary, and quaternary structure will unravel.

Consider albumin, a protein in the white of an egg. When you cook eggs, the heat does not disrupt the covalent bonds of albumin's primary structure. But it destroys albumin's weaker hydrogen bonds, and so the protein unfolds. When the translucent egg white turns opaque, we know albumin has been altered. For a few proteins, denaturation might be reversed if and when normal conditions return, but albumin is not one of them. There is no way to uncook an egg.

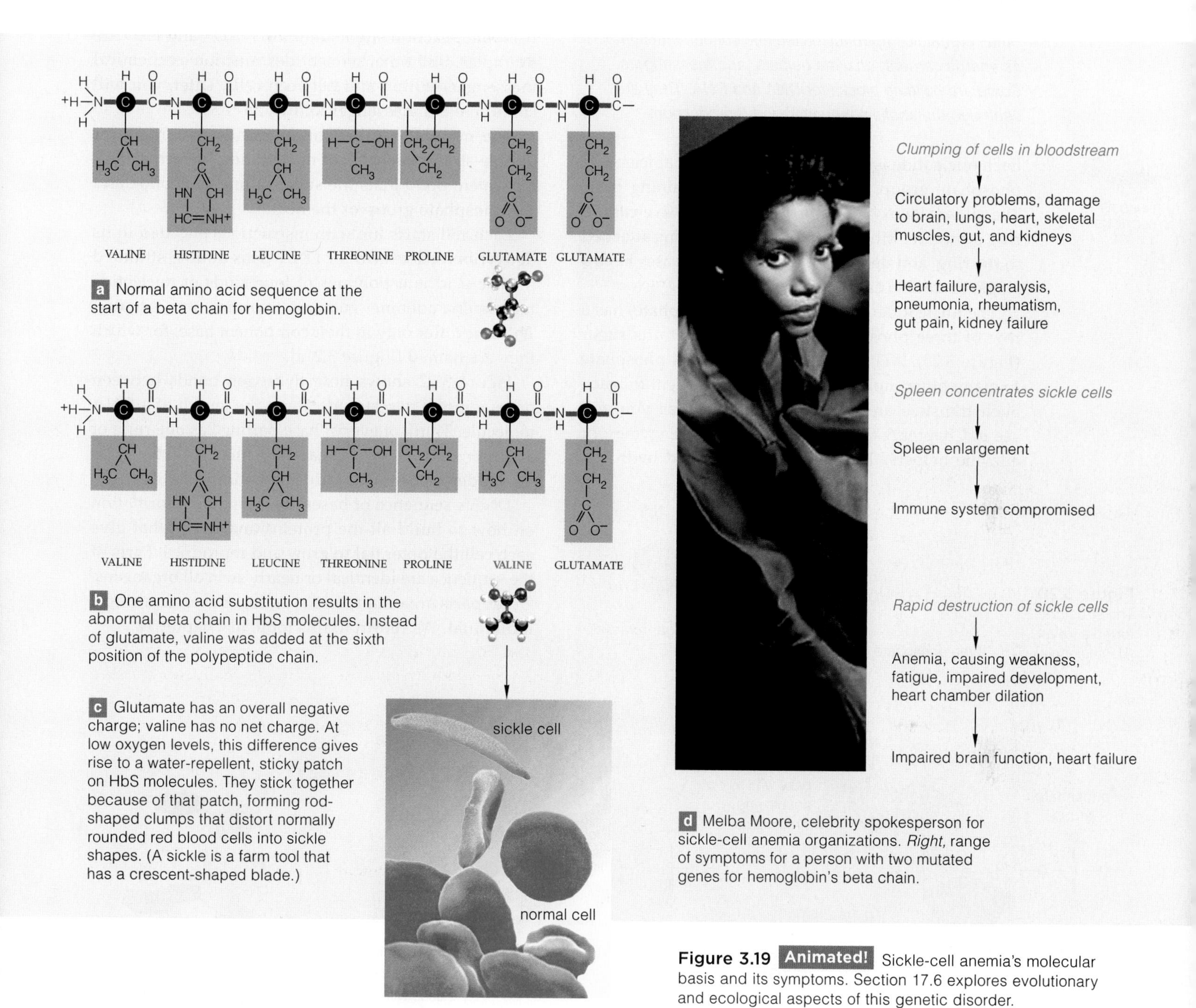

Figure 3.19 Animated! Sickle-cell anemia's molecular basis and its symptoms. Section 17.6 explores evolutionary and ecological aspects of this genetic disorder.

What is the take-home lesson? *A protein's structure dictates its function.* Hemoglobin, hormones, enzymes, transporters—such proteins help us survive. Twists and folds in their polypeptide chains form anchors, or membrane-spanning barrels, or jaws that grip foreign proteins in the body. Mutations can alter the chains enough to block or enhance an anchoring, transport, or defense function. Sometimes the consequences are awful. Yet structural and functional domain changes also give rise to variation in traits—the raw material for evolution. *Learn about protein structure and function and you are on your way to understanding life in its richly normal and abnormal expressions.*

The structure of a protein dictates its function. Mutations that alter a protein's structure can also alter its function. Occasionally, such mutations have dramatic consequences for the health of organisms that bear them.

3.6 Nucleotides, DNA, and the RNAs

Small organic compounds called nucleotides function as energy carriers, enzyme helpers, and messengers. Some are building blocks for DNA and RNA. They are central to metabolism, survival, and reproduction.

LINKS TO SECTIONS 1.2, 1.4, 2.4

Each nucleotide is composed of one sugar, at least one phosphate group, and one nitrogen-containing base. The sugar—deoxyribose or ribose—has a five-carbon ring structure. Ribose has two oxygen atoms attached to the ring, and deoxyribose has one. The bases have a single or double carbon ring structure.

The nucleotide ATP (adenosine triphosphate) has a row of three phosphate groups attached to its sugar (Figure 3.20). ATP transfers the outermost phosphate to many other molecules and so primes them to react. Such transfers are vital for metabolism, as you will see in Chapter 5. Some nucleotides are coenzymes, or enzyme helpers. They move electrons and hydrogen from one reaction site to another. NAD^+ and FAD are examples. Still other nucleotides function as chemical messengers within and between cells. Later, you will come across a messenger known as cAMP.

The molecules called nucleic acids are single- or double-stranded chains of nucleotides. In such chains, a covalent bond joins the sugar of one nucleotide and the phosphate group of the next.

Each cell starts life with instructions encoded in its deoxyribonucleic acid, or DNA. This double-stranded nucleic acid is a polymer of four kinds of nucleotide monomers: adenine, guanine, thymine, and cytosine. The four differ only in their component base, for which they are named (Figure 3.21*a*).

Figure 3.22 shows how hydrogen bonds between bases join the two strands along the length of a DNA molecule. Think of every "base pairing" as one rung of a ladder, and the two sugar–phosphate backbones as the ladder's posts. The ladder coils into a helix.

DNA's sequence of bases has heritable information on how to build all the proteins and RNAs that give each cell the potential to grow and reproduce. Parts of the sequence are identical or nearly so in all organisms. Other parts are unique to a species, or even to a single individual. We return to this topic in Chapter 12.

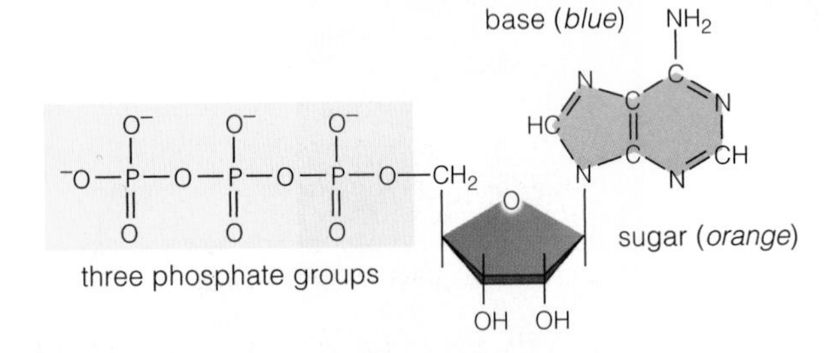

Figure 3.20 The structural formula for an ATP molecule.

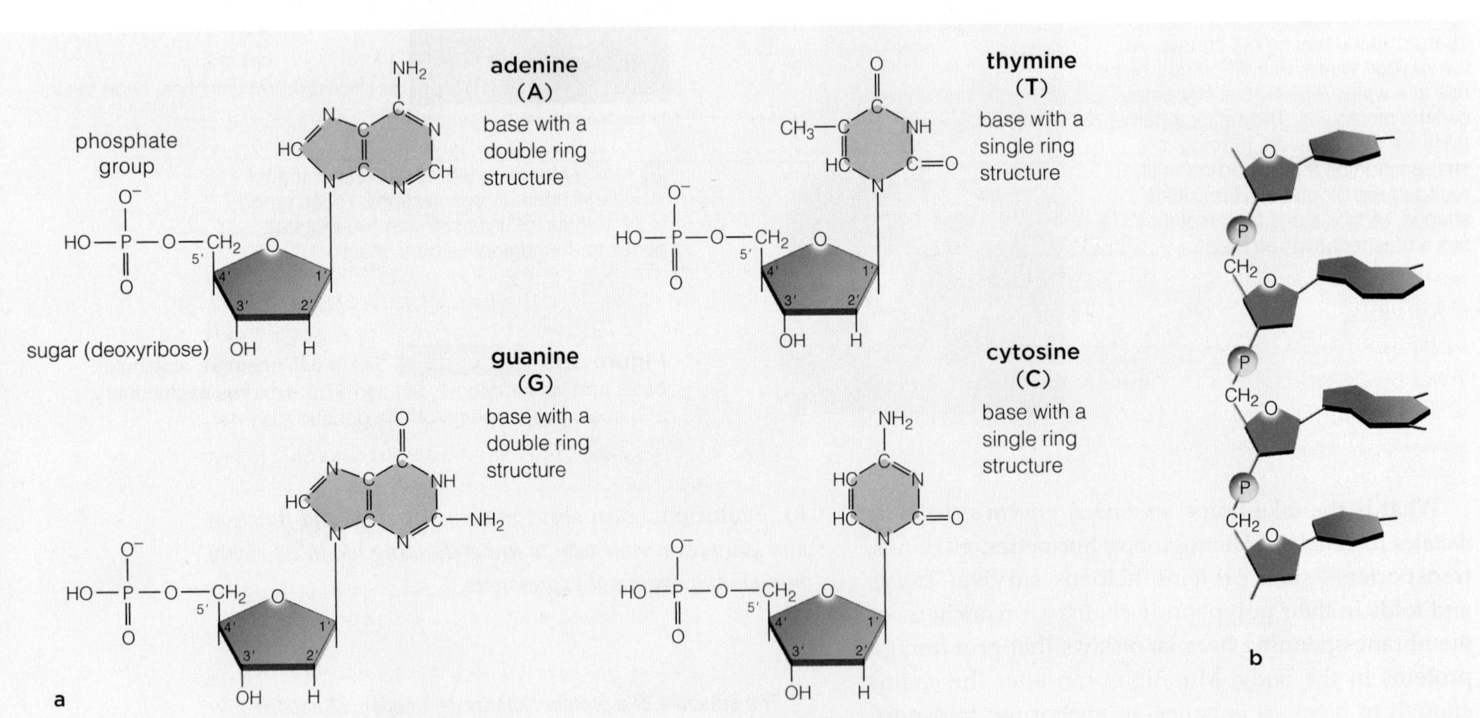

Figure 3.21 Animated! (**a**) Nucleotides of DNA. Two nucleotide bases, adenine and guanine, have a double-ring structure. Two others, thymine and cytosine, have a single-ring structure. (**b**) Bonding pattern between successive bases in nucleic acids.

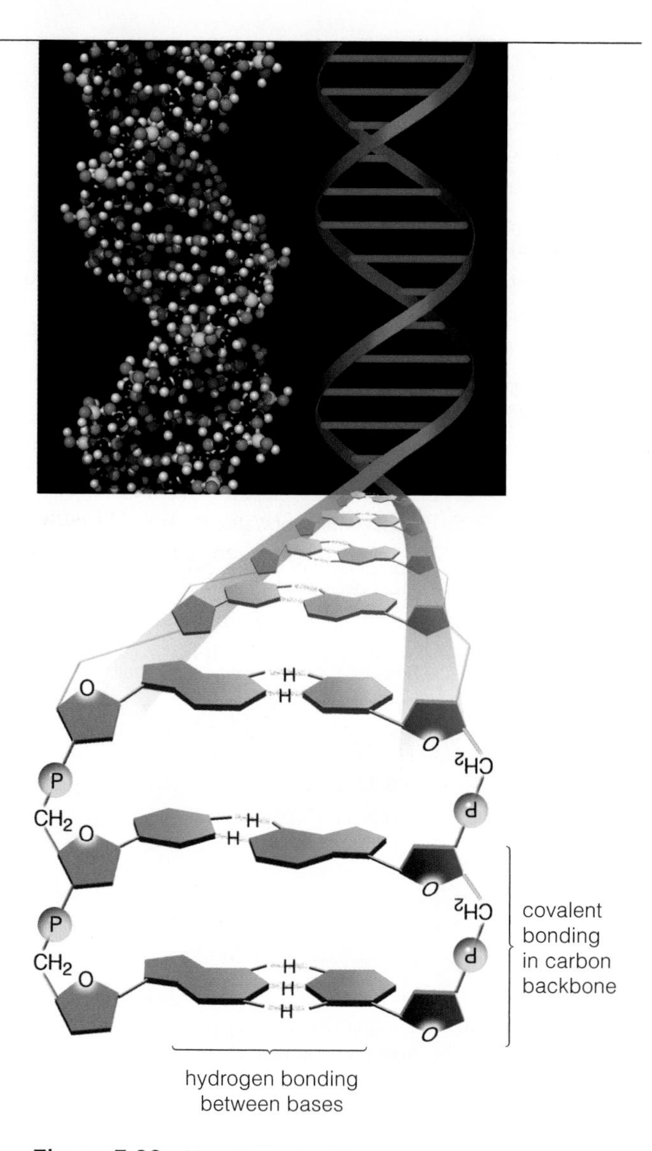

Figure 3.22 Models for the DNA molecule.

Like DNA, RNA (ribonucleic acid) has four kinds of nucleotide monomers. Unlike DNA, most RNAs are single stranded, and RNA contains uracil instead of thymine. One type of RNA is a messenger that carries eukaryotic DNA's protein-building instructions out of the nucleus and into the cytoplasm, where different RNAs translate their genetic messages into proteins. Chapter 13 returns to protein synthesis.

Nucleotides function as coenzymes, energy carriers such as ATP, chemical messengers, and monomers of the nucleic acids DNA and RNA.

DNA consists of two nucleotide strands joined by hydrogen bonds and twisted as a double helix. Its nucleotide sequence encodes heritable information.

RNA usually is a single-stranded nucleic acid. RNAs have roles in the processes by which a cell retrieves and uses genetic information in DNA to build proteins.

Summary

Section 3.1 Under present-day conditions in nature, only living cells can synthesize complex carbohydrates and lipids, proteins, and nucleic acids—the molecules of life. These molecules differ in their three-dimensional structure and function, starting with a carbon backbone and the functional groups attached to it. Looking at their structures gives us clues to how they function.

Organic compounds consist primarily of carbon and hydrogen atoms. Carbon atoms can bond covalently with as many as four other atoms, often in long chains or rings. Functional groups attached to a carbon backbone influence an organic compound's properties. Cells build large molecules from simple sugars, fatty acids, amino acids, and nucleotides. Table 3.2 (next page) summarizes these molecules.

By the process of metabolism, cells use energy to grow and maintain themselves. Enzyme-driven reactions that build, rearrange, and split organic molecules are the basis of metabolism.

- *Use the animation on ThomsonNOW to explore functional groups, condensation, and hydrolysis.*

Section 3.2 The main carbohydrates are the simple sugars, oligosaccharides, and polysaccharides. Cells use carbohydrates as instant energy sources, transportable or storable forms of energy, and structural materials.

- *Use the animation on ThomsonNOW to see how sucrose forms by condensation of glucose and fructose.*

Section 3.3 Lipids are greasy or oily molecules that tend not to dissolve in water but dissolve easily in nonpolar substances, such as other lipids. The neutral fats (triglycerides), phospholipids, waxes, and sterols are lipids. Cells use lipids as major sources of energy and as structural materials, as in cell membranes.

- *Use the animation on ThomsonNOW to see how a triglyceride forms by condensation.*

Section 3.4 Structurally and functionally, proteins are the most diverse molecules of life. Protein structure begins as a linear sequence of amino acids, a polypeptide chain (primary structure). The chains form sheets and coils (secondary structure), which pack into functional domains (tertiary structure). Many proteins, including most enzymes, consist of two or more chains (quaternary structure). Fibrous proteins form aggregates.

- *Use the animation on ThomsonNOW to explore amino acid structure and learn about peptide bond formation.*
- *Read the InfoTrac article "Protein Folding and Misfolding," David Gossard,* American Scientist, *September 2002.*

Section 3.5 A protein's structure dictates its function. Sometimes a mutation in DNA results in an amino acid substitution that alters a protein's structure enough to compromise its function. Genetic diseases such as sickle-cell anemia may result. Shifts in pH or temperature, and exposure to detergent or to salts may disrupt the many

Table 3.2 Summary of the Main Organic Molecules in Living Things

Category	Main Subcategories	Some Examples	Their Functions
CARBOHYDRATES ... contain an aldehyde or a ketone group, and one or more hydroxyl groups	**Monosaccharides** (simple sugars)	Glucose	Energy source
	Oligosaccharides (short-chain carbohydrates)	Sucrose (a disaccharide)	Most common form of sugar; the form transported through plants
	Polysaccharides (complex carbohydrates)	Starch, glycogen	Energy storage
		Cellulose	Structural roles
LIPIDS ... are mainly hydrocarbon; generally do not dissolve in water but do dissolve in nonpolar substances, such as alcohols and other lipids	**Glycerides** Glycerol backbone with one, two, or three fatty acid tails (e.g., triglycerides)	Fats (e.g., butter), oils (e.g., corn oil)	Energy storage
	Phospholipids Glycerol backbone, phosphate group, another polar group, and often two fatty acids	Lecithin	Key component of cell membranes
	Waxes Alcohol with long-chain fatty acid tails	Waxes in cutin	Conservation of water in plants
	Sterols Four carbon rings; the number, position, and type of functional groups differ among sterols	Cholesterol	Component of animal cell membranes; precursor of many steroids and vitamin D
PROTEINS ... are one or more polypeptide chains, each with as many as several thousand covalently linked amino acids	**Mostly fibrous proteins** Long strands or sheets of polypeptide chains; often strong, water-insoluble	Keratin	Structural component of hair, nails
		Collagen	Structural component of bone
		Myosin, actin	Functional components of muscles
	Mostly globular proteins One or more polypeptide chains folded into globular shapes; many roles in cell activities	Enzymes	Great increase in rates of reactions
		Hemoglobin	Oxygen transport
		Insulin	Control of glucose metabolism
		Antibodies	Immune defense
NUCLEIC ACIDS ... are chains of units (or individual units) that each consist of a five-carbon sugar, phosphate, and a nitrogen-containing base	**Adenosine phosphates**	ATP	Energy carrier
		cAMP	Messenger in hormone regulation
	Nucleotide coenzymes	NAD^+, $NADP^+$, FAD	Transfer of electrons, protons (H^+) from one reaction site to another
	Nucleic acids Chains of nucleotides	DNA, RNAs	Storage, transmission, translation of genetic information

hydrogen bonds and other molecular interactions that are responsible for the protein's shape. If a protein unfolds and loses its three-dimensional shape (denatures), it also loses its function.

■ *Use the animation on ThomsonNOW to learn more about hemoglobin structure and sickle-cell mutation.*

Section 3.6 Nucleotides are small organic molecules consisting of a sugar, a phosphate group, and a nitrogen-containing base. Different kinds have essential roles in metabolism, survival, and reproduction. ATP energizes many kinds of molecules by phosphate-group transfers. Other nucleotides function as coenzymes or as chemical messengers. DNA and the RNAs are nucleic acids, each composed of four kinds of nucleotide subunits.

DNA encodes information about the primary structure of all of a cell's proteins. Different kinds of RNA molecules interact with DNA and with one another to bring about protein synthesis.

■ *Use the animation on ThomsonNOW to explore DNA.*

Self-Quiz

Answers in Appendix III

1. Name the molecules of life and the families of small organic compounds from which they are built.

2. Each carbon atom can share pairs of electrons with as many as ______ other atom(s).
   a. one b. two c. three d. four

3. Sugars are a class of ______, which have one or more ______ groups.
   a. proteins; amino c. alcohols; hydroxyl
   b. acids; phosphate d. carbohydrates; carboxyl

4. ______ is a simple sugar (a monosaccharide).
   a. Glucose c. Ribose e. both a and b
   b. Sucrose d. Chitin f. both a and c

5. Unlike saturated fats, the fatty acid tails of unsaturated fats incorporate one or more ______.
   a. single covalent bonds b. double covalent bonds

6. Is this statement true or false? Unlike saturated fats, all of the unsaturated fats are beneficial to health because their fatty acid tails bend and do not pack together.

7. Sterols are among the lipids with no ______ .
   a. saturation c. hydrogens
   b. fatty acids tails d. carbons

8. Which of the following is a class of molecules that encompasses all of the other molecules listed?
   a. triglycerides c. waxes e. lipids
   b. fatty acids d. sterols f. phospholipids

9. ______ are to proteins as ______ are to nucleic acids.
   a. Sugars; lipids c. Amino acids; hydrogen bonds
   b. Sugars; proteins d. Amino acids; nucleotides

10. A denatured protein has lost its ______ .
   a. hydrogen bonds c. function
   b. shape d. all of the above

11. ______ consists of nucleotides.
   a. ATP b. DNA c. RNA d. all are correct

12. Which of the following nucleotides is *not* found in DNA?
   a. adenine b. uracil c. thymine d. guanine

13. Match each molecule with its most suitable description.
   ___ long chain of amino acids a. carbohydrate
   ___ energy carrier in cells b. phospholipid
   ___ glycerol, fatty acids, phosphate c. polypeptide
   ___ two strands of nucleotides d. DNA
   ___ one or more sugar monomers e. ATP

■ *Visit ThomsonNOW for additional questions.*

Critical Thinking

1. In the following list, identify the carbohydrate, the fatty acid, the amino acid, and the polypeptide:

a. $^{+}NH_3$—CHR—COO^{-} c. $(glycine)_{20}$
b. $C_6H_{12}O_6$ d. $CH_3(CH_2)_{16}COOH$

2. In 1976, researchers were developing new insecticides by modifying sugars with chlorine (Cl_2) and other toxic gases. One young member of the team misunderstood instructions to "test" a new molecule. He thought he was supposed to "taste" it. Luckily, the molecule was not insecticidal, but it was sweet. It became the food additive sucralose.

Sucralose has three chlorine atoms substituted for three hydroxyl groups of sucrose. The highly electronegative chlorine atoms make sucralose strongly electronegative (Section 2.3). Sucralose binds so strongly to sweet-taste receptors on the tongue that our brain perceives it as 600 times sweeter than sucrose. The body does not recognize sucralose as a carbohydrate. Researchers fed sucralose labeled with ^{14}C to volunteers, then analyzed radioactive molecules in their urine and feces. 92.8 percent of the sucralose passed unaltered through the body. Nonetheless, many are worried that the chlorine atoms impart toxicity to sucralose. How would you respond to that concern?

HO HO HO O O OH O HO OH HO OH
sucrose

Cl HO Cl O O OH O Cl OH HO OH
sucralose

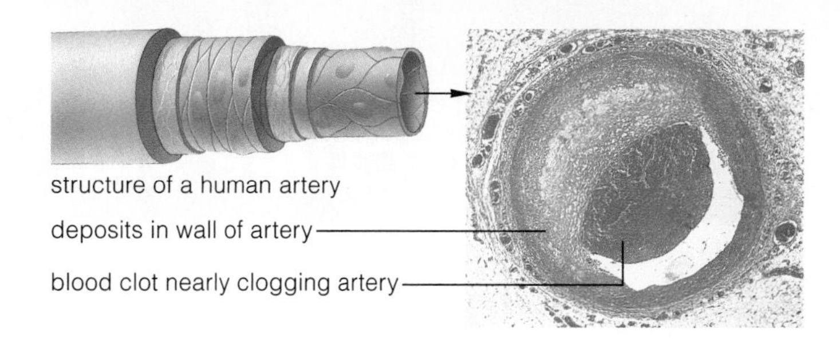

Figure 3.23 Blood clot in a patient affected by atherosclerosis.

3. Lipoproteins are relatively large, spherical clumps of protein and lipid molecules that circulate in the blood of mammals. They are like suitcases that move cholesterol, fatty acid remnants, triglycerides, and phospholipids from one place to another in the body. Given what you know about the insolubility of lipids in water, which of the four kinds of lipids would you predict to be on the outside of a lipoprotein clump, bathed in the fluid portion of blood?

4. Cholesterol from food or synthesized in the liver is too hydrophobic to circulate in blood; complexes of protein and lipids ferry it around. Low density lipoprotein, or *LDL*, transports cholesterol out of the liver and into cells. High density lipoprotein, or *HDL*, ferries the cholesterol that is released from dead cells back to the liver.

High LDL levels are implicated in atherosclerosis or clogged arteries (Figure 3.23), heart problems, and strokes. The main protein in LDL is called ApoA1. A mutant form of ApoA1 has a different amino acid (cysteine instead of arginine) at one place in its primary sequence. Carriers of this LDL mutation have very low levels of HDL, which is usually predictive of cardiovascular (heart or blood vessel) disorders. However, the carriers of the mutation have no such disorders.

Injected into patients with cardiovascular disease, the mutant LDL acted like a drain cleaner. It quickly reduced the size of cholesterol deposits in their arteries. Such a treatment may reverse years of damage. Still, many researchers caution that a low-fat, low-cholesterol diet is the best bet for long-term health. Would you choose artery-cleansing treatments over a healthy diet?

5. Protein shapes often are clues to functions. Figure 3.24 is a model for one of the HLAs, a recognition protein on the surface of vertebrate body cells. Some immune system cells use HLAs to distinguish self (the body's cells) from nonself. Each HLA has a jawlike region that can bind bits of an invader or some other threat and sound the alarm. Speculate on what may happen if a mutation makes the jawlike region misfold.

where the molecule binds and displays "enemies" (*arrow*)

one of the two chains anchors the molecule in the cell membrane

Figure 3.24 From structure to function—a protein that helps your body defend itself against bacteria and other foreign agents. HLA-A2 is composed of two polypeptide chains that are like jaws. Another protein anchors it to the plasma membrane.

4 CELL STRUCTURE AND FUNCTION

IMPACTS, ISSUES Animalcules and Cells Fill'd With Juices

Do you ever think of yourself as being about 1/1000 of a kilometer tall? Probably not, yet that is how we measure cells. Use the scale bars in Figure 4.1 like a ruler and you see that the cells shown are a few micrometers "tall." A micrometer is one-thousandth of a millimeter, which is one-thousandth of a meter, which is one-thousandth of a kilometer. The cells shown are one of the bacterial species. Bacteria are among the smallest and structurally simplest cells on Earth. Cells that make up your body are generally larger and more complex than bacteria. Your body cells are descendants of ancient species in which a nucleus and other types of internal compartments first evolved.

Nearly all cells are invisible to the naked eye. No one knew about them until the seventeenth century, when a few scholars created the first microscopes in Italy, then in France and England. Those microscopes were not much to speak of. For instance, Galileo Galilei simply put two glass lenses inside a cylinder, but the arrangement was good enough to magnify an insect's eyes.

At midcentury, Robert Hooke magnified a bit of thinly sliced cork from a mature tree and saw tiny compartments (Figure 4.2). He named them *cellulae*—a Latin word for small rooms—and thus coined the biological term "cell." Actually they were dead plant cell walls, which is what cork is made of, but Hooke did not think of them as being dead because neither he nor anyone else knew cells could be alive. He observed cells "fill'd with juices" in green plant tissues but did not realize they were alive, either.

Given the simplicity of their instruments, it is amazing that the pioneers in microscopy observed as much as they did. Antoni van Leeuwenhoek, a Dutch shopkeeper, had exceptional skill in constructing lenses and possibly the keenest vision. In the late 1600s, he was spying on sperm, protists, bacteria, and scrapings of tartar from his teeth—in which he saw "many very small animalcules, the motions of which were very pleasing to behold."

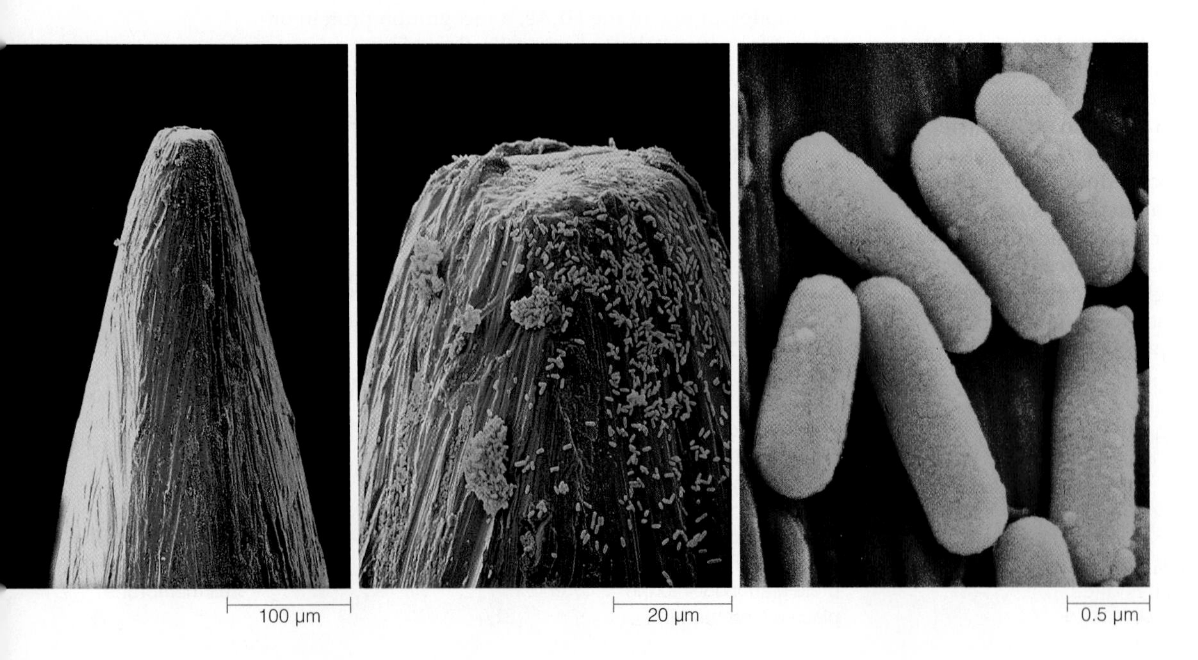

Figure 4.1 How small are cells? This example will give you an idea. The "µm" is an abbreviation for micrometer. This population of rod-shaped bacterial cells is attached to the tip of a household pin and is shown at three increasingly higher magnifications.

How would you vote? Nanoparticles deliver chemicals of many new cosmetics, sunblocks, and other products into the skin. Some scientists worry about their effects inside the body. Should products that contain nanoparticles say so on the label? See ThomsonNOW for details, then vote online.

By the 1820s, improved lenses brought cells into sharper focus. Robert Brown, a botanist, was the first to identify a plant cell nucleus. Matthias Schleiden, another botanist, hypothesized that a plant cell develops as an independent unit even though it is part of the plant. By 1839, after years of research, zoologist Theodor Schwann reported that the tissues of animals as well as plants are made of cells and their products. He realized that cells have an individual life of their own even when part of a multicelled body. Another insight emerged after the physiologist Rudolf Virchow had studied how cells grow and reproduce—that is, how they divide into daughter cells. Every cell, he realized, had descended from another living cell.

These early microscopic observations yielded three generalizations that came to be known as the cell theory: First, every organism consists of one or more cells and their products. Second, the cell is the smallest unit of organization having the properties of life. Third, the continuity of life arises directly from the growth and division of single cells.

Use this chapter as your overview of the defining features of cells and as a guide for interpreting images from microscopy. It can help you think critically about issues that affect your own health.

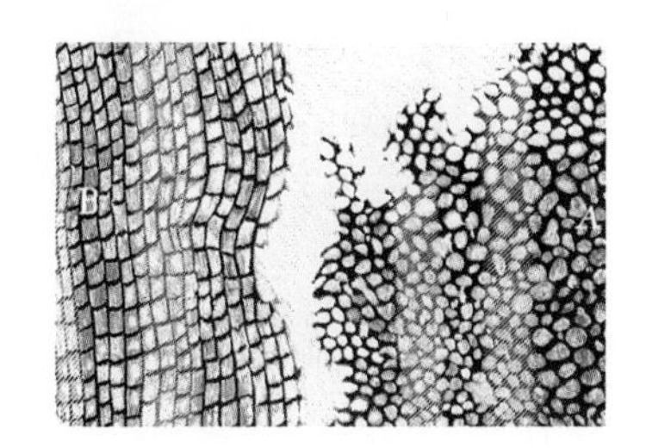

Figure 4.2 Robert Hooke's microscope and part of one of his sketches of cell walls from cork tissue.

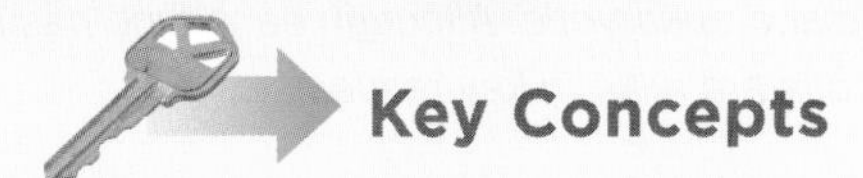

Key Concepts

WHAT ALL CELLS HAVE IN COMMON

Each cell has a plasma membrane, a boundary between its interior and the outside environment. The interior consists of cytoplasm and an innermost region of DNA. **Section 4.1**

MICROSCOPES

Microscopic analysis supports three generalizations of the cell theory: Each organism consists of one or more cells and their products, a cell has a capacity for independent life, and each new cell is descended from a living cell. **Section 4.2**

COMPONENTS OF CELL MEMBRANES

All cell membranes are mostly a lipid bilayer—two layers of lipids—and a variety of proteins. The proteins have diverse tasks, including control over which water-soluble substances cross the membrane at any given time. **Section 4.3**

PROKARYOTIC CELLS

Archaeans and bacteria are prokaryotic cells, which have few, if any, internal membrane-enclosed compartments. In general, they are the smallest and structurally the simplest cells. **Sections 4.4, 4.5**

EUKARYOTIC CELLS

Cells of protists, plants, fungi, and animals are eukaryotic; they have a nucleus and other membrane-enclosed compartments. They differ in internal parts and surface specializations. **Sections 4.6–4.11**

A LOOK AT THE CYTOSKELETON

Diverse protein filaments reinforce a cell's shape and keep its parts organized. As some filaments lengthen and shorten, they move chromosomes or other structures to new locations. **Section 4.12**

Links to Earlier Concepts

Reflect on the Section 1.1 overview of levels of organization in nature. You will see how the properties of cell membranes emerge from the organization of lipids and proteins (3.3, 3.4). You will consider the cellular location of DNA (3.6) and the sites where carbohydrates are built and broken apart (3.1, 3.2). You will also expand your understanding of the vital roles of proteins in cell functions (3.4, 3.5), and see how a nucleotide helps control cell activities (3.6).

4.1 What Is a Cell?

There are two fundamental categories of cells in nature—prokaryotic and eukaryotic. Although each type has unique traits, all cells are alike in key respects.

THE BASICS OF CELL STRUCTURE

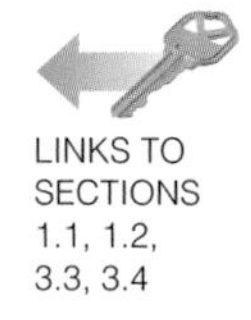

LINKS TO SECTIONS 1.1, 1.2, 3.3, 3.4

The cell is the smallest unit that shows the properties of life, which means it has a capacity for metabolism, homeostasis, growth, and reproduction. The interior of a eukaryotic cell is divided into various functional compartments, including a nucleus. Prokaryotic cells are usually smaller and simpler; none has a nucleus. Cells differ in size, shape, and activities. Yet, as Figure 4.3 suggests, all cells are similar in three respects. All cells start out life with a plasma membrane, a DNA-containing region, and cytoplasm:

1. A plasma membrane is the cell's outer membrane. It separates metabolic activities from events outside of the cell, but does not isolate the cell's interior. Water, carbon dioxide, and oxygen can cross it freely. Other substances cross only with the assistance of membrane proteins. Still others are kept out entirely.

2. All eukaryotic cells start life with a nucleus. This double-membraned sac holds a eukaryotic cell's DNA. The DNA inside prokaryotic cells is concentrated in a region of cytoplasm called the nucleoid.

3. Cytoplasm is the semifluid mixture of water, ions, sugars, and proteins between the plasma membrane and the region of DNA. Cell compartments and other components are suspended in cytoplasm. For instance, cytoplasm holds many ribosomes, structures on which proteins are built.

PREVIEW OF CELL MEMBRANES

The structural foundation of all cell membranes is the lipid bilayer, a double layer of lipids organized so that their hydrophobic tails are sandwiched between their hydrophilic heads (Figure 4.4). Phospholipids are the most abundant type of lipid in a cell membrane. Many different proteins embedded in a bilayer or attached to one of its surfaces carry out membrane functions. For example, some proteins form channels through a bilayer; others pump substances across it. In addition to a plasma membrane, many cells also have internal membranes that form channels or enclose sacs. These membranous structures compartmentalize tasks such as building, modifying, and storing substances. Section 4.3 offers a closer look at membrane structure.

CELL SIZES AND SHAPES

Are any cells big enough to be seen without the help of a microscope? A few. They include "yolks" of bird eggs, cells in watermelon tissues, and amphibian and fish eggs. These cells can be large because they are not doing too much, metabolically speaking, at maturity. Most of their volume simply acts as a warehouse.

A physical relationship, the surface-to-volume ratio, strongly influences cell size and shape. By this ratio, an object's volume increases with the cube of its diameter, but its surface area increases only with the square.

Apply this constraint to a round cell. As Figure 4.5 shows, *when a cell expands in diameter during growth, its volume increases faster than its surface area does.* Imagine that a round cell expands four times in diameter. The volume of the cell increases 64 times (4^3), but its surface

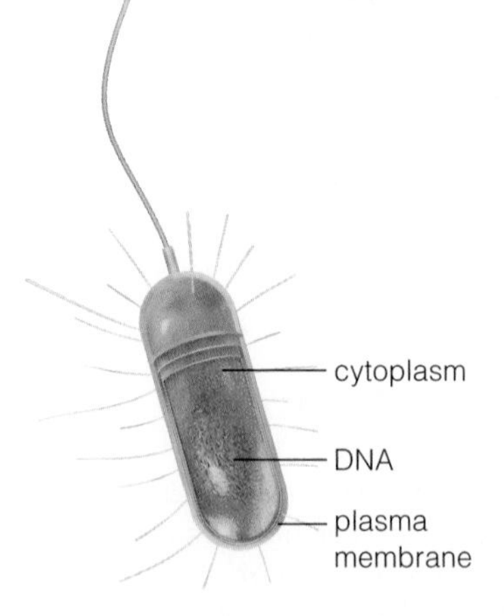

a Bacterial cell (prokaryotic)

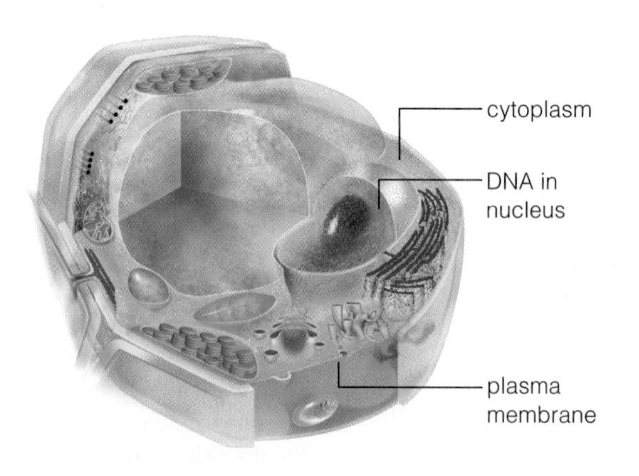

b Plant cell (eukaryotic)

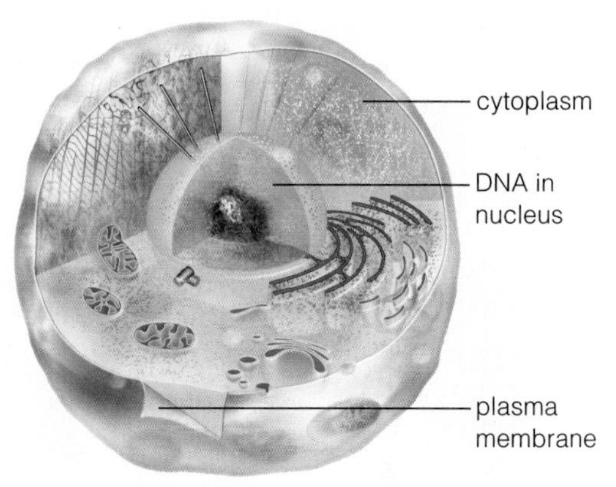

c Animal cell (eukaryotic)

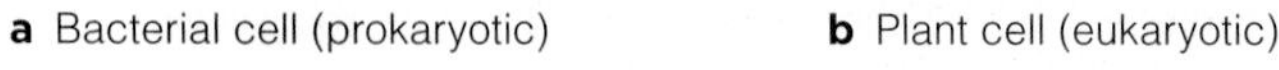

Figure 4.3 Overview of the general organization of prokaryotic cells and eukaryotic cells. The three examples are not drawn to the same scale.

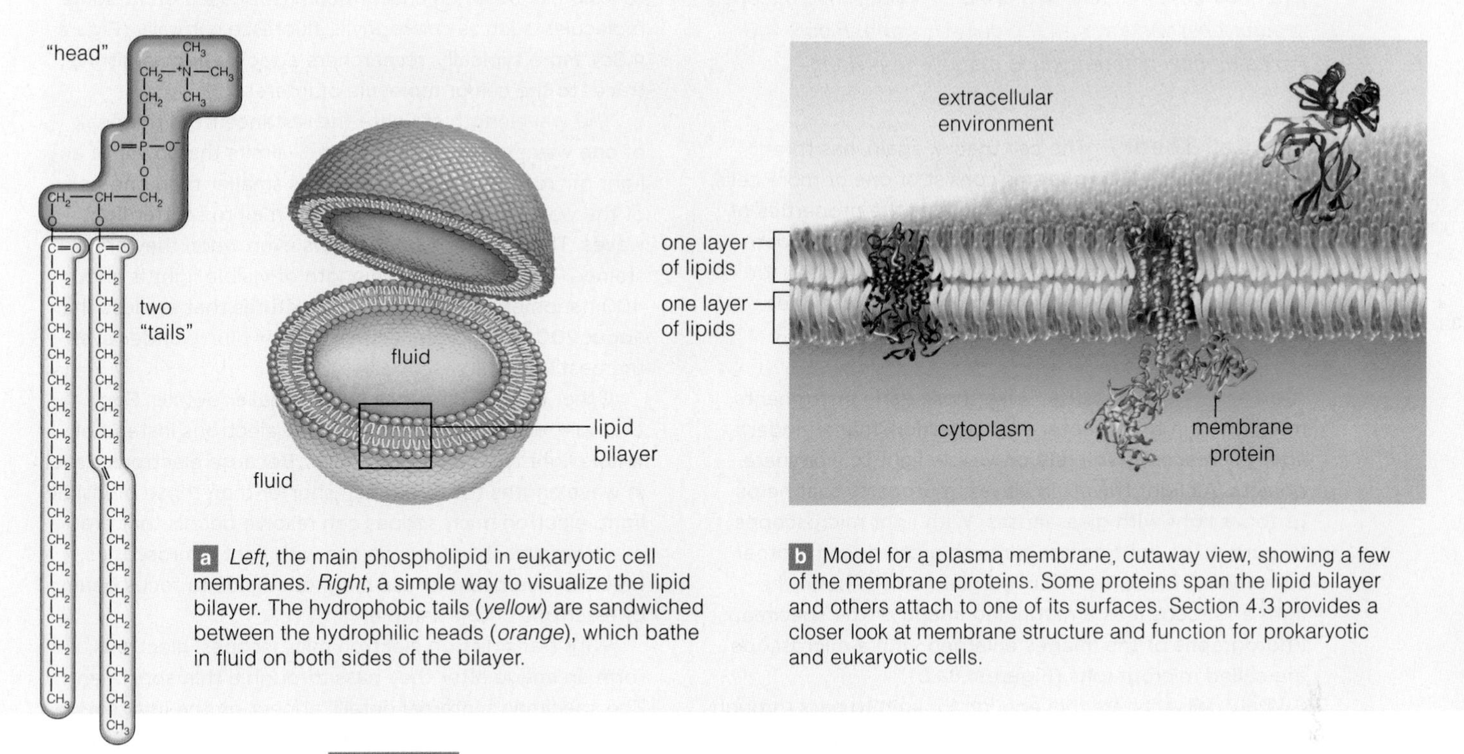

a *Left*, the main phospholipid in eukaryotic cell membranes. *Right*, a simple way to visualize the lipid bilayer. The hydrophobic tails (*yellow*) are sandwiched between the hydrophilic heads (*orange*), which bathe in fluid on both sides of the bilayer.

b Model for a plasma membrane, cutaway view, showing a few of the membrane proteins. Some proteins span the lipid bilayer and others attach to one of its surfaces. Section 4.3 provides a closer look at membrane structure and function for prokaryotic and eukaryotic cells.

Figure 4.4 **Animated!** Typical components of cell membranes.

Figure 4.5 **Animated!** Three examples of the surface-to-volume ratio. This physical relationship between increases in volume and surface area constrains cell size and shape.

Diameter (cm)	2	3	6
Surface area (cm^2)	12.6	28.2	113
Volume (cm^3)	4.2	14.1	113
Surface-to-volume ratio	3:1	2:1	1:1

area increases only 16 times (4^2). Each unit of plasma membrane now handles exchanges with four times as much cytoplasm. If a cell's circumference gets too big, the inward flow of nutrients and outward flow of wastes would not be fast enough to keep the cell alive.

A big, round cell would also have trouble moving substances through its cytoplasm. Molecules disperse by their own random motions, but they move only so quickly. Nutrients or wastes would not be distributed fast enough to keep up with a large, round, active cell's metabolism. That is why many cells are long and thin, or frilly surfaced with folds that increase surface area. The surface-to-volume ratio of such cells is enough to sustain their metabolism. The amount of raw materials that cross the plasma membrane, and the speed with which they are distributed through cytoplasm, satisfy the cell's needs. Wastes are also removed fast enough to keep the cell from getting poisoned.

Surface-to-volume constraints also affect the body plans of multicelled species. For example, small cells attach end to end in strandlike algae, so each interacts directly with its surroundings. Muscle cells in your thighs are as long as the muscle in which they occur, but each is thin, so it exchanges substances efficiently with fluids in the tissue surrounding it.

All cells start life with a plasma membrane, cytoplasm, and a region of DNA.

The DNA of eukaryotic cells is enclosed by a nucleus. In prokaryotic cells, the DNA is concentrated in a region of cytoplasm called the nucleoid.

A lipid bilayer forms the structural framework of all cell membranes. Proteins in the bilayer or attached to one of its surfaces carry out diverse membrane functions.

The surface-to-volume ratio limits cell size.

4.2 How Do We See Cells?

Like their centuries-old forerunners, modern microscopes are our best windows on the cellular world. Figure 4.6 indicates how different kinds magnify that world.

LINK TO SECTION 2.2

The Cell Theory The cell theory, again, has three generalizations: All organisms consist of one or more cells, the cell is the smallest unit that retains the properties of life, and each new cell arises from another cell. Modern microscopy still supports the theory. Microscopists, for instance, have yet to find a living organism that does not consist of one or more cells.

Modern Microscopes Like those early instruments mentioned in this chapter's introduction, many modern *light microscopes* still rely on visible light to illuminate objects. All light travels in waves, a property that helps us focus light with glass lenses. With light microscopes (Figure 4.7*a*), light passes through a cell or some other specimen. One or more curved glass lenses bend the light and focus it as a magnified image of the specimen. Photographs of the images enlarged with a microscope are called micrographs (Figure 4.8*a,b*).

Only cells that are thin enough for light to pass through will be visible with a *light microscope*. Even so, most cells are nearly colorless. They may look featureless under a light microscope unless they are stained, which means exposed to dyes that only some cell parts will soak up. The parts that absorb the most dye appear darkest.

With a *fluorescence microscope*, a cell or a molecule is the light source; it fluoresces, or emits energy in the form of visible light, when a laser beam is focused on it. Some molecules, such as chlorophylls, fluoresce naturally (Figure 4.8*c*). More typically, researchers attach a light-emitting tracer to the cell or molecule of interest.

The wavelength of light—the distance from the peak of one wave to the peak behind it—limits the power of any light microscope. Why? Structures smaller than one-half of the wavelength of light are too small to scatter light waves. They do not cast shadows even when they are stained. The smallest wavelength of visible light is about 400 nanometers. That is why structures that are less than about 200 nanometers across appear blurry under even the best light microscopes.

Other microscopes can reveal smaller details. For example, *electron microscopes* use electrons instead of visible light to illuminate samples. Because electrons travel in wavelengths that are much shorter than those of visible light, electron microscopes can resolve details that are much smaller than you can see with light microscopes. Electron microscopes use magnetic fields to focus beams of electrons onto a sample.

With transmission electron microscopes, electrons form an image after they pass through a thin specimen. The specimen's internal details appear on the image as shadows (Figure 4.8*d*). Scanning electron microscopes direct a beam of electrons back and forth across a surface of a specimen, which has been coated with a thin layer of gold or another metal. The metal emits both electrons and x-rays, which are converted into an image of the surface (Figure 4.8*e*). Both types of electron microscopes can resolve structures as small as 0.2 nanometer.

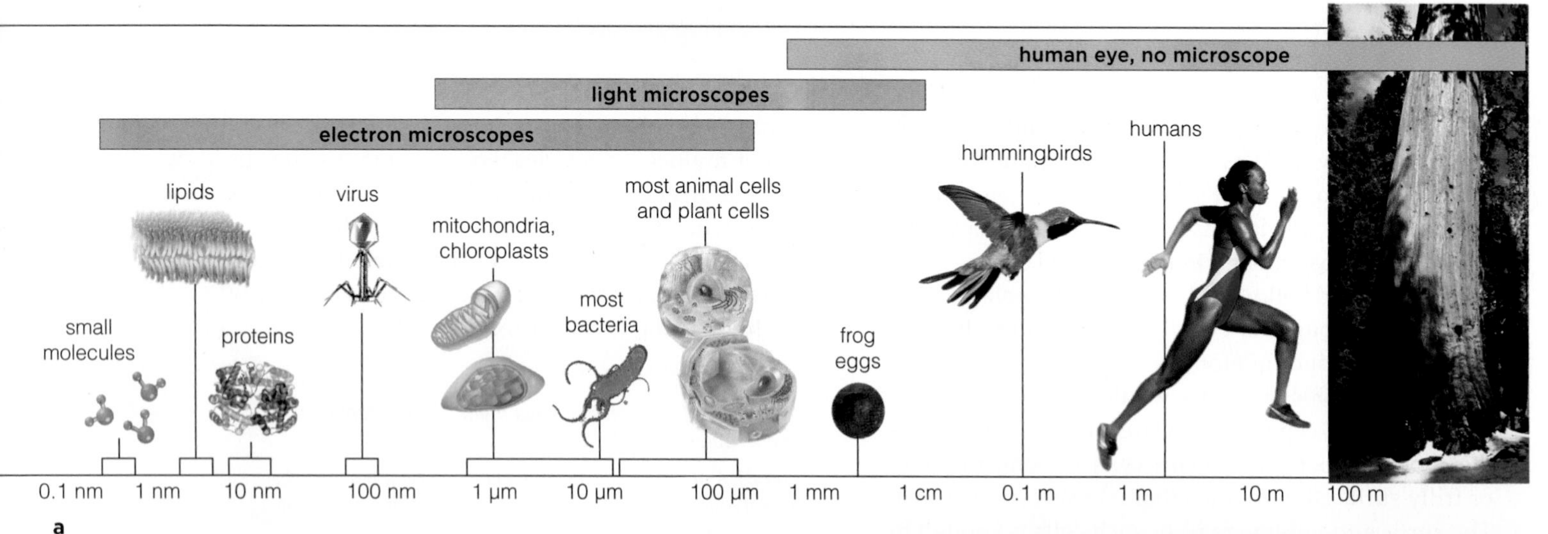

1 centimeter (cm)	=	1/100 meter, or 0.4 inch
1 millimeter (mm)	=	1/1000 meter
1 micrometer (µm)	=	1/1,000,000 meter
1 nanometer (nm)	=	1/1,000,000,000 meter

b 1 meter = 10^2 cm = 10^3 mm = 10^6 µm = 10^9 nm

Figure 4.6 (**a**) Relative sizes of typical molecules, cells, and multicelled organisms. The scale shown here is exponential, not linear; each unit of measure is ten times larger than the unit preceding it. The diameter of most cells is in the range between 1 and 100 micrometers. Frog eggs, one of the exceptions, are 2.5 millimeters across. (**b**) Units of measure. See also Appendix IX.

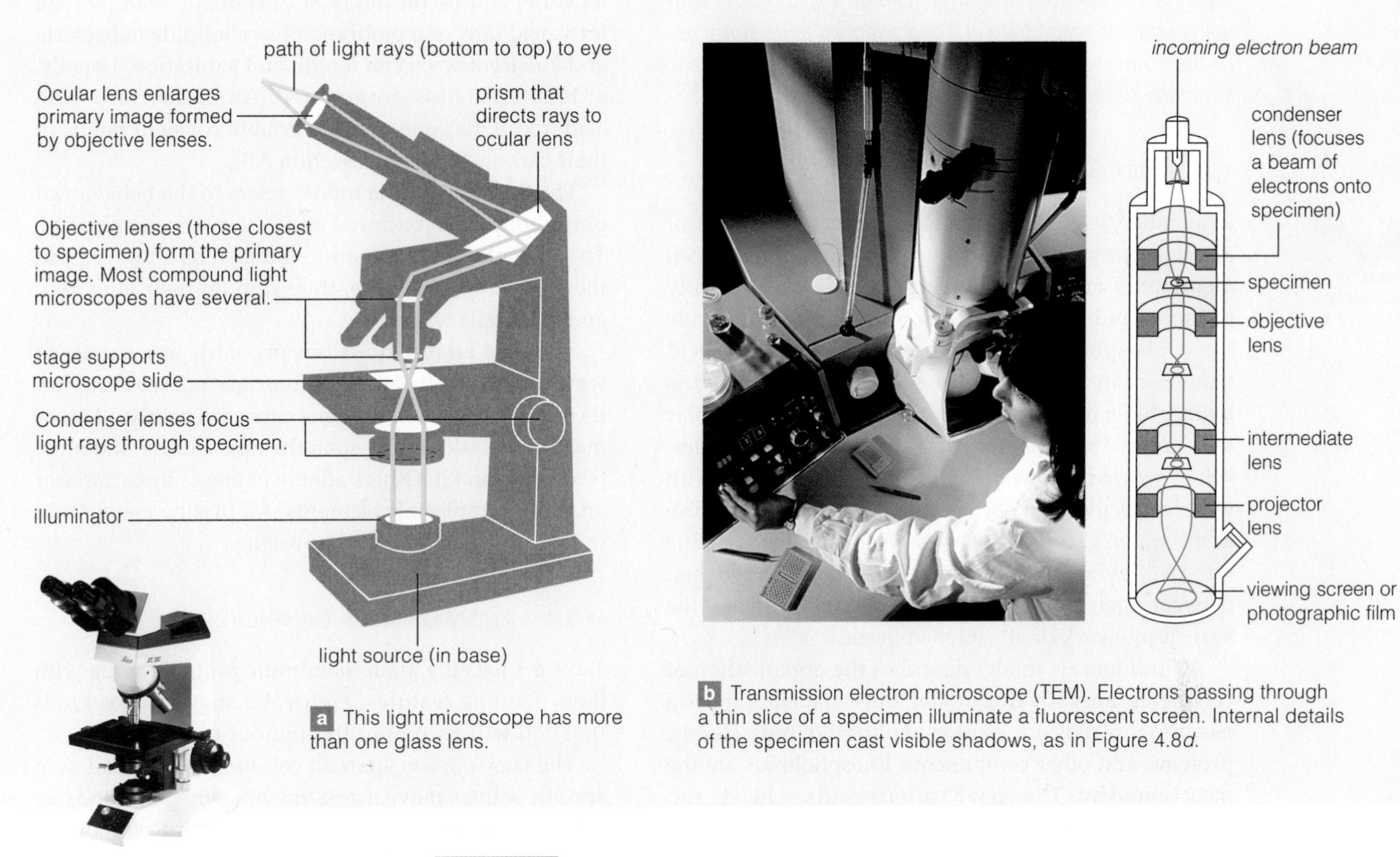

a This light microscope has more than one glass lens.

b Transmission electron microscope (TEM). Electrons passing through a thin slice of a specimen illuminate a fluorescent screen. Internal details of the specimen cast visible shadows, as in Figure 4.8*d*.

Figure 4.7 **Animated!** Examples of microscopes.

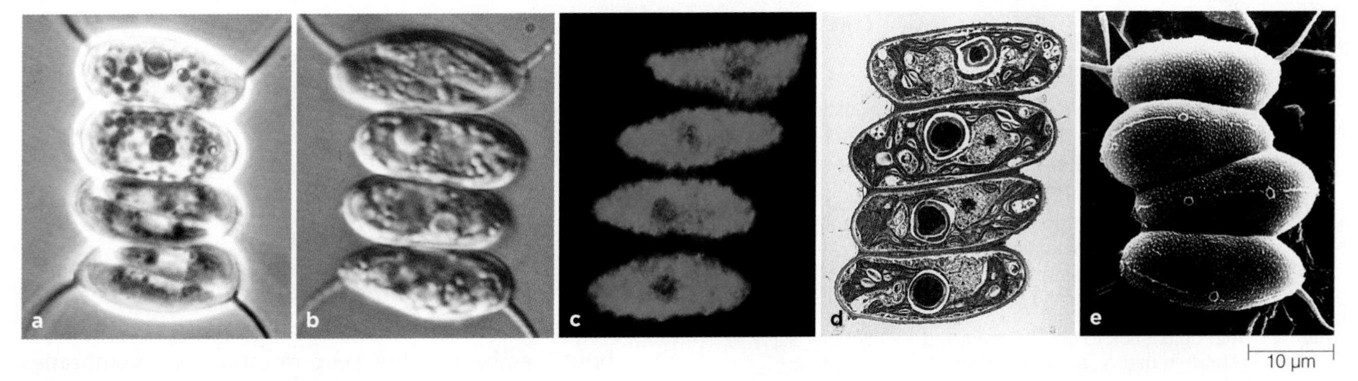

Figure 4.8 Different microscopes can reveal different characteristics of the same aquatic organism—a green alga (*Scenedesmus*). (**a**,**b**) Two types of light micrograph,

(**c**) Fluorescence micrograph. The cells are not visible in this micrograph. What you see is red light that fluoresced naturally from molecules of chlorophyll inside them. If a molecule or structure does not emit light on its own, researchers can attach a light-emitting tracer to it.

(**d**) Transmission electron micrographs reveal fantastically detailed images of internal structures. (**e**) Scanning electron micrographs show surface details of cells and structures. Often they are artificially colorized for clarity, but recent methods let natural colors show through. Try estimating the size of one of these algal cells by using the scale bar shown below like a ruler.

4.3 Membrane Structure and Function

Again, a cell membrane is organized as a lipid bilayer with many proteins embedded in it and attached to its surfaces. Focus now on the connection between that membrane's structural organization and its numerous functions.

THE FLUID MOSAIC MODEL

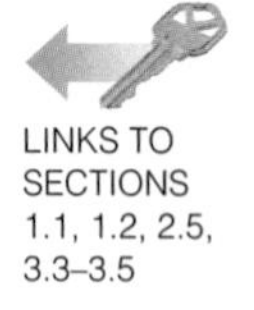

LINKS TO SECTIONS 1.1, 1.2, 2.5, 3.3–3.5

Remember Section 1.1, which introduced the idea of emergent properties? Functions that are unique to cell membranes emerge when certain molecules—mainly phospholipids—interact. Each phospholipid molecule has a phosphate-containing head and two fatty acid tails (Section 3.3 and Figure 4.4*a*). The polar head is hydrophilic; it attracts water molecules. The nonpolar tails are hydrophobic, so they repel water molecules. When swirled into water, phospholipids interact with water molecules and one another. They cluster as a film at the water's surface and spontaneously self-assemble into two layers, with their nonpolar tails sandwiched between their polar heads. Such lipid bilayers are the basic framework of all cell membranes.

A fluid mosaic model describes the organization of cell membranes. By this model, the cell membrane is a *mosaic*—a mixed composition of phospholipids, sterols, proteins, and other components. Phospholipids are the most abundant. They have various kinds of heads and tails, depending on the type of cell. For example, the fatty acid tails of membrane phospholipids in bacteria and eukaryotes vary in length and saturation. Usually, at least one tail is unsaturated. An unsaturated fatty acid, recall, has one or more double covalent bonds in their carbon backbone (Section 3.3).

The *fluid* part of the model refers to the behavior of phospholipids in bacterial and eukaryotic membranes. The phospholipids remain organized as a bilayer, but they also drift sideways, they spin on their long axis, and their tails wiggle.

The lipid bilayer functions primarily as a barrier to water-soluble substances; water does not readily enter its nonpolar interior. Proteins carry out nearly all other membrane tasks. Many span the lipid bilayer and often project beyond it. Others adhere to one of its surfaces or attach to cytoplasmic elements. All plasma membranes incorporate many kinds of proteins.

Table 4.1 Common Types of Membrane Proteins

Category	Function	Examples
Passive transporters	Allow ions or small molecules to cross a membrane to the side where they are less concentrated. Open or gated channels.	Porins; glucose transporter
Active transporters	Pump ions or molecules through membranes to the side where they are more concentrated. Require energy input, as from ATP.	Calcium pump; serotonin transporter
Receptors	Initiate change in a cell activity by responding to an outside signal (e.g., by binding a signaling molecule or absorbing light energy).	Insulin receptor; B cell receptor
Cell adhesion molecules (CAMs)	Help cells stick to one another and to protein matrixes that are part of tissues.	Integrins; cadherins
Recognition proteins	Identify cells as self (belonging to one's own body or tissue) or nonself (foreign to the body).	Histocompatibility molecules
Communication proteins	Join together and form cytoplasm-to-cytoplasm junctions through which ions and small molecules pass freely and quickly between adjacent cells.	Connexins in gap junctions

THE MAIN CATEGORIES OF PROTEINS

Table 4.1 lists the main membrane proteins along with their defining features. Figure 4.9 shows major kinds that you will encounter throughout the book.

The transporters span all cell membranes and help specific solutes move across the bilayer. Some kinds let solutes flow through a channel in their interior; others pump them across. Receptors trigger changes in cell activities by responding to signals or stimuli from the outside. Different cells have different combinations of receptors. Recognition proteins identify a cell as self (belonging to one's body) or as nonself (foreign to the body). Adhesion proteins help cells migrate to certain regions, and then stay there. Communication proteins form channels that allow substances or signals to flow freely across the plasma membranes of adjacent cells. Such proteins are abundant in heart muscle and other tissues in which cells interact fast as a unit.

VARIATIONS ON THE MODEL

Differences in Membrane Composition The fluid mosaic model is a good starting point for thinking about membranes, but keep in mind that membranes differ in molecular composition. The differences reflect their functions in cells. Even the two surfaces of a lipid bilayer are different. For example, carbohydrate chains attached to certain proteins and lipids project outward from a plasma membrane but not into the cell.

Differences in Fluidity Researchers once thought that all proteins of the membrane coated its surface.

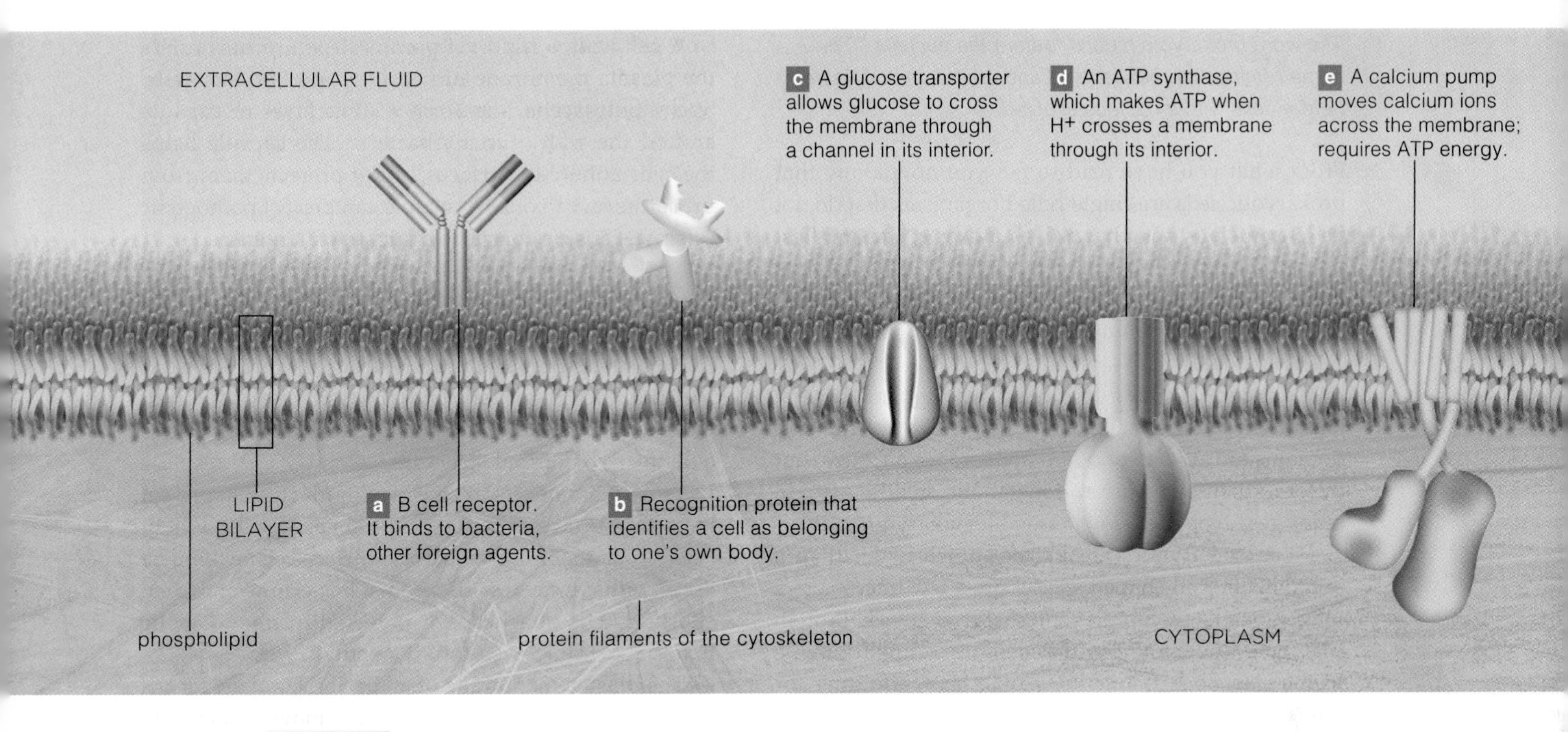

Figure 4.9 Animated! Icons for typical proteins associated with the plasma membrane of animal cells.

Then they learned to split a frozen plasma membrane down the middle of its bilayer. They discovered many proteins embedded within the bilayer (Figure 4.10*a*). Other researchers showed that membrane proteins can drift sideways. They isolated a single human cell and a single mouse cell, then induced the two cells to fuse. When the plasma membranes merged, the human and mouse proteins did not stay on their respective sides of the hybrid cell. Instead, they quickly intermingled throughout the membrane (Figure 4.10*b*).

As we now know, many proteins do drift through the bilayer, but others stay put. Some cluster as rigid pores. In eukaryotes, protein filaments that make up a cellular skeleton, or cytoskeleton, lock others in place.

Each cell membrane is a boundary that selectively controls exchanges between the cell and the surroundings. It is a mosaic of different kinds of lipids and proteins.

Two layers of lipid molecules—mainly phospholipids—are the membrane's primary structure. The hydrophobic parts of the lipids are sandwiched between hydrophilic parts, which face cytoplasmic fluid or extracellular fluid.

Many different kinds of proteins are embedded in a lipid bilayer or positioned at one of its surfaces. Some of the proteins function as transporters or receptors. Others function in cell adhesion, communication, and recognition.

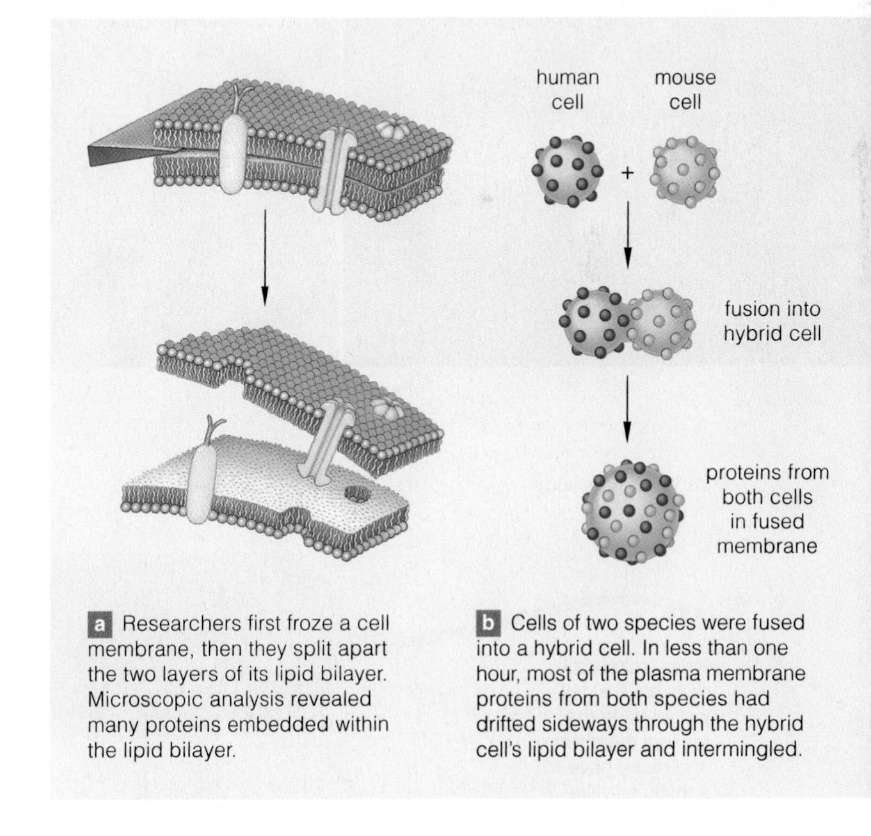

Figure 4.10 Animated! Results from two studies of membrane structure. (**a**) An observational test and (**b**) an experimental test.

4.4 Introducing Prokaryotic Cells

The word prokaryote means "before the nucleus." The name reminds us that bacteria and archaeans originated before cells with a nucleus evolved.

LINKS TO SECTIONS 1.1, 1.3, 3.1, 3.4

From what you have read so far, you now know that prokaryotic cells are single-celled organisms that do not have a nucleus and that are, as a group, the smallest and most metabolically diverse forms of life we know about (Figures 4.11 and 4.12). Prokaryotic cells live in nearly all Earth's environments, including some very hostile places.

We group all prokaryotic cells into domains Bacteria and Archaea (Sections 1.3 and 19.3). Cells of the two domains are alike in outward appearance and size, but differ in their structure and metabolic details. Also, in many respects, archaeans resemble eukaryotic cells.

Most prokaryotic cells are not much wider than a micrometer. Rod-shaped species are a few micrometers long. None has a complex internal framework. Protein filaments under the plasma membrane impart shape to the cell. Such filaments also act as scaffolding for internal structures, such as ribosomes.

A cell wall, a rigid yet porous structure, surrounds the plasma membrane of nearly all prokaryotic cells. Sticky polysaccharides form a slime layer or capsule around the wall of many bacteria. The capsule helps the cells adhere to surfaces, and it protects them from predators and toxins. A capsule can protect pathogenic (disease-causing) bacteria against host defenses.

Projecting past the wall of many prokaryotic cells are one or more flagella (singular, flagellum): slender cellular structures used for motion. Bacterial flagella move like a propeller that drives the cell through fluid habitats, such as a host's body fluids. They differ from eukaryotic flagella, which bend like a whip and have a distinctive internal structure.

Protein filaments called pili (singular, pilus) project from the surface of some bacterial species. They help cells cling to or move across surfaces. One kind, a "sex" pilus, attaches to another bacterium and then shortens. The attached cell is reeled in, then genetic material is transferred into it (Section 19.1).

The plasma membrane of all bacteria and archaeans selectively controls which substances move to and from

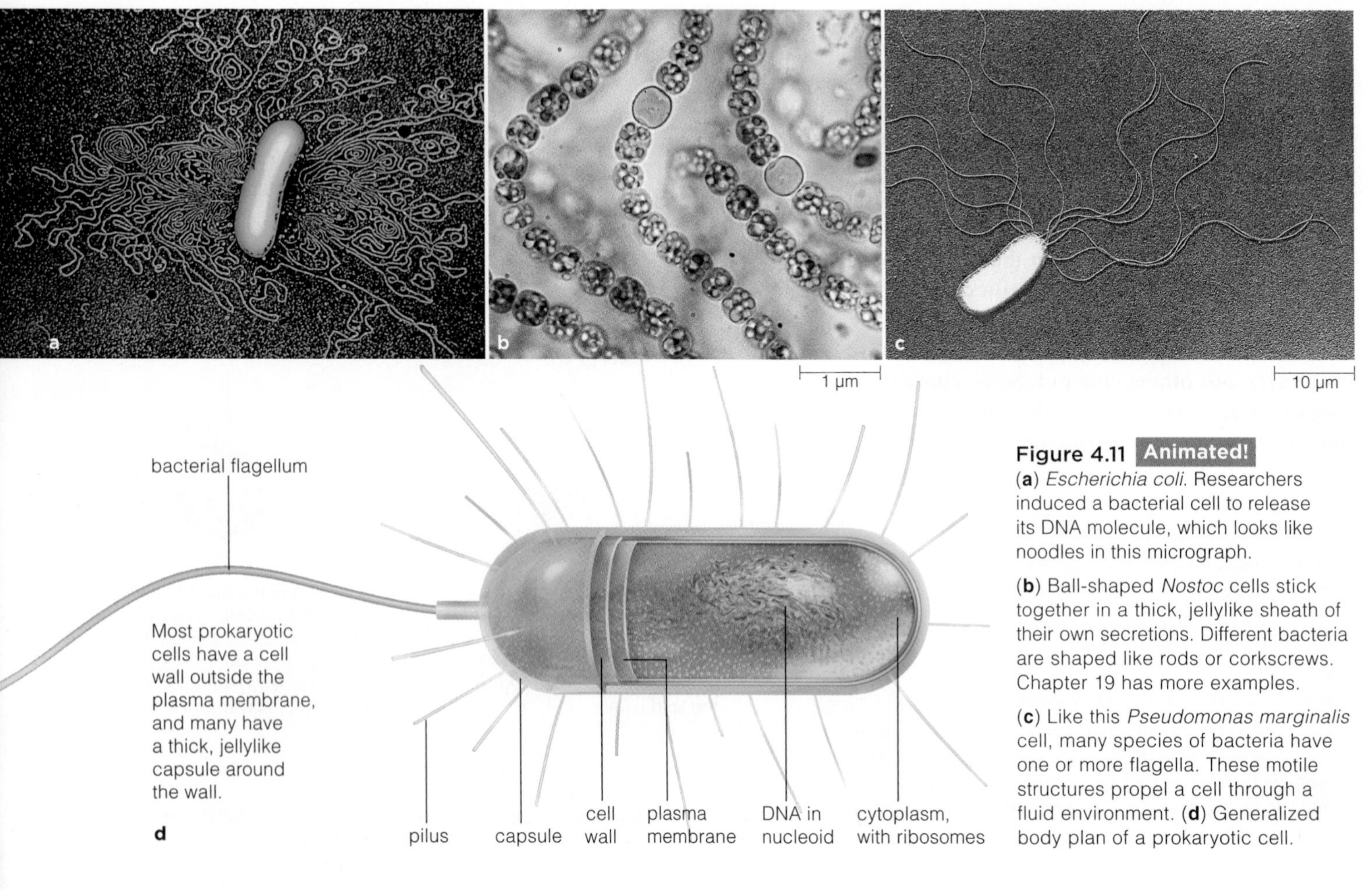

Figure 4.11 Animated! (**a**) *Escherichia coli.* Researchers induced a bacterial cell to release its DNA molecule, which looks like noodles in this micrograph.

(**b**) Ball-shaped *Nostoc* cells stick together in a thick, jellylike sheath of their own secretions. Different bacteria are shaped like rods or corkscrews. Chapter 19 has more examples.

(**c**) Like this *Pseudomonas marginalis* cell, many species of bacteria have one or more flagella. These motile structures propel a cell through a fluid environment. (**d**) Generalized body plan of a prokaryotic cell.

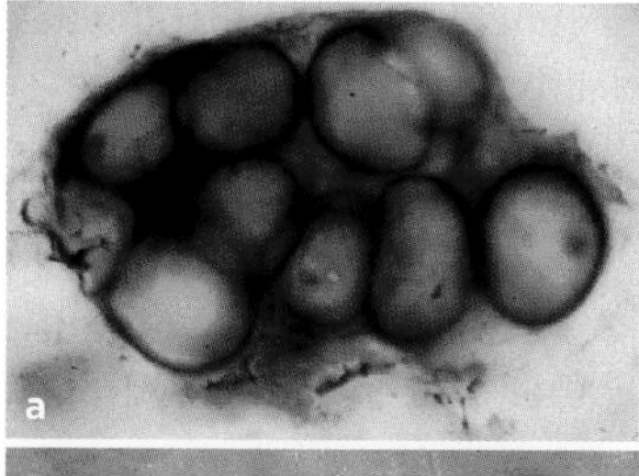

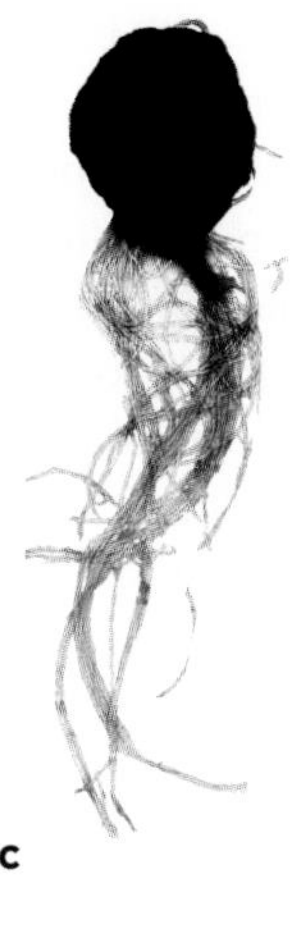

Figure 4.12 Prokaryotes. From Bitter Springs, Australia, fossilized bacterial cells that lived about 850 million years ago: (**a**) Colonial bacteria, probably *Myxococcoides minor,* and (**b**) filamentous *Palaeolyngbya*. (**c**) *Methanococcus jannaschii,* a methane producer, lives in near-boiling water spewing from hydrothermal vents in the seafloor. Many other living archaeans inhabit extreme environments.

the cytoplasm, as it does for eukaryotic cells. A plasma membrane bristles with transporters, receptors, and proteins that carry out important metabolic processes. Similar metabolic processes occur in eukaryotes, but they take place at specialized internal membranes.

The cytoplasm contains many ribosomes on which polypeptide chains are assembled. Continuous with the cytoplasm is the nucleoid, an irregularly shaped region where DNA is located but is not enclosed in a membrane. A prokaryotic cell's single chromosome is a circular DNA molecule. Some prokaryotes also have plasmids in the cytoplasm. These far smaller circles of DNA carry a few genes (units of inheritance) that can confer advantages, such as antibiotic resistance.

One more intriguing point: protists, plants, fungi, and animals all may have evolved from a few ancient lineages of prokaryotes. Section 18.4 looks at a theory about this possibility.

Bacteria and archaeans are the only prokaryotic cells. They do not have a nucleus. Most kinds have a cell wall around their plasma membrane. The wall is permeable, and it reinforces and imparts shape to the cell body.

Although structurally simple, prokaryotic cells as a group show the most metabolic diversity.

Some metabolic processes occur at the plasma membrane of prokaryotic cells. They are similar to complex processes that occur at certain internal membranes of eukaryotic cells.

4.5 Microbial Mobs

FOCUS ON THE ENVIRONMENT

Even though prokaryotes are single-celled organisms, few kinds live in isolation. Typically, they have company—lots of it—as when they form biofilms.

Bacterial cells often live so close together that an entire group shares a layer of secreted polysaccharides and glycoproteins. Such communal living arrangements, in which single-celled organisms live in a shared mass of slime, are called biofilms. In nature, a biofilm typically consists of multiple species, all entangled in their own mingled secretions. It may include bacteria, algae, fungi, protists, and other microbes. Such associations allow cells to linger in a particular spot rather than be swept away by currents in a fluid habitat (Figure 4.13).

When free-living cells encounter a biofilm that offers a favorable accommodation, their metabolism changes to support a less active, more communal lifestyle. Flagella disassemble, sex pili form, and they join the biofilm community. The inhabitants of such communities are interdependent. The rigid or netlike secretions of some species serve as permanent scaffolding for others. Still other types busily maneuver themselves through this matrix. Species that break down toxic chemicals allow more sensitive ones to thrive in polluted habitats that they could not withstand on their own. Waste products of some serve as raw materials for others.

Like a busy metropolitan city, a biofilm is organized as "neighborhoods," each with a distinct microenvironment that stems from its location within the biofilm and the particular species that inhabit it. For example, cells that reside near the middle of a biofilm are very crowded and do not divide often. Those at the edges divide repeatedly and expand the biofilm.

Wherever there is water, biofilms form. Slippery goo on river rocks, in water pipes, and in sewage treatment ponds are biofilms. So is the slime on teeth. Biofilms form on contact lenses, in airways to the lungs, and in the digestive tract. They are implicated in gum disease, middle-ear and urinary tract infections, and cystic fibrosis.

Figure 4.13 Biofilm formation. Bacteria and other microbes settle on surfaces, such as the lining of your airways, and then reproduce. Slime holds them together in a biofilm. When conditions become unfavorable, a biofilm's inhabitants may revert to a flagellated form and disperse.

4.6 Introducing Eukaryotic Cells

All cells synthesize, store, degrade, and move substances about, but eukaryotic cells in particular carry out many of these tasks inside membrane-enclosed organelles.

Again, all eukaryotic cells start out life with a nucleus. *Eu–* means true; and *karyon*, meaning kernel, is taken to mean a nucleus. A nucleus is a type of organelle: a structure that carries out a specialized function inside a cell. Many organelles, particularly in eukaryotic cells, are bounded by membranes. Like all cell membranes, those around organelles control the types and amounts of substances that cross them. Such control maintains a special internal environment that allows an organelle to carry out its particular function. That function may be isolating a toxic or sensitive substance from the rest of the cell, transporting some substance through the cytoplasm, maintaining fluid balance, or providing a favorable environment for a reaction that could not occur in the cytoplasm. For example, organelles called mitochondria make ATP after concentrating hydrogen ions inside their membrane system.

Much as interactions among organ systems keep an animal body running, interactions among organelles keep a cell running. Substances shuttle from one kind of organelle to another, and to and from the plasma membrane. Some metabolic pathways take place in a series of different organelles.

Table 4.2 lists common components of eukaryotic cells. These cells all start out life with certain kinds of organelles such as a nucleus and ribosomes. They also have a cytoskeleton, a dynamic "skeleton" of proteins (*cyto–* means cell). Specialized cells contain additional kinds of organelles and structures. Figure 4.14 shows two typical eukaryotic cells.

All eukaryotic cells start out life with a nucleus and other organelles, which are structures that carry out specific functions inside a cell. Many types of eukaryotic organelles are enclosed by membranes.

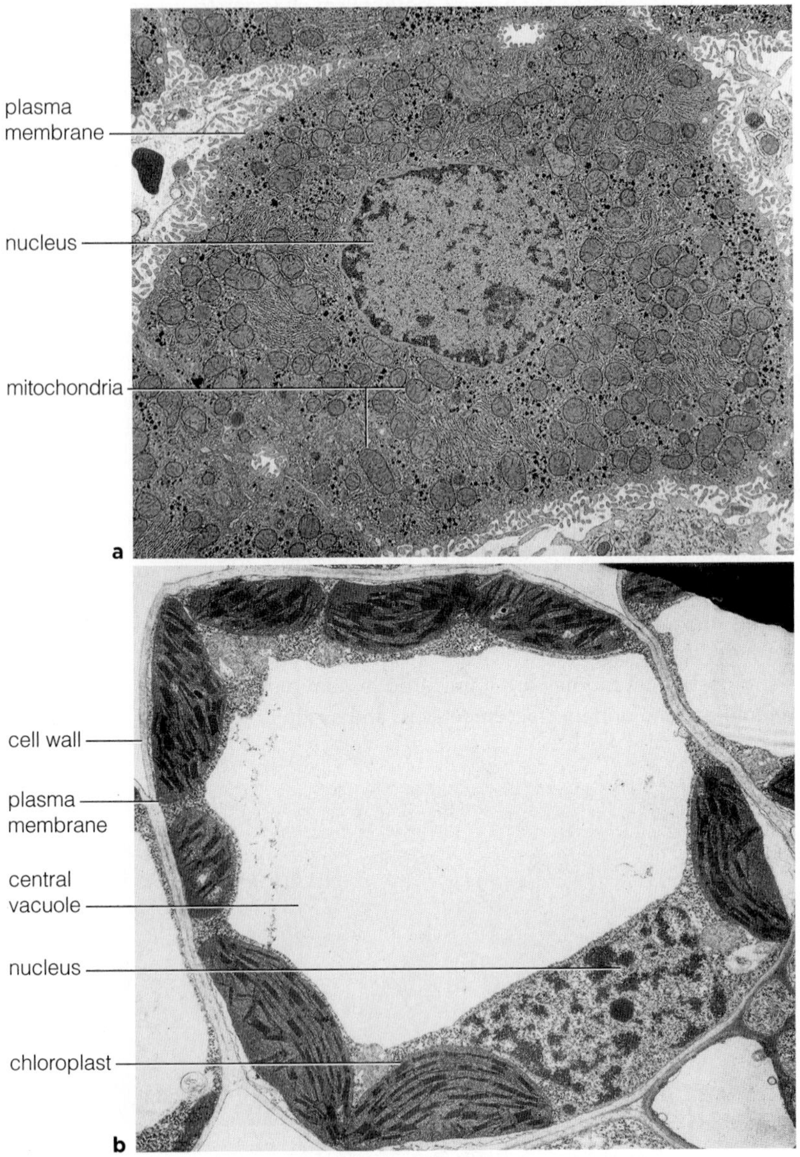

Figure 4.14 Transmission electron micrographs of eukaryotic cells. (**a**) Rat liver cell. (**b**) Photosynthetic cell from a blade of timothy grass.

Table 4.2 Components of Eukaryotic Cells

Organelles with membranes	
Nucleus	Protecting and controlling access to DNA
Endoplasmic reticulum (ER)	Routing, modifying new polypeptide chains; synthesizing lipids; other tasks
Golgi body	Modifying new polypeptide chains; sorting, shipping proteins and lipids
Vesicles	Transporting, storing, or digesting substances in a cell; other functions
Mitochondrion	Making ATP by glucose breakdown
Chloroplast	Making sugars in plants, some protists
Organelles without membranes	
Ribosomes	Assembling polypeptide chains
Other structures	
Cytoskeleton	Contributing to cell shape, internal organization, movement

4.7 The Nucleus

The nucleus keeps eukaryotic DNA separated from the cytoplasm and isolates it from potentially damaging reactions. It also controls when the DNA is accessed.

Let's zoom in on the components of the cell nucleus, as listed in Table 4.3. The outer boundary, or nuclear envelope, consists of two lipid bilayers. As Figure 4.15 shows, the outer bilayer of the nucleus is continuous with the membrane of another organelle, the ER.

Receptors, pores, and transporters that span the two lipid bilayers of the nuclear envelope control when and how much the information encoded in DNA gets used. The controls also protect the cell's all-important hereditary material from reactions in the cytoplasm that might damage it.

The nuclear envelope encloses a semifluid matrix called nucleoplasm. Embedded in the matrix is at least one nucleolus (plural, nucleoli), an irregularly shaped region where ribosome subunits are assembled from proteins and RNA. The subunits pass through nuclear pores into the cytoplasm, where they join and become active in protein synthesis.

Each eukaryotic chromosome is a double-stranded molecule of DNA with attached proteins. The genetic material inside eukaryotic cells is distributed among a number of chromosomes that differ in length and shape. For instance, there are twelve chromosomes in oak tree cells, forty-six in human body cells, and 208 in king crab cells. Chromatin is the name for all of the chromosomal DNA and proteins in a nucleus.

Chromosomes change in appearance. When a cell is not dividing, its chromosomes can appear grainy in transmission electron micrographs (Figure 4.15*a*). Just before a cell divides, each chromosome gets duplicated. During cell division, all of the chromosomes condense. As they do, they become visible in micrographs first as threadlike forms, then as rodlike forms.

A cell nucleus has an envelope of two lipid bilayers.

Pores, receptors, and transport proteins in the nuclear envelope control the passage of substances between the nucleus and cytoplasm. The control protects the cell's DNA and regulates its use.

Each chromosome in a nucleus consists of a duplicated or unduplicated DNA molecule and many attached proteins.

Table 4.3 Components of the Nucleus

Nuclear envelope	Pore-riddled double-membrane that controls which substances enter and leave the nucleus
Nucleoplasm	Semifluid interior portion of the nucleus
Nucleolus	Rounded mass of proteins and copies of genes for ribosomal RNA used to construct ribosomal subunits
Chromatin	Total collection of all DNA molecules and associated proteins in the nucleus; all of the chromosomes
Chromosome	One DNA molecule, duplicated or not, and the many proteins associated with it:

one chromosome (one dispersed DNA molecule + proteins; not duplicated)	one chromosome (threadlike and now duplicated; two DNA molecules + proteins)	one chromosome (duplicated and also condensed tightly)

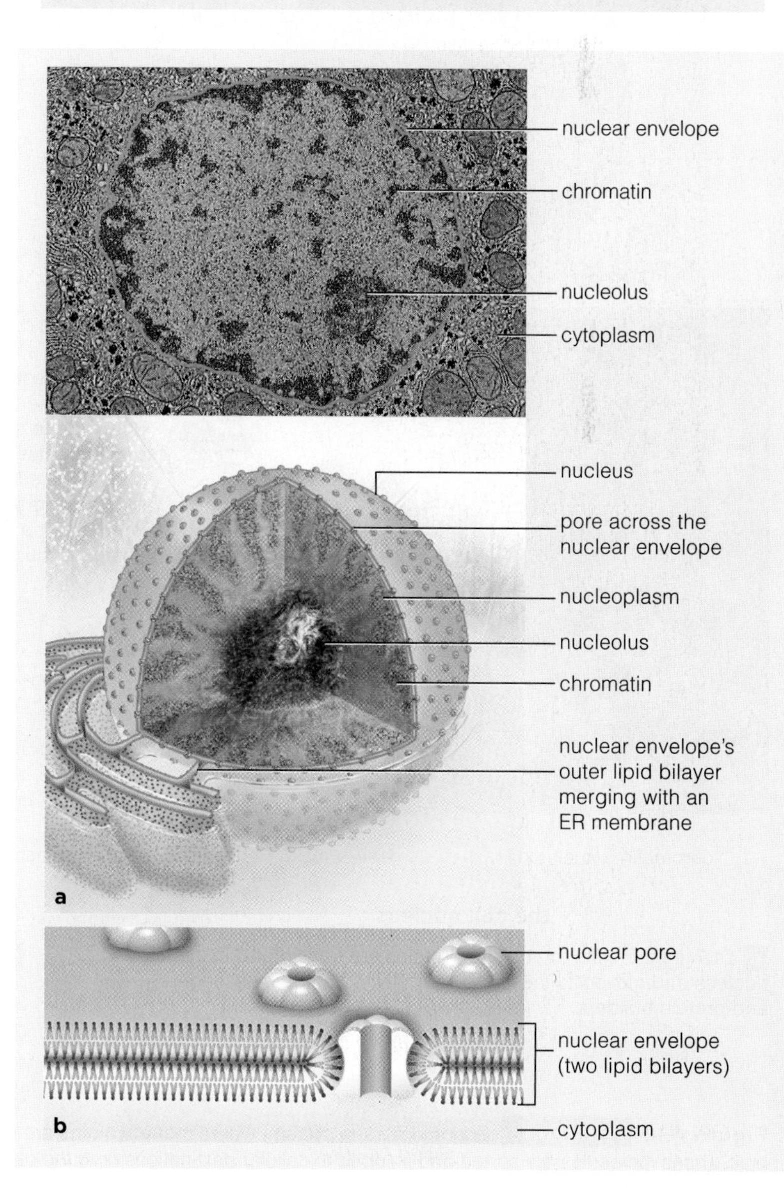

Figure 4.15 **Animated!** (**a**) Micrograph (TEM) and sketch of the nucleus of a rat liver cell. (**b**) Closer view of the nuclear envelope's structure. Each nuclear pore is an organized cluster of membrane proteins.

4.8 The Endomembrane System

The endomembrane system is a set of organelles in the cytoplasm of eukaryotic cells. It includes endoplasmic reticulum, Golgi bodies, vesicles that bud from both, and vesicles that bud from the plasma membrane. These organelles have different but interconnected functions.

ENDOPLASMIC RETICULUM

LINK TO SECTION 3.4

Endoplasmic reticulum, or ER, is an extension of the nuclear envelope. It forms a continuous compartment that folds over and over into flattened sacs and tubes (Figure 4.16). The space inside the compartment is the site where many new polypeptide chains are modified.

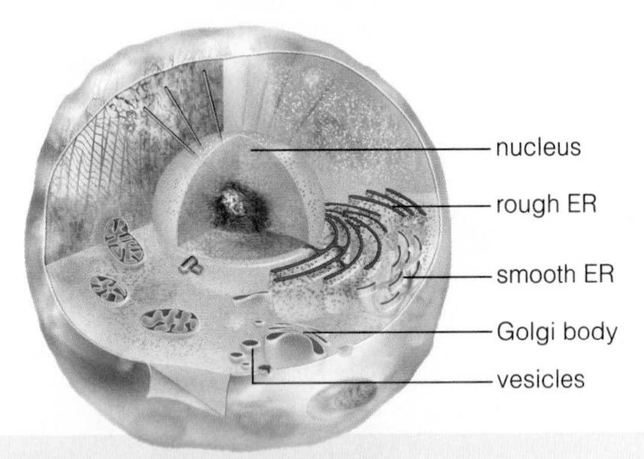

Two kinds of ER are named for their appearance in electron micrographs. *Rough* ER has many thousands of ribosomes attached to its outer surface. Some of the polypeptide chains assembled on ribosomes enter the ER, where enzymes modify them. Other polypeptide chains become part of the ER membrane itself.

Cells that make, store, and secrete proteins have a lot of rough ER (Figure 4.16*c*). For example, ER-rich gland cells in the pancreas make and secrete enzyme molecules that help digest meals in the small intestine.

Smooth ER (Figure 4.16*d*) does not have ribosomes. It makes most of the lipids that get incorporated into cell membranes. It has roles in carbohydrate and fatty acid breakdown, and in detoxifying some drugs and poisons. In skeletal muscle cells, one type of smooth ER stores calcium ions and has a role in contraction.

GOLGI BODIES

Patches of rough ER membrane bulge and break away as vesicles, a name that refers to a variety of small, sac-shaped organelles. Many vesicles fuse with and empty their contents into a Golgi body. This organelle has a

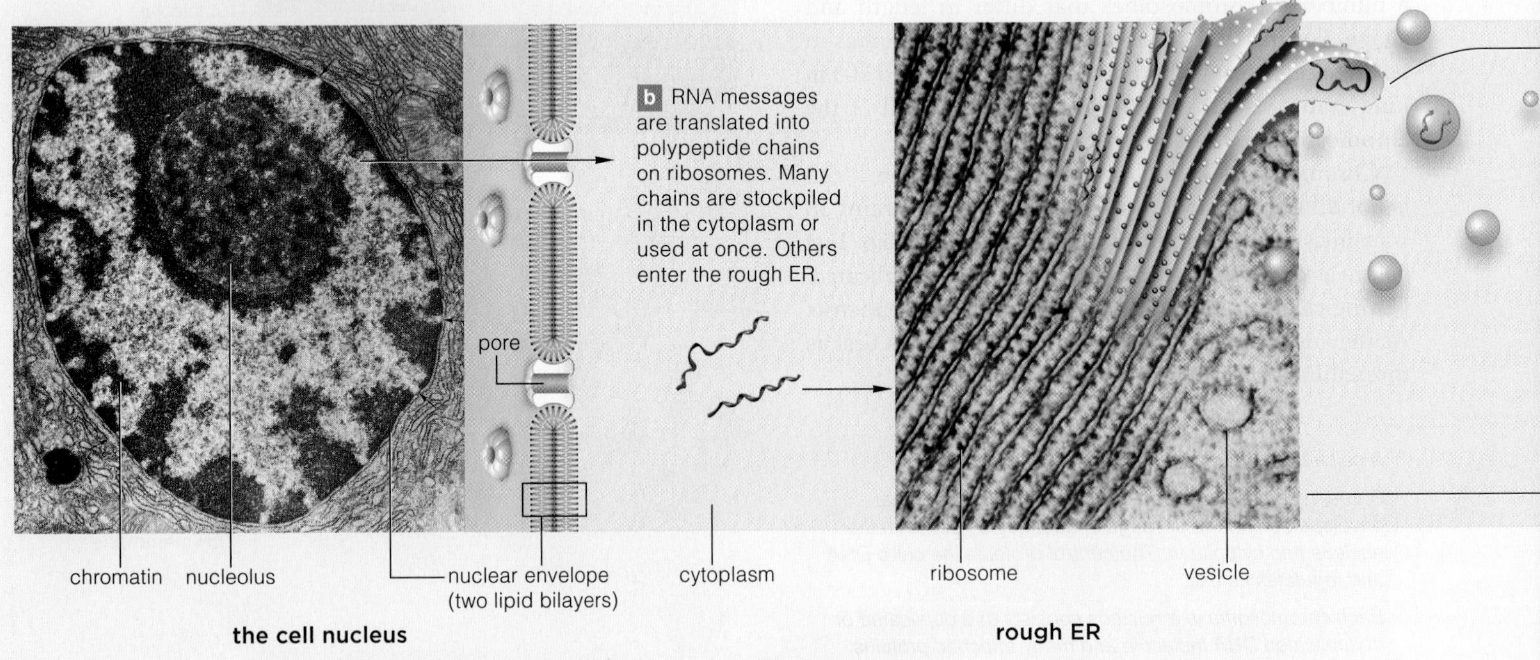

a DNA instructions for making proteins are transcribed in the nucleus and moved to the cytoplasm. RNAs are the messengers and protein builders.

c Flattened sacs of rough ER form one continuous channel between the nucleus and smooth ER. Polypeptide chains that enter the channel undergo modification. They will be inserted into organelle membranes or will be secreted from the cell.

Figure 4.16 **Animated!** Endomembrane system, where many proteins are modified and lipids are built. These molecules are sorted and shipped to cellular destinations or to the plasma membrane for export.

folded membrane that typically looks like a stack of pancakes (Figure 4.16*e*). Enzymes inside a Golgi body modify polypetide chains and lipids that vesicles have delivered from the ER. They attach phosphate groups or sugars, and cleave certain polypeptide chains. The finished products are sorted and packaged into new vesicles that transport them through the cell.

A VARIETY OF VESICLES

The endomembrane system includes diverse vesicles. *Endocytic* kinds form as a patch of plasma membrane sinks into the cytoplasm; *exocytic* kinds bud from the ER or Golgi membranes and transport substances to the plasma membrane for export (Figure 4.16*f–h*). Still other vesicles form on their own in the cytoplasm.

Lysosomes are vesicles that bud from Golgi bodies and take part in intracellular digestion. They contain enzymes that can break down carbohydrates, proteins, nucleic acids, and some lipids. Other types of vesicles deliver ingested bacteria, cell parts, and other materials to lysosomes for destruction. This happens after white blood cells or predatory amoebas engulf their targets.

As another example, small vesicles fuse into larger sacs called vacuoles. Amino acids, sugars, toxins, and ions accumulate in the fluid-filled interior of a plant cell's central vacuole (Figure 4.14*b*). Fluid pressure in the central vacuole keeps plant cells—and structures such as stems and leaves—firm.

In plants and animals, vesicles called peroxisomes form and divide on their own. They contain enzymes that digest fatty acids and amino acids. Peroxisomes have a variety of functions. For instance, they break down hydrogen peroxide, a toxic by-product of fatty acid metabolism. Enzymes convert hydrogen peroxide to water and oxygen, or they use it in reactions that break down alcohol and other toxins.

Many new polypeptide chains are modified and lipids are assembled in the endomembrane system's ER. They may be further modified in Golgi bodies.

The endomembrane system includes a variety of vesicles. Different kinds of vesicles help integrate cell activities by storing substances or transporting them through the cell. Enzymes in some vesicles break down substances.

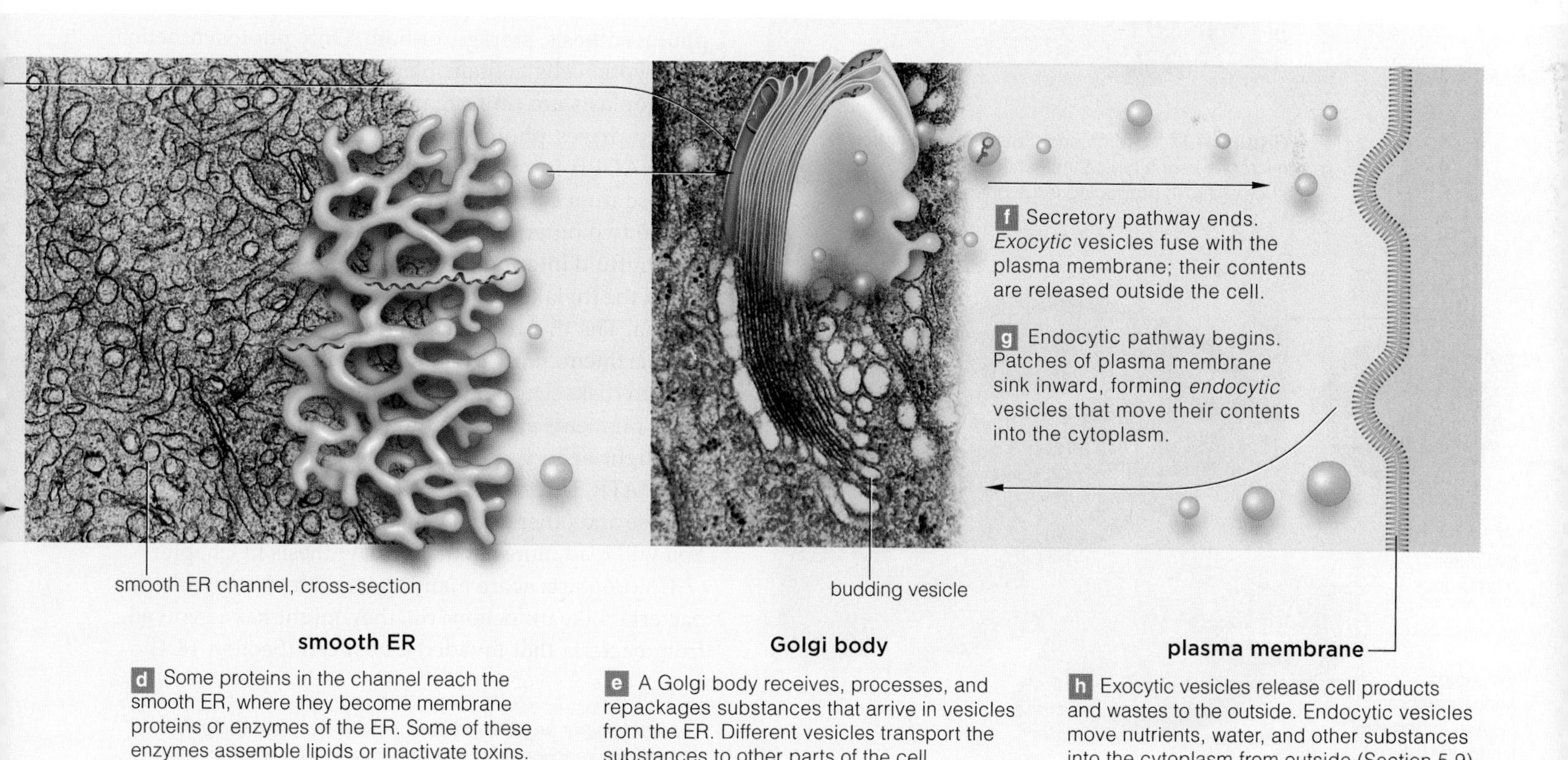

d Some proteins in the channel reach the smooth ER, where they become membrane proteins or enzymes of the ER. Some of these enzymes assemble lipids or inactivate toxins.

e A Golgi body receives, processes, and repackages substances that arrive in vesicles from the ER. Different vesicles transport the substances to other parts of the cell.

h Exocytic vesicles release cell products and wastes to the outside. Endocytic vesicles move nutrients, water, and other substances into the cytoplasm from outside (Section 5.9).

4.9 Mitochondria and Chloroplasts

Recall, from Section 3.6, that ATP is an energy carrier. It delivers energy, in the form of phosphate-group transfers, that drives many of the reactions in cells. Eukaryotic cells make ATP inside mitochondria. Some plant and protist cells also make ATP inside chloroplasts.

LINK TO SECTION 3.6

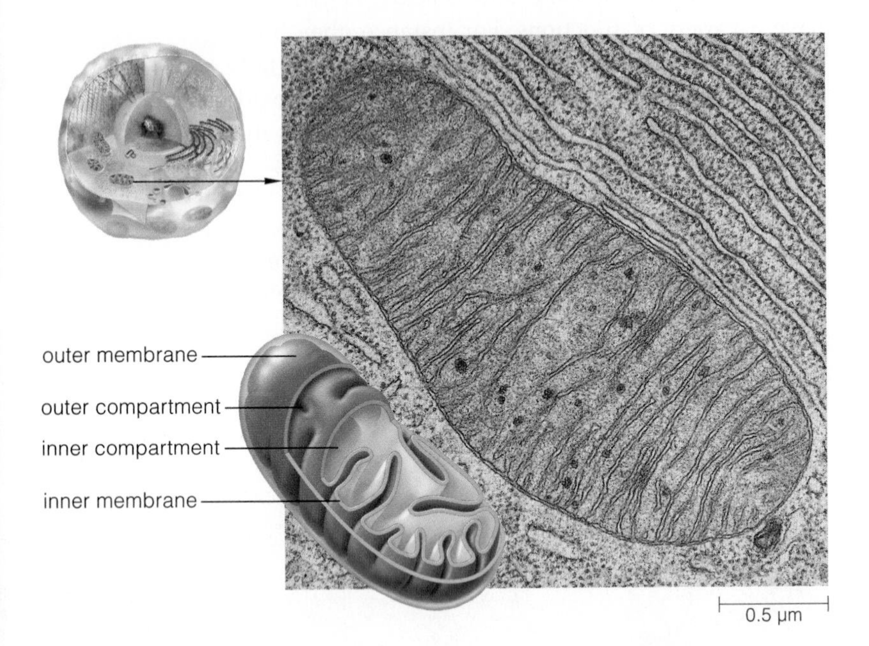

Figure 4.17 Sketch and transmission electron micrograph of a typical mitochondrion. This organelle specializes in producing large quantities of ATP.

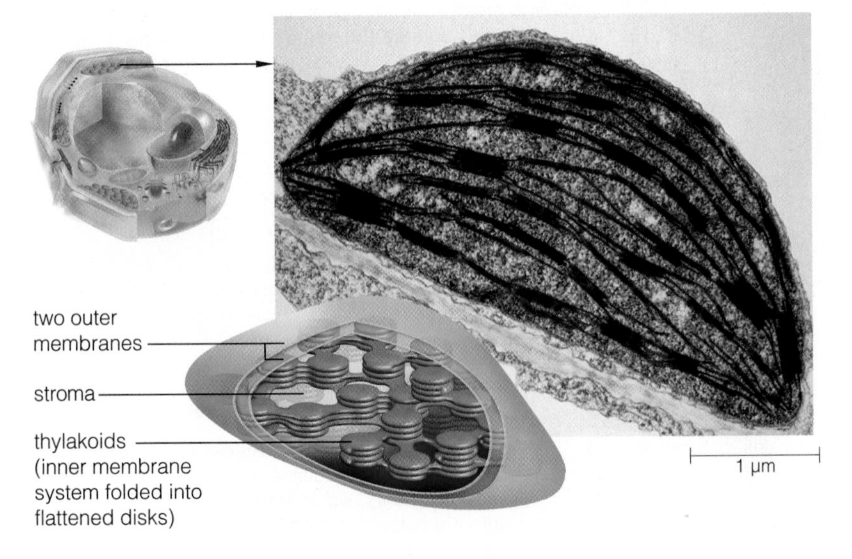

Figure 4.18 The chloroplast, a defining character of all photosynthetic eukaryotic cells. This transmission electron micrograph shows a chloroplast from a tobacco leaf (*Nicotiana tabacum*), thin section.

MITOCHONDRIA

The mitochondrion (plural, mitochondria) specializes in aerobic respiration, an oxygen-requiring metabolic pathway that produces many ATP by breaking down organic molecules. With each breath, you are taking in oxygen mainly for the mitochondria in your trillions of cells. Cells of your liver, heart, skeletal muscles, and other tissues that demand a lot of energy may contain a thousand or more mitochondria.

A mitochondrion has two membranes, and one is highly folded inside the other (Figure 4.17). Hydrogen ions accumulate between the two membranes. Then the ions flow across the inner membrane, through the interior of ATP synthases of the sort shown in Figure 4.9. That flow drives the formation of ATP.

Mitochondria are found in the cytoplasm of nearly all eukaryotic cells. Curiously, they share many traits with bacteria. Mitochondria might have evolved from bacteria that took up permanent residence inside a host cell (Section 18.4).

CHLOROPLASTS

Many plant cells have plastids, which are organelles of photosynthesis, storage, or both. Only photosynthetic eukaryotic cells contain plastids called chloroplasts. Chloroplasts are tiny sugar factories. Inside, sunlight energy drives photosynthesis. In this pathway, ATP and NADPH form, and then both are used to produce glucose from carbon dioxide.

The two outer membranes of a chloroplast enclose its semifluid interior, the stroma (Figure 4.18). A third, called the thylakoid membrane, is folded up inside the stroma. The thylakoid membrane forms a continuous compartment. Parts of it are organized as channels or stacked disks. During photosynthesis, chlorophyll and other pigments embedded in the thylakoid membrane trap light energy, which other molecules then use to make ATP. The energy of ATP drives the assembly of glucose and other organic molecules inside the stroma. You will read more about photosynthesis in Chapter 6.

Chloroplasts share many traits with photosynthetic bacteria. Like mitochondria, they might have evolved from bacteria that invaded a host cell (Section 18.4).

Mitochondria are ATP-producing powerhouses. Oxygen-requiring reactions that release energy from organic compounds occur inside them.

Chloroplasts in photosynthetic eukaryotic cells are sugar-producing factories that run on energy from the sun.

4.10 Visual Summary of Eukaryotic Cell Components

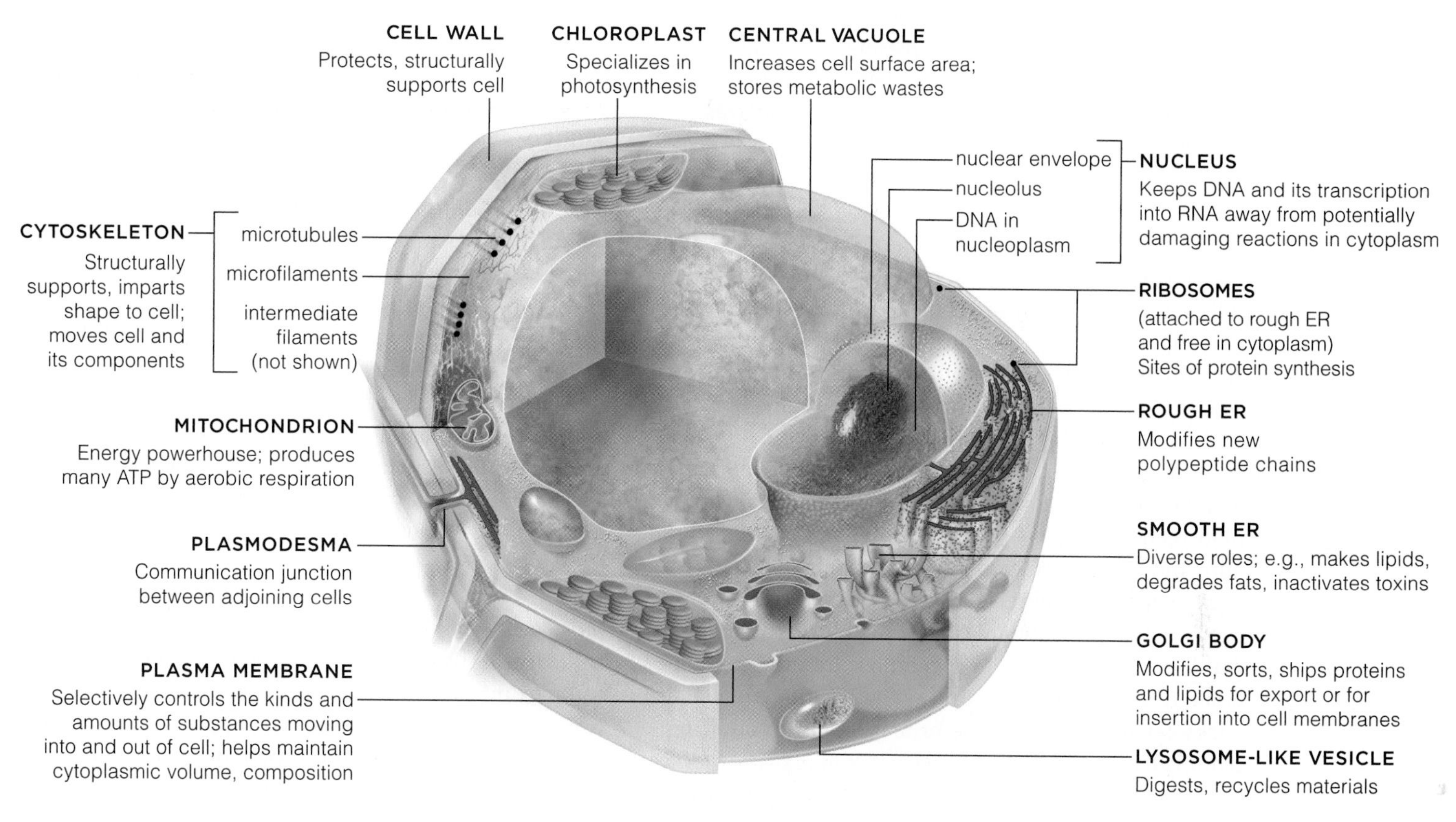

a Typical plant cell components.

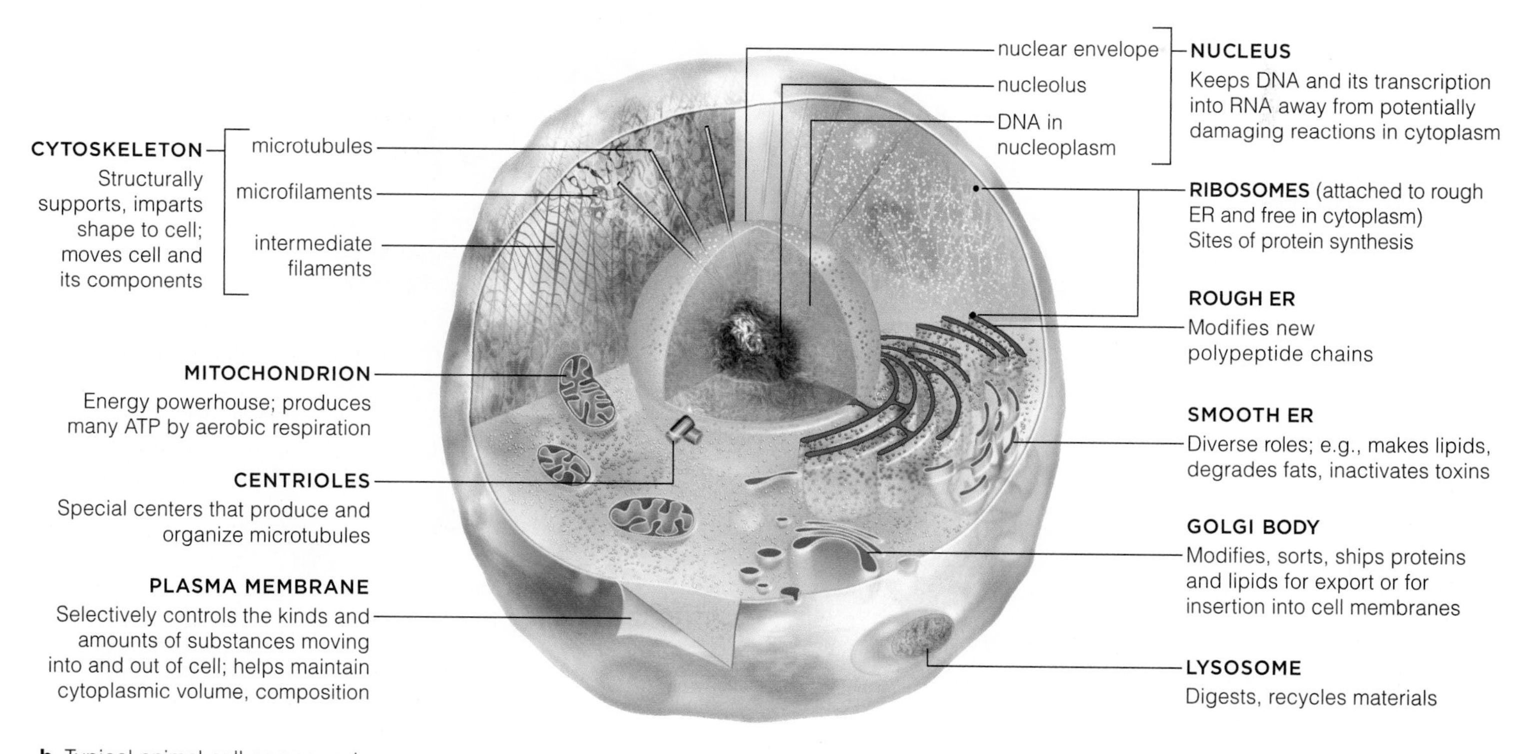

b Typical animal cell components.

Figure 4.19 Animated! Organelles and structures typical of (**a**) plant cells and (**b**) animal cells.

4.11 Cell Surface Specializations

For many eukaryotic cells, a porous wall or protective covering intervenes between the plasma membrane and the surroundings. In animal tissues, one plasma membrane may press right up to another, and matrixes and junctions usually form between them.

EUKARYOTIC CELL WALLS

Like most prokaryotic cells, many types of eukaryotic cells have a cell wall around the plasma membrane. The wall is a porous structure that protects, supports, and imparts shape to the cell. Water and solutes easily cross it on the way to and from the plasma membrane. Cells could not live without such exchanges.

Animal cells do not have walls, but plant cells and many protist and fungal cells do. For example, a young plant cell secretes pectin and other polysaccharides onto the outer surface of its plasma membrane. The sticky coating is shared between adjacent cells, and it cements them together. Each cell then forms a primary wall by secreting strands of cellulose into the coating. Some of the coating remains as the middle lamella, a sticky layer in between the primary walls of abutting plant cells (Figure 4.20*a*,*b*).

Being thin and pliable, the primary wall allows the growing plant cell to enlarge. Plant cells with only a thin primary wall can change shape as they develop. At maturity, cells in some plant tissues stop enlarging and begin to secrete material onto the primary wall's inner surface. These deposits form a firm secondary wall, of the sort shown in Figure 4.20*b*. One of the materials deposited is lignin. This organic compound makes up as much as 25 percent of the secondary wall of cells in older stems and roots. Lignified plant parts are stronger, more waterproof, and less susceptible to plant-attacking organisms than younger tissues.

A cuticle is a protective body covering made of cell secretions. In plants, a semitransparent cuticle helps protect exposed surfaces of soft parts and limits water loss on hot, dry days (Figure 4.21).

MATRIXES BETWEEN ANIMAL CELLS

Most cells of multicelled organisms are surrounded and organized by extracellular matrix. This nonliving, complex mixture of substances is secreted by cells, and varies with the type of tissue. It supports and anchors cells, separates tissues, and functions in cell signaling.

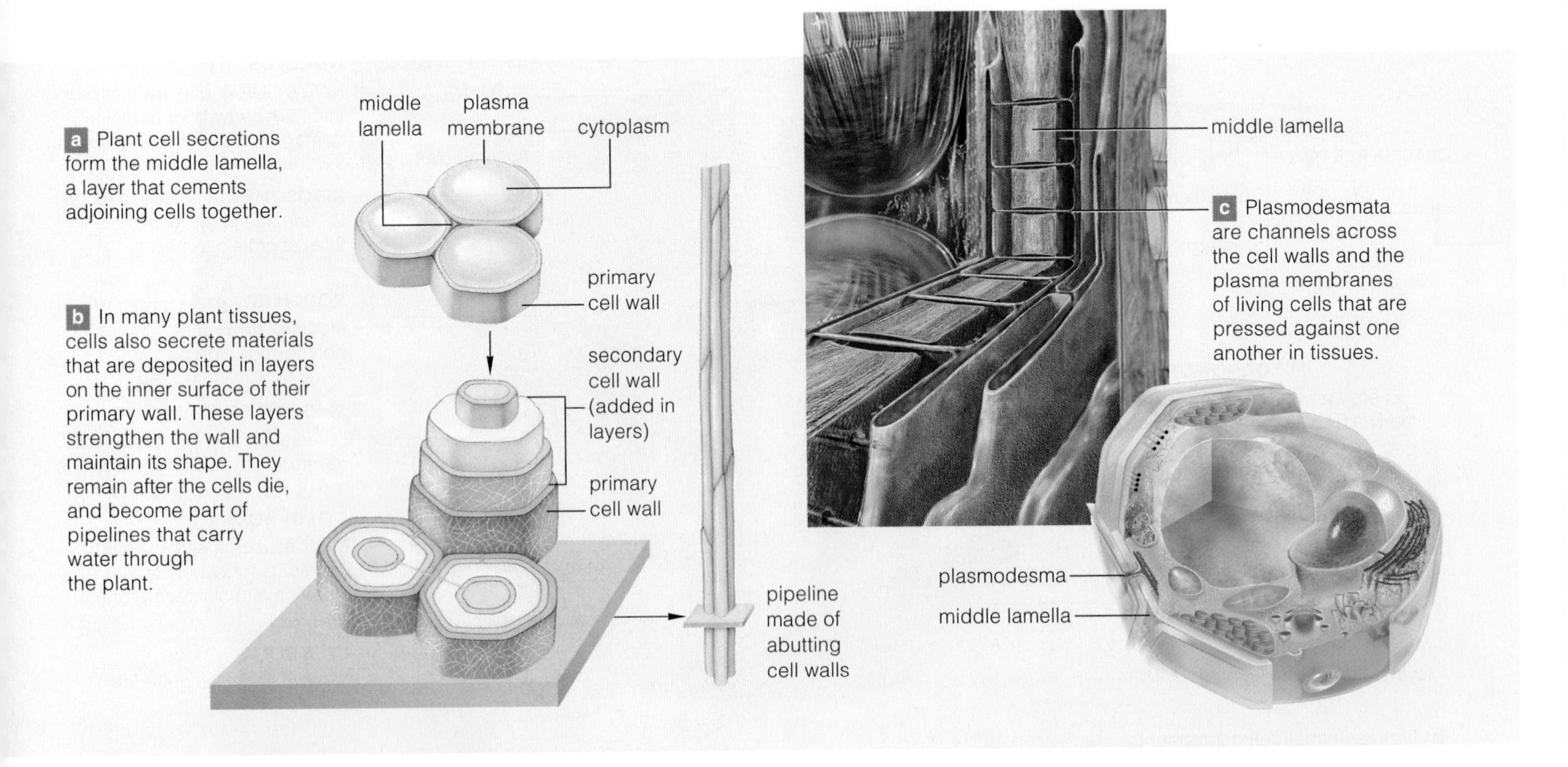

Figure 4.20 **Animated!** Some characteristics of plant cell walls.

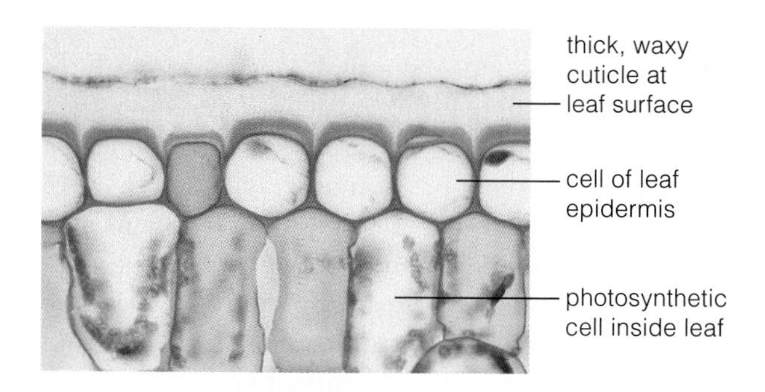

Figure 4.21 Section through a plant cuticle, a protective covering made of deposits secreted from living cells.

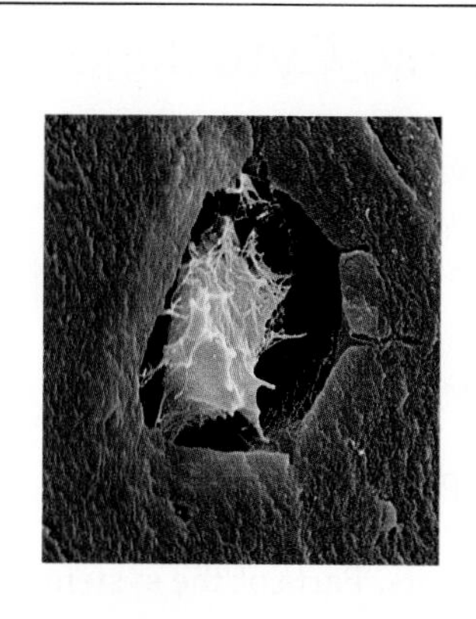
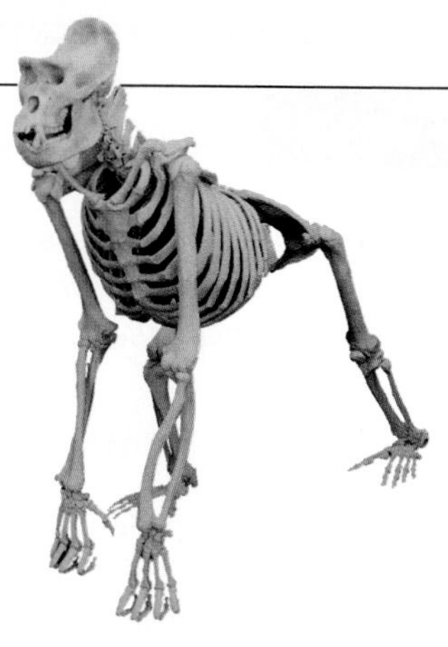

Figure 4.22 A living cell imprisoned in hardened bone tissue, the main structural material in the skeleton of most vertebrates.

Primary cell walls are a type of extracellular matrix, which in plants is mostly cellulose. The extracellular matrix of fungi is mainly chitin (Section 3.2). In most animals, extracellular matrix consists of various kinds of carbohydrates and proteins; it is the basis of tissue organization, and it provides structural support. For example, bone is largely an extracellular matrix that supports and protects body parts (Figure 4.22). The matrix is mostly a fibrous protein—collagen—and it is hardened by deposits of calcium and other minerals.

CELL JUNCTIONS

Cells imprisoned by a wall or other secretions can still interact with other cells and with their surroundings. In multicelled species, such interaction occurs by way of cell junctions, structures that connect a cell to other cells and to the environment. Cells can send or receive signals or materials through some types of junctions. Other types of junctions help cells recognize and stick to each other or to extracellular matrix.

In plants, channels called plasmodesmata (singular, plasmodesma) extend across the primary wall of two adjoining cells and connect their cytoplasm (Figure 4.20*c*). Substances flow quickly from cell to cell across these junctions.

In most tissues of animals, three types of cell-to-cell junctions are common (Figure 4.23). *Tight* junctions link cells in most body tissues, especially those that line the outer surfaces, internal cavities, and organs. These junctions seal cells together—tightly—so water-soluble substances cannot pass between them. Tight junctions are the reason why gastric fluid does not leak out of your stomach and damage your internal tissues. *Adhering* junctions anchor cells to one another and also to the extracellular matrix. *Gap* junctions connect the cytoplasm of adjacent cells. Substances flow quickly from one cell to another through these open channels. Gap junctions allow entire regions of cells to respond simultaneously to a stimulus, as when a signal causes cells in heart muscle to contract as a unit.

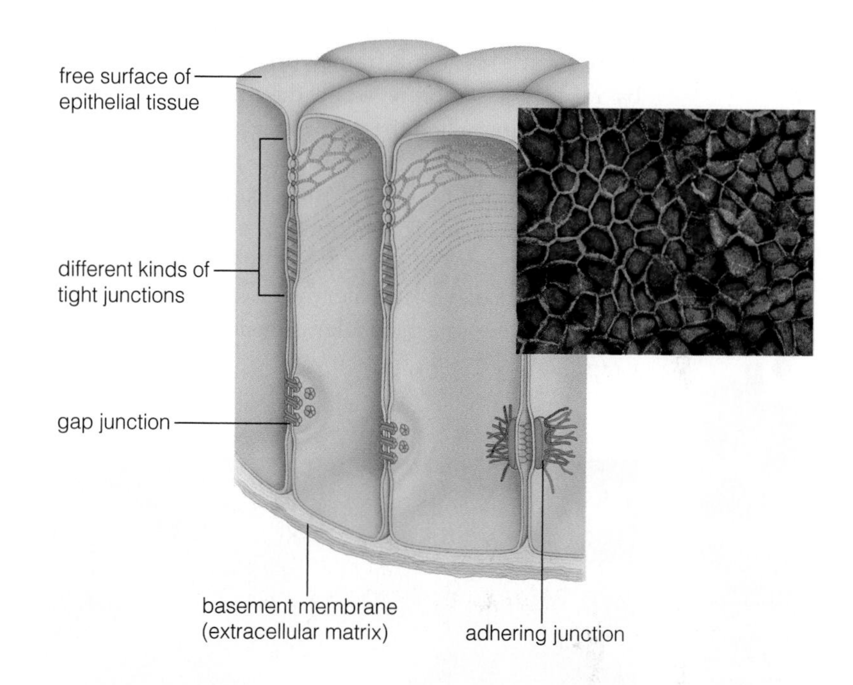

Figure 4.23 Animated! Three types of cell junctions in animal tissues: tight junctions, gap junctions, and adhering junctions. In the micrograph, a profusion of tight junctions (*green*) seals the abutting surfaces of kidney cell membranes and forms a waterproof tissue. DNA in each cell nucleus appears *red*.

Many protist and fungal cells, and all plant cells, have a porous wall around the plasma membrane. Animal cells do not have walls.

Cell secretions form a semitransparent cuticle that helps protect the exposed surfaces of soft plant parts.

Secretions form extracellular matrixes between cells in many kinds of animal tissues.

Cells make structural and functional connections with one another and with extracellular matrix in tissues.

4.12 The Dynamic Cytoskeleton

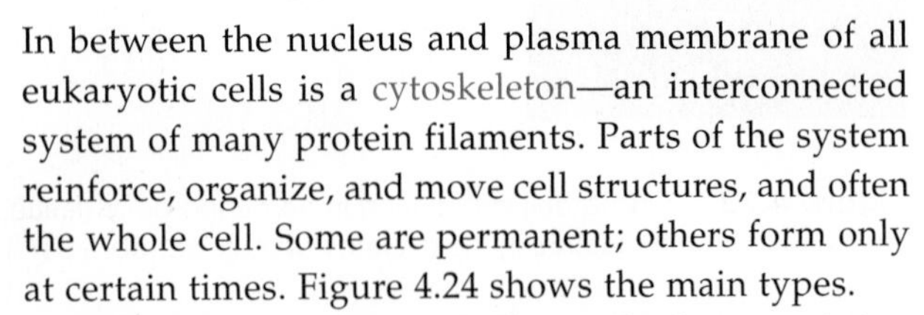

Like you, all eukaryotic cells have an internal structural framework—a skeleton. Unlike your skeleton, theirs has elements that assemble and disassemble quickly.

LINKS TO SECTIONS 3.4, 3.6

COMPONENTS OF THE CYTOSKELETON

In between the nucleus and plasma membrane of all eukaryotic cells is a cytoskeleton—an interconnected system of many protein filaments. Parts of the system reinforce, organize, and move cell structures, and often the whole cell. Some are permanent; others form only at certain times. Figure 4.24 shows the main types.

Microtubules are long, hollow cylinders consisting of many subunits of the protein tubulin (Figure 4.23*a*). They form a dynamic framework for many activities by rapidly assembling when they are required, then disassembling when they are not. For example, before a dividing eukaryotic cell splits in two, microtubules assemble, attach to the cell's duplicated chromosomes, and separate them. The microtubules then disassemble. As another example, the microtubules that form in the growing end of a young nerve cell support and guide its lengthening in a particular direction (Figure 4.24*d*).

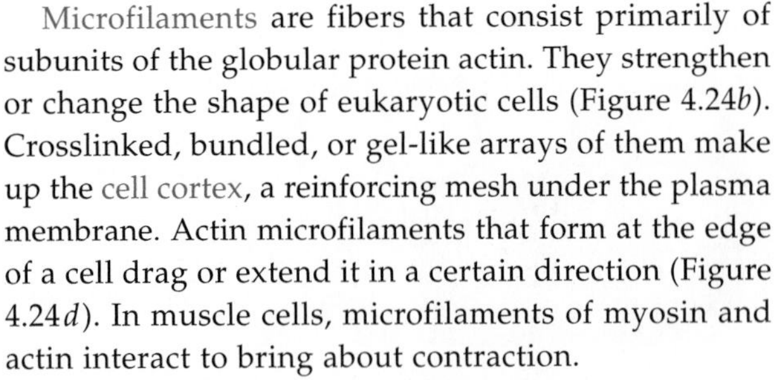

Microfilaments are fibers that consist primarily of subunits of the globular protein actin. They strengthen or change the shape of eukaryotic cells (Figure 4.24*b*). Crosslinked, bundled, or gel-like arrays of them make up the cell cortex, a reinforcing mesh under the plasma membrane. Actin microfilaments that form at the edge of a cell drag or extend it in a certain direction (Figure 4.24*d*). In muscle cells, microfilaments of myosin and actin interact to bring about contraction.

Intermediate filaments are the most stable parts of a cell's cytoskeletons (Figure 4.24*c*). They strengthen and maintain cell and tissue structures. For example, certain lamins form a layer that structurally supports the inner surface of the nuclear envelope.

All eukaryotic cells have similar microtubules and microfilaments. Despite the uniformity, both kinds of elements play diverse roles. How? They interact with accessory proteins, such as the motor proteins that can move cell parts in a sustained direction when they are repeatedly energized by ATP.

A cell is like a train station during a busy holiday, with molecules being transported through its interior. Microtubules and microfilaments are like dynamically assembled train tracks. Motor proteins are the freight engines that move along those tracks (Figure 4.25).

Some motor proteins move chromosomes. Others slide one microtubule over another. Some chug along tracks in nerve cells that extend from your spine to your toes. Many engines are organized in series, each moving some vesicle partway along the track before giving it up to the next in line. From dawn to dusk, kinesins inside plant cells drag chloroplasts to new positions, where they can intercept the most light as the angle of the sun changes overhead.

CILIA, FLAGELLA, AND FALSE FEET

Organized arrays of microtubules occur in eukaryotic flagella (singular, flagellum) and cilia (cilium). These whiplike structures propel cells such as sperm through

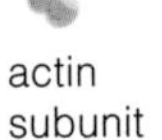

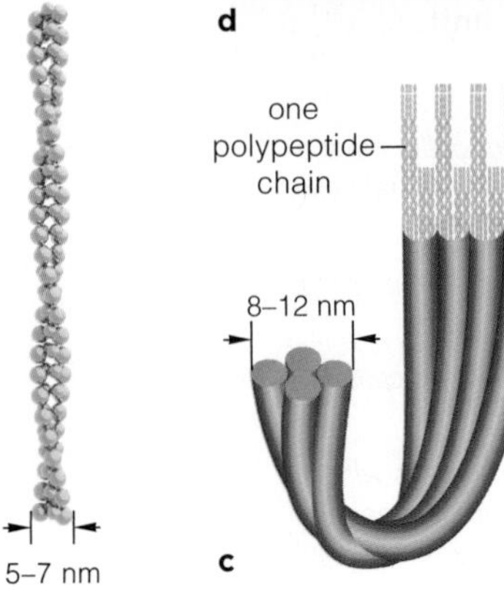

Figure 4.24 Components of the cytoskeleton. The structural arrangement of subunits in (**a**) microtubules, (**b**) microfilaments, and (**c**) intermediate filaments. Fluorescence microscopy yields startling images of cytoskeletal elements. (**d**) The microtubules (*yellow*) and actin microfilaments (*blue*) in the growing end of a nerve cell support and guide the cell's lengthening.

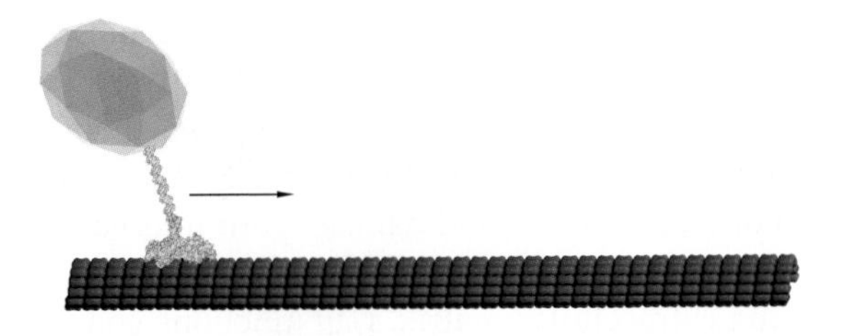

Figure 4.25 **Animated!** Kinesin (*tan*), a motor protein dragging cellular freight (in this case, a *pink* vesicle) as it inches along a microtubule.

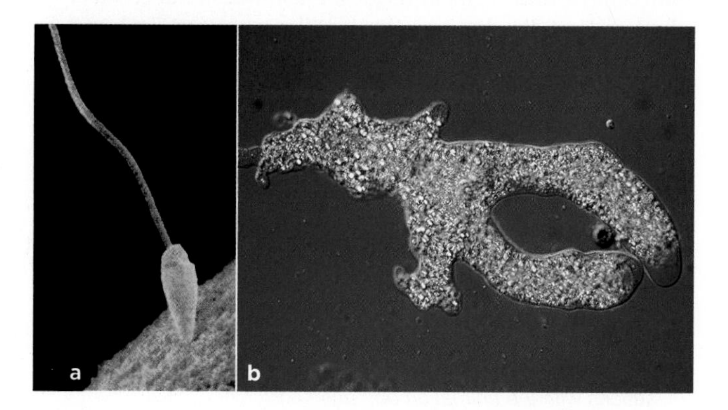

Figure 4.26 (**a**) Flagellum of a human sperm, which is about to penetrate an egg. (**b**) A predatory amoeba (*Chaos carolinense*) extending two pseudopods around its hapless meal: a single-celled green alga (*Pandorina*).

fluid (Figure 4.26*a*). Flagella tend to be longer and less profuse than cilia. The coordinated beating of cilia can propel motile cells through fluid and it can stir fluid around stationary ones. For example, many thousands of ciliated cells line the airways to your lungs. The cilia beat in a coordinated motion that sweeps bacteria and other particles away from the lungs.

Extending the length of a flagellum or cilium is a 9+2 array: nine pairs of microtubules ringing another pair in the center. Protein spokes and links stabilize the array. A barrel-shaped structure, the centriole, gives rise to the microtubules, then it remains below the finished array as a basal body. Figure 4.27 shows this internal structure and the sliding mechanism that moves it.

Amoebas and some other types of eukaryotic cells form temporary, irregular lobes called pseudopods, or "false feet" (Figure 4.26*b*). As they grow and bulge outward, pseudopods move the cell and engulf prey or some other target. Microfilaments inside elongate; they force the lobe to advance in a steady direction. Motor proteins that are attached to the microfilaments drag the plasma membrane with them.

A cytoskeleton of protein filaments is the basis of eukaryotic cell shape, internal structure, and movement.

Microtubules have roles in organizing the cell and moving its parts. Microfilament networks reinforce the cell surface. Intermediate filaments strengthen and maintain the shape of most cells and tissues of multicelled organisms.

When energized by ATP, motor proteins move along tracks of microtubules and microfilaments, delivering molecules and structures to new locations in a cell. As part of cilia, flagella, and pseudopods, they can move the whole cell.

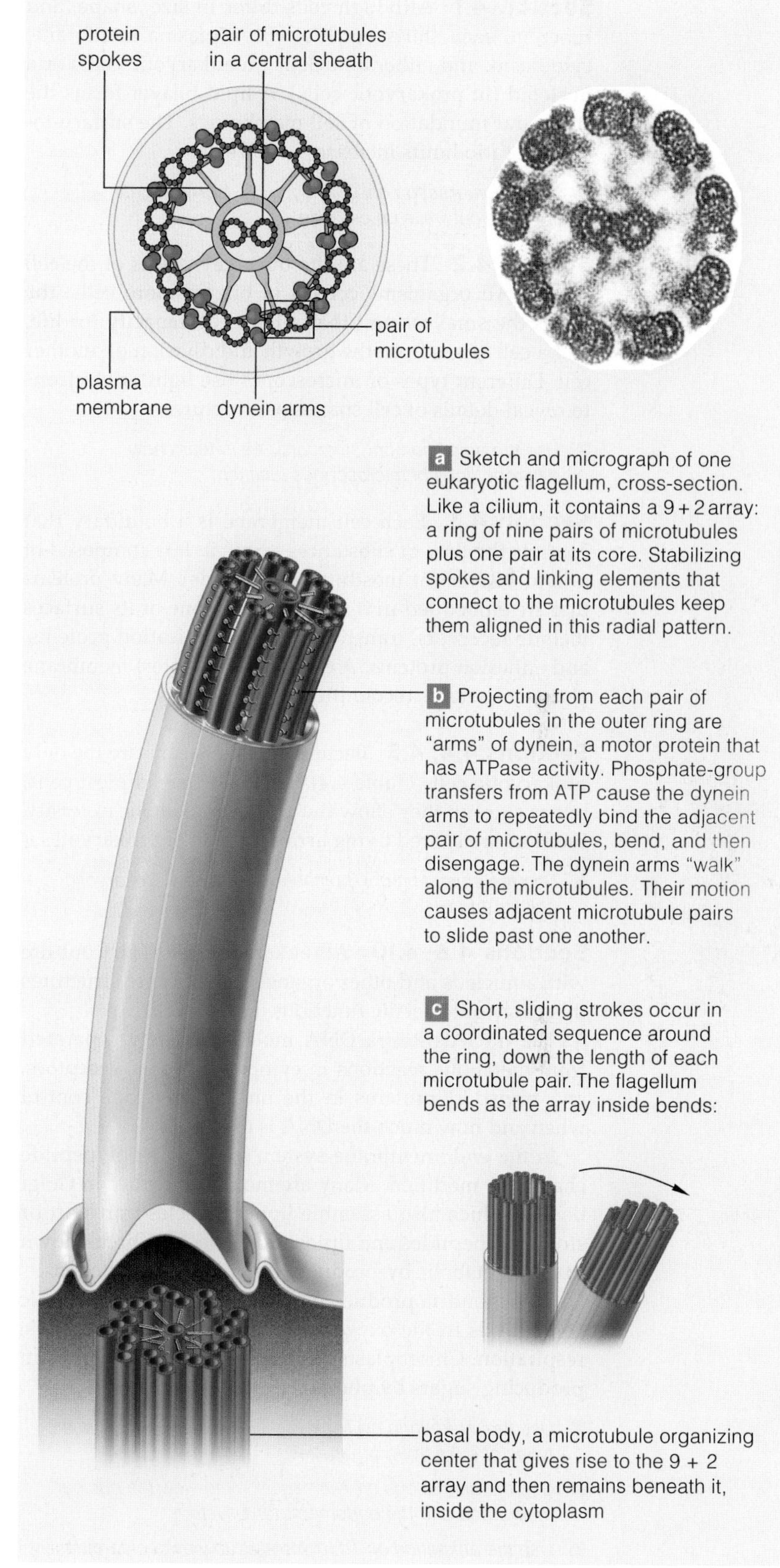

Figure 4.27 Animated! Eukaryotic flagella and cilia.

www.thomsonedu.com ThomsonNOW!

Summary

Section 4.1 Although cells differ in size, shape, and function, each starts out life with a plasma membrane, cytoplasm, and either a nucleus (in eukaryotic cells) or a nucleoid (in prokaryotic cells). A lipid bilayer forms the structural foundation of cell membranes. The surface-to-volume ratio limits increases in cell size.

■ *Use the interaction on ThomsonNOW to investigate the physical limits on cell size.*

Section 4.2 These are the three key points of the cell theory: All organisms consist of one or more cells, the cell is the smallest unit that retains the capacity for life, and a cell arises from the growth and division of another cell. Different types of microscopes use light or electrons to reveal details of cell shapes or structures.

■ *Use the animation on ThomsonNOW to learn how different types of microscopes function.*

Section 4.3 Each cell membrane is a boundary that controls the flow of substances across it. It is composed of a lipid bilayer (of mostly phospholipids). Many proteins that are embedded in it or attached to one of its surfaces include receptors, transporters, communication proteins, and adhesion proteins. A cell's plasma (outer) membrane also incorporates recognition proteins.

Sections 4.4, 4.5 Bacteria and archaeans are the only prokaryotic cells (Table 4.4). They are the simplest cells, but as a group they show the greatest metabolic diversity. Biofilms are shared living arrangements of prokaryotes.

■ *Use the animation on ThomsonNOW to view prokaryotic cell structure.*

Sections 4.6–4.10 All eukaryotic cells start out life with a nucleus and other organelles, which are structures that carry out specific functions inside a cell.

The nucleus keeps DNA molecules safely separated from metabolic reactions in cytoplasm. Pores, receptors, and transport proteins in the nuclear envelope control when and how much the DNA is used.

In the endomembrane system's ER, new polypeptide chains are modified. Many are modified further in Golgi bodies, which also assemble lipids. Vesicles transport or store polypeptides and lipids. Other vesicles break down toxins, metabolic by-products, and ingested pathogens.

Mitochondria produce ATP by breaking down organic compounds in the oxygen-requiring pathway of aerobic respiration. Chloroplasts are organelles that specialize in producing sugars by photosynthesis.

■ *Use the interaction on ThomsonNOW to survey the major types of eukaryotic organelles.*

■ *Use the animations on ThomsonNOW to view the nuclear membrane and the endomembrane system.*

■ *Use the animation on ThomsonNOW to view a chloroplast.*

Section 4.11 Cells of most prokaryotes, protists, and fungi and all plant cells have a porous wall around their plasma membrane. Many eukaryotic cells have surface specializations such as a wall or a cuticle. Extracellular matrix surrounds cells in tissues. Plasmodesmata connect plant cells. Adhering junctions, tight junctions, and gap junctions connect cells of animals.

■ *Study the structure of cell walls and junctions with the animation on ThomsonNOW.*

Section 4.12 Eukaryotic cells have a cytoskeleton of microtubules, microfilaments, and (in most) intermediate filaments. A cytoskeleton organizes and moves cell parts (and sometimes the whole cell). It reinforces cell shape. Interactions between motor proteins and microtubules in cilia, flagella, and pseudopods can move the whole cell.

■ *Learn more about cytoskeletal elements and their actions with the animation on ThomsonNOW.*

Self-Quiz

Answers in Appendix III

1. Is this statement true or false? The cytoplasm includes all of the cell's components except the plasma membrane and cell wall (if present).

2. Cell membranes consist mainly of a ______ .
   a. carbohydrate bilayer and proteins
   b. protein bilayer and phospholipids
   c. lipid bilayer and proteins

3. Unlike eukaryotic cells, prokaryotic cells ______ .
   a. have no plasma membrane
   b. have RNA but not DNA
   c. have no nucleus
   d. all of the above

4. Organelles ______ .
   a. are often enclosed by membranes
   b. are typical of eukaryotic cells
   c. have specific functions
   d. All of the above are features of organelles.

5. In the nucleolus, ______ .
   a. DNA remains organized
   b. ribosome subunits are built
   c. lysosomes operate
   d. digestion occurs

6. You will not observe an animal cell with ______ .
   a. mitochondria
   b. a plasma membrane
   c. ribosomes
   d. a cell wall

7. Is this statement true or false? The plasma membrane is the outermost component of all cells. Explain.

8. Match each cell component with its function.

___ mitochondrion	a. protein synthesis
___ chloroplast	b. initial modification of new polypeptide chains
___ ribosome	c. final modification of proteins; lipid assembly; shipping tasks
___ smooth ER	d. photosynthesis
___ Golgi body	e. production of ATP
___ rough ER	f. assembles lipids; other tasks

■ *Visit ThomsonNOW for additional questions.*

Critical Thinking

1. In a classic episode of *Star Trek*, a gigantic cell engulfs an entire starship. Spock blows the cell to bits before it reproduces and eats the universe. Think of at least one problem a biologist would have with this scenario.

Table 4.4 Summary of Typical Components of Prokaryotic and Eukaryotic Cells

Cell Component	Main Functions	Prokaryotic	Eukaryotic			
		Bacteria, Archaea	Protists	Fungi	Plants	Animals
Cell wall	Protection, structural support	✓*	✓*	✓	✓	*None*
Plasma membrane	Control of substances moving into and out of cell	✓	✓	✓	✓	✓
Nucleus	Physical separation of DNA from cytoplasm	*None*	✓	✓	✓	✓
DNA	Encoding of hereditary information	✓	✓	✓	✓	✓
RNA	Transcription, translation of DNA messages into polypeptide chains of specific proteins	✓	✓	✓	✓	✓
Nucleolus	Assembly of ribosome subunits	*None*	✓	✓	✓	✓
Ribosome	Protein synthesis	✓	✓	✓	✓	✓
Endoplasmic reticulum (ER)	Initial modification of many of the newly forming polypeptide chains of proteins; lipid synthesis	*None*	✓	✓	✓	✓
Golgi body	Final modification of proteins, lipid assembly, and packaging of both for use inside cell or export	*None*	✓	✓	✓	✓
Lysosome	Intracellular digestion	*None*	✓	✓*	✓*	✓
Mitochondrion	ATP formation	**	✓	✓	✓	✓
Photosynthetic pigments	Light–energy conversion	✓*	✓*	*None*	✓	*None*
Chloroplast	Photosynthesis; some starch storage	*None*	✓*	*None*	✓	*None*
Central vacuole	Increasing cell surface area; storage	*None*	*None*	✓*	✓	*None*
Bacterial flagellum	Locomotion through fluid surroundings	✓*	*None*	*None*	*None*	*None*
Flagellum or cilium with 9+2 microtubular array	Locomotion through or motion within fluid surroundings	*None*	✓*	✓*	✓*	✓
Cytoskeleton	Cell shape; internal organization; basis of cell movement and, in many cells, locomotion	*Rudimentary****	✓*	✓*	✓*	✓

* Known to be present in cells of at least some groups.
** Many groups use oxygen-requiring (aerobic) pathways of ATP formation, but mitochondria are not involved.
*** Protein filaments form a simple scaffold that helps support the cell wall in at least some species.

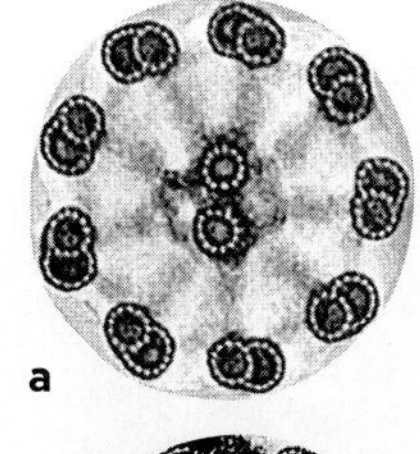

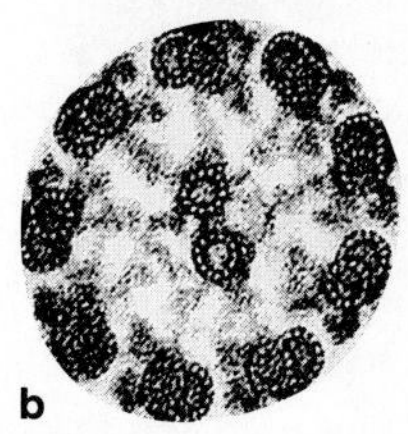

Figure 4.28 Cross-section of the flagellum of a sperm cell from (**a**) a human male affected by Kartagener syndrome and (**b**) an unaffected male. Check out the dynein arms projecting from the microtubule pairs.

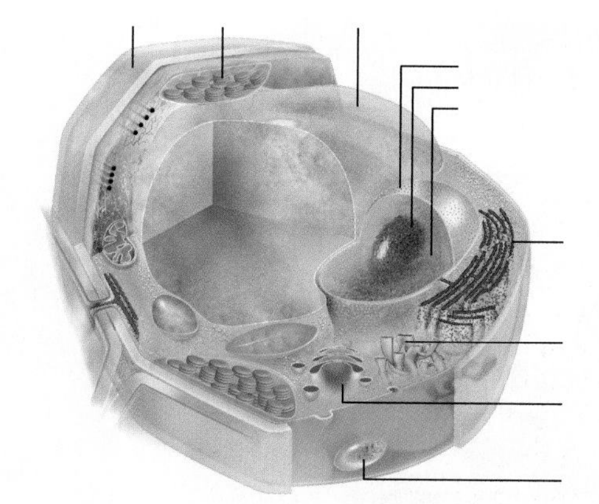

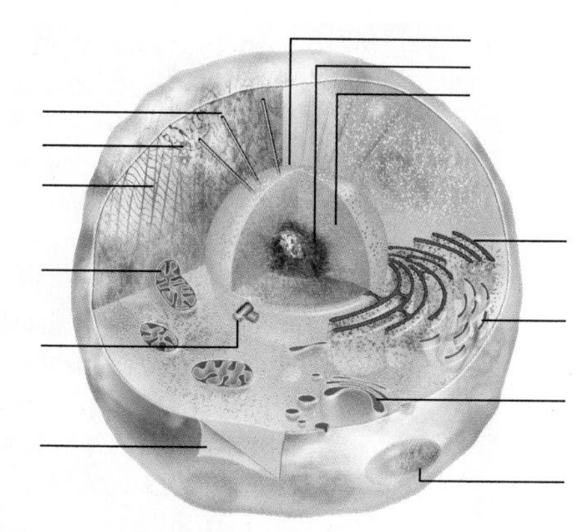

2. A mutated form of the protein dynein causes *Kartagener syndrome*. People affected by this genetic disorder have chronically irritated sinuses, and thick mucus collects in airways to the lungs. Biofilms form in the mucus, and the resulting bacterial activities and inflammation damage tissues. Also, affected men can produce sperm but are infertile (Figure 4.28). Some have become fathers after a doctor injects their sperm cells directly into eggs. Review Figure 4.27, and then explain how an abnormal dynein molecule could cause the observed effects.

3. Sketches of a plant cell and an animal cell are shown above. Identify as many of their components as you can.

4. Your professor shows you an electron micrograph of a cell with many mitochondria, Golgi bodies, and a lot of rough ER. What kinds of cellular activities would require such an abundance of the three kinds of organelles?

5. Many plant cells form a secondary wall on the inner surface of their primary wall. Speculate on the reason why the secondary wall does not form on the outer surface.

5 GROUND RULES OF METABOLISM

IMPACTS, ISSUES Alcohol, Enzymes, and Your Liver

The next time someone asks you to have a drink or two, or three, stop for a moment and think about the cells that keep alcohol from killing you. It makes no difference whether you drink a bottle of beer, a glass of wine, or 1-1/2 ounces of eighty-proof vodka. Each holds the same amount of alcohol or, more precisely, ethanol (CH_3CH_2OH). Ethanol molecules have polar and nonpolar regions, so they move very quickly from the stomach and small intestine into the bloodstream. Almost all of the alcohol someone drinks ends up in the liver, which has impressive numbers of alcohol-metabolizing enzymes. One of those enzymes, catalase, helps rid the body of harmful toxins, including ethanol (Figure 5.1).

As ethanol in blood circulates through the liver, catalase and other enzymes convert it to acetaldehyde (CH_3CHO), an organic compound even more toxic than ethanol and the most likely source of various hangover symptoms. A different enzyme in liver cells converts acetaldehyde to acetic acid (CH_3COOH), a less toxic compound.

A human liver can convert only so much acetaldehyde to acetic acid in any given hour, which is why people risk a hangover when they drink more than one alcoholic beverage in any two-hour interval.

Detoxifying alcohol is hard on liver cells. It causes a slowdown in protein and glucose synthesis, and disrupts lipid and carbohydrate breakdown. Mitochondria use up oxygen in ethanol metabolism—oxygen that normally would take part in the breakdown of fatty acids. Fatty acids accumulate as large fat globules in the tissues of heavy drinkers. Many liver cells die of oxygen starvation.

The liver is the largest gland in the human body, and its functions impact everything else. Without it, your body would have a hard time digesting fats or regulating its

See the video! **Figure 5.1** Is this a celebration of catalase, an enzyme with four hemes (*red*) tucked into its four polypeptide chains? It should be. Catalase helps detoxify many substances that can damage the body, such as the alcohol in beer.

How would you vote? Some people damage their liver because they drink too much alcohol. Others have liver problems from infection. There are not enough liver donors for all the people waiting for liver transplants. Should life-styles be a factor in deciding who gets a transplant? See ThomsonNOW for details, then vote online.

blood sugar level. The liver rids the body of many toxic compounds, not just acetaldehyde. It also makes some of the plasma proteins that circulate freely in blood. Plasma proteins are essential for blood clotting, immune function, and keeping the volume of the internal environment within a range that favors cell survival.

Given the liver's essential role in alcohol metabolism, heavy drinkers gamble with alcohol-induced liver diseases. In time, they have fewer and fewer cells for detoxification. One possible outcome is alcoholic hepatitis, a common disease characterized by inflammation and destruction of liver tissue. Alcoholic cirrhosis, another possibility, leaves the liver permanently scarred. Eventually the liver just stops working, with devastating effects.

Also think about a self-destructive behavior known as binge drinking. The idea is to consume large amounts of alcohol in a brief period. Binge drinking is now the most serious drug problem on college campuses throughout the United States. For example, here is one finding from a 1999 study: Almost half of the 14,138 students surveyed at 119 colleges and universities are binge drinkers, meaning they consumed five or more alcoholic drinks in a row at least once during the two weeks prior to the survey. One-fourth of those students binged at least three times during the same two-week period.

Binge drinking can do far more than damage the liver. Aside from the related 500,000 injuries from accidents, the 600,000 assaults by intoxicated students, 100,000 cases of date rape, and 400,000 cases of (whoops) unprotected sex among students in an average year, binge drinking can kill before you know what hit. Drink too much, too fast, and your heart may stop beating.

With this example we turn to metabolism, the cell's capacity to acquire energy and use it to build, degrade, store, and release substances in controlled ways. At times, the activities of your cells may be the last thing you want to think about, but they help define who you are and what you will become, liver and all.

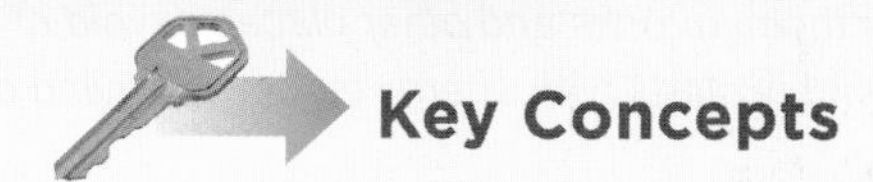

Key Concepts

THE NATURE OF ENERGY FLOW

Energy tends to disperse spontaneously. Each time energy is transferred, some of it disperses. Organisms maintain their complex organization only by continually harvesting energy. **Section 5.1**

ENERGY, ATP, AND ENZYMES

ATP couples metabolic reactions that release usable energy with reactions that require energy input. On their own, those reactions proceed too slowly to sustain life. Enzymes increase reaction rates. Environmental factors influence enzyme activity. **Sections 5.2–5.4**

THE NATURE OF METABOLISM

Metabolic pathways are energy-driven sequences of enzyme-mediated reactions. They concentrate, convert, or dispose of materials in cells. Controls over enzymes that govern key steps in these pathways can shift cell activities fast. **Section 5.5**

MEMBRANES AND METABOLISM

Concentration gradients drive the directional movements of ions and molecules into and out of cells. Transport proteins raise and lower water and solute concentrations across the plasma membrane and internal cell membranes. Other mechanisms move larger cargo across the plasma membrane. **Sections 5.6–5.9**

METABOLISM EVERYWHERE

Knowledge about metabolism, including how enzymes work, can help you interpret what you see in nature. **Section 5.10**

Links to Earlier Concepts

Reflect again on the road map for life's organization (Section 1.1). Here you will gain insight into how organisms tap into a grand, one-way flow of energy to maintain that organization (1.2). You will start thinking about how cells use the chemical behavior of electrons (2.3). You will see how pH (2.6) affects enzyme activity. You will consider some examples of how specific functions arise from protein structure (3.4). You will apply your knowledge of the properties of water molecules to the movement of water across membranes (2.5). You will see how the endomembrane system (4.8) helps cycle the lipid and protein components of membranes.

5.1 Energy and the World of Life

The molecules of life do not form spontaneously or we would see them in rocks and other places devoid of life. Their assembly starts with energy inputs into living cells.

ENERGY DISPERSES

LINKS TO SECTIONS 1.1, 1.2, 2.3, 3.2, 3.6, 4.12

Although we define energy as a capacity to do work, we can understand energy only by thinking about the way it behaves. For example, energy does not appear from nowhere, and it does not vanish into nothing. *Energy cannot be created or destroyed.* This concept is called the first law of thermodynamics.

There is an unchanging amount of energy in the universe that is distributed in many forms, such as sunlight, motion, and heat. Energy can be converted from one form to different forms, but the total amount of energy in the universe is constant.

Energy is either in play or in storage. When you see something change—say, when a ball is bouncing —energy is in play. A ball perched on the top shelf of your closet has energy in storage. *Potential* energy is a capacity to cause change because of where an object is located or how its parts are arranged.

As one example, ATP has potential energy because of the chemical bonds that hold its atoms in certain arrangements. In muscle cells, the chemical energy of ATP is transferred to myosin, a protein with a role in contraction (Section 4.12). That energy makes muscles contract. The chemical energy of ATP is converted to *kinetic* energy, the energy of motion, as in muscles that help you bounce (Figure 5.2). In each conversion, a bit of energy also is converted to *thermal* energy, or heat.

Here is another characteristic of energy: It tends to disperse spontaneously. For example, heat flows from a hot pan to air in a cool kitchen until the temperature of both is the same. You never see cool air raise the temperature of a hot pan. Each form of energy—not just heat—tends to disperse until no part of a system holds more than another part. Entropy is a measure of how much the energy in the universe, or any given part of it, has been dispersed. This is the second law of thermodynamics: *Entropy tends to increase.* If entropy decreases in one place, there will be a corresponding increase somewhere else.

Energy flows throughout the world of life by the making and breaking of bonds in chemical reactions. Entropy changes are part of the reason some reactions are spontaneous and others require an energy input.

ENERGY IN, ENERGY OUT

All chemical bonds hold potential energy. The amount in any bond depends on which elements are taking part in it. We measure bond energy in kilocalories. A kilocalorie is the amount of energy it takes to heat 1,000 grams of water by 1°C at a standard pressure. For instance, the amount of energy it takes to break one covalent bond between a hydrogen atom and an oxygen atom is 1.8×10^{-22} kilocalories. That also is the amount released when the bond forms.

Figure 5.2 Organisms converting the potential energy of molecules in their muscles to kinetic energy.

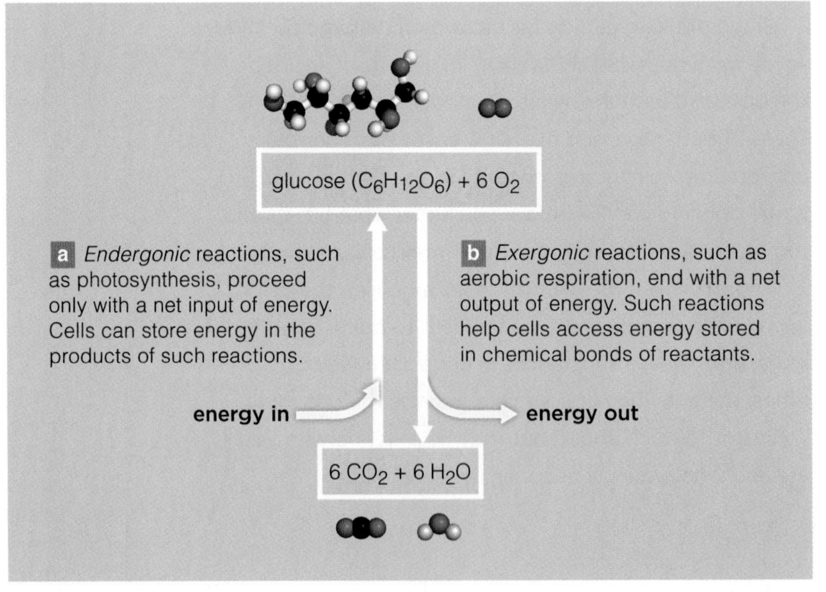

Figure 5.3 Animated! Two categories of energy change that occur during chemical work—in this case, building and breaking apart a glucose molecule.

Molecules that enter a reaction are reactants, and those remaining at the end of the reaction are products. Reactant bonds break and product bonds form at the same time in metabolic reactions. A reaction is a single event in which the reactant bonds are broken by the energy of product bond formation. We predict whether a reaction uses or releases energy by comparing bond energies of the reactants with those of the products.

Take a look at Figure 5.3*a*. In photosynthesis, six molecules of carbon dioxide (CO_2) and six of water (H_2O) become converted to one molecule of glucose ($C_6H_{12}O_6$) and six of oxygen (O_2). It takes more energy to break the bonds of the reactants than is released by formation of the products, so photosynthesis cannot occur without an energy boost. Reactions that will not happen without a net input of energy are said to be *endergonic*, meaning energy in.

Cells store energy by way of endergonic reactions. For example, energy input in the form of light drives the reactions of photosynthesis. Unlike light, glucose is a storable form of energy.

In other reactions, energy released when products form is greater than the energy required to break the bonds of the reactants. Such reactions end with a net release of energy; they are *exergonic*, meaning energy out. Exergonic reactions give cells access to energy in chemical bonds. In aerobic respiration, such reactions convert a glucose molecule to six molecules of CO_2 and six of H_2O, for a net gain of energy (Figure 5.3*b*).

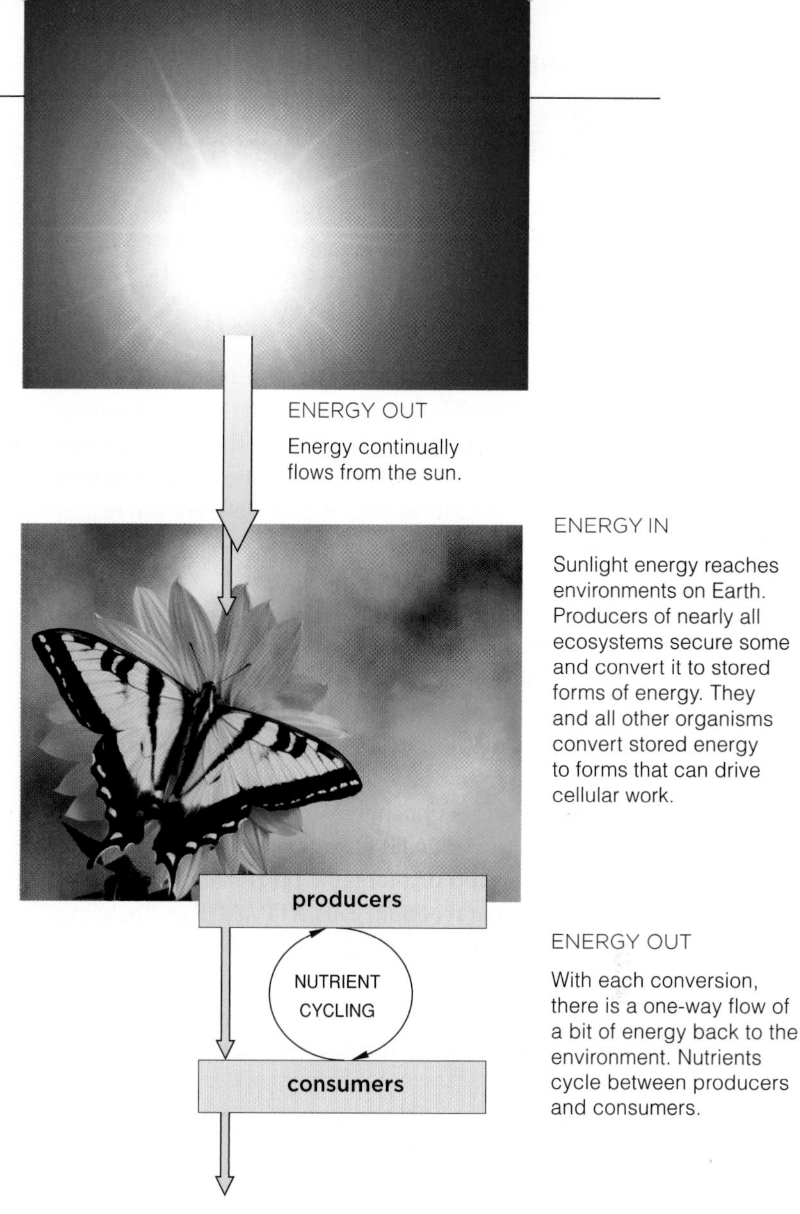

Figure 5.4 Example of how a one-way flow of energy into living organisms compensates for a one-way flow of energy out. Energy inputs drive a cycling of materials among producers and consumers.

THE GREAT ONE-WAY FLOW OF ENERGY

The tendency of energy to disperse gives us our sense of "time's arrow." Remember, a hot pan gives off heat when it cools, and that heat will not spontaneously diffuse back from the air to the pan. Heat lost from all of the reactions that converted one fertilized egg into your adult body will never be gathered up again. No matter what the example, released energy tends to flow in one direction—toward maximum dispersal.

Yet how can that be, given that an organized body is hardly dispersed? Living things constantly use energy to maintain themselves—to grow, to move, to acquire nutrients, to reproduce, and so on. Organisms replace energy they use for such purposes with energy they gain from the environment. They can stay alive only as long as they continue to resupply themselves with energy they get from someplace else.

Energy that enters the web of life comes mainly from the sun—which has been losing energy since it formed 5 billion years ago. Photosynthetic producers such as plants convert sunlight energy to bond energy when they build glucose and other compounds from simple raw materials. Consumers as well as producers access energy stored in those compounds by breaking and rearranging bonds. Nearly all organisms survive by tapping into the one-way energy flow from the sun. This flow organizes the world of life, from molecules through the biosphere (Figure 5.4).

Energy is the capacity to do work, and it cannot be created or destroyed. It can be converted from one form to another.

Energy tends to spread out, or disperse, spontaneously through any system and its surroundings.

Organisms maintain their complex organization as long as they replace their inevitable energy losses, mainly in the form of heat, with energy from someplace else.

5.2 ATP in Metabolism

ATP is the currency in a cell's economy. Cells spend it in energy-requiring reactions and invest it in energy-releasing reactions that help keep them alive. That is why we use a cartoon coin to symbolize ATP.

Next time you watch an animal or a person running, think of all the ATP that thousands of cells are using. Adenosine triphosphate, or ATP, is the main energy carrier in cells. This means ATP accepts energy released by exergonic reactions, and it also delivers energy to endergonic reactions.

All cells make ATP, which is a nucleotide that has three phosphate groups attached (see Section 3.6 and Figure 5.5*a*). When ATP donates a phosphate group to another molecule, it transfers energy that primes the recipient molecule to react. Phosphorylation is simply the formal name for such a phosphate-group transfer.

Cells have many ways of renewing ATP. When ATP gives up one of its phosphate groups, ADP (adenosine diphosphate) forms. ATP forms again when ADP binds a phosphate group or inorganic phosphate (P_i) by way of an endergonic reaction. This ATP/ADP cycle drives most metabolic reactions (Figure 5.5*b*).

ATP, the main energy carrier in all cells, couples reactions that release energy with reactions that require energy.

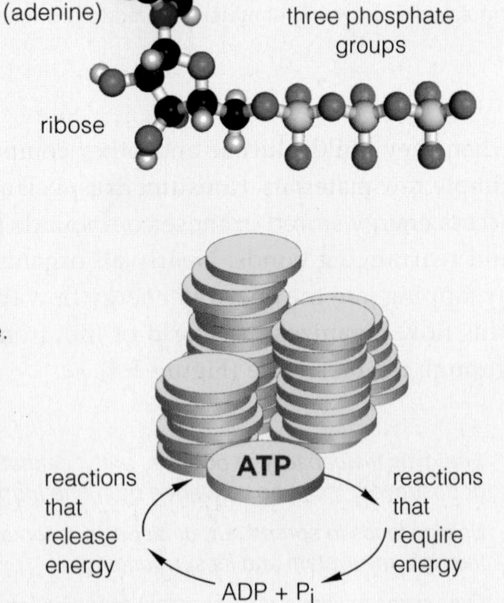

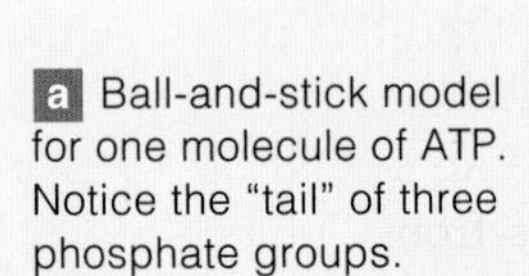
a Ball-and-stick model for one molecule of ATP. Notice the "tail" of three phosphate groups.

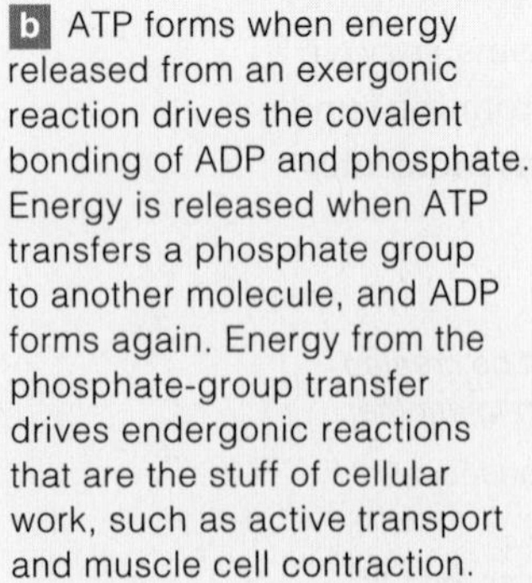
b ATP forms when energy released from an exergonic reaction drives the covalent bonding of ADP and phosphate. Energy is released when ATP transfers a phosphate group to another molecule, and ADP forms again. Energy from the phosphate-group transfer drives endergonic reactions that are the stuff of cellular work, such as active transport and muscle cell contraction.

Figure 5.5 **Animated!** ATP, the energy currency of cells.

5.3 Enzymes in Metabolism

If you left a cup of sugar out in the open, centuries might pass before all of it would break down to carbon dioxide and water. Yet that same conversion takes just a few seconds in your body. Enzymes make the difference.

ACTIVATION ENERGY—WHY THE WORLD DOES NOT GO UP IN FLAMES

The molecules of life are not particularly stable in the presence of oxygen. For example, think of how a spark ignites tinder-dry wood in a campfire. Wood is mostly cellulose. Cellulose, remember, consists of chains of many repeating units of glucose (Section 3.2). A spark converts cellulose to water and carbon dioxide. This reaction is highly exergonic, and it releases enough energy to keep itself going fast.

Earth is rich in oxygen—and in potential exergonic reactions. Why doesn't it burst into flames? Luckily, it takes energy to break the chemical bonds of reactants. Activation energy is the minimum amount of energy that will get a chemical reaction going. Some reactions require a lot of activation energy; others do not. For instance, nitrocellulose is made from cellulose. So little activation energy is required to change molecules of nitrocellulose into water, carbon dioxide, and nitrogen gas that this substance tends to explode spontaneously.

ENZYME STRUCTURE AND FUNCTION

Enzymes are catalysts: molecules that make chemical reactions occur much faster than they would on their own. Nearly all enzymes are proteins, but a few kinds of RNAs also show enzymatic activity. Enzymes are not consumed or altered by participating in a reaction; they can work again and again. Each kind recognizes, binds, and alters only specific reactants. For instance, the enzyme thrombin cleaves peptide bonds, but only a specific peptide bond in a protein called fibrinogen. Thrombin converts fibrinogen to fibrin, a protein that helps clot blood.

Reactants must gain a minimum amount of energy before a reaction proceeds. That amount, the activation energy, is like a hill that reactants must climb before they can run down the other side to products (Figure 5.6). Enzymes lower that energy hill by providing a microenvironment that is more favorable for a reaction than the surroundings.

Most enzymes are larger than their substrates, or the specific reactants that an enzyme recognizes and acts upon. An enzyme's polypeptide chains are folded into one or more chemically stable active sites. The sites are pockets or crevices where substrates bind and

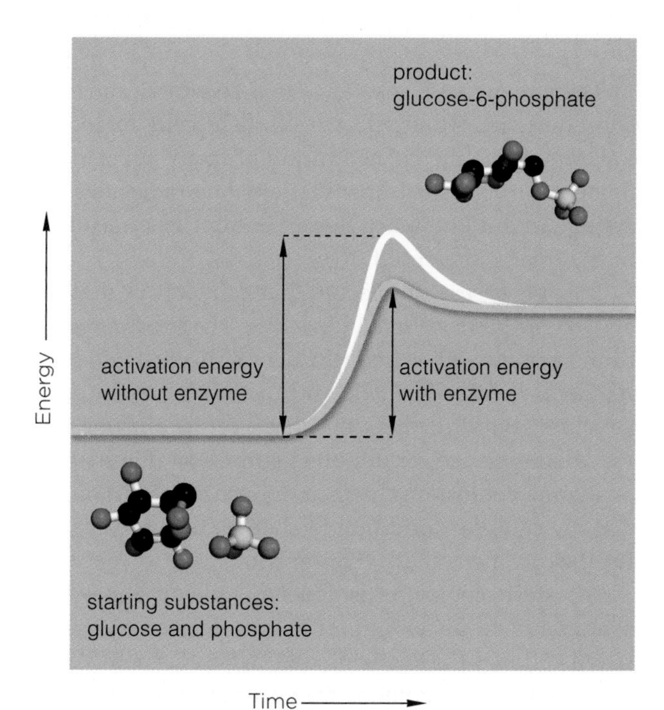

Figure 5.6 **Animated!** Activation energy, with or without an enzyme. Activation energy is the minimum amount of energy that will get a reaction started. An enzyme enhances the rate of reaction by lowering the energy hill.

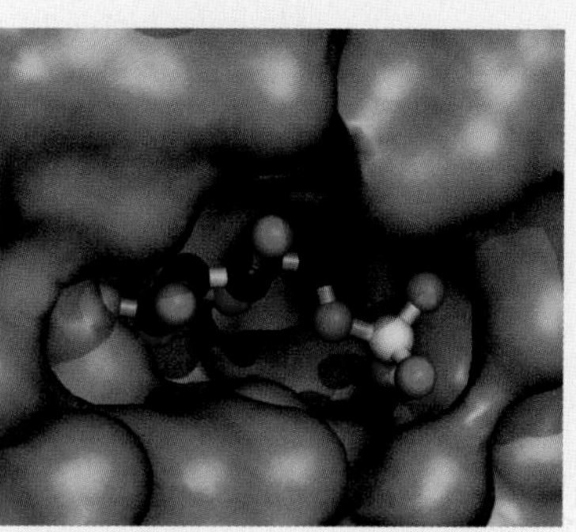

a A glucose molecule and a phosphate group meet inside hexokinase's active site, which has a microenvironment that favors binding of the substrates.

b The glucose molecule has bonded with the phosphate group. The product of this reaction, glucose-6-phosphate, is shown leaving the active site.

Figure 5.7 Models of the active site of hexokinase. This enzyme attaches phosphate groups to glucose and some other sugars with the help of ATP (not shown). The inset above shows the whole hexokinase molecule. Look at the size of the active site relative to the enzyme.

where reactions proceed (Figure 5.7). All or part of a substrate "fits" the active site; it is complementary in shape, size, polarity, and charge. Because of that fit, each enzyme acts only on specific substrates.

Think back on the main types of enzyme-mediated reactions (Section 3.1). With *functional group transfers*, one molecule gives up a functional group to another. With *electron transfers*, electrons from one molecule are accepted by another. With *rearrangements*, one kind of molecule is converted to another. With *condensation*, two or more molecules covalently bind and become a larger molecule. With *cleavage* reactions, one molecule splits into two smaller ones.

When we talk about activation energy, we really are talking about the energy it takes to align reactive chemical groups, destabilize electric charges, and break bonds. Such events bring on the transition state, when the substrate's bonds reach the breaking point and the reaction can run spontaneously to product.

The following four mechanisms work alone or in combination during enzyme-mediated reactions:

Helping substrates get together. When molecules of substrates are far apart, they rarely react. Binding at an active site is as effective as bringing substrates 10 millionfold closer together.

Orienting substrates in positions that favor reaction. On their own, substrates collide from random directions. By contrast, in an active site, individually weak but extensive bonds closely align reactive groups.

Inducing a fit between the enzyme and its substrate. By the induced-fit model, the substrate is almost but not quite complementary to the active site. The enzyme restrains the substrate and stretches or squeezes it into a shape that often puts it next to a reactive group or to another molecule. By forcing a substrate to fit into the active site, the enzyme ushers in the transition state.

Shutting out water molecules. Because of its capacity to form hydrogen bonds so easily, water can interfere with the breaking and formation of chemical bonds in reactions. Certain active sites repel water and keep it away from the reactions.

LINKS TO SECTIONS 3.1, 3.2, 3.4

Enzymes greatly enhance reaction rates. They lower the minimum amount of energy required to get a reaction to the point where it will run to completion, with no further energy input. Activation energy differs for different reactions.

In an enzyme's active site, substrates move to a transition state, when their bonds are at the breaking point and the reaction can run spontaneously to completion.

Four mechanisms move substrates to the transition state. They concentrate substrate molecules, orient them in positions that favor reaction, induce a substrate to fit the active site, and exclude water from the active site.

5.4 Enzymes Don't Work Alone

Many factors influence what an enzyme molecule does at any given time or whether it is built in the first place. Here we highlight a few of the major factors.

CONTROLS OVER ENZYMES

LINKS TO SECTIONS 2.6, 3.4, 3.5

What happens when one or another of the thousands of substances in cells becomes too abundant or scarce? Feedback mechanisms and other controls that activate or inhibit enzymes allow cells to conserve energy and resources. These adjustments help cells produce what they require—no more, no less—at any given moment.

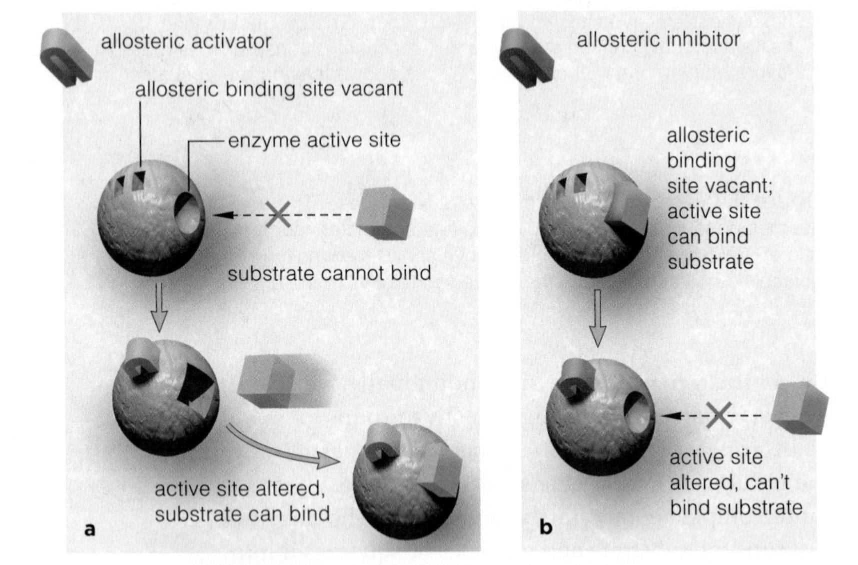

Figure 5.8 **Animated!** Examples of allosteric control. (**a**) An active site is unblocked when an activator binds to an allosteric site. (**b**) An active site is blocked when an inhibitor binds to an allosteric site.

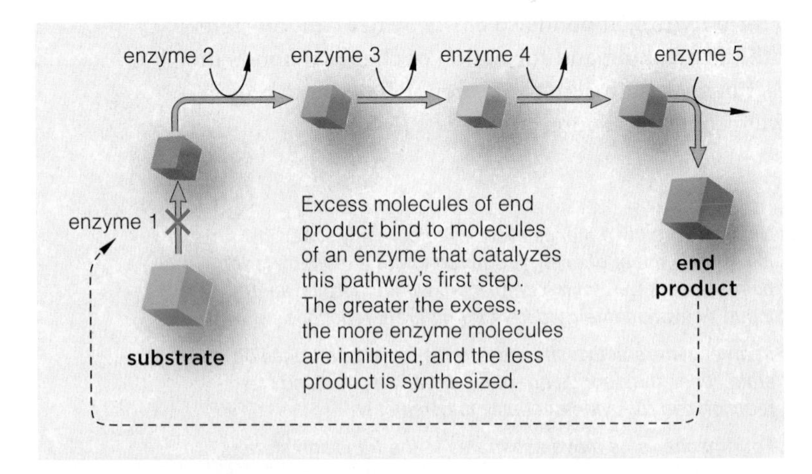

Figure 5.9 **Animated!** Feedback inhibition. In this example, five kinds of enzymes act in sequence to convert a substrate to a product, which inhibits the activity of the first enzyme.

The amount of a substance in a given volume is its concentration. Different controls help cells maintain, lower, or raise the concentration of many substances. Some controls adjust how fast enzyme molecules are synthesized. Others activate or inhibit enzymes that have already been built.

In some cases, a molecule that activates or inhibits an enzyme binds to an *allosteric* site. The site is a region of the enzyme other than the active site that can bind regulatory molecules (*allo–* means other; *steric* means structure). Binding alters the shape of the enzyme in a way that enhances or inhibits its function (Figure 5.8).

By one control mechanism, feedback inhibition, an activity causes a condition to change, then the change itself stops the activity (Figure 5.9).

Allosteric control of isoleucine biosynthesis is one example of feedback inhibition. Such control ensures that a cell makes isoleucine only when needed. While cells are making proteins, they are also making amino acid building blocks of proteins, including isoleucine. When protein synthesis slows, isoleucine is no longer being incorporated into proteins, and it accumulates. The unused isoleucine binds to an allosteric site on an enzyme in its own biosynthesis pathway. The binding changes the enzyme's shape, so less isoleucine forms. When the cell starts to make proteins again, it uses up accumulated isoleucine until the allosteric site is free. The isoleucine synthesis enzyme begins to work, and isoleucine is produced again.

EFFECTS OF TEMPERATURE, pH, AND SALINITY

Conditions in the environment affect enzyme activity. Factors such as temperature, pH, and salinity (or salt concentration) influence reaction rates.

Temperature, remember, is a measure of molecular motion (Section 2.6). As it rises, it boosts reaction rates both by increasing the likelihood that a substrate will bump into an enzyme and by increasing a substrate molecule's internal energy. The more internal energy a reactant molecule has, the closer it is to jumping the activation energy hill and taking part in a reaction.

Above the range of temperatures that an enzyme can tolerate, weak bonds are broken. The shape of the enzyme changes, so substrates no longer can bind to the active site. The reaction rate falls sharply (Figure 5.10). For example, temperatures above 42°C (107.6°F) adversely affect your body's enzymes, which is why such severe fevers are dangerous.

Also, remember that the pH of solutions can vary (Section 2.6). In the human body, most enzymes work best at pH 6–8. For instance, the hexokinase molecule in Figure 5.7 is active in the small intestine, where the pH is around 8. One exception, the enzyme pepsin, is activated only in gastric fluid in the stomach, where it digests any protein. The fluid is very acidic, with a pH of 2. If any activated pepsin were to leak out of your stomach, it would digest the proteins in your tissues instead of those in your food. Figure 5.11 shows the effect of pH on pepsin and other kinds of enzymes.

Also, an enzyme's activity is influenced by shifts in the amount of salt in the surrounding fluid. Too much or too little salt can interfere with the hydrogen bonds that hold an enzyme in its three-dimensional shape. If an enzyme loses its shape, it will lose its function.

HELP FROM COFACTORS

Cofactors are atoms or molecules other than proteins that associate with enzymes and are necessary for their function. Some are metal ions. Organic cofactors are called coenzymes. Almost all vitamins are coenzymes or precursors of them.

We can use catalase as an example of how cofactors work. Catalase has four hemes: organic ring structures with an iron cofactor at their center (Figure 5.1). The iron helps catalase speed the breakdown of hydrogen peroxide to water. How? Like other metal ions, iron affects the electron arrangement of nearby molecules and so helps bring on the transition state.

Catalase is an antioxidant. The iron in each heme helps catalase neutralize other strong oxidizers such as free radicals. *Free radicals* are atoms or molecules with at least one unpaired electron. These dangerous leftovers of metabolic reactions attack the structure of biological molecules. Free radicals accumulate as we age, in part because the body makes fewer and fewer catalase molecules.

Some coenzymes are tightly bound to an enzyme. Others, such as NAD+ and NADP+, can diffuse freely through the cytoplasm. Unlike enzymes, many kinds of coenzymes become modified during a reaction, but they are regenerated elsewhere.

Controls maintain, lower, and raise concentrations of many substances by enhancing or inhibiting the activity of specific enzymes. These adjustments help cells produce only what they require in any given interval.

Enzymes work best within limited ranges of temperature, pH, and salt concentration. Ranges of tolerance differ from one type of enzyme to the next.

Cofactors are metal ions or coenzymes that associate with enzymes and are necessary for their function.

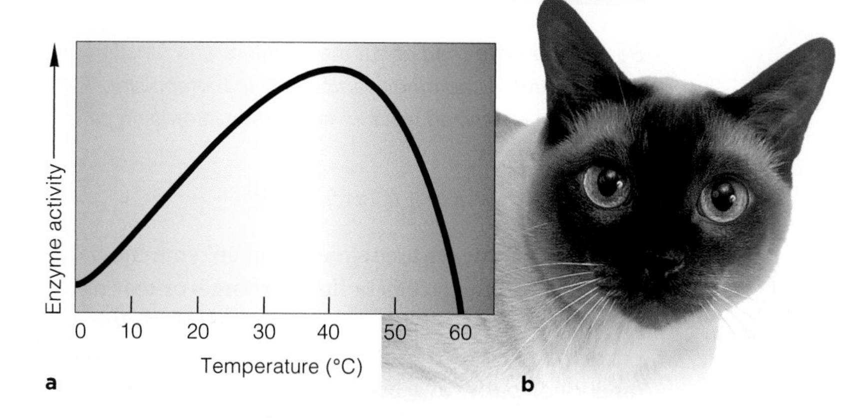

Figure 5.10 Enzymes and the environment. (**a**) Graph that shows how increases in temperature affect one enzyme's activity.

(**b**) The air temperature outside the body affects the fur color of Siamese cats. Epidermal cells that give rise to the cat's fur produce a brownish-black pigment, melanin. Tyrosinase, an enzyme in the melanin production pathway, is heat-sensitive in the Siamese. It becomes less active in warmer parts of the cat's body, which end up with less melanin, and lighter fur.

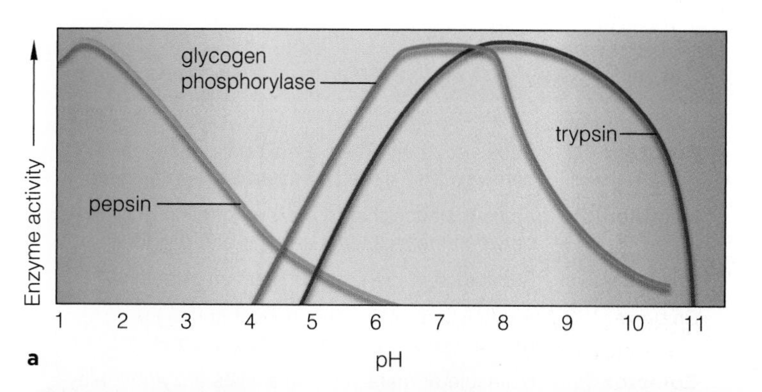

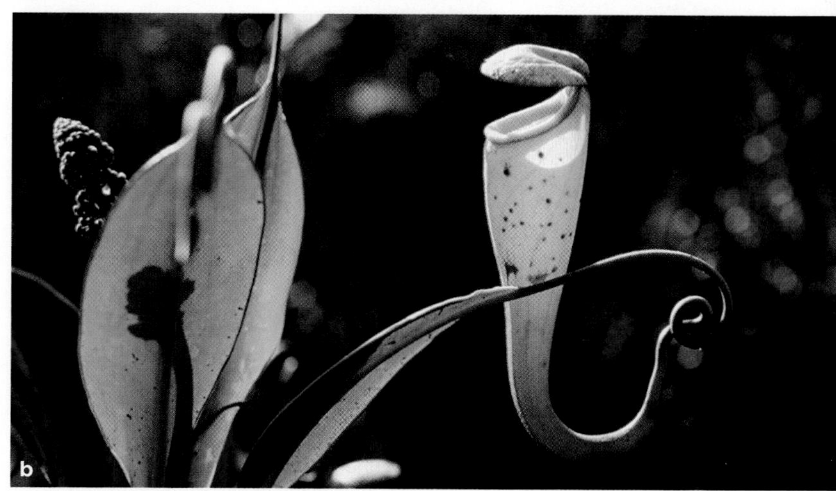

Figure 5.11 Enzymes and the environment. (**a**) How pH values affect three enzymes. The *orange* graph line tracks the activity for pepsin. (**b**) Carnivorous plants of genus *Nepenthes* grow in nitrogen-poor habitats. They secrete acids and protein-digesting enzymes into fluid in a cup made of a modified leaf. The enzymes release nitrogen from small prey, such as insects, that are attracted to odors from the fluid and then drown in it. One of these pepsin-like enzymes functions best at pH 2.6.

5.5 Metabolism—Organized, Enzyme-Mediated Reactions

So far, you have read about the structure and function of ATP, enzymes, and other participants in metabolism. Turn now to the organized ways in which they interact in cells.

TYPES OF METABOLIC PATHWAYS

LINKS TO SECTIONS 2.3, 4.3

Metabolic pathways are sequences of enzyme-mediated reactions by which cells build, rearrange, or tear down substances. You already are acquainted with the key participants, which are listed in Table 5.1. *Biosynthetic* (anabolic) pathways build organic compounds from small molecules, and they require a net energy input. By contrast, *degradative* (catabolic) pathways—which end with a net release of usable energy—break down organic compounds to smaller products.

We can categorize all organisms according to how they get the carbon required for metabolic pathways. The producers, or autotrophs, get carbon directly from carbon dioxide in their environment. Many kinds use that carbon in photosynthesis, the main biosynthetic pathway in the biosphere (Figure 5.12). Consumers are heterotrophs; they get their carbon from organic compounds that autotrophs have already assembled. Both autotrophs and heterotrophs extract energy from organic compounds. Most do so by aerobic respiration, the main degradative pathway in the biosphere.

Many metabolic pathways are linear, a straight line from reactants to products. Others are branched, with reactants or some intermediates funneled into more than one sequence of reactions. Still others are cyclic; the last step regenerates the reactant that is required for the first step. For example, such a pathway occurs during the second stage of photosynthesis. The entry point for the reactions is a molecule called RuBP; the last reaction of the pathway converts an intermediate to another molecule of RuBP.

Table 5.1 Key Players in Metabolic Reactions

Reactant	Substance that enters a metabolic reaction or pathway; also called a substrate of an enzyme
Intermediate	Any substance that forms in a reaction or pathway, between the reactants and the end products
Product	Substance at the end of a reaction or pathway
Enzyme	A catalytic protein, one that enhances the rate of a reaction; a few RNAs are catalytic
Cofactor	Molecule or metal ion that assists enzymes. May carry electrons, hydrogen, or functional groups to other reaction sites. NAD^+ is an example
Energy carrier	Mainly ATP; couples reactions that release energy with different reactions that require energy
Transport protein	Protein that passively assists or actively pumps specific solutes across a cell membrane

THE DIRECTION OF METABOLIC REACTIONS

Bear in mind, metabolic reactions do not always run from reactants to products. Most also run in reverse to some extent, with products being converted back to reactants. Reactions tend toward chemical equilibrium, when the concentrations of reactants and products no longer change because the rate of reaction is about the same in either direction. At equilibrium, the number of reactant and product molecules are not necessarily equal (Figure 5.13). Like a party in which people drift between two rooms, the number in each room may stay the same even as people move back and forth.

Why bother to think about this? Big changes can occur in cellular activities through control of enzymes that mediate a few steps of reversible pathways.

For example, membrane transport proteins called ATP synthases (Section 4.3) in bacteria, mitochondria,

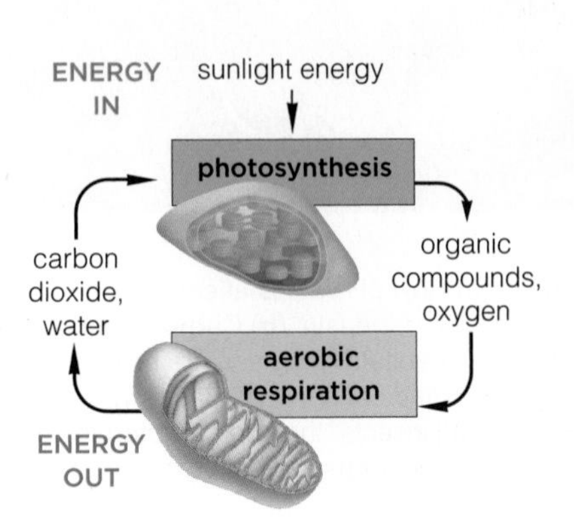

Figure 5.12 Main metabolic pathways. In plants, energy from the sun drives the synthesis of glucose from carbon dioxide and water. In plants, animals, and many other kinds of organisms, aerobic respiration uses oxygen in the complete breakdown of glucose and releases usable energy. The two pathways are now linked on a global scale, as Section 7.8 explains.

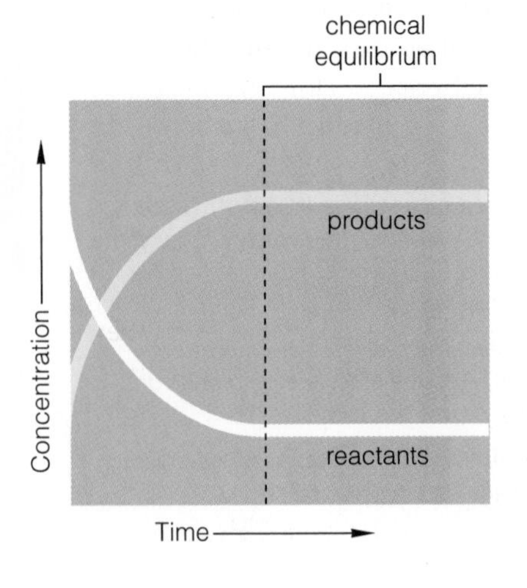

Figure 5.13 Chemical equilibrium, when there is no more net change in the concentrations of reactants and products for a reaction.

The forward and reverse reaction rates change over time until chemical equilibrium is reached.

Figure 5.14 **Animated!** Uncontrolled versus controlled energy release. (**a**) Glucose and oxygen react when exposed to a spark. Energy is released in an explosive burst as carbon dioxide and water form. (**b**) The very same reaction occurs in small steps with an electron transfer chain. The steps release energy in amounts that cells can harness for cellular work.

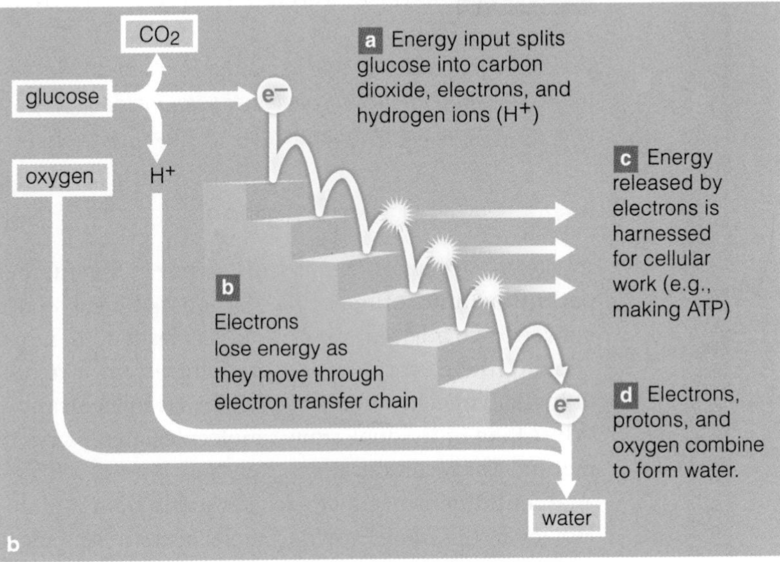

and chloroplasts serve a critical role in cellular energy conversion and transfers. Each one is a channel across the membrane through which hydrogen ions can flow; in one direction, that flow drives the synthesis of ATP from ADP and inorganic phosphate. Depending on the conditions in the cell or organelle, ATP synthases can also function in reverse: They can use energy released from ATP hydrolysis to pump hydrogen ions across the membrane in the opposite direction.

REDOX REACTIONS

If a glucose molecule were to break apart and give up all of its stored energy at once, the energy would be released explosively (Figure 5.14*a*). Explosions are not good for cells. The only way cells can capture energy from glucose is to break down the molecule in small, manageable steps. Most of these steps are oxidation–reduction reactions, or electron transfers. In each of these "redox" reactions, a molecule accepts electrons (it becomes *red*uced) from another molecule (which becomes *ox*idized). Coenzymes are among the many molecules that accept electrons in redox reactions.

In the next two chapters, you will see how redox reactions also occur in electron transfer chains. Such chains are membrane-bound arrays of enzymes and other molecules that accept and give up electrons in an organized series of steps. Electrons are at a higher energy level when they enter a chain than when they leave. Think of the electrons as descending a staircase and losing a bit of energy at each step (Figure 5.14*b*).

In photosynthesis and aerobic respiration, many coenzymes deliver electrons to electron transfer chains. Energy released at certain steps in those chains helps drive the synthesis of ATP. Figure 5.15 is an overview of how ATP and coenzymes connect energy-releasing with energy-requiring pathways. These pathways will occupy our attention in chapters to come.

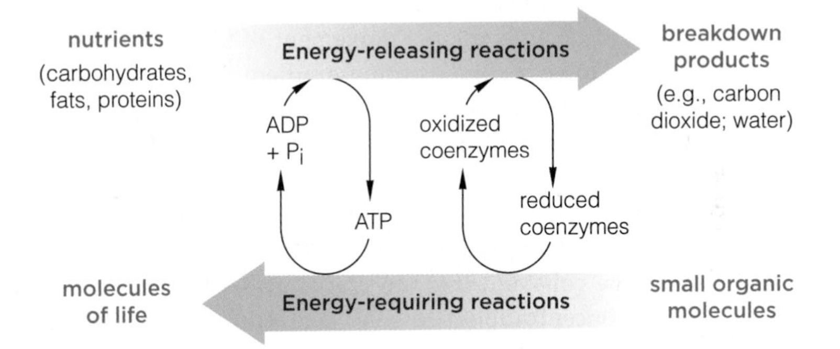

Figure 5.15 ATP, coenzymes, and metabolic pathways. Molecules of ATP form in many different energy-releasing reactions, then they deliver chemical energy to energy-requiring ones. The oxidized coenzymes NAD^+, $NADP^+$, and FAD accept electrons and hydrogen from energy-releasing reactions. The reduced coenzymes (NADH, NADPH, and $FADH_2$) deliver their cargo to energy-requiring reactions.

Metabolic pathways are orderly, enzyme-mediated reaction sequences. Many are biosynthetic; others are degradative.

Control over a key step of a metabolic pathway can bring about rapid shifts in cell activities.

Many aspects of metabolism involve electron transfers, or oxidation–reduction reactions. Redox reactions often occur in electron transfer chains. The chains are important sites of energy exchange in both photosynthesis and aerobic respiration.

5.6 Diffusion, Membranes, and Metabolism

Cells can't run to the grocery store for refills, or to the trash can when metabolic wastes pile up. Instead, they take in or expel substances across their membranes.

WHAT IS A CONCENTRATION GRADIENT?

LINKS TO SECTIONS 2.5, 2.6, 4.1, 4.3

A concentration gradient is a difference in the number per unit volume of molecules (or ions) of a substance between two adjacent regions. Molecules tend to move "down" their concentration gradient—from a region of higher concentration to one of lower concentration. Why? Like individual atoms, molecules are always in motion. They collide at random and bounce off one another millions of times each second in both regions. However, the more crowded molecules are, the more often they collide. During any interval, more molecules are knocked out of a region of higher concentration than are knocked into it.

Diffusion is the net movement of like molecules or ions down a concentration gradient. It is an essential way in which substances move into, through, and out of cells. In multicelled species, diffusion can also move substances between body regions or between the body and its environment. For instance, photosynthetic cells inside a leaf produce oxygen. The oxygen diffuses out of the cells and into air spaces inside the leaf, where its concentration is lower. Then it diffuses into the air outside the leaf, where its concentration is lower still.

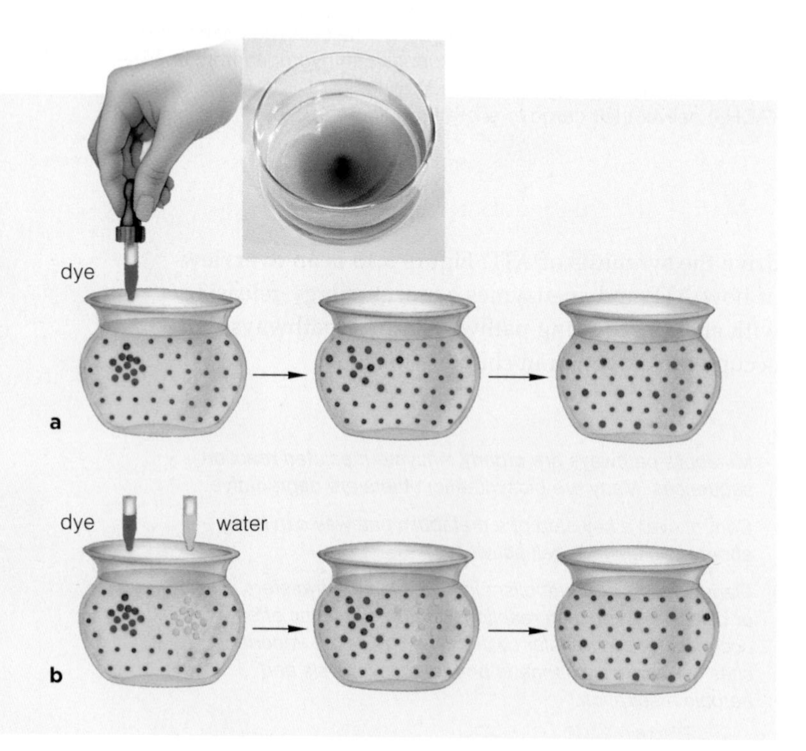

Any substance tends to diffuse in a direction set by its own concentration gradient, not by the gradients of other solutes that may be sharing the same space. You can observe this tendency by squeezing a drop of dye into water. Dye molecules diffuse slowly into the region where they are less concentrated. At the same time, water molecules move the other way, into the region where they are less concentrated (Figure 5.16).

DIFFUSION RATES

How quickly a solute diffuses depends on its size, the steepness of its concentration gradient, temperature, and electric or pressure gradients (if any).

First, small molecules diffuse faster than large ones do. Second, the diffusion rate is higher with steeper gradients, because more molecules are moving out of a region of greater concentration compared with the number moving into it. Third, more heat energy makes molecules move faster. Thus, solutes collide more often in warmer regions.

Fourth, an electric gradient can affect the rate and direction of diffusion. An electric gradient is simply a difference in electric charge between adjoining regions. For example, each ion dissolved in a fluid contributes to the fluid's overall electric charge. Opposite charges attract. A fluid with an overall negative charge attracts positively charged substances, such as sodium ions, dissolved in the fluid of an adjoining region. Later, you will see how electric and concentration gradients help drive activities such as ATP formation and the transmission of signals along nerve cells.

Fifth, diffusion also may be affected by a pressure gradient—a difference in pressure per unit volume (or area) between two adjoining regions.

DIFFUSION AND MEMBRANE PERMEABILITY

Many different substances are dissolved in cytoplasm and in extracellular fluid, but the kinds and amounts of solutes in the two fluids differ. Cells maintain these differences with a membrane property called selective

Figure 5.16 **Animated!** Two examples of diffusion. (**a**) A drop of dye enters a bowl of water. Gradually, the dye molecules become evenly dispersed among the water molecules. (**b**) The same thing happens with water molecules. Here, dye (*red*) and water (*yellow*) are added to the same bowl. Each substance shows a net movement down its own concentration gradient.

permeability: The membrane allows some substances but not others to cross it. This property helps the cell control which substances and how much of them can diffuse across it in a given time (Figure 5.17).

Membrane barriers and crossings are vital, because metabolism depends on the cell's capacity to increase, decrease, and maintain concentrations of substances required for reactions. That capacity supplies the cell with raw materials, removes wastes, and maintains the volume and pH within tolerable ranges. It also serves these functions for membrane-enclosed sacs in cells.

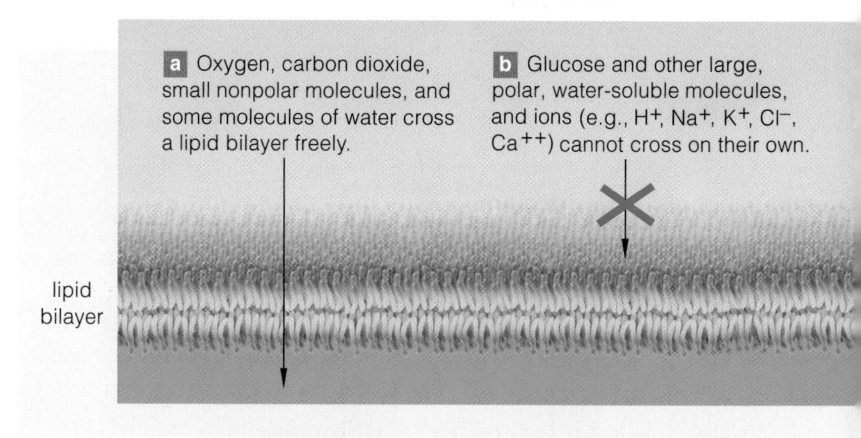

Figure 5.17 Animated! The selectively permeable nature of cell membranes. Small, nonpolar molecules, gases, and some water molecules freely cross the lipid bilayer. Large, polar molecules, ions, and water cross with the help of proteins that span the bilayer.

HOW SUBSTANCES CROSS MEMBRANES

Selective permeability arises from the structure of the membrane. Remember, a lipid bilayer has a nonpolar interior. It lets gases and small, nonpolar molecules cross. Water molecules are polar, but some slip through gaps that open when the hydrophobic tails of lipids flex and bend (Section 4.1).

The lipid bilayer is impermeable to ions and large, polar molecules, including glucose. These substances cross a membrane by diffusing through the interior of transport proteins that span the bilayer.

Passive transporters are channels through which a specific solute follows its concentration gradient across a membrane. The solute moves across the membrane simply by diffusing through the channel; this process, which is called *passive transport* or facilitated diffusion, requires no energy input.

The active transporters help specific solutes diffuse across membranes, but they are not passive about it. They move solutes against a gradient with the help of energy inputs. We call this mechanism *active transport*.

Other mechanisms move large particles into or out of cells. In *endocytosis*, a vesicle forms around particles when a patch of plasma membrane sinks inward and seals back on itself. In *exocytosis*, a vesicle that formed in the cytoplasm fuses with the plasma membrane, so that its contents are dumped outside (Section 4.8).

Before getting into the details of these mechanisms, you may wish to study the overview in Figure 5.18.

Diffusion is the net movement of molecules or ions into an adjoining region where they are not as concentrated.

The steepness of a concentration gradient as well as temperature, molecular size, and electric and pressure gradients affect the rate of diffusion.

Different substances can move across cell membranes by the mechanisms of diffusion, passive and active transport, endocytosis, and exocytosis.

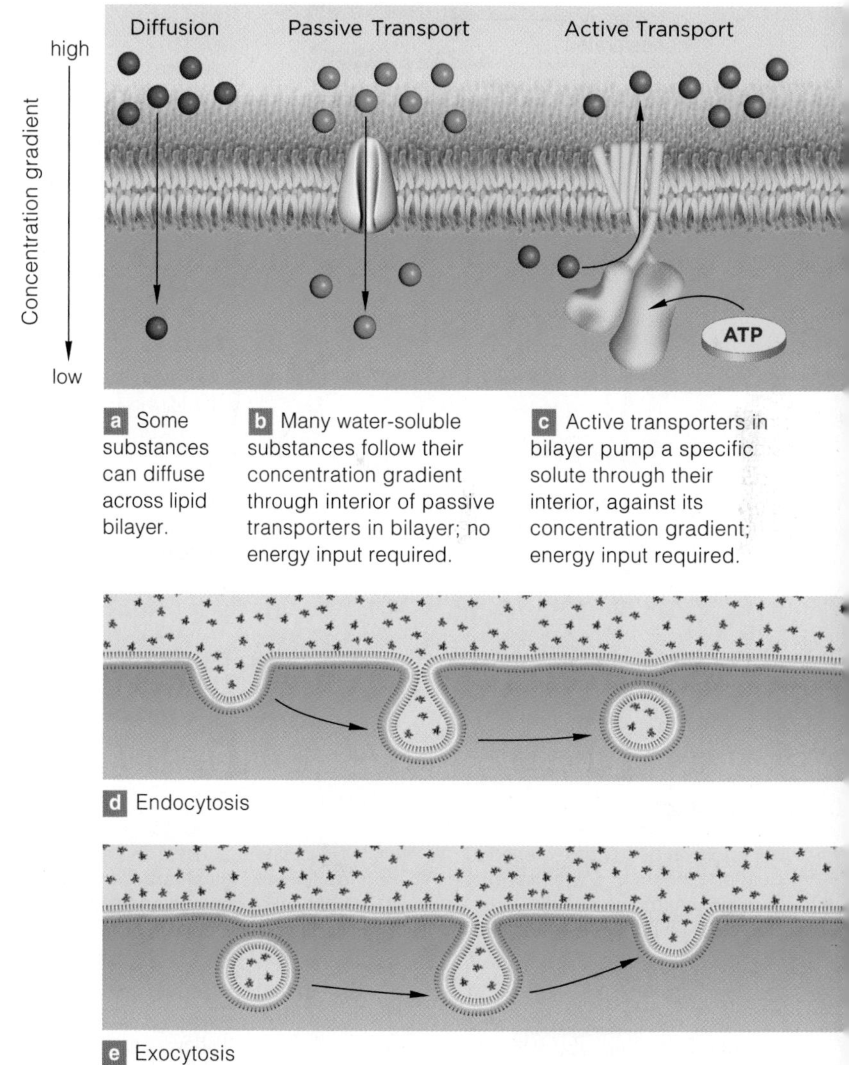

Figure 5.18 Overview of membrane-crossing mechanisms.

5.7 Working With and Against Gradients

Polar molecules and ions cannot diffuse across a lipid bilayer. They require the help of transport proteins.

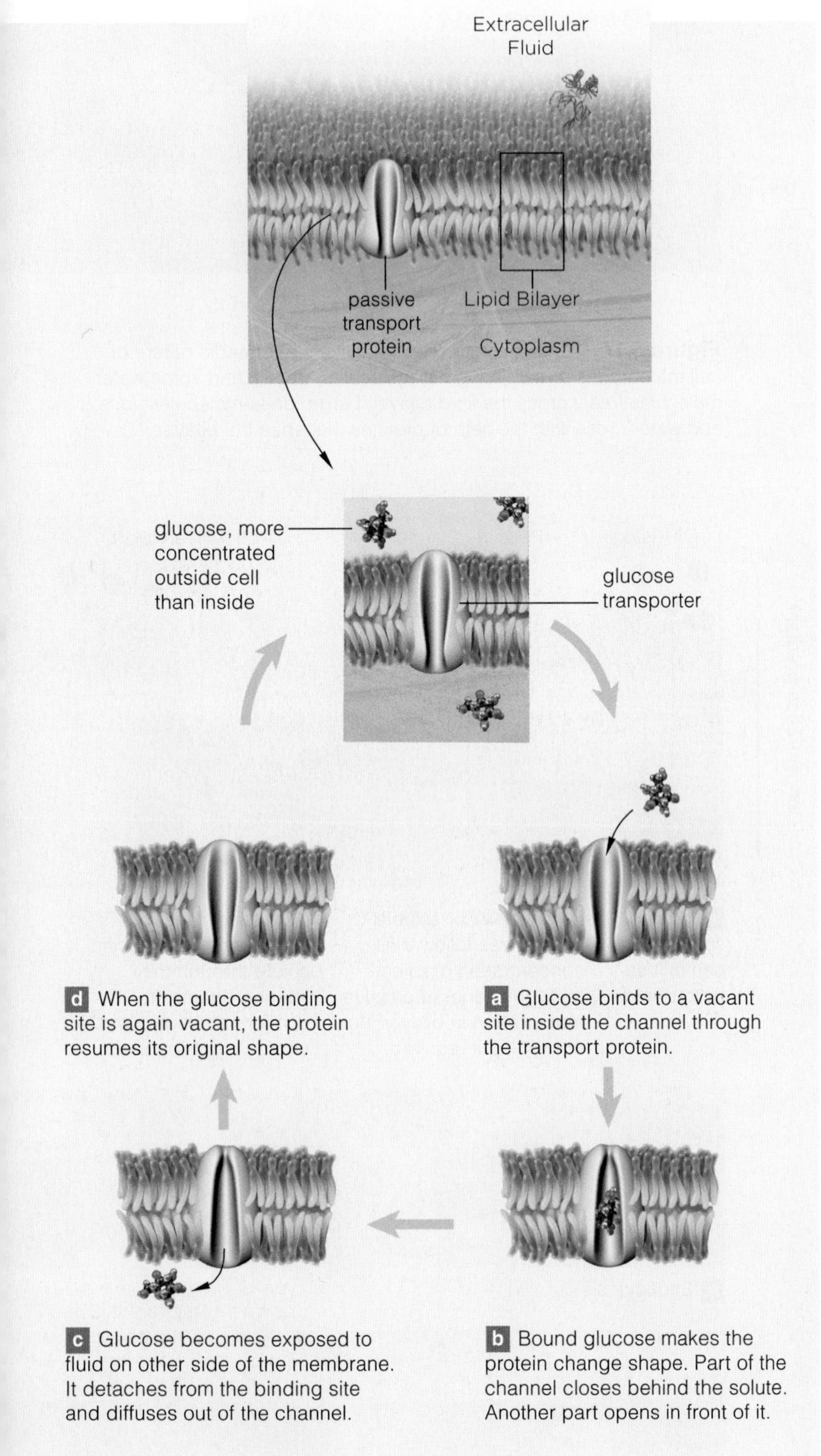

Many kinds of solutes cross a membrane by diffusing through a channel or tunnel inside transport proteins. When one solute molecule or ion enters the channel and binds to the protein, the protein's shape changes. The channel closes behind the solute and also opens in front of it. The solute leaves the channel, and then the protein reverts to its original shape.

Extracellular Fluid

passive transport protein

Lipid Bilayer

Cytoplasm

glucose, more concentrated outside cell than inside

glucose transporter

a Glucose binds to a vacant site inside the channel through the transport protein.

b Bound glucose makes the protein change shape. Part of the channel closes behind the solute. Another part opens in front of it.

c Glucose becomes exposed to fluid on other side of the membrane. It detaches from the binding site and diffuses out of the channel.

d When the glucose binding site is again vacant, the protein resumes its original shape.

Figure 5.19 Animated! Passive transport. This model shows one of the glucose transporters that span the plasma membrane. Glucose crosses in both directions. The *net* movement of this solute is to the side of the membrane where it is less concentrated.

PASSIVE TRANSPORT

In passive transport, a concentration gradient, electric gradient, or both drive diffusion of a substance across a cell membrane, through a channel inside a transport protein. The protein does not require an energy input to assist the directional movement. That is why this mechanism is also known as facilitated diffusion.

Some passive transporters are open channels; others have molecular gates that open or close as conditions change. Figure 5.19 shows how a glucose transporter works. When one end of its channel is shut, the other is open and glucose can diffuse in. The channel closes behind glucose and opens in front of it, on the other side of the membrane.

The net movement of a particular solute through passive transporters tends to be toward the side of the membrane where the solute is less concentrated. This is because molecules or ions simply collide with the transporters more often on the side of the membrane where they are more concentrated.

If nothing else were happening, passive transport might continue until concentrations on both sides of the membrane were equal. However, such equilibrium rarely occurs in a living system. For example, glucose diffuses into your cells through glucose transporters, but cells tend to use it up as fast as they get it. As soon as a glucose molecule enters a cell, it is broken down for energy or it is used to build other molecules. In this case, a cell's use of glucose helps maintain a gradient that favors the uptake of more glucose.

ACTIVE TRANSPORT

Solute concentrations shift constantly in cytoplasm and extracellular fluid. Maintaining a solute's concentration at a level necessary for a metabolic reaction to proceed often means moving the solute against its gradient, to the side of a membrane where it is *more* concentrated. Such pumping does not occur without energy inputs, usually from ATP.

In active transport, a transport protein uses energy to pump a solute across a cell membrane, against its gradient. Only specific solutes can bind to the interior channel of an active transporter. Energy, often in the form of a phosphate-group transfer from ATP, changes the transporter's shape. This makes the transporter release the solute to the other side of the membrane.

For example, calcium pumps are active transporters that move calcium ions across muscle cell membranes (Figure 5.20). Muscle cells contract when the nervous system causes calcium ions to flood out from a special compartment wrapped around the muscle fiber. The flood clears out binding sites on motor proteins that make muscles contract (Section 4.12). Contraction ends after calcium pumps have moved most of the calcium ions back into the compartment, against their gradient. Calcium pumps can keep the concentration of calcium in that compartment 1,000 to 10,000 times higher than it is in muscle cell cytoplasm.

The sodium–potassium pump is a *co*transporter—it moves two substances at the same time. Nearly all of your body's cells have these pumps, which maintain gradients by pumping sodium and potassium ions in opposite directions across the membrane. Sodium ions (Na^+) in the cytoplasm diffuse into the pump's open channel and bind to its interior. The pump changes shape after it receives a phosphate group from ATP. Its channel opens to the extracellular fluid, and it releases the Na^+. Then, potassium ions (K^+) from extracellular fluid diffuse into the channel and bind to its interior. The transporter releases the phosphate group and then reverts to its original shape. The channel opens to the cytoplasm, and the K^+ is released there.

Bear in mind, the membranes of all cells, not just those of animals, have membrane pumps. In Section 27.5, for example, you will learn how sugars made in a plant's leaves are pumped into specialized tubes that distribute them through the plant body.

Membrane transport proteins act as open or gated channels across cell membranes. They undergo reversible changes in shape that assist solutes across the membrane.

In passive transport, a transporter allows a solute to cross a cell membrane simply by diffusing through its interior.

In active transport, the net diffusion of a specific solute is against its gradient. The transporter must be activated, usually by an energy input from ATP, which counters the force inherent in the gradient.

Passive and active transport continually help lower or raise gradients across a membrane, which helps the cell respond to signals and to chemical changes.

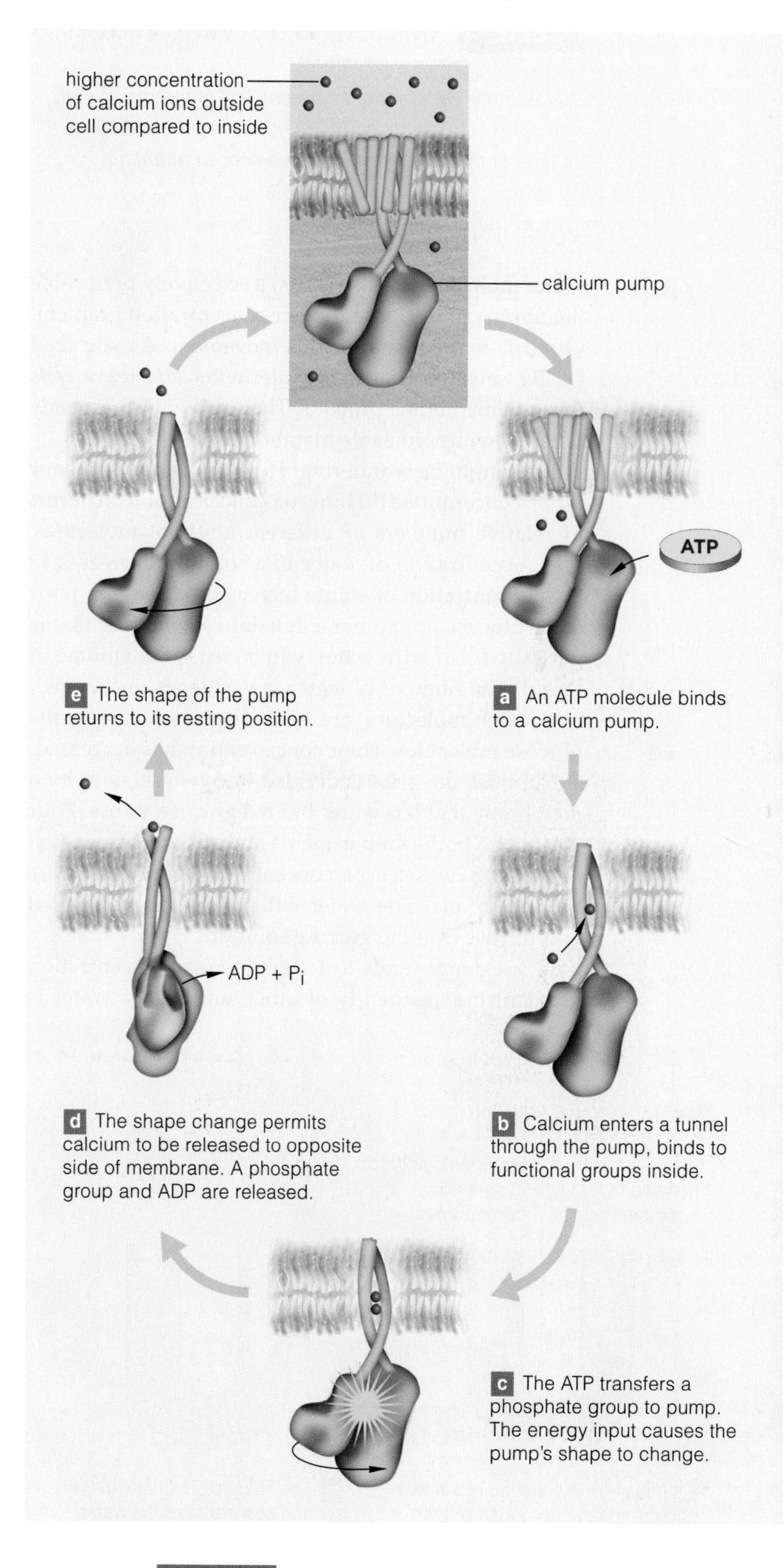

Figure 5.20 **Animated!** Active transport. In this example, you can see the channel for transport of calcium ions through a calcium pump that spans the plasma membrane. After two calcium ions bind to the pump, ATP transfers a phosphate group to it, thus providing energy that drives the movement of calcium *against* a concentration gradient across the cell membrane.

5.8 Which Way Will Water Move?

By far, more water diffuses across cell membranes than any other substance, so the main factors that influence its directional movement deserve special attention.

MOVEMENT OF WATER

LINKS TO SECTIONS 2.5, 4.11

Water molecules diffuse across a selectively permeable membrane in response to their concentration gradient. Osmosis is the name for this movement. As you read earlier, most molecules of water enter and leave cells through membrane proteins. They also will cross many types of semipermeable membranes.

You might be wondering: How can water be less or more concentrated? Think of concentration in terms of relative numbers of different kinds of molecules. The concentration of water in a solution decreases as the concentration of solute increases. When you pour some glucose or another solute into a container that is partially filled with water, you increase the volume of liquid. The number of water molecules is unchanged, but water molecules are now dispersed among the glucose molecules. Their concentration has decreased.

Suppose the glass is divided into two sections by a membrane that lets water but not glucose across. Pour water into both compartments and add glucose to just one. You have set up a concentration gradient across the membrane. The water will follow its gradient; it will diffuse into the glucose solution.

A substance tends to follow its own concentration gradient independently of other substances. Water is no exception. The concentration of water is influenced by the total number of molecules or ions dissolved in a particular volume, not by the kinds of solutes.

hypotonic solution in first compartment

hypertonic solution in second compartment

a Initially, the volumes of the two compartments are equal, but the solute concentration across the membrane differs.

b The fluid volume rises in the second compartment as water follows its concentration gradient and diffuses into it.

Figure 5.21 **Animated!** Experiment showing a change in fluid volume as an outcome of osmosis. A semipermeable membrane separates two regions.

EFFECTS OF TONICITY

Tonicity refers to the relative concentrations of solutes in two fluids that are separated by a semipermeable membrane. When the solute concentrations differ, the fluid with the lower concentration of solutes is called the hypotonic fluid. The other one, with the higher solute concentration, is the hypertonic fluid. Water tends to diffuse from hypotonic to hypertonic fluid. Isotonic fluids have the same solute concentration.

Later chapters have a variety of examples of how osmosis and tonicity affect the concentration of water and solutes inside plants and animals. For now, focus on tonicity as it applies to solutions in general. What happens when fluid in a cell, an organelle, or another membrane-bound compartment is hypotonic? Solutes cannot cross a plasma membrane to follow their own gradient, so cell volume will decrease as water diffuses out of it. If fluid in a cell is hypertonic, cell volume will increase, because water flows into it (Figure 5.21).

To visualize these osmotically induced changes in volume, imagine that you construct three bags from a selectively permeable membrane that water—but not sucrose—can cross. You fill each bag with 1 liter of a solution that has 2 percent sucrose. Drop the first bag into pure water (a hypotonic solution), the second into a solution that has 10 percent sucrose (a hypertonic solution), and the third into a solution with 2 percent sucrose (an isotonic solution). Figure 5.22*a* shows what would happen. Tonicity would dictate the direction of water movement, if any, across the membrane.

Most free-living cells can counter shifts in tonicity by selectively transporting solutes across the plasma membrane. Most cells of multicelled species cannot. Figure 5.22*b–d* shows what happens to red blood cells when tonicity is experimentally manipulated. Fluid in these cells normally is isotonic with tissue fluid. If the fluid were to become hypotonic, far too much water would diffuse into the cells, which would burst apart. If tissue fluid were to become hypertonic, the cells would lose water, and they would shrivel and die.

EFFECTS OF FLUID PRESSURE

Hydrostatic pressure, or as botanists say, turgor, often counters osmosis. Both terms refer to pressure that a volume of fluid exerts against a cell wall, membrane, tube, or any other structure that holds it. Cell walls in

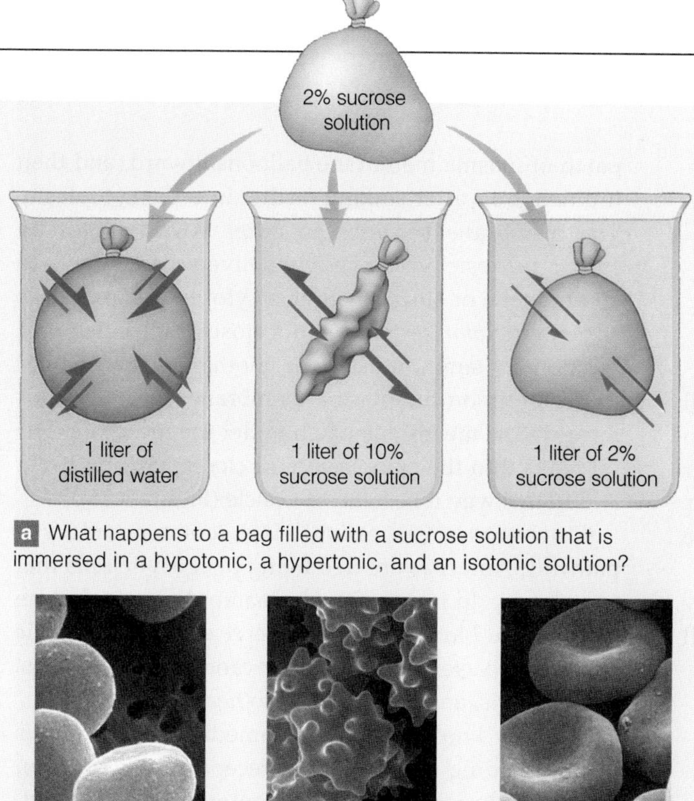

a What happens to a bag filled with a sucrose solution that is immersed in a hypotonic, a hypertonic, and an isotonic solution?

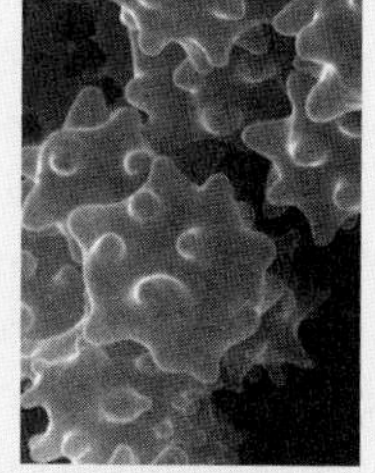

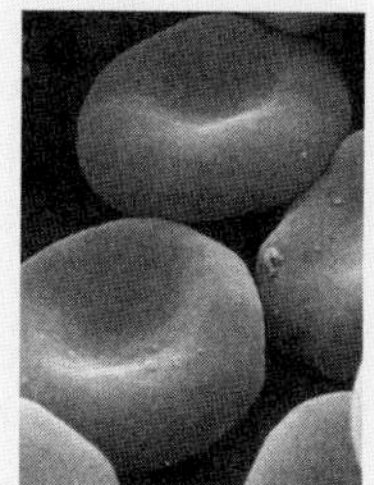

b Red blood cells in a hypotonic solution swell because water diffuses into them.

c Red blood cells in a hypertonic solution shrivel because water diffuses out of them.

d Red blood cells in an isotonic solution do not change in volume; no net water flow.

Figure 5.22 Animated! (**a**) A tonicity experiment. (**b–d**) The micrographs show human red blood cells that were immersed in fluids of different tonicity.

Figure 5.23 (**a**) A tomato plant undergoing osmotically induced wilting within thirty minutes after salty water was added to the soil in the pot. (**b**) Cells from an iris petal, plump with water. Their cytoplasm and central vacuole extend to the cell wall. (**c**) Cells from a wilted iris petal. Their cytoplasm and central vacuole shrank, and the plasma membrane moved away from the wall.

plants and many protists, fungi, and bacteria resist increases in the volume of cytoplasm. Blood vessel walls resist increases in blood volume. The amount of hydrostatic pressure that can stop water from diffusing into cytoplasmic fluid or other hypertonic solution is called osmotic pressure.

As one example, growing plant cells are hypertonic relative to water in soil (the cytoplasmic fluid usually has more solutes than soil water). Water diffusing into a young plant cell by osmosis exerts fluid pressure on the primary wall. The thin, pliable wall expands under pressure, which lets the cytoplasmic volume increase (Section 4.11). Expansion of the wall—and the cell—ends when the osmotic pressure inside the cell builds up enough to prevent the uptake of additional water.

Hydrostatic pressure also supports soft plant parts. When a plant with soft green leaves is growing well and has enough water, hydrostatic pressure keeps the cells plump—and the plant erect. If the soil dries out for too long, it becomes salty; the soil water becomes hypertonic with respect to the cytoplasmic fluid. Then, hydrostatic pressure in the young plant cells falls as water diffuses out of them. Their cytoplasm shrinks, and the plant wilts. Adding too much salt to the soil has the same effect. Figure 5.23 shows what happens when you pour salty water into soil around a tomato plant's roots. Within thirty minutes, the plant droops.

The greater the number of molecules and ions dissolved in a given amount of water, the higher the solute concentration of the solution.

Osmosis is a net diffusion of water between two solutions that differ in solute concentration and are separated by a selectively permeable membrane.

Water tends to move osmotically to regions of greater solute concentration (from hypotonic to hypertonic solutions). There is no net diffusion between isotonic solutions.

Fluid pressure that a solution exerts against a membrane or wall influences the osmotic movement of water.

5.9 Membrane Traffic To and From the Cell Surface

By the processes of exocytosis and endocytosis, vesicles move substances to and from the plasma membrane. The vesicles help cells take in and expel items that are too big for transport proteins.

ENDOCYTOSIS AND EXOCYTOSIS

LINKS TO SECTIONS 3.4, 4.8

Think back on the membrane traffic to and from a cell surface (Figure 5.18). By exocytosis, a vesicle moves to the cell surface, and the protein-studded lipid bilayer of its membrane fuses with the plasma membrane. As this exocytic vesicle loses its identity, its contents are released to the surroundings (Figures 5.24 and 5.25).

There are three pathways of endocytosis, but they all take up substances near the cell's surface. A small patch of plasma membrane balloons inward, and then it pinches off after sinking farther into the cytoplasm. The membrane becomes an outer boundary for an endocytic vesicle. The vesicle delivers its contents to an organelle or stores them in a cytoplasmic region.

With *receptor-mediated* endocytosis, molecules of a hormone, vitamin, mineral, or another substance bind to receptors on the plasma membrane. A shallow pit forms in the membrane patch under the receptors. The pit sinks into the cytoplasm and closes back on itself, and in this way it becomes a vesicle (Figure 5.25).

Phagocytosis ("cell eating") is a common endocytic pathway. Amoebas are free-living phagocytic cells that engulf prey. In many animals, macrophages and some other white blood cells engulf and digest pathogenic viruses or bacteria, worn-out or cancerous body cells, tissue debris, and other threats to health.

We now know that receptor-mediated endocytosis is a misleading name, because receptors also function in phagocytosis. When these receptors bind to a target, they cause microfilaments to assemble in a mesh under the plasma membrane. The microfilaments contract, forcing some cytoplasm and plasma membrane above it to bulge outward as a lobe, or pseudopod (Figures 4.26*b* and 5.26). Pseudopods engulf a target and merge as a vesicle, which sinks into the cytoplasm and fuses with a lysosome. Enzymes inside the lysosome break down the vesicle's contents.

Bulk-phase endocytosis is not as selective. A vesicle forms around a small volume of the extracellular fluid regardless of the kinds of substances dissolved in it.

Endocytosis
Exocytosis
a Molecules get concentrated inside coated pits at the plasma membrane.
coated pit
b The pits sink inward and become endocytic vesicles.
c Vesicle contents are sorted.
d Many of the sorted molecules cycle to the plasma membrane.
e Some vesicles are routed to the nuclear envelope or ER membrane. Others fuse with Golgi bodies.
f Some vesicles and their contents are delivered to lysosomes.

Figure 5.24 **Animated!** Endocytosis and exocytosis.

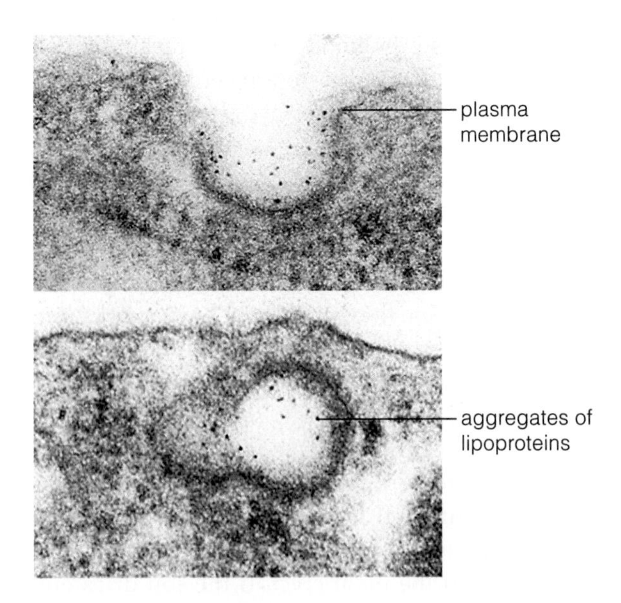

Figure 5.25 Endocytosis of clumps of lipoproteins.

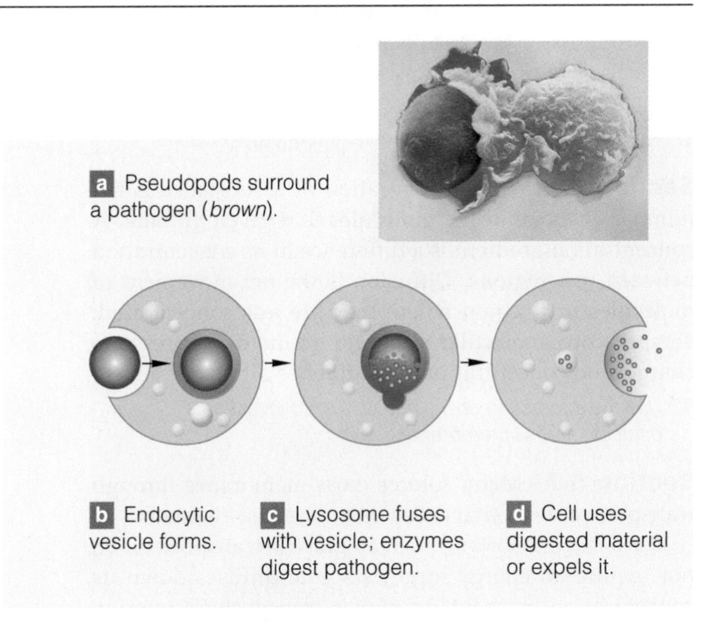

Figure 5.26 Animated! Phagocytosis. A phagocytic cell's pseudopods (extending lobes of cytoplasm) surround a pathogen. The plasma membrane above the bulging lobes fuse and form an endocytic vesicle. Inside the cytoplasm, the vesicle fuses with a lysosome, which digests its contents.

MEMBRANE CYCLING

As long as a cell is alive, exocytosis and endocytosis are continually replacing and withdrawing patches of its plasma membrane, as in Figure 5.24.

The composition of a plasma membrane begins in the ER (Section 4.8). Proteins are modified and lipids are synthesized, and both are packaged in vesicles that transport them to Golgi bodies for final modification. The finished proteins and lipids are packaged in new vesicles that transport them to the plasma membrane. The vesicles and their cargo then fuse with the plasma membrane. This is how new plasma membrane forms.

In a cell that is no longer growing, the total area of the plasma membrane remains more or less constant. Membrane is lost as a result of endocytosis, but it is replaced by membrane arriving as exocytic vesicles.

Exocytosis and endocytosis move materials in bulk across a plasma membrane.

By exocytosis, a cytoplasmic vesicle fuses with the plasma membrane. By endocytosis, a patch of plasma membrane sinks inward and forms a vesicle in the cytoplasm.

Receptor-mediated endocytosis and phagocytosis are two endocytic pathways that occur when specific substances bind to receptors. Bulk-phase endocytosis is not specific.

Plasma membrane lost during endocytosis is replaced by membrane that surrounds exocytic vesicles.

METABOLISM EVERYWHERE

5.10 Night Lights

FOCUS ON SCIENCE

Leave the chapter with this thought: Everything that organisms do starts with metabolism. Night flashers offer immediate and vivid evidence of it.

At night, in the warm waters of tropical seas or in the summer air above gardens and fields, you may catch sight of abrupt shimmerings or flashes of bioluminescent light. Bioluminescence refers to light emitted from metabolic reactions in living organisms. In different species, it helps attract mates or prey, or confuse predators.

Many species, including assorted bacteria, protists, fungi, insects, jellyfishes, and fishes, are flashers. The effect is especially startling when something disturbs masses of plankton: aquatic communities of tremendous numbers of organisms, most of which are microscopic.

Bioluminescent organisms emit light when enzymes called luciferases convert chemical bond energy to light energy. Figure 5.27 shows a model for firefly luciferase. A light-emitting reaction occurs when luciferase transfers a phosphate group from ATP and an oxygen atom to a type of pigment molecule called luciferin. Energized by the phosphate-group transfer, the modified luciferin releases its excess energy in the form of light. Different luciferins emit colors across the spectrum of visible light—from red to orange, yellow, green, blue, and purple. Some even emit infrared or ultraviolet light.

Figure 5.27 Bioluminescence. The inset shows a ribbon model for firefly luciferase, a light-releasing enzyme. To the *right*, a North American firefly (*Photinus pyralis*) emits a flash from its light organ, which contains peroxisomes packed with luciferase molecules. Firefly flashes may help potential mates find each other in the dark.

www.thomsonedu.com 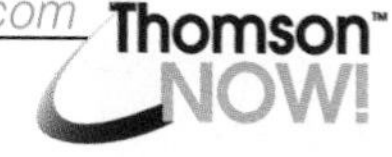

Summary

Section 5.1 Energy is often defined as the capacity to do work. Energy cannot be created from nothing and it cannot vanish, but it can be converted from one form to another and thus transferred between objects or systems. Energy tends to disperse spontaneously. A bit disperses at each energy transfer, usually in the form of heat.

Energy flows in one direction, starting mainly from the sun, then into and out of ecosystems. Producers and then consumers use that energy to assemble, rearrange, and dispose of substances. Throughout ecosystems, many substances cycle among organisms over time.

Energy itself is not cycled. All living things maintain their organization only as long as they harvest energy from someplace else.

Energy-requiring (endergonic) reactions require a net energy input. Energy-releasing (exergonic) reactions end with a net release of energy.

Section 5.2 ATP is the main energy carrier between reaction sites in cells. Phosphate-group transfers to and from ATP couple metabolic reactions that release usable energy to metabolic reactions that require energy.

- *Use the animation on ThomsonNOW to learn about energy changes in chemical reactions and the role of ATP.*

Section 5.3 Enzymes are catalysts. They enormously enhance reaction rates by lowering the activation energy, the minimum amount of energy needed to get a reaction going. Nearly all enzymes are proteins. Some RNAs are catalytic. Reactions occur in a small cleft in an enzyme's surface, the active site, which offers a microenvironment more favorable for reaction than the surroundings.

Enzymes lower activation energy by boosting local concentrations of the substrates, orienting substrates in positions that favor reaction, inducing the fit between a substrate and the active site, and excluding water.

- *Use the animation and interaction on ThomsonNOW to investigate how enzymes facilitate reactions.*

Section 5.4 Each type of enzyme functions best within a characteristic range of temperature, salt concentration, and pH. Most enzymes require the assistance of cofactors, which are metal ions or organic coenzymes. Controls over enzyme activity, such as negative feedback mechanisms, adjust the types and amounts of substances in cells.

- *Use the animation on ThomsonNOW to observe mechanisms that exert control over enzymes.*

Section 5.5 Cells concentrate, convert, and dispose of most substances in orderly, enzyme-mediated reaction sequences called metabolic pathways.

Biosynthetic pathways construct large molecules from smaller ones. They require energy. Photosynthesis is the main biosynthetic pathway in the biosphere. Degradative pathways break down molecules to smaller products, with a release of usable energy. The main degradative pathway in the biosphere is aerobic respiration. Electron transfers known as oxidation–reduction (redox) reactions are typical of many metabolic pathways. Coenzymes and electron transfer chains take part in organized sequences of reactions in photosynthesis and aerobic respiration.

- *Use the animation on ThomsonNOW to compare the effects of controlled and uncontrolled energy release.*

Section 5.6 The concentration of a substance is the number of its atoms or molecules in a given volume. A concentration gradient is a difference in its concentration between two regions. Diffusion is the net movement of molecules to a region where they are less concentrated. Temperature, molecular size, and gradients of pressure, charge, and concentration, all influence diffusion rates.

- *Use the interaction on ThomsonNOW to investigate diffusion across membranes.*

Section 5.7 Many solutes cross membranes through transport proteins that act as open or gated channels.

Passive transporters, such as glucose transporters, do not require an energy input; a solute diffuses down its concentration gradient through a transporter's interior. Active transporters require ATP energy to move a specific solute against its concentration gradient. Different kinds help maintain gradients across cell membranes.

- *Use the animation on ThomsonNOW to compare the processes of passive and active transport.*

Section 5.8 Osmosis is the diffusion of water across a selectively permeable membrane. The water molecules follow their concentration gradient, which is influenced by solute concentration.

- *Use the interaction and animation on ThomsonNOW to explore the effects of osmosis.*

Section 5.9 By exocytosis, a cytoplasmic vesicle fuses with the plasma membrane, and its contents are released outside. By endocytosis, a patch of the plasma membrane forms a vesicle that sinks into cytoplasm. Some single cells use an endocytic process called phagocytosis.

- *Use the animation on ThomsonNOW to see phagocytosis and how membrane components are cycled.*

Section 5.10 Bioluminescence is fluorescent light that is released by enzyme-mediated reactions in organisms.

Self-Quiz

Answers in Appendix III

1. ______ is life's primary source of energy.
   a. Food b. Water c. Sunlight d. ATP

2. If we liken a chemical reaction to an energy hill, then an ______ reaction is an uphill run.
   a. endergonic c. ATP-assisted
   b. exergonic d. both a and c

3. Energy ______.
   a. cannot be created or destroyed
   b. can change from one form to another
   c. tends to disperse spontaneously
   d. all of the above

4. Enzyme function is influenced by ______.
   a. temperature c. salt
   b. pH d. all of the above

5. Enzymes ______ .
 a. are proteins, except for a few RNAs
 b. lower the activation energy of a reaction
 c. are destroyed by the reactions they catalyze
 d. a and b

6. Diffusion is the movement of ions or molecules from one region to another where they are less concentrated. The rate of diffusion is affected by ______ .
 a. temperature
 b. electric gradients
 c. solute size
 d. all of the above

7. Transporters that require an energy boost help sodium ions across a cell membrane. This is a case of ______ .
 a. passive transport
 b. active transport
 c. facilitated diffusion
 d. a and c

8. Immerse a living human cell in a hypotonic solution, and water will tend to ______ .
 a. diffuse into the cell
 b. diffuse out of the cell
 c. show no net movement
 d. move in by endocytosis

9. Vesicles form by way of ______ .
 a. endocytosis
 b. exocytosis
 c. phagocytosis
 d. halitosis
 e. a through c
 f. all of the above

10. Match each term with its most suitable description.

___ passive transporter	a. assists enzymes
___ enzyme	b. works against a gradient
___ phagocyte at work	c. substance that enters a reaction
___ product	d. energy currency
___ ATP, mostly	e. enhances reaction rate
___ active transporter	f. a basis of diffusion
___ concentration gradient	g. no energy to help solutes across membrane
___ cofactor	h. cell engulfing a cell
___ reactant	i. there at reaction's end

■ *Visit ThomsonNOW for additional questions.*

Critical Thinking

1. Often, beginning physics students are taught the basic concepts of thermodynamics with two phrases: First, you can't win. Second, you can't break even. Explain.

2. Why does applying lemon juice to sliced apples keep them from turning brown?

3. Hydrogen peroxide bubbles if dribbled on an open cut but does not bubble on unbroken skin. Explain why.

4. Free radicals are atoms or molecules that are like ions with the wrong number of electrons. They form in many enzyme-catalyzed reactions, such as the digestion of fats and amino acids. They slip out of electron transfer chains. They form when x-rays and other kinds of ionizing radiation strike water and other molecules. Free radicals react easily with many molecules, and can damage the molecules of life.

Hydrogen peroxide, an oxygen-containing molecule, can easily become a radical. It forms in most organisms as a by-product of aerobic respiration. Hydrogen peroxide is toxic, so cells must dispose of it fast or risk being damaged. One molecule of catalase can inactivate about 6 million hydrogen peroxide molecules per minute by combining them two at a time. Catalase also inactivates other toxins, including alcohol. Given that its active site is specific for hydrogen peroxide, how can catalase act on other substances?

5. Superoxide dismutase (SOD) is an important enzyme in nearly all organisms that live around oxygen. Like catalase, SOD is an antioxidant that protects cells from damage by free radicals (Figure 5.28). Two metal cofactors, copper and zinc ions, help SOD convert free radicals to hydrogen peroxide, which other enzymes such as catalase break down to water.

Figure 5.28 Model of the enzyme superoxide dismutase.

SOD is one of the most well-studied enzymes, and researchers thought they had it all figured out. Then they discovered that a mutated version of the enzyme causes amyotrophic lateral sclerosis, also known as Lou Gehrig's disease. With this disease, nerve cells that control voluntary movement degenerate and eventually die.

Researchers are scrambling to figure out how mutated SOD could cause a fatal neurodegenerative disease. At first, they assumed that loss of enzyme function would result in the accumulation of free radicals, which in turn would kill nerve cells. That hypothesis was shown to be wrong: Mice that are engineered to lack SOD entirely do not get the disease.

Researchers now think the mutated SOD may have gained some toxic activity. They know the mutated enzyme has lost its ability to hold zinc ions but do not know how the loss would make the enzyme toxic. Look up current research articles to see how close researchers have come to answering their question since this book was printed.

6. Water moves osmotically into *Paramecium*, a single-celled aquatic protist. If unchecked, the influx would bloat the cell and burst it, but contractile vacuoles expel excess water (Figure 5.29). Water enters each vacuole's tubelike extensions and collects inside. A full vacuole contracts and squirts water out of the cell through a pore. Are this cell's surroundings hypotonic, hypertonic, or isotonic?

7. Is the white blood cell in Figure 5.30 disposing of a worn-out red blood cell by endocytosis or phagocytosis? And yes, this is a trick question.

Figure 5.29 Light micrograph of *Paramecium*, a ciliated protozoan.

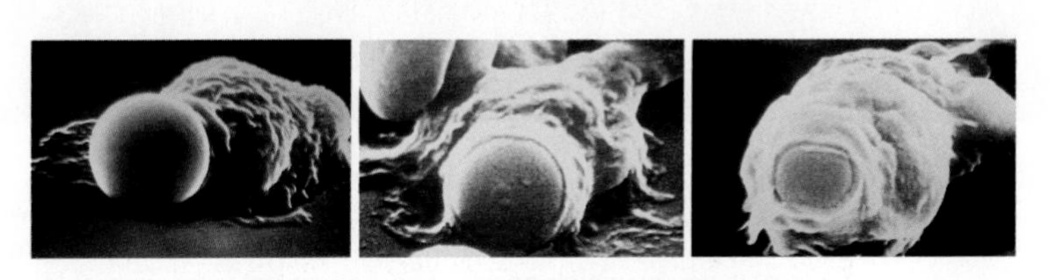

Figure 5.30 Name the mystery membrane mechanism.

6 WHERE IT STARTS—PHOTOSYNTHESIS

IMPACTS, ISSUES Sunlight and Survival

Think about the last bit of apple, lettuce, chicken, pizza, or other food you put in your mouth. Where did it come from? Look beyond the refrigerator, the market or restaurant, and the farm. Look to plants, the starting point for nearly all food—the carbon-based compounds—you eat. Plants, recall, are autotrophs, or "self-nourishing" organisms. Like other autotrophs, they make their own food by securing energy and carbon directly from the environment. Most bacteria, many protists, and all fungi and animals are heterotrophs. They get energy and carbon by feeding on autotrophs, one another, or organic wastes or remains. *Hetero-* means other, as in "being nourished by others."

Plants are a kind of *photo*autotroph. By the process of photosynthesis, they make sugars using energy from sunlight and carbon atoms from carbon dioxide. Each year, plants collectively produce about 220 billion tons of sugar, enough to make 300 quadrillion sugar cubes. That is a *lot* of sugar. They also release oxygen in the process.

It was not always this way. The first cells on Earth did not have the option of tapping into sunlight. They were *chemo*autotrophs that extracted energy and carbon from simple inorganic and organic compounds in the environment, such as hydrogen sulfide and methane.

Ways of securing food did not change much for about a billion years. Then light-trapping metabolic pathways evolved in the first photoautotrophs, and sunlight offered them an essentially unlimited supply of energy. Not long afterward, a photosynthetic pathway became modified in some species. The new pathway split water molecules apart into hydrogen and oxygen, and it harvested their electrons for cellular work. Over millions of years, the oxygen diffused out of uncountable numbers of cells

See the video! **Figure 6.1** Then and now—a view of how our atmosphere was irrevocably altered by photosynthesis. For most of the past as well as the present, photosynthesis has been the main pathway by which energy and carbon enter the web of life. Plants in this orchard are producing oxygen and carbon-rich parts—apples—at the Jerzy Boyz organic farm in Chelan, Washington.

How would you vote? Ethanol and other fuels manufactured from crops cost more than gasoline, but are renewable energy sources and have fewer emissions. Would you pay a premium to drive a vehicle that uses biofuels? If so, how much? See ThomsonNOW for details, then vote online.

and accumulated in the atmosphere. From that time on, the world of life would never be the same (Figure 6.1).

Oxygen enrichment of the early atmosphere exerted selection pressure on life all over the world. Oxygen reacts with metals, and free radicals can form during the reactions. Free radicals, remember, are toxic to cells (Section 5.4). Many species were not able to neutralize the oxygen radicals, so they became extinct or persisted only in oxygen-free habitats. In others, new pathways evolved. One of the new pathways, aerobic respiration, put oxygen's reactive properties to use. Oxygen accepts electrons at the end of electron transfer chains in the ATP-forming reactions of aerobic respiration.

Meanwhile, high in the ancient atmosphere, oxygen molecules were combining into ozone (O_3). An ozone layer slowly formed and helped shield life against the sun's lethal ultraviolet radiation. Aerobic species could now emerge from the deep ocean, out from sediments or mud, and diversify under the open sky.

What does the evolution of metabolic pathways have to do with you right now? The emergence and continuity of photosynthesis are big reasons why you can exist, and breathe, and think about what it takes to stay alive. Photosynthesis is one reason you can travel about in cars or buses, planes or trains. Three hundred million years ago, photosynthesis fueled the growth of vast swamp forests. In time, successive forests slowly decayed, compacted, and became the fossil fuels that we now extract from the earth.

Recently, fossil fuels have become commodities on a dangerous geopolitical stage. A sense of urgency surrounds the search for alternative sources of fuel energy. Photosynthesis is far more efficient at getting and storing energy than any current technology. Why is it so hard for us? Perhaps we can take a lesson from the biochemistry of photosynthesis.

Key Concepts

THE RAINBOW CATCHERS

A great one-way flow of energy through the world of life starts after chlorophylls and other pigments absorb the energy of visible light from the sun's rays. In plants, some bacteria, and many protists, that energy ultimately drives the synthesis of glucose and other carbohydrates. **Sections 6.1, 6.2**

OVERVIEW OF PHOTOSYNTHESIS

Photosynthesis proceeds through two stages in the chloroplasts of plants and many types of protists. First, pigments embedded in a membrane inside the chloroplast capture light energy, which is then converted to chemical energy. Next, that chemical energy drives the synthesis of carbohydrates. **Section 6.3**

MAKING ATP AND NADPH

In the first stage of photosynthesis, sunlight energy is converted to the chemical bond energy of ATP. The coenzyme NADPH forms in a pathway that also releases oxygen. **Sections 6.4, 6.5**

MAKING SUGARS

The second stage is the "synthesis" part of photosynthesis. Enzymes speed the assembly of sugars from carbon and oxygen atoms, both obtained from carbon dioxide. The reactions use ATP and NADPH that form in the first stage of photosynthesis. ATP delivers energy, and NADPH delivers electrons and hydrogens to the reaction sites. Details of the reactions vary among organisms. **Sections 6.6, 6.7**

PHOTOSYNTHESIS, CO_2, AND GLOBAL WARMING

Photosynthesis by autotrophs removes carbon dioxide from the atmosphere; metabolism by all organisms puts it back in. Human activitities have disrupted this balance, and so have contributed to global warming. **Section 6.8**

Links to Earlier Concepts

Before considering the chemical basis of photosynthesis, you may wish to review the nature of electron energy levels (Section 2.3). You will be using your knowledge of carbohydrate structure (3.2), chloroplasts (4.9), active transport proteins (4.3, 5.7), and concentration gradients (5.6).

Remember the concepts of energy flow and the underlying organization of life (5.1)? They help explain how energy flows through photosynthesis reactions. You also will expand your understanding of how ATP links reactions that release energy with reactions that require them (5.2), and see an example of how cells harvest energy with electron transfer chains (5.5). An example of nutrient cycling (1.2) illustrates one of the connections between the biosphere and its inhabitants through photosynthesis.

6.1 Sunlight as an Energy Source

Remember how energy flows in one direction through the world of life? In nearly all ecosystems, the flow starts when photoautotrophs intercept energy from the sun.

PROPERTIES OF LIGHT

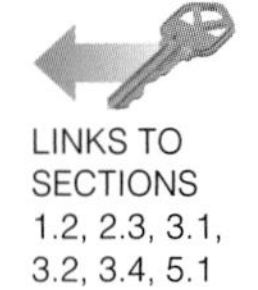

LINKS TO SECTIONS 1.2, 2.3, 3.1, 3.2, 3.4, 5.1

So far, we have mentioned photosynthesis in passing (Sections 1.2 and 5.1). Turn now to the process itself, starting with the energy inputs that drive it.

Visible light is part of a spectrum of electromagnetic energy radiating from the sun. Such radiant energy travels in waves, undulating across space as waves move across a sea. The distance between crests of two successive waves of light is called wavelength, which we measure in nanometers (nm).

Although light travels in waves, it is organized as photons, or packets of electromagnetic energy. All of the photons at a particular wavelength have the same amount of energy. Photons that have the least energy travel in longer wavelengths. Those having the most energy travel in shorter wavelengths.

Photoautotrophs, again, use light energy to build organic molecules from inorganic raw materials. They capture only light of wavelengths between 380 and 750 nanometers. Together, light of all of these wavelengths appears white, but humans and many other organisms perceive certain wavelengths in this range as different colors. White light passing through a prism separates into individual colors (Figure 6.2). A prism bends the longer wavelengths more than the shorter ones, and a rainbow of colors forms.

Figure 6.2 also indicates where visible light falls in the electromagnetic spectrum: a chart of the range of all wavelengths of radiant energy. Wavelengths of UV (ultraviolet) light, x-rays, and gamma rays are shorter than about 380 nanometers. They are energetic enough to alter or break chemical bonds in DNA and proteins, so they threaten life. That might explain why the first forms of life evolved away from sunlight, deep in the ocean or hidden under rocks or in mud. Organisms did not live out in the open until after an ozone layer formed and started to absorb much of the UV light.

THE RAINBOW CATCHERS

Certain pigments are the molecular bridges between sunlight and photosynthesis. In general, a pigment is an organic compound that selectively absorbs light of specific wavelengths. The wavelengths of light that are not absorbed are reflected. That reflected light gives each pigment its characteristic color.

Collectively, photosynthetic pigments absorb nearly all wavelengths of visible light. Different kinds cluster together in photosynthetic membranes. Together, they

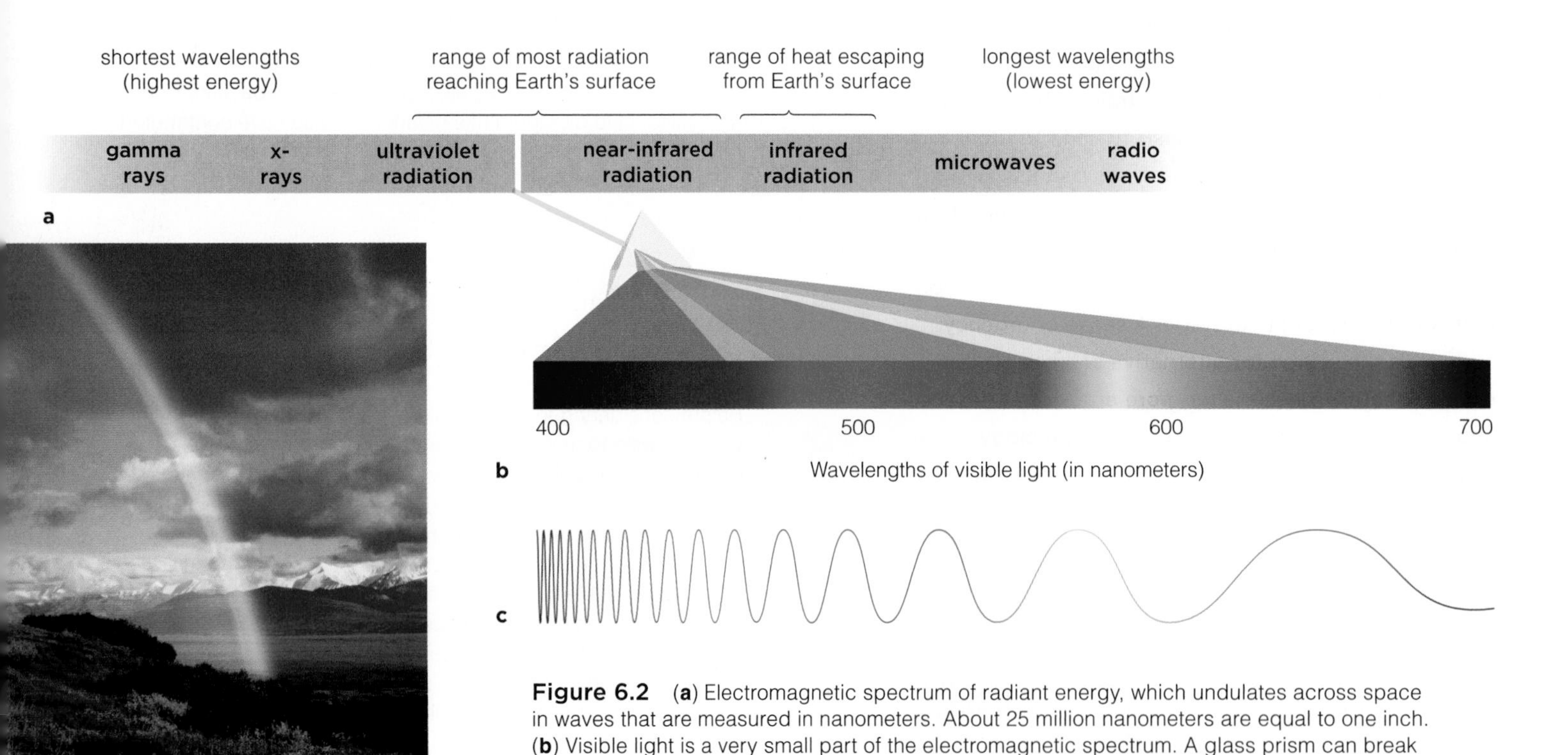

Figure 6.2 (**a**) Electromagnetic spectrum of radiant energy, which undulates across space in waves that are measured in nanometers. About 25 million nanometers are equal to one inch. (**b**) Visible light is a very small part of the electromagnetic spectrum. A glass prism can break it into the bands we see in a rainbow. (**c**) The shorter the wavelength, the higher the energy.

can absorb a broad range of wavelengths, like a radio antenna that can pick up different stations.

Chlorophyll *a* is the main photosynthetic pigment in plants, algae, and cyanobacteria. It absorbs violet and red light, so it appears green. This large molecule has a light-trapping ring structure and a hydrocarbon tail that anchors it in a membrane (Figure 6.3*a*,*b*).

Carotenoids and other accessory pigments extend the range of wavelengths usable for photosynthesis (Table 6.1). Accessory pigments are often masked by an abundance of chlorophylls. In autumn, chlorophyll synthesis lags behind its breakdown in many kinds of leafy plants. Accessory pigments do not break down as fast—their structure is more stable—so the leaves appear red, orange, and yellow (Figure 6.3). Tourists in New England spend about a billion dollars a year to see a glorious blaze of accessory pigments.

Certain accessory pigments have additional roles. For instance, carotenes and lycopenes (two kinds of carotenoids) are antioxidants; they neutralize oxygen radicals. Their colors attract animal pollinators to some flowers, and seed-dispersing animals to many fruits.

The light-trapping part of any pigment is an array of atoms in which single bonds alternate with double bonds (Figure 6.3*b*,*c*). Photon absorption excites the electrons of those atoms. Remember, excited electrons can move to a higher energy level (Section 2.3). They return quickly to a lower energy level by emitting the extra energy. That energy is not necessarily lost, as you will see in the next section. Arrays of pigments can use it to jump-start the reactions of photosynthesis.

Energy radiating from the sun travels through space in waves and is organized as packets called photons. The shorter the wavelength of light, the greater its energy.

We can perceive radiant energy of certain wavelengths as light of different colors.

Chlorophylls, carotenoids, and other photosynthetic pigments absorb specific wavelengths of visible light.

Collectively, photosynthetic pigments can harvest energy from the entire spectrum of visible light.

Table 6.1 Major Photosynthetic Pigments

Pigments	Reflected Colors	Present In
Chlorophyll *a*	Yellow–green	Main pigment in all plants, algae, cyanobacteria
Chlorophyll *b*	Blue–green	Plants, green algae, cyanobacteria
Carotenoids		
Carotenes, lycopenes	Yellow, orange, red	Plants, algae, cyanobacteria
Xanthophylls	Yellow, brown	Plants, algae, cyanobacteria
Phycobilins	Red	Red algae, cyanobacteria

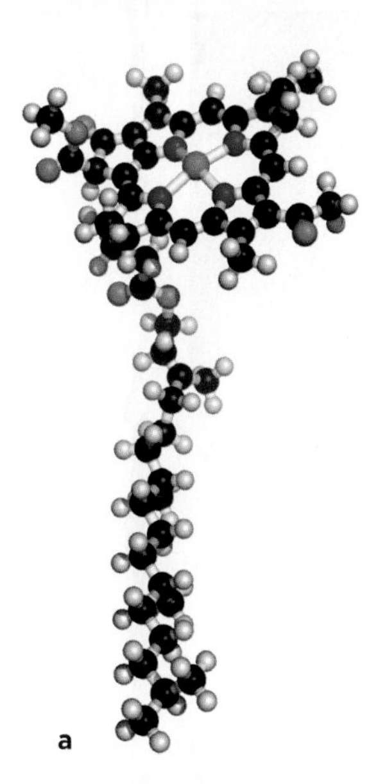

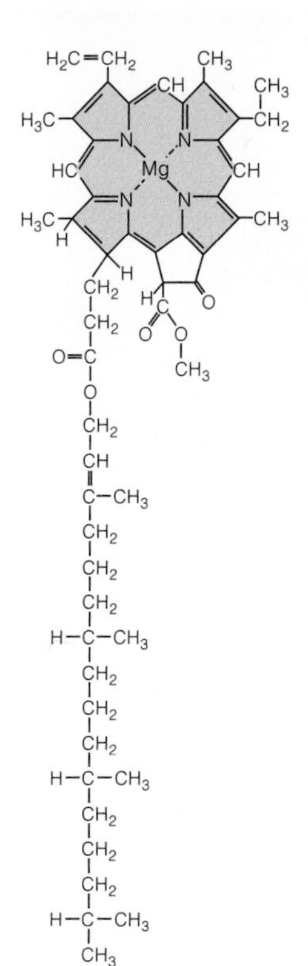

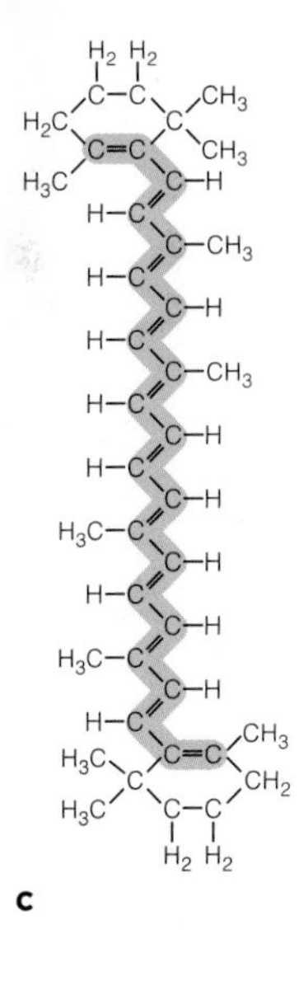

Figure 6.3 (**a**) Ball-and-stick model for chlorophyll *a*. (**b**,**c**) Structural formulas for chlorophyll *a* and beta-carotene, an accessory pigment that shows up in many autumn leaves. The light-catching part is tinted the color that each pigment reflects. The hydrocarbon tail is embedded in the lipid bilayer of photosynthetic membranes.

6.2 Exploring the Rainbow

FOCUS ON SCIENCE

Different photosynthetic pigments work together. How efficient are the pigments at harvesting light of different wavelengths in the sun's rays?

alga

a Outcome of T. Engelmann's experiment.

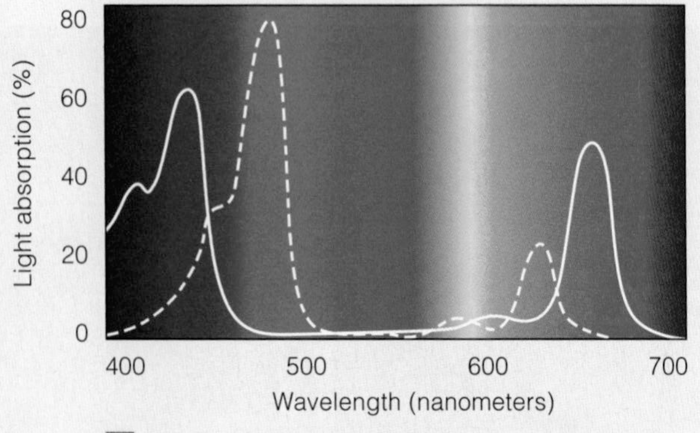

b Absorption spectra for chlorophyll *a* (solid graph line) and chlorophyll *b* (dashed line). Compare these graphs with the clustering of bacteria shown in (**a**).

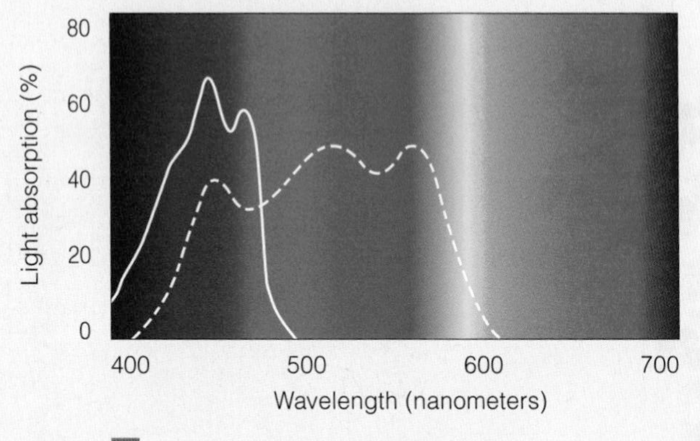

c Absorption spectra for beta-carotene (solid line) and one of the phycobilins (dashed line).

Figure 6.4 Animated! (**a**) One of the earliest photosynthesis experiments. T. Engelmann directed a ray of light through a prism so that bands of colors crossed a water droplet on a microscope slide. The water held aerobic bacterial cells and a strand of a photosynthetic alga. Bacteria quickly clustered around the algal cells that were releasing the most oxygen. The clusters were bathed in red and violet light which, for this algal species, is most efficient at driving photosynthesis.

(**b**,**c**) As later research revealed, all photosynthetic pigments combined absorb most of the wavelengths of visible light. Absorption spectra of chlorophylls *a* and *b*, beta-carotene, and a phycobilin reveal the efficiency with which these pigments absorb different wavelengths of visible light.

At one time, people thought that plants used substances in soil to make food. By 1882, a few chemists had an idea that plants use sunlight, water, and something in the air. The botanist Theodor Engelmann wondered: *What parts of sunlight are being used?* Engelmann already knew that photosynthesis releases free oxygen. He came up with a hypothesis: If the color of light affects photosynthesis, then photosynthetic cells will release different amounts of oxygen when they are illuminated by different colors.

Engelmann also knew that certain bacteria require oxygen for aerobic respiration. He predicted they would accumulate in places where photosynthetic cells were releasing the most oxygen. He directed a spectrum of visible light across a water droplet. The water held many bacterial cells and one strand of a photosynthetic alga (*Cladophora*), as shown in Figures 6.4 and 6.5.

Within minutes, most of the bacterial cells gathered where violet and red light fell across the algal strand. Engelmann concluded that the algal cells illuminated by this light were releasing the most oxygen—a sign that violet and red light are best at driving photosynthesis.

Engelmann's experiment allowed him to correctly identify the colors of light most efficient for *Cladophora* photosynthesis. The result of his experiment, shown in Figure 6.4*a*, was an early absorption spectrum—a graph that shows which wavelengths of light are absorbed by the substance of interest. Peaks in the graph indicate wavelengths of light that the substance absorbs best.

We now use more sophisticated techniques to measure absorption spectra of individual photosynthetic pigments. As Figure 6.4*b* shows, chlorophylls are best at absorbing red and violet light, and they reflect green light. What if you combined absorption spectra for chlorophylls and all of the accessory pigments, including those in Figure 6.4*c*? You would see that, collectively, they absorb almost the full spectrum of visible light wavelengths.

Figure 6.5 Light micrograph of photosynthetic cells in a strand of the green alga *Chladophora*.

6.3 Overview of Photosynthesis

Chloroplasts are organelles of photosynthesis in plants and all other eukaryotic species that harness sunlight. Their structure gives rise to their specialized function.

Let us now zoom in on the chloroplast, the organelle that specializes in photosynthesis in plants and many protists (Figure 6.6*a,b*). Plant chloroplasts have three membranes. Two enclose a semifluid matrix called the stroma. The third, the thylakoid membrane, is folded up in the stroma. Commonly, the folds are organized as stacks of disks (thylakoids), with channels between the stacks. The space inside all the disks and channels is continuous; it forms a single compartment (Figure 6.6*b*). Sugars are built outside of it, in the stroma.

The thylakoid membrane contains clusters of light-harvesting pigments. The clusters are "antennas" that absorb photons of different energies (Section 6.1). The membrane also has photosystems, groups of hundreds of pigments and other molecules that work as a unit to begin the reactions of photosynthesis. Chloroplasts contain two kinds of photosystems, type I and type II, which were named in the order of their discovery. Both types can convert light energy into chemical energy.

Often, photosynthesis is summarized by this simple equation, from reactants to products:

$$6H_2O + 6CO_2 \xrightarrow[\text{enzymes}]{\text{light energy}} 6O_2 + C_6H_{12}O_6$$

water — carbon dioxide — oxygen — glucose

However, the equation does not show that there are many reactions, which unfold in two stages. In the first stage, called light-dependent reactions, light energy is converted to the chemical bond energy of ATP. Water molecules are split apart, and typically the coenzyme $NADP^+$ accepts the released hydrogen and electrons, thus becoming NADPH. Oxygen released from water molecules escapes into the surroundings. The second stage, the light-independent reactions, runs on energy delivered by the ATP and NADPH. That energy drives the synthesis of glucose and other carbohydrates from carbon dioxide and water (Figure 6.6*c*).

In chloroplasts, photosynthesis has two reaction stages. The first stage occurs at the thylakoid membrane. Sunlight energy drives ATP and NADPH formation, and oxygen is released. The second stage occurs in the stroma. Energy from ATP and NADPH drives the synthesis of sugars from water and carbon dioxide.

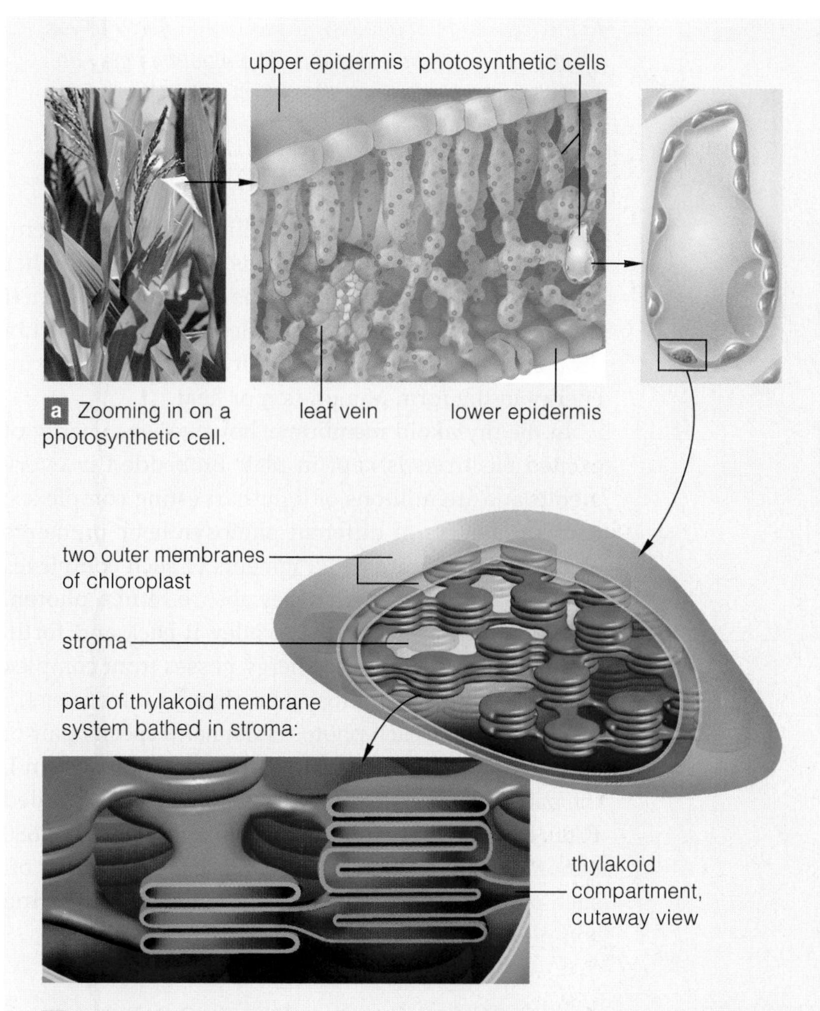

a Zooming in on a photosynthetic cell.

b Chloroplast structure. No matter how highly folded, its thylakoid membrane system forms a single, continuous compartment in the stroma.

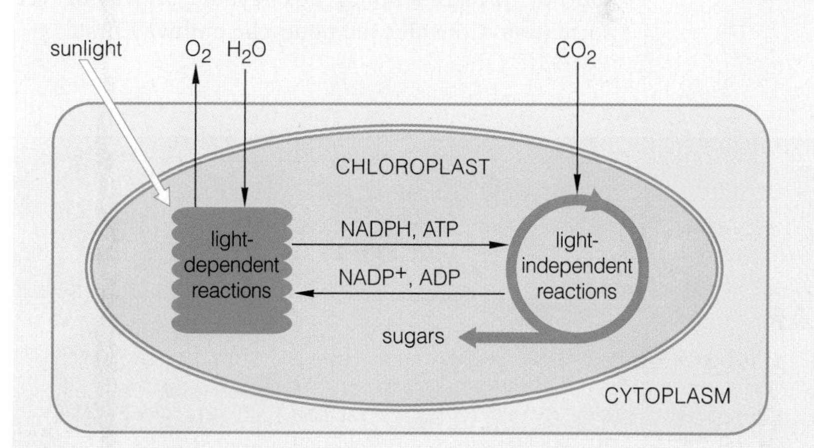

c In chloroplasts, ATP and NADPH form in the light-dependent stage of photosynthesis, which occurs at the thylakoid membrane. The second stage, which produces sugars and other carbohydrates, proceeds in the stroma.

Figure 6.6 **Animated!** Sites of photosynthesis in a typical leafy plant.

6.4 Light-Dependent Reactions

In the first stage of photosynthesis, light energy drives electrons out of photosystems. The electrons may be used in a noncyclic or cyclic pathway of ATP formation.

CAPTURING ENERGY FOR PHOTOSYNTHESIS

LINKS TO SECTIONS 2.3. 4.3, 5.5

Visualize a lone photon as it collides with a pigment. One of the pigment's electrons is boosted to a higher energy level after it absorbs the photon's energy. If nothing else were to occur, the electron would quickly drop back to its unexcited state. It would lose the extra energy in the form of a photon or heat.

In the thylakoid membrane, however, the energy of excited electrons is kept in play. Embedded in every membrane are millions of light-harvesting complexes: circular clusters of different photosynthetic pigments and proteins (Figure 6.7). Pigments of such complexes can hold on to the energy they absorb from a photon. Like a volleyball team, they volley it back and forth. At the speed of light, the energy passes from complex to complex until a photosystem absorbs it for keeps.

The center of each photosystem has a special pair of chlorophyll *a* molecules (Figure 6.3). In photosystem I, the pair absorbs energy of 700 nanometers and is called P700. In photosystem II, the pair absorbs energy of 680 nanometers and is called P680. Electrons pop right off such pairs when energy passes from a light-harvesting complex to a photosystem.

ELECTRON FLOW IN A NONCYCLIC PATHWAY

The electrons that light drives out of a photosystem journey through a noncyclic or cyclic pathway of ATP formation. Consider the noncyclic pathway first.

Using Electrons To Make ATP Electrons lost from a photosystem immediately enter an electron transfer chain embedded in the thylakoid membrane (Figure 6.8*a*). Remember, electron transfer chains are organized arrays of enzymes, coenzymes, and other proteins that accept and donate electrons in turn (Section 5.5). The entry of electrons from a photosystem into an electron transfer chain is the first step in the light-dependent reactions. With this step, light energy is converted to chemical energy. Light does not take part in chemical reactions, but electrons do.

Electrons release energy bit by bit as they "bounce" down an electron transfer chain (Figure 6.8*b*). Various components of the chain use that released energy for cellular work—in this case, they move hydrogen ions (H^+) across the thylakoid membrane, from the stroma to the thylakoid compartment. As more electrons flow through these transfer chains, a hydrogen ion gradient builds up across the membrane.

This gradient attracts hydrogen ions back toward the stroma, but H^+ cannot diffuse across a lipid bilayer without assistance. Hydrogen ions cross the thylakoid membrane only by flowing through the interior of ATP synthases (Section 4.3). The flow causes these proteins to attach phosphate groups to ADP. By this process, ATP forms in the stroma (Figure 6.8*c*).

Replacing Lost Electrons As long as electrons flow through transfer chains, the cell can continue to produce ATP. However, the number of electrons in a photosystem is not unlimited, so where do they come from? Photosystem II pulls replacement electrons from water molecules, which then dissociate into hydrogen ions and oxygen inside the thylakoid compartment. The hydrogen ions contribute to gradients that drive ATP formation, and the oxygen diffuses out of the cell (Figure 6.8*a*). The process by which the energy of light breaks down a molecule is called photolysis.

The noncyclic pathway of photosynthesis is named because the electrons that leave photosystem II do not return to it; they end up in NADPH. After the electrons from photosystem II move through an electron transfer

Figure 6.7 The thylakoid membrane surface facing the stroma. Arrays of light-harvesting complexes and photosystems are embedded in the membrane's lipid bilayer. Many electron transfer chains and ATP synthases are also present, but not shown for clarity.

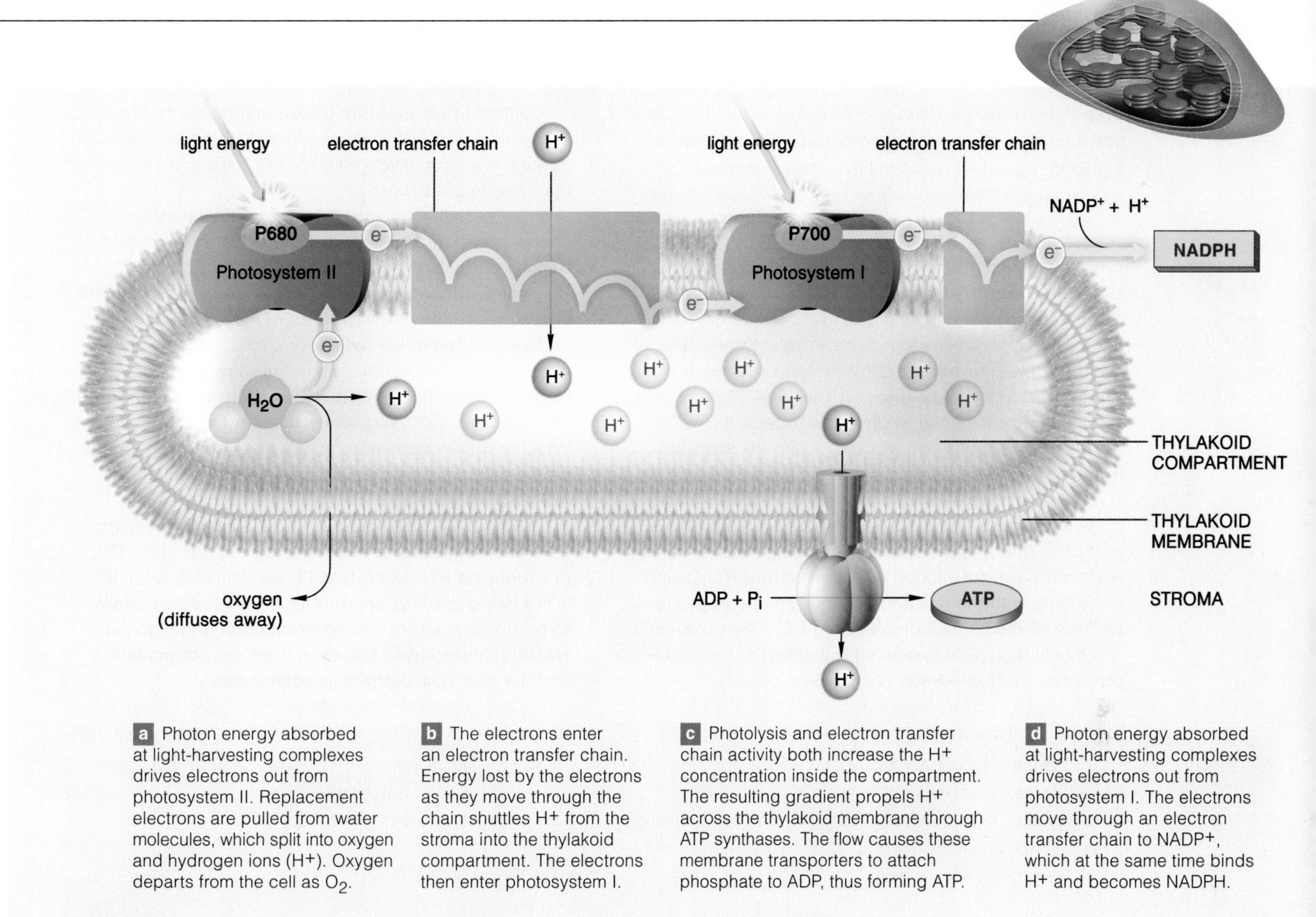

Figure 6.8 **Animated!** Noncyclic ATP-forming pathway of photosynthesis. This art shows a cross-section through a disk-shaped fold in the thylakoid membrane. Electrons that left photosystem II are not cycled back to it. They end up in NADPH, a reduced coenzyme that delivers them to sites where sugars are built. Appendix VI shows individual components of the electron transfer chains.

chain, they enter a type I photosystem. Energy from a light-harvesting complex then causes the P700 of this photosystem to release electrons, which enter another electron transfer chain. At the end of the chain, $NADP^+$ accepts electrons and H^+, thereby becoming NADPH (Figure 6.8*d*).

ELECTRON FLOW IN A CYCLIC PATHWAY

The noncyclic pathway is the dominant mode of ATP formation in chloroplasts. However, when NADPH is not being used, it accumulates in the stroma and the noncyclic pathway backs up and stalls. Then, a cyclic pathway runs independently in type I photosystems and allows cells to continue making ATP. In this case, electrons lost from photosystem I get cycled back to it. On their return journey, the electrons pass through an electron transfer chain that shuttles hydrogen ions into the thylakoid compartment. An H^+ gradient builds up and drives ATP formation, just as in the noncyclic pathway, but NADPH does not form.

In chloroplasts, ATP forms during light-dependent reactions of photosynthesis. There are two reaction pathways.

In the noncyclic pathway, electrons flow from water molecules, through two photosystems and two electron transfer chains, and end up in the coenzyme NADPH. This pathway releases oxygen.

In the cyclic pathway, electrons lost from photosystem I return to it after moving through an electron transfer chain. ATP forms, but NADPH does not. Oxygen is not released.

6.5 Energy Flow in Photosynthesis

FOCUS ON SCIENCE

One of the recurring themes in biology is that organisms use energy to drive cellular processes. Energy flow in the light-dependent reactions is a classic example of how organisms harvest energy from their environment.

LINK TO SECTION 5.5

Any light-driven reaction that attaches phosphate to a molecule is known as photophosphorylation. Figure 6.9 compares the energy flow of the cyclic and noncyclic photophosphorylations that occur in chloroplasts. The cyclic pathway evolved first, in anaerobic species. It was simpler and less energy efficient. It yields ATP alone, and it still operates in nearly all photoautotrophs. No NADPH forms, and oxygen is not released. Again, electrons lost from photosystem I cycle back to it (Figure 6.9*a,b*).

Later, the photosynthetic machinery in some kinds of photoautotrophs was modified. Photosystem II became part of it. That was the start of a combined sequence of reactions powerful enough to strip electrons from water molecules, with the release of oxygen and hydrogen ions. Remember redox reactions (Section 5.5)? Photosystem II is the only biological system strong enough to oxidize—pull electrons from—water (Figure 6.9*c–e*).

Electrons that leave photosystem II do not return to it. They end up in NADPH, a powerful reducing agent (electron donor). NADPH delivers electrons to the sugar-producing reactions in the stroma.

In both cyclic and noncyclic photophosphorylation, energy associated with electrons flowing through electron transfer chains shuttles H^+ across the thylakoid membrane. Hydrogen ions accumulate in the thylakoid compartment and form a gradient that powers ATP synthesis.

Today, different bacteria have either type I or type II photosystems in their plasma membrane. Cyanobacteria, plants, and all photoautotrophic protists have both types, and both the cyclic and noncyclic pathway can operate. Which pathway dominates at any given time depends on the organism's immediate metabolic demands for ATP and NADPH.

Having these alternate pathways is energy efficient. Cells can direct energy to producing NADPH and ATP or producing ATP alone. NADPH accumulates when it is not being used up, and this causes the cyclic pathway to predominate. When sugar production is in high gear, NADPH is being used quickly. It does not accumulate, and the noncyclic pathway predominates.

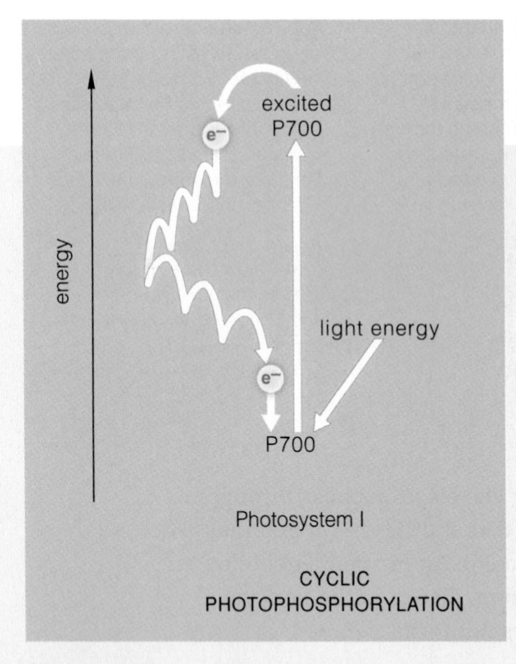

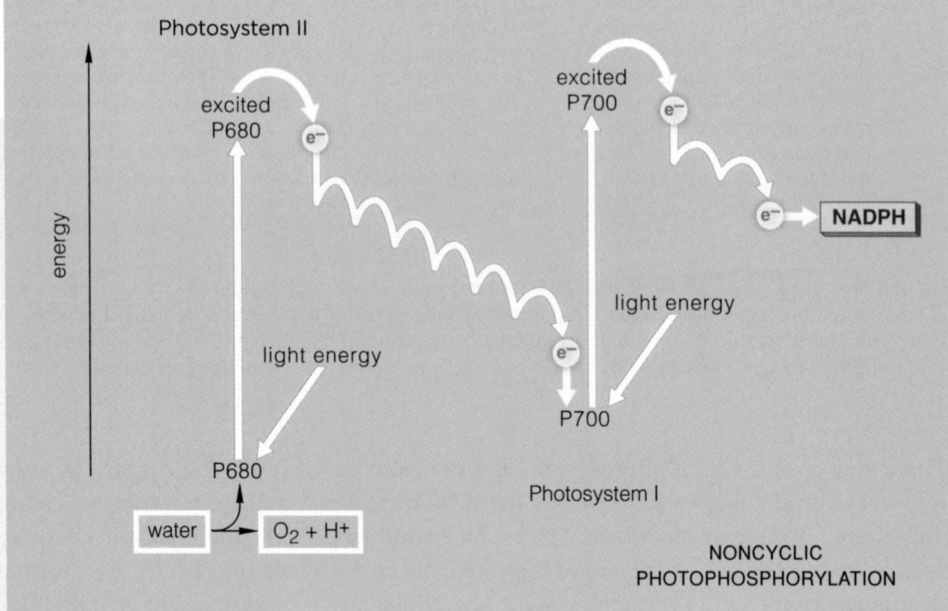

a Energy from light-harvesting complexes causes photosystem I to lose electrons.

b Electrons give up energy as they pass through an electron transfer chain. The energy drives H^+ across the thylakoid membrane, against its gradient. The electrons reenter photosystem I.

c Energy from a light-harvesting complex drives electrons out of photosystem II. Then, the photosystem pulls replacement electrons from water molecules.

d Electrons from photosystem II pass through an electron transfer chain. Energy lost at each step moves H^+ across the thylakoid membrane. At the end of the chain, the electrons enter photosystem I.

e $NADP^+$ combines with hydrogen and with electrons driven from photosystem II by energy from a light-harvesting complex. The resulting NADPH delivers electrons and hydrogen to the next stage of reactions.

Figure 6.9 **Animated!** Energy flow in the light-dependent reactions of photosynthesis. The P700 in photosystem I absorbs photons of a 700-nanometer wavelength. The P680 of photosystem II absorbs photons of a 680-nanometer wavelength. Energy inputs boost P700 and P680 to an excited state in which they lose electrons.

6.6 Light-Independent Reactions: The Sugar Factory

The chloroplast is a sugar factory operated by enzymes of the Calvin–Benson cycle. The cyclic, light-independent reactions are the "synthesis" part of photosynthesis.

The enzyme-mediated reactions of the Calvin–Benson cycle build sugars in the stroma of chloroplasts. These are light-independent reactions because light does not power them. Instead, they run on the bond energy of ATP and the reducing power of NADPH—molecules that formed in the light-dependent reactions.

Light-independent reactions use carbon atoms from CO_2 to make glucose. Carbon fixation is the process of extracting carbon atoms from an inorganic source and incorporating them into an organic molecule. In most plants, rubisco fixes carbon. This enzyme attaches the carbon atom of CO_2 to a five-carbon molecule of RuBP, or ribulose biphosphate (Figure 6.10*a*).

The six-carbon intermediate that forms is unstable. It splits right away into two PGA (phosphoglycerate) molecules, each with a three-carbon backbone (Figure 6.10*a*). Next, ATP transfers a phosphate group to each PGA, and NADPH donates hydrogen and electrons to it (Figure 6.10*b*). Thus, ATP energy and the reducing power of NADPH convert each PGA molecule into a molecule of PGAL (phosphoglyceraldehyde), a different three-carbon compound.

LINK TO SECTION 3.2

Glucose, remember, has six carbon atoms. For each glucose molecule to form, six CO_2 are fixed into twelve PGAL molecules (Figure 6.10*c*). Two PGAL combine to form a six-carbon glucose molecule with an attached phosphate group. The remaining PGAL regenerate the starting compound of the cycle, RuBP (Figure 6.10*d–f*).

Plants can use the glucose they make as an energy source, and also as building blocks for other organic molecules. Most of the glucose is converted at once to sucrose or starch by other pathways that conclude the light-independent reactions. Sucrose is a transportable carbohydrate in plants. Excess glucose is stored in the form of starch grains inside the stroma of chloroplasts. When sugars are needed in other parts of the plant, the starch is converted to sucrose and exported.

Driven by ATP energy, the light-independent reactions of photosynthesis use hydrogen and electrons (from NADPH), and carbon and oxygen (from CO_2) to build sugars.

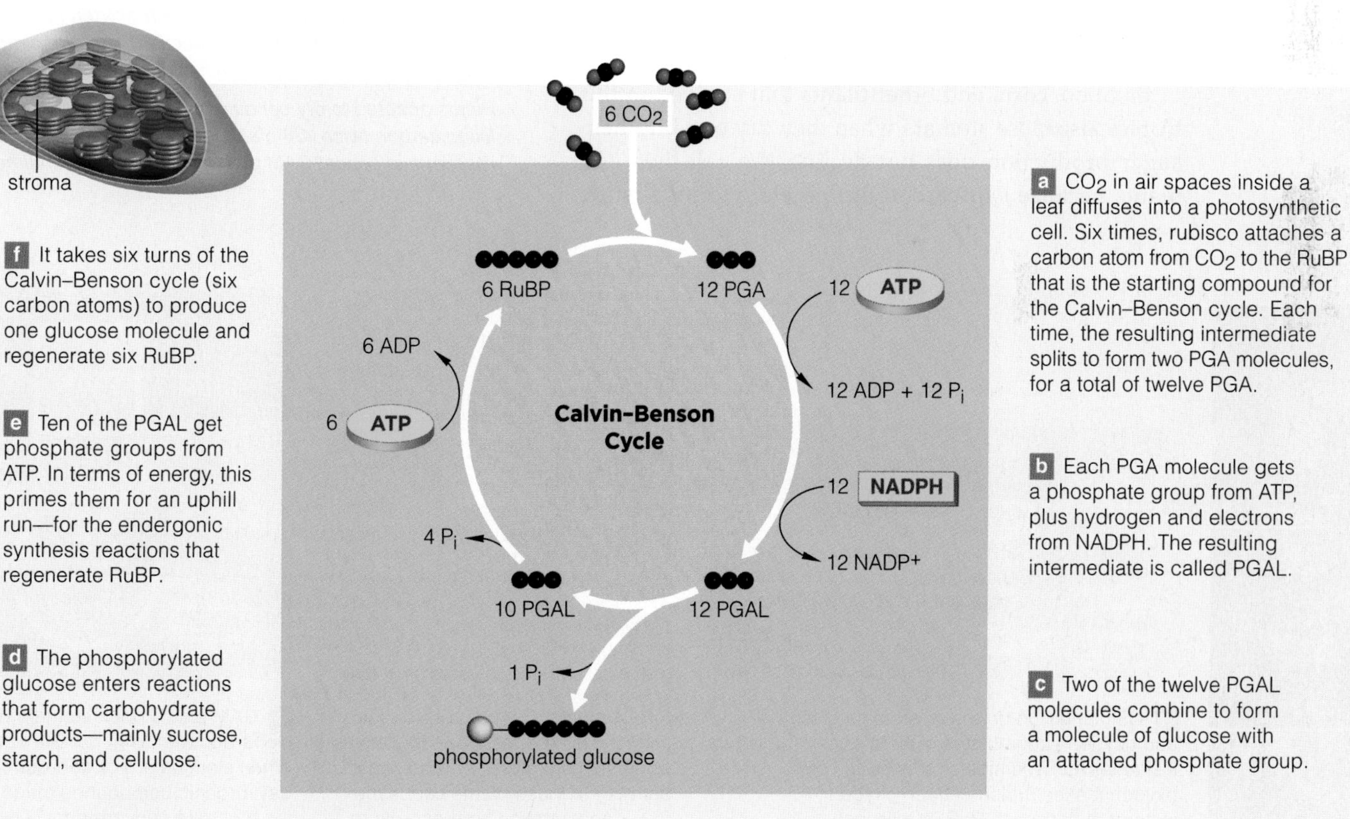

Figure 6.10 **Animated!** Light-independent reactions of photosynthesis which, in chloroplasts, occur in the stroma. The sketch is a summary of six turns of the Calvin–Benson cycle and their product, one glucose molecule. *Black* balls signify carbon atoms. Appendix VI details the reaction steps.

6.7 Adaptations: Different Carbon-Fixing Pathways

If sunlight intensity, air temperature, rainfall, and soil composition never varied, photosynthesis might be the same in all plants. But environments differ, and so do details of photosynthesis.

LINK TO SECTION 5.5

The only way for gases to diffuse into or out of a plant is at stomata (singular, stoma), small openings across the surface of leaves and green stems. Stomata close on dry days. Water stays in the plant, but O_2 produced by light-dependent reactions cannot diffuse out, and CO_2 for the light-independent reactions cannot diffuse in.

That is why 85 percent of all plant species do not grow well in drier climates without steady irrigation. We call them C3 plants because they fix carbon with the Calvin–Benson cycle, in which *three*-carbon PGA is the first stable intermediate. When photosynthetic reactions run with stomata closed, oxygen builds up in leaves. At high O_2 levels, rubisco attaches oxygen (not carbon) to RuBP in a pathway called photorespiration. CO_2 is a product; a cell loses carbon instead of fixing it. ATP and NADPH are used to shunt intermediates back to the Calvin–Benson cycle. So, sugar production in C3 plants is inefficient on dry days (Figure 6.11*a*).

Photorespiration can limit growth; plants compensate for rubisco's inefficiency by making a lot of it. Rubisco is the most abundant protein on Earth.

Bamboo, corn, and other plants that evolved in the tropics also close stomata when they dry out, but their sugar production does not decline. We call them C4 plants because *four*-carbon oxaloacetate forms first in carbon-fixation reactions that run through two types of cells (Figure 6.11*b*). In *mesophyll* cells, the enzyme that catalyzes carbon fixation will not use oxygen even if its concentration is high. Another intermediate moves into *bundle-sheath cells,* where it gets converted to CO_2. Carbon is fixed for the second time as the CO_2 enters the Calvin–Benson cycle.

The C4 cycle keeps the CO_2 level near rubisco high enough to minimize photorespiration. Because they fix carbon twice, C4 plants use more ATP than C3 plants do, but they can make more sugar on dry days.

We see additional adaptations to dry conditions in CAM plants (Crassulacean Acid Metabolism). In these plants, the two carbon fixing reactions are separated in time, rather than in space. CAM plants open their stomata at night, when mesophyll cells use a C4 cycle to fix carbon from CO_2 in the air. The products of the cycle are stored until the next day. Then, the stomata close. The C4 cycle's products are converted to CO_2, which enters the Calvin–Benson cycle (Figure 6.11*c*).

When stomata are closed, oxygen builds up inside leaves of C3 plants. Rubisco then can attach oxygen (instead of carbon dioxide) to RuBP. This reaction, photorespiration, reduces efficiency of sugar production, so it can limit growth.

Plants adapted to dry conditions limit photorespiration by fixing carbon twice. C4 plants separate the two sets of reactions in space; CAM plants separate them in time.

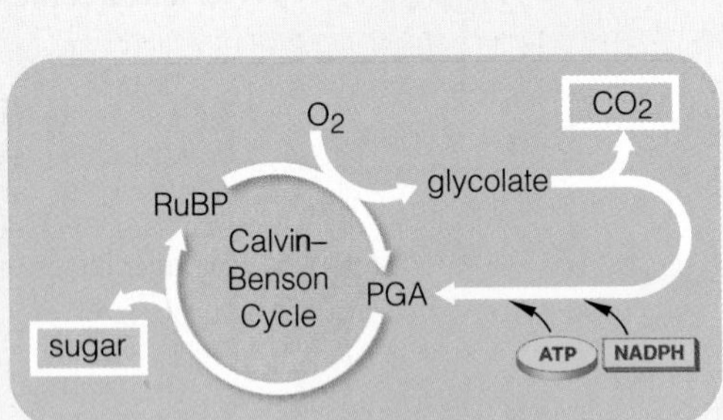

a C3 plants. On dry days, stomata close and oxygen accumulates in air spaces inside leaves. The high concentration of oxygen makes rubisco attach oxygen instead of carbon to RuBP. Cells lose carbon and energy as they make sugars.

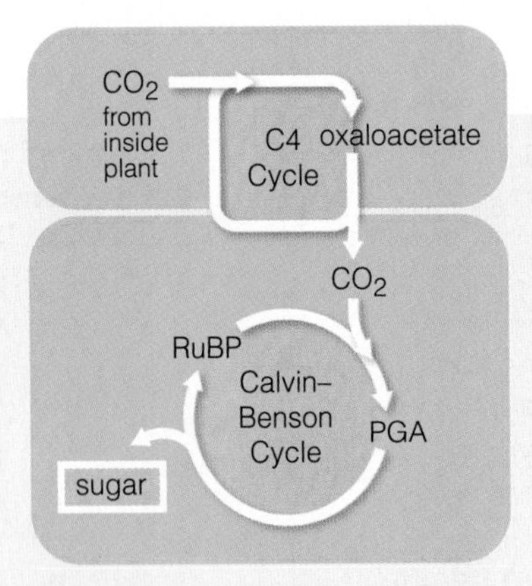

b C4 plants. Oxygen also builds up in the air spaces inside the leaves when stomata close. An additional pathway in these plants keeps the CO_2 concentration high enough to prevent rubisco from using oxygen.

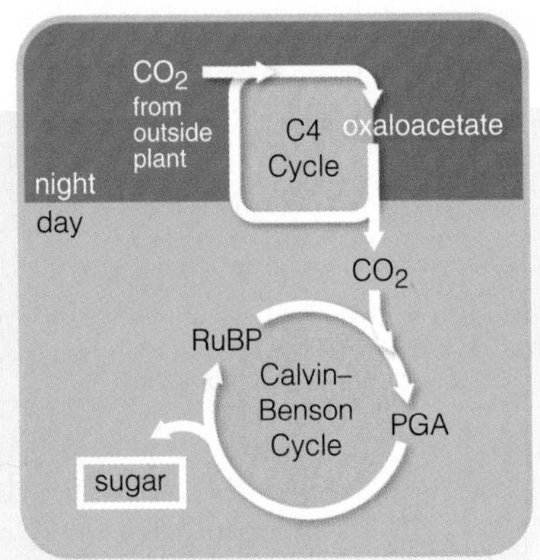

c CAM plants open stomata and fix carbon with a C4 pathway at night. When stomata are closed during the day, organic compounds made during the night are converted to CO_2 that enters the Calvin–Benson cycle.

Figure 6.11 Carbon-fixing adaptations in three kinds of plants.

6.8 A Burning Concern

FOCUS ON THE ENVIRONMENT

Earth's natural atmospheric cycle of carbon dioxide is out of balance, mainly as a result of human activity.

Have you ever wondered where all of the atoms in your body came from? Think about just the carbon atoms. You eat other organisms to get the carbon atoms your body uses for energy and for raw materials. Those atoms may have passed through other heterotrophs before you ate them, but at some point they were part of photoautotrophic organisms. Photoautotrophs strip carbon from carbon dioxide, then use the atoms to build organic compounds. Your carbon atoms—and those of most other organisms—came from carbon dioxide.

Photosynthesis removes carbon dioxide from the atmosphere, and locks its carbon atoms inside organic compounds. When photosynthesizers and other aerobic organisms break down the organic compounds for energy, carbon atoms are released in the form of CO_2, which then re-enters the atmosphere. Since photosynthesis evolved, these two processes have constituted a balanced cycle of the biosphere. You will learn more about the carbon cycle in Section 42.7. For now, know that the amount of carbon dioxide that photosynthesis removes from the atmosphere is roughly the same amount that organisms release back into it—at least it was, until humans came along.

As early as 8,000 years ago, humans began burning forests to clear land for agriculture. When trees and other plants burn, most of the carbon locked in their tissues is released into the atmosphere as carbon dioxide. Fires that occur naturally release carbon dioxide the same way, but they do not occur nearly as often.

Today, we are burning a lot more than our ancestors ever did. In addition to wood, we are burning fossil fuels—coal, petroleum, and natural gas—to satisfy our greater and greater needs for energy. As you will see in Section 21.4, fossil fuels are the organic remains of ancient organisms. When we burn these fuels, we release the carbon that has been locked inside them for hundreds of millions of years back into the atmosphere—as carbon dioxide (Figure 6.12).

Researchers find pockets of our ancient atmosphere in Antarctica. Snow and ice have been accumulating in layers there, year after year, for the last 15 million years. Air and dust trapped in each layer reveal the composition of the atmosphere that prevailed when the layer formed. Thus, we now know that the atmospheric CO_2 level had been relatively stable for about 10,000 years before 1850—about the time the industrial revolution was underway. Since then, the CO_2 level has been steadily rising. In 2006, it was higher than it had been in *23 million years*.

Our activities have put Earth's atmospheric cycle of carbon dioxide out of balance. We are adding far more CO_2 to the atmosphere than photosynthetic organisms are removing from it. Today, we release around 26 billion tons of carbon dioxide into the atmosphere each year, more than ten times the amount we released in the year 1900. Most of it comes from burning fossil fuels. How do we know? Researchers can tell how long ago the carbon atoms in a sample of CO_2 were part of a living organism by measuring the ratio of different carbon isotopes in it. From that ratio, they can calculate the number of carbon atoms that came from the breakdown of ancient organic molecules (such as those in fossil fuels), and the number that came from the breakdown of more recent ones. You will read more about radiocarbon dating in Section 16.5.

Figure 6.12 Visible evidence of fossil fuel emissions in the atmosphere: the sky over New York City on a sunny day.

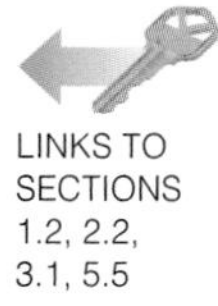

LINKS TO SECTIONS 1.2, 2.2, 3.1, 5.5

The increase in atmospheric carbon dioxide is having dramatic effects on climate. CO_2 contributes to global warming, as you will read in Section 42.8. We are seeing a warming trend that mirrors the increase in CO_2 levels; Earth is now the warmest it has been for 12,000 years. The climate change is affecting biological systems everywhere. Life cycles are changing: Birds are laying eggs earlier; plants are flowering at the wrong times; mammals are hibernating for shorter periods. Migration patterns and habitats are also changing. For many species, the changes are too fast, and extinctions are now occurring at an unprecedented rate.

Under normal circumstances, extra carbon dioxide stimulates photosynthesis, which means additional uptake of CO_2. However, the changes we are already seeing in temperature and moisture patterns as a result of global warming are offsetting this benefit. Such changes are proving harmful to plants and other photosynthesizers.

Much research today targets development of energy sources that are not based on fossil fuels. For example, photosystem II catalyzes photolysis, the most efficient oxidation reaction in nature. Researchers are working to duplicate its catalytic function in artificial systems. If they are successful, then perhaps we too might be able to use light to split water into hydrogen, oxygen, and electrons—all of which can be used as clean sources of energy. Other research is focused on ways to remove carbon dioxide from the atmosphere—for example, by improving the efficiency of rubisco.

www.thomsonedu.com Thomson NOW!

Summary

Introduction Before photosynthesis evolved, Earth's atmosphere held very little free oxygen. Oxygen released during photosynthesis changed the atmosphere. It was a selective force that favored the evolution of new metabolic pathways, including aerobic respiration.

Sections 6.1, 6.2 Visible light is a very small part of a spectrum of electromagnetic energy radiating from the sun. That energy travels in waves, and it is organized as photons. Visible light drives photosynthesis, which begins when photons are absorbed by photosynthetic pigment molecules. Pigments are molecules that absorb light of particular wavelengths only; photons not captured by a pigment are reflected as its characteristic color. The main photosynthetic pigment—chlorophyll *a*—absorbs violet and red light, so it appears green.

Accessory pigments, including chlorophyll *b* and the carotenoids, absorb additional wavelengths. Collectively, photosynthetic pigments efficiently absorb almost all of the wavelengths of visible light.

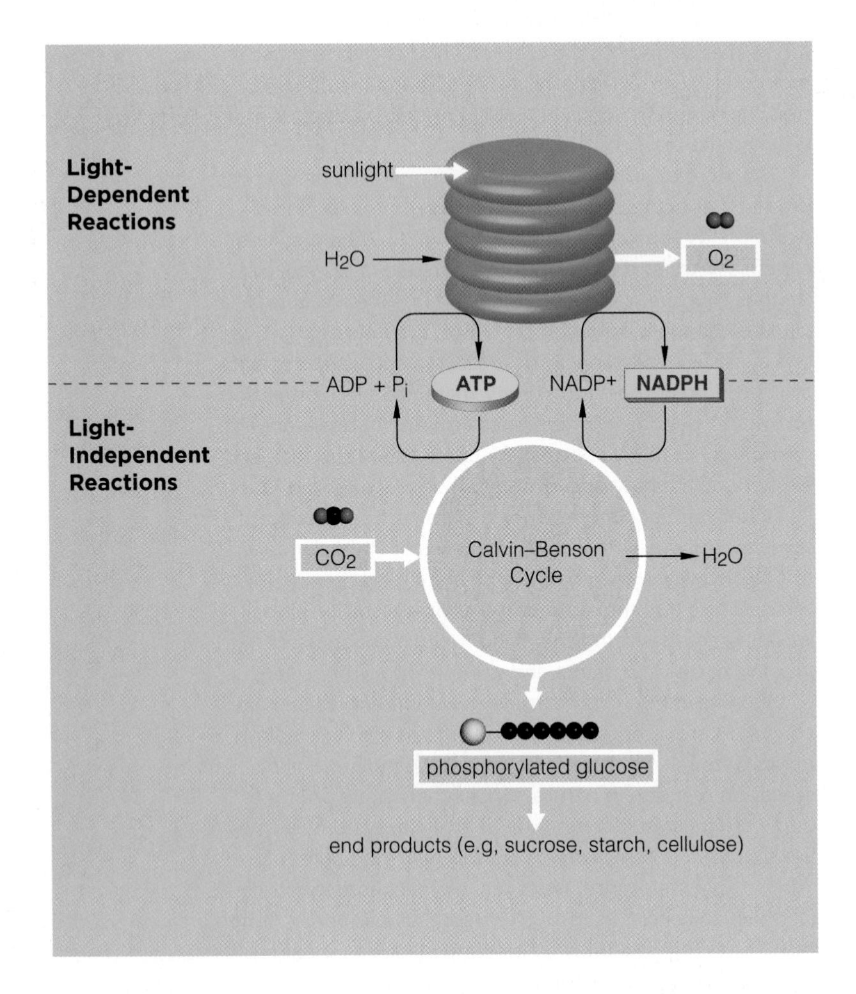

Figure 6.13 Visual summary of photosynthesis.

Section 6.3 Photosynthesis proceeds in two stages: light-dependent and light-independent reactions. Figure 6.13 and the following equation summarize the process:

$$6H_2O + 6CO_2 \xrightarrow[\text{enzymes}]{\text{light energy}} 6O_2 + C_6H_{12}O_6$$

water, carbon dioxide → oxygen, glucose

In chloroplasts, the light-dependent reactions occur at a much-folded thylakoid membrane. The membrane forms a single, continuous compartment inside the stroma, the chloroplast's semifluid interior. The light-independent reactions occur in the stroma. Typically, sunlight energy drives the formation of ATP and NADPH, and oxygen is released from the chloroplast (and the cell).

- *Use the animation on ThomsonNOW to view the sites where photosynthesis takes place.*

Sections 6.4, 6.5 The thylakoid membrane contains two types of photosystems and many light-harvesting complexes—arrays of pigments and other proteins that as a unit absorb light energy and pass it to photosystems, which then release electrons. The electrons enter the light-dependent reactions of photosynthesis.

In noncyclic photophosphorylation, electrons released from photosystem II flow through an electron transfer chain. At the end of the chain, they enter photosystem I. Photon energy causes photosystem I to release electrons, which end up in NADPH. Photosystem II replaces lost electrons by pulling them away from water, which then dissociates into H^+ and O_2.

In cyclic photophosphorylation, the electrons released from photosystem I enter an electron transfer chain, then cycle back to photosystem I. NADPH does not form.

In both pathways, electron flow through the electron transfer chains causes H^+ to accumulate in the thylakoid compartment, and so a hydrogen ion gradient builds up across the thylakoid membrane. H^+ flows back across the membrane through ATP synthases. This flow results in the formation of ATP in the stroma.

- *Use the animation on ThomsonNOW to review pathways by which light energy is used to form ATP.*

Section 6.6 The light-independent reactions proceed in the stroma. The enzyme rubisco attaches carbon from CO_2 to RuBP to start the Calvin–Benson cycle. This cyclic pathway uses energy from ATP, carbon and oxygen from CO_2, and hydrogen and electrons from NADPH to make phosphorylated glucose, which quickly enters reactions that form photosynthetic products (e.g., sucrose, starch, and cellulose). With six turns of the Calvin–Benson cycle, the six carbons required to build a glucose molecule are fixed from CO_2.

- *Read the InfoTrac article "Robust Plants' Secret? Rubisco Activase!" Marcia Wood,* Agricultural Research, *November 2002.*

Section 6.7 Environments differ, and so do details of sugar production in the light-independent reactions. On dry days, plants conserve water by closing their stomata, but O_2 from photosynthesis cannot escape. In C3 plants, the resulting high O_2 level in leaves causes rubisco to attach O_2 instead of CO_2 to RuBP. This pathway, called photorespiration, reduces the efficiency of sugar production. In C4 plants, carbon fixation occurs twice. The first reactions release CO_2 near rubisco, and thus limit photorespiration when stomata are closed. CAM plants open their stomata and fix carbon at night.

■ *Read the InfoTrac article "Light of Our Lives," Norman Miller,* Geographical, *January 2001.*

Section 6.8 Photoautotrophs remove CO_2 from the atmosphere; metabolic activity of every organism puts it back. Human activities disrupt this cycle by adding more CO_2 to the atmosphere than photoautotrophs can remove from it. The imbalance contributes to global warming.

Self-Quiz

Answers in Appendix III

1. Photosynthetic autotrophs use _______ from the air as a carbon source and _______ as their energy source.

2. Chlorophyll *a* absorbs mainly violet and red light, and it reflects mainly _______ light.
   a. violet and red
   b. yellow and green
   c. green only
   d. white and orange

3. Light-*dependent* reactions in plants occur at the _______.
   a. thylakoid membrane
   b. plasma membrane
   c. stroma
   d. cytoplasm

4. In the light-*dependent* reactions, _______.
   a. carbon dioxide is fixed
   b. ATP forms
   c. CO_2 accepts electrons
   d. sugars form

5. What accumulates inside the thylakoid compartment during the light-*dependent* reactions?
   a. glucose
   b. RuBP
   c. hydrogen ions
   d. CO_2

6. When a photosystem absorbs light, _______.
   a. sugar phosphates are produced
   b. electrons are transferred to ATP
   c. RuBP accepts electrons
   d. light-dependent reactions begin

7. Light-*independent* reactions proceed in the _______.
   a. cytoplasm
   b. plasma membrane
   c. stroma

8. The Calvin–Benson cycle starts when _______.
   a. light is available
   b. carbon dioxide is attached to RuBP
   c. electrons leave photosystem II

9. What substance is *not* part of the Calvin–Benson cycle?
   a. ATP
   b. NADPH
   c. RuBP
   d. PGAL
   e. O_2
   f. CO_2

10. Match each event with its most suitable description.
   ___ ATP formation only
   ___ CO_2 fixation
   ___ photolysis
   ___ PGAL formation
   a. rubisco required
   b. water molecules split
   c. ATP, NADPH required
   d. electrons cycled back to photosystem I

■ *Visit ThomsonNOW for additional questions.*

Figure 6.14 (**a**) Red alga from a tropical reef. (**b**) Coastal green alga (*Codium*).

Critical Thinking

1. About 200 years ago, Jan Baptista van Helmont did experiments on the nature of photosynthesis. He wanted to know where growing plants get the materials necessary for increases in size. He planted a tree seedling weighing 5 pounds in a barrel filled with 200 pounds of soil and then watered the tree regularly. After five years, the tree weighed 169 pounds, 3 ounces, and the soil weighed 199 pounds, 14 ounces. Because the tree had gained so much weight and the soil had lost so little, he concluded the tree had gained all of its additional weight by absorbing the water he had added to the barrel. Given what you know about biological molecules, why was he misguided? What really happened?

2. A cat eats a bird, which earlier ate a caterpillar that chewed on a weed. Which organisms are autotrophs? Which are heterotrophs?

3. While gazing into an aquarium, you observe bubbles coming from an aquatic plant (*right*). What is happening?

4. Krishna exposes pea plants to a carbon radioisotope ($^{14}CO_2$), which they absorb. In which compound will the labeled carbon appear first in C3 plants? In C4 plants?

5. Most pigments respond to only part of the rainbow of visible light. If acquiring energy is so vital, then why doesn't each kind of photosynthetic pigment absorb the whole spectrum? *Why isn't each one black?*

If early photoautotrophs evolved in the seas, then so did their pigments. Ultraviolet and red wavelengths do not penetrate water as deeply as green and blue wavelengths do. Possibly natural selection favored the evolution of different pigments at different depths. Many relatives of the red alga in Figure 6.14*a* live deep in the sea. Some are nearly black. Green algae, such as the one in Figure 6.14*b*, live in shallow water. Their chlorophylls absorb red wavelengths. Their accessory pigments harvest others, and some also function as shields against ultraviolet radiation.

Speculate on how natural selection may have favored the evolution of different pigments at different depths.

6. Only about eight classes of pigment molecules are known, but this limited group gets around. For example, photoautotrophs make carotenoids, which move through food webs, as when tiny aquatic snails graze on green algae and then flamingos eat the snails. Flamingos modify the carotenoids. Their cells split beta-carotene to form two molecules of vitamin A. This vitamin is the precursor of retinol, a visual pigment that converts light energy to electric signals in the eyes. Beta-carotene gets dissolved in fat under the skin. Cells that give rise to bright pink feathers take it up. Research another organism to identify sources for pigments that color its surfaces.

7 HOW CELLS RELEASE CHEMICAL ENERGY

IMPACTS, ISSUES When Mitochondria Spin Their Wheels

In the early 1960s, a Swedish physician, Rolf Luft, mulled over a patient's odd symptoms. The young woman felt weak and too hot all the time. Even on the coldest winter days she could not stop sweating, and her skin was always flushed. She was thin, yet had a huge appetite. Luft inferred that his patient's symptoms pointed to a metabolic disorder. Her cells seemed to be spinning their wheels. They were very active, but much of their activity was being lost as metabolic heat. Luft checked the patient's basal rate of metabolism, the amount of energy her body was expending when resting. Even at rest, her oxygen consumption was the highest ever recorded!

As examination of a tissue sample revealed, the patient's skeletal muscles had plenty of mitochondria, the cell's ATP-producing powerhouses. But there were too many of them, and they were abnormally shaped. Further studies showed that her mitochondria were making very little ATP despite working at top speed.

The disorder, now called Luft's syndrome, was the first to be linked to defective mitochondria. The cells of someone with the disorder are like cities that are burning tons of coal in many power plants but not getting much energy output. Skeletal and heart muscles, the brain, and other hardworking body parts with high energy demands are hit the hardest.

More than forty other disorders related to defective mitochondria are now known. One of the heritable types, Friedreich's ataxia, causes loss of coordination (ataxia), weak muscles, and serious heart problems. Many of those affected die when they are young adults. Figure 7.1 shows a girl and a boy affected by this disorder.

See the video! **Figure 7.1** Sister, brother, and broken mitochondria. Both of these individuals show symptoms of Friedreich's ataxia, a genetic disorder that prevents them from making enough ATP to keep their body structurally and functionally sound.

Leah started to lose her sense of balance and coordination at age five. Six years later she was in a wheelchair; now she is diabetic and partially deaf. Her brother Joshua could not walk by the time he was eleven, and is now blind. Both have heart problems; both had spinal fusion surgery. Special equipment allows them to attend school and work part-time. Leah is a professional model.

How would you vote? Developing new drugs is costly, which is why pharmaceutical companies tend to ignore Friedreich's ataxia and other disorders that affect relatively few people. Should governments fund private companies that research potential treatments? See ThomsonNOW for details, then vote online.

Remember how ATP forms at a membrane folded up inside chloroplasts? It also forms at a membrane folded up in mitochondria by the process of aerobic respiration. Operation of electron transfer chains embedded in the inner membrane of both organelles sets up hydrogen ion gradients that power ATP formation.

In Luft's syndrome, the transfer chains work overtime, but too little ATP forms. In Friedreich's ataxia, the gene for a protein called frataxin (*below*) is mutated, so the protein does not work properly. This protein helps build some of the iron-containing enzymes of electron transfer chains. When it malfunctions, iron atoms that are supposed to be incorporated into those enzymes accumulate in mitochondria instead.

Molecular oxygen is present in mitochondria. As you know, toxic free radicals form when oxygen reacts with metals. Too much iron inside oxygen-rich mitochondria means too many free radicals, which destroy the molecules of life faster than the cell can repair or replace them. Soon, the mitochondria stop functioning, and the cell dies.

You already have a sense of how cells harvest energy in electron transfer chains. Even prokaryotic cells make ATP with electron transfer chains built into their plasma membrane. Details of the reactions vary from one type of organism to the next. Even so, such variations cannot mask life's universal reliance on ATP-forming machinery. When you consider mitochondria in this chapter, do not assume they are too remote from your interests. Without them, you would not make enough ATP even to read about how they do it.

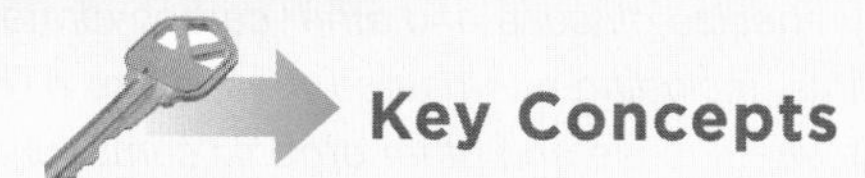

Key Concepts

ENERGY FROM CARBOHYDRATE BREAKDOWN

All organisms produce ATP by various degradative pathways that extract chemical energy from glucose and other organic compounds. Aerobic respiration yields the most ATP from each glucose molecule. In eukaryotes, it is completed inside mitochondria. **Section 7.1**

GLYCOLYSIS

Glycolysis is the first stage of aerobic respiration and of anaerobic routes, such as the fermentation pathways. As enzymes break down glucose to pyruvate, the coenzyme NAD^+ picks up electrons and hydrogen atoms. The net energy yield is two ATP. **Section 7.2**

HOW AEROBIC RESPIRATION ENDS

Aerobic respiration has two more stages. In the Krebs cycle and a few reactions before it, pyruvate is broken down to carbon dioxide, and many coenzymes pick up electrons and hydrogen atoms. In electron transfer phosphorylation, coenzymes deliver electrons to transfer chains that set up conditions for ATP formation. Oxygen accepts electrons at the end of the chains. **Sections 7.3, 7.4**

HOW ANAEROBIC PATHWAYS END

Fermentation pathways start with glycolysis. Substances other than oxygen are the final electron acceptor. Compared with aerobic respiration, the net yield of ATP is small. **Sections 7.5, 7.6**

OTHER METABOLIC PATHWAYS

Molecules other than glucose are common energy sources. Different pathways convert lipids and proteins to substances that may enter glycolysis or the Krebs cycle. **Section 7.7**

PERSPECTIVE AT UNIT'S END

Life shows unity in its molecular and cellular organization and in its dependence on a one-way flow of energy. **Section 7.8**

Links to Earlier Concepts

This chapter expands the picture of life's dependence on energy flow (Section 5.1). It focuses on metabolic pathways (3.1, 5.5) that make ATP (5.1, 5.2) by degrading glucose and other molecules. The reactions occur in the cytoplasm or in mitochondria (4.1, 4.9). You will reflect on a global connection between one of these pathways and photosynthesis (6.3). You may wish to review the structure of glucose and other carbohydrates (3.2). You will come across more examples of what electron transfer chains do (5.5).

7.1 Overview of Carbohydrate Breakdown Pathways

Photoautotrophs make ATP during photosynthesis and use it to synthesize glucose and other carbohydrates. All organisms, including photoautotrophs, make ATP by breaking down glucose and other organic compounds.

LINKS TO SECTIONS 5.1, 5.2, 5.5

Organisms stay alive only as long as they get more energy to replace the energy they use up (Section 5.1). Plants and all other photoautotrophs get energy from the sun; heterotrophs get energy by eating plants and one another. Regardless of its source, energy must be converted to a form that can drive the thousands of diverse reactions necessary to sustain life. That form is adenosine triphosphate—ATP—the common currency for energy expenditures in all living cells.

COMPARISON OF THE MAIN PATHWAYS

The first metabolic pathways were operating billions of years before Earth had an oxygen-rich atmosphere, so the ATP-forming reactions probably were *anaerobic;* they did not use oxygen. Anaerobic reactions are still common among many prokaryotes and protists that live where oxygen is absent. **Fermentation pathways**, which produce ATP under anaerobic conditions, are among them. Certain kinds of eukaryotic cells can use one of those pathways; however, nearly all eukaryotic cells engage in **aerobic respiration**. This ATP-forming pathway uses oxygen. Every breath you take provides your aerobically respiring cells with a fresh supply of oxygen. Some types of prokaryotes also use aerobic respiration, but we will focus on this pathway mainly as it occurs in eukaryotic cells.

Both fermentation pathways and aerobic respiration begin with the exact same reactions in the cytoplasm. In the initial reactions, enzymes convert one molecule of six-carbon glucose into two molecules of **pyruvate**, an organic compound with a three-carbon backbone. After the initial reactions, the pathways diverge. The fermentation pathways end in the cytoplasm, where a molecule other than oxygen accepts electrons at the end of electron transfer chains. Aerobic respiration ends in mitochondria, where oxygen accepts electrons at the end of electron transfer chains (Figure 7.2).

Aerobic respiration can extract energy from glucose far more efficiently than the anaerobic pathways. The fermentation pathways end with a net yield of two ATP per glucose. Aerobic respiration typically yields thirty-six. You and all other complex organisms would die without the higher yield of aerobic respiration.

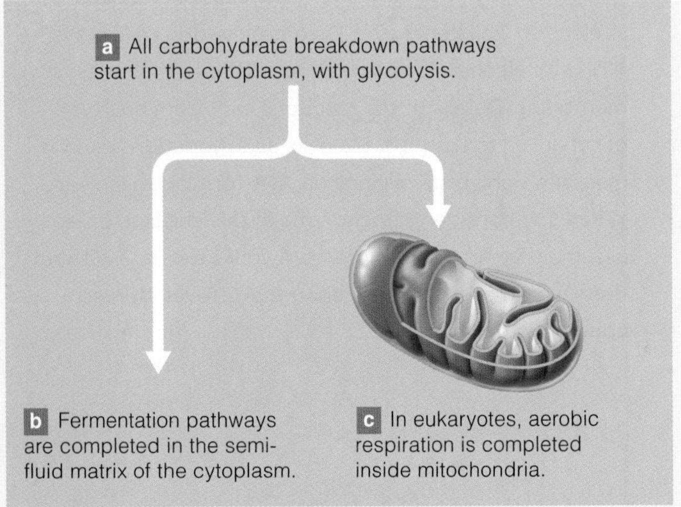

Figure 7.2 **Animated!** Where the different pathways of carbohydrate breakdown start and end. Aerobic respiration alone can deliver enough ATP to sustain large multicelled organisms such as people, redwoods, and Canada geese.

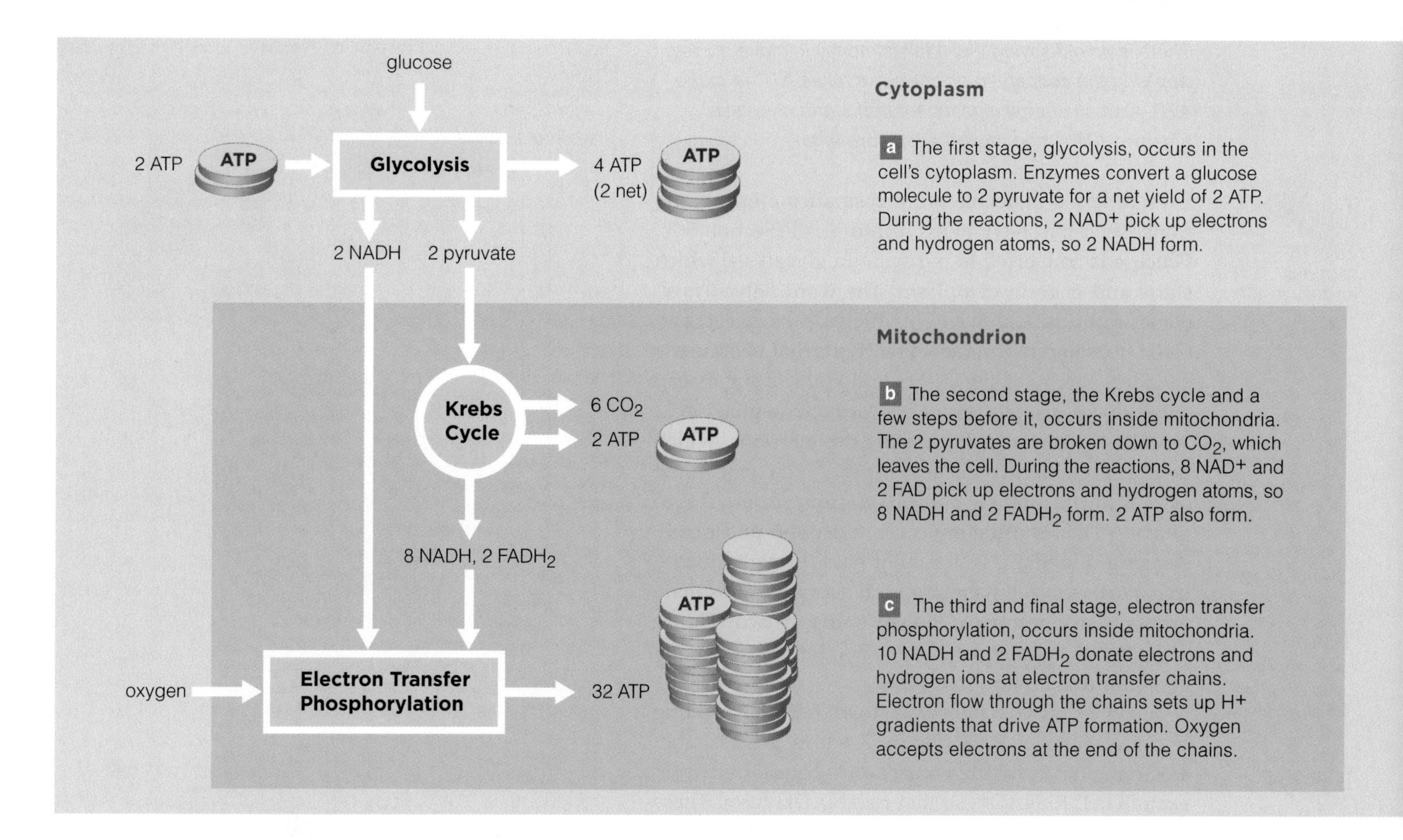

Figure 7.3 Animated! Overview of aerobic respiration. The reactions start in the cytoplasm and end inside mitochondria. A typical net energy yield from one glucose molecule is thirty-six ATP.

OVERVIEW OF AEROBIC RESPIRATION

This simple equation summarizes aerobic respiration:

$$\underset{\text{glucose}}{C_6H_{12}O_6} + \underset{\text{oxygen}}{6O_2} \longrightarrow \underset{\text{carbon dioxide}}{6CO_2} + \underset{\text{water}}{6H_2O}$$

This equation only shows the substances at the start and end of the pathway but not those at three stages in between (Figure 7.3). In the first stage, glycolysis, enzymes convert glucose to pyruvate. In the second stage of reactions, enzymes break down the pyruvate to carbon dioxide. We call these second stage reactions acetyl-CoA formation and the Krebs cycle.

During the first two stages, electrons and hydrogen atoms are picked up by two coenzymes called NAD^+ (nicotinamide adenine dinucleotide) and FAD (flavin adenine dinucleotide). When they carry electrons and hydrogen, we refer to them as NADH and $FADH_2$.

Few ATP form during the first two stages. The big payoff occurs in the third stage after coenzymes give up electrons and hydrogen to electron transfer chains —the machinery of electron transfer phosphorylation. Operation of the transfer chains sets up hydrogen ion (H^+) gradients that drive ATP formation. Many ATP form during electron transfer phosphorylation, which ends when oxygen in mitochondria accepts electrons at the end of the transfer chains. The oxygen also picks up H^+ at the same time and forms water, which is a by-product of aerobic respiration.

Cells convert the chemical energy of carbohydrates and other organic compounds to the chemical energy of ATP, which drives nearly all life-sustaining reactions.

Carbohydrates can be degraded by aerobic respiration and fermentation pathways. These degradative pathways start in the cytoplasm, with glycolysis.

Fermentation pathways end in the cytoplasm. They do not use oxygen. The net yield per glucose molecule is two ATP.

Aerobic respiration ends in mitochondria. It uses oxygen, and the net yield per glucose molecule is thirty-six ATP.

7.2 Glycolysis—Glucose Breakdown Starts

There is an old saying that it takes money to make money. Applying the concept to glycolysis, it takes ATP to make ATP. A small energy investment that starts glycolysis nets two ATP—and maybe a lot more later.

LINKS TO SECTIONS 3.1, 5.5, 5.7

Let's follow a molecule of glucose after a membrane transporter passively helps it into a cell (Section 5.7). Glucose is converted to pyruvate in glycolysis, which starts and ends in cytoplasm. The word "glycolysis" comes from the Greek *glyk–* (which means sweet) and *–lysis* (loosening). It refers to reactions that liberate the chemical energy of sugars. Several kinds of sugars can enter glycolysis, but let's focus for now on glucose.

Cells invest two ATP to start the reactions, which do not run without an energy input. An enzyme in the cytoplasm transfers a phosphate group from ATP to glucose. The resulting molecule, glucose-6-phosphate, stays right where it is; it cannot pass through glucose transporters in cell membranes. It accepts a phosphate group from another ATP, then splits in two (Figure 7.4*b*). PGAL is a simple abbreviation for the two three-carbon intermediates that result.

Enzymes attach a phosphate to each PGAL, forming two molecules of PGA (Figure 7.4*c*). In this reaction, electrons and a hydrogen atom are transferred from each PGAL to NAD^+, so that two NADH form. These reduced coenzymes will give up their cargo during reactions that follow glycolysis.

Two ATP form after enzymes transfer a phosphate group from each PGA to ADP. Two more ATP form when a phosphate is transferred from another pair of intermediates to two ADP. Both of these reactions are substrate-level phosphorylations—the direct transfer of phosphate groups from a substrate to ADP.

Remember, two ATP were invested to initiate the reactions of glycolysis. A total of four ATP form, so the *net* yield is two ATP per molecule of glucose that enters glycolysis (Figure 7.4*f*).

Glycolysis ends with the formation of two pyruvate molecules, each with three carbon atoms. These end products may now enter a second stage of reactions by which cells break down sugars.

Glycolysis is the first stage of aerobic respiration and some fermentation pathways. These pathways convert chemical energy of carbohydrates to the chemical energy of ATP.

In glycolysis, one molecule of glucose is converted to two molecules of pyruvate, with a net energy yield of two ATP. Two NADH also form. The reactions occur in cytoplasm.

Pyruvate and NADH that formed in glycolysis may enter the second stage of carbohydrate breakdown pathways.

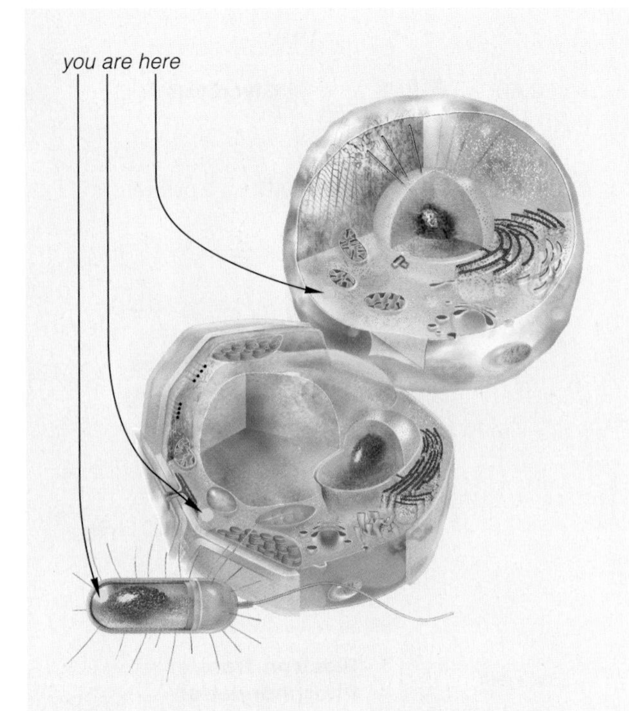

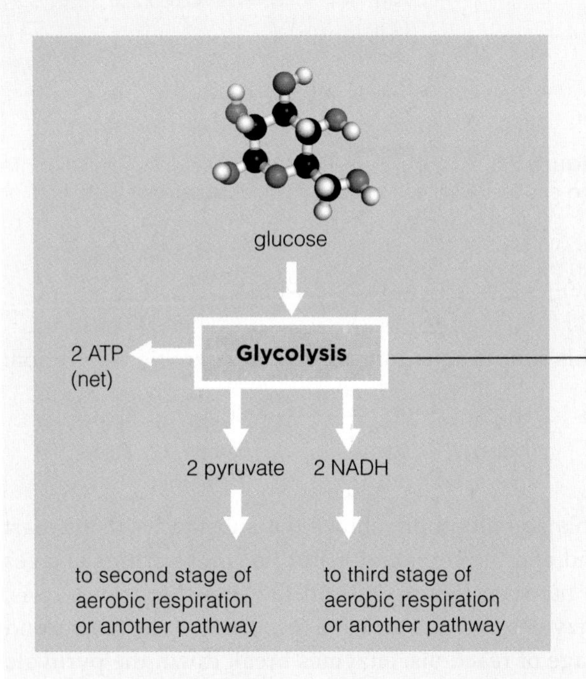

Figure 7.4 **Animated!** Glycolysis. This first stage of carbohydrate breakdown starts and ends in the cytoplasm of all prokaryotic and eukaryotic cells. Glucose is the reactant in this example; we track only its six carbon atoms (*black circles*). For interested students, Appendix VI has structural formulas of intermediates and products.

Cells invest two ATP to start glycolysis, so the *net* energy yield from one glucose molecule is two ATP. Two NADH also form, and two pyruvate molecules are the end products.

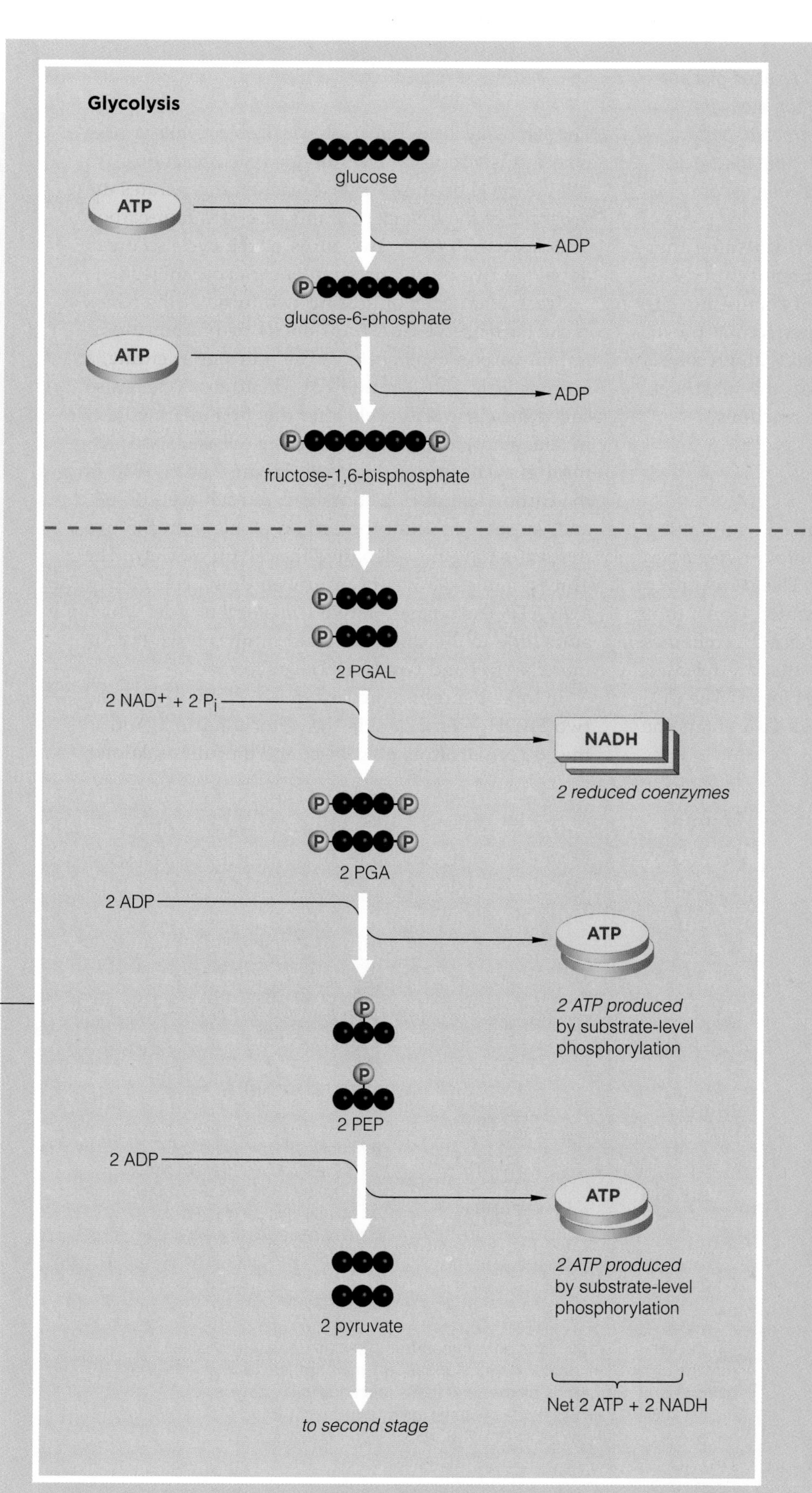

ATP-Requiring Steps

a An enzyme transfers a phosphate group from ATP to glucose, forming glucose-6-phosphate.

b A phosphate group from a second ATP is transferred to the glucose-6-phosphate. The resulting molecule is unstable, and it splits into two three-carbon molecules. The molecules are interconvertible, so we will call them both PGAL (phosphoglyceraldehyde).

So far, two ATP have been invested in the reactions.

ATP-Generating Steps

c Enzymes attach a phosphate to the two PGAL, and transfer two electrons and a hydrogen atom from each PGAL to NAD^+. Two PGA (phosphoglycerate) and two NADH are the result.

d Enzymes transfer a phosphate group from each PGA to ADP. Thus, *two ATP have formed by substrate-level phosphorylation.*

The original energy investment of two ATP has now been recovered.

e Enzymes transfer a phosphate group from each of two intermediates to ADP. *Two more ATP have formed by substrate-level phosphorylation.*

Two molecules of pyruvate form at this last reaction step.

f Summing up, glycolysis yields two NADH, two ATP (net), and two pyruvate for each glucose molecule.

Depending on the type of cell and environmental conditions, the pyruvate may enter the second stage of aerobic respiration. Or it may be used in other ways, as in fermentation pathways.

7.3 Second Stage of Aerobic Respiration

The second stage of aerobic respiration finishes glucose breakdown. Enzymes split off all six carbon atoms in the two pyruvate from glycolysis. The carbon departs, in CO_2. The reactions yield only two ATP, but the big payoff is the formation of many reduced coenzymes.

LINKS TO SECTIONS 3.5, 4.9, 5.2

Let's follow the two molecules of pyruvate after they leave the cytoplasm and enter the premier producer of ATP, a mitochondrion. Figure 7.5*a* zooms into the structure of this organelle. In its inner compartment, the dismantling of glucose that started with glycolysis will be finished. Figure 7.5*b* introduces the substrates, intermediates, and products of these reactions.

ACETYL–CoA FORMATION

The second-stage reactions start as an enzyme splits a molecule of three-carbon pyruvate into a two-carbon acetyl group and a molecule of CO_2. The acetyl group combines with coenzyme A (abbreviated CoA), thus forming acetyl–CoA. The coenzyme NAD^+ combines with hydrogen atoms and electrons that are released from pyruvate breakdown, so NADH forms. The CO_2 diffuses out of the cell, and the acetyl–CoA enters the Krebs cycle (Figure 7.6*a*).

THE KREBS CYCLE

The Krebs cycle is not a physical object, like a bike wheel. It is a series of enzyme-mediated reactions. We call it a cycle because the last reaction regenerates the substrate of the first step; in this case, it is four-carbon oxaloacetate. It takes two turns of the cycle to break down the two pyruvates from one glucose molecule.

Track what goes on during two turns of the Krebs cycle. Each acetyl–CoA transfers its two carbon atoms to four-carbon oxaloacetate. The outcome is citrate, a form of citric acid (Figure 7.6*b*). The Krebs cycle is also called the citric acid cycle after this first intermediate. In later steps, four carbon atoms are released; two CO_2 form in each turn of the cycle (Figure 7.6*c,d*). Add in the carbon lost during formation of each acetyl–CoA, and we see that all of the carbon atoms from the two pyruvates have now departed the cell (in six CO_2). The glucose molecule has been broken down completely. Two ATP also formed in two turns of the cycle, which adds little to the small net yield of glycolysis. But six more NADH and two $FADH_2$ also formed.

In total, ten reduced coenzymes—eight NADH and two $FADH_2$—form inside the mitochondrion. Add in the two NADH from glycolysis, and the full breakdown

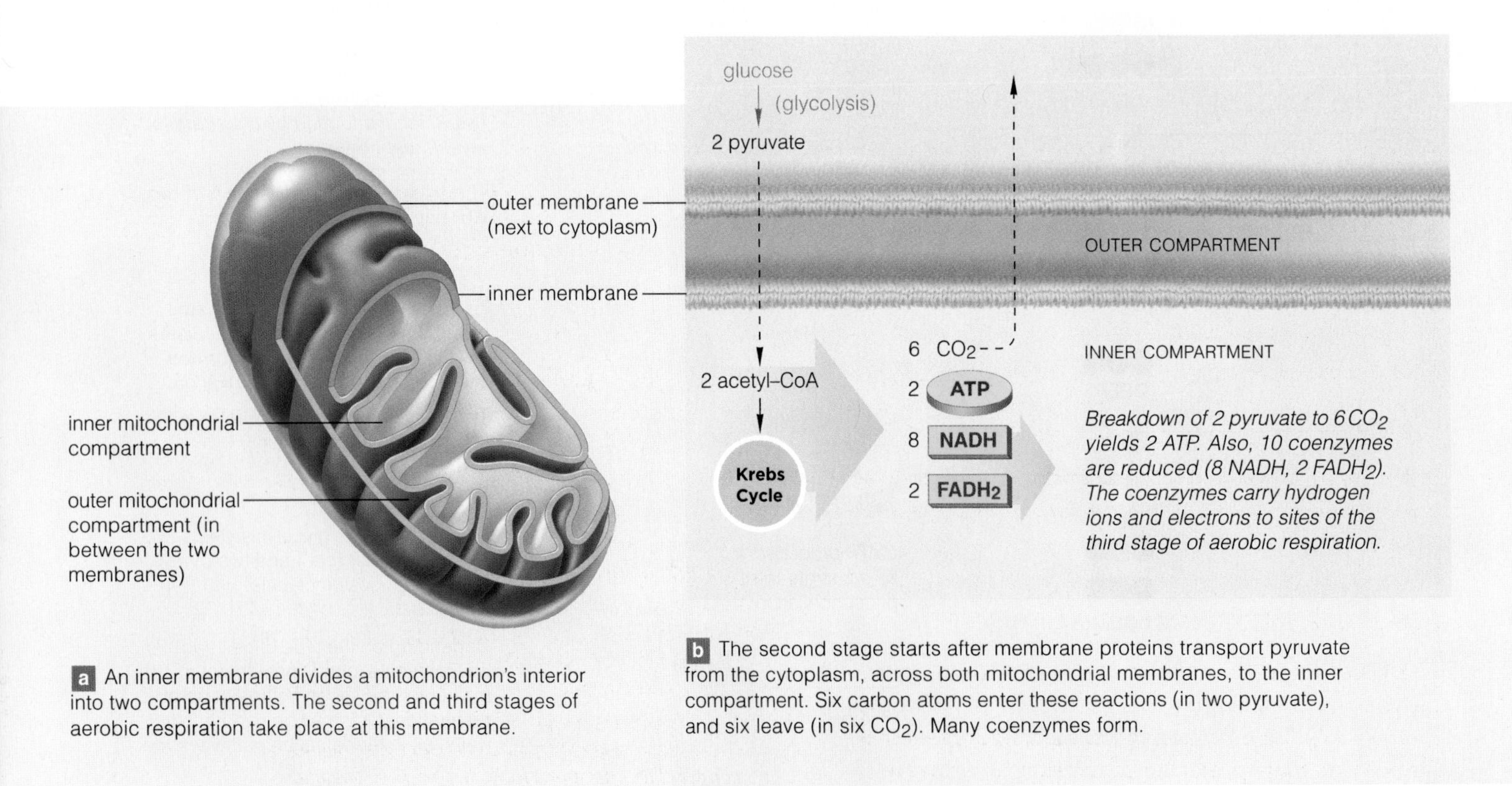

a An inner membrane divides a mitochondrion's interior into two compartments. The second and third stages of aerobic respiration take place at this membrane.

b The second stage starts after membrane proteins transport pyruvate from the cytoplasm, across both mitochondrial membranes, to the inner compartment. Six carbon atoms enter these reactions (in two pyruvate), and six leave (in six CO_2). Many coenzymes form.

Figure 7.5 Animated! Zooming in on aerobic respiration inside a mitochondrion.

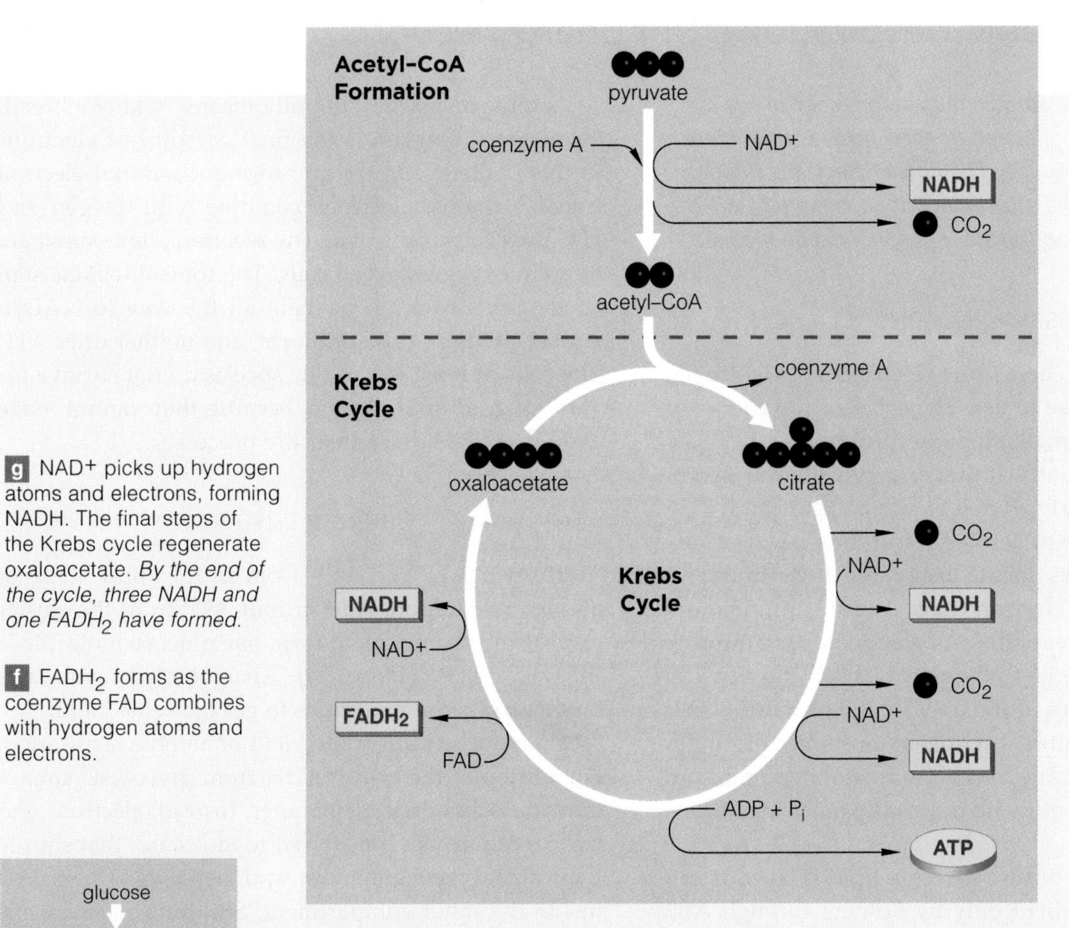

a A pyruvate molecule binds to coenzyme A. One carbon atom is released in CO_2, and two-carbon acetyl–CoA remains. NADH forms as NAD^+ picks up the released hydrogen atoms and electrons.

b The Krebs cycle starts as one carbon is transferred from acetyl–CoA to oxaloacetate. Citrate forms. Coenzyme A is regenerated.

c For simplicity, intermediates are not shown. Another carbon from an intermediate is released, in CO_2. NAD^+ picks up hydrogen atoms and electrons released by the reaction, so NADH forms.

d A carbon atom from another intermediate is released, in CO_2, and another NADH forms. *The three carbon atoms of pyruvate that entered the second-stage reactions have now been released in the form of CO_2.*

e A phosphate group becomes attached to ADP. One ATP has now formed by substrate-level phosphorylation.

f $FADH_2$ forms as the coenzyme FAD combines with hydrogen atoms and electrons.

g NAD^+ picks up hydrogen atoms and electrons, forming NADH. The final steps of the Krebs cycle regenerate oxaloacetate. *By the end of the cycle, three NADH and one $FADH_2$ have formed.*

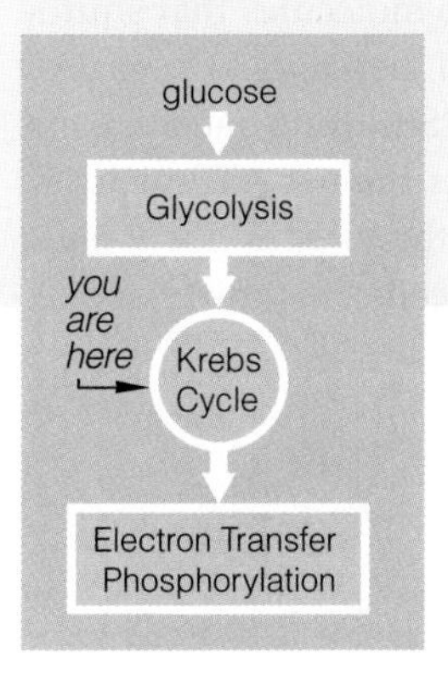

Figure 7.6 **Animated!** Aerobic respiration's second stage: formation of acetyl–CoA and the Krebs cycle. The reactions occur in the mitochondrion's inner compartment. *It takes two cycles to break down the two pyruvates from one glucose molecule.* After two cycles, all six carbons that entered glycolysis in one glucose molecule have left the cell, in six CO_2. Two ATP, eight NADH, and two $FADH_2$ form during the two cycles. See Appendix VI for details.

of glucose has a big potential payoff. *All electrons carry energy, which twelve coenzymes will now deliver to sites of the third stage of aerobic respiration.*

Figure 7.6 is a simple version of the second stage of aerobic respiration. For interested students, Appendix VI offers more details.

ABNORMAL STRUCTURE, ALTERED FUNCTION

The Krebs cycle gives us another opportunity to see the connection between protein structure and function (Section 3.5). For example, pyruvate dehydrogenase is the enzyme that speeds the conversion of pyruvate to acetyl–CoA. Rare mutations that change the structure of this enzyme impair or destroy its function. When the enzyme is mutated, citrate cannot form during the Krebs cycle. Pyruvate dehydrogenase malfunction is correlated with severe disorders, including Alzheimer's disease, Parkinson's disease, and certain cancers.

Aerobic respiration's second stage occurs in the inner compartment of mitochondria.

During the reactions, pyruvate is converted to acetyl–CoA, which enters the Krebs cycle. For two pyruvates, two ATP and ten coenzymes (eight NADH, two $FADH_2$) form. All of pyruvate's carbons depart, in the form of carbon dioxide.

In sum, the first two stages of aerobic respiration net four ATP and twelve reduced coenzymes for every molecule of glucose. The coenzymes will deliver electrons and hydrogen to the third stage of reactions.

7.4 Aerobic Respiration's Big Energy Payoff

In the third stage of aerobic respiration, coenzymes deliver electrons to electron transfer chains in the inner mitochondrial membrane. The flow of electrons through the chains drives the attachment of phosphate to ADP. That is what electron transfer phosphorylation means.

ELECTRON TRANSFER PHOSPHORYLATION

LINKS TO SECTIONS 4.3, 5.5

Electron transfer chains put electrons and hydrogen atoms from glucose to use. They release the energy of the electrons in small, stepwise increments that cells can harvest efficiently. If that energy were released in one big burst, nearly all of it would be lost.

The third stage of aerobic respiration begins when reduced coenzymes donate their cargo of electrons and hydrogen ions to electron transfer chains in the inner mitochondrial membrane. As electrons pass through the chains, they give up energy bit by bit (Section 5.5). Some molecules of the transfer chains harness that energy to shuttle hydrogen ions into the outer compartment. The ions accumulate there; thus, an H^+ gradient builds up across the inner membrane (Figure 7.7).

glucose
Glycolysis
Krebs Cycle
you are here
Electron Transfer Phosphorylation

H^+ cannot diffuse across a lipid bilayer. It can follow its gradient only by flowing through ATP synthases that span the inner membrane (Section 4.3 and Figure 7.7*c*). The flow drives attachment of phosphate to ADP, thus forming ATP.

Aerobic respiration literally means "taking a breath of oxygen." Oxygen is the final acceptor of electrons in this pathway. At the end of mitochondrial electron transfer chains, electrons combine with oxygen and H^+, thus forming water. The electrons have nowhere to go in oxygen-starved cells. The transfer chains stall as electrons back up in them all the way to NADH. An H^+ gradient does not form, and neither does ATP. The cells of most eukaryotic species do not survive for very long without oxygen, because they cannot make enough ATP to drive their life processes.

SUMMING UP: THE ENERGY HARVEST

Thirty-two ATP typically form in the third stage of aerobic respiration. Add in four ATP from the earlier stages, and the net yield from one glucose molecule is thirty-six ATP (Figure 7.8). Anaerobic pathways use eighteen glucose molecules to get the same yield.

Many factors affect the yield of aerobic respiration. For example, the two NADH from glycolysis cannot cross mitochondrial membranes. Instead, electrons and hydrogen ions are transferred to molecules that shuttle them across the membranes, and then to NAD^+ or FAD inside the inner compartment. Shuttling mechanisms differ among cells, which affects the net ATP yield. In brain and skeletal muscle cells, the yield is thirty-eight ATP. In liver, heart, and kidney cells, it is thirty-six.

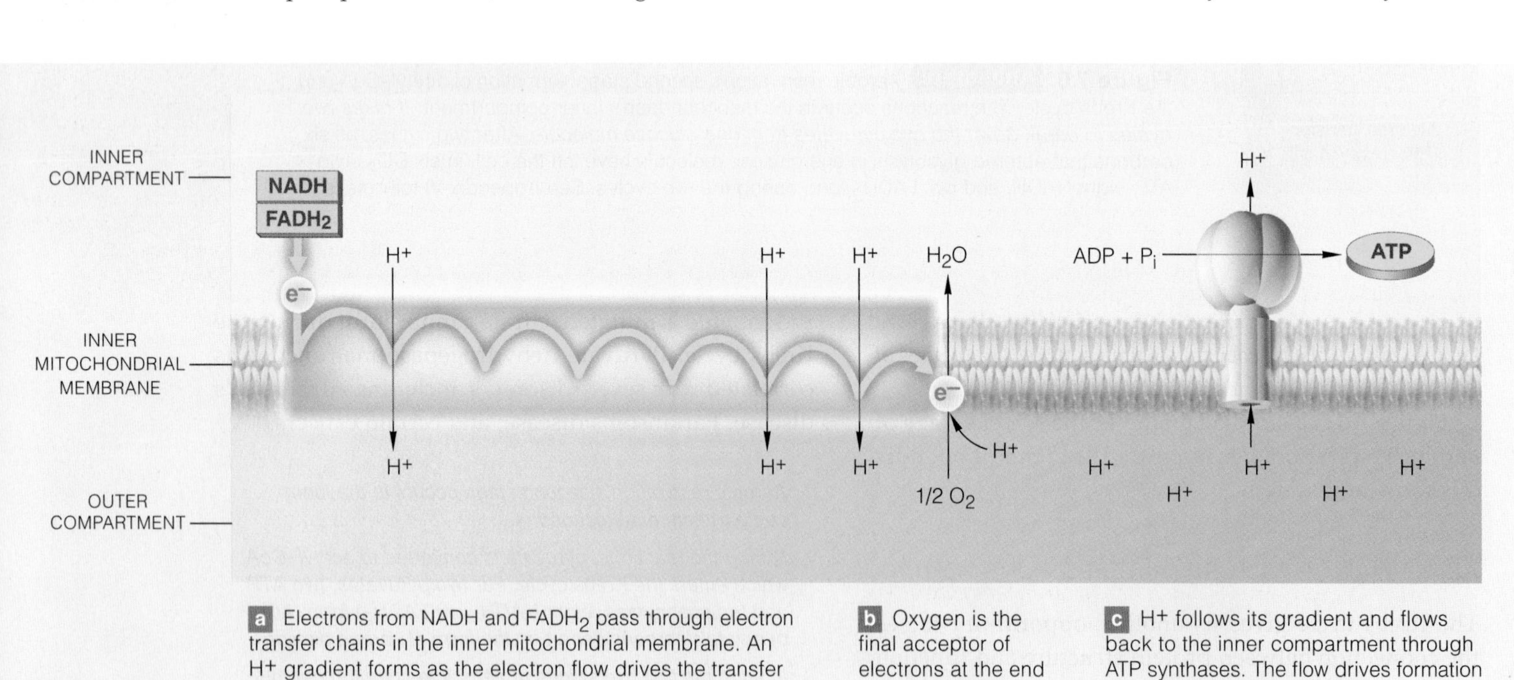

a Electrons from NADH and $FADH_2$ pass through electron transfer chains in the inner mitochondrial membrane. An H^+ gradient forms as the electron flow drives the transfer of H^+ from the inner to the outer compartment.

b Oxygen is the final acceptor of electrons at the end of the transfer chains.

c H^+ follows its gradient and flows back to the inner compartment through ATP synthases. The flow drives formation of ATP from ADP and phosphate (P_i).

Figure 7.7 Electron transfer phosphorylation, the third and final stage of aerobic respiration.

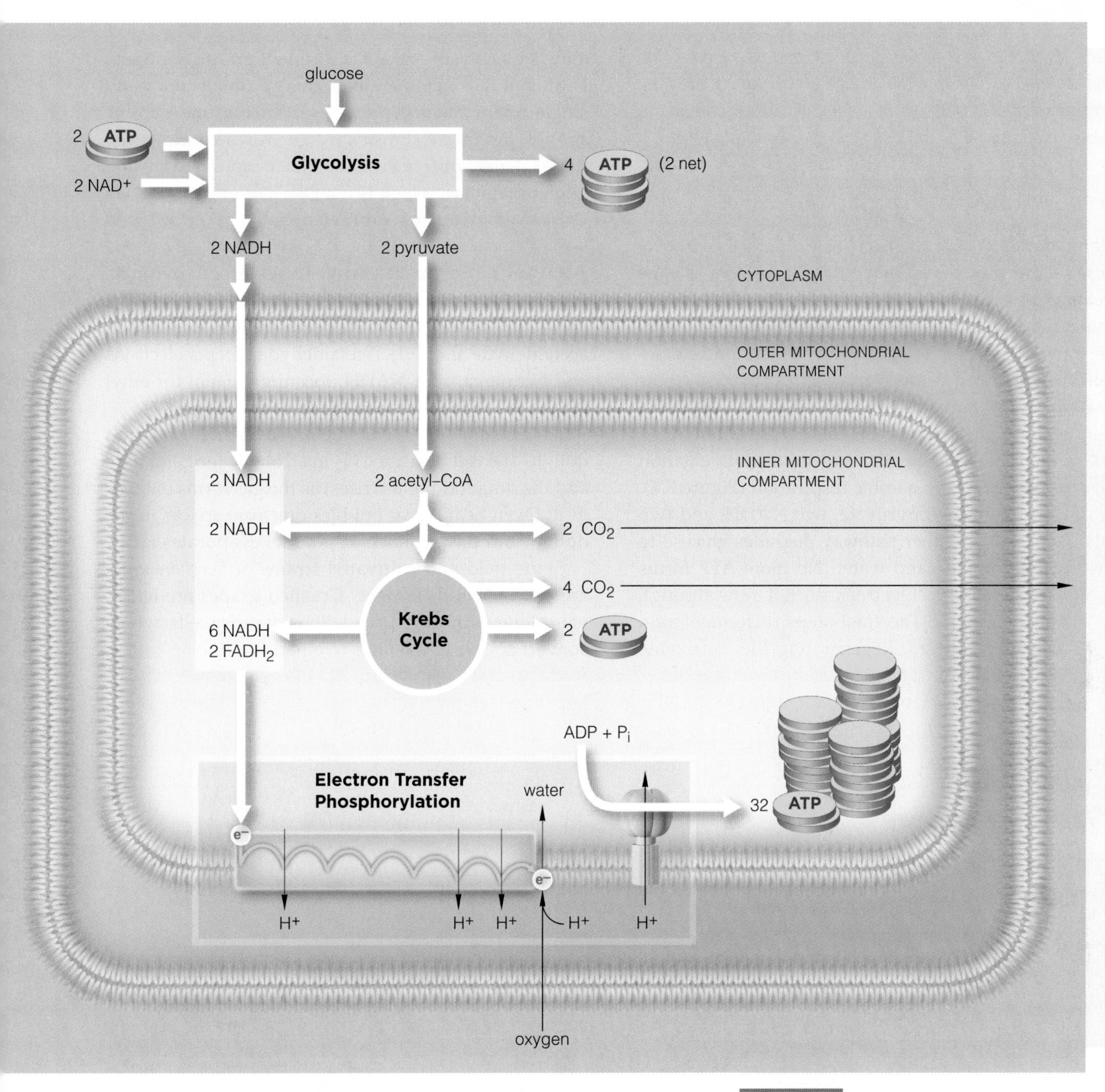

a Glucose is broken down to 2 pyruvate; 2 NADH and 4 ATP form. The net energy yield is only 2 ATP (because an up-front energy investment of 2 ATP was required).

b 12 coenzymes (10 NAD^+, 2 FAD) accept electrons and hydrogen atoms during glycolysis and the second stage reactions (including acetyl–CoA formation and the Krebs cycle). All six carbons of glucose leave the cell, as 6 CO_2. 2 ATP form.

c Coenzymes give up electrons and hydrogen ions to electron transfer chains. Operation of the chains pumps H^+ from the inner to the outer compartment. The resulting gradient causes H^+ to flow back into the inner compartment, through ATP synthases. The flow drives synthesis of ATP. Typical yield of last stage: 32 ATP per glucose.

Figure 7.8 Animated! Summary of the steps in aerobic respiration.

Remember, ATP is the energy currency in all cells. Cells can harvest its energy with a phosphate group transfer, and it moves between reaction sites. Energy from ATP drives many cell activities, such as assembly of sugars, fats, nucleotides, and proteins; movement of the cell or parts of it; and active transport.

Also remember this: With each transfer of energy in a reaction, some energy is lost (Section 5.1). Although aerobic respiration is the most efficient pathway, about 60 percent of the energy in glucose is still lost, as heat.

Aerobic respiration's third stage occurs when coenzymes deliver electrons and hydrogen ions to electron transfer chains in the inner mitochondrial membrane.

Energy released by electrons as they pass through electron transfer chains pumps hydrogen ions from the inner to the outer compartment; thus, an H^+ gradient forms.

The gradient drives H^+ flow through ATP synthases, which in turn drives ATP formation. The overall yield of aerobic respiration is typically thirty-six ATP per glucose molecule.

7.5 Anaerobic Energy-Releasing Pathways

Turn now to the ATP-producing pathways of fermentation. Like aerobic respiration, they break down sugars, starting with glycolysis. Unlike aerobic respiration, they do not use oxygen, and their final steps only regenerate NAD+.

FERMENTATION PATHWAYS

Many fermenters are single-celled bacteria and protists in sea sediments, the animal gut, canned food, sewage treatment ponds, mud, and other oxygen-free places. Some of them, such as the bacteria that cause botulism, die when exposed to oxygen. Others, including single-celled fungi called yeasts, can switch between aerobic respiration and fermentation. Animal muscle cells rely on both fermentation and aerobic respiration.

glucose
Glycolysis
you are here
Fermentation Pathway

Glycolysis is the first stage of the fermentation routes, as it is in aerobic respiration (Figure 7.4). Here again, two pyruvate, two NADH, and two ATP form. Neither pathway degrades glucose to carbon dioxide and water. No more ATP forms after glycolysis. Electrons do not flow through transfer chains. The final steps of fermentation regenerate NAD+. Regenerating this coenzyme allows glycolysis—and ATP production—to continue. Fermentation yields enough energy to sustain many single-celled anaerobic species. It helps some aerobic species, produce ATP when oxygen is scarce. However, the ATP yield of fermentation is not enough to sustain large organisms, which is why we can predict that we will never see an anaerobic elephant.

Alcoholic Fermentation Three-carbon pyruvate is broken down in alcoholic fermentation (Figures 7.9*a* and 7.10). An enzyme splits pyruvate into two-carbon acetaldehyde and CO_2. Acetaldehyde accepts electrons and hydrogen from NADH, forming ethanol (or ethyl alcohol). The last reaction step regenerates NAD+.

Bakers mix one yeast, *Saccharomyces cerevisiae,* into dough. The cells release CO_2 in alcoholic fermentation, and the dough expands (rises) as the gas forms bubbles in it. Oven heat forces bubbles out from spaces in the dough, and the alcohol end product evaporates away.

Some wild and cultivated strains of *Saccharomyces* are used to produce wine. Crushed grapes are left in vats along with large populations of yeast cells, which convert sugars in the juice to ethanol.

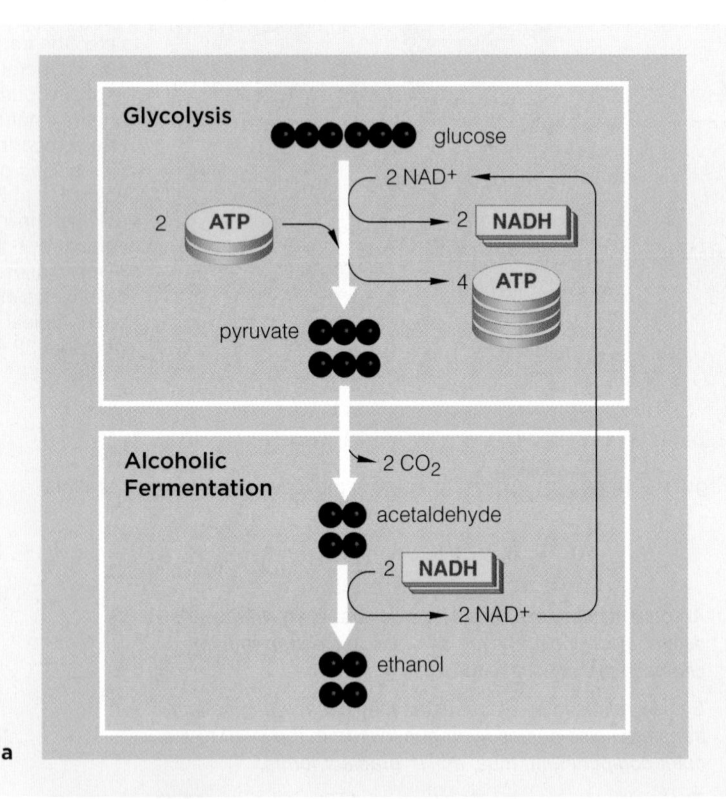

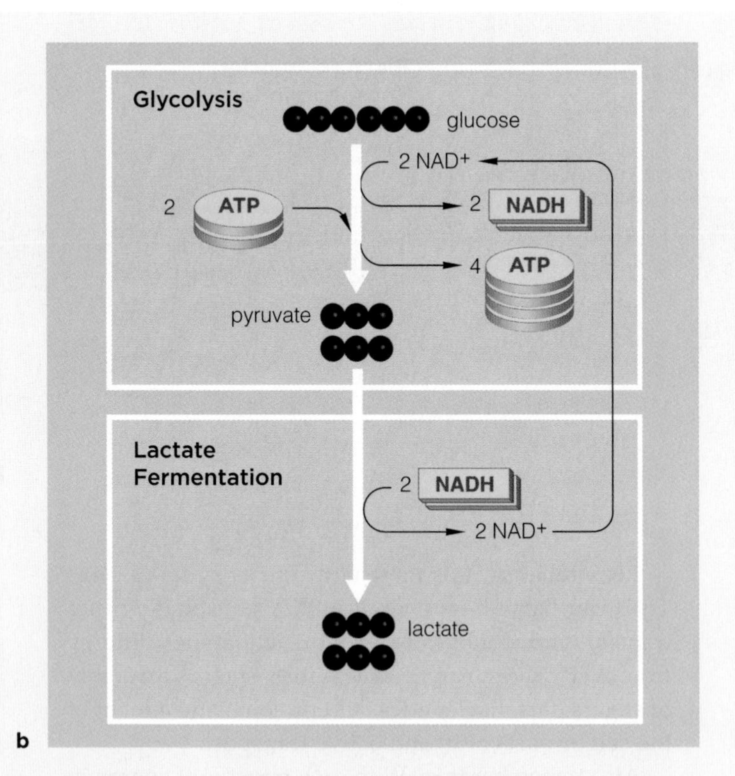

Figure 7.9 Animated! (**a**) Alcoholic fermentation. (**b**) Lactate fermentation. In both pathways, the final steps do not produce ATP. They regenerate NAD+. The net yield is two ATP per molecule of glucose (from glycolysis).

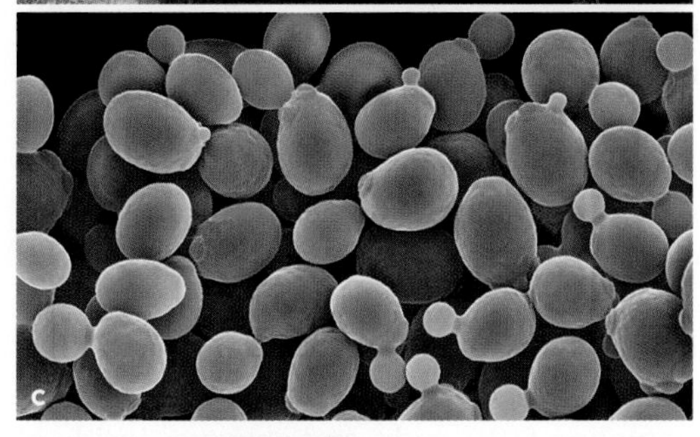

Figure 7.10 Animated! Alcoholic fermentation in action. (**a**) A vintner examines the color and clarity of one fermentation product of *Saccharomyces*. (**b**) A commercial vat of yeast dough rising with the help of live cultures of yeast cells. (**c**) Scanning electron micrograph of yeast cells. Some are reproducing by budding.

Lactate Fermentation In lactate fermentation, an enzyme transfers electrons and hydrogen from NADH to pyruvate. The reaction converts pyruvate to three-carbon lactate (lactic acid), and also regenerates NAD^+ (Figure 7.9*b*).

Some fermenters can preserve food. For instance, we use *Lactobacillus acidophilus*, which digests lactose in milk, to ferment dairy products such as buttermilk, cheese, and yogurt. Other species of yeast ferment and preserve pickles, corned beef, sauerkraut, and kimchi. Still others can spoil food (Section 22.3).

ATP can form by carbohydrate breakdown in fermentation pathways, which are anaerobic.

The end product of lactate fermentation is lactate. The end product of alcoholic fermentation is ethanol.

Both pathways have a net yield of two ATP per glucose molecule. The ATP forms during glycolysis.

Fermentation reactions regenerate the coenzyme NAD^+, without which glycolysis and ATP production would stop.

7.6 The Twitchers

FOCUS ON HEALTH

Lactate fermentation as well as aerobic respiration yields ATP for muscles that are partnered with bones.

Skeletal muscles, which move bones, consist of cells fused as long fibers. The fibers differ in how they make ATP.

LINK TO SECTION 3.5

Slow-twitch muscle fibers have many mitochondria and produce ATP by aerobic respiration. They dominate during prolonged activity, such as long runs. Slow-twitch fibers are red because they have an abundance of myoglobin, a pigment related to hemoglobin (Section 3.5). Myoglobin stores oxygen in muscle tissue.

Fast-twitch muscle fibers have few mitochondria and no myoglobin; they are pale. The ATP they make by lactate fermentation sustains only short bursts of activity, such as sprints or weight lifting (Figure 7.11). The pathway makes ATP quickly but not for long; it cannot support sustained activity. That is one reason you will never see migrating chickens. The flight muscles of a chicken are mostly fast-twitch fibers, which make up the "white" breast meat. Chickens fly only in short bursts, which they do to escape predators. More often, a chicken walks or runs. Its leg muscles are mostly slow-twitch muscle, the "dark meat."

Would you expect to find more light or dark breast muscles in a migratory duck? An ostrich? An albatross that can skim the ocean surface for months?

Most human muscles are a mix of fast-twitch and slow-twitch fibers, but the proportions vary among muscles and among individuals. Great sprinters tend to have more fast-twitch fibers. Great marathon runners tend to have more slow-twitch fibers. Section 33.5 offers a closer look at energy-releasing pathways in skeletal muscle.

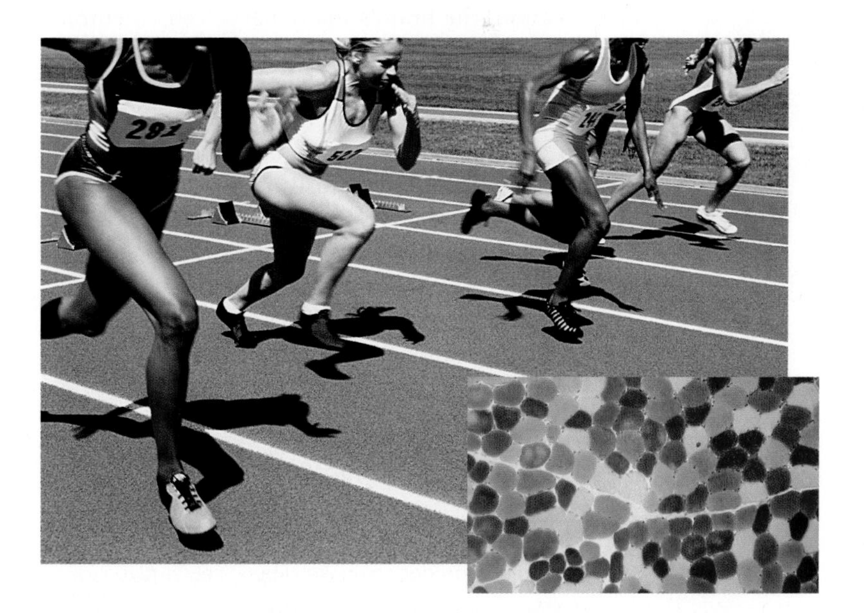

Figure 7.11 Sprinters and lactate fermentation. The micrograph, a cross-section through a human muscle, reveals three types of fibers. Light and medium colored fibers sustain short, intense bursts of speed; they make ATP by lactate fermentation. Dark fibers contribute to endurance; they make ATP by aerobic respiration.

7.7 Alternative Energy Sources in the Body

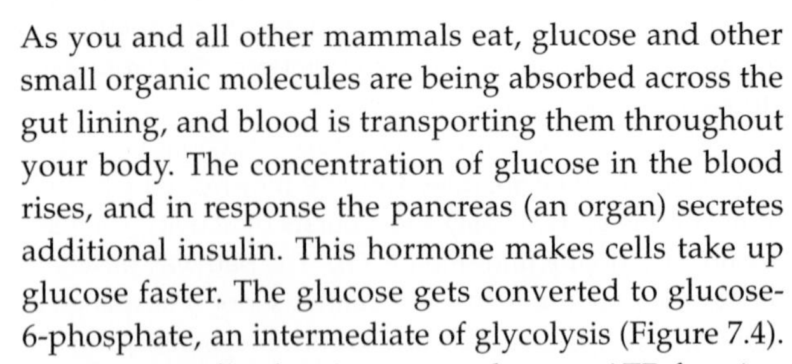

So far, you have looked at what happens when glucose is broken down in the main pathways of ATP formation. However, many other degradative pathways also help keep organisms alive.

THE FATE OF GLUCOSE AT MEALTIME AND BETWEEN MEALS

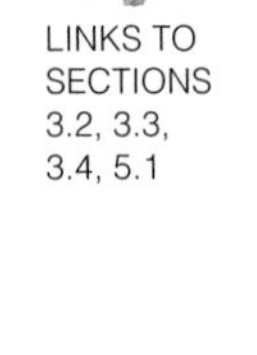

LINKS TO SECTIONS 3.2, 3.3, 3.4, 5.1

As you and all other mammals eat, glucose and other small organic molecules are being absorbed across the gut lining, and blood is transporting them throughout your body. The concentration of glucose in the blood rises, and in response the pancreas (an organ) secretes additional insulin. This hormone makes cells take up glucose faster. The glucose gets converted to glucose-6-phosphate, an intermediate of glycolysis (Figure 7.4).

When a cell takes in excess glucose, ATP-forming machinery goes into high gear. Unless the ATP is used quickly, its concentration rises in the cytoplasm. This causes glucose-6-phosphate to be diverted away from glycolysis and into a biosynthesis pathway. Glycogen, a polysaccharide, forms in this pathway (Section 3.2). Liver and muscle cells especially favor the conversion of glucose to glycogen. These cells maintain the body's largest stores of glycogen (Table 7.1).

Between meals, the blood level of glucose declines. If the decline were not countered, that would be bad news for the brain, your body's glucose hog. At any time, your brain is taking up more than two-thirds of the freely circulating glucose. Why? Except in times of starvation, the brain's many nerve cells (neurons) use only this sugar. They cannot store it.

The pancreas responds to low glucose levels in the blood by secreting glucagon. The hormone causes liver cells to convert stored glycogen to glucose, which is released from cells and diffuses into blood. The level of blood glucose rises, so brain cells keep on working. Thus, *hormones control whether your cells use glucose as an energy source or tuck it away.*

Don't let this explanation lead you to believe your cells store huge amounts of glycogen. Glycogen makes up about 1 percent of an average adult's total energy reserves, an energy equivalent of two cups of cooked pasta. Unless you eat regularly, you will deplete your liver's glycogen stores in less than twelve hours.

Of the total energy reserves in, say, a typical adult who eats well, 78 percent (about 10,000 kilocalories) is concentrated in body fat and 21 percent in proteins.

Table 7.1 Disposition of Organic Compounds

During meals	Excess glucose converted to glycogen or fat
Between meals	Glycogen degraded, glucose subunits enter glycolysis
	Fats degraded to fatty acids; some fragments enter glycolysis, others converted to acetyl–CoA
	Proteins degraded to amino acids, fragments become intermediates in Krebs cycle

ENERGY FROM FATS

How does a human body access its fat reservoir? A fat molecule, recall, has a glycerol head and one, two, or three fatty acid tails (Section 3.3). The body stores most fats as triglycerides, which have three fatty acid tails. Triglycerides accumulate in fat cells of adipose tissue. This tissue is an energy reservoir. It also insulates and pads the buttocks and other body regions.

When the blood glucose level falls, triglycerides are tapped as an energy alternative. Enzymes in fat cells cleave bonds between glycerol and fatty acids, which both enter the blood. Enzymes in the liver convert the glycerol to PGAL, which, remember, is an intermediate of glycolysis (Figure 7.4). Nearly all cells of your body take up circulating fatty acids, and enzymes inside of them cleave the fatty acid backbones. The fragments become converted to acetyl–CoA, which can enter the Krebs cycle (Figures 7.6 and 7.12).

Compared to carbohydrate breakdown, fatty acid breakdown yields more ATP for each atom of carbon. Between meals or during steady, prolonged exercise, fatty acid breakdown supplies about half of the ATP that muscle, liver, and kidney cells require.

What happens if you eat too many carbohydrates? Aerobic respiration converts the glucose subunits to pyruvate, then to acetyl–CoA, which enters the Krebs cycle. When too much glucose is circulating through the body, acetyl–CoA is diverted to a pathway that synthesizes fatty acids. *Too much glucose ends up as fat.*

ENERGY FROM PROTEINS

Some enzymes in your digestive system split dietary proteins into their amino acid subunits, which are then absorbed into the bloodstream. Cells use amino acids to build proteins or other molecules. Even so, when you eat more protein than your body needs, the amino acids are broken down further. Their NH_3^+ group is removed, and it becomes ammonia (NH_3). Depending on the amino acid, their carbon backbone is split, and acetyl–CoA, pyruvate, or an intermediate of the Krebs cycle forms. Your cells can divert any of these organic molecules into the Krebs cycle (Figure 7.12).

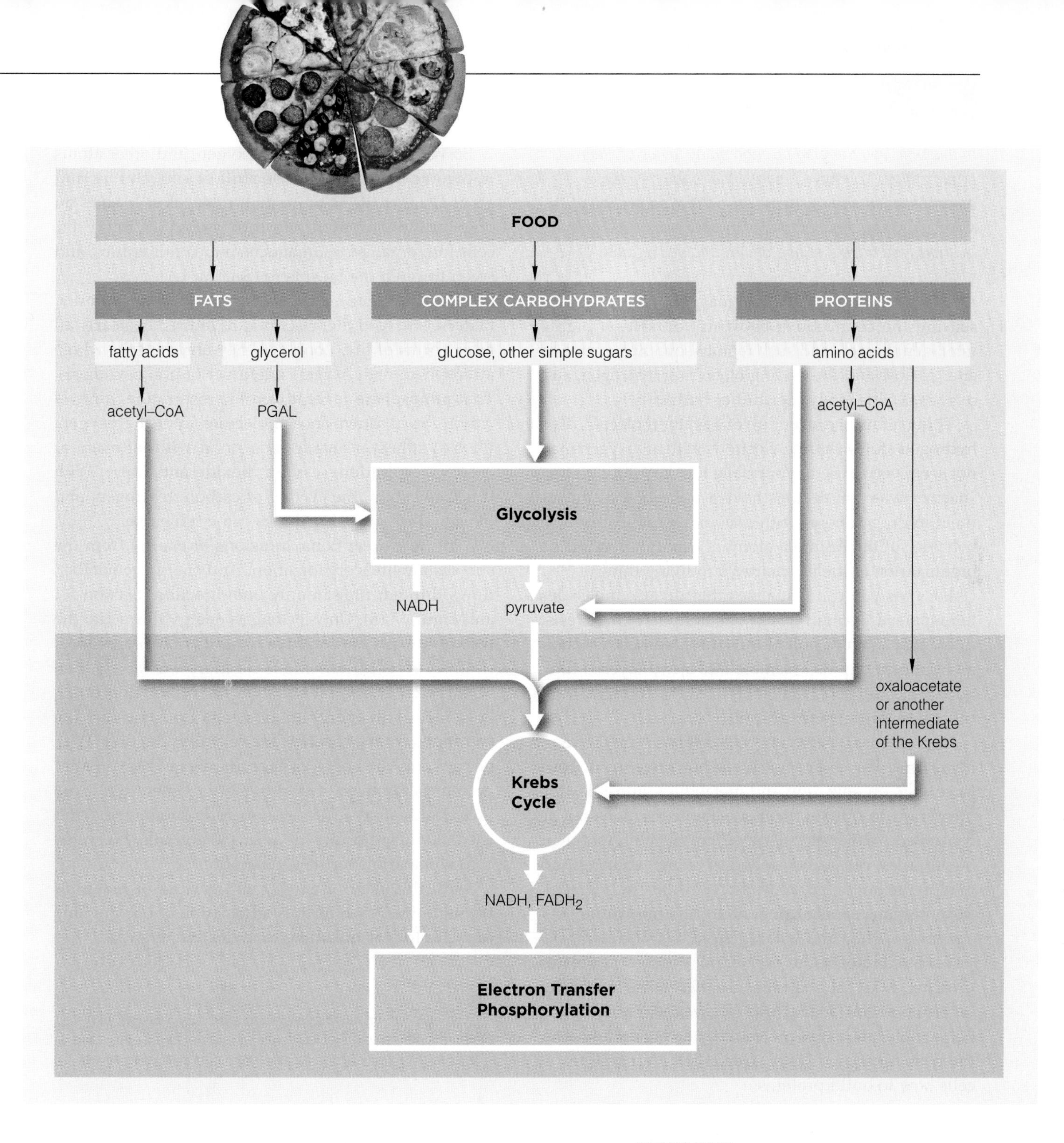

As you can see, maintaining and accessing energy reserves is complicated business. Controlling the use of glucose is special because it is the fuel of choice for the brain. However, providing all of your cells with energy starts with the kinds of food you eat.

In humans and other mammals, the entrance of glucose or other organic compounds into an energy-releasing pathway depends on the kinds and proportions of carbohydrates, fats, and proteins in the diet.

Figure 7.12 **Animated!** Reaction sites where a variety of organic compounds enter the reactions of aerobic respiration. Such compounds are alternative energy sources in the human body.

In humans and other mammals, complex carbohydrates, fats, and proteins from food do not enter the pathway of aerobic respiration directly. First, the digestive system, and then individual cells, break apart all of these molecules into simpler subunits. We return to this topic in Chapter 37.

7.8 Reflections on Life's Unity

CONNECTIONS

In this unit, you traveled through many levels of life's organization. You have a sense that each new life emerges when energy inputs drive the organization of molecules, and their interactions, into units called cells. In short, you have a sense of life's molecular unity.

LINKS TO SECTIONS 1.1, 2.5, 3.3, 4.3, 5.1, 5.6, 6.3

At this point in the book, you may still have difficulty sensing the connections between yourself—a highly intelligent being—and such remote-sounding events as energy flow and the cycling of carbon, hydrogen, and oxygen. Is this really the stuff of humanity?

Think about the structure of a water molecule. Two hydrogen atoms sharing electrons with an oxygen may not seem very close to your daily life. Yet, through that sharing, water molecules have a polarity that makes them hydrogen-bond with one another. The chemical behavior of three simple atoms is a foundation for the organization of lifeless matter into living things.

For now you can visualize other diverse molecules interspersed through water. The nonpolar kinds resist interacting with it; polar kinds dissolve in it. On their own, phospholipids assemble into a two-layered film. Such lipid bilayers, remember, are the framework of cell membranes, hence all cells.

From the very beginning, the cell has been the basic *living* unit. The essence of life is not some mysterious force. It is organization and metabolic control. With a membrane to contain them, metabolic reactions can be controlled. With molecular mechanisms built into their membranes, cells can respond to energy changes and to shifts in solute concentrations in the environment. Response mechanisms operate by "telling" proteins—enzymes—when and what to build or tear down.

And it is not some mysterious force that creates proteins. DNA, the double-stranded encyclopedia of inheritance, has a structure—*a chemical message*—that helps molecules copy molecules, one generation after the next. Your own DNA strands tell your trillions of cells how to build proteins.

So yes, carbon, hydrogen, oxygen, and other atoms of organic molecules are the stuff of you, and us, and all of life. Yet life is more than molecules. It takes an ongoing flow of energy to turn molecules into cells, cells into organisms, organisms into communities, and so on through the biosphere (Section 1.1).

Photosynthesizers use energy from the sun and raw materials to feed themselves and, indirectly, nearly all other forms of life. Long ago they enriched the whole atmosphere with oxygen, a leftover of photosynthesis. That atmosphere favored aerobic respiration, a novel way to break down food molecules by using oxygen. Photosynthesizers made more food with leftovers of aerobic respiration—carbon dioxide and water. With this connection, the cycling of carbon, hydrogen, and oxygen through living things came full circle.

With few exceptions, infusions of energy from the sun sustain life's organization. And energy, remember, flows through time in only one direction (Section 5.1 and Figure 7.13). Only as long as energy flows into the web of life can life continue in all its rich expressions.

In short, each new life is no more *and no less* than a marvelously complex system for prolonging order. Sustained with energy transfusions from the sun, life continues by its capacity for self-reproduction. With energy and the codes of inheritance in DNA, matter becomes organized, generation after generation. Even as individuals die, life elsewhere is prolonged. With each death, molecules are released and may be cycled as raw materials for new generations.

With this flow of energy and cycling of materials through time, each birth is affirmation of our ongoing capacity for organization, each death a renewal.

Through biology, we have gained a profound insight into nature: The diversity of life, and its continuity through time, arises from unity at the bioenergetic and molecular levels.

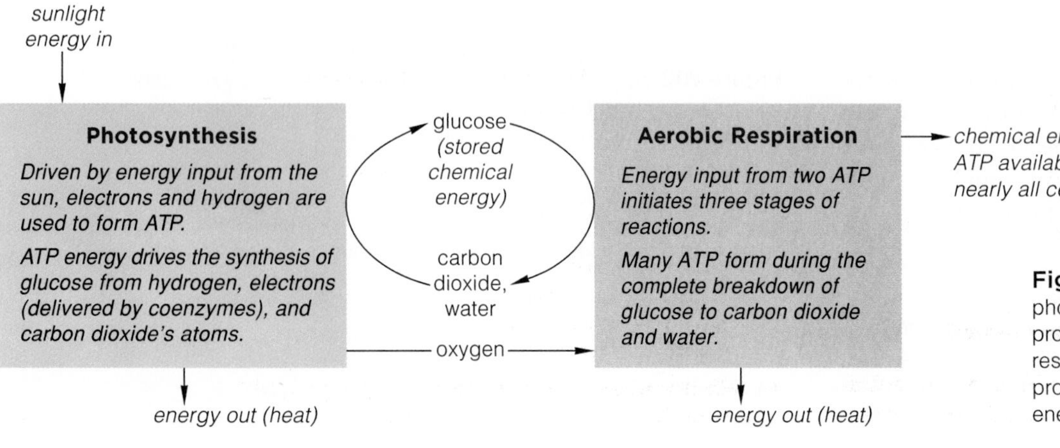

Figure 7.13 Summary of links between photosynthesis—the main energy-requiring process in the biosphere—and aerobic respiration, the main energy-releasing process. Notice the one-way flow of energy and the cycling of materials.

www.thomsonedu.com Thomson NOW!

Summary

Section 7.1 All organisms, including photosynthetic types, convert the chemical energy of carbohydrates and other organic compounds to the chemical energy of ATP. ATP is a common energy currency that drives thousands of metabolic reactions in cells.

Different pathways of carbohydrate breakdown start in the cytoplasm with glycolysis, a series of reactions that convert glucose and other sugars to pyruvate.

Fermentation pathways end in the cytoplasm, do not use oxygen, and yield two ATP per molecule of glucose. Aerobic respiration ends in mitochondria, uses oxygen, and yields much more ATP than fermentation—typically thirty-six ATP per glucose molecule.

■ *Use the animation on ThomsonNOW for an overview of aerobic respiration.*

Section 7.2 Enzymes of glycolysis use two ATP to convert one molecule of glucose or another six-carbon sugar to two molecules of three-carbon pyruvate. In the reactions, electrons and hydrogen atoms are transferred to two NAD^+, which are thereby reduced to NADH. Four ATP also form, by substrate-level phosphorylation.

The net yield of glycolysis is two pyruvate, two ATP, and two NADH per glucose molecule. The pyruvate may continue in fermentation pathways in the cytoplasm, or it may enter mitochondria and be broken down further in the next steps of aerobic respiration.

■ *Use the animation on ThomsonNOW for a step-by-step journey through glycolysis.*

Section 7.3 The second stage of aerobic respiration takes place in the inner compartment of mitochondria. It starts with acetyl–CoA formation and proceeds through the Krebs cycle.

The first steps convert two pyruvates from glycolysis to 2 acetyl–CoA. Two CO_2 leave the cell, and the acetyl–CoA enters the Krebs cycle. Each time the cycle runs, one acetyl–CoA is converted to two molecules of CO_2. After two cycles, two pyruvates are dismantled; the glucose molecule that entered glycolysis is fully broken down.

During these reactions, electrons and hydrogen atoms are transferred to NAD^+ and FAD, which are reduced to NADH and $FADH_2$. ATP also forms by substrate-level phosphorylation, the direct transfer of a phosphate group from a reaction intermediate to ADP.

In total, the second stage of aerobic respiration results in the formation of six CO_2, two ATP, eight NADH, and two $FADH_2$ for every two pyruvates. Adding the yield from glycolysis, the total tally for the first two stages of aerobic respiration is twelve reduced coenzymes and four ATP for each glucose molecule. The coenzymes deliver electrons and hydrogen to the third stage of reactions.

■ *Use the animation on ThomsonNOW to explore a mitochondrion and observe the reactions inside it.*

Section 7.4 Aerobic respiration ends in mitochondria. In the third stage of reactions, coenzymes deliver electrons and hydrogen ions to electron transfer chains in the inner mitochondrial membrane. Energy released by electrons flowing through the transfer chains moves H^+ from the inner to the outer compartment.

Hydrogen ions accumulate in the outer compartment, and a gradient forms across the inner membrane. The ions follow their gradient back to the inner compartment only through ATP synthases. Hydrogen ion flow through these transport proteins drives the synthesis of ATP.

Oxygen combines with electrons and H^+ at the end of the transfer chains, thus forming water.

Overall, aerobic respiration typically yields thirty-six ATP for each glucose molecule.

■ *Use the animation on ThomsonNOW to see how each step in aerobic respiration contributes to a big energy harvest.*

Sections 7.5, 7.6 Different fermentation pathways begin with glycolysis and end in the cytoplasm. They do not use oxygen or electron transfer chains. The final steps do not produce ATP. They only regenerate the oxidized NAD^+ required for glycolysis to continue.

The end product of lactate fermentation is lactate. The end product of alcoholic fermentation is ethyl alcohol, or ethanol. Both pathways have a net yield of two ATP per glucose (from glycolysis).

Slow-twitch and fast-twitch skeletal muscle fibers can support different activity levels. Aerobic respiration and lactate fermentation proceed in different fibers that make up these muscles.

■ *Use the animation on ThomsonNOW to compare alcoholic and lactate fermentation.*

Section 7.7 In humans and other mammals, simple sugars from carbohydrates, glycerol and fatty acids from fats, and carbon backbones of amino acids from proteins may enter aerobic respiration at various steps.

■ *Use the interaction on ThomsonNOW to follow the breakdown of different organic molecules.*

Section 7.8 Photosynthesis and aerobic respiration are interconnected on a global scale. In its organization, diversity, and continuity through generations, life shows unity at the bioenergetic and molecular levels.

Self-Quiz

Answers in Appendix III

1. Is the following statement true or false? Unlike animals, which make many ATP by aerobic respiration, plants make all of their ATP by photosynthesis.

2. Glycolysis starts and ends in the ______ .
   a. nucleus
   b. mitochondrion
   c. plasma membrane
   d. cytoplasm

3. Which of the following metabolic pathways require oxygen?
   a. aerobic respiration
   b. lactate fermentation
   c. alcoholic fermentation
   d. all of the above

4. Which molecule does not form during glycolysis?
   a. NADH b. pyruvate c. $FADH_2$ d. ATP

5. In eukaryotes, aerobic respiration is completed in the ______ .
   a. nucleus c. plasma membrane
   b. mitochondrion d. cytoplasm

6. The following reactions are part of the second stage of aerobic respiration:
   a. substrate-level phosphorylation c. Krebs cycle
   b. acetyl–CoA formation d. all of the above

7. After the Krebs reactions run through ______ cycle(s), one glucose molecule has been completely oxidized.
   a. one b. two c. three d. six

8. In the third stage of aerobic respiration, ______ is the final acceptor of electrons from glucose.
   a. water b. hydrogen c. oxygen d. NADH

9. In alcoholic fermentation, ______ is the final acceptor of electrons stripped from glucose.
   a. oxygen c. acetaldehyde
   b. pyruvate d. sulfate

10. Fermentation makes no more ATP beyond the small yield from glycolysis. The remaining reactions ______ .
   a. regenerate FAD c. regenerate NADH
   b. regenerate NAD^+ d. regenerate $FADH_2$

11. Your body cells can use ______ as an alternative energy source when glucose is in short supply.
   a. fatty acids c. amino acids
   b. glycerol d. all of the above

12. Fill in the blanks in the diagram below.

13. Match the event with its most suitable description.
   ___ glycolysis a. ATP, NADH, $FADH_2$, CO_2, and water form
   ___ fermentation b. glucose to two pyruvates
   ___ Krebs cycle c. NAD^+ regenerated, little ATP
   ___ electron transfer phosphorylation d. H^+ flows via ATP synthases

■ *Visit ThomsonNOW for additional questions.*

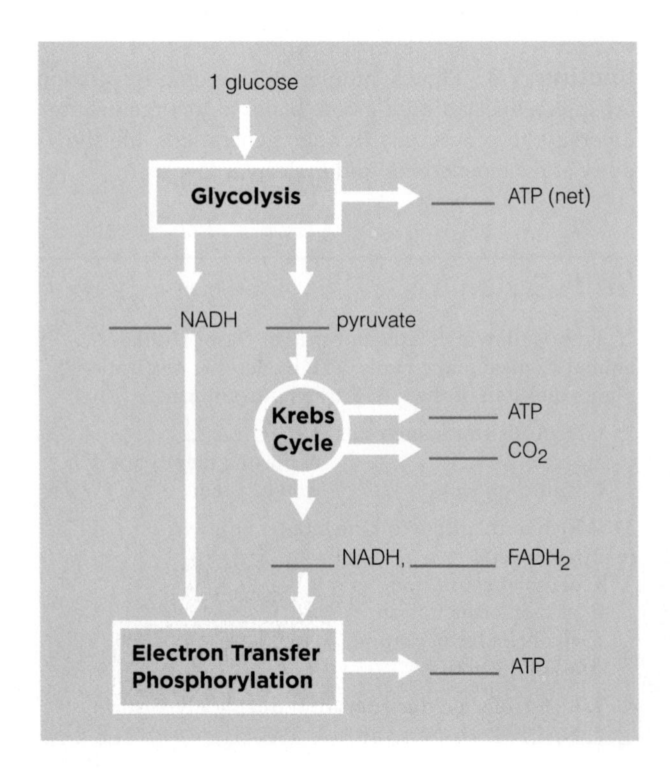

Critical Thinking

1. Cells of your body never use their nucleic acids as an energy source. Why not?

2. Suppose you start a body-building program. A qualified nutritionist recommends that you start a protein-rich diet that includes protein supplements. Speculate on how extra dietary proteins will be put to use, and in which tissues.

3. Each year, Canada geese lift off from their northern breeding grounds. They head south to spend the winter in warmer climates, then return in spring. As is the case for other migratory birds, their flight muscle cells efficiently use fatty acids as an energy source. Remember, the carbon backbone of fatty acids can be cleaved into small fragments, which are converted to acetyl–CoA for the Krebs cycle.

Suppose a goose has been steadily flapping along for about 3,000 kilometers. It looks down and notices a coyote chasing a rabbit. With a stunning burst of speed, the rabbit reaches the safety of its burrow.

Which energy-releasing pathway predominated in muscle cells in the rabbit's legs? Which pathway was the goose relying on for most of its journey? Why wouldn't the pathway used by goose flight muscle cells be helpful to a rabbit making a mad dash from a hungry coyote?

4. At high altitudes, oxygen levels are low. Mountain climbers risk altitude sickness, which is characterized by shortness of breath, weakness, dizziness, and confusion.

Oddly, early symptoms of cyanide poisoning resemble altitude sickness. Cyanide, a highly toxic poison, binds tightly to cytochrome c oxidase, a protein complex that is the last component of mitochondrial electron transfer chains. Cytochrome c oxidase with bound cyanide can no longer transfer electrons. Explain why cyanide poisoning starts with the same symptoms as altitude sickness.

5. As you learned, membranes impermeable to hydrogen ions are required for electron transfer phosphorylation. Membranes in mitochondria serve this function in eukaryotes. Prokaryotes do not have this (or any other) organelle, but they can make ATP by electron transfer phosphorylation. How do you think they do it, given that they have no mitochondria to help them generate H^+ gradients?

6. ATP forms in mitochondria. In warm-blooded animals, so does a lot of heat, which is circulated in ways that help control body temperature. Cells of brown adipose tissue make a protein that disrupts the formation of electron transfer chains in mitochondrial membranes. H^+ gradients are affected, so fewer ATP form; electrons in the transfer chains give up more of their energy as heat. Because of this, some researchers hypothesize that brown adipose tissue may not be like white adipose tissue, which is an energy (fat) reservoir. Brown adipose tissue may function in thermogenesis, or heat production.

Mitochondria, recall, contain their own DNA, which may have mutated independently in human populations that evolved in the Arctic and in the hot tropics. If that is so, then mitochondrial function may be adapted to climate.

How do you suppose such a mitochondrial adaptation might affect people living where the temperature range no longer correlates with their ancestral heritage? Would you expect people whose ancestors evolved in the Arctic to be more or less likely to put on a lot of weight than those whose ancestors lived in the tropics? See *Science*, January 9, 2004: 223–226 for more information.

II Principles of Inheritance

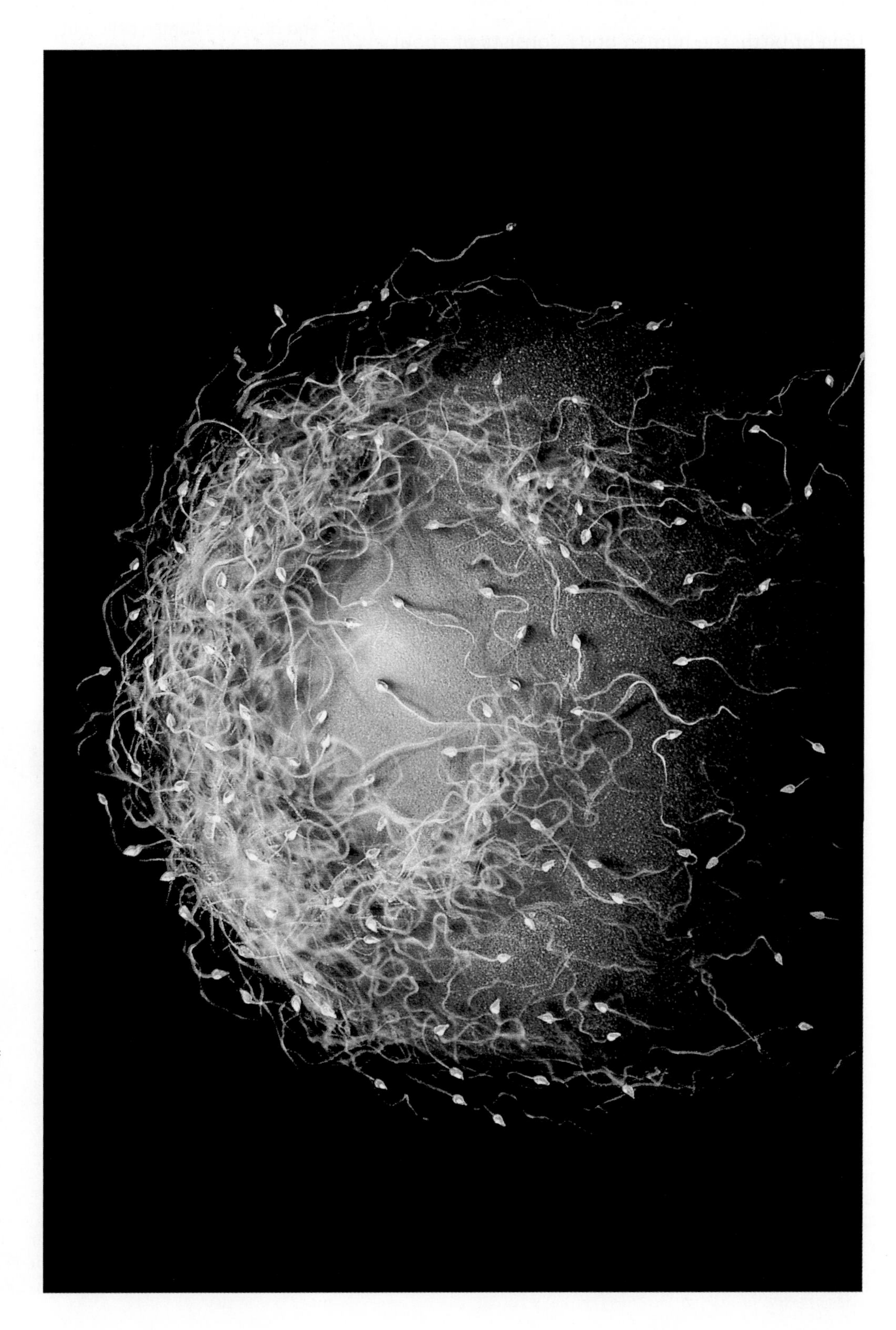

Human sperm, one of which will penetrate this mature egg and so set the stage for the development of a new individual in the image of its parents. This exquisite art is based on a scanning electron micrograph.

8 HOW CELLS REPRODUCE

IMPACTS, ISSUES Henrietta's Immortal Cells

Each human starts out as a fertilized egg. By the time of birth, the human body consists of about a trillion cells, all descended from that single cell. Even in an adult, billions of cells divide every day and replace their damaged or aged predecessors. At one time, researchers did not know how to keep human cells dividing outside of the body. They kept trying, because they knew "immortal" cell lineages could help them study cancer and other diseases, as well as normal life processes, without experimenting on patients and risking their already vulnerable lives. At Johns Hopkins University, George and Margaret Gey were among those persistent researchers.

For almost thirty years, the Geys tried to grow normal and diseased human cells, but they could not stop the cellular descendants from dying within a few weeks. Mary Kubicek, their lab assistant, tried again and again to establish a lineage of cultured human cancer cells. She was about to give up, but she prepared one last sample. She named them *HeLa* cells, after the first and last names of the patient from whom the cells were taken.

The HeLa cells began to divide. They divided again and again. Four days later, there were so many cells that the researchers had to transfer part of the population to more culture tubes. The cell populations increased at a phenomenal rate; cells were dividing every twenty-four hours and coating the inside of the tubes within days.

Sadly, cancer cells in the patient were dividing just as fast. Six months after she had been diagnosed with cancer, malignant cells had invaded tissues throughout her body. Two months after that, Henrietta Lacks, a young woman from Baltimore, was dead.

Although Henrietta passed away, her cells lived on in the Geys' laboratory (Figure 8.1). In time, HeLa cells were

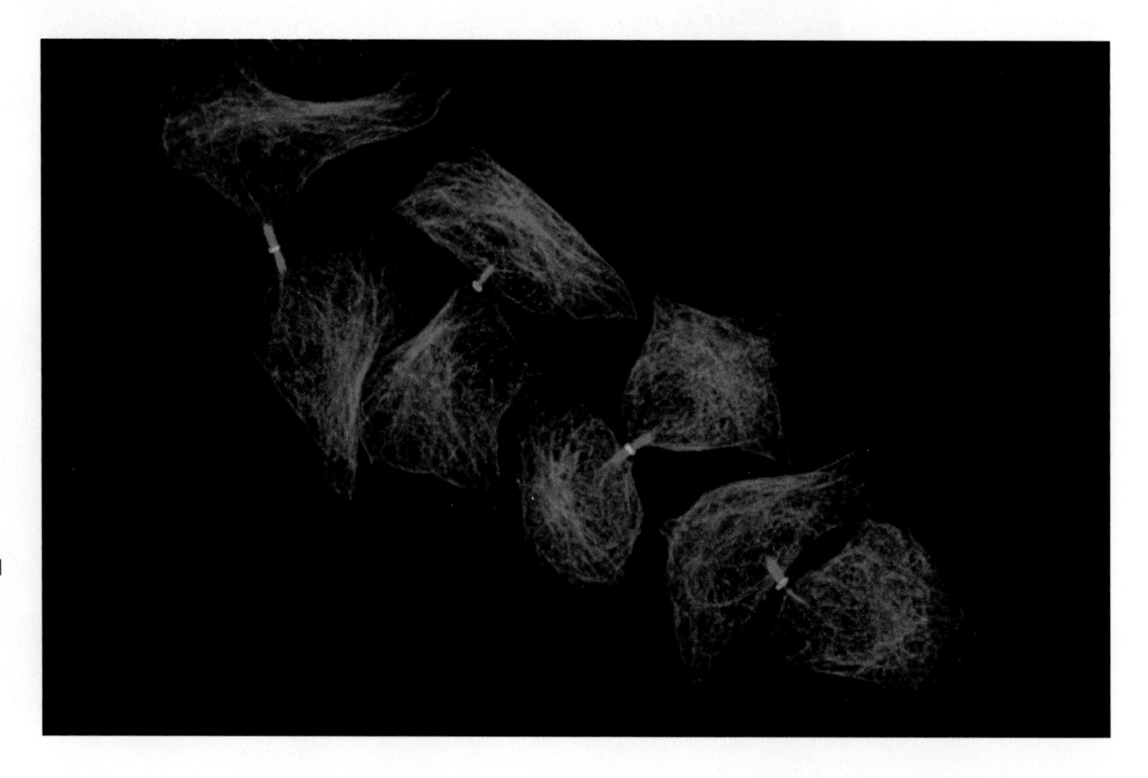

See the video! Figure 8.1 The terrifying beauty of dividing HeLa cells—the cellular legacy of Henrietta Lacks, who was a young casualty of cancer. Her contribution to science is still helping others every day.

How would you vote? No one asked Henrietta Lacks' permission to culture her cells or to use them for research. Her family did not find out about them until twenty-five years after she died. HeLa cells are still being sold worldwide by cell culture firms. Should the family of Henrietta Lacks share in the profits? See ThomsonNOW for details, then vote online.

Figure 8.2 Henrietta Lacks.

shipped to research laboratories all over the world. The Geys used HeLa cells to identify viral strains that cause polio, which at the time was epidemic. They also used the cells to test the newly developed polio vaccine. Other researchers used HeLa cells to investigate cancer, viral growth, protein synthesis, the effects of radiation on cells, and more. Some HeLa cells even traveled into space for experiments on the *Discoverer XVII* satellite. Even now, hundreds of important research projects make use of Henrietta's immortal cells.

Figure 8.2 shows a photograph of Henrietta. She was thirty-one, a wife and mother of four, when runaway cell divisions killed her. Decades later, her legacy continues to help humans all around the world, through her cells that are still dividing day after day.

Understanding cell division—and, ultimately, how new individuals are put together in the image of their parents—starts with answers to three questions. *First*, what kind of information guides inheritance? *Second*, how is that information copied inside a parent cell before being distributed to each of its daughter cells? *Third*, what kinds of mechanisms parcel out the information to daughter cells?

We will require more than one chapter to survey the nature of cell reproduction and other mechanisms of inheritance. In this chapter, we introduce the structures and mechanisms that cells use to reproduce.

Key Concepts

CHROMOSOMES AND DIVIDING CELLS

Individuals of a species have a characteristic number of chromosomes in each of their cells. The chromosomes differ in length and shape, and they carry different portions of the cell's hereditary information. Division mechanisms parcel out the information to each daughter cell, along with enough cytoplasm for that cell to start up its own operation. **Section 8.1**

WHERE MITOSIS FITS IN THE CELL CYCLE

A cell cycle starts when a daughter cell forms and ends when that cell completes its own division. A typical cycle goes through interphase, mitosis, and cytoplasmic division. In interphase, a cell increases its mass and number of components, and copies its DNA. **Section 8.2**

STAGES OF MITOSIS

Mitosis divides the nucleus, not the cytoplasm. It has four sequential stages: prophase, metaphase, anaphase, and telophase. A microtubular spindle forms. It moves the cell's duplicated chromosomes into two parcels, which end up in two genetically identical nuclei. **Section 8.3**

HOW THE CYTOPLASM DIVIDES

After nuclear division, the cytoplasm divides and typically puts a nucleus in each daughter cell. The cytoplasm of an animal cell is simply pinched in two. In plant cells, a cross-wall forms in the cytoplasm and divides it. **Section 8.4**

THE CELL CYCLE AND CANCER

Built-in mechanisms monitor and control the timing and rate of cell division. On rare occasions, the surveillance mechanisms fail, and cell division becomes uncontrollable. Tumor formation and cancer are the outcome. **Section 8.5**

Links to Earlier Concepts

Before you start reading, think back on the changing appearance of chromosomes in the nucleus of eukaryotic cells (Section 4.7). You may also wish to review the introduction to microtubules and motor proteins (4.12). Doing so will help you understand the nature of the mitotic spindle and the potential value of cancer research. A review of plant cell walls (4.10, 4.11) will help give you a sense of why plant cells do not divide by pinching their cytoplasm into two parcels, as animal cells do.

8.1 Overview of Cell Division Mechanisms

Section 1.2 introduced the idea that the continuity of life depends on reproduction. By this process, parents produce cells or multicelled individuals like themselves. Cell division is the bridge between generations.

LINKS TO SECTIONS 1.2, 4.7

A dividing cell faces challenges. Each of its daughter cells must get information encoded in the parental DNA and enough cytoplasm to start up its own operation. DNA "tells" a cell which proteins to build. Some of the proteins are structural materials; others are enzymes that speed construction of organic molecules. If a new cell does not inherit all of the information required to build proteins, it will not grow or function properly.

In addition, the parent cell's cytoplasm already has enzymes, organelles, and other metabolic machinery. When a daughter cell inherits what looks like a blob of cytoplasm, it actually is getting start-up metabolic machinery that will keep it running until it can use the information in its DNA for growing on its own.

MITOSIS, MEIOSIS, AND THE PROKARYOTES

In general, a eukaryotic cell cannot simply split in two, because only one of its daughter cells would get the nucleus—and the DNA. A cell's cytoplasm splits only after its DNA has been copied and packaged into more than one nucleus by way of mitosis or meiosis.

Mitosis is a nuclear division mechanism that occurs in somatic cells (body cells) of multicelled eukaryotes. Mitosis is the basis of increases in body size during growth, replacements of worn-out or dead cells, and tissue repair. Many plants, animals, fungi, and single-celled protists can also make copies of themselves, or reproduce asexually, by way of mitosis (Table 8.1).

Meiosis is a different nuclear division mechanism. It precedes the formation of gametes or spores, and it is the basis of sexual reproduction. In humans and other mammals, the gametes called sperm and eggs develop from immature reproductive cells. Spores, which form during the life cycle of fungi, plants, and many kinds of protists, protect and disperse new generations.

As you will discover in this chapter and the next, meiosis and mitosis have much in common. Even so, their outcomes differ.

What about prokaryotes—bacteria and archaeans? Such cells reproduce asexually by an entirely different mechanism known as prokaryotic fission. We consider prokaryotic fission later, in Section 19.1.

KEY POINTS ABOUT CHROMOSOME STRUCTURE

Remember, in eukaryotic species, genetic information is distributed among a certain number of chromosomes of different lengths and shapes (Section 4.7). Before a eukaryotic cell enters nuclear division, each of those chromosomes is one double-stranded DNA molecule. Each chromosome gets duplicated; each becomes *two* double-stranded DNA molecules. The two molecules of DNA stay attached as one chromosome until late in nuclear division (Figure 8.3). Until they separate, they are called sister chromatids.

During the early stages of mitosis and meiosis, each duplicated chromosome coils back on itself again and again, into a highly condensed form. Figure 8.4*a* has

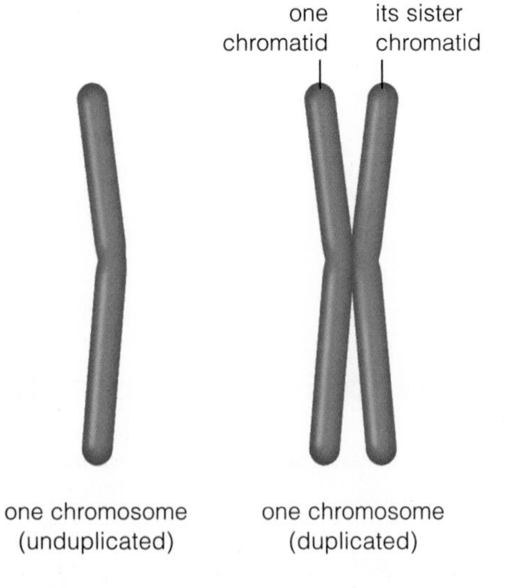

Figure 8.3 Simple way to visualize a eukaryotic chromosome in the unduplicated state and duplicated state. Eukaryotic chromosomes are duplicated before mitosis or meiosis. Each becomes two sister chromatids.

Table 8.1 Comparison of Cell Division Mechanisms

Mechanisms	Functions
Mitosis, cytoplasmic division	In *all* multicelled eukaryotes, the basis of three processes: 1. Increases in body size during growth 2. Replacement of dead or worn-out cells 3. Repair of damaged tissues In single-celled and many multicelled species, also the basis of asexual reproduction
Meiosis, cytoplasmic division	In single-celled and multicelled eukaryotes, the basis of sexual reproduction; precedes gamete formation or spore formation (Chapter 9)
Prokaryotic fission	In bacteria and archaeans alone, the basis of asexual reproduction (Section 19.1)

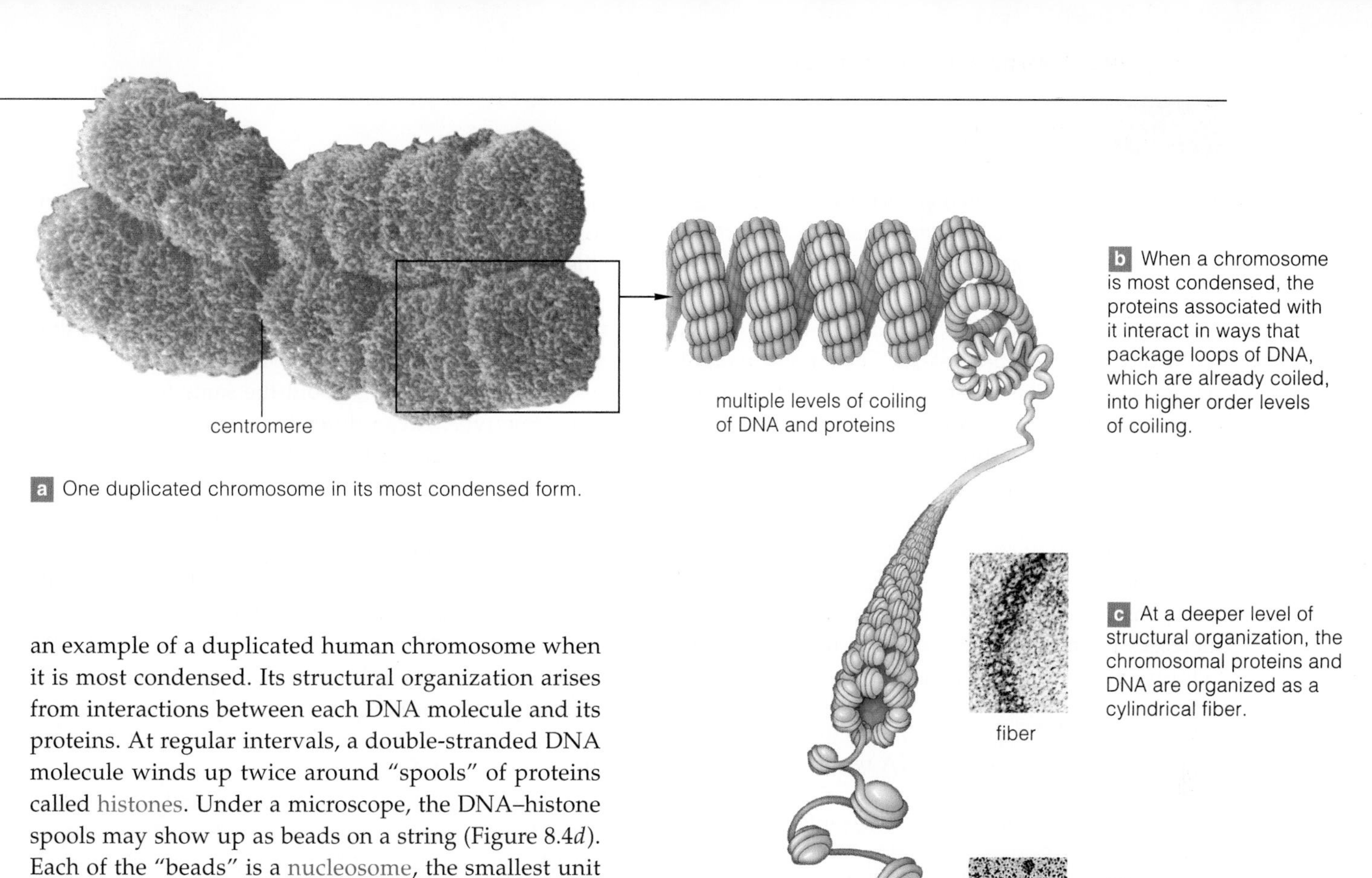

an example of a duplicated human chromosome when it is most condensed. Its structural organization arises from interactions between each DNA molecule and its proteins. At regular intervals, a double-stranded DNA molecule winds up twice around "spools" of proteins called histones. Under a microscope, the DNA–histone spools may show up as beads on a string (Figure 8.4*d*). Each of the "beads" is a nucleosome, the smallest unit of structural organization in eukaryotic chromosomes (Figure 8.4*e*).

As a duplicated chromosome condenses, its sister chromatids constrict where they attach to one another. This constricted region is called a centromere (Figure 8.4*a*). The location of a centromere differs for each type of chromosome. During nuclear division, a kinetochore forms at the centromere. Kinetochores are binding sites for microtubules that tether the chromatids.

What is the point of all this structural organization? Tight packaging probably keeps the chromosomes from tangling as they are moved and sorted out into parcels during nuclear division. Also, between cell divisions, nucleosome packaging loosens selectively, so enzymes access specific units of information in DNA at specific times. The rest of the DNA stays tightly packed, and information that is not required is not used.

When a cell divides, each daughter cell receives a required number of chromosomes and some cytoplasm. Eukaryotic cells divide their nucleus first, then the cytoplasm.

A nuclear division mechanism called mitosis is the basis of increases in body size during growth, cell replacements, and tissue repair of multicelled eukaryotes. Mitosis also is the basis of asexual reproduction in single-celled and some multicelled eukaryotes.

In eukaryotes, a nuclear division mechanism called meiosis precedes the formation of gametes and, in many species, spores. It is the basis of sexual reproduction.

Figure 8.4 **Animated!** (**a**) Scanning electron micrograph of a duplicated human chromosome in its most condensed form. (**b**,**c**) Proteins package many loops of coiled DNA into a cylindrical fiber. (**d**,**e**) The smallest unit of structural organization is the nucleosome: part of a DNA molecule looped twice around a core of histone molecules.

8.2 Introducing the Cell Cycle

The sequence of stages through which individuals of a species pass during their lifetime is called a life cycle. A cell's life passes through a different set of stages.

LINKS TO SECTIONS 4.7, 4.12

A cell cycle is a series of events from one cell division to the next (Figure 8.5). It starts when a new daughter cell forms by mitosis and cytoplasmic division. It ends when that cell divides. Mitosis, cytoplasmic division, and interphase constitute one turn of this cycle. For most cells, interphase is the longest interval.

THE WONDER OF INTERPHASE

Interphase consists of three stages during which a cell increases its mass, roughly doubles the number of its cytoplasmic components, and duplicates its DNA:

G1 Interval ("Gap") of cell growth and activity before the onset of DNA replication

S Time of "Synthesis" (DNA replication)

G2 Second interval (Gap), after DNA replication when the cell prepares for division

It is important to understand that a nucleus houses a lot of DNA. For instance, if you stretched out all the DNA from just one of your body cells, it would extend about 2 meters (6.5 feet)! The wonder is, enzymes and other proteins in cells can selectively access, activate, and silence information in all that DNA. They copy it, base by base, before cells divide. Most of this cellular work is completed during interphase.

G1, S, and G2 of interphase have distinct patterns of biosynthesis. Most of your cells remain in G1 while they are building proteins, carbohydrates, and lipids. Cells destined to divide enter S, when they copy their DNA and the proteins attached to it. During G2, they make the proteins that will drive mitosis.

The cycle's length is about the same for all cells of the same type. It can differ from one cell type to the next. For example, 2 to 3 million new red blood cells replace old ones circulating in your blood each second. The parent of a red blood cell divides every 12 hours. Cells in the tips of a bean plant root divide every 19 hours. In a sea urchin embryo, which develops rapidly from a fertilized egg, the cells divide every 2 hours. At the other end of the spectrum, the neurons (nerve cells) in most parts of your brain remain permanently in G1 of interphase; once they mature, they will never divide again. Experimentally driving them out of G1 causes them to die, not divide.

Once S begins, DNA replication usually proceeds at a predictable rate and ends before the cell prepares to divide. The rate holds for all cells of a species.

Control mechanisms operate at certain points in the cycle. Like built-in molecular brakes, they work when parts of the road become slippery, so to speak. Apply some brakes that are supposed to work in G1, and the cell cycle will stall in G1. Lift the brakes, and it will run to completion.

Imagine a car losing its brakes as it starts down a steep road. As you will see shortly, cancer begins this way. Crucial controls over the cell cycle are lost, and the cell cycle spins out of control.

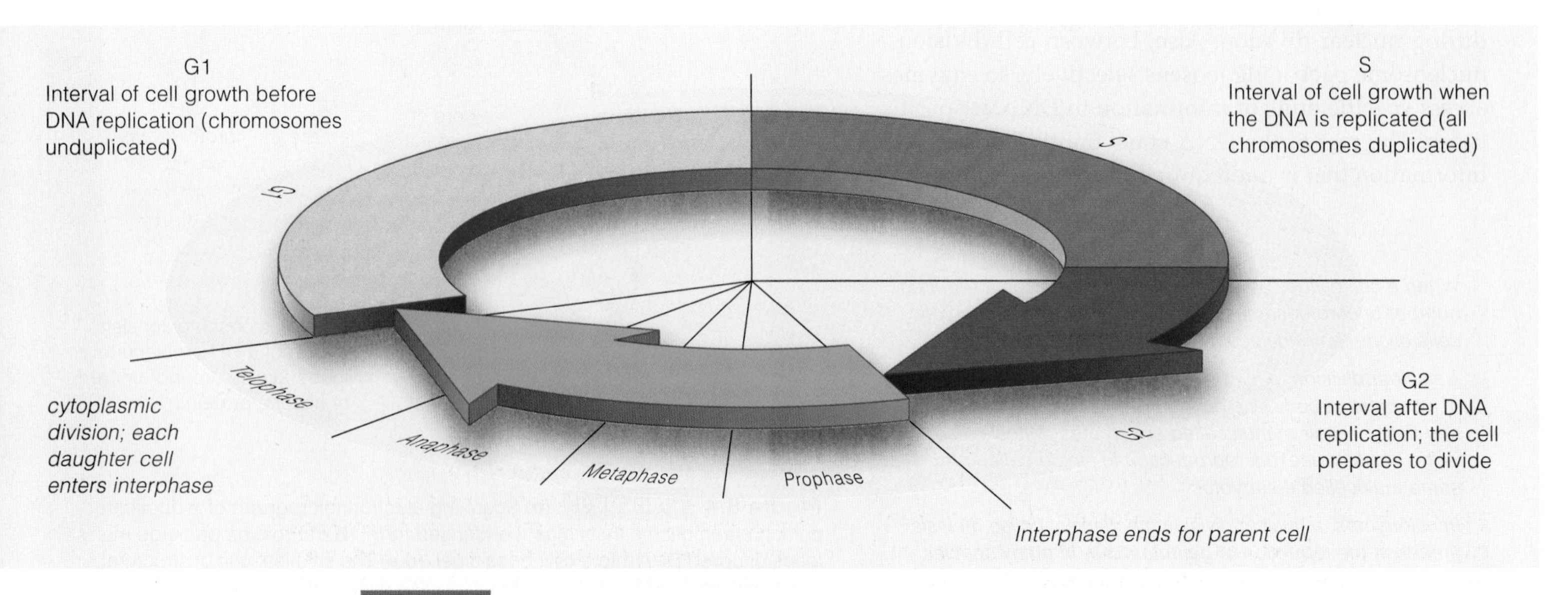

Figure 8.5 **Animated!** Eukaryotic cell cycle. The length of each interval differs among cells.

MITOSIS AND THE CHROMOSOME NUMBER

After G2, a cell divides by mitosis. Identical daughter cells result, each with the same number and kind of chromosomes as the parent. The chromosome number is the sum of all chromosomes in a cell of a given type. The body cells of gorillas have 48, those of human cells have 46. Pea plant cells have 14.

Actually, human body cells have two of each type of chromosome: their chromosome number is diploid (2*n*). The 46 are like two sets of books numbered from 1 to 23 (Figure 8.6*a*). You have two volumes of, say, chromosome 22—*a pair*. Except for a sex chromosome pair (XY), both have the same length and shape, and they hold information about the same heritable traits.

Think of them as two sets of books on how to build a house. Your father gave you one set. Your mother had her own ideas about wiring, plumbing, and so on. She gave you an alternate edition on the same topics, but it says slightly different things about many of them.

With mitosis, a diploid parent cell can produce two diploid daughter cells. It is not just that each daughter gets forty-six or forty-eight or fourteen chromosomes. If only the total mattered, then one cell might get, say, two pairs of chromosome 22 and no pairs whatsoever of chromosome 9. Neither cell could function like its parent without two of each type of chromosome.

As the next section explains, mitosis consists of four stages: prophase, metaphase, anaphase, and telophase. During prophase, a dynamic network of microtubules called the bipolar spindle grows from opposite poles of the cell. Some of the spindle's microtubules attach to the duplicated chromosomes:

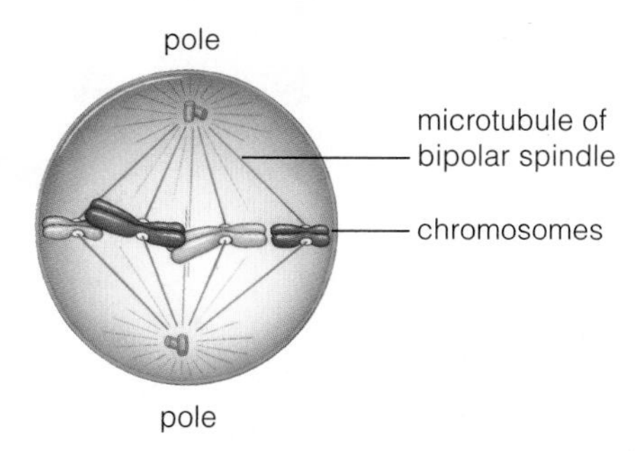

Microtubules from one pole connect to one chromatid of each chromosome, and microtubules from the other pole connect to its sister. In anaphase, the microtubules separate sister chromatids and move them to opposite poles. In telophase, two nuclei form around the two parcels of unduplicated chromosomes. The cytoplasm then divides, and two daughter cells form. Figure 8.6*b* shows a preview of how mitosis maintains the parental chromosome number in both daughter cells.

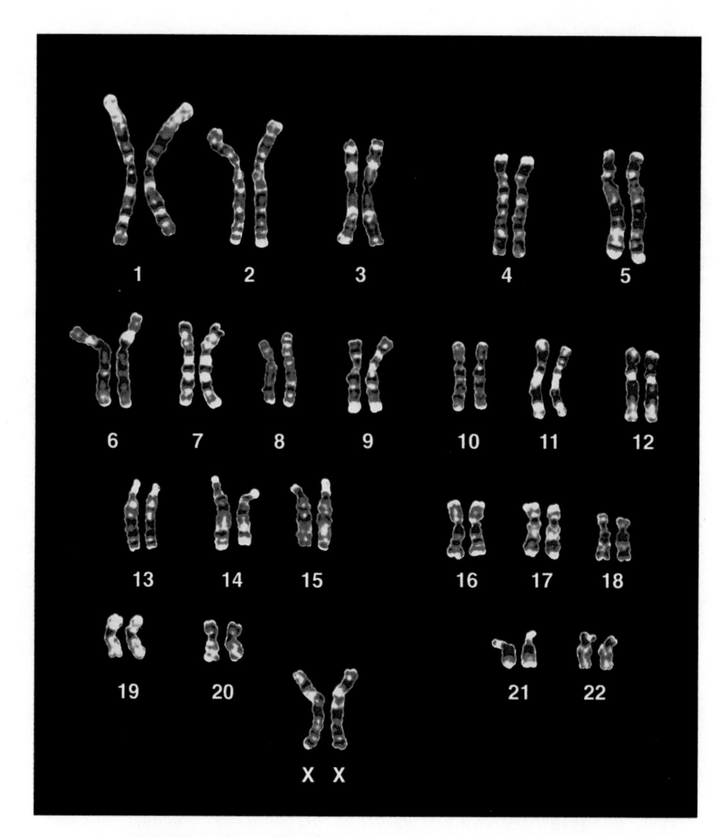

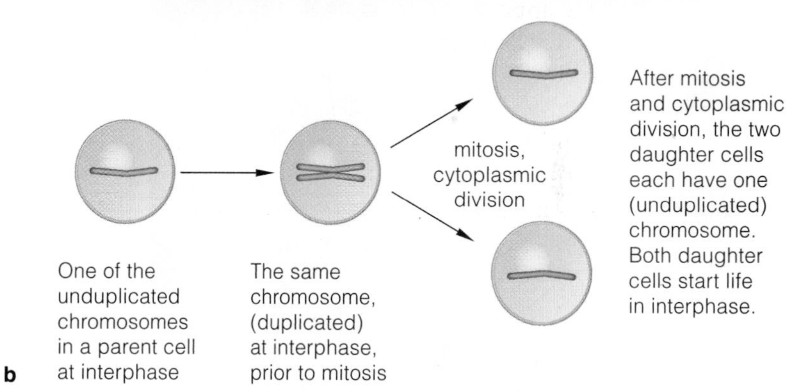

Figure 8.6 Preview of how mitosis maintains the parental chromosome number, one generation to the next. (**a**) Human diploid cells have twenty-three pairs of chromosomes, for a total of forty-six. The last ones in this lineup of metaphase chromosomes are a pair of sex chromosomes. Here, they are XX, so these chromosomes are from a female. In males, they are XY.

(**b**) What happens to each one of the forty-six chromosomes? Each time a human somatic cell undergoes mitosis and cytoplasmic division, its daughter cells end up with a set of forty-six chromosomes.

Interphase, mitosis, and cytoplasmic division constitute one cell cycle. During interphase, a new cell increases its mass, doubles the number of its cytoplasmic components, and duplicates its chromosomes. The cycle ends after the cell undergoes mitosis and then divides its cytoplasm.

8.3 A Closer Look at Mitosis

Focus now on a "typical" animal cell to see how a spindle of microtubules separates the chromosomes in a nucleus into equivalent parcels.

LINKS TO SECTIONS 4.7, 4.12

We know that a cell is in prophase, the first stage of mitosis, when its chromosomes become visible in light microscopes as threadlike forms. "Mitosis" is from the Greek *mitos,* meaning thread. Each chromosome was duplicated earlier, during interphase; each is two sister chromatids joined at the centromere. During prophase, the duplicated chromosomes condense. They become threadlike, and then rod-shaped (Figure 8.7*a–c*). New microtubules also assemble during prophase.

Most animal cells have a centrosome—a region near the nucleus that will organize microtubules while they are forming. The centrosome, which usually includes two barrel-shaped centrioles, is duplicated just before prophase. Remember, basal bodies are centrioles that give rise to cilia and flagella (Section 4.12). In prophase, one of the two centrosomes (with its pair of centrioles) moves to the opposite pole of the nucleus (Figure 8.7*d*). Microtubules that grow from both centrosomes are the start of the bipolar spindle.

As prophase ends, the nuclear envelope breaks up into flattened vesicles. The microtubules growing from each centrosome can now penetrate the nuclear region; the spindle forms as they interact with chromosomes and with one another. The ends of some microtubules attach to the chromosomes; others keep growing until they overlap midway between the centrosomes. Motor proteins traveling on the microtubules help the spindle grow in the proper direction. Remember, motor protein movement is driven by phosphate-group transfers from ATP (Section 4.12).

Again, some microtubules extending from one pole tether one chromatid of each chromosome, and some from the opposite pole tether the sister chromatid. The opposing sets of microtubules engage in a tug-of-war. They add and lose tubulin subunits, so they grow and shrink until they are the same length.

All of the duplicated chromosomes are now aligned midway between the two spindle poles (Figure 8.7*e*).

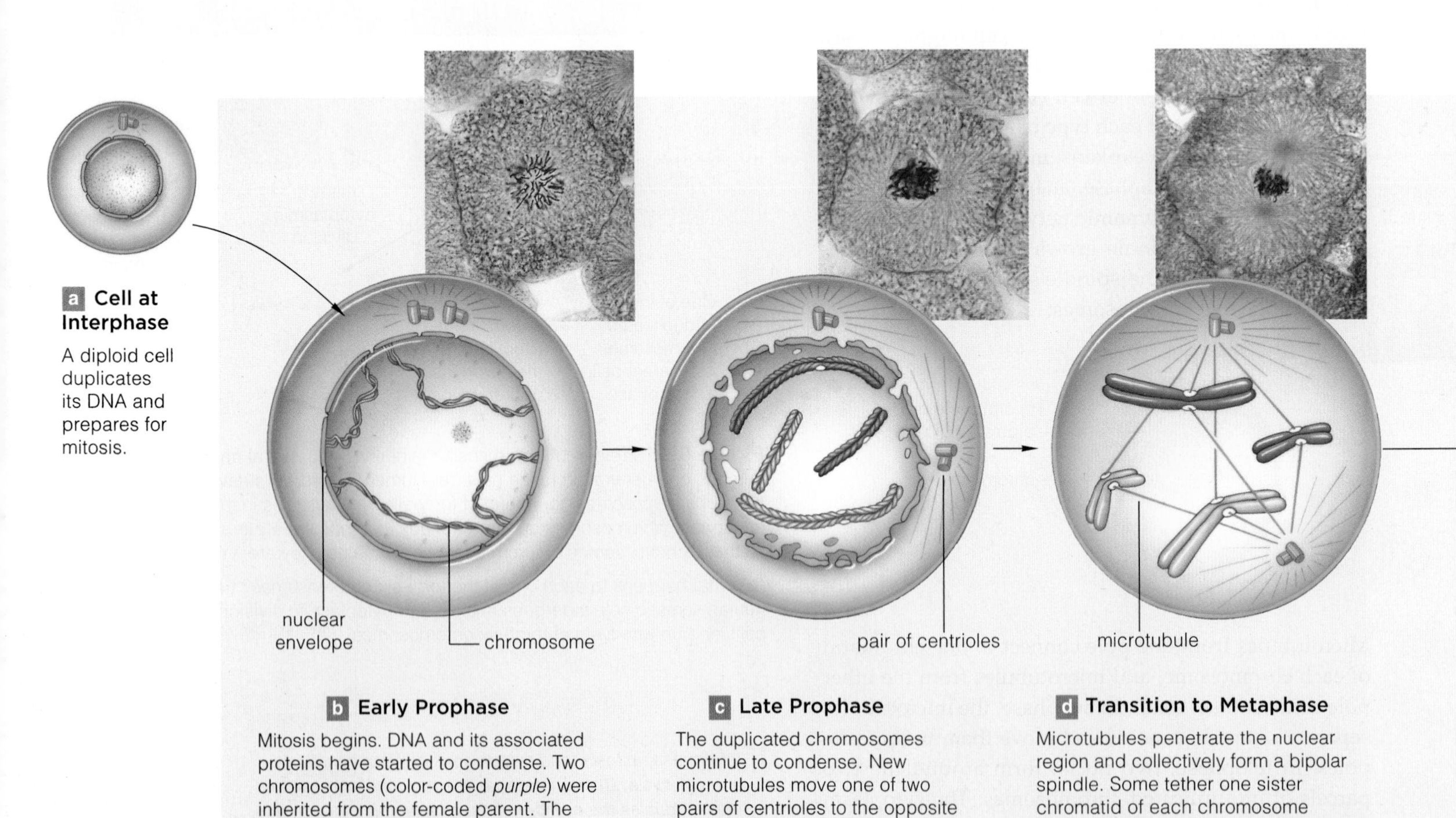

a Cell at Interphase
A diploid cell duplicates its DNA and prepares for mitosis.

b Early Prophase
Mitosis begins. DNA and its associated proteins have started to condense. Two chromosomes (color-coded *purple*) were inherited from the female parent. The other two (*blue*) are their counterparts, inherited from the male parent.

c Late Prophase
The duplicated chromosomes continue to condense. New microtubules move one of two pairs of centrioles to the opposite side of the nucleus. The nuclear envelope starts to break up.

d Transition to Metaphase
Microtubules penetrate the nuclear region and collectively form a bipolar spindle. Some tether one sister chromatid of each chromosome to a spindle pole. Others overlap at the spindle equator (*not shown*).

The alignment marks metaphase (*meta–*, midway), and it is crucial for the next stage of mitosis.

Anaphase is the interval when sister chromatids of each chromosome separate and move toward opposite spindle poles (Figure 8.7*f*). Three cell activities bring this about. First, the spindle microtubules attached to each chromatid shorten. Second, motor proteins drag chromatids along the shrinking microtubules toward the spindle poles. Third, the microtubules that overlap midway between spindle poles slide past one another. Motor proteins drive the movement, which pushes the two spindle poles farther apart.

As anaphase ends, each original chromosome and its duplicate are heading to opposite spindle poles, as part of two full sets of unduplicated chromosomes.

Telophase gets under way when one of each type of chromosome reaches a spindle pole. Two genetically identical clusters of chromosomes are now located at opposite "ends" of the cell. All of the chromosomes decondense and become threadlike. Vesicles derived from old nuclear envelope fuse and form patches of membrane around each cluster. Patch joins with patch until a new nuclear envelope encloses each cluster. Thus, two nuclei form (Figure 8.7*g*). In our example, the parent cell had a diploid number of chromosomes; so does each nucleus. Once two nuclei have formed, telophase is over—and so is this round of mitosis.

Prior to mitosis, each chromosome in a cell's nucleus is duplicated, so it consists of two sister chromatids.

In prophase, chromosomes condense and microtubules form a bipolar spindle. The nuclear envelope breaks up. Some microtubules harness the chromosomes.

At metaphase, all duplicated chromosomes are aligned midway between the spindle's poles, at its equator.

In anaphase, microtubules move the sister chromatids of each chromosome apart, to opposite spindle poles.

In telophase, a new nuclear envelope forms around each of two clusters of decondensing chromosomes.

Thus two daughter nuclei form. Each one has the same chromosome number as the parent cell's nucleus.

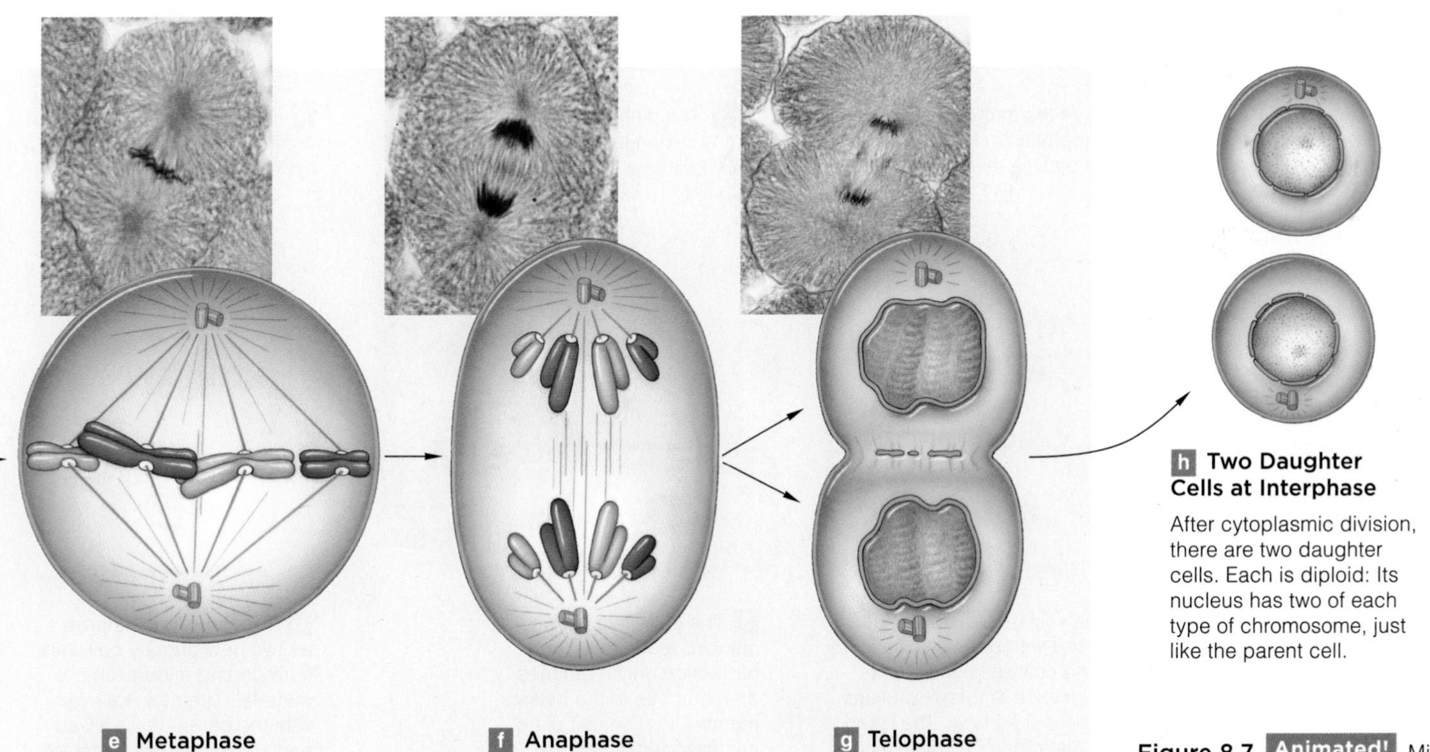

e Metaphase

All of the chromosomes have become lined up midway between the spindle poles. At this stage of mitosis, the chromosomes are in their most tightly condensed form.

f Anaphase

Sister chromatids separate as motor proteins moving along spindle microtubules drag them to opposite spindle poles. Other microtubules push the poles farther apart.

g Telophase

There are two clusters of chromosomes, which now decondense. Patches of new membrane fuse to form a new nuclear envelope. Mitosis is over.

Figure 8.7 **Animated!** Mitosis. For clarity, we track only two pairs of chromosomes from a diploid (2n) animal cell. Cells of nearly all eukaryotes have more pairs. The micrographs show mitosis in a whitefish. Appendix X compares mitosis in plant and animal cells.

8.4 Cytoplasmic Division Mechanisms

In most kinds of eukaryotic cells, the cytoplasm divides between late anaphase and the end of telophase. The mechanism of cytoplasmic division—or, more formally, cytokinesis—differs among species.

HOW DO ANIMAL CELLS DIVIDE?

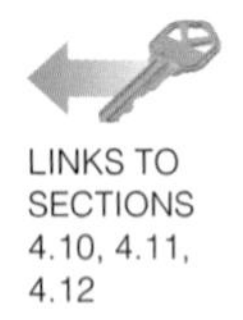

LINKS TO SECTIONS 4.10, 4.11, 4.12

Animal cells partition their cytoplasm by pinching it in two. Most often, the plasma membrane starts to sink inward as a thin indentation about halfway between the cell's poles (Figure 8.8*a*). The indentation is called a cleavage furrow, and it is the first visible sign that the cytoplasm is dividing. The furrow advances until it extends around the cell. As it does so, it deepens along a plane that corresponds to the former spindle equator.

What is happening? Part of the cell cortex, the mesh of cytoskeletal elements under the plasma membrane, is a thin band of actin and myosin filaments wrapped around the cell's midsection. Just as it does in muscle, ATP hydrolysis causes these filaments to interact, and the interaction results in contraction (Section 4.11). The band, which is called a contractile ring, is anchored to the plasma membrane. As it shrinks, the ring drags the plasma membrane inward until the cytoplasm—and

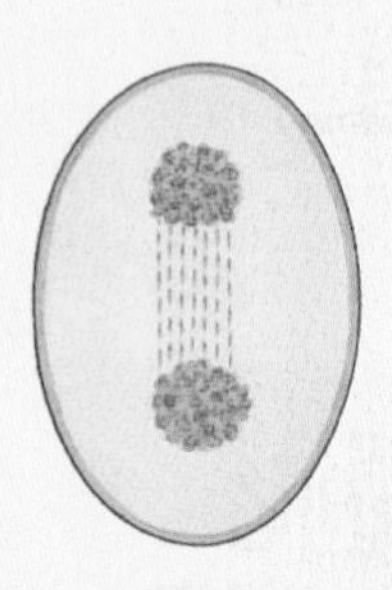

1 Mitosis is completed, and the bipolar spindle is starting to disassemble.

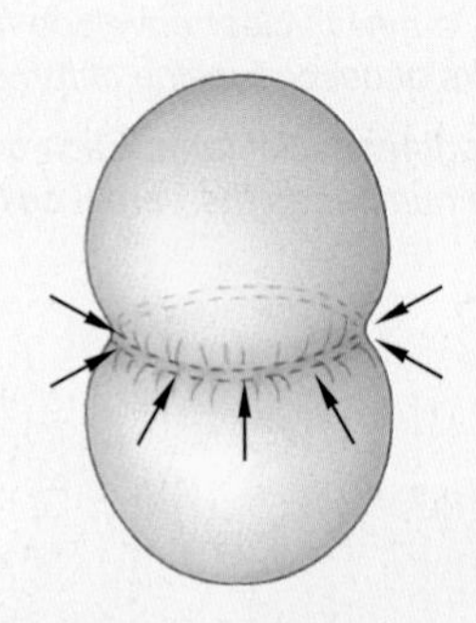

2 At the former spindle equator, a ring of actin filaments attached to the plasma membrane contracts.

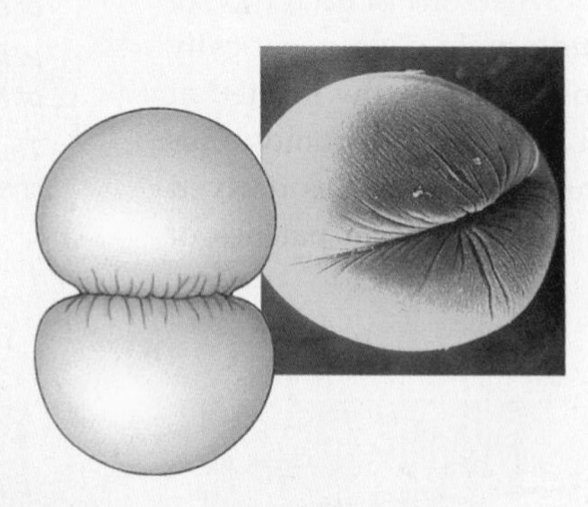

3 The diameter of the contractile ring continues to shrink and pull the cell surface inward.

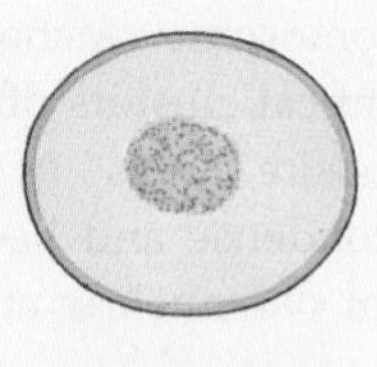

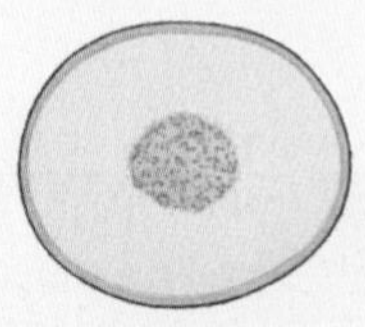

4 The contractile mechanism continues to operate until the cytoplasm is partitioned.

a Contractile Ring Formation

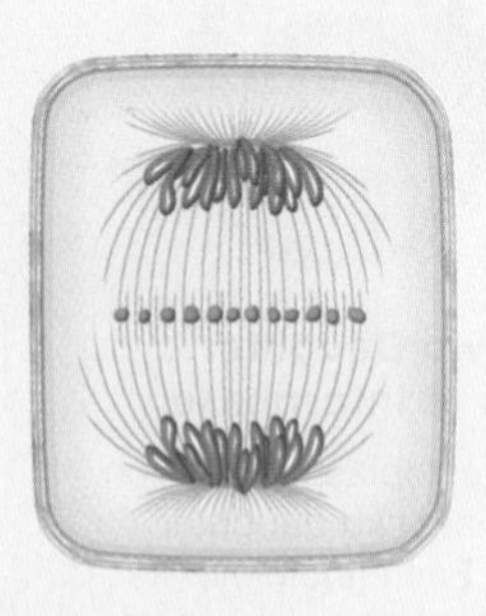

1 The plane of division and of a future cross-wall was established by a band of microtubules and actin filaments that formed and broke up before mitosis. Vesicles cluster here when mitosis ends.

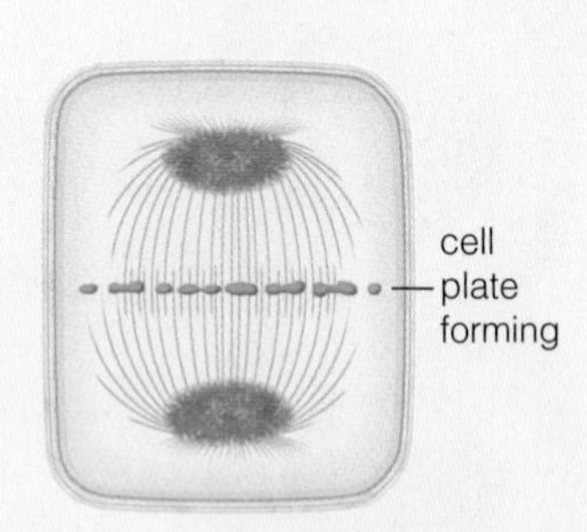

2 The vesicles fuse with each other and with endocytic vesicles bringing cell wall components and plasma membrane proteins from the cell surface. The fused materials form a cell plate along the plane of division.

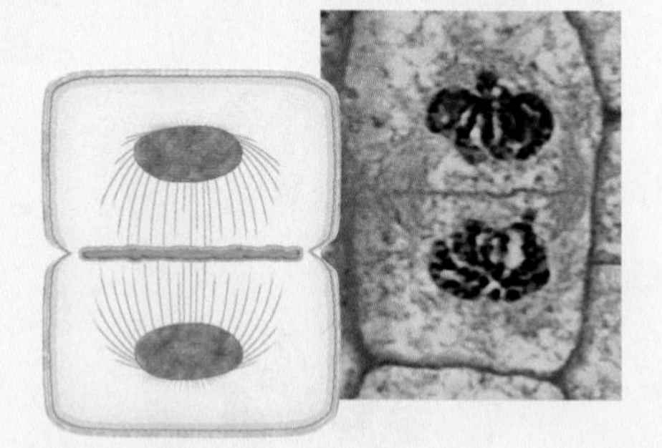

3 The cell plate grows outward along the plane of division until it reaches and attaches to the plasma membrane. The cell plate has now partitioned the cell's cytoplasm.

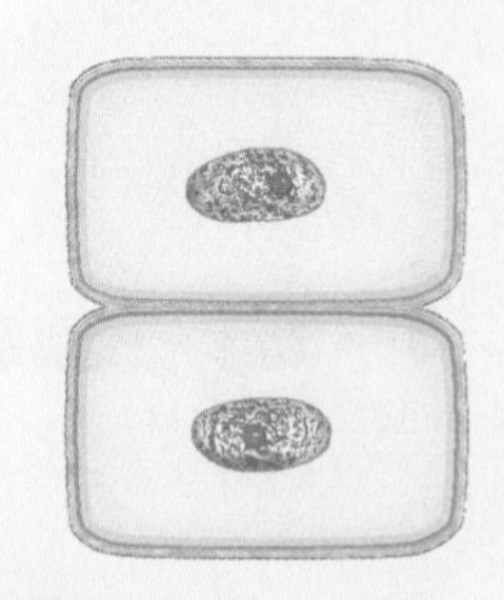

4 The cell plate matures as two new primary cell walls surrounding middle lamella material. The new walls join with the parent cell wall, so each daughter cell becomes enclosed by its own wall.

b Cell Plate Formation

Figure 8.8 **Animated!** Cytoplasmic division of an animal cell (**a**) and a plant cell (**b**).

the cell—is pinched in two (Figure 8.8*a*). Two daughter cells form this way. Each has a nucleus and some of the parent cell's cytoplasm, and each is enclosed in its own plasma membrane.

HOW DO PLANT CELLS DIVIDE?

The contractile ring mechanism that works for animal cells would not work for a plant cell. The contractile force is not strong enough to pinch through plant cell walls, which are stiff with cellulose and often lignin. Microtubules under a plant cell's plasma membrane orient cellulose fibers in the cell wall. Before prophase, these microtubules disassemble; new ones assemble in a narrow band around the nucleus. The band includes actin filaments. As other microtubules of the bipolar spindle form, the narrow band disappears; an actin-depleted zone is left behind. The zone marks the plane in which cytoplasmic division will occur (Figure 8.8*b*).

By the end of anaphase, a set of short microtubules has formed on either side of the division plane. These microtubules now guide vesicles from Golgi bodies and the cell's surface to the division plane. There, the vesicles and their wall-building contents start to fuse into a disk-shaped cell plate.

The plate grows outward until its edges reach the plasma membrane. It attaches to the membrane, and so partitions the cytoplasm. In time, the cell plate will develop into a primary cell wall that merges with the parent cell wall. Thus, by the end of division, each of the daughter cells will be enclosed by its own plasma membrane and its own cell wall.

APPRECIATE THE PROCESS!

Take a moment to look closely at your hands. Visualize the cells making up your palms, thumbs, and fingers. Now imagine the mitotic divisions that produced all of the cell generations that preceded them while you were developing, early on, inside your mother (Figure 8.9). Be grateful for the precision of the mechanisms that led to the formation of your hands and other body parts at the right times, in the proper numbers, for the alternatives can be terrible indeed.

Why? Good health and survival itself depend on the proper timing and completion of cell cycle events. Some genetic disorders arise as the result of errors in duplication or distribution of even one chromosome. In other cases, unchecked cell divisions may destroy the surrounding tissues and, ultimately, the individual. Problems can start in body cells. They can also start in cells that give rise to sperm and eggs, although rarely.

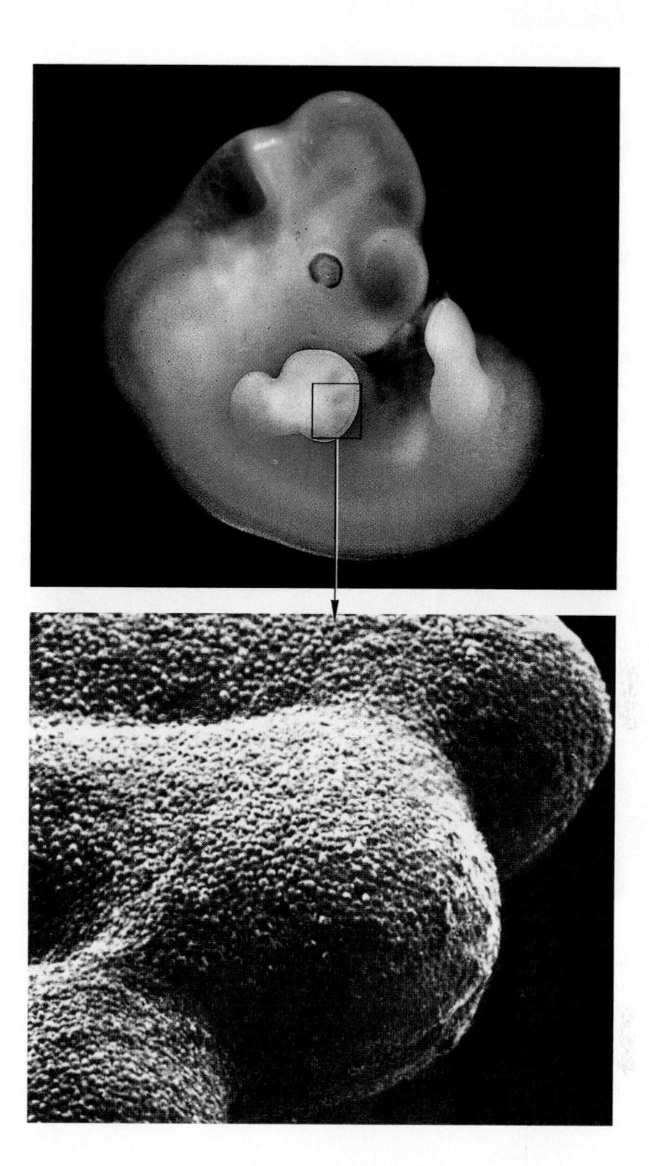

Figure 8.9 The paddlelike structure of a human embryo that develops into a hand by mitosis, cytoplasmic divisions, and other processes. The scanning electron micrograph reveals individual cells.

After mitosis, a separate mechanism partitions the cytoplasm of the parent cell into two daughter cells, each with its own nucleus.

In animal cells, a contractile ring partitions the cytoplasm. A band of actin filaments that rings the cell midsection contracts and pinches the cytoplasm in two.

In plant cells, a mechanism called cell plate formation partitions the cytoplasm. Vesicles deposit material at a plane of cytoplasmic division to form a cross-wall, which connects to the parent cell wall.

8.5 When Control Is Lost

Controls over cell division are central to growth and reproduction. On rare occasions, something goes wrong in a somatic cell or reproductive cell. Cancer may be the outcome.

LINKS TO SECTIONS 5.4, 6.1

The Cell Cycle Revisited Every second, millions of cells in your skin, bone marrow, gut lining, liver, and elsewhere are dividing and replacing worn-out, dead, and dying predecessors. They do not divide at random. Many mechanisms control cell growth, DNA replication, and when cell division begins and ends.

What happens when something goes wrong? Suppose sister chromatids do not separate as they should during mitosis. As a result, one daughter cell ends up with too many chromosomes and the other with too few. Or suppose the wrong nucleotide gets added to a growing strand of DNA when a chromosome is being duplicated. Suppose free radicals, peroxides, or ultraviolet radiation disrupts a cell's DNA (Sections 5.4 and 6.1). Such problems are frequent but inevitable, and a cell may not function properly unless they are quickly countered.

The cell cycle has built-in checkpoints that keep such errors from getting out of hand. Certain proteins, the products of checkpoint genes, monitor whether a cell's DNA is fully replicated, whether it is damaged, and even whether nutrient concentrations are sufficient to support cell growth. Such proteins interact to advance, delay, or stop the cell cycle (Figure 8.10).

For example, kinases are a class of enzymes that can activate other molecules by transferring a phosphate group to them. When DNA is broken or incomplete, kinases can activate certain proteins in a cascade of signaling events that ultimately stop the cell cycle or induce cell death. As another example, the checkpoint proteins called growth factors activate genes that stimulate cells to grow and divide. One epidermal growth factor activates a kinase by binding to receptors on target cells in epithelial tissues. The binding is a signal to start mitosis.

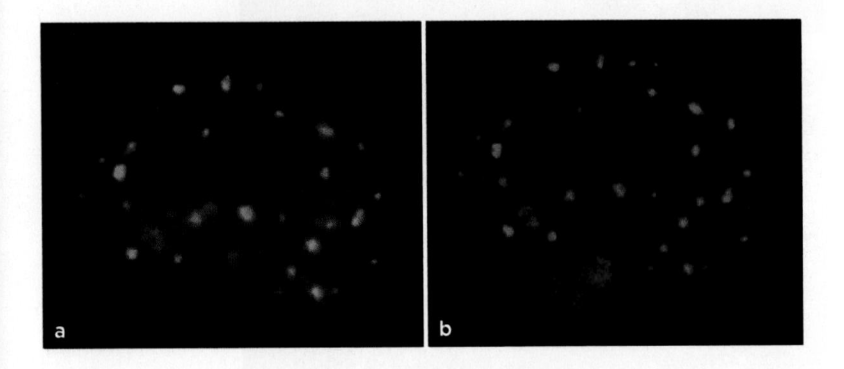

Figure 8.10 Protein products of checkpoint genes in action. A form of radiation damaged the DNA inside this nucleus. (**a**) *Green* dots pinpoint the location of a protein called *53BP1*, and (**b**) *red* dots pinpoint the location of another protein, *BRCA1*. Both proteins have clustered around the same chromosome breaks in the same nucleus. The integrated action of these proteins and others blocks mitosis until the DNA breaks are fixed.

Checkpoint Failure and Tumors Sometimes a checkpoint gene mutates and its protein product no longer works properly. When all checkpoint mechanisms fail, the cell loses control over the cell cycle. Figures 8.11 through 8.13 show a few of the outcomes.

In some cases, the cycle skips interphase, and the cell divides again and again without resting. In other cases, DNA that has been damaged is replicated. In still other cases, signaling mechanisms that can make an abnormal cell commit suicide are disabled. You will read more about this mechanism in Section 25.5. Regardless of the cause, the cell's continually dividing descendants form a tumor—an abnormal mass—in the surrounding tissue.

Usually, one or more checkpoint proteins are absent in tumor cells. That is why checkpoint gene products that inhibit mitosis are called *tumor suppressors*. Checkpoint genes encoding proteins that stimulate mitosis are known as *proto-oncogenes*. Mutations that alter their products or the rate at which they are synthesized help transform a normal cell into a tumor cell. Mutant checkpoint genes are associated with an increased risk of tumor formation, and sometimes they run in families.

Moles and other tumors are neoplasms, or abnormal masses of cells that lost control over how they grow and divide. Ordinary skin moles are among the noncancerous, or *benign*, neoplasms. They grow very slowly, and their cells retain the surface recognition proteins that keep them in their home tissue (Figure 8.12). Unless a benign neoplasm grows too large or becomes irritating, it poses no threat to the body.

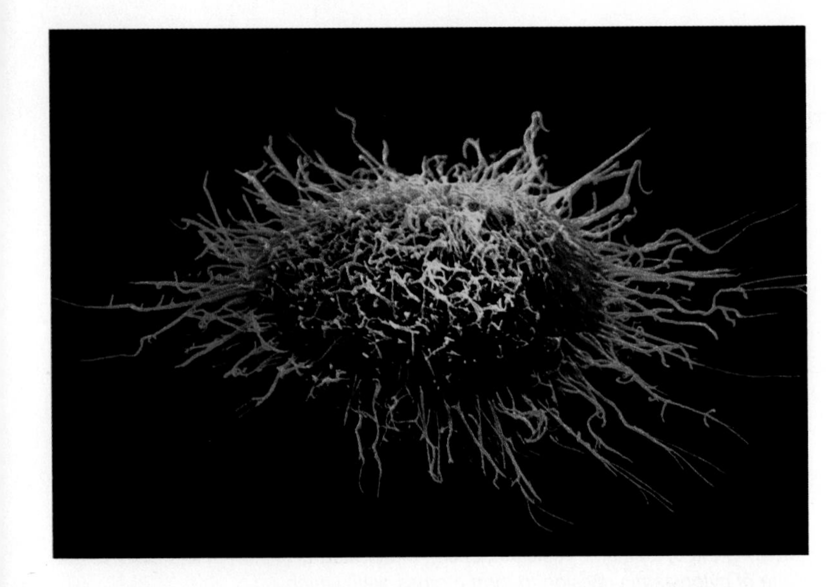

Figure 8.11 Scanning electron micrograph of the surface of a cervical cancer cell, the kind of malignant cell that killed Henrietta Lacks.

Characteristics of Cancer All cancers are abnormally growing and dividing cells of a *malignant* neoplasm. They physically and metabolically disrupt surrounding tissues. Cancer cells may be disfigured. They can break loose from home tissues, slip into and out of blood vessels and lymph vessels, and invade other tissues where they do not belong (Figure 8.12). They typically display these characteristics:

First, cancer cells grow and divide abnormally. The controls that keep cells from getting overcrowded in tissues are lost and cell populations reach extremely high densities. The number of small blood vessels, or capillaries, that transport blood to the growing cell mass also increases abnormally.

Second, both the cytoplasm and plasma membrane of cancer cells become altered. The membrane gets leaky and has altered or missing proteins. The whole cytoskeleton shrinks, becomes disorganized, or both. Enzyme action shifts, as in amplified reliance on ATP formation by glycolysis.

Third, cancer cells often have a weakened capacity for adhesion. Because their recognition proteins are altered or lost, they cannot stay anchored in proper tissues. They break away and may establish growing colonies in distant tissues. *Metastasis* is the name for this process of abnormal cell migration and tissue invasion.

Unless chemotherapy, surgery, or another procedure eradicates them, cancer cells can put the individual on a painful road to death. Each year in developed countries alone, cancers cause 15 to 20 percent of all human deaths. Cancers are not just a human problem. They are known to occur in most of the animal species studied to date.

Cancer is a multistep process. Researchers already know about many mutant genes that contribute to it. They are working to identify drugs that target and destroy cancer cells or stop them from dividing.

HeLa cells, for instance, were used in early tests of taxol, a drug that keeps microtubules from disassembling and so hampers mitosis. Frequent divisions of cancer cells make them more vulnerable to this poison than normal cells. Such research may yield drugs that put the brakes on cancer. We return to this topic in later chapters.

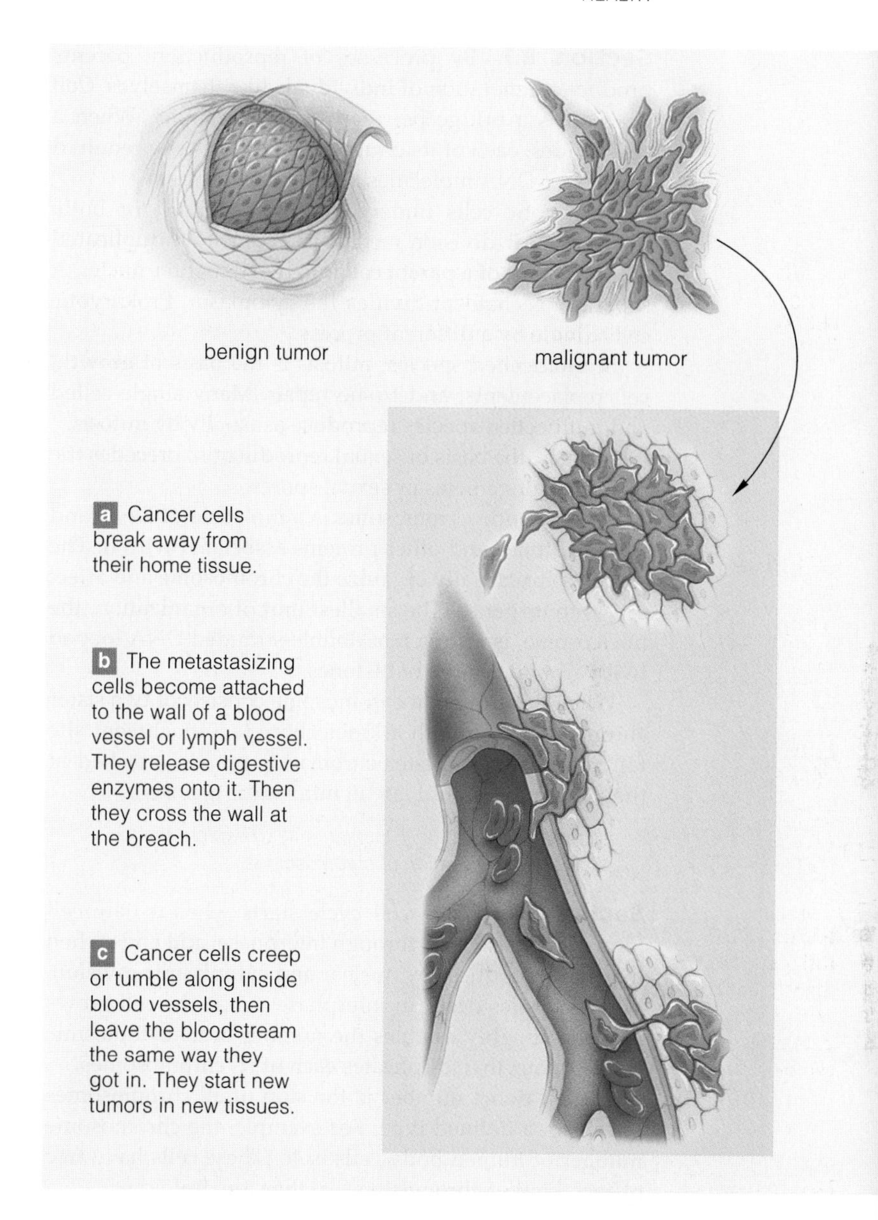

Figure 8.12 Animated! Comparison of benign and malignant tumors. Benign tumors typically are slow-growing and stay in their home tissue. Cells of a malignant tumor migrate abnormally through the body and establish colonies even in distant tissues.

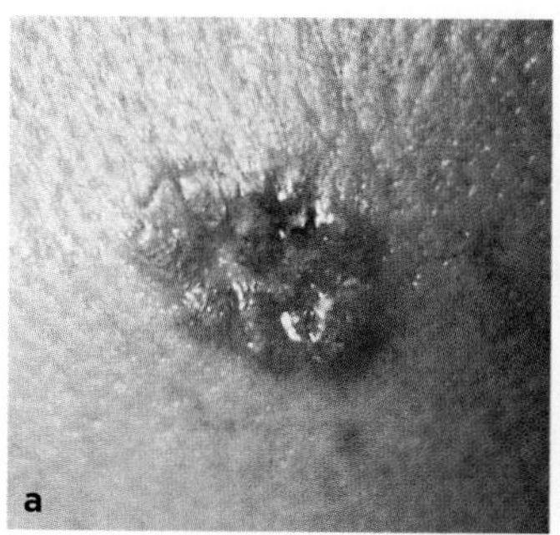

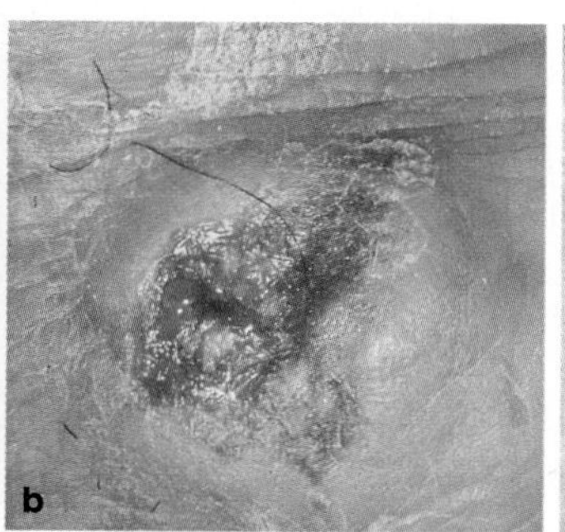

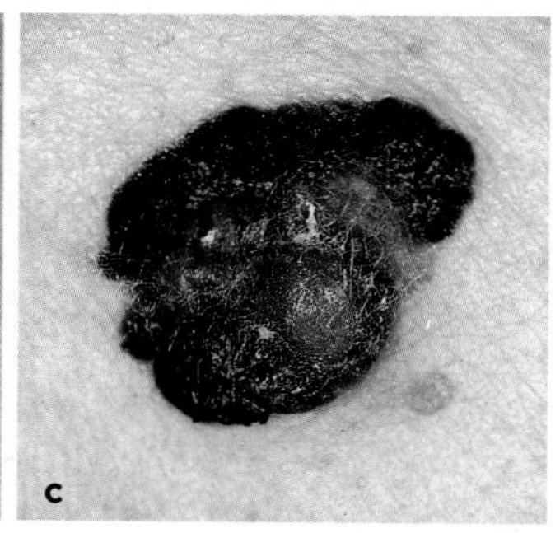

Figure 8.13 Skin cancers. (**a**) A *basal cell carcinoma* is the most common type. This slow-growing, raised lump is typically uncolored, reddish-brown, or black.

(**b**) The second most common form of skin cancer is a *squamous cell carcinoma*. This pink growth, firm to the touch, grows fast under the surface of skin exposed to the sun.

(**c**) *Malignant melanoma* spreads fastest. Cells form dark, encrusted lumps. They may itch like an insect bite or bleed easily.

www.thomsonedu.com Thomson NOW!

Summary

Section 8.1 By processes of reproduction, parents produce a generation of individuals like themselves. Cell division is a bridge between two generations. When a cell divides, each of its daughter cells receives a required number of DNA molecules and some cytoplasm.

Eukaryotic cells undergo mitosis, meiosis, or both. These nuclear division mechanisms partition duplicated chromosomes of a parent cell into two daughter nuclei. A separate mechanism divides the cytoplasm. Prokaryotic cells divide by a different process.

In multicelled species, mitosis is the basis of growth, cell replacements, and tissue repair. Many single-celled and multicelled species reproduce asexually by mitosis.

Meiosis, the basis of sexual reproduction, precedes the formation of gametes or sexual spores.

A eukaryotic chromosome is a molecule of DNA and many histones and other proteins associated with it. The proteins structurally organize the chromosome and affect access to its genes. The smallest unit of organization, the nucleosome, is a stretch of double-stranded DNA looped twice around a spool of histones.

When duplicated, a chromosome consists of two sister chromatids, each with a kinetochore (an attachment site for microtubules). Sister chromatids remain attached at their centromere until late in mitosis (or meiosis).

■ *Use the animation on ThomsonNOW to explore the structural organization of chromosomes.*

Section 8.2 Each cell cycle starts when a new cell forms. The cycle runs through interphase, and ends when that cell reproduces by nuclear and cytoplasmic division. Most activities occur in interphase, when the cell grows in mass, roughly doubles the number of its cytoplasmic components, then duplicates each of its chromosomes.

Chromosome number is the sum of all chromosomes in cells of a defined type. For example, the chromosome number of human body cells is 46. These cells have two of each kind of chromosome, so they are diploid.

■ *Use the interaction on ThomsonNOW to investigate the stages of the cell cycle.*

Section 8.3 Mitosis is a nuclear division mechanism that maintains the chromosome number. It proceeds in four sequential stages, which we summarize as follows:

Prophase. Duplicated chromosomes become threadlike as they start to condense. Microtubules form a bipolar spindle, and the nuclear envelope starts to break apart. Some of the microtubules that extend from one spindle pole harness one chromatid of each chromosome; others that extend from the opposite spindle pole tether its sister chromatid. Other microtubules extend from both poles and grow until they overlap at the spindle's midpoint.

Metaphase. All chromosomes have become aligned at the spindle's midpoint.

Anaphase. The sister chromatids of each chromosome detach from each other, and some spindle microtubules start moving them toward opposite spindle poles. Other microtubules that overlap at the spindle's midpoint slide past each other in a way that pushes the poles farther apart. Motor proteins drive the movements.

Telophase. Two identical clusters—each consisting of two chromosomes of each type—have reached opposite spindle poles. A nuclear envelope forms around each one. Both new nuclei have the parental chromosome number.

■ *Use the animation on ThomsonNOW to see how mitosis proceeds.*

Section 8.4 The mechanisms of cytoplasmic division differ. In animal cells, a microfilament ring that is part of the cell cortex contracts and pulls the cell surface inward until the cytoplasm is partitioned. In plant cells, a band of microtubules and microfilaments forms around the nucleus before mitosis starts. It marks the site where a cell plate will form. The cell plate will become a cross-wall that partitions the cytoplasm.

■ *Compare the cytoplasmic division of plant and animal cells with the animation on ThomsonNOW.*

Section 8.5 Checkpoint gene products exert control over the cell cycle. Mutant checkpoint genes can cause tumors by disrupting the normal controls. Cancer is a multistep process involving altered cells that grow and divide abnormally. Malignant cells may metastasize, or break loose and colonize distant tissues.

■ *Use the animation on ThomsonNOW to see how cancers spread through the body.*

Self-Quiz

Answers in Appendix III

1. Mitosis and cytoplasmic division function in ______ .
   a. asexual reproduction of single-celled eukaryotes
   b. growth and tissue repair in multicelled species
   c. gamete formation in prokaryotes
   d. both a and b

2. A duplicated chromosome has ______ chromatid(s).
   a. one b. two c. three d. four

3. The basic unit that structurally organizes a eukaryotic chromosome is the ______ .
   a. higher order coiling c. nucleosome
   b. bipolar mitotic spindle d. microfilament

4. The chromosome number is ______ .
   a. the sum of all chromosomes in a cell of a given type
   b. an identifiable feature of each species
   c. maintained by mitosis
   d. all of the above

5. A somatic cell having two of each type of chromosome has a(n) ______ chromosome number.
   a. diploid b. haploid c. tetraploid d. abnormal

6. Interphase is the part of the cell cycle when ______ .
   a. a cell ceases to function
   b. a cell forms its spindle apparatus
   c. a cell grows and duplicates its DNA
   d. mitosis proceeds

7. After mitosis, the chromosome number of a daughter cell is ______ the parent cell's.
   a. the same as c. rearranged compared to
   b. one-half d. doubled compared to

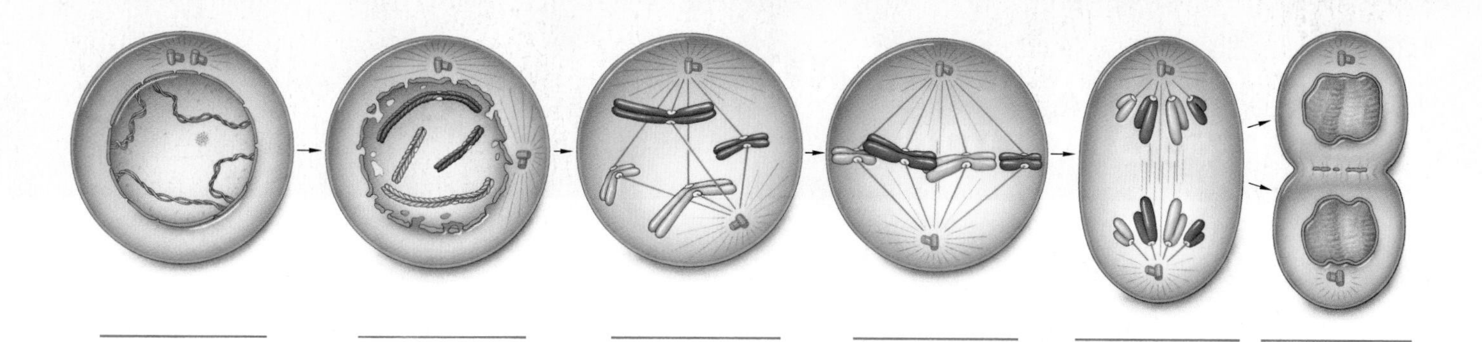

8. In the above diagram of mitosis, fill in the blanks with a descriptive name for each interval.

9. Only ______ is not a stage of mitosis.
   a. prophase b. interphase c. metaphase d. anaphase

10. Match each stage with the events listed.
   ___ metaphase a. sister chromatids move apart
   ___ prophase b. chromosomes start to condense
   ___ telophase c. daughter nuclei form
   ___ anaphase d. all duplicated chromosomes are aligned at the spindle equator

■ *Visit ThomsonNOW for additional questions.*

Critical Thinking

1. Figure 8.14 shows a cell going through stages of mitosis. Think about the dense vertical arrays of short microtubules midway between the two clusters of chromosomes at the stage of telophase. Does this clue suggest that the cell is from a plant or an animal?

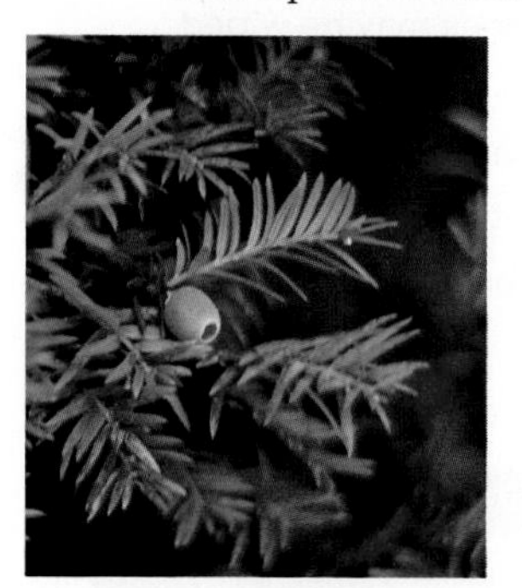

2. The anticancer drug taxol was first isolated from Pacific yews (*Taxus brevifolia*), which are slow-growing trees. Bark from about six yew trees provided enough taxol to treat one patient, but removing the bark killed the trees. Fortunately, taxol is now produced using plant cells that grow and divide in big vats rather than in trees. What challenges do you think had to be overcome to get plant cells to grow and divide in laboratories?

3. X-rays emitted from some radioisotopes damage DNA. Humans exposed to high levels of x-rays face *radiation poisoning*. Speculate about why hair loss and damage to the lining of the gut are early symptoms. Also speculate about why exposure to radioactivity is used as a therapy to treat some kinds of cancers.

4. Suppose you have a way to measure the amount of DNA in one cell during the cell cycle. You first measure the amount at the G1 phase. At what points in the rest of the cycle will you see a change in the amount of DNA per cell?

5. The cervix is part of the uterus, a chamber in which embryos develop. The *Pap smear* is a screening procedure that can detect *cervical cancer* in its earliest stages.

Treatments include freezing precancerous cells or killing them with a laser beam, and hysterectomy (removal of the uterus). The treatments are more than 90 percent effective when this cancer is detected early. However, the chances of survival plummet to less than 9 percent after it spreads.

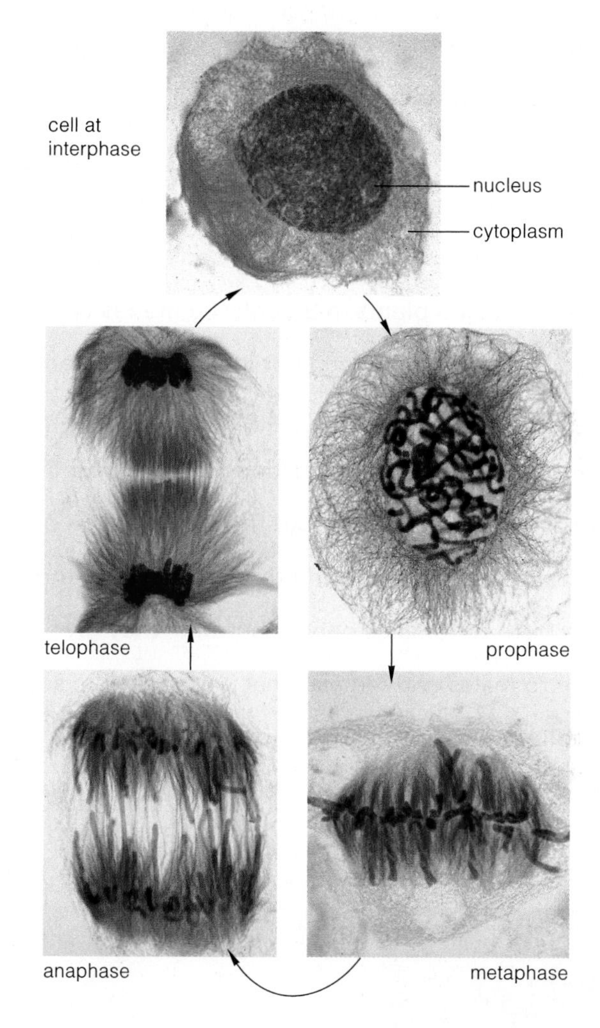

Figure 8.14 Is this the nucleus of a plant cell or animal cell undergoing mitosis?

Most cervical cancers develop slowly. Unsafe sex increases the risk. A key risk factor is infection by human papillomaviruses (HPV), which cause genital warts. Viral genes coding for the tumor-inducing proteins get inserted into the DNA of cervical cells. Of one group of cervical cancer patients, 91 percent had been infected with HPV.

Not all women request Pap smears. Many wrongly believe the procedure is costly. Many do not recognize the importance of abstinence or "safe" sex. Others don't want to think about whether they have cancer. Knowing about the cell cycle and cancer, what would you say to a woman who falls into one of these groups?

9 MEIOSIS AND SEXUAL REPRODUCTION

IMPACTS, ISSUES Why Sex?

Think about the shared traits that make all of us "human," such as two forward-directed eyes, two legs, and so on. Now think about how much we differ in the details, such as the color of our eyes and how long our legs are. Variation in heritable traits is typical of sexual reproducers. Sexual reproduction dominates the life cycle of nearly all animals and plants, although the modes of producing offspring vary. For instance, the plant-sucking insects called aphids alternate between asexual and sexual reproduction. Females bear live offspring asexually—without the participation of males (Figure 9.1*a*). Spring through summer, they give birth to females only. In autumn, they give birth to males and females that will engage in sex. Fertilized eggs form. The next spring, females hatch from the eggs, and the cycle begins again.

Like aphids, the females of a few fishes, reptiles, and birds can propagate themselves naturally. The females of mammals do not make fatherless offspring except in laboratories (Figure 9.1*d*).

If the function of reproduction is the perpetuation of one's genes—heritable units of information about traits—then an asexual reproducer would seem to win. In asexual reproduction, all of an individual's genes are passed to all of its offspring. By contrast, sexual reproduction mixes up genetic information from two parents. Only about half of each parent's genes are passed to offspring.

So why sex? Recall, from Section 1.4, that some forms of traits are more adaptive than others to conditions in the environment. If conditions change in a big way, at least some of the diverse offspring of sexual reproducers may have forms of traits that help them to survive the change; they may have just the ticket to go on the new ride. By contrast, all offspring of asexual reproducers are adapted the same way to the environment—and equally vulnerable to changes in it. Entire lineages may be wiped out at once.

Other organisms are part of the environment, and they, too, can change. Think of a predator and its prey—say, a fox and a rabbit. If one rabbit is better than others

See the video! **Figure 9.1** Reproductive moments. (**a**) Aphid giving birth. As in a few other species, the females of this kind of insect reproduce asexually in spring, then switch to sexual reproduction in autumn. (**b**) Poppy plant being helped by a beetle, which makes pollen deliveries for it. (**c**) Mealybugs mating. (**d**) Kaguya, the first fatherless mouse.

How would you vote? Japanese researchers successfully created a "fatherless" mouse with genetic material from the eggs of two females. The mouse is healthy and fertile. Should researchers be allowed to try the same process with human eggs? See ThomsonNOW for details, then vote online.

at running away from a fox, it has a better chance of reproducing and passing on the genetic basis for the trait. If one fox is better than others at outrunning the faster rabbit, the genetic basis for that trait will be passed on, also. As one species changes, so does the other.

According to the Red Queen hypothesis, two species locked in a predator–prey or parasite–host interaction continually adapt to one another or face extinction. The hypothesis gives a nod to Lewis Carroll's book *Through the Looking Glass*. The Queen of Hearts tells Alice, "Now here, you see, it takes all the running you can do, to keep in the same place."

Compared to all other reproductive modes, sexual reproduction is most responsive to changing conditions because it gives rise most quickly to variation in traits. Does this advantage mean that sexual reproducers win? You might think so, given the natural human tendency to focus on ourselves first, other vertebrates second, and everything else last. However, in terms of numbers of individuals and how long their lineages have endured, the most successful organisms on Earth are bacteria, which usually just copy their DNA and divide. *Successful reproductive modes are as diverse as the environments to which they are adapted.*

With that humbling thought in mind, turn now to the mechanisms of sexual reproduction. Three connected events—meiosis, gamete formation, and fertilization—are hallmarks of this mode of reproduction. The outcome is the production of offspring with novel combinations of traits. That outcome has contributed immensely to the range of diversity, past and present, but it is not the only route to reproductive success.

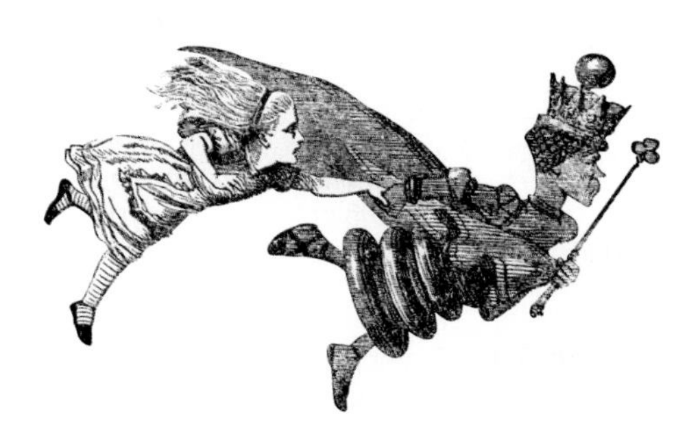

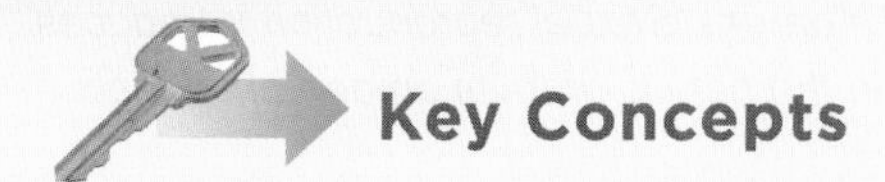

Key Concepts

SEXUAL VERSUS ASEXUAL REPRODUCTION

By asexual reproduction, one parent alone transmits its genetic information to offspring. By sexual reproduction, offspring typically inherit information from two parents that differ in their alleles. Alleles are different forms of the same gene; they specify different versions of a trait. **Section 9.1**

STAGES OF MEIOSIS

Diploid cells have a pair of each type of chromosome, one maternal and one paternal. Meiosis, a nuclear division mechanism, reduces the chromosome number. It occurs only in cells set aside for sexual reproduction. Meiosis sorts out a reproductive cell's chromosomes into four haploid nuclei, which are distributed to daughter cells by way of cytoplasmic division. **Sections 9.2, 9.3**

CHROMOSOME RECOMBINATIONS AND SHUFFLINGS

During meiosis, each pair of maternal and paternal chromosomes swaps segments and exchanges alleles. The pairs get randomly shuffled, so forthcoming gametes end up with different mixes of maternal and paternal chromosomes. Chance also governs which gametes combine during fertilization. All three events contribute to variation in traits among offspring. **Section 9.4**

SEXUAL REPRODUCTION IN THE LIFE CYCLES

In animals, gametes form by different mechanisms in males and females. In most plants, spore formation and other events intervene between meiosis and gamete formation. **Section 9.5**

MITOSIS AND MEIOSIS COMPARED

Recent molecular evidence suggests that meiosis originated through mechanisms that already existed for mitosis and, before that, for repairing damaged DNA. **Section 9.6**

Links to Earlier Concepts

For this chapter, reflect on the simple overview of growth and reproduction in Section 1.2. You will be revisiting the microtubules that move chromosomes about (4.12, 8.2, 8.3). Be sure you have a clear picture of the structural organization of chromosomes (8.1) and can define chromosome number (8.2). You will draw on your understanding of cytoplasmic division (8.4) and checkpoint gene products that monitor and repair chromosomal DNA during the cell cycle (8.5).

SEXUAL VERSUS ASEXUAL REPRODUCTION

9.1 Introducing Alleles

Asexual reproduction produces genetically identical copies of a parent. Sexual reproduction introduces variation in the details of traits among offspring.

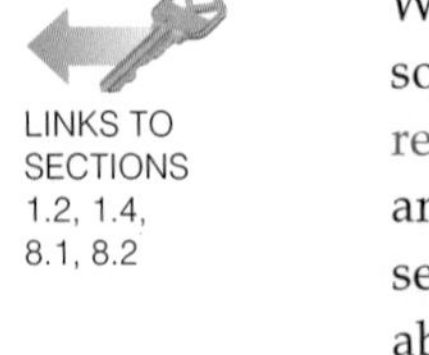

LINKS TO SECTIONS 1.2, 1.4, 8.1, 8.2

When an orchid or aphid reproduces by itself, what sort of offspring does it get? By the process of asexual reproduction, all offspring inherit the same number and kinds of genes from a single parent. Genes are sequences of DNA that encode heritable information about traits. Each species has a unique set of genes that collectively encodes all the information necessary to make a new individual. Rare mutations aside, then, asexually produced individuals are usually clones, or genetically identical copies of the parent.

Inheritance gets much more interesting with sexual reproduction, a process involving meiosis, formation of mature reproductive cells called gametes, and then fertilization—a union of two mature reproductive cells. In most sexual reproducers, such as humans, the first cell of a new individual holds pairs of genes, on pairs of chromosomes. Usually, one chromosome of a pair is maternal and the other paternal (Figure 9.2).

If information in all pairs of genes were identical down to the last detail, sexual reproduction would also produce clones. Just imagine—you, every person you know, the entire human population might be a clone, with everybody looking alike. But the two genes of a pair might *not* be identical. Why not? The molecular structure of any gene can change permanently; it can mutate. As one outcome, the two genes that happen to be paired in an individual's cells might "say" slightly different things about a trait. Each unique molecular form of the same gene is called an allele.

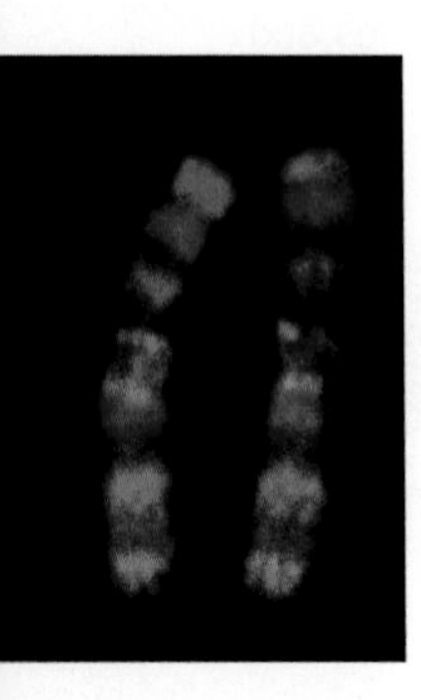

Figure 9.2 A maternal and a paternal chromosome pair. They seem identical at this magnification, but any gene on one might differ slightly from its partner on the other.

Differences in alleles affect thousands of traits. For instance, whether your chin has a dimple depends on which pair of alleles you inherited at one chromosome location. One allele at that location says "put a dimple in the chin." A different allele says "no dimple in the chin." Variant alleles are one reason individuals of a species do not all look alike. By sexual reproduction, offspring inherit new combinations of alleles, which lead to variations in the details of traits.

This chapter gets into the cellular basis of sexual reproduction. More importantly, it starts you thinking about far-reaching effects of its gene shufflings. The process introduces variations in traits among offspring. That variation is the foundation for evolution.

Sexual reproduction introduces variation in traits by bestowing novel combinations of alleles on offspring, which is a foundation for evolution.

STAGES OF MEIOSIS

9.2 What Meiosis Does

Meiosis is a nuclear division mechanism that precedes cytoplasmic division of immature reproductive cells. It occurs only in sexually reproducing eukaryotic species.

THINK "HOMOLOGUES"

Meiosis is a nuclear division process that divides the parental chromosome number in half. Meiosis differs from mitosis, a nuclear division mechanism you read about in the previous chapter. Unlike mitosis, meiosis sorts chromosomes into parcels not once but *twice*. Also unlike mitosis, meiosis occurs only in specialized reproductive cells. In animals, the mature reproductive cells called gametes form by meiosis of germ cells; in plants, spores form. A sperm is an example of a male gamete; an egg, a female gamete. At fertilization, an egg and a sperm fuse to form a zygote, the first cell of a new individual. In most multicelled species, gametes form from cells in reproductive structures or organs, such as the three examples in Figure 9.3.

As you know, the chromosome number is the total number of chromosomes in a cell of a given type. If the cell has a diploid number ($2n$), it has a *pair* of each type of chromosome, often from two parents. Except for two

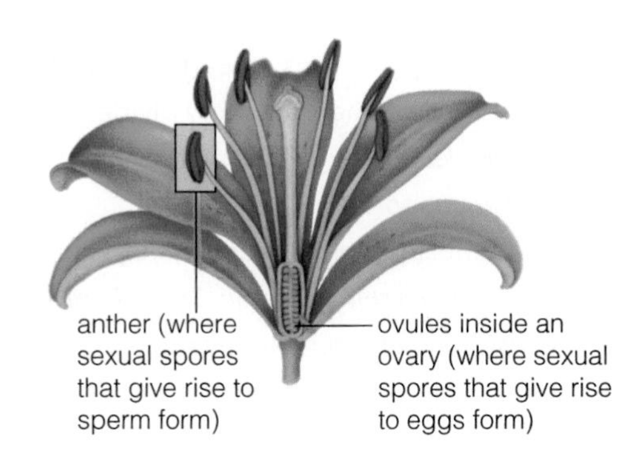

a Flowering plant

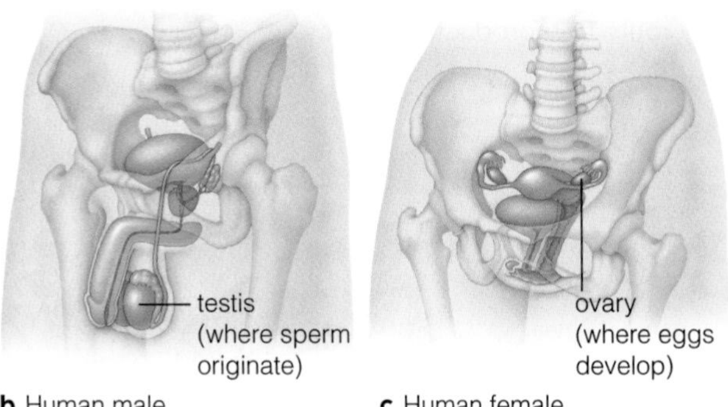

b Human male **c** Human female

Figure 9.3 Examples of reproductive organs, where cells that give rise to gametes originate.

nonidentical sex chromosomes, the chromosomes of a pair have the same length, shape, and assortment of genes. These homologous chromosomes briefly zipper together during meiosis (*hom*– means alike).

Human body cells are diploid; they have 23 pairs of homologous chromosomes (Figure 9.4). Meiosis halves the parental chromosome number, to a haploid number (n). After meiosis, every gamete normally ends up with 23 chromosomes—one of each pair.

TWO DIVISIONS, NOT ONE

Bear in mind, meiosis is similar to mitosis in certain respects. A cell duplicates its DNA before the division process starts. The two DNA molecules and associated proteins stay attached at the centromere, the notably constricted region along their length. For as long as they remain attached, we call them sister chromatids:

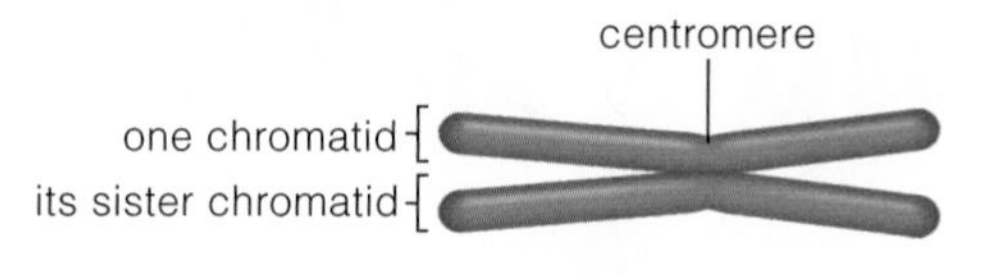

As in mitosis, the microtubules of a spindle move the chromosomes to opposite poles of the cell. However, with meiosis, two consecutive nuclear divisions form four haploid nuclei. There is no interphase between the two divisions, which are called meiosis I and II:

	Meiosis I		Meiosis II
Interphase (DNA is replicated prior to meiosis I)	Prophase I Metaphase I Anaphase I Telophase I	*No* interphase (DNA is *not* replicated prior to meiosis II)	Prophase II Metaphase II Anaphase II Telophase II

In meiosis I, each duplicated chromosome aligns with its partner, homologue to homologue. After they are sorted and arranged this way, each homologous chromosome is pulled away from its partner:

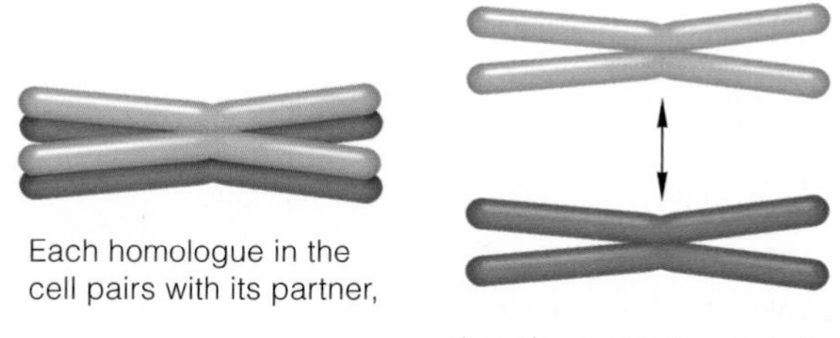

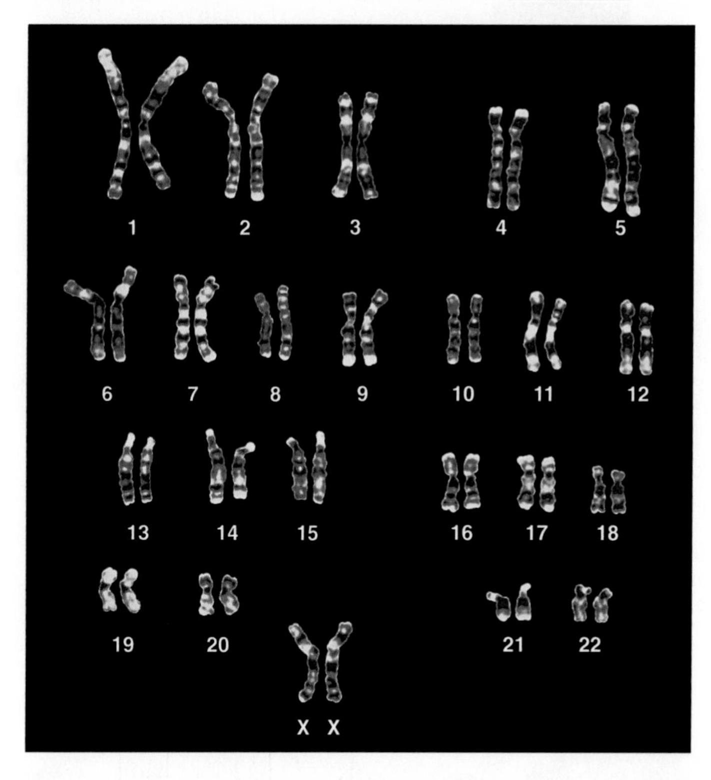

Figure 9.4 Twenty-three homologous pairs of human chromosomes. This example is from a human female, with two X chromosomes. Human males have a different pairing of sex chromosomes (XY).

All homologues move apart from their partners and end up in two new nuclei. When the cytoplasm divides, there are two daughter cells with one of each type of chromosome. The chromosomes are still duplicated. Then, during meiosis II, the two sister chromatids of each chromosome separate. They become individual, unduplicated chromosomes:

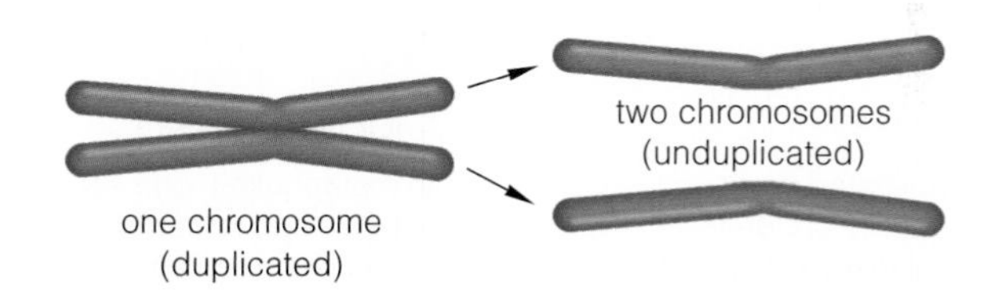

There are now four parcels of chromosomes; each contains one unduplicated chromosome of each type. New nuclear envelopes begin to enclose them as four nuclei inside two cells. Typically, four haploid (n) cells form after a second cytoplasmic division. Figure 9.5 in the next section puts these chromosomal movements into the context of the sequential stages of meiosis.

Meiosis is a nuclear division mechanism that occurs in immature reproductive cells of eukaryotes. It halves the parental cell's chromosome number, to the haploid number (n), before gametes form.

LINKS TO SECTIONS 4.12, 8.2, 8.3

9.3 Visual Tour of Meiosis

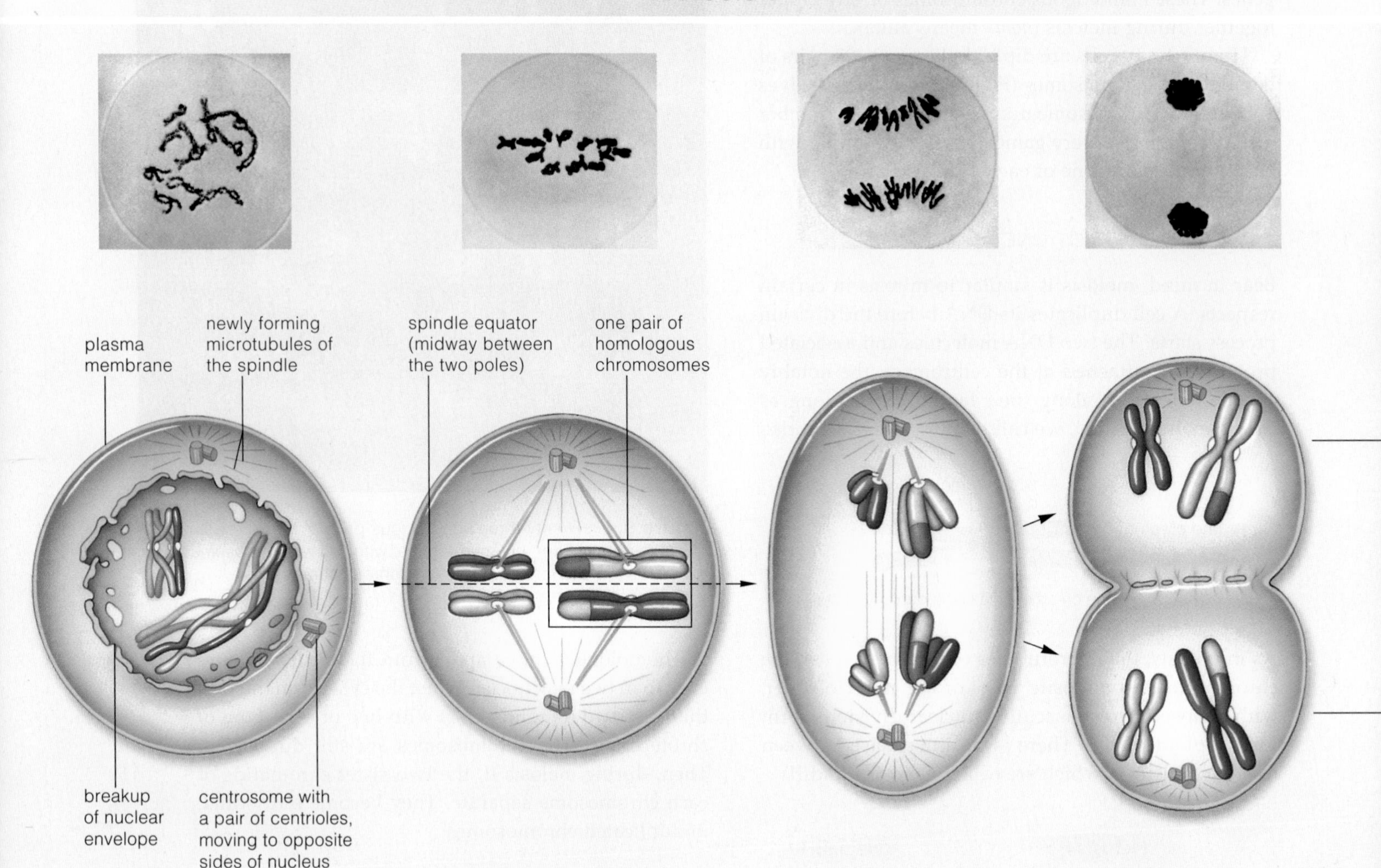

a Prophase I

Chromosomes were duplicated earlier, in interphase. Now they start to condense. Each pairs with its homologue and swaps segments with it, as indicated by color breaks in the large chromosomes. New microtubules are forming a bipolar spindle (Section 8.2). One of the two centrosomes moves to the opposite side of the nuclear envelope, which begins to break up.

b Metaphase I

Prior to metaphase I, one set of microtubules had tethered one chromosome of each type to one spindle pole and another set tethered its homologue to the other spindle pole. A tug-of-war between the two sets has now aligned the chromosomes midway between the two spindle poles.

c Anaphase I

One of each duplicated chromosome, maternal or paternal, moves to a spindle pole; its homologue moves to the opposite pole. Motor proteins that interact with microtubules bring about the movement, as explained in Section 4.12.

d Telophase I

One of each type of chromosome has arrived at a spindle pole. In most species, the cytoplasm divides at this time. All of the chromosomes are still duplicated.

Figure 9.5 **Animated!** Meiosis in one type of animal cell. This nuclear division mechanism halves the parental chromosome number in immature reproductive cells, to the haploid number, for forthcoming gametes. To keep things simple, we track only two pairs of homologous chromosomes. Maternal chromosomes are shaded *purple* and paternal chromosomes *blue*. Of the four haploid cells that form by meiosis and cytoplasmic divisions, one or all may develop into gametes and function in sexual reproduction. In plants, the cells that form may develop into spores, a stage that precedes gamete formation in the life cycle.

The light micrographs above each sketch show corresponding stages of meiosis in a lily plant (*Lilium regale*). These steps precede the formation of pollen grains.

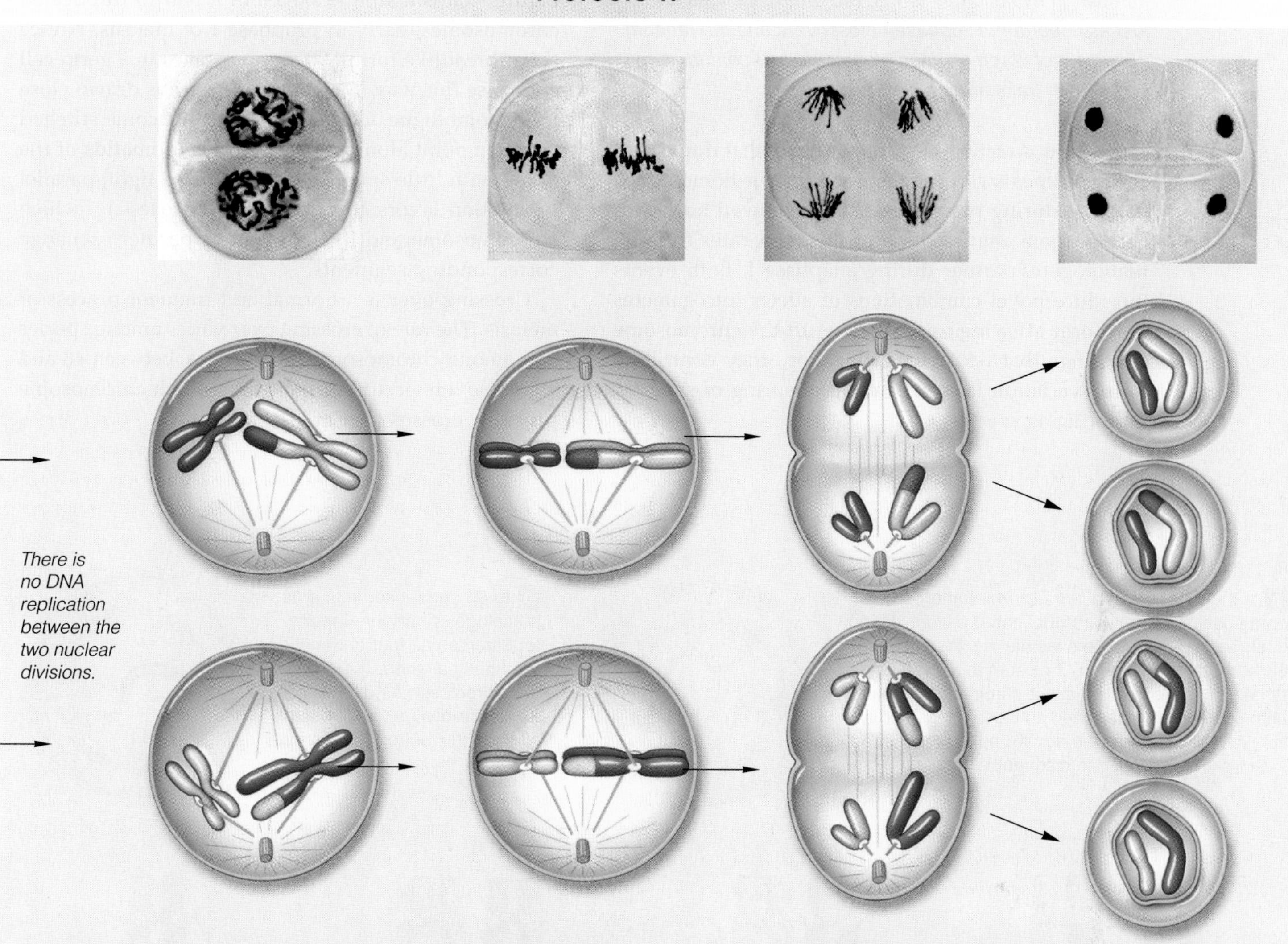

e Prophase II

In each cell, one of two centrioles moves to the opposite side of the cell, and a new bipolar spindle forms. Some spindle microtubules harness one chromatid of each chromosome to a spindle pole; other microtubules harness its sister chromatid to the other pole.

f Metaphase II

By now, microtubules from both spindle poles have finished a tug-of-war. They have aligned all of the still-duplicated chromosomes midway between the poles.

g Anaphase II

The sister chromatids of each chromosome move apart and are now individual, unduplicated chromosomes. Microtubules pull them toward opposite spindle poles as other microtubules push the poles apart. A parcel of one of each type of chromosome will end up near each pole.

h Telophase II

A new nuclear envelope encloses each parcel of chromosomes, so there are now four nuclei. The cytoplasm divides. Each of the daughter cells now has a haploid number (n) of unduplicated chromosomes.

9.4 How Meiosis Introduces Variations in Traits

In Sections 9.2 and 9.3, you read that meiosis functions to halve the parental chromosome number, to the haploid number. In evolutionary terms, two other functions are just as important: Prophase I crossovers and the random alignment of chromosomes at metaphase I contribute to variation in traits among offspring.

LINKS TO SECTIONS 8.1, 8.2

The previous section mentioned briefly that duplicated chromosomes swap segments with their homologous partners during prophase I. It also showed how each chromosome aligns with and then separates from its homologous partner during anaphase I. Both events introduce novel combinations of alleles into gametes that form after meiosis. Along with the chromosome shufflings that occur in fertilization, they contribute to the variation in traits among offspring of sexually reproducing species.

CROSSING OVER IN PROPHASE I

Figure 9.6*a* is a simple sketch of a pair of duplicated chromosomes, early in prophase I of meiosis. Notice their threadlike form. All chromosomes in a germ cell condense this way. When they do, each is drawn close to its homologue. Chromatids of one become stitched point by point along their length to chromatids of the other, with little space in between. This tight, parallel orientation favors crossing over—a process by which a chromosome and its homologous partner exchange corresponding segments.

Crossing over is a normal and frequent process of meiosis. The rate of crossing over varies among species and among chromosomes; in humans, between 46 and 95 crossovers occur per meiosis, so each chromosome probably crosses over at least once.

a A maternal chromosome (*purple*) and paternal chromosome (*blue*) were duplicated earlier, during interphase. They become visible in microscopes early in prophase I, when they start to condense to threadlike form. The two sister chromatids of each chromosome are positioned so closely together that they look like a single thread. We pulled them apart in this sketch so you can distinguish between them.

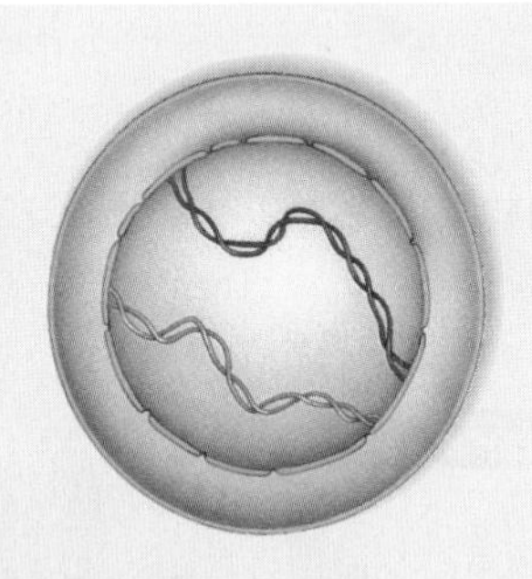

b Each chromosome and its homologous partner zipper together, so all four chromatids are tightly aligned. If the two sex chromosomes differ (such as X paired with Y), they still align tightly, but only in a small region at their ends.

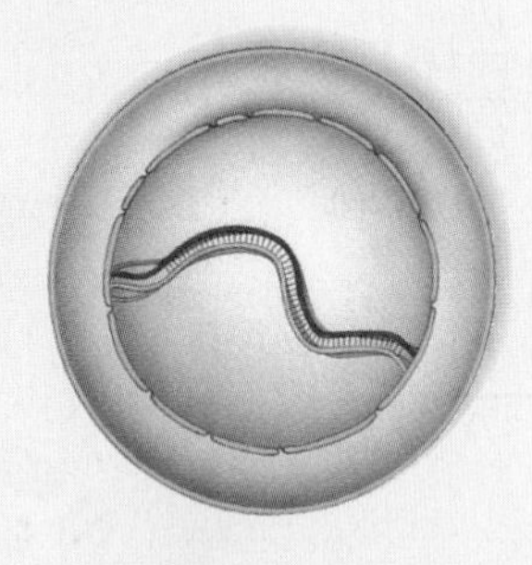

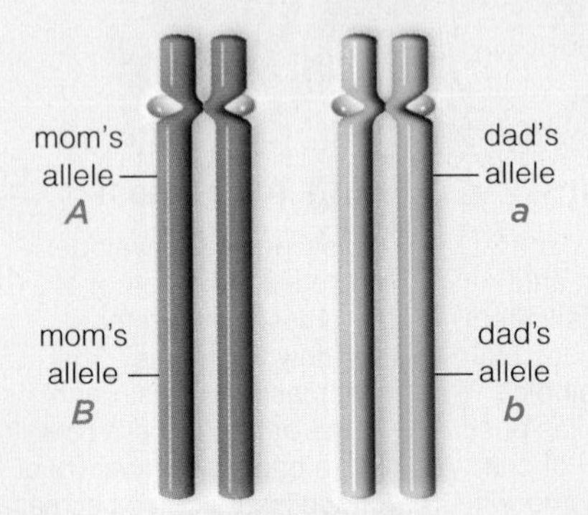

c Here is a simple way to think about crossing over. (Chromosomes are still condensed and threadlike, and each is tightly aligned with its homologous partner.)

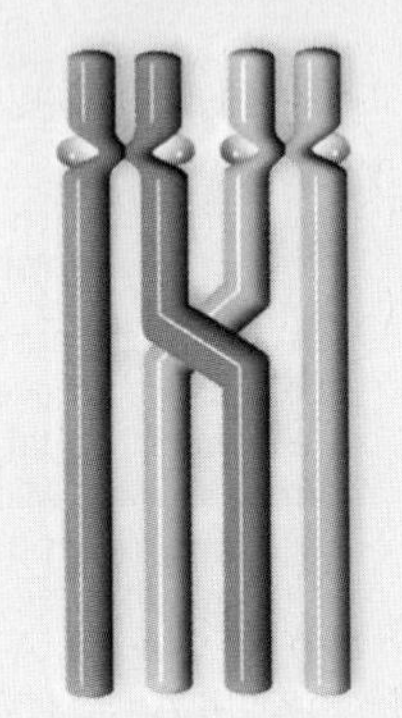

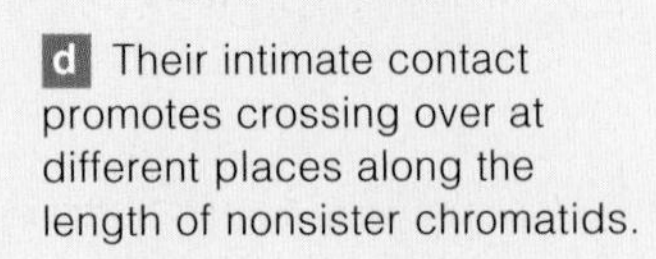

d Their intimate contact promotes crossing over at different places along the length of nonsister chromatids.

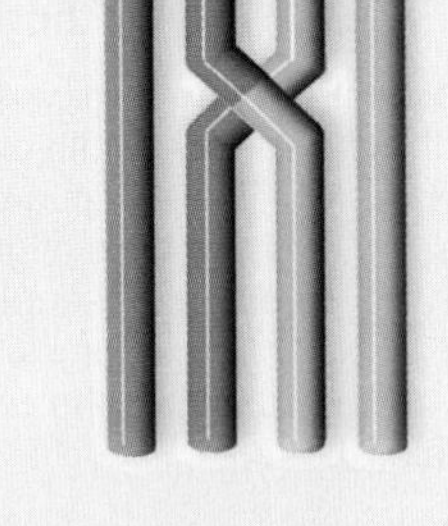

e At the crossover site, paternal and maternal chromatids exchange corresponding segments.

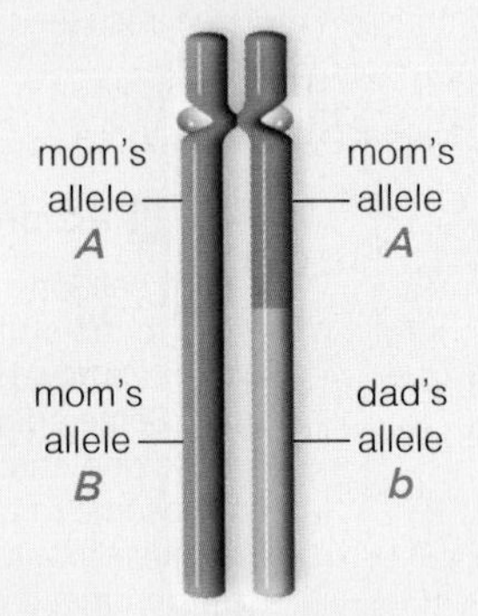

f Crossing over mixes up maternal and paternal alleles on homologous chromosomes.

Figure 9.6 **Animated!** Key events of prophase I, the first stage of meiosis. For clarity, we show only one pair of homologous chromosomes and one crossover. More than one crossover may occur in each chromosome pair. *Blue* signifies a paternal chromosome, and *purple*, its maternal homologue.

Gene swapping would be pointless if each type of gene never varied. But remember, a gene can come in slightly different forms—alleles. You can predict that a number of the alleles on one chromosome will *not* be identical to their partner alleles on the homologous chromosome. That is why each crossover event is an opportunity to exchange slightly different versions of heritable information.

We will be returning to the impact of crossing over in later chapters. For now, just remember this point: *Crossing over between a pair of homologous chromosomes puts novel combinations of alleles in both, which results in novel variations in forms of traits among offspring.*

METAPHASE I ALIGNMENTS

Major shufflings of whole chromosomes start during the transition from prophase I to metaphase I. Suppose this is happening right now in one of your germ cells. Crossovers have already made genetic mosaics of the chromosomes, but put this aside to simplify tracking. Just call the twenty-three chromosomes you inherited from your mother the maternal chromosomes, and the twenty-three chromosomes you inherited from your father the paternal chromosomes.

At metaphase I, microtubules from both poles have now aligned all of the duplicated chromosomes at the spindle equator (Figure 9.5*b*). Have they attached all maternal chromosomes to one pole and all paternal chromosomes to the other? Most likely not. All of the microtubules that grow out from a spindle pole latch onto the first chromosome they contact, regardless of whether it is maternal or paternal. As a result, there is no pattern to the attachment of maternal or paternal chromosomes to either pole during metaphase I. *Either partner can end up at a spindle pole.*

Then, in anaphase I, each duplicated chromosome separates from its homologous partner and is pulled toward the pole to which it is attached.

Think of the possibilities while tracking just three pairs of homologues. By metaphase I, these three pairs may be arranged in any one of four possible positions (Figure 9.7). This means that eight combinations (2^3) are possible for forthcoming gametes.

Cells that give rise to human gametes have twenty-three pairs of homologous chromosomes, not three. Thus, every time a human sperm or egg forms, there is a total of 8,388,608 (or 2^{23}) possible combinations of maternal and paternal chromosomes! Moreover, in a sperm or an egg, many hundreds of alleles inherited from the mother might not "say" the exact same thing about hundreds of different traits as alleles inherited from the father. Are you getting an idea of why such fascinating combinations of traits show up among the generations of your own family tree?

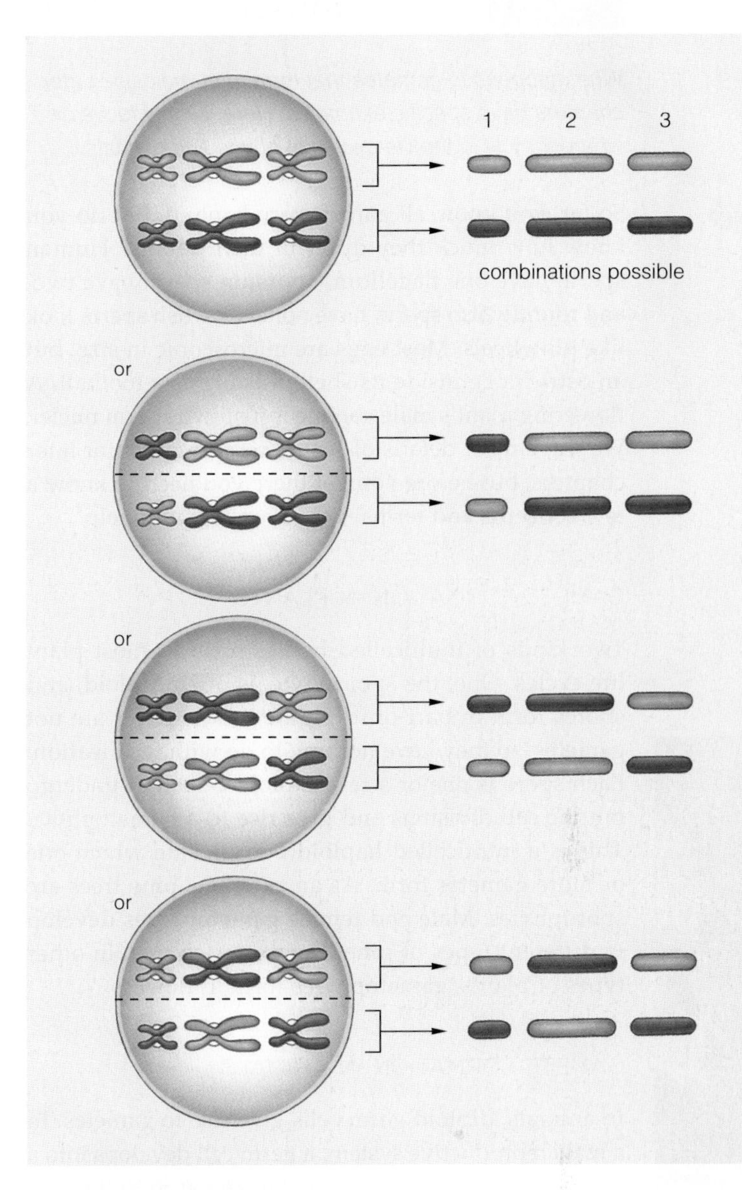

Figure 9.7 Animated! Possible outcomes for random alignment of three pairs of homologous chromosomes at metaphase I. The three types of chromosomes are labeled 1, 2, and 3. With four alignments, eight combinations of maternal chromosomes (*purple*) and paternal chromosomes (*blue*) are possible in gametes.

Crossing over, or recombination between homologous chromosomes, breaks up old combinations of alleles and puts new ones together during prophase I of meiosis.

The random tethering and subsequent positioning of each pair of maternal and paternal chromosomes at metaphase I lead to different combinations of maternal and paternal traits in each new generation.

9.5 From Gametes to Offspring

What happens to gametes that form after meiosis? Later chapters have specific examples. Here, simply focus on where they fit in the life cycles of plants and animals.

LINK TO SECTION 8.4

So far, you know all gametes are haploid, but do you know how much they differ in their details? Human sperm have one flagellum, opossum sperm have two, and roundworm sperm have none. Crayfish sperm look like pinwheels. Most eggs are microscopic in size, but an ostrich egg inside its shell is as big as a football. A flowering plant's male gamete is just two sperm nuclei. We leave most details of sexual reproduction for later chapters, but before you get there you need to know a few concepts and terms. This overview will help.

GAMETE FORMATION IN PLANTS

Two kinds of multicelled bodies form in most plant life cycles. One, the sporophyte, is often diploid, and spores form in part of it (Figure 9.8*a*). Spores are not gametes, so they have nothing to do with fertilization. Each spore is one or a few haploid cells that undergo mitotic cell divisions and give rise to a gametophyte. This is a multicelled haploid body inside which one or more gametes form. As an example, pine trees are sporophytes. Male and female gametophytes develop in different types of pine cones on each tree. In other types of plants, gametophytes form in flowers.

GAMETE FORMATION IN ANIMALS

In animals, diploid germ cells give rise to gametes. In a male reproductive system, a germ cell develops into a primary spermatocyte. This large, immature cell enters meiosis and its cytoplasm divides. Four haploid cells result and develop into spermatids (Figure 9.9). Each cell develops one or more flagella. It becomes a sperm, a type of mature male gamete.

In female animals, a germ cell becomes a primary oocyte, which is an immature egg. Unlike sperm, the primary oocyte increases in size and stockpiles many cytoplasmic components. Also, its four daughter cells differ in size and function (Figure 9.10).

When the primary oocyte divides after meiosis I, one daughter cell—the secondary oocyte—gets nearly all of the cytoplasm. The other cell, a first polar body, is quite small. Later, both haploid cells enter meiosis II, then the cytoplasm divides. One of the secondary oocyte's daughter cells becomes a second polar body. The secondary oocyte's other daughter cell gets most of the cytoplasm and develops into a gamete. A mature female gamete is an ovum (plural, ova), or egg.

Polar bodies are not nutrient-rich or plump with cytoplasm, and generally do not function as gametes. In time they will degenerate. Their formation simply ensures that the egg will have a haploid chromosome number and will get enough metabolic machinery to support early divisions of the new individual.

MORE SHUFFLINGS AT FERTILIZATION

The chromosome number characteristic of parents is restored at fertilization: fusion of the haploid nuclei of two gametes. If meiosis did not precede fertilization, the chromosome number would double with every generation. Such doublings in chromosome number change the individual's hereditary information, which is a fine-tuned set of blueprints that must be followed exactly, page by page, to build a normal body. Changes in those blueprints can have serious consequences.

Fertilization also adds to variation among offspring. Reflect on the possibilities for humans alone. During prophase I, every human chromosome undergoes an average of one or two crossovers. In addition to those crossovers, the random positioning of pairs of paternal and maternal chromosomes at metaphase I results in

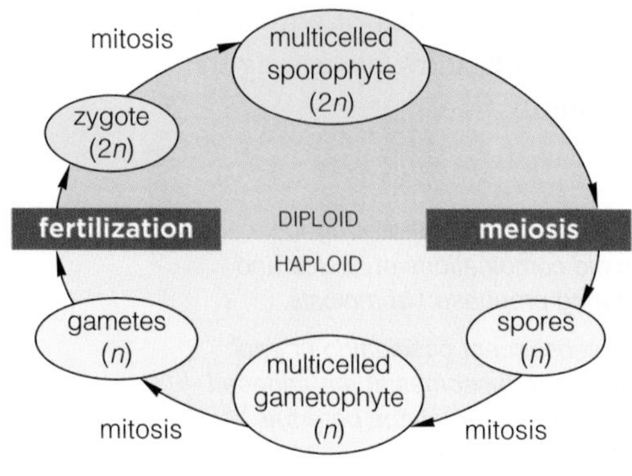

a Plant life cycle

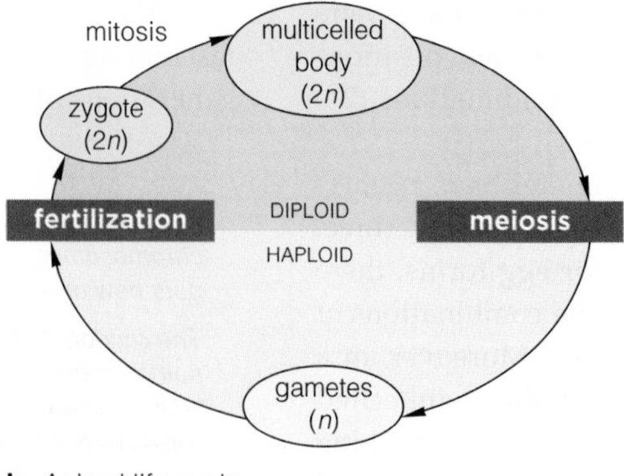

b Animal life cycle

Figure 9.8 (**a**) Generalized life cycle for most plants. A pine tree is a typical sporophyte.
(**b**) Generalized life cycle for animals. The zygote is the first cell to form when the nuclei of two gametes, such as a sperm and an egg, fuse at fertilization.

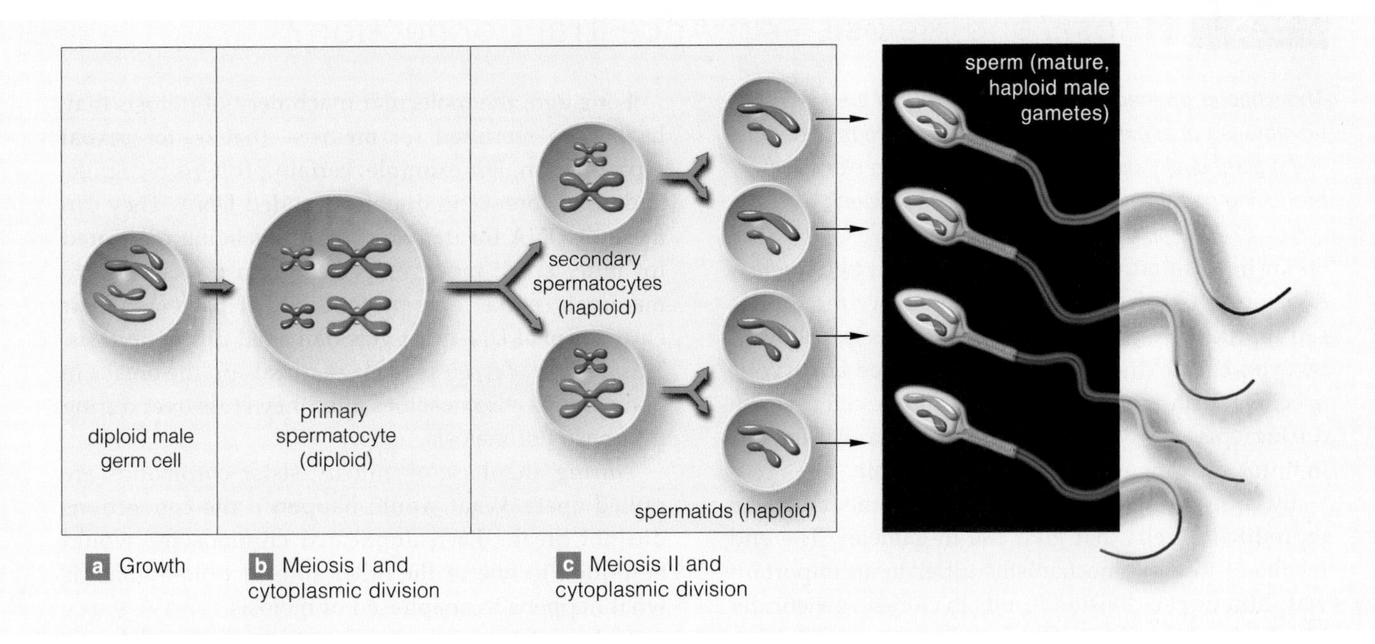

Figure 9.9 **Animated!** Generalized sketch of sperm formation in animals. Figure 39.13 shows a specific example: how sperm form in human males.

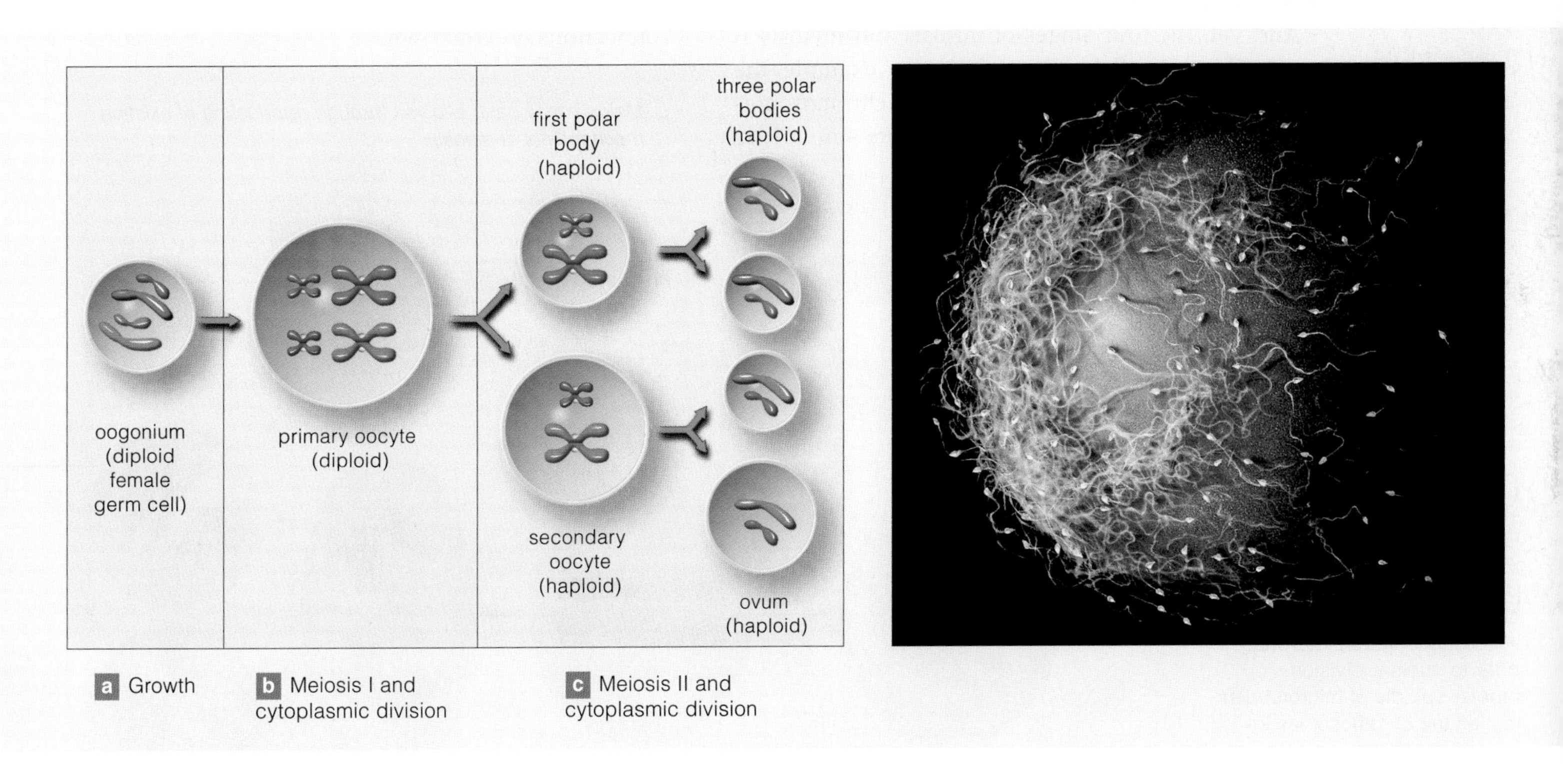

Figure 9.10 **Animated!** Animal egg formation. Eggs are far larger than sperm and larger than the three polar bodies. The painting, based on a scanning electron micrograph, depicts human sperm surrounding an ovum. Figure 39.17 shows how human eggs form.

one of millions of possible chromosome combinations in each gamete. And of all male and female gametes that form, *which* two actually get together is a matter of chance. The sheer number of combinations that can exist at fertilization is staggering!

Meiosis and cytoplasmic division precede the development of haploid gametes in animals and spores in plants. The eventual union of two haploid gametes at fertilization results in a diploid zygote that inherited alleles from both parents.

9.6 Mitosis and Meiosis—An Ancestral Connection?

This chapter opened with hypotheses about the survival advantages of asexual and sexual reproduction. It seems like a giant evolutionary step from producing clones to producing genetically varied offspring. But was it?

LINKS TO SECTIONS 8.3, 8.5

So far in this unit, our focus has been on two nuclear division mechanisms. Single-celled eukaryotic species can reproduce asexually by way of mitosis, followed by cytoplasmic division. Many multicelled eukaryotic species switch to reproduction by mitotic cell division during episodes of asexual reproduction. All engage in mitosis during growth and tissue repair.

By contrast, meiosis happens only in the immature reproductive cells that give rise to gametes. The end results of the two mechanisms differ in an important way. Mitotic cell division results in clones—genetically identical copies of a parent cell. Meiotic cell division and fertilization promotes variation among offspring. Remember, such variation in traits contributes to life's immense diversity.

And yet, the four stages of mitosis and meiosis II have striking parallels (Figure 9.11). For example, the same kind of bipolar spindle separates chromosomes into parcels during both processes. More similarities occur at the molecular level.

Long ago, the molecular machinery of mitosis may have been recruited for meiosis—that is, for sexual reproduction. For example, certain proteins recognize and repair breaks in double-stranded DNA. They can monitor DNA for damage while it is being replicated for mitosis. All modern species, from prokaryotes to mammals, make these proteins. Other proteins repair chromosomal DNA that gets damaged during mitosis. *This same set of repair proteins* also seals up the breaks in homologous chromosomes after they cross over during prophase I of meiosis.

During anaphase of mitosis, sister chromatids are pulled apart. What would happen if the connections did not break? Each duplicated chromosome would be pulled to one or the other spindle pole—which is what happens in anaphase I of meiosis.

Did sexual reproduction originate through chance mutations in mechanisms that come into play before and during mitosis? We invite you to think about this possibility as you read later chapters or explore the connections on your own.

Meiosis may have evolved through remodeling of existing mechanisms of mitosis.

Meiosis I

Figure 9.11 Comparative summary of key features of mitosis and meiosis, starting with a diploid cell. Only two paternal and two maternal chromosomes are shown. Both were duplicated in interphase, prior to nuclear division. A bipolar spindle of microtubules moves the chromosomes in mitosis as well as meiosis.

Mitosis maintains the parental chromosome number. Meiosis halves it, to the haploid number.

Mitotic cell division is the basis of asexual reproduction among eukaryotes. It also is the basis of growth and tissue repair of multicelled eukaryotic species.

Meiotic cell division is a required step before the formation of gametes or sexual spores.

Prophase I
In a diploid ($2n$) germ cell, duplicated chromosomes now condense. The bipolar spindle forms and tethers the chromosomes. Crossovers occur between homologues.

Metaphase I
Each maternal chromosome and its paternal homologue are randomly aligned midway between the two spindle poles. Either one may get attached to either pole.

Anaphase I
Homologous partners separate and move to opposite poles.

Telophase I
There are two clusters of chromosomes. New nuclear envelopes may form and the cytoplasm may divide before meiosis II begins.

Right, a bipolar spindle at metaphase, anaphase, and telophase of mitosis in a mouse cell. The *green* stain identifies the microtubules of the spindle. The *blue* stain identifies DNA in the cell's chromosomes. Neither meiosis nor mitosis can proceed without a spindle.

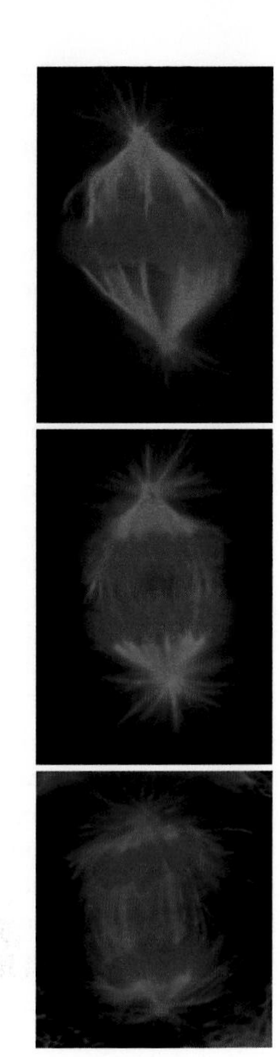

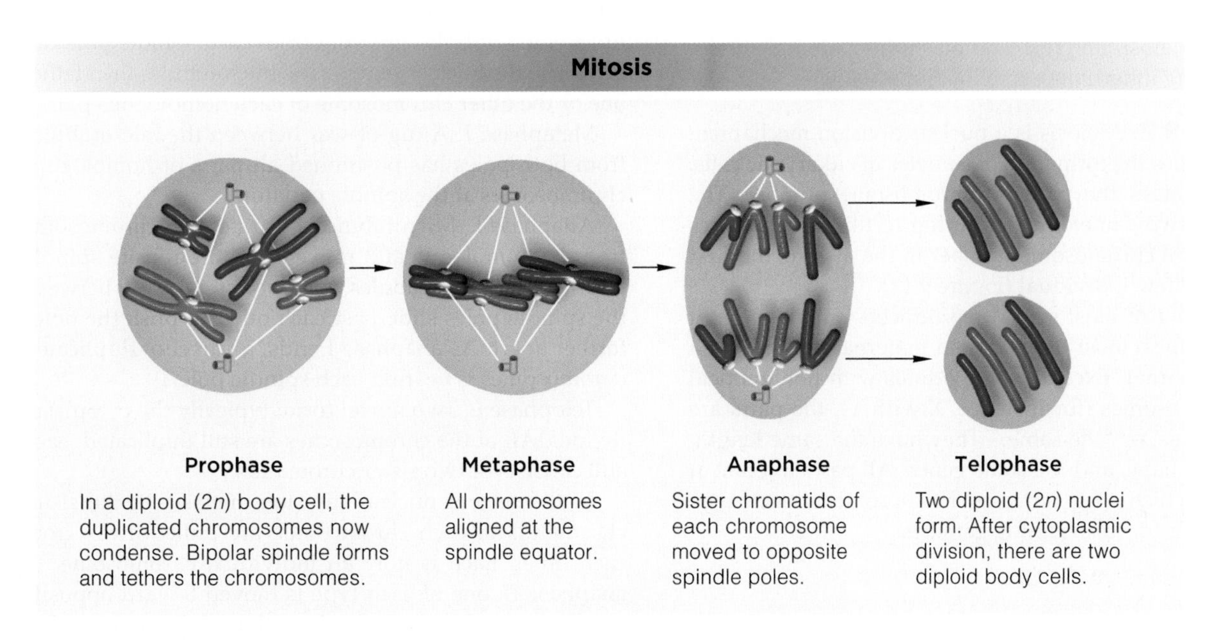

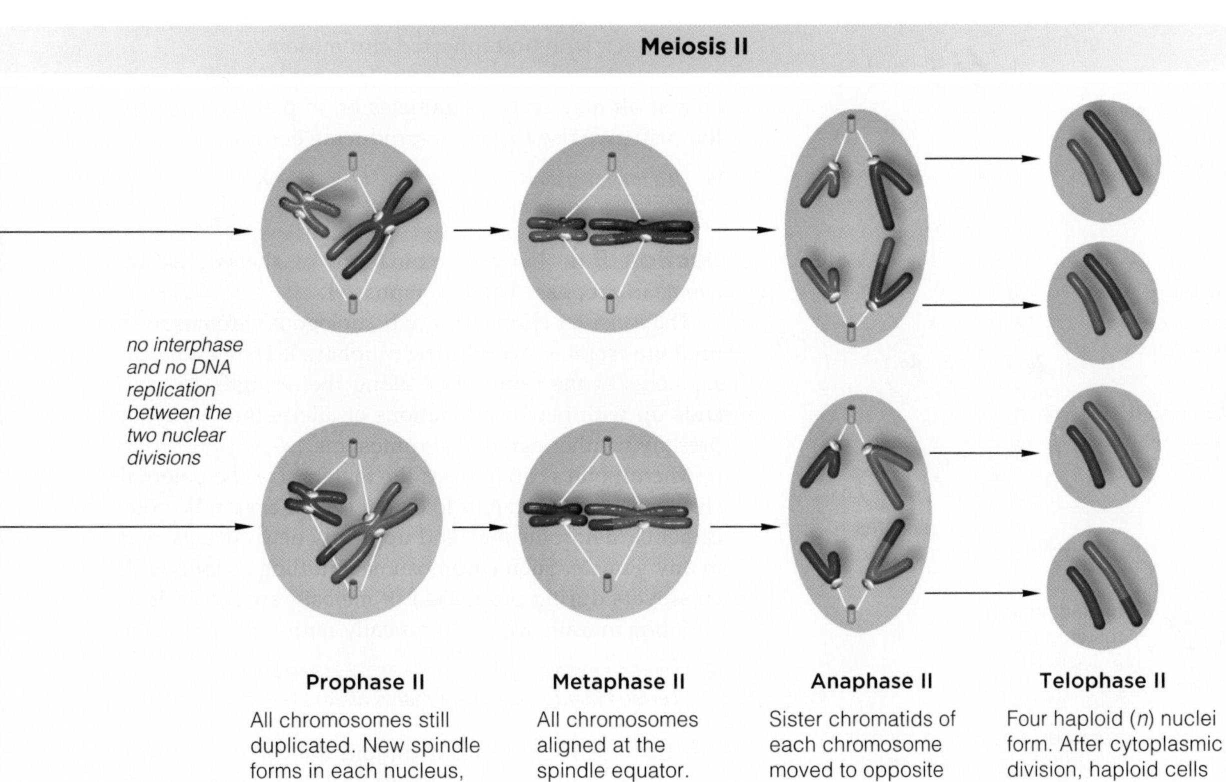

www.thomsonedu.com Thomson NOW!

Summary

Section 9.1 Many eukaryotic life cycles have asexual and sexual phases. In asexual reproduction, offspring are genetically identical to their one and only parent. In sexual reproduction, offspring differ from their two parents, and often from one another, in their combinations of alleles and in the details of their shared traits. Meiosis, gamete formation, and fertilization occur in sexual reproduction.

Alleles are slightly different molecular forms of the same gene. Each specifies a different version of the gene product. Meiosis and fertilization shuffle parental alleles, so offspring inherit new combinations of alleles.

Section 9.2 Meiosis is a nuclear division mechanism that precedes the formation of gametes in eukaryotic cells. Meiosis halves the parental chromosome number. The fusion of two gamete nuclei during fertilization restores the parental chromosome number in the zygote, the first cell of the new individual (Figure 9.12).

Offspring of most sexual reproducers inherit pairs of chromosomes, most often from a maternal parent and a paternal parent. Except in individuals with nonidentical sex chromosomes (for instance, X with Y), the pairs are homologous, or "the same." They have the same length, the same shape, and the same genes. All pairs interact at meiosis, which sorts out one chromosome of each type into parcels for forthcoming gametes.

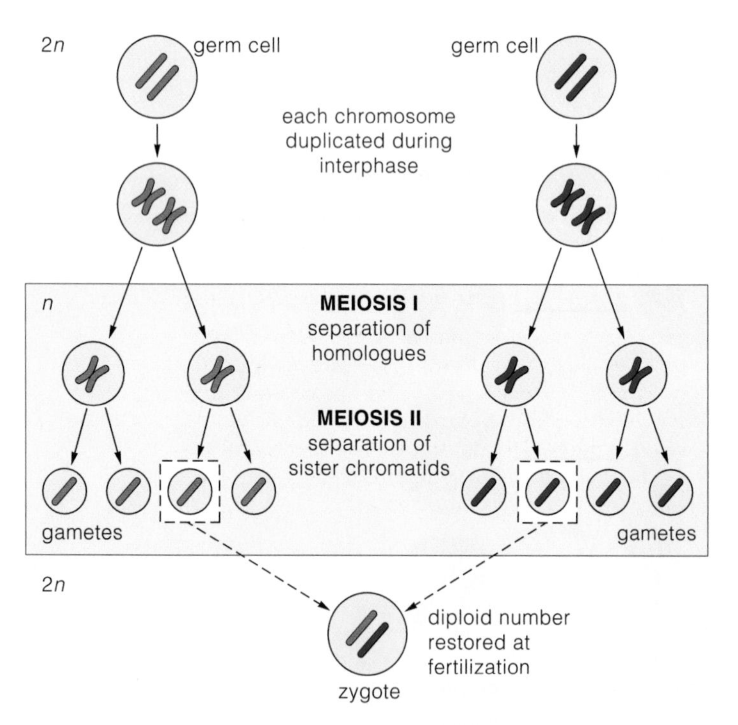

Figure 9.12 Summary of changes in chromosome number at different stages of sexual reproduction, using two diploid ($2n$) germ cells as the example. During two nuclear divisions, meiosis halves the chromosome number, and results in the formation of haploid (n) gametes. Later, the union of haploid nuclei of two gametes at fertilization restores the diploid number.

Section 9.3 All chromosomes are duplicated during interphase, before meiosis. Two divisions, meiosis I and II, divide the parental chromosome number by two so that each forthcoming gamete will be haploid (n).

In the first nuclear division, meiosis I, each duplicated chromosome lines up with its homologous partner; then the two move apart, toward opposite spindle poles.

Prophase I. Chromosomes condense and align tightly with their homologues. Each pair of homologues typically undergoes crossing over. Microtubules form the bipolar spindle. One of two pairs of centrioles is moved to the other side of the nucleus. The nuclear envelope breaks up, so microtubules growing from each spindle pole can penetrate the nuclear region. The microtubules then tether one or the other chromosome of each homologous pair.

Metaphase I. A tug-of-war between the microtubules from both poles has positioned all pairs of homologous chromosomes at the spindle equator.

Anaphase I. Microtubules separate each chromosome from its homologue and move both to opposite spindle poles. Other microtubules that overlap midway between the spindle poles slide past each other to push the poles farther apart. As anaphase I ends, a parcel of duplicated chromosomes is nearing each spindle pole.

Telophase I. Two nuclei form; typically the cytoplasm divides. All of the chromosomes are still duplicated; each still consists of two sister chromatids.

In the second nuclear division, meiosis II, the sister chromatids of each chromosome are pulled away from each other. Each is now an individual chromosome. In anaphase II, one of each type is moved toward opposite spindle poles. This happens in both nuclei that formed in meiosis I. By the end of telophase II, there are four haploid nuclei, each with unduplicated chromosomes.

When the cytoplasm divides, four haploid cells result. One or all may serve as gametes or, in plants, as spores that will give rise to gamete-producing bodies.

- *Use the animation on ThomsonNOW to explore what happens in the stages of meiosis.*

Section 9.4 Novel combinations of alleles arise by events in prophase I and metaphase I.

The *non*sister chromatids of homologous chromosomes undergo crossing over during prophase I. They exchange segments at the same place along their length, so each ends up with new combinations of alleles that were not present in either parental chromosome.

Microtubules can harness the maternal *or* the paternal chromosome of each pair to one or the other spindle pole. Either chromosome may end up in any new nucleus, and in any gamete. Such chromosome shufflings, along with crossovers during prophase I of meiosis, are the basis of variation in traits we see in sexually reproducing species.

- *Use the animation on ThomsonNOW to see how crossing over and metaphase I alignments affect allele combinations.*

Section 9.5 Multicelled diploid and haploid bodies are typical in life cycles of plants and animals. A diploid sporophyte is a multicelled plant body that makes haploid spores. Spores give rise to gametophytes, or multicelled plant bodies in which haploid gametes form. Germ cells in the reproductive organs of most animals give rise to sperm or eggs. Fusion of a sperm and egg at fertilization results in a zygote, the first cell of a new individual.

Three events give rise to novel combinations of alleles: crossing over during prophase I, the random alignment of maternal and paternal chromosomes at metaphase I, and the chance meeting of gametes at fertilization. All three contribute to variation in traits among offspring.

■ *Use the animation on ThomsonNOW to see how gametes form.*

Section 9.6 Like mitosis, meiosis requires a bipolar spindle to move and sort duplicated chromosomes, but meiosis occurs only in immature cells set aside for sexual reproduction. Mitosis maintains the parental chromosome number. Meiosis halves the chromosome number, and it introduces new combinations of alleles into offspring.

Some mechanisms of meiosis resemble those of mitosis, and may have evolved from them. DNA repair enzymes that function in both are an example.

Self-Quiz

Answers in Appendix III

1. Meiosis and cytoplasmic division function in ______ .
   a. asexual reproduction of single-celled eukaryotes
   b. growth and tissue repair
   c. sexual reproduction
   d. both b and c

2. Sexual reproduction requires ______ .
   a. meiosis c. spore formation
   b. fertilization d. a and b

3. Generally, a pair of homologous chromosomes ______ .
   a. carry the same genes c. are the same length, shape
   b. interact at meiosis d. all of the above

4. Meiosis ______ the parental chromosome number.
   a. doubles c. maintains
   b. halves d. mixes up

5. Meiosis ends with the formation of ______ .
   a. two cells c. eight cells
   b. two nuclei d. four nuclei

6. Sister chromatids of each duplicated chromosome separate during ______ .
   a. prophase I d. anaphase II
   b. prophase II e. both b and c
   c. anaphase I

7. ______ contributes to variation in traits among the offspring of sexual reproducers.
   a. Crossing over c. Fertilization
   b. Metaphase I random orientations d. both a and b
   e. All of the above are factors

8. The cell in the diagram below is in anaphase I, not anaphase II. I know this because ______ .

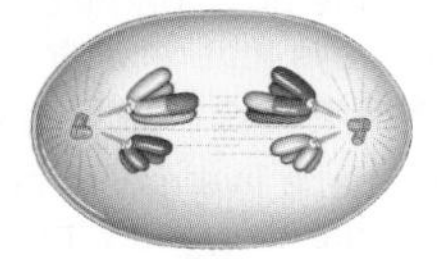

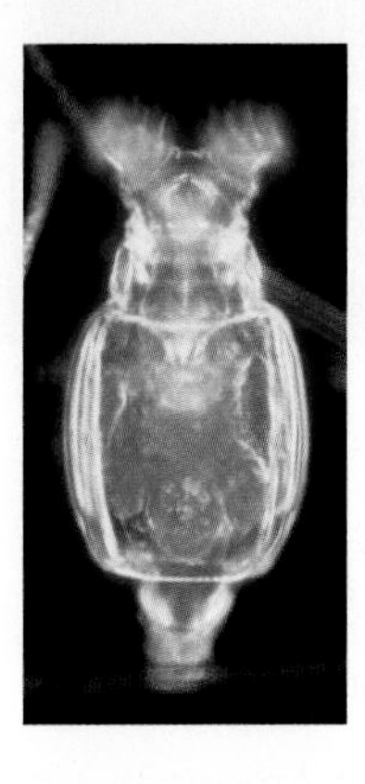

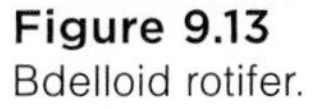

Figure 9.13 Bdelloid rotifer.

Figure 9.14 Viggo Mortensen (**a**) with and (**b**) without a chin dimple.

9. Match each term with its description.
   ___ chromosome number a. different molecular forms of the same gene
   ___ alleles b. none between meiosis I, II
   ___ metaphase I c. all chromosomes aligned at spindle equator
   ___ interphase d. all chromosomes in a given type of cell

■ *Visit ThomsonNOW for additional questions.*

Critical Thinking

1. Explain why you can predict that meiosis gives rise to genetic differences between parent cells and daughter cells in fewer cell divisions than mitosis does.

2. The bdelloid rotifer lineage started at least 40 million years ago (Figure 9.13). These tiny aquatic animals show tremendous genetic diversity: There are about 360 known species worldwide. Every bdelloid rotifer is female. What is the most probable cause of their great genetic diversity?

3. Actor Viggo Mortensen inherited a gene that makes his chin dimple. Figure 9.14*b* shows what he might have looked like if he inherited a different form of that gene. What is the name for alternative forms of the same gene?

4. Assume you can measure the amount of DNA in the nucleus of a primary oocyte, and then in the nucleus of a primary spermatocyte. Each gives you a mass *m*. What mass of DNA would you expect to find in the nucleus of each mature gamete (each egg and sperm) that forms after meiosis? What mass of DNA will be (1) in the nucleus of a zygote that forms at fertilization and (2) in that zygote's nucleus after the first DNA duplication?

5. The diploid chromosome numbers for the somatic cells of several eukaryotic species are listed at right. What is the number of chromosomes that normally ends up in gametes of each species? What would that number be after three generations if meiosis did not occur before gamete formation?

Species	Number
Fruit fly, *Drosophila melanogaster*	8
Garden pea, *Pisum sativum*	14
Frog, *Rana pipiens*	26
Earthworm, *Lumbricus terrestris*	36
Human, *Homo sapiens*	46
Amoeba, *Amoeba*	50
Dog, *Canis familiaris*	78
Vizcacha rat, *Tympanoctomys barrerae*	102
Horsetail, *Equisetum*	216

10 OBSERVING PATTERNS IN INHERITED TRAITS

IMPACTS, ISSUES Menacing Mucus

Cystic fibrosis (CF) is a debilitating and ultimately fatal genetic disorder. It affects about one in 3,200 newborns of European descent. The American College of Obstetricians and Gynecologists set off a wave of protest in 2001 when it advocated that obstetricians should advise all prospective parents to be tested for mutations associated with CF. Such testing would let parents know if they were at risk for having an affected child. If they were, they could choose a prenatal test to detect CF in a fetus. If the test was positive, parents could choose to end the pregnancy or to address the special needs of a child with CF.

The gene associated with CF encodes a membrane transport protein called CFTR (Figure 10.1). The protein forms channels across the plasma membrane of epithelial cells that line the passageways and ducts of the lungs, liver, pancreas, intestines, reproductive tract, and skin. CFTR actively transports chloride ions out of these cells; water follows the ions and forms a thin film on the free surface of the epithelial linings. Mucus, which lubricates the tissues and helps prevent infection, slides freely on the watery film.

More than 10 million people in the U.S. inherited one normal and one abnormal copy of the CFTR gene. Some of them have sinus problems, but no other symptoms develop. Most do not know they carry the mutated gene.

CF develops in anyone who inherits two copies of the mutated gene—one from each parent. Without a working copy of CFTR, transport of chloride ions across epithelial cell membranes is disrupted, and so is the movement of water. The result is thick, dry mucus that accumulates on epithelial linings.

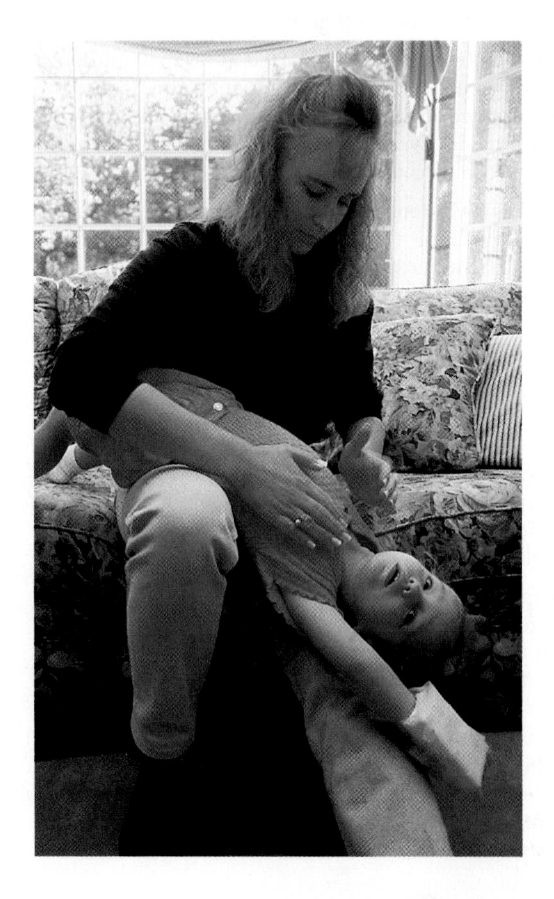

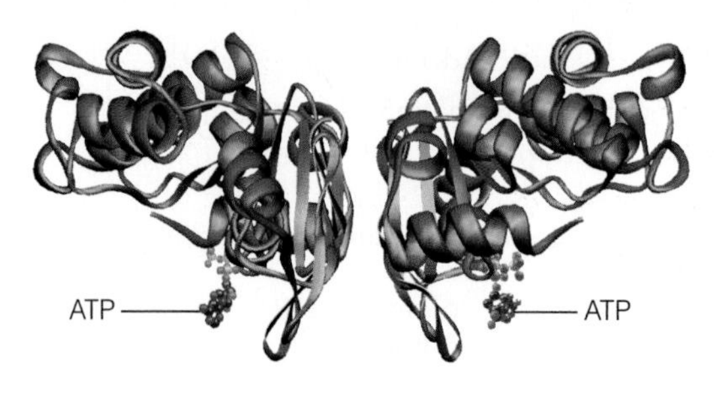

Figure 10.1 *Left*, child affected by cystic fibrosis, or CF, who each day endures thumps on the chest and back to dislodge the thick mucus that collects in the airways to the lungs.

Above, model for part of an ABC transporter, a category of membrane proteins that includes CFTR. The parts shown here are ATP-driven motors that can widen an ion channel across the plasma membrane. Symptoms vary from one affected individual to the next, partly because the CFTR gene can mutate in more than 500 ways that result in CF. Environmental factors and a person's genetic makeup also affect the outcome.

Opposite, a sample of lung tissue from a five-month-old patient diagnosed with cystic fibrosis. The white areas are plugged with mucus.

How would you vote? The ability to detect mutations that cause severe disorders raises bioethical questions. Should we encourage screening for alleles that cause cystic fibrosis? Should society encourage women to give birth only if their child will not develop severe medical problems? How severe? See ThomsonNOW for details, then vote online.

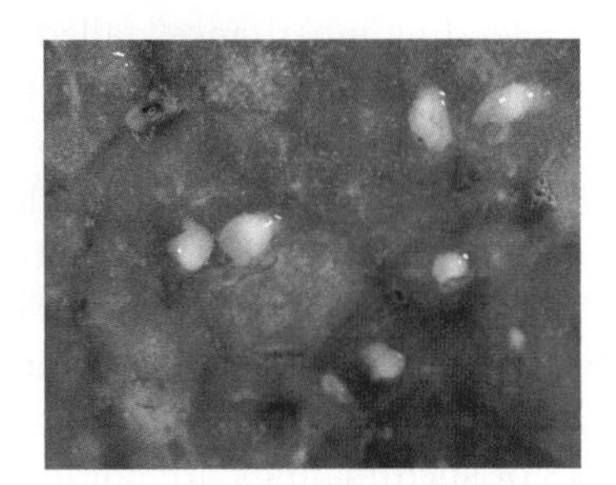

In the respiratory tract, the mucus clogs airways to the lungs and makes breathing difficult. It is too thick for the ciliated cells lining the airways to sweep out, and bacteria thrive in it. Low-grade infections may persist for years. Even with a lung transplant, most patients do not live longer than thirty years, at which time their lungs usually fail. At present, there is no cure.

We now know of hundreds of mutations that cause CF, but when prospective parents find out that they carry one of those mutations, what are they to do with the information? Such questions raise even larger ethical issues. What is the goal of mass screening for carriers of harmful mutations? Is it the welfare of individuals and families? Reduced medical costs? A healthy gene pool?

So here we are, working our way through the ethical consequences of understanding human genetics. It started long before DNA, genes, or chromosomes were known, in a small garden, with a monk named Gregor Mendel. By breeding many generations of pea plants in monastery garden plots, Mendel discovered evidence of how parents bestow units of hereditary information upon their offspring.

This chapter starts out with the methods and some representative results of Mendel's experiments. His work remains a classic example of how a scientific approach can pry open important secrets about the natural world. To this day, it is the foundation of modern genetics.

Key Concepts

WHERE MODERN GENETICS STARTED

Gregor Mendel gathered the first indirect, experimental evidence of the genetic basis of inheritance. His meticulous work tracking traits in many generations of pea plants gave him clues that heritable traits are specified in units. The units, which are distributed into gametes in predictable patterns, were later identified as genes. **Section 10.1**

INSIGHTS FROM MONOHYBRID EXPERIMENTS

Some experiments yielded evidence of gene segregation: When one chromosome separates from its homologous partner during meiosis, the pairs of alleles on those chromosomes also separate and end up in different gametes. **Section 10.2**

INSIGHTS FROM DIHYBRID EXPERIMENTS

Other experiments yielded evidence of independent assortment: During meiosis, the members of a pair of homologous chromosomes are distributed into gametes independently of how all other pairs are distributed. **Section 10.3**

VARIATIONS ON MENDEL'S THEME

Not all traits have clearly dominant or recessive forms. One allele of a pair may be fully or partially dominant over its nonidentical partner, or codominant with it. Two or more gene pairs often influence the same trait, and some single genes influence many traits. The environment also influences variation in gene expression. **Sections 10.4–10.7**

Links to Earlier Concepts

Before starting this chapter, be sure you can generally define genes, alleles, and diploid versus haploid chromosome numbers (Sections 9.1 and 9.2). As you read, you may wish to refer to the earlier introduction to natural selection (1.4) and to the visual road map for the four stages of meiosis (9.3). You will be considering experimental evidence of two major topics that were introduced earlier—the effects that crossing over and metaphase I alignments have on inheritance (9.4). You may wish to scan the introductions to experimental design (1.6, 1.7), protein structure (3.4), and pigments (6.1).

10.1 Mendel, Pea Plants, and Inheritance Patterns

We turn now to recurring inheritance patterns among humans and other sexually reproducing species. You already know meiosis halves the parental chromosome number, which is restored at fertilization. Here the story picks up with some observable outcomes of these events.

LINKS TO SECTIONS 1.6, 9.1

More than a century ago, people wondered about the basis of inheritance. Most had an idea that two parents contribute hereditary material to their offspring, but few suspected that the material is organized as units, or genes. According to the prevailing view, hereditary material must be fluid, with fluids from both parents blending at fertilization like milk into coffee.

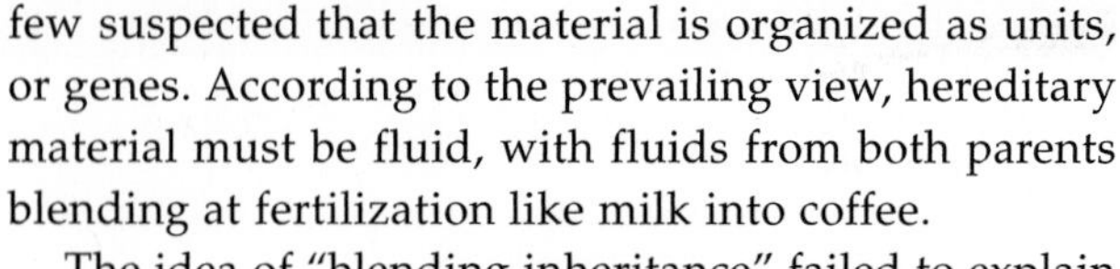

The idea of "blending inheritance" failed to explain the obvious. For example, many children who differ in their eye or hair color have the same two parents. If parental fluids blended, then the color would be some blended shade of the parental colors. If neither parent had freckles, freckled children would never pop up. A white mare bred with a black stallion should always give birth to gray offspring, but as all horse breeders knew, offspring of such matings are not always gray. Blending inheritance did not explain the variation in traits that people could see with their own eyes.

Charles Darwin did not accept the idea of blending inheritance. However, though inheritance was central to his theory of natural selection, he could not quite see how it works. By his theory, forms of traits often vary among individuals in a population. Variations that help an individual survive and reproduce tend to show up among more offspring over generations. Less helpful kinds become less frequent and, in time, may vanish. But neither he nor anyone else knew that hereditary material is divided into discrete units (genes), and that insight is crucial to explaining natural selection.

Even before Darwin presented his theory, someone was gathering evidence that would support it. Gregor Mendel, a monk (Figure 10.2), had already formulated a hypothesis that sperm and eggs carry separate units of information about heritable traits. By documenting the traits of his pea plants, generation after generation, he had been collecting indirect but *observable* evidence of how parents transmit those units to offspring.

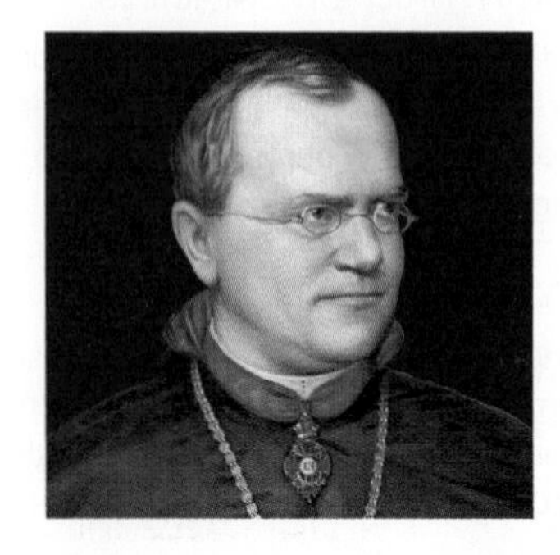

Figure 10.2 Gregor Mendel, the founder of modern genetics.

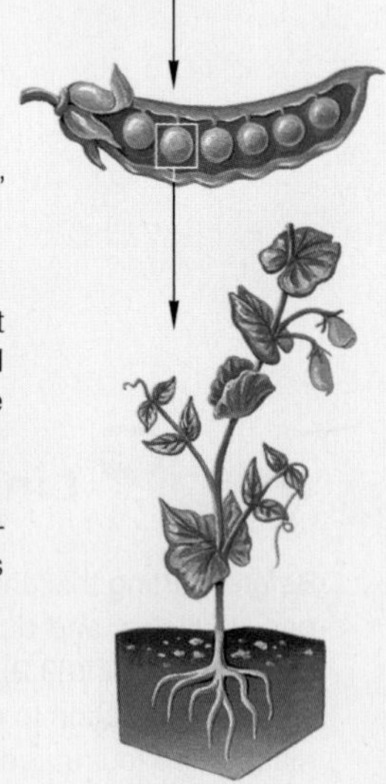

a Garden pea flower, cut in half. Sperm form in pollen grains, which originate in male floral parts (stamens). Eggs develop, fertilization takes place, and seeds mature in female floral parts (carpels).

b Pollen from a plant that breeds true for purple flowers is brushed onto a floral bud of a plant that breeds true for white flowers. The white flower had its stamens snipped off. This is one way to assure that a plant will not self-fertilize.

c Later, seeds develop inside pods of the cross-fertilized plant. An embryo in each seed develops into a mature pea plant.

d Each new plant's flower color is indirect but observable evidence that hereditary material has been transmitted from the parent plants.

Figure 10.3 **Animated!** Garden pea plant (*Pisum sativum*), which can self-fertilize or cross-fertilize. Experimenters can control the transfer of its hereditary material from one flower to another.

MENDEL'S EXPERIMENTAL APPROACH

Mendel spent most of his adult life in Brno, a city near Vienna that is now part of the Czech Republic. He was not a man of narrow interests who just stumbled onto dazzling principles. He lived in a monastery close to European cities that were centers of scientific inquiry. Having been raised on a farm, Mendel was aware of agricultural principles and their applications. He kept abreast of current literature on breeding experiments. He was a dedicated member of an agricultural society, and he won awards for developing improved varieties of fruits and vegetables.

Just after Mendel entered the monastery at Brno, he took courses in mathematics, physics, and botany at the University of Vienna. Few scholars of his time were trained in both plant breeding and mathematics.

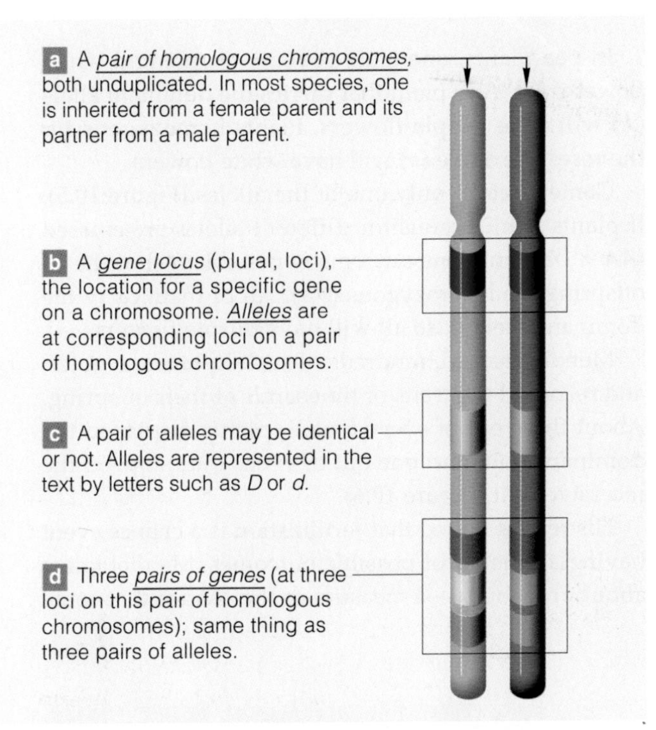

Figure 10.4 **Animated!** A few genetic terms. Like other species with a diploid chromosome number, garden pea plants have pairs of genes, on pairs of homologous chromosomes. Most genes come in slightly different molecular forms called alleles. Different alleles specify different versions of the same trait. An allele at any given location on a chromosome may or may not be identical with its partner on the homologous chromosome.

Shortly after his university training, Mendel started to study *Pisum sativum*, the garden pea plant (Figure 10.3). This plant is self-fertilizing. Its flowers produce both male and female gametes—call them sperm and eggs—that can come together and give rise to a new plant. One lineage of pea plants can "breed true" for certain traits. This means successive generations will be identical in one or more traits, as when all offspring grown from the seeds of self-fertilized, white-flowered parent plants also have white flowers.

Also, plant breeders cross-fertilize plants when they transfer pollen from one plant to the flower of another plant. Breeders open a floral bud of a plant that bred true for white flowers or some other trait and snip out its anthers. (Pollen grains, structures in which sperm develop, start forming in anthers.) The buds can be brushed with pollen from a plant that bred true for a different form of the trait.

Mendel predicted that such observable differences could help him follow a certain trait through many generations. If there were patterns in its inheritance, they might tell him something about heredity itself.

TERMS USED IN MODERN GENETICS

In Mendel's time, no one knew about genes, meiosis, or chromosomes. As we follow his thinking, we will clarify the picture by substituting some modern terms used in inheritance studies, as stated here and in Figure 10.4:

1. Genes are heritable units of information about traits. Parents transmit genes to offspring. Each gene occurs at a specific location (locus) on a specific chromosome.

2. Cells with a diploid chromosome number ($2n$) have pairs of genes, on pairs of homologous chromosomes.

3. A mutation is a permanent change in a gene and in the information it carries. It may cause a trait to change, as when a gene for flower color specifies purple and a mutated form specifies white. All molecular forms of the same gene are known as alleles.

4. When offspring inherit a pair of *identical* alleles for a trait generation after generation, we say they belong to a true-breeding lineage. The offspring of a cross, or mating, between two individuals that breed true for different forms of a trait are called hybrids. A hybrid inherited two *nonidentical* alleles for the trait.

5. A *heterozygous* condition occurs when there are two different alleles at the same gene locus on homologous chromosomes. A *homozygous* condition occurs when the homologous chromosomes carry the same allele.

6. An allele is dominant when its effect on a trait masks the effect of any recessive allele paired with it. Capital letters signify dominant alleles, and lowercase letters signify recessive ones. *A* and *a* are examples.

7. A homozygous dominant individual has a pair of dominant alleles (*AA*). A heterozygous individual has a pair of nonidentical alleles (*Aa*). Heterozygotes are hybrids. A homozygous recessive individual has a pair of recessive alleles (*aa*).

8. Gene expression is the process by which a gene's information is converted to a structural or functional part of a cell. Expressed genes determine traits.

9. Two terms help keep the distinction clear between genes and the traits they specify: Genotype refers to the particular alleles that an individual carries; phenotype refers to an individual's traits.

10. F_1 stands for the first-generation offspring, and F_2 for the second-generation offspring of self-fertilized or intercrossed F_1 individuals.

By tracking observable traits through generations of pea plants, Mendel collected evidence of how inheritance works.

10.2 Mendel's Theory of Segregation

Mendel used monohybrid experiments to test this hypothesis: Garden pea plants inherit two "units" of information (genes) for a trait, one from each parent.

LINK TO SECTION 1.8

Monohybrid experiments test for dominant or recessive alleles at one locus. Individuals with different alleles of one gene are crossed (or self-fertilized); traits of offspring are clues about the alleles. Monohybrid experiments can be crosses between individuals homozygous for different alleles (*AA* × *aa*), or between heterozygous individuals (*Aa* × *Aa*). Mendel used such monohybrid experiments to track seven pea plant traits through two generations. In one set of experiments, he crossed plants that bred true for purple flowers with plants that bred true for white flowers. All of the F_1 offspring had purple flowers. When he crossed the F_1 offspring, he saw that some F_2 offspring had white flowers! What was going on?

In pea plants, one gene governs purple and white flower color. Any plant that carries the dominant allele (*A*) will have purple flowers. Plants homozygous for the recessive allele (*a*) will have white flowers.

Gametes carry only one of the alleles (Figure 10.5). If plants homozygous for different alleles are crossed (*AA* × *aa*), only one outcome is possible: All of the F_1 offspring are heterozygous (*Aa*). All of them carry the dominant allele *A*, so all will have purple flowers.

Mendel crossed hundreds of such F_1 heterozygotes, and recorded the traits of thousands of their offspring. About three out of every four F_2 plants displayed the dominant trait, and one out of every four showed the recessive trait (Figure 10.6).

His results hinted that fertilization is a chance event having a number of possible outcomes. Mendel knew about probability—a measure of the chance that some

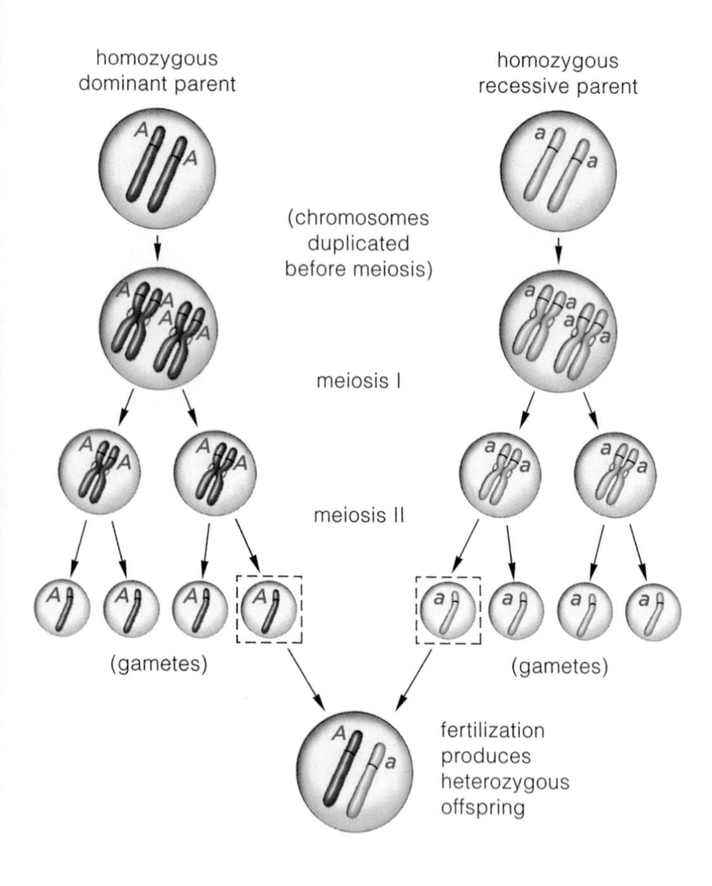

Figure 10.5 Segregation of a pair of alleles at a gene locus.

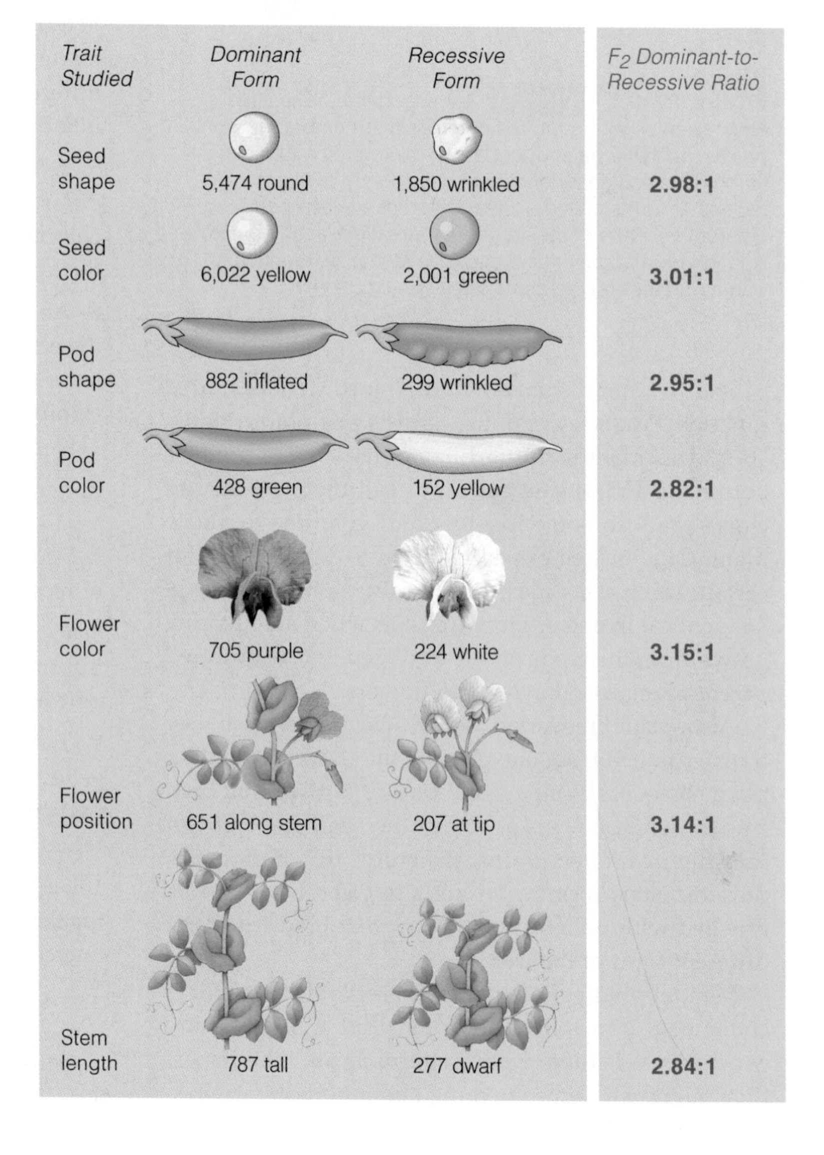

Trait Studied	*Dominant Form*	*Recessive Form*	*F_2 Dominant-to-Recessive Ratio*
Seed shape	5,474 round	1,850 wrinkled	**2.98:1**
Seed color	6,022 yellow	2,001 green	**3.01:1**
Pod shape	882 inflated	299 wrinkled	**2.95:1**
Pod color	428 green	152 yellow	**2.82:1**
Flower color	705 purple	224 white	**3.15:1**
Flower position	651 along stem	207 at tip	**3.14:1**
Stem length	787 tall	277 dwarf	**2.84:1**

Figure 10.6 From some of Mendel's monohybrid experiments with pea plants, counts of F_2 offspring with dominant or recessive hereditary "units" (alleles). On average, a 3:1 phenotypic ratio held for these traits.

particular outcome will occur. That chance depends on the number of possible outcomes. For example, if you cross two *Aa* heterozygotes, the two types of gametes (*A* and *a*) can meet four different ways in fertilization:

Possible Event	Probable Outcome
sperm *A* meets egg *A*	1 out of 4 offspring *AA*
sperm *A* meets egg *a*	1 out of 4 offspring *Aa*
sperm *a* meets egg *A*	1 out of 4 offspring *Aa*
sperm *a* meets egg *a*	1 out of 4 offspring *aa*

Each of the offspring of this cross has 3 chances in 4 of inheriting at least one dominant *A* allele (and purple flowers). It has 1 chance in 4 of inheriting two recessive *a* alleles (and white flowers). Thus, the probability that an offspring of this cross will have allele *A* (and purple flowers) is 3:1. We use grids called Punnett squares to calculate the probability of genotypes (and phenotypes) that will occur in offspring (Figure 10.7).

Mendel's observed ratios were not exactly 3:1, but he knew that deviations can arise from sampling error (Section 1.8). For example, if you flip a coin, it is just as likely to end up heads as tails (a probability of 1:1). But often it ends up heads, or tails, several times in a row. If you flip the coin only a few times, the observed ratio might differ greatly from the predicted ratio. Flip it many times, and you are more likely to see that ratio. Mendel minimized his sampling error by maximizing his sample sizes.

TESTCROSSES

A testcross is a method of determining genotype. One individual of an unknown genotype is crossed with another that is homozygous recessive; the results may show if the individual is heterozygous or homozygous for a dominant trait. For example, Mendel crossed his F_1 purple-flowered plants with true-breeding white-flowered plants. If the F_1 plants were homozygous dominant, then all the F_2 offspring should be purple-flowered. If the F_1 plants were heterozygous, then half of the F_2 offspring should be purple-flowered. About half of the F_2 offspring had purple flowers and half of them had white, so the F_1 plants were heterozygous.

The results from Mendel's monohybrid experiments became the basis of his theory of segregation, which we state here in terms of modern genetics:

MENDEL'S THEORY OF SEGREGATION *Diploid cells have pairs of genes, on pairs of homologous chromosomes. The two genes of each pair are separated from each other during meiosis, so they end up in different gametes.*

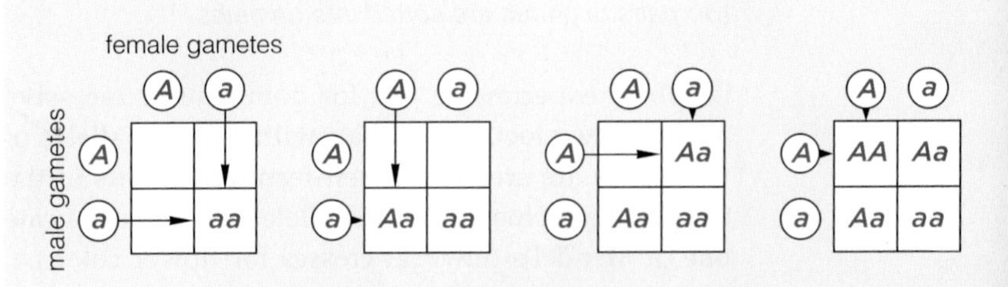

a From left to right, step-by-step construction of a Punnett square. Circles signify gametes. *A* stands for a dominant allele and *a* for a recessive allele at the same gene locus. Offspring genotypes are indicated inside the squares.

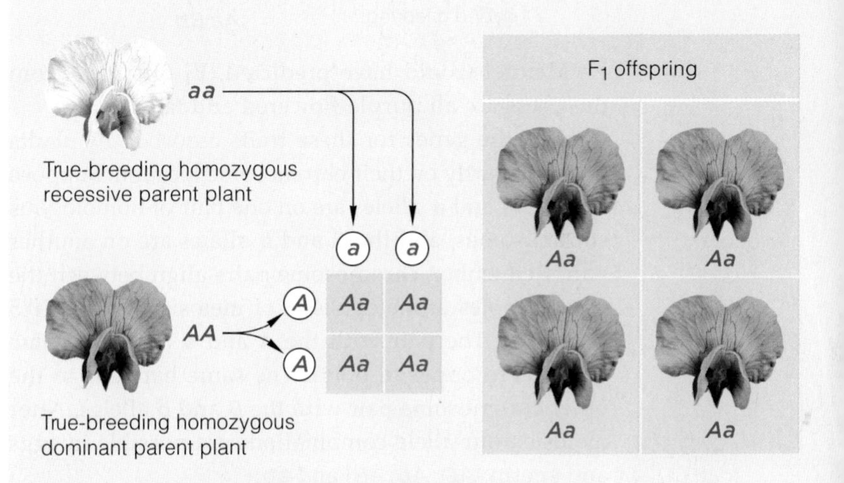

b Cross between two plants that breed true for different forms of a trait.

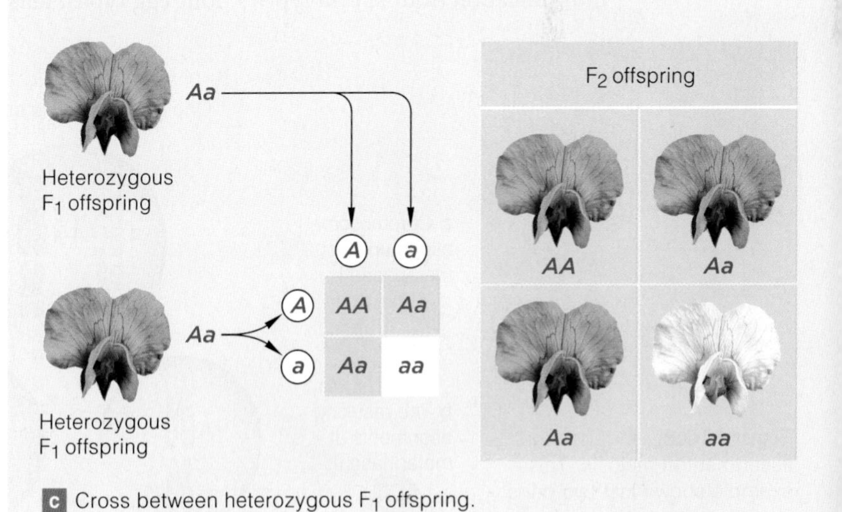

c Cross between heterozygous F_1 offspring.

Figure 10.7 **Animated!** (**a**) Punnett-square method of predicting probable outcomes of genetic crosses. (**b**,**c**) Results from one of Mendel's monohybrid experiments. On average, the ratio of dominant-to-recessive phenotypes among second-generation (F_2) plants was 3:1.

10.3 Mendel's Theory of Independent Assortment

Mendel used dihybrid experiments to explain how two pairs of genes are sorted into gametes.

Dihybrid experiments test for dominant or recessive alleles at two loci. Individuals with different alleles of two genes are crossed (or self-fertilized); traits of the offspring are clues about the alleles. We can duplicate one of Mendel's dihybrid crosses for flower color (*A*, purple; *a*, white) and height (*B*, tall; *b*, short):

True-breeding parents: *AABB* × *aabb*

Gametes: *AB* *AB* *ab* *ab*

F_1 hybrid offspring: *AaBb*

As Mendel would have predicted, F_1 offspring from this cross are all purple-flowered and tall (*AaBb*).

How the genes for these traits assort in F_1 plants depends partly on their chromosome location. Suppose that the *A* and *a* alleles are on one pair of homologous chromosomes, and the *B* and *b* alleles are on another pair. Remember, chromosome pairs align between the spindle poles at metaphase I of meiosis (Figures 10.5 and 10.8). The pair with the *A* and *a* alleles become tethered to opposite poles. The same happens to the other chromosome pair with the *B* and *b* alleles. After meiosis, four allele combinations are possible in eggs and sperm: *AB*, *Ab*, *aB*, and *ab*.

Given the alternative metaphase I alignments, many allelic combinations can result at fertilization. Simple multiplication (four sperm types × four egg types) tells us that sixteen allele combinations are possible among the F_2 offspring of a dihybrid cross (Figure 10.9).

Sorting all possible phenotypes gives us a 9:3:3:1 ratio, or 9/16 tall purple-flowered, 3/16 dwarf purple-flowered, 3/16 tall white-flowered, and 1/16 dwarf white-flowered F_2 plants. Results of Mendel's dihybrid experiment were close to this predicted ratio.

Mendel analyzed the numerical results from such experiments, but he could only hypothesize that two units for flower color were sorted out into gametes independently of the two units for height. He did not know that seven pairs of homologous chromosomes carry a pea plant's "units" of inheritance.

In time, his hypothesis became known as the theory of independent assortment. In modern terms, after meiosis ends, the genes on each pair of homologous chromosomes are sorted into gametes independently of how genes on other pairs of homologues are sorted out. Independent assortment and segregation give rise to genetic variation. With a monohybrid cross for one gene pair, three genotypes are possible: *AA*, *Aa*, and *aa*. We represent this as 3^n, where n is the number of gene pairs. With more gene pairs, more combinations are possible. If the parents differ in, say, twenty gene pairs, 3.5 billion allele combinations are possible!

In 1866 Mendel published his work. Apparently his article was read by few and understood by no one. In 1871 he became monastery abbot, and his pioneering experiments ended. He died in 1884, never to know that his experiments would be the starting point for modern genetics. Mendel's theory of segregation still

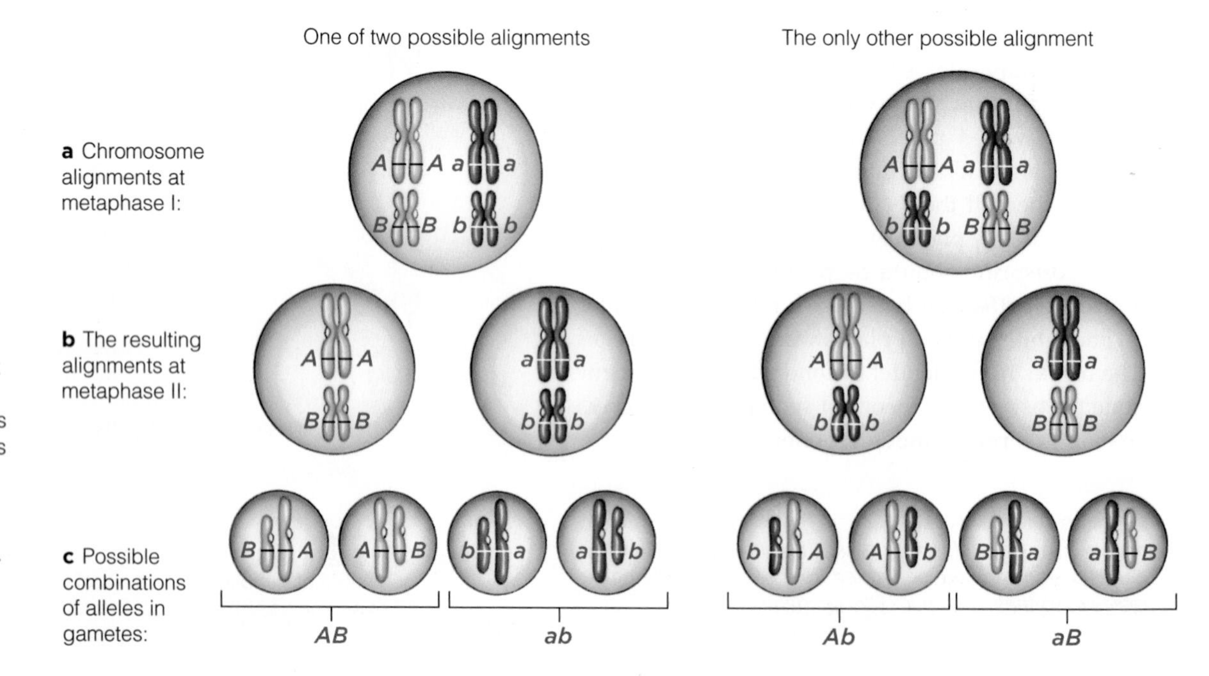

Figure 10.8 Independent assortment at meiosis. This example shows just two pairs of homologous chromosomes in the nucleus of a diploid ($2n$) reproductive cell. Either chromosome of a pair may get harnessed to either pole. When two pairs are tracked, two different metaphase I alignments are possible.

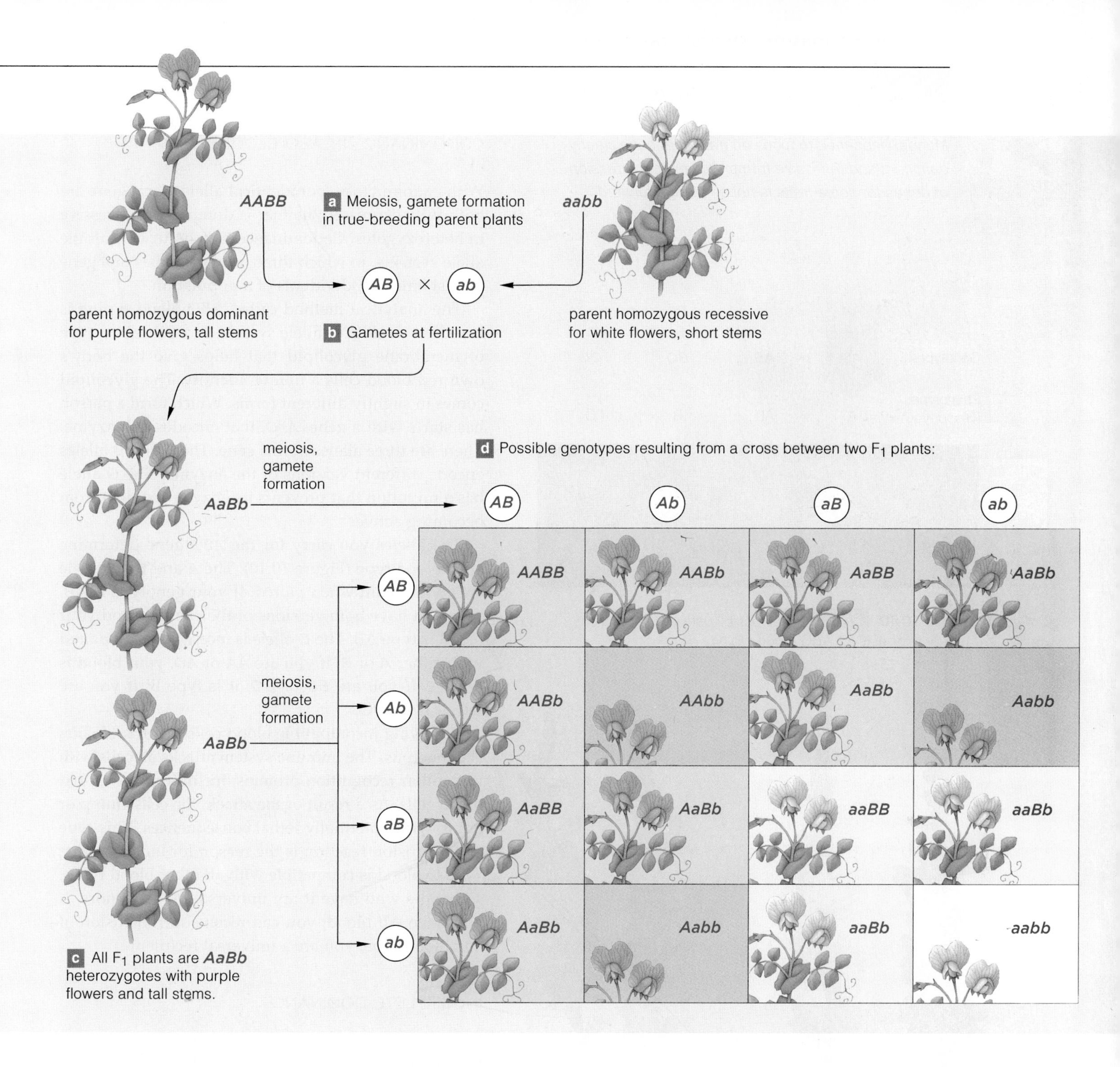

holds for most genes in most organisms: The units of hereditary material (genes) do retain their identity all through meiosis. However, his theory of independent assortment requires qualification, because the alleles of gene pairs do not always assort independently into gametes, as Section 10.5 explains.

MENDEL'S THEORY OF INDEPENDENT ASSORTMENT *As meiosis ends, genes on pairs of homologous chromosomes have been sorted out for distribution into one gamete or another, independently of gene pairs on other chromosomes.*

Figure 10.9 **Animated!** Results from one of Mendel's dihybrid experiments with the garden pea plant. Parent plants were true-breeding for different versions of two traits: flower color and plant height. ***A*** and ***a*** represent the dominant and recessive alleles for flower color. ***B*** and ***b*** represent dominant and recessive alleles for height. This Punnett square shows all allelic combinations possible in the F_2 generation. Adding up the corresponding F_2 phenotypes, we get:

- 9/16 or 9 purple-flowered, tall
- 3/16 or 3 purple-flowered, dwarf
- 3/16 or 3 white-flowered, tall
- 1/16 or 1 white-flowered, dwarf

VARIATIONS ON MENDEL'S THEME

10.4 Beyond Simple Dominance

LINK TO SECTION 4.3

Mendel happened to focus on traits that have clearly dominant and recessive forms. However, expression of genes for some traits is not as straightforward.

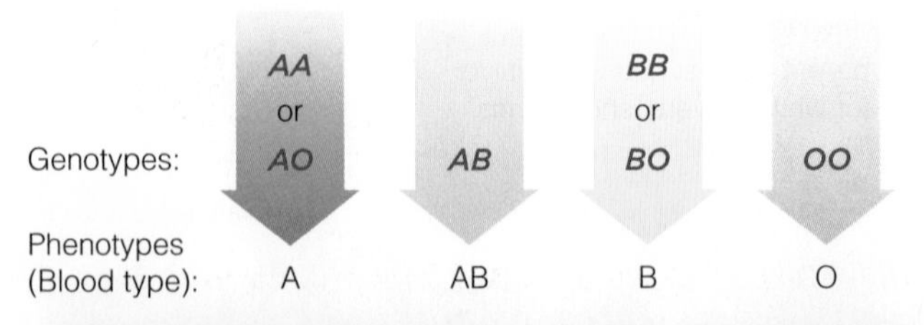

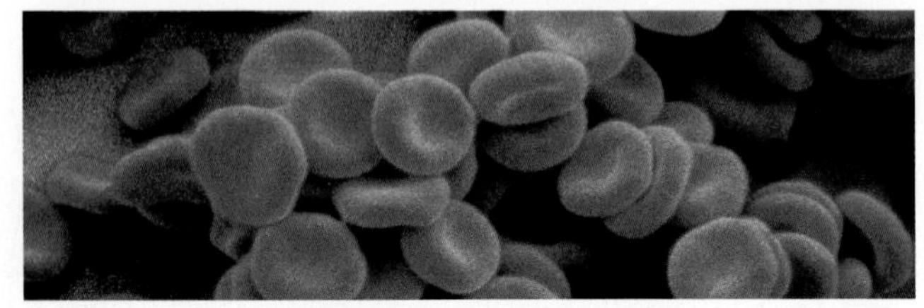

Figure 10.10 **Animated!** Possible allelic combinations that are the basis for ABO blood typing.

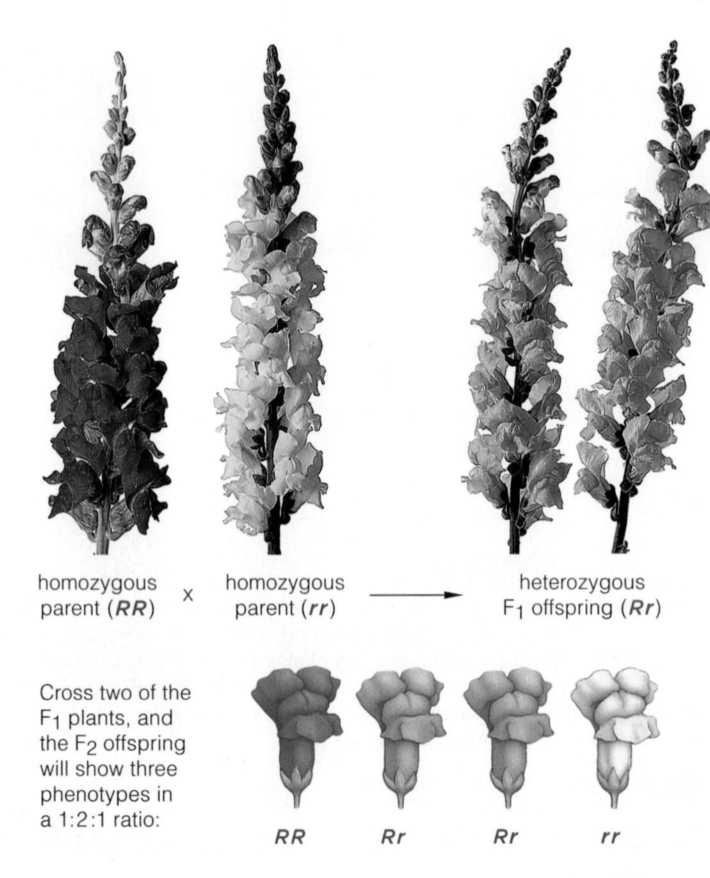

Figure 10.11 Incomplete dominance in heterozygous (*pink*) snapdragons. An allele that affects red pigment is paired with a "white" allele.

CODOMINANCE IN ABO BLOOD TYPES

With codominance, nonidentical alleles for a gene are both fully expressed; neither is dominant or recessive in heterozygotes. Codominance may occur in multiple allele systems, in which three or more alleles of a gene persist among individuals of a population.

The analytical method called ABO blood typing is based on a multiple allele system. It checks for a type of membrane glycolipid that helps give the body's own red blood cells a unique identity. The glycolipid comes in slightly different forms. Which form a person has starts with a gene, *ABO*, that encodes an enzyme. There are three alleles of this gene. The ***A*** and ***B*** alleles encode different versions of the enzyme. The ***O*** allele has a mutation that prevents its enzyme product from becoming active.

The alleles you carry for the *ABO* gene determine your blood type (Figure 10.10). The ***A*** and the ***B*** allele are codominant when paired. If your genotype is ***AB***, then you have both versions of the enzyme, and your blood is type AB. The ***O*** allele is recessive when paired with either ***A*** or ***B***. If you are ***AA*** or ***AO***, your blood is type A. If you are ***BB*** or ***BO***, it is type B. If you are ***OO***, it is type O.

Receiving incompatible blood cells in a transfusion is dangerous. The immune system attacks all cells with unfamiliar recognition proteins, including foreign red blood cells. As a result of the attack, the cells clump or burst, with potentially lethal consequences. This type of transfusion reaction is the reason for blood typing. Type O blood is compatible with all other blood types, so people who have it are universal blood donors. If you have AB blood, you can receive a transfusion of any blood type; you are a universal recipient.

INCOMPLETE DOMINANCE

In incomplete dominance, one allele of a pair is not fully dominant over its partner, so the heterozygote's phenotype is *somewhere between* the two homozygotes.

A cross between two true-breeding snapdragons, one red and one white, is a case in point. The F_1 offspring are pink-flowered. Cross two F_1 plants and you can expect to see red, white, and *pink* flowers in a certain ratio (Figure 10.11). Why? Red snapdragons have two alleles that let them make a lot of molecules of a red pigment. White snapdragons have two mutant alleles; they do not make any pigment at all and so appear colorless. Pink snapdragons have a "red" allele and a "white" allele; these heterozygotes make only enough pigment to color the flowers pink, not red.

EPISTASIS

Traits also arise through epistasis: interactions among products of two or more gene pairs. Two alleles might mask expression of another gene's alleles, and some expected phenotypes might not appear at all. As an example, in chickens, interactions between products of alleles of the *R* and *P* genes cause variations in combs, the fleshy, red crest on the head (Figure 10.12).

As another example, several genes affect coat color of Labrador retrievers, which can be black, yellow, or brown (Figure 10.13). A dog's coat color depends on how enzymes and other products of alleles at more than one locus make a dark pigment, melanin, and deposit it in tissues. Allele *B* (black) is dominant to *b* (brown). At a different locus, allele *E* promotes the deposition of melanin in fur but two recessive alleles (*ee*) reduce it. A dog with two *e* alleles has yellow fur regardless of which alleles it has at the *B* locus.

SINGLE GENES WITH A WIDE REACH

One gene may influence two or more traits, an effect called pleiotropy. Genes with products that carry out basic tasks in all or most cells are the most likely to be pleiotropic. For example, *Marfan syndrome* is a genetic disorder that arises by mutations in the fibrillin gene. Long, thin fibers of fibrillin protein impart elasticity to many tissues—heart, skin, blood vessels, skeleton, and tendons.

In Marfan syndrome, tissues form with defective fibrillin or none at all. The largest blood vessel leading from the heart, the aorta, is particularly vulnerable. Muscle cells in the aorta's thick wall do not function well, and the aortic wall itself is not as elastic as it should be. Being under pressure, the aorta expands, and it becomes thinner and leaky. Calcium deposits accumulate inside. Inflamed, weakened, and thinned, the aorta might rupture suddenly during exercise.

Marfan syndrome can be very difficult to diagnose. Affected people are often tall, thin, and loose-jointed, but there are plenty of tall, thin, loose-jointed people without the disorder. Symptoms may not be apparent; many affected people are not aware that they have it. Until recently, it killed most of them before the age of fifty. Olympian Flo Hyman was one (Figure 10.14).

An allele may be fully dominant, incompletely dominant, or codominant with its partner on a homologous chromosome.

In epistasis, two or more gene products influence a trait. In pleiotropy, one gene product influences two or more traits.

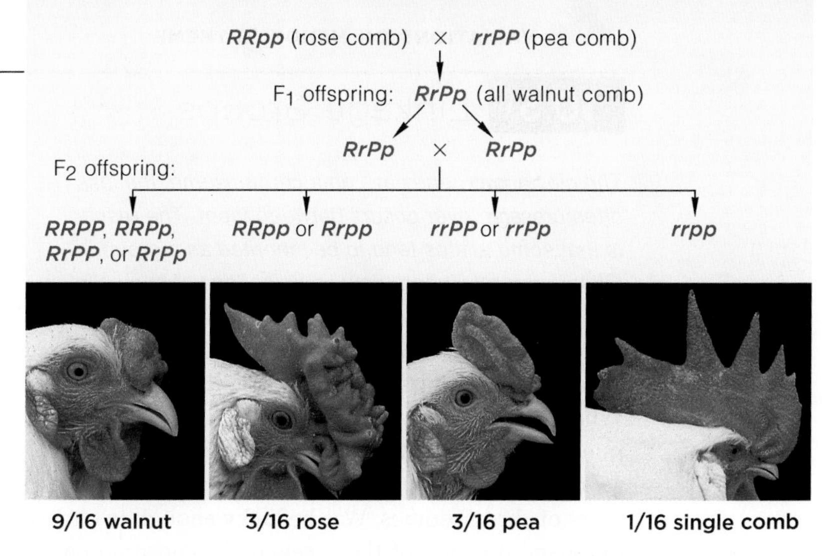

Figure 10.12 Variation in chicken combs, an outcome of Interactions among the products of alleles at two gene loci.

	EB	*Eb*	*eB*	*eb*
EB	*EEBB* black	*EEBb* black	*EeBB* black	*EeBb* black
Eb	*EEBb* black	*EEbb* chocolate	*EeBb* black	*Eebb* chocolate
eB	*EeBB* black	*EeBb* black	*eeBB* yellow	*eeBb* yellow
eb	*EeBb* black	*Eebb* chocolate	*eeBb* yellow	*eebb* yellow

Figure 10.13 *Left to right*, black, chocolate, and yellow Labrador retrievers. Epistatic interactions among products of two gene pairs affect the coat color trait.

Figure 10.14 Flo Hyman, left, captain of the United States volleyball team that won an Olympic silver medal in 1984. Two years later, at a game in Japan, she slid to the floor and died. A dime-sized spot in her aorta's wall had burst.

10.5 Linkage Groups

The closer two genes are on a chromosome, the less often crossing over occurs between them. The result is that some alleles tend to be inherited as a group. Offspring rarely inherit new combinations of such alleles.

LINKS TO SECTIONS 3.5, 5.4, 9.3, 9.4

As you learned in Section 10.3, the genes on different chromosomes all assort independently into gametes. What about the genes on the same chromosome? This question cast a shadow on Mendel's work. He studied seven genes in pea plants, which have exactly seven pairs of chromosomes. Was he lucky enough to choose one gene on each of those seven chromosome pairs? Some surmised that if he had studied just one more gene, he would have discovered an exception to his theory of independent assortment.

As it turns out, some of the genes Mendel studied *are* on the same pea plant chromosome. As long as two genes are far enough apart on a chromosome, crossing over occurs between them so often that they assort as if they were on different chromosomes. That was the case with the pea plant genes Mendel studied. Other genes are so close together that crossing over between them does not happen very often; such genes tend to stay together during meiosis. Genes that do not assort independently are said to be linked (Figure 10.15).

All of the genes on a chromosome form one linkage group. Thus, humans have 23 linkage groups, and peas have 7. Genes of a linkage group tend to stay together, but they also can be separated by crossovers.

Some linked genes stay together more often than others. They are closer together on the chromosome, so they are separated less frequently by crossovers. For example, if genes *A* and *B* are twice as far apart as genes *C* and *D* on a chromosome, then we can expect crossovers to disrupt the linkage between genes *A* and *B* more often than between genes *C* and *D*:

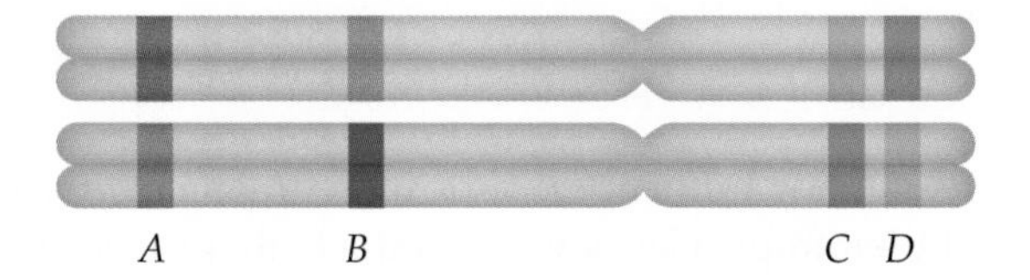

Generalizing from this example, the probability that a crossover event will disrupt a linkage between two genes is proportional to the distance between them. Two genes are tightly linked if the distance between them is relatively small. The parental combinations of these alleles nearly always end up in the same gamete. Linkage is more vulnerable to crossover events if the distance between two gene loci is greater. When two gene loci are relatively far apart, crossing over occurs so frequently between them that they almost always assort independently into gametes.

Human gene linkages were identified by tracking inheritance in families over several generations. One thing became clear: Crossovers are not at all rare. In many eukaryotes, at least two crossover events occur between each pair of homologous chromosomes when prophase I of meiosis is under way.

Mendel's blind luck in the genetics game?

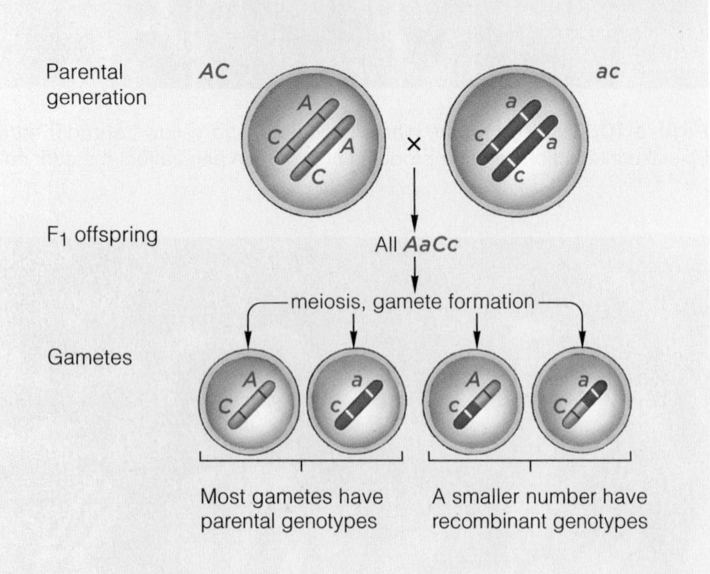

Figure 10.15 **Animated!** Linkage and crossing over. Alleles of two genes on the same chromosome stay together when there is no crossover between them, and recombine when there is a crossover between them.

All genes on a chromosome are part of one linkage group.

Crossing over between a pair of homologous chromosomes disrupts gene linkages. The outcome is recombination of alleles between homologous chromosomes.

The farther apart two genes are on a chromosome, the more often crossing over occurs between them. Genes that are very close to each other do not assort independently into gametes; those that are very far apart do not.

10.6 Genes and the Environment

The environment often influences gene expression, as a few classic examples demonstrate.

In Section 5.4, you read about a heat-sensitive enzyme, tyrosinase, that affects the coat color of Siamese cats. The enzyme catalyzes one step in the synthesis of the brown-black pigment melanin. It only works in cooler body regions, such as the legs, tail, and ears. It also affects the coat color of Himalayan rabbits. The rabbits are homozygous for the c^h allele, which encodes a form of tyrosinase that does not work if temperatures in cells exceed 33°C, or 91°F. Metabolic heat keeps the main body mass warm enough to stop the enzyme from working, so the fur is light there. Ears and other slender appendages lose metabolic heat faster and are cooler, so melanin darkens them. Figure 10.16 shows an experiment that tested the impact of temperature on rabbit fur color.

As another example, yarrow plants can grow from cuttings, so they are a useful experimental organism. All cuttings from a plant have the same genotype, so experimenters know that genes are not the basis for any phenotypic differences among them. In one study, researchers grew cuttings from each of several yarrow plants at three elevations (Figure 10.17). The cuttings grew differently at the different altitudes. The yarrow plants were genetically identical, but their phenotypes differed in the different habitats.

Invertebrates, too, show phenotypic variation with environmental conditions. For instance, daphnias are microscopic freshwater relatives of shrimps. Aquatic insects prey on them. *Daphnia pulex* living in ponds with few predators have rounded heads, but those in ponds with many predators have more pointed heads (Figure 10.18). *Daphnia's* predators emit chemicals that trigger the different phenotype.

The environment affects human genes, too. One of our genes encodes a protein that transports serotonin across the membrane of brain cells. Serotonin lowers anxiety and depression during traumatic times. Some mutations in the serotonin transporter gene can reduce our ability to cope with stress. It is as if some of us are bicycling through life without an emotional helmet. Only when we fall does the mutation's phenotypic effect—depression—appear. Other human genes affect emotional state, but mutations in this one reduce our capacity to snap out of it when bad things happen.

Variation in traits arises not only from gene mutations and interactions, but also in response to variations in environmental conditions that each individual faces.

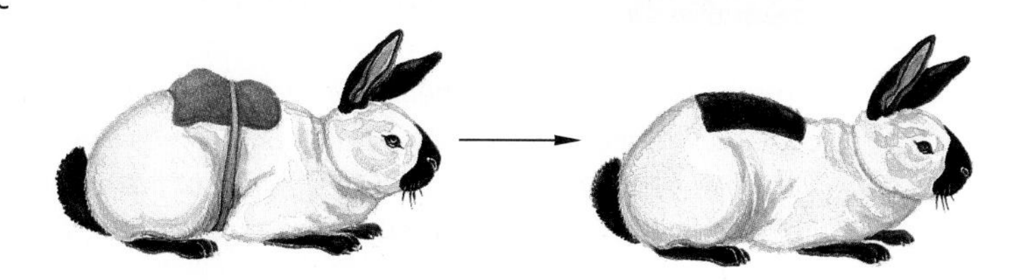

Figure 10.16 Animated! Observable effect of the environment on gene expression. A Himalayan rabbit is homozygous for an allele that encodes a heat-sensitive form of an enzyme required for melanin synthesis. Cooler body parts, such as ears, are dark. The main body mass is warmer, and light. A patch of one rabbit's white fur was shaved off. An ice pack was tied above the fur-free patch. Fur grew back, but it was dark. The ice pack had cooled the patch enough for the enzyme to work, and melanin was produced.

Figure 10.17 Experiment showing environmental effects on phenotype in yarrow (*Achillea millefolium*). Cuttings from the same parent plant were grown in the same kind of soil at three different elevations.

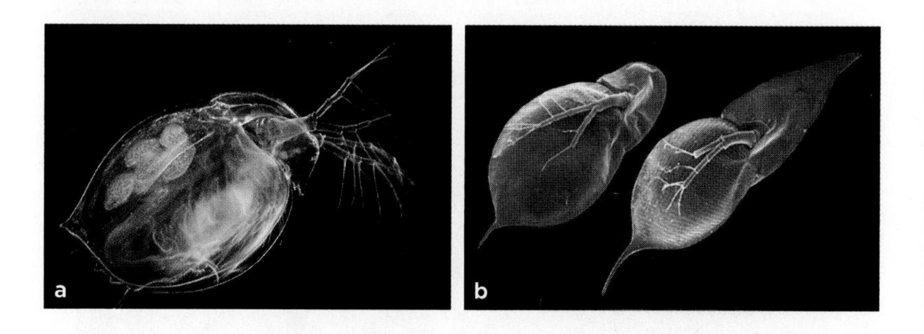

Figure 10.18 (**a**) Light micrograph of a living daphnia. (**b**) Phenotypic effects of the presence of insects that prey on daphnias. The body form at the *left* develops when predators are absent or few. The form at the *right* develops when water contains chemicals emitted by the daphnia's insect predators. It has a longer tail spine and a pointed spine at the head.

10.7 Complex Variations in Traits

For most populations or species, individuals show rich variation for many of the same traits. Sometimes the phenotypes cannot be predicted, and most of the time they are part of a continuous range of variation.

CONTINUOUS VARIATION IN POPULATIONS

Individuals of populations generally show a range of small differences in most traits. This feature of natural populations is known as continuous variation. It arises by polygenic inheritance: the inheritance of multiple genes that affect a trait. The distribution of all forms of a trait becomes more continuous as greater numbers of genes and environmental factors are involved.

Consider your eye color. The colored part is the iris, a doughnut-shaped, pigmented structure just under the cornea. Several gene products contribute to that color. They make and distribute different amounts and kinds of melanins like the light-absorbing pigment that colors mammalian fur. Irises that are nearly black have dense melanin deposits, which absorb most light. Deposits are not as extensive in brown eyes, so some unabsorbed light is reflected. Light brown or hazel eyes have even less melanin.

Green, gray, or blue eyes have lesser amounts of the pigments. Some or most of the blue light that reaches the iris is simply reflected.

How do we show continuous variation in a trait—say, height? First, divide the total range of phenotypes into measurable categories, such as inches. Next, count how many individuals fall into each category; this will give you relative frequencies of phenotypes across the range of measurable values. Finally, plot out the data as a bar chart, such as the one in Figure 10.19*a*. In this chart, the shortest bars are categories with the fewest individuals, and the tallest bar is the category with the most. The graph line around the top of these bars will be a bell-shaped curve. Such bell curves are typical of any trait that shows continuous variation. Two real-life examples are given in Figure 10.19*b*,*c*.

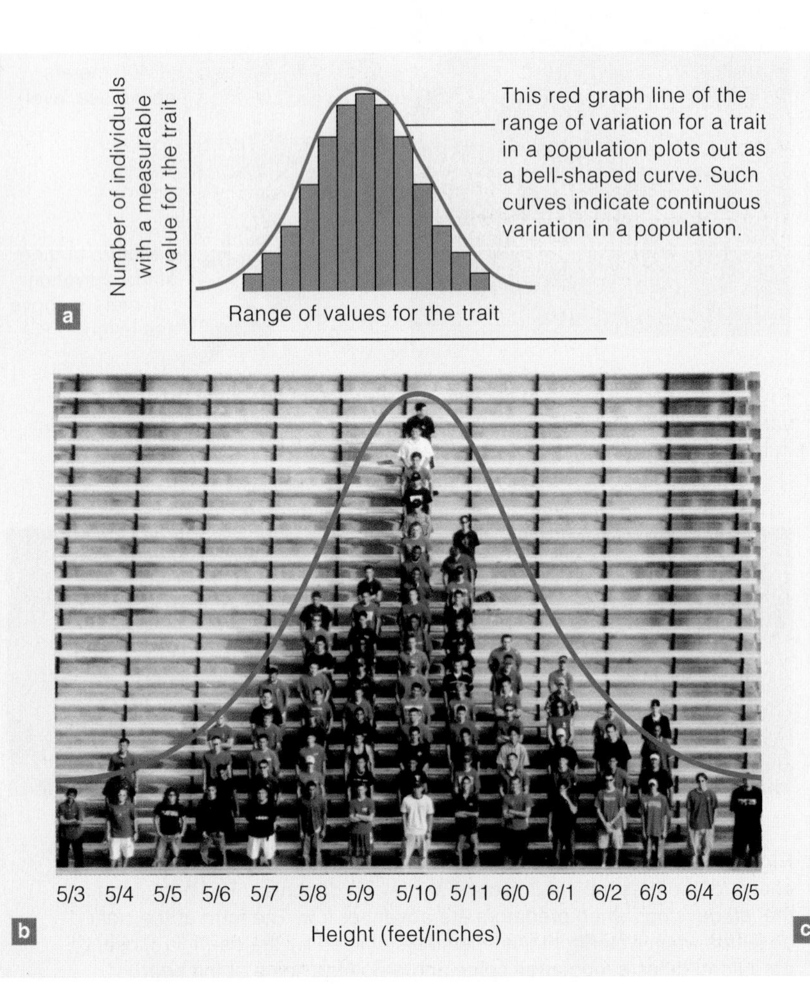

REGARDING THE UNEXPECTED PHENOTYPE

Think back on Mendel's dihybrid crosses. Nearly all of the traits he tracked showed up in predictable ratios because the gene pairs happened to be on different chromosomes or far apart on the same chromosome. They tended to segregate cleanly. More often, there is far more variation in phenotypes, and not all of it is a result of tight linkage or crossing over.

For example, *camptodactyly* affects finger shape and movement. People who carry a mutant allele for this rare heritable abnormality have immobile, bent fingers on both hands or immobile, bent fingers on the left or right hand only, or no obvious difference at all.

What causes complex variation? Remember, most organic compounds are synthesized by a sequence of metabolic steps. *Different enzymes, each a gene product, control different steps.* A gene may mutate in a number of ways. Its altered product may block a pathway or make it run nonstop or not long enough. Perhaps poor nutrition or another environmental variable influences an enzyme's activity. Such factors can introduce big or small variations even in expected phenotypes.

So here is the take-home lesson: Phenotype results from complex interactions among genes, enzymes and other gene products, and the environment. Chapter 17 considers some of the evolutionary consequences.

Individuals of populations and species show continuous variation—a range of small differences in a trait. Usually, the more genes and environmental factors that influence a trait, the more continuous the distribution of phenotypes.

Enzymes and other gene products control steps of most metabolic pathways. Mutations, interactions among genes, and environmental conditions can affect one or more steps, and so contribute to variation in phenotypes.

Figure 10.19 **Animated!** Continuous variation. (**a**) A bar graph can reveal continuous variation in a population. The number of individuals in each measured category is plotted against the range of phenotypes.

Two examples of continuous variation: many biology students organized into rows on the basis of height. Professors Jon Reiskind and Greg Pryor wanted to illustrate the frequency distribution for height among their biology students at the University of Florida. They divided all of the students into two groups: (**b**) males and (**c**) females. In both groups, they subdivided the range of possible heights, measured the students, and assigned each to the appropriate category.

Summary

www.thomsonedu.com ThomsonNOW!

Section 10.1 Genes are heritable units of information about traits. Each has its own locus, or location, along the length of a chromosome. Different molecular forms of the same gene are alleles. By experimenting with pea plants, Mendel was the first to gather evidence of patterns by which parents transmit genes to offspring.

Offspring of a cross between two individuals that breed true for different forms of a trait are hybrids; each inherited nonidentical alleles for a trait being studied.

An individual with two dominant alleles for a trait (*AA*) is homozygous dominant. A homozygous recessive has two recessive alleles (*aa*). A heterozygote has two nonidentical alleles (*Aa*). A dominant allele may mask the effect of a recessive allele partnered with it on the homologous chromosome. The alleles at any or all gene loci constitute an individual's genotype. Phenotype refers to an individual's observable traits.

- *Learn how Mendel crossed garden pea plants, and the definitions of important genetic terms, on ThomsonNOW.*

Section 10.2 Crossing two true-breeding parents of different genotypes yields hybrid offspring. All the F_1 offspring are identically heterozygous for the same gene, and can be used in monohybrid experiments. Mendel's monohybrid experiments gave him indirect evidence that some forms of a gene are dominant over others.

All F_1 offspring of a parental cross *AA* × *aa* were *Aa*. Crosses between F_1 monohybrids resulted in these allelic combinations among the F_2 offspring:

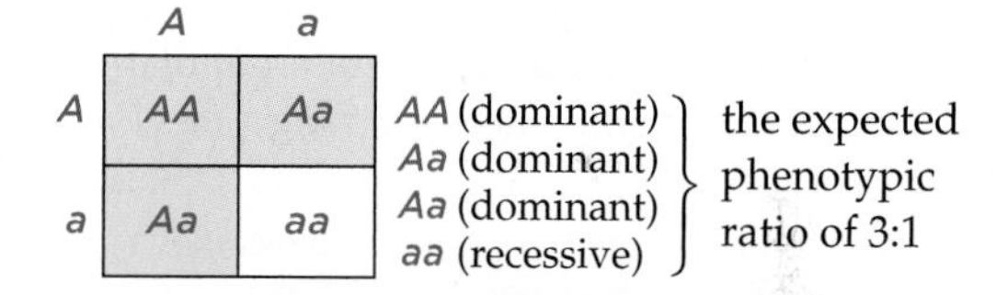

Mendel's monohybrid experiment results led to his theory of segregation (stated in modern terms): Diploid organisms have pairs of genes, on pairs of homologous chromosomes. During meiosis, the genes of each pair separate, so each gamete gets one or the other gene.

- *Use the interaction on ThomsonNOW to carry out monohybrid experiments.*

Section 10.3 Dihybrid experiments start with a cross between true-breeding heterozygous parents that differ for alleles of two genes (*AABB* × *aabb*). All F_1 offspring are heterozygous for both genes (*AaBb*). In Mendel's dihybrid experiments, phenotypes of the F_2 offspring of F_1 hybrids were close to a 9:3:3:1 ratio:

9 dominant for both traits
3 dominant for *A*, recessive for *b*
3 dominant for *B*, recessive for *a*
1 recessive for both traits

The results led to his theory of independent assortment (stated in modern terms): Meiosis assorts gene pairs of homologous chromosomes independently of gene pairs on all other chromosomes. The random alignment of all pairs of homologous chromosomes at metaphase I is the basis of this outcome.

■ *Use the interactions on ThomsonNOW to observe the results of a dihybrid cross.*

Section 10.4 Inheritance patterns often vary. With incomplete dominance, an allele is not fully dominant over its partner on a homologous chromosome, and both are expressed. The combination of alleles gives rise to a phenotype between the two homozygous conditions.

Some alleles are codominant and are both expressed at the same time in heterozygotes, as in a multiple allele system underlying ABO blood typing. Also, interacting products of one or more genes often affect the same trait, and a single gene may have effects on two or more traits.

■ *Use the interactions on ThomsonNOW to explore patterns of non-Mendelian inheritance.*

Section 10.5 The farther apart two genes are on a chromosome, the greater the frequency of crossing over between them. Genes that are relatively close together on a chromosome tend to stay together during meiosis; few crossover events occur between them. Genes that are relatively far apart tend to assort independently into gametes. All of the genes on a chromosome constitute a linkage group.

Section 10.6 Environmental factors may affect gene expression in individuals.

■ *Use the interactions on ThomsonNOW to see how the environment can affect phenotype.*

Section 10.7 Gene interactions and environmental factors affect most phenotypes. When products of many genes influence a trait, individuals of a population show small, incremental differences—a range of continuous variation for the trait.

■ *Use the interaction on ThomsonNOW to plot the continuous distribution of height for a class.*

Figure 10.20 Two albino organisms. By not posing his subjects as objects of ridicule, the photographer of human albinos is attempting to counter the notion that there is something inherently unbeautiful about them.

Self-Quiz

Answers in Appendix III

1. Alleles are ______.
   a. different molecular forms of a gene
   b. different phenotypes
   c. self-fertilizing, true-breeding homozygotes

2. A heterozygote has a ______ for a trait being studied.
   a. pair of identical alleles
   b. pair of nonidentical alleles
   c. haploid condition, in genetic terms

3. The observable traits of an organism are its ______.
   a. phenotype c. genotype
   b. sociobiology d. pedigree

4. Second-generation offspring of a cross between parents who are homozygous for different alleles are the ______.
   a. F_1 generation c. hybrid generation
   b. F_2 generation d. none of the above

5. F_1 offspring of the cross *AA* × *aa* are ______.
   a. all *AA* c. all *Aa*
   b. all *aa* d. 1/2 *AA* and 1/2 *aa*

6. Refer to Question 5. Assuming complete dominance, the F_2 generation will show a phenotypic ratio of ______.
   a. 3:1 b. 9:1 c. 1:2:1 d. 9:3:3:1

7. Assuming complete dominance, crosses between two dihybrid F_1 pea plants, which are offspring from a cross *AABB* × *aabb*, result in F_2 phenotype ratios of ______.
   a. 1:2:1 b. 3:1 c. 1:1:1:1 d. 9:3:3:1

8. The probability of a crossover occurring between two genes on the same chromosome ______.
   a. is unrelated to the distance between them
   b. decreases with the distance between them
   c. increases with the distance between them

9. Two genes that are close together on the same chromosome are ______.
   a. linked c. homologous e. all of the above
   b. identical alleles d. autosomes

10. Match each example with the most suitable description.
   ___ dihybrid experiment a. *bb*
   ___ monohybrid experiment b. *AABB* × *aabb*
   ___ homozygous condition c. *Aa*
   ___ heterozygous condition d. *Aa* × *Aa*

■ *Visit ThomsonNOW for additional questions.*

Genetics Problems

Answers in Appendix III

1. One gene encodes the second enzyme in a melanin-synthesizing pathway. An individual who is homozygous for a recessive mutant allele of this gene cannot make or deposit melanin in body tissues. *Albinism*, the absence of melanin, is the result. Humans and many other organisms can have this phenotype. Figure 10.20 shows examples. In the following situations, what are the probable genotypes of the father, the mother, and their children?

a. Both parents have normal phenotypes; some of their children are albino and others are unaffected.

b. Both parents are albino and have albino children.

c. The woman is unaffected, the man is albino, and they have one albino child and three unaffected children.

dominant

recessive

dominant

recessive

Figure 10.21 (**a**) A climbing rose and (**b**) a shrub rose. (**c**) Urn-shaped buds versus (**d**) globe-shaped buds.

2. Several alleles affect traits of roses, such as plant form and bud shape (Figure 10.21). Alleles of one gene govern whether a plant will be a climber (dominant) or shrubby (recessive). All F_1 offspring from a cross between a true-breeding climber and a shrubby plant are climbers. If an F_1 plant is crossed with a shrubby plant, about 50 percent of the offspring will be shrubby; 50 percent will be climbers. Using symbols *A* and *a* for the dominant and recessive alleles, make a Punnett-square diagram of the expected genotypes and phenotypes in F_1 offspring and in offspring of a cross between an F_1 plant and a shrubby plant.

3. One gene has alleles *A* and *a*. Another has alleles *B* and *b*. For each genotype, what type(s) of gametes will form, assuming independent assortment during meiosis?

a. *AABB* b. *AaBB* c. *Aabb* d. *AaBb*

4. Refer to Problem 3. Determine the frequencies of each genotype among offspring from the following matings:

a. *AABB* × *aaBB*
b. *AaBB* × *AABb*
c. *AaBb* × *aabb*
d. *AaBb* × *AaBb*

5. Refer to Problem 3. Assume a third gene has alleles *C* and *c*. For each genotype listed, what allele combinations will occur in gametes, assuming independent assortment?

a. *AABBCC*
b. *AaBBcc*
c. *AaBBCc*
d. *AaBbCc*

6. Certain alleles are vital for normal development. When mutated, they are lethal in homozygous recessives. Even so, heterozygotes can perpetuate these recessive, lethal alleles in a population. The allele *Manx* (M^L) in cats is an example. Homozygous cats (M^LM^L) die before birth. In heterozygotes (M^LM), the spine develops abnormally, and the cats end up with no tail (Figure 10.22).

Two M^LM cats mate. What is the probability that any one of their surviving kittens will be heterozygous?

7. In one experiment, Mendel crossed a true-breeding garden pea plant with green pods and a true-breeding plant with yellow pods. All the F_1 plants had green pods. Which color is recessive?

8. Suppose you identify a new gene in mice. One of its alleles specifies white fur, another specifies brown. You want to see if the two interact in simple or incomplete dominance. What sorts of genetic crosses would give you the answer? What types of observations would you require to form conclusions?

9. In sweet pea plants, an allele for purple flowers (*P*) is dominant to an allele for red flowers (*p*). An allele for long pollen grains (*L*) is dominant to an allele for round pollen grains (*l*). Bateson and Punnett crossed a plant having purple flowers/long pollen grains with one having white flowers/round pollen grains. All F_1 offspring had purple flowers and long pollen grains. Among the F_2 generation, the researchers observed the following phenotypes:

296 purple flowers/long pollen grains
19 purple flowers/round pollen grains
27 red flowers/long pollen grains
85 red flowers/round pollen grains

What is the best explanation for these results?

10. Red-flowering snapdragons are homozygous for allele R^1. White-flowering snapdragons are homozygous for a different allele (R^2). Heterozygous plants (R^1R^2) bear pink flowers. What phenotypes should appear among first-generation offspring of the crosses listed? What are the expected proportions for each phenotype?

a. $R^1R^1 \times R^1R^2$
b. $R^1R^1 \times R^2R^2$
c. $R^1R^2 \times R^1R^2$
d. $R^1R^2 \times R^2R^2$

(Incompletely dominant alleles are usually designated by superscript numerals, as shown, not by uppercase letters for dominance and lowercase letters for recessiveness.)

11. A single mutant allele gives rise to an abnormal form of hemoglobin (Hb^S, not Hb^A). Homozygotes (Hb^SHb^S) develop sickle-cell anemia (Section 3.5). Heterozygotes (Hb^AHb^S) show few symptoms. A couple who are both heterozygous for the Hb^S allele plan to have children. For each of the pregnancies, state the probability that they will have a child who is:

a. homozygous for the Hb^S allele
b. homozygous for the Hb^A allele
c. heterozygous; Hb^AHb^S

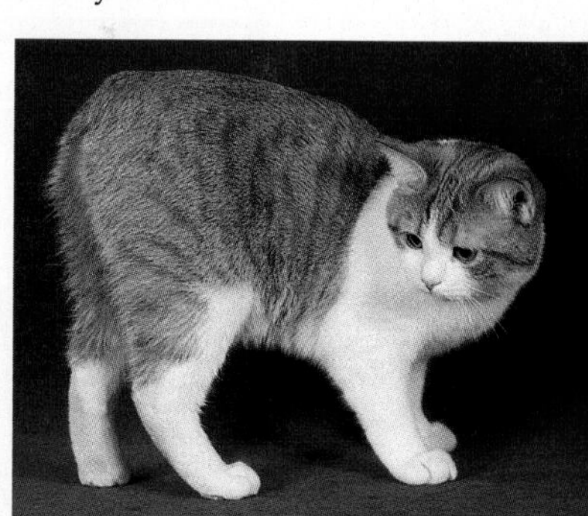
Figure 10.22 The Manx, a breed of cat that has no tail.

12. Watermelon plants (*Citrullus*) that are homozygous for recessive allele *e* have fruit with rinds that explosively split when cut. Plants with genotype *EE* have a nonexplosive rind that ships to market better. Plants with genotype *ff* have an unappealing furrowed rind. Those bearing dominant allele *F* have a smooth rind. A testcross is performed in which a dihybrid plant with melons that have a smooth, nonexplosive rind is crossed with a plant that has melons with a furrowed, explosive rind. Make a Punnett square of these results: 118 smooth, nonexplosive; 112 smooth, explosive; 109 furrowed, nonexplosive; and 121 furrowed, explosive. What is the smooth rind/furrowed rind ratio among the testcross offspring? What is the ratio of nonexplosive rind/explosive rind? Are the two gene loci assorting independently of each other?

11 CHROMOSOMES AND HUMAN INHERITANCE

IMPACTS, ISSUES Strange Genes, Tortured Minds

"This man is brilliant." That was the extent of a letter of recommendation from Richard Duffin, a professor of mathematics at Carnegie Mellon University. Duffin wrote the line in 1948 on behalf of John Forbes Nash, Jr. (Figure 11.1). Nash was twenty years old at the time and applying for admission to Princeton University's graduate school. Over the next ten years, Nash made his reputation as one of the foremost mathematicians. He was socially awkward, but so are many highly gifted people. Nash showed no symptoms warning of the paranoid schizophrenia that eventually would debilitate him.

Full-blown symptoms emerged in his thirtieth year. Nash had to abandon his position at the Massachusetts Institute of Technology. Two decades passed before he was able to return to his pioneering work in mathematics.

Of every 100 people, 1 is affected by schizophrenia. This neurobiological disorder (NBD) is characterized by delusions, hallucinations, disorganized speech, and abnormal social behavior. Exceptional creativity often accompanies schizophrenia. It also accompanies other NBDs, including autism, chronic depression, and bipolar disorder, which manifests itself as jarring swings in mood and social behavior.

Compared to the general population, highly intelligent individuals are *less* likely to develop NBDs—unless they also happen to be outside-the-box creative thinkers. Disturbingly, creative writers alone are eighteen times more suicidal, ten times more likely to be depressed, and twenty times more likely to have bipolar disorder. Virginia Woolf's suicide after a prolonged mental breakdown is a tragic example.

See the video! **Figure 11.1** *Left*, John Forbes Nash, Jr., a prodigy who solved problems that had baffled some of the greatest minds in mathematics. His early work in economic game theory won him a Nobel Prize. He is shown here at a premiere of *A Beautiful Mind*, a film based on his battle with schizophrenia. His neural disorder places him in the ranks of other highly creative, distinguished, yet troubled individuals, including Abraham Lincoln, Virginia Woolf (*right*), and Pablo Picasso (*next page*).

How would you vote? Tests for predisposition to neurobiological disorders will be available soon. Insurance companies and employers may use the information to discriminate against otherwise healthy individuals. Would you support legislation governing the tests? See ThomsonNOW for details, then vote online.

Emotionally healthy, creatively brilliant people have more personality traits in common with people affected by NBDs than they do with individuals closer to the norm. For instance, both are hypersensitive to environmental stimuli. Some may be on a razor's edge between mental stability and instability. Those who do develop NBDs become part of a crowd that includes names such as Newton, Socrates, Beethoven, Darwin, Lincoln, Dickens, Tolstoy, van Gogh, Freud, Churchill, Einstein, Picasso, Poe, Hemingway, Woolf, and Nash.

We have not yet identified all the interactions among genes and environment that might tip such individuals one way or the other. But we do know about several mutations that predispose them to develop NBDs.

Creatively gifted people, as well as those affected by NBDs, often turn up in the same family tree—which points to a genetic basis for their special traits. Also, those affected by bipolar disorder and schizophrenia show altered gene expression in certain brain regions. Cells of these regions make too many or too few of the enzymes that carry out electron transfer phosphorylation. Remember, this stage of aerobic respiration yields the bulk of the body's ATP. Does its disruption alter brain cells in ways that boost creativity but also invite illness?

With this intriguing connection, we invite you to reflect on how far you have come in this unit of the book. You first surveyed mitotic and meiotic cell divisions. You looked at how chromosomes and genes become shuffled during meiosis and then during fertilization. You also became acquainted with Gregor Mendel's discovery of major patterns of inheritance. This knowledge is your portal to the chromosomal basis of human inheritance.

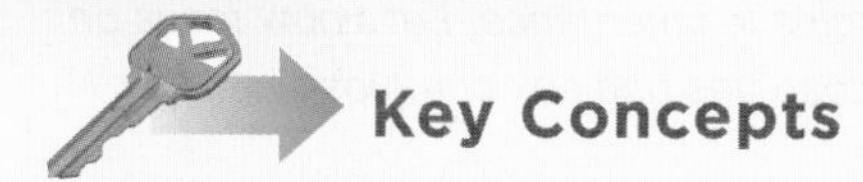

Key Concepts

AUTOSOMES AND SEX CHROMOSOMES

All animals have pairs of autosomes—chromosomes that are identical in length, shape, and which genes they carry. Sexually-reproducing species also have a pair of sex chromosomes. The members of this pair differ between females and males. A gene on one of the human sex chromosomes dictates the male sex. Karyotyping, a diagnostic tool, reveals changes in the structure or number of an individual's chromosomes. **Section 11.1**

AUTOSOMAL INHERITANCE

Many genes on autosomes are expressed in Mendelian patterns of simple dominance. **Sections 11.2, 11.3**

SEX-LINKED INHERITANCE

Some traits are affected by genes on the X chromosome. Inheritance patterns of such traits differ in males and females. **Section 11.4**

CHANGES IN CHROMOSOME STRUCTURE

On rare occasions, a chromosome may undergo permanent change in its structure, as when a segment of it is deleted, duplicated, inverted, or translocated. **Section 11.5**

CHANGES IN CHROMOSOME NUMBER

On rare occasions, the number of autosomes or sex chromosomes changes. In humans, the change usually results in a genetic disorder. **Section 11.6**

HUMAN GENETIC ANALYSIS AND OPTIONS

Various analytical and diagnostic procedures often reveal genetic disorders. Ethical questions are associated with what individuals as well as society at large do with the information. **Sections 11.7, 11.8**

Links to Earlier Concepts

You will be drawing upon your knowledge of chromosome structure (Sections 8.1, 8.2), meiosis (9.3, 9.4), and gamete formation (10.5). Be sure you understand dominant and recessive genes, and the homozygous and heterozygous conditions (10.1). Remember that environmental factors influence gene expression (10.6). Glycolysis will turn up again (7.2), this time in the context of a genetic disorder.

11.1 Human Chromosomes

You already know quite a bit about chromosomes and their roles in inheritance. Let's now focus on human autosomes and sex chromosomes.

AUTOSOMES AND SEX CHROMOSOMES

LINKS TO SECTIONS 5.8, 8.1, 8.3, 9.3

Most animals, including humans, normally are female or male. They have a diploid chromosome number ($2n$), and pairs of homologous chromosomes in their body cells. All except one pair are autosomes—chromosomes with the same length, shape and centromere location. Autosomes carry the same genes in both females and males. Two sex chromosomes form the last pair. These chromosomes differ between males and females, and the differences determine an individual's sex.

Human females inherit two X chromosomes (XX). Males inherit one X and one Y (XY). These two sex chromosomes differ in length, shape, and which genes they carry. They zipper up together in a small region along their length and interact as homologues during prophase I of meiosis.

XX females and XY males are the rule among fruit flies, mammals, and many other animals, but there are other patterns. In butterflies, moths, birds, and certain fishes, males have two identical sex chromosomes, not females. Environmental factors (not sex chromosomes) determine sex in some species of invertebrates, turtles, and frogs. As an example, the temperature of the sand in which sea turtle eggs are buried determines the sex of the hatchlings.

SEX DETERMINATION

In humans, a new individual inherits a combination of sex chromosomes that dictates whether it will become a male or a female. All eggs made by a human female have one X chromosome. One-half of the sperm cells made by a male carry an X chromosome; the other half carry a Y chromosome. If an X-bearing sperm fertilizes an X-bearing egg, the resulting zygote will develop into a female. If the sperm carries a Y chromosome, the zygote will develop into a male (Figure 11.2*a*).

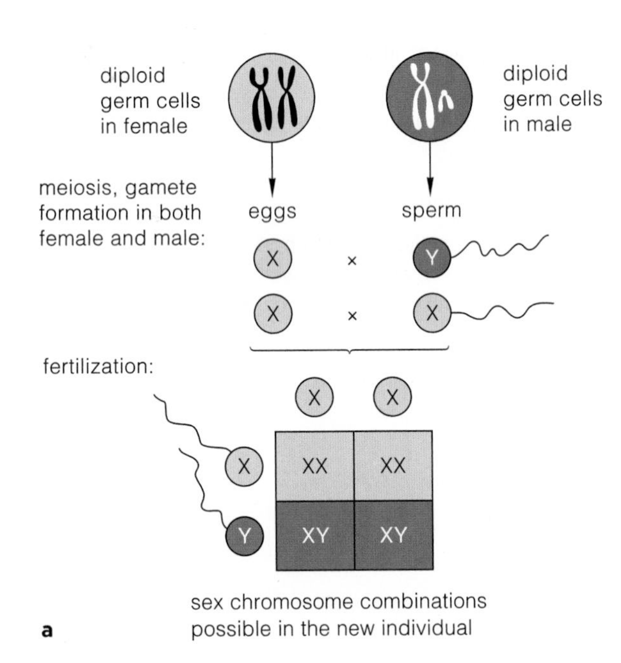

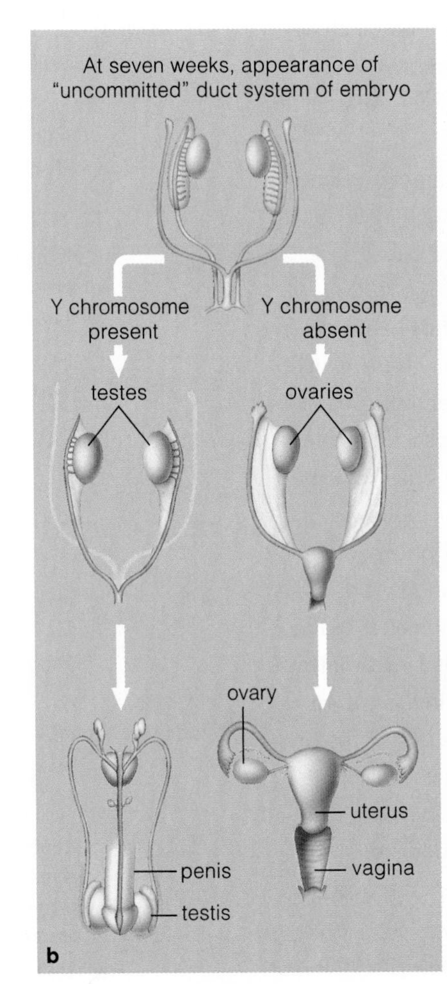

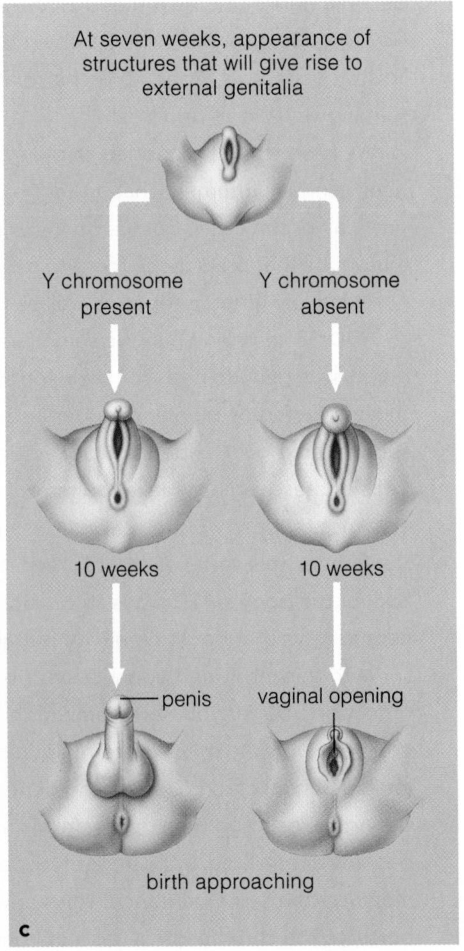

Figure 11.2 Animated! (**a**) Punnett-square diagram showing the sex determination pattern in humans.

(**b**) An early human embryo is neither male nor female. Then tiny ducts and other structures that can develop into male *or* female reproductive organs start forming. In an XX embryo, ovaries form in the absence of the Y chromosome and its *SRY* gene. In an XY embryo, the gene product triggers formation of testes, which secrete a hormone that initiates development of other male traits. (**c**) External reproductive organs in human embryos.

The human Y chromosome carries only 307 genes, but one of them is the *SRY* gene—the master gene for male sex determination. Its expression in XY embryos triggers the formation of testes, which are male gonads (Figure 11.2*b*). Some of the cells in these primary male reproductive organs make testosterone, a sex hormone that controls the emergence of male secondary sexual traits. How do we know *SRY* is the male sex master gene? Mutations that disable it cause individuals with an XY chromosome pair to develop as females.

An XX embryo has no Y chromosome, no *SRY* gene, and much less testosterone. Therefore, primary female reproductive organs—ovaries—form instead. Ovaries make estrogens and other sex hormones that govern the development of female secondary sexual traits.

The human X chromosome carries 1,336 genes. Some of those genes are associated with sexual traits, such as the distribution of body fat and hair. However, most of the genes on the X chromosome govern nonsexual traits such as blood clotting. Such genes are expressed in males as well as in females. Males, remember, also inherit one X chromosome.

KARYOTYPING

Sometimes the structure of a chromosome can change during mitosis or meiosis. Chromosome number can change also. Both events have variable consequences. A diagnostic tool called karyotyping helps us analyze an individual's diploid complement of chromosomes (Figure 11.3). With this procedure, a sample of cells taken from an individual is put into a growth medium that stimulates them to start mitosis. The medium also contains colchicine, a poison that binds tubulin and so interferes with assembly of mitotic spindles. The cells enter mitosis, but the colchicine prevents them from dividing, so their cell cycle stops at metaphase.

The cells and the medium are transferred to a tube. Then, the cells are separated from the liquid medium with a centrifuge. Like a washing machine spin cycle, the centrifuge whirls the tubes around a center post. The force of spinning moves the cells to the bottom of the tube. The medium is removed, and a hypotonic solution is added (Section 5.8). The cells swell up, so the chromosomes inside of them move apart. The cells are spread on a microscope slide and stained so the chromosomes become visible with a microscope.

The microscope reveals metaphase chromosomes in every cell. A micrograph of a single cell is digitally rearranged so the images of all the chromosomes are lined up by their centromere location, and arranged according to size, shape, and length. The final array is the individual's karyotype, which is compared with a normal standard. The karyotype shows whether there are extra or missing chromosomes. Some other kinds of structural abnormalities are also visible.

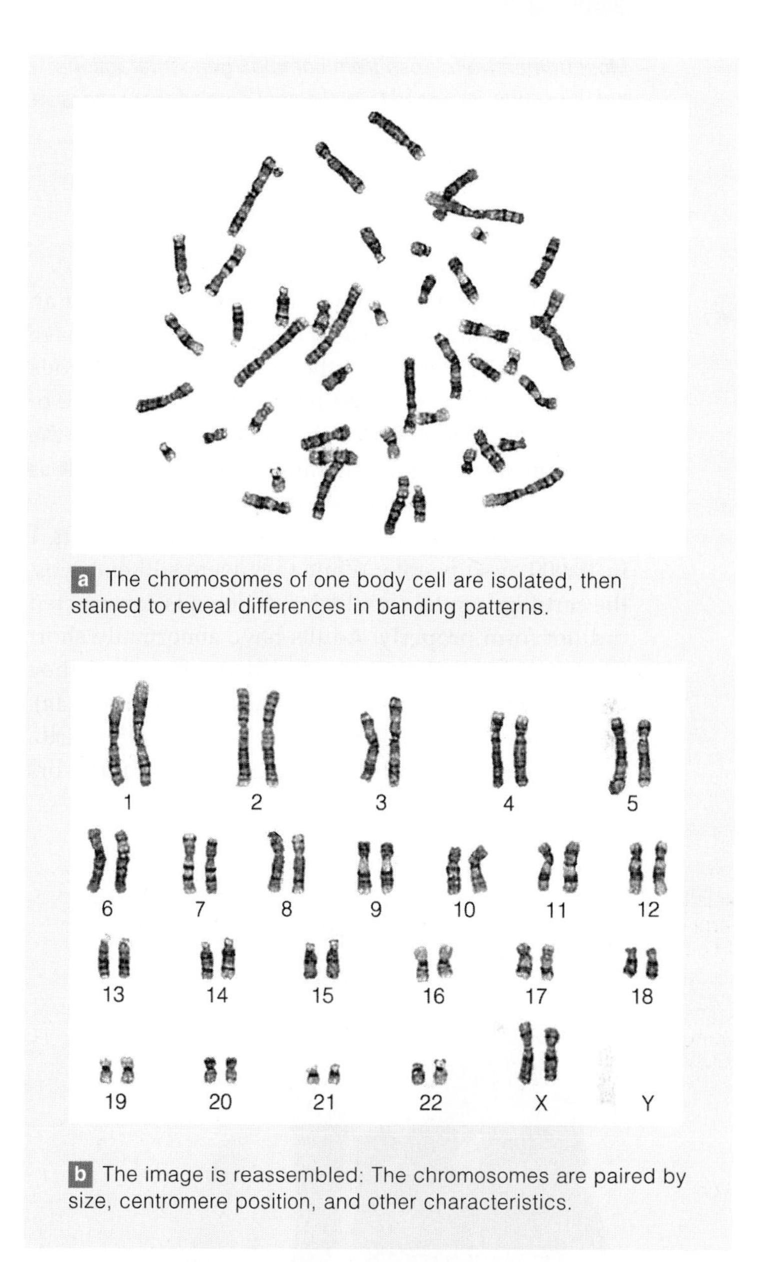

a The chromosomes of one body cell are isolated, then stained to reveal differences in banding patterns.

b The image is reassembled: The chromosomes are paired by size, centromere position, and other characteristics.

Figure 11.3 **Animated!** Karyotyping, a diagnostic tool that reveals an image of a single cell's diploid complement of chromosomes. This human karyotype shows 22 pairs of autosomes and a pair of X chromosomes. Was the cell taken from a male or a female?

Autosomes are pairs of chromosomes that are the same in males and females of a species. Members of a pair of sex chromosomes differ between males and females.

11.2 Examples of Autosomal Inheritance Patterns

Most human traits arise from complex gene interactions, but many can be traced to autosomal dominant or recessive alleles that are inherited in simple Mendelian patterns. Some of these alleles cause genetic disorders.

AUTOSOMAL DOMINANT INHERITANCE

LINKS TO SECTIONS 3.2, 4.12, 7.7

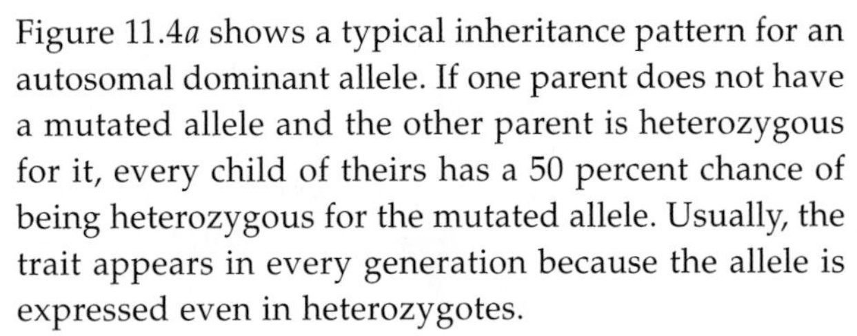

Figure 11.4*a* shows a typical inheritance pattern for an autosomal dominant allele. If one parent does not have a mutated allele and the other parent is heterozygous for it, every child of theirs has a 50 percent chance of being heterozygous for the mutated allele. Usually, the trait appears in every generation because the allele is expressed even in heterozygotes.

One autosomal condition, *achondroplasia*, affects 1 in 10,000 or so people. While they were still embryos, the cartilage model on which a skeleton is constructed did not form properly. Adults have abnormally short arms and legs relative to other body parts and they are only about four feet, four inches tall (Figure 11.4*a*). Most homozygotes die before or not long after birth. The allele does not affect the capacity of the survivors to reproduce.

In *Huntington's disease*, the nervous system slowly deteriorates, and involuntary muscle action increases. Symptoms often do not start until past age thirty, and those affected die during their forties or fifties. Many unknowingly transmit the mutated allele to children before then. The mutation causing the disorder alters a protein necessary for normal brain cell development. It is an expansion mutation. In such mutations, three nucleotides become duplicated multiple times in a row. Hundreds of thousands of expansion repeats occur in and between genes on the human chromosomes. This one alters a gene product's function.

A few dominant alleles that cause severe problems persist in populations because expression of the allele may not interfere with reproduction, or affected people reproduce before the symptoms become severe. Also, spontaneous mutations reintroduce some of them.

AUTOSOMAL RECESSIVE INHERITANCE

Inheritance patterns may point to a recessive allele on an autosome. First, if both parents are heterozygous for a recessive allele, each of their children will have a 50 percent chance of being heterozygous for the allele and a 25 percent chance of being homozygous for it (Figure 11.4*b*). Second, if both parents are homozygous, all of their children will be homozygous too.

Galactosemia is a heritable metabolic disorder that affects about 1 in every 50,000 newborns. This case of autosomal recessive inheritance involves an allele for an enzyme that helps digest the lactose in milk or in milk products. The body normally converts lactose to glucose and galactose. Then, a series of three enzymes converts the galactose to glucose-6-phosphate (Figure 11.5). This intermediate can enter glycolysis or it can be converted to glycogen (Sections 3.2 and 7.7).

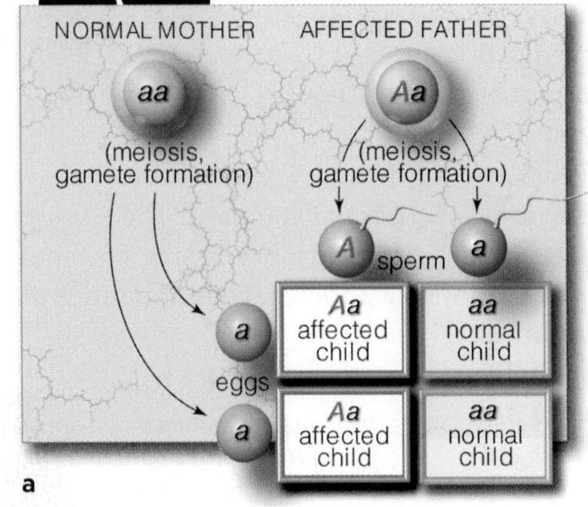

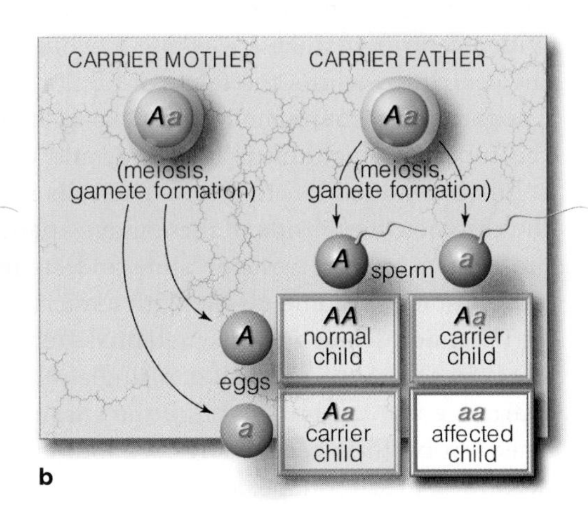

Figure 11.4 Animated! (**a**) Example of autosomal dominant inheritance. One dominant allele (*red*) is fully expressed in carriers. Achondroplasia, an autosomal dominant disorder, affects the three males shown above. At center, Verne Troyer (or Mini Me in the Mike Myers spy movies), stands two feet, eight inches tall.

(**b**) An autosomal recessive pattern. In this example, both of the parents are heterozygous carriers of the recessive allele (*red*).

galactose

ATP → ADP (enzyme action)

galactose-1-phosphate

enzyme action

glucose-1-phosphate

enzyme action

glucose-6-phosphate

Figure 11.5 How galactose is normally converted to a form that can enter the breakdown reactions of glycolysis. A mutation that affects the second enzyme in the conversion pathway gives rise to galactosemia.

People with galactosemia do not make one of these three enzymes; they are homozygous recessive for a mutated allele. Galactose-1-phosphate accumulates to toxic levels in their body, and it can be detected in the urine. The condition leads to malnutrition, diarrhea, vomiting, and damage to the eyes, liver, and brain.

When they do not receive treatment, galactosemics typically die young. When they are quickly placed on a diet that excludes all dairy products, the symptoms may not be as severe.

WHAT ABOUT NEUROBIOLOGICAL DISORDERS?

Most of the neurobiological disorders mentioned in the chapter introduction do not follow simple patterns of Mendelian inheritance. In most cases, mutations in one gene do not give rise to depression, schizophrenia, or bipolar disorder. Multiple genes and environmental factors contribute to the outcome. Nonetheless, it is useful to search for mutations that make some people more vulnerable to NBDs.

For example, researchers who conducted extensive family and twin studies have predicted that mutated alleles in specific regions of autosomes 1, 3, 5, 6, 8, 11 through 15, 18, and 22 increase an individual's chance of developing schizophrenia. Similarly, some mutated alleles are linked to bipolar disorder and depression.

Some traits can be traced to dominant or recessive alleles on autosomes because they are inherited in Mendelian patterns. Certain alleles on these chromosomes give rise to genetic abnormalities and genetic disorders.

11.3 Young, Yet Old

FOCUS ON HEALTH

Sometimes textbook examples of the human condition seem a bit abstract, so take a moment to think about two boys who were too young to be old.

Imagine being ten years old with a mind trapped in a body that is getting a bit more shriveled, more frail—*old*—every day. You are barely tall enough to peer over the top of a table. You weigh less than thirty-five pounds. Already you are bald and have a wrinkled nose. Possibly you have a few more years to live. Would you, like Mickey Hays and Fransie Geringer, still be able to laugh?

On average, of every 8 million newborn humans, one will grow old far too soon. On one of its autosomes, that rare individual carries a mutated allele that gives rise to Hutchinson-Gilford progeria syndrome. While that new individual was still an embryo inside its mother, billions of DNA replications and mitotic cell divisions distributed the information encoded in that gene to each newly formed body cell. Its legacy will be an accelerated rate of aging and a sharply reduced life span.

The disorder arises by spontaneous mutation of a gene for lamin, a protein that normally makes up intermediate filaments in the nucleus (Section 4.12). The altered lamin is not processed properly. It builds up on the inner nuclear membrane and distorts the nucleus. How this buildup causes the symptoms of progeria is not yet known.

Those symptoms start before age two. Skin that should be plump and resilient starts to thin. Skeletal muscles weaken. Limb bones that should lengthen and grow stronger soften. Premature baldness is inevitable (Figure 11.6). Affected people do not usually live long enough to reproduce, so progeria does not run in families.

Most progeriacs can expect to die in their early teens as a result of strokes or heart attacks. These final insults are brought on by a hardening of the wall of arteries, a condition typical of advanced age. Fransie was seventeen when he died. Mickey died at age twenty.

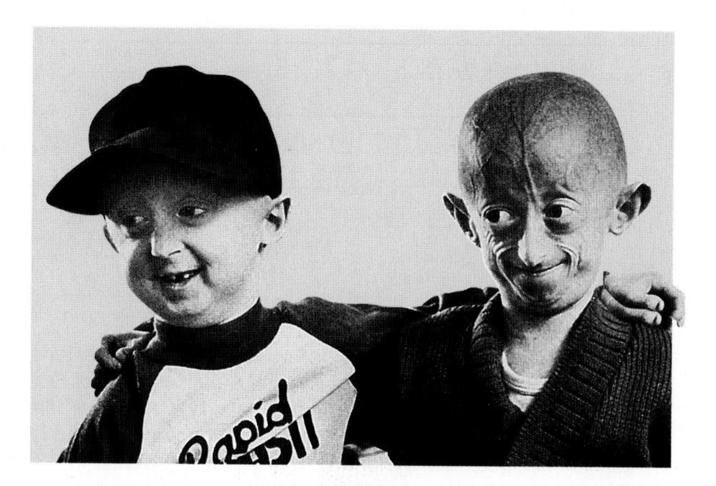

Figure 11.6 Mickey (*left*) and Fransie (*right*) met at a gathering of progeriacs at Disneyland, California. They were not yet ten years old.

11.4 Examples of X-Linked Inheritance Patterns

Alleles on an X chromosome give rise to phenotypes that also reflect simple Mendelian patterns of inheritance. Many of the recessive ones cause problems.

LINKS TO SECTIONS 4.12, 6.1

A recessive allele on an X chromosome often leaves certain clues when it causes a genetic disorder. First, more males than females are affected. Heterozygous females still have a dominant allele on their other X chromosome that masks the recessive allele's effects. Males are not protected, because they inherit only one X chromosome along with one Y chromosome (Figure 11.7). Second, an affected father cannot pass his one X-linked recessive allele to a son (children who inherit an X chromosome from their fathers are female). Thus, a heterozygous female must be the bridge between an affected male and his affected grandson.

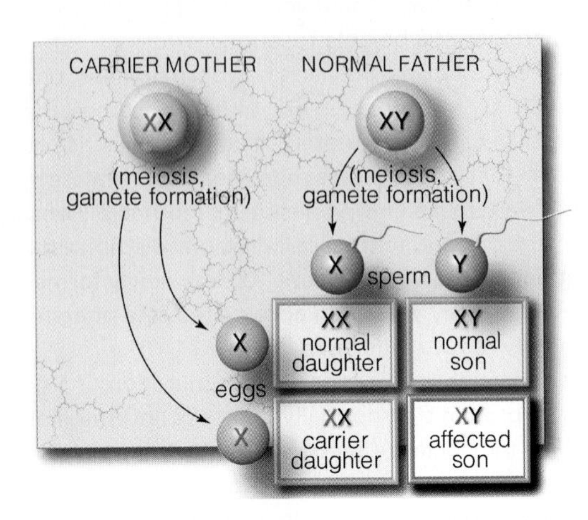

Figure 11.7 **Animated!** X-linked recessive inheritance. In this case, the mother carries a recessive allele on one of her X chromosomes (*red*).

Recently, researchers mapped the genes on the X chromosome. They discovered two things. First, a bit more than 6 percent of all of the genes we have reside on this sex chromosome. Second, mutated alleles that cause or contribute to more than 300 known genetic disorders occur on this chromosome.

HEMOPHILIA A

Hemophilia A, a type of blood clotting disorder, is a case of X-linked recessive inheritance. Most of us have a clotting mechanism that quickly stops bleeding from minor injuries (Section 34.2). The mechanism involves the protein products of genes on the X chromosome. Bleeding is prolonged in males who carry a mutated form of one of these X-linked genes, or in females who are homozygous for a mutation. Affected people bruise easily, and internal bleeding causes problems in their

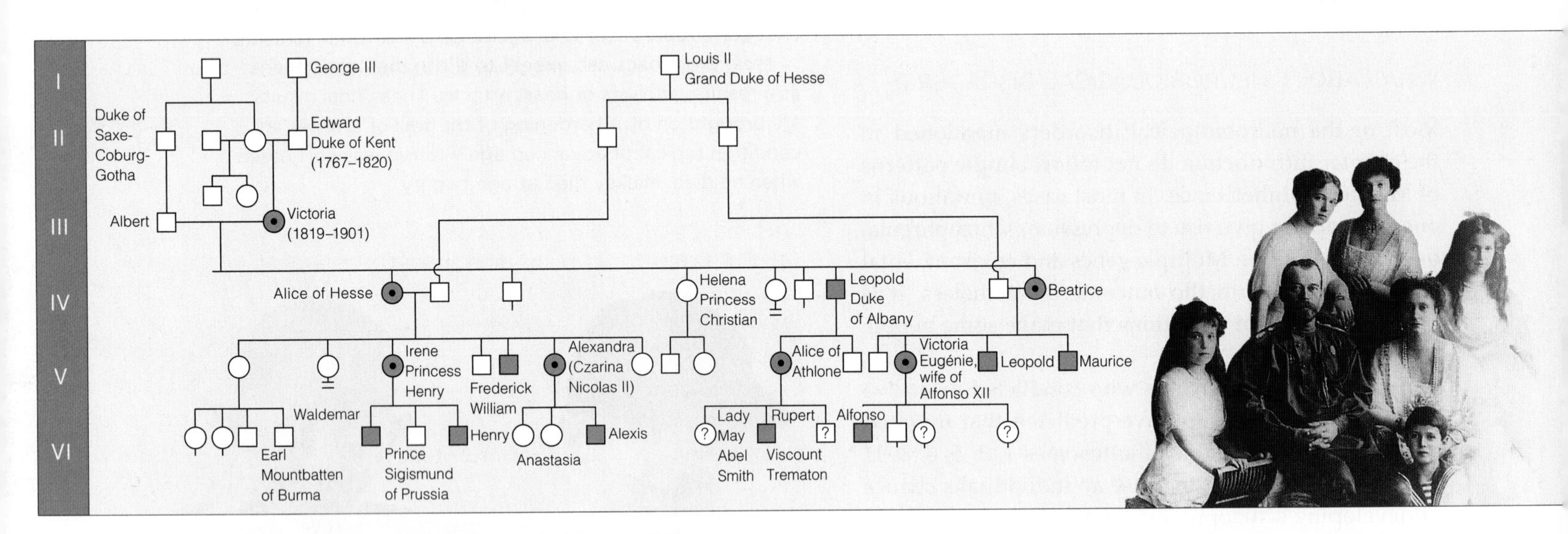

Figure 11.8 A classic case of X-linked recessive inheritance. This is a partial pedigree, or a chart of genetic connections, among descendants of Queen Victoria of England. It focuses on the male line of descent. Females affected by hemophilia A are indicated by *green* circles; affected males, *red* squares. At one time, the recessive X-linked allele was present in eighteen of Victoria's sixty-nine descendants, who sometimes intermarried. Of the Russian royal family members shown, the mother was a carrier. Through her obsession with the vulnerability of her son Alexis, a hemophiliac, she became involved in political intrigue that helped trigger the Russian Revolution of 1917.

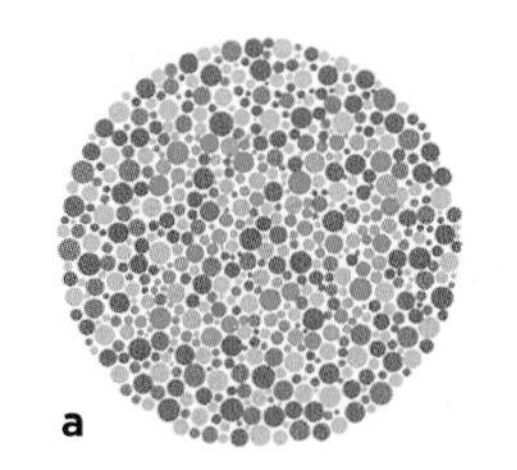

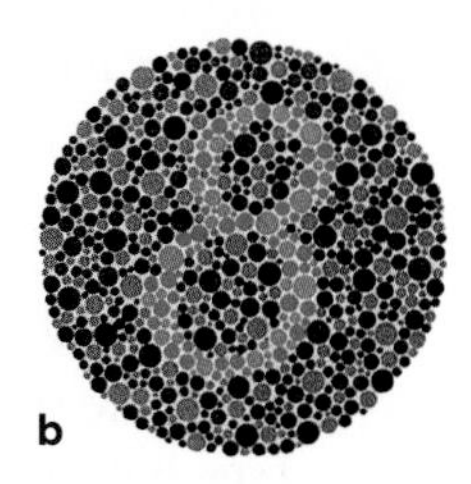

Figure 11.9 *Left*, what red–green color blindness means, using ripe red cherries on a green-leafed tree as an example. In this case, the perception of blues and yellows is normal, but the affected individual has difficulty distinguishing red from green.

Above, two of many Ishihara plates, which are standardized tests for different forms of color blindness. (**a**) You may have one form of red–green color blindness if you see the numeral "7" instead of "29" in this circle. (**b**) You may have another form if you see a "3" instead of an "8."

muscles and joints. In heterozygous females, clotting time is close to normal.

This disorder affects 1 in 7,500 people, on average, but because it is now treatable that number may rise. The disorder's frequency was relatively high in royal families of Europe and Russia in the nineteenth century, probably because the common practice of inbreeding kept the harmful allele in the family tree (Figure 11.8).

RED–GREEN COLOR BLINDNESS

The pattern of X-linked recessive inheritance shows up among individuals who have some degree of *color blindness*. The term refers to a range of conditions in which an individual cannot distinguish among some or all colors in the spectrum of visible light. Mutated genes result in altered function of the photoreceptors (light-sensitive receptors) in the eyes.

Normally, humans can sense the differences among 150 colors. A person who is red–green color blind sees fewer than 25 colors: Some or all of the receptors that respond to red and green wavelengths are weakened or absent. Some people confuse red and green colors. Others see green as shades of gray, but perceive blues and yellows quite well (Figure 11.9). Two sections of a standard set of tests for color blindness are shown in Figure 11.9*a*,*b*.

Color blindness is more common in men, who are about twelve times more likely than women to develop the condition. Heterozygous women show symptoms as well. Can you explain why?

DUCHENNE MUSCULAR DYSTROPHY

Duchenne muscular dystrophy (DMD) is one of a group of X-linked recessive disorders characterized by rapid degeneration of muscles, starting early in life. About 1 in 3,500 boys is affected.

A recessive allele encodes dystrophin, a protein that structurally supports the fused cells in muscle fibers by anchoring the cell cortex to the plasma membrane. When dystrophin is abnormal or absent, the cell cortex weakens and muscle cells die. The cell debris that remains in the tissues triggers chronic inflammation.

Most boys with DMD are diagnosed between the ages of three and seven. The progression of this disorder cannot be stopped. When the affected boy is about twelve years old, he will begin to use a wheelchair. His heart muscles will start to break down. Even with the best care, he will probably die before he is thirty, from a heart disorder or from respiratory failure (suffocation).

Diverse recessive alleles on the human X chromosome are implicated in more than 300 genetic disorders.

A female heterozygous for one of those alleles may not show symptoms.

Males (XY) transmit an X-linked allele only to daughters, not to their sons.

11.5 Heritable Changes in Chromosome Structure

On rare occasions, a chromosome's structure changes. Many of the alterations have severe or lethal outcomes.

LINKS TO SECTIONS 3.5, 9.3

MAIN CATEGORIES OF STRUCTURAL CHANGE

Large-scale changes in the structure of a chromosome may give rise to a genetic disorder. Such changes are rare, but they do occur spontaneously in nature. Some also are induced by exposure to certain chemicals or radiation. Either way, such alterations may be detected by karyotyping. Large-scale changes in chromosome structure include duplications, deletions, inversions, and translocations.

Duplication Even normal chromosomes have DNA sequences that are repeated two or more times. These are called duplications:

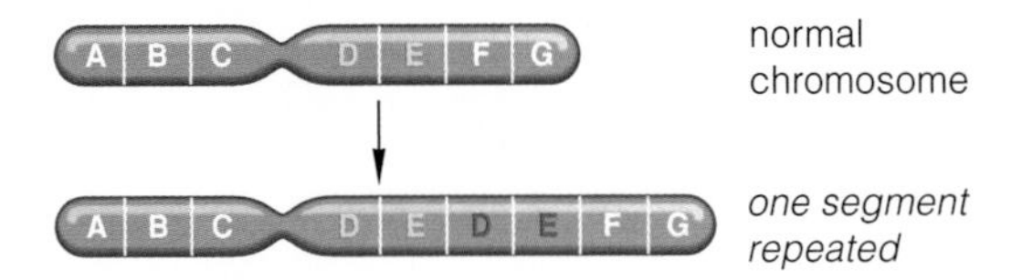

Duplications can occur through unequal crossovers at prophase I. Homologous chromosomes align side by side, but their DNA sequences misalign at some point along their length. The probability of misalignment is greater in regions where DNA has long repeats of the same sequence of nucleotides. A stretch of DNA gets deleted from one chromosome and is spliced into the partner chromosome. Some duplications cause neural problems and physical abnormalities. As you will see, others apparently were important in the evolution of primates that were ancestral to humans.

Deletion A deletion is the loss of some portion of a chromosome:

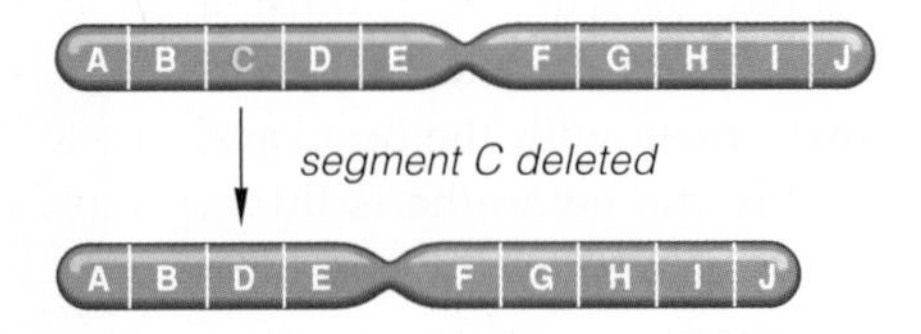

In mammals, deletions usually cause serious disorders and are often lethal. The loss of genes results in the disruption of growth, development, and metabolism. For instance, a small deletion in chromosome 5 causes mental impairment and an abnormally shaped larynx. Affected infants tend to make a sound like the meow of a cat, hence the name of the disorder, cri-du-chat, which is French for "cat's cry" (Figure 11.10).

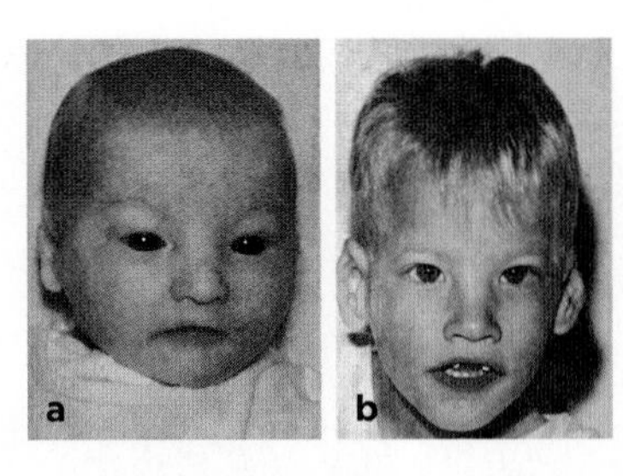

Figure 11.10 Cri-du-chat syndrome. (**a**) This infant's ears are low relative to his eyes. (**b**) Same boy, four years later. The high-pitched monotone of cri-du-chat children may persist into their adulthood.

Inversion With an inversion, part of the sequence of DNA within the chromosome becomes oriented in the reverse direction, with no molecular loss:

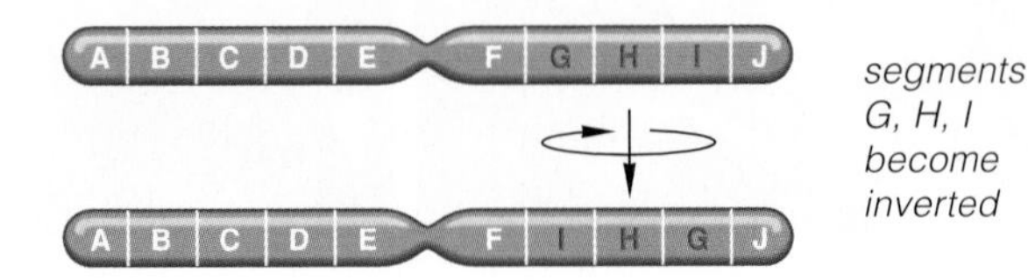

An inversion may not affect a carrier's health if it does not disrupt a gene region. However, it may affect an individual's fertility. Crossovers in an inverted region during meiosis may result in deletions or duplications that affect the viability of forthcoming embryos. Some carriers do not know that they have an inversion until they are diagnosed with infertility and their karyotype is tested.

Translocation If a chromosome breaks, the broken part may get attached to a different chromosome, or to a different part of the same one. This structural change is a translocation. Most translocations are reciprocal, or balanced; two chromosomes exchange broken parts:

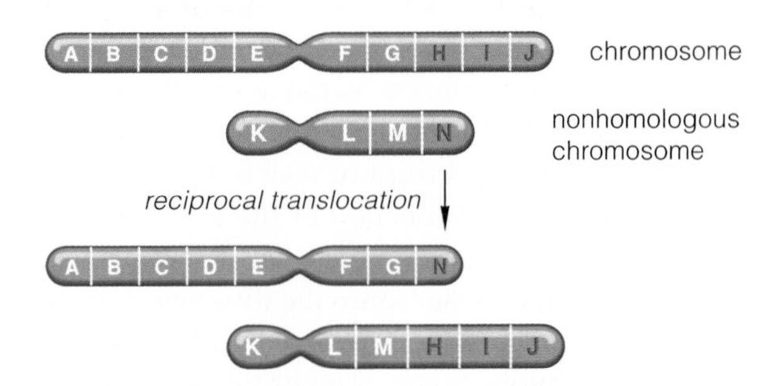

A reciprocal translocation that does not disrupt a gene region may have no adverse effect on its bearer. Many people do not realize they carry a translocation until they have difficulty with fertility. The two translocated chromosomes pair in a different way with their non-translocated counterparts. They segregate improperly about half of the time, so about half of the resulting gametes will have major duplications or deletions. If one of these gametes unites with a normal gamete, the resulting embryo almost always dies.

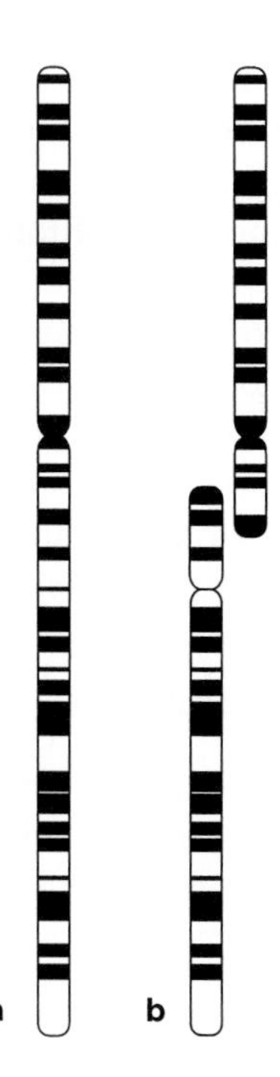

Figure 11.11 Banding patterns of human chromosome 2 (**a**), compared with two chimpanzee chromosomes (**b**). Bands appear because different regions of the chromosomes take up stain differently.

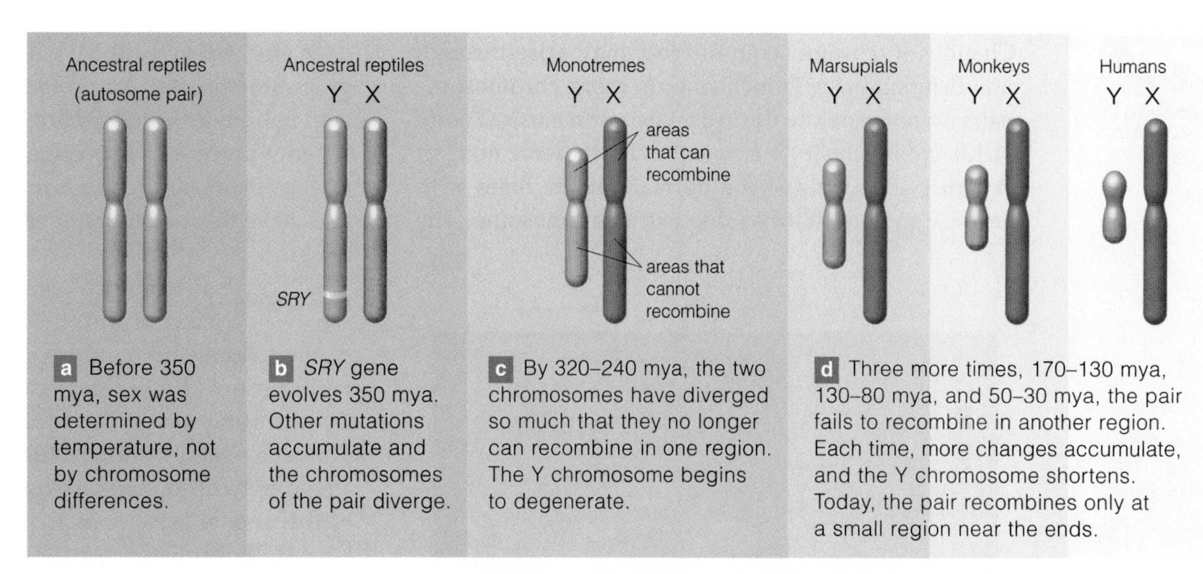

Figure 11.12 Evolution of the Y chromosome.

DOES CHROMOSOME STRUCTURE EVOLVE?

As you see, alterations in chromosome structure may reduce fertility; individuals who are heterozygous for multiple changes may not be able to produce offspring at all. However, accumulation of multiple alterations in homozygous individuals can be the start of a new species. It may seem as if this outcome would be rare, but it can and does occur over generations in nature. Karyotyping studies show that structural alterations have been built into the DNA of nearly all species.

For example, certain duplications have allowed one copy of a gene to mutate while a different copy carries out its original function. The globin genes of primates appear to have evolved like this. Reflect on the two globin chains that make up hemoglobin (Section 3.5). Humans and other primates have several different but strikingly similar genes for globin chains. These genes apparently evolved by duplications and mutations. The globin chains encoded by the genes differ in their oxygen-binding capacity under a range of conditions.

Some chromosome structure alterations contributed to differences among closely related organisms, such as apes and humans. There are twenty-three pairs of chromosomes in the body cells of humans, but twenty-four pairs in the body cells of chimpanzees, gorillas, and orangutans. Thirteen out of the twenty-three pairs of human chromosomes are almost identical with the corresponding pairs of chimpanzee chromosomes. Nine pairs differ by inversions, and one human chromosome arose from two chromosomes that remained separate in chimpanzees and the other great apes. During human evolution, those two chromosomes fused end to end and formed our chromosome 2 (Figure 11.11). In the fused region, researchers discovered the remnants of a telomere—a special DNA sequence that caps the *ends* of all chromosomes.

As another example, X and Y chromosomes were once homologous autosomes in reptile-like ancestors of mammals. In those organisms, ambient temperature probably determined sex, as it still does in turtles and some other modern reptiles. Then, about 350 million years ago, one of the two chromosomes underwent a structural alteration that interfered with homologous recombination in meiosis. Eventually, the homologues became so different that they no longer recombined at all in the changed region—which by that time held the *SRY* gene on the Y chromosome. A gene on the X chromosome is similar to *SRY*; they probably diverged from a common ancestral gene (Figure 11.12).

A segment of a chromosome may be duplicated, deleted, inverted, or moved to a new location. Such changes can be harmful or lethal. Others have been conserved over time; they confer advantages or have had neutral effects.

11.6 Heritable Changes in the Chromosome Number

Occasionally, abnormal events occur before or during cell division, and gametes and new individuals end up with the wrong chromosome number. Consequences range from minor to lethal changes in form and function.

LINKS TO SECTIONS 1.8, 8.3, 9.3

Changes in chromosome number may arise through nondisjunction, in which one or more chromosome pairs do not separate during mitosis or meiosis (Figure 11.13). Nondisjunction affects the chromosome number at fertilization. Suppose a normal gamete fuses with an $n+1$ gamete that has one extra chromosome. The new individual will be trisomic ($2n+1$), with three of one type of chromosome and two of every other type. If an $n-1$ gamete and a normal n gamete fuse, the new individual will be $2n-1$, or monosomic. Mitotic divisions perpetuate such mistakes.

In aneuploidy, cells have too many or too few copies of a chromosome. Autosomal aneuploidy is typically fatal in humans, and it is linked to many miscarriages. Seventy percent of flowering plant species, and some insects, fishes, and other animals are polyploid—their cells have three or more of each type of chromosome.

AUTOSOMAL CHANGE AND DOWN SYNDROME

A few trisomic humans are born alive, but only those with trisomy 21 will reach adulthood. A newborn with three chromosomes 21 will develop *Down syndrome*. This autosomal disorder is the most common type of aneuploidy in humans; it occurs once in 800 to 1,000 births and affects more than 350,000 people in the United States. Figure 11.13*a* shows a karyotype for a trisomic 21 female. About 95 percent of all cases arise through nondisjunction. The affected individuals have upward-slanting eyes, a fold of skin that starts at the inner corner of each eye, a deep crease across the sole of each palm and foot, one (instead of two) horizontal furrows on their fifth fingers, slightly flattened facial features, and other symptoms.

Not all of the outward symptoms develop in every individual. That said, trisomic 21 individuals tend to have moderate to severe mental impairment and heart problems. Their skeleton grows and develops abnormally; older children have shorter body parts, loose joints, and misaligned bones of the fingers, toes, and hips. Muscles and reflexes are weak, and motor skills, including speech, develop slowly. With medical care, trisomy 21 individuals can live about fifty-five years. Early training can help affected individuals learn to take part in normal activities. As a group, they tend to be cheerful.

The incidence of Down syndrome, and nondisjunction in general, rises with the increasing age of the mother (Figure 11.14). Nondisjunction may occur in the father, although far less frequently. Trisomy 21 is just one of hundreds of conditions that can be detected through prenatal diagnosis (Section 11.8).

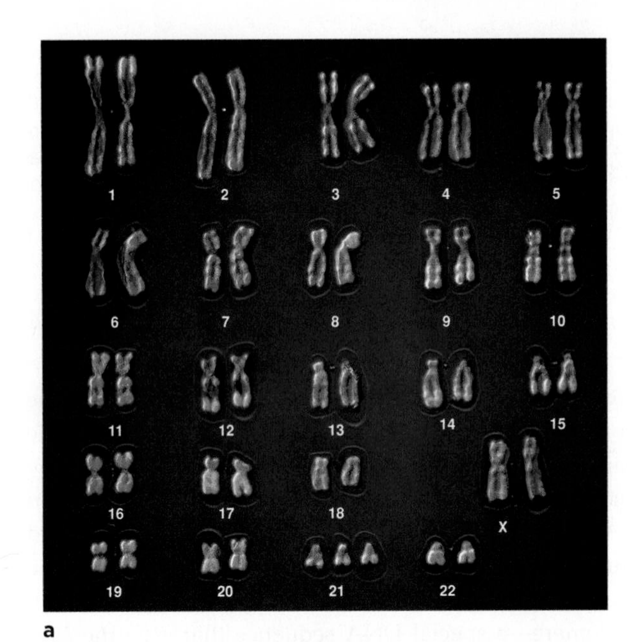

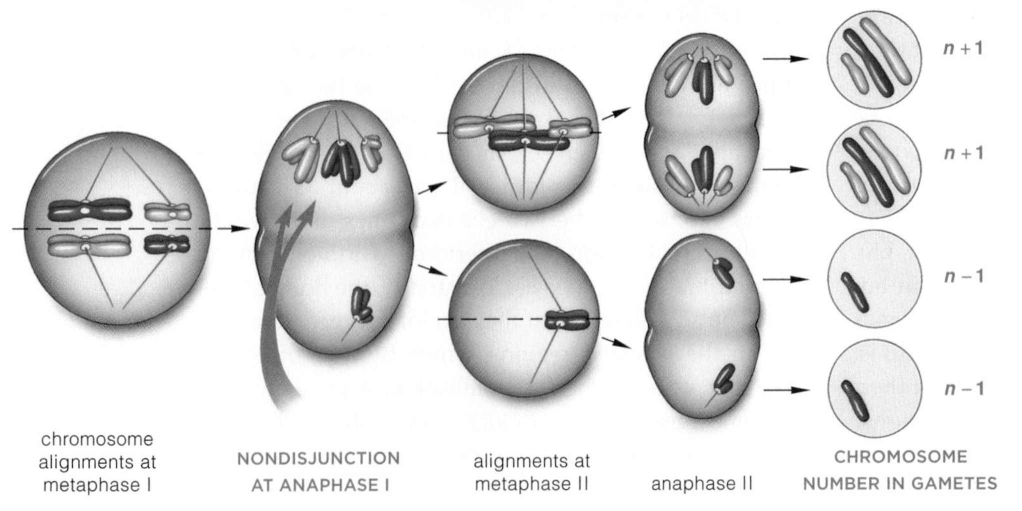

Figure 11.13 (**a**) A case of nondisjunction. This karyotype reveals the trisomic 21 condition of a human female. (**b**) One example of how nondisjunction arises. Of the two pairs of homologous chromosomes shown here, one fails to separate during anaphase I of meiosis. The chromosome number is altered in the gametes that form after meiosis.

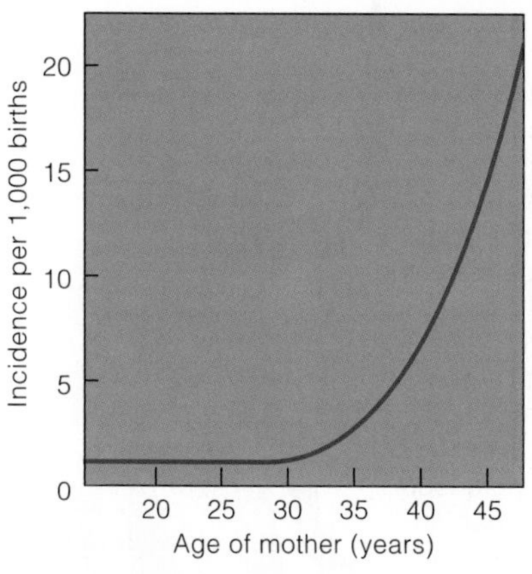

Figure 11.14 Relationship between the frequency of Down syndrome and mother's age at childbirth. The data are from a study of 1,119 affected children. The risk of having a trisomic 21 baby rises with the mother's age. About 80 percent of trisomic 21 individuals are born to women under thirty-five, but these women have the highest fertility rates, and they have more babies.

Figure 11.15 A 6-year-old with Turner's syndrome. Affected girls tend to be shorter than average, but daily hormone injections can help them reach normal height.

CHANGE IN THE SEX CHROMOSOME NUMBER

Nondisjunction also causes alterations in the number of X and Y chromosomes, with a frequency of about 1 in 400 live births. Most often, such alterations lead to difficulties in learning and impaired motor skills such as a speech delay, but problems may be so subtle that the underlying cause is never diagnosed.

Female Sex Chromosome Abnormalities *Turner syndrome* individuals have an X chromosome and no corresponding X or Y chromosome (XO). About 1 in 2,500 to 10,000 newborn girls are XO (Figure 11.15). Nondisjunction originating with the father accounts for 75 percent of these cases. Yet there are fewer cases compared with other sex chromosome abnormalities: At least 98 percent of XO embryos will spontaneously abort early in pregnancy.

Despite the near lethality, the XO survivors are not as disadvantaged as other aneuploids. They grow up well proportioned but short (with an average height of four feet, eight inches). Most do not have functional ovaries, so they do not make enough sex hormones to mature sexually. The development of secondary sexual traits such as breasts is affected.

A few females inherit three to five X chromosomes. The resulting *XXX syndrome* occurs in about 1 of 1,000 births. Only one X chromosome is typically active in female cells, so having extra X chromosomes usually does not result in any physical or medical problems.

Male Sex Chromosome Abnormalities About 1 of every 500 males has an XXY karyotype. Most cases are an outcome of nondisjunction during meiosis. The disorder that results, *Klinefelter syndrome*, develops at puberty. XXY males tend to be overweight, tall, and within a normal range of intelligence. They make more estrogen and less testosterone than normal males, and these hormones have feminizing effects. The testes and prostate gland are smaller than average. Sperm count is low, hair is sparse, the voice is high-pitched, and the breasts are somewhat enlarged. Testosterone injections during puberty can reverse the feminized traits.

About 1 in 500 to 1,000 males has an *XYY condition*. They tend to be taller than average and have mild mental impairment, but most are otherwise normal. XYY men were once thought to be predisposed to a life of crime. This misguided view was based on sampling error (too few cases in narrowly chosen groups such as prison inmates) and bias (researchers who gathered the karyotypes also took the personal histories).

In 1976 a Danish geneticist reported results from his study of 4,139 tall males, all twenty-six years old, who had registered at their draft board. Besides their data from physical examinations and intelligence tests, the draft records offered clues to social and economic status, education, and any criminal convictions. Only twelve of the males studied were XYY, which meant that the "control group" had more than 4,000 males. The only findings? Mentally impaired, tall males who engage in criminal deeds are just more likely to get caught—irrespective of karyotype.

Nondisjunction can change the number of autosomes or the number of sex chromosomes an individual has. Such changes usually cause genetic disorders.

Nondisjunction at meiosis causes most sex chromosome abnormalities, which can lead to difficulties with learning, and delays in speech and other motor skills.

11.7 Human Genetic Analysis

Some organisms, including pea plants and fruit flies, are ideal for genetic analysis. They have few chromosomes, and can reproduce fast in small spaces under controlled conditions. It does not take long to track a trait through many generations. Humans, however, are another story.

LINKS TO SECTIONS 1.8, 10.4

Unlike flies grown in laboratory bottles, we humans live under varying conditions in diverse environments, and we live as long as geneticists who study us. Most of us select our own mates and reproduce if and when we want to. Most human families are not large, which means that there are not enough offspring available for researchers to make easy inferences.

To reduce sampling error, geneticists often gather information from several generations (Section 1.8). If a trait follows a simple Mendelian inheritance pattern, geneticists can predict the probability of its recurrence in future generations. Some inheritance patterns are clues to past events (Figure 11.16).

Inheritance patterns are often displayed as charts of genetic connections called pedigrees. Standardized methods, definitions, and symbols are used to make these charts. You already came across one in Figure 11.8. Figures 11.17 and 11.18 show two more.

Figure 11.16 An intriguing pattern of inheritance. Eight percent of the men in Central Asia carry nearly identical Y chromosomes, which implies descent from a shared ancestor. If so, then 16 million males living between northeastern China and Afghanistan—close to 1 of every 200 men alive today—may be part of a lineage that started with the warrior and notorious womanizer Genghis Khan. In time, his offspring ruled an empire that stretched from China all the way to Vienna.

Those who analyze pedigrees use their knowledge of probability and patterns of Mendelian inheritance that may yield clues to a trait. Such researchers have traced many genetic abnormalities and disorders to a dominant or recessive allele and often to its location on an autosome or a sex chromosome. Table 11.1 is a list of some that are used as examples in this book.

As individuals and as members of society, what do we do with genetics information? The next section gets into options. When considering them, keep in mind some distinctions. First, a genetic *abnormality* is a rare or uncommon version of a trait, such as when a person is born with six digits on each hand or foot instead of the usual five (Figure 11.17). Such abnormalities are not inherently life-threatening, and how you view them is a matter of opinion. By contrast, a genetic *disorder* is an inherited condition that sooner or later causes mild to severe medical problems. A genetic disorder is characterized by a specific set of symptoms—a syndrome.

A disease—an illness caused by infection or environmental factors—also has a characteristic set of symptoms. It is appropriate to use the term *genetic disease* when environmental factors modify

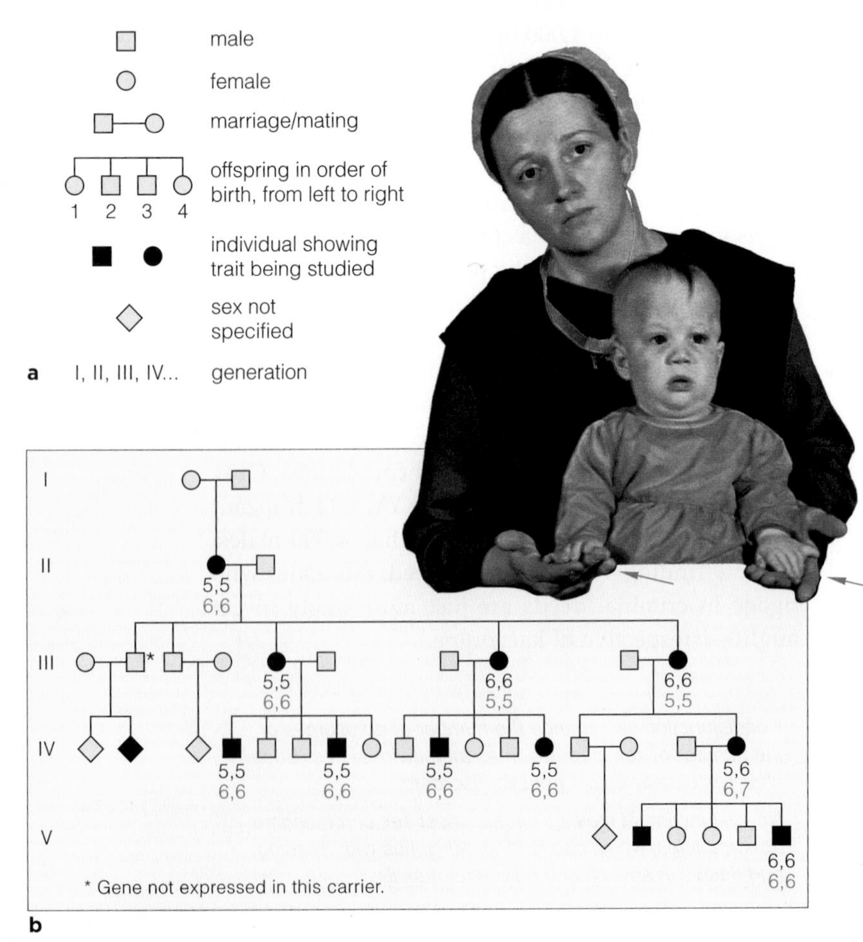

Figure 11.17 Animated! (**a**) Some standard symbols used in pedigrees. (**b**) A pedigree for *polydactyly*, which is characterized by extra fingers, toes, or both. The *black* numbers signify the number of fingers on each hand; the *blue* numbers signify the number of toes on each foot. Though it occurs on its own, polydactyly is also one of several symptoms of Ellis–van Creveld syndrome.

Table 11.1 Examples of Human Genetic Disorders and Genetic Abnormalities

Disorder or Abnormality	Main Symptoms
Autosomal recessive inheritance	
Albinism	Absence of pigmentation
Hereditary methemoglobinemia	Blue skin coloration
Cystic fibrosis	Abnormal glandular secretions leading to tissue, organ damage
Ellis–van Creveld syndrome	Dwarfism, heart defects, polydactyly
Fanconi anemia	Physical abnormalities, bone marrow failure
Galactosemia	Brain, liver, eye damage
Phenylketonuria (PKU)	Mental impairment
Sickle-cell anemia	Adverse pleiotropic effects on organs throughout body
Autosomal dominant inheritance	
Achondroplasia	One form of dwarfism
Camptodactyly	Rigid, bent fingers
Familial hypercholesterolemia	High cholesterol levels in blood; eventually clogged arteries
Huntington's disease	Nervous system degenerates progressively, irreversibly
Marfan syndrome	Abnormal or no connective tissue
Polydactyly	Extra fingers, toes, or both
Progeria	Drastic premature aging
Neurofibromatosis	Tumors of nervous system, skin
X-linked recessive inheritance	
Androgen insensitivity syndrome	XY individual but having some female traits; sterility
Red–green color blindness	Inability to distinguish among some or all shades of red and green
Fragile X syndrome	Mental impairment
Hemophilia	Impaired blood clotting ability
Muscular dystrophies	Progressive loss of muscle function
X-linked anhidrotic dysplasia	Mosaic skin (patches with or without sweat glands); other effects
Changes in chromosome structure	
Chronic myelogenous leukemia (CML)	Overproduction of white blood cells in bone marrow; organ malfunctions
Cri-du-chat syndrome	Mental impairment; abnormally shaped larynx
Changes in chromosome number	
Down syndrome	Mental impairment; heart defects
Turner syndrome (XO)	Sterility; abnormal ovaries, abnormal sexual traits
Klinefelter syndrome	Sterility; mild mental impairment
XXX syndrome	Minimal abnormalities
XYY condition	Mild mental impairment or no effect

previously workable genes in a way that disrupts some body function.

One more point to keep in mind: Alleles that give rise to severe genetic disorders are generally rare in populations, because they put their bearers at risk. Why don't they disappear entirely? Rare mutations can reintroduce them. A normal allele may mask the harmful effects of expression of a mutated allele in heterozygotes. Or, heterozygotes with a codominant mutated allele may have an advantage in a particular environment. You will encounter an example of the latter case in Section 17.6.

Pedigree analysis may reveal simple patterns of Mendelian inheritance. From such patterns, geneticists can infer the probability that offspring will inherit certain alleles.

A genetic abnormality is a rare or less common version of a heritable trait. A genetic disorder is a heritable condition that results in mild to severe medical problems.

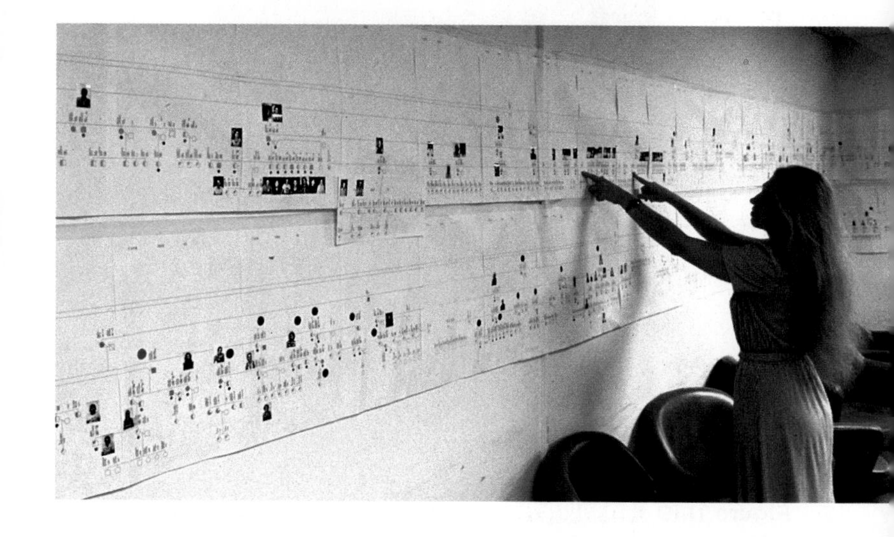

Figure 11.18 Pedigree for Huntington's disease, a progressive degeneration of the nervous system. Researcher Nancy Wexler and her team constructed this extended family tree for nearly 10,000 Venezuelans. Their analysis of unaffected and affected individuals revealed that a dominant allele on human chromosome 4 is the culprit. Wexler has a special interest in the disorder; it runs in her family.

11.8 Prospects in Human Genetics

With the first news of pregnancy, parents-to-be typically wonder if their baby will be normal. Quite naturally, they want their baby to be free of genetic disorders, and most babies are. What are the options when they are not?

Many prospective parents have difficulty coming to terms with the possibility that a child of theirs might develop a severe genetic disorder. What are their options?

Genetic Counseling Genetic counseling starts with diagnosis of parental genotypes, pedigrees, and genetic testing for known disorders. Using information gained from the tests, genetic counselors can predict a couple's probability of having a child with a genetic disorder.

Parents-to-be commonly ask genetic counselors to compare the risks associated with diagnostic procedures against the likelihood that their future child will be affected by a severe genetic disorder. At the time of counseling, they also should compare the small overall risk (3 percent) that complications during the birth process can affect *any* child. They should talk about how old they are—the older either prospective parent is, the greater the risk.

As a case in point, suppose a first child or a close relative has a severe disorder. A genetic counselor will evaluate the pedigrees of the parents, and the results of any genetic tests. Using this information, counselors can predict risks for disorders in future children. The same risk applies to each pregnancy.

Prenatal Diagnosis Doctors and clinicians commonly use methods of *prenatal diagnosis* to determine the sex of embryos or fetuses and to screen for more than 100 known genetic problems. *Prenatal* means before birth. *Embryo* is a term that applies until eight weeks after fertilization, after which *fetus* is appropriate.

Suppose a forty-five-year-old woman is pregnant and worries about Down syndrome. Between fifteen and twenty weeks after conception, she might opt for *amniocentesis* (Figure 11.19). By this diagnostic procedure, a clinician uses a syringe to withdraw a small sample of fluid from the amniotic cavity. The "cavity" is a fluid-filled sac, bounded by a membrane—the amnion—that encloses the fetus. The fetus normally sheds some cells into the fluid. Cells suspended in the fluid sample can be analyzed for many genetic disorders, including Down syndrome, cystic fibrosis, and sickle-cell anemia.

Chorionic villi sampling (CVS) is a similar diagnostic procedure. A clinician withdraws a few cells from the chorion, a membrane that surrounds the amnion and helps form the placenta—an organ that allows substances to be exchanged between mother and embryo. However, unlike amniocentesis, CVS can be performed as early as eight weeks into pregnancy.

It is now possible to see a live, developing fetus with the aid of an endoscope, a fiber-optic device. In *fetoscopy*, sound waves are pulsed across the mother's uterus. Images of parts of the fetus, umbilical cord, or placenta show up

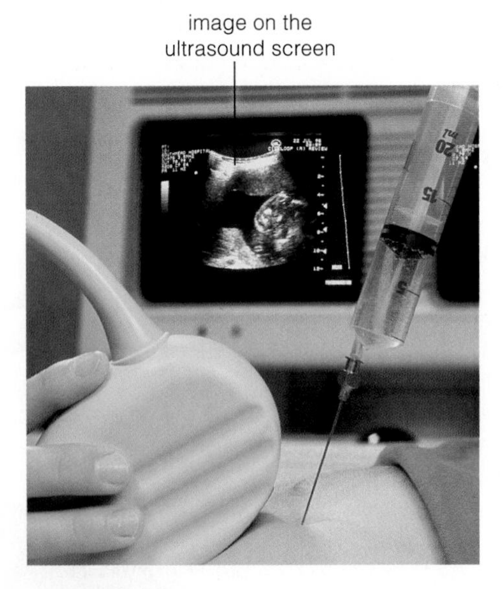

Figure 11.19 **Animated!** Amniocentesis, a prenatal diagnostic tool. A pregnant woman's doctor holds an ultrasound emitter against her abdomen while drawing a sample of amniotic fluid into a syringe. He monitors the path of the needle with an ultrasound screen in the background. Then he directs the needle into the amniotic sac that holds the developing fetus and withdraws 20 milliliters or so of amniotic fluid. The fluid contains fetal cells and wastes that can be analyzed for genetic disorders.

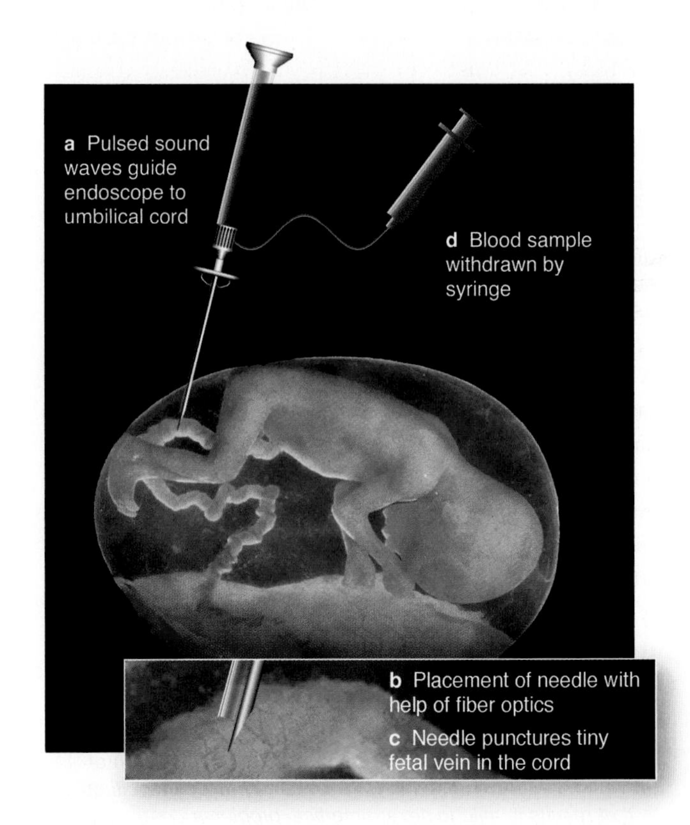

Figure 11.20 Fetoscopy for prenatal diagnosis.

on a computer screen that is connected to the endoscope (Figure 11.20). A sample of fetal blood is often drawn at the same time, in order to diagnose blood cell disorders such as sickle-cell anemia and hemophilia.

There are risks to a fetus associated with all three procedures, including punctures or infections. If the amnion does not reseal itself quickly, too much fluid may leak out of the amniotic cavity. Amniocentesis increases the risk of miscarriage by 1 to 2 percent. CVS occasionally disrupts the placenta's development and thus causes underdeveloped or missing fingers and toes in 0.3 percent of newborns. Fetoscopy raises the miscarriage risk by 2 to 10 percent.

Preimplantation Diagnosis This procedure relies on *in vitro fertilization*. Sperm and eggs from prospective parents are mixed in a sterile culture medium. One or more eggs may become fertilized. Then, mitotic cell divisions can turn the fertilized egg into a ball of eight cells within forty-eight hours (Figure 11.21).

According to one view, the tiny, free-floating ball is a pre-pregnancy stage. Like all of the unfertilized eggs that a woman's body discards monthly during her reproductive years, it has not attached to the uterus. All of its cells have the same genes, but they are not yet committed to being specialized one way or another. Doctors can remove one of these undifferentiated cells and analyze its genes. If it has no detectable genetic defects, the ball is inserted into the uterus. The withdrawn cell will not be missed. Many of the resulting "test-tube babies" are born in good health. Some couples who are at risk of passing on the alleles for cystic fibrosis, muscular dystrophy, or some other genetic disorder have opted for this procedure.

Phenotypic Treatments Surgery, prescription drugs, hormone replacement therapy, and often dietary controls can minimize and in some cases eliminate the symptoms of many genetic disorders.

For instance, strict dietary controls work in cases of *phenylketonuria*, or PKU. Individuals affected by this genetic disorder are homozygous for a recessive allele on an autosome. They cannot make a functional form of an enzyme that catalyzes the conversion of one amino acid (phenylalanine) to another (tyrosine). Because the conversion is blocked, phenylalanine accumulates and is diverted into other metabolic pathways. The outcome is an impairment of brain function.

Affected people who restrict phenylalanine intake can lead essentially normal lives. They must avoid soft drinks and other products that are sweetened with aspartame, a compound that contains phenylalanine.

Regarding Abortion What happens after prenatal diagnosis reveals a severe problem? Some prospective parents opt for an induced abortion. An *abortion* is an expulsion of a pre-term embryo or fetus from the uterus. We can only say here that individuals must weigh their awareness of the severity of the genetic disorder against their ethical and religious beliefs. Worse, they must play out their personal tragedy on a larger stage dominated by a nationwide battle between highly vocal "pro-life" and "pro-choice" factions. We return to this very volatile topic in Section 39.12, after explaining the stages of human embryonic development.

Figure 11.21 Eight-cell and multicelled stages of human development.

Genetic Screening Genetic screening detects alleles associated with genetic disorders, provides information on reproductive risks, and helps families that are already affected. Often, carriers may be detected early enough to start countermeasures for minimizing the damage before symptoms develop.

A few large-scale screening programs are operational. Besides helping individuals, the information they generate is being used to estimate the prevalence and distribution of harmful alleles in populations. In the United States, for instance, most hospitals routinely screen newborns for PKU. Those affected receive early treatment, so we now see fewer individuals with symptoms of the disorder.

There are social risks that must be considered. How would you feel if you were labeled as someone that carries a "bad" allele? Would the knowledge invite anxiety? If you become a parent even though you know you have a "bad" allele, how would you feel if your child ends up affected by a genetic disorder? No easy answers here.

www.thomsonedu.com Thomson NOW!

Summary

Section 11.1 Of the twenty-three pairs of homologous chromosomes in human body cells, one is a pairing of sex chromosomes. All the others are autosomes. In both sexes, the two autosomes of a pair are the same in length and shape, have the same centromere location, and carry the same genes along their length.

Human females have identical sex chromosomes (XX) and males have nonidentical ones (XY). The *SRY* gene on the Y chromosome is the basis of male sex determination. Its expression initiates the synthesis of testosterone, the hormone that causes a human embryo to develop into a male. If an embryo has no Y chromosome (no *SRY* gene), it develops into a female.

Karyotyping is a diagnostic tool that reveals missing or extra chromosomes and some structural changes in an individual's chromosomes. With this technique, a person's metaphase chromosomes are prepared for microscopy, then imaged. Images of the chromosomes are arranged in sequence on the basis of their defining features.

- *Use the interaction on ThomsonNOW to see how sex is determined in humans.*
- *Use the animation on ThomsonNOW to learn how to create a karyotype.*

Sections 11.2, 11.3 Some of the alleles on autosomes are inherited in simple Mendelian patterns that can be predictably associated with specific phenotypes. Certain mutated forms of alleles give rise to genetic abnormalities or genetic disorders.

- *Use the interaction on ThomsonNOW to investigate autosomal inheritance.*

Section 11.4 Certain dominant and recessive alleles on the X chromosome are inherited in simple Mendelian patterns. Mutated alleles on the X chromosome contribute to more than 300 known genetic disorders. Males cannot transmit a recessive X-linked allele to their sons; a female passes such alleles to male offspring.

- *Use the interaction on ThomsonNOW to investigate X-linked inheritance.*

Section 11.5 On occasion, a chromosome's physical structure may undergo alteration. Part of it is duplicated, deleted, inverted, or translocated (a piece moves to a new location in the same chromosome or a different one).

Most alterations are harmful or lethal. Even so, many have accumulated in the chromosomes of all species over evolutionary time. Either they had neutral effects or they later proved to be useful. Many duplications, inversions, and translocations are built into primate chromosomes. They are strikingly similar among human, chimpanzee, gorilla, orangutan, and gibbon chromosomes, which is evidence of divergences from a common ancestor.

Section 11.6 The chromosome number of a parental cell can change permanently. Most often, such a change is an outcome of nondisjunction, which is the failure of one or more pairs of duplicated chromosomes to separate from each other during meiosis. In aneuploidy, cells have too many or too few copies of a chromosome. In humans, the most common aneuploidy, trisomy 21, causes Down syndrome. Most other human autosomal aneuploids die before birth.

Polyploid individuals inherit three or more of each type of chromosome from their parents. About 70 percent of all flowering plants, and some insects, fishes, and other animals are polyploid.

A change in the number of sex chromosomes usually results in learning and motor skill impairment. Problems can be so subtle that the underlying cause may not ever be diagnosed, as among XXY, XXX, and XYY children.

Sections 11.7, 11.8 Geneticists construct pedigrees, or charts of genetic connections among individuals, to estimate the chance that a couple's offspring will inherit a certain trait. Potential parents who may be at risk of transmitting a harmful allele to their offspring have the choice of several options that include genetic counseling, prenatal diagnosis, preimplantation diagnosis, phenotypic treatments, abortions, or genetic screening.

- *Use the animation on ThomsonNOW to examine a human pedigree.*
- *Use the animation on ThomsonNOW to explore amniocentesis.*

Self-Quiz

Answers in Appendix III

1. The ______ of chromosomes in a cell are compared to construct karyotypes.
   a. length and shape
   b. centromere location
   c. gene sequence
   d. both a and b

2. The ______ determines sex in humans.
   a. X chromosome
   b. *Dll* gene
   c. *SRY* gene
   d. both a and c

3. If one parent is heterozygous for a dominant allele on an autosome and the other parent does not carry the allele, any child of theirs has a ______ chance of being heterozygous.
   a. 25 percent
   b. 50 percent
   c. 75 percent
   d. no chance; it will die

4. Expansion mutations occur ______ within and between genes in human chromosomes.
   a. only rarely
   b. frequently
   c. not at all
   d. only in multiples of ten

5. Galactosemia is a case of ______ inheritance.
   a. autosomal dominant
   b. autosomal recessive
   c. X-linked dominant
   d. X-linked recessive

6. Is this statement true or false? A son can inherit an X-linked recessive allele from his father.

7. Color blindness is a case of ______ inheritance.
   a. autosomal dominant
   b. autosomal recessive
   c. X-linked dominant
   d. X-linked recessive

8. A(an) ______ can alter chromosome structure.
   a. deletion
   b. duplication
   c. inversion
   d. translocation
   e. all of the above

9. Nondisjunction may occur during ______ .
   a. mitosis   c. fertilization
   b. meiosis   d. both a and b

10. Is this statement true or false? Body cells may inherit three or more of each type of chromosome characteristic of the species, a condition called polyploidy.

11. The karyotype for Klinefelter syndrome is ______ .
   a. XO   c. XXY
   b. XXX   d. XYY

12. A recognized set of symptoms that characterize a specific disorder is a ______ .
   a. syndrome   b. disease   c. pedigree

13. Match the chromosome terms appropriately.

___ polyploidy	a. number and defining features of an individual's metaphase chromosomes
___ deletion	b. segment of a chromosome moves to a nonhomologous chromosome
___ aneuploidy	c. extra sets of chromosomes
___ translocation	d. gametes with the wrong chromosome number
___ karyotype	e. a chromosome segment lost
___ nondisjunction during meiosis	f. one extra chromosome

■ *Visit ThomsonNOW for additional questions.*

Genetics Problems *Answers in Appendix III*

1. Human females are XX and males are XY.
   a. Does a male inherit the X from his mother or father?
   b. With respect to X-linked alleles, how many different types of gametes can a male produce?
   c. If a female is homozygous for an X-linked allele, how many types of gametes can she produce with respect to that allele?
   d. If a female is heterozygous for an X-linked allele, how many types of gametes might she produce with respect to that allele?

2. In Section 10.4, you read about a mutation that causes a serious genetic disorder, *Marfan syndrome*. A mutated allele responsible for the disorder follows a pattern of autosomal dominant inheritance. What is the chance that any child will inherit it if one parent does not carry the allele and the other is heterozygous for it?

3. Somatic cells of individuals with Down syndrome usually have an extra chromosome 21; they contain forty-seven chromosomes.
   a. At which stages of meiosis I and II could a mistake alter the chromosome number?
   b. A few individuals with Down syndrome have forty-six chromosomes, two of which are normal-appearing chromosomes 21 and a longer-than-normal chromosome 14. Speculate on how this chromosome abnormality may have arisen.

4. As you read earlier, *Duchenne muscular dystrophy* is a genetic disorder that arises through the expression of a recessive X-linked allele. Usually, symptoms start to appear in childhood. Gradual, progressive loss of muscle function leads to death, usually by age twenty or so. Unlike color blindness, the disorder is nearly always restricted to males. Suggest why.

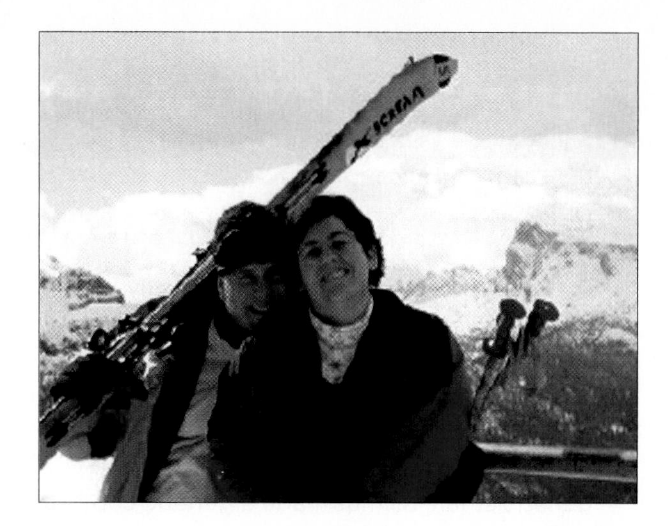

Figure 11.22 A case of Klinefelter syndrome. Until his teenage years, Stefan was shy, reserved, and prone to rage for no apparent reason. Psychologists and doctors discovered he had learning disabilities that affected comprehension, auditory processing, memory, and abstract thinking. One told Stefan he was stupid and lazy, and would be lucky to graduate from high school. In time, Stefan graduated from college with degrees in business administration and sports management. He never discussed his learning disabilities. Instead, he took pride in doing the work on his own and not being treated differently.

Stefan was twenty-five years old before laboratory tests as well as karyotyping revealed a 46XY/47XXY mosaic condition. That same year, he started a job as a software engineer. Having a full-time position helped him open doors to volunteer work with the Klinefelter syndrome network. During his volunteer work, he met his future fiancée, whose son also has the syndrome.

5. In the human population, mutation of two genes on the X chromosome causes two types of X-linked *hemophilia* (A and B). In a few cases, a woman is heterozygous for both mutated alleles (one on each of the X chromosomes). All of her sons should have either hemophilia A or B.
   However, on very rare occasions, one of these women gives birth to a son who does not have hemophilia, and his one X chromosome does not have either mutated allele. Explain how such an X chromosome could arise.

6. Does the phenotype indicated by red circles and squares in this pedigree show a Mendelian inheritance pattern that is autosomal dominant, autosomal recessive, or X-linked?

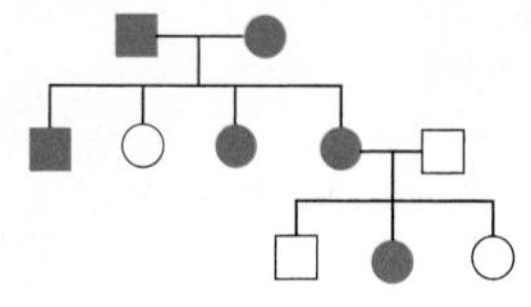

7. When it comes to acceptance of an unusual genetic condition, people tend to be subjective. As an example, consider the individual described in Figure 11.22. How would you have categorized him without knowing the genetic basis of his early behavior? How would you categorize him now in terms of what we as a society consider to be "ideal" phenotypes?

12 DNA STRUCTURE AND FUNCTION

IMPACTS, ISSUES Here, Kitty, Kitty, Kitty, Kitty, Kitty

By now, you have been told repeatedly that DNA holds heritable information. Has anybody actually demonstrated that it does? Well, yes. One jarring demonstration occurred in 1997, when Scottish geneticist Ian Wilmut made a genetic copy—a clone—of a fully grown sheep. His team removed the nucleus from an unfertilized egg and replaced it with the nucleus from an adult sheep cell. The egg with its new nucleus became an embryo, and then a lamb, which the researchers named Dolly. At first, Dolly looked and acted like a normal sheep, but five years later something was wrong. She was as fat and arthritic as a much older sheep, and a year later a lung disease that is typical of advanced age set in.

Dolly's telomeres indicated that she developed health problems because she was a clone. Telomeres are short, repeated DNA sequences at the ends of chromosomes. They become shorter and shorter as an animal ages. When Dolly was only two years old, her telomeres were as short as those of a six-year-old sheep—the exact age of the adult animal that had been her genetic donor.

Since Dolly was born, mice, rabbits, pigs, cattle, goats, mules, deer, horses, cats, and a dog have been cloned, but cloning mammals is far from routine. Not very many clonings end successfully. It takes hundreds of attempts to produce one embryo, and most embryos that do form die before birth or shortly after. About 25 percent of the clones that do survive have health problems. For example, cloned pigs tend to limp and have heart problems. One never did develop a tail or, even worse, an anus.

What causes the problems? Even though all cells of an individual *inherit* the same DNA, an adult cell *uses* only a fraction of it compared to an embryonic cell. To make a clone from an adult cell, researchers must reprogram its DNA to function like the DNA of an egg. As Dolly's story reminds us, we still have a lot to learn about that.

Why do geneticists keep at it? The potential benefits are enormous. Replacement cells and organs may help people with presently incurable degenerative diseases.

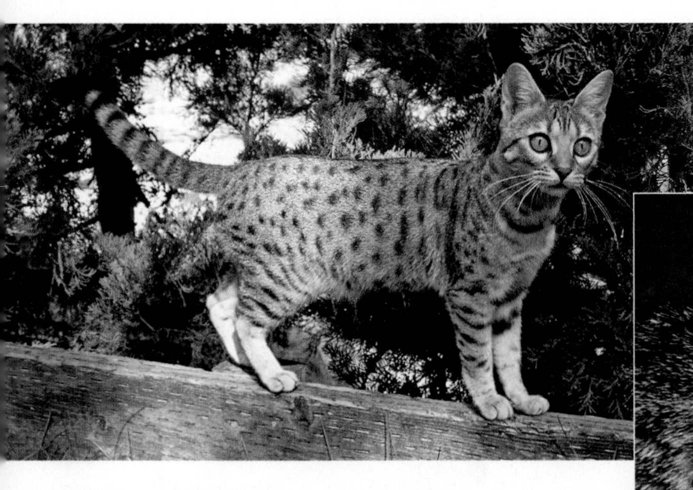

See the video! **Figure 12.1** Demonstration that DNA holds heritable information—cloning of an adult down to the last detail. Compare the markings on the top of the head, face, legs, and tail of Tahini, a Bengal cat (*above*) with those of Tabouli and Baba Ganoush, two of her clones (*right*). Eye color changes as a Bengal cat matures; both clones will have the same eye color as Tahini's. All three cats are household pets of the founder of a company called Genetic Savings and Clone.

How would you vote? Some view sickly or deformed clones as unfortunate but acceptable casualties of animal cloning research that may also result in new drugs and organ replacements for human patients. Should animal cloning be banned? See ThomsonNOW for details, then vote online.

Figure 12.2 Watson, Crick, and their model of DNA.

Endangered animals might be saved from extinction; extinct animals may be brought back into existence. Livestock animals with desirable traits are already being cloned for breeding; some day, their clones may be mass-produced. A few companies now clone pets (Figure 12.1).

Controversy swirls around adult cloning. Perfecting the methods to make healthy animal clones brings us closer to the possibility of cloning humans, technically and also ethically. For example, if cloning a lost cat for a grieving pet owner is acceptable, would it be acceptable to clone a lost child for a grieving parent?

With this chapter, we delve more deeply into the basis of inheritance. We turn to the investigations and models that led to our current understanding of DNA (Figure 12.2). As you will see, the chapter is more than a march through details of how the molecular structure of DNA encodes hereditary information. *It also reveals how ideas are generated in science.*

On the one hand, having a shot at fame and fortune quickens the pulse of men and women in any profession, and scientists are no exception. On the other hand, science is a community effort. Individuals share not only what they can explain but also what they do not understand. Even if an experiment fails to produce the anticipated results, it may turn up information that others can use, or lead to questions that others can answer. Unexpected results, too, may be clues to something important about the natural world.

Key Concepts

DISCOVERY OF DNA'S FUNCTION

In all living cells, DNA molecules store information that governs heritable traits. **Section 12.1**

THE DNA DOUBLE HELIX

A DNA molecule consists of two chains of nucleotides, hydrogen-bonded together along their length and coiled into a double helix. Four kinds of nucleotides make up the chains: adenine, thymine, guanine, and cytosine.

The order in which one kind of nucleotide base follows the next along a DNA strand encodes heritable information. The order in some regions of DNA is unique for each species. **Section 12.2**

THE FRANKLIN FOOTNOTE

Like any race, the one that led to the discovery of DNA's structure had its winners—and its losers. **Section 12.3**

HOW CELLS DUPLICATE THEIR DNA

Before a cell divides, enzymes and other proteins copy its DNA. Newly forming DNA strands are monitored for errors, most of which are corrected. Uncorrected errors are mutations. **Section 12.4**

DNA AND THE CLONING CONTROVERSIES

Knowledge about the structure and function of DNA is the basis of several methods of cloning. **Section 12.5**

Links to Earlier Concepts

This chapter builds on your understanding of hydrogen bonding (Section 2.4), condensation reactions (3.1), and the overview of DNA structure and function (3.6). Knowledge of chromosomes, mitosis, and meiosis will help you understand cloning procedures (8.2, 8.3, 9.3). Keep the image of the eight-cell stage of human development (11.8) in mind when you read about embryo cloning, because the cells used are no more developed than this.

12.1 The Hunt for Fame, Fortune, and DNA

About the time Gregor Mendel was born, a Swiss medical student, J. Miescher, was ill with typhus. Miescher became partially deaf, so he could not be a doctor. He switched to organic chemistry instead, and made a major discovery.

EARLY AND PUZZLING CLUES

LINKS TO SECTIONS 2.2, 3.4

In 1869, Johann Miescher was studying the composition of the nucleus. He had collected white blood cells from pus-filled bandages and sperm from fish. Such cells do not contain much cytoplasm, so it is easy to isolate the substances in their nucleus. Miescher found an acidic substance that contains nitrogen and phosphorus. Later it would be called deoxyribonucleic acid, or DNA.

Now let's jump to 1928. Frederick Griffith, a British medical officer, wanted to make a pneumonia vaccine. He isolated two strains of *Streptococcus pneumoniae*, a bacterium that can cause pneumonia. He named one strain *R*, because colonies of it have a Rough surface when grown in petri dishes. He named the other strain *S*; its colonies have a Smooth surface. Griffith used both strains in four experiments that unfortunately did not result in a vaccine. However, the experiments revealed a clue about inheritance (Figure 12.3).

First, he injected mice with live *R* cells. The mice did not develop pneumonia. *The* R *strain was harmless.*

Second, he injected other mice with live *S* cells. The mice died. Blood samples from them teemed with live *S* cells. *The* S *strain was pathogenic; it caused the disease.*

Third, he killed *S* cells by exposing them to high temperature. *Mice injected with dead* S *cells did not die.*

Fourth, he mixed live *R* cells with heat-killed *S* cells and injected the mixture into mice. The mice died—*and blood samples drawn from them teemed with live* S *cells!*

What happened in the fourth experiment? Maybe the heat-killed *S* cells in the mix were not really dead. But if that were so, then mice injected with the heat-killed *S* cells in experiment 3 would have died. Or maybe the harmless *R* cells had mutated into a killer form. If that were so, then the mice injected with the *R* cells only in experiment 1 would have died.

The simplest explanation was that heat killed the *S* cells, but did not destroy their hereditary material—including whatever part that specified "infect mice." Somehow, that material had been transferred from the dead *S* cells into live *R* cells, which put it to use.

The transformation was permanent and heritable. Even after a few hundred generations, descendants of transformed *R* cells were infectious, also. What caused the transformation? In broader terms, *which substance encodes the information about traits that parents pass on to offspring?* Researchers knew of Miescher's discovery, but few gave DNA much thought. Most were thinking PROTEINS! After all, heritable traits are tremendously diverse, and proteins are the most structurally diverse molecules. Other molecules seemed too uniform.

Still, Griffith's results intrigued Oswald Avery, who read Griffith's paper and set out to determine which substance could transform *R* cells into *S* cells. He and his colleagues found that only DNA transformed the *R* cells into *S* cells. Their experiments confirmed that DNA was the substance of heredity in these bacteria.

CONFIRMATION OF DNA FUNCTION

By the 1950s, researchers trying to find the hereditary material were doing experiments with bacteriophage, a type of virus that infects bacteria. Like all viruses, these

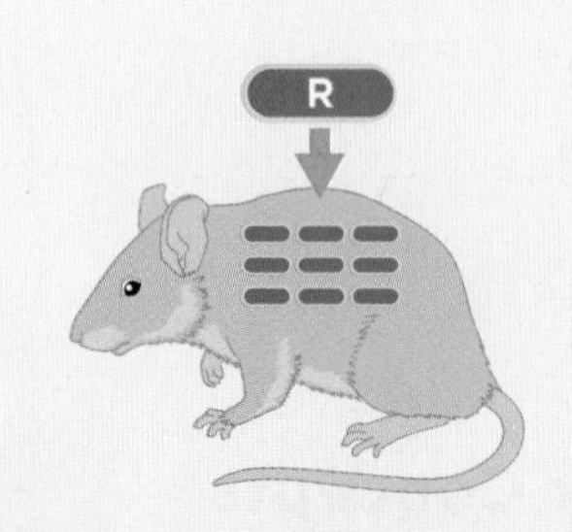

a Mice injected with live cells of harmless strain *R* do not die. Live *R* cells in their blood.

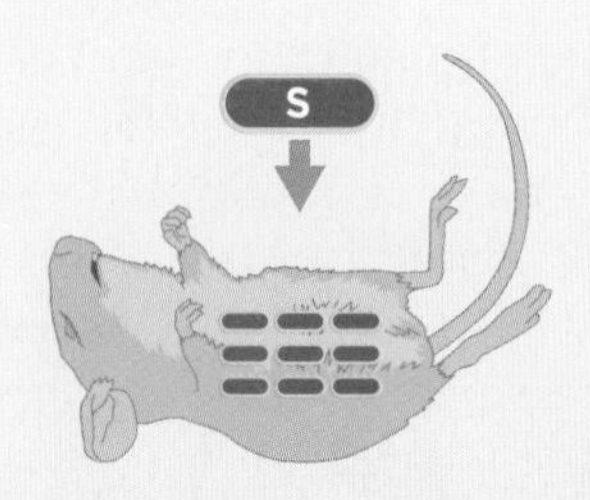

b Mice injected with live cells of killer strain *S* die. Live *S* cells in their blood.

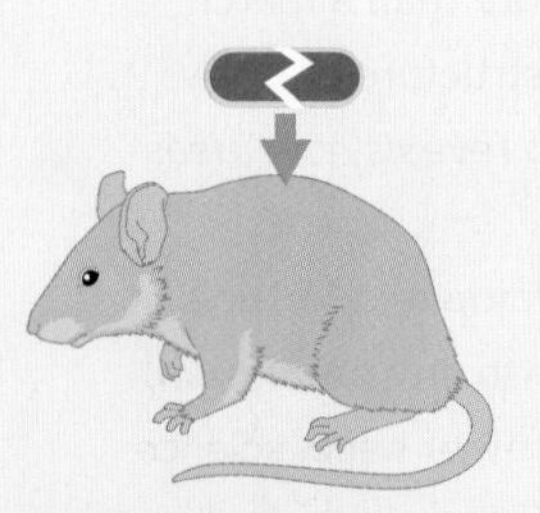

c Mice injected with heat-killed *S* cells do not die. No live *S* cells in their blood.

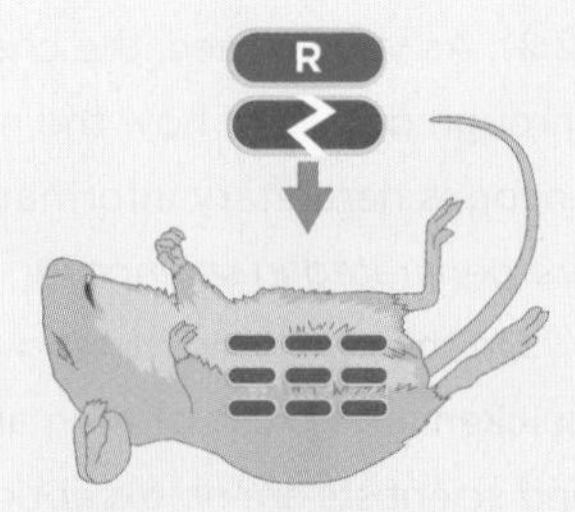

d Mice injected with live *R* cells plus heat-killed *S* cells die. Live *S* cells in their blood.

Figure 12.3 **Animated!** Summary of results from Fred Griffith's experiments. The hereditary material of harmful *Streptococcus pneumoniae* cells transformed harmless cells into killers.

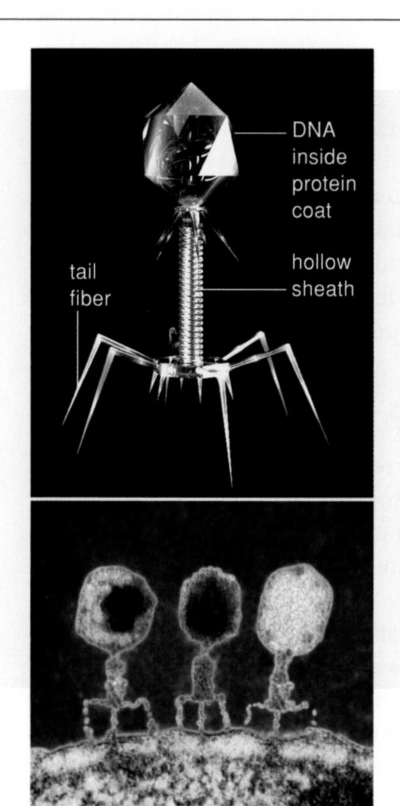

Top, model of a bacteriophage. *Bottom*, micrograph of three viruses injecting DNA into an *E. coli* cell.

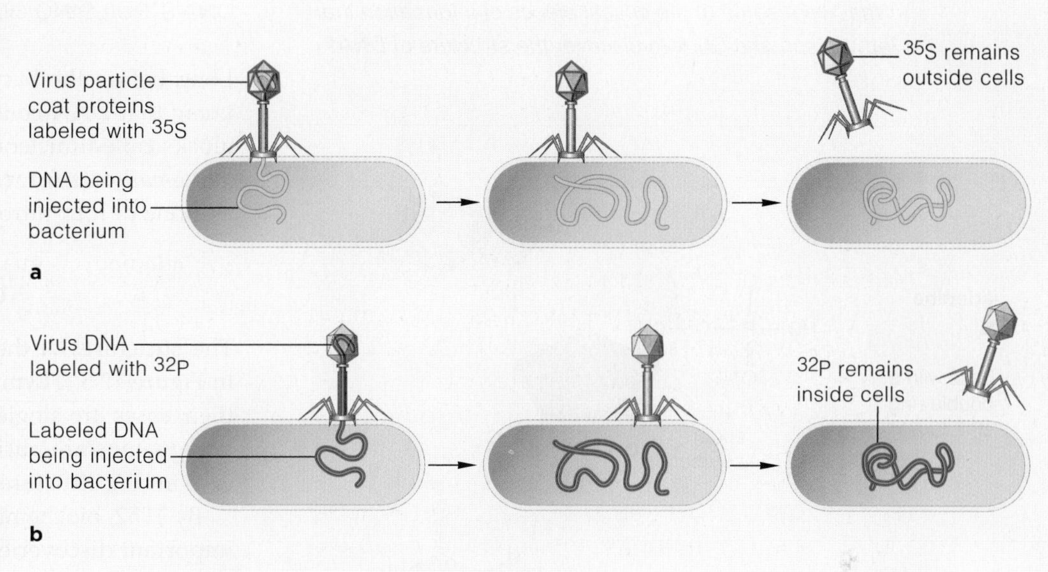

Figure 12.4 **Animated!** Landmark experiments that tested whether genetic material resides in bacteriophage DNA, proteins, or both. As Alfred Hershey and Martha Chase knew, sulfur (S) but not phosphorus (P) is present in proteins, and phosphorus but not sulfur is present in DNA.

(**a**) In one experiment, bacteria were infected with virus particles labeled with a radioisotope of sulfur (^{35}S). The sulfur had labeled only viral proteins. The mixture was whirled in a kitchen blender, which dislodged the viruses from the bacterial cells. Radioactive sulfur was detected mainly in the solution, not inside the cells. The viruses had not injected protein into the bacteria.

(**b**) In another experiment, bacteria were infected with virus particles labeled with a radioisotope of phosphorus (^{32}P). The phosphorus had labeled only viral DNA. When the viruses were dislodged from the bacteria, the radioactive phosphorus was detected mainly inside the bacterial cells. The viruses had injected DNA into the cells—evidence that DNA is the genetic material of this virus.

infectious particles carry hereditary information about how to make new viruses. After a virus infects a cell, the cell starts making virus particles.

As researchers knew, some bacteriophages consist only of DNA and protein. The viruses were injecting genetic material into host cells, but was that material DNA, protein, or both? Figure 12.4 describes just two landmark experiments that supported the hypothesis that DNA, not protein, is the material of heredity.

Then, in 1951, the biochemists Linus Pauling, Robert Corey, and Herman Branson described the details of a coiled pattern that occurs in many proteins (Section 3.4). The discovery was electrifying. If someone could pry open the secrets of proteins, why not DNA? And if the structural details of DNA could be worked out, then wouldn't they be clues to its function? *Someone would go down in history as having discovered the secret of life!*

ENTER WATSON AND CRICK

Scientists all over the world started to reevaluate their data and sift through everyone else's. James Watson, a scientist from Indiana University, was among them. In 1951, he began work at Cambridge University, where he met the brilliant physicist Francis Crick. The two spent many hours arguing about the size, shape, and bonding requirements of the four kinds of nucleotides that make up DNA. They pestered chemists to help them identify bonds they might have overlooked. They fiddled with cardboard cutouts, and made models from bits of metal connected by suitably angled "bonds" of wire.

In 1953, Watson and Crick built a model consistent with every pertinent biochemical constraint and insight known at the time. They had discovered the double-helix structure of DNA. As you will see in the following section, the structure is striking in its simplicity. That simplicity is the basis of why life can have such unity at the level of molecules, and such diversity at the level of organisms.

The cumulative work of many scientists over more than a century led to our knowledge that the DNA molecule, with its double-helix structure, encodes heritary information.

12.2 The Discovery of DNA's Structure

What were some of the critical pieces of information that led Watson and Crick to discover the structure of DNA?

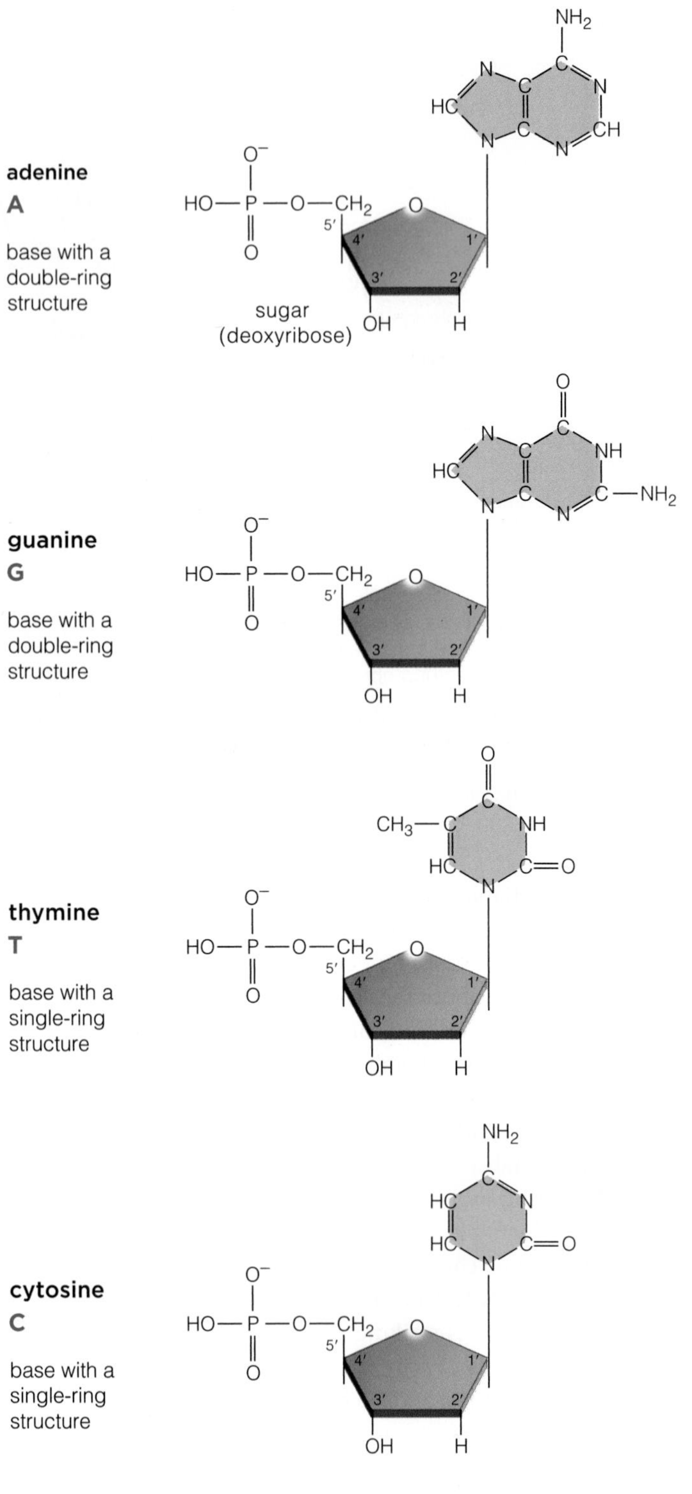

adenine
A

base with a double-ring structure

guanine
G

base with a double-ring structure

thymine
T

base with a single-ring structure

cytosine
C

base with a single-ring structure

Figure 12.5 Four kinds of nucleotides in the DNA molecule. Each is named after its component base (*blue*). Biochemist Phoebus Levene identified the structure of these bases and how they are connected in DNA in the early 1900s. He worked with DNA for almost 40 years.

DNA'S BUILDING BLOCKS

Long before the bacteriophage research, biochemists knew that DNA contains only four kinds of building blocks called nucleotides. Each nucleotide consists of a five-carbon sugar (deoxyribose), a phosphate group, and one of four nitrogen-containing bases:

adenine	guanine	thymine	cytosine
A	G	T	C

The structures of these four nucleotides are depicted in Figure 12.5. Thymine and cytosine are pyrimidines; their bases are single carbon ring structures. Adenine and guanine are purines; their larger bases are double carbon ring structures.

By 1952, biochemist Erwin Chargaff had made two important discoveries about the composition of DNA. First, the amounts of thymine and adenine in DNA are the same, as are the amounts of cytosine and guanine. Second, the proportion of adenine and guanine differs among species. We may show Chargaff's rules as:

$$A = T \quad \text{and} \quad G = C$$

The symmetrical proportions had to mean something important. They did. The proportions were one more clue to how nucleotides are arranged in DNA.

The first convincing evidence of the arrangement came from Rosalind Franklin, a researcher in Maurice Wilkins's laboratory in London, England. Franklin had made exceptional x-ray diffraction images of DNA. Such images form after x-rays are beamed through a crystalline sample of purified molecules. Atoms in the molecules scatter the x-rays in a pattern that can be captured on film. Researchers use the positions of dots and streaks to calculate the size, shape, and spacing between any repeating elements of the molecule—all details of its structure.

DNA is a large molecule, and hard to crystallize. Also, as Franklin discovered, "wet" and "dry" DNA samples have different shapes. She made the first clear image of the wet form—which is how DNA occurs in cells. She used the image to calculate that DNA is very long compared to its 2-nanometer diameter. She also identified one repeating pattern every 0.34 nanometer along its length, and another every 3.4 nanometers. It was the first glimpse into DNA's actual structure.

Wilkins reviewed Franklin's diffraction image with Watson. Watson and Crick now had all the information they needed to build their model of the DNA helix—one with two sugar–phosphate chains running in opposite directions, and paired bases inside (Figure 12.6).

Pauling, the discoverer of protein helixes, already was thinking "helix." So was everyone else, including Franklin, Wilkins, Watson, and Crick. Later on, Watson wrote, "We thought, why not try it on DNA? We were worried that Pauling would say, why not try it on DNA? Certainly he was a clever man. He was a hero of mine. But we beat him at his own game. I still can't figure out why."

Pauling, like most other people, had assumed that the nucleotide bases were on the outside of the helix, so they would be accessible to enzymes that copy DNA. Franklin's data showed otherwise.

PATTERNS OF BASE PAIRING

From all the clues that had been accumulating, Watson and Crick proposed that DNA consists of two chains (strands) of nucleotides, running in opposite directions and coiled up into a double helix. Hydrogen bonds between the internally positioned bases hold the two strands together. Only two kinds of base pairings form: **A—T** and **G—C**.

How do two kinds of simple base pairings give rise to such stunning diversity in traits? The *order* in which one pair follows the other—the DNA base sequence—can be tremendously variable. For instance, a small piece of DNA from a petunia, a human, or any other organism might be:

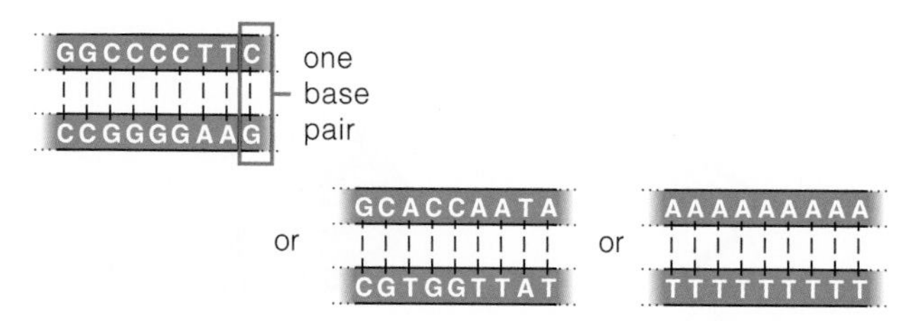

Notice how the two strands of DNA match up; each base on one is suitably paired with a partner base on the other. This bonding pattern (**A—T** and **G—C**) is the same in all molecules of DNA. However, which base pair follows the next in line differs among species, and among individuals of the same species. Thus, *DNA, the molecule of inheritance in every cell, is the basis of life's unity. Variations in its base sequence from one individual or one species to the next is the basis for life's diversity.*

The pattern of base pairing between the two strands in DNA is constant for all species—A with T, and G with C. However, each species has a number of unique base pair sequences along the length of its DNA molecules.

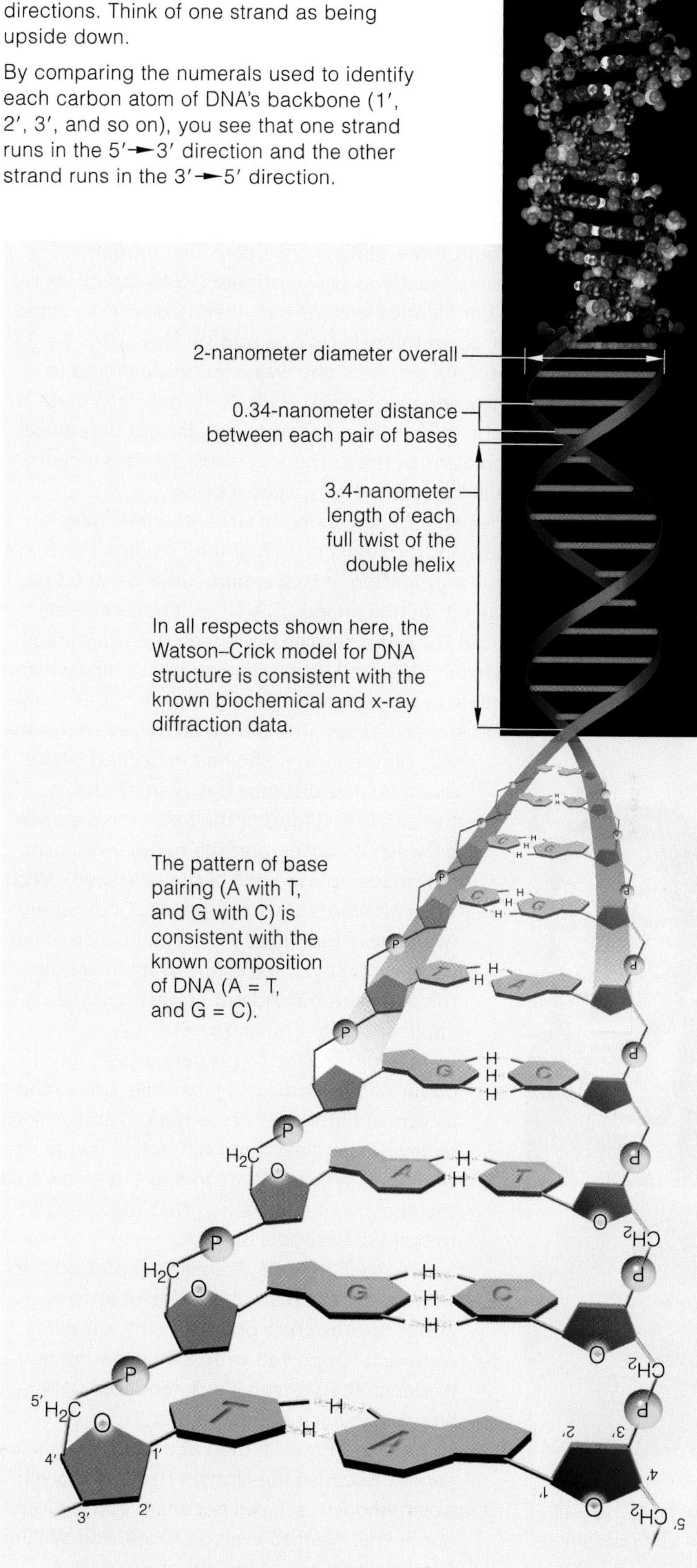

Figure 12.6 **Animated!** Composite of three different ways to represent the DNA double helix. DNA's two sugar–phosphate backbones run in parallel but opposite directions. Think of one strand as being upside down.

By comparing the numerals used to identify each carbon atom of DNA's backbone (1′, 2′, 3′, and so on), you see that one strand runs in the 5′→3′ direction and the other strand runs in the 3′→5′ direction.

12.3 Fame and Glory

FOCUS ON BIOETHICS

There is a saying among researchers in any discipline—publish or perish. As soon as Watson and Crick's model of DNA fell into place, they published a brief paper that dazzled the world. Others who had helped fill in pieces of the puzzle did not quite share the same limelight.

LINKS TO SECTIONS 8.2, 8.5, 10.1

Rosalind Franklin was an expert at x-ray crystallography. She had solved the structure of coal, which is complex and unorganized (as are large biological molecules such as DNA), and she took a new mathematical approach to interpreting x-ray diffraction images. Like Pauling, she had built three-dimensional molecular models.

Her assignment was to investigate DNA's structure. No one told her Maurice Wilkins was already doing the same thing just down the hall. No one told Wilkins about her assignment; he assumed she was a technician hired to do his x-ray crystallography work. And so a clash began. He thought she displayed an appalling lack of deference that technicians of the era usually accorded researchers. To her, Wilkins seemed inexplicably prickly.

Wilkins had a prized cache of DNA which he gave to his "technician" for x-ray crystallography studies. Franklin's meticulous preparation of that sample gave her the first clear x-ray diffraction image of DNA as it occurs inside cells (Figure 12.7), and she gave a research presentation on this work in 1952. DNA, she said, had two chains twisted into a double helix, with phosphate groups in a backbone on the outside, and bases arranged in an as-yet unknown way on the inside. She had measured DNA's diameter, the distance between its chains, the pitch (or angle) of the helix, the distance between its bases, and the number of bases that made up each coil of the helix (ten). With his crystallography background, Crick surely would have recognized the significance of her work—if he had been there. Watson was in the audience but did not understand the implications of Franklin's data.

Franklin started to prepare a research paper on her findings. Meanwhile, Crick read a copy of her earlier presentation, and Wilkins reviewed her clear x-ray diffraction image of DNA with Watson. Watson and Crick now had the final bit of information that they needed to build their model of DNA.

On April 25, 1953, Franklin's report on her work appeared third in a series of articles about the structure of DNA in the journal *Nature*. It supported with solid experimental evidence the Watson-Crick model of DNA structure, which appeared first.

Rosalind Franklin died at age 37, of ovarian cancer. Because the Nobel Prize is not given posthumously, she did not share in the 1962 honor that went to Watson, Crick, and Wilkins for the discovery of the structure of DNA.

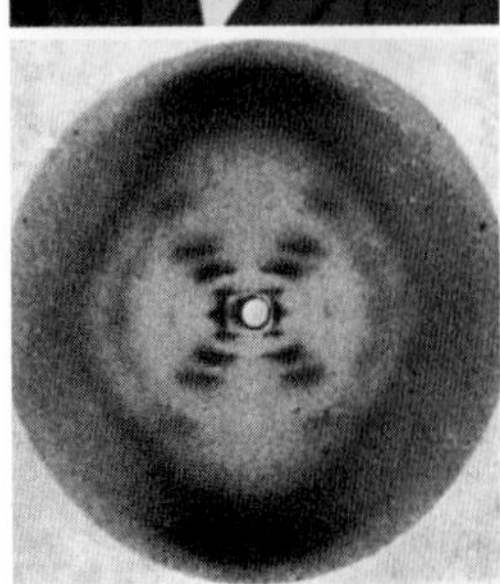

Figure 12.7 Rosalind Franklin and her famous x-ray diffraction image.

12.4 Replication and Repair

The discovery of DNA structure was a turning point in studies of inheritance. Suddenly it became clear how cells duplicate their DNA before they divide.

HOW DNA GETS COPIED

Until Watson and Crick presented their model, no one could explain DNA replication, or how the molecule of inheritance is duplicated before a cell divides. The process requires a team of molecular workers that are activated in interphase of the cell cycle (Section 8.2).

Both strands of a DNA double helix are replicated. Enzymes break the hydrogen bonds that hold the helix together, and the two DNA strands unwind. Each one serves as a template, or guide, for DNA polymerase to assemble a new strand of DNA. This enzyme bonds nucleotides together into a new strand according to the order of nucleotides of a template. For example, it pairs an **A** on a parent strand with a free **T**; it pairs a **G** with a **C**, and so on. Each nucleotide provides energy for its own attachment to the end of a growing strand of DNA, when the polymerase removes two of its three phosphate groups. Another enzyme, DNA ligase, seals any gaps in the new DNA strands.

As it is forming, a new DNA strand twists up with its template strand into a double helix. Thus, half of each double-stranded DNA molecule is "old" and half is "new." Figures 12.8 and 12.9 show the details of this process, which is called semiconservative replication.

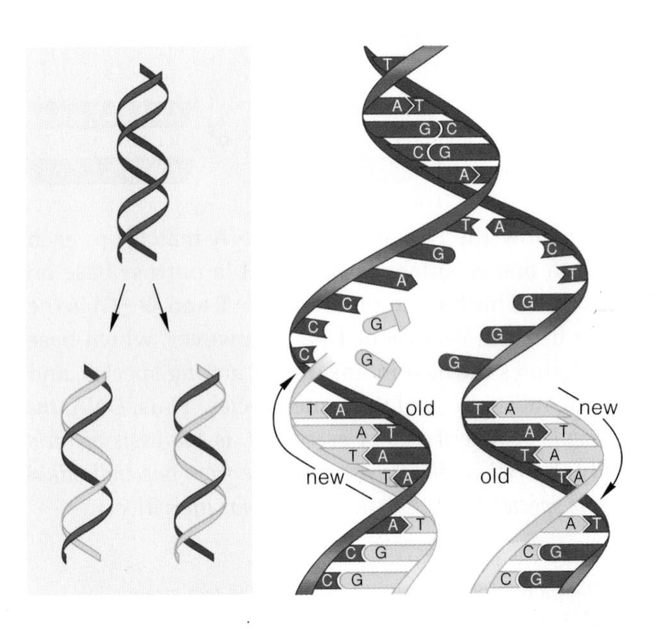

Figure 12.8 Semiconservative DNA replication. Each parent strand (*blue*) stays intact. A new strand (*gold*) is assembled on each parent strand in the direction shown by the arrow.

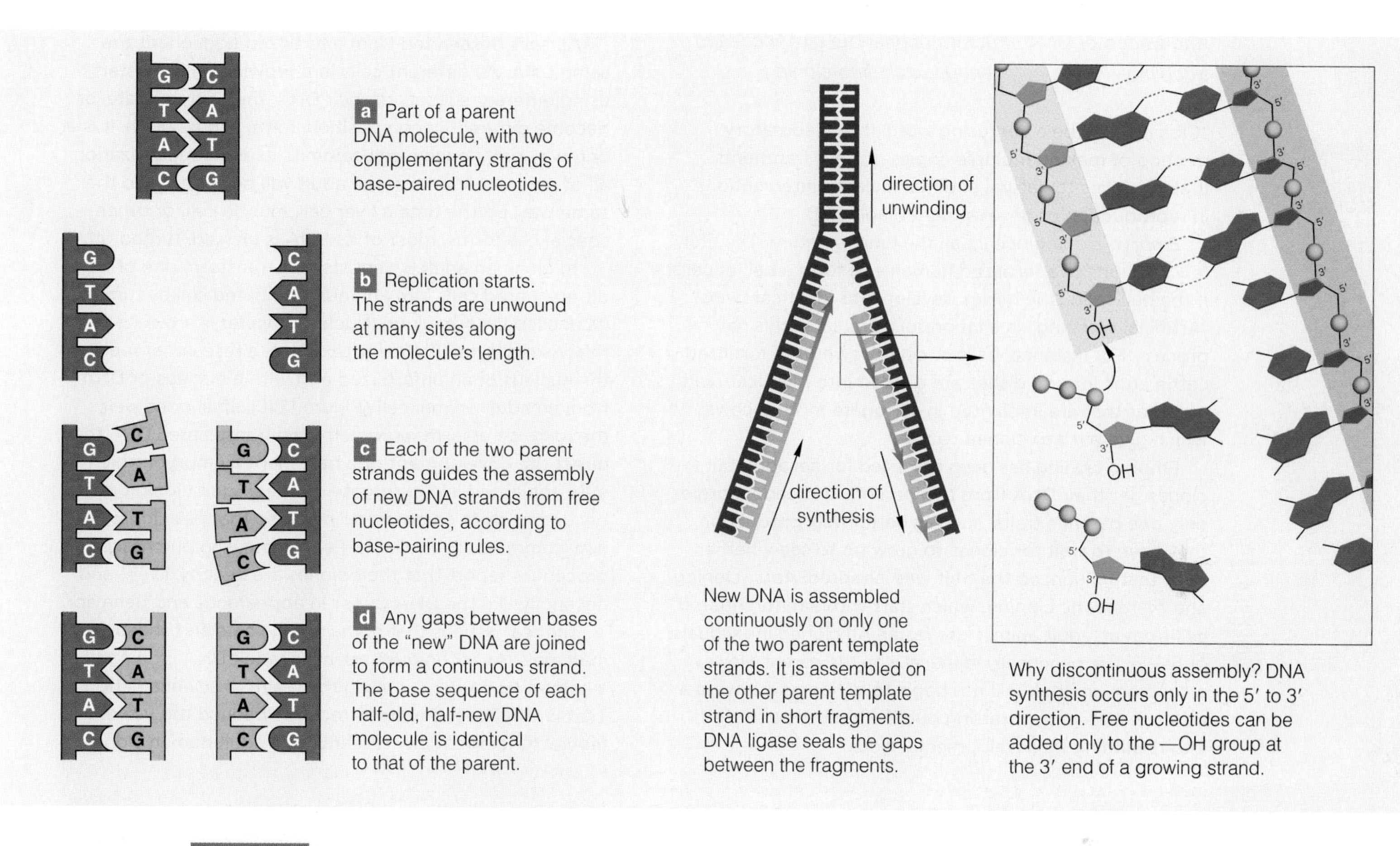

Figure 12.9 **Animated!** A closer look at how new strands form during DNA replication.

CHECKING FOR MISTAKES

Errors occasionally are introduced during replication. The wrong base may be added to a growing strand of DNA. One or more bases might be lost, or extra ones might be added. Either way, the new DNA strand will no longer match up perfectly with its parent strand.

Some of these errors occur after the DNA becomes damaged by exposure to radiation or toxic chemicals. DNA polymerases do not copy damaged DNA very well. In most cases, DNA repair mechanisms can fix damaged DNA. Replication proceeds with DNA that has been fully repaired.

Most DNA replication errors occur because DNA polymerases catalyze a huge number of reactions very fast. Mistakes are inevitable; some polymerases make many. Fortunately, most DNA polymerases proofread their own base pairings. They correct mismatches by immediately reversing the reaction and removing the mismatched nucleotide. DNA synthesis then resumes. If an error remains uncorrected, controls may pause the cell cycle (Sections 8.2 and 8.5).

When proofreading and repair mechanisms fail, a mistake becomes a mutation—a permanent alteration in DNA. Over the long term, mutations give rise to variations in traits that help define a species; over a lifetime, a particular individual or its offspring may not survive them. Mutations in body cells can cause cancer; those in cells that form eggs or sperm can lead to genetic disorders in offspring.

DNA is replicated during interphase, before a cell divides.

Enzymes unwind the DNA double helix at several sites. Both strands serve as templates for DNA synthesis.

DNA polymerases join free nucleotides into a new DNA strand on a template according to base-pairing rules. Each new strand and its parent twist into a double helix.

DNA repair mechanisms fix nearly all DNA damage, and proofreading corrects nearly all replication errors. They help maintain the integrity of a cell's genetic information. Errors that remain unrepaired are mutations.

12.5 Using DNA To Duplicate Existing Mammals

FOCUS ON SCIENCE

Knowledge of DNA structure opened up exciting—and troubling—research avenues, including cloning.

LINKS TO SECTIONS 9.1, 9.5, 11.8

"Cloning" can be a confusing word. It is a laboratory method of making multiple copies of DNA fragments. It also applies to natural and manipulated interventions in reproduction or development (Table 12.1).

Embryo cloning occurs all the time in nature. The first few divisions of a fertilized human egg form a ball of cells. If the ball splits, the halves develop into identical twins. "Artificial twinning" in a laboratory simulates this natural process. For instance, balls of cells grown from fertilized cattle eggs in petri dishes are divided into identical-twin embryos that are implanted in surrogate mother cows, which give birth to cloned calves.

Embryo cloning has been practiced for decades, but the clones get their DNA from two parents. If breeders prefer only one parent's traits, such as better milk production, they have to wait for clones to grow up to see whether DNA that influenced the trait was inherited. *Adult* cloning and *therapeutic* cloning, which start with a differentiated cell from an adult animal, are faster. All cloned individuals or tissues are genetically identical with the parent. However, a cell from an adult will not begin to divide as if it were a newly fertilized egg starting out life. It must be tricked into rewinding its developmental clock.

All cells descended from a fertilized egg inherit the same DNA. As different cells in a growing embryo start using different subsets of their DNA, they differentiate, or become different in composition, form, and function. It is a one-way path; once a cell commits itself to specialization, all of its descendants in the adult will be specialized the same way. By the time a liver cell, muscle cell, or other special cell forms, most of its DNA is unused, turned off.

To clone an adult, scientists first transform one of its differentiated cells into an undifferentiated cell by turning its unused DNA back on. Nuclear transfer is a way to do this. As explained in the introduction, a researcher replaces the nucleus of an unfertilized egg with a nucleus or DNA from an adult animal cell (Figure 12.10). If all goes well, the egg's cytoplasm reprograms the transplanted DNA to direct the development of a ball-shaped embryo, which gets implanted in a surrogate mother. Adult cloning is becoming quite common, so much so that pet cloning is now commercially available. Pet owners who purchased the procedure report that their clones are healthy, lively, and uncannily like the DNA donor in appearance and behavior.

The real issue is that *humans*—like cats and sheep—are mammals. As techniques advance, adult DNA cloning of a human no longer seems in the realm of science fiction. That is why most countries recently banned the use of federal funding for any research into adult human cloning.

Table 12.1 Some Cloning Methods Compared

Method	Starting DNA	→	→	Result
Sexual Reproduction (basis of comparison)	Maternal DNA, Paternal DNA	fertilized egg	early embryo	Individual with mix of parental traits
Embryo Cloning	Maternal DNA, Paternal DNA	fertilized egg	artificial splitting of early embryo	Identical twins with same mix of parental traits
Adult Cloning (reproductive cloning)	Nucleus or DNA from adult cell	ball of cytoplasm (unfertilized egg stripped of nucleus)	early embryo	Clone: individual that is genetically identical to parent
Therapeutic Cloning	Nucleus or DNA from adult cell	ball of cytoplasm (unfertilized egg stripped of nucleus)	early embryo	Stem cells that are genetically identical to donor

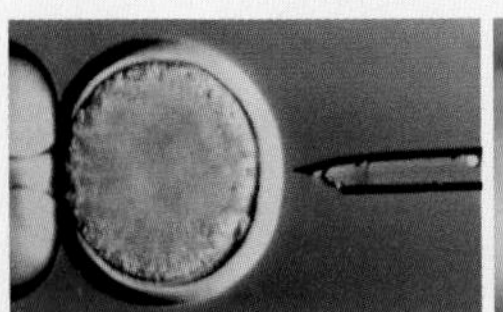

a A microneedle is about to penetrate an unfertilized sheep egg.

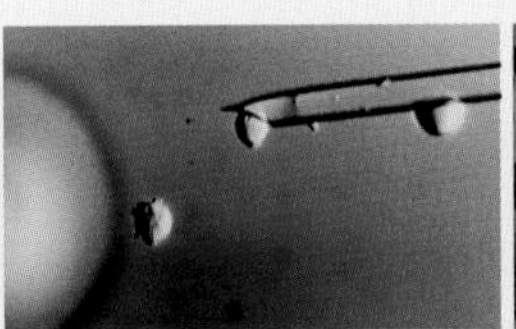

b The microneedle has now removed the sheep egg's nucleus.

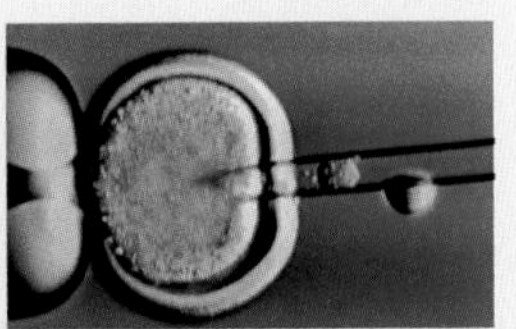

c A nucleus from a donor cell is about to be inserted into the enucleated egg.

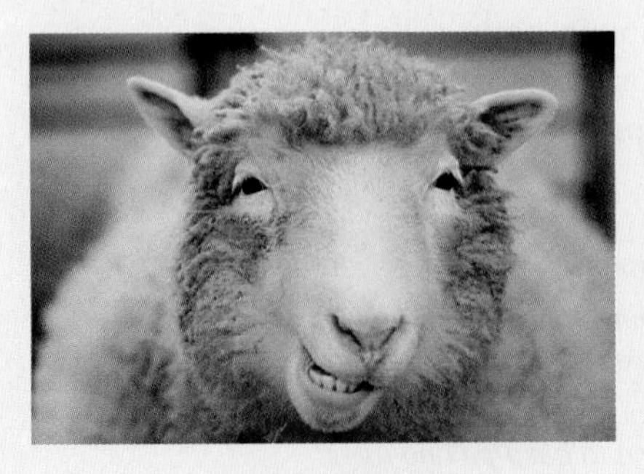

d An electric current will stimulate the egg to enter mitotic cell division. After a few rounds of divisions, the ball of cells will be implanted in the womb of a female sheep (ewe). *Left*, Dolly, the first sheep cloned from adult DNA.

Figure 12.10 Animated! Nuclear transfer in sheep cells. In the series of micrographs, a microneedle replaces the nucleus of a sheep egg with a nucleus from an adult sheep cell. Newer cloning methods involve the direct transfer of nuclear DNA that has been first treated with extracts of mitotic cells to condense the chromosomes.

www.thomsonedu.com Thomson NOW!

Summary

Section 12.1 Experimental tests that used bacteria and bacteriophages offered the first solid evidence that DNA is the hereditary material in living organisms.

■ *Use the animation on ThomsonNOW to learn about experiments that revealed the function of DNA.*

Sections 12.2, 12.3 DNA consists of two strands of nucleotides, coiled into a double helix. Each nucleotide has a five-carbon sugar (deoxyribose), a phosphate group, and one of four nitrogen-containing bases for which it is named: adenine, thymine, guanine, or cytosine.

Bases of the two DNA strands pair in a constant way. Adenine pairs with thymine (**A—T**), and guanine with cytosine (**G—C**). Which base follows another (the DNA sequence) varies among species and among individuals. The DNA of each species has unique sequences that set it apart from the DNA of all other species.

■ *Use the animation on ThomsonNOW to investigate the structure of DNA.*

Section 12.4 A living cell replicates its DNA before dividing. Enzymes unwind the double helix at several sites along its length. DNA polymerases use each strand as a template to assemble new, complementary strands of DNA from free nucleotides. Two double-stranded DNA molecules result. One strand of each is new.

DNA repair mechanisms can fix DNA damaged by chemicals or radiation. Proofreading by DNA polymerases corrects most base-pairing errors. Uncorrected errors can lead to mutations.

■ *Use the animation on ThomsonNOW to see how a DNA molecule is replicated.*

Section 12.5 Artificial twinning and nuclear transfers are methods that produce clones, or genetically identical individuals. To clone an adult animal, a cell's DNA must be reprogrammed to function like an embryonic cell and direct the development of a new individual.

■ *Use the animation on ThomsonNOW to observe the procedure used to create Dolly and other clones.*

Self-Quiz

Answers in Appendix III

1. Which is *not* a nucleotide base in DNA?
   a. adenine c. uracil e. cytosine
   b. guanine d. thymine f. All are in DNA.

2. What are the base-pairing rules for DNA?
   a. A–G, T–C c. A–U, C–G
   b. A–C, T–G d. A–T, G–C

3. One species' DNA differs from others in its ______ .
   a. sugars c. base sequence
   b. phosphates d. all of the above

4. When DNA replication begins, ______ .
   a. the two DNA strands unwind from each other
   b. the two DNA strands condense for base transfers
   c. two DNA molecules bond
   d. old strands move to find new strands

5. Show the complementary strand of DNA that forms on this template DNA fragment during replication:

   5′—GGTTTCTTCAAGAGA—3′

6. Match the terms appropriately.

___ bacteriophage	a. nitrogen-containing base, sugar, phosphate group(s)
___ clone	b. copy of an organism
___ nucleotide	c. nucleotide base with one carbon ring
___ purine	d. only DNA and protein
___ DNA ligase	e. fills in gaps, seals breaks in a DNA strand
___ DNA polymerase	f. nucleotide base with two carbon rings
___ pyrimidine	g. adds nucleotides to a growing DNA strand

■ *Visit ThomsonNOW for additional questions.*

Critical Thinking

1. Matthew Meselson and Franklin Stahl's experiments supported the semiconservative model of replication. These researchers obtained "heavy" DNA by growing *Escherichia coli* with ^{15}N, a radioactive isotope of nitrogen. They also prepared "light" DNA by growing *E. coli* in the presence of ^{14}N, the more common isotope. An available technique helped them identify which of the replicated molecules were heavy, light, or hybrid (one heavy strand and one light). Use different colored pencils to draw the heavy and light strands of DNA. Starting with a DNA molecule having two heavy strands, show the formation of daughter molecules after replication in a ^{14}N-containing medium. Show the four DNA molecules that would form if the daughter molecules were replicated a second time in the ^{14}N medium. Would the resulting DNA molecules be heavy, light, or mixed?

2. Mutations, remember, are permanent changes in DNA base sequences—the original source of genetic variation and the raw material of evolution. How can mutations accumulate, given that cells have repair systems that fix changes or breaks in DNA strands?

3. There may be millions of woolly mammoths frozen in the ice of Siberian glaciers. These huge elephant-like mammals have been extinct for about 10,000 years, but a team of privately funded Japanese scientists is planning to resurrect one of them by cloning DNA isolated from frozen remains. What are some of the pros and cons, both technical and ethical, of cloning an extinct animal?

4. *Xeroderma pigmentosum* is an autosomal recessive disorder that is characterized by rapid formation of skin sores that can develop into cancers (Figure 12.11). Affected individuals have no mechanism for dealing with the damage that ultraviolet (UV) light can inflict on skin cells. They must avoid all forms of radiation—including sunlight and fluorescent lights. These people lack a particular repair mechanism. When the nitrogen-containing bases in DNA absorb UV light, a covalent bond can form between two thymine bases in the same strand of DNA. The resulting thymine dimer puts a kink in the strand. Propose what consequences might occur (a) during interphase, when most of the cell's proteins are synthesized, and (b) during DNA replication.

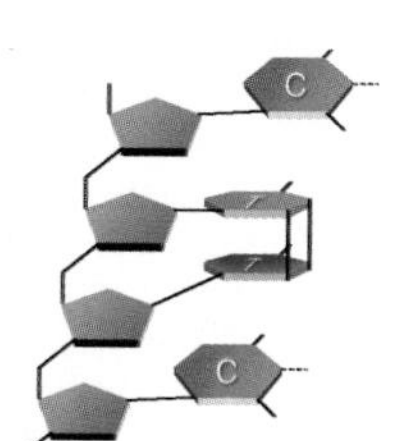

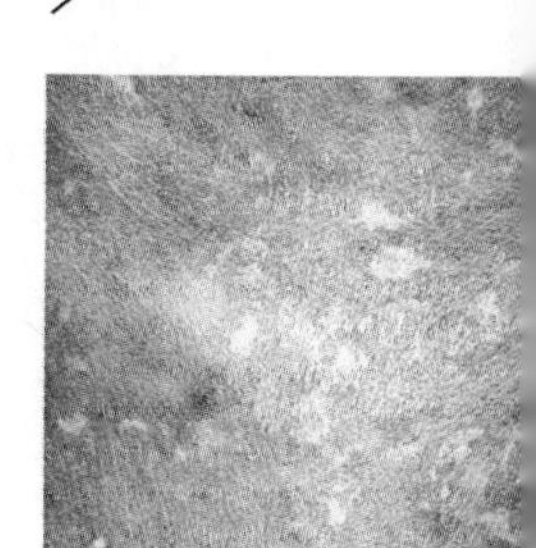

Figure 12.11 Thymine dimer (*top*), which can lead to xeroderma pigmentosum (*bottom*).

13 FROM DNA TO PROTEIN

IMPACTS, ISSUES Ricin and Your Ribosomes

In 2003, police acted on an intelligence tip and stormed a London apartment, where they found laboratory glassware and castor-oil beans (Figure 13.1). They made a few arrests and reminded the world that terrorists view ricin as a bioweapon. A dose of ricin as small as a grain of salt can kill you; only plutonium and botulism toxin are more deadly. Ricin is present in all of the tissues of the castor-oil plant (*Ricinus communis*), but the plant's oil is an ingredient in many plastics, cosmetics, paints, textiles, and adhesives. The oil—and ricin—is most concentrated in castor-oil seeds (beans), but the ricin is discarded when the oil is extracted.

Researchers knew about ricin's lethal effects as long ago as 1888. Though using ricin as a chemical weapon is now banned by most countries under the Geneva Protocol, the toxin is easily isolated. Controlling its production is impossible, so it crops up periodically in the news as a weapon wielded by terrorists.

For example, at the height of the Cold War between Russia and the West, Georgi Markov, a Bulgarian writer, had defected to England and had started working as a journalist for the BBC. As he made his way to a bus stop on a London street, an assassin used the tip of a modified umbrella to jam a small, ricin-laced ball into Markov's leg. Markov died in agony three days later.

More recently, traces were found in a United States Senate mailroom and State Department building, and also in an envelope addressed to the White House. In 2005, the FBI arrested a man who had castor-oil beans and an assault rifle stashed in his Florida home. Jars of banana baby food laced with ground castor-oil beans also made the news in 2005.

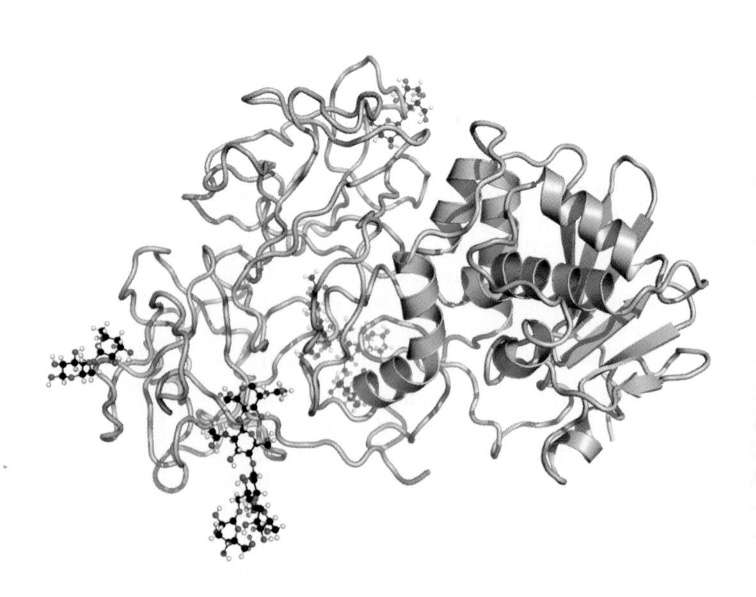

See the video! **Figure 13.1** *Left*, model for ricin. One of its polypeptide chains (*green*) helps ricin penetrate a living cell. The other chain (*tan*) destroys the cell's capacity for protein synthesis. Ricin is a glycoprotein; sugars attached to the protein are shown. *Right*, seeds of the castor-oil plant, source of ribosome-busting ricin.

How would you vote? Ricin is difficult to disperse through air and it breaks down quickly, so it is not likely to be used in a large-scale terrorist attack. However, terrorists may try to disperse it in confined spaces. Scientists have developed a vaccine against ricin. Do you want to be vaccinated? See ThomsonNOW for details, then vote online.

How does ricin exert its deadly effects? *It inactivates ribosomes, the protein-building machinery of all cells.*

Ricin is a protein with two polypeptide chains. One chain helps ricin insert itself into cells. The other chain functions as an enzyme. Its catalytic action wrecks part of the ribosome where amino acids are assembled into proteins. It yanks subunits of adenine out of a type of RNA molecule that is a crucial structural component of ribosomes. Once that happens, a ribosome's shape unravels, and protein synthesis stops. Protein synthesis is critical to all life processes: Cells that cannot synthesize proteins die, and so does the individual.

Start with what you know about DNA, the book of protein-building information in all cells. The alphabet used to write the book seems simple enough: A, T, G, and C, for the four nucleotide bases adenine, thymine, guanine, and cytosine. How does a cell make all of its proteins from this simple alphabet? The answer starts with the order, or sequence, of the four nucleotide bases in DNA.

Genetic information is encoded in the sequence itself as sets of three bases. These "genetic words" follow one another along the length of each chromosome. Like the words of a sentence, a linear series of genetic words can form a meaningful parcel of information. Such parcels are called genes, and cells use them to make RNA (ribonucleic acid) or protein products.

Most gene products are proteins. It takes two steps, transcription and translation, to go from a sequence of bases in a gene to a sequence of amino acids in a protein:

$$\text{DNA} \xrightarrow{\textit{transcription}} \text{RNA} \xrightarrow{\textit{translation}} \text{PROTEIN}$$

Transcription enzymes use the base sequence of a gene as a template to make a strand of RNA. The genetic information the RNA carries is then decoded—translated—into a sequence of amino acids. The result is a polypeptide chain that twists and folds into a protein.

This chapter details how proteins are synthesized using directions encoded in DNA. Though it is unlikely that your ribosomes will ever encounter ricin, protein synthesis is nevertheless worth appreciating for how it keeps you—and all other organisms—alive.

Key Concepts

INTRODUCTION

Life depends on enzymes and other proteins. All proteins consist of polypeptide chains. The chains are sequences of amino acids that correspond to sequences of nucleotide bases in DNA called genes. The path leading from genes to proteins has two steps: transcription and translation.

TRANSCRIPTION

During transcription, the two strands of the DNA double helix are unwound in a gene region. Exposed bases of one strand become the template for assembling a single strand of RNA (a transcript). Messenger RNA is the only type of RNA that carries DNA's protein-building instructions. **Section 13.1**

CODE WORDS IN THE TRANSCRIPTS

The nucleotide sequence in RNA is read three bases at a time. Sixty-four base triplets that correspond to specific amino acids represent the genetic code, which has been highly conserved over time. **Section 13.2**

TRANSLATION

During translation, amino acids become bonded together into a polypeptide chain in a sequence specified by base triplets in messenger RNA. Transfer RNAs deliver amino acids one at a time to ribosomes. Ribosomal RNA catalyzes the formation of peptide bonds between the amino acids. **Sections 13.3, 13.4**

MUTATIONS IN THE CODE WORDS

Mutations in genes may result in changes in protein structure, protein function, or both. The changes may lead to variation in traits among individuals. **Section 13.5**

Links to Earlier Concepts

Once again you will meet up with the nucleic acids DNA and RNA (Section 3.6). Gene transcription has features in common with DNA replication, so you may wish to review Sections 12.2 and 12.4 before you start. You will again consider protein primary structure (3.4), this time in the context of how it arises. The last section of this chapter will expand your knowledge of DNA repair mechanisms (12.4) and mutation (1.4, 3.5, 11.2, 11.5).

13.1 Transcription

LINKS TO SECTIONS 3.6, 4.1, 12.2, 12.4

Protein synthesis requires three types of RNA molecules: mRNA, tRNA, and rRNA. An mRNA is a disposable version of a gene; it carries DNA's protein-building information to the other two types of RNA. The three RNAs interact to translate the information into a protein.

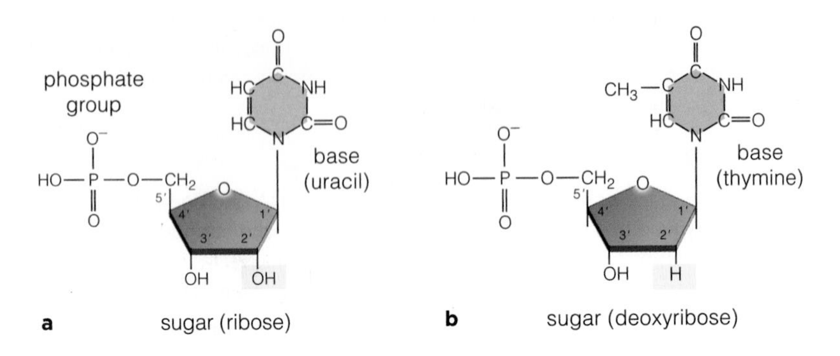

c Example of base pairing between a DNA strand and a new RNA strand assembled on it during *transcription*

G A C T — DNA template
C U G A — **RNA transcript**

d Example of base pairing between an old DNA strand and a new strand assembled on it during *DNA replication*

G A C T — DNA template
C T G A — New DNA strand

Figure 13.2 (**a**) Uracil, one of four nucleotides in RNA. Three others—adenine, guanine, and cytosine—differ only in their bases. (**b**) Compare uracil with thymine, a DNA nucleotide. (**c**) DNA–RNA base pairing during transcription, compared with (**d**) DNA–DNA base pairing during DNA replication.

During transcription, the first step in protein synthesis, a sequence of nucleotide bases becomes exposed in an unwound region of a DNA strand. That sequence acts as a template upon which a single strand of RNA—a transcript—is synthesized from free nucleotides. The transcription of genes that encode proteins results in messenger RNA (mRNA), the only kind of RNA that carries protein-building codes. Transcription of other genes results in other kinds of RNA. Ribosomal RNA (rRNA) becomes part of ribosomes, the structures upon which polypeptide chains are assembled. Transfer RNA (tRNA) delivers amino acids one by one to ribosomes, in the order specified by mRNA.

An RNA molecule is almost—but not quite—like a single strand of DNA (Table 13.1). It has four kinds of nucleotides, each with the five-carbon sugar ribose, a phosphate group, and one base. Three bases (adenine, cytosine, and guanine) are the same as those in DNA. In RNA, the fourth base is uracil, not thymine. Uracil, too, can pair with adenine; thus, RNA follows the same base-pairing rule as DNA (Figure 13.2).

Transcription differs from DNA replication in three respects. Only part of one DNA strand, not the whole molecule, is unwound and used as the template. The enzyme RNA polymerase, not DNA polymerase, adds the nucleotides one at a time to the end of a growing transcript. Also, transcription results in a single strand of RNA, not two DNA double helixes.

Transcription starts after RNA polymerase attaches to its binding site in the DNA, a nucleotide sequence called a promoter. In eukaryotes, other proteins that

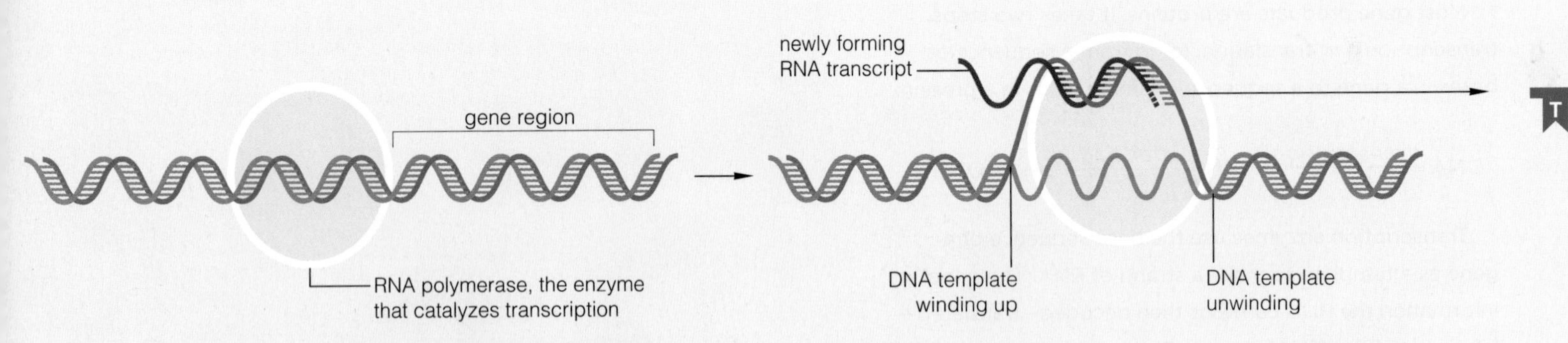

a RNA polymerase binds to a promoter region in the DNA, then initiates transcription at a nearby gene. The polymerase begins to link the nucleotides adenine, cytosine, guanine, and uracil into a strand of RNA, in the order specified by the base sequence of the DNA.

b All through transcription, the DNA double helix becomes unwound in front of the RNA polymerase. The newly forming RNA strand briefly winds up with its DNA template. As the RNA strand lengthens, it unwinds from the DNA (and the two DNA strands wind up again).

Figure 13.3 **Animated!** Gene transcription. By this process, an RNA molecule is assembled on a DNA template. (**a**) Gene region of DNA. The base sequence along one of DNA's two strands (not both) is used as the template. (**b–d**) Transcribing that region produces a molecule of RNA.

Table 13.1 RNA and DNA Compared

	RNA	DNA
Structure		
sugar	ribose	deoxyribose
bases	adenine, guanine, cytosine, uracil	adenine, guanine, cytosine, thymine
form	single-stranded; some have specific shapes	double-stranded helix
Function	disposable copies of heritable information; some are catalytic	permanent storage of heritable information

unit of transcription in DNA strand
exon intron exon intron exon
transcription into pre-mRNA
cap
poly-A tail
5′
3′
snipped out
snipped out
mature mRNA transcript

Figure 13.4 Animated! Post-translational RNA modification in the nucleus. Introns are removed, exons may be spliced. An mRNA gets a poly-A tail and modified guanine "cap."

help with transcription bind to a promoter along with RNA polymerase. Binding to a promoter positions a polymerase at a transcription start site close to a gene. Transcription then begins at that site. The polymerase joins RNA nucleotides, using the DNA sequence as a template (Figure 13.3). Transcription ends when the polymerase passes the end of the gene.

In prokaryotes, transcription and translation occur simultaneously, in the cytoplasm. In eukaryotic cells, transcription occurs in the nucleus, where the resulting RNA is modified before it is shipped to the cytoplasm for translation. Just as a dressmaker may snip off some loose threads or add bows to a dress before it leaves the shop, so do eukaryotic cells tailor their RNA before it leaves the nucleus.

For example, most eukaryotic genes contain one or more introns—nucleotide sequences that are removed from a new RNA. Introns intervene between exons, or sequences that stay in the RNA (Figure 13.4). Introns are transcribed along with the exons, but are removed before the RNA leaves the nucleus. Either all exons remain in the mature RNA, or some are removed and the rest are spliced in various combinations. By such alternative splicing, one gene can encode different but related proteins. A tail of 50 to 300 adenines also gets added to the 3′ end of each new mRNA; thus the name, poly-A tail. Enzymes begin disassembling an mRNA as soon as it arrives in the cytoplasm, starting with the tail. How long each mRNA—and its protein-building message—lasts depends on the length of its tail.

In transcription, RNA polymerase uses a gene in DNA as a template to assemble a strand of RNA from nucleotides.

Before leaving the nucleus of eukaryotes, new RNAs are modified into their final form.

direction of transcription
3′
5′
T T C T A C G G A C T C C T C T T C
A U G C C U G A G G A G A A G
5′
3′
growing RNA transcript

c What happened in the gene region? RNA polymerase catalyzed the covalent bonding of nucleotides to one another to form an RNA strand. The sequence of nucleotides in a new RNA molecule is complementary to the base sequence of its DNA template. Many other proteins assist in transcription; compare Section 12.4.

d At the end of the gene region, the last stretch of the new transcript is unwound and released from the DNA template. Shown below it is a model for a transcribed strand of RNA.

A U G C C U G A G G A G A A G

13.2 The Genetic Code

The bridge between genes and proteins takes the form of protein-building "words" in mRNA transcripts. Three nucleotide bases make up each three-letter word.

LINKS TO SECTIONS 3.4, 3.6

An mRNA is a linear sequence of genetic "words" that are spelled with an alphabet of just four nucleotides. Like sentences, mRNA transcripts can be understood by those who know the language. Researchers Marshall Nirenberg, Heinrich Matthaei, Philip Leder, Har Gobind Khorana, and Severo Ochoa deciphered the language of mRNA; they discovered the correspondence between genes and proteins.

An mRNA transcript encodes a sequence of amino acids. To translate its base sequence into an amino acid sequence, you have to know that mRNA bases are read *three at a time*, as triplets. Each base triplet in mRNA is known as a codon. Which bases are in the first, second, and third positions of a triplet determines which amino acid the codon specifies (Figure 13.5*a*).

There are a total of sixty-four codons, many more than are necessary to specify all twenty kinds of amino acids found in proteins. Most amino acids are encoded by more than one codon. For instance, the codons GAA and GAG both call for the amino acid glutamate.

The linear order of codons in an mRNA transcript determines the order of amino acids in a polypeptide chain (Figure 13.5*b*). Some codons signal the beginning and end of a gene. In most species, the first AUG is a signal to start translation. AUG is also the codon for methionine, so methionine is the first amino acid in all new polypeptide chains of these organisms. UAA, UAG, and UGA do not specify an amino acid. They are STOP signals that block further additions of amino acids to a new chain.

The set of sixty-four mRNA codons used in protein synthesis is the genetic code. This code has been highly conserved over time. Prokaryotes, a few organelles, and some protists of ancient lineages have a few slightly variant codons. For instance, a few unique codons give mitochondria their own "mitochondrial code," which was one clue that led to the theory of how eukaryotic organelles arose (Section 18.4).

The genetic code is a set of sixty-four different codons, which are nucleotide bases in mRNA that are "read" in sets of three during protein synthesis. Different codons (base triplets) specify different amino acids.

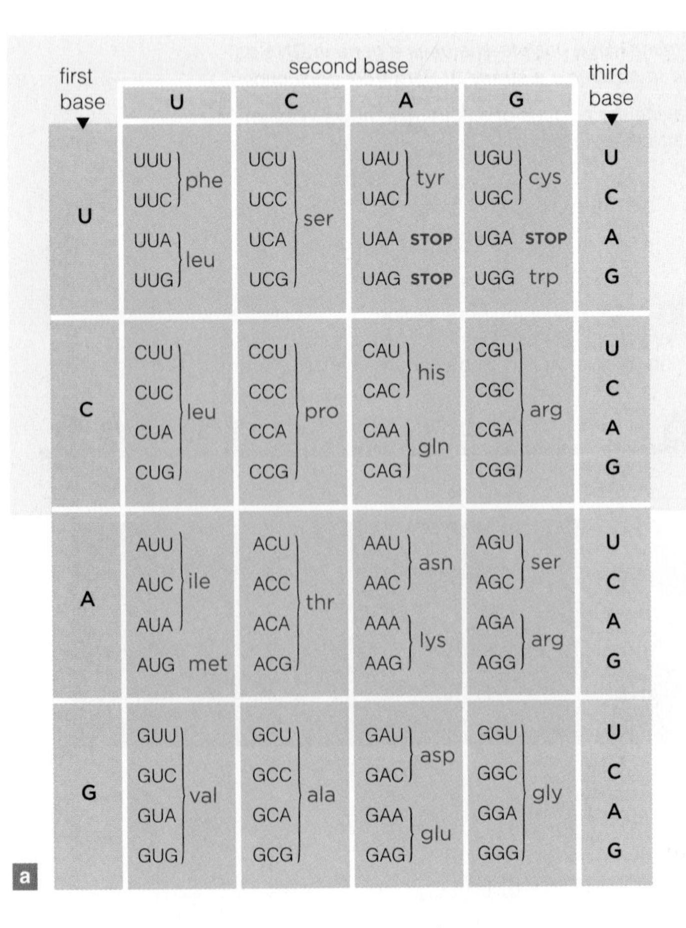

first base	second base U	second base C	second base A	second base G	third base
U	UUU phe	UCU ser	UAU tyr	UGU cys	U
U	UUC	UCC	UAC	UGC	C
U	UUA leu	UCA	UAA STOP	UGA STOP	A
U	UUG	UCG	UAG STOP	UGG trp	G
C	CUU leu	CCU pro	CAU his	CGU arg	U
C	CUC	CCC	CAC	CGC	C
C	CUA	CCA	CAA gln	CGA	A
C	CUG	CCG	CAG	CGG	G
A	AUU ile	ACU thr	AAU asn	AGU ser	U
A	AUC	ACC	AAC	AGC	C
A	AUA	ACA	AAA lys	AGA arg	A
A	AUG met	ACG	AAG	AGG	G
G	GUU val	GCU ala	GAU asp	GGU gly	U
G	GUC	GCC	GAC	GGC	C
G	GUA	GCA	GAA glu	GGA	A
G	GUG	GCG	GAG	GGG	G

a

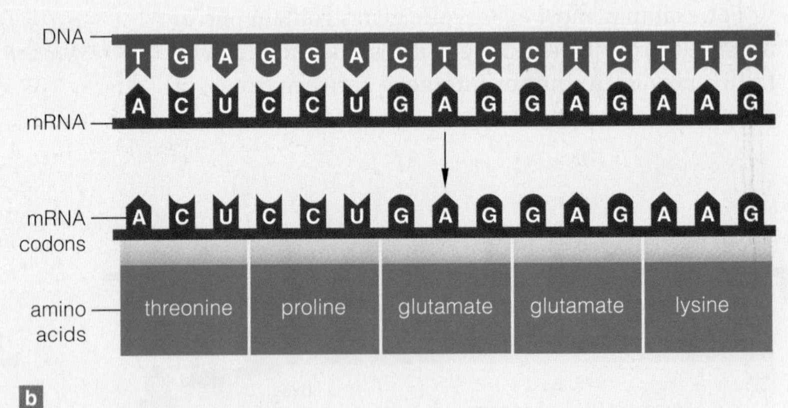

Figure 13.5 **Animated!** The genetic code. (**a**) Each codon in mRNA is a set of three nucleotide bases. Sixty-one of the triplets encode amino acids. Three are signals that stop translation. The amino acid names that correspond to these abbreviations are listed in Appendix V.

The *left* vertical column lists a codon's first base. The horizontal row on *top* lists the second base. The *right* vertical column lists the third. Some examples: reading left to right, the triplet UGG codes for tryptophan (trp); both UUU and UUC code for phenylalanine (phe).

(**b**) Example of the correspondence between genes and proteins. First, a DNA strand is transcribed into mRNA. Notice how the mRNA's nucleotide sequence is complementary to the gene sequence in the DNA. The codons of this mRNA specify a chain of amino acids: threonine–proline–glutamate–glutamate–lysine.

13.3 tRNA and rRNA

The codons in an mRNA are words in a protein-building message. Two other classes of RNA—rRNA and tRNA—translate those words into a polypeptide chain.

DNA is the molecule that stores heritable information about building proteins. In eukaryotic cells, the DNA stays protected in the nucleus. Gene by gene, mRNAs carry protein-building messages from the DNA to the protein synthesis machinery in the cytoplasm. There, rRNAs and tRNAs interact to translate mRNA codons into polypeptide chains.

The rRNAs are part of ribosomes. A ribosome has two subunits, one large and one small, each of which consists of rRNA and structural proteins (Figure 13.6). In eukaryotic cells, the subunits are assembled in the nucleus and then moved to the cytoplasm, where they converge as intact ribosomes.

The tRNAs ferry one amino acid after the next to ribosomes. Each tRNA has two attachment sites. One is an anticodon: a triplet of nucleotides that base-pairs with a complementary mRNA codon (Figure 13.7). The tRNA's other attachment site binds to a specific kind of amino acid—the one specified by that mRNA codon.

Some tRNAs can base-pair with different codons. In codon–anticodon interactions, the base-pairing rule is loose for the third base in a codon. Freedom in codon–anticodon pairing is called "wobble." For example, the codons AUU, AUC, and AUA all specify isoleucine. All of them base-pair with a tRNA that carries isoleucine.

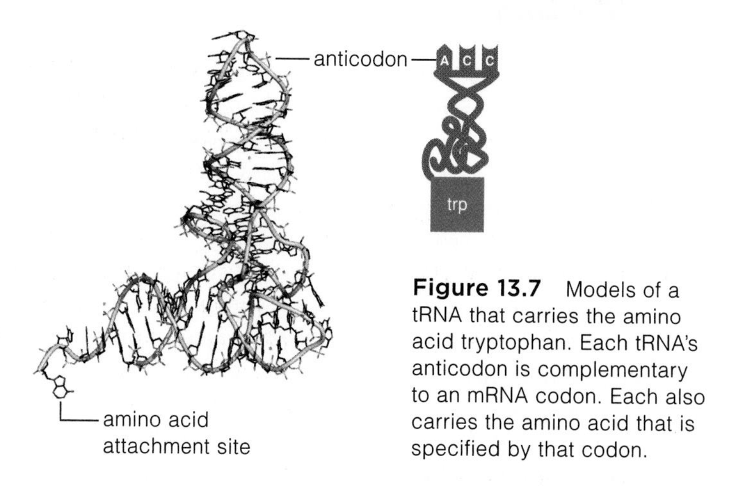

Figure 13.7 Models of a tRNA that carries the amino acid tryptophan. Each tRNA's anticodon is complementary to an mRNA codon. Each also carries the amino acid that is specified by that codon.

A ribosome, an mRNA, and tRNAs converge during protein synthesis. The order of codons in the mRNA is the order in which the tRNAs deliver their amino acid cargo to the ribosome. As amino acids are delivered, the ribosome joins them into a new polypeptide chain (Section 3.4). Thus, the order of codons in an mRNA—DNA's protein-building message—is translated into a polypeptide chain.

Ribosomes, which have two subunits that consist of rRNA and proteins, link amino acids into polypeptide chains.

tRNAs have an anticodon complementary to an mRNA codon, and also a binding site for the amino acid specified by that codon. tRNAs deliver amino acids to ribosomes.

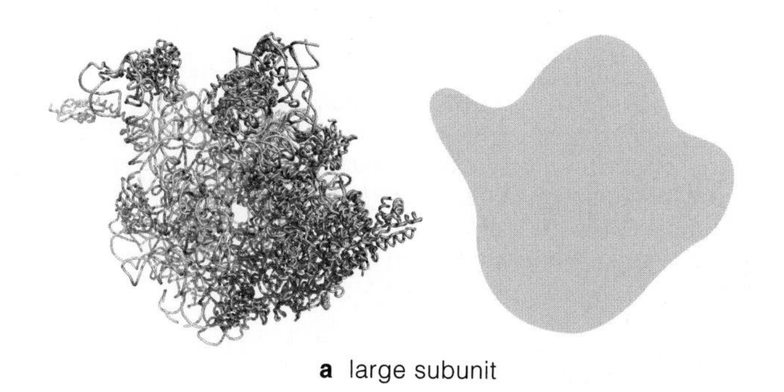

a large subunit

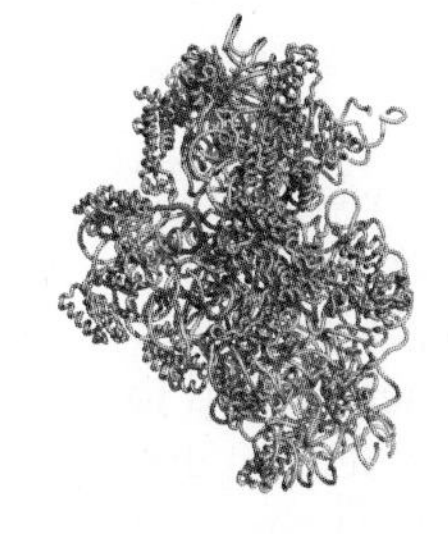

b small subunit

c intact ribosome

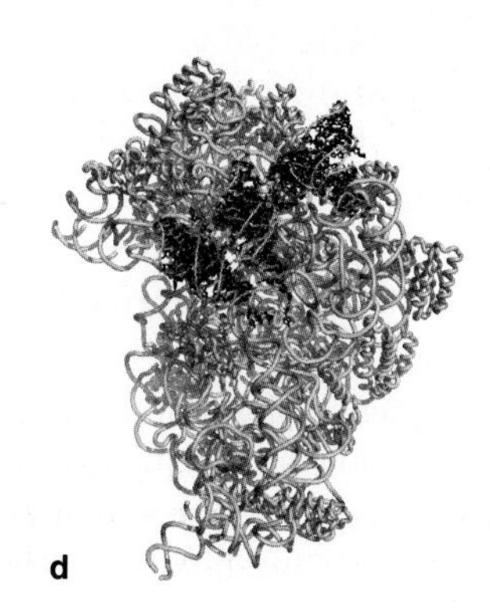

d

Figure 13.6 Models of the bacterial ribosome. This ancient, highly conserved structure is so vital that the eukaryotic ribosome, which is larger, is expected to be quite similar.

(**a**,**b**) The large and small subunits of a ribosome consist of rRNA molecules (*tan*) and structural proteins (*green*). Notice the tunnel through the interior of the large subunit. The rRNA component of the ribosome catalyzes assembly of polypeptide chains, which thread through this tunnel as they form. An mRNA (*red*) is shown attached to the small subunit.

(**c**) One large and one small ribosomal subunit fit together for translation.

(**d**) During translation, tRNAs dock at an intact ribosome. Here, three tRNAs (*brown*) are docked at the small ribosomal subunit (the large subunit is not shown, for clarity). The anticodons of the tRNAs line up with complementary codons in an mRNA (*red*).

13.4 The Three Stages of Translation

In eukaryotes, hereditary information is stored intact, safe, and in one place: in DNA, in the nucleus. mRNA is the intermediary that delivers a genetic message from DNA to a ribosome in the cytoplasm. There, tRNA and rRNA interact to translate the message into a polypeptide chain.

LINKS TO SECTIONS 3.4, 4.8

The second part of protein synthesis, translation, is a process that converts genetic information encoded in an mRNA transcript into a new polypeptide chain. Translation occurs in the cytoplasm, and it has three stages: initiation, elongation, and termination.

Only initiator tRNAs can begin the *initiation* stage. The anticodon of an initiator tRNA base-pairs with an AUG codon, and it carries the amino acid methionine. In initiation, an initiator tRNA binds a small ribosomal subunit, and then to the first AUG codon of an mRNA. Then a large ribosomal subunit joins the group. When an initiator tRNA, an intact ribosome, and an mRNA are clustered together, they form an initiation complex (Figure 13.8*a,b*).

In the *elongation* stage of translation, a polypeptide chain is assembled as the mRNA threads between the two ribosomal subunits. Remember, the initiator tRNA carries methionine, so the first amino acid of the chain will be methionine. In elongation, other tRNAs bring amino acids to the ribosome. The tRNAs bind to the ribosome one after the next, in the order determined by the successive codons in the mRNA template.

In turn, each amino acid is added to the end of the growing polypeptide chain by way of a peptide bond (Figure 13.8*c–e*). The rRNA component of the ribosome catalyzes formation of the bond (Section 3.4). Thus, the order of codons in an mRNA translates to the order of amino acids in a polypeptide chain.

During *termination*, the last stage of translation, the mRNA's STOP codon enters the ribosome. No tRNA has an anticodon that base-pairs with this codon, so translation ends. Proteins called release factors bind to the ribosome. The binding triggers enzyme activity that detaches the mRNA and the polypeptide chain from the ribosome (Figure 13.8*f*).

In cells that are quickly using or secreting proteins, there are often many clusters of ribosomes (polysomes) on an mRNA transcript, all translating it at the same time. In such cases, an mRNA forms a new initiation complex before other ribosomes finish translating it.

Many newly formed polypeptide chains carry out their functions in the cytoplasm. Others have a special sequence of amino acids that allows them to enter the ribosome-studded, flattened sacs of rough ER (Section 4.8). Inside these organelles, polypeptide chains take on final form before shipment to other parts of the cell.

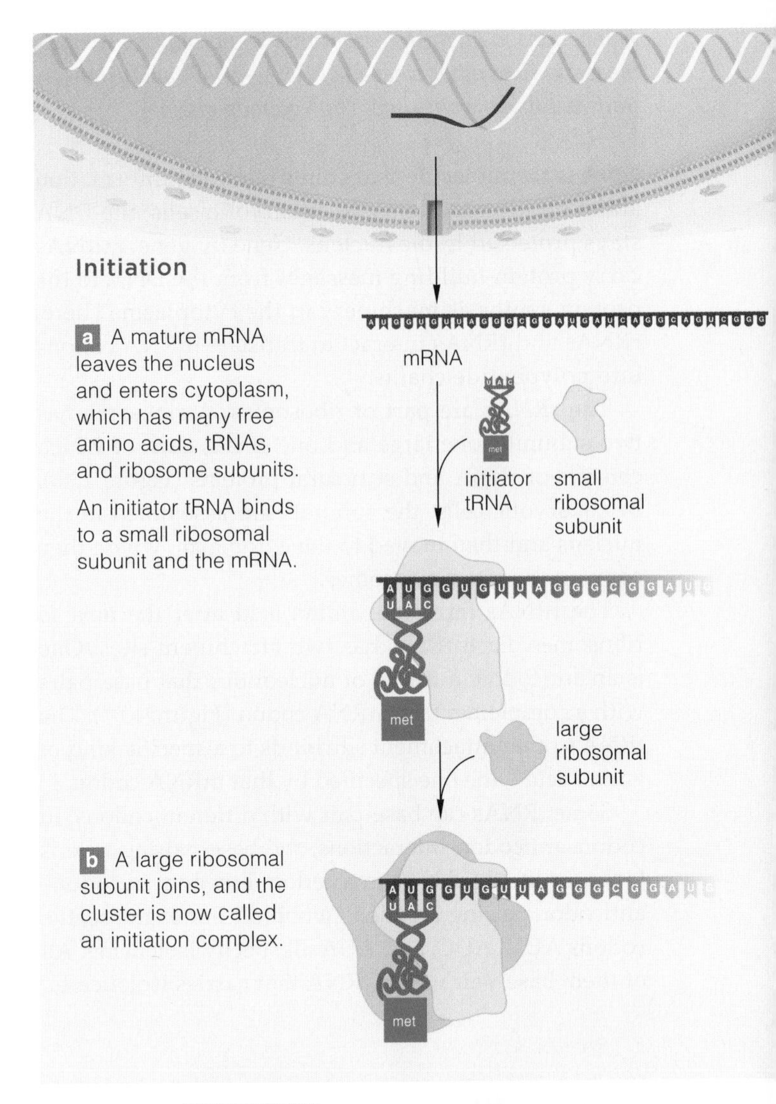

Figure 13.8 **Animated!** Stages of translation, the second step of protein synthesis. (**a**,**b**) In initiation, an mRNA, an intact ribosome, and an initiator tRNA form an initiation complex. (**c**–**e**) In elongation, the new polypeptide chain grows as the ribosome catalyzes the formation of peptide bonds between amino acids delivered by tRNAs. (**f**) In termination, the mRNA and the new polypeptide chain are released, and the ribosome disassembles.

Translation is initiated when an initiator tRNA binds the first AUG of an mRNA and a small ribosomal subunit, then a large ribosomal subunit joins them.

tRNAs deliver specific amino acids to a ribosome in the order dictated by the linear sequence of mRNA codons. A polypeptide chain lengthens as peptide bonds form between the amino acids.

Translation ends when a STOP codon triggers events that cause the polypeptide chain and the mRNA to detach from the ribosome.

Elongation

c An initiator tRNA carries the amino acid methionine, so the first amino acid of the new polypeptide chain will be methionine. A second tRNA binds the second codon of the mRNA (here, that codon is GUG, so the tRNA that binds carries the amino acid valine).

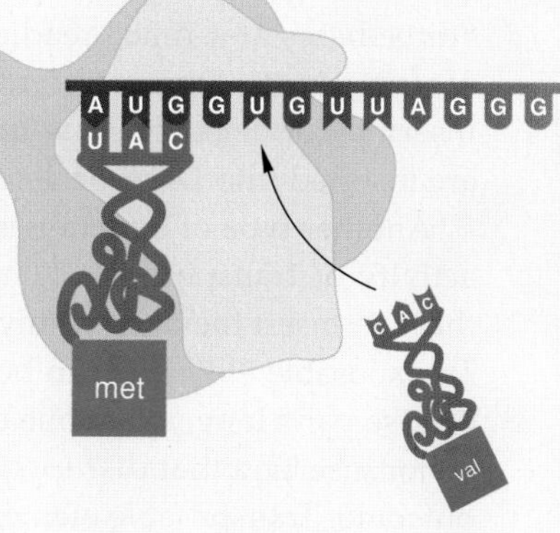

A peptide bond forms between the first two amino acids (here, methionine and valine).

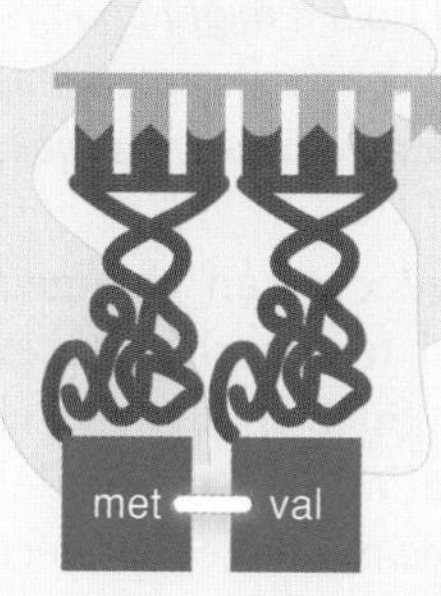

d The first tRNA is released and the ribosome moves to the next codon in the mRNA. A third tRNA binds to the third codon of the mRNA (here, that codon is UUA, so the tRNA carries the amino acid leucine).

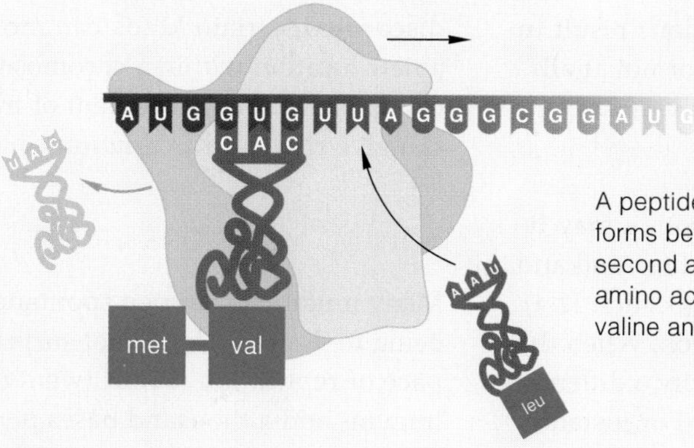

A peptide bond forms between the second and third amino acids (here, valine and leucine).

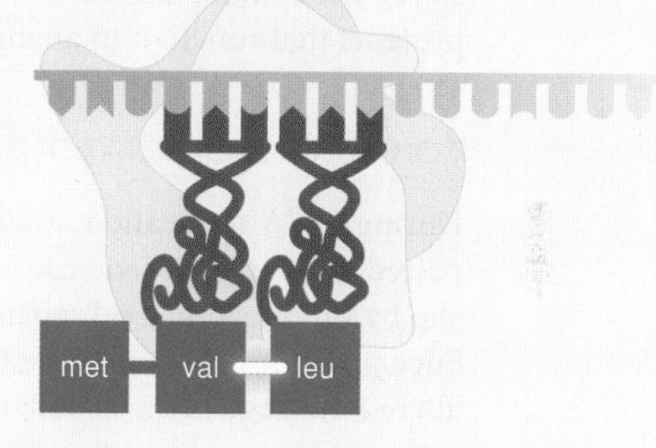

e The second RNA is released and the ribosome moves to the next codon. A fourth tRNA binds the fourth mRNA codon (here, that codon is GGG, so the tRNA carries the amino acid glycine).

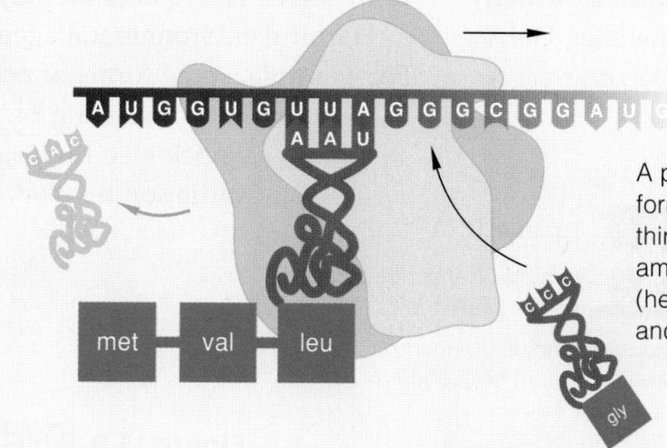

A peptide bond forms between the third and fourth amino acids (here, leucine and glycine)

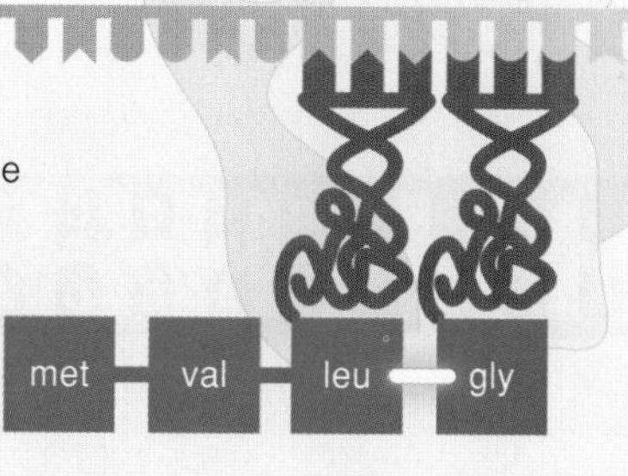

Termination

f Steps **d** and **e** are repeated over and over until the ribosome encounters a STOP codon in the mRNA. The mRNA transcript and the new polypeptide chain are released from the ribosome. The two ribosomal subunits separate from each other. Translation is now complete. Either the chain will join the pool of proteins in the cytoplasm or it will enter rough ER of the endomembrane system (Section 4.8).

13.5 Mutated Genes and Their Protein Products

When a cell accesses its genetic code, it makes RNAs and proteins that keep it alive. If a gene changes, an mRNA transcribed from it might change so that it specifies an altered protein. If the protein has a crucial role, the result may be a dead or abnormal cell.

LINKS TO SECTIONS 2.3, 3.5, 5.4, 6.1, 8.5, 12.4

Gene sequences can change. Sometimes one base is substituted for another, an extra base is inserted, or one is lost. Such small-scale changes in the nucleotide sequence of a DNA molecule are mutations, and they can alter the message that becomes encoded in mRNA. Cells have some leeway, because more than one codon can specify the same amino acid. For example, if UCU replaced UCC in an mRNA transcript, this might not be bad, because both codons specify serine. However, as the next examples show, many mutations result in proteins that function in an altered way or not at all.

COMMON GENE MUTATIONS

During DNA replication, a wrong nucleotide may be paired with an exposed base on a template strand and slip by proofreading and repair enzymes (Section 12.4). Such a mutation is a base-pair substitution. When the altered message is translated, it may specify a different amino acid or a premature STOP codon. For instance, if adenine replaces a thymine in the hemoglobin beta chain gene, sickle-cell anemia results (Figure 13.9*b*).

Figure 13.9*c* depicts a different mutation, in which a single base—thymine—was *deleted*. Remember, DNA polymerases read base sequences in blocks of three. A deletion is one of the *frameshift* mutations. Frameshift mutations garble the genetic message by shifting the "three-bases-at-a-time" reading frame, and an altered protein results. Frameshift mutations may result from insertions or deletions, in which one or more base pairs are inserted into DNA or deleted from it.

Another type of insertion mutation is caused by the activity of transposable elements—segments of DNA that can insert themselves anywhere in a chromosome. Transposable elements can be hundreds or thousands of base pairs long; when one enters a gene sequence, a major insertion that disrupts the gene's product is the outcome. Transposable elements occur in all genomes; about 45 percent of human DNA consists of them or remnants of them. As geneticist Barbara McClintock discovered, certain kinds can move spontaneously to a new location within a chromosome, or to a different chromosome. The movement of transposable elements can give rise to odd variations in traits (Figure 13.10).

HOW DO MUTATIONS ARISE?

Many mutations happen spontaneously while DNA is being replicated. That is not surprising, given the swift pace of replication (about twenty bases per second in humans and a thousand bases per second in bacteria). DNA polymerases make mistakes at predictable rates, but most fix errors when they occur (Section 12.4). Any uncorrected errors may be perpetuated as mutations.

Harmful environmental agents can cause mutations. For example, some forms of energy such as x-rays are called ionizing radiation because they can knock electrons right out of atoms. Such radiation breaks chromosomes apart into

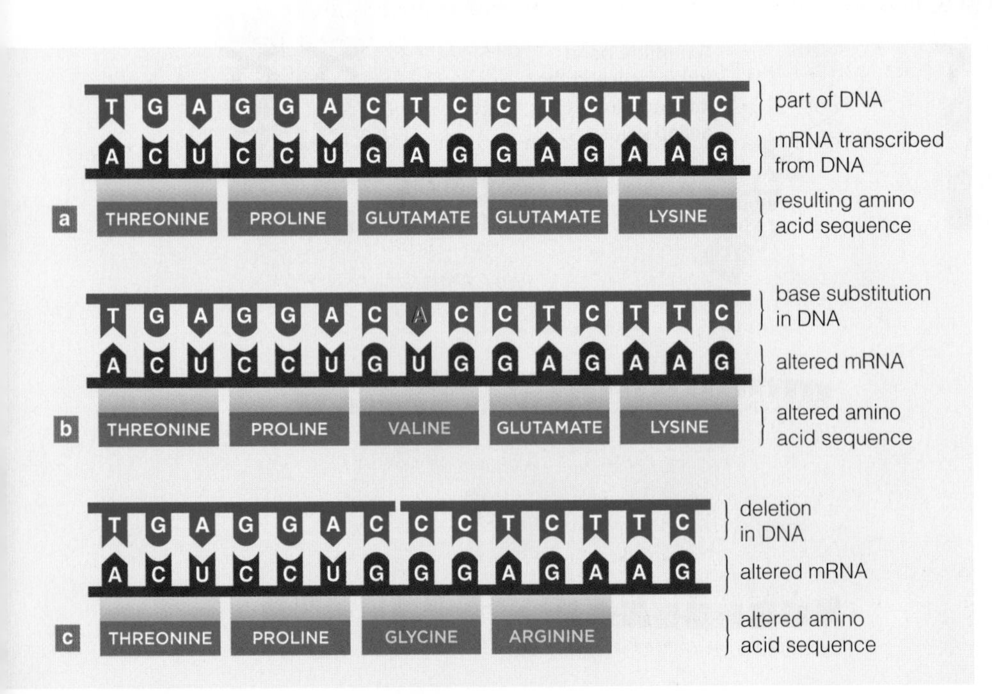

Figure 13.9 **Animated!** Examples of gene mutation. (**a**) Part of a gene, the mRNA, and the amino acid sequence of the beta chain of hemoglobin.

(**b**) A base-pair substitution in DNA replaces a thymine with an adenine. When the altered mRNA transcript is translated, valine replaces glutamate as the sixth amino acid of the new polypeptide chain. Sickle-cell anemia is the eventual outcome.

(**c**) Deletion of the same thymine causes a frameshift mutation. The reading frame for the rest of the mRNA shifts, and a different protein product forms. This mutation results in thalassemia, a type of red blood cell disorder.

Figure 13.10 Barbara McClintock, who won a Nobel Prize for her research. She found that transposable elements can slip into and out of different locations in DNA. The curiously nonuniform coloration of individual kernels in Indian corn (*Zea mays*) sent her on the road to discovery.

Several genes govern pigment formation and deposition in corn kernels, which are a type of seed. Interactions among these genes result in yellow, white, red, orange, blue, or purple kernels. McClintock realized that unstable mutations in these genes cause streaks or spots of color in individual kernels.

The same pigment genes occur in all of the cells of a kernel, but those near a transposable element are inactive. Transposable elements move while a kernel's tissues are forming, so they can end up in different locations in the DNA of different cell lineages. Streaks and spots on the kernels are evidence of transposable element movement that inactivated and reactivated different pigment genes in different cell lineages.

pieces that may be lost during DNA replication (Figure 13.11). That is how some deletions arise.

Ionizing radiation can also damage DNA indirectly. It penetrates living tissue, leaving in its wake a trail of destructive free radicals. Free radicals, remember, also damage DNA. That is why doctors and dentists use the lowest possible doses of x-rays on their patients.

Other forms of radiant energy boost electrons to a higher energy level but not enough to knock them out of an atom; they are forms of nonionizing radiation. DNA absorbs the kind called ultraviolet (UV) light. Exposure to UV light can cause two adjacent thymine bases to bond covalently to one another. This bond, a thymine dimer, kinks the DNA (Figure 12.11). During replication, DNA polymerase often copies the kinked part incorrectly, and so introduces a mutation into the DNA. Mutations that cause certain kinds of cancers arise from thymine dimers. They are the reason that exposing your unprotected skin to sunlight increases your risk of skin cancer. At least seven gene products interact as a DNA repair mechanism to remove them.

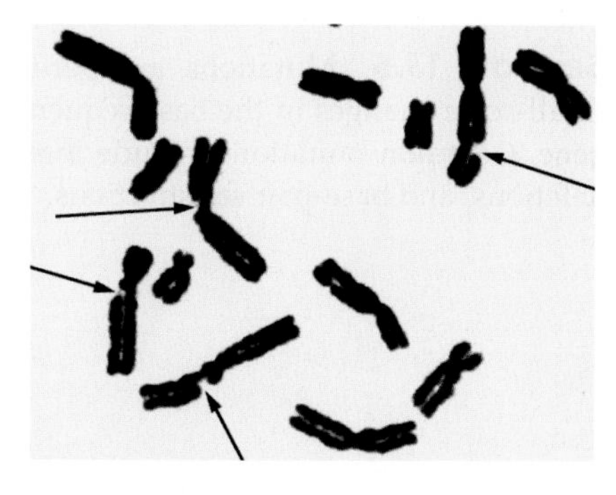

Figure 13.11 Chromosomes from a human cell after exposure to gamma rays, a form of ionizing radiation. We can expect such broken pieces (*arrows*) to be lost during interphase, when DNA is being replicated. The extent of the chromosome damage in an exposed cell typically depends on how much radiation it absorbed.

Some natural or synthetic chemicals also can cause mutations. For instance, certain chemicals in cigarette smoke transfer small hydrocarbon groups to the bases in DNA. The altered bases mispair during replication or stop further replication.

THE PROOF IS IN THE PROTEIN

When a mutation arises in a somatic cell of a sexually reproducing individual, its good or bad effects will not endure; it is not passed on to offspring. If it arises in a germ cell or a gamete, however, it may enter the evolutionary arena. It also may do so when it is passed on to offspring by asexual reproduction. Either way, a protein product of a mutated gene can have harmful, neutral, or beneficial effects on an individual's capacity to function in its prevailing environment. The effects of uncountable mutations in millions of species have had spectacular evolutionary consequences—and that is a topic of later chapters.

A mutation is a permanent small-scale change in the nucleotide sequence of DNA. The most common types are base-pair substitutions, insertions, and deletions.

Most mutations arise during DNA replication as a result of unrepaired DNA polymerase errors. Some mutations occur after exposure to harmful radiation or chemicals.

A protein specified by a mutated gene may have harmful, neutral, or beneficial effects on an individual's capacity to function in its environment.

www.thomsonedu.com ThomsonNOW!

Summary

Introduction Enzymes and all other proteins consist of polypeptide chains. Each chain, a linear sequence of amino acids, corresponds to the nucleotide base sequence of a gene. The path from genes to proteins has two steps: transcription and translation (Figure 13.12).

Section 13.1 Transcription and translation occur in the cytoplasm of prokaryotic cells, which have no nucleus. Genes are transcribed in the nucleus of eukaryotic cells, and the resulting mRNA is translated in the cytoplasm.

In transcription, DNA's two strands unwind only in a specific region. RNA polymerase assembles a new RNA strand by covalently bonding the nucleotides adenine, guanine, cytosine, and uracil in the order dictated by the nucleotide sequence of the exposed gene.

RNA of eukaryotes is modified before it leaves the nucleus. Introns are removed. Some of the exons may be removed along with the introns, and the remaining exons spliced together in different combinations. A poly-A tail is added to the 3′ end of a new mRNA. The longer its poly-A tail, the more time an mRNA transcript (and its protein-building message) will remain intact in the cytoplasm.

■ *Use the animation on ThomsonNOW to learn how genes are transcribed and transcripts are processed.*

Sections 13.2, 13.3 Only messenger RNA (mRNA) carries DNA's protein-building information to ribosomes for translation. Its genetic message is written in codons, or sets of three nucleotides along an mRNA strand. Sixty-four codons constitute a highly conserved genetic code. A few codons are signals that stop translation; the rest specify different amino acids. Variations in this genetic code occur among the prokaryotes, prokaryote-derived organelles (such as mitochondria), and in a few ancient lineages of single-celled eukaryotes.

Two other types of RNA function in translation. Each transfer RNA (tRNA) molecule has an anticodon that can bind to an mRNA codon. Each also binds the amino acid specified by that mRNA codon. Different tRNAs carry different amino acids.

tRNAs deliver free amino acids to ribosomes during protein synthesis. Ribosomal RNA (rRNA) and proteins that stabilize it make up the two subunits of ribosomes.

■ *Use the interaction on ThomsonNOW to explore the genetic code.*

Section 13.4 Genetic information carried by an mRNA transcript directs the synthesis of a polypeptide chain during translation.

Translation has three stages. In initiation, one initiator tRNA, two ribosomal subunits, and one mRNA come together as an initiation complex. In elongation, tRNAs deliver amino acids to the ribosome in the order specified by the mRNA codons. rRNA that is part of the ribosome catalyzes the formation of a peptide bond between the amino acids. In termination, translation ends when the RNA polymerase encounters a STOP codon in the mRNA. Then, the new polypeptide chain and the mRNA are released, and the ribosome's subunits separate from each other.

■ *Use the animation on ThomsonNOW to see the translation of an mRNA transcript.*

Section 13.5 Mutations are permanent, small-scale changes in the base sequence of a gene. Common mutations include insertions, deletions, and base-pair substitutions.

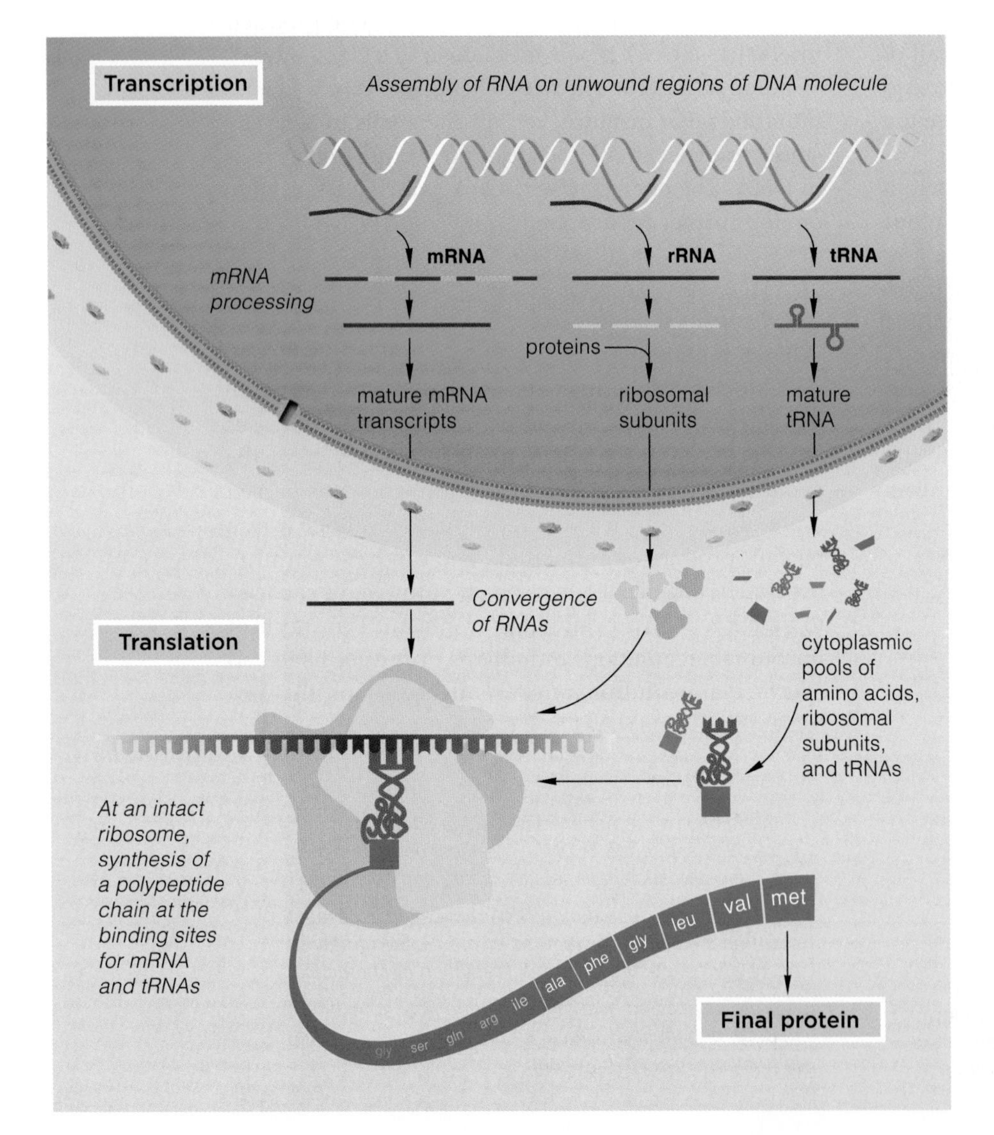

Figure 13.12 **Animated!** Summary of protein synthesis in eukaryotic cells. DNA is transcribed into RNA in the nucleus. RNA is translated in the cytoplasm. Prokaryotic cells have no nucleus; transcription and translation occur simultaneously in their cytoplasm.

Most mutations arise spontaneously as an outcome of DNA polymerase errors during DNA replication; others can occur after a cell is exposed to ionizing radiation or harmful chemicals in the environment. Transposable elements are segments of DNA that can insert themselves into a gene sequence and thus cause a major mutation. Any change in a gene's DNA sequence may lead to changes in the gene's product.

■ *Use the animation on ThomsonNOW to investigate the effects of mutation.*

Self-Quiz

Answers in Appendix III

1. DNA contains many different gene regions that are transcribed into different ______.
   a. proteins c. mRNAs, tRNAs, rRNAs
   b. mRNAs only d. all of the above

2. An RNA molecule is typically ______.
   a. a double helix c. double-stranded
   b. single-stranded d. triple-stranded

3. An mRNA molecule is assembled by ______.
   a. replication c. transcription
   b. duplication d. translation

4. ______ remain in new mRNA transcripts.
   a. Introns c. Telomeres
   b. Exons d. Amino acids

5. Each codon specifies a(n) ______.
   a. protein c. amino acid
   b. polypeptide d. mRNA

6. Each amino acid is specified by a set of ______ bases in an mRNA transcript.
   a. 3 b. 20 c. 64 d. 120

7. ______ different codons constitute the genetic code.
   a. 3 b. 20 c. 64 d. 120

8. Anticodons pair with ______.
   a. mRNA codons c. RNA anticodons
   b. DNA codons d. amino acids

9. ______ can cause mutations.
   a. Replication errors d. Non-ionizing radiation
   b. Transposons e. b and c are correct
   c. Ionizing radiation f. all of the above

10. Match the terms with the most suitable description.

___ cigarette smoke	a. protein-coding regions of mRNA
___ chain elongation	b. base triplet in mRNA
___ exons	c. second stage of translation
___ genetic code	d. base triplet; pairs with codon
___ anticodon	e. chemicals that can induce mutation in DNA
___ introns	f. set of 64 codons
___ codon	g. noncoding part of mRNA transcript, removed before translation

■ *Visit ThomsonNOW for additional questions.*

Critical Thinking

1. *Antisense drugs* help us fight some types of cancer and viral diseases. The drugs consist of short mRNA strands that are complementary to mRNAs linked to the diseases. Speculate on how antisense drugs work.

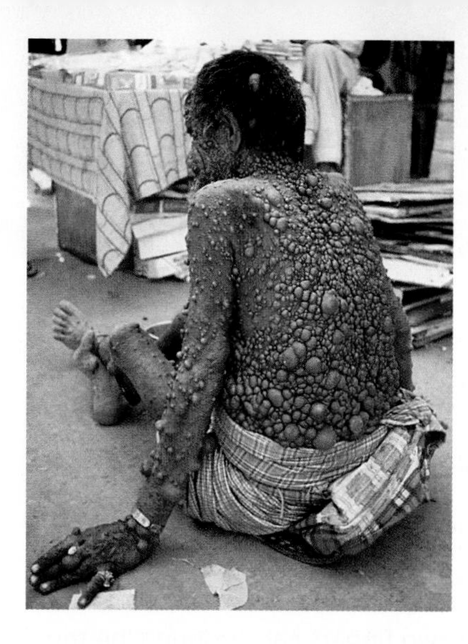

Figure 13.13 Soft skin tumors on an individual affected by the autosomal dominant disorder called neurofibromatosis.

2. *Neurofibromatosis* is a disorder caused by mutations in the *NF1* gene. Soft, fibrous tumors form in the skin and nervous system (Figure 13.13). Muscles, bones, and internal organs also show abnormalities.

NF1 is an autosomal dominant gene, so children that are affected by the disease usually have an affected parent. However, a boy developed the disorder in 1991 even though neither one of his parents did. Both copies of the boy's *NF1* gene were analyzed. The *NF1* gene on the chromosome he inherited from his father contained a transposable element, but neither his father nor his mother had a transposable element in any of their *NF1* genes. Explain the cause of neurofibromatosis in the boy and how it arose.

3. Using Figure 13.5, translate this nucleotide sequence into an amino acid sequence, starting at the first base:

5′—GGUUUCUUCAAGAGA—3′

4. Review Section 12.4. Suppose DNA polymerase made a wrong base pairing while a crucial gene region of DNA was being replicated. DNA repair mechanisms did not fix the mistake. Here is the part of the DNA with the error:

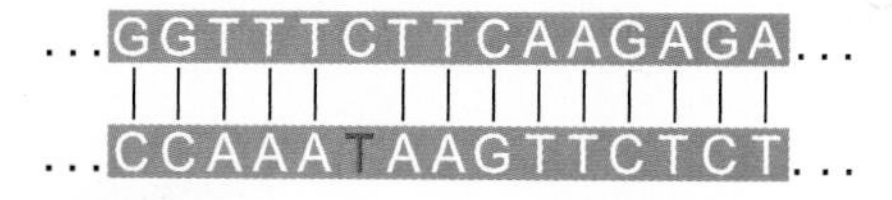

After the DNA molecule is replicated, the cell undergoes mitosis and two daughter cells form. One daughter cell dies and the other cell is normal. Develop a hypothesis to explain this observation.

5. Cigarette smoke is mostly carbon dioxide, nitrogen, and oxygen. The rest contains at least fifty-five different chemicals identified as carcinogenic, or cancer-causing, by the International Agency for Research on Cancer (IARC). When these carcinogens enter the bloodstream, enzymes convert them to a series of chemical intermediates that are easier to excrete. Some of the intermediates bind irreversibly to DNA. Propose one mechanism by which such binding can cause cancer.

6. Termination of prokaryotic DNA transcription often depends on the structure of a newly forming RNA. Transcription stops where the mRNA folds back on itself, forming a hairpin-looped structure such as the one at right. Why do you think this structure stops transcription?

14 CONTROLS OVER GENES

IMPACTS, ISSUES Between You and Eternity

You are in college, your whole life ahead of you. Your risk of developing cancer is as remote as old age, an abstract statistic that is easy to forget. "There is a moment when everything changes—when the width of two fingers can suddenly be the total distance between you and eternity." Robin Shoulla wrote those words after being diagnosed with breast cancer. She was seventeen. At an age when most young women are thinking about school, parties, and potential careers, Robin was dealing with radical mastectomy—the removal of a breast, all lymph nodes under the arm, and skeletal muscles in the chest wall under the breast. She was pleading with her oncologist not to use her jugular vein for chemotherapy and wondering if she would survive through the next year (Figure 14.1).

Robin's ordeal became part of a statistic—one of more than 200,000 new cases of breast cancer diagnosed in the United States each year. About 5,700 of those cases occur in women and men under thirty-four years of age.

Mutations in some genes predispose individuals to develop certain kinds of cancer. Tumor suppressor genes are so named because tumors are more likely to occur when these genes mutate. Two examples are *BRCA1* and *BRCA2*. A mutated version of one or both is often found in breast and ovarian cancer cells. If a *BRCA* gene mutates in one of three especially dangerous ways, a woman has an 80 percent chance of developing breast cancer before reaching age seventy.

Tumor suppressors are part of a system of stringent controls over gene expression that keeps the cells of multicelled organisms functioning normally. For example, BRCA proteins promote the transcription of genes that encode some of the DNA repair enzymes. Any mutations that alter this function also alter a cell's capacity to repair damaged DNA. Other mutations are likely to accumulate, and that sets the stage for cancer.

BRCA proteins also bind to receptors for the hormones estrogen and progesterone, which are abundant in breast and ovarian tissues. Binding regulates the transcription

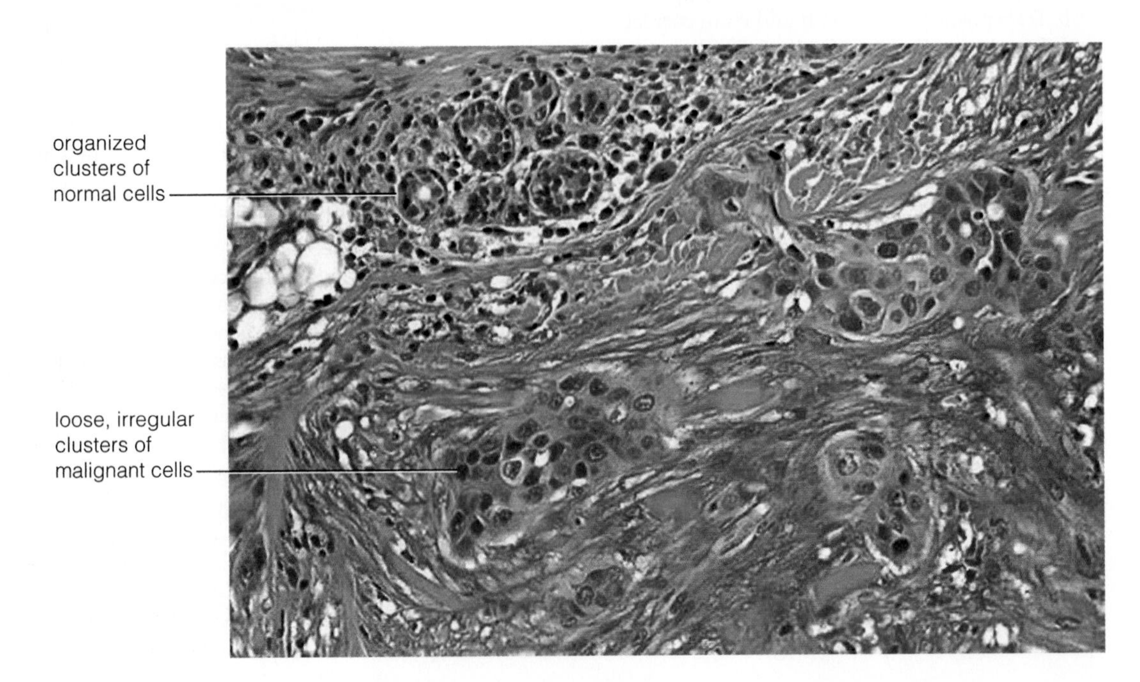

See the video! Figure 14.1 A case of breast cancer. This light micrograph revealed irregular clusters of carcinoma cells that infiltrated milk ducts in breast tissue. *Facing page*, Robin Shoulla. Diagnostic tests revealed abnormal cells such as these in her body.

How would you vote? Some females at high risk of developing breast cancer opt for prophylactic mastectomy, the surgical removal of one or both breasts even before cancer develops. Many never would have developed cancer. Should the surgery be restricted to cancer treatment? See ThomsonNOW for details, then vote online.

of growth factor genes. Among other things, growth factors stimulate cells to divide during normal, cyclic renewals of breast and ovarian tissues. When a mutation results in a BRCA protein that cannot bind to hormone receptors, the growth factors are overproduced. Cell division goes out of control, and tissue growth becomes disorganized. In other words, cancer develops.

Robin Shoulla survived. Although radical mastectomy is rarely performed today—a modified procedure is less disfiguring—it is the only option when cancer cells invade muscles under the breast. It was Robin's only option. She may never know which mutation caused her cancer. Now, sixteen years later, she has what she calls a normal life—career, husband, children. Her goal is to grow very old with gray hair and spreading hips, smiling.

Robin's story lends immediacy to the world of gene controls. By these molecular mechanisms, cells control when and how fast specific genes will be transcribed and translated, and whether gene products will be switched on or silenced. You will be considering the impact of such controls in chapters throughout the book—and in many chapters of your life.

Key Concepts

OVERVIEW OF THE CONTROLS

Control mechanisms govern when, how, and to what extent a cell's genes are expressed. They alter gene expression in response to changing conditions both inside and outside the cell.

Diverse controls govern every step between gene transcription and delivery of the final gene product to a targeted location.

In multicelled species, master genes guide the stage-by-stage development of new individuals. Selective gene expression results in cell differentiation. By this process, different cell lineages become specialized in composition, structure, and function. **Section 14.1**

EXAMPLES FROM EUKARYOTES

The orderly, localized expression of certain genes is the basis of the body plan of complex multicelled organisms. In female mammals, most of the genes on one of the two X chromosomes are inactivated in every cell. **Section 14.2**

CASE STUDY: FRUIT FLY DEVELOPMENT

Drosophila research revealed how a complex body plan emerges. All cells in a developing embryo inherited the same genes, but they selectively activate or suppress different fractions of those genes. **Section 14.3**

WHAT ABOUT THE PROKARYOTES?

Prokaryotic gene controls govern responses to short-term changes in nutrient availability and other aspects of the environment. The main gene controls bring about fast adjustments in rates of transcription. **Section 14.4**

Links to Earlier Concepts

In this chapter you will learn about some of the key mechanisms by which organisms respond to their environment and maintain homeostasis (Section 1.2). You will be applying your knowledge of chromosomal DNA organization (8.1), and post-transcriptional processing of mRNA (13.1). You also will revisit mechanisms of enzyme control (5.4). Your understanding of sex determination in humans (11.1) and the basis of autosomal recessive inheritance (11.2) will come in handy.

14.1 Gene Expression in Eukaryotic Cells

Many types of controls influence the kinds and amounts of substances in a cell in any given interval. Such controls are the basis for growth, differentiation, and homeostasis in all cells and multicelled organisms.

WHICH GENES GET TAPPED?

LINKS TO SECTIONS 1.2, 4.7, 7.3, 8.1, 13.1

All of the cells in your body are descended from the same fertilized egg, so they all contain the same DNA. All of them transcribe many of the same genes in that DNA, and they are alike in most aspects of structure and in common housekeeping activities.

In other ways, however, nearly all of your body cells are specialized in structure, composition, and function. Cell differentiation, the process by which cells become specialized, happens as different cell lineages begin to use different subsets of their genes. Which genes a cell uses determines the molecules it will produce, which in turn determines what kind of cell it will be.

For example, most of your body cells use genes that encode enzymes of glycolysis, but only immature red blood cells use the genes that encode hemoglobin. Your liver cells use genes for enzymes that neutralize toxins, but they are the only ones that do. While your eyes were forming, certain cells in them began to use genes to make crystallin. No other cells in your body use the genes for this protein, which makes up the transparent fibers of the lens in each eye.

A cell rarely uses more than 10 percent of its genes at once. Which genes it uses at any given time depends on many factors, such as conditions in the cytoplasm and extracellular fluid, and the type of cell. The factors affect gene expression—a multistep process by which cells convert information encoded in a gene into an RNA or protein product.

Controls govern all of the steps of gene expression, starting with transcription and continuing on through the delivery of an RNA or a protein product to its final destination. The controls start, enhance, slow, or stop gene expression. You will encounter specific examples of gene controls when you study how a body functions in its environment; that is, its physiology (Section 25.1 and Units V and VI). For now, just be familiar with the main types of controls, as summarized in Figure 14.2.

Control of Transcription Several controls affect attachment of RNA polymerase to DNA. For example, controls that keep RNA polymerase from attaching to a gene also prevent transcription of that gene. Other kinds of controls help RNA polymerase bind to DNA and thereby speed up transcription.

Certain base sequences in DNA are binding sites for molecules that affect transcription. For example, promoters are binding sites for RNA polymerase, as Section 13.1 explains. Enhancers are binding sites for other proteins that can speed transcription. Whether a gene is transcribed rapidly or slowly depends on what protein binds to these sites. When an activator binds to a promoter or enhancer, it speeds up transcription. When a repressor binds to a promoter, it hinders RNA

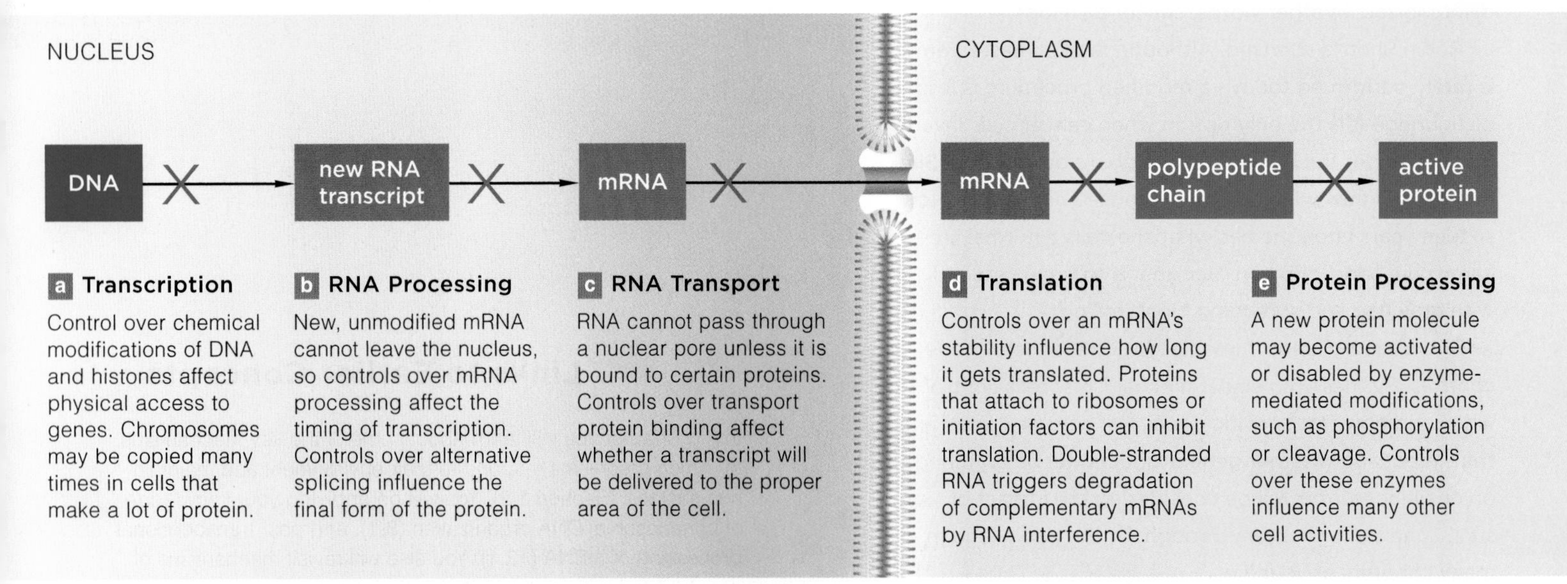

Figure 14.2 **Animated!** When controls over eukaryotic gene expression come into play.

polymerase binding, so it slows or stops transcription. Regulatory proteins such as activators and repressors are transcription factors.

Some factors that influence the interaction of DNA with histone proteins (Section 8.1) can also influence transcription. RNA polymerase cannot bind to DNA that is coiled tightly around histones, but it can bind when the coils loosen. Attachment of methyl groups (—CH_3) to DNA is a control that stops transcription (Figure 14.3*a*). As you will see, methylation underlies differences in expression that occur between alleles on homologous X chromosomes.

Controls even affect how a gene is transcribed. For instance, some gene sequences can be rearranged or multiplied. For example, in immature amphibian eggs and the gland cells of some insect larvae, chromosomes are copied repeatedly during interphase. The resulting *polytene chromosomes* consist of hundreds to thousands of side-by-side copies of genes. Polytene chromosomes allow cells to make copious amounts of mRNA, which is translated quickly into copious amounts of proteins (Figure 14.3*b*).

mRNA Processing and Transport As you know, before eukaryotic mRNAs leave the nucleus, they are modified—spliced, capped, and their poly-A tails are tailored (Section 13.1). Controls over the modifications affect the form of a protein product and when it will appear in the cell. For example, controls that determine which exons get spliced out of an mRNA affect which alternative form of a protein will be built.

Processing controls also affect mRNA transport. In eukaryotes, mRNA is transcribed in the nucleus and translated in the cytoplasm. However, a new transcript can pass through the pores of a nuclear envelope only after it has been processed appropriately. Controls that delay processing thus delay an mRNA's appearance in the cytoplasm for translation.

Controls also guide RNA to localized destinations. A short base sequence near an mRNA's poly-A tail is like a zip code. Depending on what the base sequence is, transport proteins deliver the mRNA to one place or another in the cell. Proteins that attach to a zip code region prevent an mRNA from being translated before it reaches its destination. mRNA localization allows a cell to move or grow in a certain direction, and it is the basis of pattern formation in embryonic development.

Translational Control Most of the controls over eukaryotic gene expression affect translation. Many of them govern the production or function of the different molecules that carry out translation. The stability of an mRNA is another type of translational control: The longer an mRNA lasts, the more protein is made from it. Enzymes in the cytoplasm can disassemble an mRNA within a few minutes; how fast it happens depends on the mRNA's base sequence, poly-A tail length, and attached proteins. Such fast turnover means that cells have the capacity to adjust protein synthesis quickly in response to their changing needs.

Certain kinds of RNAs also exert control over translation. For example, transcription of microRNAs can inhibit translation of other RNAs. A microRNA folds back on itself and forms a small double-stranded region. By the process of RNA interference, double-stranded RNA (such as microRNA) is cut up into small bits, which are taken up by special enzyme complexes. The complexes destroy all mRNA that is complementary in sequence to the bits.

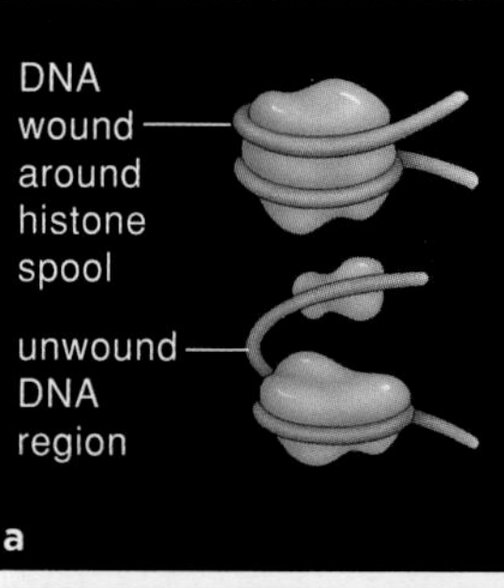

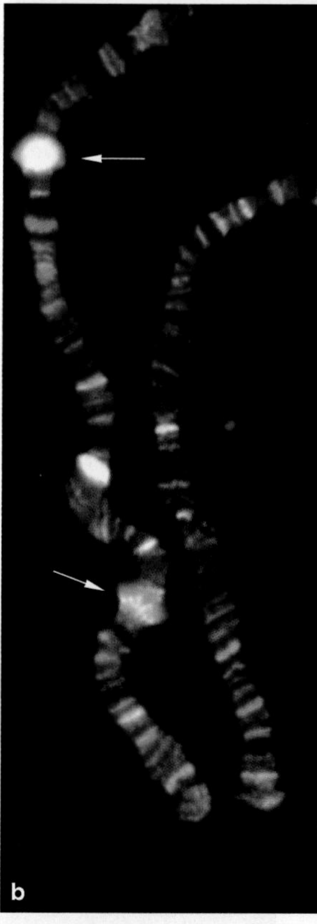

Figure 14.3 Two examples of gene controls. (**a**) Methylation of DNA causes enzymes to remove acetyl groups from histones, which in turn causes histones to wind the DNA tightly. RNA polymerase cannot bind to DNA wound tightly around histones. Enzymes that add acetyl groups to histones make them loosen their grip on DNA to be translated.

(**b**) *Drosophila* polytene chromosomes. To sustain a rapid growth rate, *Drosophila* larvae eat continuously and use a lot of saliva. In their salivary gland cells, giant polytene chromosomes form by repeated DNA replication. Each of these chromosomes consists of hundreds or thousands of the same DNA molecule, aligned side by side. Transcription is visible as puffs, where the DNA packing has loosened (*arrows*).

Controls After Translation Modifications after translation influence the expression of many proteins. For instance, some enzymes and other proteins become active only after enzymes attach a phosphate group to them. Such modifications activate, inhibit, or stabilize many different molecules, including the ones that are required for transcription and translation.

Most cells of multicelled organisms become specialized as they begin to use a unique subset of their genes.

Which genes a cell uses depends on the type of organism, its stage of development, and environmental conditions.

Gene expression is a multistep process by which a cell converts information encoded in a gene into a structural or functional part of a cell. Controls govern each step between gene and gene product.

14.2 A Few Outcomes of Gene Controls

The preceding section introduced an important idea. All differentiated cells in a complex, multicelled body carry the same genes, but each cell type uses a unique fraction of them. Such selective gene expression gives rise to certain traits. Consider some examples of controls that guide the selections during embryonic development.

X CHROMOSOME INACTIVATION

LINKS TO SECTIONS 9.5, 10.1, 11.1, 11.4, 11.6

Diploid cells have two sets of chromosomes, one from each parent. Among mammals, cells of female embryos have two X chromosomes, and one of them is always tightly condensed even in interphase (Figure 14.4*a*). We call condensed X chromosomes Barr bodies, after Murray Barr, who discovered them. The condensation is a transcriptional control that keeps RNA polymerase from accessing most of the genes on the chromosome. Such X chromosome inactivation ensures that only one of the two X chromosomes in a female's cells is active.

X chromosome inactivation occurs when an embryo is a ball of about 200 cells. In humans, cats, and many other mammals, it occurs independently in each cell of a female embryo: *Either* chromosome can be condensed. The maternal X chromosome may become inactivated in one cell; the paternal or maternal X chromosome may become inactivated in a cell next to it. Once the molecular selection is made in a cell, all of that cell's descendants make the same selection as they continue dividing to form tissues. What is the outcome? *A fully developed female has patches of tissue where genes of the maternal X chromosome are expressed and patches of tissue where genes of the paternal X chromosome are expressed.* She is a "mosaic" for expression of X-linked genes.

The homologous X chromosomes of most females have at least some alleles that are not identical. Thus, most females have variations in traits among patches of tissue. A female's mosaic tissues are visible if she is heterozygous for certain X chromosome mutations.

For example, incontinentia pigmenti is an X-linked disorder that affects the skin, teeth, nails, and hair. In heterozygous human females, mosaic tissues show up as lighter and darker patches of skin. The darker skin consists of cells in which the active X chromosome has the mutated allele; the lighter skin consists of cells in which the active X chromosome has the normal allele (Figure 14.4*c*).

Mosaic tissues are visible in other female mammals as well. For example, a gene on the X chromosomes of cats influences fur color. The expression of an allele (*O*) results in orange fur, and expression of another allele (*o*) results in black fur. Heterozygous cats (*Oo*) may be calico. They have patches of orange and black fur that arise from patches of cells in which allele *O* or *o* is on the active X chromosome (Figure 14.5).

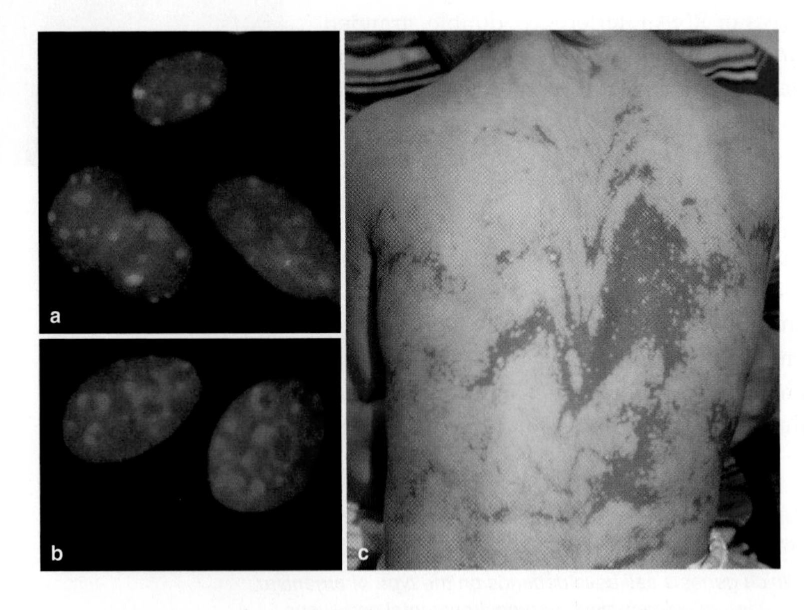

Figure 14.4 X chromosome inactivation. (**a**) Barr bodies (*red*) in the nucleus of four XX cells. (**b**) Compare the nucleus of two XY cells. (**c**) Mosaic tissues show up in human females who are heterozygous for mutations that cause incontinentia pigmenti. In darker patches of this girl's skin, the X chromosome with the mutation is active. In lighter skin, the X chromosome with the normal allele is active.

Figure 14.5 **Animated!** Why is this cat "calico"? When she was an embryo, one or the other X chromosome was inactivated in each of her cells. The descendants of the cells formed mosaic patches of tissue. Orange or black fur results from expression of different alleles on the active X chromosome. (White patches are the outcome of a different gene, the product of which blocks synthesis of all pigment.)

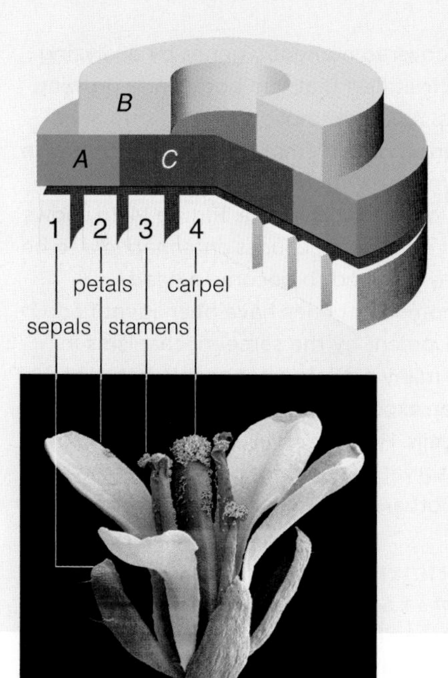

a The pattern in which the floral identity genes *A*, *B*, and *C* are expressed affects differentiation of cells growing in whorls in the plant's tips. Their gene products guide expression of other genes in cells of each whorl; a flower results.

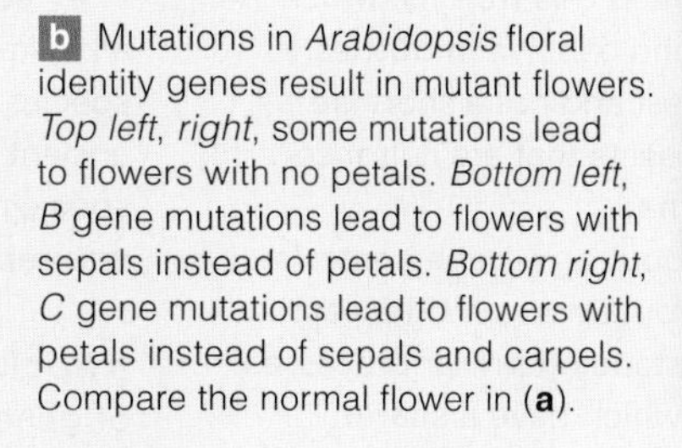

b Mutations in *Arabidopsis* floral identity genes result in mutant flowers. *Top left, right*, some mutations lead to flowers with no petals. *Bottom left*, *B* gene mutations lead to flowers with sepals instead of petals. *Bottom right*, *C* gene mutations lead to flowers with petals instead of sepals and carpels. Compare the normal flower in (**a**).

Figure 14.6 **Animated!** Control of flower formation, revealed by mutations in *Arabidopsis thaliana*.

According to the theory of dosage compensation, X chromosome inactivation equalizes the expression of X chromosome genes between the sexes. The body cells of male mammals (XY) have one set of X chromosome genes. Body cells of female mammals (XX) have two sets, but only one is expressed. Normal development of female embryos depends on this type of control.

How does just one of the two X chromosomes get inactivated? An X chromosome gene called *XIST* does the trick. This gene is transcribed on only one of the two X chromosomes. The gene's product, a large RNA, sticks to the chromosome that expresses the gene. The RNA coats the chromosome and causes it to condense into a Barr body. Thus, transcription of the *XIST* gene keeps the chromosome from transcribing other genes. Only the chromosome that expresses the *XIST* gene becomes inactivated; the other chromosome does not express the *XIST* gene, so it does not get coated with RNA. Its genes remain available for transcription.

GENE CONTROL OF FLOWER FORMATION

When it is time for a plant to flower, populations of cells that would otherwise give rise to leaves instead differentiate into floral parts—sepals, petals, stamens, and carpels. How does the switch happen? Studies of mutations in the common wall cress plant, *Arabidopsis thaliana*, support the ABC model. This model explains how the specialized parts of a flower develop. Three sets of master genes—*A*, *B*, and *C*—guide the process. Master genes encode products that affect expression of many other genes. The expression of a master gene initiates cascades of expression of other genes, with the outcome being the completion of an intricate task —such as the formation of a flower.

The master genes that control flower formation are switched on by environmental cues such as daylength, as you will see in Section 28.9. At the tip of a floral shoot (a modified stem), cells form whorls of tissue, one over the other like layers of an onion. Cells in each whorl give rise to different tissues depending on which of their *ABC* genes get activated. In the outer whorl, only the *A* genes are switched on, and their products trigger events that cause sepals to form. Cells in the next whorl express both *A* and *B* genes; they give rise to petals. Cells farther in express *B* and *C* genes; they give rise to male floral structures called stamens. The cells of the innermost whorl express only the *C* genes; they give rise to female floral structures called carpels (Figure 14.6*a*). Studies of the phenotypic effects of *ABC* gene mutations support this model (Figure 14.6*b*).

X chromosome inactivation is a gene control mechanism in eukaryotes. By the dosage compensation theory, most genes on one X chromosome in female mammals (XX) are inactivated, which balances gene expression with males (XY). The balance is vital for development of female embryos.

Gene control also guides flower formation. ABC master genes are expressed differently in tissues of floral shoots.

14.3 There's a Fly in My Research

As an embryo develops, the parts of its body take shape in places where we expect them to be. Researchers have correlated many of the patterns with expression of specific genes at particular times, in particular tissues. Tiny fruit flies yielded big clues to the connection.

fruit fly, actual size

For about a hundred years, *Drosophila melanogaster* has been the subject of choice for many research experiments. Why? It costs almost nothing to feed this fruit fly, which is only about 3 millimeters long and can live in bottles. *D. melanogaster* also reproduces fast and has a short life cycle. As well, experimenting on insects that are nuisance pests presents few ethical dilemmas.

Many important discoveries about how gene controls guide development have come from *Drosophila* research. The discoveries are clues to understanding similar processes in humans and other organisms, which have a shared evolutionary history.

Discovery of Homeotic Genes We now know of about 13,767 genes on *Drosophila's* four chromosomes. As in most other eukaryotic species, some are homeotic genes: master genes that control formation of specific body parts (eyes, legs, segments, and so on) during the development of embryos. All homeotic genes encode transcription factors with a homeodomain, a region of about sixty amino acids that can bind to a promoter or some other sequence in DNA.

Localized expression of homeotic genes in tissues of a developing embryo gives rise to details of the adult body plan. The body begins to develop as different master genes are expressed in different areas of the early embryo. The products of these master genes—transcription factors that can turn homeotic genes on or off—form in concentration gradients that span the entire embryo. Depending on where they are located within the gradients, embryonic cells begin to transcribe different homeotic genes. Products of these homeotic genes form in specific areas of the embryo. The different products cause cells to differentiate into tissues that will form specific structures such as wings or a head.

Researchers discovered homeotic genes by analyzing the DNA of mutant fruit flies that had body parts growing in the wrong places. As an example, the homeotic gene *antennapedia* is transcribed in embryonic tissues that give rise to a thorax, complete with legs. Normally, it is never transcribed in cells of any other tissue. Figure 14.7*b* shows what happens after a mutation causes *antennapedia* to be transcribed in tissue destined to become a head.

More than 100 homeotic genes have been identified. They control development by the same mechanisms in all eukaryotes, and many are interchangeable between species. Thus, we can expect that they evolved in the most ancient eukaryotic cells. Homeodomains often differ among species only in conservative substitutions—one amino acid was replaced by another with similar chemical properties.

Knockout Experiments By controlling expression of genes in *Drosophila* one at a time, researchers have made other important discoveries about how embryos of many organisms develop. In knockout experiments, researchers mutate a gene in a way that prevents its transcription or translation. Then they observe how an organism that carries the mutation differs from normal individuals. Differences are clues to the function of the missing gene product.

Researchers tend to name homeotic genes based on what happens in their absence. For instance, flies that have had their *eyeless* gene knocked out develop with no eyes. *Dunce* is required for learning and memory. The *wingless*, *wrinkled*, and *minibrain* genes are self-explanatory. *Tinman* is necessary for development of the heart, and flies with a mutated *groucho* gene have too many bristles above their eyes. Figure 14.7 shows examples of mutant *Drosophila*.

Humans, squids, mice, and many other animals have a homologue of the *eyeless* gene called *PAX6*. In humans, mutations in *PAX6* cause eye disorders such as aniridia—underdeveloped or missing irises. Altered expression of the *eyeless* gene causes eyes to form not only on a fruit fly's head but also on its wings and legs (Figure 14.7*c*). *PAX6* works the same way in frogs—it causes eyes to form wherever it is expressed in tadpoles.

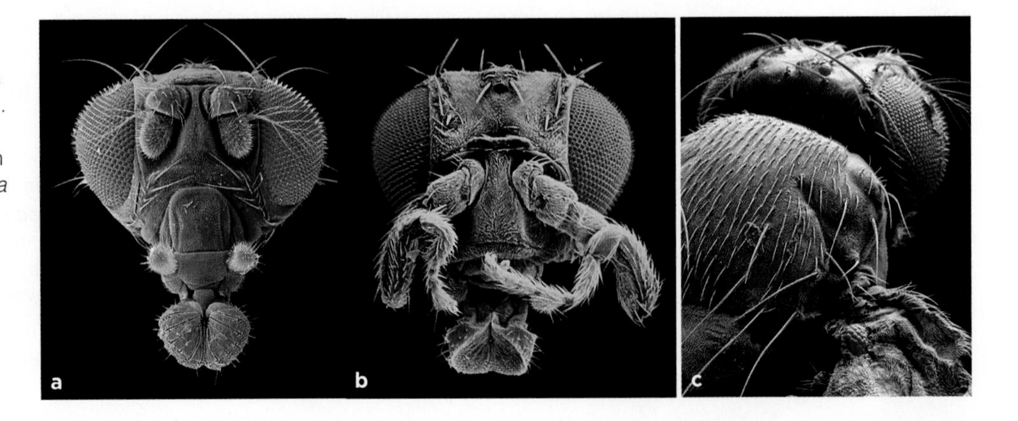

Figure 14.7 Experimental evidence of controls over embryonic development. (**a**) Normal fly head. (**b**) Transcription of the *antennapedia* gene in the embryonic tissues of the thorax causes legs to form on the body. A mutation that causes *antennapedia* to be transcribed in the embryonic tissues of the head causes legs to form there too.

(**c**) A mutation that causes the *eyeless* gene to be transcribed in the wrong tissues causes eyes to form on the wrong body parts—here, on a wing.

(**d**) *Facing page*, more *Drosophila* mutations that yielded clues to homeotic gene function.

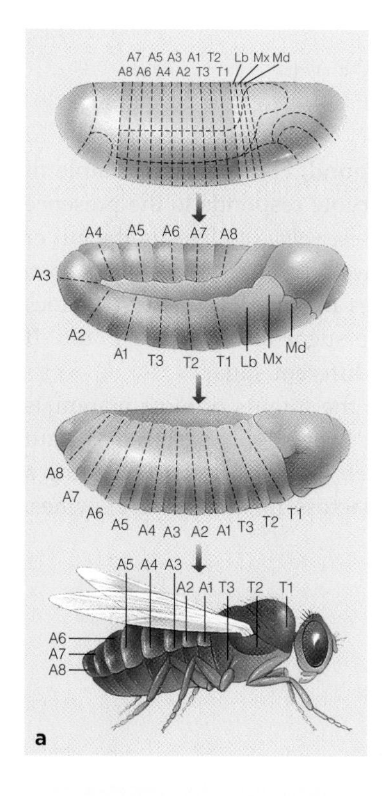

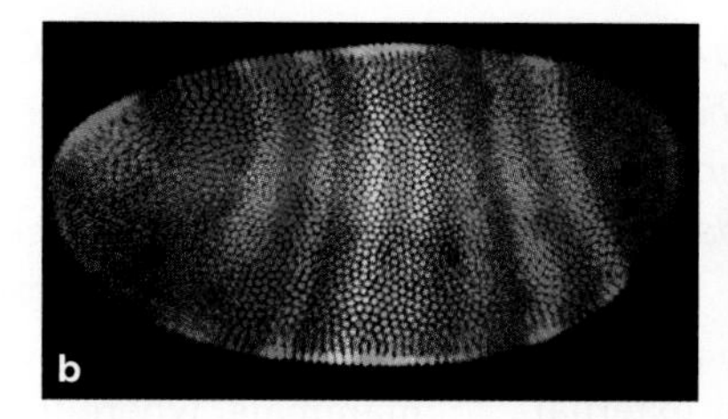

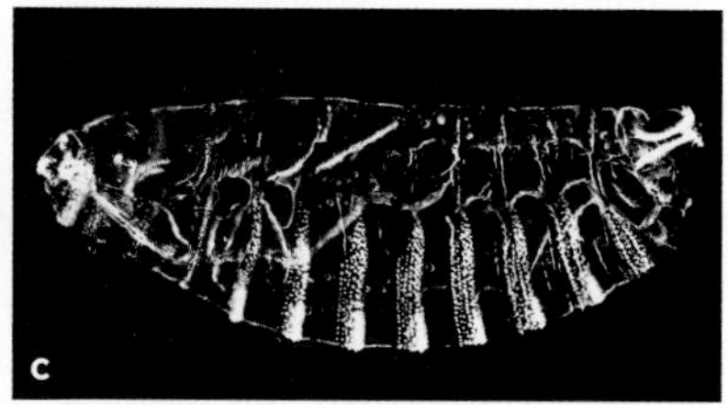

Figure 14.8 Cascades of gene expression that result in the segmented body plan of *Drosophila*. (**a**) Fate map for the surface of a *Drosophila* zygote shows where descendants of each cell end up in the adult. The pattern starts as maternal mRNAs are delivered to opposite ends of an unfertilized egg as it forms (Section 14.1). This polar distribution of maternal mRNA will dictate the future body axis.

The localized maternal mRNAs are translated right after fertilization. Their protein products diffuse away in gradients that span the entire embryo. Depending on where they fall within those gradients, cells of the embryo translate master genes called *gap* genes. Overlapping gradients of the *gap* gene products then form.

Gap gene products that accumulate in various tissues of the embryo influence translation of different master genes called *pair-rule* genes. (**b**) Here, two *gap* gene products that suppress the expression of other genes are shown in *green* and *blue*. Both constrain the expression of a pair-rule gene (*red*) to certain areas of the developing embryo. Note its expression in seven *red* stripes. (**c**) One day later, seven segments develop that correspond to the position of the stripes.

Researchers also discovered that *PAX6* is one of the homeotic genes that works across different species. If *PAX6* from a human, mouse, or squid is inserted into an *eyeless* mutant fly, it has the same effect as the *eyeless* gene: An eye forms wherever it is expressed. Such studies are evidence of a shared ancestor among evolutionarily distant animals.

Filling In Details of Body Plans Let's take stock. As an embryo develops, cells divide and differentiate in different body regions. They migrate or stick to other cells, and thus form tissues. These events fill in details of the body, and all are driven by cascades of expression of master genes. The genes are transcribed only in specific tissues at specific stages of development. Such regional gene expression during development results in a dynamic three-dimensional map of an embryo. The map consists entirely of overlapping concentration gradients of the products of master genes, and it spans the developing embryo at all times. Which master genes are active at a given time changes, and so does the map.

Some master gene products cause undifferentiated cells to differentiate; specialized tissues are the outcome. Which kinds of tissues form depends on where particular embryonic cell lineages fall on the map at any given time. The formation of body segments in a fruit fly embryo is an example of how spatial mapping works. In fruit flies, segmentation is orchestrated by master genes that are called *gap genes* and *pair-rule* genes (Figure 14.8).

Pattern formation is the process by which a complex body forms from local processes in an embryo. Section 39.4 returns to the topic of pattern formation in animals.

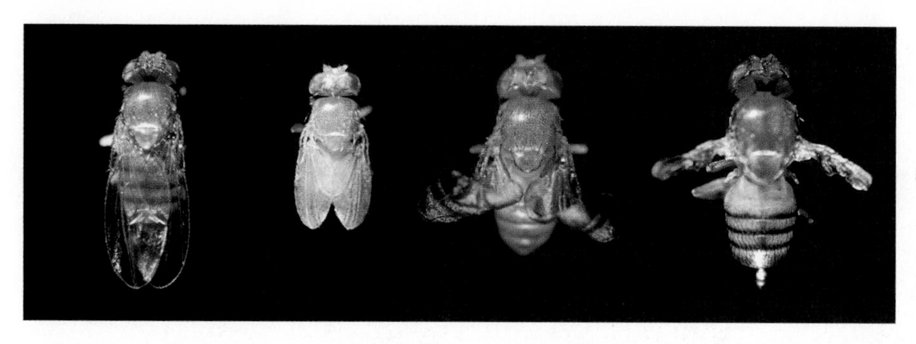

d *Left to right,* normal fly; yellow miniature; curly wings; vestigial wings; and a double thorax.

14.4 Prokaryotic Gene Control

In prokaryotic cells, controls deal mainly with slowing and increasing transcription as rapid responses to short-term shifts in environmental conditions. Long-term controls are not required, because no prokaryote gradually develops into a multicelled form.

LINKS TO SECTIONS 3.2, 5.4, 7.2

When nutrients are plentiful and when other external conditions are favorable for growth and reproduction, prokaryotic cells rapidly transcribe genes that specify the enzymes for nutrient absorption and other growth-related tasks. Genes used at the same time often occur one after the other on the chromosome. All of them are transcribed together into a single RNA, so their transcription is controllable in one step (Table 14.1).

NEGATIVE CONTROL OF THE LACTOSE OPERON

With this bit of background, consider an example of how one kind of prokaryote responds to the presence or absence of lactose. *Escherichia coli* lives in the gut of mammals, where it dines on nutrients traveling past. Milk typically nourishes mammalian infants. It does not contain glucose, the sugar of choice for *E. coli*. It does contain lactose, a different sugar.

After being weaned, the infants of most mammals drink little (if any) milk. Even so, *E. coli* cells in the gut can still use lactose when it shows up by activating a set of three genes for lactose-metabolizing enzymes.

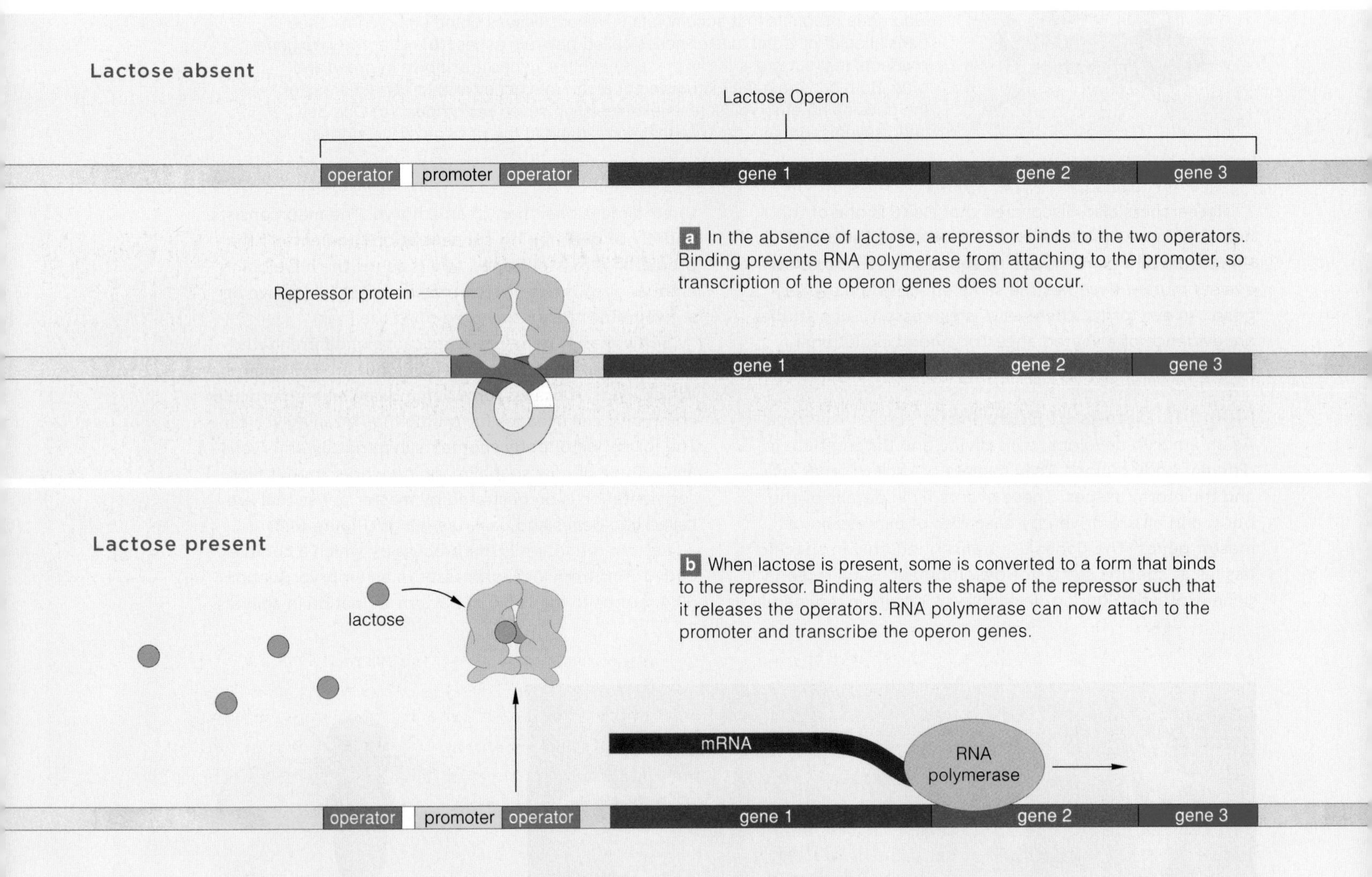

Figure 14.9 **Animated!** Negative control of the lactose operon on a bacterial chromosome. The operon consists of a promoter flanked by two operators, and three genes for lactose-metabolizing enzymes.

In *E. coli* DNA, a promoter precedes all three genes. Two operators flank the promoter. Each operator is a DNA sequence that is a binding site for a repressor; binding stops transcription. An arrangement in which a single promoter and one or more operators control access to multiple genes is called an operon.

In the absence of lactose, a repressor molecule binds to both operators. Binding causes the DNA region with the promoter to twist into a loop, as in Figure 14.9*a*. RNA polymerase cannot bind to the promoter when it is looped up, so it cannot transcribe the operon genes. Thus, the operon genes are not used when they are not required (Section 5.4).

When lactose *is* in the gut, *E. coli* converts some of it to allolactose. This sugar binds to the repressor and changes its shape. The altered repressor can no longer bind to the operators. The looped DNA unwinds, RNA polymerase attaches to the promoter, the operon genes are transcribed, and lactose-degrading enzymes are produced (Figure 14.9*b*).

POSITIVE CONTROL OF THE LACTOSE OPERON

E. coli cells pay far more attention to glucose than to lactose. Even when lactose is present, they do not use their lactose operon much—unless there is no glucose. Then, an activator called CAP (or catabolite activator protein) makes the lactose operon promoter far more inviting to RNA polymerase. However, CAP will not issue the invitation until it has already become bound to a chemical messenger called cAMP (cyclic adenosine monophosphate). A CAP–cAMP complex helps RNA polymerase bind to the promoter. By doing so, it speeds transcription of the operon genes about tenfold.

When glucose is plentiful, ATP forms by glycolysis (Section 7.2), but synthesis of an enzyme necessary to synthesize cAMP is blocked. The blocking ends when glucose is scarce and lactose becomes available. cAMP accumulates, the CAP–cAMP complexes form, and the operon genes are transcribed quickly. Thus, the gene products are produced quickly as well. They convert lactose to glucose, the preferred sugar of *E. coli*.

LACTOSE INTOLERANCE

Unlike *E. coli*, many humans are lactose intolerant. Lactase is secreted into the small intestine by cells in the intestinal lining. When there is not enough of this lactose-digesting enzyme, lactose in food cannot be broken down, and it accumulates in the colon, or large intestine. There, huge populations of resident bacteria use lactose for various fermentation reactions. Carbon dioxide, methane, and other gaseous products of these reactions accumulate quickly in the colon, distending its wall and causing pain. Undigested carbohydrates disrupt the solute–water balance inside the colon and so invite severe diarrhea.

Table 14.1 Prokaryotic Versus Eukaryotic Gene Controls

Prokaryotic Gene Controls

1. Control mechanisms adjust enzyme-mediated reactions in response to short-term changes in nutrient availability or other environmental factors.
2. Operons permit control of the expression of more than one gene at a time.
3. Transcriptional controls are reversibly inhibited in response to conditions in the environment.
4. Transcription and translation occur simultaneously; prokaryotic RNAs have no introns and are not processed.

Eukaryotic Gene Controls

1. Some control mechanisms adjust enzyme-mediated reactions in response to short-term changes in conditions inside and outside the cell.
2. In multicelled species, products of master genes activate other genes at different times, in different tissues. They induce generally irreversible events that are part of a long-term program of growth and development.
3. Diverse controls regulate every step of gene expression, starting with transcription and ending with the targeted delivery of its final product.

Prokaryotic cells are small, fast reproducers, so they do not require extensive controls over long-term development. The main controls guide transcription of enzyme-coding genes in response to short-term shifts in nutrient availability and other outside conditions.

www.thomsonedu.com Thomson NOW!

Summary

Section 14.1 Gene expression is a multistep process by which a cell converts information encoded in a gene into a gene product. Which genes a cell uses depends on the type of organism, the type of cell, factors inside and outside the cell, and, in complex multicelled species, the organism's stage of development.

Gene controls are the basis of embryonic development in multicelled eukaryotes. All cells of an embryo share the same genes, but different lineages use different subsets of them during development. The outcome of selective gene expression is cell differentiation, by which different cell lineages become unique in composition, structure, and function. Differentiated cells give rise to specialized tissues and organs.

Different controls over transcription, RNA processing and transport, translation, and post-translational protein processing collectively govern whether, when, and how a gene is expressed.

■ *Use the animation on ThomsonNOW to review the control points for gene expression.*

Section 14.2 In female mammals, most genes on one of the two X chromosomes are permanently inaccessible. This X chromosome inactivation balances gene expression between the sexes. It arises because the *XIST* gene gets transcribed on only one of the two X chromosomes. The gene's RNA product shuts down the chromosome that transcribes it. Which chromosome becomes inactivated in any cell is random.

Studies of mutations in *Arabidopsis thaliana* support an ABC model for flower formation. Three sets of master genes (*A*, *B*, *C*) guide cell differentiation in the whorls of a floral shoot; sepals, petals, stamens, and carpels form.

■ *Use the animation on ThomsonNOW to see how controls over gene expression affect eukaryotic development.*

Section 14.3 Experimental mutations in *Drosophila melanogaster* revealed controls over gene expression that govern the embryonic development of all animals. As the embryos of these fruit flies and other complex eukaryotes develop, master genes are transcribed in different tissues at different times. The gene products affect expression of other master genes, which affect the expression of others, and so on. These cascades of master gene products form overlapping concentration gradients. The gradients are a dynamic spatial map that spans the entire embryo body. Cells differentiate according to their location on the map. As their descendants form particular tissues and organs, they fill in the details of the body plan.

Section 14.4 Prokaryotic cells do not have great structural complexity and do not undergo development. Most of their gene controls reversibly adjust transcription rates in response to environmental conditions, especially nutrient availability.

Bacterial operons are examples of prokaryotic gene controls. The lactose operon governs expression of three genes, the three products of which digest lactose. Two operators that flank the promoter are binding sites for a repressor that blocks transcription.

■ *Use the animation on ThomsonNOW to explore the structure and function of the lactose operon.*

Self-Quiz

Answers in Appendix III

1. The expression of a given gene depends on ______ .
   a. the type of organism — c. the type of cell
   b. environmental conditions — d. all of the above

2. Gene expression in multicelled eukaryotes changes in response to ______ .
   a. external conditions — c. master genes
   b. operons — d. a and c

3. At ______ in DNA, regulatory protein binding can increase transcription of specific genes.
   a. promoters — c. operators
   b. enhancers — d. both a and b

4. Eukaryotic gene controls govern ______ .
   a. transcription — e. translation
   b. RNA processing — f. protein modification
   c. RNA transport — g. a through e
   d. mRNA degradation — h. all of the above

5. Eukaryotic gene expression controls guide ______ .
   a. natural selection — c. development
   b. nutrient availability — d. all of the above

6. Cell differentiation ______ .
   a. occurs in all complex multicelled organisms
   b. requires unique genes in different cells
   c. involves selective gene expression
   d. both a and c
   e. all of the above

7. During X chromosome inactivation ______ .
   a. female cells shut down — c. pigments form
   b. RNA coats chromosomes — d. both a and b

8. A cell with a Barr body is ______ .
   a. prokaryotic — c. from a female mammal
   b. from a male mammal — d. infected by Barr virus

9. Homeotic gene products ______ .
   a. are binding sites that flank a bacterial operon
   b. map out the overall body plan in embryos
   c. control the formation of specific body parts

10. Knockout experiments ______ genes.
    a. delete — c. express
    b. inactivate — d. either a or b

11. A(n) ______ is a promoter and a set of operators that control access to two or more prokaryotic genes.
    a. lactose molecule — c. dosage compensator
    b. operon — d. both b and c

12. Match the terms with the most suitable description.

___ ABC genes	a. a big RNA is its product
___ *XIST* gene	b. binding site for repressor
___ operator	c. cells become specialized in composition, function, etc.
___ Barr body	d. inactivated X chromosome
___ process of cell differentiation	e. guide flower development
___ methylation	f. —CH_3 additions to DNA

■ *Visit ThomsonNOW for additional questions.*

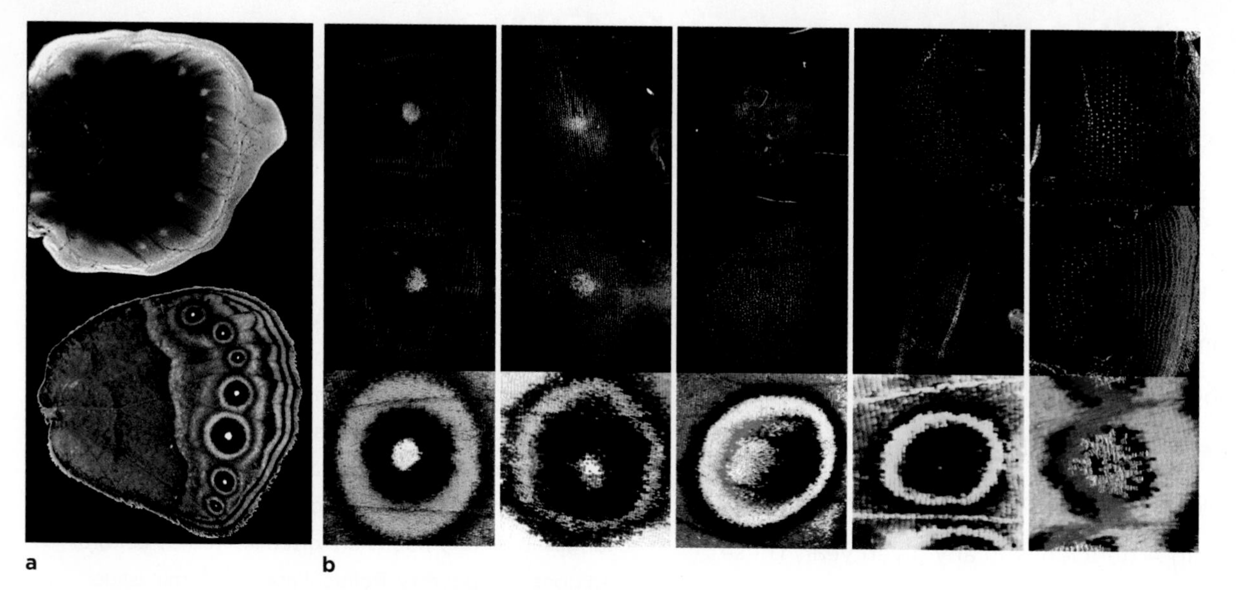

Figure 14.10 Examples of how homeotic gene expression influences the development of body details. (**a**) *Top*, seven *green* spots in the embryonic wing of a moth show expression of a homeotic gene that results in seven "eyespots" in the wing of the adult (*bottom*). (**b**) Expression of the same four homeotic genes in embryonic wings of five species of moth is shown by fluorescent *purple, green, red*, and *yellow* (*top* and *middle* rows). The diverse eyespot patterns on the adult moth wings (*bottom* row) result from differences in the patterns of expression of these and other homeotic genes.

Critical Thinking

1. Do all transcriptional controls operate in prokaryotic cells as well as eukaryotic cells? Why or why not?

2. Unlike most rodents, guinea pigs are well developed at the time of birth. Within a few days, they can eat grass, vegetables, and other plant material.

Suppose a breeder decides to separate baby guinea pigs from their mothers three weeks after they were born. He wants to raise the males and the females in different cages. However, he has trouble identifying the sex of young guinea pigs. Suggest how a quick look through a microscope can help him identify the females.

3. Calico cats are almost always female. Male calico cats are rare, and usually they are sterile. Why?

4. Small changes in master genes brought about the structural changes in the mutant *Arabidopsis thaliana* flowers in Figure 14.6. Would you predict that such changes figured in the evolution of more than 295,000 kinds of plants, each with distinctive flowers?

Reflect on Figure 14.10 and the *Drosophila* mutants shown in Figure 14.7. Then formulate a hypothesis about how homeotic gene mutations figured in the evolution of the more than 1.5 million known species of animals.

5. Geraldo isolated an *E. coli* strain in which a mutation has hampered the capacity of CAP to bind to a region of the lactose operon, as it would do normally. How will this mutation affect transcription of the lactose operon when the *E. coli* cells are exposed to the following conditions? Briefly state your answers:

a. Lactose and glucose are both available.

b. Lactose is available but glucose is not.

c. Both lactose and glucose are absent.

6. Duchenne muscular dystrophy, a genetic disorder, affects boys almost exclusively. Muscles begin to atrophy (waste away) in affected children, who typically die in their teens or early twenties (Section 11.4).

Muscle biopsies of a few women who carry an allele that is associated with the disorder identified some body regions of atrophied muscle tissue. They also showed that muscles adjacent to a region of atrophy were normal or even larger and more chemically active, as if they were compensating for the weakness of the adjoining region.

Form a hypothesis about the genetic basis of Duchenne muscular dystrophy that includes an explanation of why the symptoms might appear in some body regions but not others.

7. The mechanism by which *XIST* RNA localizes to one of the two X chromosomes in a mammalian cell is only partially understood. Two groups of researchers, one at the Dana-Farber Cancer Institute at Harvard, the other at the University of Milan, recently found that *XIST* localization is abnormal in breast cancer cells. In those cells, both X chromosomes are active.

We can sense intuitively that having two active X chromosomes in cells might have something to do with the abnormal gene expression of breast and ovarian cancer cells, but why unmutated *XIST* RNA does not localize properly in such cells remains a mystery.

Mutations in the *BRCA1* gene may be part of the answer. Remember from the introduction, a mutated *BRCA1* or *BRCA2* gene is often found in breast and ovarian cancer cells. The Harvard researchers found that the *BRCA1* protein physically associates with *XIST* RNA. They were able to restore proper *XIST* localization—and proper X chromosome inactivation—by rescuing *BRCA1* function in breast cancer cells.

Suggest another experiment that might demonstrate a connection between *BRCA1* function and *XIST*.

15 STUDYING AND MANIPULATING GENOMES

IMPACTS, ISSUES Golden Rice, or Frankenfood?

Vitamin A is necessary for good vision, growth, and immune system function. A small child can get enough of it just by eating a carrot every few days, yet about 140 million children under the age of six suffer from serious health problems due to vitamin A deficiency in any given year. These children do not grow as they should, and they succumb easily to infection. Between 250,000 and 500,000 become blind every year, and half of them die within a year of losing their sight.

It is no coincidence that populations with the highest incidence of vitamin A deficiency also are the poorest. Most people in such populations subsist mainly on rice. They tend to eat few animal products, vegetables, or fruits—all foods that are rich in vitamin A. Correcting and preventing vitamin A deficiency can be as simple as supplementing the diet with these foods, but changes in dietary habits are often limited by cultural traditions and poverty. Political and economic issues hamper long-term vitamin supplementation programs.

Geneticists Ingo Potrykus and Peter Beyer wanted to help. As they knew, beta-carotene, a yellow pigment in plant leaves, is a precursor for vitamin A. They transferred the genes for two enzymes, one from corn and one from bacteria, into rice plants. The rice plants transcribed the genes and began to make beta-carotene in their seeds—in the grains of gold-tinted rice (Figure 15.1). One cup of Golden Rice has enough beta-carotene to satisfy a child's daily recommended amount of vitamin A.

Why rice? Rice is the dietary staple for 3 billion people in impoverished countries around the world. Economies, traditions, and cuisines are based on growing and eating rice. Therefore, growing and eating rice that happens to contain enough vitamin A to prevent disease would be compatible with prevailing methods of agriculture and traditional dietary preferences.

See the video! **Figure 15.1** Making food crops better. (**a**) Vitamin A deficiency is common in Southeast Asia and other regions where people subsist mainly on rice. (**b**) Rice with artificially inserted genes make and store beta-carotene in their seeds, or rice grains. The grains of Golden Rice may help prevent vitamin A deficiency in developing countries. *Facing page*, a big kernel from a modern strain of corn next to tiny kernels of an ancestral corn species discovered in a prehistoric cave in Mexico.

How would you vote? All packaged food in the United States must have a nutrition label, but there is no requirement that genetically modified foods be labeled as such. Should food distributors be required to identify products made from genetically modified plants or livestock? See ThomsonNOW for details, then vote online.

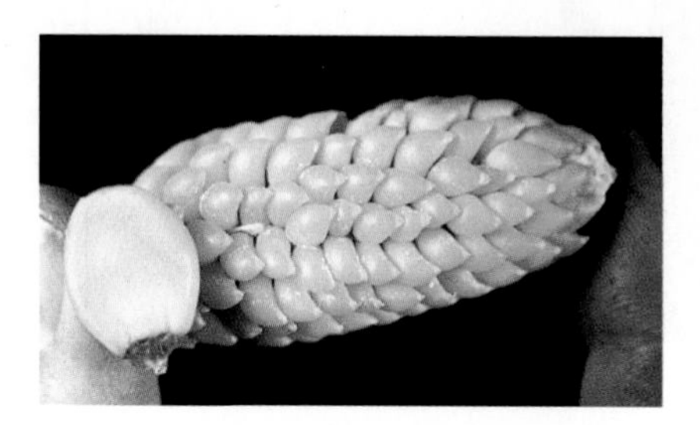

However, getting rice grains to make beta-carotene was beyond the scope of conventional methods of plant breeding. Even the best gardener could not induce rice plants to breed with corn plants. Potrykus and Beyer used recombinant DNA techniques, so Golden Rice is a genetically modified organism (GMO). Like many other GMOs, Golden Rice is transgenic—it carries genes from a different species. GMOs are made in laboratories, not on farms, but they are an extension of artificial selection and cross-breeding practices used for many thousands of years to coax new plants and new breeds of animals from wild ancestral stocks.

No one wants children to suffer or die. However, many people oppose the idea of any GMO. Some are unsettled because the pace of our genetic tinkering has picked up, hugely. Some worry because we cannot predict the long-term impact of GMOs that escape into the environment.

Should we be more cautious? Two people discovered a way to keep millions of children from suffering. How much of a risk should we take to help those children?

Take stock of how far you have come in this unit. You started with cell division mechanisms that allow parents to pass on DNA to new generations. You moved to the chromosomal and molecular basis of inheritance, then on to the gene controls that guide life's continuity. The sequence parallels the history of genetics. Now you have arrived at a point in time when geneticists hold molecular keys to the kingdom of inheritance. As you will see, what they are unlocking is already having an impact on life in the biosphere.

Key Concepts

MAKING RECOMBINANT DNA

Researchers routinely make recombinant DNA by isolating, cutting, and joining DNA from different species. Plasmids and other vectors can carry foreign DNA into host cells. **Section 15.1**

ISOLATING AND AMPLIFYING DNA FRAGMENTS

Researchers isolate and make many copies of DNA in order to study it. PCR copies particular fragments of DNA in sufficient quantity for experiments and for other practical purposes. **Section 15.2**

DECIPHERING DNA FRAGMENTS

Sequencing reveals the linear order of nucleotides in a fragment of DNA. A DNA fingerprint is an individual's unique array of DNA sequences. **Sections 15.3, 15.4**

MAPPING AND ANALYZING WHOLE GENOMES

Genomics is the study of genomes. Comparisons of the genomes of different species offer practical benefits. **Section 15.5**

USING THE NEW TECHNOLOGIES

Genetic engineering—the directed modification of an organism's genes—is now used routinely in research and medical applications. It continues to raise many ethical questions. **Sections 15.6–15.10**

Links to Earlier Concepts

This chapter builds on earlier explanations of DNA's molecular structure (Sections 3.6, 12.2), and replication (12.4). You may wish to review transcription (13.1) and the controls that govern it (14.1), as well as mRNA codons (13.2) and the effects of mutation (13.5). You will see an example of how researchers use bioluminescence (5.10) as a tracer (2.2), as well as another application for knockout experiments (14.3).

15.1 A Molecular Toolkit

Analysis of genes starts with manipulation of DNA. Researchers use molecular tools to cut up DNA from different sources and splice the fragments together.

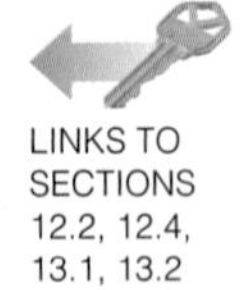

LINKS TO SECTIONS 12.2, 12.4, 13.1, 13.2

CUT AND PASTE

In the 1950s, excitement over the discovery of DNA's structure gave way to frustration. No one could figure out how to identify the sequential order of bases in DNA; the molecule was just too big. Identifying one base among millions of others was a huge technical challenge. Researchers had to divide a DNA molecule into manageable, identifiable pieces before they could determine its base sequence. But how?

Then, in 1970, Hamilton Smith and his colleagues noticed how cells of *Haemophilus influenzae*, a bacterial species, protected themselves against viral infection. Enzymes in the cells were chopping up invading viral DNA before it slipped into the bacterial chromosome. The chopping was not random; the bacterial enzymes were cutting viral DNA at specific base sequences.

In time, hundreds of strains of bacteria and a few eukaryotic cells yielded thousands of other enzymes like the kinds Smith discovered. Each is a restriction enzyme: It cuts any double-stranded DNA wherever a specific base sequence occurs. As one example, the enzyme *Eco*RI cuts between G and A in the sequence GAATTC. Like many other restriction enzymes, *Eco*RI makes staggered cuts that leave single-stranded tails, or "sticky ends," on the cut fragments (Figure 15.2*a*,*b*). Each tail can base-pair with the tail of any other DNA fragment cut by the same enzyme, because the sticky ends match (Figure 15.2*c*).

Remember DNA ligase (Section 12.4)? Researchers use it to form covalent bonds between matching sticky ends of different fragments. Recombinant DNA is the result (Figure 15.2*d*). Recombinant DNA is a molecule composed of DNA from two or more organisms.

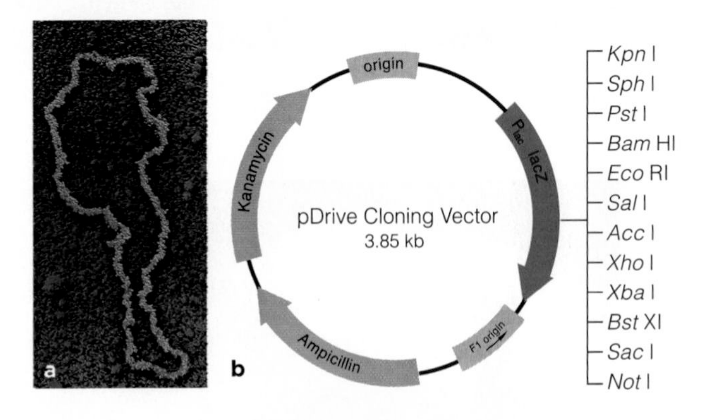

Figure 15.3 Cloning vectors. (**a**) Micrograph of a plasmid. (**b**) A commercial cloning vector. Restriction enzyme cutting sites useful for cloning are listed at *right*. Antibiotic resistance genes (*purple*) and the lactose operon (*red*) help researchers identify host cells that take up a vector with inserted DNA.

CLONING VECTORS

Bacterial cells have one circular chromosome. Many also have plasmids: small circles of DNA with just a few genes (Figure 15.3*a*). Bacteria usually can survive without plasmids, but the extra genes come in handy at times, as when they confer resistance to antibiotics.

Before a bacterium divides, its replication enzymes copy the chromosome *and* any plasmids, so that every daughter cell gets one of each. A plasmid gets copied and distributed to daughter cells even when it holds a fragment of foreign DNA. That is why bacteria can be used in DNA cloning, a set of laboratory procedures that uses living cells to make many identical copies of DNA. Researchers clone DNA to get enough material for experiments.

Like other cloning vectors, plasmids can be used to carry foreign DNA into host cells (Figure 15.3*b*). Extra restriction enzyme sites and bacterial genes are often added to vectors. Researchers insert foreign DNA into the restriction sites, and the extra bacterial genes help

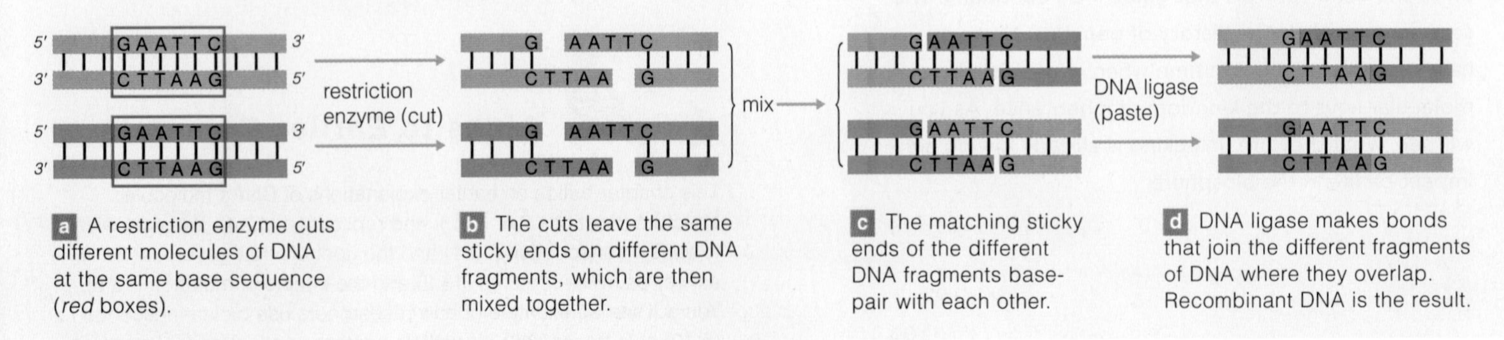

a A restriction enzyme cuts different molecules of DNA at the same base sequence (*red* boxes).

b The cuts leave the same sticky ends on different DNA fragments, which are then mixed together.

c The matching sticky ends of the different DNA fragments base-pair with each other.

d DNA ligase makes bonds that join the different fragments of DNA where they overlap. Recombinant DNA is the result.

Figure 15.2 Making recombinant DNA.

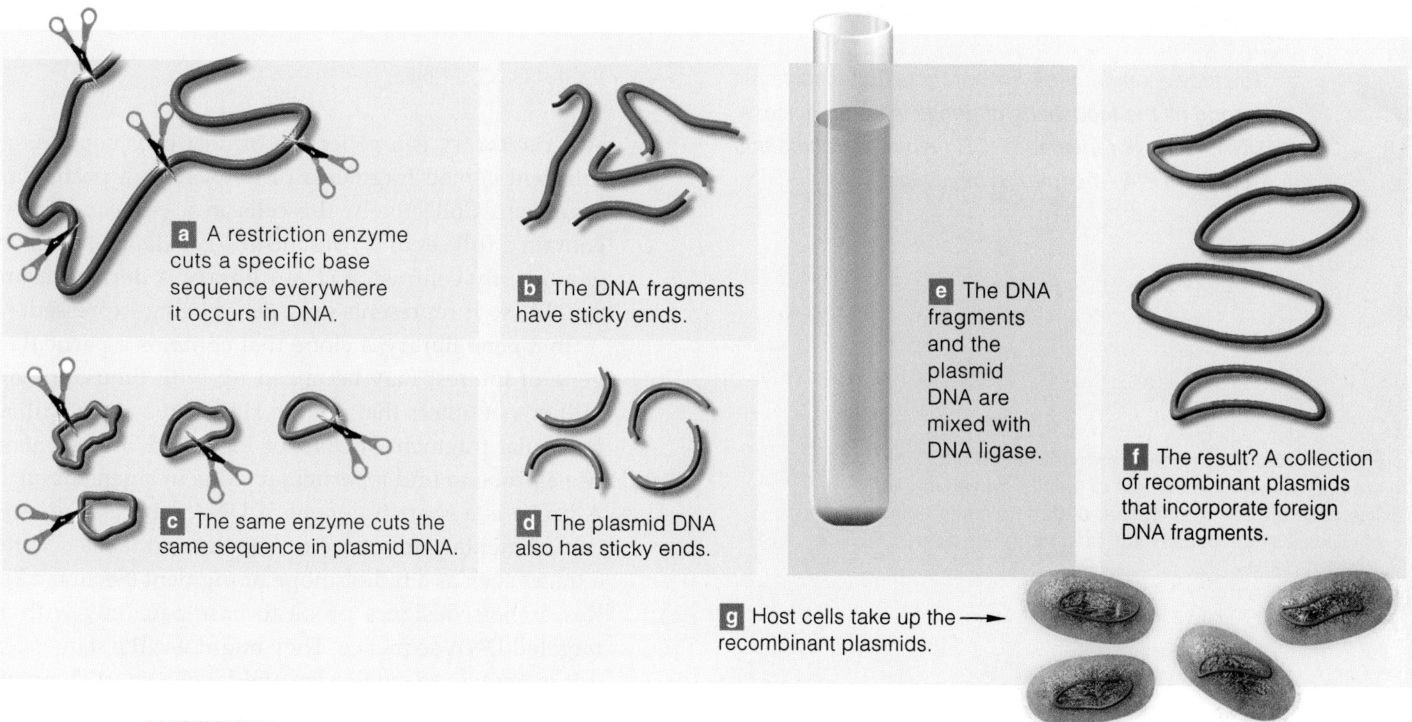

Figure 15.4 Animated! (**a–f**) Formation of recombinant DNA—in this case, a collection of DNA fragments sealed into bacterial plasmids. (**g**) Recombinant plasmids are inserted into host cells that multiply and so make multiple copies of them—and the foreign DNA they carry.

them select those host cells that take up recombinant plasmids. A host cell that takes up a cloning vector can be cultured so that it gives rise to a huge population of genetically identical daughter cells, or clones. Each clone contains a copy of the vector—and the foreign DNA incorporated into it (Figure 15.4).

cDNA CLONING

Eukaryotic DNA, remember, contains introns (Section 13.1). Unless you happen to be a eukaryotic cell, it is not an easy task to find the genes. Researchers who study eukaryotic genes and gene expression work with mRNA, because the introns have already been snipped out of its base sequence. All that remains are codons and a few signal sequences.

Restriction enzymes can only cut double-stranded DNA, so mRNA cannot be cloned directly. However, it can be converted to double-stranded DNA first and then cloned. Researchers can use reverse transcriptase, a replication enzyme isolated from certain viruses, to transcribe mRNA to DNA in a test tube. This enzyme assembles a strand of complementary DNA, or cDNA, from free nucleotides, using an mRNA as a template (Figure 15.5). The outcome is a hybrid molecule—one strand of mRNA base-paired with one strand of cDNA.

Next, DNA polymerase is added to the mixture. The polymerase strips the RNA from the hybrid molecule as it copies the cDNA into a second strand of DNA. The outcome is a double-stranded DNA copy of the original mRNA, and it may be used for cloning.

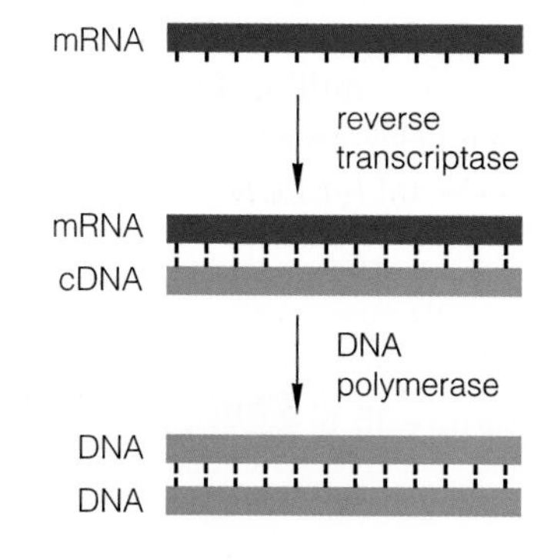

Figure 15.5 How to make cDNA. Reverse transcriptase catalyzes the assembly of a single DNA strand on an mRNA template, so that an mRNA–cDNA hybrid molecule forms. DNA polymerase replaces the mRNA with another DNA strand. The result is double-stranded DNA.

DNA cloning uses living cells to make identical copies of a particular fragment of DNA. Restriction enzymes cut DNA into fragments, then DNA ligase seals the fragments into cloning vectors. Recombinant DNA molecules result.

A cloning vector that holds foreign DNA can slip into a living cell such as a bacterium. A host cell divides and gives rise to huge populations of genetically identical daughter cells (clones), each of which contains a copy of the foreign DNA.

15.2 From Haystacks to Needles

LINKS TO SECTIONS 2.2, 3.6, 12.2

To study a single gene, researchers first must find it among all the thousands of genes in an organism's DNA. Gene libraries and PCR help researchers find genes in DNA—needles in haystacks.

a Individual cells from a library of bacterial cell clones are spread over the surface of a solid growth medium. The cells divide repeatedly and form colonies—clusters of millions of genetically identical daughter cells.

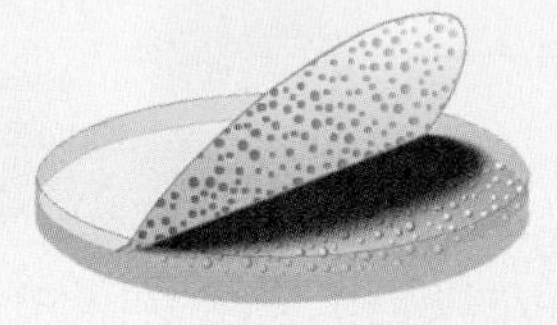

b A piece of special paper pressed onto the surface of the growth medium will bind some cells from each colony.

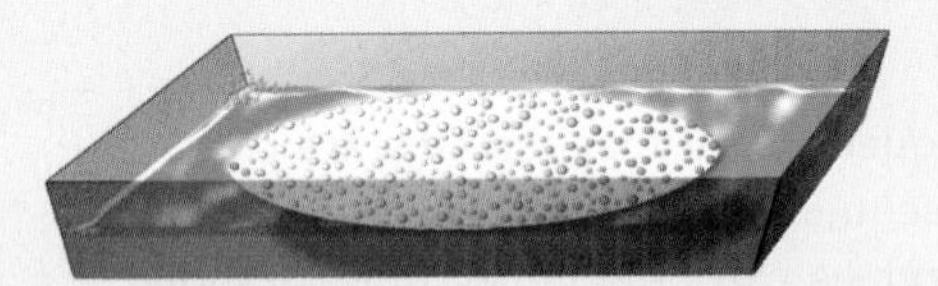

c The paper is soaked in a solution that ruptures the cells and releases their DNA. The DNA clings to the paper in spots mirroring the distribution of the colonies.

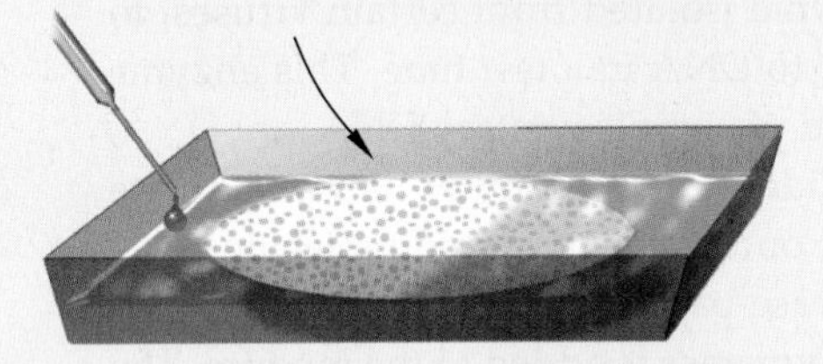

d A probe is added to the liquid bathing the paper. The probe hybridizes with (sticks to) only the spots of DNA that contain complementary base sequences.

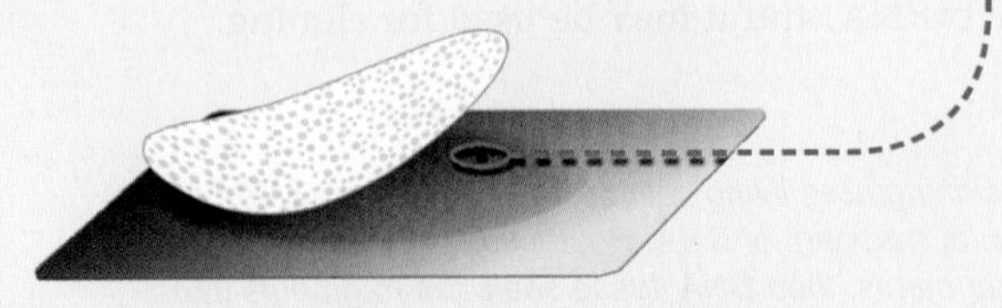

e The bound probe makes a spot. Here, one radioactive spot darkens x-ray film. The position of the spot on the film is compared to the positions of all the original bacterial colonies. Cells from the colony that made the spot are cultured, and the DNA they contain is harvested.

Figure 15.6 **Animated!** Nucleic acid hybridization. In this example, a radioactive probe helps identify a bacterial colony that contains a targeted sequence of DNA.

ISOLATING GENES

A gene library is a collection of host cells containing different cloned fragments of DNA from a particular organism. Collectively, the cells in a *genomic* library contain a full set of an organism's genetic material, or genome. By contrast, a *cDNA* library is derived from mRNA, so it represents only genes being expressed.

In a gene library, a clone that contains a particular gene of interest may be mixed up with thousands or millions of others that do not. How can a clone with a particular fragment of DNA be identified? Researchers use a probe to find a particular clone in a gene library. A probe is a short fragment of DNA with a sequence complementary to a gene of interest and labeled with a tracer such as a radioisotope or pigment (Section 2.2). Researchers design a probe to base-pair only with a targeted DNA sequence. They might use the sequence, if it is known, to synthesize and label a short chain of nucleotides. They might make a cDNA probe to screen a genomic library, or make a genomic DNA probe to screen a cDNA library.

Nucleic acid hybridization is base pairing between DNA (or DNA and RNA) from more than one source. Researchers use nucleic acid hybridization to locate clones that contain a targeted DNA sequence among all other clones in a library. When a probe is mixed with DNA from the library, it hybridizes with (sticks to) only DNA that is complementary to it (Figure 15.6). Researchers pinpoint a clone that contains the targeted DNA sequence by detecting the label on the probe. The clone is cultured, and a huge population of genetically identical cells forms. The targeted DNA can then be extracted from the cells in large quantities.

BIG-TIME AMPLIFICATION: PCR

Researchers can isolate and mass-produce a particular DNA fragment without cloning. They do so with the Polymerase Chain Reaction, or PCR. This hot-and-cold cycled reaction uses a heat-tolerant DNA polymerase to copy a fragment of DNA by the billions.

PCR can transform one needle in a haystack, that one-in-a-million DNA fragment, into a huge stack of needles with a little hay in it (Figure 15.7). The starting material for PCR is a sample of DNA with at least one molecule of a target sequence. It might be DNA from a mixture of 10 million different clones, one sperm, a hair left at a crime scene, or a mummy. Essentially any sample that has DNA in it can be used for PCR.

First, the starting material is mixed with a special DNA polymerase, nucleotides, and primers. Primers are synthetic single strands of DNA, usually between ten and thirty bases long. They are designed to base-pair only with nucleotide sequences on either end of the DNA sequence to be amplified.

Researchers expose the reaction mixture to repeated cycles of high and low temperature. High temperature disrupts the hydrogen bonds that hold the two strands of a DNA double helix together (Section 12.2). During a high temperature cycle, every molecule of double-stranded DNA unwinds and becomes single-stranded. During a low temperature cycle, single DNA strands hybridize with complementary partners, and double-stranded DNA forms again.

Most DNA polymerases are destroyed by the high temperatures required to separate DNA strands. The kind that is used in PCR reactions, *Taq* polymerase, is from *Thermus aquaticus*. This bacterial species lives in superheated springs (Chapter 18), so its polymerase is heat-tolerant. Like other DNA polymerases, this one recognizes primers as places to start DNA synthesis. During low temperature cycles, the polymerase starts synthesizing DNA wherever primers are hybridized with template. Synthesis proceeds along the template strand until the temperature rises again and the DNA separates into single strands. Any newly synthesized DNA is a copy of the target DNA sequence.

When the mixture cools, primers rehybridize, and DNA synthesis begins again. With each temperature cycle, the number of copies of target DNA can double. After about thirty PCR cycles, the number of template molecules may be amplified by about a billionfold.

Probes may be used to help identify a clone that hosts a DNA fragment of interest among the many other clones present in a gene library. The clone is cultured, and then its DNA is harvested.

PCR, the polymerase chain reaction, is a technique that rapidly increases the number of molecules of (amplifies) a particular DNA fragment.

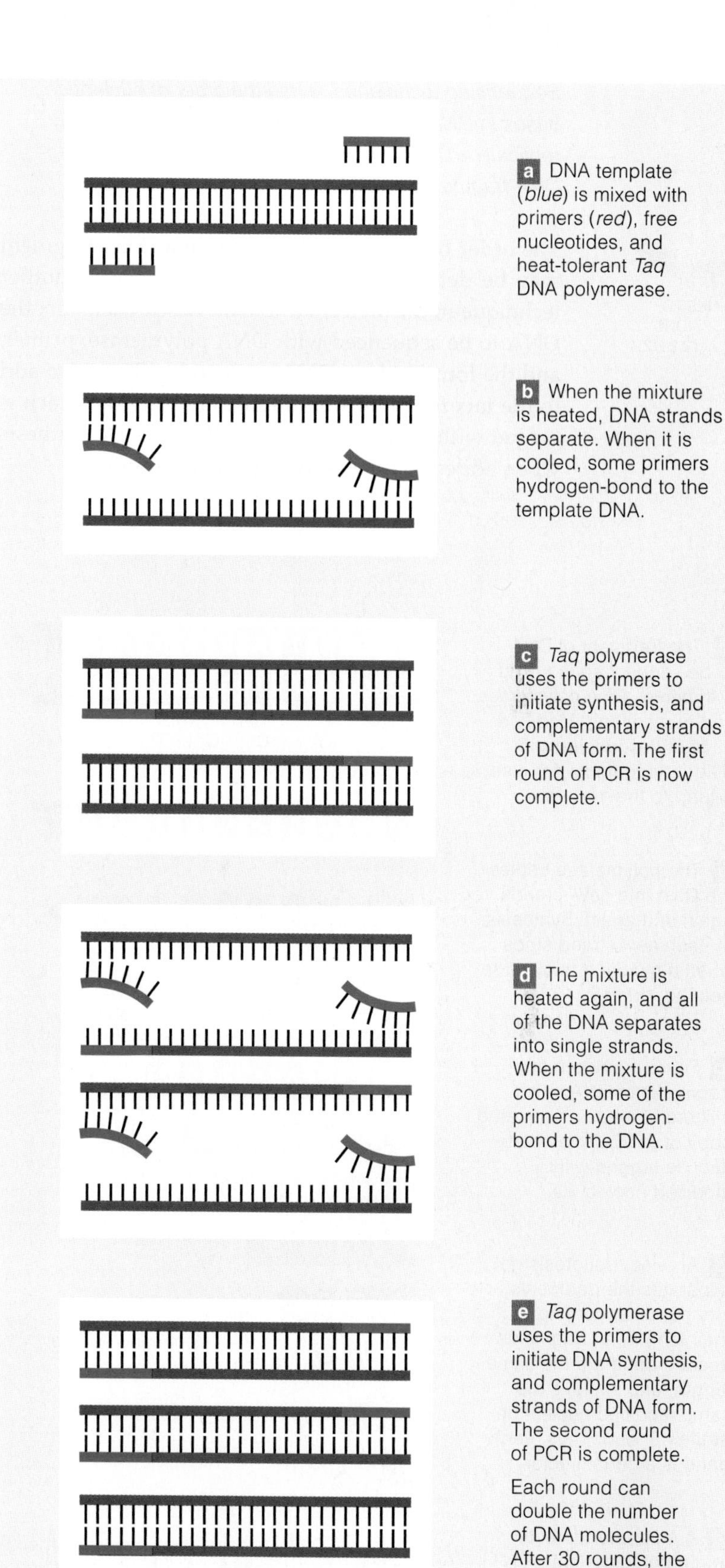

Figure 15.7 **Animated!** Two rounds of PCR. Thirty cycles of this polymerase chain reaction may increase the number of starting DNA template molecules a billionfold.

15.3 DNA Sequencing

Sequencing techniques reveal the order of nucleotide bases in DNA. They use DNA polymerase to partially replicate a DNA template, then separate the resulting DNA fragments by size.

LINKS TO SECTIONS 2.2, 12.2, 12.4

The order of the nucleotide bases in a DNA fragment may be determined with DNA sequencing—another technique in the molecular toolkit. Researchers mix the DNA to be sequenced with DNA polymerase, primer, and the four kinds of DNA nucleotides. They also add to the mix four kinds of modified nucleotides: Each is tagged with a tracer, and each will halt DNA synthesis when added to the end of a growing strand.

In one method, each kind of modified nucleotide is labeled with a different colored pigment. As in DNA replication, the polymerase constructs a new strand of DNA by joining together nucleotide bases according to the sequence of a template strand. The polymerase can add either a standard or a modified nucleotide to the 3′ end of a growing strand. If it attaches one of the modified nucleotides, synthesis of that strand stops.

After about 10 minutes, there are millions of DNA fragments of all different lengths; most are incomplete copies of the template DNA. All of the copies end with one of the four modified nucleotides (Figure 15.8*a–c*). For example, there will be many ten base-pair long copies of the template in the mixture. If the tenth base in the template was A, every one of those fragments will end with a modified A.

The fragments are separated by gel electrophoresis. With this technique, an electric field pulls all the DNA fragments through a semisolid gel. DNA fragments of different sizes move through the gel at different rates. The shorter the fragment, the faster it moves, because shorter fragments slip through the tangled molecules of the gel faster than longer fragments do. By analogy, tigers running through a forest in India slip between trees much faster than elephants do.

All fragments of the same length move through the gel at the same speed. They gather into bands. All of the fragments in a given band have the same modified nucleotide at their ends; the tracer on that nucleotide imparts its distinct color to the band (Figure 15.8*d*).

A computer records the color of all of the bands. Because each color designates one of the four kinds of nucleotides, the order of colored bands represents the DNA sequence, which is recorded by the computer.

Figure 15.8*e* shows the partial results from a DNA sequence run. Each peak in the tracing represents one color—one type of modified nucleotide. The sequence is shown beneath the graph line.

a The fragment of DNA to be sequenced is mixed with primer, DNA polymerase, and nucleotides. Modified nucleotides labeled with different pigments are also added to the mixture.

b The polymerase copies the DNA into new strands again and again. Synthesis of each new strand stops when a modified nucleotide gets added to it.

c There are now many fragments of DNA in the mixture. Each is a truncated copy of the DNA template; each is tagged with a modified nucleotide.

d An electrophoresis gel separates the fragments into bands according to their length. All fragments in each band are the same length, and all have the same modified nucleotide at their 3′ end. Thus, each band is a certain color.

e A computer detects and records the color of each band on the gel. The order of colors of the bands represents the sequence of the template DNA.

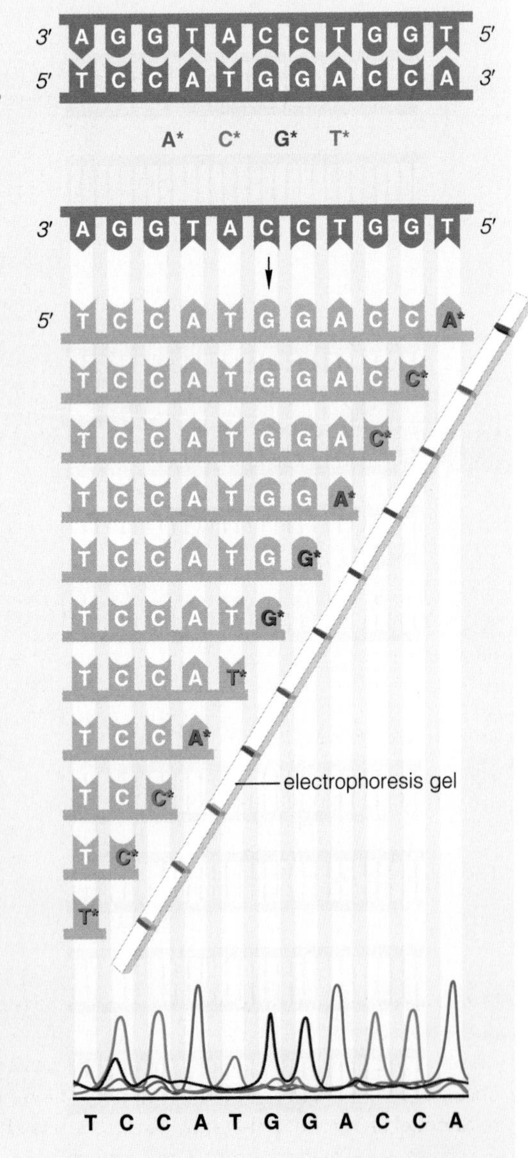

DNA sequencing reveals the order of nucleotides in DNA. The technique uses gel electrophoresis to separate partially replicated fragments of DNA according to their length.

Figure 15.8 **Animated!** DNA sequencing. Researchers sequence a sample of DNA by using it as a template for DNA synthesis. Any modified nucleotide, here represented as **T***, **C***, **A***, or **G***, stops replication when DNA polymerase attaches it to a growing DNA strand. The result is a collection of fragments of different length that can be separated by gel electrophoresis.

15.4 Analyzing DNA Fingerprints

FOCUS ON SCIENCE

Except for identical twins, no two people have exactly the same sequence of bases in their DNA. One individual can be distinguished from all others on the basis of this molecular fingerprint.

Each human has a unique set of fingerprints. In addition, like other sexually reproducing species, each also has a DNA fingerprint—a unique array of DNA sequences. More than 99 percent of the DNA in all humans is the same, but the other fraction of 1 percent is unique to each individual. Some of these unique sequences are sprinkled throughout the human genome as short tandem repeats—many copies of the same 2 to 10 base pair sequences, positioned one after the next along the length of a chromosome.

For example, one person's DNA might contain fifteen repeats of the bases TTTTC in a certain location. Another person's DNA might have TTTTC repeated two times in the same location. One person might have ten repeats of CGG; another might have fifty. Such repetitive sequences slip spontaneously into DNA during replication, and their numbers grow or shrink over generations. The mutation rate is relatively high around tandem repeat regions.

DNA fingerprinting reveals differences in the tandem repeats among individuals. With this technique, PCR is used to amplify a region of a chromosome known to have tandem repeats of 4 or 5 nucleotide bases. The size of the resulting PCR fragment differs among most individuals, because the number of tandem repeats in that region also differs.

Thus, the genetic differences between individuals can be detected by electrophoresis. As in DNA sequencing, the fragments form bands according to length as they migrate through a gel. Several regions of chromosomal DNA are typically tested. The resulting banding patterns on the electrophoresis gel constitute an individual's DNA fingerprint—which, for all practical purposes, is unique. Unless two people are identical twins, the chances that they have identical tandem repeats in even three regions of DNA is 1 in 1,000,000,000,000,000,000—or one in a quintillion—which is far more than the number of people that live on Earth.

A few drops of blood, semen, or cells from a hair follicle at a crime scene or on a suspect's clothing yield enough DNA to amplify with PCR for DNA fingerprinting (Figure 15.9). DNA fingerprints have been established as accurate and unambiguous, and are often used as evidence in court. For example, DNA fingerprints are now routinely submitted as evidence in paternity disputes. The technique is being widely used not only to convict the guilty, but also to exonerate the innocent: As of this writing, DNA fingerprinting evidence has helped release more than 160 innocent people from prison.

DNA fingerprint analysis has many applications. For instance, DNA fingerprinting was used to identify the remains of the individuals who died in the World Trade Center on September 11, 2001. It confirmed that human bones exhumed from a shallow pit in Siberia belonged to five individuals of the Russian imperial family, all shot to death in secrecy in 1918. Short tandem repeats on the Y chromosome are now often used to determine genetic relationships among male relatives and descendants, as well as to trace ancient ethnic heritage.

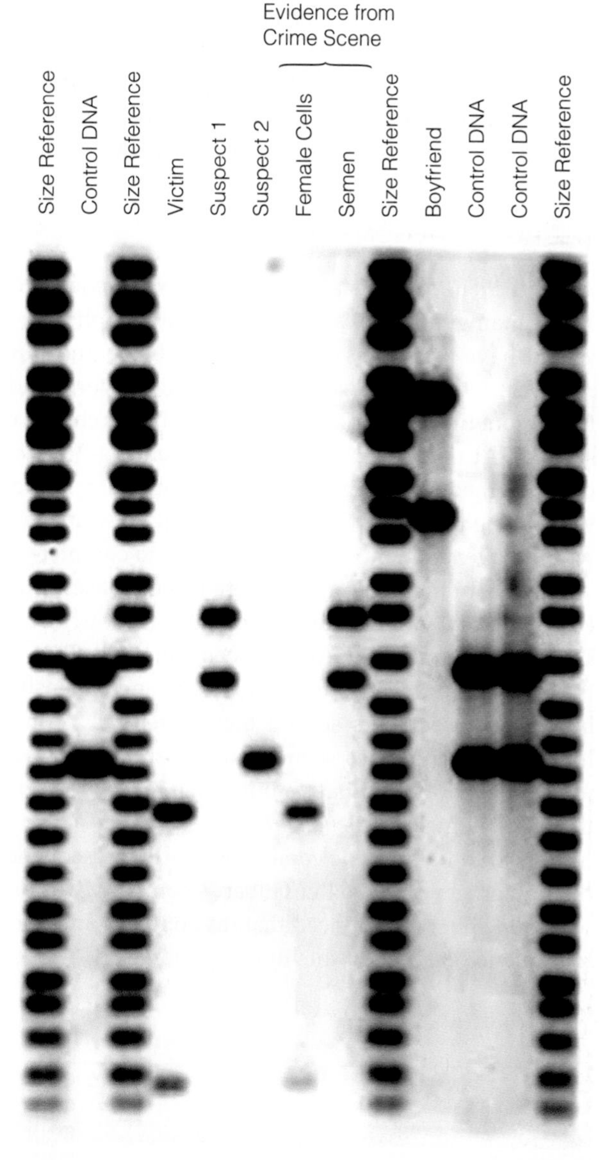

Figure 15.9 Damning comparison of DNA fingerprints. The crime: sexual assault. A single short tandem repeat region was amplified from biological evidence found at the crime scene—the perpetrator's semen and the victim's cells. The two samples were compared with the same tandem repeat region amplified from DNA of the victim, her boyfriend, and two suspects (1 and 2). Can you point out which suspect is guilty?

This is an image of an electrophoresis gel from a forensics laboratory. Note the three samples of control DNA (to confirm that the assay was working correctly), and the four size reference samples (to compare fingerprints that are not next to each other on the gel).

15.5 The Rise of Genomics

It took almost twenty years of concerted effort to finish the sequence of the human genome. Now that the task is complete, its benefits beyond pure research are apparent. Comparing our genome with those of other species is giving us new insights into how the human body works.

THE HUMAN GENOME PROJECT

LINKS TO SECTIONS 3.3, 12.1, 14.3

Around 1986, people started arguing about sequencing the human genome. Many insisted that deciphering it would have enormous payoffs for medicine and pure research. Others said sequencing would divert funds from more urgent work that also had a better chance of success. At that time, the task of sequencing 3 billion bases seemed daunting. It would take at least 6 million sequencing reactions. Given the techniques available, it would have taken more than fifty years to complete.

But techniques kept getting better; more bases could be sequenced in less time. Automated (robotic) DNA sequencing and PCR had just been invented. Both of these techniques were still cumbersome and expensive, but many researchers sensed their potential. Waiting for faster technologies seemed the most efficient way to sequence 3 billion bases, but when, exactly, would the technology be fast *enough*?

In 1987, a few private companies started to sequence the human genome. Walter Gilbert's company was one of them. Gilbert announced that his company intended to sequence *and* patent the human genome.

This announcement provoked widespread outrage, but it also spurred commitments in the public sector. In 1988, the National Institutes of Health (NIH) effectively annexed the entire project by hiring the venerable James Watson to head the official Human Genome Project, and providing $200 million a year to fund it. A consortium formed between the NIH and international institutions sequencing different parts of the genome. Watson set aside 3 percent of the funding for studies of any ethical and social issues arising from the research. He resigned later over a disagreement about patenting.

Amid ongoing squabbles over patent issues, Craig Venter announced in 1998 that his new company, Celera Genomics, would be the first to sequence and patent the genome (Figure 15.10). Then, in 2000, President Clinton and British Prime Minister Tony Blair jointly declared that the sequence of the human genome could not be patented. Celera kept on sequencing anyway, and the competition motivated the public consortium to move its efforts into high gear.

In 2001, Celera and the public consortium separately published about 90 percent of the sequence. By 2003, fifty years after the discovery of DNA, the sequence of the human genome was completed. To date, about 99 percent of its coding regions—28,976 genes—have been identified. Researchers have not discovered what all of the genes encode, only where they are in the genome.

What do we do with this vast amount of data? The next step is to find out just what the sequence means.

Figure 15.10 Some of the bases of the human genome—and a few of the supercomputers used to sequence it—at Venter's Celera Genomics in Maryland.

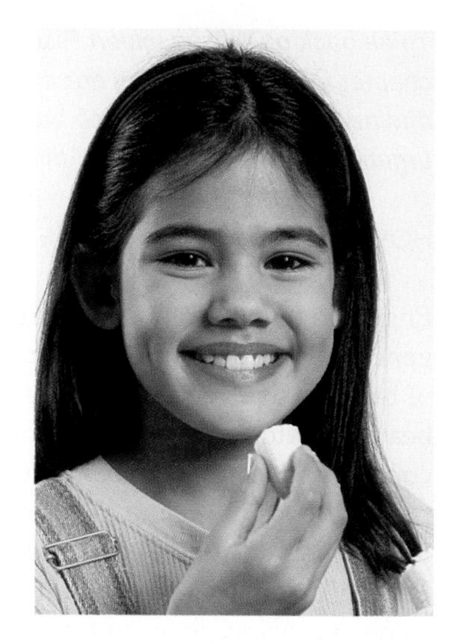

Figure 15.11 One sequencing surprise—just how much you have in common with a banana.

GENOMICS

Investigations into the genomes of humans and other species have converged into the new research field of genomics. *Structural* genomics focuses on determining the three-dimensional structure of the proteins encoded by a genome. *Comparative* genomics compares genomes of different species; similarities and differences reflect evolutionary relationships.

The human genome sequence is a massive collection of seemingly cryptic data. Currently, the only way we are able to decipher it is by comparing it to genomes of other organisms, the premise being that all organisms are descended from shared ancestors, so all genomes are related to some extent. We see evidence of such genetic relationships just by comparing the sequence data. For example, the human and mouse sequences are about 78 percent identical; the human and banana sequences are about 50 percent identical (Figure 15.11).

Intriguing as such percentages may be, gene-by-gene comparisons offer practical benefits. We have learned about the function of many human genes by studying their counterpart genes in other species. For example, researchers studying a human gene might disable the same gene in mice. The effects of the gene's absence on mice are clues to its function in humans. These types of knockout experiments (Section 14.3) are revealing the function of many human genes. Recently, researchers comparing the human and mouse genomes discovered a human version of the mouse gene *APOA5*. This gene encodes a protein that carries lipids in the blood. Mice with an *APOA5* knockout have four times the normal level of triglycerides in the blood. The researchers then looked for—and found—a correlation between *APOA5* mutations and high triglyceride levels in humans. High triglycerides are risk factor for coronary artery disease.

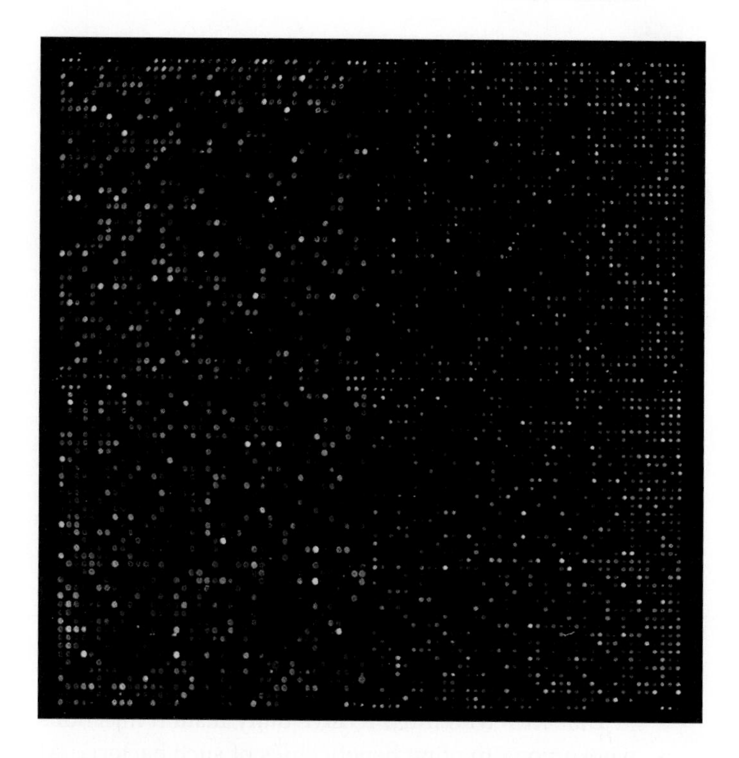

Figure 15.12 Complete yeast genome arrayed on a DNA chip about 19 millimeters (3/4 inch) across. *Green* spots indicate genes that are active during fermentation, and *red* spots indicate genes active during aerobic respiration. *Yellow* spots indicate genes active in both pathways.

DNA CHIPS

Analysis of the human genome is now advancing very quickly. Researchers are using DNA chips to pinpoint which genes are silent and which are expressed. DNA chips are microscopic arrays (microarrays) of hundreds of thousands of short DNA fragments that collectively represent an entire genome, all stamped onto a few glass plates no bigger than business cards.

DNA chips are being used to compare patterns of gene expression among cells—perhaps different types of cells from one individual, or the same cells grown under different conditions (Figure 15.12). RNA from one set of cells is used to make cDNA labeled with a green tracer, and RNA from another set of cells is used to make cDNA labeled with a red tracer. The probes are mixed, and a DNA chip is soaked in them. Green and red spots on the chip indicate expression of genes in different cells.

Now that the sequence of the human genome is complete, analysis of the data is yielding tremendous amounts of new information about the function of human genes.

15.6 Genetic Engineering

Traditional methods of producing genetic modifications only work if the organisms that carry the desired traits can cross-breed. Genetic engineering takes these gene-swapping manipulations to an entirely new level.

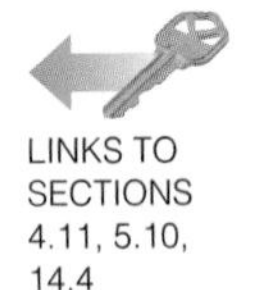

LINKS TO SECTIONS 4.11, 5.10, 14.4

Genetic engineering is a laboratory process by which deliberate changes are introduced into an individual's genome. Genes from one species may be transferred to another to make a transgenic organism, or a gene may be altered and reinserted into an individual of the same species. Both techniques result in GMOs.

Genetic engineering started with bacteria, so let's consider them first. Remember, bacteria can take up recombinant plasmids. Like you, these cells have the metabolic machinery to make many complex organic compounds. They transcribe and translate any foreign genes carried by their plasmids, and so can be used to produce proteins of other species. Bacteria that have been engineered to produce foreign proteins are now widely used in research and for practical applications.

Diabetics who must receive daily insulin injections were among the first beneficiaries of such bacteria. At one time, the insulin for their injections was extracted from pigs and cattle, but that is no longer the case. In 1982, a synthetic gene that codes for human insulin was transferred into the intestinal bacteria *Escherichia coli* (*E. coli* for short). The engineered cells produced human insulin, and became the first large-scale, cost-effective "factory" for the production of a medically important protein.

Engineered microbes produce other proteins that are useful in medicine, agriculture, and various food industries. For example, cheese is traditionally made with an extract of calf stomach linings that contains an enzyme called chymotrypsin. Today in the United States, most cheese manufacturers use chymotrypsin made by genetically engineered bacteria. Other food industry products that are made by microbes include enzymes that can improve the taste and clarity of beer and fruit juice, slow bread staling, or modify fats.

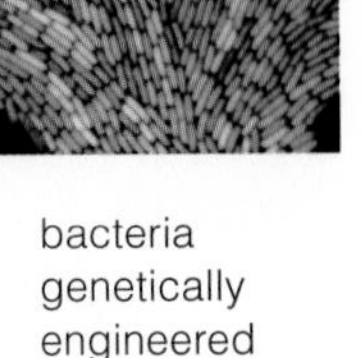

bacteria genetically engineered to glow green

Engineered bacteria also hold potential for industry and environmental remediation, or cleaning up messes in the environment. Some modified types digest spilled crude oil into less harmful compounds. Others sponge up high concentrations of phosphates, heavy metals and other pollutants, or even radioactive wastes.

Genetic engineering refers to the directed alteration of an individual's genome. Genetically modified microbes now produce a range of protein products that are useful in medicine and other practical applications.

15.7 Designer Plants

Think back on those Golden Rice plants described in the chapter introduction. They are a prime example of genetic engineering that can produce transgenic plants. There is urgency surrounding much of this work.

HOW PLANTS GET ENGINEERED

Engineering plant cells starts with vectors that can carry genes into them. *Agrobacterium tumefaciens* is a species of bacteria that can infect many plants, including peas, beans, potatoes, and other important crops. Its plasmid contains genes that cause tumors to form on infected plants; hence the name Ti plasmid (*T*umor-*i*nducing). Researchers use the Ti plasmid as a vector to transfer foreign or modified genes into plants. They take the tumor-inducing genes out of the plasmid, then insert desired genes into it. Whole plants can be grown from cultured plant cells that take up the modified plasmid (Figure 15.13).

Modified *A. tumefaciens* bacteria are used to deliver genes into some food crop plants, including soybeans, squash, and potatoes. Researchers also transfer genes into plants by way of electric shocks or chemicals, or by blasts of microscopic DNA-coated bullets.

GENETICALLY ENGINEERED PLANTS

As crop production expands to keep pace with human population growth, it places unavoidable pressure on ecosystems everywhere. Irrigation leaves mineral and salt residues in soils. Tilled soil erodes, taking topsoil with it. Runoff clogs rivers, and fertilizer in it causes algae to grow so much that fish suffocate. Pesticides harm humans, other animals, and beneficial insects.

Pressured to produce more food at lower cost and with less damage to the environment, many farmers have begun to rely on genetically modified crop plants. Some of these crops offer improved yields, such as a strain of transgenic wheat that has double the yield of unmodified wheat. Other GMO crops can help farmers use smaller amounts of toxic pesticides. For example, organic farmers often spray their crops with spores of *Bacillus thuringiensis*, a bacterial species that produces a protein toxic only to insect larvae. Genetic engineers transferred the gene encoding the *Bt* protein into crop plants. The engineered plants produce the *Bt* protein, but otherwise they are identical to unmodified plants. Insect larvae die shortly after eating their first (and only) GMO meal. Farmers use much less pesticide on crops that make their own (Figure 15.14*a*).

Transgenic crop plants are also being developed for regions that are affected by severe droughts, such as

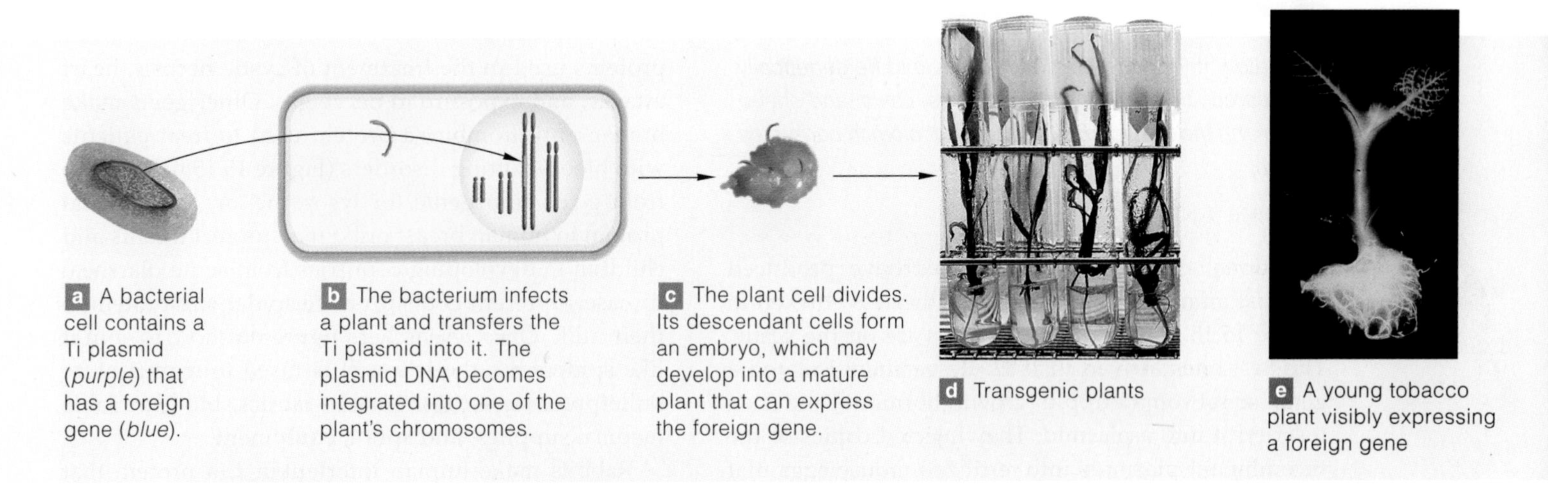

Figure 15.13 Animated! (**a**–**d**) Ti plasmid transfer of an *Agrobacterium tumefaciens* gene to a plant cell. (**e**) Transgenic plant expressing a gene for the firefly enzyme luciferase (Section 5.10).

Figure 15.14 (**a**) Some GMO crops help farmers use less insecticide. *Top*, a bacterial gene conferred insect resistance to the genetically modified plants that produced this corn. *Bottom*, unmodified corn is more vulnerable to pests.

(**b**) Control plant (*left*) and three genetically engineered aspen seedlings (*right*). Vincent Chiang and coworkers suppressed a control gene involved in a lignin synthesis pathway. Lignin, an organic compound, strengthens the secondary cell walls of many kinds of woody plants. Before paper can be made from wood, the lignin must be extracted from wood pulp.

Chiang's modified plants synthesized normal lignin, but not as much. Lignin synthesis fell by as much as 45 percent, yet cellulose production increased 15 percent. Root, stem, and leaf growth were greatly enhanced with no structural loss. Paper products and clean-burning fuels such as ethanol may be easier to manufacture from the wood of such trees.

Africa. Genes that confer drought tolerance and insect resistance are being transferred into crop plants such as corn, beans, sugarcane, cassava, cowpeas, banana, and wheat. Such crops may help people that rely on agriculture for food and income survive in drought-stricken, impoverished regions of the world.

At this writing, the United States Department of Agriculture (USDA) has approved seventy genetically modified plants for general crop use. Hundreds more are pending approval. Today, the most widely planted GMO crops include corn, sorghum, cotton, soy, canola, and alfalfa engineered for resistance to glyphosate, an herbicide. Rather than tilling the soil to control weeds, farmers can drench their fields with glyphosate, which kills the weeds but not the engineered crops. However, weeds are becoming resistant to glyphosate, so farmers are starting to use different, more toxic herbicides. The engineered gene is also cropping up in wild plants as well as nonengineered crops, which means transgenes can—and do—escape into the environment.

Such controversy invites you to read the research and form your own opinions. The alternative is to be swayed by media hype (the term "Frankenfood," for instance), or by reports from potentially biased sources (such as herbicide manufacturers).

Various methods deliver modified or foreign genes into plants. Genetically engineered plants are now commonly planted as farm crops.

15.8 Biotech Barnyards

Laboratory mice were the first mammals to be genetically engineered. Today, goats, rabbits, pigs, cows, and sheep are among the transgenic inhabitants of biotech barnyards.

OF MICE AND MEN

LINK TO SECTION 14.3

Traditional cross-breeding practices have produced unusual animals, including the featherless chicken in Figure 15.15. Now transgenic types are on the scene. The first ones arrived in 1982. Researchers isolated a gene for rat somatotropin (growth hormone) and then inserted it into a plasmid. They injected copies of the recombinant plasmids into fertilized mouse eggs that were later implanted into female mice. A third of the offspring of the surrogate mothers grew much larger than their littermates (Figure 15.16). The rat gene had become integrated into the host DNA and was being expressed in the transgenic mice.

Genetically modified animals are used routinely in medical research. The functions of many human genes and how they are controlled have been discovered by inactivating their counterparts in mice (Section 14.3). Such information also leads to better animal models of disease. For example, molecules that control glucose metabolism have been inactivated in mice; research into the effects of these knockouts has given us new insights into diabetes in humans. Genetically modified animals such as these mice allow researchers to study human diseases—and their potential cures—without experimenting on humans.

Other genetically engineered animals are sources of many proteins that have medical and industrial applications. Various transgenic goats produce proteins used in the treatment of cystic fibrosis, heart attacks, and exposure to nerve gas. Other goats make human antithrombin, a protein used to treat patients with blood-clotting disorders (Figure 15.15*b*). The milk from goats transgenic for lysozyme, an antibacterial protein in human breast milk, may protect infants and children in developing countries from acute diarrheal disease. Different goats produce spider silk protein in their milk. Once researchers figure out how to spin it like spiders do, the silk will be used to manufacture bulletproof vests, fashionable fabrics, biodegradable medical supplies, and sports equipment.

Rabbits make human interleukin-2, a protein that triggers divisions of immune cells. Genetic engineering has also given us dairy goats with heart-healthy milk, low-fat pigs, pigs with environmentally friendly low-phosphate manure, extra-large sheep, and cows that are resistant to mad cow disease.

Tinkering with the genes of animals raises ethical questions. For example, is transgenic animal research just an extension of thousands of years of acceptable breeding practices? The techniques have changed, but not the intent. Humans still have a vested interest in improving their livestock.

KNOCKOUT CELLS AND ORGAN FACTORIES

Each year, about 75,000 people are on waiting lists for an organ transplant, but human donors are in short supply. There is talk of harvesting organs from pigs; pig and human organs are about the same size, and function in the same way. Transferring an organ from one species into another is called xenotransplantation.

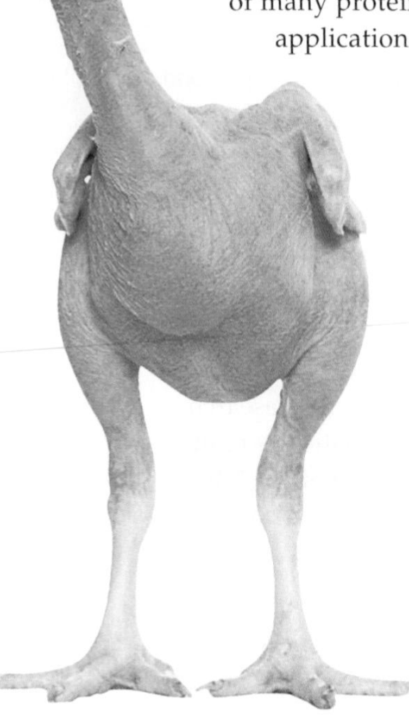

Figure 15.15 Genetically modified animals. (**a**) Featherless chicken developed by traditional cross-breeding methods in Israel. Such chickens survive in hot deserts where cooling systems are not an option. Chicken farmers in the United States lose millions of feathered chickens in hot weather. (**b**) Mira, a goat transgenic for human antithrombin III, an anticlotting factor. (**c**) The pig on the *left* is transgenic for a yellow fluorescent protein; its nontransgenic littermate is on the *right*.

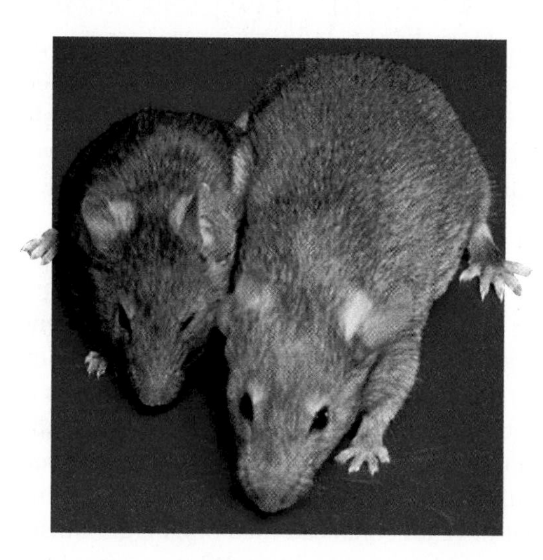

Figure 15.16 Evidence of a successful gene transfer. Two ten-week-old mouse littermates. *Left*, this mouse weighed 29 grams. *Right*, this one weighed 44 grams. It grew from a fertilized egg into which a gene for rat somatotropin had been inserted.

The human immune system battles anything that it recognizes as nonself. It rejects a pig organ at once, because it recognizes as foreign a glycoprotein on the plasma membrane of pig cells. Antibodies circulating in human blood quickly latch on to the glycoprotein and initiate an immune response. Within a few hours, blood inside the organ's vessels coagulates massively, and this dooms the transplant. Drugs can suppress the immune response but have a serious side effect: They make organ recipients vulnerable to infections.

Researchers have engineered knockout pigs that do not have the offending glycoprotein on their cells. The human immune system may not reject tissues or organs of such pigs. The idea behind this research is to help the millions of patients whose own organs or tissues are damaged beyond repair.

Critics of xenotransplantation are concerned that, among other things, pig-to-human transplants would invite pig viruses to cross the species barrier and infect humans, perhaps catastrophically. Their concerns are not unfounded. In 1918, an influenza pandemic killed 50 million people. It started with a bird flu virus that jumped to humans. This story is continued in *Critical Thinking* question 5 on page 236.

Animals that would be impossible to produce by traditional breeding methods are being created by genetic engineering. Such animals are used in research, medicine, and industry.

15.9 Safety Issues

Many years have passed since the first transfer of foreign DNA into bacteria. That transfer ignited an ongoing debate about potential dangers of transgenic organisms that may enter the environment, either accidentally or intentionally.

In 1953, James Watson and Francis Crick presented their model of the DNA double helix and ignited a global blaze of optimism about genetic research. The very book of life seemed to be open for scrutiny. In reality, no one could read it. Scientific breakthroughs are not very often accompanied by the simultaneous discovery of the tools to study them. New techniques would have to be invented before that book would become readable.

Twenty years later, Paul Berg and his coworkers discovered how to make recombinant organisms by fusing DNA from two species of bacteria. By isolating DNA in manageable subsets, researchers now had the tools to be able to study its sequence in detail. They began to clone and analyze DNA from many different organisms. The technique of genetic engineering was born, and suddenly everyone was worried about it.

Researchers knew that DNA itself was not toxic, but they could not predict with certainty what would happen every time they fused genetic material from different organisms. Would they accidentally make a superpathogen? Could they make a new, dangerous form of life by fusing DNA of two normally harmless organisms? What if that new form escaped from the laboratory and transformed other organisms?

In a remarkably quick and responsible display of self-regulation, scientists reached a consensus on new safety guidelines for DNA research. Adopted at once by the NIH, these guidelines included precautions for laboratory procedures. They covered the design and use of host organisms that could survive only under the narrow range of conditions inside the laboratory. Researchers stopped using DNA from pathogenic or toxic organisms for recombination experiments until proper containment facilities were developed.

Now, all genetic engineering research is done under these laboratory guidelines. Testing, improving, and importing genetically modified organisms is carefully regulated by the USDA. Such regulations are our best effort to minimize any risk involved in the research or as a result of it, but they are not a guarantee.

Rigorous safety guidelines for DNA research have been in place for decades in the United States. Researchers are expected to comply with these stringent standards.

15.10 Modified Humans?

CONNECTIONS

We as a society continue to work our way through ethical implications of applying the new DNA technologies. Even as we are weighing the risks and benefits, however, the manipulation of individual genomes has begun.

WHO GETS WELL?

LINK TO SECTION 11.7

Human gene therapy is often cited as one of the most compelling reasons for embracing genetic engineering research. Gene therapy is the transfer of one or more normal or modified genes into an individual's body cells to fix a genetic defect or boost disease resistance.

We already know about more than 15,500 serious genetic disorders. Many are rare, but collectively they show up in 3 to 5 percent of all newborns and cause 20 to 30 percent of all infant deaths a year. They account for about half of all mentally impaired patients and nearly a fourth of all hospital admissions. They also contribute to many age-related disorders.

Gene therapies deliver cells that carry recombinant DNA into a patient's tissues, or use viruses as vectors that inject genes into a person's cells. In many cases, such therapy can alleviate a patient's symptoms.

Rhys Evans, shown below, was born with a severe immune deficiency called SCID-X1, which stems from mutations in the gene *IL2RG*. Children affected by this disorder can survive only in germ-free isolation tents, because they cannot fight infections.

In 1998, a viral vector was used to insert unmutated copies of *IL2RG* into cells taken from the bone marrow of eleven boys with SCID-X1. Each child's modified cells were infused back into his bone marrow. Months later, ten of the boys left their isolation tents for good. Their immune systems had been repaired by the gene therapy. Since then, gene therapy has freed many other SCID-X1 patients from life in an isolation tent. Rhys is one of them.

However, manipulating a gene within the context of a living individual is unpredictable even when we know its sequence and where it is within the genome. No one, for example, can predict where a virus-injected gene will insert into chromosomes. Its insertion might disrupt other genes. If it interrupts a gene that is part of the controls over cell division, then cancer might be the outcome.

To the shock of researchers, three children from the 1998 SCID-X1 clinical trial have since developed leukemia, and one of them died. The researchers had wrongly anticipated that cancerous transformation related to the gene therapy would be extremely rare.

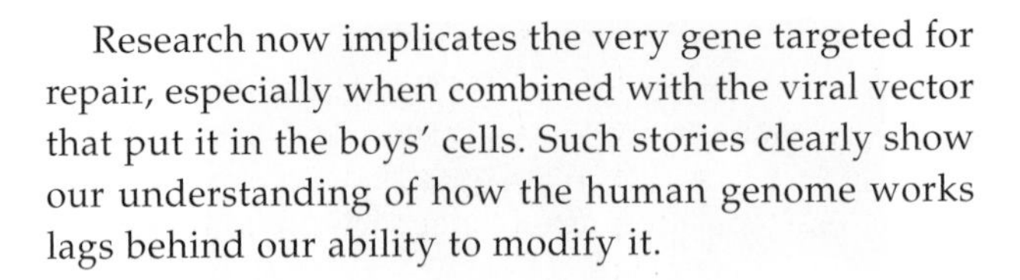

Research now implicates the very gene targeted for repair, especially when combined with the viral vector that put it in the boys' cells. Such stories clearly show our understanding of how the human genome works lags behind our ability to modify it.

WHO GETS ENHANCED?

The idea of using human gene therapy to cure genetic disorders seems like a socially acceptable goal to most people. However, go one step further. Would it also be acceptable to modify genes of an individual who is within a normal range in order to minimize or enhance a particular trait? Researchers have already produced mice that have an enhanced memory, bigger muscles, and improved learning abilities. Why not people?

The idea of selecting the most desired human traits is referred to as eugenic engineering. Yet who decides which forms of traits are most desirable? Realistically, cures for many severe but rare genetic disorders will not be found, because the financial payback will not cover the research. Eugenics, however, might just turn a profit. How much would potential parents pay to be sure that their child will be tall or blue-eyed? Would it be okay to engineer "superhumans" with breathtaking strength or intelligence? How about a treatment that can help you lose that extra weight? The line between interesting and abhorrent is not the same for everyone.

In a survey conducted in the United States, more than 40 percent of those interviewed said it would be fine to use gene therapy to make smarter and cuter babies. In one poll of British parents, 18 percent would be willing to use genetic enhancement to keep their child from being aggressive, and 10 percent would use it to keep a child from growing up to be homosexual.

Some argue that we must never alter the DNA of anything. The concern is that we just do not have the wisdom to bring about any genetic changes without causing irreparable damage to ourselves and nature.

One is reminded of our peculiar human tendency to leap before we look. And yet, something about the human experience allows us to dream of such things as wings of our own making, a capacity that carried us to the frontiers of space. In this brave new world, the questions before you are these: What do we stand to lose if serious risks are not taken? And, do we have the right to impose the consequences of taking such risks on those who would choose not to take them?

Be engaged; our understanding of the meaning of the human genome is changing even as you read this.

Summary

www.thomsonedu.com ThomsonNOW!

Section 15.1 Researchers make recombinant DNA by combining the DNA from different organisms. They use restriction enzymes to cut DNA into pieces, then DNA ligase to splice the pieces into plasmids or other cloning vectors. The resulting recombinant DNA can be cloned by inserting it into host cells such as bacteria. When a host cell divides, it forms huge populations of genetically identical daughter cells, or clones. Each clone has a copy of the foreign DNA. Researchers clone DNA in order to harvest enough for experiments.

RNA cannot be cloned directly. Reverse transcriptase, a viral enzyme, is used to convert single-stranded RNA into cDNA for cloning.

- *Use the animation on ThomsonNOW to survey the tools of researchers who make recombinant DNA.*

Section 15.2 A gene library is a collection of hundreds or millions of cells that host different fragments of DNA. Researchers use probes to identify cells in a library that host a specific fragment of DNA. These short sequences of DNA base-pair with the DNA of interest and are also labeled with a tracer such as a radioisotope or pigment. Base-pairing between nucleic acids from different sources is called nucleic acid hybridization.

The polymerase chain reaction (PCR) is a technique for rapidly increasing the number of molecules of a DNA fragment. The temperature-cycled reaction uses a special heat-resistant DNA polymerase.

- *Use the interaction on ThomsonNOW to learn how researchers isolate and copy genes.*

Section 15.3 DNA sequencing reveals the order of bases in a particular piece of DNA. DNA polymerase is used to partially replicate a DNA template. The reaction produces a mix of DNA fragments of all different lengths; the synthesis of each molecule has been terminated with one of four different pigment-labeled nucleotides.

Electrophoresis separates the fragments into bands by their length. Each band is the color of the pigment that labeled the terminal nucleotide. The order of the colored bands as they migrate through the gel reflects the base sequence of the template DNA.

- *Use the animation on ThomsonNOW to investigate the technique of DNA sequencing.*

Section 15.4 Tandem repeats are multiple copies of a short DNA sequence that follow one another along a chromosome. The number and distribution of tandem repeats, unique in each individual, can be revealed by gel electrophoresis. They form a DNA fingerprint.

- *Use the animation on ThomsonNOW to observe the process of DNA fingerprinting.*

Section 15.5 The genomes of several organisms have been completely sequenced. Genomics, or the study of genomes, is providing insights into the function of the human genome. Comparative genomics uses similarities and differences between genomes of different organisms to clarify evolutionary relationships. Similarities can also be used as a predictive tool in research.

DNA chips (also called DNA microarrays) are used to study gene expression.

Sections 15.6–15.8 Recombinant DNA technology and genome analysis are the basis of genetic engineering: directed modification of an organism's genetic makeup. Genes from one species are inserted into an individual of a different species to make a transgenic organism, or a gene is modified and reinserted into an individual of the same species. The result of either process is a genetically modified organism (GMO).

Genetically engineered bacteria are used in research, medicine, agriculture, industry, and ecology. Transgenic crop plants help farmers produce food more efficiently.

Genetically engineered animals are producing valued proteins on a commercial scale. Some kinds are useful in research of human diseases and their potential cures, and others are valuable as livestock.

- *Use the animation on ThomsonNOW to see how the Ti plasmid is used to genetically engineer plants.*

Section 15.9 Rigorous safety procedures minimize potential risks to researchers in genetic engineering labs. Although these and other strict government regulations limit the release of genetically modified organisms into the environment, such laws are not guarantees against accidental releases or unforseen environmental effects.

Section 15.10 The goal of human gene therapy is to transfer normal or modified genes into body cells to correct genetic defects. As with any new technology, the benefits must be weighed against potential risks.

Self-Quiz

Answers in Appendix III

1. Researchers can cut DNA molecules at specific sites by using ______ .
   a. DNA polymerase c. restriction enzymes
   b. DNA probes d. reverse transcriptase

2. Fill in the blank: A ______ is a small circle of bacterial DNA that contains only a few genes and is separate from the bacterial chromosome.

3. By reverse transcription, ______ is assembled on a(n) ______ template.
   a. mRNA; DNA c. DNA; ribosome
   b. cDNA; mRNA d. protein; mRNA

4. For each species, all ______ in the complete set of chromosomes is the ______ .
   a. genomes; phenotype c. mRNA; start of cDNA
   b. DNA; genome d. cDNA; start of mRNA

5. PCR can be used to ______ .
   a. amplify the number of specific DNA fragments
   b. make DNA fingerprints
   c. sequence DNA
   d. a and b are correct

6. By gel electrophoresis, fragments of DNA can be separated according to ______ .
   a. sequence b. length c. species

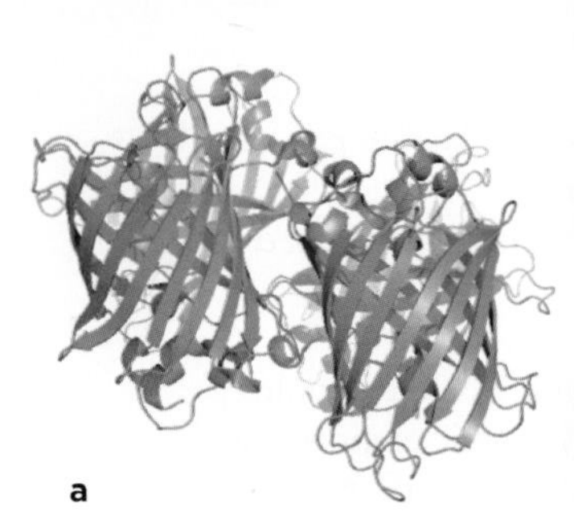

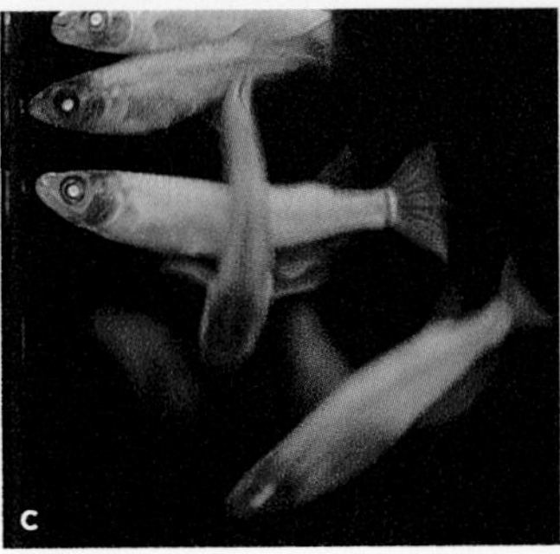

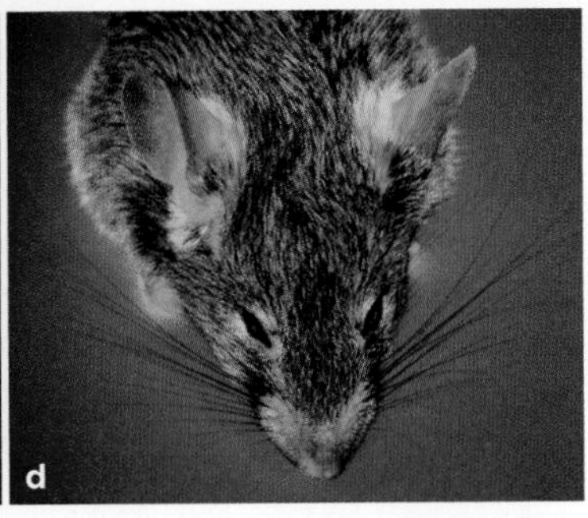

Figure 15.17 Green fluorescent animals. (**a**) Model for green fluorescent protein, the product of a jellyfish gene. (**b**) ANDi, the first transgenic primate; his cells carry the gene. The product, but not ANDi, fluoresces under blue light. The same gene was transferred into (**c**) zebrafish, and (**d**) a mouse.

7. DNA sequencing relies on ______ .
   a. standard and labeled nucleotides
   b. primers and DNA polymerase
   c. gel electrophoresis
   d. all of the above

8. ______ can be used to insert genes into human cells.
   a. PCR
   b. Modified viruses
   c. Xenotransplantation
   d. DNA microarrays

9. Match the terms with the most suitable description.

___ DNA fingerprint	a. selecting "desirable" traits
___ Ti plasmid	b. used in some gene transfers
___ nucleic acid hybridization	c. a person's unique collection of tandem repeats
___ eugenic engineering	d. base pairing of nucleotide sequences from different DNA or RNA sources

■ *Visit ThomsonNOW for additional questions.*

Critical Thinking

1. Lunardi's Market put out a bin of tomatoes having vine-ripened redness, flavor, and texture. A sign identified them as genetically engineered produce. Most shoppers selected unmodified tomatoes in the adjacent bin even though those tomatoes were pale pink, mealy textured, and tasteless. Which tomatoes would you buy? Why?

2. The sequencing of the human genome is completed, and knowledge about many genes is being used to detect genetic disorders. Many insurance companies will pay for their female subscribers to take advantage of genetic testing for breast cancer and are willing to allow them to keep the results confidential.

Explain how a health insurance company may benefit financially if it were to encourage its subscribers to take confidential tests for breast cancer susceptibility.

3. At this writing, 61 percent of corn and 89 percent of soybean crops in the United States are genetically modified. These and many other genetically modified crops are intended for human consumption and have successfully passed safety tests.

Untested genetically modified organisms are appearing in the environment despite efforts by the industries that produce them. StarLink, a strain of genetically modified corn, makes the insect larvae-killing *Bacillus thuringensis* protein. In 1999, StarLink corn somehow appeared in almost every harvest of corn in the United States, even though it was not approved for human consumption. In 2003, researchers found that weeds growing in test fields had acquired herbicide resistance genes from neighboring transgenic plants. In 2006, most commercial supplies of rice were contaminated with Liberty Link, an herbicide resistant GMO that was not intended to be a food crop.

These and many other genetically modified organisms have been inadvertently introduced into the environment. Propose why their escape may or may not be a problem.

4. Scientists at Oregon Health & Science University produced ANDi, the first transgenic primate, by inserting a jellyfish gene into the fertilized egg of a rhesus monkey. The gene encodes green fluorescent protein—a rather unimaginative name for a protein that fluoresces green (Figure 15.17). The long-term goal of this project is not to make monkeys that glow green, but the transfer of human genes into organisms with genomes most like ours.

Would be okay to transfer genes that govern distinctly human traits into chimpanzees or other primates? What would be some of the implications if the engineered animals developed a human trait such as speech?

5. Animal viruses can mutate so that they infect humans, occasionally with disastrous results. In 1918, an influenza pandemic that apparently originated with a strain of avian flu killed 50 million people.

Researchers recently isolated samples of that virus, the influenza A(H1N1) strain, from bodies of infected people that had been preserved in Alaskan permafrost since 1918. From the samples, the researchers reconstructed the DNA sequence of the entire viral genome, then reconstructed the actual virus. Being 39,000 times more infectious than modern influenza strains, the reconstructed A(H1N1) virus proved to be 100 percent lethal in mice.

Understanding how the A(H1N1) strain works in a live body can help us defend ourselves against other strains that may be like it. For example, researchers are using the reconstructed virus to discover which of its mutations made it so infectious and deadly in humans. Their work is urgent. A deadly new strain of avian influenza in Asia shares some mutations with the A(H1N1) strain. Even now, researchers are working to test the effectiveness of antiviral drugs and vaccines on the reconstructed virus, and to develop new ones.

Critics of the A(H1N1) reconstruction are concerned. If the virus escapes the containment facilities (even though it has not done so yet), it might cause another pandemic. Worse, terrorists could use the published DNA sequence and methods to make the virus for horrific purposes. Do you think this research makes us more or less safe?

III Principles of Evolution

Two male frigate birds (*Fregata minor*) in the Galápagos Islands, far from the coast of Ecuador. Each male inflates a gular sac, a balloon of red skin at his throat, in a display that may catch the eye of a female. The males lurk together in the bushes, sacs inflated, until a female flies by. Then they wag their head back and forth and call out to her. Like other structures that males use only in courtship, the gular sac probably is an outcome of sexual selection—one of the topics you will read about in this unit.

16 EVIDENCE OF EVOLUTION

IMPACTS, ISSUES Measuring Time

How do you measure time? Is your comfort level with the past limited to your own generation? Probably you can relate to a few hundred years of human events, but how about a few million? Understanding the distant past requires a huge intellectual leap from the familiar to the unknown. Perhaps the idea of an asteroid slamming into Earth will help you make that leap. Asteroids are minor planets hurtling through space. They range in size from 1 to 1,500 kilometers across. Millions of them orbit around the sun between Mars and Jupiter—cold, stony leftovers from the formation of our solar system.

Asteroids are hard to spot because they do not emit light. Many cross Earth's orbit, but most of those pass us by before we know about them. Some have passed too close for comfort.

The mile-wide Barringer Crater in Arizona is hard to miss (Figure 16.1*a*). A 330,000-ton asteroid made this impressive pock in the sandstone when it slammed into Earth 50,000 years ago. The impact was 150 times more powerful than the bomb that leveled Hiroshima.

No human witnessed the impact, so how do we know what happened? Sometimes we have physical evidence of events that occurred before we were around to see them. In this case, geologists were able to infer the most probable cause of the Barringer Crater by analyzing tons of meteorites, pulverized sand, and other rocky clues.

Similar evidence points to even larger asteroid impacts. For example, an unusual layer of clay formed all around Earth 65 million years ago (Figure 16.1*b*). The clay is rich in iridium, an element that is rare on Earth's surface, but

See the video! Figure 16.1 From evidence to inference. (**a**) The Barringer Crater in Arizona is 1.6 kilometers (1 mile) wide, and 170 meters (550 feet) deep. What made it? Rocky evidence points to a 300-ton asteroid that collided with Earth 50,000 years ago. (**b**) Bands that are part of an iridium-rich layer of rock that formed 65 million years ago, worldwide. The layer correlates with an abrupt transition in the fossil record that marks the time of a mass extinction. The red pocketknife gives you an idea of scale.

How would you vote? A huge asteroid could wipe out civilization and much of Earth's biodiversity. Should nations around the world contribute resources to locating and tracking asteroids? See ThomsonNOW for details, then vote online.

common in asteroids. After finding this iridium layer, researchers looked for evidence of an asteroid big enough to cover the entire Earth with its debris. Buried under Mexico's Yucatán Peninsula, they found a crater about 65 million years old. It is so big—273.6 kilometers (170 miles) across and 1 kilometer (3,000 feet) deep—that no one had even noticed it before. The crater is evidence of an asteroid impact *40 million times* more powerful than the one at Barringer.

Such impacts would have influenced the history of life in a big way. For example, long before the discovery of the Yucatán impact crater, fossil hunters knew about a mass extinction—a loss of major groups of organisms, including the last of the dinosaurs. It occurred 65 million years ago, at what is known as the K–T boundary. There are plenty of dinosaur fossils below the iridium layer, which marks the boundary. Above it, there are none, anywhere. Coincidence? Most scientists say no. Some catastrophe of global proportion occurred 65 million years ago—and the evidence points to a colossal asteroid impact.

Did asteroid impacts influence human history as well? As fossils tell us, in the 5 million years before modern humans appeared, several humanlike species arose in Africa. Most of them vanished, but not our ancestors. We know that at least twenty asteroids collided with Earth during that time span. Were our own ancestors just plain lucky? For instance, about 2.3 million years ago, a huge object plunged into the ocean west of what is now Chile. If it had struck the rotating Earth just a few hours earlier, it would have landed in Africa instead of the Pacific Ocean, and our ancestors might have been incinerated.

You are about to make an intellectual leap through time, to places that were not even known about a few centuries ago. We invite you to launch yourself from this premise: *Natural phenomena that occurred in the past can be explained by the same physical, chemical, and biological processes that operate today.*

That premise is the foundation for scientific research into the history of life. The research represents a shift from experience to inference—from the known to what can only be surmised. It gives us astonishing glimpses into the past.

Key Concepts

EMERGENCE OF EVOLUTIONARY THOUGHT

Long ago, Western scientists started to catalog previously unknown species and think about their global distribution. They discovered similarities and differences among major groups, including those represented as fossils in layers of sedimentary rock. **Sections 16.1, 16.2**

A THEORY TAKES FORM

Evidence of evolution, or changes in lines of descent, gradually accumulated. Charles Darwin and Alfred Wallace independently developed a theory of natural selection to explain how heritable traits that define each species evolve. **Section 16.3**

EVIDENCE FROM FOSSILS

The fossil record offers physical evidence of past changes in lines of descent. **Sections 16.4, 16.5**

EVIDENCE FROM BIOGEOGRAPHY

Correlating evolutionary theories with geologic history helps explain the distribution of species, past and present. **Section 16.6**

EVIDENCE FROM COMPARATIVE MORPHOLOGY

Species of different lineages often have similar body parts that may be evidence of descent from a shared ancestor. **Sections 16.7, 16.8**

EVIDENCE FROM COMPARATIVE BIOCHEMISTRY

Molecular comparisons help us discover and confirm relationships among species and lineages. **Section 16.9**

ORGANIZING THE EVIDENCE

Species are identified, named, and classified. Evolutionary tree diagrams are based on the premise that all species interconnect through shared ancestors—some remote, others recent. A current tree groups all organisms into three domains: Bacteria, Archaea, and Eukarya. **Sections 16.10–16.12**

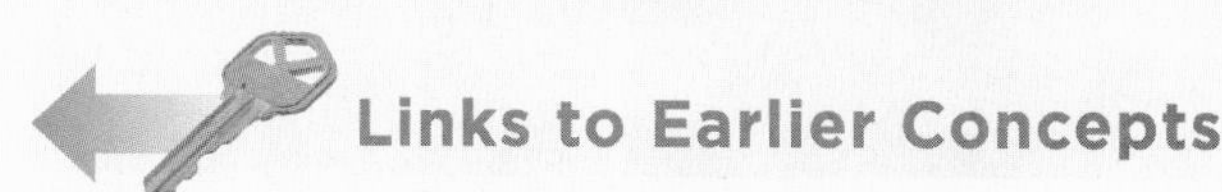

Links to Earlier Concepts

Section 1.4 sketched out the key premises of the theory of natural selection. Now you will consider evidence that led to its formulation. Remember that science does not deal with the supernatural, and it cannot answer subjective questions about nature (1.5, 1.6). You may wish to review radioisotope decay (2.2), genetic change and its effects (11.5, 13.5), and how gene controls work (14.1, 14.2, 14.3). You will build on an early introduction to species classification (1.3), and will see how biochemical and molecular tools (15.2, 15.3, 15.4) are clarifying evolutionary relationships.

16.1 Early Beliefs, Confounding Discoveries

Prevailing beliefs can influence how we interpret clues to natural processes and their observable outcomes.

QUESTIONS FROM BIOGEOGRAPHY

LINK TO SECTION 1.3

The seeds of biological inquiry were taking hold in the West more than two thousand years ago, with Aristotle the philosopher. There were no books or instruments to guide him, and yet he was more than a collector of random observations. In his writings we see evidence that he was connecting observations in an attempt to explain the order of things. He, like few others of his time, saw nature as a continuum of organization, from lifeless matter through complex forms of plants and animals. Aristotle was one of the first naturalists.

By the fourteenth century, Aristotle's ideas had been changed into a rigid view of life. A Chain of Being was seen as extending from the "lowest" forms, to humans, to spiritual beings. Each kind of being, or species, was an individual link in the chain. All of those links had been designed and forged at the same time, and they had not changed since then. Once the naturalists had discovered and described every link, the meaning of life would be revealed.

Then Europeans embarked on their globe-spanning explorations. They soon discovered that the world is a lot bigger than Europe. Tens of thousands of unique plants and animals from Asia, Africa, the New World, and the Pacific Islands were brought back home to be carefully catalogued as more links in the chain.

Later, Alfred Wallace and a few other naturalists moved beyond cataloguing species just to name them. They started to identify patterns in where species live and how species might or might not be related. These naturalists were pioneers in biogeography: the study of patterns in the geographic distribution of species. They were the first to think about the ecological and evolutionary forces that shape life on Earth.

Some patterns were intriguing. For example, globe-trotting naturalists discovered animals and plants that live only on certain islands in the middle of the ocean or in other remote places. Some of the isolated species are strikingly similar to animals and plants that live far across vast expanses of open ocean, or on the opposite side of impassible mountain ranges.

Take the three long-necked, long-legged, flightless birds shown in Figure 16.2*a–c*. Each lives on a different continent, but all sprint about in flat, open grasslands that are about the same distance from the equator. All raise their long necks to watch out for predators. These birds appear to be related (and they are), but how did they end up in geographically isolated environments?

Though the two plant species in Figure 16.2*d,e* are found on different continents, they also appear to be

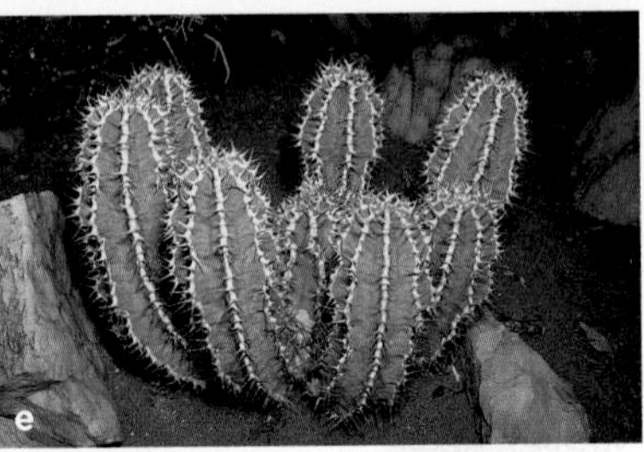

Figure 16.2 Species that resemble one another, strikingly so, even though they are native to distant geographic realms.

(**a**) South American rhea, (**b**) Australian emu, and (**c**) African ostrich. All three types of birds live in similar habitats. They are unlike most birds in several notable traits such as their long, muscular legs and their inability to fly.

(**d**) A spiny cactus native to the hot deserts of the American Southwest. (**e**) A spiny spurge native to southwestern Africa.

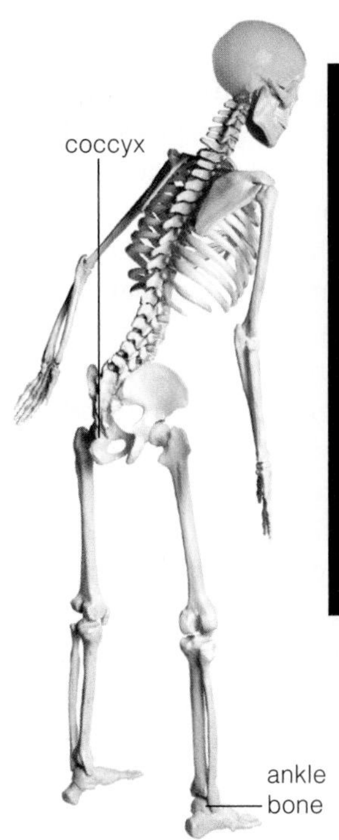

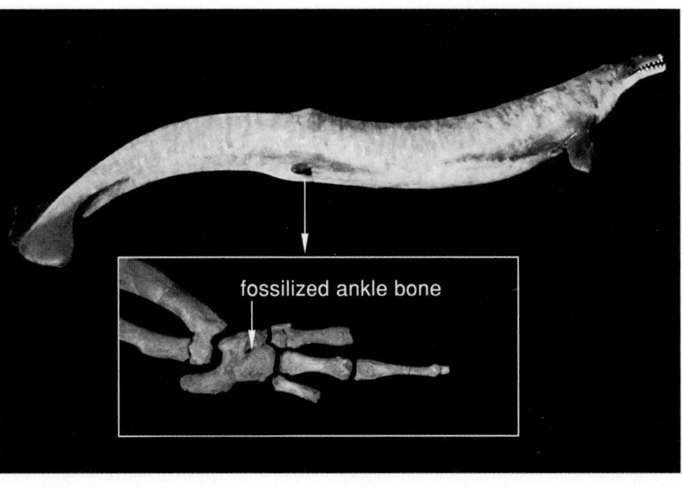

Figure 16.3 Body parts with no apparent function. *Above*, reconstruction of an ancient whale (*Basilosaurus*), with a head as long as a sofa. This marine predator was fully aquatic, so it did not use its hindlimbs to support body weight as you do—yet it had ankle bones. We use our ankles, but not our coccyx bones.

Figure 16.4 Fossilized ammonites. These large marine predators lived hundreds of millions of years ago. Their shell is similar to the shell of a modern chambered nautilus.

related. Each lives about the same distance from the equator in the same kind of environment—hot deserts—where water is seasonally scarce. Both species have rows of sharp spines that deter herbivores, and both store water in their thick, fleshy stems. However, their reproductive parts are very different, so they cannot be very closely related. How can two plant species that are at best distant cousins on different continents end up looking so much alike?

QUESTIONS FROM COMPARATIVE MORPHOLOGY

The similarities and differences among species raised questions that gave rise to comparative morphology: the study of body plans and structures among groups of organisms. Some organisms that are outwardly very similar may be quite different internally; think of fishes and porpoises. Others differ in outward appearance but have very similar body plans. As Section 16.7 explains, the bones inside a human arm, porpoise flipper, and bat wing are comparable. They develop the same way in embryos. Why do species that differ so much in some features look so much alike in other features?

By one hypothesis, body plans are so perfect there was no need for each organism to have a new design at the time of creation. Yet if that were so, then why do some organisms have body parts with no function? For instance, an ancient aquatic whale had ankle bones but it did not walk (Figure 16.3). Why the bones? Our coccyx is like the tailbone in many other mammals. We do not have a tail. Why do we have the parts of one?

QUESTIONS ABOUT FOSSILS

Geologists had been mapping layers of rock exposed by erosion or quarrying. As you will read later, they added to the confusion when they discovered identical sequences of rock layers in different parts of the world. Fossils in the layers came to be recognized as stone-hard evidence of earlier forms of life. Figures 16.3 and 16.4 show examples.

Some of the fossils were puzzling. As one example, many deep layers of rock had fossils of simple marine life. Rock layers above them held fossils that looked very similar but were more intricate. In higher layers, fossils that were similar but even more intricate looked like modern species. If every species had been created in a perfect state, then what could those sequences in complexity mean? As well, explorers were unearthing fossils of huge animals with no living representatives. If the animals had been perfect, why were they extinct?

Taken as a whole, the findings from biogeography, comparative morphology, and geology did not fit with prevailing beliefs of the nineteenth century. Scholars started thinking about new hypotheses. If species had not been created in a perfect state (and fossil sequences and "useless" body parts implied they had not), then perhaps species had changed over time.

Awareness of biological evolution emerged over centuries, through the cumulative observations of many naturalists, biogeographers, comparative anatomists, and geologists.

16.2 A Flurry of New Theories

Nineteenth-century naturalists found themselves trying to reconcile the evidence of change with a traditional conceptual framework that simply did not allow for it.

SQUEEZING NEW EVIDENCE INTO OLD BELIEFS

LINKS TO SECTIONS 1.5, 1.6

Around 1800, a respected anatomist, Georges Cuvier, was among those trying to make sense of the growing evidence of evolution—change that occurs in a line of descent. For years he had compared fossils with living organisms. He knew about abrupt changes in the fossil record and was the first to recognize that they marked times of mass extinctions.

The prevailing belief was that no species had ever become extinct, because every one had been created in a perfect state. This belief was inconsistent with many newly discovered fossils of animals that had no living counterpart. Cuvier attempted to explain how species could become extinct even if they were perfect. By his hypothesis, global catastrophes shaped Earth's surface suddenly and violently. Such catastrophes would have caused the extinction of many species. The hypothesis that Earth's surface was shaped by sudden, worldwide geologic forces very different from those operating in the present was later called catastrophism.

Other scholars kept at the puzzle. For example, Jean-Baptiste Lamarck thought that offspring inherit traits a parent *acquired during its lifetime*. By his hypothesis, environmental pressures and internal needs promote permanent changes in an individual's body form and function, which offspring can inherit. By this proposed process, life was created in a simple state. It gradually improved because it had a drive toward perfection, up the Chain of Being. Lamarck thought the drive directed an unknown "fluida" into body parts needing change.

Try using Lamarck's hypothesis to explain why the giraffe's neck is very long. We might predict that some short-necked ancestor of the modern giraffe stretched its neck to browse on leaves beyond the reach of other animals. The stretches may have even made its neck a bit longer. By Lamarck's hypothesis, that individual's offspring would inherit a longer neck, and after many generations that strained to reach ever loftier leaves, the modern giraffe would have been the result.

Like catastrophism, Lamarck's hypothesis was not supported by observations or experiments, so it faded from scientific study. The environment *does* influence change in lines of descent, but not in the way Lamarck thought. An individual's phenotype may change, as when a person develops large muscles after strength training. However, a child of a muscle-bound parent is not born muscle-bound. Offspring inherit genes, but not a parent's phenotypic changes.

VOYAGE OF THE *BEAGLE*

In 1831, in the midst of the confusion, Charles Darwin was twenty-two years old and wondering what to do

Figure 16.5 (**a**) Charles Darwin. (**b**) A replica of the *Beagle* sails off a rugged South American coastline. During one of his voyages, Darwin ventured into the Andes, where he found fossils of marine organisms in rock layers 3.6 kilometers above sea level.

(**c–e**) The Galápagos Islands are isolated in the ocean, far to the west of Ecuador. They arose by volcanic action on the seafloor about 5 million years ago. Winds and currents carried organisms to the once-lifeless islands. All of the native species are descended from those travelers. At far right, a blue-footed booby, one of many species Darwin observed during his voyage.

with his life. Ever since he was eight, he had wanted to hunt, fish, collect shells, or just watch insects and birds—anything but sit in school. Later, at his father's insistence, he did attempt to study medicine in college. The crude, painful procedures being used on patients in that era sickened him. His exasperated father urged him to become a clergyman, and so Darwin packed for Cambridge, where he earned a degree in theology. Yet he spent most of his time with faculty members who embraced natural history.

John Henslow, a botanist, perceived Darwin's real interests. He hastily arranged for Darwin to become ship's naturalist aboard the *Beagle*, which was about to embark on an extended survey expedition to South America. The young man who had hated school and had no formal training in science quickly became an enthusiastic naturalist.

The *Beagle* set sail for South America in December, 1831 (Figure 16.5). During the ship's Atlantic crossing, Darwin read Henslow's parting gift, the first volume of Charles Lyell's *Principles of Geology*. What he learned allowed him insights into the geological history of the regions he would encounter.

During the *Beagle*'s five-year voyage, Darwin found many unusual fossils. He saw diverse species living in environments that ranged from sandy shores of remote islands to plains high in the Andes.

He also started to mull over a radical theory. For many years, early geologists had been chipping away at sandstones, limestones, and other rocks that form after sediments slowly accumulate in lakebeds, river bottoms, and ocean floors. The rocks held evidence that gradual processes of geologic change operating in the present had also operated in the distant past. Lyell proposed that strange catastrophes were not necessary to explain Earth's surface. Over great spans of time, gradual, everyday geologic processes such as erosion could have sculpted the current landscape.

The idea that gradual, repetitive change had shaped Earth became known as the theory of uniformity. It challenged the prevailing belief that Earth was 6,000 years old. According to traditional scholars, people had recorded everything that happened in those 6,000 years—and in all that time, no one mentioned seeing a species evolve. Even so, by Lyell's calculations, it must have taken millions of years to sculpt Earth's surface. Was that not enough time for species to evolve? Darwin thought that it was. But *how* did they evolve? He would end up devoting the rest of his life to that burning question.

The underlying cause of some process in the natural world may not be obvious. Prevailing beliefs can influence how we interpret clues about how such processes work.

Darwin's observations during a global voyage helped him think about how species evolve.

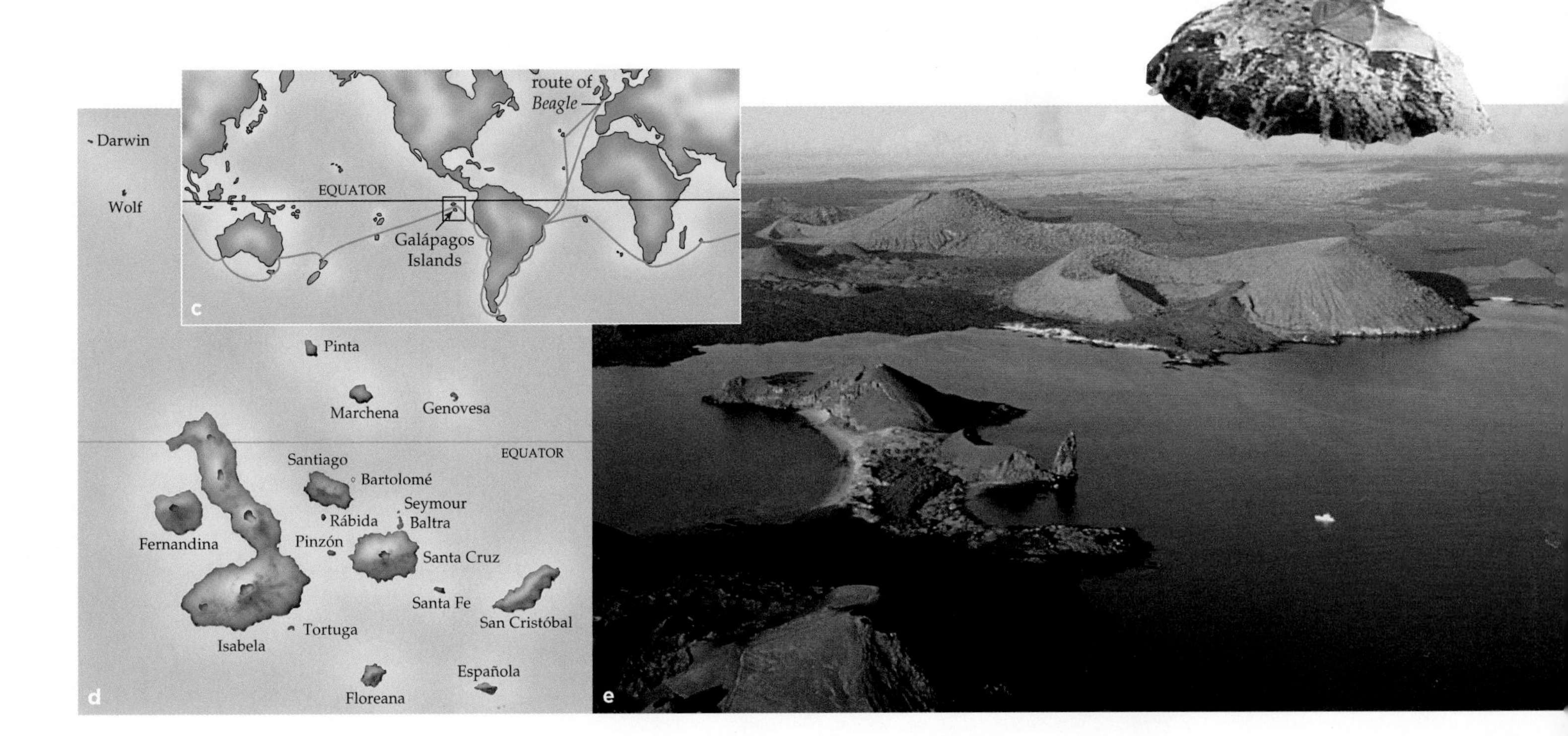

16.3 Darwin, Wallace, and Natural Selection

Darwin's observations of species in different parts of the world helped him perceive how species might evolve.

OLD BONES AND ARMADILLOS

LINK TO SECTION 1.4

Darwin sent to England the thousands of specimens he collected on his voyage. Although he had page after page of notes, he had been careless about recording where each species lived and what its habitat was like. His colleagues helped him fill in some of the blanks, and in time he was able to make enough connections to understand how species might evolve.

Among the specimens were fossils of glyptodonts from Argentina. Modern armadillos are much like the now-extinct glyptodonts (Figure 16.6). Armadillos live only in the places where glyptodonts once lived.

If these animals had been created at the same time, lived in the same place, and were so alike in certain odd traits—such as body armor made of overlapping scales—then why is one extinct and one still around? What if the glyptodonts were the ancient relatives of armadillos? What if traits of a common ancestor had changed in the line of descent that led to armadillos? *Descent with modification*—it did seem possible. If so, what was the driving force behind it?

A KEY INSIGHT—VARIATION IN TRAITS

While Darwin assessed his notes, an essay by Thomas Malthus, a clergyman and economist, made him reflect upon a topic of social interest. Malthus had correlated population size with famine, disease, and war. He said that humans run out of food, living space, and other resources because they tend to reproduce beyond the capacity of their environment to sustain them. After that happens, a population's individuals must compete with one another for resources, or develop methods to increase productivity.

Darwin deduced that *any* population has a capacity to produce more individuals than the environment can support. Even one sea star can produce 2,500,000 eggs per year, but the seas do not fill with sea stars.

Darwin also reflected on species he had observed during his voyage. Individuals of those species were not alike in their details. They varied in size, in color, and in other traits. *It dawned on Darwin that variations in traits can influence an individual's ability to secure limited resources—and thus to survive and reproduce.*

He thought about the Galápagos Islands, separated from South America by 900 kilometers of open ocean. He knew that most of their finch species live nowhere else, yet they share traits with mainland species. Could fierce storms have blown a few mainland birds far out to sea? If winds and currents had carried the birds as far as the Galápagos, then those modern species may be island-hopping descendants of the colonizers.

As he knew, different species live in diverse habitats near coasts, in dry lowlands, and in mountain forests. One strong-billed type is better than others at cracking open hard seeds (Figure 16.7). During droughts, soft seeds are scarce. A strong-billed individual might have a better chance of surviving and reproducing than a weak-billed individual during a drought. Its bill size has a heritable basis, so its offspring might have the same advantage. Eventually, strong-billed birds would predominate in a drought-stricken population. Thus, conditions in the prevailing environment "select" those individuals that are best suited to that environment.

Figure 16.6 (**a**) From Texas, a modern armadillo, about a foot long excluding the tail. (**b**) A Pleistocene glyptodont, which was about as big as a Volkswagen Beetle, and now extinct. Glyptodonts shared unusual traits and a restricted distribution with the existing armadillos. Yet the two kinds of animals are widely separated in time. Their similarities were a clue that helped Darwin develop a theory of evolution by natural selection.

Figure 16.7 Three of the thirteen species of finches that evolved on the Galápagos Islands.

(**a**) A big-billed seed cracker, *Geospiza magnirostris*.

(**b**) *G. scandens* eats cactus fruit and insects in cactus flowers.

(**c**) *Camarhynchus pallidus* uses cactus spines and twigs to probe for wood-boring insects.

NATURAL SELECTION DEFINED

Let's now put Darwin's observations and conclusions into the context and terms of modern genetics:

1. *Observation:* Natural populations have an inherent reproductive capacity to increase in size over time.

2. *Observation:* No population can expand indefinitely, because its individuals will run out of food, living space, and other resources.

3. *Inference:* Sooner or later, individuals will end up competing for dwindling resources.

4. *Observation:* Individuals of a population have the same genes—heritable information about traits.

5. *Observation:* Variations in traits begin with alleles, slightly different molecular forms of genes that arise through mutations.

6. *Inferences:* Certain forms of traits prove better than others at helping an individual compete for resources, survive, and reproduce. Over the generations, alleles for such traits become more common in a population. They increase an individual's fitness—its adaptation to an environment, as measured by its relative genetic contribution to future generations.

7. *Conclusions:* Environmental factors act on the range of variation in traits that are shared by individuals of a population. The differential survival and reproduction of such individuals is called natural selection.

Darwin kept on looking for patterns in his data and filling in gaps in his reasoning. He also wrote out his theory but let ten years pass without publishing it. He waited too long. Alfred Wallace, a naturalist who was at the time exploring wildlife in the Amazon Basin and the Malay Archipelago, wrote an essay and sent it to Darwin for advice. Wallace's essay outlined Darwin's theory! Wallace had written earlier letters to Lyell and Darwin about patterns in the geographic distribution of species; he too, had connected the dots. Wallace is now called the father of biogeography (Figure 16.8).

Figure 16.8 Alfred Wallace. For one account of the Darwin–Wallace story, read David Quammen's *Song of the Dodo*.

In 1858, just weeks after Darwin received Wallace's essay, their similar theories were presented jointly at a scientific meeting. Wallace was still in the field and knew nothing about the meeting, which Darwin did not attend. The next year, Darwin published *On the Origin of Species*, which laid out detailed evidence in support of his theory.

You may have heard that Darwin's book fanned an intellectual firestorm, but most scholars were quick to accept descent with modification, or evolution. They did fiercely debate the idea that evolution occurs by natural selection. Decades passed before experimental evidence from genetics research led to its widespread acceptance in the scientific community.

As Darwin and Wallace perceived, natural selection is the differential in survival and reproduction among individuals of a population that vary in the details of their shared traits.

Natural selection can lead to increased fitness, which is an individual's adaptation to the environment as measured by its relative genetic contribution to future generations.

16.4 Fossils—Evidence of Ancient Life

Fossils are remnants or traces of organisms that lived in the past. The fossil record holds clues to life's evolution.

LINK TO SECTION 2.3

About 500 years ago, Leonardo da Vinci was puzzled by seashells entombed in the rocks of northern Italy's high mountains, hundreds of kilometers from the sea. How did they get there? The prevailing belief was that water from a stupendous, divinely invoked flood had surged up into the mountains, where it deposited the shells. But many shells were thin, fragile, and intact. If they had been swept across such great distances, then wouldn't they be battered to bits?

Leonardo also brooded about the rocks. They were stacked like cake layers. Some layers had shells; others had none. Then he remembered how large rivers swell with spring floodwaters and deposit silt in the sea. Did such depositions happen in ancient seasons? If so, then shells in the mountains could be evidence of layered communities of organisms that once lived in the seas!

In the 1700s, fossils were accepted as remains and impressions of organisms that lived in the past. (*Fossil* comes from a Latin word for "something that was dug up.") However, people were still interpreting fossils through the prism of cultural beliefs, as when a Swiss naturalist discovered a fossil of a giant salamander and excitedly announced it was the skeleton of a man who had drowned in the great flood.

By midcentury, naturalists were questioning such interpretations. Excavations for canals, quarries, and mines were unearthing similar rock layers with similar sequences of fossils in distant places, such as the cliffs on both sides of the English Channel. If the layers had been deposited over time, then a vertical sequence of fossils embedded in them might be a record of past life—*a fossil record*.

HOW DO FOSSILS FORM?

Most fossils are remains of bones, teeth, shells, seeds, spores, or other hard parts (Figure 16.9*a*,*b*). *Trace* fossils are indirect evidence of an organism—impressions of its body, tracks, burrows, nests, and so on. Coprolites are fossilized feces (Figure 16.9*c*).

Fossilization is a slow process that starts when an organism or traces of it become covered by sediments or volcanic ash. Water slowly infiltrates the remains, and metal ions and other inorganic compounds that are dissolved in it replace the minerals in bones and other hardened tissues. Sediments that accumulate on top of the remains exert increasing pressure on them. In time, the pressure and mineralization transform the remains into stony hardness.

Remains that become buried quickly are less likely to be obliterated by scavengers. Preservation is also favored when a burial site stays undisturbed. Usually, however, erosion and other geologic assaults deform, crush, break, or scatter the fossils. This is one reason fossils are relatively rare.

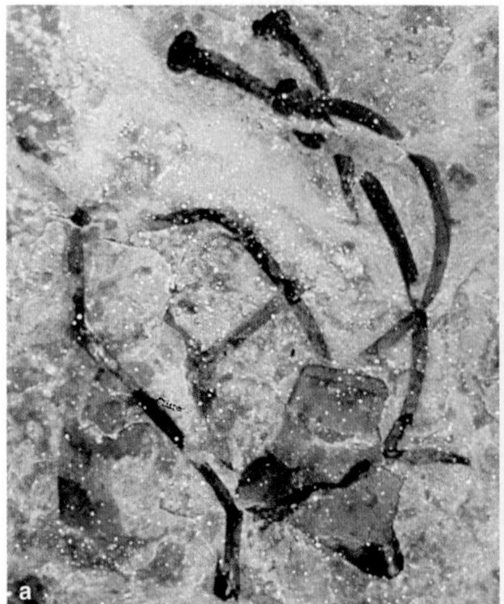

Figure 16.9 Fossils. (**a**) Fossil of one of the oldest known land plants (*Cooksonia*). Its stems were about as long as a toothpick. (**b**) Fossilized skeleton of an ichthyosaur. This marine reptile lived about 200 million years ago. (**c**) Coprolite. Fossilized food remains and parasitic worms inside such fossilized feces tell us about the diet and health of extinct species. A foxlike animal excreted this one.

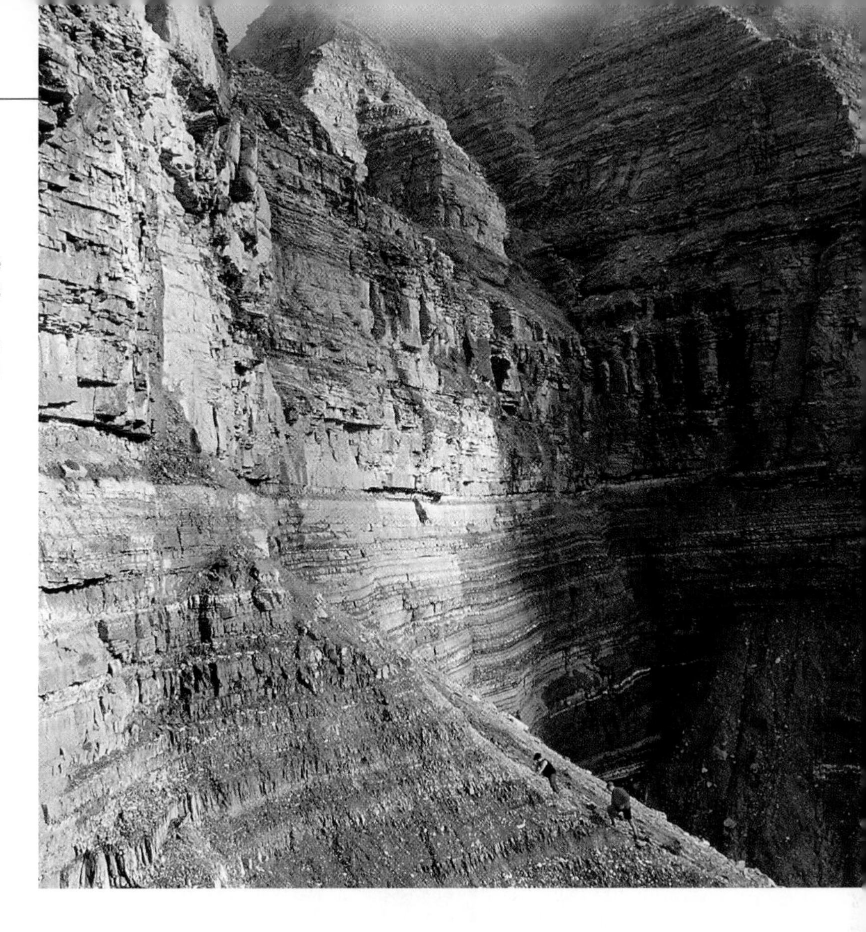

Figure 16.10 A slice through time—Butterloch Canyon, Italy, once at the bottom of a sea. Sediments piled up on the seafloor year after year. After hundreds of millions of years, they became compacted into rock-hard layers. Geologic forces lifted the layers above sea level; then, river water carved the canyon walls and exposed the layers. Scientists Cindy Looy and Mark Sephton are climbing to the Permian–Triassic boundary layer, where they will look for fossilized fungal spores.

Other factors affect preservation. Organic materials cannot decompose in the absence of free oxygen, for instance. Remains may endure encased in sap, tar, ice, mud, or other air-excluding substance. A 40,000-year-old woolly mammoth frozen in arctic ice is an example.

FOSSILS IN SEDIMENTARY ROCK LAYERS

Most fossils are found in layers of sedimentary rock. Understand how the layers formed, and you can find clues about the distant past in them. As da Vinci had suspected, sedimentary rock began forming long ago when rivers washed silt, sand, volcanic ash, and other particles from land to sea. The particles settled on the seafloors in horizontal layers that varied in thickness and composition. After hundreds of millions of years, the layers of sediments became compacted into layers of rock. The formation of layered sedimentary rock is called stratification.

Researchers study the layers in order to understand the historical context of the fossils they find in them. The deeper layers were the first to form; those closest to the surface were the last. Thus, the deeper the layer, the older the fossils it contains. A layer's composition may hold clues to local or global events that occurred as it formed; the iridium layer is an example. Relative thicknesses of different layers may reflect changes in climate. For instance, sedimentation patterns changed in ice ages. Rivers dried up as tremendous volumes of water froze and became locked in glaciers. When the glaciers melted, sedimentation resumed.

INTERPRETING THE FOSSIL RECORD

We have fossils for more than 250,000 known species. Judging from the current range of biodiversity, there must have been many, many millions more. Yet the fossil record will never be complete. Why is this so?

The odds are against finding evidence of an extinct species. Why? At least one specimen had to be buried before it decomposed or something ate it. The burial site had to escape geologic assaults and end up where someone can find it. For instance, many fossils become exposed only after a river or glacier has slowly carved out steep canyon walls (Figure 16.10).

Most ancient species had no hard parts to fossilize, so we do not find much evidence of them. Unlike the bony fishes or the hard-shelled mollusks, for example, jellyfishes and soft worms do not show up much in the fossil record, although they were probably as common.

Also think about population density and body size. One plant population might release millions of spores in a single season. The earliest humans lived in small bands and raised few offspring. What are the odds of finding even one fossilized human bone compared to the odds of finding a fossilized plant spore?

Finally, imagine one line of descent, a lineage, that vanished when its habitat on a remote volcanic island sank into the sea. Or imagine two lineages, one lasting only briefly and the other for billions of years. Which is more likely to be represented in the fossil record?

Fossils are physical evidence of organisms that lived in the remote past, a stone-hard historical record of life. In general, the oldest are in the deepest sedimentary rocks.

The fossil record is incomplete. Geologic events obliterated much of it. The record is slanted toward species that had hard parts, dense populations with wide distribution, and that persisted a long time.

Even so, the fossil record is substantial enough to help us reconstruct patterns and trends in the history of life.

16.5 Dating Pieces of the Puzzle

How do we assign fossils to a place in time? In other words, how do we know how old fossils really are?

RADIOMETRIC DATING

LINK TO SECTION 2.2

At one time, people could assign only *relative* ages to their fossil treasures, not absolute ones. For instance, a fossilized mollusk embedded in a layer of rock was said to be younger than fossils below it and older than fossils above it, and so forth.

Fossil dating became more accurate when the nature of radioisotope decay was discovered. A radioisotope, remember, is a form of an element with an unstable nucleus (Section 2.2). Atoms of a radioisotope become atoms of other elements as their nucleus disintegrates. Radioisotope decay is not influenced by temperature, pressure, chemical bonding state, or moisture; it is only influenced by time. Like the ticking of a perfect clock, each type of radioisotope decays at a constant rate into predictable products called daughter elements.

For example, radioactive uranium 238 decays into thorium 234, which decays into something else, and so on until it becomes lead 206. The time it takes for half of a radioisotope's atoms to decay into a stable product is the radioisotope's half-life (Figure 16.11*a*). The half-life of uranium 238 is 4.5 billion years.

The predictability of radioactive decay can be used to find the age of a sample, such as a rock or a fossil. In radiometric dating, the ratio of parent to daughter elements in a sample is used to calculate the sample's age. For example, the ratio of uranium 238 to lead 206 in meteorites was used to show that the solar system —and Earth—formed 4.55 billion years ago.

Recent fossils can be dated by their ratio of carbon isotopes (Figure 16.11*b–d*). After 60,000 years, most of the carbon 14 in a fossil will have decayed, so carbon isotope ratios cannot be used to date older fossils. Such fossils are dated by isotope ratios of other elements in volcanic rocks in the same sedimentary layer.

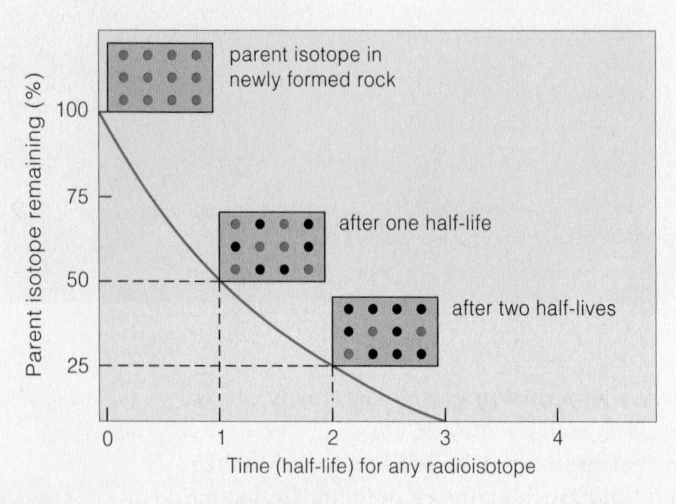

a A simple way to think about the decay of a radioisotope to a more stable isotope, as plotted against time.

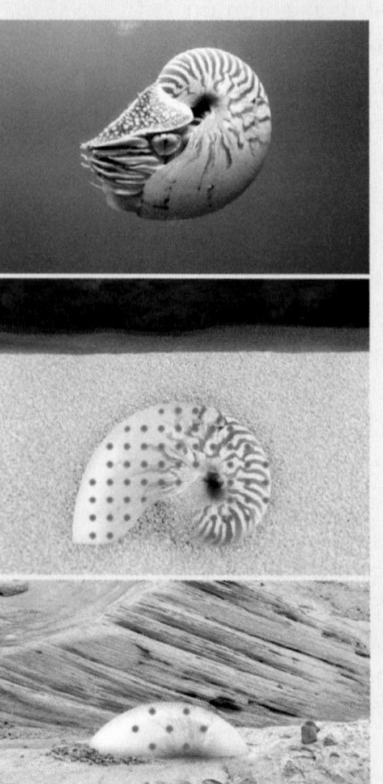

b Long ago, trace amounts of ^{14}C and a lot more ^{12}C were incorporated into tissues of a mollusk. The carbon atoms were part of organic molecules in the mollusk's food. As long as it was alive, the mollusk replenished its own tissues with carbon gained from food. Thus, the proportion of ^{14}C to ^{12}C in its tissues stayed the same.

c After the mollusk died, it no longer gained carbon. Over long spans of time, the proportion of ^{14}C to ^{12}C in its remains fell because of radioactive decay of ^{14}C. Half the ^{14}C decayed in 5,370 years, half of what was left was gone after another 5,370 years, and so on.

d Fossil hunters discover the fossil. They measure its ^{14}C to ^{12}C ratio to determine half-life reductions since death. In this example, the ratio turns out to be one-eighth of the ^{14}C to ^{12}C ratio in living organisms. Thus, the mollusk lived about 16,000 years ago.

PLACING FOSSILS IN GEOLOGIC TIME

Early geologists carefully counted backward through layers of sedimentary rock, each one a slice of geologic time. They discovered the same transitions in fossil sequences in the same rock layers around the world. The transitions became the boundaries of eons, eras, periods, and epochs—all great intervals of time in the geologic time scale, or chronology of Earth's history.

Figure 16.11 Animated! (**a**) Decay of radioisotopes at a fixed rate to more stable isotopes. The half-life of each radioisotope is the time it takes for 50 percent of a given sample to decay. After two half-lives, 75 percent of the sample has decayed, and so on.

(**b–d**) Radiometric dating of a fossil. Carbon 14 (^{14}C) forms in the atmosphere and combines with oxygen to become carbon dioxide. Along with far greater quantities of the stable carbon isotope ^{12}C, trace amounts of ^{14}C enter food chains by way of photosynthesis. All living organisms incorporate carbon in their body tissues.

Eon	Era	Period	Epoch	Millions of Years Ago	Major Geologic and Biological Events
PHANEROZOIC	CENOZOIC	QUATERNARY	Recent		
			Pleistocene	0.01	Modern humans evolve; major extinction event is now underway.
		TERTIARY	Pliocene	1.3	
			Miocene	6.3	
			Oligocene	22.0	Tropics, subtropics extend poleward. Climate cools; dry woodlands and grasslands emerge. Adaptive radiations of mammals, insects, birds.
			Eocene	33.7	
			Paleocene	65.5	
				66 ◀	Major extinction event, perhaps precipitated by asteroid impact. Mass extinction of all dinosaurs and many marine organisms.
	MESOZOIC	CRETACEOUS	Late		
				99	Climate very warm. Dinosaurs continue to dominate. Important modern insect groups appear (bees, butterflies, termites, ants, and herbivorous insects including aphids and grasshoppers). Flowering plants originate and become dominant land plants.
			Early		
				145	Age of dinosaurs. Lush vegetation; abundant gymnosperms and ferns. Birds appear. Pangea breaks up into North America, Eurasia, and Gondwana.
		JURASSIC			
				213 ◀	Major extinction event.
		TRIASSIC			Recovery from the major extinction at end of Permian. Many new groups appear, including turtles, dinosaurs, pterosaurs and mammals.
				249 ◀	Major extinction event.
	PALEOZOIC	PERMIAN			Supercontinent Pangea and world ocean form. Adaptive radiation of conifers. Cycads and ginkgos appear. Relatively dry climate leads to drought-adapted gymnosperms and insects such as beetles and flies.
				286	High atmospheric oxygen level fosters giant arthropods. Spore-releasing plants dominate. Age of great lycophyte trees; vast coal forests form. Ears evolve in amphibians; penises evolve in early reptiles (vaginas evolve later, in mammals only).
		CARBONIFEROUS			
				360 ◀	Major extinction event.
		DEVONIAN			Land tetrapods appear. Explosion of plant diversity leads to tree forms, forests, and many new plant groups including lycophytes, ferns with complex leaves, seed plants.
				410	Radiations of marine invertebrates. First appearances of land fungi, vascular plants, bony fish, and perhaps terrestrial animals (millipedes, spiders).
		SILURIAN			
				440 ◀	Major extinction event.
		ORDOVICIAN			Major period for first appearances. The first land plants, fish, and reef-forming corals appear. Gondwana moves toward the South Pole and becomes frigid.
				505	Earth thaws. Explosion of animal diversity. Most major groups of animals appear (in the oceans). Trilobites and shelled organisms evolve.
		CAMBRIAN			
				544	
PROTEROZOIC					Oxygen accumulates in atmosphere. Origin of aerobic metabolism. Origin of eukaryotic cells, then protists, fungi, plants, animals. Evidence that Earth mostly freezes over in a series of global ice ages between 750 and 600 mya.
				2,500	
ARCHAEAN AND EARLIER					3,800–2,500 mya. Origin of prokaryotes. 4,600–3,800 mya. Origin of Earth's crust, first atmosphere, first seas. Chemical, molecular evolution leads to origin of life (from proto-cells to anaerobic prokaryotic cells).

Figure 16.12 Animated! Geologic time scale. *Blue* triangles mark times of great mass extinctions. "First appearance" refers to appearance in the fossil record, not necessarily the first appearance on Earth. We often discover significantly older fossils. The spans are not to the same scale (compare Figure 16.13).

Life originated in the Archaean eon. Protists, fungi, plants, and animals all arose during the next eon, the Proterozoic. Figure 16.12 correlates the geologic time scale with macroevolution: major patterns, trends, and rates of change among lineages.

Researchers use the predictability of radioisotope decay to estimate the age of rocks and fossils.

A geologic time scale shows the chronology of life on Earth. Its dates were determined by radiometric dating.

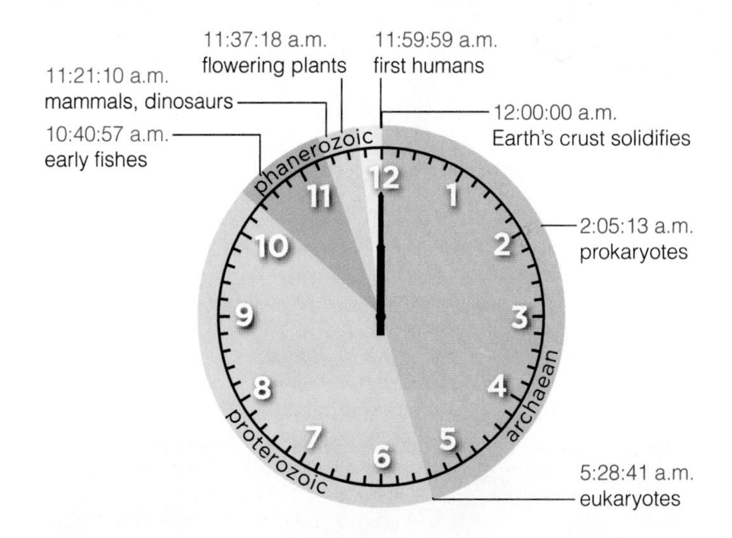

Figure 16.13 A geologic time clock. Think of the spans as minutes on a clock that runs from midnight until noon. The recent epoch started after the last 0.1 second before noon. Where does that put you?

16.6 Drifting Continents, Changing Seas

Turn now to evidence of the connection between the evolution of Earth's surface and the evolution of life.

When geologists first started to map vertical stacks of sedimentary rock, the theory of uniformity prevailed. The geologists knew that water, wind, fire, and other natural factors were continuously altering the surface of Earth. Eventually it became clear to them that such factors were part of a big picture of geologic change. Like life, Earth also changes dramatically.

AN OUTRAGEOUS HYPOTHESIS

For instance, the Atlantic coasts of South America and Africa seemed to "fit" like jigsaw puzzle pieces. One model suggested that all continents were once part of a bigger supercontinent—Pangea—that had split into fragments and drifted apart. The model explained why the same types of fossils occur in sedimentary rock on both sides of the vast Atlantic Ocean.

At first, most scientists did not accept the model, which was called continental drift. To them, continents drifting about Earth seemed to be an outrageous idea. They preferred to think that continents did not move.

However, evidence that supported the model kept turning up. For instance, molten rock deep in the Earth wells up and solidifies on the surface. Some iron-rich minerals become magnetic as they solidify, and their magnetic poles align with Earth's poles when they do. If continents never moved, then all of these ancient rocky magnets would be aligned north-to-south, like compass needles. Indeed, the magnetic poles of rock formations on different continents are aligned, but not with Earth's poles. The most likely explanation is that the poles wander, the continents wander, or both do.

Then, deep-sea explorers discovered that the ocean floors are not as static and featureless as was assumed. Immense ridges stretch out thousands of kilometers across the seafloor. Molten rock spewing from these ridges pushes old seafloor outward in both directions, then cools and hardens into new seafloor. Elsewhere, older seafloor plunges into deep trenches.

Such discoveries swayed the skeptics. Finally, there was a plausible mechanism for continental drift, which by then was named the plate tectonics theory. By this theory, Earth's relatively thin outer layer of rock is cracked into immense plates, like a gigantic cracked eggshell. The ridges and trenches on the ocean floor

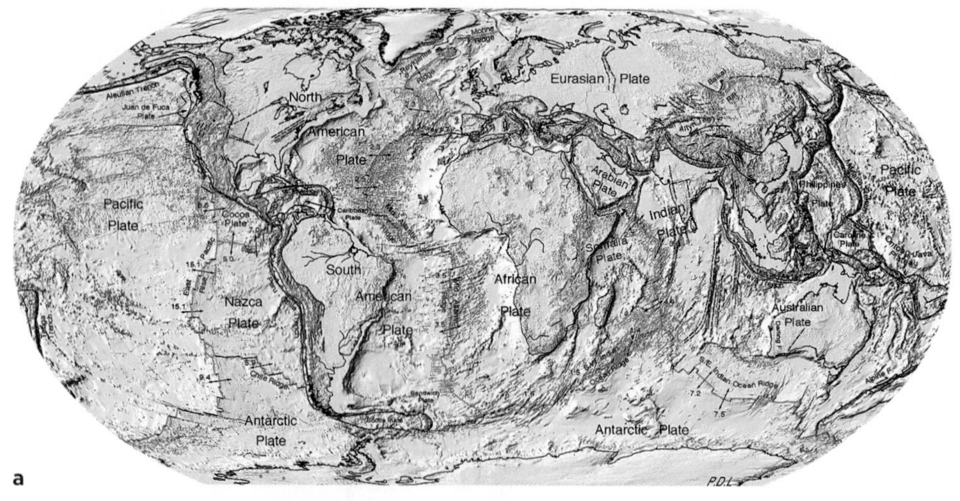

Figure 16.14 Forces of geologic change.

(**a**) Present configuration of tectonic plates. These immense, rigid portions of Earth's outer layer of rock split, drift apart, and collide, all at a rate of less than 10 centimeters a year. This map is shown enlarged in Appendix VIII.

(**b**) Huge plumes of molten rock welling up from Earth's interior drive the movement. The rock spreads laterally as it cools and hardens into new seafloor. The spreading seafloor forces tectonic plates away from the ridges and into trenches elsewhere. As the plates move, they raft continents around the globe.

The advancing edge of one plate can plow under an adjacent plate and lift it up. The Cascades, Andes, and other great mountain ranges paralleling the coasts of continents formed this way. When 2004 drew to a close, the Indian Plate lurched violently under the Eurasian Plate and caused huge tsunamis. These earthquake-generated ocean waves traveled 600 miles per hour across the Indian Ocean and killed more than 240,000 people.

Besides these forces, superplumes rupture the tectonic at what are now called "hot spots" in the mantle. The Hawaiian Archipelago has been forming this way. Continents also can rupture in their interior. Deep rifting and splitting are happening now in Missouri, at Lake Baikal in Russia, and in eastern Africa.

Figure 16.15 Animated! A series of reconstructions of drifting continents. (**a**) The supercontinent Gondwana (*yellow*) had begun to break up by the Silurian. (**b**) The supercontinent Pangea formed during the Triassic, then began to break up in the Jurassic (**c**). (**d**) K–T boundary. (**e**) The continents reached their modern configuration in the Miocene.

About 260 million years ago, seed ferns and other plants lived nowhere except on the area of Pangea that had once been Gondwana. So did mammal-like reptiles named therapsids. *Right*, fossilized leaf of one of the seed ferns (*Glossopteris*). *Far right*, a therapsid (*Lystrosaurus*) about 1 meter (3 feet) long. This tusked herbivore fed on fibrous plants in dry floodplains.

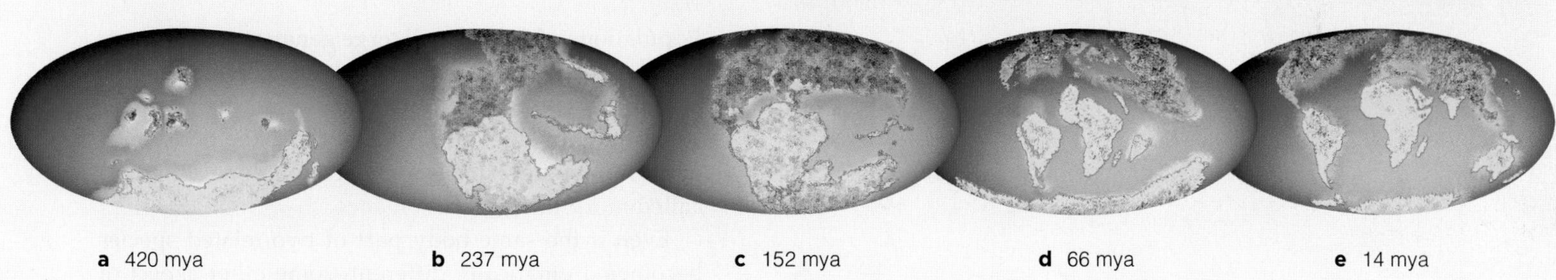

are the edges of these plates. The plates grow from the ridges and sink into the trenches. As they do, they raft land masses to new locations (Figure 16.14). The plates move no more than 10 centimeters a year, but after 40 million years or so, that is enough to carry a continent all the way around the world.

Researchers soon applied the theory to some long-standing puzzles. For example, fossils of *Glossopteris*, a seed fern, and of *Lystrosaurus*, an early reptile, occur in similar geologic formations in Africa, India, South America, and Australia (Figure 16.15). The seeds of *Glossopteris* were too heavy to float or to be blown over an ocean. The stocky, heavy reptile was not built to swim between the continents. Researchers suspected that both organisms evolved about 300 million years ago on a supercontinent called Gondwana.

Antarctica formed after Gondwana broke up. If it had once been connected to the other continents, then it would have the same geologic formations as well as *Glossopteris* and *Lystrosaurus* fossils. This prediction—and the plate tectonics theory—were supported when Antarctic explorers found the formations and fossils. Many modern species, such as the birds in Figure 16.2, live only in places that were once part of Gondwana.

A BIG CONNECTION

We now know that Gondwana drifted south, across the South Pole, then north until it merged with other land masses to form Pangea. We know that continents are always on the move. They collide, split into new continents, then collide all over again. Earth's outer layer of rock solidified 4.55 billion years ago. At least five times since then, a single supercontinent formed, with one ocean lapping at its coastline. All the while, the erosive forces of water and wind resculpted the land. So did the impacts and aftermaths of asteroids.

Such changes on land and in the ocean and atmosphere influenced life's evolution. Imagine early life in shallow, warm waters along continents. Shorelines vanished as continents collided and wiped out many lineages. Even as old habitats disappeared, new ones opened up for survivors—and evolution took off in new directions.

For billions of years, movements of Earth's outer layer as well as catastrophic events changed the land, atmosphere, and world ocean. The changes have had profound effects on the evolution of life.

16.7 Divergences From a Shared Ancestor

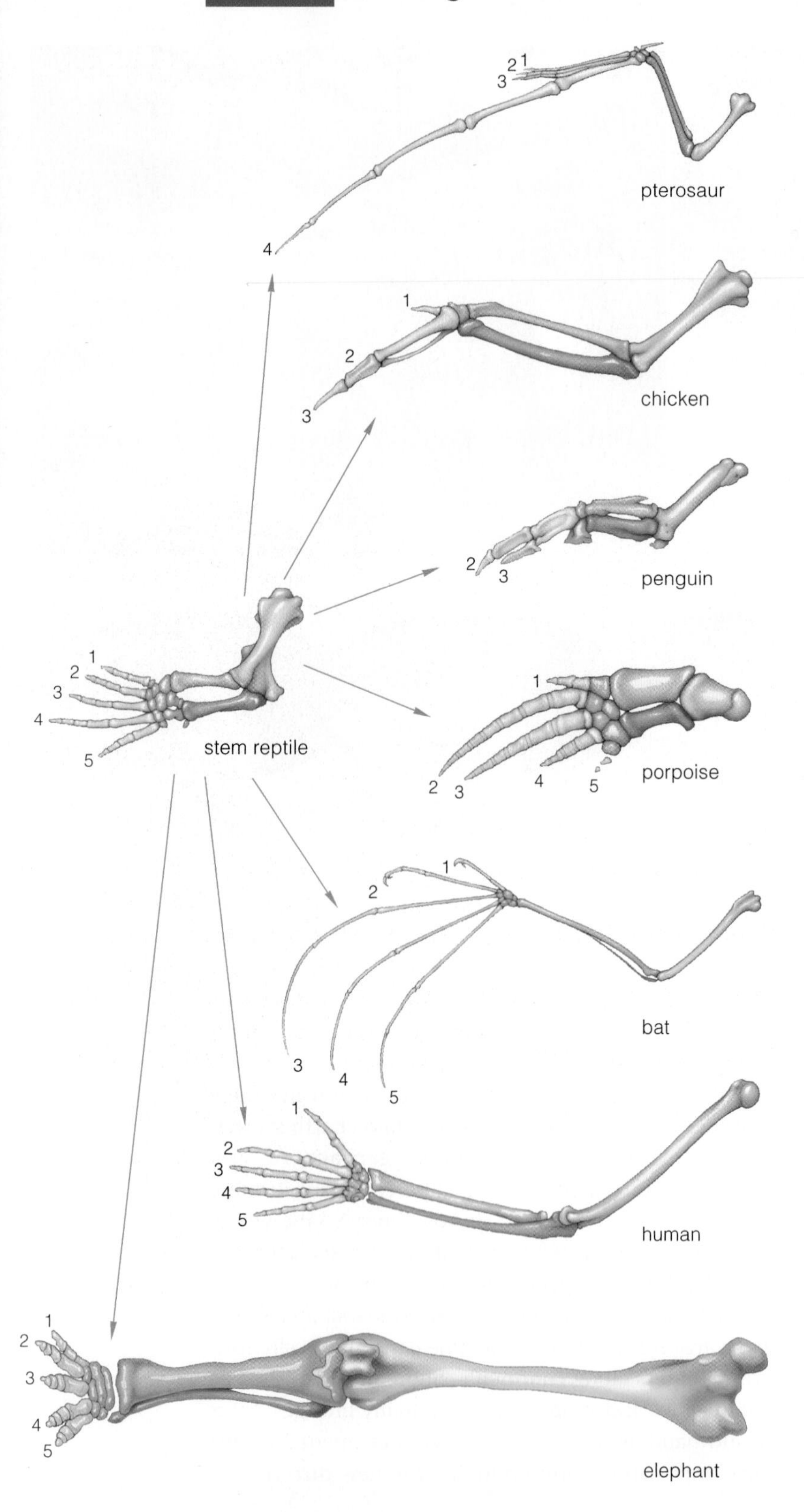

Figure 16.16 Morphological divergence among vertebrate forelimbs, starting with the bones of a stem reptile (a cotylosaur). The number and position of skeletal elements were preserved when these diverse forms evolved; notice the bones of the forearms, here colored *pink* and *orange*. Certain bones were lost over time in some of the lineages (compare the digits numbered 1 through 5). The drawings are not to the same scale.

To biologists, remember, evolution means heritable change in a line of descent. Comparisons of the body form and structures of major groups of organisms may yield clues to evolutionary relationships.

Comparative morphology, again, is the study of body forms and structures of major groups of organisms, such as vertebrates and flowering plants. (The Greek *morpho–* means body form.) Sometimes, similarities in the internal structure of one or more body parts may be evidence of a common ancestor. Similar body parts that reflect shared ancestry are homologous structures (*homo–* means the same). Such structures may be used differently in different groups, but the very same genes direct their development.

MORPHOLOGICAL DIVERGENCE

Populations of a species diverge genetically after gene flow ends between them (Chapter 17). In time, some of the morphological traits that define their species commonly diverge, also. Change from the body form of a common ancestor is a macroevolutionary pattern called morphological divergence.

Even if the same body part of two related species becomes dramatically different, some other aspect of form may remain alike. A careful look beyond unique modifications can reveal shared heritage. For instance, all land vertebrates are descended from the first four-legged amphibians. Divergences led to what we call reptiles, then to birds and mammals. We know about "stem reptiles" that probably were ancestral to these groups. Fossilized, five-toed limb bones tell us that the ancestral species crouched low to the ground (Figure 16.16*a*). Their descendants diversified into many new habitats on land. A few descendants that had become adapted for walking on land even returned to the seas after environmental conditions changed.

A five-toed limb was evolutionary clay. It became molded into different kinds of limbs having different functions (Figure 16.16). In the lineages that led to the penguins and porpoises, it became flippers used for swimming. In the lineage leading to modern horses, it became long, one-toed limbs suitable for running fast. Among the elephants, it became strong and pillarlike, suitable for supporting a great deal of weight. Among extinct reptiles called pterosaurs, most birds, and bats, it became modified for flight.

The five-toed limb also became modified into the human arm and hand. Later on, a thumb evolved in opposition to the four fingers of the hand. It was the basis of stronger and more precise motions.

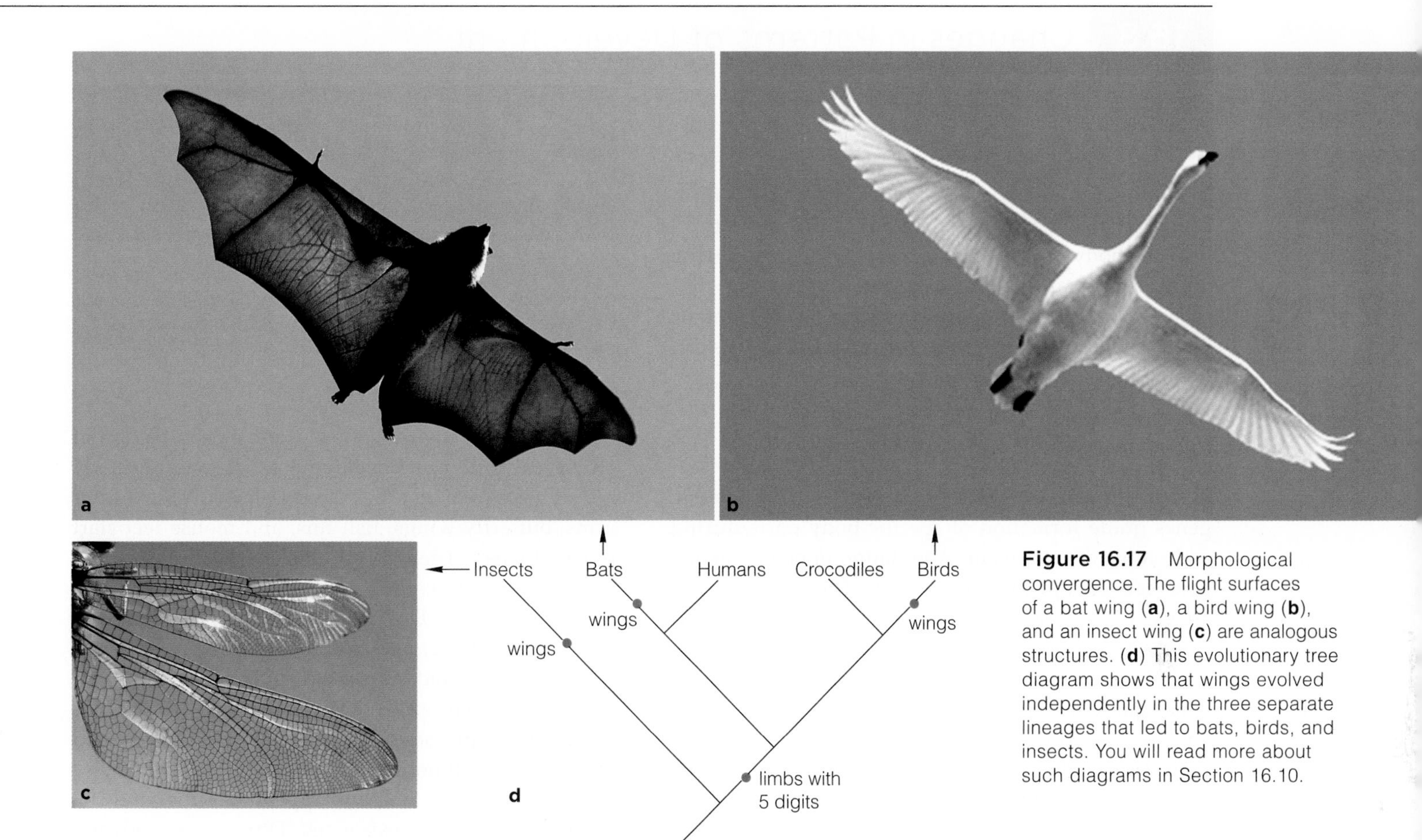

Figure 16.17 Morphological convergence. The flight surfaces of a bat wing (**a**), a bird wing (**b**), and an insect wing (**c**) are analogous structures. (**d**) This evolutionary tree diagram shows that wings evolved independently in the three separate lineages that led to bats, birds, and insects. You will read more about such diagrams in Section 16.10.

Even though vertebrate forelimbs are not the same in size, shape, or function from one group to the next, they clearly are alike in the structure and positioning of bony elements. They also are alike in the patterns of nerves, blood vessels, and muscles that develop inside them. In addition, comparisons of the early embryos of different vertebrates reveal strong resemblances in patterns of bone development. Such similarities point to a shared ancestor.

MORPHOLOGICAL CONVERGENCE

Similar body parts are not always homologous; they may have evolved independently in separate lineages as adaptations to the same environmental pressures. In this case, such parts are called analogous structures. Analogous structures look alike in different lineages but did not evolve in a shared ancestor; they evolved independently after the lineages diverged. Evolution of similar body parts in different lineages is known as morphological convergence.

We can sometimes identify analogous structures by studying their underlying form. For example, bird, bat, and insect wings all perform the same function—flight. However, several clues tell us that the flight surfaces of the wings are not homologous structures. The wing surfaces are adapted to the same physical constraints that govern flight, but the adaptations are different.

For example, although the limbs of birds and bats are homologous, the structures that make those limbs useful for flight are very different. The surface of a bat wing is a thin membrane, an extension of the animal's skin. The surface of a bird wing is a sweep of feathers, which are specialized structures derived from skin.

An insect wing forms as a saclike extension of the body wall. Except at forked veins, the sac flattens and fuses into a thin membrane. The veins are reinforced with chitin, which structurally support the wing.

The unique adaptations for flight are evidence that wing surfaces of birds, bats, and insects are analogous structures—they evolved after the ancestor of these modern groups diverged (Figure 16.17).

With morphological divergence, a body part inherited from a common ancestor becomes modified differently in different lines of descent. Such parts are homologous structures.

With morphological convergence, body parts that appear alike evolved independently in different lineages, not in a common ancestor. Such parts are analogous structures.

16.8 Changes in Patterns of Development

Comparing the patterns of embryonic development often yields evidence of evolutionary relationships.

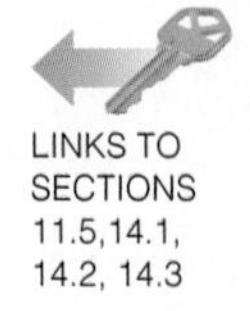

LINKS TO SECTIONS 11.5,14.1, 14.2, 14.3

In land plants and animals, multicelled embryos form from a fertilized egg. There are built-in constraints on how the body develops, step by step. Most mutations are selected against because they disrupt the steps, but other mutations may move a lineage beyond one or more of the constraints. Mutations in master genes can do just that.

Figure 16.18 Effects of mutation in one homeotic gene. (**a**) Normal and (**b**) abundantly stamened mutant flowers of wild cabbage (*Brassica oleracea*). (**c**) Normal and (**d**) petal-less mutant flowers of common wall cress (*Arabidopsis thaliana*).

GENES AND VARIATION IN PLANTS

Remember from Sections 14.2 and 14.3 that homeotic genes guide formation of specific body parts during embryonic development. A mutation in one homeotic gene can disrupt the body's form. The consequences are usually severe, but sometimes an altered body plan proves to be advantageous.

You already saw some examples of how homeotic genes guide the formation of flowers. When mutation has inactivated the floral identity gene *Apetala1*, male reproductive structures (stamens) form where petals are supposed to be in flowers of wild cabbage (Figure 16.18*a,b*). At least in the laboratory, such abundantly stamened mutants are exceptionally fertile. Mutations in *Apetala1* cause other plant species to make flowers with no petals (Figure 16.18*c,d*). A gene that functions in similar ways across different lineages is evidence of a common ancestor.

GENES AND VARIATION IN ANIMALS

How Many Legs? We have evidence of the impact of mutations on animal evolution. For example, the embryos of many vertebrate species develop in similar ways. Their tissues form the same way, as embryonic cells divide, differentiate, and interact. Development of the gut, heart, bones, muscles, and other parts is strikingly similar among these lineages (Figure 16.19).

How, then, did the adult forms of these lineages get to be so different? Part of the answer may lie with heritable changes in the onset, rate, or completion of crucial early steps in development. For example, body appendages as diverse as crab legs, beetle legs, sea star arms, butterfly wings, fish fins, and mouse feet start out as clusters of cells that bud from the surface of the embryo. Such buds form wherever the homeotic gene *Dlx* is expressed. *Dlx* encodes a transcription factor that signals clusters of embryonic cells to "stick out from the body" and give rise to an appendage.

A master gene called *Hox* helps sculpt the body's details. It suppresses *Dlx* expression in all parts of an embryo that will not have appendages. Wherever *Hox* is expressed, *Dlx* is not, and appendages do not form. Wherever *Hox* is not expressed, *Dlx* is expressed, and appendages form.

The *Dlx/Hox* gene control system operates across many phyla, which is strong evidence that it evolved long ago. *Dlx* probably came first; in some Cambrian fossils, it looks like it was not suppressed at all (Figure 16.20*a*). *Hox* gene control over *Dlx* evolved later. The variation in numbers and positioning of appendages among all complex animals may be the result (Figure 16.20*b–d*).

Chimps and Humans The last shared ancestor of chimpanzees and humans lived somewhere between 6 and 4 million years ago. Today, more than 98 percent of our DNA is identical with chimp DNA. What, then, accounts for the major morphological and behavioral differences between the two primates? Differences in genes that control growth rates may account for some

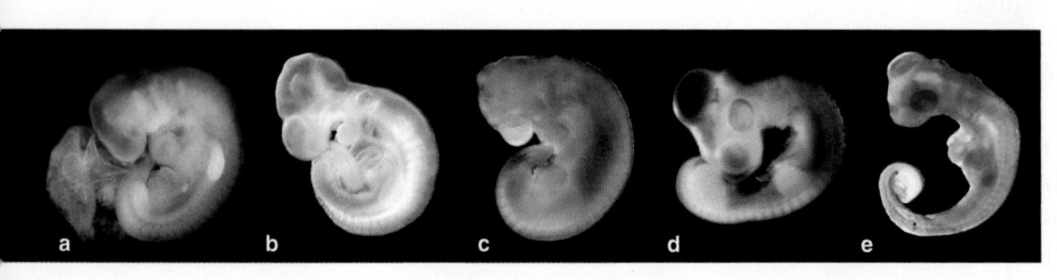

Figure 16.19 From comparative embryology, evidence of evolutionary relationships among vertebrates. Adult vertebrates are diverse, yet their embryos are similar in the early stages of development. Compare the segmented backbone, four limb buds, and tail of these early embryos of a (**a**) human, (**b**) mouse, (**c**) bat, (**d**) chicken, and (**e**) alligator.

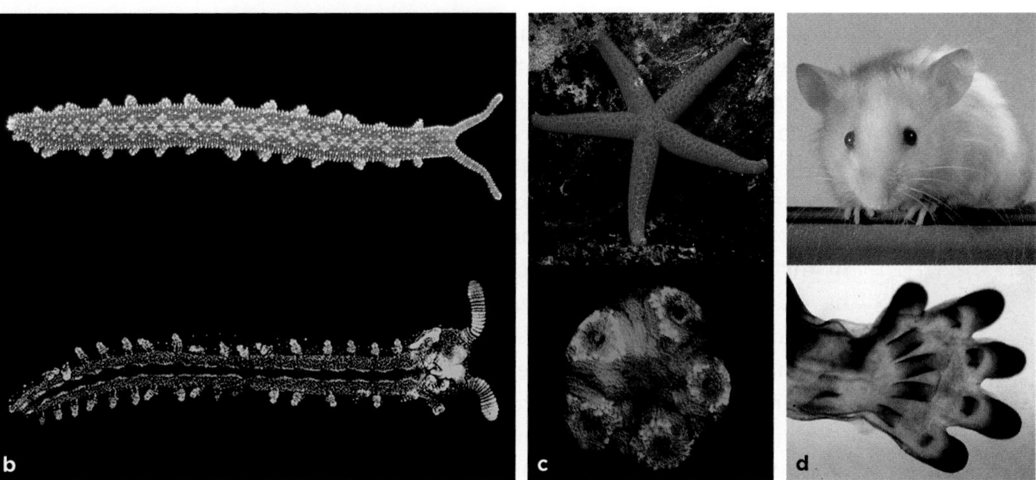

Figure 16.20 How many legs? Mutations in master genes may explain why animals differ in the number of appendages. *Dlx*, a homeotic gene, makes limbs form wherever it is expressed. (**a**) Fossil animal that may be a case of unrestricted expression of the *Dlx* gene in Cambrian times.

Variations in control over *Dlx* expression, revealed by *green* fluorescence in embryonic appendages of (**b**) a velvet walking worm and (**c**) a sea star, and (**d**) *blue* dye in a mouse embryo's foot.

proportional differences—for example, between skull bones (Figure 16.21). For humans, the bones of the face and brain chamber increase in size at about the same rate, from infant to adult. In chimps, the facial bones grow faster than the brain chamber bones. As a result, humans and chimps have notably different faces.

Many genes have been duplicated, by transposons or other mechanisms. The globin gene is an example (Section 11.5). Some genes were duplicated a different number of times in different species. The more copies of a gene, the more of its product forms during gene expression. Enhanced expression may be a factor in evolution. For example, humans have 212 copies of the gene *MGC8902*; chimps have 37, and other primates have even fewer. Mammals that are not primates have 1, and all other animals have none. The gene encodes a protein with an as-yet unknown function, but we do know that it is expressed mainly in regions of the brain associated with thinking, reasoning, and perception—all hallmarks of intelligence.

Similarities in patterns of development are often clues to shared ancestry.

Mutations in master genes are capable of launching body plans in new evolutionary directions. Changes in these and other genes that affect development may result in structural differences among related lineages.

Gene duplications may account for some major functional differences among species that have nearly identical genes.

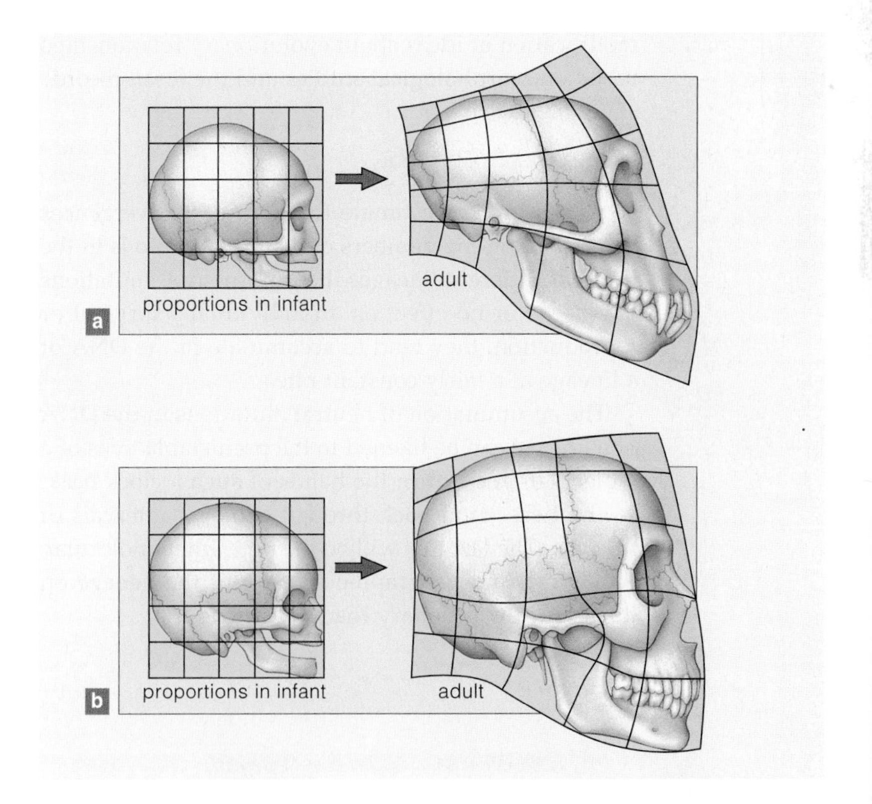

Figure 16.21 **Animated!** Thinking about the morphological differences between two primates. The skulls are depicted as paintings on a rubber sheet divided into a grid. Stretching the sheets deforms the grid. Differences in how they are stretched are analogous to different growth patterns. Shown here, proportional changes in (**a**) a chimpanzee skull and (**b**) a human skull.

16.9 Clues in DNA, RNA, and Proteins

All species have a mixture of ancestral and novel traits, including biochemical ones. The kind and number of traits that species share are clues to relationships.

LINKS TO SECTIONS 12.2, 13.2, 15.3, 15.4

Each species has its own DNA base sequence, which encodes instructions for making RNAs and proteins (Sections 12.2 and 13.2). We can expect genes to mutate in any line of descent. The more recently two lineages have diverged, the less time each one will have had to accumulate unique mutations. That is why the RNA, and the proteins, of closely related species have more similarities than those of more distantly related ones.

Identifying biochemical similarities and differences among species is now very fast and precise, thanks to DNA sequencing (Section 15.3). The sequences of new genes and proteins from many genomes are compiled continually into internationally accessible databases. With such data, we know (for example) that 31 percent of the 6,000 genes of yeast cells have counterparts in the human genome. So do 50 percent of the fruit fly genes and 40 percent of 19,023 roundworm genes.

Comparative analyses may either reinforce or invite modification of ideas about evolutionary relationships based on morphological studies and the fossil record.

MOLECULAR CLOCKS

Some researchers estimate the timing of divergences by comparing the numbers of neutral mutations in the DNA of different lineages. Because neutral mutations have little or no effect on an individual's survival or reproduction, they tend to accumulate in the DNA of a lineage at a fairly constant rate.

The accumulation of neutral mutations in the DNA of a lineage can be likened to the predictable ticks of a molecular clock. Turn the hands of such a clock back, so the ticks wind back through geologic intervals of the past. The last tick will be the time when molecular, ecological, and geographic events put the lineage on its unique evolutionary road.

How are molecular clocks calibrated? The number of differences in DNA base sequences or amino acid sequences between species is often plotted against a series of branch points that researchers have inferred from the fossil record. Section 16.11 shows how this is done. Such diagrams may reflect the relative times of divergences among species or groups of them.

PROTEIN COMPARISONS

Comparisons of protein primary structure (sequences of amino acids) can be used to decipher connections between species, and to study why some proteins have not changed as much as others. Two species with many identical proteins are likely to be close relatives. Two species with very few similar proteins probably have not shared an ancestor for a long time—long enough for many mutations to have accumulated in the DNA of their separate lineages.

A few essential genes have evolved very little; they are highly *conserved* across diverse species. One such gene encodes cytochrome *c*. This protein is a crucial component of electron transfer chains of species that range from aerobic bacteria to humans. In humans, its primary structure consists of only 104 amino acids. Figure 16.22 shows the striking similarity between the entire amino acid sequences for cytochrome *c* from a yeast, a plant, and an animal. And think about this: The *entire* amino acid sequence of human cytochrome *c* is identical with that of chimpanzee cytochrome *c*. It differs by merely 1 amino acid in rhesus monkeys, 18 in chickens, 19 in turtles, and 56 in yeasts. With this biochemical information in hand, would you predict that humans are more closely related to chimpanzees or to rhesus monkeys? To chickens or to yeast?

NUCLEIC ACID COMPARISONS

Most phenotypic differences between any two species boil down to differences in the nucleotide sequences

$^{+}NH_3$-gly asp val glu lys gly lys lys ile phe ile met lys cys ser gln cys his thr val glu lys gly gly lys his lys thr gly pro asn leu his gly leu phe gly arg lys thr gly gln ala pro gly tyr ser

$^{+}NH_3$-ala ser phe ser glu ala pro pro gly asn pro asp ala gly ala lys ile phe lys thr lys cys ala gln cys his thr val asp ala gly ala gly his lys gln gly pro asn leu his gly leu phe gly arg gln ser gly thr thr ala gly tyr ser

$^{+}NH_3$-thr glu phe lys ala gly ser ala lys lys gly ala thr leu phe lys thr arg cys leu gln cys his thr val glu lys gly gly pro his lys val gly pro asn leu his gly ile phe gly arg his ser gly gln ala glu gly tyr ser

Figure 16.22 **Animated!** Comparison of the amino acid sequence of cytochrome *c* protein from three species: yeast (*top row*), wheat (*middle*), and primate (*bottom*). *Gold* highlights the amino acids that are identical in all three sequences. The probability that such a pronounced molecular resemblance resulted by chance is extremely low. Cytochrome *c* is a vital component of electron transfer chains in cells. Its amino acid sequence has been highly conserved even in these three evolutionarily distant lineages.

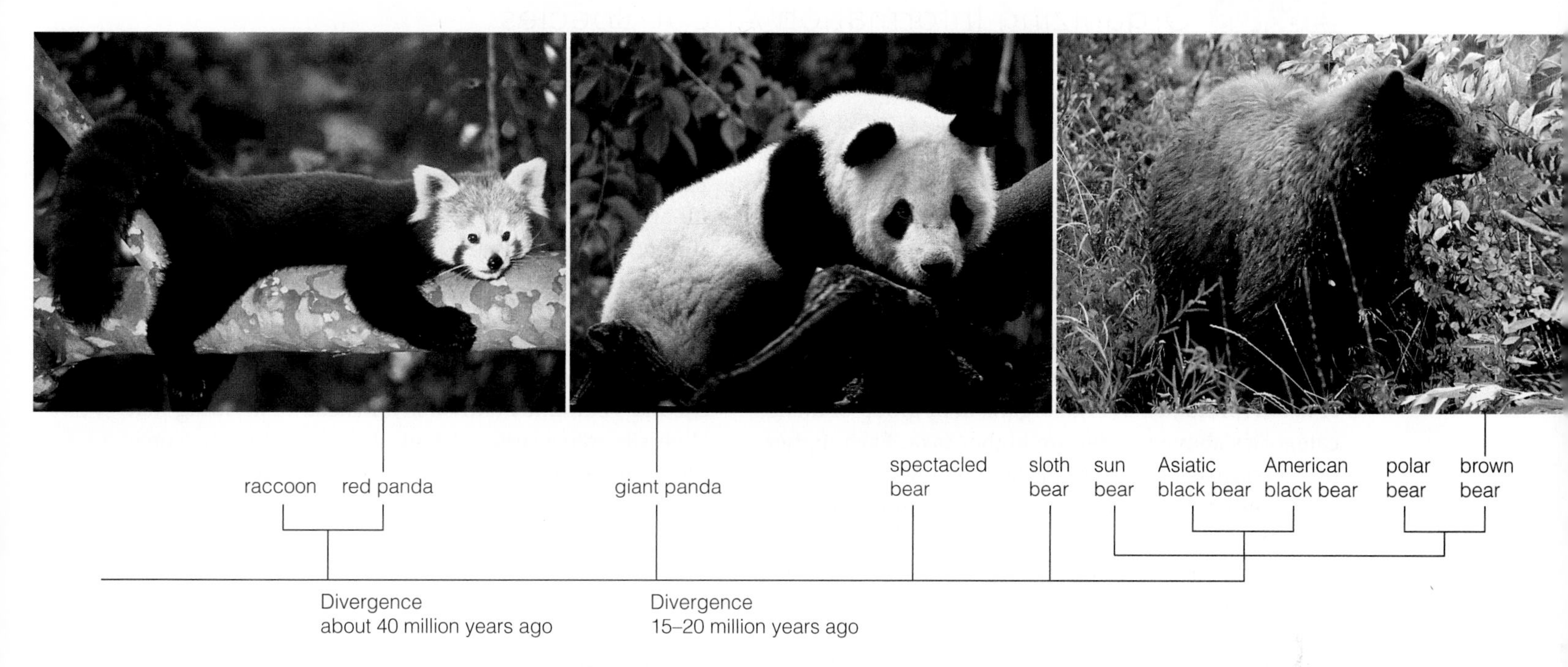

Figure 16.23 An example of how biochemical comparisons can help us construct and refine evolutionary trees. This tree for red pandas, giant pandas, and brown bears was constructed using sequence comparisons of mitochondrial and nuclear DNA. Recently, researchers found that red pandas may be more closely related to skunks, weasels, and otters than to raccoons.

of their genomes. If you find the number of bases that differ between species, you can estimate the relative evolutionary distance between them (Figure 16.23). The DNA from nuclei, mitochondria, and chloroplasts can be compared. Genes that encode ribosomal RNA are especially informative.

Mitochondrial DNA (mtDNA) is used to compare different individuals of the same sexually reproducing animal species. Mitochondria are inherited intact from a single parent, usually the mother. They contain their own DNA; thus, any differences in mtDNA sequences between maternally related individuals are the result of mutation, not genetic recombination.

DNA sequencing is the most common method of comparing DNA among different species, but nucleic acid hybridization is still used. DNA from two species is mixed and heated to unwind the double helixes. When the mixture cools, hybrid DNA molecules form. The more base pairs that match, the more hydrogen bonds form between the two strands. The amount of heat that is needed to separate the two strands of a hybrid DNA molecule can be used as a measure of their similarity. Why? More heat is needed to disrupt more extensive hydrogen bonding.

The DNA of related species can also be compared by using DNA fingerprinting techniques (Section 15.4).

Biochemical similarity is greatest among the most closely related species and smallest among the most remote.

ala ala asn lys asn lys gly ile ile trp gly glu asp thr leu met glu tyr leu glu asn pro lys lys tyr ile pro gly thr lys met ile phe val gly ile lys lys lys glu glu arg ala asp leu ile ala tyr leu lys lys ala thr asn glu-COO^-

ala ala asn lys asn lys ala val glu trp glu glu asn thr leu tyr asp tyr leu leu asn pro lys lys tyr ile pro gly thr lys met val phe pro gly leu lys lys pro gln asp arg ala asp leu ile ala tyr leu lys lys ala thr ser ser-COO^-

asp ala asn ile lys lys asn val leu trp asp glu asn asn met ser glu tyr leu thr asn pro lys lys tyr ile pro gly thr lys met ala phe gly gly leu lys lys glu lys asp arg asn asp leu ile thr tyr leu lys lys ala cys glu-COO^-

16.10 Organizing Information About Species

Connections among species, from ancient to recent, are evidence of evolution. We put species into groups based on what we know about their evolutionary relationships.

THE HIGHER TAXA

LINK TO SECTION 1.3

Taxonomy deals with naming and classifying species. Every kind of organism is assigned a unique two-part scientific name: The first part is the genus name, and together with the second part it designates the species. A taxon (plural, taxa) is an organism or group of them; categories above species are higher taxa. Each higher taxon consists of a group of the next lower taxon. An eighteenth century naturalist, Carolus Linnaeus, came up with a classification system that we still use. In his system, the higher taxa are genus, family, order, class, phylum, and kingdom (Figure 16.25). A monophyletic group is an ancestor and all of its descendants.

A six-kingdom classification system assigns all of the prokaryotic species to kingdoms Eubacteria and Archaea. It puts all single-celled eukaryotes and all of the most evolutionarily ancient multicelled species in kingdom Protista. Plants, fungi, and animals have their own kingdoms. The newer three-domain system sorts the six kingdoms into three higher taxa—the domains Bactèria, Archaea, and Eukarya (Figure 16.25). The next unit details some of the mechanisms that underlie the origin and modification of those traits.

A CLADISTIC APPROACH

Classifying species for the sake of sorting them into categories began when the prevailing belief was that all organisms were created in a perfect state. Then we discovered evolution—which is dynamic, extravagant, messy, and ongoing. Evolution can be challenging for those who like neat categories.

Our understanding of evolutionary relationships has led to a major, ongoing overhaul of the way biologists see life's extreme biodiversity. Instead of dividing that diversity into a series of ranks based on traits, we now try to organize it into a system of relationships. In this system, each species is not a member of a type or class, but is instead part of a bigger picture of evolution.

Cladistics is a method for determining evolutionary relationships, which may then be used in phylogenetic classification schemes. It groups species on the basis of derived traits—those that did not appear in the most recent ancestor. Species that share a set of derived traits are grouped into a clade (from *klados*, a Greek word for twig or branch).

A clade may correspond to a Linnaean group, but not every Linnaean group corresponds to a clade. The data used to define a clade are updated frequently, in part because techniques such as DNA sequencing keep getting cheaper and easier. Cladistic categories often change along with new data. A newly discovered fossil

KINGDOM	Bacteria	Plantae	Plantae	Animalia	Animalia
PHYLUM	Proteobacteria	Pinophyta	Magnoliophyta	Arthropoda	Chordata
CLASS	Epsilonproteobacteria	Coniferopsida	Monocotyledonae	Insecta	Mammalia
ORDER	Campylobacterales	Coniferales	Asparagales	Diptera	Primates
FAMILY	Helicobacteraceae	Cupressaceae	Orchidaceae	Muscidae	Hominidae
GENUS	*Helicobacter*	*Juniperus*	*Vanilla*	*Musca*	*Homo*
SPECIES	*H. felis*	*J. occidentalis*	*V. planifolia*	*M. domestica*	*H. sapiens*
COMMON NAME	none	western juniper	vanilla orchid	housefly	human

Figure 16.24 Linnaean classification of five species. Each species has been assigned to ever more inclusive taxa—in this case, from genus to kingdom.

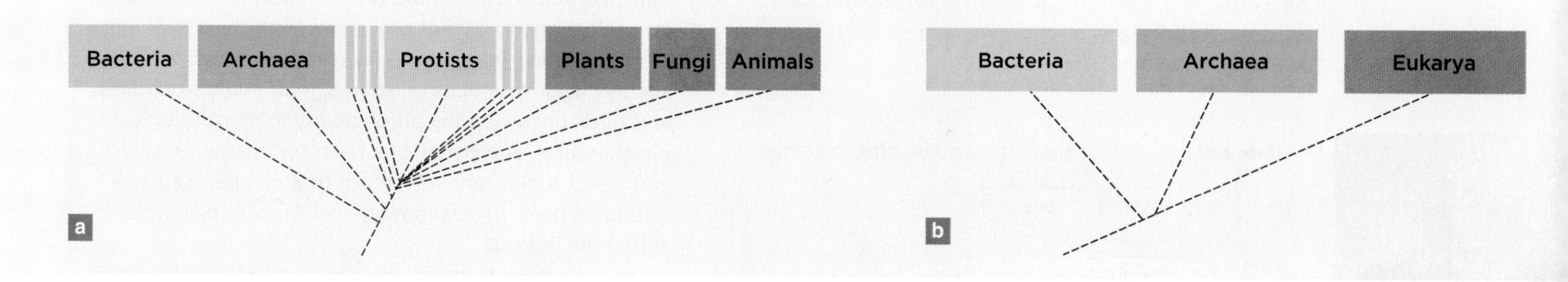

Figure 16.25 Animated! (**a**) Six-kingdom system of classification. Protists are now being divided into several new kingdoms. (**b**) Three-domain system of classification. The traditional kingdoms of protists, plants, fungi, and animals are subsumed in domain Eukarya.

or biochemical similarity may completely change our understanding of how one species is related to another.

Evolutionary trees are diagrams that summarize our best data-supported hypotheses about the pattern of evolution of a group of species. Each line represents one lineage; each branch point (or node) represents a common ancestor. A dashed line means that we know something about the lineage but not exactly where it fits in the tree (Figure 16.25).

Evolutionary tree diagrams called cladograms show "who is most closely related to whom." One clade is at the end of every branch of a cladogram, and each node has two branches. The two lineages that emerge from a node are sister groups. By definition, these two lineages are of equal age. We may not know what that age is, but we can compare any two sister groups and say something about their relative rates of evolution.

A cladogram can depict just a few clades or many of them. One that represents all of the species on Earth is called the tree of life. It looks deceptively simple, like a stick drawing. However, each stick is a connection made after detailed analysis of many derived traits.

The derived traits that define clades also help us visualize different monophyletic groups as *sets within sets*. Figure 16.26 shows an example. Cycads, ginkgos, conifers, gnetophytes, and flowering plants form a set. Only they have a common ancestor that was the first species to make seeds. Seed plants are nested inside a larger set—plants with leaves—that includes horsetails and ferns but excludes the lycophytes and bryophytes (mosses, hornworts, and liverworts). Bryophytes are not nested with the vascular plants; they do not have tubelike tissues that carry water and solutes.

The section that follows shows how to construct a cladogram. It provides a closer look at the advantages and some of the pitfalls of a cladistic approach.

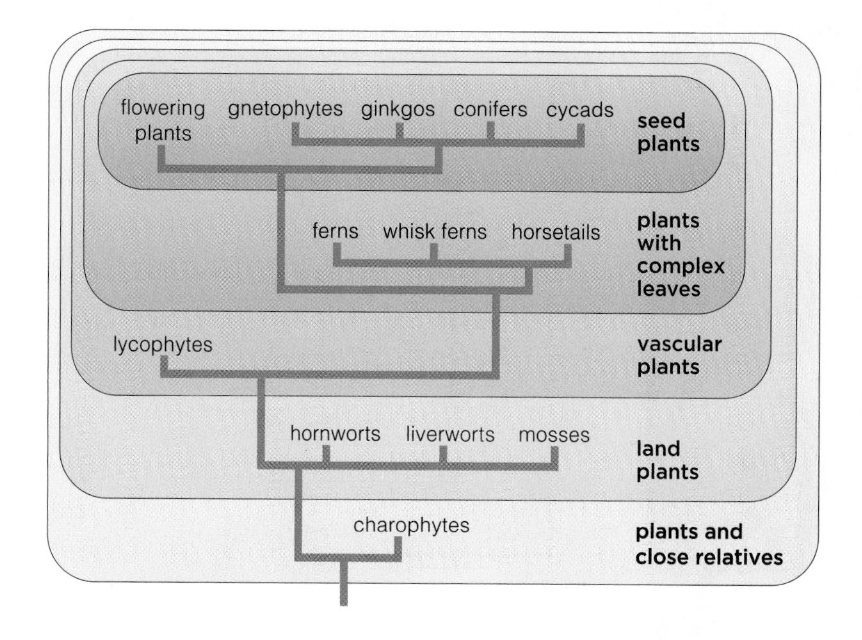

Figure 16.26 Example of sets within sets: a cladogram that reflects evolutionary relationships among major groups of plants. All groups have living representatives.

Taxonomy is a set of rules for naming organisms and classifying them into a series of ranks based on their traits.

Taxonomic classifications do not necessarily reflect true evolutionary relationships (phylogeny).

The most recent taxonomic hierarchy sorts all species into three domains: the Archaea, Bacteria, and Eukarya.

Cladistics is a set of methods by which we can determine evolutionary relationships. A clade is a group of species that share a set of derived traits.

Evolutionary trees summarize our best understanding of the pattern of evolution for a group of organisms.

16.11 How To Construct a Cladogram

In case you would like to know how a cladogram can be constructed, here is a step-by-step approach.

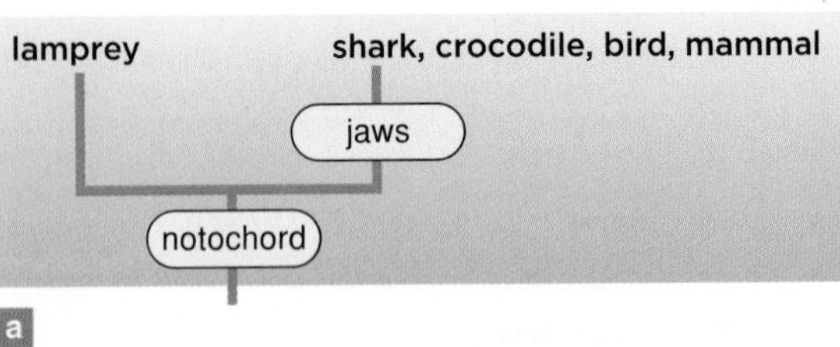

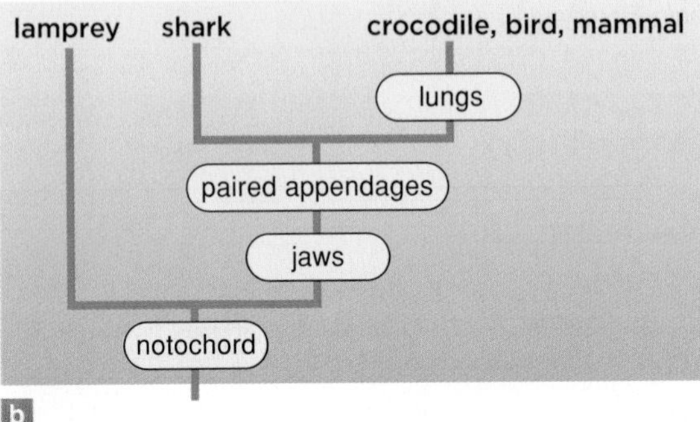

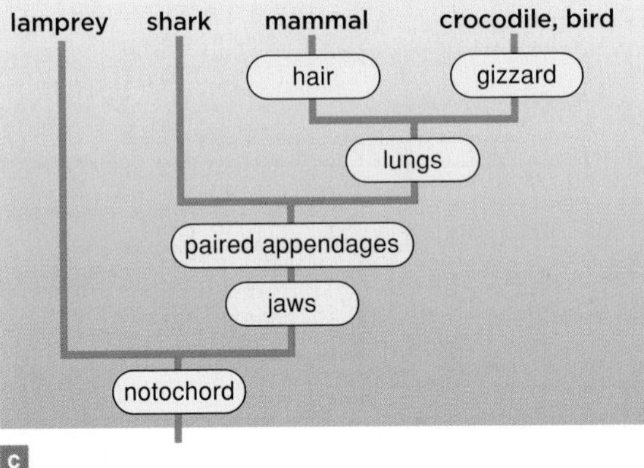

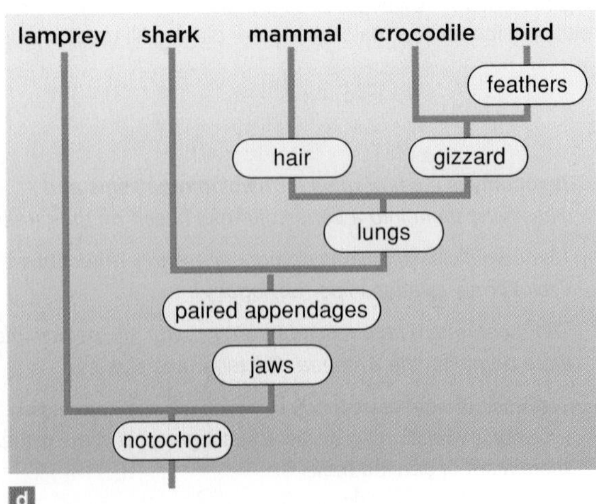

Figure 16.27 Using a selection of traits among groups to construct a simple cladogram, step by step.

Suppose you want to make a cladogram for vertebrates. You select an *ingroup* of organisms with traits that hint at relatedness—in this case, jaws and paired appendages, such as legs. You focus on sharks, mammals, crocodiles, and birds because they differ clearly in morphological, physiological, and behavioral traits, or characters. Next, you select a different vertebrate that can be used as a reference point for estimating evolutionary distances within the ingroup.

To keep things simple, you check for the presence or absence of only seven traits. After scanning Chapter 24, you decide that lampreys are only distantly related to the ingroup. For instance, although a tubular structure called a notochord forms in its embryos, as it does for all other vertebrates, lampreys have no jaws or paired appendages. Here is your *outgroup*, the one with the fewest derived traits compared to the others. Derived traits, remember, are evidence of morphological divergence.

Figure 16.27 shows how a cladogram takes shape as you add information about derived traits. Start with the presence or absence of jaws and paired appendages, two traits that the ingroup, not lampreys, got from a common ancestor. What about lungs? Like lampreys, sharks do not have lungs, so you find another branch point in vertebrate evolution. Past that branch point, only mammals have hair. Only crocodiles and birds have some form of gizzard. Only birds and their immediate ancestors have feathers.

How do you "read" your final cladogram? Remember, such charts are only an estimate of *relative* relatedness; they only imply common ancestry. Thus, birds are more closely related to crocodiles than they are to mammals. Crocodiles are not the ancestor of birds—they are modern organisms, too—but both share a more recent common ancestor than either does with mammals. Finally, birds, crocodiles, and mammals are closer to one another evolutionarily than they are to the shark.

The higher up a branch point is on a family tree, the more derived traits are shared. The lower the position of the branch point between two groups in the diagram, the fewer derived traits are shared.

A few words of caution: Interpretations of evolutionary relationships are more reliable when many traits are used, and there must be strong evidence that shared traits are derived. Using many traits helps to minimize problems that arise from a bad choice, such as including a trait that is a result of morphological convergence (Section 16.7).

The facing page shows a cladogram for major branches in the tree of life. All groups have living representatives, so ideally all should be listed at the "top"of the tree in a horizontal row, as in Figure 16.27. If we did so, however, the chart would run right off that page and on through the next two pages. We opted for compact clarity instead.

16.12 Preview of Life's Evolutionary History

Figure 16.28, the tree of life, shows macroevolutionary links among major groups of organisms, as described in the next unit. Each set of organisms (taxon) has living representatives. Each branch point represents the last common ancestor of the set above it. The small boxes within domains Archaea and Eukarya highlight taxa that are currently being recognized as the equivalent of kingdoms in earlier classification systems.

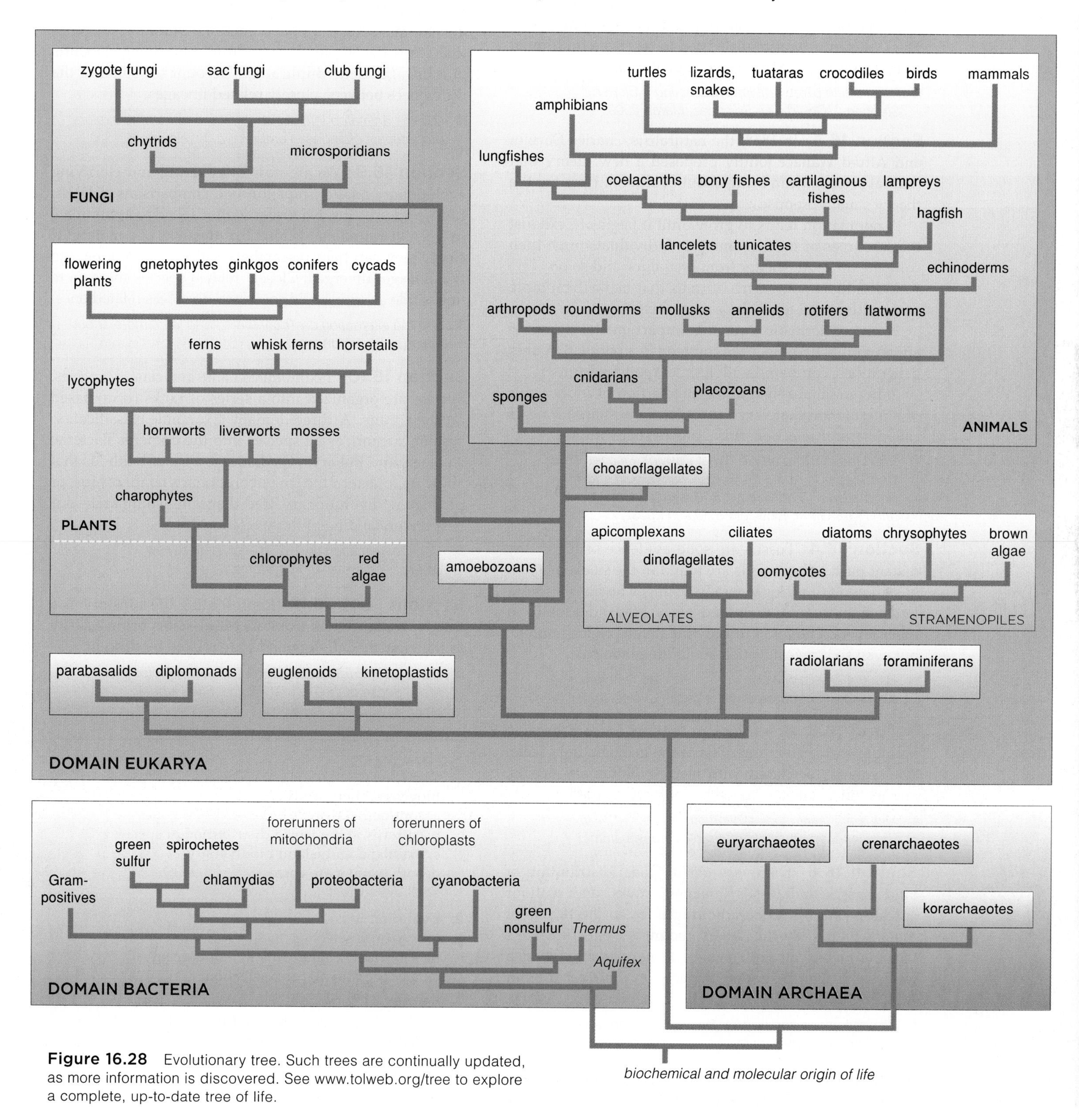

Figure 16.28 Evolutionary tree. Such trees are continually updated, as more information is discovered. See www.tolweb.org/tree to explore a complete, up-to-date tree of life.

www.thomsonedu.com ThomsonNOW!

Summary

Section 16.1 By the 19th century, advances in geology, biogeography, and morphology had resulted in a dawning awareness of change in lines of descent.

Section 16.2 Prevailing belief systems may influence interpretation of the underlying cause of a natural event. The 19th-century naturalists tried to reconcile traditional beliefs with physical evidence of evolution.

- *Read the InfoTrac article "Typecasting a Bit Part," Stephen J. Gould,* The Sciences, *March 2000.*

Section 16.3 In 1858, the naturalists Charles Darwin and Alfred Wallace jointly proposed a new theory that natural selection can bring about evolution. Here are the theory's main premises:

A population tends to grow until it begins to exhaust the resources of its environment. Individuals must then compete for food, shelter from predators, and so on.

Individuals with forms of traits that make them more competitive tend to produce more offspring.

Forms of heritable traits that impart greater fitness to an individual become more common in a population over generations, compared with less competitive forms.

Differential survival and reproduction of individuals of a population that vary in the details of shared traits is called natural selection. It is a process of evolution.

- *Read the InfoTrac article "What Darwin's Finches Can Teach Us About the Evolutionary Origin and Regulation of Biodiversity," B. Rosemary Grant and Peter Grant,* Bioscience, *March 2003.*

Section 16.4 Fossils are solid evidence of life in the distant past. Many fossils are found in the stacked layers of sedimentary rock. Generally, younger fossils are found in the more recently deposited layers, atop older fossils in older layers. The fossil record will always be incomplete, but even so it reveals much about life in the past.

- *Use the animations on ThomsonNOW to learn more about fossil formation and the geologic time scale.*

Section 16.5 Transitions in the fossil record became boundaries for great intervals of the geologic time scale. The scale is correlated with macroevolutionary events, and includes dates obtained by radiometric dating.

- *Use the animated interaction on ThomsonNOW to learn more about the half-life of a radioisotope's atoms.*

Section 16.6 Discovery of the global distribution of land masses and fossils, magnetic rocks, and seafloor spreading from mid-oceanic ridges led to the theory of plate tectonics. By this theory, the movements of Earth's tectonic plates raft land masses to new positions. Such movements had profound impacts on the directions of life's evolution.

- *Use the interaction on ThomsonNOW to learn more about drifting continents.*

Section 16.7 Comparative morphology can reveal the evolutionary connections among lineages. Homologous structures are similar body parts that, by the process of morphological divergence, became modified differently in different lineages. Such parts are evidence of a common ancestor. Analogous structures are body parts that look alike in different lineages but did not evolve in a common ancestor. By the process of morphological convergence, they evolved separately after the lineages diverged.

Section 16.8 Similarities in patterns of embryonic development suggest shared ancestry. Mutations in genes that affect development may cause morphological shifts in a lineage. Gene duplications account for some of the differences between closely related lineages.

- *Use the animated interaction on ThomsonNOW to explore proportional changes in embryonic development.*

Section 16.9 We are now discovering and clarifying evolutionary relationships through comparisons of DNA, RNA, and proteins. Neutral mutations tend to accumulate in DNA at a predictable rate; like the ticks of a molecular clock, they can help researchers estimate how long ago two lineages diverged. Closely related species share more nucleotide or amino acid sequences than less related ones.

- *Use the interaction on ThomsonNOW to learn more about amino acid comparisons.*

Section 16.10 Taxonomists name and classify species. Species are organized into a series of ranks (taxa) based on their traits. A three-domain taxonomic classification system categorizes all species into the domains Bacteria, Archaea, and Eukarya. Cladistics is a set of methods that allow us to determine true evolutionary relationships, or phylogeny. Evolutionary tree diagrams summarize our best understanding of those relationships.

- *Use the animation on ThomsonNOW to review biological classification systems.*

Sections 16.11, 16.12 Representing life's history as a tree with branchings from ancestral stems brings clarity to the view that all organisms are related by descent.

- *Read the InfoTrac article "How Taxonomy Helps Us Make Sense of the Natural World," Sue Hubbell,* Smithsonian, *May 1996.*

Self-Quiz

Answers in Appendix III

1. Biogeographers study ______.
   a. continental drift
   b. patterns in the world distribution of species
   c. mainland and island biodiversity
   d. both b and c are correct
   e. all are correct

2. Evolution ______.
   a. is natural selection
   b. is heritable change in a line of descent
   c. may occur by natural selection
   d. b and c are correct

3. ______ has/have influenced the fossil record.
   a. Sedimentation and compaction
   b. Tectonic plate movements
   c. Radioisotope decay
   d. All of the above

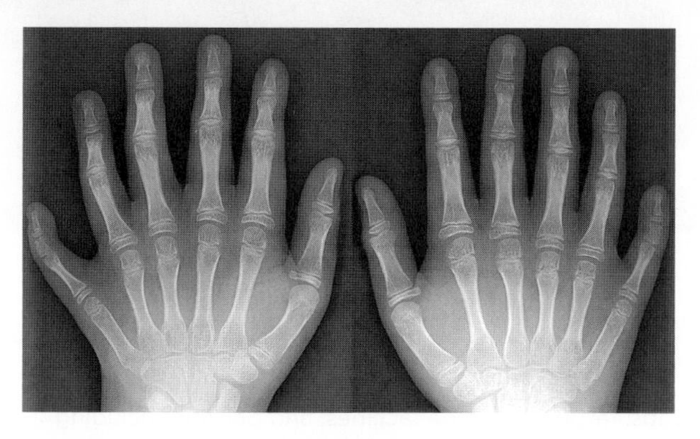

Figure 16.29 Polydactyly. Certain types of mutations result in extra fingers or toes. Most often, extra digits are duplicates.

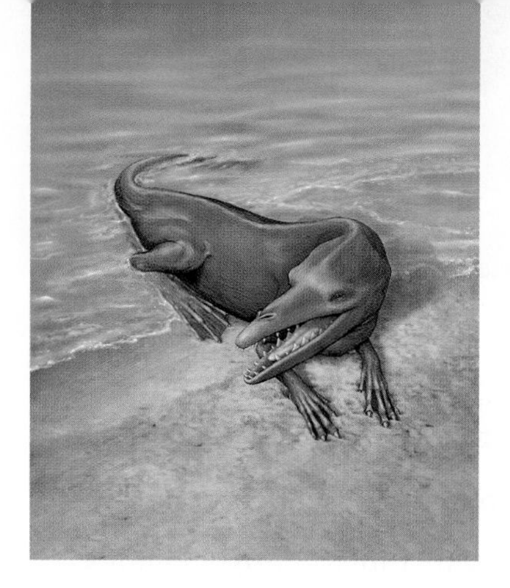

Figure 16.30 Reconstruction of *Rodhocetus* based on fossils found in Pakistan. This mammal, a cetacean, lived 47 million years ago along the shores of the Tethys Sea. Its ankle bones have details that point to a close evolutionary link between early whales and hoofed mammals on land. Compare Figure 16.3.

4. Evidence suggests that life originated in the ______ .
   a. Archaean c. Phanerozoic
   b. Proterozoic d. Cambrian

5. Did Pangea or Gondwana form first?

6. Homologous structures among major groups of organisms may differ in ______ .
   a. size c. function
   b. shape d. all of the above

7. Through ______ , a body part of an ancestor became modified differently in different lines of descent.
   a. morphological convergence
   b. morphological divergence
   c. analogous structures
   d. homologous structures

8. By altering steps in the program by which embryos develop, ______ may lead to major differences between adults of related lineages.
   a. automated gene sequencing c. transposons
   b. homeotic gene mutations d. b and c

9. Molecular clocks are based on comparisons of the number of ______ mutations between species.
   a. gene c. lethal
   b. neutral d. conserved

10. In evolutionary trees, each branch point represents a(n) ______ .
   a. single species c. divergence
   b. extinction d. adaptive radiation

11. Match the terms with the most suitable description.

___ stratification	a. evidence of life in distant past
___ fossils	b. geologic change has occurred repetitively in Earth's history
___ homeotic genes	c. human arm and bird wing
___ half-life	d. big role in development
___ homologous structures	e. layers of sedimentary rock
___ uniformity	f. insect wing and bird wing
___ analogous structures	g. time it takes half of a quantity of a radioisotope's atoms to decay into a stable product

■ *Visit ThomsonNOW for additional questions.*

Critical Thinking

1. At one time, all species were ranked in a great Chain of Being, from lowly forms to Man, then to spiritual beings. Scientists who do not understand phylogeny still refer to species as "primitive" or "advanced." For example, they may say mosses are primitive and flowering plants are advanced, or crocodiles are primitive and mammals are advanced. Why is it incorrect to refer to a modern taxon as primitive?

2. Think about the species living around you. From the evolutionary perspective, which ones are most successful in terms of sheer numbers, geographic distribution, and how long their lineage has endured on Earth?

3. At the end of your backbone is a coccyx, a few small, fused-together bones (Figure 16.3). Is the human coccyx a vestigial structure—all that is left of the tail of some distant vertebrate ancestors? Or is it the start of a newly evolving structure? Formulate a hypothesis, then design a way to test predictions you make based on the hypothesis.

4. Polydactyly is an inherited disorder characterized by extra digits on the hands or feet (Section 11.7 and Figure 16.29). Mutations in certain genes cause this disorder. In what family of genes do you think the mutations occur?

5. Comparative biochemistry can help us estimate evolutionary relationship and approximate times for divergences from ancestral stocks. DNA base sequence comparisons and amino acid comparisons yield good estimates. Reflect on the genetic code (Section 13.2), then suggest why it may be a useful measure of mutations, mutation rates, and biochemical relatedness.

6. For some time, evolutionists accepted that the ancestors of whales were four-legged animals that walked on land, then took up life in water about 55 million years ago. Fossils reveal gradual changes in skeletal features that made an aquatic life possible. But which four-legged mammals were the ancestors?

The answer recently came from Philip Gingerich and Iyad Zalmout. While digging for fossils in Pakistan, they found remains of early aquatic whales. Intact, sheeplike ankle bones *and* archaic whale skull bones were in the same fossilized skeletons (Figures 16.3 and 16.30).

Ankle bones of fossilized, early whales from Pakistan have the same form as the unique ankle bones of extinct and modern artiodactyls. Modern cetaceans no longer have even a remnant of an ankle bone. Here is evidence of an evolutionary link between certain aquatic mammals and a major group of mammals on land.

The radiometrically dated fossils are real. Yet no one was around to witness this transitional time. Because we did not see ancient life evolving, do you think there can be absolute proof of evolution in the distant past? Is the circumstantial evidence of fossil morphology enough to convince you that the theory is not wrong?

17 PROCESSES OF EVOLUTION

IMPACTS, ISSUES Rise of the Super Rats

Slipping in and out of the pages of human history are rats—*Rattus*—by far the most notorious of mammalian pests. Rats thrive in urban centers, where garbage is plentiful and natural predators are not. The average U.S. city sustains about one rat for every ten people. Part of their success stems from an ability to reproduce very quickly; rat populations expand within weeks to match the amount of garbage available for them to eat. Unfortunately for us, rats carry pathogens and parasites that cause bubonic plague, typhus, and other infectious diseases. They chew their way through walls and wires, and eat or foul 20 to 30 percent of our total food production (Figure 17.1). Rats cost us about $19 billion per year.

For years, people have been fighting back with dogs, traps, ratproof storage facilities, and poisons, including arsenic and cyanide. During the 1950s, they used baits laced with warfarin, a synthetic organic compound that interferes with blood clotting. Rats that ate the poisoned baits died within days after bleeding internally or losing blood through cuts or scrapes.

Warfarin was extremely effective. Compared to other rat poisons, it had a lot less impact on harmless species. It quickly became the rodenticide of choice.

In 1958, however, a Scottish researcher reported that warfarin did not work against some rats. Similar reports from other European countries followed. About twenty years later, 10 percent of the urban rats caught in the United States were warfarin resistant.

What happened? To find out, researchers compared warfarin-resistant rat populations with still-vulnerable rats. They traced the difference to a gene on one of the rat chromosomes. Certain mutations in the gene were common among warfarin-resistant rat populations but rare among vulnerable ones. Warfarin binds to the

See the video!

Figure 17.1 *Right*, rats infesting 80,000 hectares of the rice fields in the Philippine Islands ruin more than 20 percent of the crops. Rice is the main food for people in Southeast Asia. *Opposite*, rats thrive wherever people do. Dousing buildings and soil with rat poisons does not usually exterminate their populations, which recover quickly. Rather it selects for rats that are resistant to the poisons.

How would you vote? Antibiotic-resistant strains of bacteria are now widespread. One standard animal husbandry practice includes continually dosing healthy animals with the same antibiotics prescribed for people. Should this practice stop? See ThomsonNOW for details, then vote online.

gene's product—an enzyme involved in vitamin K-dependent synthesis of blood clotting factors. The mutations made the enzyme insensitive to warfarin.

"What happened" was evolution by natural selection. As warfarin exerted pressure on rat populations, the populations changed. The previously rare mutations suddenly proved to be adaptive. Unlucky rats that had an unmutated gene died after eating warfarin. The lucky ones that had one of the mutations survived and passed it to their offspring. Their populations recovered quickly, and a higher proportion of rats in the new populations now carried mutations. With each warfarin onslaught, the frequency of the mutation in rat populations increased.

Selection pressures can and often do change. When warfarin resistance increased in rat populations, people stopped using warfarin. The frequency of the mutation in rat populations declined—probably because rats with the mutation are mildly vitamin K-deficient, so they are not as healthy as normal rats. Now, savvy exterminators in urban areas know that the best way to control a rat infestation is to exert another kind of selection pressure: remove their source of food, which is usually garbage. Then the rats will eat each other.

The point is, when you hear someone question whether life evolves, remember this: With respect to life, evolution simply means heritable change is occurring in some line of descent. The mechanisms that can bring about such change are the focus of this chapter.

Key Concepts

MICROEVOLUTION

Populations evolve. Individuals of a population differ in which alleles they inherit, and thus in phenotypes. Over generations, any allele may increase or decrease in frequency in a population. Such shifts occur by the microevolutionary processes of mutation, natural selection, genetic drift, and gene flow. **Sections 17.1–17.8**

HOW SPECIES ARISE

Sexually reproducing species consist of one or more populations of individuals that interbreed successfully under natural conditions, produce fertile offspring, and are reproductively isolated from other species. The origin of new species varies in details and duration. Typically, it starts after gene flow ends between parts of a population. Microevolutionary events occur independently and lead to genetic divergence of the subpopulations. Such divergences are reinforced as reproductive isolation mechanisms evolve. **Sections 17.9–17.11**

PATTERNS IN LIFE'S HISTORY

Genetic change above the population level is called macroevolution. Recurring patterns of macroevolution include preadaptation, adaptive radiation, coevolution, and extinction. **Sections 17.12, 17.13**

ADAPTATION AND THE ENVIRONMENT

An evolutionary adaptation is a heritable aspect of form, function, behavior, or development that increases an individual's capacity to survive and reproduce in a particular environment. **Section 17.14**

Links to Earlier Concepts

Before starting this chapter, review the premises of the theory of natural selection (Sections 1.4 and 16.3), and the basic principles and terms of genetics (10.1, 10.2).

This chapter builds on evidence for evolution and processes of macroevolution introduced in the previous chapter (16.4,16.6). You will be drawing upon your knowledge of the genetic basis of complex traits (10.4, 10.6, 10.7), the effects of genetic change (11.6, 11.7, 13.5, 16.8), and how molecular comparisons work (16.9), as you evaluate experiments (1.6, 1.7) that provide evidence of microevolution.

17.1 Individuals Don't Evolve, Populations Do

Evolution starts with mutations in individuals. Mutation is the source of new alleles, and sexual reproduction spreads them through a population.

VARIATION IN POPULATIONS

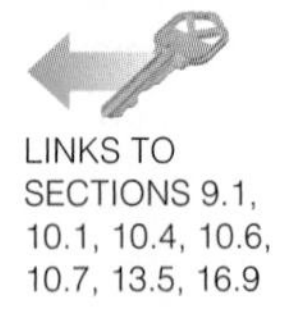

LINKS TO SECTIONS 9.1, 10.1, 10.4, 10.6, 10.7, 13.5, 16.9

A population is one group of individuals of the same species in a specified area. All individuals of a species share certain features. Giraffes have four hooved feet, very long necks, brown spots on white coats, and so on. These are examples of morphological traits (*morpho–*, form). Individuals of a species also share physiological traits such as basic metabolic activities. In addition, they respond the same way to certain stimuli, as when pigeons peck at food. These are behavioral traits.

However, the individuals of a population also show variation in the details of their shared traits. You know this just by thinking about the variations in the color and patterning of dog fur or snail shells. Figure 17.2 hints at the range of variations in human skin and eye color, and distribution, color, texture, and amount of hair. Almost every trait of any species may vary, and the variation can be dramatic.

Many traits show *qualitative* differences; they have two or more different forms, or morphs, such as the purple or white pea plant flowers that Gregor Mendel studied. In addition, for many traits, individuals of a population show *quantitative* differences, or a range of incrementally small variations in a trait (Section 10.7).

THE GENE POOL

Genes encode heritable information about traits. The individuals of a population inherit the same number and kind of genes (except for genes on nonidentical sex chromosomes). Together, the genes of a population comprise a gene pool—a pool of genetic resources.

In a sexually reproducing population, most genes in the pool have slightly different forms called alleles (Section 9.1). An individual carries two copies of each autosomal gene, and those copies may or may not be identical (Section 10.1). An individual's complement of alleles is its *genotype*. Alleles are the main source of variation in *phenotype*—the observable characteristics of an individual. For example, the color of your eyes is determined by the alleles you carry.

Polymorphism (*polymorphos*, or many forms) occurs when a gene has three or more alleles that persist in a population at a frequency of more than 1 percent. The ABO alleles that determine human blood type are one example. Some traits have two distinct forms, such as the female and male sex. This is a case of dimorphism.

You learned about patterns of inheritance in earlier chapters. Here we summarize the key events involved:

Gene mutation (source of new alleles)

Crossing over at meiosis I (introduces new combinations of alleles into chromosomes)

Independent assortment at meiosis I (mixes maternal and paternal chromosomes)

Fertilization (combines alleles from two parents)

Change in chromosome number or structure (loss, duplication, or transposition)

Mutation is the source of new alleles. Other events shuffle existing alleles into different combinations, but what a shuffle that is! There are $10^{116,446,000}$ possible combinations of human alleles. Not even 10^{10} people are alive today. Unless you are an identical twin, it is unlikely that another person with your precise genetic makeup has ever lived or ever will.

One other point about the nature of the gene pool: Offspring inherit a genotype, not a phenotype. Section 10.6 describes how environmental conditions can bring about variation in the range of phenotypes, but such effects last no longer than the individual.

MUTATION REVISITED

Being the original source of new alleles, mutations are worth another look—this time in the context of their impact on populations. We cannot predict when or in which individual a particular gene will mutate. We *can* predict an average mutation rate, the probability that a mutation will occur in a given interval, for a species. In humans, that rate is about 175 mutations per person per generation.

Many mutations give rise to structural, functional, or behavioral alterations that reduce an individual's chances of surviving and reproducing. Even a single biochemical change may be devastating. For instance, skin, bones, tendons, lungs, blood vessels, and many other vertebrate organs incorporate collagen. When the collagen gene mutates in a way that alters collagen function, the entire body may be affected. A mutation that drastically changes phenotype usually results in death; it is called a lethal mutation.

A neutral mutation, recall, alters the base sequence in DNA, but the change has no effect on survival or reproduction (Section 16.9). It neither helps nor hurts the individual. For instance, if you carry a mutation that keeps your earlobes attached to your head instead of swinging freely, this in itself should not stop you from

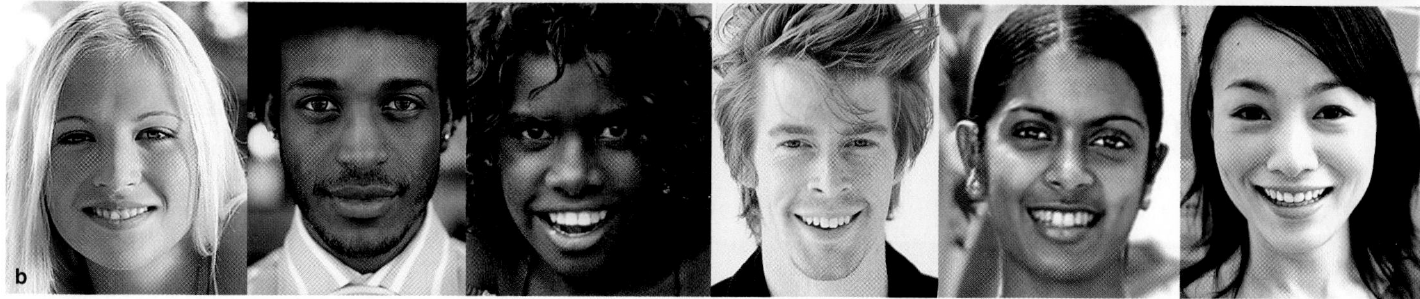

Figure 17.2 A sampling of the phenotypic variation in populations of (**a**) a type of snail found on islands in the Caribbean and (**b**) humans. Such variation among individuals of a population is an outcome of variations in alleles.

surviving and reproducing as well as anybody else. So, natural selection does not affect the frequency of this particular mutation in a population.

Every so often, a mutation proves useful. A mutant gene product that affects growth might make a corn plant grow larger or faster and thereby give it the best access to sunlight and nutrients. Even if the mutation bestows only a slight advantage, natural selection will favor its transmission to the next generation. Even a neutral mutation might prove helpful if conditions in the environment change.

Mutations have been altering genomes for billions of years. Cumulatively, they have given rise to Earth's staggering biodiversity. Think about it. The reason you do not look like a bacterium or avocado or earthworm or even your neighbor began with the mutations that arose in different lines of descent.

STABILITY AND CHANGE IN ALLELE FREQUENCIES

Researchers typically track allele frequencies, or the relative abundances of alleles of a given gene among all individuals of a population. They can start from a theoretical reference point, genetic equilibrium, when a population is *not* evolving with respect to that gene.

Genetic equilibrium can only occur if all five of the following conditions are met: Mutations never occur; the population is infinitely large; the population stays isolated from all other populations of the same species; all individuals mate at random; and all individuals of the population survive and produce the same number of offspring.

All five conditions are never met in nature; thus, natural populations are never in genetic equilibrium. Microevolution, small-scale change in a population's allele frequencies, occurs constantly. Four processes of microevolution—mutation, natural selection, genetic drift, and gene flow—prevent all natural populations from reaching equilibrium. The next section offers a closer look at genetic equilibrium.

We partly characterize a natural population or species by shared morphological, physiological, and behavioral traits. Most traits have a heritable basis.

Different alleles may give rise to variations in phenotypes—to differences in the details of shared structural, functional, and behavioral traits.

The alleles of all individuals in a population comprise a pool of genetic resources—that is, a gene pool.

Mutation is the source of new alleles. Natural selection, genetic drift, and gene flow affect the frequencies of various alleles at a given gene locus in the population.

Natural populations are always evolving, which means that allele frequencies in their gene pool are changing over the generations.

17.2 When Is A Population *Not* Evolving?

How do researchers know whether or not a population is evolving? They can start by tracking deviations from the baseline of genetic equilibrium.

The Hardy-Weinberg Formula Early in the twentieth century, Godfrey Hardy (a mathematician) and Wilhelm Weinberg (a physician) independently applied the rules of probability to sexually reproducing populations. They perceived that gene pools can remain stable only when five conditions are being met:

1. Mutations do not occur.
2. The population is infinitely large.
3. The population is isolated from all other populations of the species (no gene flow).
4. Mating is random.
5. All individuals survive and produce the same number of offspring.

These conditions never occur all at once in nature. Thus, allele frequencies for any gene in the shared pool always change. However, we can think about a hypothetical situation in which the five conditions are being met and a population is not evolving.

Hardy and Weinberg developed a simple formula that can be used to track whether a population of any sexually reproducing species is in a state of genetic equilibrium. Consider tracking a hypothetical pair of alleles that affect butterfly wing color. A protein pigment is specified by dominant allele *A*. If a butterfly inherits two *AA* alleles, it will have dark-blue wings. If it inherits two recessive alleles (*aa*), it will have white wings. If it inherits one of each (*Aa*), the wings will be medium-blue (Figure 17.3).

At genetic equilibrium, the proportions of the wing-color genotypes are

$$p^2(AA) + 2pq(Aa) + q^2(aa) = 1.0$$

where p and q are the frequencies of alleles *A* and *a*. This is what became known as the *Hardy-Weinberg equilibrium equation*. It defines the frequency of a dominant and a recessive allele for a gene that controls a particular trait in a population.

The frequencies of *A* and *a* must add up to 1.0. To give a specific example, if *A* occupies half of all the loci for this gene in the population, then *a* must occupy the other half (0.5 + 0.5 = 1.0). If *A* occupies 90 percent of all the loci, then *a* must occupy 10 percent (0.9 + 0.1 = 1.0). No matter what the proportions,

$$p + q = 1.0$$

At meiosis, remember, paired alleles are assorted into different gametes. The proportion of gametes with the *A* allele is p, and the proportion with the *a* allele is q. The Punnett square on the next page shows the genotypes possible in the next generation (*AA*, *Aa*, and *aa*). Note that the frequencies of the three genotypes add up to 1.0:

$$p^2 + 2pq + q^2 = 1.0$$

Figure 17.3 **Animated!** Finding out whether a population is evolving. The frequencies of wing-color alleles among all of the individuals in this hypothetical population of morpho butterflies are not changing; the population is not evolving.

	p (A)	q (a)
p (A)	AA (p^2)	Aa (pq)
q (a)	Aa (pq)	aa (q^2)

Suppose that the population has 1,000 individuals and that each one produces two gametes:

490 *AA* individuals make 980 *A* gametes
420 *Aa* individuals make 420 *A* and 420 *a* gametes
90 *aa* individuals make 180 *a* gametes

The frequency of alleles *A* and *a* among 2,000 gametes is

$$A = \frac{980 + 420}{2{,}000 \text{ alleles}} = \frac{1{,}400}{2{,}000} = 0.7 = p$$

$$a = \frac{180 + 420}{2{,}000 \text{ alleles}} = \frac{600}{2{,}000} = 0.3 = q$$

At fertilization, gametes combine at random and start a new generation. If the population size stays constant at 1,000, there will be 490 *AA*, 420 *Aa*, and 90 *aa* individuals. The frequencies of the alleles for dark-blue, medium-blue, and white wings are the same as they were in the original gametes. Thus, dark-blue, medium-blue, and white wings occur at the same frequencies in the new generation.

As long as the assumptions that Hardy and Weinberg identified continue to hold, the pattern persists. If traits show up in different proportions from one generation to the next, though, one or more of the five assumptions is not being met. The hunt can begin for the evolutionary forces driving the change.

Applying the Rule How does the Hardy-Weinberg formula work in the real world? Researchers can use it to estimate the frequency of carriers of alleles that cause genetic traits and disorders.

As an example, hereditary hemochromatosis (HH) is the most common genetic disorder among people of Irish ancestry. Affected individuals absorb too much iron from food. The symptoms of this autosomal recessive disorder include liver problems, fatigue, and arthritis. A study in Ireland found the frequency for one allele that causes HH to be 0.14. If $q = 0.14$, then p is 0.86. Based on this study, the carrier frequency ($2pq$) can be calculated to be about 0.24. Such information is useful to doctors and to public health professionals.

Another example: A mutation in the *BRCA2* gene has been linked to breast cancer in adults. A deviation from the birth frequencies predicted by the Hardy-Weinberg formula suggests that this mutation can also have effects even before birth. In one study, researchers looked at the mutation's frequency among newborn girls. They found fewer homozygotes than expected, based on the number of heterozygotes and the Hardy-Weinberg formula. Thus, it seems that in homozygous form the mutation impairs the survival of female embryos.

17.3 Natural Selection Revisited

Natural selection, introduced in earlier chapters, is a most profoundly influential microevolutionary event.

LINKS TO SECTIONS 1.4, 16.2, 16.3

Natural selection, again, is differential survival and reproduction among individuals of a population that vary in the details of shared traits. Some traits prove more adaptive than others in prevailing environmental conditions. Natural selection influences all levels of biological organization, which is the reason you were introduced to it early on, in Chapter 1. You also came across examples in other chapters, and Sections 16.2 and 16.3 looked at the history behind its discovery. Turn now to major categories of natural selection, as introduced in Figure 17.4.

With *directional* selection, the range of variation for a trait shifts in a consistent direction; individuals at one end of the range of variation are selected against and those at the other end are favored. With *stabilizing* selection, the forms at one or both ends of the range are selected against. With *disruptive* selection, forms at the extremes of the range of variation are favored; intermediate forms are selected against.

Diverse selection pressures acting on a population might favor forms at one end in the range of variation for a trait, or intermediate forms within that range, or extreme forms at both ends of the range.

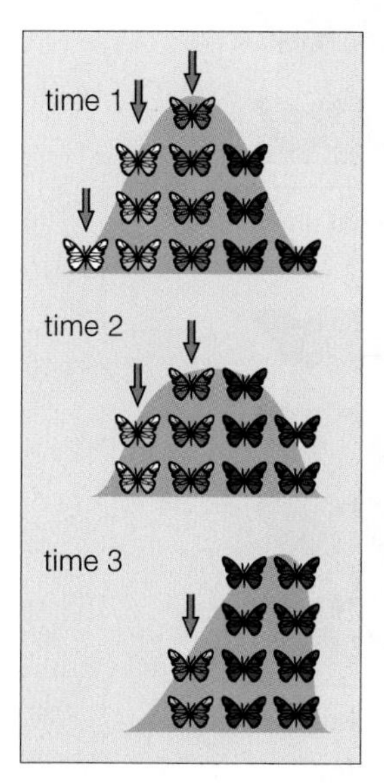

Directional selection
Extreme form at one end of the range of phenotypes favored

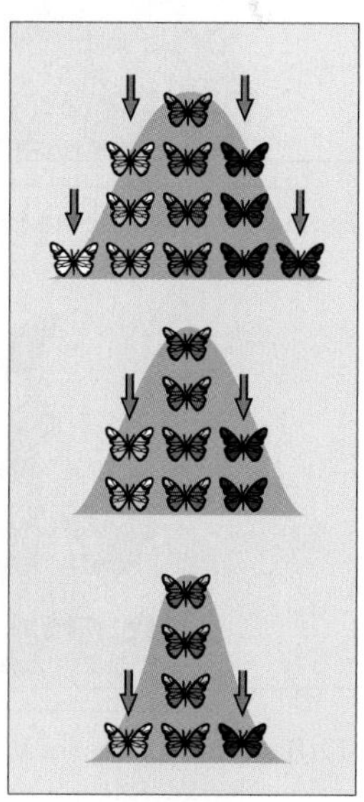

Stabilizing selection
Intermediate form within the range of phenotypes favored

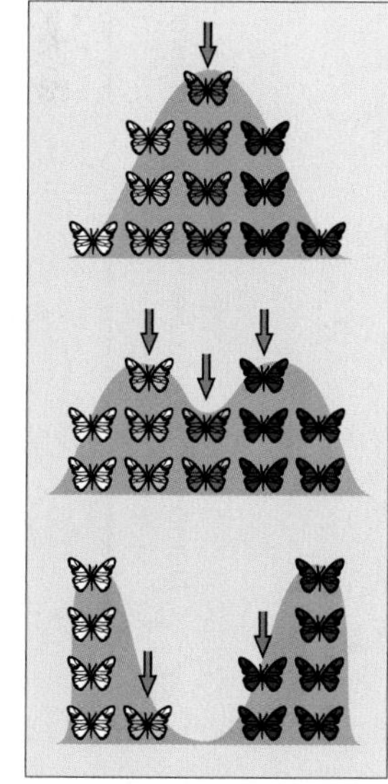

Disruptive selection
Extreme forms at both ends of the range of phenotypes favored

Figure 17.4 Overview of the outcomes of three modes of natural selection.

17.4 Directional Selection

Existing or novel environmental conditions can cause a shift in allele frequencies in a consistent direction.

LINK TO SECTION 1.7

With directional selection, allele frequencies shift in a consistent direction, so forms at one end of the range of phenotypic variation become more common than midrange forms. Figure 17.5 is a simple way to think about this microevolutionary process. The examples that follow demonstrate how observations and field experiments provide evidence of directional selection.

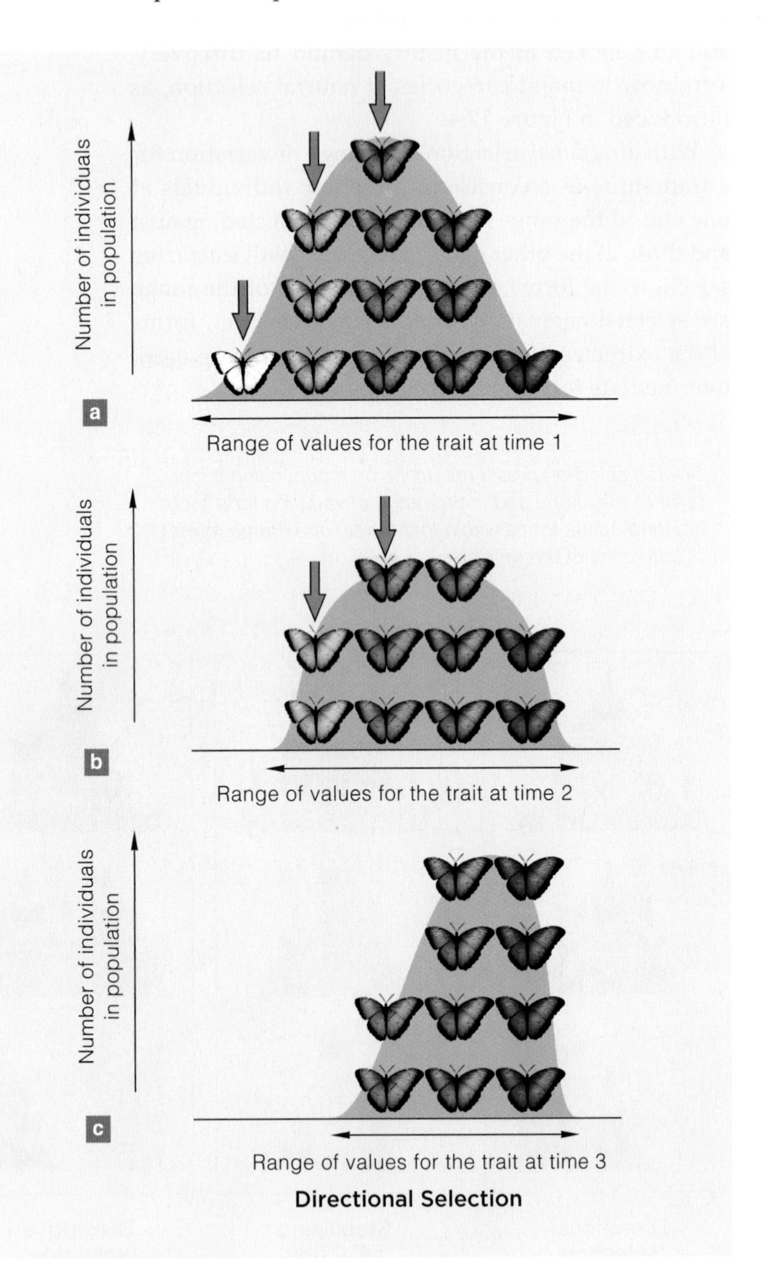

Figure 17.5 **Animated!** Directional selection. These bell-shaped curves signify a range of continuous variation in a butterfly wing-color trait. *Medium-blue* is between two phenotypic extremes—*white* and *dark purple*. *Orange* arrows signify which forms are being selected against over time.

EFFECTS OF PREDATION

The Peppered Moth The history of the peppered moth (*Biston betularia*) is a classic case of directional selection. The moths feed and mate at night, and rest motionless on trees during the day. Their behavior and coloration (mottled gray to nearly black) camouflage them from day-flying, moth-eating birds.

Light colored moths were the most common form in preindustrial England. A dominant allele that resulted in the dark color was rare. The air was clean, and light-gray lichens grew on the trunks and branches of most trees. Light moths were camouflaged when they rested on the lichens, but dark moths were not (Figure 17.6*a*).

By the 1850s, the dark moths were appearing more frequently. Why? The industrial revolution had begun, and smoke from coal-burning factories was beginning to change the environment. Air pollution was killing the lichens. Researchers hypothesized that dark moths were now better camouflaged from predators on the soot-darkened trees than the light moths (Figure 17.6*b*).

In the 1950s, H. B. Kettlewell used a mark–release–recapture method to test the hypothesis. He bred both moth forms in captivity and marked hundreds so that they could be easily identified after being released in the wild. He released them near highly industrialized areas around Birmingham and near an unpolluted part of Dorset. His team recaptured more dark moths in the polluted area and more light ones near Dorset:

	Near Birmingham (pollution high)	Near Dorset (pollution low)
Light-Gray Moths		
Released	64	393
Recaptured	16 (25%)	54 (13.7%)
Dark-Gray Moths		
Released	154	406
Recaptured	82 (53%)	19 (4.7%)

They also observed predatory birds eating more light moths in Birmingham, and more dark moths in Dorset.

Pollution controls went into effect in 1952. Lichens made a comeback, and tree trunks became largely free from soot. Phenotypes shifted in the reverse direction: Wherever pollution decreased, the frequency of dark moths decreased as well. Many other researchers since Kettlewell have confirmed the rise and fall of the dark colored form of the peppered moth.

Pocket Mice Directional selection has been in play among populations of rock pocket mice (*Chaetodipus intermedius*) in Arizona's Sonoran Desert. Rock pocket mice are small mammals that spend the day sleeping

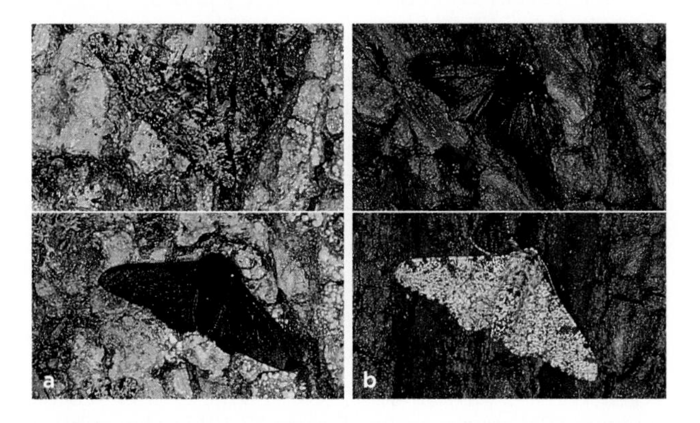

Figure 17.6 Natural selection of two forms of the same trait, body surface coloration, in two settings. (**a**) Light moths (*Biston betularia*) on a nonsooty tree trunk are hidden from predators. Dark ones stand out. (**b**) The dark color is more adaptive in places where soot darkens tree trunks.

Figure 17.7 Visible evidence of directional selection in populations of rock pocket mice.

(**a**) Rock pocket mice that have dark fur are more common in these areas of dark basalt rock.

(**b**,**c**) The two color types of rock pocket mice, each posed on the dark and light rocks of the area.

in underground burrows. They emerge to forage for seeds at night. The mice differ in coat color; some are light brown, and others are dark gray. Out of eighty or so genes known to affect coat color in this species, researchers found one gene that is responsible for the difference. One allele of the gene results in light fur; the other, dark.

The Sonoran Desert is dominated by outcroppings of light brown granite. There are also patches of dark basalt, the remains of ancient lava flows (Figure 17.7*a*). Most of the mice in populations that inhabit the light brown granite have light brown coats (Figure 17.7*b*). Most of the mice in populations that inhabit the dark rock have dark gray coats (Figure 17.7*c*).

Why? In each habitat, the individuals that match the rock color are camouflaged from predators. We can expect that night-flying owls more easily see mice that do not match the rocks, so they preferentially eliminate nonmatching mice from each population. The owls are selective agents that directionally shift the frequency of coat color alleles in rock pocket mice populations.

RESISTANCE TO ANTIBIOTICS

Our attempts to control the environment can result in directional selection, as is the case with the super rats. The use of antibiotics is another example. Antibiotics are toxins that kill bacteria. Streptomycins, for instance, block protein synthesis in certain bacteria. Penicillins disrupt the formation of covalent bonds that link the glycoproteins in bacterial cell walls. The resulting cell walls are weak, and they rupture.

When your grandparents were still young, scarlet fever, tuberculosis, and pneumonia caused one-fourth of the annual deaths in the United States alone. Since the 1940s, we have been relying on antibiotics to fight these and other dangerous bacterial diseases. We also use them in other, less dire circumstances. Antibiotics are used preventively, both in humans and in livestock. They are part of the daily rations of millions of cattle, pigs, chickens, fish, and other animals that are raised in extremely crowded conditions on factory farms.

Overuse of antibiotics places tremendous selection pressure on bacteria. Bacteria divide very quickly, so they evolve at a greatly accelerated rate compared to humans. *Escherichia coli* can divide every 17 minutes, so an average two-week course of antibiotics exerts selection pressure on over a thousand generations of this common intestinal bacterium.

Dividing bacteria quickly form huge populations, and the gene pools of such populations show great genetic variation. Thus, it is very likely that some cells can survive antibiotic treatment. Resistant strains are becoming the norm, particularly in hospitals, so they are posing a greater threat to human health each year. Even as doctors and researchers scramble to find new antibiotics, this trend is bad news for the millions of people who contract cholera, tuberculosis, or another dangerous bacterial disease each year.

With directional selection, allele frequencies underlying a range of variation tend to shift in a consistent direction in response to some environmental factor.

17.5 Selection Against or in Favor of Extreme Phenotypes

Consider now two more categories of natural selection. One works against phenotypes at the fringes of a range of variation; the other favors them.

STABILIZING SELECTION

With stabilizing selection, intermediate forms of a trait in a population are favored, and extreme forms are not. This mode of selection tends to preserve intermediate phenotypes in the population, so it does not result in evolutionary change (Figure 17.8*a*). For example, human babies who weigh 3.4 kilograms (7.5 pounds) at birth are more likely to survive than babies who weigh much more or much less (Figure 17.9). Smaller babies tend to have serious health problems. Larger babies are more difficult to deliver because of the tighter fit through the mother's pelvic bones. However, since the mid-1900s, large babies have been routinely delivered by cesarean section, so the selection effect at the top of the weight spectrum may be changing.

As a different example, the body mass of sociable weavers (*Philetairus socius*, Figure 17.10) is also subject

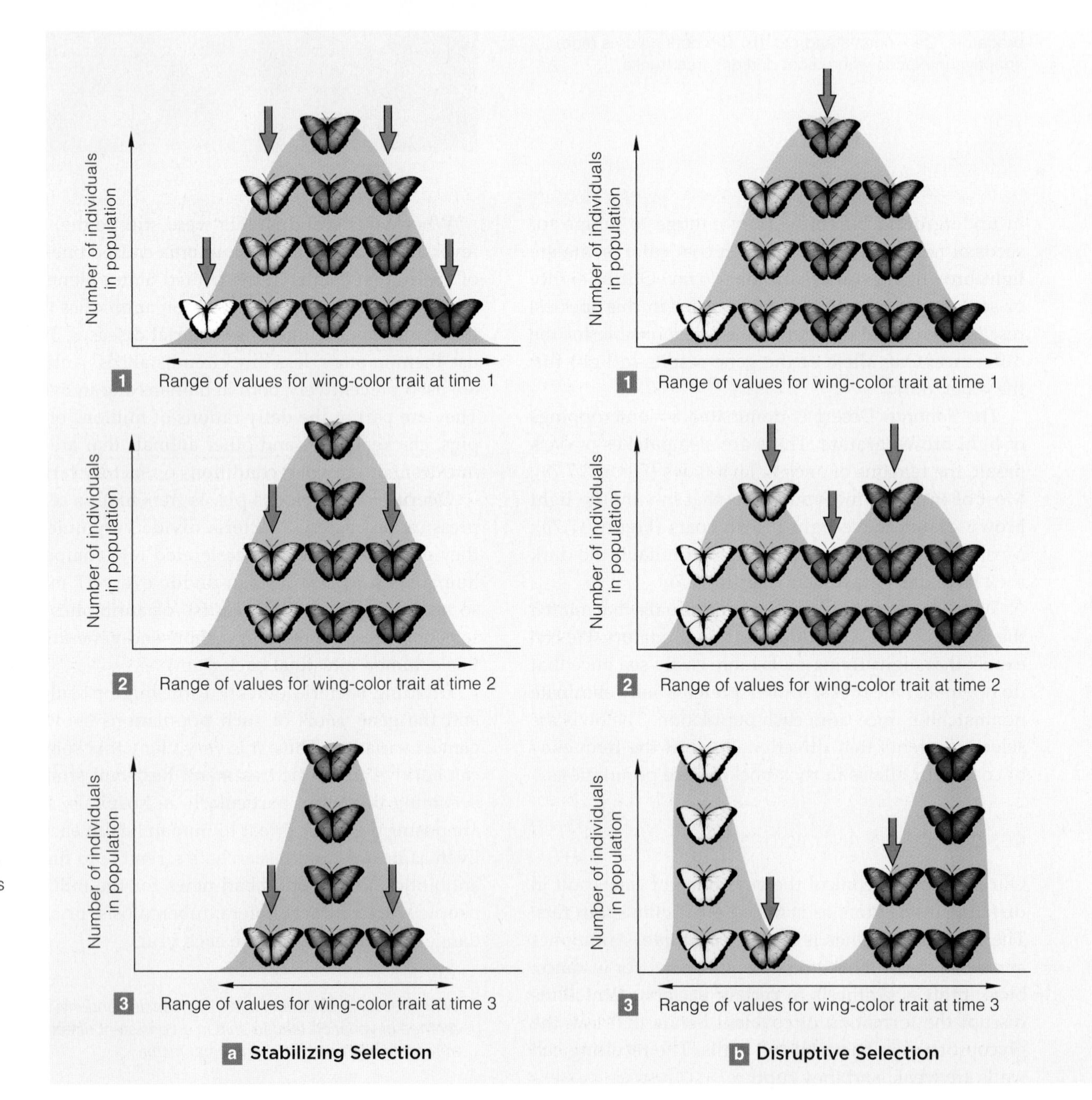

Figure 17.8 **Animated!** Selection in favor of and against extreme phenotypes. *Orange* arrows indicate forms of the trait being selected against.

(**a**) Stabilizing selection eliminates extreme forms.

(**b**) Disruptive selection eliminates midrange forms.

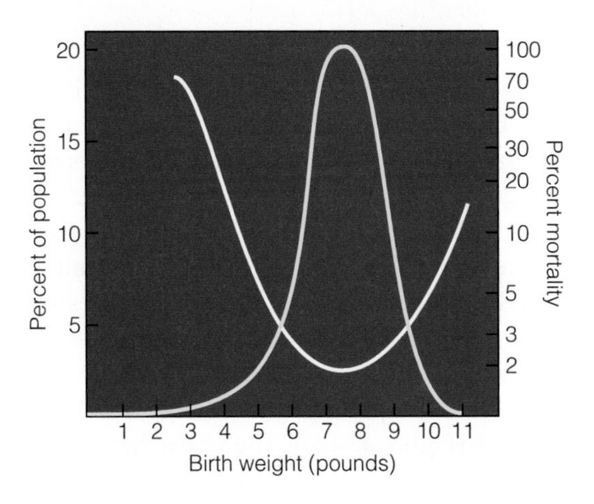

Figure 17.9 Weight distribution for 13,730 human newborns (*yellow* curve) correlated with death rate (*white* curve). This graph represents data collected in the mid-1900s. More recent advancements in medical care have likely affected this curve.

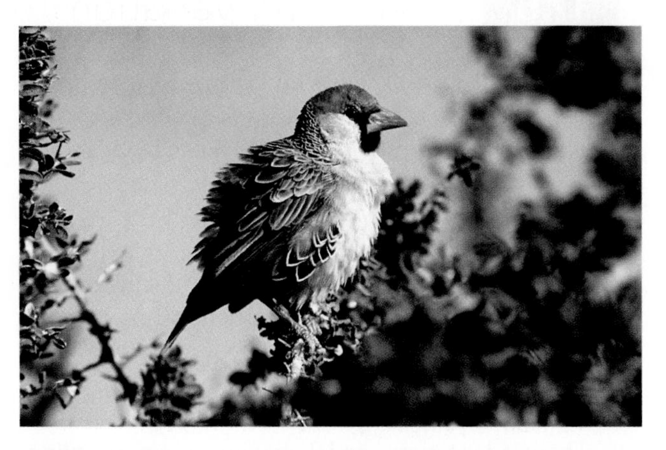

Figure 17.10 Adult sociable weaver (*Philetairus socius*). The body mass of these birds is subject to stabilizing selection; birds with intermediate body mass have the highest survival rate. Smaller birds tend to starve in the harsh environment of the African savanna, and larger birds are more attractive to predators.

to stabilizing selection. These birds cooperate in the construction and use of large communal nests in areas of the African savanna. Between 1993 and 2000, Rita Covas and her colleagues captured, measured, tagged, released, and recaptured 70 to 100 percent of the birds living in communal nests during the breeding season. Their field studies supported a prediction that body mass is a trade-off between the risks of starvation and predation. Intermediate-mass birds have the selective advantage. Foraging is not easy in this habitat, and lean birds do not store enough fat to avoid starvation. Fatter birds may be more attractive to predators, and not as good at escaping.

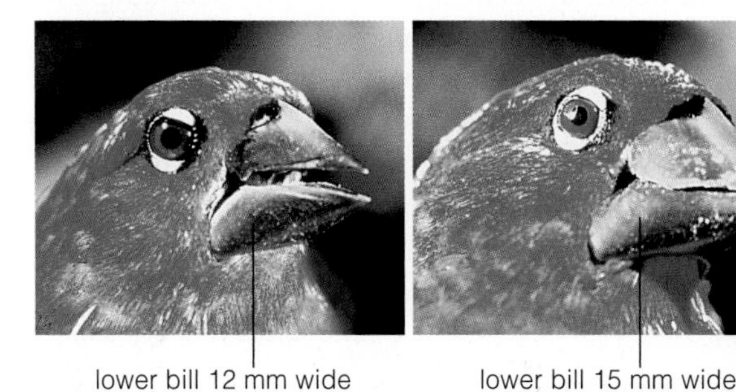

Figure 17.11 Disruptive selection in African finch populations. Selection pressures favor birds with bills that are about 12 *or* 15 millimeters wide. The difference is correlated with competition for scarce food resources during the dry season.

DISRUPTIVE SELECTION

With disruptive selection, forms at both ends of a range of variation are favored; intermediate forms are selected against (Figure 17.8*b*).

Consider the black-bellied seedcracker (*Pyrenestes ostrinus*) of Cameroon, Africa. In these finches, there is a genetic basis for bills of a particular size. Females and males have large or small bills—but no sizes in between (Figure 17.11). Both forms occur in the same geographic range, and breed randomly with respect to bill size. It is as if every human adult were four feet *or* six feet tall, with no one between.

Factors that affect feeding performance maintain the dimorphism. The birds feed mainly on the seeds of two types of grasslike plants. One plant produces hard seeds and the other, soft seeds. The small-billed seedcrackers are better at opening the soft seeds, but the birds with large bills are better at cracking the hard ones. During Cameroon's wet seasons, seeds are quite abundant and both forms of birds feed on both types of seeds. However, all seeds become scarce during dry seasons; then, each type of bird focuses on eating the seeds it opens most efficiently. The birds with the large bills feed mainly on the hard seeds, and birds with the small bills feed mainly on the soft seeds. Birds with intermediate-sized bills are not able to open either type of seed as efficiently, so they are less likely to survive the dry seasons.

With stabilizing selection, intermediate phenotypes are favored, and phenotypes at both ends of the range of variation are eliminated.

With disruptive selection, intermediate forms of traits are selected against, and phenotypes at both ends of the range of variation are favored.

17.6 Maintaining Variation

Natural selection theory helps explain diverse aspects of nature, including male–female differences and the relationship between sickle-cell anemia and malaria.

SEXUAL SELECTION

LINKS TO SECTIONS 3.5,10.1

The individuals of many sexually reproducing species show a distinct male or female phenotype, or sexual dimorphism (*dimorphos*, having two forms). One of the sexes (often the male) is often more colorful or larger than the other, and tends to be more aggressive.

These adaptations and behaviors seem puzzling. They take energy and time away from an individual's survival activities, and some may even be maladaptive because they attract predators. Why do they persist? The answer is sexual selection, in which the genetic winners outreproduce others of a population because they are better at securing mates. The most adaptive forms of a trait are those that help individuals defeat same-sex rivals for mates, or are the ones that are the most attractive to the opposite sex.

By choosing among mates, a male or female acts as a selective agent on its own species. For example, the females of some species will shop for a mate among a congregation of males, which vary in appearance and courtship behavior. Selected males pass the alleles for their attractive traits to the next generation of males. Females pass alleles that influence mate preference to the next generation of females.

Gerald Wilkinson and his colleagues demonstrated female preference for an exaggerated male trait in the stalk-eyed fly, *Cyrtodiopsis dalmanni*. The eyes of this Malaysian species form on long, horizontal eyestalks that provide no obvious adaptive advantage to their bearers, other than perhaps provoking sexual interest in other flies (Figure 17.12*a*). The researchers predicted that if eyestalk length is a sexually selected trait, then males with longer eyestalks would be more attractive to female flies than males with shorter eyestalks. They bred males with extra-long eyestalks, and found those males were indeed preferred by the female flies. Such experiments show how exaggerated traits can arise by sexual selection in nature.

Females of many species raise offspring with little help from males. In such species, males typically mate with any female; females choose males that display species-specific cues, which often include flashy body parts or behaviors (Figure 17.12*b*). Flashy traits can be a physical hindrance and they may attract predators. However, a flashy male's survival despite his obvious handicap implies health and vigor—traits that are likely to improve a female's chances of bearing healthy, vigorous offspring.

The males of species in which both sexes share parenting responsibilities usually are not overly flashy. Courtship behavior in such species may include demonstrations of the male's nurturing ability, such as offering nesting materials or food to the female.

Figure 17.12 Sexual selection. (**a**) Oh, what sexy eyestalks! Stalk-eyed flies of the species *Cyrtodiopsis dalmanni* cluster on aerial roots to mate, and females cluster preferentially around males with the longest eyestalks. This photo, taken in Kuala Lumpur, Malaysia, shows a male with very long eyestalks (*top*) that has captured the interest of the three females below him.

(**b**) This male bird of paradise (*Paradisaea raggiana*) is engaged in a flashy courtship display. He caught the eye (and, perhaps, the sexual interest) of the smaller, less colorful female. The males of this species of bird compete fiercely for females, which are selective. (Why do you suppose the females are drab-colored?)

BALANCED POLYMORPHISM

With *balancing* selection, two or more alleles of a gene persist at high frequencies in a population. We may see such persistence, which is balanced polymorphism, when environmental conditions favor heterozygotes. In some way, having nonidentical alleles is more adaptive than having identical ones.

For example, environmental pressures can favor the Hb^A/Hb^S heterozygous condition in humans. The *Hb* gene codes for hemoglobin, the oxygen-transporting protein in the blood. Hb^A is the normal allele. The Hb^S allele carries a mutation that causes homozygotes to develop sickle-cell anemia (Section 3.5). Individuals homozygous for the Hb^S allele often die in their teens or early twenties. Despite being so harmful, the allele persists at very high frequency among populations in tropical and subtropical regions of Asia and Africa.

Why? A clue to the answer is that populations with the highest frequency of the Hb^S allele also have the highest incidence of malaria (Figure 17.13). Mosquitoes transmit the parasitic agent of malaria, *Plasmodium*, to human hosts. The protozoan multiplies in the liver and then in red blood cells. The cells rupture and release new parasites during severe, recurring bouts of illness.

It turns out that Hb^A/Hb^S heterozygotes are more likely to survive malaria than people who make only normal hemoglobin. Several mechanisms are possible. For example, infected cells of heterozygotes take on a sickle shape. The abnormal shape brings infected cells to the attention of the immune system, which destroys them—along with the parasites they harbor. Infected cells of Hb^A/Hb^A homozygotes do not sickle, and the parasite may remain hidden from the immune system.

The persistence of the Hb^S allele may be a matter of relative evils. Malaria and sickle cell anemia are both potentially deadly. Hb^A/Hb^S heterozygotes are more likely to survive and reproduce in areas where malaria is common than Hb^A/Hb^A homozygotes. They are not completely healthy, but they do make enough normal hemoglobin to survive. Malaria or not, they are more likely to live long enough to reproduce than Hb^S/Hb^S homozygotes. The result is that nearly one-third of individuals that live in the most malaria-ridden regions of the world are Hb^A/Hb^S heterozygotes.

With sexual selection, some version of a trait gives an individual an advantage over others in securing mates. Sexual dimorphism is one outcome of sexual selection.

Balanced polymorphism is a state in which natural selection maintains two or more alleles at relatively high frequencies.

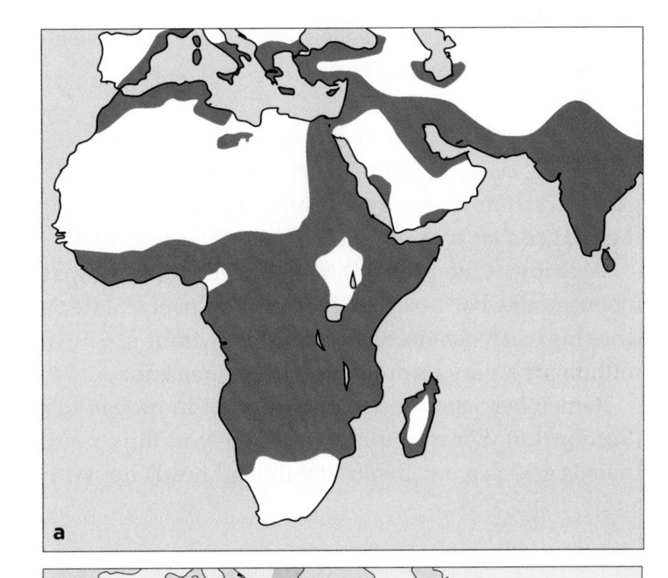

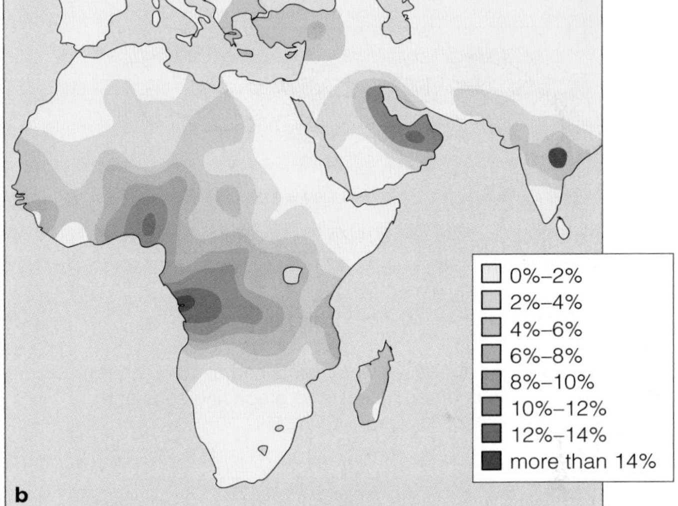

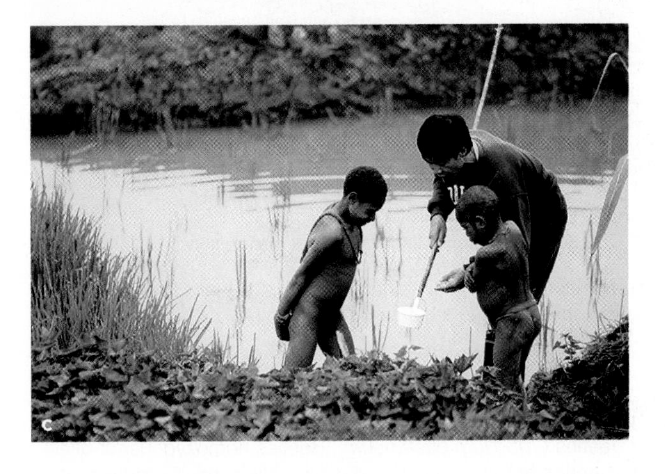

Figure 17.13 (**a**) Distribution of malaria cases reported in Africa, Asia, and the Middle East in the 1920s, before the start of programs to control mosquitoes, which transmit *Plasmodium*. (**b**) Distribution (in percent) of people that carry the sickle-cell allele. Notice the close correlation between the maps. (**c**) Physician searching for mosquito larvae in Southeast Asia.

17.7 Genetic Drift—The Chance Changes

Especially in small populations, random changes in allele frequencies can lead to a loss of genetic diversity.

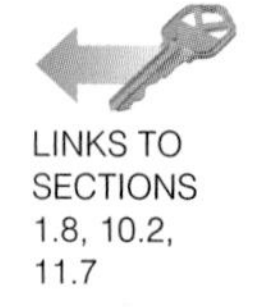
LINKS TO SECTIONS 1.8, 10.2, 11.7

Genetic drift is a random change in allele frequencies over time, brought about by chance. We can explain genetic drift in terms of probability rules. *Probability* is the chance that some event will occur (Section 10.2).

We express the probability of an event occurring as a percentage. For instance, if 10 million people enter a drawing, each has the same chance of winning: 1 in 10 million, or a very improbable 0.00001 percent.

Remember, sample size is important in probability (Section 1.8). For example, every time you flip a coin, there is a 50 percent chance it will land heads up. With 10 flips, the proportion of times heads actually land up may be very far from 50 percent. With 1,000 flips, that proportion is more likely to be near 50 percent.

We can apply the same rule to populations. Because population sizes are not infinite, there will be random changes in allele frequencies. These random changes have a minor impact on large populations. However, such changes can lead to dramatic shifts in the allele frequencies of small populations.

Imagine two human populations. Population I has 10 individuals, and population II has 100. Say an allele *b* occurs in both populations at a 10 percent frequency. Only one person carries the allele in population I. If that person dies before reproducing, allele *b* will be lost from population I. However, ten people in population II carry the allele. All ten would have to die before they reproduce for the allele to be lost from population II. Thus, the chance that population I will lose allele *b* is greater than that for population II. Steven Rich and his colleagues demonstrated this effect in populations of flour beetles (Figure 17.14).

Random change in allele frequencies can lead to a loss of genetic diversity and to the homozygous condition. This is a possible outcome of genetic drift in all populations; it simply happens faster in small ones. When all of the individuals of a population have become homozygous for one allele, we say that fixation has occurred. Once an allele is fixed, its frequency will not change again unless mutation or gene flow introduces new alleles.

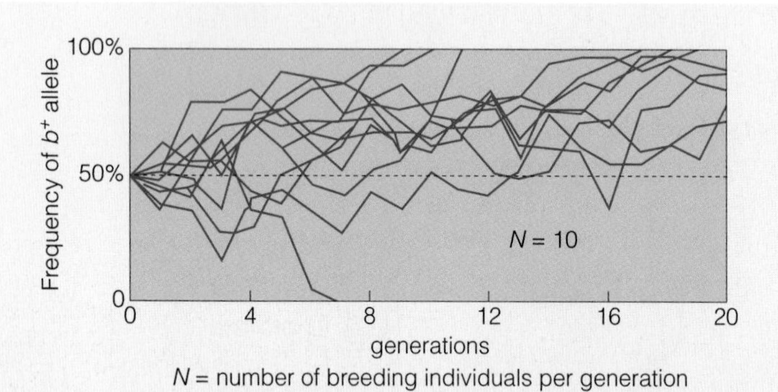

a The size of these populations of beetles was maintained at 10 breeding individuals. Allele b^+ was lost in one population (one graph line ends at 0).

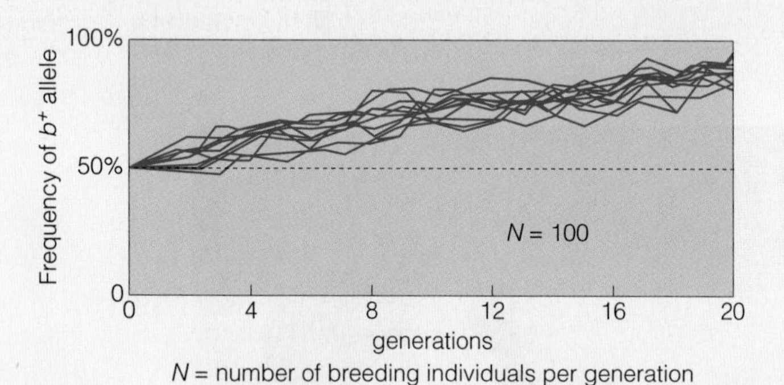

b The size of these populations was maintained at 100 individuals. Drift in these populations was less than the small populations in (**a**).

Figure 17.14 **Animated!** Experiment showing the effect of population size on genetic drift in flour beetles (*Tribolium castaneum*). Beetles homozygous for allele *b* were crossed with beetles homozygous for wild-type allele b^+. F_1 individuals (b^+b) were divided into populations of (**a**) 10 or (**b**) 100 randomly selected male and female beetles; population sizes were maintained for 20 generations.

Graph lines in (**b**) are smoother than in (**a**), indicating that drift was greatest in the sets of 10 beetles and least in the sets of 100. Notice that the average frequency of allele b^+ rose at the same rate in both groups. This means natural selection was at work too; allele b^+ was weakly favored.

BOTTLENECKS AND THE FOUNDER EFFECT

Genetic drift is pronounced when a few individuals rebuild a population or start a new one, such as occurs after a bottleneck—a drastic reduction in population size brought about by severe pressure. Suppose that a contagious disease, habitat loss, or overhunting nearly wipes out a population. Even if a moderate number of individuals do survive, allele frequencies will have been altered at random.

By the 1890s, northern elephant seals were on the edge of extinction, with only twenty known survivors. Hunting restrictions implemented since have allowed the population to recover to about 170,000 individuals. Each is homozygous for all the genes analyzed to date.

Genetic drift can also occur when a small group of individuals founds a new population. If the group is not representative of the original population in terms of allele frequencies, the new population will not be representative of it either. This is the founder effect. If the founding group was very small, a new population's genetic diversity may be quite reduced (Figure 17.15).

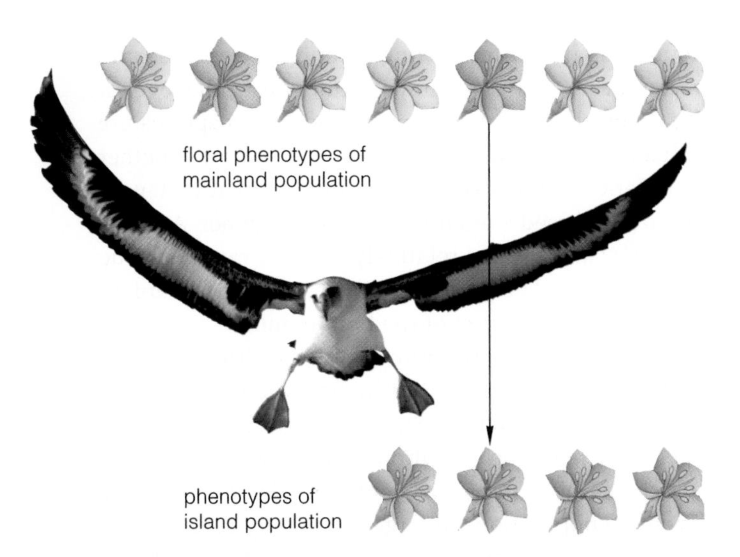

Figure 17.15 Founder effect. A wandering albatross carries seeds, stuck to its feathers, from the mainland to a remote island. By chance, most of the seeds carry an allele for flower color that is rare in the original population. Without further gene flow or selection for color, genetic drift will fix the allele on the island.

INBRED POPULATIONS

The effects of genetic drift are pronounced in inbred populations. Inbreeding is mating or breeding among close relatives. Close relatives share many of the same alleles, so inbreeding increases the frequency of the homozygous state. Thus, inbreeding lowers the genetic diversity of a population. Most societies discourage or forbid incest (breeding between parents and children or between siblings), but more distant relatives such as cousins do often mate in some societies.

As an example, the Old Order Amish in Lancaster County, Pennsylvania are moderately inbred. Amish people marry only other Amish people; intermarriage with other groups is not permitted. This population shows a high frequency of an allele associated with *Ellis–van Creveld syndrome*. Among other more serious problems, individuals affected by this syndrome have extra toes or fingers (Section 11.7). The particular allele that affects the Lancaster population has been traced to one man and his wife, who were among a group of 400 Amish who immigrated to the United States in the mid-1700s. As a result of inbreeding since then, about 1 of 8 people in the population is heterozygous for the allele, and 1 in 200 is homozygous for it.

Genetic drift is a random change in allele frequencies over generations. The magnitude of its effect is greatest in small populations, such as one that endures a bottleneck.

17.8 Gene Flow

Individuals, and their alleles, move into and away from populations. The flow of alleles counters genetic change that occurs within a population.

Individuals of the same species do not always stay in the same geographic area, or in the same population. A population can lose alleles when individuals leave it permanently, an act called *emigration*. A population gains alleles when individuals permanently move in, an act called *immigration*. Gene flow, the movement of alleles among populations, occurs in both cases. Gene flow is a microevolutionary process that counters the diversifying effects of mutations, natural selection, and genetic drift in a population.

Consider the acorns that blue jays disperse when they gather nuts for the winter. Every fall, jays visit acorn-bearing oak trees repeatedly, then bury acorns in the soil of home territories that may be as much as a mile away (Figure 17.16). The jays transfer acorns—and the alleles inside them—among populations of oak trees that would otherwise be genetically isolated.

Figure 17.16 Blue jay, a mover of acorns that helps keep genes flowing between separate oak populations.

Or think of the millions of people from politically explosive, economically bankrupt countries who seek a more stable home. The scale of their emigrations is unprecedented, but the flow of genes is not. Human history is rich with cases of gene flow that minimized many of the genetic differences among geographically separate groups. Remember Genghis Khan? His genes flowed from China to Vienna (Section 11.7). Similarly, the armies of Alexander the Great brought alleles for green eyes from Greece all the way to India.

Gene flow is the physical movement of alleles into and out of a population, by way of immigration and emigration. It tends to counter the effects of mutation, natural selection, and genetic drift.

17.9 Reproductive Isolation

Speciation differs in its details, but reproductive isolating mechanisms (Figure 17.17) are always part of the process.

LINKS TO SECTIONS 10.6, 16.10

SPECIES AND SPECIATION

Species is a Latin word for "kind," as in "one kind of plant." Earth's biodiversity is so great that we easily can identify many organisms that belong to different species—petunias, whales, monkeys, bees, and so on. In general, such organisms shared an ancestor so long ago that many changes accumulated in their separate lineages. Organisms that have a more recent common ancestor may be difficult to tell apart. For instance, how do we tell whether two forms of, say, arrowhead plant, are separate species (Figure 17.18)?

Evolutionary biologist Ernst Mayr defined a species as one or more groups of individuals that potentially can interbreed, produce fertile offspring, and do not interbreed with other groups. This biological species concept is useful for distinguishing species of sexual reproducers such as mammals, but only at the level of local populations: not all populations of a species are actually interbreeding. We may never know whether populations of a species separated by great distances could interbreed even if they did get together. As well, species that diverged relatively recently may continue to interbreed. The point is, a "species" is a convenient but artificial construct of the human mind.

In nature, sexually reproducing species attain and maintain separate identities by *reproductive isolation*—the end of gene exchanges between populations. New species arise by the evolutionary process of speciation, which begins as gene flow ends between populations. Then the populations diverge genetically as mutation, natural selection, and genetic drift operate in each one independently. Speciation may occur after a very long period of divergence, or after one generation (as often occurs among flowering plants by polyploidy).

As two populations diverge, reproductive isolating mechanisms arise (Figure 17.17). These are heritable aspects of body form, function, or behavior that often prevent interbreeding between different species. Such mechanisms reinforce differences between diverging populations. *Prezygotic* isolating mechanisms prevent successful pollination or mating. *Postzygotic* isolating mechanisms result in weak or infertile hybrids.

Different species!

Prezygotic isolating mechanisms

a

Temporal isolation: Individuals of different species reproduce at different times.

Mechanical isolation: Individuals cannot mate or pollinate because of physical incompatibilities.

Behavioral isolation: Individuals of different species ignore or do not get the required cues for sex.

Ecological isolation: Individuals of different species live in different places and never do meet up.

They interbreed anyway.

b

Gamete incompatibility: Reproductive cells meet up, but no fertilization occurs.

Zygotes form, but . . .

Postzygotic isolating mechanisms

c

Hybrid inviability: Hybrid embryos die early, or new individuals die before they can reproduce.

Hybrid sterility: Hybrid individuals or their offspring do not make functional gametes.

No offspring, sterile offspring, or weak offspring that die before reproducing

Figure 17.17 **Animated!** Reproductive isolating mechanisms that prevent interbreeding: barriers to (**a**) getting together, mating, or pollination; (**b**) successful fertilization; and (**c**) survival, fitness, or fertility of hybrid embryos or offspring.

PREZYGOTIC ISOLATING MECHANISMS

Temporal Isolation Diverging populations cannot interbreed when their timing of reproduction differs. Three species of periodical cicadas mature underground as they feed on roots. Every 17 years, they emerge to reproduce. Each has a sibling species with nearly identical form and behavior. However, the siblings emerge on a 13-year cycle. A species and its sibling might interbreed—except they only get together once every 221 years!

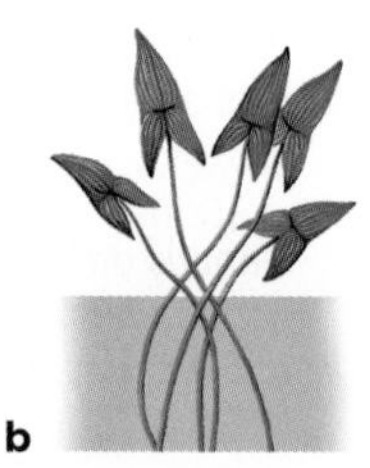

Figure 17.18 One species or two? The leaves of the arrowhead plant (*Sagittaria sagittifolia*) look very different depending on whether the plant is growing (**a**) underwater, or (**b**) partially submerged.

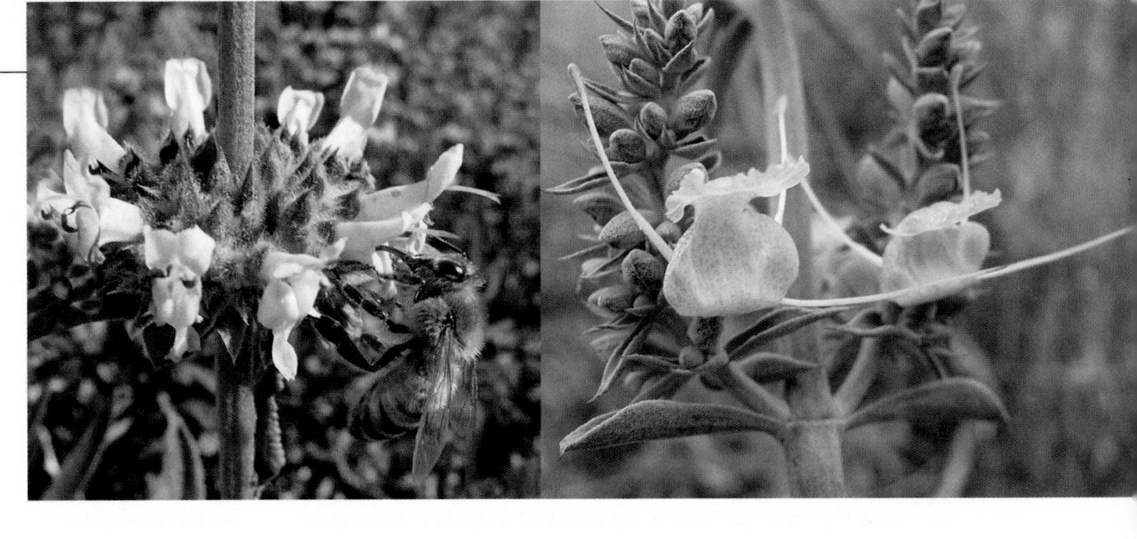

Figure 17.19 Mechanical isolation. *Left*, the flowers of black sage, *Salvia mellifera*, are too small to support bumblebees; they are pollinated by smaller honeybees. *Right*, the pollen-bearing anthers of white sage flowers (*S. apiana*) are at the tips of filaments that project high above the petals. Honeybees that land on this flower are too small to reach the anthers, so only larger bees pollinate white sage.

Mechanical Isolation Body parts of a species may not physically match with those of species that could otherwise be mates or pollinators. For example, *Salvia mellifera* (black sage) and *S. apiana* (white sage) grow in the same areas, but hybrids rarely form because the flowers of the two species have become specialized for distinct pollinators (Figure 17.19). Pollen-bearing parts (anthers) of white sage flowers are at the end of long filaments (stamens) that extend far above the petals. The anthers are too high above the flower to brush small bees that land on the petals. Thus, small bees cannot pollinate white sage. This flower is pollinated only by bumblebees and other large bees. Large bees have difficulty finding footing on the tiny flowers of black sage; this species is pollinated only by small bees.

Behavioral Isolation Behavioral differences stop gene flow between related species. For instance, males and females of some bird species engage in courtship displays before sex (Figure 17.20). A female recognizes the singing, wing spreading, or bobbing of a male of her species as an overture to sex. Females of different species usually ignore this behavior.

Ecological Isolation Two populations in different microenvironments may be ecologically isolated. Two manzanita species live in seasonally dry foothills of the Sierra Nevada, one at elevations of 600 to 1,850 meters (2,000 to 6,000 feet), the other at elevations of 750 to 3,350 meters (2,500 to 11,000 feet). They hybridize only rarely. Both species conserve water, but one is adapted to sites where water stress is not as intense. The other species lives in drier, exposed sites on rocky hillsides, so cross-pollination is unlikely.

Gamete Incompatibility The reproductive cells of different species have molecular incompatibilities, so fertilization does not occur. This may be the primary speciation route of animals that fertilize their eggs by releasing free-swimming sperm in an aquatic habitat.

POSTZYGOTIC ISOLATING MECHANISMS

Postzygotic reproductive isolating mechanisms occur as an outcome of unsuitable interactions among genes or gene products. They can cause early death, weak hybrids with low survival rates, sterility, or sterile F_2 offspring. Mules, which are offspring of female horses and male donkeys, are infertile hybrids.

Evolution is a dynamic and ongoing process that is changing the landscape of Earth's biodiversity even as we discuss it.

The individuals of a species can potentially interbreed and produce fertile offspring. They are reproductively isolated from individuals of other species.

Speciation is an evolutionary process by which a new species forms. It varies in its details and duration, but reproductive isolation is always part of the process.

Figure 17.20 **Animated!** Behavioral isolation. Courtship displays precede sex among many birds, including these albatrosses. Individuals recognize tactile, visual, and acoustical signals, such as a prancing dance, back arching, bill pointed skyward, with throat exposed and wings spread.

17.10 Allopatric Speciation

In the most common mode of speciation, a physical barrier arises and ends gene flow between populations.

LINKS TO SECTIONS 16.3,16.6

Every species is a unique outcome of its own history and environment. Thus, speciation is not a predictable process such as a metabolic reaction; it happens in a unique way every time it happens. However, we can identify some underlying principles.

Genetic changes leading to a new species usually begin with physical separation between populations, so allopatric speciation may be the most common way that new species form. (*Allo–* means different; *patria* is taken to mean homeland.) By this speciation route, a physical barrier separates two populations and ends gene flow between them. Then, reproductive isolating mechanisms arise, so even if the populations meet up again their individuals could not interbreed.

Whether a geographic barrier can block gene flow depends on an organism's means of travel (deliberate or accidental), how fast it can travel, and whether it is inclined to disperse. Populations of most species are separated by distance, and gene flow is intermittent. Barriers may arise abruptly and end the flow entirely. In the 1800s, a major earthquake buckled part of the Midwest and the Mississippi River changed course. It cut through the habitats of populations of insects that could not swim or fly, and thus interrupted gene flow in those populations.

The fossil record suggests that geographic isolation also happens slowly. For example, it happened after vast glaciers advanced into North America and Europe during the ice ages and cut off populations of plants and animals. After glaciers retreated, descendants of related populations met up again. Genetic divergences were not great between some separated populations; their descendants could still interbreed. Descendants of some other populations could no longer interbreed. Reproductive isolation had led to speciation.

Also, remember how Earth's crust is fractured into gigantic plates? Slow, colossal movements inevitably alter the configurations of land masses (Section 16.6). As Central America formed, part of an ancient ocean basin was uplifted, and it became a land bridge—now called the Isthmus of Panama. Some camelids crossed the bridge into South America. Geographic separation led to new species: llamas and vicunas (Figure 17.21).

THE INVITING ARCHIPELAGOS

An archipelago is an island chain some distance from a continent. Many chains are so close to the mainland that gene flow is more or less unimpeded, so there is little if any speciation. The Florida Keys are like this. As you read earlier, the Hawaiian Islands, Galápagos Islands, and other remote, isolated archipelagos favor adaptive radiations and speciation (Section 16.3). The islands are the tops of volcanoes that started building up on the seafloor, and, in time, broke the surface of the ocean. We can assume that their fiery surfaces were initially barren, with no life.

Winds or ocean currents carry a few individuals of some mainland species to such islands, as in Figure

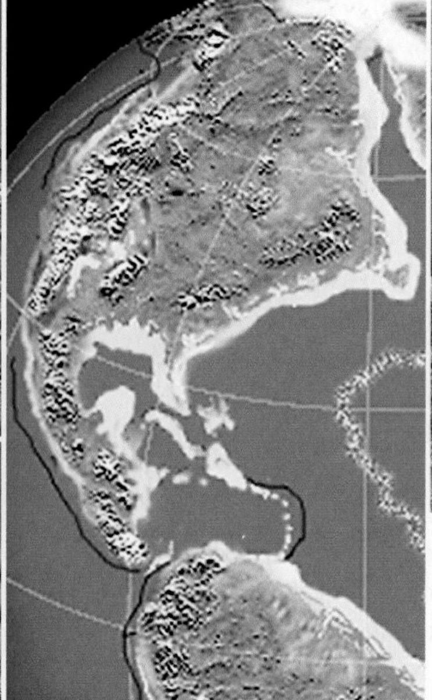

Figure 17.21 Allopatric speciations. The earliest camelids, no bigger than a jackrabbit, evolved in the Eocene grasslands and deserts of North America. By the end of the Miocene, they included the now-extinct *Procamelus*. The fossil record and comparative studies indicate that *Procamelus* may have been the common ancestral stock for llamas (**a**), vicunas (**b**), and camels (**c**). One descendant lineage dispersed into Africa and Asia and evolved into modern camels. A different lineage, ancestral to the llamas and vicunas, dispersed into South America after a land bridge formed between the two continents.

Late Eocene paleomap, before a land bridge formed between North and South America. At that time, North America and Eurasia were still connected by a land bridge.

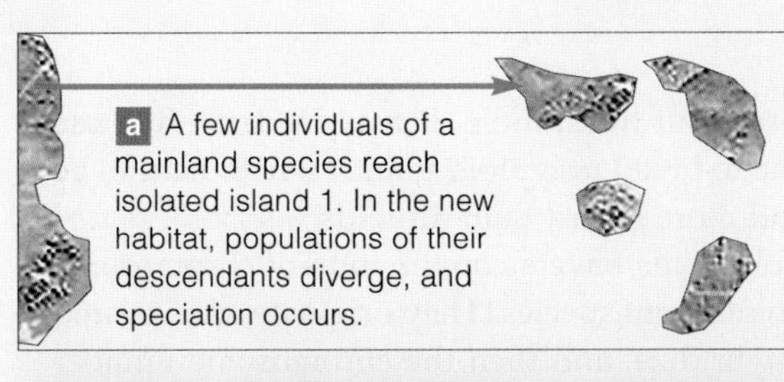

b Later, a few individuals of a new species colonize nearby island 2. Speciation follows genetic divergence in the new habitat.

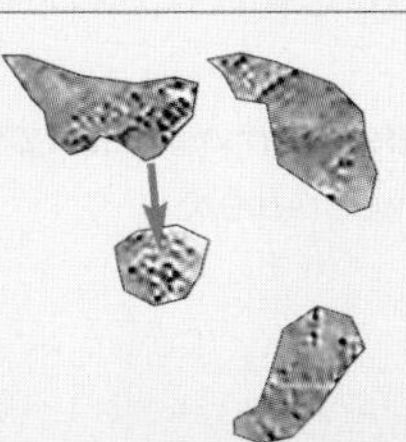

c Genetically different descendants of the ancestral species may colonize islands 3 and 4 or even invade island 1. Genetic divergence and speciation may follow.

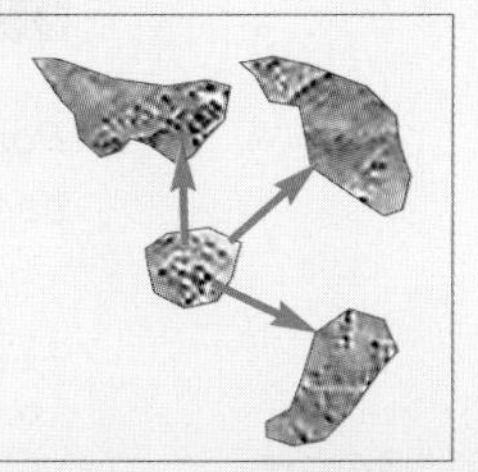

Akepa (*Loxops coccineus*)
Insects, spiders from buds twisted apart by bill; some nectar; high mountain rain forest

Akekee (*L. caeruleirostris*)
Insects, spiders, some nectar; high mountain rain forest

Nihoa finch (*Telespiza ultima*)
Insects, buds, seeds, flowers, seabird eggs; rocky or shrubby slopes

Palila (*Loxioides bailleui*)
Mamane seeds ripped from pods; buds, flowers, some berries, insects; high mountain dry forests

Maui parrotbill (*Pseudonestor xanthophrys*)
Rips dry branches for insect larvae, pupae, caterpillars; mountain forest with open canopy, dense underbrush

Apapane (*Himatione sanguinea*)
Nectar, especially of ohia-lehua flowers; caterpillars and other insects; spiders; high mountain forests

Poouli (*Melamprosops phaeosoma*)
Tree snails, insects in understory; last known male died in 2004

Maui Alauahio (*Paroreomyza montana*)
Bark or leaf insects, some nectar; high mountain rain forest

Kauai Amakihi (*Hemignathus kauaiensis*)
Bark-picker; insects, spiders; nectar; high mountain rain forest

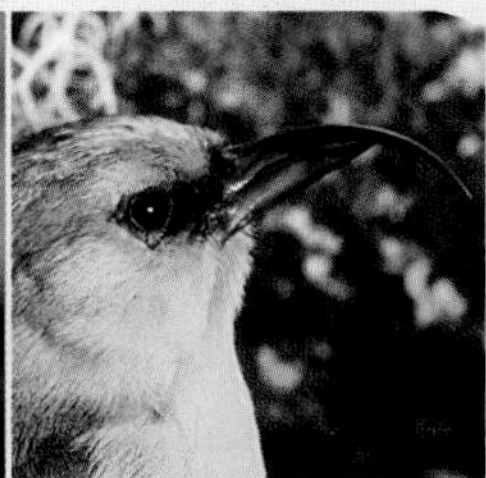

Akiapolaau (*H. munroi*)
Probes, digs insects from big trees; high mountain rain forest

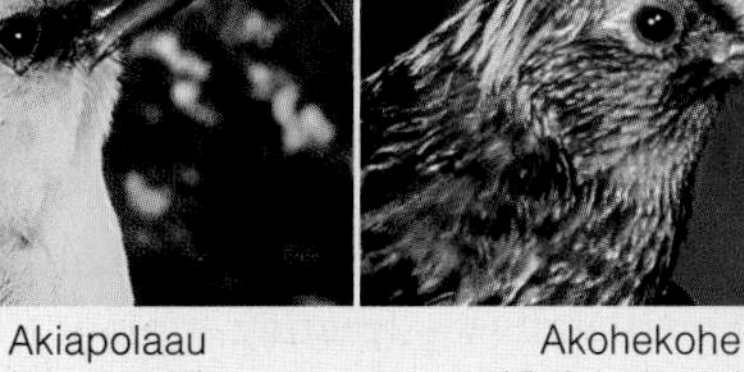

Akohekohe (*Palmeria dolei*)
Mostly nectar from flowering trees, some insects, pollen; high mountain rain forest

Iiwi (*Vestiaria coccinea*)
Mostly nectar (ohia flowers, lobelias, mints), some insects; high mountain rain forest

Figure 17.22 **Animated!** (**a–c**) Allopatric speciation on an isolated archipelago. *Above*, 12 of 57 known species and subspecies of Hawaiian honeycreepers, with some dietary and habitat preferences. Honeycreeper bills are adapted to feed on insects, seeds, fruits, nectar in floral cups, and other foods. Morphological studies, and comparisons of chromosomal and mitochondrial DNA sequences for proteins (such as cytochrome *b*) suggest that the ancestor of all Hawaiian honeycreepers resembled the housefinch (*Carpodacus*) shown at *left*.

17.22*a*. Their descendants colonize other islands that formed in the chain. Habitats and selection pressures differ within and between these islands, so allopatric speciation proceeds by way of divergences. Later, new species may return to islands that were colonized by their ancestors. Distances between the islands in many archipelagos are enough to favor divergence but not enough to stop the occasional colonizers.

The big island of Hawaii formed less than 1 million years ago. Its habitats range from old lava beds, rain forests, and grasslands to snow-capped volcanoes. The first birds to colonize it found a buffet of fruits, seeds, nectars, tasty insects—and few competitors for them. The near absence of competition in an abundance of vacant habitats spurred rapid speciation. Figure 17.22 hints at the variation that arose among the Hawaiian honeycreepers. Like thousands of other species, they are unique to Hawaii.

As another example of their speciation potential, the Hawaiian Islands combined make up only about 0.01 percent of the world's total land mass. Yet 40 percent of the 1,450 known *Drosophila* (fly) species arose there.

In allopatric speciation, a physical barrier intervenes between populations or subpopulations of a species and prevents gene flow among them. Gene flow ends, and genetic divergences give rise to daughter species.

17.11 Other Speciation Models

Populations sometimes speciate even without a physical barrier that bars gene flow between them.

SYMPATRIC SPECIATION

LINKS TO SECTIONS 4.12, 8.3, 9.3, 11.1, 11.6, 16.10

In sympatric speciation, new species form *within* the home range of an existing species, in the absence of a physical barrier. *Sym*– means together, as in "together in the homeland."

Polyploidy Speciation can occur in an instant with a change in chromosome number. Sometimes, somatic cells duplicate their chromosomes but do not divide during mitosis. Or, nondisjunction in meiosis results in gametes with an unreduced chromosome number. Cells that result from such events are polyploid: They have three or more sets of chromosomes characteristic of their species (Section 11.6).

In plants, one individual can give rise to an entire population. Polyploid cells that arise in a parent plant may proliferate to form polyploid shoots and flowers. If the flowers self-fertilize, a new species may result. The new plant species will be *auto*polyploid—it arose by chromosome multiplication in one parent species.

Autopolyploids occur spontaneously in nature, but can be induced artificially by treating a plant's seeds or buds with colchicine. This toxin prevents spindle microtubules from assembling (Sections 8.3 and 9.3). Chromosomes cannot separate during either mitosis or meiosis without the spindle. Plant breeders often use colchicine to breed polyploid plants, which tend to be larger and more robust than diploids.

*Allo*polyploids have a combination of chromosome sets from different species. They originate after related species hybridize, and then the chromosome number multiplies in the offspring. For example, the ancestor of common bread wheat was a wild species, *Triticum monococcum*, that hybridized about 11,000 years ago with another wild species of *Triticum* (Figure 17.23). A spontaneous chromosome doubling in the resulting hybrid gave rise to *T. turgidum*, an allopolyploid species with two sets of chromosomes. Another hybridization led to the common bread wheat, *T. aestivum*.

Often, polyploids do not produce fertile offspring by breeding with the parent species; the mismatched chromosomes pair abnormally during meiosis. Some polyploid plants are crossed with diploid parents to make sterile offspring that are valued for agriculture. Seedless watermelons are produced this way.

About 95 percent of ferns originated by polyploidy; 70 percent of flowering plants are now polyploid, as well as a few conifers, insects and other arthropods, mollusks, fishes, amphibians, and reptiles.

Other Examples Speciation with no physical barrier to gene flow may occur with no chromosome number change. For example, sister species *Howea forsteriana* (thatch palm) and *H. belmoreana* (curly palm) diverged

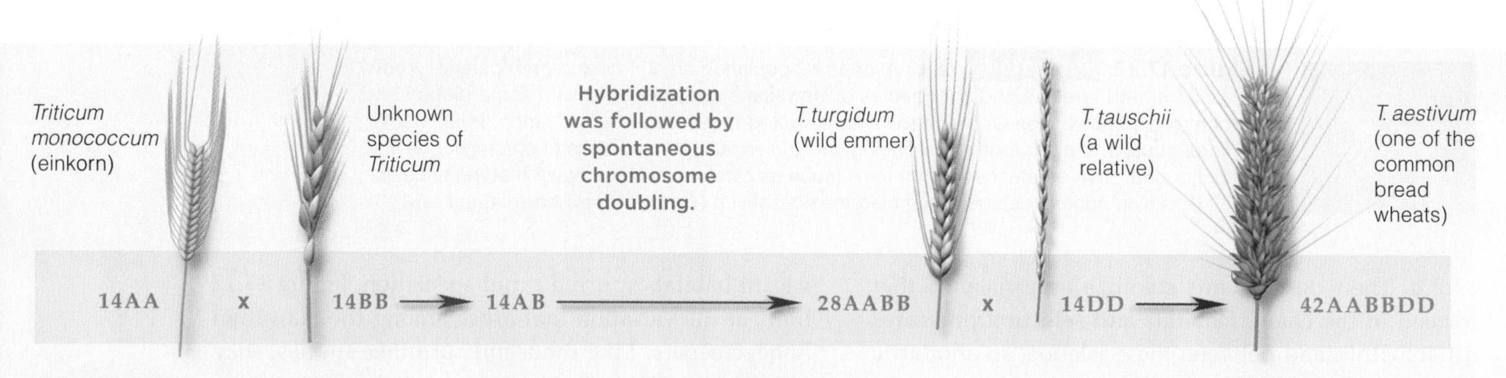

a By 11,000 years ago, humans were cultivating wild wheats. Einkorn has a diploid chromosome number of 14 (two sets of 7). It probably hybridized with another wild wheat species having the same number of chromosomes.

b About 8,000 years ago, the alloploid wild emmer originated from an AB hybrid wheat plant in which the chromosome number doubled. Wild emmer is tetraploid, or AABB; it has two sets of 14 chromosomes. There is recently renewed culinary interest in emmer, also called farro.

c AABB emmer probably hybridized with *T. tauschii*, a wild relative of wheat. Its diploid chromosome number is 14 (two sets of 7 DD). Common bread wheats have a chromosome number of 42 (six sets of 7 AABBDD).

Figure 17.23 Animated! Presumed sympatric speciation in wheat. Wheat grains 11,000 years old and diploid wild wheats have been found in the Middle East, and chromosome analysis indicates that they hybridized. Later, in a self-fertilizing hybrid, homologous chromosomes failed to separate at meiosis, and it produced fertile polyploid offspring. A polyploid descendant hybridized with a wild species. We make bread from grains of their hybrid descendants.

Figure 17.24 Sympatric speciation in palms. (**a**) Lord Howe Island is so small that geographic isolation of native wind-pollinated palm species is impossible. (**b**) The thatch palm *Howea forsteriana* and (**c**) the curly palm *H. belmoreana* may have speciated in sympatry by reproductive isolation.

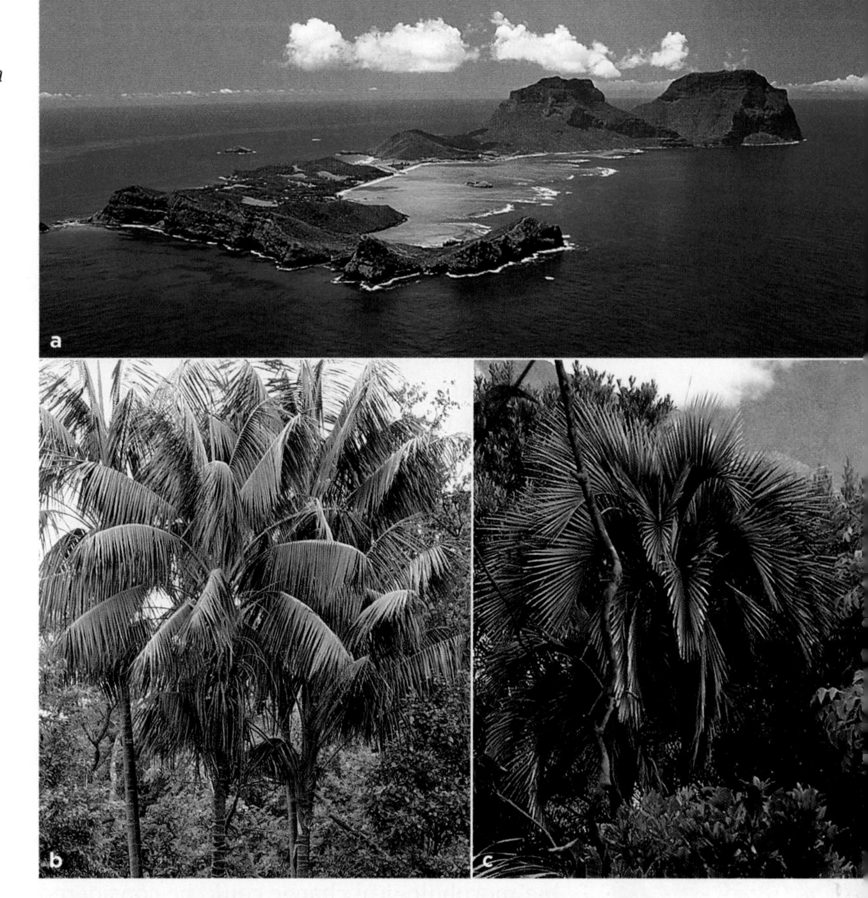

about 2 million years ago (Figure 17.24). The palms are still abundant in their native habitat, tiny Lord Howe Island. The island is so small that geographic isolation of the wind-pollinated palms is not possible, so their speciation was probably not allopatric. What ended gene flow between the species?

The answer lies in the pH of the island soils. Most of the thatch palms grow in the low-lying parts of the island, where the soil's pH is basic. Curly palms grow in volcanic soils, which are more acidic. Thatch palms growing in the basic soil flower six weeks earlier than palms of either species growing in acidic soil.

We can expect that some individuals of an ancestral palm species colonized the lower-lying regions of the island, and they began to flower earlier. If the island then was as small as it is now, reproductive isolation (temporal) occurred with no physical barrier to gene flow. Disruptive selection would have reinforced the divergence of the two populations.

Sympatric speciation has also occurred in greenish warblers of Siberia (*Phylloscopus trochiloides*). A chain of populations of this bird forms a ring around Tibet. The adjacent populations interbreed, but small genetic differences between each add up to major divergences between the populations at the ends. Such *ring species* present one of those paradoxes for those who like neat categories: Populations at the ends of the chain cannot interbreed; thus, they are technically distinct species. However, gene flow occurs continuously all around the ring; where should the line that divides the two species be drawn?

ISOLATION AT HYBRID ZONES

Parapatric speciation may occur when one population extends across a broad region encompassing diverse habitats. The different habitats exert distinct selection pressures on parts of the population, and the result may be divergences that lead to speciation. Hybrids that form in a contact zone between habitats are less fit than individuals on either side of it (Figure 17.25).

By a sympatric speciation model, daughter species arise from a population even in the absence of a physical barrier. Polyploid flowering plants probably formed this way.

By a parapatric speciation model, populations maintaining contact along a common border evolve into distinct species.

Figure 17.25 Example of parapatric speciation. The habitats of two rare species, (**a**) the giant velvet walking worm, *Tasmanipatus barretti*, and (**b**) the blind velvet walking worm, *T. anophthalmus*, overlap in a hybrid zone on the island of Tasmania (**c**). Hybrid offspring are sterile, which may be the main reason these two species are maintaining separate identities in the absence of a physical barrier between their habitats.

17.12 Macroevolution

Microevolution describes genetic changes within a species or population. Macroevolution is our name for large-scale patterns such as one species giving rise to several others, the origin of major groups, and major extinction events.

THE RATE OF EVOLUTIONARY CHANGE

LINKS TO SECTIONS 16.4–16.8, 16.10, 16.12

Models of speciation make no direct predictions about rate. How fast does evolution occur? The fossil record shows a stable rate of change in about 99 percent of most lineages; episodes of abrupt change occur in the remaining 1 percent. A model called gradualism holds that evolution occurs by slight changes over long time spans. This model fits with many sequences of fossils. For example, vertical sequences of fossil foraminiferan shells reflect gradual changes in morphology in these shelled protists (Figure 17.26).

A different model, punctuated equilibrium, holds that evolutionary change tends to occur over a brief time span, followed by very long periods of little or no change. Remember, we are discussing geologic time in this context. "Brief" means that change in a lineage appeared in few sedimentary rock layers relative to the lineage's entire existence in the fossil record. A big morphological change could be considered "abrupt" even if it occurred in gradual steps over 100,000 years.

The two models are not necessarily mutually exclusive. The same genetic changes may be at the root of all evolution—fast or slow, large- or small-scale. Big jumps in morphology, if they are not just artifacts of gaps in the fossil record, may be the result of mutations in homeotic or other regulatory genes (Section 16.8).

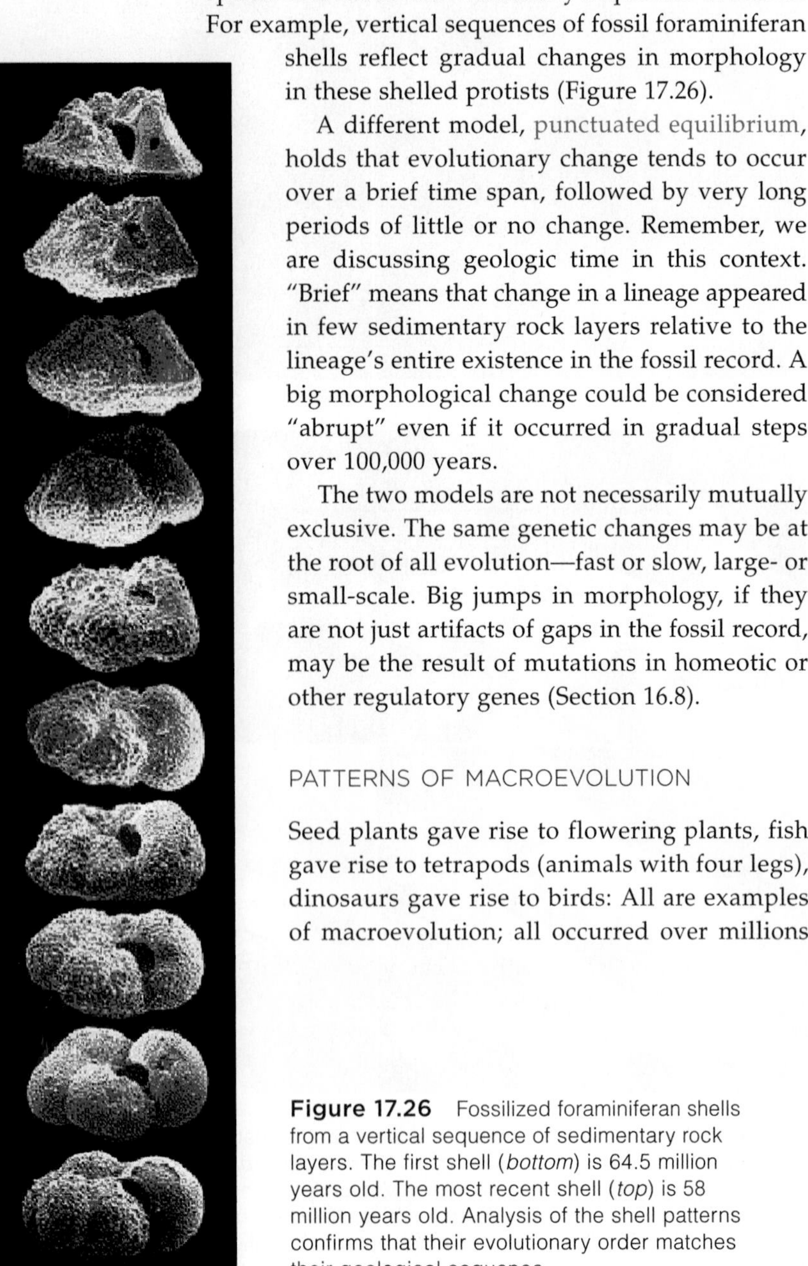

Figure 17.26 Fossilized foraminiferan shells from a vertical sequence of sedimentary rock layers. The first shell (*bottom*) is 64.5 million years old. The most recent shell (*top*) is 58 million years old. Analysis of the shell patterns confirms that their evolutionary order matches their geological sequence.

PATTERNS OF MACROEVOLUTION

Seed plants gave rise to flowering plants, fish gave rise to tetrapods (animals with four legs), dinosaurs gave rise to birds: All are examples of macroevolution; all occurred over millions of years. Using cladistics, we are able to reconstruct such large-scale evolutionary events with far greater clarity than we can with microevolutionary patterns of change at the population level.

Preadaptation Typically, major evolutionary novelty stems from adaption of existing structures. Bird wings, which are derived from the forelimbs of their reptilian ancestors, are one example (Section 16.7). Sometimes, such complex traits held a different adaptive value in an ancestral lineage; in other words, they may have been used for a very different purpose. For example, the feathers that allow birds to fly are derived from the feathers that first evolved in certain dinosaurs. Those dinosaurs did not use their feathers for flight, but they probably did use them for insulation.

Adaptive Radiation A burst of divergences from a single lineage is called adaptive radiation, and it leads to many new species. This is the evolutionary pattern that gave rise to the Hawaiian honeycreepers (Figure 17.22). Adaptive radiation only occurs when a lineage encounters a set of new niches. Genetic divergences then give rise to new species that fill vacant niches, or outcompete resident species for occupied ones. Think of a *niche* as a certain way of life, such as "burrowing in seafloor sediments" or "catching winged insects in the air at night." In Chapter 41, you will learn more about niches in the context of community structure.

A lineage may encounter a set of new niches when some of its individuals gain physical access to a new habitat by migrating to a different region. Geologic or climatic events may also change an existing habitat. For example, mammals were once distributed through tropical regions of Pangea. That supercontinent broke up into huge land masses, which drifted apart over millions of years (Section 16.6). Changes in habitats and resources set the stage for adaptive radiations.

Genetic change may allow individuals of a lineage to enter niches within their existing habitat that had been unavailable before the change. A key innovation is a structural or functional modification that bestows upon its bearer the opportunity to exploit a habitat more efficiently or in a novel way. For example, new niches opened up for the ancestors of birds after they began to use their feathered forelimbs for flight.

Coevolution The process by which close ecological interactions among species cause them to evolve jointly is coevolution. Each species adapts to changes in the other; over time the two may become interdependent. Some coevolved species can no longer survive without

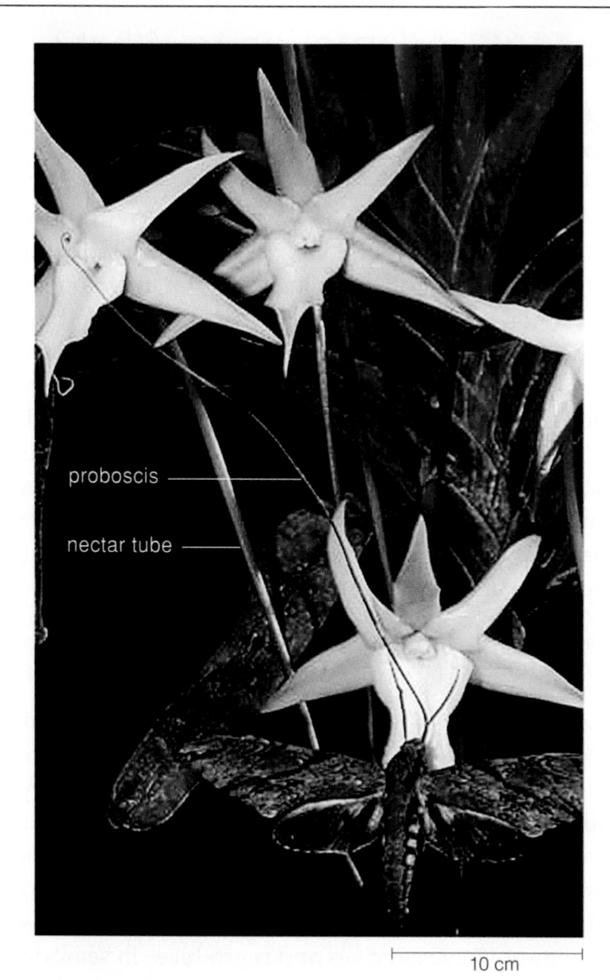

Figure 17.27 Example of two coevolved species, the orchid *Angraecum sesquipedale* and hawkmoth *Xanthopan morgani praedicta*. The orchid, which was discovered in Madagascar in 1852, stores its nectar at the base of a 30-centimeter-long floral tube. Darwin predicted that there had to be an insect in Madagascar that had a proboscis long enough to reach the nectar—and pollinate the flower. Decades later, the moth was discovered in Madagascar. Its proboscis is 30–35 cm long.

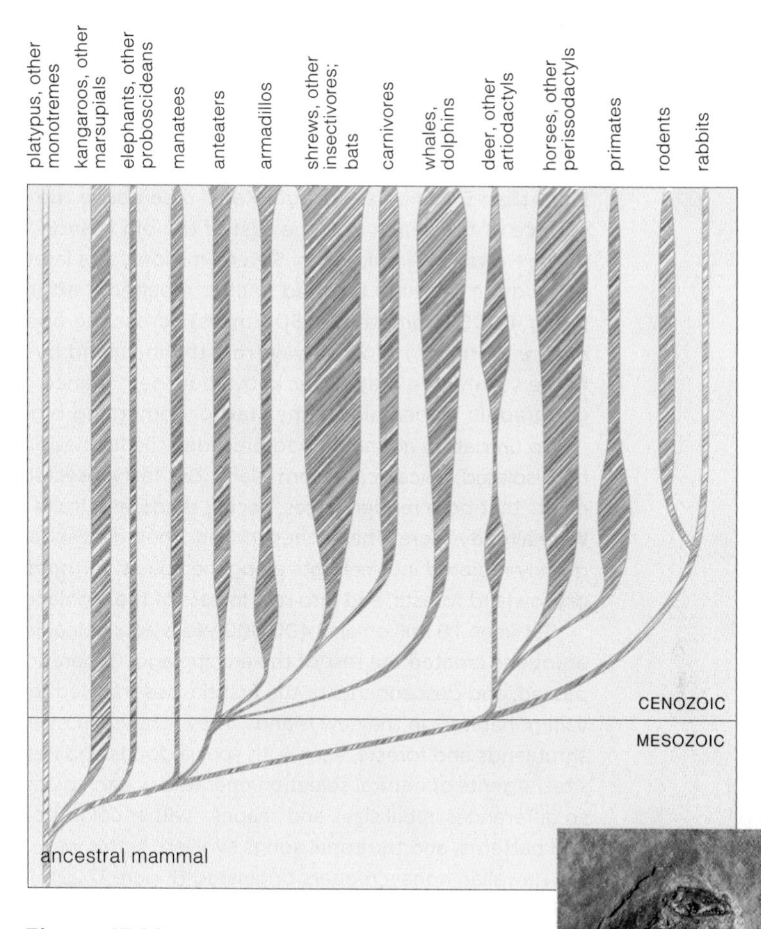

Figure 17.28 Adaptive radiation of mammals that followed the K–T extinction event. Branch widths indicate the range of biodiversity at different times. We show only a sampling of the mammalian lineage, which includes 4,000 modern species.

The photograph shows a fossil of *Eomaia scansoria* (Greek for ancient mother climber). About 125 million years ago, this insect-eater crawled on low branches. It is thought to be in the lineage that led to mammals.

one another. We know of many coevolved species of predator and prey, host and parasite, pollinator and flower (Figures 17.19 and 17.27). Later chapters detail specific examples.

Extinction An extinction is the irrevocable loss of a species. By current estimates, more than 99 percent of all species that ever lived are now extinct. In addition to continuing small-scale extinctions, the fossil record indicates that there have been more than twenty mass extinctions, which are losses of many lineages. These include five catastrophic events in which the majority of species on Earth disappeared (Section 16.5). After each event, new species evolved that filled the vacated niches. Biodiversity recovered very slowly, over tens of millions of years (Figure 17.28).

EVOLUTIONARY THEORY

Biologists do not doubt that macroevolution occurs, but many disagree about *how* it occurs. Macroevolution may include more processes than microevolution—or not. It may be an accumulation of microevolution, or it may be a very different process. Though evolutionary biologists disagree about such hypotheses, they all are trying to explain the same thing—the tree of life, which connects all species by ancestry (Section 16.12).

Macroevolution includes patterns of evolutionary change above the population level. The patterns represent the shape of the tree of life, based on the origin of species and major groups, and loss through extinction.

17.13 For the Birds

FOCUS ON EVOLUTION

At one time, Hawaiian honeycreepers flourished. Half of the known species are now extinct.

LINKS TO SECTIONS 3.5, 5.8, 16.6

More than 5 million years ago, Kauai rose above the surface of the sea. It was the first of the big islands of the Hawaiian Archipelago. Several million years later, a few quite possibly terrified finches reached it after flying 4,000 kilometers (2,500 miles) across the open ocean. Were they blown away from the mainland by a fierce storm? We may never know, but their chance geographic dispersal was the start of something big.

No predatory mammals had preceded the finches onto that isolated, volcanically born island. But tasty insects and plants that bore tender leaves, nectar, seeds, and fruits were already there. The finches thrived. Their descendants quickly radiated into habitats along the coasts, through dry lowland forests, and into rain forests of the highlands.

Between 1.8 million and 400,000 years ago, volcanic eruptions created the rest of the archipelago. Generations passed, and descendants of the first finches traveled to vacant habitats in the new islands. They foraged in many shrublands and forests, each with special foods and nest sites. Agents of natural selection operated in each place, so differences in bill sizes and shapes, feather coloration and patterns, and territorial songs evolved. In this way, the Hawaiian honeycreepers originated (Figure 17.22).

Ironically, the very isolation that favored specialized adaptations to conditions in unique habitats made these birds vulnerable to extinction. When conditions changed, they had nowhere to go. They had no built-in defenses against predatory mammals and avian diseases of the mainland, against humans who coveted cloaks made of their eye-catching feathers, or against climate change.

Accompanying humans to the islands were rats, cats, and other voracious predators. People also imported chickens and other birds that happened to be infected with disease agents. Over time, people cleared more and more of the forests. Imported crop plants and plant-eating mammals became established. The Hawaiian honeycreeper habitats shrank. Today, with a long-term increase in global temperatures—global warming—forests at higher elevations are not as cool as they once were. They have been infiltrated by mosquitoes, which thrive in warm climates. Mosquitoes transmit avian malaria and other diseases.

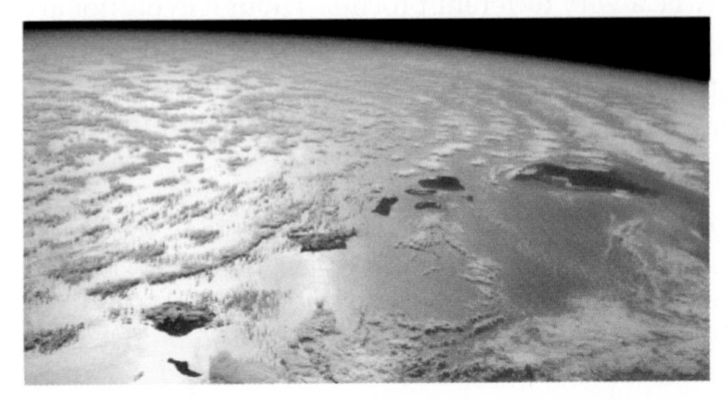

Hawaiian Islands, in the vast Pacific Ocean

17.14 Adaptation to What?

Observable traits are not always easy to correlate with conditions in an organism's environment.

"Adaptation" is one of those words that have different meanings in different contexts. Often, an individual plant or animal quickly adjusts its form, function, and behavior. Junipers in windy places grow less tall than junipers of the same species in more sheltered places. This is an example of a *short-term* adaptation: It is not heritable, and it lasts only as long as the individual.

An evolutionary adaptation is a trait—a heritable aspect of form, function, or behavior—that improves an individual's odds of surviving and reproducing in its prevailing environment. Such adaptations are the outcome of microevolutionary processes.

SALT-TOLERANT TOMATOES

As an example of long-term adaptation, compare how tomato species handle salty water. Tomatoes evolved in Ecuador, Peru, and the Galápagos Islands. The type sold most often in markets, *Lycopersicon esculentum*, has eight close relatives in the wild. Even if soil has only 2,500 parts per million of salt, it grows poorly. However, the Galápagos tomato (*L. cheesmanii*) shown in Figure 17.29*a* survives and reproduces in seawater-washed soils. We know its salt tolerance is a heritable adaptation. Crossing *L. cheesmanii* with a commercial variety yielded a small, edible hybrid. The hybrid can tolerate irrigation water salty enough to wither most plants is scarce and where salts have built up in croplands.

It may take modification of only a few traits to get new salt-tolerant plants. Revving up just one gene for a sodium–hydrogen ion transporter helps the tomato plants use salty water and still bear edible fruits.

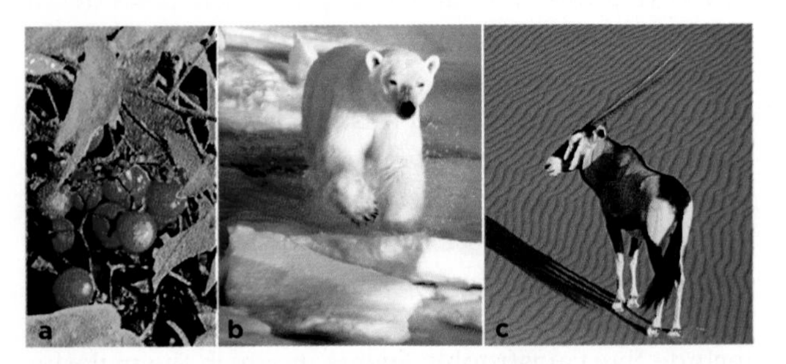

Figure 17.29 Long-term adaptations. (**a**) *Lycopersicon cheesmanii.* (**b**,**c**) Make a list of adaptations that may help the polar bear (*Ursus maritimus*) and the oryx (*Oryx beisa*) survive in their extreme environments. After you finish Unit Six, see how you can expand the list.

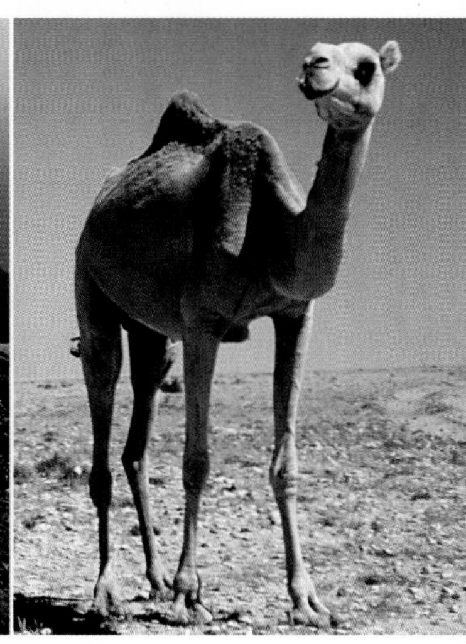

Figure 17.30 Adaptation to what? A heritable trait is an adaptation to specific environmental conditions. The hemoglobin of llamas (*left*), which live at high altitudes, has high oxygen-binding affinity. However, so does the hemoglobin of camels (*right*), which live at much lower elevations.

NO POLAR BEARS IN THE DESERT

You can safely bet that a polar bear (*Ursus maritimus*) is finely adapted to the icy Arctic, and that its form and function would be a flop in a desert (Figure 17.29*b*,*c*). You might be able to guess why that is so. However, knowledge of a polar bear's anatomy and physiology might give you a different, more respectful perspective about this species—or any other. How does the polar bear maintain its internal temperature when it sleeps on ice? How can its muscles function in frigid water? As another example, how can an oryx walk all day in the blistering heat of an African desert? How does it get enough water when there is no water to drink? You will find some answers, or at least ideas about how to look for them, in the next three units of this book.

ADAPTATION TO WHAT? A WORD OF CAUTION

A relationship between adaptation and environment is not always obvious. A prevailing environment may be different from the one in which a trait arose. Consider the llama, native to cloud-piercing peaks of the South American Andes (Figure 17.30). It lives 4,800 meters (16,000 feet) above sea level. Compared to humans at lower elevations, its lungs have more blood vessels and air sacs. Its heart has larger chambers, so it pumps larger volumes of blood. Llamas need not make extra blood cells, as people do after they move permanently from lowlands to high elevations. (Extra cells make the blood "stickier," so the heart has to pump harder.) The best known adaptation of llamas to high altitude is that their hemoglobin is better than ours at binding oxygen in the lungs. Did the higher oxygen affinity of llama hemoglobin evolve as an adaptation to thin air at high altitudes? Apparently not.

Llamas are in the same family as dromedary camels. Their shared camelid ancestors evolved in the Eocene grasslands and deserts of North America. Later on, the ancestors of camels and llamas went separate ways. The ancestors of camels reached Asia's low-elevation grasslands and deserts by a land bridge, which later submerged when the sea level rose. Llama forerunners moved down the Isthmus of Panama to South America.

Intriguingly, a dromedary camel's hemoglobin also shows a high oxygen-binding capacity. So if the trait arose in a shared ancestor, then in what respect is it adaptive at low elevations if it is also adaptive at high elevations? Camels and llamas did not just happen to evolve to be similar. They are close kin, and their most recent ancestors lived in very different environments with different oxygen concentrations.

The high oxygen affinity of llama hemoglobin may be a preadaptation. The Eocene climate fluctuated, and hemoglobin binds less oxygen as the temperature rises. What if the allele for more efficient hemoglobin became fixed in an ancestral population by chance? Was it a neutral mutation at first? Or did it have an adaptive value that we do not yet understand? Use such "what-if" questions as a reminder to think critically about the connections between a trait and a given environment. Identifying the correct connections often takes more than intuition.

An evolutionary adaptation is a heritable trait that enhances the fit between an individual and its environment.

An adaptive trait improves the odds of surviving and reproducing—at least under conditions that prevailed when the trait first evolved.

www.thomsonedu.com ThomsonNOW!

Summary

Section 17.1 Individuals of one population generally have the same number and kinds of genes for the same traits. Individuals who inherit different combinations of alleles vary in the details of one or more traits. Mutations are the original source of new alleles.

Microevolution, or changes in the allele frequencies of a population, occurs by processes of mutation, natural selection, genetic drift, and gene flow.

Section 17.2 Genetic equilibrium is a state in which a population is not evolving. It never occurs in nature. We use deviations from genetic equilibrium to study how a population is evolving.

- *Use the interaction on ThomsonNOW to investigate gene frequencies and genetic equilibrium.*

Section 17.3 Natural selection is differential survival and reproduction among individuals of a population that show variations in the details of their shared traits. Allele frequencies are maintained by stabilizing selection, and shifted by directional or disruptive selection.

Section 17.4 Directional selection shifts the range of variation in traits in one direction. Individuals at one end of the range are favored; those at the other end are not.

- *Use the animation on ThomsonNOW to see how directional selection works.*
- *Read the InfoTrac article "AIDS in Africa Has Potential to Affect Human Evolution,"* AIDS Weekly, *June 2001.*

Section 17.5 Stabilizing selection works against both extremes in the range of phenotypic variation; it favors intermediate forms. Disruptive selection favors forms at the extremes of the range of variation.

- *Use the animation on ThomsonNOW to see how disruptive and stabilizing selection work.*
- *Read the InfoTrac article "Portraits of Evolution: Studies of Coloration in Hawaiian Spiders," Geoffrey S. Oxford, Rosemary G. Gillespie,* Bioscience, *July 2001.*

Section 17.6 By sexual selection, a female or a male acts as an agent of selection on its own species. Sexual selection leads to forms of traits that favor reproductive success. Sexual dimorphism is one outcome.

In balanced polymorphism, nonidentical alleles for a trait are maintained at relatively high frequencies.

- *Read the InfoTrac article "High-Risk Defenses," Gregory Cochran, Paul W. Ewald,* Natural History, *February 1999.*

Section 17.7 Genetic drift is a random change in a population's allele frequencies over time due to chance. It can lead to the loss of genetic diversity. Its effect is most pronounced in small or inbred populations.

- *Use the interaction on ThomsonNOW to explore genetic drift.*

Section 17.8 Gene flow is the movement of alleles into or out of a population by immigration or emigration. It helps keep populations of the same species similar. Gene flow can counter mutation, natural selection, and genetic drift, all processes that cause populations to diverge.

Section 17.9 The individuals of a sexually reproducing species produce fertile offspring under natural conditions, and are reproductively isolated.

Typically, when gene flow between populations stops, reproductive isolating mechanisms evolve. Divergences may lead to new species (Table 17.1).

- *Use the animation on ThomsonNOW to explore how species become reproductively isolated.*
- *Read the InfoTrac article "Tracking the Red-Eyed, Sluggish, and Ear-Splitting," Tabitha M. Powledge,* American Scientist, *July 2004.*

Section 17.10 In allopatric speciation, a geographic barrier stops gene flow between two or more populations of a species. Then, genetic divergence and reproductive isolation give rise to new species.

- *Use the animation on ThomsonNOW to learn more about speciation on an archipelago.*

Section 17.11 In sympatric speciation, populations in physical contact diverge into separate species. Polyploid species of many plants (and a few animals) originated by chromosome doublings and hybridizations.

In parapatric speciation, selection pressures across a broad region act on populations of a species in contact along a common border. Unfit hybrids form in the zone.

- *Use the animation on ThomsonNOW to explore the effects of sympatric speciation in wheat.*

Sections 17.12, 17.13 Macroevolution refers to large-scale patterns of evolution such as one species giving rise to others, origin of major groups, and major extinctions.

A lineage may use a structure for a different purpose than its ancestor did. Adaptive radiations occur when a species encounters a set of new niches, and the lineage diverges rapidly into new species that occupy the niches. A key innovation is some modification that allows an organism to exploit its environment in a new or more efficient way.

Coevolution occurs when close ecological interactions cause two species to act as agents of selection upon one

Table 17.1 Different Speciation Models

	Allopatric	Sympatric	Parapatric
Original population			
Initiating event	barrier arises	genetic change	new niche entered
Reproductive Isolation	in isolation	within population	in new niche
New species			

another. Most species that ever existed are now extinct. Mass extinctions and recoveries have occurred several times in the history of life.

Section 17.14 Evolutionary adaptations are heritable traits that improve an individual's chance of surviving and reproducing, or at least did so under conditions that prevailed when genes for the trait first evolved.

Self-Quiz

Answers in Appendix III

1. Individuals don't evolve, ______ do.

2. Biologists define evolution as ______.
   a. purposeful change in a lineage
   b. heritable change in a line of descent
   c. acquiring traits during the individual's lifetime

3. ______ is the original source of new alleles.
   a. Mutation
   b. Natural selection
   c. Genetic drift
   d. Gene flow
   e. All are original sources of new alleles

4. Natural selection can only occur in a population when there are ______.
   a. differences in forms of heritable traits
   b. different selection pressures
   c. both a and b

5. Directional selection ______.
   a. eliminates common forms of alleles
   b. shifts allele frequencies in one direction only
   c. does not favor intermediate forms of a trait
   d. works against adaptive traits

6. Disruptive selection ______.
   a. eliminates uncommon forms of alleles
   b. shifts allele frequencies in one direction only
   c. does not favor intermediate forms of a trait
   d. both b and c

7. Sexual selection, such as competition between males for access to fertile females, frequently influences aspects of body form and leads to ______.
   a. violence
   b. sexual behavior
   c. sexual dimorphism
   d. both b and c

8. The persistence of malaria and sickle-cell anemia in a population is a case of ______.
   a. bottlenecking
   b. balanced polymorphism
   c. natural selection
   d. artificial selection
   e. both b and c

9. ______ tends to keep populations similar to one another.
   a. Genetic drift
   b. Gene flow
   c. Mutation
   d. Natural selection

10. Match the evolution concepts.

___ gene flow	a. source of new alleles
___ natural selection	b. changes in a population's allele frequencies due to chance alone
___ mutation	c. allele frequencies change owing to immigration, emigration, or both
___ genetic drift	d. survival of the fittest
___ adaptive radiation	e. burst of divergences from one lineage into a set of niches

■ *Visit ThomsonNOW for additional questions.*

Figure 17.31 Rama the cama, a llama–camel hybrid, displays his short temper.

Critical Thinking

1. About 50,000 years ago, humans began domesticating wild dogs. By 14,000 years ago, they started to favor new varieties (breeds) by way of artificial selection. Individual dogs having desirable forms of traits were selected from each new litter and, later, encouraged to breed. Those with undesired forms of traits were passed over.

After favoring the pick of the litter for hundreds or thousands of generations, we ended up with sheep-herding border collies, badger-hunting dachshunds, bird-fetching retrievers, and sled-pulling huskies. At some point we began to delight in odd, extraordinary forms.

In practically no time at all, evolutionarily speaking, we picked our way through the pool of variant dog alleles and came up with such extreme breeds as chihuahuas and Great Danes. Explain what aspects of artificial selection processes are like, and unlike, natural selection processes.

2. Rama the cama, a llama–camel hybrid, was born in 1997 (Figure 17.31). The idea was to breed an animal that has the camel's strength and endurance and the llama's gentle disposition. However, instead of being large, strong, and sweet, Rama is smaller than expected and has a camel's short temper. He has his eye on Kamilah, a female cama born in early 2002. The question is, will any offspring from such a match be fertile? What might the offspring look like?

What does Rama's story tell you about the genetic changes required for irreversible reproductive isolation in nature? Explain why a biologist might not view Rama as evidence that llamas and camels are the same species.

3. An inherited form of methemoglobinemia recurs in a few families in a remote region of Kentucky. People who are homozygous for a mutant allele lack an enzyme that keeps the hemoglobin protein in its normal form. Without it, a brownish form of hemoglobin that accumulates in blood makes the skin appear bright blue. Formulate a hypothesis to explain why this trait recurs among a cluster of families but is rare in the human population at large.

4. Some theorists have hypothesized that many of our uniquely human traits arose by sexual selection. Over many thousands of years, women attracted to charming, witty men perhaps prompted the development of human intellect far beyond what was necessary for mere survival. Men attracted to women with juvenile features may have shifted the species as a whole to be less hairy and softer featured than any of our simian relatives. How would it be possible to test this hypothesis?

18 LIFE'S ORIGIN AND EARLY EVOLUTION

IMPACTS, ISSUES Looking for Life in All the Odd Places

In the 1960s, microbiologist Thomas Brock was looking for signs of life in hot springs and pools at Yellowstone National Park (Figure 18.1). He found a simple ecosystem of microscopically small cells, including *Thermus aquaticus*. This single-celled bacterium extracts energy from simple carbon compounds dissolved in water. It is known as one of the thermophiles, or "heat lovers," for good reason. *T. aquaticus* withstands temperatures on the order of 80°C (176°F).

Brock's work had two unexpected results. First, it put researchers on investigative paths that identified the boundaries of a great domain of life—Archaea. Second, it led to a faster way to copy DNA and end up with useful amounts of it. How? *T. aquaticus* happens to have heat-resistant enzymes that help it survive and replicate its DNA in a superheated environment. Researchers learned to use one of its enzymes to catalyze the polymerase chain reaction—PCR (Section 15.2). Synthetic versions of the enzyme helped trigger a revolution in biotechnology.

Bioprospecting became the new game in town. Many companies started to look closely at thermal pools and other extreme environments for species that might yield valuable products. They found archaeans, bacteria, and protists adapted to extraordinary levels of temperature, acidity, alkalinity, salinity, and pressure.

To extreme thermophiles on the seafloor, Yellowstone's hot water would be too cool. They are adapted to the superheated, mineral-rich water near hydrothermal vents; one species even grows and reproduces at 121°C (249°F)! Highly alkaline lakes contain hardy prokaryotes. So do acidic springs where the pH approaches zero. In polar regions, still other prokaryotes cling to life in salt ponds that never freeze and glacial ice that never melts.

Extreme environments also support some eukaryotic species of ancient lineages. Populations of snow algae tint mountain glaciers red. Another red alga, *Cyanidium caldarium*, lives in acidic hot springs. A variety of single-celled protists survive in waters tainted by toxic metals. Free-living photosynthetic cells called diatoms live in extremely salty lakes, where the hypertonicity would make cells of most organisms shrivel and die.

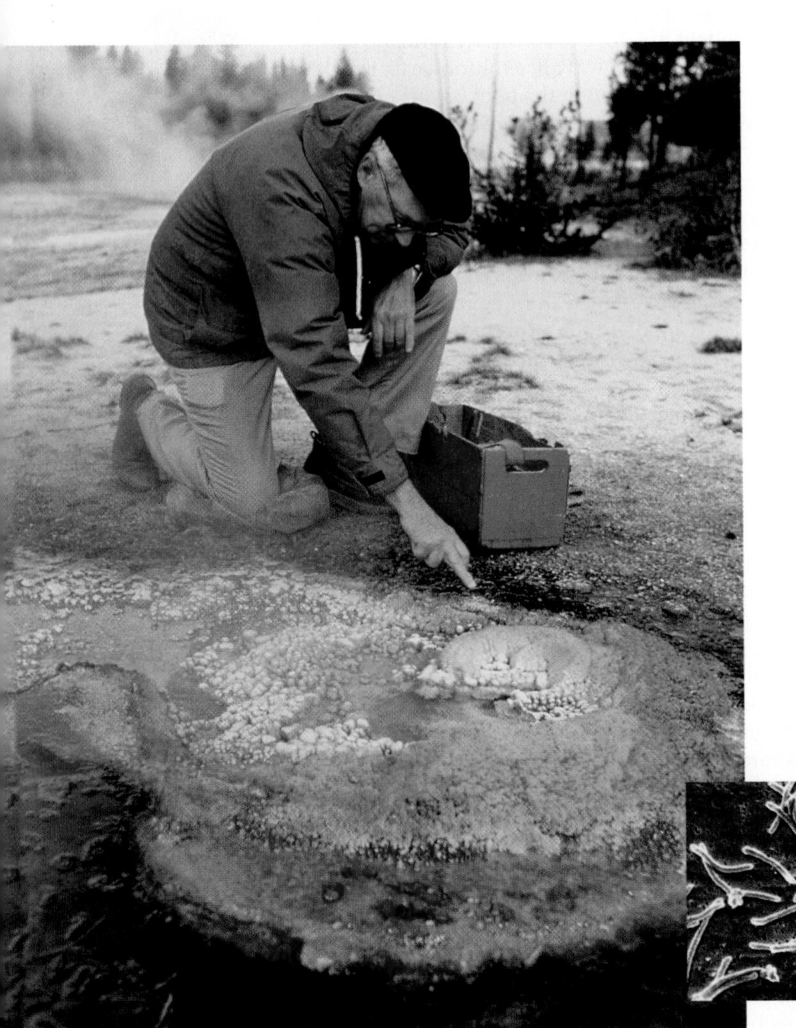

See the video! Figure 18.1 Thomas Brock looking for life in a thermal pool at Yellowstone National Park. The micrograph shows cells of one of the resident bacterial species—*Thermus aquaticus*. Recombinant DNA researchers make great use of its heat-resistant enzymes.

How would you vote? Private companies make millions of dollars selling an enzyme first isolated from cells in Yellowstone National Park. Should private companies be allowed to bioprospect in national parks? See ThomsonNOW for details, then vote online.

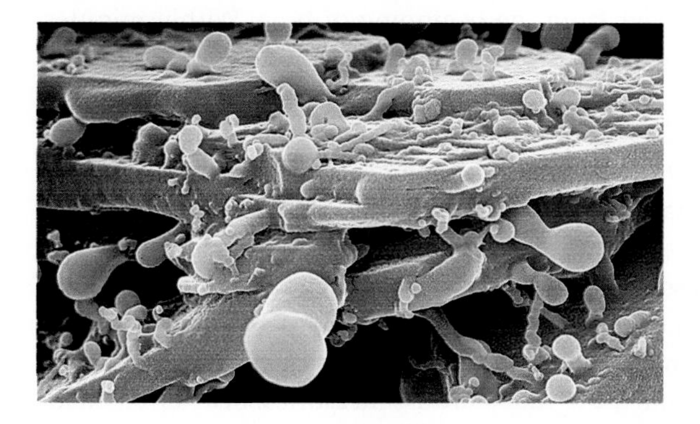

Figure 18.2 Nanobes—possible models for the first proto-cells on Earth. Typical nanobes are only 15 to 20 nanometers across, but they have DNA and other organic compounds enclosed within a membrane, and they seem to grow. This image was magnified 20,000 times.

What could top that? Nanobes! Australian researchers discovered these odd, threadlike and bloblike structures in hot rocks nearly 4 kilometers (2.5 miles) below Earth's surface. The rock temperatures reached 170°C (338°F).

At one-tenth the size of most bacteria, nanobes are visible only with electron microscopes (Figure 18.2). They may be too small to hold all of the metabolic machinery necessary to sustain life. The smallest ones discovered to date are about the size of a ribosome. Yet they contain DNA. They appear to grow and take up substances from their surroundings. Are nanobes more like proto-cells, the presumed forerunners of the first living cells? Perhaps.

Why think about life in such extreme environments? These examples show that life can take hold almost any place, as long as there are sources of carbon and energy.

This chapter is your introduction to a sweeping slice through time, one that cuts back to Earth's formation and to life's chemical origins. The picture we paint here sets the stage for the next unit, which will take you along lines of descent to the present range of biodiversity.

The picture is incomplete. Even so, evidence now converging from many avenues of research leads to a concept that can help you organize information about an immense journey: *Life is a magnificent continuation of the physical and chemical history of the universe, and of the planet Earth.*

Key Concepts

ABIOTIC SYNTHESIS OF ORGANIC COMPOUNDS

When Earth first formed about 4 billion years ago, conditions were too harsh to support life. Over time, its crust cooled, seas formed, and organic compounds of the sort now found in living cells may have formed spontaneously or arrived in meteorites. **Section 18.1**

ORIGIN AND EARLY EVOLUTION OF CELLS

Laboratory experiments and advanced computer simulations support the hypothesis that forerunners of living cells arose through known physical and chemical processes, such as the tendency of lipids to assemble into membrane-like structures when mixed with water.

The first cells probably were anaerobic prokaryotes. Some gave rise to bacteria, others to archaeans and to ancestors of eukaryotic cells. Photosynthetic bacteria started releasing free oxygen into the atmosphere. Oxygen accumulated over time and became a global selection pressure. **Sections 18.2, 18.3**

HOW THE FIRST EUKARYOTIC CELLS EVOLVED

A nucleus, ER, and other membrane-enclosed organelles are among the defining features of eukaryotic cells. Some organelles may have evolved from infoldings of the plasma membrane. Mitochondria and chloroplasts probably are descendants of bacterial cells that became modified after taking up residence in host cells. **Section 18.4**

VISUAL PREVIEW OF THE HISTORY OF LIFE

A time line for milestones in the history of life offers insight into shared connections among all organisms. **Section 18.5**

Links to Earlier Concepts

This chapter starts your survey of the sweep of biodiversity, as introduced in Section 1.3. This is where details of cell metabolism, genetics, and evolutionary theory converge and help explain life's fabulous journey. Here, you will correlate prokaryotes (4.4) and eukaryotes (4.5) with a time line of Earth history (16.5).

You will use your knowledge of organic compounds in general (3.1), and of amino acids (3.4), membranes (4.3), and enzymes (5.3). You will return to connections between photosynthesis and aerobic respiration (Chapter 6). You also will consider how the nucleus, ER, mitochondria, and chloroplasts (4.6–4.9) might have originated. You may wish to scan the sections on DNA replication (12.4), the genetic code, and protein synthesis (13.1–13.3).

18.1 In the Beginning . . .

Life originated when Earth was a thin-crusted inferno, so we may never find evidence of the first cells. Even so, answers to some questions yield clues to their origins. What were conditions like? Can the building blocks of cells assemble spontaneously under those conditions? Can experiments disprove that they did? Let's take a look.

LINKS TO SECTIONS 2.1, 2.5, 3.2, 3.4, 4.1, 4.3

Some clear evening, look up at the moon. Five billion trillion times farther away from you are galaxies at the edge of the known universe. Many billions of years ago, those distant systems of stars gave off light that is only now reaching Earth.

By all known measures, all near and distant galaxies are moving away from one another, which is taken to mean that the universe is expanding. A theory of this colossal expansion and its aftermath might account for every bit of matter in every living thing.

Think about rewinding a videotape on a VCR. Now imagine rewinding the universe. Galaxies move closer together until, after 12 to 15 billion years of rewinding, all galaxies—all matter and space—are compressed in a hot, dense volume. You have arrived at time zero.

In a big bang, all of that matter and energy became simultaneously distributed throughout the universe in a single instant. Radio telescopes have detected a relic of the big bang—cooled, diluted background radiation left over from the beginning of time. Within minutes, the temperature dropped by a billion degrees. Then the simplest elements formed by nuclear reactions. One of them, helium, still is the most abundant element.

During the next billion years, uncountable numbers of gaseous particles collided and condensed under the forces of gravity into the first stars. When stars grew massive enough, nuclear reactions ignited inside them, and heavier elements formed. Stars have a life history, from birth to an often explosive death. In what might be called the first stardust memories, heavier elements released from dying stars were swept up as new stars formed, and even heavier elements formed.

When explosions of dying stars ripped through our galaxy, they formed a dusty, gaseous cloud trillions of kilometers across. The cloud cooled, and bits of matter gravitated toward its center. By 5 billion years ago, the star of our solar system—the sun—was born.

CONDITIONS ON THE EARLY EARTH

Figure 18.3 shows part of one of the vast clouds in the universe. That cloud is mostly hydrogen gas, water, iron, silicates, hydrogen cyanide, ammonia, methane, formaldehyde, and other small inorganic and organic substances. Most likely, the cloud that formed our solar system had a similar composition. Clumps of ice and minerals near the cloud's edges condensed into planets (Figure 18.4*a*). One such clump became Earth.

Four billion years ago, gases blanketed patches of Earth's thin, fiery crust (Figure 18.4*b*). The atmosphere most likely was a mix of gaseous hydrogen, nitrogen, carbon monoxide, and carbon dioxide. It held little or no free oxygen. How do we know oxygen levels were low? Free oxygen binds to the iron in rocks and forms rust. The geologic record shows that rust did not start to form until much later in Earth history.

If free oxygen had been abundant, then the organic compounds necessary for life could not have formed and persisted. This gas is highly reactive. Had oxygen been present, it would have reacted with and disabled organic compounds as quickly as they formed.

Figure 18.3 Part of the Eagle nebula, a hotbed of star formation. Each pillar is wider than our solar system. New stars shine on the tips of gaseous streamers.

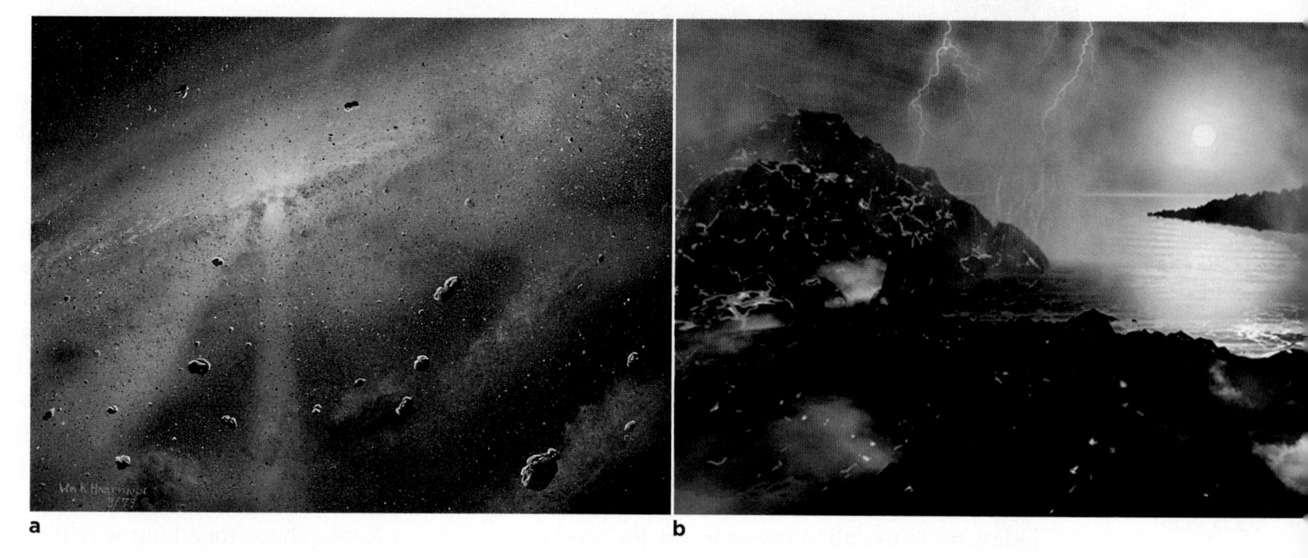

Figure 18.4 (**a**) What the cloud of dust, gases, rocks, and ice around the early sun may have looked like. (**b**) Within 500 million years, Earth was a thin-crusted inferno.

At first, any water falling on Earth's molten surface evaporated fast. As the surface cooled, rocks formed and erosion began. Rains washed mineral salts out of the rocks and salty runoff pooled in shallow seas. As you know, organisms consist largely of water (Section 2.5) and the lipid bilayers of cell membranes cannot form without water (Section 4.3). Think it through: No water, no membranes, no cells—and no life.

ORIGIN OF THE MOLECULES OF LIFE

Besides liquid water, it takes complex carbohydrates, lipids, proteins, and nucleic acids to build living cells. As you know, simple sugars, fatty acids, amino acids, and nucleotides are the main building blocks for the molecules of life (Chapter 3). By one theory, all of these small organic compounds formed spontaneously on the early Earth. Atmospheric gases and mineral salts from rocks supplied the reactants. The reactions could have run on energy inputs from the sun, lightning, or heat from hydrothermal vents on the sea floor.

The idea that small organic compounds once formed spontaneously was first tested in the 1950s. At that time, scientists thought that Earth's early atmosphere had contained water vapor, hydrogen, and ammonia. Stanley Miller put these gases into a reaction chamber (Figure 18.5). He kept the mix circulating and zapped it with sparks to simulate lightning. Within a week, he had amino acids and other small organic compounds.

Since Miller's experiment, researchers have revised their ideas about the first atmosphere's composition. They ran many simulations with different gases and energy sources, such as ultraviolet (UV) light. Even then, small organic compounds formed, including all twenty kinds of the amino acids found in living cells.

By another hypothesis, simple organic compounds formed in outer space. Supporting this hypothesis is the presence of amino acids in interstellar clouds and in some carbon-rich meteorites that landed on Earth.

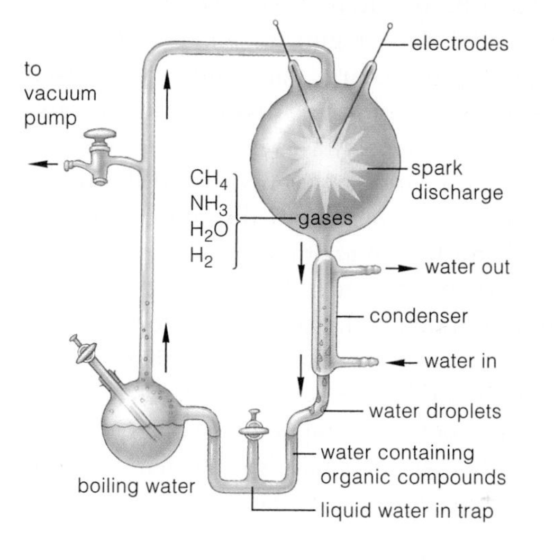

Figure 18.5 **Animated!** Simple diagram of the type of apparatus that Stanley Miller used to test whether organic compounds could have spontaneously self-assembled on the early Earth. Miller circulated water vapor, hydrogen gas (H_2), methane (CH_4), as well as ammonia (NH_3) in a glass chamber to simulate the first atmosphere. Sparks from an electrode simulated lightning.

Regardless of whether the small organic compounds formed on Earth or arrived with meteorites, questions remain. Where and how did these small subunits get assembled into complex carbohydrates, proteins, and phospholipids? Such polymers cannot form from low concentrations of subunits—so what concentrated the subunits? Besides that, the complex carbohydrates and proteins are water soluble. Even if they had formed in the open seas, they would have dissolved and fallen apart quickly. We consider some possible answers to these questions in the next section.

Earth, and its early atmosphere and salty seas, formed by physical and chemical processes that govern the universe.

Results from many laboratory experiments designed to simulate conditions on the early Earth support this theory: All of the small organic compounds that now function as the building blocks of life once self-assembled spontaneously.

18.2 How Did Cells Emerge?

All cells now have a plasma membrane with a lipid bilayer. They all translate heritable information stored in DNA into RNA, then into proteins. Their similarities in structure, metabolism, and genetic replication point to a common origin. We may never know how cells originated, but we can test hypotheses about the event.

ORIGIN OF AGENTS OF METABOLISM

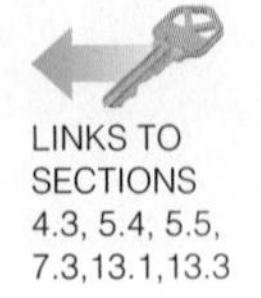

LINKS TO SECTIONS 4.3, 5.4, 5.5, 7.3, 13.1, 13.3

Metabolism and genetic replication boil down to this: One interacting group of molecules makes copies of itself over and over again, using concentrated sources of different molecules as "food" for the reactions.

Today, proteins called enzymes are the workhorses of metabolism. By one hypothesis, they first formed on clay-rich tidal flats (Figure 18.6*a*). Clay particles carry a slight negative charge that would have caused positively-charged amino acids dissolved in seawater to stick to them. If the clay was exposed during low tide, evaporation would have concentrated the amino acids even more, and heat from the sun could have made them bond to one another. Some experimental results support this clay-template hypothesis. When exposed to conditions designed to simulate tidal flats, amino acids bond together into proteinlike chains. Similarly, nucleotides bond and form RNA.

By another hypothesis, simple metabolic pathways evolved at hydrothermal vents. Superheated, mineral-rich water is forced out from these seafloor fissures under high pressure. Sulfides of iron, hydrogen, and other minerals settle out of the water and build up as deposits near the vents (Figure 18.6*b*). The idea is that, long ago, these minerals behaved like cofactors. They promoted the formation of organic compounds from carbon dioxide and other gases dissolved in the water. Remember the Krebs cycle (Section 7.3)? Among some ancient prokaryotic lineages, the cycle runs in reverse. Carbon dioxide's atoms get tacked on to two-carbon acetate, then to four-carbon oxaloacetate, and so on until six-carbon citrate—which splits into oxaloacetate and acetate. This metabolic cycle is self-replicating; it forms building blocks that allow it to turn again.

Researchers are a long way from simulating such a self-replicating cycle, but they keep at it. For example, some constructed cell-sized chambers to simulate the ones discovered in rocks at hydrothermal vents (Figure 18.6*c*). The iron sulfide in the chamber walls acted like a cofactor. It promoted reactions between compounds, which became concentrated within the chambers and started to interact in simple chemical pathways.

ORIGIN OF THE PLASMA MEMBRANE

All modern cells have a plasma membrane composed of lipids and proteins (Section 4.3). We do not know whether the first agents of metabolism were enclosed in a membrane. At some point, however, a membrane did become the outermost boundary of protocells. We

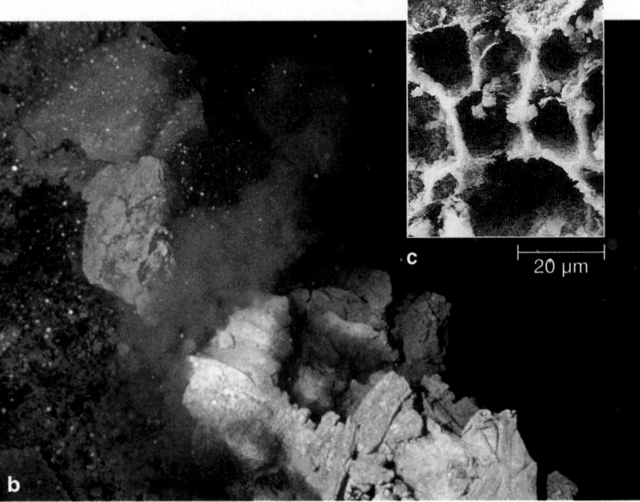

Figure 18.6 Where did the first complex organic compounds form? Two candidates are (**a**) clay templates in tidal flats, and (**b**) iron-sulfide rocks at hydrothermal vents on the deep ocean floor. (**c**) Laboratory simulations of conditions near the vents produce rocks riddled with cell-sized chambers. By one hypothesis, such chambers could have served as protected environments in which organic compounds accumulated and reactions took place. Iron-sulfide cofactors in living cells may be a legacy of such events.

Figure 18.7 Laboratory-formed models for proto-cells that preceded the emergence of the first cells. (**a**) Selectively permeable vesicles with an outer membrane made of proteins were formed by heating amino acids, then wetting the resulting protein chains. (**b**) RNA-coated clay (stained *red*) enclosed in a simple membrane of fatty acids and alcohols (*green*). The mineral-rich clay catalyzes the formation of single RNA strands. It also promotes assembly of vesicle-like forms.

(**c**) Model for the steps that might have taken place on the chemical road to living cells.

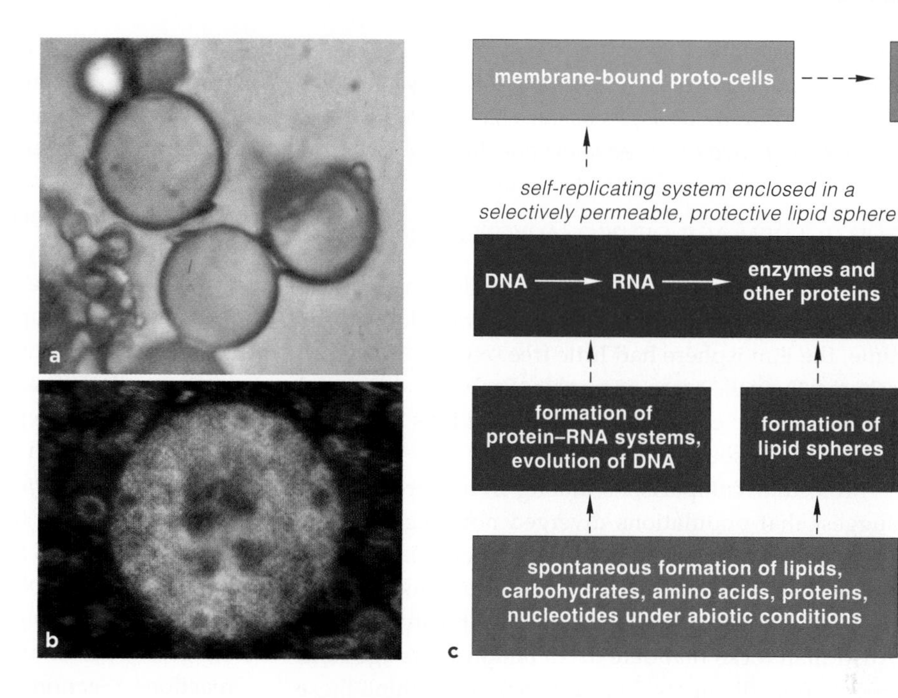

define a protocell as any membrane-enclosed sac of molecules that captures energy, concentrates materials, engages in metabolism, and replicates itself.

Experiments show that membranous sacs can form spontaneously. In simulations of sunbaked tidal flats, amino acids formed long chains. When moistened, the chains assembled as vesicle-like structures with fluid inside (Figure 18.7*a*). A mix of fatty acids and alcohols will self-assemble into sacs around bits of clay (Figure 18.7*b*). Finally, remember that when phospholipids and water mix, a lipid bilayer forms. Such a bilayer is the structural basis of all cell membranes.

ORIGIN OF SELF-REPLICATING GENETIC SYSTEMS

Cells replicate their own genetic information, which is encoded in DNA. As you know from Chapter 13, cells also build proteins based on the instructions in DNA. RNA, enzymes, and additional molecules take part in this process. You also know RNA structurally resembles DNA. The sugar in RNA—ribose—differs from deoxyribose by only one hydroxyl group. Three of four nitrogenous bases in RNA (adenine, cytosine, and guanine) also occur in DNA. RNA's fourth base, uracil, is structurally similar to DNA's thymine.

We have indirect evidence of an early RNA world—a time before DNA evolved, when RNA both stored genetic information and functioned like an enzyme in protein synthesis. Remember, certain RNAs in living cells do serve as enzymes. One rRNA in ribosomes catalyzes formation of peptide bonds during protein synthesis (Figure 10.4). Ribosomes have not changed much over evolutionary time. Ribosomes in eukaryotic cells closely resemble those in prokaryotic cells. This suggests that RNA's role in protein synthesis evolved early in life's history.

The discovery of other catalytic RNAs, known as ribozymes, also supports the RNA world hypothesis. Natural ribozymes cut up and splice RNAs as part of transcript processing (Section 13.1). In the laboratory, researchers made synthetic, self-replicating ribozymes that copy themselves by assembling free nucleotides.

If the earliest self-replicating genetic systems were RNA-based, then what was the advantage of a switch to DNA? The structure of DNA may hold the answer. Compared to a double-stranded DNA molecule, single-stranded RNA breaks apart more easily and mutates more often. Thus, a switch from RNA to DNA would make larger, more stable genomes possible. By another hypothesis, the switch might have protected the early replicating systems from RNA viruses. If such viruses attacked RNA-based genomes, then DNA-based ones would afford a powerful selective advantage.

Living cells show a capacity for metabolism and genetic self-replication. The picture of how this capacity emerged is incomplete; no human was there to witness it.

However, laboratory experiments and computer simulations demonstrate this: The large organic compounds unique to life can assemble spontaneously under suitable conditions and start to interact as complex systems.

18.3 The First Cells

The first cells apparently emerged during the Archaean, an eon that lasted from 3.8 billion to 2.5 billion years ago. Not long afterward, divergences gave rise to three great lineages that have persisted to the present.

THE GOLDEN AGE OF PROKARYOTES

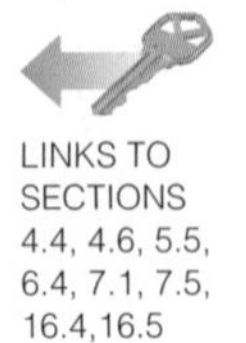

LINKS TO SECTIONS 4.4, 4.6, 5.5, 6.4, 7.1, 7.5, 16.4, 16.5

Fossils indicate that early cells resembled the modern prokaryotes; they had no nucleus (Section 4.4). At the time, the atmosphere had little free oxygen, so we can assume the cells were anaerobic. Probably they used simple organic compounds or mineral ions dissolved in the seas as sources of energy.

Molecular comparisons among living prokaryotes suggest that populations diverged not long after life originated. One lineage gave rise to bacteria, another to the shared ancestors of archaeans and eukaryotes.

What is the earliest fossil evidence for living cells? Australian rocks that date to 3.5 billion years ago hold microscopic filaments (Figure 18.8*a*). Some think these strands may be fossils of early cells. Others argue that they could have formed by geologic processes.

Fossils from another Australian site offer evidence that bacteria lived near deep-sea hydrothermal vents by about 3.2 million years ago. These bacteria might have had pigments that detected thermal radiation (heat) emitted at hydrothermal vents. Detecting high temperature would have helped the cells avoid boiling water. Pigments function this way for some existing species that live around hydrothermal vents.

In at least one lineage, gene mutations in radiation-sensitive pigments and other cellular machinery led to a new mechanism of nutrition—the cyclic pathway of photosynthesis (Section 6.4). The mutated cells were the Earth's first photoautotrophs and they tapped into sunlight, an unlimited source of energy.

As those self-feeding cells reproduced, populations grew on top of one another. They formed dense mats that trapped mineral ions and sediments. Over time, layer after layer accumulated and formed the dome-shaped structures called stromatolites (Figure 18.9).

Stromatolites became abundant in the Proterozoic. A noncyclic pathway of photosynthesis had evolved in cyanobacteria, one of their resident bacterial lineages. Cyanobacterial populations increased, and so did the pathway's waste product: free oxygen. Oxygen started to accumulate, first in the seas, then in the air. Sound familiar? Here we pick up the Chapter 6 storyline.

An oxygen-rich atmosphere had three irreversible outcomes. First, *life no longer could arise spontaneously*. The self-assembly of complex organic compounds had come to an end; as fast as they formed, oxygen attacked them. Second, *aerobic respiration evolved and became the main energy-releasing pathway*. This pathway, recall, uses oxygen as the final electron acceptor for ATP-forming reactions (Section 7.1). It became the most efficient energy-releasing pathway of all, a key innovation that foreshadowed the emergence of complex, multicelled eukaryotes. Third, *with oxygen enrichment of the global atmosphere, an ozone layer formed and shielded Earth from harmful UV radiation*. Without that protective layer, life could never have moved onto land.

THE RISE OF EUKARYOTES

In the Proterozoic, the forerunners of eukaryotic cells split from prokaryotic lineages. Rocks 2.8 billion years old have traces of lipids like those made by modern eukaryotes. The oldest eukaryotic fossils discovered so far date to 2.1 billion years ago (Figure 18.8*b*,*c*).

As you know, organelles are the defining features of eukaryotic cells (Section 4.6). Where did they come from? The next section presents a few plausible ideas. Another question: Where do the early eukaryotes fit in evolutionary trees? The earliest species that we can assign to a modern group is *Bangiomorpha pubescens*, a

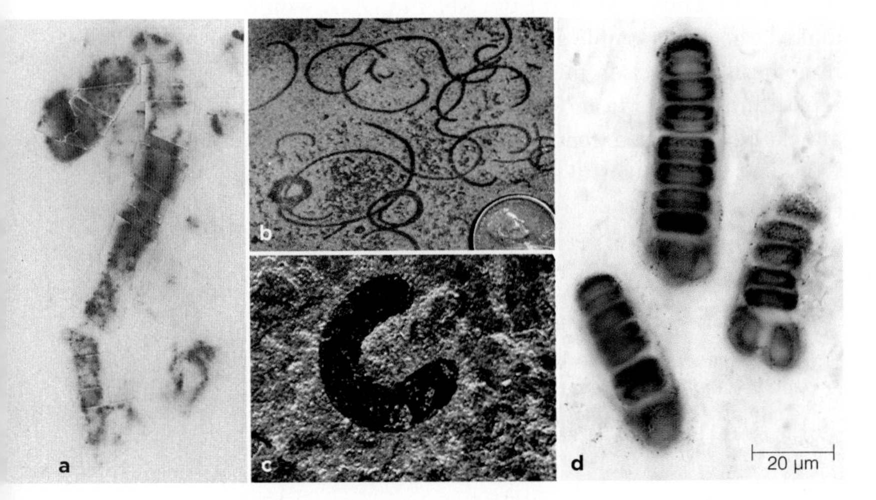

Figure 18.8 A sampling of early life. (**a**) A strand of what may be walled prokaryotic cells dates back 3.5 billion years. (**b**) One of the oldest known eukaryotic species, *Grypania spiralis*, which lived about 2.1 billion years ago. Its fossilized colonies are large enough to observe without a microscope. (**c**) Fossil of *Tawuia*, another early eukaryotic species that lived during the Proterozoic era. (**d**) Fossils of a red alga, *Bangiomorpha pubescens*. This multicelled species lived 1.2 billion years ago, and it reproduced sexually.

Figure 18.9 A glimpse of early forms of life and the environmental stage on which they evolved.

(**a**) From Christopher Scotese's Paleomap Project, a reconstruction of the supercontinent Rodinia about 5.7 billion years ago. (The Russian *rodinia* means homeland.) Most of the original land mass was a shallow floodplain, exposed to fierce storms that swept in from a vast world ocean and to ultraviolet (UV) radiation from the sun.

(**b**) Artist's view of the cushion-shaped stromatolites in a shallow sea near Rodinia. At the time, Earth's atmosphere held little ozone to deflect UV radiation and less than 5 percent of the present-day concentration of oxygen. You can observe beautifully preserved stromatolites and multicolored sedimentary rocks from Rodinia in the Okanogan Highlands of eastern Washington and on into Idaho and Montana.

(**c**) In Shark Bay, Australia, mounds that are 2,000 years old. In their structure, they are similar to stromatolites that formed 3.4 billion years ago. (**d**) Longitudinal cut through a stromatolite. Notice the many layers of fine sediments and mineral deposits. Many mats contain individual cyanobacterial cells that became fossilized.

(**e**) Fossil of *Dickinsonia*, a few centimeters thick. This animal lived with stromatolites 570 million years ago.

red alga that lived 1.2 billion years ago (Figure 18.8*d*). This multicelled alga had specialized structures. Some of its cells helped anchor it. Others produced two types of sexual spores. Apparently, *B. pubescens* was one of the earliest sexually reproducing organisms.

Stromatolites started forming 3.4 billion years ago. By 1.1 billion years ago, they dotted the vast shoreline of the supercontinent Rodinia. About 400 million years later, they began to decline. Newly evolved bacteria-eating animals may have contributed to their decline.

We have fossils of complex animals from about 570 million years ago (Figure 18.9*e*). They shared the seas with many bacteria, archaeans, fungi, and protists—including the lineage of green algae that would later give rise to land plants.

By the dawn of the Cambrian, 543 million years ago, a great adaptive radiation of animals was under way. At the end of this period, all major animal lineages—including vertebrates—were represented in the seas.

The first living cells evolved by 3.8 billion years ago, in the Archaean eon. All were prokaryotic, and they obtained energy by anaerobic pathways. Not long afterward, the ancestors of archaeans and eukaryotic cells diverged from the lineage that led to modern bacteria.

After the noncyclic pathway of photosynthesis evolved, free oxygen accumulated in the atmosphere and ended the further spontaneous chemical origin of life. The stage was set for the evolution of eukaryotic cells.

18.4 Where Did Organelles Come From?

Fossil evidence tells us that eukaryotic cells evolved by about 2 billion years ago. Many lines of evidence support the hypothesis that eukaryotic cells have a composite ancestry, with different cellular components descended from different prokaryotic species.

ORIGIN OF THE NUCLEUS, ER, AND GOLGI BODY

LINKS TO SECTIONS 4.4–4.9

A nucleus, remember, houses eukaryotic DNA (Section 4.7). Its outer boundary is a double membrane, which is continuous with the membrane of the endoplasmic reticulum (ER). Some modern prokaryotes may hold clues about how the inner membrane systems evolved. Their plasma membrane folds into the cytoplasm, and it contains enzymes and other structures with roles in metabolic reactions (Figure 18.10*a*). Similar infoldings could have evolved in the ancestor of eukaryotic cells and become more elaborate over time.

What advantages could such infoldings offer? They might have served as channels that could concentrate specific substances. Also, a membrane with a greater surface area—a physical platform—could hold more metabolic machinery as well as transport proteins.

Internal compartments formed by the membrane of the ER and Golgi bodies may be derived from such infoldings. Perhaps they initially protected metabolic machinery from uninvited guests. We know that, from time to time, parasitic foreign cells enter and feed on the cytoplasm of living prokaryotic cells.

Infoldings that extended around the DNA could have evolved into a nuclear envelope (Figure 18.10*b*). By one hypothesis, the nuclear envelope was favored because it kept genes safe from foreign DNA. Modern bacteria take up DNA from the surroundings; viruses inject it into bacterial cells. An alternative, explained shortly, is that the nuclear envelope evolved after two prokaryotic cells fused. The nuclear envelope might have helped to keep their two genomes separate.

THE ROLE OF ENDOSYMBIOSIS

Early in the history of life, cells became food for one another. Heterotrophs engulfed autotrophs and other heterotrophs. Intracellular parasites dined inside their hosts. In some cases, the engulfed prey or parasites struck an uneasy balance with host cells. They were protected, they withdrew nutrients from their host's cytoplasm, and—like their hosts—they continued to divide and reproduce.

Such partnerships are the premise of a hypothesis of endosymbiosis, championed by Lynn Margulis and others. (*Endo–* means within; *symbiosis* means living together.) The symbiont lives out its life inside a host, and the interaction benefits one or both of them.

Most likely, mitochondria evolved after an aerobic bacterium entered and survived in a host cell (Figure 18.11*a*). The host cell could have been either an early eukaryote or an archaean. If it was an archaean, then a nuclear membrane might have evolved after the two cells fused. The membrane would have kept bacterial enzymes and genes from interfering with expression of the archaean host's genes.

In any case, the host began to use ATP produced by its aerobic symbiont while the symbiont began to rely on the host for raw materials. Over time, genes that specified the same or similar proteins in both the host and its symbiont were free to mutate. If a gene lost its function in one partner, a gene from the other could take up the slack. Eventually, the host and symbiont both became incapable of living independently.

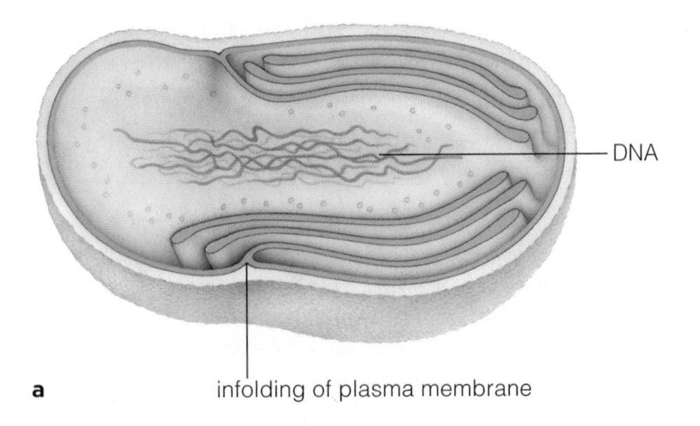

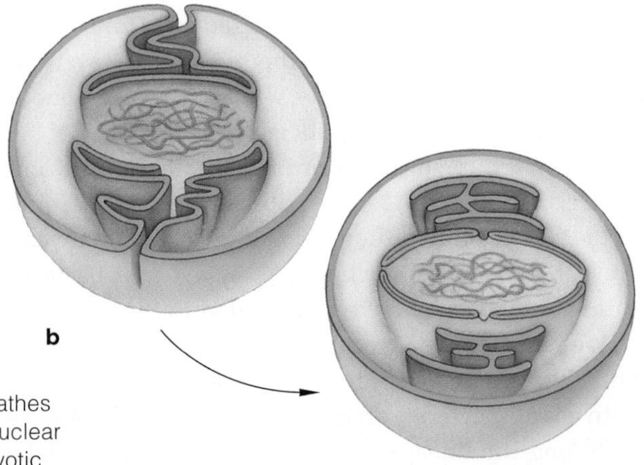

Figure 18.10 (**a**) Sketch of a soil bacterium, *Nitrobacter*. Cytoplasmic fluid bathes permanent infoldings of the plasma membrane. (**b**) Model for the origin of the nuclear envelope and the endoplasmic reticulum. In the prokaryotic ancestors of eukaryotic cells, infoldings of the plasma membrane may have evolved into these organelles.

EVIDENCE OF ENDOSYMBIOSIS

A chance discovery made by microbiologist Kwang Jeon supports the hypothesis that bacteria can evolve into organelles. In 1966, Jeon was studying *Amoeba proteus*, a species of single-celled protist. By accident, one of his cultures became infected by a rod-shaped bacterium. Some infected amoebas died right away. Others kept growing, but only slowly. Intrigued, Jeon maintained those infected cultures to see what would happen. Five years later, the descendant amoebas were host to many bacterial cells, yet they seemed healthy. When those amoebas received bacteria-killing drugs that usually do not harm amoebas, they died.

Experiments confirmed that the amoebas had come to rely on the bacteria. When amoebas in uninfected cultures were stripped of their nucleus and given one from an infected amoeba, they died. Yet, when bacteria were included with the nuclear transplant, most cells survived. Other studies showed that infected amoebas had lost the ability to make an essential enzyme. They depended on bacterial invaders to make it for them! The bacterial cells had become vital endosymbionts.

Mitochondria in living cells do resemble bacteria in size and structure (Figure 18.11*a*). A mitochondrion's inner membrane is like a bacterial plasma membrane. Like a bacterial chromosome, mitochondrial DNA is a circle with few noncoding regions between genes, and few or no introns. A mitochondrion does not replicate its DNA or divide at the same time as the cell.

Did chloroplasts, too, originate by endosymbiosis? By one scenario, predatory aerobic bacteria engulfed photosynthetic cells. Those cells escaped digestion and continued to function by absorbing nutrients from the host cytoplasm. The cells released oxygen and sugars into their aerobically respiring hosts, which may have had a selective edge as a result.

In their metabolism and their overall nucleic acid sequence, existing chloroplasts resemble cyanobacteria. Also, chloroplast DNA replicates itself independently of cellular DNA. Chloroplasts and the cells in which they reside divide independently of each other.

The glaucophytes offer more clues. These protists have unique photosynthetic organelles that resemble cyanobacteria. Like cyanobacteria, the organelles even have a cell wall (Figure 18.11*b*).

However they arose, early eukaryotic cells had a nucleus, endomembrane system, mitochondria, and—in certain lineages—chloroplasts. These cells were the first protists. Over time, their many descendants came to include the modern protist lineages, as well as the plants, fungi, and animals. The next section provides a time frame for these pivotal evolutionary events.

A nucleus and other organelles are defining features of eukaryotic cells. The nucleus and ER may have evolved from infoldings of the plasma membrane. Mitochondria and chloroplasts may have evolved by endosymbiosis between cells that were the prey or parasites of heterotrophic cells.

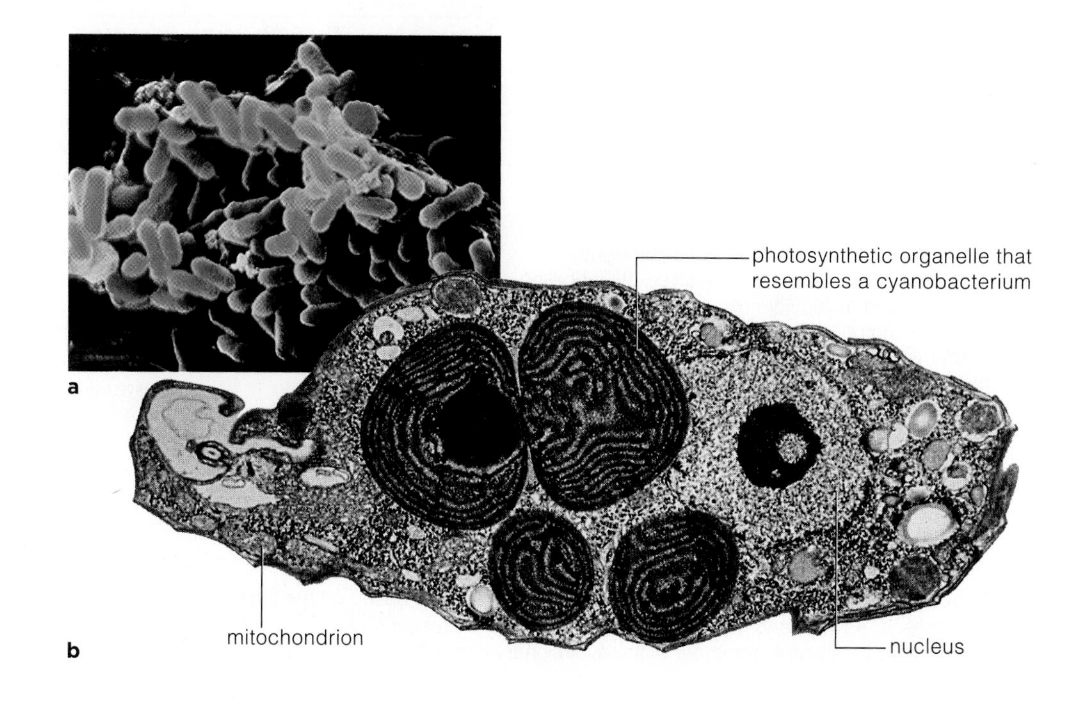

Figure 18.11 Examples of the many clues to ancient endosymbiotic interactions.

(**a**) What the ancestors of mitochondria may have looked like. Genes of the structurally simplest mitochondria we know about are similar to genes of *Rickettsia prowazekii*, a parasitic bacterium that causes typhus. Like mitochondria, *R. prowazekii* divides only inside the cytoplasm of eukaryotic cells. Enzymes in the host's cytoplasm catalyze the partial breakdown of organic compounds—a task that is completed inside aerobically respiring mitochondria.

(**b**) *Cyanophora paradoxa* is one of the flagellated protists called glaucophytes. Its mitochondria resemble aerobic bacteria in size and structure. Its photosynthetic structures resemble cyanobacteria. They even have a wall similar in composition to the wall around a cyanobacterial cell.

18.5 Time Line for Life's Origin and Evolution

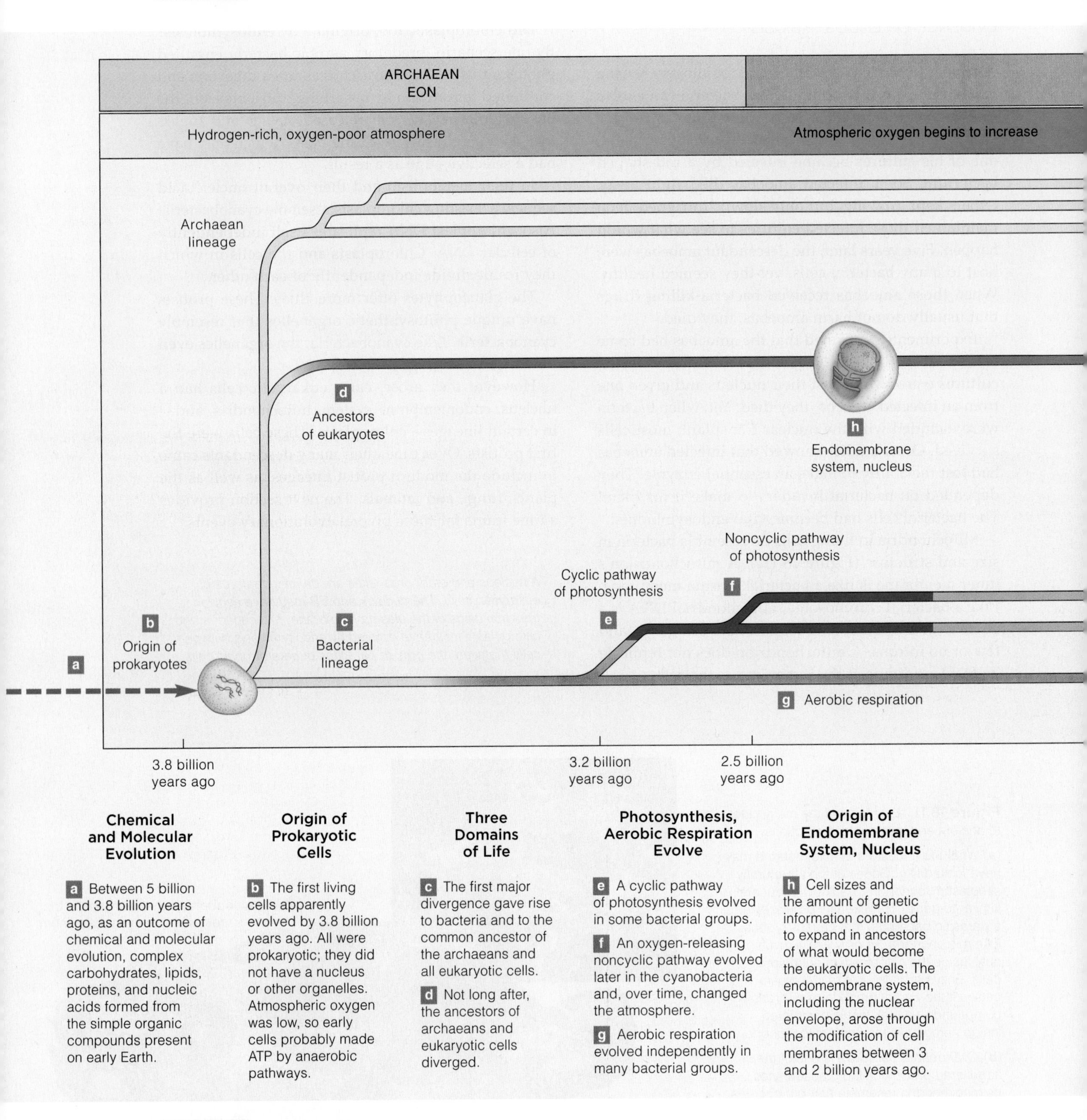

Figure 18.12 **Animated!** Milestones in the history of life, based on the most widely accepted hypotheses. As you read the next unit on life's past and present diversity, refer to this visual overview. It can serve as a simple reminder of the evolutionary connections among all groups of organisms.

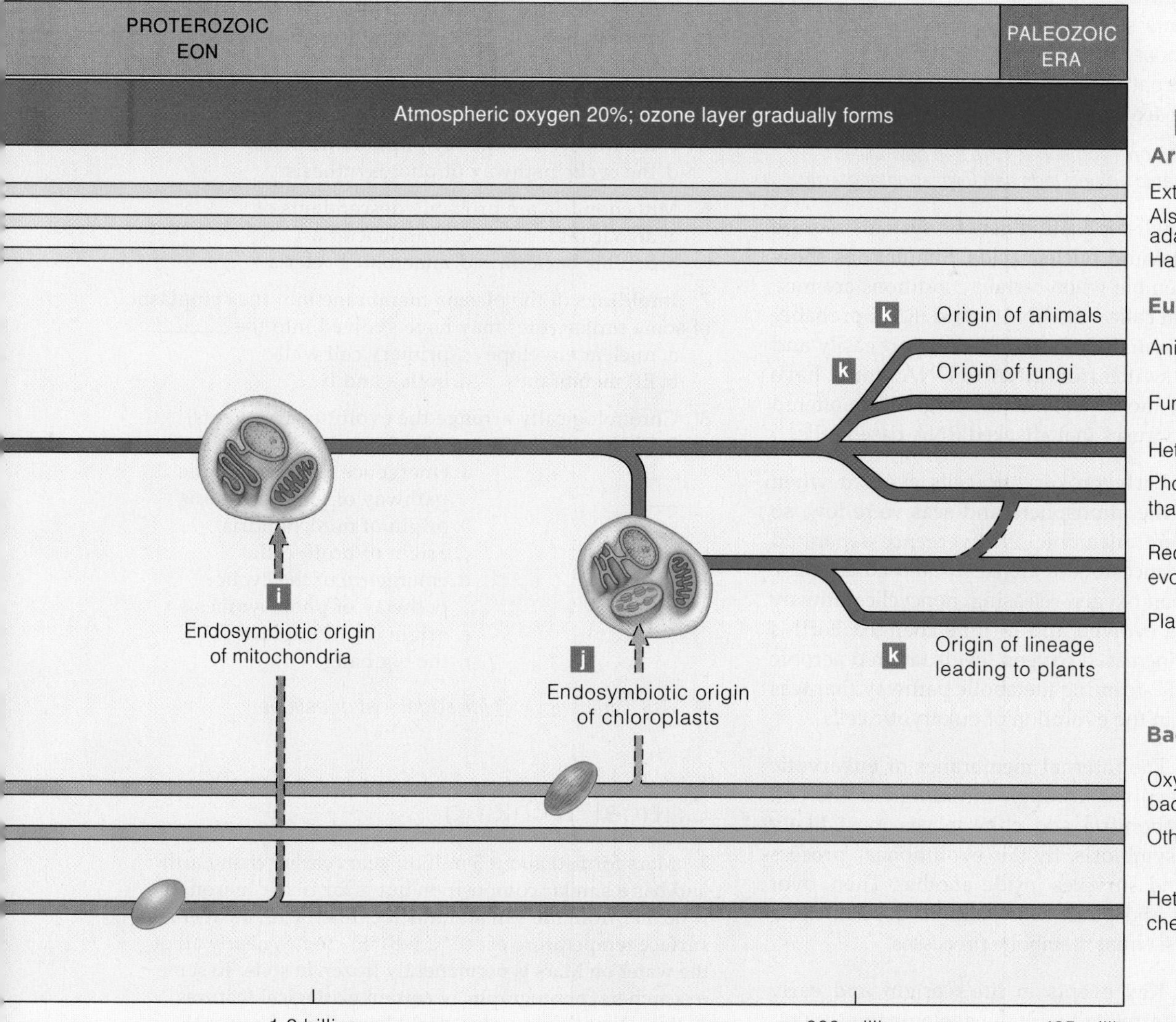

Endosymbiotic Origin of Mitochondria

i Before about 1.2 billion years ago, an aerobic bacterium entered an anaerobic eukaryotic cell. Over generations, the two species established a symbiotic relationship. Descendants of the bacterial cell became mitochondria.

Endosymbiotic Origin of Chloroplasts

j Between about 1 and 1.5 billion years ago, a cyanobacterium entered into a symbiotic interaction with a protist. Its descendants evolved into chloroplasts. Later, some photosynthetic protists would evolve into chloroplasts inside other protist hosts.

Plants, Fungi, and Animals Evolve

k By 900 million years ago, all major lineages—including fungi, animals, and the algae that would give rise to plants—had evolved along shorelines of Rodinia, the first known supercontinent.

Lineages That Have Endured to the Present

l Today, organisms live in nearly all regions of Earth's waters, crust, and atmosphere. They are related by descent and share certain traits. However, each lineage encountered different selective pressures, and unique traits evolved in each one.

Summary

Section 18.1 Earth formed more than 4 billion years ago. Laboratory simulations offer indirect evidence that organic compounds self-assemble spontaneously under conditions like those thought to have prevailed on the early Earth. Alternatively, the compounds might have formed in deep space and reached Earth in meteorites.

■ *Use the animation on ThomsonNOW to see how Miller showed that organic compounds can form spontaneously.*

Section 18.2 Self-replicating genetic systems require proteins (enzymes) and nucleic acids. Simulations show that both self-assemble when certain conditions are met. During an interval called the RNA world, RNA probably stored genetic information, but it breaks apart easily and mutates often. A switch from RNA to DNA would have made the genome more stable. It also might have offered a defense against viruses that attacked RNA-based cells.

Section 18.3 Early prokaryotic cells evolved when oxygen levels in the atmosphere and seas were low, so they probably were anaerobic. A divergence separated bacteria from the ancestors of archaeans and eukaryotes. In cyanobacteria, an oxygen-releasing, noncyclic pathway of photosynthesis evolved and in time changed Earth's atmosphere. The increased oxygen levels favored aerobic respiration, an ATP-forming metabolic pathway that was a key innovation in the evolution of eukaryotic cells.

Section 18.4 The internal membranes of eukaryotic cells may have evolved through infoldings of the cell membrane. Mitochondria and chloroplasts most likely evolved by endosymbiosis. By this evolutionary process one cell enters and survives inside another. Then, over generations, host and guest cells come to depend upon one another for essential metabolic processes.

Section 18.5 Key events in life's origin and early evolution can be correlated with the geologic time scale.

■ *Use the animated interaction on ThomsonNOW to investigate the history of life on Earth.*

Self-Quiz

Answers in Appendix III

1. An abundance of ______ in the atmosphere would have prevented the spontaneous assembly of organic compounds on early Earth.
   a. hydrogen b. methane c. oxygen d. nitrogen

2. The prevalence of iron-sulfide cofactors in living organisms may be evidence that life arose ______ .
   a. in outer space c. near deep-sea vents
   b. on tidal flats d. in the upper atmosphere

3. RNA in ribosomes can catalyze formation of peptide bonds. This supports the hypothesis that ______ .
   a. RNA was the first template for protein synthesis
   b. RNA can hold more information than DNA
   c. the first protists had RNA as their genetic material
   d. all of the above

4. Simulations of conditions on early tidal flats support the hypothesis that ______ .
   a. clay can facilitate the assembly of polypeptides
   b. oxygen levels in the atmosphere increased over time
   c. the first enzymes were RNAs
   d. all of the above

5. The evolution of ______ resulted in an increase in the levels of atmospheric oxygen.
   a. sexual reproduction
   b. aerobic respiration
   c. the noncyclic pathway of photosynthesis
   d. the cyclic pathway of photosynthesis

6. Mitochondria are probably descendants of ______ .
   a. archaeans c. cyanobacteria
   b. aerobic bacteria d. anaerobic bacteria

7. Infoldings of the plasma membrane into the cytoplasm of some prokaryotes may have evolved into the ______ .
   a. nuclear envelope c. primary cell wall
   b. ER membranes d. both a and b

8. Chronologically arrange the evolutionary events, with 1 being the earliest and 6 the most recent.

___ 1	a. emergence of the noncyclic pathway of photosynthesis
___ 2	b. origin of mitochondria
___ 3	c. origin of proto-cells
___ 4	d. emergence of the cyclic pathway of photosynthesis
___ 5	e. origin of chloroplasts
___ 6	f. the big bang

■ *Visit ThomsonNOW for additional questions.*

Critical Thinking

1. Mars formed about 5 million years earlier than Earth and has a similar composition but is far richer in iron. It is farther from the sun and much chillier, with an average surface temperature of −63°C (−81°F). Today, nearly all of the water on Mars is permanently frozen in soils. To some researchers, photographs of certain geological features indicate that liquid water might have flowed across the planet's surface during an earlier and warmer time. The Martian atmosphere is now richer in carbon dioxide than Earth's but quite low in nitrogen and oxygen. Based on this information, would you rule out the possibility that life could have existed on Mars or that simple life forms could currently exist there? Explain your reasoning.

2. Researchers looking for fossils of the earliest life forms face many hurdles. For example, few sedimentary rocks date back more than 3 billion years. Review what you learned about plate tectonics (Section 16.6). Explain why so few remaining samples of these early rocks remain.

3. Craig Venter and Claire Fraser are working to create a "minimal organism." They are starting with *Mycoplasma genitalium*, a bacterium that has 517 genes. By disabling its genes one at a time, they discovered that 265–350 of them code for essential proteins. The scientists are synthesizing the essential genes and inserting them, one by one, into an engineered cell consisting only of a plasma membrane and cytoplasm. They want to see how few genes it takes to build a new life form. What properties would such a cell have to exhibit for you to conclude that it was alive?

IV Evolution and Biodiversity

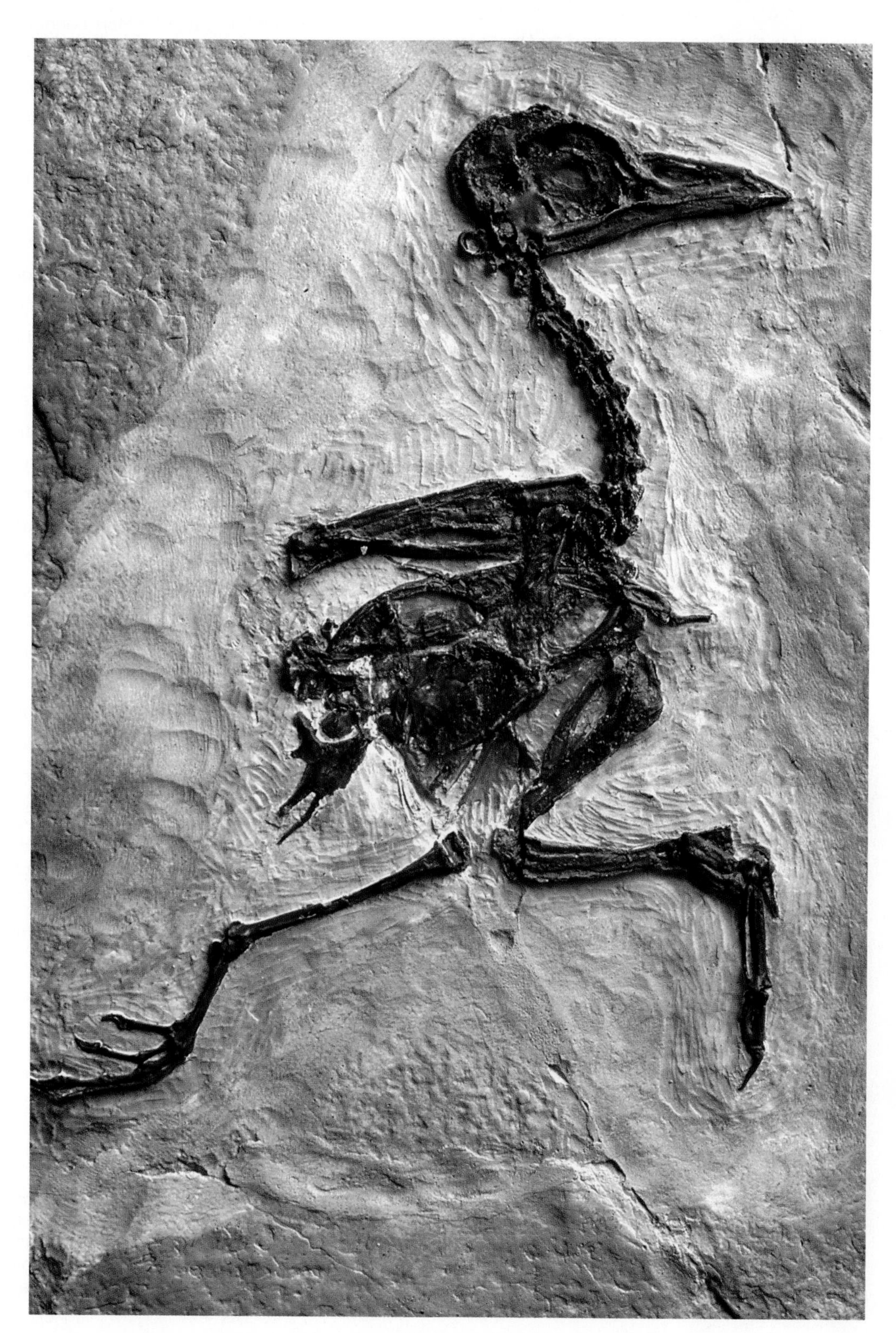

From the Green River formation near Lincoln, Wyoming, the stunning fossilized remains of a bird trapped in time. During the Eocene, some 50 million years ago, sediments that had been gradually deposited in layers at the bottom of a large inland lake became its tomb. In this same formation, fossilized remains of sycamore, cattails, palms, and other plants suggest that the climate was warm and moist when the bird lived. Fossils from places all around the world yield clues to life's early history.

19 PROKARYOTES AND VIRUSES

IMPACTS, ISSUES West Nile Virus Takes Off

In 336 B.C., when he was twenty years old, Alexander the Great ascended to the throne of Macedonia (Figure 19.1). During his twelve-year reign, he carved out an empire that stretched across the Middle East and into northern India. He died after entering Babylon, now the site of Baghdad. According to one recorded account, a flock of ravens announced Alexander's arrival in the city. The birds behaved strangely, and some fell dead at his feet. Soon thereafter, Alexander was bedridden with severe back pain, fever, chills, and muscle weakness. He quickly spiraled down through delirium and paralysis, to death.

An infectious disease expert, Charles Calisher, and epidemiologist John Marr connected the dots between the birds' behavior and Alexander's reported symptoms. They hypothesize that Alexander died as an outcome of West Nile encephalitis, a severe inflammation of the brain that follows a viral infection. The virus that causes this disease has been common in Africa, Western Asia, and the Middle East through the ages. Was there an outbreak of West Nile encephalitis during Alexander's time? Possibly.

Until the summer of 1999, no one had ever reported the presence of the virus in the Western Hemisphere. Then some people in and around New York City could not help but notice the dead and dying crows—as well as mysteriously sickened horses and humans. Sixty-two people became ill that summer, and seven died. Tissue samples from them contained West Nile virus.

During the next seven years, infected birds spread the virus across the United States and into Mexico, Canada, and the Caribbean islands. By 2006, virus-infected birds had been detected in every one of the continental United States. About 20,000 cases have been reported in this region, and 770 of those infections proved fatal.

West Nile virus is one of the known flaviviruses. Other infectious agents in this group cause the deadly tropical diseases yellow fever and dengue fever. All flaviviruses move from host to host with the assistance of mosquitoes. As a mosquito feeds on an infected individual, it sucks up virus particles along with blood. When the insect feeds again, it delivers virus particles to a new host.

Humans cut their risk of infection by controlling mosquito populations and by applying repellents. Domestic horses now can be protected by a vaccine. Wildlife does not have these protections.

See the video! **Figure 19.1** A clue to viral history? Alexander the Great may have died of West Nile encephalitis. Ravens and crows are highly susceptible to this viral disease. Biologists now monitor birds to assess the spread of the virus and its effects.

How would you vote? Spraying pesticides where mosquitoes can breed offers protection against West Nile virus. Spraying might disrupt ecological balances and adversely affect human health. Would you support spraying in your community? See ThomsonNOW for details, then vote online.

West Nile virus now infects more than 220 species of North American birds. Some, such as robins, are not sickened. They act as viral reservoirs that can put less resistant species in the same habitat at risk.

Birds can also carry avian influenza strain H5N1, which causes "bird flu." The disease can kill birds and people who have contact with birds. Discovered in Asia during 2003, bird flu spread to the Pacific, Africa, Europe, and the Middle East by 2006. Smuggling of infected poultry or the migration of infected wild birds could introduce the virus to North America. So far, the H5N1 virus does not seem to spread from person to person. If mutations give it that capacity, the result could be a health disaster. The virus kills nearly all of the humans it infects.

This chapter can start you thinking about microbes, the single cells and noncellular infectious particles that can be seen only through a microscope. Viruses, viroids, and prions hover near the boundary between living and nonliving things. Prokaryotic cells are just inside it. Most of us tend to judge microbes through the prism of human interests. We view them as dangerous, or beneficial, or simply of no concern to us. However, in the evolutionary view, microbes compete to persist and make copies of themselves, just as all the bigger and more conspicuous organisms do.

Key Concepts

DISTINCTLY PROKARYOTIC FEATURES

We divide all prokaryotic cells into domains Bacteria and Archaea. These single-celled organisms are structurally simple; they do not have a nucleus or the diverse cytoplasmic organelles found in most eukaryotic cells. Collectively, they show great metabolic diversity.

Bacteria and archaeans alone reproduce by prokaryotic fission. Exchanges of chromosomal DNA and plasmid DNA often occur within and among species. **Section 19.1**

THE PROKARYOTIC LINEAGES

Bacteria are the most studied prokaryotic species. They are the most abundant and widely distributed organisms. Archaeans, discovered more recently, are less well known. Many are adapted to extreme environments. **Sections 19.2, 19.3**

NONCELLULAR INFECTIOUS PARTICLES

Viruses are noncellular particles that cannot replicate themselves without taking over the metabolic machinery of a host cell. A protein coat encloses their DNA or RNA and a few enzymes. Some viruses become enveloped in membrane when they bud from host cells.

Viroids and prions are even simpler than viruses. Viroids are short sequences of infectious RNA. Prions are infectious misfolded versions of normal proteins. **Sections 19.4, 19.5**

THE BAD BUNCH

An immense variety of pathogens, or disease-causing agents, infect human hosts. Pathogens and their hosts have been coevolving by way of natural selection. **Section 19.6**

Links to Earlier Concepts

This chapter picks up where Section 4.4—your introduction to prokaryotic cells—left off. It offers more detailed glimpses into the bacterial and archaean branches of the tree of life (17.12).

You will be drawing on theories of the physical and chemical conditions under which life originated (18.1–18.3). You will be reflecting on how certain bacteria evolved into chloroplasts and mitochondria (4.9, 18.4). You also will compare prokaryotic and eukaryotic cell division mechanisms (10.6) and look once more at plasmids (15.2).

19.1 Characteristics of Prokaryotic Cells

Prokaryotic cells are the smallest, most abundant, and most widely dispersed organisms. Billions of these single cells live in a handful of rich soil; those living in your gut and on your skin outnumber your body cells.

LINKS TO SECTIONS 4.4, 5.5, 15.1, 18.3

The first cells, remember, had no nucleus; they were all prokaryotic. (*Pro*– means before, *karyon* is taken to mean nucleus.) None had the profusion of organelles typical of eukaryotes. Their prokaryotic descendants, too, are structurally simple, but simplicity does not mean they are inferior to eukaryotic cells. Prokaryotic lineages have endured for more than 3 billion years. All eukaryotes are their descendants. Also, in terms of reproductive success, the prokaryotes are unparalleled. University of Georgia biologists once estimated that 5,000,000,000,000,000,000,000,000,000,000 bacterial cells were alive at that moment on Earth.

CELL SIZES, SHAPES, AND STRUCTURES

Section 4.4 introduced the prokaryotic cells. Table 19.1 and Figure 19.2 recap their main features. The average prokaryotic cell is no longer or wider than 0.5 to 1 micrometer. The most common cell shapes are a sphere (coccus; plural cocci), rod (bacillus; bacilli), and spiral (spirillum; spirilla), but shapes vary. A cell might be oval or square, a curvy rod, or a tight corkscrew. Also, cells often stick together in chains after division.

Nearly all prokaryotes have a semirigid, porous cell wall around their plasma membrane (Figure 19.2*b*). A sticky, secreted layer of slime coats the wall of many bacteria and helps them adhere to surfaces. A capsule around the wall of some parasitic bacteria helps them slip past an animal host's defensive cells.

Many prokaryotic cells have one or more flagella (Figure 19.3*a*). Prokaryotic flagella are built differently than eukaryotic flagella (Section 4.12). They have no microtubules and do not bend side to side, but instead rotate like a propeller.

METABOLIC DIVERSITY

Like all other organisms, prokaryotes require energy to make ATP, and carbon atoms to make sugars and other organic compounds. Some prokaryotes are autotrophs, or self-feeders. They get carbon from carbon dioxide. The *photo*autotrophs get energy from sunlight. Like plants, some release oxygen during photosynthesis and help maintain the atmosphere. The *chemo*autotrophs get energy by releasing electrons from a great variety of mostly inorganic substances, such as sulfides.

Heterotrophs are not self-feeders. They get carbon from organic compounds made by other organisms and use it to make organic compounds of their own. A few bacteria are the only *photo*heterotrophs; they get energy from the sun. *Chemo*heterotrophs get carbon and energy from tissues, wastes, or remains of other

Table 19.1 Prokaryotic Cell Characteristics

1. No nucleus; chromosome in nucleoid
2. Generally a single chromosome (a circular DNA molecule); many species also contain plasmids
3. Cell wall present in most species
4. Reproduction mainly by prokaryotic fission
5. Collectively, great metabolic diversity among species

coccus

bacillus

a spirillum

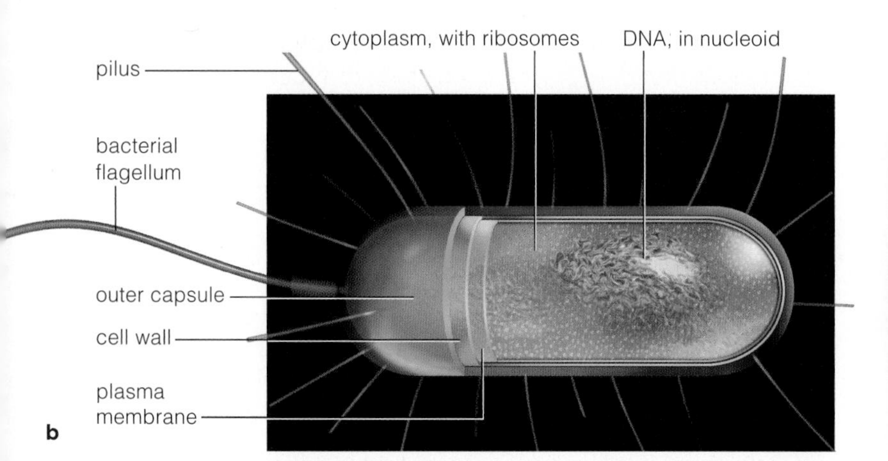

Figure 19.2 (**a**) The three most common shapes among prokaryotic cells: spheres, rods, and spirals. (**b**) Generalized body plan of a prokaryotic cell.

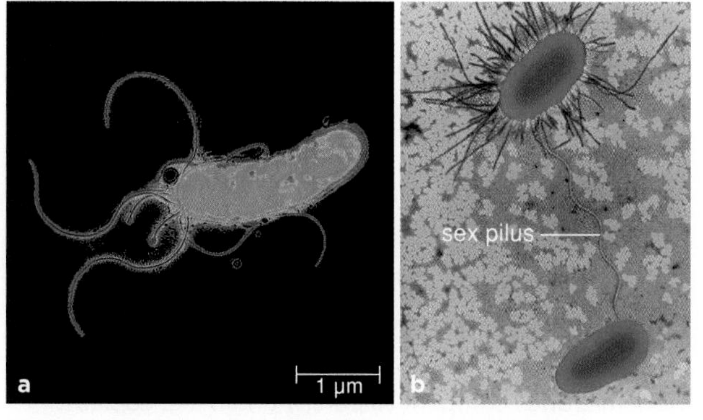

Figure 19.3 (**a**) *Helicobacter pylori*, a flagellated bacterium that can cause stomach ulcers. (**b**) A sex pilus connects two *Escherichia coli* cells. It will shorten and draw them together.

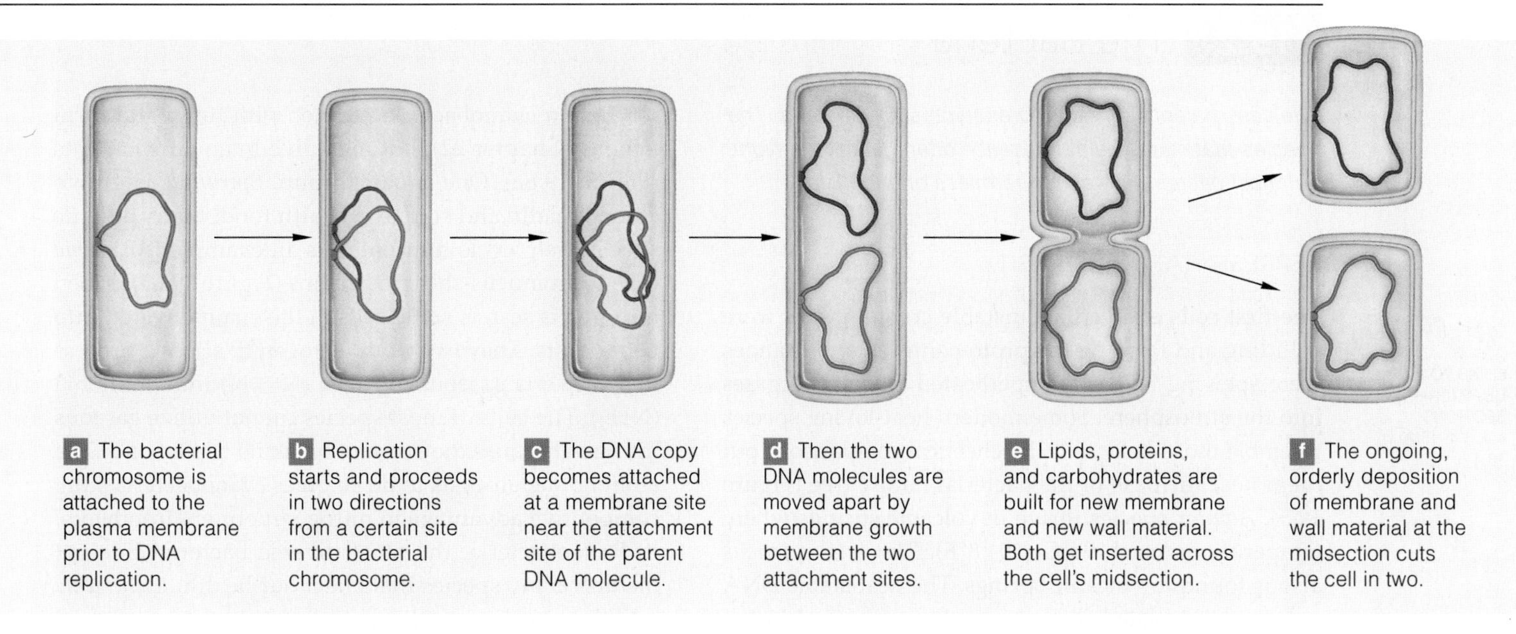

Figure 19.4 Animated! Prokaryotic fission, the reproductive mode of bacteria and archaeans.

organisms. Most chemoheterotrophs act as parasites; they feed on a living host. Others are saprobes; they break down wastes or remains, then absorb some of the released nutrients. By serving as decomposers, the saprobes help supply other organisms with nutrients.

REPRODUCTION AND GENE TRANSFERS

Prokaryotes have staggering reproductive potential. Some types can divide every twenty minutes. One cell becomes two, the two become eight, the eight become sixteen, and so on. Before each cell divides, it nearly doubles in size. After division, each daughter cell has one copy of the prokaryotic chromosome—a circular double-stranded molecule of DNA with a few proteins.

In some species, a daughter cell buds from a parent cell. More commonly, a cell reproduces by prokaryotic fission (Figure 19.4). Briefly, a parent cell replicates its single chromosome, and this DNA replica attaches to the plasma membrane adjacent to the parent molecule. The addition of more membrane material moves the two DNA molecules apart. Eventually, the membrane and cell wall extend across the cell's midsection and so divide the parent cell into two daughter cells.

Besides inheriting DNA "vertically" from a parent cell, prokaryotes engage in horizontal gene transfers: they pick up genes from cells of the same or different species. One mechanism of horizontal gene transfer, conjugation, requires contact with a cell that makes a sex pilus. A pilus (plural, pili) is a hairlike extension from a microbe and it functions in attachment. After one cell attaches to another cell, its sex pilus shortens and draws the cells together (Figure 19.3*b*). The donor cell puts a copy of a plasmid and perhaps a few of its chromosomal genes into the recipient. The plasmid is an independently replicated circle of DNA with a few genes (Section 15.1). Both bacteria and archaeans have plasmids and both can engage in conjugation.

CLASSIFICATION AND PHYLOGENY

Prokaryotic fossils are scarce except for stromatolites (Section 18.3), so numerical taxonomy has dominated classification efforts. This process compares traits of modern groups, such as shape and wall structure. The more traits two groups share, the closer is the inferred relatedness. Signs of horizontal gene transfers in the past cloud the picture. However, gene sequencing and other tools of comparative biochemistry are helping the classification efforts. For instance, comparisons of genes for ribosomal RNA (rRNA) were key factors in the reclassification of all prokaryotic species into two domains—Archaea and Bacteria:

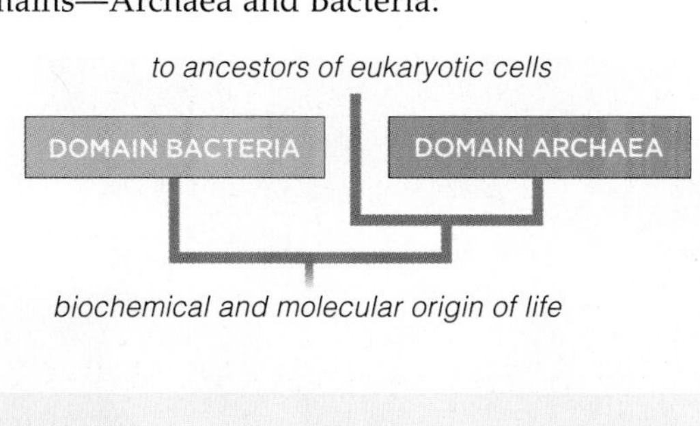

Despite their small size and relatively simple structure, prokaryotes are the most abundant and widely dispersed organisms. Diverse modes of nutrition and a capacity for rapid reproduction contribute to the group's success.

19.2 The Bacteria

The vast majority of named prokaryotes are bacteria. The species that cause human disease often get the spotlight, but most bacteria are either harmless or beneficial.

REPRESENTATIVE DIVERSITY

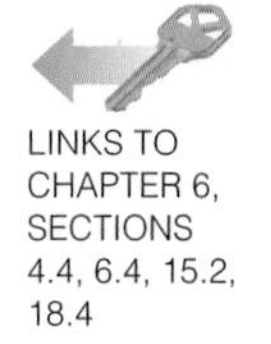

LINKS TO CHAPTER 6, SECTIONS 4.4, 6.4, 15.2, 18.4

The first cells evolved as unstable crustal plates were colliding and forming the proto-continents. Volcanoes were spewing lava into superheated water, and gases into the atmosphere. Some modern heat-loving species resemble those early cells. Biochemical comparisons put them near the base of the bacterial family tree (Figure 19.5). *Aquifex* species thrive in volcanic springs where temperatures reach 96°C (204.8°F). *Thermus aquaticus* also is found in volcanic springs. The heat-stable DNA polymerase isolated from this species was used in the first PCR reactions (Section 15.2).

Photosynthesis evolved in many bacterial lineages. However, only cyanobacteria release free oxygen by a noncyclic pathway, as plants do. Chloroplasts evolved from an ancient cyanobacteria (Section 18.4). Thus we have cyanobacteria and their chloroplast descendants to thank for nearly all the oxygen in the atmosphere.

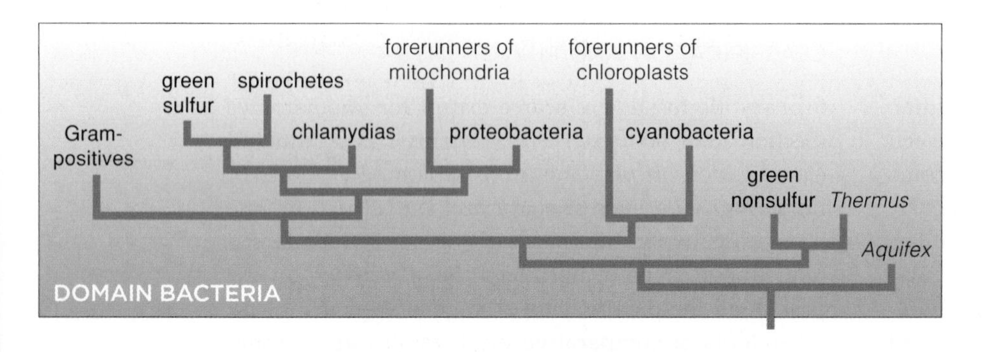

Figure 19.5 Domain Bacteria. This tree shows the probable relationships among bacterial groups mentioned in this book. For more information on the major groups see Appendix I.

Certain cyanobacteria partner with fungi and form lichens (Chapter 22), but most live in aquatic habitats (Figure 19.6*a*). One aquatic group, *Spirulina,* is grown commercially and sold as a health food. Many aquatic species help cycle nutrients. As an example, *Anabaena* cells form mucus-sheathed chains (Figure 19.6*b*). When nitrogen is scarce, some cells in the chain develop into heterocysts. They carry out nitrogen fixation, a process that converts gaseous nitrogen (N≡N) into ammonia (NH_3). The cells of most species cannot utilize gaseous nitrogen because they cannot break its triple bond. The ability of heterocysts to do so gives cyanobacteria a big competitive advantage in nitrogen-poor environments.

Proteobacteria, the most diverse bacterial lineage, include many species that affect our health. *Escherichia coli,* the best known, lives in the mammalian gut and synthesizes vitamin K. Other species in this group are pathogens. A pathogen is a disease-causing agent that infects a host organism and lives in or on it. Disease occurs when a pathogen's metabolic activities disrupt host function. For example, *Neisseria gonorrhoeae* is the cause of gonorrhea, *Vibrio cholerae* causes cholera, and *Helicobacter pylori* (Figure 19.3*a)*, causes stomach ulcers. Rickettsias cause typhus and Rocky Mountain spotted fever. They are the closest living relatives of the bacteria that gave rise to mitochondria.

Most proteobacteria are photoautotrophs or free-living heterotrophs. They include the largest prokaryote, *Thiomargarita namibiensis,* which cycles nitrogen and sulfur in the seas (Figure 19.6*c*). *Rhizobium* species live inside the roots of certain plants (legumes), where they fix nitrogen. Their activity benefits host plants and also enriches the soil.

Chlamydias are intracellular parasites of animals. They cannot make ATP, but obtain it from host cells. Every year, *C. trachomatis*

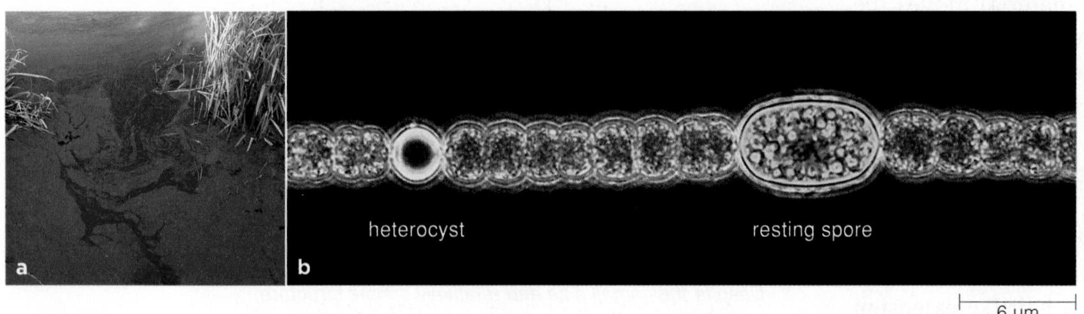

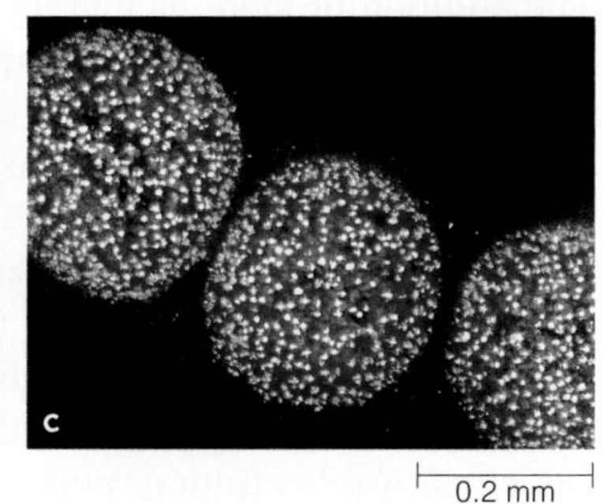

Figure 19.6 Aquatic bacteria. (**a**) Cyanobacteria in a nutrient-rich pond. (**b**) Some cyanobacterial cells become resting spores or heterocysts when conditions restrict growth. (**c**) *Thiomargarita namibiensis,* the largest bacterium; it can be seen with the naked eye. The cells are mostly a nitrate-filled vacuole.

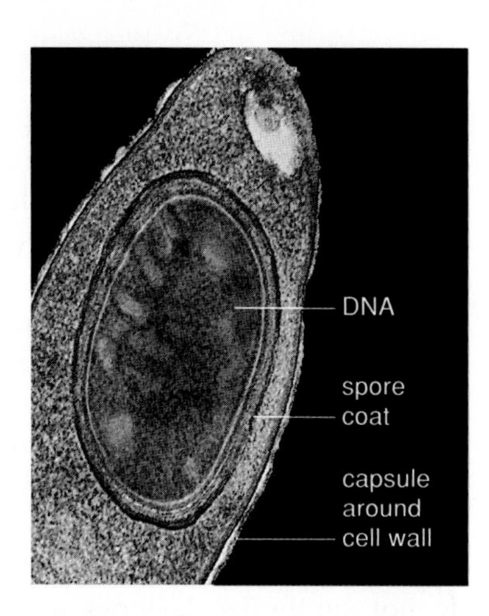

Figure 19.7 Transmission electron micrograph of an endospore forming inside the soil bacterium *Clostridium tetani*. The spore coat encloses the DNA.

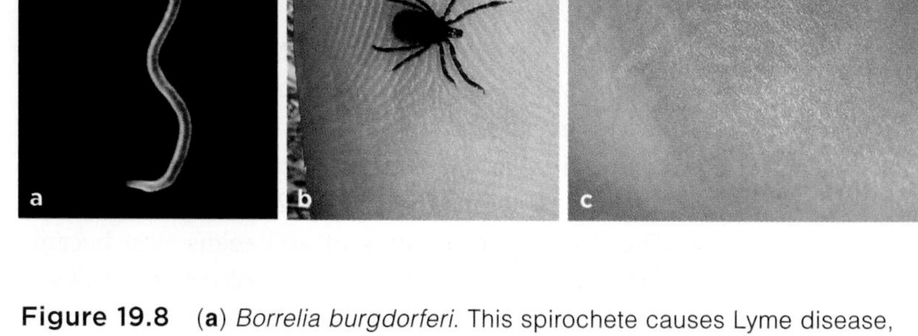

Figure 19.8 (**a**) *Borrelia burgdorferi*. This spirochete causes Lyme disease, now the most prevalent tick-borne disease in the United States. (**b**) Deer ticks (*Ixodes*) carry this bacterium between vertebrate hosts. Bottom to top: a female, male, and nymph. (**c**) Bull's-eye rash, the first visible sign of Lyme disease.

causes nearly a million cases of sexually transmitted disease in the United States. Many infected people display few or no symptoms. Without diagnosis and treatment, they may become sterile.

Gram-positive bacteria include diverse lineages. All have a thick wall that retains purple dye if prepared for microscopy using the Gram-staining technique. In contrast, thin walls of most bacteria lose dye and turn pink with this method; these cells are Gram-negative.

Most Gram-positive bacteria are chemoheterotrophs. *Lactobacillus* species carry out fermentation reactions used to produce yogurt and other foods. *L. acidophilus* lives on skin and in intestinal and vaginal mucus. Its presence lowers the pH, which helps keep pathogenic bacteria in check. Antibiotics can disturb this balance and cause diarrhea and vaginitis as side effects.

The Gram-positive *Clostridium* and *Bacillus* species form endospores when conditions do not favor growth. These resting structures hold a bacterial chromosome and a bit of cytoplasm (Figure 19.7). Endospores resist heat, boiling, irradiation, acids, and disinfectants. As living conditions improve, an endospore germinates, and a single bacterium emerges.

Endospores can germinate inside a human body. In 2001, someone put *Bacillus anthracis* endospores into envelopes and mailed them. A few people breathed in the powdery contents and died of anthrax. *Clostridium tetani* endospores that enter the body through wounds can cause tetanus. This disease can make muscles lock in irreversible contraction. Endospores of one relative, *C. botulinum*, sometimes taint canned food and cause botulism, a dangerous type of food poisoning.

Spirochetes resemble stretched-out springs (Figure 19.8*a*). Some are free-living; others live inside a host organism. Ticks can carry spirochetes that cause Lyme disease (Figure 19.8*b*). Untreated, this disease can harm the heart, nervous system, and joints. A bull's-eye rash (Figure 19.8*c*) is often the earliest symptom.

REGARDING THE "SIMPLE" BACTERIA

Bacteria are small. Their insides are not elaborate. *But bacteria are not simple.* They sense and move toward areas where nutrients are more plentiful and where other conditions also favor growth. Aerobes move to oxygen; anaerobes move away from it. Photosynthetic types move toward light but away from light that is too intense. Many species avoid toxins or predators.

Magnetotactic bacteria migrate in the ocean, and they grow best in oxygen-poor seawater. They contain a compass—chains of magnetic particles that respond to Earth's magnetic field (Figure 1.8*a*). For instance, in the Northern Hemisphere, geomagnetic north points down at a slight angle from the equator to the North Pole. Cells responding to the angle move farther down from surface waters, to greater depths where oxygen concentrations are lower.

Myxobacteria show collective behavior and undergo differentiation. Thousands of cells form a colony that moves as a predatory unit. They secrete enzymes that digest prey, such as other bacteria. As food dwindles, the cells mass together, interact chemically, and form spore-bearing structures. Some cells differentiate and form a stalk; others become spores. Germination of a spore releases a single cell that can start a new colony.

Bacteria are the most abundant of prokaryotic cells. This group includes photoautotrophs, chemoautotrophs that cycle nutrients in habitats, and chemoheterotrophs. The chemoheterotrophs are the most diverse; some of these species are dangerous parasites and pathogens.

19.3 The Archaeans

Archaeans are prokaryotic, but in some ways they are as similar to eukaryotes as they are to bacteria. Many survive in extreme environments, but new species are turning up almost everywhere on and in the planet.

LINKS TO CHAPTER 3 INTRODUCTION, SECTIONS 6.1, 13.1, 16.10

THE THIRD DOMAIN

All prokaryotes were once put into the same kingdom, and it is easy to see why. Archaeans and bacteria look alike in size and shape. Neither has a nucleus. Both have one circular chromosome, with some of the genes arranged as operons (Section 14.2).

The distinctive features of archaeans first became apparent in the 1970s. Molecular biologist Carl Woese began comparing the ribosomal RNAs of prokaryotes to find out how they are related to one another. Genes for rRNA are essential for protein synthesis (Section 13.1). However, certain sequences in those genes can mutate a bit without any loss of function. The longer two lineages have traveled on separate evolutionary roads, the more their genes for rRNA will differ.

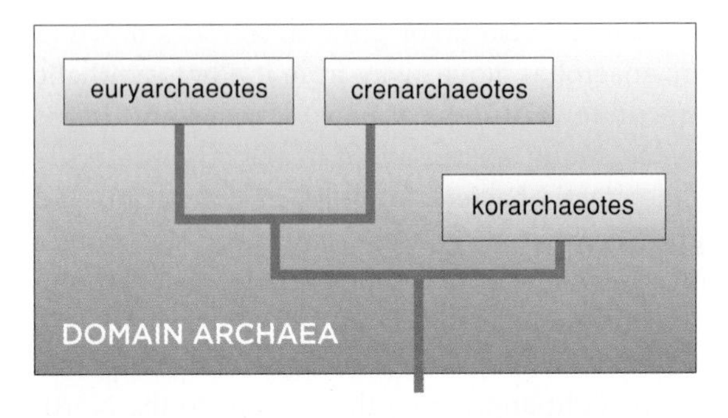

Figure 19.9 The major groups of domain Archaea. See Appendix I for descriptions of each group.

Woese discovered that some prokaryotes fell into a distinct group. Their rRNA gene sequences positioned them between the bacteria and the eukaryotes. On the basis of this evidence, Woese proposed a three-domain classification system (Section 19.1).

The three-domain system is now widely accepted, and evidence that supports it is accumulating. It turns out that the archaeans and bacteria have different cell wall components. Their membrane phospholipids are mirror images of each other and are synthesized in different ways. Like eukaryotic cells, archaeans spool their DNA around proteins called histones. Bacteria do not synthesize histones or structurally organize their DNA to the same extent.

Woese compares the discovery of archaeans to the discovery of a new continent, which he and others are now exploring. Already the explorers have identified three subgroups: Euryarchaeota, Crenarchaeota, and Korarchaeota (Figure 19.9). Their work also has shown that, as a group, archaeans are diverse, widespread, and ecologically important.

HERE, THERE, EVERYWHERE

In their physiology, most archaeans are methanogens (methane producers), extreme halophiles (salt lovers), and extreme thermophiles (heat lovers). These three informal designations are not phylogenetic groupings. Some bacterial species also are methanogens, extreme halophiles, and extreme thermophiles. Archaeans and bacteria often coexist, and they often exchange genes.

Chapter 3 introduced the methanogenic archaeans. Species have been discovered in marshes, Antarctic ice, oceans, and deep in Earth's crust. A few are symbionts inside the gut of termites and other animals (Figure 19.10*b*). All of the methanogens are strict anaerobes; free oxygen kills them. When forming ATP, they pull electrons away from hydrogen gas or acetate. Methane (CH_4) forms as a product of the reactions.

By their metabolic activity, methanogens produce 2 billion tons of methane annually. The release of this

Figure 19.10 Examples of methanogenic archaeans. (**a**) *Methanococcus jannaschii* grows in water heated to 85°C (185°F) near hydrothermal vents. (**b**) Researchers have isolated several kinds of methanogenic archaeans from the cattle gut.

Figure 19.11 Life in extreme environments. (**a**) In salty evaporation ponds in Utah's Great Salt Lake, extreme halophilic archaeans and red algae tint the water pink. (**b**) Parasitic *Nanoarchaeum equitans* (smaller *blue* spheres) grows as a parasite on another archaean, *Ignicoccus* (larger spheres). Both were isolated from 100°C water near a hydrothermal vent. (**c**) Crenarchaeote membrane lipids have been discovered in the Three Buddhas, a hot spring in Gerlach, Nevada. (**d**) Yellowstone's hot springs and pools are home to many extreme thermophiles.

carbon-containing gas into the air has a major impact on the global carbon cycle. As Section 42.8 explains, methane also contributes to global warming.

Extreme halophilic archaeans live in the Dead Sea, the Great Salt Lake, saltwater evaporation ponds, and other highly salty habitats (Figure 19.11*a*). Most make ATP by aerobic reactions but switch to photosynthesis when oxygen is low. A unique purple pigment called bacteriorhodopsin (Section 6.1) is embedded in their plasma membrane. When excited by light, this protein pumps protons (H^+) out of the cell. H^+ flows back in through an ATP synthase and drives ATP formation.

Some extreme thermophilic archaeans live beside hydrothermal vents, where temperatures can exceed 110°C (230°F). These archaeans use hydrogen sulfide escaping from the vents as an electron source for ATP-forming reactions. Their existence is cited as evidence that life could have originated on the seafloor.

Researchers came across *Nanoarchaeum equitans* as they were exploring hydrothermal vents near Iceland. Only 400 nanometers across, *N. equitans* is among the smallest known cells. It is a parasite, and its host is another, slightly larger archaean (Figure 19.11*b*).

Other heat-loving archaeans thrive in mineral-rich hot springs (Figure 19.11*c,d*). *Sulfolobus* species grow in well-oxygenated water at 80°C (176°F) and a pH of 3. They metabolize sulfur or switch to a heterotrophic mode and feed on carbon compounds. They are among the few archaeans that can be cultured in laboratories.

Archaeans are common in the seas—and not just at vents. There may be as many archaeans in deep water as there are bacteria in water near the ocean's surface. Biologists are finding archaeans just about everywhere they look. Some kinds thrive in the human gut, vagina, and mouth. A few that may cause dental problems are the only archaeans suspected to be human pathogens.

Like bacteria, archaeans are prokaryotic cells, but in some respects they resemble eukaryotes. Their characteristics are distinctive enough to assign the archaeans to their own domain, separate from bacteria and eukaryotes.

Archaeans were once thought to be confined to extreme environments. Many are now being discovered alongside bacteria in more hospitable habitats.

19.4 The Viruses

In ancient Rome, virus meant "poison" or "venomous secretion." In the 1800s, this nasty word was bestowed on newly discovered pathogens, smaller than bacteria. Viruses deserve the name; think of any organism, and there is probably a virus that can infect it.

THE STRUCTURE OF VIRUS PARTICLES

LINKS TO SECTIONS 12.1, 15.1

A virus is a noncellular infectious particle consisting of DNA or RNA, a few enzymes, and a protein coat. A virus cannot multiply by itself. In an act of molecular piracy, its genes direct a host cell to make building blocks that can be assembled into new virus particles.

The viral coat protects the genetic material as the virus moves between hosts. It consists of a rodlike or many-sided array of protein subunits (Figures 19.12 and 19.13). Some viral coat proteins bind to receptors on host cells. Coats of complex viruses have sheaths, tail fibers, and other accessory structures attached. In some classes of viruses, a bit of membrane envelops the coat when a virus buds from an infected cell.

Each kind of virus multiplies only in specific hosts. Researchers can grow infected host cells in culture to study a virus. For instance, bacteriophages infect only bacteria. These viruses and their hosts were used in studies of the nature of inheritance (Section 12.1).

Figure 19.12*a* shows part of a tobacco mosaic virus. Like most plant viruses, its genetic material is RNA. Plant cells have a thick wall, so viruses usually enter after insects or pruning cut through the plant surface.

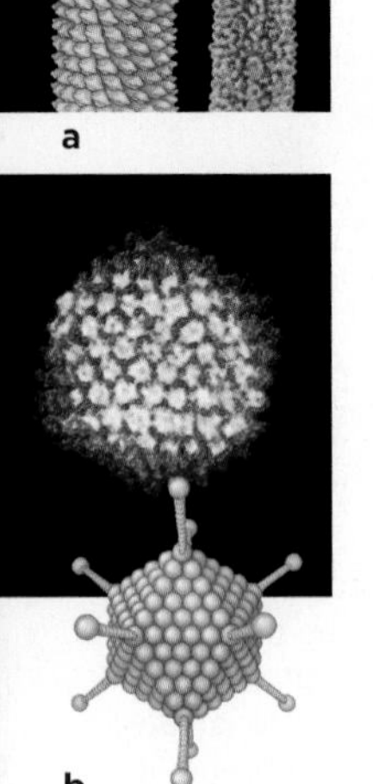

Figure 19.12 Examples of viral structure. (**a**) Tobacco mosaic virus, a helical virus. Subunits of its rod-shaped protein coat coil helically around RNA. (**b**) Adenovirus. Like other polyhedral viruses, it has a many-sided coat. *Facing page*, bacteriophages and other complex viruses have a coat with accessory parts. *Herpes simplex*, HIV, and other enveloped viruses have a membrane-wrapped coat.

Table 19.2 Steps in Most Viral Multiplication Cycles

Attachment	Molecular groups on virus particle chemically recognize and lock onto molecular groups of receptors at the host cell surface.
Penetration	Either the virus particle or its genetic material (DNA or RNA) crosses the plasma membrane of a host cell and enters the cytoplasm.
Replication and synthesis	Viral DNA or RNA directs host transcription and translation mechanisms to make viral nucleic acids and viral proteins.
Assembly	New infectious particles are built.
Release	The new virus particles are released from the cell.

Viruses that infect animals range in size from tiny parvoviruses (18 nanometers) to "big" poxviruses (350 nanometers). Adenoviruses cause most common colds (Figure 19.12*b*). Other viral diseases include measles, influenza, and warts. Human immunodeficiency virus (HIV) infects certain white blood cells, weakens body defenses, and causes AIDS. We discuss AIDS in detail in Chapters 35 and 39. Human papilloma virus (HPV) and a few other viruses can cause cancer.

VIRAL MULTIPLICATION

Viral multiplication cycles are varied, but nearly all go through the five steps listed in Table 19.2. The virus attaches to an appropriate host cell. The virus or just its genetic material enters the cell. Viral genes direct the cell to replicate the viral DNA or RNA and to build viral proteins. New virus particles are assembled. The new virus particles are released from the infected cell by way of lysis or budding, as explained below.

Figure 19.13 shows two bacteriophage multiplication pathways. In a lytic pathway, the steps of attachment through assembly occur fast, and lysis follows. Here, lysis refers to the disintegration of a host cell plasma membrane, wall, or both, which lets the cytoplasm—and virus particles—dribble out. The host synthesized a viral enzyme that initiated lysis, and its own death.

In a lysogenic pathway, a virus enters a latent state that extends the multiplication cycle. Viral genes get integrated into the host chromosome. The recombinant molecule is passed along to all descendants of the host cell. Like miniature time bombs, the virus inside these descendants awaits a signal to enter the lytic pathway.

Latency is typical of many viruses, such as type I *Herpes simplex*, which causes cold sores (fever blisters). The virus infects nerve cells. When stress or sunburn reactivates it, the virus moves to the skin and causes painful lesions. Like other enveloped viruses, it enters a host cell by endocytosis, and new particles form by budding (Figure 19.14). Eventually the host cell lyses.

RNA viruses have a different multiplication cycle. Viral RNA functions as a template for making mRNA or DNA. One of the enzymes that the retrovirus HIV carries into host cells catalyzes the formation of DNA from RNA by reverse transcription (Section 15.1).

A virus is a noncellular infectious particle that consists of nucleic acid enclosed in a protein coat and sometimes an outer envelope. It cannot multiply without taking over the metabolic machinery of a specific type of host cell.

Nearly all organisms are targets of specific viruses.

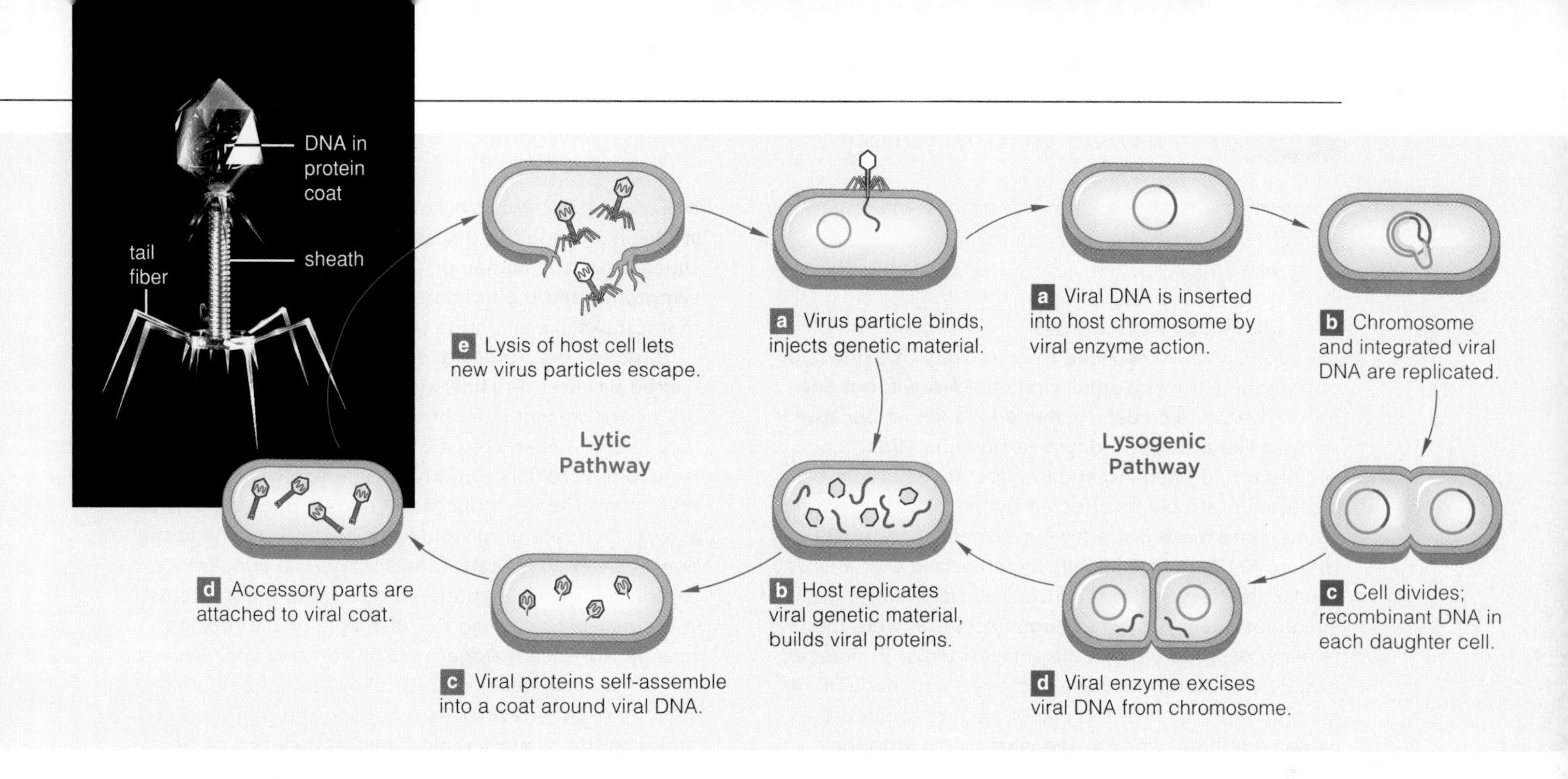

Figure 19.13 **Animated!** Pathways in the multiplication cycle of a bacteriophage, a complex virus.

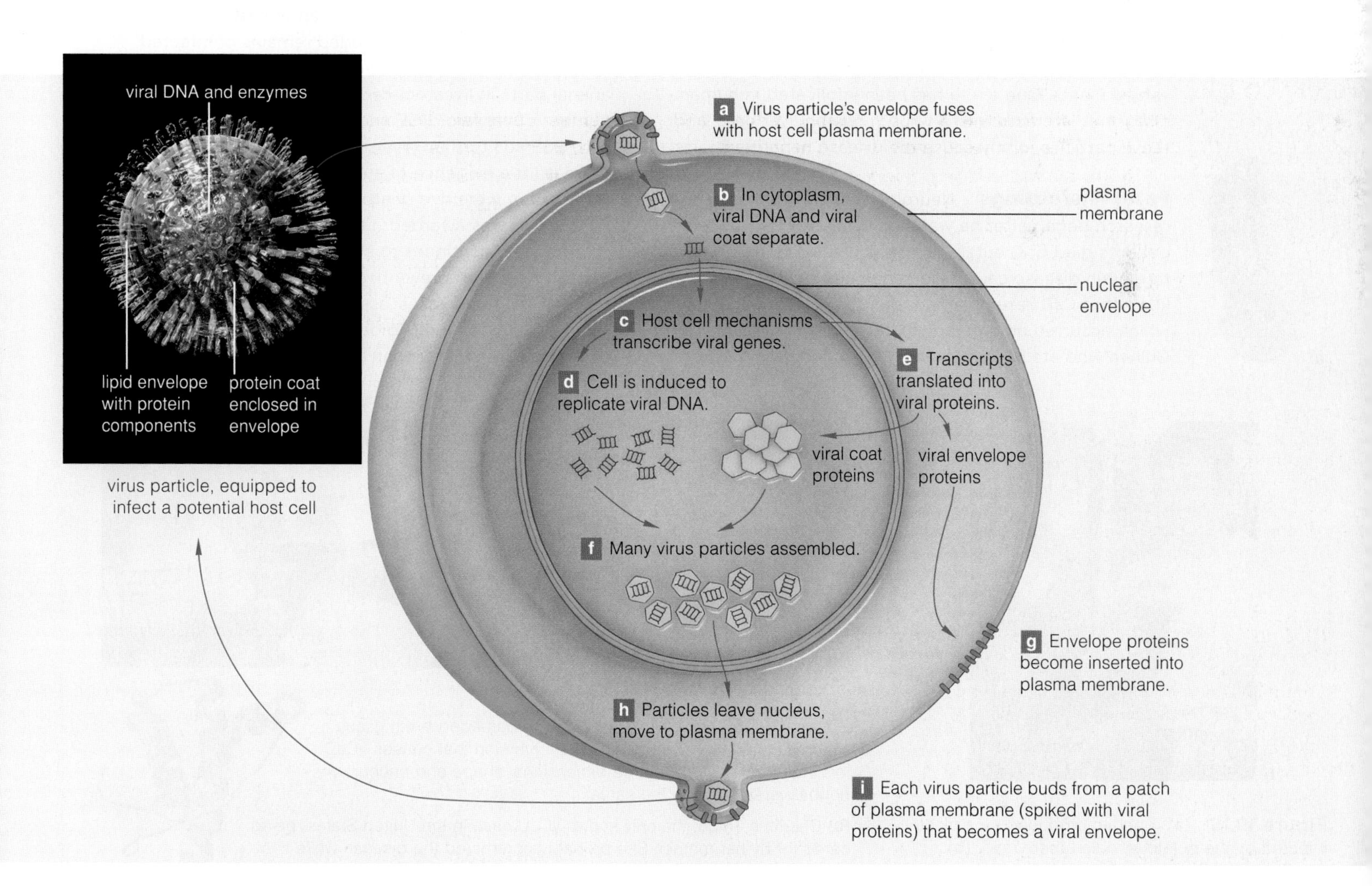

Figure 19.14 **Animated!** Multiplication cycle of *Herpes simplex*, an enveloped DNA virus.

19.5 Viroids and Prions

FOCUS ON SCIENCE

Some infectious particles are only a single molecule—a small circle of RNA or an oddly folded protein.

The Smallest Pathogens In 1971, plant pathologist Theodor Diener announced the discovery of a new type of pathogen. It was a small circle of RNA without any protective protein coat. He named it a viroid, because it seemed like a stripped-down version of a virus.

Diener had been investigating potato spindle tuber disease. Potato plants affected by this disease become stunted and make just a few small, deformed tubers (Figure 19.15). Diener initially hypothesized that a virus causes the disease. He reconsidered after realizing the infectious agent passed through filters that were too fine to allow passage of even the smallest virus. To find out exactly what this minuscule pathogen was made of, he made extracts of infected plants. He then treated the extracts with enzymes to see what would stop their ability to infect plants. Extracts treated with enzymes that digested DNA or protein still infected plants. Only RNA-digesting enzymes made the extracts harmless.

Plant pathologists have now described about thirty viroids, many of which cause disease in commercially valued plants. One viroid has been implicated in human disease. It interacts with a virus in human liver cells, and these particles jointly cause the disease hepatitis D.

Fatal Misfoldings Neurologist Stanley Prusiner's research began after he watched helplessly as one of his patients died of Creutzfeldt-Jakob disease (CJD). This rare brain disease causes dementia and death. Prusiner knew that CJD resembled another rare disease, kuru, which occurred among individuals of a tribe in New Guinea who ate human brains. What agent could cause these diseases? Prusiner approached the question by studying still another disease—scrapie—that affects sheep. As in the human diseases, it causes neurological symptoms, and the brain becomes so riddled with holes that it looks like a sponge (Figure 19.16*a*).

Based on his studies of scrapie and his knowledge of related diseases, Prusiner proposed that proteins called prions are present in a normal nervous system, where they fold in a characteristic way. Disease develops after some prions fold incorrectly. In an unknown manner, their altered shape induces the normal prions to misfold as well. Deposits of misfolded prions accumulate in the brain, kill cells, and cause the sponge-like appearance.

Prusiner's prion hypothesis generated great interest in the mid-1980s during a British epidemic of bovine spongiform encephalopathy (BSE) or mad cow disease. A rise in human cases of CJD followed that epidemic in cattle (Figure 19.16*b*). Prusiner showed that a prion similar to the one in scrapie-infected sheep could be isolated from cows with BSE and from humans affected by the variant Creutzfeldt-Jakob disease (vCJD).

How did a prion get from sheep to cattle to people? The cattle ate feed that included remains of infected sheep, then the infected beef sickened humans. Use of animal parts in livestock feed is now banned, and the number of cases of BSE and cVJD has declined. Cattle with BSE still turn up, even in the United States, but there is little evidence of a major threat to humans. In 2005, there were five deaths from vCJD, all in Britain.

Prusiner was awarded a Nobel Prize for his discovery of prions. He continues to study prion diseases, and he hopes to develop preventive treatments and cures. His research also addresses issues raised by skeptics. Some scientists still think misfolded proteins are a symptom of the diseases rather than their ultimate cause.

a

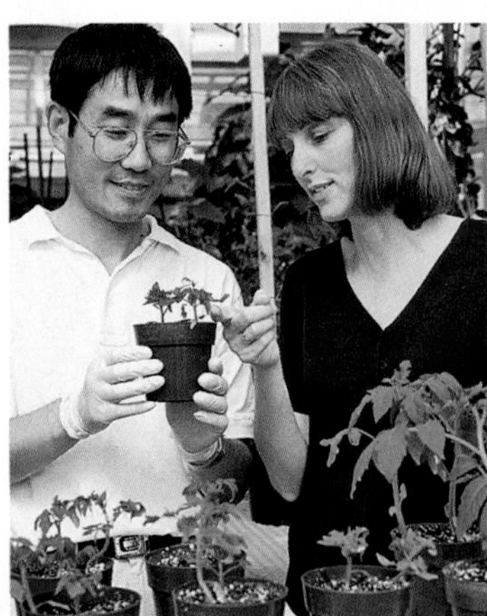

b

Figure 19.15 (**a**) A potato produced by a plant infected by the potato spindle tuber viroid. (**b**) Yan Zhao and Rosemarie Hammond study plants infected by this viroid. They hope to learn how it gets into the nucleus of an infected plant cell.

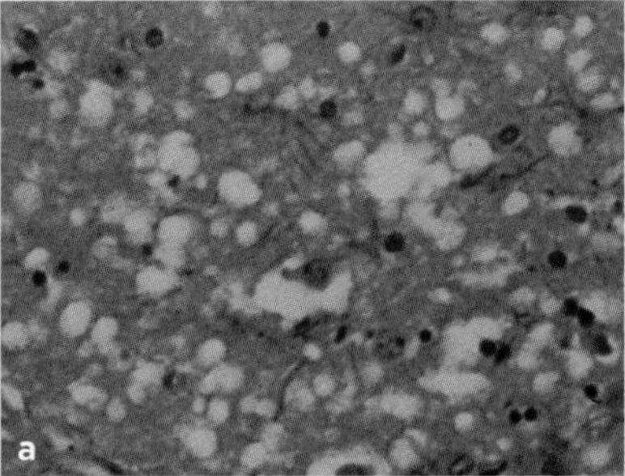

a

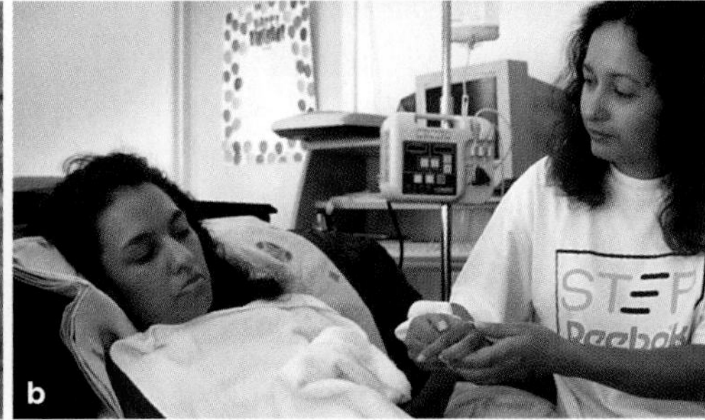

b

Figure 19.16 (**a**) Perforations in BSE-damaged brain tissue. *Right*, a model for a normal prion. The version that causes vCJD misfolds into a different three-dimensional shape and encourages normal versions to do the same.

(**b**) Charlene Singh, the only known vCJD case in the United States, being cared for by her mother. She probably contracted the disease while she was growing up in Britain. A gradual loss of memory and sense of balance led to her diagnosis in 2001. She became confined to bed and could not control her body functions or communicate. She died in 2004.

19.6 Evolution and Disease

CONNECTIONS

Just by being human, you are a potential host for a huge variety of pathogenic bacteria, viruses, fungi, protozoans, and parasitic invertebrates (Table 19.3).

THE NATURE OF DISEASE

During an infection, a pathogen multiplies in cells or tissues of its host. Unless defenses halt this process, disease results. Contagious diseases such as common colds spread by contact with mucus, blood, or other fluids from an infected individual. Sporadic diseases such as whooping cough break out from time to time and affect few people. Endemic diseases persist at a low level. Tuberculosis is an example.

During an *epidemic*, a disease spreads through part of a population for a limited time, then subsides. A *pandemic* occurs if epidemics of the same disease break out in different geographic regions at the same time.

AIDS is a pandemic with no end in sight. A 2003 outbreak of SARS (short for Severe Acute Respiratory Syndrome) was a brief viral pandemic that started out in China. Travelers carried it around the world. Before government-ordered quarantines stopped the spread, thousands were sickened, and hundreds died.

Think of disease in terms of a pathogen's prospects for survival. A pathogen survives only for as long as it can access energy and raw materials. For many kinds of disease agents, the host's body is the jackpot. Any pathogen that overcomes host defenses has access to an abundance of resources, and it can multiply to very impressive numbers. Evolutionarily, the organism that has the most descendants wins.

Two barriers prevent pathogens from taking over the world. First, any species with a history of being attacked by a specific pathogen has coevolved with it and has built-in defenses. The vertebrate immune system is an example. Second, a pathogen that kills an individual host too fast can vanish along with it. Most pathogens have less-than-fatal effects. Table 19.3 lists the diseases with the highest death rates.

Usually, an individual will die only if it becomes host to multiple pathogens, if it is a novel host with no coevolved defenses, or if a mutant pathogenic strain emerged and breached current defenses. *Ebola*, one of the viruses that causes a hemorrhagic fever, coevolved with monkeys in Africa. Humans are a novel host and *Ebola* kills 70 to 90 percent of those that it infects. The H5N1 strain of avian influenza (bird flu) has infected and killed more than 170 people. Nearly all the victims had direct contact with infected birds. It may take no more than a few mutations to make H5N1 contagious and usher in a pandemic.

Table 19.3 The Eight Deadliest Infectious Diseases

Disease	Main Agents	Estimated New Cases per Year	Deaths per Year
Acute respiratory infections*	Bacteria, viruses	1 billion	4.7 million
Diarrheas**	Bacteria, viruses, protists	1.8 billion	3.1 million
Tuberculosis	Bacteria	9 million	3.1 million
Malaria	Protists	110 million	2.5–2.7 million
AIDS	Virus (HIV)	5.6 million	2.6 million
Measles	Viruses	200 million	1 million
Hepatitis B	Virus	200 million	1 million
Tetanus	Bacteria	1 million	500,000

* Includes pneumonia, influenza, and whooping cough.
** Includes amoebic dysentery, cryptosporidiosis, and gastroenteritis.

DRUG-RESISTANT PATHOGENS

As explained in Section 16.6, using antibiotics to treat any infectious disease results in directional selection. Individuals in a population of pathogens that are least affected by the drug have a selective advantage. The frequency of drug-resistant individuals increases over generations—which, for most pathogens, are short.

The first antibiotic drug manufactured synthetically and put to wide use was penicillin. It saved many lives during World War II. At the time, there was discussion of eliminating threats of bacterial diseases around the world. Optimism about a disease-free future began to fade as penicillin-resistant bacteria evolved.

For example, *Streptococcus pneumoniae* is commonly transmitted among children at day care centers. It can cause pneumonia, meningitis, and chronic middle-ear infections. Penicillin-resistant strains of *S. pneumoniae* first appeared in 1967. Today, about half of all strains of this bacterium are resistant.

Penicillin-resistant strains of *S. pneumoniae* arise by spontaneous mutations or horizontal gene transfers. Some of their cell wall proteins differ from those in nonresistant strains. As we now know as a result of gene sequencing studies, genes for wall proteins that confer resistance were transferred to an *S. pneumoniae* cell from the related species *S. mitis*.

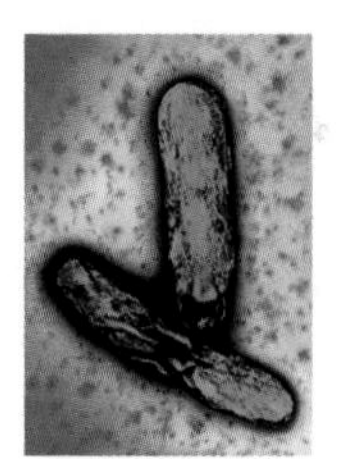

Mycobacterium tuberculosis

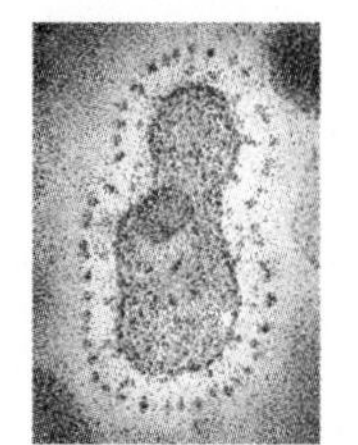

SARS virus

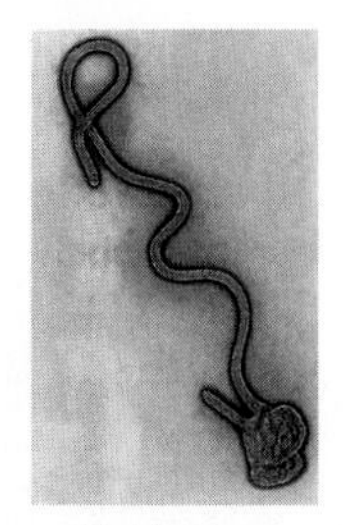

Ebola virus

Pathogens evolve by natural selection. Selection favors the species or strains that do not kill off their host too quickly. It also favors the ones that are resistant to antibiotics.

Summary

Section 19.1 Bacteria and archaeans are single-celled prokaryotes. They do not have a nucleus or the typically diverse array of cytoplasmic organelles that characterize eukaryotic cells. As a group, they are the smallest, and the most widely distributed, numerous, and metabolically diverse organisms (Table 19.4). Cell shapes vary. Typical surface structures are a cell wall, an outermost protective capsule or slime layer, one or more flagella, and pili.

Only bacteria and archaeans reproduce by prokaryotic fission: replication of a single, circular chromosome and division of a parent cell into two genetically equivalent daughter cells. Horizontal gene transfers can move genes between prokaryotes, as when conjugation puts a plasmid and some chromosomal genes into another cell.

- *Use the animation on ThomsonNOW to explore prokaryotic structure.*
- *Use the animation on ThomsonNOW to observe prokaryotic fission and conjugation.*

Section 19.2 The bacteria are the most common and diverse prokaryotes. Cyanobacteria are oxygen-releasing photoautotrophs. Chloroplasts probably evolved from ancient cyanobacteria by endosymbiosis. Gram-positive bacteria have thick walls. Most are chemoheterotrophs, and a number are human pathogens. A pathogen is an agent that infects a host and causes disease.

The proteobacteria, the most diverse bacterial group, include autotrophs and heterotrophs, free-living species, beneficial symbionts, and pathogens. Chlamydias are all intracellular parasites. Spring-shaped spirochetes live on their own or in hosts and some are pathogens.

Section 19.3 Archaeans are prokaryotic, but they are like eukaryotic cells in certain features. Comparisons of structure, function, and genetic sequences position them in a separate domain, between eukaryotes and bacteria. Ongoing research is showing that they are more diverse and widely distributed than was previously thought.

Comparative rRNA studies have helped identify three archaean groups—euryarchaeotes, crenarchaeotes, and korarchaeotes. In their physiology, most are methanogens (methane makers), halophiles (salt lovers), and extreme thermophiles. Archaeans coexist with bacteria in many habitats and apparently exchange genes with them.

Table 19.4 Prokaryotic Nutritional Modes

Mode of Nutrition	Energy Source	Carbon Source
Photoautotrophic	Sunlight	CO_2
Photoheterotrophic	Sunlight	Organic compounds
Chemoautotrophic	Inorganic substances	CO_2
Chemoheterotrophic	Organic compounds	Organic compounds

Section 19.4 All viruses are noncellular infectious particles that cannot reproduce on their own. They infect a host cell, and their genes and enzymes take over the host's mechanisms of replication and protein synthesis.

Each virus particle consists of a core of DNA or RNA and a protein coat. In some viruses, the coat is enveloped in a bit of an infected cell's plasma membrane. The outer envelope forms as each new virus particle is released by budding or lysis. In bacteriophages and other complex viruses, the coat has a sheath and other structures.

Nearly all viral multiplication cycles have five steps. The virus attaches to a suitable host cell. The whole virus or just its genetic material enters the host cell. Viral genes and enzymes direct host mechanisms to replicate viral genetic material and synthesize viral proteins. New virus particles are assembled. The particles are released.

Multiplication pathways vary greatly. Two are common among bacteriophages. In a lytic pathway, multiplication is rapid, and the new virus particles are released by lysis. In a lysogenic pathway, the virus enters a latent state that extends the cycle; the host cell is not killed outright. Viral nucleic acid gets integrated into the host chromosome. All of the host cell's descendants inherit these miniature time bombs, which may be reactivated many generations later, causing the cell to enter the lytic pathway.

- *Use the animation on ThomsonNOW to learn how a bacteriophage and an enveloped virus can multiply.*

Section 19.5 Viroids and prions are small infectious agents. Viroids are circles of RNA without a protein coat. Many cause disease in plants. Prions are proteins that occur naturally in the vertebrate nervous system but can cause fatal disease when they misfold.

Section 19.6 Hosts coevolve with pathogens. Hosts evolve defenses. Pathogens evolve so they do not kill a host before they can infect other host individuals. Use of antibiotics favors antibiotic-resistant bacteria. Genes that convey drug resistance can arise by mutation and may spread among members of the same or different species by conjugation.

Diseases can be fatal if an individual becomes host to multiple pathogens, if it has no coevolved defenses, or if a pathogen mutates into a different form that can breach current defenses. *Ebola* and the H5N1 strain of bird flu are two of the deadly emerging pathogens.

- *Read the InfoTrac article "Origins of HIV: The Interrelationship Between Nonhuman Primates and the Virus," Myra Watanabe,* Bioscience, *September 2004.*

Self-Quiz

Answers in Appendix III

1. Only ______ are prokaryotic.
   a. archaeans
   b. bacteria
   c. prions
   d. both a and b

2. Bacteria transfer plasmids by ______.
   a. prokaryotic fission
   b. endospore formation
   c. conjugation
   d. the lytic pathway

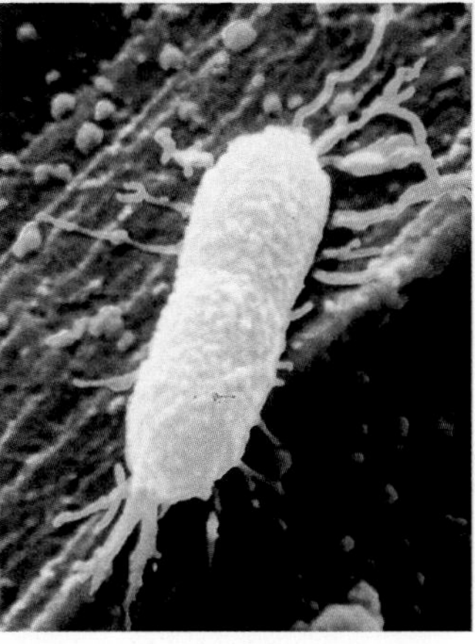

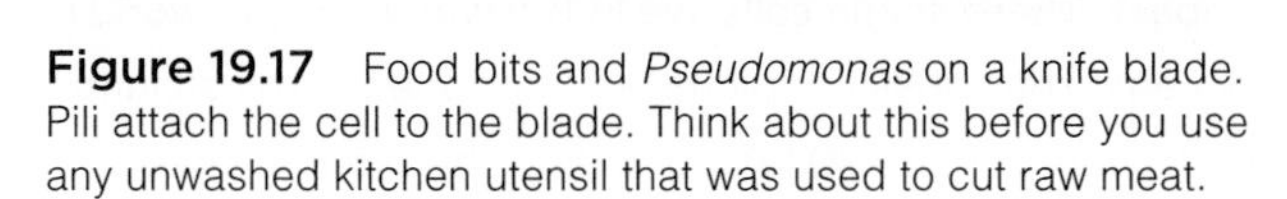

Figure 19.17 Food bits and *Pseudomonas* on a knife blade. Pili attach the cell to the blade. Think about this before you use any unwashed kitchen utensil that was used to cut raw meat.

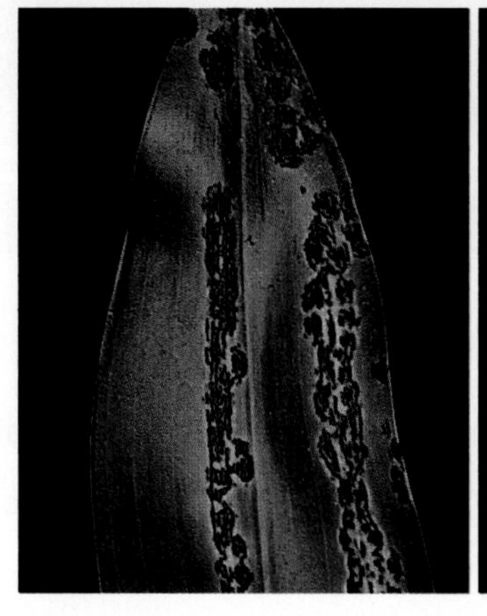

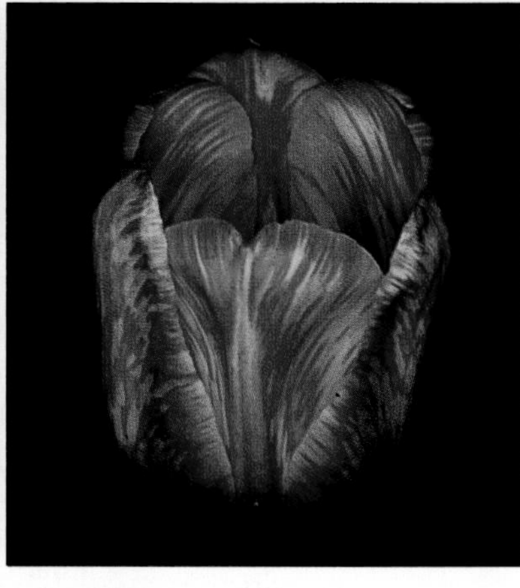

Figure 19.18 Effects of plant viruses. *Left*, orchid leaf infected by a rhabdovirus. *Right*, streaking in tulip petals. In the lighter areas, a virus disrupted pigment formation.

3. All ______ are oxygen-releasing photoautotrophs.
   a. spirochetes c. cyanobacteria
   b. chlamydias d. proteobacteria
4. Normally harmless *E. coli* cells in your gut are ______ .
   a. spirochetes c. cyanobacteria
   b. chlamydias d. proteobacteria
5. All ______ are intracellular parasites of vertebrates.
   a. spirochetes c. cyanobacteria
   b. chlamydias d. proteobacteria
6. Some Gram-positive bacteria (e.g., *Bacillus anthracis*) survive harsh conditions by forming a(n) ______ .
   a. pilus c. endospore
   b. heterocyst d. plasmid
7. Only ______ reproduce by prokaryotic fission.
   a. viruses c. bacteria
   b. archaeans d. both b and c
8. DNA or RNA may be the genetic material of ______ .
   a. a bacterium b. a prion c. a virus d. an archaean
9. A viroid consists entirely of ______ .
   a. a DNA b. RNA c. protein d. lipids
10. Bacteriophages can multiply by ______ .
   a. prokaryotic fission c. a lysogenic pathway
   b. a lytic pathway d. both b and c
11. Match the terms with their most suitable description.
   ___ archaean a. infectious protein
   ___ bacteria b. nonliving infectious particle; nucleic acid core, protein coat
   ___ virus c. involved in conjugation
   ___ plasmid d. prokaryotes that most closely resemble eukaryotes
   ___ extreme halophile e. most common prokaryotic cells
   ___ prion f. small circle of bacterial DNA
   ___ sex pilus g. salt lover

■ *Visit ThomsonNOW for additional questions.*

Critical Thinking

1. Annual costs of treating known cases of food poisoning are between 5 billion and 22 billion dollars, and pathogens in kitchens may cause at least one-half of the cases. Electron microscopes show microbes on wood or plastic cutting boards, even stainless steel knives (Figure 19.17). Carlos Enriquez and his colleagues at the University of Arizona sampled 75 dishcloths and 325 sponges from homes. They found *Salmonella*, *E. coli*, *Pseudomonas*, and *Staphylococcus* colonies in most of them. Sponges and dishcloths can be sanitized by a few minutes in a microwave or a cycle in a dishwasher. If your class has access to a light microscope, sample your kitchen sponges or dishcloths and compare what is living with you.

2. Food irradiation (exposing food to gamma rays that kill pathogenic bacteria) can prevent bacterial food poisoning, slow spoilage, and prolong the shelf life of food. It does not make food radioactive or alter its nutritional value. Some people worry that irradiation can cause harmful chemicals to form, and it does not kill the endospores that can cause botulism. In their view, a better way to minimize the risk of food poisoning is to tighten and enforce existing food safety standards.

Irradiated meat, poultry, and fruits are already available in many supermarkets. By law, they must be marked with the symbol shown at right. Would seeing this symbol on a package make you any more or less likely to purchase the product? Explain your answer.

3. When planting bean seeds, gardeners are advised to inoculate them first with powder that contains nitrogen-fixing *Rhizobium* cells, which infect plant roots. How could the presence of these bacteria benefit the plants?

4. Curtis Suttle at the University of British Columbia studies interactions among viruses, bacteria, and algae in seawater. He found that selectively removing viruses from seawater made algae in the water stop growing. Algal growth depended on nutrients released by virus-infected bacteria as they died. Make a list of some other ways that viruses might serve important functions in natural ecosystems.

5. Like other organisms, plants suffer from viral diseases (Figure 19.18). Recall from Section 4.11 that plant parts are usually protected by a waxy cuticle and that each plant cell has a cellulose wall around its plasma membrane. Plant viruses often get into a host plant cell with assistance from plant-eating insects. Once inside a cell, the virus causes modifications in the cell's plasmodesmata. Explain how altering these structures might benefit the virus.

20 PROTISTS—THE SIMPLEST EUKARYOTES

IMPACTS, ISSUES Tiny Critters, Big Impacts

Sample water from just about anywhere in nature and you will find multitudes of organisms known informally as protists. Like their most ancient ancestors, nearly all of these eukaryotic species live in fresh water or the seas. They range from producers, to predators, to decomposers. Many species release oxygen into the atmosphere, take up carbon, and otherwise play important roles in cycling of nutrients. Some live inside the cells or tissues of fungi, plants, and animals, where they benefit their host or have no effect one way or another. Others are harmful pathogens.

Structurally, single-celled protists are the simplest of all eukaryotes. Although most are microscopically small, they often have a big collective impact on the world.

Foraminiferans and coccolithophores have such an impact. These single cells live in the waters of the world's oceans. Their shells or plates are hardened with calcium carbonate (Figure 20.1*a*). Long ago, the calcium-rich remains of their ancient relatives started to pile up on the seafloor. Over great spans of time, the deposits became compressed into stone. Limestone and chalk are the legacy of those uncountable numbers of single cells.

About 5,200 years ago on Malta, foraminiferan-rich limestone was quarried to build the Hagar Qim temple, now a World Heritage site (Figure 20.1*b*). A few thousand years later, other limestones were used to construct the Egyptian pyramids and Sphinx. Look closely at the walls of the Empire State Building, the Pentagon, or Chicago's Tribune Tower, and you will see foraminiferan shells.

Dover, England, is renowned for its white chalk cliffs (Figure 20.1*c*). The cliffs are a monument to individual

See the video!

Figure 20.1 (**a**) What shells of foraminiferans, including *Spiroloculina* and *Peneroplis* species, look like through a magnifying lens. Foraminiferans have contributed to architectural marvels, such as (**b**) the ancient Hagar Qim temple on Malta, an island south of Sicily, and (**c**) the white chalk cliffs near Dover, England, which tower 91 meters (about 300 feet) above the English Channel.

How would you vote? The pathogen that causes sudden oak death infects some plants commonly sold as nursery stock in the United States. Should states where the pathogen is absent be allowed to prohibit shipping of all plants from affected states? See ThomsonNOW for details, then vote online.

coccolithophores that lived inside chalky plates, died, and then drifted to the seafloor. Each year, their remains added 0.5-millimeter to the chalky sediments. They did so for many millions of years.

Protists have big effects on land, too. *Phytophthora* species infect many plants. One caused a terrible famine during the mid-1800s by destroying potatoes that were the staple crop in Ireland. Today, a related species that causes sudden oak death threatens forests in the western United States. Tens of thousands of oaks have already died. Other trees and shrubs also are sickened by the pathogen. As a result, many forest animals have been left scrambling for food and shelter.

These two contrasting examples can help you keep the larger perspective in mind. Some protists do cause disease or damage, and we tend to assign value to them in terms of their direct impact on human lives. There is nothing wrong with battling the harmful species and admiring the beneficial ones. Even so, do not lose sight of how protists fit into nature's larger picture.

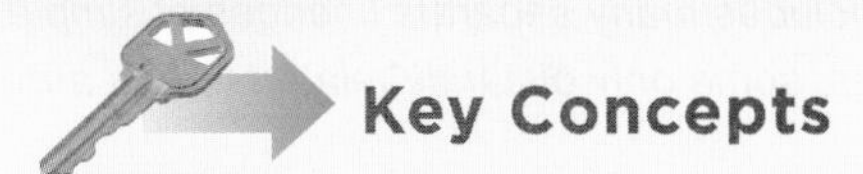

Key Concepts

SORTING OUT THE PROTISTS

Protists include many lineages of single-celled eukaryotic organisms and their closest multicelled relatives. Gene sequencing and other methods are clarifying how protist lineages are related to one another and to plants, fungi, and animals. **Section 20.1**

GROUPS WITH ANCIENT ROOTS

Two of the earliest lineages of eukaryotic cells are known informally as flagellated protozoans. The foraminiferans and radiolarians are another ancient lineage. **Sections 20.2, 20.3**

THE ALVEOLATES

Ciliated protozoans, dinoflagellates, and apicomplexans are single-celled photoautotrophs, predators, and parasites with a unique layer of tiny sacs under their plasma membrane. **Sections 20.4, 20.5**

THE STRAMENOPILES

Chrysophytes, diatoms, and brown algae are stramenopiles, most of which are photoautotrophs. The colorless, parasitic oomycotes are an offshoot of the stramenopile lineage. **Sections 20.6, 20.7**

THE CLOSEST RELATIVES OF LAND PLANTS

Red algae and green algae are photosynthetic single cells and multicelled forms. One lineage of multicelled green algae is the closest living relatives of land plants. **Sections 20.8, 20.9**

DISTANT RELATIVES OF FUNGI AND ANIMALS

A great variety of amoeboid species formerly classified as separate lineages are now united as the amoebozoans. They are the closest living protistan relatives of fungi and animals. **Section 20.10**

Links to Earlier Concepts

This chapter continues the journey through life's diversity, so reflect on the evolutionary road map in Section 16.12 and on the logic behind constructing cladograms (16.10,16.11). You will revisit DNA sequencing and other biochemical methods of comparing groups of organisms (15.3,15.5,16.9). You consider again the effects of osmosis on cells living in fresh water (5.8). You will build on your knowledge of life cycles (9.5), think about the origin of organelles (18.4), and consider another case of bioluminescence (5.10).

20.1 An Evolutionary Road Map

Genetic analysis has shown that the traditional kingdom Protista includes many separate lineages of simple eukaryotes, some only distantly related to one another.

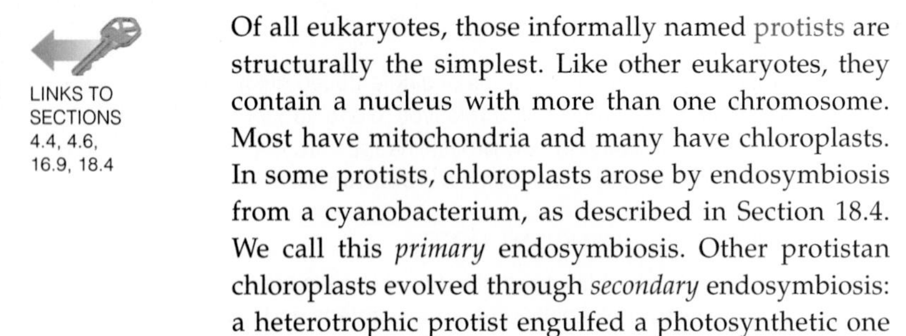

LINKS TO SECTIONS 4.4, 4.6, 16.9, 18.4

Of all eukaryotes, those informally named protists are structurally the simplest. Like other eukaryotes, they contain a nucleus with more than one chromosome. Most have mitochondria and many have chloroplasts. In some protists, chloroplasts arose by endosymbiosis from a cyanobacterium, as described in Section 18.4. We call this *primary* endosymbiosis. Other protistan chloroplasts evolved through *secondary* endosymbiosis: a heterotrophic protist engulfed a photosynthetic one that lived inside it and evolved into an organelle.

Most protists are microscopic single cells that live independently or in colonies. Some multicelled species have big, complex bodies. Life cycles typically include haploid and diploid stages, but the details vary (Figure 20.2). It can be difficult to categorize protists in terms of their mode of nutrition. Some protist lineages include both heterotrophic and autotropic species, and members of some species can switch from one mode to the other.

Until recently, protists were defined mainly in terms of what they are not, as in "not bacteria, not plants, not fungi, and not animals." They were lumped into a kingdom that represented an evolutionary crossroads between prokaryotes and "higher" forms of life.

As you read in earlier chapters, researchers are in the process of carving up kingdom Protista into many smaller groups. By using gene sequencing and other methods, they have begun to understand evolutionary connections. They have discovered that many protist groups are only distantly related to one another. Some lineages are more closely related to plants, fungi, or animals than to other protists. To keep track of these relationships, use the evolutionary tree in Figure 20.3 as a road map through this chapter.

Many kinds of single-celled and multicelled eukaryotes, both heterotrophs and autotrophs, were once lumped together in the kingdom, Protista. Gene comparisons can now identify distinct lineages and show how they relate to one another, and to other eukaryotes.

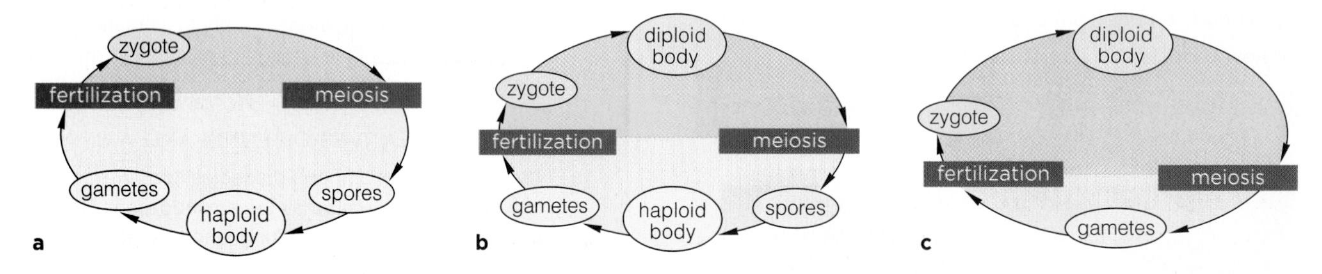

Figure 20.2 Generalized protist life cycles. Haploid and diploid bodies may be single-celled, multicelled, or both. (**a**) A haploid body dominates some cycles; only the zygote is diploid. (**b**) Other cycles show alternation of generations—a diploid body alternates with a haploid one, as in plants. (**c**) A diploid body dominates other cycles. As in animals, only gametes are haploid, and no spores form.

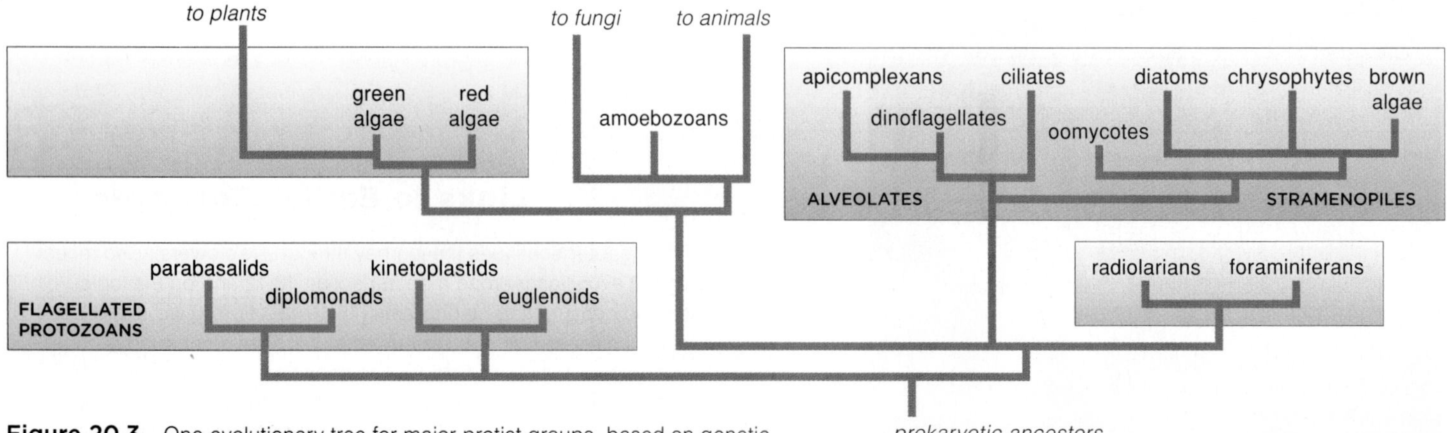

Figure 20.3 One evolutionary tree for major protist groups, based on genetic and morphological comparisons. All groups shown have living representatives.

20.2 Evolutionarily Ancient Flagellates

Some time ago, zoologists thought that motile, single-celled heterotrophs were ancient relatives of animals. They named the organisms "protozoans" (proto–, first; and –zoa, animal). We now know that most protozoans are not at all close to animals in the tree of life.

Flagellated protozoans are single-celled heterotrophs with one or more flagella. A pellicle, a layer of elastic proteins just under the plasma membrane, helps these unwalled cells retain their shape.

Gene sequence comparisons put diplomonads and parabasalids close to the base of the eukaryotic family tree. These flagellated cells do not have mitochondria, but they have organelles that seem to be remnants of them. The common ancestor of both groups may have had more typical mitochondria.

Diplomonads alone have two more or less identical nuclei. Some species are free-living in ponds. *Giardia lamblia* is a parasite that attacks humans, cattle, and some other mammals (*left*). Like many other protists, *G. lamblia* forms cysts. A cyst is a thick-walled resting stage. Drink from a stream that is contaminated with *G. lamblia*, cysts, and you may end up with giardiasis. Symptoms of this disease range from mild cramping to severe diarrhea that can last for months.

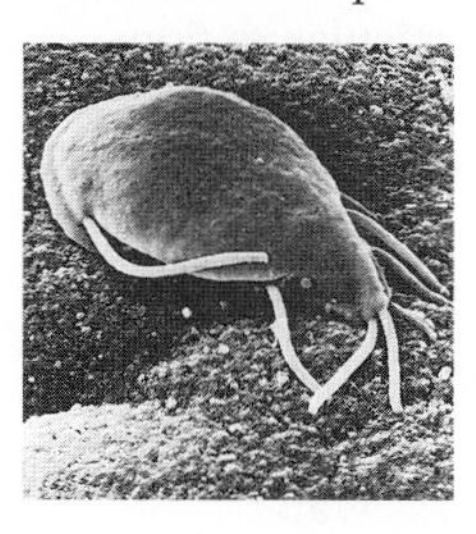

Parabasalids live in or on animals. Some live in the termite gut and help their hosts by enabling them to digest cellulose. *Trichomonas vaginalis* (Figure 20.4*a*), one of the parasitic types, attaches to the lining of the human vagina or male reproductive tract and sucks out nutrients. It causes trichomoniasis, one of the most common sexually transmitted diseases.

Kinetoplastids are flagellated protozoans with one large mitochondrion that contains a clump of DNA. Trypanosomes, the main subgroup, are parasites with their flagellum attached to the cell body in a way that forms an undulating membrane (Figure 20.4*b*).

Biting insects are vectors for many kinetoplastids that parasitize humans. In microbiology, a vector is an insect or some other animal that carries a pathogen between hosts. Tsetse flies spread *Trypanosoma brucei*, which harms the nervous system and causes African sleeping sickness. Blood-sucking bugs carry *T. cruzi*, which causes Chagas disease and damages the heart. Desert sandflies spread *Leishmania donovani*, the cause of leishmaniasis. Skin sores, inflamed nasal linings, and enlargement of the liver are symptoms.

Euglenoids are flagellated, free-living cells common in freshwater habitats. Genetic comparisons show they share a recent common ancestor with kinetoplastids. Most euglenoids are colorless heterotrophs. About a third have chloroplasts that evolved after a green alga was engulfed by a heterotrophic ancestor. An eyespot near the base of a long flagellum helps the cell detect light. Euglenoids have a unique pellicle made up of translucent strips of protein that spiral around the cell (Figure 20.5). Reproduction is asexual; the cell divides in two along its long axis.

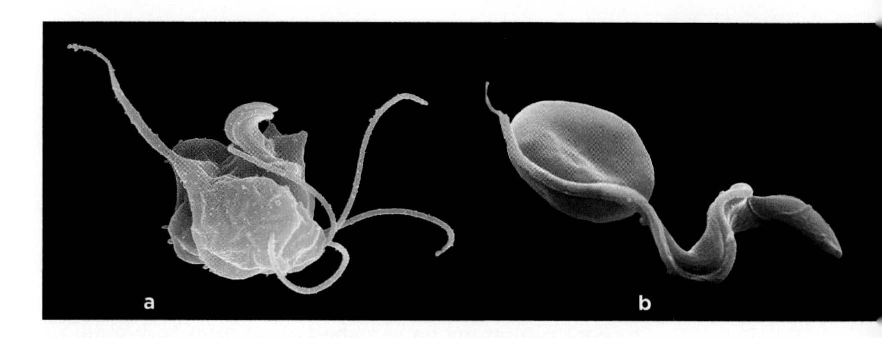

Figure 20.4 (**a**) *Trichomonas vaginalis*, cause of a common sexually transmitted disease. (**b**) *Trypanosoma brucei* with a red blood cell. It swims through the human bloodstream using a flagellum that forms an undulating membrane along the cell.

A euglenoid is hypertonic relative to freshwater. As in the other freshwater protists, one or more contractile vacuoles counter the tendency of water to diffuse into the cell. Excess water collects in contractile vacuoles, which contract and expel it to the outside.

Parabasalids and diplomonads are heterotrophs that lack mitochondria. Some are important human pathogens.

Trypanosomes are parasites with a large mitochondrion. Biting insects transmit some that cause human disease.

Euglenoids include heterotrophs and photoautotrophs that have chloroplasts that evolved from green algae.

Figure 20.5 Animated! Scanning electron micrograph and sketch of a euglenoid (*Euglena*), which has many mitochondria.

20.3 Shelled Amoebas

Radiolarians and foraminiferans are soft cells inside hard shells. Like the amoebas we will discuss in Section 20.10, they engulf their prey with cytoplasmic extensions called pseudopods, but they are not close relatives of amoebas.

LINKS TO SECTIONS 4.12, 5.6, 5.10, 16.10

Foraminiferans and radiolarians are heterotrophs that have no pellicle to give them shape. They live enclosed in a secreted shell. Food-gathering pseudopods radiate like a starburst through pores that perforate the shell.

Foraminiferans have a shell of calcium carbonate or silica (Figure 20.6*a*). Most live on the ocean floor where their pseudopods continually probe the water and the sediments for prey. Others drift in sunlit upper waters. These often have smaller photosynthetic protists such as diatoms or golden algae as symbionts inside them.

Foraminiferan shells have been accumulating on the seafloor since Cambrian times. The shells entombed in sedimentary layers yield clues about previous shifts in global climate, such as the temperature of ancient seas.

Radiolarians secrete an intricate, glassy silica shell (Figure 20.6*b*). Most are part of the marine plankton, the mostly microscopic organisms that drift or swim in seawater. Dense populations of radiolarians occur from surface waters to 5,000-meter depths. Numerous vacuoles in an outer zone of cytoplasm fill with air and keep radiolarians afloat. The vacuoles also assist in digesting food, including other protists.

Foraminiferans and radiolarians are both soft-bodied, heterotrophic cells that have intricate mineralized shells. Their hundreds of pseudopods extend outward through pores in the shell, probing aquatic habitats for food.

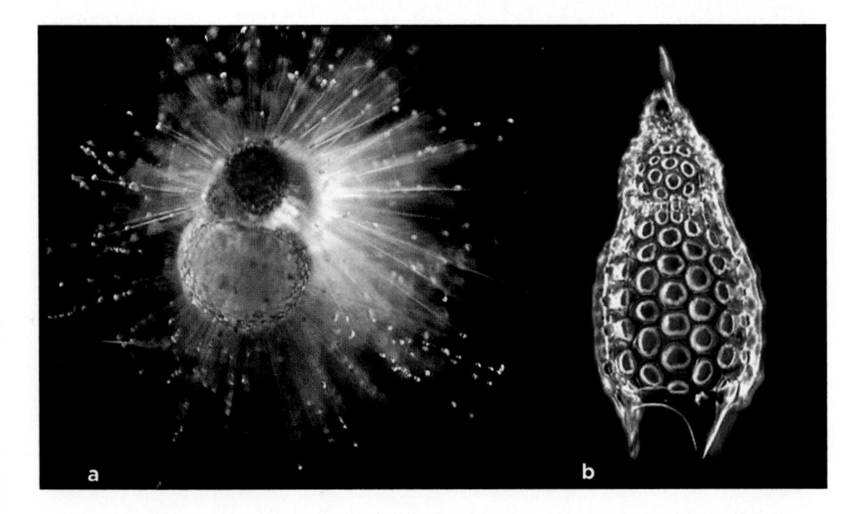

Figure 20.6 (**a**) Living foraminiferan extending thin pseudopods past its shell. The yellow "dots" are golden algae. (**b**) A radiolarian's silica shell.

20.4 Alveolates

Ciliates, dinoflagellates, and apicomplexans may seem to have little in common. Some are autotrophs, others are predators or parasites. But all share a unique trait, an array of tiny sacs just under the plasma membrane.

Three major groups of protists have sacs beneath their plasma membrane, as shown at right. *Alveolus* means "sac," and these protists are called alveolates.

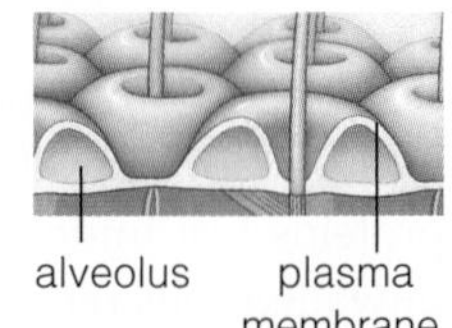

CILIATES

Ciliates, also known as ciliated protozoans, live in the seas and in fresh water, including water in soil. Most of the 8,000 or so species are free-living heterotrophs. About a third live in the gut or tissues of animals, and a few are colonial. Some species have rows of cilia, but others have only tufts of cilia in particular regions.

Paramecium is a common ciliate (Figure 20.7). Cilia cover its entire surface and beat in synchrony; they resemble a field of grass swaying in the wind. Starting at an oral depression in the cell surface, cilia sweep water laden with bacteria, algae, and other food into a gullet. Food is digested inside enzyme-filled vesicles. Contractile vacuoles squirt out excess water.

Ciliates reproduce asexually through binary fission: a cell divides in two, forming identical daughter cells. Ciliates also reproduce sexually. A cell has two kinds of nuclei. A macronucleus governs daily functions and a micronucleus undergoes meiosis and serves in sexual reproduction. Sexual partners swap micronuclei.

DINOFLAGELLATES

About half of the 4,000 or so species of dinoflagellates are predators and parasites. The rest have chloroplasts that most likely evolved from red algal cells. Some of the photosynthetic dinoflagellates live as symbionts in cells of reef-forming corals. The corals cannot survive without the sugars that the dinoflagellates produce.

Dinoflagellates deposit cellulose inside their alveoli. In some species, the cellulose forms thick plates. Most species have two flagella. One runs like a belt through a groove in the middle of the cell, and the other extends out from the cell's base (Figure 20.8).

Cells reproduce asexually until conditions do not favor growth. Then two cells fuse and form a resting stage that later undergoes meiosis.

Dinoflagellates abound in seawater and fresh water, especially in the tropics. Some of the tropical species are bioluminescent (Figure 20.8*a*). As in fireflies, the

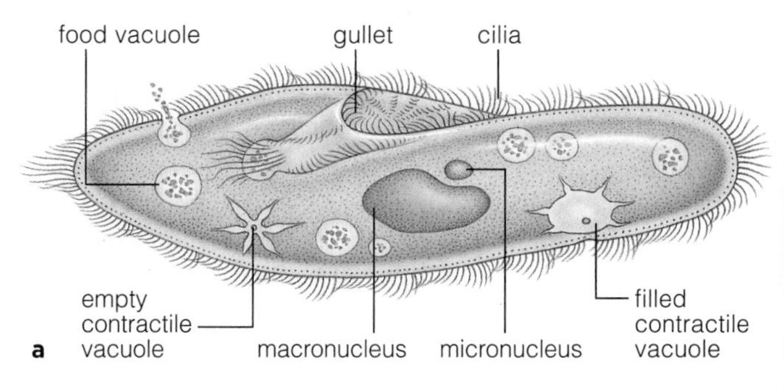

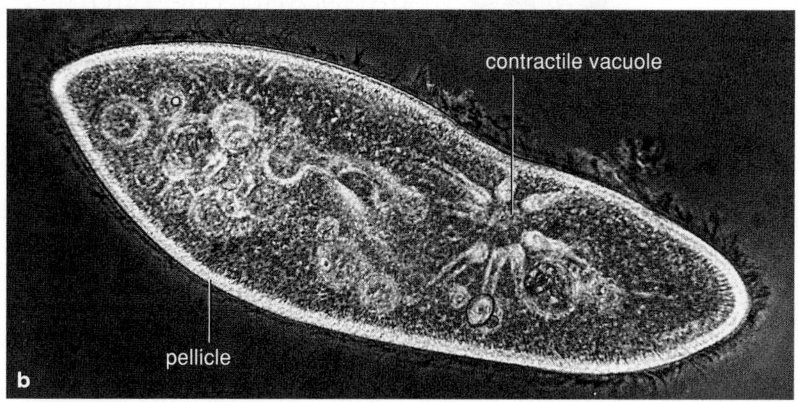

Figure 20.7 Animated! (**a**) Sketch and (**b**) light micrograph of *Paramecium*, a typical ciliate. See also Figure 5.29.

enzyme luciferase converts ATP energy to light energy (Section 5.6). The light may protect a cell by startling a small predator about to eat it, or by attracting larger predators that chase would-be dinoflagellate eaters.

Photosynthetic dinoflagellates sometimes undergo algal blooms: dramatic increases in population sizes of single-celled photoautotrophs. In habitats enriched with nutrients, as from agricultural runoff, each liter of water may hold millions of cells. Blooms of certain species, including *Karenia brevis,* cause red tides; they tint water red, especially near coasts.

Metabolic wastes from cells that are part of an algal bloom kill off other aquatic organisms (Figure 20.8*b*). Aerobic bacteria decompose algal remains and deplete the oxygen dissolved in water, so that aquatic animals suffocate. Some dinoflagellate toxins also kill directly. *Karenia brevis* produces a toxin that binds to sodium channels in the plasma membrane of nerve cells. Eat shellfish contaminated with this toxin and you might end up dizzy and nauseated from neurotoxic shellfish poisoning. Symptoms usually develop hours after the meal and persist for a few days. Blooms of *K. brevis* occur almost every year in the Gulf of Mexico, but the severity and effects vary.

APICOMPLEXANS

Apicomplexans are parasitic alveolates with a unique structure made of microtubules. They use it to pierce and enter a host cell. About 5,000 species are parasites of animals, from worms and insects to humans. Even though the apicomplexans are heterotrophs, they have organelles similar to chloroplasts. They probably had a photosynthetic ancestor. Gene sequences show that this group is closely related to the dinoflagellates.

Apicomplexans reproduce asexually and sexually inside their hosts. Only the gametes have flagella. The next section describes the life cycle of *Plasmodium,* an apicomplexan that causes malaria.

Figure 20.8 Two effects of dinoflagellates. (**a**) Tropical waters light up when *Noctiluca scintillans*, a bioluminescent dinoflagellate, is disturbed by something moving through the water. (**b**) In 2003, blooms of *Karenia brevis* released toxins into waters near Naples, Florida. Poisoned fishes died and were washed ashore.

Alveolates alone are single cells with a continuous array of membrane-bound sacs beneath the plasma membrane.

The ciliates are heterotrophs. Cilia cover all or part of the cell surface and function in motility, feeding, or both.

Dinoflagellates are common in plankton. These flagellated heterotrophs or photoautotrophs have cellulose plates.

Apicomplexans, such as Plasmodium *species, are parasites with a host-piercing device.*

20.5 Malaria and the Night-Feeding Mosquitoes

FOCUS ON HEALTH

About 1 million people develop malaria each year, mainly in the tropics and subtropics. More than 300 million are now infected by the apicomplexan agents of the disease.

LINK TO SECTION 17.6

Malaria got its name in the seventeenth century. Italians connected the disease with swamps near Rome, where noxious gases and mosquitoes filled the air (*mal*, bad; *aria*, air). So many people became ill that the disease accelerated the decline of ancient empires.

With her bite, a female *Anopheles* mosquito transmits a motile infective stage of *Plasmodium* to a human host (Figure 20.9*a*). The motile stage, a sporozoite, travels in blood vessels to liver cells and then reproduces asexually (Figure 20.9*b*). Some offspring—merozoites—enter red blood cells and liver cells, reproduce in them, and kill them (Figure 20.9*c,d*). Others develop into immature gametes, or gametocytes (Figure 20.9*e*). They mature into gametes only after another blood-sucking mosquito takes them up (Figure 20.9*f,g*).

Gametocytes cannot mature and form gametes in a human host, because *Plasmodium* is highly sensitive to temperature and oxygen levels. Humans maintain a high internal body temperature, and their blood has little free oxygen; most is bound to hemoglobin. A female *Anopheles* mosquito has a lower body temperature, than a human and she takes in oxygen from the air along with blood. Gametes mature and fuse inside her gut. The resulting zygotes give rise to new sporozoites that migrate to the insect's salivary glands and await transmission to a new vertebrate host.

Malaria symptoms usually start a week or two after a bite, when the infected liver cells rupture and release merozoites, metabolic wastes, and cellular debris into blood. Shaking, chills, a burning fever, and drenching sweats follow. After the initial episode, symptoms may subside for a few weeks or even months. Infected people often feel healthy. However, ongoing infection damages the liver, spleen, and kidneys. Debris clogs blood vessels and cuts blood flow to the brain. The result is convulsions, coma, and eventual death.

Historically, malaria has been most prevalent in tropical and subtropical Africa (Section 17.6). It was common in much of the United States until extensive swamp drainage and DDT spraying programs in the mid-1940s eliminated nearly all populations of infected mosquitoes.

Many *Plasmodium* strains are now resistant to older antimalarial drugs. Artemisinin, a synthetic version of a compound discovered in *Artemisia* plants, is now favored. It reduces fever and the number of gametocytes in blood, so it can slow the course of infection.

Efforts to design a vaccine are ongoing. Vaccines are preparations that induce an individual's immune system to recognize a specific pathogen. Vaccines developed for malaria are not effective against all stages of *Plasmodium* life cycles. So far, no one has come up with an effective vaccine against any parasite with a multistage life cycle. However, researchers armed with biotechnology have come up with promising vaccines that are being tested in Africa and elsewhere. In the meantime, every thirty seconds, one African child dies of malaria.

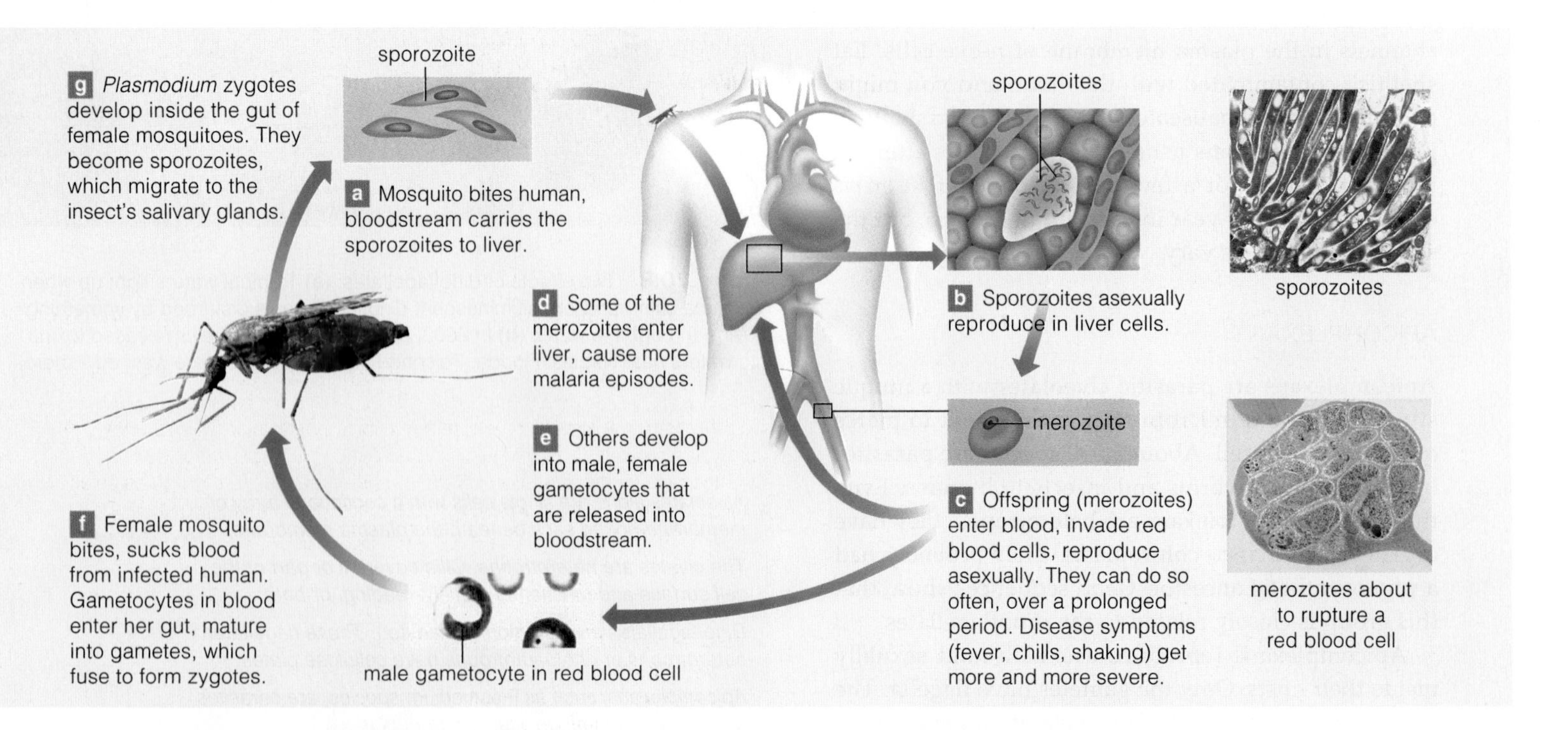

Figure 20.9 Animated! Life cycle of one of the four *Plasmodium* species that cause malaria.

20.6 Single-Celled Stramenopiles

Stramenopile means "straw-haired" and refers to a flagellum with short, hairlike filaments that occurs on cells during the life cycle of most members of this group. The group includes the single cells introduced here, and the brown algae, the next section's focus.

The water molds and downy mildews are examples of oomycotes. Like fungi, they are heterotrophs that live as a mesh of nutrient-absorbing filaments. But they are not closely related to fungi. Most act as decomposers. Some parasitize plants or animals (Figure 20.10).

Many oomycotes are plant pathogens. *Phytophthora ramorum* causes sudden oak death, mentioned earlier. *P. infestans* causes late blight in potatoes (*below left*). In the mid-1800s, a major outbreak of *P. infestans* caused a widespread famine in Ireland. Resulting starvation, diseases, and emigration cut Ireland's population by one third. As another example, in the 1870s, a downy mildew that infects grapes, was accidentally brought from North America to France on imported vines. This mildew spread and destroyed nearly all of France's vineyards before it was finally brought under control.

Single-celled photosynthetic stramenopiles include diatoms, coccolithophores, and golden algae. All are members of the *phyto*plankton—a collection of mostly microscopic producers in aquatic habitats. A pigment in their chloroplasts—fucoxanthin—also occurs in red algae. This suggests that the stramenopile chloroplasts evolved by secondary endosymbiosis from a red alga.

Diatoms have a two-part, perforated shell made of silica (Figure 20.11a,*b*). They live in oceans, streams, rivers, lakes, and damp soils. Their shells piled up on the ocean floor over millions of years and now form vast deposits of material called diatomaceous earth. We use this silica-rich material in filters and cleaners, and as an insecticide that is nontoxic to humans and other vertebrates.

Coccolithophores have plates of calcium carbonate beneath their plasma membrane (Figure 20.11*c*). Over time, their remains helped form vast chalk deposits. Modern species still tie up carbon in their shells and so play a major role in the global carbon cycle. Like dinoflagellates, they can undergo sudden population explosions (algal blooms) in nutrient-enriched waters.

Most golden algae live as single cells in lakes and other freshwater habitats. Some colonial species have a silica covering (Figure 20.11*d*). Many golden algae are not full-time autotrophs. They switch to predation in waters that are too dark for photosynthesis.

Figure 20.10 Effects of oomycotes. (**a**) Dead and dying trees near Big Sur, California, where *Phytophthora ramorum* is epidemic. (**b**) *Saprolegnia*, sometimes called cotton mold, growing on a fish. (**c**) Spores that have two flagella, one of them "hairy," spread oomycote diseases.

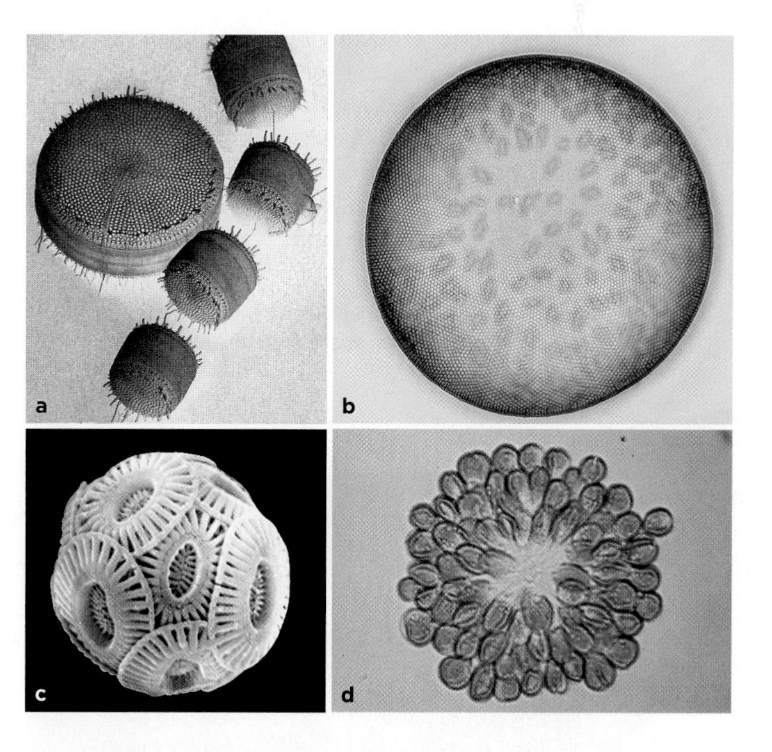

Figure 20.11 Stramenopiles of the phytoplankton. (**a**) Diatoms form by mitotic divisions inside parent cells and get smaller with each division. They do not grow afterward; when they reach a certain small size, they reproduce sexually. The resulting zygote develops into a full-sized cell. (**b**) Fucoxanthin-tinted chloroplasts of a diatom are visible through its silica shell. (**c**) Calcium carbonate plates lie beneath the plasma membrane of a coccolithophore. (**d**) Cells of a colonial freshwater golden alga (*Synura*).

Stramenopiles have a flagellum with hairlike filaments. Oomycotes are single-celled decomposers or parasites. The photosynthetic species include coccolithophores, diatoms, and golden algae—all phytoplankton.

20.7 Brown Algae

What do giant seaweeds have in common with single-celled stramenopiles? Fucoxanthin in their chloroplasts and a filament-covered flagellum on their sperm.

LINKS TO SECTIONS 2.2, 4.11, 6.1, 6.2, 8.4, 9.5

All 1,500 species of brown algae are multicelled. Most live in cool or temperate seawater, from the intertidal zone to the open oceans. Besides chlorophylls *a* and *c*, they contain an abundance of fucoxanthin and other pigments that color them gold, olive-green, or brown. Most brown algal life cycles alternate between diploid and haploid multicelled stages (Figure 20.2).

Brown algae range in size from microscopic strands to giant kelps that are the largest protists; some stand thirty meters tall (Figure 20.12). A kelp bed is a highly productive ecosystem, an underwater forest teeming with life. Kelp beds serve as nurseries for many fishes. Sea urchins that feed on kelp often become food for a sea otter. A decline in sea otter numbers can put kelp beds at risk for overgrazing by sea urchins.

Macrocystis and some other species are harvested commercially. We use algins extracted from kelps to thicken many foods, beverages, cosmetics, and other products. In certain parts of the world, especially the Far East, people harvest kelps for food and mineral salts, and as fertilizer for crops.

Brown algae are multicelled, photosynthetic stramenopiles. Different kinds are ecologically and commercially important.

Figure 20.12 Giant kelps (*Macrocystis*) form dense underwater forests. The life cycle has an alternation of generations. A diploid spore-forming stage (sporophyte) has stemlike parts called stipes, leaflike blades, and anchoring holdfasts. Gas-filled bladders impart buoyancy to the stipes and blades and keep the kelp upright in water. Haploid spores formed on blades are released into the water. They develop into the microscopic gamete-forming stage (the gametophyte).

20.8 Green Algae

Green algae are photoautotrophs with chloroplasts that evolved from cyanobacteria. One lineage gave rise to the plants, but all green algae have certain plantlike traits.

"Green algae" is the informal name for about 7,000 species belonging mainly to two groups: chlorophytes and charophytes. Most green algae are chlorophytes. The charophytes are the closest relatives of plants. Like plants, green algae have chloroplasts with chlorophylls *a* and *b*, store their excess sugars as starch, and deposit cellulose in their cell wall.

Green algae range in size from microscopic cells to multicelled filamentous species several meters long. Most live in fresh water; some live in the ocean, in the soil, or on surfaces. A few chlorophytes are symbionts inside invertebrates or with fungi in lichens.

Green algal life cycles alternate between diploid and haploid stages, but details vary (Figure 20.2). In some species, the diploid stage is a big, multicelled body and the haploid stage is microscopic. In others, the opposite occurs or both stages are about the same size.

CHLOROPHYTES

Haploid flagellated spores dominate the life cycle of *Chlamydomonas*, a single-celled, freshwater chlorophyte (Figure 20.13). When nutrients and light are plentiful, spores reproduce asexually. When conditions worsen, gametes develop and fuse, forming a zygote enclosed by a thick, protective wall. When conditions improve, a zygote undergoes meiosis and germinates, releasing the next generation of haploid, flagellated spores.

Volvox, a colonial chlorophyte, lives in freshwater ponds. The flagellated cells resemble *Chlamydomonas*, but hundreds to thousands form a whirling, spherical colony (Figure 20.14*a*). Daughter colonies form inside the parental sphere, which eventually ruptures and releases them.

Wispy sheets of *Ulva*, a multicelled species, cling to coastal rocks (Figure 20.14*b*). They grow longer than your arm, but are usually less than forty microns thick.

Some chlorophytes help out in the lab. Theodor Engelmann used *Cladophora* filaments in his studies of photosynthesis (Section 6.2). Melvin Calvin used cells of *Chlorella*, a single-celled species, to study the light-independent photosynthetic reactions (Section 2.2).

Figure 20.14 *Facing page:* (**a**) *Volvox* colony, with many flagellated cells joined by thin strands of cytoplasm. It ruptured and is releasing new colonies. (**b**) Sea lettuce (*Ulva*). (**d**) Two desmid cells, formed by binary fission. (**c**) *Chara*, known as muskgrass or stinkweed for its strong odor.

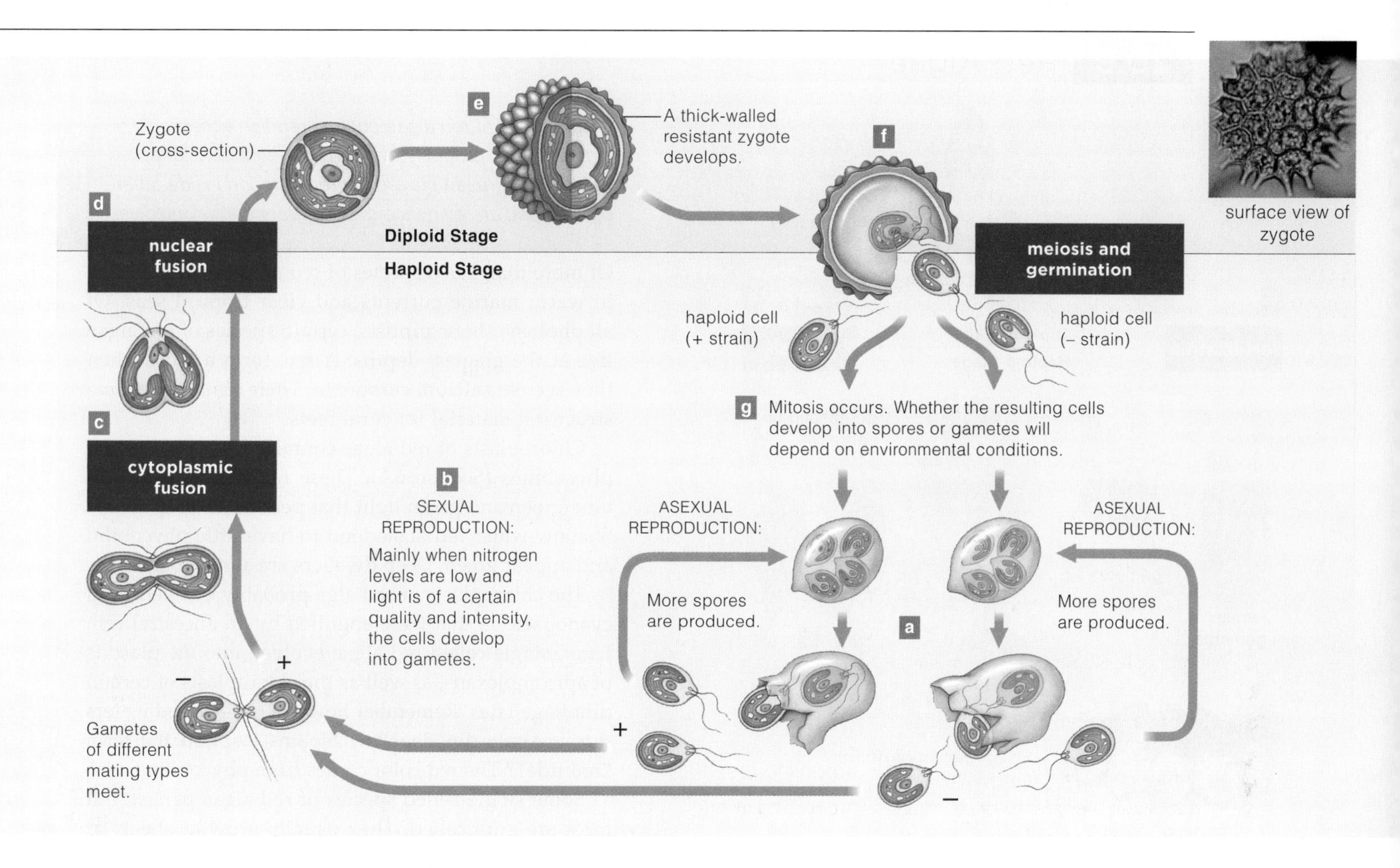

Figure 20.13 Animated! Life cycle of *Chlamydomonas*, a single-celled, freshwater green alga.

CHAROPHYTES

Like chlorophytes, charophytes can be single-celled or multicelled. Figure 20.14*c* shows desmid cells formed by a recent cell division. Most desmids are freshwater species, but some live in seawater, or on snow and ice. *Chara,* a multicelled charophyte native to Florida, can be an invasive nuisance in lakes and ponds (Figure 20.14*d*). It releases spores from ball-like structures on its branches. Like plants, and unlike most other green algae, *Chara* cells divide their cytoplasm by cell plate formation (Section 8.4), and plasmodesmata connect cytoplasm of neighboring cells (Section 4.11).

Chlorophytes and charophytes—the closest relatives of land plants—are two lineages of green algae. All are photosynthetic, but form and life cycles vary greatly.

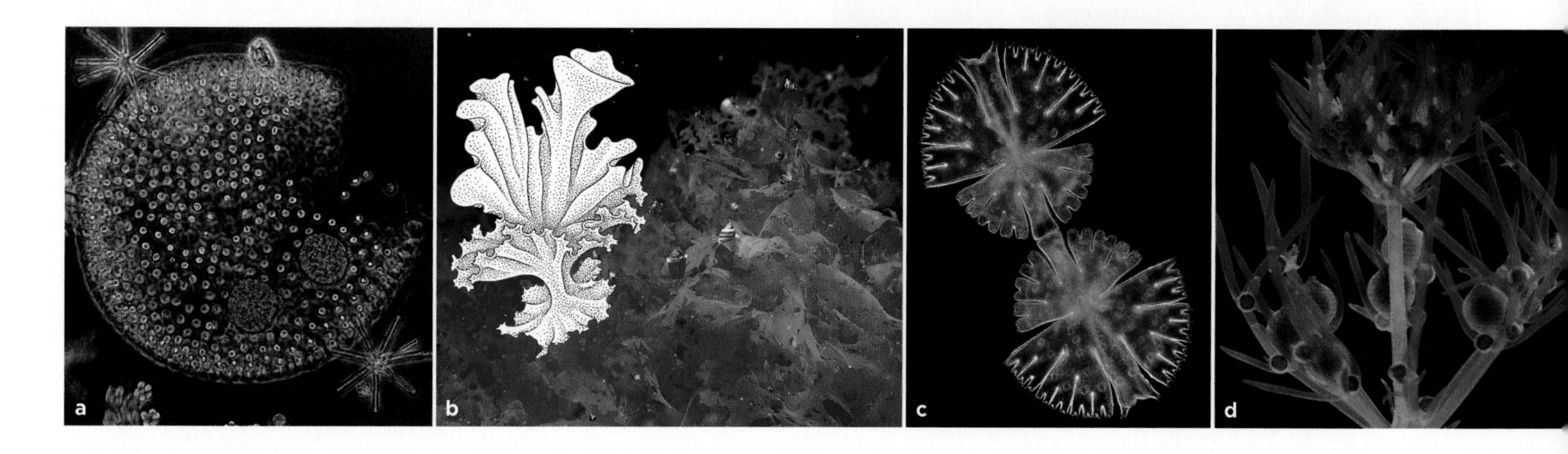

20.9 Red Algae

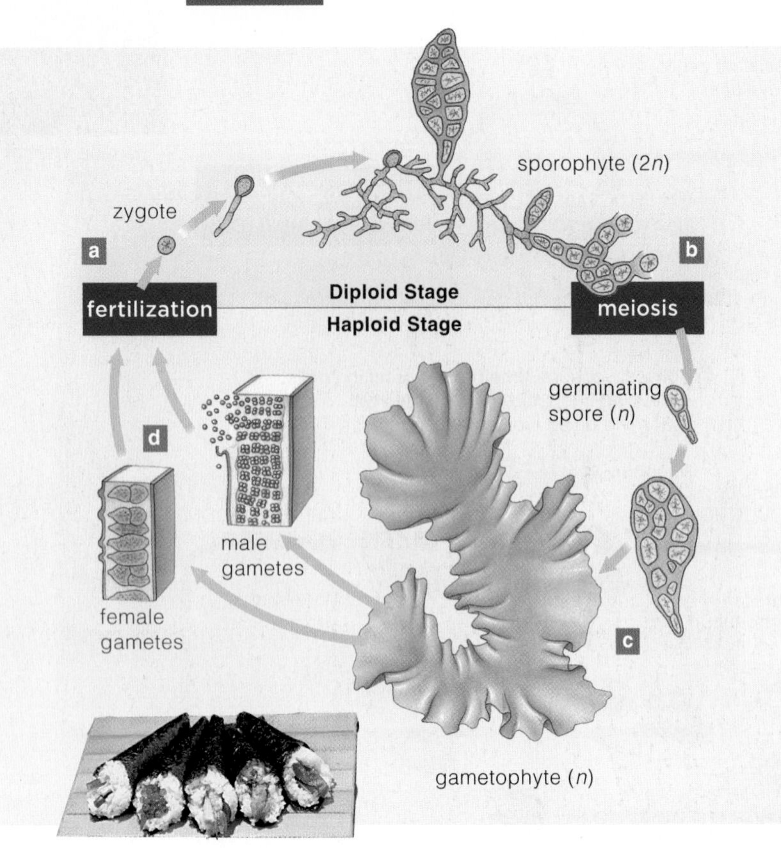

Figure 20.15 **Animated!** Life cycle of a red alga (*Porphyra*). For centuries, Japanese fishermen cultivated and harvested a red alga in early fall. The rest of the year, it seemed to vanish. Kathleen Drew-Baker studied its sheetlike form in the laboratory. She saw gametes forming in packets near the sheet margins. She also studied gametes in a petri dish. After zygotes formed, individuals developed into tiny, branching filaments on bits of shell in the dish. That was how the alga spent most of the year!

People had known about the pinkish growths on shells, but no one figured out the growths were algal sporophytes. *Porphyra* species could be cultivated on shells or other calcium-rich surfaces! Within a few years, researchers worked out the life cycle of *P. tenera*, a species used for seasoning or as a wrapper for sushi. By 1960, cultivation of *P. tenera* had become a billion-dollar industry.

Figure 20.16 The red alga *Antithamnion plumula*. The filamentous, branching growth pattern is common among red algae.

Chloroplasts of red algae contain reddish accessory pigments and probably evolved from cyanobacteria. Plastids that resemble red algae are found inside other protists and are evidence of secondary endosymbiosis.

Of more than 4,000 species of red algae, nearly all live in warm marine currents and clear tropical seas. Of all photosynthetic protists, certain species of red algae live at the greatest depths. A few form a crust when they secrete calcium carbonate. Their remains become structural material for coral reefs.

Chloroplasts of red algae contain chlorophyll *a* and phycobilins (Section 6.1). These red pigments absorb blue-green and green light that penetrates deep water. Shallow-water red algae tend to have little phycobilin and appear green. Deep dwellers are almost black.

The chloroplasts of red alga probably evolved from cyanobacteria that were engulfed by an ancestral cell. Later, single-celled red algae evolved into the plastids of apicomplexans, as well as the chloroplasts of certain dinoflagellates. Remember how the red-colored waters during some dinoflagellate blooms inspired the term "red tide"? The red color comes from phycobilins.

Some single-celled species of red algae persist, but most are multicelled. They usually grow as sheets or in a branching pattern (Figures 20.15 and 20.16). Life cycles vary and are often complex, with both asexual and sexual phases. There is no flagellated stage.

Red algae have many commercial uses. Agar is a polysaccharide extracted from cell walls of some red algae. It is used to keep baked goods and cosmetics moist, to set jellies, as a culture medium, and as soft capsules that hold medicines. Carrageenan, another polysaccharide, is added to soy milk, dairy foods, and fluid sprayed onto airplanes to prevent ice formation. *Porphyra* is now cultivated worldwide as food (Figure 20.15). More than 130,000 tons are harvested annually.

Most red algae are multicelled. Some survive at great depths because their phycobilins can absorb light of wavelengths that penetrate deep waters. Some red algae are cultivated and used in commercial products.

20.10 Amoeboid Cells at the Crossroads

FOCUS ON SCIENCE

The amoebas and their relatives are shape-shifting heterotrophs. Many are solitary, but some display communal behavior and cell differentiation that hint at complexities to come in fungi and animals.

Amoebozoa is one of the monophyletic groups now being carved out of the former hodgepodge kingdom Protista. Few amoebozoans have a cell wall, shell, or pellicle; nearly all undergo dynamic changes in shape. A compact blob of a cell can quickly send out pseudopods, move about, and capture food (Section 4.12).

The amoebas live as single cells. Figure 20.17*a* shows *Amoeba proteus*. Like most amoebas, it is a predator in freshwater habitats. Some amoebas can live in the gut of humans and other animals. Each year, about 50 million people are affected by amoebic dysentery after drinking water contaminated with *Entamoeba histolytica* cysts.

Slime molds are "social amoebas." The *plasmodial* kinds spend most of their life cycle as a plasmodium, a slimy, multinucleated mass. It forms from a diploid cell that undergoes mitosis many times without cytoplasmic division. A plasmodium streams out along the forest floor feeding on microbes and organic matter (Figure 20.17*b*). When stressed, as by dwindling food, a plasmodium gives rise to spore-bearing fruiting bodies (Figure 20.17*c*).

Cellular slime molds, such as *Dictyostelium discoideum,* spend the bulk of their existence as individual amoeboid (amoeba-like) cells (Figure 20.18). Each cell eats bacteria and reproduces by mitosis. Thousands of cells aggregate when food runs out. Environmental gradients induce them to crawl as a "slug." When the slug reaches a suitable spot, it becomes a fruiting body: A stalk forms, lengthens, and nonmotile spores form at its tip. Germination of a spore releases a diploid amoeboid cell that starts the cycle anew.

Dictyostelium and other amoebozoans provide clues to how signaling pathways of multicelled organisms evolved. Coordinated behavior—an ability to respond to stimuli as a unit—is a hallmark of multicellularity. It requires cell-to-cell communication, which may have originated in amoeboid ancestors. In *Dictyostelium*, a nucleotide called cyclic AMP is the signal that induces solitary amoeboid cells to stream together. It also triggers changes in gene expression. The changes cause some cells to differentiate into components of a stalk or into spores. Cyclic AMP also functions in the signaling pathways among cells of multicelled organisms.

Intriguingly, molecular comparisons suggest that the fungi and animals are a monophyletic group. They also suggest that animals and fungi descended from an ancient amoebozoan-like ancestor.

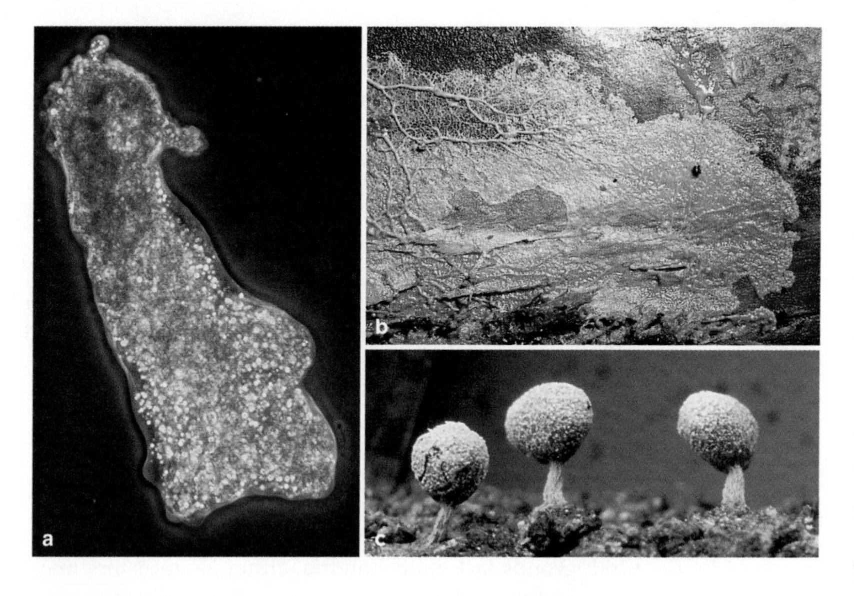

Figure 20.17 (**a**) *Amoeba proteus*, a free-living, freshwater amoeba. (**b**) *Physarum* plasmodium streaming across a rotting log. (**c**) *Physarum* fruiting bodies, which release haploid motile spores. When two spores fuse, they form a diploid cell. That cell undergoes repeated rounds of mitosis without cytoplasmic division, which forms a new plasmodium.

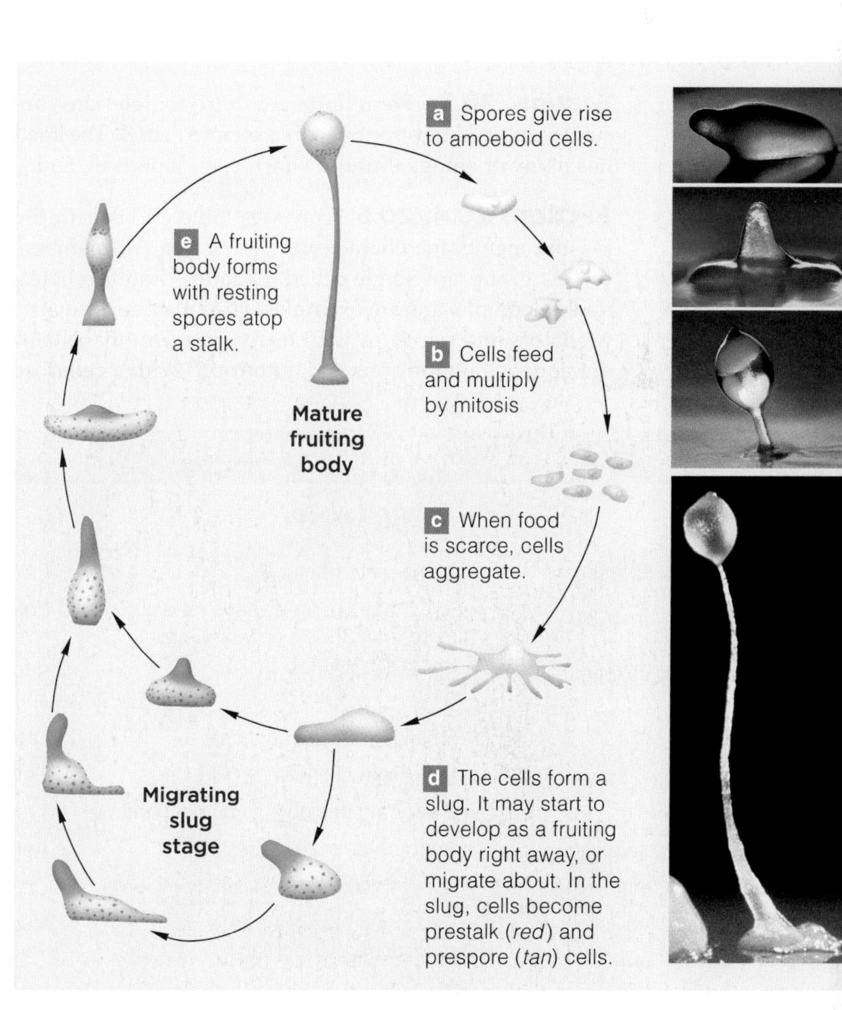

Figure 20.18 Animated! Life cycle of *Dictyostelium discoideum*, a cellular slime mold. During aggregation, the cells secrete and respond to cyclic AMP.

www.thomsonedu.com ThomsonNOW!

Summary

In this chapter and the preceding one, you learned about prokaryotes and the simplest eukaryotes: protists. Table 20.1 summarizes their similarities and differences.

Section 20.1 Protists are a collection of mostly single-celled eukaryotes, although some are multicelled and big. As in other eukaryotes, cells have a nucleus. Most also have one or more mitochondria. Many have chloroplasts that evolved from cyanobacteria or from another protist. The dominant stage of the life cycle may be haploid or diploid. Gene sequencing has shown that protists are not a monophyletic group, but rather a collection of lineages, some only distantly related. The evolutionary connections among the groups are gradually becoming clear.

Section 20.2 Single-celled heterotrophs traditionally known as flagellated protozoans include the most ancient eukaryotic lineages: the diplomonads and parabasalids. Both groups lack mitochondria, although ancestors may have had them. Both include species that infect humans.

Euglenoids and kinetoplastids also are single-celled flagellates. Most euglenoids live in freshwater. Some have chloroplasts that arose by secondary endosymbiosis from a green alga. Trypanosomes and most other kinetoplastids are parasites. They have a single giant mitochondrion.

- *Use the animation on ThomsonNOW to explore the structure of a euglenoid.*

Section 20.3 Foraminiferans and radiolarians are single-celled heterotrophs with a secreted shell. The shell has many openings through which pseudopods extend.

Sections 20.4, 20.5 Tiny sacs (alveoli) beneath the plasma membrane characterize alveolates. All members of this group are single-celled. Ciliates, dinoflagellates, and apicomplexans are examples. The ciliates are aquatic predators and parasites with many cilia. Dinoflagellates are aquatic heterotrophs and autotrophs with a cellulose covering. In nutrient-rich water, photosynthetic protists undergo population explosions, known as algal blooms. Apicomplexans are parasites that live in cells of animals. *Plasmodium* species cause malaria.

- *Use the animation and video on ThomsonNOW to explore ciliate structure and function.*
- *Use the animation on ThomsonNOW to explore the life cycle of a malaria-causing* Plasmodium *species.*

Sections 20.6, 20.7 Stramenopiles typically have two flagella, one with hairlike filaments. Oomycotes are heterotrophs—decomposers and parasites—that grow as a mesh of absorptive filaments. Some parasitic species are important plant pathogens.

Diatoms, coccolithophores, and golden algae often are part of the phytoplankton. They are photosynthetic cells and, like brown algae, contain the pigment fucoxanthin. Coccolithophores have calcium carbonate plates; diatoms have silica shells, as do some of the golden algae. Over time, these parts accumulated as vast deposits of chalk, limestone, and diatomaceous earth. Brown algae include microscopic strands and giant kelps, the largest protists.

- *Use the video on ThomsonNOW to visit a kelp forest.*

Section 20.8 Most green algae are chlorophytes. The smaller group of charophytes are the closest relatives of plants. Like plants, green algae have chloroplasts with chlorophylls *a* and *b*, and store carbohydrates as starch grains. Some complex charophytes have plasmodesmata and undergo cell plate formation during cell division.

- *Use the animation on ThomsonNOW to learn about the life cycle of a green alga.*

Section 20.9 Most red algae are multicelled. They can survive in deeper water than most photoautotrophs because their chloroplasts have phycobilins. Chloroplasts of red algae evolved from cyanobacteria. Red algae-like plastids occur in dinoflagellates and apicomplexans.

- *Use the animation on ThomsonNOW to explore the life cycle of a red alga.*

Table 20.1 Comparison of Prokaryotes With Eukaryotes

	Prokaryotes	Eukaryotes
Organisms represented:	Archaeans, bacteria	Protists, plants, fungi, and animals
Ancestry:	Two major lineages that evolved more than 3.5 billion years ago	Equally ancient prokaryotic ancestors gave rise to forerunners of eukaryotes, which evolved more than 1.2 billion years ago
Level of organization:	Single-celled	Single-celled or multicelled species with a division of labor among specialized cells; complex types have tissues and organ systems
Typical cell size:	Small (1–10 micrometers)	Large (10–100 micrometers)
Cell wall:	Many with cell wall	Cellulose or chitin; none in animal cells
Organelles:	Rarely; no nucleus; no mitochondria	Typically profuse; nucleus present; mitochondria in most
Modes of metabolism:	Anaerobic, aerobic, or both	Aerobic modes predominate
Genetic material:	One chromosome; plasmids in some	Chromosomes of DNA and many associated proteins; in a nucleus
Mode of cell division:	Prokaryotic fission, mostly; some reproduce by budding	Nuclear division (mitosis, meiosis, or both) associated with one of various modes of cytoplasmic division, including binary fission

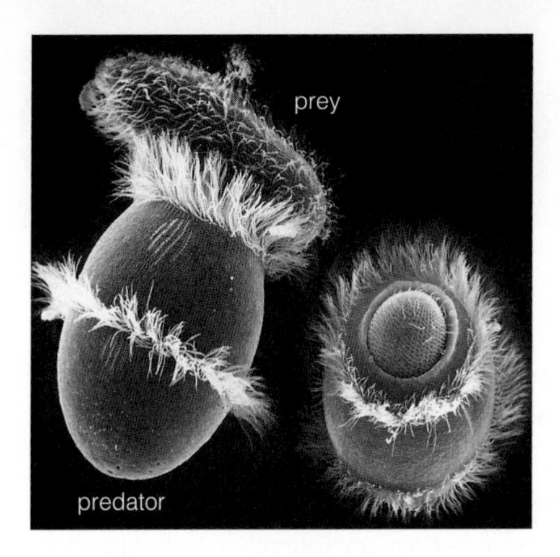

Figure 20.19 A composite photograph of (*left*) a larger protist catching a smaller one, and then (*right*) engulfing it. The cells are in a drop of pond water.

Section 20.10 Amoebozoans are heterotrophic free-living amoebas and slime molds. The plasmodial slime molds feed as a multinucleated mass. Amoeba-like cells of cellular slime molds aggregate when food is scarce and form multicelled fruiting bodies that disperse resting spores. Fungal and animal signaling mechanisms may have started in amoebozoan ancestors.

■ *Use the animation on ThomsonNOW to learn about the life cycle of a cellular slime mold.*

Self-Quiz

Answers in Appendix III

1. Trypanosomes cause which disease(s)?
   a. giardiasis
   b. Chagas disease
   c. amoebic dysentery
   d. African sleeping sickness
   e. malaria
   f. both b and d

2. Foraminiferans and radiolarians have ______ .
   a. pseudopods
   b. chloroplasts
   c. cilia

3. Which of the following is not an alveolate?
   a. a ciliate
   b. an apicomplexan
   c. a dinoflagellate
   d. a water mold

4. Diatoms, golden algae, and brown algae are most closely related to the ______ .
   a. dinoflagellates
   b. oomycotes
   c. green algae
   d. red algae

5. The chloroplasts of ______ evolved by primary endosymbiosis after a protist engulfed a cyanobacterium.
   a. green algae
   b. brown algae
   c. dinoflagellates
   d. red algae
   e. euglenoids
   f. both a and d

6. ______ are the protists that are most closely related to the fungi and animals.
   a. Stramenopiles
   b. Radiolarians
   c. Apicomplexans
   d. Amoebozoans

7. Match each term with its most suitable description.

___ diplomonad	a. protist population explosion
___ apicomplexan	b. silica shelled producer
___ algal bloom	c. multinucleated motile mass
___ diatom	d. no mitochondria, two nuclei
___ brown alga	e. closest relatives of land plants
___ red alga	f. multicelled, with fucoxanthin
___ green alga	g. agent of malaria
___ slime mold	h. deep dwellers with phycobilins

■ *Visit ThomsonNOW for additional questions.*

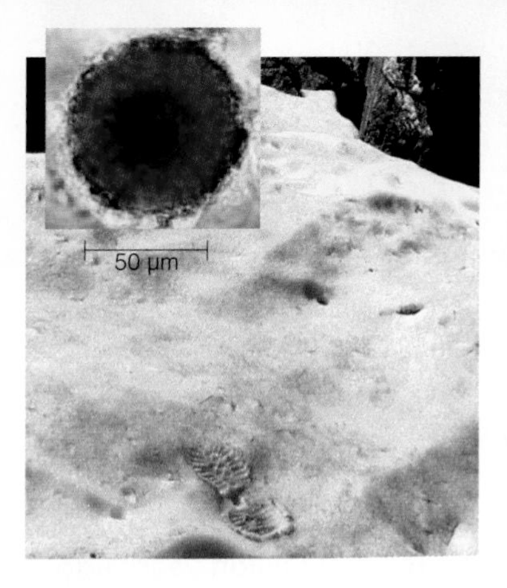

Figure 20.20 *Chlamydomonas nivalis* and other snow algae tinting a snowdrift.

Figure 20.21 *Postelsia*, a brown alga along coasts exposed to heavy surf from Vancouver Island to central California.

Critical Thinking

1. Suppose you vacation in a developing country where sanitation is poor. Having read about parasitic flagellates in water and damp soil, what would you consider safe to drink? What foods might be best to avoid or which food preparation methods might make them safe to eat?

2. You use a light microscope to look at a drop of pond water and see the scene in Figure 20.19. A living cell of one protist species is engulfing another. Both lack chloroplasts. To which group do they belong? What kind of information would help you decide among the possibilities?

3. The apicomplexan parasite *Toxoplasma gondii* infects birds and mammals and causes the disease toxoplasmosis. Cats are primary hosts. About 60 million people in the United States are infected. Some have flu-like symptoms, but most infected people are symptom-free. Dining on undercooked meat invites infection. Exposure to cats that eat raw meat or wildlife is also a risk factor, as is exposure to feces from such cats.

Toxoplasmosis during pregnancy increases chances of miscarriage or a premature delivery. A child infected prior to birth may have nervous disorders, such as seizures and mental retardation. Pregnant women are advised to avoid cats and litter boxes. Do you think keeping a cat around poses an acceptable risk during pregnancy? How could you minimize the risk of problems?

4. Runoff from highly fertilized cropland promotes algal blooms that can result in massive kills of aquatic species, birds, and other forms of wildlife. Are the massive kills an unfortunate but inevitable side effect of our lives? If you find the environmental cost unacceptable, how would you stop the pollution? Bear in mind that we now absolutely depend on high-yield (and heavily fertilized) crops.

5. The most common "snow alga," *Chlamydomonas nivalis*, lives on glaciers (Figure 20.20). It is a green alga, but it has so many carotenoid pigments that it appears red. Think about the solar radiation striking its icy habitats. Besides their role in photosynthesis, what other function might the carotenoids be serving? (*Hint:* Review Section 6.1.)

6. *Postelsia* (Figure 20.21) glues itself to wave-swept rocks along coasts. Few organisms can withstand the often violent conditions in this habitat. Research and report on the most distinctive structural adaptations that help it survive.

21 PLANT EVOLUTION

IMPACTS, ISSUES Beginnings, And Endings

Change is the way of life. As long ago as the Devonian, stalked plants were popping up on the margins of continents. By 300 million years ago, tree-sized ancestors of modern club mosses and horsetails dominated vast swamp forests. Then things changed. The global climate became cooler and drier, and moisture-loving plants declined. Hardier plants—cycads, ginkgos, and then conifers—rose to dominance. They were the gymnosperms, and they had a novel trait: They made seeds.

Later, one branch of the gymnosperm lineage gave rise to flowering plants and things changed again. The newcomers spread and soon became dominant in most regions (Figure 21.1). However, the conifers retained their competitive edge in certain environments, including the high latitude forests of the Northern Hemisphere.

Things changed yet again. About 11,000 years ago, humans learned to cultivate plants. Planting cereal crops provided a foundation for human population growth. Human populations grew spectacularly, and in time it took more resources than crops to sustain them. Conifers had the bad luck to become premier sources of lumber, paper, and other products. Deforestation—the removal of all trees from large tracts of land—was under way.

Only about 4 percent of California's original forest of coast redwoods remains intact. In Maine, an area the size of Delaware was deforested in just the past fifteen years. Many of the harvested logs get exported to mills around the world. At the same time, the United States imports timber from Canada's coniferous forests and from New Zealand's tropical rain forests.

DEVONIAN SWAMP

How would you vote? Demand for paper is a big factor in deforestation. However, processing costs make recycled paper expensive. Are you willing to pay more for papers, books, and magazines printed on recycled paper? See ThomsonNOW for details, then vote online.

Deforestation disrupts ecosystems. Exposed soils lose nutrients and fine sediments, which clog streams. Herbicides sprayed to kill plants other than timber trees slow the ecosystem's recovery and may push some plant species toward extinction. Almost 750 plant species in the United States alone are on the endangered list.

With this bit of perspective on change, we turn to the origins and adaptations of the land plants. With few exceptions, they are photoautotrophs. These metabolic wizards make organic compounds by absorbing energy from the sun, carbon dioxide from the air, and water and dissolved minerals from soil. By the noncyclic pathway of photosynthesis, they split water molecules and release oxygen. Their oxygen output and carbon uptake sustains the atmosphere. Think of it—every molecule of carbon in a redwood tree that stands a hundred meters high and weighs thousands of tons was taken up from the air.

We know of at least 295,000 kinds of plants. Their existence makes our lives possible. Without them, we humans and other land-dwelling animals never would have made it onto the evolutionary stage.

See the video! **Figure 21.1** *From left to right*, plants through time. By Devonian times, simple stalked plants were established, then conifers and other gymnosperms rose to dominance. *Near left*, this coniferous forest is being clear-cut on a mountain in British Columbia. Such forests once cloaked much of the world, but their decline started long before human populations crossed evolutionary paths with them. Flowering plants arose and coexisted with gymnosperms during the age of dinosaurs. Their descendants, including the orchids shown above, became spectacularly successful.

Key Concepts

MILESTONES IN PLANT EVOLUTION

The earliest known plants date from 475 million years ago. Ever since then, environmental changes have triggered divergences, adaptive radiations, and extinctions. Structural and functional adaptations of lineages are responses to some of the changes. **Section 21.1**

NONVASCULAR PLANTS

Bryophytes are nonvascular, with no internal pipelines to conduct water and solutes through the plant body. A gamete-producing stage dominates their life cycle, and sperm reach the eggs by swimming through droplets or films of water. **Section 21.2**

SEEDLESS VASCULAR PLANTS

Lycophytes, whisk ferns, horsetails, and ferns have vascular tissues but do not produce seeds. A large spore-producing body that has internal vascular tissues dominates the life cycle. As with bryophytes, sperm swim through water to reach eggs. **Sections 21.3, 21.4**

SEED-BEARING VASCULAR PLANTS

Gymnosperms and, later, angiosperms radiated into higher and drier environments. The packaging of male gametes in pollen grains and embryo sporophytes in seeds contributed to the expansion of these groups into new habitats.

Angiosperms alone make flowers, which wind, water, and animals help pollinate. In distribution and diversity, angiosperms are the most successful group of plants. Nearly all plant species that we rely upon for food are angiosperms. **Sections 21.5–21.9**

Links to Earlier Concepts

Section 20.8 introduced you to charophytes, a green algal group that is the closest relative of plants. Section 9.5 introduced you to plant life cycles. Now you will see how the cycles became modified in different lineages. You will get a sense of environmental changes that favored the evolution of waxy cuticles, lignin-reinforced cell walls (4.11), stomata (6.7), and other structures.

Much of what we know about plant evolution is based on the fossil record (16.4). The record offers clues to adaptive radiations and extinctions (17.12) on a changing geologic stage (16.6).

21.1 Evolutionary Trends Among Plants

In the past, changes in structural traits and life cycles helped plant lineages endure and often thrive while geological events were reshaping the world stage.

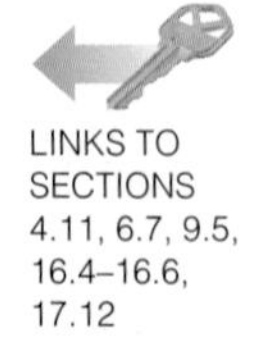

LINKS TO SECTIONS 4.11, 6.7, 9.5, 16.4–16.6, 17.12

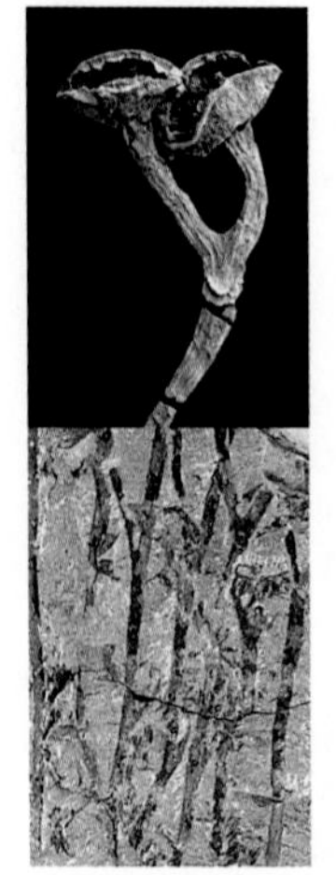

Fossil *Cooksonia* and *Psilophyton*, both extinct

About 500 million years ago, the invasion of land was under way. Green algae grew at the water's edge, and one lineage, the charophytes, gave rise to land plants. *Cooksonia*, a simple branching plant a few centimeters high, evolved by 430 million years ago. *Psilophyton*, a taller, more complex plant, appeared about 60 million years later. From that time forward, plants colonized most land habitats through adaptive radiations. How did they manage it? Natural selection favored adaptive changes in their life cycles, structure, and function.

FROM HAPLOID TO DIPLOID DOMINANCE

All plants show an alternation of haploid and diploid multicelled stages (Figure 21.2*a*). The haploid body is the gametophyte, which produces gametes by mitosis. Fusion of male and female gametes produces a zygote, which develops into the sporophyte: a diploid body that produces spores by meiosis. Each plant spore is a haploid cell that can remain dormant until conditions favor its growth. After a spore germinates, it divides by mitosis and develops into a new gametophyte.

The history of plants reveals a gradual shift from gametophyte dominance of the life cycle to sporophyte dominance (Figure 21.2*b*). Gametophytes dominate in many green algal life cycles. In some cases, there is no multicelled diploid stage; the zygote produces spores by meiosis. In the oldest lineages of land plants, there is a multicelled sporophyte, but it is small relative to the gametophyte. Spore production took on additional importance when plants colonized drier habitats. Here, selection favored larger, longer lived sporophytes that could produce more spores. It also favored new traits that helped plants nourish, protect, and disperse each new generation of sporophytes.

ROOTS, STEMS, AND LEAVES

Once plants moved onto land, selection favored traits that prevented water loss. Early on, the aboveground parts of plants became covered by cuticle, a secreted waxy layer that restricts evaporation. Openings across the cuticle, called stomata, became control points for balancing water conservation with the need to obtain carbon dioxide for photosynthesis (Section 6.7).

Early plants had structures that held them in place, but true roots evolved later. Roots not only anchored sporophytes, they also took up water with dissolved mineral ions from the soil. Fungal symbionts in or on roots assisted in these tasks, as they still do today.

Moving substances taken up by roots to other body regions required vascular tissues, a system of internal pipelines. Xylem is the vascular tissue that distributes water and mineral ions. Phloem is the vascular tissue that distributes sugars made in photosynthetic cells.

Of 295,000 or so modern plant species, more than 90 percent have xylem and phloem. These plants are members of the vascular plant lineage. The remainder are nonvascular plants, or bryophytes, a more ancient lineage. Mosses are among the modern bryophytes.

What made the vascular plants successful? For one thing, their vascular tissues are reinforced by lignin, an organic compound that is deposited in cell walls and lends structural support (Section 4.11). Thus, vascular

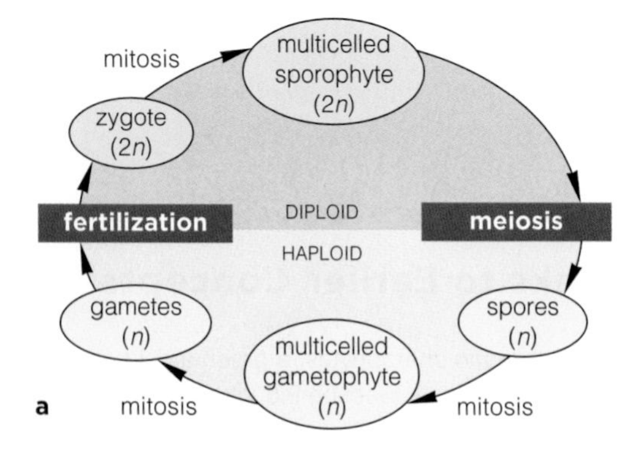

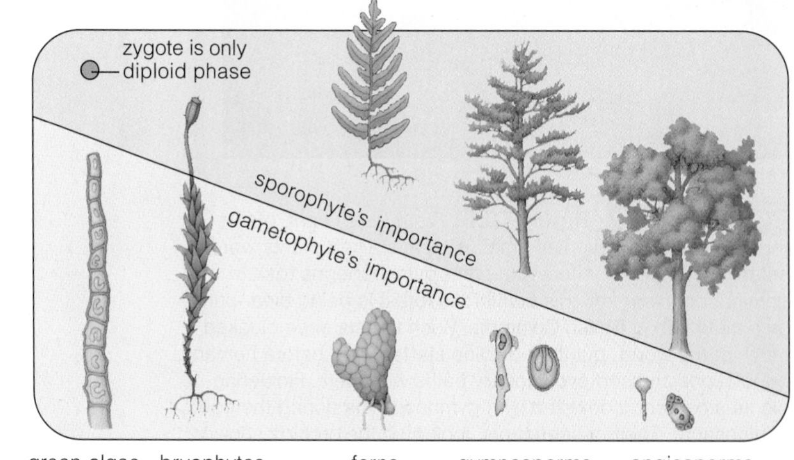

Figure 21.2 **Animated!** (**a**) Generalized life cycle for land plants, as explained in Section 9.5. (**b**) One evolutionary trend in plant life cycles. Algae and bryophytes put the most energy into making gametophytes. Groups in seasonally dry habitats put the most energy into making sporophytes, which retain, nourish, and protect the new generation through harsh times.

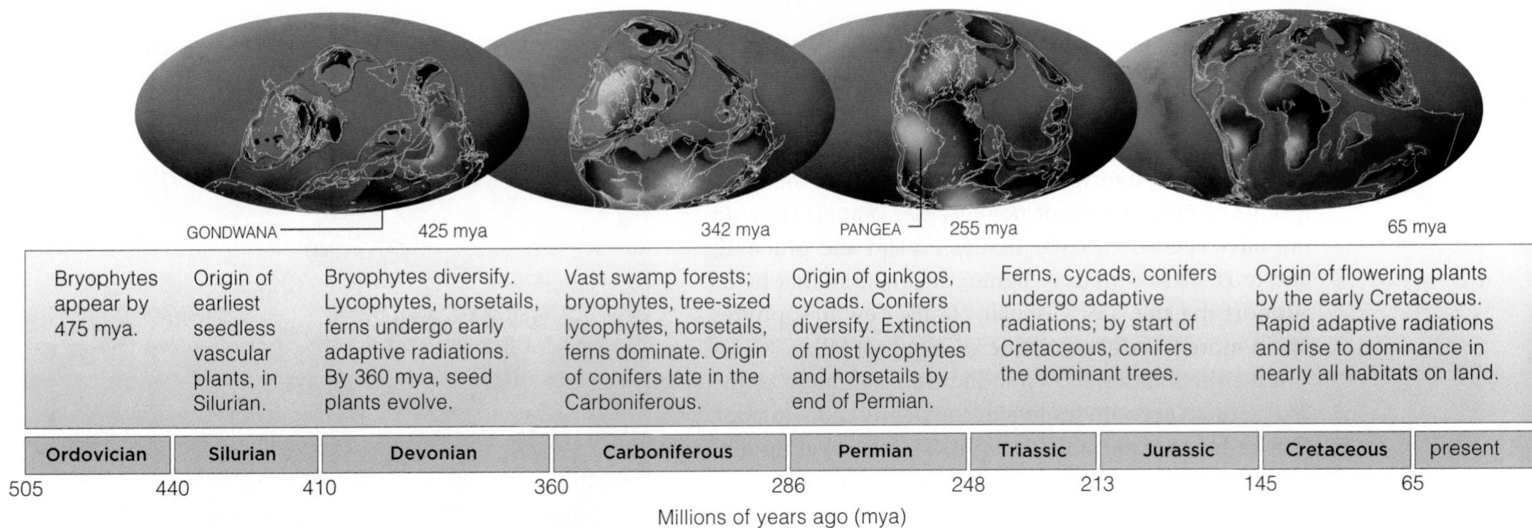

Figure 21.3 Animated! *Above*, overview of milestones in plant evolution. *Right*, nested monophyletic groups of plants. Each branching in this cladogram represents the origin of unique traits in response to novel or changing conditions in the environment.

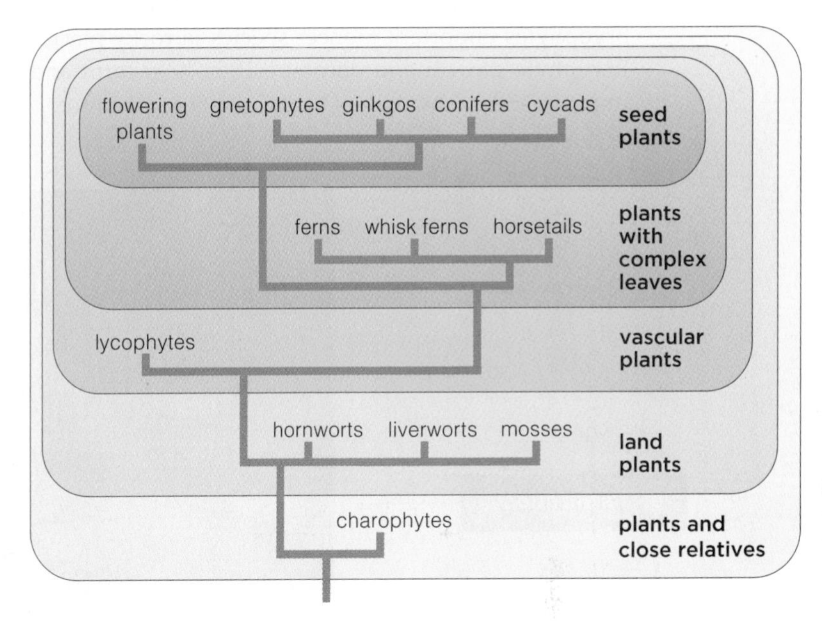

tissues not only helped a plant to distribute materials inside its body, they also helped it stand upright. This gave vascular plants an advantage in competition for sunlight. The most successful vascular plant lineage evolved leaves, which increased their surface area for intercepting sunlight and for gas exchange.

POLLEN AND SEEDS

Reproductive traits also gave some vascular plants a competitive edge. All nonvascular plants—and some vascular ones such as ferns—release spores. However, seed-bearing vascular plants hold and protect spores. They make two kinds of spores. One kind gives rise to an egg-producing female gametophyte. The other kind develops into a pollen grain: a walled, immature male gametophyte that will give rise to sperm.

Pollen grains are released. They travel to eggs on the same or another plant with the help of air currents or animals, especially insects. Production of pollen put seed plants at an advantage in dry environments. Plants that do not produce pollen need water to allow their sperm to swim to eggs. Pollen allowed the seed plants to reproduce even when water was scarce.

Seeds also helped survive dry times. A seed is an embryo sporophyte and some nutritive tissue enclosed inside a waterproof coat. It is no coincidence that seed plants rose to dominance during the Permian, when all land masses combined as Pangea and warm, dry climates prevailed (Figure 21.3).

One of the first seed-bearing lineages gave rise to gymnosperms; pine trees and other conifers are among their descendants. Another lineage, the angiosperms, arose later. They alone make flowers, and they are the most widely distributed and diverse plant group.

Plants became structurally adapted to life on land. Most have root and shoot systems, a waxy cuticle, stomata, and lignin-reinforced vascular tissues.

Sporophytes with well-developed roots, stems, and leaves came to dominate the life cycle of most plants.

Seed plants are the most recently evolved plant lineage. Their male gametes can be dispersed without standing water and their embryos are protected inside seeds.

21.2 The Bryophytes—No Vascular Tissues

The first plants were bryophytes. Like modern species, they were low-growing plants that required the presence of water droplets to reproduce sexually.

LINK TO SECTION 16.10

Today, mosses, liverworts, and hornworts are the main groups of bryophytes, or nonvascular plants. They do not have special vascular tissues (xylem and phloem), and most do not produce lignin, the material that helps support the stems of vascular plants. Few bryophytes stand more than 20 centimeters (8 inches) tall.

Like the mosses shown in Figure 21.4, many of the 24,000 or so bryophytes live in constantly moist places. Others tolerate periodic drought. They shrivel up and become dormant when water gets scarce, then resume growth when rains return.

Drought tolerance and wind-dispersed spores make bryophytes important pioneer species in rocky places. Bryophyte growth and decomposition help form and improve soil, allowing less hardy plants to take root.

Figure 21.4 Moss plants growing on damp rocks. Most bryophytes live in continually or seasonally moist habitats.

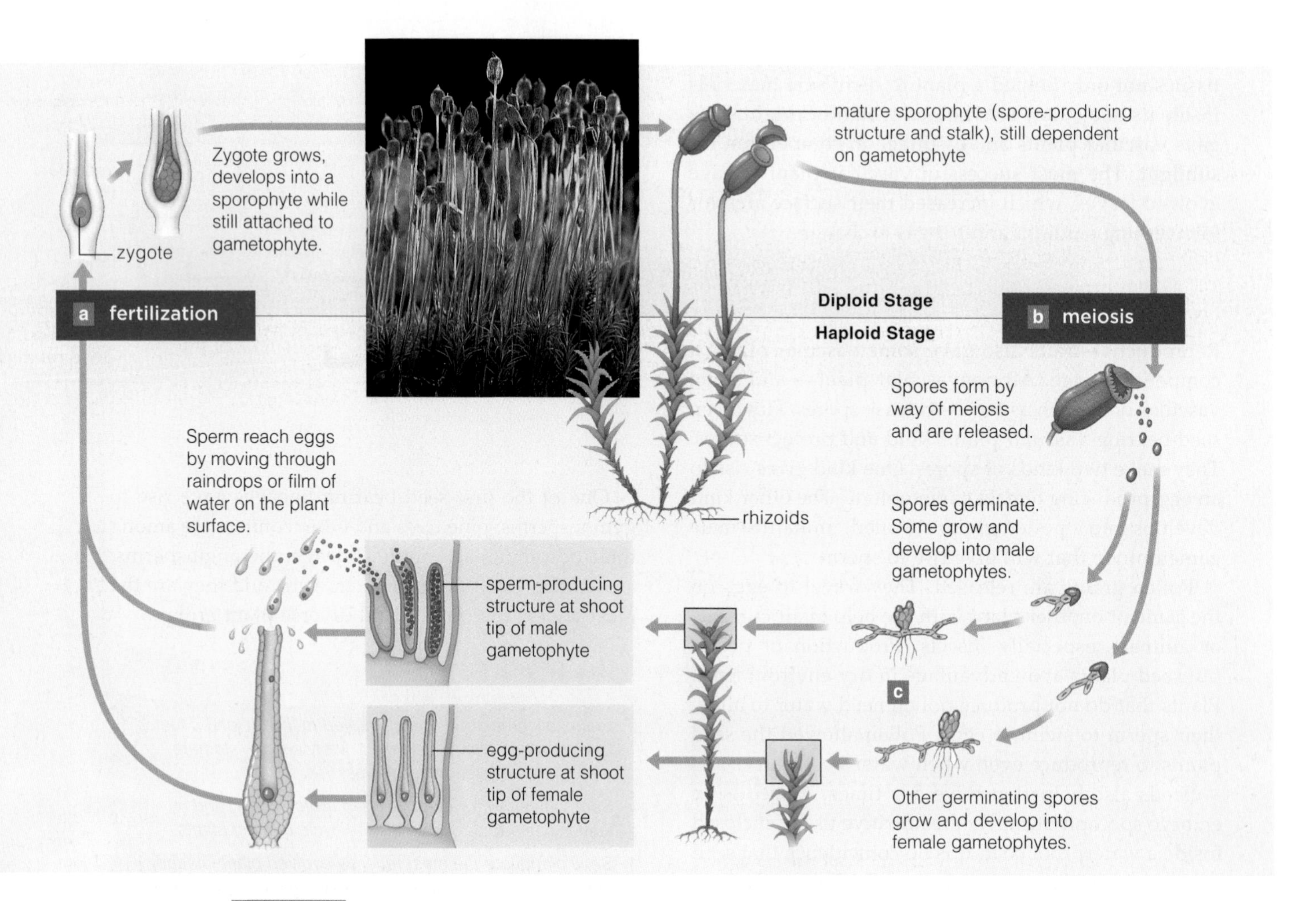

Figure 21.5 Animated! Life cycle of a moss (*Polytrichum*). The sporophyte remains attached to the gametophyte, which provides it with nutrients and absorbed water.

A haploid gametophyte is the largest, longest lived, and most structurally complex phase of the bryophyte life cycle. For example, the gametophytes of hair cap moss (*Polytrichum*) have leaflike, photosynthetic parts attached to a stemlike structure (Figure 21.5). Rootlike parts called rhizoids soak up and store moisture while anchoring the gametophytes to soil or surfaces.

Bryophytes cannot reproduce sexually unless water splashes or coats them. Sperm form inside structures at the tip of a male gametophyte, then swim to eggs in chambers at the tip of a nearby female gametophyte. As in all other plants, eggs and embryos form inside multicelled structures. This is the defining trait of the embryophytes, the clade of land plants (Section 16.10).

After the sporophyte forms, it remains attached to the gametophyte, which usually continues nourishing it. A moss sporophyte consists of a nonphotosynthetic stalk and a thin-walled capsule. Spores form inside the capsule by way of meiosis. Wind disperses the spores, they germinate, and the life cycle starts again.

The 350 or so species of peat mosses (*Sphagnum*) are of great ecological and commercial importance. In cold and temperate regions, their remains can pile up for decades forming vast peat bogs (Figure 21.6). The soil in a peat bog can be as acidic as vinegar. Only acid-tolerant plants, such as cranberries, survive alongside the mosses. Most bacterial and fungal decomposers do not grow well in this acidic habitat. That is why well-preserved, 2,000-year-old human remains occasionally turn up in European peat bogs. The acidity kept the bodies from decomposing.

Peat moss is acidic enough to serve as an antiseptic, and it is highly absorbent. When World War I shut down shipping to Europe, cotton was scarce and peat moss was used to make bandages. It is now a common ingredient in planting mixes. Cells of the dead moss absorb water and slowly release it to plant roots.

Liverworts and hornworts are less well known than mosses. Their gametophytes are ribbonlike or leaflike (Figure 21.7). Some liverworts reproduce asexually by dispersing clumps of cells that develop inside cups on the gametophyte's surface.

Bryophytes are small, structurally simple plants with no vascular tissues. Most species grow in fully or seasonally moist habitats. Their flagellated sperm reach nonmotile eggs by swimming through water films or droplets.

A sporophyte grows from gametophyte tissues and draws nutritional support from it. Only among bryophytes does the sporophyte remain attached to a larger gametophyte.

Figure 21.6 (**a**) Peat bog in Ireland. This family is cutting blocks of peat and stacking them to dry as a home fuel source. Most bryophytes are slow growing. Peat mosses grow fast enough to yield 13 tons of organic matter per hectare annually, about twice the yield from corn plants. Nova Scotia alone exports 200,000 bales of peat per year to Japan and the United States. Nearly all harvested peat is burned to generate electricity in power plants, which releases fewer pollutants compared to coal burning.

(**b**) Peat moss (*Sphagnum*). You can clearly see four sporophytes (the brown, jacketed structures on the white stalks) attached to the pale gametophyte.

Figure 21.7 (**a**) One of the hornworts. A hornlike sporophyte is attached to a ribbonlike gametophyte. (**b**) *Marchantia*, a liverwort. Like other liverwort species, this nonvascular plant reproduces sexually. Its gametophytes are male or female. (**c**) *Marchantia* also reproduces asexually by way of gemmae, clumps of cells that are produced in cups on the leaflike body. If raindrops transport a gemma to a suitable site, it grows and develops into a new plant.

21.3 Seedless Vascular Plants

Early seedless vascular plants were rootless, branching, and leafless. Many leafy and treelike forms evolved in a great adaptive radiation during the Carboniferous.

LINKS TO SECTIONS 4.10, 16.9

Like bryophytes, seedless vascular plants are species that require water during the sexual phase of their life cycle; sperm swim to eggs. Also like bryophytes, they release spores directly to the environment. Unlike the bryophytes, their sporophytes include vascular tissues and produce lignin-reinforced cell walls.

Molecular comparisons point to two major lineages of modern seedless vascular plants. Club mosses and spike mosses are lycophytes. Whisk ferns, horsetails, and ferns are now grouped as pteridophytes, a term once used exclusively for the ferns.

CLUB MOSSES AND SPIKE MOSSES

Tree-sized lycophytes grew during the Carboniferous, and club mosses and spike mosses are among their far smaller modern descendants. Despite their common names, these are vascular plants, not mosses.

All lycophytes share a unique trait: aboveground stems have microphylls, or tiny leaves with a single unbranched vein. Stems and roots extend out from a rhizome, an underground stem. Spores form within small, cone-shaped structures called strobili (singular, strobilus). As you will learn later, some gymnosperms also produce spore-bearing strobili.

Most of the 1,200 or so lycophyte species are club mosses. They typically grow on moist forest floors. You many have heard of "ground pines," a type of club moss that grows in North America's hardwood forests. They are species of *Lycopodium* (Figure 21.8*a*). Most spike mosses (*Selaginella*) live in tropical rain forests. Some live in deserts and are the most drought-tolerant vascular plants. People call them resurrection plants. When water gets scarce, the plants curl up and turn brown. When the rains return, they uncurl, turn green with new chlorophylls, and resume growth.

WHISK FERNS AND HORSETAILS

Whisk ferns (*Psilotum*) are native to the southeastern United States. They have underground rhizomes and their branching photosynthetic stems appear leafless (Figure 21.8*b*). Spores form inside sporangia (singular, sporangium), multicelled chambers on the stems. You may have noticed whisk ferns in bouquets. There is a brisk commercial market for their unusual branches.

Tree-sized sphenopytes also grew in Carboniferous swamp forests. About 25 smaller species of *Equisetum* persist. You may know them as horsetails or scouring rushes (Figure 21.8*c–e*). They grow best in acidic, wet soil, but some thrive in soggy patches of meadows, fields, and grasslands, or in well-drained gravelly or sandy soils along railroad tracks, roads, and streams.

A horsetail has rhizomes, hollow stems, and tiny nonphotosynthetic leaves. Photosynthesis occurs in the stems and leaflike branches. Silica deposits inside cell walls support the plant and give stems a sandpapery texture. When American pioneers were moving west, they used scouring rush stems as pot scrubbers.

Depending on the species, *Equisetum* strobili either form at tips of photosynthetic stems or colorless ones that do not make chlorophylls. Each spore gives rise to a gametophyte not much bigger than a pin head.

Figure 21.8 Seedless vascular plants. (**a**) A club moss (*Lycopodium*), one of the modern lycophytes. (**b**) A whisk fern (*Psilotum*). Horsetails: (**c**) The vegetative stem of *Equisetum arvense*. Its profuse rhizomes make it a vexing weed in parts of North America and Europe. (**d**) Stems and (**e**) strobilus of scouring rushes (*E. hyemale*).

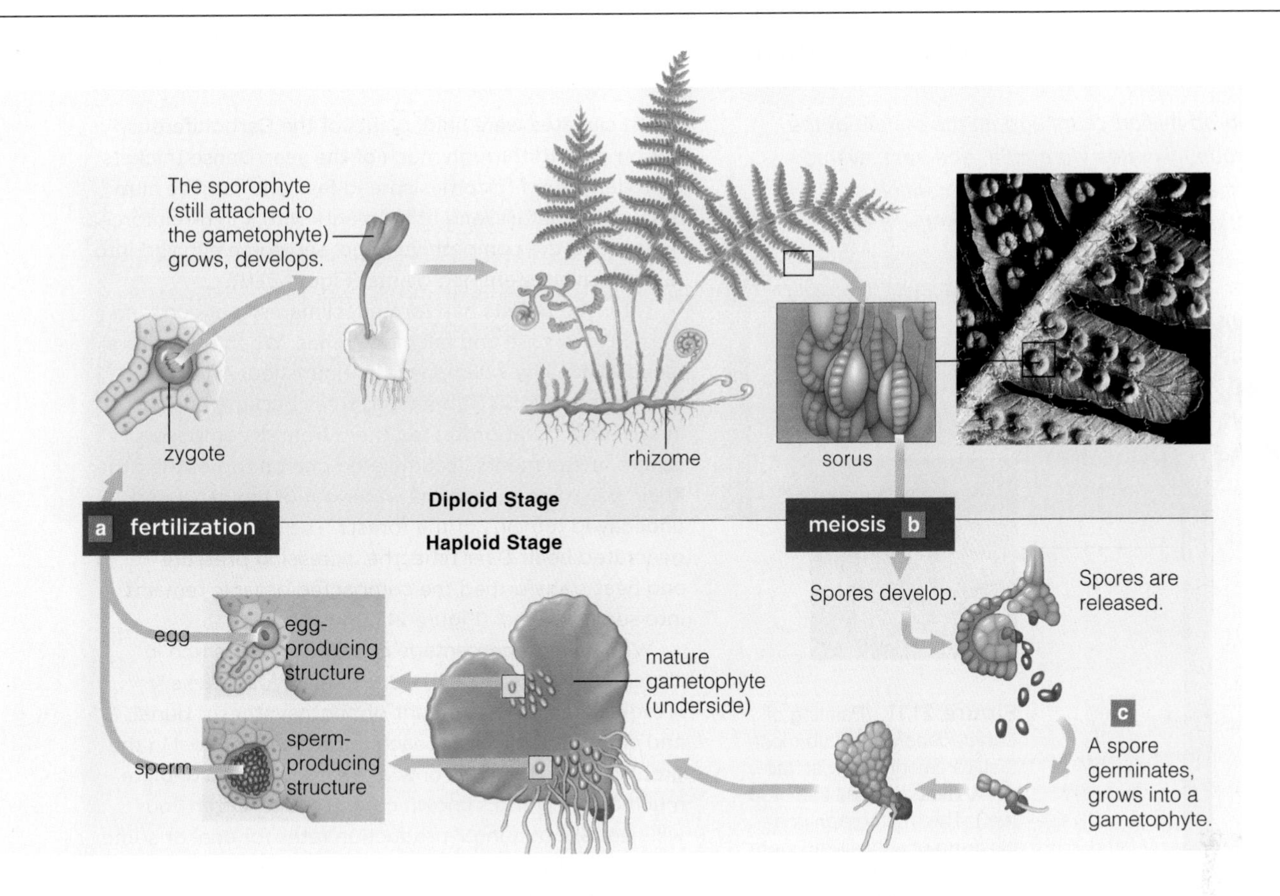

Figure 21.9 **Animated!** Chain fern (*Woodwardia*) life cycle.

(**a**) After swimming sperm reach eggs, fertilization results in a diploid zygote. The zygote is the start of a sporophyte with much-admired fronds.

(**b**) Many sori form on the underside of fronds. Each sorus is a cluster of small chambers in which spores form by way of meiosis.

(**c**) After the spores are released, they germinate and develop into small gametophytes that have a distinctive heart shape.

FERNS—NO SEEDS, BUT MUCH DIVERSITY

With 12,000 or so species, ferns are the most diverse seedless vascular plants. All but 380 or so species live in the tropics. Most fern sporophytes have leaves and roots that grow out from rhizomes (Figure 21.9). Fern leaves, also known as fronds, often start out in a tight coil known as a "fiddlehead" before unfurling.

Sori (singular, sorus) are clusters of spore-forming chambers that form only on the lower surface of fern fronds. They spring open and catapult haploid spores through the air. After germination, a spore develops into a heart-shaped gametophyte that is no more than a few centimeters across (Figure 21.9*c*).

Fern sporophytes vary greatly in structure and size (Figure 21.10). Some floating ferns have fronds only 1 millimeter long, but tree ferns can be 25 meters high. Fern fronds can be swordlike or divided into leaflets. Many tropical ferns are epiphytes. Such plants attach to and grow on a trunk or branch of another plant but do not withdraw any nutrients from it.

Large, independent sporophytes with xylem, phloem, and leaves dominate the life cycle of seedless vascular plants.

Tiny, independent gametophytes produce sperm that swim to eggs through films or droplets of water.

Ferns are the most diverse seedless vascular plants.

Figure 21.10 A sampling of fern diversity (**a**) The floating fern *Azolla pinnata*. The whole plant is not as wide a finger. Chambers in the leaves shelter nitrogen-fixing cyanobacteria. Southeast Asian farmers grow this species in rice fields as a natural alternative to chemical fertilizers. (**b**) Bird's nest fern (*Asplenium nidus*), one of the epiphytes. (**c**) Lush forest of tree ferns (*Cyathea*) in Australia's Tarra-Bulga National Park. A tree fern's "trunk" is an enlarged rhizome.

21.4 Ancient Carbon Treasures

FOCUS ON THE ENVIRONMENT

LINK TO SECTION 16.4

Three hundred million years ago, in the middle of the Carboniferous, climates were mild, and vast swamp forests formed in the wet lowlands of continents. They were the start of the coal we dig out of the ground.

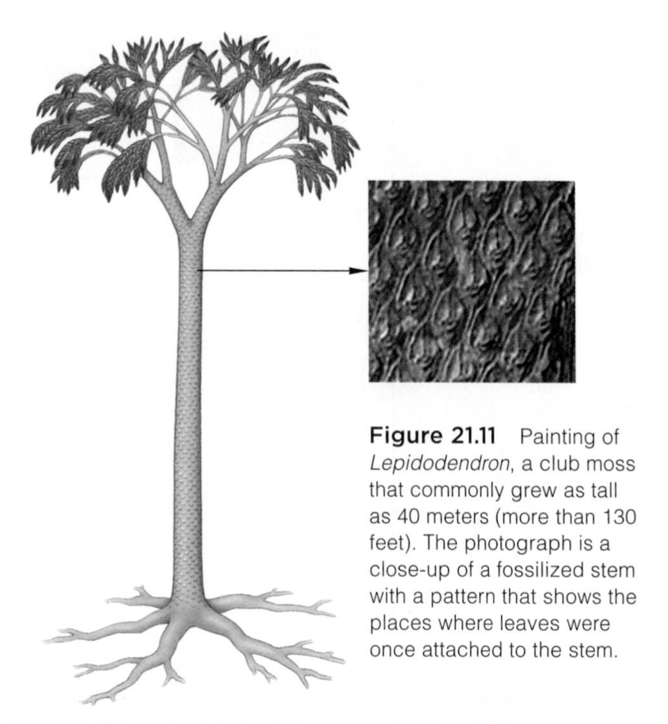

Figure 21.11 Painting of *Lepidodendron*, a club moss that commonly grew as tall as 40 meters (more than 130 feet). The photograph is a close-up of a fossilized stem with a pattern that shows the places where leaves were once attached to the stem.

When climates were mild, plants of the Carboniferous, put on growth through much of the year. Dense thickets of underground rhizomes spread fast, and far. The club mosses, horsetails, and other plants with lignin-reinforced tissues had the competitive edge, and some evolved into tall, massively stemmed giants (Figure 21.11).

After the forests had formed, climates changed, and the sea level rose and fell many times. When the waters receded, steamy swamp forests flourished. After the sea moved back in, submerged trees became buried in sediments that protected them from decomposers. Layers of sediments accumulated one on top of the other. Their weight squeezed the water out of the saturated, undecayed remains of the forests. The compaction generated heat. Over time, the increasing pressure and heat transformed the compacted organic remains into seams of coal (Figure 21.12).

With its high percentage of carbon, coal is rich in stored energy and one of our premier "fossil fuels." It took a staggering amount of photosynthesis, burial, and compaction to form each major seam of coal in the ground. It has taken us only a few centuries to deplete much of the world's known coal deposits. Often you will hear about annual production rates for coal or some other fossil fuel. How much do we really produce each year? We produce nothing. We simply extract it from the ground. Coal is a nonrenewable source of energy.

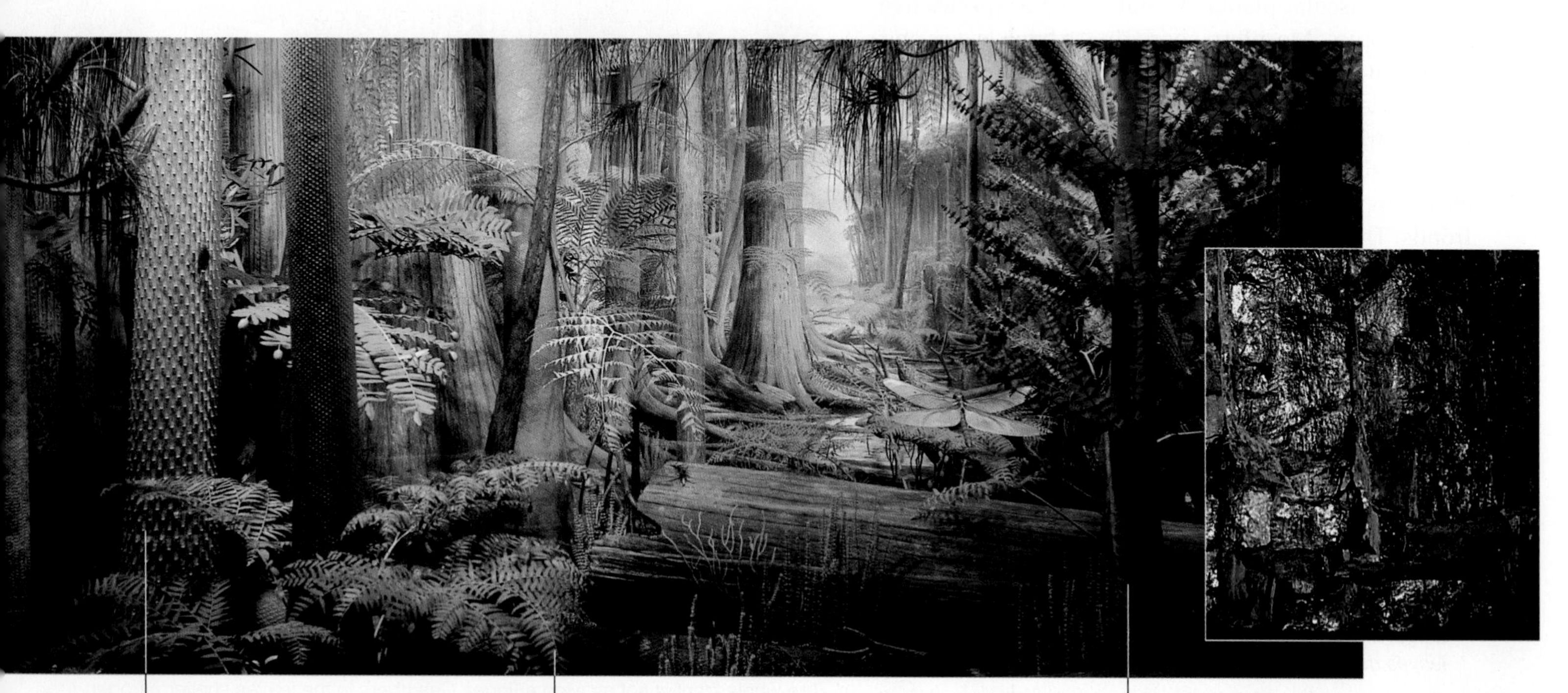

Figure 21.12 Reconstruction of a Carboniferous forest. *Right*, photograph of part of a seam of coal.

21.5 The Rise of Seed-Bearing Plants

In diversity, numbers, and distribution, seed producers became the most successful groups of the plant kingdom.

Seed-bearing plants arose late in the Devonian. Early species were vinelike or shrubby. Most did not make it past the Carboniferous (Figure 21.12). Others gave rise to cycads and other gymnosperms. (What some people call the Age of Dinosaurs, botanists call the Age of Cycads.) Later, by about 120 million years ago, flowering plants arose and started down the road to supremacy in most environments.

Think back on the factors that promoted the rise of seed plants. Structural modifications, such as a water-conserving cuticle, were one reason seed plants did well even in seasonally dry habitats. Just as important were their two kinds of spores, formed by meiosis.

The microspores become pollen grains. Again, these walled male gametophytes consist of a few cells, one of which produces sperm. As they journey on their own to eggs, pollen grains can withstand drought and cold. Air currents or animals such as insects easily disperse them. Pollination is the arrival of pollen grains on the female reproductive parts of a seed plant.

After pollen grains evolved, it no longer mattered whether water was splashing around or forming films on the plant. Sperm reached eggs without it.

Unlike microspores, megaspores form inside ovules and stay attached to a parent plant. An ovule starts as a small mass of sporophyte tissue. A megaspore inside the mass gives rise to a female gametophyte that has an egg cell. Fertilization occurs and an embryo forms inside the ovules. *Each seed is a mature ovule.* The seed coat derived from ovule tissues protects the embryo sporophyte after it has been released from a parent plant. Nutrient-rich tissue inside the seed jump-starts the embryo's renewed growth as soon as conditions favor germination.

A footnote to this story: Many seed plants have our help in dispersing seeds. Some had recruited humans by 500,000 years ago; fossils show that *Homo erectus* was roasting seeds and stashing nuts and rose hips in caves. By 11,000 years ago, modern humans had started to domesticate seed plants as reliable sources of food.

We now recognize 3,000 or so species as edible and grow about 150 as crops (Figure 21.13). We also grow thousands more to decorate our homes and gardens. We grow others as sources of oils for perfumes and medicines and balms. Juices of *Aloe vera* leaves soothe rough skin, digitalin from foxgloves makes the heart contract harder, and chemicals from periwinkle and yew trees slow the growth of certain cancers. People grow tobacco, marijuana, opium poppies, and coca (the source of cocaine) for their mind-altering properties. They use flax, cotton, and hemp to create fabrics and carpets and ropes. In such ways, humans promote the evolutionary success of many seed-bearing plants.

In diversity, numbers, and distribution, seed plants are the most successful plants. Structural modifications, pollen grains, and seeds helped gymnosperms and, later, flowering plants radiate into dry habitats.

pine pollen grains

Figure 21.13 Edible treasures from flowering plants. (**a**) Fruits, which function in seed dispersal. (**b**) Mechanized harvesting of bread wheat, *Triticum*. (**c**) Indonesians picking shoots of tea plants (*Camellia sinensis*). Leaves of plants on hillsides in moist, cool regions have the best flavor. Only the terminal bud and the two or three youngest leaves make the best teas. (**d**) In Hawaii, a field of sugarcane, *Saccharum officinarum*. We make sugar and syrup by boiling down sap extracted from its stems.

21.6 Gymnosperms—Plants With Naked Seeds

Gymnosperms are one of the two modern lineages of seed-bearing plants. Those conifers mentioned in the chapter opening are the most well-known gymnosperms.

Gymnosperms are vascular seed plants that have no ovaries, as flowering plants do. Ovaries are chambers in which eggs form, get fertilized, and become seeds. (*Gymnos* means naked; *sperma* is taken to mean seed.)

CONIFERS

The 600 or so species of conifers are woody trees and shrubs, most with needlelike or scalelike leaves. Most shed some leaves steadily but stay *evergreen*, or leafy. A few *deciduous* kinds shed their leaves in one season. The tallest trees (redwoods), the oldest (bristlecone pine), and one of the most abundant kinds (pines) are conifers. So are firs, junipers, spruces, podocarps, and cypresses. Figures 21.14*a* and 21.15 show two species.

Conifer sporophytes make cones, microspore- or megaspore-bearing structures with woody, papery, or fleshy scales around a central axis. In between female cone scales, ovules mature into seeds. Male cones are much smaller, and most are not woody.

LESSER KNOWN GYMNOSPERMS

Cycads and ginkgos were diverse in dinosaur times. Now they are the only seed plants that have flagellated sperm. Sperm emerge from pollen grains then swim in fluid produced by the plant's ovule.

About 130 species of cycads made it to the present, mainly in the dry tropics and subtropics. Cycads look like palms or ferns but are not close relatives of them (Figure 21.14*b*). The "sago palms" commonly used in landscaping and as houseplants are actually cycads.

The only living ginkgo species is *Ginkgo biloba*, the maidenhair tree (Figure 21.14*c–e*). It is one of the few deciduous gymnosperms. *G. biloba* is native to China, but its attractive fan-shaped leaves and resistance to insects, disease, and air pollution make it a popular street tree in urban areas. Usually only male trees are planted because the seeds produced by female trees give off a strong, unpleasant odor when they decay. Some studies indicate that dietary supplements made from ginkgo leaves may slow memory loss in people who have Alzheimer's disease.

Gnetophytes include tropical trees, leathery vines, and desert shrubs. Extracts from the photosynthetic stems of *Ephedra* (Figure 21.14*g*) were sold as an herbal stimulant and a weight loss supplement until 2004. The FDA banned sale of ephedra after a few users, including baseball player Steve Bechler, died.

Welwitschia is an odd-looking gnetophyte that lives in African deserts. It has a taproot and a woody stem with strobili. Two straplike leaves can reach 5 meters in length. These leaves split lengthwise repeatedly as the plant matures (Figure 21.14*h*).

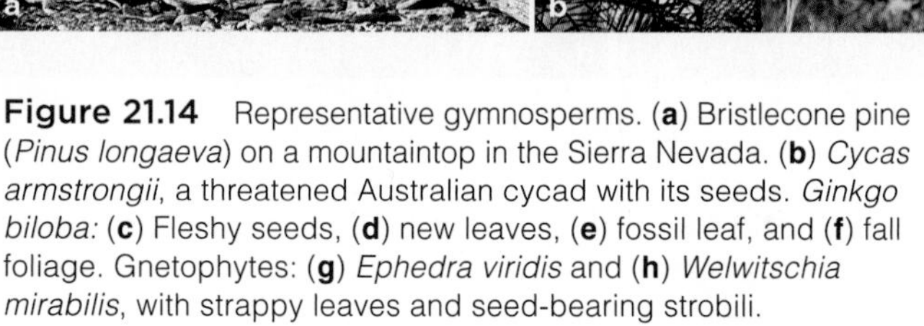

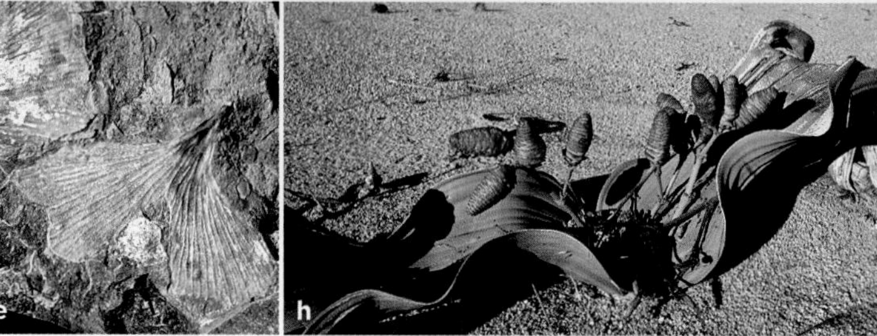

Figure 21.14 Representative gymnosperms. (**a**) Bristlecone pine (*Pinus longaeva*) on a mountaintop in the Sierra Nevada. (**b**) *Cycas armstrongii*, a threatened Australian cycad with its seeds. *Ginkgo biloba:* (**c**) Fleshy seeds, (**d**) new leaves, (**e**) fossil leaf, and (**f**) fall foliage. Gnetophytes: (**g**) *Ephedra viridis* and (**h**) *Welwitschia mirabilis*, with strappy leaves and seed-bearing strobili.

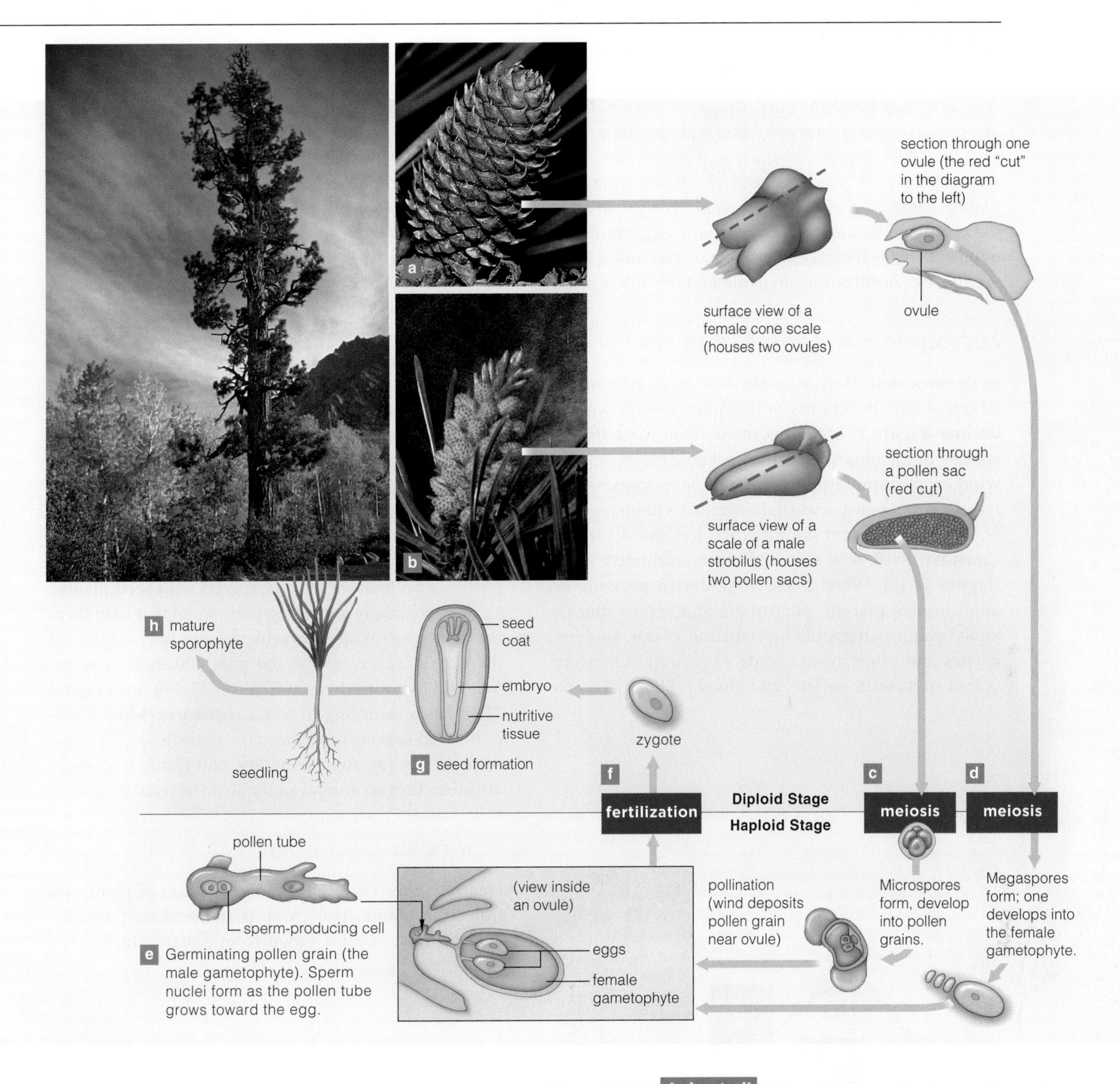

Figure 21.15 Animated! Life cycle of a conifer, the ponderosa pine.

A REPRESENTATIVE LIFE CYCLE

A pine tree is a sporophyte, and its life cycle is typical of conifers (Figure 21.15). Ovules form on the upper surfaces of scales in female cones. An egg-producing female gametophyte develops in each ovule. In male cones, microspores become winged pollen grains.

Millions of tiny pollen grains are released and drift on the winds. Pollination occurs when pollen lands on an ovule. The pollen grain germinates and some cells of the male gametophyte begin to grow, forming the pollen tube (Figure 21.15*e*). Fertilization occurs when the nucleus of a sperm cell in the tube fuses with the egg nucleus, forming a zygote. The zygote develops into an embryo sporophyte, which, along with ovule tissues, becomes the seed. The entire process, from pollination to seed release, takes about one year.

Gymnosperms are seed-bearing plants. Their eggs and seeds do not form in ovaries, as they do in flowering plants.

All gymnosperms produce pollen. In the cycads and ginkos, sperm emerge from pollen and swim through fluid released by the ovule. In other groups, the sperm are nonmotile.

21.7 Angiosperms—The Flowering Plants

Angiosperms are seed plants that produce flowers. Their seeds form inside an ovary that later matures into a fruit.

Angiosperms are vascular seed plants, and the only plants that make flowers. Their name refers to ovaries, the chambers that enclose one or more egg-producing ovules. (*Angio*– means enclosed chamber; and *sperma*, seed.) After fertilization, an ovule matures into a seed.

LINKS TO SECTIONS 16.10, 17.12

COEVOLUTION WITH POLLINATORS

In the Mesozoic, flowering plants began a spectacular adaptive radiation even as other plant groups were in decline (Figure 21.16). Now there are at least 260,000 species in meadows, forests, parched deserts, and on windswept mountaintops. A few live in streams and lakes, salt marshes, and shallow marine habitats.

What accounts for their distribution and diversity? Consider the flower, a specialized reproductive shoot (Figure 21.17). When plants first started growing on land, insects that ate plant parts and spores quickly joined them. After pollen-producing plants evolved, beetles and other insects made a connection between "plant parts with pollen" and "food." Plants did give up some pollen but gained a reproductive edge. How? Insects crawling on flowers were dusted with pollen, which they brushed directly onto female plant parts. Beetles made more reliable deliveries than winds did.

Later still, flowering plants evolved in ways that made them more attractive to insects, as with vibrant colors, strong fragrances, and plentiful nectar. Instead of randomly looking for food, some insects specialized on particular floral species. These insects made more direct pollen deliveries. Not coincidentally, the great adaptive radiation of flowering plants coincided with an adaptive radiation of insects.

Coevolution refers to two or more species jointly evolving because of their close ecological interactions. Heritable changes in one exert selection pressure on the other, which also evolves. That is how pollinators evolved. Pollinators are animals that move pollen of one plant species onto female reproductive structures of the same species. Insects, some bats, and many birds are pollinators that coevolved with seed plants.

By successfully recruiting pollinators that help them reproduce sexually, flowering plants have remained the dominant group for the past 100 million years. You will focus on them in Chapter 27. For now, Figure 21.18 has a sampling of floral structures. Most wind-pollinated species have inconspicuous flowers. Flashy, colorful flowers, sugary nectars, and heady fragrances are clues that an animal pollinates the plant.

FLOWERING PLANT DIVERSITY

Nearly 90 percent of all modern species of plants are flowering plants. The group is tremendously diverse even in size. Species range from aquatic duckweeds

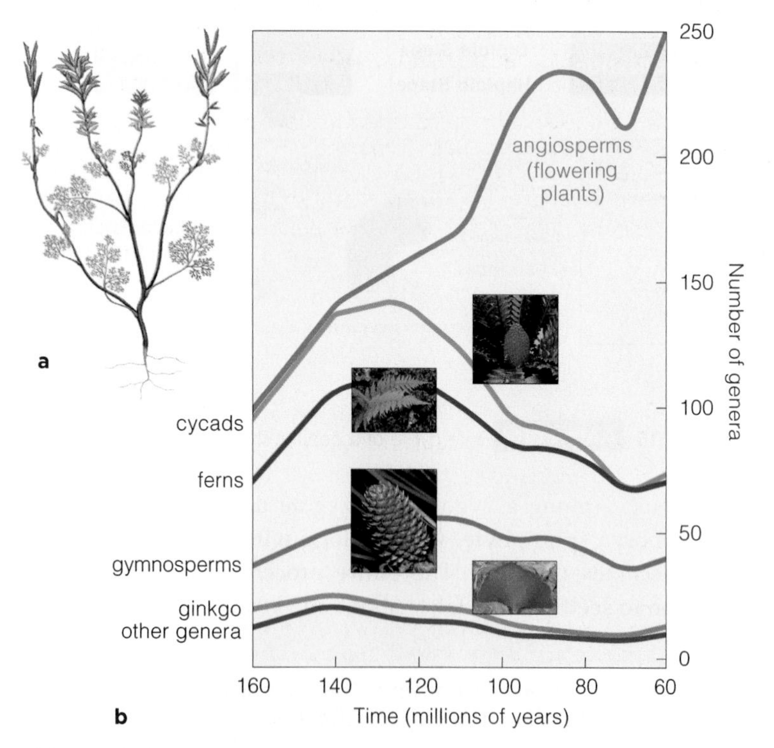

Figure 21.16 (**a**) Sketch of *Archaefructus sinensis*, one of the earliest known flowering plants. Apparently it grew in shallow lakes. (**b**) Diversity of vascular plants in Mesozoic times. Conifers and other gymnosperms started to decline even before flowering plants started their major adaptive radiation.

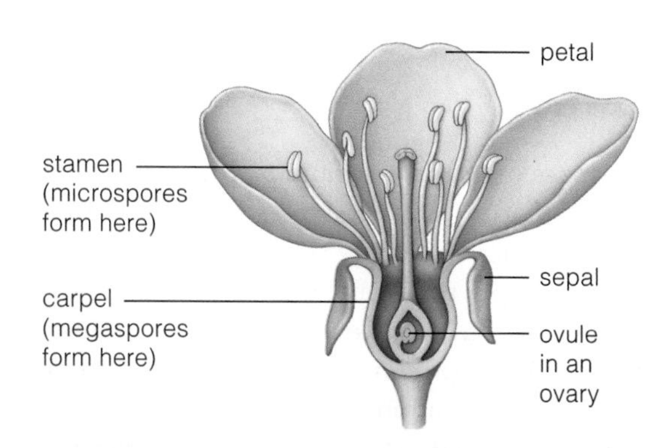

Figure 21.17 Structures of a typical modern flower. Like the earliest flowering plants, it has male and female parts (stamens and carpels). Unlike them, it also has petals and sepals.

1 millimeter long to *Eucalyptus* trees 100 meters tall. A few are parasites that withdraw nutrients from other plants (Figure 21.18*e*). Pitcher plants and some others in nitrogen-poor habitats attract, trap, and dissolve insects, then absorb the nutrients (Figure 5.11).

Flowering plants were once divided into only two groups based on their number of cotyledons, or seed leaves, that form on embryo sporophytes. Plants with one cotyledon were named monocots; those with two were named dicots. It now appears that the monocots branched from a more ancient dicot lineage.

Researchers have now identified the oldest lineages of flowering plants. There are three, as represented by their modern descendants: water lilies, star anise, and *Amborella* (Figure 21.18*f*).

Genetic divergences gave rise to other groups that became dominant: magnoliids, eudicots (true dicots), and monocots.

Among the 9,200 or so magnoliids are magnolias and avocado trees. The most diverse group, eudicots, has about 170,000 species. It includes most herbaceous (nonwoody) plants such as lettuces, cabbages, crocus, and daisies, and cacti. Most flowering shrubs or trees, such as roses, maples, oaks, elms, and fruit trees are eudicots. Among the 80,000 named monocot species are palms, lilies, grasses, and orchids, such as the one shown in Figure 21.1. Sugarcane and cereal grasses—especially rice, wheat, corn, oats, and barley—are the most important monocot crop plants.

Figure 21.18 (**a**) Representative flowers: reproductive structures with roles in pollination and seed formation. Their colors, patterns, and shapes attract pollinators. (**b**) This hummingbird pollinator has a long bill that fits the long, delicate nectar tube of the flower of a columbine (*Aquilegia*).

(**c**) Sacred lotus (*Nelumbo nucifera*), an aquatic species. The radial pattern of petals is typical of ancient eudicot lineages. (**d**) The more recent eudicot lineages, including pansies (*Viola*), have a bilaterally symmetrical flower with roughly equivalent left and right parts. (**e**) Dwarf mistletoe (*Arceuthobium*) is a highly specialized, nonphotosynthetic eudicot that parasitizes conifers. Colorless flowers produce droplets of nectar that lure insect pollinators.

(**f**) Evolutionary tree diagram for flowering plants.

Angiosperms are the most successful plants. They alone produce flowers. These specialized reproductive structures have pollen-producing male parts and female parts with ovaries, protective chambers in which seeds develop.

Flowers coevolved with pollinators, which have contributed greatly to the reproductive success of angiosperms.

Magnoliids, eudicots, and monocots are the main lineages of flowering plants. The largest group, eudicots, includes 170,000 named species.

21.8 Focus on a Flowering Plant Life Cycle

A flower-producing sporophyte dominates the life cycle of flowering plants.

Chapters in Unit Five offer a closer look at flowering plant structure and function. For now, begin thinking about the flower-producing sporophyte that dominates angiosperm life cycles. Figure 21.19 shows a monocot life cycle. Section 28.2 shows the life cycle of one of the eudicots. The seeds of both kinds of angiosperms contain endosperm, a nutrient-rich tissue that will be tapped during early growth of the new sporophyte. A seed forms in an ovary, which matures into a fruit. Only flowering plants make fruits. Fragrant, colorful, hard-shelled, spiny, winged, sticky—no matter what their features, all fruits function in dispersing the new generation of sporophytes, as Chapter 28 explains.

Flowering plant life cycles involve formation of pollen and eggs, fertilization, and the packaging of embryo sporophytes in seeds. Ovaries mature into fruits that function in seed dispersal.

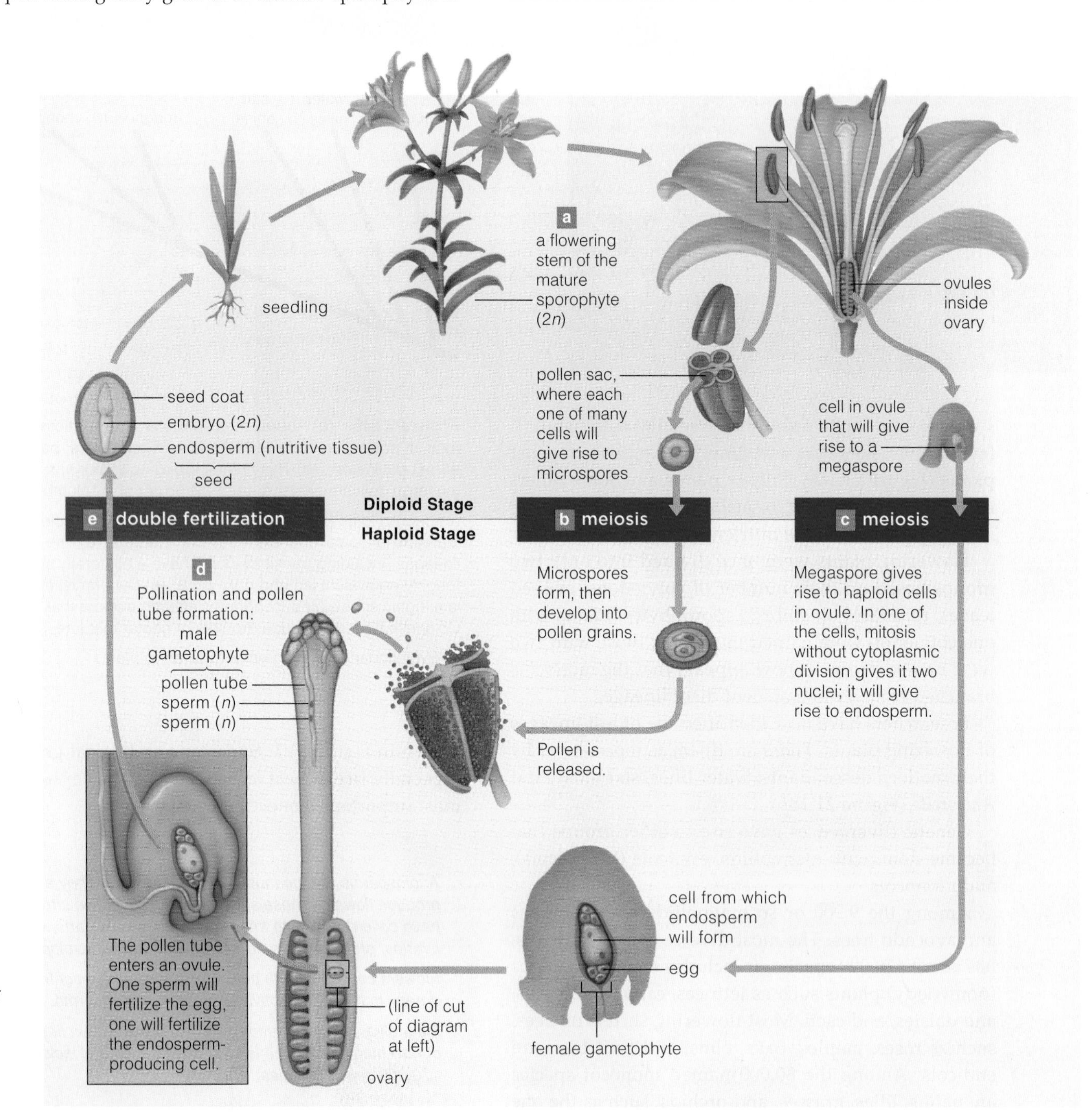

Figure 21.19 Animated! Life cycle of lily (*Lilium*), one of the monocots.

(**a**) The sporophyte dominates this life cycle. (**b**) Pollen forms in pollen sacs. (**c**) Eggs develop in an ovule within an ovary. (**d**) Pollination, occurs and a tube grows from the pollen grain into the ovule, delivering two sperm.

(**e**) Double fertilization occurs in all flowering plant life cycles. One sperm fertilizes the haploid egg. The other fertilizes a diploid cell that gives rise to the endosperm, a tissue that will nourish the embryo sporophyte.

21.9 The World's Most Nutritious Plant

FOCUS ON SCIENCE

This chapter introduced you to the tremendous range of plant diversity. Despite the diversity, only about 150 species are grown commercially as crops. Of those, only 12 are the main food source for human populations around the world. Some botanists are putting knowledge of plant genetics to use to help counter malnutrition.

Alejandro Bonifacio was raised in poverty in rural Bolivia. As a child he spoke a language that predates the Incas. He learned Spanish before college. He earned a bachelor's degree in agriculture and became a plant breeder for Bolivia's department of agriculture.

His research interest is *Chenopodium quinoa*, a plant that originated in the Andes. Quinoa (pronounced keen-wa) is a leafy eudicot, a distant relative of spinach and beets. Its nutritious seeds are not a cereal grain, but for many thousands of years, they have been a staple of Latin American diets. Together with corn and potatoes, quinoa helped feed the great Inca civilization.

Quinoa seeds are 16 percent protein, on average. Some contain more. Wheat seeds are about 12 percent protein, and rice seeds, 8 percent. Quinoa has all amino acids that humans require, wheat and rice proteins are deficient in lysine. Quinoa also has more iron than most cereal grains, and a lot of calcium, phosphorus, and many B vitamins.

Just as important, quinoa plants are easy to grow. They are highly resistant to drought, frost, and salty soils. Quinoa is the only food crop that can grow in the arid salt deserts that prevail in much of Bolivia.

Far to the north, even before Alejandro received his college scholarship, Daniel Fairbanks became a botanist at Brigham Young University (BYU). He, too, saw quinoa's potential to feed millions in Peru and Bolivia, the most impoverished places in Latin America. Most families are subsistence farmers, and kwashiorkor is common.

Kwashiorkor is one form of malnutrition caused by protein-deficient diets, and it is common in developing countries. Fatigue, drowsiness, and irritability are early symptoms. Ongoing protein deficiency stunts growth. Tissues swell with excess fluid, muscles lose mass, and the stomach often protrudes. The immune system starts to weaken. Epidermal conditions, are common. In severe cases, mental and physical problems lead to coma and death.

In 1991, Alejandro and Daniel met at a conference on Andean crops and became friends. Later, the World Bank awarded Alejandro a fellowship to study in the United States, and Dan became his advisor at BYU. Alejandro earned his PhD and learned a third language—English.

The two are now codirectors of an international research program with a holistic approach to improving quinoa production for farmers locked in poverty. They collect quinoa strains and look for ways to conserve, improve, and use genetic diversity. They are identifying the traits of each strain and researching the best way of preserving sample seeds for future study. They are developing a genetic map of quinoa.

Today, more than twenty scientists in four countries take part in this program. They research the economic impact of new strains, agricultural technologies, nutrition, hygiene, health, and community programs. They look for substitutes for chemical pesticides to control quinoa moths. They keep in mind cultural preferences for seeds of particular sizes and colors. Farmers and home cooks help evaluate new varieties.

Thousands of Bolivian families now grow more food, thanks to the new quinoa strains. Children who would have died from kwashiorkor are now attending school.

In a recent letter, Dan told us that he learned more from Alejandro than Alejandro learned from him. He appended a photograph of his colleague in a research field, standing next to one of his new quinoa varieties, so that we can put a face with the name (Figure 21.20).

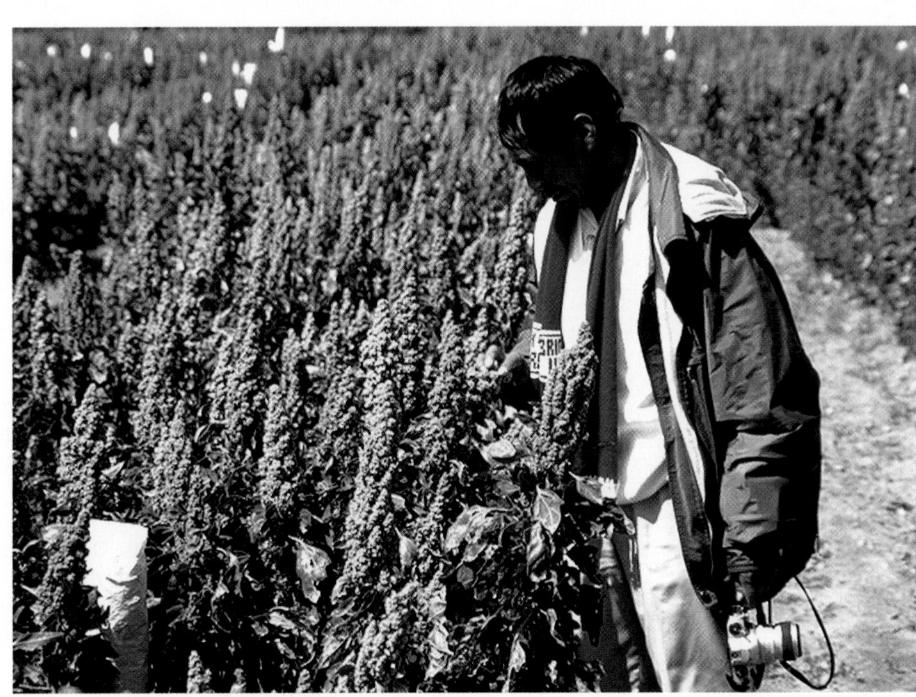

Figure 21.20 Alejandro Bonifacio checks genetically improved quinoa plants.

www.thomsonedu.com Thomson NOW!

Summary

Section 21.1 Plants evolved about 475 million years ago from charophytes, a group of green algae. Nearly all modern plants are photoautotrophs on land. The groups listed in Table 21.1 and Figure 21.21 reflect these trends:

Most groups adapted to dry and often cold habitats through structural modifications, such as stomata across epidermal surfaces, a waterproof cuticle, lignin-reinforced tissues, and xylem and phloem (vascular tissues).

A haploid body dominates life cycles of many algae and nonvascular plants. Diploid dominance emerged in most plants. Complex sporophytes evolved that retained, nourished, and protected the new generations through seasons that did not favor growth.

Production of two types of spores paved the way for the evolution of pollen grains and seeds in two lineages.

- *Use the animation on ThomsonNOW to investigate plant life cycles and plant classification.*

Section 21.2 Mosses, liverworts, and hornworts are bryophytes. They are nonvascular (no xylem or phloem). Their sperm swim through water droplets to eggs. The sporophyte remains attached to a larger gametophyte.

- *Use the animation on ThomsonNOW to observe the life cycle of a moss.*

Sections 21.3, 21.4 Lycophytes, horsetails, whisk ferns, and true ferns are seedless vascular plants. Their life cycle is dominated by the sporophyte. Spore-bearing structures include the strobili of horsetails and the sori of ferns. Sperm swim through water to reach eggs. Energy-rich, compacted remains of Carboniferous swamp forests that were dominated by giant lycophytes became coal.

Section 21.5 The gymnosperms and flowering plants (angiosperms) are seed-bearing vascular plants.

Seed plants produce microspores that become pollen grains in which sperm-producing male gametophytes develop. They make megaspores that give rise to female gametophytes (with eggs) inside ovules. Part of the ovule forms nutritive tissue and a seed coat. A seed is a mature ovule. Its tough outer coat protects the embryo sporophyte inside from harsh conditions.

Section 21.6 Gymnosperms include conifers, cycads, ginkgos, and gnetophytes. Many are well adapted to dry climates. Their ovules form on the exposed surfaces of strobili or, in the case of conifers, female cones.

Table 21.1 Comparison of Major Plant Groups

Bryophytes	24,000 species. No roots, leaves; moist, humid habitats.
Lycophytes	1,100 species. Simple leaves; mainly wet or shady habitats.
Whisk ferns	7 species, sporophytes with no obvious roots or leaves.
Horsetails	25 species of one genus. Swamps, disturbed habitats.
Ferns	12,000 species. Wet, humid, tropical or temperate regions.
Conifers	600 species. Most evergreen, woody trees. Widespread.
Cycads	130 species. Slow-growing; tropics.
Ginkgo	1 species, a tree with fleshy-coated seeds.
Gnetophytes	70 species. Limited to some deserts, tropics.
Flowering plants	
Monocots	80,000 species. Floral parts often arranged in threes or in multiples of three; one seed leaf; parallel leaf veins common; pollen with one furrow.
Eudicots (true dicots)	170,000 species at least. Floral parts often in fours, fives, or multiples of these; two seed leaves; often net-veined leaves; pollen with three or more pores and/or furrows.
Magnoliids and older groups	9,200 species. Many or few spirally arranged floral parts, arrayed in threes; two seed leaves; net-veined leaves common; pollen with one furrow.

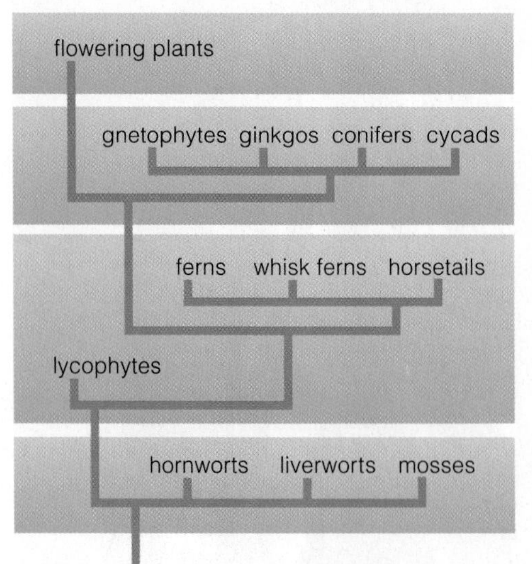

Angiosperms. Only seed-bearing vascular plants with flowers. Male gametophytes dispersed as pollen grains. Eggs and seeds develop from ovules in floral ovaries, which mature into fruits. Most flowers coevolved with animal pollinators.

Gymnosperms. Seed-bearing vascular plants without ovaries. Male gametophytes dispersed as pollen grains. Eggs, seeds develop from ovules on upper surfaces of cone scales or other reproductive structures of sporophyte.

Seedless vascular plants; xylem and phloem present. Sporophytes sometimes lack well-developed leaves. Spores that give rise to gametophytes are wind-dispersed. Sperm swim through water droplets or film of water to eggs.

Bryophytes. Nonvascular plants. Only plants with haploid dominance; sporophytes stay attached to larger gametophytes. Spores that give rise to gametophytes are wind-dispersed. Sperm swim through water droplets or film of water to eggs.

Figure 21.21 Summary of plant evolutionary trends. All groups shown have living representatives.

Figure 21.22 From clear-cut forests to new houses—where many conifers end up.

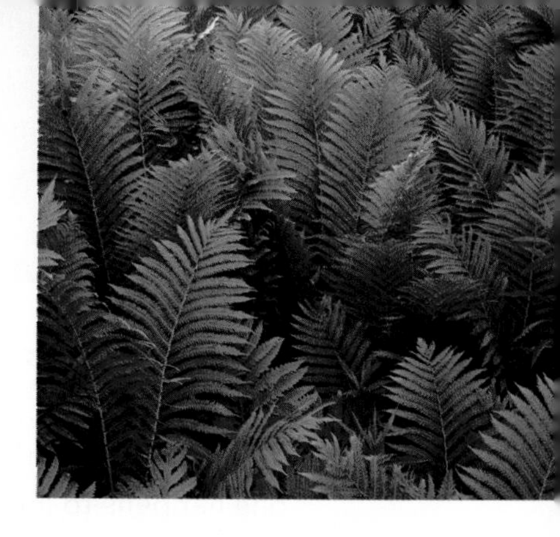

Figure 21.23 Ostrich ferns.

■ *Use the animation on ThomsonNOW to explore the life cycle of a conifer.*

Section 21.7 Angiosperms alone have flowers, and many coevolved with birds, bees, bats, and other animal pollinators. They are the most widely distributed and diverse plant group. The two largest classes are eudicots and monocots. Magnoliids are one of the early groups that preceded eudicots and monocots.

■ *Read the InfoTrac Article "Flower Power," Barbara Eaglesham, Odyssey, March 2003.*

Section 21.8 A monocot life cycle is one example of sexual reproduction in many flowering plants. Double fertilization is the start of an embryo sporophyte and of the nutritive tissue that will support it. Seeds form in ovaries. Outer ovary tissues later develop into fruits.

■ *Use the animation on ThomsonNOW to explore the life cycle of a flowering plant.*

Section 21.9 Botanists are now working to improve crops for human populations around the world.

Self-Quiz

Answers in Appendix III

1. The first plants were ______ .
   a. gnetophytes c. bryophytes
   b. pteridophytes d. lycophytes

2. Which of the following statements is false?
   a. Ferns produce seeds inside strobili.
   b. Bryophytes do not have xylem or phloem.
   c. Gymnosperms and angiosperms produce seeds.
   d. Only angiosperms produce flowers.

3. Bryophytes alone have independent ______ as well as attached, dependent ______ .
   a. sporophytes; gametophytes
   b. gametophytes; sporophytes
   b. rhizoids; stalked sporangia

4. Lycophytes, horsetails, and ferns are ______ plants.
   a. multicelled aquatic c. seedless vascular
   b. nonvascular seed d. seed-bearing vascular

5. Coal consists primarily of compressed remains of the ______ that dominated Carboniferous swamp forests.
   a. seedless vascular plants c. flowering plants
   b. seed-bearing plants d. a and c

6. The ______ produce flagellated sperm.
   a. bryophytes d. monocots
   b. lycophytes e. a and b
   c. conifers f. a through c

7. A seed is a (an) ______ .
   a. female gametophyte c. mature pollen tube
   b. mature ovule d. immature microspore

8. Which does *not* apply to gymnosperms or angiosperms?
   a. vascular tissues c. single spore type
   b. diploid dominance d. cuticle with stomata

9. Match the terms appropriately.
   ___ gymnosperm a. gamete-producing body
   ___ sporophyte b. help control water loss
   ___ lycophyte c. "naked" seeds
   ___ ovary d. only plant group that produces flowers
   ___ bryophyte e. spore-producing body
   ___ gametophyte f. nonvascular land plant
   ___ stomata g. seedless vascular plant
   ___ angiosperm h. ovules form in it

■ *Visit ThomsonNOW for additional questions.*

Critical Thinking

1. Figure 21.22 shows a Nahmint Valley forest in British Columbia after logging and wood frames of new homes. Many species that supply lumber for builders can be grown in managed tree farms. Investigate the ecological and economic impacts of tree farming. Do you think it is a better alternative than logging existing natural forests?

2. Fiddleheads (young fronds) of the ostrich fern (Figure 21.23) are sold as food, but sometimes raw or undercooked ones sicken people. Many similar looking ferns have highly toxic fiddleheads. How might production of a toxin by developing fronds benefit the fern plant?

3. Fern life cycles puzzled botanists of the 1700s. They propagated ferns from what looked like dustlike "seeds" on the undersides of fronds but could not find the pollen source. Write a note, addressed to them, to clear up their confusion.

4. The 2004 Nobel Peace Prize was awarded to Wangari Maathai, who founded the Green Belt Movement in 1977 (Figure 21.24). Since then, group members—mostly poor rural women—have planted more than 25 million trees. Reforestation efforts help halt soil erosion, and provide sources of fruits, food, and firewood. Do some research to find out if any nonprofit groups are working to protect forests in your area.

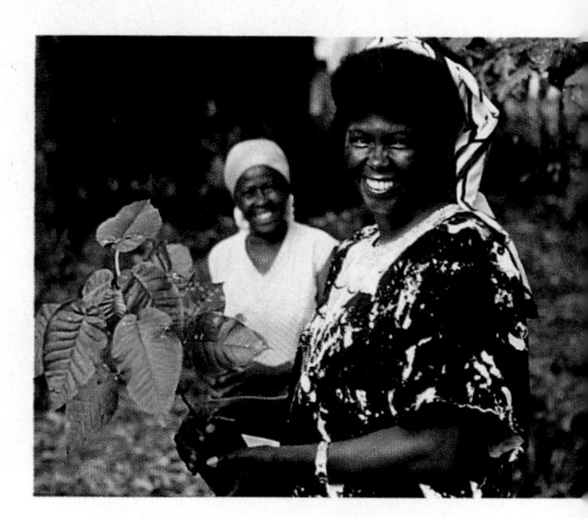

Figure 21.24 Wangari Maathai.

22 FUNGI

IMPACTS, ISSUES Food, Forests, and Fungi

Most of us do not think much about fungi unless one happens to infect us or end up on a dinner plate. However, the many members of this group have global impact. Most get energy and carbon by breaking down the wastes and remains of other organisms. Some products of this process escape into the soil, where they enhance plant growth. Fungi also feed animals. For example, the lichen known as reindeer moss (*Cladonia rangiferina*) is the primary winter food for caribou (Figure 22.1). Like all lichens, it consists of a fungus together with single-celled photosynthetic species.

During summer, caribou (*Rangifer tarandus caribou*) fatten up on plants and lichens of the arctic tundra. In fall, herds of up to 500,000 migrate to more protected forests inland and to the south. Once the caribou reach winter foraging grounds, a keen sense of smell helps them locate lichens buried in snow on the forest floor.

Humans started herding caribou about 3,000 years ago. We call the domesticated kinds reindeer. From Russia through Scandinavia, people have accompanied herds during migrations. But now northern Scandinavian forests have been cleared to make way for industry. The reindeer are herded into big trailers that can move them past cities to the remaining wintering grounds.

Whether natural or human-assisted, these migrations are testimony to connections among organisms. Reindeer are food for humans, and lichens are food for reindeer. A lichen itself is an example of a symbiosis, an enduring close association between two organisms, a fungus and alga or cyanobacterium.

Figure 22.1 Luscious lichens (*Cladonia rangiferina*), at least to caribou (*facing page*) that have migrated long distances to their wintering grounds.

How would you vote? Pollutants emitted by coal-fired power plants contribute to the decline of ecologically important lichens and fungi. Reducing emissions raises the cost of energy. Should pollution standards for power plants be tightened? See ThomsonNOW for details, then vote online.

Lichens survive in nearly all sunlit land environments, even at high latitudes, where the growing season is brief. Compared with plants, lichens are more tolerant of dry conditions. They colonize rocks, fence posts, tree trunks, and other seemingly inhospitable places.

Lichens help form soil by secreting acids that break down bedrock. The cyanobacteria of many lichens also improve soil nutrition. They capture gaseous nitrogen from the air and convert it to forms that plants can take up and use. After lichens improve the soil, plant species can move in. The plants often crowd out the hardy lichen pioneers that made their entry possible.

Collectively, lichens help trees of forest ecosystems secure nutrients. As an example, lichens of the genus *Lobaria* secure 20 percent of the nitrogen required by trees of old-growth forests in the Pacific Northwest.

Air pollution can kill lichens. They absorb harmful chemicals from airborne particles that alight on them and cannot rid themselves of the poisons. This is why environmental scientists often use lichens as natural indicators of air quality.

Lichens invite you to start thinking about the fungi, which associate in beneficial or harmful ways with many organisms. Although a minority of fungi are pathogens that infect crops and animals, webs of life would fall apart without the collective activities of the group.

Key Concepts

TRAITS AND CLASSIFICATION

Fungi are single-celled and multicelled heterotrophs more closely related to animals than to plants. They feed by secreting digestive enzymes onto organic matter in their surroundings, then absorbing the released nutrients. Multicelled species form a mesh of absorptive filaments, some of which intertwine as spore-producing structures.

Zygote fungi, club fungi, and sac fungi are three major groups. Others include chytrids, an ancient lineage with flagellated spores, and the intracellular parasites called microsporidians. **Section 22.1**

THE MAJOR GROUPS

Spore formation dominates fungal life cycles. Zygote fungi make thick-walled zygotes that give rise to sexual spores. Many sac fungi and club fungi make complex spore-bearing structures. In sac fungi, sexual spores form in a sac. In club fungi, they form at the tip of a club-shaped cell. **Sections 22.2–22.4**

LIVING TOGETHER

Many fungi live on, in, or with other species. Some live inside plant leaves, stems, or roots. Others form lichens by living with algae or cyanobacteria. **Section 22.5**

ABOUT THE NOTORIOUS ONES

A minority of fungi are parasites. Certain species cause diseases in crop plants and humans. **Section 22.6**

Links to Earlier Concepts

Before starting, review Section 16.12 to get a sense of where fungi fit in the evolutionary history of life. You will put your understanding of nutrient cycling in ecosystems to use (1.1, 1.2). You will become reacquainted with fermentation pathways (7.5), cyanobacteria (19.2), and green algae (20.8).

22.1 Characteristics of Fungi

Fungi evolved from a flagellated, amoeboid ancestor more than 900 million years ago. They are more closely related to animals than to plants.

FUNGAL GROUPS

LINKS TO SECTIONS 3.2, 16.12, 20.1

Fungi are heterotrophs that secrete digestive enzymes onto a food source, then absorb the released nutrients. Most kinds are saprobes; they feed on and decompose organic wastes and remains. As decomposers, they help cycle nutrients in ecosystems. Other kinds live in or on living organisms, with varying effects on their hosts.

Most fungi are multicelled, but some live as single cells informally called yeasts. Chitin, a polysaccharide, reinforces fungal cell walls. Most of the about 100,000 species are placed into one of the three major groups: zygote fungi (zygomycetes), sac fungi (ascomycetes), and club fungi (basidiomycetes). They are the primary focus of this chapter. Chytrids and microsporidians are more ancient fungal lineages (Figure 22.2).

Chytrids are the only modern fungi in which the life cycle includes flagellated cells. Most chytrids are freshwater saprobes or parasites. Some parasitic ones are endangering amphibians (Section 23.5). Chytrids also live in the stomach of sheep and cattle and assist their host by breaking down cellulose.

Microsporidians were once viewed as protists, but genetic comparisons place them among the fungi. All live as parasites inside animal cells. Microsporidians commonly infect people with a weak immune system, such as AIDS patients. They cause chronic diarrhea, eye infections, and urinary tract infections.

OVERVIEW OF FUNGAL LIFE CYCLES

In fungi, as in some protists, the diploid stage is the least conspicuous part of the life cycle. Depending on the fungal group, a haploid stage or dikaryotic stage dominates the cycle. "Dikaryotic" means that each cell contains two genetically different nuclei.

Most fungi disperse by producing spores. A fungal spore is a cell or cluster of cells, often with a thick wall that allows it to survive harsh conditions. Spores may form by mitosis (asexual spores) or by meiosis (sexual spores). Each fungal group produces sexual spores in a distinctive structure, a trait that aids in classification.

In multicelled species, germinating spores give rise to a mesh of absorptive filaments called a mycelium (plural, mycelia). Each filament is one hypha (plural, hyphae), a string of cells attached end to end. In some multicelled fungi, many hyphae become interwoven into spore-producing structures, as in Figure 22.3.

zygote fungi | sac fungi | club fungi | chytrids | microsporidians | FUNGI | amoeboid ancestors

Figure 22.2 Family tree for the fungi.

Fungi are single-celled or multicelled saprobes, parasites, or symbionts. All digest organic material outside their body, then individual cells absorb the released nutrients.

The haploid stage or dikaryotic stage dominates the life cycle. Dispersal occurs by way of sexual and asexual spores.

In multicelled fungi, some spores give rise to hyphae. These absorptive filaments form a mycelium, a mesh that grows in or on food. In some groups, many hyphae also become interwoven and develop into spore-producing structures.

a PURPLE CORAL FUNGUS *Clavaria* **b** RUBBER CUP FUNGUS *Sarcosoma*

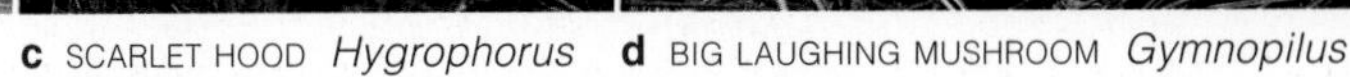

c SCARLET HOOD *Hygrophorus* **d** BIG LAUGHING MUSHROOM *Gymnopilus*

Figure 22.3 A sampling of fungal spore-producing structures on the floor of a Virginia forest.

22.2 Zygomycetes—The Zygote Fungi

Have you noticed how berries and other foods mold fast? The molds are among the zygote fungi, which make up about 1 percent of the known species of fungi.

Zygote fungi are fungi that make zygospores during sexual reproduction. Unlike club fungi and sac fungi, they do not have cross walls between the cells making up a hypha. Most zygote fungi are saprobes, but some parasitize animals, protists, or other fungi. Still others associate with plant roots and benefit the plants.

Rhizopus is a common genus. *R. nigricans* spoils our fruits and vegetables. *R. oryzae* can infect people who have diabetes or a weak immune system. It invades blood vessels and causes zygomycosis, an often fatal disease. "Mycosis" is the general term for any disease caused by a fungus. *R. stolonifer*, black bread mold, is a nuisance in homes and bakeries.

Consider the life cycle of *R. stolonifer*, which has two mating strains, plus (+) and minus (–). If hyphae of these two strains meet, they form gamete-making structures (gametangia). Gametes (+ and – nuclei) fuse and a diploid zygospore with a thick, protective wall forms (Figure 22.4*a*). Meiosis occurs as the zygospore germinates and a hypha emerges bearing a spore sac at its tip. Haploid spores form in the spore sac. After release, the spores germinate and each gives rise to a haploid mycelium. This mycelium grows rapidly and forms asexual spores on special hyphae (Figure 22.4*b*).

Zygote fungi alone produce thick-walled zygospores when they reproduce sexually. The fast-growing molds that spoil food are often zygote fungi. A few species are dangerous pathogens that infect humans who are already weakened.

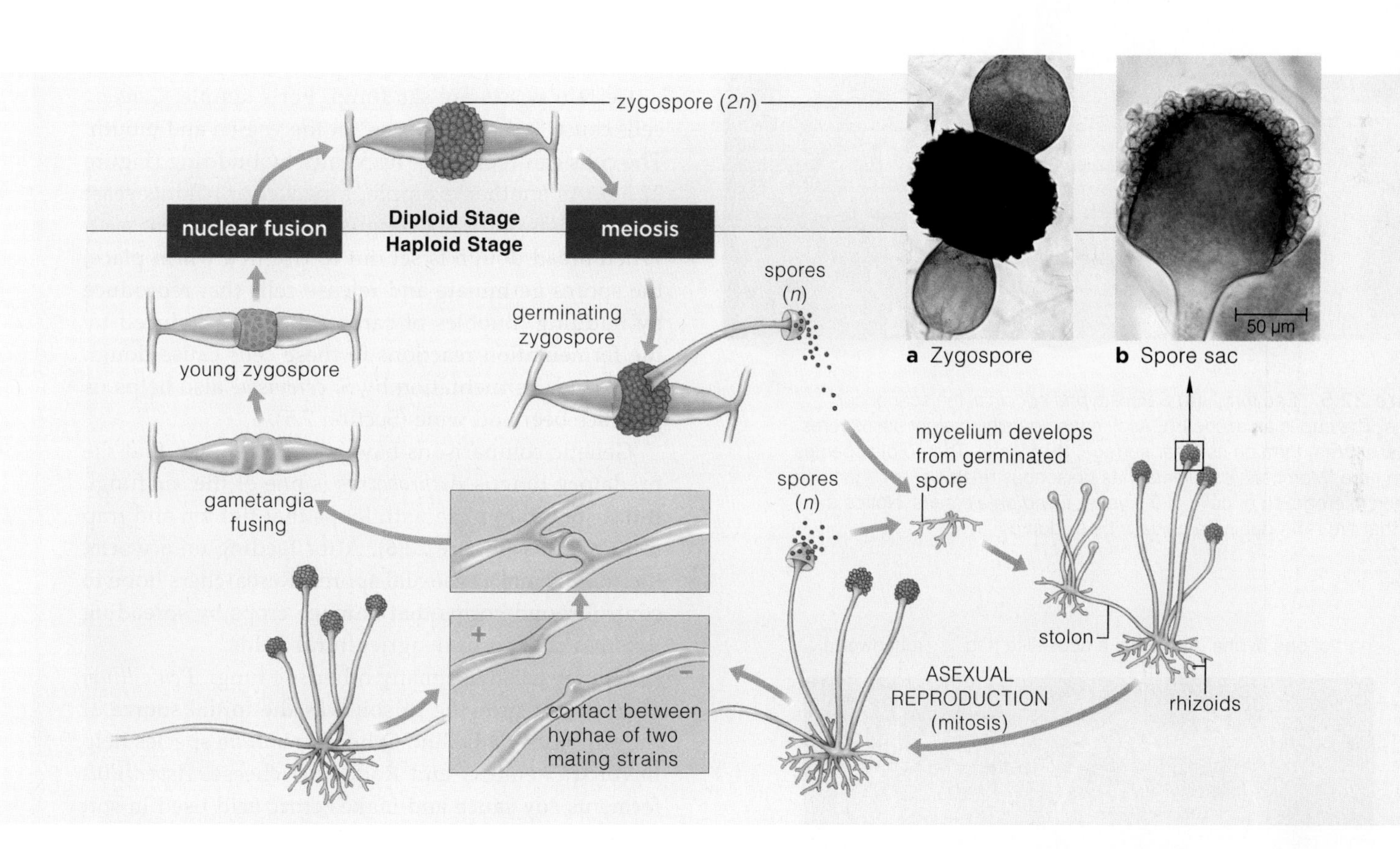

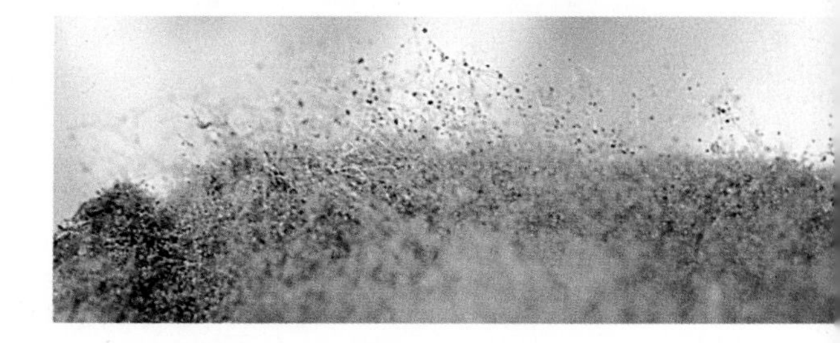

Figure 22.4 **Animated!** Life cycle of *Rhizopus stolonifer*, a black bread mold.

Asexual reproduction occurs frequently. Different mating strains (+ and –) can also reproduce sexually. Either way, haploid spores form and then give rise to mycelia. Attraction between a + hypha and – hypha causes them to fuse and form a thick-walled, diploid zygospore (**a**) that can remain dormant for months. Cells within the zygospore undergo meiosis, then the zygospore germinates. A hypha emerges with a spore sac at its tip. Haploid spores germinate and give rise to new mycelia. As long as conditions are favorable, a mycelium grows fast and spore sacs form at the ends of specialized hyphae (**b**), as on the slice of moldy bread shown at right.

22.3 Ascomycetes—The Sac Fungi

Hyphae with cross-walls evolved in the common ancestor of sac fungi and club fungi. Cross-walls made the hyphae stronger and allowed formation of a large spore-producing body. They also divided the cytoplasm so that damage to one part of a hypha did not cause the whole hypha to dry out and die. That is one reason why the sac fungi and club fungi do better than zygote fungi in drier environments.

LINK TO SECTION 7.5

KEY CHARACTERISTICS

Sac fungi form sexual spores inside a baglike structure called an ascus (plural, asci). Inside an ascus, meiosis and one mitotic division produce eight haploid spores. Most multicelled species produce many asci at a time as part of an ascocarp, a reproductive structure made of specialized, interwoven hyphae (Figure 22.5*a*,*b*).

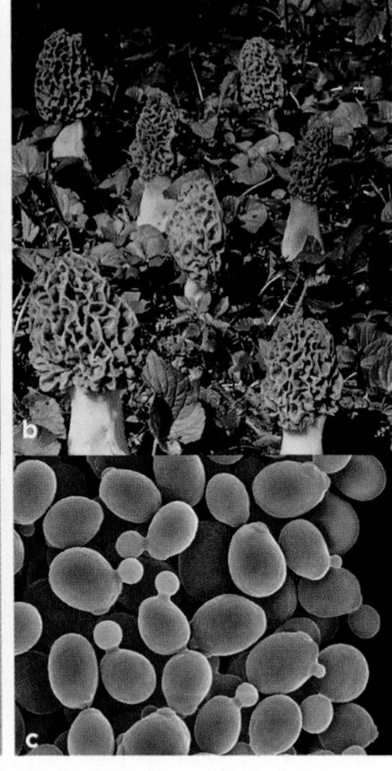

Figure 22.5 Sac fungi. (**a**) *Sarcoscypha coccinia*, the scarlet cup fungus. The cup is an ascocarp. Asci, each containing eight ascospores (sexual spores) form on its inner surface. (**b**) Morels. The edible species shown here, *Morchella esculenta*, has poisonous relatives. (**c**) Scanning electron micrograph of cells of the yeast *Candida albicans*. Notice the cells that are reproducing asexually by budding.

A SAMPLING OF DIVERSITY

With more than 32,000 species, sac fungi are the most diverse fungi. Many partner with algal cells in lichens. Others associate with tree roots and form underground ascocarps known as "truffles." A mature truffle emits some of the same chemicals as an amorous male pig. Excited female pigs disperse truffle spores as they dig up soil in search of the seemingly subterranean suitor. Some truffles are an expensive delicacy. Humans use sows and trained dogs to snuffle out these truffles.

Certain yeasts are sac fungi. For example, *Candida* cells cause "yeast infections" of the vagina and mouth. The cells can reproduce asexually by budding (Figure 22.5*c*). As another example, a packet of baking yeast holds spores of the sac fungus *Saccharomyces cerevisiae*. When bread dough is set out to rise in a warm place, the spores germinate and release cells that reproduce by budding. Bubbles of carbon dioxide produced by the fermentation reactions in these cells cause dough to expand. Fermentation by *S. cerevisiae* also helps us produce beer and wine (Section 7.5).

Genetic comparisons have recently shown that the predatory fungus *Arthrobotrys* is one of the sac fungi. It has special hyphae with loops that tighten and trap roundworms (Figure 22.6). After feeding on a worm, the fungus makes asexual spores. Researchers hope to control roundworms that damage crops by spreading *Arthrobotrys* spores in agricultural fields.

Humans also use many other sac fungi. *Penicillium chrysogenum* growing in soil was the initial source of the antibiotic penicillin. Other *Penicillium* species help flavor Camembert and Roquefort cheese. *Aspergillus* ferments soy sauce and makes citric acid used in soft drinks. The red bread mold, *Neurospora crassa*, serves as an experimental organism for geneticists. It grows fast, results of crosses can be easily observed, and its genome has been sequenced.

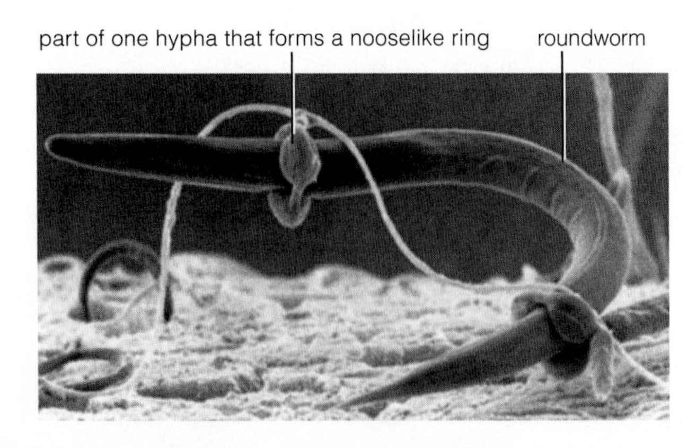

Figure 22.6 Animated! A predatory fungus (*Arthrobotrys*) that captures and feeds on roundworms. Rings that form on the hyphae constrict and entrap the worms, then hyphae grow into the captive and digest it.

Sac fungi make sexual spores inside sac-shaped asci. They include multicelled species with hyphae divided by internal walls, and yeasts that can reproduce asexually by budding.

22.4 Basidiomycetes—The Club Fungi

Most of the 30,000 species of club fungi are multicelled, with cross-walled hyphae. A few are single-celled yeasts.

Club fungi alone produce club-shaped cells that bear four sexual spores (Figure 22.7). The familiar club fungi include puffballs, shelf fungi, and mushroom-forming species. Each mushroom is a short-lived reproductive structure made of specialized hyphae.

Those button mushrooms in markets and on pizzas are spore-bearing parts of *Agaricus bisporus*. Haploid hyphae of *A. bisporus* grow in the soil. When hyphae of two mating strains meet, their cytoplasm fuses. The resulting dikaryotic cell holds two genetically different nuclei (Figure 22.7*a*,*b*). This cell divides and produces a dikaryotic mycelium. Mushrooms pop up from the mycelium when conditions favor sexual reproduction. Hanging beneath each mushroom's cap are thin tissue sheets (gills) fringed with club-shaped cells. The two nuclei in these dikaryotic cells fuse and form a diploid zygote (Figure 22.7*c*,*d*). Meiosis in the zygote results in four haploid spores, which disperse in air, germinate, and start a new cycle (Figure 22.7*e*–*g*).

Club fungi play an important role as decomposers of forest plants; they are the only fungi that can break down the lignin that stiffens stems (Section 21.1). Some club fungi are long-lived giants. In one Oregon forest, the mycelium of a honey mushroom (*Armillaria ostoyae*) extends through 2,200 acres of soil. By one estimate, this fungus is 2,400 years old.

Club fungi produce their sexual spores at the tips of club-shaped cells. These cells form on a short-lived reproductive body, such as a mushroom.

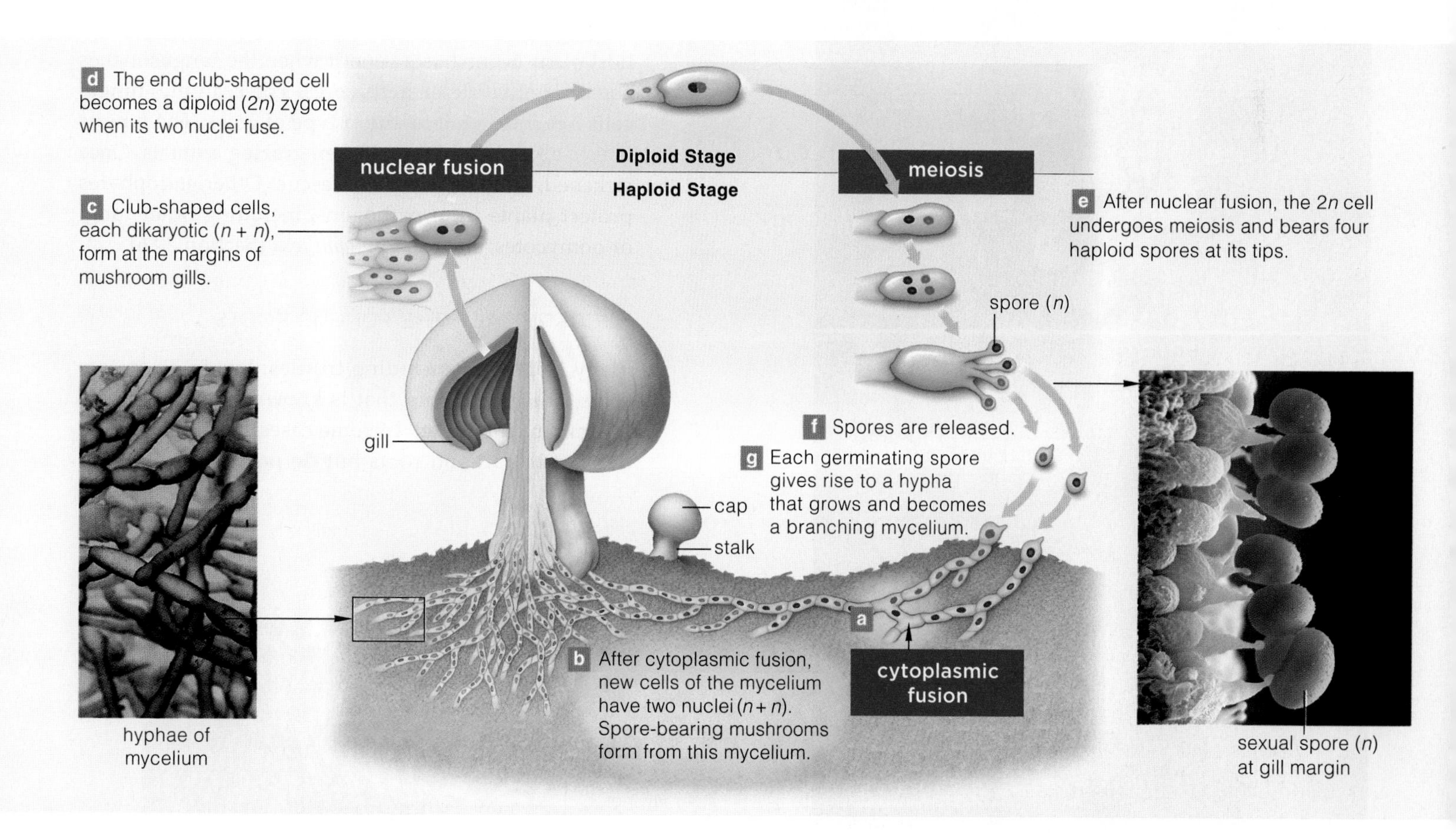

Figure 22.7 **Animated!** Life cycle typical of a club fungus having two mating strains of hyphae. (**a**) Haploid hyphal cells of two compatible strains meet. Their cytoplasm fuses, but the nuclei do not (**b**) Mitotic cell divisions form a mycelium in which each cell has two nuclei. Under favorable conditions, many hyphae of the mycelium intertwine and form a mushroom. (**c**,**d**) Club-shaped structures develop on the mushroom gills. The end cell on the "club" becomes diploid when two nuclei fuse. (**e**) Meiosis results in four haploid spores, which migrate into four short cytoplasmic extensions at the tip of the club. (**f**) The spores drift away from the gills. (**g**) Each may germinate and give rise to a new mycelium.

22.5 The Fungal Symbionts

Symbiosis, recall, refers to two organisms that interact closely throughout their life cycle. Many fungal species are symbiotic with photosynthetic cells in lichens, while others are symbionts that live in or on plants.

LICHENS REVISITED

LINKS TO SECTIONS 19.2, 20.6

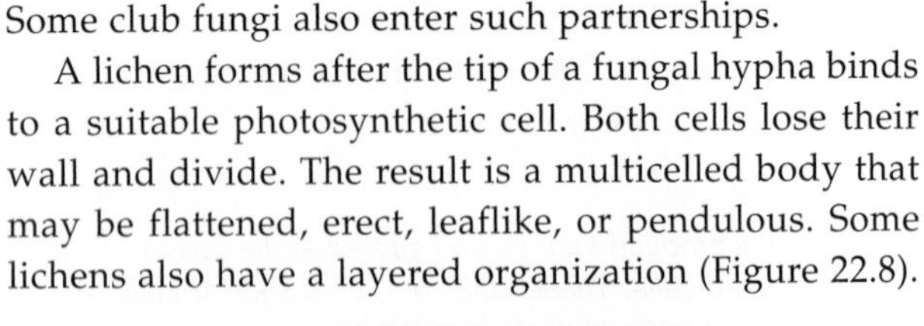

Most lichens are a symbiotic interaction between a sac fungus and one or more cyanobacteria or green algae. Some club fungi also enter such partnerships.

A lichen forms after the tip of a fungal hypha binds to a suitable photosynthetic cell. Both cells lose their wall and divide. The result is a multicelled body that may be flattened, erect, leaflike, or pendulous. Some lichens also have a layered organization (Figure 22.8).

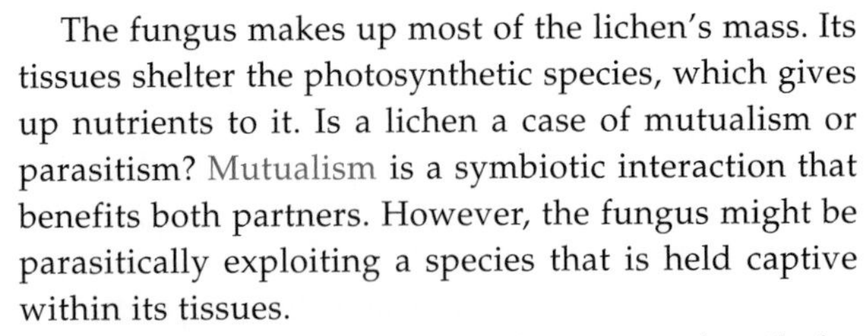

The fungus makes up most of the lichen's mass. Its tissues shelter the photosynthetic species, which gives up nutrients to it. Is a lichen a case of mutualism or parasitism? Mutualism is a symbiotic interaction that benefits both partners. However, the fungus might be parasitically exploiting a species that is held captive within its tissues.

Lichens can colonize places that are too hostile for most organisms. For example, when a glacier retreats, lichens colonize newly exposed bedrock. By releasing acids and holding water that freezes and thaws, they break down the rock. When soil conditions improve, plants move in and take root. Long ago, lichens may have preceded the invasion of plants onto land.

FUNGAL ENDOPHYTES

Endophytic fungi are mostly sac fungi that reside in the leaves and stems of the vast majority of plants. Usually, the interaction neither helps nor harms the host plant. Some hosts benefit when the fungus makes chemicals that deter herbivores. For example, fungal cells live inside tall fescue, a type of grass. The fungus makes alkaloids that can sicken grazing animals. Once sickened, animals avoid tall fescue. Other endophytes protect plants from pathogens, including other fungi or oomycotes, such as *Phytophthora* (Section 20.6).

MYCORRHIZAE—THE FUNGUS ROOTS

Many soil fungi, including truffles, live in or on tree roots in a partnership that is known as a mycorrhiza (plural, mycorrhizae). In some cases, the hyphae form a dense net around roots but do not penetrate them.

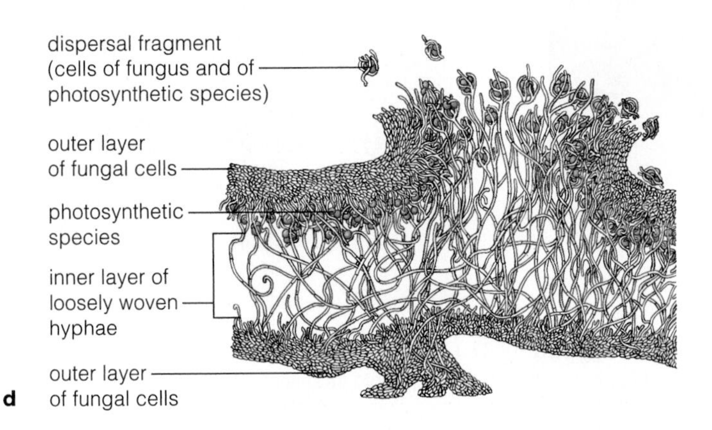

Figure 22.8 (**a**) Leaflike lichen on a birch tree. (**b**) *Usnea*, or old man's beard, is one of the pendant lichens. (**c**) Encrusting lichens on granite. (**d**) Organization of one stratified lichen, as it would look in cross-section.

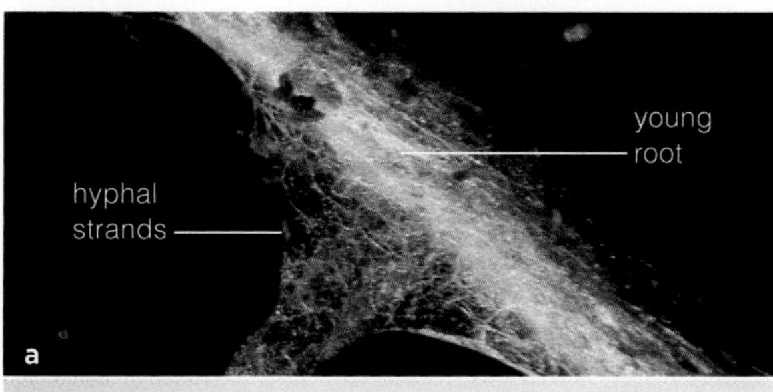

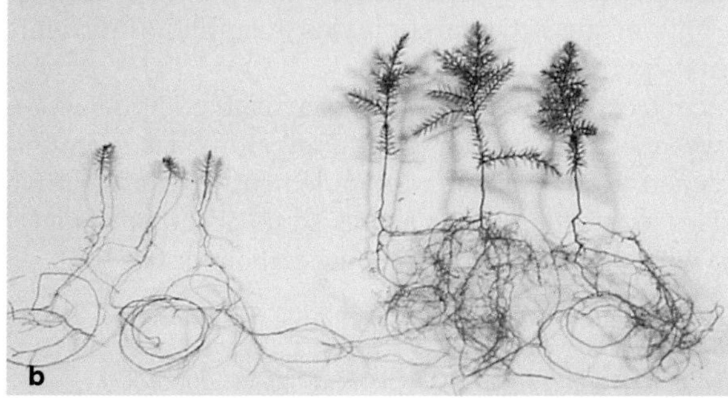

Figure 22.9 (**a**) Mycorrhiza formed by a fungus and its partner, a young hemlock tree. (**b**) Effects of the presence or absence of mycorrhizae on plant growth in phosphorus-poor, sterilized soil. The juniper seedlings at left were controls; they grew without a fungus. The six-month-old seedlings at right, the experimental group, were grown with a partner fungus.

Club fungi often take part in mycorrhizae with tree roots in temperate forests. Most forest mushrooms are reproductive structures of these fungi. In other cases, hyphae of a zygote fungus penetrate root cells. About 80 percent of vascular plants form such a partnership with a zygote fungus.

Hyphae of both kinds of mycorrhizae grow through soil and functionally increase the absorptive surface area of their partner. Fungal hyphae are thinner and better at growing in between soil particles, compared to even the thinnest plant roots. The fungus absorbs and concentrates nitrogen, phosphorus, and other ions near the plant's roots. A plant gives up some sugars to the fungus. It is an advantageous trade; many plants do poorly without mycorrhizae (Figure 22.9).

In lichens, a fungus shelters one or more photoautotrophs and shares carbon dioxide and mineral ions with them, while receiving some carbohydrates in return. Often the photosymbionts are nitrogen fixers, such as cyanobacteria.

Endophytes are fungal symbionts living in leaves or stems. Some protect plants from grazing animals or pathogens.

In mycorrhizae, a fungus living in or on a plant's young roots increases the plant's uptake of water and dissolved mineral ions and helps protect it from pathogens. The fungus withdraws some nutrients from its partner.

22.6 An Unloved Few

FOCUS ON HEALTH

You know you are a serious student of biology when you view organisms objectively in terms of their place in nature. You salute saprobic fungi as decomposers and praise parasitic fungi that keep populations of harmful insects and weeds in check. The true test is when you meet a fungus that feeds on you or your resources.

What thoughts run through your head when you open the refrigerator to grab some fruit and discover a mold beat you to it? How appreciative would you be if a fungus was feeding on warm, damp tissues in your groin, under your nails, or between your toes (Figure 22.10*a*)?

Which home gardeners sing praises of black spot or powdery mildew on roses? Which farmers happily hand over millions of dollars a year to sac fungi that attack corn, wheat, peaches, and apples? Who rejoices that a certain sac fungus, *Cryphonectria parasitica*, killed off all the chestnut trees in eastern North America?

Who willingly inhales *Histoplasma capsulatum* spores? In moist lung tissues, spores of this sac fungus germinate and release cells that cause histoplasmosis, a respiratory disease. The body fights back, but the battle results in calcified lung tissue.

Spores of *Penicillium*, *Aspergillus*, and other molds can cause sinus congestion, a raw throat, and chronic sneezing. If you are susceptible to asthma, exposure to these mold spores may increase the frequency, intensity, and duration of attacks.

Some fungi even have a place in human history. One notorious species, *Claviceps purpurea*, parasitizes rye and other cereal grains (Figure 22.10*c*). Give it credit; we use some of its alkaloid products to relieve pain of migraine headaches and to stop bleeding after childbirth by shrinking the uterus. Still, the alkaloids can be toxic. Eat a lot of bread made with tainted rye flour and you end up with ergotism. Vomiting, diarrhea, hallucinations, hysteria, and convulsions are typical symptoms. Unless the disease is treated, it can turn limbs gangrenous and bring on death.

Ergotism epidemics were common in Europe in the Middle Ages, when rye was a major crop. They thwarted Peter the Great, a Russian czar who became obsessed with conquering ports along the Black Sea for his otherwise landlocked empire. Soldiers laying siege to the ports ate mostly rye bread and fed rye to horses. The soldiers went into convulsions and the horses into "blind staggers." Remember those witch hunts in early American colonies? The behavior of the supposedly "bewitched" resembled symptoms of ergotism.

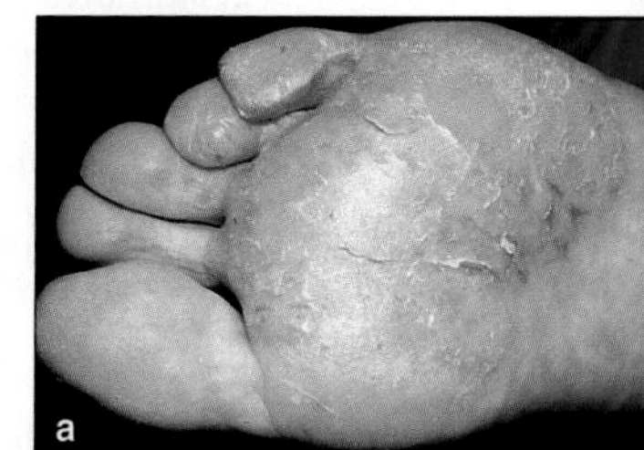

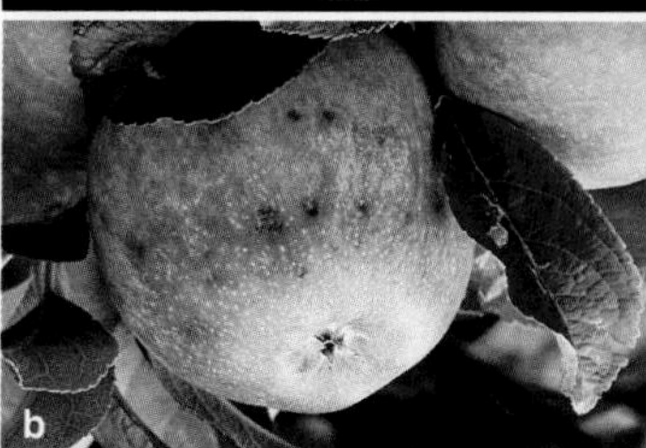

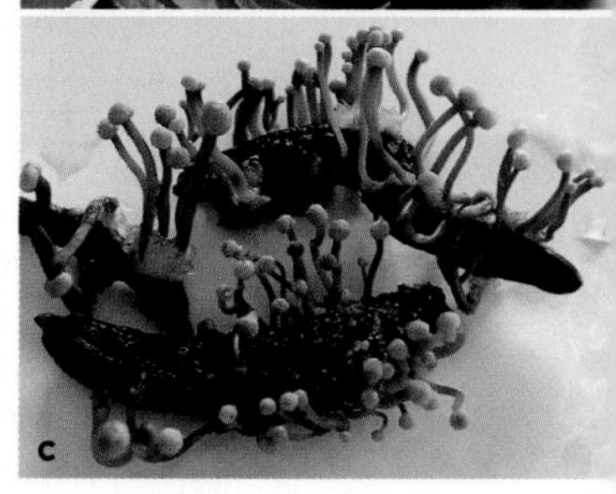

Figure 22.10 Love those fungi! (**a**) Athlete's foot, thanks to *Epidermophyton floccosum*. (**b**) Apple scab, the trademark of *Venturia inaequalis*. (**c**) Spores of *Claviceps purpurea*, on an infected rye plant.

www.thomsonedu.com ThomsonNOW!

Summary

Section 22.1 All fungi are heterotrophs that secrete digestive enzymes on organic matter and absorb released nutrients. Most are saprobes that feed on organic remains; they are major decomposers in ecosystems. Other fungi are harmless or beneficial symbionts or parasites.

Fungi and animals share an amoebozoan-like relative. Fungi include single-celled yeasts and large multicelled species. Zygote fungi, sac fungi, and club fungi are the major groups. Microsporidians are intracellular parasites. Chytrids are the only fungi with flagellated spores.

Sexual as well as asexual spore production dominates the life cycles. In multicelled species, spores germinate and give rise to filaments called hyphae. The filaments typically grow as an extensive mesh called a mycelium. When conditions favor sexual reproduction, numerous hyphae may weave together and form short-lived spore-producing reproductive structures, such as mushrooms.

Section 22.2 Zygote fungi include many familiar molds that grow on fruits, breads, and other foods. Their hyphae are continuous tubes that have no cross-walls. A zygote forms after two compatible hyphae meet and then fuse. It develops into a thick-walled zygospore that can survive harsh conditions. Meiosis occurs as a zygospore germinates and a specialized hypha grows out from it. Haploid spores that form at the tip of this hypha can give rise to a haploid mycelium. Mycelia also produce asexual spores atop specialized hyphae.

■ *Use the animation on ThomsonNOW to observe the life cycle of black bread mold* (Rhizopus)*, a zygote fungus.*

Section 22.3 Sac fungi are the most diverse group. They include single-celled yeasts and multicelled species. Many sac fungi produce asexual spores at the tip of a specialized type of hypha. Sexual spores are produced in asci. In multicelled species, these saclike structures form on an ascocarp, a reproductive structure. Ascocarps can become large because their hyphae contain reinforcing cross-walls. Such hyphae help many sac fungi survive in habitats that are far drier than zygote fungi can tolerate.

■ *View the video on ThomsonNOW and watch a nematode-trapping fungus in action.*

Section 22.4 Like sac fungi, multicelled club fungi have hyphae with cross-walls and can produce complex reproductive structures, such as mushrooms. Many club fungi are major decomposers in forest habitats.

Typically, a dikaryotic mycelium dominates the life cycle of multicelled species. It grows by mitosis and, in some species, extends through a vast volume of soil. The mycelium forms when two haploid hyphae of different mating strains meet and their cytoplasm fuses but their nuclei remain separate.

When conditions favor reproduction, a reproductive structure, also made up of dikaryotic hyphae, develops. A mushroom is an example. Nuclei fuse in club-shaped cells at the edges of sheets of tissue called gills. Meiosis of the resulting diploid cell produces four spores at the tips of the club-shaped cells.

■ *Use the the animation on ThomsonNOW to learn about the life cycle of a club fungus.*

Section 22.5 Many fungi are symbionts that spend all or part of their life cycle in or on another species. Endophytes are fungi that live in many stems and leaves without harming the host plant. Some protect their hosts from herbivores or from plant pathogens.

A lichen is a composite organism that consists of a fungal symbiont and one or more photoautotrophs, such as green algae and cyanobacteria. The fungus makes up the bulk of the lichen and obtains a supply of nutrients from its photosynthetic partner.

A mycorrhiza (fungus-root) is a symbiotic interaction between a fungus and a plant. Fungal hyphae surround or penetrate the roots and supplement their absorptive surface area. The fungus shares some absorbed mineral ions with the plant and gets some carbohydrates back.

■ *Read the InfoTrac article "The Fungus Among Us: Tiny But Ubiquitous, Fungi Form Vital Connections Underground," Janet Wallace,* Alternatives Journal, *December 2004.*

Section 22.6 A number of pathogenic fungi can destroy crops, spoil food, and cause diseases in humans.

Self-Quiz

Answers in Appendix III

1. All fungi ______ .
   a. are multicelled c. are heterotrophs
   b. form flagellated spores d. all of the above

2. Saprobic fungi derive nutrients from ______ .
   a. nonliving organic matter c. living animals
   b. living plants d. both b and c

3. In ______ a hypha lacks cross-walls.
   a. chytrids c. sac fungi
   b. zygomcyetes d. club fungi

4. A slice of white bread contains the remains of many yeast cells, one type of ______ .
   a. chytrid c. sac fungus
   b. zygote fungus d. club fungus

5. In many ______ an extensive dikaryotic mycelium is the most conspicuous phase of the life cycle.
   a. chytrids c. sac fungi
   b. zygote fungi d. club fungi

6. A mushroom is ______ .
   a. the food-absorbing part of a club fungus
   b. the only part of the fungal body not made of hyphae
   c. a reproductive structure that releases sexual spores
   d. produced by meiosis in a zygote

7. Spores released from a mushroom's gills are ______ .
   a. club-shaped c. haploid
   b. dikaryotic d. both a and c

8. Nitrogen-fixing cyanobacteria often partner with a fungus as a(n) ______ .
   a. endophyte c. mycorrhiza
   b. lichen d. mycosis

9. Histoplasmosis is an example of a(n) ______ .

a. endophyte
b. lichen
c. mycorrhiza
d. mycosis

10. Match the terms appropriately.

___ hypha
___ chitin
___ chytrid
___ sac fungus
___ club fungus
___ lichen
___ mycorrhiza

a. produces flagellated spores
b. component of fungal cell walls
c. partnership between a fungus and one or more photoautotrophs
d. filament of a mycelium
e. fungus–root partnership
f. forms sexual spores in an ascus
g. many form mushrooms

■ *Visit ThomsonNOW for additional questions.*

Critical Thinking

1. *Pilobolus* is a type of fungus that often lives on horse dung. A dark cluster of asexual spores sits on the bulging tip of stalked hyphae (Figure 22.11). Inside the bulge is a fluid-filled central vacuole. Sunlight makes the stalk bend, which puts pressure on the central vacuole that causes it to rupture. The forceful blast can propel the cluster of spores as far away as 2 meters (about 6.5 feet) onto grass. Spore-dusted grass passes through a horse gut. Then the spores exit in feces, and the life cycle turns again. Reflect on the examples of fungi in this chapter. Would you say *Pilobolus* is a zygote fungus, a club fungus, or a sac fungus? What other information would allow you to say for sure?

2. The fungus *Fusarium oxysporum* is a plant pathogen. Strains of the fungus attack and kill only specific plants, which make it a candidate for a natural mycoherbicide. Proposals to spray *F. oxysporum* to kill marijuana plants in Florida were abandoned after public outcries. What concerns might you have about spraying fungal spores as a way to destroy plants that are sources of the drugs favored by substance abusers?

3. Figure 22.12 shows a sulfur shelf fungus (*Laetiporus sulphureus*), one of the club fungi. Devise a study or experiment to determine if this fungal species benefits, harms, or has no effect on its host tree.

4. Chances are, a dermatophytic (skin-living) fungus has taken up residence in you or in someone you know. *Trichophyton, Microsporum, Epidermophyton* are the main culprits. They cause diseases known as tineas, and health professionals refer to each kind according to which body tissues are infected. As Table 22.1 shows, dermatophytes thrive on almost all body surfaces. They feed on the outer, dead layers of skin by secreting enzymes that dissolve keratin, the main skin protein, and other skin components. Infected areas commonly become raised, red, and itchy.

Dermatophyte diseases are persistent. Ointments and creams may not reach the deepest infected skin layers. Oral antifungal drugs are far less common than antibacterials and often have bad side effects. Reflect on the evolutionary relationships among bacteria, fungi, and humans. Explain why it is harder to create drugs against fungi than bacteria.

Table 22.1 Common Dermatophyte Diseases

Disease	Infected Body Parts
Tinea corporis (ringworm)	Trunk, limbs
Tinea pedis (athlete's foot)	Feet, toes
Tinea capitis	Scalp, eyebrows, eyelashes
Tinea cruris (jock itch)	Groin, perianal area
Tinea barbae	Bearded areas
Tinea unguium	Toenails, fingernails

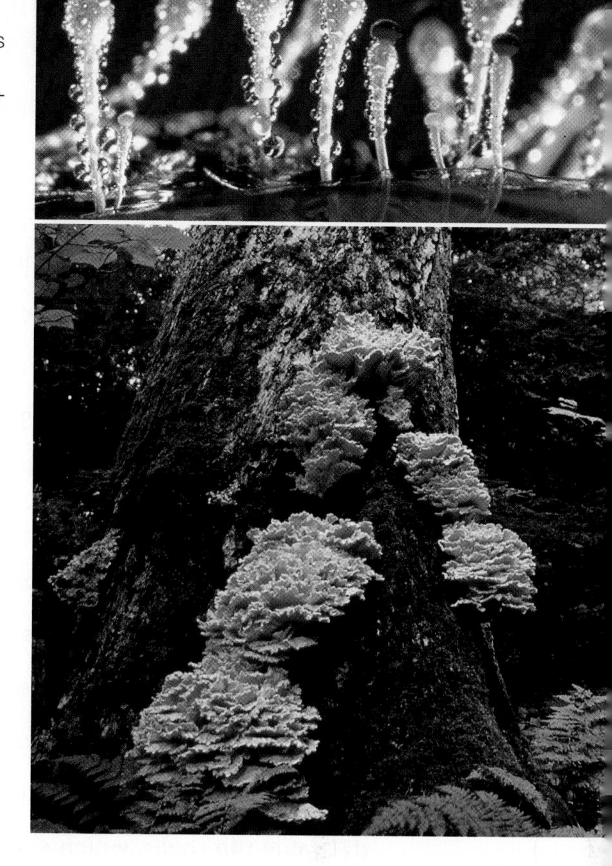

Figure 22.11 Reproductive structures of *Pilobolus*, named for the Greek word for "hat-thrower." Each "hat" is a dark-walled cluster of asexual spores.

Figure 22.12 Sulfur shelf fungus growing on a tree in Virginia.

5. Certain toxic mushrooms have bright, distinctive colors that mushroom-eating animals learn to recognize (Figure 22.13). Once sickened, day-feeding animals tend to avoid them. Other toxic mushrooms are as dull-looking as edible ones, but they have a much stronger odor. Some scientists think the strong odors aid in defense against predators that are active at night. Explain their reasoning.

Figure 22.13 Two kinds of poisonous mushrooms. (**a**) Fly agaric mushroom (*Amanita muscaria*), a hallucinogenic species with vivid coloration. (**b**) The death cap mushroom (*A. phalloides*) is a poisonous relative that resembles edible mushrooms. It is the most common cause of fatal mushroom poisoning in North America, which brings up an old saying: There are old mushroom hunters and bold mushroom hunters—but no old, bold mushroom hunters.

23 ANIMAL EVOLUTION— THE INVERTEBRATES

IMPACTS, ISSUES Old Genes, New Drugs

East of Australia, small, reef-fringed islands dot the vast expanse of the South Pacific Ocean. Shelled animals abound in the warm, nearshore waters of the islands, which include Samoa, Fiji, Tonga, and Tahiti. Among them are more than 500 kinds of predatory mollusks called cone snails (*Conus*), which have endured for millions of years. Humans find them beautiful as well as tasty (Figure 23.1). Archaeologists studying ancient cultures often discover bracelets, rings, and ritual objects made with the elaborately patterned shells. Collectors still value the shells, which end up in jewelry or as decorations on coffee tables, walls, and shelves.

Cone snails fascinate biologists for different reasons. All paralyze prey by injecting them with conotoxins. These toxic secretions can paralyze a small fish within seconds by disrupting the signals flowing through its nervous system. Some can kill larger animals. If you were to step on a cone snail and it managed to pump toxins into you, your chest muscles could freeze up. Without prompt treatment, you would stop breathing.

Each *Conus* species makes a unique mix of 100 to 300 conotoxins that target different receptors on cells of the nervous system. The broad range of specific targets makes the snails potential sources of many new drugs.

One *C. magnus* conotoxin blocks calcium transporters in cell membranes. Its action stops cells from releasing signaling molecules that contribute to the sense of pain. Ziconotide, a synthetic version of the toxin, relieves pain for people who have severe back problems, inoperable cancers, and AIDS. The nonaddictive drug is 1,000 times more potent than morphine and patients do not develop tolerance to it; they do not require increasing doses over time to achieve the same result.

See the video! **Figure 23.1** (**a**) The mollusk *Conus geographicus*, engulfing a small fish. The tubelike structure extended straight up in this photograph is a siphon. It can detect small pressure changes in water, as when small fishes and other prey swim within range. *C. geographicus* impaled this fish with a harpoon-like device. Then it pumped paralyzing conotoxins into it. (**b**) A small sampling of the diverse patterns of *Conus* shells.

How would you vote? Most cone snails are highly vulnerable to extinction. No one monitors how many are taken from their habitats. Trading in endangered species is regulated. Should the rules be extended to cover any species captured in the wild? See ThomsonNOW for details, then vote online.

C. geographicus, shown in Figure 23.1, secretes a toxin that one day might help epileptics. While studying this species, University of Utah researchers discovered a gene that has persisted in the DNA of different species for a long time. That highly conserved gene encodes an enzyme, gamma-glutamyl carboxylase (GGC), that catalyzes a step in the conotoxin synthesis pathway.

The gene for GGC has been around for at least 500 million years. It also is in the genome of fruit flies and humans, which means it arose before divergences gave rise to snails, insects, and vertebrates. The gene mutated independently in those separate lineages, and its product diverged in its functions. GGC helps repair damaged blood vessels in humans. We have yet to figure out what it does in fruit flies.

This example supports an organizing principle in the study of life. *Look back through time, and you discover that all organisms interconnect.* At each branch point in the animal family tree, microevolutionary processes gave rise to novel changes in biochemistry, body plans, or behavior. They were the source of unique traits that help define each species.

Tens of millions of animal species have appeared and disappeared over time. Of more than 2 million named species, only 50,000 kinds of fishes, amphibians, reptiles, birds, and mammals have an internal backbone; they are vertebrates. The vast majority of animal species are known informally as the invertebrates. They have many fabulous features, but a backbone is not one of them.

This chapter focuses mainly on the invertebrates. The next chapter continues the story, focusing mainly on the vertebrates and their closest invertebrate relatives. Even with two chapters, we can scarcely do justice to so many animal species. The best we can do is consider the scope of diversity and the evolutionary trends.

One final note: Do not assume that the most ancient lineages are somehow stunted or "primitive." Even the structurally simplest animals are exquisitely adapted to their environment.

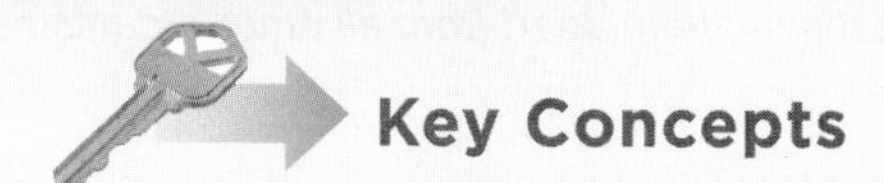

Key Concepts

INTRODUCING THE ANIMALS

Animals are multicelled heterotrophs that ingest other organisms, grow and develop through a series of stages, and actively move about during all or part of the life cycle. Cells of most animals form tissues and extracellular matrixes.

The earliest animals were small and structurally simple. Their descendants evolved larger bodies with a more complex structure and greater integration among specialized parts.

Animals' body plans vary. Bodies may or may not show symmetry. There may or may not be an internal body cavity, a head, or division into segments. An early divergence gave rise to two major branches: protostomes and deuterostomes. **Sections 23.1, 23.2**

STRUCTURALLY SIMPLE INVERTEBRATES

Sponges and placozoans have no body symmetry or true tissues. Cnidarians are radially symmetrical, with two tissue layers and a gelatinous matrix between the two. **Sections 23.3, 23.4**

BILATERAL INVERTEBRATES

Most animals show bilateral symmetry. Bilateral animals have tissues, organs, and organ systems. All adult tissues arise from two or three simple layers that form in early embryos. **Sections 23.5–23.8**

THE MOST SUCCESSFUL ANIMALS

In diversity, numbers, and distribution, arthropods are the most successful animals. In the seas, crustaceans are the dominant arthropod lineage; on land, insects rule. **Sections 23.9–23.13**

ON THE ROAD TO CHORDATES

Echinoderms are on the same branch of the animal family tree as the chordates. They are invertebrates with bilateral ancestors, but adults now have a decidedly radial body plan. **Section 23.14**

Links to Earlier Concepts

This chapter builds on your understanding of levels of organization in nature (Section 1.1), the surface-to-volume ratio (4.1), membrane proteins (4.3), and extracellular matrixes (4.9). You will consider more mutations that altered key steps in embryonic development (11.5, 14.3, 16.8). You will see how drifting and colliding continents influenced animal evolution (16.5, 16.6). You will reflect again on pesticide resistance (Chapter 17 Introduction) and the vectors for human diseases (19.6).

23.1 Animal Traits and Trends

No one trait defines all animals. It takes a list of features to see what sets them apart from all other organisms.

WHAT IS AN ANIMAL?

LINKS TO SECTIONS 1.1, 4.1, 4.11, 16.4, 16.12, 18.3

Animals are multicelled heterotrophs that ingest the tissues, juices, or wastes of other organisms. Nearly all reproduce sexually; many also do so asexually. All develop in a series of stages, and most form tissues. Animals actively move about during at least part of the life cycle. Their body cells are usually diploid and always unwalled. Table 23.1 lists the main groups we will survey. Again, most are invertebrates, or animals that do not have a backbone. Animals with a backbone, the vertebrates, will be discussed in Chapter 24.

Animal Tissues From simple origins, millions of animal species evolved by way of a key innovation: *Their cells had a capacity to stick together and interact in functional units called epithelium and connective tissues.*

An epithelium (plural, epithelia) is a sheet of cells that lines a body surface, cavity, or tube. Cells of this tissue serve in absorption, secretion, and other tasks. Connective tissues consist of cells scattered within an extracellular matrix of their own secretions. They bind tissues together and structurally support a body. Other tissues evolved later, but these were the first.

All tissues that make up the adult animal develop from simpler layers that form when embryos are still tiny balls of cells: surface ectoderm, inner endoderm, and often mesoderm between the two.

Body Size Revisited Physical constraints affect animal sizes and shapes. The surface-to-volume ratio is a case in point: As an animal body increases in size, its volume increases faster than its surface area (Section 4.1). If a body is more than a few cells thick, diffusion alone cannot supply its cells with gases and nutrients (and remove wastes) fast enough to keep them alive.

In most animals, an energy-demanding circulatory system distributes materials. Additional energy cost is offset by the advantages of larger size. In a large body, more cells are shielded from outside threats. The body survives even if it loses a few cells.

Advantages multiply when different cells, tissues, and the internal compartments called organs split up essential tasks, such as food distribution and defense. Each cell lineage can specialize in one task instead of having to make every type of molecule required to perform all of them.

Table 23.1 Animal Groups Surveyed in Chapters 23 and 24

Group	Representatives	Named Species
Poriferans	Barrel sponges, encrusting sponges	8,000
Placozoans	*Trichoplax*	1?
Cnidarians	Sea anemones, jellyfishes, corals, siphonophores	11,000
PROTOSTOMES		
Flatworms	Turbellarians, tapeworms, flukes	15,000
Rotifers	Bdelloids	2,000
Annelids	Polychaetes, earthworms, leeches	15,000
Mollusks	Snails, slugs, clams, squids, octopuses	110,000
Roundworms	Pinworms	20,000
Tardigrades	Water bears	800
Arthropods	Horseshoe crabs, crabs, spiders, insects	1,113,000*
DEUTEROSTOMES		
Echinoderms	Sea stars, sea urchins, sand dollars, sea lilies	6,000
Chordates	Invertebrate chordates	2,100
	Vertebrates:	
	Jawless fishes	84
	Jawed fishes	21,000
	Amphibians	4,900
	Reptiles	8,200
	Birds	8,600
	Mammals	4,500

* Named species only; arthropods probably number in the tens of millions.

OVERVIEW OF ANIMAL BODY PLANS

How can you get a handle on animals as different as spiders and humans? *You might organize them by body symmetry (if any), and whether they have some type of head, gut, main body cavity, and repeating body units.*

With few exceptions, the animal body shows some symmetry. It has a ventral surface (underside) and a dorsal one (upper side). With radial symmetry, parts are arranged around a central axis, like spokes of a wheel (Figure 23.2*a*). With bilateral symmetry, many appendages and organs are paired, one to each side of the front-to-back body axis (Figure 23.2*b*). Nearly all animals are bilateral, and most are cephalized—two legacies from early animals that crept forward. Cephalization means that many sensory cells became concentrated at the body's leading end, or head. In many lineages, the process led to a complex brain.

Most animals have a gut: a digestive sac or tube that opens at the body surface. A saclike gut is an *incomplete* digestive system; food enters and wastes must leave through the same opening. A tubular gut is a *complete* digestive system, with a mouth at one end and an anus at the other.

A complete digestive system has advantages. Parts of the tube became regionally specialized for taking

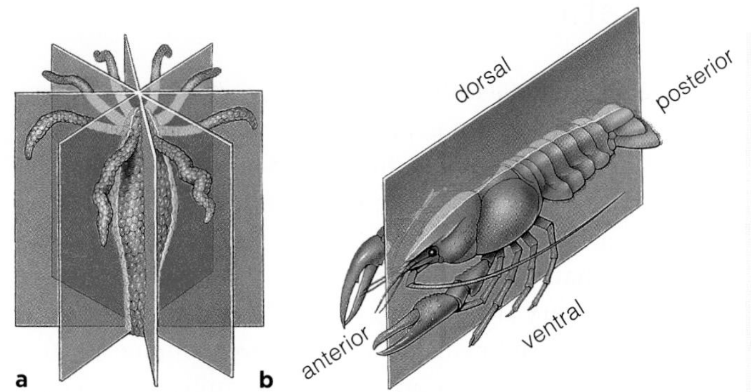

Figure 23.2 (**a**) Radial body symmetry of *Hydra*, a cnidarian. (**b**) Bilateral symmetry of a lobster, an arthropod.

in food, breaking food down, absorbing nutrients, or compacting wastes. Unlike a saclike gut, a complete gut can carry out these tasks simultaneously.

Animals also differ in what is in between the gut and body wall. Some small invertebrates have a more or less solid mass of tissues and organs (Figure 23.3*a*). Other invertebrates have a pseudocoel: a fluid-filled, unlined cavity surrounding the gut (Figure 23.3*b*). Most bilateral animals have a coelom. They have a fluid-filled cavity between the gut and body wall, and it has a lining derived from mesoderm (Figure 23.3*c*). Like other vertebrates, you have a coelom with a lining called peritoneum. Fluid inside the coelomic cavity cushions organs from shocks, and it allows organs to form and move independently of the body wall.

We define two major lineages of bilateral animals in part by the differences in how the digestive system and coelom form. In protostomes, the first opening that appears on the embryo develops into the mouth. In deuterostomes, the first opening becomes the anus.

Segmentation is a repetition of the same or similar units along a body's front-to-back axis. Fossils show that a segmented body evolved early in some lineages (Figure 23.4). Repetition allowed specialization. When many segments performed the same task, some were free to change, fuse, or add a variety of appendages.

A human embryo yields clues to our segmented ancestry. During early human development, some of the embryo's mesoderm forms a series of segments along the main body axis. The segments later develop into the ribs, backbone, and back muscles.

Animals are multicelled heterotrophs with unwalled cells. They develop in stages and move about during the life cycle. Most have epithelia and other tissues. Animals differ in body size, complexity, symmetry, cephalization, type of gut and main body cavity, and segmentation.

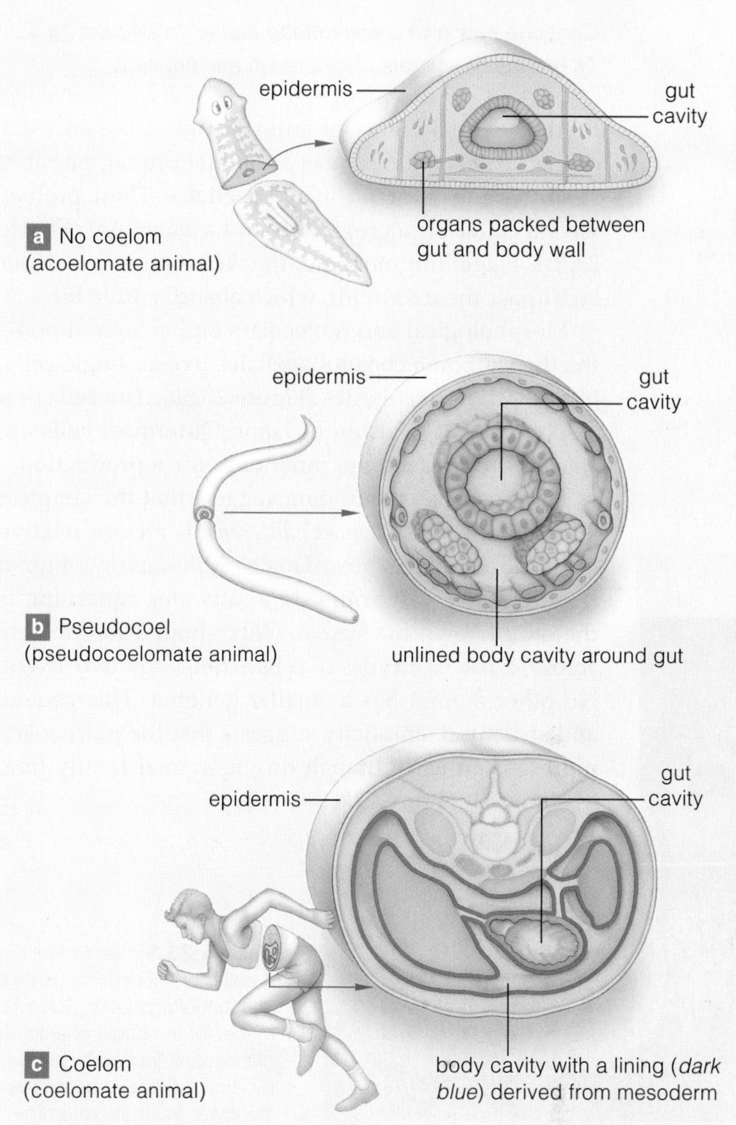

Figure 23.3 Animated! Main cavity (if any) between the gut and body wall of animals. (**a**) A flatworm is acoelomate; it has no body cavity. (**b**) A roundworm has an unlined cavity (a pseudocoel). (**c**) All vertebrates are coelomate. The main body cavity is lined with tissue (*dark blue*).

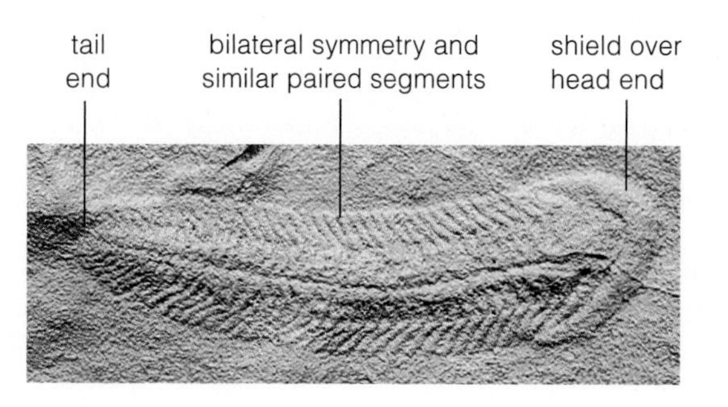

Figure 23.4 Fossil of a cephalized, segmented animal (*Spriggina floundersi*) that was alive 570 million years ago. For more fossils, go online to PaleoBase.com as one portal to the past.

23.2 Animal Origins and Early Radiations

Continue now with a point made earlier, in Section 18.3. All groups of animals originated in ancient seas.

LINKS TO SECTIONS 5.9, 14.3, 16.5, 16.8, 18.3, 20.10

By the colonial theory of animal origins, the ancestor of the earliest animals was a colonial protist, possibly similar to the modern choanoflagellates. These protists have a collar of microvilli around a flagellum (Figure 23.5*a*). Flagellum movement makes food-laden water swirl past the microvilli, which absorb edible bits.

Morphological and molecular comparisons support the theory. Some choanoflagellates live as single cells, but others form colonies (Figure 23.5*b*). The cells of a colony show a division of labor. Outermost cells are specialized for feeding; inner ones for reproduction.

Also, gene sequence data suggest that the simplest animal known, *Trichoplax adhaerens*, is a close relative of the choanoflagellates. This bloblike marine animal (Figure 23.5*c*), discovered in a saltwater aquarium, is the only known placozoan. Only about 5 millimeters across, it has four types of cells arranged in two layers. No other animal has a smaller genome. This genetic and structural simplicity suggests that the placozoans represent an early branch on the animal family tree, as shown in Figure 23.6.

By about 600 million years ago, a diverse collection of multicelled organisms that may have been early animals were living in the seas. Known as Ediacarans because their fossils were first found in Australia's Ediacaran hills, they included tiny blobs and frondlike forms that stood two meters tall. Most were probably evolutionary dead ends, but early representatives of some modern lineages may have been among them.

When the Cambrian started 543 million years ago, novel predators and prey with hard, protective parts began to emerge. Within about 40 million years, all the major animal lineages had evolved in the seas. What explains this Cambrian explosion in diversity? Shifting sea levels, drifting continents, and changes in ocean circulation patterns may have played a role. The evolution of master genes that affect body plans could also be a factor (Section 16.8). Mutations in these genes would have allowed adaptive changes in body form in response to predation or altered habitat conditions.

Ancestors of modern animals originated in the Cambrian seas. Perhaps they were colonies of flagellated cells similar to the simplest modern animal, the placozoan.

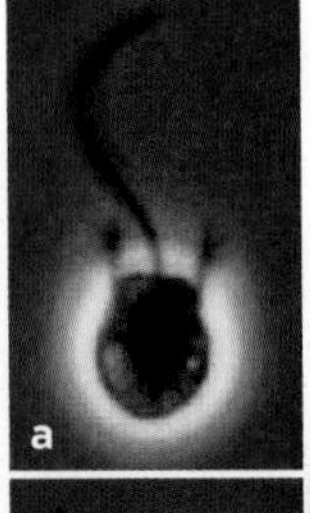

Figure 23.5 (**a**) Free-living choanoflagellate. A collar of microvilli rings its flagellum. (**b**) Colony of choanoflagellates. Some researchers view it as a model for the origin of animals. (**c**) The only named placozoan, *Trichoplax adhaerens*. Its asymmetrical, two-layered body humps over food. Gland cells in the lower layer secrete digestive enzymes.

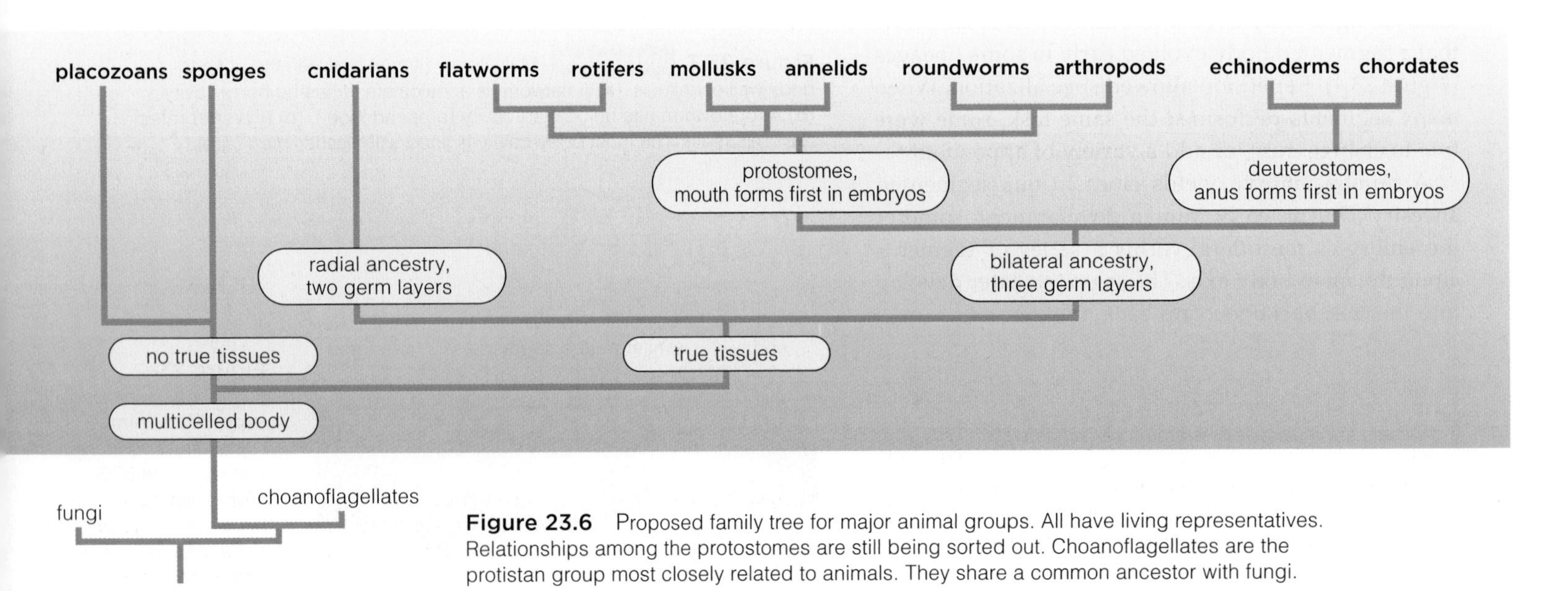

Figure 23.6 Proposed family tree for major animal groups. All have living representatives. Relationships among the protostomes are still being sorted out. Choanoflagellates are the protistan group most closely related to animals. They share a common ancestor with fungi.

23.3 Sponges—Success in Simplicity

Yes, it is possible to endure through time with simplicity, as demonstrated by the sponges.

GENERAL CHARACTERISTICS

Sponges (phylum Porifera) are aquatic animals with no symmetry, tissues, or organs. They are like a colony of choanoflagellates but with more kinds of cells and a greater division of labor. Most live in tropical seas, others in frigid ocean water or fresh water. A few are large enough to sit in; others would fit on a fingertip. Different kinds range in shape from sprawling and flat to lobed, compact, or tubelike (Figure 23.7).

A typical sponge attaches to a surface and has a body riddled with pores. Flat, nonflagellated cells form the outer surface, flagellated collar cells line the inner one, and a jellylike matrix lies in between (Figure 23.8).

Most sponges feed on bacteria filtered from water. As in choanoflagellates, movement of flagella drives food-laden water past a collar of microvilli. As water flows through pores in the body wall, collar cells trap bits of food and engulf them by phagocytosis (Section 5.9). All digestion is intracellular. Amoeba-like cells in the matrix receive food from the collar cells and then distribute it to other cells throughout the body.

Sponges cannot run away from predators, but they have other defenses. In many species, cells inside the matrix secrete fibrous proteins or glassy spikes called spicules (Figure 23.7*c*). These defenses make sponges unappetizing to most predators. Some sponges secrete slime or chemicals that repel predators. The chemicals also help fend off competitors for living space.

SPONGE REPRODUCTION

A typical sponge is a hermaphrodite; it produces eggs and sperm. Sperm are released (Figure 23.7*b*). Eggs are held until after fertilization. A zygote develops into a ciliated larva. A larva (plural, larvae) is a free-living, sexually immature stage in an animal life cycle. Sponge larvae swim briefly, then settle and become adults.

Many sponges reproduce asexually as small buds or fragments break away and grow into new sponges. Some freshwater species survive oxygen-poor water, drying out, or freezing by producing gemmules: tiny clumps of cells encased in a hardened coat. In better times, gemmules grow into a new sponge.

Sponges are filter feeders with no symmetry, tissues, or organs. Fibers and sharp spicules in the body wall, as well as chemical secretions, help deter most predators.

Figure 23.7 (**a**) Vase-shaped sponges. (**b**) Barrel sponge releasing a cloud of sperm. (**c**) Framework for a Venus's flower basket (*Euplectella*). Many silica spicules fuse together as a rigid framework. A thin layer of flattened cells stretches over its outer surface. (**d**) A bright-red encrusting sponge on a ledge in a temperate sea.

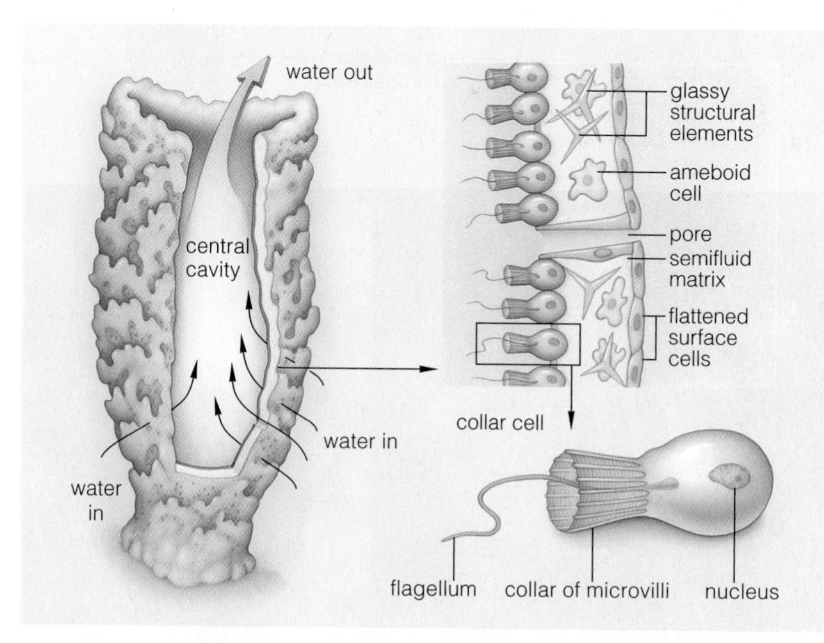

Figure 23.8 **Animated!** Body plan of a simple sponge. Flattened cells cover the outer surface and line pores. Flagellated collar cells line inner canals and chambers. Microvilli of these cells act like a sieve that strains food from the water. Cells in the matrix distribute nutrients and secrete structural elements.

23.4 Cnidarians—Simple Tissues, No Organs

Nearly all cnidarians live in the seas. They have radial body plans with true epithelial tissues.

GENERAL CHARACTERISTICS

The cnidarians (phylum Cnidaria) are radial, tentacled animals. All but a few of the 10,000 kinds are marine. A major subgroup, the *anthozoans*, includes soft corals, reef-forming corals, and sea anemones. *Medusozoans*, the other major subgroup, includes hydras, jellyfishes, and siphonophores (Figures 23.9 through 23.12).

Medusae (singular, medusa) and polyps are typical body forms (Figure 23.9*a,b*). In both, a tentacle-ringed mouth opens onto a gastrovascular cavity that serves in digestion and gas exchange. Medusae are shaped like a bell or an umbrella, and most swim about in the water column. Polyps are tubular. They usually attach to some surface and remain there.

Cnidarians have two epithelia: an outer epidermis and inner gastrodermis that secretes digestive enzymes from gland cells. Between the two is mesoglea (middle glue), a jellylike matrix with a few cells scattered in it.

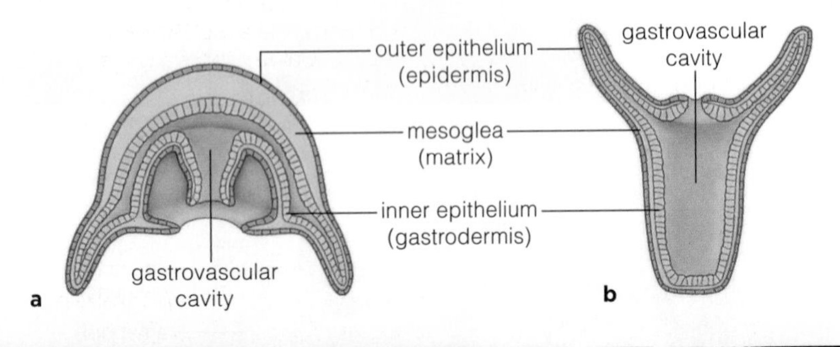

Figure 23.9 **Animated!** Two cnidarian body plans: (**a**) medusa and (**b**) polyp, cutaway views. (**c**) Sea anemone and (**d**) part of a colony of a reef-building coral, with interconnected hard walls. See also Sections 30.1, 41.2, and 43.14.

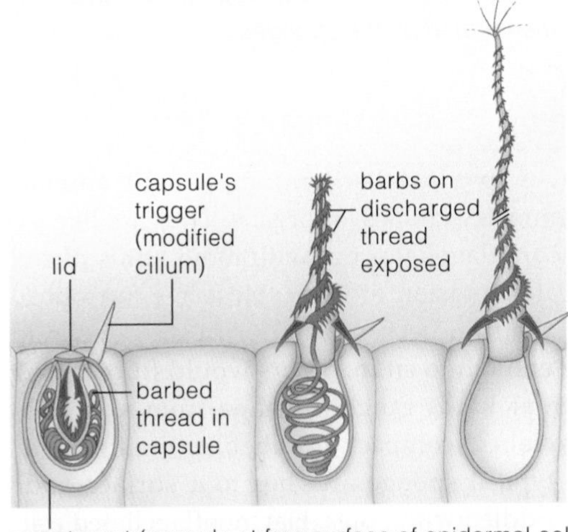

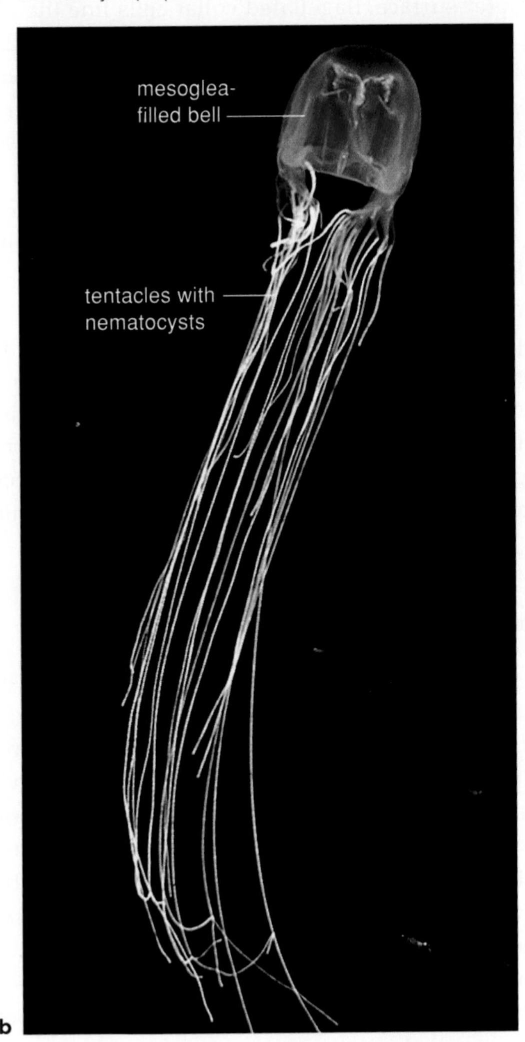

Figure 23.10 (**a**) Nematocyst before and after prey (not shown) touched its trigger and made the capsule "leakier." Water diffused in, pressure built up, and the thread was forced to turn inside out. The barbed tip pierces prey. (**b**) Sea wasp (*Chironex*), a box jelly with nematocyst-studded tentacles as long as five meters. Its toxin can kill you within minutes.

Interconnecting nerve cells thread through epithelia, forming a nerve net, a simple nervous system. Body parts move when these nerve cells signal contractile cells. The resulting contractions redistribute mesoglea, just as a water-filled balloon changes shape when you squeeze it. Any fluid-filled cavity or cellular mass that contractile cells act upon is a hydrostatic skeleton.

UNIQUE CNIDARIAN WEAPONS

Cnidarians sometimes are called the "flowers of the seas," but most are carnivores. Cnidarians alone have cells that contain nematocysts, thick-walled capsules used in prey capture and as a defense (Figures 23.10 and 23.11). Like a jack-in-the box, the capsules hold a coiled thread, often spiny or barbed, under a hinged lid. When something touches the capsule's trigger, the top pops open and the thread delivers stinging barbs or toxins into whatever it hits. Nematocysts irritate human skin and, rarely, can be deadly.

LIFE CYCLES

Cnidarian life cycles vary. For example, anthozoan cycles do not include a medusa stage. Gamete-making organs (gonads) form on polyps. In medusozoan life cycles, asexually-reproducing polyps usually alternate with sexually-reproducing medusae. As one example, consider *Obelia*, a colonial hydroid (Figure 23.12). The colony grows asexually when new polyps branch from older ones. Sexual reproduction occurs after medusae form on special polyps. Medusae release gametes that fuse, forming a zygote. As in most cnidarians, bilateral ciliated larvae develop. They swim briefly, then settle and take on the adult form.

Many cnidarians are colonial. Siphonophores such as a Portuguese man-of-war (*Physalia*) are not a single animal, but a colony of many specialized polyps and medusae (Figure 23.11*b*). As another example, coral reefs are large colonies of calcium carbonate-secreting polyps. The polyps have dinoflagellates living inside their tissues. These photosynthetic protists get shelter and carbon dioxide from the coral, and the coral gets sugars and oxygen from them. If a reef-building coral loses its symbionts, an event called "coral bleaching," it may die. We will discuss this again in Section 43.14.

Cnidarians are radially-symmetrical polyps or medusae. Both body forms have tentacles with nematocysts around the mouth, which opens into a gastrovascular cavity. A nerve net signals contractile cells during movements.

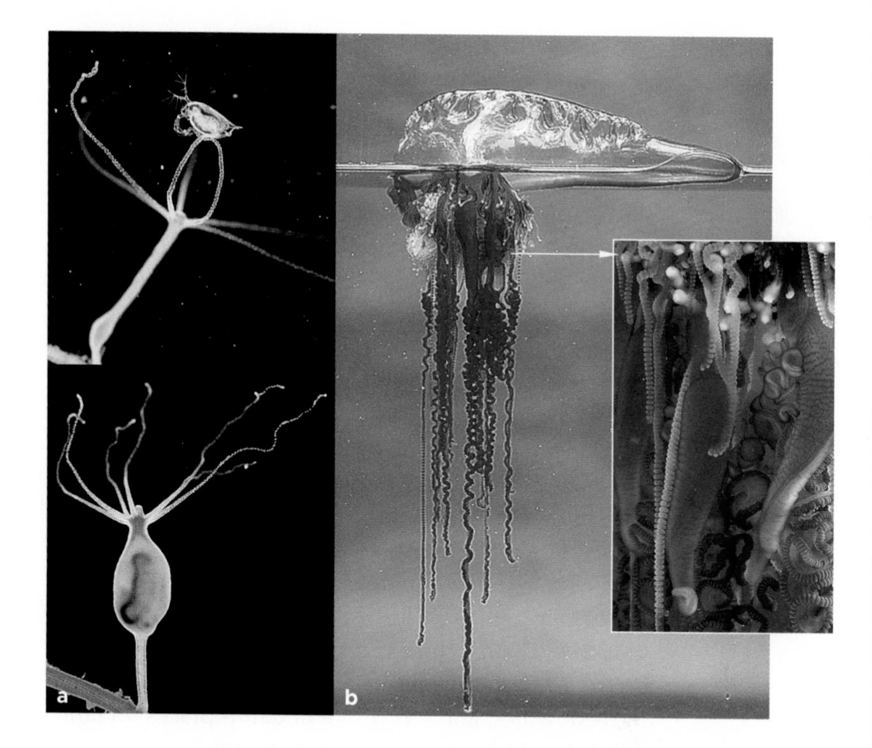

Figure 23.11 (**a**) One of the few freshwater cnidarians: a hydroid (*Hydra*), capturing a water flea (*above*) and digesting it (*below*). (**b**) Portuguese man-of-war (*Physalia utriculus*), a siphonophore. A purplish-blue, air-filled float—a modified polyp—keeps the colony at the surface of warm seas. Beneath it are gamete-producing medusae and polyps that specialize in prey capture, defense, and digestion. Tentacles of some polyps can be ten meters long.

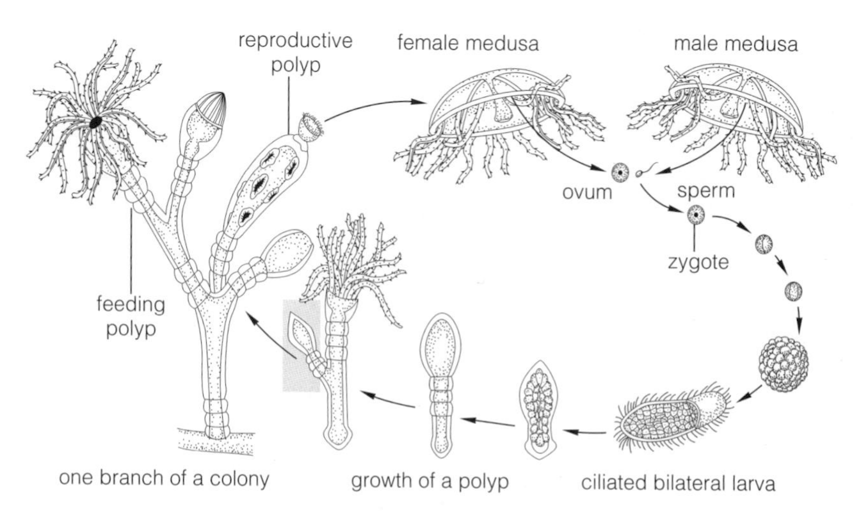

Figure 23.12 **Animated!** Life cycle of a colonial hydroid (*Obelia*). Reproductive polyps release medusae, which make eggs and sperm. Zygotes develop into ciliated, bilateral larvae that crawl briefly, then settle and grow into polyps. More polyps form asexually. A colony may have thousands of feeding and reproductive polyps.

23.5 Flatworms—Simple Organ Systems

Beyond the cnidarians are animals ranging from worms to humans. Most are bilateral, with a distinct head, an anterior–posterior body axis, and organ systems.

LINKS TO SECTIONS 1.1, 8.1, 14.1

Although cnidarians form tissues and some sensory structures, they do not have organs. Organs, recall, are structural units of two or more tissues that develop in predictable patterns and that interact in one or more tasks (Section 1.1). Each organ system consists of two or more organs interacting chemically, physically, or both as they carry out specialized tasks.

Structurally, flatworms (phylum Platyhelminthes) are the simplest animals with a three-layered embryo. They are bilateral and cephalized. Some of their genes are similar to those that map out body segments and a coelom in other animals, but they have no segments or coelom. By one hypothesis, their ancestors had both features, which were lost as the lineage evolved.

Turbellarians, flukes, and tapeworms are the main groups. Most turbellarians are marine, but some live in fresh water, and a few live in damp places on land. All flukes and tapeworms are parasites.

STRUCTURE OF A FREE-LIVING FLATWORM

Planarians are free-living turbellarians commonly seen in ponds and streams. A muscular tube, the pharynx, connects the mouth with the gut. The pharynx sucks in food and expels waste; a planarian digestive system is incomplete (Figure 23.13*a*).

A pair of nerve cords, each a communication line, run the length of the body (Figure 23.13*b*). Clusters of nerve cell bodies, called ganglia (singular, ganglion), serve as a simple brain. The head also has chemical receptors and light-detecting eyespots.

Planarians have female and male sex organs; they are hermaphrodites (Figure 23.13*c*). Some species can also reproduce asexually. The body splits in two near the midsection, then each piece regrows the missing part. Regrowth also occurs if a planarian is cut in two.

A system of tubes regulates water and solute levels. Flame cells with a tuft of "flickering" cilia drive any excess water into these tubes, which open to the body surface at a pore (Figure 23.13*d–f*).

FLUKES AND TAPEWORMS—THE PARASITES

Flukes and tapeworms show specializations for their parasitic life style. Immature stages often live in one or more intermediate hosts, then reproduction occurs in the main host. For example, Figure 23.14 shows the life cycle of a blood fluke (*Schistosoma*). Aquatic snails are the intermediate host, but reproduction can only take place inside a mammal, such as a human.

Ancestors of tapeworms probably had a gut and mouth but lost both as they evolved in the vertebrate

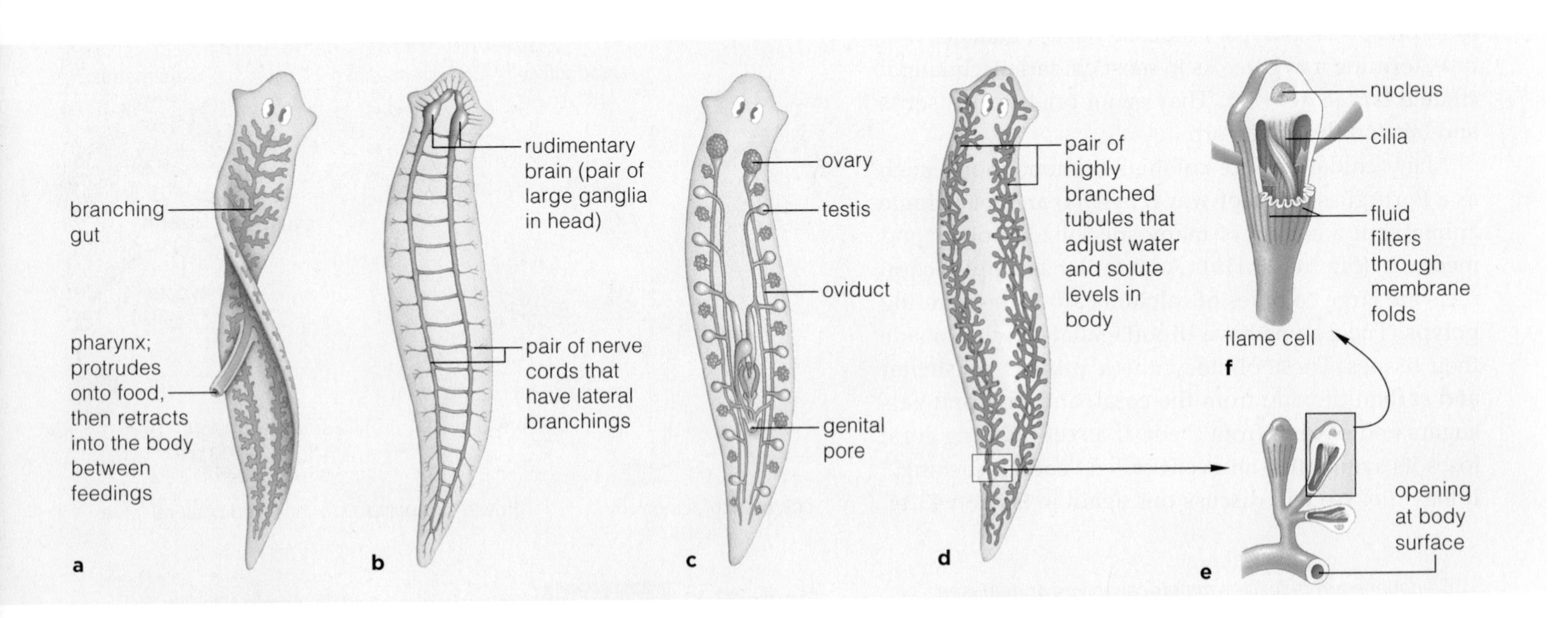

Figure 23.13 **Animated!** Organ systems of a planarian, one of the flatworms. (**a**) Digestive system. This branching saclike gut is connected to a pharynx, which protrudes onto food, then retracts into a chamber between feedings. (**b**) Nervous system. (**c**) Reproductive system. (**d–f**) Water-regulating system.

gut, a habitat rich in predigested food. Modern species latch onto the intestinal wall with a scolex, a structure with hooks or suckers at the head end. Nutrients reach cells by diffusing across the tapeworm body wall.

A tapeworm body consists of proglottids. It grows as these repeating body units form and bud from the region behind the scolex. The tapeworm can fertilize itself, because each proglottid is hermaphroditic. The sperm from one can fertilize eggs in another. Older proglottids (farthest from the scolex) contain fertilized eggs. The oldest units break off, then exit the body in feces. Fertilized eggs can survive for months on their own before reaching an intermediate host.

Some tapeworms parasitize humans. Larvae enter the body when a person eats undercooked meat or fish that contains larvae. Figure 23.15 shows the life cycle of a beef tapeworm.

Flatworms are among the simplest bilateral, cephalized animals with organ systems. Organs arise from three primary tissue layers that form in embryos.

Free-living turbellarians and some notorious parasitic flukes and tapeworms are in this group.

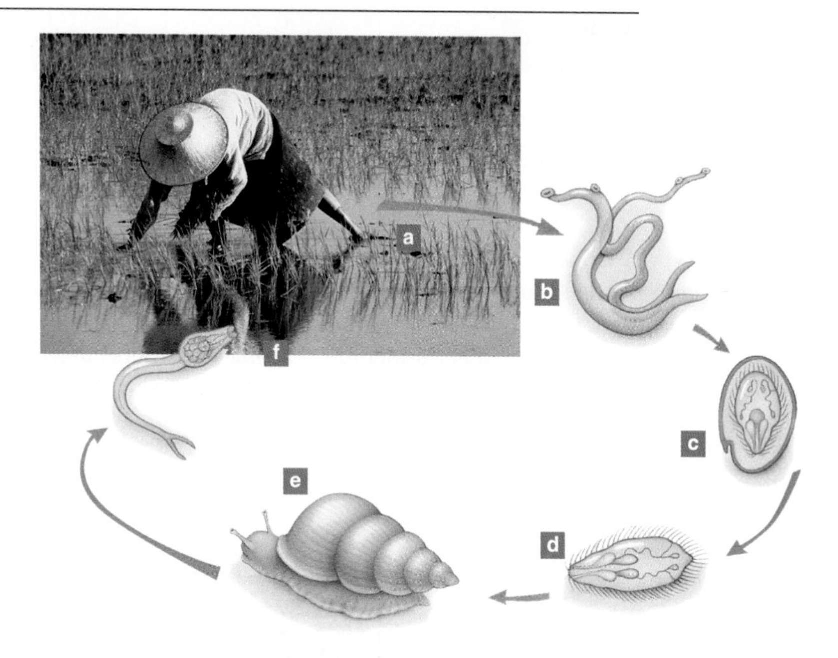

Figure 23.14 Life cycle of *Schistosoma japonicum*, found mainly in China, Indonesia, and the Philippines. (**a**) This blood fluke grows, matures, and mates in human hosts. (**b**,**c**) Fertilized eggs exit in feces and hatch as ciliated larvae. (**d**) Larvae burrow into an aquatic snail and multiply asexually. (**e**,**f**) Fork-tailed, swimming larvae develop, leave the snail, and bore into a human host. They enter intestinal veins and start a new cycle. Early symptoms of the resulting disease, schistosomiasis, are not obvious. Later on, side effects of immune responses to fluke eggs damage internal organs. Worldwide, an estimated 200 million people are currently infected by some kind of schistosome.

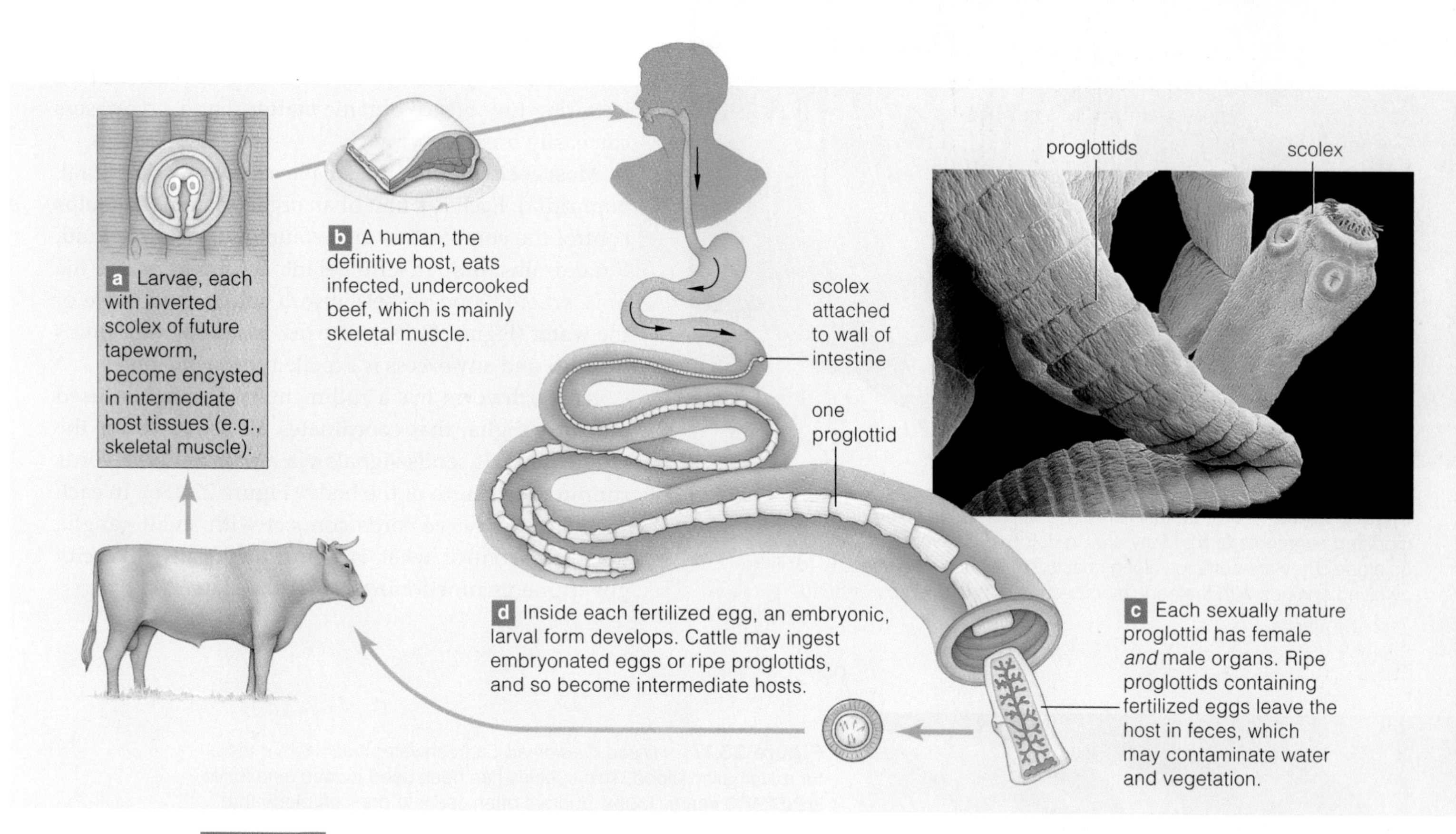

Figure 23.15 **Animated!** Life cycle of a beef tapeworm (*Taenia saginata*). Adult worms can grow 7 meters (22 feet) long. The photograph shows the pork tapeworm, *T. solium*.

23.6 Annelids—Segments Galore

Most annelids are segmented inside and out. In certain groups, many segments became highly modified.

SIMPLE TO HIGHLY MODIFIED SEGMENTS

LINK TO SECTION 16.7

Annelids are bilateral worms with a coelom and a body that is impressively segmented, inside and out. The 12,000 or so species include polychaetes, oligochaetes such as earthworms, and leeches. Except in leeches, nearly all segments bear chaetae, or chitin-reinforced bristles. Hence the names polychaete and oligochaete (*poly–*, many; *oligo–*, few). Bristles shoved into marine sediments or soil give the worm traction for crawling or burrowing.

All segments were similar in early annelids, but the repetition invited modification. Many polychaetes have a variety of appendages sticking out of their head and pairs of fleshy, lobed structures along their body. Some lobes handle food, but most act like feet (Figure 23.16). Besides having feeding structures on the head, leeches have a sucker at both ends (Figure 23.17).

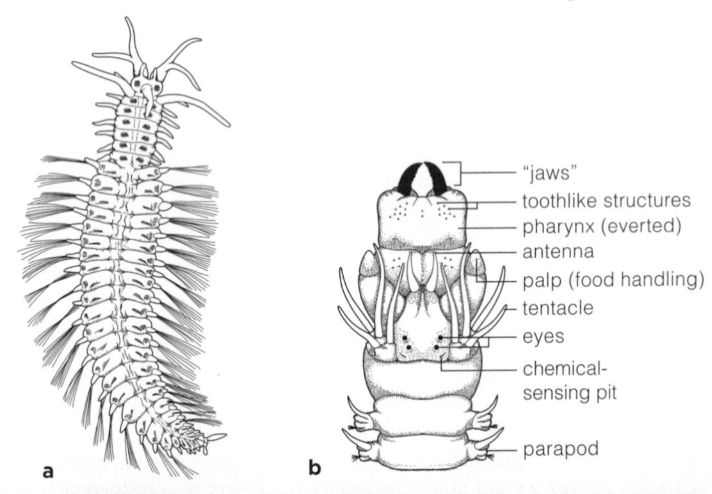

Figure 23.16 Polychaetes—annelids with often stunning arrays of highly modified segments (**a**,**b**). Many are predators or scavengers; others graze on algae. They are common along coasts. (**c**) A burrowing species. (**d**) Tube-dwelling species with feathery, mucus-coated structures on its head.

ANNELID ADAPTATIONS—A CASE IN POINT

An earthworm body has many similar segments, each containing a coelomic chamber (Figure 23.18*a*). Gas exchange occurs across the moist body surface, which is covered by a secreted cuticle. Running through one coelomic chamber after another is a closed circulatory system. Such systems confine the blood and pump it through the interconnected vessels that run through the body (Figure 23.18*b*).

A complete digestive system also extends through all coelomic chambers (Figure 23.18*c*). Earthworms are scavengers that ingest soil, then digest organic debris. The worms improve soil by loosening its particles and excreting tiny bits of organic matter that decomposers can easily break down.

Most coelomic chambers have a nephridium (plural, nephridia). Each is a unit of an organ system that helps control the composition and volume of coelomic fluid. Fluid drains through a funnel into a tubular part of the unit, where blood vessels absorb solutes and some of the water (Figure 23.18*d*). The rest is stored in the unit's bladder, and any excess is expelled through a pore.

An earthworm has a rudimentary "brain," a fused pair of ganglia, that coordinates all activities for the whole body. It sends signals via a pair of nerve cords running the length of the body (Figure 23.18*e*). In each segment, the nerve cords connect with small ganglia that help control what is going on locally, like city governments functioning across the state.

Figure 23.17 *Hirudo medicinalis*, a freshwater leech with a taste for mammalian blood. This species has been used in medicine for at least 2,000 years. Today, doctors often use it to draw off blood that pools after a severed ear, lip, or fingertip has been reattached. The body can't do that until the blood circulation has been reestablished. Most leech species prey on invertebrates or act as scavengers.

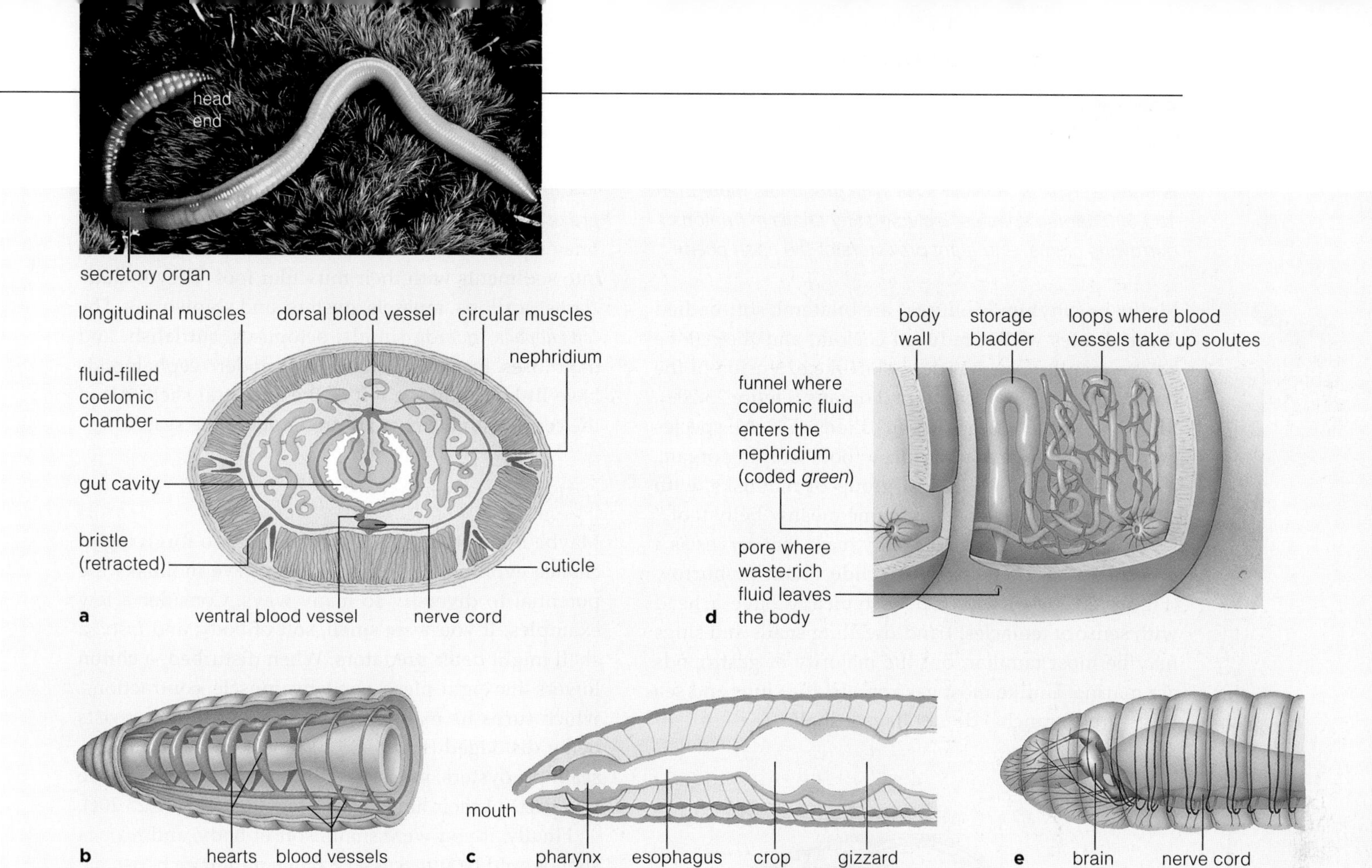

Figure 23.18 Animated! Earthworm body plan. (**a**) Transverse section. (**b**) Closed circulatory system. (**c**) Digestive system. (**d**) A nephridium, one of many functional units that balance the volume and composition of body fluids. (**e**) Nervous system.

An earthworm has a hydrostatic skeleton. Muscle contractions put pressure on fluid trapped inside each segment. As circular muscle of a segment contracts, longitudinal muscle relaxes, and the segment gets long and thin. Then circular muscle relaxes, longitudinal muscle contracts, and the segment gets shorter and wider. Coordinated waves of contraction propel the worm through the soil (Figure 23.19). Bristles of each segment alternately extend outward to anchor it, then retract as the segment is pulled forward.

Earthworms are hermaphrodites. A secretory organ produces mucus that glues two worms together while they swap sperm. Later, that organ secretes a case that protects the fertilized eggs.

Annelids are bilateral, coelomate, segmented worms that have digestive, nervous, excretory, and circulatory systems. Some have simple segments; the variety of appendages on others is testimony to the potential of segmentation.

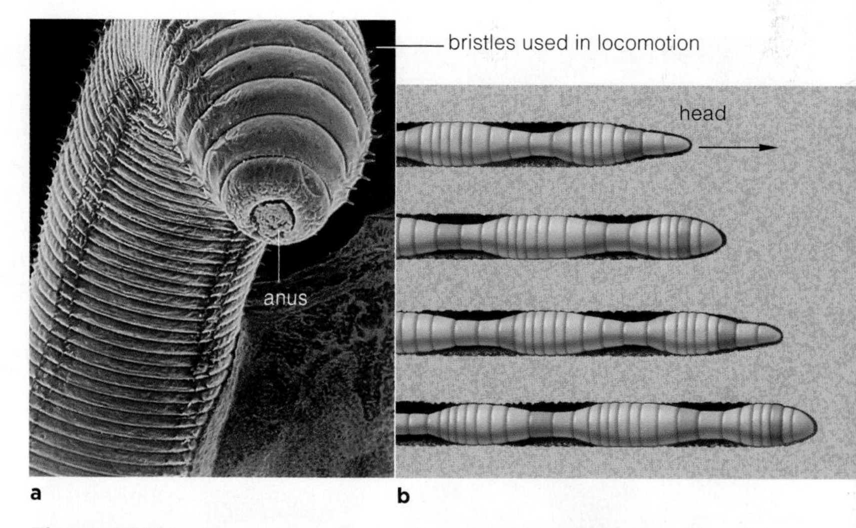

Figure 23.19 How earthworms move through soil. (**a**) Bristles on sides of the body extend and withdraw as muscle contractions act on coelomic fluid inside each segment. (**b**) Bristles are extended when a segment's diameter is at its widest (when circular muscle is relaxed and longitudinal muscle is contracted). They retract as the segment gets long and thin. A worm's front end is pushed forward, then bristles anchor it and the back of the body is pulled up behind it.

23.7 The Pliable Mollusks

Is there a "typical" mollusk? No. The group has more than 100,000 named species, including tiny snails in treetops, burrowing clams, and giant predators of the open ocean.

LINK TO SECTION 16.5

Mollusks (phylum Mollusca) are bilateral, soft-bodied invertebrates with a reduced coelom, and they alone have a mantle. The mantle, a skirtlike extension of the body wall, drapes over internal organs (Figure 23.20*a*). Most have a respiratory organ called a gill. All species except bivalves have a radula, a food-scraping organ.

Gastropods are the largest group of mollusks, with at least 60,000 species. Their name means "belly foot." Many species have a lower body mass that serves as a muscular foot; it helps them glide about or burrow (Figure 23.20*a–c*). Gastropods typically have a head with sensory tentacles. Land-dwelling snails and slugs may be most familiar, but the majority of gastropods are marine. Unlike most gastropods, the slugs and sea slugs (nudibranchs) do not have a shell.

Chitons have a shell of eight plates. They mostly graze on algae. *Bivalves* have a two-part, hinged shell. They siphon and filter food from water and often dig into sediments with their muscular foot. They include clams, scallops, mussels, oysters, and shipworms. The *cephalopods* include squids, octopuses, cuttlefish, and nautiluses. Except for nautiluses, modern cephalopods have little or nothing left of the ancestral shell. All of the cephalopods are swift predators of the seas.

HIDING OUT, OR NOT

Maybe it was their fleshy, soft bodies, so forgiving of chance evolutionary changes, that gave mollusks the potential to diversify so many ways. Consider a few examples. If you were small, soft of body, and *tasty*, a shell might deter predators. When disturbed, a chiton lowers its eight-piece shell by muscle contractions, which turns its mantle into a suction cup that resists being dislodged (Figure 23.20*d*). When danger looms, scallops, oysters, mussels, and other bivalves snap the two parts of their hinged shell together (Figure 23.20*e*).

Finally, if you were small, soft of body, and *toxic*, a shell would be unnecessary. Some nudibranchs secrete nasty-tasting substances, such as sulfuric acid. Others eat cnidarians and store undischarged nematocysts in their tissues. Spanish shawl nudibranchs do this. They store nematocysts inside red respiratory organs (Figure 23.20*f*). The sting of nematocysts teaches predators to avoid this colorful, easily recognized gastropod.

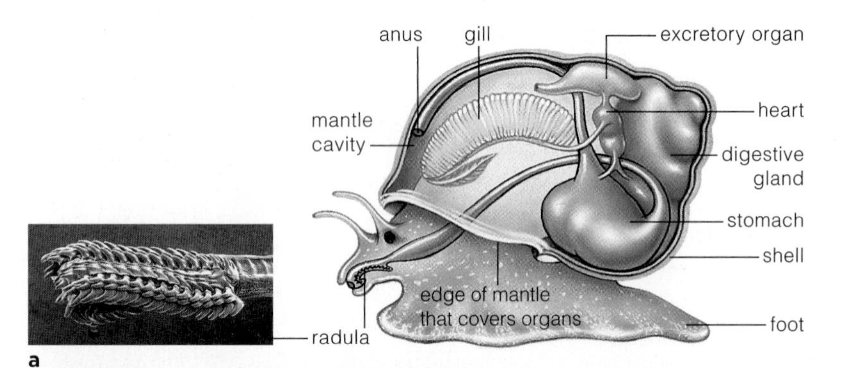

Figure 23.20 **Animated!** (**a**) Body plan of an aquatic snail. The micrograph shows the radula of a land snail. (**b**) An aquatic snail using its big "foot" to crawl on the glass wall of an aquarium. (**c**) A land snail. (**d**) Chiton, with a shell made of eight plates. (**e**) Scallop, a bivalve with many eyes (*blue* dots). (**f**) Two Spanish shawl nudibranchs (*Flabellina iodinea*).

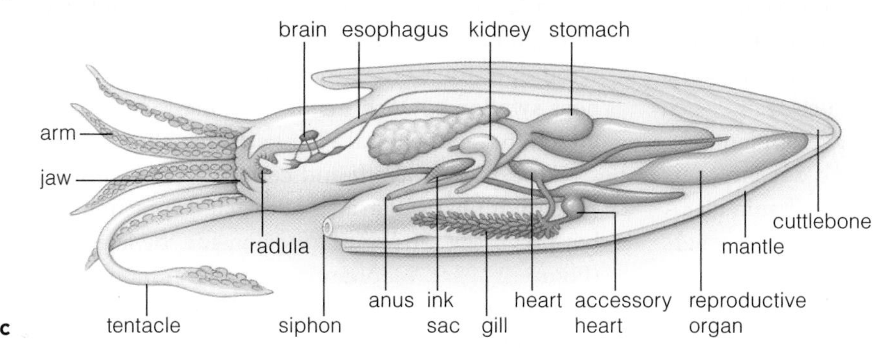

Figure 23.21 **Animated!** Cephalopods. (**a**) Painting of an Ordovician sea. In the foreground are two nautiloids, now extinct. The chambered nautilus (**b**) is their living relative. (**c**) Cuttlefish body plan. A "cuttlebone" is all that is left of an ancestral shell. (**d**) Human diver and giant squid (*Dosidicus*) inspecting each other. (**e**) An octopus.

ON THE CEPHALOPOD NEED FOR SPEED

Five hundred million years ago, in the Ordovician, cephalopods were the top predators of the open seas (Figure 23.21*a*). All lived inside a shell with multiple chambers. Except for a few species of nautiluses, their modern descendants have a highly reduced shell or none at all (Figure 23.21*b–e*).

What happened? Jawed fishes started an adaptive radiation in the Devonian, about 400 million years ago (Section 16.5). Many fishes that hunted cephalopods or competed with them for prey became swifter and larger. In what appears to have been a long-term race for speed, most cephalopods left their external shell behind. They became fast, highly active, streamlined, and—for invertebrates—smart.

For cephalopods, jet propulsion became the name of the game. They moved faster by shooting a jet of water out from the mantle cavity, through a funnel-shaped siphon. All modern cephalopods do the same. The brain controls siphon activity and the direction in which the body moves. Increased speed correlates with increasingly complex eyes and respiratory and circulatory systems. Of all groups of mollusks, only cephalopods have a closed circulatory system. Blood pumped by a heart gives up carbon dioxide and picks up oxygen in two gills, then two accessory hearts pump it to all body tissues.

Cephalopods include the fastest (squids), biggest (giant squid), and smartest (octopuses) invertebrates. Of all mollusks, the octopuses have the largest brain relative to body size, and the most complex behavior.

Mollusks are bilateral, soft-bodied, coelomate animals, the only ones with a mantle over the body mass. Some are filter feeders without a head; others are brainy predators.

23.8 Roundworms

Roundworms are among the most abundant animals. Sediments in shallow water may hold a million per square meter; a cupful of topsoil teems with them.

LINKS TO SECTIONS 15.5, 22.3

Roundworms, or nematodes (phylum Nematoda), are bilateral, unsegmented animals that have a false coelom and a cuticle-covered, cylindrical body (Figure 23.22). Nearly all of the 22,000 or so species are less than five millimeters long; but one that lives inside the sperm whale can grow thirteen meters long. A roundworm's cuticle is collagen-rich and pliable, and it is repeatedly molted. Molting is the shedding of a body covering (or hair, or feathers) between growth spurts.

One free-living species, *Caenorhabditis elegans*, is a favorite for genetic experiments. It has the same tissue types as complex animals. Yet it is transparent, it has only 959 body cells, and it reproduces fast. Its genome is about 1/30 the size of ours. With such traits, each cell's fate is easy to monitor during development.

Although most species are free-living decomposers in soil or water, some parasitize plants and animals. A fungus that preys on roundworms may help farmers control the kinds that destroy crops (Section 22.3).

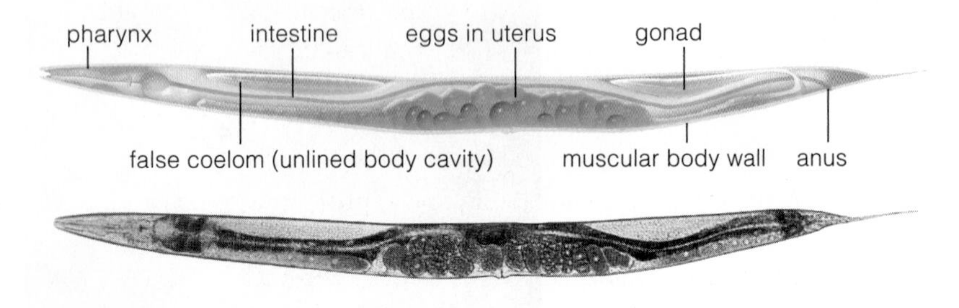

Figure 23.22 **Animated!** Body plan and micrograph of *Caenorhabditis elegans*, a free-living roundworm. Sexes are separate and this is a female.

Some parasitic roundworms impair human health, especially in developing countries in tropical zones. Eating undercooked pork or wild game can lead to an infection by *Trichinella spiralis*. The roundworm moves from the intestines, through blood, and into muscles, where it forms cysts (Figure 23.23*a*). Infection causes trichinosis, a painful disease that can be fatal.

Ascaris lumbricoides, a large roundworm, currently infects more than 1 billion people, primarily in Asia and in Latin America (Figure 23.23*b*). People become infected when they eat its eggs, which survive in soil and get on hands and into food. When enough adults occupy a host, they can clog the digestive tract.

Hookworms, too, infect more than 1 billion people. Juveniles in the soil cut into human skin and migrate through blood vessels to the lungs. They climb up the windpipe, then enter the digestive tract when their host swallows. Once inside the small intestine, they attach to the intestinal wall and suck blood.

Wuchereria bancrofti and certain other roundworms cause lymphatic filariasis. Repeated infections injure lymph vessels, so lymph pools inside the legs and feet (Figure 23.23*c*). Elephantiasis, the common name for this disease, refers to fluid-filled, "elephantlike" legs. Mosquitoes carry larval roundworms to new hosts.

Pinworms (*Enterobius vermicularis*) commonly infect children. Female worms less than a millimeter long leave the rectum at night and lay eggs near the anus. The migration causes itching, and scratching puts eggs under fingernails. From there, they get into food and onto toys. If swallowed, the eggs spread the infection.

Roundworms are bilateral worms with a cuticle that they molt as they grow. Organ systems almost fill their false coelom. Some are agricultural pests or human parasites.

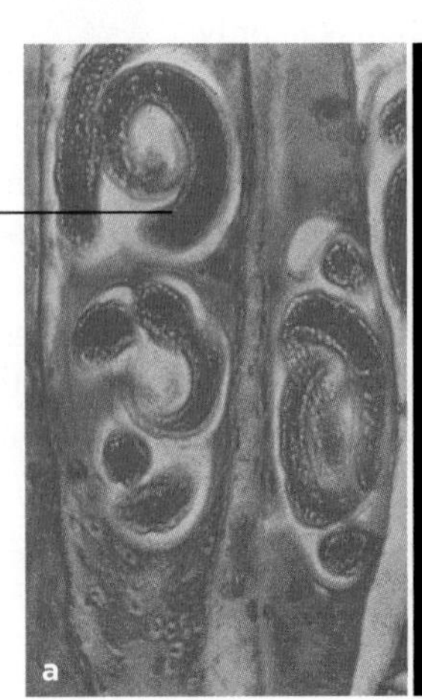

Figure 23.23 (**a**) *Trichinella spiralis* larvae in muscle tissue of a host animal. (**b**) Live roundworms (*Ascaris lumbricoides*). These intestinal parasites cause stomach pain, vomiting, and appendicitis. (**c**) A case of elephantiasis that resulted from an infection by the roundworm *Wuchereria bancrofti*.

23.9 Why Such Spectacular Arthropod Diversity?

Evolutionarily speaking, "success" means having the most species in the most habitats, fending off competition and threats, exploiting the most food, and producing the most offspring. These features characterize arthropods.

MAJOR GROUPS

Arthropods (phylum Arthropoda) are bilateral animals with a hardened, jointed external skeleton, a complete gut, and a reduced coelom. Major lineages are *trilobites* (now extinct, Figure 23.24*a*), *chelicerates* (spiders and close relatives), *crustaceans* (such as lobsters, shrimps and crabs) and *uniramians* (millipedes, centipedes, and insects). We survey the modern groups in sections to follow, so start thinking about the six key adaptations that contributed to their evolutionary success.

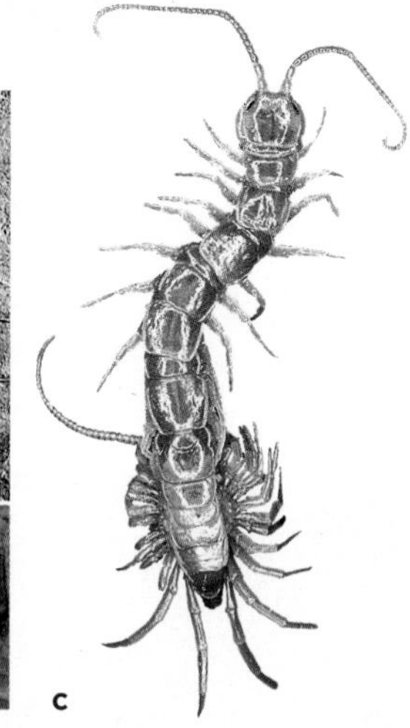

Figure 23.24 Structurally simple arthropods. (**a**) Fossil of a trilobite, an extinct arthropod. (**b**) A millipede, a type of scavenger. Each fused segment has two pairs of legs. (**c**) Centipede molting its old exoskeleton (*gray*). Each segment has a pair of legs.

KEY ARTHROPOD ADAPTATIONS

Hardened Exoskeleton Arthropods have a cuticle of chitin, proteins, and waxes. It is an exoskeleton, an external skeleton. It protected aquatic species against predators. It took on additional functions when some groups invaded land. An exoskeleton helps the body conserve water and support its weight on land.

A hard exoskeleton does not restrict size increases, because arthropods shed it after each growth spurt, as roundworms do. Hormones control molting. They call for the formation of a soft, new cuticle under the old one, which is shed (Figure 23.24*c*). Before the new one has a chance to harden, the body mass increases by the uptake of air or water and mitotic cell divisions.

Jointed Appendages If a cuticle were uniformly hard and thick like a plaster cast, it would prevent movement. Arthropod cuticles thin at joints: regions where two hardened body parts meet. Contraction of muscles that span the joints makes the cuticle bend, so abutting parts move relative to one another. The jointed exoskeleton was a key innovation; it led to the evolution of jointed legs, wings, and other specialized appendages. *Arthropod* means jointed leg.

Highly Modified Segments In early arthropods, body segments were more or less alike. In many of their descendants, segments became fused together or modified for specialized tasks. For example, among insects, thin extensions of the wall of some segments evolved into wings—appendages used in flight.

Respiratory Structures Aquatic arthropods have a type of gill for gas exchange. Land dwellers have air-conducting tubes that start at surface pores, branch into finer tubes, and get oxygen to all internal tissues. Flight and other activities that use a great deal of ATP require the rapid uptake of oxygen in all aerobically respiring tissues, such as muscle tissue (Section 7.1).

LINK TO SECTION 7.1

Sensory Specializations Most arthropods have one or more pairs of eyes: organs that sample the visual world. Most also have paired antennae that can detect touch and waterborne or airborne odors.

Specialized Stages of Development Especially among insects, the tasks of surviving and reproducing are divided among different stages of development. Individuals of many species undergo metamorphosis: Tissues get remodeled as embryonic stages make the transition to the adult form. A typical immature stage specializes in eating and growing rapidly, whereas the adult specializes in reproduction and dispersal. For instance, caterpillars chew leaves and metamorphose into butterflies, which fly about, mate, then deposit eggs. Differences among stages of development are adaptations to specific conditions in the environment, including seasonal shifts in food, water, and shelter.

As a group, arthropods are abundant and widespread. A hardened, jointed exoskeleton, modified body segments, and specialized appendages, respiratory structures, and sensory structures underlie their evolutionary success.

In many species—most notably insects—success also arises from a division of labor among different stages of the life cycle, such as larvae, juveniles, and adults.

23.10 Spiders and Their Relatives

Arthropods called chelicerates arose in shallow seas early in the Paleozoic. Their name refers to their first pair of feeding appendages. Horseshoe crabs have changed little and still live in the seas. Of familiar species on land—spiders, scorpions, ticks, and chigger mites—we might say this: Never have so many been loved by so few.

Figure 23.25 (**a**) Horseshoe crab (*Limulus*). (**b**) Scorpion. It has a venom-delivering stinger on its last segment. Its modified pedipalps function as claws that seize prey. (**c**) Tarantula with a fearless female. Like other spiders, it keeps insect populations in check. Its bite does not harm humans. Section 23.13 shows spiders that can be dangerous. (**d**) Dust mite, less than 0.5 millimeter in length. Some people have allergic reactions to mite feces and corpses.

Horseshoe crabs have a hardened, dorsal shield over a segmented body with five pairs of walking legs. A spikelike tail helps them to steer or to right themselves after a big wave flips them (Figure 23.25*a*). Like other chelicerates, the horseshoe crabs do not have antennae. They prey on clams and worms, and their eggs serve as a seasonal food source for migratory shore birds.

Spiders, scorpions, ticks, and mites are arachnids, and each has four pairs of legs. Scorpions and spiders are predators that subdue prey with venom (Figure 23.25*b,c*). Scorpions dispense venom through a stinger on their last body segment. Spiders deliver it with a bite; their fanglike chelicerae have poison-producing glands (Figures 23.26). Of 38,000 species, only about 30 can be dangerous to humans (Section 23.13). Most spiders indirectly help us by eating harmful or pesky insects and keeping their populations in check.

Spider segments fuse into a forebody and hindbody. Legs, chelicerae, and sensory pedipalps are forebody appendages. The hindbody has paired spinners that eject silk for webs. An open circulatory system in both allows blood to mingle with tissue fluids. Malpighian tubules move excess water and nitrogen-rich wastes from the tissues to the gut for disposal. Gas exchange occurs at leaflike "book lungs" in many species.

Of all arachnids, ticks and mites are the smallest, most diverse, and most widespread. Ticks are parasites with skin-piercing chelicerae (Section 19.8). Mites are parasites, predators, and scavengers on animals and in water, soil, and our homes (Figure 23.25*d*).

Spiders, scorpions, and their relatives are predators or parasites. Their first pair of appendages, chelicerae, are structurally and functionally unique feeding structures.

chelicerae

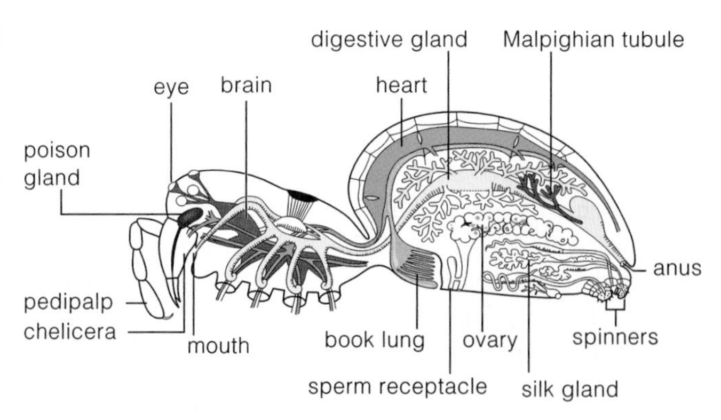

Figure 23.26 A spider's internal organization.

23.11 A Look at the Crustaceans

Tens of thousands of known species of lobsters, crabs, shrimps, and other crustaceans live in the ocean, where they are nicknamed "insects of the seas." Crustaceans also live in freshwater habitats. Pillbugs, sandhoppers, and others do well on land.

Crustaceans have two pairs of antennae that provide sensory information about their habitat, which is most often the sea. Krill swim in upper ocean waters (Figure 23.27*a*). These animals, typically just a few centimeters long, eat phytoplankton and serve as food for squids, fishes, penguins, and baleen whales.

About a third of all crustacean species are copepods (Figure 23.27*b*). Most are tiny swimmers, but some that parasitize fish or whales are as long as your forearm.

Larval barnacles swim, but adults are enclosed in a calcified shell and live attached to piers, rocks, and even whales (Figure 23.27*c*). They filter food from the water with feathery legs. As adults, they cannot move about, so you would think that mating might be tricky. But barnacles tend to settle in groups, and most are hermaphrodites. An individual extends a penis, often several times its body length, out to neighbors.

The lobsters, crabs, and shrimps are close relatives. All are bottom feeders that have five pairs of walking legs (Figure 23.28). In some of the lobsters and crabs, the first pair has become modified as a pair of claws. Like other arthropods, crabs repeatedly molt as they grow (Figure 23.29). Certain spider crabs grow quite a bit. With legs that can be more than a meter in length, they are the largest living arthropods.

Impressive numbers of and kinds of crustaceans live in the seas; others live in fresh water and on land. Krill and copepod species are major links in aquatic food webs.

Figure 23.27 Representative crustaceans (**a**) Antarctic krill (*Euphausia superba*), no more than six centimeters long. Their populations reach densities of 10,000 individuals per cubic meter of seawater and may extend for kilometers. (**b**) Free-living male copepod (*Cyclops*) about one millimeter long. (**c**) Goose barnacle. Adults cement themselves to one spot. You might mistake them for mollusks until they open their hinged shell and you see jointed legs. For more crustacean photos, see Figures 10.18, 43.34, and 44.9.

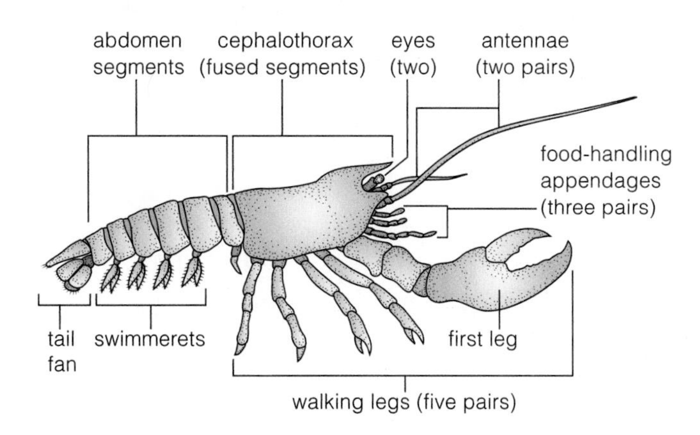

Figure 23.28 Body plan of a clawed lobster (*Homarus americanus*) showing its specialized appendages.

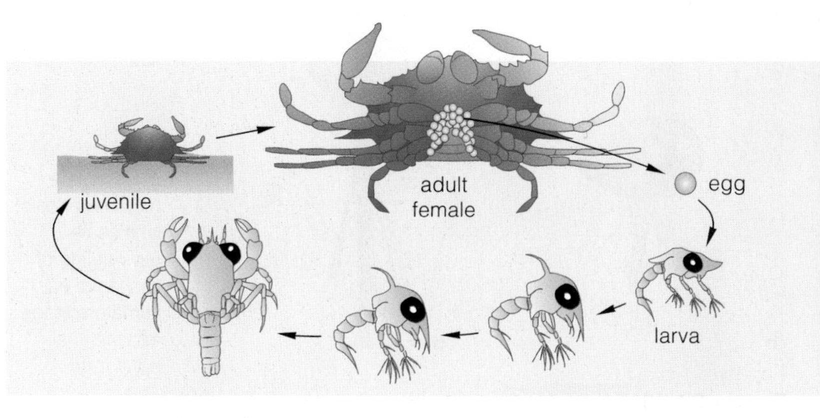

Figure 23.29 Animated! Crab life cycle. Larval and juvenile stages molt repeatedly and grow in size before they are mature adults. Adults continue to molt. A female carries her fertilized eggs under her abdomen until they hatch.

23.12 A Look at Insect Diversity

Insects are the most diverse group of animals. There are more species of dragonflies alone than there are species of mammals. Many insects produce staggering numbers of offspring. Ants and termites account for as much as one-third of the biomass of all animals on land.

LINKS TO SECTIONS 5.3, 17.4, 19.6, 21.7

In most adult insects, the segmented body is divided into three distinct parts: a head, thorax, and abdomen. Their head has paired sensory antennae and paired mouthparts that are specialized for biting, chewing, or other functions (Figure 23.30). The thorax has three pairs of legs and, usually, two pairs of wings. Insects are the only winged invertebrates.

Insects have a complete digestive system divided into a foregut, a midgut where food is digested, and a hindgut, where water is reabsorbed. As in spiders and other land-dwelling arthropods, Malpighian tubules function in excretion. Nitrogen-rich wastes produced by digestion of proteins diffuse from blood into these tubes. There, enzymes convert the waste to crystals of uric acid, which an insect excretes. Malpighian tubules help insects eliminate toxic metabolic wastes without losing precious water.

Larvae, nymphs, and pupae are immature stages of many insect life cycles (Figure 23.31). Like a human child, some insect nymphs have the form of an adult, although in miniature. Unlike children, however, these immature stages molt. Most insect larvae are not like the adults and must undergo metamorphosis. By this process, tissues are reorganized and remodeled in the transition to the adult form (Figure 23.31*b*,*c*).

Figure 23.32 shows a few of the more than 800,000 species. If numbers and distribution are the measure, the most successful insects are small and reproduce often. Great numbers grow and reproduce on a plant that would be just an appetizer for another animal. By one estimate, if all the offspring of a single female fly survived and reproduced for six more generations, she would have more than 5 trillion descendants!

The most successful insects are winged. They move among food sources that are too widely scattered to be exploited by other invertebrates. The ability to fly contributed hugely to insect success on land.

The very features that contribute to insect success also make them our most aggressive competitors for crops and forest products such as paper and lumber.

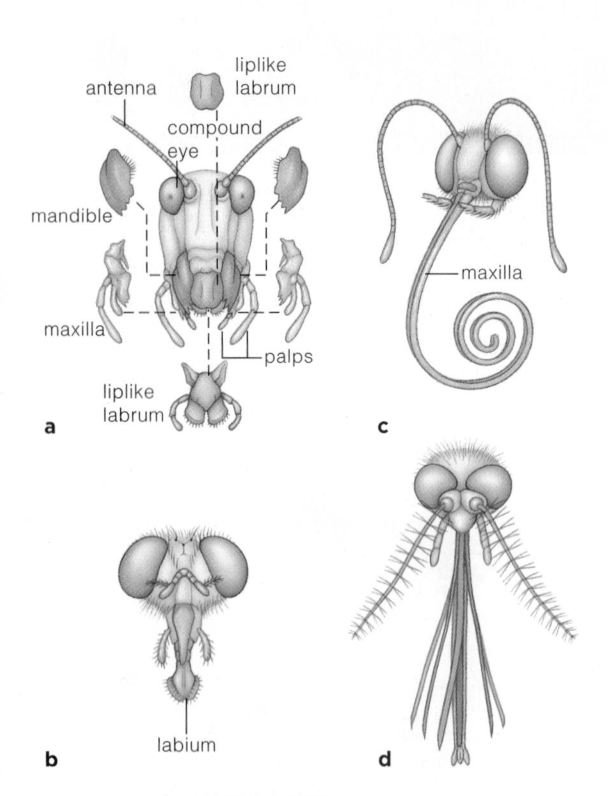

Figure 23.30 **Animated!** Examples of insect appendages. Head parts typical of (**a**) grasshoppers, which chew food; (**b**) flies, which sponge up nutrients; (**c**) butterflies, which siphon nectar; and (**d**) mosquitoes, which pierce hosts and suck up blood.

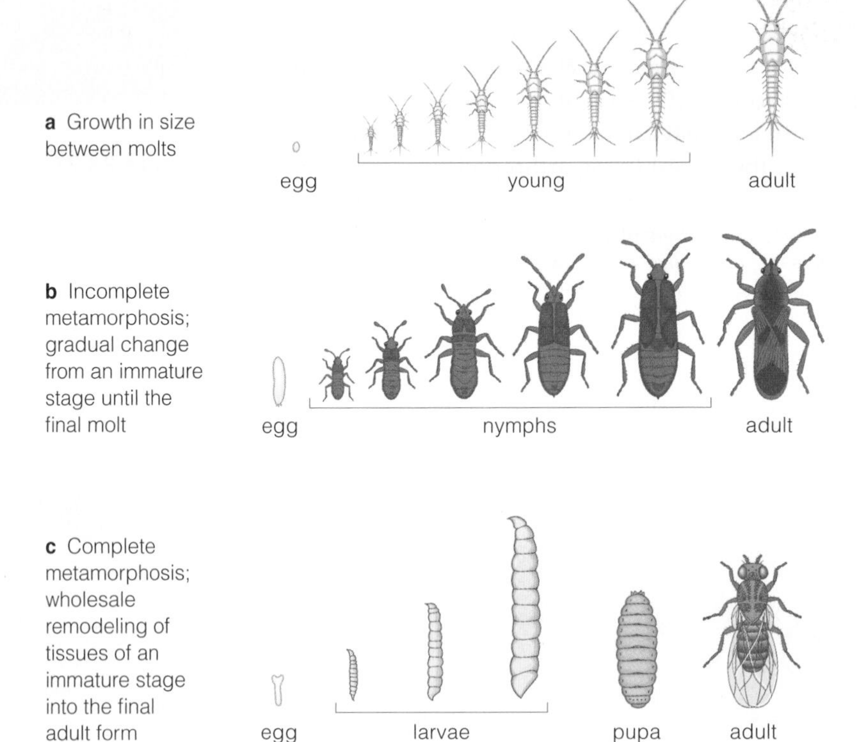

Figure 23.31 Insect development. (**a**) Young silverfish change mostly in size and proportion as they mature into adults. (**b**) True bugs show *incomplete* metamorphosis, or slow change from an immature form until the last molt. (**c**) Fruit flies show *complete* metamorphosis. Tissues of immature stages are remodeled into the adult form.

Figure 23.32 Examples of insects. (**a**) Mediterranean fruit fly. Its larvae destroy citrus fruit and other crops. (**b**) Duck louse. It eats bits of feathers and skin. (**c**) European earwig, a common household pest. Curved pincers at the tail end indicate this is a male. In females, pincers are straight.

(**d**) Flea, with strong legs good for jumping onto and off of its animal hosts. (**e**) Stinkbugs, newly hatched. (**f**) In the center of this group of honeybee workers, one bee is dancing in a way that communicates the position of a food source to her hive mates, as explained in Section 44.4.

With more than 300,000 species of beetles, Coleoptera is the largest order in the animal kingdom. (**g**) Ladybird beetles swarming. (**h**) Staghorn beetle (*Cyclommatus*) from New Guinea.

(**i**) Adult western corn rootworm (*Diabrotica virgifera*). This beetle has been making the rounds in cornfields as one of our major competitors for food. Its larvae feed on corn roots, killing the plant. Each year, rootworms cause about a billion dollars' worth of crop losses. These enormous losses compel farmers to spread about 30 million pounds of pesticides annually through cornfields—about half the total for all row crops in the United States. Many rootworms are now pesticide resistant as a result of directional selection (Section 17.4).

(**j**) Swallowtail butterfly and (**k**) luna moth. (**l**) An adult damselfly, which preys on flying insects. Its aquatic larvae are voracious predators of other invertebrates and small fishes.

Some insects transmit pathogens that sicken us. We fight "bad" insects by spraying pesticides. Yet most are essential decomposers or pollinators of flowering plants—including many major crop plants. Also, the "good" insects are predators or parasites of the ones we would rather do without. Think about it.

Like other arthropod groups, insects have a hardened exoskeleton, jointed appendages, modified body segments, efficient respiratory structures, and specialized sensory organs. Many kinds also compartmentalize tasks among different stages in the life cycle. The most successful species of insects are winged.

In terms of distribution, number of species, population sizes, competitive adaptations, and exploitation of diverse foods, insects are the most successful animals.

23.13 Unwelcome Arthropods

FOCUS ON HEALTH

Few of us encounter arthropods that cause real pain. We collectively do more harm to more species than, say, spiders, most of which do good by eating so many insect pests. But the "good" ones have a few "bad" relatives, and it doesn't hurt to know them when we see them.

LINKS TO SECTIONS 19.2, 20.5

Harmful Spiders Recluse spiders (*Loxosceles*) hide out in cracks and crevices and hunt at night. There are eleven native species and one South American import. Sometimes they enter attics, basements, and garages. Their cephalothorax has a violin-shaped marking (Figure 23.33*a*). Most spiders have eight eyes; they have six. Their toxin is dangerous enough to kill small children. Tissues around a bite blister, then blacken as cells die. Some bitten people have required skin grafts.

The paralytic neurotoxin of the black widow spider (*Latrodectus*) kills about five people a year in the United States; treatment must be prompt (Figure 23.33*b*).

Terrible Ticks Ticks do not fly or jump. They crawl on low plants, then onto animals that brush past. They move to the host's groin or other hair-covered regions, and gorge on blood (Figure 23.33*c*). When walking in tick habitats, wear light-colored clothes, use a repellent, and check your clothes and body carefully afterward.

The deer tick (*Ixodes scapularis*) can transmit *Borrelia burgdorferi,* the bacterial agent of Lyme disease (Figure 19.8). Biting nymphs cause most of the infections. They are as small as a pin head. Often they feed undetected for several days, which gives bacteria time to move from the tick into its host. Adult ticks are easier to spot and more likely to be removed fast.

Symptoms of Lyme disease include fever, a stiff neck, headache, listlessness, muscle and joint pain, and blurred vision. They may subside and then show up years later. Early diagnosis and treatment are critical. In untreated people, infection can damage many organs.

The most common tick-borne disease in the United States is Rocky Mountain spotted fever. The American dog tick usually transmits the bacterial disease agent, *Rickettsia rickettsii*, to humans. Cases of this disease turn up throughout the continental United States.

Other ticks transmit pathogens that cause tularemia, babesiosis, scrub typhus, and encephalitis. Also, saliva of some female ticks contains a toxin that disrupts nerve function. The resulting paralysis affect the feet first, then moves up the body. No infectious agent is transmitted, so removal of the feeding tick leads to a speedy recovery.

The Mosquito Menace Which animals pose the greatest threat to human health? Mosquitoes. Give them credit; they are integral parts of many food webs that sustain birds and fishes. But mosquitoes also transmit pathogens (Figure 23.33*d*).

Each year, mosquito-borne diseases kill about 3 million people worldwide. Many species of mosquitoes act as intermediate hosts for pathogens that cause Rift Valley fever, dengue fever, yellow fever, and different forms of encephalitis as well as malaria and West Nile virus. They also transmit some parasitic roundworms (Section 23.8).

Only female mosquitoes are bloodsuckers, and most will feed on any warm-blooded animal that they come across. The males feed on plant nectar and juices. After a female has a blood meal, she produces a batch of eggs, which must be laid in standing water. Repellents can help prevent bites, but draining or treating bodies of stagnant water is the best way to control mosquito populations.

Mosquitoes thrive in warm, humid climates. With the current long-term warming trend, they may become active for more of the year, in ever expanding habitats.

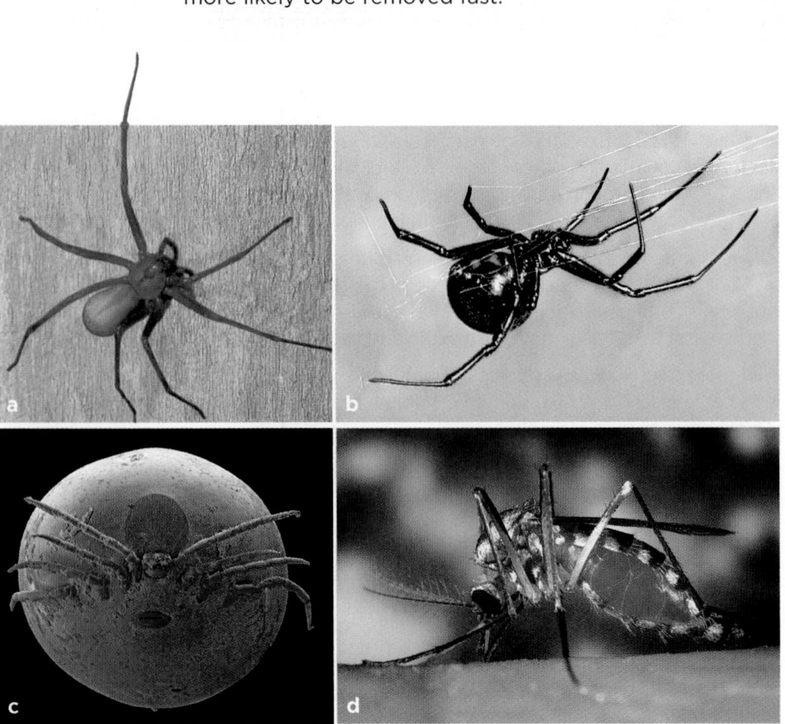

Figure 23.33 Dangerous spiders. (**a**) Brown recluse, with a violin-shaped mark on its forebody. Most bites heal without complication, but some have had severe to fatal consequences. (**b**) Female black widow, which has a red, hourglass-shaped marking on the underside of her shiny black abdomen. Males are smaller and do not bite.

Blood suckers. (**c**) Like spiders, this blood-engorged tick is an arachnid and has eight legs. (**d**) Female mosquito (*Aedes triseriatus*) siphoning blood from a human. This species is one of the vectors for the West Nile virus. Male mosquitoes sip flower nectar, not blood.

23.14 The Spiny-Skinned Echinoderms

Reflect on Figure 23.6. Genetic divergences among early bilateral animals gave rise to roundworms, arthropods, annelids, mollusks, and other protostomes. They put deuterostomes on a separate evolutionary journey, with echinoderms and chordates being major travelers.

An echinoderm is one of 6,000 marine invertebrates with a covering of spines and plates, all interlocking and stiffened with calcium carbonate. (Its name means spiny-skinned.) Figure 23.34 shows a few echinoderms. The group includes sea stars, brittle stars, feather stars, sea urchins, sand dollars, sea cucumbers, sea pens, sea biscuits, and sea lilies (crinoids). Although most adults are radial, they have a few bilateral traits. The larvae *are* bilateral. Did the radial traits evolve secondarily in this lineage? DNA sequencing studies say yes.

All echinoderms are brainless. Their decentralized nervous system responds to food, danger, and mates that can arrive from any direction.

Some sea lilies live attached to the seafloor. But most adult echinoderms glide about on little tube feet. Echinoderm tube feet are fluid-filled, muscularized structures, often with suckerlike disks (Figure 23.34). They help their owners walk, burrow, cling to rocks, and grip prey. Tube feet are part of a water–vascular system unique to echinoderms. Figure 23.34*e* shows how the system starts out as fluid-filled canals in each arm of a sea star. Side canals get water into a bulb in each tube foot (Figure 23.34*e*). The foot extends as the bulb contracts and forces fluid into it. Contractions redistribute fluid among hundreds of tube feet. All of the feet lift up, swing forward, reattach, swing back, and let go in coordination, so gliding is very smooth.

Most sea stars are active predators. Some swallow prey whole. Others push part of their stomach out of their mouth and around prey, then start digesting the prey before swallowing it. All expel coarse, undigested residues through their mouth, not their small anus.

In sea urchins, calcium carbonate plates form a stiff, rounded cover from which spines protrude (Figure 23.34*a*). Sea urchin roe (eggs) are used in some sushi. Overharvesting for markets in Asia threatens species that produce the most highly-prized roe.

Sea cucumbers, too, are heavily harvested for Asian markets. They have microscopic plates embedded in a soft body (Figure 23.34*b*). Some filter food from the seawater; most feed on organic sediments. They can expel organs through the anus to distract a predator. If a sea cucumber escapes, its missing parts grow back.

The spiny-skinned echinoderms with their radial and bilateral traits, make a case for this point: There are exceptions to the major trends in animal evolution.

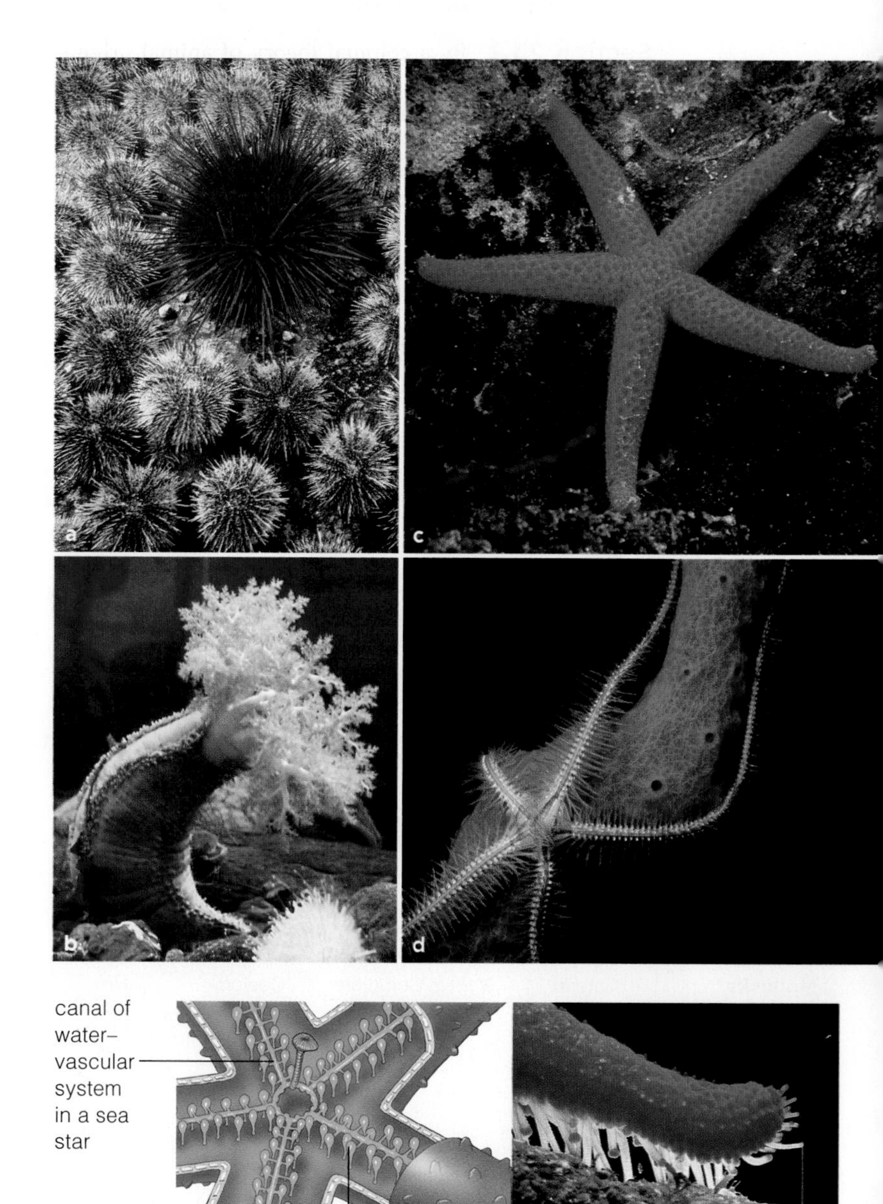

Figure 23.34 **Animated!** Some echinoderms. (**a**) Underwater "forest" of sea urchins, which move on spines and a few tube feet. (**b**) Sea cucumber, with rows of tube feet. (**c**) Sea star. (**d**) Brittle star. (**e**) A water–vascular system which, in combination with many tube feet, is the basis of locomotion.

www.thomsonedu.com ThomsonNOW!

Summary

Section 23.1 Animals are multicelled heterotrophs that move about for at least part of their life cycle. They develop in a series of stages. Ectoderm, endoderm, and often mesoderm form as layers in the early embryo.

Comparing key groups reveals a trend toward greater size, compartmentalization (the division of labor among cells, tissues, and organs), and integration of specialized activities that keep the organism alive. Most animals are bilateral and cephalized (Table 23.2).

Two major branches of bilateral animals, protostomes and deuterostomes, have a coelom and a complete gut. The groups differ in several respects. For example, in protostomes, the first opening on the embryo becomes the mouth. In deuterostome embryos, an anus forms first.

- *Use the animation on ThomsonNOW to familiarize yourself with terms necessary to understand animal body plans.*

Section 23.2 By a colonial theory of animal origins, animals arose from a colony of flagellated protist cells, similar to modern choanoflagellates. Placozoans are the simplest known modern animals, and gene sequencing data show placozoans are close to choanoflagellates.

The oldest animal fossils, called the Ediacarans, date back about 600 million years. A great adaptive radiation during the Cambrian gave rise to most modern lineages.

Section 23.3 Sponges do not have tissues or organs. Flattened cells line the body wall, which has many pores and is stiffened with spikes of silica, proteins, or both. Flagellated collar cells of the inner lining absorb food from water; amoeboid cells digest and distribute it.

- *Use the animation on ThomsonNOW to explore the body plan of a sponge.*

Table 23.2 Comparative Summary of Animal Body Plans

Group	Body Symmetry	Germ Layers	Digestive System	Main Cavity	Segmented
Placozoans	None	None	None	None	No
Sponges	None	None	None	None	No
Cnidarians	Radial	2	Gut cavity	None	No
PROTOSTOMES					
Flatworms	Bilateral	3	Incomplete	None*	No
Annelids	Bilateral	3	Complete	Coelom	Yes
Mollusks	Bilateral	3	Complete	Coelom	No
Roundworms	Bilateral	3	Complete	False coelom	No
Arthropods	Bilateral	3	Complete	Coelom	Yes
DEUTEROSTOMES					
Echinoderms	Radial**	3	Complete	Coelom	No
Chordates	Bilateral	3	Complete	Coelom	Yes

* An ancestor may have had a coelom.
** Radial with some bilateral features; probably had a bilateral ancestor.

Section 23.4 Cnidarians, such as jellyfishes, corals, and sea anemones, are radial carnivores. They have true epithelial tissues with a jellylike layer in between. They alone make nematocysts, which they use to capture prey and in defense. A gastrovascular cavity functions in both respiration and digestion.

- *Use the animation on ThomsonNOW to compare cnidarian body plans and life cycles.*

Section 23.5 Flatworms, the simplest animals with organ systems, are bilateral protostomes. They include free-living turbellarians (such as planarians) as well as the parasitic tapeworms and flukes.

- *Use the animation on ThomsonNOW to learn about planarian organ systems and a tapeworm life cycle.*

Section 23.6 Annelids are segmented worms (such as earthworms and polychaetes) and leeches. Circulatory, digestive, solute-regulating, and nervous systems extend through all coelomic chambers. Muscles and fluid in the chambers act as a hydrostatic skeleton.

- *Use the animation on ThomsonNOW to investigate the body plan of an earthworm.*

Section 23.7 With 100,000 named species, mollusks are one of the largest animal groups. They alone have a mantle, a sheetlike part of the body mass that is draped back on itself. Examples are gastropods (such as snails), bivalves (such as scallops) and cephalopods. The fastest (squids), largest (giant squids), and smartest (octopuses) invertebrates are cephalopods.

- *Use the animation on ThomsonNOW to compare molluscan body plans.*

Section 23.8 The roundworms (nematodes) have a cylindrical body with bilateral features, a cuticle that is molted, a complete gut, and a false coelom. More than 22,000 kinds are free-living decomposers or parasites.

- *Use the animation on ThomsonNOW to learn about the roundworm body plan.*

Sections 23.9–23.13 There are more than 1 million arthropod species, such as trilobites (extinct), chelicerates, crustaceans, centipedes, millipedes, and insects.

The success of arthropods as a group is attributable to a hardened exoskeleton, jointed appendages, specialized and fused segments, efficient respiratory and sensory structures, and often a division of labor among stages of development that face different environmental conditions.

Chelicerates include horseshoe crabs and arachnids (spiders, scorpions, ticks, and mites). They are predators, parasites, or scavengers. The mostly marine crustaceans include crabs, lobsters, barnacles, krill, and copepods.

Insects are the most successful of all animal groups, and include the only winged invertebrates. They are our major competitors for food.

- *Use the animation on ThomsonNOW to learn about the crab life cycle and insect mouthpart specializations.*

Section 23.14 Echinoderms, such as sea stars, are invertebrates of the deuterostome lineage. They have an exoskeleton with spines, spicules, or plates of calcium carbonate. A water–vascular system with tube feet helps most glide about. Adults are radial, but bilateral ancestry is evident in their larval stages and other features.

■ *Look into the body plan of a sea star and watch tube feet in action on ThomsonNOW.*

Self-Quiz

Answers in Appendix III

1. All animals ______ .
   a. consist of tissues arranged as organs
   b. are motile for at least some stage in the life cycle
   c. can reproduce asexually as well as sexually
   d. both a and b

2. A coelom is a ______ .
   a. type of bristle
   b. resting stage
   c. sensory organ
   d. lined body cavity

3. Cnidarians alone have ______ .
   a. nematocysts
   b. a mantle
   c. a hydrostatic skeleton
   d. Malpighian tubules

4. Flukes are most closely related to ______ .
   a. tapeworms
   b. roundworms
   c. arthropods
   d. echinoderms

5. Which group has the most named species?
   a. crustaceans
   b. insects
   c. mollusks
   d. roundworms

6. Nephridia have the same functional role as ______ .
   a. gemmules of sponges
   b. mandibles of insects
   c. flame cells of planarians
   d. tube feet of echinoderms

7. A spider's chelicerae ______ .
   a. detect light
   b. inject venom
   c. produce silk
   d. eliminate excess water

8. Barnacles are shelled ______ .
   a. gastropods
   b. cephalopods
   c. arthropods
   d. copepods

9. The ______ include the only winged invertebrates.
   a. cnidarians
   b. echinoderms
   c. arthropods
   d. placozoans

10. The ______ have a coelom and are radial as adults.
   a. cnidarians
   b. echinoderms
   c. roundworms
   d. both a and b

11. Match the organisms with their descriptions.

___ choanoflagellates	a. complete gut, false coelom
___ placozoan	b. sister group to animals
___ sponges	c. simplest organ systems
___ cnidarians	d. no tissues, no organs
___ flatworms	e. jointed exoskeleton
___ roundworms	f. mantle over body mass
___ annelids	g. segmented worms
___ arthropods	h. tube feet, spiny skin
___ mollusks	i. nematocyst producers
___ echinoderms	j. simplest known animal

■ *Visit ThomsonNOW for additional questions.*

Critical Thinking

1. A colony of soft coral is a branching form covered with feathery polyps (Figure 23.35). Unlike reef-building corals, which occur only in sunlit tropical waters, soft corals also can live in cooler middle to high latitude seas and survive in the dark depths. Some species often colonize the interior of wrecked ships. Suggest a hypothesis to explain how the feeding strategies of these two types of coral might differ. How might you test your hypothesis?

Figure 23.35 A soft coral (*Carijoa*).

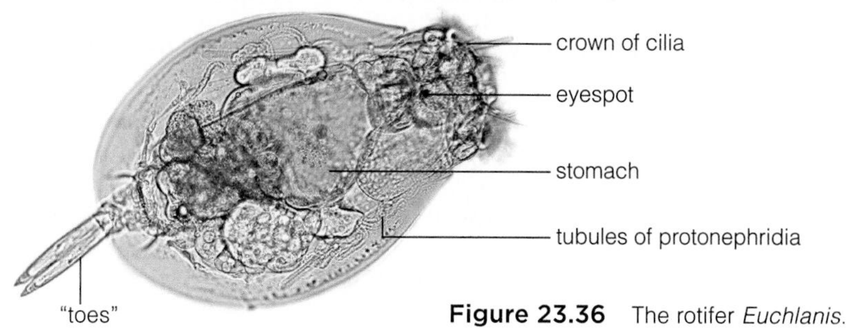

Figure 23.36 The rotifer *Euchlanis*.

2. There are many invertebrate groups besides those introduced in this chapter. For example, more than 2,100 rotifer species live mostly in freshwater lakes, ponds, and streams. Some swim; others glue themselves to substrates by exuding sticky substances from their toes. All are less than 2 millimeters long (Figure 23.36). Cilia on the head end are in continuous motion and direct food to the mouth. The body has a false coelom and complete digestive system. Protonephridia like those of flatworms control water and solute levels and eliminate wastes. A cluster of nerve cells in the head integrates body activities. Rotifers do not have a circulatory system or respiratory organs. Explain how essential gases and nutrients reach their cells and why you will never see a rotifer as big as a fish.

3. Many different species of flatworms, roundworms, and annelids are parasites of mammals. There are no such parasites among sponges, cnidarians, mollusks, and echinoderms. Propose a plausible explanation for this difference.

4. Figure 23.37 shows a roundworm being devoured by a tardigrade (phylum Tardigrada), an invertebrate known as the water bear. Do research to find out why tardigrades are often said to be the "toughest" of all living animals.

5. In 2000, there was a massive die-off of lobsters in the Long Island Sound. Many lobstermen blamed pesticides sprayed to control the mosquitoes that cause West Nile virus. Speculate on why a chemical designed to kill insects also may harm lobsters.

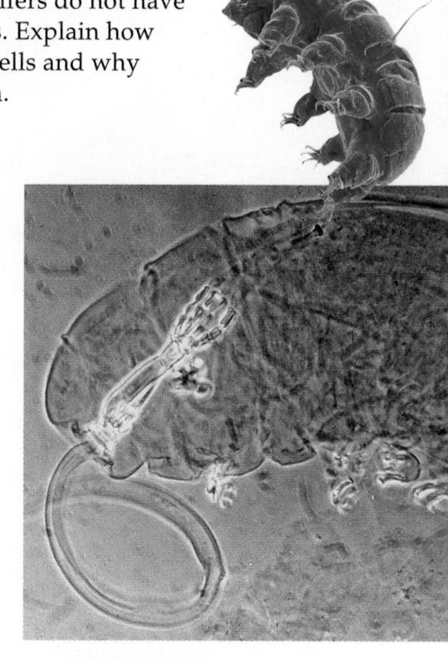

Figure 23.37 A tardigrade eating a roundworm (*lower left*).

24 ANIMAL EVOLUTION—THE VERTEBRATES

IMPACTS, ISSUES Interpreting and Misinterpreting the Past

By Charles Darwin's time, all major groups of organisms had been identified. A big obstacle to accepting his theory of evolution by natural selection was the seeming lack of transitional forms between groups. If new species evolve from older ones, then where were the "missing links" in the fossil record, the species with traits intermediate between two groups?

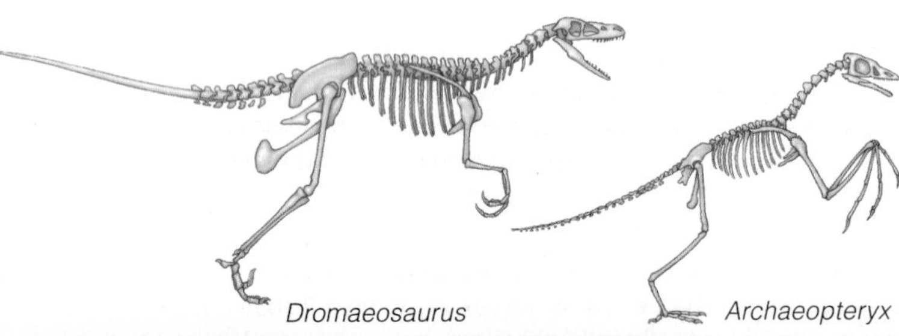

Ironically, workmen at a limestone quarry in Germany had already unearthed one link. The pigeon-sized fossil looked like a small dinosaur (*Dromaeosaurus*). It had short spiky teeth, three long clawed fingers on a pair of forelimbs, and a long bony tail. Later, diggers found another specimen. Later still, someone noticed feathers. If they were fossilized birds, then why did they have teeth and a bony tail? If dinosaurs, what were they doing with *feathers*? The specimen type was named *Archaeopteryx*, meaning ancient winged one (Figure 24.1).

Between 1860 and 1988, six *Archaeopteryx* specimens and a fossilized feather were found. Anti-evolutionists tried to dismiss them as forgeries. Someone, they said, pressed modern bird bones and feathers against wet plaster; the imprints only looked like fossils. Microscopic examination confirmed that the fossils are authentic.

See the video! **Figure 24.1** Placing *Archaeopteryx* in time. This painting is based on fossils of plants and animals that lived in a Jurassic tropical forest. Center foreground, two gliding *Archaeopteryx*. Background, left to right, two *Stegosaurus* and an *Apatosaurus* (herbivores), a *Saurophaganax* ("king of reptile eaters"), and three *Camptosaurus* (beaked herbivores). The small-brained *Apatosaurus* grew as long as 27 meters (90 feet) and weighed 33 to 38 tons; it was one of the largest land animals that ever lived. Far right, a climbing mammal on a tree. Facing page, one *Archaeopteryx* fossil.

How would you vote? Private collectors find and protect rare vertebrate fossils, but the private trade raises purchase costs for museums and encourages theft from protected fossil beds. Should private collecting of vertebrate fossils be banned? See ThomsonNOW for details, then vote online.

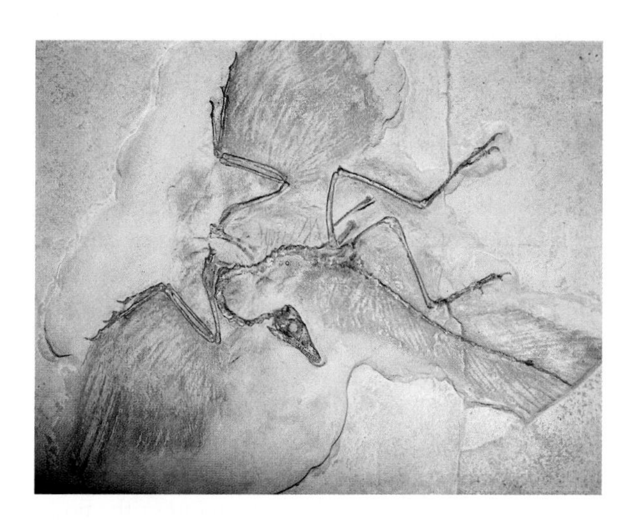

Later, radiometric dating (Section 16.5) revealed that *Archaeopteryx* lived 150 million years ago. Why were the remains preserved so well? This early bird lived near a lagoon. When bodies of organisms fell into the lagoon, fine sediments quickly covered them. There was little wave action, so skeletons often remained intact. Over time, the sediments compacted and hardened. They became a limestone tomb for more than 600 species, including *Archaeopteryx*.

No human witnessed the major transitions that led to modern animal diversity. However, fossils are physical evidence of changes, and radiometric dating assigns the fossils to places in time. The structure, biochemistry, and genetic makeup of living organisms provide information about branchings. Biological evolution is not a "theory" in the common sense of the word—an idea off the top of someone's head. It is a *scientific* theory. Its predictive power has been tested many times in the natural world, and the theory has withstood these tests (Section 1.6).

Evolutionists often argue among themselves. They argue over how to interpret data and which of the known mechanisms can best explain life's history. At the same time, they eagerly look to new evidence to support or disprove hypotheses. As you will see, fossils and other evidence form the foundation for this chapter's account of vertebrate evolution, including our own origins.

Key Concepts

CHARACTERISTICS OF CHORDATES

A unique set of four traits characterizes the chordates: a supporting rod (notochord), a dorsal nerve cord, a pharynx with gill slits in the wall, and a tail extending past an anus. Certain invertebrates and all vertebrates belong to this group. **Section 24.1**

TRENDS AMONG VERTEBRATES

In some vertebrate lineages, a backbone replaced the notochord as the partner of muscles used in motion. Jaws evolved, sparking the evolution of novel sensory organs and brain expansions. On land, lungs replaced gills, and more efficient blood circulation enhanced gas exchange. Fleshy fins with skeletal supports evolved into limbs, which are now typical of vertebrates on land. **Section 24.2**

TRANSITION FROM WATER TO LAND

Vertebrates first evolved in the seas, where lineages of cartilaginous and bony fishes persist. Of all vertebrates, modern bony fishes show the most diversity. Mutations in master genes that control body plans were pivotal in the rise of aquatic tetrapods and their move onto dry land. **Sections 24.3–24.5**

THE AMNIOTES

As a group, the amniotes—known informally as the reptiles, birds, and mammals—are vertebrate lineages that radiated into nearly all habitats on land. **Sections 24.6–24.10**

EARLY HUMANS AND THEIR ANCESTORS

Primates that were ancestral to the human lineage became physically and behaviorally adapted to changes in global climate and available resources. Behavioral and cultural flexibility helped humans disperse from Africa throughout the world. **Sections 24.11–24.13**

Links to Earlier Concepts

Think back on the branchings that led to the emergence of animals by reviewing the evolutionary tree in Section 16.12. Also check the geologic time scale (16.5) to see where you are heading from here.

You will come across more examples of how small changes in master genes changed the course of evolution (14.3, 16.8). You will draw on your understanding of speciation, adaptive radiation, and extinctions (17.9–17.12). You will revisit the dinosaurs's demise (Chapter 16). You will come across more examples of biochemical and molecular tools that are unlocking secrets of the past (16.9).

24.1 The Chordate Heritage

As you consider vertebrates and their chordate heritage, remember this: Each kind of organism is a mosaic of traits, many conserved from remote ancestors and others unique to one branch of the family tree.

CHORDATE CHARACTERISTICS

LINKS TO SECTIONS 23.1, 23.2, 23.9

The preceding chapter ended with echinoderms, one of the earliest deuterostome lineages. The chordates dominate this branch of the animal family tree (Figure 24.2). Chordates are bilateral, coelomate animals that show these four defining traits in their embryos: (1) A rod of stiff but flexible connective tissue, a notochord, extends the length of the body and supports it. (2) A dorsal, hollow nerve cord parallels the notochord. (3) Gill slits open across the wall of the pharynx (the throat region). (4) A muscular tail extends beyond the anus. Depending on the group, some, none, or all of these defining traits persist in the adult.

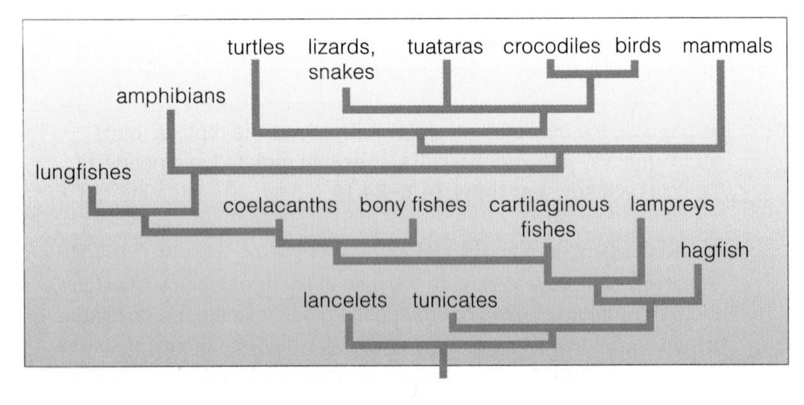

Figure 24.2 Family tree for the chordates which, together with echinoderms and some lesser known groups, belong to the deuterostome lineage. Appendix I includes a classification system for vertebrate groups and subgroups.

Most of the 50,000 or so chordates are vertebrates—animals with a backbone—and the bulk of this chapter is devoted to their characteristics and evolution. Here we begin our survey with tunicates and lancelets, the two groups of invertebrate chordates. Both are marine.

INVERTEBRATE CHORDATES

Tunicates The tunicates are invertebrate chordates that briefly swim about as larvae, then metamorphose into an adult that has a secreted covering or "tunic" (Figure 24.3*a*). Most adult tunicates cling to rocks and other undersea surfaces. The only chordate trait that adult tunicates retain is a muscular pharynx with gill slits (Figure 24.3*b*).

The adult is a filter feeder; it strains pieces of food out of the watery surroundings. Water flows in at an oral opening, past gill slits (where the food sticks to mucus and gets sent on to a gut), then out of the body by way of another opening. At the same time, oxygen diffuses from water into blood vessels next to the gill slits. Carbon dioxide diffuses out, into water.

Tunicate larvae are free-swimming and display all the characteristic chordate traits (Figure 24.3*c*). Their notochord interacts with muscles when they swim. It bends one way as the muscles contract, then it springs back as muscles relax. The side-to-side motion propels the larva forward. A larva swims about briefly before attaching to a surface. Then it metamorphoses; the tail breaks down and other parts become rearranged into the adult body form.

Lancelets The lancelets are invertebrate chordates with a fishlike body 3 to 7 centimeters long. Unlike the tunicates, adults retain all characteristic chordate features (Figure 24.4). The dorsal nerve cord extends

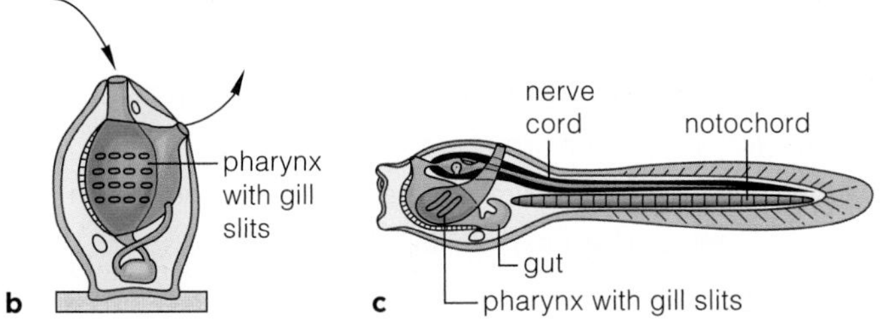

Figure 24.3 (**a**) Cluster of adult tunicates (*Rhopalaea crassa*). (**b**) Adult body plan. Arrows indicate the direction of water flow: In one opening, into pharynx and out through gill slits, then out of the body through another opening. (**c**) Tunicate larva. It swims briefly, then glues its head to a surface and metamorphoses. Tissues of its tail, notochord, and most of the nervous system are remodeled.

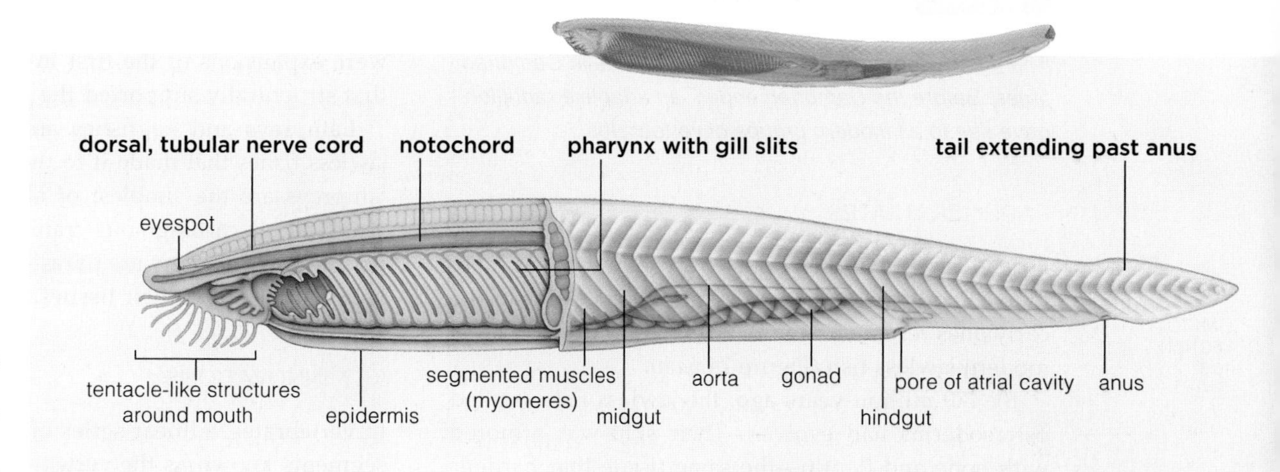

Figure 24.4 **Animated!** Body plan of a lancelet, a small filter-feeder. It clearly shows the set of four traits that define chordates. The traits develop in all chordate embryos but do not always persist in adults.

into the head, where a group of nerve cells serves as a simple brain. An eyespot at the end of the nerve cord detects light, but the head has no brain case or paired sensory organs like those of fishes.

A lancelet wiggles backward into sediments until it is buried up to its mouth. Movement of cilia lining its pharynx draws water into the pharynx and through gill slits. As in tunicates, gill slits filter food particles out of the water and also function in gas exchange.

Like vertebrates, lancelets have segmented muscles. Contractile units in muscle cells run parallel with the body's long axis. The force that muscles direct against the notochord produces a side-to-side motion that lets a lancelet burrow and swim short distances. As you will see, that is how fish swim and how the first land vertebrates walked about.

Until very recently, lancelets were thought to be the closest invertebrate relatives of vertebrates. A lancelet certainly looks more like a fish than a tunicate does. But appearances can be deceiving. Gene sequence data show that tunicates are the sister group of chordates. Keep in mind that neither tunicates nor lancelets are ancestors of vertebrates. These groups share a recent common relative, but each has unique traits that put it onto a separate branch of the animal family tree.

A BRAINCASE BUT NO BACKBONE

Fishes, amphibians, reptiles, birds, and mammals are craniates. A cranium, a braincase of cartilage or bone, encloses and protects their brain.

Hagfishes are the only modern animals that have a cranium, but no backbone (Figure 24.5). Like lancelets, these soft-bodied, jawless fishes have a notochord that supports the body. Sensory tentacles help them detect food—soft invertebrates and dead or dying fish. Like other craniates, a hagfish has paired ears that detect vibrations and a pair of eyes. However, the eye has no lens and vision is poor to nonexistent.

When threatened, a hagfish can secrete a gallon of slimy mucus. Fishermen who catch a hagfish consider this behavior disgusting. However, slime has been a useful defense for the otherwise vulnerable hagfishes; it has helped them survive since the Carboniferous.

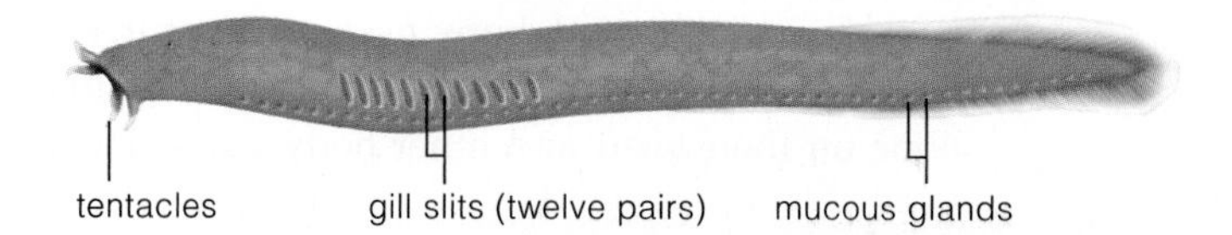

Figure 24.5 Hagfish body plan. The two photographs show a hagfish before and after it coated its body with slimy mucous secretions.

All chordate embryos have a notochord, a dorsal tubular nerve cord, a pharynx with gill slits in its wall, and a tail that extends past the anus. These traits are legacies from a shared ancestor, an early invertebrate chordate.

Modern filter-feeding lancelets and tunicates are among the groups closest to the most ancient chordate lineages.

Hagfishes are at the next level of complexity among the chordates. Some cartilage protects part of their brain.

24.2 Evolutionary Trends Among the Vertebrates

The first vertebrates—jawless fishes—arose in Cambrian times. Before the Cambrian ended, an adaptive radiation gave rise to all modern groups of vertebrates.

EARLY CRANIATES

LINK TO SECTION 16.5

Fossils reveal that the craniates arose in the seas more than 530 million years ago. In body form, some of the early ones resembled the larvae of lampreys, a type of modern jawless fish (Figure 24.6*a*,*b*).

By 470 million years ago, the jawless fishes called ostracoderms had evolved. Their skin was armored with bone and dentin—the same tissue that hardens your teeth. Armor defended them from the pincers of giant, predatory sea scorpions but was useless against jaws. Jaws are hinged, bony feeding structures.

Early jawed fishes called placoderms had plates of bone on their head and other body parts. Their jaws were expansions of the first in a series of hard parts that structurally supported the gill slits (Figure 24.7).

Lampreys and hagfishes are the only lineages of jawless fishes that made it to the present. Structurally, lampreys are the simplest of all modern vertebrates. They have a cartilaginous cranium and backbone but no fins or jaws. Many are parasites that attach to other fishes and feed on their tissues (Figure 24.6*c*).

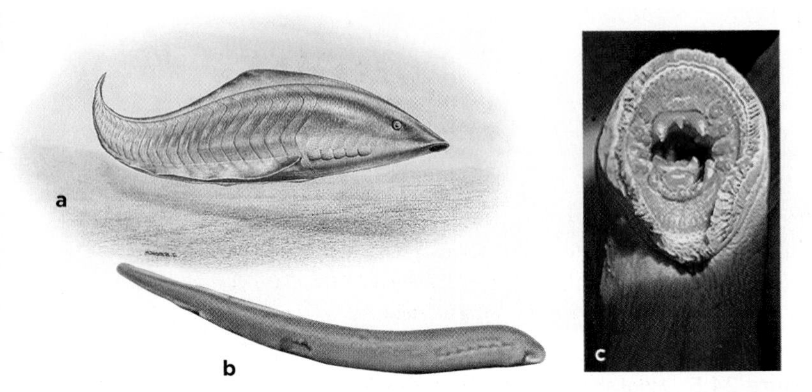

Figure 24.6 (**a**) Painting of an early craniate (*Myllokunmingia*) based on fossils discovered in China. (**b**) Larva of a lamprey, a modern jawless fish. Many lampreys are parasites that attach to other fishes with a suckerlike oral disk (**c**).

KEY INNOVATIONS

In vertebrates, a linear series of cartilaginous or bony segments known as the vertebrae (singular, vertebra) replaced the notochord. They formed a backbone, or vertebral column, that was part of an internal skeleton. The segmented muscles attached to the backbone and resulting changes increased both maneuverability and strength. The vertebral column was a key innovation. *It opened the way for an adaptive radiation of strong, fast-moving, agile predators and prey in the seas.*

Jaws sparked another trend. Early chordates fed by filtering, sucking, and rasping food. Evolution of jaws in predatory fishes selected for new defenses in prey. New defenses then gave predators able to overcome them an advantage. For example, fishes with a bigger brain that could better plan pursuit had the edge. *A trend toward more complex sensory organs and nervous systems began among ancient lineages of fishes and, later, continued among vertebrates on land.*

A shift in respiration set another trend in motion. In lancelets and tunicates, some gas exchange occurs at gill slits, but most gases just diffuse across the body wall. Paired gills evolved in early vertebrates. Gills are respiratory organs with moist, thin folds that are richly supplied with blood vessels. Gills enhance the

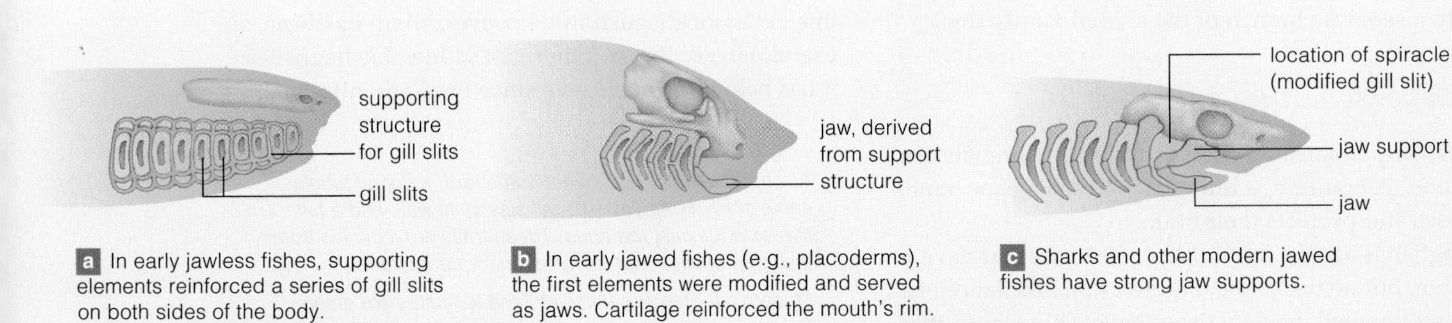

a In early jawless fishes, supporting elements reinforced a series of gill slits on both sides of the body.

b In early jawed fishes (e.g., placoderms), the first elements were modified and served as jaws. Cartilage reinforced the mouth's rim.

c Sharks and other modern jawed fishes have strong jaw supports.

Figure 24.7 **Animated!** Comparison of gill-supporting structures.

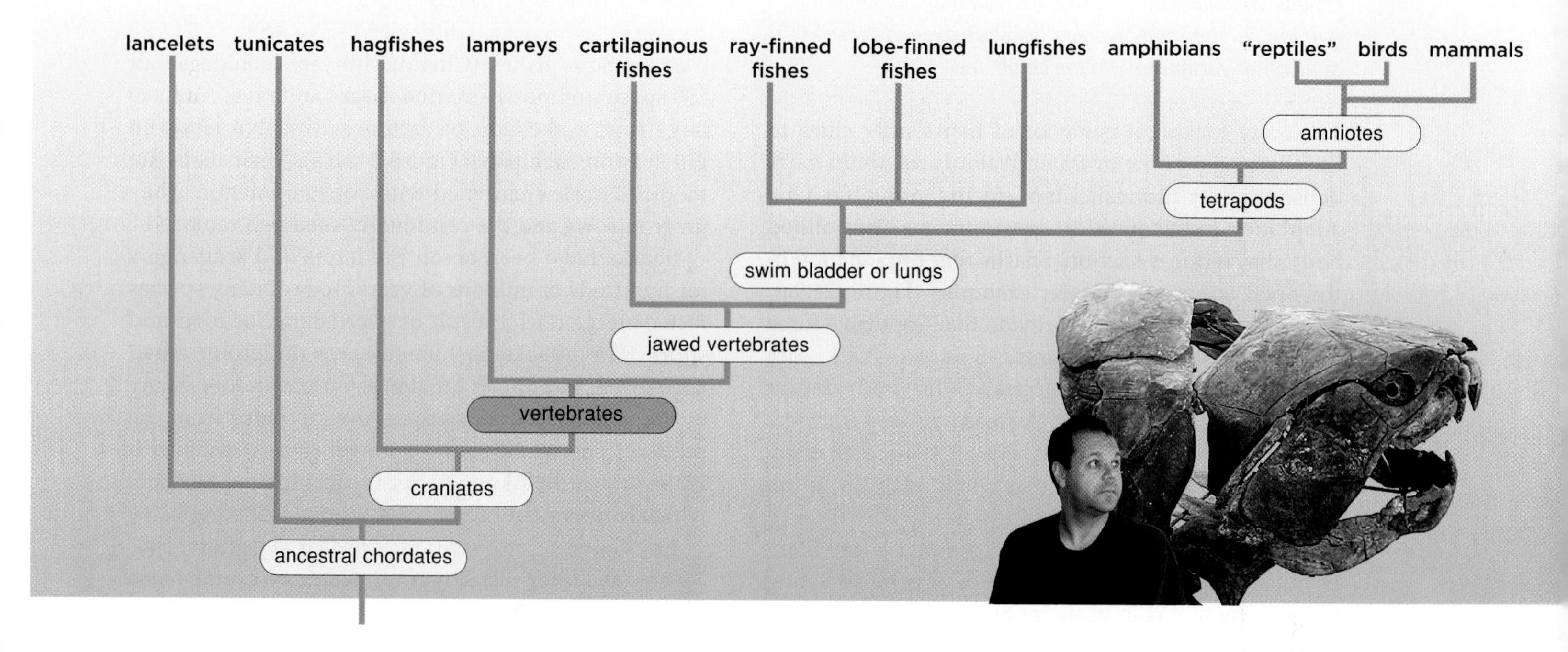

Figure 24.8 Where living vertebrates fit in the chordate family tree. All groups listed are multicelled, bilateral, coelomate animals. Compare the size of this human with an extinct jawed fish, a placoderm (*Dunkleosteus*).

exchange of gases and thereby support higher levels of activity. As water flows over gills, oxygen diffuses from the water into blood vessels in each gill. Carbon dioxide diffuses in the opposite direction, from blood vessels into the water. Blood flows through vessels in fish gills driven by the force of a beating heart.

Gills became more efficient in larger, more active fishes. But gills cannot function out of water; their thin surfaces stick together unless flowing water moistens them. In fishes ancestral to land vertebrates, two small outpouchings on the side of the gut wall evolved into lungs: moist, internal sacs that serve in gas exchange.

Lungs became the main organs of gas exchange in most vertebrates that moved onto land. In a related trend, changes to the circulatory system increased the rate of blood flow. *The evolution of a pair of lungs and modifications to the circulatory system accompanied the move to land.*

Another trend started as pairs of fins evolved. Fins are appendages that help propel, stabilize, and guide a fish body through water. They go by these names:

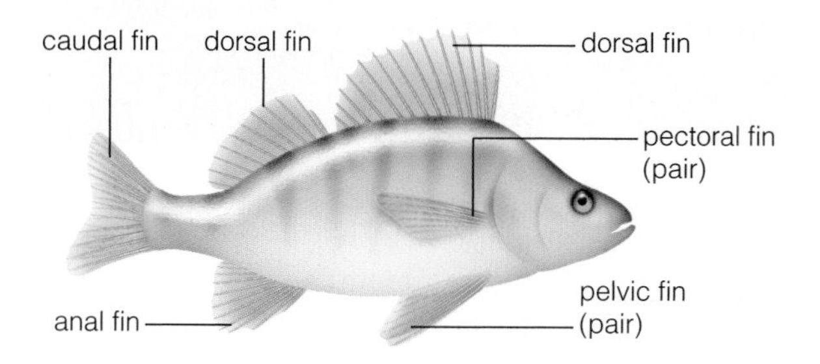

Some Devonian fishes had fleshy pelvic and anal fins with internal skeletal supports. *Paired, fleshy fins were a starting point for all legs, arms, and wings that evolved among amphibians, reptiles, birds, and mammals.*

In the Carboniferous, agile predators with bigger brains evolved and the placoderms died out. Through adaptations that proved to be key innovations—and luck—other descendants of some jawless and jawed groups made it to the present.

MAJOR VERTEBRATE GROUPS

Figure 24.8 shows the vertebrate family tree. Besides lampreys, there are cartilaginous fishes, bony fishes (ray-finned fishes, lobe-finned fishes, and lungfishes), amphibians, reptiles, birds, and mammals. The rest of the chapter is devoted to their traits and evolution.

Vertebrates arose in the Cambrian. All major modern groups evolved in an adaptive radiation of jawed fishes.

Jaws, a vertebral column, paired fins, and lungs proved to be key innovations in the evolution of lineages that gave rise to amphibians, reptiles, birds, and mammals.

24.3 Jawed Fishes and the Rise of Tetrapods

Unless you study underwater life, you may not know that the number and diversity of fishes exceed those of all other vertebrate groups combined.

LINKS TO SECTIONS 11.5,16.5

The body form and behavior of fishes offer clues to the challenges of life in water. Water is 800 times more dense than air and resists movements through it. One adaptation to this physical constraint is a streamlined body that reduces friction. Sharks that pursue prey in the open ocean are premier examples (Figure 24.9*a*). Strong muscles in the tail provide the propulsive force for a shark's rapid movements.

Scales and a hard skeleton make a fish body denser and more prone to sinking. As many fishes swim, the shape of their fins and body help lift them. The effect is something like the way that wings help lift up an airplane and stabilize it in air.

Some bony fishes have a swim bladder: a gas-filled chamber that serves as a flotation device. By adjusting the volume of gas in its swim bladder, a fish can stay suspended in water at different depths. For example, a trout fills its swim bladder by gulping in air at the water surface, then burps when it wants to descend.

CARTILAGINOUS FISHES

Cartilaginous fishes (Chondrichthyes) include about 850 species of mostly marine sharks and rays. All have large fins, a skeleton of cartilage, and five to seven gill slits on each side (Figure 24.9*a*,*b*). Their teeth are modified scales hardened with bone and dentin. They grow in rows and are continually shed and replaced.

Sharks have been ocean predators and scavengers for hundreds of millions of years. Today, many species face extinction as a result of overfishing for food and sport. Rare attacks on humans give the group a bad reputation, but not all are dangerous predators. Many of the smaller sharks suck up invertebrates from the seafloor. The whale shark is longer than a city bus. It is the largest fish known, yet it eats mostly plankton, which it filters out of seawater with modified gills.

Rays have a flattened body with large pectoral fins. Manta rays filter plankton from water and some reach 6 meters across (Figure 24.9*b*). Stingrays are bottom feeders. They have a barbed tail with a venom gland. Some rays have electric organs that help them locate food and mates, and in some cases, stun prey.

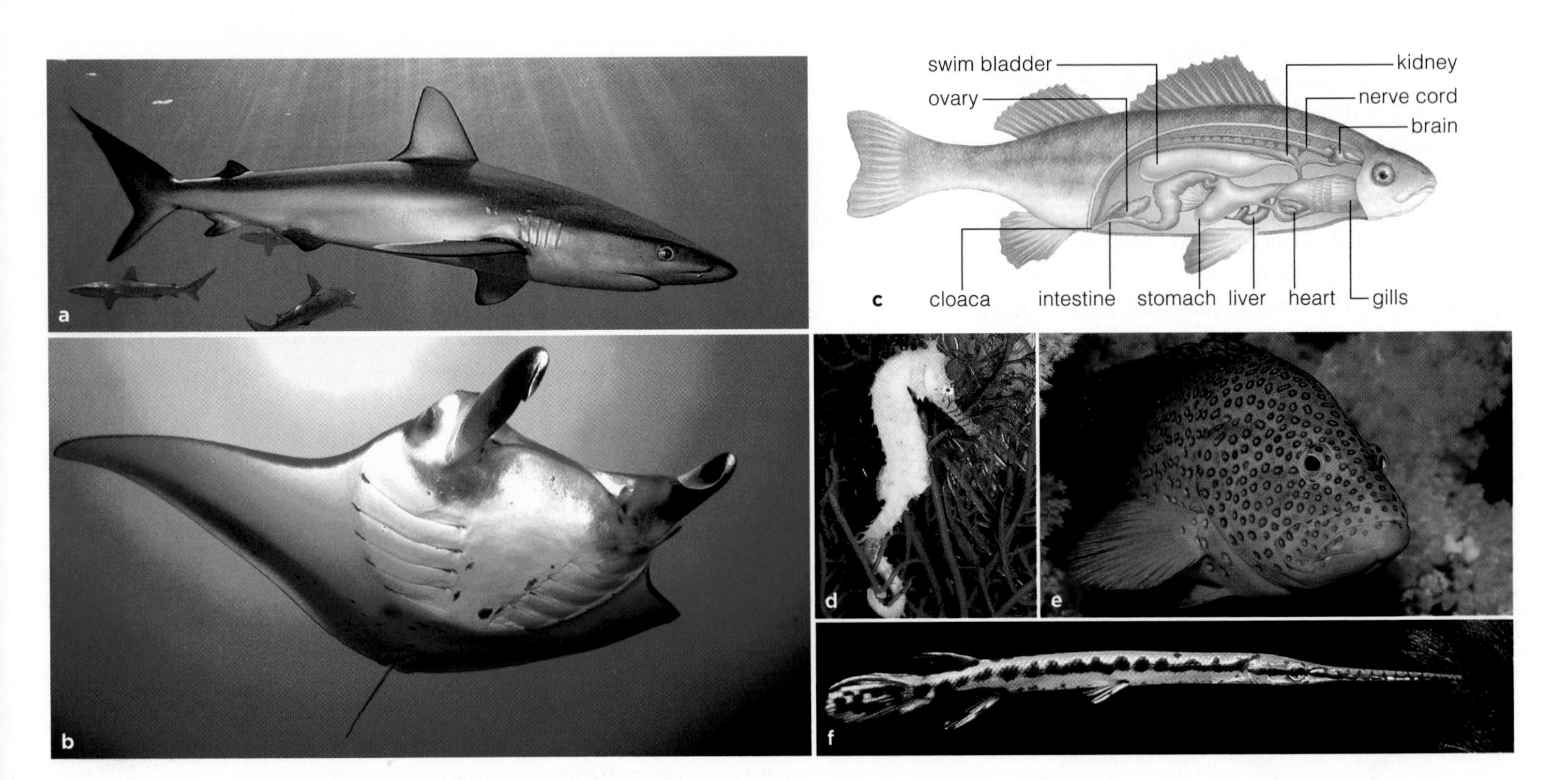

Figure 24.9 Cartilaginous fishes. (**a**) Galápagos sharks. (**b**) Manta ray. Two fleshy projections on its head unfurl and waft plankton to its mouth, in between them. Notice the gill slits on both fishes.

Ray-finned body fishes. (**c**) Body plan of a perch. (**d**) Sea horse. (**e**) Coral grouper, one of many species of bony fishes that fishermen are overharvesting. (**f**) Long-nose gar, a fast-moving predator.

BONY FISHES

Jawed fishes with a bony endoskeleton evolved 400 million years ago. The ray-finned fishes have flexible fin supports, derived from skin, and thin scales. With more than 21,000 freshwater and marine species, they are the most diverse vertebrate group. Figure 24.9*c–f* shows a few ray-finned fishes.

Paddlefishes and sturgeons belong to one lineage of ray-finned fishes. Sturgeon eggs are highly prized as caviar. Another lineage, gars, are long, streamlined predators of North American freshwater habitats. The largest ray-finned lineage, teleosts, includes eels, sea horses, and puffers, as well as salmon, bass, tuna, and most other fishes that end up on dinner plates. Long ago, the whole teleost genome underwent duplication. Subsequent mutations in the copied genes may have helped this group to adapt to diverse environments.

Coelacanths (*Latimeria*) are the only modern group of lobe-finned fishes. The two populations we know about may be separate species. Their ventral fins are fleshy extensions of the body wall and have internal skeletal elements that support them (Figure 24.10*a*). Like ray-finned fishes, coelacanths have gills.

Lungfishes (Figure 24.10*b*) have gills and lunglike sacs—modified outpouchings of the gut wall. They fill the sacs by gulping air, then oxygen diffuses from the sacs into the blood. When a pond dries out seasonally, lungfishes can encase themselves in slimy secretions and mud. They remain inactive until the rains return.

THE FIRST TETRAPODS

By the time the Devonian ended, some aquatic species were walking (Figure 24.10*c*). Evidence suggests that lungfishes and lobe-finned fishes are the closest living relatives of tetrapods—the four-legged walkers. Like tetrapods, a lungfish has a separate blood circuit to its "lungs." Its skullbones are arranged in a similar way. Also, bony elements in the pectoral and pelvic fins of lobe-finned fishes and lungfishes resemble the bones of early tetrapod limbs. Fossils reveal how skeletons changed from fish to early tetrapods (Figure 24.10*d–e*).

Fishes have an endoskeleton of cartilage, bone, or both. Bony fishes include ray-finned fishes, now the most diverse and abundant vertebrates, lobe-finned fishes (with fleshy ventral fins reinforced with skeletal parts), and lungfishes, with lunglike sacs that supplement respiration by gills.

Walking probably started under water during the Devonian, among the aquatic forerunners of tetrapods on land.

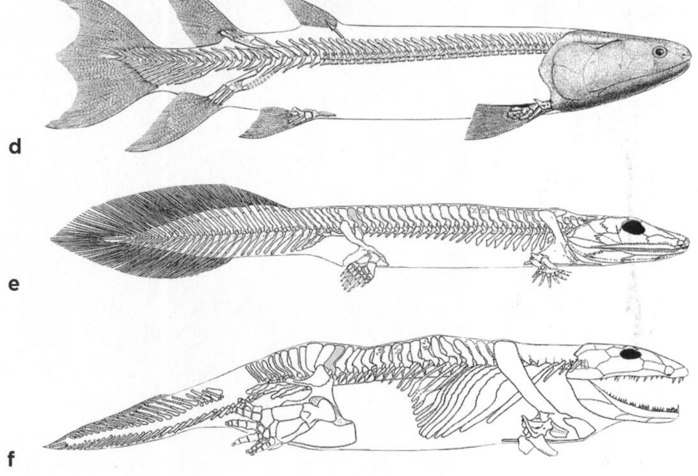

Figure 24.10 (**a**) Living coelacanth (*Latimeria*), of the only modern lineage of lobe-finned fishes. (**b**) Living Australian lungfish. (**c**) Painting of Devonian tetrapods: *Acanthostega*, submerged, and *Ichthyostega* crawling onto land. Compare the Devonian lobe-finned fish *Eusthenopteron* (**d**), with two early tetrapods *Acanthostega* (**e**), and *Ichthyostega* (**f**) to get a sense of how the skeleton became modified as tetrapods moved onto land.

24.4 Amphibians—First Tetrapods on Land

The preceding section introduced you to the Devonian forerunners of amphibians. Turn now to their descendants, which include the salamanders, frogs, and toads.

EVOLUTIONARY HIGH POINTS

LINKS TO SECTIONS 15.3, 17.8

Amphibians are a lineage of thin-skinned vertebrates, the first to evolve from aquatic Devonian tetrapods. Adults are carnivores and may live on land, but nearly all return to the water to reproduce (Figures 24.11 and 24.12). In body plans and reproductive modes, they are somewhere between fishes and reptiles.

What rewards did the early tetrapods ancestral to amphibians achieve by dragging themselves onto land with their modified fins? Clambering onto land might have helped them escape from an aquatic predator or move from an evaporating puddle to another home. It might have helped them chase land-dwelling insects, a bountiful food source.

Over time, descendants of early tetrapods evolved into the earliest amphibians. Structural modifications let them spend more time out of water and move in more coordinated ways on land. Tetrapods with lungs had an advantage; they could get oxygen by gulping air. These developments served as the foundation for a great adaptive radiation at the end of the Devonian.

Figure 24.12 One of the caecilians, a legless tropical amphibian.

Life in the new, drier habitats was both dangerous and promising. Temperatures fluctuated more on land than in the water, air did not support the body as well as water did, and drying out became a risk. On the other hand, air holds more oxygen than water. Larger lungs improved gas exchange and the heart became divided into three chambers. These changes supported more active life-styles.

Natural selection led to changes in sensory systems that help amphibians detect and pursue prey, or detect and escape predators. Changes to components of the inner ear improved the detection of airborne sounds and a sense of balance. Vision sharpened; eyes became

Figure 24.11 Familiar amphibians. (**a**) Red-spotted salamander (*Notophthalmus*). Salamander and newt forelimbs and hindlimbs are about the same size, and most project at right angles from the body. Coordinated controls over the legs move the body the same way that controls over fins move fishes. (**b**) A jumping frog displaying its tetrapod heritage. (**c**) American toad.

protected by eyelids. At the same time, those brain regions that receive, process, and respond to sensory input expanded.

No modern amphibian has fully escaped its aquatic heritage. Even the species that have gills or lungs also exchange oxygen and carbon dioxide across their thin skin. But respiratory surfaces must be kept moist at all times, and skin dries out easily in air. Only a few of the modern amphibians remain fully aquatic. Most release eggs in water, where larvae develop. Species adapted to dry habitats usually lay their eggs in moist places. In a few lineages, the embryos develop inside moist tissues of the adult body.

THE MAIN MODERN GROUPS

Frogs and toads belong to the most diverse amphibian lineage; there are more than 5,000 living species. Most live in the tropics. Muscular, elongated hindlegs allow the stocky, tailless adults to swim, hop, and make leaps that can be spectacular, given their body size (Figure 24.11*b*). In many species, venom or mucus secreted by the skin repels predators. The tongue attaches at the front of the mouth, not at the back as yours does. The animal unrolls its sticky tongue to catch prey, usually insects. Larvae, commonly known as tadpoles, have gills and a tail, but no limbs. During metamorphosis, hormones cause major remodeling; the tail and gills disappear, and legs form.

About 535 salamander species and their relatives, the newts, live mainly in North America, Europe, and Asia. Adults have a tail, and limbs extend outward from the sides of the body. As salamanders walk, their body bends from side to side, just like the body of a swimming fish (Figure 24.11*a*). The first tetrapods to move onto land probably walked this way.

Some salamanders, such as axolotls, retain external gills and other juvenile traits as adults. Their bones just stop growing at an early stage. Some are sexually precocious; they breed before getting the adult form.

The caecilian lineage includes about 165 limbless, blind species (Figure 24.12). Most burrow into moist soil, using their senses of touch and smell to pursue insects. A few aquatic, predatory types detect weak electric currents produced by prey moving in the water.

Amphibians show adaptations to life on land, such as lungs and a three-chambered heart, but nearly all still must have access to water to complete their life cycle.

Frogs, toads, and salamanders are the best known modern amphibians.

24.5 Vanishing Acts

FOCUS ON THE ENVIRONMENT

Amphibians are survivors. Their lineage originated before the dinosaurs, and it outlasted them. These tetrapods have been around a thousand times longer than humans, but now human activities are putting many kinds at risk throughout the world.

LINKS TO SECTIONS 22.1, 23.5

There is no question that amphibians are in trouble. Of about 5,500 known species, population sizes of at least 200 are plummeting. The alarming declines have been well documented in North America and Europe, but the changes are happening worldwide.

At this writing, six frog species, four toad species, and eleven salamander species are listed as threatened or endangered in the United States and Puerto Rico. One, the California red-legged frog (*Rana aurora*), inspired Mark Twain's well-known short story, "The Celebrated Jumping Frog of Calaveras County." This species is the largest frog native to the western United States.

Researchers correlate many declines with shrinking or deteriorating habitats. Developers and farmers commonly fill in low-lying ground that once collected seasonal rains and formed pools of standing water. Nearly all species of amphibians have trouble breeding unless they deposit eggs in water, where their larvae can develop.

Also contributing to declines are introductions of new species in amphibian habitats, long-term shifts in climate, increases in ultraviolet radiation, and the spread of fungal and parasitic diseases into different habitats (Section 22.1 and Figure 24.13). Chemical pollution of aquatic habitats also causes problems. We will consider the effects of one agricultural chemical in detail in Chapter 32.

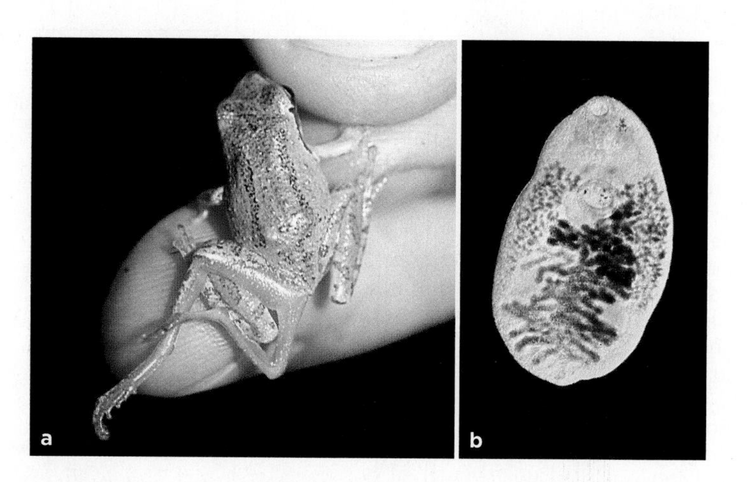

Figure 24.13 (**a**) Example of frog deformities. (**b**) A parasitic fluke (*Ribeiroia*). It burrows into limb buds of frog tadpoles and physically or chemically alters individual cells. Infected tadpoles grow extra legs or none at all. Where *Ribeiroia* populations are most dense, the number of tadpoles that successfully complete metamorphosis is low. Nutrient enrichment of water by fertilizers and pesticide contamination may make frogs more easily infected.

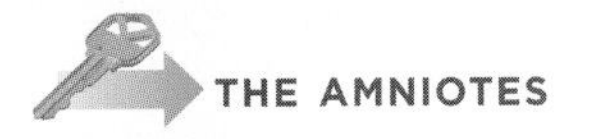

24.6 The Rise of Amniotes

Amniotes were the first vertebrates to adapt fully to dry land. They did so through modifications in their organ systems, behavior, and eggs, which have four membranes that conserve water and support embryonic development.

LINKS TO SECTIONS 16.5, 16.6

In the late Carboniferous, one amphibian lineage gave rise to the "stem" reptiles, the first amniotes. Amniotes make eggs having four unique membranes that allow embryos to develop away from water (Figure 24.14*a,b*). They also have a pair of highly efficient kidneys and a waterproof skin. Nearly all fertilize eggs inside the female's body. These traits adapt them to life on land.

One early branching of the amniote lineage led to synapsids: mammals and extinct mammal-like species (Figure 24.14*c*). A now extinct synapsid subgroup, the therapsids, included ancestors of mammals, as well as *Lystrosaurus*, a tusked herbivore shown in Section 16.6.

Three other branches of the amniote lineage have survived. One branch led to turtles, another to lizards and snakes, and a third to crocodilians and birds. The common term "reptile," is not a formal taxon because it places birds and crocodilians, both the descendants of archosaurs, into separate groups. Nevertheless, the term persists as a way to refer to amniotes that have none of the defining traits of birds or mammals. That is how we use the term in this book.

The earliest reptiles had a heavy, lizardlike body but slender limbs. With well-muscled jaws and sharp teeth, they could seize and crush insect prey with more force than amphibians. Their waterproof skin was far more suitable in drier habitats, but gases no longer could be exchanged across it. Compared to amphibians, early reptiles had larger, more efficient lungs. They also had larger brains and showed more complex behavior.

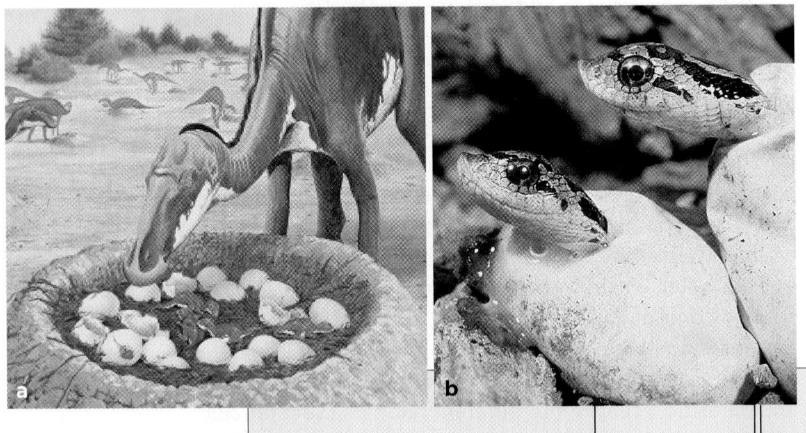

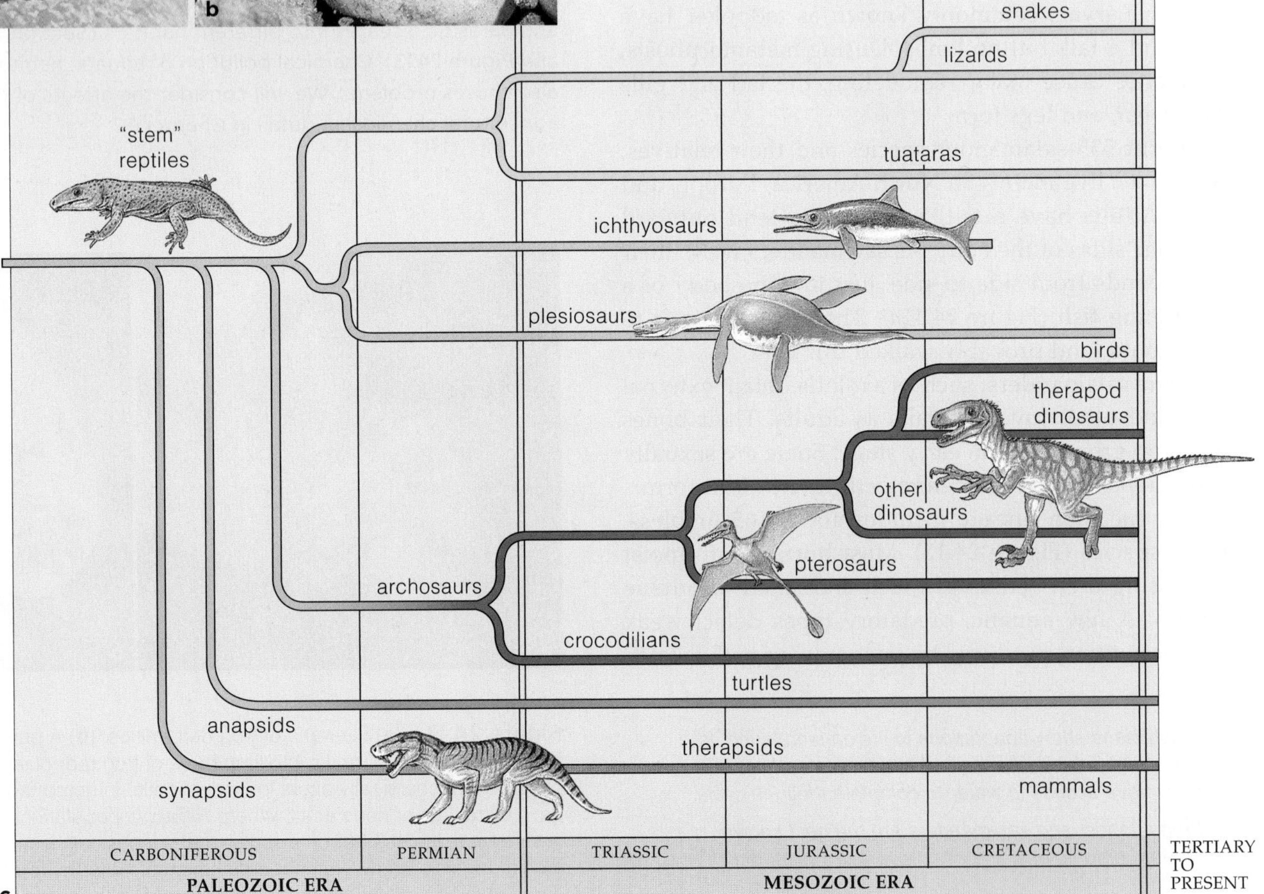

Figure 24.14 Amniote eggs and phylogeny.

(**a**) A painting of a nest of a duck-billed dinosaur (*Maiasaura*) that lived about 80 million years ago in what is now Montana. Like the modern crocodilians and birds, this dinosaur protected its eggs in a nest and may have cared for the hatchlings. (**b**) Two eastern hognose snakes emerging from amniote eggs.

(**c**) Family tree for amniotes. Snakes, lizards, tuataras, birds, crocodilians, turtles, and mammals are modern amniote groups.

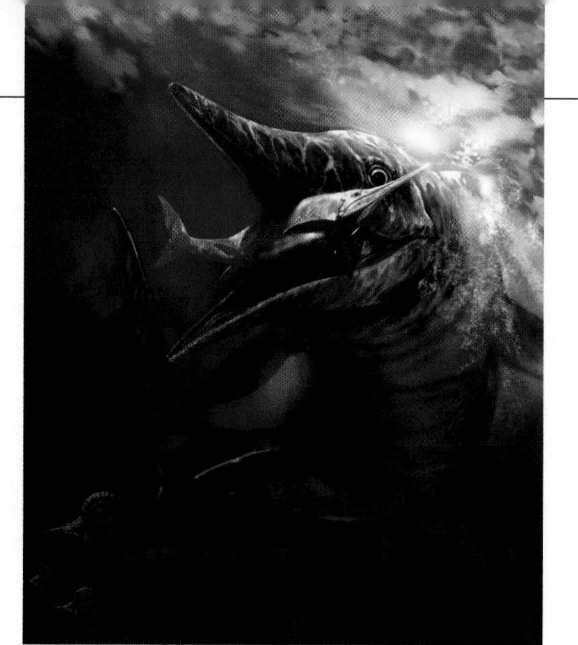

Figure 24.15 *Temnodontosaurus.* This ichthyosaur hunted large squids, ammonites, and other prey in the warm, shallow seaways of the early Jurassic. Fossils that measure 30 feet (9 meters) long have been found in England and Germany.

Biologists define the reptiles known as dinosaurs by skeletal features, such as the shape of their pelvis and hips. Dinosaurs evolved by the late Triassic. Early species were the size of a turkey and ran on two legs. Adaptive zones opened up for this lineage about 213 million years ago, after fragments from an asteroid or comet fell on the spinning Earth and hit what is now France, Quebec, Manitoba, and North Dakota. Nearly all the animals that survived the impacts were small, had high rates of metabolism, and could tolerate big changes in temperature.

Surviving dinosaur groups, such as those shown in the chapter introduction, became the "ruling reptiles." For 125 million years they dominated the land even as other groups, including the ichthyosaurs, flourished in the seas (Figure 24.15). Many kinds of dinosaurs were lost in a mass extinction that ended the Jurassic. Others died off in the Cretaceous.

The Cretaceous ended as cosmic bad luck wiped out nearly all dinosaurs. Feathered kinds, known as birds, survived. So did the ancestors of modern reptiles, the crocodilians, turtles, tuataras, snakes, and lizards.

Eggs with four membranes that help embryos develop, water-conserving skin and kidneys, and active life-styles helped early amniotes adapt fully to dry land habitats.

One amniote group, the synapsids, includes mammals and now-extinct mammal-like species. Another group, the sauropsids, includes reptiles and birds.

Much of amniote history was a matter of luck, of being in the right or wrong places on a changing geologic stage.

24.7 So Long, Dinosaurs

FOCUS ON EVOLUTION

Reflect for a moment on the dinosaurs. Did they reign supreme? No question about it. Their lineage dominated the land for 140 million years. In the end, did it matter? Not a bit. Sixty-five million years ago, the remaining members of their lineage perished. Why? Bad luck.

Chapter 16 made passing reference to a mass extinction that defines the Cretaceous–Tertiary (K–T) boundary. After methodically analyzing iridium levels in soils, maps of gravitational fields, and other evidence from around the world, Walter Alvarez and Luis Alvarez developed a hypothesis: A direct hit by an enormous asteroid caused the K–T extinction event. This came to be known as the K–T asteroid impact hypothesis.

LINK TO CHAPTER 16 INTRODUCTION

Later, researchers discovered an enormous impact crater on the sea floor in the Gulf of Mexico. Known as the Chicxulub crater, it is 9.6 kilometers deep and 300 kilometers across. By one estimate, to make a crater that big, an asteroid had to hit Earth at 160,000 kilometers per hour. It blasted at least 200,000 cubic kilometers of dense gases and debris into the sky.

Did this impact cause the K–T extinction event? Many researchers think so. However, Gerta Keller and others hold that the Chicxulub crater was formed 300,000 years before the K–T extinction. They think a series of asteroid impacts occurred, and that the crater formed at the K–T boundary is yet to be discovered.

Researchers also debate the mechanism by which an asteroid impact could have caused the known extinctions. Some argue that atmospheric debris must have blocked sunlight for months. They are trying to explain the fossil record, which shows that many plant and animal species on land died off. Others think that the volume of debris blasted aloft would not have been great enough to have this dramatic and widespread effect.

An alternative scenario was proposed after the comet Shoemaker–Levy 9 slammed into Jupiter in 1994. Debris blasted into the Jovian atmosphere and triggered intense heating. That event led Jay Melosh and his colleagues to propose that an enormous asteroid impact raised Earths atmospheric temperatures by thousands of degrees. In one terrible glowing hour, the world erupted in flames. Any animals out in the open—including nearly all dinosaurs—were broiled alive.

Not every living thing disappeared. Snakes, lizards, crocodiles, and turtles survived. So did birds and mammals. The proponents of Melosh's hypothesis argue that smaller species may have missed the firestorm by burrowing underground. Critics point out that most birds are ill-equipped to burrow. Also, protists of the open ocean and invertebrates on the ocean floor disappeared too. How could they have "broiled"?

In short, one or more asteroids are implicated in the K–T extinctions. Where they hit and exactly what happened next remains an open question.

An asteroid impact ended the golden age of dinosaurs.

24.8 Portfolio of Modern "Reptiles"

All modern amniotes that are not birds or mammals belong to diverse reptilian lineages.

GENERAL CHARACTERISTICS

LINK TO SECTION 16.5

"Reptile" is derived from the Latin *repto*, which means to creep. Some reptiles do creep. Others swim or race or lumber about. Modern reptiles include more than 8,160 species of turtles, lizards, tuataras, snakes, and crocodilians. Figure 24.16 shows one reptile body plan. Like amphibians and fishes, reptiles are cold-blooded and have a cloaca, an opening that expels digestive and urinary waste and also functions in reproduction. All male reptiles, except tuataras, have a penis and fertilize eggs inside the female body. Females lay eggs on land.

MAJOR GROUPS

Turtles About 305 modern species of turtles have a shell attached to their skeleton. When threatened, they pull the head and limbs inside (Figure 24.17*a*,*b*). Their body plan has endured since Triassic times. The shell is reduced in sea turtles and other highly active types.

Instead of teeth, turtles have horny plates that can grind food. Strong jaws and a thick shell help deter predation on adults. However, many predators savor turtle eggs. Construction of resorts and homes along beaches that were previously sea turtle nest sites has helped push these species toward extinction.

Lizards With 4,710 species, the lizards are the most diverse reptiles. The smallest lizard can fit on a dime (*left*). The largest, the Komodo dragon, grows up to 3 meters (10 feet) long. It is an ambush predator that lurches out of bushes to attack deer, wild pigs, and a few unfortunate humans. Like most predatory lizards, it snags prey with peglike teeth. Chameleons catch prey with the flick of a sticky tongue, which can be longer than their body. Iguanas are herbivores.

Many lizards serve as tasty prey for other animals. Some try to outrun a predator or startle it by flaring a throat fan (Figure 24.17*c*,*d*). Many detach their tail when a predator grabs it. The tail wriggles for a bit and may distract a predator from its fleeing owner.

Tuataras Two species of tuatara are found on small islands near the coast of New Zealand. They are the remnants of a lineage that thrived during the Triassic. Tuatara means "peaks on the back" in the language of New Zealand's native Maori people. This name refers to the spiny crest (Figure 24.17*e*). Tuataras are reptiles but walk like salamanders and have some amphibian-like brain structures. Also, a third eye develops under the skin of the forehead. It becomes covered by scales in adults and its function, if any, is unknown.

Snakes During the Cretaceous, short-legged, long-bodied lizards gave rise to the limbless snakes. Some of the 2,995 modern species still have bony remnants of hindlimbs. All are carnivores. Many have flexible jaws that help them swallow prey whole. All snakes have teeth; not all have fangs. Rattlesnakes and other fanged types bite and subdue prey with venom produced in modified salivary glands (Figure 24.17*f*). On average, about 2 of the 7,000 snake bites reported annually in the United States are fatal. Worldwide, snake bites kill 30,000 to 40,000 people each year.

Snakes help keep rodent populations in check and also are important as prey. Hawks, herons, and other birds prey on adult snakes, as do foxes and coyotes. Snake eggs are vulnerable, too. Often, a female stores sperm from a mating, then lays small clutches of eggs.

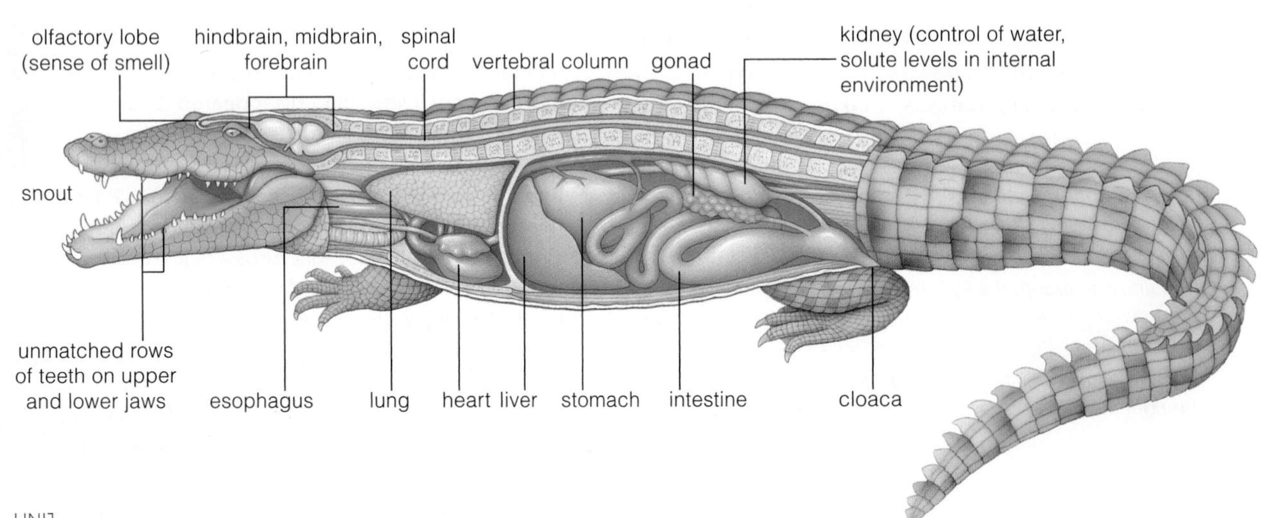

Figure 24.16 **Animated!** Body plan of a crocodile which, like its lifestyle, has changed very little for nearly 200 million years. The group's future is not rosy. Human habitats encroach on many crocodilian habitats. Alligator belly skin is in high demand for wallets, handbags, belts, and shoes.

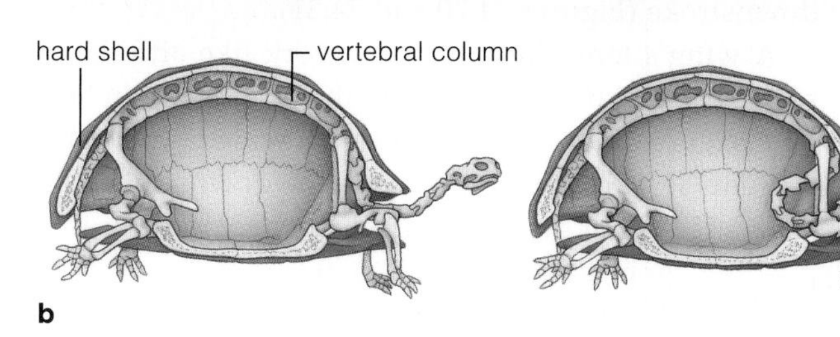

Figure 24.17 (**a**) Galápagos tortoise. (**b**) Turtle shell and skeleton. (**c**) Lizard fleeing and (**d**) lizard confronting. (**e**) Tuatara (*Sphenodon*). (**f**) Rattlesnake. (**g**) Spectacled caiman, a crocodilian. Its upper and lower rows of teeth do not align, as mammalian teeth do.

Laying her eggs at different times in different places improves odds that some will survive until hatching. Other snakes, including all sea snakes, brood eggs in the body and give birth to well-developed young.

Crocodilians Almost a dozen species of crocodiles, alligators, and caimans are the closest living relatives of birds. All are predators in or near water. They have powerful jaws, a long snout, and sharp teeth (Figure 24.17*g*). They clench prey, drag it under water, tear it apart as they spin around, and gulp the chunks. The feared saltwater crocodile and Nile crocodile weigh up to a 1,000 kilograms (2,200 pounds).

Crocodilians are the only reptiles that have a four-chambered heart, as mammals and birds do. Also like most birds, the crocodilians display complex parental behavior. For example, they make and guard a nest, then feed and care for the hatchlings.

Body plans of turtles and tuataras have changed little since the Triassic. The most diverse reptilians are lizards and snakes. The ones with the most complex brain and heart structure are crocodilians, the closest relatives of birds.

24.9 Birds—The Feathered Ones

A capacity for flight evolved in pterosaurs, which are now extinct, and in insects, birds, and bats. A few dinosaurs had feathers; today, birds alone have them.

LINKS TO SECTIONS 16.3, 17.5, 17.10

Birds are the only living animals that have feathers. These lightweight structures have roles in insulation, flight, and courtship displays. They vary in their size, shape, and coloration (Figure 24.18).

As an adaptive radiation of reptiles got under way in the Mesozoic, the ancestors of birds branched from a lineage of small, carnivorous dinosaurs that roamed about on two legs. *Archaeopteryx* (Figure 24.1) was in or close to the lineage. *Confuciusornis sanctus*, another early bird, is shown in Figure 24.18*b*. Both species had feathers, which evolved from reptilian scales.

Like some beaked dinosaurs, all birds have a horny beak, or bill. Differences in the size and shape of the bill correlate with differences in food preferences, as Sections 16.3, 17.5, and 17.10 show.

Unlike reptiles, birds have mechanisms that closely control their core temperature. They gain most of their heat not from outside sources but from metabolism. When cold, they fluff their insulating feathers. When they become too hot, they breathe rapidly. This helps dissipate excess metabolic heat to the outside air.

Like the dinosaurs and crocodilians—their closest relatives—birds have scales on their legs and some of the same organs, such as a cloaca. Birds have amniote eggs (Figure 24.19) and fertilization occurs internally. But several modifications to the body plan gave birds the capacity for flight.

Highly efficient respiratory and circulatory systems support the energy demands of flight. Birds alone have a respiratory system in which air sacs connect to lungs and keep air flowing through them. Most reptiles have a three-chambered heart, but in crocodilians, birds, and mammals, the heart has four chambers. Such a heart allows blood to travel through two separate circuits; oxygen-rich blood and oxygen-poor blood never mix. A bird also has a rapid heartbeat and its heart is large relative to its body size.

Flight also demands an airstream, low weight, and a powerful downstroke that can provide lift: a force at right angles to an airstream. Each wing, a modified forelimb, consists of feathers and lightweight bones attached to powerful muscles. Its bones do not weigh much, owing to profuse air cavities in the bone tissue. Flight muscles are attached to an enlarged breastbone (sternum) and to the upper limb bones attached to it. When the muscles contract, they produce a powerful downstroke (Figures 24.20 and 24.21).

A wing's long flight feathers work like airfoils on planes. The bird normally folds these feathers slightly on each upstroke, which lessens the wing surface and presents less resistance to air. On the downstroke, the bird spreads out the feathers, which increases the area pushing against air (Figure 24.20*a*).

Approximately 9,000 named species of birds vary in their size, proportions, coloration, and capacity for flight. The bee hummingbird weighs 1.6 grams (about half an ounce). An ostrich, a flightless sprinter, weighs about 150 kilograms (330 pounds). Perching birds are the largest subgroup, with more than half of all named species, including familiar songbirds. Bird song and other behaviors are discussed in Chapter 44.

We see one of the most dramatic forms of behavior among birds that migrate with the changing seasons. Migration is a recurring pattern of movement from one region to another in response to environmental rhythms. Seasonal change in daylength is one cue for internal timing mechanisms called "biological clocks." It triggers physiological and behavioral changes that

Figure 24.18 (**a**) The brilliant feathers of a male northern cardinal. (**b**) Artist's depiction of *Confuciusornis sanctus*, which lived about the same time as *Archaeopteryx*. This is the earliest known bird with a toothless bill. It wings show how grasping limbs became modified for flight. Each wing had three feathered digits, one with a small claw at its tip.

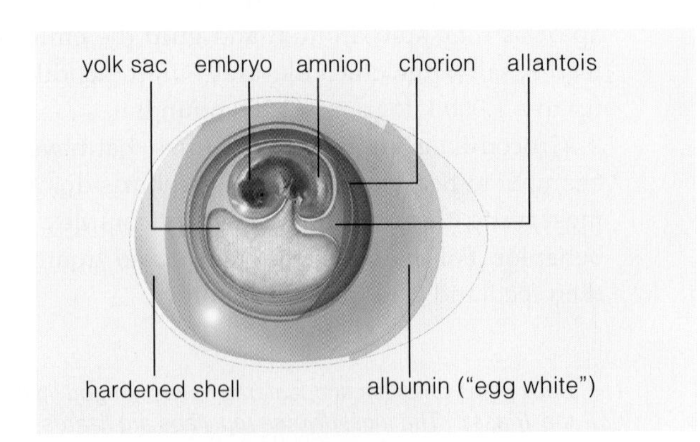

Figure 24.19 **Animated!** Components of a bird egg, a type of amniote egg.

Figure 24.20 *Above*, bird flight. (**a**) Of all the modern vertebrates, only birds and bats can fly by flapping a pair of wings. A bird wing consists of feathers and lightweight bones. Section 16.4 addresses the evolution of wing bones. (**b**) The inner down feathers serve as insulation. (**c**). A hollow central shaft and an interlocked lattice of barbs and barbules strengthen the flight feathers and increase a wing's surface area without adding much weight.

(**d**) Laysan albatross, a seabird with a wingspan of more than two meters across, weighs less than 10 kilograms (22 pounds). It is so at home in the air that it sleeps while riding the winds. One bird monitored by researchers flew 39,980 kilometers in ninety days as it searched for food and brought it back to its nestling. That is the equivalent of a trip around the globe.

induce migratory birds to make round-trips between breeding grounds and wintering grounds. Arctic terns migrate the farthest. They spend summers in the arctic regions and winters in the antarctic. One lifted off from an island near Great Britain and landed three months later in Melbourne, Australia. It racked up more than 22,000 kilometers (14,000 miles) on that one journey.

Of all animals, birds alone make feathers, which function in flight, conservation of body heat, and socially significant communication displays. Feathers, lightweight bones, and highly efficient respiratory and circulatory systems are vital components of bird flight.

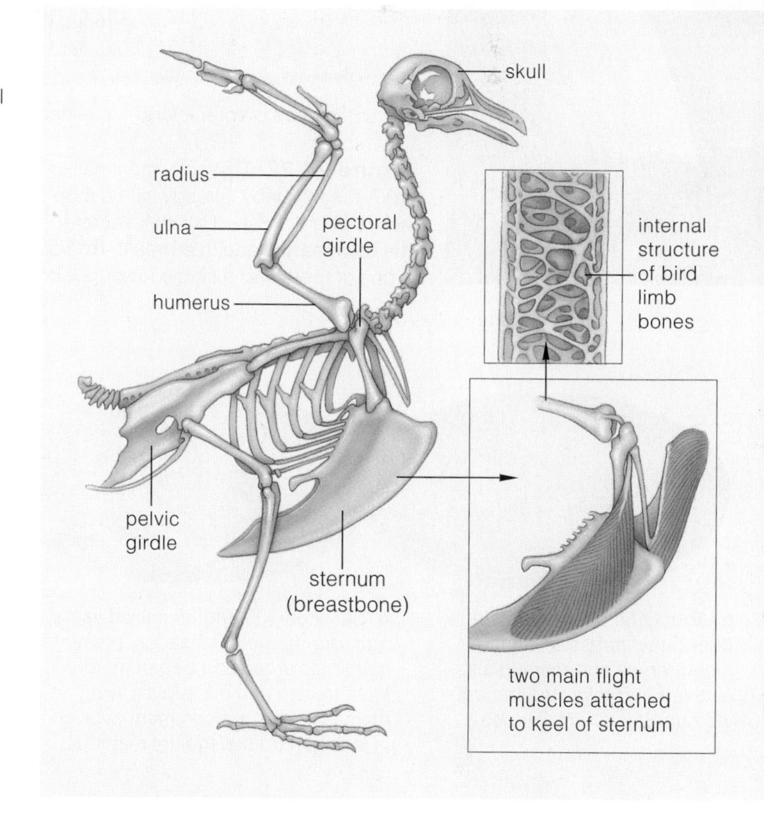

Figure 24.21 Animated! Bird skeleton and main flight muscles.

24.10 The Rise of Mammals

Christine Janis put it this way: In a world dominated by dinosaurs, mammals were artful dodgers that survived by being small, efficient at exploiting food, and flexible in behavior. They originated before dinosaurs, survived the K–T catastrophe, then underwent an adaptive radiation.

LINKS TO SECTIONS 16.5–16.7

Mammals are animals in which the females have milk-producing mammary glands (Figure 24.22*a*). The term is from the Latin *mamma*, meaning breast. Other traits unique to mammals include the presence of hair or fur, three bones in the middle ear, and a jaw with a single lower jawbone and four kinds of teeth. Section 30.5 explains how the middle ear bones function in hearing. Figure 24.22*b* illustrates the arrangement of the four kinds of mammalian teeth.

Again, as Figure 24.14 shows, synapsid ancestors of mammals branched from stem reptiles during the Carboniferous. Mammals are the only synapsids that survived to the present. Their ancestors lived through a mass extinction that ended the Permian. Like most lizards, their legs extended out to their sides and their belly was close to the ground. By the Triassic, legs of one group became positioned under the body, as in dogs. This raised the trunk higher above the ground, making the body more maneuverable, but less stable. Expansion of the brain region that controls balance and coordination accompanied the changes in posture.

Modern mammals arose in the Jurassic. Three major lineages had evolved by the Cretaceous. Monotremes (egg-laying mammals) arose first, then the marsupials (pouched mammals) and finally eutherians (placental mammals). All the while, the continents continued to move. For example, after Pangea broke up, immense fragments rafted certain monotremes and marsupials away from the placental mammals that were radiating elsewhere (Figure 24.23). When dinosaurs vanished 65 million years ago, mammals began their great adaptive radiation. Climates were mild; forests and woodlands

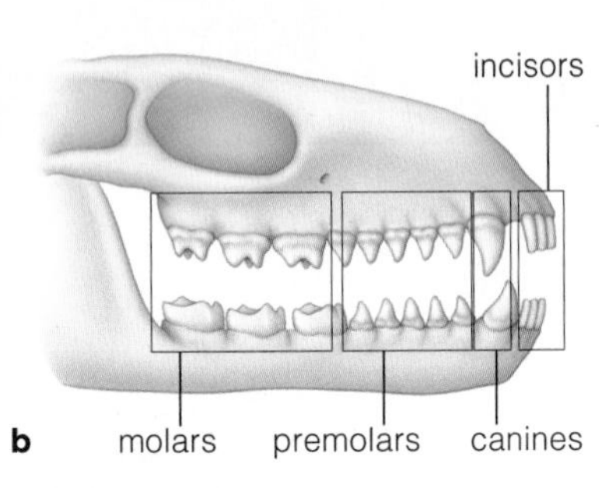

Figure 24.22 Distinctly mammalian traits. (**a**) A human baby, already with a mop of hair, being nourished by milk secreted from the mammary gland in a breast. (**b**) Four types of teeth and a single lower jaw bone.

a About 150 million years ago, during the Jurassic, the first monotremes and marsupials evolved and migrated through the supercontinent Pangea.

b Between 130 and 85 million years ago, during the Cretaceous, placental mammals arose and began to spread. Monotremes and marsupials living on the southern supercontinent evolved in isolation from placental mammals.

c Starting about 65 million ago, mammals expanded in range and diversity. On Antarctica, mammals vanished. Marsupials and early placental mammals displaced monotremes in South America.

d About 5 million years ago, in the Pliocene, advanced placental mammals invaded South America. They drove most marsupials and the early placental species to extinction.

Figure 24.23 **Animated!** Adaptive radiations of mammals. The painting shows mammals that lived about 60 million years ago in sequoia forests. Fossils of them have been discovered in Wyoming.

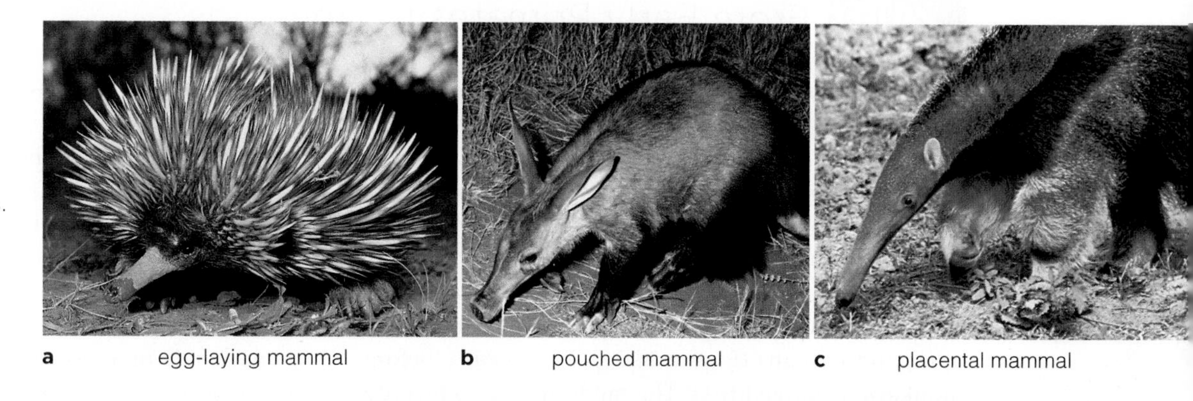

Figure 24.24 **Animated!** One outcome of convergent evolution. (**a**) Australia's spiny anteater, one of only three modern species of monotremes. (**b**) Africa's aardvark and (**c**) South America's giant anteater. Compare the specialized ant-snuffling snouts.

formed even in polar regions. Carnivores, omnivores, and herbivores evolved. Their teeth could nip, shred, or grind food. Then, about 20 million years ago, great mountain ranges formed and altered air circulation and rainfall. This climate change favored grasses, and grazing mammals underwent an adaptive radiation. In the Pliocene, which started 5.3 million years ago, ice ages alternated with warmer interglacial periods. Mammalian diversity peaked just before the onset of the Pliocene and continues to decline.

Often, species of separate lineages became adapted to very similar habitats on separate continents. Over time, they came to resemble one another in body form and function. For example, Australia's spiny anteater, South America's giant anteater, and Africa's aardvark eat ants and have a similar snout (Figure 24.24). They show morphological convergence (Section 16.7).

Modern monotremes include spiny anteaters and the duck-billed platypus. Females incubate their eggs and the young hatch in a relatively undeveloped state. They cling to the mother or are held in a skin fold on her belly. Milk feeds them as they grow and develop.

Most of the 240 modern species of marsupials live in Australia and nearby islands. They include koalas and kangaroos. The opposum is the only kind in North America. Marsupials are born tiny, hairless, and blind. They crawl along their mother to a permanent pouch on her ventral surface, where they suckle and grow.

Placental mammals are named for the placenta, an organ that allows materials to pass between a mother and an embryo developing inside her uterus (Figure 24.25). This innovation gave the placental mammals a competitive advantage over other mammals; placental embryos grow faster. Also the offspring are born more fully formed and thus are less vulnerable to predation. Placental mammals are now the dominant mammals in most land habitats, and the only ones living in the seas. Figure 24.25 shows a few of the more than 4,000 species. Appendix I lists major groups.

The body form, function, behavior, and ecology of mammals will occupy our attention later in the book.

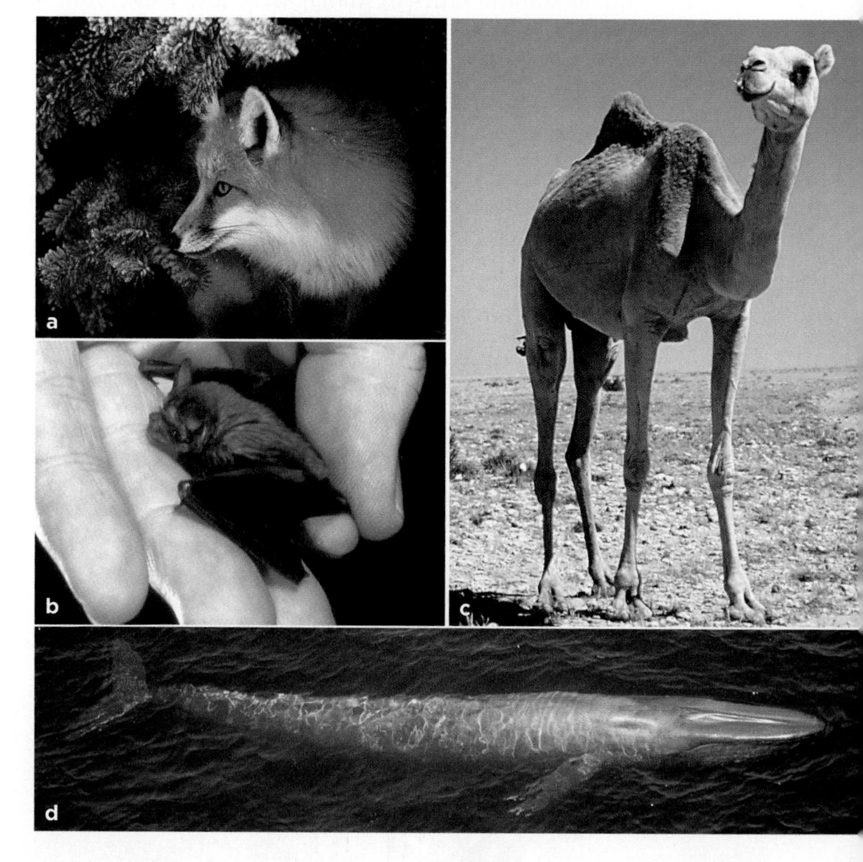

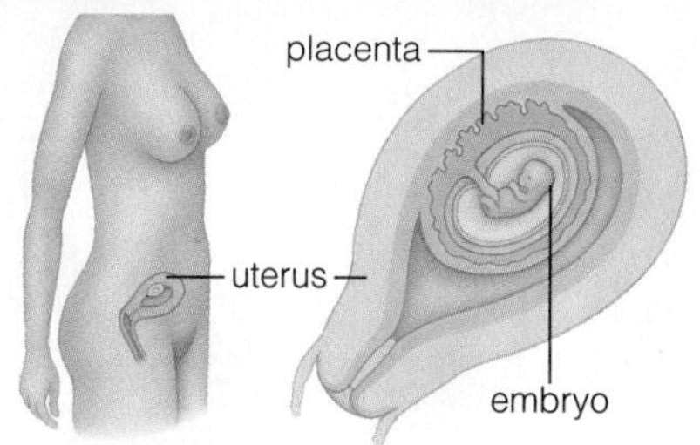

Figure 24.25 Placental mammals, which have radiated into nearly all habitats, from icy polar regions, to hot deserts, into the seas. (**a**) Red fox. (**b**) Kitti's hog-nosed bat, the smallest mammal. (**c**) Camel. (**d**) Blue whale. At 200 tons, it is the largest living animal. *Right*, the location of the uterus and placenta in a pregnant human female.

Mammals are the only animals that have hair or fur. Only female mammals produce milk from mammary glands.

The egg-laying monotremes, pouched marsupials, and placental mammals are three major groups. Placental mammals are now the dominant group in most regions.

24.11 From Early Primates to Hominids

So far, you have traveled 570 million years through time, from flattened invertebrates to craniates, then on to the vertebrates and early tetrapods with limbs that evolved from fleshy lobed fins. You moved on to the amniotes—reptiles, birds, and mammals. You are about to travel on the evolutionary roads that lead to humans.

LINK TO SECTION 16.5

Primates include 240 species of prosimians, monkeys, apes, and humans (Figure 24.26). *Prosimians* ("before monkeys") evolved first. The modern species include tarsiers and lemurs in Madagascar and New Guinea (Figure 1.2). *Anthropoids* include monkeys, apes, and humans, all widely dispersed. *Hominoids* include apes and humans, who are biochemically and structurally similar. Our closest living relatives are chimpanzees and bonobos. Humans and extinct humanlike species are hominids. Table 24.1 summarizes these groupings.

Table 24.1 Primate Classification

Prosimians	Lemurs, tarsiers
Anthropoids	New World monkeys (e.g., spider monkeys)
	Old World monkeys (e.g., baboons, macaques)
	Hominoids:
	Hylobatids (gibbons, siamangs)
	Pongids (orangutans, gorillas, chimpanzees, bonobos)
	Hominids (humans, extinct humanlike species)

OVERVIEW OF KEY TRENDS

Most primates live in the trees of forests, woodlands, and savannas, or grasslands with a scattering of trees. Five trends that led to uniquely human traits began in early tree-dwelling species. They came about through modifications to the eyes, bones, teeth, and brain.

Enhanced daytime vision. Early primates had an eye on each side of a mouse-shaped head. Later on, some had a more upright, flattened face, with eyes up front. The ability to focus both eyes on an object improved depth perception. Also, eyes became more sensitive to variations in light intensity and in color. During this time, the sense of smell declined in importance.

Upright walking. Humans display bipedalism; their skeleton and muscles are adapted for upright walking. For example, an S-shaped backbone keeps their head and torso centered over the feet, and arms are shorter than legs. By contrast, prosimians and monkeys walk on four legs of about the same length. Gorillas walk on two legs while leaning on the knuckles of longer arms (Figure 24.26*c*,*e*). As in other vertebrates, a primate's brain connects with its spinal cord through a hole in the skull: the foramen magnum. In animals that walk on all fours, the hole is located at the back of the skull. In upright walkers, it is near the center of the skull's base, as you can see from Figures 24.26*d–f* and 24.27.

Better grips. Early mammals spread their toes apart to support their weight as they walked or ran on four

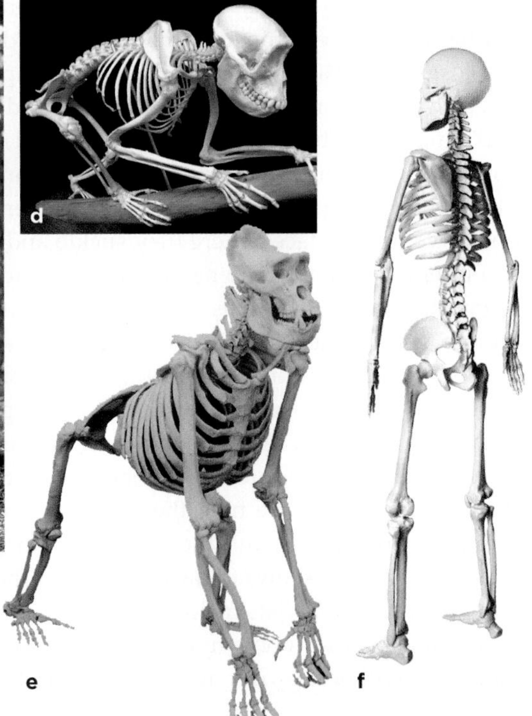

Figure 24.26 Primates. (**a**) Tarsier, a prosimian climber and leaper. (**b**) Spider monkey, an agile climber. (**c**) Male gorilla, using its forearms to support its weight as it walks on its knuckles and two legs. Comparisons of the skeletal structure of (**d**) a monkey, (**e**) a gorilla, and (**f**) a human. Skeletons are not to the same scale.

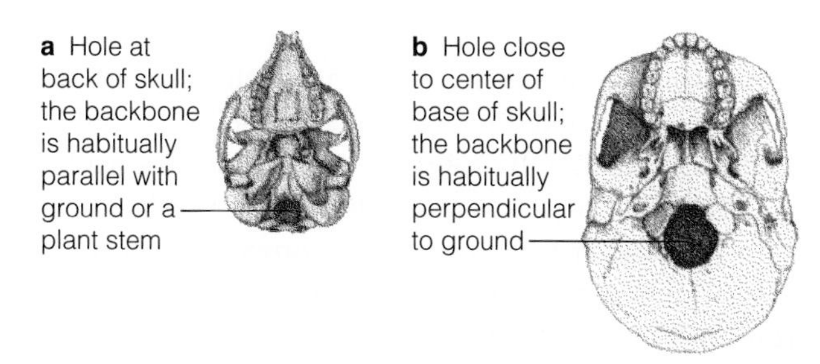

Figure 24.27 A hole in the head—the foramen magnum—in (**a**) a four-legged walker and (**b**) an upright walker. The foramen magnum is diagnostic trait for interpreting fossils.

legs. In ancient tree-dwelling primates, hands became modified. Fingers could curl around things (*prehensile* movements), and the thumb could touch all fingertips (*opposable* movements). In time, hands were freed from load-bearing functions and were modified in ways that allowed powerful or precision gripping of objects:

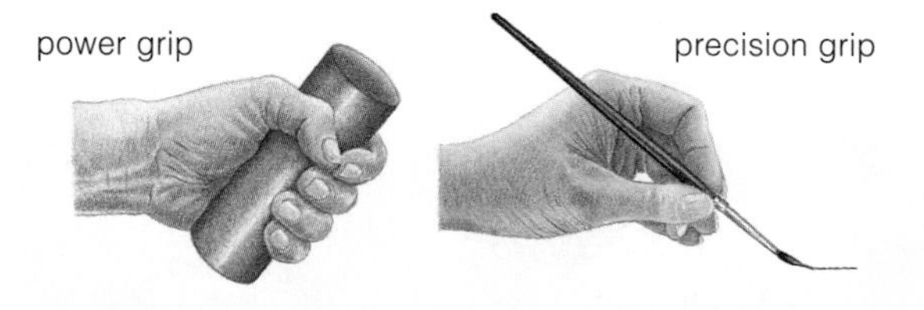

Having the capacity for versatile hand positions gave the ancestors of humans the capacity to make and use tools. Refined prehensile and opposable movements led to the development of technologies and culture.

Modified jaws and teeth. Modifications to the jaws correlate with a shift from eating insects, to fruits and leaves, to a mixed, or omnivorous, diet. Rectangular jaws and long canine teeth evolved in monkeys and apes. A bow-shaped jaw and smaller, more uniformly sized teeth evolved in the early hominids.

Brain, behavior, and culture. The braincase and the brain increased in size and complexity. As brain size increased, so did the length of pregnancy and extent of maternal care. Compared to early primates, later ones had fewer offspring and invested more in each one.

Early primates were solitary. Later, some started to live in small groups. Social interactions and cultural traits started to affect reproductive success. Culture is the sum of all learned behavioral patterns transmitted among members of a group and between generations.

ORIGINS AND EARLY DIVERGENCES

The first primates arose in the tropical forests of East Africa by about 65 million years ago (mya). The early species resembled modern tree shrews (Figure 24.28*a*). They foraged at night under trees and in low branches

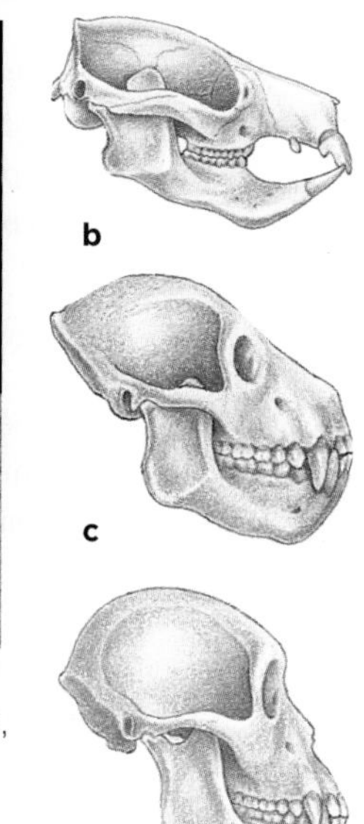

Figure 24.28 (**a**) Southeast Asian tree shrew (*Tupaia*), a close relative of modern primates. Skull comparisons: (**b**) *Plesiadapis*, an early, shrewlike primate. (**c**) Monkey-sized *Aegyptopithecus*, one of the Oligocene anthropoids. (**d**) *Proconsul africanus*. This early hominoid was no larger than an average four-year-old child.

for insects and seeds. They had a long snout and an excellent sense of smell.

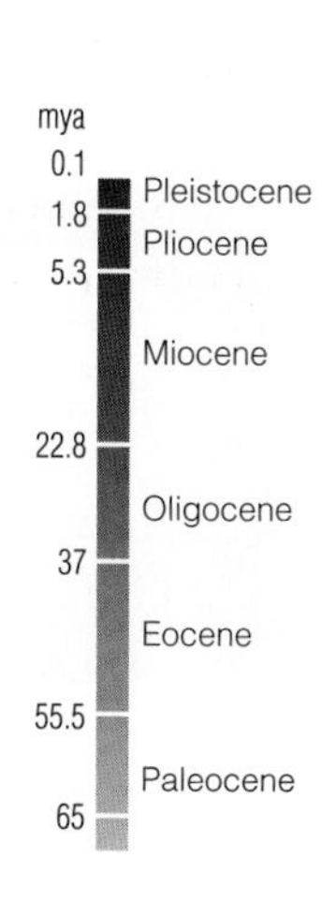

By the Eocene, prosimians had evolved. They were climbing higher, where odor-sensitive snouts did not help much; breezes disperse odors. Skeletal changes, front-facing eyes, and a brain that assessed motion, depth, shape, and color were favored. Climbers and leapers had to estimate body weight, distance, wind speed, and suitable destinations. Adjustments had to be quick for a body in motion far above the ground.

By 36 million years ago, tree-dwelling anthropoids arose (Figure 24.28*c*). Between 23 and 18 million years ago, in tropical rain forests, they gave rise to the first hominoids: the early apes (Figure 24.28*d*). Hominoids spread through Africa, Asia, and Europe as climates were changing due to shifts in land masses and ocean circulation patterns. During this time, Africa became cooler and drier. Tropical forests, with their abundance of edible soft fruits and leaves, were replaced by open woodlands and, later, grasslands. Food became drier, harder, and more difficult to find. Hominoids that had evolved in moist forests had two options: move into the new adaptive zones or die out. Most species died out, but not the shared ancestor of apes and humans. By 6 million years ago, hominids had emerged.

Complex, forward-directed vision; bipedalism, refined hand movements, generalized teeth, and interlocked elaboration of brain regions, behavior, and culture were key adaptations on the road from arboreal primates to modern humans.

24.12 Emergence of Early Humans

Fossils from central, eastern, and southern Africa show that the Miocene through the Pliocene was a "bushy" time of hominid evolution. In other words, many forms were rapidly evolving. We still do not know how they are related.

EARLY HOMINIDS

LINKS TO SECTIONS 16.8, 16.9

Genetic comparisons indicate that hominids diverged from apelike ancestors about 6 to 8 million years ago. Fossils that may be hominids are about 6 million years old. *Sahelanthropus tchadensis* had a hominid-like flat face, prominent brow, and small canines, but its brain was the size of a chimpanzee's (Figure 24.29*a*). *Orrorin tugenensis* and *Ardipithecus ramidus* also had hominid-like teeth. Some researchers suspect that those species stood upright; others disagree. More fossils will have to be discovered to clarify the picture.

An indisputably bipedal hominid, *Australopithecus afarensis*, was walking in Africa by about 3.9 million years ago. Remarkably complete skeletons reveal that it habitually walked upright (Figure 24.30). About 3.7 million years ago, two *A. afarensis* individuals walked across a layer of newly deposited volcanic ash. A light rain fell, and it transformed the powdery ash into a fast-drying cement, which preserved their footprints (Figure 24.30*c,d*).

A. afarensis was one of the australopiths ("southern apes"). This informal group includes *Australopithecus* and *Paranthropus* species. *Australopithecus* species were petite; they had a small face and teeth (Figure 24.29*b*). One or more species are suspected to be ancestral to modern humans. In contrast, *Paranthropus* species had a stockier build, a wider face, and larger molars. Jaw muscles attached to a pronounced bony crest at the top of their skull (Figure 24.29*c*). The big molars and strong jaw muscles indicate that fibrous, difficult-to-chew plant parts accounted for a large part of the diet. *Paranthropus* died out about 1.2 million years ago.

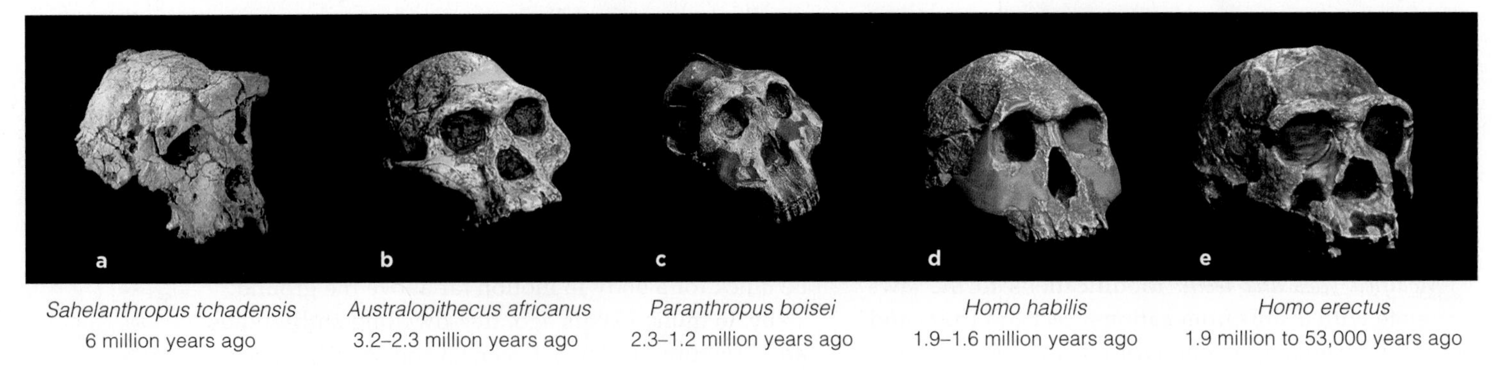

Sahelanthropus tchadensis	*Australopithecus africanus*	*Paranthropus boisei*	*Homo habilis*	*Homo erectus*
6 million years ago	3.2–2.3 million years ago	2.3–1.2 million years ago	1.9–1.6 million years ago	1.9 million to 53,000 years ago

Figure 24.29 A sampling of fossilized hominid skulls from Africa, all to the same scale.

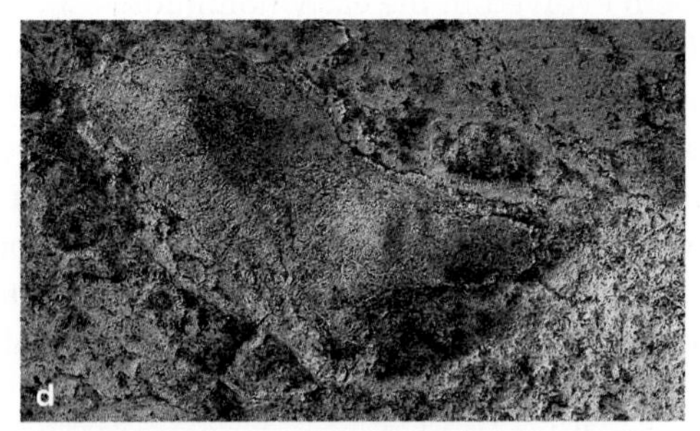

Figure 24.30 (**a**) Fossilized bones of Lucy, a female australopith (*Australopithecus afarensis*). This bipedal hominid lived in Africa 3.2 million years ago.

Chimpanzees and other apes have a splayed-out big toe (**b**). Early hominids did not. How do we know? (**c,d**) At Laetoli in Tanzania, Mary Leakey discovered footprints made in soft, damp, volcanic ash 3.7 million years ago. The arch, big toe, and heel marks of these footprints are signs of bipedal hominids.

EARLY HUMANS

What do the fossilized fragments of early hominids tell us about human origins? The record is still too sketchy for us to interpret how all the diverse forms were related, let alone which might have been our ancestors. Besides, exactly which traits should we use to define humans—members of the genus *Homo*?

Well, what about brains? Our brain is the basis of unsurpassed analytical skills, verbal skills, complex social behavior, and technological innovations. Early hominids did use simple tools, but so do chimpanzees and bonobos. So how did an early hominid make the evolutionary leap to becoming human?

Comparing the brains of modern primates can give us clues. As explained in Section 16.8, genes for some brain proteins underwent repeated duplication as the primate lineage evolved. Further studies of how these proteins function may provide additional insight into how our uniquely human traits arose.

Until then, we are left to speculate on the evidence of physical traits among diverse fossils. They include a skeleton that permitted bipedalism, a smaller face, larger cranium, and smaller teeth with more enamel. These traits emerged during the late Miocene and can be observed in *Homo habilis*. The name of this early human means "handy man" (Figure 24.31).

Most of the early known forms of *Homo* are from the East African Rift Valley. Fossil teeth indicate that these early humans ate hard-shelled nuts, dry seeds, soft fruits and leaves, and insects. *H. habilis* may have enriched its diet by scavenging carcasses left behind by carnivores such as saber-tooth cats, but it did not have teeth adapted to a diet rich in meat.

Our close relatives, the chimpanzees and bonobos, use sticks and other natural objects as tools (Section 44.6). Early hominids most likely did the same. They smashed nuts open with rocks and used sticks to dig into termite nests and capture insects. By 2.5 million years ago, some hominids were modifying rocks in ways that made them better tools. Pieces of volcanic rock chipped to a sharp edge were found with animal bones that look like they were scraped by such tools.

The layers of Tanzania's Olduvai Gorge document refinements in toolmaking abilities (Figure 24.32). The layers that date to about 1.8 million years ago hold crudely chipped pebbles. More recent layers contain more complex tools, such as knifelike cleavers.

Olduvai Gorge also held hominid fossils. At the time of their discovery, these fossils were classified as *Homo erectus*. The name means "upright man." Today, some researchers reserve that name for fossils in Asia, and they assign the African fossils to *H. ergaster*. In our discussions, we will adopt a traditional approach using "*H. erectus*" in reference to African populations and to descendant populations who, over generations, migrated into Europe and Asia.

Figure 24.31 Painting of a band of *Homo habilis* in an East African woodland. Two australopiths are shown in the distance at the left.

Figure 24.32 A sample of stone tools from Olduvai Gorge in Africa. From left to right, crude chopper, more refined chopper, hand ax, and cleaver.

H. erectus adults averaged about 5 feet (1.5 meters) tall, and had a larger brain than *H. habilis*. Improved hunting skills may have helped *H. erectus* get the food needed to maintain a large body and brain. Also, *H. erectus* built fires. Cooking probably broadened their diet by softening previously inedible hard foods.

Australopiths and certain hominids that preceded them walked upright. Some Australopithecus *species were close to or on the lineage that led to humans.*

Like some hominids that preceded them, Homo habilis, *the earliest known human species, walked upright.* Homo erectus *had a larger brain and dispersed out of Africa.*

24.13 Emergence of Modern Humans

Judging from the fossil record, the earliest members of the human lineage emerged about 2.5 million years ago, in the great East African Rift Valley.

BRANCHINGS OF THE HUMAN LINEAGE

LINKS TO SECTIONS 4.1, 16.4, 17.9

By 1.7 million years ago *Homo erectus* populations had become established in places as far away from Africa as the island of Java and eastern Europe. At the same time, African populations continued to thrive. Over thousands of generations, genetic divergences led to adaptations to local conditions. Some populations became so different from parental *H. erectus* groups that we call them new species: *H. neanderthalensis*—Neandertals—*H. floresiensis*, and *H. sapiens*, or fully modern humans (Figure 24.33).

A fossil from Ethiopia shows that *Homo sapiens* had evolved by 195,000 years ago. Compared to *H. erectus, H. sapiens* had smaller teeth, facial bones, and jawbones. Many fossils reveal a new feature—a prominent chin. Compared to earlier hominids, *H. sapiens* had a higher, rounder skull, a larger brain, and the capacity for spoken language.

From 200,000 to 30,000 years ago, Neandertals lived in Africa, the Middle East, Europe, and Asia. They were stocky enough to endure colder climates. Remember the surface-to-volume ratio (Section 4.1)? A stocky body has a lower surface area-to-volume ratio than a thin one and loses heat less quickly.

Neandertals had a big brain. Did they believe in an afterlife? Did they have a spoken language? We don't know. They vanished when *H. sapiens* entered the same regions. Did the new arrivals drive them to extinction through direct warfare or indirectly, by outcompeting Neandertals for the same resources?

Figure 24.33 Recent *Homo* species. (**a**) *H. neanderthalensis*, (**b**) *H. sapiens* (modern human), (**c**) *H. floresiensis*.

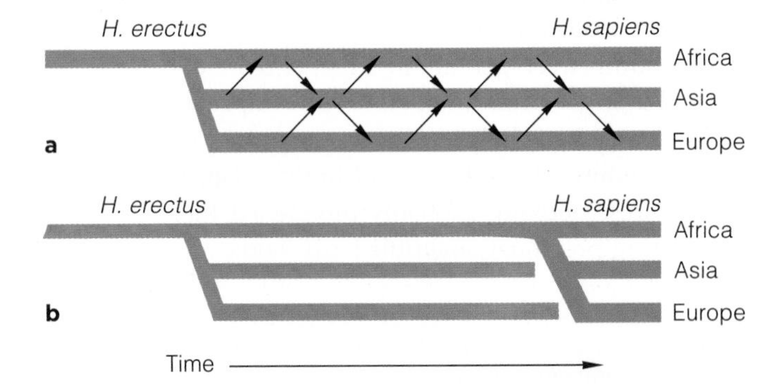

Figure 24.34 Two models for the origin of *H. sapiens*. (**a**) Multiregional model. *H. sapiens* slowly evolves from *H. erectus* in many regions. Arrows represent ongoing gene flow among populations. (**b**) Replacement model. *H. sapiens* rapidly evolves from one *H. erectus* population in Africa, then disperses and replaces *H. erectus* populations in all regions.

We do not know. Even if a few matings between the species did take place, sequencing data indicates that Neandertal genes did not contribute to the gene pool of modern humans. Even so, because of our common ancestry, modern humans share about 99.5 percent of their genes with Neandertals.

In 2003, human fossils about 18,000 years old were discovered on the Indonesian island of Flores. Like *H. erectus*, they had a heavy brow and a relatively small brain for their body size. Adults would have stood a meter tall. Scientists who found the fossils assigned them to a new species, *H. floresiensis*. Not everyone is convinced. Some think the fossils belong to *H. sapiens* individuals who had a disorder of some sort.

WHERE DID MODERN HUMANS ORIGINATE?

Neandertals evolved from *H. erectus* populations in Europe and western Asia. *H. floresiensis* evolved from *H. erectus* in Indonesia. Where did *H. sapiens* originate? Two major models agree that *H. sapiens* evolved from *H. erectus* but differ over where and how fast. Both attempt to explain the distribution of *H. erectus* and *H. sapiens* fossils, as well as genetic differences among modern humans who live in different regions.

Multiregional Model By the multiregional model, populations of *H. erectus* in Africa and other regions evolved into populations of *H. sapiens* gradually, over more than a million years. Gene flow among them maintained the species through the transition to fully modern humans (Figure 24.34*a*).

By this model, some of the genetic variation now seen among modern Africans, Asians, and Europeans began to build up soon after their ancestors branched from an ancestral *H. erectus* population. The model is based on interpretation of fossils. For example, faces of *H. erectus* fossils from China are said to look more like modern Asians than those of *H. erectus* that lived in Africa. The idea is that much variation seen among modern *H. sapiens* evolved long ago, in *H. erectus*.

Replacement Model By the more widely accepted replacement model, *H. sapiens* arose from a single *H. erectus* population in sub-Saharan Africa within the past 200,000 years. Later, bands of *H. sapiens* entered regions already occupied by *H. erectus* populations, and drove them all to extinction (Figure 24.34*b*). If this model is correct, then the regional variations observed among modern *H. sapiens* populations arose relatively recently. This model emphasizes the enormous degree of genetic similarity among all living humans.

Fossils support the replacement model. *H. sapiens* fossils date back to 195,000 years ago in East Africa and 100,000 years ago in the Middle East. In Australia, the oldest date to 60,000 years ago and, in Europe, to 40,000 years ago. Global comparisons of markers in mitochondrial DNA, and in the X and Y chromosomes, place the modern Africans closest to the root of the family tree. They also reveal that the most recent common ancestor of all modern humans was alive in Africa 60,000 years ago.

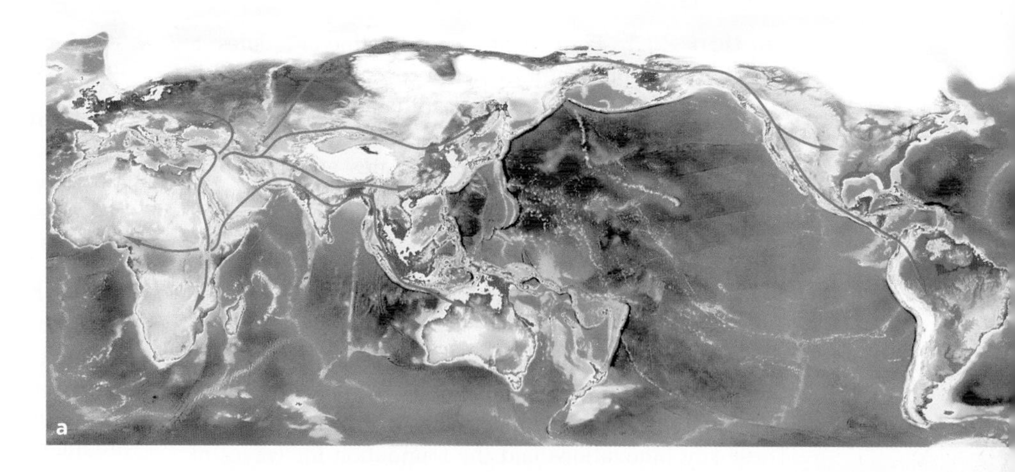

LEAVING HOME

Long-term shifts in the global climate drove human bands away from Africa (Figure 24.35). About 120,000 years ago, Africa's interior was becoming cooler and drier. As patterns and amounts of rainfall changed, so did the distribution of herds of grazing animals and the humans who hunted them. A few hunters may have journeyed north from East Africa and into Israel, where fossils 100,000 years old were found in a cave. They have no living descendants. Eruption of Mount Toba in Indonesia may have killed them, along with other ancient travelers. An enormous eruption 73,000 years ago released 10,000 times more ash than the 1981 eruption of Mount St. Helens in Washington State. The debris had a devastating impact on the global climate.

Later waves of travelers had better luck, as some individuals left established groups and ventured into new territory. Successive generations continued along the coasts of Africa, then Australia and Eurasia. In the Northern Hemisphere, much of Earth's water became locked in vast ice sheets, which lowered the sea level by hundreds of meters. Previously submerged land was drained off between some regions. About 15,000 years ago, one small band of humans crossed such a land bridge from Siberia into North America.

Deserts and mountains also influenced the routes available (Figure 24.35*b–e*). Until about 100,000 years ago, enough rain fell across northern Africa to sustain plants and herds of grazing animals. By 45,000 years ago, blazing hot sand stretched for more than 3,218 kilometers (2,000 miles). Humans whose ancestors had passed through this region no longer had the option of moving back to the grasslands of central Africa.

Instead, they moved east into central Asia, where the towering Himalayas and other peaks of the Hindu Kush forced some to detour north, into western China, and others south, into India. Descendants of humans that moved into Asia eventually reached Siberia, and some traveled on into North America. Colonists from central Asia moved west, across dry, cold grasslands. Some crossed the mountains in the Balkans, and then continued on into Europe.

With each step of their journey, humans faced and overcame extraordinary hardships. During this time, they devised *cultural* means to survive in inhospitable environments. Unrivaled capacities for modifying the habitat and for language served them well. Cultural evolution is ongoing. Hunters and gatherers persist in a few parts of the world, but others moved on from "stone-age" technology to the age of "high tech." This coexistence of such diverse groups is a tribute to the deep behavioral plasticity of the human species.

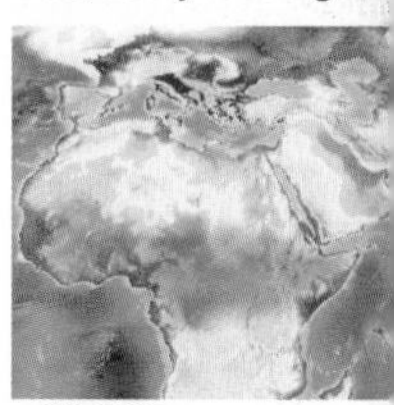

Figure 24.35 (**a**) Some dispersal routes for small bands of *Homo sapiens*. This map shows ice sheets and deserts that prevailed about 60,000 years ago. It is based on clues from sedimentary rocks and ice core drillings. Fossils, and studies of genetic markers in Y chromosomes and mitochondrial DNA from 10,000 individuals around the world, indicate the times when modern humans appeared in these regions:

Africa	by 195,000 years ago
Israel	100,000
Australia	60,000
China	50,000
Europe	40,000
North America	11,000

(**b**) Global climate changes caused expansion and contraction of deserts in Africa and the Middle East. Resulting changes in food sources may have encouraged migrations of small groups out of Africa. Locations of ice sheets, deserts, and tall mountain ranges influenced migration routes.

Fossils and genetic evidence indicate that modern humans, H. sapiens, *evolved from a* H. erectus *population in Africa. Modern humans dispersed out of Africa at a time when long-term shifts in climate influenced their options.*

www.thomsonedu.com Thomson NOW!

Summary

Section 24.1 Four features help define chordates: a notochord, a dorsal hollow nerve cord, a pharynx with gill slits, and a tail extending past the anus. All features form in embryos and may or may not persist in adults. Invertebrate chordates include tunicates and lancelets, both marine filter-feeders. Craniates are chordates with a braincase of cartilage or bone. Structurally, a jawless fish called the hagfish is the simplest modern craniate.

- *Use the animation on ThomsonNOW to examine the body plan and chordate features of a lancelet.*

Section 24.2 The craniates called vertebrates have a vertebral column of cartilaginous or bony segments. Jaws, paired fins, and lungs evolved in some lineages. These key innovations laid the foundation for waves of adaptive radiations of vertebrates.

- *Use the animation on ThomsonNOW to explore the chordate family tree.*

Section 24.3 Jawed fishes include the cartilaginous fishes and bony fishes. Their body plans are adapted to life in water. For example, a shark's streamlined shape reduces drag and the swim bladder of some bony fishes adjusts buoyancy. The most diverse vertebrates are bony fishes. Late in the Devonian, bony fishes that had lobed fins gave rise to tetrapods, or four-legged vertebrates.

Sections 24.4, 24.5 Amphibians such as the frogs, toads, and salamanders are carnivorous vertebrates, the first to evolve from aquatic Devonian tetrapods. Most are adapted to life on land, but nearly all return to the water to reproduce. Many now face extinction.

Sections 24.6, 24.7 Amniotes, the first vertebrates able to complete their life cycle on dry land, have water-conserving skin and kidneys, and amniote eggs. Reptiles (including extinct dinosaurs) and birds are one amniote lineage; modern mammals are another. A huge asteroid impact caused the extinction of the last of the dinosaurs but spared the earliest birds and mammals.

- *Use the animation on ThomsonNOW to explore the body plan of a crocodile.*

Section 24.8 Modern reptilian groups live on land or in water. Eggs are fertilized in the body and usually laid on land. Lizards and snakes are the most diverse group. Tuataras have some amphibian-like traits and a third eye. Crocodilians are the closest relatives of birds.

Section 24.9 Birds are warm-blooded amniotes and the only modern animals with feathers. The body plan of most species has been highly modified for flight.

- *Use the animation on ThomsonNOW to see what is inside a bird egg and how birds are adapted for flight.*

Section 24.10 Mammals are the only animals with hair and with females that nourish the young with milk secreted from mammary glands. Three lineages are the egg-laying mammals (monotremes), pouched mammals (marsupials), and placental mammals (eutherians), which are the most diverse and widespread of the three.

- *Use the animation on ThomsonNOW to see how the current distribution of mammalian groups arose.*

Section 24.11 Primates include prosimians (such as tarsiers), anthropoids (such as monkeys and apes), and hominids (humans and extinct humanlike forms). Early primates were shrewlike. Better daytime vision, upright walking, more refined hand movements, smaller teeth, bigger brains, social complexity, extended parental care, and, later, culture evolved in some lineages.

Sections 24.12, 24.13 All hominoids and hominids originated in Africa. The human lineage (*Homo*) arose by 2 million years ago (Figure 24.36), with *H. habilis* as an early toolmaking species. Bands of *H. erectus* dispersed into Europe and Asia. Extinct Neandertals and modern humans are close relatives, but have distinct gene pools.

Modern humans, *H. sapiens*, evolved by 195,000 years ago. By the multiregional model, *H. erectus* populations in far-flung regions evolved into *H. sapiens*. The African emergence model has modern humans evolving from *H. erectus* in Africa, then dispersing into regions already occupied by *H. erectus* and driving them to extinction. Most data now support the African emergence model. Its underlying premise is that regional variations among human groups evolved very recently.

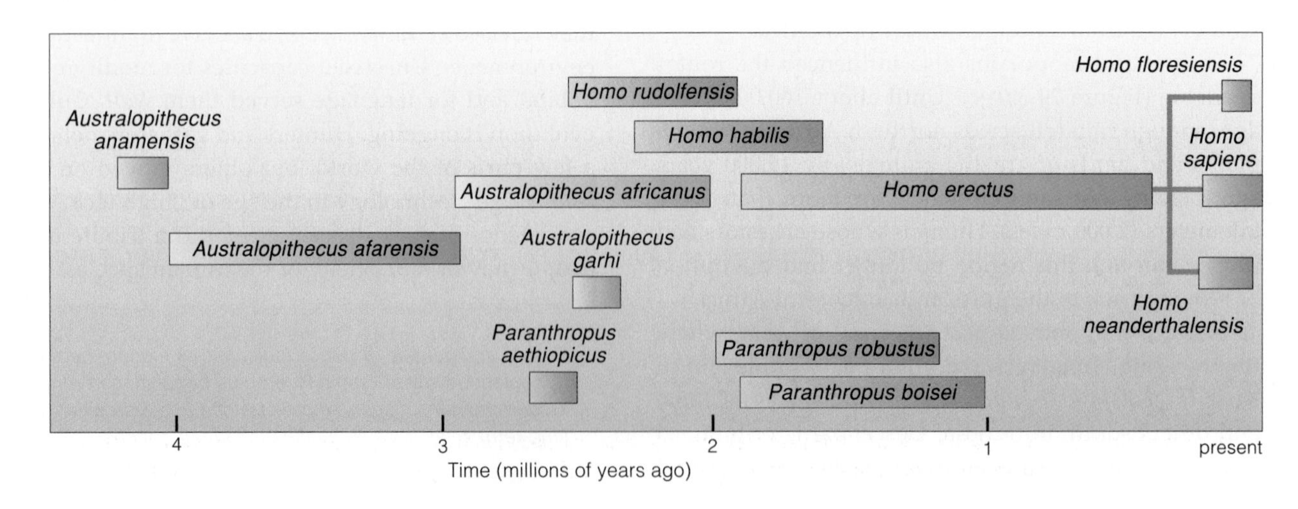

Figure 24.36 Summary of the estimated dates for the origin and extinction of three hominid genera. Purple lines are one view of how human species are related. The number of species, which fossils belong to each group, and how the species relate to one another are being debated.

Self-Quiz

Answers in Appendix III

1. All chordates have (a) ______ .
   a. backbone b. jaws c. notochord d. both b and c

2. The lancelet pharynx functions in ______ .
   a. respiration c. reproduction
   b. filter-feeding d. both a and b

3. Vertebrate jaws evolved from ______ .
   a. gill supports b. ribs c. scales d. teeth

4. Lampreys and sharks both have ______ .
   a. jaws d. a swim bladder
   b. a bony skeleton e. a four-chambered heart
   c. a cranium f. lungs

5. A divergence from ______ gave rise to tetrapods.
   a. ray-finned fishes c. cartilaginous fishes
   b. lizards d. lobe-finned fishes

6. Reptiles and birds belong to one major lineage of amniotes, and ______ belong to another.
   a. sharks c. mammals
   b. frogs and toads d. salamanders

7. Reptiles are adapted to life on land by ______ .
   a. tough skin d. amniote eggs
   b. internal fertilization e. both a and c
   c. good kidneys f. all of the above

8. The closest modern relatives of birds are ______ .
   a. crocodilians c. prosimians b. tuataras d. lizards

9. Only birds have ______ .
   a. a cloaca c. feathers
   b. a four-chambered heart d. amniote eggs

10. An australopith is a ______ .
    a. craniate d. amniote
    b. vertebrate e. placental mammal
    c. hominoid f. all of the above

11. *Homo erectus* ______ .
    a. was the earliest member of the genus *Homo*
    b. was one of the australopiths
    c. evolved in Africa and dispersed to many regions
    d. disappeared as the result of an asteroid impact

12. Match the organisms with the appropriate description.

___ lancelets	a. pouched mammals
___ fishes	b. invertebrate chordates
___ amphibians	c. feathered amniotes
___ reptiles	d. egg-laying mammals
___ birds	e. humans and close relatives
___ monotremes	f. cold-blooded amniotes
___ marsupials	g. first land tetrapods
___ placental mammals	h. most successful mammal lineage
___ hominids	i. most diverse vertebrates

■ *Visit ThomsonNOW for additional questions.*

Critical Thinking

1. The genome of *Ciona savignyi* (Figure 24.37) is being sequenced. Preliminary data show that this tunicate has about half as many genes as the average vertebrate. What might this easily cultured invertebrate—one with a small genome that displays all chordate traits—tell researchers about the development of other, more complex chordates?

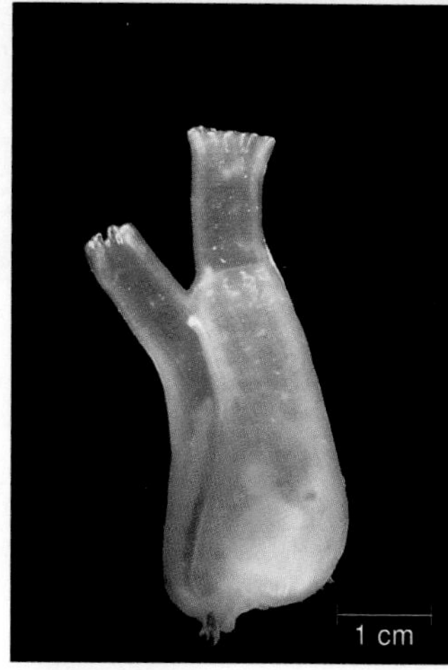

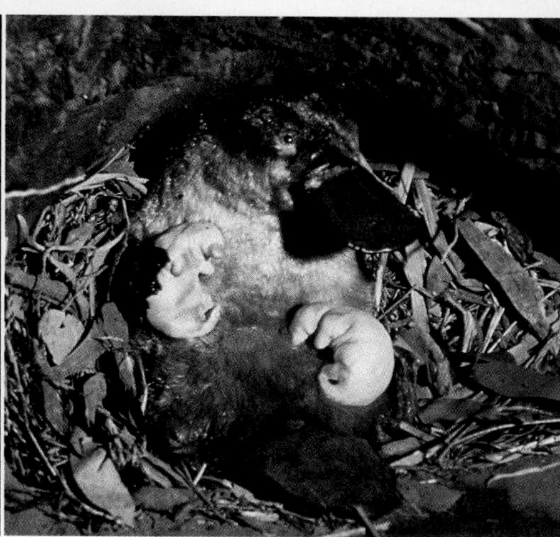

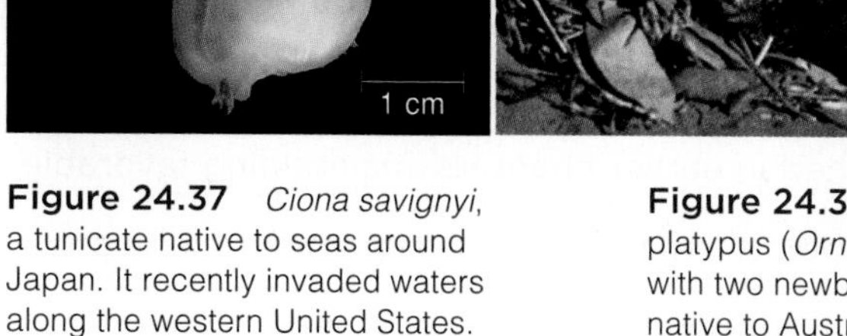

Figure 24.37 *Ciona savignyi*, a tunicate native to seas around Japan. It recently invaded waters along the western United States.

Figure 24.38 Duck-billed platypus (*Ornithorhynchus anatinus*) with two newborns. This mammal is native to Australia and Tasmania.

2. In 1798, a stuffed platypus specimen was delivered to the British Museum. Many thought it was a fake. It had brown fur, a beaverlike tail, a ducklike bill, and webbed feet (Figure 24.38). Reports that it laid eggs added to the debate. We now know platypuses burrow in riverbanks and hunt prey under water. They are the only modern venom-producing mammals. Webbing on the feet retracts and exposes claws. Sensory receptors in the bill help the animal detect prey under water, when its eyes and ears are tightly shut.

To modern biologists, a platypus is clearly a mammal. Like other mammals, it has fur and the females produce milk. Young animals have more typical mammalian teeth that are replaced by hardened pads as the animal matures. Why do you think modern biologists can easily accept that a mammal can have some reptilian traits?

3. During the Triassic, hundreds of species of tuatara-like reptiles roamed Pangea. Only two species have endured on a few small, cold islands off the coast of New Zealand. Speculate on why this relic population has survived there despite disappearing from the rest of its former range.

4. The cranial volume of early *H. sapiens* averaged 1,200 cubic centimeters. It now averages 1,400 cubic centimeters. By one hypothesis, females chose the cleverest mates, the advantage being offspring with genes that favorably affect intelligence. What types of data might a researcher gather to test this sexual selection hypothesis?

5. Figure 24.39 compares five primate lineages, from most ancient to most recent. It graphs life spans and years spent as infants and juveniles, when individuals are most vulnerable and still learning survival skills from adults. What overall trends can you identify with respect to life spans and the time it takes to reach adulthood? What selective pressures may have contributed to differences in life histories?

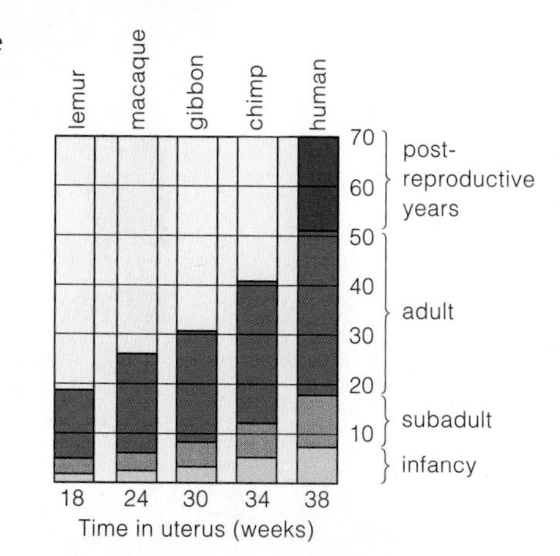

Figure 24.39 Trend toward longer life spans and greater dependency of offspring on adults for five primate lineages.

25 PLANTS AND ANIMALS—COMMON CHALLENGES

IMPACTS, ISSUES A Cautionary Tale

Even when outside temperatures change, your body works to keep its inside temperature within a range that individual cells can tolerate. As you learned in earlier chapters, maintaining favorable temperatures in the internal environment is one aspect of homeostasis. Homeostasis requires the coordinated activities of many types of cells that make up different organ systems. To understand homeostasis, you have to know how the body is organized and how its parts function.

Sometimes terrible things happen when someone does not recognize the threat of internal imbalances. In the summer of 2001, football player Korey Stringer collapsed after morning practice (Figure 25.1). On that day his team, the Minnesota Vikings, was working out in full uniform on a field where temperatures were high. The humidity put the heat index above 100°F (38°C).

Stringer's internal body temperature soared to 108.8°F, and his blood pressure was too low even to record. He was rushed to the hospital, where doctors immersed him in an icewater bath, then wrapped him in cold, wet towels. It was too late.

Stringer's blood clotting mechanism shut down and he started to bleed internally. His kidneys faltered and he was placed on dialysis. He stopped breathing on his own and was attached to a respirator, but his heart gave out. Less than twenty-four hours after football practice had started, Stringer was pronounced dead. He was twenty-seven years old.

All organisms function best within a limited range of internal operating conditions. For humans, "best" is when the body's internal temperature remains between about 97°F and 100°F. Above 104°F (40°C), metabolism is in an uproar, and controls over blood transport divert heat from the brain and other internal organs to the skin. Skin can transfer heat to the outside environment, as long as

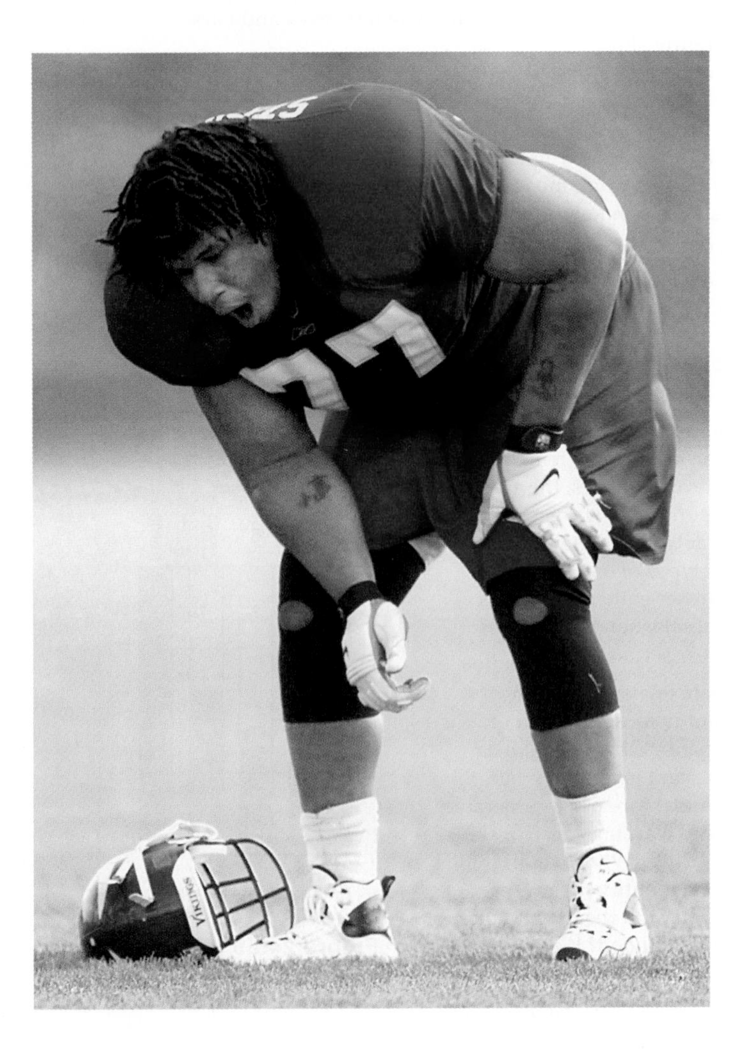

See the video! **Figure 25.1** *Left*, Korey Stringer, during his last practice with his team. When the body's temperature rises, profuse sweating increases evaporative cooling. Also, blood is directed to capillaries of the skin (*facing page*), which radiate heat into the air. In Stringer's case, homeostatic control mechanisms were no match for strenuous activity on a hot, humid day.

How would you vote? Some states have laws that protect outdoor workers from heat illness by requiring periodic breaks with access to shade and water. Such rules can be costly for employers. Do you think that they are necessary? See ThomsonNOW for details, then vote online.

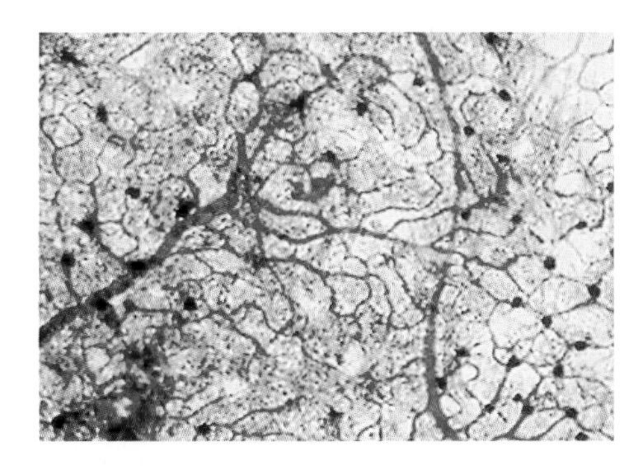

it is not too hot outside. Profuse sweating can dissipate more heat, but not on hot, humid days.

Past 105°F (40.6°), normal cooling processes fail. The body cannot sweat as much, and its core temperature starts to climb rapidly. The heart beats faster; fainting or confusion follow. These symptoms of heat stroke arise because a high internal temperature disrupts enzyme function. When heat stroke is not countered fast enough, brain damage or death is the expected outcome.

We use this sobering example as our passport to the world of anatomy and physiology. *Anatomy* is the study of body form—that is, its morphology. *Physiology* is the study of patterns and processes by which an individual survives and reproduces in the environment. It deals with how the body's parts are put to use and how metabolism and behaviors are adjusted when conditions change. It also deals with how, and to what extent, physiological processes can be controlled.

On a personal level, this information can help you monitor what is going on inside your own body. More broadly, it can also help you identify commonalities among animals and plants. Regardless of the species, *the structure of any given body part almost always has something to do with a present or past function.* Most aspects of body form and function are long-standing adaptations that evolved as responses to particular environmental challenges.

That said, nothing in the evolutionary history of the human species suggests that organ systems are fine-tuned to handle the combination of intense exercise, high temperatures, and high humidity.

Key Concepts

MANY LEVELS OF STRUCTURE AND FUNCTION

Anatomy is the study of body form at one or more levels of structural organization, from molecules to cells, tissues, and organ systems. Physiology is the study of patterns and processes by which the body functions in its environment. The structure of most body parts correlates with current or past functions, and it emerges during growth and development. **Section 25.1**

SIMILARITIES BETWEEN ANIMALS AND PLANTS

Animals and plants must exchange gases with their environment, transport materials throughout a body, maintain the volume and composition of their internal environment, and coordinate cell activities. They must also respond to threats and to variations in available resources. **Section 25.2**

HOMEOSTASIS

Extracellular fluid bathes all living cells in the multicelled body. Cells, tissues, organs, and organ systems contribute to maintaining this internal environment within a range that individual cells can tolerate.

Homeostasis is the process of maintaining favorable operating conditions in the body's internal environment. Negative and positive feedback systems contribute to homeostasis. **Sections 25.3, 25.4**

CELL COMMUNICATION IN MULTICELLED BODIES

Cells of tissues and organs communicate by secreting chemical molecules into extracellular fluid, and by selectively responding to signals secreted by other cells. **Section 25.5**

Links to Earlier Concepts

Look back at the levels of biological organization in Section 1.1. You are about to see how new properties of life emerge through interactions among the tissues, organs, and organ systems. You will also explore mechanisms that ensure homeostasis (1.2).

You will draw on your understanding of the constraints on multicelled body plans—the surface-to-volume ratio (4.1, 23.1), diffusion and demands for gas exchange (5.6), and internal transport (5.7, 21.1, 24.2). You will find examples of how signaling molecules can help control growth, development, day-to-day activities, and reproduction (4.3, 20.10). You also will consider specific cases of evolutionary adaptation (17.14).

25.1 Levels of Structural Organization

We introduce important concepts in this chapter. They have broad application across the next two units of the book. Becoming familiar with them can deepen your sense of how plants and animals function in stressful as well as favorable environments.

FROM CELLS TO MULTICELLED ORGANISMS

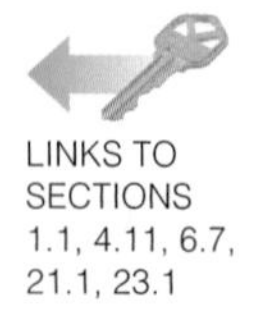

LINKS TO SECTIONS 1.1, 4.11, 6.7, 21.1, 23.1

Depending on its size, a plant or animal body consists of hundreds to hundreds of trillions of cells. In all but the simplest bodies, cells are organized into tissues, organs, and organ systems, each capable of specialized functions. Said another way, the plant or animal body typically shows a division of labor (Section 23.1).

A tissue is a community of cells and extracellular substances that interact in one or more tasks. As an example, bone and wood are two tissues that provide structural support.

An organ is a structural unit made up of tissues arranged in proportions and patterns that allow it to carry out a specific task or tasks. For example, a leaf is an organ of gas exchange and photosynthesis (Figure 25.2); lungs are organs of gas exchange (Figure 25.3).

Organs that interact in one or more tasks form an organ system. Leaves and stems are components of a plant's gas exchange system. Lungs and airways are organs of the respiratory system of land vertebrates.

GROWTH VERSUS DEVELOPMENT

A plant or animal becomes structurally organized as it grows and develops. For any multicelled species, growth refers to an increase in the number, size, and volume of cells. We measure it in *quantitative* terms. Development is a series of stages in which specialized tissues, organs, and organ systems form in heritable patterns. We measure it in *qualitative* terms; that is, whether and how well the patterns are followed.

STRUCTURAL ORGANIZATION HAS A HISTORY

The structural organization of a plant or animal body has an evolutionary history. It is the result of natural selection, of differences in survival and reproduction among individuals who varied in their anatomy and physiology. For example, Section 21.1 surveyed how plants adapted to life on dry land. As their ancestors ventured out of the aquatic cradle, they faced a new challenge—how to keep from drying out in air.

Think about that challenge when micrographs show the internal structure of plant roots, stems, and leaves (Figure 25.2). Xylem conduct columns of water from soil to leaves. Stomata, the small gaps across a leaf's epidermis, open and close in ways that conserve water. Collectively, lignin-strengthened cell walls support the growth of tall stems. Remember that challenge as well when you consider root systems. A patch of soil that

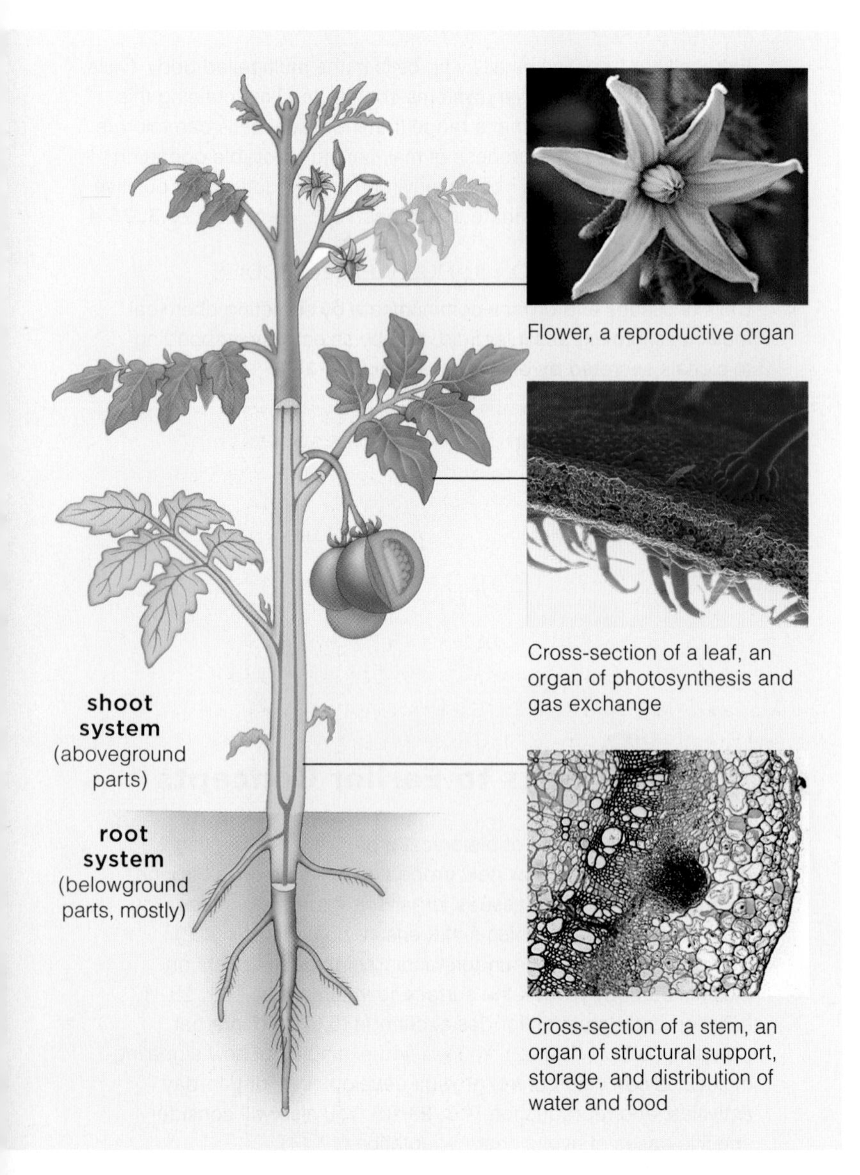

Figure 25.2 **Animated!** Anatomy of a tomato plant. Its vascular tissues (*purple*) conduct water, dissolved mineral ions, and organic compounds. Another tissue makes up the bulk of the plant body. A third covers all external surfaces. Organs such as flowers, leaves, stems, and roots are each made up of all three tissue types in specific proportions.

Figure 25.3 Parts of the human respiratory system. Cells making up the tissues of this system carry out specialized tasks. Airways to paired lungs are lined with ciliated cells that whisk away bacteria and other airborne particles that might cause infections. Lungs are organs of gas exchange. Inside them are tubes (blood capillaries) filled with blood (a fluid connective tissue), and thin-walled air sacs (epithelial tissue). All of these components indirectly or directly assist the flow of oxygen into the internal environment and carbon dioxide out of it.

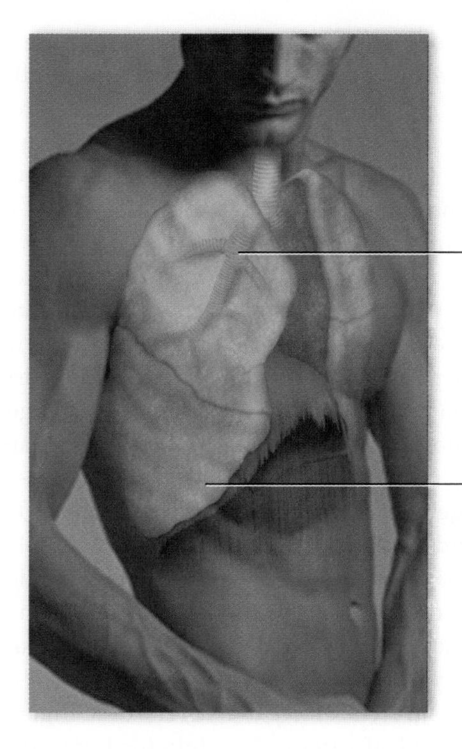

Organs (lungs), part of an organ system (the respiratory tract) of a whole organism

Ciliated cells and mucus-secreting cells of a tissue that lines respiratory airways

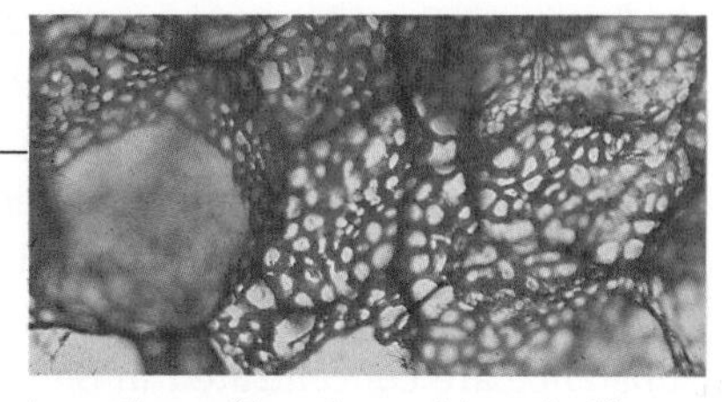

Lung tissue (tiny air sacs) laced with blood capillaries—one-cell-thick tubular structures that hold blood, which is a fluid connective tissue

has plenty of water and dissolved mineral ions will stimulate root growth in its direction. The branchings of root systems are the response.

Similarly, respiratory systems of land animals are adaptations to life in air. Gases only move into and out of an animal by diffusing across a moist surface. That is not a problem for aquatic organisms, but moist surfaces dry out in air. Animals on dry land contain moist sacs for gas exchange *inside* their body (Figure 25.3), which brings us to the internal environment.

THE BODY'S INTERNAL ENVIRONMENT

Plant and animal cells must be bathed in a fluid that delivers nutrients and carries away metabolic wastes. In this respect they are no different from free-living single cells. But each plant or animal has thousands to many trillions of living cells. All must draw nutrients from the fluid bathing them and release wastes into it.

The body fluid not inside cells is extracellular fluid (ECF). It acts as an internal environment. Changes in its composition and volume affect cell activities. The type and number of ions in the ECF must be kept at concentrations compatible with metabolism. It makes no difference whether the plant or animal is simple or complex. *The body requires a stable fluid environment for its cells*. This concept is central to understanding how plants and animals work.

START THINKING "HOMEOSTASIS"

The next two units describe how a plant or an animal carries out the following essential functions: The body maintains favorable conditions for all its living cells. It acquires water, nutrients, and other raw materials, distributes them, and disposes of wastes. It passively and actively defends itself against attacks. It has the capacity to reproduce. Specialized parts of it nourish and protect gametes, and, in most species, embryonic stages of the next generation.

Each living cell engages in metabolic activities that will favor its own survival. Collectively, however, the activities of cells in tissues, organs, and organ systems sustain the body as a whole. Their interactions keep the operating conditions of the internal environment within tolerable limits—a process we call homeostasis.

The structural and functional organization of plants and animals emerges in stages of growth and development.

Cells, tissues, and organs require a favorable internal environment, and they collectively maintain it. All body fluids not contained in cells make up that environment, which is known as the extracellular fluid (ECF).

Acquiring materials and distributing them to cells, getting rid of wastes, protecting cells and tissues, reproducing, and often nurturing offspring are basic body functions.

25.2 Recurring Challenges to Survival

Plants and animals have such diverse body plans that we sometimes forget how much they have in common.

GAS EXCHANGE IN LARGE BODIES

LINKS TO SECTIONS 4.1, 5.7, 6.7, 7.1

Think for a moment about the ways that Tiger Woods, the famous golfer, is like a tulip (Figure 25.4). In both bodies, cells release energy by aerobic respiration and require oxygen. Both have to get rid of waste gases. Tiger has to dispose of carbon dioxide. The tulip has to get rid of excess oxygen formed by photosynthesis. As Section 6.7 explains, excess oxygen interferes with the sugar-producing reactions of photosynthesis.

All multicelled species respond, structurally and functionally, to this common challenge: Quickly move gaseous molecules to and from individual cells.

Remember diffusion? When ions or molecules of a substance are concentrated in one place, they tend to move to a place where they are not as concentrated. Plants and animals keep gases diffusing in directions most suitable for metabolism and cell survival. How? That question will lead you to stomata at leaf surfaces (Section 27.4) and to the circulatory and respiratory systems of animals (Chapters 34 and 36).

INTERNAL TRANSPORT IN LARGE BODIES

Metabolic reactions happen fast. If reactants take too long to diffuse through the body or to and from the surface, reactions slow or stop. That possibility helps explain the sizes and shapes of cells and multicelled bodies. As cells and multicelled organisms grow, their volume increases in three dimensions—length, width, and depth—while surface area increases in only two. Thus, surface area-to-volume ratio decreases (Section 4.1). If a body were to develop as a dense mass of cells, diffusion alone would not be able to move materials into, out of, and through it fast enough to sustain life.

When a body or some body part is thin, as it is for the flatworm and lily pads in Figure 25.5, substances easily diffuse among body cells and between cells and the external environment. In a body, systems of rapid internal transport service individual cells that are far from an exchange point with the environment.

Most plants and animals have vascular tissues, or systems of tubes through which substances move to and from cells. In land plants, xylem distributes soil water and dissolved minerals. Phloem distributes the sugars made in leaves. Each leaf vein consists of long strands of xylem and phloem (Figure 25.5*c*).

In most animals, the vascular tissues extend from a surface exposed to the environment to the cells inside. Every time Tiger takes a breath, he is moving oxygen into his lungs and carbon dioxide out of them. Blood vessels thread through lung tissue where the gases are exchanged. Large vessels transport oxygen and branch into small capillaries in all tissues—where interstitial fluid and cells exchange gases (Figure 25.5*d*).

In plants and animals, a vascular system transports diverse substances, including water, nutrients, and the signalling molecules called hormones. Animal blood has infection fighting white blood cells. Plant phloem transports chemicals made in response to injury.

Figure 25.4 What do these organisms have in common besides good looks?

MAINTAINING THE WATER–SOLUTE BALANCE

Plants and animals continually gain and lose water. On land, gas exchange invariably causes water losses. Plants and animals also lose and gain solutes. Given all the inputs and outputs, how does the volume and composition of their internal environment stay within a tolerable range? Plants and animals differ hugely in this respect, yet you can still find common responses by zooming down to the level of molecules.

Substances tend to follow concentration gradients. At the surface of a body or an organ, cells in sheets of tissue carry out active and passive transport. In active transport, recall, a protein pumps one specific solute from a region of low concentration to one of higher concentration (Section 5.7).

Active transport mechanisms in roots help control which solutes can move into the plant. In leaves, they help control water loss and gas exchange by closing and opening stomata at different times. For animals,

Figure 25.5 (a) Flatworm gliding in water, (b) water-lily leaves floating on water, (c) veins in a decaying dicot leaf, (d) human veins and capillaries. Which constraint has influenced these body plans and structures?

such mechanisms occur in kidneys and other organs. Active transport mechanisms sustain metabolism and maintain the internal environment by their effects on the kinds, amounts, and movements of solutes. In the next two units, you will learn about many examples.

CELL-TO-CELL COMMUNICATION

Plants and animals show another crucial similarity: They depend on communication among cells. Many types of specialized cells release signaling molecules that help coordinate and control events in the body as a whole. Signaling mechanisms guide how the plant or animal body grows, develops, and maintains itself, and also reproduces. Section 25.5 has a good example.

ON VARIATIONS IN RESOURCES AND THREATS

A habitat is the place where individuals of a species typically live. Each habitat has different resources and poses a unique set of challenges. What are its physical characteristics? Is water plentiful, with the right kinds and amounts of solutes? Is the habitat rich or poor in nutrients? Is it brightly lit, shady, or dark? Is it warm or cool, windy or still? How much does temperature vary from daytime to night? How much do conditions change with the seasons?

Biotic (living) components of the habitat vary as well. Which producers, predators, prey, pathogens, or parasites live there? Is competition for resources and reproductive partners minimal or fierce? Variation in these factors promotes diversity in form and function.

Even with all the diversity, we may still see similar responses to similar challenges. Sharp cactus spines or porcupine quills deter most animals that might eat a cactus or porcupine (Figure 25.6). Modified epidermal cells give rise to both spines and quills. Deliveries from vascular tissues and other body parts keep such cells alive. In return, by forming spines or quills and other structures that defend the body at its surface, those cells help protect vascular tissues and other internal parts against environmental threats.

Figure 25.6 Protecting body tissues from predation: (a) Cactus spines. (b) Quills of a porcupine (*Erethizon dorsatum*).

Plant and animal cells function in ways that help ensure survival of the body as a whole. At the same time, tissues and organs of a body function in ways that contribute to the survival of individual living cells.

Each cell and the multicelled body interconnect through their requirements for—and contributions to—gas exchange, nutrition, internal transport, stability in the internal environment, and defense.

25.3 Homeostasis in Animals

In preparation for your trek through the next two units, keep focused on what homeostasis means to survival.

LINKS TO SECTIONS 1.2, 5.4

In an adult human body, about 65 trillion living cells withdraw nutrients from and dump wastes into about 4 gallons (15 liters) of extracellular fluid. Most of the extracelluar fluid is interstitial fluid, which fills spaces between cells and tissues. Plasma, the fluid portion of blood, is a lesser component. Cells and blood exchange solutes with interstitial fluid, and fluid moves between blood and interstitial fluid.

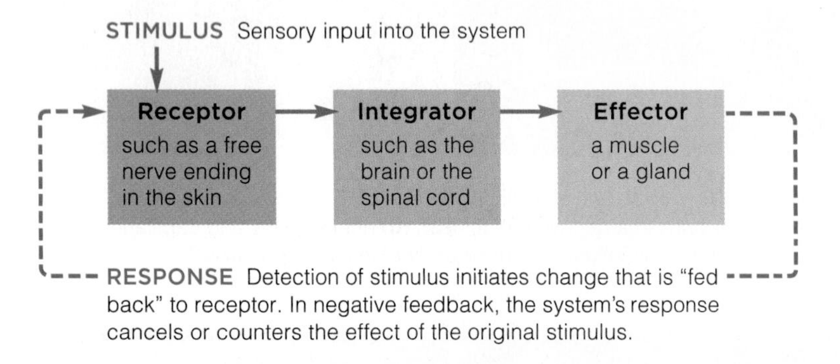

Figure 25.7 Three essential components of negative feedback systems in multicelled organisms. Figure 25.8 shows a specific example.

Homeostasis, again, is a state in which the internal environment is being maintained within a range that cells can tolerate. Sensory receptors, integrators, and effectors are in charge of it. Collectively, they detect, process, and respond to information about *how things are* compared to preset points for *how things should be*.

Sensory receptors are cells or cell parts that detect stimuli, which are specific forms of energy. A kiss, for example, is a form of mechanical energy that changes pressure on the lips. Receptors in lip tissues translate a kiss into signals that reach the brain. The brain is an integrator: a central command post that receives and processes information about stimuli. It issues signals to effectors—muscles, glands, or both—that carry out suitable responses to the stimulation. Figures 25.7 and 25.8 are simple ways to think about these systems.

NEGATIVE FEEDBACK

Feedback mechanisms are major homeostatic controls over what goes on in cells and the multicelled body (Section 5.4). With the negative feedback mechanisms, an activity changes a specific condition in the internal environment, and when the condition changes past a certain point, a response reverses the change.

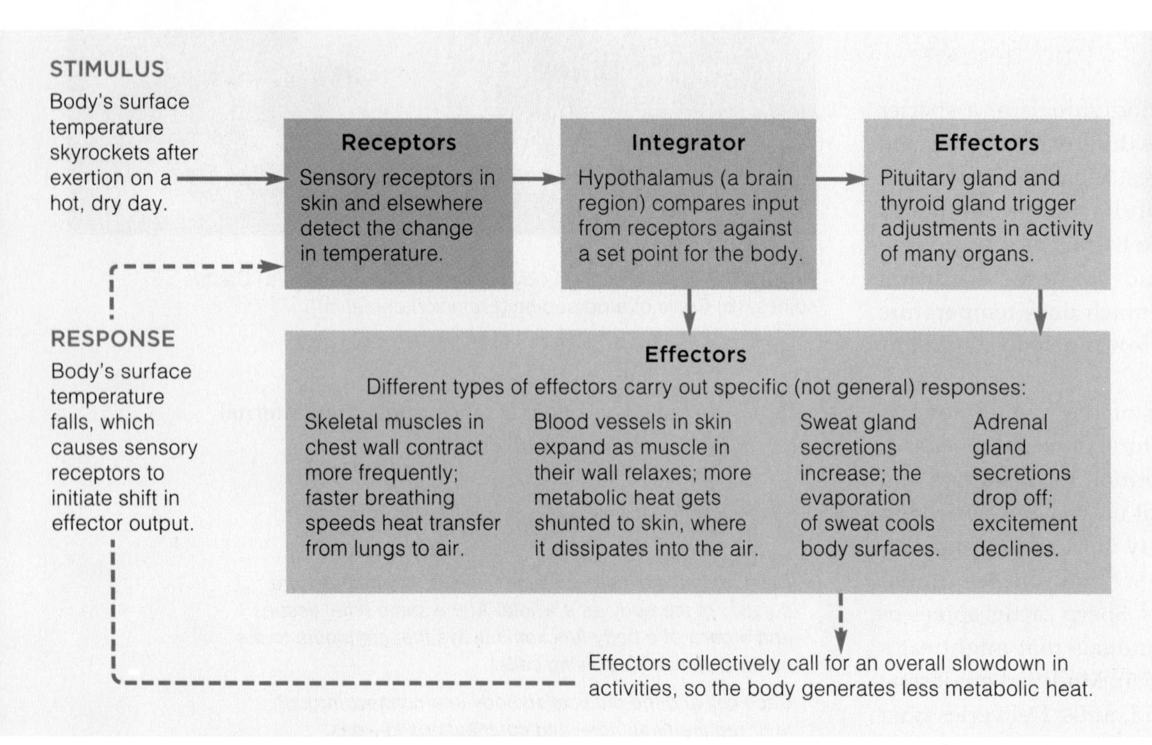

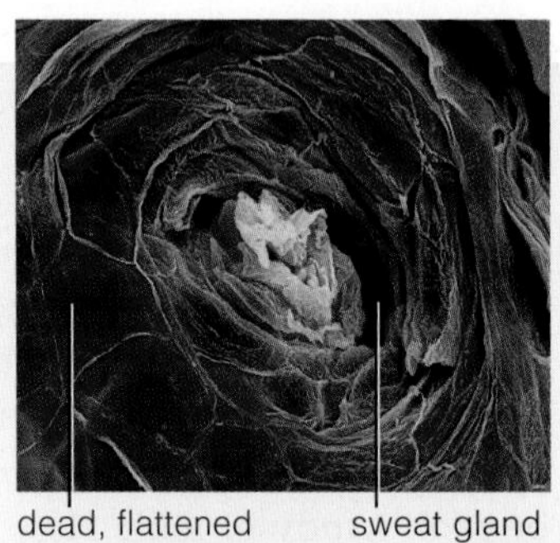

Scanning electron micrograph of a sweat gland pore at the skin surface. Such glands are among the effectors for this control pathway.

Figure 25.8 **Animated!** Major homeostatic controls over a human body's internal temperature. *Solid* arrows signify the main control pathways. *Dashed* arrows signify the feedback loop.

Think of a furnace with a thermostat. A thermostat senses the surrounding air temperature relative to a preset point on a thermometer built into the furnace's control system. When the temperature falls below the preset point, it signals a switching mechanism, which turns on the furnace. When the air gets warm enough to match the preset level, the thermostat senses that match and signals the switch to turn off the furnace.

Similarly, feedback mechanisms can help keep the internal body temperature of humans and many other mammals near 98.6°F (37°C) even during hot or cold weather. When someone runs on a hot summer day, the body becomes hot. Receptors trigger changes that affect activities of the whole body and individual cells (Figure 25.8). Controls normally prevent overheating. They curb the events that generate metabolic heat and stimulate events that deliver excess heat to the skin, for its release to the surrounding air.

As another example, the brain receives signals from receptors that monitor carbon dioxide concentration in blood. It compares the signals against a set point—the desired carbon dioxide level. When your muscles are highly active, lots of carbon dioxide enters blood. Receptors detect the change and alert the brain, which calls for adjustments in breathing that move the blood level of carbon dioxide closer to the set point.

Another set of receptors monitors oxygen levels. If you climb above 3,300 meters (10,000 feet), oxygen is scarce, and its concentration in blood declines (Figure 25.9). Receptors sensitive to oxygen signal the brain, which calls for faster, deeper breathing. If the oxygen level remains too low, shortness of breath, changes in heartbeat, headaches, nausea, and vomiting can follow. These symptoms of altitude sickness are messages that cells desperately need oxygen.

POSITIVE FEEDBACK

In some situations, positive feedback mechanisms will operate. The mechanisms spark a chain of events that *intensify* change from an original condition. In a living organism, intensification eventually leads to a change that ends feedback. Positive feedback mechanisms are usually associated with instability in a system.

For instance, a neuron can help move information through a nervous system because of the way sodium ions flow across its plasma membrane. A signal causes a few sodium channels to open in the membrane. The disturbance causes more channels to open, then more, until sodium ions flood inside. Ion flows disturb the membrane all the way down to an output zone, where the neuron may send information to another cell.

Figure 25.9 A climber near the summit of Mount Everest, where normal breathing is not enough to sustain the body's oxygen needs.

As another example, during childbirth, a fetus puts pressure on the wall of the surrounding chamber, the uterus. Pressure induces the production and secretion of a hormone, oxytocin, that makes muscle cells in the wall contract. Contractions exert pressure on the fetus, which puts more pressure on the wall to expand, and so on until the fetus leaves the mother's body.

We have been introducing ways in which the body detects, evaluates, and responds to certain signals in internal and external environments. As you read the next two units, ask these questions about how organ systems interact during all of this activity:

1. Which physical or chemical aspects of the internal environment are organ systems working to maintain?
2. How are organ systems kept informed of change?
3. By what means do they process the information?
4. What mechanisms are set in motion in response?

Discovering answers is likely to give you an abiding appreciation of how your own body works.

The internal environment is all the fluid not inside the body's cells. It consists of interstitial fluid and plasma, the fluid portion of blood.

Homeostatic controls, such as negative and positive feedback mechanisms, help maintain physical and chemical aspects of the body's internal environment within ranges that its individual cells can tolerate.

25.4 Does Homeostasis Occur in Plants?

Plants differ from animals in important respects. Even so, they, too, exert controls over their internal environment.

LINKS TO SECTIONS 21.6, 22.5

Directly comparing plants and animals is not always possible. As a plant grows, new tissues form only at specific sites in roots and shoots. In animal embryos, tissues form all through the body. Plants do not have the equivalent of an animal brain. But they do have some decentralized mechanisms that influence the internal environment and help keep their body functioning. Two simple examples will make the point; chapters to follow include more.

WALLING OFF THREATS

Unlike people, trees consist mostly of dead and dying cells. Also unlike people, trees cannot run away from attacks. When a pathogen infiltrates their tissues, trees cannot unleash infection-fighting white blood cells in response, because they have none. However, plants do have system acquired resistance: a defense response to infections and injured tissues. Cells in an affected tissue release signaling molecules that call for the synthesis and release of organic compounds that will protect the plant against attacks for days or months to come. The compounds are so effective that synthetic versions are being developed and used to boost disease resistance in crop plants and ornamental plants.

An additional response helps most trees protect the internal environment. When wounded, such trees wall off the damaged tissue, release phenols and other toxic compounds, and often secrete resins. A heavy flow of gooey compounds saturates and protects the bark and wood at the wound. It also seeps into the soil around roots. Some of these toxins are so potent that they can kill cells of the tree itself. Compartments form around injured, infected, or poisoned tissues, and new tissues grow right over them. This plant response to wounds is called compartmentalization.

Drill holes into a tree species that makes a strong compartmentalization response and the wound gets walled off fast (Figure 25.10). In a species that makes a moderate response, decomposers cause the decay of more wood surrounding the holes. Drill into a weak compartmentalizer, and decomposers cause massive decay deep into the trunk.

Even strong compartmentalizers live only so long. If too much tissue gets walled off, flow of water and solutes to living cells slows and the tree begins to die. What about the bristlecone pine, which grows high in mountain regions (Section 21.6)? One tree we know of is almost 5,000 years old. These trees live under harsh conditions in remote habitats where pathogens are few. The trees spend most of each year dormant beneath a blanket of snow, and grow slowly during a short, dry summer. This slow growth makes a bristlecone pine's wood so dense that few insects can bore into it.

Figure 25.10 **Animated!** Results of an experiment in which holes were drilled into living trees to test compartmentalization responses. From top to bottom, decay patterns (*green*) in trunks of three species of trees that made strong, moderate, and weak compartmentalization responses, respectively.

SAND, WIND, AND THE YELLOW BUSH LUPINE

Anybody who has tiptoed barefoot across sand near the coast on a hot, dry day has a tangible clue to why few plants grow in it. One exception is the yellow bush lupine, *Lupinus arboreus* (Figure 25.11).

L. arboreus is native to warm, dry areas of central and southern California. It is a hardy colonizer of soil exposed by fires or abandoned after being cleared for agriculture. Like all other legumes, this species has nitrogen-fixing symbionts in its young roots (Section 22.5). The interaction provides a competitive edge in

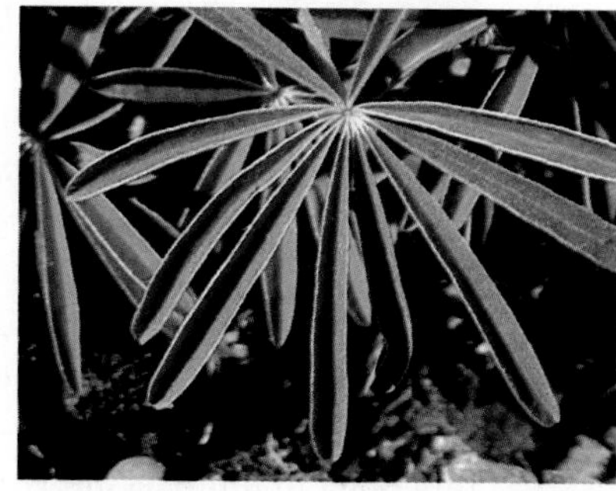

Figure 25.11 Yellow bush lupine, *Lupinus arboreus*, in a sandy shore habitat. On hot, windy days, its leaflets fold up longitudinally along the crease that runs down their center. This helps minimize evaporative water loss.

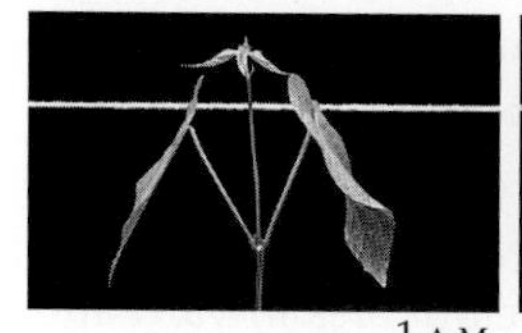

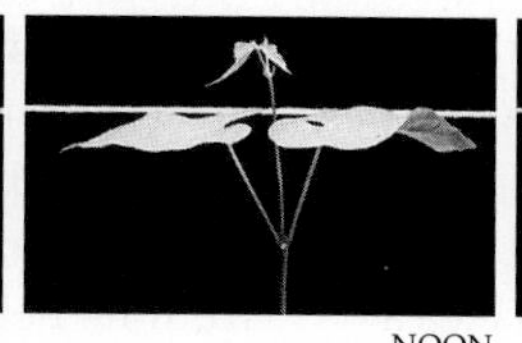

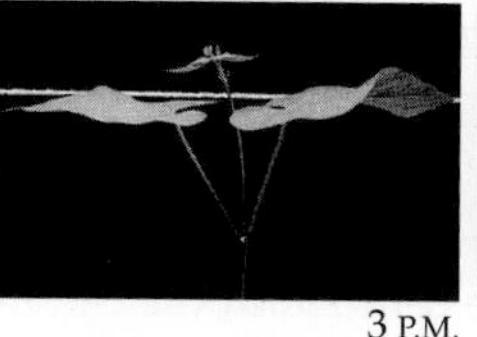

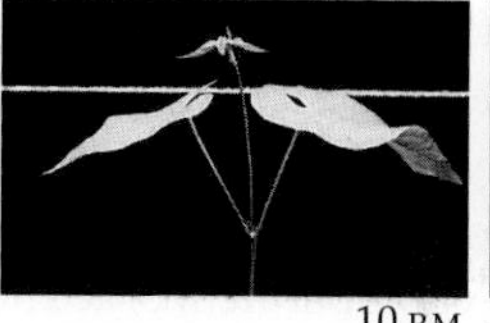

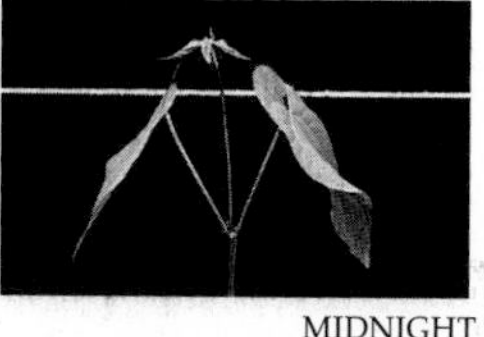

Figure 25.12 **Animated!** Observational test of rhythmic leaf movements by a young bean plant (*Phaseolus*). Physiologist Frank Salisbury kept the plant in darkness for twenty-four hours. Its leaves kept on folding and unfolding independently of sunrise (6 A.M.) and sunset (6 P.M.).

nitrogen-deficient soil. In the early 1900s, this species was planted in northern California to stabilize coastal dunes. Unfortunately, it is too successful in this new habitat. It outgrows and displaces native plants.

One big environmental challenge near the beach is the lack of freshwater. Leaves of a yellow bush lupine are structurally adapted for water conservation. Each leaf has a surprisingly thin cuticle, but a dense array of fine epidermal hairs projects above it, particularly on lower leaf surfaces. Collectively, all of the hairs trap much of the moisture escaping from stomata. Trapped moist air slows evaporation and helps maintain water levels inside the leaf at levels that favor metabolism.

The leaves also make a homeostatic response under conditions when water loss could be a problem. They fold in half lengthwise (Figure 25.11). This shelters the stomata from evaporative effects of the wind and thus slows water loss.

Leaf folding by *L. arboreus* is a controlled response to changing conditions. When winds are strong and the potential for water loss is greatest, the leaves fold tightly. The least-folded leaves are close to the plant's center or on the side most sheltered from the wind. Folding is a response to heat as well as to wind. When air temperature is highest during the day, leaves fold at an angle that helps reflect the sun's rays from their surface. The response minimizes heat absorption.

ABOUT RHYTHMIC LEAF FOLDING

In case you think leaf folding could not possibly be a coordinated response, take a look at the bean plant in Figure 25.12. Like some other plants, it holds its leaves horizontally during the day but folds them closer to a stem at night. Keep the plant in full sun or darkness for a few days and it will continue to move its leaves in and out of the "sleep" position, independently of sunrise and sunset. The response might help reduce heat loss at night, when air cools, and so maintain the plant's internal temperature within tolerable limits.

Rhythmic leaf movements are just one example of a circadian rhythm: a biological activity pattern that recurs with an approximately 24-hour cycle. Circadian means "about a day." A molecule called phytochrome, described in Section 28.9, helps control this and other rhythmic activity in plants.

Control mechanisms that help maintain homeostasis are at work in plants, although they are not centrally controlled by a central command center as they are in most animals.

System acquired resistance, compartmentalization, and rhythmic leaf movements in response to environmental challenges are examples of these mechanisms.

25.5 How Cells Receive and Respond to Signals

Signal reception, transduction, and response—this is a fancy way of saying cells chatter among themselves in ways that bring about changes in their activities.

LINKS TO SECTIONS 4.3, 4.11, 8.4, 20.10

Cells in any multicelled body communicate with their neighbors and often with cells farther away. Section 4.11 described how plasmodesmata in plants and gap junctions in animals allow substances to pass quickly between adjoining cells. Communication among more distant cells involves amino acids, peptides, proteins, lipids, and gases. Some signals diffuse from one cell to another through interstitial fluid. Others travel in blood vessels or a plant's vascular tissues.

Molecular mechanisms by which cells "talk" to one another evolved early in the history of life. They often have three steps. *First*, a specific receptor is activated, as by reversibly binding a signaling molecule. *Second*, the signal is transduced—it is converted into a form that can operate inside the cell. *Third*, the cell makes a response to the signal, as by altering its metabolism or which genes it expresses (Figure 25.13*a*).

Many receptors that respond to molecular signals are membrane proteins of the sort shown in Section 4.3. Once activated, some receptors will activate one enzyme, which in turn activates many molecules of a different kind of enzyme. These activate molecules of another kind, and so on. Cascading reactions amplify the response to that one signal many times over.

In the next two units, you will come across diverse cases of signal reception, transduction, and response. For now, how cells commit suicide is a fine example.

The first cell of a new multicelled individual holds marching orders that guide its descendants through growth, development, reproduction, and often death. As part of that program, many cells heed calls to self-destruct at a particular time. Apoptosis is the process of programmed cell death. It often starts when certain molecular signals bind to receptors at the cell surface (Figure 25.13*b*). A chain of reactions leads to activation of self-destructive enzymes. Some chop up structural proteins, such as cytoskeleton proteins and histones that organize DNA. Others snip apart nucleic acids.

An animal cell undergoing apoptosis shrinks away from its neighbors. Its membrane bubbles inward and outward. The nucleus and then the whole cell breaks apart. Phagocytic white blood cells that patrol tissues engulf the dying cells and their remnants. Lysosomes inside the phagocytes digest the engulfed bits.

Many cells committed suicide as your hands were developing (Section 8.4). Each hand starts forming as a paddlelike structure. Normally, apoptosis in vertical rows of cells divides the paddle into fingers within a few days (Figure 25.14). When the cells do not die on cue, the paddle does not split properly (Figure 25.15).

Signal Reception — Signal binds to a receptor, usually at the cell surface. → **Signal Transduction** — Binding brings about changes in cell properties, activities, or both. → **Cellular Response** — Changes alter cell metabolism, gene expression, or rate of division.

a

b

Figure 25.13 (**a**) Generalized signal transduction pathway. A signaling molecule docks at a receptor. The signal activates enzymes or other cytoplasmic components that cause changes in metabolism, gene expression, or the cell's rate of division. (**b**) Apoptosis. Body cells self-destruct after they finish their functions or if they become infected or damaged in ways that could threaten the body as a whole.

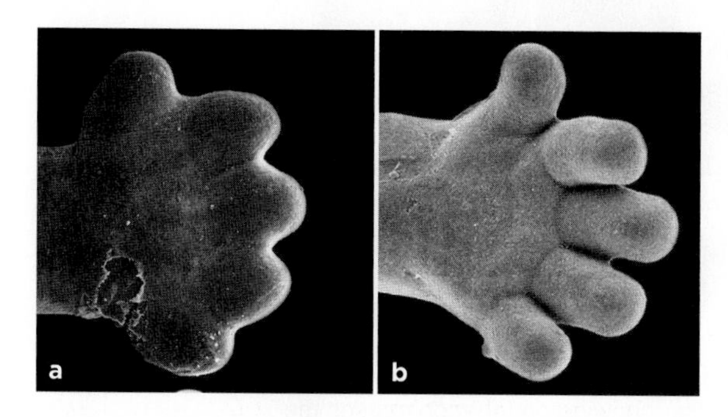

Figure 25.14 Animated! Formation of human fingers. (**a**) Forty-eight days after fertilization, tissue webs connect embryonic digits. (**b**) Three days later, after apoptosis by cells making up the tissue webs, the digits are separated.

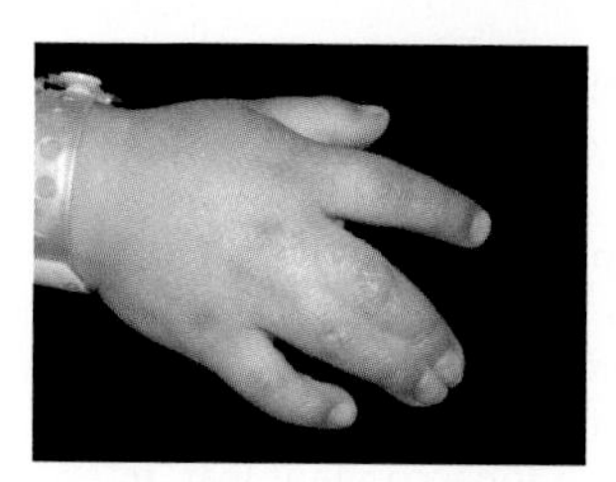

Figure 25.15 Digits that remained attached when embryonic cells did not commit suicide on cue.

Besides helping to sculpt certain developing body parts, apoptosis also removes aged or damaged cells from a body. For example, keratinocytes are the main cells in your skin. Normally they live for three weeks or so, then undergo apoptosis. Formation of new cells balances out the death of old ones, so your skin stays uniformly thick. If you spend too much time in the sun, cells enter apoptosis ahead of schedule, so your skin peels. Peeling is bad news for individual cells but helps protect your body. Cells exposed to excess UV radiation often end up with damaged DNA and might turn cancerous.

Some walled plant cells also die on cue. They get emptied of cytoplasm, and the walls of abutting ones act as pipelines for water. Cells that attach leaves to a stem die in response to seasonal change or stress, and leaves are shed. When a plant tissue is wounded or attacked by a pathogen, signals may trigger the death of nearby cells, which form a wall around the threat, as described in the previous section.

Cells within a plant or animal body communicate with one another. They exchange materials through cell junctions and secrete signaling molecules into extracellular fluid.

Membrane proteins serve as receptors for signals. Once, received, a signal is transduced into a form that initiates a change in the target cell's activity.

Summary

Section 25.1 Anatomy is the study of body form at one or more levels of structural organization. Physiology is the study of patterns and processes by which the body functions in its environment. Structural and functional organization emerges as each new individual grows and develops according to heritable patterns.

Each cell carries out metabolic tasks that keep it alive. At the same time, the individual cells maintain the body. They interact in tissues, organs, and often organ systems that acquire and distribute necessary materials, remove wastes, protect the body, and allow reproduction.

- *Use the animation on ThomsonNOW to investigate the structural organization of a tomato plant.*

Section 25.2 Plants and animals adapted in similar ways to environmental challenges. They exchange gases with the outside environment, transport materials to and from cells, and maintain water and solute concentrations inside their bodies at levels that favor cell survival. They both evolved mechanisms for integrating and controlling body parts in ways that favor survival of the organism. They both have mechanisms that allow them to respond to signals from other cells, as well as to signals or cues from the outside.

Section 25.3 In multicelled organisms, homeostasis is the state in which the body's internal environment is kept within the range that cells can tolerate. Interacting sensory receptors, integrators, and effectors maintain the tolerable conditions, often by feedback mechanisms.

With negative feedback, the detection of a change in internal conditions causes a response that brings about reversal of that change.

With positive feedback mechanisms, a change in the internal environment leads to a response that intensifies the condition that caused it.

- *Use the animation on ThomsonNOW to observe the effects of negative feedback on temperature control in humans.*

Section 25.4 Plants have decentralized mechanisms of homeostasis, such as systemic resistance and walling off a wound (compartmentalization). Plants also make homeostatic responses to changes in their environment when they fold leaves closer to their stem. Rhythmic leaf folding is one type of circadian rhythm, an event that is repeated on a 24-hour cycle.

- *Use the animation on ThomsonNOW to learn about plant defense mechanisms.*

Section 25.5 Communication between cells involves signal reception, signal transduction, and a response by a target cell. Many signals are transduced by membrane proteins that trigger reactions in the cell. Reactions may alter gene expression or metabolic activities. An example is a signal that unleashes the protein-cleaving enzymes of apoptosis, the programmed self-destruction of a cell.

- *Use the animation on ThomsonNOW to see how a human hand forms.*

Figure 25.16 (**a**) Consuelo De Moraes studies plant stress responses. (**b,c**) A caterpillar chewing on tobacco causes the plant to secrete chemicals that attract a wasp. (**d**) The wasp grabs the caterpillar and lays an egg inside it.

Self-Quiz

Answers in Appendix III

1. An increase in the number, size, and volume of plant cells or animal cells is called ______ .
   a. growth c. differentiation
   b. development d. all of the above

2. The internal environment consists of ______ .
   a. all body fluids c. all body fluids outside cells
   b. all fluids in cells d. interstitial fluid

3. ______ influences the concentrations of water and solutes in the internal environment.
   a. Diffusion c. Passive transport
   b. Active transport d. all are correct

4. Both plants and animals ______ .
   a. release energy by aerobic respiration
   b. require oxygen
   c. require carbon dioxide
   d. a and b

5. A plant's xylem and phloem are ______ tissues.
   a. vascular c. respiratory
   b. sensory d. digestive

6. An animals muscles and glands are ______ .
   a. integrators c. effectors
   b. receptors d. all are correct

7. With ______ , a change in conditions triggers a response that intensifies that change.
   a. positive feedback c. negative feedback
   b. differentiation d. compartmentalization

8. Cell communication typically involves signal ______ .
   a. reception c. response
   b. transduction d. all are correct

9. Match the terms with their most suitable description.
   ___ circadian rhythm a. programmed cell death
   ___ homeostasis b. 24-hour or so cyclic activity
   ___ apoptosis c. central command center
   ___ integrator d. stable internal environment
   ___ effectors e. muscles and glands
   ___ negative feedback f. an activity changes some condition, then the change triggers its own reversal

■ *Visit ThomsonNOW for additional questions.*

Critical Thinking

1. Many plants protect themselves with thorns or toxins or nasty-tasting chemicals that deter plant-eating animals. Some get help from wasps.

Consuelo De Moraes, currently at Pennsylvania State University, studies interactions among plants, caterpillars, and parasitoid wasps. *Parasitoids* are a special class of parasites; their larvae eat a host from the inside out.

When a caterpillar chews on a tobacco plant leaf, it secretes a lot of saliva. Some chemicals in its saliva are an external signal that triggers a chemical response from leaf cells. The cells release certain molecules that diffuse through the air. Parasitoid wasps follow the molecules down their concentration gradient to the stressed leaves. They attack a caterpillar, and each wasp deposits one egg inside it. When wasp eggs hatch they release caterpillar-munching larvae (Figure 25.16).

As De Moraes discovered, plant responses are highly specific. Leaf cells release different chemicals in response to different caterpillar species. Each chemical attracts only the wasps that parasitize the particular kind of caterpillar that triggers the chemical's release.

Are the plants "calling for help"? Not likely. Give a possible explanation for this plant–wasp interaction in terms of cause, effect, and natural selection theory.

2. The Arabian oryx (*Oryx leucoryx*), an endangered antelope, evolved in the harsh deserts of the Middle East. Most of the year there is no free water, and temperatures routinely reach 117°F (47°C). The most common tree in the region is the umbrella thorn tree (*Acacia tortilis*). List common challenges that the oryx and the acacia face. Also research and report on the morphological, physiological, and behavioral responses of both organisms.

3. In the summer of 2003, a record heat wave lasted for weeks and caused the death of more than 5,000 people in France. The elderly and very young were at greatest risk. High humidity increases the chance of heat-related illness because it reduces the rate of evaporative cooling. So does tight clothing that does not "breathe," like the uniform Korey Stringer wore during his last practice. Other risk factors are obesity, poor circulation, dehydration, and alcohol intake. Using Figure 25.8 as a reference, briefly suggest how these factors may overwhelm homeostatic controls over the body's internal temperature.

V How Plants Work

The sacred lotus, *Nelumbo nucifera*, busily doing what its ancestors did for well over 100 million years—flowering spectacularly during the reproductive phase of its life cycle.

26 PLANT TISSUES

IMPACTS, ISSUES Droughts Versus Civilization

The more we dig up records of past climates, the more we wonder about what is happening now. In any given year, places around the world have severe droughts—far less rainfall than we expect to see. In themselves, droughts are not unusual. However, they are getting worse in North America, Australia, Africa, and elsewhere. Some have been severe enough to cause mass starvation, cripple economies, and invite conflicts over dwindling resources. What is the long-term forecast? More of the same. Global warming is changing Earth's weather patterns. Heat waves are expected to be more intense, and droughts more frequent.

Humans built the whole of modern civilization on a vast agricultural base. Today we reel from droughts that last two, five, seven years or so. Imagine one lasting *200* years! It happened. About 3,400 years ago, rainfall dried up and brought an end to the Akkadian civilization in northern Mesopotamia. We know this from deep core drillings. Researchers drill a long pipe down through a glacial ice sheet that can be more than 3,000 meters (9,800 feet) thick, then pull it out. The ice core inside the pipe holds air bubbles trapped in layers of snow that fell year in, year out, over the past 200,000 years. Such air holds clues to past atmospheric conditions, and it points to recurring climate changes that may have brought an end to many societies all around the world.

A drought lasting more than 150 years contributed to the collapse of the Mayan civilization centuries ago (Figure 26.1). More recently, Afghanistan was scorched by seven years of drought—the worst in the past century. The vast majority of Afghans are subsistence farmers. The drought wiped out harvests, dried up wells, and killed livestock. Despite relief efforts, starvation was rampant. Desperate rural families sold their land, their possessions, and their

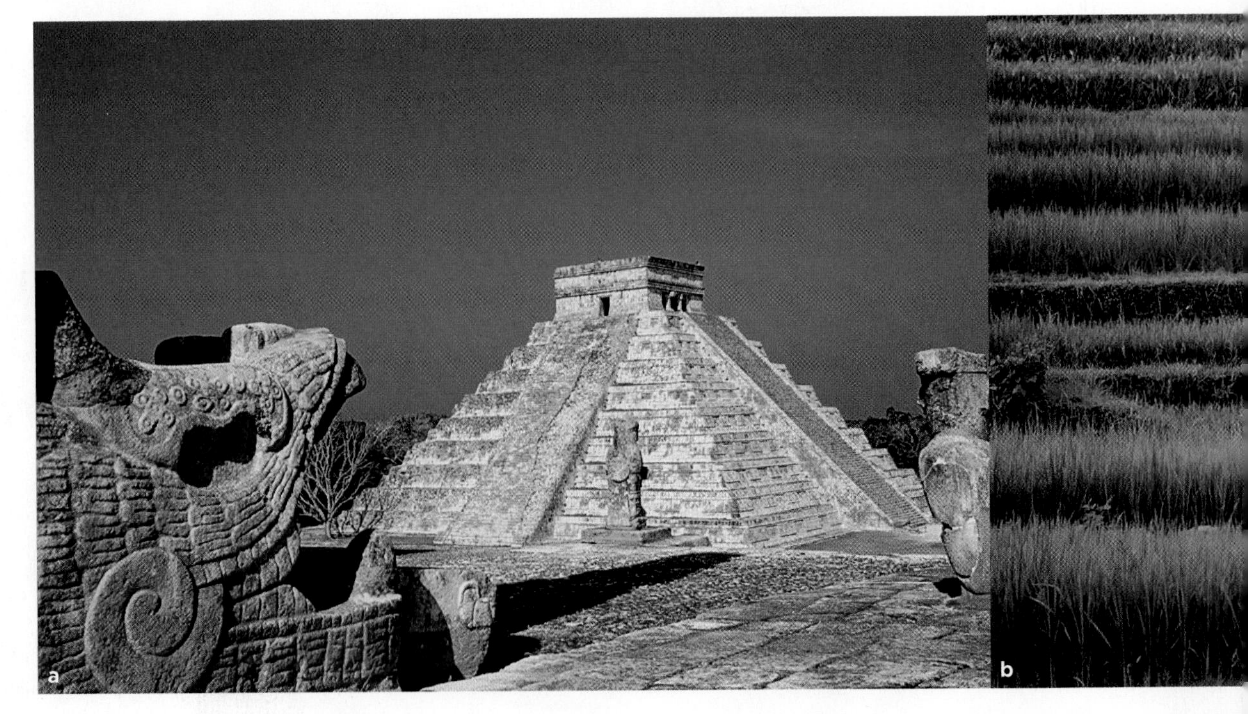

See the video! Figure 26.1 Why study plants? We absolutely depend on adaptations by which they get and use environmental resources, which include liquid water. Directly or indirectly, plants make the food that sustains nearly all forms of life on Earth.

(**a**) Mute reminder of the failed Mayan civilization. (**b**) Rice crop, which must grow in shallow water of continuously flooded fields. (**c**) From a Guatemalan field, an atrophied corncob—a reminder of a prolonged drought and widespread crop failures.

How would you vote? Large-scale farms and large cities compete for clean, fresh water. Should cities restrict urban growth? Should farming be restricted to areas with sufficient rainfall to sustain agriculture? See ThomsonNOW for details, then vote online.

daughters. What will happen to people that depend on wheat and other food crops exported by countries that also face long-term drought?

Even a short drought reduces photosynthesis and crop yields. Like other plants, crop plants conserve water by closing stomata, which of course also stops carbon dioxide from moving in. No carbon dioxide, no sugars. Drought-stressed flowering plants make fewer flowers or stunted ones. Even if flowers get pollinated, fruits may fall off the plant before ripening.

This unit focuses on seed-bearing vascular plants, especially the flowering types that are part of our lives. You will be looking at how these plants function and at their patterns of growth, development, and reproduction. You will consider how they are adapted to withstand a variety of stressful conditions and why prolonged water deprivation kills them.

The vulnerability of the agricultural base for societies around the world will impact your future. Which nations will stumble during long-term climate change? Which ones will make it through a severe drought that does not end any time soon?

Key Concepts

OVERVIEW OF PLANT TISSUES

Seed-bearing vascular plants have a shoot system, which includes stems, leaves, and reproductive parts. Most also have a root system. Such plants consist mostly of ground tissues. Their vascular tissues distribute water, nutrients, and products of photosynthesis. Their dermal tissues cover all surfaces exposed to the environment.

Plants lengthen or thicken only at active meristems: zones where undifferentiated cells are dividing rapidly. Meristems near the tips of young shoots and roots drive primary growth, or the lengthening of plant parts. **Sections 26.1, 26.2**

ORGANIZATION OF PRIMARY SHOOTS

The ground, vascular, and dermal tissue systems of monocot and eudicot stems and leaves have characteristic patterns of organization. Patterns of leaf growth and internal leaf structure maximize sunlight interception, water conservation, and gas exchange. **Sections 26.3, 26.4**

ORGANIZATION OF PRIMARY ROOTS

The ground, vascular, and dermal tissue systems in the roots of monocots and eudicots are organized differently. Roots absorb water and mineral ions, and anchor the plant. **Section 26.5**

THE WOODY PLANTS

In many plants, older branches or roots put on secondary growth; they thicken during successive growing seasons. Extensive secondary growth is known as wood. **Sections 26.6, 26.7**

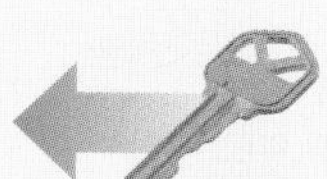

Links to Earlier Concepts

In this chapter, you will build on Sections 21.1 and 25.1, which introduced the nature of plant structure and correlated it with present and past functions. You will deepen your understanding of how the fine structure of plant cells (4.11) and photosynthetic pathways (6.3) are adaptations to environmental conditions (6.7).

26.1 Components of the Plant Body

LINKS TO SECTIONS 21.1, 21.7, 25.1

Think back on Chapter 21, which surveyed the plant kingdom. With 260,000 species, the flowering plants are the dominant group. They are the main focus of this introductory chapter on plant cells and tissues.

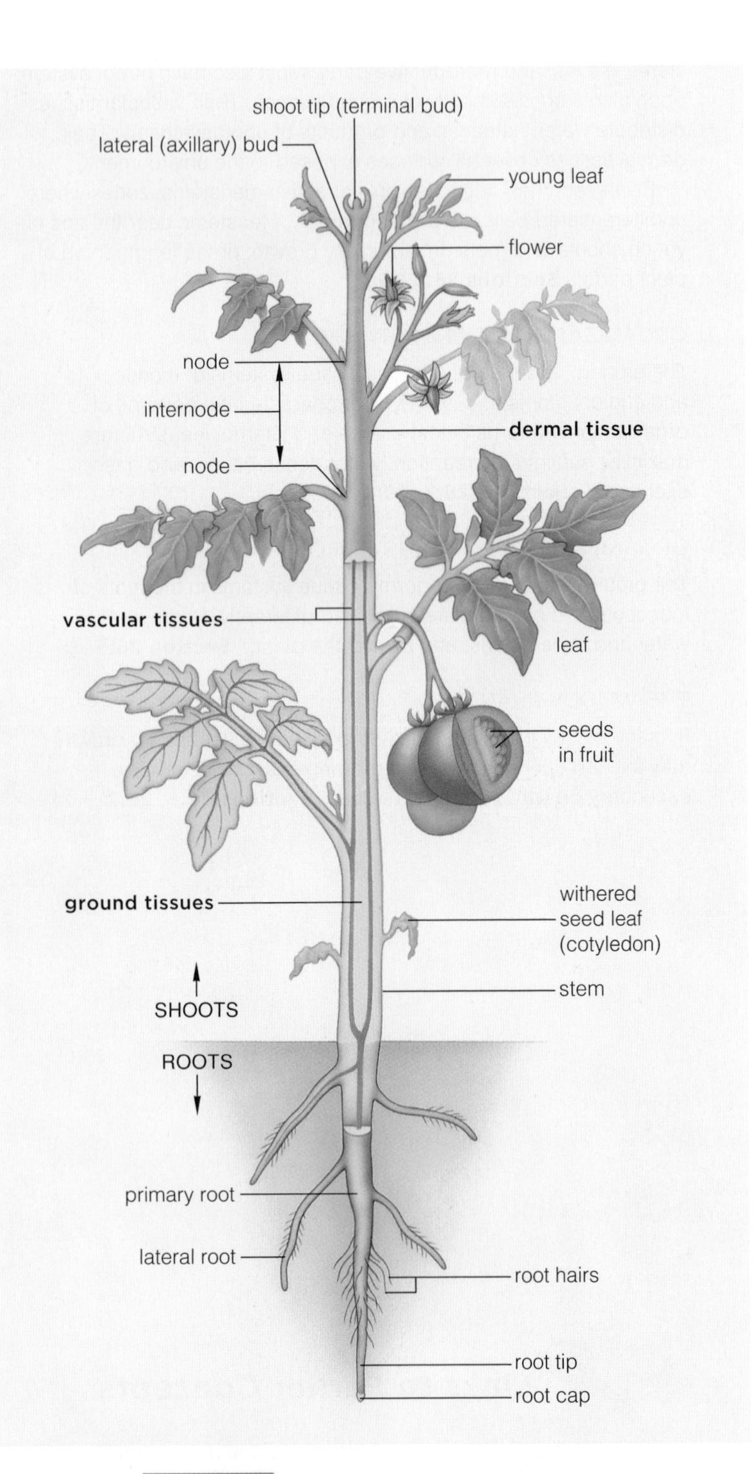

Figure 26.2 **Animated!** Body plan for a tomato plant (*Lycopersicon esculentum*). Its vascular tissues (*purple*) conduct water, dissolved minerals, and organic substances. They thread through ground tissues that make up most of the plant. Epidermis, a type of dermal tissue, covers root and shoot surfaces.

THE BASIC BODY PLAN

Figure 26.2 shows the body plan of a typical flowering plant. It has shoots: aboveground parts such as stems, leaves, and flowers. Stems support upright growth, a boon for cells that intercept energy from the sun. The plant also has roots—structures that absorb water and dissolved minerals as they grow down and outward in the soil. Roots often anchor the entire plant. All root cells store food for their own use, and some types also store it for the rest of the plant body.

OVERVIEW OF THE TISSUE SYSTEMS

Three plant tissue systems—the ground, vascular, and dermal systems—make up all shoots and roots. The ground tissue system functions in many basic tasks, such as photosynthesis, storage, or structural support of other tissues. The vascular tissue system takes up and then distributes water and dissolved ions from the environment. It also carries sugars from photosynthetic cells to the rest of the plant. The dermal tissue system covers and protects all exposed surfaces.

All three systems incorporate simple and complex tissues. Parenchyma, collenchyma, and sclerenchyma are simple tissues, each with one cell type only. Xylem, phloem, and epidermis are complex tissues, composed of two or more cell types. You will learn more about all of these tissues in Section 26.2.

EUDICOTS AND MONOCOTS—SAME TISSUES, DIFFERENT FEATURES

Although the same tissue systems form in all flowering plants, they do so in different patterns. Think about eudicots and monocots, the main classes of flowering plants (Section 21.7). Most shrubs and trees, such as roses, maples, and beans, are eudicots. Lilies, orchids, and corn are typical monocots. These plants are alike in structure and function but differ in what they do with their tissues. For instance, tissues make up two cotyledons in eudicot seeds but only one in monocot seeds. Cotyledons are leaflike structures that store or absorb food for the embryo attached to them. These "seed leaves" wither after a seed germinates and new leaves start to produce food by photosynthesis. Other differences are listed in Figure 26.3 and Table 21.1.

INTRODUCING MERISTEMS

All plant tissues arise at meristems, each a region of undifferentiated cells that can divide rapidly. Portions

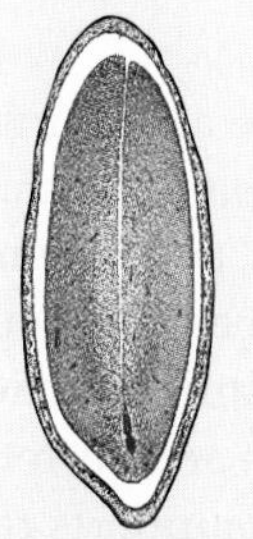

In seeds, two cotyledons (seed leaves of embryo)

Flower parts in fours or fives (or multiples of four or five)

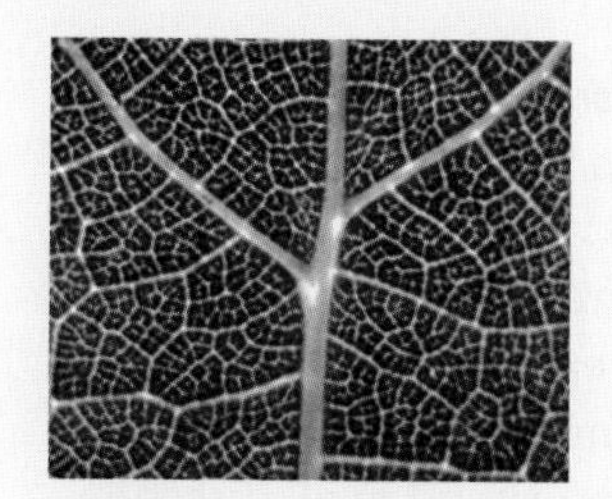

Leaf veins usually forming a netlike array

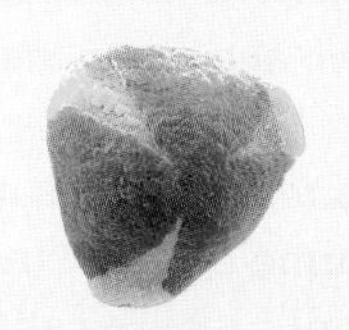

Pollen grain with three pores or furrows (or furrows with pores)

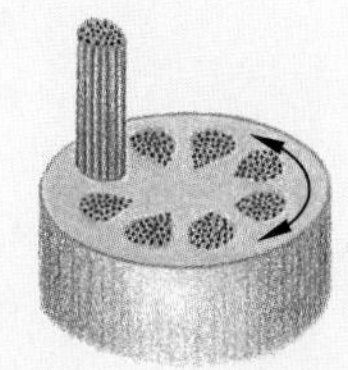

Vascular bundles organized in a ring in ground tissue

a Eudicots

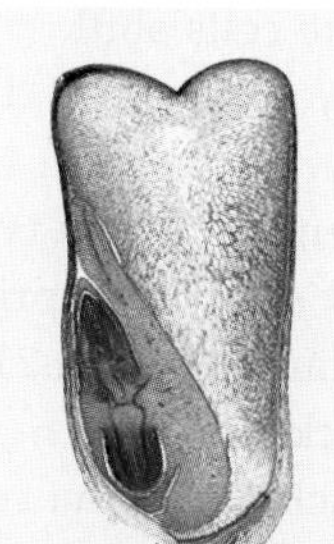

In seeds, one cotyledon (seed leaf of embryo)

Flower parts in threes (or multiples of three)

Leaf veins usually running parallel with one another

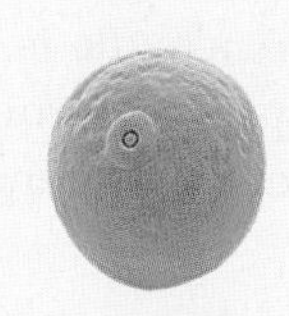

Pollen grain with only one pore or furrow on its surface

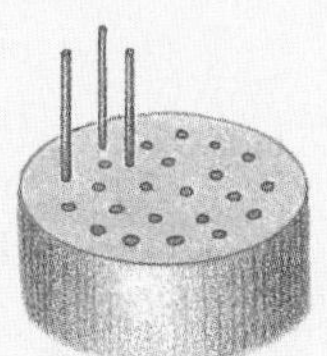

Vascular bundles throughout ground tissue

b Monocots

Figure 26.3 **Animated!** Comparison of eudicots and monocots.

Figure 26.4 *Right,* locations of apical and lateral meristems.

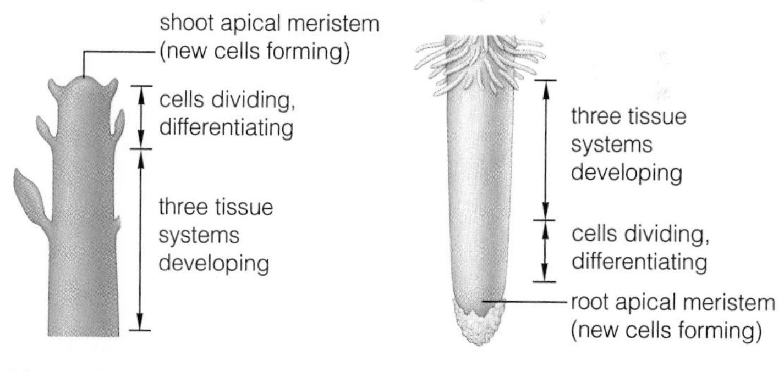

a Many cellular descendants of *apical meristems* are the start of lineages of differentiated cells that grow, divide, and lengthen shoots and roots.

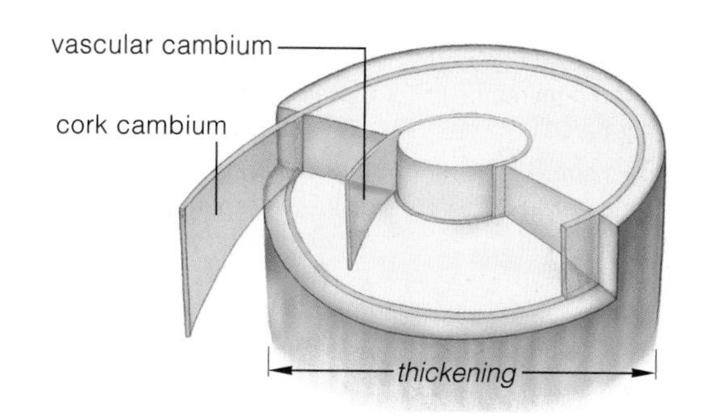

b In woody plants, the activity of two *lateral meristems—vascular cambium* and *cork cambium*—result in secondary growth that thickens older stems and roots.

of the descendant cells differentiate and mature into specialized tissues. New, soft plant parts lengthen only by activity at *apical* meristems, in shoot and root tips. The seasonal lengthening of young shoots and roots is called primary growth (Figure 26.4*a*).

In many plants, activity at *lateral* meristems results in secondary growth: a thickening of older stems and roots. A lateral meristem forms as a cylindrical sheet a few cells thick. The cylinder is not far below the stem or root surface and runs parallel with it (Figure 26.4*b*).

Plants typically have aboveground shoots, such as stems, leaves, and flowers. All have ground, vascular, and dermal tissue systems. The patterns in which these tissues are organized differ between eudicots and monocots.

Plants lengthen, or put on primary growth, only at soft shoot and root tips. Many plants put on secondary growth; older stems and roots thicken over successive growing seasons.

26.2 Components of Plant Tissues

Plant tissues form just behind shoot and root tips, and on older stem and root parts that are no longer growing.

LINKS TO SECTIONS 4.11, 6.7, 25.1

Growth, remember, is an increase in the number, size, and volume of cells. In flowering plants, new growth occurs only where cells formed by meristems divide, differentiate, then lengthen or expand in all directions. Figure 26.5 will help you interpret micrographs of the resulting tissues, which are listed in Table 26.1.

CELLS OF SIMPLE TISSUES

Figures 26.6 and 26.7 show examples of parenchyma, collenchyma, and sclerenchyma, which are all simple tissues. Parenchyma is the tissue that makes up most of the soft primary growth of roots, stems, leaves, and flowers. Cells of parenchyma tend to be thin-walled, pliable, and many-sided; they are alive at maturity and can still divide. Mitotic divisions of parenchyma cells repair plant wounds. Photosynthetic parenchyma, or mesophyll, has many small air spaces that enhance gas exchange between cells and the air inside leaves. Parenchyma also functions in storage and secretion.

Collenchyma, a stretchable tissue, supports rapidly growing plant parts, including young stems and leaf stalks (Figure 26.7*a*). Collenchyma cells are elongated and alive at maturity. Pectin, a pliable polysaccharide, imparts flexibility to their primary cell wall, which is thickened where three or more of the cells abut.

Cells of sclerenchyma are variably shaped and dead at maturity, but the lignin-rich walls that remain help this tissue resist compression. Remember, lignin is the organic compound that structurally supports upright plants, and helped them evolve on land (Section 21.1). Lignin also deters some fungal attacks.

Fibers and sclereids are typical sclerenchyma cells. *Fibers* are long, tapered cells that structurally support the vascular tissues in some stems and leaves (Figure 26.7*b*). They flex and twist, but resist stretching. We use certain fibers as materials for cloth, rope, paper, and other commercial products. The far stubbier and often branched *sclereids* strengthen hard seed coats, such as peach pits, and make pear flesh gritty (Figure 26.7*c*).

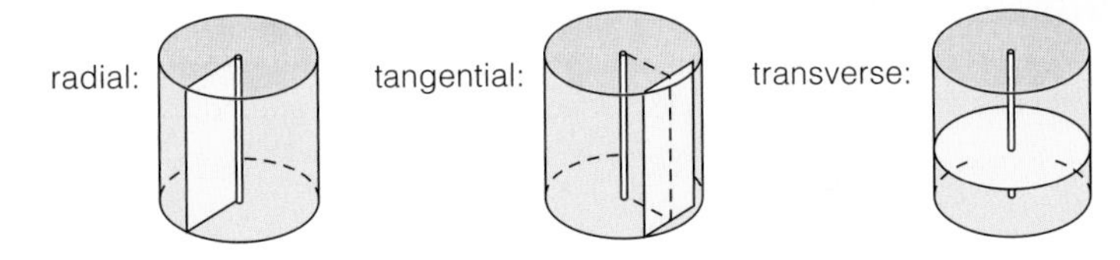

Figure 26.5 Terms that identify how tissue specimens are cut from a plant. Longitudinal cuts along a stem or root radius give *radial* sections. Cuts at right angles to the radius give *tangential* sections. Cuts perpendicular to the long axis of a stem or root give *transverse* sections—that is, cross-sections.

Table 26.1 Overview of Flowering Plant Tissues

Tissue Type	Main Components	Main Functions
Simple		
Parenchyma	Parenchyma cells	Photosynthesis (mesophylls), storage, secretion, tissue repair, other diverse tasks
Collenchyma	Collenchyma cells	Pliable structural support
Sclerenchyma	Fibers or sclereids	Structural support
Complex		
Vascular		
Xylem	Tracheids, vessel members, or both); parenchyma cells; sclerenchyma cells	Tubes for conducting water, dissolved mineral ions, plus reinforcing components
Phloem	Sieve-tube members, parenchyma cells; sclerenchyma cells	Tubes of living cells that distribute organic compounds; a cast of supporting cells
Dermal		
Epidermis	Undifferentiated as well as specialized cells (e.g., guard cells)	Secretion of protective cuticle; protection, control of gas exchange, water loss
Periderm	Cork cambium; cork cells; parenchyma	Formation of protective cover over older stems and roots

CELLS OF COMPLEX TISSUES

Vascular Tissues Xylem and phloem are vascular tissues. Both have elongated conducting cells that are often sheathed in fibers and parenchyma.

Xylem conducts water and dissolved mineral ions. The two types of cells that form xylem, *vessel members* and tracheids, are dead at maturity (Figure 26.8*a*,*b*). The secondary walls of these cells, which are stiffened

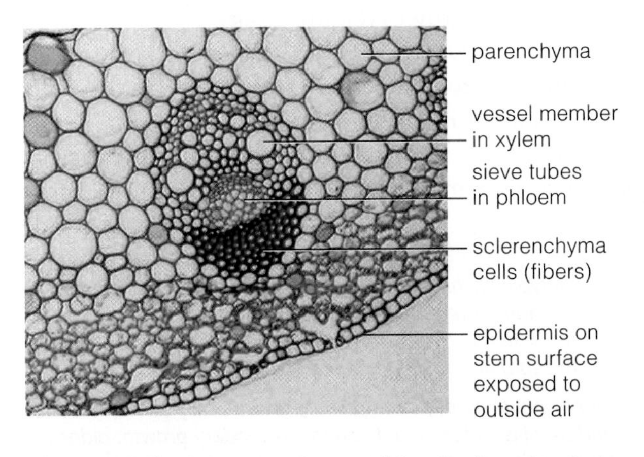

Figure 26.6 Locations of some of the simple and complex tissues of a buttercup (*Ranunculus*) stem, transverse section.

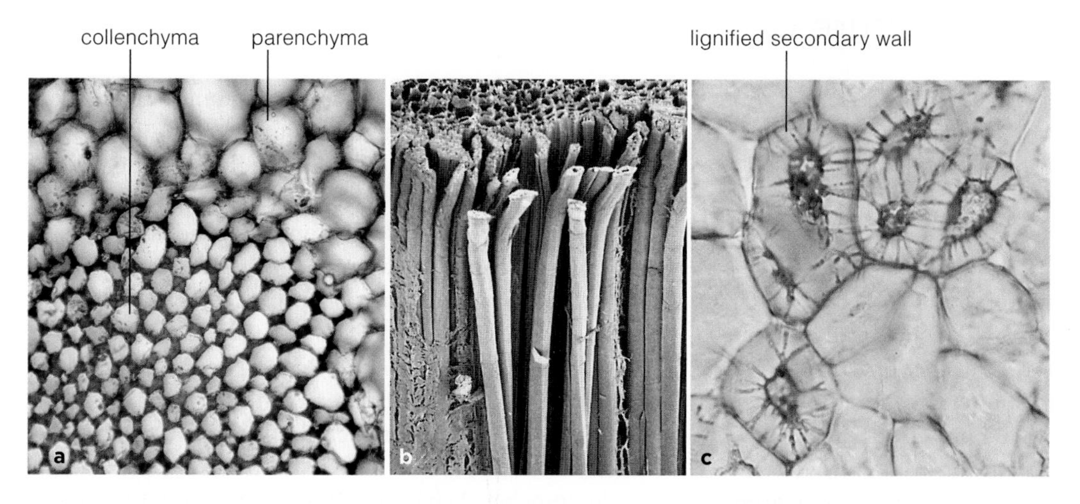

Figure 26.7 Simple tissues. (**a**) Collenchyma from one of the supporting strands inside a celery stem, transverse section.

Sclerenchyma: (**b**) Fibers from a strong flax stem, tangential view. (**c**) Stone cells, a type of sclereid in pears, transverse section.

and waterproofed with lignin, interconnect to form sturdy conducting tubes that also lend structural support to the plant. Perforations in adjoining cell walls match up, so fluid can move laterally between the tubes as well as upward through them.

Phloem conducts sugars and other organic solutes. Its main cells, *sieve-tube members*, are alive at maturity. They connect end to end, at *sieve plates*, to form sieve tubes that carry sugars to all parts of the plant (Figure 26.8*c*). Companion cells, a type of living parenchyma, load products of photosynthetic cells into the tubes.

Dermal Tissues The first dermal tissue to form on a plant is epidermis, which usually is a single layer of cells. Secretions deposited on the outward-facing cell walls form a cuticle. In plants, this surface covering is rich in deposits of cutin, a waxy substance. It helps the plant conserve water and wards off attacks by certain pathogens (Figure 26.9). Periderm, a different tissue, replaces the epidermis in woody stems and roots.

The epidermis of leaves and young stems also has specialized cells. Remember, each stoma is a tiny gap across the epidermis that opens when a pair of guard cells around it swells up (Section 6.7). The diffusion of water vapor, oxygen, and carbon dioxide gases across the epidermis is controlled at stomata.

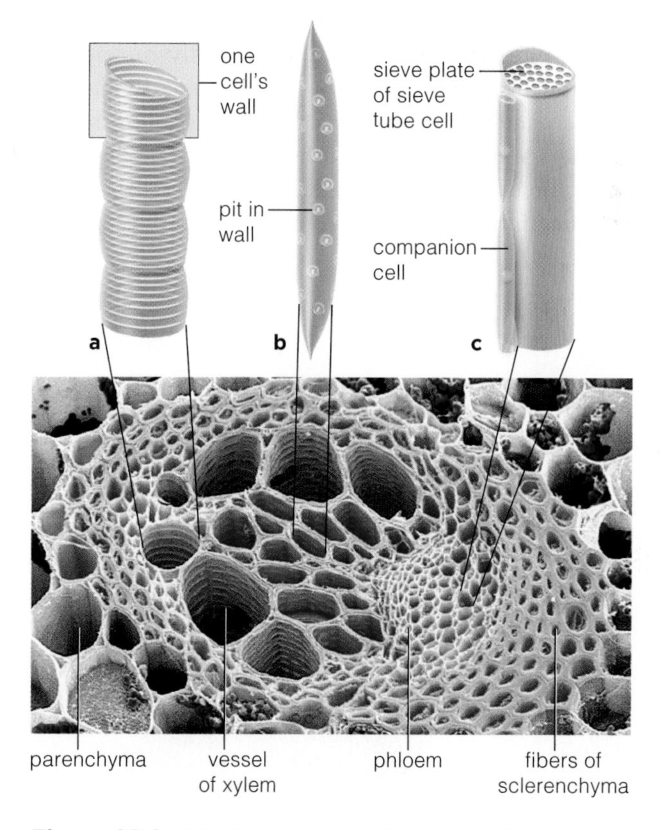

Figure 26.8 Simple and complex tissues in a stem. In xylem, (**a**) part of a column of vessel members, and (**b**) a tracheid. (**c**) One of the living cells that interconnect as sieve tubes in phloem.

Cells of parenchyma have diverse roles, such as secretion, storage, photosynthesis, and tissue repair. Collenchyma and sclerenchyma support and strengthen plant parts.

Xylem and phloem are vascular tissues that thread through the ground tissue. In xylem, water and dissolved ions flow through tubes of dead tracheid and vessel member cells. In phloem, sieve tubes of living cells distribute sugars.

Epidermis covers all young plant parts exposed to the surroundings. Periderm that forms on older stems and roots replaces epidermis of younger stems.

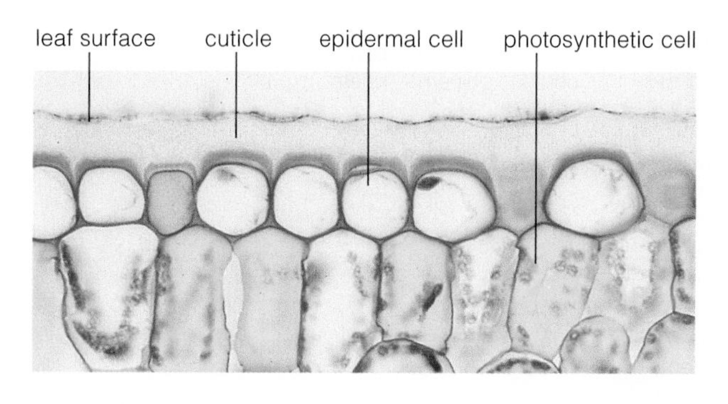

Figure 26.9 A typical plant cuticle, with many epidermal cells as well as photosynthetic cells beneath it.

26.3 Primary Structure of Shoots

Inside the soft, young stems and leaves of both eudicots and monocots, the ground, vascular, and dermal tissue systems are organized in predictable patterns.

BEHIND THE APICAL MERISTEM

The structural organization of a new flowering plant has become mapped out by the time it is an embryo sporophyte inside a seed coat. As you will read later, a tiny primary root and shoot have already formed as part of the embryo. Both are poised to resume growth and development as soon as the seed germinates.

Terminal buds are a shoot's main zone of primary growth. Just beneath a terminal bud's surface, cells of shoot apical meristem divide continually during the growing season. Some of the descendants divide and differentiate into specialized tissues. Each descendant cell lineage divides in particular directions, at different rates, and the cells go on to differentiate in size, shape, and function. Figure 26.10 shows an example.

Buds may be naked or encased in modified leaves called bud scales. Small regions of tissue bulge out near the sides of a bud's apical meristem. Each is the start of a new leaf. As the stem lengthens, leaves form and mature in orderly tiers, one after the next. A stem region where one or more leaves form is a node; the region between two successive nodes is an internode (Figure 26.2).

Lateral buds, or *axillary buds*, are dormant shoots of mostly meristematic tissue. Each one forms inside a leaf axil, the point at which the leaf is attached to the stem. Different kinds of axillary buds are the start of side branches, leaves, or flowers. A hormone secreted by a terminal bud can keep lateral buds dormant, as Section 28.7 explains.

INSIDE THE STEM

In most flowering plants, cells of primary xylem and phloem are bundled up in the same cylindrical sheath

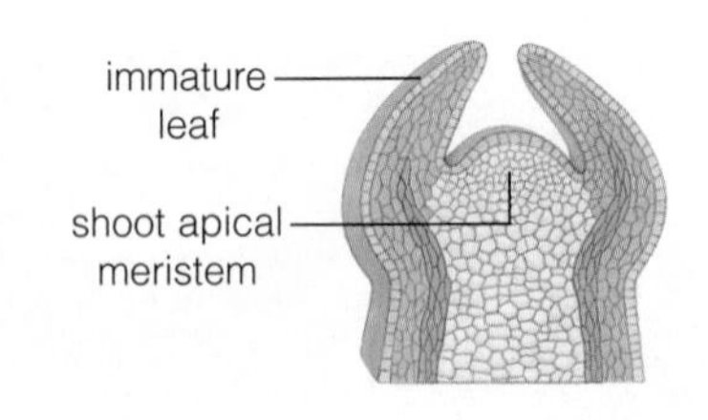

a Sketch of the shoot tip in the micrograph at right, tangential cut. The descendant meristematic cells are color-coded *orange*.

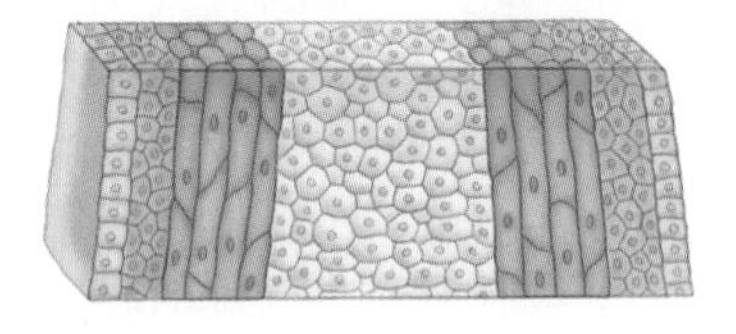

b Same tissue region later on, after the shoot tip lengthened above it

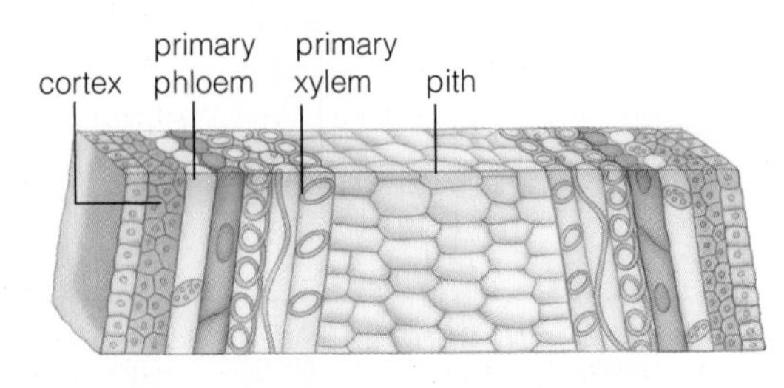

c Same tissue region later still, with lineages of cells lengthening and differentiating

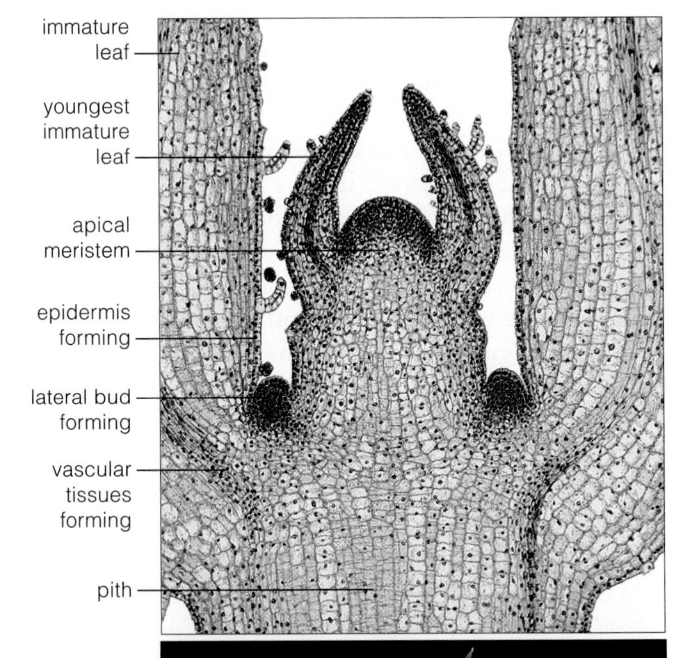

Figure 26.10 Stem of *Coleus*, a eudicot. (**a–c**) Successive stages of the stem's primary growth, starting with the shoot apical meristem. (**d**) The light micrograph is a longitudinal cut through the stem's center. The tiers of leaves in the photograph below it formed in this linear pattern of development.

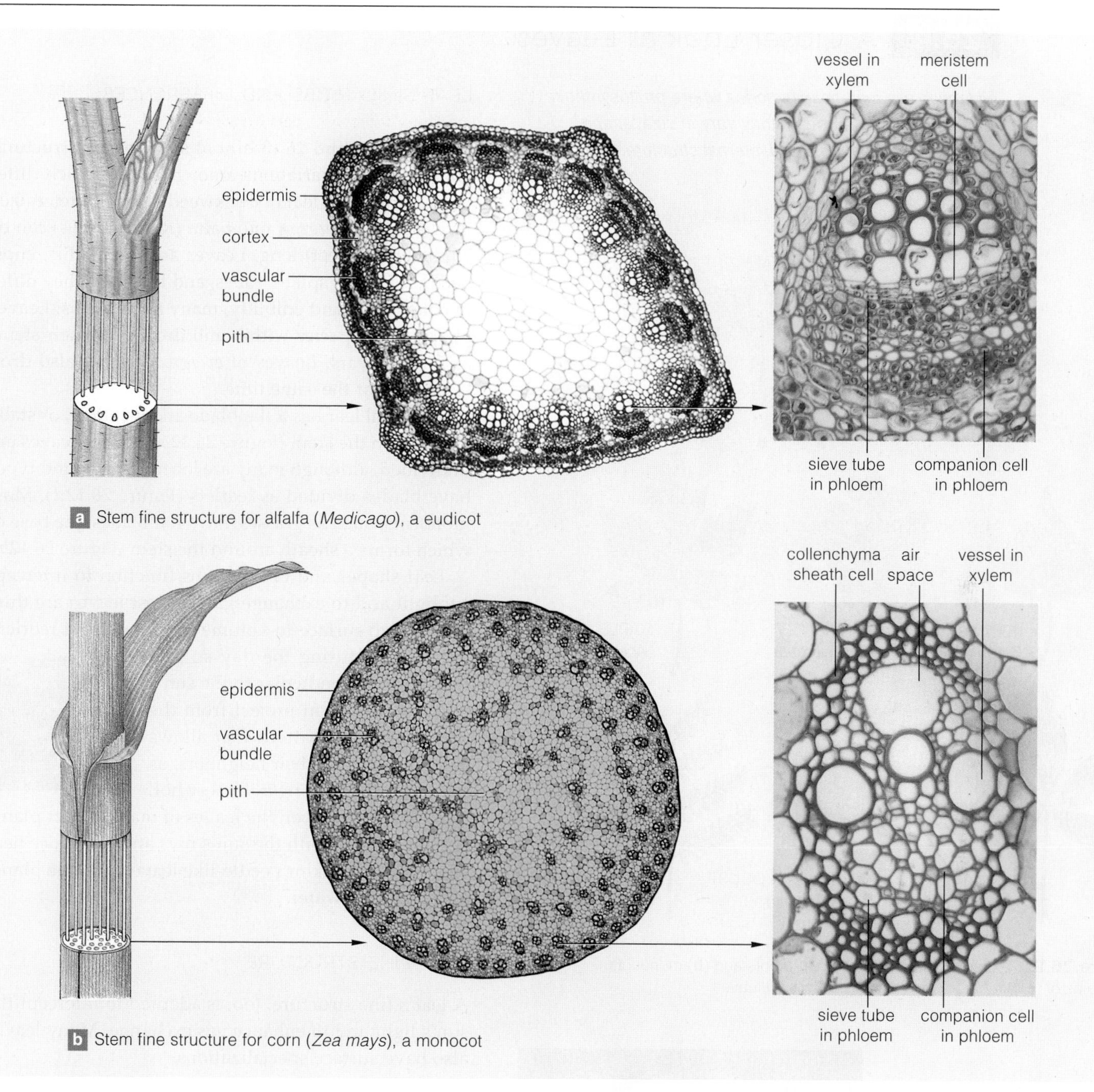

Figure 26.11 **Animated!** Primary structure of a eudicot and a monocot stem.

of cells, as long, multistranded cords. These cords are called vascular bundles, and they thread lengthwise through the ground tissue system of all shoots.

Vascular bundles form in two distinct patterns. In most eudicots, long vascular bundles form a ring that runs parallel with the long axis of the shoot. The ring divides the parenchyma of ground tissue into cortex and pith (Figure 26.11*a*). The *cortex* is the parenchyma between the vascular bundles and the epidermis. *Pith* is the parenchyma inside the ring of vascular bundles.

Most monocot and some magnoliid stems have a different arrangement. Their vascular bundles are not arranged in a ring. They are distributed all through the ground tissue (Figure 26.11*b*). In the next chapter, you will see how these vascular tissues take up, conduct, and give up water and solutes throughout the plant.

Buds are the main zones of primary growth in shoots. Ground, vascular, and dermal tissues form in organized patterns. The arrangement of vascular bundles, which are multistranded cords of vascular tissue, differs between eudicot and monocot stems.

26.4 A Closer Look at Leaves

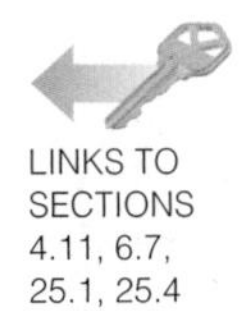

LINKS TO SECTIONS 4.11, 6.7, 25.1, 25.4

All leaves are metabolic factories where photosynthetic cells churn out sugars, but they vary in size, shape, surface specializations, and internal structure.

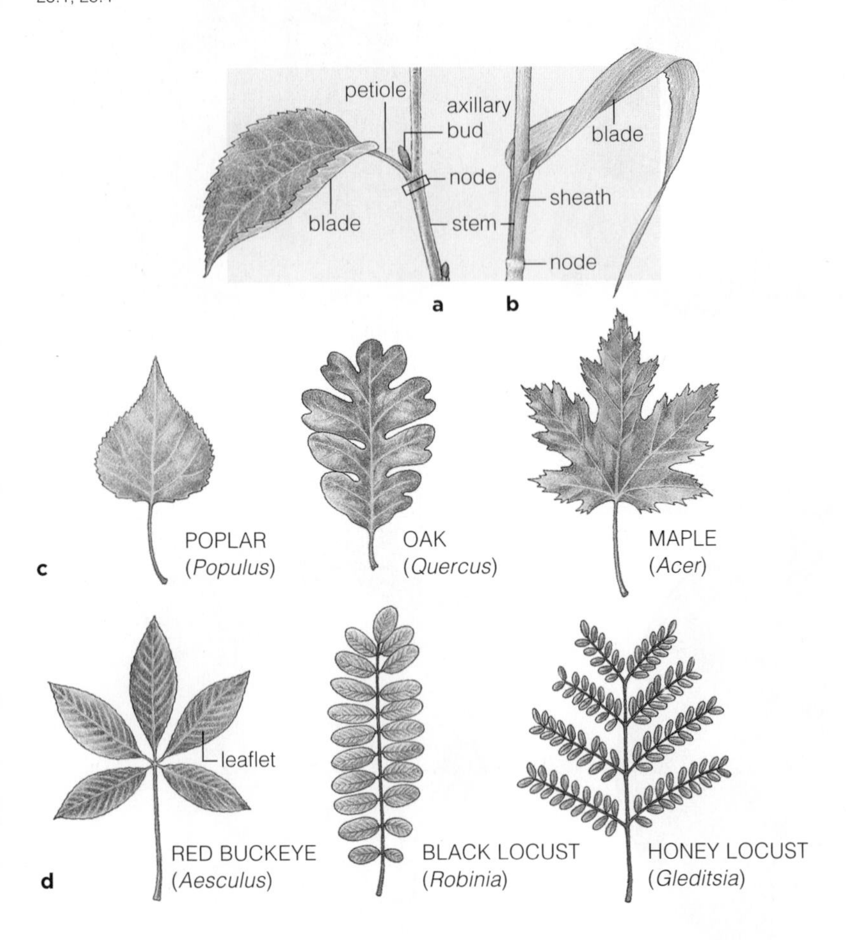

Figure 26.12 Common leaf forms of (**a**) eudicots and (**b**) monocots. Examples of (**c**) simple leaves and (**d**) compound leaves.

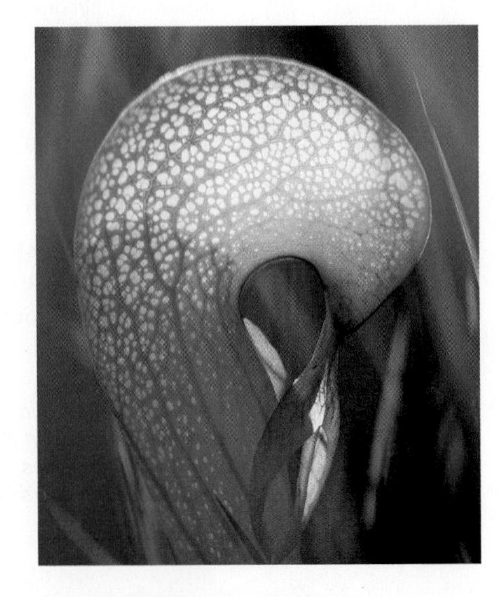

Figure 26.13 Example of leaf specialization. The leaves of the cobra lily (*Darlingtonia californica*) form a "pitcher." The pitcher becomes partially filled with plant secretions that hold digestive enzymes. It also gives off chemical odors that some insects find irresistible. Insects lured in often cannot find the way back out; light that is shining through the patterned dome of the pitcher's leaves confuses them. They just wander around and down, adhering to downward-pointing leaf hairs—which are slickened with wax above the potent vat.

LEAF SIMILARITIES AND DIFFERENCES

Figures 26.12 and 26.13 hint at some of the structural and functional variations among leaves, which differ in size as well. A leaf of duckweed is 1 millimeter (0.04 inch) across; leaves of one palm (*Raphia regalis*) can be 25 meters (82 feet) long. Leaves are shaped like cups, needles, blades, spikes, tubes, and feathers. They differ in color, odor, and edibility; many form toxins. Leaves of *deciduous* species wither and drop from their stems as winter nears. Leaves of *evergreen* plants also drop but not all at the same time.

A typical leaf has a flat blade and a petiole, or stalk, attached to the stem (Figure 26.12*a*). Simple leaves are undivided, although many are lobed. Compound types have blades divided as leaflets (Figure 26.12*d*). Most monocots, such as grasses, have flat blades, the base of which forms a sheath around the stem (Figure 26.12*b*).

Leaf shapes and orientations function to intercept sunlight and to exchange gases. Most leaves are thin, with a high surface-to-volume ratio, and most reorient themselves during the day so that they stay perpendicular to the sun's rays. Leaves often project from the same stem in a pattern that allows sunlight to reach their neighbors, as in the photo to the *right*. In very hot, arid places, however, the leaves of many desert plants remain parallel with the sun's rays and so reduce heat absorption. Thick or needle-like leaves of some plants also conserve water.

LEAF FINE STRUCTURE

A leaf's fine structure, too, is adapted to intercept the sun's light and to enhance gas exchange. Many leaves also have surface specializations.

Leaf Epidermis Epidermis covers every leaf surface exposed to the air. This surface tissue may be smooth, sticky, or slimy, with hairs, scales, spikes, hooks, and other specializations. A cuticle coats the sheetlike array of epidermal cells; it restricts water loss (Figures 26.9 and 26.14). Most leaves have far more stomata on the lower surface. In arid or cold habitats, stomata and thickly coated epidermal hairs often are positioned in depressions in the leaf surface. Both adaptations help conserve water (Section 25.4).

Mesophyll—Photosynthetic Ground Tissue Each leaf has mesophyll, a photosynthetic parenchyma with air spaces between cells (Section 6.7 and Figure 26.14).

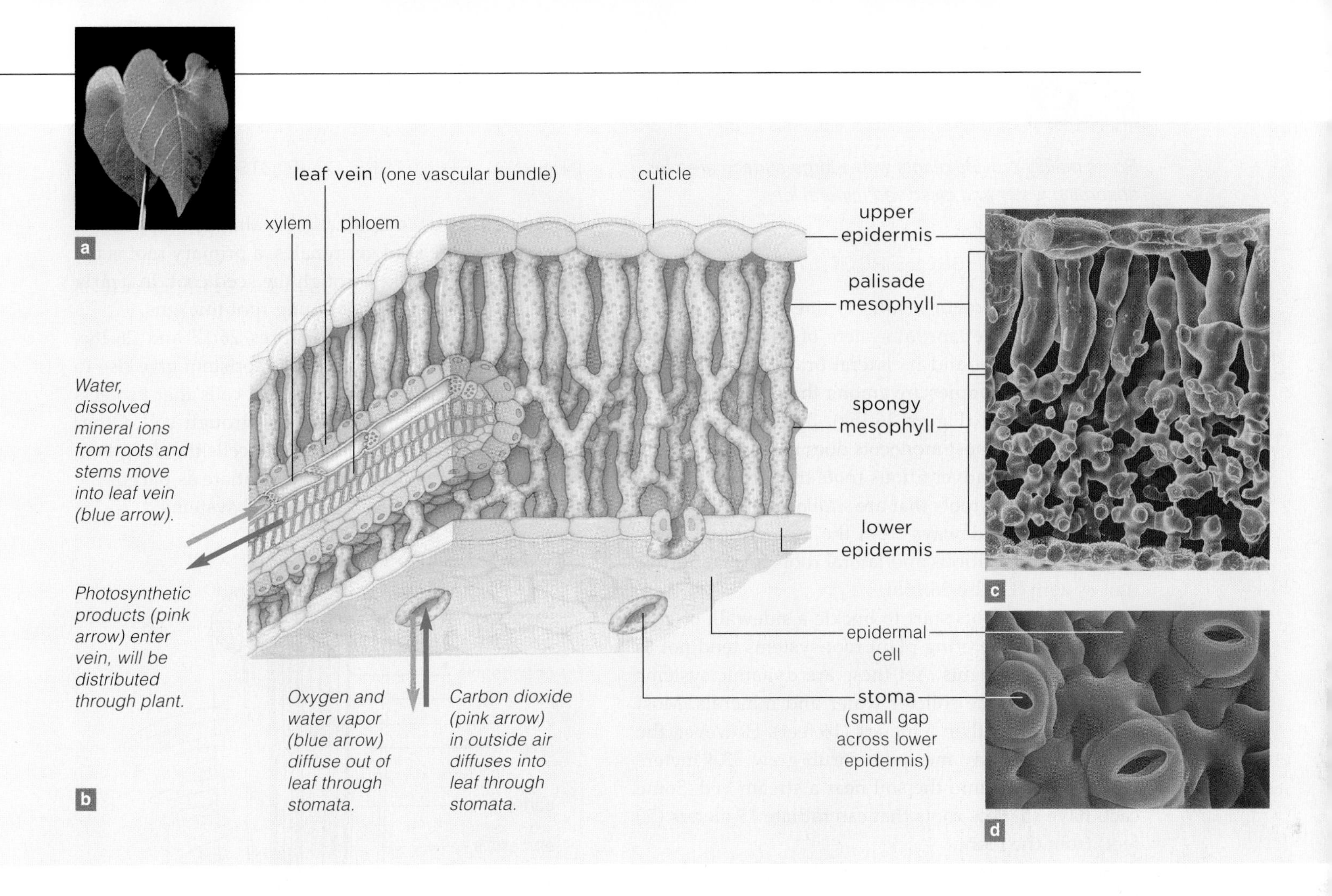

Figure 26.14 Animated! Leaf organization for *Phaseolus*, a bean plant. (**a**) Foliage leaves. (**b–d**) Leaf fine structure.

Carbon dioxide reaches the cells by diffusing into the leaf at stomata; the oxygen released by photosynthesis diffuses out the same way. Plasmodesmata connect the cytoplasm of adjacent cells. Remember, substances flow rapidly through these cell junctions across the walls of adjoining cells (Section 4.11).

Leaves oriented perpendicular to the sun have two mesophyll regions. Attached to the upper epidermis is *palisade* mesophyll. These elongated parenchyma cells have more chloroplasts (and photosynthetic potential) than cells of the *spongy* mesophyll layer below (Figure 26.14). In grass blades and other monocot leaves, the mesophyll is not divided into two layers. Such leaves grow vertically and intercept light from all directions.

Veins—The Leaf's Vascular Bundles Leaf veins are vascular bundles, usually strengthened with fibers. Their continuous strands of xylem rapidly move water and dissolved nutrients to all mesophyll cells, and the continuous strands of phloem transport photosynthetic products, especially sugars, away from them. In most eudicots, veins branch into minor veins embedded in the mesophyll. In most monocots, all of the veins are similar in length and run parallel with the leaf's long axis (Figure 26.15).

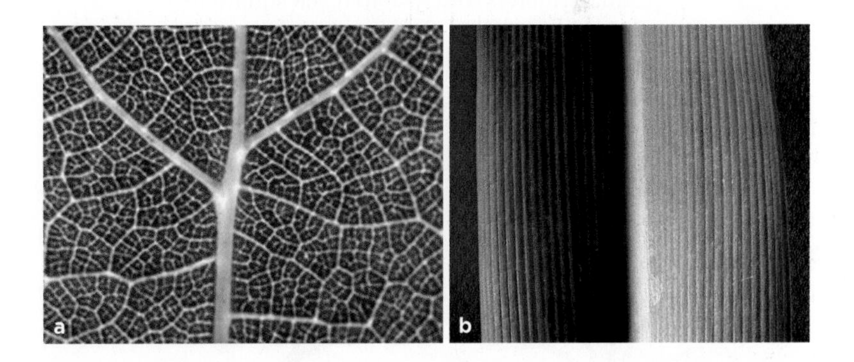

Figure 26.15 Typical vein patterns in flowering plants. (**a**) The netlike array in this grape leaf is common among eudicots. A stiffened midrib runs from the petiole to the leaf tip. Ever smaller veins branch from it. (**b**) The strong parallel orientation of veins in an *agapanthus* leaf is typical of monocots. Like umbrella ribs, stiffened veins help maintain leaf shape.

A leaf's shape, orientation, and structure typically function in sunlight interception, gas exchange, and distribution of water and solutes to and from living cells. Its epidermis encloses mesophyll and veins.

26.5 Primary Structure of Roots

Roots mainly provide plants with a large surface area for absorbing water and dissolved mineral ions.

TAPROOT AND FIBROUS ROOT SYSTEMS

LINKS TO SECTIONS 25.1, 25.2

Root primary growth results in one of two kinds of root systems. The taproot system of eudicots consists of a primary root and its lateral branchings. Carrots, oak trees, and poppies are among the plants that have a taproot system (Figure 26.16*a*). By comparison, the primary root of most monocots does not last very long; it is replaced by adventitious roots that grow out from the stem. Lateral roots that are similar in diameter and length branch sideways from the adventitious roots. Together, adventitious and lateral roots form a fibrous root system (Figure 26.16*b*).

Unless tree roots start to buckle a sidewalk or clog a sewer line, flowering plant root systems tend not to occupy our thoughts. Yet these are dynamic systems that actively mine soil for water and minerals. Most grow no deeper than 5 meters (16 feet). However, the roots of one hardy mesquite shrub grew 53.4 meters (175 feet) down into the soil near a stream bed. Some cacti have shallow roots that can radiate 15 meters (50 feet) from the plant.

Someone measured the roots of a young rye plant that had been growing for four months in 6 liters (16 gallons) of soil. If the surface area of that root system were laid out as one sheet, it would occupy over 600 square meters, or close to 6,500 square feet!

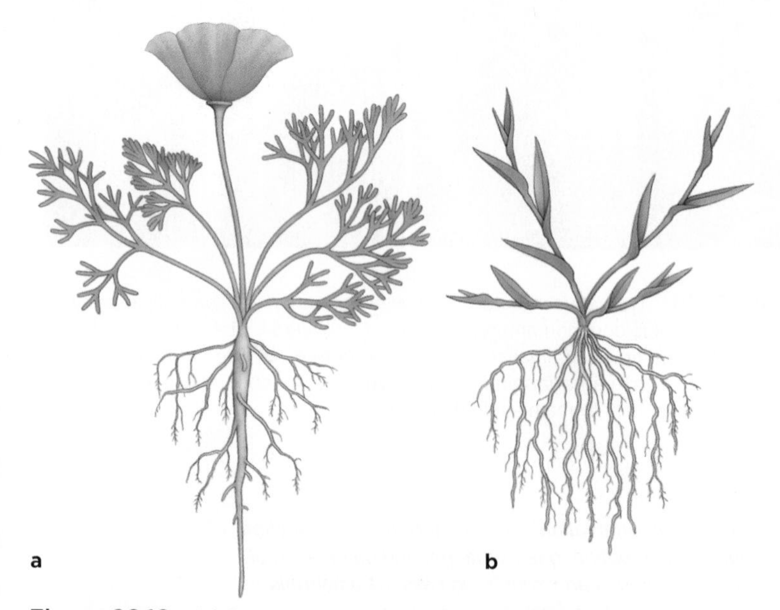

Figure 26.16 (**a**) Taproot system of the California poppy, a eudicot. (**b**) Fibrous root system of a grass plant, a monocot.

INTERNAL STRUCTURE OF ROOTS

A root's structural organization is already mapped out in a seed. As the seed germinates, a primary root is the first structure to poke through the seed coat. In nearly all eudicot seedlings, that young root thickens.

Look at the root tip in Figures 26.17 and 26.18*a*. Some descendants of root apical meristem give rise to a root cap, a dome-shaped mass of cells that protects the soft, young root as it grows through soil. Other descendants give rise to lineages of cells that lengthen, widen, or flatten when they differentiate as part of the dermal, ground, and vascular tissue systems.

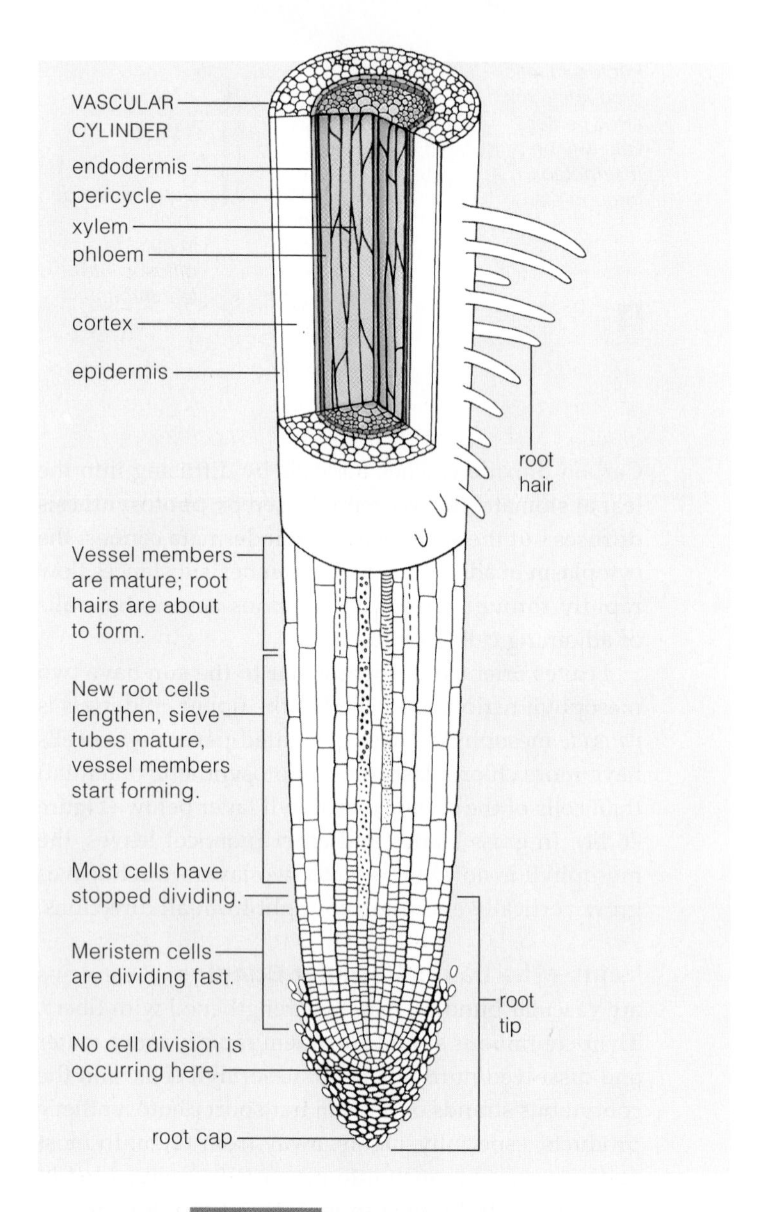

Figure 26.17 Animated! Organization of a typical primary root. The zones where cells divide, lengthen, and differentiate into primary tissues are shown here.

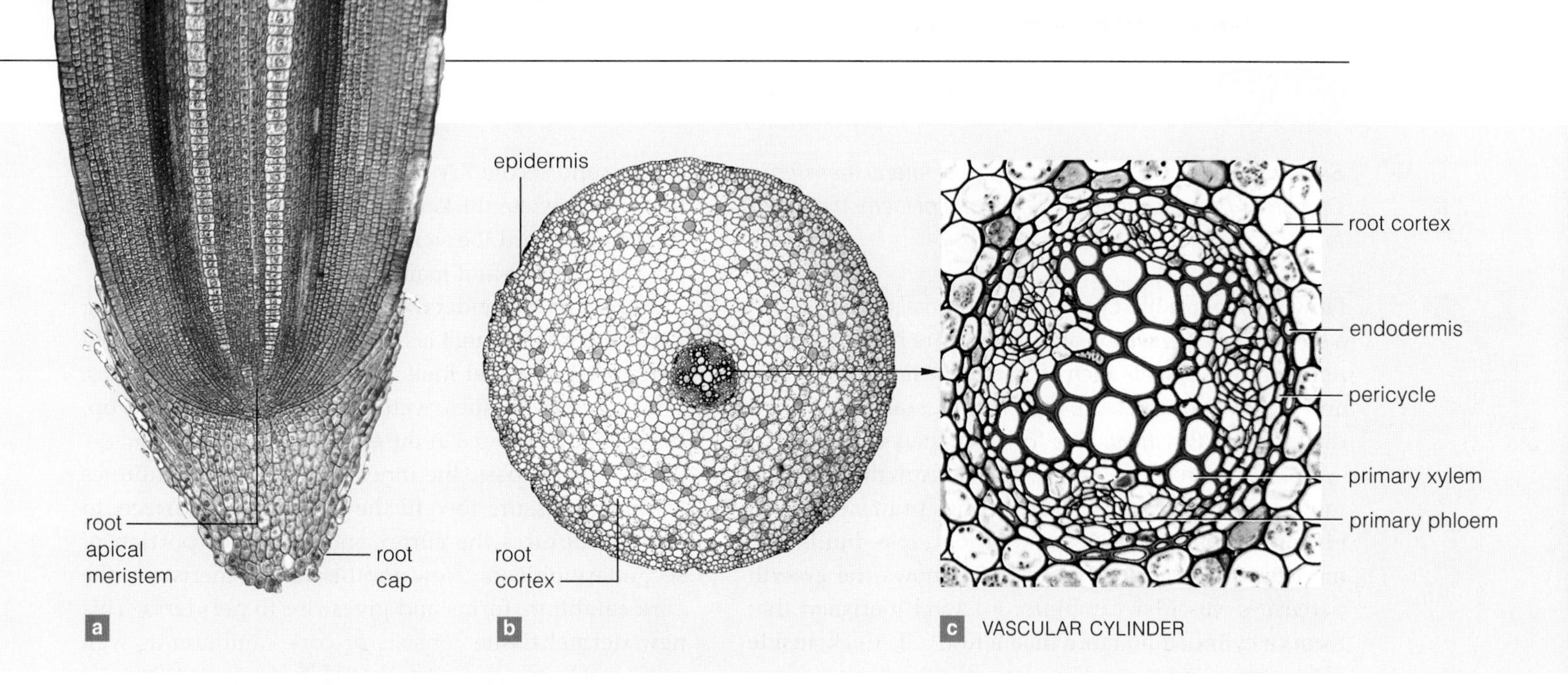

Figure 26.18 Animated! (**a**) Root tip of corn (*Zea mays*), a monocot. The oldest cells are farthest from the apical meristem, which a root cap protects. (**b**,**c**) Transverse sections of root and vascular cylinder from buttercup (*Ranunculus*), a dicot.

Ongoing divisions push cells away from the active root apical meristem. Some of their descendants form epidermis. The root epidermis is the plant's absorptive interface with soil. Many of its specialized cells send out fine extensions called root hairs, which collectively increase the surface area available for taking up soil water, dissolved oxygen, and mineral ions. Chapter 27 looks at the role of root hairs in plant nutrition. In the meantime, just be aware that root cells, like all other living cells in the plant, use oxygen as they make ATP by aerobic respiration.

Descendants of the meristematic cells also form a vascular cylinder. A vascular cylinder consists of the root's primary xylem and phloem (Figure 26.18*c*). In the roots of corn and some other monocots, it divides ground tissue into cortex and pith (Figure 26.19).

The vascular cylinder is sheathed by a pericycle, an array of parenchyma cells one or more layers thick (Figure 26.18*c*). Pericycle cells are differentiated, but they still divide, repeatedly. They do so in a direction perpendicular to the root axis. Masses of cells erupt through the cortex and epidermis as the start of new, lateral roots (Figure 26.20).

As you will see in Chapter 27, water entering a root moves from cell to cell until it reaches the endodermis, a layer of cells that encloses the pericycle. Wherever endodermal cells abut, their walls are waterproofed. Water must pass through the cytoplasm of endodermal cells to reach the vascular cylinder. Transport proteins in the plasma membrane control the uptake of water and dissolved substances.

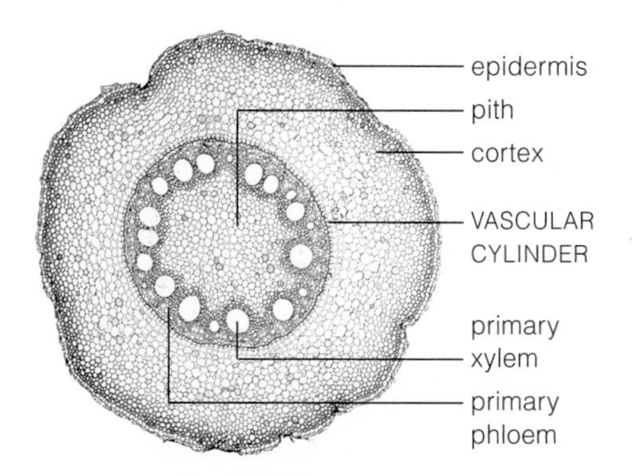

Figure 26.19 Transverse section of root from corn (*Zea mays*), a monocot. Its vascular cylinder divides the ground tissue into two zones, cortex and pith.

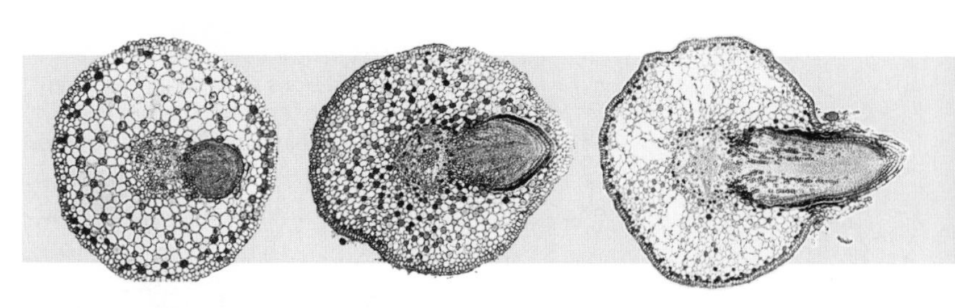

Figure 26.20 Lateral root formation from the pericycle, a cylindrical sheet of cells inside the endodermis. Transverse sections.

Roots provide a plant with a tremendous surface area for absorbing water and solutes. Inside each is a vascular cylinder, with long strands of primary xylem and phloem.

Taproot systems consist of a primary root and lateral branchings. Fibrous root systems consist of adventitious and lateral roots that replace the primary root.

26.6 Accumulated Secondary Growth—The Woody Plants

Secondary growth starts at two types of lateral meristem. One gives rise to secondary xylem and phloem, the other to a sturdier surface covering.

LINKS TO SECTIONS 25.1, 25.4

Flowering plant life cycles differ. *Annuals* live only for a single growing season and usually are herbaceous, or nonwoody. *Biennials* such as parsley form roots, stems, and leaves in one season, then flower, make seeds, and die the next. *Perennials* live for more than two years.

Woody plants put on secondary growth in two or more growing seasons. Each spring, primary growth resumes at buds on twigs and in root tips—but lateral meristems also form. Figure 26.21 shows the growth pattern at vascular cambium, a lateral meristem that forms a cylinder, no more than a few cells thick, inside older stems and roots. Divisions of vascular cambium cells produce secondary xylem on the cylinder's inner surface and secondary phloem on its outer surface. As the core of xylem thickens, it also displaces the vascular cambium toward the stem surface. Some meristem cells divide sideways and maintain the widening cylinder.

Part of the secondary xylem and phloem forms long vertical cords through a stem or root. Other parts form horizontal rays that look a bit like bike wheel spokes. Through these tissues, water and solutes can travel up, down, and sideways in the enlarging woody stem.

As seasons pass, the inner core of xylem continues to direct pressure toward the stem or root surface. In time it ruptures the cortex and the outer portion of secondary phloem. Now another lateral meristem, the cork cambium, forms and gives rise to periderm. This new dermal tissue consists of cork cambium as well as parenchyma and cork that form from it. What we call bark is secondary phloem and periderm.

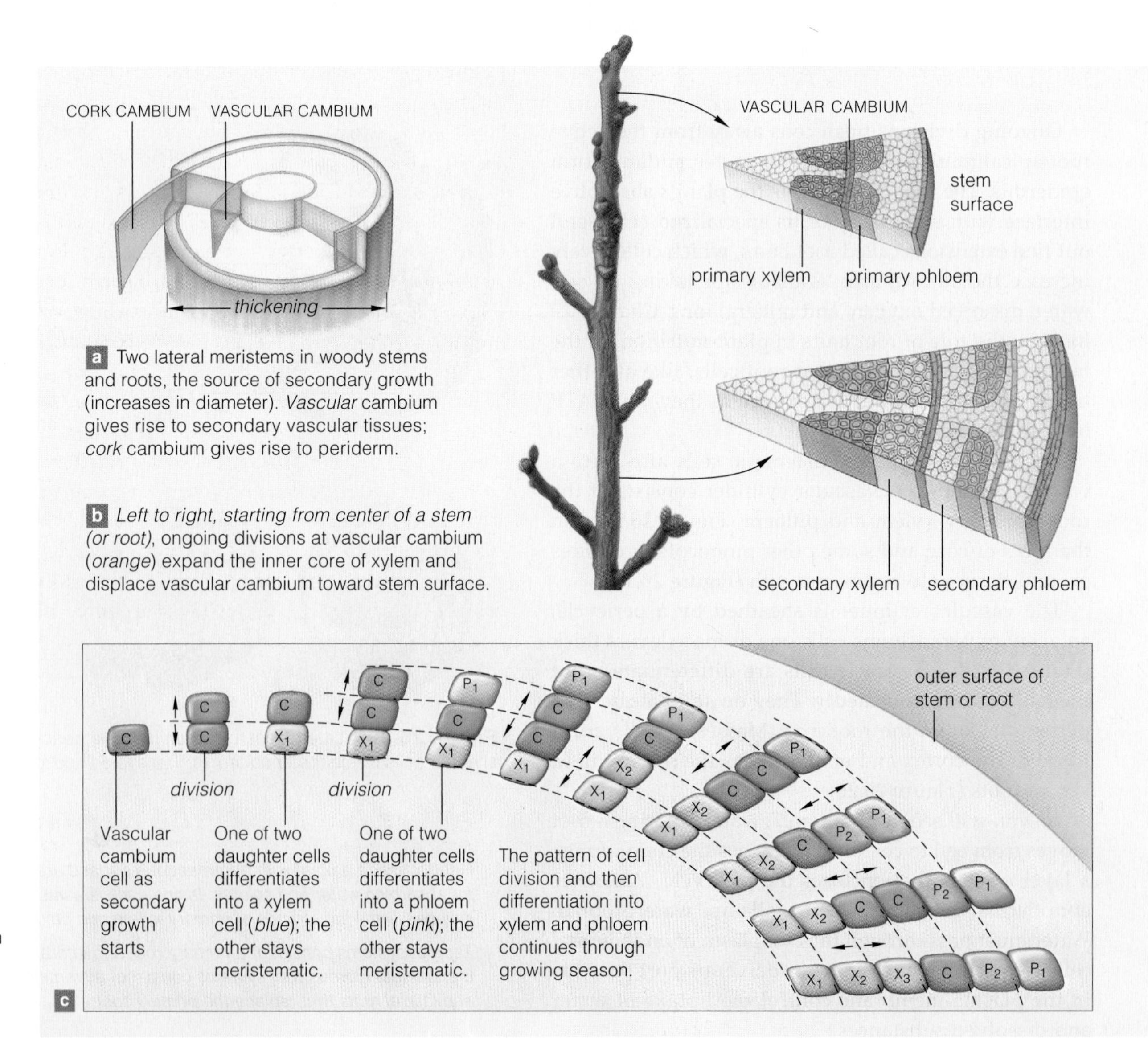

Figure 26.21 **Animated!** (**a**) Two meristems that become active during secondary growth. (**b**) Twig from a walnut tree in winter, after leaves dropped. In spring, *primary* growth resumes at terminal and lateral buds. *Secondary* growth resumes at vascular cambium. (**c**) Overall pattern of growth at vascular cambium.

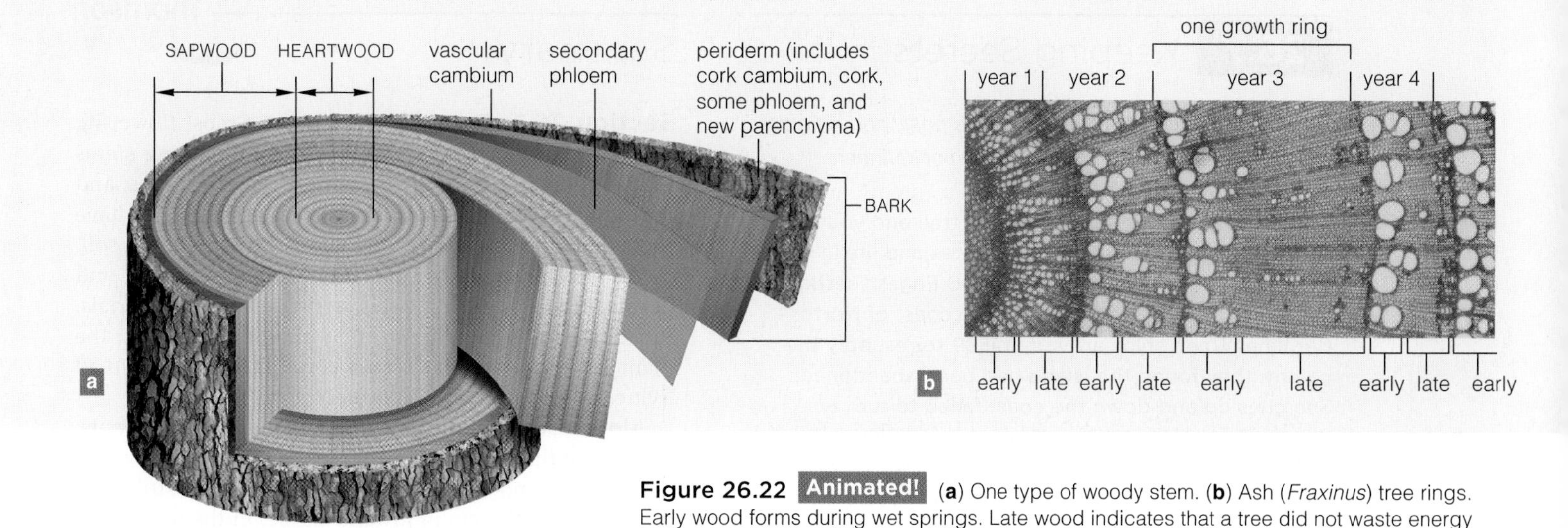

Figure 26.22 Animated! (**a**) One type of woody stem. (**b**) Ash (*Fraxinus*) tree rings. Early wood forms during wet springs. Late wood indicates that a tree did not waste energy making large-diameter xylem cells for water uptake during a dry summer or drought.

Cell divisions at cork cambium produce cork. This tissue has densely packed rows of dead cells, the walls of which are thickened with a fatty substance called suberin. Cork protects, insulates, and waterproofs the stem or root surface. Cork also forms over wounded tissues. When leaves drop from the plant, cork forms at the places where petioles had attached to stems.

Wood's appearance and function change as a stem or root ages. Metabolic wastes such as resins, tannins, gums, and oils, clog and fill the oldest xylem so much that it no longer is able to transport water and solutes. These substances often darken and strengthen wood, which becomes more aromatic and prized by furniture builders as heartwood.

Sapwood is all of the still-functional xylem between the vascular cambium and heartwood (Figure 26.22*a*). Sapwood is wet, typically pale, and not as strong as heartwood. In spring, dissolved sugars that had been stored in roots flow through secondary xylem to buds. You may have tasted the sugar-rich fluid, or sap, that New Englanders tap from maple trees.

Vascular cambium is inactive over cool winters or long dry spells. In wet springs, it gives rise to *early* wood, with large-diameter, thin-walled cells. In dry summers, it forms *late* wood, with smaller diameter, thick-walled xylem cells. A transverse cut from older trunks reveal alternating bands of early and late wood. Each band of early and late wood is one growth ring, known informally as a "tree ring" (Figure 26.22*b*).

Where seasonal change is pronounced, trees may add one growth ring a year. In deserts, thunderstorms rumble through at different times, and trees may add more than one ring of early wood in the same season. In the tropics, seasonal change is almost nonexistent, so growth rings are not a feature of tropical trees.

Oak, hickory, and other eudicot trees that evolved in temperate and tropical zones are hardwoods, with vessels, tracheids, and fibers in xylem. Pines and other conifers are softwood trees. Their xylem has tracheids and parenchyma rays but no vessels or fibers, so they are weaker and less dense than the hardwoods.

Some tree species, such as redwoods and bristlecone pines, lay down wood over centuries. Most die sooner from old age and environmental assaults. Remember compartmentalization? This stress response counters threats but in time blocks the vital flow of water and solutes through the vascular system (Section 25.4).

We are left to ask: *What is the point of laying down so much wood?* Wood offers big selective advantages. Like other organisms, plants compete for resources. Plants with taller stems or broader canopies that defy the pull of gravity can intercept more light energy streaming from the sun. By tapping a greater supply of energy for photosynthesis, they have the metabolic means to form large root and shoot systems. With larger root and shoot systems, they can be more competitive than their neighbors in acquiring resources.

In woody plants, secondary vascular tissues form at a ring of vascular cambium inside older stems and roots. Wood is accumulated secondary growth, xylem especially.

Pressure from an ever expanding core of xylem ruptures the cortex and outer layers of secondary phloem. Another lateral meristem, cork cambium, forms and gives rise to a sturdier covering, periderm. Together, periderm and the intact secondary phloem make up bark.

A plant that can secure more energy than its neighbors to drive photosynthesis has advantages in terms of metabolic capabilities, growth, and reproductive success.

26.7 Keeping Secrets

FOCUS ON SCIENCE

Ancient trees hold secrets about the past, and a field of study called dendroclimatology unlocks them.

Count the rings in a section from a tree and you have clues to the tree's age and to climates and life in the past. For instance, In 1587, about 150 English settlers arrived at Roanoke Island off of the coast of North Carolina. When ships arrived in 1589 to resupply the colony, they found the island had been abandoned. Searches up and down the coast failed to turn up the missing colonists. About twenty years later, the English established a colony at Jamestown, Virginia. Although this colony survived, the initial years were hard. In the summer of 1610 alone, more than 40 percent of the colonists died, many of starvation.

Scientists examined wood cores from bald cypress (*Taxodium distichum*) that had been growing at the time the Roanoke and Jamestown colonies were founded. Growth rings revealed the colonists were in the wrong place at the wrong time (Figure 26.23). They arrived at Roanoke just in time for the worst droughts in eight hundred years. Nearly a decade of severe drought struck Jamestown.

We know the corn crop of the Jamestown colony failed totally. Drought-related crop failures probably occurred at Roanoke. The settlers also had difficulty finding fresh water. Jamestown was established at the head of an estuary; when river levels dropped, the fresh water mixed with ocean water and became salty.

If the secrets of trees intrigue you, find out what growth layers can reveal about shifts in climate where you live. See whether you can correlate them with human events at the time the trees were busily recording climate changes.

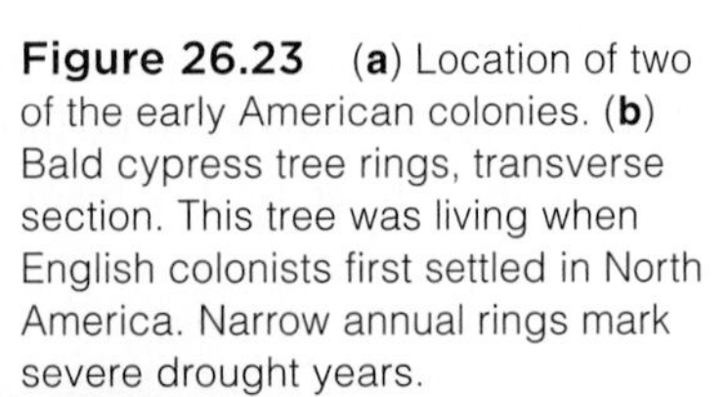

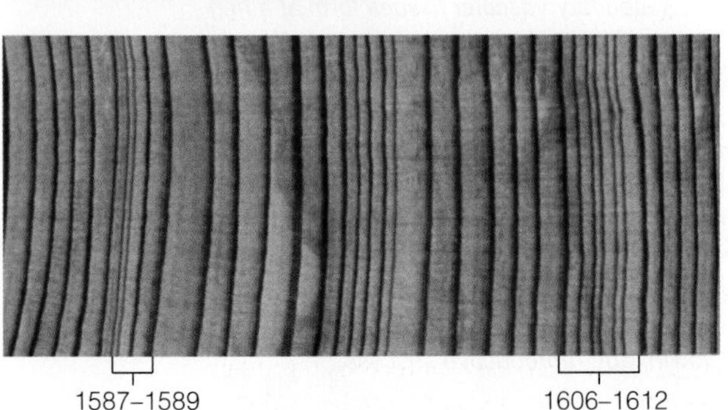

Figure 26.23 (**a**) Location of two of the early American colonies. (**b**) Bald cypress tree rings, transverse section. This tree was living when English colonists first settled in North America. Narrow annual rings mark severe drought years.

www.thomsonedu.com Thomson NOW!

Summary

Section 26.1 The basic body plan of most flowering plants consists of aboveground shoots, including stems that support upright growth, photosynthetic leaves, and reproductive shoots called flowers. Most kinds also have roots that typically grow downward and outward in soil.

All shoots and roots consist of ground, vascular, and dermal tissue systems. Ground tissues store materials, function in photosynthesis, and structurally support the plant. Tubes in vascular tissues conduct substances to all living cells. Dermal tissues protect plant surfaces.

Monocots and eudicots are the main flowering plants. They have the same tissues organized in different ways. For example, monocots and eudicots differ in distribution of xylem and phloem in ground tissue, in the number of petals in flowers, and in the number of cotyledons.

All tissues originate at meristems, which are regions of undifferentiated cells that retain a capacity for division. Apical meristems are the sources of primary growth, or lengthening of young plant parts. Lateral meristems are the sources of secondary growth, or thickening of older plant parts.

- *Use the animation on ThomsonNOW to explore a plant body plan and to compare monocot and eudicot tissues.*

Section 26.2 Simple plant tissues are parenchyma, collenchyma, and sclerenchyma. The living, thin-walled cells of parenchyma have diverse roles in ground tissue. In collenchyma, living cells with sturdy, flexible walls support fast-growing plant parts. Cells of sclerenchyma die at maturity, but their lignin-reinforced walls remain and support the plant.

The vascular tissues (xylem and phloem) and dermal tissues (epidermis and periderm) are examples of complex plant tissues. Vessel members and tracheids of xylem are dead at maturity; their perforated, interconnected walls conduct water and dissolved minerals. Phloem's sugar-conducting cells, sieve-tube members, are alive at maturity and form tubes that conduct sugars. Companion cells load sugars into the sieve tubes. Both xylem and phloem also incorporate parenchyma and sclerenchyma cells.

Epidermis covers and protects the outer surfaces of primary plant parts. Periderm replaces epidermis on the woody plants, which show extensive secondary growth.

Section 26.3 Stems of most species serve in upright growth, which favors interception of sunlight. Vascular bundles (xylem and phloem) thread through them.

In most herbaceous and young woody eudicot stems, a ring of bundles divides ground tissue into cortex and pith. In woody eudicot stems, the ring of bundles becomes bands of different tissues in the stem. Monocot stems often have vascular bundles distributed through ground tissue.

- *Use the animation on ThomsonNOW to look inside stems.*

Section 26.4 Leaves are photosynthesis factories. Between the upper and lower epidermis of each leaf are veins (vascular bundles) and mesophyll (photosynthetic parenchyma). Air spaces around mesophyll cells enhance gas exchange. Water vapor, oxygen, and carbon dioxide cross the cuticle-covered epidermis at stomata.

■ *Use the animation on ThomsonNOW to explore the structure of a leaf.*

Section 26.5 Roots absorb water and mineral ions for distribution to aboveground parts of the plant. Most store food, and support aboveground parts of the plant. Most eudicots have a taproot system; many monocots have a fibrous, side-branching root system.

■ *Use the animation on ThomsonNOW to learn about root structure and function.*

Sections 26.6, 26.7 Activity at lateral meristems thickens many older stems and roots. Wood is classified by location and function, as in heartwood or sapwood. Bark consists of secondary phloem and periderm.

■ *Use the animation on ThomsonNOW to learn about the structure of wood.*

Figure 26.24 *Above*, flower of St. John's wort (*Hypericum*) and *below*, of an iris (*Iris*).

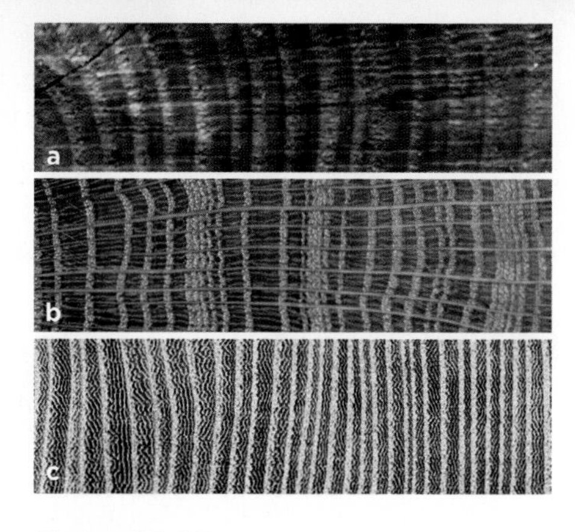

Figure 26.25 Examples of tree rings from three kinds of wood, transverse sections.

Self-Quiz

Answers in Appendix III

1. Which of the following two distribution patterns for vascular tissues is common among eudicots? Which is common among monocots?

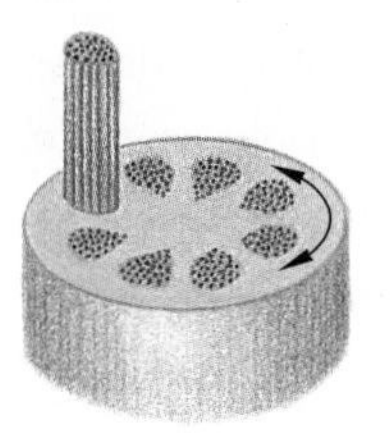

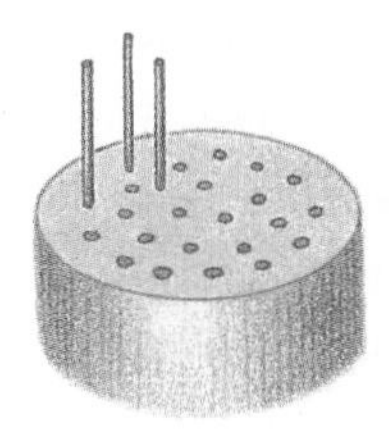

2. Roots and shoots lengthen through activity at ______ .
   a. apical meristems c. vascular cambium
   b. lateral meristems d. cork cambium

3. In many plant species, older roots and stems thicken by activity at ______ .
   a. apical meristems c. vascular cambium
   b. cork cambium d. both b and c

4. Is the plant with the yellow flower in Figure 26.24 a eudicot or a monocot? What about the plant with the purple flower?

5. ______ conducts water and ions; ______ conducts food.
   a. Phloem; xylem c. Xylem; phloem
   b. Cambium; phloem d. Xylem; cambium

6. Mesophyll consists of ______ .
   a. waxes and cutin c. photosynthetic cells
   b. lignified cell walls d. cork but not bark

7. In phloem, organic compounds flow through ______ .
   a. collenchyma cells c. vessels
   b. sieve tubes d. tracheids

8. Xylem and phloem are ______ tissues.
   a. ground b. vascular c. dermal d. both b and c

9. In early wood, cells have ____ diameters, ____ walls.
   a. small; thick c. large; thick
   b. small; thin d. large; thin

10. Match each plant parts with a suitable description.
   ___ apical meristem a. massive secondary growth
   ___ lateral meristem b. source of primary growth
   ___ xylem c. distribution of sugars
   ___ phloem d. source of secondary growth
   ___ vascular cylinder e. distribution of water
   ___ wood f. central column in roots

■ *Visit ThomsonNOW for additional questions.*

Critical Thinking

1. Think about the conditions that might prevail in a hot desert in New Mexico or Arizona; on the floor of a shady, moist forest in Georgia or Oregon; and in the arctic tundra in Alaska. Then "design" a flowering plant that could do well in one of those places, and explain why.

2. Fitzgerald, widely known for his underdeveloped sense of nature, sneaks into a forest preserve at night and then maliciously girdles an old-growth redwood. Girdling a tree means cutting away the bark and cambium in a ring around its girth, and can kill it by interrupting the circulation of water and nutrients. He settles down and watches the tree, waiting for it to die. But the tree will not die that night or any time soon. Explain why.

3. Sylvia lives in Santa Barbara, where droughts are common. She replaced most of her lawn with drought-tolerant plants and waters the remaining lawn twice a week after the sun goes down, soaking it to a depth of several inches. Why is her strategy good for lawn grasses?

4. Oscar and Lucinda meet in a tropical rain forest and fall in love, and he carves their initials into the bark of a tiny tree. They never do get together, though. Ten years later, still heartbroken, Oscar searches for the tree. Given what you know about primary and secondary growth, will he find the carved initials higher relative to ground level? If he goes berserk and chops down the tree, what kinds of growth rings will he see?

5. Figure 26.25 shows tree rings for (**a**) oak and (**b**) elm. Both hardwood eudicots are durable and strong. Tree rings for pine, a softwood, are shown in (**c**). Pine is lightweight, resists warping, and grows faster than hardwoods. It is commercially farmed as a source of relatively inexpensive lumber. Pine also is less costly. Why is it considered less desirable for furniture, wall panels, and kitchen cabinets?

27 PLANT NUTRITION AND TRANSPORT

IMPACTS, ISSUES Leafy Clean-Up Crews

From World War I until the 1970s, the United States Army tested and disposed of weapons at J-Field, Aberdeen Proving Ground in Maryland (Figure 27.1*a*). Obsolete chemical weapons and explosives were burned in open pits, together with plastics, solvents, and other wastes. Lead, arsenic, mercury, and other metals heavily contaminated the soil and groundwater. So did trichloroethylene (TCE) and dozens of other highly toxic organic compounds. TCE damages the nervous system, lungs, and liver, and it can lead to coma and death. Today, toxic groundwater is seeping toward nearby marshes and the Chesapeake Bay, part of which is already so polluted from other sources that it is a dead zone, where no marine life survives.

With the Environmental Protection Agency, the Army turned to phytoremediation: the use of plants to take up and concentrate or degrade environmental contaminants. They could not dig up and cart off the contaminated soil at J-Field; there is too much of it. Instead, they planted hybrid poplars (*P. trichocarpa* x *deltoides*). The trees can cleanse groundwater by taking up TCE and other organic compounds (Figure 27.1*b*).

The roots of the hybrid poplars take up water from the soil. Along with the water come dissolved nutrients and chemical contaminants, including TCE. The trees break down some of the TCE, and release some of it into the atmosphere. Airborne TCE is the lesser of two evils: TCE persists for a long time in groundwater, but it breaks down quickly in polluted air.

In other kinds of plants used for phytoremediation, groundwater contaminants accumulate in tissues. The plants are harvested for safer disposal elsewhere.

The best plants for phytoremediation take up many contaminants, grow fast, and grow big. Not many species tolerate toxic substances, but genetically engineered

See the video! **Figure 27.1** (**a**) J-Field, once a weapons testing and disposal site. (**b**) Today, hybrid poplars are helping to remove substances that contaminate the field's soil and groundwater. (**c**) Researcher Kuang-Yu Chen analyzing zinc and cadmium levels in plants that can tolerate these and other contaminants.

How would you vote? Genetically engineered plants may be more efficient at cleaning up contaminated sites. Do you support using genetically engineered plants for phytoremediation? See ThomsonNOW for details, then vote online.

ones may expand the range of choices. For instance, alpine pennycress (*Thlaspi caerulescens*) absorbs zinc, cadmium, and other potentially toxic minerals dissolved in soil water. Unlike most plants, its cells store zinc and cadmium out of harm's way, inside a central vacuole. Researchers are working to transfer a gene that confers this toxin-storing capacity to other plants.

Phytoremediation invites you into the world of plant physiology, the study of the mechanisms and processes by which plants function. Many adaptations that help the toxin-busters cleanse contaminated areas are the same ones that absorb and distribute water and solutes through the plant body.

When considering these adaptations, remember a key point. In nature, plants rarely have unlimited supplies of the resources they require to nourish themselves. The soils in most natural habitats are frequently dry. Nowhere except in overfertilized gardens does soil water contain lavish amounts of dissolved minerals. Many aspects of plant structure and function are adaptations to limited environmental resources.

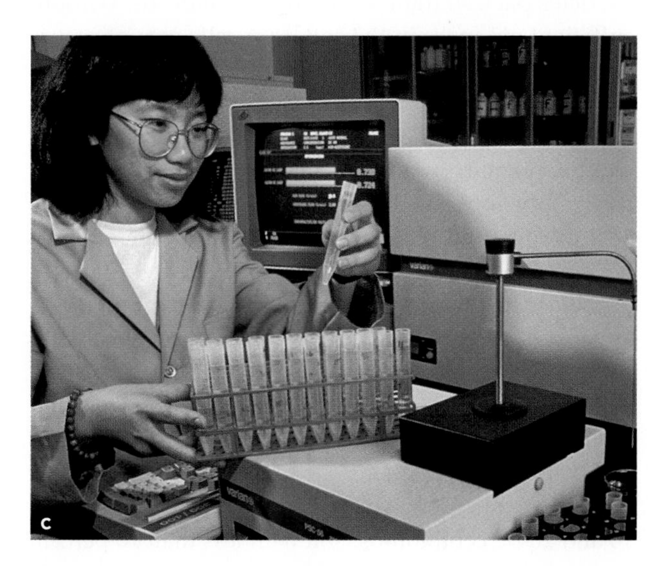

c

Key Concepts

UPTAKE OF NUTRIENT-LADEN WATER

Many plant structures and functions are adaptations to limited amounts of water and dissolved mineral ions. The root systems of vascular plants absorb water and mine the soil for nutrients, and many have symbionts that help them do so.

The amount of water and nutrients available for plants to take up depends on the composition of soil. Soil is vulnerable to leaching and erosion. **Sections 27.1, 27.2**

WATER MOVEMENT THROUGH PLANTS

Xylem distributes absorbed water and solutes from roots to leaves. Evaporation from leaves pulls up water molecules that are hydrogen-bonded to one another in long columns inside xylem. New molecules entering leaves replace the ones evaporating away. **Section 27.3**

WATER LOSS VERSUS GAS EXCHANGE

A cuticle and stomata help plants conserve water, a limited resource in most land habitats. Closed stomata stop water loss but also stop gas exchange. Some plant adaptations are trade-offs between water conservation and photosynthesis. **Sections 27.4, 27.5**

SUGAR DISTRIBUTION THROUGH PLANTS

Phloem distributes sucrose and other organic compounds from photosynthetic cells in leaves to living cells throughout the plant. Organic compounds are actively loaded into conducting cells, then unloaded in growing tissues or storage tissues. **Section 27.6**

Links to Earlier Concepts

You will be taking a closer look at the adaptations of C3, C4, and CAM plants (Section 6.7). You will revisit ionization and hydrogen bonding (2.4), the cohesive properties of water (2.5), membrane transport mechanisms (5.7), and osmosis and turgor (5.8). You will use your knowledge of vascular tissues (26.2), primary stems and roots (26.3, 26.5), and the fine structure of leaves (26.4). You will gain more insight into the close interactions between plants and their fungal symbionts (22.5).

27.1 Plant Nutrients and Availability in Soil

We've mentioned nutrients in passing. But exactly what are they, and which ones are available to plants?

THE REQUIRED NUTRIENTS

LINKS TO SECTIONS 1.2, 2.5

A nutrient, remember, is an element or molecule that has some essential role in an organism's growth and survival. Sixteen nutrients—all elements—are vital for plant growth. They are all ionized in water (Section 2.5). Examples include calcium ions (Ca^{++}) and potassium ions (K^{+}). Nine are *macro*nutrients; they are required in amounts above 0.5 percent of the plant's dry weight (its weight after all of the water has been removed). Seven other elements are *micro*nutrients, which make up traces—typically a few parts per million—of the plant's dry weight. A deficiency in any one of these nutrients may visibly affect the growth of a plant (Tables 27.1 and 27.2).

Table 27.1 Plant Nutrients and Deficiency Symptoms

Type of Nutrient	Deficiency Symptoms
MACRONUTRIENT	
Carbon, Oxygen, Hydrogen	None; all are available in abundance from water and carbon dioxide
Nitrogen	Stunted growth; chlorosis (leaves turn yellow and die because of insufficient chlorophyll)
Potassium	Reduced growth; curled, mottled, or spotted older leaves, burned leaf edges; weakened plant
Calcium	Terminal buds wither; deformed leaves; stunted roots
Magnesium	Chlorosis; drooped leaves
Phosphorus	Purplish veins; stunted growth; fewer seeds, fruits
Sulfur	Light-green or yellowed leaves; reduced growth
MICRONUTRIENT	
Chlorine	Wilting; chlorosis; some leaves die
Iron	Chlorosis; yellow, green striping in leaves of grasses
Boron	Buds die; leaves thicken, curl, become brittle
Manganese	Dark veins, but leaves whiten and fall off
Zinc	Chlorosis; mottled or bronzed leaves; abnormal roots
Copper	Chlorosis; dead spots in leaves; stunted growth
Molybdenum	Pale green, rolled or cupped leaves

Table 27.2 What To Ask If Home Gardening Hits a Wall

1. What are symptoms? (e.g., brown, yellow, curled, wilted, chewed leaves)
2. What is the species? Is all or part of one or more plants affected?
3. Is soil loose or compact? Amended or fertilized? How often?
4. Is watering by hand, hose, sprinklers, drip system? When, how often?
5. Is the plant indoors? Outdoors, in full sun or partial or full shade? In wind?
6. Gently expose a few small feeder roots. Are they white, crisp? Or black, mushy (overwatering) or brown, dry (not enough water)?
7. Do you see insects, or insect droppings, webs, cast skins, or slime?

PROPERTIES OF SOIL

Soil consists of mineral particles mixed with variable amounts of decomposing organic material, or humus. The particles form by the weathering of hard rocks. Humus forms from dead organisms and organic litter: fallen leaves, feces, and so on. Water and air occupy spaces between the particles and organic bits.

Soils differ in their proportions of mineral particles and how compacted they are. The particles, which vary in size, are primarily sand, silt, and clay. The biggest sand grains are 0.05 to 2 millimeters across. You can see individual grains by sifting beach sand through your fingers. Individual particles of silt are too small to see; they are only 0.002 to 0.05 millimeter across. Particles of clay are even smaller.

Each clay particle consists of thin, stacked layers of negatively charged crystals. Sheets of water molecules alternate between layers of clay. Because of its negative charge, clay can temporarily bind positively charged mineral ions dissolved in the soil water. This chemical behavior is the reason that plants grow in soil. Clay latches onto dissolved nutrients that would otherwise trickle past roots too quickly to be absorbed.

Although they do not bind mineral ions as well as clay, sand and silt are necessary for growing plants too. Without enough intervening sand and silt, the tiny clay particles pack so tightly that they exclude air. Without sufficient air space in the soil, root cells cannot secure enough oxygen for aerobic respiration.

Soils and Plant Growth How suitable is a given soil for plant growth? The answer depends partly on the proportions of sand, silt, and clay. Soils with the best oxygen and water penetration are loams, which have roughly equal proportions of the three types of particle. Most plants grow best in loams. Humus also affects plant growth because it releases nutrients, and its negatively charged organic acids trap the positively charged mineral ions in soil water. Humus swells and shrinks as it absorbs and releases water, which aerates soil by opening spaces for air to penetrate.

Most plants grow well in soils that contain between 10 and 20 percent humus. Soil with less than 10 percent humus may be nutrient-poor. Soil with more than 90 percent humus stays so heavily saturated with water that air (and the oxygen in it) is excluded. Swamps and bogs are like this; few kinds of plants grow in them.

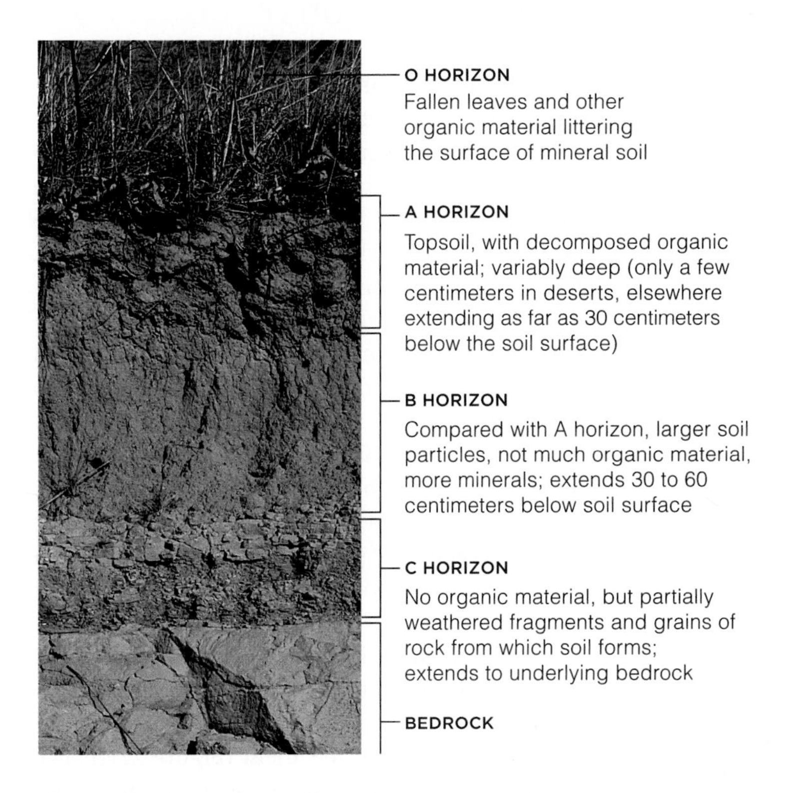

Figure 27.2 From a habitat in Africa, an example of soil horizons.

Figure 27.3 *Right:* (**a**) Erosion. In Chiapas State, Mexico, forests that once sponged up water from soil were cleared. Runoff from rains cut deep, wide gullies and carried away topsoil. (**b**) Why the Great Plains of North America was known as the Dust Bowl. Drought and strong winds prevail in this region. In the 1850s, native prairies were plowed under for farms. In time, a third of the once-deep topsoil and half the nutrients were blown away. By the 1930s, dust clouds were monstrous. They spurred erosion control practices.

How Soils Develop Soils develop over thousands of years. They are in different stages of development in different places. Most form in layers, or horizons, that are distinct in color and other properties (Figure 27.2). The layers help us profile soil in a given place. For instance, the A horizon is topsoil. Topsoil, the layer most essential for plant growth, is far deeper in some places than in others. Section 43.5 shows soil profiles for some of the major classes of ecosystems on land.

LEACHING AND EROSION

Water absorbs salts, mineral ions, and other nutrients from soil as it filters through. Leaching is the removal of nutrients from soil by the downward percolation of water. Leaching is fastest in sandy soils, which do not bind nutrients as well as clay soils. During heavy rains and the resulting runoff, more leaching occurs in forests than in grasslands. Why? Grass plants grow faster than trees, so they absorb water more quickly.

Soil erosion is a loss of soil under the force of wind and water. Strong winds, fast-moving water, and poor vegetation cover cause the greatest losses (Figure 27.3). For example, each year, about 25 billion metric tons of topsoil eroded from croplands enters the Mississippi River, which then dumps it in the Gulf of Mexico. The nutrient losses affect plants as well as other organisms that depend on plants for survival.

Nutrients are elements or molecules essential for an organism's survival and growth. Plants require nine macronutrients and seven micronutrients.

Soil is mainly particles ranging from large-grained sand to silt and fine-grained clay. Clay attracts and reversibly binds dissolved mineral ions.

Soil contains humus, a reservoir of organic material rich in organic acids. Most plants grow best in loams (soils with equal proportions of sand, silt, and clay) and 10 to 20 percent humus.

27.2 How Do Roots Absorb Water and Mineral Ions?

Where soil composition and texture change, new roots form, replace old ones, and infiltrate new areas. The roots are not "exploring" soil. Rather, their growth is simply greater in patches of soil with higher concentrations of water and minerals.

SPECIALIZED ABSORPTIVE STRUCTURES

LINKS TO SECTIONS 5.6, 22.5, 26.5

A mature corn plant can absorb as much as 3 liters of water per day. Plants in general use a lot of water and mineral ions. Mycorrhizae, root nodules, and root hairs of the plant itself enhance the uptake.

Mycorrhizae As Section 22.5 explains, a mycorrhiza (plural, mycorrhizae) is a form of mutualism between a young root and a fungus, in which both species gain benefits. The fungal hyphae grow as a velvety covering around the root or penetrate its cells. Collectively, the hyphae have a far larger surface area and can absorb scarce minerals from a larger volume of soil than the root can do on its own. The root's cells give up some sugars and nitrogen-rich compounds to the fungus, which gives up some minerals to the plant.

Root Nodules Certain bacteria in soil are mutualists with clover, peas, and other legumes which, like other plants, require nitrogen for growth. Gaseous nitrogen ($N{\equiv}N$, or N_2) is plentiful in air, but plants do not have enzymes that can break it apart. The bacteria do. They can convert nitrogen gas to ammonia (NH_3). The metabolic conversion of gaseous nitrogen to ammonia is called nitrogen fixation. Other soil bacteria convert ammonia to nitrate (NO_3), the form of nitrogen that plants can use most easily. You will read more about nitrogen fixation in Section 42.9.

Root nodules are swollen masses of root cells and cells of bacteria that infected them (Figure 27.4). The bacteria (*Rhizobium* and *Bradyrhizobium*) are nitrogen-fixing symbionts. They fix nitrogen and share it with the plant. In return, the bacteria take a small amount of photosynthetically-produced sugars from the plant.

Root Hairs As most plants put on primary growth, their root system may develop billions of root hairs (Figure 27.5). Collectively, these thin extensions of root epidermal cells enormously increase the surface area available for absorption. These are fragile structures that do not become roots. They grow no more than a few millimeters through soil and die after a few days. New ones form just behind the root tip (Section 26.5).

a Root nodule of a soybean plant

Figure 27.4 (**a**) Nutrient uptake at root nodules of a soybean plant, Like other legumes, soybean plants are symbionts with nitrogen-fixing bacteria (*Rhizobium* and *Bradyrhizobium*).

(**b**) Bacterial cells infect root cells. The bacterial cells and plant cells divide repeatedly, forming the swollen mass called a root nodule. Bacteria start fixing nitrogen. The plant takes up some of the nitrogen; the bacteria get some photosynthetic compounds in return.

(**c**) Soybean plants growing in nitrogen-poor soil show the effect of root nodules on growth. Only the plants in the rows at *right* were inoculated with *Rhizobium* bacteria and formed nodules.

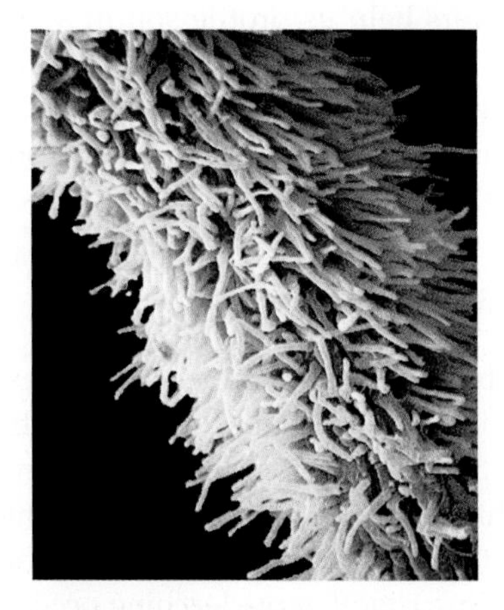

Figure 27.5 Root hairs, extensions of young root epidermal cells that specialize in absorbing water and dissolved ions.

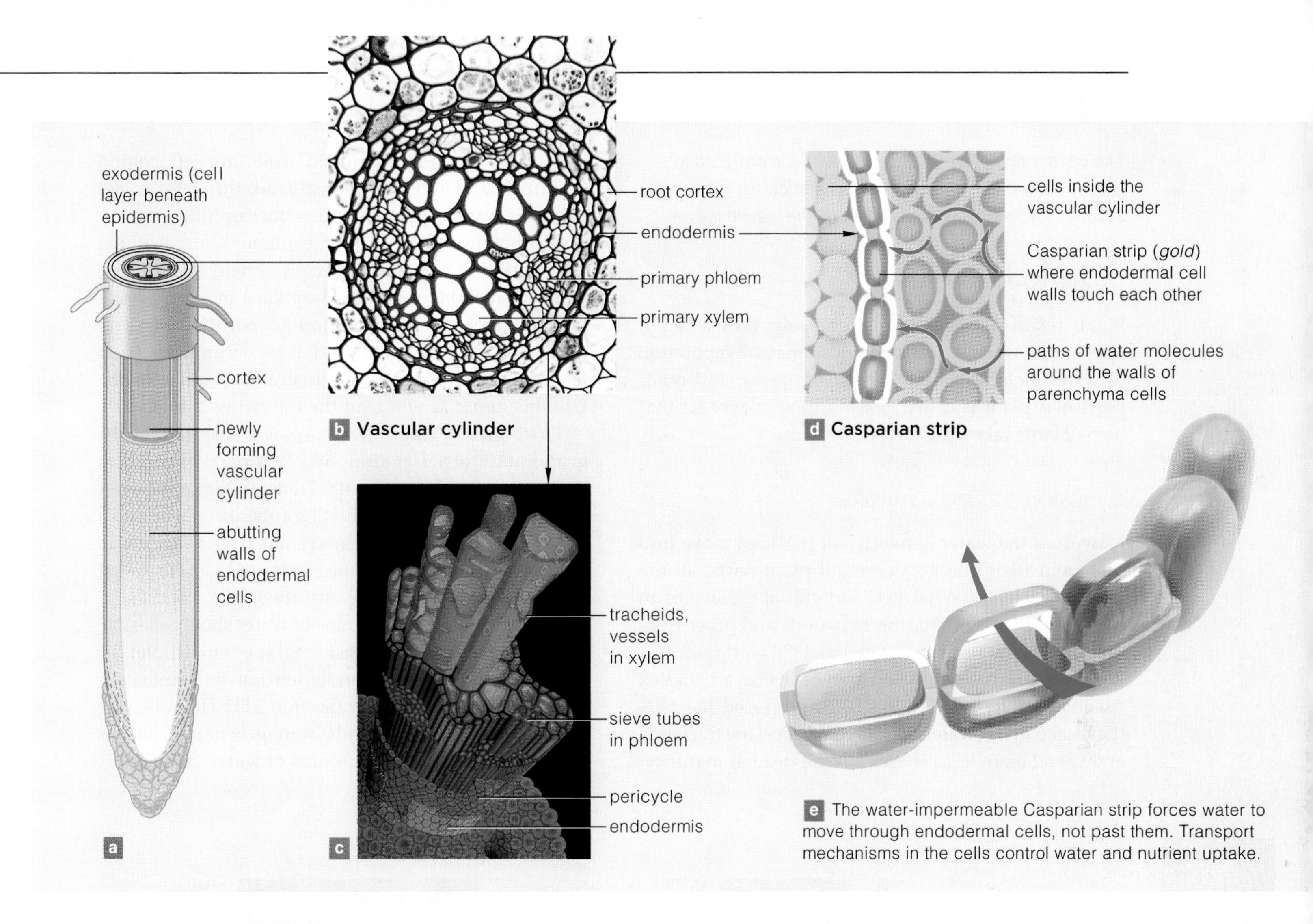

Figure 27.6 **Animated!** Where roots of most flowering plants control uptake of nutrient-laden water. (**a–c**) Cells of the endodermis (and exodermis) have a waxy Casparian strip sealing their abutting walls. The strip forces water to move through the cytoplasm of the cells. (**d**,**e**) Transport proteins in the plasma membrane of the cells control water and nutrient uptake (see Section 5.6).

HOW ROOTS CONTROL WATER UPTAKE

Section 26.5 introduced you to the tissue organization of typical roots. Turn now to how roots carry out their absorptive function. Water molecules in soil are only weakly bound to clay particles, so they readily move across the root epidermis. They move on to a vascular cylinder, a column of vascular tissue inside the root. A cylindrical sheet of endodermal cells separates the root cortex from the cylinder's xylem and phloem, which are pipelines to the rest of the plant (Figure 27.6*a–c*).

Endodermal cells deposit a Casparian strip, a waxy band, wherever their walls touch each other (Figure 27.6*d*,*e*). The band stops the unrestricted flow of water and solutes into the vascular cylinder. It forces them to cross only at unwaxed wall regions, then *through* the cell. Water must flow across the plasma membrane facing the cortex, through the cytoplasm, then across the plasma membrane facing the vascular cylinder.

Transport proteins let some solutes but not others cross the plasma membrane of endodermal cells. These proteins are the points at which a plant can adjust the amount and types of solutes it absorbs from soil water.

Roots of many plants also have an exodermis, a cell layer just beneath their surface (Figure 27.6*a*). Walls of exodermal cells commonly have a Casparian strip that functions like the one next to the vascular cylinder.

Root hairs, root nodules, and mycorrhizae greatly enhance a plant's uptake of water and dissolved nutrients.

Roots control the type and amount of solutes that enter their vascular cylinder. A layer of endodermal cells wraps around the cylinder, and a waxy Casparian strip seals the abutting walls of the cells. Membrane transport proteins of these endodermal cells control water and nutrient uptake.

27.3 How Does Water Move Through Plants?

The distribution of water and dissolved mineral ions to cells is central to plant growth and functioning. How, exactly, does water get from roots all the way to leaves?

TRANSPIRATION DEFINED

LINKS TO SECTIONS 2.4, 2.5, 26.2–26.5

Plants retain just a fraction of the water they take up. Most of the water is lost through stomata. Evaporation of water molecules from the leaves, stems, and other parts of a plant is called transpiration, a process that helps plants take up water.

COHESION–TENSION THEORY

How does the water between soil particles move into roots and then into aboveground plant parts, all the way into leaves? What gets individual molecules to the top of plants, including redwoods and other trees that may be more than 100 meters (330 feet) tall?

In vascular plants, water moves inside a complex tissue called xylem. Section 26.2 introduced the cells that make up its water-conducting tubes, the tracheids and vessel members. These cells are dead at maturity; only their lignin-impregnated walls are left behind (Figure 27.7). Obviously, being dead, the cells are not expending any energy to pull water "uphill."

The botanist Henry Dixon explained how water is transported in plants. By his cohesion–tension theory, water inside xylem is pulled upward by air's drying power, which creates a continuous negative pressure called tension. The tension extends continuously from leaves to roots. Figure 27.8 illustrates Dixon's theory. Use this figure as you read the following points:

First, air's drying power causes transpiration: the evaporation of water from all plant parts exposed to the air, but mostly at stomata. Transpiration puts water molecules inside the conducting tubes of xylem into a state of tension. The tension extends from veins inside the leaves, down through the stems, and on into young roots where water is being absorbed.

Second, continuous columns of water show cohesion, which means that they resist breaking into droplets as they are being pulled up under tension. Remember the cohesive property of water (Section 2.5)? The collective strength of hydrogen bonds among water molecules imparts cohesion to the columns of water in xylem.

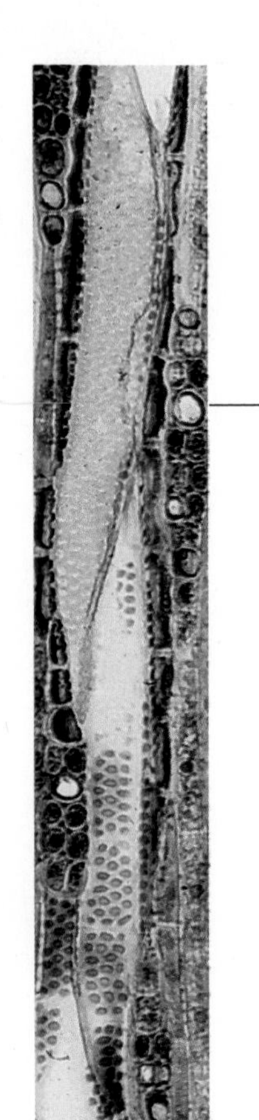

a Tracheids have tapered, unperforated end walls. Perforations in the side walls of adjoining tracheids match up.

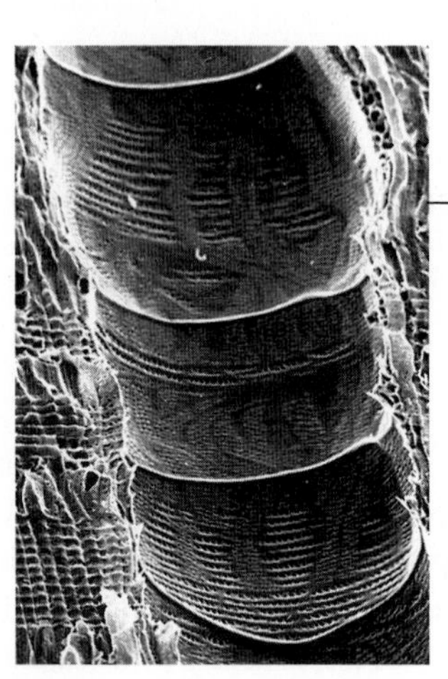

b Three adjoining vessel members. The thick, finely perforated end walls of dead cells connect to make long tubes that conduct water through xylem.

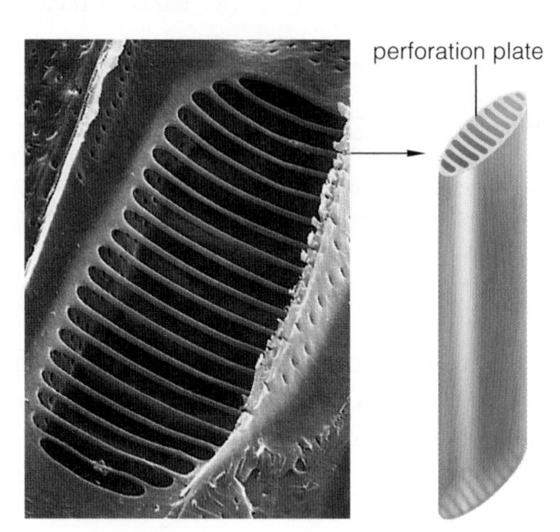

c Perforation plate at the end wall of one type of vessel member. The perforated ends allow water to flow freely through the tube.

Figure 27.7 Tracheids and vessel members from xylem. Interconnected, perforated walls of dead cells form these water-conducting tubes. The pectin-coated perforations may help control water distribution to specific regions. When hydrated, the pectins swell and stop the flow. During dry periods, they shrink, and water moves freely through open perforations toward leaves.

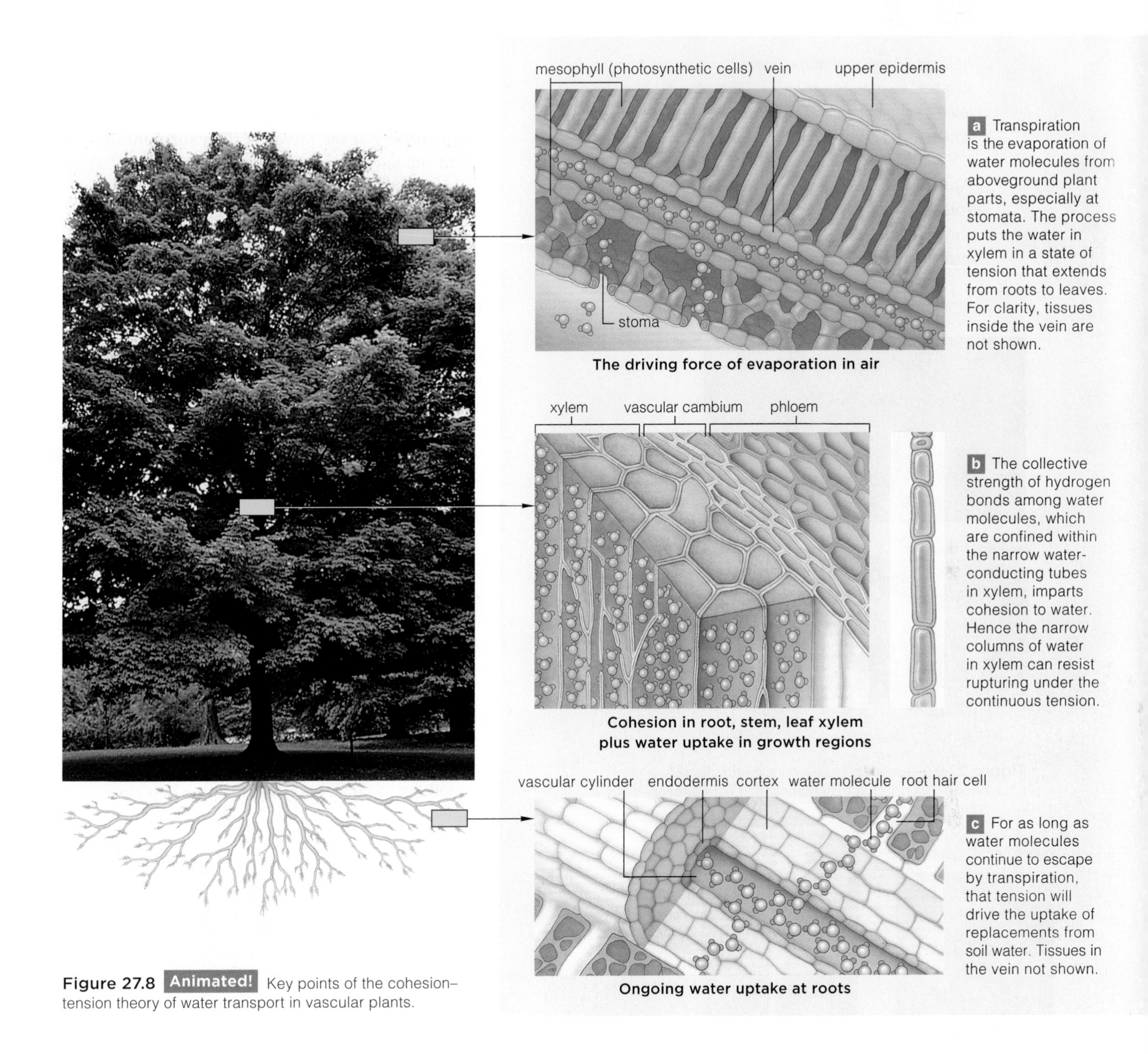

Figure 27.8 Animated! Key points of the cohesion–tension theory of water transport in vascular plants.

Third, for as long as individual molecules of water escape from a plant, the continuous tension inside the xylem permits more molecules to be pulled up from the roots to replace them.

Hydrogen bonds are strong enough to hold water molecules together in the conducting tubes of xylem. However, the bonds are not strong enough to stop the molecules from breaking away from one another and escaping the plant, during transpiration.

Transpiration is the evaporation of water from leaves, stems, and other plant parts.

By a cohesion–tension theory, transpiration puts water in xylem in a state of tension from leaf veins down to roots where water is being absorbed.

Tension pulls columns of water upward through the plant. The collective strength of many hydrogen bonds (cohesion) keeps the water from breaking into droplets as it rises.

27.4 How Do Stems and Leaves Conserve Water?

At least 90 percent of the water transported from roots to a leaf evaporates right out. Only about 2 percent gets used in photosynthesis, growth, membrane functions, and other events, but that amount must be maintained.

LINKS TO SECTIONS 4.11, 5.8, 6.3, 6.7, 21.1, 25.5, 26.4

Water is an essential resource for plants. Think of a young plant cell, with a soft primary wall. As it grows, water diffuses in, fluid pressure increases on the wall and causes it to expand, and the cell grows in volume. Expansion ends when there is enough internal fluid pressure against the wall to balance water uptake. If soil dries out, that balance can end. However, plants are not entirely at the mercy of their surroundings, at least in the short term (Sections 4.11 and 21.1). Their cuticle and stomata help them conserve water.

THE WATER-CONSERVING CUTICLE

Even mildly water-stressed plants would wilt and die without a cuticle. This water-impermeable layer coats all cell walls exposed to air (Figure 27.9). It consists of epidermal cell secretions—a mixture of waxes, pectin, and cellulose fibers embedded in cutin, an insoluble lipid polymer. The cuticle is translucent, so it does not prevent light from reaching photosynthetic tissues. It restricts water loss. At the same time, it restricts the inward diffusion of carbon dioxide for photosynthesis and the outward diffusion of the oxygen by-product.

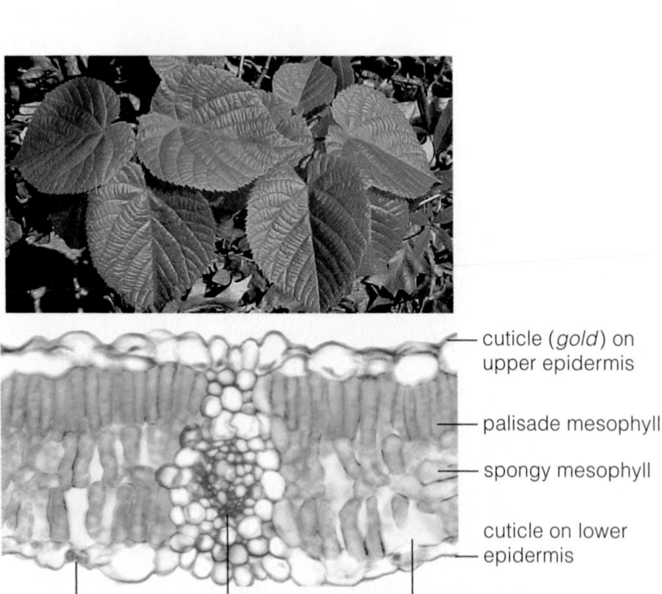

Figure 27.9 Upper and lower cuticle and one stoma of a basswood (*Tilia*) leaf, transverse section. See Section 26.4.

CONTROLLED WATER LOSS AT STOMATA

Photosynthesis in plants requires carbon dioxide and releases oxygen (Section 6.3). Both gases diffuse across the epidermis at stomata. Water also moves out, but if soil holds enough water, roots can replace the losses.

A pair of specialized epidermal cells defines each stoma (Figure 27.10). When these two guard cells swell with water, they bend slightly so a gap forms between them. The gap is the stoma. When the cells lose water, they collapse against each other, so the gap closes.

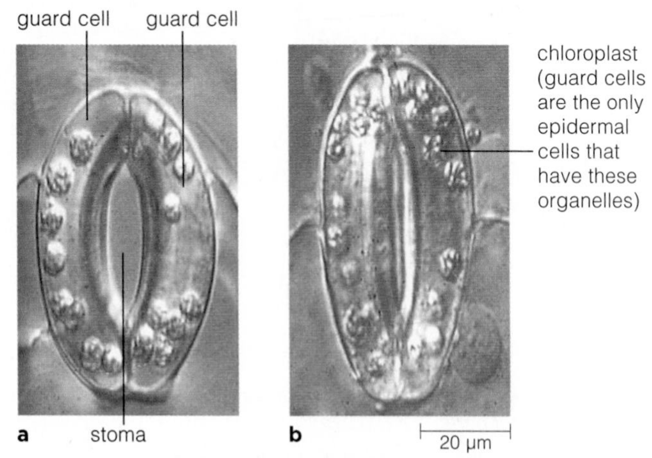

Figure 27.10 Stomata in action. Whether a stoma is open or closed depends on the shape of two guard cells, one on each side of this small gap across a plant's cuticle-covered epidermis. (**a**) This stoma is open. The two guard cells are swollen with water, which caused them to bend so that a gap opened between them. (**b**) This stoma is closed. Water has diffused out of the two guard cells, which caused them to collapse against each other so that the gap between them closed.

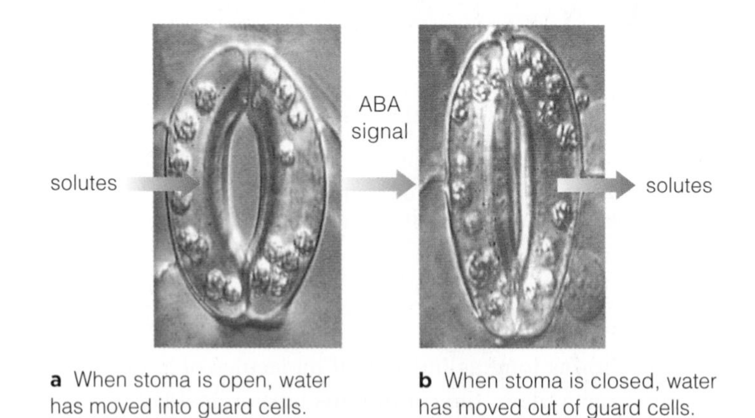

Figure 27.11 Hormonal control of stomatal closure. (**a**) When a stoma is open, the guard cell cytoplasm is maintaining a relatively high concentration of solutes. Water diffusing into the hypertonic cytoplasm is keeping the cells plump. (**b**) When water is scarce, a hormone (ABA) activates a pathway that lowers the concentrations of solutes in guard cell cytoplasm. Water follows its gradient and diffuses out of the cells, and the stoma closes.

Figure 27.12 Stomata at the leaf surface of a holly plant growing in a smoggy, industrialized region. Gritty airborne pollutants clog stomata and prevent sunlight from reaching photosynthetic cells in the leaf.

Environmental cues such as water availability, the level of carbon dioxide inside the leaf, and light, affect whether stomata remain open or closed. For example, photosynthesis starts after the sun comes up. Carbon dioxide levels fall in all photosynthetic cells, including guard cells. This decrease provokes stomata to open.

The stomata of most plants close at night. Water is conserved, and carbon dioxide builds up in leaves as cells make ATP by aerobic respiration.

CAM plants, including most cacti, open stomata at night, when they take in and fix carbon from carbon dioxide. During the day, they close their stomata and use the fixed carbon for photosynthesis (Section 6.7).

During water shortages, abscisic acid (ABA) signals stoma to close. ABA is a hormone (Section 25.5), and guard cells have receptors for it. When ABA binds to these receptors, solutes flow out from the cytoplasm. Water follows its gradient and moves out of the guard cells. The guard cells collapse against each other, and the stoma closes (Figure 27.11).

Plant survival depends on stomatal function. Think about it on a smoggy day (Figure 27.12).

A waxy cuticle covers all epidermal surfaces of the plant exposed to air. It restricts water loss from plant surfaces.

Plants conserve water mainly by closing their many stomata. Closed stomata also prevent gas exchanges necessary for photosynthesis and aerobic respiration.

A stoma stays opens when the guard cells that define it are plump with water. It closes when the cells lose water and collapse against each other.

27.5 The Bad Ozone

FOCUS ON THE ENVIRONMENT

Smog is bad for plants as well as people. Plants close stomata not only to conserve water, but also to keep out the harmful stuff in smog. Closed stomata mean reduced sugar production and reduced growth.

Remember how ozone (O_3) forms a wonderful shield against harmful ultraviolet radiation high above Earth? Ozone also forms near the ground when sunlight acts on emissions from motor vehicles and industries. Especially on hot, windless days, it becomes part of a nasty type of polluted air called smog.

LINK TO SECTION 18.3

A little ozone makes your eyes smart. A lot of it invites shortness of breath and coughing. It makes the lungs and immune system work harder, which is not good for people with asthma and other chronic respiratory problems.

Given that breathing ozone is bad for us, is it hard to imagine what ozone near the ground does to crop plants and ornamental plants? Ozone crosses stomata during gas exchange. It is reactive, a strong oxidant. Figure 27.13 shows a plant exposed to ozone. In plants, the symptoms of ozone exposure include chlorosis and flecked, stippled, bronzed, or reddened leaves. Plants damaged by ozone exposure are weak and stumpy. Grow plants in air filtered through activated charcoal to remove the ozone, and none of these symptoms develop.

In a study at the University of Illinois, botanist Stephen Long looked at the impact of controlled concentrations of ozone on soybean crops. With a 20 percent increase in ozone exposure, the crop yield plummeted by 20 percent.

Here are some estimated annual crop losses due to ozone: western and eastern Europe combined, $7.5 billion; India $5 billion; China $2.5 billion; and the United States, $2–3 billion. By 2050, ground-level ozone is expected to increase by at least 25 percent in China. Like China, India is undergoing fast industrial expansion, and ground-level ozone is expected to rise greatly there also.

As Daniel Gallie at the University of California, Riverside, found out, plants use vitamin C to detoxify ozone that enters leaves. Plants with higher levels of vitamin C fared better than plants with lower levels, regardless of whether their stomata were open or closed. Finding a way to raise the vitamin C level may help plants tolerate ozone, which would translate into protection and better crop yields.

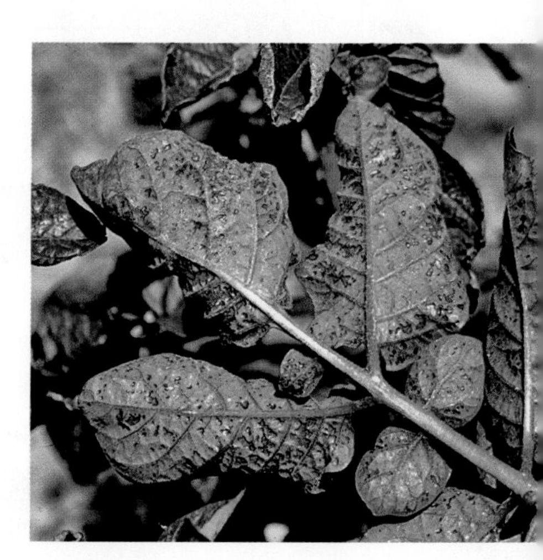

Figure 27.13 Leaves of an ozone-stressed potato plant.

27.6 How Do Organic Compounds Move Through Plants?

Xylem distributes water and minerals through plants. The vascular tissue called phloem distributes organic products of photosynthesis.

CONDUCTING TUBES IN PHLOEM

LINKS TO SECTIONS 3.2, 5.6, 5.8, 6.6, 26.2

Phloem is a vascular tissue having organized arrays of conducting tubes, fibers, and strands of parenchyma cells. Unlike xylem, it has sieve tubes through which organic compounds rapidly flow. Living cells form the long tubes, the cells of which are positioned side by side and end to end. Their abutting end walls, called sieve plates, are porous (Figure 27.14*a,b*). Companion cells are pressed against the tubes. These cells help load organic compounds into neighboring sieve tubes by active transport mechanisms.

Some organic products of photosynthesis are used in leaf cells that make them. The rest move to roots, stems, buds, flowers, and fruits (Section 6.6). Starch is the main carbohydrate storage form. Starch molecules are too big for transport across the plasma membrane of cells and too insoluble for transport. Cells convert them to sucrose, which is more easily transportable.

Experiments with insects show that sucrose is the main form of carbohydrate being transported inside phloem. Aphids feeding on the juices in conducting tubes of phloem were anesthetized with high levels of carbon dioxide (Figure 27.15). Then their bodies were detached from the mouthparts, which were attached to the plant. Researchers collected and analyzed fluid exuded from the mouthparts. For most of the plants studied, sucrose was the most abundant carbohydrate in the fluid being forced out of the tubes.

TRANSLOCATION

Translocation is the formal name for the process that moves sucrose and other organic compounds through phloem of vascular plants. High fluid pressure drives the movement (Section 5.8). The pressure in phloem's conducting tubes can be five times higher than the air pressure inside an automobile tire.

Phloem translocates photosynthetic products along declining pressure and solute concentration gradients. The *source* of the flow is any region of the plant where organic compounds are being loaded into sieve tubes.

a

one of a series of living cells that abut, end to end, and form a sieve tube

companion cell (in the background, pressed tightly against sieve tube)

perforated end plate of sieve-tube cell, of the sort shown in (**b**)

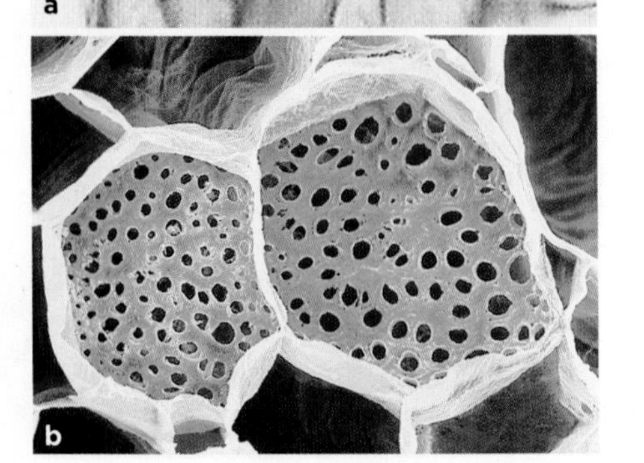

Figure 27.14 (**a**) Part of a sieve tube inside phloem. Arrows point to perforated ends of individual tube members. (**b**) Scanning electron micrograph of the sieve plates on the ends of two side-by-side sieve-tube members.

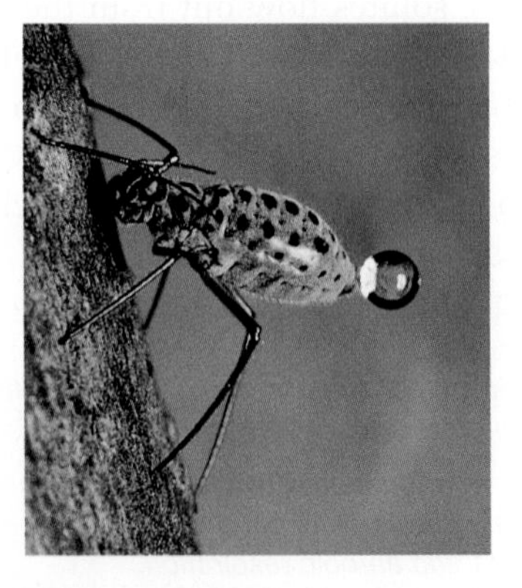

Figure 27.15 Honeydew exuding from an aphid after this insect's mouthparts penetrated a sieve tube. High pressure in phloem forced this droplet of sugary fluid out through the terminal opening of the aphid gut.

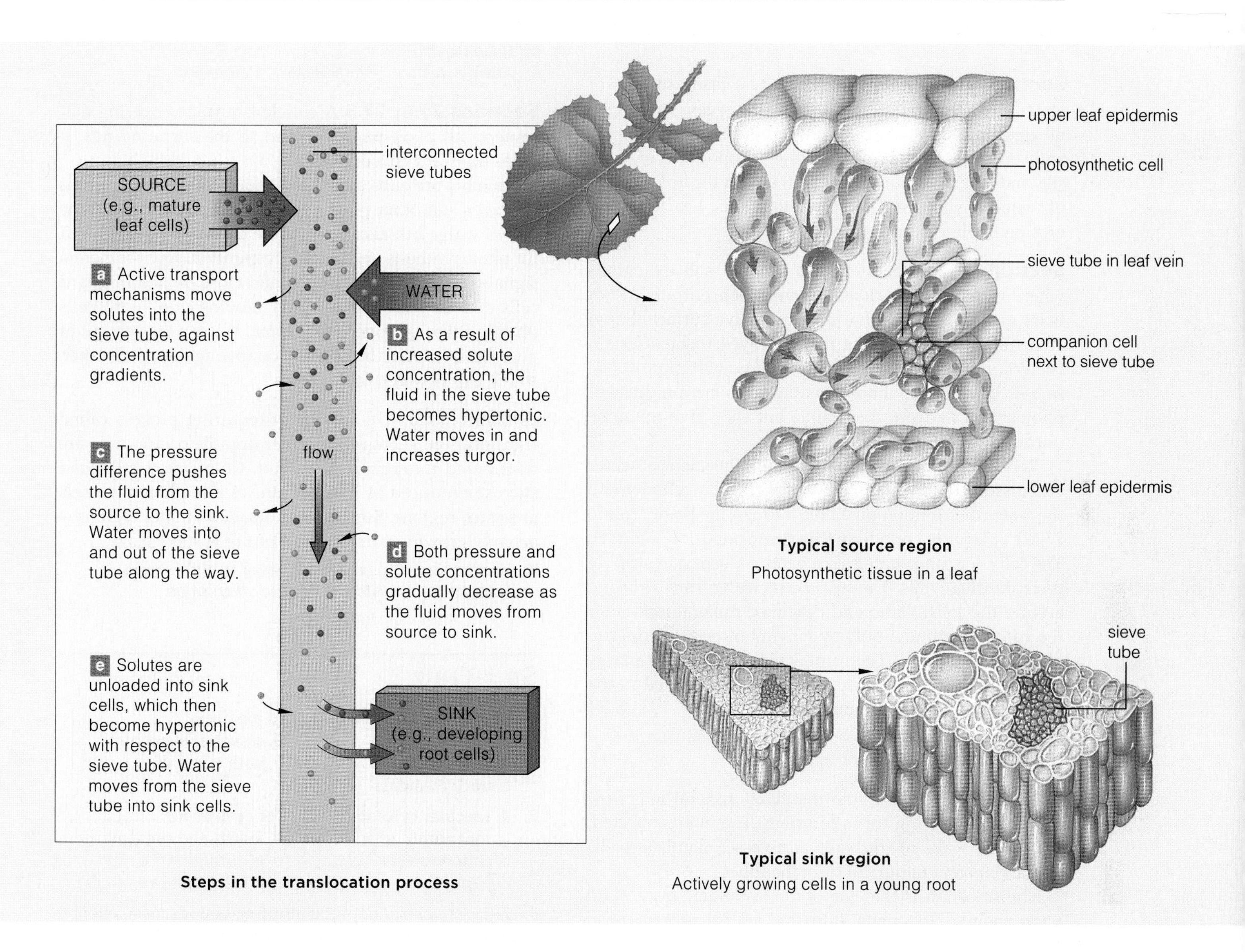

Figure 27.16 Animated! Translocation of organic compounds. Review Section 6.7 to get an idea of how translocation relates to photosynthesis in vascular plants.

Common sources are mesophylls—the photosynthetic tissues in leaves. The flow ends at a *sink*, which is any plant region where products are being used or stored. For instance, while flowers and fruits are forming on the plant, they are sink regions.

Why do organic compounds flow from a source to a sink? According to the pressure flow theory, internal pressure builds up at the source end of the sieve tube system and *pushes* the solute-rich solution on toward any sink, where solutes are being removed.

Use Figure 27.16 to track what happens to sucrose as it moves from the photosynthetic cells into small leaf veins. By energy-requiring reactions, companion cells in veins load sucrose into sieve-tube members. When the sucrose concentration increases in the tubes, water also moves into them, by osmosis. Rising fluid volume exerts more pressure on the wall of the sieve tubes. With sufficient turgor, the sucrose-laden fluid in the tubes is forced out of the leaf, into the stem, and on toward the sink.

Plants store carbohydrates as starch but distribute them in the form of sucrose and other small, water-soluble units.

Translocation is the distribution of organic compounds to different plant regions. It depends on concentration and pressure gradients in the sieve-tube system of phloem.

Gradients last as long as companion cells load compounds into sieve tubes at sources, such as mature leaves, and as long as compounds are unloaded at sinks, such as roots.

Summary

www.thomsonedu.com ThomsonNOW!

Section 27.1 Plant growth requires steady sources of water and nutrients, such as carbon, oxygen, hydrogen, nitrogen, and phosphorus from carbon dioxide in the air and soil water. A soil's composition (proportions of sand, silt, and clay; and humus content) affects the accessibility of water, oxygen, and nutrients to plants. Leaching and erosion deplete nutrients in soil.

Section 27.2 Roots expand through soil to regions where water and nutrients are most concentrated. Root hairs greatly increase the root absorptive surface. Fungi are symbionts with young roots in mycorrhizae. Certain bacteria in root nodules are symbionts with plant roots. In both cases, the symbionts withdraw some products of photosynthesis from the plants, but they give up some nutrients in return.

Roots exert some control over the movement of water and dissolved mineral ions into the vascular cylinder and the water distribution pipelines through the plant (Figure 27.17). A layer of endodermal cells surrounds the cylinder. The cells deposit a waterproof band, a Casparian strip, in their abutting walls. The strip keeps water from diffusing around the cells. Water and dissolved mineral ions enter the vascular cylinder only by moving through cytoplasm of endodermal cells. Their uptake is controlled to a large extent by the active transport proteins embedded in the endodermal cell membranes.

- *Use the animation on ThomsonNOW to see how vascular plant roots control nutrient uptake.*

Section 27.3 Water and dissolved mineral ions flow through conducting tubes of xylem. The interconnected, perforated walls of tracheids and vessel members (cells that are dead at maturity) form the tubes.

Transpiration is the evaporation of water from plant parts, mainly at stomata, into air. By a cohesion–tension theory, transpiration pulls water upward through xylem by causing continuous negative pressure (tension) from leaves to roots. Collectively, the hydrogen bonds among water molecules resist rupturing; they impart cohesion, so water is pulled upward as continuous fluid columns. The bonds break and water molecules diffuse into the air during transpiration.

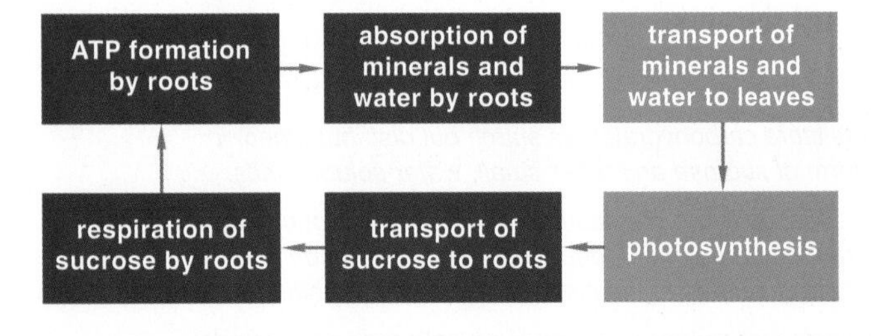

Figure 27.17 Summary of interdependent processes that sustain plant growth. All living cells in plants require at least sixteen nutrients. They all produce ATP, which drives metabolic activities.

- *Use the animation on ThomsonNOW to learn about water transport in vascular plants.*
- *Read the InfoTrac article "How Plants Get High," Adam Summers,* Natural History, *March 2005.*

Sections 27.4, 27.5 A cuticle is a waxy covering that protects all plant parts exposed to the surroundings. It helps the plant conserve water.

Stomata are gaps across the cuticle-covered epidermis of leaves and other plant parts. Closed stomata limit the loss of water, but also prevent the gas exchange required for photosynthesis and aerobic respiration. Environmental signals cause stomata to open and close. A pair of guard cells defines each stoma. Water moving into guard cells plumps them and opens the stoma. Water diffusing out of guard cells causes the cells to collapse against each other, so the gap between them closes.

Section 27.6 By an energy-requiring process called translocation, sucrose and other organic compounds are distributed throughout the plant. Companion cells load sucrose produced by photosynthesis into phloem vessels at source regions. Sugars are unloaded at sink regions—actively growing parts of the plant or storage parts.

- *Use the animation on ThomsonNOW to observe how vascular plants distribute organic compounds.*

Self-Quiz

Answers in Appendix III

1. Carbon, hydrogen, and oxygen are plant ______.
   a. macronutrients
   b. micronutrients
   c. trace elements
   d. essential elements
   e. both a and d

2. A vascular cylinder consists of cells of the ______.
   a. root cortex
   b. endodermis
   c. pericycle
   d. xylem and phloem
   e. b through d
   f. all of the above

3. A ______ strip between abutting endodermal cell walls forces water and solutes to move through these cells rather than around them.
   a. cutin b. Casparian c. lignin d. cellulose

4. The nutrition of some plants depends on a root–fungus association known as a ______.
   a. root nodule
   b. mycorrhiza
   c. root hair
   d. root hypha

5. Water evaporation from plant parts is called ______.
   a. translocation
   b. expiration
   c. transpiration
   d. tension

6. Water transport from roots to leaves occurs by ______.
   a. pressure flow
   b. differences in source and sink solute concentrations
   c. the pumping force of xylem vessels
   d. cohesion–tension among water molecules

7. Stomata open in response to light and ______.
   a. ion influx into guard cells
   b. ion influx in air spaces inside leaf

8. During water shortages, ______ acts on guard cells and initiates closure of stomata.
   a. air temperature
   b. humidity
   c. abscisic acid
   d. oxygen

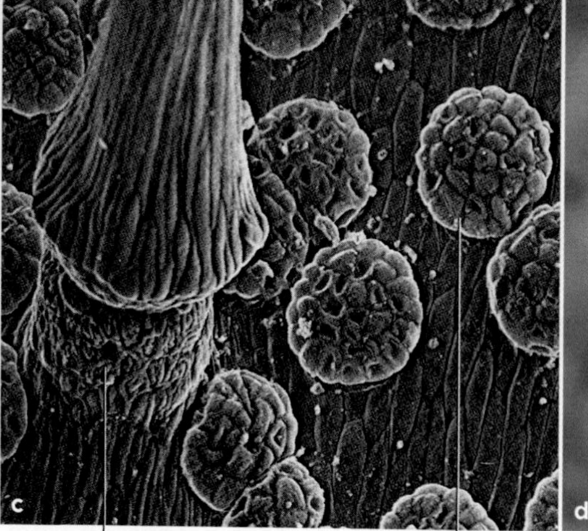

Figure 27.18 (**a**) Venus flytrap (*Dionaea muscipula*), one of the carnivorous plants. (**b**) A fly stuck in sugary goo on a lobed leaf. (**c**) It brushes against hairlike triggers that activate the leaf, which snaps shut in half a second. (**d**) While a trap is open, mesophyll cells below the epidermis are compressed. Spring the trap and turgor abruptly decompresses the cells. Whoosh!

9. Match the concepts of plant nutrition and transport.

___ stomata	a. evaporation from plant parts
___ nutrient	b. harvesting soil nutrients
___ sink	c. balance water loss with carbon dioxide requirements
___ root system	d. cohesion in water transport
___ hydrogen bonds	e. sugars unloaded from sieve tubes
___ transpiration	f. organic compounds distributed through the plant body
___ translocation	g. element with roles in metabolism that no other element can fulfill

■ *Visit ThomsonNOW for additional questions.*

Critical Thinking

1. Successful home gardeners, like farmers, make sure that their plants get enough nitrogen from either nitrogen-fixing bacteria or fertilizer. Which biological molecules incorporate nitrogen? Nitrogen deficiency stunts plant growth; leaves yellow and then die. How would nitrogen deficiency cause these symptoms?

2. You just returned home from a three-day vacation. Your severely wilted plants tell you they weren't watered before you left. Being aware of the cohesion–tension theory of water transport, explain what happened to them.

3. When moving a plant from one location to another, it helps to include some native soil around the roots. Explain why, in terms of mycorrhizae and root hairs.

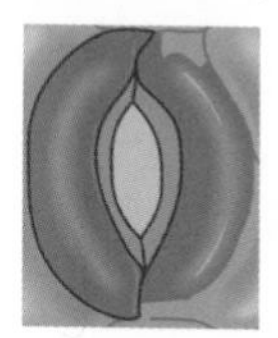

4. In the sketch at *left*, label the stoma. Now think about Cody, who discovered a way to keep all of a plant's stomata open at all times. He also figured out how to keep those of another plant closed all the time. Both plants died. Explain why.

5. Allen is studying the rate of transpiration from tomato plant leaves. He notices that several environmental factors, including wind and relative humidity, affect the rate. Explain how they might do so.

6. The Venus flytrap (*Dionaea muscipula*) is a flowering plant native to bogs of North and South Carolina. Its two-lobed, spine-fringed leaves open and close like a steel trap (Figure 27.18*a–d*). Like all other plants, it cannot grow well without nitrogen and other nutrients, which are scarce in bogs. Plenty of insects fly in from places around the bogs.

Epidermal glands on leaf surfaces secrete sticky sugars that attract insects. As insects land, they brush against hairlike structures—the triggers for the trap. When an insect touches two hairs at the same time or the same hair twice in rapid succession, the two lobes of the leaf snap shut. Digestive juices pour out from certain leaf cells, pool around the insect, and dissolve the prey. This plant makes its own nutrient-rich water, which it proceeds to absorb!

The Venus flytrap is only one of several species of "carnivorous" plants. Their mode of nutrient acquisition is a form of extracellular digestion and absorption. Not all carnivorous plants actively spring traps. Some have fluid-filled traps into which prey slip, slide, or fall and then simply drown. Given the variety and numbers of insects and other animals that attack plants, you can just imagine how endearing the carnivorous plants are to botanists. With their plucky modes of nutrition, these plants also can give you glimpses into the diversity of adaptations by which plants function in their environment.

All carnivorous plants evolved in habitats where nitrogen and other nutrients are hard to come by. They take hold even in shallow freshwater lakes and streams, which have only dilute concentrations of dissolved minerals. Insects and other small invertebrates are fair game, and so is the occasional tiny amphibian.

As a class activity, divide into research groups, each focusing on one genera of the carnivorous plants listed below. Gather data on the number of known species, their distribution, abiotic and biotic conditions in their habitats, the type and numbers of prey captured in a given interval, and mechanisms by which they capture prey. Note whether one or more species in the genera are threatened or endangered. If so, note possible causes. Present an oral or written report of your findings.

Byblis Rainbow plant
Cephalotus Western Australian pitcher plant
Darlingtonia Cobra lily
Dionaea Venus flytrap
Drosera Sundews
Drosophyllum Dewy pine
Nepenthes Monkey cup (Tropical pitcher plants)
Pinguicula Butterworts
Sarracenia North American pitcher plants
Utricularia Bladderworts

28 PLANT REPRODUCTION AND DEVELOPMENT

IMPACTS, ISSUES Imperiled Sexual Partners

Imagine a world with no chocolates, no chocolate ice cream or candy, no cocoa, no brownies. Too far-fetched? Chocolate is processed from seeds of *Theobroma cacao* (Figure 28.1), a tree that grows in shady tropical rain forests. It also grows in sun-drenched plantations carved out of rain forest habitats. There is just one problem: Trees in sun-drenched plantations do not form many seeds.

As plantation owners found out, *T. cacao* does not get pollinated without midges, a flying insect smaller than a pinhead. Midges cart its pollen from flower to flower. They live and breed only in the damp, deep shade of tropical rain forests. No forests, no midges. No midges, no chocolate.

Growers have now started planting cacao trees in rain forests. Besides keeping trees shaded and awash in midges, they are saving otherwise threatened habitats for endangered birds, orchids, and other species.

The pollinator-plant connection is commanding more attention from growers all over the world. About three-fourths of all crop plants set more fruit and complete their life cycle when pollinators—most notably bees, butterflies, flies, moths, and beetles—help them out.

Honeybees (*Apis mellifera*) pollinate most crop plants in the United States. Beekeepers pack up and transport artificial hives with thousands of bees from field to field. The problem is that pesticides, diseases, parasitic mites, and other threats are wiping out the honeybees.

Other bee species can pollinate some crop plants. Squash bees pollinate squash and pumpkins. The alfalfa leaf-cutting bee pollinates alfalfa plants. The southeastern blueberry bee helps blueberry plants. But blueberry bees are dwindling because of pesticide applications, habitat losses, and (whoops) accidentally imported ants that eat blueberry bee larvae.

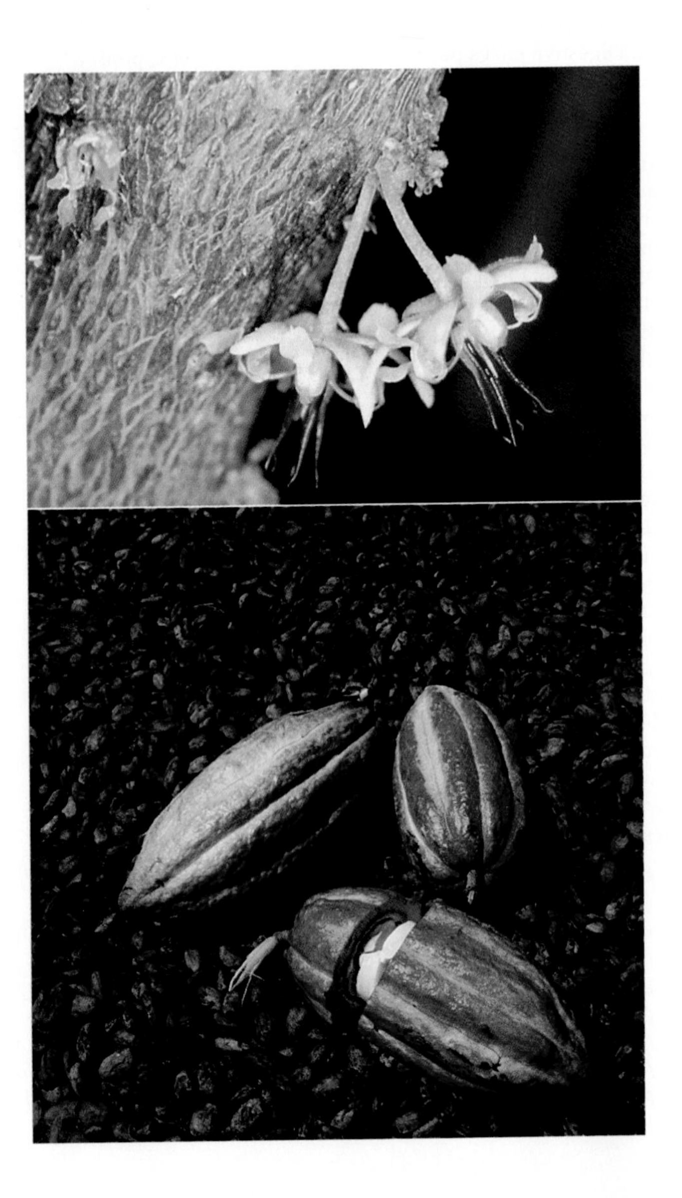

See the video! **Figure 28.1** Flowers, fruits, and seeds of *Theobroma cacao*, a plant pollinated exclusively by midges. Each fruit holds as many as forty seeds—the "cocoa beans."

We process the seeds into cocoa butter and essences of chocolate, which are mildly addictive. The average American, including the chocolate ice-cream lovers shown on the facing page, buys 8 to 10 pounds of chocolate annually. The average Swiss citizen craves a whopping 22 pounds. With that kind of demand, *T. cacao* trees are in luck; many humans are intensely interested in promoting their survival and reproduction.

How would you vote? Microencapsulated pesticides are easy to apply and effective for long periods. They also are about the size of pollen grains and are tempting but toxic to certain pollinators. Should we restrict their use? See ThomsonNOW for details, then vote online.

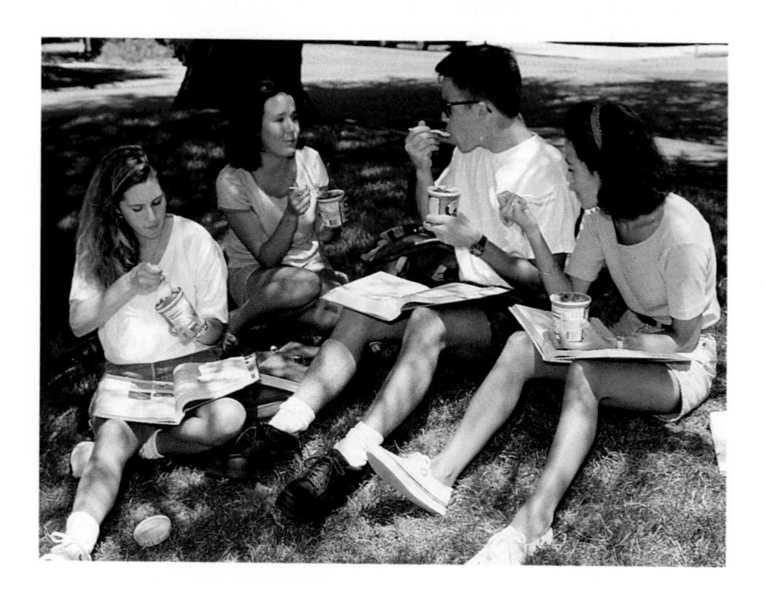

Many hundreds of flowering plant species are listed as endangered or threatened, so many pollinators that coevolved with them are at risk, too. They may disappear along with the plants or the habitats. Migrating from one place to another won't help. Monarch butterflies, rufous hummingbirds, and lesser long-nosed bats are among the pollinators that go back and forth between breeding grounds in the United States and wintering grounds in Mexico. When flowering plants that sustain them dwindle in either habitat or anywhere along the migratory route, so do their populations.

This chapter puts you squarely in the middle of what it takes for flowering plants to make more of themselves and why you might want to care about it. You already have an overview of their tissues and body plans. You know how a plant secures mineral-laden water, makes food, and distributes both through its tissues. Step now into the life cycles, starting with reproductive modes. Along the way, you will view controls over how a new plant develops. You will come across environmental cues that turn specific controls on or off at different times, in different seasons.

Key Concepts

STRUCTURE AND FUNCTION OF FLOWERS

Sexual reproduction is the dominant reproductive mode of flowering plant life cycles, which typically depend on pollinators. Spores and gametes form in the specialized reproductive shoots called flowers. **Section 28.1**

LIFE CYCLE OF A EUDICOT

Sperm-bearing male gametophytes develop within pollen grains and female gametophytes that bear eggs inside ovules form in reproductive parts of flowers. After fertilization, ovules mature into seeds, each an embryo sporophyte and tissues that nourish and protect it.

As seeds develop, tissues of the ovary and often other floral parts mature into fruits, which function in seed dispersal. Air currents and animals are the main dispersal agents. **Sections 28.2–28.4**

ASEXUAL REPRODUCTIVE MODES

Many species of flowering plants also reproduce asexually by vegetative growth and other mechanisms. **Section 28.5**

CONTROLS OVER GROWTH AND DEVELOPMENT

Interactions among hormones and other signaling molecules control plant growth and development. Hormone synthesis starts at seed germination and guides all events of the life cycle, such as root and shoot development, flowering, fruit formation, and dormancy. **Sections 28.6, 28.7**

RESPONSES TO ENVIRONMENTAL CUES

Plant cell receptors that govern hormone synthesis respond to environmental cues, such as gravity, sunlight, and seasonal shifts in night length and temperatures. **Sections 28.8–28.10**

Links to Earlier Concepts

This chapter builds on reproductive adaptations that contributed to the evolutionary success of flowering plants (Sections 21.1, 21.5, 21.7). You will dig deeper into plant life cycles (9.5, 21.8). You may wish to review where meristems are located (26.1) and how primary roots and shoots are organized (26.3, 26.5). You will see more outcomes of signaling pathways (25.5) and controls over development, including flower formation (14.2, 16.8). You will draw upon your understanding of wavelengths of light (6.1), pH (2.6), cell walls (4.11), active transport (5.7), turgor (5.8), stomata (27.4), and transport through phloem (27.6).

28.1 Reproductive Structures of Flowering Plants

Flowering plants coevolved in richly varied ways with pollination vectors that assist in sexual reproduction.

LINKS TO SECTIONS 6.1, 9.5, 14.2, 21.1, 21.5, 21.7, 26.3

Sporophytes, recall, are spore-producing plant bodies that grow by mitotic cell divisions from a fertilized egg (Sections 9.5 and 21.1). Flowers are reproductive shoots of flowering plant sporophytes. Sexual spores that form inside them give rise to haploid gametophytes: structures in which male or female haploid gametes form. Male and female gametes meet at fertilization, and the resulting diploid zygote gives rise to a new sporophyte (Figure 28.2*a*).

FLORAL STRUCTURE AND FUNCTION

Some of the lateral buds along the stem of a sporophyte differentiate as floral shoots. Four whorls of modified leaves form on the shoot tip, the receptacle. The outer whorl becomes a ring of sepals—a calyx. A whorl just inside it becomes a ring of nonfertile petals called the flower's corolla (Figure 28.2*b*). Two innermost whorls form the fertile parts called stamens and carpels.

Stamens are male reproductive structures of floral shoots. They usually consist of an anther at the tip of a thin filament (Figure 28.2*b*). Inside a typical anther are two to six pairs of pouches called pollen sacs. In each sac, meiosis of diploid cells gives rise to haploid, walled spores. The spores then differentiate into pollen grains, structures that contain the male gametophytes. A pollen grain is a suitcase for the journey to an egg.

Carpels are the female reproductive parts of floral shoots. Their upper region, a sticky or hairlike stigma, traps pollen grains. Often a slender stalk called a style elevates the stigma (Figure 28.2*b*). Many flowers have one carpel; others have two or more that are fully or partly fused (Figure 28.2*c,d*). A carpel's lower region is an ovary, a chamber in which one or more ovules form and eggs are fertilized. An ovule is a structure in which a haploid, egg-producing female gametophyte forms. After fertilization, the ovules mature into seeds; their wall becomes the seed coat. Thus, each seed is a mature ovule (Section 21.5).

REVISITING THE POLLINATORS

Pollination vectors are any agents that deliver pollen grains to structures that house female gametophytes. Winds are a vector for a species that grows densely in

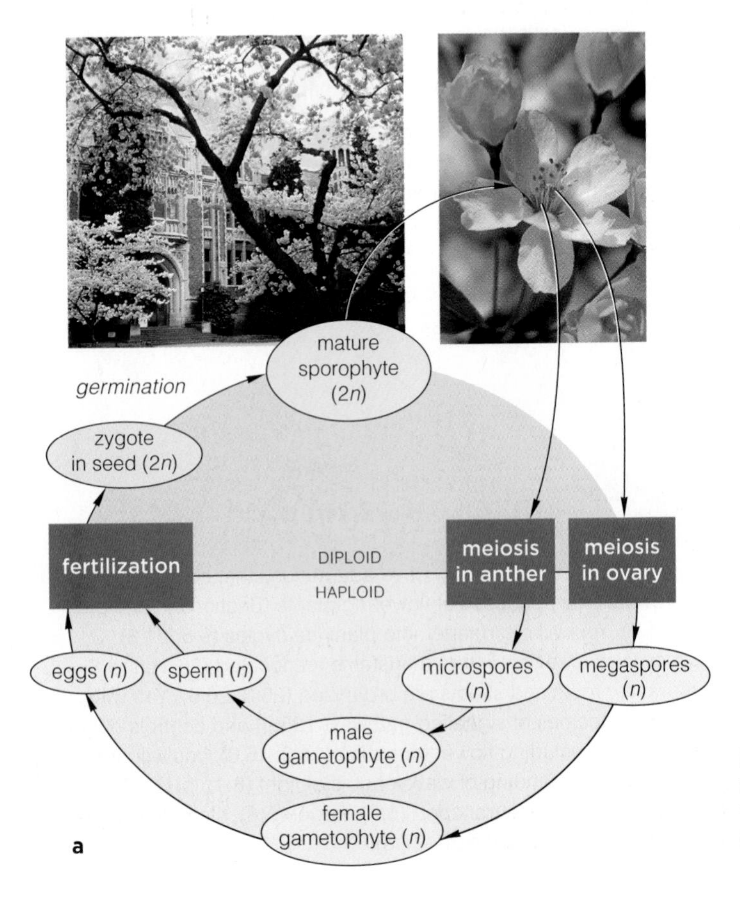

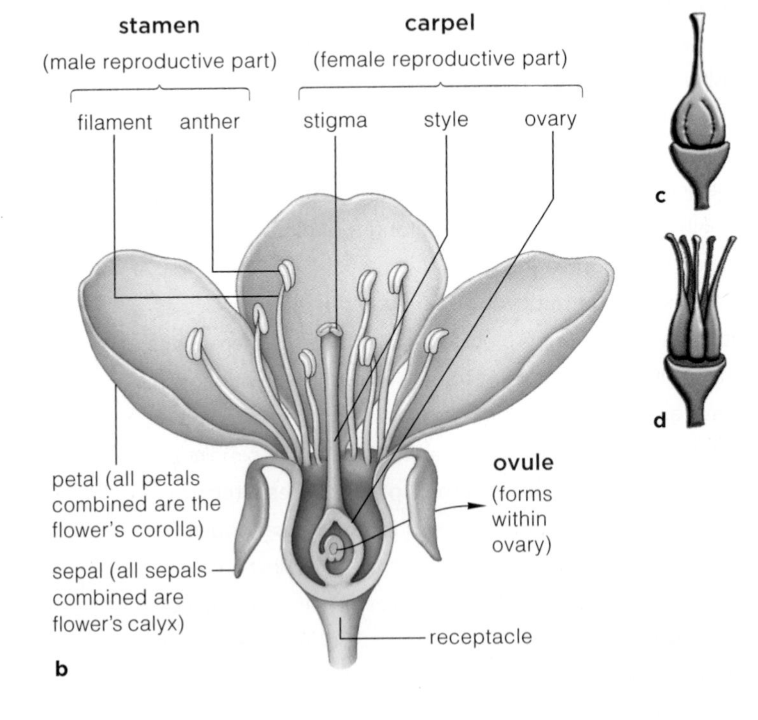

Figure 28.2 **Animated!** (**a**) Typical flowering plant life cycle. A cherry tree is the example. (**b**) Structure of a cherry blossom (*Prunus*). Like many flowers, it has a single carpel and stamens. (**c**,**d**) Flowers of other species have two or more fully or partially fused carpels, as in these examples.

grasslands; they easily shake clouds of pollen grains (Figure 28.3). However, dozens or hundreds of species often grow together, where wind can deposit pollen on the wrong flowers. Thus, many flowering plant species are pollinated only by specific insects, birds, or other animals. About 90 percent of the 295,000 named species of flowering plants coevolved with specific pollinators (Section 21.7). A plant's flower structure, color, pattern, fragrance, and nectar attracts its coevolved pollinator.

Fragrant, colorful flowers with sturdy petals attract day-flying butterflies and moths. Some pollinators have an excellent sense of smell, and can track concentration gradients of volatile chemicals that some flowers give off. Night-flying bats and moths pollinate flowers with intensely sweet odors and white or pale colored petals that are easy to see in the dark (Figure 28.4*a*). Odors like dung or rotting fruit beckon some beetles and flies.

Bee-pollinated flowers have yellow, blue, or purple petals, typically with pigments that reflect ultraviolet light. These pigments are distributed in patterns that bees recognize as visual guides to nectar (Figure 28.4*b*). We see these floral patterns only with special camera filters; our eyes do not have receptors that respond to ultraviolet light.

Nectar and pollen are the rewards for revisiting a particular species. Nectar, a sucrose-rich secretion, is dilute enough for pollinators to sip easily. It is the only food for most butterflies, as well as the fuel of choice for hummingbirds. In beehives, nectar is converted to honey, which helps feed the bees through the winter. Pollen is an even richer food, with more vitamins and mineral ions than nectar. Stamens are adapted to brush or lob pollen onto a pollinator's body.

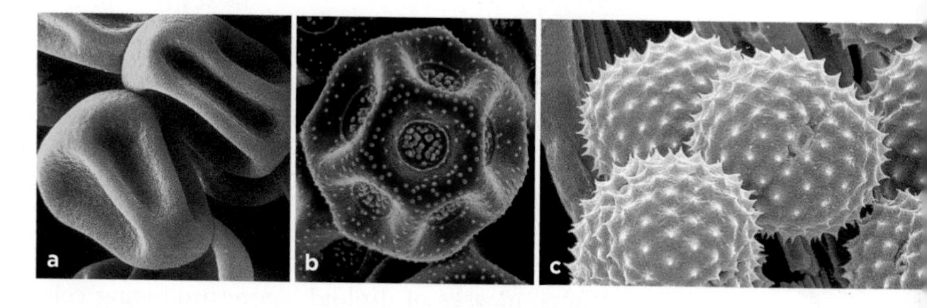

Figure 28.3 Pollen grains of (**a**) grass, (**b**) chickweed, and (**c**) ragweed plants. Pollen grains of most plants differ in size, sculpturing, and number of furrows or pores.

Floral sizes and shapes do not promote visits from animals that are not efficient at delivering that plant's pollen. They fit best with coevolved partners. Nectar-filled floral tubes or spurs are the same length as their coevolved pollinator's feeding device (Figure 17.27). "Flowers" such as sunflowers actually are composites of many flowers, each with a nectar tube of no interest to, say, finches or bats. Flowers with tall, thin stems and flimsy landing platforms exclude heavy beetles.

Sporophytes alternate with gametophytes in flowering plant life cycles. Flowers form on sporophytes. In floral stamens, male gametophytes develop in pollen grains. Inside the ovaries of carpels, female gametophytes with egg cells form as part of ovules.

An ovule is a structure in which a haploid, egg-producing female gametophyte forms. Ovules mature into seeds.

The vast majority of flowering plants coevolved with animal pollinators. Their flowers attract specific pollinators and reward them with pollen and nectar.

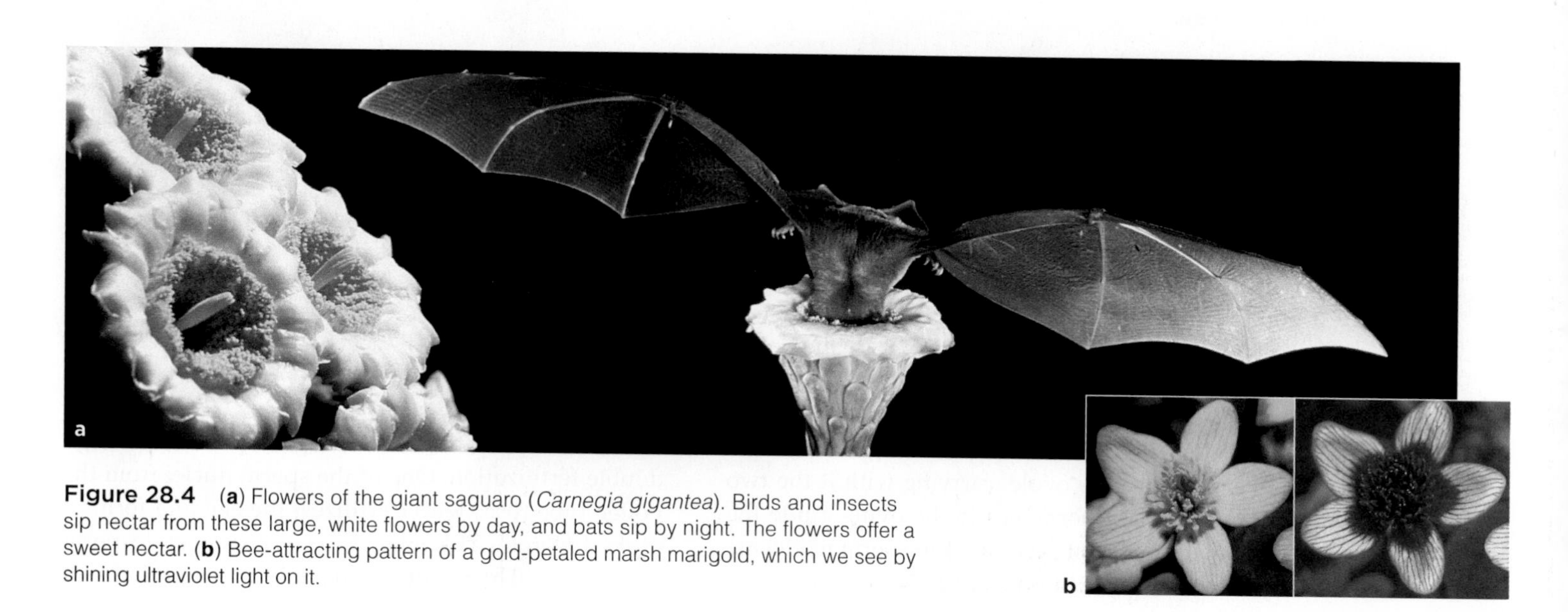

Figure 28.4 (**a**) Flowers of the giant saguaro (*Carnegia gigantea*). Birds and insects sip nectar from these large, white flowers by day, and bats sip by night. The flowers offer a sweet nectar. (**b**) Bee-attracting pattern of a gold-petaled marsh marigold, which we see by shining ultraviolet light on it.

28.2 A New Generation Begins

In flowering plants, fertilization has two outcomes: It results in a zygote, and it is the start of endosperm—a nutritious tissue that nourishes the embryo sporophyte.

MICROSPORE AND MEGASPORE FORMATION

LINK TO SECTION 21.8

Figure 28.5 zooms in on a flowering plant life cycle. In the anthers, masses of diploid, spore-producing cells form by mitosis. There are typically four masses; walls that develop around each mass form four pollen sacs. Each spore-producing cell inside the sacs undergoes meiosis and forms four haploid microspores. The cell wall of a microspore has species-specific recognition proteins on it. The proteins will bind only to receptors for them on a stigma of the same plant species.

Mitosis and differentiation of microspores produce pollen grains, which contain the male gametophytes. After a period of dormancy, the pollen sacs split open, and the pollen grains are released from the anther to be taken up by pollinators.

Meanwhile, in the ovary of a carpel, different kinds of spores are forming inside ovules (Figure 28.5*e*). An ovule starts as a cell mass on the ovary's inner wall. One of the cells gives rise to four haploid spores by meiosis and cytoplasmic division. Spores that form in ovules are called megaspores. While they are forming, protective cell layers, or integuments, develop around them. Three of the megaspores typically disintegrate. The remaining cell undergoes three rounds of mitosis but not cytoplasmic division. The outcome is one cell with eight haploid nuclei. The cytoplasm then divides and forms a seven-celled embryo sac that is the mature female gametophyte. One cell, the endosperm mother cell, has two nuclei ($n + n$). Another cell in the embryo sac is the egg.

POLLINATION AND FERTILIZATION

Back at the anthers, pollen grains have been released. Pollination refers to the arrival of pollen grains on a receptive stigma. In response to the stigma's chemical cues, the pollen grain germinates; it resumes metabolic activity. A pollen tube, which is a tubular outgrowth of a pollen grain, develops from the gametophyte. Two sperm nuclei, the male gametes, form inside the pollen tube. Taken together, a pollen tube and its contents of male gametes are a mature male gametophyte.

The pollen tube grows down through the carpel and ovary toward the ovule, carrying with it the two sperm nuclei. Plant hormones in the ovule guide the tube's growth to the embryo sac. The sperm nuclei are then released into the sac (Figure 28.5*i*).

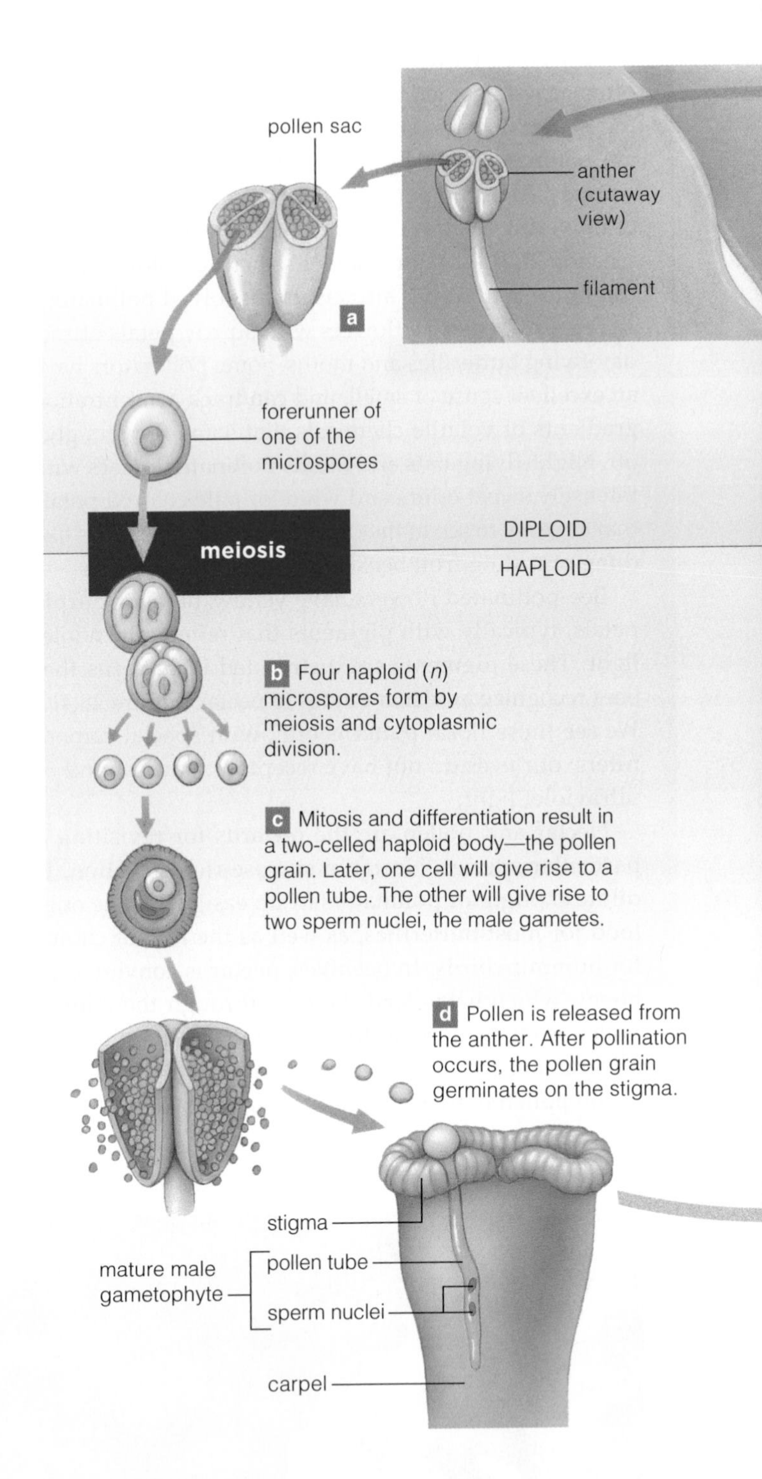

Flowering plants (and a few gnetophytes) undergo double fertilization. One of the sperm nuclei from the pollen tube fuses with (fertilizes) the egg and forms a diploid zygote. The other fuses with the endosperm mother cell. The result is a triploid ($3n$) cell that will

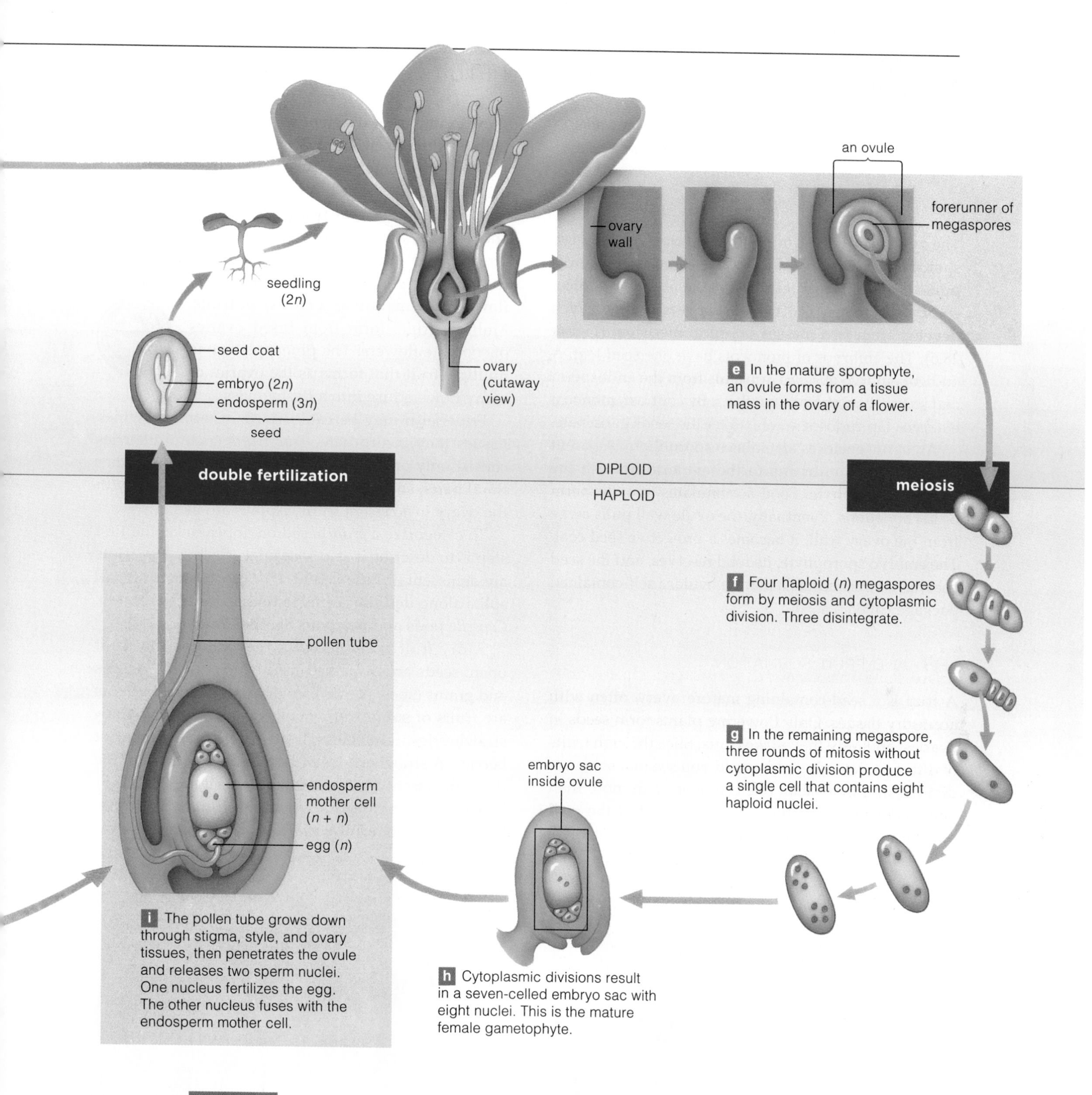

Figure 28.5 **Animated!** Life cycle of a cherry (*Prunus*) tree, a eudicot. Compare Figure 21.19.

give rise to endosperm, a nutritious tissue that forms only in seeds of flowering plants. Right after a seed germinates, endosperm will sustain the rapid growth of the sporophyte seedling until true leaves form and photosynthesis gets under way.

In flowering plants, male gametophytes form inside pollen grains. After pollination, a pollen tube containing the male gametes grows into carpel tissues. One nucleus fuses with the egg, the other with a cell that gives rise to endosperm.

28.3 From Zygotes to Seeds and Fruits

After zygotes form inside the ovules of flowering plants, tissues around them expand and form seeds and fruits.

THE EMBRYO SPOROPHYTE

The zygotes of flowering plants develop into embryo sporophytes in different ways. We can use shepherd's purse (*Capsella*) as an example. Like all other eudicot embryos, it has two cotyledons, or seed leaves, which develop from two masses of apical meristem (Figure 28.6). The embryos of monocots have one seed leaf. A eudicot embryo absorbs nutrients from the endosperm and stores them in its cotyledons. By contrast, monocot embryos tap endosperm only after the seeds germinate.

An ovule encloses an embryo sporophyte. A parent plant transfers nutrients to the ovule to nourish the embryo sporophyte. Food accumulates in endosperm or in cotyledons. Eventually, the ovule wall pulls away from the ovary wall; it becomes a protective seed coat. The embryo sporophyte, its food reserves, and the seed coat have now become a mature ovule, a self-contained package called a seed.

SEED AND FRUIT FORMATION

A fruit is a seed-containing mature ovary, often with accessory tissues. Only flowering plants form seeds in ovaries, and only they make fruits. Slice through fruits of different flowering plants and you see that some are divided into more than one chamber, with more than one seed attached to the ovary wall. Part of the wall may be a fleshy mass of tissue in the ovary's center. We show examples of different seed positions with the three sketches at right.

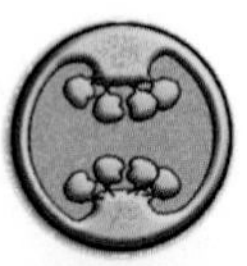

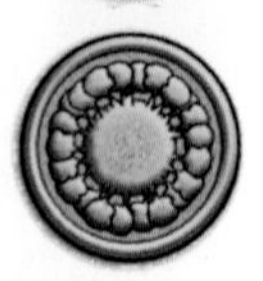

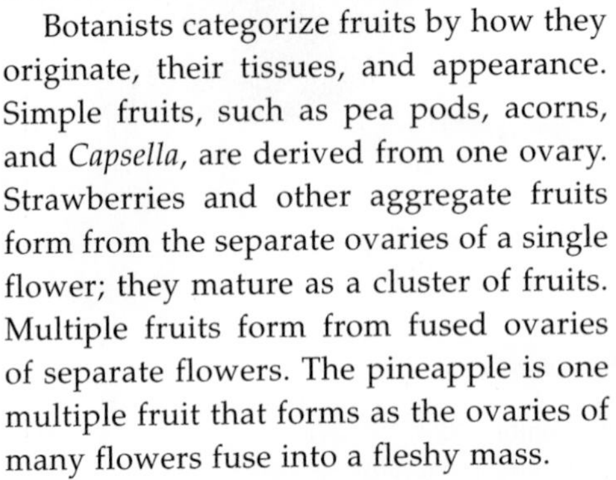

Botanists categorize fruits by how they originate, their tissues, and appearance. Simple fruits, such as pea pods, acorns, and *Capsella*, are derived from one ovary. Strawberries and other aggregate fruits form from the separate ovaries of a single flower; they mature as a cluster of fruits. Multiple fruits form from fused ovaries of separate flowers. The pineapple is one multiple fruit that forms as the ovaries of many flowers fuse into a fleshy mass.

Fruits also may be categorized in terms of which tissues they incorporate. *True* fruits such as cherries consist only of the ovary wall and its contents. Other floral parts, such as the receptacle, expand along with the ovary in *accessory* fruits. Apples are like this.

To categorize a fruit based on appearance, the first step is to describe it as dry or juicy (fleshy). Dry fruits are dehiscent or indehiscent. If *dehiscent*, the fruit wall splits along definite seams to release the seeds inside. *Capsella* pods and pea pods are like this.

A dry fruit is *indehiscent* if the wall does not split open; seeds are dispersed inside intact fruits. Acorns and grains (such as corn) are dry indehiscent fruits, as are fruits of sunflowers, maples, and—surprisingly—strawberries. Strawberry fruits are not juicy and not berries. A strawberry's red flesh is an accessory to the dry indehiscent fruits on its surface (Figure 28.7*e–h*).

Three major categories of fleshy fruits are drupes, berries, and pomes. *Drupes* have a pit—a stone-hard

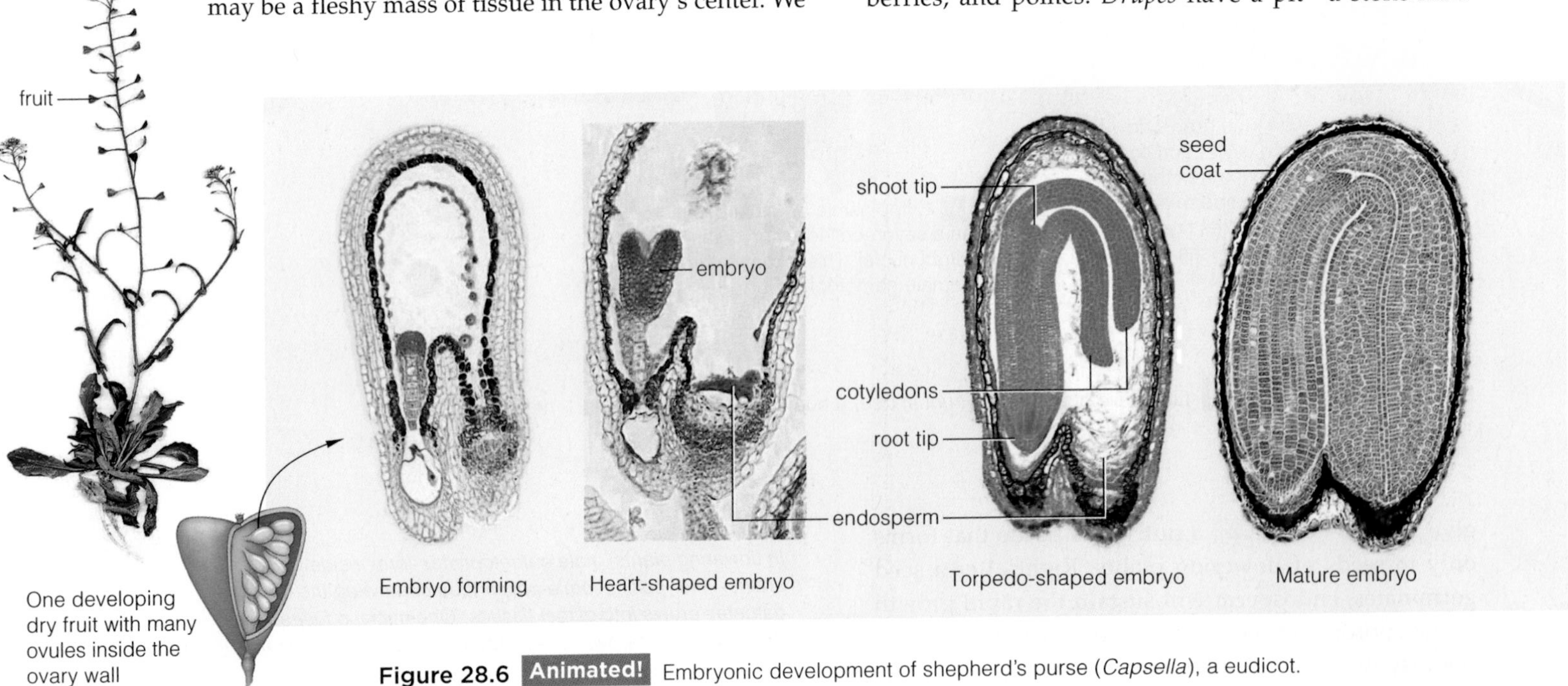

Figure 28.6 **Animated!** Embryonic development of shepherd's purse (*Capsella*), a eudicot.

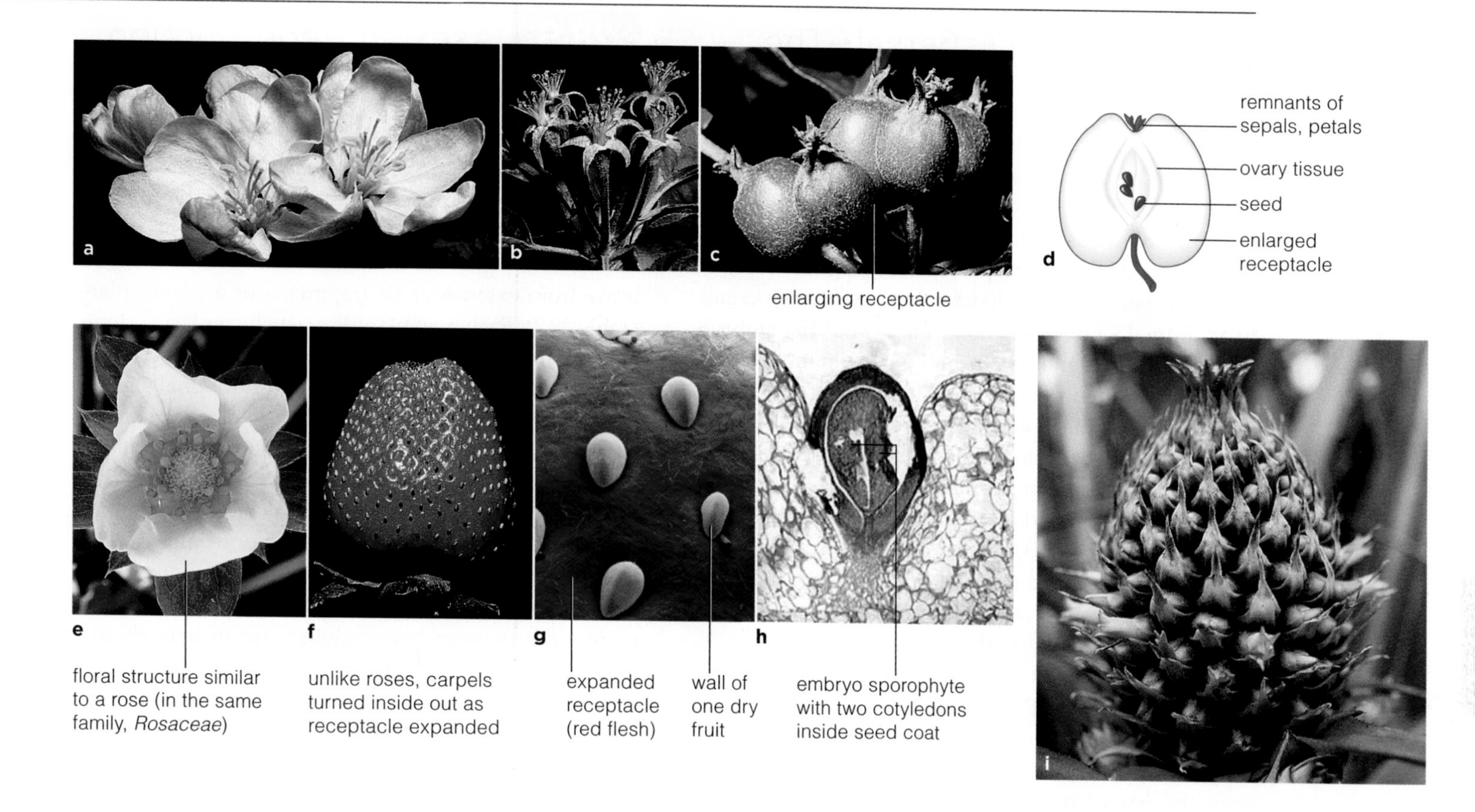

Figure 28.7 From flowers to fruits. (**a–d**) Fruit formation on an apple tree (*Malus*). Petals fall after eggs are fertilized. (**e–h**) Strawberry (*Fragaria*). Many ovaries are at the surface of the mature receptacle; each will have a hard fruit wall around an embryo sporophyte. (**i**) Pineapple (*Ananas*), a multiple fruit that became a symbol for hospitality. Native Americans cultivated pineapples and used the juice as a base for an alcoholic beverage. Christopher Columbus thought a pineapple looked like a pinecone, hence the name.

jacket around one seed (sometimes more), and fleshy fruit that encloses the pit. Cherries, peaches, apricots, almonds, and olives are examples of drupes.

A *berry* has one to many seeds, no pit, and fleshy fruit. Grapes and tomatoes are in this category. So are lemons, oranges, grapefruits, and other citrus fruits; a pithy layer encloses the berries, and an oily, leathery rind encloses the pith. Each "section" of a citrus fruit started out as one ovary of a fused carpel. Pumpkins, watermelons, and cucumbers are still another type of berry in which a hard rind of accessory tissues forms over the somewhat slippery true fruit.

A *pome* has seeds in a somewhat elastic core tissue and fleshy accessory tissues that enclose its core. Two familiar pomes are apples (Figure 28.7*a*) and pears.

Let's use an apple as an example. You could call it a simple fruit, because it originates from one flower. Or you could call it an accessory fruit, because a fleshy receptacle expands around five carpels. Or you could call it a pome; the carpels form an elastic core in the fleshy accessory tissue (Table 28.1).

A seed is a mature ovule, which consists of an embryo sporophyte, food reserves, and a protective seed coat. A mature ovary, with or without accessory tissues, is a fruit.

We can categorize a fruit in terms of how it originated, its composition, and whether it is dry or fleshy.

Table 28.1 Three Ways To Classify Fruits

How did the fruit originate?	
1. Simple fruit	One flower, single or fused carpels
2. Aggregate fruit	One flower, several unfused carpels; becomes cluster of several fruits
3. Multiple fruit	Individually pollinated flowers grow and fuse together
What is the fruit's tissue composition?	
1. True fruit	Only ovarian wall and its contents
2. Accessory fruit	Ovary as well as other floral parts, such as receptacle
Is the fruit dry or fleshy?	
1. Dry:	
a. Dehiscent	Dry fruit wall splits on definite seam to release seeds
b. Indehiscent	Seeds dispersed from parent plant in intact, dry fruit wall
2. Fleshy:	
a. Drupe	Fleshy fruit around hard pit with usually one seed inside
b. Berry	Fleshy fruit, no pit, one to many seeds
	Pepo: Hard rind on ovary wall
	Hesperidium: Leathery rind on ovary wall
c. Pome	Fleshy accessory tissues, seeds in elastic core

LIFE CYCLE OF A EUDICOT

28.4 Seed Dispersal—The Function of Fruits

Turn now to some of the remarkable adaptations by which fruits help disperse seeds from parent plants.

LINK TO SECTION 8.1

Tasty, fleshy fruits or nuts invite many insects, birds, and mammals to help disperse seeds (Figure 28.8*a*). A seed coat becomes abraded by the digestive enzymes in an animal's gut—not enough to expose the embryo sporophyte, but enough to make germination easier after the seed departs in feces. The seeds of cocklebur, bur clover, and many other plants have hooks, spines, hairs, and surfaces that stick to feathers, feet, fur, and clothing (Figure 28.8*b*). Human explorers have carried seeds all over the world, although most countries now restrict such imports. Cacao, oranges, corn, and other plants were encouraged to grow in new habitats, where they now have more reproductive success than they would have achieved on their own.

Many plant species, such as maples (*Acer*), use wind as a dispersal agent. Part of a maple fruit extends like a pair of thin, lightweight wings. It breaks in half and drops, then air currents spin it sideways and away from the parent tree (Figure 28.8*c*). Thistles, cattails, dandelions, and milkweed have lightweight fruits with a pluming "parachute" that may be blown as far as 10 kilometers from the parent plant. Great numbers of orchid seeds, fine as dust particles, drift through air. A fruit capsule of *Impatiens capensis* propels its seeds through the air by popping open explosively.

Water-dispersed types, such as coconut palm fruits that may drift hundreds of kilometers across the ocean, have thick, water-repellent layers that help them float.

Fruits help disperse seeds with their structural adaptations to air currents, water currents, and diverse animal species.

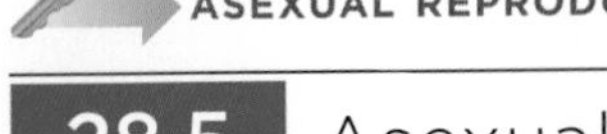

ASEXUAL REPRODUCTIVE MODES

28.5 Asexual Reproduction of Flowering Plants

Many plants also reproduce asexually, which permits rapid production of genetically identical offspring.

Many flowering plants reproduce asexually by modes of vegetative growth, in which new roots and shoots grow from extensions or fragments of a parent plant (Table 28.2). Each new plant is a genetic replica, a clone, of the parent. The clones of the plant world are vivid demonstrations of this concept: An individual's genes hold all of the information required to duplicate itself.

For example, strawberry plants send out runners—aboveground horizontal stems; new roots and shoots grow from nodes on the runners. As another example, "forests" of the quaking aspen (*Populus tremuloides*) are clones that arise from vegetative reproduction. Figure 28.9 shows shoots of an individual plant. That parent plant's root system keeps giving rise to new shoots. One aspen forest in Colorado stretches for hundreds of acres; it consists of about 47,000 shoots.

No one knows how old those aspen clones are. As long as conditions in the environment favor growth, such clones are as close as any organism gets to being immortal. One very old clone is a creosote bush (*Larrea divaricata*) that has been growing in the Mojave Desert for about 11,700 years.

Most houseplants, woody ornamentals, and orchard trees are clones from cuttings or fragments of shoot systems. Gardeners cut off African violet leaves and induce them to form a callus from which adventitious roots grow. A callus is a mass of meristematic tissue.

Twigs or buds of a plant may be grafted, or joined, to a plant of a related species. In 1862, the plant louse *Phylloxera* was accidentally introduced into France via imported American grapevines. European grapevines had little resistance to this tiny insect, which attacks

Figure 28.8 Examples of fruit dispersal. (**a**) Mountain ash berries, a simple fleshy accessory fruit, attracts cedar waxwings.

(**b**) Cocklebur, a prickly accessory fruit, sticks to the fur of animals and clothing of humans that brush past.

(**c**) Simple, dry fruits of maple (*Acer*) use winds to lift their "wings" and spin the seeds inside away so that they will not have to compete with the parent plant for water, mineral ions, and sunlight.

Figure 28.9 Quaking aspen (*Populus tremuloides*). By reproducing asexually by runners, a single plant gave rise to this stand of genetically identical shoots. In Colorado, one clone with about 47,000 shoots stretches for hundreds of acres. Rare mutations aside, trees at the north end are identical with those to the south. Water travels from roots near a lake to shoot systems in drier soil. Dissolved ions travel in the opposite direction.

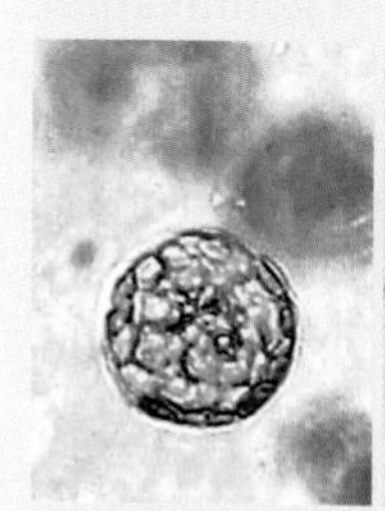

a Protoplast: a living plant cell without a wall, which enzymes digested away.

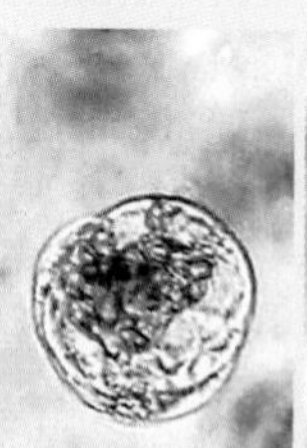

b Protoplast regenerates missing wall, mitotic cell divisions start.

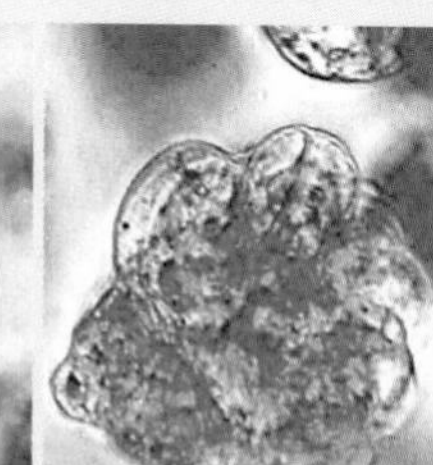

c Divisions yield a small clump of living cells, not yet differentiated.

d Ongoing divisions yield a callus, a mass of undifferentiated cells that are induced to give rise to a plant embryo.

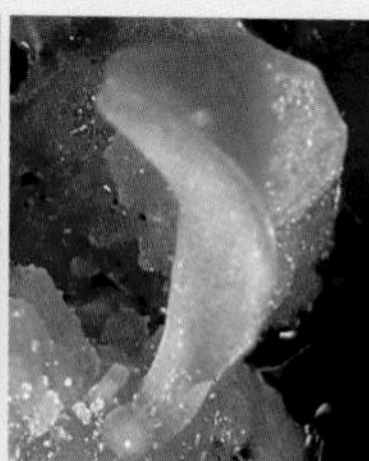

e A newly developing somatic embryo.

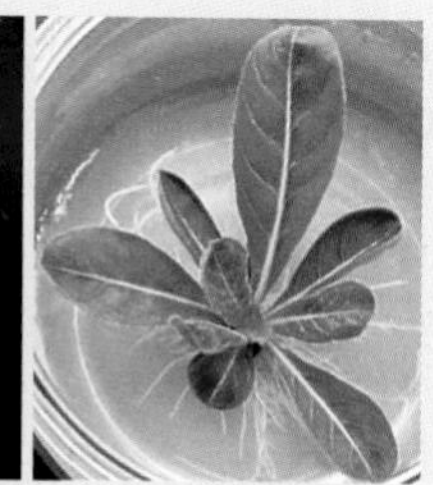

f Young plant that grew from a somatic embryo.

Figure 28.10 One method of tissue culture propagation. Many asexually formed embryos can be produced at the same time in tanks. With such micropropagation methods, plants grown from the somatic embryos can be cloned by the thousands.

and kills the root systems of vines. By 1900, *Phylloxera* had destroyed two thirds of the vineyards in Europe, devastating the wine-making industry. Today, French vintners routinely graft their prized grapevines onto the roots of *Phylloxera*-resistant American vines.

Researchers now use shoot tips, leaf or stem pieces, and other parts of individual plants for tissue culture propagation, in which an entire plant is cloned from a single cell (Figure 28.10). Tissue culture propagation can yield millions of genetically identical plants from a single specimen. The technique is being employed in efforts to improve food crops, including corn, wheat, rice, and soybeans. It also is used to propagate rare or hybrid ornamental plants such as orchids and lilies.

Individual flowering plants reproduce asexually in nature by modes of vegetative growth. All of the offspring are genetically identical with the parent plant.

The world's largest known individual plants, and the oldest, are the result of vegetative growth.

Tissue culture propagation is a laboratory procedure for cloning a plant from a single somatic cell.

Table 28.2 Asexual Reproduction in Flowering Plants

Mode	Examples	Characteristics
Vegetative reproduction on modified stems		
Runner	Strawberry	Plants form at aboveground horizontal stem nodes.
Rhizome	Cordgrass	Plants form at underground horizontal stem nodes.
Corm	Gladiolus	Plants form from axillary buds on underground, short, carbohydrate-storing stems.
Tuber	Potato	Shoots arise from axillary buds (tubers are the enlarged tips of slender underground rhizomes).
Bulb	Onion, lily	Bulbs arise from buds on short underground stems.
Vegetative growth		
	Jade plant, African violet	New plant develops from tissue or structure (a leaf, for instance) that drops from the parent plant or gets separated from it.
Tissue culture propagation		
	Orchid, lily, wheat, rice	New plant induced to arise from a parent plant cell that is not irreversibly differentiated.

28.6 Overview of Plant Development

In Section 28.4, we left the embryo sporophyte after its dispersal from the parent plant. Its growth idles inside the seed coat until conditions favor germination. In response to genetically prescribed programs and environmental cues, growth and development resume. The sporophyte matures, then enters a phase of sexual reproduction, and the life cycle turns again.

SEED GERMINATION

LINKS TO SECTIONS 1.2, 14.1, 14.3, 26.1

Plants, recall, consist of ground, vascular, and dermal tissues (Section 26.1). The meristems that give rise to all of these tissues form while an embryo sporophyte is still in its seed coat. Figure 28.11 shows where they are located in a corn grain, a type of dry fruit. A corn grain germinates in response to seasonal factors, such as the temperature, oxygen level, and moisture of the surrounding soil. For plants, germination is a process by which an embryo sporophyte resumes growth after an interval of arrested development.

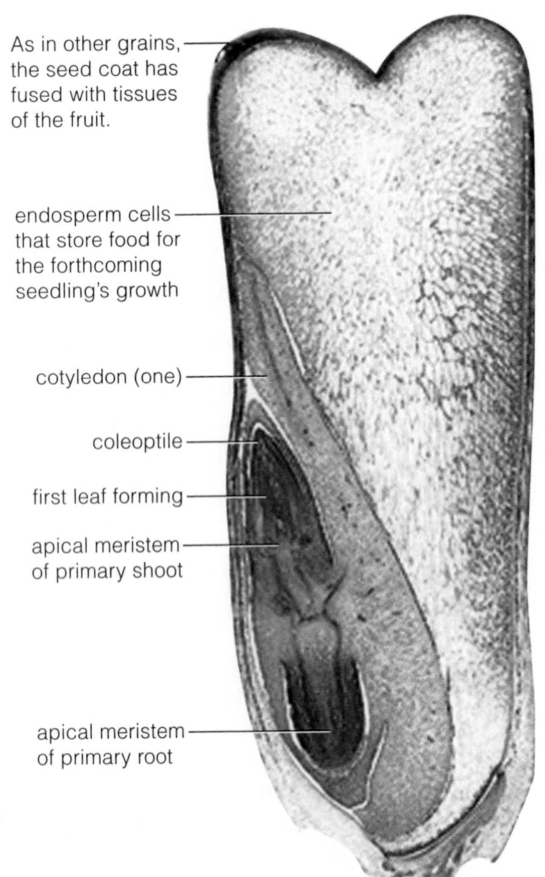

Figure 28.11 Embryo sporophyte and food reserves inside a grain of corn (*Zea mays*).

Mature seeds do not hold enough water to support cell expansion or metabolism. Where water is scarce for much of the year, germination coincides with the rainy season. Water molecules infiltrate seeds, being attracted by hydrophilic groups of storage proteins in endosperm. The seed swells with water, its coat splits, and more oxygen enters. Meristematic cells inside the seed begin to consume oxygen in aerobic respiration as they start dividing rapidly. Germination ends when a primary root breaks out of the seed coat.

PATTERNS OF EARLY GROWTH

Figures 28.12 and 28.13 show patterns of germination, growth, and development. The patterns have a genetic basis. All cells in a seedling inherited the same genes. Now, as cells in different tissues use different subsets of genes, they begin to differentiate (Section 14.1).

Growth, again, refers to an increase in the number, size, and volume of cells. For plants, the mitotic cell divisions that increase the number of cells occur only at meristems. Some meristem cells never differentiate; they keep dividing and making more meristem. Other meristem cells differentiate as they begin to selectively express a subset of their genes. Differentiated meristem cells are the basis of development in plants—they form roots, stems, leaves, and other parts of the plant that differ in size, shape, location, and function.

You may be surprised that development in plants depends on extensive coordination among cells, just as animal development does (Section 14.3). As occurs in animals, transcription of master genes guides the formation of body details in plants. Plant cells also communicate with other plant cells that are often some distance away from one another. As occurs in animals, hormones are the primary signals in such cell-to-cell communication. For instance, cells of apical meristem make a hormone called auxin that diffuses downward into new tissues below. This hormone stimulates cells in the tissues to divide and elongate in a direction that lengthens the shoot.

In plants, environmental cues, including seasonal shifts in daylength, availability of water, temperature, and gravity, trigger the production of hormones. Thus, such cues influence plant growth and development.

The body plan of flowering plants starts taking form while the embryo sporophyte is still part of a seed.

Interactions among genes, hormones, and the environment govern how each plant grows and develops.

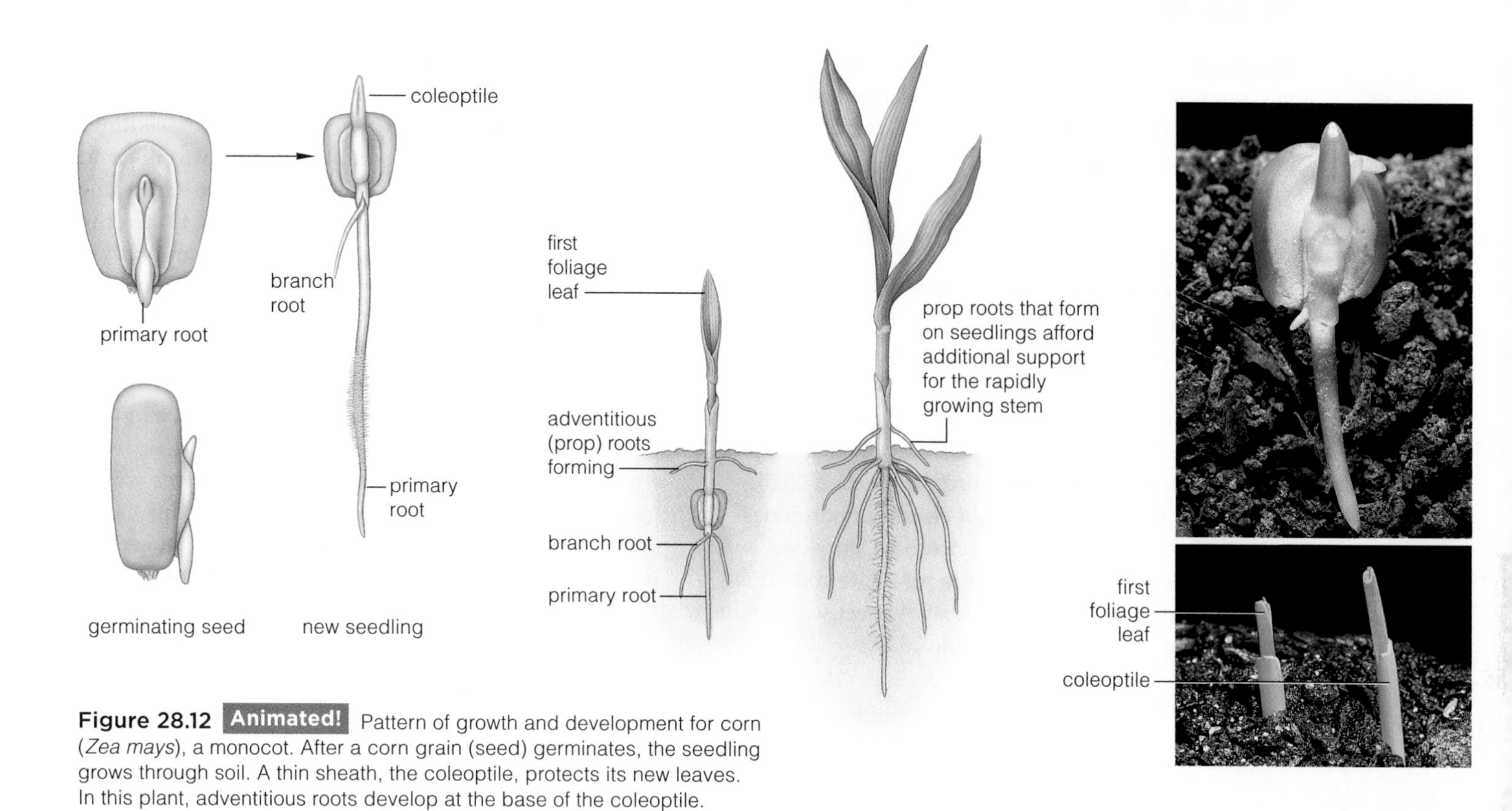

Figure 28.12 **Animated!** Pattern of growth and development for corn (*Zea mays*), a monocot. After a corn grain (seed) germinates, the seedling grows through soil. A thin sheath, the coleoptile, protects its new leaves. In this plant, adventitious roots develop at the base of the coleoptile.

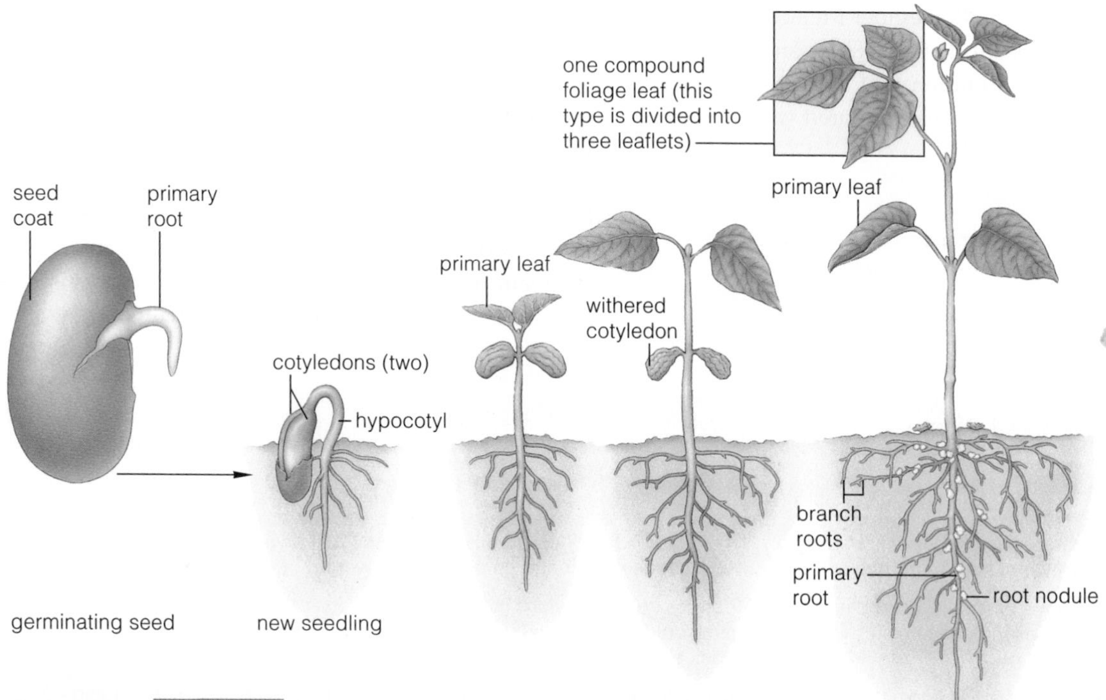

one compound foliage leaf (this type is divided into three leaflets)
seed coat
primary root
primary leaf
primary leaf
withered cotyledon
cotyledons (two)
hypocotyl
branch roots
primary root
root nodule
germinating seed
new seedling

Figure 28.13 **Animated!** Pattern of growth and development for the common bean plant (*Phaseolus vulgaris*), a eudicot. After a bean seed germinates, its primary shoot bends in the shape of a hook just below the cotyledons. Sunlight makes this "hypocotyl" straighten, which forces a channel through soil for the cotyledons. Photosynthetic cells in the cotyledons make food for several days, then foliage leaves take over the task. The cotyledons wither and fall off.

28.7 Plant Hormones and Other Signaling Molecules

As a plant grows and develops, its diverse cells increase in number, size, and volume, and specialized tissues form. These events require cell communication, as introduced in Chapter 25.

MAJOR TYPES OF PLANT HORMONES

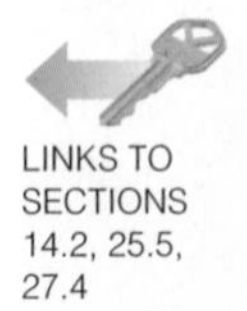

LINKS TO SECTIONS 14.2, 25.5, 27.4

Hormones, recall, are signaling molecules that alter an activity of a target cell. Hormone binding often causes a receptor to change shape, which in turn may affect gene expression, enzyme activity, ion concentrations, or activation of second messengers in the cytoplasm. The process by which a cell converts one signal into another is called signal transduction (Section 25.5).

In plants, five major classes of hormones—gibberellins, auxins, cytokinins, ethylene, and abscisic acid—interact to stimulate and inhibit growth and development at specific times in the plant life cycle (Table 28.3).

Gibberellins The gibberellins are growth hormones in flowering plants, ferns, mosses, gymnosperms, and some fungi. Gibberellins stimulate cells to divide and elongate, and so cause stems to lengthen. They also help seeds germinate and induce flowering in biennials and other plants. Remember Mendel's dwarf pea plants (Section 10.3)? Their short stems result from a mutation that reduces the rate of gibberellin synthesis compared with that of normal plants.

Figure 28.14 Demonstration of the effect of gibberellins. The three tall cabbage plants were treated with gibberellins. Next to them are two untreated cabbage plants.

Gibberellins were discovered in Japan in the 1930s. It had long been known that infection with the fungus *Gibberella fujikuroi* makes the stems of rice seedlings lengthen so much that the plants eventually topple and die. Researchers discovered they could cause the same effect by applying extracts of the fungus to healthy plants (Figure 28.14). They identified the compound in the extracts that triggered stem elongation, and named it after the fungus.

Auxins The auxins are produced in all meristems, young leaves, and seeds, and they promote or inhibit growth in different tissues. As an auxin diffuses away from its source, the resulting gradient dictates which genes are transcribed in which tissues. Auxins cause stems to bend toward light, leaves to grow in certain patterns, and roots to grow down through soil.

Auxins produced in apical meristems of shoots and coleoptiles cause both of these structures to elongate by cell divisions (Figure 28.15). Auxins also induce cell division and differentiation in the vascular cambium, and stimulate fruit formation.

Auxins also have inhibitory effects. For example, secretion of auxin by a shoot tip prevents the growth of lateral buds along the lengthening stem, an effect called apical dominance. Gardeners routinely pinch off shoot tips to make a plant bushier. Pinching the tips ends the supply of auxin in a main stem, so lateral buds are free to give rise to branches. Auxins also prevent abscission, the dropping of leaves, flowers, and fruits.

Cytokinins Unlike other plant hormones, cytokinins also occur in animal cells. They induce cells to divide rapidly. Most plant cytokinins form in roots and travel via xylem to shoots. There, they induce cell divisions in apical meristems and in maturing fruits. They also

Table 28.3 Major Classes of Plant Hormones and Their Main Effects

Hormone	Source and Mode of Transport	Stimulatory or Inhibitory Effects
Gibberellins	Young tissues of shoots, seeds, possibly roots. May travel in xylem and phloem	Make stems lengthen greatly (stimulates cell division, elongation). Help seeds germinate; help induce flowering in some plants
Auxins	Apical meristems of shoots, coleoptiles, embryos. Diffuses through shoot cells toward base of roots	Promote cell elongations that lengthen shoots, coleoptiles. Start vascular cambium activity, vascular tissue differentiation, and fruit formation. Block abscission and lateral bud formation
Cytokinins	Mainly root tip. Travels in xylem to shoots	Stimulate roots and shoot growth, and leaf expansion. Inhibit leaf aging
Ethylene	Most ripening, aging, or stressed tissues. A gas; diffuses in all directions	Promotes or inhibits cell division in different directions. Promotes abscission and senescence (aging and death)
Abscisic acid (ABA)	Root cells. Travels in xylem to leaves	Makes stomata close. Stimulates sugar transport from leaves to seeds and embryo's development. Induces and maintains dormancy in some species

release lateral buds from apical dominance, and they can stop leaves from aging prematurely.

Ethylene The only gaseous hormone, ethylene, can promote or inhibit cell division in different directions as tissues expand. It can also induce fruit to ripen. Its concentration is highest in maturing apples, bananas, avocados, and other fruits in which aerobic respiration increases greatly. Its concentration is high in stressed plants, as happens in autumn or near the end of the life cycle. Then, ethylene induces dropping of leaves and fruits, and often the death of the whole plant.

Abscisic Acid During water shortages, root cells make more abscisic acid (ABA), which xylem moves to leaves. ABA is part of a stress response that causes stomata to close so that the plant can minimize water loss (Section 27.4). Also, when a growing season ends, ABA's effect overrides the growth-promoting effects of other hormones. It causes photosynthetic products to be diverted from the leaves to seeds, where it helps embryos mature. ABA induces dormancy in buds and seeds of some plants. This growth-inhibiting hormone was misnamed; it has little to do with abscission.

OTHER SIGNALING MOLECULES

As we now know, other signaling molecules have roles in plant growth and development. *Brassinosteroids* help promote cell division and elongation; stems stay short in their absence. *Jasmonates*, derived from fatty acids, help other hormones control germination, root growth, and defense of tissues. *FT protein* is part of a signaling pathway in flower formation. *Salicylic acid* is similar to aspirin. It interacts with *nitric oxide* in calling for gene products that resist pathogenic attacks. *Systemin* forms as insects feed on plant tissues. It activates transcription of genes for substances that cripple an insect's capacity to digest proteins.

Figure 28.16 and Table 28.4 will give you a sense of how natural and synthetic plant hormones are put to use for commercial applications.

Plant hormones are signaling molecules secreted by plant cells; they alter activities of other plant cells.

The five main classes of plant hormones are gibberellins, auxins, cytokinins, abscisic acid, and ethylene.

Interactions among hormones and other kinds of signaling molecules stimulate or inhibit cell division, elongation, differentiation, and other events.

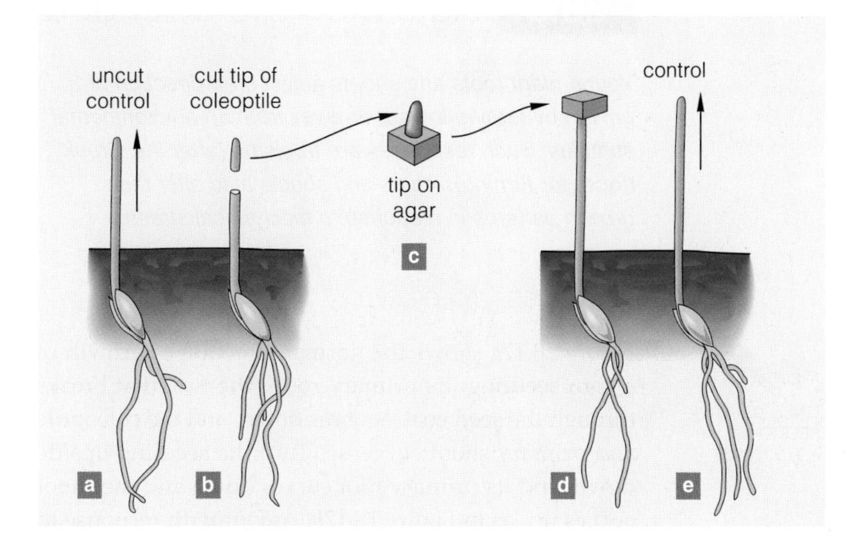

Figure 28.15 Animated! Auxin diffusing down from a coleoptile tip makes a seedling's cells lengthen. (**a**,**b**) A seedling with a cut tip will not lengthen as much as an uncut control. (**c**) A block of agar positioned under a cut tip for a few hours will absorb auxin from the tip. (**d**) If the block is placed on a de-tipped coleoptile, cells below it will lengthen as fast as an uncut seedling (**e**).

Figure 28.16 Effect of rooting powders that contain auxin. Cuttings of winter honeysuckle (*Lonicera fragrantissima*) that were treated with a lot of auxin (*left*), some auxin (*middle*), and no auxin (*right*).

Table 28.4 Some Applications of Plant Hormones
Gibberellins Increase fruit size; delay citrus fruit ripening; synthetic forms can make some dwarf mutants grow tall
Synthetic auxins Promote root formation in cuttings; induce seedless fruit production before pollination; keep mature fruit on trees until harvest time; widely used as herbicides against broad leaf weeds in agriculture
Cytokinins Tissue culture propagation; biotechnology; prolong shelf life of cut flowers
Ethylene Allows shipping of green, still-hard fruit (minimizes bruises and rotting). Carbon dioxide application stops ripening of fruit in transit to market, then ethylene is applied to ripen distributed fruit fast
ABA Induces nursery stock to enter dormancy before shipment to minimize damage during handling

28.8 Adjusting the Direction and Rates of Growth

Young plant roots and shoots adjust their direction of growth by turning toward or away from an environmental stimulus. Such responses are tropisms (after the Greek trope, *for turning). Roots and shoots also alter their growth patterns in response to mechanical stress.*

RESPONSES TO GRAVITY

LINKS TO SECTIONS 4.9, 6.1, 26.5

Figure 28.17*a* shows the normal direction of growth of a corn seedling. Its primary root—the first that breaks through the seed coat—curves down, and the coleoptile and primary shoot curve up. Turn the seedling upside down, and its primary root curves down and the shoot curves up, as in Figure 28.17*b*. Any growth response to Earth's gravitational force is a form of gravitropism.

Turn the seedling on its side in a dark room and it still makes a gravitropic response. Its shoot curves up even in the absence of light. What makes the plant do this? In that horizontally oriented stem, cells do not elongate much on the side facing up, but those on the lower side elongate rapidly. The different elongation rates result in an upward-bent shoot.

Auxin, and auxin transporters in cell membranes, function in gravitropism. How do we know? Position seedlings perpendicular to Earth's surface and expose them to a substance that inhibits auxin transporters. Unlike untreated control seedlings, the seedlings will not bend (Figure 28.17*c*,*d*).

Gravity-sensing mechanisms of many organisms are based on statoliths. Plant statoliths are modified plastids stuffed with clusters of heavy starch grains. In response to gravity, statoliths in specialized cells drift downward into the lowest region of cytoplasm (Figure 28.18). If the plant is reoriented with respect to gravity, shifting statoliths (and possibly other organelles) cause the redistribution of auxins, which in turn initiates a gravitropic response.

RESPONSES TO LIGHT

When light streams in from one direction only, a stem or leaf adjusts its rate and direction of growth so that it grows toward the light source. This response is called phototropism. Phototropism orients certain parts of a

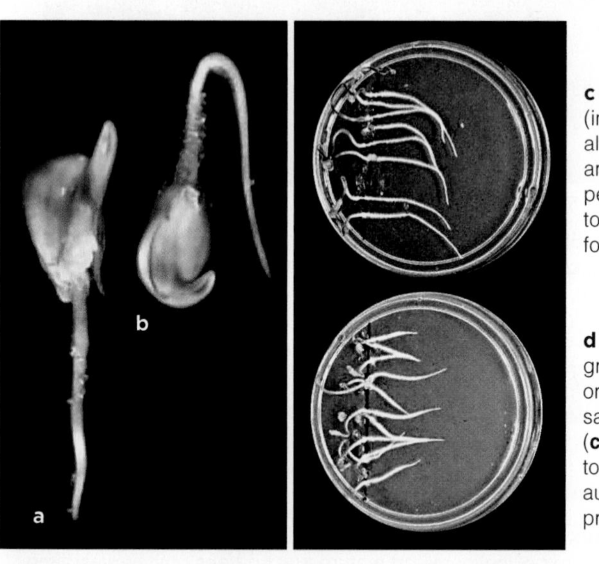

c Control group (in a petri dish, all seedlings are oriented perpendicular to gravitational force of Earth)

d Experimental group (seedlings oriented the same way as in (**c**) but exposed to an inhibitor of auxin transport protein)

Figure 28.17 Gravitropic responses of a corn primary root and shoot growing in a normal orientation (**a**), and turned upside down (**b**).

(**c**) Experimental test of whether gravitropism requires auxin transport proteins in root cell membranes. In primary roots turned on their side, auxin is transported to the down-facing side, where it stops cells from lengthening. Cells of the up-facing side continue to lengthen, so the roots bend downward.

(**d**) Seedlings exposed to inhibitors of auxin transport do not bend in response to gravity. Mutations in genes that encode auxin transport proteins have the same effect.

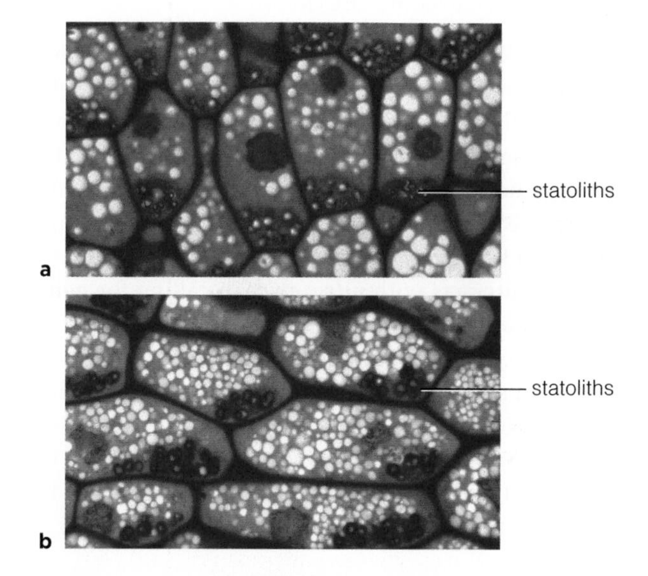

Figure 28.18 Animated! Gravity, the distribution of statoliths in root cells, and auxin.

(**a**) Starch-packed statoliths have settled to the bottom of the gravity-sensing cells in a corn root cap. (**b**) The same root, ten minutes after it was turned sideways. The statoliths have already settled to the new "bottom" of the cells.

Such redistribution of statoliths causes the redistribution of auxin in the root, which in turn causes the root tip to curve downward.

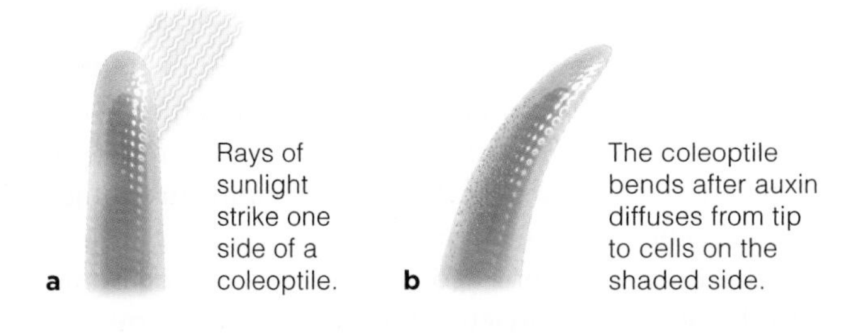

plant in a direction that maximizes the amount of light intercepted by photosynthetic cells.

Phototropism in plants occurs in response to blue light. Nonphotosynthetic pigments called *flavoproteins* absorb blue light, and transduce its energy into signals that cause auxin to be redistributed to the shaded side of a shoot or coleoptile. As a result, cells on the shaded side elongate faster than cells on the illuminated side. Differences in growth rates between cells on opposite sides of a shoot or coleoptile causes the entire structure to bend toward the light (Figure 28.19*a,b*).

You can observe a phototropic response by putting seedlings of sun-loving plants next to a sunlit window. In time, the seedlings will begin to curve toward the light (Figure 28.19*c*).

Figure 28.19 **Animated!** Phototropism. (**a**,**b**) Hormone-mediated differences in rates of cell elongation along its length induce a coleoptile to bend toward light. (**c**) Flowering shamrock (*Oxalis*) responding to light.

RESPONSES TO CONTACT

In some plants, contact with a solid object results in the redistribution of auxin and ethylene, which changes the direction of the plant's growth. This response is called thigmotropism. As an example, when a vine's shoot tip touches an object, the cells near the area of contact stop elongating. The cells on the opposite side of the shoot keep elongating. The unequal growth rates of cells on opposite sides of the shoot cause it to curl around the object (Figure 28.20). A similar mechanism causes roots to grow away from contact.

Figure 28.20 Passion flower (*Passiflora*) tendril twisting thigmotropically.

RESPONSES TO MECHANICAL STRESS

Mechanical stress, as inflicted by prevailing winds and grazing animals, inhibits stem lengthening. Trees high up in windswept mountains are stubbier than sheltered trees of the same species at lower elevations. Similarly, plants grown outdoors commonly have shorter stems than the same kinds of plants grown in a greenhouse. Briefly shake a plant every day and you will inhibit its overall growth (Figure 28.21).

Plants adjust the direction and rate of growth in response to environmental stimuli that include gravity, light, contact, and mechanical stress.

Figure 28.21 Effect of mechanical stress on tomato plants. (**a**) This plant, the control, grew in a greenhouse. (**b**) Each day for twenty-eight days, this plant was mechanically shaken for thirty seconds. (**c**) This one had two shakings each day.

28.9 Sensing Recurring Environmental Changes

Seasonal shifts in night length, temperature, and other environmental cues trigger plant responses.

BIOLOGICAL CLOCKS

LINKS TO SECTIONS 6.1, 6.2, 14.2, 25.4

Most organisms have a biological clock—an internal mechanism that governs the timing of rhythmic cycles of activity. Section 25.4 showed a bean plant changing the position of its light-intercepting leaves over twenty-four hours—even when it was kept in the dark.

Any cycle of activity that starts anew every twenty-four hours or so is known as a circadian rhythm. The Latin *circa* means about; *dies* means day. In a circadian response called solar tracking, a leaf or flower shifts position in response to the changing angle of the sun through the day. For example, the stem of a sunflower plant swivels so the flower on top of it always faces the sun (Figure 28.22). Unlike a phototropic response, solar tracking does not involve redistribution of auxin and differential growth. Instead, the absorption of blue light by photoreceptor proteins causes increased fluid pressure in cells on the sunlit side of a stem or petiole. The cells change shape in ways that bend the stem.

Similar mechanisms cause flowers of some plants to open only at certain times. For example, the flowers of many bat-pollinated flowers unfurl, secrete nectar, and release fragrance only at night. Closing flowers during the day keeps delicate reproductive parts tucked away when the likelihood of pollination is low.

SETTING THE CLOCK

Like a mechanical clock, a biological one can be reset. Sunlight resets biological clocks in plants by activating and inactivating photoreceptors called phytochromes. These blue-green pigments are sensitive to red light (660 nanometers) and far-red light (730 nanometers). The relative amounts of these wavelengths in sunlight that reaches a given environment vary during the day and with the season. Red light causes phytochromes to change from an inactive form to an active form. Far-red light causes them to change back to their inactive form (Figure 28.23).

Active phytochromes bring about transcription of many genes, including some that encode components of rubisco, photosystem II, ATP synthase, and other proteins used in photosynthesis. Flowering, tropisms such as gravitropism and phototropism, germination, and other processes are under phytochrome control.

WHEN TO FLOWER?

Photoperiodism is an organism's response to changes in the length of night relative to day. Unless you live at the equator, the night length varies with the season. Nights are longer in winter than in summer, and the difference increases with latitude.

You have probably noticed that different species of plants flower at different times of the year. In such plants, flowering is photoperiodic. Chrysanthemums and other *short-day* plants flower only when the hours of darkness are greater than some critical value (Figure 28.24*a,c*). *Long-day* plants such as irises flower only when the hours of darkness fall below a critical value (Figure 28.24*b*). Sunflowers and other *day-neutral* plants flower when they mature, regardless of night length.

Phytochrome has a role in photoperiodism. Figure 28.25*a* shows results from an experiment that exposed

Figure 28.22 Sunflowers (*Helianthus*), which track the sun.

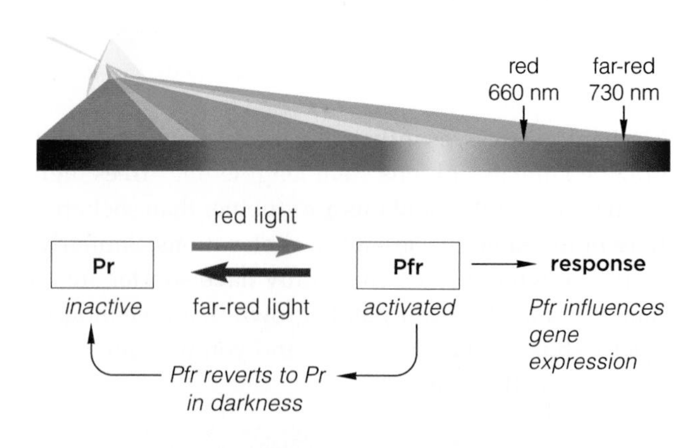

Figure 28.23 Animated! Phytochromes. Red light changes the structure of a phytochrome from inactive to active form; far-red light changes it back to the inactive form. Activated phytochromes control important processes such as germination and flowering.

Figure 28.24 Animated! (**a**) Flowering response of chrysanthemum, a short-day plant, to (*left*) long-day and (*right*) short-day exposure.

Experiments showing that different plant species flower in response to different night lengths. Each horizontal bar represents 24 hours. *Yellow* signifies daylight; *blue-gray* signifies night. (**b**) This long-day plant, an iris, flowered only when hours of darkness were *less* than the value that is critical for flowering of its species. (**c**) This short-day plant, a chrysanthemum, flowered only when the hours of darkness were *more* than a critical value for its species.

plants to long "nights," interrupted by a brief pulse of red light. The pulse of light activated phytochrome and made both long-day and short-day plants respond as if the nights were short. In a related experiment, a pulse of far-red light followed the pulse of red light. In this case, plants reacted as if the nights were long (Figure 28.25*b*). The far-red light counteracted the effect of the earlier pulse of red light; it deactivated phytochrome.

Leaves detect night length and produce signals that travel through the plant. In one experiment, a single leaf was left on a cocklebur, a short-day plant. The leaf was shielded from light for 8–1/2 hours every day, which is the threshold amount of darkness required for flowering. The plant flowered. Later, the leaf was grafted onto another cocklebur plant that had *not* been exposed to long hours of darkness. Then, the recipient plant flowered, too.

How does a compound produced by leaves cause flowering? In response to night length and other cues, leaf cells transcribe more or less of a flowering gene. The transcribed mRNA migrates from leaves to shoot tips, where it is translated. Its protein product helps activate the master genes that control the formation of floral structures, as Section 14.2 explains.

Night length is not the only cue for flowering. Some biennials and perennials flower only after exposure to low winter temperatures (Figure 28.26). This response is known as vernalization (from *vernalis*, meaning "to make springlike").

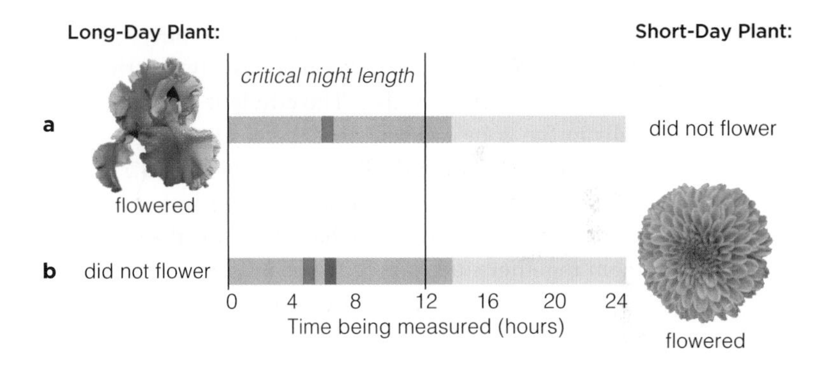

Figure 28.25 Two experiments that pointed to phytochrome as a trigger for flowering in long-day and short-day plants. (**a**) An intense red flash interrupted a long night. Both plants responded as if it were a short night (long day). Only the long-day plant flowered. (**b**) A short pulse of far-red light after the red flash canceled the effect of the red flash by inactivating phytochrome. Only the short-day plant flowered.

Like all other organisms, flowering plants have internal mechanisms that regulate recurring cycles of growth, development, and reproduction, in response to cues from the environment.

The main environmental cue for flowering is the length of night relative to the length of day, which varies by the season in most places.

Low winter temperatures stimulate the flowering of many plant species in spring.

Figure 28.26 Localized effect of cold temperature on dormant buds of lilac (*Syringa*). For this experiment, a single branch was positioned to protrude from a greenhouse through a cold winter. The rest of the plant was kept inside and exposed only to warm temperatures. Only buds exposed to the low outside temperatures resumed growth and flowered in springtime.

28.10 Entering and Breaking Dormancy

Many plants live in places where growth conditions vary seasonally. During part of the year it is too dry or too cold for optimal growth. Plants may drop their leaves and slow their growth during these unfavorable intervals.

ABSCISSION AND SENESCENCE

LINK TO SECTION 4.11

The process by which plant parts are shed is called abscission. It occurs in deciduous plants as a response to shortening daylight hours, and in evergreen plants year-round. Injury, water or nutrient deficiencies, or high temperatures can also induce abscission.

Let's use deciduous plants as an example. As leaves and fruits grow in early summer, their cells produce auxin, which moves into stems. Auxin helps maintain growth. By midsummer, the nights are getting longer. Plants begin to divert nutrients away from their leaves, stems, and roots, and into flowers, fruits, and seeds. As the growing season comes to an end, nutrients are routed to twigs, stems, and roots, and auxin production declines in leaves and fruits.

The auxin-deprived structures release ethylene that diffuses into nearby abscission zones—twigs, the base of leaves, and fruit stalks. The ethylene is a signal for cells in the zone to produce enzymes that digest their own walls and the middle lamella (Section 4.11). The cells bulge as their walls soften, and separate from one another as the middle lamella—the layer that cements them together—dissolves. Tissue in the zone weakens, and the structure above it drops (Figure 28.27).

If the seasonal diversion of nutrients into flowers, seeds, and fruits is interrupted, leaves and stems stay on a deciduous plant longer (Figure 28.28). Senescence refers to the phase from full maturity until the death of plant parts or the whole plant. It is a normal part of the plant life cycle.

DORMANCY

For many species, growth stops in autumn as a plant enters dormancy, a period of arrested growth caused by—and later ended by—environmental cues. Long nights, cold temperatures, and dry, nitrogen-poor soil are strong cues for dormancy.

Dormancy-breaking cues usually operate between fall and spring. Dormant plants do not resume growth until certain conditions in the environment occur. A few species require exposure of the dormant plant to many hours of cold temperature, but more typical cues include the return of milder temperatures and plentiful water and nutrients. With the return of such favorable conditions, life cycles begin to turn once more. Seeds germinate, buds resume growth, and new leaves form.

Multiple cues from the environment influence processes of plant growth, senescence, and dormancy. Changes in daylength, nutrient and moisture availability, and temperature are such cues.

Figure 28.27 Horse chestnut (*Aesculus hippocastanum*) leaves changing color in autumn. The leaf scar at *right* is all that remains of an abscission zone. Before the leaf detached from the tree, a tissue formed in a horseshoe-shaped zone (hence the tree's name).

control (pods not removed) experimental plant (pods removed)

Figure 28.28 The observable results from an experiment in which seed pods were removed from a soybean plant as soon as they formed. Removal delayed senescence.

www.thomsonedu.com ThomsonNOW!

Summary

Section 28.1 Flowering plants make sexual spores in stamens and carpels of floral shoots (Figure 28.29). Male and female gametophytes develop from the spores. Most flowering plants coevolved with particular pollinators, such as insects, that transfer pollen from stamens to the carpels of flowers of the same species.

- *Use the animation on ThomsonNOW to investigate a flowering plant life cycle and floral structure.*

Section 28.2 Pollen sacs form in anthers of stamens. In these chambers, haploid microspores form by meiosis of diploid spore-producing cells. A microspore develops into a sperm-bearing male gametophyte, which is housed in a pollen grain.

A carpel's base has one or more ovaries. Ovules form from the inner ovary wall. One cell in the ovule, a haploid megaspore, gives rise to the mature female gametophyte. One cell of the gametophyte becomes the egg.

Pollination is the arrival of pollen grains on a receptive stigma. When a pollen grain germinates, it forms a pollen tube (with two sperm nuclei inside) that grows through ovary tissues, to the egg. In double fertilization, one of the sperm nuclei fertilizes the egg, forming a zygote; the other fuses with the endosperm mother cell.

- *Use the animation on ThomsonNOW to take a closer look at the life cycle of a eudicot.*
- *Read the InfoTrac article "What's So Special About Flowers?" Karl Nicklas,* Natural History, *May 1999.*

Sections 28.3, 28.4 A seed is a mature ovule: an embryo sporophyte and endosperm inside a seed coat. Eudicot embryos have two cotyledons; monocot embryos have one. The ovary wall and sometimes other tissues develop into fruits around seeds.

- *Use the animation on ThomsonNOW to see how an embryo sporophyte develops in a eudicot seed.*

Section 28.5 Many species of flowering plants can reproduce asexually by vegetative growth. A laboratory method, tissue culture propagation, can produce millions of clones from a single somatic cell.

Section 28.6 Embryo sporophytes remain dormant in seeds until conditions favor germination, or resumption of growth. Genes, signaling molecules, and various cues from the environment guide growth and development.

- *Use the animation on ThomsonNOW to compare monocot and eudicot growth and development.*

Section 28.7 Like animal hormones, plant hormones secreted by one cell alter the activity of a different cell. Plant hormones can promote or arrest growth of a plant by stimulating or inhibiting cell division, differentiation, elongation, and reproduction.

Gibberellins help make stems lengthen, help seeds and buds break dormancy, and often stimulate flowering.

Auxins help make coleoptiles and shoots lengthen, and function in phototropism and gravitropism (adjustments in the direction of elongation during primary growth). Cytokinins stimulate cell division, release lateral buds from apical dominance, and retard leaf aging.

Ethylene promotes fruit ripening and abscission (the dropping of leaves, fruits, and other plant parts).

Abscisic acid promotes bud and seed dormancy, and it limits water loss by promoting stomatal closure.

- *Use the animation on ThomsonNOW to observe the effect of auxin on plant growth.*

Section 28.8 Environmental cues activate hormones that adjust the direction and rate of growth.

In gravitropism, roots grow down and stems grow up in response to gravity. In phototropism, stems and leaves bend toward or away from light. Blue light is the trigger for a phototropic response. In some plants, the direction of growth changes in response to contact with an object. Growth may also be affected by mechanical stress.

- *Use the animation on ThomsonNOW to investigate plant tropisms.*

Sections 28.9, 28.10 Internal timing mechanisms respond to daily and seasonal cycles. Photoperiodism is a response to changes in length of night. Phytochromes are photoreceptors that control germination, flowering, and tropisms. Short-day plants flower mainly in spring or fall, when nights are shorter. Long-day plants flower in summer, when nights are longer. Day-neutral plants flower whenever they are mature enough to do so.

Dormancy is a period of arrested growth that does not end until specific environmental cues occur. Senescence is the part of the plant life cycle between maturity and death of the plant or plant parts.

- *Use the animation on ThomsonNOW to learn how plants respond to night length.*

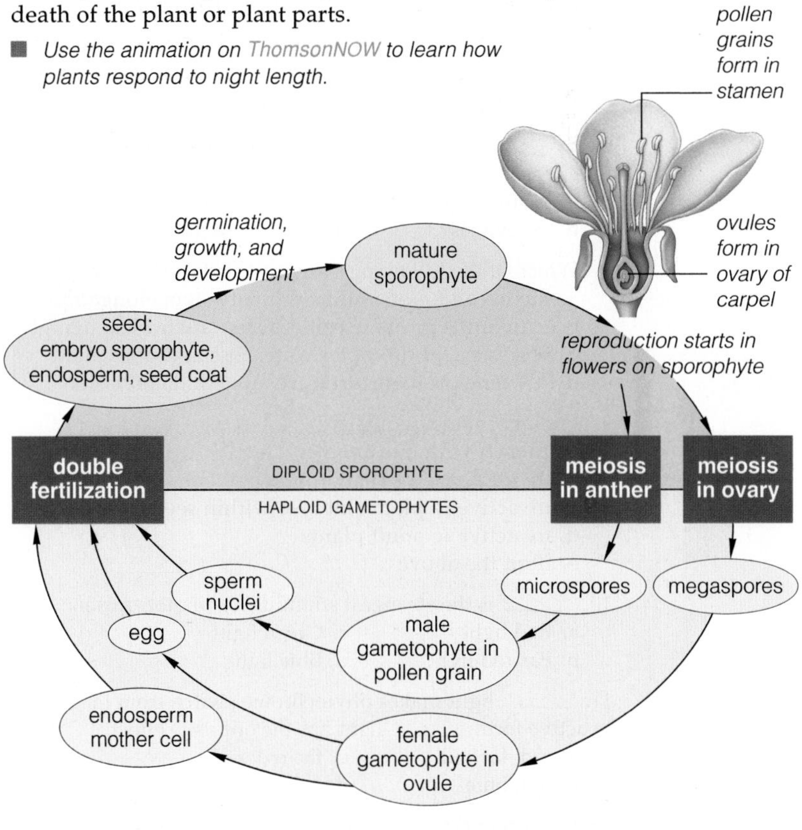

Figure 28.29 Summary of stages of growth, development, and reproduction in the life cycle of a typical eudicot.

Figure 28.30 (**a**) Vivid colors and patterns of the *Gazania* hybrid Fiesta Red, a composite flower. Tangential sections that reveal seeds of two mature fruits: (**b**) papaya (*Carica papaya*) and (**c**) peaches (*Prunus*).

Self-Quiz

Answers in Appendix III

1. The ______ of a flower contains one or more ovaries in which eggs develop, fertilization occurs, and seeds mature.
   a. pollen sac c. receptacle
   b. carpel d. sepal

2. Seeds are mature ______; fruits are mature ______.
   a. ovaries; ovules c. ovules; ovaries
   b. ovules; stamens d. stamens; ovaries

3. After meiosis within pollen sacs, haploid ______ form.
   a. megaspores c. stamens
   b. microspores d. sporophytes

4. After meiosis in an ovule, ______ megaspores form.
   a. two b. four c. six d. eight

5. The seed coat forms from the ______.
   a. ovule wall c. endosperm
   b. ovary d. residues of sepals

6. Cotyledons develop as part of ______.
   a. carpels c. embryo sporophytes
   b. accessory fruits d. petioles

7. By ______, a new plant forms from a tissue or structure that drops or is separated from the parent plant.
   a. parthenogenesis c. vegetative growth
   b. exocytosis d. nodal growth

8. Which of the following statements is false?
   a. Auxins and gibberellins promote stem elongation.
   b. Cytokinins promote cell division, retard leaf aging.
   c. Abscisic acid promotes water loss and dormancy.
   d. Ethylene promotes fruit ripening and abscission.

9. Plant hormones ______.
   a. interact with one another
   b. are influenced by environmental cues
   c. are active in plant embryos within seeds
   d. are active in adult plants
   e. all of the above

10. ______ is the strongest stimulus for phototropism.
    a. Red light c. Green light
    b. Far-red light d. Blue light

11. ______ light makes phytochrome switch from inactive to active form; ______ light has the opposite effect.
    a. red; far-red c. far-red; red
    b. red; blue d. far-red; blue

12. Flowering is a ______ response.
    a. phototropic c. photoperiodic
    b. gravitropic d. thigmotropic

13. Match the terms with the most suitable description.

___ ovule	a. pollen tube together with its contents
___ receptacle	b. embryo sac of seven cells, one with two nuclei
___ double fertilization	c. starts out as cell mass in ovary; may become a seed
___ anther	d. female reproductive part
___ carpel	e. pollen sacs inside
___ mature female gametophyte	f. base of floral shoot
___ mature male gametophyte	g. formation of zygote and first cell of endosperm

■ *Visit ThomsonNOW for additional questions.*

Critical Thinking

1. Would you expect winds, bees, birds, bats, butterflies, or moths to pollinate the flower in Figure 28.30*a*? Pick one. Explain why you picked it and rejected the other candidates.

2. All but one species of large-billed birds native to New Zealand's tropical forests are now extinct. Numbers of the surviving species, the kereru, are declining rapidly due to the habitat loss, poaching, predation, and interspecies competition that wiped out the other native birds. The kereru remains the sole dispersing agent for several native trees that produce big seeds and fruits. One tree, the puriri (*Vitex lucens*), is New Zealand's most valued hardwood. Explain, in terms of natural selection, why we might expect to see no new puriri trees in New Zealand.

3. Wanting to impress friends with her sophisticated botanical knowledge, Dixie Bee prepares a plate of tropical fruits for a party and cuts open a papaya (*Carica papaya*) for the first time. Soft skin and soft fleshy tissue enclose many seeds in a slimy tissue (Figure 28.30*b*). Knowing her friends will ask how to categorize this fruit, she panics, runs to her biology book, and opens it to Section 28.3. What does she find out?

4. Having succeeded spectacularly in her papaya research, Dixie Bee prepares a platter of peaches (Figure 28.30*c*) for her next party. How will she categorize the peach?

5. Before cherries, apples, peaches, and many other fruits ripen and the seeds inside them mature, their flesh is bitter or sour. Only later does it become tasty to animals that help disperse its seeds. Develop a hypothesis about how this feature can help the plant's reproductive success.

6. Reflect on Chapter 26. Would you expect hormones to influence primary growth only? What about secondary growth in, say, a hundred-year-old oak tree?

7. Photosynthesis sustains plant growth, and inputs of sunlight sustain photosynthesis. Why, then, do seedlings that germinated in a fully darkened room grow taller than different seedlings of the same species that germinated in the full sun?

8. Belgian scientists isolated a mutant gene in common wall cress (*Arabidopsis thaliana*) that leads to excess auxin production. Predict the impact on the plant's phenotype.

9. Beef cattle typically are given somatotropin, an animal hormone that makes them grow bigger (the added weight means greater profits). There is concern that such hormones may have unforeseen effects on beef-eating humans. Can plant hormones affect humans? Why or why not?

VI How Animals Work

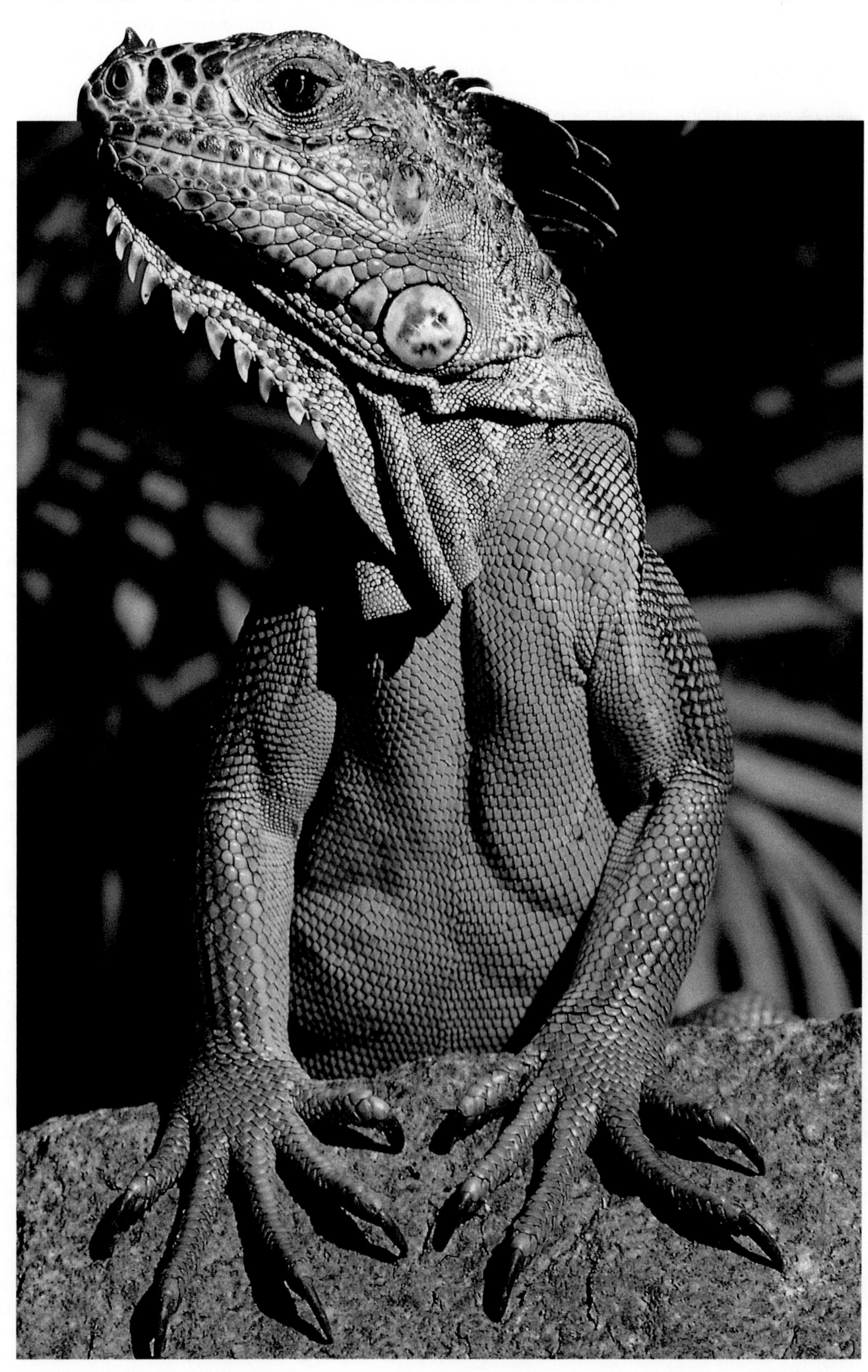

How many and what kinds of body parts does it take to function as a lizard in a tropical forest? Make a list of what comes to mind as you start reading Unit VI, then see how resplendent the list can become at the unit's end.

29 ANIMAL TISSUES AND ORGAN SYSTEMS

IMPACTS, ISSUES Open or Close the Stem Cell Factories?

Imagine being able to grow new body parts to replace lost or diseased ones. Some animals do this. A sea star that gets one of its arms lopped off grows another. A salamander can lose its tail and then replace it. Making new body parts starts with stem cells. These unspecialized cells are self-renewing; they produce more stem cells. They also produce cells that differentiate and form specific tissues. Early in human development, a fertilized egg divides and forms a ball of cells that includes embryonic stem cells (Figure 29.1). Descendants of the stem cells divide, differentiate, and give rise to all tissues and organs of the human body.

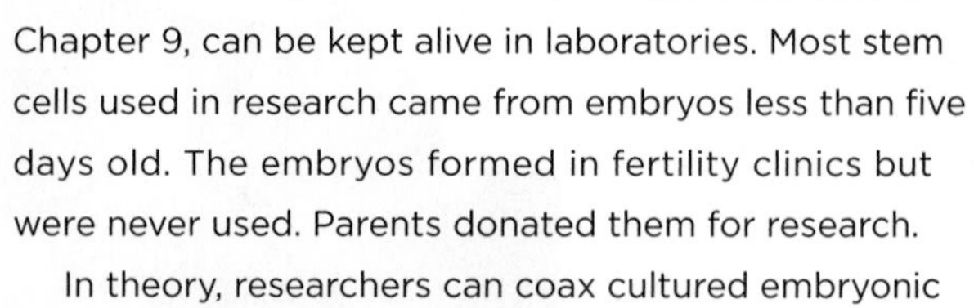

Human embryonic stem cells, like the HeLa cells in Chapter 9, can be kept alive in laboratories. Most stem cells used in research came from embryos less than five days old. The embryos formed in fertility clinics but were never used. Parents donated them for research.

In theory, researchers can coax cultured embryonic stem cells to be the start of any tissue. The idea is to replace a patient's cells that faltered or died because of injuries, Parkinson's disease, diabetes, heart attacks, and other severe problems.

Christopher Reeve, an actor best known for his role as Superman, stoked public awareness of stem cells. In 1995, he was thrown from his horse and landed on his head. The fall paralyzed him; he could not even breathe on his own. Medical advances kept him alive and gave him hope. He became a champion for the disabled and for human embryonic stem cell research (Figure 29.2).

Injecting embryonic stem cells into rats with spinal cord injuries can restore some neural function. The first trial of a human embryonic stem cell treatment may clarify whether a similar approach can restore some function to people who have a crushed spinal cord.

Adults have some stem cells too. For example, stem cells in adult bone renew themselves and also give rise to blood cells. But adult stem cells have already become somewhat specialized and produce a limited set of cell types. For example, stem cells in adult bone give rise to blood cells, but not to muscle cells or nerve cells.

Stem cell research is a fitting introduction to this unit, which deals with animal anatomy (how a body is put together) and physiology (how a body works). The chapter invites you to reflect on who we are, where we came from, and where medical technology is taking us.

In this chapter, you will start with the four basic types of animal tissues—epithelial, connective, muscle, and nervous tissues. A tissue, recall, is a community of

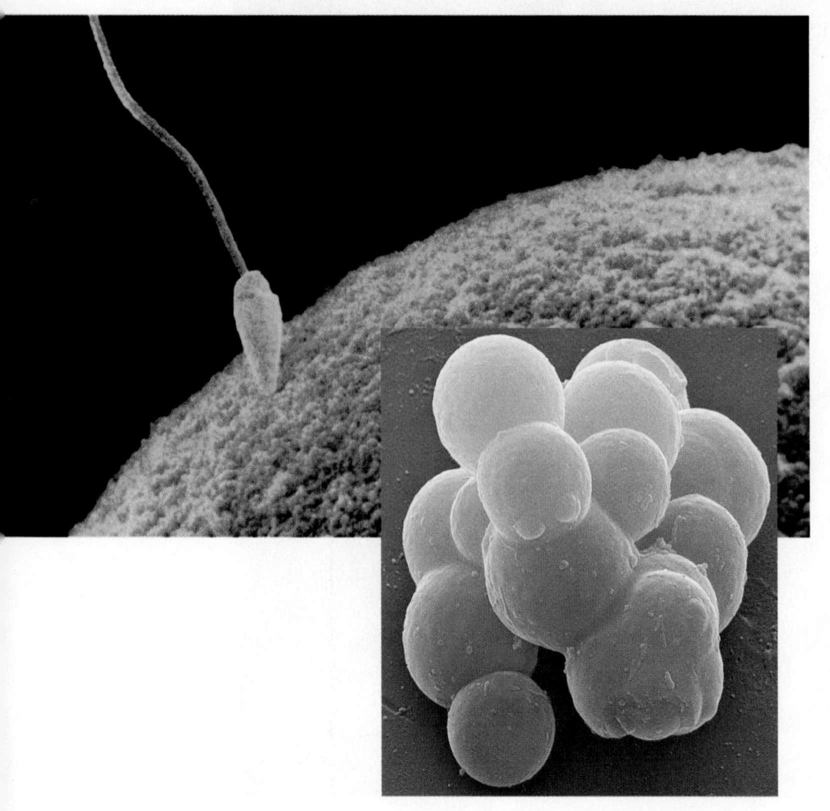

See the video! **Figure 29.1** *Left*, moment before the union of two cells— sperm and egg. *Right*, after fertilization and early divisions, a cluster of human stem cells. Each cell has the capacity to give rise to all of the body's specialized cells, tissues, and organ systems.

How would you vote? Human embryonic stem cells have potential medical benefits, but some people object to using them. Should researchers be allowed to start embryonic stem cell lines from human embryos that were frozen but never used for *in vitro* fertilization? See ThomsonNOW for details, then vote online.

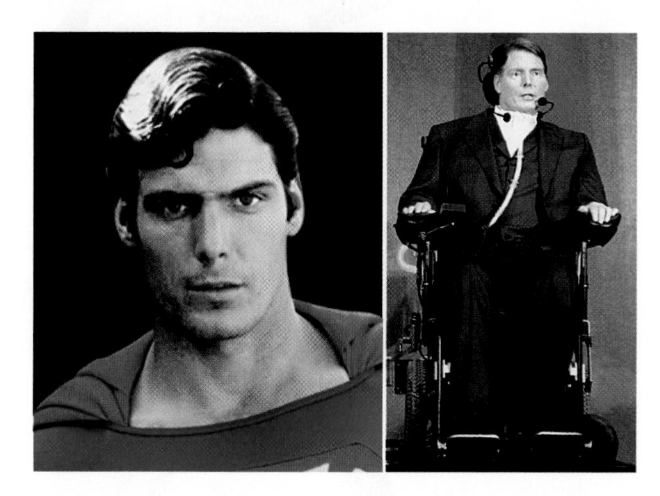

Figure 29.2 Actor Christopher Reeve before and after a spinal cord injury left him paralyzed. He died in 2004.

cells and intercellular substances that carry out one or more tasks, such as muscle tissue contraction. An organ is a structural unit of two or more tissues organized in proportions and patterns necessary to carry out specific tasks. Your heart is an organ made of all four types of tissues in certain proportions and arrangements. In organ systems, two or more organs and other components interact physically, chemically, or both in a common task, as when a beating heart forces blood through interconnected vessels.

In this unit, you will return repeatedly to a concept outlined in Chapter 25. The body's cells, tissues, and organs interact smoothly when its internal environment is maintained within a range that individual cells can tolerate. This process, remember, is called homeostasis. In most kinds of animals, blood and interstitial fluid act as an internal environment. Whether your focus is on the body of a flatworm or salmon, a bird or human, you will discover that it must perform the following tasks:

1. *Maintain homeostasis through coordination and control of the activities of its individual parts.*

2. *Acquire and distribute raw materials to individual cells and dispose of wastes.*

3. *Protect tissues against injury or attack.*

4. *Reproduce and, in many species, nourish and protect offspring through early growth and development.*

Key Concepts

BASIC TYPES OF ANIMAL TISSUES

Epithelial, connective, muscle, and nervous tissues are the basic categories of tissues in nearly all animals.

Epithelia line the body surface and its internal cavities and tubes. They have protective and secretory functions.

Connective tissues bind, support, strengthen, protect, and insulate other tissues. They include soft connective tissues, cartilage, bone, blood, and adipose tissue.

Muscle tissues help move the body and its parts. The three kinds are skeletal, cardiac, and smooth muscle tissue.

Nervous tissue provides local and long-distance lines of communication among cells. Its cellular components are neurons and neuroglia. **Sections 29.1–29.4**

INTRODUCING ANIMAL ORGAN SYSTEMS

Vertebrate organ systems compartmentalize the tasks of survival and reproduction for the body as a whole; they show a division of labor. Different systems arise from ectoderm, mesoderm, and endoderm, the primary tissue layers that form in the early embryo. **Section 29.5**

CASE STUDY: AN INTEGUMENTARY SYSTEM

Human skin is an example of an organ system. It has epithelial layers, connective tissue, adipose tissue, glands, blood vessels, and sensory receptors. It helps protect the body from injury and some pathogens, conserve water and control body temperature, excrete wastes, and detect some external stimuli. **Section 29.6**

Links to Earlier Concepts

With this unit, you have now arrived at the tissue and organ system levels of biological organization for animals (Section 1.1). This chapter expands on the nature of multicelled body plans (24.1, 25.1–25.3). It builds on your knowledge of the origin and evolution of animal tissues (24.1). You may wish to review the introductions to cell junctions (4.11), the surface-to-volume ratio (4.1), diffusion and transport proteins (5.6, 5.7), aerobic respiration (7.1), and energy conversion pathways (7.6).

29.1 Epithelial Tissue

Most of what you see when you look in a mirror—skin, hair, and nails—is epithelial tissue or structures derived from it. Epithelium also lines internal tubes and cavities, such as your blood vessels and gut.

GENERAL CHARACTERISTICS

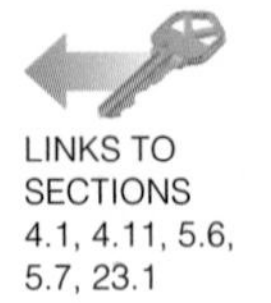

LINKS TO SECTIONS 4.1, 4.11, 5.6, 5.7, 23.1

Epithelium (plural, epithelia) is a sheetlike tissue of cells with little extracellular material between them. One free surface is exposed to the environment or to some body fluid (Figure 29.3). At the opposite surface, epithelial cell secretions form a basement membrane that glues the epithelium to another tissue.

Most epithelial cells have a squamous (flattened), cuboidal, or columnar shape, as in Figure 29.3*c*. Some epithelial cells are ciliated, whereas others specialize in secretion. For instance, the epithelial lining of your airways has ciliated and mucus-secreting cells (Figure 25.3). Other epithelia, such as those in your kidneys and gut, function in absorption. Fingerlike projections called microvilli at their free surface increase the area across which substances are absorbed.

Simple epithelium has one layer of cells; *stratified* epithelium has two or more layers. The outer layer of your skin is mostly stratified squamous epithelium.

GLANDULAR EPITHELIUM

Only epithelial tissue contains gland cells. These cells produce and secrete substances that function outside

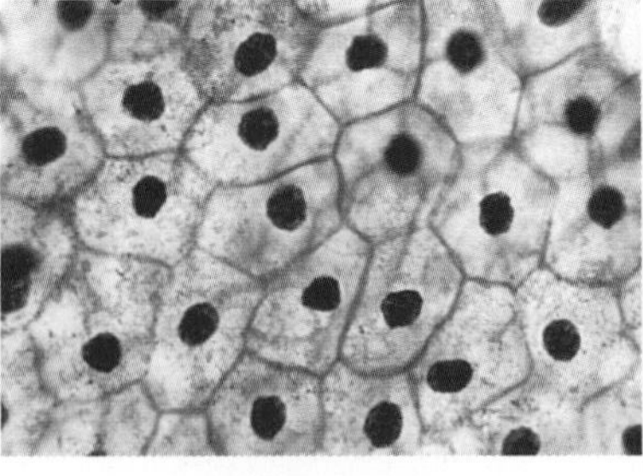

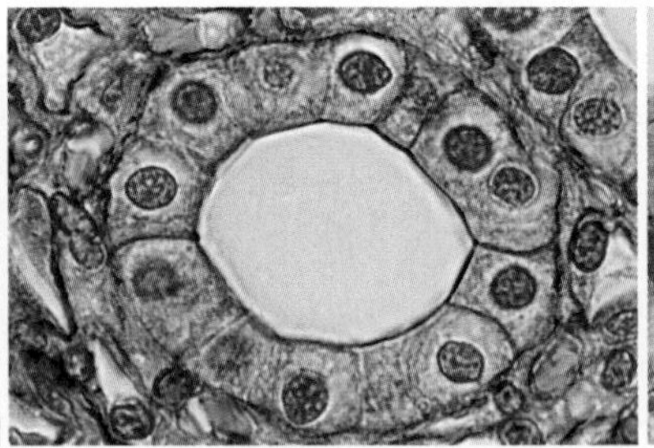

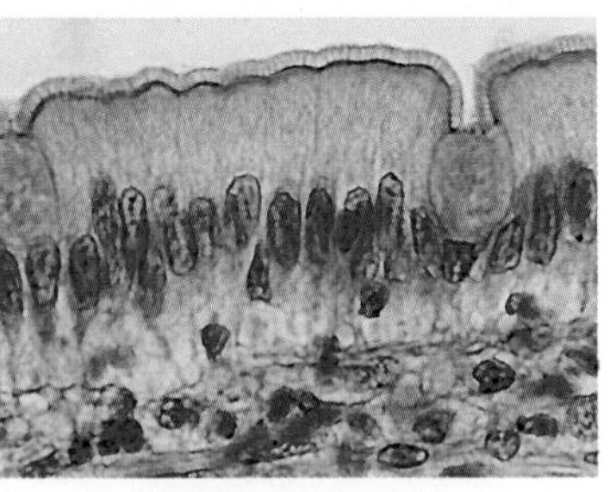

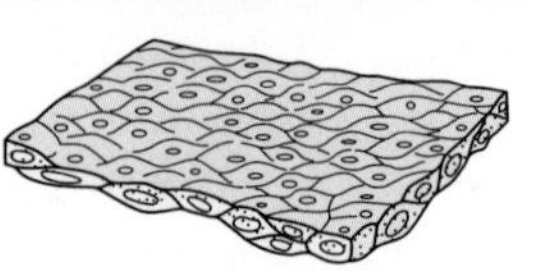

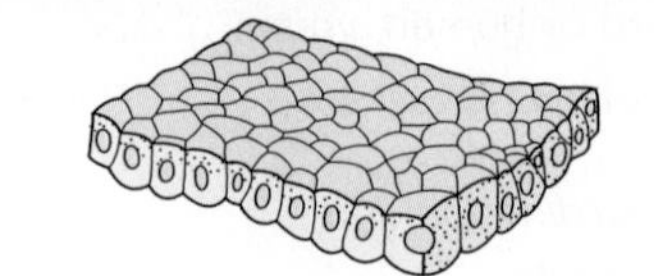

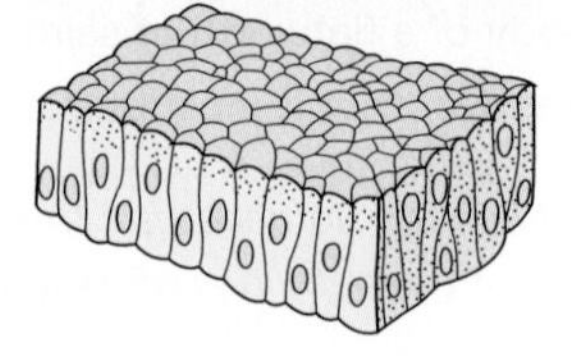

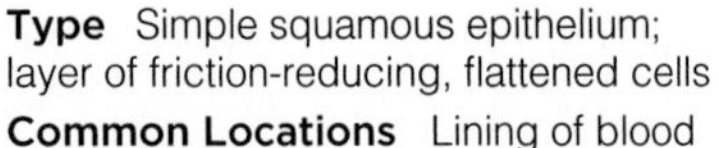

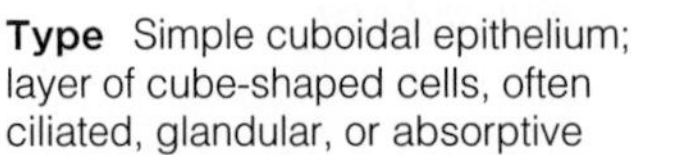

Figure 29.3 Some of the characteristics of epithelium. This animal tissue has a free surface that faces either the outside environment or some internal body fluid.

(**a**) Squamous epithelium of skin, a section starting at the surface. It consists of multiple layers of cells that flatten as they near the free surface.

(**b**) An epithelium attaches to tissue below it by means of the basement membrane that it secretes.

(**c**) Light micrographs and illustrations of three simple epithelia, showing the most common cell shapes.

c

Type Simple squamous epithelium; layer of friction-reducing, flattened cells
Common Locations Lining of blood and lymph vessels, heart, air sacs in lungs, abdominal cavity
Functions Diffusion of nutrients and gases, filtration of fluids

Type Simple cuboidal epithelium; layer of cube-shaped cells, often ciliated, glandular, or absorptive
Common Locations Lining of kidney tubules and bronchioles; ducts and secretory part of glands
Functions Secretion, absorption

Type Simple columnar epithelium; layer of elongated cells, often ciliated, secretory, or absorptive
Common Locations Lining of most of the gut, cervix, oviducts; glands and their ducts
Functions Secretion, absorption

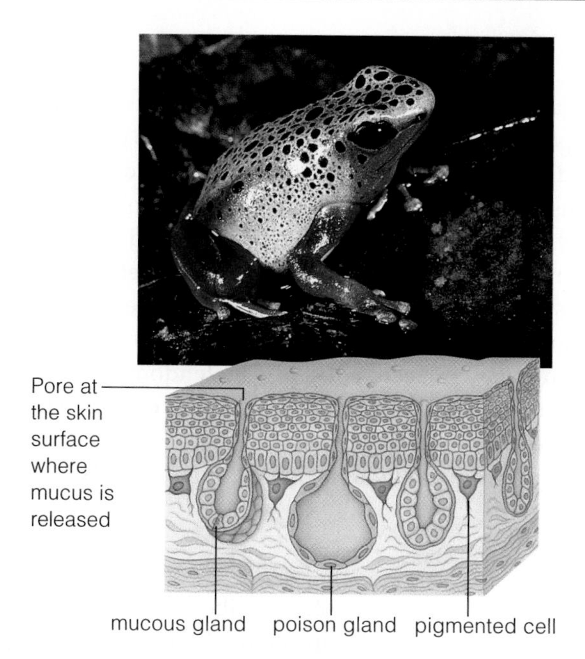

Figure 29.4 Glandular epithelium of a frog (*Dendrobates azureus*) that secretes a paralyzing poison. The hunters in one South American tribe tip their blowgun darts with this poison. The pigment-rich skin of all poisonous frogs has vivid colors and patterns that evolved as a warning signal. In essence, it says to predators, "Don't even think about it."

the cell. In most animals, secretory cells are clustered inside glands, organs that secrete substances onto the skin, or into a body cavity or the interstitial fluid.

Exocrine glands have ducts or tubes that deliver secretions onto a free epithelial surface (Figure 29.4). Exocrine secretions include mucus, saliva, tears, milk, digestive enzymes, and earwax.

Endocrine glands have no ducts. They secrete their products, hormones, directly into interstitial fluid. The hormone diffuses into the bloodstream, which delivers it to target cells. The plasma membrane of a target cell has receptors that bind the hormone. Target cells often are some distance from the source of the hormone.

CELL JUNCTIONS

Cell junctions connect adjoining cells in most tissues. The rows of proteins in tight junctions seal epithelial cell membranes together. These junctions, which occur only in epithelia, prevent fluids from seeping through the spaces in between cells (Figure 29.5*a*). This means that for fluid to cross the tissue, it must pass through the cells, not around them. Thus, transport proteins in cell membranes can control which ions and molecules cross the epithelium (Section 5.7).

An abundance of tight junctions in the lining of the stomach normally keeps acidic fluid from leaking out. If a bacterial infection damages this lining, acid and

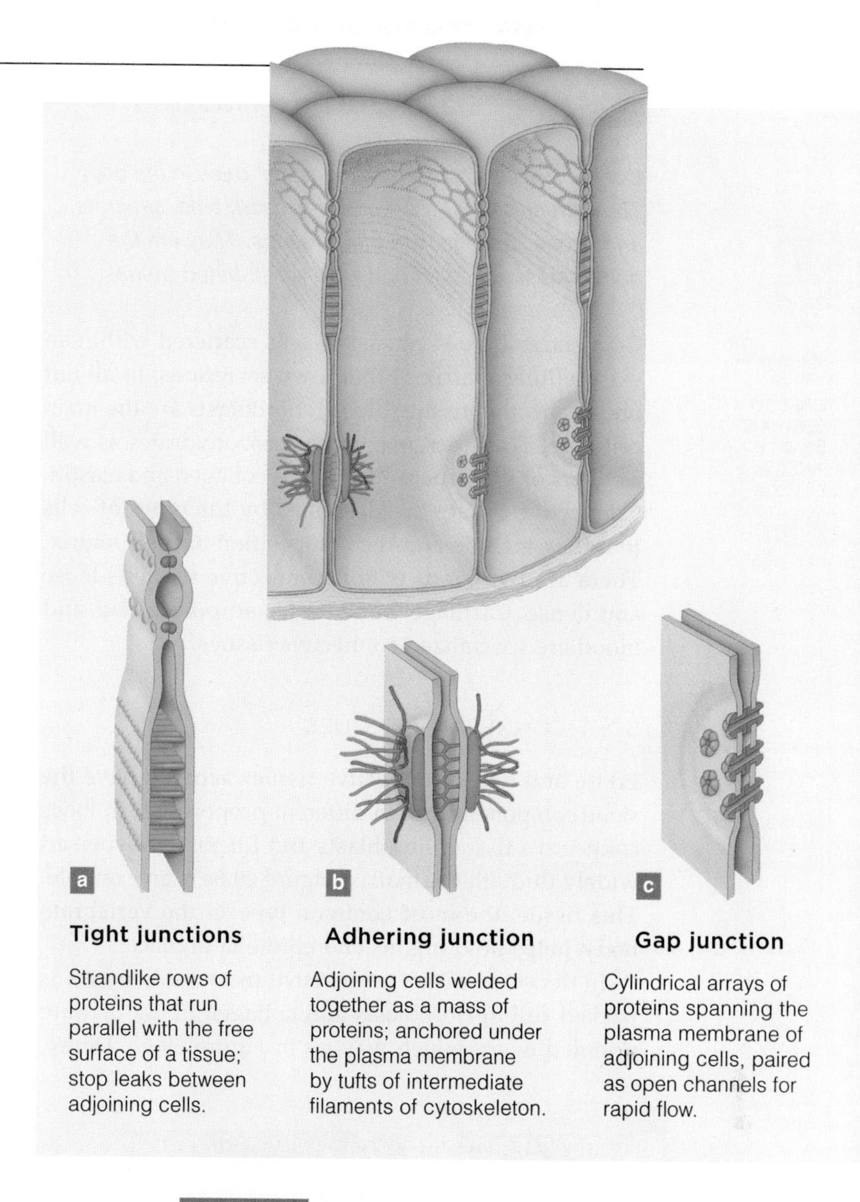

Figure 29.5 Animated! Examples of cell junctions in animal tissues.

enzymes can erode the underlying connective tissue and muscle layers. The result is a painful peptic ulcer.

Adhering junctions hold cells together at distinct spots, like buttons holding a shirt closed (Figure 29.5*b*). Skin and other tissues that are subject to abrasion or stretching have an abundance of adhering junctions.

Gap junctions permit ions and small molecules to pass freely from the cytoplasm of one cell to another (Figure 29.5*c*). These communication channels abound in heart muscle and other tissues in which the cells perform some coordinated action.

Epithelia are sheetlike tissues that line the body's surface and its cavities, ducts, and tubes. Epithelia have one free surface exposed to the outside environment or a body fluid. Glands are secretory organs derived from epithelium. Cell junctions structurally and functionally link adjoining cells.

29.2 Connective Tissues

Connective tissues have "connecting" roles in the body. They structurally or functionally support, bind, separate, and in one case insulate other tissues. They are the body's most abundant and widely distributed tissues.

LINKS TO SECTIONS 3.3, 4.11, 7.7, 24.3

Connective tissues consist of cells scattered within an extracellular matrix of their own secretions. In all but one connective tissue (blood), fibroblasts are the main cell type. They secrete complex carbohydrates as well as fibers of the structural proteins collagen and elastin. Connective tissues are described by the types of cells that they include and the composition of their matrix. There are two kinds of soft connective tissues—loose and dense. Cartilage, bone tissue, adipose tissue, and blood are specialized connective tissues.

SOFT CONNECTIVE TISSUES

Loose and dense connective tissues actually have the same components but in different proportions. In loose connective tissue, fibroblasts and fibers are dispersed widely through the matrix. Figure 29.6*a* is an example. This tissue, the most common type in the vertebrate body, helps hold organs and epithelia in place.

In dense, irregular connective tissue, the matrix is packed full of fibroblasts and collagen fibers that are oriented every which way, as in Figure 29.6*b*. Dense, irregular connective tissue makes up deep skin layers. It supports intestinal muscles and also forms capsules around organs that do not stretch, such as kidneys.

Dense, regular connective tissue has fibroblasts in orderly rows between parallel, tightly packed bundles of fibers (Figure 29.6*c*). This organization helps keep the tissue from being torn apart when placed under mechanical stress. Tendons and ligaments are mainly dense, regular connective tissue. The tendons connect skeletal muscle to bones. Ligaments attach one bone to another and are stretchier than tendons. Elastic fibers in their matrix facilitate movements around joints.

SPECIALIZED CONNECTIVE TISSUES

Cartilage has a matrix of collagen fibers and rubbery, compression-resistant glycoproteins. Cells secrete the matrix, which imprisons them (Figure 29.6*d*). Sharks have a cartilage skeleton. When you were an embryo, cartilage formed a model for your developing skeleton, then bone replaced most of it. Cartilage still supports the outer ears, nose, and throat. It cushions joints and is a shock absorber between vertebrae. Blood vessels do not extend through cartilage, as they do in other connective tissues. Substances diffuse from vessels in nearby tissues. Also unlike cells of other connective tissues, cartilage cells do not divide often in adults.

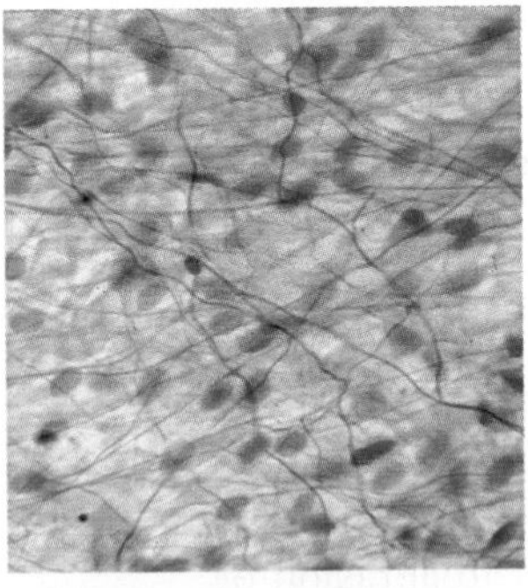

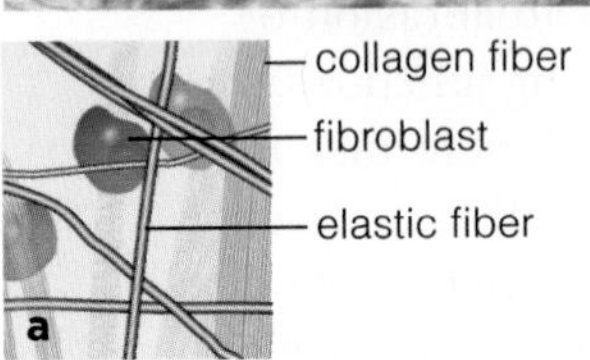

Type Loose connective tissue; fibroblasts, other cells scattered in a squishy matrix with relatively few fibers
Common Locations Beneath skin and most epithelia
Functions Elasticity, diffusion

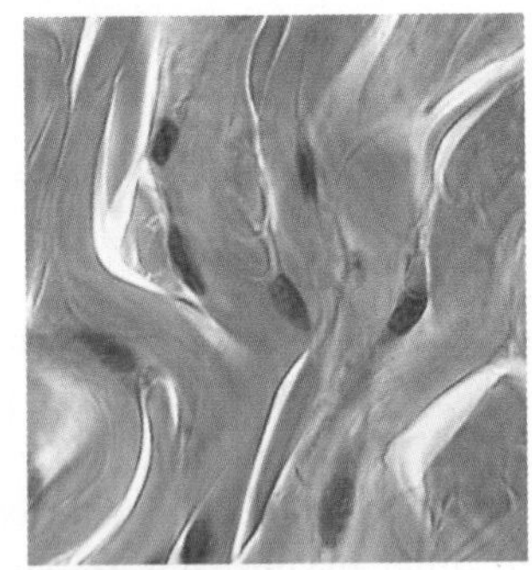

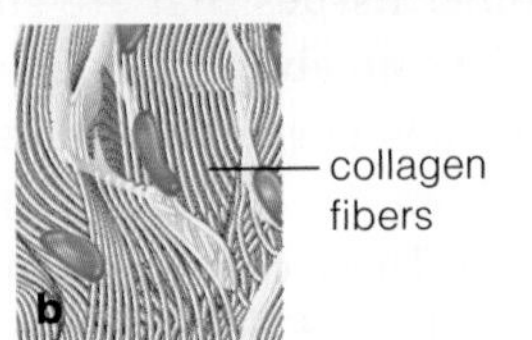

Type Dense, irregular connective tissue; fibroblasts in a semisolid matrix with many loosely interwoven collagen fibers
Common Locations In skin and in capsules around some organs
Function Structural support

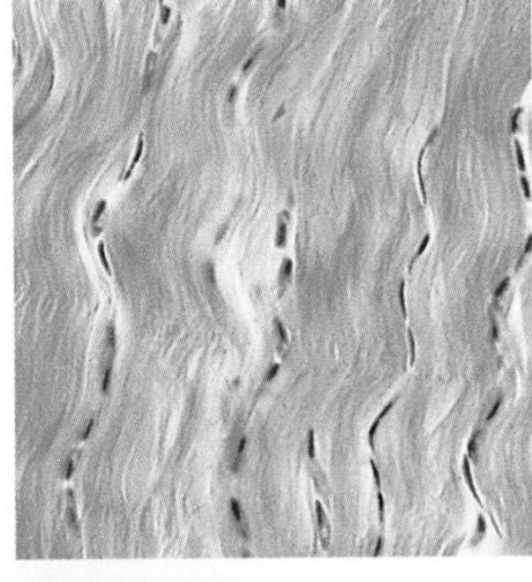

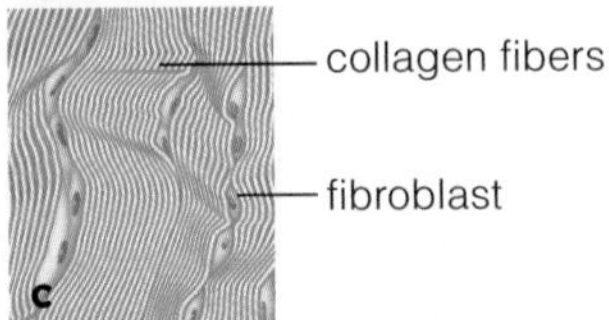

Type Dense, regular connective tissue; fibroblasts in rows between tight parallel bundles of many collagen fibers
Common Locations Tendons, ligaments
Functions Strength, elasticity

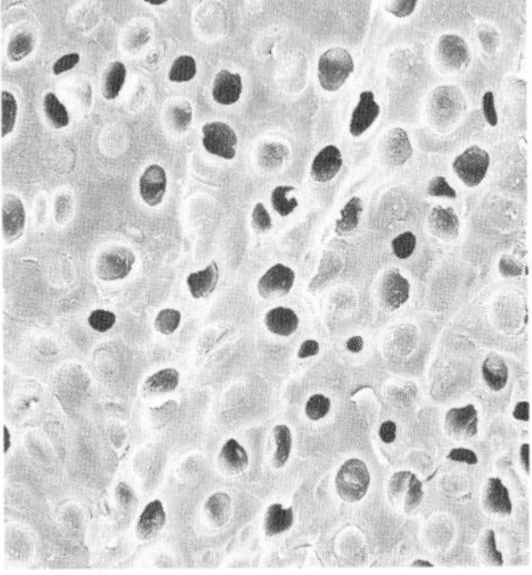

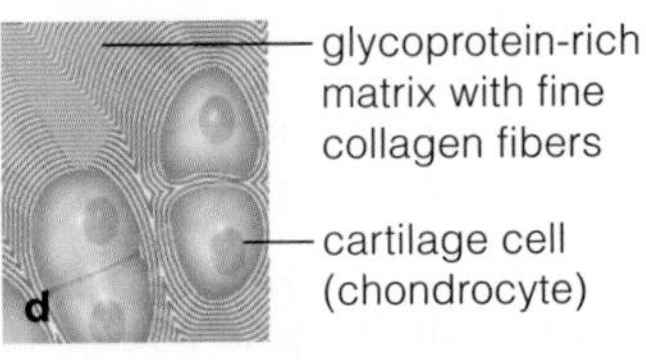

Type Cartilage; chondrocytes and collagen fibers in a rubbery matrix
Common Locations Nose, ends of long bones, airways, skeleton of cartilaginous fish, vertebrate embryo
Functions Support, protection, low-friction surface for joint movements

Figure 29.6 Characteristics of connective tissue.

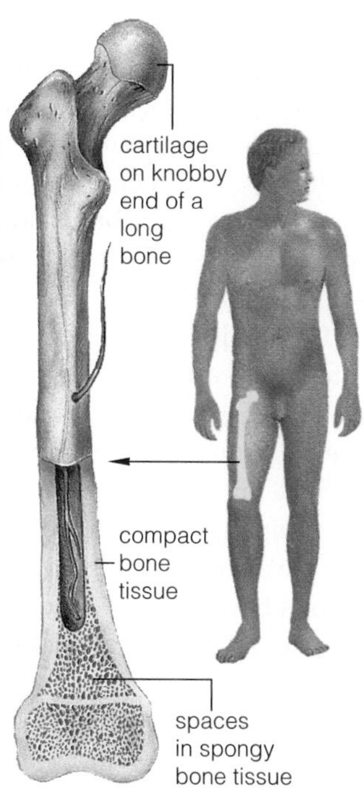

Figure 29.7 Cartilage and bone tissue. Spongy bone tissue has needlelike hard parts with spaces between. Compact bone tissue is more dense. Bone, a load-bearing tissue, resists being compressed and gives giraffes and other big animals selective advantages. They can ignore most predators and roam farther for food and water. Also, they have a lower surface-to-volume ratio than smaller animals do, so they lose or gain heat more slowly.

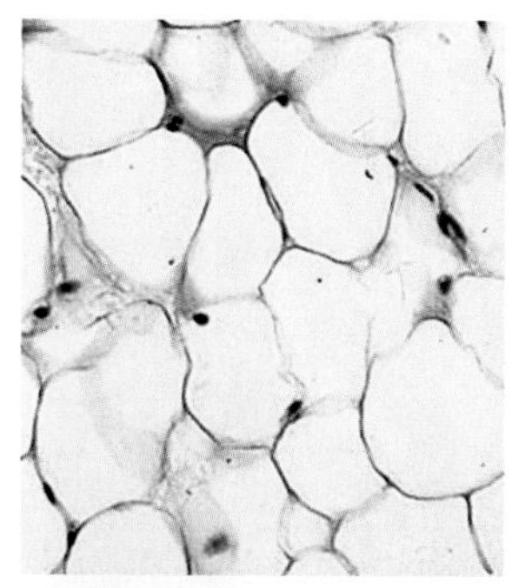

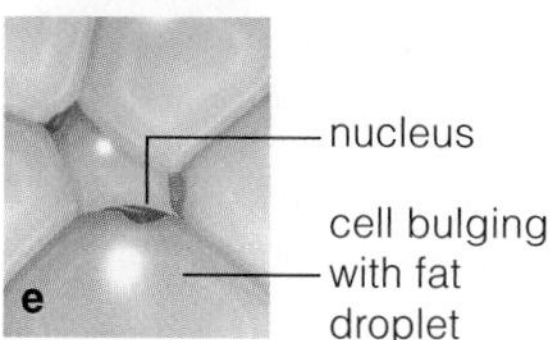

Type Adipose tissue; large, tightly packed fat cells with little extracellular matrix

Common Locations Under skin, around the heart and kidneys

Functions Energy storage, insulation, padding

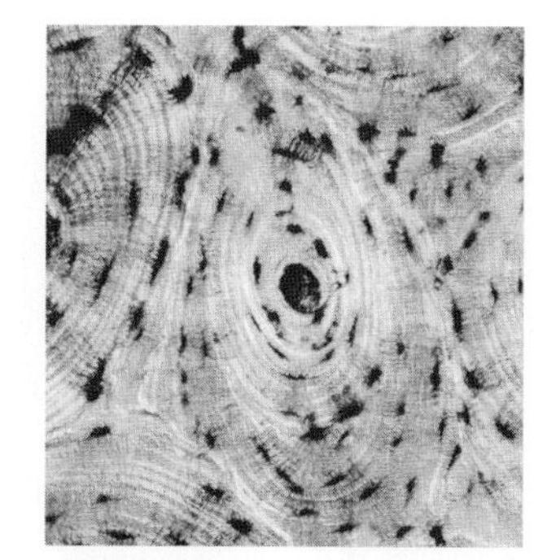

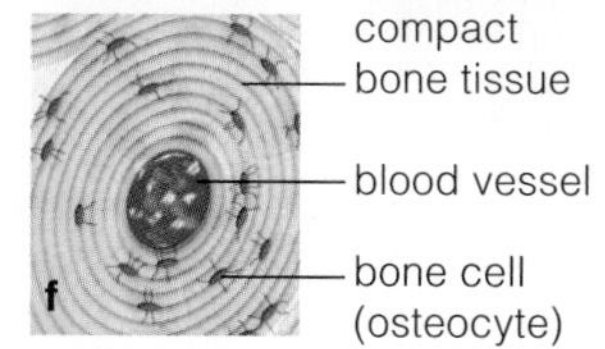

Type Bone tissue; collagen fibers, osteocytes in chambers inside an extensive, calcium-hardened extracellular matrix

Location All bony vertebrate skeletons

Functions Movement, support, protection

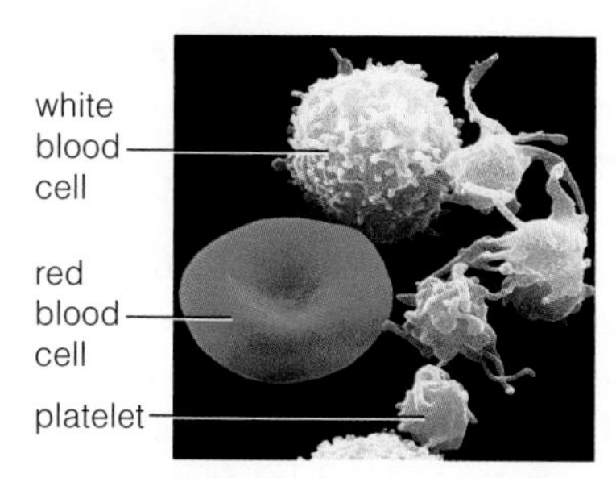

Figure 29.8 Cellular components of human blood. Many diverse proteins, nutrients, oxygen and carbon dioxide, and other substances also are dissolved in plasma, blood's straw-colored fluid portion.

Adipose tissue is the body's main energy reservoir. Most cells can convert excess sugars and lipids into droplets of fat (Section 7.7). However, only the cells of adipose tissue bulge with so much stored fat that the nucleus gets pushed to one side and flattened (Figure 29.6*e*). Adipose cells have little matrix between them. Small blood vessels run through the tissue and carry fats to and from cells. In addition to its energy-storage role, adipose tissue cushions and protects body parts, and a layer of adipose tissue under the skin functions as insulation; it helps keep internal body temperature within the optimal range.

Bone tissue is a connective tissue with living cells imprisoned in their own calcium-hardened secretions (Figure 29.6*f*). This is the main tissue of bones—organs that interact with muscles to move a body. Bones also support and protect the internal organs. Figure 29.7 shows a femur, a leg bone that is structurally adapted to bear weight. Blood cells form in the spongy interior of some bones.

Blood is considered a connective tissue because its cellular components (red blood cells, white blood cells, and platelets) are descended from stem cells in bone (Figure 29.8). Red blood cells filled with hemoglobin transport oxygen (Section 3.5). White blood cells help defend the body against pathogens. Platelets function in clot formation after a vessel is injured. The cellular components of blood drift along in the plasma, a fluid extracellular matrix consisting mostly of water, with dissolved nutrients, gases, and other substances.

Connective tissues support, protect, organize, or insulate other tissues. They consist of cells within an extracellular matrix. Except for blood, each contains fibroblasts.

The matrix of soft connective tissues contains characteristic proportions and arrangements of fibroblasts and fibers.

Cartilage, bone, adipose tissue, and blood are specialized connective tissues. Cartilage and bone are both structural materials. Adipose tissue is a reservoir of stored energy. Blood, a fluid connective tissue, has transport functions.

29.3 Muscle Tissues

Vertebrates have three types of muscle tissue: skeletal, cardiac, and smooth muscle tissues. Each type has unique properties that reflect its functions.

LINKS TO SECTIONS 4.12, 7.1

In muscle tissues, cells *contract*, or forcefully shorten in response to stimulation, then they relax and passively lengthen. These tissues consist of many cells arranged in parallel with one another, in tight or loose arrays. Coordinated contractions of layers or rings of muscles move the whole body or its component parts. Muscle tissue occurs in most animals, but we focus here on the kinds found in vertebrates.

SKELETAL MUSCLE TISSUE

Skeletal muscle tissue, the functional partner of bone (or cartilage), helps move and maintain the positions of the body and its parts. Skeletal muscle tissue has parallel arrays of long, cylindrical *muscle fibers* (Figure 29.9*a*). The fibers are not single cells. While embryos are developing, groups of cells fuse together and form each fiber, which ends up with multiple nuclei. Inside the fiber are myofibrils—long strands with row after row of contractile units. These rows are so regular that skeletal muscle has a striated, or striped, appearance.

Each unit, a sarcomere, is contractile. It has parallel, interacting arrays of the contractile proteins actin and myosin (Section 4.12). The structure and function of skeletal muscle is the focus of Section 33.4.

Skeletal muscle tissue makes up 40 percent or so of the weight of an average human. Reflexes activate it, but we also make it contract simply by thinking about it, as is happening in Figure 29.9*a*. That is why skeletal muscles are commonly called "voluntary" muscles.

CARDIAC MUSCLE TISSUE

Cardiac muscle tissue occurs only in the heart wall (Figure 29.9*b*). Like skeletal muscle tissue, it contains sarcomeres and looks striated. Unlike skeletal muscle tissue, it has branching cells. Cardiac muscle cells abut at their ends, where adhering junctions prevent them from being ripped apart during forceful contractions. Signals to contract pass swiftly from cell to cell at gap

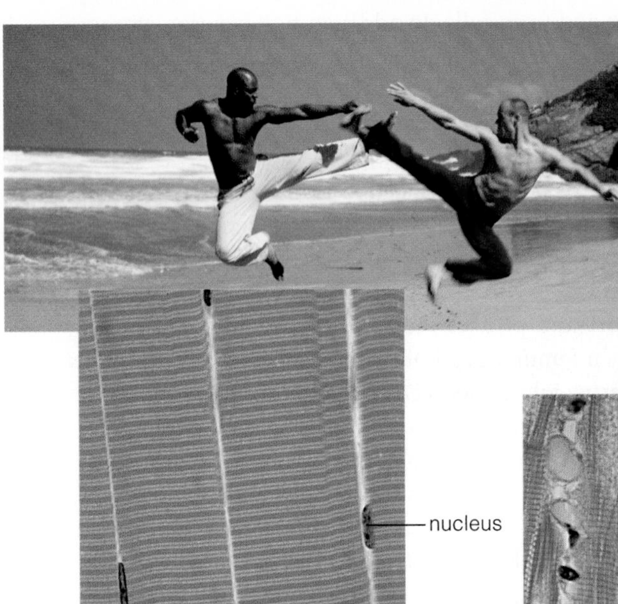

Type Skeletal muscle; bundles of long, cylindrical, striated muscle fibers, many mitochondria; often reflex-activated but under conscious control
Location Partner of skeletal bones, against which it exerts great force
Functions Locomotion; posture; head, limb movements

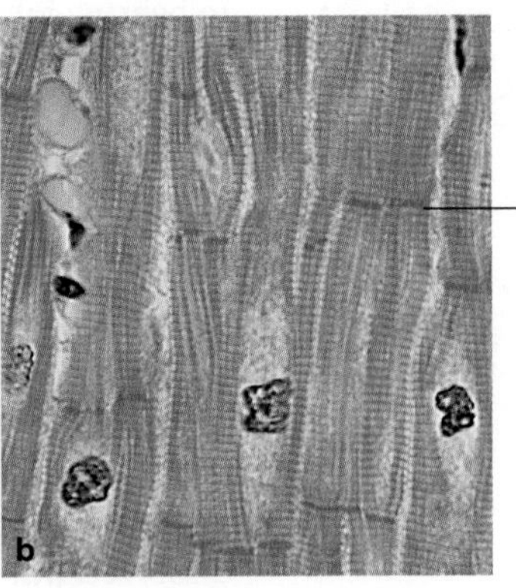

Type Cardiac muscle; cylindrical, unevenly striated muscle fibers that abut at their ends; signal flow through gap junctions makes them contract rapidly as a unit
Location Heart wall only
Function Forcefully pump blood through circulatory system

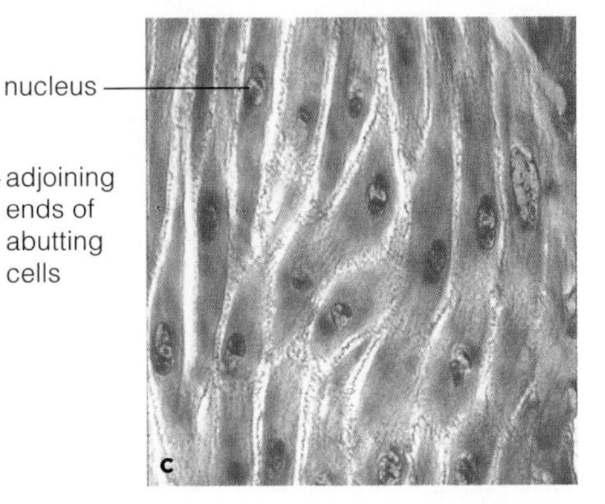

Type Smooth muscle; not striated, with contractile cells tapered at both ends
Locations Wall of arteries, veins, sphincters, stomach, urinary bladder, many other internal organs
Functions Controlled constriction, motility (as in gut), arterial blood flow

Figure 29.9 (**a**) Skeletal muscles. These functional partners of bones move the vertebrate body. Rows of contractile units give skeletal muscle a striated, or striped, appearance. Multiple nuclei press against each fiber's periphery. (**b**) Cardiac muscle tissue. Adhering junctions in horizontal bands hold its striated cells together. (**c**) Smooth muscle tissue, with tapered cells that have no striations.

junctions along their length. The rapid signals make all cells in cardiac muscle tissue contract as a unit.

Compared to other muscle tissues, cardiac muscle has far more mitochondria, which provide the beating heart with a dependable supply of ATP from aerobic respiration. Unlike skeletal muscle, cardiac muscle has little stored glycogen. If the blood flow to cardiac cells is interrupted, cells run out of glucose and oxygen fast. This happens during a heart attack. ATP production plummets, and cardiac muscle dies as a result.

Cardiac muscle and smooth muscle tissue occur in "involuntary" muscle, so named because we usually cannot make it contract just by thinking about it.

SMOOTH MUSCLE TISSUE

We find layers of smooth muscle tissue in the wall of many soft internal organs, such as the stomach, uterus, and bladder. This tissue's unbranched cells contain a nucleus at their center and are tapered at both ends (Figure 29.9*c*). Contractile units are not arranged in an orderly repeating fashion, as they are in skeletal and cardiac muscle tissue, so smooth muscle tissue is not striated. Even so, cells of this tissue contain actin and myosin filaments, which are anchored to the plasma membrane by intermediate filaments.

Smooth muscle tissue contracts more slowly than skeletal muscle, but its contractions can be sustained much longer. Contractions drive many internal events, as when they propel material through the gut, shrink the diameter of arteries, and close sphincters.

Muscle tissue, which functions in movement, contracts in response to stimulation.

Skeletal muscle is the functional partner of bones. Cardiac muscle is present only in the heart wall. Smooth muscle tissue is present in many soft internal organs.

29.4 Nervous Tissue

Of all animal tissues, the one with the communication lines made of neurons exerts the most control over how the body senses and responds to changing conditions.

Nervous tissue is composed of neurons and a variety of cells, collectively called neuroglia, that structurally and functionally support them. Neurons are a kind of excitable cell that makes up communication lines in most nervous systems. Figure 29.10 shows a common type, a motor neuron. A motor neuron consists of a cell body with a nucleus and typically long extensions that receive and transmit signals.

All cells respond to stimulation, but a neuron has a plasma membrane that can become excited in a specific way. When a neuron is stimulated, a message travels along its plasma membrane to the end of one of its cytoplasmic extensions. Here, the message triggers the release of signaling molecules, which diffuse across a small gap and convey information to another cell.

Your nervous system contains more than 100 billion neurons, and half of its volume consists of neuroglial cells that keep neurons positioned and functioning as they should. Sensory neurons detect specific stimuli, such as light and pressure. Neurons in the spinal cord and brain are interneurons. They receive and integrate sensory information. They also store information and coordinate responses to stimuli. Motor neurons relay commands from the brain and spinal cord to muscle cells, as in Figure 29.11, and to glands. Such signals can stimulate or inhibit activity in target cells, which is a topic of later chapters.

Figure 29.10 Motor neuron, which relays signals from the nervous system to muscle cells.

Neurons are the basic units of communication in nervous tissue. Different kinds detect specific stimuli, integrate information, and issue or relay commands to other tissues.

Nervous tissue also contains neuroglia. Diverse cells in this category structurally and functionally support the neurons.

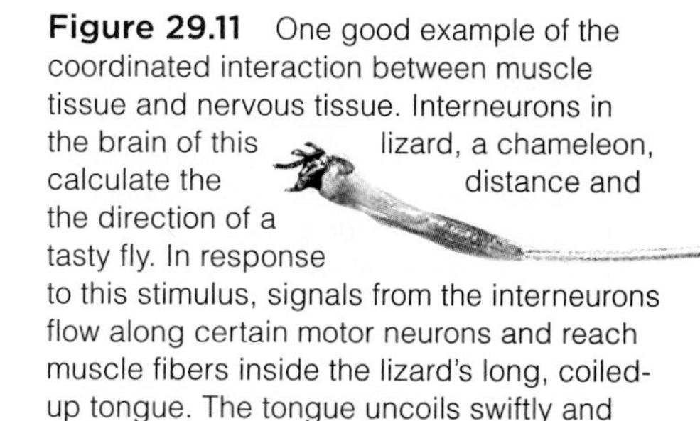

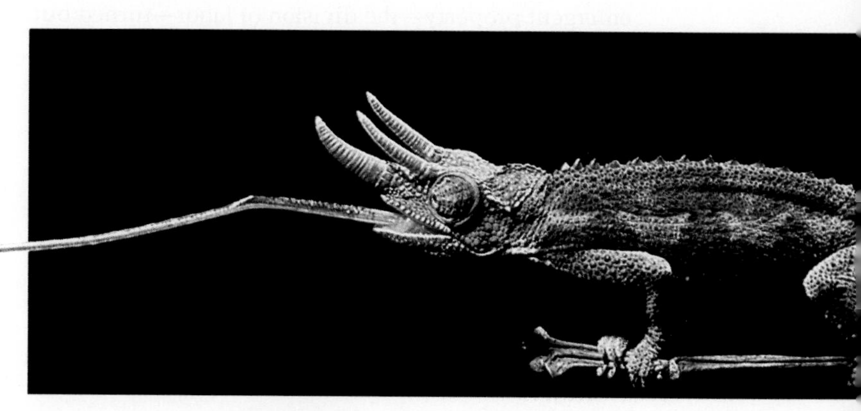

Figure 29.11 One good example of the coordinated interaction between muscle tissue and nervous tissue. Interneurons in the brain of this lizard, a chameleon, calculate the distance and the direction of a tasty fly. In response to this stimulus, signals from the interneurons flow along certain motor neurons and reach muscle fibers inside the lizard's long, coiled-up tongue. The tongue uncoils swiftly and precisely in the direction of the fly.

29.5 Overview of Major Organ Systems

Organ systems perform compartmentalized functions, such as gas exchange, blood circulation, and locomotion.

DEVELOPMENT OF TISSUES AND ORGANS

LINKS TO SECTIONS 1.1, 9.5, 23.1, 24.2

How do tissues of a vertebrate body develop? Recall that fertilization unites an egg and a sperm and forms a zygote. Mitotic cell divisions form a ball of cells that arrange themselves into three germ layers, or primary tissue layers. Growth and differentiation of the germ layers produce all of the adult tissues. Ectoderm, the outermost germ layer, becomes the nervous tissue and the epithelium of skin. Mesoderm, the middle germ layer, gives rise to muscle, connective tissue, and the lining of body cavities derived from the coelom. The innermost germ layer, endoderm, forms epithelium of the gut and also organs—such as lungs—that evolved from outpocketings of the gut (Section 24.2).

VERTEBRATE ORGAN SYSTEMS

Like other vertebrates, humans are bilateral and have a lined body cavity known as a coelom (Section 23.1). A sheet of smooth muscle, the diaphragm, divides the coelom into an upper thoracic cavity and a cavity that has abdominal and pelvic regions (Figure 29.12*a*). The heart and lungs are in the thoracic cavity. The stomach, intestines, and liver lie inside the abdominal cavity. The bladder and reproductive organs are in the pelvic cavity. A cranial cavity in the head and spinal cavity in the back are not derived from the coelom.

Figure 29.13 introduces the eleven organ systems of the human body. Other vertebrates have the same systems, which carry out the same functions. Figure 29.12*b*,*c* introduces anatomical terms used to describe the vertebrate body.

Collectively, organ systems of the vertebrate body show a division of labor—a compartmentalization of functions. Each helps a body survive in ways that no one tissue can. In Chapter 23, you glimpsed how this emergent property—the division of labor—turned out to be a key innovation that allowed the evolution of large-bodied animals. In the rest of this unit, you will come across examples of its extraordinary potential.

Animal organ systems compartmentalize many specialized tasks which, taken together, contribute to the survival and reproduction of the whole body.

In vertebrate embryos, organ systems arise from three germ layers: ectoderm, mesoderm, and endoderm.

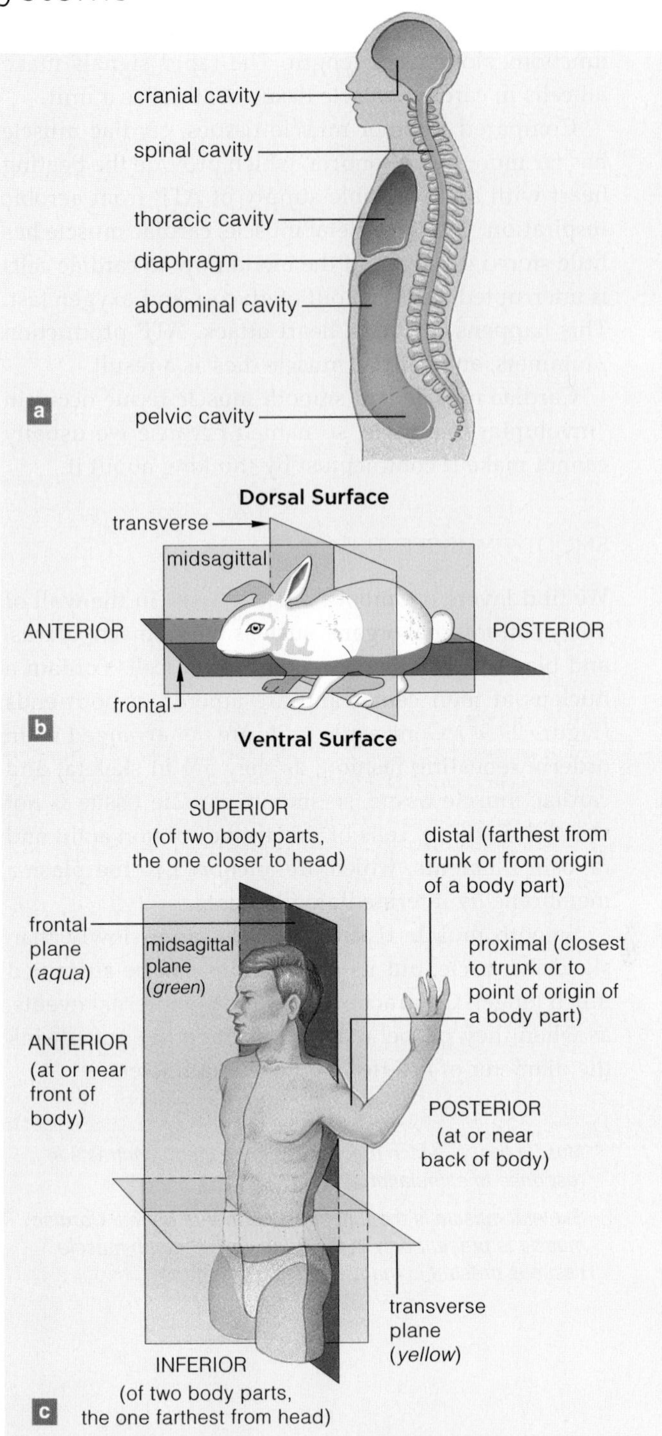

Figure 29.12 **Animated!** (**a**) Main body cavities in humans. (**b**,**c**) Directional terms and planes of symmetry for the body. For vertebrates that keep their main body axis parallel with Earth's surface, *dorsal* refers to the upper surface (back) and *ventral* to the lower surface. For upright walkers, *anterior* (the front) corresponds to ventral and *posterior* (the back) to dorsal.

Figure 29.13 **Animated!** *Facing page,* human organ systems and their functions.

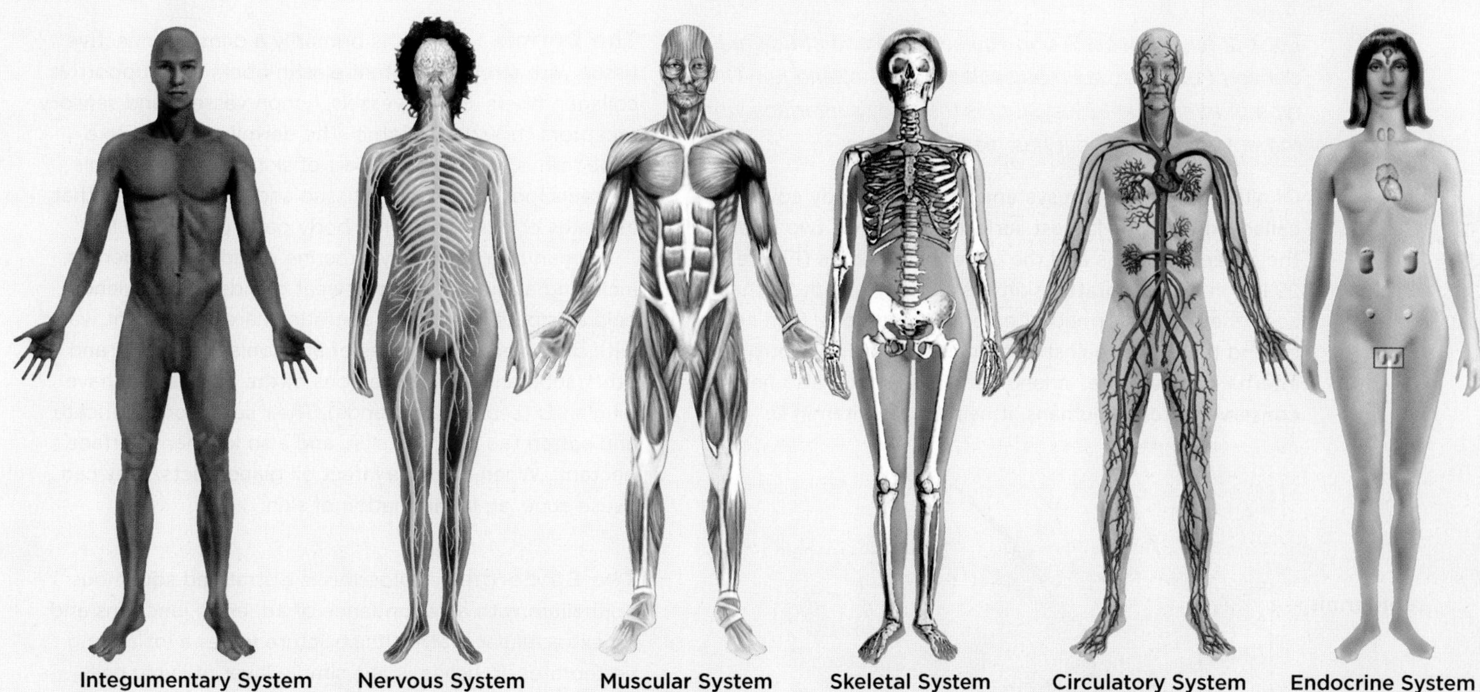

Integumentary System
Protects body from injury, dehydration, and some pathogens; controls its temperature; excretes certain wastes; receives some external stimuli.

Nervous System
Detects external and internal stimuli; controls and coordinates the responses to stimuli; integrates all organ system activities.

Muscular System
Moves body and its internal parts; maintains posture; generates heat by increases in metabolic activity.

Skeletal System
Supports and protects body parts; provides muscle attachment sites; produces red blood cells; stores calcium, phosphorus.

Circulatory System
Rapidly transports many materials to and from interstitial fluid and cells; helps stabilize internal pH and temperature.

Endocrine System
Hormonally controls body functioning; with nervous system integrates short- and long-term activities. (Male testes added.)

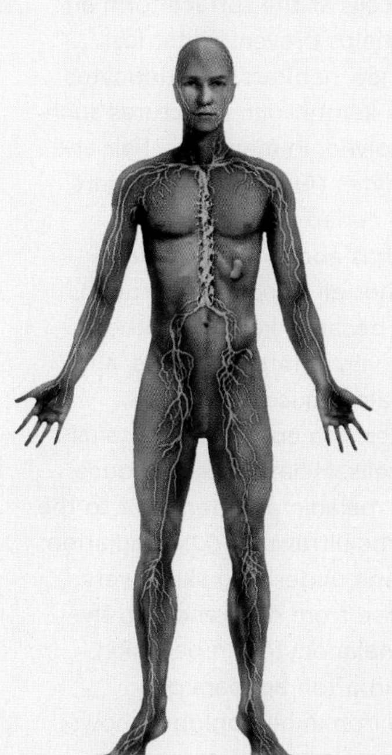

Lymphatic System
Collects and returns some tissue fluid to the bloodstream; defends the body against infection and tissue damage.

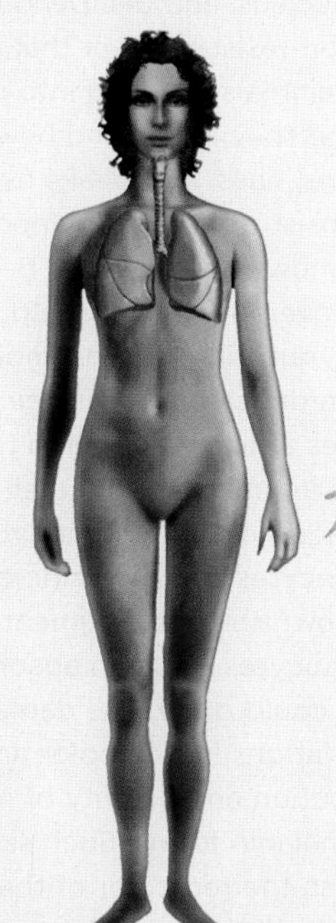

Respiratory System
Rapidly delivers oxygen to the tissue fluid that bathes all living cells; removes carbon dioxide wastes of cells; helps regulate pH.

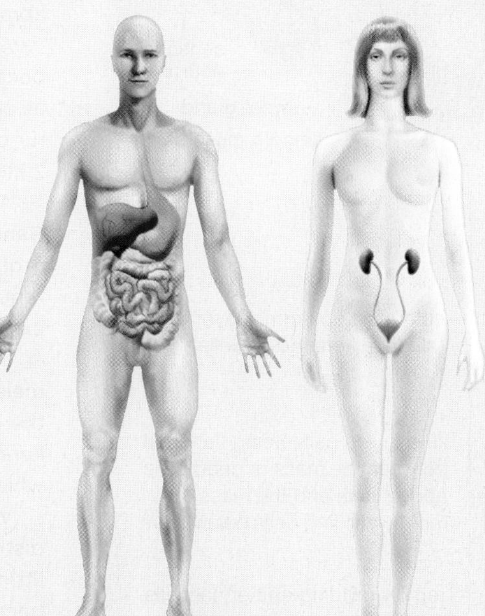

Digestive System
Ingests food and water; mechanically, chemically breaks down food and absorbs small molecules into internal environment; eliminates food residues.

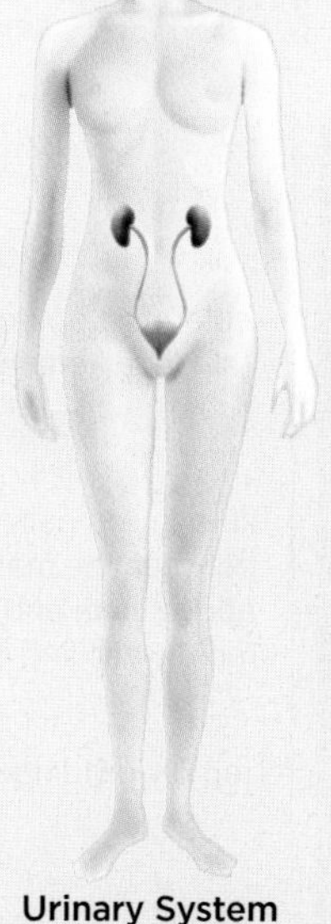

Urinary System
Maintains the volume and composition of internal environment; excretes excess fluid and bloodborne wastes.

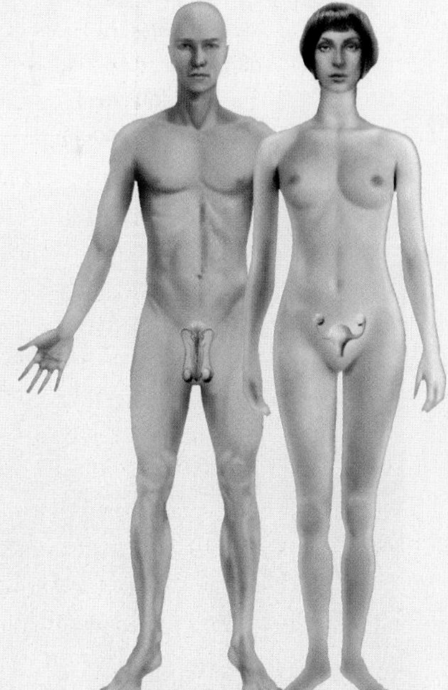

Reproductive System
Female: Produces eggs; after fertilization, affords a protected, nutritive environment for the development of new individuals. *Male:* Produces and transfers sperm to the female. Hormones of both systems also influence other organ systems.

29.6 Vertebrate Skin—Example of an Organ System

Look at any vertebrate and you see skin and structures derived from skin, such as scales, feathers, and fur. This type of integumentary system is the body's interface with the environment, and it has many functions.

LINKS TO SECTIONS 8.5, 24.10, 24.13

Of all vertebrate organ systems, the outer body covering called skin has the largest surface area. It has two layers, the outer epidermis and the underlying dermis (Figure 29.14). In all vertebrates, skin has sensory receptors that keep the brain informed of external conditions. Skin helps defend the body against pathogens and helps control its internal temperature. In land vertebrates, skin also helps conserve water. In humans, it helps make vitamin D.

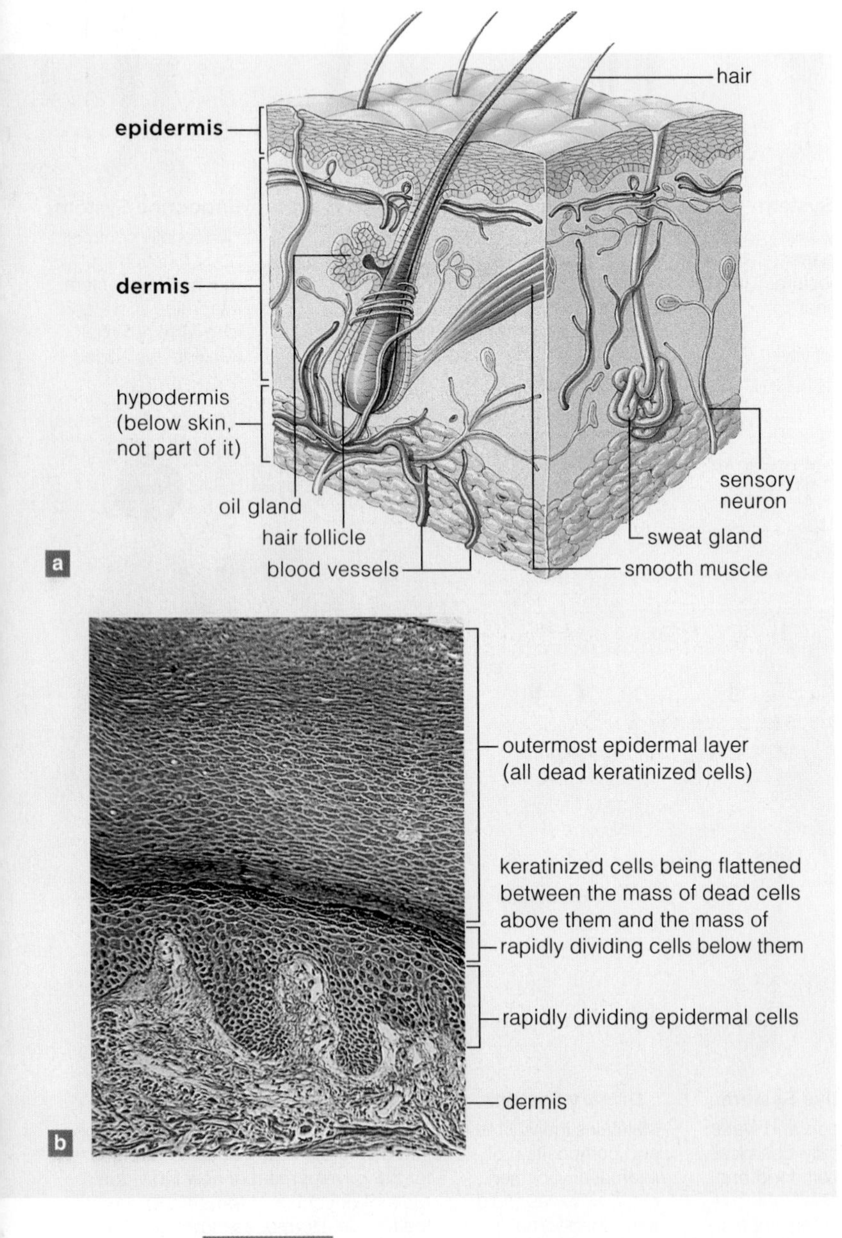

Figure 29.14 Animated! (**a**) Skin structure. Its components, such as glands, differ from one body region to the next. (**b**) Section through human skin.

The Dermis Dermis is primarily a dense connective tissue with stretch-resistant elastin fibers and supportive collagen fibers. Blood vessels, lymph vessels, and sensory receptors thread through it. The dermis attaches to a hypodermis, which is not part of skin. The hypodermis contains loose connective tissue and adipose tissue that insulates or cushions some body parts (Figure 29.14).

Human skin has many exocrine glands in the dermis, including about 2.5 million sweat glands. Sweat glands help dissipate heat. Their secretions are 99 percent water, with dissolved salts, traces of ammonia, vitamin C, and other substances. Most regions of the dermis also have oil glands (sebaceous glands). Their secretions lubricate and soften the hair and skin, and also kill many surface bacteria. When bacteria infect oil gland ducts, they can cause acne, an inflammation of skin.

The Epidermis Epidermis is a stratified squamous epithelium with an abundance of adhering junctions and no extracellular matrix. Its structure varies a lot among vertebrates. In fishes, it is a simple layer of living cells. The evolution of a thick outer layer of keratinocytes—cells that make the tough, waterproof protein keratin—accompanied the move onto land. In tetrapods, ongoing mitotic cell divisions in deep epidermal layers push newly formed keratinocytes toward the skin's surface. As cells move toward the surface, they become flattened, lose their nucleus, and die. Dead cells at the surface form an abrasion-resistant layer that helps prevent water loss.

As animals radiated into new habitats, keratinocytes became specialized and new keratin-rich structures such as claws, nails, and beaks evolved. In mammals, hair and fur consist of dead keratinocytes (Figure 29.15). Figure 29.16 shows how hair can be curled or straightened.

An average human scalp has about 100,000 hairs. Genes, nutrition, and hormones all affect their growth. Protein deficiency thins hair because keratin synthesis requires amino acids. High fever, emotional stress, and excess vitamin A in the diet also cause hair loss.

In addition to keratinocytes, the epidermis contains melanocytes and dendritic cells. Melanocytes produce the brownish-black pigment melanin and donate it to the keratinocytes. Melanin absorbs ultraviolet (UV) radiation, which could otherwise damage underlying skin layers.

Variations in skin color arise from differences in the distribution and activity of melanocytes. In pale skin, little melanin forms. Such skin often appears pink because the red color of the iron in hemoglobin shows through thin-walled blood vessels and epidermis. An orange pigment, carotene, also contributes to skin color.

Dendritic cells migrate through epidermis. These phagocytic white blood cells engulf bacteria or viruses and alert the immune system to the threat. Ultraviolet radiation damages them. Skin becomes more vulnerable to viral outbreaks, such as painful cold sores caused by the *Herpes simplex* virus (Section 19.4).

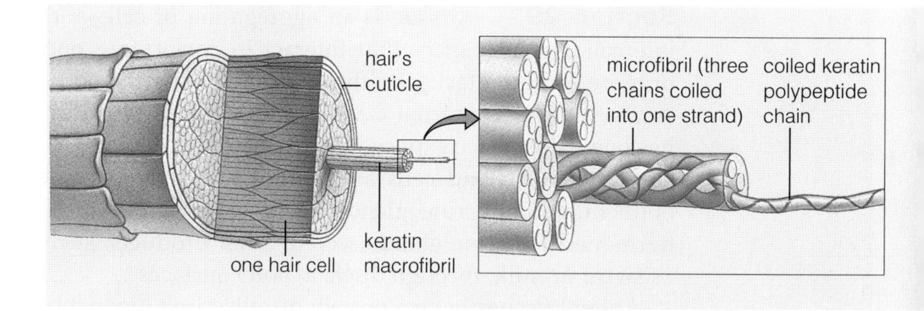

Figure 29.15 Animated! Hair fine structure. Dead, flattened hair cells accumulate and form a tubelike cuticle around a hair's shaft. They are derived from modified skin cells that synthesize polypeptide chains of the protein keratin. Disulfide bridges link three chains into thin fibers, which become bundled into larger fibers. The fibers almost completely fill the cells, which in time die off. Many people play with disulfide bridges (Figure 29.16).

Lab-grown epidermis is often used to protect tissues and aid wound healing. One company grows it from cells of foreskins that were discarded earlier from circumcised male infants. The resulting product is missing some cell types, such as melanocytes, and it has no glands.

Suntans and Shoe-Leather Skin Sunlight can burn unprotected light skin, sometimes severely. UV light stimulates melanocytes in skin to make melanin, which gives skin the "tan" that many light-skinned people covet (Figure 29.17). Melanin production accelerates, then peaks about ten days after tanning starts.

Dark-skinned people are better protected than light-skinned ones. But in anyone, prolonged or repeated UV exposure damages collagen and causes elastin fibers to clump. Chronically tanned skin gets less resilient and starts to look like shoe leather. UV harms DNA, and the damage can lead to skin cancer (Section 8.5).

As we age, epidermal cells divide less often. Skin thins and becomes less elastic as collagen and elastin fibers become sparse. Glandular secretions that kept it soft and moist dwindle. Wrinkles deepen. Many people needlessly accelerate the aging process by indulging in tanning or smoking, which shrinks the skin's blood supply.

Evolution of Skin Color A bit of UV radiation is a good thing; it stimulates melanocytes to make a molecule that the body later converts to vitamin D. We need this vitamin to absorb calcium ions from food. At the same time, UV exposure causes the breakdown of folate, one of the B vitamins. Among other problems, a deficiency in folate during development damages the nervous system.

By one hypothesis, variations in skin color among human populations evolved as adaptations to differences in sunlight exposure. Humans arose in Africa, where the sun's rays are intense (Section 24.13). There, melanin-rich skin protected folate and still made enough vitamin D. Later, some humans moved to colder regions, where sunlight is less intense, winter days are short, and more time is spent indoors or bundled in clothing. There, skin with fewer melanocytes was advantageous; it was easier to make enough vitamin D during cold, dark winters.

Researchers tested the hypothesis by comparing annual UV levels and skin color of people native to more than fifty countries. They found people in areas where the annual UV level is high tend to be darker skinned than those in regions where UV exposure is lower.

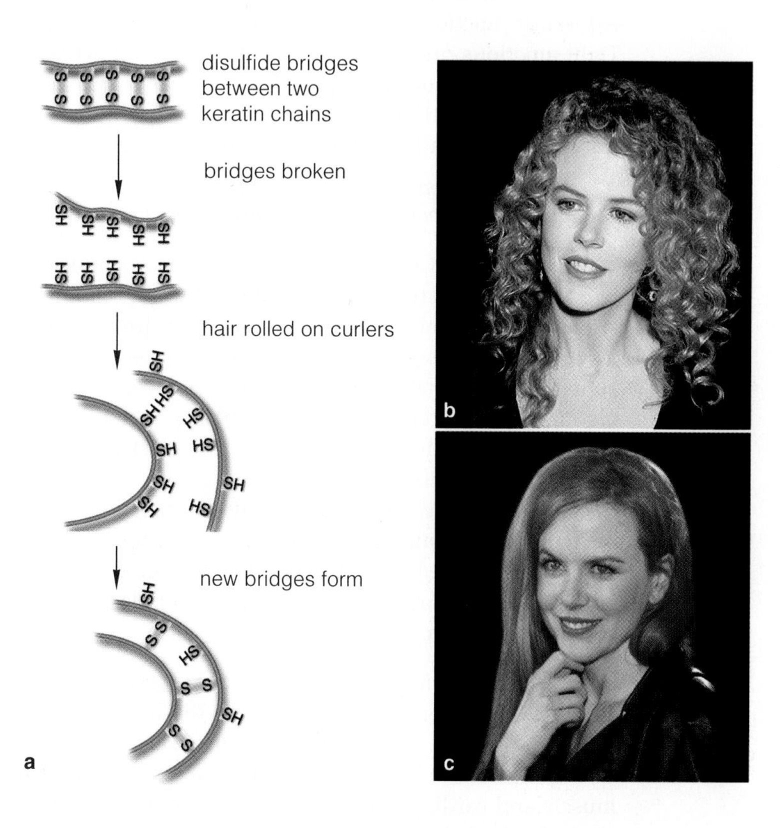

Figure 29.16 (**a**) Curly or straight hair? Hair's long keratin chains are bonded together by disulfide bridges, which break apart when exposed to chemicals. Ironing (flattening out) hair or rolling it around curlers holds the unbonded chains in new positions. Exposure to a different chemical makes new disulfide bridges form between different sulfur-bearing amino acids. The displaced bonding locks chains in new positions. That is how many women, including actress Nicole Kidman (**b**,**c**), straighten their naturally curly hair.

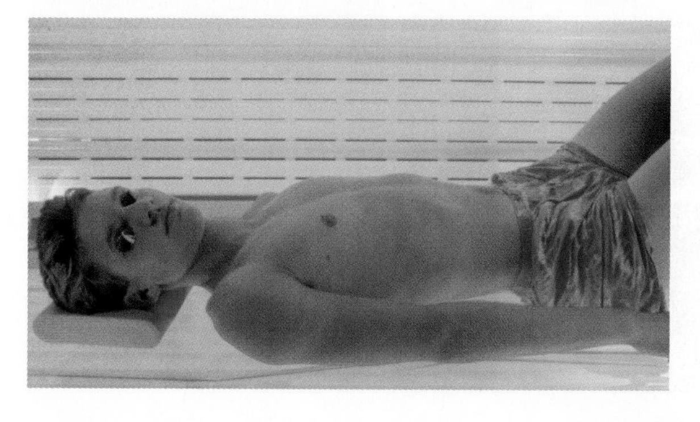

Figure 29.17 Demonstration of how to get shoe-leather skin, in this case, in a tanning salon.

Summary

Section 29.1 A tissue is an aggregation of cells and intercellular substances that interact in performing one or more common tasks. Epithelial tissues cover the body surface and line internal cavities and tubes. Epithelium has one free surface exposed to either a body fluid or the environment. Gland cells and glands are derived from epithelium. Endocrine glands are ductless and secrete hormones. Exocrine glands secrete other products, such as sweat or milk, through ducts to body surfaces.

Animal tissues have a variety of cell-to-cell junctions. Adhering junctions cement neighboring cells together. Tight junctions prevent substances from leaking across a tissue. Gap junctions are open channels that connect the cytoplasm of abutting cells. These junctions permit rapid transfer of ions and small molecules between cells.

- *Use the animation on ThomsonNOW to compare the structure and function of the main animal cell junctions.*

Section 29.2 Connective tissues "connect" tissues to one another, both functionally and structurally. Different types bind, organize, support, strengthen, protect, and insulate other tissues. All contain cells scattered within an extracellular matrix of their own secretions. Except in blood, the main cell type is the fibroblast. It makes and secretes structural fibers of collagen and elastin into the extracellular matrix.

Loose connective tissue and dense connective tissue have the same components but differ in the proportions. They are classified as soft connective tissues. Cartilage, bone tissue, adipose tissue, and blood are classified as specialized connective tissues.

Section 29.3 Muscle tissues contract (shorten), then passively lengthen. They help move the body and its component parts. The three types are skeletal muscle, cardiac muscle, and smooth muscle tissue. Only skeletal muscle and cardiac muscle tissues appear striated. Only skeletal muscle is under voluntary control.

Section 29.4 Neurons in nervous tissue make up communication lines through the body. Different kinds detect, integrate, and assess stimuli about internal and external conditions, and deliver commands to muscles and glands that carry out responses. Nervous tissue also contains a diverse collection of cells known as neuroglia. Neuroglia protect and support the neurons.

Section 29.5 An organ system consists of two or more organs that interact chemically, physically, or both in tasks that help keep individual cells as well as the whole body functioning. Most vertebrate organ systems contribute to homeostasis; they help maintain conditions in the internal environment within tolerable limits and so benefit individual cells and the body as a whole.

All tissues and organs of an adult animal arise from three primary tissue layers that form in early embryos: ectoderm, mesoderm, and endoderm. Ectoderm develops first and its cellular descendants become the epidermis and parts of the nervous system. The other two primary tissues give rise to other internal tissues and organs.

Collectively, the organ systems of a multicelled body demonstrate a division of labor, a compartmentalization of functions—that help the body survive in ways that no one tissue can offer.

- *Use the animation on ThomsonNOW to investigate the function of vertebrate organ systems and learn terms that describe their locations.*

Section 29.6 An organ system called skin functions in protection, temperature control, detection of shifts in external conditions, vitamin D production, and defense.

- *Use the animation on ThomsonNOW to explore the structure of human skin.*

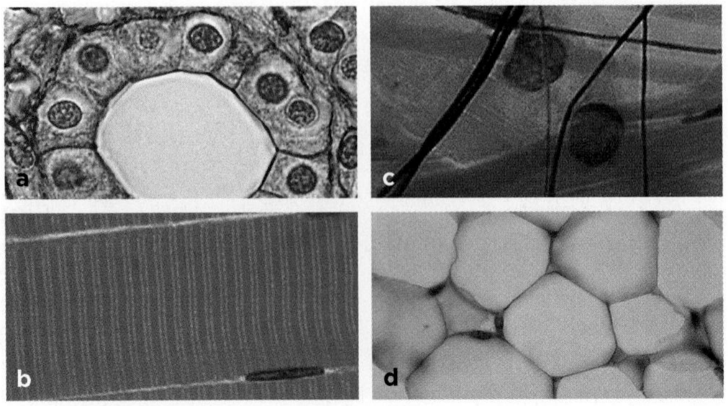

Self-Quiz

Answers in Appendix III

1. The four light micrographs at lower left show four types of animal tissues. Identify each type and write out a brief description of its defining features.

2. ______ tissues are sheetlike with one free surface.
   a. Epithelial c. Nervous
   b. Connective d. Muscle

3. ______ function in cell-to-cell communication.
   a. Tight junctions c. Gap junctions
   b. Adhering junctions d. all of the above

4. In most animals, glands are located in ______ tissue.
   a. epithelial c. muscle
   b. connective d. nervous

5. Most ______ have many collagen and elastin fibers.
   a. epithelial tissues c. muscle tissues
   b. connective tissues d. nervous tissues

6. ______ is mostly plasma.
   a. Adipose tissue c. Cartilage
   b. Blood d. Bone

7. Your body converts excess carbohydrates and proteins to fats. ______ specializes in storing the fats.
   a. Epithelial tissue c. Adipose tissue
   b. Dense connective tissue d. both b and c

8. In your body, cells of ______ can shorten (contract).
   a. epithelial tissue c. muscle tissue
   b. connective tissue d. nervous tissue

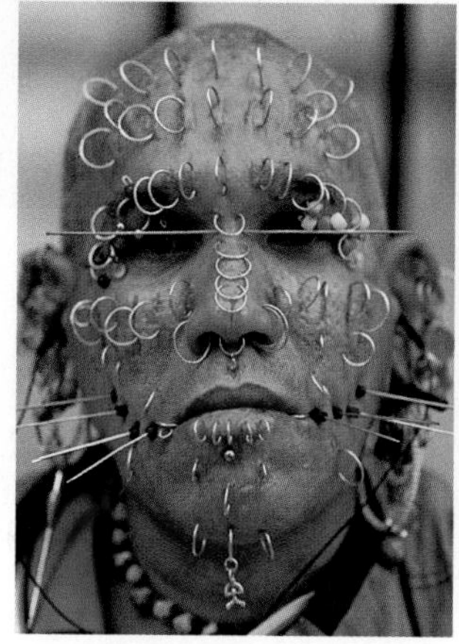

Figure 29.18 Assaults on the integument: skin piercings.

Figure 29.19 In the Kalahari Desert, gray meerkats (*Suricata suricatta*). Each morning, they come out of their burrows and face the sun's warming rays.

9. Only ______ muscle tissue has a striated appearance.
   a. skeletal c. cardiac
   b. smooth d. a and c

10. ______ detects and integrates information about changes and controls responses to those changes.
   a. Epithelial tissue c. Muscle tissue
   b. Connective tissue d. Nervous tissue

11. Match the terms with the most suitable description.

___ exocrine gland	a. strong, pliable; like rubber
___ endocrine gland	b. secretion through duct
___ cartilage	c. outermost primary tissue
___ ectoderm	d. contracts, not striated
___ smooth muscle	e. cements cells together
___ blood	f. fluid connective tissue
___ adhering junction	g. ductless secretion

■ *Visit ThomsonNOW for additional questions.*

Critical Thinking

1. The nose, lips, tongue, navel, nipples, and genitals are often targets for *body piercing:* cutting holes into the body so jewelry can be threaded through them (Figure 29.18). *Tattooing*, or using permanent dyes to make patterns in skin, is another fad. Besides being painful, both skin invasions can invite bacterial infections, chronic viral hepatitis, AIDS, and other diseases if the piercers and tattooers reuse unsterilized needles, dye, razors, gloves, swabs, and trays. Months may pass before any problems develop, so the cause-and-effect connection isn't always obvious. If despite the risks you think tissue invasions are okay, how can you be sure the equipment used is sterile?

2. Adipose tissue and blood are often said to be atypical connective tissues. Compared to other types of connective tissues, which of their features are *not* typical?

3. Many people oppose the use of animals for testing the safety of cosmetics. They say alternative test methods are available, such as the use of *lab-grown tissues* in some cases. Given what you learned in this chapter, speculate on the advantages and disadvantages of tests that use specific lab-grown tissues as opposed to living animals.

4. After a cold night in Africa's Kalahari Desert, animals small enough to fit inside a coat pocket emerge stiffly from burrows. These meerkats are a type of mongoose. They stand on their hind legs and face east, exposing their chilled bodies to the warm rays of the morning sun (Figure 29.19). Once meerkats warm up, they fan out and search for food. Into the meerkat gut go insects and the occasional lizard. These are pummeled, dissolved, and digested into glucose and other nutritious bits small enough to move across the gut wall, into blood, and on to the body's cells.

Name as many tissues as you can that might have roles in (1) keeping the meerkat body warm, (2) moving the body, as during foraging and heart-thumping flights from predators, and (3) digestion and absorption of nutrients, and elimination of the residues.

5. *Porphyria* is a name for a set of rare genetic disorders. Affected people lack one of the enzymes in the metabolic pathway that forms heme, the iron-containing group of hemoglobin. As a result, intermediates of heme synthesis (porphyrins) accumulate in the body. In some forms of the disorder, porphyrins pile up in the bones, skin, and teeth. When porphyrins in the skin are exposed to sunlight, they absorb energy and release energized electrons. Electrons careening around the cell can break bonds and cause free radicals to form. The most notable result is the formation of lesions and scars on skin (Figure 29.20). In the most extreme cases, gums and lips can recede, which makes some front teeth—the canines—look more fanglike.

Affected individuals must avoid sunlight, and garlic can exacerbate their symptoms. By one hypothesis, people who were affected by the most extreme forms of porphyria may have been the source for vampire stories. Would you consider this hypothesis plausible? What other kinds of historical data might support or disprove it?

Figure 29.20 Skin lesions of an individual affected by a severe form of porphyria.

30 NEURAL CONTROL

IMPACTS, ISSUES In Pursuit of Ecstasy

Ecstasy, a street drug, can make you feel socially accepted, less anxious, and more aware of your surroundings while giving you a mild high. Ecstasy also can leave you dying in a hospital, foaming at the mouth and bleeding from all orifices as your temperature skyrockets. It can send your family and friends spiraling into horror and disbelief as they watch you stop breathing. Lorna Spinks ended life that way. She was nineteen years old (Figure 30.1).

Her anguished parents released the photograph at far right, taken minutes after her death. They wanted others to know what their daughter did not: Ecstasy can kill.

Ecstasy is a psychoactive drug; it alters brain function. Its active ingredient, MDMA, is a type of amphetamine, or "speed." As one effect, it makes neurons release too much serotonin, which is a signaling molecule. The serotonin saturates receptors on target cells and cannot be cleared away, so cells cannot be released from overstimulation.

The surge of serotonin promotes feelings of energy, empathy, and euphoria. But the unrelenting stimulation calls for rapid breathing, dilated eyes, restricted urine formation, and a racing heart. Blood pressure soars, and the body's internal temperature can spiral out of control. Spinks became dizzy, flushed, and incoherent after taking just two Ecstasy tablets. She died because her soaring temperature shut down her organ systems.

Few Ecstasy overdoses end in death. Panic attacks and fleeting psychosis are more common short-term effects. We do not know much about its long-term effects; users are unwitting guinea pigs for unscripted experiments.

We know that neurons do not rebound fast after Ecstasy depletes the brain's store of serotonin. Multiple doses of MDMA given to laboratory animals altered the structure and number of serotonin-secreting neurons. In humans, lower serotonin levels contribute to a loss of concentration, problems with memory, and depression. As the frequency

See the video! **Figure 30.1** Nervous systems and a psychoactive drug. Ecstasy use at all-night raves and other parties caused a rise in Ecstasy-related visits to emergency rooms. Sociology major Lorna Spinks (*facing page*) died in a hospital thirty-six hours after taking two Ecstasy tablets. If you suspect someone is having a bad reaction to Ecstasy or any other drug, get medical help fast and be honest about the cause of the problem. Immediate, informed medical action may save a life.

How would you vote? Should people caught using illegal drugs enter mandatory drug rehabilitation programs as an alternative to jail? Or does the threat of jail make some think twice before experimenting with possibly dangerous drugs? See ThomsonNOW for details, then vote online.

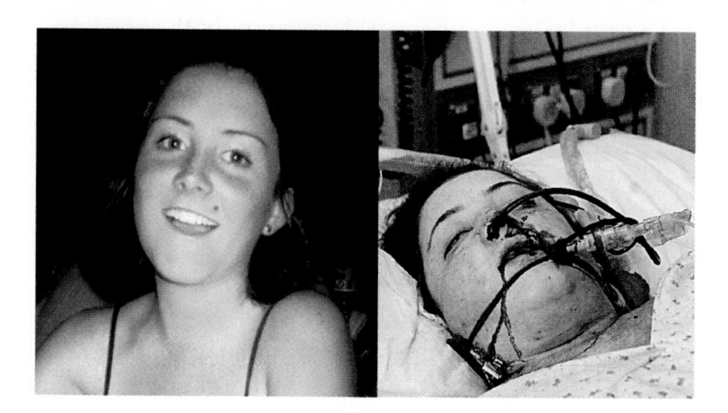

of drug intake rises, so does memory loss. At least over the short term, the capacity for memory seems to be restored when Ecstasy use stops. Undoing the neural imbalances often takes many months.

No one is regulating the manufacture of Ecstasy. The tablets vary in their concentration of MDMA. Often they incorporate a hodgepodge of other chemicals, such as methamphetamine, mescaline, ephedrine, and ketamine. Sometimes a drug called PMA is passed off as Ecstasy because it is easier and cheaper to make. PMA is far more potent and more likely to cause deadly reactions.

Think about it. The nervous system evolved as a way to sense and respond fast to changing conditions inside and outside the body. Sights, odors, hunger and passion, fear and rage—awareness of stimulation starts with a flow of information along communication lines of the nervous system. Even before you were born, excitable cells called neurons started organizing themselves into gridworks in newly forming tissues. They began to chatter among themselves. All through your life, in moments of danger or reflection, excitement or sleep, their chattering has continued and will continue for as long as you do.

Each of us possesses a complex nervous system, a legacy of millions of years of evolution. Its architecture and its functions give us an unparalleled capacity for learning and sharing experiences with others. Perhaps the saddest consequence of drug abuse is the implicit denial of this legacy—the denial of self when we choose not to assess threats, or cease to care.

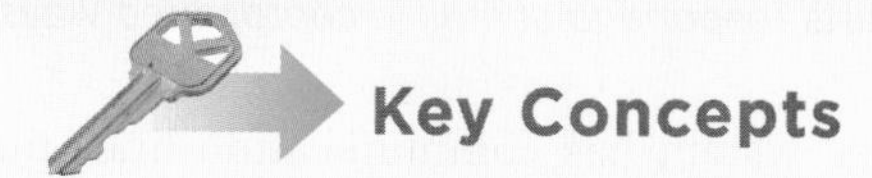

Key Concepts

OVERVIEW OF NERVOUS SYSTEMS

Excitable cells called neurons interconnect and form communication lines of animal nervous systems. In the radially symmetrical animals, neurons connect as a nerve net. Bilaterally symmetrical animals have a concentration of neurons at their head end; one or more nerve cords run the length of the body. **Section 30.1**

HOW NEURONS WORK

Messages flow along a neuron's plasma membrane, from input to output zones. The messages are brief, self-propagating reversals in the distribution of electric charge across the membrane. At an output zone, they are transduced to a chemical signal that may stimulate or inhibit activity in another cell. Psychoactive drugs interfere with the information flow between cells. **Sections 30.2–30.6**

PATHS OF INFORMATION FLOW

Nerves are long-distance communication lines, with the extensions of many neurons bundled together. Electrically insulating sheaths that wrap around extensions of many neurons greatly enhance the rate at which messages are propagated. Reflex arcs are the simplest routes of information flow. **Section 30.7**

VERTEBRATE NERVOUS SYSTEM

The central nervous system of vertebrates consists of the brain and spinal cord. The peripheral nervous system includes many pairs of nerves that connect the brain and spinal cord to the rest of the body. **Sections 30.8, 30.9**

CLOSER LOOK AT THE HUMAN BRAIN

The cerebral cortex is the most recently evolved part of the brain. In humans, it governs conscious behavior and interacts with the limbic system in forming and retrieving memories. **Section 30.10**

THE SUPPORTING CAST

A great variety of cells called neuroglia make up more than half the volume of vertebrate nervous systems. Neurons cannot function without them. **Section 30.11**

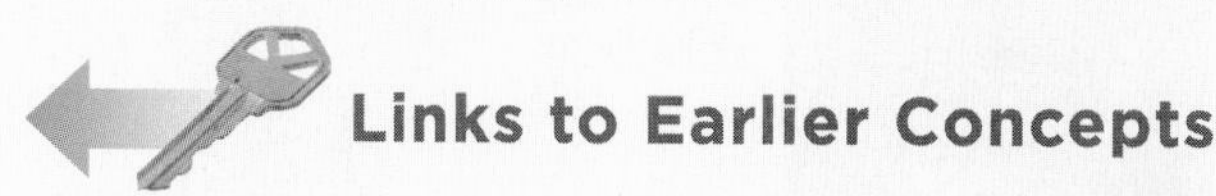

Links to Earlier Concepts

In this chapter, you will draw upon your knowledge of diffusion, concentration gradients (Section 5.6), and mechanisms of passive and active transport (5.7). You might wish to review an earlier explanation of how positive feedback mechanisms work (25.3). You will be revisiting the body plans of a few invertebrates (23.4, 23.6) and major trends in vertebrate evolution (24.2).

30.1 Evolution of Nervous Systems

A nervous system gives an animal a capacity to make its various parts respond to stimuli in coordinated ways.

LINKS TO SECTIONS 25.1–25.7, 26.2, 29.4

A nervous system has communication lines that can detect, process, and call for responses to information about conditions inside and outside the body. The lines consist of neurons: one type of excitable cell (Section 29.4). Collectively, diverse cells that structurally and metabolically support the neurons are called neuroglia.

There are three kinds of neurons. Sensory neurons detect stimuli, then signal other cells in response. Interneurons integrate sensory signals and activate or suppress motor neurons. Motor neurons signal effectors, the muscles or glands that can carry out responses (Figure 30.2).

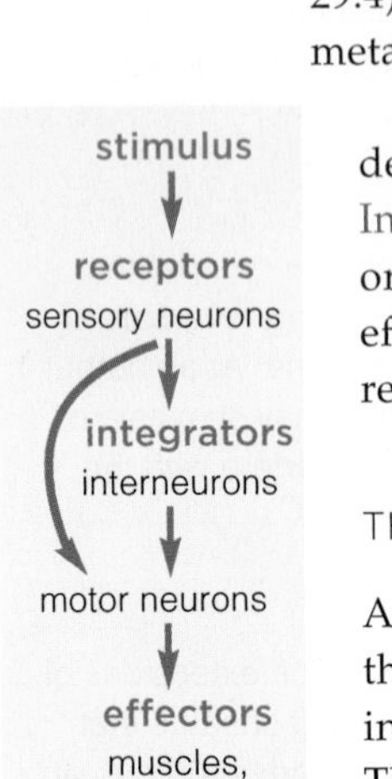

Figure 30.2 The line of communication.

THE CNIDARIAN NERVE NET

Animals first evolved in the seas, and the kinds with the simplest nervous system still live in water. They include hydras, jellyfishes, and all other cnidarians. These radial animals contain a nerve net: a mesh of neurons that detect stimuli and control contractile cells (Sections 23.1 and 23.4). As an example, certain neurons detect mechanical stimulation. In response to touch, they signal some contractile cells in the body wall to contract and others to relax. The coordinated interactions among these cells change the body shape or move tentacles toward or away from a stimulus.

The cnidarians will never dazzle you with speed or acrobatics. Even so, their nerve net helps them capture food that brushes against them and move it into their gut. A radially distributed net responds equally well to tidbits that drift into tentacles from any direction.

BILATERAL, CEPHALIZED SYSTEMS

As you know, most animals are bilateral, with paired organs arranged on either side of the main body axis (Section 23.1). Evolution of these bilateral body plans was accompanied by cephalization: the concentration of neurons that detect and process information at the body's anterior end, or "head."

Planarians and related flatworms are the simplest animals having a bilateral, cephalized nervous system. They have nerves: long, cordlike communication lines. Two nerve cords extend the length of their body and interconnect through a ladderlike array of branching nerves (Figure 30.3*b*). Both also connect with pairs of ganglia in the head. A ganglion (plural, ganglia) is a

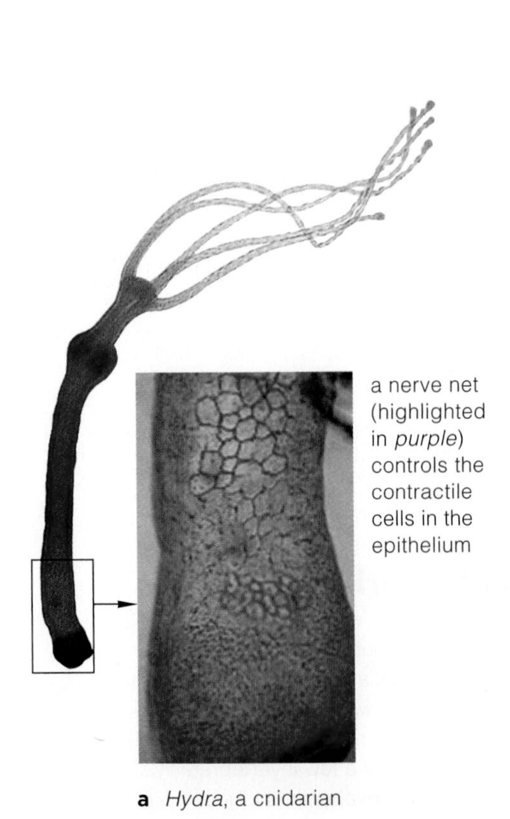

a *Hydra*, a cnidarian

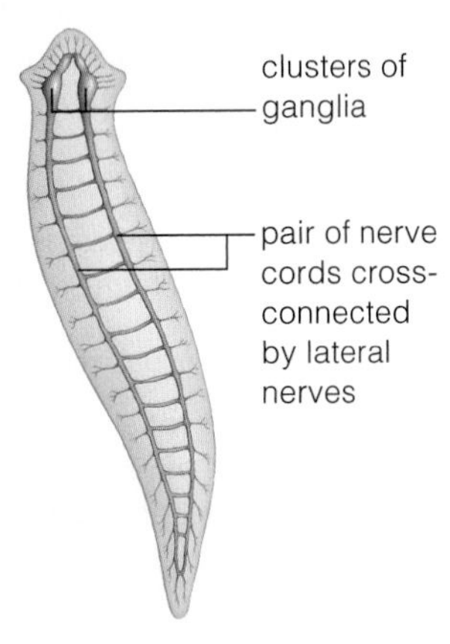

b Planarian, a flatworm

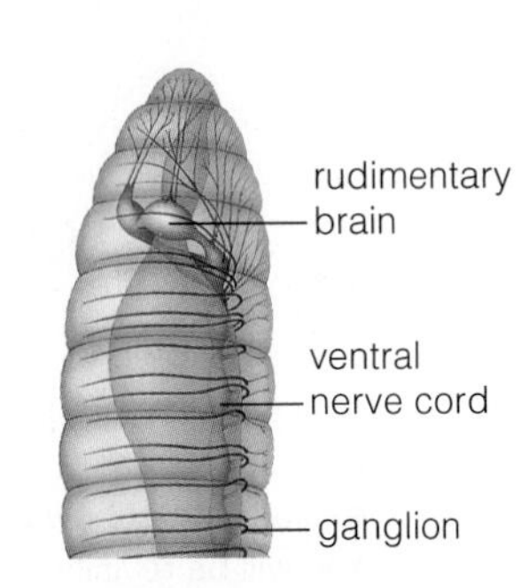

c Earthworm, an annelid

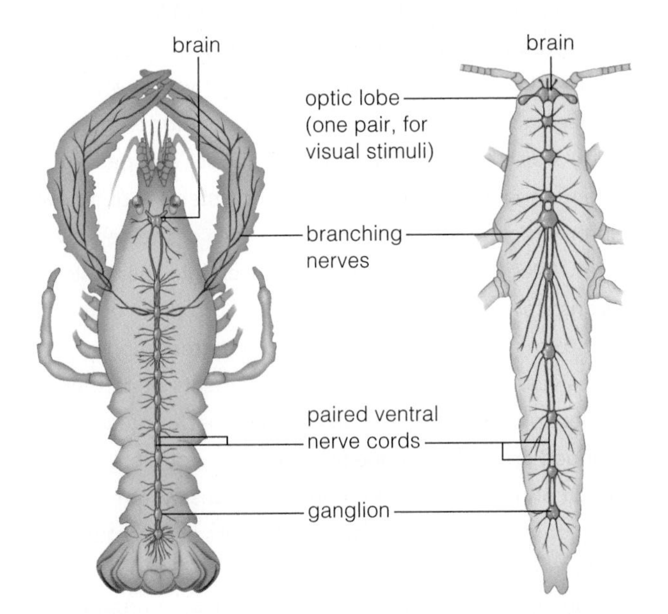

d Crayfish, a crustacean (a type of arthropod)

e Grasshopper, an insect (a type of arthropod)

Figure 30.3 Gallery of invertebrate nervous systems. Hydras and other radial animals have a nerve net. The other representatives shown have a bilateral, cephalized nervous system.

cluster of nerve cell bodies that functions as a local information-processing center. Ganglia in a planarian head accept signals from light-sensitive eye spots and from sensory neurons elsewhere in the body.

Annelids and arthropods have a simple brain with paired ganglia (Figure 30.3*c–e*). In addition, a pair of ganglia in each body segment provides local control over a segment's muscles and connects to one or two ventral nerve cords, near the body's lower surface.

Chordates have one dorsal nerve cord (Section 24.1). In vertebrates, the anterior region of the cord evolved into a complex brain. Bigger brains gave individuals a competitive edge in assessing and responding to food and danger. Among the first vertebrates that invaded land, agents of selection favored expansion of sensory structures and different brain centers that coordinate, process, and direct responses to novel stimuli.

THE VERTEBRATE NERVOUS SYSTEM

The nervous system of vertebrates has two functional divisions (Figure 30.4). Most interneurons are located in the central nervous system—the brain and spinal cord. Nerves that extend through the rest of the body make up the peripheral nervous system. These nerves are further classified as autonomic or somatic, based on which parts of the body they deal with.

Figure 30.5 shows the location of the human brain, spinal cord, and some peripheral nerves. As you will see, each nerve contains long extensions, or axons, of sensory neurons, motor neurons, or both. *Afferent* axons relay sensory signals into the central nervous system; *efferent* ones relay commands for response out of it. For instance, you have a sciatic nerve in each leg. It swiftly relays signals from sensory receptors in leg muscles, joints, and skin into the spinal cord even as it relays signals from the spinal cord to the leg's muscles.

In sections to follow, you will consider the kinds of messages that flow along these communication lines.

Radial animals have a meshlike array of neurons called a nerve net. Bilateral animals have paired nerves that service both sides of the body. Most are cephalized, with ganglia or a brain inside the head end.

Functionally, the nervous system of vertebrates has two divisions. The brain and spinal cord are the central nervous system. Threading through the rest of the body are spinal and cranial nerves, the major communication lines of the peripheral nervous system.

Each nerve delivers signals into the brain and spinal cord and also carries signals away from them.

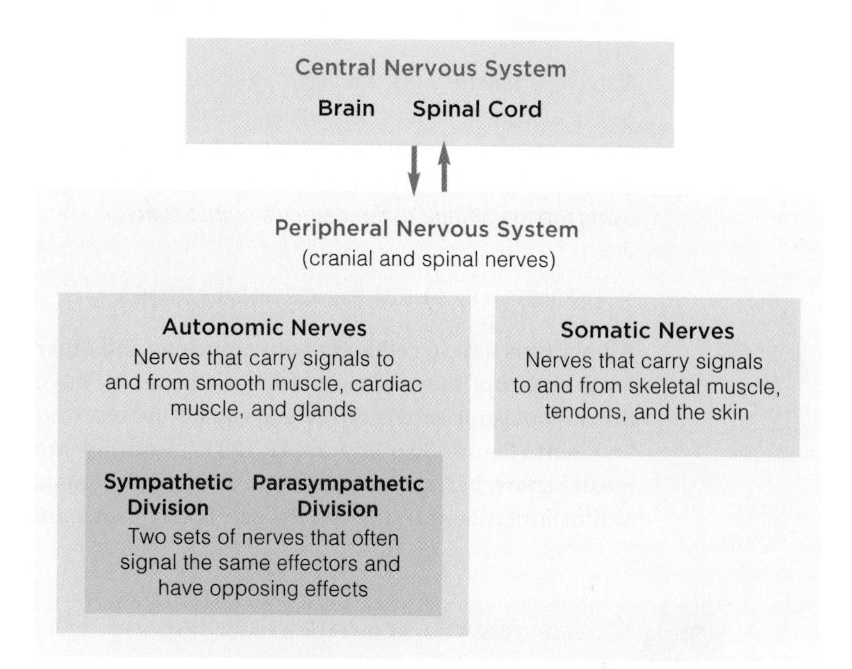

Figure 30.4 Functional divisions of vertebrate nervous systems. The spinal cord and brain are its central portion. The peripheral nervous system includes spinal nerves, cranial nerves, and their branchings, which extend through the rest of the body. They carry signals to and from the central nervous system. Section 30.8 explains the functional divisions of the peripheral system.

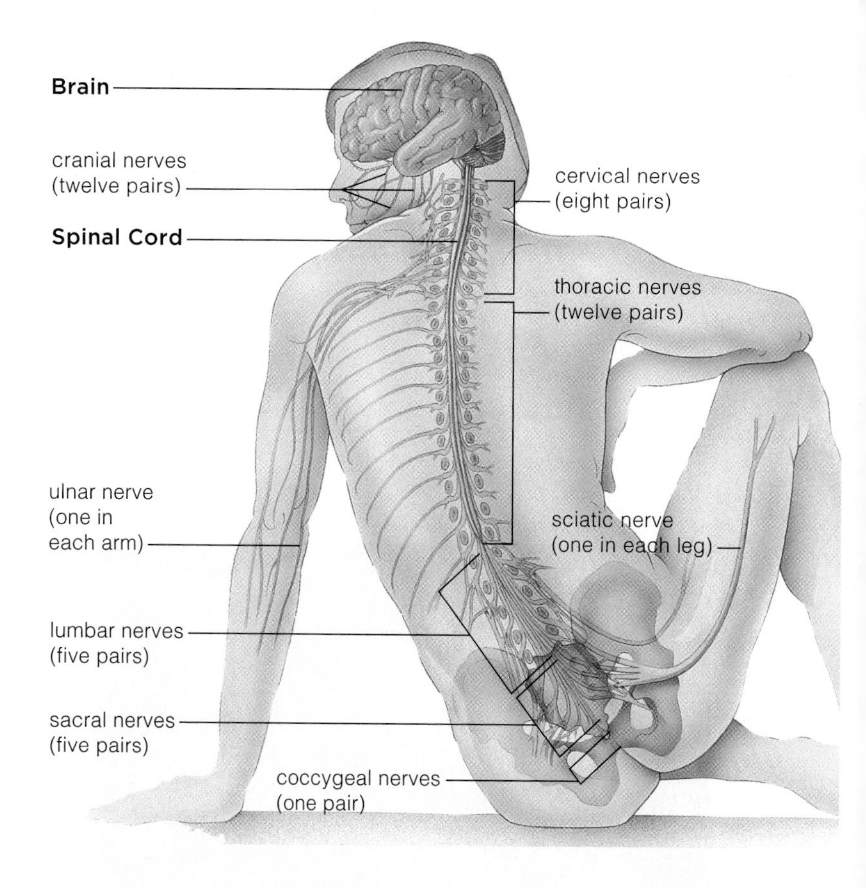

Figure 30.5 Some of the major nerves of the human nervous system.

30.2 Neurons—The Great Communicators

Neurons are among the "excitable" cells. Although all cells maintain an electric gradient across the plasma membrane, only in excitable cells does the gradient abruptly and briefly reverse. The reversal is conducted along the membrane to the neuron's output zone.

NEURONS AND THEIR FUNCTIONAL ZONES

LINKS TO SECTIONS 4.2, 5.1, 5.6, 5.7, 5.8

All neurons have a cell body with a nucleus and other organelles, but they differ in size and shape. All have thin cytoplasmic extensions where signals are received and sent (Figure 30.6 and Table 30.1). Dendrites are mostly short, branching extensions that accept signals and conduct them in toward the cell body. Axons are mostly long extensions, as mentioned in Section 30.1. They conduct signals away from the cell body.

Sensory neurons typically have one dendrite, and it has receptor endings that detect a specific stimulus. Motor neurons and interneurons typically have many dendrites. Nearly all neurons have only one axon. The axons in some nerves have a fatty sheath that helps speed information flow.

The cell body and dendrites are *input* zones, where arriving signals can alter electric gradients across the plasma membrane (Figure 30.6). A large disturbance spreads to the *trigger* zone, a patch of membrane with special properties. From here, the disturbance travels along the neuron's axon, its *conducting* zone, to axon endings. At these *output* zones, electrical disturbances get transduced; their arrival causes release of chemical signals that diffuse to neighboring cells.

Table 30.1 Extensions of a Neuron's Cell Body

Dendrites	Axons
Carry information toward cell body	Carry information away from cell body
One per cell in most sensory neurons; many per cell in most other neurons	One per cell in most neurons
Branch close to cell body	Branch farther from cell body
No insulating sheath	Insulating sheath for most

MEMBRANE GRADIENTS AND POTENTIALS

When a neuron is "at rest," or not being stimulated, it maintains a slight electric gradient across its plasma membrane. The cytoplasmic fluid near the membrane has a slight negative charge with respect to the fluid outside. As in a battery, these separated charges have potential energy, which can be measured in millivolts (thousandths of a volt). The steady voltage difference across the neuron's plasma membrane is known as the resting membrane potential. In many neurons, resting potential is about −70 millivolts.

All living cells maintain an electric gradient across the plasma membrane. Only in neurons, muscle cells, and a few other excitable cells does stimulation cause an action potential—an abrupt reversal in the electric gradient across a plasma membrane. A signal travels along an axon when the gradient reverses in one axon region after another, from trigger zone to output zone.

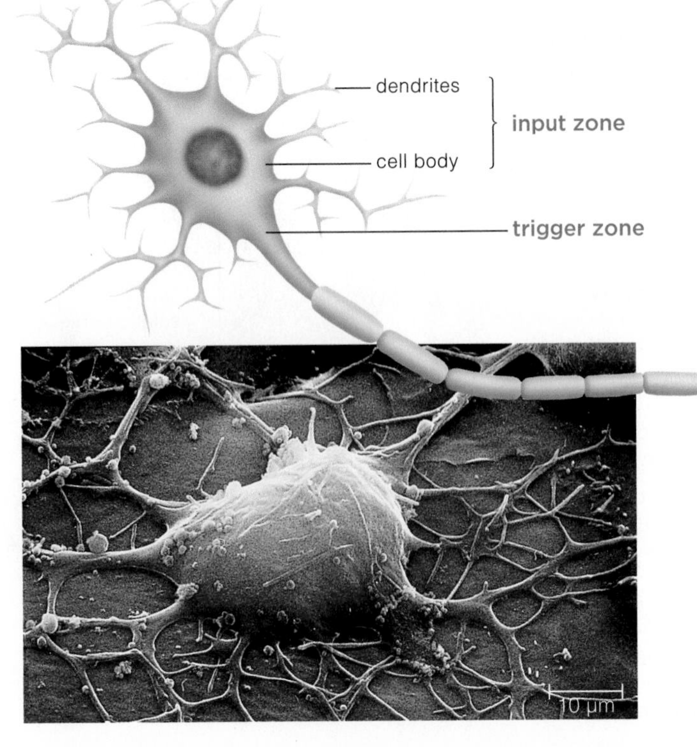

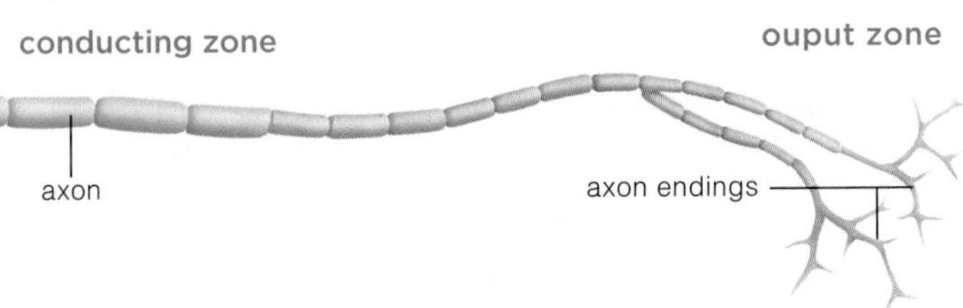

Figure 30.6 **Animated!** Scanning electron micrograph and sketch of a motor neuron. Dendrites receive information and relay it to the cell body. Signals that spread to the trigger zone may be conducted along the axon to its endings. From here, signals flow to another cell—in the case of a motor neuron, a muscle cell.

Figure 30.7 **Animated!** Icons for the main types of protein channels and pumps that span the lipid bilayer of a neuron's plasma membrane. (**a**) Sodium–potassium pumps (Na^+/K^+ pumps) and (**b**) open K^+ channels contribute to the membrane's resting potential. (**c**) Voltage-gated channels are required for action potentials.

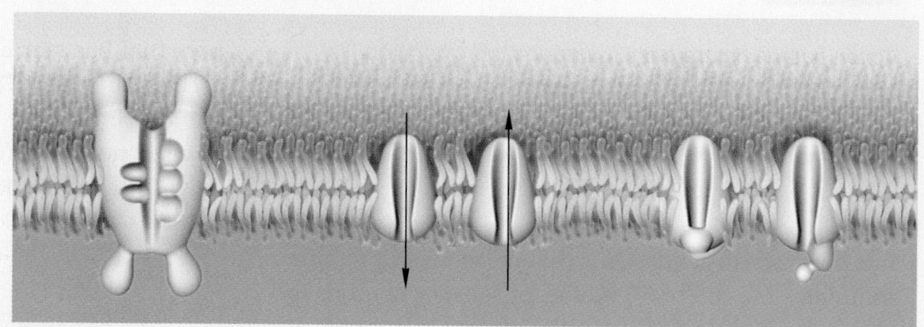

a Sodium–potassium pumps actively transport Na^+ and K^+ against their gradients. They pump three Na^+ out of the neuron for every two K^+ they pump in.

b Passive transporters with open channels allow K^+ ions to leak across the plasma membrane, down their concentration gradient.

c In a resting neuron, gates of voltage-sensitive channels are shut (*left*). During an action potential, gates open (*right*). Na^+ or K^+ flow through them, down concentration gradients.

A neuron is able to maintain resting potential and undergo action potentials because of the properties of its plasma membrane. As Section 5.7 explained, ions and proteins do not diffuse across the lipid bilayer of a cell membrane. Negatively-charged proteins trapped in the cytoplasm give the interior of a resting neuron a negative charge. Potassium ions (K^+) and sodium ions (Na^+) move in and out of the neuron through the interior of transport proteins (Figure 30.7*a–c*).

Sodium–potassium pumps are busy even when the neuron is at rest. Each time ATP transfers a phosphate group to a pump, two K^+ ions move into the cell and three Na^+ move out (Figure 30.7*a*). The membrane of a neuron at rest is largely impermeable to Na^+, so very little of the Na^+ that gets pumped out leaks back in.

The membrane crossings of K^+ are more complex. Even as K^+ is being pumped in, some K^+ is passively leaking out, through open channels for that ion. There *is* an outward-directed K^+ concentration gradient, but negatively charged proteins in the cytoplasm attract some K^+ back in.

As a result of the pumping and leaking processes, an unstimulated neuron has fewer Na^+ and more K^+ than the interstitial fluid. To give you a sense of the size of the ion gradients across a resting membrane, for every 3 Na^+ inside, there are 30 outside; and for every 30 K^+ inside, there is 1 outside:

30 Na^+ 1 K^+ interstitial fluid

plasma membrane

3 Na^+ 30 K^+ neuron's cytoplasm

Without those ion gradients, an action potential cannot arise. As Section 5.6 explained, an electrical gradient is a form of potential energy. Neurons put this energy to work by altering their membrane permeability to K^+ and Na^+. The brief change in permeability allows a reversal in the electrical gradient. Gated ion channels for K^+ and Na^+ are central to the reversal.

Sodium–potassium pumps and open K^+ channels abound in all zones of a neuron. The trigger zone and conducting zone membranes also have channels with molecular gates that open and close in response to the voltage difference across the neuron membrane. Some of these voltage-gated channels let K^+ diffuse across the membrane through their interior. Others let Na^+ move across. Voltage-gated channels for K^+ and Na^+ stay shut in a neuron at rest, then swing open during an action potential (Figure 30.7*c*).

With this bit of background on membrane proteins and ion gradients, you are ready to look at how an action potential arises at a neuron's trigger zone and propagates itself, undiminished, to an output zone.

A neuron has a cell body, an input zone consisting of one or more dendrites, and usually one axon that has a signal-conducting zone and an output zone.

In a resting neuron, transport proteins maintain ion concentration gradients by assisting or restricting the diffusion of ions across the plasma membrane.

The resting membrane potential is a steady voltage difference associated with those ion gradients.

An action potential is an abrupt, fleeting reversal in the voltage difference across the plasma membrane of any excitable cell. Sodium–potassium pumps contribute to the ion gradients required for a reversal.

30.3 A Look at Action Potentials

Action potentials are easy to understand when you remember that ions can cross cell membranes only through the interior of transport proteins.

LINKS TO SECTIONS 5.7, 5.8, 25.3

APPROACHING THRESHOLD

Tap your wrist gently, and this mechanical pressure stimulates the receptor endings of sensory neurons in your skin. Pressure deforms the plasma membrane at the input zone of these neurons and ions slip across. This ion flow causes a local, graded potential—a slight shift in the voltage difference across the membrane. "Graded" means the size of this voltage shift varies. With a stronger or longer lasting stimulus, more ions cross the membrane, so the voltage shift is larger.

When a stimulus is intense or long-lasting, graded potentials can spread into a trigger zone. Unlike input zones, the trigger zone is richly endowed with gated Na^+ channels (Figure 30.8*a*). Remember, the gates are voltage-sensitive. When a stimulus shifts the voltage difference across the membrane by a certain amount —the threshold level—all of the gated Na^+ channels in the trigger zone open and start an action potential.

As the gates open, Na^+ streams in, from interstitial fluid into the axon (Figure 30.8*b*). The influx makes the cytoplasm more positive, which causes more gates to open and more ions to enter. This inward flow of more and more ions demonstrates positive feedback, whereby some activity intensifies because of its own occurrence. Once threshold is reached and an action potential has started, gated Na^+ channels open in an accelerating way. The stimulus strength no longer has any bearing on the outcome. The rush of Na^+ into the neuron, not the diffusion of ions from the input zone, drives the feedback cycle:

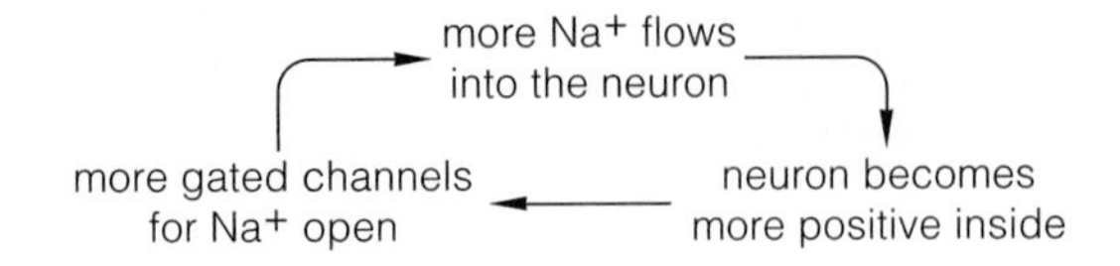

AN ALL-OR-NOTHING SPIKE

Researchers can record action potentials. They insert one electrode into an axon and position another just outside of it, then connect both to a recording device. Figure 30.9*f* shows what a recording looks like before, during, and after an action potential. Once threshold has been reached, the change in membrane potential in any neuron always spikes with the same intensity, within milliseconds. We call it an all-or-nothing event.

The voltage difference across the membrane quickly rises. Once it reaches a certain level, the Na^+ inflow ends as gates on Na^+ channels shut. About the same time, the gates on K^+ channels open, so K^+ flows out (Figure 30.8*c*). The outflow of positive charge makes the cytoplasm slightly negative relative to interstitial fluid. The sodium–potassium pumps restore the ion gradients—and the resting membrane potential—by pumping Na^+ out and K^+ in (Figures 30.7*a* and 30.8*d*).

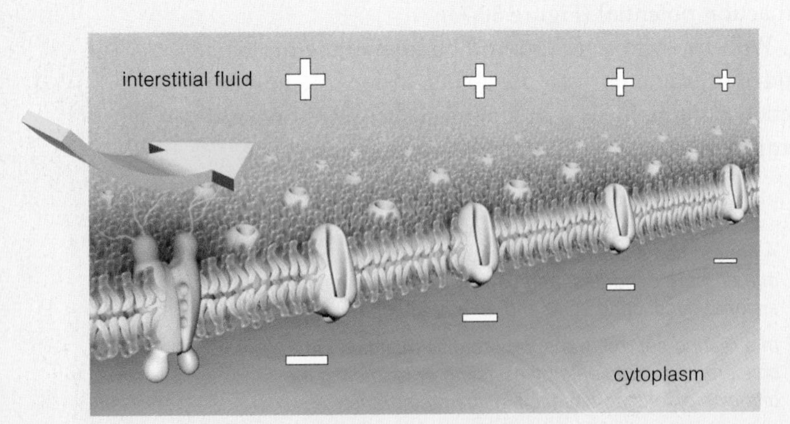

a The cytoplasm of this axon at rest is negative relative to interstitial fluid. Gated channels for ions of sodium (Na^+) and potassium (K^+) are shut. An electrical disturbance (*yellow* arrow) from an adjacent trigger zone will shift the voltage difference in this region toward threshold.

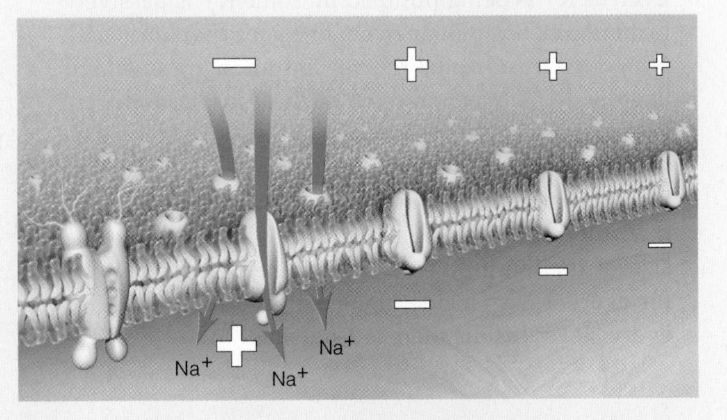

b The disturbance causes a few gated Na^+ channels to open. The inflow of Na^+ causes more gates to open until threshold is reached. Then, all gated Na^+ channels open abruptly. As a result, the voltage difference across this patch of membrane suddenly reverses.

Figure 30.8 **Animated!** Propagation of an action potential along part of a motor neuron's axon.

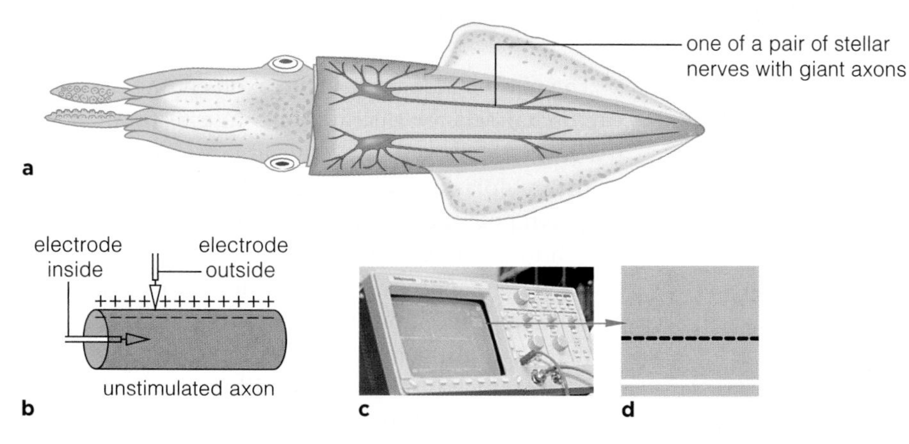

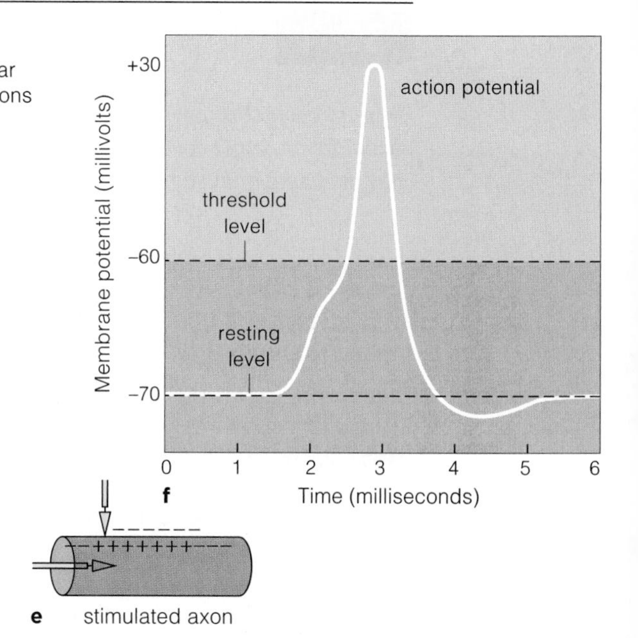

Figure 30.9 Action potential recordings. (**a**) A squid (*Loligo*) yielded the first recordings. (**b**) Researchers put electrodes inside and outside one of the squid's "giant" axons and connected them to a monitoring device called an oscilloscope (**c**). (**d**) The resting membrane potential showed up as a horizontal beam of light (the *white* line) across the oscilloscope's screen. (**e**,**f**) Strong stimulation of the axon deflected the beam. The resulting waveform in (**f**) is typical of action potentials.

DIRECTION OF PROPAGATION

Each action potential is self-propagating. Some of the Na^+ that enters one region of an axon diffuses into an adjoining region, driving that region to threshold and opening Na^+ gates. As Na^+ gates open in one region after the next, the action potential travels along toward the axon endings without weakening.

Action potentials travel only toward axon endings. Gated Na^+ channels are briefly inactivated after they close, so diffusion of Na^+ from a region undergoing an action potential can only cause more gated Na^+ channels to open in a region farther down the axon.

An action potential begins in the neuron's trigger zone. Here, a strong stimulus decreases the voltage difference across the membrane. This causes gated Na^+ channels to open and the voltage reverses. Self-propagating voltage reversals, which do not weaken with distance, occur at consecutive patches of membrane toward axon endings.

At each patch of membrane, an action potential ends when K^+ ions flow out of the neuron, and the voltage difference across the membrane is restored. Action potentials move in one direction, toward axon endings, because gated Na^+ channels are briefly inactivated after an action potential.

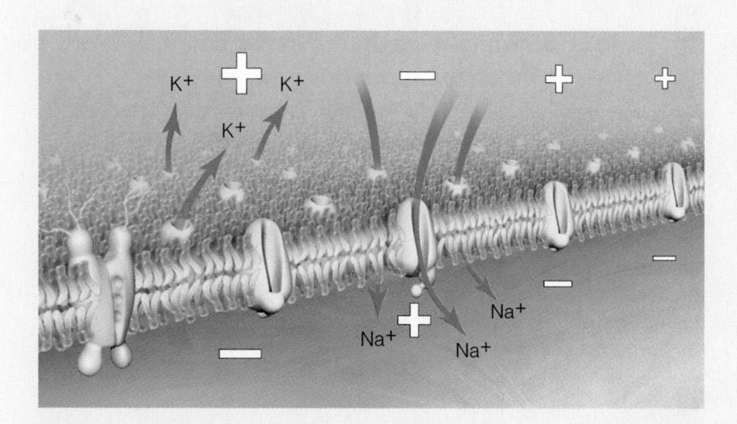

c The charge reversal makes gated Na^+ channels shut and gated K^+ channels open. The K^+ outflow restores the voltage difference across the membrane. The action potential is propagated along the axon as positive charges spreading from one region push the next region to threshold.

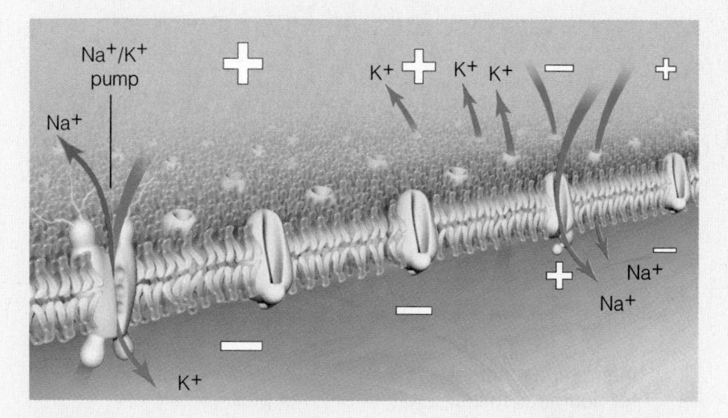

d After an action potential, gated Na^+ channels are briefly inactivated, so the action potential moves one way, toward axon endings. Also after an action potential, K^+ and Na^+ gradients are rebuilt. Sodium–potassium pumps restore the gradients by pumping Na^+ out and K^+ in.

30.4 How Neurons Send Messages to Other Cells

Neurons release signaling molecules that bind to specific receptors on target cells. Once transduced, the signals may stimulate or inhibit the target cell's activities.

CHEMICAL SYNAPSES

LINKS TO SECTIONS 5.6–5.9, 25.5

Recall, from Section 25.5, that most pathways of cell-to-cell communication proceed through three steps:

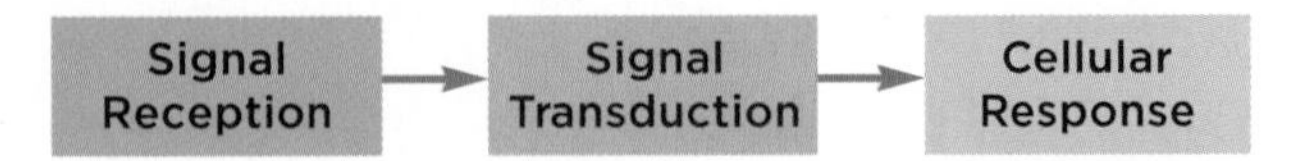

A chemical synapse is a region where signals from a neuron's axon endings are received and transduced by another cell (Figure 30.10*a*,*b*). The neuron sending the signal is the *pre*synaptic cell. A fluid-filled cleft about 20 nanometers wide separates it from the input zone of a *post*synaptic cell—a neuron, a muscle fiber, or a gland cell that receives the signal.

Neurotransmitters, a type of signaling molecule, relay messages between presynaptic and postsynaptic cells. Neurons synthesize and store neurotransmitter molecules in vesicles in their axon endings. An action potential stimulates secretion of those molecules.

The plasma membrane of an axon ending has many gated channels for calcium ions. In a resting neuron, the gates stay closed, and there are fewer calcium ions in the cytoplasm than in interstitial fluid. Arrival of an action potential unlocks the gates, and calcium ions flow into the axon. Calcium inflow causes exocytosis (Section 5.9); neurotransmitter-filled vesicles move to the plasma membrane and fuse with it, releasing their contents into the synaptic cleft.

Neurotransmitter molecules diffuse across the cleft and bind to receptors on the postsynaptic cell. Some receptors respond to neurotransmitter by changing in a way that allows ions to travel through them (Figure 30.10*c*). Other receptors alter membrane permeability indirectly. When activated by neurotransmitter they set in motion events that open or close specific ion channels elsewhere in the postsynaptic membrane.

For example, a neuromuscular junction is a synapse between a motor neuron and one skeletal muscle fiber that receives its signals (Figure 30.11). Arrival of an action potential induces the motor neuron to release acetylcholine (ACh). Binding of this neurotransmitter to receptors on a skeletal muscle fiber opens channels for Na^+ in the muscle fiber's plasma membrane. Like

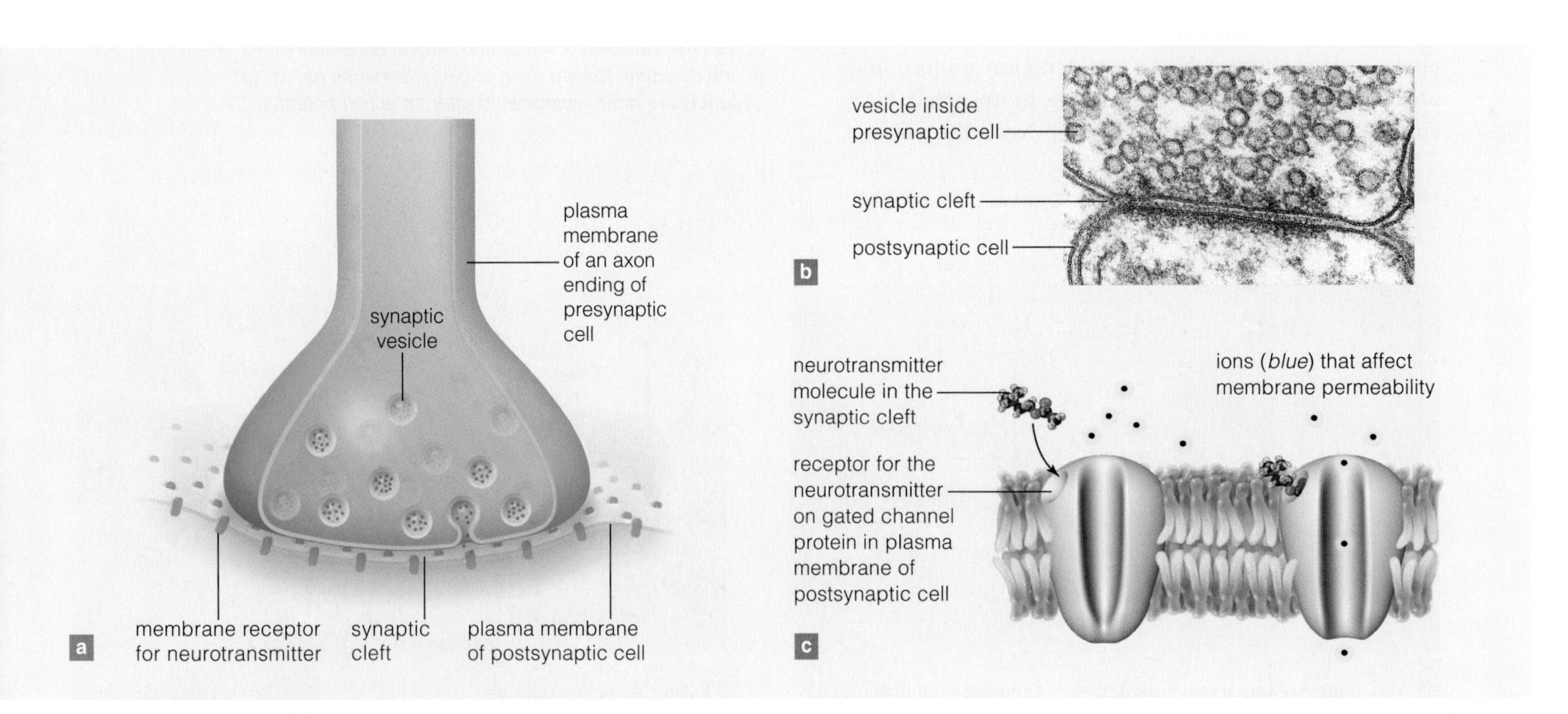

Figure 30.10 Animated! A chemical synapse, part of a signaling pathway in the nervous system. (**a**,**b**) An action potential arriving at an axon ending of a presynaptic cell causes neurotransmitter to be released. (**c**) Neurotransmitter diffuses across a cleft to the postsynaptic cell membrane. When it binds to a receptor, a passage opens through the receptor interior. The signal is transduced to ion flow into the cell (a form of electrochemical energy). The signal may stimulate or inhibit the target cell's activity.

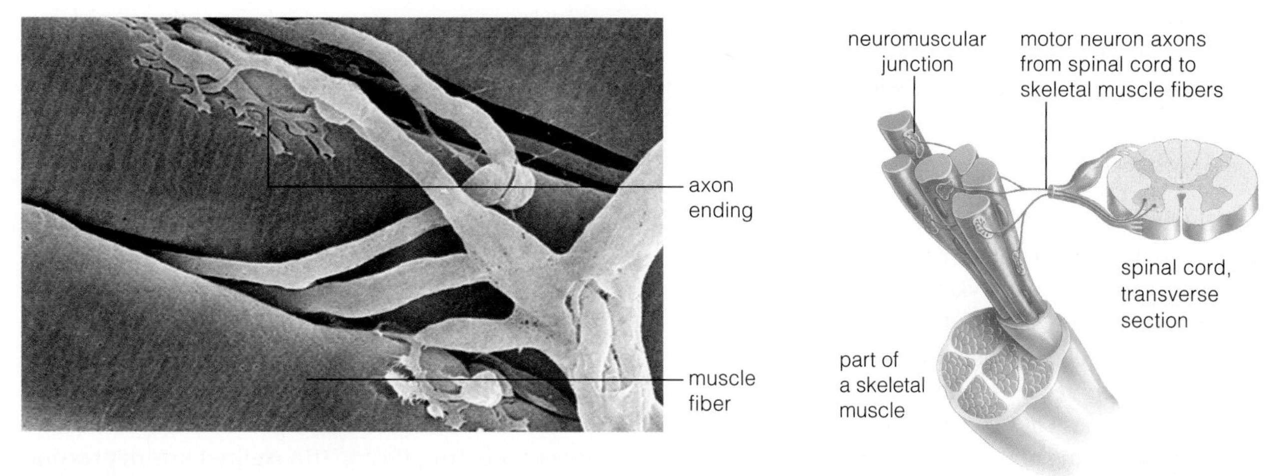

Figure 30.11 A common chemical synapse: a neuromuscular junction. A micrograph reveals some of these synapses between the axon endings of one motor neuron and the muscle fibers in the skeletal muscle that it controls.

neurons, muscle fibers are excitable. Na^+ inflow drives the fiber's membrane toward threshold. The resulting action potentials stimulate muscle contraction by way of a process described in detail in Section 33.5.

Some neurotransmitters bind to more than one type of postsynaptic cell, causing a different result in each. For example, ACh stimulates contraction in skeletal muscle but it slows contraction in cardiac muscle.

SYNAPTIC INTEGRATION

Typically, a neuron or effector cell gets messages from many neurons at the same time. Certain interneurons in the brain are on the receiving end of synapses with 10,000 neurons! An incoming signal may be excitatory and push the membrane potential closer to threshold. Or it may be inhibitory and nudge the potential away from threshold.

How does a postsynaptic cell respond to all of this information? Through synaptic integration, a neuron sums all inhibitory and excitatory signals arriving at its input zone. For example, the signals from different presynaptic cells amplify, dampen, or cancel the effect of another. Figure 30.12 shows what happens when an excitatory signal and an inhibitory signal arrive at the same time.

Competing signals cause the membrane potential at the postsynaptic cell's input zone to rise and fall. When excitatory signals outweigh inhibitory ones, ions diffuse from the input zone into the trigger zone and drive the postsynaptic cell to threshold. Gated Na^+ channels swing open, and an action potential occurs as described in the preceding section.

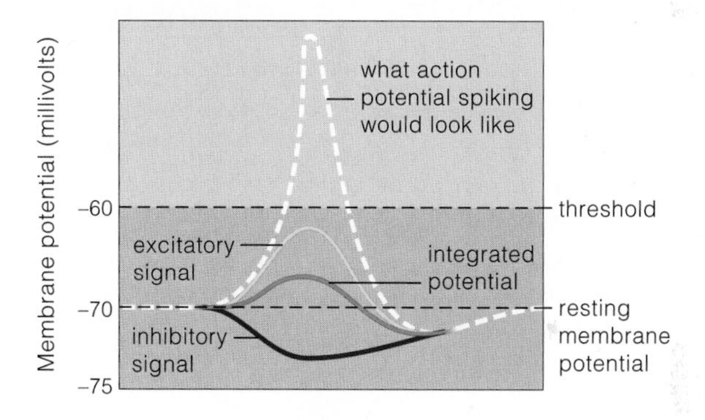

Figure 30.12 Synaptic integration. Excitatory and inhibitory signals arrive at a postsynaptic neuron's input zone at the same time. The graph lines show a postsynaptic cell's response to an excitatory signal (*yellow*), to an inhibitory signal (*purple*) and to both at once (*red*). In this example, summation of the two signals did not lead to an action potential (*white* waveform).

Neurons also integrate signals that arrive in quick succession from a single presynaptic cell. An ongoing stimulus can trigger a series of action potentials in a presynaptic cell, which will bombard a postsynaptic cell with waves of neurotransmitter.

Action potentials travel to a neuron's output zone. There, these signals typically are transduced to a chemical form—neurotransmitter—that can be sent onward to another cell.

Neurotransmitters are signaling molecules secreted into a synaptic cleft from a neuron's output zone. They may have excitatory or inhibitory effects on a postsynaptic cell.

Synaptic integration is the summation of all excitatory and inhibitory signals arriving at a postsynaptic cell's input zone at the same time.

30.5 A Smorgasbord of Signals

Controlled release and removal of neurotransmitters and other signaling molecules by neurons are essential for movement, memory, mood, and other functions.

NEUROTRANSMITTER DIVERSITY

LINK TO SECTION 6.4

Acetylcholine (ACh) was the first neurotransmitter to be discovered. As explained in the preceding section, it stimulates contraction of skeletal muscle, but it has the opposite effect on the heart. In myasthenia gravis, an autoimmune disease, the body mistakenly attacks its skeletal muscle receptors for ACh. Eyelids start to droop, and weakness in other muscles follows.

Interneurons in the brain also release and bind ACh. A lower ACh level in the brain contributes to memory loss in Alzheimer's disease. Affected people often can recall long-known facts, such as a childhood address, but have trouble remembering recent events.

There are many other neurotransmitters (Table 30.2). Norepinephrine and epinephrine (adrenaline), prime the body to respond to stress or to excitement. They are synthesized from the amino acid tyrosine. So is dopamine, which influences reward-based learning and acts in fine motor control. In Parkinson's disease, dopamine-secreting neurons in a certain brain region are destroyed (Figure 30.13). Most often, symptoms start with minor tremors of the hands. Later, sense of balance may be affected, so walking can be difficult.

As you read earlier, serotonin influences mood and memory. The drug Prozac (fluoxetine) lifts depression by increasing serotonin levels. GABA (gamma aminobutyric acid) inhibits the release of neurotransmitters from other neurons. Valium and Xanax are drugs that lower anxiety by boosting GABA's effects.

Table 30.2 Some Neurotransmitters and Their Effects

Neurotransmitter	Examples of Effects
Acetylcholine (ACh)	Induces skeletal muscle contraction, slows cardiac muscle contraction rate, affects mood and memory
Epinephrine and norepinephrine	Speed heart rate; dilate the pupils and airways to lungs; slow gut contractions; increase anxiety
Dopamine	Dampens excitatory effects of other neurotransmitters; roles in memory, learning, fine motor control
Serotonin	Elevates mood; role in memory
GABA	Inhibits release of other neurotransmitters

Figure 30.13 Battling Parkinson's disease. (**a**) This neurological disorder affects former heavyweight champion Muhammad Ali, actor Michael J. Fox, and about half a million other people in the United States. PET scans from an unaffected individual (**b**) and an affected person (**c**). *Red* and *yellow* indicate high metabolic activity in dopamine-secreting neurons.

THE NEUROPEPTIDES

Some neurons also make neuropeptides that serve as neuromodulators, molecules that influence the effects of neurotransmitters. One neuromodulator, substance P, enhances pain perception. Neuromodulators called enkephalins and endorphins are natural painkillers. They are secreted in response to strenuous activity or injuries and inhibit release of substance P. Endorphins also are released when people laugh, reach orgasm, or get a comforting hug or a relaxing massage.

CLEANING THE CLEFT

After they do their work, signaling molecules must be removed from synaptic clefts to make way for new signals. Some diffuse away. Membrane pumps move others back into presynaptic cells or neuroglial cells. Secreted enzymes break down certain kinds, as when acetylcholinesterase breaks down ACh.

When neurotransmitter accumulates inside the cleft, it disrupts the signaling pathways. That is how nerve gases, such as sarin, exert their deadly effects. After being inhaled, they bind to acetylcholinesterase and inhibit ACh breakdown. ACh accumulates and causes skeletal muscle paralysis, confusion, headaches, and, when the dosage is high enough, death.

ACh, norepinephrine, epinephrine, dopamine, serotonin, GABA, and other neurotransmitters have diverse effects.

Endorphins, enkephalins, and other neuromodulators magnify or reduce the effects of neurotransmitters.

30.6 Drugs and Disrupted Signaling

FOCUS ON HEALTH

Psychoactive drugs affect neurotransmitter function. Some induce a neurotransmitter's release. Others prevent its breakdown or uptake. Still others block receptors where a neurotransmitter normally binds.

Many people take psychoactive drugs, both legal and illegal, to alleviate pain, relieve stress, or feel pleasure. Many drugs are habit-forming, and users often develop tolerance; it takes larger or more frequent doses of the drug to get the desired effect.

Habituation and tolerance can lead to drug addiction, by which a drug takes on a vital biochemical role. Table 30.3 lists the main warning signs of addiction. Three or more signs may be cause for concern.

All major addictive drugs stimulate the release of dopamine, a neurotransmitter involved in the sense of pleasure. For example, most people think smoking is just a habit, but people smoke to experience a spike in the release of dopamine when nicotine hits neurons. All addicts crave the spike; most cannot stop the craving.

Stimulants Stimulants make users feel alert but also anxious, and they can interfere with fine motor control. Nicotine is a stimulant, as is the caffeine in coffee, tea, and many soft drinks. It blocks brain receptors for ACh. Caffeine blocks receptors for adenosine, which can act as a signaling molecule to suppress brain cell activity.

Cocaine, a powerful stimulant, is inhaled or smoked. Users feel elated and aroused, then become depressed and exhausted. Cocaine stops the uptake of dopamine, norepinephrine, and serotonin from synaptic clefts. When norepinephrine is not cleared away, blood pressure soars. Overdoses may cause strokes or heart attacks that can end in death. Cocaine is highly addictive. Long-term addiction to it remodels the brain until only cocaine can bring about a sense of pleasure (Figure 30.14).

Amphetamines increase dopamine, serotonin, and norepinephrine secretion in the brain, which reduces appetite and energizes users. Various forms are ingested, smoked, or injected. The chapter introduction focused on the synthetic amphetamine MDMA, found in Ecstasy. At present, crystal meth is a widely abused amphetamine. As with cocaine, users require more and more to get high or just to feel okay. Long-term use shrinks the brain areas involved in memory and emotions.

Depressants Depressants slow motor responses by inhibiting ACh output. Examples are alcohol (ethyl alcohol) and barbiturates. Alcohol stimulates release of endorphins and GABA, so users experience a brief euphoria followed by depression. Combining alcohol with barbiturates can be deadly. As the introduction to Chapter 5 explains, alcohol abuse damages the brain, liver, and other organs. Alcoholics deprived of the drug undergo tremors, seizures, nausea, and hallucinations.

Analgesics Analgesics mimic the body's natural painkillers—endorphins and enkephalins. The narcotic analgesics, such as morphine, codeine, heroin, fentanyl, and oxycontin, suppress pain. They cause a rush of euphoria and are highly addictive. Ketamine and PCP (phencyclidine) are a different type of analgesic. Both drugs give users an out-of-body experience and numb the extremities, because they slow down the clearing of synapses. Both have caused seizures, kidney failure, and fatal heat stroke. PCP sometimes induces a violent, agitated psychosis that lasts more than a week.

Hallucinogens Hallucinogens distort sensory perception and bring on a dreamlike state. LSD (lysergic acid diethylamide) resembles serotonin and binds to receptors for it. Tolerance develops, but LSD is not addictive. However, users can get hurt, and even die, because they do not perceive and respond to hazards, such as oncoming cars. Flashbacks, or brief distortions of perceptions, may occur years after the last intake of LSD. Two related drugs, mescaline and psilocybin, have weaker effects.

Marijuana consists of parts of *Cannabis* plants. Smoking a lot of it can cause hallucinations. More often, users become relaxed and sleepy as well as uncoordinated and inattentive. The active ingredient is THC (delta-9-tetrahydrocannabinol). It disrupts dopamine, serotonin, norepinephrine, and GABA levels. Chronic use impairs short-term memory and decision-making processes.

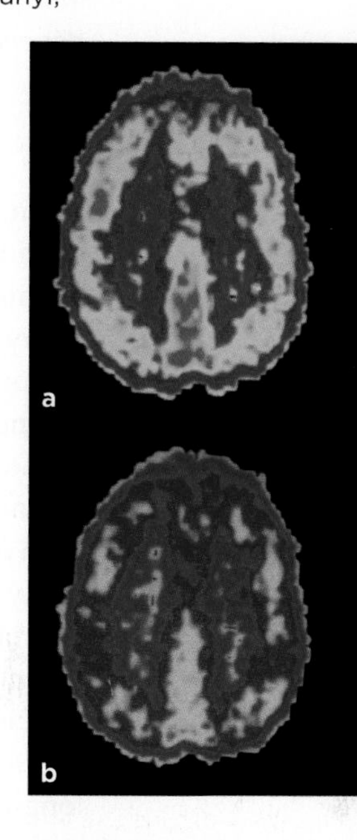

Figure 30.14 PET scans revealing (**a**) normal brain activity and (**b**) cocaine's long-term effect. *Red* shows areas of most activity, and *yellow*, *green*, and *blue* for successively reduced activity.

Psychoactive drugs affect mood, perception, memory, and other processes by interfering with neurotransmitter function. In some cases, they cause permanent damage.

Table 30.3 Warning Signs of Drug Addiction

1. Tolerance; takes increasing amounts of the drug to get the same effect.
2. Habituation; takes continued drug use over time to maintain the self-perception of functioning normally.
3. Inability to stop or curtail drug use, even if desire to do so persists.
4. Concealment; not wanting others to know of the drug use.
5. Extreme or dangerous actions to get and use a drug, as by stealing, by asking more than one doctor for prescriptions, or by jeopardizing employment by using drugs at work.
6. Deterioration of professional and personal relationships.
7. Anger and defensiveness if someone suggests there may be a problem.
8. Drug use preferred over previous customary activities.

30.7 Organization of Neurons in Nervous Systems

Through synaptic integration, messages arriving at a neuron may be reinforced and sent on to its neighbors. In which direction will a given message travel? That depends on how neurons are organized in the body.

BLOCKS AND CABLES OF NEURONS

Again, information about what is going on inside and outside the body usually flows from sensory receptors to interneurons, to motor neurons, then to effectors. Many signals also loop about in astonishing ways.

For instance, billions of interneurons in your brain belong to circuits that integrate and organize responses to messages. In *diverging* circuits, dendrites and axons extend from one block of neurons and communicate with other blocks. In *converging* circuits, signals from many neurons zero in on just a few. In other circuits, neurons synapse back on themselves, repeating signals like gossip that just won't go away. Such *reverberating* circuits make eye muscles twitch while you sleep.

Information also flows rapidly through nerves, the long-distance cables between regions (Section 30.1). In each nerve, axons of sensory neurons, motor neurons, or both are bundled inside connective tissues. Figure 30.15*a* shows the structure of one kind of nerve.

The neuroglial cells called Schwann cells wrap like jelly rolls around axons of most peripheral nerves, one after another. They form a myelin sheath: an insulator that makes action potentials flow faster. Ions cannot cross the neural membrane where it is sheathed. Ion disturbances associated with an action potential must spread down the axon's cytoplasm until they reach a node, a small, exposed gap between two Schwann cells (Figure 30.15*b*). At each node, the membrane is loaded with gated Na^+ channels. When these gates open, the voltage difference reverses abruptly. By jumping from node to node in long axons, a signal can move as fast as 120 meters per second. In unmyelinated axons, the maximum speed is about 10 meters per second.

In the disease multiple sclerosis (MS), white blood cells wrongly identify a type of protein in the myelin sheaths as foreign. They destroy it. This autoimmune response leads to inflammation of axons in the brain and spinal cord. The glial cells called oligodendrocytes die. Some people are genetically predisposed to the disorder, but a viral infection might set it in motion. Either way, information flow is disrupted. Dizziness, numbness, muscle weakness, fatigue, visual problems, and other symptoms follow. MS affects about 500,000 people in the United States.

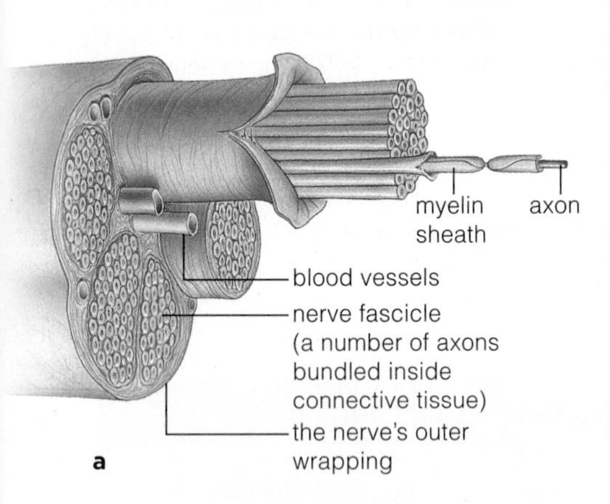

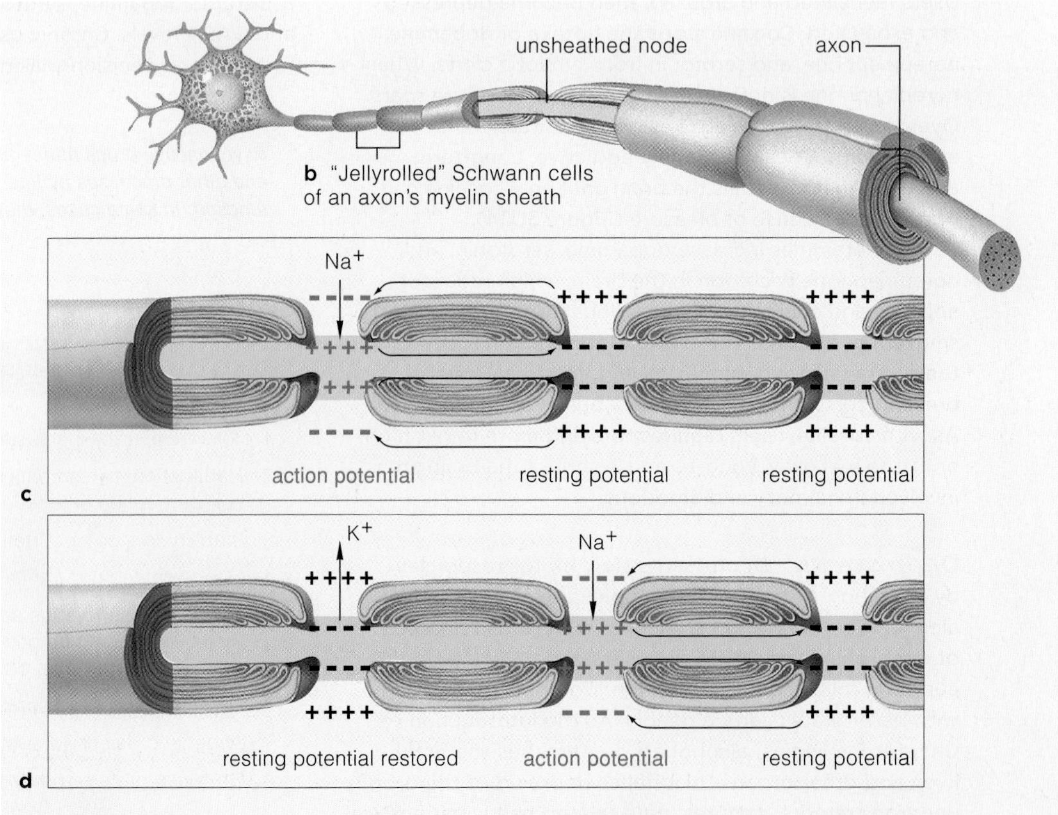

Figure 30.15 **Animated!** (**a**) Structure of one type of nerve. (**b–d**) In axons with a myelin sheath, ions flow across the neural membrane at nodes, or small gaps between the cells that make up the sheath. Many gated Na^+ channels for sodium ions are exposed to extracellular fluid at the nodes. When excitation caused by an action potential reaches a node, the gates open and start a new action potential. Excitation spreads fast to the next node, where it triggers a new action potential, and so on down the line to the output zone.

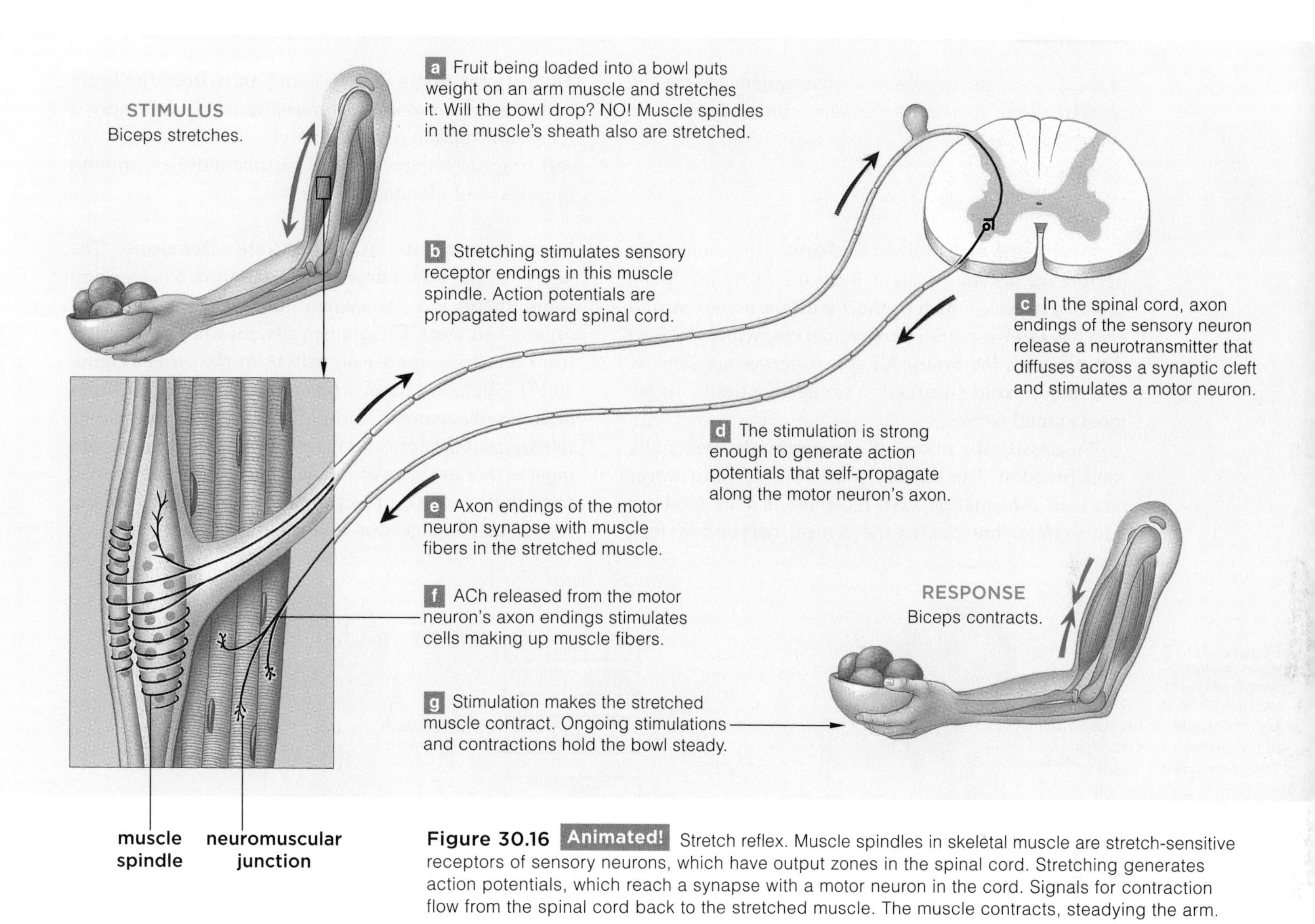

Figure 30.16 Animated! Stretch reflex. Muscle spindles in skeletal muscle are stretch-sensitive receptors of sensory neurons, which have output zones in the spinal cord. Stretching generates action potentials, which reach a synapse with a motor neuron in the cord. Signals for contraction flow from the spinal cord back to the stretched muscle. The muscle contracts, steadying the arm.

REFLEX ARCS

Reflexes are the simplest and most ancient paths of information flow. A reflex is an automatic response to a stimulus, a movement or other action that does not require thought. In the simplest reflex arcs, sensory neurons synapse directly on motor neurons. In more complex reflexes, sensory neurons interact with one or more interneurons. These interneurons then signal motor neurons, which call for the response.

The stretch reflex, a simple arc, causes a muscle to contract after gravity or some other force stretches it. Suppose you hold a bowl as someone drops peaches into it. The load makes your hand drop a bit, which stretches an arm muscle called a biceps. As a biceps is stretched, receptor endings of muscle spindles inside it are stretched. These sensory organs house sheathed receptors of sensory neurons (Figure 30.16).

The more the biceps muscle stretches, the greater the frequency of action potentials along axons of the muscle spindle neurons. Inside the spinal cord, these axons synapse with motor neurons—axons of which lead right back to the muscle. Action potentials reach axon endings of the motor neurons, where they cause the release of ACh. Remember, this neurotransmitter can make muscles contract. Contraction of the biceps helps steady the arm against the added load.

The knee-jerk reflex is another stretch reflex. A tap just below the knee stretches the thigh muscle. Signals flow to the spinal cord, and the leg jerks in response.

In complex animals, neurons are organized in blocks and in cables, some of which arc back to the point of stimulation.

Nerves are long-distance cables between body regions. They contain bundles of long axons of sensory neurons, motor neurons, or both. These axons are wrapped in a myelin sheath, which speeds signal propagation.

In reflex arcs, the simplest paths of information flow, sensory neurons synapse directly with motor neurons.

30.8 What Are the Major Expressways?

In vertebrates, the peripheral nervous system and the spinal cord interconnect as the main expressways for information flow.

PERIPHERAL NERVOUS SYSTEM

Somatic and Autonomic Systems In humans, the peripheral nervous system includes thirty-one pairs of *spinal* nerves, which connect with the spinal cord. It also has twelve pairs of *cranial* nerves, which connect directly with the brain. All spinal nerves are sensory and motor axons sheathed in connective tissue. So are most cranial nerves.

We classify the nerves of the peripheral system by their function. The sensory portion of somatic nerves conducts information from receptors in skin, tendons, and skeletal muscles to the central nervous system. Their motor axons deliver commands from the brain and spinal cord to skeletal muscles. Autonomic nerves relay information to and from viscera. *Viscera* refers to soft internal organs, such as cardiac muscles, smooth muscles, and glands.

Sympathetic and Parasympathetic Divisions The nerves of the autonomic system fall in two categories: sympathetic and parasympathetic. Both service most organs and work antagonistically, meaning the signals from one type oppose signals from the other (Figure 30.17). Sympathetic neurons are most active in times of stress, excitement, and danger. Their axon endings release norepinephrine. Parasympathetic neurons are most active in times of relaxation. The release of ACh from their axon endings promotes daily housekeeping tasks, such as digestion and urine formation.

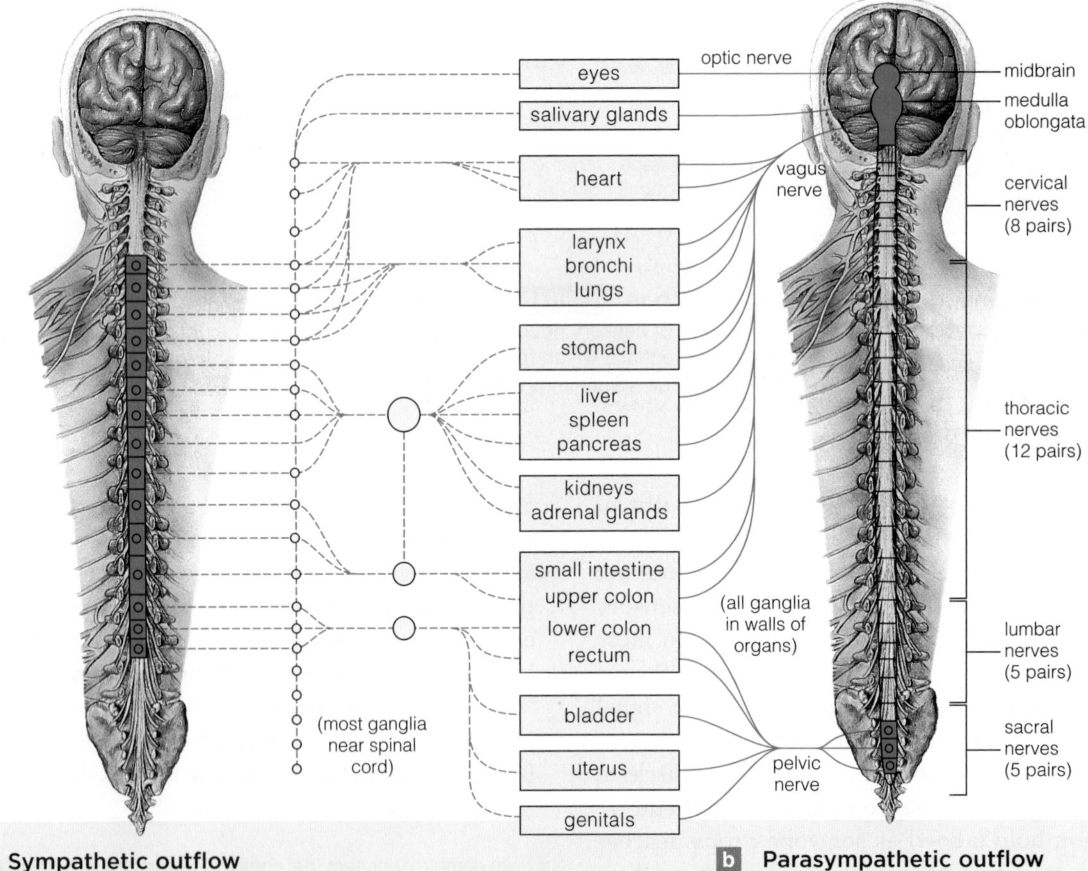

Figure 30.17 Animated! (**a**) Sympathetic and (**b**) parasympathetic nerves of the autonomic system. These nerves are paired; the body's right and left halves each have one. Ganglia, remember, are the clusters of nerve cell bodies that act as local control centers. Their axons, too, are bundled together inside nerves.

a Sympathetic outflow from spinal cord

Some responses to sympathetic outflow

- Heart rate increases
- Pupils of eyes dilate (widen, let in more light)
- Glandular secretions decrease in airways to lungs
- Salivary gland secretions thicken
- Stomach and intestinal movements slow down
- Sphincters contract

b Parasympathetic outflow from spinal cord and brain

Some responses to parasympathetic outflow

- Heart rate decreases
- Pupils of eyes constrict (keep more light out)
- Glandular secretions increase in airways to lungs
- Salivary gland secretions become more watery
- Stomach and intestinal movements increase
- Sphincters relax

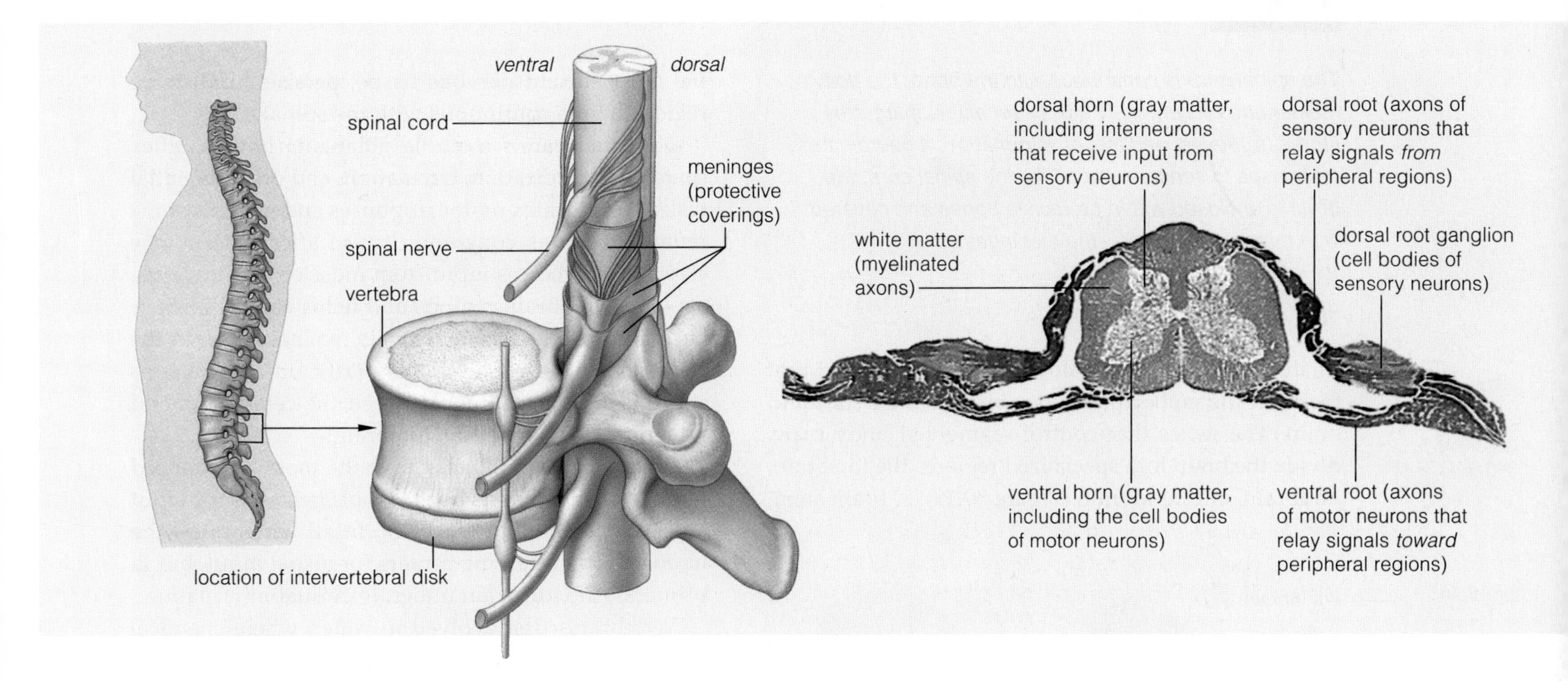

Figure 30.18 Animated! Location and organization of the spinal cord. Axons enclosed in myelin sheaths make up the white matter and function in rapid signal transmission. Cell bodies, dendrites, and unmyelinated axons of neurons, along with neuroglia, make up the gray matter.

What happens when something startles or scares you? Parasympathetic input falls. Sympathetic signals increase. When unopposed, sympathetic signals raise your heart rate and blood pressure, make you sweat more and breathe faster, and induce adrenal glands to secrete epinephrine. The signals put you in a state of intense arousal, so you are primed to fight or make a fast getaway. Hence the term fight–flight response.

Opposing sympathetic and parasympathetic signals control most organs. For instance, both act on smooth muscle cells in the gut wall. As sympathetic neurons are releasing norepinephrine at synapses with these cells, parasympathetic neurons are releasing ACh at other synapses with the same muscle cells. One signal tells the gut to slow down contractions; the other calls for increased activity. The outcome is finely adjusted through synaptic integration.

SPINAL CORD

The spinal cord threads through the vertebral column and connects peripheral nerves with the brain (Figure 30.18). Three layers of connective tissue, the meninges, cover and protect the spinal cord and brain. Viral or bacterial infection causes meningitis, an inflammation of the coverings. Symptoms may include headaches, a stiff neck, back pain, fever, and nausea.

The inner portions of the spinal cord and brain are gray matter: cell bodies, dendrites, and unsheathed axons of neurons, and neuroglia. The outer portions are white matter: myelin-sheathed axons organized as tracts, as well as neuroglia.

The spinal cord controls simple reflexes, including the stretch reflex (Section 30.7). Each spinal nerve has motor and sensory components, and each connects to the spinal cord by two "roots." Sensory information travels to the spinal cord through a dorsal root. Motor commands travel away from it through a ventral root.

An injury that disrupts the signal flow through the spinal cord can cause a loss of sensation and paralysis. Symptoms depend on where the cord was damaged. Nerves to and from the upper body lie higher in the cord than nerves that affect the lower body. An injury to the lumbar region of the cord often paralyzes the legs. An injury to higher cord regions can paralyze all the limbs as well as the muscles used in breathing. Christopher Reeves's injury, described in the Chapter 29 introduction, was like this. The top of his vertebral column and spinal cord were crushed.

Nerves of the peripheral nervous system connect the brain and spinal cord with the rest of the body.

The somatic division of the peripheral nervous system deals with skeletal muscle movements. Its autonomic division deals with smooth muscle, cardiac muscle, and glands.

The spinal cord is a vital expressway for signals between the brain and the peripheral nerves.

30.9 The Vertebrate Brain

The spinal cord is continuous with the brain, the body's master control center. The brain receives, integrates, stores, retrieves, and issues information. It coordinates responses to sensory input. Like the spinal cord, the brain is enclosed within protective bones and covered with three membranes—the meninges.

THE BRAIN'S SUBDIVISIONS

LINKS TO SECTIONS 23.1, 24.2, 24.11

In all vertebrates, the hollow, tubular nerve cord that forms in the embryo develops into a spinal cord and brain. The genes that control segmented body plans divide the brain into specialized regions: the forebrain, midbrain, and hindbrain (Figure 30.19). A brain stem, the most ancient nervous tissue, persists in all three regions and is continuous with the spinal cord.

The hindbrain's medulla oblongata houses reflex centers for respiration, circulation, and other essential tasks. It integrates motor responses and governs some reflexes, such as coughing. It also affects sleep. The cerebellum receives input from muscle spindles, eyes, ears, and forebrain regions and helps control posture and motor skills. Axons from its two halves reach the pons (meaning bridge). Like a traffic officer, the pons controls signals flowing between the cerebellum and integrating centers in the forebrain.

Fishes and amphibians have the most pronounced midbrain, which sorts out most of their sensory input and initiates motor responses. In all vertebrates, the midbrain has receiving centers for visual input, but in primates, the forebrain integrates visual information.

Vertebrates first evolved in water, where chemical odors diffusing from predators, prey, and mates were vital cues. They relied heavily on olfactory lobes and paired outgrowths from the brain stem that integrated olfactory input and responses to it. Especially among land vertebrates, the outgrowths expanded into two halves of the cerebrum, the two cerebral hemispheres.

The thalamus became a forebrain center for sorting out sensory input and relaying it to the cerebrum. The hypothalamus ("under the thalamus") began serving as the center for homeostatic control of the internal environment. It assesses and regulates all behaviors related to internal organ activities, such as thirst, sex, and hunger. It also governs related responses, such as sweating with passion and vomiting from fear.

PROTECTION AT THE BLOOD–BRAIN BARRIER

The neural tube's lumen—the space inside it—persists in adult vertebrates as a system of cavities and canals filled with cerebrospinal fluid. This clear fluid forms inside brain cavities called ventricles, then seeps out and bathes the brain and spinal cord. It helps cushion both against potentially jarring movements.

A blood–brain barrier protects the spinal cord and brain from harmful substances. It exerts some control

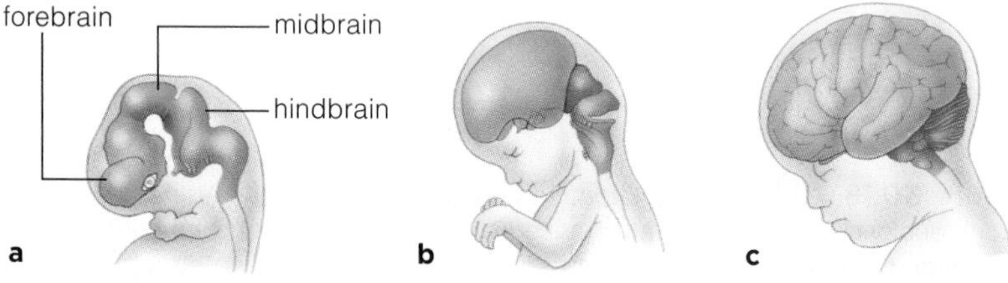

Region	Component	Function
FOREBRAIN	Cerebrum	Localizes, processes sensory inputs; initiates, controls skeletal muscle activity; governs memory, emotions, abstract thought in the most complex vertebrates
	Olfactory lobe	Relays sensory input from nose to olfactory centers of cerebrum
	Thalamus	Has relay stations for conducting sensory signals to and from cerebral cortex; has role in memory
	Hypothalamus	With pituitary gland, a homeostatic control center. Adjusts volume, composition, temperature of internal environment; governs organ-related behaviors (e.g., sex, thirst, hunger), and expression of emotions
	Limbic system	Governs emotions; has roles in memory
	Pituitary gland (Chapter 32)	With hypothalamus, provides endocrine control of metabolism, growth, development
	Pineal gland (Chapter 32)	Helps control some circadian rhythms; also has role in mammalian reproduction
MIDBRAIN	Roof of midbrain (tectum)	In fishes and amphibians, its centers coordinate sensory input (as from optic lobes), motor responses. In mammals, its reflex centers swiftly relay sensory input to forebrain
HINDBRAIN	Pons	Tracts bridge cerebrum and cerebellum, also connect spinal cord with forebrain. With the medulla oblongata, controls rate and depth of respiration
	Cerebellum	Coordinates motor activity for moving limbs and maintaining posture, and for spatial orientation
	Medulla oblongata	Tracts relay signals between spinal cord and pons; its reflex centers help control heart rate, adjustments in blood vessel diameter, respiratory rate, vomiting, coughing, other vital functions

Figure 30.19 Neural tube to brain. The human neural tube at (**a**) 7 weeks. The brain at (**b**) 9 weeks, and (**c**) at birth. The chart lists and describes major components in each of the three main regions of the adult vertebrate brain.

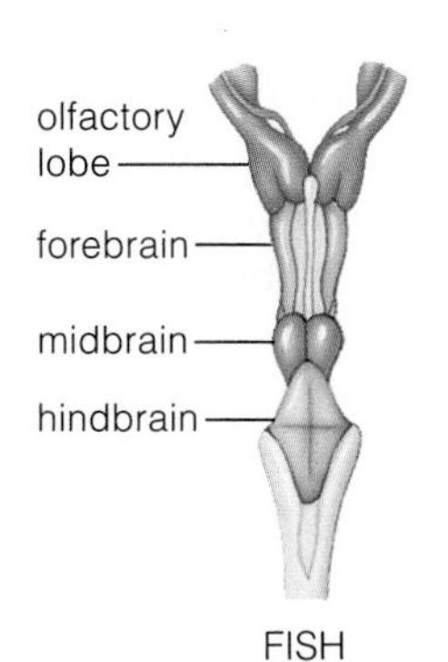

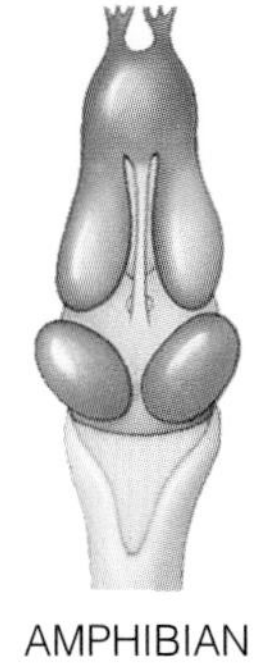

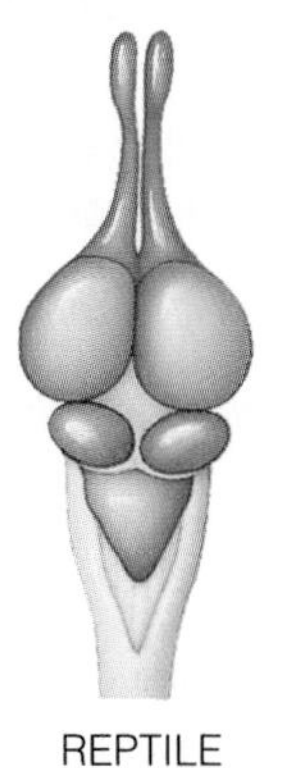

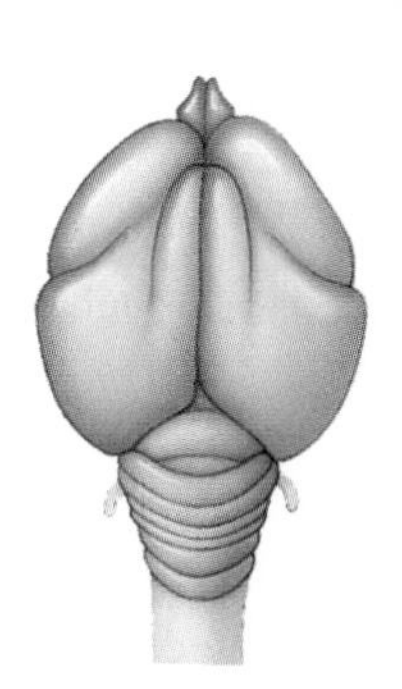

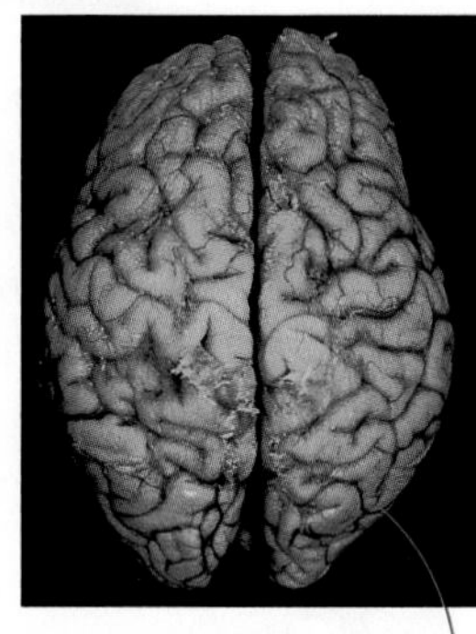

Figure 30.20 **Animated!** (**a**) Major brain regions for five vertebrates, dorsal views. The sketches are not to the same scale. (**b**) Right half of a human brain in sagittal section, showing the locations of the major structures and regions. Meninges around the brain were removed for this photograph.

over which solutes enter cerebrospinal fluid. No other portion of extracellular fluid has solute concentrations maintained within such narrow limits. Even changes brought on by eating and exertion are limited. Why? Hormones and other chemicals in blood affect neural function. Also, changes in ion concentrations can alter the threshold for action potentials.

The barrier works at the wall of blood capillaries that service the brain. In most parts of the brain, tight junctions form a seal between adjoining cells of the capillary wall, so water-soluble substances must pass through the cells to reach the brain. Transport proteins in the plasma membrane of these cells allow essential nutrients to cross. Oxygen and carbon dioxide diffuse across the barrier, but most waste urea cannot breach it. The blood-brain barrier is not perfect; some toxins such as nicotine, alcohol, caffeine, and mercury slip across. Also, inflammation or a traumatic blow to the head can damage it and compromise neural function.

THE HUMAN BRAIN

The average human brain weighs 1,300 grams, or 3 pounds. Again, it has about 100 billion interneurons, and neuroglia makes up more than half of its volume. The human midbrain is relatively smaller than that of other vertebrates (Figure 30.20). A human cerebellum is the size of a fist and has more interneurons than all other brain regions combined. As in other vertebrates, the cerebellum plays a role in the sense of balance, but it took on added functions as humans evolved. It affects learning of motor and mental skills, such as language.

A deep fissure divides the forebrain's cerebrum into two halves, the cerebral hemispheres (Figure 30.20). Each half deals mainly with input from the opposite side of the body. For instance, signals about pressure on the right arm reach the left hemisphere. Activities of both halves are coordinated by signals that flow both ways across the corpus callosum, a thick band of nerve tracts. The next section focuses on the cerebral cortex, the thin outer layers of the cerebrum.

A pineal gland is located near the hypothalamus. The hypothalamus receives signals about light sources from the retina and relays them to this light-sensitive endocrine gland. Chapter 32 explores the functions of these forebrain regions in the context of connections between the nervous and endocrine systems.

The vertebrate brain develops from a hollow neural tube, the lumen of which persists in adults as a system of cavities and canals filled with cerebrospinal fluid. The fluid cushions nervous tissue from sudden, jarring movements.

Nervous tissue is subdivided into a hindbrain, forebrain, and midbrain. The brain stem is the most ancient tissue. The forebrain has the most complex integrating centers.

30.10 The Human Cerebrum

Our "humanness" starts in an outer layer of gray matter, the cerebral cortex. This brain region processes and coordinates responses to sensory input. It governs conscious behavior. It interacts with the limbic system, which governs emotions and contributes to memory.

FUNCTIONS OF THE CEREBRAL CORTEX

LINKS TO SECTIONS 5.3, 24.2, 24.4, 24.10

Each half of the cerebrum, or cerebral hemisphere, is divided into frontal, temporal, occipital, and parietal lobes. The cerebral cortex, the outermost gray matter on each lobe, contains distinct areas that receive and process diverse signals. *Motor* areas govern voluntary motor activity. *Sensory* areas assist in perceptions of specific sensations. Various *association* areas integrate information that brings about conscious actions.

The hemispheres overlap in function, but there are differences. For instance, the left hemisphere's cortex deals mainly with analytical skills, mathematics, and speech. The right hemisphere's cortex interprets music, judges spatial relations, and assesses visual inputs.

The body is spatially mapped out in the primary motor cortex of each frontal lobe, which controls and coordinates the movements of skeletal muscles on the opposite side of the body. Much of the motor cortex is devoted to finger, thumb, and tongue muscles. Figure 30.21 depicts the proportions of the motor cortex that are devoted to controlling different body parts.

The premotor cortex of each frontal lobe regulates complex movements and learned motor skills. Swing a golf club, play the piano, or type on a keyboard, and the premotor cortex coordinates the activity of many different muscle groups.

Broca's area helps translate thoughts into speech. It controls the tongue, throat, and lip muscles and gives us our capacity to create complex sentences. In most individuals, Broca's area is in the frontal cortex of the left hemisphere (Figure 30.22*a*). Damage to it prevents normal speech, although an affected individual is still able to understand language.

The primary somatosensory cortex is at the front of the parietal lobe. Like the motor cortex, it is organized as a map that corresponds to body parts. It functions as a receiving center for sensory input from the skin and joints, and one part of it deals with taste perception (Section 31.2). A primary visual cortex at the back of

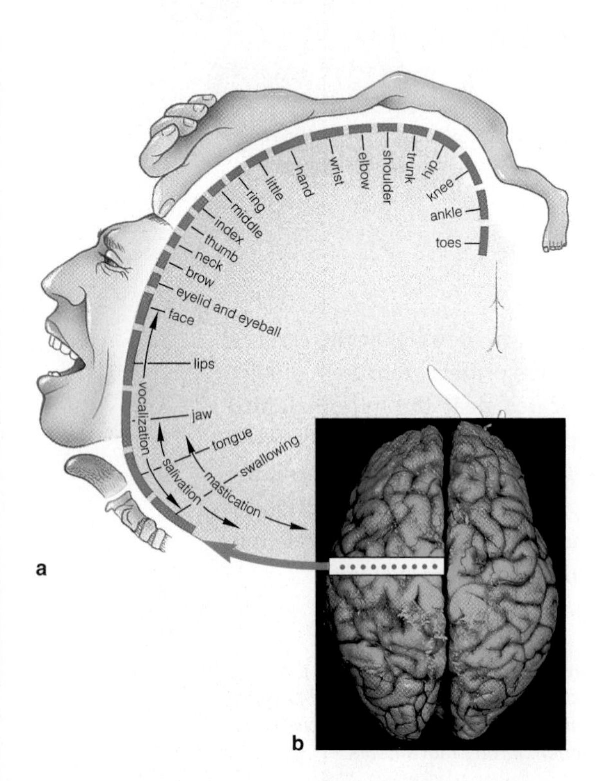

Figure 30.21 (**a**) A slice of the primary motor cortex through the region identified in (**b**). Sizes of body parts draped over the artful slice are distorted to indicate which ones get the most precise control.

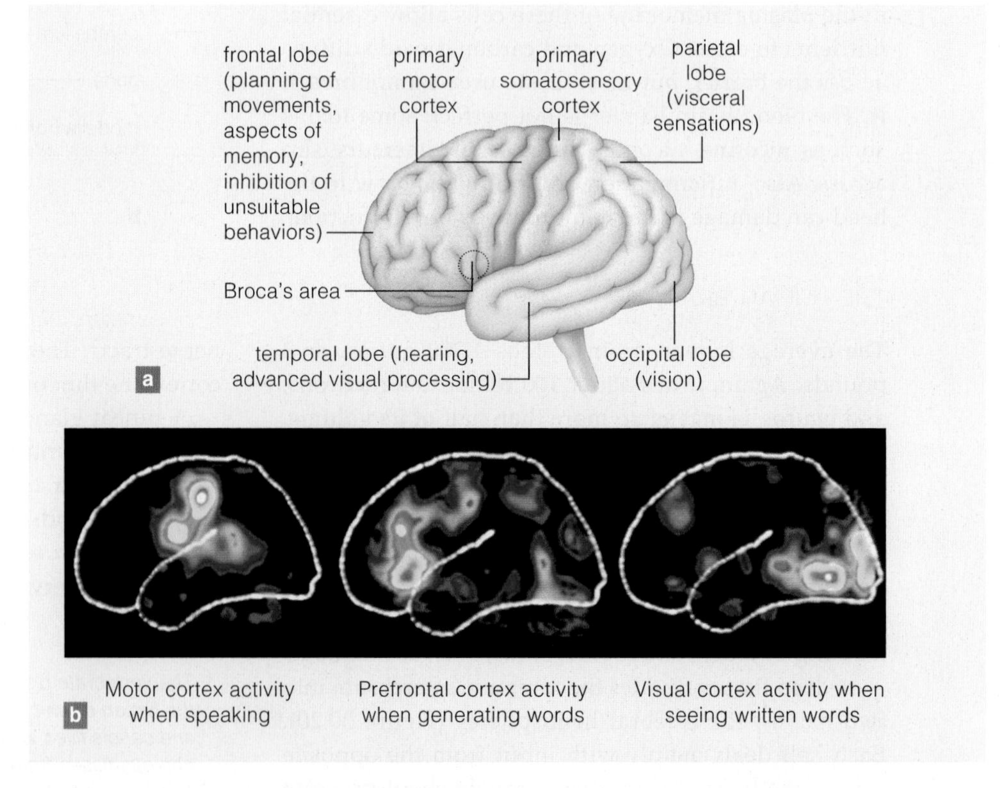

Figure 30.22 (**a**) Primary receiving and integrating centers of the human cerebral cortex. Association areas coordinate and process sensory input from diverse receptors. (**b**) Three PET scans identifying which areas were active when a person performed three tasks. *Yellow* and *orange* indicate high activity.

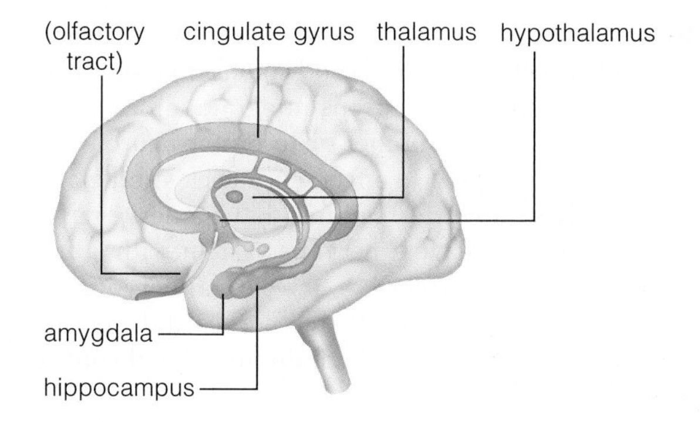

Figure 30.23 Limbic system components.

each occipital lobe receives input from both eyes. The perceptions of sound and odor arise in sensory areas of each temporal lobe.

Association areas are scattered through the cortex, but not in the primary motor and sensory areas. Each integrates diverse inputs (Figure 30.22*b*). For instance, one visual association area around the primary visual cortex compares what we see with visual memories. The most recently evolved area, the prefrontal cortex, is the basis of our personality and intellect, of abstract thought, judgment, planning, and concern for others.

CONNECTIONS WITH THE LIMBIC SYSTEM

The limbic system encircles the upper brain stem. It governs emotions, assists in memory, and correlates organ activities with self-gratifying behavior such as eating and sex. That is why the limbic system is called our emotional-visceral brain. It can put a heart on fire with passion and a stomach on fire with indigestion. These and other "gut reactions" can be overridden by signals from the prefrontal cortex.

The hypothalamus, hippocampus, amygdala, and adjacent structures make up the limbic system (Figure 30.23*b*). The hypothalamus is the major control center for homeostatic responses and it correlates emotions with visceral activities. The hippocampus helps store memories and access memories of earlier threats. The almond-shaped amygdala helps interpret social cues, and contributes to the sense of self. It is highly active during episodes of fear and anxiety, and often it is overactive in people with panic disorders.

Evolutionarily, the limbic system is related to the olfactory lobes. Olfactory input causes signals to flow to the hippocampus, amygdala, and hypothalamus as well as to the olfactory cortex. That is one reason why you feel warm and fuzzy when you recall the scent of someone special. Signals about taste also travel to the limbic system and can call up emotional responses.

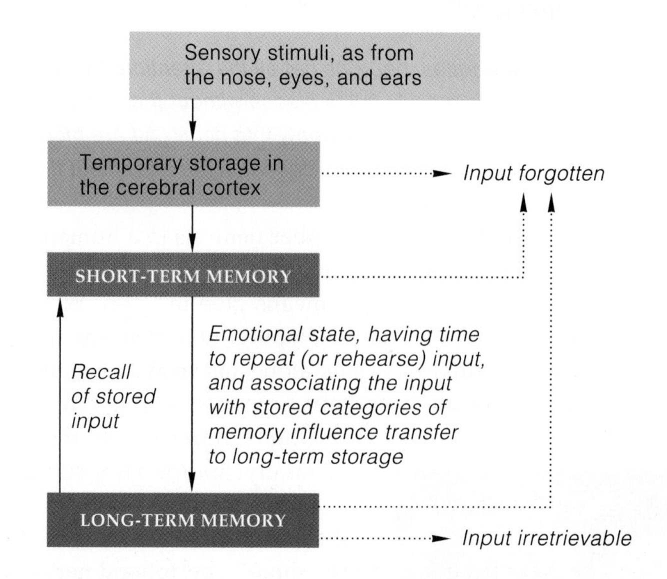

Figure 30.24 Stages in memory processing.

MAKING MEMORIES

The cerebral cortex receives information continually, but only a fraction of it becomes memories. Memory forms in stages. *Short-term memory* lasts a few seconds or hours. This stage holds a few bits of information—a set of numbers, words of a sentence, and so forth. In *long-term memory*, larger chunks of information become stored more or less permanently (Figure 30.24).

Different types of memories are stored and brought to mind by different mechanisms. Repetition of some motor task creates skill memories, which are highly persistent. Once you learn how to drive a car, dribble a basketball, or play the accordion, you seldom forget how, even if you rarely do it again. Skill memories involve the cerebellum, which controls motor activity.

Declarative memory stores facts and impressions of events, as when it helps you remember how a lemon smells or that a quarter is worth more than a dime. It starts when the sensory cortex signals the amygdala, a gatekeeper to the hippocampus. A memory will be retained only if signals loop repeatedly in the sensory cortex, hippocampus, basal ganglia, and thalamus.

Emotions influence memory retention. For instance, epinephrine released in times of stress helps shuffle short-term memories into long-term storage.

The cerebral cortex, each hemisphere's outermost layer of gray matter, contains motor, sensory, and association areas that interact to govern conscious behavior.

The cerebral cortex also interacts with the limbic system, which affects emotions and contributes to memory.

30.11 Neuroglia

Like celebrities, neurons get all the attention. Even so, they would quickly fall to pieces without their supporting staff. Similarly, we now know that neuroglia are more than bit players; neurons cannot act at all without them.

LINK TO SECTION 8.2

Neuroglial cells outnumber neurons in a human brain by about 10 to 1. Many act as a framework that holds neurons in place; *glia* means glue in Latin. While the nervous system is developing, new neurons migrate to their final positions along highways of neuroglial axons, which extend out from the forming brain.

Oligodendrocytes in the brain make myelin, the fatty substance in the insulating sheaths of long-distance axons. Schwann cells do the same in peripheral nerves.

Star-shaped astrocytes are the most abundant cells in the brain (Figure 30.25). Astrocytes serve in diverse roles. For example, they control the local concentrations of ions and neurotransmitters. They also function in immune defense, make lactate that fuels activities of neurons, and synthesize nerve growth factor. Growth factors are signaling molecules secreted from one cell that bind to receptors on target cells, which divide, mature, or differentiate in response.

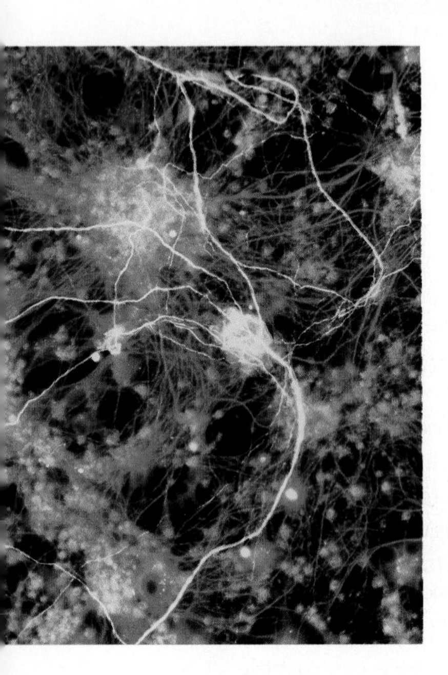

Figure 30.25 Astrocytes (*orange*) and a neuron (*yellow*) in brain tissue. The cells in this light micrograph were made visible by immunofluorescence. This procedure attaches fluorescent dye molecules to antibodies that bind to specific molecules on a cell.

Nerve growth factor does not cause cell division; most mature neurons remain in G1 of the cell cycle and do not undergo mitosis. Instead, it stimulates a neuron to form additional synapses with its neighbors. Such new connections are required for memory storage.

Microglia hang out in the brain, in resting form. When a tissue is injured, they become active, motile cells. They prowl the brain, engulf dead or dying cells and debris, and also issue chemical signals that alert immune cells to the threat.

Most cells of the vertebrate nervous system are neuroglia. These diverse cells help organize neurons as the nervous system is developing. They structurally support and nourish neurons, and promote synapse formation and immunity.

Astrocytes, oligodendrocytes, and microglia are major components of the brain. Myelin-producing Schwann cells sheathe many axons of peripheral nerves.

Summary

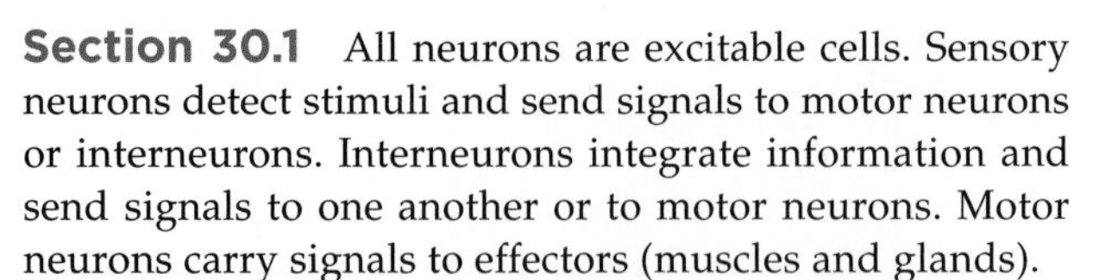

Section 30.1 All neurons are excitable cells. Sensory neurons detect stimuli and send signals to motor neurons or interneurons. Interneurons integrate information and send signals to one another or to motor neurons. Motor neurons carry signals to effectors (muscles and glands).

Radially symmetrical animals have a nerve net. Most bilateral animals have a nervous system with a brain at the head end. In vertebrates, the central nervous system is a brain and spinal cord. Paired nerves of the peripheral nervous system connect the brain and spinal cord to the rest of the body.

Section 30.2 Dendrites of a neuron receive signals and axon endings transmit signals. Neurons maintain a slight voltage difference across their plasma membrane.

- *Use the animation on ThomsonNOW to explore a neuron's structure and its membrane properties.*

Section 30.3 An action potential is a brief reversal in the voltage difference across the plasma membrane. It occurs only if membrane potential reaches the threshold level. Gated Na^+ channels open and Na^+ rushes into the axon. Then gated K^+ channels open and K^+ rushes out. Sodium–potassium pumps maintain ion gradients that are required for the resting membrane potential.

- *Use the animation on ThomsonNOW to view an action potential step by step.*

Section 30.4 Neurons signal other neurons, muscle fibers, or gland cells at chemical synapses. The arrival of an action potential at a presynaptic cell's axon endings triggers the release of neurotransmitter, which diffuses to receptors on the postsynaptic cell and binds to them. A postsynaptic cell's response is determined by synaptic integration of messages arriving at the same time.

- *Use the animation on ThomsonNOW and learn about a synapse between a motor neuron and a muscle cell.*

Sections 30.5, 30.6 A postsynaptic cell membrane has receptors for neurotransmitter molecules, which have different effects on different cell types. Neuromodulators alter neurotransmitter effects. Psychoactive drugs mimic neurotransmitters or disrupt their release or uptake.

Section 30.7 Nerves are long-distance cables, with bundles of axons that carry signals through the body. Myelin sheaths enclose most of the axons and increase signal conduction rates. Reflexes are simple, automatic responses to stimulation. In the simplest reflex arcs, a sensory neuron synapses directly on a motor neuron.

- *Use the animation on ThomsonNOW to see what happens during a stretch reflex.*

Section 30.8 The somatic nerves of the peripheral nervous system control skeletal muscles. The autonomic nerves control soft internal organs. Sympathetic signals dominate during danger or heightened awareness, and parasympathetic signals dominate at less stressful times. The spinal cord connects peripheral nerves to the brain.

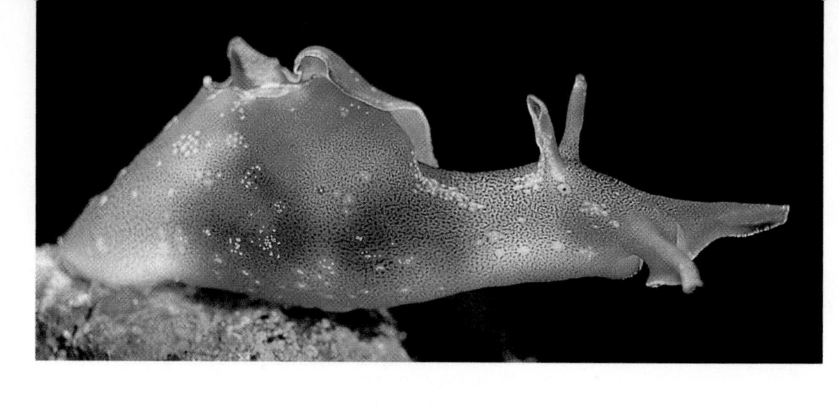

Figure 30.26 Sea hare (*Aplysia*). This brainless mollusk is a boon for studies into the neural basis of behavior in a simple organism.

■ *Use the animation on ThomsonNOW to explore the structure of the spinal cord and compare the effects of sympathetic and parasympathetic stimulation.*

Sections 30.9, 30.10 A vertebrate embryo's neural tube develops into the spinal cord and brain. The brain stem is the most evolutionarily ancient neural tissue. It governs reflex centers for breathing and other tasks. A blood–brain barrier stops many harmful substances from reaching the brain. The cerebral cortex, the most recently evolved region, governs the most complex functions. It receives and integrates motor and sensory information and controls conscious behavior. It also interacts with the limbic system in emotions and memory.

■ *Use the animation on ThomsonNOW to learn about the structure and function of the human cerebral cortex.*

Section 30.11 Neuroglial cells make up the bulk of the brain. They insulate and support the neurons.

Self-Quiz

Answers in Appendix III

1. _____ relay messages from the brain and spinal cord to muscles and glands.
   a. Motor neurons c. Interneurons
   b. Sensory neurons d. Neuroglia

2. When a neuron is at rest, _____.
   a. it is at threshold potential
   b. gated sodium channels are open
   c. the sodium–potassium pump is operating
   d. both a and c

3. Action potentials occur when _____.
   a. a neuron receives adequate stimulation
   b. more and more sodium gates open
   c. sodium–potassium pumps kick into action
   d. both a and b

4. Neurotransmitters are released by _____.
   a. axon endings c. dendrites
   b. the cell body d. the myelin sheath

5. The most abundant cells in the brain are _____.
   a. Schwann cells c. astrocytes
   b. microglia d. neurons

6. Skeletal muscles are controlled by _____.
   a. sympathetic signals c. somatic nerves
   b. parasympathetic signals d. both a and b

7. When you sit quietly on the couch and read, output from the _____ system prevails.
   a. sympathetic c. Both prevail.
   b. parasympathetic d. Neither prevails.

8. Skeletal muscles contract in response to _____.
   a. ACh c. serotonin
   b. dopamine d. all of the above

9. The cerebrum is part of the _____.
   a. forebrain c. hindbrain
   b. midbrain d. brain stem

10. Match each item with its description.
   ___ muscle spindle a. start of brain, spinal cord
   ___ neurotransmitter b. connects the hemispheres
   ___ limbic system c. protects brain and spinal cord from some toxins
   ___ corpus callosum d. type of signaling molecule
   ___ cerebral cortex e. support team for neurons
   ___ neural tube f. stretch-sensitive receptor
   ___ neuroglia g. roles in emotion, memory
   ___ gray matter h. most complex integration
   ___ blood–brain barrier i. unmyelinated axons and cell bodies

■ *Visit ThomsonNOW for additional questions.*

Critical Thinking

1. The sea hare *Aplysia californica* is a marine mollusk that can be as long as your forearm (Figure 30.26). Its nervous system only has about 20,000 neurons, some with a large cell body into which scientists can easily insert electrodes or inject dyes. Its limited behaviors include responses to food and to touch. Researchers have found that dopamine influences its feeding behavior. This neurotransmitter also affects human reward-seeking behavior, such as addiction to drugs. How might studies of dopamine's effects on sea hares help shed light on treating drug addiction?

2. In human newborns, especially premature ones, the blood–brain barrier is not yet fully developed. Why is this one reason to pay careful attention to the diet of infants?

3. In humans, the axons of some motor neurons extend from the base of the spinal cord to the big toe, a distance of more than a meter. In a giraffe, the longest axons are several meters long. What are some of the functional challenges involved in the development and maintenance of such impressive cellular extensions?

4. Some survivors of disastrous natural events, violent assaults, terrible accidents, and military combat develop post-traumatic stress disorder (PTSD). Symptoms include nightmares about the experience and suddenly feeling as if the event is recurring. Brain imaging studies of soldiers with PTSD showed that their hippocampus was abnormally small and their amygdala unusually active when compared to normal controls. Given these changes, what other brain functions might be disrupted in PTSD?

5. Colorful frogs of the genus *Dendrobates* secrete a toxin that is similar in structure to the neurotransmitter ACh. Hunters in some South American tribes dip their arrows in the toxin. What effect would an injection of an ACh-like chemical have on prey? How might the effect increase the hunters' success?

31 SENSORY PERCEPTION

IMPACTS, ISSUES A Whale of a Dilemma

Imagine yourself in the sensory world of a whale, 200 meters below the ocean surface where little sunlight penetrates. You cannot see well at all as you move through water. Many fishes detect your motion with a lateral line system, which responds to differences in water pressure. All of them use chemicals dissolved in the water as navigational cues. However, a whale has no lateral line, and it has a very poor sense of smell. How, then, does it know where it is going?

All whales use sounds—acoustical cues. Water is an ideal medium for transmitting sound waves, which move five times faster in water than in air. Whales have many neurons that collect and integrate auditory information.

Unlike humans, whales do not have a pair of ear flaps that collect sound waves. Some do not even have a canal leading to ear components inside their head. Others have ear canals packed with wax. How, then, do whales hear? Their *jaws* pick up vibrations traveling through water. The vibrations are transmitted from the jaws, through a layer of fat, to a pair of pressure-sensitive middle ears.

Whales use sound to communicate, locate food, and find their way around underwater. Killer whales and some other species of toothed whales use echolocation. The

See the video! **Figure 31.1** A few children drawn to one of the whales that beached itself during military testing of a new sonar system. Of sixteen stranded whales, six died on the beach. Volunteers pushed the others out to sea. Their fate is unknown.

How would you vote? Maritime activities such as shipping cause an underwater ruckus. Would you support a ban on activities that generate excessive noise levels from territorial waters of the United States and other nations? See ThomsonNOW for details, then vote online.

whale emits high-pitched sounds and then listens as the echoes bounce off objects, including prey. Its ears are especially sensitive to sounds of high frequencies. Baleen whales, including the humpback whale, make very low-pitched sounds that can travel across an entire ocean basin. Their ears are adapted to respond to those sounds.

The ocean is becoming a lot noisier, and the superb acoustical adaptations of whales now put them at risk. In 2001, for example, some whales beached themselves near an area where the United States Navy was testing a sonar system (Figure 31.1). This sonar system uses echoes of low-frequency sounds to detect a new generation of missile-launching submarines that can run in near-silence. Humans cannot hear the intense sounds. Whales can.

As autopsies later revealed, the beached whales had blood in their ears and in acoustic fat. Apparently the intense sounds made them race to the surface in fear. The rapid change in pressure damaged internal tissues.

Public outcry halted the deployment of the new sonar system. Testing continues, however, because the threat of stealth attacks against the United States is real.

Besides, noise pollution from commercial shipping may be a worse problem for whales. Massive tankers generate low-frequency sounds that frighten whales or drown out acoustical cues. Realistically, global shipping of oil and other resources that nations require is not going to stop. If research shows that whales are at risk, will those same nations be willing to design and deploy new tankers that minimize the damage?

In this chapter, we turn to sensory systems. With these organ systems, animals receive signals from inside and outside the body, decode them, and become aware of touches, sounds, sights, odors, and other sensations. As you will see, animals differ in the type and number of sensory receptors that sample the environment, and differ in their perception of it.

Key Concepts

HOW SENSORY PATHWAYS WORK

Sensory systems are front doors of the nervous system. Each has sensory receptors, nerve pathways to the brain, and brain regions that receive and process the sensory input. A stimulus is a form of energy that activates a specific type of sensory receptor. Information becomes encoded in the number and frequency of action potentials sent to the brain along particular nerve pathways. **Section 31.1**

SOMATIC SENSES

Touch, pressure, pain, temperature, and muscle sense are somatic sensations. These sensations start at mechanoreceptors in skin, muscles, and the wall of internal organs. **Section 31.2**

CHEMICAL SENSES

The senses of smell and taste require chemoreceptors, which bind molecules of specific substances that have become dissolved in the fluid bathing them. **Section 31.3**

BALANCE AND HEARING

The sense of balance starts at mechanoreceptors, which detect gravity, velocity, acceleration, and other forces that influence the position and motion of the body or specific parts of it.

In vertebrates, the sense of hearing starts with structures that collect, amplify, and sort out pressure variations caused by sound waves. The variations trigger action potentials in mechanoreceptors that send signals to the brain. **Sections 31.4, 31.5**

VISION

Most organisms have light-sensitive pigments, but vision requires eyes with a dense array of photoreceptors and image formation in the brain. Paired, camera-like eyes of cephalopoda and vertebrates collect and process information on distance, brightness, shape, position, and movement of visual stimuli. A sensory pathway starts at the retina and ends in the visual cortex. **Sections 31.6–31.8**

Links to Earlier Concepts

Sensory systems are composed of nervous tissue, and cross-references will guide you to specific sections in Chapter 30. You may wish to review the properties of light (6.1). You will come across more examples of signal transduction pathways (25.5). You will reflect again on the evolution of fast-moving cephalopoda (23.7) and the challenges that faced the first vertebrates that moved onto land (24.4).

31.1 Overview of Sensory Pathways

Nervous systems include communication pathways in which sensory neurons detect stimulus energy and transduce it to a form that can trigger a response.

LINKS TO SECTIONS 1.2, 30.1–30.3, 30.7

A sensory system is a portion of the nervous system. In vertebrates, it consists of sensory neurons that can detect stimuli, nerves that carry information about the stimulus to the brain, and brain regions that process the information. A stimulus (plural, stimuli) is a form of energy that activates receptor endings of a sensory neuron. The energy is transduced to electrochemical energy of action potentials—the messages that travel along the plasma membrane of neuron (Section 30.2).

Processing of sensory signals in the brain gives rise to a sensation: awareness of a stimulus. Perception is understanding what the sensation means. The sensory neuron that start in tissues throughout the body are the start of somatic sensations. The types confined to particular organs, such as eyes, are the start of special senses—taste, smell, balance, hearing, and vision.

For instance, stretch receptors in a gymnast's arm and leg muscles keep the brain informed of changes in muscle length (Section 30.7 and Figure 31.2*a*). The brain integrates the sensory input with other signals, then issues commands that cause muscles to adjust their length and help maintain balance and posture.

In this book, you will come across examples from five classes of sensory receptors:

Mechanoreceptors detect mechanical energy, such as a change in pressure, position, or acceleration.

Pain receptors detect tissue damage.

Thermoreceptors detect a temperature change.

Chemoreceptors detect chemical energy of specific substances dissolved in the fluid that bathes them.

Osmoreceptors detect change in the concentration of solutes in a body fluid, such as blood.

Photoreceptors detect forms of light energy.

Regardless of the differences, all sensory receptors transduce stimulus energy into action potentials. But action potentials, remember, are always the same size. How, then, does the brain assess stimulus location and intensity? It takes into account *which* nerve pathways are carrying action potentials, the *frequency* of action potentials traveling on each axon in the pathway, and the *number* of axons recruited by the stimulus.

First, an animal's brain is prewired, or genetically programmed, to interpret action potentials in certain ways. That is why you may "see stars" after an eye gets poked, even in a dark room. Many photoreceptors in the eye were mechanically disturbed and sent signals along one of two optic nerves to the brain. The brain interprets any signals from an optic nerve as "light."

Second, a strong signal makes receptors fire action potentials more often and longer. The same receptors are stimulated by a whisper and a whoop. Your brain interprets differences between these stimuli through variations in the frequency of signals (Figure 31.2*b*).

Third, a strong stimulus recruits far more sensory receptors, compared to a weak stimulus. A gentle tap on the arm activates far fewer receptors than a sudden slap. A larger disturbance results in action potentials in many more sensory axons, and the brain interprets all the activity as an increase in stimulus intensity.

In sensory adaptation, sensory neurons stop firing despite continued stimulation. Put on a sock and you briefly feel it pressing on your skin, then quickly lose awareness of it. Mechanoreceptors in the skin adapt to this stimulation, allowing you to focus on other things.

Sensory receptors detect specific stimuli. Nerves carry signals to brain regions that assess a stimulus according to which nerve pathway is signaling it, the frequency of action potentials, and the number of axons the stimulus excited.

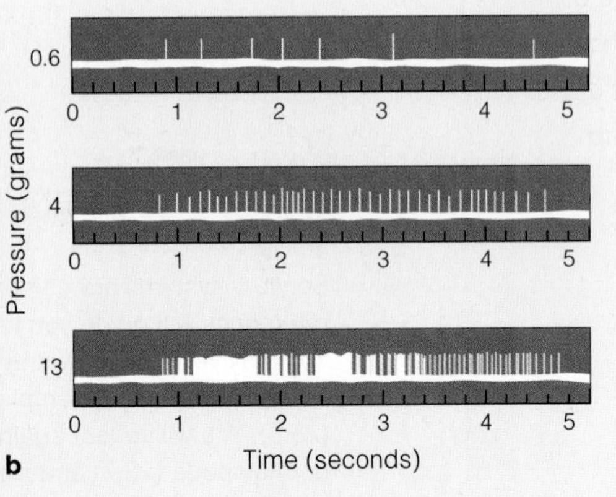

Figure 31.2 Animated! (**a**) Gymnast benefitting from the sensory input of stretch receptors and other sensory receptors. (**b**) Recordings of action potentials from a pressure receptor with endings in a human hand. The graphs chart the variations in stimulus strength. A thin rod was pressed against skin with the amount of pressure indicated to the left of each diagram. Vertical bars above each thick horizontal line record individual action potentials. Increases in the frequency of action potentials correspond to increases in the stimulus strength.

31.2 Somatic Sensations

Different kinds of somatic sensory receptors are widespread in skin and in soft internal tissues.

LINKS TO SECTIONS 30.5, 30.10

Somatic sensations are responses to receptors near the body surface and in skeletal muscle, joints, and walls of soft internal organs. Signals reach the spinal cord and the somatosensory cortex of the cerebrum. In the somatosensory cortex, interneurons are arrayed like a map of the body surface (Figure 31.3). The largest map regions correspond to the most sensitive parts, such as lips. Fewer neurons represent less sensitive parts.

Free nerve endings, or branched endings of sensory neurons, are the most abundant receivers of somatic information. Pressure, temperature changes, and pain activate them. Free nerve endings coiled around the roots of hairs in the dermis detect even the slightest pressure (Figure 31.4). Others can detect temperature changes or damage to tissues. Free nerve endings also occur in skeletal muscles, tendons, joints, and walls of internal organs. Here, they give rise to sensations that range from itching, to a dull ache, to sharp pain.

A capsule surrounds the endings of certain kinds of receptors, many of which are named for the scientist who first described them. Meissner's corpuscles and Pacinian corpuscles are the main receptors for touch and pressure in fingertips, palms, soles of the feet, and other hairless skin regions. Meissner's corpuscles in the upper dermis detect light touches. Deeper in the dermis, bigger Pacinian corpuscles respond to strong pressure. Onionlike layers of connective tissue wrap around their one sensory ending. Pacinian corpuscles also occur near joints and in the wall of some organs.

Either pressure or a specific temperature can trigger the response of other encapsulated receptors. Ruffini endings in skin adapt more slowly than Meissner's and Pacinian corpuscles. If you hold a stone in your hand, Ruffini endings tell your brain that the stone is still there even after other receptors have adapted and stopped responding. Ruffini endings also fire when temperature exceeds 45°C (113°F). The bulb of Krause, an encapsulated receptor in skin and certain mucous membranes, responds to both touch and cold.

Remember the stretch receptors in muscle spindle fibers (Section 30.7)? The more a muscle stretches, the more frequently these receptors fire. In concert with receptors in skin and near movable joints, they inform the brain about positions of the body's limbs.

Pain, the perception of tissue injury, occurs after the brain receives signals from free nerve endings that are pain receptors. A sense of *somatic* pain starts with activation of receptors in skin, skeletal muscle, joints, and tendons. *Visceral* pain starts with nerve endings of internal organs, such as the heart and stomach.

When signals from pain receptors enter the spinal cord, they stimulate secretion of substance P (Section 30.5). Signals are relayed to the sensory cortex, which evaluates the intensity and type of pain. Signals also stimulate the release of endorphins and enkephalins, natural opiates that can diminish the sense of pain by inhibiting the release of substance P.

Sensory receptors near the body surface and in internal organs detect touch, pressure, temperature, pain, motion, and positional changes of body parts.

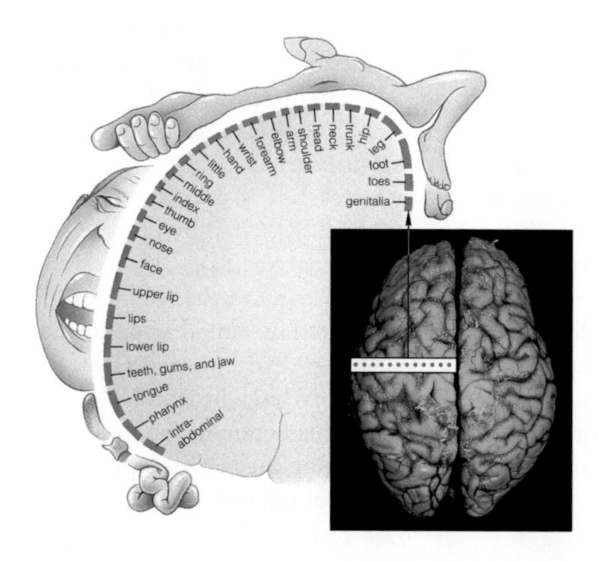

Figure 31.3 A map of where different body regions are represented in the human primary somatosensory cortex, a narrow strip of the cerebral cortex that runs from the top of the head to just above the ear. Compare Section 30.10.

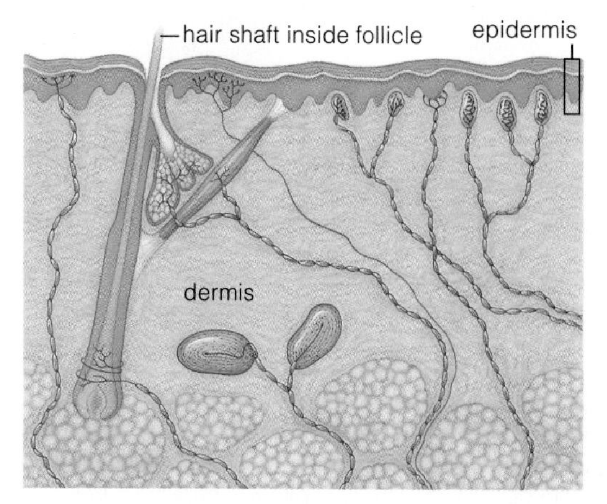

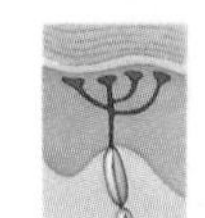

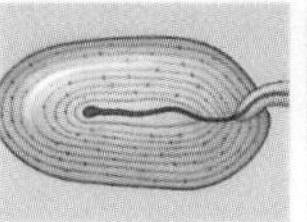
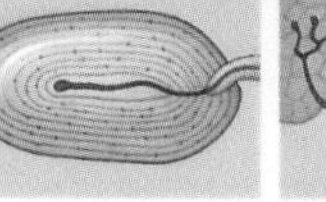

Figure 31.4 Animated! A sampling of sensory receptors in human skin.

31.3 Sampling the Chemical World

The rest of this chapter samples organs that give rise to the senses of smell, taste, balance, hearing, and vision.

SENSE OF SMELL

LINK TO SECTION 30.10

Olfaction, a sense of smell, starts with chemoreceptors that bind substances that become dissolved in fluid around them. A stimulus can trigger action potentials that olfactory nerves transmit to the cerebral cortex, where coarse perceptions are refined. Messages also reach the limbic system, which integrates them with emotional state and stored memories (Section 30.10).

Olfactory receptors detect water-soluble or volatile (easily vaporized) chemicals. A human nose has about 5 million of them; a bloodhound nose has 200 million. Receptor axons lead into one of two olfactory bulbs. In these small brain structures, neurons sort out the components of a scent, then signal the cerebrum for further processing (Figure 31.5).

Many animals use olfactory cues to navigate, find food, and communicate socially, as with pheromones. As later chapters explain, pheromones are signaling molecules secreted by one individual that change the social behavior of other individuals of its species. As one example, olfactory receptors on the antennae of a male silk moth help him find a pheromone-secreting female that may be more than a kilometer upwind.

In the nasal cavity of reptiles and most mammals, a cluster of sensory neurons forms a vomeronasal organ that responds to pheromones. Humans have a reduced version of this organ; it is only 12.7 millimeters (0.5 inch) long. Some studies suggest that humans make and respond to certain pheromones (Chapter 44).

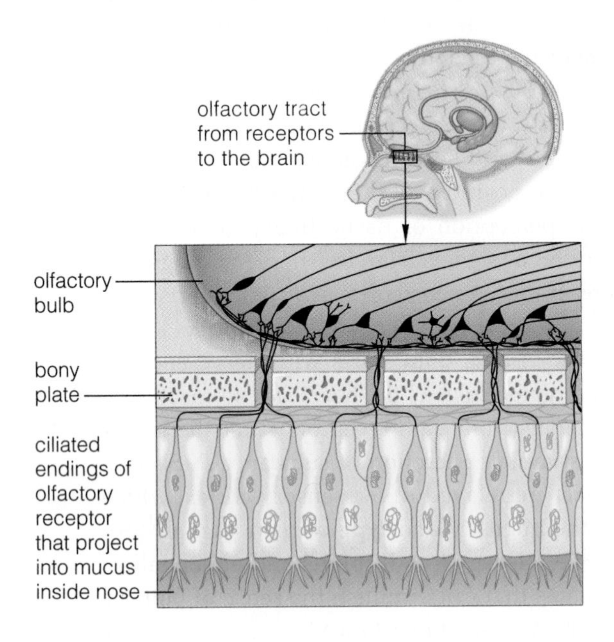

Figure 31.5 Pathway from the sensory endings of olfactory receptors in the human nose to the cerebral cortex and limbic system. Axons of these sensory receptors pass through holes in a bony plate between the lining of the nasal cavities and the brain.

SENSE OF TASTE

Taste receptors are other chemoreceptors that detect chemicals dissolved in fluid, but they have a different structure and location than olfactory receptors. Taste receptors help animals locate food and avoid poisons. An octopus "tastes" with receptors in suckers on its tentacles; a fly does this with receptors in its antennae and feet. In humans, many taste buds are embedded in the upper surface of the tongue (Figure 31.6). These sensory organs are located in specialized epithelial structures, or papillae, that look like raised bumps or red dots on the tongue surface.

You perceive many tastes, but all are a combination of five main sensations: *sweet* (elicited by glucose and the other simple sugars), *sour* (acids), *salty* (sodium chloride or other salts), *bitter* (plant toxins, including alkaloids), and *umami* (elicited by amino acids such as glutamate which, like aged cheese and aged meat, has a savory taste). You probably have heard of MSG, or monosodium glutamate. This common food additive enhances flavor by stimulating the taste receptors that contribute to the sensation of umami.

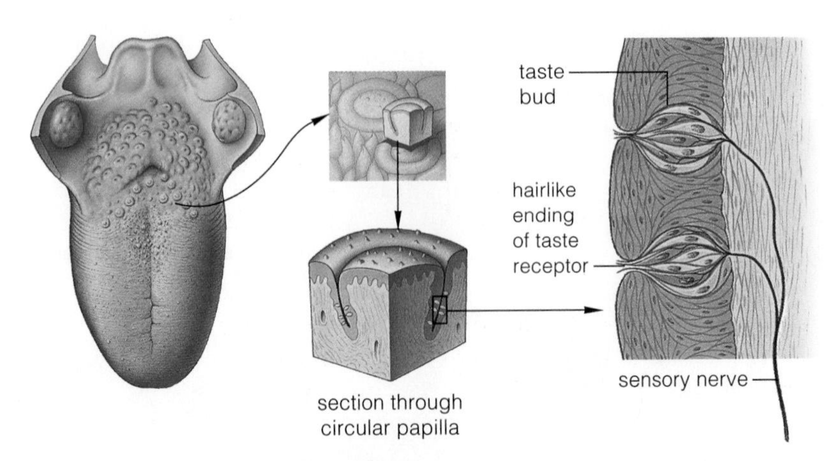

Figure 31.6 Taste receptors in the human tongue. Taste buds are clusters of receptor cells and supporting cells inside special epithelial papillae. One type, a circular papilla, is shown in section here. The tongue has about 5,000 taste buds, each enclosing as many as150 taste receptor cells.

The senses of smell and taste start at chemoreceptors. Both involve sensory pathways that lead to processing regions in the cerebral cortex and in the limbic system.

31.4 Keeping the Body Balanced

Even brainless jellyfishes keep their body balanced, as by righting themselves when they are turned upside down.

Organs of equilibrium are parts of sensory systems that monitor the body's positions and motions. Each vertebrate ear has such organs inside a fluid-filled sensory structure called the vestibular apparatus. The organs are located in three semicircular canals and in two sacs, the saccule and utricle (Figure 31.7*a*).

Organs of the vestibular apparatus have hair cells, a type of mechanoreceptor with modified cilia at one end. Fluid pressure inside the canals and sacs makes the cilia bend. That mechanical energy deforms the hair cell plasma membrane just enough to let ions slip across and stimulate an action potential. A vestibular nerve carries the sensory input to the brain. As you will see, hair cells also function in hearing.

The three semicircular canals are oriented at right angles to one another, so rotation of the head in any combination of directions—front/back, up/down, or left/right—moves the fluid inside them. An organ of equilibrium rests on the bulging base of each canal. The cilia of its hair cells are embedded in a jellylike mass (Figure 31.7*b*). When fluid moves in the canal, it pushes against the mass and generates the pressure required for initiating action potentials.

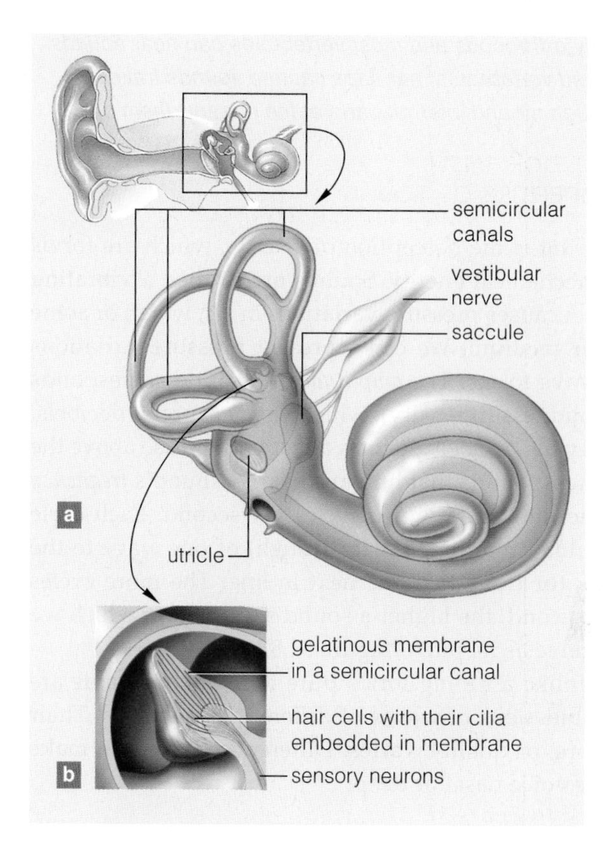

Figure 31.7 **Animated!** (**a**) The vestibular apparatus inside a human ear. Organs of equilibrium in its fluid-filled sacs and canals contribute to a sense of balance. (**b**) Components of one of the organs in a semicircular canal.

The brain receives signals from semicircular canals on both sides of the head. By comparing the frequency of action potentials and the number of sensory neurons responding on nerves from both sides, the brain can sense *dynamic* equilibrium: any angular movement and rotation of the head. Among other things, you can use this sense to keep your eyes locked on an object even as your head is swiveling or nodding about.

Organs in the saccule and utricle function in *static* equilibrium. They help the brain monitor the head's position and how fast the head is moving in a straight line. They also help keep the head upright and maintain posture. A jellylike mass weighted with calcite crystals lies just above hair cells in these organs. Tilt your head, or start or stop moving, and gravity causes this mass to shift. As the mass shifts, the hair cells bend and alter their firing pattern.

The brain also takes into account information from the eyes, and from receptors in the skin, muscles, and joints. Integration of the signals provides awareness of the body's position and motion in space, as shown by figure skater Sarah Hughes.

A stroke, an inner ear infection, or loose particles in the semicircular canals cause vertigo, a sensation that the world is moving or spinning around. Vertigo also arises from conflicting sensory inputs, as when you stand at a height and look down. The vestibular apparatus reports that you are motionless, but your eyes report that your body is floating in space.

Mismatched signals can cause motion sickness. On a curvy road, passengers in a car experience changes in acceleration and direction that scream "motion" to their vestibular apparatus. At the same time, signals from their eyes about objects inside the car tell their brain that the body is at rest. Driving can minimize motion sickness because the driver focuses on sights outside the car such as scenery rushing past, so the visual signals are consistent with vestibular signals.

Organs of equilibrium help keep the body balanced in relation to gravity, velocity, acceleration, and other forces that influence its position and movement.

31.5 Collecting, Amplifying, and Sorting Out Sounds

Many arthropods and most vertebrates can hear sounds. In land vertebrates, ear flaps capture sounds traveling through air and internal parts of the ear sort them out.

LINKS TO SECTIONS 24.4, 25.5

PROPERTIES OF SOUND

Hearing is the perception of sounds, which are forms of mechanical energy. Sounds arise when a vibrating object causes pressure variations in air, water, or some other medium. We can represent pressure variations as wave forms. The *amplitude* of a sound corresponds to loudness (intensity), which we measure in decibels. Every ten decibels denotes a tenfold increase above the faintest sounds that humans hear. A sound's *frequency* is the number of wave cycles per second. Each cycle extends from the peak (or trough) of one wave to the peak (or trough) of the next in line. The more cycles per second, the higher a sound's frequency, which we perceive as its pitch (Figure 31.8).

Unlike a tuning fork's pure tone, most sounds are combinations of waves of different frequencies. Their timbre, or quality, varies. Differences in timbre make your voice nasal or deep.

THE VERTEBRATE EAR

Water readily transfers vibrations to body tissues, so fishes do not require fancy ears to detect them. When some vertebrates left water for land, the capacity to collect and *amplify* vibrations evolved in response to a new environmental challenge: transfer of sound waves from air to body tissues is inefficient. The structure of human ears helps maximize the efficiency of transfer.

As Figure 31.9*a* indicates, the *outer* ear of humans and most other mammals is adapted to gathering sounds from the air. The pinna, a skin-covered flap of cartilage projecting from the side of the head, collects sound waves and directs them into the auditory canal. The canal conveys sounds to the middle ear.

The *middle* ear amplifies and transmits air waves to the inner ear. An eardrum evolved in early reptiles as a shallow depression on each side of the head. The drum, a thin membrane, vibrates fast in response to pressure waves. Behind it is an air-filled cavity and a hammer, anvil, and stirrup, which are three small bones (Figure 31.9*b*). Together, the bones transmit the force of sound waves from the eardrum onto the smaller surface of an elastic membrane called the oval window, which is positioned in front of the inner ear.

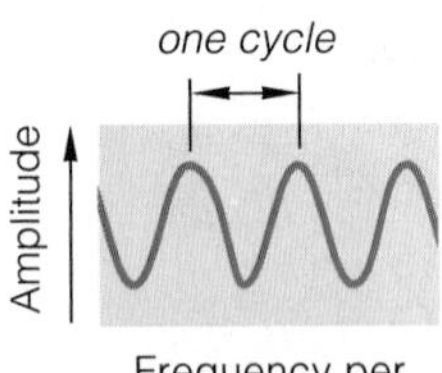

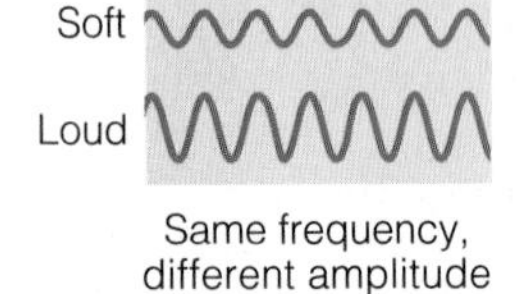

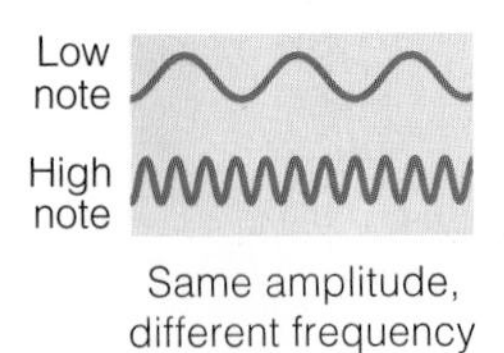

Figure 31.8 **Animated!** Wavelike properties of sound.

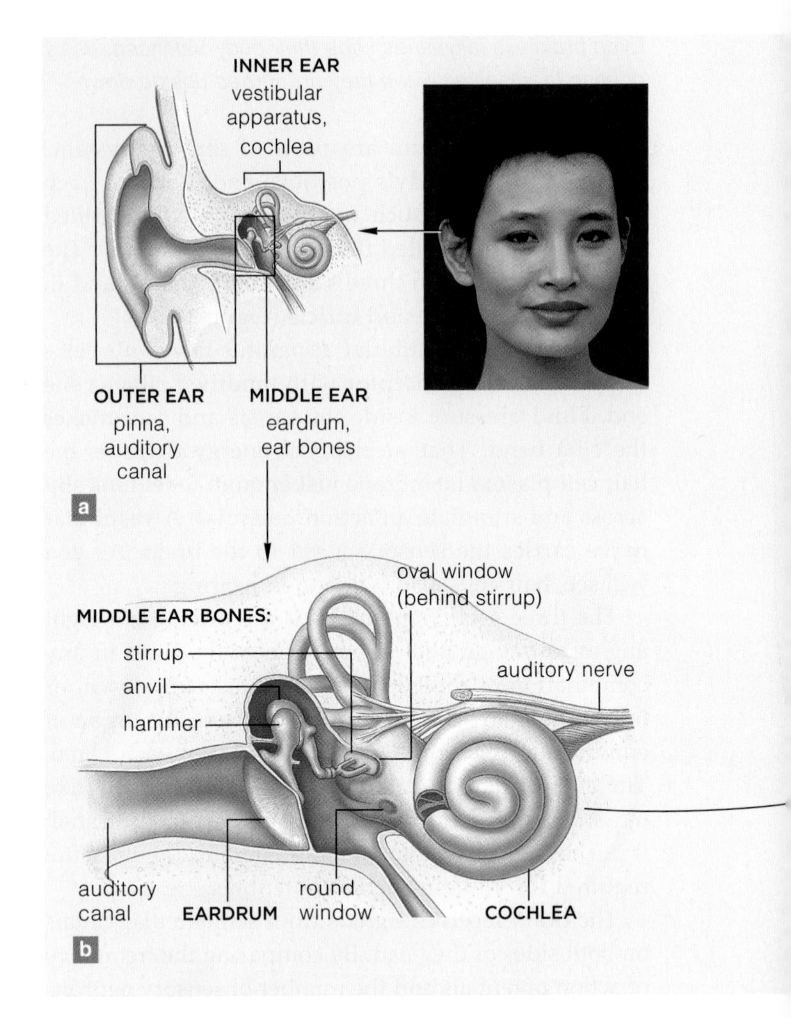

Figure 31.9 **Animated!** Components of the human ear.

The *inner* ear, remember, has a vestibular apparatus (Section 31.4). It also has a cochlea, which in humans is a pea-sized, fluid-filled structure that resembles a coiled snail shell (the Greek *koklias* means snail). The transduction of waves of sound into action potentials takes place in the cochlea (Figure 31.9*c*).

Sound waves make the stirrup vibrate, causing this middle ear bone to push against the oval window. The pushing transmits pressure waves to the fluid inside two cochlear ducts (scala vestibuli and scala tympani). The waves move through fluid to the round window, which bows inward and outward in response.

A third cochlear duct sorts out the pressure waves. Its wall, a basilar membrane, is stiff and narrow near the oval window, then it broadens and becomes more

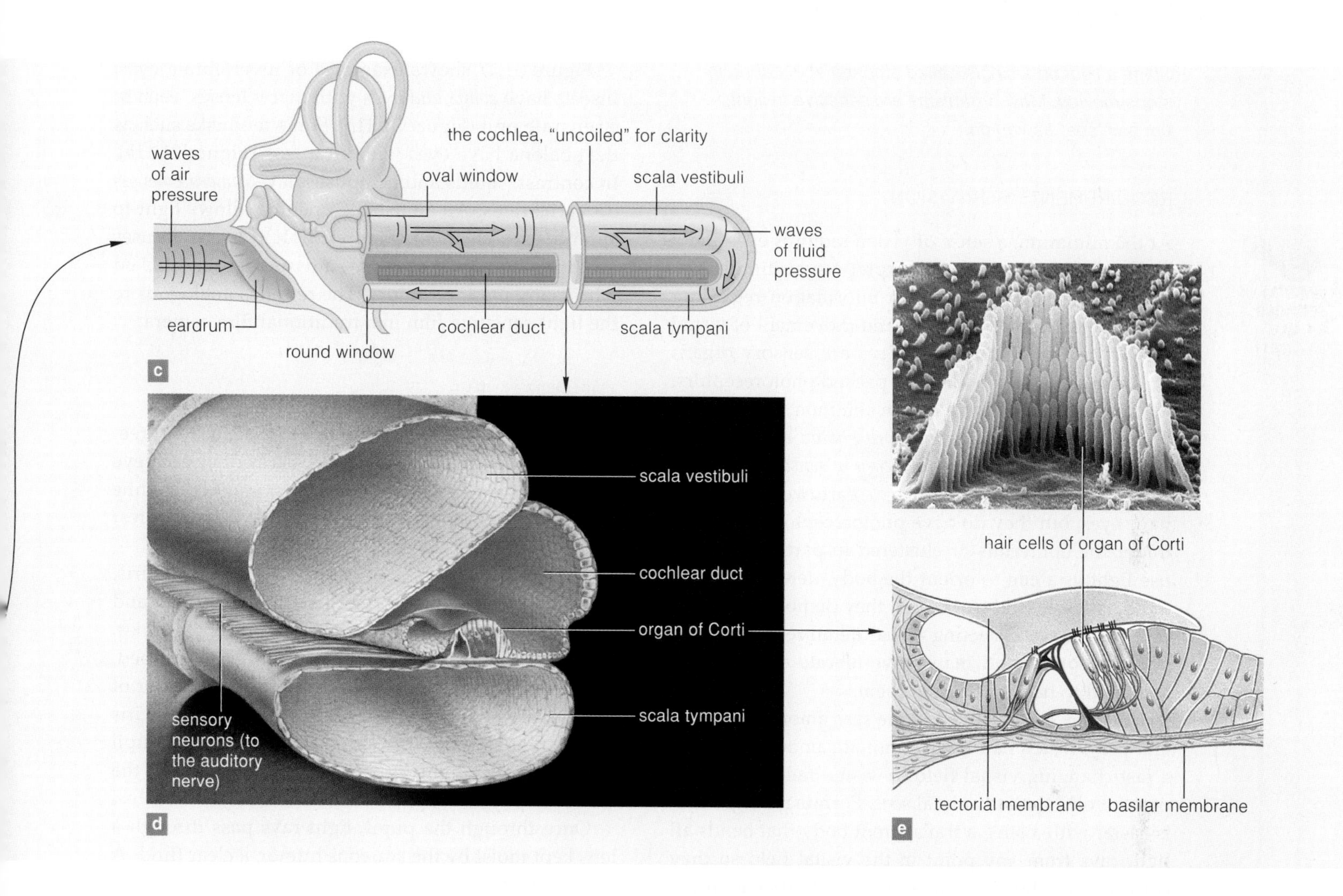

flexible deeper in the coil. Because of the differences, frequency variations have different effects along the length of the basilar membrane. High-pitched sounds make the stiff, narrow part of the cochlear duct vibrate; low-pitched sounds make the flexible part vibrate.

Attached to the top of the basilar membrane is an organ of Corti, an acoustical organ with arrays of hair cells—mechanoreceptors with a tuft of modified cilia at one end. The cilia project into a tectorial membrane that drapes over them (Figure 31.9*d*,*e*). Movement of the basilar membrane pushes cila against the tectorial membrane and causes hair cells to fire. Signals travel along an auditory nerve to the brain.

Exposure to loud sound can destroy hair cells and cause hearing loss (Figure 31.10). Certain antibiotics such as streptomycin also can harm the hair cells.

Ears of land vertebrates collect, amplify, and sort out sound waves. In the inner ear, sound waves cause fluid pressure variations and trigger action potentials in hair cells.

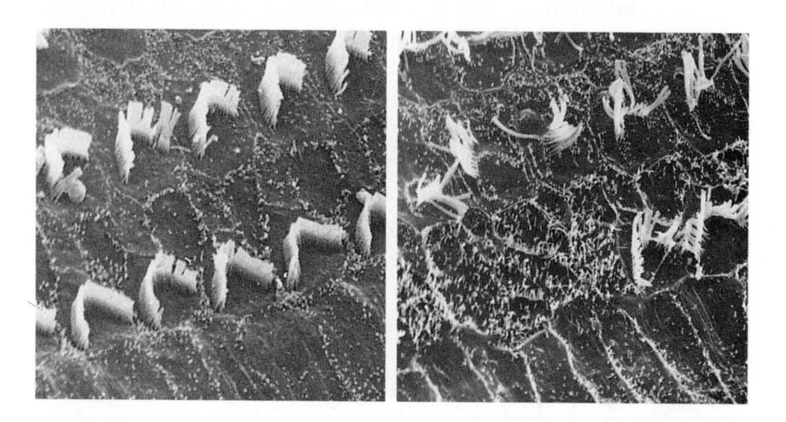

Figure 31.10 Results of an experiment on the effect of intense sound on the inner ear. *Left,* from a guinea pig ear, two rows of hair cells that normally project into the tectorial membrane in the organ of Corti. *Right,* hair cells inside the same organ after twenty-four hours of exposure to noise levels comparable to extremely loud music.

To give you a sense of how sound intensity is measured, a ticking watch measures 10 decibels. Normal conversation is about 60 decibels, a food blender operating at high speed is about 90 decibels, and an amplified rock concert is about 120 decibels. The perceived loudness of a sound depends not only on its intensity but also on its frequency.

31.6 Do You See What I See?

Shine a light on a single-celled amoeba and it abruptly stops moving. Most organisms are sensitive to light, but few "see" as you do.

REQUIREMENTS FOR VISION

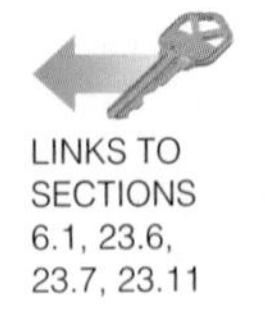

LINKS TO SECTIONS 6.1, 23.6, 23.7, 23.11

At the minimum, a sense of vision requires eyes and a brain with the capacity to interpret visual stimulation. Image perception arises when information regarding shapes, brightness, positions, and movement of visual stimuli becomes integrated. Eyes are sensory organs that contain a layer of densely packed photoreceptors. All photoreceptors have this in common: *They contain pigment molecules that can absorb photon energy, which can be converted to excitation energy in sensory neuron.*

Certain invertebrates, such as earthworms, do not have eyes, but they do have photoreceptors dispersed under the epidermis or clustered in parts of it. They use light as a cue to orient the body, detect shadows, or adjust biological clocks, but they do not have a true sense of vision. Detecting visual detail requires many photoreceptors, and many invertebrate eyes are just too small to have enough of them.

Complex eyes evolved in the seas among predators and prey, which had to discriminate among objects in a fast-changing visual field. A visual field is the part of the world that an animal sees. Forming images of it is easier with a lens, a transparent body that bends all light rays from any point in the visual field so they converge on photoreceptors. Rays of light bend at the boundaries between substances of different densities. Bending sends them off in new directions.

Figure 31.11 shows examples of invertebrate eyes. Insects have *compound* eyes with many lenses, each in a separate unit (Figure 31.11*a*). Some mollusks such as the abalone have eyes without a lens (Figure 31.11*b*). In contrast, squids and octopuses have *camera* eyes, as do vertebrates. An adjustable opening allows light to enter a dark chamber (Figure 31.11*c*). A lens focuses incoming light onto a retina—a tissue densely packed with many photoreceptors. The retina is analogous to the light-sensitive film in a traditional film camera.

THE HUMAN EYE

As in most vertebrates, a human eyeball has a three-layered structure (Figure 31.12). The front of each eye is covered by a cornea made of transparent crystalline proteins. A dense, white, fibrous sclera covers the rest of the eye's outer surface.

The middle layer has a choroid, ciliary body, iris, and pupil. Blood vessels run through the choroid and pigments in it absorb light that photoreceptors miss. Suspended behind the cornea is a doughnut-shaped, pigmented iris. The pupil, an opening at the center of the iris, lets light into the eye. Bright light causes iris muscle that encircles the pupil to contract, so the pupil shrinks. In dim light, spoke-like radial muscle of the iris contracts, so the pupil dilates (widens).

Once through the pupil, light rays pass through a lens kept moist by the aqueous humor, a clear fluid. A jellylike vitreous body fills the space behind the lens. The retina is at the back of this chamber.

The cornea and lens both bend the incoming light rays so that these rays converge at the back of the retina. The image formed on the retina is upside-down and the mirror image of the real world (Figure 31.13*a*). The brain makes necessary adjustments so you perceive the correct orientation of things.

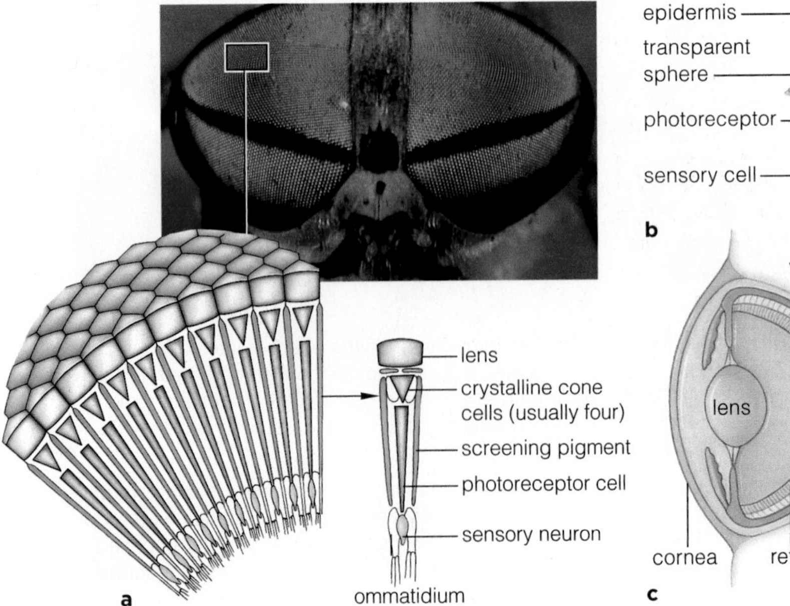

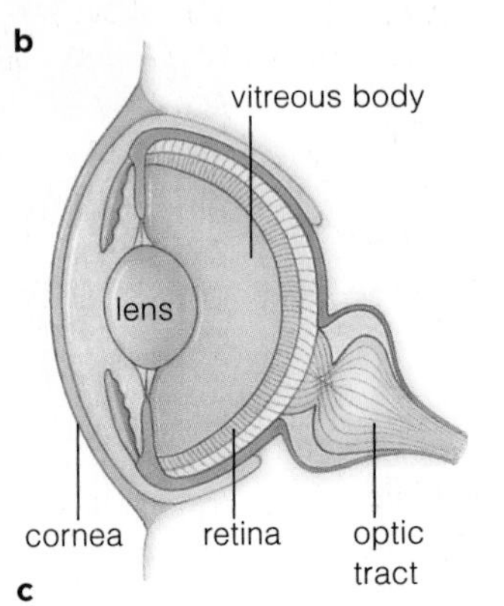

Figure 31.11 Invertebrate eyes. There are far more photoreceptors than can be shown in the simple diagrams. (**a**) Compound eye of a deerfly. A lens in each of many units directs light onto a crystalline cone, which focuses it onto one photoreceptor cell. (**b**) Eye of an abalone (a gastropod). (**c**) Camera eye of squid (a cephalopod).

Wall of eyeball (three layers)

Outer layer	*Sclera.* Protects eyeball
	Cornea. Focuses light
Middle layer	*Pupil.* Serves as entrance for light
	Iris. Adjusts diameter of pupil
	Ciliary body. Its muscles control the lens shape; its fine fibers hold lens in place
	Choroid. Its blood vessels nutritionally support wall cells; its pigments stop light scattering
	Start of optic nerve. Carries signals to brain
Inner layer	*Retina.* Absorbs, transduces light energy

Interior of eyeball

Lens	Focuses light on photoreceptors
Aqueous humor	Transmits light, maintains fluid pressure
Vitreous body	Transmits light, supports lens and eyeball

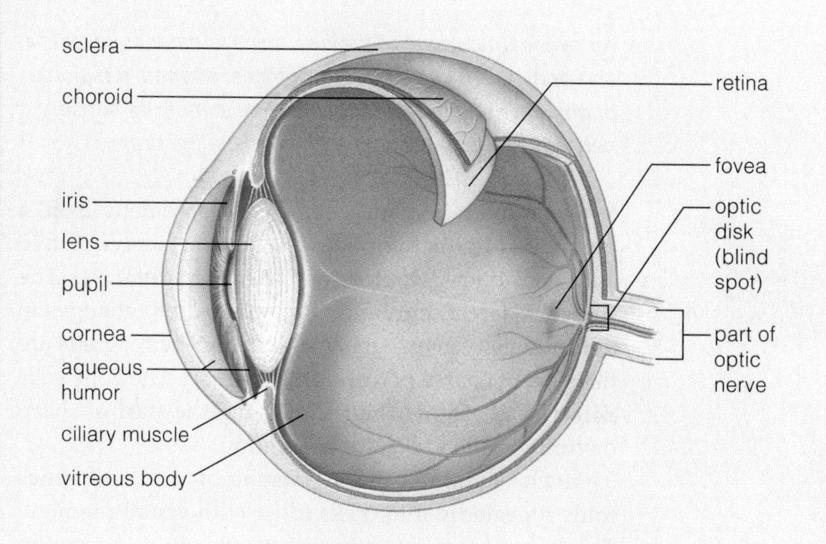

Figure 31.12 **Animated!** Components and structure of the human eye.

VISUAL ACCOMMODATION

With visual accommodation, mechanisms adjust the shape or the position of a lens in ways that help focus incoming light rays precisely on the retina, not in front of it or some place past it. Without these adjustments, only objects at a fixed distance would stimulate retinal photoreceptors in a focused pattern. Objects closer or farther away would appear fuzzy.

In fish and reptilian eyes, the lens is rather like a magnifying glass. The lens can move forward or back, but its shape stays the same. Extending or shrinking the distance between the lens and retina keeps light focused on the retina.

In birds and mammals, the lens is elastic; pulling on it can change its shape. A ciliary muscle encircles the lens and attaches to it by short fibers. Contraction of the ciliary muscle adjusts the shape of the lens for proper focus (Figure 31.13*b*). When the ciliary muscle is relaxed, the lens is under tension, and it stretches out and flattens. When the ciliary muscle contracts, fibers around the edge of the lens slacken. This allows the lens to become more round.

A relatively flat lens will focus light from a distant object onto the retina, but the lens must be rounder to focus light from closer objects. When you read a book, ciliary muscle contracts and fibers between the lens and this muscle slacken. The decreased tension on the lens allow it to round up enough to focus light from the page onto your retina. Gaze into the distance and muscle around the lens relaxes, so the lens flattens. To reduce eyestrain when looking at a book or computer screen, take breaks and focus on more distant objects.

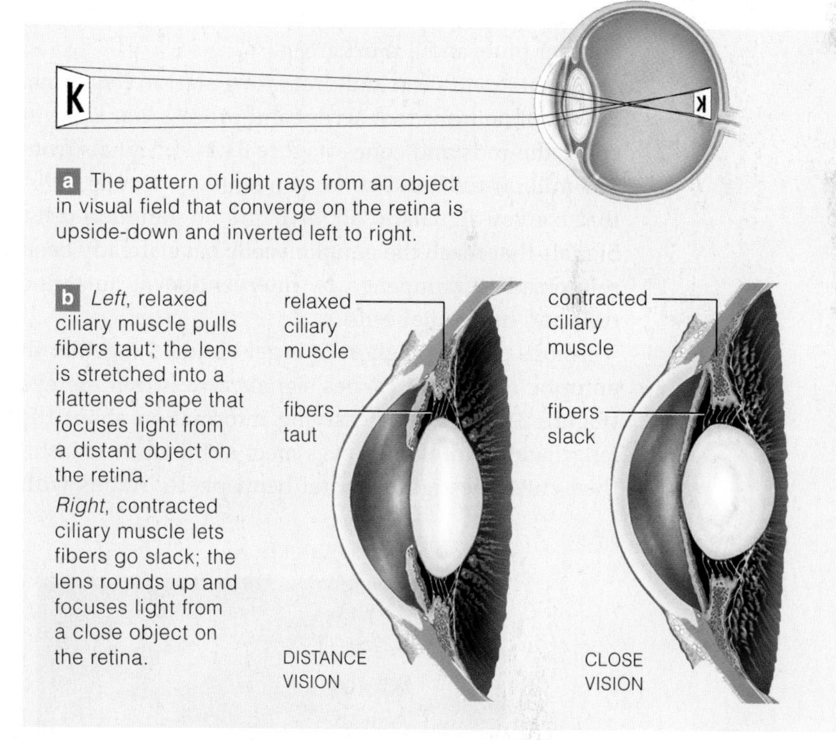

Figure 31.13 **Animated!** (**a**) Pattern in which light stimulates the human retina. (**b**) Ciliary muscle adjustments focus light from near or distant sources on the retina by adjusting the tension of fibers that ring the lens.

Vision requires eyes with a dense array of photoreceptors and image formation in the brain. Brain centers receive and process patterns of information about brightness, shapes, positions, and movement in the visual field.

31.7 From the Retina to the Visual Cortex

An eye is an outpost of the nervous system that collects and analyzes signals about distances, shapes, positions, brightness, and movements of visual stimuli. Its sensory pathway starts at the retina and ends in the brain.

LINKS TO SECTIONS 6.1, 25.5, 30.9, 30.10

Between the retina and choroid is a basement layer, a pigmented epithelium. Densely packed arrays of two classes of photoreceptors rest on it (Figure 31.14). The rod cells detect dim light. They respond to changes in light intensity across the visual field. Their signals are the start of coarse perception of motion. The cone cells detect bright light. Their signals are the start of sharp daytime vision and color perception.

Much of the plasma membrane of rods and cones folds repeatedly into disks filled with visual pigment. The visual pigment in rods is rhodopsin. Absorption of a photon causes it to change shape, which initiates reactions that result in an action potential. There are three types of cone cells, and each contains a different pigment. One pigment is best at absorbing red light, another blue, and a third green.

Signal integration and processing start in the retina. Layers of neurons that accept and process visual input cover the rods and cones (Figure 31.14*b*). Signals from 125 million rods and cones converge on bipolar cells, that convey signals to about 1 million ganglion cells. Signals that reach the ganglion cells have already been reinforced or dampened by intervention of amacrine cells and horizontal cells.

Axons of the ganglion cells get bundled together in an optic nerve that carries signals away from the eye (Figure 31.15). Axons carrying information about the left visual field of both eyes meet at the optic chiasm, then enter the right cerebral hemisphere. Axons with signals about the right visual field of each eye deliver information to the left hemisphere. Each optic nerve ends in a brain center (lateral geniculate nucleus) that processes signals. From there, information flows along different pathways to parts of the visual cortex. Final integration of signals produces visual sensations.

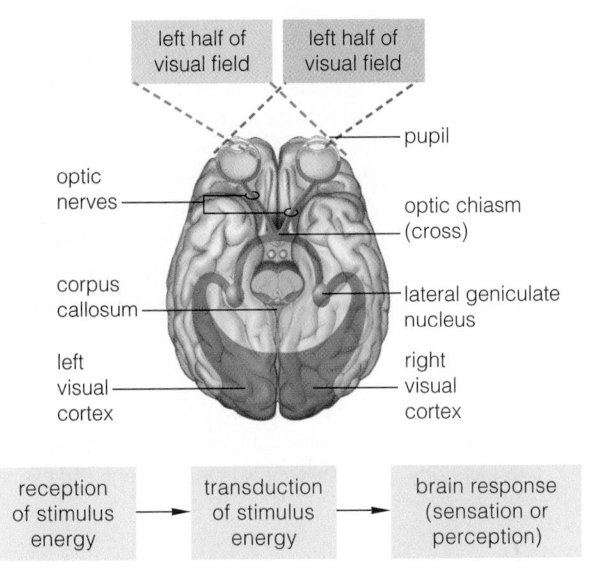

Figure 31.15 Sensory pathway from the retina into brain centers. In the retina, photon energy is transduced into action potentials. Signals travel along an optic nerve from each eye to a lateral geniculate nucleus, which relays them to the visual cortex. Synaptic integration here results in a sense of vision.

Integration and processing of visual input starts at receptive fields in the retina. The input is processed further in a brain center and finally integrated in the visual cortex.

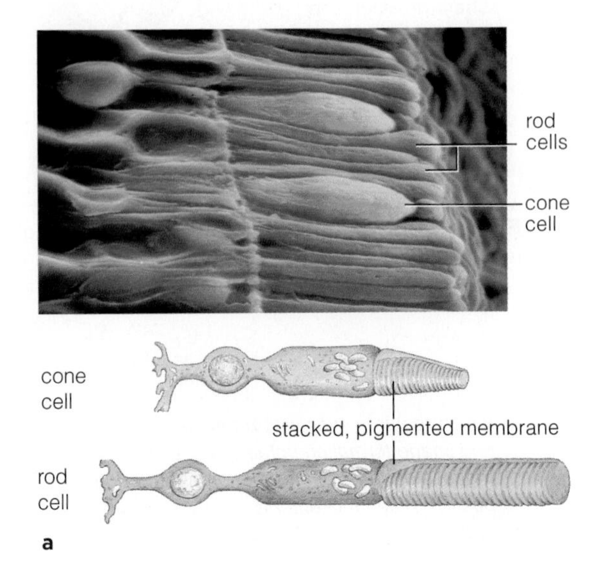

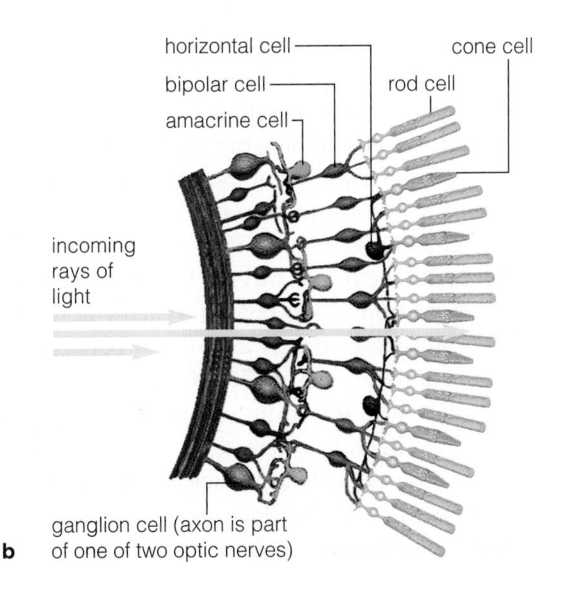

Figure 31.14 Animated! (**a**) Scanning electron micrograph and sketch of the photoreceptors called rods and cones. (**b**) Organization of photoreceptors as well as other sensory neurons in the retina.

31.8 Visual Disorders

FOCUS ON HEALTH

Vision is impaired when light is not focused properly, when photoreceptors do not respond as they should, or when some aspect of visual processing breaks down.

Color Blindness Sometimes one or more types of cones fail to develop or work properly. The outcome is one form or another of color blindness. In red–green color blindness, a person has trouble distinguishing reds from greens. This X-linked recessive trait affects about 7 percent of males in the United States. As is the case for other X-linked traits, it shows up more in males; only 0.4 percent of women are affected (Section 11.4).

LINK TO SECTION 11.4

Lack of Focus About 150 million Americans have disorders in which light rays do not converge as they should. Astigmatism results from an unevenly curved cornea, which cannot properly focus incoming light on the lens. Nearsightedness occurs when the distance from the front to the back of the eye is longer than normal or when ciliary muscles react too strongly. With either disorder, images of distant objects get focused in front of the retina instead of on it (Figure 31.16*a*).

In farsightedness, the distance from front to back of the eye is unusually short or ciliary muscles are too weak. Either way, light rays from nearby objects get focused behind the retina (Figure 31.16*b*). Also, the lens loses its flexibility as a person ages. That is why most people who are over age forty have relatively impaired close vision.

Glasses, contact lenses, or surgery can correct most focusing problems. About 1.5 million Americans undergo laser surgery (LASIK) annually. Typically, LASIK can eliminate the need for glasses during most activities, although some older adults still need reading glasses. Chronic eye irritation is a common complication.

Age-Related Disorders In the United States, an estimated 13 million people have age-related macular degeneration (AMD). The macula, a small spot in the retina, is the basis of clear central vision. All cone cells are positioned at its center, the fovea (Figure 31.17). Destruction of photoreceptors in the macula clouds the center of the visual field more than the periphery.

Some mutant genes increase the risk of AMD. So do smoking, obesity, and high blood pressure. A vegetable-rich diet seems to protect against it. Damage caused by AMD usually cannot be reversed, but drug injections and laser therapy can slow its progression.

Glaucoma results when too much aqueous humor builds up inside the eyeball. The increased fluid pressure damages blood vessels and ganglion cells. It also can interfere with peripheral vision and visual processing. Although we often associate chronic glaucoma with old age, the conditions that give rise to the disorder start to develop long before symptoms show up. When doctors detect the increased fluid pressure before the damage becomes severe, they can manage the disorder with medication, surgery, or both.

A cataract is a clouding of the lens that develops slowly. It reduces the amount and focusing of light that reaches the retina. Early symptoms are poor night vision and blurred vision. Vision ends after the lens becomes fully opaque. Excessive exposure to UV radiation, steroid use, and diabetes promote the onset and development of cataracts. An artificial implant can replace a badly clouded lens. Millions of people in developed countries undergo cataract surgery each year. Worldwide, about 16 million are currently blind as a result of cataracts.

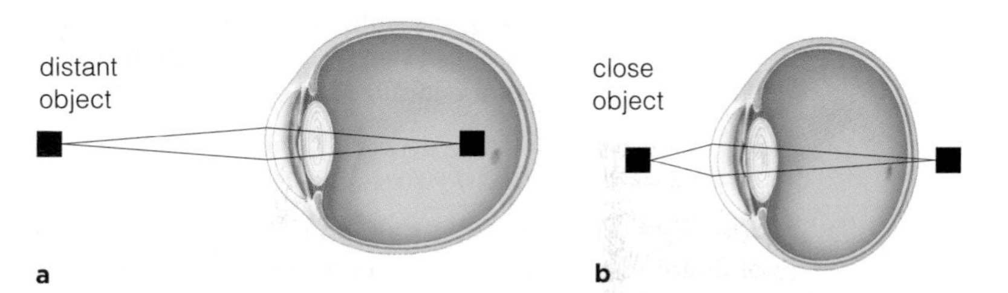

Figure 31.16 Focusing problems. (**a**) In nearsightedness, light rays from distant objects converge in front of the retina. (**b**) In farsightedness, light rays from close objects have not yet converged when they arrive at the retina.

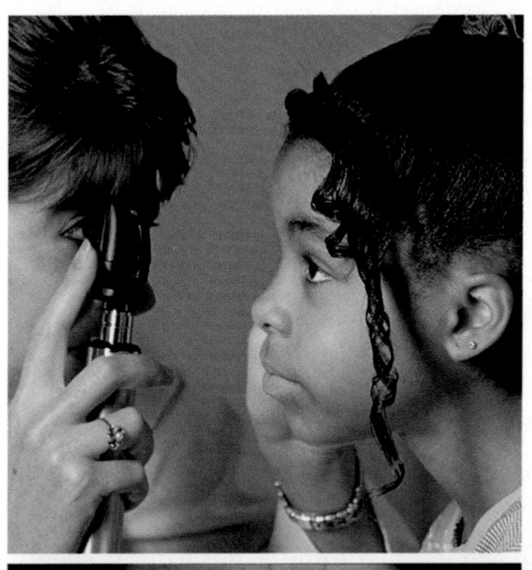

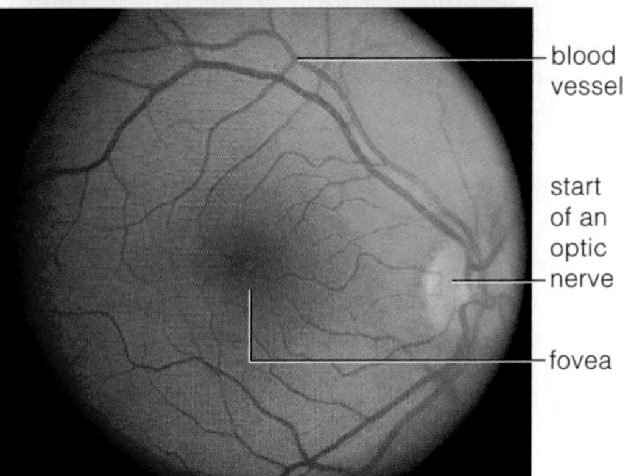

Figure 31.17 A doctor using a lighted magnifying instrument as she examines a patient's eye. The photograph below shows what she sees: part of the retina. The fovea has cone cells only. The start of an optic nerve rings the entry point for blood vessels that service the eye. In an older person who has macular degeneration, photoreceptors of the fovea die off.

Summary

Section 31.1 Sensory pathways are examples of cell communication, from signal reception and transduction to a response. Sensory receptors transduce the energy of a specific stimulus, such as pressure, heat, or light, into the electrochemical energy of action potentials. Types of sensory receptors include pain receptors, chemoreceptors, thermoreceptors, mechanoreceptors, osmoreceptors, and photoreceptors.

The brain evaluates action potentials from sensory receptors based on which of the body's nerves delivers them, their frequency, and the number of axons firing in any given interval. Continued stimulation of a receptor may lead to a diminished response (sensory adaptation).

Receptors for somatic sensations, such as touch and warmth, occur in many locations, rather than in a specific organ. Receptors for special senses: taste, smell, hearing, balance, and vision—are localized in specific organs.

■ *Use the animation on ThomsonNOW to see how stimulus intensity affects action potential frequency.*

Section 31.2 Somatic sensations start at free nerve endings, encapsulated receptors, and stretch receptors in the skin, in the walls of internal organs, and in skeletal muscles. Signals reach the sensory areas of the cerebral cortex, where interneurons are organized like maps for individual parts of the body surface.

■ *Use the animation on ThomsonNOW to learn about the sensory receptors in human skin.*

Section 31.3 The senses of taste and smell involve pathways from chemoreceptors to processing regions in the cerebral cortex and limbic system. In humans, taste receptors are concentrated in taste buds of the tongue and mouth. Olfactory receptors line the nasal passages. Pheromones are chemical signals that act as social cues among many animals that have the means to detect them.

Section 31.4 The vestibular apparatus, a system of fluid-filled sacs and canals in the inner ear, houses organs of equilibrium. These organs detect gravity, acceleration, and other forces that affect body positions and motions.

■ *Use the animation on ThomsonNOW to explore static and dynamic equilibrium.*

Section 31.5 Sound is a form of mechanical energy. A vibrating object generates pressure waves, which can vary in amplitude and frequency. Humans have a pair of ears with three regions: the outer, middle, and inner ear. The outer ear collects sound waves. The middle ear amplifies sound waves and transmits them to the inner ear, which includes the vestibular apparatus and the cochlea: a coiled, fluid-filled structure with three ducts.

Pressure waves traveling through the fluid inside the cochlea bend mechanoreceptors called hair cells, which are embedded in one of the cochlear membranes. Sounds get sorted out according to their frequency. Mechanical energy (pressure) is transduced to action potentials that are relayed along auditory nerves to the brain.

■ *Use the animation on ThomsonNOW to learn about the properties of sound and the human sense of hearing.*

Sections 31.6–31.8 Most organisms can respond to light, but vision requires eyes and brain centers capable of processing the visual information. An eye is a sensory organ that contains a dense array of photoreceptors.

Like squids and octopuses, humans have camera eyes, each with an iris that adjusts incoming light and a lens that focuses light on the retina in the back of the eyeball chamber. The retina has densely packed photoreceptors.

Rods detect dim light, while cones detect bright light and colors. These photoreceptor cells interact with other cells in the retina that start processing visual information before sending it to the brain. Visual signals travel to the cerebral cortex along two optic nerves.

Abnormalities in eye shape, in the lens, and in cells of the retina often impair vision.

■ *Use the animation on ThomsonNOW to investigate the structure, function, and organization of the retina.*

Self-Quiz

Answers in Appendix III

1. A stimulus is a specific form of energy in the outside environment that is detected by ______ .
   a. a sensory receptor c. the brain
   b. nerves d. all of the above

2. ______ is defined as a decrease in the response to an ongoing stimulus.
   a. Perception c. Sensory adaptation
   b. Visual accommodation d. Somatic sensation

3. Which is a somatic sensation?
   a. taste c. touch e. a through c
   b. smell d. hearing f. all of the above

4. Chemoreceptors play a role in the sense of ______ .
   a. taste c. touch e. both a and b
   b. smell d. hearing f. all of the above

5. In the ______ , neurons are arranged like maps that correspond to different parts of the body surface.
   a. cerebral cortex c. basilar membrane
   b. retina d. all of the above

6. Mechanoreceptors in the ______ send signals to the brain about the body's position relative to gravity.
   a. eye b. ear c. tongue d. nose

7. The middle ear functions in ______ .
   a. detecting shifts in body position
   b. amplifying and transmitting sound waves
   c. sorting sound waves out by frequency

8. Label the parts of the human eye in this diagram:

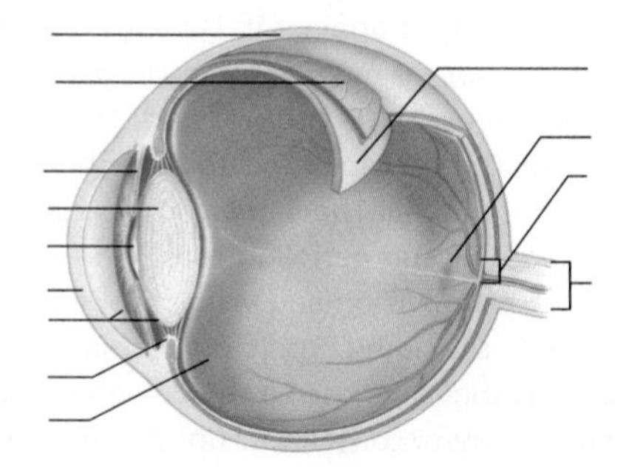

9. Match each structure with its description.

___ rod cell	a. sensitive to vibrations
___ cochlea	b. functions in balance
___ lens	c. detects color
___ hair cell	d. detects dim light
___ cone cell	e. contains chemoreceptor
___ taste bud	f. focuses rays of light
___ vestibular apparatus	g. sorts out sound waves
___ free nerve ending	h. helps brain assess heat, pressure, pain

■ *Visit ThomsonNOW for additional questions.*

Critical Thinking

1. Laura loves to eat broccoli and brussels sprouts. Lionel cannot stand them. Everyone has the same five kinds of taste receptors, so what is going on? Is Lionel just being difficult? Perhaps not. The number and distribution of receptors that respond to bitter substances vary among individuals of a population—and studies now indicate that some of this variation is heritable.

People who have the greatest number of receptors for bitter substances find many fruits and vegetables highly unpalatable. These supertasters make up about 25 percent of the general population. They tend to be slimmer than average but are more likely to develop colon polyps and colon cancer. How might Lionel's highly sensitive taste buds put him at increased risk for colon cancer?

2. Above and below a python's mouth are rows of pits that contain thermoreceptors (Figure 31.18). They detect body heat, or infrared energy, of nearby prey. Name the type of prey organisms the receptors can detect. Which kinds of otherwise edible animals would they miss so that the snake would slither on by?

3. Nearly all of the mammals called bats sleep during the day and spread webbed wings at dusk. Different kinds take to the air in search of nectar, fruit, frogs, or insects. Many sensory receptors in their eyes, nose, ears, mouth, and skin are not all that different from yours. Others are very different. They help bats navigate and capture flying insects swiftly in the dark.

As one of these bats flies, it emits a steady stream of about ten clicking sounds per second. You cannot hear a bat's clicks. They are ultrasounds, of an intensity beyond the range of sound waves that receptors in human ears can detect. When a bat hears a pattern of distant echoes from, say, an airborne mosquito, it increases the rate of ultrasonic clicks to as many as 200 per second. That is faster than a machine gun can fire bullets. In the few milliseconds of silence between the clicks, sensory receptors in the bat's pair of inner ears detect the echoes. The bat brain swiftly constructs a "map" of the sounds, which the bat follows during its maneuvers through the night world.

Which part of the cochlear duct inside the bat ear do the ultrasounds stimulate?

4. Are organs of dynamic equilibrium, static equilibrium, or both activated during a heart-thumping roller coaster ride, as in Figure 31.19?

5. In humans, photoreceptors are most concentrated at the very back of the eyeball. In birds of prey, such as owls and hawks, the greatest density of photoreceptors is in a region closer to the eyeball's roof. When these birds are on the ground, they cannot see objects even slightly above them unless they turn their head almost upside down (Figure 31.20). What is the adaptive advantage of this peculiar type of retinal organization?

6. The strength of Earth's magnetic field and its angle relative to the surface vary with latitude. Diverse species sense these differences and use them as cues for assessing their location and direction of movement. Behavioral experiments have shown that sea turtles, salamanders, and spiny lobsters use information from Earth's magnetic field during their migrations. Whales and some burrowing rodents also seem to have a magnetic sense. Evidence about humans is contradictory. Is it likely that humans have such a sense? Suggest an experiment that might support or disprove the possibility.

Figure 31.18 Thermoreceptors in pits above and below a python's mouth detect body heat, or infrared energy, of nearby prey.

Figure 31.19 Giving organs of equilibrium a workout.

Figure 31.20 Here's looking at you! In owls and some other birds of prey, photoreceptors are concentrated more on top of the inner eyeball, not back. Such birds look down more than up when they fly and scan the ground for a meal. When they are on the ground, they cannot see something overhead very easily unless they turn their head almost upside down.

32 ENDOCRINE CONTROL

IMPACTS, ISSUES Hormones in the Balance

Atrazine has been a widely used herbicide for more than forty years. Each year in the United States, about 76 million pounds are sprayed out, mostly in fields of corn and other monocot crops. From there, it gets into soil and water. Atrazine molecules break down within a year but they still turn up in ponds, wells, groundwater, and rain. Atrazine contamination is common in soils and water of the American Midwest. Does it have bad effects? Tyrone Hayes, a University of California biologist, thinks so. His studies suggest that atrazine is an endocrine disruptor: a synthetic compound that alters the action of natural hormones and adversely affects health and development (Figure 32.1).

Hayes studied atrazine's effects on African clawed frogs (*Xenopus laevis*) and leopard frogs (*Rana pipiens*). He found that exposing male tadpoles to atrazine in the laboratory caused some to develop both female and male reproductive organs. This effect occurred even at atrazine levels far below those allowed in drinking water.

Does atrazine have similar effects in the wild? Hayes collected native leopard frogs from ponds and ditches across the Midwest. Male frogs from every contaminated pond had abnormal sex organs. In the pond with the most atrazine, 92 percent of males had ovary tissue.

Later, other scientists reported that atrazine causes or contributes to frog deformities. The Environmental Protection Agency found the data intriguing. Among other tasks, this agency regulates chemical applications in agriculture. It called for further study of atrazine's effects on amphibians and is encouraging farmers to minimize atrazine-laden runoff from their fields.

Other hormone disruptors infiltrate aquatic habitats. For instance, the estrogens in birth control pills are excreted in urine and cannot be removed by standard

See the video! **Figure 32.1** Benefits and costs of herbicide applications. *Left*, atrazine can keep cornfields nearly weed-free; no need for constant tilling that causes soil erosion. Tyrone Hayes (*right*) suspects that the chemical scrambles amphibian hormonal signals.

How would you vote? Some widely used agricultural chemicals may disrupt hormone action in untargeted species. Should potentially harmful chemicals be kept on the market while researchers investigate them? See ThomsonNOW for details, then vote online.

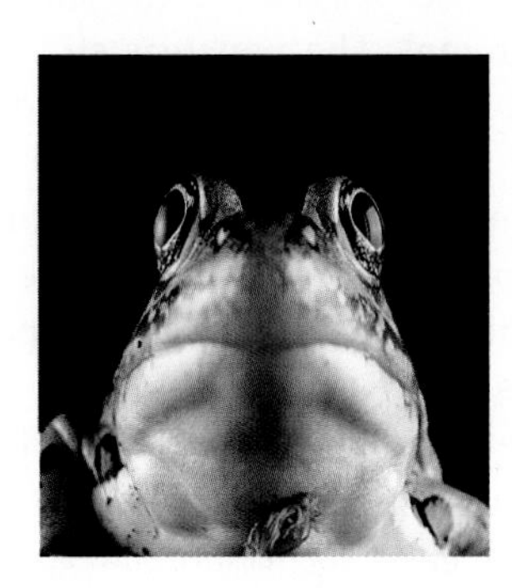

not a good time to be a frog

wastewater treatments. In streams or rivers, estrogen-tainted water causes male fish to develop female traits. So do estrogen-mimicking pollutants from industries.

In some pesticide-laden Florida lakes, male alligators have low testosterone levels and an abnormally small penis. In New York, salamanders exposed to herbicide-laden runoff from a new golf course in the neighborhood developed strange coloration patterns.

Are amphibians, fishes, and alligators too removed from your concerns? Endocrine disruptors may be acting on humans, too. Between 1938 and 1990, sperm counts of males in Western countries fell by about 40 percent. Males in agricultural regions may have the lowest counts. As University of Missouri researchers discovered, semen samples from men in Columbia, Missouri, contained only about half as many motile sperm as samples from men in New York City and Los Angeles.

An excess of estrogen-like chemicals may lower sperm counts. Estrogen is a sex hormone. Both men and women produce it and have receptors for it, although the females make much more. In males, estrogen docks at receptors on target cells in reproductive organs and helps sperm to mature. Other chemicals, including kepone and DDT, can bind to the estrogen receptors in place of estrogen. Both insecticides are now banned in the United States.

This chapter focuses on hormones—their sources, targets, and interactions. All vertebrates have similar hormone-secreting glands and systems. If the details start to blur, remember the endocrine disruptors. What you learn here can help you evaluate research that may influence your health and life-style.

Key Concepts

THE VERTEBRATE ENDOCRINE SYSTEM

Hormones and other signaling molecules regulate the pathways that control metabolism, growth, development, and reproduction. Nearly all vertebrates have an endocrine system composed of the same hormone sources. **Section 32.1**

SIGNALING MECHANISMS

A hormone signal acts on any cell that has receptors for it. Receptor activation leads to transduction of the signal and a response in the targeted cell. **Section 32.2**

A MASTER INTEGRATING CENTER

In vertebrates, the hypothalamus and pituitary gland are connected structurally and functionally. Together, they coordinate activities of many other glands. **Section 32.3**

OTHER HORMONE SOURCES

Negative feedback loops to the hypothalamus and pituitary control secretions from many glands. Other glands secrete hormones in response to local changes in the internal environment. **Sections 32.4–32.8**

INVERTEBRATE HORMONES

Hormones control molting and other events in invertebrate life cycles. Vertebrate hormones and receptors for them first evolved in ancestral lineages of invertebrates. **Section 32.9**

THE HORMONAL SYMPHONY

Mammalian organ systems are a focus of many chapters in this unit. This section is an overview of tissue-specific responses to hormones, including the impact of variations in receptors. **Section 32.10**

Links to Earlier Concepts

This chapter builds on your knowledge of membrane proteins (Section 4.3), gene controls (14.1), feedback loops (25.3), and cell-to-cell signaling (25.5). You will draw on your understanding of energy-releasing pathways (7.7), nervous system functioning (30.8, 30.9), and sexual development (11.1). You will revisit the nature of molting (23.8, 23.9).

An earlier section introduced glandular epithelium (29.1). All hormone-secreting cells are embedded in this kind of tissue, which may form a gland or a sheetlike lining for some organs.

32.1 Introducing the Vertebrate Endocrine System

Throughout their life, cells respond to changing conditions by taking up and releasing various chemical substances. In vertebrates, responses of millions to trillions of cells become integrated in ways that help keep the whole body alive and functioning.

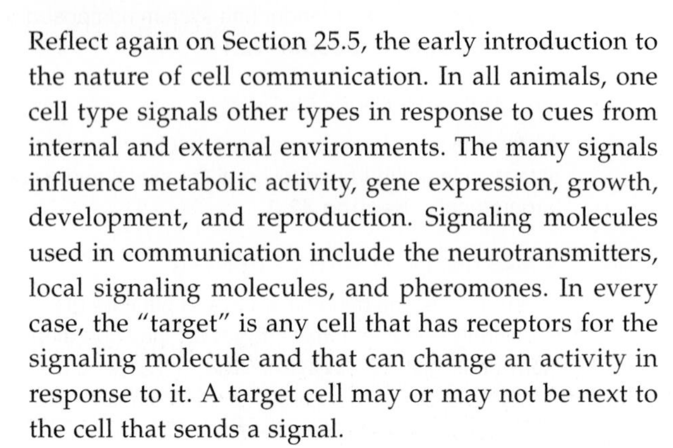

CATEGORIES OF SIGNALING MOLECULES

LINKS TO SECTIONS 25.5, 29.1, 30.4, 30.5

Reflect again on Section 25.5, the early introduction to the nature of cell communication. In all animals, one cell type signals other types in response to cues from internal and external environments. The many signals influence metabolic activity, gene expression, growth, development, and reproduction. Signaling molecules used in communication include the neurotransmitters, local signaling molecules, and pheromones. In every case, the "target" is any cell that has receptors for the signaling molecule and that can change an activity in response to it. A target cell may or may not be next to the cell that sends a signal.

Animal hormones are cell products secreted from endocrine glands, endocrine cells, and a few neurons. In most cases, the bloodstream circulates hormones to target cells that are some distance away.

Neurotransmitters are secreted by neurons into the tiny synaptic cleft between a neuron and a target cell (Section 30.4). Local signaling molecules are secreted by many cell types into the extracellular fluid. They are broken down quickly; molecules persist only long enough to affect the nearby tissues. For example, cells in a damaged tissue secrete prostaglandins that affect smooth muscle in the walls of adjacent blood vessels. The vessels dilate, and more of the infection-fighting proteins in blood flow to the tissue.

Pheromones are signals that diffuse through water or air to target cells in other individuals of the same species. Pheromones help integrate social behavior, as when a female moth releases her sex pheromone and attracts a mate. Many vertebrates have a vomeronasal organ in their nose that responds to pheromones. We consider pheromones in Chapter 44, in the context of social behavior. Here, hormones are the focus.

OVERVIEW OF THE ENDOCRINE SYSTEM

The word "hormone" dates back to the early 1900s. Physiologists W. Bayliss and E. Starling were trying to determine what triggers the secretion of pancreatic juices when food travels through a dog's gut. As they knew, acids mix with food in the stomach. Arrival of the acidic mixture inside the small intestine triggers pancreatic secretions that reduce the acidity. Was the nervous system stimulating this pancreatic response, or was some other signaling mechanism at work?

To find an answer, Bayliss and Starling blocked the nerves—but not blood vessels—to the small intestine of a laboratory animal. The pancreas still responded when acidic food from the stomach entered the small intestine. It even responded to extracts of cells from the intestinal lining. That lining is a type of *glandular* epithelium (Section 29.1). Apparently, some substance produced by glandular cells signaled the pancreas to start its secretions.

That substance is now called secretin. Identifying its mode of action supported a hypothesis that dated back centuries: *The blood carries internal secretions that influence the activities of the body's organs.*

Starling coined the term "hormone" for glandular secretions (the Greek *hormon* means to set in motion). Later on, researchers identified many other hormones and their sources. Figure 32.2 surveys major sources of human hormones. In general, vertebrate hormones are secreted by the following sources:

- Pituitary gland
- Adrenal glands (*two*)
- Pancreatic islets (*numerous cell clusters*)
- Thyroid gland
- Parathyroid glands (*in humans, four*)
- Pineal gland
- Thymus gland
- Gonads (*two*)
- Endocrine cells of the hypothalamus, stomach, small intestine, liver, kidneys, heart, placenta, skin, adipose tissue, and other organs

The term endocrine system refers to all sources of hormones in the animal body. That system is linked structurally and functionally with the nervous system in intercellular communication. The two systems share a few structures and issue a few of the same signals. Most organs accept and respond to signals from both.

In all animals, signaling molecules integrate cell activities. Each type of signal acts on all cells that have receptors for it. The targeted cells may alter their activities in response.

Most vertebrates have the same types of hormones produced by similar sources. Collectively, their hormone-secreting glands and cells make up an endocrine system.

Integrated interactions between the nervous system and nearly all endocrine glands coordinate many different functions for the body as a whole.

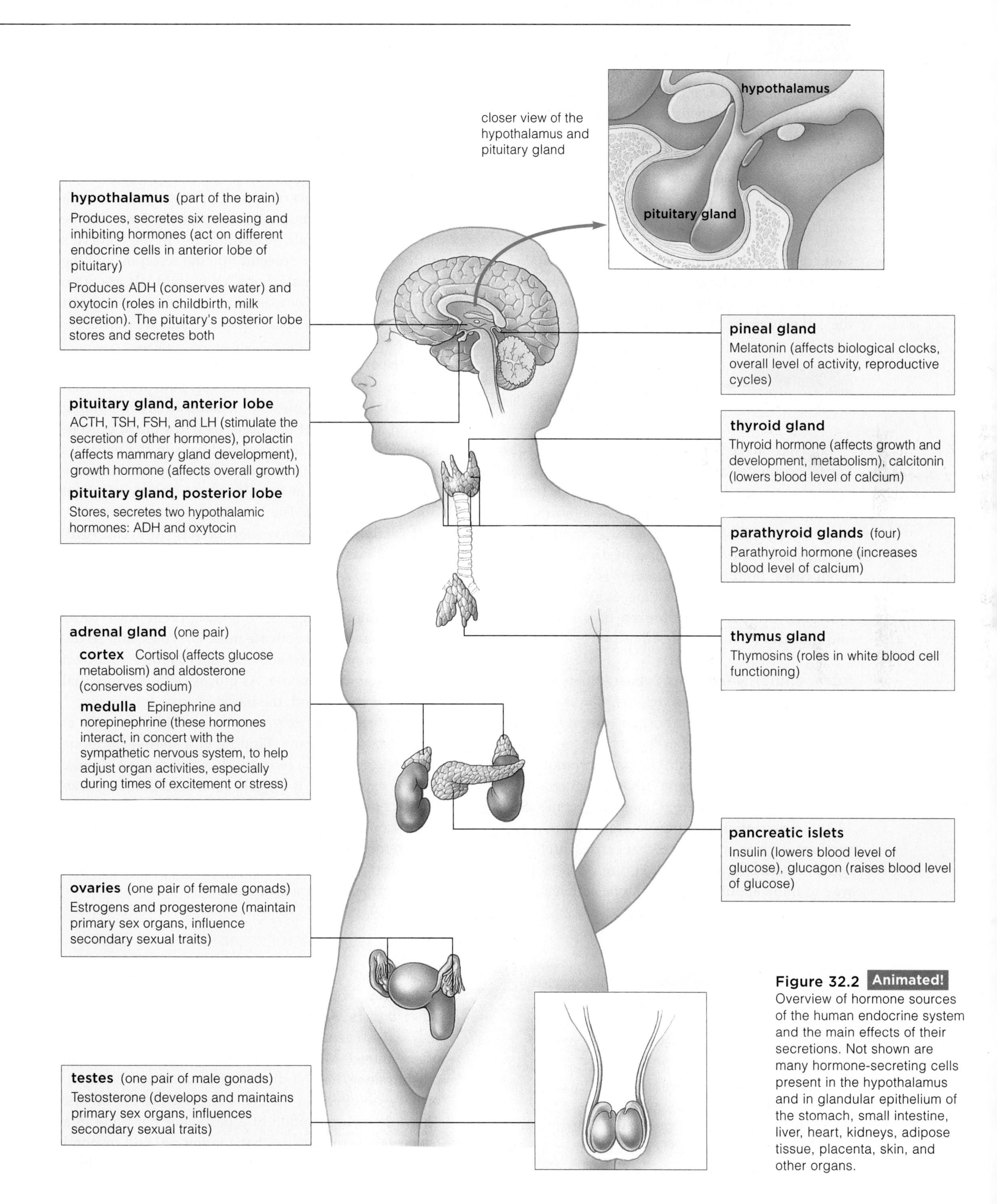

Figure 32.2 Animated! Overview of hormone sources of the human endocrine system and the main effects of their secretions. Not shown are many hormone-secreting cells present in the hypothalamus and in glandular epithelium of the stomach, small intestine, liver, heart, kidneys, adipose tissue, placenta, skin, and other organs.

32.2 The Nature of Hormone Action

Hormones can induce target cells to express specific genes, make more proteins or fewer, increase glucose uptake, step up or slow down secretions, grow, divide, or even commit suicide—all in the service of programs of growth, maintenance, and reproduction.

SIGNAL RECEPTION, TRANSDUCTION, RESPONSE

LINKS TO SECTIONS 4.3, 11.1, 13.1, 25.3, 25.5

Cell communication has three steps (Section 25.5). A signal activates a target cell receptor, it is transduced to a molecular form that acts in the receiving cell, and the cell may make a functional response:

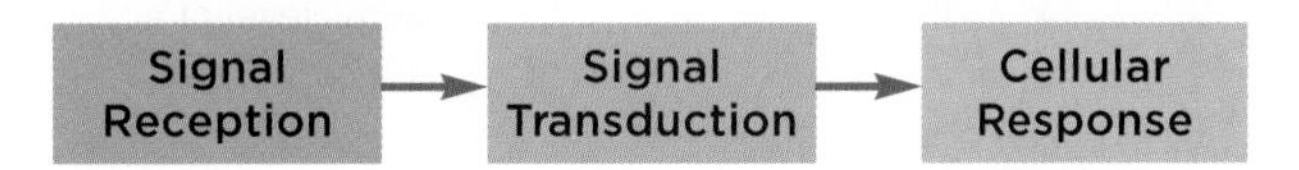

Enzymes make hormones from a variety of sources. *Steroid* hormones are derived from cholesterol. *Amine* hormones are modified amino acids. *Peptide* hormones are short chains of amino acids; *protein* hormones are longer chains. Table 32.1 lists a few examples of each.

Hormones initiate responses in different ways, but in all cases, binding to a receptor is reversible. Also, the responses decline over time. Why? The hormone molecules are eventually broken down by the body and thus can no longer elicit a response.

Intracellular Receptors Being lipids, the steroid hormones easily diffuse across the bilayer of a plasma membrane. They form a hormone–receptor complex by binding to a receptor in the cytoplasm or nucleus. Most often, the complex binds to a promoter near a hormonally regulated gene. RNA polymerase, recall, binds to promoters before it transcribes genes (Section 13.1). Transcription and translation result in a protein product, such as an enzyme, that carries out the target cell's response to the signal. Figure 32.3*a* is a simple way to think about steroid hormone action.

Table 32.1 Categories and Examples of Hormones

Steroid hormones	Testosterone and other androgens, estrogens, progesterone, aldosterone, cortisol
Amines	Melatonin, epinephrine, thyroid hormone
Peptides	Glucagon, oxytocin, antidiuretic hormone, calcitonin, parathyroid hormone
Proteins	Growth hormone, insulin, prolactin, follicle-stimulating hormone, luteinizing hormone

Signal response also requires functional receptors. As an example, typical male genitals cannot form in an XY embryo without testosterone, one of the steroid hormones (Section 11.1). XY individuals affected by androgen insensitivity syndrome secrete testosterone, but a mutation alters receptors for this hormone. With total androgen insensitivity, receptors for testosterone do not respond at all. Testes develop in the embryo's abdomen but never descend into the scrotum, and the genitals appear female. Such XY individuals are often raised as females. They develop breasts at puberty, but have no ovaries, do not menstruate, and are sterile.

Receptors at the Plasma Membrane Peptide and protein hormones bind to receptor proteins that span a target cell's plasma membrane. Often, binding sets off a cascade of reactions, as when blood glucose level falls below a set point and cells in the pancreas respond by secreting glucagon. This peptide hormone binds to receptors in the plasma membrane of target cells (Figure 32.3*b*). The binding activates an enzyme that catalyzes the conversion of ATP to cyclic AMP (short for cyclic adenosine monophosphate). The cyclic AMP serves as a second messenger: a molecule that is formed in response to an external signal and causes more cellular changes. In this case, cyclic AMP turns on enzymes, which activate more enzymes, and so on. The last enzyme to be activated speeds the breakdown of glycogen into glucose. As glucose enters interstitial fluid, and then blood, the blood glucose level rises.

Some cells have receptors for steroid hormones at the plasma membrane. Binding of a steroid hormone to such a membrane receptor does not cause a change in gene expression. It triggers a faster response by way of a second messenger or by altering a property of the membrane. As an example, when the steroid hormone aldosterone binds to membrane receptors on its target cells inside kidneys, these target cells quickly become more permeable to sodium ions.

VARIATIONS IN THE RESPONSES

Cells do not always switch some activity on or off in response to hormones. Their sensitivity is affected by other factors. *First,* cells ignore any hormone unless they have receptors for it. *Second,* different hormones often interact. Most cells have receptors for more than one hormone, and binding of one may block, enhance, or have no effect on the cell's response to the others. *Third,* the concentration of a particular hormone in a tissue affects the chance that a target cell will respond to that hormone. The more hormone molecules there

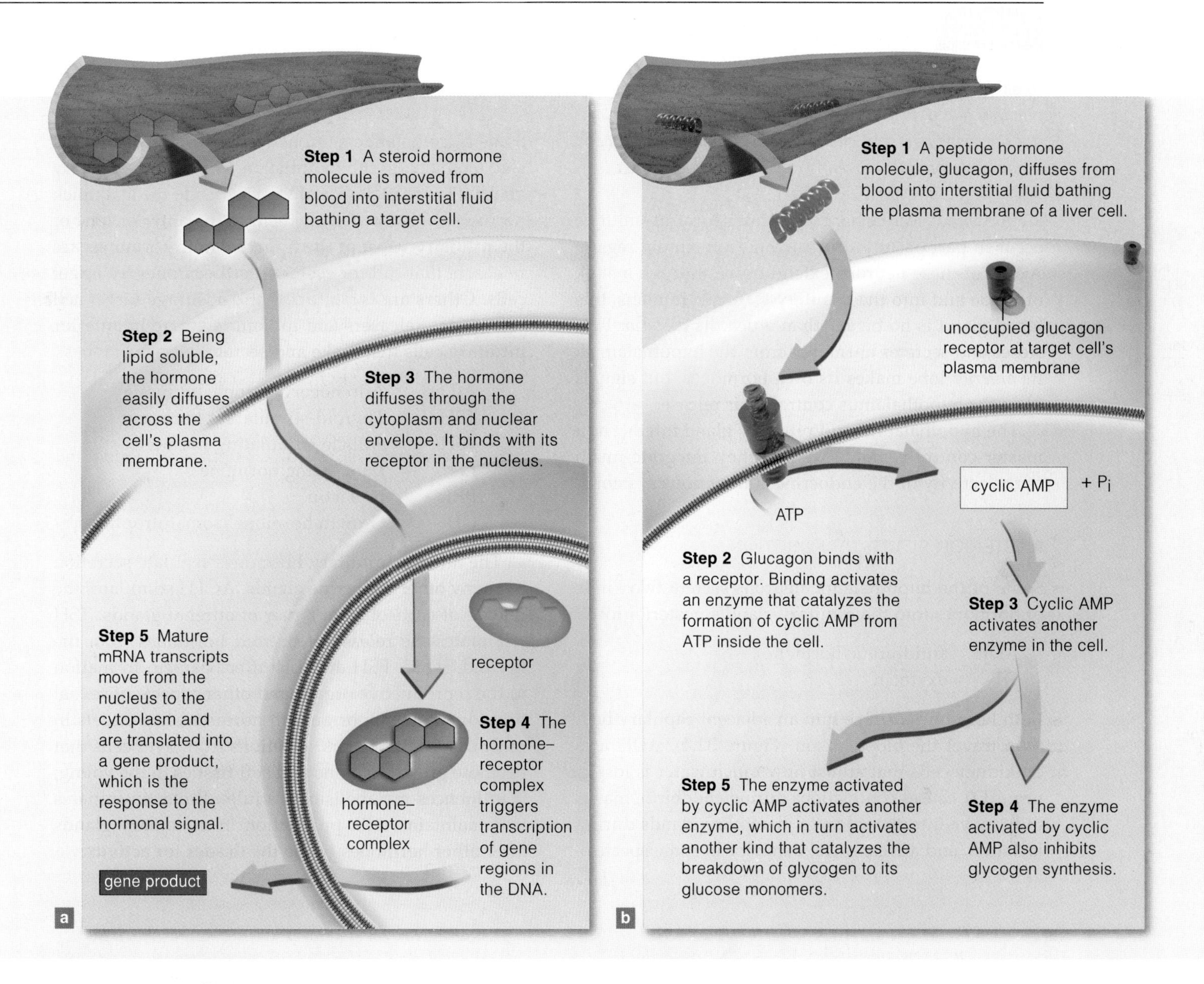

Figure 32.3 Animated! (**a**) Example of *steroid* hormone action inside a target cell. (**b**) Example of *peptide* hormone action. Cyclic AMP, a second messenger, relays a signal from a plasma membrane receptor into the cell.

are in interstitial fluid around a target cell, the greater the likelihood that some will bind with the receptors. *Fourth,* a target cell's metabolic and nutritional state affect responses. For instance, a target cell can only respond to a signal that calls for synthesis of a protein if it has plenty of amino acid building blocks on hand. *Fifth,* specific environmental cues such as temperature and day length can influence a target cell's capacity to respond to a hormone.

You will come across many examples of hormone action in this unit. They will give you a sense of the factors that can cause variations in cellular responses.

Hormone molecules reversibly bind with receptors on or in target cells. The hormonal signal is transduced into a form that elicits a functional response.

Most steroid and thyroid hormones bind with receptors inside cells and change gene expression. Some bind to receptors at the cell surface and trigger fast responses.

Peptide and protein hormones bind to membrane receptors. Often, a second messenger in the cytoplasm relays the signal into the cell's interior.

The type and state of target cell receptors, interventions by other hormones, feedback mechanisms, environmental cues, and other factors influence hormone action.

32.3 The Hypothalamus and Pituitary Gland

In all vertebrates, the hypothalamus interacts with the pituitary gland as a major integrating center. Together, the hypothalamus and pituitary control the activities of many other organs, including other endocrine glands.

LINKS TO SECTIONS 30.1, 30.9

Neurons that secrete hormones, not neurotransmitters, occur in part of the hypothalamus, a forebrain region. Axons of these neurons extend down through a stalk of tissue and into the pituitary gland. In humans, this lobed gland is no bigger than a pea. Its *posterior* lobe stores and secretes hormones from the hypothalamus. Its *anterior* lobe makes its own hormones, but signals from the hypothalamus control their release.

The hypothalamus and pituitary gland interact as a master control center. Together, they integrate much of the activity of the endocrine and nervous systems.

POSTERIOR PITUITARY FUNCTION

Some of the hypothalamic neurons secrete two kinds of hormones into the pituitary gland's posterior lobe:

ADH	antidiuretic hormone
OT	oxytocin

Both hormones diffuse into an adjacent capillary bed, then travel the bloodstream (Figure 32.4). ADH acts on kidney cells that adjust how much water is lost in urine. OT causes contractions during childbirth, makes milk move into the ducts of mammary glands during lactation, and affects social behavior in some species.

ANTERIOR PITUITARY FUNCTION

Some hypothalamic neurons secrete hormones into the stalk connecting the hypothalamus with the pituitary gland (Figure 32.5). Blood vessels inside the stalk pick hormones up and deliver them to the anterior lobe of the pituitary. Most of the hypothalamic hormones are releasers that call for secretion of hormones by target cells. Others are inhibitors that discourage target cell secretions. Releasers and inhibitors act upon anterior pituitary cells that make and secrete these hormones:

ACTH	Adrenocorticotropic hormone
TSH	Thyroid-stimulating hormone
FSH	Follicle-stimulating hormone
LH	Luteinizing hormone
PRL	Prolactin
GH	Growth hormone (somatotropin)

The anterior pituitary hormones regulate secretion by many other endocrine glands. ACTH stimulates the release of cortisol from a pair of adrenal glands. TSH stimulates the release of thyroid hormones from the thyroid gland. FSH and LH affect gamete formation in the reproductive organs and other aspects of sexual reproduction. GH, or growth hormone, has targets in most tissues. It triggers secretions from liver cells that promote growth of bone and soft tissues in the young; it influences metabolism in adults. Prolactin initiates and maintains milk production in mammary glands after other hormones prime the tissues for activity.

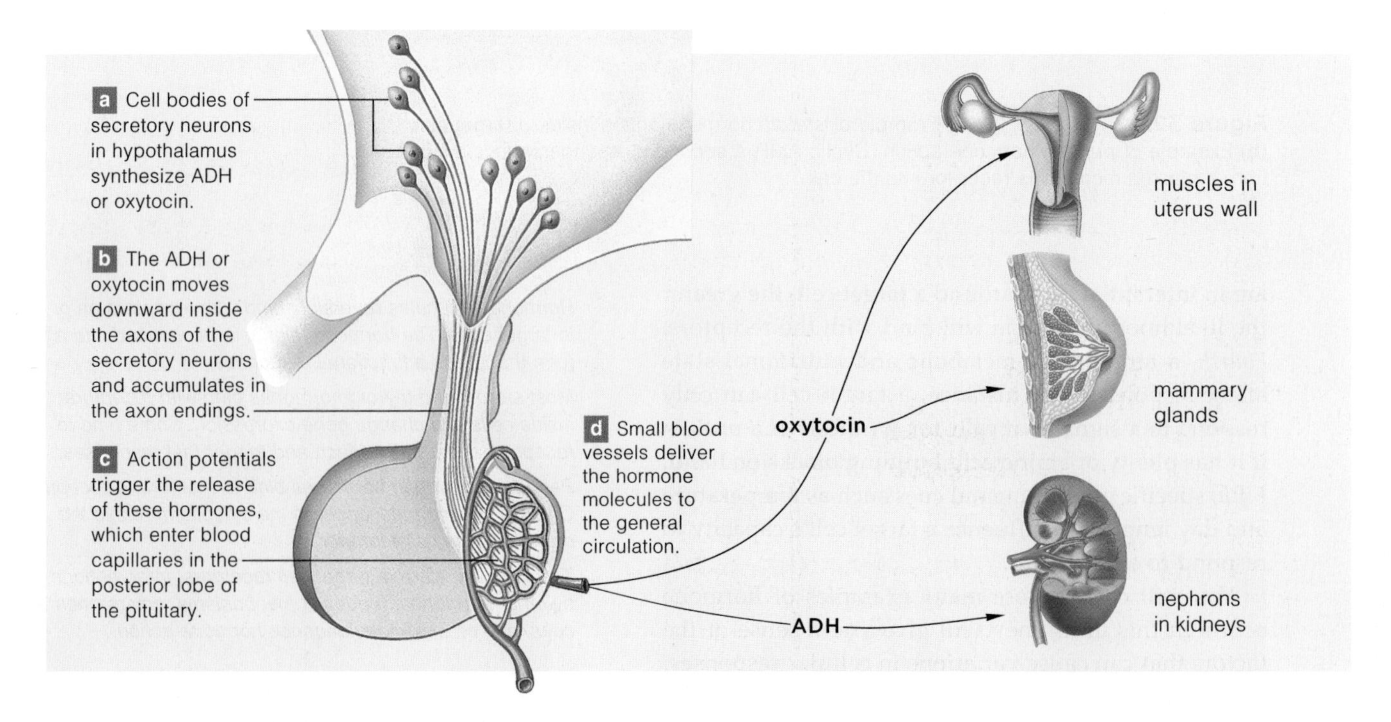

Figure 32.4 Animated! Functional links between the hypothalamus and the pituitary gland's posterior lobe. Targets of posterior lobe secretions are also shown.

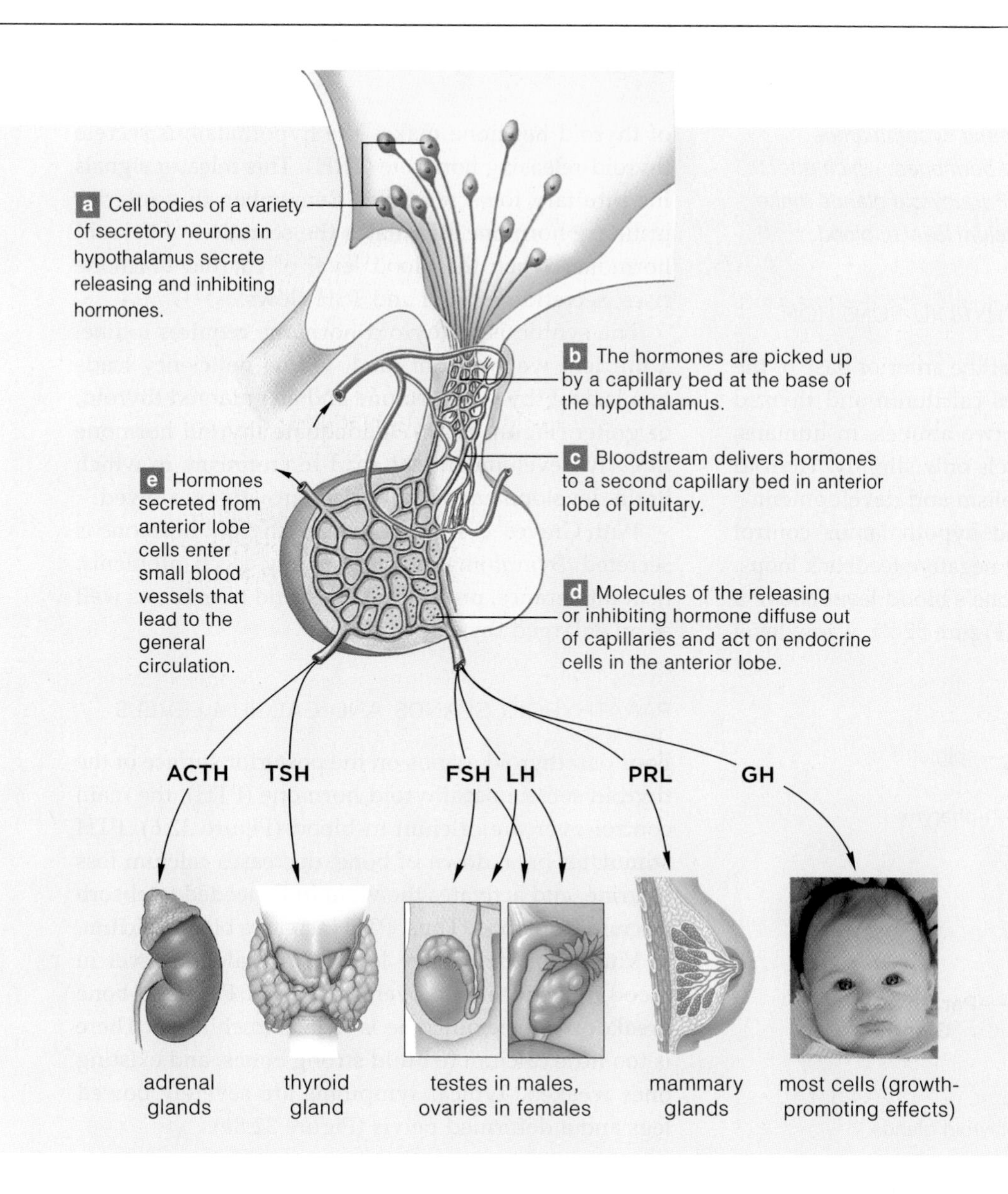

Figure 32.5 Animated! Functional links between the hypothalamus and the pituitary gland's anterior lobe. Also shown are the main targets of anterior lobe secretions. The photograph above shows the result of abnormal output of an anterior pituitary hormone. This boy is twelve years old and is affected by pituitary gigantism, which resulted from overproduction of GH. At 6 feet, 5 inches tall, he towers over his mother.

ABNORMAL PITUITARY OUTPUTS

The vertebrate body does not churn out vast amounts of any hormone. Roger Guillemin and Andrew Schally started with half a ton of hypothalamic tissue to purify a single milligram of the releasing hormone for TSH. Even so, tiny amounts of hormones have big impacts. When control of their secretion fails, the body's form, function, or both become altered.

Oversecretion of growth hormone in children leads to pituitary gigantism. An affected adult has a normal body form, but is unusually tall, as in Figure 32.5. Low GH secretion in childhood causes pituitary dwarfism. Affected adults have the form of an average person but are smaller. High GH secretion during adulthood causes acromegaly. Long bones cannot lengthen, but the hands and feet become enlarged and swollen. Too much cartilage and bone form, which distorts facial features. Skin thickens; internal organs may enlarge.

Another example: One cause of diabetes insipidus is a hard blow to the head that damages the posterior pituitary enough to slow or stop ADH secretion. In this condition, the body loses too much water in urine. The resulting dehydration can be life threatening.

The functional links between the nervous and endocrine systems are most evident in the close interaction between the hypothalamus and pituitary gland.

The hypothalamus produces ADH and oxytocin. Both hormones are stored in and secreted from the pituitary's posterior lobe. Other hypothalamic hormones stimulate or inhibit secretion of anterior pituitary hormones.

The anterior pituitary makes and secretes ACTH, TSH, FSH, LH, PRL, and GH. These hormones trigger secretion of hormones from different glands and have diverse effects.

32.4 Thyroid and Parathyroid Glands

A feedback loop to the pituitary and hypothalamus controls the secretion of thyroid hormones, which affect development and metabolism. Parathyroid glands make a hormone that controls the calcium level in blood.

FEEDBACK CONTROL OF THYROID FUNCTION

LINKS TO SECTIONS 7.7, 25.3, 30.8

The human thyroid gland is at the anterior base of the neck (Figure 32.6). It secretes calcitonin and thyroid hormone, which is a mix of two amines. In humans, calcitonin affects calcium levels only slightly. Thyroid hormone is central to metabolism and development.

The anterior pituitary and hypothalamus control thyroid hormone secretion by negative feedback loops. In this case, a rise in a hormone's blood level above a set point slows its secretion (Figure 32.7). A low level of thyroid hormone makes the hypothalamus secrete thyroid-releasing hormone (TRH). This releaser signals the pituitary to secrete TSH. Remember, this anterior pituitary hormone stimulates the secretion of thyroid hormone. When the blood level of thyroid hormone rises, secretion of TRH and TSH slows.

The synthesis of thyroid hormone requires iodine, a nutrient we get from food. Iodine deficiency leads to a lack of thyroid hormone and an enlarged thyroid, or goiter (Figure 32.8*a*). Inadequate thyroid hormone in early development can lead to cretinism, in which brain development and physical growth are slowed.

With Graves' disease, too much thyroid hormone is secreted. Symptoms include anxiety, sleep problems, heat intolerance, protruding eyes, and tremors, as well as an enlarged thyroid.

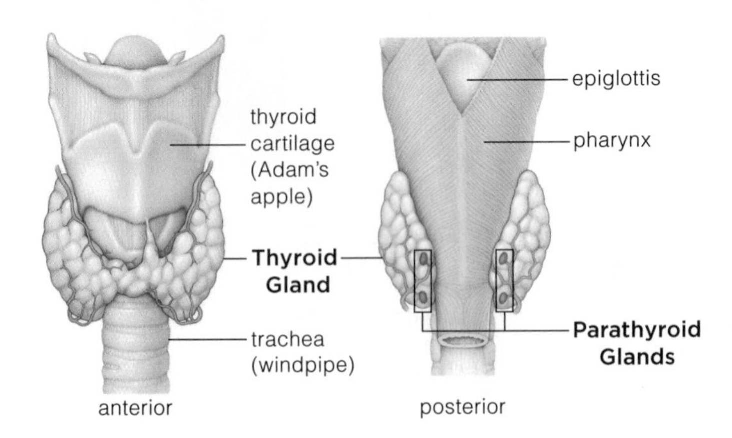

Figure 32.6 Location of human thyroid and parathyroid glands.

PARATHYROID GLANDS AND CALCIUM LEVELS

Four parathyroid glands on the posterior surface of the thyroid secrete parathyroid hormone (PTH), the main control over the calcium in blood (Figure 32.6). PTH stimulates breakdown of bone, decreases calcium loss in urine, and activates the vitamin D needed to absorb calcium from food. Thus, PTH increases blood calcium.

Vitamin D deficiency lowers the calcium level in blood, which leads to oversecretion of PTH and bone breakdown. One outcome is rickets in children. There is too little calcium to build strong bones, and existing ones weaken. Typical symptoms are severely bowed legs and a deformed pelvis (Figure 32.8*b*).

Secretion of thyroid hormone is governed by negative feedback and requires dietary iodine. Parathyroid glands secrete PTH, the main regulator of calcium levels in blood.

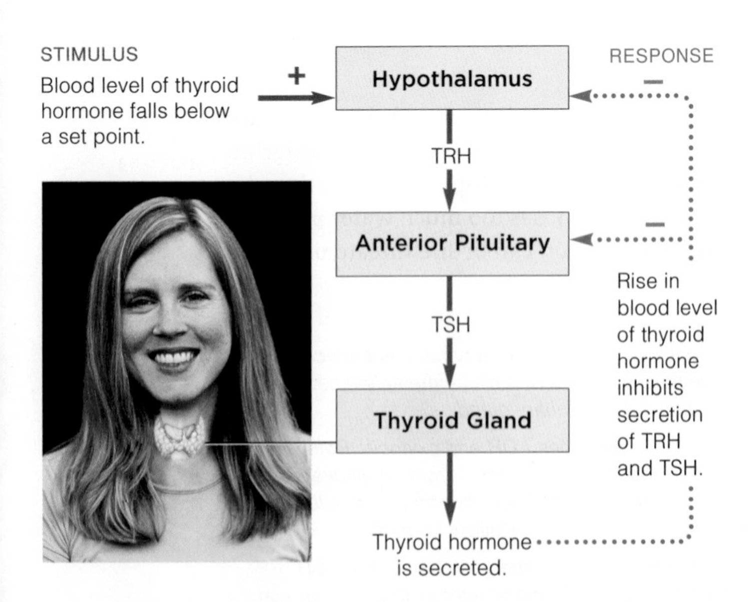

Figure 32.7 Negative feedback loop to the hypothalamus and the pituitary's anterior lobe that governs thyroid hormone secretion.

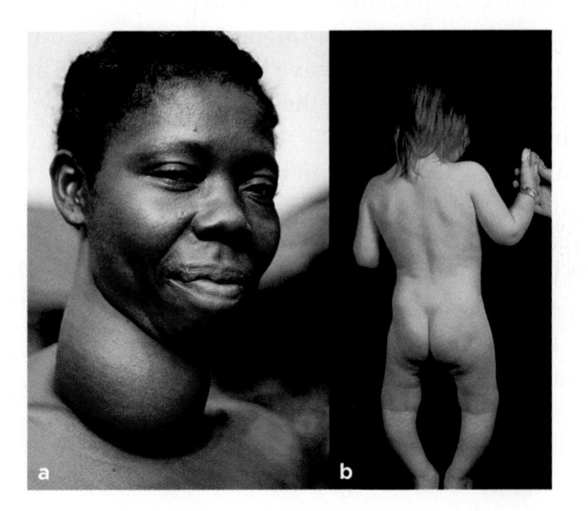

Figure 32.8 (**a**) A goiter caused by a dietary iodine deficiency. (**b**) A child with rickets caused by a lack of vitamin D has characteristic bowed legs.

32.5 The Adrenal Glands

An adrenal gland has two functional zones that secrete different substances. Its outer cortex secretes steroid hormones. Its inner medulla releases molecules that function as neurotransmitters.

THE ADRENAL CORTEX

Humans have a pair of adrenal glands, one on top of each kidney (Figure 32.2). The outermost part of each gland—the adrenal cortex—releases steroid hormones. One of these, aldosterone, controls sodium and water reabsorption in the kidneys. Chapter 38 explains this process. The adrenal cortex also produces and secretes small amounts of sex hormones, which we discuss in Section 32.8. Here we will consider cortisol, a hormone that affects metabolism and immune responses.

A negative feedback loop governs the cortisol level in blood (Figure 32.9). A decrease in cortisol triggers secretion of CRH (corticotropin-releasing hormone) by the hypothalamus. CRH then stimulates the secretion of ACTH, an anterior pituitary hormone that causes the adrenal cortex to release cortisol. The blood level of cortisol keeps increasing until it reaches a set point. Then the hypothalamus and anterior pituitary secrete less CRH and ACTH, and cortisol secretion slows.

Cortisol helps maintain the blood level of glucose. Most notably, it induces liver cells to break down their store of glycogen, and it suppresses uptake of glucose by other cells. Cortisol also induces adipose cells to degrade fats, and skeletal muscles to degrade proteins. The breakdown products—fatty acids and amino acids—function as alternative energy sources (Section 7.7). Cortisol also makes immune responses wind down.

The nervous system overrides the cortisol feedback loop in times of injury, illness, or anxiety, so cortisol levels soar. Over the short term, this cortisol stress response is adaptive; it helps get glucose to the brain.

THE ADRENAL MEDULLA

Stress, excitement, or danger also triggers secretion of norepinephrine and epinephrine by an adrenal gland's inner part, the adrenal medulla. These hormones have the same effects on targets that sympathetic nerves do, they bring about a fight–flight response (Section 30.8).

STRESS AND HEALTH

A short-term stress response can be advantageous, but ongoing stress can lead to health problems. Elevated cortisol levels disrupt the production and release of other hormones, suppress immune function, and also impair memory. We can observe the impact of excess cortisol in people with Cushing syndrome. Symptoms include a puffy, round "moon face" and too much fat on the trunk. Blood sugar levels and blood pressure are too high. White blood cell counts are low, and the risk of infection is great. In women, menstrual cycles are erratic or nonexistent. Men often are impotent.

Long-term use of synthetic corticosteroids, such as cortisone and prednisone, has similar effects. Both are prescribed to relieve pain and suppress inflammation. They work as cortisol does inside the body.

Negative feedback loops usually control secretion of cortisol from the adrenal cortex, but signals from the nervous system override this loop in times of stress.

The adrenal medulla contains sympathetic neurons that release signaling molecules into the blood.

Continual stress can lead to health problems.

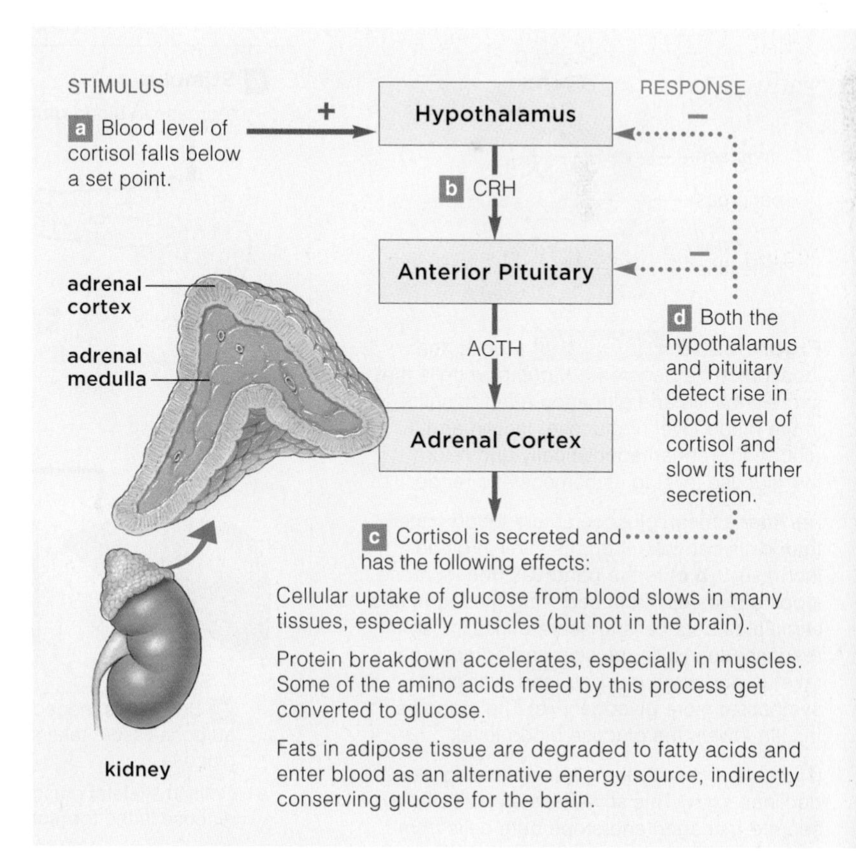

Figure 32.9 Animated! Structure of the human adrenal gland. An adrenal gland rests on top of each kidney. The diagram shows a negative feedback loop that governs cortisol secretion.

32.6 Pancreatic Hormones

Many hormones work antagonistically: The action of one opposes the action of another. Insulin and glucagon act this way. Controls over secretion of these two pancreatic hormones help maintain the glucose level in blood.

LINKS TO SECTIONS 7.7, 25.3

The pancreas lies in the abdominal cavity, behind the stomach (Figure 32.10). It has exocrine and endocrine functions. Its exocrine cells secrete digestive enzymes into a duct to the small intestine. Its endocrine cells are grouped in clusters called pancreatic islets. Each islet contains three types of hormone-secreting cells.

Alpha cells secrete the peptide hormone glucagon. Between meals, all cells take up glucose from blood. When the glucose level falls below a set point, alpha cells secrete glucagon. Glucagon binds to cells in the liver and causes the activation of enzymes that break glycogen into glucose subunits (Figure 32.3*b*). By its action, *glucagon raises the level of glucose in blood.*

Beta cells, the most abundant cells in the pancreatic islets, secrete insulin—the only hormone that causes target cells to take up and store glucose. After a meal, the blood level of glucose rises, which stimulates beta cells to release insulin. The main targets are liver, fat, and skeletal muscle cells. Insulin especially stimulates muscle and fat cells to take up glucose. In all target cells, it activates enzymes that function in protein and fat synthesis, and it inhibits the enzymes that catalyze protein and fat breakdown. As a result of its actions, *insulin lowers the level of glucose in the blood.*

Delta cells secrete somatostatin. This hormone helps control digestion and nutrient absorption. Also, it can inhibit the secretion of insulin and glucagon.

Some shifts in the blood glucose level are typical of all animals that show a discontinuous eating pattern. By working in opposition, glucagon and insulin from the pancreas maintain that level within a homeostatic range that keeps cells functioning (Figure 32.10).

Glucagon triggers the breakdown of glycogen; it raises the blood level of glucose. Insulin helps cells take up and store more glucose; it lowers the blood level of glucose.

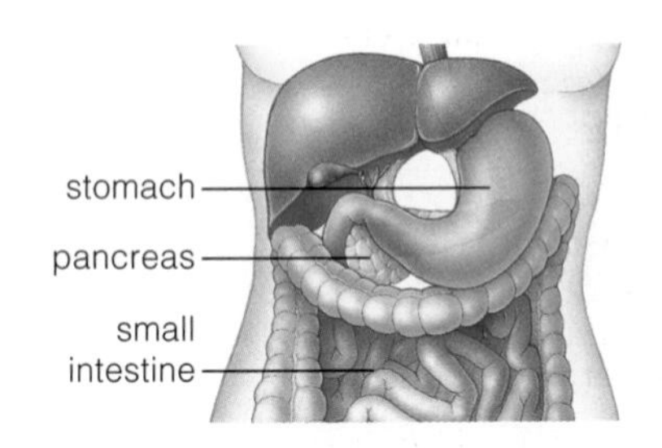

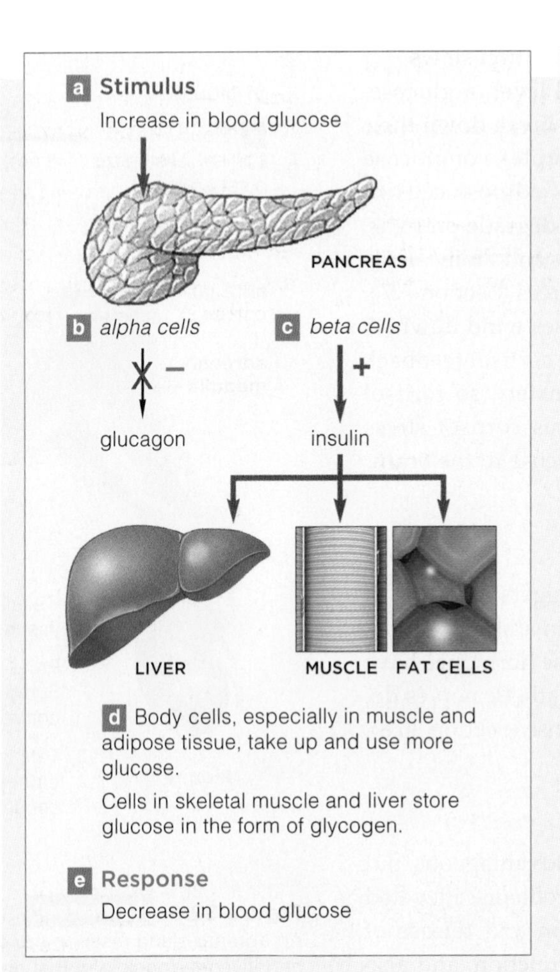

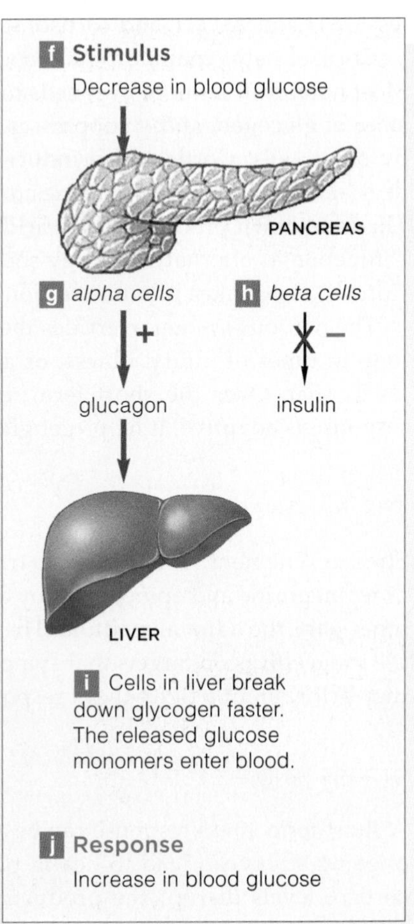

Figure 32.10 **Animated!** *Above*, the location of the pancreas. *Right*, how cells that secrete insulin and glucagon react to shifts in the blood level of glucose. Insulin and glucagon work antagonistically and return the glucose level to its homeostatic range.

(**a**) *After* a meal, glucose enters blood faster than cells can take it up. Its level in blood increases. (**b**,**c**) In the pancreas, the increase stops alpha cells from secreting glucagon and stimulates beta cells to secrete insulin. (**d**) In response to insulin, muscle and adipose cells take up and store glucose, and liver cells synthesize more glycogen. (**e**) The outcome? Insulin *lowers* the glucose blood level.

(**f**) *Between* meals, the glucose level in blood declines. (**g**,**h**) This stimulates alpha cells to secrete glucagon and stops beta cells from secreting insulin. (**i**) In the liver, glucagon causes cells to break glycogen down into glucose, which enters the blood. (**j**) The outcome? Glucagon *raises* the amount of glucose in blood.

32.7 Blood Sugar Disorders

FOCUS ON HEALTH

All body cells use glucose, the preferred energy source for brain cells and the only one for red blood cells. The importance of a stable glucose level becomes clear when we study people who cannot produce enough insulin or whose target cells do not respond properly.

Diabetes mellitus is a metabolic disorder in which cells do not take up glucose as they should. As a result, sugar accumulates in blood and in urine. Complications develop throughout the body (Table 32.2). Excess sugar in the urine encourages growth of pathogenic bacteria, and it damages small blood vessels in the kidneys. Diabetes is the single greatest cause of permanent kidney failure.

Uncontrolled diabetes also damages blood vessels and nerves elsewhere, especially in the arms, hands, legs, and feet. Diabetics account for more than 60 percent of the lower limb amputations in the population at large.

Type 1 Diabetes Of two main types of diabetes mellitus, type 1 develops after the body has mounted an autoimmune response against its insulin-secreting beta cells. Certain white blood cells wrongly identify the cells as foreign (nonself) and destroy them. Environmental factors add to a genetic predisposition to the disorder. Symptoms usually start to appear during childhood and adolescence, which is why this metabolic disorder also is known as juvenile-onset diabetes. All affected individuals require injections of insulin, and they must monitor their blood sugar level carefully (Figure 32.11).

Type 1 diabetes accounts for only 5 to 10 percent of all reported cases, but it is the most dangerous in the short term. In the absence of a steady supply of glucose, the body of affected people uses fats and proteins as energy sources (Section 7.7). Two outcomes are weight loss and ketone accumulation in the blood and urine. Ketones are normal acidic products of fat breakdown, but when too many build up, the result is ketoacidosis. The altered acidity and solute levels can interfere with normal brain function. Extreme cases may lead to coma or death.

LINK TO SECTION 7.7

Type 2 Diabetes Type 2 diabetes is by far the most common form of the disorder. Insulin levels are normal or even high. However, target cells do not respond to the hormone as they should, and blood sugar levels remain elevated. Symptoms typically start to develop in middle age, when insulin production declines. Genetics also is a factor, but obesity increases the risk.

Diet, exercise, and oral medications can control most cases of type 2 diabetes. Even so, if glucose levels are not lowered, pancreatic beta cells continue to receive continual stimulation. Eventually they will falter, and so will insulin production. When that happens, a type 2 diabetic may require insulin injections.

Worldwide, rates of type 2 diabetes are soaring. By one estimate, more than 150 million people are now affected. Western diets and sedentary life-styles are contributing factors. The prevention of diabetes and its complications is acknowledged to be among the most pressing public heath priorities around the world.

Hypoglycemia In hypoglycemia, blood glucose level falls low enough to disrupt normal body functions. Rare insulin-secreting tumors can cause it, but most cases occur after an insulin-dependent diabetic miscalculates and injects a bit too much insulin to balance food intake. The result is insulin shock. The brain stalls as its fuel dwindles. Common symptoms are dizziness, confusion, and difficulty speaking. Insulin shock can be lethal, but an injection of glucagon quickly reverses the condition.

Table 32.2	Some Complications of Diabetes
Eyes	Changes in lens shape and vision; damage to blood vessels in retina; blindness
Skin	Increased susceptibility to bacterial and fungal infections; patches of discoloration; thickening of skin on the back of hands
Digestive system	Gum disease; delayed stomach emptying that causes heartburn, nausea, vomiting
Kidneys	Increased risk of kidney disease and failure
Heart and blood vessels	Increased risk of heart attack, stroke, high blood pressure, and atherosclerosis
Hands and feet	Impaired sensations of pain; formation of calluses, foot ulcers; possible amputation of a foot or leg because of necrotic tissue that formed owing to poor circulation

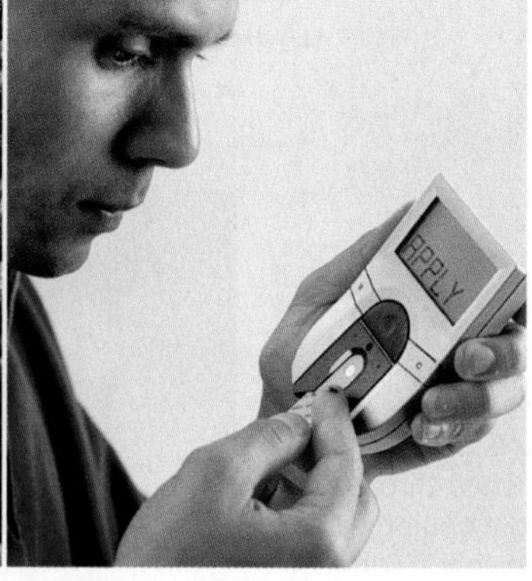

Figure 32.11 A diabetic checks his blood glucose by placing a blood sample into a glucometer. Compared with Caucasians, Hispanics and African Americans are about 1.5 times more likely to be diabetic. Native Americans and Asians are at even greater risk. Proper diet helps control blood sugar, even in type 1 diabetics.

32.8 Other Endocrine Glands

Environmental factors often influence the timing and rates of secretion for hormones that help orchestrate certain aspects of the life cycle. In addition, hormonal secretions may vary over the course of the lifetime.

THE GONADS

LINK TO SECTION 31.6

Gonadal hormones are part of the feedback loops that control gonads, or primary reproductive organs. The gonads of male vertebrates are testes (singular, testis). Females have ovaries. Gonads produce gametes. They also secrete sex hormones, including estrogens and the androgens, such as testosterone.

Puberty is a post-embryonic stage of development when the reproductive organs and structures mature. At puberty, a female mammal's ovaries increase their estrogen production, which causes breasts and other female secondary sexual traits to develop. Estrogens and progesterone control egg formation and ready the uterus for pregnancy. In males, a rise in testosterone output triggers the onset of sperm formation and the development of secondary sexual traits.

Both sexes make most of the hormones. Females, for instance, make a small amount of testosterone. It affects libido, or the desire for sex. Removal of the ovaries reduces female libido. Males produce a small amount of estrogen; sperm cannot develop without it. We return to the sex hormones in Chapter 39.

THE PINEAL GLAND

Deep inside the vertebrate brain is the pineal gland (Figure 32.2). This small gland secretes the hormone melatonin, which serves as part of an internal timing mechanism—a biological clock. Its secretion declines when the brain responds to signals about light that are sent from the retina (Section 31.6). The amount of light varies seasonally and from day to night, and so does melatonin production.

Variations in melatonin levels affect the gonads in many species. For instance, during long winter nights, an elevated level of melatonin in blood dampens the sexual activity of a male songbird. In spring, there are fewer hours of darkness, so the melatonin level falls. Freed from melatonin's suppressing effect, the bird's gonads secrete hormones that affect singing and other behaviors associated with courtship (Figure 32.12).

Melatonin may affect human gonads. A decline in the production of this hormone starts at puberty and may help trigger it. Some pineal gland disorders are known to accelerate or delay puberty.

Melatonin also targets neurons that can lower body temperature and make us drowsy in dim light. The blood level of melatonin peaks in the middle of the night. Exposure to bright light sets a biological clock that controls sleeping versus arousal. Travelers who cross many time zones are advised to spend time in the sun after reaching a destination. If they reset their biological clock, they minimize jet lag.

In winter, seasonal affective disorder (SAD) affects some people. These "winter blues" depress them and make them binge on carbohydrates and crave sleep. Bright artificial light in the morning decreases pineal gland activity and can improve outlooks.

Exposure to bright light in the middle of the night disrupts melatonin production and may increase the risk of cancer. Nurses who typically work night shifts have a lower melatonin level and higher breast cancer rates than day-shift nurses.

THE THYMUS

The thymus gland lies beneath the breastbone (Figure 32.2). The hormones it secretes help infection-fighting white blood cells called T cells to mature. The thymus grows until a person reaches puberty, when it is about the size of an orange. Then, the surge in sex hormones causes it to shrink, and its secretions decline. However, they enhance immune function even in adults.

Figure 32.12 Male white-throated sparrow belting out a song that began, indirectly, with an environmentally induced decline in melatonin secretion from the pineal gland.

Gonads make hormones with roles in sexual reproduction and in the development of secondary sexual traits. Testes produce androgens, including testosterone. The ovaries secrete progesterone and estrogens.

Melatonin from the pineal gland may affect the onset of puberty. It controls a biological clock, and its production is inhibited by time spent in a well-lit environment.

32.9 A Comparative Look at a Few Invertebrates

This chapter focused on vertebrates, but most animals produce hormones of one sort or another. How did the splendid array of these signaling molecules originate?

LINKS TO SECTIONS 23.8, 23.9

EVOLUTION OF RECEPTOR DIVERSITY

How did vertebrates evolve so many diverse hormones and hormone receptors? Molecular evidence points to gene duplications and subsequent divergences by way of mutations. Genetic analysis reveals the invertebrate beginnings of some hormone receptors. For example, sea anemones have receptors at the plasma membrane that are structurally similar to vertebrate receptors for TSH, LH, FSH, and other signaling molecules. Also, the genes that encode them have similar nucleotide sequences in both kinds of animals, and they have the same number and type of introns in similar regions. Presumably, the gene for this receptor protein arose millions of years ago in a common ancestor.

CONTROL OF MOLTING

Do not assume invertebrate hormones are somehow primitive in structure and function just because they evolved first. To get a sense of what they do, think back on molting, the periodic shedding of a too-small body covering (or hair, feathers, or horns) during the life cycle. In arthropods, remember, a soft new cuticle forms beneath an old one, which is then shed (Section 23.9). Before the new cuticle hardens, the body mass increases by the rapid uptake of air or water and by continuous mitotic cell divisions. Details vary among groups. In all cases, however, molting is largely under the control of ecdysone, a steroid hormone.

In arthropods, a molting gland produces and stores ecdysone and releases it for distribution through the body at molting time. Hormone-secreting neurons in the brain seem to control its release. They respond to a combination of internal signals and environmental cues, including light and temperature.

Figure 32.13 is an example of the control steps in crabs and other crustaceans. Right before and during an episode of molting, coordinated interactions among ecdysone and other hormones bring about structural and physiological changes. The interactions make the old cuticle separate from the epidermis and muscles. They induce changes that dissolve inner layers of the cuticle and recycle the remnants. These interactions trigger changes in metabolism and in the composition of the internal environment. They promote rapid cell divisions, secretions, and pigment formation that help make a new cuticle. At the same time, the hormonal interactions govern heart rate, muscle action, changes in body coloration, and other processes.

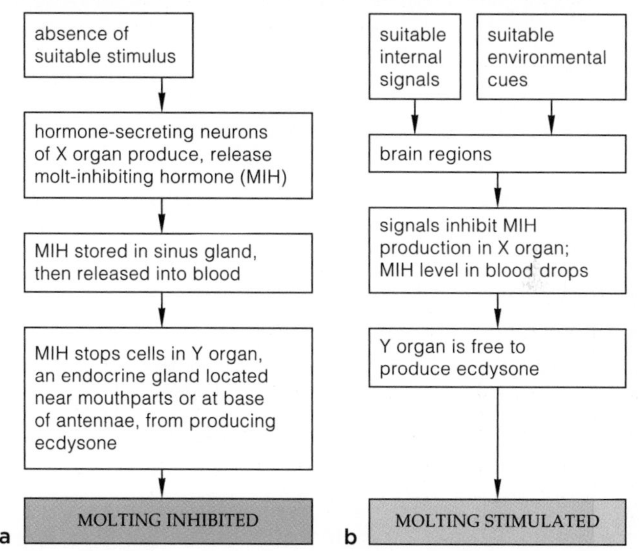

Figure 32.13 Steps in hormonal control of molting in crabs and other crustaceans. The hormone ecdysone stimulates molting. (**a**) In the absence of environmental cues for molting, an X organ in each crab eye stalk produces a hormone that inhibits ecdysone synthesis. (**b**) Right before and during molts, signals from the brain turn off cell activities in the X organ, and so ecdysone is produced and secreted.

The steps differ a bit in insects, which do not have a molt-inhibiting hormone. Rather, stimulation of the insect brain sets in motion a cascade of signals that trigger the production of molt-inducing ecdysone.

Chemicals that mimic or interfere with ecdysone and other hormones can be used as insecticides. Also, nematode pests of crop plants are controlled with the help of ecdysone inhibitors.

A diverse array of animal hormones and receptors arose through gene mutations and divergences.

Ecdysone, a steroid hormone, controls the molting process in nematodes and in arthropods, such as insects. Ecdysone secretion is influenced by environmental cues.

32.10 The Hormone Connection

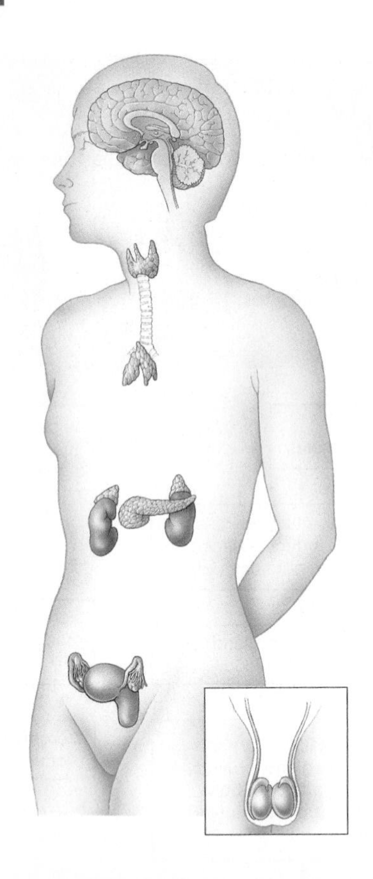

If we liken hormones to violins, harps, and other musical instruments, then the whole body is attuned to a sweeping symphony, and the hypothalamus and pituitary gland are its premier conductors.

Take a moment to reflect on Tables 32.3 and 32.4. This chapter introduced many of the hormones listed. Some other hormones are included for reference purposes, and still others are topics of later chapters.

The point is not to memorize this information but rather to get a sense of the importance of hormones. No tissue or organ is beyond their reach. The cells in most tissues have receptors for different hormones that may compete for or reinforce a cellular response. Every skeletal muscle fiber has receptors for glucagon, insulin, cortisol, epinephrine, estrogen, testosterone, growth hormone, somatostatin, and thyroid hormone, as well as others. What happens at a given moment in each muscle fiber depends in part on blood levels of these hormones, on interactions among them, and on the impact of other signaling molecules.

Remember, a hormone often can bind to receptors in many different tissues. These receptors may differ and binding may summon up a different response in

Table 32.3 Primary Actions of Hypothalamic and Pituitary Hormones

Pituitary Lobe	Secretions	Designation	Main Targets	Primary Actions
Posterior Nervous tissue (extension of hypothalamus)	Antidiuretic hormone (or vasopressin)	ADH	Kidneys	Induces water conservation as required during control of extracellular fluid volume and solute concentrations
	Oxytocin	OT	Mammary glands	Induces milk movement into secretory ducts
			Uterus	Induces uterine contractions during childbirth
Anterior Glandular tissue, mostly	Adrenocorticotropic hormone	ACTH	Adrenal cortex	Stimulates release of cortisol, an adrenal steroid hormone
	Thyroid-stimulating hormone	TSH	Thyroid gland	Stimulates release of thyroid hormones
	Follicle-stimulating hormone	FSH	Ovaries, testes	In females, stimulates estrogen secretion, egg maturation; in males, helps stimulate sperm formation
	Luteinizing hormone	LH	Ovaries, testes	In females, stimulates progesterone secretion, ovulation, corpus luteum formation; in males, stimulates testosterone secretion, sperm release
	Prolactin	PRL	Mammary glands	Stimulates and sustains milk production
	Growth hormone (or somatotropin)	GH	Most cells	Promotes growth in young; induces protein synthesis, cell division; roles in glucose, protein metabolism in adults
Intermediate* Glandular tissue, mostly	Melanocyte-stimulating hormone	MSH	Pigmented cells in skin and other integuments	Induces color changes in response to external stimuli; affects some behaviors

* An intermediate pituitary lobe is present in most vertebrates, but not in humans. MSH is associated with the human anterior lobe.

each tissue. For example, in Chapter 38, you will see how ADH (antidiuretic hormone) from the posterior lobe of the pituitary acts on kidney cells in ways that helps maintain solute concentrations in the internal environment. ADH is sometimes called vasopressin, because it also binds to receptors in the wall of blood vessels and causes these vessels to narrow. In many mammals, ADH helps maintain the blood pressure. It also can influence sexual and social behavior (Section 44.1) by binding to brain cells. This great diversity in the responses to a single hormone is an outcome of variations in ADH receptors. Each kind of receptor summons a different cellular response.

During any interval, vertebrate cells are exposed to and are selectively responding to a complex mix of hormones, the secretions of which are under endocrine and neural control.

Table 32.4 Sources and Primary Actions of Additional Vertebrate Hormones

Source	Examples of Secretion(s)	Main Targets	Primary Actions
Adrenal cortex	Glucocorticoids (including cortisol)	Most cells	Promote breakdown of glycogen, fats, and proteins as energy sources; thus help raise blood level of glucose
	Mineralocorticoids (including aldosterone)	Kidney	Promote sodium reabsorption (sodium conservation); help control the body's salt–water balance
Adrenal medulla	Epinephrine (adrenaline)	Liver, muscle, adipose tissue	Raises blood level of sugar, fatty acids; increases heart rate and force of contraction
	Norepinephrine	Smooth muscle of blood vessels	Promotes constriction or dilation of certain blood vessels; thus affects distribution of blood volume to different body regions
Thyroid	Thyroid hormone	Most cells	Regulates metabolism; has roles in growth, development
	Calcitonin	Bone	Lowers calcium level in blood
Parathyroids	Parathyroid hormone	Bone, kidney	Elevates calcium level in blood
Gonads			
Testes (in males)	Androgens (including testosterone)	General	Required in sperm formation, development of genitals, maintenance of sexual traits, growth, and development
Ovaries (in females)	Estrogens	General	Required for egg maturation and release; preparation of uterine lining for pregnancy and its maintenance in pregnancy; genital development; maintenance of sexual traits; growth, development
	Progesterone	Uterus, breasts	Prepares, maintains uterine lining for pregnancy; stimulates development of breast tissues
Pancreatic islets	Insulin	Liver, muscle, adipose tissue	Promotes cell uptake of glucose; thus lowers glucose level in blood
	Glucagon	Liver	Promotes glycogen breakdown; raises glucose level in blood
	Somatostatin	Insulin-secreting cells	Inhibits digestion of nutrients, hence their absorption from gut
Thymus	Thymopoietin, thymosin	T lymphocytes	Poorly understood regulatory effect on T lymphocytes
Pineal	Melatonin	Gonads (indirectly)	Influences daily biorhythms, seasonal sexual activity
Stomach, small intestine	Gastrin, secretin, etc.	Stomach, pancreas, gallbladder	Stimulate activities of stomach, pancreas, liver, gallbladder; required for food digestion, absorption
Liver	Somatomedins	Most cells	Stimulate cell growth and development
Kidneys	Erythropoietin	Bone marrow	Stimulates red blood cell production
	Angiotensin*	Adrenal cortex, arterioles	Helps control secretion of aldosterone (indirectly affects sodium reabsorption and blood pressure)
	1,25-hydroxyvitamin D_6* (calcitriol)	Bone, gut	Enhances calcium reabsorption from bone and calcium absorption from gut
Heart	Atrial natriuretic hormone	Kidney, blood vessels	Increases sodium excretion; lowers blood pressure

* Kidneys produce enzymes that modify precursors of this substance, which enters blood as an activated hormone.

www.thomsonedu.com 

Summary

Section 32.1 The hormones, neurotransmitters, local signaling molecules, and pheromones are chemicals that are secreted by one cell type and adjust the behavior of other, target cells. Any cell is a target if it has receptors for a signaling molecule at its plasma membrane or in the cytoplasm or nucleus.

All vertebrates have an organ system of endocrine glands and cells. In most cases, the hormonal secretions travel through the bloodstream to nonadjacent targets.

- *Use the animation on ThomsonNOW to learn about the main sources of hormones in the human body.*

Section 32.2 Steroid hormones are lipid soluble and derived from cholesterol. Some kinds enter a target cell and interact directly with DNA. Others bind to the cell's plasma membrane and alter the membrane properties.

The peptide and protein hormones bind to plasma membrane receptors. Binding may lead to the formation of a second messenger, such as cyclic AMP, that relays the signal into the cytoplasm. There, the transduced signal causes a cascade of enzyme activations.

- *Use the animation on ThomsonNOW to compare the mechanisms of steroid and protein hormone action.*

Section 32.3 The hypothalamus, a forebrain region, is structurally and functionally linked with the pituitary gland. They are a major center for homeostatic control.

Some hypothalamic neurons make ADH or oxytocin, two hormones that the posterior pituitary gland secretes. ADH acts in kidneys. Oxytocin acts on the uterus and milk ducts. Other hypothalamic neurons make releasers and inhibitors that enhance or slow secretion of anterior pituitary hormones.

The anterior pituitary makes and secretes hormones of its own. ACTH targets the adrenal cortex, TSH targets the thyroid, FSH and LH act on male and female gonads, and PRL influences mammary glands and the uterus. GH promotes growth throughout the body. Gigantism and dwarfism result from mutations that affect GH secretion or response to this hormone.

- *Use the animation on ThomsonNOW to study how the hypothalamus and pituitary interact.*

Section 32.4 A negative feedback loop involving the anterior pituitary gland and the hypothalamus governs thyroid hormone secretion. Iodine deficiency or immune system disorders can disrupt thyroid gland function.

The parathyroid glands are the main regulators of calcium levels in the blood. They release parathyroid hormone (PTH) in response to low calcium levels. PTH acts on bone cells and kidney cells in ways that raise calcium levels in the blood.

Section 32.5 The adrenal medulla secretes two of the steroid hormones: aldosterone and cortisol. Cortisol secretion is governed by a negative feedback loop to the anterior pituitary gland and hypothalamus. In times of stress, the central nervous system overrides the feedback controls so that cortisol levels rise.

Norepinephrine and epinephrine released by neurons of the adrenal medulla influence organs as sympathetic stimulation does; they cause a fight–flight response.

- *Watch the animation on ThomsonNOW to see how cortisol levels are maintained by negative feedback.*

Sections 32.6, 32.7 Pancreatic islet cells secrete the hormones insulin and glucagon in response to changes in the blood concentration of glucose. Beta cells secrete insulin when the glucose level is high. Insulin stimulates glucose uptake by muscle and liver cells, which lowers blood glucose level. Alpha cells secrete glucagon, which stimulates the release of glucose, when the blood level is too low. The two hormones work in opposition to keep blood glucose within the optimal range.

Diabetes occurs when the body does not make insulin or its cells do not respond to it. Diabetes has adverse effects on organs throughout the body.

- *Use the animation on ThomsonNOW to see how the actions of insulin and glucagon regulate blood sugar.*

Section 32.8 Environmental factors affect secretion of some hormones, including sex hormones such as the estrogens, progesterone, and testosterone. Sex hormones control gamete formation and regulate the development of secondary sexual traits.

Light suppresses secretion of melatonin by the pineal gland in the brain. Melatonin functions in internal timing mechanisms—biological clocks. In humans, it influences the onset of puberty and the sleep/wake cycle.

The thymus in the chest produces hormones that help some white blood cells (T cells) mature.

Section 32.9 Vertebrate hormone receptor proteins often resemble similar receptor proteins in invertebrates and probably evolved from them. The steroid hormone ecdysone affects molting in arthropods and roundworms. Environmental cues such as day length affect its secretion.

Section 32.10 The big picture is this: All cells inside the vertebrate body are bathed in an ever-changing array of hormones and selectively respond to them according to the types of receptors in different tissues, competing or enhancing hormone interactions, and other factors.

Self-Quiz

Answers in Appendix III

1. ______ are signaling molecules released from one type of cell that can alter target cell activities.
   a. Hormones
   b. Neurotransmitters
   c. Pheromones
   d. Local signaling molecules
   e. both a and b
   f. a through d

2. ADH and oxytocin are hormones produced in the hypothalamus but distributed from the ______.
   a. anterior lobe of pituitary
   b. posterior lobe of pituitary
   c. pancreas
   d. pineal gland

3. Overproduction of ______ causes gigantism.
   a. growth hormone
   b. ADH
   c. insulin
   d. melatonin

4. Which do not stimulate hormone secretions?
   a. neural signals
   b. local chemical changes
   c. hormonal signals
   d. environmental cues
   e. All of the above can stimulate secretion.

5. ______ lowers blood sugar levels; ______ raises it.
   a. Glucagon; insulin
   b. Insulin; glucagon
   c. Gastrin; insulin
   d. Gastrin; glucagon

6. The pituitary detects a rising hormone concentration in blood and inhibits the gland secreting the hormone. This is a ______ feedback loop.
   a. positive b. negative c. long-term d. b and c

7. The ______ has endocrine and exocrine functions.
   a. hypothalamus
   b. pancreas
   c. pineal gland
   d. parathyroid gland

8. Secretion of ______ stimulates breakdown of bone.
   a. glucagon
   b. melatonin
   c. thyroid hormone
   d. parathyroid hormone

9. Exposure to bright light lowers blood ______ levels.
   a. glucagon
   b. melatonin
   c. thyroid hormone
   d. parathyroid hormone

10. Match the term listed at left with the most suitable description at right.

___ adrenal medulla	a. affected by day length
___ thyroid gland	b. has potent local effects
___ parathyroid glands	c. raises blood calcium level
___ pancreatic islets	d. source of epinephrine
___ pineal gland	e. secretes insulin, glucagon
___ prostaglandin	f. hormones require iodide

■ *Visit ThomsonNOW for additional questions.*

Critical Thinking

1. In the late 1990s, Marcia Herman-Giddens asked this question of pediatricians throughout the United States: At what age are female patients developing secondary sexual traits? The compiled answers showed that girls are developing breasts six months to one year earlier than they did in earlier decades.

Herman-Giddens's data were criticized because the girls were not a random sample; they had appointments to see pediatricians. Nevertheless, many researchers suspect that her findings reflect a real trend. What might be the cause of this precocious breast development?

Evidence that chemicals may play a role comes from the island of Puerto Rico, which has the world's highest incidence of premature breast development. A recent study found that most affected girls started to develop breasts when they were between six and twenty-four months old (Figure 32.14*a*). Sixty-eight percent had high blood levels of phthalates—chemicals used in the manufacture of some plastics and pesticides. By comparison, none of the normal girls that researchers examined had high phthalate levels.

What type of information might be used to test the hypothesis that exposure to phthalates is contributing to the early breast development among young girls in the mainland United States?

2. Many abnormally small but normally proportioned people live in a remote village in Pakistan. On average, the men are about 130 centimeters (a little over 4 feet) tall,

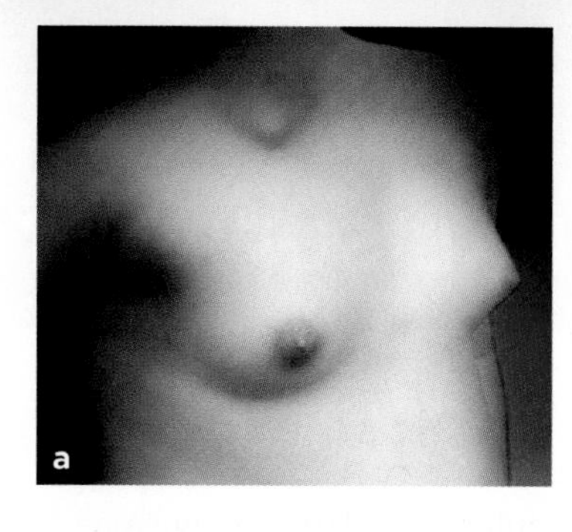

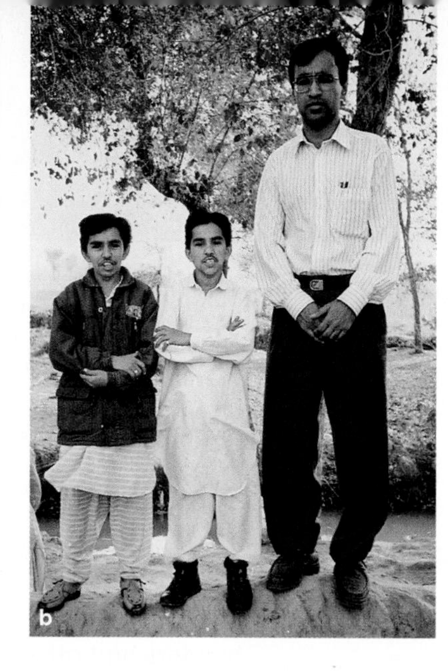

Figure 32.14 Two examples of hormone effects in humans. (**a**) At *twenty-three months*, this Puerto Rican girl already shows breast development. Industrial chemicals that mimic estrogen may be a factor. (**b**) Researcher Hiralal Maheshwari, with two men who have a heritable form of dwarfism.

and the women about 115 centimeters (3 feet, 6 inches) tall. Northwestern University researchers interviewed the people. They found out that all were part of an extended family. The abnormality, a type of dwarfism, is one case of autosomal recessive inheritance (Figure 32.14*b*).

Biochemical and genetic analysis show that affected individuals have a low level of growth hormone but a normal gene that codes for this hormone. The abnormality starts with a mutant gene that specifies a receptor for one of the hypothalamic releasing hormones. That receptor is on cells in the anterior lobe of the pituitary gland. Explain how a mutant receptor could cause the abnormality.

3. Tuberculosis and other infectious diseases damage the adrenal glands, and slow or halt cortisol secretion. The result is Addison's disease. In developed countries, this hormonal disorder most often arises after autoimmune attacks on the adrenal gland. President John F. Kennedy showed symptoms of this form of the disorder. Emotional or physical stress can push an affected person into a crisis mode in which many organ systems spin out of control. Explain why an absence of cortisol has this outcome.

4. The blue crab (*Callinectes sapidus*) lives in estuaries along the Gulf and Atlantic coasts of the United States. Like other arthropods, the crabs molt as they grow (Figure 32.15). After molting, it takes about twelve hours for the new shell to harden. During this time, the crab is vulnerable to natural predators and human crabbers who can sell it as an expensive "soft-shelled" crab. Blue crab populations have declined in some areas. Chemical pollutants block ecdysone receptors. Could a chemical that affects ecdysone action interfere with a crab's life cycle? If so, how?

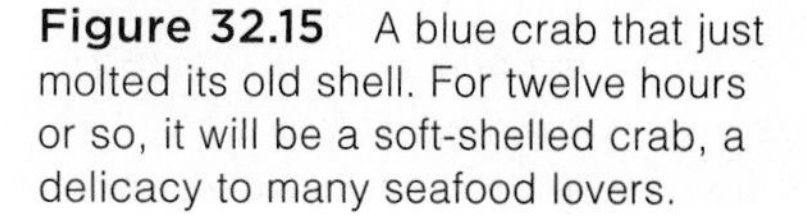

Figure 32.15 A blue crab that just molted its old shell. For twelve hours or so, it will be a soft-shelled crab, a delicacy to many seafood lovers.

33 STRUCTURAL SUPPORT AND MOVEMENT

IMPACTS, ISSUES Pumping Up Muscles

Want to be more muscular and stronger, with a lot more endurance? Just use our powders or pills and you'll soon look like the guy in Figure 33.1. That is the message in advertisements for many dietary supplements that target body builders and other athletes. The supplements are easily purchased from health-food stores or through the Internet. The U. S. Food and Drug Administration (FDA) does not classified dietary supplements as drugs, so testing to verify their effectiveness, or even that they contain the ingredients listed on the label, is negligible.

In 1998, sales of androstenedione, or "andro," got a big boost after Mark McGwire admitted he used it during his successful attempt to break Major League Baseball's home-run record. Andro forms naturally in the body as an intermediate in the synthesis of testosterone, a sex hormone. Testosterone has tissue-building, anabolic effects that are well documented. Andro supplements supposedly raise the blood level of testosterone, which in turn boosts the rate of protein synthesis in muscles.

Does andro work? Probably not. In some controlled studies, males of an experimental group used an andro supplement. They did not gain any more muscle mass or strength than males of a control group who were given a placebo. At most, the andro supplement raised the testosterone level in blood for a few hours.

Andro also forms as an intermediate in the synthesis of estrogen. This sex hormone has feminizing effects on

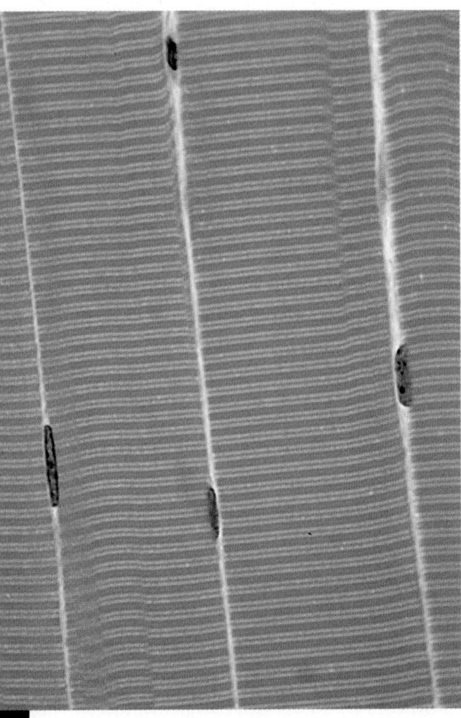

See the video! **Figure 33.1** *Left*, overabundance of skeletal muscle tissue and its parallel rows of muscle fibers (*above*).

How would you vote? Dietary supplements are largely unregulated. Should they be placed under the jurisdiction of the Food and Drug Administration, which could subject them to more stringent testing for effectiveness and safety? See ThomsonNOW for details, then vote online.

males, including shrunken testicles, formation of female-like breasts, and hair loss. Common effects on females include masculinized patterns of hair growth, deepening of the voice, and disruption of menstrual cycles. Users in both sexes risk acne, liver damage, and a lower blood level of "good" cholesterol, or HDL.

In 2004, the FDA announced that andro supplements have serious side effects. Companies were ordered to stop distributing them immediately.

Creatine, a short chain of amino acids, is touted as another performance-enhancing supplement. The body makes some creatine and gets more from food. When muscles must contract hard and fast, they normally use phosphorylated creatine as an instant energy source.

Unlike andro, creatine supplements might work. In controlled studies, they improved performance during brief, high-intensity exercise. Clinical trials are under way to evaluate whether creatine may benefit people affected by muscle disorders, including muscular dystrophy.

Nevertheless, excessive creatine intake puts a strain on the kidneys. The strain is risky, given that kidneys cleanse the blood and maintain the body's fluid balance around the clock. It is too soon to know whether creatine supplements have any long-term side effects. Also, no regulatory agency is checking to see how much creatine is actually present in any commercial product.

With this chapter, we turn to the advantage of having muscles in the first place. All animals move from place to place during their life cycle. In adults, the movement of body parts requires contractile cells and some enclosed fluid or skeletal element against which contractile force can be applied. Skeletal and muscular systems have a long evolutionary history, one that can help you evaluate how far both systems can be pushed in the pursuit of enhanced performance.

Key Concepts

SKELETAL SYSTEMS

Contractile force exerted against some type of skeleton moves the animal body. Many invertebrates have a hydrostatic skeleton, which is a fluid-filled body cavity. Others have an exoskeleton of hardened structures at the body surface. Vertebrates have an endoskeleton, an internal skeleton of cartilage, bone, or both. **Section 33.1**

VERTEBRATE SKELETONS

Bones are collagen-rich organs that help the body move. They also protect and support soft organs, and store minerals. Blood cells form in some bones. Cartilage or ligaments connect bones at joints. **Section 33.2**

THE MUSCLE–BONE PARTNERSHIP

Skeletal muscles are bundles of muscle fibers that interact with bones and with one another. Some cause movements by working as pairs or groups. Others oppose or reverse the action of a partner muscle. Tendons attach skeletal muscles to bones. **Section 33.3**

HOW SKELETAL MUSCLE CONTRACTS

A muscle fiber contains many myofibrils, each divided crosswise into sarcomeres, the basic units of contraction. Sarcomeres contain many parallel arrays of actin and myosin filaments. ATP-driven interactions between the arrays shorten sarcomeres, which collectively accounts for contraction of a whole muscle. **Section 33.4**

PROPERTIES OF WHOLE MUSCLES

Muscle fibers in a muscle are organized in motor units that contract in response to signals from one motor neuron. Cross-bridges form in all the sarcomeres and collectively exert tension. A muscle contracts only when the tensile force exceeds other, opposing forces. Exercise enhances the properties of whole muscles, and aging and disease diminish them. **Sections 33.5, 33.6**

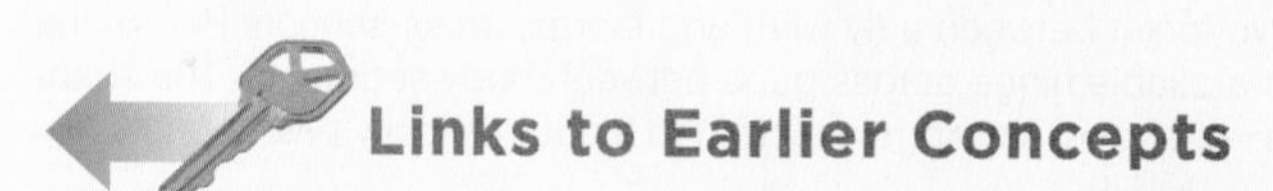

Links to Earlier Concepts

In this chapter you will return to the contractile proteins actin and myosin (Section 4.11). You will take a closer look at the skeletons of invertebrates (23.4, 23.9) and vertebrates (16.7, 24.2, 24.3, 24.9, 24.11), and at the fine structure of bone tissue and muscle tissues (29.2, 29.3). You will draw on your knowledge of active transport (5.7), ATP function (5.2), and pathways of organic metabolism (7.1, 7.6). You will gain more insight into the hormonal control of calcium levels (32.4) and how nervous signals lead to contraction (30.4). You will revisit a bacterial species that can severely disrupt muscle function (19.2).

33.1 Animal Skeletons

Most animal bodies are supported by a hydrostatic skeleton, an exoskeleton, or an endoskeleton.

LINKS TO SECTIONS 17.9, 23.4, 23.9, 24.2, 24.3, 24.9, 24.11

Soft-bodied invertebrates typically have a hydrostatic skeleton: a closed chamber or chambers containing a fluid that muscles exert pressure against. Recall that an earthworm moves when waves of contraction alter the shape of its fluid-filled coelomic chambers (Section 23.6). An anemone elongates when it closes its mouth and contracts muscles that ring its body wall (Figure 33.2). Pressure on water enclosed inside the gut cavity alters the body shape, just as squeezing a water-filled balloon causes the balloon to elongate.

An exoskeleton is a stiff body covering to which the muscles attach. For example, clams have a hinged shell that muscles can close up tight. In insects and other arthropods, the cuticle serves as an exoskeleton. Contracting muscles bend the cuticle at hinge points and the resulting lever-like actions move body parts, such as a fly's wings (Figure 33.2).

An endoskeleton is an internal framework of hard elements, to which the muscles attach. All vertebrates have an endoskeleton of cartilage, bone, or both. For instance, the human skeleton has 206 bones. Its *axial* portion (head and trunk) includes jaws and other skull bones, twelve pairs of ribs, the sternum (breastbone), and twenty-six vertebrae (singular, vertebra), or bony segments that form a column-like backbone. Between each vertebra, intervertebral disks of flexible cartilage act as shock absorbers and flex points. Paired muscles attach to this vertebral column, and a spinal cord runs through the canal inside it. The skeleton's *appendicular* portion consists of paired arms and legs, and the sites where limbs attach to the vertebral column: a pectoral (shoulder) girdle and a pelvic (hip) girdle.

Human jaws and skull bones can be traced back to aquatic ancestors that lived in Cambrian seas (Section 24.1). Pairs of fleshy, lobed fins of certain fishes were forerunners of limb bones. Skeletal elements evolved inside the fins and became attached to the pelvic and pectoral girdles of the first tetrapods, or four-legged walkers (Section 24.3). In reptilian lineages on land, both girdles became larger and sturdier. They helped support the body, which was now deprived of water's buoying effects (Figure 33.3).

Your pectoral girdle and pelvic girdle are legacies of early reptiles (Figures 33.3 and 33.4). So is the sternum. It became attached to a protective rib cage that helped hold the heart, lungs, and other soft organs in place inside the thorax, now positioned above the ground.

Chapter 24 introduced the evolutionary history of humans, from lobe-finned fishes, through amphibians, reptiles, then early primates. By 4 million years ago, some of our primate ancestors had become bipedal; they walked upright on two legs. Remember how an upright posture resulted in an S-shaped curve in the backbone? Vertebrae and intervertebral disks became

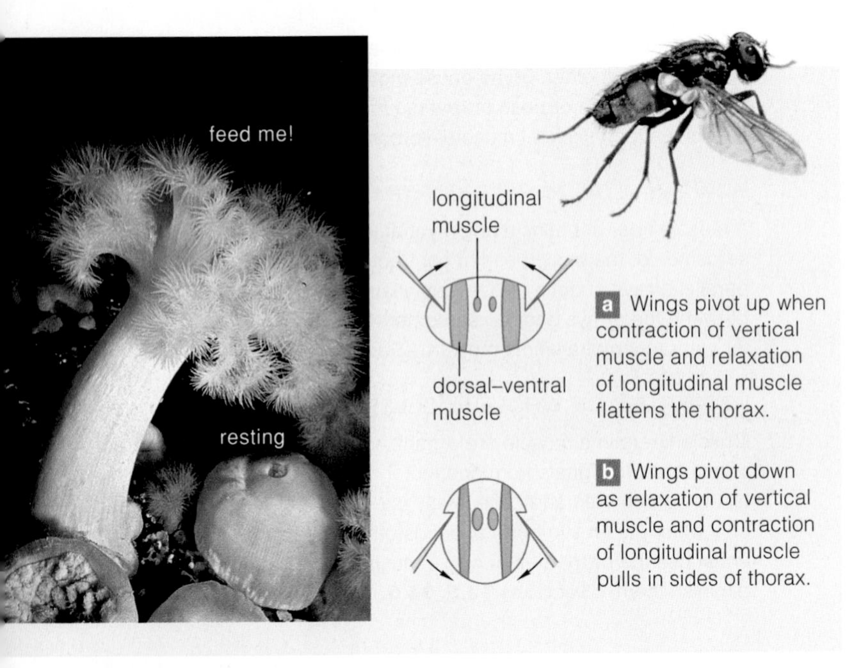

Figure 33.2 **Animated!** Examples of invertebrate skeletons. *Left*, in sea anemones, muscles ringing the fluid-filled gut cavity contract and change the body's shape from its relaxed, resting position to its upright feeding position. *Right*, pivot point between a fly wing and thorax, cross-section. Part of the cuticle is a pliable hinge across gaps between body segments. The shape of the thorax changes, so wings attached to it automatically pivot up and down.

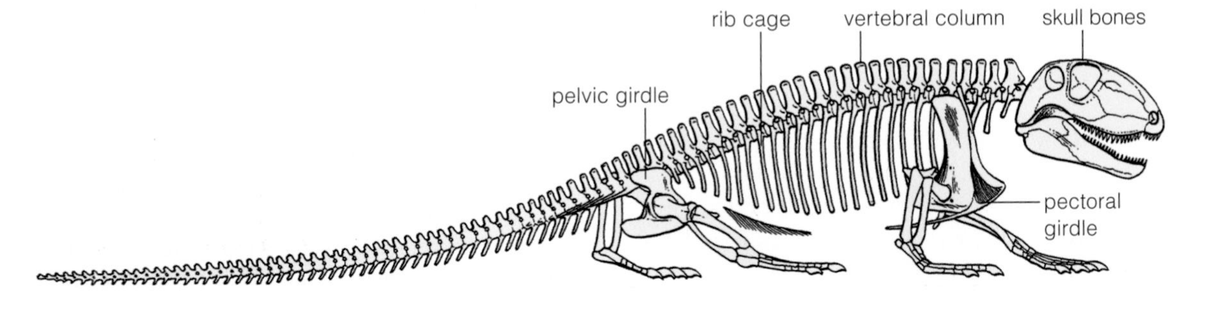

Figure 33.3 Skeletal elements typical of early reptiles. Compare Figures 24.10, 24.21, and 24.25.

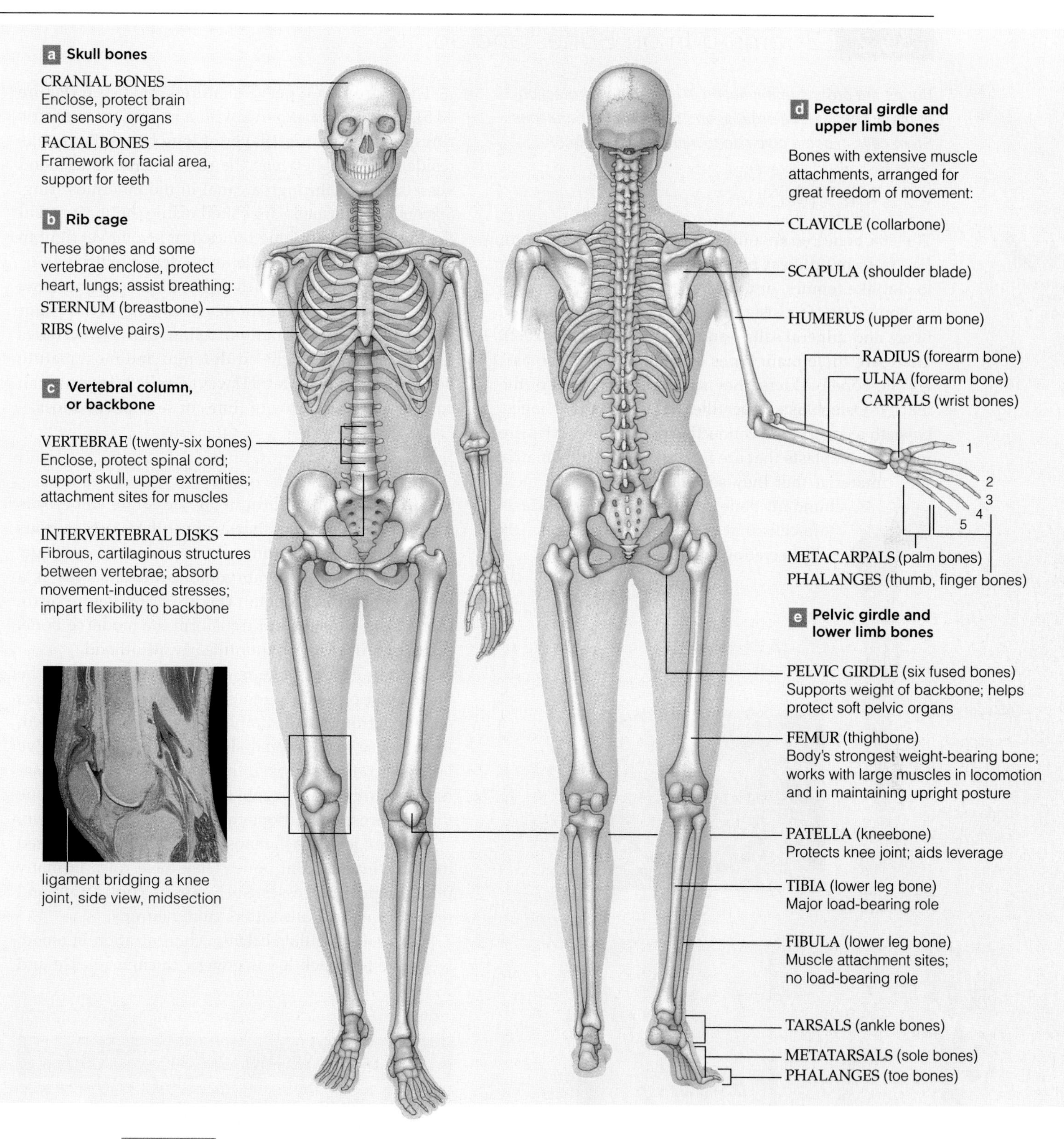

Figure 33.4 **Animated!** Major bone (*tan*) and cartilage (*light blue*) elements of the human skeleton: axial portion (*left*) and appendicular portion (*right*). Ligaments bridge joints.

stacked on top of each other. These disks sometimes slip out of place or rupture, causing chronic back pain. No longer aquatic and no longer tetrapods, modern humans live with costs as well as benefits of walking upright through the world.

Animals move when muscles interact with an enclosed fluid, or with hard external or internal parts. Vertebrates have an internal skeleton. The human skeleton shows modifications for life on land and for upright walking.

33.2 Zooming In on Bones and Joints

Bones are organs that function in movement, protection of many soft internal organs, and storage of mineral ions. Stem cells in some give rise to blood cells (Table 33.1).

BONE STRUCTURE AND FUNCTION

LINKS TO SECTIONS 29.2, 32.4

The 206 bones of an adult human's skeleton range in size from middle ear bones as small as a grain of rice to clublike femurs, or thighbones.

Bone tissue consists of a matrix—mainly collagen fibers and mineral salts—and bone cells (Section 29.2). There are three main types of bone cells. Osteoblasts are the bone builders; they secrete components of the matrix. Osteoblasts cover the surface of adult bones, beneath a covering of connective tissue. Osteocytes are former osteoblasts that are now surrounded by matrix material that they secreted. They are the most abundant bone cells in adults. Osteoclasts are cells that can break down the matrix by secreting enzymes and acids.

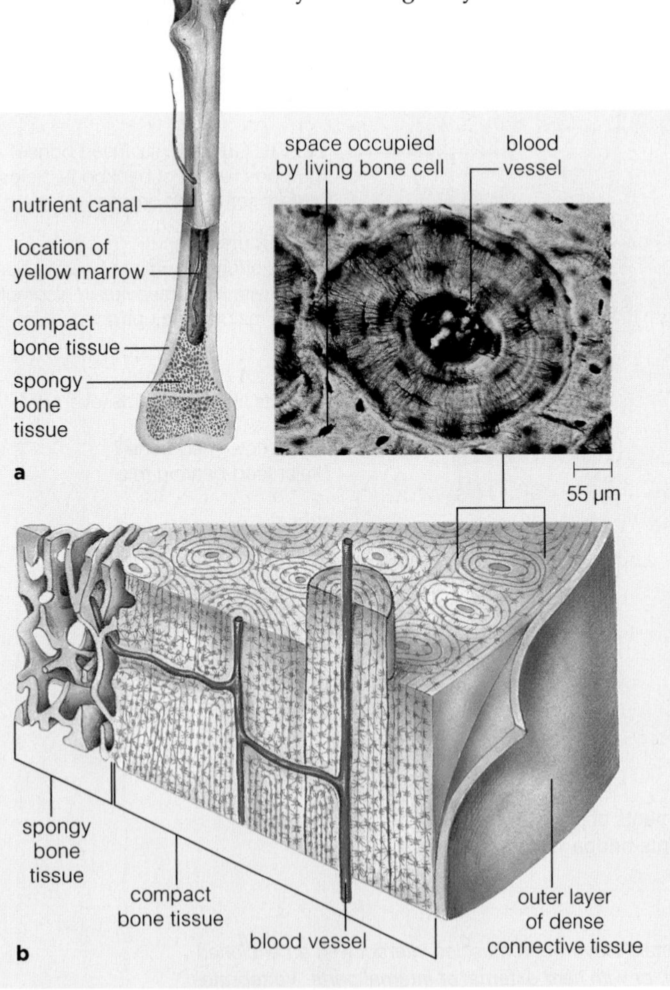

Figure 33.5 **Animated!** (**a**) Structure of a human femur, or thighbone, and (**b**) a section through its spongy and compact bone tissues.

There are two types of mature bone tissue (Figure 33.5). Dense *compact* bone, with a matrix of concentric rings, is the outermost layer of long bones. Bone cells reside in spaces between the rings. Nerves and blood vessels extend through a canal in the innermost ring. *Spongy* bone tissue is deposited inside the shaft and at the knobby ends of long bones. It is strong yet light in weight; open spaces riddle its hardened matrix.

Marrow fills the cavities inside a bone. Red marrow occupies the spaces in spongy bone and is the major site of blood cell formation. Yellow marrow occupies the central cavity of the adult femur and most mature bones and is mostly fat. However, yellow marrow can convert to red marrow in times of severe blood loss.

BONE FORMATION AND REMODELING

The first skeleton to form in all vertebrate embryos is made of cartilage. In sharks and other cartilaginous fishes, the adult skeleton consists entirely of cartilage. In other vertebrates, embryonic cartilage serves as a model for a bony skeleton (Figure 33.6). Before birth, osteoblasts move in and transform the model to bone. Bones continue to grow until early adulthood.

Even in adults, bone is a dynamic tissue, that the body must constantly remodel. Microscopic fractures resulting from small, daily movements are repaired. In response to hormonal signals, osteoclasts dissolve portions of the matrix so that stored mineral ions are released to blood. Osteoblasts make new bone tissue that replaces the portions that osteoclasts break down.

Calcium ions are the most abundant mineral stored in and released from bone. They have roles in many physiological processes, such muscle contraction and release of neurotransmitters from neurons.

Hormones regulate calcium concentration in blood. Negative feedback loops govern calcium release and

Table 33.1 Functions of Bone

1. *Movement.* Bones interact with skeletal muscle and change or maintain positions of the body and its parts.
2. *Support.* Bones support and anchor muscles.
3. *Protection.* Many bones form hardened chambers or canals that enclose and protect soft internal organs.
4. *Mineral storage.* Bones are a reservoir for calcium and phosphorus ions. Deposits and withdrawals of these ions help maintain their concentrations in body fluids.
5. *Blood cell formation.* Only certain bones contain the tissue where blood cells form.

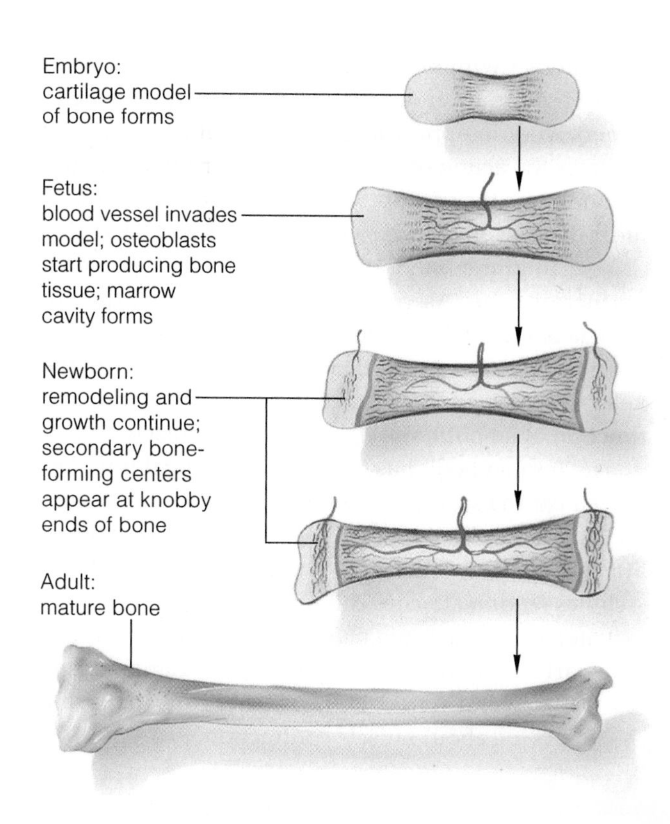

Figure 33.6 Long bone formation, starting with osteoblast activity in a cartilage model formed earlier in the embryo. The bone-forming cells are active first in the shaft region, then at the knobby ends. In time, cartilage is left only at the ends.

uptake (Section 32.4). When the blood calcium level is too high, the thyroid gland secretes calcitonin. This hormone slows the release of calcium into blood by inhibiting osteoclast action. When blood has too little calcium, the parathyroid glands release parathyroid hormone, or PTH. This hormone stimulates osteoclast actions. PTH also decreases calcium loss in urine and helps activate vitamin D, which stimulates cells in the gut lining to absorb calcium.

Other hormones also affect bone turnover. The sex hormones estrogen and testosterone encourage bone deposition. Cortisol, the stress hormone, slows it.

Until humans are about twenty-four years old, the osteoblasts secrete more matrix than osteoclasts break down, so bone mass increases. Bones become denser and stronger. Later in life, the osteoblasts become less active, and bone mass gradually declines.

Osteoporosis is a disease in which bone loss speeds up and bones become brittle (Figure 33.7). Vitamin D deficiency, thyroid or parathyroid problems, and lack of exercise increase risk. Declining sex hormone levels after menopause, smoking, and the prolonged use of synthetic steroids are other contributing factors.

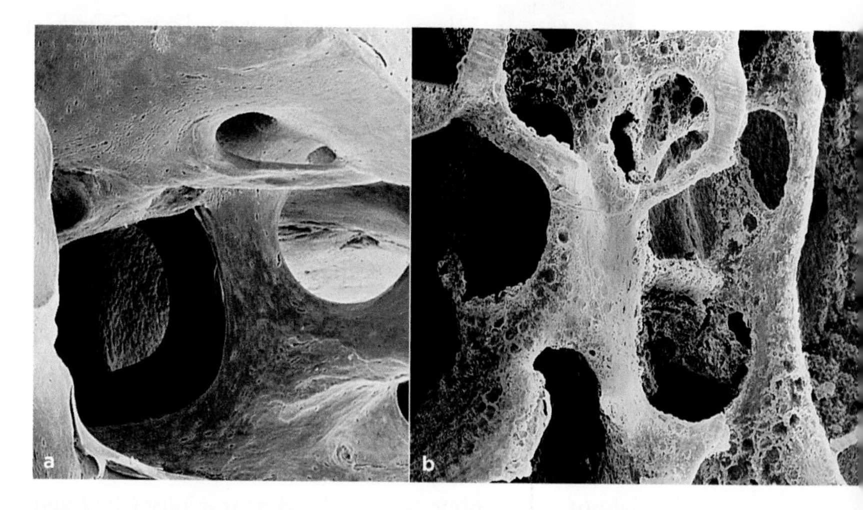

Figure 33.7 (**a**) Normal bone tissue. (**b**) Bone weakened by osteoporosis.

WHERE BONES MEET—SKELETAL JOINTS

A joint is an area of contact or near contact between bones. Bones at a *fibrous* joint, such as skull bones, are fixed in place. At *cartilaginous* joints (pads or disks of cartilage), bones move a bit and are cushioned from shocks. Cartilaginous joints occur between vertebrae and between ribs and the breastbone.

Most common are *synovial* joints, which allow the bones to move freely at knees, shoulders, ankles, and elsewhere. Smooth cartilage covers the end of bones at these joints and reduces friction during movements. Ligaments are stretchy cords of connective tissue that connect bone to bone in a synovial joint.

Different synovial joints allow different movements. For example, joints at the shoulders and hips allow a ball-and-socket motion. Joints at the elbows and knees function like a hinge; they allow the bones to move back and forth in one plane.

Joint injuries and disorders are common. Tearing or overstretching a ligament results in a sprain. Knee, wrist, and ankle joints are most frequently sprained. Arthritis means "inflammation of a joint." There are two types. Osteoarthritis occurs when cartilage wears down at a frequently used joint. This disorder usually appears in old age and affects only certain joints. For instance, only the knees or even just one knee may be affected. In rheumatoid arthritis, the immune system mistakenly attacks all movable joints. It can happen at any age and is more common in females.

Bones are collagen-rich, mineralized organs that function in movement, support, protection, storage of mineral ions, and blood cell formation. They are constantly remodeled.

Joints are regions where bones meet. Most joints allow bones to move, and ligaments help to stabilize them.

33.3 Skeletal–Muscular Systems

Skeletal muscles are the functional partners of bones. Unlike smooth muscle or cardiac muscle, they are under voluntary control; you can consciously make them move.

LINKS TO SECTIONS 25.9, 29.3

Skeletal muscles consist of bundles of muscle fibers sheathed in dense connective tissue. A muscle fiber is a long, cylindrical cell with multiple nuclei that holds contractile filaments. It has several nuclei because it is descended from a group of cells that fused together in the developing embryo.

Most muscles and bones interact as a lever system, in which a rod is attached to a fixed point and moves about it. The bone is a rigid rod near a joint (the fixed point). Muscle contraction transmits force to the bone and makes it move, as in Figure 33.8.

Fully extend your right arm, place your left hand over the upper arm, and slowly bend your elbow, as in Figure 33.9*a*. Can you feel the muscle contracting? By causing this muscle to shorten a bit, you allowed the bone attached to the muscle to move a big distance. Besides interacting with bone, skeletal muscles also interact with one another. Some skeletal muscles work in pairs or groups to bring about a movement. Others work in opposition; the action of one resists or reverses the action of another. The biceps in the upper arm opposes the triceps. Such pairings are the case for most muscles in the limbs (Figures 33.8 and 33.9).

Bear in mind, only *skeletal* muscle is the functional partner of bone. As you read in Section 25.3, smooth muscle is mainly a component of soft internal organs, such as the stomach. Cardiac muscle forms only in the heart wall. Later chapters consider the structure and function of smooth muscle and cardiac muscle.

The human body has close to 700 skeletal muscles, some near the surface, others in the body wall (Figure 33.10). A straplike tendon made of connective tissue attaches skeletal muscles to bone. As an example, the Achilles tendon attaches calf muscles to the heel bone.

Later chapters explain how skeletal muscles also take part in respiration and in blood circulation. Here, we turn next to the mechanisms that bring about muscle contraction.

Cordlike or straplike tendons of dense connective tissue attach skeletal muscles to bones.

Skeletal muscles transmit contractile force to bones. Small muscle movements can bring about large movements of bones. Many muscle groups have opposing actions.

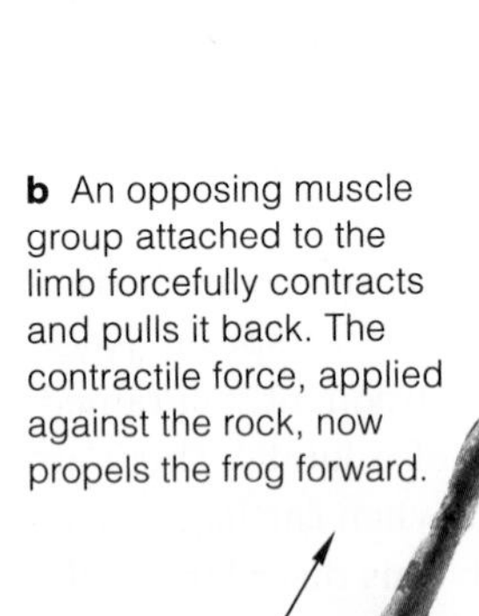

Figure 33.8 A frog on a rock demonstrating how small contractions and the action of opposing muscles can cause big movements.

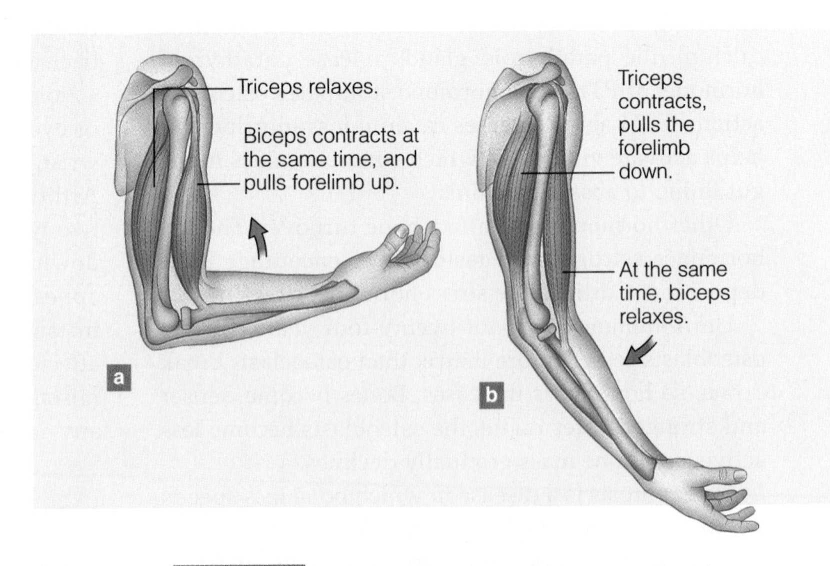

Figure 33.9 **Animated!** Two opposing muscle groups in human arms. (**a**) When the triceps relaxes and its opposing partner (biceps) contracts, the elbow joint flexes and the forearm is pulled upward. (**b**) When the triceps contracts and the biceps relaxes, the forearm is extended down.

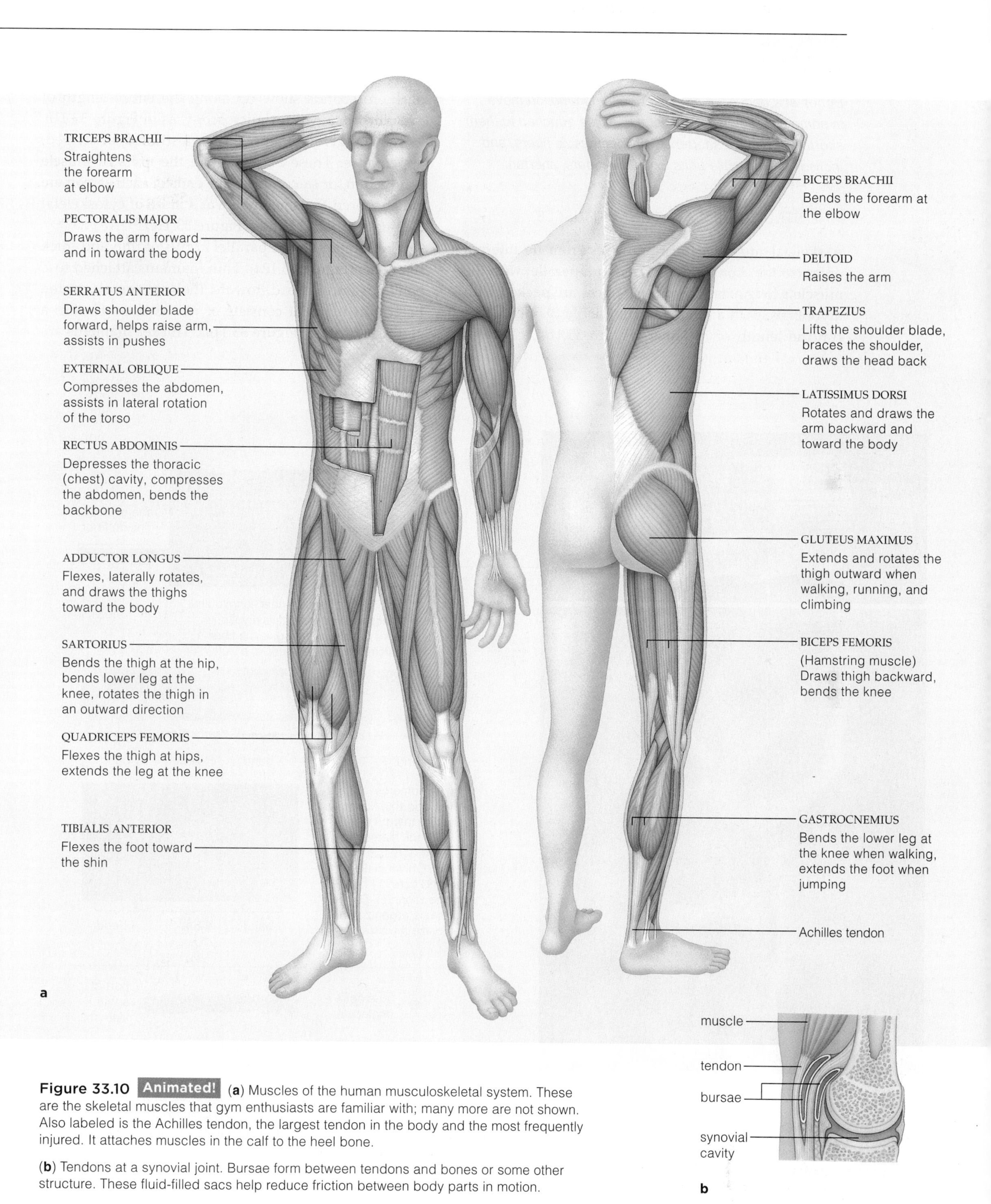

Figure 33.10 Animated! (**a**) Muscles of the human musculoskeletal system. These are the skeletal muscles that gym enthusiasts are familiar with; many more are not shown. Also labeled is the Achilles tendon, the largest tendon in the body and the most frequently injured. It attaches muscles in the calf to the heel bone.

(**b**) Tendons at a synovial joint. Bursae form between tendons and bones or some other structure. These fluid-filled sacs help reduce friction between body parts in motion.

33.4 How Does Skeletal Muscle Contract?

Bones of a dancer or any other human in motion move in some direction when skeletal muscles attached to them shorten. A muscle shortens when its muscle fibers, and individual contractile units inside the fibers, shorten.

FINE STRUCTURE OF SKELETAL MUSCLE

LINKS TO SECTIONS 4.10, 5.2, 25.3

A skeletal muscle's function arises from its internal organization. Long muscle fibers run parallel with the muscle's long axis. The muscle fibers are packed with myofibrils, each a bundle of contractile filaments that run the length of the fiber (Figure 33.11*a*). Light-to-dark crossbands show up along the entire length of myofibrils stained for microscopy, as in Figure 33.11*b*. The bands give the muscle fiber a striated, or striped, appearance. These bands define the units of muscle contraction, or sarcomeres. The ends of each sarcomere are anchored to its neighbors at a mesh of cytoskeletal elements called Z bands (Figure 33.11*c*).

The sarcomere has parallel arrays of thin and thick filaments (Figure 33.12*a*). Thin filaments attached to Z bands extend inward, toward the sarcomere's center. Each thin filament consists of two chains of actin, a globular protein (Figure 33.11*d*). Thicker filaments are

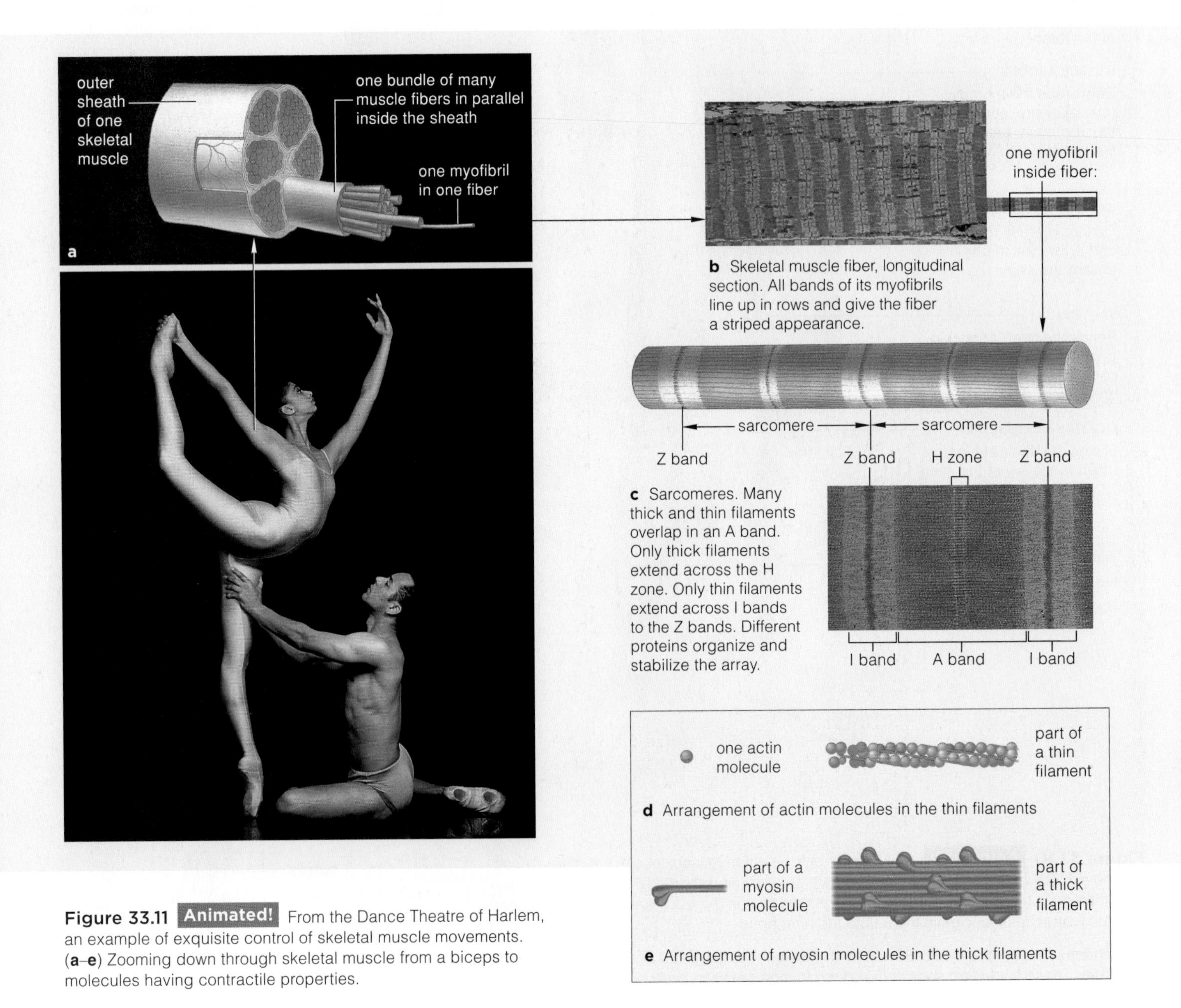

b Skeletal muscle fiber, longitudinal section. All bands of its myofibrils line up in rows and give the fiber a striped appearance.

c Sarcomeres. Many thick and thin filaments overlap in an A band. Only thick filaments extend across the H zone. Only thin filaments extend across I bands to the Z bands. Different proteins organize and stabilize the array.

d Arrangement of actin molecules in the thin filaments

e Arrangement of myosin molecules in the thick filaments

Figure 33.11 **Animated!** From the Dance Theatre of Harlem, an example of exquisite control of skeletal muscle movements. (**a**–**e**) Zooming down through skeletal muscle from a biceps to molecules having contractile properties.

centered in the sarcomere. A thick filament consists of myosin, a motor protein with a clublike head (Figure 33.11*e*). The head is positioned just a few nanometers away from a thin filament.

Muscle fibers, myofibrils, thin filaments, and thick filaments all have the same orientation; they all run parallel with a muscle's long axis. What function does this repetitive orientation serve? It focuses the force of contraction; all sarcomeres in all fibers of a muscle work together and pull a bone in the same direction.

SLIDING-FILAMENT MODEL FOR CONTRACTION

The sliding-filament model explains how interactions between thick and thin filaments bring about muscle contraction. *Filaments do not change length and myosin filaments do not change position.* Myosin heads bind to actin filaments and slide them toward the center of a sarcomere. As the actin filaments are pulled inward, Z bands attached to them are drawn closer together, and the sarcomere shortens (Figure 33.12*a,b*).

Part of the myosin head can bind ATP and break it into ADP and phosphate. The binding readies myosin for action (Figure 33.12*c*). It's like what happens when you pull back the rubber band of a slingshot. Muscle contraction occurs when nervous signals cause a rise in calcium level, allowing myosin heads to form cross-bridges with actin filaments (Figure 33.12*d*). ADP and phosphate bound to myosin earlier are now released, and each myosin head tilts—like a slingshot snapping back to its unstretched position. As a myosin head tilts, it pulls an actin filament toward the sarcomere center, drawing the Z bands inward (Figure 33.12*e,f*).

Binding of a new ATP frees the myosin head from actin, and the head goes back to its original position (Figure 33.12*g*). The head attaches to another binding site on the actin, tilts in another stroke, and so on. The contraction of a sarcomere occurs when hundreds of myosin heads perform a series of repeated strokes all along the length of the actin filaments.

Sarcomeres are the basic units of contraction in skeletal muscle. They follow one another in myofibrils, which run parallel with the muscle fibers that house them. In turn, muscle fibers run parallel with the whole muscle.

Collectively, this parallel orientation of the components of a skeletal muscle focuses contractile force on a bone that is to be moved.

By energy-driven interactions between myosin and actin filaments, the many sarcomeres of a muscle cell shorten and bring about the muscle's contraction.

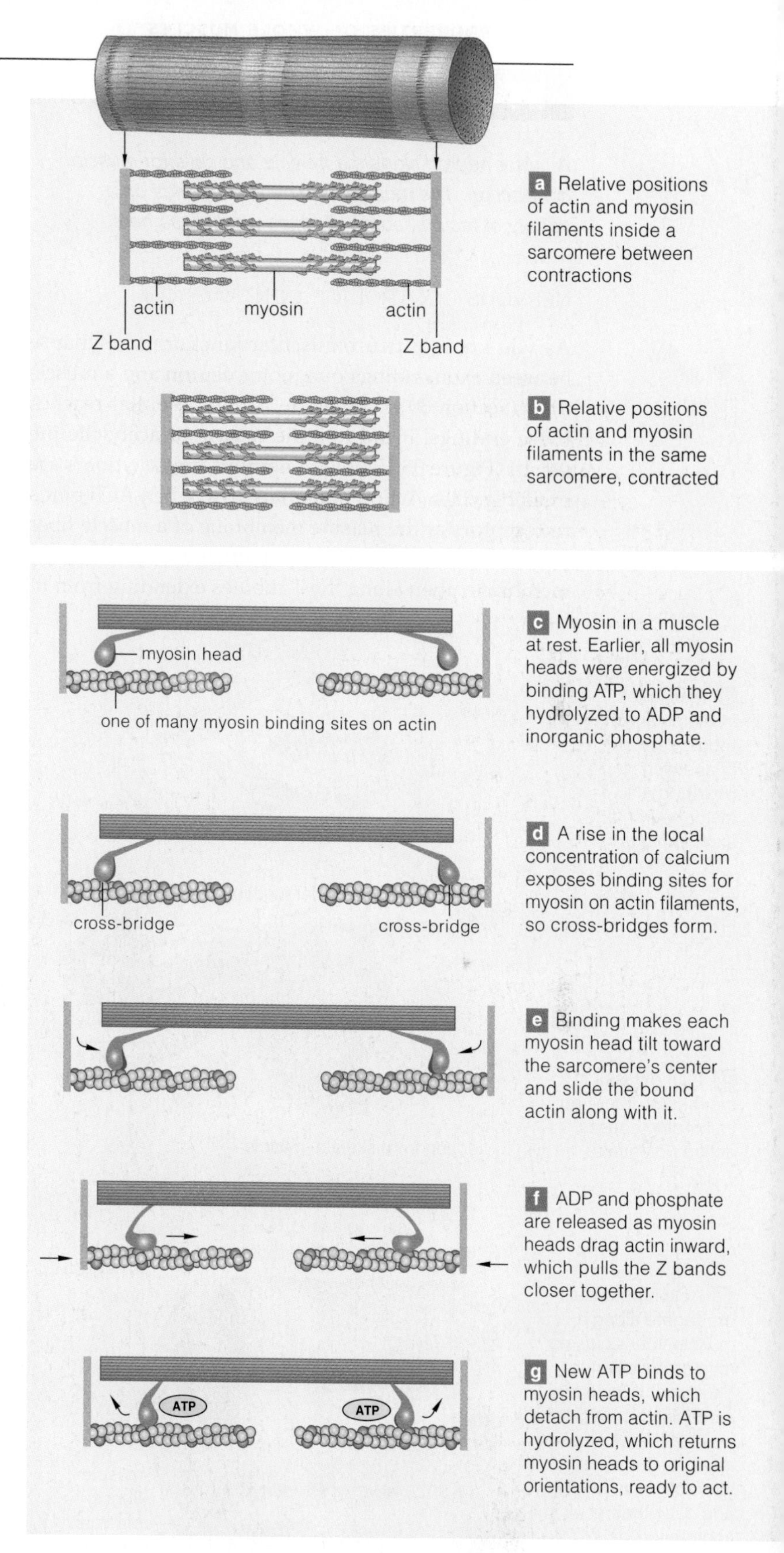

Figure 33.12 **Animated!** A sliding-filament model for the contraction of a sarcomere in skeletal muscle. (**a**,**b**) Organized, overlapping arrays of actin and myosin filaments interact and reduce each sarcomere's width. (**c**–**g**) For clarity, we show the action of two myosin heads only. Each head binds repeatedly to an actin filament and slides it toward the center of the sarcomere. The collective action makes the sarcomere shorten (contract).

33.5 From Signal to Responses

A motor neuron signals a muscle and calls for muscle contraction. The muscle's response depends on a variety of factors, some of which we discuss here.

NERVOUS CONTROL OF CONTRACTION

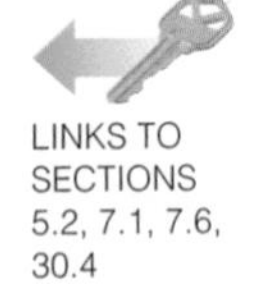

LINKS TO SECTIONS 5.2, 7.1, 7.6, 30.4

As you know, a neuromuscular junction is a synapse between axon endings of a motor neuron and a muscle fiber (Section 30.4). When an action potential reaches those endings, it induces the release of acetylcholine (ACh) (Figure 33.13*a*). Like neurons, muscle fibers are excitable cells. Action potentials arise when ACh binds to receptors at the plasma membrane of a muscle fiber (Figure 33.13*b*). The action potentials travel along the membrane, then along the T tubules extending from it.

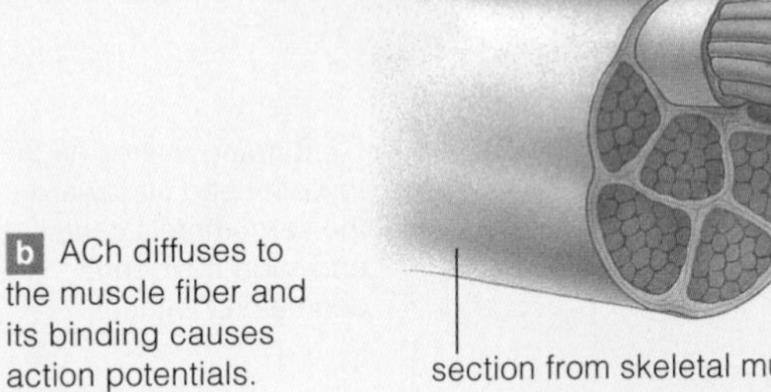

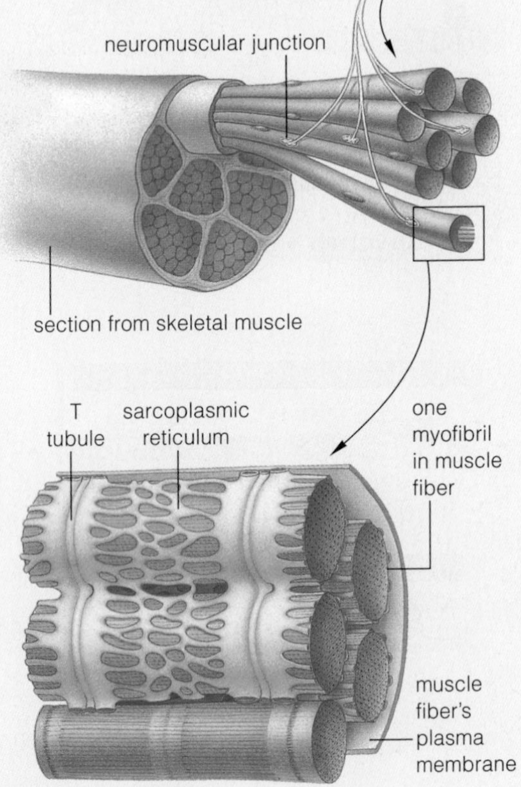

Figure 33.13 Animated! Pathway by which the nervous system controls skeletal muscle contraction. A muscle fiber's plasma membrane encloses many individual myofibrils. Tubelike extensions of the membrane connect with part of the sarcoplasmic reticulum, which wraps lacily around the myofibrils.

The T tubules lie next to the sarcoplasmic reticulum: a system of membranous chambers that wraps around the myofibrils. The sarcoplasmic reticulum stores and releases calcium ions (Figure 33.13*c*).

Arrival of action potentials causes the sarcoplasmic reticulum to release calcium ions. Sites on actin where myosin heads can bind are blocked in resting muscle, but an influx of calcium ions clears these binding sites. Rising calcium concentration allows actin and myosin to interact, and muscle contraction gets underway.

When contraction ends, membrane proteins actively transport the calcium ions back into the sarcoplasmic reticulum. The muscle fiber is ready for another signal.

MOTOR UNITS AND MUSCLE TENSION

A motor neuron has many axon endings that synapse on different fibers in a muscle. One motor neuron and all of the muscle fibers it synapses with constitute one motor unit. Briefly stimulate a motor neuron, and the fibers of its motor unit contract for a few milliseconds. That contraction is a muscle twitch (Figure 33.14*a*).

A new stimulus that hits before a response ends makes the fibers twitch again. Repeatedly stimulating a motor unit during a short interval makes all of the twitches run together and causes tetanus, a sustained contraction (Figure 33.14*c*). Force generated by tetanus is three or four times the force of a single twitch.

Muscle tension is the mechanical force exerted by a muscle. The more motor units stimulated, the greater the muscle tension. Opposing muscle tension is a load, either the weight of an object or gravity's pull on the muscle. Only when muscle tension exceeds opposing forces does a stimulated muscle shorten. *Isotonically* contracting muscles shorten and move some load, as when you lift a light object. *Isometrically* contracting muscles develop tension, yet do not shorten, as when you try to lift an object but fail because it is too heavy.

EXERCISE AND AGING

When unrelenting stimulation keeps a skeletal muscle in tetanus, muscle fatigue follows. Muscle fatigue is a decrease in a muscle's capacity to generate force; the muscle tension declines despite ongoing stimulation. After a few minutes of rest, the fatigued muscle will contract again in response to stimulation.

In humans, all muscle fibers form before birth and exercise does not stimulate the addition of new ones. Aerobic exercise—low intensity, but long duration—makes muscles more resistant to fatigue by increasing their blood supply and number of mitochondria.

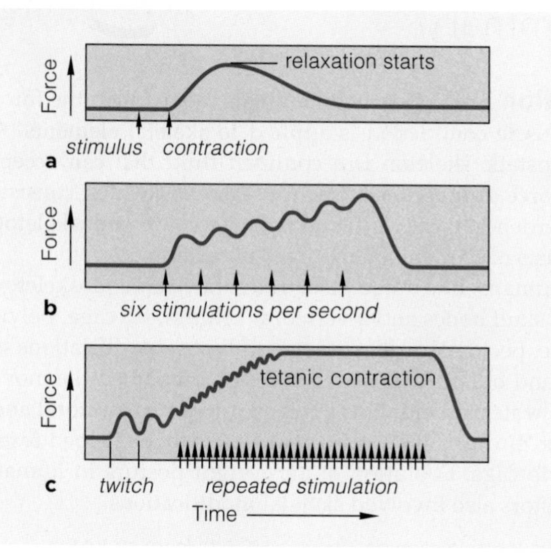

Figure 33.14 Animated! Recordings of twitches in a muscle exposed to artificial stimulation. (**a**) A single twitch. (**b**) Summation of twitches after six stimulations per second. (**c**) Tetanic contraction after twenty stimulations per second.

Brief, intense exercise such as weight lifting causes actin and myosin synthesis. This helps a muscle exert more tension but does not improve endurance.

As people age, the number and size of their muscle fibers decline. The tendons that attach muscle to bone stiffen and are more likely to tear. Older people may exercise intensely for long periods, but their muscle mass no longer can increase as much. Even so, aerobic exercise does improve blood circulation, and modest strength training can slow the loss of muscle tissue. Exercise can also help prevent or lift depression. It is good for the mind as well as the muscles.

ENERGY FOR CONTRACTION

The availability of ATP affects whether and how long a muscle can contract. ATP is the first energy source a muscle uses, but its stores of ATP are limited. Once its ATP gets used up, the muscle turns to stored creatine phosphate. ATP forms when ADP accepts a phosphate from creatine phosphate (Figure 33.15). The transfers keep the muscle going until other pathways increase ATP output. This is why taking creatine supplements, as described in the introduction, may be helpful.

Aerobic respiration yields most of the ATP during prolonged, moderate activity. Glucose derived from stored glycogen fuels five to ten minutes of activity, then glucose and fatty acids that the blood delivers to muscle fibers are burned. Fatty acids are the main fuel for activities that last more than half an hour.

Not all fuel is burned aerobically. Even in resting muscle, some pyruvate is converted to lactate by the fermentation pathway. Lactate production rises with exercise. Remember, this pathway does not yield much ATP, but it can operate even in the absence of oxygen.

MUSCULAR DYSTROPHIES

Muscular dystrophies are a class of genetic disorders in which skeletal muscles progressively weaken. With Duchenne muscular dystrophy, symptoms appear in childhood. Myotonic muscular dystrophy is the most common kind in adults.

A single mutant gene on the X chromosome causes Duchenne muscular dystrophy. This gene encodes a protein (dystrophin) found in the plasma membrane of muscle fibers. A mutant form of dystrophin allows foreign material to enter a muscle fiber, and the fiber deteriorates. Muscular dystrophy arises in about 1 in 3,500 males. The affected individuals usually require a wheelchair in their teens and die in their twenties. There is no cure, but treatments that use stem cells or gene therapy are under investigation.

In muscle fibers, signals from motor neurons initiate action potentials that cause the release of calcium ions from storage. Contraction cannot proceed without the release.

A muscle's response to stimulation varies in strength and duration. Recent stimulation, prior exercise, and age are factors that influence the response.

Three metabolic pathways supply energy for contraction.

Muscular dystrophies are disorders in which muscles are progressively and irreversibly weakened.

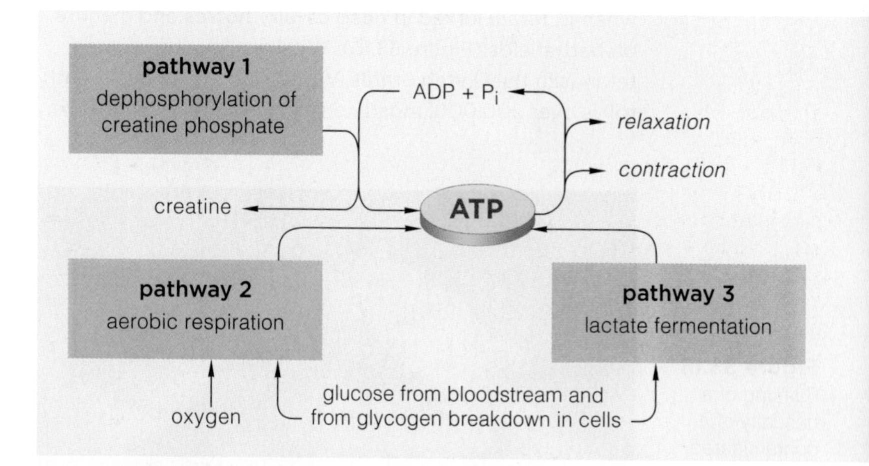

Figure 33.15 Animated! Three metabolic pathways by which muscles obtain the ATP molecules that fuel their contraction.

33.6 Oh, *Clostridium!*

FOCUS ON HEALTH

We conclude this chapter with a look at two bacterial infections that interfere with how muscles work.

LINKS TO SECTIONS 19.2, 30.4, 30.5

Section 19.2 introduced you to *Clostridium botulinum*. This anaerobic soil bacterium can cause disease in humans. Food stored in unsterilized cans or jars may contain its endospores. When the spores germinate, they start making botulinum, an odorless toxin.

Botulinum enters motor neurons and stops the release of acetylcholine (ACh). Muscles cannot contract without this neurotransmitter, so they become limp and paralyzed. These are symptoms of botulism, a type of deadly food poisoning. Affected people must be treated quickly with antitoxin before cardiac muscle and the skeletal muscles with roles in breathing are affected.

A related bacterium, *C. tetani*, lives in the gut of cattle, horses, and other grazing animals, and often people. Its endospores can last for years in soil—manure-rich soils especially—as long as sunlight and oxygen do not reach them. The endospores resist strong disinfectants, heat, and boiling water. When they enter the body through a deep puncture or cut, they can germinate in necrotic (dead) tissues. Bacterial cells do not spread out from anaerobic, dead tissues, but they produce a toxin that blood or nerves deliver to the spinal cord and brain.

In the spinal cord, the bacterial toxin blocks release of neurotransmitters that exert inhibitory control over motor neurons. Without the controls, nothing puts the brakes on signals to contract, so symptoms of the disease known as tetanus begin.

After four to ten days, overstimulated muscles stiffen and cannot be released from contraction; they go into spasm. Prolonged, spastic paralysis follows. Fists and the jaws may stay clenched; lockjaw is a common name for the disease. The backbone may become locked in an abnormally arching curve. When respiratory and cardiac muscles become paralyzed, death nearly always follows.

Vaccines were not available for soldiers of early wars, when *C. tetani* lurked in dead cavalry horses and manure on battlefields (Figure 33.16). Vaccines have eradicated tetanus in the United States. Worldwide, the annual death toll is over 200,000, mostly due to unsanitary childbirths.

Figure 33.16 Painting of a casualty of a contaminated battle wound as he lay dying of tetanus in a military hospital.

Summary

www.thomsonedu.com ThomsonNOW!

Section 33.1 Nearly all animals move when the force of muscle contraction is applied to skeletal elements. A hydrostatic skeleton is a confined fluid that can accept the force of muscle contraction. An exoskeleton consists of hardened parts at the body surface. An endoskeleton consists of hardened parts inside the body.

Humans, like other vertebrates, have an endoskeleton with skull bones and a vertebral column, rib cage, pelvic girdle, pectoral girdle, and paired limbs. Modifications to fins and other skeletal structures accompanied the move from water to land. Fins evolved into limbs, pectoral and pelvic girdles got sturdier, and a breastbone helped form the rib cage. Evolution of an upright posture in human ancestors also involved skeletal modifications.

- *Use the animation on ThomsonNOW to learn about the skeletal systems of invertebrates and humans.*

Section 33.2 Bones are organs rich in collagen and minerals. Bones function in mineral storage, movement, protection and support of internal organs, and blood cell formation. In human embryos, the bones develop from a cartilage model. Hormonally regulated bone remodeling helps maintain blood levels of mineral ions. Ligaments and cartilage hold bones together at joints.

- *Use the animation on ThomsonNOW to study the structure of a human femur.*

Section 33.3 A muscle fiber is a long, cylindrical cell with multiple nuclei. It forms from groups of embryonic cells that fuse before they differentiate and mature. In a skeletal muscle, muscle fibers are bundled inside a dense connective tissue sheath that extends beyond the fibers. Tendons are extensions of this sheath. They attach most skeletal muscles to bones.

When skeletal muscles contract, they transmit force to bones and move them. Some muscles work together and others work as opposing pairs.

- *Use the animation on ThomsonNOW to review the location and function of human skeletal muscles.*

Section 33.4 The internal organization of a skeletal muscle promotes a strong, directional contraction. Many myofibrils make up a skeletal muscle fiber. A myofibril consists of sarcomeres, basic units of muscle contraction, lined up along its length. Each sarcomere has parallel arrays of actin and myosin filaments. ATP-driven sliding of actin filaments past myosin filaments shortens the sarcomere. Shortening of all sarcomeres in all myofibrils of all muscle fibers brings about muscle contraction.

- *Use the animation on ThomsonNOW to explore muscle structure and observe muscle contraction.*

Section 33.5 Signals from motor neurons result in action potentials in muscle fibers that cause the release of calcium stored in the sarcoplasmic reticulum. Calcium floods out and allows actin and myosin heads to interact and muscle contraction occurs.

A motor neuron and all the muscle fibers it controls are one motor unit. Repeated stimulation of a motor unit causes a sustained contraction, or tetanus. A muscle will shorten when muscle tension exceeds an opposing load.

Muscle fibers produce the ATP needed for contraction by way of three pathways: dephosphorylation of creatine phosphate, aerobic respiration, and lactate fermentation. Age and muscular dystrophy can weaken muscles.

- *Use the animation on ThomsonNOW to observe how the nervous system controls muscle contraction, the effects of stimulation on a muscle cell, and how a muscle gets the energy needed for contraction.*

Section 33.6 Nervous system signals that stimulate contraction can be disrupted by toxins, such as those that cause deadly food poisoning and the disease tetanus.

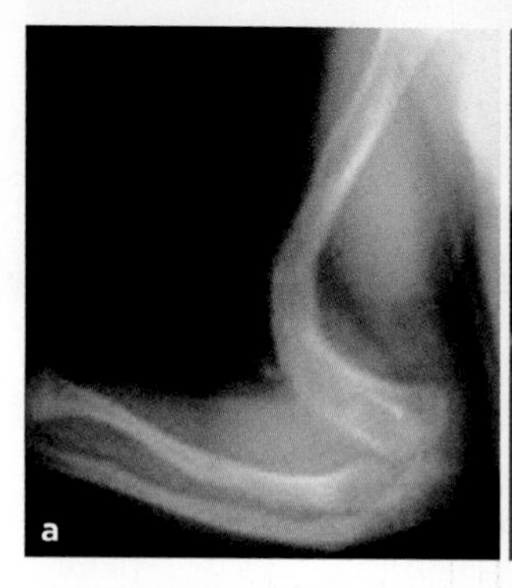

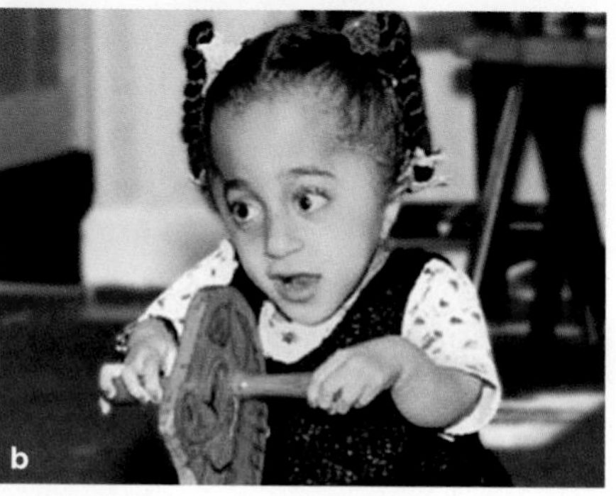

Figure 33.17 (**a**) X-ray of an arm bone deformed by osteogenesis imperfecta (OI). (**b**) Tiffany is affected by OI. She was born with multiple fractures in her arms and legs. By age six, she had undergone surgery to correct more than 200 bone fractures and to place steel rods in her legs. Every three months she receives intravenous infusions of an experimental drug that may help strengthen her bones.

Self-Quiz

Answers in Appendix III

1. A hydrostatic skeleton consists of ______.
   a. a fluid in an enclosed space
   b. hardened plates at the surface of a body
   c. internal hard parts
   d. none of the above

2. Bones are ______.
   a. mineral reservoirs
   b. skeletal muscle's partners
   c. sites where blood cells form (some bones only)
   d. all of the above

3. Bones move when ______ muscles contract.
   a. cardiac
   b. skeletal
   c. smooth
   d. all of the above

4. A ligament connects ______.
   a. bones at a joint
   b. a muscle to a bone
   c. a muscle to a tendon
   d. a tendon to bone

5. Bone breakdown is stimulated by ______.
   a. parathyroid hormone
   b. estrogen
   c. vitamin D
   d. cortisol and calcitonin

6. The ______ is the basic unit of contraction.
   a. osteoblast
   b. sarcomere
   c. muscle fiber
   d. myosin filament

7. In sarcomeres, phosphate-group transfers from ATP activate ______.
   a. actin b. myosin c. both d. neither

8. A sarcomere shortens when ______.
   a. thick filaments shorten
   b. thin filaments shorten
   c. both thick and thin filaments shorten
   d. none of the above

9. ATP for muscle contraction can be formed by ______.
   a. aerobic respiration
   b. lactate fermentation
   c. creatine phosphate breakdown
   d. all of the above

10. A motor unit is ______.
    a. a muscle and the bone it moves
    b. two muscles that work in opposition
    c. the amount a muscle shortens during contraction
    d. a motor neuron and the muscle fibers it controls
    e. none of the above

11. Match the words with their defining feature.

___ osteoblast	a. stores and releases calcium
___ muscle twitch	b. all in the hands
___ muscle tension	c. blood cell production
___ joint	d. decline in tension
___ myosin	e. bone-forming cell
___ red marrow	f. motor unit response
___ metacarpals	g. force exerted by cross-bridges
___ myofibrils	h. area of contact between bones
___ muscle fatigue	i. muscle fiber's threadlike parts
___ sarcoplasmic reticulum	j. actin's partner

- *Visit ThomsonNOW for additional questions.*

Critical Thinking

1. Osteogenesis imperfecta (OI) is a genetic disorder that arises from a mutant form of collagen. As bones develop, this protein forms a scaffold for deposition of mineralized bone tissue. The scaffold forms improperly in children with OI, who end up with fragile, brittle, easily broken bones (Figure 33.17). OI is rare; most doctors never come across it. Parents are often accused of child abuse when they take an OI child with multiple bone breaks to an emergency room. Prepare a handout about OI that could be used to alert emergency room staffs to the condition. You may wish to start with resources on the Osteogenesis Imperfecta Foundation web site at www.oif.org.

2. Lydia is training for a marathon. While training, she plans to take creatine supplements because she heard that they give athletes extra energy. She asks you whether you think this is a good idea. What is your response?

3. Compared to most people, long-distance runners have far more mitochondria in skeletal muscles. In sprinters, skeletal muscle fibers have more of the enzymes required for glycolysis but not as many mitochondria. Suggest why.

4. Zachary's younger brother Noah had Duchenne muscular dystrophy and died at the age of 16. Zachary is now 26 years old, healthy, and planning to start a family of his own. However, he worries that his sons might be at a high risk for muscular dystrophy. His wife's family has no history of this genetic disorder. Review Section 11.4 and decide whether Zachary's concerns are well founded.

34 CIRCULATION

IMPACTS, ISSUES And Then My Heart Stood Still

In the nineteenth century, physiologist Augustus Waller put the paws of his pet bulldog Jimmie in bowls of salty water. Electrical signals emitted by Jimmie's heart were transmitted to his skin, then to the salt water, and then picked up by a simple recording device. The device scratched out the world's first graph of a heartbeat—the very first electrocardiogram (Figure 34.1).

Fast-forward to the present. A graph of your heart's normal activity would look much the same as Jimmie's. The pattern emerged a few weeks after you started to grow from a fertilized egg and develop into a multicelled embryo. Early on, some embryonic cells differentiated into cardiac muscle cells and started to contract on their own. These spontaneously firing cells set the pace for all other cardiac muscle cells. They function as a natural pacemaker for the beating heart. If all goes well, those contracting cells will make your heart beat about seventy times per minute until the day you die.

In cases of sudden cardiac arrest, the heart's electrical signaling becomes disrupted, the heart stops beating, and blood flow through the circulatory system halts. In the United States alone, cases exceed 250,000 per year.

Tammy Higgins became one of those cases before she was thirty years old. Just after leaving a church service, she collapsed and stopped breathing. Her husband, trained as a lifeguard, could not find her pulse. He knew the flow of oxygen-rich blood to her brain had stopped and had to resume fast. He started cardiopulmonary resuscitation—CPR. He alternated mouth-to-mouth respiration with repetitive compression of her chest. He kept Tammy alive until an ambulance arrived.

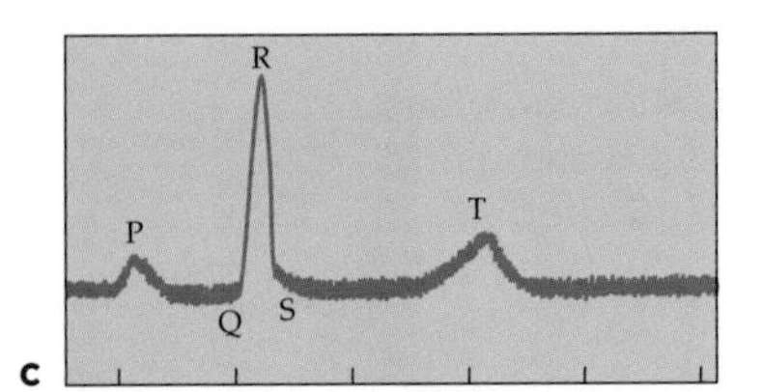

See the video! **Figure 34.1** A bit of history in the making. (**a**) Jimmie the bulldog standing in saltwater that conducted electrical signals emitted by his heart to wires of a recording device. (**b**) Physiologist Augustus Waller relaxes with Jimmie after the experiment, which yielded a recording of Jimmie's heartbeat (**c**). The labels P, QRS complex, and T mark three waves of electrical activity caused by the spread of action potentials across cardiac muscle, then recovery of the normal resting potential. *Facing page:* (**d**) Defibrillator used to shock a heart into beating after cardiac arrest.

How would you vote? Should public high schools in your state require all students to take a course in CPR? Is such a course worth diverting time and resources from the basic curriculum? See ThomsonNOW for details, then vote online.

The emergency crew used a portable defibrillator. This device often jump-starts the heart into beating again by delivering an electric shock to the chest (Figure 34.1*d*). Tammy was lucky; her brain escaped damage and she did not die. In the hospital, she found out that her heartbeat is abnormal. Surgeons implanted a small defibrillator in her chest wall. The implant continually monitors her heart and shocks it when it stops beating. Since then, the implant has saved her twice, once during a pregnancy.

The chance of surviving sudden cardiac arrest rises by 50 percent when CPR starts within four to six minutes. For every minute that passes without defibrillation, the chance of survival decreases by about 7 to 10 percent. Automated external defibrillators (AEDs) are about the size of a laptop computer. Many senior centers, shopping malls, hotels, and other public places have them on hand. An AED's electronic voice commands trained bystanders to check for a heartbeat and, if required, shock the heart. The American Heart Association, American Red Cross, community groups, and adult education programs offer CPR and AED training to the public. We have come a long way from Waller's experiments with Jimmie to checking out the heart and helping hearts that falter.

With this bit of history, we turn to the structure and function of the circulatory system. In humans, the durable muscular pump of this organ system generates pressure that drives blood through two elaborate loops of blood vessels. As you will see, blood functions in homeostasis, and the lymphatic system supports many of its functions.

d

Key Concepts

OVERVIEW OF CIRCULATORY SYSTEMS

Many animals have either an open or a closed circulatory system that transports substances to and from all body tissues. **Section 34.1**

BLOOD COMPOSITION AND FUNCTION

Vertebrate blood is a fluid connective tissue. It consists of red blood cells, white blood cells, platelets, and diverse substances dissolved in plasma, the transport medium. Red blood cells function in gas exchange; white blood cells and platelets help defend the tissues. **Sections 34.2–34.4**

THE HUMAN HEART AND TWO FLOW CIRCUITS

The human heart has four chambers. Blood flows into its two atria and then into two ventricles, which pump it into two separate circuits of blood vessels. One circuit extends through all body regions, the other through lung tissue only. Both circuits loop back to the heart. **Sections 34.5, 34.6**

BLOOD VESSEL STRUCTURE AND FUNCTION

The heart pumps blood rhythmically, on its own. Blood pressure is highest in the heart's ventricles, drops as it flows through arteries, and is lowest in the atria. Adjustments at arterioles regulate how much blood volume is distributed to tissues. Exchange of gases, wastes, and nutrients between the blood and tissues takes place at capillaries. **Sections 34.7, 34.8**

WHEN THE SYSTEM BREAKS DOWN

Ruptured or clogged blood vessels or abnormal heart rhythms cause problems. Some problems have a genetic basis; most are related to age or life-styles. **Section 34.9**

LINKS WITH THE LYMPHATIC SYSTEM

A lymph vascular system delivers excess fluid that collects in tissues to the blood. Lymphoid organs cleanse blood of infectious agents and other threats to health. **Section 34.10**

Links to Earlier Concepts

This chapter expands on earlier introductions to circulatory systems (Sections 23.1, 23.6, 24.1, 24.4, 24.9), cardiac muscle (29.3), and muscle contraction (33.4). You will draw upon your knowledge of hemoglobin (3.5), diffusion (5.6), osmosis (5.8), endocytosis (5.9), action potentials (30.3), and cell junctions (29.1). You will revisit ABO blood typing (10.4). You will be invited to reflect on sickle-cell anemia (3.5, 17.5) and on connections between cardiovascular function and cholesterol in blood (3.3).

34.1 The Nature of Blood Circulation

Imagine an earthquake closing off your neighborhood's highway. Grocery trucks cannot enter; wastes cannot be taken away. Food supplies dwindle and the garbage piles up. Similar problems would occur if something were to disrupt the flow along the body's circulatory highways.

FROM STRUCTURE TO FUNCTION

LINKS TO SECTIONS 5.6, 23.1, 23.6, 24.1, 24.4, 24.9, 25.1

A circulatory system moves substances into and out of cellular neighborhoods. Blood, its transport medium, typically flows inside tubular vessels under pressure generated by a heart, a muscular pump. Blood makes exchanges with interstitial fluid—fluid in tissue spaces between cells—which exchanges substances with cells.

Blood and interstitial fluid are the body's internal environment. The body works to keep the composition, volume, and temperature of that environment within ranges that cells can tolerate (Section 25.1).

Structurally, there are two main kinds of circulatory systems. Arthropods and most mollusks have an *open* circulatory system. Their blood moves through hearts and large vessels but also mixes with interstitial fluid (Figure 34.2*a*). Annelids and vertebrates have a *closed* circulatory system. Their blood remains inside a heart or blood vessel at all times (Figures 34.2*b* and 34.3).

In a closed system, the total volume of blood moves continually away from the heart, through vessels, and back to the heart. Blood moves fastest when confined to a few large vessels. It slows in blood capillaries, the vessels with the smallest diameter. This slowdown in capillaries gives the blood and interstitial fluid time to exchange substances by diffusion (Section 5.6).

Blood slows in capillaries not only because these vessels are small, but also because of their numbers. Your body has billions. Their collective cross-sectional area is much greater than that of the far fewer, larger vessels that deliver blood to them. When blood enters capillaries, its speed declines, as if a narrow river (the few larger vessels) were delivering water into a wide lake (the many capillaries). Figure 34.3*d* illustrates the concept. The velocity picks up after blood flows out of capillary beds into the larger diameter vessels that return it to the heart. Similarly, water picks up speed when it flows from a wide lake into a narrow river.

EVOLUTION OF CIRCULATION IN VERTEBRATES

All vertebrates have a closed circulatory system, but fishes, amphibians, birds, and mammals differ in their pumps and plumbing. These differences evolved over

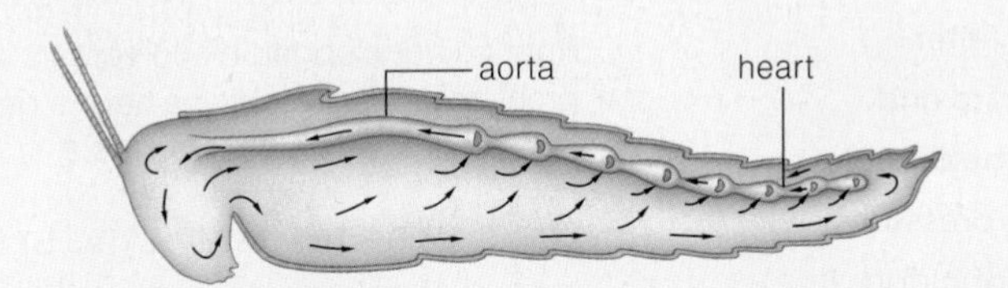

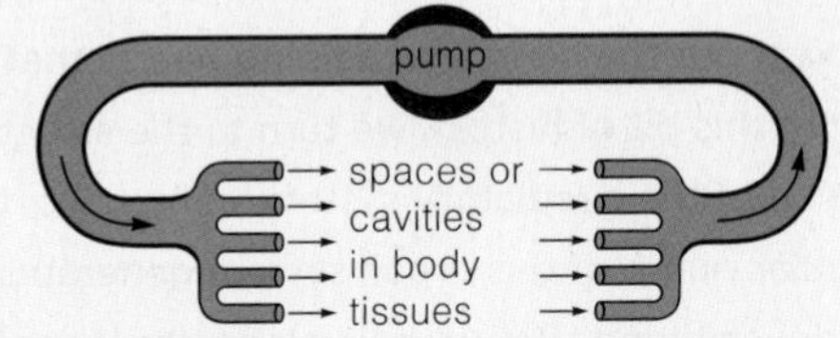

a In a grasshopper's open system, a heart (not like yours) pumps blood through a vessel, a type of aorta. From there, blood moves into tissue spaces, mingles with interstitial fluid, then reenters the heart at openings in the heart wall.

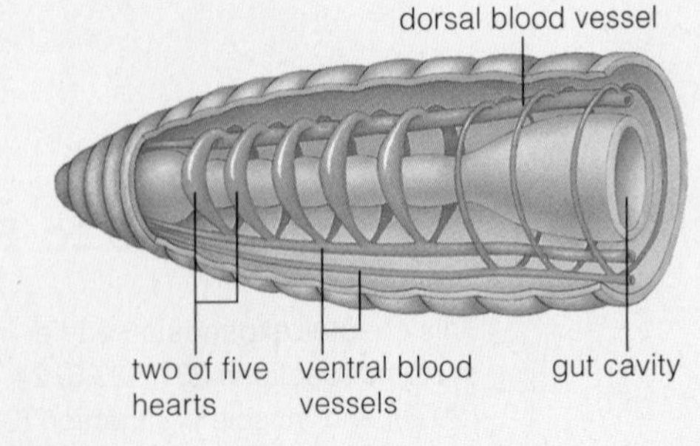

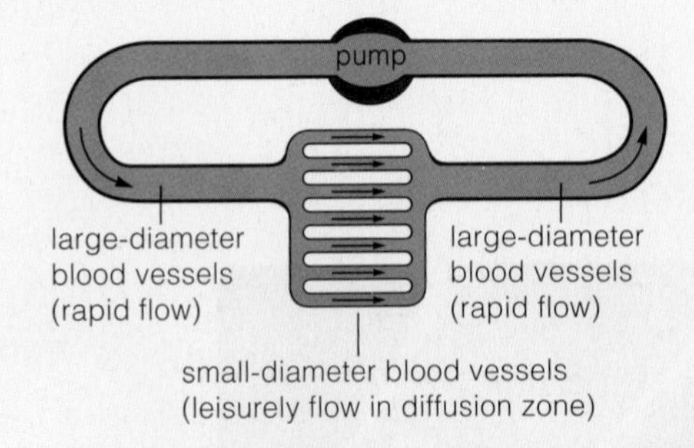

b The closed system of an earthworm confines blood inside pairs of muscular hearts near the head end and inside many blood vessels.

Figure 34.2 **Animated!** Comparison of open and closed circulatory systems.

hundreds of millions of years after some vertebrates left the water for land. The earliest vertebrates, recall, had gills. Like other respiratory structures, gills have a thin, moist surface, which oxygen and carbon dioxide diffuse across. In time, *internally moistened* sacs called lungs evolved and supported the move to dry land. Other modifications helped blood flow more quickly in a loop between the heart and lungs (Section 24.4).

In most fishes, blood flows in one circuit (Figure 34.3*a*). The contractile force of a two-chambered heart drives it through a capillary bed inside each gill. From there, blood flows into a large vessel, then through capillary beds in body tissues and organs, and back to the heart. The blood is not under much fluid pressure when it leaves the gill capillaries, so it moves slowly through the single circuit back to the heart.

In amphibians, the heart became divided into three chambers, with two atria emptying into one ventricle. Oxygenated blood flowed from the lungs to the heart in one circuit, and a forceful contraction pumped it through the rest of the body in a second circuit. Still, the oxygenated blood and oxygen-poor blood mixed a bit in the ventricle (Figure 34.3*b*).

In birds and mammals, the heart has fully separate right and left halves, each with two chambers, and it pumps blood in two separate circuits (Figure 34.3*c*). In the pulmonary circuit, oxygen-poor, carbon dioxide–rich blood flows from the *right* half of the heart to the lungs. There, blood picks up oxygen, gives up carbon dioxide, and flows into the left half of the heart.

In the longer systemic circuit, the heart's *left* half pumps oxygenated blood to tissues where oxygen is used and carbon dioxide forms in aerobic respiration. Blood gives up oxygen and picks up carbon dioxide at tissues, then flows to the heart's right half.

With two fully separate circuits, blood pressure can be regulated in each one. Strong contraction of the heart's left ventricle can propel blood fast through the long systemic circuit. Because its right ventricle contracts less strongly, delicate lung capillaries are protected; they would be blown apart under higher pressure.

Circulatory systems move substances to and from the interstitial fluid that surrounds the body's living cells.

Blood and interstitial fluid mix in open circulatory systems. By contrast, closed systems confine blood inside a heart and blood vessels. Substances are exchanged across the wall of capillaries, the vessels with the smallest diameter.

In fishes and amphibians, oxygen-poor and oxygenated blood mix a bit. In birds and mammals, blood flows in two separate circuits, and fluid pressure is optimized in each.

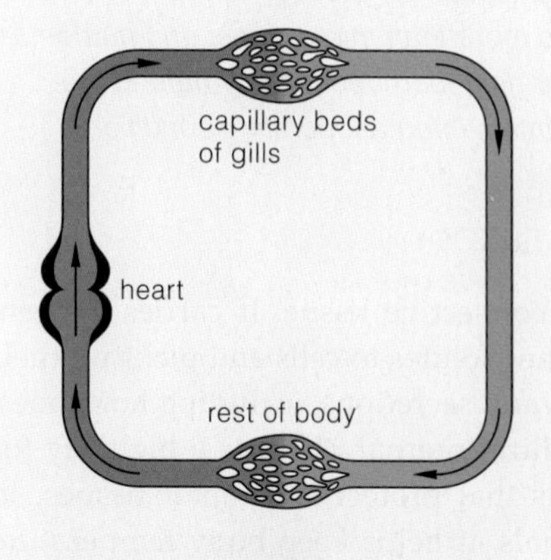

a In fishes, the heart has two chambers: one atrium and one ventricle. Blood flows through one circuit. It picks up oxygen in the capillary beds of the gills, and delivers it to capillary beds in all body tissues. Oxygen-poor blood then returns to the heart.

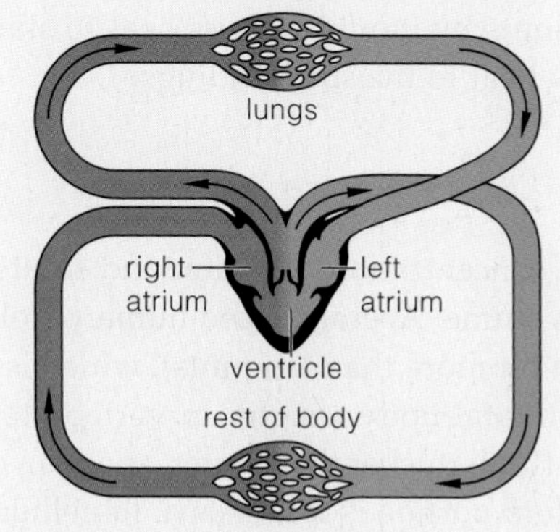

b In amphibians, the heart has three chambers: two atria and one ventricle. Blood flows along two partially separated circuits. The force of one contraction pumps blood from the heart to the lungs and back. The force of a second contraction pumps blood from the heart to all body tissues and back to the heart.

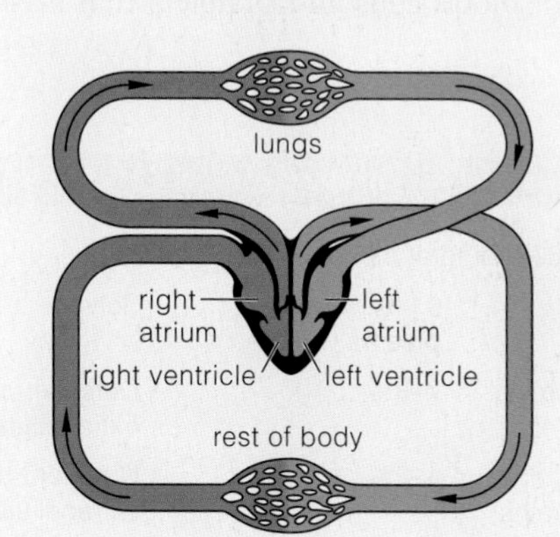

c In birds and mammals, the heart has four chambers: two atria and two ventricles. The blood flows through two fully separated circuits. In one circuit, blood flows from the heart to the lungs and back. In the second circuit, blood flows from the heart to all body tissues and back.

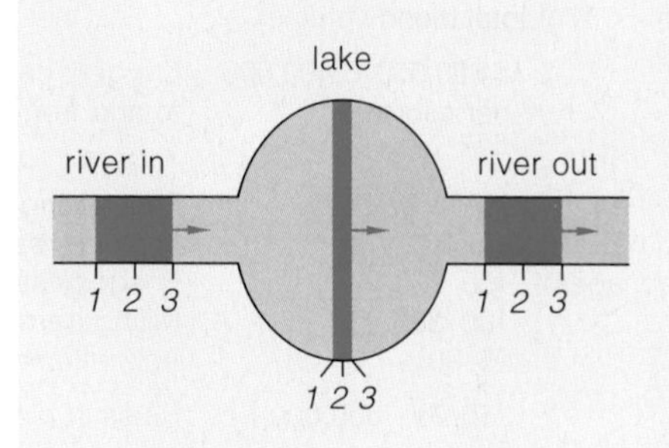

d Why flow slows in capillaries. Picture a volume of water in two fast rivers flowing into and out of a lake. The flow *rate* is constant, with an identical volume moving from points *1* to *3* in the same interval. However, flow *velocity* decreases in the lake. Why? The volume spreads out through a larger cross-sectional area and flows forward a shorter distance during the specified interval.

Figure 34.3 **Animated!** Comparison of flow circuits in the closed circulatory systems of fishes, amphibians, birds, and mammals.

34.2 Characteristics of Blood

Tumbling along in a vertebrate bloodstream are cells and substances that do more than move gases and nutrients to and from cellular neighborhoods. Many defend the body and help maintain internal operating conditions.

FUNCTIONS OF BLOOD

LINKS TO SECTIONS 3.5, 5.9, 29.2, 33.2

Blood is a fluid connective tissue. It carries oxygen, nutrients, and other solutes to cells and picks up their metabolic wastes and secretions, including hormones. Blood helps stabilize internal pH. It is a highway for cells and proteins that protect and repair tissues. In birds and mammals, it helps keep body temperature within tolerable limits by moving excess heat to skin, which can give up heat to the surroundings.

BLOOD VOLUME AND COMPOSITION

Body size and the concentrations of water and solutes dictate the blood volume. Average-sized humans hold about five liters (a bit more than ten pints), which is 6 to 8 percent of the total body weight. In vertebrates, blood is a viscous fluid, thicker than water, and slower flowing. Blood's fluid portion is the plasma. Its cellular portion consists of blood cells and platelets that arise from stem cells in bone marrow (Section 29.2). A stem cell is an unspecialized cell that retains a capacity for mitotic cell division. Some portion of its daughter cells divide and differentiate into specialized cell types.

Plasma About 50 to 60 percent of the blood's total volume is plasma (Figure 34.4). Plasma is 90 percent water. Besides being the transport medium for blood cells and platelets, it acts as a solvent for hundreds of different plasma proteins, other molecules, and ions. Some of the proteins transport lipids and fat-soluble vitamins, and some kinds function in blood clotting. Other solutes include glucose, oxygen, carbon dioxide, gaseous nitrogen, and a variety of lipids, amino acids, vitamins, and hormones.

Red Blood Cells Erythrocytes, or red blood cells, transport oxygen from lungs to aerobically respiring cells and carry carbon dioxide wastes from them. In all mammals, red blood cells lose their nucleus when they mature. Mature red blood cells are flexible disks with a depression at their center (Figure 34.5).

Most oxygen that diffuses into your blood binds to hemoglobin in red blood cells. You learned about this protein in Section 3.5. Stored hemoglobin fills about

Components	Amounts	Main Functions
Plasma Portion (50–60% of total blood volume)		
1. Water	91–92% of total plasma volume	Solvent
2. Plasma proteins (albumins, globulins, fibrinogen, etc.)	7–8%	Defense, clotting, lipid transport, extracellular fluid volume controls
3. Ions, sugars, lipids, amino acids, hormones, vitamins, dissolved gases, etc.	1–2%	Nutrition, defense, respiration, extracellular fluid volume controls, cell communication, etc.
Cellular Portion (40–50% of total blood volume)		
1. Red blood cells	4,600,000–5,400,000 per microliter	Oxygen, carbon dioxide transport to and from lungs
2. White blood cells:		
Neutrophils	3,000–6,750	Fast-acting phagocytosis
Lymphocytes	1,000–2,700	Immune responses
Monocytes (macrophages)	150–720	Phagocytosis
Eosinophils	100–380	Killing parasitic worms
Basophils	25–90	Anti-inflammatory secretions
3. Platelets	250,000–300,000	Roles in blood clotting

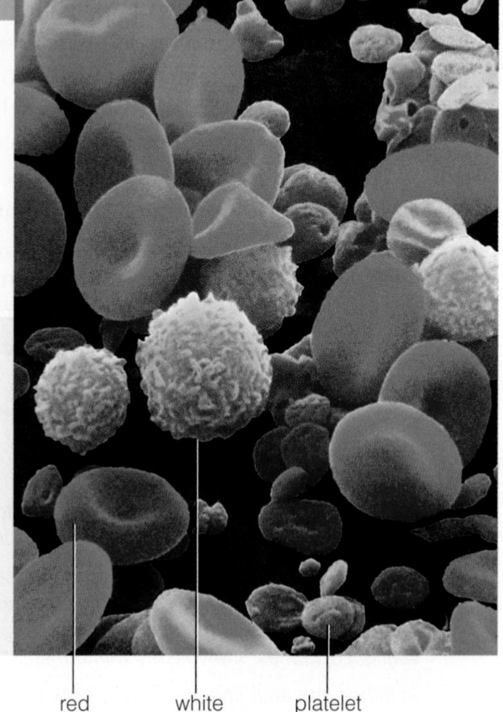

Figure 34.4 Typical components of human blood. The sketch of a test tube shows what happens when you prevent a blood sample from clotting. The sample separates into straw-colored plasma, which floats on a reddish cellular portion. Blood makes up 6 to 8 percent of total body weight. The scanning electron micrograph shows some cellular components.

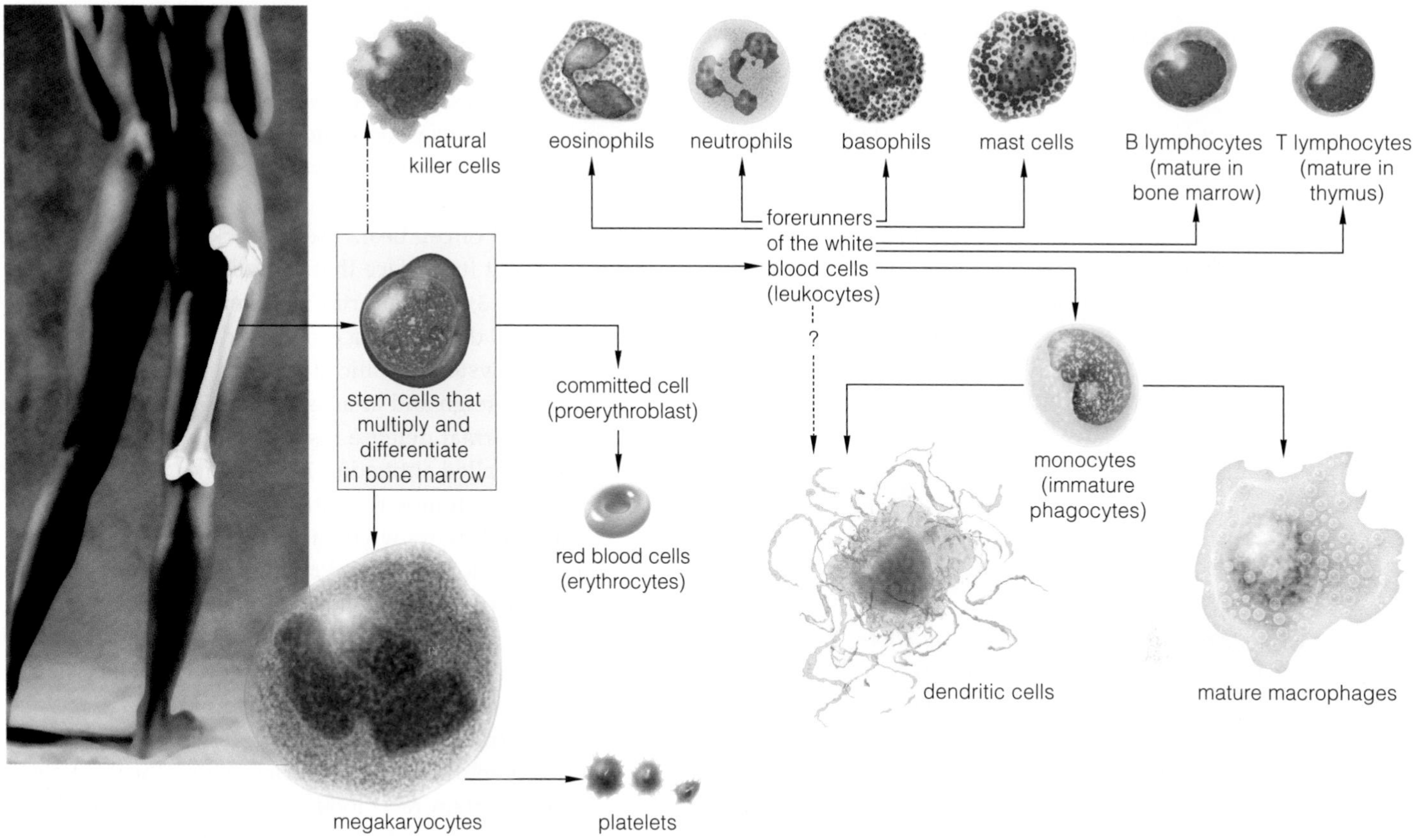

natural killer cells
eosinophils
neutrophils
basophils
mast cells
B lymphocytes (mature in bone marrow)
T lymphocytes (mature in thymus)
forerunners of the white blood cells (leukocytes)
?
stem cells that multiply and differentiate in bone marrow
committed cell (proerythroblast)
red blood cells (erythrocytes)
monocytes (immature phagocytes)
dendritic cells
mature macrophages
megakaryocytes
platelets

Figure 34.5 Cellular components of mammalian blood and where they originate.

98 percent of the interior of human red blood cells. It makes the cells and oxygenated blood appear bright red. Oxygen-poor blood is dark red, but it looks blue through blood vessel walls near the body surface.

In addition to hemoglobin, a mature red blood cell has enough stored sugars, RNAs, and other molecules to live about 120 days. In a healthy person, ongoing replacements keep red blood cell numbers at a fairly stable level. A cell count is a measure of the quantity of cells of one type per cubic millimeter of blood. Men tend to have a higher red blood cell count than women, who lose blood during menstruation.

White Blood Cells Leukocytes, or white blood cells, function in daily housekeeping activities and in defense. Some are phagocytes that engulf damaged, dead, or dying cells, as well as any materials that they chemically recognize as "nonself." Some types sound the alarm when viruses, bacteria, or other pathogens attack the body. As Sections 34.10 and 35.4 explain, armies of white blood cells congregate in some organs of the lymphatic system.

White blood cells differ in size, nuclear shape, and staining traits (Figure 34.5). Their numbers fluctuate with levels of activity and state of health. We consider their roles in detail in the next chapter, but here is a preview. The mast cells, neutrophils and basophils are phagocytes with roles in inflammation. Macrophages and dendritic cells are phagocytes that can stimulate immune responses to particular threats. Two types of lymphocytes, the B cells and T cells, are specialized for immune responses. Natural killer cells directly kill infected or ailing body cells.

Platelets Megakaryocytes are ten to fifteen times bigger than other blood cells that form in bone marrow. They break up into platelets, which are no more than membrane-wrapped fragments of cytoplasm. After a platelet forms, it lasts five to nine days. Hundreds of thousands of platelets are always circulating in blood. When activated, they release substances that initiate blood clotting.

Blood, a fluid connective tissue, functions as a transport medium and solvent. Diverse proteins, gases, sugars, and other substances are dissolved in it. Its cellular components include red blood cells, white blood cells, and platelets.

34.3 Blood Disorders

FOCUS ON HEALTH

The body quickly and continually replaces its blood cells for good reason. Besides aging and dying off regularly, blood cells are targets of some parasites and pathogens, and their formation or functioning may be disrupted as a result of rare mutations.

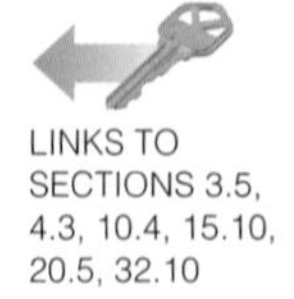

LINKS TO SECTIONS 3.5, 4.3, 10.4, 15.10, 20.5, 32.10

Red Blood Cell Disorders Too few red blood cells or deformed ones result in the disorders collectively called anemias. Oxygen delivery slows and metabolism falters. Shortness of breath, fatigue, and chills follow. Hemorrhagic anemias result from a sudden blood loss, as from a wound; chronic anemias result from low red blood cell production or a slight but persistent blood loss, as happens from a bleeding ulcer.

Bacteria and protozoans that replicate in blood cells cause some hemolytic anemias; they kill infected cells as they escape by lysis. Insufficient iron in the diet causes iron deficiency anemia; red blood cells cannot make enough of the iron-containing groups that carry oxygen in hemoglobin. Sickle-cell anemia arises from a mutation that alters hemoglobin and allows cells to change shape (Section 3.5). In thalassemias, mutations disrupt or stop synthesis of globin chains of hemoglobin. Too few red blood cells form. Those that do form are thin and fragile.

Polycythemia is a condition in which there are too many red blood cells. It makes blood more viscous and elevates blood pressure. So does blood doping, in which an athlete withdraws and stores red blood cells and injects them into the body just before competing in an event. Withdrawal triggers the formation of new red blood cells. When stored cells are injected, red cell count shoots up. The idea is to raise blood's oxygen-carrying capacity. Some athletes inject erythropoietin, a hormone that stimulates red cell production, for the same reason.

White Blood Cell Disorders Epstein-Barr virus causes infectious mononucleosis, a highly contagious disease in which too many monocytes and lymphocytes form. Most people recover after a few weeks of fatigue, muscle aches, low-grade fever, and a chronic sore throat.

Recovery is less certain for leukemias. These cancers originate in bone marrow and interfere with white blood cell formation (Figure 34.6). Radiation therapy or chemotherapy kills the cancer cells, but side effects can be severe. Symptom-free periods, or remissions, sometimes last months or years. Experimental gene therapies can put some leukemias into remission (Section 15.10).

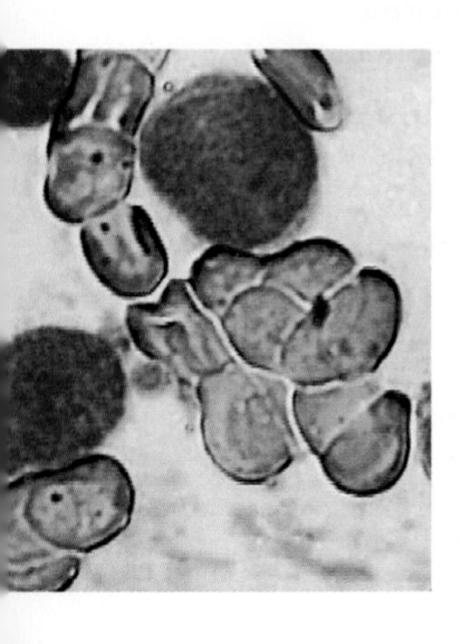

Figure 34.6 Light micrograph of a blood sample that is typical of chronic myelogenous leukemia. Abnormal, immature white blood cells are starting to crowd out normal cells.

34.4 Blood Typing

Blood from donors can be transfused into patients who are affected by severe blood loss or a blood disorder. Such blood transfusions cannot be hit-or-miss. Red blood cells of a potential donor and recipient must bear the same kinds of recognition proteins at their surface.

Each kind of cell bears "self" markers, or recognition proteins at its surface that provide its unique identity (Section 4.3). When certain markers on red blood cells of a blood donor and a recipient differ, the recipient's immune system chemically recognizes those donated cells as foreign and attacks them.

In a normal defense response called agglutination, proteins called antibodies bind foreign cells and make them form clumps that attract phagocytes. However, Figure 34.7 shows what happens when the blood from incompatible donors and recipients intermingles. Free antibodies circulating in the recipient's plasma bind to the nonself markers on the introduced cells and make them clump. There are so many foreign cells that the clumps clog small blood vessels and damage tissues. Without treatment, death may follow. The same thing can happen during a pregnancy if a mother and child differ in certain red blood cell markers. The mother's immune system will attack the child's cells.

Blood typing—the analysis of the surface markers on red blood cells—can help prevent such problems.

ABO BLOOD TYPING

ABO blood typing is a method of analyzing variations in one type of self marker on red blood cells. Section 10.4 describes the genetics of these variations. People who have one form of the marker have type A blood. Those with a different form have type B blood. When they have both forms of the marker on their red blood cells, their blood is type AB. When they have neither form, their blood is type O.

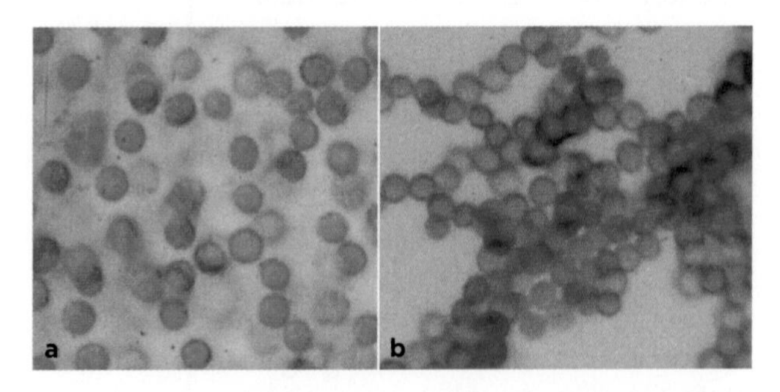

Figure 34.7 Light micrographs showing (**a**) an absence of agglutination in a mixture of two different yet compatible blood types and (**b**) agglutination in a mixture of incompatible types.

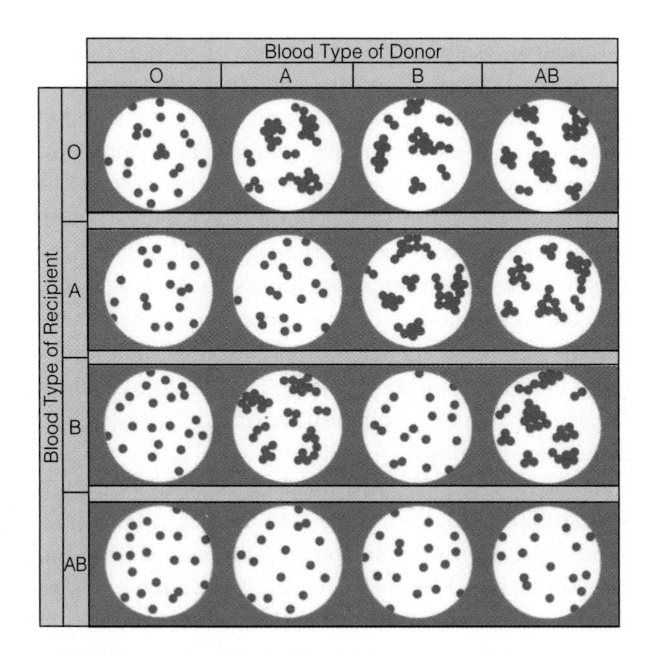

Figure 34.8 Animated! Responses in blood types A, B, AB, and O when mixed with samples of the same and different blood types.

Look at Figure 34.8. If you are blood type A, your immune system will recognize blood cells with type B markers as foreign. If type B, it will react against cells with type A markers. If you are blood type AB, your immune system treats both markers as "self," so you can receive blood from anyone. If you are blood type O, your immune system will recognize both A and B markers as foreign. You can accept blood only from people who are type O. However, lacking both A and B markers, you can donate blood to anyone.

Rh BLOOD TYPING

Rh blood typing is based on the presence or absence of the Rh marker (first identified in blood of *Rh*esus monkeys). If you are type Rh+, your blood cells bear this marker. If you are type Rh⁻, they do not.

Normally, Rh⁻ individuals do not have antibodies against the Rh marker. However, they will produce such antibodies if they are exposed to Rh+ blood. This can happen during some pregnancies. If an Rh+ man impregnates an Rh⁻ woman, the resulting fetus may be Rh+. The first time that an Rh⁻ woman carries an Rh+ fetus, she will not have antibodies against the Rh marker. However, fetal red blood cells may slip into her blood during childbirth and she may form such antibodies (Figure 34.9*a*). If the woman gets pregnant again, anti-Rh+ antibodies cross the placenta and get into fetal blood. If a fetus is Rh+, the antibodies attack

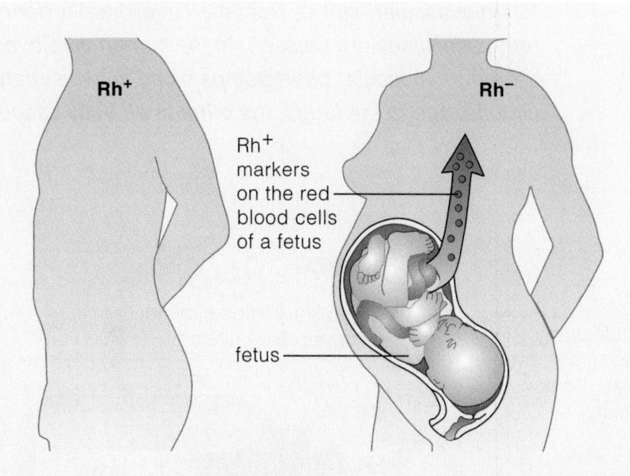

a An Rh+ man and an Rh⁻ woman carrying his Rh+ child. This is the mother's first Rh+ pregnancy, so she has no anti-Rh+ antibodies. But during birth, some of the child's Rh+ marked cells get into her blood.

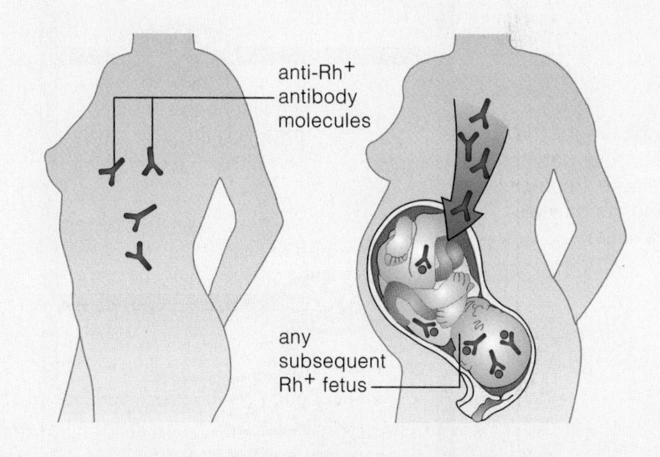

b The foreign marker stimulates antibody formation. If this woman gets pregnant again and if her second fetus (or any other) carries the Rh+ marker, her anti-Rh+ antibodies may attack the fetal red blood cells.

Figure 34.9 Animated! Maternal production of antibodies in response to Rh+ markers on red blood cells of her fetus.

its red blood cells and can kill the fetus (Figure 34.9*b*). To prevent any problems, an Rh⁻ mother who has just given birth to an Rh+ child should be injected with a drug that blocks production of antibodies that could cause problems during future pregnancies.

ABO blood types do not cause a similar condition because maternal antibodies for A and B markers do not cross the placenta and attack the fetal cells.

ABO and Rh blood typing methods analyze self markers on the surface of red blood cells. Incompatible blood types cause problems in transfusions and in some pregnancies.

34.5 Human Cardiovascular System

"Cardiovascular" comes from the Greek kardia *(for heart) and Latin* vasculum *(vessel). In the human cardiovascular system, a muscular heart pumps blood in two separate circuits: one to the lungs, the other to all body tissues.*

In humans, as in all mammals, the heart is a double pump that propels blood through two cardiovascular circuits. Each circuit extends from the heart, through arteries, arterioles, capillaries, venules, and veins, and reconnects with the heart (Figures 34.10 and 34.11). A short loop, the pulmonary circuit, oxygenates blood. Again, it leads from the heart's right half to capillary beds in each of two lungs, then back to the heart's left half. The systemic circuit is a longer loop. The heart's left half pumps oxygenated blood into the main artery in the body: the aorta. That blood gives up oxygen in all tissues, then the oxygen-poor blood flows back to the heart's right half.

In the systemic circuit, most blood flows through one capillary bed, then returns to the heart. However, blood from capillaries in the small intestine flows on to a capillary bed in the liver. This allows the blood to pick up glucose and other substances absorbed from the gut, and deliver them to the liver. The liver stores glucose as glycogen (Section 7.7). It also neutralizes absorbed toxins, including alcohol (Chapter 5).

As Figure 34.12 shows, the cardiovascular system distributes oxygen, nutrients, and other substances that enter the body by way of the digestive system and respiratory system. It moves carbon dioxide and other metabolic wastes to the respiratory and urinary systems for disposal. These are the main systems that keep operating conditions of the internal environment within tolerable ranges, a process we call homeostasis (Sections 25.1 and 25.3).

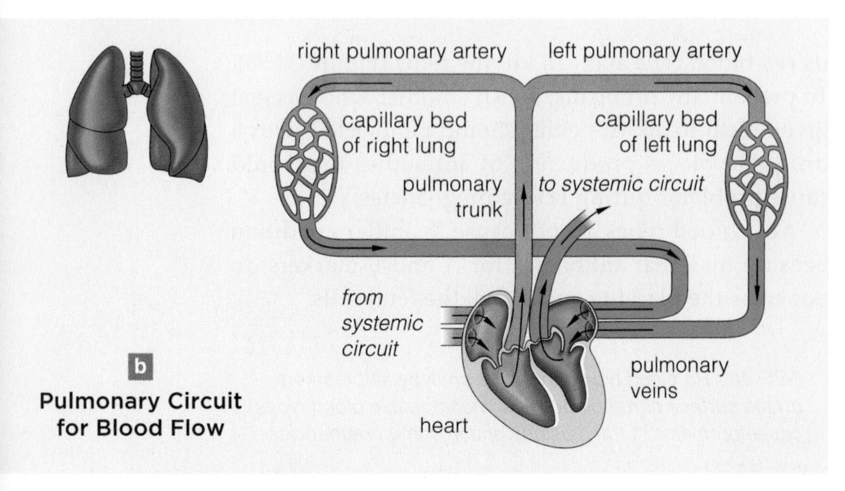

In the human cardiovascular system's pulmonary circuit, oxygen-poor blood flows from the heart's right half, through a pair of lungs, then back to the heart. It takes up oxygen and gives up carbon dioxide in the lungs.

In the systemic circuit, oxygenated blood flows from the heart's left half and aorta to capillary beds of all tissue regions. There it gives up oxygen and takes up carbon dioxide, then flows into the heart's right half.

Figure 34.10 **Animated!** (**a**,**b**) Systemic and pulmonary circuits for blood flow through the human cardiovascular system. Blood vessels carrying oxygenated blood are color-coded *red*. Those carrying oxygen-poor blood are color-coded *blue*.

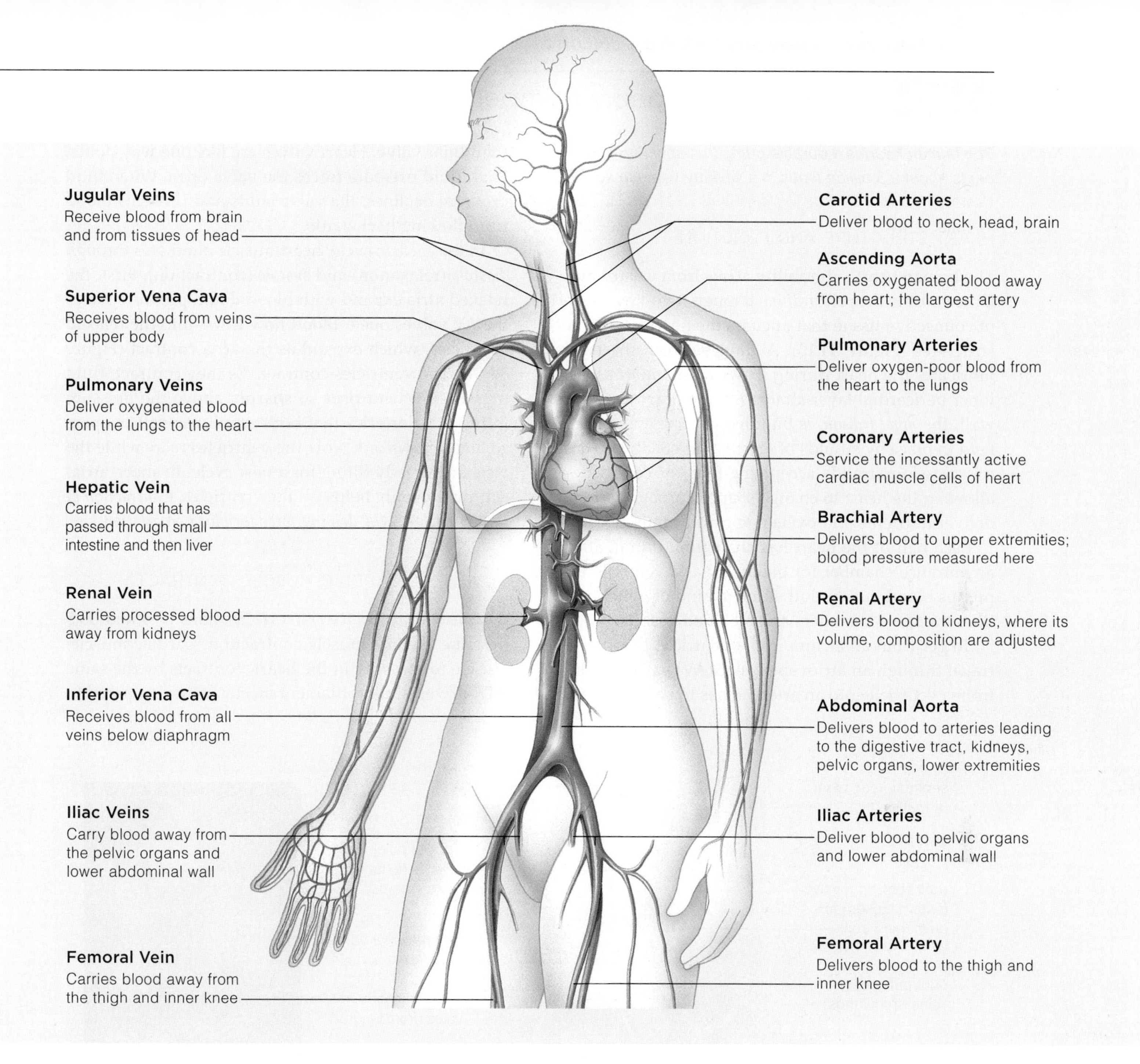

Figure 34.11 Animated! Major blood vessels of the human cardiovascular system. This art is simplified for clarity. For example, each arm has the arteries and veins listed; so does each leg. Humans, remember, show bilateral symmetry (Section 23.1).

Carotid bodies are located at the first branching of carotid arteries. Aortic bodies are in the aorta, where it arches above the heart. Both sensory receptors monitor chemical changes in blood. Also in these locations, pressure receptors monitor blood pressure. In response to signals from all three kinds of receptors, the brain adjusts the heart's strength and rate of contractions, as well as the diameter of arterioles.

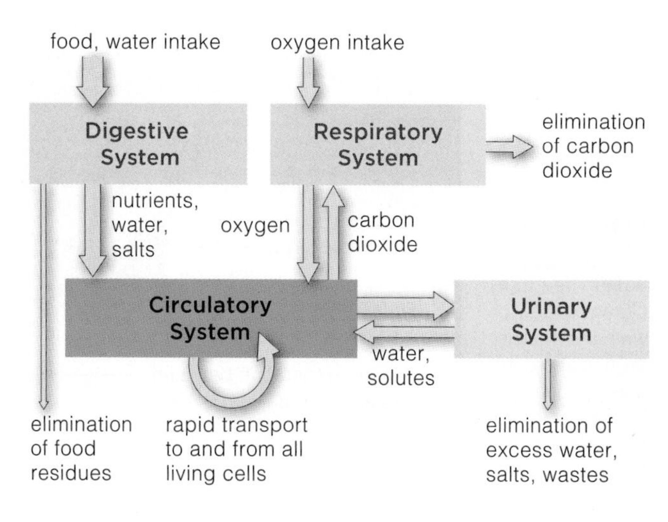

Figure 34.12 Functional links between the circulatory system and other organ systems with big roles in maintaining the internal environment.

34.6 The Heart Is a Lonely Pumper

The human heart is a durable pump that spontaneously beats about 2.5 billion times in a seventy-year life span.

LINKS TO SECTIONS 29.1, 29.3, 30.3, 33.4, 33.5

HEART STRUCTURE AND FUNCTION

The human heart's durability arises from its structure. Outermost is the *pericardium*, a tough, two-layered sac of connective tissue that anchors the heart to adjacent structures (Figure 34.13). A fluid between the layers lubricates the heart during its wringing motions. The inner pericardial layer attaches to the heart wall. This wall, the *myocardium*, is bundled cardiac muscle cells held in place by strands of elastin and collagen. These crisscrossing strands accept the force of contraction, allowing the heart to change shape. Coronary arteries deliver nutrients and oxygen to cardiac muscle cells.

Each half of the heart has an atrium (plural, atria), an entrance chamber for blood. It has a ventricle that pumps blood out. Endothelium, a kind of epithelium, lines the heart chambers and all blood vessels.

To get from an atrium into a ventricle, blood must travel through an atrioventricular (AV) valve. To flow from a ventricle into an artery, it has to pass through a semilunar valve. Heart valves are like one-way doors. High fluid pressure forces the valve open. When fluid pressure declines, the valve shuts and prevents blood from flowing backwards.

In the cardiac cycle, heart muscle alternates through *diastole* (relaxation) and *systole* (contraction). First, the relaxed atria expand with blood. Fluid pressure forces the AV valves open. Blood now flows into the relaxed ventricles, which expand as the atria contract (Figure 34.14). The ventricles contract. As they contract, fluid pressure in them rises so sharply above the pressure in the great arteries that both semilunar valves open, so blood flows out. Now the ventricles relax while the atria are already filling for a new cycle. In short, atrial contraction only helps fill the ventricles. *Contraction of the ventricles is the driving force for blood circulation.*

HOW DOES CARDIAC MUSCLE CONTRACT?

Cardiac Muscle Revisited Sections 33.4 and 33.5 describe skeletal muscle contraction. Cardiac muscle, muscle found only in the heart, contracts by the same ATP-driven sliding-filament mechanism. Compared to

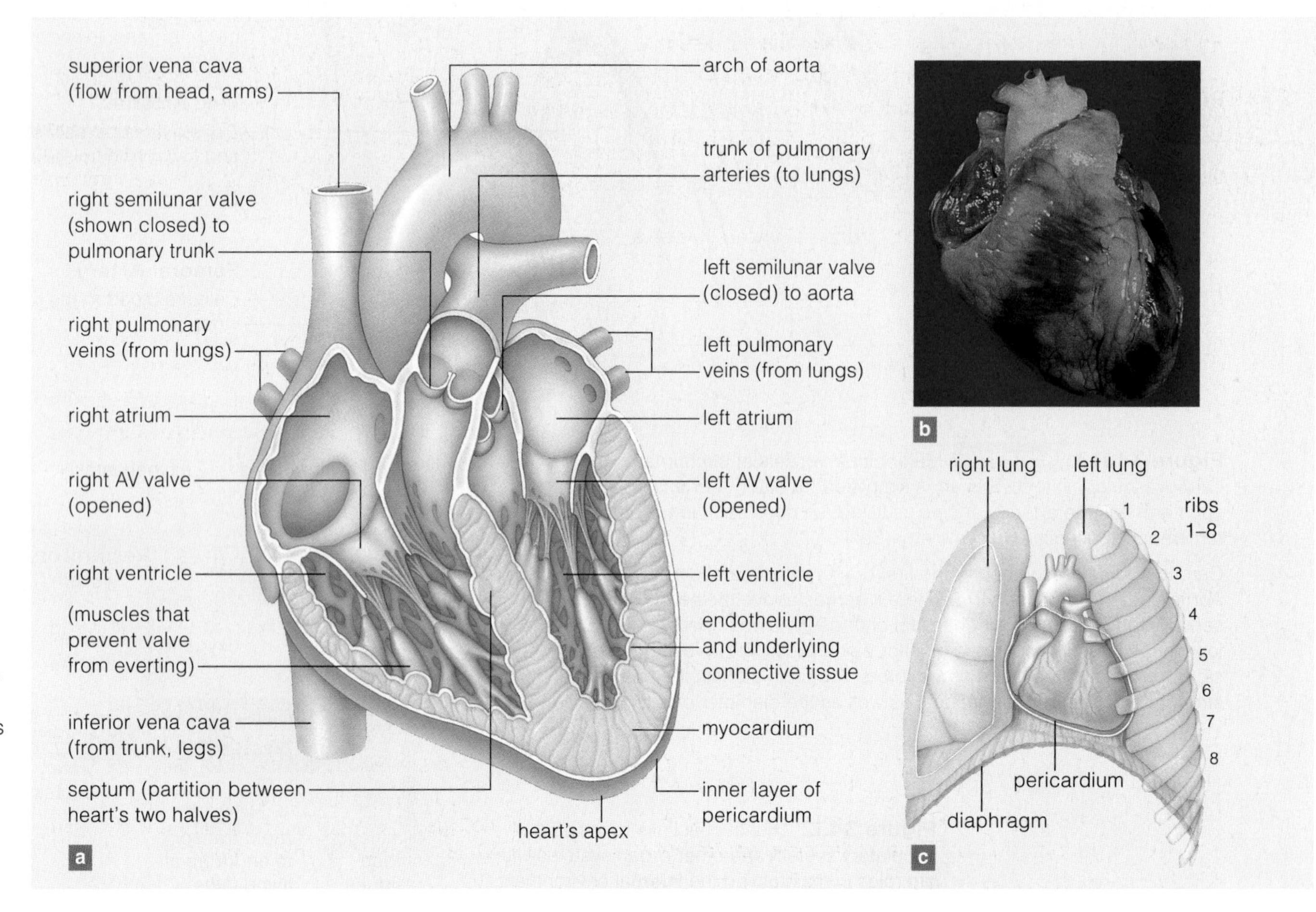

Figure 34.13 **Animated!** The human heart. (**a**) Cutaway view of its wall and internal organization, (**b**) outer appearance, and (**c**) its location between lungs in the thoracic cavity.

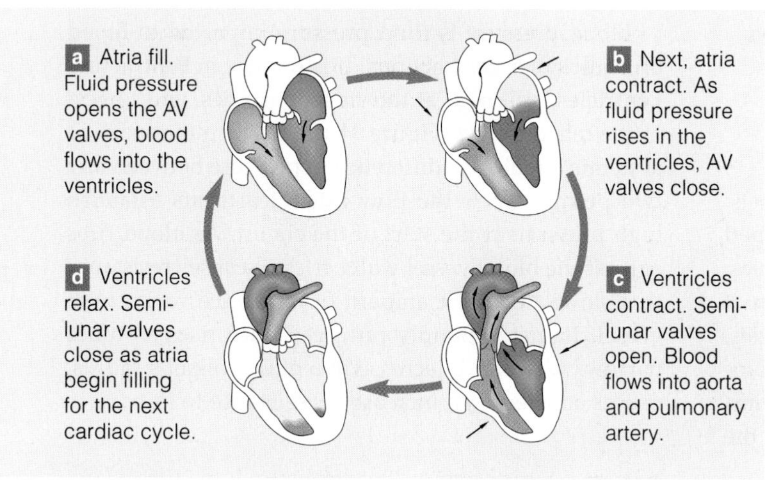

Figure 34.14 Animated! Cardiac cycle. You can hear the cycle through a stethoscope as a "lub-dup" near the chest wall. At each "lub," the heart's AV valves are closing as its ventricles are contracting. At each "dup," the heart's semilunar valves are closing as its ventricles are relaxing.

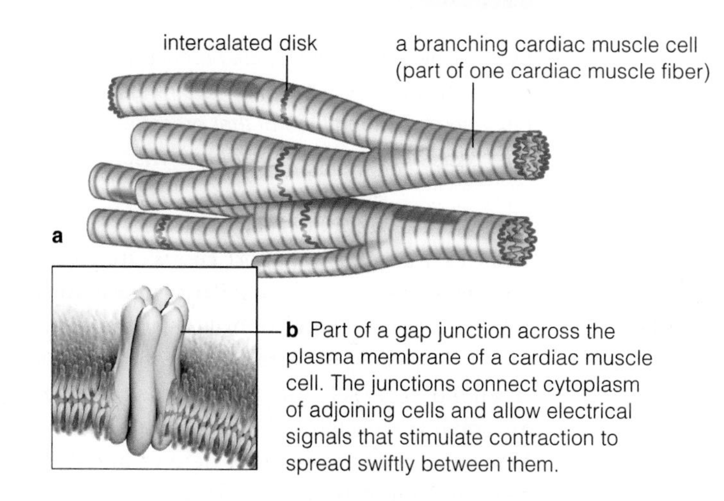

Figure 34.15 (**a**) Cardiac muscle cells. Compare Figure 29.9*b*. Many adhering junctions in intercalated disks at the ends of cells hold adjacent cells together, despite the mechanical stress caused by the heart's wringing motions. (**b**) The sides of cardiac muscle cells are subject to less mechanical stress than at the ends. The sides have a profusion of gap junctions across the plasma membrane.

other muscle, cardiac muscle has more mitochondria that supply ATP.

Sarcomeres arranged along the length of each cell give cardiac muscle a striated appearance. Cells branch and attach end-to-end at intercalated disks, regions abundant in adhering junctions (Figure 34.15*a*). Gap junctions connect the cytoplasm of neighboring cells. They allow waves of excitation to wash swiftly over the entire heart (Section 29.1 and Figure 34.15*b*).

How the Heart Beats In cardiac muscle, some specialized cells do not contract. Instead, they are part of the cardiac conduction system, which initiates and distributes signals that tell other cardiac muscle cells to contract. As Figure 34.16 shows, the system consists of a sinoatrial (SA) node and an atrioventricular (AV) node, functionally linked by junctional fibers. These fibers are bundles of long, thin cardiac muscle cells.

The SA node, a clump of noncontracting cells in the right atrium's wall, is the cardiac pacemaker. Its cells have specialized membrane channels that let them fire action potentials about seventy times per minute. The brain does not have to direct the SA node to fire; this natural pacemaker has spontaneous action potentials. Nervous signals only adjust the rate and strength of contractions. Even if a heart is removed from the body, it will keep beating for a short time. The defibrillators you read about earlier work by resetting the SA node.

A signal from the SA node starts the cardiac cycle. The signal spreads through the atria, causing them to contract. Simultaneously, the signal excites junctional fibers, which conduct it to the AV node. This clump of cells is the only electric bridge to the ventricles. The time it takes for a signal to cross this bridge is enough to keep ventricles from contracting before they fill.

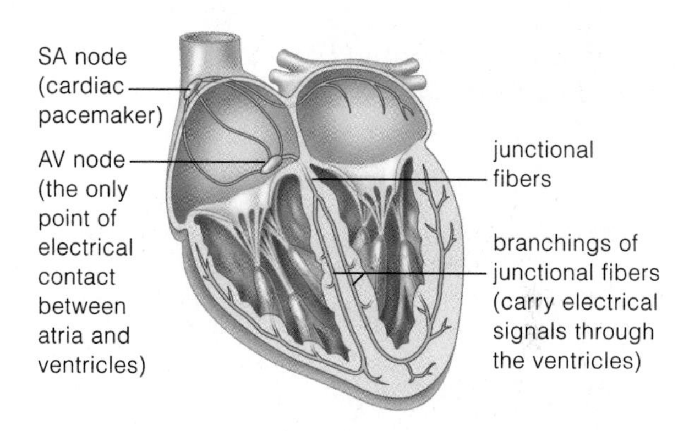

Figure 34.16 The cardiac conduction system.

From the AV node, a signal travels along a bundle of fibers. These junctional fibers branch in the septum, between the heart's left and right ventricles. Branching fibers extend down to the heart's lowest point and up the ventricle walls. Starting at the atria, cardiac muscle responds by contracting in a twisting motion, which ejects blood into the aorta and pulmonary arteries.

The four-chambered heart is partitioned into two halves, each with an atrium and a ventricle. Forceful contraction of its ventricles is the driving force for blood circulation.

The SA node is the cardiac pacemaker. Its spontaneous, rhythmic signals make cardiac muscle fibers of the heart wall contract in a coordinated fashion.

34.7 Pressure, Transport, and Flow Distribution

During any given interval, fluid pressure is driving blood through your body. The rate and strength of heartbeats influence the magnitude of that pressure. So does the total resistance to flow through the vascular system.

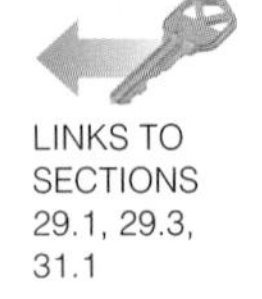

LINKS TO SECTIONS 29.1, 29.3, 31.1

Figure 34.17 compares the structure of blood vessels. Arteries are rapid-transport vessels for blood pumped out of the heart's ventricles. They supply the arterioles: smaller vessels where controls over the distribution of blood flow operate. Arterioles branch into capillaries, small blood vessels that form diffusion zones. Venules are small vessels located between capillaries and veins. Veins are large vessels that deliver blood back to the heart and serve as blood volume reservoirs.

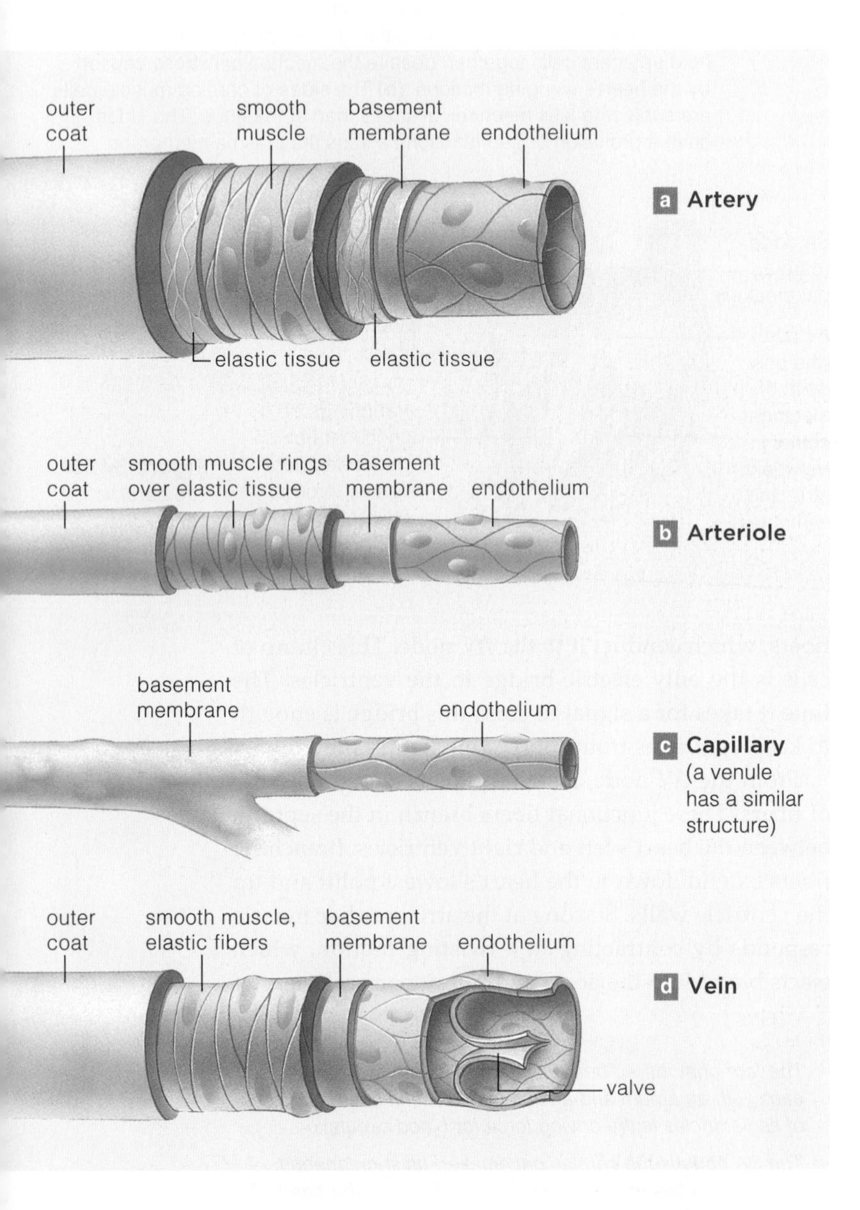

Figure 34.17 Structural comparison of human blood vessels. The drawings are not to the same scale.

Blood pressure is fluid pressure imparted to blood by ventricular contractions. It is highest in contracting ventricles, still high at the start of arteries, and lowest in the relaxed atria (Figure 34.18). In the pulmonary or systemic circuit, the difference in pressure between any two points affects the flow rate. Heartbeats establish high pressure at the start of the circuit. As blood rubs against the blood vessel walls, friction causes resistance that slows flow. The amount of resistance varies with vessel diameter. Simply put, resistance rises as tubes narrow. A twofold decrease in a blood vessel's radius, causes a sixteenfold increase in resistance to flow.

RAPID TRANSPORT IN ARTERIES

With their large diameter and low resistance to flow, arteries are efficient rapid transporters of oxygenated blood. They also are pressure reservoirs that smooth out pressure differences during every cardiac cycle. Their thick, muscular, elastic wall bulges whenever a heartbeat forces a large volume of blood into them. Between contractions, the wall recoils, which makes the blood move faster and farther down the circuit.

FLOW DISTRIBUTION AT ARTERIOLES

No matter how active you are, all blood from the right half of your heart flows to your lungs, and all blood from the left half is distributed to other tissues along the systemic circuit. The brain gets a constant supply of blood, but flow to other organs varies with activity. When you are resting, the blood flow is distributed as shown in Figure 34.19.

When you exercise, less blood flows to the kidneys and gut, and more flows to skeletal muscles in your

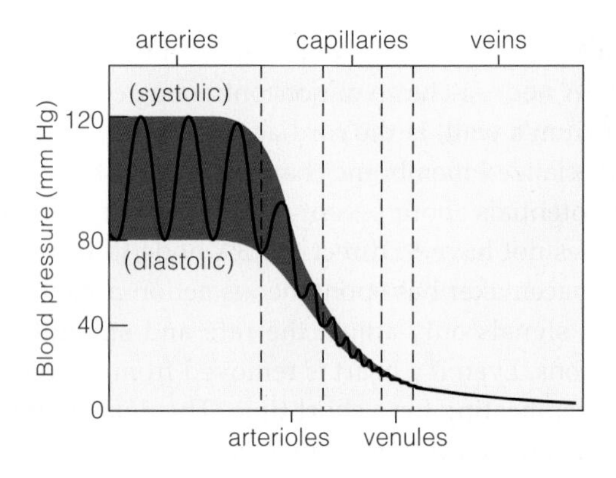

Figure 34.18 Plot of the decline in fluid pressure for a volume of blood that flowed through the systemic circuit.

legs. Like traffic cops, your arterioles guide the flow under orders from command posts: the nervous and endocrine systems. Signals from both act on rings of smooth muscle cells in arteriole walls (Figure 34.17*b*). Some signals cause dilation, or widening, of a blood vessel by causing the smooth muscle cells in its wall to relax. Other signals decrease blood vessel diameter by causing the smooth muscle in its wall to contract. When arterioles that supply a particular organ widen, more blood flows to that organ.

Arterioles also respond to shifts in concentrations of substances in a tissue. Such local chemical changes are "selfish," in that they invite or divert blood flow in response to the tissue's metabolic activities. As an example, when you exercise, your skeletal muscle cells use up oxygen, and concentrations of carbon dioxide, hydrogen and potassium ions, and other solutes rise. Arterioles in skeletal muscle tissue widen in response to the local chemical changes. More oxygenated blood flows through the tissue, and more metabolic wastes and products are carted away. When skeletal muscles relax, they require less oxygen. The concentration of oxygen rises locally, and the arterioles narrow.

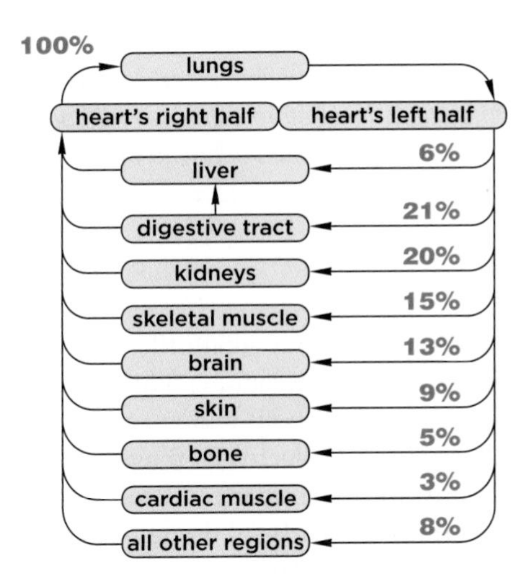

Figure 34.19 Distribution of the heart's output in people napping. How much blood flows through a given tissue is adjusted by selective vasodilation and vasoconstriction of arterioles all along the systemic circuit.

CONTROLLING BLOOD PRESSURE

We generally measure blood pressure at the brachial artery in an upper arm (Figure 34.20). In each cardiac cycle, *systolic* (peak) pressure occurs when contracting ventricles force blood into arteries. *Diastolic* pressure, the lowest pressure, occurs when ventricles are most relaxed. Blood pressure is measured in millimeters of mercury (mm Hg) and recorded as "systolic pressure over diastolic pressure," as in 120/80 mm Hg.

Blood pressure depends on the total blood volume, how much blood the ventricles pump out (the cardiac output), and whether the arterioles are constricted or dilated. Receptors in the aorta and in carotid arteries of the neck alert a control center in the medulla when blood pressure increases or decreases (Section 31.1). In response, this brain region calls for changes in cardiac output and arteriole diameter. This reflex response is a short-term control over blood pressure. The kidneys exert longer term control by adjusting fluid loss and thus altering the total blood volume.

The rate and strength of heartbeats and resistance to flow through blood vessels dictates blood pressure. Pressure is greatest in contracting ventricles and at the start of arteries.

How much blood flows to specific tissues varies over time and is altered by adjustments to the diameter of arterioles.

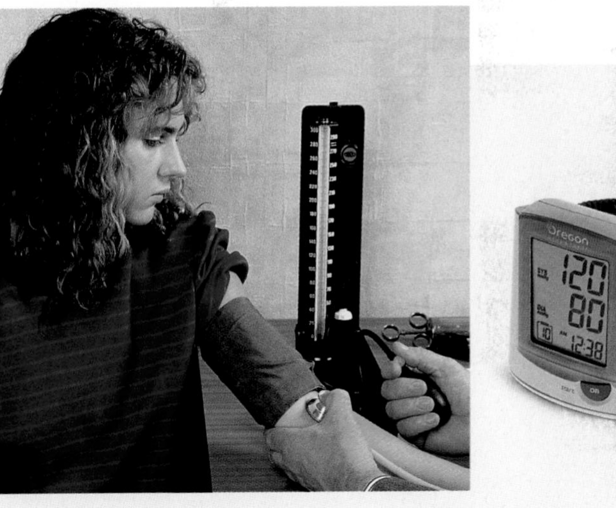

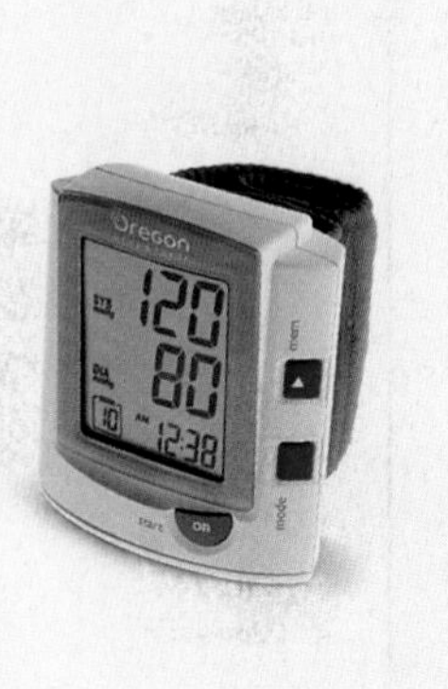

Figure 34.20 **Animated!** Measuring blood pressure. *Left*, a hollow inflatable cuff attached to a pressure gauge is wrapped around the upper arm. A stethoscope is placed over the brachial artery, just below the cuff.

The cuff is inflated with air to a pressure above the highest pressure of the cardiac cycle, when ventricles contract. Above this pressure, you will not hear sounds through the stethoscope, because no blood is flowing through the vessel.

Air in the cuff is slowly released until the stethoscope picks up soft tapping sounds. Blood flowing into the artery under the pressure of the contracting ventricles—the systolic pressure—causes the sounds. When these sounds start, a gauge typically reads about 120 mm Hg. That amount of pressure will force mercury (Hg) to move up 120 millimeters in a glass column of a standardized diameter.

More air is released from the cuff. Eventually the sounds stop. Blood is now flowing continuously, even when the ventricles are the most relaxed. The pressure when the sounds stop is the lowest during a cardiac cycle, the diastolic pressure, is usually about 80 mm Hg.

Right, compact monitors are now available that automatically record the systolic/diastolic blood pressure.

34.8 Diffusion at Capillaries, Then Back to the Heart

A capillary bed is a diffusion zone, where blood exchanges substances with interstitial fluid before veins carry it back to the heart. Living cells die fast when deprived of the exchanges. Brain cells start dying within four minutes.

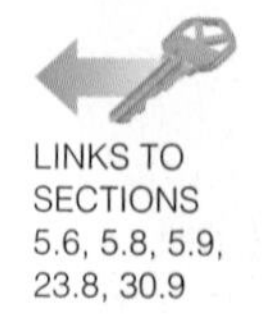

LINKS TO SECTIONS 5.6, 5.8, 5.9, 23.8, 30.9

CAPILLARY FUNCTION

A capillary is a cylinder of endothelial cells, one cell thick, wrapped in a basement membrane. Figure 34.21 shows a few of the 10 billion to 40 billion capillaries that service a human body. Collectively, they offer an enormous surface area for the exchange of substances with the interstitial fluid. In nearly all tissues, cells are very close to one or more capillaries. The proximity is essential. Diffusion distributes molecules and ions so slowly that it is effective only over small distances.

Red blood cells are 8 micrometers in diameter. They have to squeeze single file through the capillaries. The squeeze puts oxygen-transporting red blood cells and solutes in the plasma in direct or near contact with the exchange surface—the capillary wall.

To move between the blood and interstitial fluid, a substance must cross a capillary wall. Oxygen, carbon dioxide, and small lipid-soluble molecules, can diffuse across endothelial cells of a capillary. Proteins are too big to diffuse across plasma membranes, but some do enter endothelial cells by endocytosis, diffuse through the cell, then escape by exocytosis on the opposite side. Also, fluid containing small solutes and ions leaks out of capillaries through narrow spaces between adjacent endothelial cells.

Compared to other capillaries in the body, those in the brain are much less leaky. Brain endothelial cells adhere so tightly to one another, that plasma does not leak between them. This property of brain capillaries creates the blood–brain barrier (Section 30.9).

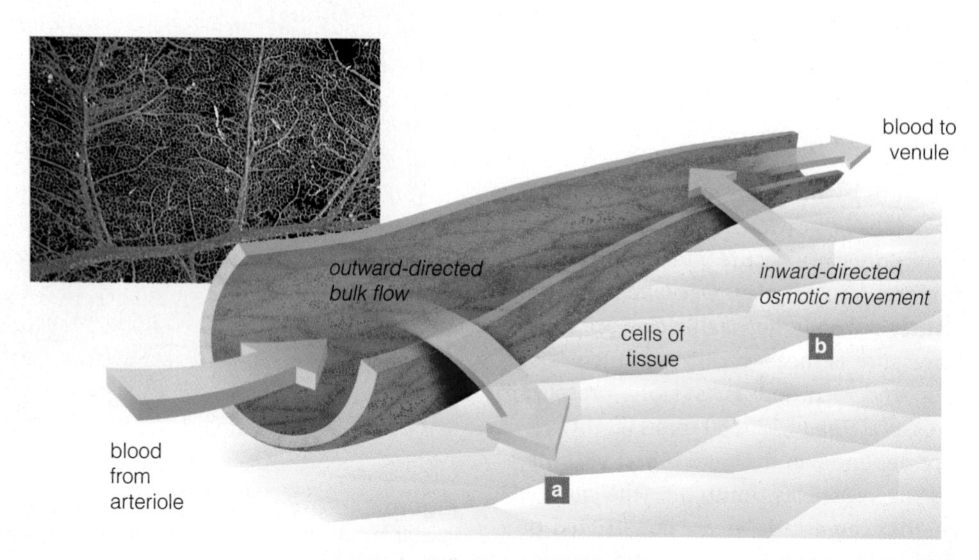

Arteriole end of capillary bed	
Outward-Directed Pressure:	
Hydrostatic pressure of blood in capillary:	35 mm Hg
Osmosis due to interstitial proteins:	28 mm Hg
Inward-Directed Pressure:	
Hydrostatic pressure of interstitial fluid:	0
Osmosis due to plasma proteins:	3 mm Hg
Net Ultrafiltration Pressure:	
(35 – 0) – (28 – 3) =	10 mm Hg
Ultrafiltration favored	

Venule end of capillary bed	
Outward-Directed Pressure:	
Hydrostatic pressure of blood in capillary:	15 mm Hg
Osmosis due to interstitial proteins:	28 mm Hg
Inward-Directed Pressure:	
Hydrostatic pressure of interstitial fluid:	0
Osmosis due to plasma proteins:	3 mm Hg
Net Reabsorption Pressure:	
(15 – 0) – (28 – 3) =	–10 mm Hg
Reabsorption favored	

Figure 34.21 Fluid movement at a capillary bed. Fluid crosses a capillary wall by way of ultrafiltration and reabsorption. (**a**) At the capillary's arteriole end, a difference between blood pressure and interstitial fluid pressure forces out plasma, but few plasma proteins, through clefts between endothelial cells of the capillary wall. Ultrafiltration is the outward flow of fluid across the capillary wall as a result of hydrostatic pressure.

(**b**) Reabsorption is the osmotic movement of some interstitial fluid *into* the capillary. It happens when the water concentration between interstitial fluid and the plasma differs. Plasma, with its dissolved proteins, has a greater solute concentration and therefore a lower water concentration. Reabsorption near the end of a capillary bed tends to balance ultrafiltration at the start of it. Normally, there is only a small *net* filtration of fluid, which the lymphatic system returns to blood.

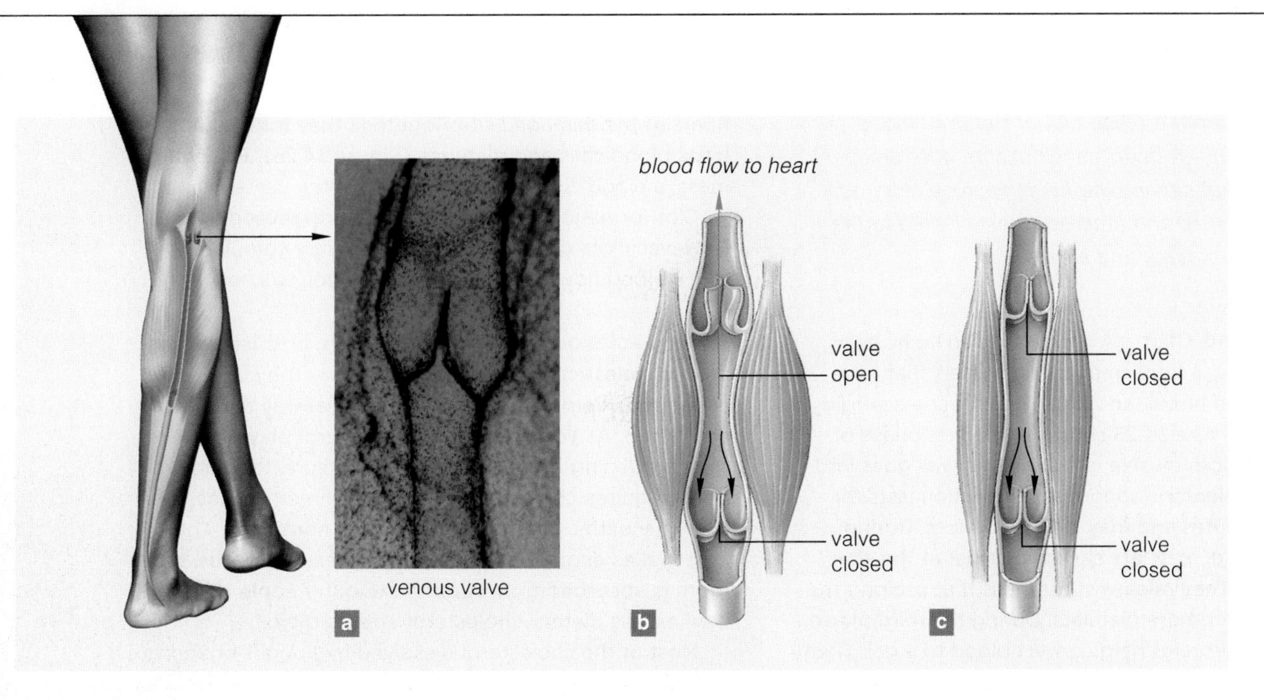

Figure 34.22 Vein structure and function.

(**a**) Valves in medium-sized veins prevent the backflow of blood.

Adjacent skeletal muscles helps raise fluid pressure inside a vein. (**b**) These muscles bulge into a vein as they contract. Pressure inside the vein rises and helps keeps blood flowing forward. (**c**) When muscles relax, the pressure that they exerted on the vein is lifted. Venous valves shut and cut off backflow.

As blood flows through a typical capillary bed, it is subject to two opposing forces. Hydrostatic pressure, an outward-directed force, results from contraction of the ventricles. Osmotic pressure, an inward-directed force, results from differences in solute concentration between blood and interstitial fluid.

At the arterial end of a capillary bed, hydrostatic pressure is high. It forces fluid out between cells of the capillary wall, into interstitial fluid (Figure 34.21*a*). This process is ultrafiltration. The fluid forced out has high levels of oxygen, ions, and nutrients, such as glucose. Ultrafiltration allows movement of large quantities of essential substances from blood into interstitial fluid.

As the blood continues on to the venous end of the capillary bed, hydrostatic pressure drops and osmotic pressure predominates (Figure 34.21*b*). Water is drawn by osmosis from interstitial fluid into the protein-rich plasma. This process is capillary reabsorption.

Normally, there is a small *net* outward flow of fluid from capillaries, which lymph vessels return to blood. If high blood pressure causes too much fluid to flow out or something slows down fluid return, interstitial fluid collects in tissues. The resulting swelling is called edema. Roundworm infections that damage the lymph vessels can cause severe edema (Section 23.8).

VENOUS PRESSURE

Blood from several capillaries flows into each venule. These thin-walled vessels join together to form veins, the large-diameter, low-resistance transport tubes that carry blood to the heart. Many veins, especially in the legs, have flaplike valves that help prevent backflow (Figure 34.22). These valves automatically shut when blood in the vein starts to reverse direction.

Sometimes venous valves lose their elasticity. Then veins become enlarged and bulge near the surface of skin. This commonly occurs in veins of the legs, and the result is varicose veins. Failure of valves in veins around the anus causes hemorrhoids. Weight control and regular exercise can help prevent both problems.

The vein wall can bulge quite a bit under pressure, much more so than an arterial wall. Thus, veins act as reservoirs for great volumes of blood. When you rest, they hold about 60 percent of the total blood volume.

During exercise, fluid pressure in veins increases, and less blood collects inside them. Veins have a bit of smooth muscle inside their wall and exercise-induced signals from the nervous system make it contract. This causes veins to stiffen so they cannot hold as much blood, and the pressure inside them rises. At the same time, skeletal muscles used in limb movements bulge and press on veins, squeezing blood toward the heart. (Figure 34.22*b*,*c*). Exercise-induced deep breathing also raises venous pressure. As the chest expands, organs get squeezed and press against veins.

Capillary beds are diffusion zones where blood exchanges substances with interstitial fluid. Bulk flow contributes to the fluid balance between blood and interstitial fluid.

Venules deliver blood from capillaries to veins. Veins are blood volume reservoirs. The amount of blood residing in the veins can be adjusted depending on activity level.

34.9 Cardiovascular Disorders

A stroke results from a blood clot or ruptured blood vessel in the brain. A heart attack occurs after small blood vessels that service the heart become clogged. High blood pressure and atherosclerosis increase the risk of both heart attack and stroke.

LINKS TO SECTIONS 3.3, 32.7

Good Clot, Bad Clot A process called hemostasis can stop blood loss from small blood vessels that have become ruptured or cut, and it can construct a scaffold for repairs. As Figure 34.23 shows, in the first phase of hemostasis, smooth muscle in a damaged wall goes into spasm. This involuntary, abnormal contraction lasts for about thirty minutes and may stop blood loss. During the second phase, platelets clump together at the site of the damage. They release substances that prolong the spasm and attract more platelets. During the third phase, certain plasma proteins help convert blood to a gel. Then fibrinogens (rodlike plasma proteins) form. They converge as long, insoluble threads that stick to exposed collagen fibers at the damaged site. Together, they form a net that traps blood cells and platelets (Figure 34.23). The entire mass, a blood clot, seals the breach in the vessel wall.

Clot formation is essential for repairing blood vessels. However, clots cause problems when they completely block blood flow through a vessel, as you will now see.

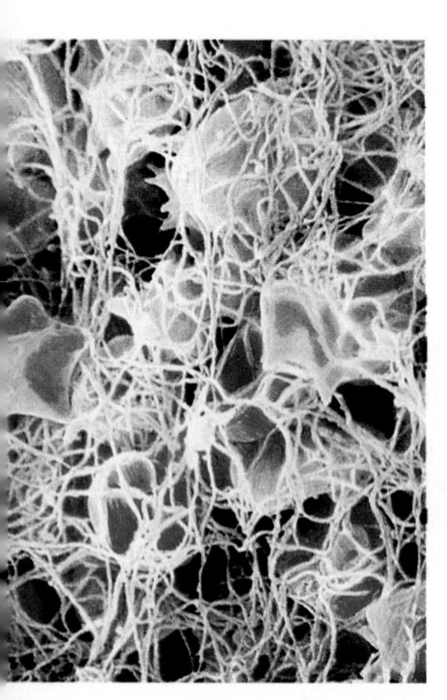

Stimulus

Blood vessel damage

Phase 1 response

A vascular spasm constricts the vessel at the site of damage, slowing blood loss.

Phase 2 response

Platelets aggregate and stick together within fifteen seconds, thus plugging the site.

Phase 3 response

Clot formation starts after thirty seconds:

1. Enzymes activate factor X; prothrombin forms.
2. Prothrombrin converts an enzyme precursor to thrombin.
3. Thrombin converts fibrinogen, a plasma protein, to insoluble protein threads (fibrin).
4. Fibrin forms a net that entangles blood cells and platelets; the entire mass is a blood clot.

Figure 34.23 Hemostasis. The photomicrograph shows a net of fibrous proteins that helps blood clots form on a damaged blood vessel wall.

Atherosclerosis In arteriosclerosis, arteries thicken and lose elasticity. With atherosclerosis, the condition worsens as lipids build up in the arterial wall and narrow the lumen. As you may know, cholesterol plays a role in this "hardening of the arteries." To be sure, the human body requires cholesterol to make cell membranes, myelin sheaths, bile salts, and steroid hormones. The liver makes enough cholesterol for these purposes, but more is absorbed from food in the gut. People differ in how excess dietary cholesterol affects them.

Most of the cholesterol dissolved in blood is bound to protein carriers. The complexes are known as low density lipoproteins, or LDLs, and most cells can take them up. A lesser amount is bound up in high density lipoproteins, or HDLs. Cells in the liver metabolize HDLs, using them in the formation of bile, which the liver secretes into the gut. Eventually, the bile leaves the body in the feces.

When the LDL level in blood rises, so does the risk of atherosclerosis. The first sign of trouble is a build-up of lipids in an artery's endothelium (Figure 34.24). Smooth muscle then inflames the arterial wall by growing into it. Fibrous connective tissue forms over the entire mass. The mass, called an atherosclerotic plaque, narrows the vessel's internal diameter.

A hardened plaque can rupture an artery wall, thereby triggering clot formation. A clot that stays put is called a thrombus. A clot that breaks loose and travels in blood is an embolus. Both can block vessels and cause problems.

A heart attack occurs when a cardiac artery becomes completely blocked, most commonly by a clot. If the blockage is not removed fast, cardiac muscle cells die. Clot-dissolving drugs can restore blood flow if they are given within an hour of the onset of an attack, so any suspected heart attack should receive prompt attention.

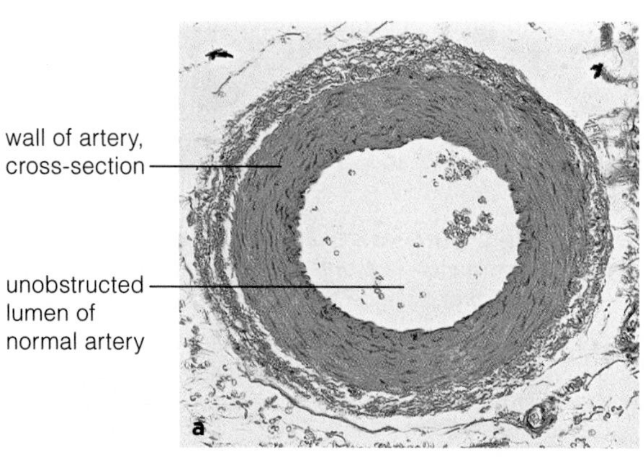

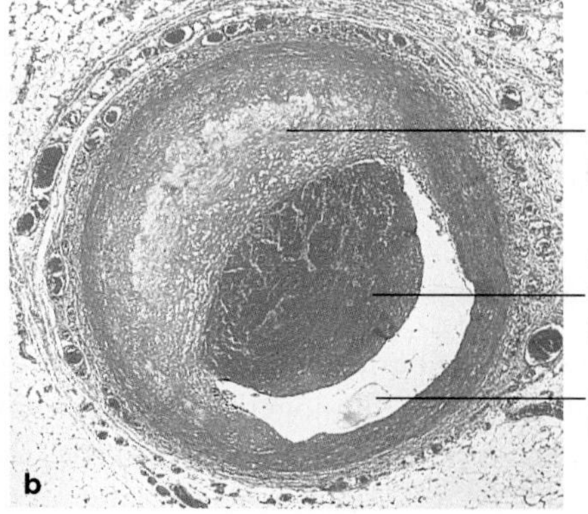

Figure 34.24 Sections from (**a**) a normal artery and (**b**) an artery with a lumen narrowed by an atherosclerotic plaque. A clot clogged this one.

With coronary bypass surgery, doctors stitch a section of a blood vessel from elsewhere in the body to the aorta and to the coronary artery below a clogged part (Figure 34.25). With laser angioplasty, laser beams vaporize the plaques. With balloon angioplasty, doctors inflate a small balloon in a blocked artery to flatten the plaques.

Hypertension—A Silent Killer Hypertension refers to chronically high blood pressure even during periods of rest. Blood pressure stays above 140/90, often for unknown reasons. We refer to this chronic condition as a silent killer, because people often are unaware they have it. Hypertension tends to run in families. Diet and lack of exercise increase risk. Also, in some people, high salt intake raises blood pressure and makes the heart pump harder. The heart may enlarge enough to pump less efficiently. High blood pressure may contribute to atherosclerosis, which interferes with the delivery of oxygen to the brain, heart, and other vital organs. Of an estimated 23 million hypertensive Americans, most do not seek treatment. About 180,000 die each year.

Rhythms and Arrhythmias As you read in Section 34.6, the SA node controls the rhythmic beating of the heart. Electrocardiograms, or ECGs, record the electrical activity of a beating heart (Figures 34.1 and 34.26*a*).

ECGs can reveal arrhythmias, which are abnormal heart rhythms (Figure 34.26*b–d*). Arrhythmias are not always dangerous. For example, endurance athletes commonly experience bradycardia, a below-average resting cardiac rate. In response to ongoing exercise, the nervous system has adjusted the firing rate of their cardiac pacemaker downward. Tachycardia, a faster than normal heart rate, can be caused by exercise, stress, or some underlying heart problem.

In atrial fibrillation, the atria do not contract normally. They quiver, which increases the risk of blood clots and stroke. Ventricular fibrillation is the most dangerous type of arrhythmia. It caused the collapse of Tammy Higgins, as described at the start of this chapter. Ventricles flutter, and their pumping action falters or stops, which causes loss of consciousness and death. A defibrillator often can restore the heart's normal rhythm.

Risk Factors In the United States, cardiovascular disorders are the leading cause of death. Each year, they affect about 40 million people, and about 1 million die. Tobacco smoking tops the list of risk factors (Section 36.7). Other factors include a hereditary predisposition to heart attack, hypertension, a high level of cholesterol in blood, obesity, and diabetes mellitus (Section 32.7). Advancing age also is a risk factor. The older you get, the greater the risk. Physical inactivity, too, increases the risk. Regular exercise lowers it, even when the exercise is not particularly strenuous. Gender is another factor; until about age fifty, males are at greater risk.

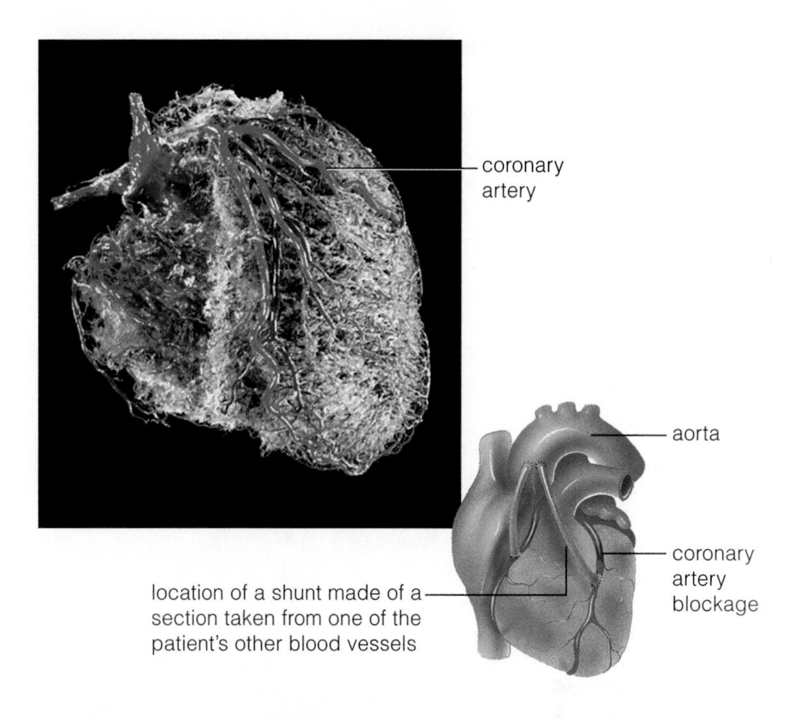

Figure 34.25 Coronary arteries and other blood vessels that service the heart. Resins were injected into them. Then the cardiac tissues were dissolved to make an accurate, three-dimensional corrosion cast. The sketch shows two coronary bypasses (color-coded *green*), which extend from the aorta past two clogged parts of the coronary arteries.

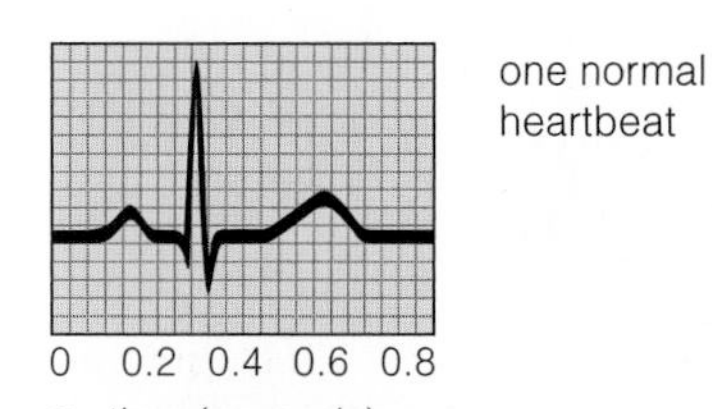

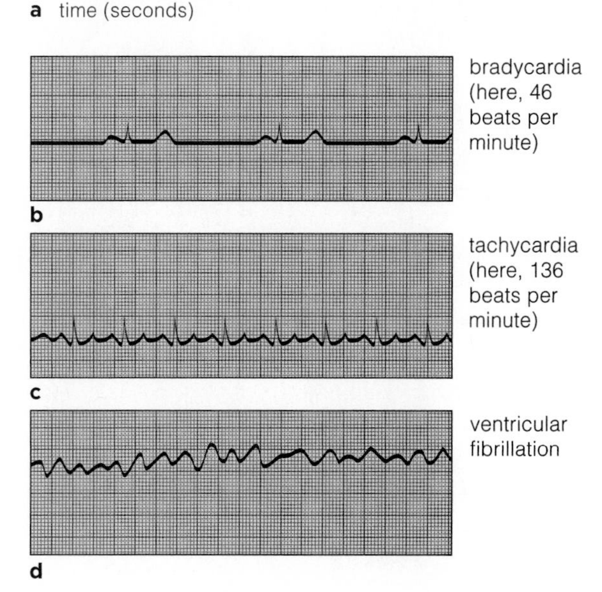

Figure 34.26 (**a**) ECG of one normal beat of the human heart. (**b–d**) Recordings that identified three types of arrhythmias.

34.10 Interactions With the Lymphatic System

LINK TO SECTION 32.10

We conclude this chapter with a brief look at how the lymphatic system and circulatory system interact. Think of this section as a bridge to the next chapter, on immunity, because the lymphatic system also helps defend the body against injury and attack.

LYMPH VASCULAR SYSTEM

A portion of the lymphatic system, called the lymph vascular system, consists of vessels that collect water and solutes from interstitial fluid, then delivers them to the circulatory system. The lymph vascular system includes lymph capillaries and vessels (Figure 34.27). Fluid that moves through these vessels is the lymph.

The lymph vascular system serves three functions. First, its vessels are drainage channels for water and plasma proteins that leaked out of capillaries and that must be returned to the circulatory system. Second, it delivers fats absorbed from food in the small intestine to the blood (Section 37.4). Third, it transports cellular debris, pathogens, and foreign cells to lymph nodes, which serve as the system's disposal sites.

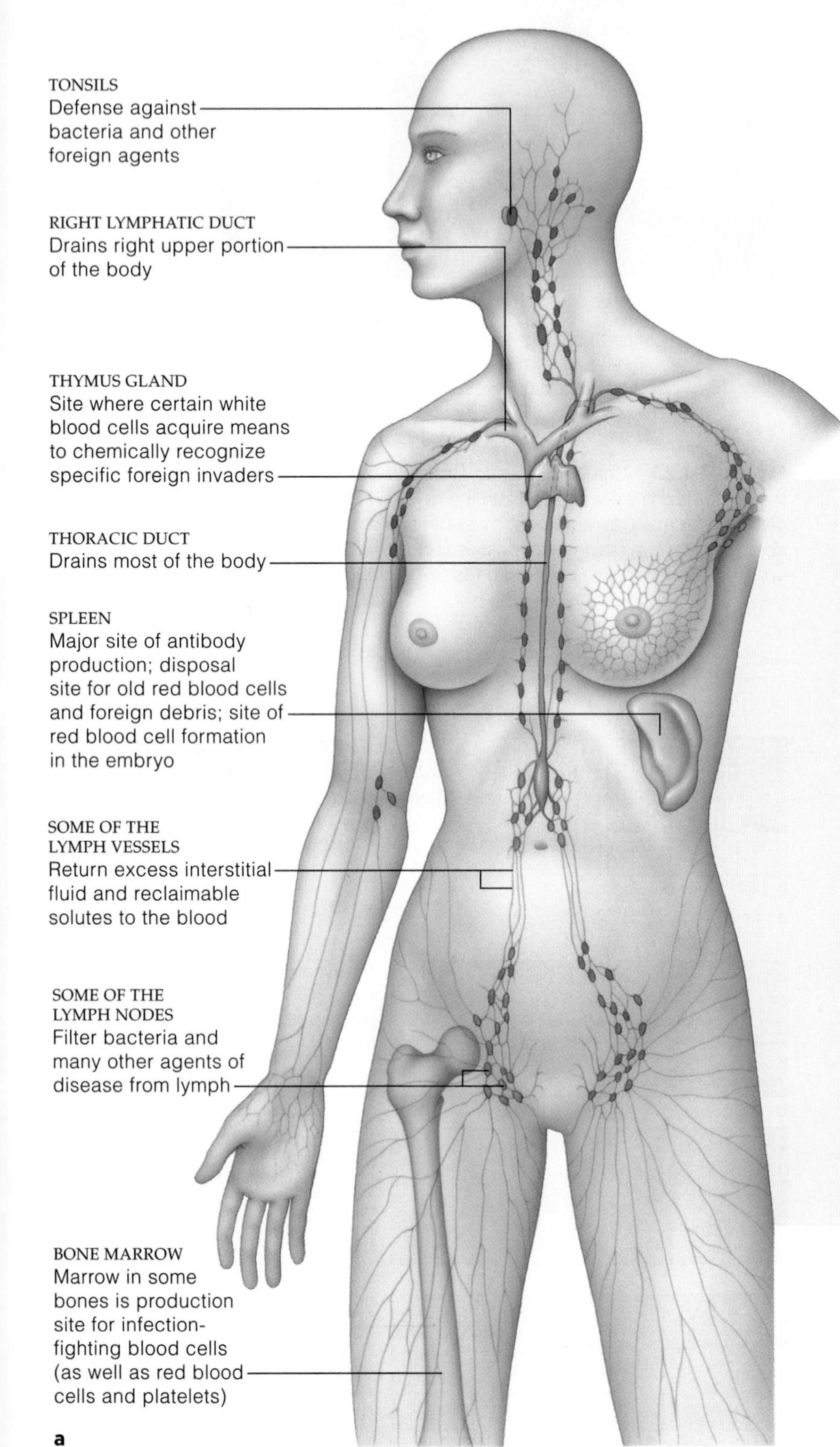

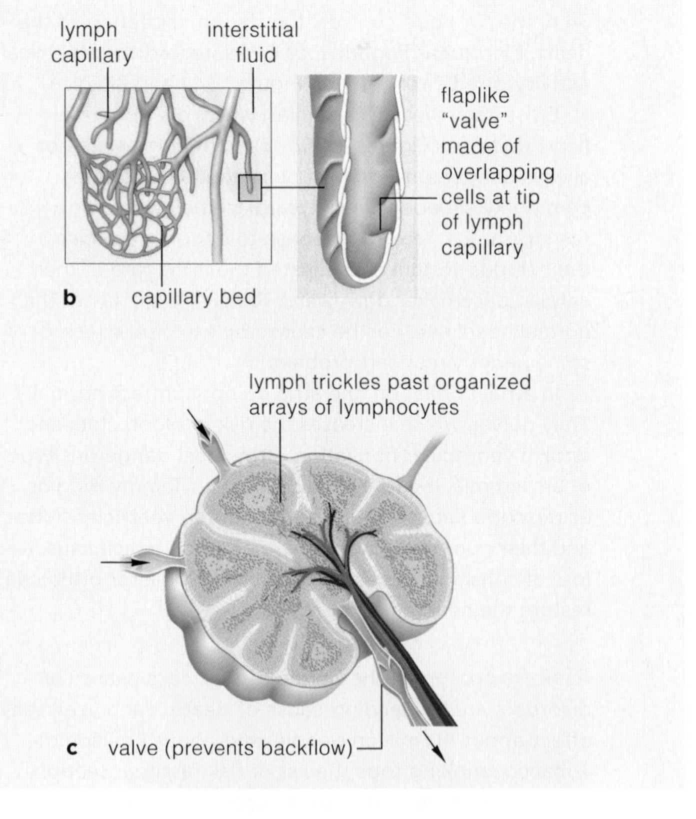

Figure 34.27 Animated! (**a**) Components of the human lymphatic system and their functions. Not shown are patches of lymphoid tissue in the small intestine and in the appendix. (**b**) Diagram of lymph capillaries at the start of a drainage network, the lymph vascular system. (**c**) Cutaway view of a lymph node. Its inner compartments are packed with organized arrays of infection-fighting white blood cells.

The lymph vascular system extends to all capillary beds. There, excess fluid enters the lymph capillaries. These capillaries have no obvious entrance; water and solutes move into clefts between cells. As you can see from Figure 34.27*b*, endothelial cells overlap, forming flaplike valves. Lymph capillaries merge into larger diameter lymph vessels, which have smooth muscle in their wall and valves that prevent backflow. Finally, lymph vessels converge onto collecting ducts, which drain into veins in the lower neck.

LYMPHOID ORGANS AND TISSUES

The other portion of the lymphatic system has roles in the body's defense responses to injury and attack. We call its components lymphoid organs and tissues. They include the lymph nodes, spleen, and thymus, as well as the tonsils, and some patches of tissue in the wall of the small intestine and appendix.

Lymph nodes are strategically located at intervals along lymph vessels (Figure 34.27*c*). Before entering blood, lymph trickles through at least one node and gets filtered. Masses of lymphocytes that formed in the bone marrow take up stations inside the nodes. When something identified as nonself reaches a node, they divide rapidly and form large armies that destroy it.

The spleen is the largest lymphoid organ, about the size of a fist in an average adult. In embryos only, it functions as a site of red blood cell formation. After birth, the spleen filters pathogens, worn-out red cells and platelets from blood vessels that branch through it. It holds phagocytic white blood cells that engulf and digest altered body cells and alert the immune system to threats. It holds antibody-producing cells. People can survive the removal of a badly damaged spleen, as often occurs in highway accidents. Lack of a spleen makes them more vulnerable to infections.

In the thymus gland, T lymphocytes differentiate and become capable of recognizing and responding to particular pathogens. The thymus gland also makes the hormones that influence these actions. It is central to immunity, the focus of the next chapter.

The lymphatic system supports the circulatory system and helps defend the body against pathogens.

The lymph vascular portion of the system consists of many tubes that collect and deliver excess water and solutes from interstitial fluid to blood. It also delivers absorbed fats to the blood, and delivers disease agents to lymph nodes.

The system's lymphoid organs, including lymph nodes, have specific roles in body defenses.

Summary

Section 34.1 A circulatory system moves substances to and from interstitial fluid faster than diffusion alone could move them. Interstitial fluid fills spaces between cells. It exchanges substances with cells and with blood.

Some invertebrates have an open circulatory system, in which blood spends part of the time mingling with tissue fluids. In vertebrates, a closed circulatory system confines blood inside a heart and blood vessels. Closed systems differ in whether blood flows through one or two circuits of blood vessels and in how many chambers divide the heart's interior. As lungs evolved in the early vertebrates on land, the circulatory system was modified in ways that made gas exchange more efficient.

- *Use the animation on ThomsonNOW to compare animal circulatory systems.*

Section 34.2 Blood is a fluid connective tissue that consists of plasma, blood cells, and platelets. Plasma is mostly water in which diverse ions and molecules are dissolved. Red blood cells, or erythrocytes, contain the hemoglobin that functions in rapid transport of oxygen and, to a lesser extent, carbon dioxide. They do not have a nucleus when mature. A variety of white blood cells, or leukocytes, have roles in day-to-day tissue maintenance and repair and in defenses against pathogens. Platelets function in blood clotting. Platelets and all blood cells arise from stem cells in bone marrow.

Section 34.3 In a blood disorder, an individual has too many, too few, or abnormal red or white blood cells.

Section 34.4 Among the recognition proteins on the surface of red blood cells are self markers that identify an individual's blood types. The body mounts an attack against any cells that bear nonself markers. ABO blood typing helps match the blood of donors and recipients to avoid blood transfusion problems. Rh blood typing and suitable treatment prevents the problems that sometimes arise when maternal and fetal Rh blood types differ.

- *Use the animation on ThomsonNOW to learn about blood types and blood transfusions.*

Section 34.5 The human heart is a four-chambered muscular pump, the contraction of which forces blood through two separate circuits.

In the pulmonary circuit, oxygen-poor blood from the heart's right half flows to the lungs, picks up oxygen, then flows to the heart's left half.

In the systemic circuit, the oxygen-rich blood flows from the heart's left half to all body tissues, then oxygen-poor blood flows to the heart's right half.

Most blood flows through only one capillary system. But blood in intestinal capillaries will later flow through liver capillaries. The liver metabolizes or stores nutrients and neutralizes some bloodborne toxins.

- *Use the animation on ThomsonNOW to explore the human cardiovascular system.*

Section 34.6 A human heart is a double pump. It is partitioned into two halves, each with two chambers: an atrium and a ventricle.

During one cardiac cycle, all heart chambers undergo rhythmic relaxation (diastole) and contraction (systole). When a cycle starts, each atrium expands as blood fills it and opens a valve to a ventricle. Both ventricles already are filling as the atria contract. When ventricles contract, they force blood into the aorta and pulmonary arteries. Contraction of the ventricles alone provides the force that powers movement of blood through blood vessels.

A cardiac conduction system is the basis of the heart's beating. It consists of an SA node in the right atrium wall that is functionally linked by long bundles of conducting fibers to an AV node.

The SA node, the cardiac pacemaker, spontaneously generates action potentials that set the pace for cardiac contractions. The nervous system does not initiate heart beats; it only adjusts their rate and strength. Waves of excitation wash over the heart's atria, down fibers in its septum, then up the walls of the ventricles.

■ *Use the animation on ThomsonNOW to learn about the structure and function of the human heart.*

Section 34.7 Blood pressure varies in the circulatory system. It is highest in contracting ventricles. It declines as blood travels through arteries, arterioles, capillaries, venules, and veins of the systemic or pulmonary circuit. It is lowest in relaxed atria. The speed of flow depends on heartbeat strength and rate, and on resistance to flow in the blood vessels. Adjusting the diameter of arterioles that supply different parts of the body redistributes the blood volume depending on conditions. In any interval, tissues that require the most metabolic support get the most blood flow.

■ *Use the animation on ThomsonNOW to see how blood pressure is measured.*

Section 34.8 Capillary beds are zones of diffusion between the blood and interstitial fluid. Ultrafiltration pushes a small amount of fluid out of capillaries. Fluid moves back in by capillary reabsorption. Normally, both processes are almost balanced, with no more than a small net outward flow of fluid from a capillary bed.

Several capillaries drain into each venule. Veins are transport vessels that serve as a blood volume reservoir where the flow volume back to the heart is adjusted.

Section 34.9 Hemostasis is the process that stops blood flow from small vessels after injury by bringing about clot formation. Blood clotting is beneficial except in certain cardiovascular disorders. Common circulatory disorders include atherosclerosis, hypertension (chronic high blood pressure), heart attacks, strokes, and certain arrhythmias. Regular exercise, maintaining normal body weight, and not smoking lower risk for these disorders.

Section 34.10 The lymphatic system interacts with the circulatory system. The lymphatic system's vascular portion includes lymph capillaries and lymph vessels. It takes up excess water from interstitial fluid, as well as fats absorbed from the gut, and delivers them to blood. It also delivers bloodborne pathogens to lymph nodes. Lymph nodes filter the lymph and white blood cells in the nodes attack any pathogens.

The spleen filters the blood and removes any old red blood cells. The thymus is a hormone-secreting organ in which T lymphocytes (a kind of white blood cell) mature.

■ *Learn about the human lymphatic system with the animation on ThomsonNOW.*

Self-Quiz

Answers in Appendix III

1. Cells directly exchange substances with ______ .
   a. blood vessels
   b. lymph vessels
   c. interstitial fluid
   d. both a and b

2. All vertebrates have ______ .
   a. an open circulatory system
   b. a closed circulatory system
   c. a four-chambered heart
   d. both b and c

3. Which are not found in the blood?
   a. plasma
   b. blood cells and platelets
   c. gases and dissolved substances
   d. All of the above are found in blood.

4. A person who has type O blood ______ .
   a. can receive a transfusion of blood of any type
   b. can donate blood to a person of any blood type
   c. can donate blood only to a person of type O
   d. cannot be a blood donor
   e. both a and b

5. In the blood, most oxygen is transported ______ .
   a. in red blood cells
   b. in white blood cells
   c. bound to hemoglobin
   d. both a and c

6. Blood flows directly from the left atrium to ______ .
   a. the aorta
   b. the left ventricle
   c. the right atrium
   d. the pulmonary arteries

7. Contraction of ______ drives the flow of blood through the aorta and pulmonary arteries.
   a. atria
   b. arterioles
   c. ventricles
   d. skeletal muscle

8. Blood pressure is highest in the ______ and lowest in the ______ .
   a. arteries; veins
   b. arterioles; venules
   c. veins; arteries
   d. capillaries; arterioles

9. At rest, the largest volume of blood is in ______ .
   a. arteries
   b. capillaries
   c. veins
   d. arterioles

10. At the start of a capillary bed (closest to arterioles), ultrafiltration moves ______ .
    a. proteins into the capillary
    b. interstitial fluid into the capillary
    c. proteins into the interstitial fluid
    d. water, ions, and small solutes into interstitial fluid

11. Which is not a function of the lymphatic system?
    a. filters out pathogens
    b. returns fluid to the circulatory system
    c. helps certain white blood cells mature
    d. distributes oxygen to the tissues

12. Match the components with their functions.

___ capillary bed	a. filters out pathogens
___ lymph node	b. cardiac pacemaker
___ blood	c. main blood volume reservoir
___ ventricle	d. largest artery
___ SA node	e. fluid connective tissue
___ veins	f. zone of diffusion
___ aorta	g. contractions drive blood circulation

13. Label the components of the heart, shown below:

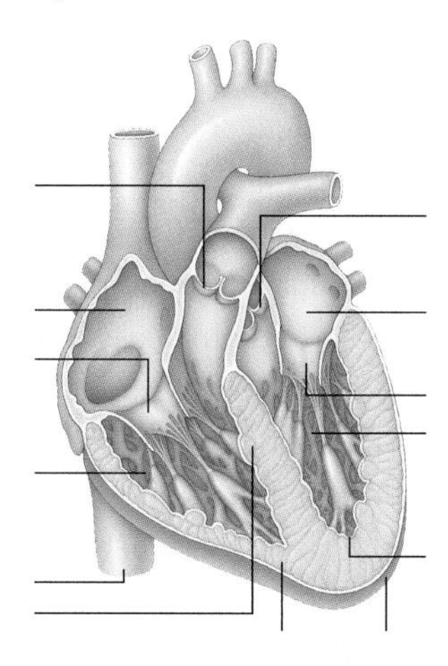

14. Identify these blood vessels and list their functions:

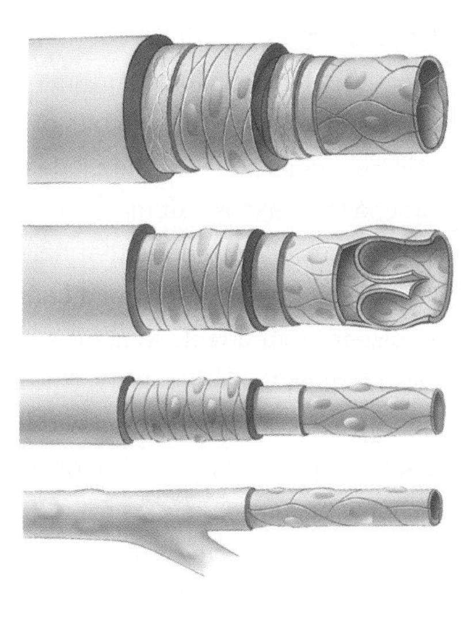

■ *Visit ThomsonNOW for additional questions.*

Critical Thinking

1. The highly publicized deaths of a few airline travelers led to warnings about economy-class syndrome. The idea is that sitting motionless for long periods on flights allows blood to pool and clots to form in legs. Low oxygen levels in airline cabins may increase clotting. If a clot gets large enough to block blood flow or breaks free and is carried to the lungs, the outcome can be deadly.

More recent studies suggest that long-distance flights cause problems in about one percent of air travelers, and that the risk is the same regardless of whether a person is in a first class seat or an economy seat. Physicians suggest that air travelers drink plenty of fluids and periodically get up and walk around the cabin. Given what you know about blood flow in veins, explain why these precautions can lower the risk of clot formation.

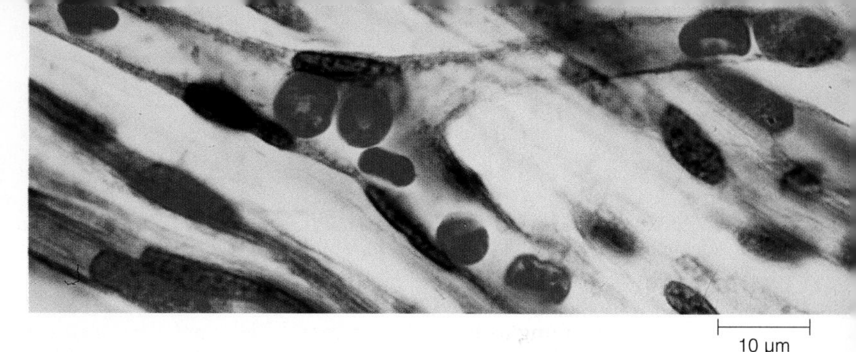

Figure 34.28 Light micrograph of a branching blood vessel.

2. Consider the micrograph in Figure 34.28. It shows red blood cells moving through a blood vessel. What type of vessel is this? Explain how you came to this conclusion.

3. Some membrane proteins of *Streptococcus pyogenes* are similar to those of cells in connective tissues throughout a human body. When this bacterium causes throat infections, the body mounts a defensive response against it. In some cases, this response also targets connective tissues of the heart, joints, and elsewhere. Chronic inflammation over the course of years or even decades leads to rheumatic heart disease. Heart valves become damaged or deformed. Explain how this disease affects the heart's function and what problems might arise as a result.

4. Mitochondria occupy about 40 percent of the volume of human cardiac muscle but only about 12 percent of the volume of skeletal muscle. Explain this difference.

5. Like other insects, the fruit fly (*Drosophila melanogaster*) has an open circulatory system. The transport medium is hemolymph, not blood. Contraction of one of the main vessels of the system drives a slow flow of hemolymph through the tiny fly body.

In 1993, researchers found that normal development of the fly "heart" requires the presence of a gene they named *tinman*. The gene's name is a reference to the character in *The Wizard of Oz* who had no heart. If the *tinman* gene is mutated, no heart forms in the embryonic fly, which dies. Genes having a similar base sequence have been found in zebrafish, the African clawed toad, chickens, mice, and humans. In all of these evolutionarily distant species, mutant versions result in abnormal heart development. In humans, mutations cause some genetic defects in the partition dividing the heart chambers and in the cardiac conduction system. Does it surprise you that a single gene would have such a major effect on the development of hearts or heartlike organs in so many different organisms? Explain your answer.

6. Miranda is nineteen, a college student, and a dedicated runner. About a month ago, a friend convinced her to go on a strict vegan diet. She now avoids all foods and other products that contain materials from animals, including eggs and milk. As she began training for a race, she found herself feeling unusually fatigued. Her friends commented that she looked pale. When she visited her doctor, she was told that the iron stores in her body were being depleted. Explain how iron deficiency would affect her blood and could cause her symptoms.

35 IMMUNITY

IMPACTS, ISSUES Viruses, Vulnerability, and Vaccines

Chedo Gowero was ten years old when both of her parents died from AIDS. She left school to support herself, her grandmother, and her seven-year-old brother. After that, Chedo spent her days working in the homes and fields of her neighbors in exchange for food, and struggling to pay her brother's tuition (Figure 35.1). Chedo and her brother are among 12 million children in sub-Saharan Africa that have been orphaned by AIDS. These two are lucky; many others are forced into prostitution because they have no other way to support themselves. Now expand this picture. Over 23 million people have died from AIDS worldwide. Almost 39 million are now infected with HIV, the virus that causes it. Among the infected are 2.3 million children.

After decades of high-priority, top-notch research all over the world, we still do not have an effective vaccine against AIDS. Why not? After all, we have been able to develop vaccines for many other dangerous diseases with far less understanding of the way they work.

For example, the very first vaccine was the result of desperate attempts to survive the smallpox epidemics that swept again and again through the world's cities. Smallpox is a severe disease that kills one-third of the people it infects (Figure 35.2). Before the 1880s, no one knew what caused infectious diseases or how to protect anyone from getting them. There were clues, however. In the case of smallpox, survivors had disfiguring scars, but they seldom contracted the disease again. They were *immune*, or protected from infection.

The idea of acquiring immunity to smallpox was so appealing that people had been risking their lives on it for almost two thousand years. Some ground up dried scabs from people that had smallpox, and then inhaled the powder. Others poked into their skin bits of scabs or threads soaked in pus. Many survived the crude practices and became immune to smallpox, but many others did not.

By 1774, it was common knowledge that dairymaids usually did not get smallpox after they had contracted cowpox (a mild disease that affects humans as well as cattle). An English farmer, Benjamin Jesty, made this prediction: If people accidentally infected with cowpox become immune to smallpox, then people deliberately infected with it should become immune too. In a desperate attempt to protect his family from a smallpox epidemic

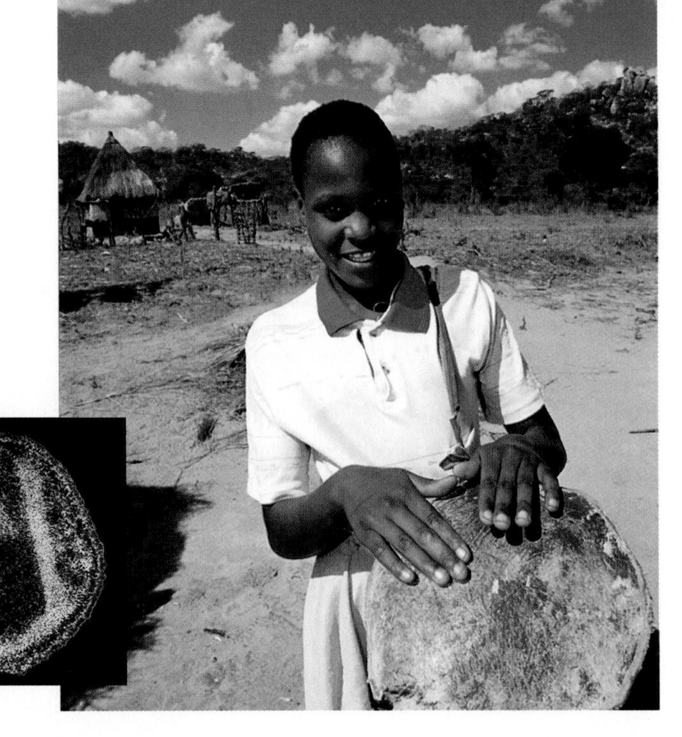

See the video! **Figure 35.1** The face of AIDS. *Left*, this color-enhanced transmission electron micrograph reveals HIV, the viral agent of AIDS. Despite decades of intense research, we still have no vaccine that protects us from HIV infection. *Right*, a social casualty of AIDS, Chedo Gowero is one of 12 million children in sub-Saharan Africa that have lost their parents to the disease. Many of these orphans struggle to survive on their own in this region, which is so impoverished that the average wage of half of the population is $0.60 per day.

How would you vote? Some clinical trials of potential AIDS vaccines are taking place in underdeveloped countries that have fewer limitations for human testing than the United States. Should clinical trials be held to the same ethical standards no matter where they take place? See ThomsonNOW for details, then vote online.

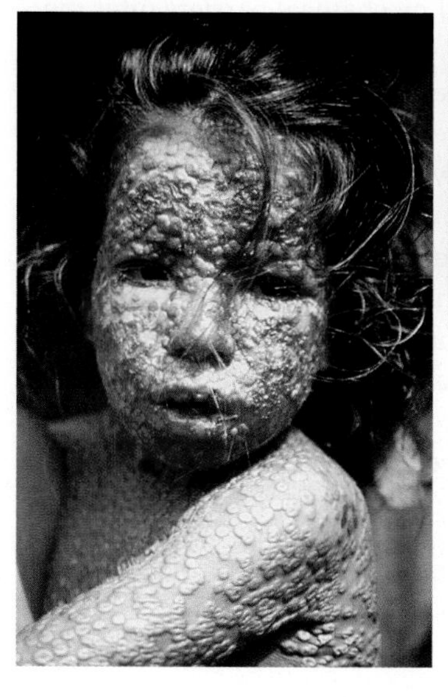

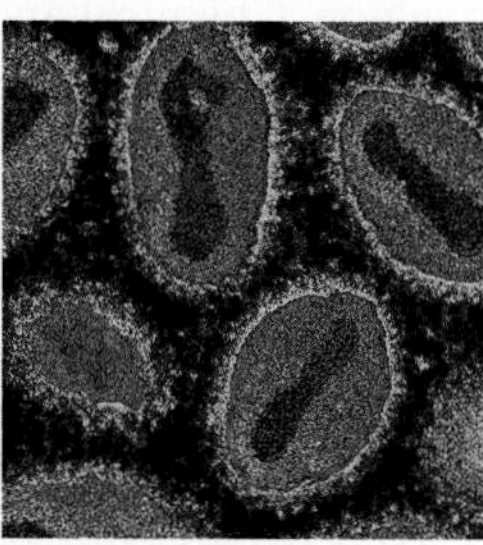

Figure 35.2 Young victim and the cause of her disease, smallpox viruses. Worldwide use of the vaccine eradicated naturally occurring cases of smallpox; vaccinations for it ended in 1972.

raging in Europe at the time, Jesty collected pus from a cowpox sore on a cow's udder, and poked it into the arm of his pregnant wife and two small children. All survived the smallpox epidemic, though they were from that time on subject to derision and rock peltings by neighbors who were convinced they would turn into cows.

Then, in 1796, Edward Jenner, an English physician, performed a definitive experiment. He injected liquid from a cowpox sore into the arm of a healthy boy. Six weeks later, Jenner tested the boy's immunity by injecting into him liquid from a smallpox sore. The boy did not get smallpox. Jenner's experiment showed directly that the agent of cowpox provokes immunity against smallpox.

Jenner named his procedure vaccination, after the Latin word for cowpox (*vaccinia*). Though it was still controversial, the use of Jenner's vaccine spread quickly throughout Europe, then to the rest of the world. The last known naturally occurring case of smallpox occurred in 1977, in Somalia. Vaccination had eradicated the disease.

Since the beginning of the twentieth century, many advances in microscopy, biochemistry, and molecular biology have increased our understanding of how the body defends itself against threats. As our difficulty making an AIDS vaccine reminds us, however, there is still much to learn. With that sobering thought, we invite you into the world of immunity.

Key Concepts

OVERVIEW OF BODY DEFENSES

The vertebrate body has three lines of defense against threats to health. Surface barriers can prevent invasion of internal tissues by ever-present pathogens. The innate immune response rids the body of most invaders before infection becomes established. Adaptive immune responses target specific pathogens and cancer cells. **Section 35.1**

SURFACE BARRIERS

Skin, mucous membranes, and secretions at the body's surfaces function as barriers that exclude most microbes. **Section 35.2**

INNATE IMMUNITY

The second line of defense, innate immunity, involves a set of general, immediate defenses against invading pathogens. Innate immunity includes phagocytic white blood cells, plasma proteins, inflammation, and fever. **Section 35.3**

ADAPTIVE IMMUNITY

Activation of innate immunity also triggers an adaptive immune response. White blood cells target and destroy specific invaders or altered cells. Some kinds make antibodies in an antibody-mediated immune response, and others directly destroy diseased body cells in a cell-mediated response. **Sections 35.4–35.7**

IMMUNITY GONE WRONG

There is a fine line between immune responses against nonself and self. The line is crossed in allergies and autoimmune disorders. Also, failed or faulty mechanisms can result in immune deficiencies. **Sections 35.8, 35.9**

Links to Earlier Concepts

In this chapter you will be integrating what you have learned about disease-causing agents and their hosts. You will be applying what you know about prokaryotic cells and viruses in Chapter 19 as you learn about their interactions with eukaryotic cells. You will use what you know about protein structure (3.4), the endomembrane system (4.8), eukaryotic tissue structure (4.11, 29.1, 29.2) and endocytosis and phagocytosis (5.9) to understand eukaryotic cellular defenses. Eukaryotic gene control (14.1) and cell signaling (25.5, 32.1) gave you the background to understand immune signaling mechanisms. You will see how body systems, including skin (29.6), the circulatory system (34.2, 34.8), and the lymphatic system (34.10), work together to fight infection. Finally, you may wish to review the coevolution of pathogens and hosts (17.12,18.4).

35.1 Integrated Responses to Threats

You continually cross paths with a staggering array of viruses, bacteria, fungi, maybe parasitic worms, and other pathogens, but you need not lose sleep over this. You and every other animal have defenses against them.

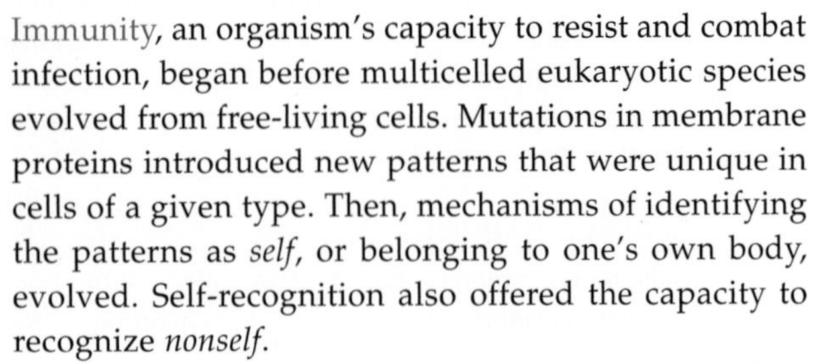

EVOLUTION OF THE BODY'S DEFENSES

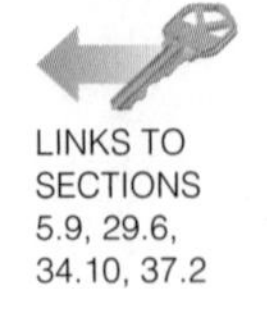

LINKS TO SECTIONS 5.9, 29.6, 34.10, 37.2

Immunity, an organism's capacity to resist and combat infection, began before multicelled eukaryotic species evolved from free-living cells. Mutations in membrane proteins introduced new patterns that were unique in cells of a given type. Then, mechanisms of identifying the patterns as *self*, or belonging to one's own body, evolved. Self-recognition also offered the capacity to recognize *nonself*.

Nonself recognition had evolved by about 1 billion years ago. Now, cells in all multicelled eukaryotes bear a set of receptors that collectively can recognize about 1,000 different nonself cues. These cues are molecular patterns that occur mainly on or in pathogens—certain components of prokaryotic cell walls, bacterial flagella and pili proteins, the double-stranded RNA of some viruses, and so on. When a cell's receptors bind to one of the patterns, they trigger a set of immediate, general defense responses. In mammals, for example, binding triggers activation of complement. Complement is a set of proteins that circulate in inactive form throughout the body. Activated complement can destroy microbes or tag them for phagocytosis (Section 5.9).

Pattern receptors and the responses they initiate are part of innate immunity, a set of fast, general defenses against infection. The defenses do not change within an individual's lifetime, but all multicelled organisms start out life with them.

Vertebrates have an additional set of defenses that involves many interacting cells, tissues, and proteins. Vertebrate adaptive immunity tailors immune defenses to a vast array of specific pathogens that an individual may encounter during its life. It is triggered by antigen: any molecule or particle that the body recognizes as nonself. Most antigens are polysaccharides, lipids, and proteins typically present on viruses, bacteria or other foreign cells, tumor cells, toxins, and allergens.

Adaptive immunity evolved within the context of innate immune systems. Researchers once thought that the systems worked independently, but we now know they function as an integrated mechanism. We describe both together in terms of three lines of defense.

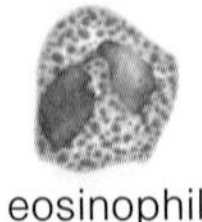

eosinophil

neutrophil

basophil

mast cell

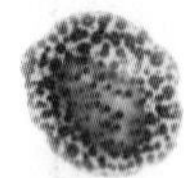

B cell

T cell

NK cell

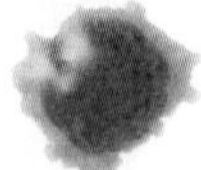

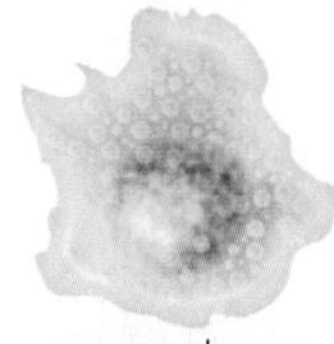

macrophage

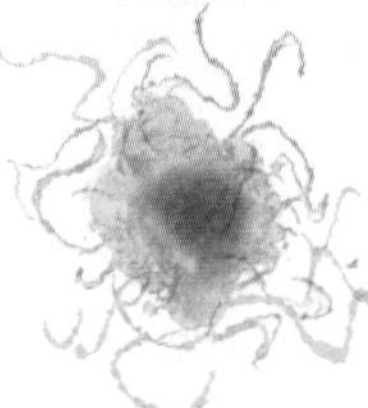

dendritic cell

Figure 35.3 White blood cells (leukocytes). Staining shows structural details such as cytoplasmic granules that contain enzymes, toxins, and signalling molecules. Compare Figure 34.5.

THREE LINES OF DEFENSE

Immune defenses begin at the skin and linings of body tubes and cavities, all of which are physical barriers to infection. Chemical and mechanical barriers offer more surface protection.

Innate immunity, the second line of defense, begins immediately after tissue is damaged, or after antigen is detected inside the body. General response mechanisms can rid the body of many different kinds of invaders before infection becomes established.

Activation of innate immunity triggers the third line of defense, adaptive immunity. White blood cells form huge populations that can target and destroy a specific antigen. Some of the cells persist long after infection ends. If the same antigen returns, these memory cells mount a secondary response. Adaptive immunity can target billions of specific antigens.

THE DEFENDERS

White blood cells participate in innate and adaptive immunity (Figure 35.3). Many kinds circulate through the body in the blood and lymph; others populate the lymph nodes, spleen, and other tissues. Some white blood cells are phagocytic, and all are secretory. Their secretions include cytokines—signaling molecules that take part in every aspect of vertebrate immunity.

The different white blood cell types are specialized for specific tasks. Neutrophils are the most abundant circulating phagocytes. Phagocytic macrophages patrol tissue fluids. Dendritic cells, also phagocytic, alert the adaptive immune system to the presence of an antigen. After their receptors bind antigen, basophils in blood and mast cells in tissues release chemicals that cause inflammation. Eosinophils target parasites too big for phagocytic cells. Lymphocytes are central to adaptive immunity. The B and T lymphocytes (or B and T cells) recognize billions of specific antigens. There are several kinds of T cells, including cytotoxic T cells that target infected or cancerous body cells. Natural killer cells (NK cells) destroy infected or cancerous body cells that are undetectable by cytotoxic T cells.

The innate immune system involves general defenses and responses that target a fixed number of pathogens. It acts immediately to prevent infection.

Adaptive immunity is a system of defenses that specifically targets billions of different pathogens.

White blood cells are central to both systems; signaling molecules integrate their activities.

35.2 Surface Barriers

A pathogen can cause infection only if it enters the internal environment by penetrating skin or other protective barriers at the body's free surfaces.

Your skin teems with about 200 species of microbial residents; there can be thousands to billions of them per square inch. Your surfaces provide their microbial inhabitants with a stable environment and a constant supply of nutrients. By contrast, the internal tissues of healthy people are typically microorganism-free. Body cavities are another matter. Tremendous populations of microbes inhabit the cavities and tubes that open out on the body's surface, including the eyes, nose, mouth, and anal and genital openings.

Keeping these microbial inhabitants on the outside of the body is critical to your health: Some of them can cause serious illness when they invade internal tissues. The bacterial agent of tetanus, *Clostridium tetani*, passes through our intestinal tract so often that we say it is indigenous. *Staphylococcus aureus*, a common resident of human skin, nasal membranes, and intestines, also is the leading cause of human bacterial disease (Figure 35.4*b*). Other common microbial inhabitants can cause or worsen pneumonia, ulcers, colitis, whooping cough, meningitis, abscesses of the lung and brain, and colon, stomach, and intestinal cancers.

Physical, chemical, and mechanical barriers usually keep microbes on the outside of your body (Table 35.1). For example, healthy, intact skin is an effective physical barrier to most microorganisms. Vertebrate skin has a tough surface layer of dead, densely packed epithelial cells (Section 29.6). Microorganisms flourish on top of this waterproof, oily environment, but rarely penetrate it. Populations of resident microbes also keep the more dangerous types from colonizing skin.

Thick, sticky mucus that coats the surfaces of many epithelial linings is a mechanical barrier that can trap bacteria. Mucus also functions as a chemical barrier because it contains lysozyme, an enzyme that cleaves polysaccharides in bacterial cell walls and so unravels their structure. When you breathe in, airborne bacteria are blown against mucus-coated airways to the lungs (Figure 35.4*b*). Coughing, another mechanical barrier, expels many bacteria. Lysozyme in the mucus ensures that the rest will not survive long enough to breach the walls of the sinuses and lower respiratory tract.

Microbes that colonize the mouth resist lysozyme in saliva. Any that enter the stomach are usually killed by gastric fluid, which is a potent brew of hydrochloric acid and protein-digesting enzymes. If any microbes survive and reach the small intestine, bile salts in the intestinal lumen usually will kill them. The few hardy ones that end up in the large intestine must compete with about 500 established resident species. If they do manage to displace the resident populations, diarrhea typically flushes them out.

Lactic acid produced by *Lactobacillus* helps keep the vaginal pH outside of the range of tolerance for most bacteria and fungi. Urination's flushing action usually stops pathogens from colonizing the urinary tract.

Table 35.1 Vertebrate Surface Barriers

Physical	Intact skin and epithelia that line the body's tubes and cavities, such as the gut and eye sockets
Mechanical	Sticky mucus; broomlike action of cilia; flushing action of tears, saliva, urination, diarrhea
Chemical	Protective secretions (sebum, other waxy coatings); low pH of urine, gastric juices, urinary and vaginal tracts; lysozyme; established populations of resident microbes

Surface barriers protect the internal environment of all vertebrates. They include intact skin, lysozyme and other secretions, and internal linings with mucus-secreting cells and often ciliated cells. Populations of normally harmless bacteria help stop pathogens from colonizing surfaces.

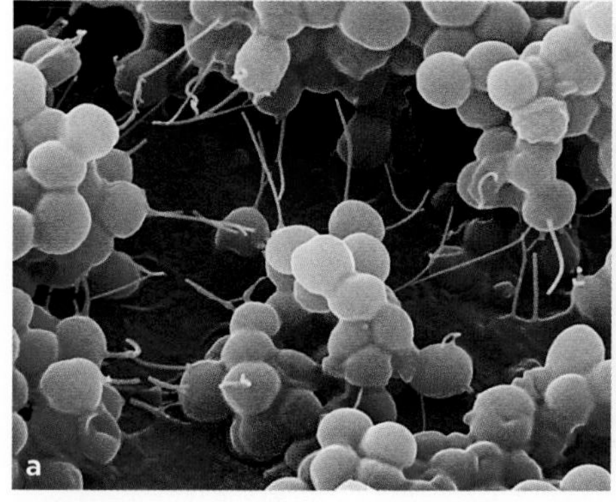

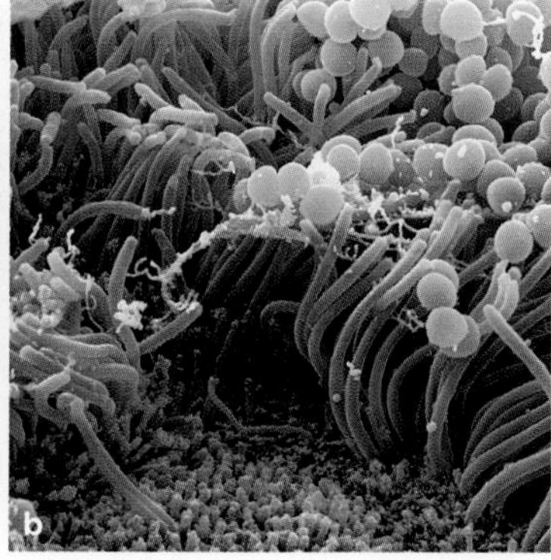

Figure 35.4 Two microbial inhabitants of human surfaces. (**a**) *Staphylococcus epidermidis*, the most common colonizer of human skin.

(**b**) *Staphylococcus aureus* cells (*yellow*) adhering to mucus-coated cilia of human nasal epithelial cells. *S. aureus* is a common inhabitant of human skin and linings of the mouth, nose, throat, and intestines. It is the leading cause of bacterial disease in humans. Antibiotic-resistant strains of *S. aureus* are now widespread. A particularly dangerous kind (MRSA) that is resistant to all penicillins is now endemic in most hospitals around the world. MRSA is called a "superbug."

35.3 Innate Immune Responses

Innate immunity includes the body's fixed, off-the-shelf mechanisms that counter threats and prevent infection. All animals are normally born with these mechanisms, which do not change much over the course of a lifetime.

PHAGOCYTES AND COMPLEMENT

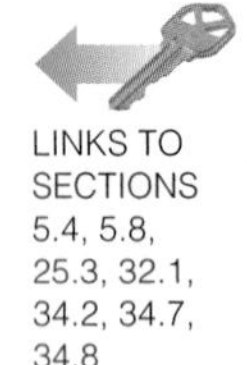

LINKS TO SECTIONS 5.4, 5.8, 25.3, 32.1, 34.2, 34.7, 34.8

Macrophages engulf and digest essentially everything except undamaged body cells. They patrol interstitial fluid, and they are among the first white blood cells to encounter an invading pathogen. When their receptors bind antigen, macrophages secrete signaling molecules (cytokines) that attract more macrophages, as well as neutrophils and dendritic cells.

Antigen also triggers complement activation. About 30 different complement proteins circulate through the blood and interstitial fluid of vertebrates. Complement becomes "activated" when it binds antigen. Activated complement proteins activate more complement, which activates more complement, and so on. The cascading reactions produce tremendous local concentrations of activated complement very quickly.

Activated complement proteins attract phagocytic cells. Like snuffling bloodhounds, phagocytes follow gradients of complement to a threatened tissue. Some complement proteins also attach directly to microbes. Phagocytes have complement receptors, so a microbe coated with the proteins gets recognized and engulfed faster than one without a coating. Other complement molecules self-assemble into complexes that puncture bacterial cell walls or plasma membranes (Figure 35.5).

The activated complement proteins also function in adaptive immunity. They help guide how immune cells mature and mediate interactions among them.

INFLAMMATION

Cytokine secretions from macrophages and activated complement both trigger inflammation, a local response to tissue invasion or damage. Its outward symptoms are redness, warmth, swelling, and pain. Inflammation begins when the pattern receptors on basophils, mast cells, or neutrophils bind antigen, or when mast cells bind complement directly. In response, the cells release

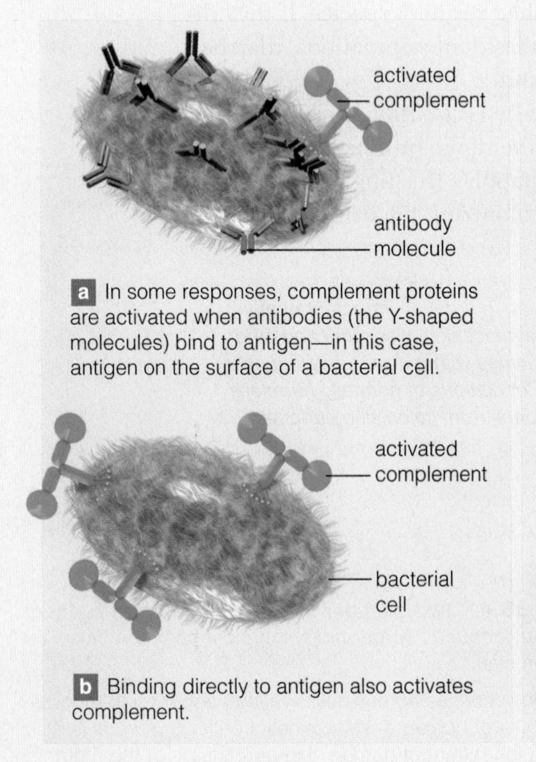

a In some responses, complement proteins are activated when antibodies (the Y-shaped molecules) bind to antigen—in this case, antigen on the surface of a bacterial cell.

b Binding directly to antigen also activates complement.

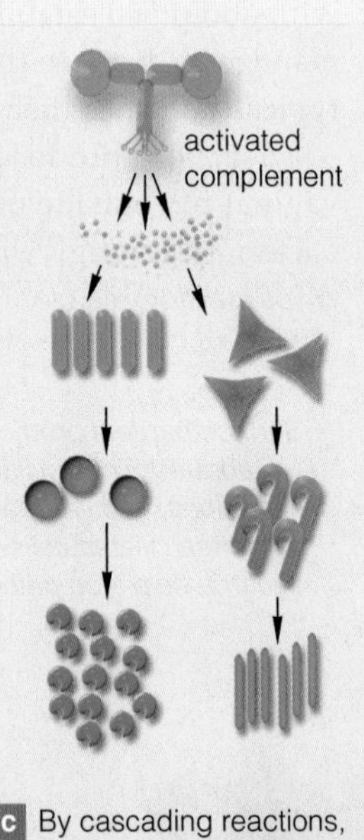

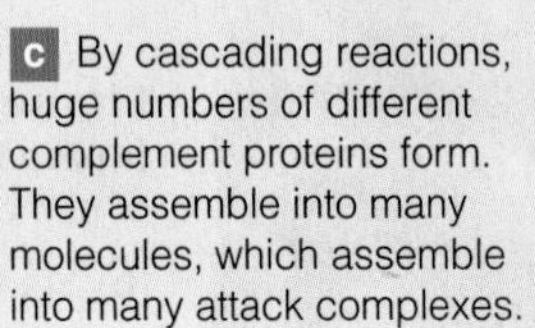

c By cascading reactions, huge numbers of different complement proteins form. They assemble into many molecules, which assemble into many attack complexes.

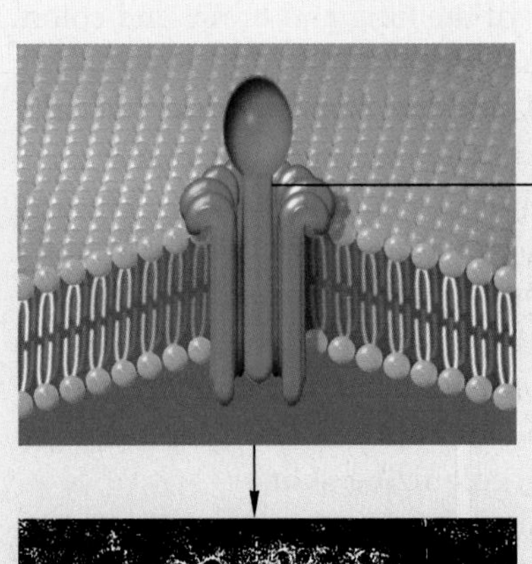

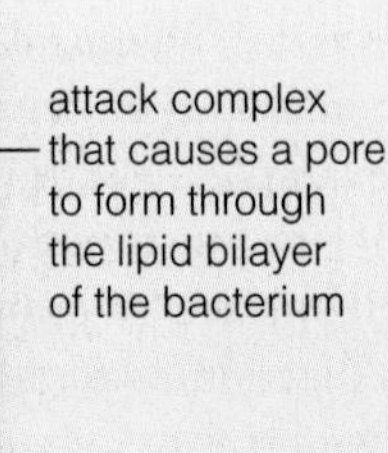

d The attack complexes become inserted into the target cell's lipid envelope or plasma membrane. Each complex makes a large pore form across it.

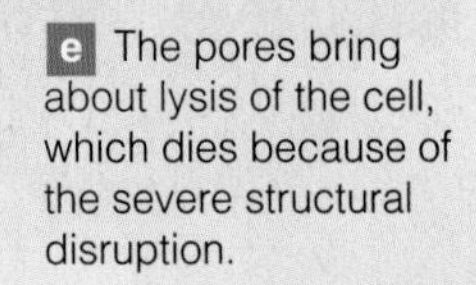

e The pores bring about lysis of the cell, which dies because of the severe structural disruption.

Figure 35.5 **Animated!** One effect of complement protein activation. Activation at step (**a**) or (**b**) causes lysis-inducing pore complexes to form. The transmission electron micrograph of a pathogen's surface shows holes made by membrane attack complexes.

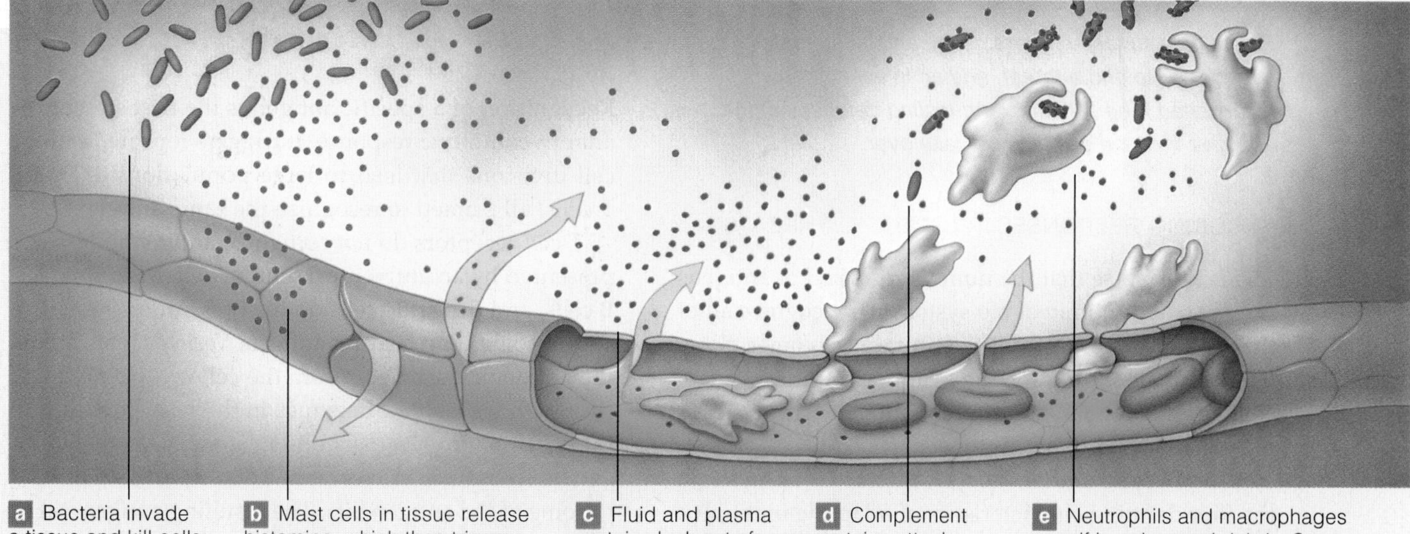

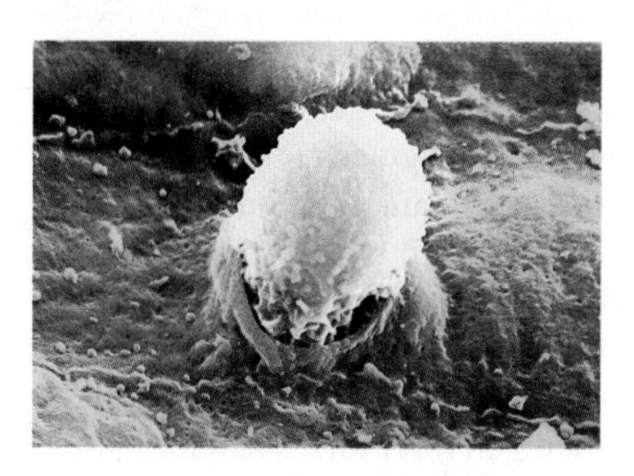

Figure 35.6 Animated! (**a–e**) A case of inflammation in response to bacterial infection. Fast-acting white blood cells and plasma proteins enter a damaged tissue. The micrograph on the *left* shows a phagocyte squeezing through a blood vessel wall.

histamine, prostaglandins, and other vasodilators into the affected tissue (Section 34.7).

These substances have two effects. First, they cause nearby arterioles to widen. As a result, blood flow to the area increases, reddening the tissue and warming it with bloodborne metabolic heat. The increased flow also speeds the arrival of more phagocytes attracted by the cytokines.

Second, the signaling molecules make capillaries in the affected tissue "leaky." Spaces open between cells in the capillary wall. Phagocytes and plasma proteins squeeze between the cells, out of the blood vessel and into interstitial fluid (Figure 35.6). The transfer changes the osmotic balance across the capillary wall, so more water diffuses from the blood into tissue. The tissue swells with fluid, and this puts pressure on free nerve endings. The pressure gives rise to sensations of pain.

FEVER

Fever is a temporary rise in body temperature above the normal 37°C (98.6°F) that often occurs in response to infection. Remember, macrophages secrete cytokines when their pattern receptors bind to an antigen. Some cytokines stimulate brain cells to make and release a prostaglandin (Section 32.1). This signaling molecule acts on the hypothalamus to elevate the body's internal temperature set point. As long as the temperature of the body is below the new set point, the hypothalamus signals effectors (Section 25.3) to give rise to a sensation of cold, to constrict blood vessels in the skin, and to trigger shivering. All of these responses help raise the internal temperature of the body.

A fever is a sign that the body is fighting something, so it should never be ignored. However, a moderate fever—39°C (102°F) or so—does not necessarily require treatment in an otherwise healthy person. Fever may enhance the body's immune defenses by increasing the rate of enzyme activity, thus speeding up metabolism, tissue repair, and formation and activity of phagocytes. Some types of microorganisms grow more slowly at elevated temperatures, so white blood cells can get a head start in the proliferation race against them.

A fever usually will not rise above 40°C (105°F), but if it does, immediate hospitalization is recommended. A fever of 42°C (107.6°F) is deadly.

Innate immunity is the body's built-in set of fixed defenses against internal threats. Complement, phagocytes, acute inflammation, and fever quickly eliminate most invaders from the body before infection is established.

35.4 Overview of Adaptive Immunity

Sometimes surface barriers, fever, and inflammation are not enough to end a threat, and an infection becomes established. The specific, long-lasting defenses of the adaptive immune system now take over.

TAILORING RESPONSES TO SPECIFIC THREATS

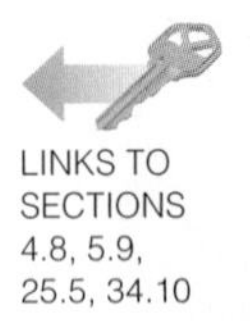

LINKS TO SECTIONS 4.8, 5.9, 25.5, 34.10

Life is so diverse that the number of different antigens is essentially unlimited. No system can recognize all of them, but vertebrate adaptive immunity comes close. Unlike innate immunity, the adaptive immune system changes: It "adapts" to different antigens encountered during an individual's lifetime.

Lymphocytes and phagocytes interact to bring about the four defining characteristics of adaptive immunity. These features are self/nonself recognition, specificity, diversity, and memory.

Self versus nonself recognition starts with molecular patterns that give each kind of cell or virus its unique identity. The plasma membranes of your cells bear self-recognition proteins known as MHC markers, after the genes that encode them. Your T cells also have T cell receptors, or TCRs, which are antigen receptors at their surfaces. Part of the TCR recognizes MHC markers as self; part of it also recognizes antigen as nonself.

Specificity means that each B cell or T cell produces receptors that bind to one—and only one—antigen.

Diversity refers to the collection of antigen receptors on all B and T cells in the body. There are potentially billions of different antigen receptors, so an individual has the potential to counter billions of different threats.

Memory refers to the immune system's capacity to "remember" an antigen. It take a few days for B and T cells to respond in force the first time they recognize antigen. If the same antigen shows up again, they make a faster, stronger response. That is why we do not get as sick when we encounter a pathogen a second time.

FIRST STEP—THE ANTIGEN ALERT

Recognition of a specific antigen is the first step of the adaptive immune response. It triggers repeated mitotic cell divisions that lead to large populations of B and T cells, all primed to recognize the same antigen.

T cell receptors do not recognize antigen unless it is presented by an antigen-presenting cell. Macrophages, B cells, and dendritic cells do the presenting. First, they engulf something bearing antigen. Vesicles that contain the antigenic particle form in the cell cytoplasm; these fuse with lysosomes. Enzymes in the lysosomes digest the particle into bits (Sections 4.8 and 5.9).

The lysosomes also contain MHC markers that bind to some of the antigen bits. The resulting antigen–MHC complexes become displayed at the cell's surface when the vesicles fuse with (and become part of) the plasma membrane (Figure 35.7). *Display of a cell's MHC markers paired with antigen fragments is a call to arms.*

Any T cell that bears a receptor for this antigen will bind the antigen–MHC complex. The T cell then starts secreting cytokines. These signals cause all other B or T cells with the same antigen receptor to divide again and again. Huge populations of B and T cells form after a few days; all of the cells recognize the same antigen. Most are *effector* cells—differentiated lymphocytes that act at once. Some are *memory* cells—long-lived B and T cells reserved for future encounters with the antigen.

TWO ARMS OF ADAPTIVE IMMUNITY

Like a boxer's one-two punch, adaptive immunity has two separate arms, called antibody-mediated and cell-mediated immune responses (Figure 35.8). They work together to eliminate diverse threats.

Not all threats present themselves in the same way. For example, bacteria, fungi, and toxins circulating in

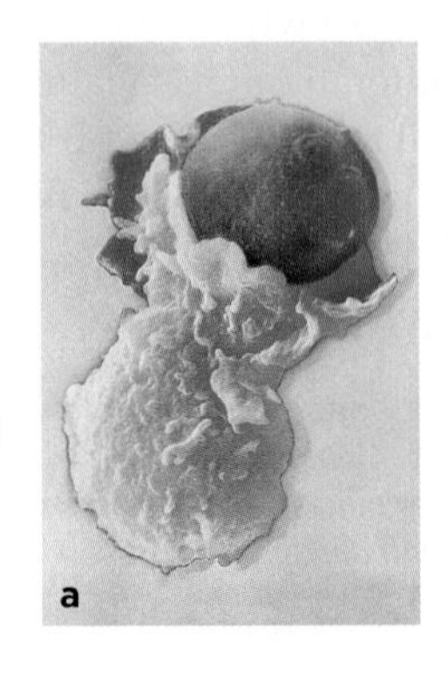

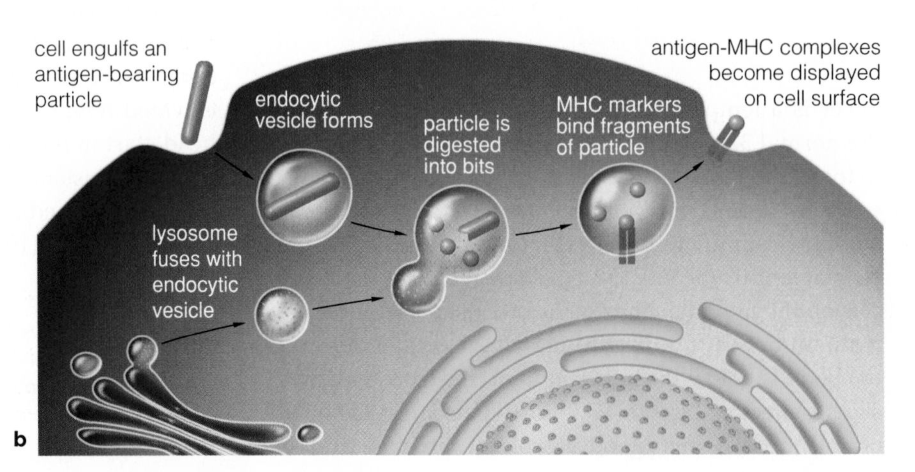

Figure 35.7 Antigen processing. (**a**) Micrograph of a macrophage ingesting a foreign cell.

(**b**) From encounter to display, what happens when a B cell, macrophage, or dendritic cell engulfs an antigenic particle—in this case, a bacterium. These cells engulf, process, and then display antigen bound to MHC markers. The displayed antigen is presented to T cells.

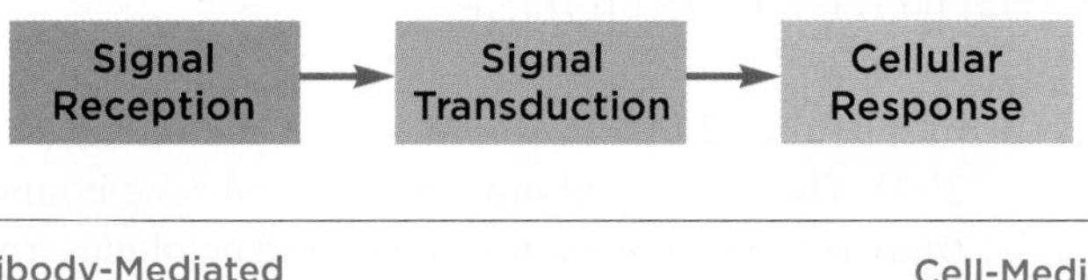

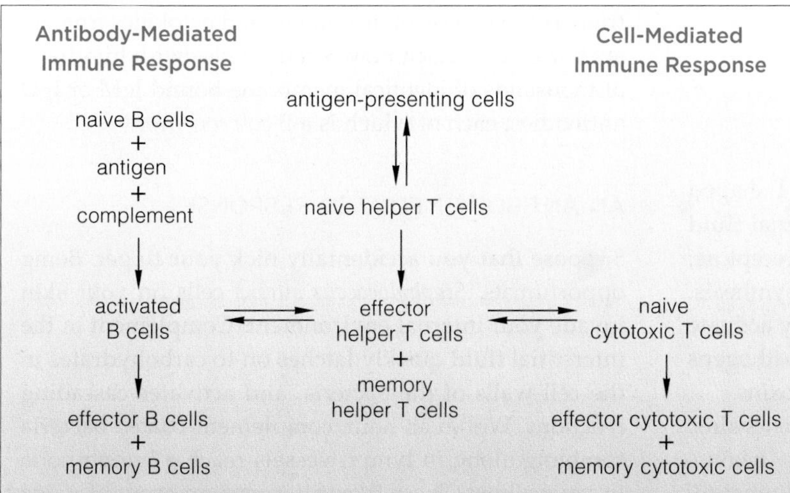

Figure 35.8 Overview of key interactions between antibody-mediated and cell-mediated responses—the two arms of adaptive immunity. Here again, cells communicate by way of signal transduction pathways (Section 25.5). A "naive" cell simply is one that has not made contact with its specific antigen.

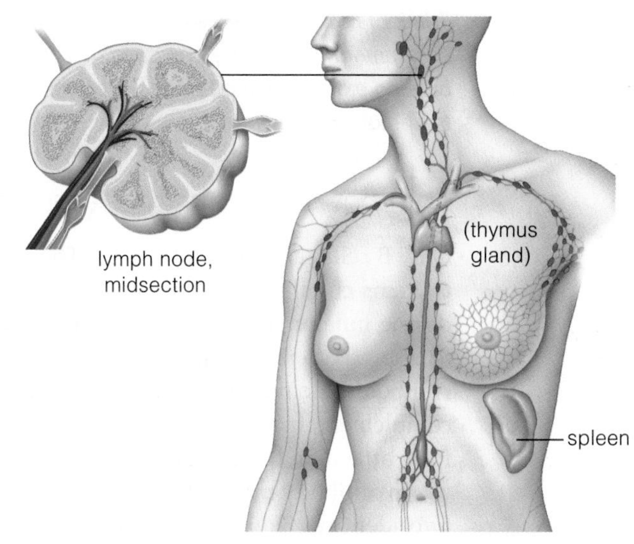

Figure 35.9 Revisiting the big battlegrounds in adaptive immunity. Lymph nodes positioned along lymph vascular highways hold macrophages, dendritic cells, B cells, and, deep in the node, T cells. The spleen filters free antigen from blood. Section 34.10 explains the structure and function of all of the lymphatic system's components.

blood or interstitial fluid are intercepted quickly by B cells and other phagocytes. These cells interact in the antibody-mediated immune response. In this response, B cells produce antibodies—proteins that can bind to specific antigen-bearing particles.

Some kinds of threats are not targeted by B cells. For example, B cells cannot detect body cells altered by cancer. As another example, some viruses, bacteria, fungi, and protists can hide and reproduce inside body cells; B cells can detect them only briefly, when they slip out of one cell to infect others. Such *intracellular* pathogens are targeted primarily by the cell-mediated immune response, which does not involve antibodies. In this response, cytotoxic T cells and NK cells detect and destroy altered or infected body cells.

INTERCEPTING AND CLEARING OUT ANTIGEN

After engulfing an antigenic particle, a dendritic cell or macrophage will migrate to a lymph node (Section 34.10). Once inside, it will present antigen to the many T cells filtering through the node (Figure 35.9). Every day, about 25 billion T lymphocytes pass through each lymph node. Those that recognize and bind to antigen presented by a phagocyte begin an adaptive response.

Free antigen in interstitial fluid ends up in a lymph vessel that leads to a lymph node. Inside the node, the antigen encounters arrays of resident B cells, dendritic cells, and macrophages. These phagocytic cells engulf, process, and present the antigen to the T cells that pass through the nodes. Antigen that manages to escape the nodes and enter blood is taken up by the spleen.

During infection, the lymph nodes swell as antigen-presenting T cells accumulate inside them. When you are sick, you may notice them as tender lumps under your jaw and elsewhere.

Immune responses subside once antigen is cleared away. The tide of battle turns as effector cells and their secretions kill most antigen-bearing agents. With less antigen present, fewer immune fighters are recruited. Complement proteins assist in the cleanup by binding antibody–antigen complexes. They form large clumps that can be quickly cleared from the blood by the liver and the spleen.

Adaptive immunity has four important characteristics: self/nonself recognition, specificity, diversity, and memory.

The two arms of adaptive immunity work together. Antibody-mediated responses target antigen in blood or interstitial fluid. Cell-mediated responses target altered body cells.

Lymph nodes house arrays of immune cells that recognize and trap most antigen-bearing particles.

35.5 The Antibody-Mediated Immune Response

Think of B cells as assassins. Each one has a genetic assignment to liquidate one particular target—an antigen-bearing extracellular pathogen or toxin. Antibodies are their molecular bullets.

ANTIBODY STRUCTURE AND FUNCTION

LINKS TO SECTIONS 3.4, 5.9, 29.1

Only B cells can make antibodies. These Y- or T-shaped proteins circulate in blood, and enter interstitial fluid during inflammation. Antibodies are antigen receptors; each binds to the antigen that prompted its synthesis. Antibodies do not kill pathogens directly. They activate complement, facilitate phagocytosis, prevent pathogens from attaching to body cells, and neutralize toxins.

An antibody molecule has four polypeptide chains: two identical "light" ones and two identical "heavy" ones (Figure 35.10). Each chain has a variable region. An antibody's variable regions form a unique array of bumps, grooves, and charge distribution. The antibody binds only to an antigen with a complementary array of grooves, bumps, and distribution of charge.

Most humans can make about 2.5 billion different antibodies. This diversity arises by random assortment of gene segments. During differentiation of each B cell, a few gene segments are selected at random from a large pool, and then spliced together into one antibody gene. A similar process generates T cell receptor genes.

In addition to a variable region, an antibody also has a constant region that determines its structural identity, or class. There are five antibody classes—IgG, IgA, IgE, IgM, and IgD (Ig stands for immunoglobulin, another name for antibody). The different classes serve different functions (Table 35.2). Most antibodies in the blood are IgG; they trigger complement cascades. IgA is found in mucus and other exocrine gland secretions (Section 29.1). The binding of antigen to membrane-bound IgE triggers the release of histamines and cytokines from an anchoring cell. Each new B cell bristles with hundreds of thousands of identical membrane-bound IgM or IgD antibodies, each of which is a B cell receptor.

AN ANTIBODY-MEDIATED RESPONSE

Suppose that you accidentally nick your finger. Being opportunists, *Staphylococcus aureus* cells on your skin invade your internal environment. Complement in the interstitial fluid quickly latches on to carbohydrates in the cell walls of the bacteria, and activates cascading reactions. Within an hour, complement-coated bacteria tumbling along in lymph vessels reach a lymph node in your elbow. There they filter past an army of naive B cells (ones that have not met antigen).

As it happens, a naive B cell bears IgM antibodies that bind polysaccharide in the bacterial cell wall. The B cell also has receptors that bind to the complement coat on the bacteria. Binding causes the B cell to engulf bacteria by receptor-mediated endocytosis (Section 5.9). The B cell is now activated (Figure 35.11*a*).

Meanwhile, more *S. aureus* cells have been secreting metabolic products into interstitial fluid around your cut. The secretions attract phagocytes. A dendritic cell engulfs several bacteria, then migrates to the lymph node in your elbow. By the time it gets there, it has digested the bacteria and is displaying their fragments bound to MHC markers on its surface (Figure 35.11*b*).

Each hour, about 500 different naive T cells travel through the lymph node, inspecting resident dendritic cells. In this case, one of those T cells has TCRs that

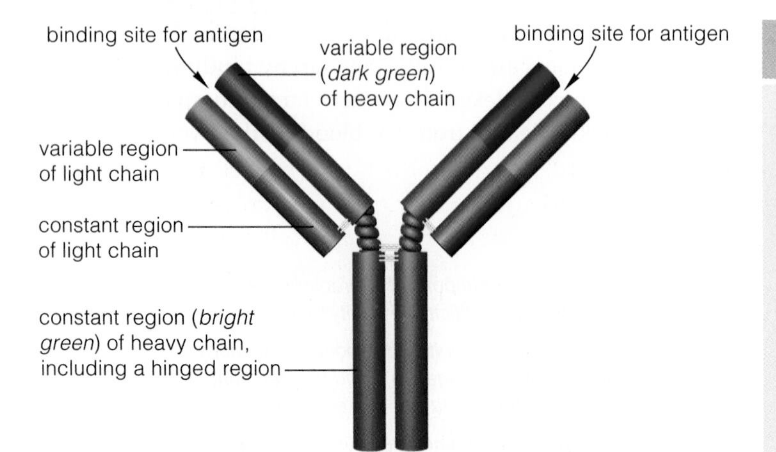

Figure 35.10 Antibody structure. An antibody molecule has four polypeptide chains, joined in a Y- or T-shaped configuration.

Table 35.2 Structural Classes of Antibodies

Secreted antibodies	
IgG	Main antibody in blood; activates complement, neutralizes toxins; protects fetus and is secreted in early milk.
IgA	Abundant in exocrine gland secretions (e.g., tears, saliva, milk, mucus). Interferes with binding of bacteria, viruses.
Membrane-bound antibodies	
IgE	Becomes anchored to surface of mast cells, basophils, dendritic cells. Binding antigen induces anchoring cell to release histamines, cytokines. Factor in allergies and AIDS.
IgM	B cell receptor; also secreted as pentamer (group of five).
IgD	B cell receptor.

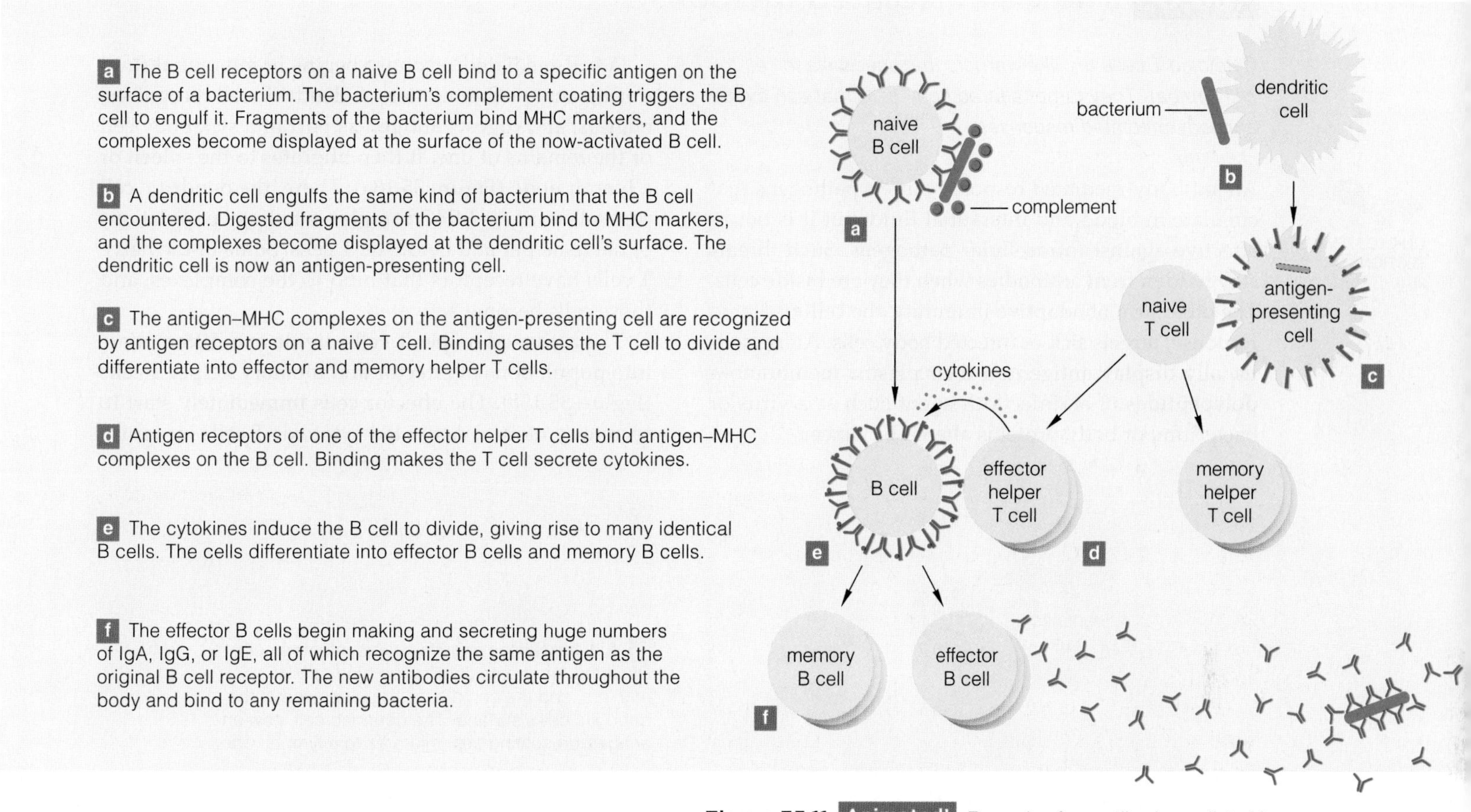

a The B cell receptors on a naive B cell bind to a specific antigen on the surface of a bacterium. The bacterium's complement coating triggers the B cell to engulf it. Fragments of the bacterium bind MHC markers, and the complexes become displayed at the surface of the now-activated B cell.

b A dendritic cell engulfs the same kind of bacterium that the B cell encountered. Digested fragments of the bacterium bind to MHC markers, and the complexes become displayed at the dendritic cell's surface. The dendritic cell is now an antigen-presenting cell.

c The antigen–MHC complexes on the antigen-presenting cell are recognized by antigen receptors on a naive T cell. Binding causes the T cell to divide and differentiate into effector and memory helper T cells.

d Antigen receptors of one of the effector helper T cells bind antigen–MHC complexes on the B cell. Binding makes the T cell secrete cytokines.

e The cytokines induce the B cell to divide, giving rise to many identical B cells. The cells differentiate into effector B cells and memory B cells.

f The effector B cells begin making and secreting huge numbers of IgA, IgG, or IgE, all of which recognize the same antigen as the original B cell receptor. The new antibodies circulate throughout the body and bind to any remaining bacteria.

Figure 35.11 Animated! Example of an antibody-mediated immune response.

bind the *S. aureus* antigen–MHC complexes displayed by the dendritic cell. For the next 24 hours, the T cell and the dendritic cell interact. When they disengage, the T cell returns to the circulatory system and begins to divide (Figure 35.11*c*). A huge population of T cells that have receptors for *S. aureus* antigen forms. These cells differentiate into helper and memory T cells.

Return to that activated B cell in the lymph node. By now, fragments of *S. aureus* bound to MHC markers are displayed at the B cell's surface. One of the new helper T cells recognizes the complexes. Like long-lost friends, the two cells exchange signals. The helper T cell secretes cytokines that signal the B cell to divide and differentiate (Figure 35.11*d*).

When the cells disengage, the B cell divides again and again to form a huge population of cells, all with receptors for *S. aureus* antigen. The cells differentiate into effector and memory B cells (Figure 35.11*e*).

The effector cells go to work immediately. They stop producing membrane-bound B cell receptors. Instead, they start making and secreting IgG, IgA, or IgE. Each of the antibody molecules recognizes the same *S. aureus* antigen. Antibodies now circulate throughout the body and bind remaining bacterial cells. Antibody binding prevents the bacteria from attaching to body cells, and also tags them for disposal (Figure 35.11*f*).

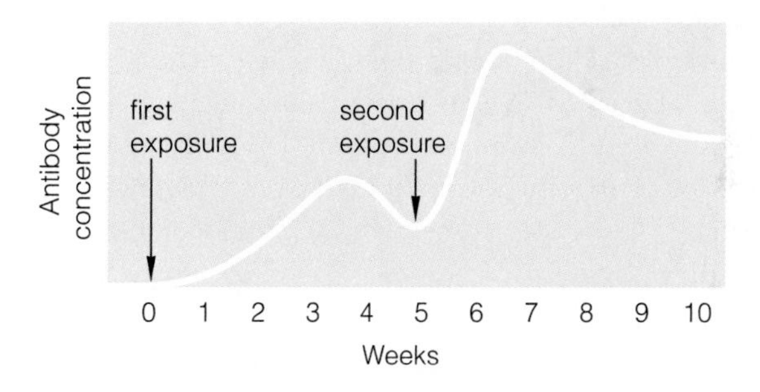

Figure 35.12 Antibody levels in a primary and secondary immune response. A secondary immune response is faster and stronger than the primary response that preceded it.

Memory B and T cells also form, but do not act right away. They persist long after the initial infection ends. If they encounter the same antigen again, these cells will initiate a secondary response that is stronger and faster than the primary one (Figure 35.12).

Antigen-presenting cells, T cells, and B cells interact in an antibody-mediated immune response targeting a specific antigen. Populations of B cells form; these make and secrete antibodies that recognize and bind the antigen.

35.6 The Cell-Mediated Response

Cytotoxic T cells are like warriors that specialize in cell-to-cell combat. They target altered body cells that can evade antibody-mediated responses.

LINK TO SECTION 25.5

An antibody-mediated response targets pathogens that circulate in blood and interstitial fluid, but it is not as effective against intracellular pathogens. Such threats stay hidden from antibodies when they are inside cells. The other arm of adaptive immunity, the cell-mediated response, targets sick or infected body cells. Ailing cells usually display antigen on their plasma membrane—polypeptides of an infectious agent such as a virus or bacterium, or body proteins altered by cancer.

A cell-mediated response begins in interstitial fluid during acute inflammation. A dendritic cell recognizes, engulfs, and digests antigen as part of a sick body cell or the remains of one. It then migrates to the spleen or a lymph node (Figure 35.13*a*). There, the dendritic cell presents antigen–MHC complexes to huge populations of naive helper and cytotoxic T cells. Some of the naive T cells have receptors that bind to the complexes, and those cells become activated.

The activated helper T cells divide and differentiate into populations of effector and memory helper T cells (Figure 35.13*b*). The effector cells immediately start to secrete cytokines. Activated cytotoxic T cells recognize

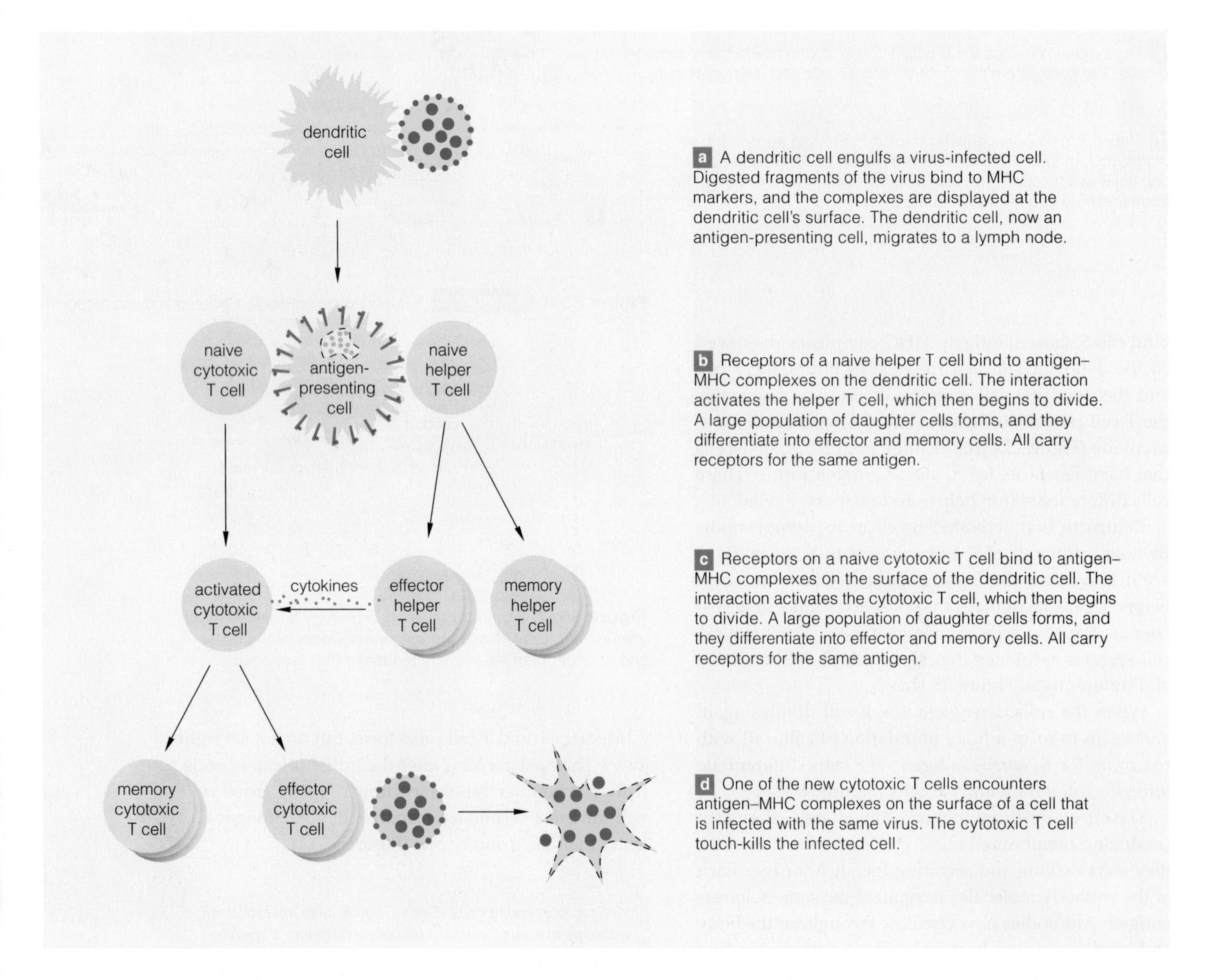

a A dendritic cell engulfs a virus-infected cell. Digested fragments of the virus bind to MHC markers, and the complexes are displayed at the dendritic cell's surface. The dendritic cell, now an antigen-presenting cell, migrates to a lymph node.

b Receptors of a naive helper T cell bind to antigen–MHC complexes on the dendritic cell. The interaction activates the helper T cell, which then begins to divide. A large population of daughter cells forms, and they differentiate into effector and memory cells. All carry receptors for the same antigen.

c Receptors on a naive cytotoxic T cell bind to antigen–MHC complexes on the surface of the dendritic cell. The interaction activates the cytotoxic T cell, which then begins to divide. A large population of daughter cells forms, and they differentiate into effector and memory cells. All carry receptors for the same antigen.

d One of the new cytotoxic T cells encounters antigen–MHC complexes on the surface of a cell that is infected with the same virus. The cytotoxic T cell touch-kills the infected cell.

Figure 35.13 **Animated!** Example of a primary cell-mediated immune response.

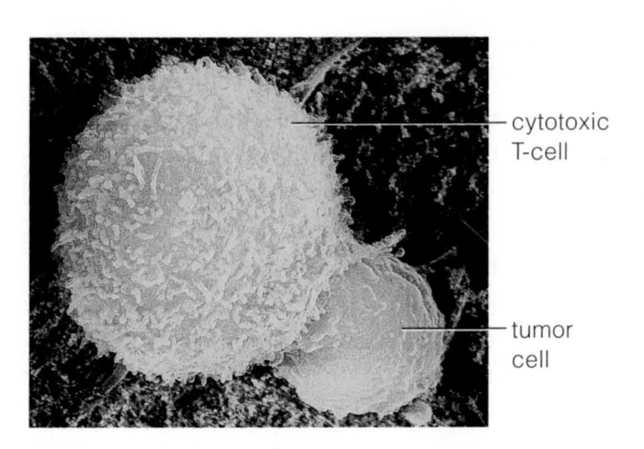

Figure 35.14 Cytotoxic T cell caught in the act of touch-killing a tumor cell.

the cytokines as signals. They respond by dividing and then differentiating into huge populations of effector and memory cytotoxic T cells (Figure 35.13*c*).

As in an antibody-mediated response, the memory cells that form in a primary cell-mediated response will mount a secondary response if the antigen returns.

Effector cytotoxic T cells now circulate throughout blood and interstitial fluid. All of them recognize and bind the same antigen—the one displayed by that first ailing cell. They bind to any other body cell displaying that antigen together with MHC markers, and inject it with perforin and proteases. These toxins poke holes in the cell and induce it to die by apoptosis (Section 25.5, Figure 35.13*d*, and Figure 35.14).

Cytotoxic T cells must recognize MHC molecules on the surface of a body cell in order to kill it. They can even recognize the MHC markers of foreign body cells (cytotoxic T cells bring about rejection of transplanted organs). However, an infected or cancerous cell's MHC markers may be altered or absent. NK cells are crucial for fighting such cells; unlike cytotoxic T cells, they can kill body cells that lack MHC markers.

Helper T cells secrete cytokines that stimulate NK cell division. These "natural killers" attack body cells tagged for destruction by antibodies. They also detect special proteins displayed by body cells that are under stress. Stressed body cells with normal MHC markers are not killed, but those with altered or missing MHC markers are destroyed.

Antigen-presenting cells, T cells, and NK cells interact in a cell-mediated immune response, which targets body cells infected or altered by cancer.

35.7 Remember to Floss

FOCUS ON HEALTH

Nine of every ten cardiovascular disease patients have serious periodontal disease. What is the connection?

LINKS TO SECTIONS 4.5, 4.11, 29.1, 34.9

Your mouth is a particularly inviting habitat for microbes, offering plenty of nutrients, warmth, moisture, and surfaces for colonization. Accordingly, it harbors huge populations of various streptococcus, lactobacillus, staphylococcus, and other bacterial species.

A few of the 400 or so species of microorganisms that normally live in the mouth cause dental plaque, a thick biofilm composed of bacteria, their extracellular products, and saliva glycoproteins. Plaque sticks tenaciously to teeth (Figure 35.15). Other bacteria that live in it are fermenters. They break down bits of carbohydrate that stick to teeth and then secrete organic acids, which etch away the tooth enamel and make cavities.

In young, healthy people, tight junctions between the gum epithelium and teeth form a barrier that keeps oral microorganisms out of the internal environment. As we age, connective tissue beneath gum epithelium thins, and the barrier becomes vulnerable. Deep pockets form between the teeth and gums, and a very nasty gang of anaerobic bacteria accumulates in these pockets. Their noxious secretions, including destructive enzymes and acids, cause inflammation of surrounding gum tissues—a condition called periodontitis.

Porphyromonas gingivalis is one of those anaerobic species. Along with every other species of oral bacteria that is associated with periodontitis, *P. gingivalis* also occurs in atherosclerotic plaque (Section 34.9). The wounds of periodontitis offer oral bacteria an open door to the circulatory system—and to the arteries.

Atherosclerosis is now known to be a disease of inflammation. Macrophages and T cells are attracted to lipid deposits in the vessel walls. Their secretions initiate inflammation that further attracts lipids, and the lesion grows as the immune cells die and become part of the deposits. What role the oral bacteria play in this scenario is not yet clear, but one thing is certain—they contribute to the inflammation that fuels coronary artery disease.

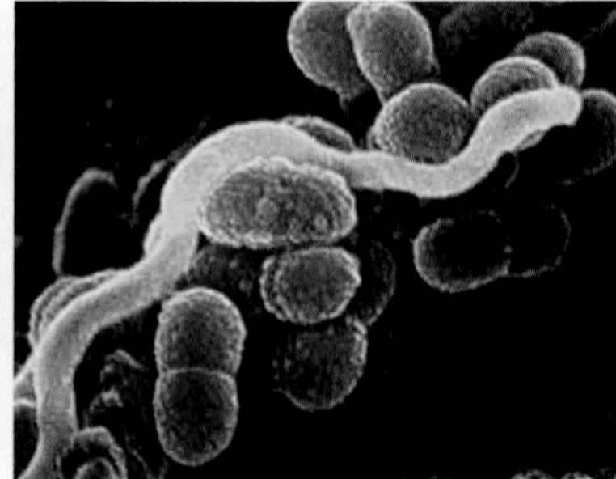

Figure 35.15 Plaque. *Left*, micrograph of toothbrush bristles scrubbing plaque on a tooth surface. *Right*, the main cause of plaque, *Streptococcus mutans*, which is actually a group of related bacteria.

35.8 Defenses Enhanced or Compromised

Sometimes immune responses are not strong enough or are misdirected or compromised. Here are examples of what we can and cannot do about it.

IMMUNIZATION

LINKS TO SECTIONS 29.2, 30.7

Immunization refers to processes designed to induce immunity. In *active* immunization, a preparation that contains antigen—a vaccine—is administered orally or injected (Table 35.3). The first immunization elicits a primary immune response, just as an infection would. A booster, or second immunization, elicits a secondary immune response for enhanced immunity.

Vaccines save millions of lives every year and are an important part of any public health program. Many vaccines consist of weakened or killed pathogens, or inactivated bacterial toxins. Some are harmless viruses that have been genetically engineered to carry genes of a pathogen. Such carrier viruses infect body cells, which then temporarily produce the gene products—antigens that provoke an immune response.

With *passive* immunization, a person who is already infected receives antibodies purified from the blood of another person. Such antibodies immediately protect someone who has been exposed to a potentially lethal agent, such as tetanus, rabies, Ebola virus, or a venom or toxin. Because the antibodies were not produced by the recipient's lymphocytes, memory cells do not form. Benefits last only as long as the injected antibodies do.

Vaccines can fail to elicit an immune response. For example, the hepatitis B vaccine does not work in one of ten people. Some have rare but serious side effects. If you are considering a vaccination, discuss the risks and benefits of the procedure with your physician.

ALLERGIES

In millions of people, exposure to harmless substances stimulates an immune response. Any substance that is ordinarily harmless yet provokes such responses is an allergen. Hypersensitivity to them is called an allergy (Figure 35.16). Drugs, foods, pollen, dust mites, fungal spores, poison ivy, and venom from bees, wasps, and other insects are among the most common allergens.

Some people are genetically predisposed to having allergies. Infections, emotional stress, and changes in air temperature can trigger reactions. A first exposure to an allergen stimulates the immune system to make IgE, which attaches to mast cells and basophils. With later exposures, antigen binds the IgE. Binding triggers the anchoring cell to secrete histamine and cytokines, which initiate inflammation. If this reaction occurs at the lining of the respiratory tract, a copious amount of mucus is secreted and the airways constrict; sneezing, stuffed-up sinuses, and a drippy nose result. Contact with an allergen that penetrates the skin's outer layers causes the skin to redden, swell, and become itchy.

Table 35.3 Recommended Immunization Schedule for Children

Vaccine	Age of Vaccination
Hepatitis B	Birth to 2 months
Hepatitis B boosters	1–4 months and 6–18 months
Rotavirus	2, 4, and 6 months
DTP: diphtheria, tetanus, and pertussis (whooping cough)	2, 4, and 6 months
DTP boosters	15–18 months, 4–6 years, and 11–12 years
HiB (*Haemophilus influenzae*)	2, 4, and 6 months
HiB booster	12–15 months
Pneumococcal	2, 4, and 6 months
Pneumococcal booster	12–15 months
Inactivated poliovirus	2 and 4 months
Inactivated poliovirus boosters	6–18 months and 4–6 years
Influenza	Yearly, 1–18 years
MMR (measles, mumps, rubella)	12–15 months
MMR booster	4–6 years
Varicella (chicken pox)	12–15 months
Varicella booster	4–6 years
Hepatitis A series	1–2 years
Human papillomavirus	11–12 years
Meningococcal	11–12 years

Source: Centers for Disease Control (CDC), 2007

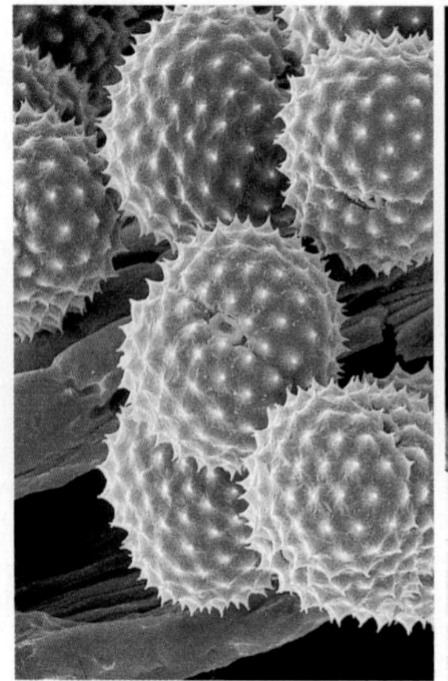

Figure 35.16 An effect of ragweed pollen (*left*) on a sensitive person (*right*).

Figure 35.17 A case of severe combined immunodeficiency (SCID). Cindy Cutshwall was born with a deficient immune system. She carries a mutated gene for adenosine deaminase (ADA). Without this enzyme, her cells could not break down adenosine, so a reaction product that is toxic to white blood cells accumulated in her body. High fevers, severe ear and lung infections, diarrhea, and an inability to gain weight were outcomes.

In 1991, when Cindy was nine years old, she and her parents consented to one of the first human gene therapies. Genetic engineers spliced the normal ADA gene into the genetic material of a harmless virus. The modified virus was the vector that delivered copies of the normal gene into her bone marrow cells. Some cells incorporated the gene in their DNA and started making the missing enzyme.

Now in her early twenties, Cindy is doing well. She still requires weekly injections to supplement her ADA production. Other than that, she is able to live a normal life. She is a strong advocate of gene therapy.

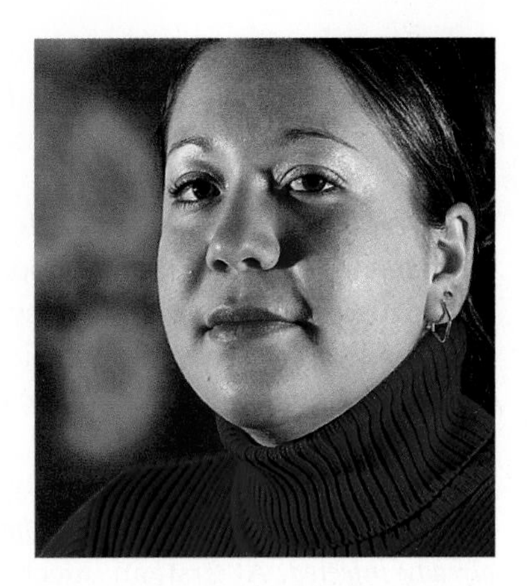

Antihistamines can relieve allergy symptoms. These drugs dampen the effects of histamines by acting on histamine receptors, and they also inhibit the release of cytokines and histamines from basophils and mast cells. Allergen desensitization can reduce or eliminate an allergy. Such programs involve injecting the patient with higher and higher doses of an allergen. With each injection, the patient's body makes less inflammation-causing IgE and more allergen-neutralizing IgG.

An allergic reaction may be annoying for some, but for others it is deadly. For instance, someone allergic to wasp or bee venom can die within minutes of a sting. Anaphylactic shock is a rapid, life-threatening allergic reaction that involves the entire body.

AUTOIMMUNE DISORDERS

Sometimes lymphocytes and antibody molecules fail to discriminate between self and nonself. When that happens, they mount an autoimmune response, or an immune response that targets one's own tissues.

For instance, rheumatoid arthritis is an autoimmune disease in which self antibodies form and bind to the soft tissue in joints. The resulting inflammation leads to eventual disintegration of the bone and cartilage in the joints.

Antibodies to self proteins may bind to hormone receptors, as in Graves' disease. Self antibodies bind stimulatory receptors on the thyroid gland, causing it to produce excess thyroid hormone. This quickens the body's overall metabolic rate. Antibodies are not part of the feedback loops that normally regulate thyroid hormone production. So, antibody binding continues unchecked, the thyroid continues to release too much hormone, and the metabolic rate spins out of control. Symptoms of Graves' disease include uncontrollable weight loss; rapid, irregular heartbeat; sleeplessness; pronounced mood swings; and bulging eyes.

A common neurological disorder, multiple sclerosis, results if self–reactive T cells attack the myelin sheaths of axons and enter cerebrospinal fluid. The symptoms include paralysis and blindness. Certain alleles of MHC genes have been linked to susceptibility, but bacterial or viral infections may trigger the disorder.

Immune responses tend to be stronger in women than in men, and autoimmunity is far more frequent in women. We know that estrogen receptors are part of gene expression controls throughout the body. T cells have receptors for estrogen. Estrogens may enhance the activation of T cells in autoimmune diseases, thus amplifying the interactions between B cells and T cells (Figure 35.8).

DEFICIENT IMMUNE RESPONSES

Loss of immune function can have a lethal outcome. *Primary* deficiencies, present at birth, are outcomes of mutant genes or abnormal developmental steps. The severe combined immunodeficiencies (SCIDs) are like this. The genetic disorder called adenosine deaminase (ADA) deficiency is one case (Figure 35.17). A *secondary* immune deficiency is the loss of immune function after exposure to an outside agent, such as a virus. Immune deficiencies make individuals vulnerable to infections by opportunistic agents that are typically harmless to those in good health.

AIDS (acquired immunodeficiency syndrome) is the most common secondary immune deficiency. The next section describes its cause and its effects.

Immunization programs are designed to boost immunity to specific diseases.

Some heritable disorders, developmental abnormalities, and attacks by viruses and other outside agents result in misdirected, compromised, or nonexistent immunity.

35.9 AIDS Revisited—Immunity Lost

Worldwide, HIV infection rates continue to skyrocket. An effective vaccine still eludes us. For now, the best protection is avoiding unsafe behavior.

LINKS TO SECTIONS 15.1, 19.4

AIDS is a constellation of disorders that develop after infection by the human immunodeficiency virus, HIV. The virus cripples the immune system and makes the body highly susceptible to infections and rare forms of cancer. Worldwide, HIV has infected an estimated 38.6 million individuals (Table 35.4).

There is no way to rid the body of the known forms of the virus, HIV-I and HIV-II. *There is no cure for those already infected.* At first, an infected person appears to be in good health, maybe fighting "a bout of flu." But symptoms emerge that foreshadow AIDS: fever, many enlarged lymph nodes, fatigue, chronic weight loss, and drenching night sweats. Then, infections by normally harmless microorganisms strike. Yeast infections of the mouth, esophagus, and vagina often occur, as well as a form of pneumonia caused by the fungus *Pneumocystis carinii*. Painless colored lesions erupt. The lesions are evidence of Kaposi's sarcoma, a type of cancer common among AIDS patients (Figure 35.18).

HIV INFECTION

HIV Revisited HIV is a retrovirus that has a lipid envelope. Remember, this type of envelope is a small bit of plasma membrane acquired when a virus particle buds from an infected cell (Section 19.4). Proteins jut from the envelope, span it, and line its inner surface. Just beneath the envelope, more viral proteins enclose two RNA strands and copies of reverse transcriptase. Figure 35.19 shows this structural arrangement.

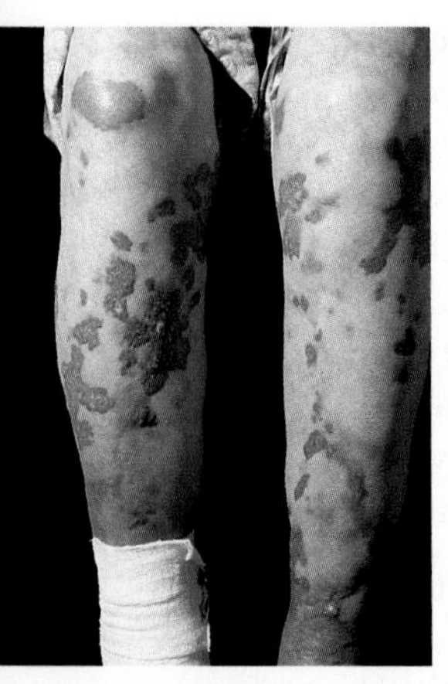

Figure 35.18 The lesions that are a sign of Kaposi's sarcoma.

Table 35.4 Global HIV and AIDS Cases

Region	AIDS Cases	New HIV Cases
Sub-Saharan Africa	24,700,000	2,800,000
South/Southeast Asia	7,800,000	860,000
Central Asia/East Europe	1,700,000	270,000
Latin America	1,700,000	140,000
North America	1,400,000	43,000
Western/Central Europe	740,000	22,000
Middle East/North Africa	460,000	68,000
Caribbean Islands	250,000	27,000
Australia/New Zealand	81,000	7,100
Worldwide total	39,500,000	4,300,000

Source: Joint United Nations Programme HIV/AIDS, 2006 data

A Titanic Struggle Begins HIV primarily infects macrophages, dendritic cells, and helper T cells. When the virus enters the body, dendritic cells engulf it. The dendritic cells migrate to the lymph nodes, where they present processed HIV antigen to naive T cells. The result is a battery of virus-neutralizing IgG antibodies and cytotoxic T cells that can kill HIV-infected cells.

We have just described a typical adaptive immune response. It rids the body of most—but not all—of the virus. In this first response, the HIV infects just a few helper T cells in a few lymph nodes. For years or even decades, the IgG antibodies and the cytotoxic T cells keep the level of HIV in the blood low.

Patients are contagious during this stage, although they might show no symptoms of AIDS. HIV viruses persist in a few of their helper T cells, in a few lymph nodes. The viruses shed proteins that end up in the bloodstream. Helper T cells bind to the viral proteins and secrete cytokines that include interleukin-4 (IL-4). IL-4 is a signal that causes B cells to make IgE instead of virus-neutralizing IgG.

IgE anchors itself to the membrane of mast cells and basophils. When IgE binds antigen, the cell to which it is anchored releases histamines and cytokines. The response effectively defends the internal tissues from invasions of worms and arthropods: Cytokines attract lymphocytes, and histamine allows the cells to enter and defend the infested tissues.

In an HIV infection, however, the IgE binds viral proteins, not worms or arthropods. The response of the anchoring cell is the same: It releases histamines and cytokines—including IL-4.

Here, a vicious cycle begins. The IL-4 signals more B cells to make IgE instead of virus-neutralizing IgG. More IgE binds viral proteins, which causes release of even more IL-4, which calls for less IgG and more IgE, and so on. Levels of virus-neutralizing IgG plummet. Rising IL-4 levels cause helper T cells to stop secreting cytokines, so cytotoxic T cell production slows.

The adaptive immune response becomes less and less effective at fighting the HIV, and the number of virus particles rises; up to 1 billion virus particles are built every day. Up to 2 billion helper T cells become infected. Half of the viruses are destroyed and half of the helper T cells are replaced every two days. Lymph nodes begin to swell with infected T cells.

Eventually, the battle tilts as the body makes fewer replacement helper T cells and the body's capacity for adaptive immunity is destroyed. Other types of viruses make more particles in a day, but the immune system wins. HIV demolishes the immune system. Secondary infections and tumors kill the patient.

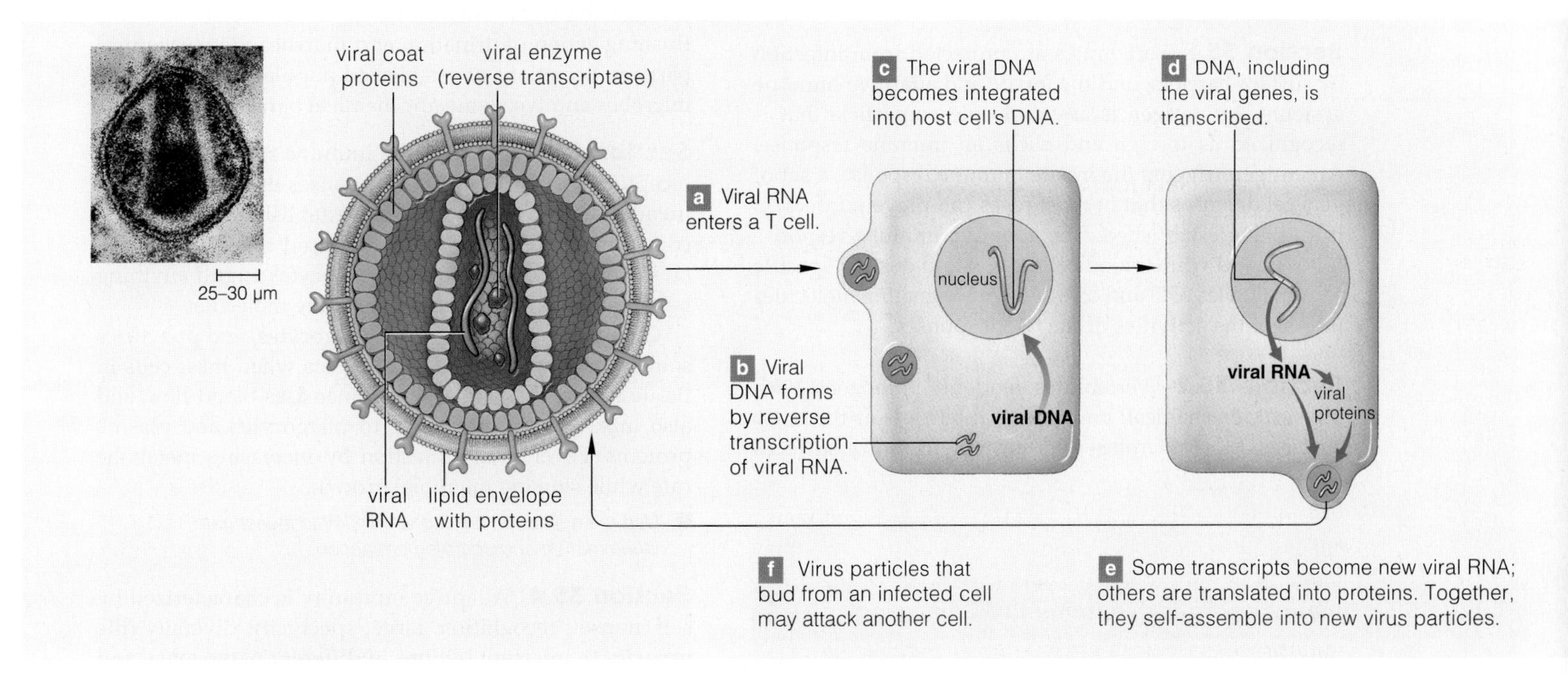

Figure 35.19 **Animated!** Replication cycle of HIV, the retrovirus that causes AIDS.

HOW IS HIV TRANSMITTED?

Most often, HIV is transmitted by having unprotected sex with an infected partner. The virus is transmitted via semen and vaginal secretions, and enters a partner through epithelial linings of the penis, vagina, rectum, and the mouth. The risk of transmission increases by the type of sexual act; for example, anal sex carries 50 times the risk of oral sex.

Infected mothers can transmit HIV to a child during pregnancy, labor, delivery, or breast-feeding. HIV also travels in tiny amounts of infected blood in the syringes shared by intravenous drug abusers, or by patients in hospitals of poor countries. HIV is not transmitted by casual contact.

WHAT ABOUT DRUGS AND VACCINES?

Drugs cannot cure HIV, they only slow its progress. Most target processes unique to retroviral replication. As an example, nucleotide phosphate analogs such as AZT interrupt HIV replication when they substitute for normal nucleotides in cDNA (Section 15.1).

At this writing, 35 HIV vaccines are undergoing human clinical trials around the world. Most involve isolated HIV proteins or peptides, and 12 use carrier viruses. None of them are live, weakened HIV virus; although such vaccines are effective in chimpanzees, the risk of HIV infection from the vaccines themselves would far outweigh their benefits in humans. Other types of HIV vaccines have been—so far—notoriously ineffective. Neutralizing IgG antibody exerts selective pressure on the virus; given the immense rate of viral replication in infected people, HIV genes have a very high mutation rate. Vaccine or no vaccine, our immune system just cannot produce antibodies fast enough to keep up with mutations in the virus.

Even so, persistent researchers are using several strategies to develop an HIV vaccine. An immediate, strong immune response to a primary HIV exposure might clear the virus before it has a chance to infect any helper T cells.

At present, our only option for halting the spread of HIV appears to be prevention, by teaching people how to avoid being infected. In most circumstances, HIV infection is the consequence of a choice—either to have unprotected sex, or to use a shared needle for intravenous drugs. Educational programs around the world are having an effect on the spread of the virus: In many—but not all—countries, the incidence of new cases of HIV each year is beginning to slow. Overall, our global battle against AIDS is not being won.

Learning about the structure and replication of viruses has practical applications. Ongoing research to understand the interaction of HIV with the immune system may one day lead to a vaccine that can save millions of lives.

www.thomsonedu.com Thomson NOW!

Summary

Section 35.1 Vertebrates are protected from infection by surface barriers, and the innate and adaptive immune systems. An antigen is any molecule or particle that is recognized as foreign and elicits an immune response. An antigen triggers the innate immune response, a set of general defenses that in most cases can prevent infection from being established. The adaptive immune response follows, and white blood cells target and destroy specific threats (Tables 35.5 and 35.6). Various signaling molecules integrate the activities of the two responses.

Section 35.2 Vertebrates fend off pathogens with physical, mechanical, and chemical barriers at the body surfaces. Most microbes that normally colonize surfaces do not cause disease unless they penetrate inner tissues. The skin and linings of the body's tubes and cavities are physical barriers. Sticky mucus, ciliated cells, and the flushing action of urination and diarrhea are mechanical barriers. Secretions of established populations of resident microbes and lysozyme are chemical barriers.

Section 35.3 In an innate immune response, pattern receptors trigger fast, preset responses that can eliminate invaders before infection can be established. Phagocytes, complement, acute inflammation, and fever are part of an innate immune response. Phagocytes engulf anything bearing antigen, then secrete signaling molecules.

Complement attracts more phagocytes, and punctures some invaders. Inflammation begins when mast cells in tissue release histamine, which increases blood flow and also makes capillaries leaky to phagocytes and plasma proteins. Fever fights infection by increasing metabolic rate while slowing microbial growth.

- *Use the animation on ThomsonNOW to investigate inflammation and complement action.*

Section 35.4 Adaptive immunity is characterized by self/nonself recognition, target specificity, diversity (the capacity to intercept billions of different pathogens), and memory. B and T cells are central to it.

Antibody-mediated and cell-mediated responses work together to rid the body of specific antigen. Macrophages, dendritic cells, and B cells are the antigen-presenting cells. They engulf and digest antigen, then display fragments of it in complex with MHC markers (self markers) at their surface. Such complexes stimulate the formation of many lymphocytes that can target the antigen.

Section 35.5 B cells carry out antibody-mediated responses, assisted by T cells and signaling molecules. B cells make antibodies: Y- or T-shaped proteins that bind to specific antigens.

Different kinds of antibodies have different functions. Secreted antibodies can neutralize an antigen or tag it for destruction. Binding of antigen to membrane-bound IgE can trigger inflammation.

The memory cells that form during a primary immune response persist long after infection ends. They initiate a faster, stronger secondary response if the same antigen is encountered at a later time.

- *Use the animation on ThomsonNOW to see an antibody-mediated immune response.*

Sections 35.6, 35.7 Antigen-presenting cells, T cells, and NK cells interact in cell-mediated responses. They target and kill body cells altered by infection or cancer.

- *Use the animation on ThomsonNOW to observe a cell-mediated immune response.*

Section 35.8 Vaccines offer protection against specific diseases. Allergens are normally harmless substances that induce immune responses; hypersensitivity to an allergen is called allergy. In autoimmunity, the body's own cells are inappropriately recognized as foreign and attacked. Immune deficiency is a weakened or nonexistent capacity to mount an immune response.

Table 35.5 Innate and Adaptive Immunity Compared

	Innate Immunity	Adaptive Immunity
Response time	Immediate	A few days
How antigen is detected	Fixed set of receptors for molecular patterns found on pathogens	Random recombinations of gene sequences generates billions of receptors
Specificity of response	None	Specific antigens targeted
Persistence	None	Long-term

Table 35.6 Summary of the Immune Fighters

Macrophage	Phagocyte. Presents antigen to helper T cells; secretes cytokines during innate and adaptive immune responses.
Neutrophil	Fast-acting and most abundant phagocyte. Takes part in inflammation; most effective against bacteria.
Eosinophil	Granules contain enzymes that target parasitic worms.
Basophil	Circulates in blood. Granules contain histamine, other substances that cause inflammation.
Mast cell	Cell in tissues. Granules contain histamine, other substances that cause inflammation; contributes to allergies.
Dendritic cell	Circulating phagocyte. Presents antigen to naive T cells.
Lymphocytes:	*Act in most immune responses. After antigen recognition, clonal populations of effector and memory cells form.*
B cell	Recognizes antigens via surface antibodies. It is the only cell that produces antibodies.
Helper T cell	Coordinates all immune responses with signaling molecules (cytokines); activates naive B cells and T cells.
Cytotoxic T cell	Recognizes antigen–MHC complexes; touch-kills infected, cancerous, or foreign cells.
Natural Killer (NK) cell	Cytotoxic; kills stressed body cells lacking MHC markers; also kills antibody-tagged cells.

Section 35.9 HIV, a retrovirus, causes AIDS. This virus destroys the immune system mainly by infecting helper T cells. At present, AIDS cannot be cured.

■ *Use the animation on ThomsonNOW to see how HIV invades a cell and replicates inside it.*

Self-Quiz

Answers in Appendix III

1. ______ is/are the first line of defense against threats.
   a. Skin, mucous membranes
   b. Tears, saliva, gastric fluid
   c. Urine flow
   d. Resident bacteria
   e. a through c
   f. all of the above

2. Complement proteins ______.
   a. form pore complexes
   b. promote inflammation
   c. neutralize toxins
   d. a and b

3. ______ trigger immune responses.
   a. Cytokines
   b. Lysozymes
   c. Immunoglobulins
   d. Antigens
   e. Histamines
   f. all of the above

4. ______ characterize innate immunity.
   a. Unchanging responses
   b. About 1,000 pattern receptors
   c. Inborn mechanisms
   d. Fast responses
   e. all of the above

5. ______ characterizes adaptive immunity.
   a. Self/nonself recognition
   b. Antigen receptor diversity
   c. Antigen specificity
   d Lasting protection
   e. all of the above

6. Antibodies are ______.
   a. antigen receptors
   b. made only by B cells
   c. proteins
   d. all of the above

7. Antibody-mediated responses work against ______.
   a. intracellular pathogens
   b. extracellular pathogens
   c. extracellular toxins
   d. both a and c
   e. both b and c
   f. all of the above

8. ______ binding antigen triggers allergic responses.
   a. IgA b. IgE c. IgG d. IgM e. IgD

9. ______ are targets of cytotoxic T cells.
   a. Extracellular virus particles in blood
   b. Virus-infected body cells or tumor cells
   c. Parasitic flukes in the liver
   d. Bacterial cells in pus
   e. Pollen grains in nasal mucus

10. Allergies occur when the body responds to ______.
    a. pathogens
    b. normally harmless substances
    c. toxins
    d. all of the above

11. ______ participate in the vicious cycle of HIV infection. Choose all that apply.
    a. IgE antibodies
    b. Helper T cells
    c. Cytotoxic T cells
    d. Mast cells
    e. B cells
    f. Cytokines

12. Match the immunity concepts.
    ___ inflammation
    ___ antibody secretion
    ___ phagocyte
    ___ immune memory
    ___ autoimmunity
    a. neutrophil
    b. effector B cell
    c. general defense
    d. immune response against own body
    e. secondary response

■ *Visit ThomsonNOW for additional questions.*

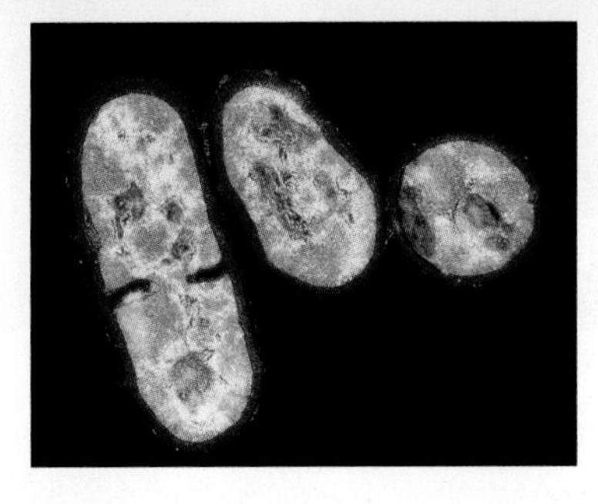

Figure 35.20 This common skin bacterium, *Propionibacterium acnes*, causes acne.

Critical Thinking

1. *Propionibacterium acnes* is a pervasive skin occupant (Figure 35.20). It feeds on sebum, a greasy mixture of fats, carbohydrates, and waxes that lubricates the skin and hair. Sebaceous glands secrete sebum at hair follicles. During puberty, the levels of sex hormones in the bloodstream increase, causing sebaceous glands to make more sebum than before. Excess sebum, along with dead skin cells that flake away at the skin surface, obstruct the openings of hair follicles. *P. acnes* can survive on the surface of the skin, but far prefer anaerobic habitats. They multiply to tremendous numbers inside the closed hair follicles. Secretions of the flourishing *P. acnes* populations leak into internal tissues. They attract neutrophils that initiate inflammation around the follicles. The resulting pustules are called acne. Is this an example of adaptive or innate immunity?

2. Pigs are considered the most likely animal candidates as future organ donors for humans. As a first step, some researchers are developing transgenic pigs—animals that received and are expressing foreign genes. Which genes might be inserted, or deleted, that would stop the human immune system from attacking a pig-to-human transplant?

3. Elena developed chicken pox when she was in first grade. Later in life, when her children developed chicken pox, she remained healthy even though she was exposed to countless virus particles daily. Explain why.

4. Before each flu season, you get a flu shot, an influenza vaccination. This year, you get "the flu" anyway. What happened? There are at least three explanations.

5. Remember, cancer arises when normal body cells lose control over the cell cycle. Researchers have been trying for decades without much success to make cancer vaccines.

New cancer treatments based on cell-therapy are promising, however. The premise is to replace defective or deficient body cells with functional ones. One new method is to culture some of a cancer patient's own naive T cells with tumor-specific antigens. The resulting "super" cytotoxic T cells are very effective at killing cancer cells when they are infused back into the patient. What do you think is a reason that traditional vaccines have not been as universally effective against cancer as the vaccine developed for smallpox? Why do you think cell therapy might be more effective?

6. Monoclonal antibodies are made by immunizing a mouse with a particular antigen, then removing its spleen. Individual B cells producing mouse antibodies specific for the antigen are isolated from the spleen and fused with myeloma cells that grow *in vitro*. The resulting hybridoma cells grow and divide indefinitely; they also produce and secrete antibodies that can bind to the antigen. Monoclonal antibodies are used for passive immunization (Section 35.8), but they tend to be effective only for a short while. Explain why.

36 RESPIRATION

IMPACTS, ISSUES Up in Smoke

Each day, 3,000 or so teenagers join the ranks of habitual smokers in the United States. Most are not even fifteen years old. When they first light up, they cough and choke on irritants in the smoke. Most become dizzy and nauseated, and develop headaches. Sound like fun? Hardly. Why, then, do they ignore signals about the threat to the body and work so hard to be a smoker? Mainly to fit in. To many adolescents, a misguided perception of social benefits overwhelms seemingly remote threats to health (Figure 36.1).

Changes that can make the threat a reality start right away. Ciliated cells keep many pathogens and pollutants that enter airways from reaching the lungs. These cells can be immobilized for hours by the smoke from a single cigarette. Smoke also kills white blood cells that patrol and defend respiratory tissues. Pathogens multiply in the undefended airways (Section 4.5). More colds, more asthma attacks, and bronchitis result.

Social pressures that invite addiction also invite clogged arteries, heart attacks, and strokes. Inhale smoke-filled air, and nicotine can travel to the circulatory system and the brain. This highly addictive stimulant constricts blood vessels, which increases blood pressure. The heart has to work harder to pump blood through the narrowed tubes. Nicotine also triggers a rise in "bad" cholesterol (LDL) and a decline in the "good" kind (HDL) in blood. It makes blood more sluggish—stickier—and encourages clots.

Do smokers know that carcinogens in cigarette smoke can induce cancers in organs throughout the body, not just the lungs? For instance, females who start smoking as teenagers are about 70 percent more likely to get breast cancer than those who do not smoke.

Families, coworkers, and friends also get unfiltered doses of carcinogens in tobacco smoke. As urine samples reveal, carcinogens also end up in tissues of nonsmokers who live with smokers. Each year in the United States, lung cancers arising from second-hand smoke kill about 3,000. Children

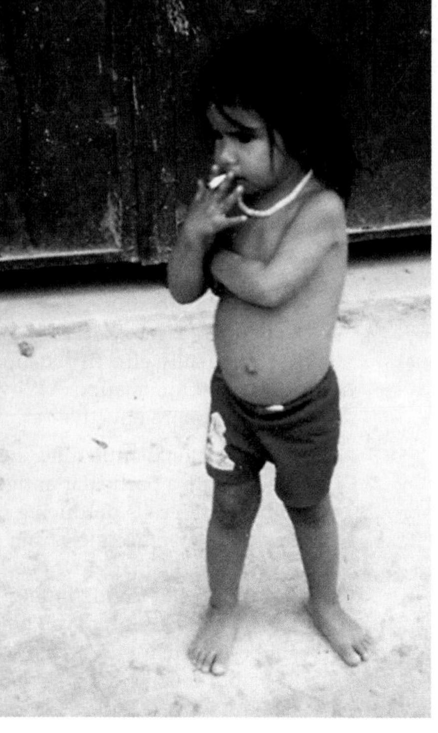

See the video! **Figure 36.1** An addiction that has become pandemic. *Left*, learning to smoke is easy, compared with trying to quit. In one survey, two-thirds of female smokers who were sixteen to twenty-four wanted to give up smoking entirely. Of those who tried to quit, only about 3 percent remained nonsmokers for an entire year.

Right, a child in Mexico City, already adept at smoking cigarettes. This behavior ultimately will endanger her capacity to breathe.

How would you vote? Tobacco is a threat to health and a profitable product for American companies. As tobacco use declines at home, should the United States encourage international efforts to reduce tobacco use around the world? See ThomsonNOW for details, then vote online.

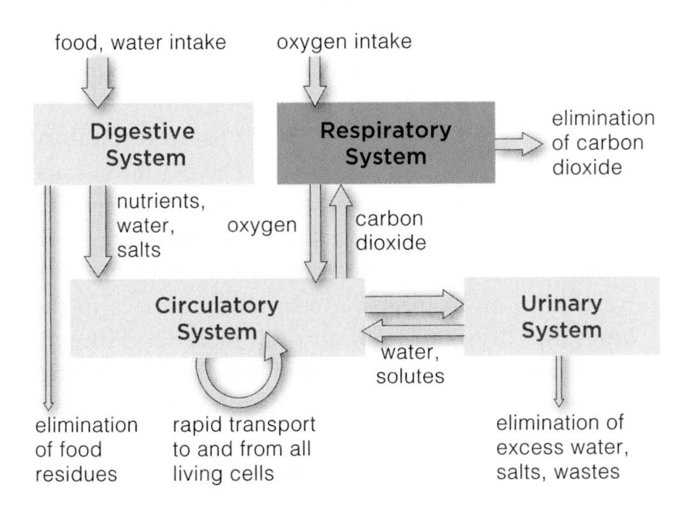

Figure 36.2 Interactions between the respiratory system and other organ systems that contribute most to homeostasis in humans and other vertebrates.

exposed to secondhand smoke also are more likely to develop chronic middle ear infections, asthma, and other respiratory problems later in life.

In the United States, smoking is banned from airline cabins and airports, and some restaurants, theaters, and other enclosed spaces. Cigarette sales to minors are prohibited. Tobacco companies are expected to restrict advertising near schools, but do they view children and women of developing countries as untapped markets? Mark Palmer, former ambassador to Hungary, thinks so. He thinks selling tobacco is the worst thing the United States does to the rest of the world.

This chapter samples a few respiratory systems. All exchange gases with the outside environment. They also contribute to homeostasis—to maintaining the body's internal operating conditions within ranges that cells can tolerate (Figure 36.2). If you or someone you know has joined the culture of smoking, you might use the chapter as a guide to smoking's impact on health. For a more graphic preview, find out what goes on every day with smokers in hospital emergency rooms or intensive care units. No glamour there. It is not cool, and it is not pretty.

Key Concepts

PRINCIPLES OF GAS EXCHANGE

Aerobic respiration uses free oxygen, and its carbon dioxide wastes are removed before the internal environment's pH shifts dangerously. Respiration is the sum of processes that move oxygen from the outside environment to all metabolically active tissues and move carbon dioxide from those tissues to the outside. **Section 36.1**

GAS EXCHANGE IN INVERTEBRATES

Gas exchange occurs across the body surface or gills of aquatic invertebrates. In large invertebrates on land, it occurs across a moist, internal respiratory surface or at fluid-filled tips of branching tubes that extend from the surface to internal tissues. **Section 36.2**

GAS EXCHANGE IN VERTEBRATES

Gills, skin, and paired lungs are gas exchange organs. Breathing ventilates lungs. Blood is a transport medium. It picks up oxygen and gives up carbon dioxide at the respiratory surface. It also exchanges gases with the interstitial fluid that bathes cells.

Gas exchange is most efficient when rates of air flow and blood flow at the respiratory surface match. Respiratory centers adjust the rate and depth of breathing. **Sections 36.3–36.6**

RESPIRATORY PROBLEMS

Respiration can be disrupted by damage to respiratory centers in the brain, physical obstructions, infectious disease, and inhalation of pollutants, including cigarette smoke. **Section 36.7**

GAS EXCHANGE IN EXTREME ENVIRONMENTS

At high altitudes, the human body makes short-term and long-term adjustments to the thinner air. Built-in respiratory mechanisms and specialized behaviors allow sea turtles and diving marine mammals to stay under water, at great depths, for long periods. **Section 36.8**

Links to Earlier Concepts

In this chapter, you will draw on your knowledge of the surface-to-volume ratio, concentration and pressure gradients, and diffusion (Sections 4.1, 5.6, 23.1). You will look at the organ systems that support aerobic respiration as well as homeostasis (7.1, 24.1, 25.1, 34.1, 34.5). You will consider how reflexes, sensory receptors, and brain centers affect breathing (30.7, 30.9, 31.1). You will see how hemoglobin and red blood cells (3.5, 34.2) function in gas exchange.

36.1 The Nature of Respiration

Aerobic respiration is the only metabolic pathway that produces enough ATP to sustain the active life-styles of animals, especially large-bodied ones (Section 7.1). It requires oxygen and produces carbon dioxide waste.

THE BASIS OF GAS EXCHANGE

LINKS TO SECTIONS 3.5, 4.1, 5.6, 7.1, 23.5, 23.6, 34.2

Respiration is the physiological process by which an animal exchanges oxygen and carbon dioxide with its environment. Respiration depends upon the tendency of gaseous oxygen (O_2) and carbon dioxide (CO_2) to diffuse down their concentration gradients—or, as we say for gases, their pressure gradients—between the external and internal environments.

The amount of O_2 dissolved in water varies among habitats and over time. More dissolves in cooler, fast-flowing water than in warm, still water. When water temperature rises or water becomes stagnant, aquatic species with high oxygen needs may suffocate.

Air-breathing animals have a more reliable source of oxygen. Earth's atmosphere is 78 percent nitrogen, 21 percent oxygen, 0.04 percent carbon dioxide, and 0.06 percent other gases. Total atmospheric pressure as measured by a mercury barometer is 760 mm at sea level (Figure 36.3). Oxygen's contribution to the total, its partial pressure, is 21 percent of 760, or 160 mm Hg.

Gases enter and leave the internal environment by crossing a respiratory surface: a moistened layer thin enough for gases to diffuse across. Gases diffuse fast only across small distances. They will diffuse across a membrane only if dissolved in fluid.

FACTORS AFFECTING DIFFUSION RATES

How much gas diffuses across the respiratory surface depends upon the surface area available for diffusion and the pressure gradient across it.

Surface-to-Volume Ratio More molecules diffuse across a large respiratory surface than a small one in any given interval. Remember, as an animal grows, its volume increases faster than its surface area does (Section 4.1). An animal without any respiratory organs usually has a small, flattened body. In such animals, diffusion occurs fast enough for gas exchange because all living cells are no more than a few millimeters away from gases outside the body.

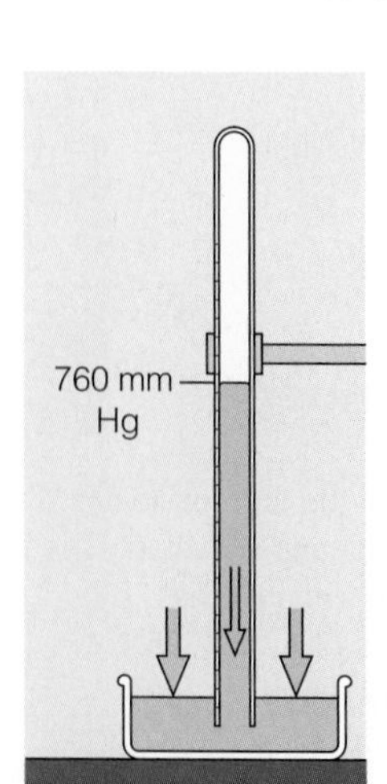

Figure 36.3 How a mercury barometer measures atmospheric pressure. That pressure makes mercury (Hg), a viscous liquid, rise or fall in a narrow tube. At sea level, it rises 760 millimeters (29.91 inches) from the tube's base. Atmospheric pressure varies with altitude. On top of Mount Everest, air pressure is only about one-third of the pressure at sea level.

Ventilation Moving air or water past a respiratory surface keeps the pressure gradient across the surface high and thus increases the rate of gas exchange. For example, frogs and humans breathe in and out, which ventilates their lungs. Breathing forces stale air with waste CO_2 away from the respiratory surface in the lungs and draws in fresh air with more O_2. Fish and other animals that live in water have mechanisms that keep water moving across their respiratory surface.

Respiratory Proteins Respiratory proteins house one or more metal ions that bind oxygen atoms when oxygen levels are high and release them when oxygen levels fall. By binding and releasing oxygen they help maintain a steep partial pressure gradient for oxygen between cells and the blood.

Hemoglobin, an iron-containing respiratory protein, occurs in vertebrate red blood cells (Sections 3.5 and 34.2). It also circulates freely in the blood of annelids, mollusks, and crustaceans, which do not have any red blood cells. An annelid hemoglobin molecule is much larger than yours. It has 200 or so globin chains with heme associated. In other invertebrates, hemerythrin (with iron associated) or hemocyanin (with copper) function in oxygen transport.

Myoglobin, a heme-containing protein in muscle of vertebrates and some invertebrates, helps stabilize the oxygen level inside cells. Oxygen diffuses from blood into muscle cells and binds to myoglobin. When cell activity increases and oxygen levels begin to decline, myoglobin releases its bound oxygen.

Respiration is the sum of processes that supply cells of the animal body with oxygen for aerobic respiration and remove this pathway's carbon dioxide wastes.

A respiratory surface is thin, moist membrane across which gases diffuse into and out of the internal environment. Its surface area influences the rate of gas exchange.

The diffusion rate of a gas is affected by its partial pressure gradient across the respiratory surface. A steeper gradient increases the diffusion rate.

Steep gradients are maintained by modes of ventilation that move water or air to and from the respiratory surface and by respiratory proteins that reversibly bind oxygen.

36.2 Invertebrate Respiration

Invertebrates differ a good deal in body sizes, shapes, and lifestyles. Their diverse respiratory organs reflect challenges of life in water and on dry land.

Small-bodied invertebrates of aquatic or continually moist habitats have the simplest forms of respiration. Gases just diffuse across the body surface covering—the integument (Figure 36.4*a*). This mode of respiration is called integumentary exchange.

Many aquatic invertebrates have thin-walled, moist respiratory organs called gills, although these are not the same in structure as fish gills. Extensively folded gill walls increase the respiratory surface area and gas exchange rates between body fluids and the outside. Figure 36.4*b* shows sea hare (*Aplysia*) gills, which help supplement integumentary exchange.

Arthropods that live on dry land have a hardened integument that helps conserve water but also blocks gas exchange. They exchange gases across an *internal* respiratory surface. For example, insects, millipedes, centipedes, and some spiders have a tracheal system that consists of repeatedly branching, air-filled tubes.

Tracheal tubes start at spiracles—small pores across the integument (Figure 36.4*c*). At the tips of the finest branches is a bit of fluid that gases dissolve in. The tips of insect tracheal tubes lie adjacent to body cells, and oxygen and carbon dioxide diffuse between these tubes and the tissues. Because insects' tracheal tubes end next to cells, they have no need for a respiratory protein such as hemoglobin to carry gases.

LINKS TO SECTIONS 23.7, 23.10, 25.2

Some insects can force air into and out of tracheal tubes. For example, when a grasshopper's abdominal muscles contract, organs press on the pliable tracheal tubes and force air out of them. When these muscles relax, pressure on tracheal tubes decreases, the tubes widen, and air rushes in.

Most spiders have one or two book lungs. In these respiratory organs, air and blood flow through spaces separated only by thin sheets of tissue (Figure 36.4*d*). Hemocyanin in the blood picks up oxygen and turns blue-green as it passes through a book lung. It gives up oxygen and becomes colorless in body tissues.

In some small invertebrates of moist habitats the body wall is the respiratory surface. Others have a gill, a thin, folded internal or external respiratory surface.

Insects and some spiders use tracheal respiration: Tubes carry air from the body surface to tissues. Other spiders have a book lung.

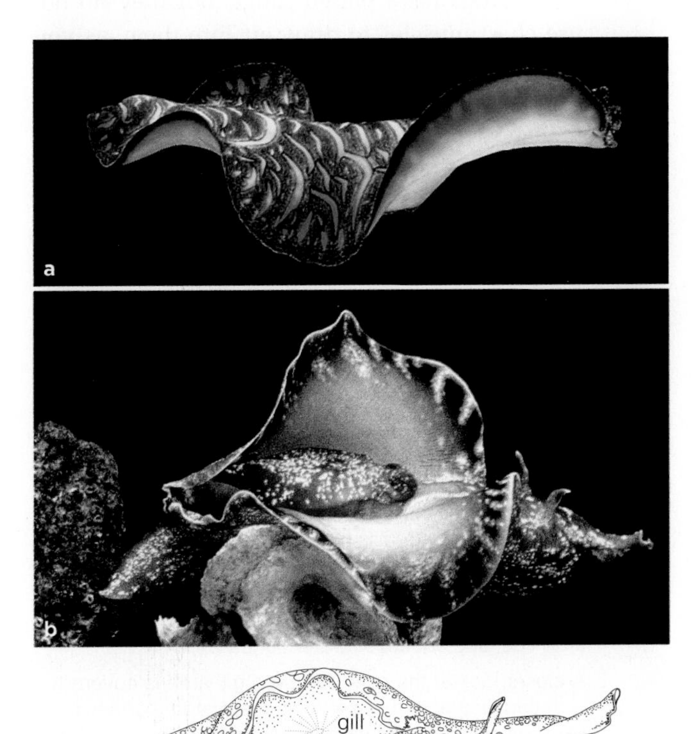

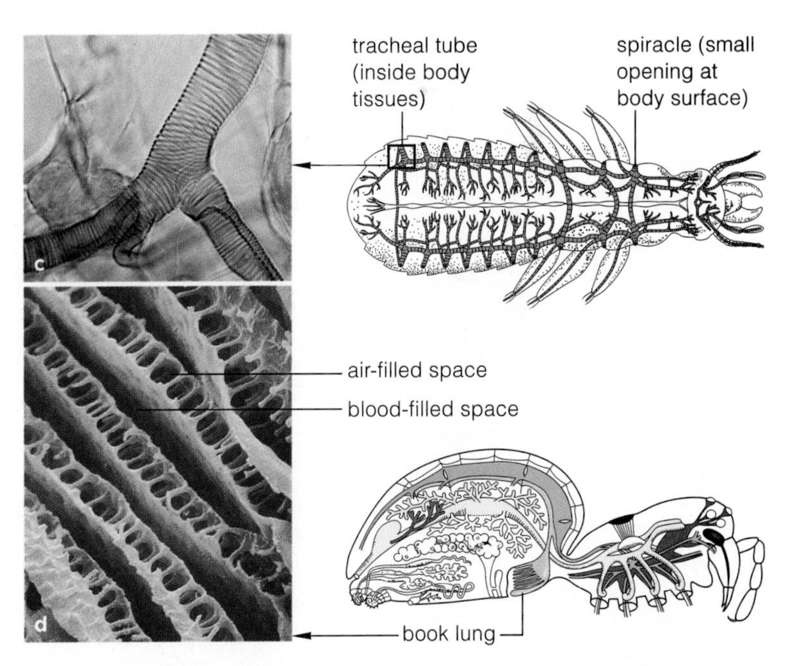

Figure 36.4 Diversity in invertebrate respiratory systems. (**a**) Flatworms are thin enough for diffusion across the body wall to meet all gas exchange need. (**b**) Gill of a sea hare (*Aplysia*), one of the sea slugs, supplement gas exchange across the body wall. Compare Sections 23.7 and 25.2. (**c**) In an insect's tracheal system, chitin rings reinforce branching tubes. (**d**) Air-filled and blood-filled layers alternate inside a spider's book lung.

36.3 Vertebrate Respiration

Depending on the species, vertebrates exchange gases across gills, skin, or the surface of paired internal lungs.

VERTEBRATE GILLS

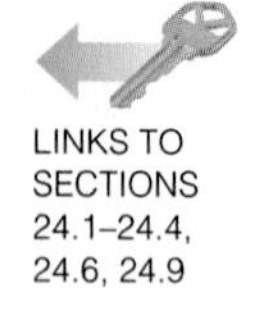

LINKS TO SECTIONS 24.1–24.4, 24.6, 24.9

Recall that gill slits—openings across a pharynx—are a defining trait of chordates (Section 24.1). Gill arches that support these openings evolved in the early fishes (Section 24.2). Today, some fish and amphibian larvae have *external* gills that project from their body. Most adult fish have *internal* gills located inside a slit or a pouch that opens to the body surface (Figure 36.5*a*).

Fish ventilate their internal gills. They draw water into the mouth, shut the mouth, then force water over gills and out of the body (Figure 36.5*b,c*). Water flows over gill filaments that attach to gill arches and have many platelike folds (Figure 36.5*d,e*). Each fold holds a capillary bed that exchanges gases with the blood.

Blood flowing in gill capillaries and water flowing past gill filaments move in opposite directions (Figure 36.5*e*). This allows countercurrent exchange, a process by which two fluids (blood and water in this example) flow in opposite directions and exchange substances. A fish extracts about 80 to 90 percent of the dissolved O_2 in water flowing past its gills. Countercurrent flow ensures that although water gives up O_2 as it flows over a capillary, the water generally has more oxygen than the adjacent blood. As a result, blood picks up more and more O_2 as it travels the length of a capillary.

EVOLUTION OF PAIRED LUNGS

A lung is a saclike respiratory organ located inside a body cavity and connected by airways to the outside air. The first lungs evolved from outpouchings of the gut wall in some bony fishes. Lungs may have helped these fishes survive short trips from pond to pond. Gills that allowed these fish to exchange gases with water would have been useless in exchanges with the air. Lungs became increasingly important as aquatic tetrapods began moving onto land (Section 24.4).

Modern amphibians typically exchange gases across gills and skin as larvae. Some salamanders lack lungs even as adults and rely on exchange across skin. Others have a pair of small lungs. A few, such as axolotls, have external gills.

All frogs have paired lungs, but they do not use chest muscles to draw air into them, as you do. Instead they suck in air through nostrils by lowering the floor of the mouth. Then they close their nostrils and lift the floor of the mouth and throat, pushing air into the lungs (Figure 36.6).

A frog makes croaking sounds by forcing air back and forth between its lungs and pouches in

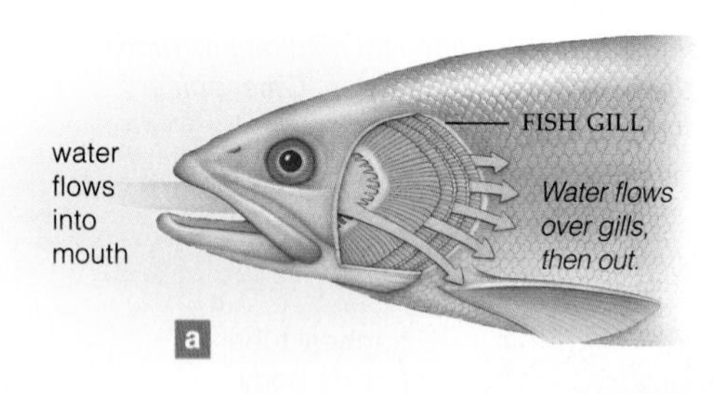

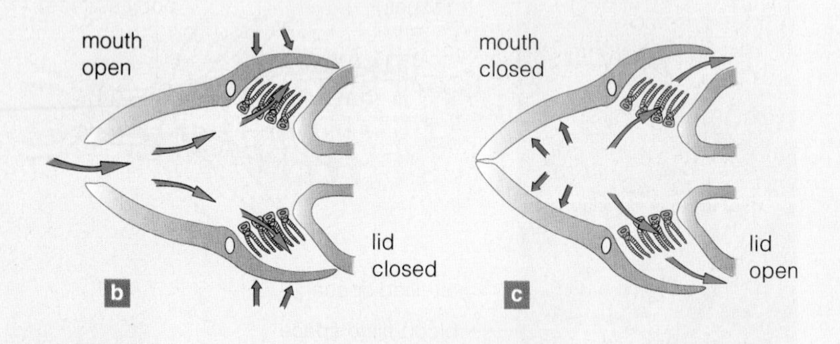

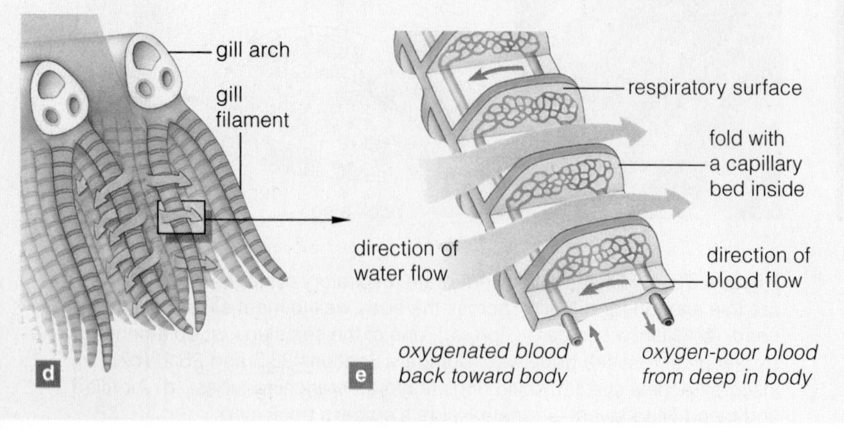

Figure 36.5 **Animated!** Fish respiration. (**a**) One of a pair of gills in a bony fish. The lid that protects and covers the delicate gills has been removed in this sketch. (**b**) Water moves into the mouth when a fish closes its gill lid, opens its mouth, and expands its oral cavity. (**c**) This water is forced out when the fish closes its mouth, opens it gill lids, and squeezes water out past the gills.

A closer look at the gills. (**d**) Each gill arch is covered with thin filaments that serve as respiratory surfaces. (**e**) A gill filament has many folds, each with a capillary bed inside it. Blood inside these capillaries exchanges gases with the water flowing in the opposite direction. Countercurrent flow favors movement of oxygen down its partial pressure gradient from water into blood.

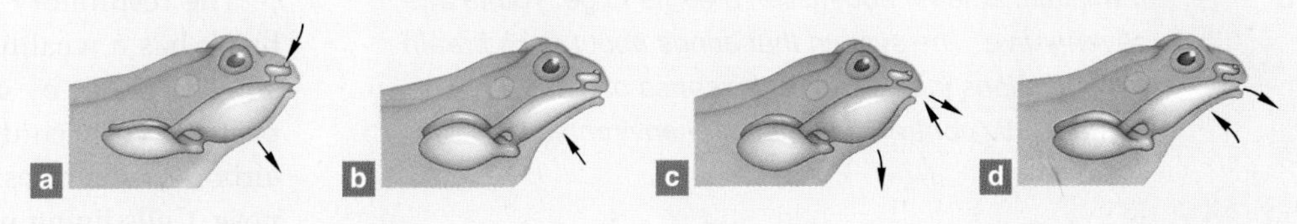

Figure 36.6 Animated! How frogs breathe. (**a**) The frog lowers the floor of its mouth and inhales air through nostrils. (**b**) It closes the nostrils, opens the glottis, and elevates the mouth's floor. This *forces* air into the lungs. (**c**) Rhythmic ventilation assists in gas exchange. (**d**) Air is forced out when muscles in the body wall above the lungs contract and lungs elastically recoil.

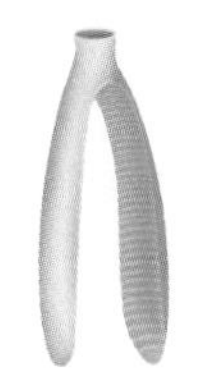

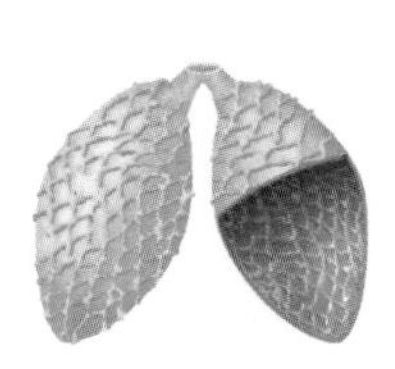

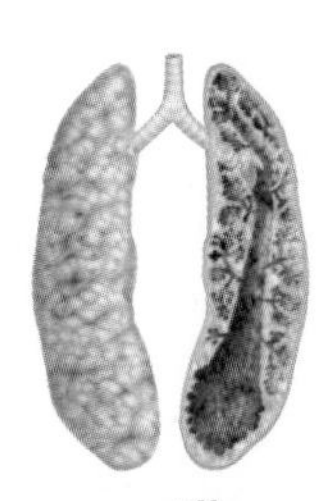

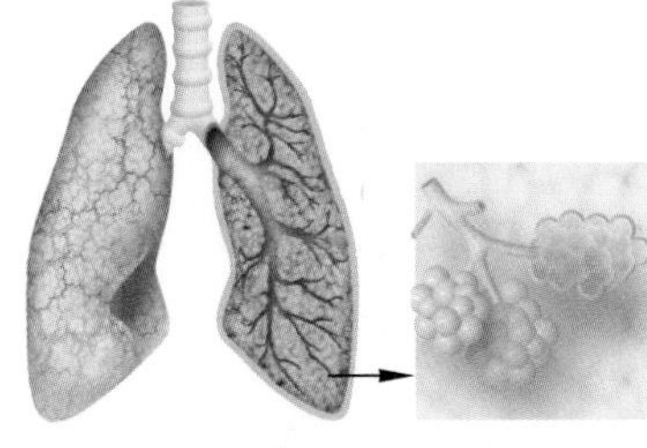

Figure 36.7 Comparison of the structure of lungs in three kinds of vertebrates. The differences suggest that there was an evolutionary trend from simple sacs for gas exchange to larger, more complex respiratory surfaces.

its mouth. Air flow causes a pair of membrane folds at the start of an airway to vibrate. As you will see, such folds also help us make sounds.

Reptiles, birds, and mammals—the amniotes—have waterproof skin and no gills as adults. They exchange gases in two well-developed lungs (Figure 36.7). Chest muscles draw air inward. The respiratory surface area is large and serviced by numerous blood capillaries.

In reptiles and mammals, gas exchange occurs at the ends of the smallest airways. In birds, there are no "dead ends" inside the lung. Tiny tubes that convey air *through* the lungs to air sacs serve as the respiratory surface (Figure 36.8). Continual movement of air past this surface increases the efficiency of gas exchange.

We turn next to the human respiratory system. Its operating principles apply to most vertebrates, even though the lungs evolved differently among them.

Countercurrent exchange in fish gills boosts the oxygen uptake from water. Gills do not work in dry land habitats, where internal air sacs—lungs—are more efficient.

Amphibians gulp air and push it into their lungs. Reptiles, birds, and mammals pull air into their lungs when they contract muscles in the chest.

Birds have a highly efficient respiratory system. Air flows through tubes in their lungs where gases are exchanged, and into air sacs.

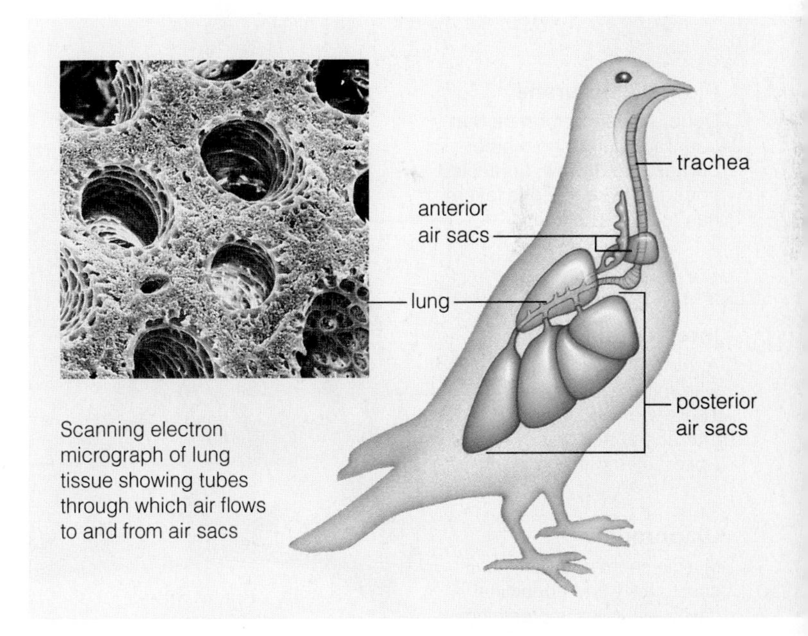

Figure 36.8 Animated! Respiratory system of a bird. Large air sacs attach to two small, inelastic lungs. Air flows in through many air tubes inside the lung, and into posterior air sacs. The lining of the tiniest of the air tubes, sometimes called air capillaries, is the respiratory surface.

The first exhalation forces air from posterior sacs, *through* air capillaries, and into anterior sacs. The next breath out forces it from the air sacs and out of the body. It takes more than one breath for air to flow through the system, but air flows continuously over the respiratory surface. This unique ventilating system supports the high metabolic rates that birds require for flight and other energy-demanding activities.

36.4 Human Respiratory System

It will take at least 300 million breaths to get you to age seventy-five. The system that brings about each breath also functions in speech, in the sense of smell, and in homeostatic control of the internal environment.

THE SYSTEM'S MANY FUNCTIONS

LINKS TO SECTIONS 31.3, 34.1

Figure 36.9 shows the human respiratory system and lists its functions. Notice the skeletal muscles attached to the rib cage. These muscles have an accessory role in respiration. Their rhythmic contraction and relaxation cause air to move into and out of the paired lungs.

The respiratory system functions in gas exchange, but it has a wealth of additional roles. We can speak, sing, or shout by controlling vibrations as air moves past our vocal cords. We have a sense of smell because airborne molecules stimulate olfactory receptors in the nose. Cells lining nasal passages and other airways of the system help defend the body; they intercept and neutralize airborne pathogens. The respiratory system contributes to the body's acid–base balance by getting rid of carbon dioxide wastes. Controls over breathing even help maintain body temperature, because water evaporating from airways has a cooling effect.

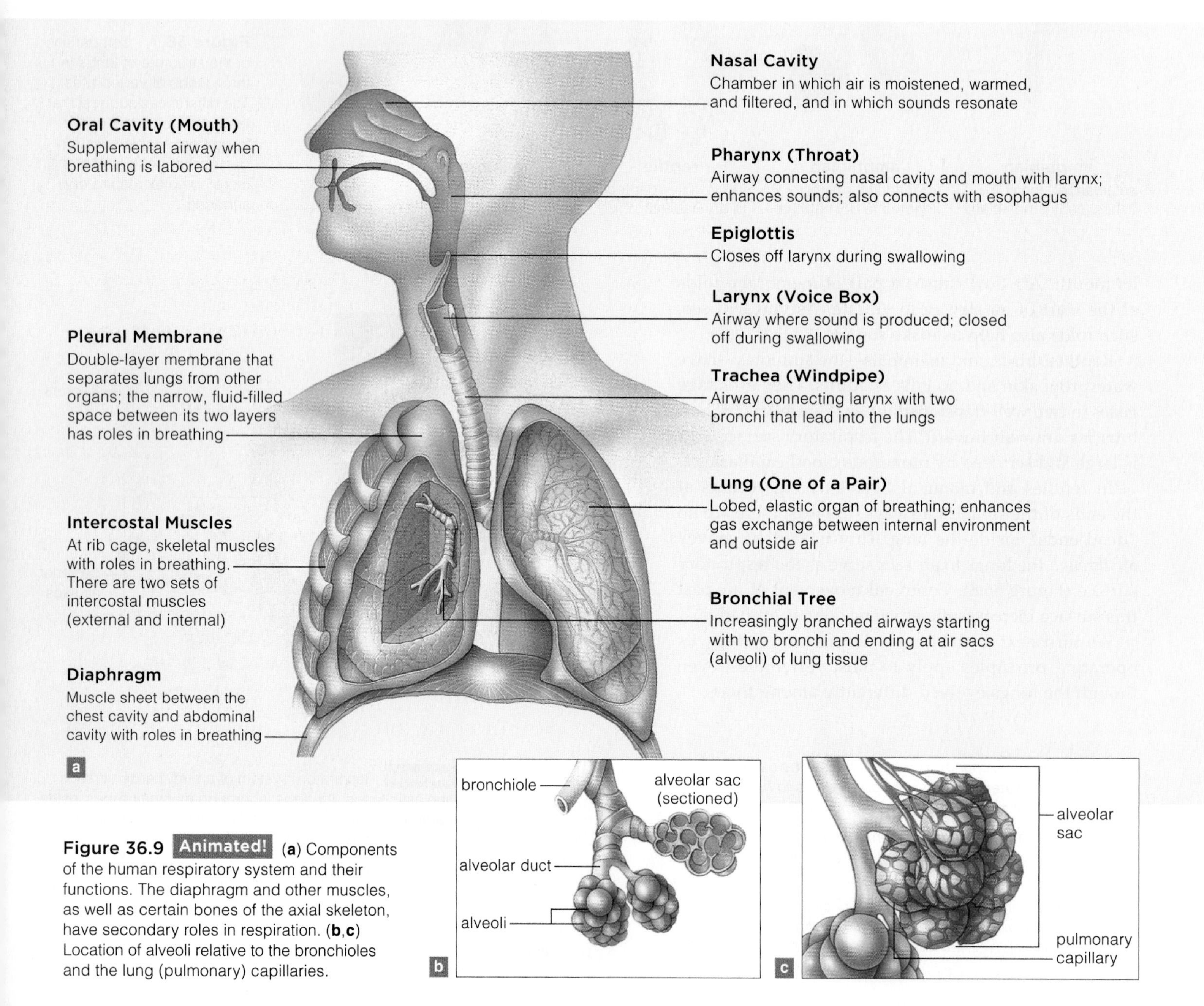

Figure 36.9 Animated! (**a**) Components of the human respiratory system and their functions. The diaphragm and other muscles, as well as certain bones of the axial skeleton, have secondary roles in respiration. (**b**,**c**) Location of alveoli relative to the bronchioles and the lung (pulmonary) capillaries.

FROM AIRWAYS TO ALVEOLI

The Respiratory Passageways Take a deep breath. Now look at Figure 36.9 to get an idea of where the air traveled in your respiratory system.

If you are healthy and sitting quietly, air probably entered through your nose, rather than your mouth. As air moves through your nostrils, tiny hairs filter out any large particles. Mucus secreted by cells of the nasal lining captures most fine particles and airborne chemicals. Ciliated cells in the nasal lining also help remove any inhaled contaminants.

Air from the nostrils enters the nasal cavity, where it gets warmed and moistened. It flows next into the pharynx, or throat. It continues to the larynx, a short airway commonly known as the voicebox because a pair of vocal cords projects into it (Figure 36.10). Each vocal cord is skeletal muscle with a cover of mucus-secreting epithelium. Contraction of the vocal cords changes the size of the glottis, the gap between them.

When the glottis is wide open, air flows through it silently. When muscle contraction narrows the glottis, outgoing air flowing through the tighter gap makes the vocal cords vibrate; it gives rise to sounds. The tension on the cords and changes in the position of the larynx change the sound's pitch. To get a feel for how that works, place one finger on your "Adam's apple," the laryngeal cartilage that sticks out most at the front of your neck. Hum a low note, then a high one. You will feel the vibration of vocal cords and how laryngeal muscles shift the position of the larynx.

In laryngitis, overuse or infection has inflamed the vocal cords. The swollen cords cannot vibrate as they should, which makes speaking difficult.

At the entrance to the larynx is an epiglottis. When this tissue flap points up, air moves into the trachea, or windpipe. When you swallow, the epiglottis flops over, points down, and covers the larynx entrance, so food and fluids enter the esophagus. The esophagus connects the pharynx to the stomach.

The trachea branches into two airways, one to each lung. Each airway is a bronchus (plural, bronchi). Its epithelial lining has many ciliated and mucus-secreting cells that fend off respiratory tract infections. Bacteria and airborne particles stick to the mucus. Cilia sweep the mucus toward the throat for expulsion.

The Paired Lungs Human lungs are cone-shaped organs in the thoracic cavity, one on each side of the heart. The rib cage encloses and protects them. A two-layer-thick pleural membrane covers each lung's outer surface and lines the inner thoracic cavity wall.

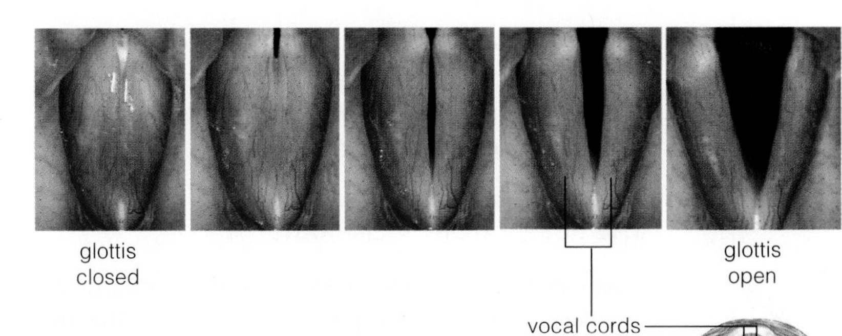

Figure 36.10 Human vocal cords, inside the larynx. Contraction of skeletal muscle in the cords changes the width of the glottis, the gap between them. The glottis closes tightly when you swallow. It is open during quiet breathing. It narrows when you speak, so that air flow causes the cords to vibrate.

Inside each lung, air flows through finer and finer branchings of a "bronchial tree." All of these branches are bronchioles. At tips of the finest bronchioles are respiratory alveoli (singular, alveolus), little air sacs where gases are exchanged (Figure 36.9*b*,*c*). The wall of each alveolus is one cell thick. Collectively, alveoli provide an extensive surface for gas exchange. If all 6 million alveoli in your lungs could be stretched out in a single layer, they would cover half a tennis court!

Air in alveoli exchanges gases with blood flowing through pulmonary capillaries (Latin *pulmo*, lung). At this point, a different organ system gets involved. The circulatory system transports oxygen to body tissues and carries carbon dioxide away from them.

Muscles and Respiration A broad sheet of smooth muscle beneath the lungs, the diaphragm, partitions the coelom into a thoracic cavity and an abdominal cavity. Of all smooth muscle, it alone can be controlled voluntarily. You can make it contract by deliberately inhaling. The diaphragm and intercostal muscles, the skeletal muscles between the ribs, interact and change the volume of the thoracic cavity during breathing.

The human respiratory system functions in gas exchange. It also has roles in sense of smell, voice production, body defenses, acid–base balance, and temperature regulation.

Air enters through the nose or mouth. It flows through the pharynx (throat) and larynx (voicebox) to a trachea that branches into two bronchioles, one to each lung.

Inside each lung, additional branching airways deliver air to the alveoli, where gases are exchanged with pulmonary capillaries. Action of the diaphragm and muscles between the ribs alter the size of the chest cavity during breathing.

36.5 Gas Exchange and Transport

You already know about the structure of hemoglobin, a respiratory protein that gives red blood cells their color. Turn now to the actual mechanisms of oxygen and carbon dioxide transport.

LINKS TO SECTIONS 3.5, 5.5, 29.1, 34.2, 34.8

THE RESPIRATORY MEMBRANE

Gases diffuse between an alveolus and a pulmonary capillary at the lung's respiratory membrane. This thin membrane is made up of alveolar epithelium, capillary endothelium, and fused basement membranes of the alveolus and capillary (Figure 36.11). Secretions keep the alveolar side of the respiratory membrane moist so that gases can cross it.

O_2 and CO_2 diffuse passively across the respiratory membrane. Therefore, the net direction of movement for these gases depends upon their partial pressure gradients across the membrane. Air flow in and out of lungs and blood flow through pulmonary capillaries keep these partial pressure gradients steep.

OXYGEN TRANSPORT

Normally, inhaled air that reaches alveoli contains a great deal of O_2, compared to the blood in pulmonary capillaries. As a result, O_2 in lungs tends to diffuse into blood plasma in the pulmonary capillaries, and then into red blood cells.

As many as 30 trillion red blood cells circulate in your blood. Each holds many millions of hemoglobin molecules. Again, the hemoglobin molecule consists of four polypeptide chains, each associated with one heme group (Figure 36.12*a*). Each heme group includes one iron atom that reversibly binds O_2. Hemoglobin with oxygen bonded to it is oxyhemoglobin, or HbO_2.

About 98.5 percent of the oxygen you inhale gets bound to heme groups of hemoglobin. The amount of HbO_2 that forms in a given interval depends on the partial pressure of O_2. The higher the partial pressure of O_2, the more HbO_2 will form.

Heme binds O_2 only weakly. It releases O_2 in places where the partial pressure of O_2 is lower than that in the alveoli. This is true in metabolically active tissues, as the boxes color-coded *pink* in Figure 36.13 show. Other factors that encourage release of O_2 from heme, including high temperature, low pH, and high CO_2 partial pressure, also are typical of these tissues.

Myoglobin, a different iron-containing respiratory protein, stores oxygen in cardiac muscle and in some skeletal muscles. Structurally, it resembles the globin in hemoglobin, but it holds more tightly onto oxygen (Figure 36.12*b*). The O_2 that hemoglobin gives up near a cardiac muscle cell diffuses into the cell and binds to myoglobin inside it. When blood flow cannot keep up with a cell's increased O_2 needs, as during periods of intense exercise, the myoglobin releases O_2, which allows mitochondria to keep on making ATP.

CARBON DIOXIDE TRANSPORT

Carbon dioxide diffuses into blood capillaries in any tissue where its partial pressure is higher than it is in blood. As the boxes color-coded *blue* in Figure 36.13 show, metabolically active tissues are such regions.

Carbon dioxide is transported to the lungs in three forms. About 10 percent remains dissolved in plasma. Another 30 percent reversibly binds with hemoglobin and forms carbaminohemoglobin ($HbCO_2$). However, most CO_2 that diffuses into the plasma—60 percent—is transported as bicarbonate (HCO_3^-).

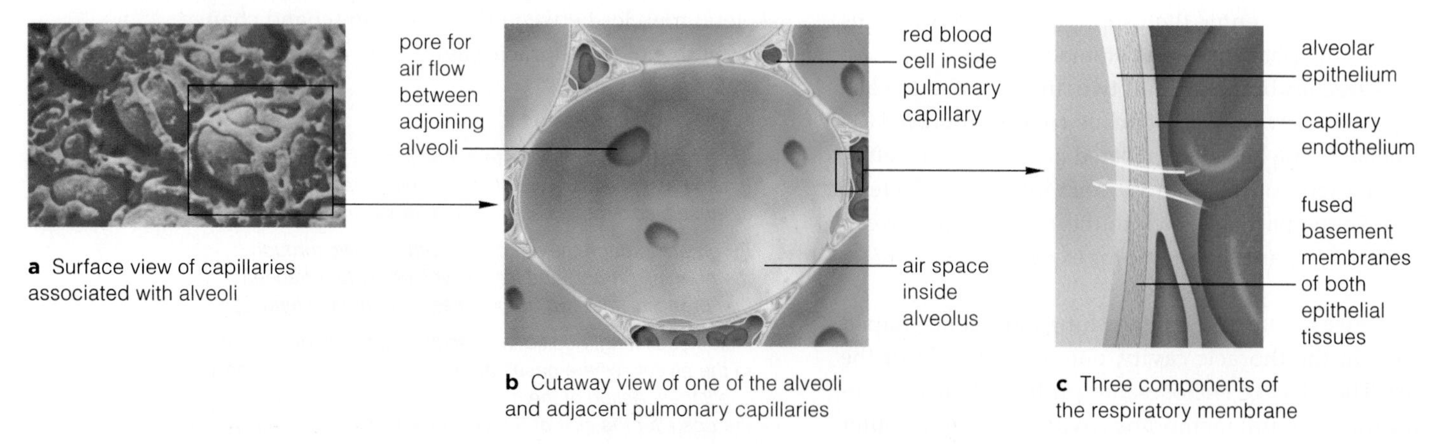

a Surface view of capillaries associated with alveoli

b Cutaway view of one of the alveoli and adjacent pulmonary capillaries

c Three components of the respiratory membrane

Figure 36.11 Zooming in on the respiratory membrane in human lungs.

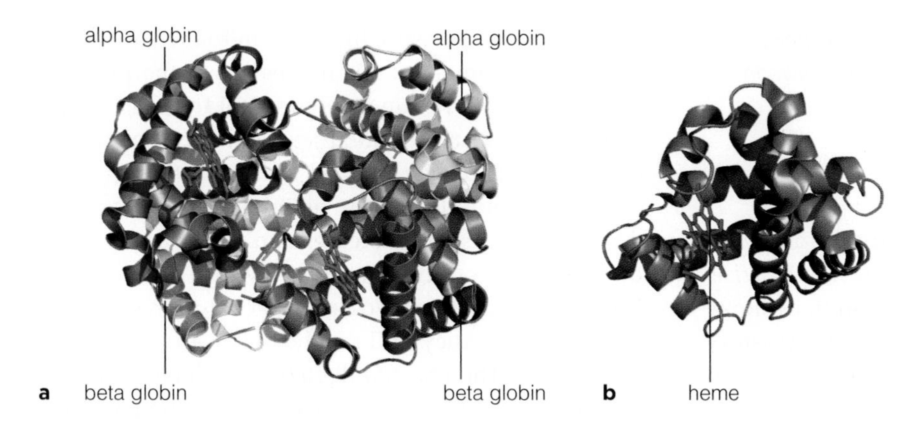

Figure 36.12 (**a**) Structure of hemoglobin, the oxygen-transporting protein of red blood cells. It consists of four globin chains, each associated with an iron-containing heme group, color-coded *red.*

(**b**) Myoglobin, an oxygen-storing protein in muscle cells. Its single chain associates with a heme group. Compared to hemoglobin, myoglobin has a higher affinity for oxygen, so it helps speed the transfer of oxygen from blood to muscle cells.

Similarities between myoglobin and hemoglobin suggest that hemoglobin arose through duplication of genes for a myoglobin-like protein, followed by mutations in some of the extra gene copies.

How does HCO_3^- form? The carbon dioxide first combines with water, forming carbonic acid (H_2CO_3). This compound separates into bicarbonate and H^+:

$$CO_2 + H_2O \rightleftharpoons \underset{\text{carbonic acid}}{H_2CO_3} \rightleftharpoons \underset{\text{bicarbonate}}{HCO_3^-} + H^+$$

Carbonic anhydrase, an enzyme inside red blood cells, speeds the reaction. HCO_3^- tends to diffuse out of red blood cells into the plasma. Most of the H^+ binds to hemoglobin. The reverse reactions occur in the alveoli, where the CO_2 partial pressure is lower than that in lung capillaries. The water and CO_2 that form inside the alveolar sacs are exhaled.

THE CARBON MONOXIDE THREAT

Carbon monoxide (CO) is a colorless, odorless gas. It is present in the smoke from cigarettes and fossil fuel combustion. Hemoglobin has a higher affinity for CO than for O_2. When CO builds up in the air, it fills O_2 binding sites on hemoglobin, preventing transport of O_2 and causing carbon monoxide poisoning. Nausea, headache, confusion, dizziness, and weakness set in as tissues are starved of oxygen. In the United States, accidental CO poisoning kills about 500 people each year. To minimize your risk, be sure that fuel-burning appliances have been properly vented to the outside, and install a carbon monoxide detector.

Driven by its partial pressure gradient, oxygen diffuses from alveoli into pulmonary capillaries. Carbon dioxide, driven by its partial pressure gradient, diffuses the opposite way.

Hemoglobin in red blood cells enormously enhances the oxygen-carrying capacity of blood. Most carbon dioxide is transported in blood in the form of bicarbonate, nearly all of which forms by enzyme action in red blood cells.

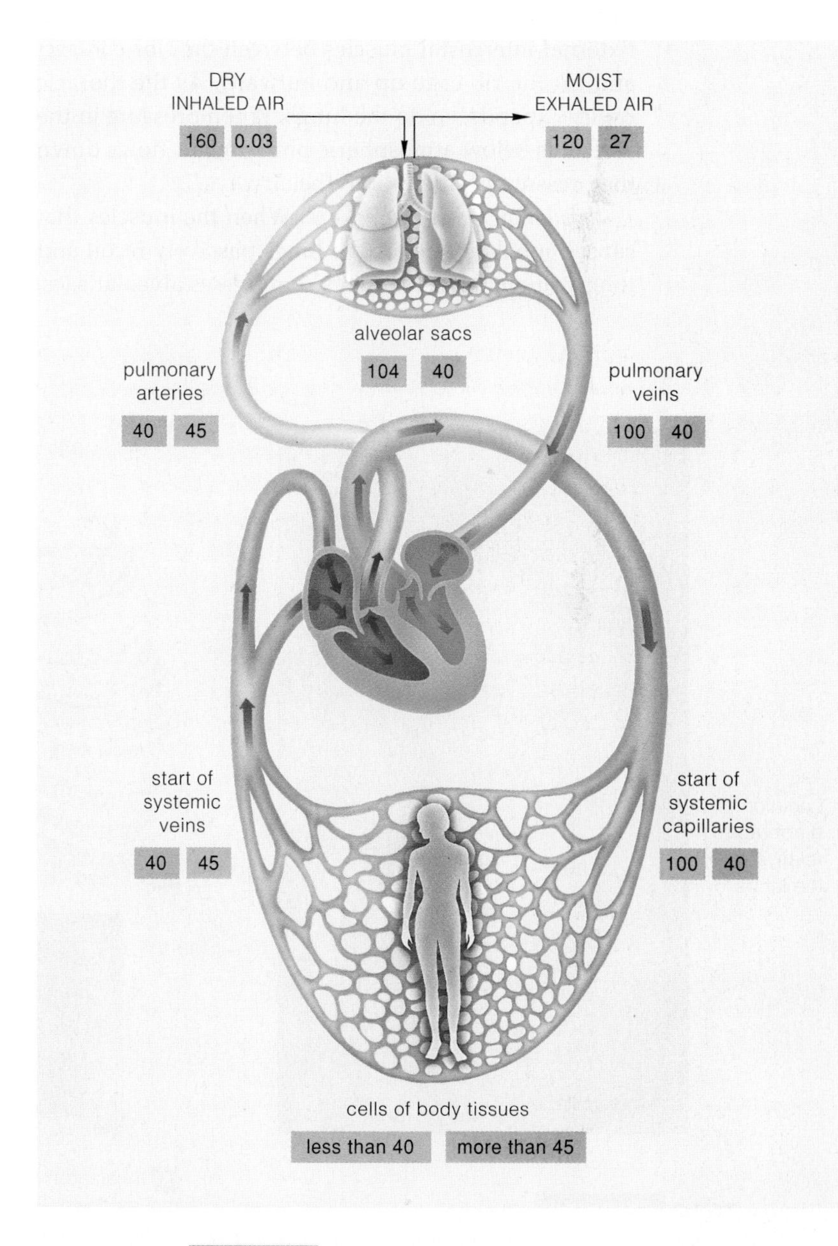

Figure 36.13 Animated! Partial pressures (in mm Hg) for oxygen (*pink* boxes) and carbon dioxide (*blue* boxes) in the atmosphere, blood, and tissues.

36.6 Cyclic Reversals in Air Pressure Gradients

Breathing ventilates both human lungs in a rhythmic, adjustable pattern. It promotes gas exchange between the atmosphere and alveolar sacs.

THE RESPIRATORY CYCLE

LINKS TO SECTIONS 29.5, 30.7

A respiratory cycle is one breath in (inhalation) and one breath out (exhalation). *Inhalation is always active, and muscle contractions drive it.* Changes in the volume of the lungs and thoracic cavity during a respiratory cycle alter pressure gradients between air inside and outside the respiratory tract.

Figure 36.14*a* shows what happens when you start to inhale. The diaphragm flattens and moves down. External intercostal muscles between the ribs contract and lift the rib cage up and outward. As the thoracic cavity expands, so do the lungs. When pressure in the alveoli is below atmospheric pressure, air flows down the pressure gradient, into the airways.

Exhalation is usually passive. When the muscles that caused inhalation relax, the lungs passively recoil and lung volume decreases. This compresses alveolar sacs, causing the air pressure inside them to increase above atmospheric pressure. Air moves down the pressure gradient, out of the lungs (Figure 36.14*b*).

Exhalation is active only when you exercise vigorously or consciously attempt to expel more air. During active exhalation, muscles of the abdominal wall contract. Abdominal pressure increases and exerts an upward-directed force on the diaphragm, which is forced up. Internal intercostal muscles contract at the same time. When they do, they pull the thoracic wall inward and downward. The chest wall flattens, which decreases the volume of the thoracic cavity. The lung volume decreases as elastic tissues passively recoil.

RESPIRATORY VOLUMES

The maximum volume of air that the lungs can hold, their vital capacity, averages 5.7 liters in healthy adult males and 4.2 liters in females. Most of the time, the lungs are about half full. Tidal volume—the volume that moves into and out of lungs during a respiratory cycle—averages about 0.5 liter (Figure 36.15). Your

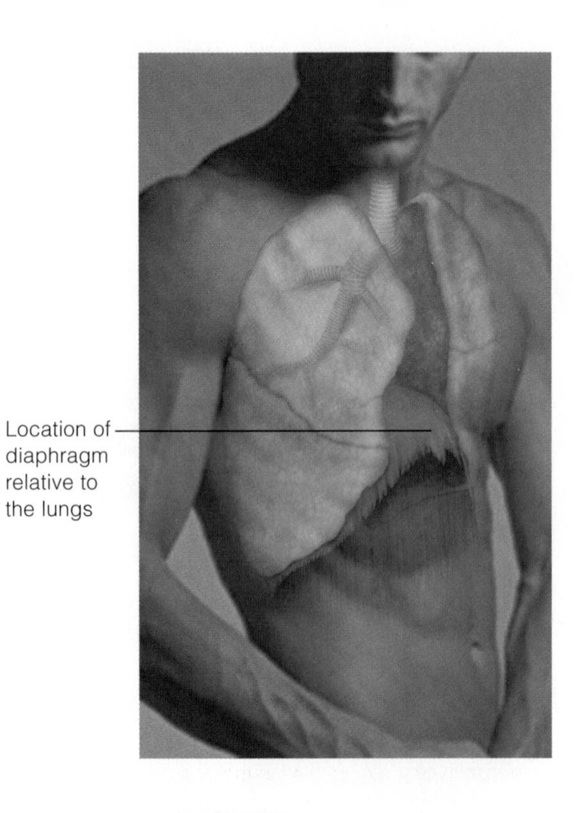

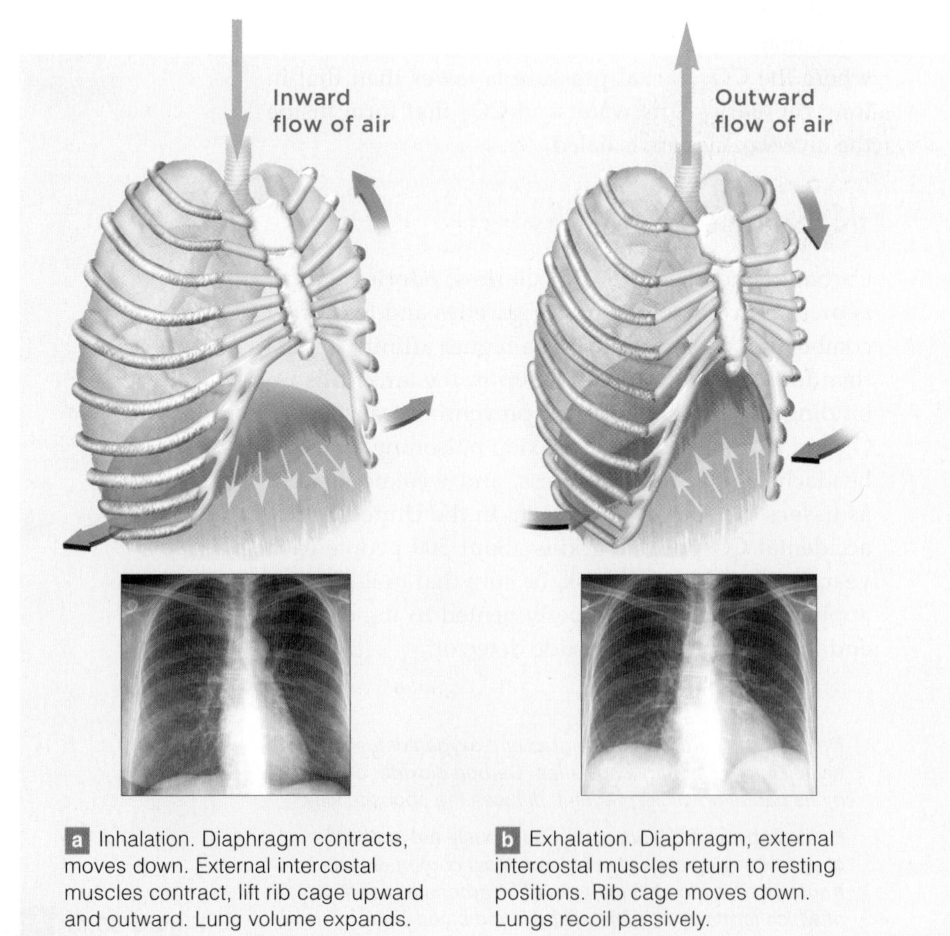

a Inhalation. Diaphragm contracts, moves down. External intercostal muscles contract, lift rib cage upward and outward. Lung volume expands.

b Exhalation. Diaphragm, external intercostal muscles return to resting positions. Rib cage moves down. Lungs recoil passively.

Figure 36.14 **Animated!** Changes in the size of the thoracic cavity during a single respiratory cycle. The x-ray images reveal how inhalation and expiration change the lung volume.

lungs never fully deflate; thus air inside them always is a mix of freshly inhaled air and "stale air" that was left behind during the previous exhalation. Even so, there is plenty of oxygen available for exchange.

CONTROL OF BREATHING

Respiratory centers in the brain stem control the rate and depth of breathing. Some neurons in the medulla oblongata are the pacemaker for inhalation. When you rest, these neurons spontaneously fire action potentials ten to fourteen times per minute. Nerves deliver these signals to the diaphragm and intercostal muscles. The contraction of these muscles causes inhalation. Between action potentials, the muscles relax and you exhale.

Breathing patterns change with activity level. When you are more active, muscle cells increase their rate of aerobic respiration and produce more CO_2. This CO_2 enters blood, where it combines with water and forms carbonic acid (Section 36.5). The acid dissociates and H^+ levels rise in the blood and in cerebrospinal fluid. Chemoreceptors inside the medulla oblongata and in carotid artery and aorta walls detect changes. These receptors signal the respiratory center, which calls for changes in the breathing pattern (Figure 36.16).

The chemoreceptors in the blood vessel walls also signal brain centers when the O_2 partial pressure in arterial blood falls below a life-threatening 60 mm Hg. Ordinarily, the O_2 partial pressure does not fall that low. This control mechanism has survival value only at high altitudes and during severe lung diseases.

Reflexes such as swallowing or coughing, briefly halt breathing. Also commands from the sympathetic nerves make you breathe faster if you are frightened (Section 30.8). Breathing patterns can be deliberately altered, as when you hold your breath during a dive, or break normal breathing rhythm to talk or sing.

Breathing reverses pressure gradients between the lungs and the air outside the body. A respiratory cycle consists of one inhalation and one exhalation.

Inhalation is always an active, energy-requiring process. It primarily requires contractions of the diaphragm and the external intercostal muscles.

During quiet breathing, exhalation is a passive process. Muscles relax, the thoracic cavity volume decreases, and the lungs recoil elastically. Forceful exhalation is an active process that requires abdominal muscle contraction.

The brain controls the rhythm and depth of breathing. A respiratory pacemaker establishes the initial rhythm, which is adjusted in response to changes in activity levels.

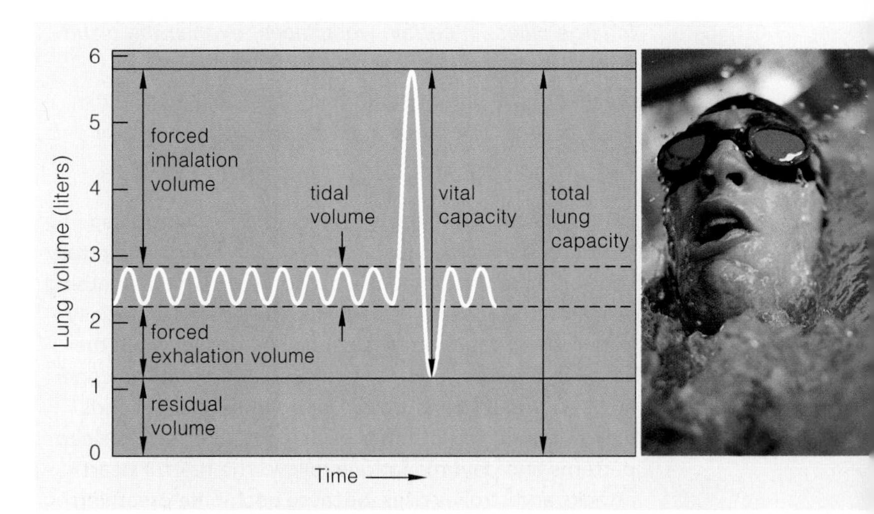

Figure 36.15 **Animated!** Respiratory volumes. In quiet breathing, the lungs hold 2.7 liters at the end of one inhalation and 2.2 liters at the end of one exhalation; the tidal volume of air entering and leaving is only 0.5 liter. Lungs never deflate completely. When air flows out and lung volume is low, the wall of the smallest airways collapses and prevents further air loss.

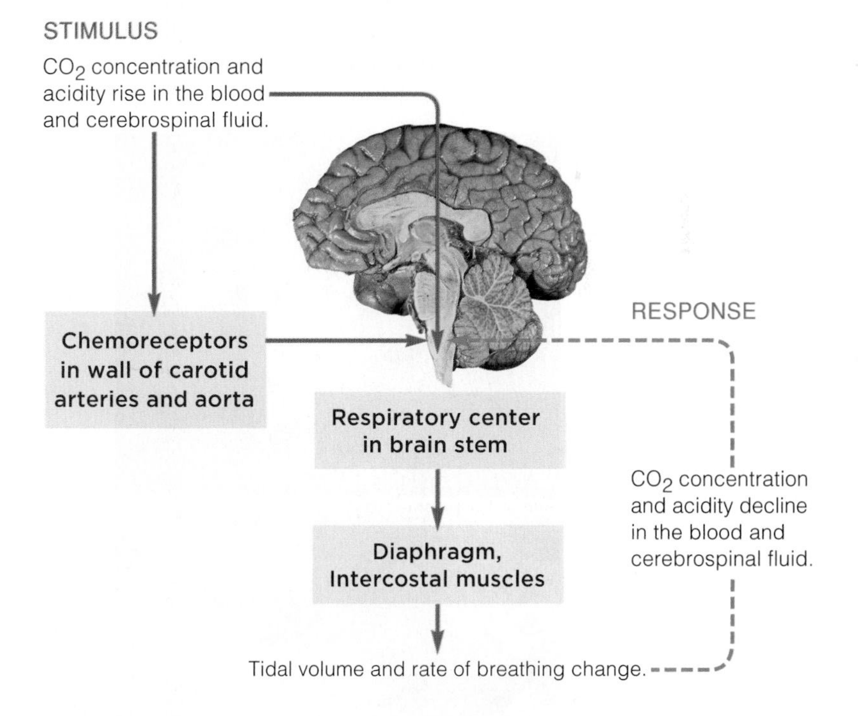

Figure 36.16 Respiratory response to increased activity levels. An increase in activity raises the CO_2 output. It also makes the blood and cerebrospinal fluid more acidic. Chemoreceptors in blood vessels and the medulla sense the changes and signal the brain's respiratory center, also in the brain stem.

In response, the respiratory center sends signals along nerves to the diaphragm and intercostal muscles. The signals call for alterations in the rate and depth of breathing. Excess CO_2 is expelled, which causes the level of this gas and acidity to decline. Chemoreceptors sense the decline and signal the respiratory center, so breathing is adjusted accordingly.

36.7 Respiratory Diseases and Disorders

In large cities, in certain workplaces, even in the cloud around a cigarette smoker, airborne particles and gases are present in abnormally high concentrations. They put extra workloads on the respiratory system.

LINKS TO SECTIONS 19.6, 30.5, 30.6

Interrupted Breathing A tumor or other damage to the brain stem's medulla oblongata can affect respiratory controls. It can cause apnea, a disorder in which breathing repeatedly stops and restarts spontaneously, especially during sleep. More often, sleep apnea occurs when the tongue, tonsils, or another soft tissue obstruct the upper airways. Breathing may stop for up to several seconds many to hundreds of times each night. Interrupted sleep patterns and daytime fatigue follow. The risk for heart attacks and strokes rises, because each time breathing stops, blood pressure soars. Changes in sleeping positions or using a mouthpiece or other kinds of devices can help mild sleep apnea. Severe cases require surgical removal of the tissues that block the airways.

Sudden infant death syndrome (SIDS) occurs when an infant does not awaken from an apneic episode. Infants who sleep on their back are less vulnerable to SIDS than stomach sleepers. They are more at risk if their mother smoked or was exposed to smoke during pregnancy.

Hannah Kinney of Harvard Medical School reported that an underlying weakness in the respiratory control center may be fatal when combined with environmental stresses. She compared brains of infants who died of SIDS with those of infants who died of other causes. The SIDS babies had fewer receptors for serotonin in their medulla oblongata. This neurotransmitter carries signals between neurons (Section 30.5). Weak signaling may impair the responses to potentially deadly respiratory stress.

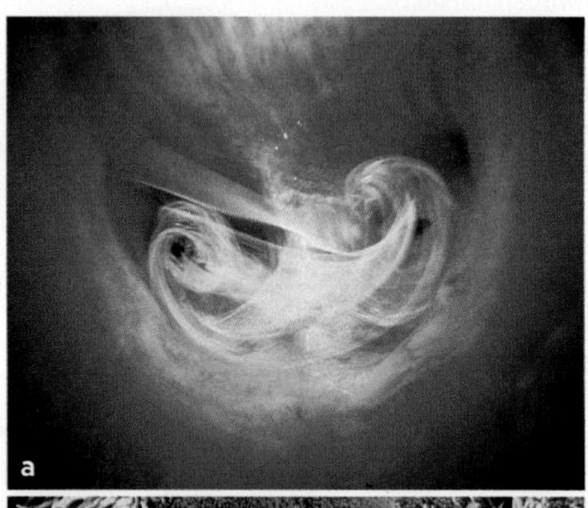

Figure 36.17 (**a**) Cigarette smoke about to enter bronchi that lead to the lungs. Smoke irritates ciliated and mucus-secreting cells that line the airways (**b**) and can encourage bronchitis.

Potentially Deadly Infections About one-third of the human population is infected by *Mycobacterium tuberculosis*, the cause of tuberculosis (Section 19.6). This bacterial species colonizes the lungs, but infection may not result in disease. Carriers can be identified by a TB skin test. About 10 percent of them will develop the disease. They start to cough and may have chest pain. When untreated, they may have trouble breathing and cough up bloody mucus. Antibiotics cure TB, but only if they are taken diligently for at least six months. An active, untreated infection can be fatal.

Lungs also get infected by bacteria, viruses, and—less commonly—fungi that cause pneumonia. Pneumonia is not one disease; it is a general term for lung inflammation caused by an infectious organism. Coughing, an aching chest, shortness of breath, and fever are usual symptoms. An x-ray can reveal infected tissues filled with fluid and immune cells instead of air. The treatment and outcome depend on the type of pathogen.

Chronic Bronchitis and Emphysema Facing the lumen of your bronchioles is a ciliated, mucus-producing epithelium (Figure 36.17). It is one of many defenses that protect you from respiratory infections. Chronic irritation of the lining may lead to bronchitis. With this respiratory disease, epithelial cells become irritated and secrete too much mucus. Thickened mucus triggers the coughing reflex and allows pathogens to grow at the lung surface.

Early attacks of bronchitis are treatable. When the aggravation continues, bronchioles become chronically inflamed. Bacteria, chemical agents, or both attack the lining of these airways. The lining's ciliated cells die, and mucus-secreting cells multiply. Fibrous scar tissue forms. Over time, scarring narrows or obstructs the airways. Breathing becomes labored and difficult.

Chronic bronchitis can lead to emphysema. With this condition, tissue-destroying bacterial enzymes digest the thin, stretchable alveolar wall. As walls deteriorate, inelastic fibrous tissue builds up around them. Alveoli enlarge, and not as many exchange gases. In time, the lungs become distended and inelastic, so the balance between air flow and blood flow is compromised. It becomes hard even to catch a breath.

About 2 million people in the United States currently have emphysema, and it causes or contributes to about 100,000 deaths every year.

A number of individuals are genetically predisposed to develop emphysema. They do not have a workable gene

Risks Associated With Smoking	Reduction in Risks by Quitting
Shortened life expectancy Nonsmokers live about 8.3 years longer than those who smoke two packs a day from midtwenties on.	Cumulative risk reduction; after 10–15 years, the life expectancy of ex-smokers approaches that of nonsmokers.
Chronic bronchitis, emphysema Smokers have 4–25 times higher risk of dying from these diseases than do nonsmokers.	Greater chance of improving lung function and slowing down rate of deterioration.
Cancer of lungs Cigarette smoking is the major cause.	After 10–15 years, risk approaches that of nonsmokers.
Cancer of mouth 3–10 times greater risk among smokers.	After 10–15 years, risk is reduced to that of nonsmokers.
Cancer of larynx 2.9–17.7 times more frequent among smokers.	After 10 years, risk is reduced to that of nonsmokers.
Cancer of esophagus 2–9 times greater risk of dying from this.	Risk proportional to amount smoked; quitting should reduce it.
Cancer of pancreas 2–5 times greater risk of dying from this.	Risk proportional to amount smoked; quitting should reduce it.
Cancer of bladder 7–10 times greater risk for smokers.	Risk decreases gradually over 7 years to that of nonsmokers.
Cardiovascular disease Cigarette smoking a major contributing factor in heart attacks, strokes, and atherosclerosis.	Risk for heart attack declines rapidly, for stroke declines more gradually, and for atherosclerosis it levels off.
Impact on offspring Women who smoke during pregnancy have more stillbirths, and the weight of liveborns is lower than the average (which makes babies more vulnerable to disease and death).	When smoking stops before fourth month of pregnancy, risk of stillbirth and lower birth weight eliminated.
Impaired immunity More allergic responses, destruction of white blood cells (macrophages) in respiratory tract.	Avoidable by not smoking.
Bone healing Surgically cut or broken bones may take 30 percent longer to heal in smokers, perhaps because smoking depletes the body of vitamin C and reduces the amount of oxygen delivered to tissues. Reduced vitamin C and reduced oxygen interfere with formation of collagen fibers in bone (and many other tissues).	Avoidable by not smoking.

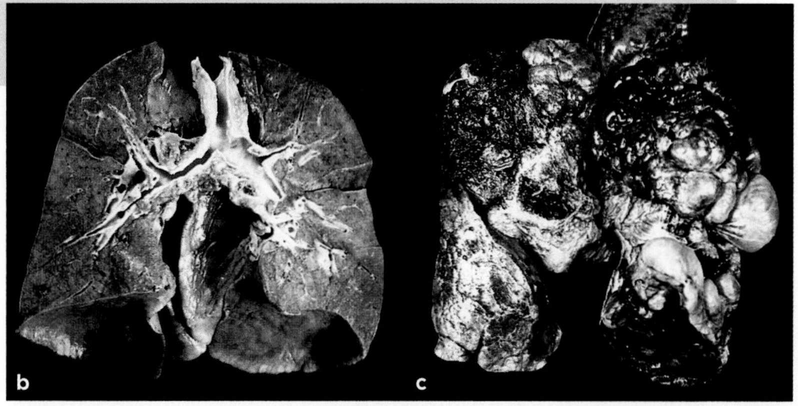

Figure 36.18 (**a**) From the American Cancer Society, a list of major risks incurred by smoking and the benefits of quitting. (**b**) Appearance of normal lung tissue in humans. (**c**) Appearance of lung tissues from someone who was affected by emphysema.

for antitrypsin, an enzyme that inhibits bacterial attacks on alveoli. Poor diet and persistent or recurring colds and other respiratory infections also invite emphysema later in life. Air pollution and chemicals in the workplace may contribute to the problem. However, tobacco smoking is by far the main risk factor for emphysema. Most of those affected are over age 50. Twenty or thirty years of smoke exposure leave lungs looking like those in Figure 36.18*c*.

Smoking's Impact Globally, cigarette smoking kills 4 million people each year. By 2030, the number may rise to 10 million, with about 70 percent of the deaths occurring in developing countries. In the United States, the direct medical costs of treating smoke-induced disorders drains $22 billion a year from the economy. As G. H. Brundtland—a medical doctor and the former director of the World Health Organization—points out, tobacco is the only legal consumer product that kills half of its regular users. If you are a smoker, you may wish to reflect on the information in Figure 36.18*a*.

Cigarettes also do more than sicken and kill smokers. Nonsmokers die of cancers and disease brought on by breathing secondhand smoke. Children who breathe cigarette smoke at home have a heightened risk for developing lung problems. Smoking while pregnant increases risk of miscarriage and low birth weight.

In 2004 in the United States, 14.6 million individuals twelve years and older were smoking pot, or marijuana (*Cannabis*), as a way to induce light-headed euphoria. Many are chronic users. The collection of toxic particles that they are inhaling, or "tar," is far more abundant in marijuana cigarettes than in tobacco cigarettes. Pot smokers also tend to inhale deeply and to smoke the cigarettes down to stubs, where tar accumulates. They deliver big doses of irritants and carcinogens into their throat and lungs. Using marijuana over the long term may lead to chronic throat problems, as well as bronchitis and emphysema. These problems are in addition to the negative effects on the nervous system (Section 30.6).

36.8 High Climbers and Deep Divers

Most humans function best within a limited range of atmospheric pressures. We run into trouble when we move out of that range too fast.

RESPIRATION AT HIGH ALTITUDES

LINKS TO SECTIONS 16.4, 25.3

Atmospheric pressure decreases with altitude. Above 5,500 meters, or about 18,000 feet, it is 380 mm Hg—half of what it is at sea level. Oxygen still is 21 percent of the total pressure, so there is about half as much oxygen as there is at sea level.

Recall, from Section 16.4, that llamas live at high altitudes. Their hemoglobin helps them survive in the "thin air," where oxygen levels are low. Compared to the hemoglobin of humans and other mammals, llama hemoglobin binds O_2 more efficiently (Figure 36.19). Also, the lungs and the heart of a llama are unusually large relative to the animal's body size.

Most people live at lower altitudes where there is plenty of oxygen. When they ascend too fast to high altitudes, the transport of oxygen to cells plummets. *Hypoxia*, or cellular oxygen deficiency, is the result.

In an acute compensatory response to hypoxia, the brain commands the heart and respiratory muscles to work harder. People breathe faster and more deeply than usual; they hyperventilate. This causes CO_2 to be exhaled faster than it forms and ion balances in the cerebrospinal fluid get skewed. Shortness of breath, a pounding heart, dizziness, nausea, and vomiting are symptoms of the resulting altitude sickness.

Compared to people at low elevations, people who grow up at high altitudes develop more alveoli and blood vessels in their lungs. The heart develops larger ventricles and pumps greater volumes of blood.

A healthy person who grew up by the sea can still move to the mountains. Through acclimatization, the acute compensatory response that the body made to a markedly different environment slowly gives way to adjustments in cardiac output, the rhythmic pattern of breathing, and the magnitude of breathing.

Within a few days, hypoxia stimulates kidney cells to secrete more erythropoietin. This hormone induces stem cells in bone marrow to divide repeatedly, and induces the descendant cells to develop as red blood cells. Under normal conditions, 2 million to 3 million red blood cells per second replace those that die off in the adult body. Under extreme stress, the stepped-up erythropoietin secretion causes six times as many to form. Increased numbers of circulating red blood cells improve the oxygen-delivery capacity of blood. When the O_2 concentration in blood returns to the normal range, erythropoietin secretion slows.

Acclimatization comes at high cost. Having many more red cells thickens the blood, so the heart has to work harder to pump blood through vessels. The risk of heart problems rises as a result.

Figure 36.19 The binding and releasing capacity for the hemoglobin of humans, llamas, and other mammals.

DEEP-SEA DIVERS

Water pressure increases with depth. Human divers using tanks of compressed air risk nitrogen narcosis, sometimes called "raptures of the deep." Starting at depths of about 45 meters (150 feet), the fluid pressure forces increased amounts of gaseous nitrogen (N_2) to dissolve in interstitial fluid. N_2 dissolves in the lipid bilayer of cell membranes. In neurons, this dissolved nitrogen can disrupt signaling, causing a diver to feel euphoric and drowsy. The deeper divers descend, the more weakened and clumsy they become.

Returning to the surface from a deep dive also has risks. As a diver ascends, pressure falls and N_2 moves from interstitial fluid into blood and is exhaled. If a diver rises too fast, N_2 bubbles form inside the body. The resulting decompression sickness, also known as "the bends," usually begins with joint pain. Bubbles of N_2 in blood can slow flow to organs. If bubbles affect the brain or lungs, the result can be fatal.

Humans who train to dive without oxygen tanks can remain submerged for about three minutes. So far, the human free diving record is 210 meters. Compare

Species	Maximum Depth
Sperm whale (*Physeter macrocephalus*)	2,200 meters
Leatherback turtle (*Dermochelys coriacea*)	1,200 meters
Southern elephant seal (*Mirounga leonina*)	1,620 meters
Weddell seal (*Leptonychotes weddelli*)	741 meters
Bottlenose dolphin (*Tursiops truncatus*)	>600 meters
Emperor penguin (*Aptenodytes forsteri*)	565 meters

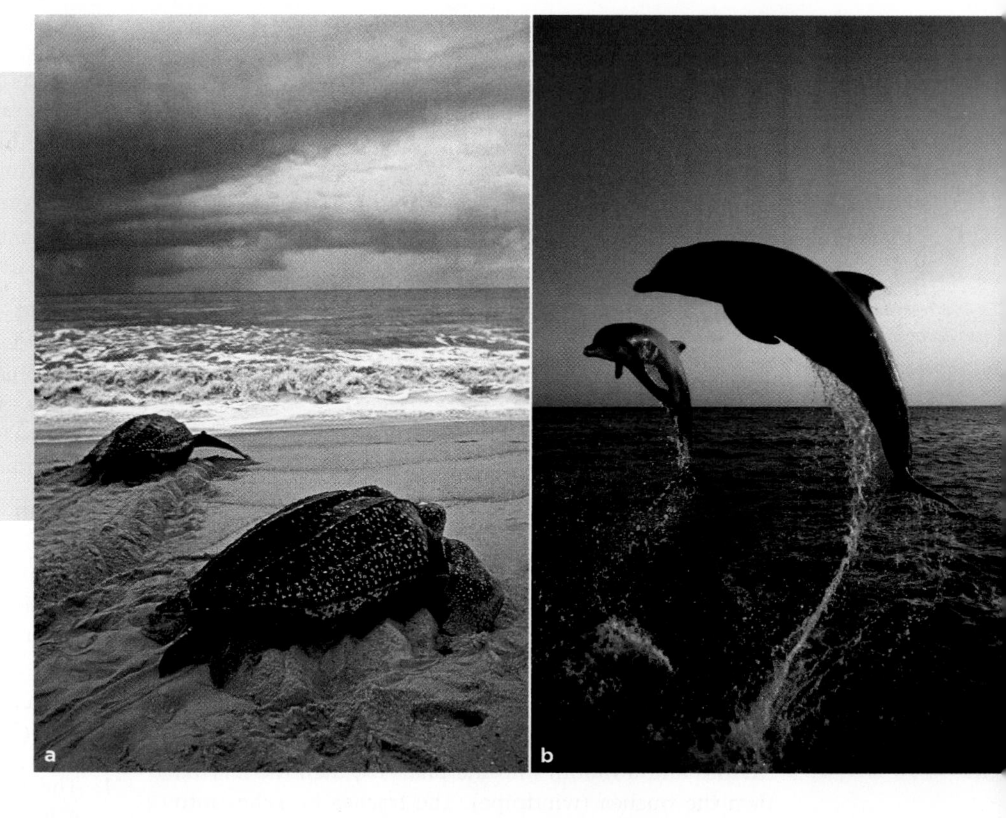

Figure 36.20 (**a**) Two Atlantic leatherback sea turtles returning to the sea after laying eggs. The leathery shell is adapted for deep diving; it bends rather than breaks under extreme pressure. (**b**) Bottlenose dolphins. The chart at *left* lists a few diving records.

that with the impressive depth records for species listed in Figure 36.20. What types of adaptations make deep dives possible?

Leatherback sea turtles leave water only to lay eggs (Figure 36.20*a*). They spend the rest of their time in open oceans diving for jellyfishes, their main prey. As a turtle or other air-breathing animal dives deeper and deeper, the weight of more and more water presses down onto the body. Lungs filled with air probably would collapse inward, but most diving animals move air out of the lungs and into cartilage-reinforced airways before they dive too deep. Also, the pressure at great depths could crack a typical turtle's hard shell, but the leatherback's soft shell bends and flexes under such pressure.

Making a deep dive means spending long intervals without access to air. The longest dive recorded for a leatherback turtle lasted for a little more than an hour. Sperm whales can stay submerged for two hours.

If a diving animal's lungs are emptied of air and if it has no access to the surface, then how does it meet its oxygen requirements? It does so in four ways.

First, before it dives, it breathes deeply. A sperm whale blows out about 80–90 percent of the air in its lungs with each exhalation; you exhale only about 15 percent. The deep breaths keep the O_2 pressure inside alveoli high, so more oxygen diffuses into the blood.

Second, diving animals can store great amounts of oxygen inside their blood and muscles. They tend to have a large blood volume relative to their body size, a high red blood cell count, and considerable amounts of myoglobin in their muscles. A skeletal muscle of a bottlenose dolphin (Figure 36.20*b*) has about 3.5 times the amount of myoglobin that a comparable skeletal muscle in a dog has. A muscle in a sperm whale has 7 times as much as the dog muscle.

Third, more O_2 gets distributed to the heart, brain, and other organs that require an uninterrupted supply of ATP energy for a deep dive. The blood volume and dissolved gases are stored and distributed efficiently with the help of valves and plexuses: meshes of blood vessels in local tissues. The metabolic rate and heart rate also decrease. So do oxygen uptake and carbon dioxide formation.

Fourth, whenever possible, a diving animal makes the most of its oxygen stores by sinking and gliding instead of actively swimming. It conserves energy by avoiding unnecessary movements.

There is less oxygen at high altitudes than lower ones. Humans make short-term and long-term adjustments in breathing patterns and cardiac output.

Unlike humans, diving sea turtles and marine mammals have built-in mechanisms for conserving oxygen during extended submergence.

www.thomsonedu.com Thomson NOW!

Summary

Section 36.1 Aerobic respiration uses O_2 and yields CO_2 as a product. Respiration is a physiological process by which, O_2 enters the internal environment and CO_2 leaves by diffusing across a respiratory surface. Each gas follows its own partial pressure gradient between the air and metabolically active tissues. Constraints imposed by the surface-to-volume ratio shaped respiratory structures and ventilation mechanisms. Respiratory proteins help maintain gradients favoring gas exchange.

Section 36.2 Small invertebrates that live in aquatic or damp habitats exchange gases mainly across the body surface. Gills, book lungs, and tracheal tubes function as organs of respiration in other invertebrates.

Section 36.3 Water and blood that flow in opposing directions exchange gases at fish gills. The countercurrent flow increases the efficiency of exchanges. Paired lungs are the main respiratory organs in the reptiles, birds, and mammals. Most amphibians also have lungs, but some have gills or exchange gases across the skin.

- *Use the animation on ThomsonNOW to compare various vertebrate respiratory systems.*

Section 36.4 In humans, air flows through two nasal cavities and a mouth into the pharynx, then the larynx, then the trachea (windpipe). The trachea branches into two bronchi that enter the lungs. These two airways and finely branching bronchioles form the bronchial tree. At the ends of the finest branches of this tree are the thin-walled alveoli. Gases are exchanged at these air sacs.

- *Use the animation on ThomsonNOW to explore the human respiratory system.*

Section 36.5 In human lungs, the alveolar wall, the wall of a pulmonary capillary, and their fused basement membranes form a thin respiratory membrane between air inside an alveolus and the internal environment. O_2 following its partial pressure gradient diffuses across the respiratory membrane, into the plasma of the blood, and finally into red blood cells.

Red blood cells are filled with hemoglobin molecules. Heme groups of hemoglobin bind O_2 that diffused into the internal environment. In metabolically active tissues throughout the body, the blood becomes warmer, the pH gets lower, and the CO_2 partial pressure increases. These conditions favor the release of O_2 from hemoglobin and its diffusion out of capillaries, through interstitial fluid, and into cells.

CO_2 follows its partial pressure gradient and diffuses from cells to interstitial fluid, to blood. Most CO_2 reacts with water in red blood cells, forming bicarbonate. These reactions are reversed in the lungs. There, CO_2 diffuses out of blood into air inside alveoli. It is expelled, along with water vapor, in exhalations.

- *Use the animation on ThomsonNOW to compare partial pressures of CO_2 and O_2 in different body regions.*

Section 36.6 Each respiratory cycle consists of one inhalation and one exhalation. Inhalation is always an active process. As muscle contractions expand the chest cavity, pressure in lungs decreases below atmospheric pressure, and air flows into the lungs. These events are reversed during exhalation, which normally is passive.

The brain stem has respiratory centers that adjust the rate and magnitude of breathing. The centers respond to an assortment of chemoreceptors that detect shifts in H^+ levels in cerebrospinal fluid and in arterial blood.

- *Use the animation on ThomsonNOW to learn about the respiratory cycle.*

Section 36.7 Respiratory disorders include apnea and sudden infant death syndrome (SIDS). Tuberculosis, pneumonia, bronchitis, and emphysema are respiratory diseases. Smoking worsens or causes many respiratory problems. Worldwide, smoking remains a leading cause of debilitating diseases and deaths.

Section 36.8 Air's oxygen concentration decreases with altitude. People acclimatize to high altitude through altered breathing patterns and other changes. Specialized mechanisms and behaviors allow sea turtles and diving marine mammals to dive deeply for long intervals.

Self-Quiz

Answers in Appendix III

1. The most abundant gas in the atmosphere is ______.
   a. nitrogen c. oxygen
   b. carbon dioxide d. hydrogen

2. Respiratory proteins such as hemoglobin ______.
   a. contain metal ions
   b. occur only in vertebrates
   c. increase the efficiency of oxygen transport
   d. both a and c

3. In insects, most gas exchange occurs at ______.
   a. the tips of tracheal tubes c. gills
   b. the body surface d. paired lungs

4. Countercurrent flow of water and blood increases the efficiency of gas exchange in ______.
   a. fishes c. birds
   b. amphibians d. all of the above

5. In human lungs, gas exchange occurs at the ______.
   a. two bronchi c. alveolar sacs
   b. pleural sacs d. both b and c

6. When you breathe quietly, inhalation is ______ and exhalation is ______.
   a. passive; passive c. passive; active
   b. active; active d. active; passive

7. During inhalation ______.
   a. the thoracic cavity expands
   b. the diaphragm relaxes
   c. atmospheric pressure declines
   d. both a and c

8. Most oxygen being transported in blood ______.
   a. is bound to hemoglobin
   b. combines with carbon to form carbon dioxide
   c. is in the form of bicarbonate
   d. is dissolved in the plasma

9. At high altitudes, ________ .
   a. nitrogen bubbles out of the blood
   b. hemoglobin has fewer oxygen-binding sites
   c. atmospheric pressure is lower than at sea level
   d. both b and c

10. Match the words with their descriptions.

___ trachea	a. muscle of respiration
___ pharynx	b. gap between vocal cords
___ alveolus	c. between bronchi and alveoli
___ hemoglobin	d. windpipe
___ bronchus	e. respiratory protein
___ bronchiole	f. site of gas exchange
___ glottis	g. airway leading to lung
___ diaphragm	h. throat

■ *Visit ThomsonNOW for additional questions.*

Critical Thinking

1. Select one of the animals shown in Figure 36.21, do some research, and explain how the animal is enhancing the rate at which O_2 diffuses into its internal environment and CO_2 diffuses out of it.

2. Using the sketch below, define and distinguish between the following respiratory system components:

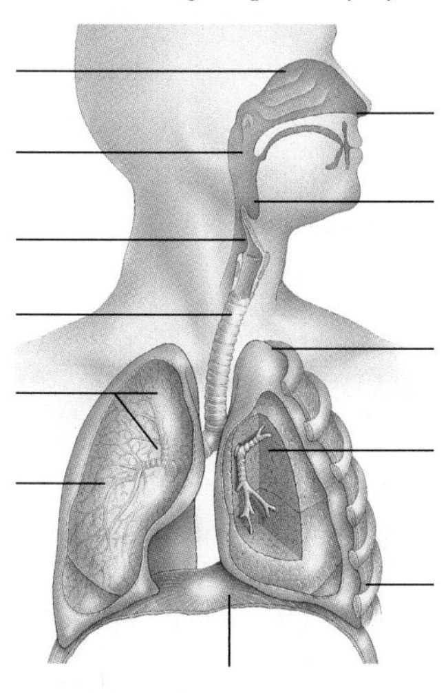

a. Epiglottis
   Glottis

b. Pharynx
   Larynx

c. Bronchiole
   Bronchus

d. Pleural sac
   Alveolar sac

3. Using the same sketch, name the types of bones and skeletal muscles that help ventilate our lungs. Do the muscles function in passive exhalation, in active inhalation, or both events?

4. A hiccup is a sudden involuntary contraction of the diaphragm. Contraction of this muscle draws air inward during normal inhalation. With hiccups, the glottis closes abruptly and stops the air flow. The cutoff makes the sound associated with a hiccup.

Humans start to hiccup long before birth, by about two months into a pregnancy. Contraction of the muscles of inhalation at the same time the glottis closes is of no use to humans. In lungfish and some amphibians that have gills and a glottis, however, a hiccuping reflex pushes water over the gills while preventing it from entering the lungs. Humans do not have functional gills at any point in their development. Speculate on why the hiccuping reflex occurs in humans.

5. Look again at Figure 36.13. Notice that the oxygen and carbon dioxide content of blood in pulmonary veins is the same as at the start of the systemic capillaries. Notice also that systemic veins and pulmonary arteries have equal partial pressures. Explain the reason for these similarities.

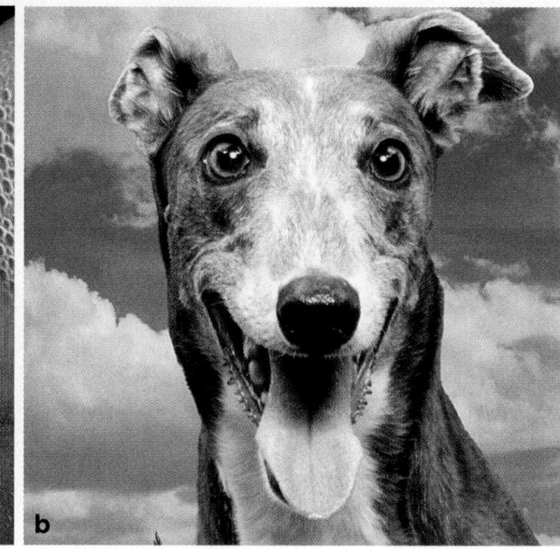

Figure 36.21 (**a**) Operculum of an Indian yellow-tail angelfish (*Apolemichthys xanthurus*). This stiffened lid over the fish gill can alternately open and close a bit. (**b**) A panting pooch.

6. When you swallow food, contracting muscles force the epiglottis down, to its closed position. This prevents food from going down the trachea and blocking air flow. Yet each year, several thousand people choke to death after food enters the trachea and blocks air flow for as little as four or five minutes.

The Heimlich maneuver—an emergency procedure only —may dislodge food stuck in the trachea. A rescuer stands behind the victim, makes a fist with one hand, and then positions the fist, thumb-side in, against the abdomen of the person who is choking (Figure 36.22). The fist must be just a bit above the navel and well below the rib cage. The fist is pressed into the abdomen with a sudden upward thrust. The thrust can be repeated if necessary.

When correctly performed, the Heimlich maneuver forcibly elevates the diaphragm. Repeated thrusts are often enough to dislodge an obstruction. When they do, a doctor must see the person at once; an inexperienced rescuer can inadvertently cause internal injuries or crack a rib.

Reflect now on our social climate in which lawsuits abound. Would you risk performing the Heimlich maneuver to save a relative's life? A stranger's life? Why or why not?

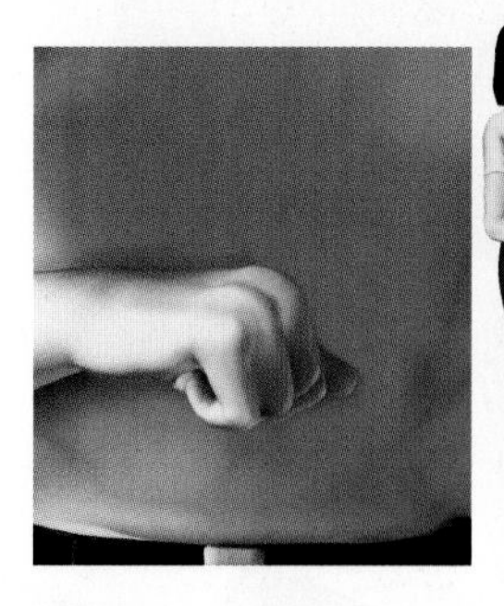

a Place a fist just above the choking person's navel, with the flat of your thumb against the abdomen.

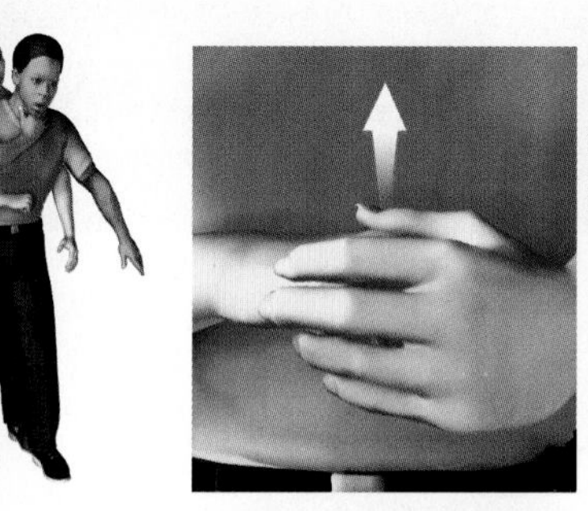

b Cover the fist with your other hand. Thrust both fists up and in with enough force to lift the person off his or her feet.

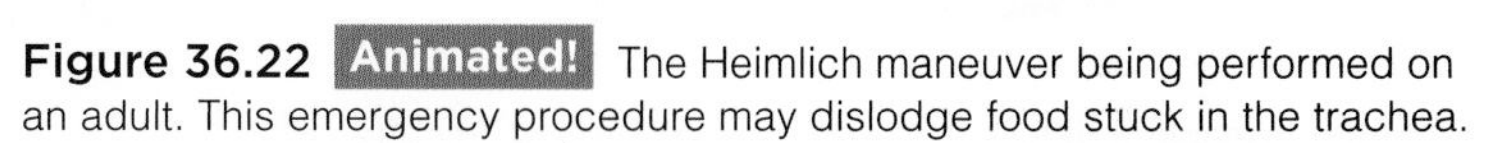

Figure 36.22 **Animated!** The Heimlich maneuver being performed on an adult. This emergency procedure may dislodge food stuck in the trachea.

37 DIGESTION AND HUMAN NUTRITION

IMPACTS, ISSUES Hominids, Hips, and Hunger

Like other mammals, humans have an abundance of fat-storing cells in adipose tissue. This energy warehouse served our early hominid ancestors well. As foragers, they could seldom be certain of where their next meal was coming from (Figure 37.1*a*). Putting on fat when food was abundant could help them survive later when it was scarce.

Lean pickings are not a problem for most Americans. With 60 percent of the adults overweight or obese, they are among the fattest people in the world. "Obesity" is an overabundance of fat in adipose tissue. It invites many severe problems, including heart disease, diabetes, and some cancers. You would think the urgency surrounding the need to lose weight is obvious, but extra pounds are tough to lose. Hormones are involved.

Eating just a bit or a lot merely changes how empty or full each adipose cell gets. When we take in more calories than we burn, the cells plump up with fat droplets and synthesize leptin. This hormone acts on a brain center that helps control appetite. Some mutant mice cannot make leptin. They eat and eat, until they look like inflated balloons (Figure 37.1*b*). Inject an obese mutant mouse with leptin, and it eats less and slims down. But lack of leptin cannot explain human weight gain. Obese people have plenty of leptin. Do they lack leptin receptors or have faulty ones? These are weighty questions.

Ghrelin, another hormone, makes you feel hungry. Some cells in the stomach lining and the brain secrete it when the stomach is empty. Secretions slow after a big meal. In one study of ghrelin's effects, a group of obese volunteers stayed on a low-fat, low-calorie diet for six months. They lost weight, but the concentration of ghrelin in their blood climbed dramatically and they were hungrier than ever!

Some extremely obese people opt for gastric bypass surgery, in which "stomach-stapling" closes off part of the stomach and most of the small intestine (Figure 37.2). They can eat only so much before feeling full, which cuts down the amount of nutrients that can be absorbed.

a

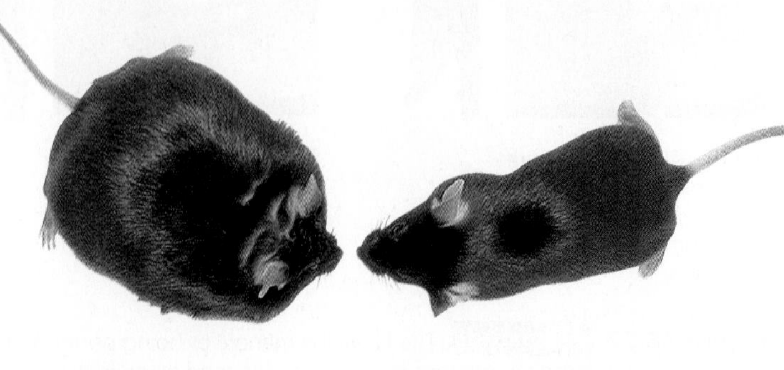

b Mutant leptin-deficient mouse **c** Normal mouse

See the video! **Figure 37.1** (**a**) No fast food or excess body fat for *Homo habilis*. Fat-storing adaptations evolved among the hominids that were ancestral to modern humans, many of whom store far too much fat. (**b**) This mutant mouse cannot synthesize leptin, a hormone that helps control an appetite center in the brain. Compared with a normal mouse (**c**), the mutant eats a lot more, and it is far heavier.

How would you vote? Rampant consumption of "fast foods" may be a big factor in the obesity epidemic in the United States. Should fast-food labels carry consumer warnings, as alcohol and cigarette labels do? See ThomsonNOW for details, then vote online.

Gastric bypass is more effective than standard weight loss methods. Patients are far less likely to regain pounds. Output from their ghrelin-producing stomach cells drops after the bypass surgery, so they feel hungry less often. Can researchers come up with a drug that can slow the synthesis or release of ghrelin? It might be a less risky alternative to surgery.

Some hormones encourage weight loss. For instance, elderly people tend to lose their appetite. Some eat so little that they endanger their health. Cholecystokinin (CCK) may be at work suppressing their appetite. A new drug that blocks secretion of this hormone might help these individuals. Conversely, a drug that promotes CCK secretion might prevent obesity.

Questions about food intake and body weight lead us into the world of nutrition. The word encompasses all the processes by which an animal ingests and digests food, then absorbs the released nutrients as energy sources and building blocks for cells. When all works well, inputs balance the outputs, and weight remains within a range that promotes good health.

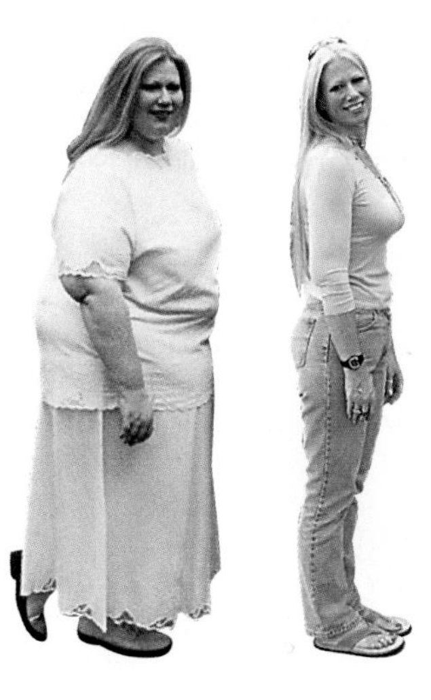

Figure 37.2 Young woman before and after gastric bypass surgery.

Key Concepts

OVERVIEW OF DIGESTIVE SYSTEMS

Some animal digestive systems are saclike, but most are a tube that extends through the body. In complex animals, a digestive system interacts with other organ systems in the distribution of nutrients and water, disposal of residues and wastes, and the maintenance of the internal environment. **Section 37.1**

HUMAN DIGESTIVE SYSTEM

In humans, digestion starts in the mouth, continues in the stomach, and is finished in the small intestine. Secretions from salivary glands, the pancreas, and the liver function in digestion. Most nutrients are absorbed in the small intestine. The large intestine, or colon, absorbs water and concentrates and stores wastes. **Sections 37.2–37.6**

ORGANIC METABOLISM AND NUTRITION

Nutrients absorbed from the gut are raw materials in the synthesis of the body's complex carbohydrates, lipids, proteins, and nucleic acids. A healthy diet normally provides all nutrients, vitamins, and minerals necessary to support metabolism. **Sections 37.7–37.9**

BALANCING CALORIC INPUTS AND OUTPUTS

Maintaining body weight requires balancing calories taken in with calories burned in metabolism and physical activity. **Section 37.10**

Links to Earlier Concepts

This chapter expands on the sampling of digestive systems in Chapters 23 and 24, on animal diversity. You will consider dietary aspects of complex carbohydrates, lipids, and proteins (Sections 3.2–3.4) and the nature of organic metabolism, especially the use and storage of glucose (7.7). You will draw on your knowledge of diffusion, transport mechanisms, and osmosis (5.6–5.8). You will revisit pH, buffer systems, and some of the cofactors (2.6, 5.4). You will look again at helpful and harmful bacteria inside the human gut (19.1, 19.2).

OVERVIEW OF DIGESTIVE SYSTEMS

37.1 The Nature of Digestive Systems

All animals ingest food, digest it, then absorb nutrients and convert them to the body's own carbohydrates, lipids, proteins, and nucleic acids. Build now on the examples given earlier, in Chapters 23 and 24.

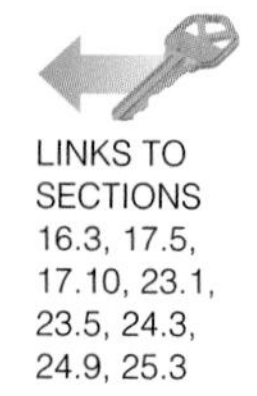

LINKS TO SECTIONS 16.3, 17.5, 17.10, 23.1, 23.5, 24.3, 24.9, 25.3

An animal's digestive system is a body cavity or a tube that mechanically and chemically breaks food down to small particles, then to molecules small enough to get absorbed into the internal environment. The digestive system also expels unabsorbed residues. By interacting with other systems, it helps maintain homeostasis for the body as a whole (Figure 37.3).

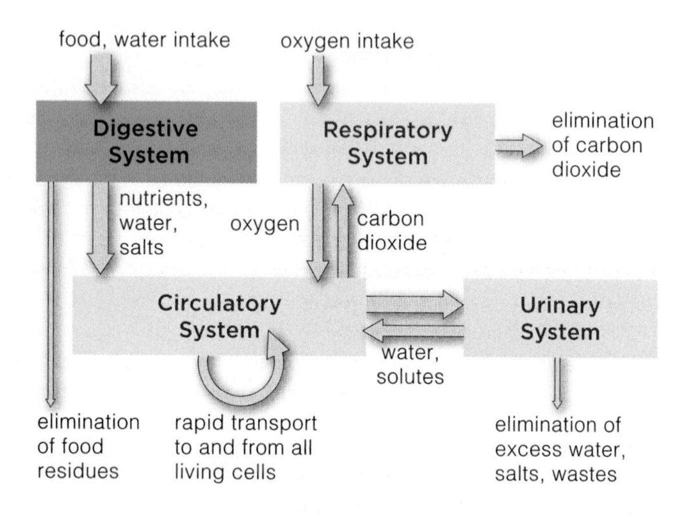

Figure 37.3 Organ systems with key roles in the uptake, processing, and distribution of nutrients and water in complex animals.

INCOMPLETE AND COMPLETE SYSTEMS

Recall, from Section 23.1, that some invertebrates have an incomplete digestive system. Food enters and wastes leave their saclike gut through a single opening at the body surface. For flatworms, a saclike, branching gut cavity opens at the start of a pharynx, a muscular tube (Figure 37.4*a*). Food that enters the sac is digested, its nutrients are absorbed, then wastes are expelled. The two-way traffic does not favor regional specialization.

Like most animals, all vertebrates have a complete digestive system: a tubular gut with two openings. A mouth is at one end and an anus at the other. Along the tube's length are specialized regions that process food, absorb nutrients, and concentrate wastes.

Figure 37.4*b* shows the complete digestive system of a frog. The tubular portion consists of the mouth, pharynx, esophagus, stomach, small intestine, large intestine, and anus. A liver, gallbladder, and pancreas are accessory organs that assist digestion by secreting enzymes and other products into the small intestine.

Regardless of the complexity, a complete digestive systems carries out five overall tasks:

1. *Mechanical processing and motility.* Movements that break up, mix, and directionally propel food material.

2. *Secretion.* Release of substances, especially digestive enzymes, into the lumen (the space inside the tube).

3. *Digestion.* Breakdown of food into particles, then to nutrient molecules small enough to be absorbed.

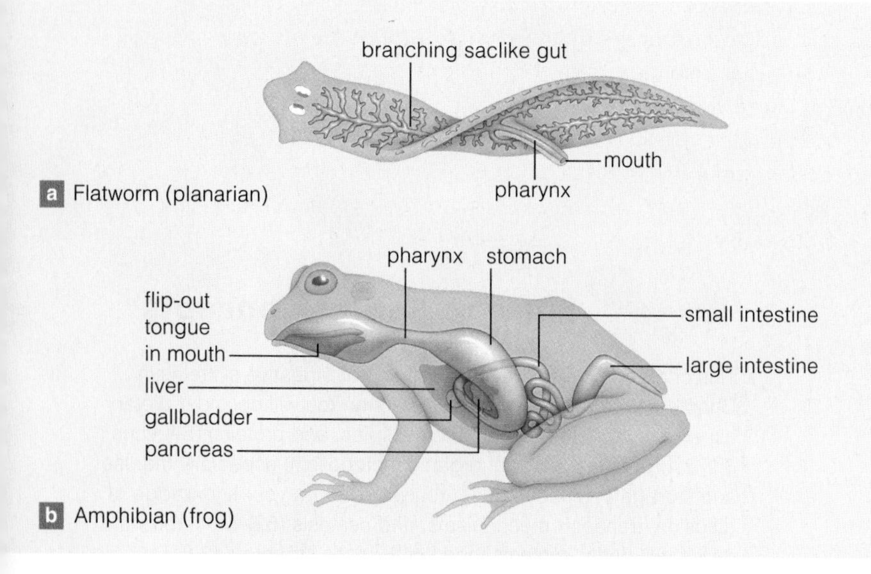

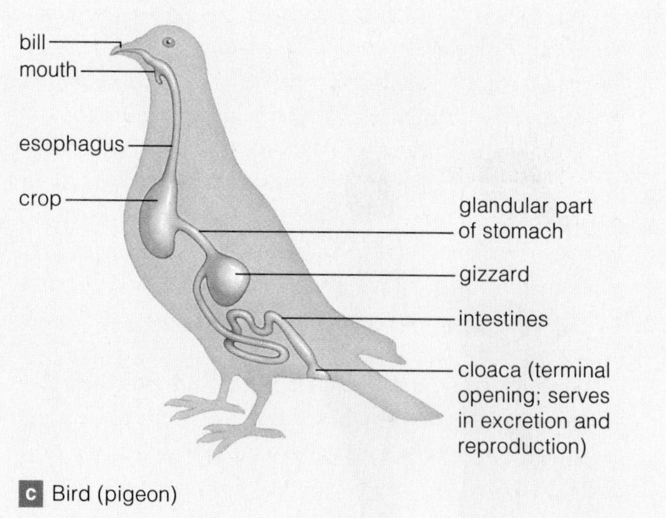

Figure 37.4 Animated! (**a**) Incomplete digestive system. (**b**,**c**) Two complete digestive systems.

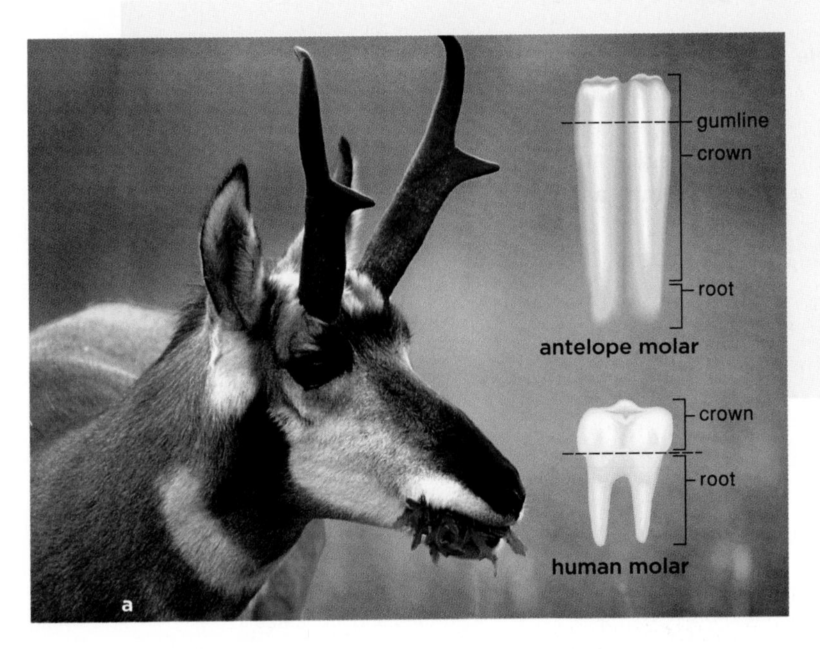

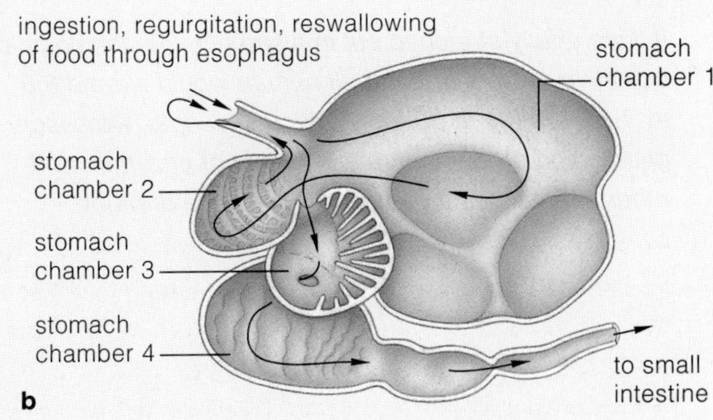

Figure 37.5 Animated! (**a**) Human and pronghorn antelope molars. (**b**) An antelope's multiple stomach chambers. In the first two, food is mixed with fluid and exposed to microbial symbionts that engage in fermentation. Some of the microbes degrade cellulose; others synthesize organic compounds, fatty acids, and vitamins. Partly digested food is regurgitated into the mouth, chewed, then swallowed. It enters the third chamber and is digested again before entering the last stomach chamber.

4. *Absorption.* Uptake of digested nutrients and water across the gut wall, into extracellular fluid.

5. *Elimination.* Expulsion of undigested or unabsorbed solid residues.

DIETARY ADAPTATIONS

In the birds, bill size and shape are diet-related traits shaped by natural selection (Sections 16.3, 17.5, and 17.10). So are other traits. Consider the pigeon (Figure 37.4*c*). It uses its bill to peck up small seeds from the ground. Like other seed eaters, a pigeon has a large crop, a saclike food-storing region above the stomach. The bird quickly fills its crop with seeds, then digests them later, in safer places. This eat-and-run strategy can reduce the amount of time that the bird is on the ground, where it is most vulnerable to predators.

Birds do not have teeth. They grind up food inside a gizzard: a stomach chamber lined with hard protein particles. Compared to hawks and other meat-eating birds, seed eaters have larger gizzards relative to their size. Also, the intestinal tract is longer; seeds require more processing time than meat. In all birds, residues collect in a cloaca before being expelled.

Mammals also display diet-related adaptations. For example, fall through winter, on the mountain ridges from central Canada on into northern Mexico, we find pronghorn antelope browsing on wild sage. In spring, they move to open grasslands and deserts, and graze on young plants. Antelope molars (cheek teeth) have a flattened crown that is a grinding platform (Section 24.10). It is proportionally much larger than the crown on your molars (Figure 37.5*a*). Why such a difference? You do not brush your mouth against dirt as you eat, but an antelope does. Abrasive soil particles mix with the animal's food, so the crown of an antelope molar gets a lot of wear. An enlarged crown is an adaptation that keeps the molars from wearing down to nubs.

The antelope's gut also shows dietary adaptations. Antelopes are ruminants, a type of hoofed mammal with multiple stomach chambers in which cellulose is slowly digested (Figure 37.5*b*). The chambers steadily accept food during long feeding periods, then release nutrients slowly when the animal rests. Compared to meat, cellulose-rich plants require more attention from digestive enzymes, so ruminants tend to have longer intestines than carnivores of the same size.

The stomach of many carnivorous and scavenging mammals expands with really big meals. A male lion that weighs 250 kilograms may eat 40 kilograms of flesh at one feeding. Gorging when food is available allows the lion to rest for days between hunts.

Digestive systems mechanically and chemically degrade food into small molecules that can be absorbed, along with water, into the internal environment. These systems also expel the undigested residues from the body.

Incomplete digestive systems are a saclike cavity with one opening. Complete digestive systems are a tube with two openings and regional specializations in between.

Digestive systems differ in diet-related adaptations, such as the type of feeding structures and length of the gut.

37.2 Overview of the Human Digestive System

LINKS TO SECTIONS 29.1, 31.3, 36.4

If it were fully stretched out in a straight line, the complete digestive system of an adult human would extend 6.5 to 9 meters (21 to 30 feet). Along its length, accessory glands and organs secrete a variety of enzymes and other substances that digest food inside the tube.

Like all vertebrates, humans have a complete digestive system; a tubular gut with two openings (Figure 37.6). Mucus-covered epithelium lines this tube, which also has regions that specialize in digesting food, absorbing nutrients, and concentrating and storing unabsorbed

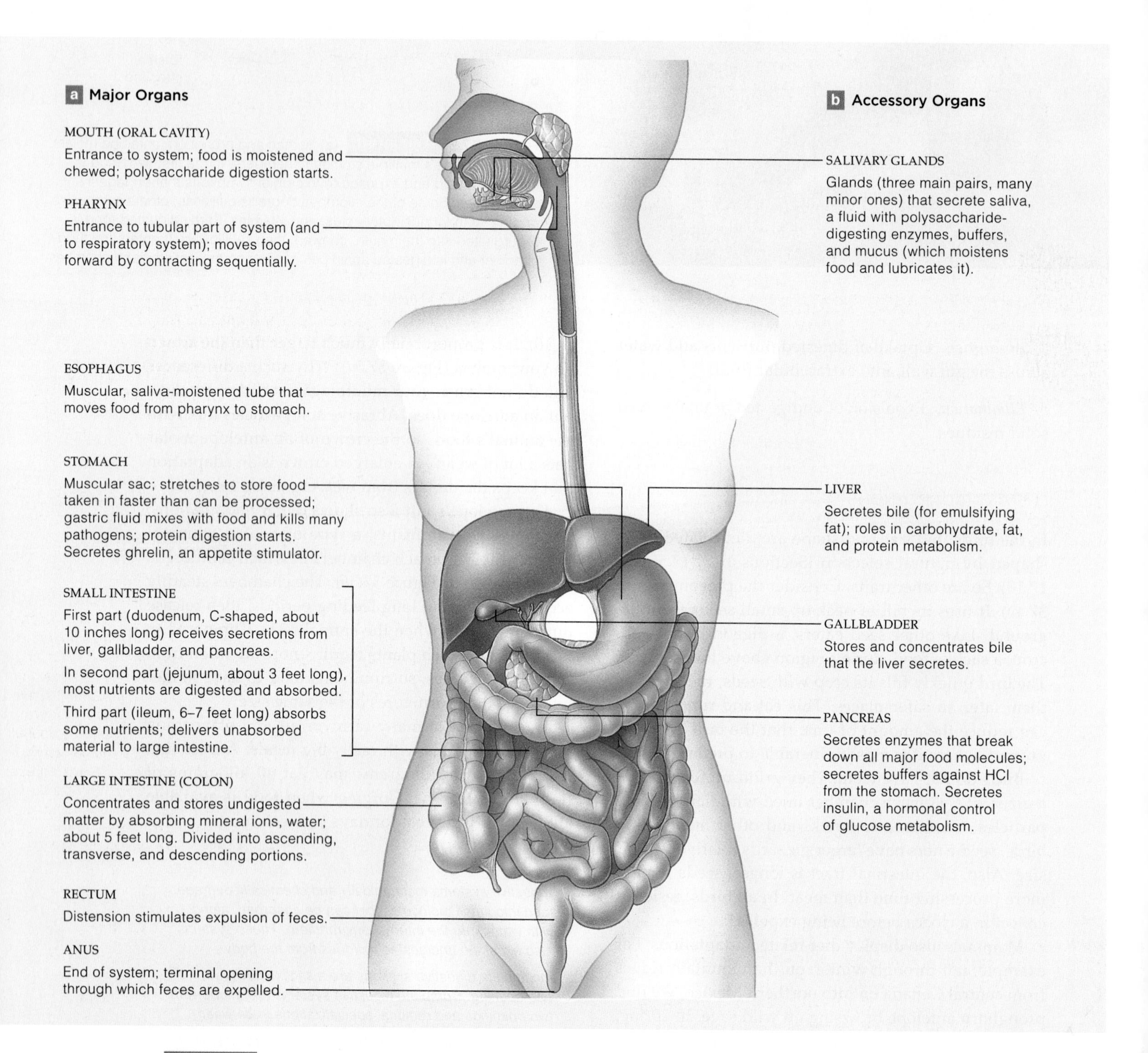

Figure 37.6 **Animated!** Overview of (**a**) major organs and (**b**) accessory organs of the human digestive system, together with a list of their primary functions.

waste. Salivary glands, the pancreas, and the liver are accessory organs that secrete substances into the tube. Food enters the mouth and goes through the pharynx, the esophagus, and finally the gut. The human gut, or gastrointestinal tract, starts at the stomach and extends through the intestines to the tube's terminal opening.

Food is partially processed inside the mouth, or oral cavity. Resting on the floor of the mouth is a tongue, an organ that consists of membrane-covered skeletal muscle. The tongue positions food for swallowing. It helps us round or flatten out sounds as we speak. The many chemoreceptors at its surface contribute to our sense of taste (Section 31.3).

Swallowing forces food into the pharynx. A human pharynx, or throat, is the entrance to the digestive *and* respiratory tracts (Section 36.4). The presence of food at the back of the throat triggers a swallowing reflex. When you swallow, the flaplike epiglottis flops down and the vocal cords constrict, so the route between the pharynx and larynx is blocked. The reflex keeps food from getting stuck in an airway and choking you.

Contractions propel food through the esophagus, a muscular tube between the pharynx and stomach. The stomach is a stretchable sac that stores food, secretes acid and enzymes, and mixes them all together.

Between the esophagus and stomach is a sphincter. Like all sphincters, this ring of smooth muscle blocks the flow of substances past it when it has contracted. Sometimes this one does not close properly after food passes through. The result is gastroesophageal reflux disease, or heartburn. Pain arises when stomach acid splashes onto esophageal tissues and irritates them. Chronic irritation raises the risk of esophageal cancer.

The stomach leads to the small intestine, the part of the gut where most carbohydrates, lipids, and proteins are digested and where most of the released nutrients and water are absorbed. Secretions from the liver and pancreas assist the small intestine in these tasks.

The colon (large intestine) absorbs water and ions, thus compacting the wastes. Wastes are briefly stored in a stretchable tube, the rectum, before being expelled from the gut's terminal opening, or anus.

Humans have a complete digestive system. Swallowing forces food and water from the mouth into the pharynx. Food continues through an esophagus to the stomach.

Food processing starts in the mouth. Most digestion and absorption occurs in the small intestine. The colon absorbs most of the remaining water and ions, which causes the wastes to compact. The rectum briefly stores the wastes before they are expelled through the anus.

37.3 Food in the Mouth

Mechanical digestion, the smashing of food into smaller pieces, begins in the mouth. So does chemical digestion, the enzymatic breakdown of food into molecular subunits.

LINKS TO SECTIONS 2.6, 3.2, 3.3, 24.10, 33.2

Mechanical breakdown of food begins with the action of teeth. Each tooth is a hard structure made mostly of bonelike dentin and embedded in the jaw at a fibrous joint (Figure 37.7*a* and Section 24.10). Dentin-secreting cells reside in a central pulp cavity. They are serviced by nerves and blood vessels that extend through the tooth's root. Enamel, the hardest material in the body, covers the tooth's exposed crown and reduces wear.

Human adults have thirty-two teeth of four types (Figure 37.7*b*). Chisel-shaped incisors shear off bits of food. Cone-shaped canines tear up meats. Premolars and molars have broad bumpy crowns and serve as platforms for grinding and crushing food.

Chewing mixes food bits with saliva, which consists mostly of water with dissolved enzymes, mucins, and bicarbonate. Bicarbonate is a buffer and it helps keep the pH in the mouth from getting too acidic (Section 2.6). The enzyme salivary amylase hydrolyzes starch, breaking it down into glucose subunits (Sections 3.2 and 3.3). Mucin proteins form mucus that binds food pieces together in easy-to-swallow clumps.

Teeth mechanically break food into particles. Salivary amylase begins the chemical digestion of carbohydrates.

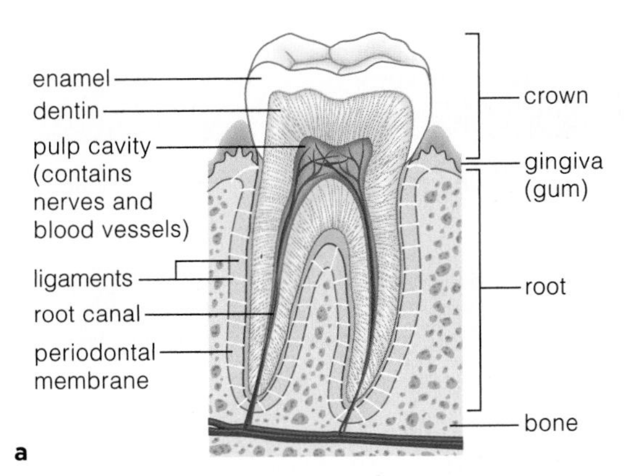

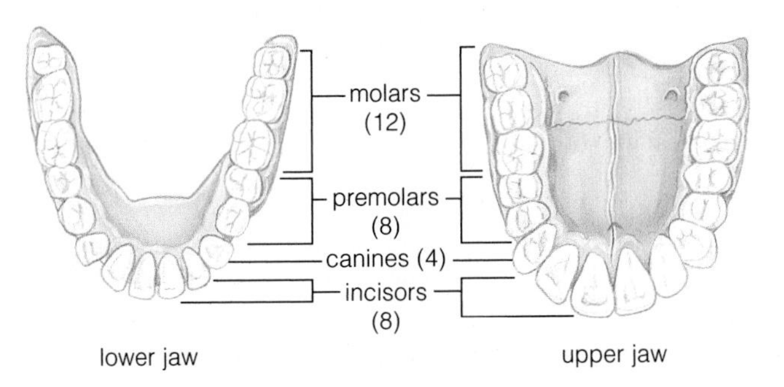

Figure 37.7 (**a**) Cross-section of one human tooth. The crown is the portion extending above the gum; the root is embedded in the jaw. Tiny ligaments attach the tooth to the jawbone. (**b**) Adult human teeth. Humans are omnivores, so some teeth are best suited to tearing meat, while others can grind up vegetable matter.

37.4 Food Breakdown in the Stomach and Small Intestine

In the stomach and small intestine, digestive enzymes and other secretions help break down nutrients into molecules small enough to be absorbed.

LINKS TO SECTIONS 2.6, 3.2–3.4, 5.4, 19.1, 30.8, 32.10

Carbohydrate breakdown, again, *starts* in the mouth. Protein breakdown *starts* in the stomach. The digestion of both is completed in the small intestine, along with lipids and nucleic acids. Smooth muscle layers in the stomach and intestinal wall mix these molecules with digestive enzymes (Figure 37.8 and Table 37.1).

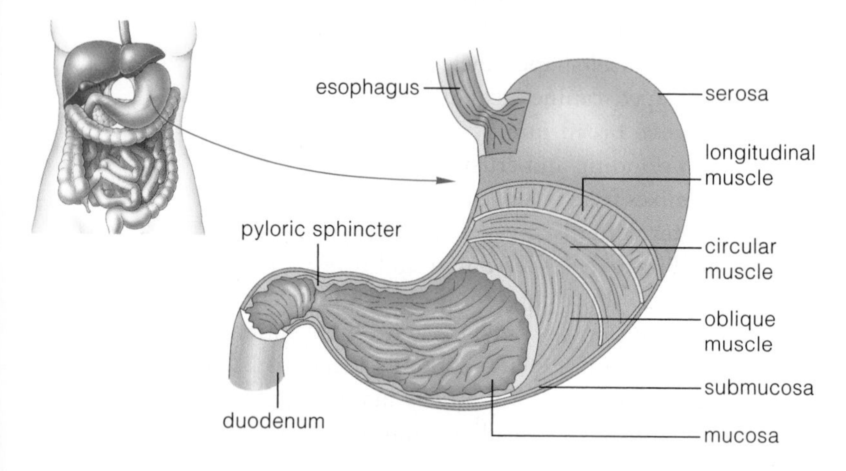

Figure 37.8 Structure of the stomach wall. Like most of the gut wall, it has an inner mucosa, which rests on a connective tissue with a mesh of nerves that exert local control of digestion. Next to this tissue are smooth muscle layers that differ in orientation and in the direction of contraction. The outer layer, the serosa, is connective tissue covered by epithelium.

DIGESTION IN THE STOMACH

The stomach, again, is a muscular, stretchable sac with three functions. It briefly stores and pummels bits of food. It secretes substances that help chemically break down food. It controls the length of time it takes for food to move on into the small intestine.

A mucus-secreting epithelium—the mucosa—lines the inner gut wall. In the stomach, cells of the mucosa secrete about two liters of gastric fluid each day. This fluid includes mucus, hydrochloric acid, and enzymes such as pepsinogens. Acid lowers the pH to about 2. When food enters the stomach, endocrine cells in the stomach lining secrete the hormone gastrin into blood. Gastrin binds to secretory cells of the mucosa, causing them to step up secretion of acid and pepsinogen.

By contracting rhythmically, smooth muscle in the stomach wall mixes the gastric fluid and food into a semiliquid mass called chyme. Chyme's acidity makes proteins unfold, which exposes their peptide bonds. It also converts pepsinogens to pepsins—enzymes that break peptide bonds. In time, the muscle contractions push chyme through the pyloric sphincter, an opening between the stomach and small intestine (Figure 37.8).

Chyme's strong acidity can kill most bacteria, but *Helicobacter pylori* occasionally infects the lining of the stomach and upper intestine (Figure 19.3*a*). Infection can damage the lining and expose tissues beneath to acid. The result is a painful ulcer. Antibiotics are now commonly used to treat such ulcers.

Table 37.1 Summary of Chemical Digestion

Location	Enzymes Present	Enzyme Source	Enzyme Substrate	Main Breakdown Products
Carbohydrate Digestion				
Mouth, stomach	Salivary amylase	Salivary glands	Polysaccharides	Disaccharides
Small intestine	Pancreatic amylase	Pancreas	Polysaccharides	Disaccharides
	Disaccharidases	Intestinal lining	Disaccharides	**Monosaccharides*** (such as glucose)
Protein Digestion				
Stomach	Pepsins	Stomach lining	Proteins	Protein fragments
Small intestine	Trypsin, chymotrypsin	Pancreas	Proteins	Protein fragments
	Carboxypeptidase	Pancreas	Protein fragments	**Amino acids***
	Aminopeptidase	Intestinal lining	**Amino acids***	
Lipid Digestion				
Small intestine	Lipase	Pancreas	Triglycerides	**Free fatty acids, monoglycerides***
Nucleic Acid Digestion				
Small Intestine	Pancreatic nucleases	Pancreas	DNA, RNA	Nucleotides
	Intestinal nucleases	Intestinal lining	Nucleotides	**Nucleotide bases, monosaccharides***

* Breakdown products small enough to be absorbed into the internal environment.

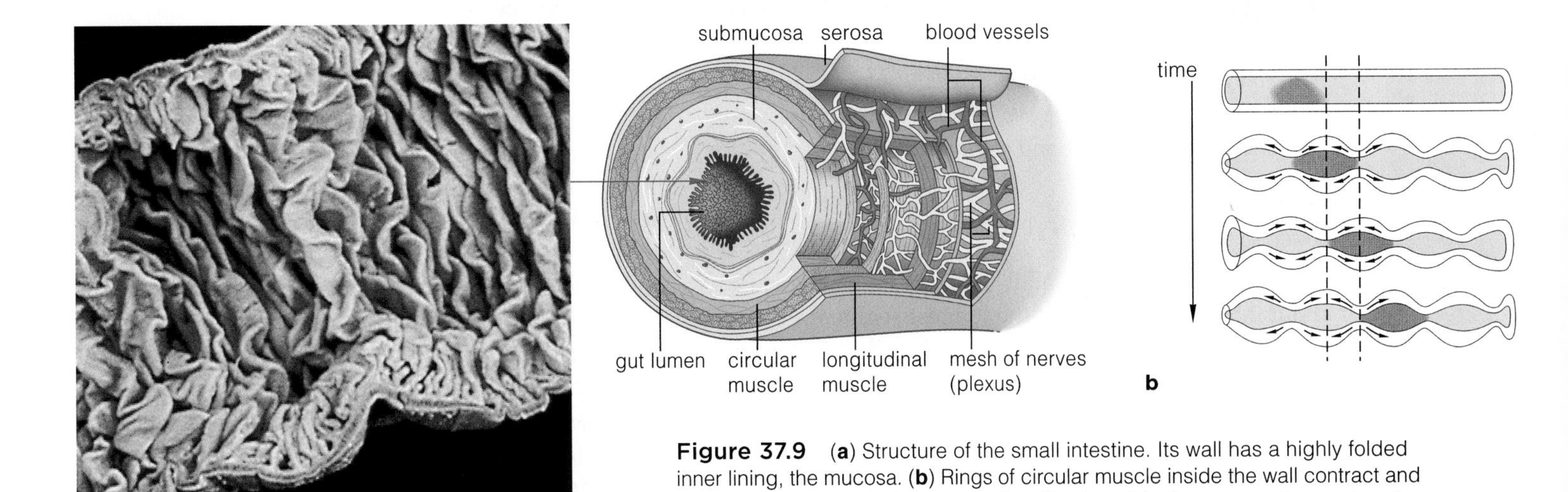

a A section of highly folded mucosa

Figure 37.9 (**a**) Structure of the small intestine. Its wall has a highly folded inner lining, the mucosa. (**b**) Rings of circular muscle inside the wall contract and relax in a pattern called segmentation. Back-and-forth movement propels, mixes, and forces chyme up against the wall, enhancing digestion and absorption.

DIGESTION IN THE SMALL INTESTINE

Chyme from the stomach and various secretions from the pancreas enter the duodenum, the first part of the small intestine. Pancreatic enzymes break down large organic compounds in chyme into monosaccharides, monoglycerides, fatty acids, amino acids, nucleotides, and nucleotide bases (Table 37.1). Bicarbonate secreted from the pancreas buffers the acids in chyme. It raises the pH enough for digestive enzymes to function.

In addition to enzymes, fat digestion requires bile. Bile is a mix of salts, pigments, cholesterol, and lipids. It is made in the liver and is concentrated and stored in the gallbladder. A fatty meal makes the gallbladder contract, forcing bile out through a duct that leads to the small intestine.

Gallstones, hard pellets of cholesterol and bile salts, can form in the gallbladder. Most are harmless. If they block the bile duct or otherwise interfere with function of the gallbladder, they can be removed surgically.

Bile salts enhance fat digestion by emulsification, a process which disperses droplets of fat in some fluid. Water-insoluble triglycerides from food tend to clump together forming fat globules. Movement of the small intestine counteracts this tendency. Rings of smooth muscle in the intestinal wall contract in an oscillating pattern known as segmentation (Figure 37.9*b*). While the contractions are mixing chyme, fat globules break up into small droplets that become coated with bile salts and remain separated. The many small droplets present a greater surface area to enzymes that break up fats into fatty acids and monoglycerides.

The breakdown products of digestion are absorbed across the epithelial lining of the small intestine, into the internal environment. How each kind gets across is the focus of the next section.

CONTROLS OVER DIGESTION

The nervous system, endocrine system, and a mesh of nerves in the gut wall control the pace of digestion. Food distends the stomach and activates receptors in its wall. Receptor signals flow along reflex pathways to gut muscles and glands, and other pathways alert the brain. In response, muscles contract and glands secrete hormones that enter blood (Table 37.2). The volume and composition of chyme influences these responses. Hearty meals cause more stretching and excite more receptors, so the gut muscles contract with more force. A high-fat meal slows stomach emptying.

With intense exercise or stress, sympathetic neurons command gut muscles to contract more slowly (Section 30.8). This is why heavy exercise after a large meal or chronic stress can cause digestive problems.

Gastric fluid, digestive enzymes, and wall contractions of the stomach and small intestine combine to break down food bits into molecules small enough to be absorbed.

Signals from the nervous system, nerves in the gut wall, and hormones exert control over digestion.

Table 37.2 Main Hormonal Controls of Digestion

Hormone	Source	Effects on Digestive System
Gastrin	Stomach	Increases acid secretion by stomach
Cholecystokinin (CCK)	Small intestine	Increases enzyme secretion by pancreas and causes contraction of gallbladder
Secretin	Small intestine	Increases bicarbonate secretion by pancreas and slows contractions in the small intestine

37.5 Absorption From the Small Intestine

The highly folded surface of the intestinal mucosa, along with microvilli on individual cells, enormously increase the surface area for water and nutrient absorption.

FROM STRUCTURE TO FUNCTION

LINKS TO SECTIONS 4.1, 5.6–5.8, 34.5, 34.8, 34.10

The small intestine is "small" only in its diameter—about 2.5 cm (1 inch). It is the longest segment of the gut. Uncoiled, it would extend for about 5 to 7 meters (16 to 23 feet). Water and nutrients cross the lining of this long tube to reach the internal environment.

Three features of the small intestine lining enhance absorption. *First*, this lining is folded (Figure 37.10*a*). *Second*, millions of multicelled, fingerlike absorptive structures called villi (singular, villus) extend out from each of the folds (Figure 37.10*b*). Each villus houses a lymph vessel and blood vessels (Figure 37.10*c*). *Third*, most cells on the villus surface are brush border cells (Figure 37.10*d*). These specialized cells have membrane extensions called microvilli (singular, microvillus) that project into the lumen. Collectively, all of these folds and projections increase the surface area of intestinal mucosa to about the size of half a tennis court!

The brush border cells function in both digestion and absorption. Digestive enzymes at the surface of the microvilli break down sugars, protein fragments, and nucleotides as listed in Table 37.1. Also at the microvillus surface are many transport proteins that act in absorption, as explained below.

In addition to brush border cells, the lining of the small intestine has a variety of secretory cells (Figure 37.10*d*). They secrete hormones, mucus, and bacteria-fighting chemicals, such as lysozyme (Section 35.1).

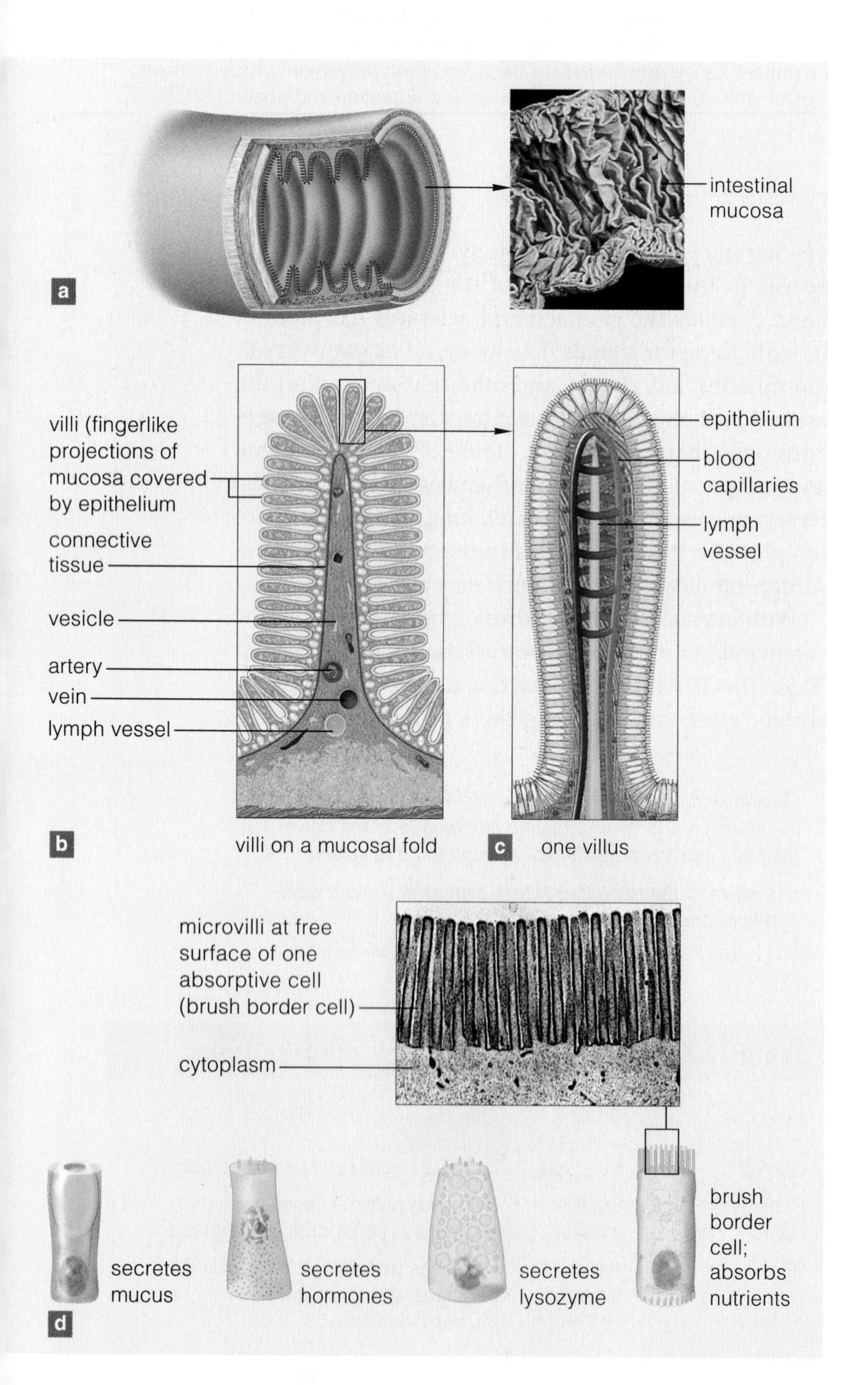

Figure 37.10 **Animated!** (**a**) Structure of the lining of the small intestine, the mucosa. The repetitive circular folds of this inner lining are permanent.

(**b**) At the free surface of each mucosal fold are many fingerlike absorptive structures called villi.

(**c**) A villus is covered with specialized epithelial cells and it contains blood capillaries and lymph vessels.

(**d**) Types of epithelial cells in the intestinal mucosa. The four shown here are color-coded enlargements of cells on the surface of the villus shown above. Absorptive brush border cells are the most abundant cells on a villus. Their crown of microvilli extends into the intestinal lumen. The small-intestinal enzymes discussed in the previous section are built into brush border cell plasma membranes. Other cells of the mucosa secrete mucus, hormones, or lysozyme, an enzyme that digests bacterial cell walls.

WHAT ARE THE ABSORPTION MECHANISMS?

Water and Solute Absorption Each day, eating and drinking puts 1 to 2 liters of fluid into the lumen of your small intestine. Secretions from your stomach, accessory glands, and the intestinal lining add another 6 to 7 liters. About 80 percent of the water that enters moves across the intestinal lining and into the internal environment by osmosis (Section 5.8). The activity of

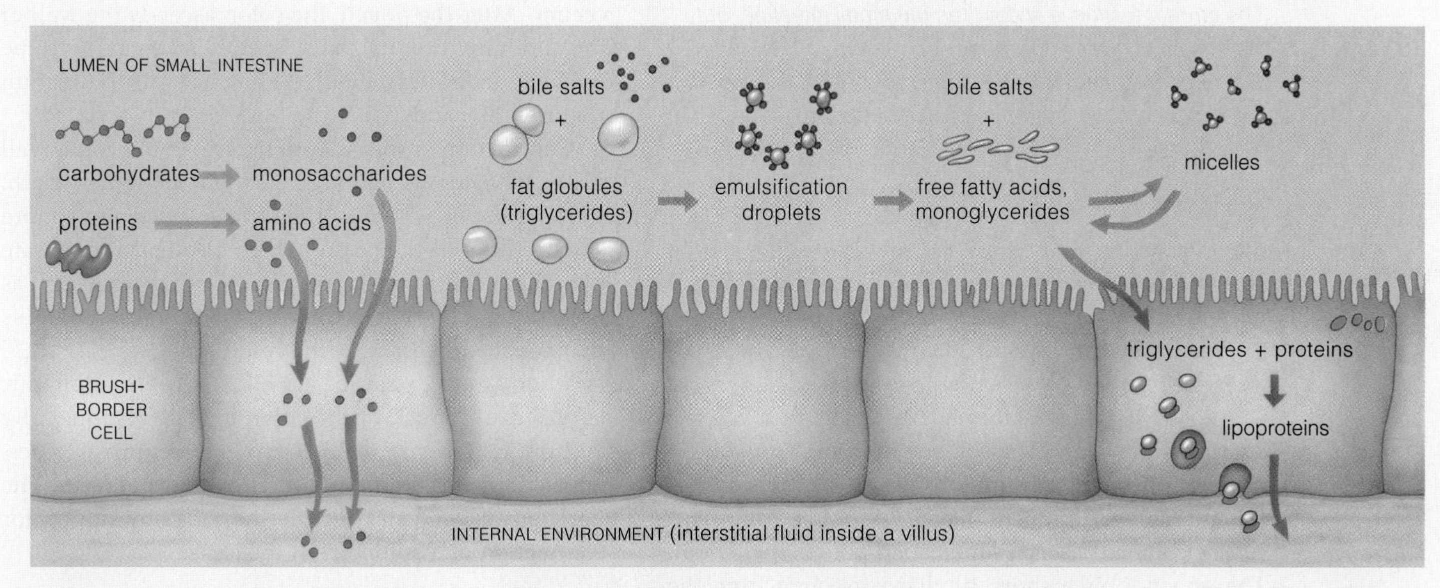

a Enzymes secreted by the pancreas and cells of the intestinal mucosa complete the digestion of carbohydrates to monosaccharides, and proteins to amino acids.

b Monosaccharides and amino acids are actively transported across the plasma membrane of brush border cells in the intestinal lining, then out of the same cells and into the internal environment.

c Movements of the intestinal wall break up fat globules into small droplets. Bile salts coat the droplets, so that globules cannot form again. Pancreatic enzymes digest the droplets to fatty acids and monoglycerides.

d Micelles form when bile salts combine with products of fat digestion: monoglycerides and fatty acids. These products slip into and out of micelles.

e Concentrating monoglycerides and fatty acids in micelles enhance diffusion of these substances into brush border cells. These lipids diffuse across the plasma membrane's lipid bilayer, into the cells.

f Inside a brush border cell, products of fat digestion form triglycerides, which become coated with proteins before being expelled by exocytosis into the interstitial fluid inside the villus (the internal environment).

Figure 37.11 **Animated!** Summary of digestion and absorption in the small intestine.

transport proteins in the plasma membrane of brush border cells moves salts, sugars, and amino acids from the intestinal lumen into these cells. Other transport proteins then move solutes out of the brush border cells and into interstitial fluid inside a villus (Figure 37.11*b*). This movement of solutes creates an osmotic gradient, so water moves in the same direction.

From the interstitial fluid, water, salts, sugars, and amino acids enter the blood capillary inside the villus. The blood then distributes them throughout the body.

Fat Absorption Being lipid soluble, the fatty acids and monoglycerides released by fat digestion enter a villus by diffusing across the lipid bilayer of brush border cells. Bile salts help out by concentrating them next to the cells (Section 37.4 and Figure 37.11*c*). By a process called micelle formation, bile salts combine with fatty acids as tiny droplets, or micelles. The fatty acids in micelles continually change places with those suspended in chyme, like people moving into and out of a room. A micelle simply holds a lot of them next to mucosal cells (Figure 37.11*d*). When concentrated enough, fatty acids and monoglycerides diffuse into brush border cells (Figure 37.11*e*). Inside these cells, they form triglycerides that get coated by proteins. The resulting lipoproteins are moved by exocytosis into interstitial fluid inside a villus (Figure 37.11*f*).

From the interstitial fluid, triglycerides enter into lymph vessels that eventually drain into the general circulation (Section 34.10).

With its richly folded mucosa, millions of villi, and many hundreds of millions of microvilli, the small intestine has a vast surface area for absorbing water and nutrients.

Substances are absorbed through the brush border cells that line the free surface of each villus. Passive and active transport mechanisms help water and solutes cross; micelle formation helps lipid-soluble products cross.

37.6 The Large Intestine

The large intestine is wider than the small intestine, but also much shorter—only about 1.5 meters (5 feet) long. The large intestine completes the process of absorption, then concentrates, stores, and eliminates wastes.

COLON STRUCTURE AND FUNCTION

LINKS TO SECTIONS 19.1, 19.2

Not everything that enters the small intestine can be or should be absorbed. Muscular contractions propel indigestible material, dead bacteria and mucosal cells, inorganic substances, and some water from the small intestine into the colon, or large intestine.

As the wastes travel through the colon, they become compacted as feces. The colon concentrates wastes by actively pumping sodium ions out of the lumen, into the internal environment. Water follows by osmosis.

The colon starts as a cup-shaped pouch, the cecum. The appendix is a short, tubular projection from the cecum. After the pouch, the colon ascends the wall of the abdominal cavity, extends across the cavity to the opposite side, descends, and connects to the rectum (Figures 37.6 and 37.12).

Contraction of the smooth muscle in the colon wall mixes its contents and propels them along its length. Compared with other gut regions, wastes move more slowly through the colon, which also has a moderate pH. These conditions favor growth of bacteria such as *Escherichia coli*. This bacterial species makes vitamins that get absorbed across the colon lining.

After a meal, gastrin and signals from autonomic nerves cause much of the colon to contract forcefully and propel feces to the rectum. The rectum stretches, which activates a defecation reflex to expel feces. The nervous system can override the reflex by calling for contraction of a sphincter at the anus.

Figure 37.12 (**a**) Cecum and appendix of the large intestine (colon). (**b**) Sketch and photo of polyps in the transverse colon.

HEALTH AND THE COLON

Healthy adults typically defecate about once a day, on average. Emotional stress, a diet low in fiber, minimal exercise, dehydration, and some medications can lead to constipation. This means defecation occurs fewer than three times a week, is difficult, and yields small, hardened, dry feces. Occasional constipation usually goes away on its own. A chronic problem should be discussed with a doctor. Diarrhea—frequent passing of watery feces—can result from bacterial infection or problems with nervous controls. If prolonged, it can cause dehydration and disrupt blood solute levels.

Appendicitis—an inflammation of the appendix—requires prompt treatment. It most often occurs after a bit of feces lodges in the appendix and infection sets in. Removing the inflamed appendix prevents it from bursting and releasing large numbers of bacteria into the abdominal cavity. Such a rupture could cause a life-threatening infection.

Some people are genetically predisposed to develop colon polyps, small growths on the colon wall (Figure 37.12*b*). Most polyps are benign, but some can become cancerous. If detected in time, colon cancer is highly curable. Blood in feces and dramatic changes in bowel habits may be symptoms of colon cancer and should be reported to a doctor. Also, anyone over the age of 50 should have a colonoscopy, in which clinicians use a camera to examine the colon for polyps or cancer.

By absorbing water and mineral ions, the colon compacts undigested residues and other wastes as feces, which are stored briefly in the rectum before expulsion.

37.7 What Happens to Absorbed Organic Compounds?

Small organic compounds—sugars, amino acids, and triglycerides—are absorbed from the small intestine. They are distributed to body cells and burned as fuel, stored, or used in the synthesis of larger organic compounds.

LINKS TO SECTIONS 7.7, 33.5

Sections 7.7 and 33.5 sketched out some mechanisms of control over organic metabolism, the disposition of glucose and other organic compounds in the body as a whole. They introduced the main pathways by which carbohydrates, fats, and proteins are broken down to forms used as intermediates in aerobic respiration, an ATP-producing pathway. Figure 37.13*a* rounds out the picture by showing all of the major routes by which organic compounds obtained from food are shuffled and reshuffled in the body as a whole.

Living cells constantly recycle some carbohydrates, lipids, and proteins by breaking them apart. They use breakdown products as energy sources and building blocks. The nervous and endocrine systems regulate this massive molecular turnover.

Reflect for a moment on a few crucial points about organic metabolism. When you eat, your body adds to its pool of raw materials. Excess carbohydrates and other organic compounds absorbed from the gut are transformed mostly into fats, which become stored in adipose tissue. Some are converted to glycogen in the liver and in all muscles. While the raw materials are being absorbed and stored, most cells use glucose as the main energy source. There is no net breakdown of protein in muscle tissue or other tissue during meals, nor is there any net fat breakdown.

In between meals, the brain takes up two-thirds of the circulating glucose, so most body cells tap fat and glycogen stores. Adipose cells degrade fats to glycerol and fatty acids, which enter blood. Liver cells break down glycogen and release glucose, which also enters blood. Most body cells take up the fatty acids as well as the glucose for ATP production.

Figure 37.13*b* highlights the liver's central role in metabolism and homeostasis. All blood that passes through intestinal capillaries enters capillary beds in the liver before returning to the heart (Section 34.5). Thus, any ingested nutrients or toxins pass through the liver, which removes many potentially dangerous substances and stores nutritious ones. As an example, ammonia (NH_3) is a potentially toxic product of amino acid breakdown. The liver converts it to urea, a much less toxic compound that gets excreted in urine.

Absorbed sugars are the human body's most accessible energy source. Between meals, the brain draws on glucose in blood; other cells tap fat and glycogen stores. Adipose cells convert and store excess carbohydrates as fats.

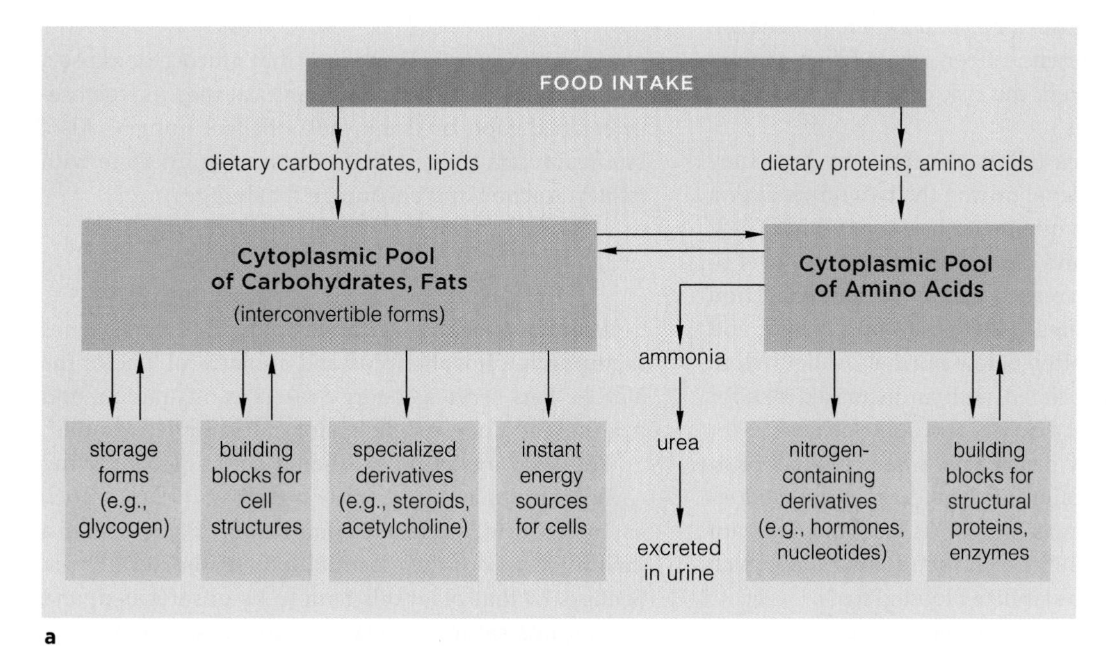

Liver Functions

- Forms bile (assists fat digestion), rids body of excess cholesterol and blood's respiratory pigments
- Controls amino acid levels in the blood; converts potentially toxic ammonia to urea
- Controls glucose level in blood; major reservoir for glycogen
- Removes hormones that served their functions from blood
- Removes ingested toxins, such as alcohol, from blood
- Breaks down worn-out and dead red blood cells, and stores iron
- Stores some vitamins

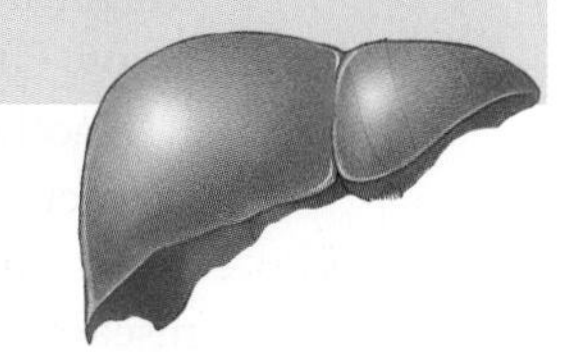

b

Figure 37.13 (**a**) Summary of major pathways of organic metabolism. Cells continually synthesize and tear down carbohydrates, fats, and proteins. Most urea forms in the liver, an organ that is at the crossroads of organic metabolism (**b**).

37.8 Human Nutritional Requirements

There is an old saying, "You are what you eat," which is meant to imply that diet profoundly affects your body's structure and function. So what do you eat?

USDA DIETARY RECOMMENDATIONS

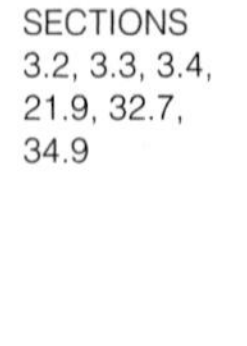

LINKS TO SECTIONS 3.2, 3.3, 3.4, 21.9, 32.7, 34.9

Scientists at the Department of Agriculture and other United States government agencies research diets that may help prevent diabetes, cancers, and other health problems. They periodically update their nutritional guidelines. Recently, they replaced a traditional "food pyramid" with a list that recommended the number of servings for different food groups (Figure 37.14).

In stark contrast to the diet of a typical American, the new guidelines recommend lowering the intake of refined grains, saturated fats, *trans*-fatty acids, added sugar or caloric sweeteners, and salt (no more than a teaspoon a day). They also recommend eating more vegetables and fruits with a high potassium and fiber content, fat-free or low-fat milk products, and whole grains. You can download the complete government report from www.health.gov/dietaryguidelines.

Also by these new guidelines, about 55 percent of daily caloric intake should come from carbohydrates.

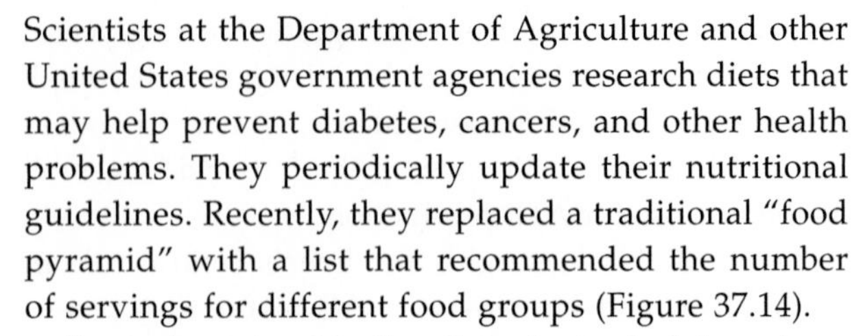

USDA Nutritional Guidelines

Food Group	Amount Recommended
Vegetables	2.5 cups/day
Dark green vegetables	3 cups/week
Orange vegetables	2 cups/week
Legumes	3 cups/week
Starchy vegetables	3 cups/week
Other vegetables	6.5 cups/week
Fruits	2 cups/day
Milk Products	3 cups/day
Grains	6 ounces/day
Whole grains	3 ounces/day
Other grains	3 ounces/day
Fish, poultry, lean meat	5.5 ounces/day
Oils	24 grams/day

Figure 37.14 Summary of nutritional guidelines as of 2006, from the United States Department of Agriculture (USDA). The recommended proportions add up to a daily 2,000-kilocalorie intake for sedentary females between ages ten and thirty. Recommended intake and serving sizes are larger for males and highly active females and less for older females. The USDA recommends varying protein choices (fish, poultry, lean meats, eggs, beans, nuts, and seeds).

ENERGY-RICH CARBOHYDRATES

Fresh fruits, whole grains, and vegetables—especially legumes, such as peas and beans—provide abundant complex carbohydrates. The body breaks the starch in these foods into glucose, its primary source of energy. These foods are also rich in fiber, which helps prevent constipation and lowers the risk of heart disease and other chronic diseases.

The glycemic index (GI) ranks foods by how they affect blood glucose level during the two hours after a meal. For example, white bread has a very high GI. It is digested quickly and causes blood glucose level to soar. This triggers increased insulin secretion (Section 32.7). Insulin stimulates cells to take up glucose, and blood glucose falls, often below normal. A diet rich in high GI foods is associated with an increased risk for type 2 diabetes, heart disease, and some cancers.

In contrast, kidney beans have a very low GI. Slow digestion and absorption of their carbohydrates does not make glucose rise as high or as fast, and does not trigger a spike in insulin secretion. Eating a diet rich in low GI foods helps stabilize blood glucose levels.

Another carbohydrate that may affect blood sugar is fructose. Use of this simple sugar in food products has risen in concert with obesity levels, causing some researchers to suspect a causal link. Fructose may not satisfy hunger in the same way that glucose does. As a result, a person can take in many calories in fructose-sweetened food or drink, but still feel hungry. Also, studies in rats suggest that fructose can interfere with insulin's action and encourage fat storage.

GOOD FAT, BAD FAT

You cannot stay alive without lipids. Cell membranes incorporate phospholipids and cholesterol, one of the sterols. Fats serve as energy reserves, insulation, and cushioning. They also help store fat-soluble vitamins.

The body makes most lipids it needs by converting carbohydrates and proteins. Essential fatty acids, such as linoleic acid, are fats it cannot make. Plant oils are a healthy source of the essential fatty acids. Recall from Section 3.3 that plant oils tend to be unsaturated, and animal fats saturated. Dairy products and meats are rich in saturated fats and cholesterol. Eating too much of these foods increases risk for heart disease, stroke, and certain cancers. *Trans*-fatty acids, or "trans fats,"

are manufactured from vegetable oils, but they have a molecular structure that makes them even worse for the heart than saturated fats (Figure 3.12). Food labels now must list the amounts of *trans* fat, saturated fat, and cholesterol per serving.

BODY-BUILDING PROTEINS

Your body also cannot function without proteins. It requires amino acid components of dietary proteins for its own protein-building programs. Of the twenty common types, eight are essential amino acids: Your cells cannot synthesize them. Such amino acids must make up a small portion of the total protein intake. Those eight are methionine (or cysteine, its metabolic equivalent), isoleucine, leucine, lysine, phenylalanine, threonine, tryptophan, and valine.

Most proteins in animal tissues are *complete*; their amino acid ratios match a human's nutritional needs. Nearly all plant proteins are *incomplete*, in that they lack one or more amino acids essential for humans. Proteins of quinoa (*Chenopodium quinoa*) are a notable exception (Section 21.9). To get required amino acids from a vegetarian diet, one must combine plant foods so the amino acids missing from one component are present in some others. As an example, rice and beans together provide all necessary amino acids, but rice alone or beans alone do not. You do not have to eat the two complementary foods at the same meal, but both should be consumed within a 24-hour period.

REGARDING THE ALTERNATIVE DIETS

Even with new dietary guidelines in place, alternative diets still abound. The *Mediterranean diet* pyramid, for instance, has a big base of grain products, then fruits and vegetables, legumes and nuts, then *olive oil* as the fat group, then cheese and yogurt. The pyramid limits weekly intakes of fish, poultry, eggs, simple sugars, and, at its tiny top, red meat. Olive oil provides about 40 percent of the diet's total energy intake. Because it is monounsaturated, olive oil is less likely to raise the blood cholesterol level, compared to saturated fats. It also is an antioxidant that helps remove free radicals. You might call this an "olive-oil-is-good" diet.

Also popular are *low-carb diets:* fewer carbohydrates and more proteins and fats. Such diets promote rapid weight loss. However, health professionals are still studying long-term effects on weight and health.

Increased protein intake causes increased ammonia production (Section 37.7). Enzymes in the liver convert ammonia to urea, which kidneys filter from blood and excrete in urine. Also, when a body burns fats—rather than carbohydrates—as its main energy source, acidic metabolic wastes called ketones enter blood. Ketones must be filtered out of blood and excreted. High-fat, high-protein diets make kidneys work harder, raising the risk of kidney stones and other kidney problems. Anyone with impaired kidney function should avoid a high-protein, high-fat diet.

You might expect a low-carb diet to increase risk of heart disease. However, studies show that following a low-carb diet for six months does not increase level of LDLs ("bad" cholesterol; Section 34.9). In some cases, they actually fell. Still, given extensive evidence that a diet high in saturated fat increases the risk of heart disease, low-carb dieters are advised to obtain protein from fish, lean meat, or vegetable sources.

In short, even the latest government guidelines do not provide all the nutritional answers.

A healthy diet provides energy and all necessary building blocks for assembling essential body components.

Nutritional guidelines are periodically revised in light of new research. Current guidelines call for most calories to come from complex carbohydrates, rather than simple sugars. They also favor fat and protein sources that are low in saturated fats.

37.9 Vitamins and Minerals

Think back on the outcomes of vitamin A deficiency, as sketched out in the Chapter 15 introduction. This is not an isolated problem. Chronically inadequate or excess amounts of most vitamins and minerals can disrupt normal development and metabolism.

LINKS TO SECTIONS 5,4, 5.5, 32.4, 36.5

Vitamins are *organic* substances that are essential in very small amounts; no other substance can carry out their metabolic functions. Unlike most plant species, which synthesize all the vitamins they require, most animals lost the ability to do so and must obtain all vitamins from their food. At a minimum, human cells require the thirteen vitamins listed in Table 37.3. Each has specific roles. For instance, the B vitamin niacin is modified to make NAD, a coenzyme (Section 5.4).

Minerals are *inorganic* substances that are essential for growth and survival because no other substance can serve their metabolic functions (Table 37.4). As an

Table 37.3 Major Vitamins: Sources, Functions, and Effects of Deficiencies or Excesses*

Vitamin	Common Sources	Main Functions	Effects of Chronic Deficiency	Effects of Extreme Excess
Fat-Soluble Vitamins				
A	Its precursor comes from beta-carotene in yellow fruits, yellow or green leafy vegetables; also in fortified milk, egg yolk, fish, liver	Used in synthesis of visual pigments, bone, teeth; maintains epithelia	Dry, scaly skin; lowered resistance to infections; night blindness; permanent blindness	Malformed fetuses; hair loss; changes in skin; liver and bone damage; bone pain
D	Inactive form made in skin, activated in liver, kidneys; in fatty fish, egg yolk, fortified milk products	Promotes bone growth and mineralization; enhances calcium absorption	Bone deformities (rickets) in children; bone softening in adults	Retarded growth; kidney damage; calcium deposits in soft tissues
E	Whole grains, dark green vegetables, vegetable oils	Counters effects of free radicals; helps maintain cell membranes; blocks breakdown of vitamins A and C in gut	Lysis of red blood cells; nerve damage	Muscle weakness, fatigue, headaches, nausea
K	Enterobacteria form most of it; also in green leafy vegetables, cabbage	Blood clotting; ATP formation via electron transport	Abnormal blood clotting; severe bleeding (hemorrhaging)	Anemia; liver damage and jaundice
Water-Soluble Vitamins				
B_1 (thiamin)	Whole grains, green leafy vegetables, legumes, lean meats, eggs	Connective tissue formation; folate utilization; coenzyme action	Water retention in tissues; tingling sensations; heart changes; poor coordination	None reported from food; possible shock reaction from repeated injections
B_2 (riboflavin)	Whole grains, poultry, fish, egg white, milk	Coenzyme action	Skin lesions	None reported
B_3 (niacin)	Green leafy vegetables, potatoes, peanuts, poultry, fish, pork, beef	Coenzyme action	Contributes to pellagra (damage to skin, gut, nervous system, etc.)	Skin flushing; possible liver damage
B_6	Spinach, tomatoes, potatoes, meats	Coenzyme in amino acid metabolism	Skin, muscle, and nerve damage; anemia	Impaired coordination; numbness in feet
Pantothenic acid	In many foods (meats, yeast, egg yolk especially)	Coenzyme in glucose metabolism, fatty acid and steroid synthesis	Fatigue, tingling in hands, headaches, nausea	None reported; may cause diarrhea occasionally
Folate (folic acid)	Dark green vegetables, whole grains, yeast, lean meats; enterobacteria produce some folate	Coenzyme in nucleic acid and amino acid metabolism	A type of anemia; inflamed tongue; diarrhea; impaired growth; mental disorders	Masks vitamin B_{12} deficiency
B_{12}	Poultry, fish, red meat, dairy foods (not butter)	Coenzyme in nucleic acid metabolism	A type of anemia; impaired nerve function	None reported
Biotin	Legumes, egg yolk; colon bacteria produce some	Coenzyme in fat, glycogen formation and in amino acid metabolism	Scaly skin (dermatitis); sore tongue; depression; anemia	None reported
C (ascorbic acid)	Fruits and vegetables, especially citrus, berries, cantaloupe, cabbage, broccoli, green pepper	Collagen synthesis; possibly inhibits effects of free radicals; structural role in bone, cartilage, and teeth; used in carbohydrate metabolism	Scurvy; poor wound healing; impaired immunity	Diarrhea, other digestive upsets; may alter results of some diagnostic tests

* Guidelines for appropriate daily intakes are being worked out by the Food and Drug Administration.

example, all of your cells use iron as a component of electron transfer chains (Section 5.5). Red blood cells require iron to make oxygen-transporting hemoglobin (Section 36.5). Iodine is essential for development of a healthy nervous system and to make thyroid hormone (Section 32.4).

Healthy people can get all vitamins and minerals they require from a well-balanced diet. In most cases, vitamin and mineral supplements are necessary only for strict vegetarians, the elderly, and people who are chronically ill or taking medicine that interferes with nutrient absorption.

Some studies suggest that supplementing vitamins A, C, and E may counter some effects of aging. They may protect the body from free radicals. A free radical, recall, is an uncharged atom or group of atoms with an unpaired electron that makes it highly reactive. When a free radical reacts with DNA or another molecule of life, it disrupts that molecule's structure and function.

Excessive amounts of many vitamins and minerals can harm individuals. Large doses of vitamins A and D are examples. Like all fat-soluble vitamins, they can accumulate in body tissues and interfere with normal metabolic activity when routinely taken in megadoses.

Normal metabolism requires the organic substances called vitamins and inorganic substances called minerals.

A balanced diet provides the required amounts of vitamins and minerals for most people. Deficiencies or excesses of either can cause health problems.

Table 37.4 Major Minerals: Sources, Functions, and Effects of Deficiencies or Excesses*

Mineral	Common Sources	Main Functions	Effects of Chronic Deficiency	Effects of Extreme Excess
Calcium	Dairy products, dark green vegetables, dried legumes	Bone, tooth formation; blood clotting; neural and muscle action	Stunted growth; possibly diminished bone mass (osteoporosis)	Impaired absorption of other minerals; kidney stones in susceptible people
Chloride	Table salt (usually too much in diet)	HCl formation in stomach; contributes to body's acid–base balance; neural action	Muscle cramps; impaired growth; poor appetite	Contributes to high blood pressure in certain people
Copper	Nuts, legumes, seafood, drinking water	Used in synthesis of melanin, hemoglobin, and some transport chain components	Anemia, changes in bone and blood vessels	Nausea, liver damage
Fluorine	Fluoridated water, tea, seafood	Bone, tooth maintenance	Tooth decay	Digestive upsets; mottled teeth and deformed skeleton in chronic cases
Iodine	Marine fish, shellfish, iodized salt, dairy products	Thyroid hormone formation	Enlarged thyroid (goiter), with metabolic disorders	Toxic goiter
Iron	Whole grains, green leafy vegetables, legumes, nuts, eggs, lean meat, molasses, dried fruit, shellfish	Formation of hemoglobin and cytochrome (transport chain component)	Iron-deficiency anemia, impaired immune function	Liver damage, shock, heart failure
Magnesium	Whole grains, legumes, nuts, dairy products	Coenzyme role in ATP–ADP cycle; roles in muscle, nerve function	Weak, sore muscles; impaired neural function	Impaired neural function
Phosphorus	Whole grains, poultry, red meat	Component of bone, teeth, nucleic acids, ATP, phospholipids	Muscular weakness; loss of minerals from bone	Impaired absorption of minerals into bone
Potassium	Diet alone provides ample amounts	Muscle and neural function; roles in protein synthesis and body's acid–base balance	Muscular weakness	Muscular weakness, paralysis, heart failure
Sodium	Table salt; diet provides ample to excessive amounts	Key role in body's salt–water balance; roles in muscle and neural function	Muscle cramps	High blood pressure in susceptible people
Sulfur	Proteins in diet	Component of body proteins	None reported	None likely
Zinc	Whole grains, legumes, nuts, meats, seafood	Component of digestive enzymes; roles in normal growth, wound healing, sperm formation, and taste and smell	Impaired growth, scaly skin, impaired immune function	Nausea, vomiting, diarrhea; impaired immune function and anemia

* Guidelines for appropriate daily intakes are being worked out by the Food and Drug Administration.

37.10 Weighty Questions, Tantalizing Answers

For many people, weight is a touchy subject. Part of the problem is that the standard of what constitutes an "ideal weight" often varies from culture to culture. Here we consider weight only as it relates to health. Thinner people really do live longer, on average.

LINK TO SECTION 30.10

A whopping 60 percent of the United States population—more than 108 million—are overweight. About 300,000 or so die each year as a result of preventable, weight-related conditions. As weight increases, so does the risk of type 2 diabetes, hypertension, heart disease, breast cancer, colon cancer, gallstones, and many other ailments.

What Is the "Right" Body Weight? Figure 37.15 shows one of the widely accepted weight guidelines for women and men. The body mass index (BMI) is another guideline. It is a measurement designed to help assess increased health risk associated with weight gains. You can calculate your body mass index with this formula:

$$\text{BMI} = \frac{\text{weight (pounds)} \times 703}{\text{height (inches)}^2}$$

Generally, individuals with a BMI of 25 to 29.9 are said to be overweight. A score of 30 or more indicates obesity: an overabundance of fat in adipose tissue that may lead to severe health problems. How body fat gets distributed also helps predict the risks. Fat deposits just above the belt, as in a "beer belly," are associated with an increased likelihood of heart problems.

Dieting alone cannot lower your BMI value. When you eat less, your body slows its metabolic rate to conserve energy. So how do you function normally over the long term while maintaining an acceptable weight? You must balance your caloric intake and energy output. For most people, this means eating only recommended portions of low-calorie, nutritious foods and exercising regularly.

Energy stored in food is expressed as kilocalories or Calories (with a capital C). One kilocalorie equals 1,000 calories, which are units of heat energy.

Here is a way to calculate how many kilocalories you should take in daily to maintain a preferred weight. First, multiply the weight (in pounds) by 10 if you are not active physically, by 15 if you are moderately active, and by 20 if you are highly active. Next, subtract one of the following amounts from the multiplication result:

Age:	*Subtract:*
25–34	0
35–44	100
45–54	200
55–64	300
Over 65	400

For example, if you are 25 years old, are highly active, and weigh 120 pounds, you will require 120 × 20 = 2,400 kilocalories daily to maintain weight. If you want to gain weight you will require more; to lose, you will require less. The amount is only a rough estimate. Other factors, such as height, must be considered. A person who is 5 feet, 2 inches tall and is active does not require as much energy as an active 6-footer whose body weight is the same.

A Question of Portion Sizes Have you noticed that portions of restaurant food are getting ever more "super-sized"? What was enough to feed two in 1977 is

Weight Guidelines for Women

Starting with an ideal weight of 100 pounds for a woman who is 5 feet tall, add five additional pounds for each additional inch of height. Examples:

Height (feet)	Weight (pounds)
5' 2"	110
5' 3"	115
5' 4"	120
5' 5"	125
5' 6"	130
5' 7"	135
5' 8"	140
5' 9"	145
5' 10"	150
5' 11"	155
6'	160

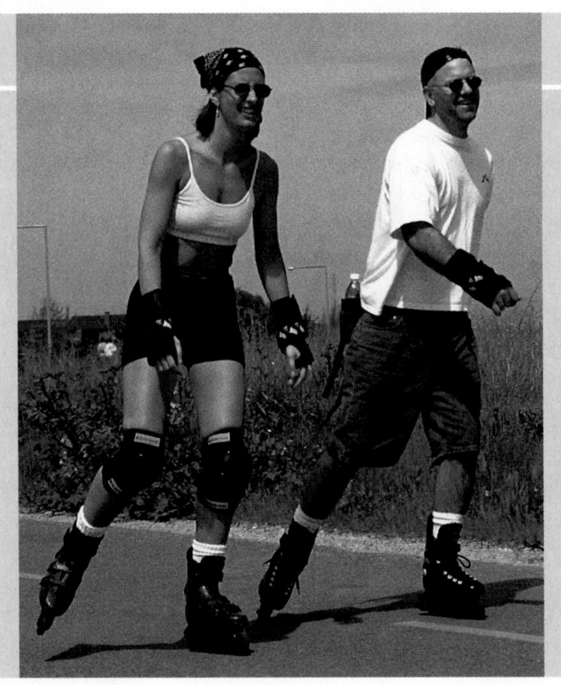

Weight Guidelines for Men

Starting with an ideal weight of 106 pounds for a man who is 5 feet tall, add six additional pounds for each additional inch of height. Examples:

Height (feet)	Weight (pounds)
5' 2"	118
5' 3"	124
5' 4"	130
5' 5"	136
5' 6"	142
5' 7"	148
5' 8"	154
5' 9"	160
5' 10"	166
5' 11"	172
6'	178

Figure 37.15 How to estimate "ideal" weights for adults. Values shown are consistent with a long-term Harvard study into the link between excess weight and risk of cardiovascular disorders. The "ideal" varies. It is influenced by specific factors such as having a small, medium, or large skeletal frame; bones are heavy.

now served on a plate for one. Correlated with bigger portions are bigger waistlines and more increases in body weight. How much we eat affects weight gain as much as what we eat and how much we exercise.

The USDA guidelines no longer say "servings" of food but rather specify amounts. Too many people have a distorted view of portion sizes. For instance, a serving of bread is two slices, not an entire loaf.

Genes, Hormones, and Obesity Numerous studies have explored the role that genetics plays in obesity. As one example, Claude Bouchard studied experimental overeating by twelve pairs of male twins. All were lean young men in their early twenties. For 100 days they did not exercise, and they adhered to a diet that provided 6,000 more kilocalories a week than usual.

They all gained weight, but some gained three times as much as others. Members of each set of twins tended to gain a similar amount, which suggests that genes affect the response to overfeeding. For another test, Bouchard put sets of obese twins on a low-calorie diet. Once again, each set of twins lost a similar amount.

As the chapter introduction indicated, we are learning more about how genes that encode hormones contribute to obesity. Figure 37.16 details how researchers uncovered the role of the appetite-suppressing hormone leptin in mice. Researchers have now also identified the leptin gene in humans; it is on chromosome 7 (Appendix VII).

Complete leptin deficiency of the sort seen in mice is extremely rare in humans. Only three cousins in a Turkish family have been shown to be entirely leptin deficient. All three were greatly obese. When UCLA researchers gave them daily leptin injections, they lost an average of 50 percent of their body weight without even trying to diet. The injections apparently caused changes in their brains. Scans showed increases in the gray matter of the cingulate gyrus, a portion of the limbic system known from other research to be involved in cravings (Section 30.10, Figure 30.23).

Another appetite-affecting hormone is a small peptide known as PYY. Glandular cells in the lining of the stomach and small intestine release it after a meal. PYY acts in the brain to suppress appetite. When volunteers were given an intravenous dose of PYY before a buffet meal, they ate less than a control group that received intravenous saline solution. A nasal spray with PYY is now being tested on volunteers to determine its safety and effectiveness.

In such ways, researchers seek to identify all steps of ancient pathways that control weight and how those contribute to other aspects of the body's physiology.

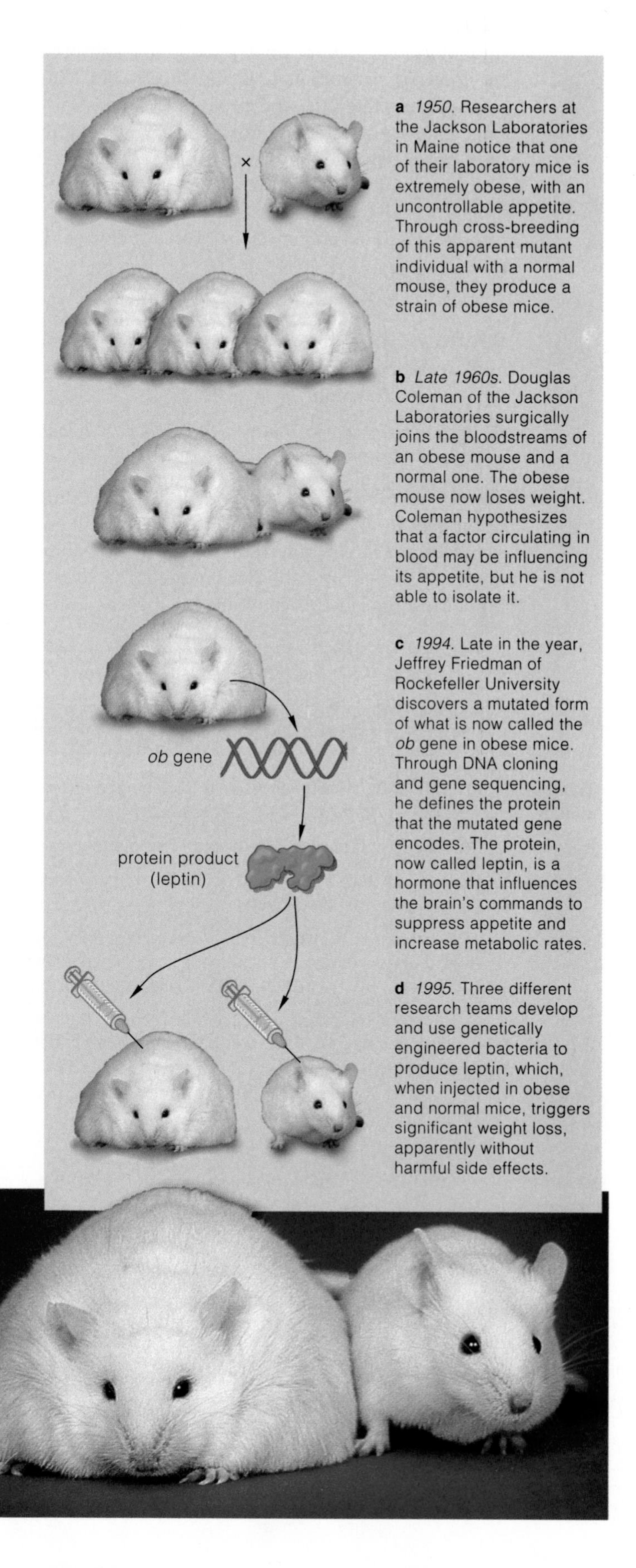

Figure 37.16 Chronology of research developments that identified leptin as a heritable factor that affects body weight.

www.thomsonedu.com ThomsonNOW!

Summary

Section 37.1 A digestive system breaks food down into molecules that are small enough to be absorbed into the internal environment. It also stores and eliminates unabsorbed materials, and promotes homeostasis by its interactions with other organ systems. Some invertebrates have an incomplete digestive system: a saclike gut with a single opening. Most animals, and all vertebrates, have a complete digestive system: a tube with two openings (mouth and anus) and specialized areas between them.

- *Use the animation on ThomsonNOW to compare vertebrate digestive systems.*

Section 37.2 Table 37.5 summarizes the components and accessory organs of the human digestive system, as well as their functions.

- *Use the animation on ThomsonNOW to tour the human digestive system.*

Sections 37.3, 37.4 Carbohydrate digestion begins in the mouth, where chewing mixes food with enzyme-rich saliva. Protein digestion begins in the stomach, a muscular sac with a glandular lining that secretes gastric fluid. Most digestion is completed in the small intestine.

Table 37.5 Summary of the Human Digestive System

Mouth (oral cavity)	With tongue, teeth, and saliva, softens and moistens food; starts polysaccharide digestion
Pharynx	Acts as entrance to digestive and respiratory tracts
Esophagus	Muscular tube, moistened by saliva, that moves food from pharynx to stomach
Stomach	Stretchable sac where food mixes with gastric fluid and protein digestion starts; stores food taken in faster than can be processed; its fluid kills most microbes
Small intestine	Receives secretions from liver, gallbladder, pancreas; digests most nutrients; delivers unabsorbed material to colon
Colon (large intestine)	Concentrates undigested food residues and other wastes by absorbing water and mineral ions
Rectum	Stores feces until defecation
Anus	Sphincter; site of control over defecation
Accessory Organs:	
Salivary glands	Secrete saliva, a fluid that contains polysaccharide-digesting enzymes, buffers, and mucus
Liver	Secretes bile; roles in carbohydrate, fat, and protein metabolism
Gallbladder	Stores and concentrates bile from the liver
Pancreas	Secretes enzymes that digest all major food molecules; buffers against HCl secretions from stomach lining

Ducts from the pancreas and gallbladder empty into the small intestine. The pancreas produces and secretes digestive enzymes. Bile, which assists in fat digestion, is made in the liver and stored in the gallbladder.

Local controls as well as the nervous and endocrine systems respond to the volume and composition of food in the gut. They cause changes in muscle activity and in secretion rates for hormones and enzymes.

Section 37.5 The mucosa that lines the small intestine is highly folded and covered with muticelled, absorptive structures called villi. Most of the cells at the surface of a villus are brush border cells that have microvilli on the surface that face the lumen. Brush border cells function in digestion and in absorption. Their many membrane proteins transport salts, simple sugars, and amino acids from the intestinal lumen into the villus interior. A blood vessel inside each villus takes up absorbed sugars and amino acids.

Monoglycerides and fatty acids diffuse into the brush border cells. Here, they recombine as triglycerides, which exocytosis moves into the interstitial fluid. From there, they enter lymph vessels that deliver them to blood.

- *Use the animation on ThomsonNOW to learn about the small intestine's structure and how it absorbs nutrients.*

Section 37.6 More water and ions are absorbed in the large intestine, or colon, which compacts undigested solid wastes as feces. Feces are stored in the rectum, the stretchable tubular region just before the anus.

Sections 37.7–37.9 The small organic compounds absorbed from the gut are stored, used in biosynthesis or as energy sources, or excreted by other organ systems. Blood that flows through the small intestine travels next to the liver, which eliminates ingested toxins and stores excess glucose as glycogen.

Dietary guidelines are in flux. Most nutritionists agree that you should minimize intake of refined carbohydrates and saturated fat. Food must provide energy and raw materials, including essential amino acids and fatty acids. It must also include two additional types of compounds needed for metabolism: vitamins, which are organic, and minerals, which are inorganic. Vegetarian diets can meet all these needs, only if foods are carefully combined.

Section 37.10 Obesity increases the risk of health problems and shortens life expectancy. To maintain body weight, energy (caloric) intake must balance with energy (caloric) output. Genetic factors can make it difficult for people to reach and maintain a healthy weight. Hormones affect both appetite and metabolic rate.

- *Use the interaction on ThomsonNOW to calculate your body mass index.*

Self-Quiz

Answers in Appendix III

1. A digestive system functions in ______ .
   a. secreting enzymes
   b. absorbing compounds
   c. eliminating wastes
   d. all of the above

Figure 37.17 One success story: In 2000, after recovering from anorexia, Dutch cyclist Leontien Zijlaard won three Olympic gold medals. Four years earlier, she was too malnourished and weak to compete.

Figure 37.18 Exceptionally Big Mac of the snake world.

2. Protein digestion begins in the ______ .
   a. mouth c. small intestine
   b. stomach d. colon

3. Most nutrients are absorbed in the ______ .
   a. mouth c. small intestine
   b. stomach d. colon

4. Bile has roles in ______ digestion and absorption.
   a. carbohydrate c. protein
   b. fat d. amino acid

5. Monosaccharides and amino acids are both absorbed from the gut ______ .
   a. at membrane proteins c. as fat droplets
   b. at lymph vessels d. both b and c

6. The largest number of bacteria thrive in the ______ .
   a. stomach c. large intestine
   b. small intestine

7. The pH is lowest in the ______ .
   a. stomach c. large intestine
   b. small intestine

8. Most water that enters the gut is absorbed across the lining of the ______ .
   a. stomach c. large intestine
   b. small intestine

9. ______ are inorganic substances with metabolic roles that no other substance can fulfill.
   a. Fats d. Vitamins
   b. Minerals e. Simple sugars
   c. Proteins f. both b and d

10. Match each organ with a digestive function.
   ___ gallbladder a. makes bile
   ___ colon b. compacts undigested residues
   ___ liver c. secretes most digestive enzymes
   ___ small intestine d. absorbs most nutrients
   ___ stomach e. secretes gastric fluid
   ___ pancreas f. stores, secretes bile

■ *Visit ThomsonNOW for additional questions.*

Critical Thinking

1. Anorexia nervosa is an eating disorder in which people starve themselves. Although the name means "nervous loss of appetite," most affected people are obsessed with food and continually hungry. The good news is that recovered anorexics can enjoy normal lives (Figure 37.17).

Anorexia nervosa has complex causes. Among them are some recently discovered genetic factors. Preliminary studies implicate mutations in *MAOA*, *SERT*, and *NET* genes. All affect the clearing of neurotransmitters away from synapses (Section 30.5). *MAOA* encodes monoamine oxidase, an enzyme that breaks down neurotransmitter. *SERT* encodes a serotonin transporter and *NET* encodes a transporter for norepinephrine. Researchers hope that better understanding of the genetic risk factors will help them devise medications to treat anorexia nervosa.

Bulimia is an eating disorder characterized by binge-eating followed by self-induced vomiting or laxative use to get rid of the excess calories. Like anorexia, it occurs mainly among women and certain genes seem to increase risk.

Reported incidence of anorexia and bulimia have risen dramatically during the past 50 years. By one hypothesis, a celebration of extreme female thinness in the media has led to more frequent triggering of these disorders among those with a genetic predisposition toward them. How might we test this hypothesis?

2. A python can survive by eating a large meal once or twice a year (Figure 37.18). When it does eat, microvilli in its small intestine lengthen fourfold and its stomach pH drops from 7 to 1. Explain the benefits of these changes.

3. A glassful of whole milk contains lactose (a sugar), proteins, butterfat, vitamins, and minerals. Explain what will happen to each component in your digestive tract.

4. A peptic ulcer is a sore in the lining of the stomach or duodenum. Acid leaking through an ulcer causes pain and may threaten life if it dissolves its way through to a blood vessel. Most ulcers form after an infection by the bacterium *Helicobacter pylori* and are treated with antibiotics. Side effects are common and include cramps and diarrhea. Women have double the normal risk of vaginal infections during treatment. Explain why these drugs could cause such side effects. What might help a person to avoid these effects?

5. Biologist Dianne Anderson presented the hypothetical nutrition label shown at right to her students. Can you name the organism that would have the kinds and proportions of ingredients listed here?

Nutrition Facts

Serving size: 1/2 cup (112g)
Servings per container: About 68

Amount per serving	
Calories	201
Calories from fat	121
Percent Daily Value	
Total Fat 13g	17%
Saturated fat 13g	78%
Total Carbohydrate 2g	
Dietary fiber 0g	0%
Sugars 0g	0%
Other carbohydrates 2g	
Protein 18g	40%

Ingredients: ___?___ , saltwater

38 THE INTERNAL ENVIRONMENT

IMPACTS, ISSUES Truth in a Test Tube

Light or dark? Clear or cloudy? A lot or a little? Asking about and examining urine is an ancient art (Figure 38.1). About 3,000 years ago in India, the pioneering healer Susruta reported that some patients formed an excess of sweet-tasting urine that attracted insects. In time, the disorder was named diabetes mellitus, which loosely translates as "passing honey-sweet water." Doctors still diagnose it by testing the sugar level in urine, although they have replaced the taste test with chemical analysis.

Today, physicians routinely check the pH and solute concentrations of urine to monitor their patients' health. Acidic urine suggests metabolic problems. Alkaline urine can indicate an infection. Damaged kidneys will produce urine high in proteins. An abundance of some salts can result from dehydration or trouble with the hormones that control kidney function. Special urine tests detect chemicals produced by cancers of the kidney, bladder, and prostate gland.

Do-it-yourself urine tests have become popular. If a woman is hoping to become pregnant, she can use one test to keep track of the amount of luteinizing hormone, or LH, in her urine. About midway through a menstrual cycle, LH triggers ovulation, the release of an egg from an ovary. Another over-the-counter urine test can reveal whether she has become pregnant. Still other tests help older women check for declining hormone levels in urine, a sign that they are entering menopause.

Not everyone is in a hurry to have their urine tested. Olympic athletes can be stripped of their medals when mandatory urine tests reveal they use prohibited drugs. Major League Baseball players agreed to urine tests only after repeated allegations that certain star players took prohibited steroids. The National Collegiate Athletic

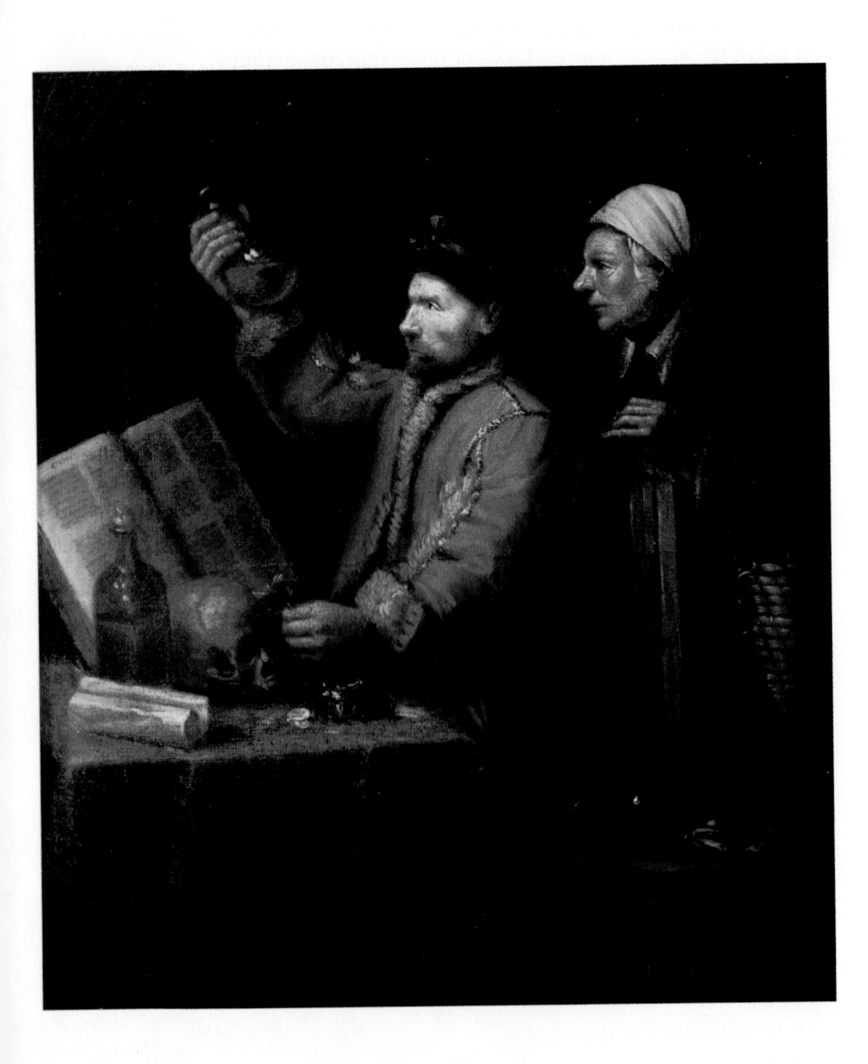

See the video! **Figure 38.1** *This page,* a seventeenth-century physician and a nurse examining a urine specimen. Urine's consistency, color, odor, and—at least in the past—taste afford clues to health problems. Urine forms inside kidneys, and it provides clues to abnormal changes in the volume and composition of blood and interstitial fluid. *Facing page,* testing for the presence of drugs in urine samples.

How would you vote? Prospective employees are sometimes screened for drug and alcohol use by testing their urine. Should an employer be allowed to require a urine test before hiring an individual, or are such tests an invasion of privacy? See ThomsonNOW for details, then vote online.

Association (NCAA) tests urine samples from about 3,300 student athletes per year for any performance-enhancing substances as well as for "street drugs."

If you use marijuana, cocaine, Ecstasy, or other kinds of psychoactive drugs, urine tells the tale. After the active ingredient of marijuana enters blood, the liver converts it to another compound. As kidneys filter blood, they add the compound to newly forming urine. It can take as long as ten days for all molecules of the compound to become fully metabolized and removed from the body. Until that happens, urine tests can detect it.

It is a tribute to the urinary system that urine is such a remarkable indicator of health, hormonal status, and drug use. Each day, a pair of fist-sized kidneys filter all of the blood in an adult human body, and they do so more than forty times. When all goes well, kidneys get rid of excess water and excess or harmful solutes, including a variety of metabolites, toxins, hormones, and drugs.

So far in this unit, you have considered several organ systems that work to keep cells supplied with oxygen, nutrients, water, and other substances. Turn now to the kinds that maintain the composition, volume, and even the temperature of the internal environment.

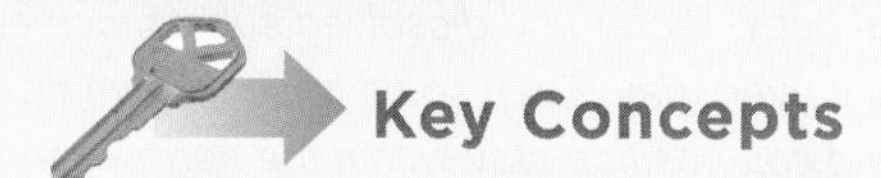

Key Concepts

MAINTAINING THE EXTRACELLULAR FLUID

Animals continually produce metabolic wastes. They continually gain and lose water and solutes. Yet the overall composition and volume of extracellular fluid must be kept within a range that individual cells can tolerate. In humans, as in other vertebrates, a urinary system interacts with other organ systems in this task. **Sections 38.1, 38.2**

THE HUMAN URINARY SYSTEM

The human urinary system consists of two kidneys, two ureters, a bladder, and a urethra. Inside a kidney, millions of nephrons filter water and solutes from the blood on an ongoing basis. Most of this filtrate is returned to the blood. Water and solutes not returned to the blood by way of kidneys leave the body as urine. **Section 38.3**

WHAT KIDNEYS DO

Urine forms by processes of filtration, reabsorption, and secretion. Its content is adjusted continually by hormonal and behavioral responses to shifts in the internal environment. The hormones ADH and aldosterone, as well as a thirst mechanism, influence whether urine becomes concentrated or dilute during any given interval. **Sections 38.4–38.6**

ADJUSTING THE CORE TEMPERATURE

Heat losses to the environment and heat gains from the environment and from metabolic activity determine an animal's body temperature. Controls over metabolism and a variety of adaptations in body form and behavior help maintain the core temperature within a tolerable range. **Sections 38.7, 38.8**

Links to Earlier Concepts

Vertebrates, recall, originated in water, and some moved onto land (Section 24.2). This chapter explores mechanisms that made the move possible, including protein-mediated transport and osmosis (5.7, 5.8). You will tap your knowledge of pH and buffer systems (2.6), and of the circulatory system (34.5), especially capillary function (34.8). You will see what happens to metabolic waste products (37.7) and reconsider the effects of pituitary hormones and adrenal glands on the internal environment (32.3, 32.5).

Thermal homeostasis will be easier to understand if you review water's temperature-stabilizing effects (2.5), the nature of energy flow (5.1), and mitochondrial function (7.4).

38.1 Gains and Losses in Water and Solutes

An internal environment, recall, bathes all living cells of the animal body. Water and dissolved substances enter and leave it every day. Even so, in healthy animals, its volume and composition stay within the narrow ranges that cells can tolerate. The inputs and outputs balance.

WATER, SALTS, AND THE ECF

LINKS TO SECTIONS 5.8, 23.5, 34.1, 34.2

By weight, all organisms consist mostly of water, with salts and other dissolved solutes. Most fluid in animal bodies is in cells. Fluid outside cells, extracellular fluid (ECF), is the body's internal environment. It consists mostly of interstitial fluid, which fills spaces between the cells that make up tissues (Figure 38.2). The rest is plasma, the solute-rich fluid of blood (Sections 34.1 and 34.2), and other fluids.

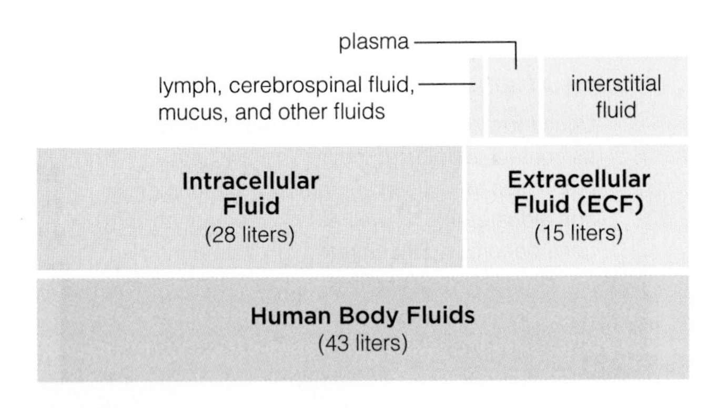

Figure 38.2 Fluid distribution in the human body. Fluid in cells (intracellular fluid) is the single greatest fluid reservoir. Fluid outside the cells (extracellular fluid, or ECF) includes interstitial fluid in the spaces between cells, plasma (the fluid portion of the blood), and a small amount of other fluids. The ECF is the body's internal environment.

The composition and volume of ECF is kept within ranges that living cells can tolerate, a process we call homeostasis. Maintaining the solute composition and volume of ECF can be a challenge because all animals continually gain and lose water and solutes as a result of normal activities.

There are many inputs into ECF. For example, most water and solutes that enter the gut become absorbed into ECF. The large amount of carbon dioxide released by aerobic respiration diffuses into ECF, as does water formed by other metabolic reactions. Breathing adds oxygen to the ECF.

What about outputs? As you know from Chapters 23 and 24, animals differ in how they remove excess water and solutes from the ECF. In this chapter, we focus on how vertebrate organ systems deal with all the inputs and outputs (Figure 38.3).

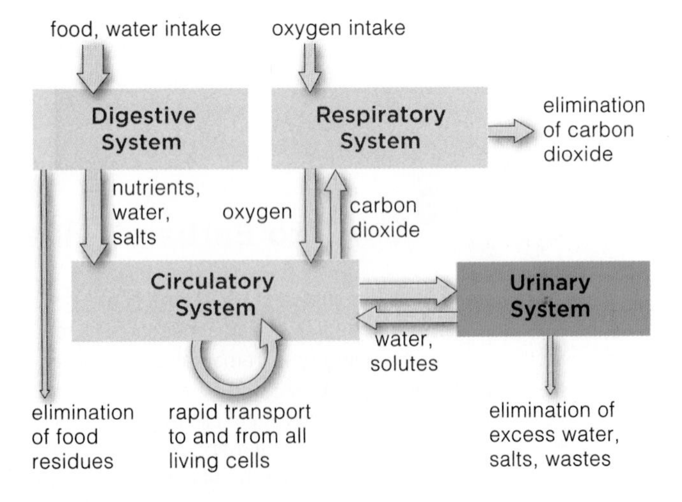

Figure 38.3 Functional links between the urinary, digestive, respiratory, and circulatory systems. Guided by the nervous and endocrine systems, these systems help maintain homeostasis.

VERTEBRATE BALANCING ACTS

Water–Solute Balance in Humans Like other vertebrates, humans have a urinary system that filters water and many solutes from blood, then reclaims or excretes them in amounts required to maintain the ECF volume and composition. A pair of organs called kidneys filter blood. Excess water and solutes collect in kidneys as urine, which is stored and then excreted by other components of the urinary system.

Urea usually is the most abundant metabolic waste in human urine. It forms when liver enzymes detoxify potentially poisonous ammonia, a product of protein metabolism (Section 37.7). Products from breakdown of hemoglobin in red blood cells gives your urine its distinctive yellow color. Human urine often contains breakdown products of synthetic compounds, such as drugs and food additives.

A Comparative Look at Other Vertebrates In vertebrates, the ECF is one-third as salty as seawater. That is why fishes in the seas lose water continually, by osmosis across the body surface (Section 5.8). They offset the losses by gulping seawater. Then cells inside their gills pump out the excess salts (Figures 36.5 and 38.4*a*). Other metabolic wastes are excreted in a small volume of urine.

The ECF of freshwater fishes and amphibians has more solutes than the surroundings. These vertebrates do not drink. Osmosis draws in enough water across the skin and the gills (Figure 38.4*b*). Solutes that leave in the dilute urine are offset by solutes absorbed from

water loss
by osmosis
gulps
water
cells in gills pump
solutes out
water loss in very small volume
of concentrated urine

a Marine bony fish; body fluids are less salty than the surrounding water; they are hypotonic.

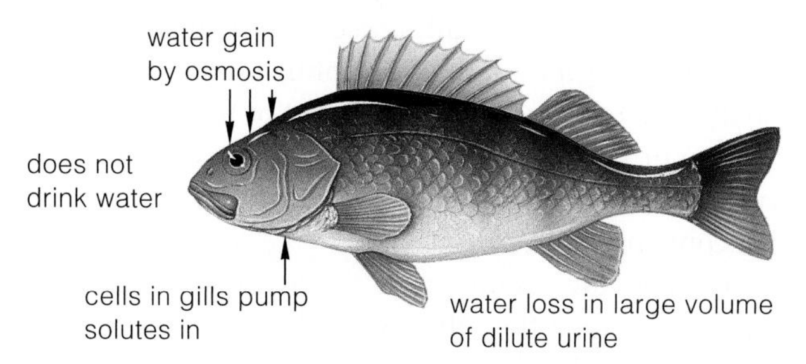

b Freshwater bony fish; body fluids are saltier than the surrounding water; they are hypertonic.

Figure 38.4 Balancing acts in fishes.

the gut, and by sodium ions pumped inward across cells of the gills and skin.

Osmosis across body surfaces became unimportant among vertebrates that colonized land. The pioneers faced intense sunlight, drying winds, wider swings in temperature, water of variable salt content, and often no water at all. These environmental factors became selective agents; they favored novel mechanisms that could control the body's water–solute balance.

Birds and reptiles convert ammonia formed during protein metabolism to crystals of uric acid. In contrast, mammals convert ammonia to urea. It takes twenty to thirty times more water to excrete one gram of urea than to excrete one gram of uric acid. Many mammals also lose water in sweat. Thus, a typical mammal has greater water requirements than a bird or reptile. Even so, some mammals have adaptations that allow them to get along with very little water, as you will learn in the next section.

In all animals in good health, daily gains in water and solutes balance the daily losses.

In all vertebrates, a urinary system counters unwanted shifts in the volume and composition of the ECF. Paired kidneys filter blood, form urine, and help maintain solute levels within tolerable limits.

38.2 Desert Rats

FOCUS ON THE ENVIRONMENT

Balancing water losses from the body with water gains is easier in some places than in others.

In a New Mexico desert, free water is scarce, except in a brief rainy season. There, a kangaroo rat (*Dipodomys deserti*) waits out the heat of the day in a burrow, then forages at night for dry seeds and bits of plants. This tiny mammal hops rapidly and far, searching for seeds and fleeing from predators. All that hopping requires ATP energy and water. Metabolic reactions release the necessary energy and water from carbohydrates and other compounds in seeds. Each day, "metabolic water" makes up a whopping 90 percent of a kangaroo rat's water intake. In contrast, metabolic water is only about 12 percent of a human's daily water gain (Figure 38.5).

A kangaroo rat conserves and recycles water when it rests in its cool burrow. It moistens and warms the air that it inhales. When it exhales, water condenses in its cooler nose, and some diffuses back into the body. Seeds emptied from a kangaroo rat's cheek pouches soak up water dripping from the nose. The kangaroo rat reclaims water when it eats dripped-on seeds.

A kangaroo rat has no sweat glands and its feces hold only half the water that human feces do, relative to body size. It loses water when urinating, but its two specialized kidneys do not let it lose much. Its kidneys excrete urine that can be as much as three to five times more concentrated than yours.

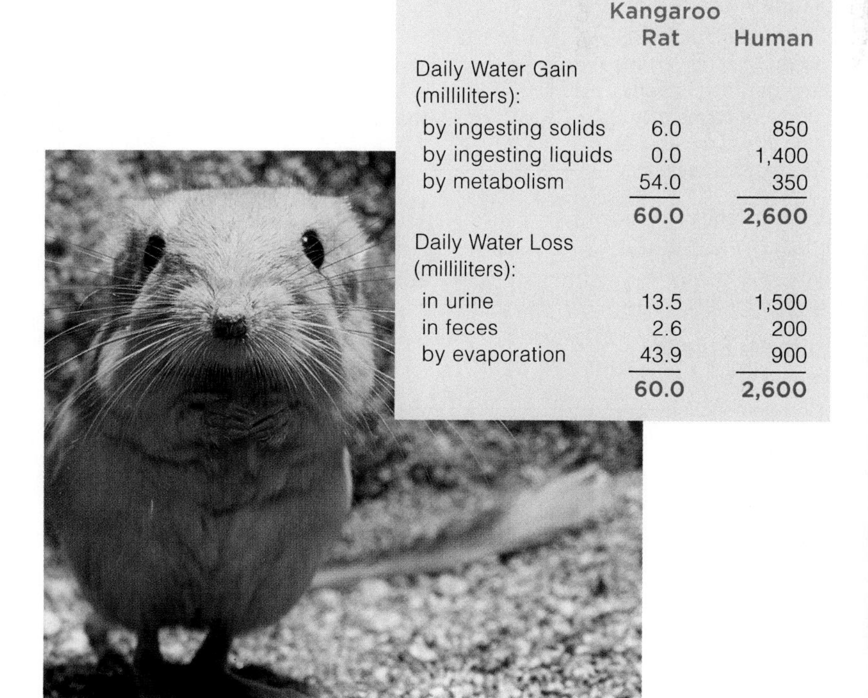

	Kangaroo Rat	Human
Daily Water Gain (milliliters):		
by ingesting solids	6.0	850
by ingesting liquids	0.0	1,400
by metabolism	54.0	350
	60.0	**2,600**
Daily Water Loss (milliliters):		
in urine	13.5	1,500
in feces	2.6	200
by evaporation	43.9	900
	60.0	**2,600**

Figure 38.5 Comparison of water gains and losses between a kangaroo rat (*Dipodomys deserti*) and a typical human.

38.3 Structure of the Urinary System

Kidneys filter water, mineral ions, organic wastes, and other substances from the blood. They adjust the volume and composition of this filtrate, and return most of it to the blood. The fluid not returned becomes urine.

COMPONENTS OF THE SYSTEM

LINKS TO SECTIONS 34.5, 34.6

Again, a urinary system helps maintain the volume and composition of the ECF. It filters water and most solutes from blood and selectively reclaims or excretes both. Drink too little water, eat a lot of salty foods, or lose sodium in your sweat, and the urinary system responds fast.

A human urinary system includes two kidneys, two ureters, one urinary bladder, and one urethra (Figure 38.6*a*). Kidneys filter blood and form urine. The other organs collect and store urine, and channel it to the body surface for excretion.

Kidneys are bean-shaped organs about the size of an adult fist. They lie just beneath the peritoneum that lines the abdominal cavity, to the left and right of the backbone (Figure 38.6*a*,*b*). The outermost kidney layer, the renal capsule, consists of fibrous connective tissue (Figure 38.6*c*). The Latin *renal* means relating to the kidneys. The bulk of tissue inside the renal capsule is divided into two zones: the outer renal cortex and the inner renal medulla. A renal artery transports blood to each kidney and a renal vein transports it out.

Urine collects in the renal pelvis, a central cavity inside each kidney. A tubular ureter conveys the fluid from a kidney into the urinary bladder. This muscular organ stores urine until a sphincter at its end opens and urine flows into the urethra.

When the bladder gets full, a reflex action kicks in. Stretch receptors signal motor neurons in the spinal cord, which signal smooth muscle in the bladder wall to contract. At the same time, sphincters that encircle the urethra relax, so urine flows out of the body. After age two or three, the brain overrides this spinal reflex and prevents urine from flowing through the urethra at inconvenient moments.

In males, the urethra runs the length of the penis. Urine and semen flow through it, but a sphincter cuts off urine flow during erections. In females, the urethra opens onto the body surface close to the vagina. It is a short tube, so infectious microbes can easily reach the urinary bladder. That is one reason why women get bladder infections more often than men do.

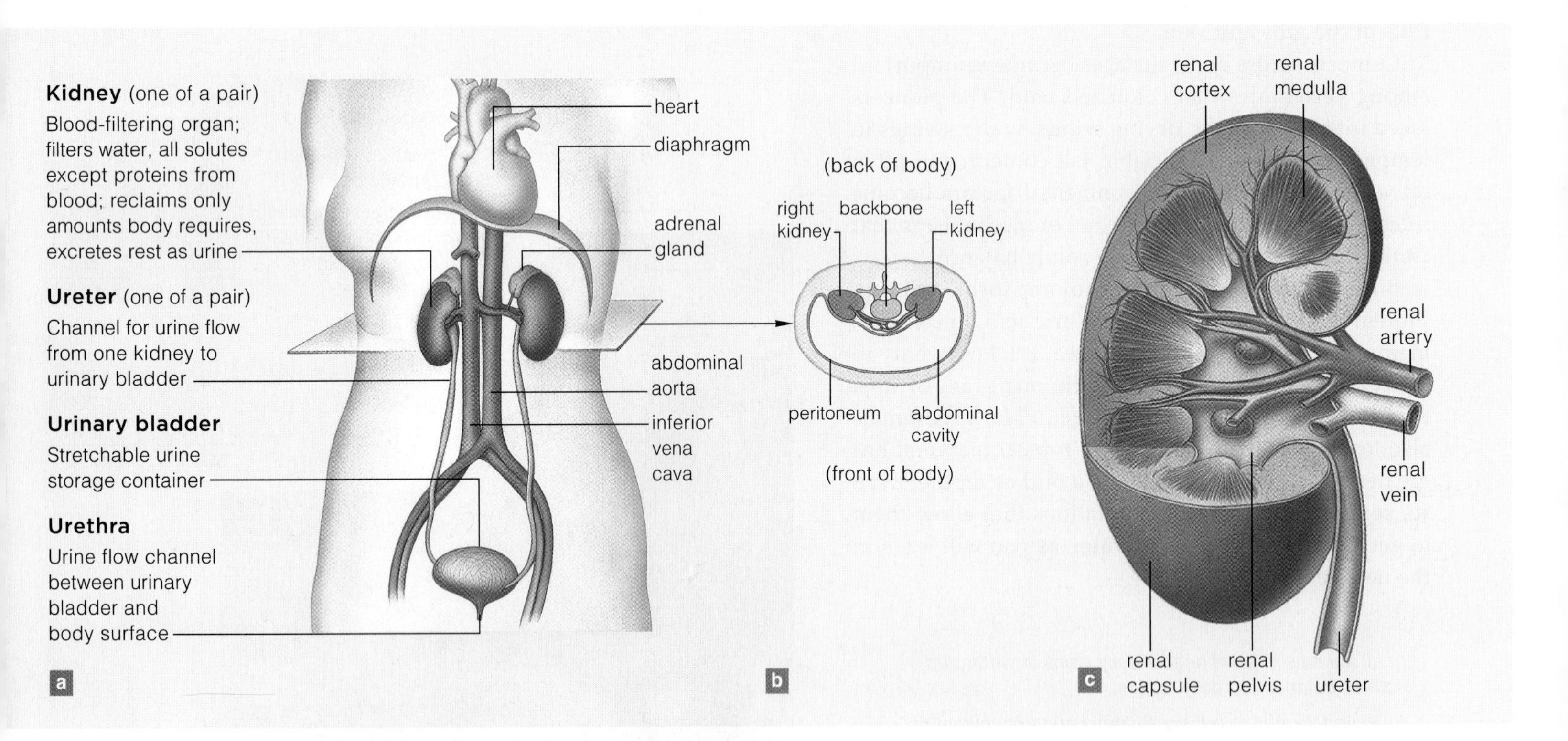

Figure 38.6 **Animated!** (**a**) Human urinary system and its functions. (**b**) The paired kidneys are located between the peritoneum, which lines the abdominal cavity, and the abdominal wall. (**c**) Structural organization of a human kidney.

Figure 38.7 **Animated!**
(**a**) The structure of a nephron. Nephrons are functional units of a kidney. They interact with neighboring blood vessels to form the urine.

(**b**) The arterioles and blood capillaries associated with each nephron. Large gaps between cells in the walls of glomerular capillaries make the capillaries about a hundred times more permeable than any others in the body. Only a thin basement membrane separates each capillary wall from cells of the innermost layer of Bowman's capsule. Cells of this inner layer have long extensions that interdigitate with one another, like interlaced fingers. Fluid flows through the narrow slits between these extensions.

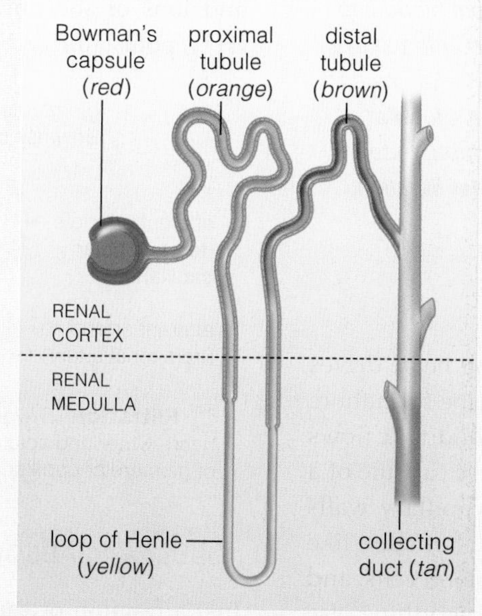

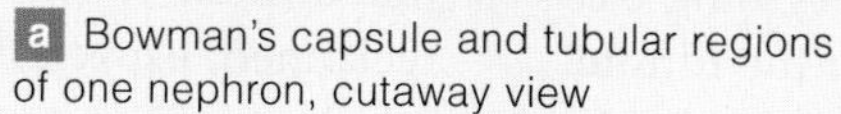

a Bowman's capsule and tubular regions of one nephron, cutaway view

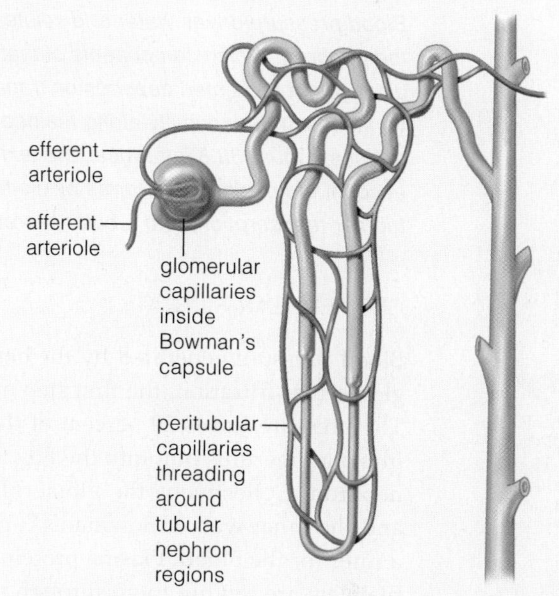

b Arterioles and the two sets of blood capillaries associated with the nephron

INTRODUCING THE NEPHRONS

In the section to follow, you will be taking a look at three processes that rid the body of excess water and solutes in the form of urine. Tracking the steps of the processes will be simpler if you first acquaint yourself with the structures that carry out these functions.

Overview of Nephron Structure A kidney has more than 1 million nephrons—microscopically small tubes with a wall only one cell thick. Each nephron begins in the cortex, where its wall balloons out and folds back on itself, to form a cup-shaped Bowman's capsule (Figure 38.7*a*). Past the capsule, the nephron twists a bit and straightens out as a proximal tubule (the part nearest the beginning of the nephron). After extending down into the renal medulla, the nephron makes a hairpin turn, the loop of Henle. It reenters the cortex and it twists again, as the distal tubule (the farthest from the start of the nephron), which drains into a collecting duct. Up to eight nephrons drain into each duct. Many collecting ducts extend through the kidney medulla and open onto the renal pelvis.

Blood Vessels Around Nephrons Inside each kidney, the renal artery branches into smaller, afferent arterioles. Each arteriole branches into a glomerulus, a capillary bed inside Bowman's capsule (Figure 38.7*b*). As the next section explains, these capillaries interact with Bowman's capsule as a blood-filtering unit.

As blood passes through the glomerulus, a portion of it is filtered into Bowman's capsule. The rest enters an efferent arteriole. This arteriole quickly branches to become peritubular capillaries, which thread lacily around the nephron (*peri–*, around). Blood inside the capillaries flows on through an efferent arteriole, and later through a vein leading out of the kidney.

Urine forms constantly by three physiological processes that involve all of the nephrons, glomerular capillaries, and peritubular capillaries. These processes are glomerular filtration, tubular reabsorption, and tubular secretion. They are the topic of the next section.

Remember, each human kidney has more than 1 million nephrons. Each minute, the nephrons of both kidneys collectively filter about 125 milliliters of fluid from blood flowing past, which amounts to 180 liters (about 47.5 gallons) per day. This means the kidneys filter the entire volume of blood about 40 times a day!

The human urinary system has two kidneys, two ureters, a urinary bladder, and a urethra. Kidneys filter water and solutes from blood. The body reclaims most of the filtrate. The rest flows as urine through ureters into a bladder that stores it. Urine flows out of the body through the urethra.

The functional unit of the human kidneys is the nephron, a microscopic tube that interacts with two systems of capillaries to filter blood and form urine.

38.4 Urine Formation

Blood pressure drives water and solutes from blood into the nephron. Which components of that filtrate leave in urine or get reclaimed depends on transport proteins and variations in permeability along the nephron's tubular regions. Figure 38.8 introduces the text's step-by-step account of how the urine forms by glomerular filtration, tubular reabsorption, and tubular secretion.

LINKS TO SECTIONS 5.6–5.8, 31.1, 32.3, 32.4, 34.7

GLOMERULAR FILTRATION

Blood pressure generated by the beating heart drives glomerular filtration, the first step of urine formation. The pressure forces 20 percent of the fluid that flows into a glomerulus out, into the Bowman's capsule of a nephron. Collectively, the glomerular capillary walls and the inner wall of Bowman's capsule function like a filter for the blood. Plasma proteins, blood cells, and platelets are too big to go through it. Everything else that filters into a nephron becomes part of the filtrate. Substances in the filtrate include glucose, amino acids, and ions of sodium (Na^+), chloride (Cl^-), hydrogen (H^+), potassium (K^+), and other solutes:

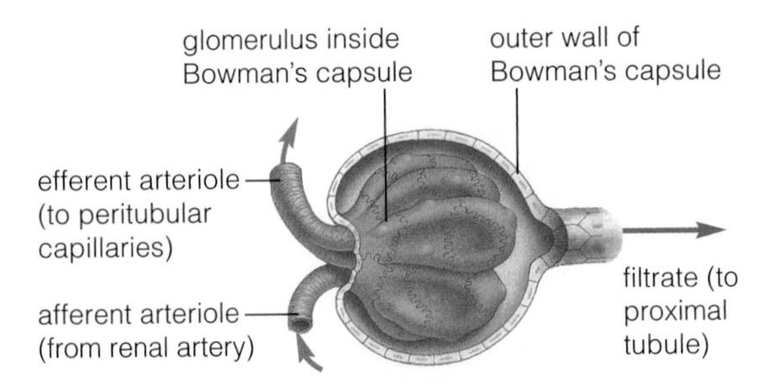

a Filtration Driven by pressure from a beating heart, water and solutes are forced across the wall of glomerular capillaries and into Bowman's capsule.

TUBULAR REABSORPTION

Just a fraction of the filtrate will be excreted. Most is reclaimed, especially at the proximal tubules. During tubular reabsorption, substances leak or get pumped out of a nephron, diffuse through interstitial fluid, and enter a peritubular capillary. Transport proteins move Na^+, Cl^-, glucose, bicarbonate, and other substances across the tubule wall and into peritubular capillaries. Water follows by osmosis:

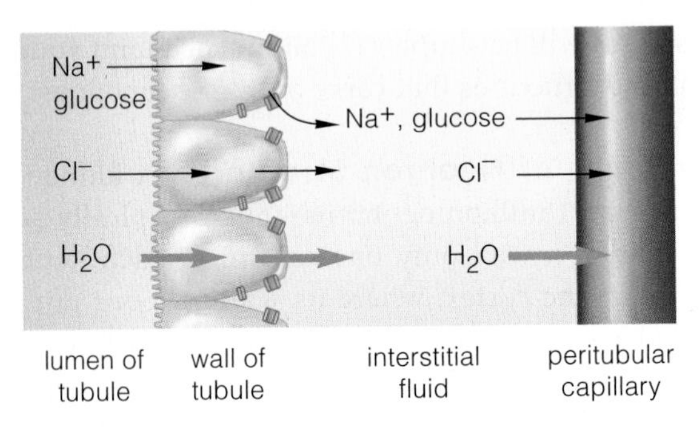

b Tubular reabsorption Filtrate flows through the proximal tubule. Ions and some nutrients are actively and passively transported into interstitial fluid. Water follows by osmosis. Cells of peritubular capillaries transport ions and nutrients into blood. Water again follows by osmosis.

Tubular reabsorption returns close to 99 percent of water in filtrate to the blood. It returns all the glucose and amino acids, most Na^+ and bicarbonate, but only 50 percent of the metabolic waste product urea.

TUBULAR SECRETION

Metabolism generates many ions, mainly H^+ and K^+. Together with urea and other wastes, the ions end up in the blood. During tubular secretion, transporters in

a Filtration Occurs at glomerular capillaries in Bowman's capsule.

b Tubular reabsorption Occurs all along a nephron's tubular parts.

c Tubular secretion Starts at proximal tubule and continues all along a nephron's tubular parts.

d Urine concentration occurs in loop of Henle and collecting duct.

proximal tubule
distal tubule
glomerular capillaries
CORTEX
MEDULLA
peritubular capillaries
loop of Henle
increasing solute concentration
urine outflow from collecting duct into renal pelvis

Figure 38.8 Animated! A nephron and associated blood capillaries. Urine forms and becomes more concentrated than interstitial fluid through glomerular filtration, tubular secretion, and tubular reabsorption.

peritubular capillary walls move ions into interstitial fluid. Then transporters in the nephron's tubular wall move them from the interstitial fluid into the filtrate, so that they may be excreted in urine:

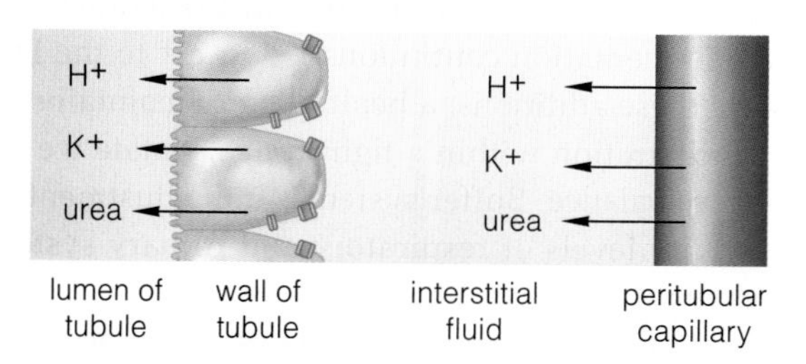

c Tubular secretion Transporters move H^+, K^+, urea, and wastes out of peritubular capillaries and into filtrate.

As Section 38.6 explains, secretion of H^+ is essential to maintenance of the body's acid–base balance.

CONCENTRATING THE URINE

Sip soda all day and your urine will be dilute; sleep eight hours and it will be concentrated. However, even the most dilute urine has far more solutes compared to plasma or most of the interstitial fluid. It gets that way by a concentrating mechanism.

For water to move out of a nephron by osmosis, the interstitial fluid must be saltier than the filtrate. Only in the renal medulla does an outward-directed solute concentration gradient form, with the interstitial fluid saltiest near the turn in the loop of Henle.

The ascending and descending arms of the loop of Henle are close together and differ in permeability, so solutes move between them. As a result, urine becomes concentrated as it moves down through the loop:

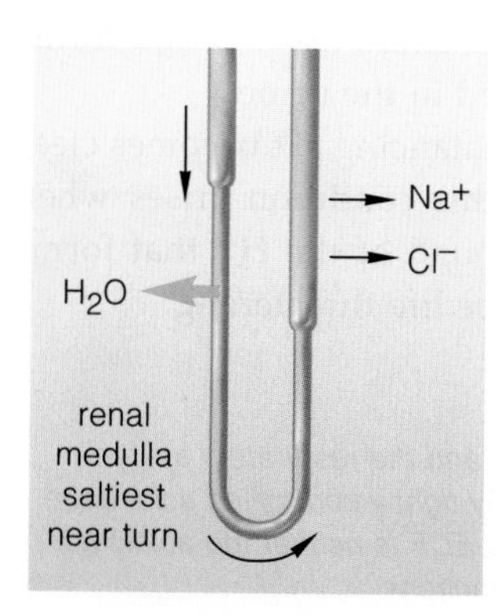

d Before the turn in the loop of Henle, water moves by osmosis out of the filtrate, which gets more and more concentrated as the loop descends. Water cannot cross the wall of the loop's ascending part, but the wall has transport proteins that pump out salt, making the interstitial fluid saltier. This attracts more water out of filtrate that is entering the descending arm of the loop.

Urine gets more concentrated as it travels through the cortex and medulla inside a collecting duct. Water moves out from the collecting duct by osmosis as the filtrate moves toward the renal pelvis.

HORMONAL CONTROLS

A nephron or collecting duct wall can be made more or less permeable to water and sodium. Adjustments in permeability can make urine more or less dilute. Two hormones and the hypothalamus, a brain region that is a major integrating center, cause the adjustments.

In the hypothalamus, osmoreceptors detect a rise in Na^+ in the ECF. A hypothalamic thirst center responds by initiating water-seeking behavior. In addition, the hypothalamus stimulates the pituitary gland to secrete ADH, or antidiuretic hormone. In response, the distal tubules and collecting ducts become more permeable to water. More water is reabsorbed, so less is lost in urine. ADH secretion slows when ECF volume rises and the Na^+ concentration returns to normal.

Cells in the walls of arterioles that carry blood to nephrons release the enzyme renin when ECF volume falls. Renin sets off a chain of reactions. The result is that the adrenal gland on top of each kidney secretes aldosterone. This hormone makes distal tubules and collecting ducts more permeable to Na^+. Thus, more Na^+ is reabsorbed. Water follows by osmosis, and the urine becomes more concentrated.

Hormonal disorders can disrupt water and sodium balance. When the pituitary gland produces too little ADH or receptors do not respond to ADH, diabetes insipidus develops. An affected person passes a large volume of dilute urine and is constantly thirsty. With overproduction of aldosterone, excess fluid retention occurs and causes high blood pressure.

The ECF also contains calcium ions (Ca^{++}). As you know from Section 32.4, parathyroid hormone (PTH) controls calcium levels in interstitial fluid, in part by its effects on kidneys. PTH binds to cells of proximal tubules and increases Ca^{++} reabsorption.

During glomerular filtration, pressure generated by the beating heart drives water and solutes out of glomerular capillaries and into nephrons, forming a filtrate.

In tubular reabsorption, membrane transporters set up a Na^+ gradient that draws water, other ions, and selected solutes out of the filtrate and into peritubular capillaries.

In tubular secretion, transporters move urea, H^+, and K^+ from peritubular capillaries into the nephron for excretion.

The deep renal medulla is very salty. The urine becomes more concentrated than interstitial fluid as it flows through a nephron's loop of Henle, which extends into this region.

Antidiuretic hormone promotes water reabsorption and aldosterone promotes reabsorption of Na^+. Parathyroid hormone promotes calcium reabsorption.

38.5 Kidney Disease

FOCUS ON HEALTH

One healthy kidney alone is enough to filter the blood and regulate fluid content for the body. Unfortunately, the failure of both kidneys is not uncommon.

LINK TO SECTION 2.6

Hard deposits called kidney stones form when uric acid, calcium, and other wastes crystallize in the renal pelvis. Inadequate fluid intake, a high-protein diet, and being male increase risk. Most stones pass without problems, but some get stuck, blocking urine flow and causing severe pain. If necessary, kidney stones can be broken up using sound waves or removed surgically. Rarely, an infection caused by kidney stones leads to kidney failure.

Kidneys fail when glomerular filtration plunges to half the normal rate. Kidney failure can kill; wastes, including excess H^+, build up and interfere with normal metabolism.

The main causes of kidney failure are diabetes mellitus and high blood pressure, which damage kidney capillaries. Some people are genetically predisposed to disorders that can damage the kidneys. Kidneys also fail after toxins enter the body. Taking aspirin or acetaminophen in high doses for long periods increases risk of kidney failure.

At one time or another, about 13 million people in the United States have experienced kidney failure. A kidney dialysis machine is used to restore proper solute balances. "Dialysis" simply refers to exchanges of solutes across any artificial membrane between two different solutions.

With *hemodialysis*, the machine is connected to a vein or an artery. It pumps blood through semipermeable tubes submerged in a solution of salts, glucose, and other substances. As blood flows through the tubes, the wastes dissolved in it diffuse out, so solute concentrations return to normal levels. The cleansed, solute-balanced blood is then allowed to flow back into the patient's body.

With *peritoneal dialysis*, fluid is pumped into a patient's abdominal cavity. It exchanges solutes with extracellular fluid, then it is drained out. In this case the peritoneum that lines the abdominal cavity serves as the membrane for dialysis (Figure 38.9).

Dialysis can keep a person alive through an episode of temporary kidney failure. When the damage is permanent, dialysis must be continued for life or until a kidney becomes available for transplant.

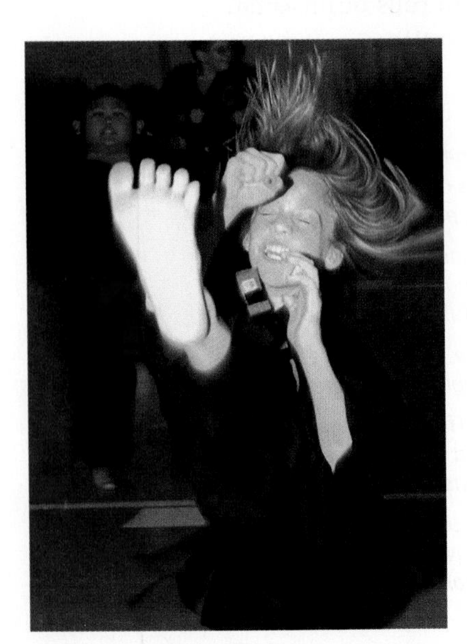

Figure 38.9 Inspirational attitude. Despite having kidney failure, Karole Hurtley became a national champion in karate at age thirteen. Each night, as she sleeps, a peritoneal dialysis machine pumps a salt solution into her abdomen. Wastes her damaged kidneys cannot remove flow into the fluid, which is then pumped out.

38.6 Acid-Base Balance

Besides maintaining the volume and composition of extracellular fluid, kidneys help keep it from getting too acidic or too basic (alkaline).

Metabolic reactions such as protein breakdown and lactate fermentation continuously add H^+ to the ECF. Despite these additions, a healthy body maintains the H^+ concentration within a tight range; a state we call acid–base balance. Buffer systems, and adjustments to the activity levels of respiratory and urinary systems are essential to this balance.

A buffer system, recall, involves substances that reversibly bind and release H^+ or OH^-. In doing so, it minimizes pH changes as acidic or basic molecules enter or leave a solution (Section 2.6).

In humans, the pH of extracellular fluid remains between 7.35 and 7.45. In the absence of a balancing mechanism, adding acids to ECF could make its pH fall. But excess hydrogen ions react with buffers, most notably a bicarbonate–carbonic acid buffer system:

$$\underset{\text{BICARBONATE}}{H^+ + HCO_3^-} \rightleftharpoons \underset{\text{CARBONIC ACID}}{H_2CO_3} \rightleftharpoons CO_2 + H_2O$$

When the pH of blood shifts, adjustments in the rate and magnitude of breathing offset that change. When the blood pH decreases, breathing quickens and CO_2 is expelled faster than it forms. As you can tell from the equation above, less CO_2 means less carbonic acid can form, so the pH rises. Slower breathing lets CO_2 accumulate, so more carbonic acid can form.

Control of bicarbonate reabsorption and secretion of H^+ can adjust the pH inside kidneys. Reabsorbed bicarbonate moves into peritubular capillaries, where it buffers excess acid. H^+ secreted into tubule cells combines with phosphate or ammonia ions and forms compounds that are excreted in the urine.

The importance of this balancing act becomes clear with metabolic acidosis. This condition arises when kidneys cannot excrete enough of the H^+ that forms during metabolism. It can be life threatening.

The kidneys, buffer systems, and the respiratory system interact to neutralize acids. By tightly controlling acid–base balance, they ensure that the ECF is neither too acidic nor too basic (alkaline) for cell functions.

A bicarbonate–carbon dioxide buffer system neutralizes excess H^+. Shifts in the rate and depth of breathing affect this buffer system, hence the pH of blood.

The kidneys also can shift the pH of blood by adjusting bicarbonate reabsorption and H^+ secretion.

38.7 Heat Gains and Losses

We turn now to another major aspect of homeostasis. How does the body maintain the core of its internal environment within a tolerable temperature range?

HOW THE CORE TEMPERATURE CAN CHANGE

The core temperature of an animal body rises when heat from the surroundings or metabolism builds up. A warm body will give up heat to cooler surroundings. The core temperature stabilizes when the rate of heat loss is equal to the rate of heat gain and production. The heat content of any complex animal depends on a balancing act between gains and losses:

$$\text{change in body heat} = \text{heat produced} + \text{heat gained} - \text{heat lost}$$

Heat is gained or lost at body surfaces. Radiation, conduction, convection, and evaporation bring about heat flow between body surfaces and the environment.

Thermal radiation is emission of heat into space around a warm object. Energy radiating from the sun heats animals. Also, metabolic activity produces heat, which radiates from the body. A typical human at rest produces about as much heat as a 100 watt light bulb.

In conduction, heat is transferred between objects in direct contact with each other. An animal loses heat when it rests on objects cooler than it is. If it contacts objects that are warmer, the animal will gain heat.

In convection, moving air or water transfers heat. Conduction plays a part; heat moves down a thermal gradient between the body and air or water next to it. Hot air rises; it constantly moves away from the body, so the thermal gradient remains steep.

In evaporation, a liquid converts to gaseous form and heat is lost in the process. Evaporation from body surfaces cools the body; water molecules carry energy away with them. Evaporative heat loss rises with dry air and breeze; high humidity and still air slow it.

ENDOTHERM? ECTOTHERM? HETEROTHERM?

Fishes, amphibians, and reptiles are warmed mostly by heat gained from the environment rather than by metabolically generated heat. These animals are called ectotherms, which means "heated from outside." They can only adjust behaviorally to rising or falling outside temperatures. Most species have low metabolic rates and little insulation. A rattlesnake (Figure 38.10*a*) is an example. When its body is cold, it basks in the sun. When hot, the snake moves into shade.

Most birds and mammals are endotherms, which means "heat from within." They have relatively high metabolic rates. Compared to, say, a foraging lizard of the same weight, a foraging mouse uses thirty times *more* energy. Metabolic heat helps endotherms remain active across a wider range of temperatures. Fur, fat, or feathers function as insulating layers and minimize heat transfers (Figure 38.10*b*).

Figure 38.10 (**a**) Sidewinder, an ectotherm. (**b**) Pine grosbeak, an endotherm, using fluffed feathers as insulation against winter cold.

LINKS TO SECTIONS 2.5, 5.1

Some birds and mammals are heterotherms. They can maintain a fairly constant core temperature some of the time but let it shift at other times. For example, hummingbirds have very high metabolic rates when foraging for nectar during the day. At night, metabolic activity decreases so much that the bird's body may become almost as cool as the surroundings.

Warm climates favor the ectotherms, which do not have to spend as much energy as endotherms do on maintaining the core temperature. In tropical regions, reptiles have the advantage, because they can spend more energy on reproduction and other tasks. There they exceed mammals in numbers and diversity. In all cool or cold regions, however, most vertebrates tend to be endotherms. About 130 species of mammals and 280 species of birds occur in the arctic, but fewer than 5 species of reptiles have ever been reported there.

The core temperature of a body is maintained when heat losses balance heat gains. Ectotherms gain heat mainly from the outside, endotherms generate most heat from within. Heterotherms do both at different times.

Animals gain and lose heat by way of thermal radiation, conduction, convection, and evaporation. They can make physiological, and behavioral adjustments in response to changes in environmental temperature.

38.8 Temperature Regulation in Mammals

Section 25.3 introduced the negative feedback controls that govern increases in human body temperature. Now compare the controls that generally come into play when mammals are subjected to heat stress and cold stress.

LINKS TO SECTIONS 2.5, 7.4, 25.3, 29.6, 31.2, 35.3

The hypothalamus, again, contains centers that help control the core temperature of the mammalian body. These hypothalamic centers receive input from many thermoreceptors inside skin and from others deep in the body (Figure 35.2). When the temperature deviates from a set point, the centers integrate the responses of skeletal muscles, smooth muscle of arterioles in the skin, and sweat glands. Negative feedback loops back to the hypothalamus inhibit the responses when the core temperature returns to the set points.

Most mammals maintain body temperature within a narrow range, but dromedary camels can adjust the hypothalamic set point (Figure 38.11). Over the course of a day, their body temperature can vary from 34°C to 41.7°C (93°F to 106°F).

RESPONSES TO HEAT STRESS

When any mammal becomes too hot, the temperature control centers in the hypothalamus issue commands for peripheral vasodilation: The diameter of the blood vessels in skin increases. More blood flows to the skin and delivers more metabolic heat that can be given up to the surroundings (Table 38.1).

Another response to heat stress, evaporative heat loss, occurs at moist respiratory surfaces and across skin. Animals that sweat lose some water this way. For instance, humans and some other mammals have sweat glands that release water and solutes through pores at the skin's surface (Section 29.6). An average-sized adult human has 2–1/2 million or more sweat glands. For every liter of sweat produced, about 600 kilocalories of heat energy leave the body by way of evaporative heat loss.

Sweat dripping from skin dissipates little heat. The body cools greatly when sweat evaporates. On humid days, the air's high water content slows evaporation rate, so sweating is less effective at cooling the body. When you exercise strenuously, sweating helps offset heat production by skeletal muscles.

Not all mammals sweat. Many drool, lick their fur, or pant to speed cooling. "Panting" refers to shallow, rapid breathing. It assists evaporative water loss from the respiratory tract, nasal cavity, mouth, and tongue.

Sometimes peripheral blood flow and evaporative heat loss cannot counter heat stress, so hyperthermia follows: Core temperature rises above normal. As you read earlier in Chapter 25, a human body temperature above 41.5°C (105°F) is dangerous.

What about a fever? Recall, from Section 35.3, that fever is not itself an illness. It is one of the responses to infection. When activated by a threat, macrophages release signaling molecules that stimulate part of the brain to secrete prostaglandins. These local signaling molecules cause the hypothalamus to allow the core temperature to rise a bit above the normal set point. The rise makes the body less hospitable for pathogens and calls up more immune responses. Generally, the hypothalamus does not let the core temperature rise above 41.5°C (105°F). When a fever exceeds that point or lasts more than a few days, the condition causing it is life threatening and medical evaluation is essential.

RESPONSES TO COLD STRESS

Selectively distributing blood flow, fluffing up hair or fur, and shivering help mammals respond to the cold.

Figure 38.11 Short-term adaptation to desert heat stress. Dromedary camels let their core temperature rise during the hottest hours of the day. A hypothalamic mechanism adjusts their internal thermostat, so to speak. By allowing their temperature set point to rise, camels minimize their sweat production and thus conserve water.

Table 38.1 Heat and Cold Stress Compared

Stimulus	Main Responses	Outcome
Heat stress	Widespread vasodilation in skin; behavioral adjustments; in some species, sweating, panting	Dissipation of heat from body
	Decreased muscle action	Heat production decreases
Cold stress	Widespread vasoconstriction in skin; behavioral adjustments (e.g., minimizing surface parts exposed)	Conservation of body heat
	Increased muscle action; shivering; nonshivering heat production	Heat production increases

Figure 38.12 Two responses to cold stress.

(**a**) Polar bears (*Ursus maritimus*, "bear of the sea"). A polar bear is active even during severe arctic winters. It does not get too chilled after swimming because the coarse, hollow guard hairs of its coat shed water quickly. Thick, soft underhair traps heat. Brown adipose tissue, about 11.5 centimeters (4–1/2 inches) thick, insulates and helps generate metabolic heat.

(**b**) In 1912, the *Titanic* collided with an immense iceberg on her maiden voyage. It took about 2–1/2 hours for the *Titanic* to sink and rescue ships arrived in less than two hours. Even so, 1,517 bodies were recovered from the calm sea. All of the dead had on life jackets; none had drowned. Hypothermia killed them.

Some thermoreceptors in skin signal the hypothalamus when things get chilly. The hypothalamus then causes smooth muscle in arterioles that deliver blood to the skin to contract. This response to cold stress is called peripheral vasoconstriction. When arterioles in skin constrict, less metabolic heat reaches the body surface. When your fingers or toes become chilled, all but 1 percent of the blood that would usually flow to skin is diverted to other regions of the body.

Also, muscle contractions make hair (or fur) "stand up." This pilomotor response creates a layer of still air next to skin. It helps reduce heat loss by convection and radiation. Minimizing exposed body surfaces can also prevent heat loss, as when polar bear cubs curl up and cuddle against their mother (Figure 38.12*a*).

With prolonged cold exposure, the hypothalamus commands skeletal muscles to contract ten to twenty times each second. Although this shivering response increases heat production, it has a high energy cost.

Long-term or severe cold exposure also leads to a an increase in thyroid activity that raises the rate of metabolism. Thyroid hormone binds to cells of *brown* adipose tissue, causing nonshivering heat production. By this process, mitochondria in cells of brown adipose tissue carry out reactions that release energy as heat, rather than storing it in ATP.

Brown adipose tissue occurs in mammals that live in cold regions and in the young of many species. In human infants, this tissue makes up about 5 percent of body weight. Unless cold exposure is ongoing, the tissue disappears after childhood ends.

Failure to protect against cold causes hypothermia, a condition in which the core temperature plummets. In humans, a decline to 95°F changes brain functions. "Stumbles, mumbles, and fumbles" are said to be the symptoms of early hypothermia. Severe hypothermia causes loss of consciousness, disrupts heart rhythm, and can be fatal (Figure 38.12*b* and Table 38.2).

Table 38.2 Impact of Increases in Cold Stress

Core Temperature	Physiological Responses
36°–34°C (about 95°F)	Shivering response; faster breathing, metabolic heat output. Peripheral vasoconstriction, more blood deeper in body. Dizziness, nausea.
33°–32°C (about 91°F)	Shivering response ends. Metabolic heat output declines.
31°–30°C (about 86°F)	Capacity for voluntary motion is lost. Eye and tendon reflexes inhibited. Consciousness lost. Cardiac muscle action becomes irregular.
26°–24°C (about 77°F)	Ventricular fibrillation sets in (Section 34.9). Death follows.

Temperature shifts are detected by thermoreceptors that send signals to an integrating center in the hypothalamus.

This center serves as the body's thermostat and calls for adjustments that maintain core temperature.

Mammals counter cold stress by vasoconstriction in skin, behavioral adjustments, increased muscle activity, and shivering and nonshivering heat production.

Mammals counter heat stress by widespread peripheral vasodilation in skin and evaporative water loss.

www.thomsonedu.com ThomsonNOW!

Summary

Sections 38.1. 38.2 Plasma and interstitial fluid are the main components of the extracellular fluid, or ECF. Maintaining ECF volume and composition is an essential aspect of homeostasis. All organisms balance solute and fluid gains with solute and fluid losses. In vertebrates, the kidneys and other components of a urinary system are central to this balancing act.

Animals in different habitats face different challenges. Those living in water lose or gain water by osmosis. On land, the main challenge is avoiding dehydration. Birds and reptiles conserve water by eliminating nitrogen-rich wastes as uric acid crystals. Mammals excrete urea, which must be dissolved in a lot of water.

Section 38.3 The human urinary system consists of a pair of kidneys, a pair of ureters, a urinary bladder, and a urethra. Urination is a reflex response, but it can be overridden by voluntary control.

Nephrons are small, tubular structures in the kidneys. They interact with capillaries as the functional units that cleanse blood and form urine.

Each nephron starts as Bowman's capsule in the renal cortex. It continues as a proximal tubule, a loop of Henle that descends into and ascends from the renal medulla, and a distal tubule that drains into a collecting duct. All collecting ducts drain into the renal pelvis.

Bowman's capsule and the glomerular capillaries it cups around are a blood-filtering unit. Most filtrate that enters Bowman's capsule is reabsorbed into peritubular capillaries around the nephron. The portion of the filtrate not returned to blood is excreted as urine.

■ *Use the animation on ThomsonNOW to explore the anatomy of the human urinary system.*

Section 38.4 Urine forms in nephrons by processes of filtration, tubular reabsorption, and tubular secretion (Table 38.3). Hormones affect reabsorption (Table 38.4). Sensory receptors detect high solute levels in the ECF or a decrease in blood volume. They signal a hypothalamic thirst center, which calls for water-seeking behavior and secretion of antidiuretic hormone (ADH) by the pituitary gland. ADH causes reabsorption of water, so urine gets more concentrated. Aldosterone, a hormone secreted by the adrenal gland, increases sodium reabsorption. Water follows sodium into the blood, so aldosterone indirectly concentrates the urine. Parathyroid hormone promotes calcium reabsorption.

■ *Use the animation on ThomsonNOW to learn about the three processes of urine formation.*

Section 38.5 When kidneys fail, frequent dialysis or a kidney transplant is required to sustain life.

■ *Read the InfoTrac article "The Kidney Swap: Adventures in Saving Lives," Denise Grady and Anahad O'Connor,* The New York Times, *October 2004.*

Section 38.6 Buffering systems and changes in the respiratory rate help maintain acid–base balance. So does the urinary system, when it eliminates H^+ in urine and reabsorbs bicarbonate.

Section 38.7 All animals produce metabolic heat. To maintain the core temperature, heat gains by metabolism and by absorption from the environment must balance heat losses to the environment.

For ectotherms such as reptiles, the core temperature depends more on heat exchanges with the environment than on any metabolic heat. Such animals control core temperatures mainly by modifications in behavior.

For endotherms (most birds and mammals), a high metabolic rate is the primary source of heat. Endotherms regulate their core temperature mainly by controlling the production and loss of metabolic heat.

Heterotherms control core temperature tightly part of the time and allow it to fluctuate with the environmental temperature at other times.

Section 38.8 In mammals, the hypothalamus is the main integrating center for the control of temperature. It receives information from thermoreceptors and calls for responses by smooth muscle in arterioles, sweat glands, and other effectors. The core temperature is maintained by behavioral, metabolic, and physiological responses.

Dilation of blood vessels in the skin, sweating, and panting are responses to heat. Mammals alone can sweat, but not all mammals have this ability.

Constriction of blood vessels in skin, shivering, and making hair (or fur) stand upright are responses to cold.

Table 38.3 Processes of Urine Formation

Process	Characteristics
Glomerular filtration	Pressure generated by heartbeats drives water and small solutes (not proteins) out of leaky glomerular capillaries and into Bowman's capsule, the entrance to the nephron.
Tubular reabsorption	Most water and solutes in the filtrate move from a nephron's tubular portions, into interstitial fluid around the nephron, then into blood inside the peritubular capillaries.
Tubular secretion	Urea, H^+, and some other solutes move out of peritubular capillaries, into interstitial fluid, then into the filtrate inside the nephron for excretion in urine.

Table 38.4 Some Hormonal Effects on Reabsorption

Hormone	Source	Target	Action
ADH (antidiuretic hormone)	Pituitary gland	Distal tubules, collecting ducts	Increases permeability to water, concentrates the urine
Aldosterone	Adrenal cortex	Distal tubules, collecting ducts	Increases reabsorption of sodium, indirectly causes concentration of urine
PTH (parathyroid hormone)	Parathyroid gland	Proximal tubules	Promotes reabsorption of calcium ions

Self-Quiz

Answers in Appendix III

1. Label the structures in the two diagrams at right and describe their functions.

2. A freshwater fish gains most of its water by ______ .
   a. drinking
   b. eating food
   c. osmosis
   d. transport across the gills

3. Bowman's capsule, the start of the tubular part of a nephron, is located in the ______ .
   a. renal cortex
   b. renal medulla
   c. renal pelvis
   d. renal artery

4. Fluid filtered into Bowman's capsule flows directly into the ______ .
   a. renal artery
   b. proximal tubule
   c. distal tubule
   d. loop of Henle

5. Water and small solutes from blood enter nephrons during ______ .
   a. filtration
   b. tubular reabsorption
   c. tubular secretion
   d. both a and c

6. Kidneys return most of the water and small solutes back to blood by way of ______ .
   a. filtration
   b. tubular reabsorption
   c. tubular secretion
   d. both a and b

7. ADH binds to receptors on distal tubules and collecting ducts, making them ______ permeable to ______ .
   a. more; water
   b. less; water
   c. more; sodium
   d. less; sodium

8. Increased sodium reabsorption at distal tubules ______ .
   a. will make urine more concentrated
   b. will make urine more dilute
   c. is stimulated by ADH and aldosterone
   d. both a and c

9. Match each structure with a function.

___ ureter	a. start of nephron
___ Bowman's capsule	b. delivers urine to body surface
___ urethra	c. carries urine from kidney to bladder
___ collecting duct	d. secretes ADH
___ pituitary gland	e. target of aldosterone

10. The main control center for maintaining the temperature of the mammalian body is in the ______ .
    a. anterior pituitary
    b. renal cortex
    c. adrenal gland
    d. hypothalamus

11. Negative feedback loops that maintain a mammal's core temperature involve ______ .
    a. the hypothalamus
    b. receptors in skin
    c. receptors in deep tissue
    d. all of the above

12. Is this statement true or false? Thirst behavior is initiated and controlled exclusively by a thirst center in the hypothalamus.

13. Match each term with the most suitable description.

___ endotherm	a. environment dictates core temperature
___ ectotherm	b. metabolism dictates core temperature
___ convection	c. heat transfer between objects that are in direct contact
___ conduction	d. water, air current transfers heat
___ thermal radiation	e. emission of radiant energy

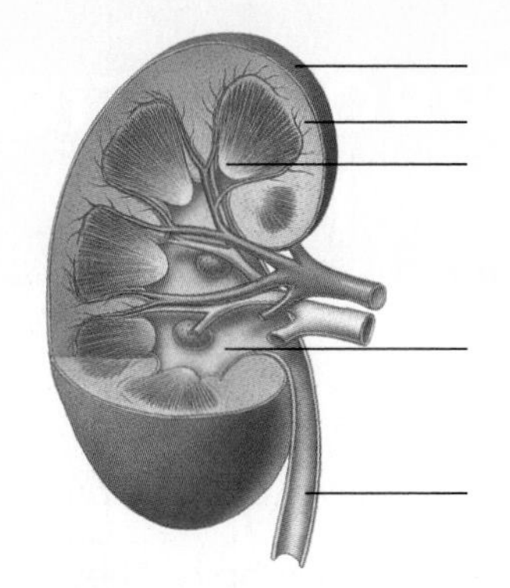

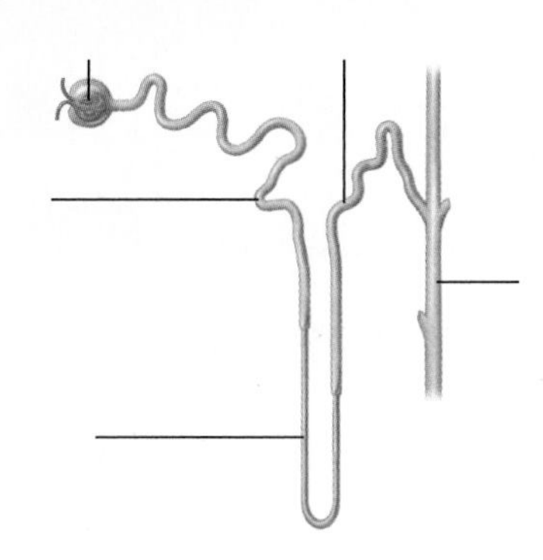

■ *Visit ThomsonNOW for additional questions.*

Critical Thinking

1. Each year in the United States, about 12,000 people are recipients of kidney transplants. More than 40,000 others remain on the waiting list because of a shortage of donors. Most transplanted kidneys are from people who donated organs after death, but some come from living donors. One healthy kidney is adequate to maintain health. The shortage of organs for transplant and high cost of dialysis have led some to suggest that people should be allowed to sell a kidney. Critics say it is unethical to tempt people to risk their health in this way. What do you think?

2. The kangaroo rat kidney efficiently excretes a very small volume of urine (Section 38.2). Compared with a human nephron, its nephrons have a loop of Henle that is proportionally much longer. Explain how this helps the kangaroo rat conserve water.

3. Drinking too much water can be a bad thing (Figure 38.13). As marathoners or other endurance athletes sweat heavily and drink lots of water, their sodium levels drop. The resulting "water intoxication" can be fatal. Why is maintaining sodium balance so important?

4. In cold habitats, ectotherms are few and endotherms often show morphological adaptations to cold. Compared to closely related species that live in warmer areas, cold dwellers tend to have smaller appendages. Also, animals adapted to cool climates tend to be larger than related species that evolved in warmer places. For example, the largest bear species is the polar bear and the largest penguin is Antarctica's emperor penguin.

Think about heat transfers between animals and their habitat, then explain why smaller appendages and larger overall body size are advantageous in very cold climates.

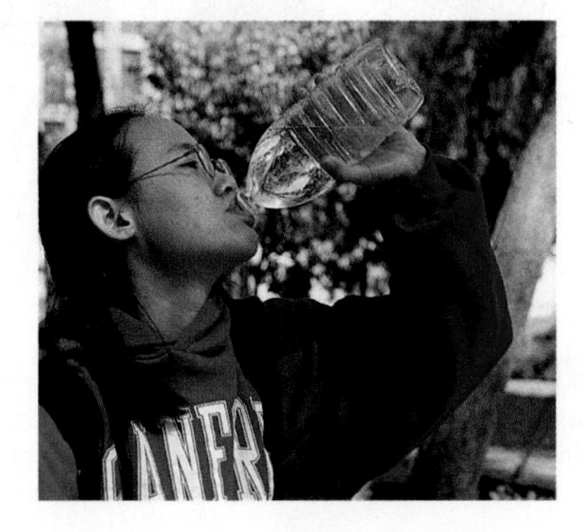

Figure 38.13 Excessive water intake during an endurance event—not a good idea.

39 ANIMAL REPRODUCTION AND DEVELOPMENT

IMPACTS, ISSUES Mind-Boggling Births

In December of 1998, Nkem Chukwu of Houston, Texas, gave birth to six girls and two boys. They were the first set of human octuplets to be born alive (Figure 39.1). The births were premature. In total, all eight newborns weighed a bit more than 4.5 kilograms (10 pounds). Odera, the smallest, weighed about 300 grams (less than 1 pound), and six days later she died when her heart and lungs gave out. Two others required surgery. All seven survivors had to spend months in the hospital before going home, but now are in good health.

Why did octuplets form in the first place? Chukwu had trouble getting pregnant. Her doctors gave her hormone injections, which caused many of her eggs to mature and be released at the same time. When the doctors realized that she was carrying a large number of embryos, they suggested reducing the number. Chukwu decided to try to carry all of them to term.

Her first child was thirteen weeks premature. The others were surgically delivered two weeks later.

Over the past two decades, the incidence of multiple births has increased by almost 60 percent. There have been four times as many higher order multiple births—triplets or more. What is going on? A woman's fertility peaks in her mid-twenties. By thirty-nine, her chance of conceiving naturally has declined by about half. Yet the number of first-time mothers who are more than forty years old doubled in the past decade. Many had turned to reproductive intervention, including fertility drugs and *in vitro* fertilization.

Weigh the rewards against risks. Carrying more than one embryo increases the risk of miscarriage, premature

See the video! **Figure 39.1** Testimony to the potency of fertility drugs—seven survivors of a set of octuplets. Besides manipulating so many other aspects of nature, humans are now manipulating their own reproduction. *Facing page*, human embryo at an early stage of development. From the top of its forming head, down a series of segments that will give rise to muscles and a backbone, to the tip of its transient chordate tail, it is less than an inch long.

How would you vote? Fertility drugs make many eggs mature at the same time and increase the odds of multiple pregnancies. Should the use of such drugs be discouraged to lower the number of high-risk pregnancies? See ThomsonNOW for details, then vote online.

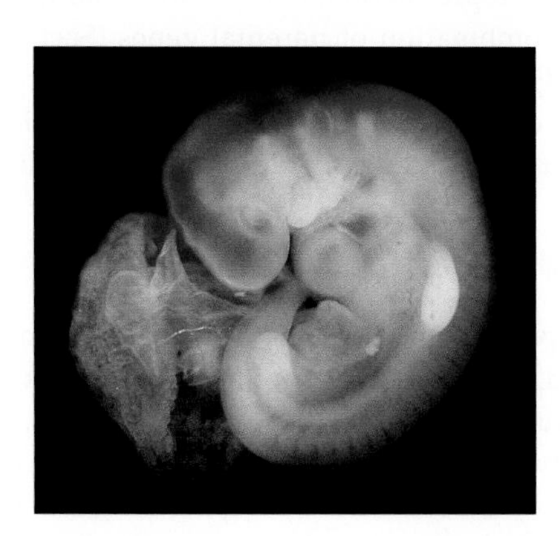

delivery, or stillbirth. Multiple-birth newborns weigh less than normal and are more likely to have birth defects, including cleft lip, heart malformations, and disorders in which the bladder or spinal cord is exposed at the body surface.

With this example, we turn to one of life's greatest dramas—the reproduction and development of complex animals in the image of their parents. *How is a fertilized egg of a human—or frog or bird or any other animal—transformed into the specialized cells and structures of the adult form?*

Answers will emerge through this chapter's survey of principles that govern animal life cycles, from gamete formation and reproduction, through embryonic and postnatal development, and on to aging and death. We focus on human sexual reproduction and development. In humans, as in other mammals, reproductive events play out in the pair of gonads—primary reproductive organs—of males and females. These organs produce gametes. They secrete sex hormones, which control reproductive functions as well as the development of many traits we associate with maleness and femaleness.

This chapter explains the developmental processes that shaped your body, how your reproductive system functions, and how you can protect those functions. What you learn might help you work through health-related problems. You might gain perspective on ethical issues, including how to control human fertility in an age of runaway population growth. These are issues we now face as individuals and as members of society at large.

Key Concepts

COSTS AND BENEFITS OF SEXUAL REPRODUCTION

Biologically, having separate sexes is more costly than asexual reproduction. However, sexual reproduction produces genetically variable offspring, thus increasing the likelihood that some will survive if the environment changes. **Section 39.1**

PRINCIPLES OF EMBRYOLOGY

Most animal life cycles have six stages: gamete formation, fertilization, cleavage, gastrulation, organ formation, and then growth and tissue specialization. **Sections 39.2–39.4**

HUMAN REPRODUCTION

The primary reproductive organs are sperm-producing testes in human males and oocyte-producing ovaries in females. Both make and release sex hormones in response to the hypothalamus and pituitary gland. The hormones guide reproductive functions and the development of secondary sexual traits. From puberty onward, human females are fertile on a cyclic basis. **Section 39.5–39.10**

HUMAN SEXUAL BEHAVIOR

Sexual intercourse leads to pregnancy, which humans may prevent or promote. Human sexual behavior transmits some pathogens. **Sections 39.11–39.13**

HUMAN EMBRYONIC DEVELOPMENT

A pregnancy starts with fertilization and implantation of a blastocyst in the uterine lining. A placenta connects the embryo with its mother—its protector and provider. **Sections 39.14–39.18**

FROM BIRTH ONWARD

Growth and proportional changes continue even after birth. Aging and death are inevitable, the result of genetic factors and ongoing environmental assaults on cells and tissues. **Sections 39.19, 39.20**

Links to Earlier Concepts

This chapter builds on earlier discussions of cleavage (Section 8.4), gamete formation, and fertilization (9.5). You will reconsider sex from an evolutionary perspective (Chapter 9 introduction). You will revisit the nature of cell differentiation and the master genes that influence it (14.1, 14.3, 16.8). You will learn more about primary tissue layers of embryos (23.1, 29.5), and see more examples of feedback controls (25.3) and cell signaling (25.5). You will be reminded of the effects of sexually transmitted pathogens (19.2, 20.2, 35.10). You will build upon your understanding of vertebrate evolution (23.1, 24.1, 24.6, 24.10), including long-term changes in development patterns (16.8).

39.1 Reflections on Sexual Reproduction

Sexual reproduction dominates the life cycle of most animals, even those that also can reproduce asexually. We therefore can expect that the benefits of sexual reproduction outweigh the costs.

SEXUAL VERSUS ASEXUAL REPRODUCTION

LINK TO SECTION 9.4

In Chapter 9, you learned about the genetic basis of sexual reproduction. Again, meiosis and the formation of gametes typically occur in two prospective parents. At fertilization, a gamete from one parent fuses with a gamete from the other and forms the first cell of the new individual—the zygote. You looked at asexual reproduction, whereby a single organism—just one parent—produces its offspring. Turn now to examples of the structural, behavioral, and ecological aspects of the two modes of animal reproduction.

Some animals can reproduce asexually by splitting into pieces. A fragment torn from a sponge or sea star can grow by mitosis into a new individual. Flatworms spontaneously divide (Figure 39.2*a*). A constriction forms in the middle. One end of the body grips a substrate and starts a tug-of-war with the other. The two parts separate a few hours later. Each incomplete worm goes off on its way and grows the missing pieces.

Mutation aside, by *asexual* reproduction, one parent has all of its genes represented in its offspring. All offspring are genetically identical with the parent and one another. This can be a good thing if the environment does not vary over time. Gene combinations that allowed the parent to reproduce can be expected to do the same for its offspring.

Most animals live where opportunities, resources, and dangers change over time. In such environments, producing offspring that differ can be advantageous. With sexual reproduction, every individual inherits a different combination of parental genes (Section 9.4). By reproducing sexually, a parent increases the odds that some offspring will have a gene combination that suits them to the changed environmental conditions.

COSTS OF SEXUAL REPRODUCTION

Sexual reproduction is costly. Only half of a parent's genes end up in each of its offspring. Resources and energy must be allocated to forming gametes. Often, reproductive structures that help deliver or receive sperm must be built. Timing of gamete formation and mating must be synchronized between the sexes. A potential mate might have to be courted.

Sexual reproduction requires careful timing. Sperm in one individual must mature at the same time as eggs of another. Each individual constructs, maintains, and uses costly neural and hormonal control mechanisms to ensure gametes mature at the right time. Cues such as daylength often signal the optimum time to produce gametes. For example, moose are sexually active only in the early autumn. As a result, offspring are born in spring, when the weather is mild and plants that they feed on are most abundant.

Finding and keeping a mate requires investment as well. Some animals produce chemical sex attractants. Many males fend off potential rival with horns, claws, or a huge body (Figure 39.2*b*,*c*). Others attract females with colorful body parts (Figure 39.2*d*). We consider courtship and mating behavior again in Chapter 44.

Producing enough offspring so that some survive can increase reproductive cost (Figure 39.3). Most of

Figure 39.2 (**a**) Example of asexual reproduction by an animal. This flatworm (*Dugesia*) can reproduce asexually by spontaneous fission. It divides into two pieces. Each piece replaces what is missing, and it is genetically identical with the original flatworm; all of the parent worm's genes are represented.

Biological costs associated with sexual reproduction. Reflect on the energy and raw materials directed into producing (**b**) sable antelope horns, (**c**) the body mass of male northern elephant seals fighting for access to the far smaller female at lower right, and (**d**) the attention-getting bill and feathers of a toucan.

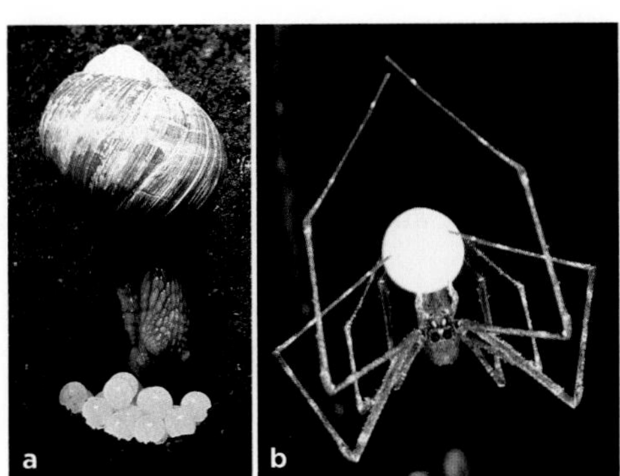

Figure 39.3 A look at where some invertebrate and vertebrate embryos develop, how they are nourished, and how (if at all) parents protect them.

(**a**) Most snails lay eggs and abandon them. (**b**) Spider eggs develop in a silk egg sac. Females often die soon after they make the sac, but some species guard the sac, then cart spiderlings about for a few days while they feed them.

(**c**) Ruby-throated hummingbirds and all other birds lay fertilized eggs with big yolk reserves. The eggs develop and hatch outside the mother. One or both parent birds typically expend energy feeding and caring for the young.

(**d**) Embryos of most sharks, most lizards, and some snakes develop in their mother, receive nourishment continuously from yolk reserves, and are born live. Shown here, live birth of a lemon shark.

Embryos of most mammals draw nutrients from maternal tissues and are born live. (**e**) In kangaroos and other marsupials, embryos are born "unfinished." They complete embryonic development inside a pouch on the mother's ventral surface. (**f**) Juveniles (joeys) continue to be nourished with milk secreted by mammary glands inside the pouch.

A human female (**g**) retains a fertilized egg inside her uterus. Her own tissues nourish the developing individual until birth.

the invertebrates and some vertebrates release sperm, eggs, or both into the water. These animals increase the odds that their gametes will contribute to the next generation by releasing lots of them. For example, a female oyster releases thousand of eggs at one time.

As another example, nearly all animals on land use internal fertilization, or the union of sperm and egg within the female body. They invest metabolic energy to construct elaborate reproductive organs, such as a penis and a uterus. A penis deposits sperm inside the female, and a uterus is a chamber in which an embryo develops inside certain mammalian females.

Most female animals also make an investment in yolk, a thick fluid containing proteins and lipids that nourishes the embryo until it can feed. Eggs of some species have more yolk than others. Sea urchins make tiny eggs that hold little yolk, but each fertilized egg grows into a self-feeding, swimming larva in less than a day. In contrast, birds funnel energy and resources into making eggs rich with yolk, the embryo's only nourishment during an extended time in an eggshell. Kiwi birds have the longest incubation time, about 11 weeks, and their eggs have an usually large amount of yolk. A typical bird's egg is about one-third yolk; a kiwi's egg is two-thirds yolk.

Your mother nourished you through nine months of development from a nearly yolkless, fertilized egg. Physical exchanges with her bloodstream supported your embryonic development (Figure 39.3*g*). You will learn more about how humans nourish their young later in the chapter.

As these examples suggest, animals show diversity in reproduction and development. Even so, as you will see, some patterns recur throughout their kingdom.

Separation into male and female sexes requires special reproductive cells and structures, neural and hormonal control mechanisms, and forms of behavior.

A selective advantage—variation in traits among offspring—offsets biological costs related to sexual reproduction.

39.2 Stages of Reproduction and Development

Animals as different as sea stars and sea otters pass through the same stages in their developmental journey from a single, fertilized egg to a multicelled adult.

LINKS TO SECTIONS 8.4, 9.5, 23.1, 29.5

Figure 39.4 shows six stages in animal reproduction and development. Most invertebrates and vertebrates pass through the early stages. Species with true tissues and organs continue past cleavage (Section 23.1).

In the first stage, gamete formation, eggs or sperm arise from germ cells in the parental body, as outlined in Section 9.5. At fertilization, the first cell of the new individual—a zygote—forms after a sperm penetrates a mature egg, when the sperm and egg nuclei fuse.

Cleavage carves up the zygote by repeated mitotic cell divisions. The number of cells increases, but the zygote's original volume does not. Cells become more numerous but smaller (Figure 39.5*a–d*). Cells formed during cleavage are called blastomeres. They typically are arranged as a blastula: a ball of cells that enclose a cavity (blastocoel) filled with their own secretions.

In the fourth stage, gastrulation, cells self-organize as an early embryo—a gastrula—that has two or three primary tissue layers. The tissues are the germ layers of the new individual. Germ layers, remember, are the forerunners of the adult animal's tissues and organs (Section 29.5).

Ectoderm, the outer germ layer, forms first. It gives rise to nervous tissue and to the outer layer of skin or other body covering. Endoderm, the inner germ layer, is the start of the respiratory tract and gut linings. In most kinds of animals, a third layer called mesoderm forms between the ectoderm and the endoderm. This "middle" layer is the source of all muscles, connective tissues, and the circulatory system.

During organ formation, interactions among cells cause body parts to form in certain places, at certain times, orchestrated by the expression of master genes. Many organs incorporate tissues from more than one germ layer. As one example, the stomach's epithelial lining is derived from endoderm whereas its smooth muscle develops from mesoderm.

Growth and tissue specialization is the final stage of animal development. The tissues and organs continue to grow, and they slowly take on their final sizes, shapes, proportions, and functions. The stage continues into adulthood.

Figure 39.5 has examples of the stages for one vertebrate, the leopard frog (*Rana pipiens*). This figure reinforces an important principle: Each stage builds on the one preceding it.

Most animal life cycles proceed through gamete formation, fertilization, cleavage, gastrulation, organ formation, and then growth and tissue specialization.

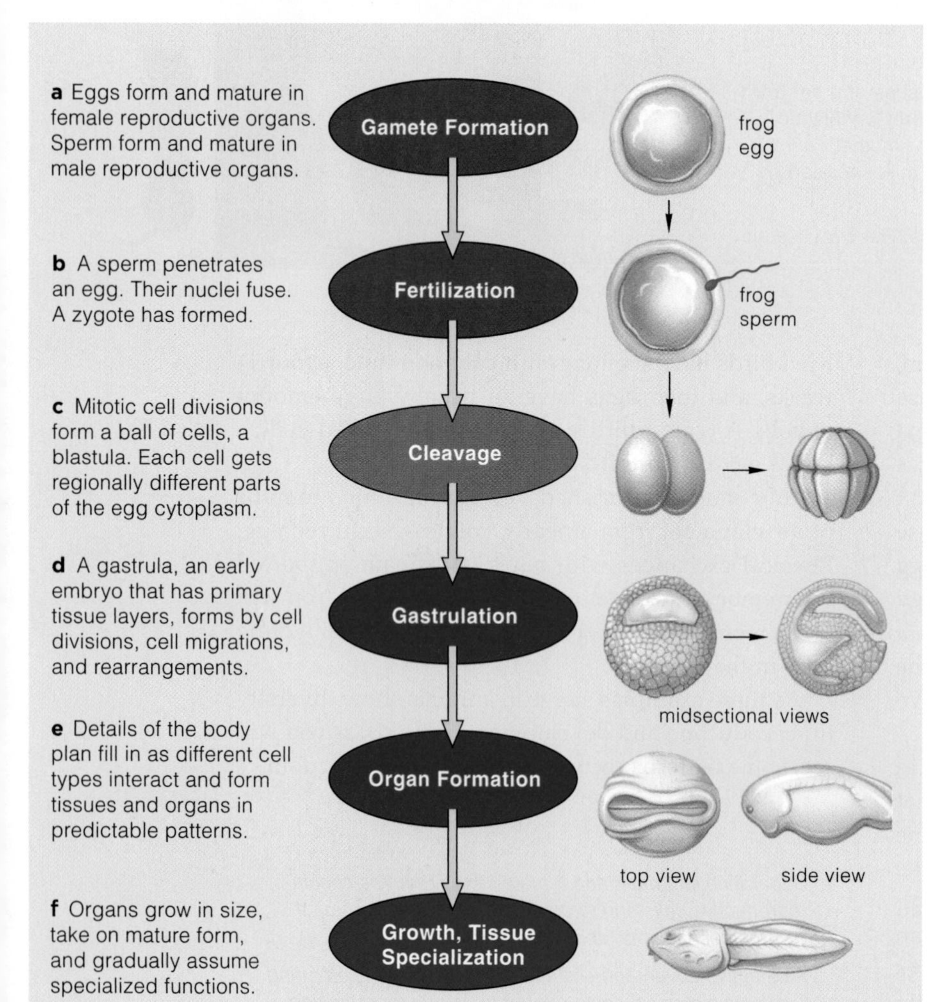

Figure 39.4 Overview of stages of animal reproduction and development. We use a few forms that appear during the frog life cycle as examples. The drawings are not to the same scale.

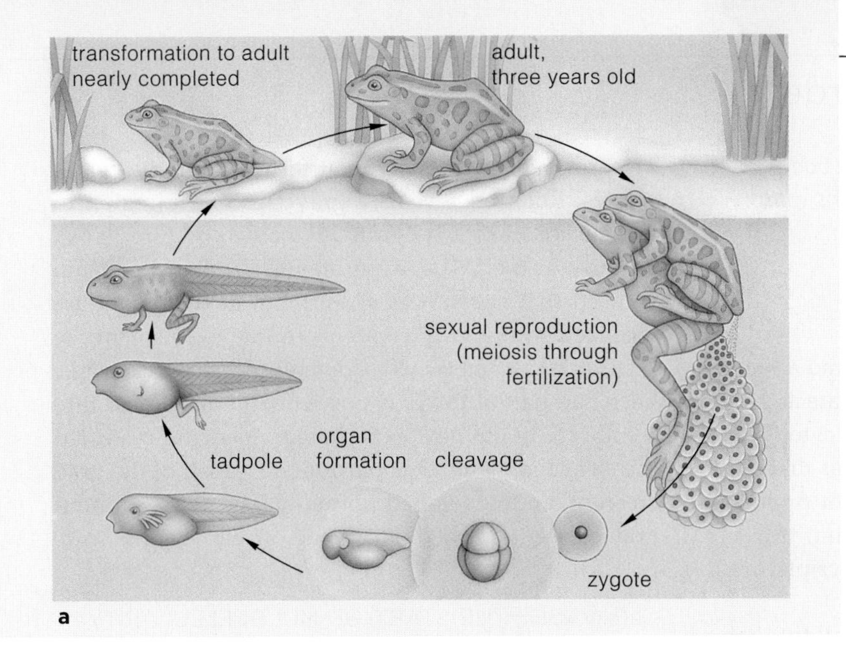

Figure 39.5 **Animated!** Reproduction and development in the life cycle of the leopard frog, *Rana pipiens.*

(**a**) We zoom in on the life cycle as a female releases her eggs into the surrounding water and a male releases sperm over the eggs. A frog zygote forms at fertilization. About one hour after fertilization, a surface feature called the gray crescent appears on this type of embryo. It establishes the frog's head-to-tail axis. Gastrulation will start here.

(**b–e**) Division planes of the first three cuts of cleavage. In this species, cleavage results in a blastula, a ball of cells with a fluid-filled cavity. (**f–i**) The blastula becomes a three-layered gastrula. At the dorsal lip, a fold of ectoderm above the first opening that appears in the blastula, cells migrate inward and start rearranging themselves. A primitive gut cavity opens up. A neural tube, then a notochord and other organs form from the primary tissue layers. (**j–l**) The embryo becomes a tadpole, which metamorphoses into an adult.

Carving up the egg cytoplasm during cleavage:

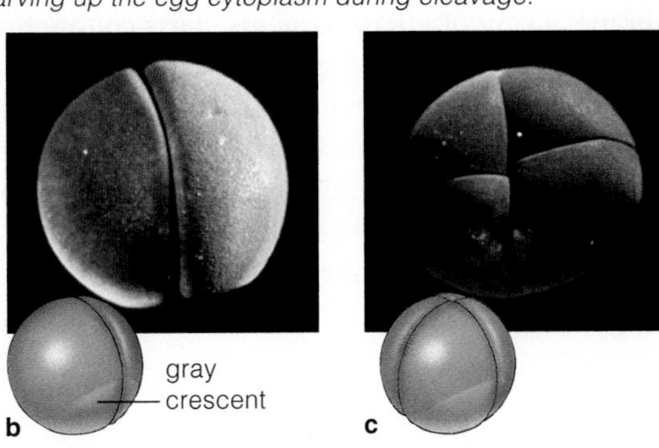

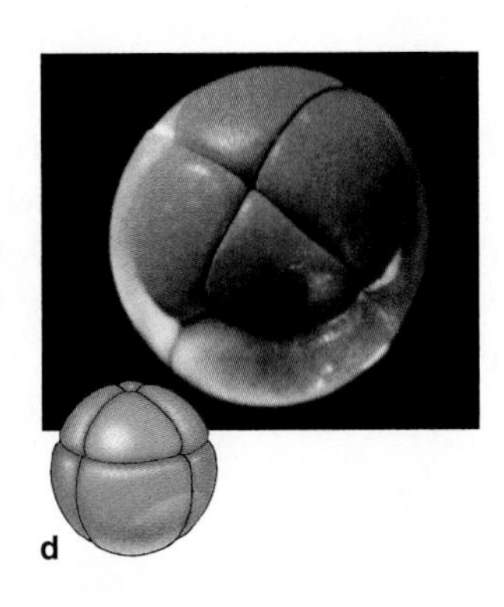

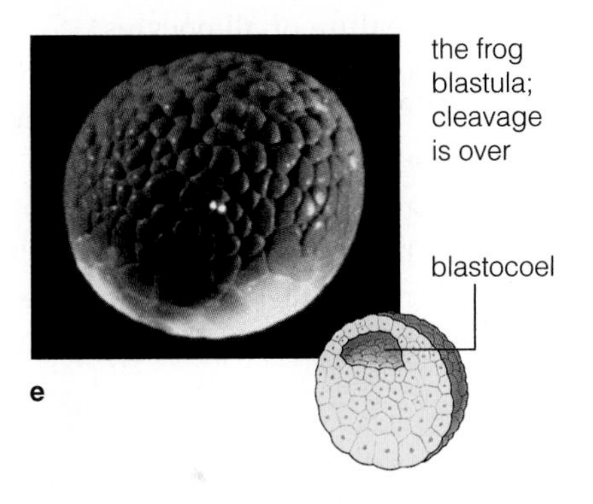

Changes going on during gastrulation and organ formation:

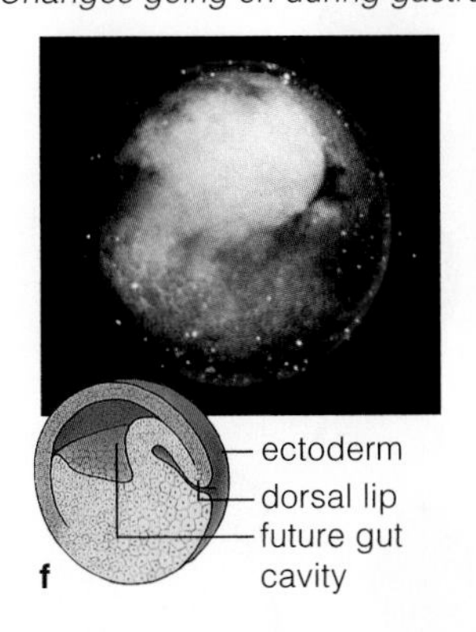

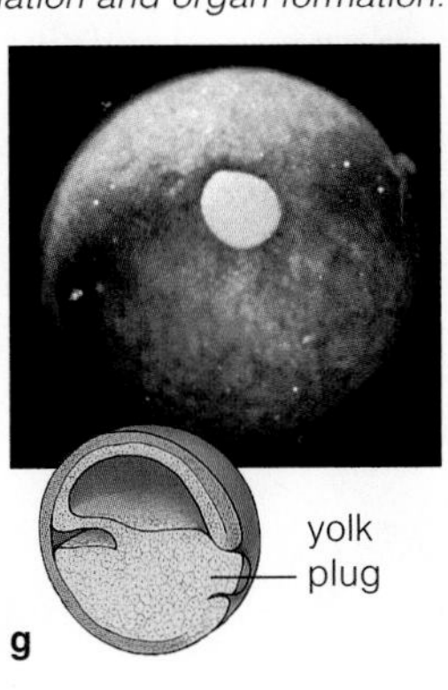

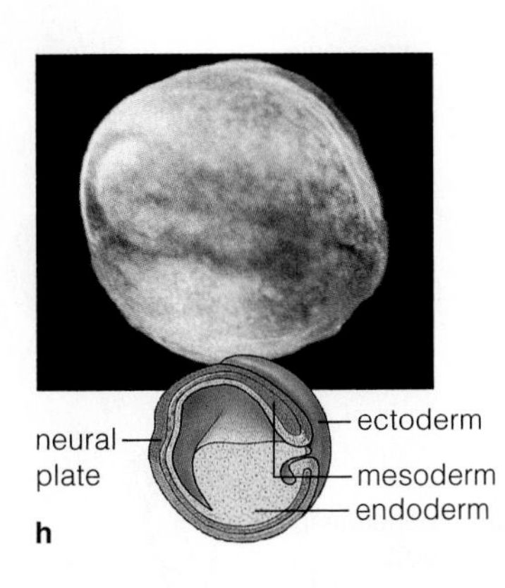

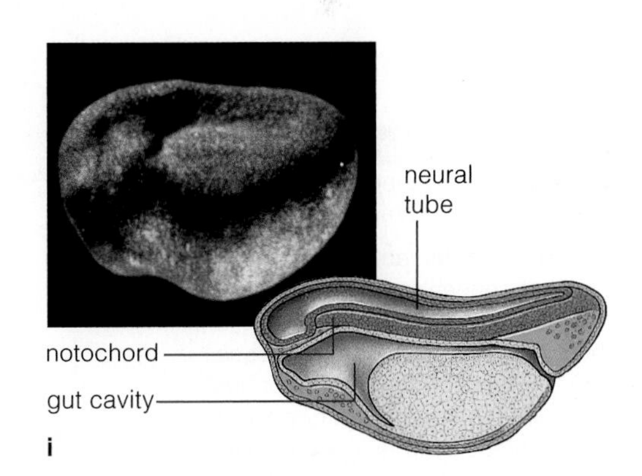

Changes in body form during growth and tissue specialization:

j Tadpole, a swimming larva with segmented muscles and a notochord extending into a tail.

k Metamorphosis to adult form under way. Limbs growing, tail tissues being resorbed.

l Sexually mature, four-legged adult leopard frog.

39.3 Early Marching Orders

How does a single fertilized egg develop into a body with specialized tissues and organs? It all begins with the positioning of material in the egg cytoplasm.

INFORMATION IN THE CYTOPLASM

LINKS TO SECTIONS 8.4, 9.5, 23.1

A sperm, recall, consists of paternal DNA and a bit of equipment that helps it swim to and penetrate an egg. An oocyte, or immature egg, has far more cytoplasm (Section 9.5). The cytoplasm has yolk proteins that will nourish a new embryo, mRNA transcripts for proteins that will be translated in early development, tRNAs and ribosomes to translate the mRNA transcripts, and proteins required to build mitotic spindles.

Certain components are not distributed all through the egg cytoplasm; they are localized in one particular region or another. This cytoplasmic localization is a feature of all oocytes.

Cytoplasmic localization gives rise to the polarity that characterizes all animal eggs. In a yolk-rich egg, the *vegetal* pole has most of the yolk and the *animal* pole has little. In some amphibian eggs, dark pigment molecules accumulate in the cell cortex, a cytoplasmic region just beneath the plasma membrane. Pigment is the most concentrated close to the animal pole. After a sperm penetrates the egg at fertilization, the cortex rotates. Rotation reveals a gray crescent, a region of the cell cortex that is lightly pigmented (Figure 39.6*a*).

Early in the 1900s, experiments by Hans Spemann showed that substances essential to development are localized in the gray crescent. In one experiment, he separated the first two blastomeres formed at cleavage. Each had half of the gray crescent and developed into a embryo. In the next experiment, Spemann modified the cleavage plane. One blastomere got all of the gray crescent, and developed normally. The other, without gray crescent, formed only a ball of cells (Figure 39.6).

CLEAVAGE—THE START OF MULTICELLULARITY

During cleavage, a furrow appears on the cell surface and defines the plane of the cut. Beneath the plasma membrane, a ring of microfilaments starts contracting, and it eventually divides the cell in two (Section 8.4). These divisions are not random. Their pattern dictates what types and proportions of materials a blastomere will get, as well as its size. *Cleavage puts different parts of the egg cytoplasm into different blastomeres.*

Each species has a characteristic cleavage pattern. Remember the branching of the coelomate lineage into

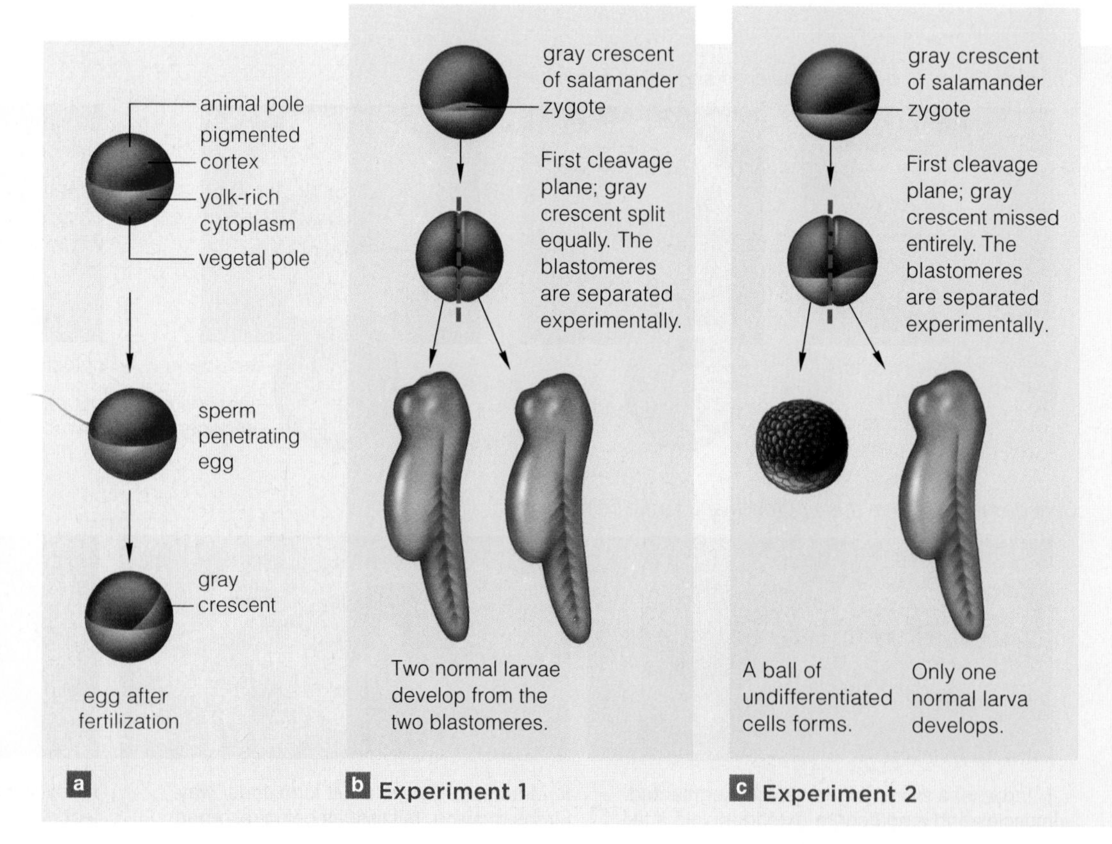

Figure 39.6 **Animated!** Experimental evidence of cytoplasmic localization in an amphibian oocyte.

(**a**) Many amphibian eggs have dark pigment concentrated in cytoplasm near the animal pole. At fertilization, the cytoplasm shifts, and exposes a gray crescent-shaped region just opposite the sperm's entry point. With normal first cleavage, each resulting cell gets half of the gray crescent.

(**b**) In one experiment, the first two cells formed by normal cleavage were physically separated from each other. Each developed into a normal larva.

(**c**) In another experiment, a zygote was manipulated so the first cleavage plane missed the gray crescent. Only one of the descendant cells received gray crescent material, and it alone developed normally. The other gave rise to an undifferentiated ball of cells.

Figure 39.7 Gastrulation in a fruit fly (*Drosophila*). In fruit flies, cleavage is restricted to the outermost region of cytoplasm; the interior is filled with yolk. The series of photographs, all cross-sections, shows sixteen cells (stained gold) migrating inward. The opening the cells move in through will become the fly's mouth. Descendants of the stained cells will form mesoderm. Movements shown in these photos occur during a period of less than 20 minutes.

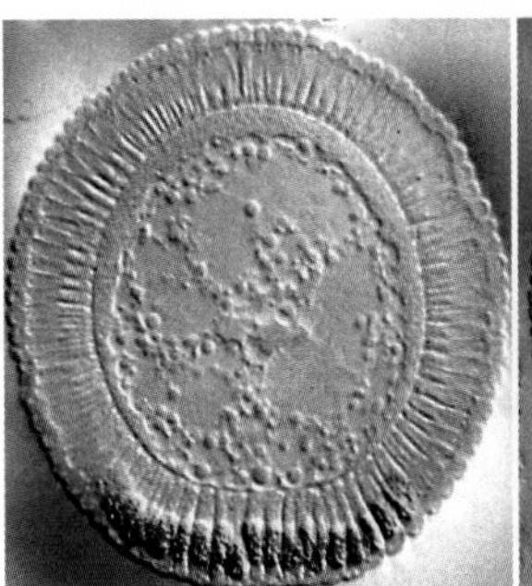

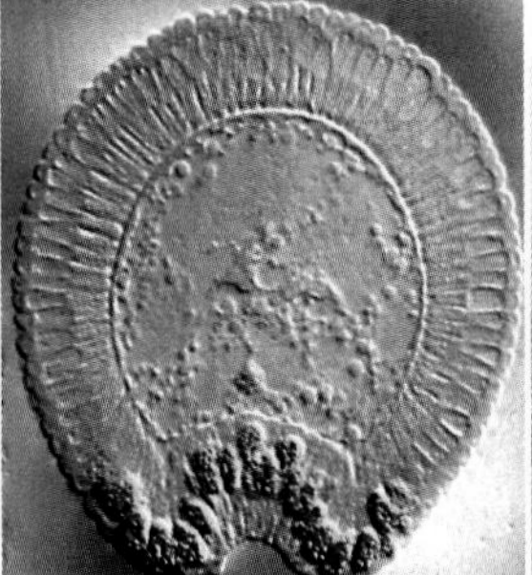

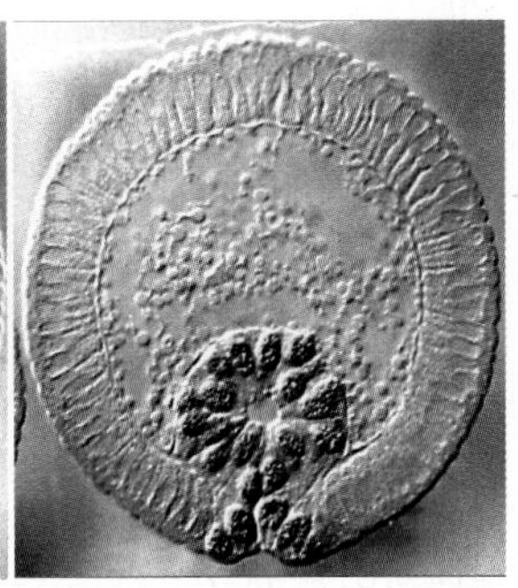

the protostomes and deuterostomes (Section 23.1)? These two groups differ in certain details of cleavage, such as the angle of divisions relative to an egg's polar axis. The amount of yolk also influences the pattern of division. Insects, frogs, fishes, and birds have yolk-rich eggs. In such eggs, a large volume of yolk slows or blocks some of the cuts, as it did for the frog blastula shown in Figure 39.5*d*,*e*. The cuts slice right through the nearly yolkless eggs of sea stars and mammals.

FROM BLASTULA TO GASTRULA

A hundred to thousands of cells may form at cleavage, depending on the species. Starting with gastrulation, cells now migrate about and rearrange themselves. A portion at the embryo's surface moves inward. Figure 39.7 is an example. In most animals, the small ball of cells formed at cleavage develops into a gastrula with three distinct germ layers: ectoderm, mesoderm, and endoderm (Table 23.2).

What initiates gastrulation? Hilde Mangold, one of Spemann's students discovered the answer. She knew that during gastrulation, cells of a salamander blastula move inward through an opening on its surface. Cells in the dorsal (upper) lip of the opening are descended from a zygote's gray crescent. Mangold hypothesized that signals from dorsal lip cells caused gastrulation. She predicted that a transplant of dorsal lip material from one embryo to another would cause gastrulation at the recipient site. She did many transplants (Figure 39.8*a*), and the results supported her prediction. Cells migrated inward at the transplant site, as well as at the usual location (Figure 39.8*b*). A salamander larva with two joined sets of body parts developed (Figure 39.8*c*). Apparently, the transplanted cells had signaled their new neighbors to develop in a novel way.

This experiment also explained the results shown in Figure 39.6*c*. Without any gray crescent cytoplasm, an embryo does not have cells that would normally form the dorsal lip. Without proper signals from these cells, development stops short.

In unfertilized eggs, enzymes, mRNAs, yolk, and other materials are localized in specific parts of the cytoplasm. This cytoplasmic localization helps guide development.

Cleavage divides a fertilized egg into a number of small cells but does not increase its original volume. The cells—blastomeres—inherit different parcels of cytoplasm that make them behave differently, starting at gastrulation.

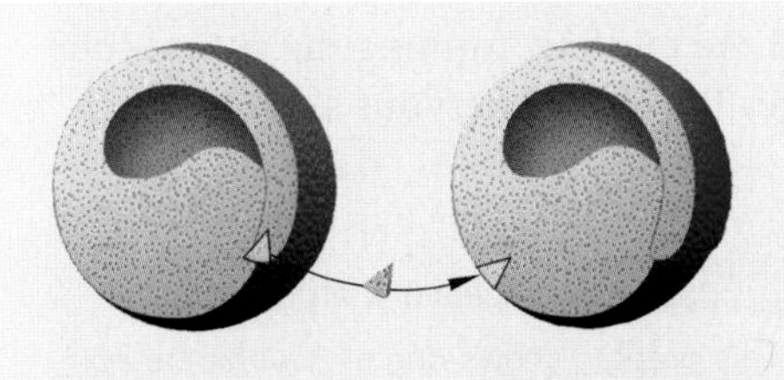

a Dorsal lip excised from donor embryo, grafted to novel site in another embryo

b Graft induces a second site of inward migration

c The embryo develops into a "double" larva, with two heads, two tails, and two bodies joined at the belly

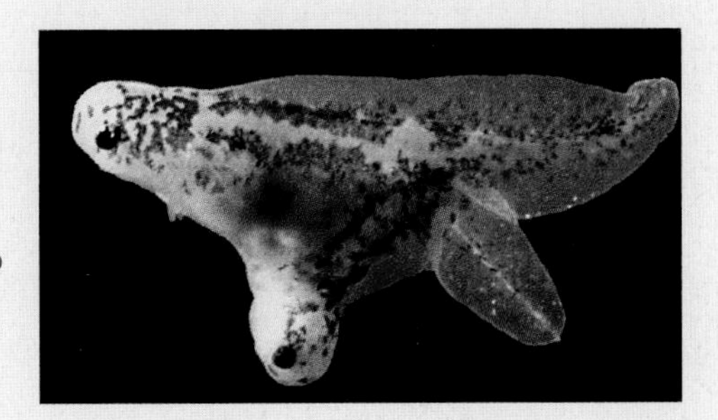

Figure 39.8 Animated! Experimental evidence that signals from dorsal lip cells start amphibian gastrulation. A dorsal lip region of a salamander embryo was transplanted to a different site in another embryo. A second set of body parts started to form.

39.4 Specialized Cells, Tissues, and Organs

Localized molecules in the egg cytoplasm guide early stages of development. Later, tissues and organs form in expected patterns when embryonic cells interact.

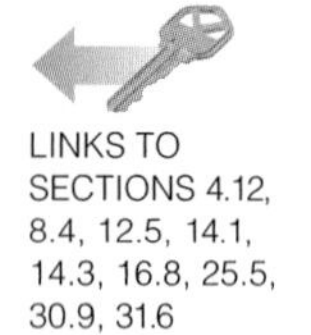

LINKS TO SECTIONS 4.12, 8.4, 12.5, 14.1, 14.3, 16.8, 25.5, 30.9, 31.6

CELL DIFFERENTIATION

All cells in an embryo are descended from the same zygote, so all have the same genes. How then, do the specialized tissue and organs arise? From gastrulation onward, selective gene expression occurs: Different cell lineages express different subsets of genes. That is the start of cell differentiation, the process by which cell lineages become specialized in composition, structure, and function (Section 14.1).

An adult human body has about 200 differentiated cell types. As your eye developed, cells of one lineage turned on genes for crystallin, a transparent protein. These differentiated cells formed the lens of your eye. No other cells in your body make crystallin.

A differentiated cell still retains the entire genome. That is why it is possible to clone an adult animal—to create a genetic copy—from one of its differentiated cells (Section 12.5). As an example, the DNA of Dolly, the first sheep clone, came from a mammary gland cell of a six-year-old female sheep.

CELL COMMUNICATION IN DEVELOPMENT

Intercellular signals can encourage differentiation. For example, certain embryonic cells secrete morphogens, molecular signals that are encoded by master genes. Morphogens diffuse out from their source and form a concentration gradient in the embryo. A morphogen's effects on target cells depends on its concentration. Cells nearest the source of a morphogen will respond differently than distant cells. Other signals operate at close range, as when cells of a salamander gastrula's dorsal lip cause adjacent cells to migrate inward and become mesoderm. This is an example of embryonic induction: Embryonic cells produce signals that alter the behavior of neighboring cells.

MORPHOGENESIS

Long-range and short range signals help bring about morphogenesis, the process by which the tissues and organs form. During morphogenesis, the body takes shape as cells migrate, entire sheets of tissue fold and bend, and specific cells die on cue.

Cells migrate to specific locations. As an example, cell migrations produce the multilayered embryo during gastrulation. Another example: Neurons in the center of the developing brain creep along extensions of glial cells or axons of other neurons until they reach their final position (Figure 39.9*a*). Once in place, they send out extensions of their own.

Sheets of cells expand and fold as cells change in shape. Controlled assembly and disassembly of microtubules and microfilaments cause these changes. For example, a neural tube develops in vertebrates (Figure 39.9*b*). Ectodermal cells at the embryo's midline elongate as microtubules inside them assemble. Then neighboring cells become wedge-shaped as microfilaments at one end constrict, forming a neural groove. Edges of the groove move inward. Eventually flaps of tissue fold over and meet at the midline, forming the neural tube that later becomes the central nervous system.

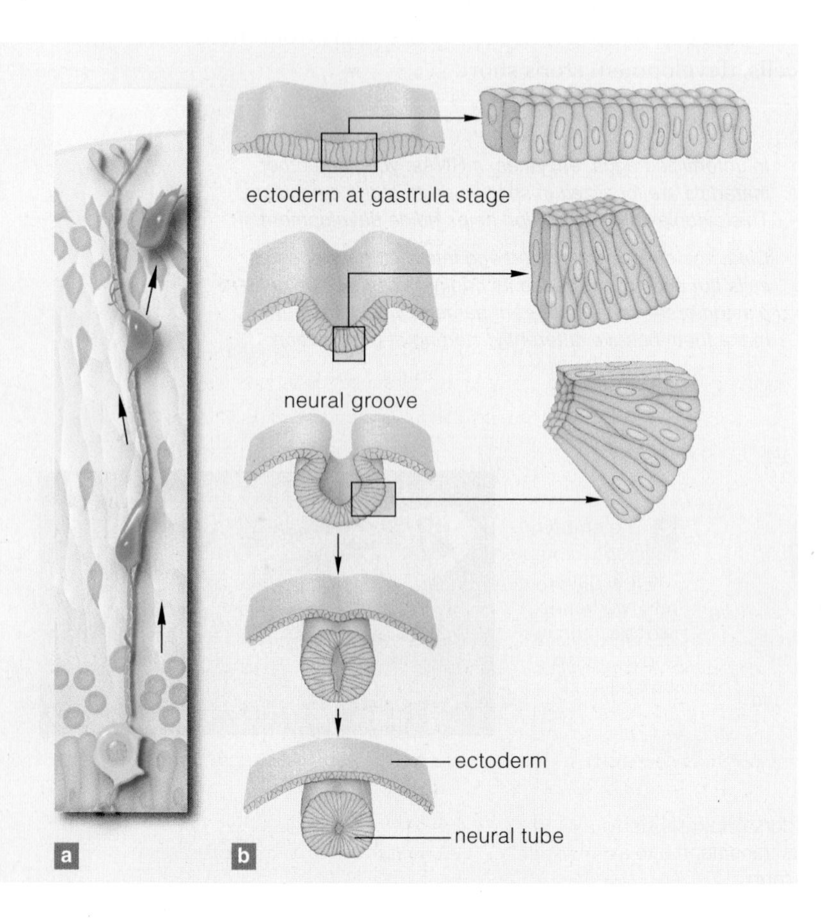

Figure 39.9 **Animated!** Examples of morphogenesis.

(**a**) Cell migration. This graphic shows one embryonic neuron (*orange*) at successive times as it migrates along a glial cell (*yellow*). Its adhesion proteins stick to glial cell proteins.

(**b**) Neural tube formation. By gastrulation's end, ectoderm is a sheet of uniform cells. Microtubules inside its cells lengthen along the axis of the future tube. Some neighboring cells become wedge-shaped as rings of microfilaments constrict inside them. They make part of the ectodermal sheet fold back on itself, then they detach from it as a separate tube.

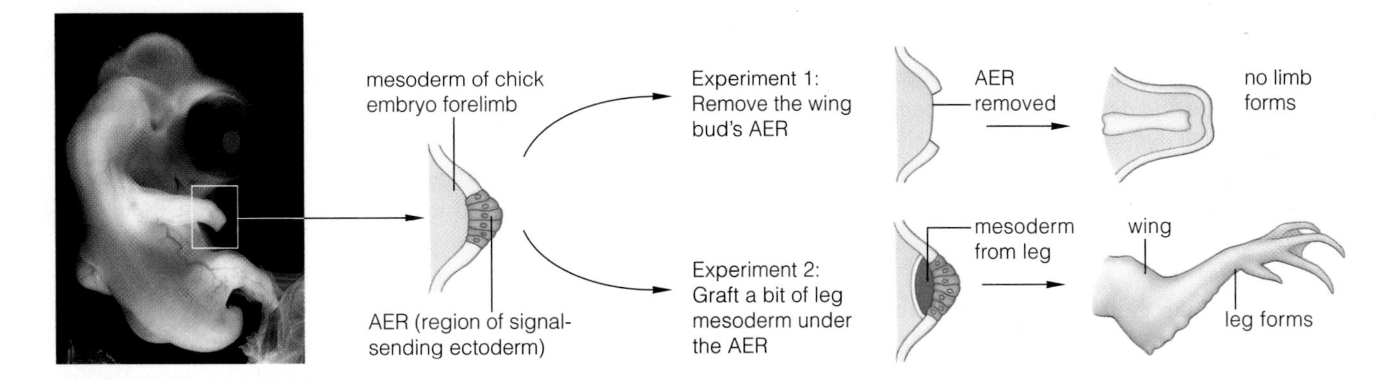

Figure 39.10 Experiments demonstrating interactions between AER (ectoderm) and mesoderm in chick wing development. AER at the tip of a limb bud tells mesoderm under it to form a limb. Whether that limb becomes a wing or a leg depends on positional signals that the mesoderm received earlier.

Programmed cell death helps sculpt body parts. By this process, apoptosis, signals from certain cells activate the tools of self-destruction in target cells. Recall how a human hand develops from a paddle-shaped part? Programmed cell death separates digits (Section 25.5). The same process shapes other body parts.

PATTERN FORMATION

Think again about that human hand. Why does a hand form at the end of an arm? Why not a foot? The answer lies in pattern formation, the process by which certain body parts form in a specific place. As one example, tissue called AER (apical ectodermal ridge) develops at the tips of limb buds in all tetrapods. AER induces mesoderm under it to form a limb. If you remove the AER from the wing bud of a chick, wing development comes to a halt (Figure 39.10*a*).

AER stimulates mesoderm to develop, but earlier positional cues have previously determined what that mesoderm will become. Implant a piece of mesoderm from a chick hindlimb under the wing AER, and a leg forms (Figure 39.10*b*). Early events set this mesoderm down the developmental road toward leg formation, and there is no turning back.

EVOLUTION AND DEVELOPMENT

Through studies of many animals, researchers have come up with a general model for development. The key point is this: *where and when particular genes are expressed determines how an animal body develops.*

First, molecules localized in different areas of an unfertilized egg induce localized expression of master genes in the zygote. Products of these genes become distributed in gradients along the front-to-back and top-to-bottom axis of the developing embryo.

Second, depending on where they fall within these concentration gradients, cells in the embryo activate or suppress other master genes. The products of these genes form gradients, and so on.

Third, this positional information affects expression of homeotic genes, genes that regulate development of specific body parts, as introduced in Section 14.3. All animals have similar homeotic genes. For example, a mouse's *eyeless* gene guides development of its eyes. Introduce this gene into a fruit fly, and eyes will form in the tissue where it is expressed.

This model helps explain similarities among animal body plans. We know that body plans are influenced by *physical* constraints (such as the surface-to-volume ratio) and *architectural* constraints (as imposed by body axes). There are also *phyletic* constraints on body plans. These constraints are imposed by interactions among genes that regulate development in a lineage. Eyes do not form at the tips of vertebrate limbs, even though this might be useful. Once master genes had evolved and started interacting, a major change in any one of them probably would have been lethal. Mutations led to a variety of forms among animal lineages. But they did so by modifying existing developmental pathways, rather than by blazing entirely new genetic trails.

All cells in an embryo have the same genes, but they express different subsets of this genome. Selective gene expression is the basis of cell differentiation. It results in cell lineages with characteristic structures and functions.

Cytoplasmic localization sets the stage for cell signaling. The signals activate sets of master genes, the products of which cause embryonic cells to form tissues and organs at certain locations, and in a certain order.

Physical, architectural, and evolutionary constraints have limited drastic changes in the basic pattern of development.

39.5 Reproductive System of Human Males

Turn now to applying some of the principles of animal reproduction and development to humans, starting with the male reproductive system.

WHEN MALE GONADS FORM AND BECOME ACTIVE

LINKS TO SECTIONS 11.1, 32.8

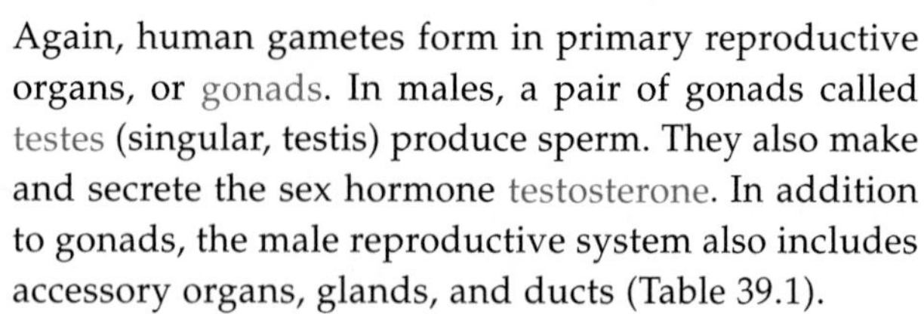

Again, human gametes form in primary reproductive organs, or gonads. In males, a pair of gonads called testes (singular, testis) produce sperm. They also make and secrete the sex hormone testosterone. In addition to gonads, the male reproductive system also includes accessory organs, glands, and ducts (Table 39.1).

Earlier, Figure 11.2 showed how two testes form on the wall of an XY embryo's abdominal cavity. Before birth, testes descend into the scrotum, a pouch of loose skin suspended below the pelvic girdle (Figure 39.11). Inside the pouch, a smooth muscle encloses the testes. Reflex contraction of this muscle pulls the testes close to the body when the temperature declines. Moderate temperatures are better for sperm formation.

A male enters puberty—the stage of development when reproductive organs mature—sometime between the ages of 11 and 16 years. Testes enlarge and sperm production begins. Secretion of testosterone increases. This leads to development of secondary sexual traits: thickened vocal cords that deepen the voice; increased growth of hair on the face, chest, armpits, and pubic region; and an altered distribution of fat and muscle. Such traits do not play a direct role in reproduction.

STRUCTURE AND FUNCTION OF THE MALE REPRODUCTIVE SYSTEM

In mammals, immature sperm travel from the testes through several ducts (Figure 39.12). First they enter an epididymis, one of a pair of coiled ducts. Secretions from glands in the epididymis wall trigger events that put finishing touches on sperm cells. The last part of the epididymis stores mature sperm. About 100 million sperm mature each day during a male's reproductive years. Unused ones are resorbed or passed in urine.

In a sexually aroused male, smooth muscle in the reproductive tract wall contracts and propels mature sperm into the vasa deferentia (singular, vas deferens), a pair of thick-walled ducts. Continued contractions propel sperm through paired ejaculatory ducts, into the urethra. The urethra extends through the penis, a male sex organ. Sperm leave the body by way of an opening at the tip of the penis.

Sperm traveling to the urethra mix with glandular secretions and form semen, a thickened fluid that gets expelled from the penis during sexual activity. Most of the semen's volume is a fructose-rich fluid secreted by seminal vesicles located behind the bladder. Sperm use fructose as an energy source. Seminal vesicles and the prostate gland also secrete prostaglandins into the semen (Section 32.1). These local signaling molecules can increase the growth rate of cervical and uterine cancers. The partners of women at heightened risk for these cancers should use a condom.

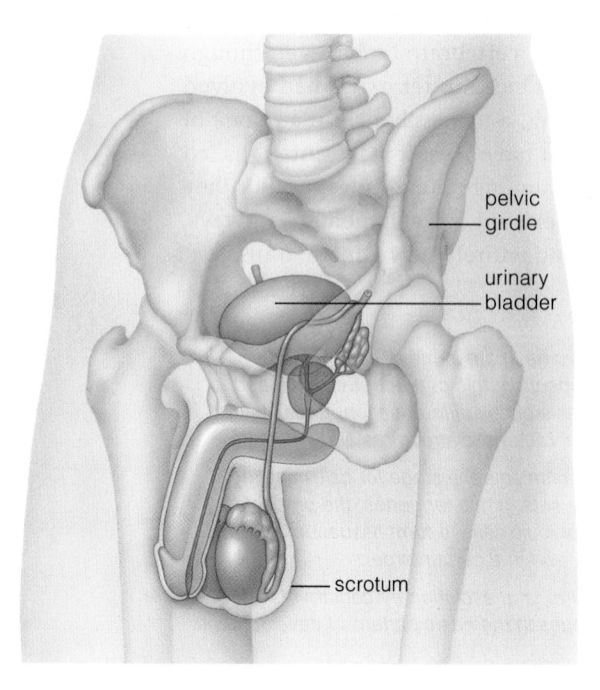

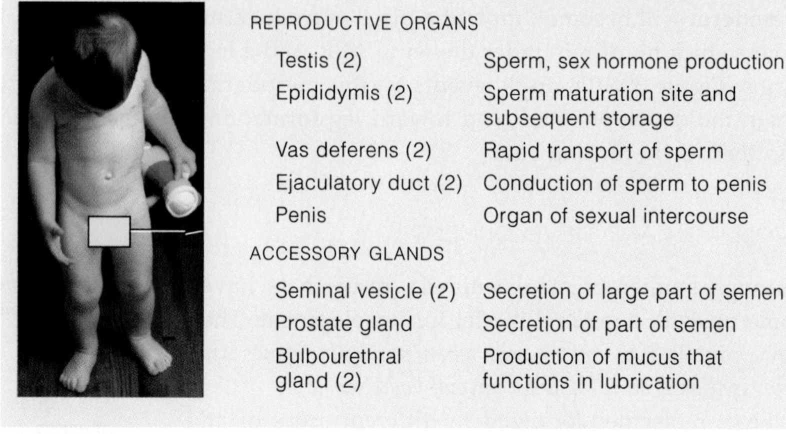

Table 39.1 Human Male Reproductive System

REPRODUCTIVE ORGANS	
Testis (2)	Sperm, sex hormone production
Epididymis (2)	Sperm maturation site and subsequent storage
Vas deferens (2)	Rapid transport of sperm
Ejaculatory duct (2)	Conduction of sperm to penis
Penis	Organ of sexual intercourse
ACCESSORY GLANDS	
Seminal vesicle (2)	Secretion of large part of semen
Prostate gland	Secretion of part of semen
Bulbourethral gland (2)	Production of mucus that functions in lubrication

Figure 39.11 Position of the human male reproductive system relative to the pelvic girdle and urinary bladder.

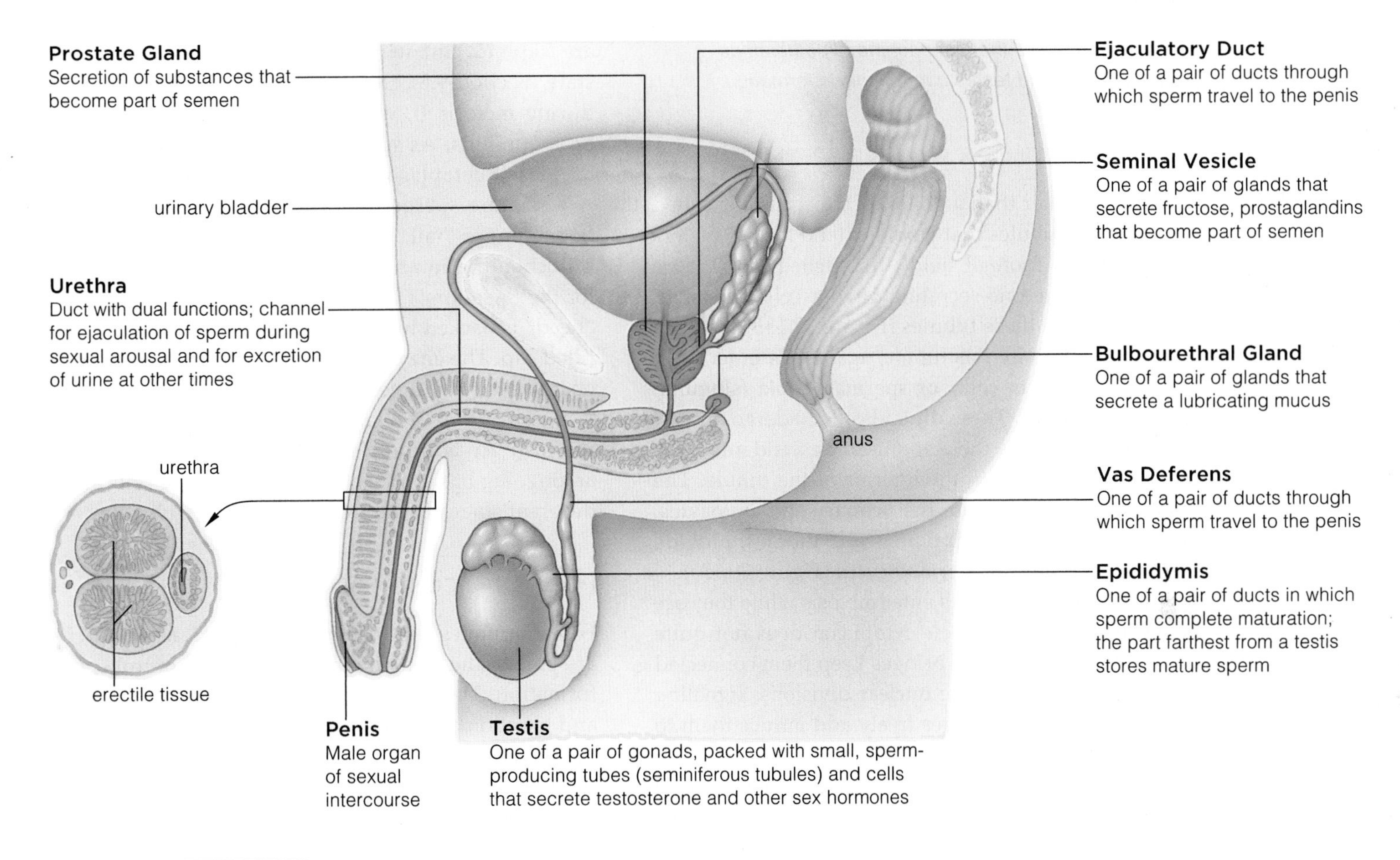

Figure 39.12 Animated! Components of the human male reproductive system and their functions.

Other prostate gland secretions can help buffer the acidic conditions in the female reproductive tract. The pH of vaginal fluid is about 3.5–4.0, but sperm swim more efficiently when the pH is 6.0. Finally, a pair of bulbourethral glands secrete mucus-rich fluid into the urethra during moments of sexual arousal.

CANCERS OF THE PROSTATE AND TESTES

Prostate cancer is a leading cause of death for men, surpassed only by lung cancers. In the United States, more than 200,000 males are diagnosed annually, and about 35,000 die. It is often said that if you examine the prostate gland of all men who died of something else, you will find many who had prostate cancer and never knew it. This saying trivializes the risk. Many prostate cancers grow slowly, but some grow fast and spread easily to other parts of the body. Risk factors for prostate cancer include advancing age, a diet rich in animal fats, smoking, and a couch-potato life-style. There is a genetic factor. Having a father or brother with prostate cancer doubles the risk.

Doctors can detect prostate cancer by blood tests for increases in prostate-specific antigen (PSA) and by physical examination. Surgery and radiation therapy can cure cancers that are detected early.

Testicular cancer is relatively rare, with 7,000 cases a year in the United States. Among men between ages fifteen and thirty-four, it is the most common cancer. Once a month, after a warm shower or bath, a male should examine his testes for lumps, enlargement, or hardening. Treating testicular cancer has a high rate of success when the cancer is detected early, before it has spread to other parts of the body.

A pair of testes, the primary reproductive organs in human males, produce sperm. They also make and secrete the sex hormone testosterone.

An epididymis leads from each testis. Sperm mature and are stored in this duct. Semen is a mix of mature sperm and noncellular secretions from accessory glands. During sexual arousal, semen is propelled through a series of ducts. It is ejaculated from the last duct, the urethra.

39.6 Sperm Formation

Signaling pathways that connect the hypothalamus, pituitary gland, and testes control sperm formation.

LINKS TO SECTIONS 9.2, 9.5, 32.3, 32.8

FROM GERM CELLS TO MATURE SPERM

Although smaller than a golf ball, a testis holds coiled seminiferous tubules that would extend for 125 meters (longer than a football field) if stretched out (Figure 39.13*a*). Testosterone-secreting Leydig cells cluster in spaces between these tubules (Figure 39.13*b*).

Inside a seminiferous tubule, against its inner wall, are the male germ cells, or spermatogonia (singular, spermatogonium). These diploid cells undergo mitosis again and again, so their youngest descendants force older ones farther into the interior of the tubule. The displaced, older cells are the primary spermatocytes. Sertoli cells, another cell type inside tubules, provide metabolic support to spermatocytes (Figure 39.13*c*).

Primary spermatocytes enter meiosis while they are being displaced—but their cytoplasm does not quite divide. Thin cytoplasmic bridges keep them connected to one another during the nuclear divisions. Signaling molecules cross the bridges freely and induce them to mature at the same rate.

By the end of meiosis I, each cell has given rise to two secondary spermatocytes (Figure 39.13*c*). The two are haploid, but with chromosomes in a duplicated state (Section 9.5). Sister chromatids will move apart during meiosis II, which produces immature sperm, or spermatids. As spermatids mature into sperm, the bridges of cytoplasm between them break down.

A mature sperm is a haploid, flagellated cell. The flagellum, or "tail," allows it to swim toward an egg. Mitochondria in an adjacent midpiece supply energy for the flagellum's motions (Figure 39.13*d*). A sperm's "head" is packed with DNA and tipped by an enzyme-filled cap. The enzymes can help a sperm penetrate an oocyte by partly digesting away its outer layer.

Sperm formation takes about 100 days, from start to finish. An adult male normally produces sperm on an ongoing basis, so that many millions of cells are in different stages of development on any given day.

HORMONAL CONTROLS

Four hormones—testosterone, LH, FSH, and GnRH—are part of the signaling pathways that control sperm formation. The testosterone also governs the structure and function of the male reproductive tract. It induces the development of male secondary sexual traits and promotes sexual and aggressive behavior. Leydig cells inside the testes secrete it.

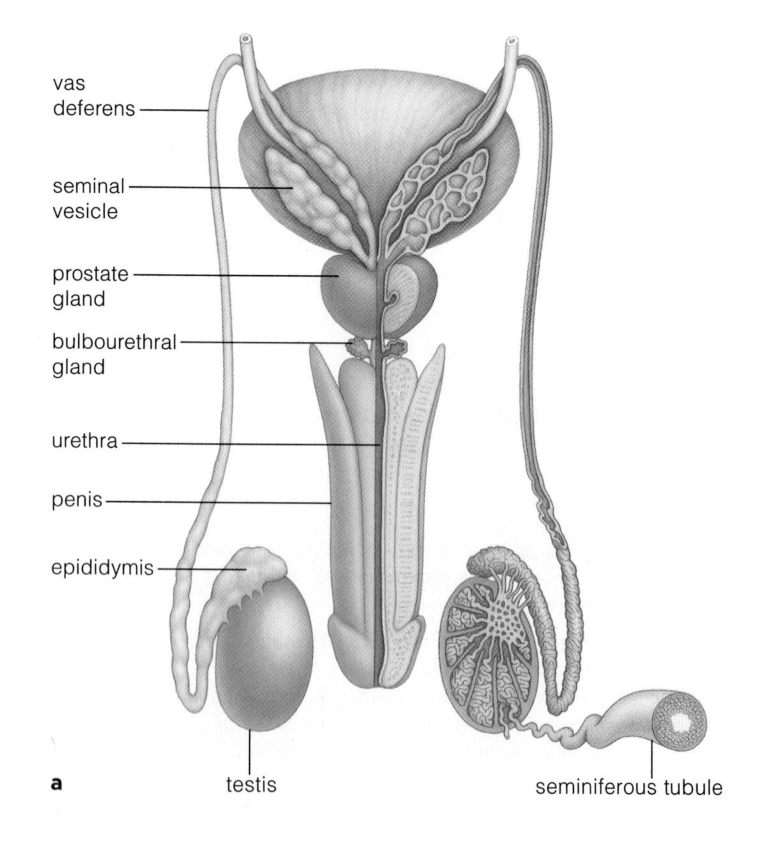

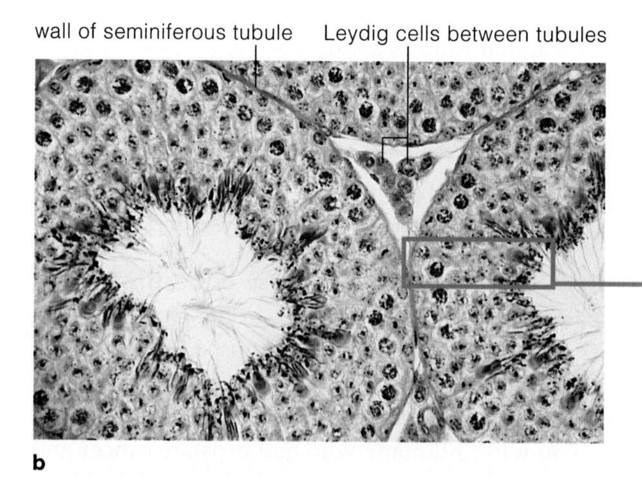

Figure 39.13 **Animated!** (**a**) Male reproductive tract, posterior view.

Sperm formation. (**b**) Light micrograph of cells in three adjacent seminiferous tubules, cross-section. Leydig cells occupy spaces between tubules. (**c**) How sperm form, starting with a diploid germ cell in a seminiferous tubule. (**d**) Structure of a mature sperm, a male gamete.

As you read in Section 32.3, the anterior lobe of the pituitary gland produces *L*uteinizing *H*ormone (LH) and FSH (*F*ollicle-*S*timulating *H*ormone). The names refer to the hormones' effects in ovaries, which were identified before it became clear that males, too, make LH and FSH.

The hypothalamus secretes GnRH (gonadotropin-releasing hormone). GnRH causes the anterior lobe of the pituitary to secrete LH and FSH (Figure 39.14). LH stimulates Leydig cells inside the testes to secrete testosterone, which promotes sperm formation. Before puberty, FSH stimulates Sertoli cells to divide. It also supports sperm production in adults.

Figure 39.14 also shows how feedback loops to the hypothalamus guide testosterone secretion and sperm formation. High testosterone in blood slows secretion of GnRH. In addition, a high sperm count encourages Sertoli cells to release inhibin, a hormone that calls for a slowdown in GnRH and FSH secretion.

In seminiferous tubules of the testes, meiosis in male germ cells produces sperm—haploid male gametes.

Sperm formation requires secretions of LH, FSH, and testosterone. Negative feedback loops from testes to the hypothalamus and pituitary control the secretions.

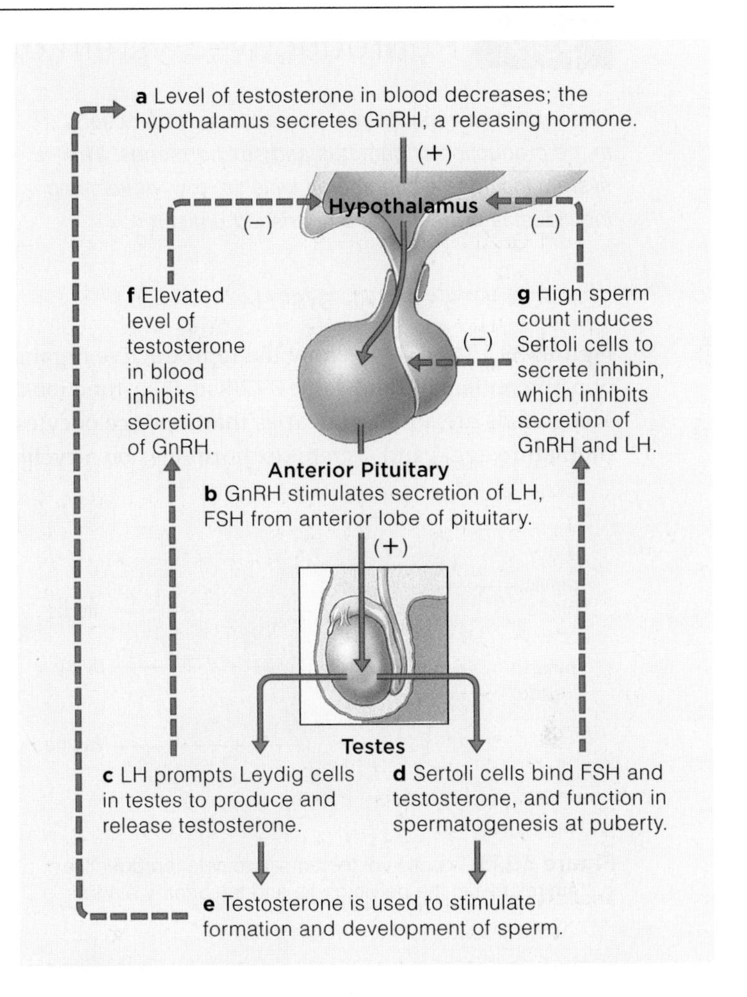

Figure 39.14 Signaling pathways in sperm formation. Negative feedback loops control hormonal secretions of the hypothalamus, the anterior lobe of the pituitary gland, and the testes.

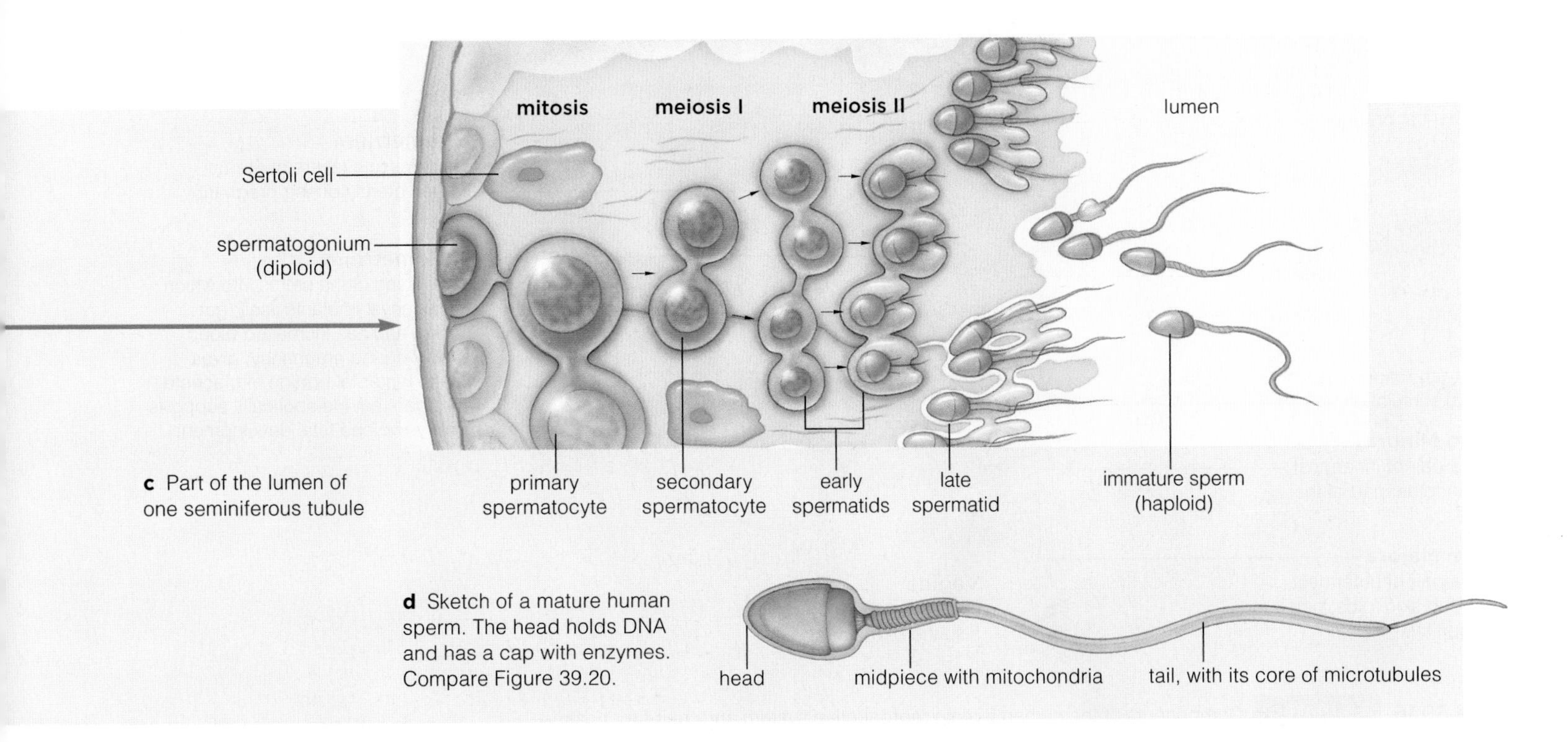

39.7 Reproductive System of Human Females

The reproductive system of human females functions in the production of gametes and sex hormones. The system includes a chamber in which a new, developing individual is protected and nourished until birth.

COMPONENTS OF THE SYSTEM

LINKS TO SECTIONS 11.1, 32.3, 32.8

Figures 39.15 and 39.16 show the reproductive organs of a human female, and Table 39.2 lists their functions. The gonads are a pair of ovaries that produce oocytes (immature eggs) and secrete sex hormones on a cyclic basis. One of the pair of oviducts, or Fallopian tubes, sweeps up an oocyte when it is released.

Most often, an oocyte gets fertilized in the oviduct. Then it tumbles into the uterus, a hollow, pear-shaped organ above the urinary bladder. A blastocyst forms and development is completed in the uterus. A thick layer of smooth muscle, the myometrium, makes up most of the uterine wall. Endometrium lines the inner wall. The lining is composed of glandular epithelium, connective tissues, and blood vessels. The narrowed-down, lowest portion of the uterus is the cervix, which opens into the vagina.

The vagina, a muscular, mucosa-lined tube, extends from the cervix to the body's surface. It is lubricated by its own mucous secretions, and it functions as the female organ of intercourse. The vagina also functions as the birth canal in childbirth. Two pairs of skin folds enclose the surface openings of both the vagina and urethra. Adipose tissue fills paired outer folds (the labia majora). Thin inner folds (the labia minora) have a rich blood supply and swell during sexual arousal.

The tip of the clitoris, a highly sensitive sex organ, is positioned between the two inner folds, just in front of the urethra. The clitoris and penis develop from the

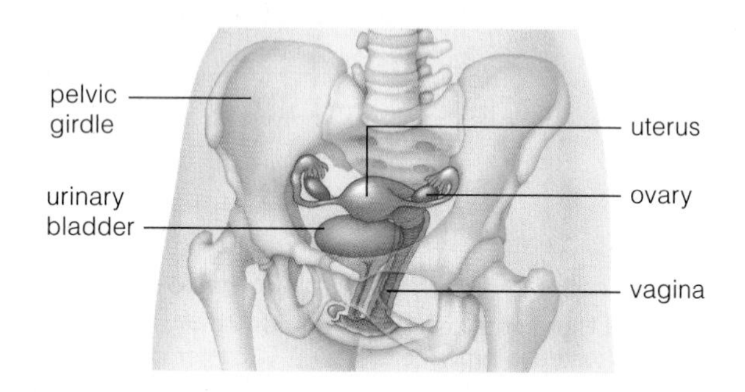

Figure 39.15 Location of the human female reproductive system relative to the pelvic girdle and the urinary bladder.

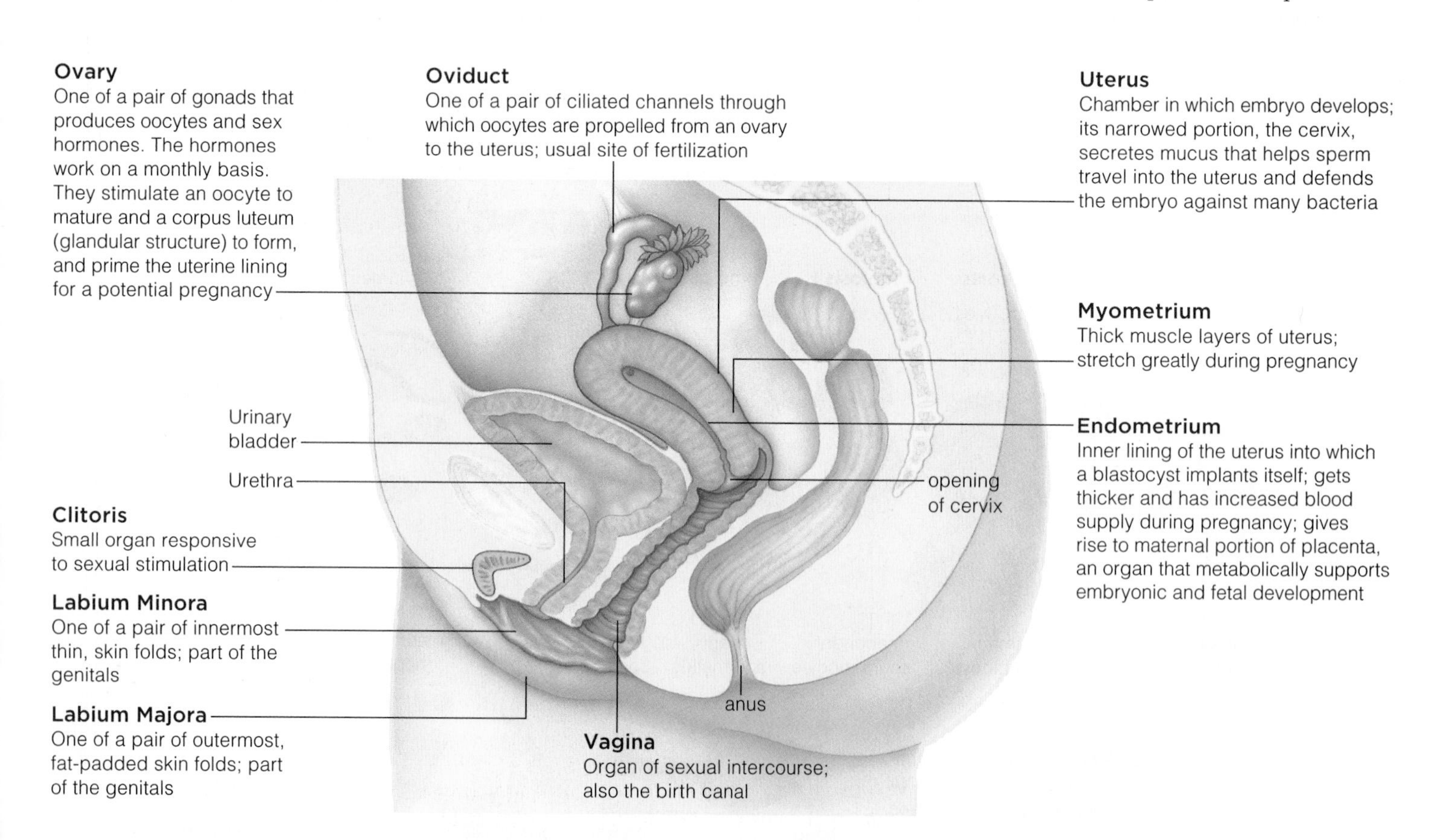

Figure 39.16 Animated! Components of the human female reproductive system and their functions.

Table 39.2 Human Female Reproductive Organs

Ovaries (two)	Oocyte production and maturation, sex hormone production
Oviducts (two)	Tubes between the ovaries and the uterus; fertilization normally takes place here
Uterus	Chamber in which new individual develops
Cervix	Entrance to the uterus; secretes mucus that enhances sperm travel into uterus and reduces embryo's risk of infection
Vagina	Organ of sexual intercourse; birth canal

same embryonic tissue. Both have an abundance of highly sensitive touch receptors, and both swell with blood and become erect during sexual arousal.

OVERVIEW OF THE MENSTRUAL CYCLE

Females of most mammalian species follow an estrous cycle, meaning they are fertile and "in heat" (sexually receptive to males) only at certain times. Females of humans and some other primates follow a menstrual cycle. Their fertile periods are cyclic, intermittent, and not tied to sexual receptivity. In other words, they can get pregnant only during certain times in their cycle but may be receptive to sex at any time.

Females typically start to menstruate at about age twelve. Section 39.9 will explain the menstrual cycle in detail, but here is an overview: Every twenty-eight days or so, an oocyte matures in an ovary, and then it is released. During a two-week interval, the uterus gets primed for pregnancy. When the oocyte does not get fertilized, blood and bits of endometrium flow out through the vagina. Menstruation indicates "there is no embryo," and it marks the start of a new cycle.

A woman goes through these monthly cycles until she reaches her late forties or early fifties. Then sex hormone secretions start to dwindle. The decline in secretions correlates with the onset of menopause, the twilight of a female's fertility.

Ovaries, the primary reproductive organs of human females, produce immature eggs and sex hormones. Endometrium lines the uterus, a chamber in which embryos develop. Other organs function in sexual arousal and intercourse.

The menstrual cycle is the cyclic growth and release of oocytes from the ovary, and the breakdown and rebuilding of the endometrium. Menstrual cycles start at puberty and end with menopause.

39.8 Female Troubles

FOCUS ON HEALTH

Hormonal changes cause premenstrual symptoms, menstrual pain, and hot flashes. Nearly all women experience one or more at some point in their life.

LINK TO SECTION 38.4

PMS Many women regularly experience discomfort a week or two before they menstruate. Some hormones released during the cycle cause milk ducts to enlarge, so breasts swell and become tender. Other tissues may swell also, because premenstrual changes influence aldosterone secretion. This hormone stimulates the reabsorption of sodium and, indirectly, water (Section 38.4). Cycle-induced changes also cause depression, irritability, or anxiety. Headaches and sleeping problems are common.

The regular recurrence of these symptoms is known as premenstrual syndrome (PMS). A balanced diet and regular exercise make PMS less likely and less severe. Taking oral contraceptives minimizes hormone swings and therefore PMS. In some cases, drugs that suppress the secretion of sex hormones entirely can help.

Menstrual Pain Prostaglandins secreted during menstruation stimulate contractions of smooth muscle in the uterine wall. Many women do not even notice the contractions, but others experience a dull ache or sharp pains. Women who secrete high levels of prostaglandins are more likely to feel uncomfortable.

Endometriosis, the growth of endometrial tissue in the wrong places in the pelvis, also can cause painful menstruation. The disorder affects about 10 percent of all women. Hormones target the mislocated tissue as well as the uterine lining. Repeated bleeding and healing cause pain and scarring. Hormone suppression methods help, but only surgery can provide a cure.

Fibroids, or benign uterine tumors, can cause long, painful menstrual periods. More than one-third of women over age thirty have fibroids, but most show no symptoms. Large fibroids that cause excessive bleeding can be treated surgically.

Hot Flashes, Night Sweats Three-fourths of the women entering menopause have hot flashes. They get abruptly and uncomfortably hot, flushed, and sweaty as blood surges to their skin. When episodes occur at night, they disrupt sleep. These symptoms can be prevented by hormone replacement therapy, but the therapy raises the risk of breast cancer and stroke, especially if used for more than a few years. Exercising, avoiding alcohol, and eating soy-based products reduces symptoms.

39.9 The Menstrual Cycle

Even before a human female is born, germ cells in her ovaries enter meiosis, but the nuclear division process in an oocyte stalls. It resumes only at fertilization.

PHASES OF THE CYCLE

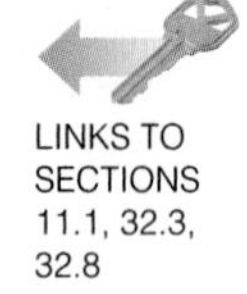

LINKS TO SECTIONS 11.1, 32.3, 32.8

At birth, a girl has about 2 million primary oocytes, immature eggs that have stopped short in prophase I of meiosis. Beginning with her first menstrual cycle, meiosis will resume in one primary oocyte at a time.

Consider a 28-day cycle. In the *follicular* phase, the menstrual flow lasts 1 to 5 days as the uterine lining breaks down. Over the next 6 to 13 days, the uterine lining is slowly rebuilt as an oocyte begins to mature. During the next phase, ovulation, one ovary releases an oocyte on around day 14. In the final *luteal* phase, a glandular corpus luteum develops and its secretions cause the endometrium to thicken in preparation for pregnancy. This last phase continues from day 15 to 28. All three phases are governed by feedback loops from the ovaries to the hypothalamus and pituitary gland.

The signals flowing in the loops are GnRH from the hypothalamus, FSH and LH from the anterior lobe of the pituitary, and estrogens and progesterone from the ovaries. Estrogens control how the female reproductive organs develop in embryos. Also, starting at puberty, they control female secondary sexual traits. Together with progesterone, they induce oocytes to mature and prime the uterus for pregnancy.

CHANGES IN THE OVARY AND UTERUS

Figure 39.17 depicts one follicle near the surface of an ovary. The follicle consists of one primary oocyte and cells that surround and nourish it. As the menstrual cycle begins, the hypothalamus secretes GnRH, which

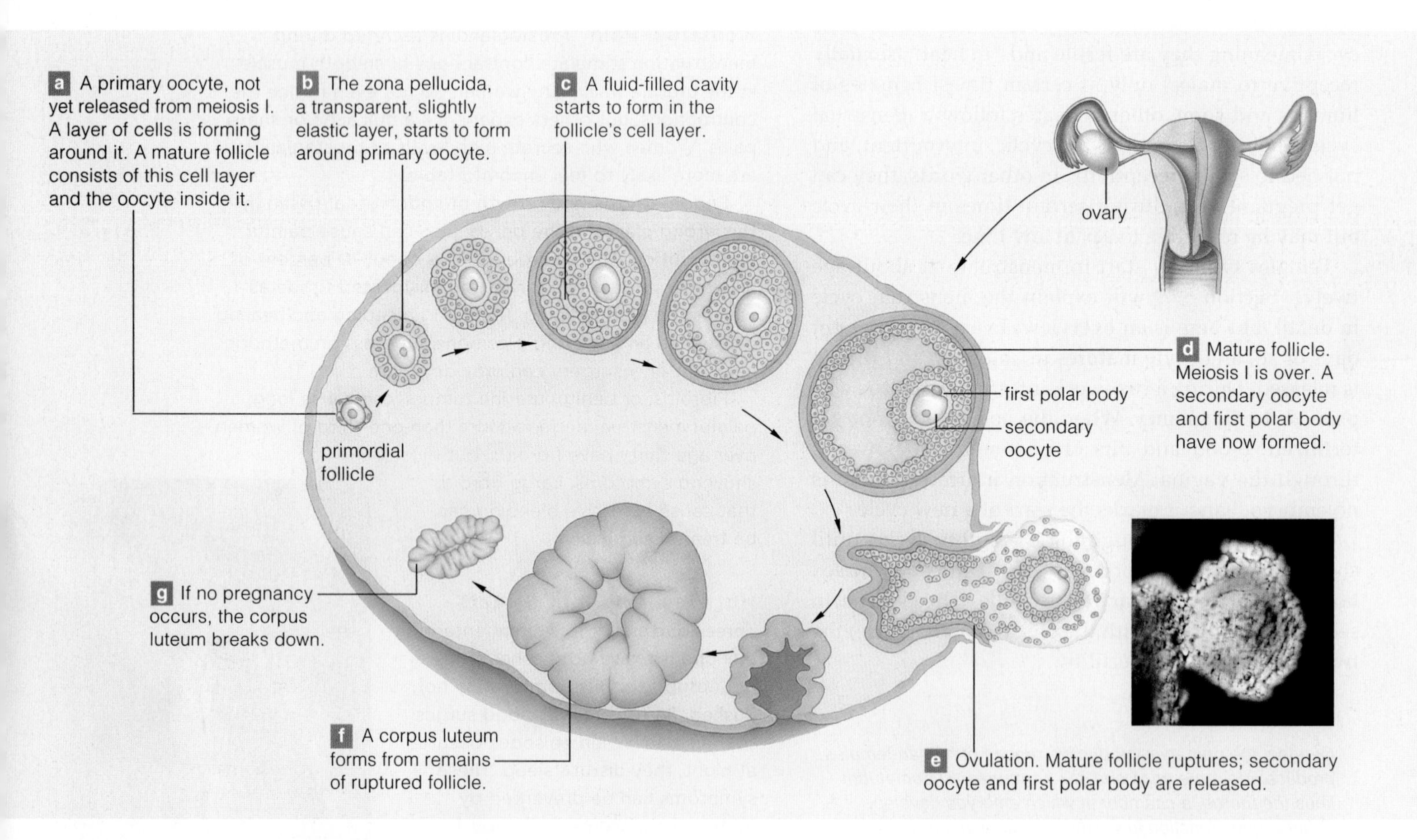

Figure 39.17 **Animated!** Cyclic events in a human ovary, cross-section. The follicle does not "move around" as in this diagram, which simply shows the *sequence* of events. All of these structures form in the same place during one menstrual cycle. In the cycle's first phase, a follicle grows and matures. At ovulation, the second phase, the mature follicle ruptures and releases a secondary oocyte. In the third phase, a corpus luteum forms from the follicle's remnants.

makes some cells in the anterior pituitary step up the secretion of FSH and LH. The rising blood levels of FSH and LH induce the oocyte and cells around it to grow. Proteins build up in a layer, the zona pellucida, between the oocyte and those cells (Figure 39.17*b*).

At the same time, FSH and LH cause the maturing follicle cells to secrete estrogens, so the blood level of this hormone starts to rise (Figure 39.18*a*).

Eight to ten hours before its release from an ovary, the oocyte completes meiosis I. Its cytoplasm divides in two. One of the resulting haploid cells, a secondary oocyte, gets most of the cytoplasm. Its chromosomes are still in the duplicated state. The other haploid cell is the first of three polar bodies (Figure 39.17*d*).

Halfway through the cycle, the pituitary detects the rising level of estrogens in blood. It responds with an outpouring of LH (Figure 39.18*a,c*). Because of that surge, the follicle swells, and enzymes are produced that digest the swollen wall. The wall ruptures, which releases fluid and the secondary oocyte (Figures 39.17*e* and 39.18*b*). *That brief midcycle surge of LH has triggered ovulation, the release of a secondary oocyte from the ovary.*

Estrogens released early in the cycle also stimulate growth of the endometrium and its glands, which sets the stage for pregnancy. Before the midcycle surge of LH, the follicle cells were busy secreting progesterone and estrogens. Blood vessels grew fast in the thickened endometrium (Figure 39.18*d*).

The midcycle surge also causes a corpus luteum to form from the remains of the ruptured follicle (Figure 39.17*f*). Progesterone and estrogens from this structure cause the endometrium to continue to thicken. When pregnancy does not occur, the corpus luteum lasts no more than 12 days or so. During the last days of the cycle, it breaks down (Figure 39.17*g*).

Without the corpus luteum, levels of progesterone and estrogens plummet, and the endometrium starts to break down. Blood and endometrial tissues make up a menstrual flow that lasts for 3 to 6 days. After menstruation, rising levels of estrogens encourage the repair and growth of the endometrium.

During a menstrual cycle, FSH and LH stimulate growth of an ovarian follicle. The primary oocyte undergoes the first meiotic cell division in the oocyte. The outcome is a secondary oocyte and the first polar body.

A midcycle surge of LH triggers ovulation—the release of the secondary oocyte and the polar body from the ovary.

Feedback loops to the hypothalamus and pituitary from the ovaries and, later, the corpus luteum control changes in the ovary and in the structure of the uterine lining.

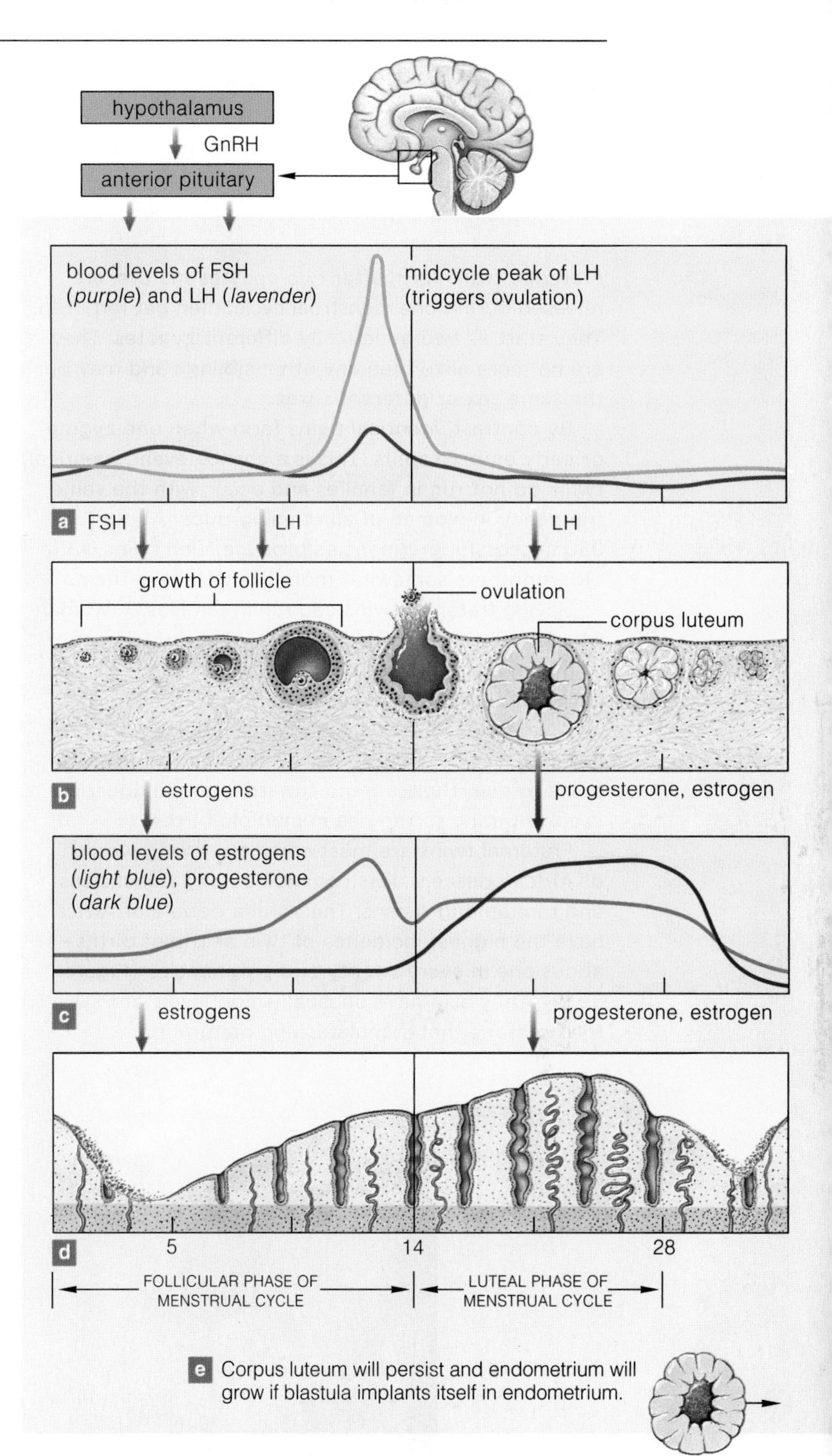

Figure 39.18 **Animated!** Changes in a human ovary and uterus correlated with changing hormone levels. We start with the onset of menstrual flow on day one of a twenty-eight-day menstrual cycle. *Green* arrows signify which hormones dominate.

(**a**,**b**) The anterior pituitary secretes FSH and LH, which stimulate a follicle to grow and an oocyte to mature in an ovary. A midcycle surge of LH triggers ovulation and the formation of a corpus luteum. A decline in FSH after ovulation stops more follicles from maturing.

(**c**,**d**) Early on, estrogen from a maturing follicle calls for repair and rebuilding of the endometrium. After ovulation, the corpus luteum secretes some estrogen and more progesterone that primes the uterus for pregnancy. (**e**) If pregnancy occurs, the corpus luteum will persist, and its secretions will stimulate maintenance of the uterine lining.

39.10 FSH and Twins

FOCUS ON HEALTH

Typically, only a single egg matures and gets released during each menstrual cycle. Abundant FSH can cause two eggs to mature and possibly lead to fraternal twins.

LINKS TO SECTIONS 32.3, 32.8

Fraternal twins form after two oocytes mature, are released during one menstrual cycle, then get fertilized. They start as two genetically different zygotes. They are no more alike than any other siblings and may be the same sex or different sexes.

By contrast, *identical* twins form when one zygote or early embryo splits. This is a chance event; identical twins do not run in families and occur with the same frequency in women of all ethnic groups. About 1 in 250 successful pregnancies produce such twins, with older mothers somewhat more likely to have them.

Having fraternal twins can run in families. A woman who is a fraternal twin has double the normal chances of giving birth to fraternal twins. If she does so once, her odds triple for a second set.

A woman's FSH levels rise from puberty through her mid-thirties, causing her odds of having fraternal twins to rise. Thus, a trend toward later childbearing is contributing to the rise in multiple births.

Fraternal twins are most common among women of African descent, less common among Caucasians, and rare among Asians. The Yoruba people of Africa have the highest incidence of twin or triplet births—about one in every twenty-two pregnancies (Figure 39.19). They also have unusually high levels of FSH, the hormone that stimulates egg maturation.

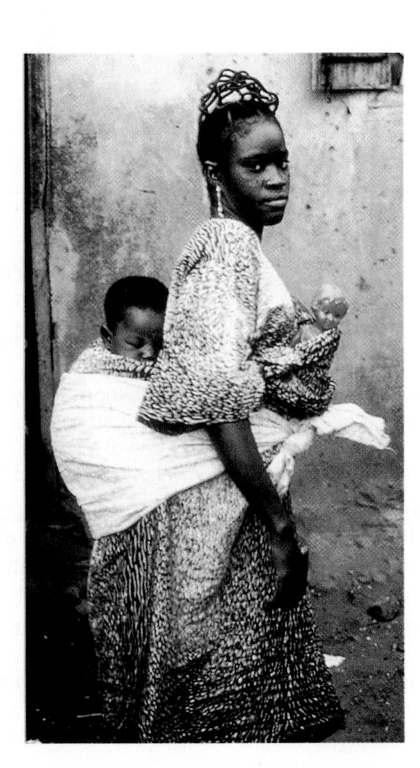

Figure 39.19 Yoruba mother. The rate of twin births among the Yoruba is the world's highest, but the mortality rate also is high; half of the twins die shortly after birth. Grieving mothers use a carving (Ere Ibeji) as a ritual point of contact with lost infants. Commercially produced plastic dolls are now being substituted for traditional wood carvings.

39.11 Pregnancy Happens

When a female and male engage in sexual intercourse, the excitement of the moment may obscure what can happen if a secondary oocyte is in an oviduct.

Internal fertilization involves coordinated changes in the physiology of two individuals and then additional interactions between their gametes. It all begins with sexual intercourse, or coitus.

SEXUAL INTERCOURSE

Physiology of Sex For males, intercourse requires an erection. A penis consists mostly of long cylinders of spongy tissue (Figure 39.12). When a male is not sexually aroused, his penis is limp, because the large blood vessels that transport blood to its spongy tissue are constricted. When he is aroused, parasympathetic signals induce these vessels to dilate. They also make the smooth muscle in the wall of chambers inside the penis relax. Inward blood flow exceeds outward flow, and the increasing fluid pressure expands the internal chambers. As a result, the penis enlarges and stiffens, so it can be inserted into a female's vagina.

During intercourse, pelvic thrusting stimulates the abundant mechanoreceptors in a male's penis, and a female's clitoris. The female's vaginal wall, skin folds around the opening of the vagina, and clitoris swell with blood.

In both partners, the heart rate and breathing rate rise. The posterior pituitary steps up its secretion of oxytocin, which inhibits signals from the brain center that controls fear and anxiety (the amygdala). When stimulation continues, oxytocin surges at orgasm.

At orgasm, oxytocin causes rhythmic contractions of muscles in the male reproductive tract and female vagina. Endorphins flood the brain and evoke feelings of pleasure. In males, orgasm is usually accompanied by ejaculation, in which contracting muscles force the semen out of the penis. You may have heard that a female will not become pregnant as long as she does not reach orgasm. Do not believe it.

Regarding Viagra The ability to get and sustain an erection peaks during the late teens. As a male grows older, he may have episodes of erectile dysfunction. With this disorder, the penis cannot stiffen enough for intercourse. Circulatory problems are a common cause. Medical conditions that harm the circulation increase the risk, as does smoking. The National Institutes of Health estimates that 30 million men are affected in the United States. Viagra and other pills prescribed for erectile dysfunction help blood enter the penis by

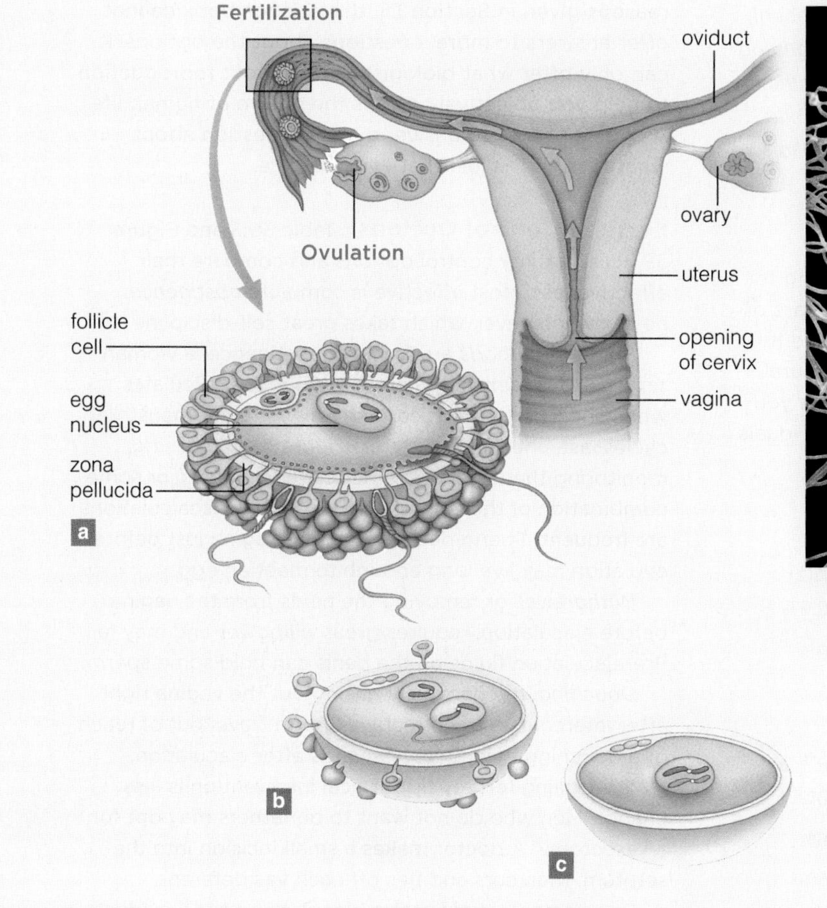

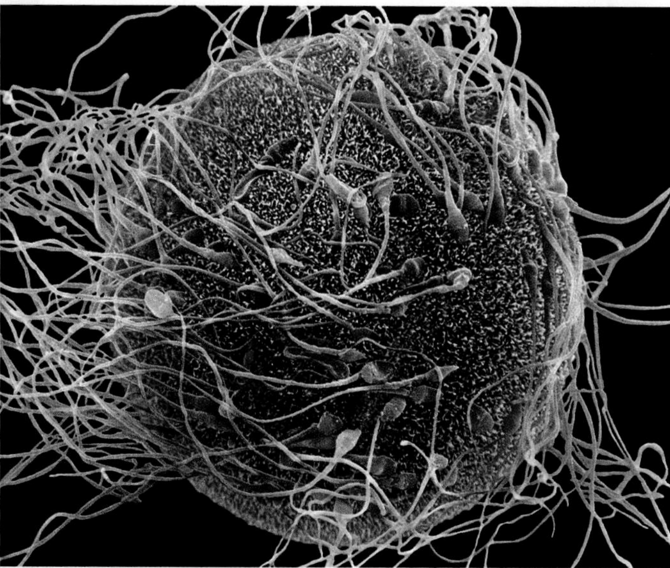

Figure 39.20 Animated! Fertilization. (**a**) Many human sperm travel swiftly through the vaginal canal into an oviduct (*blue* arrows). Inside that duct, they surround a secondary oocyte. Digestive enzymes released from the cap of each sperm clear a path through the zona pellucida.

(**b**) When a single sperm penetrates it, the secondary oocyte releases substances that prevent the other sperm from penetrating the zona pellucida. It also is stimulated to complete meiosis II in the nucleus. (**c**) The sperm's tail degenerates; its nucleus enlarges and fuses with the egg nucleus. Fertilization is over; a zygote has formed.

relaxing the smooth muscle inside it. Such pills cause headaches and other problems. They also can interact with other drugs, so no one should take them without a prescription.

FERTILIZATION

On average, an ejaculation can put 150 million to 350 million sperm in the vagina. Sperm can live for about three days after ejaculation, so fertilization may occur if they arrive a few days before or after ovulation or any time in between.

Less than thirty minutes after sperm arrive in the vagina, a few hundred of them can reach the oviducts with the help of uterine contractions. Then they swim toward the ovaries. Sperm usually meet the egg in the upper portion of the oviduct (Figure 39.20).

Sperm initially bind to the oocyte's zona pellucida. Binding triggers the release of enzymes from the cap over each sperm's head. Collectively, these digestive enzymes make a passage through the zona pellucida. Usually only one sperm enters the secondary oocyte. Its entry triggers chemical modifications in the zona pellucida that prevent other sperm from binding.

The secondary oocyte has completed only meiosis I. Signals from sperm activate the oocyte and the first polar body, causing them to complete meiosis II and undergo cytoplasmic division. This produces a single mature egg—an ovum (plural, ova)—and three polar bodies. The polar bodies degenerate. The egg nucleus and sperm nucleus fuse, forming the zygote's diploid nucleus (Figure 39.20*c*).

LINKS TO SECTIONS 11.1, 30.8, 30.10, 32.3

Sexual arousal involves nervous signals and hormones in both males and females. Ejaculation releases millions of sperm into the vagina. Uterine contractions help some of them reach the oviducts, where fertilization usually occurs. Fertilization is over when a sperm nucleus and egg nucleus fuse and form the diploid nucleus of a new individual.

39.12 Preventing or Seeking Pregnancy

As many as 500 eggs form during a woman's lifetime, and all are alive. The quarter of a billion sperm present in just one ejaculation are alive. Before a sperm and egg merge by chance, they are as much alive as any other form of life. It is scarcely tenable, then, to say that life begins at fertilization. Life began more than 3.8 billion years ago—and each gamete, zygote, and adult is a fleeting stage in the continuation of that beginning.

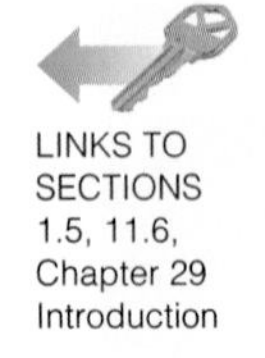

LINKS TO SECTIONS 1.5, 11.6, Chapter 29 Introduction

The motivation to engage in sex has been evolving for hundreds of millions of years. A few thousand years of moralizing about self-control have not suppressed it. How do we reconcile our biological past and cultural present? Think about the biological perspective as you assess social and economic factors that drive individuals around the world to seek fertility control options. For reasons given in Section 1.5, this science book cannot offer answers to moral questions about the options. It can only offer what biologists know about reproduction to help you objectively assess the nature of human life. A choice of how to answer a moral question about what is "right" is just that—your choice.

Fertility Control Options Table 39.3 and Figure 39.21 list fertility control options and compare their effectiveness. Most effective is complete *abstinence*: no sex whatsoever, which takes great self-discipline.

Rhythm methods are forms of abstinence; a woman simply avoids sex in her fertile period. She calculates when she is fertile by recording how long her menstrual cycles last, checking her temperature each morning, monitoring the thickness of her cervical mucus, or some combination of these methods. However, miscalculations are frequent. Sperm deposited in the vagina just before ovulation may live long enough to meet an egg.

Withdrawal, or removing the penis from the vagina before ejaculation, requires great willpower and may fail. Pre-ejaculation fluids on the penis can hold some sperm.

Douching, or chemically rinsing out the vagina right after intercourse, is ineffective. Sperm travel out of reach of a douching fluid ninety seconds after ejaculation.

Controlling fertility by surgical intervention is less chancy. Men who do not want to be fathers may opt for a *vasectomy*. A doctor makes a small incision into the scrotum, then cuts and ties off each vas deferens.

For women, *tubal ligation* blocks or cuts the oviducts. The method almost always results in permanent infertility.

Less drastic fertility control methods use physical and chemical barriers to stop sperm from reaching an egg. *Spermicidal foam* and *spermicidal jelly* poison sperm. They are not always reliable, but their use with a condom or diaphragm reduces the risk of pregnancy.

A *diaphragm* is a flexible, dome-shaped device. Prior to intercourse, it is positioned inside the vagina so that it covers the cervix. It is relatively effective when first fitted by a doctor and used correctly with a spermicide. A *cervical cap* is a similar but smaller device.

Condoms are thin, tight-fitting sheaths worn over the penis during intercourse. Good brands may be as much as 95 percent effective when used with a spermicide. Only

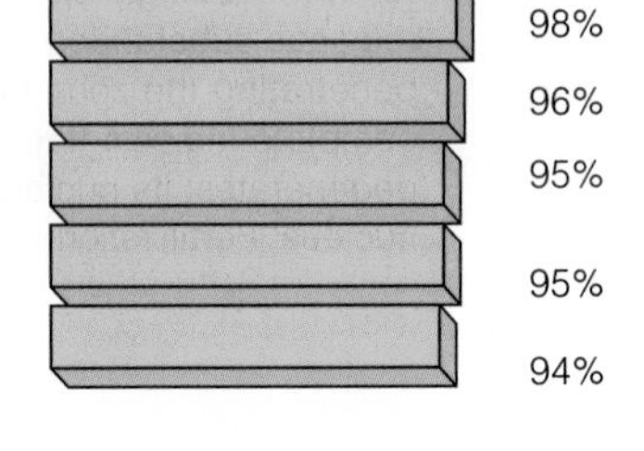

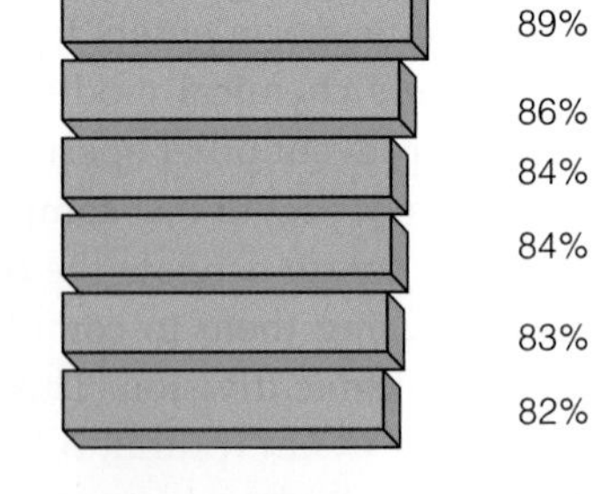

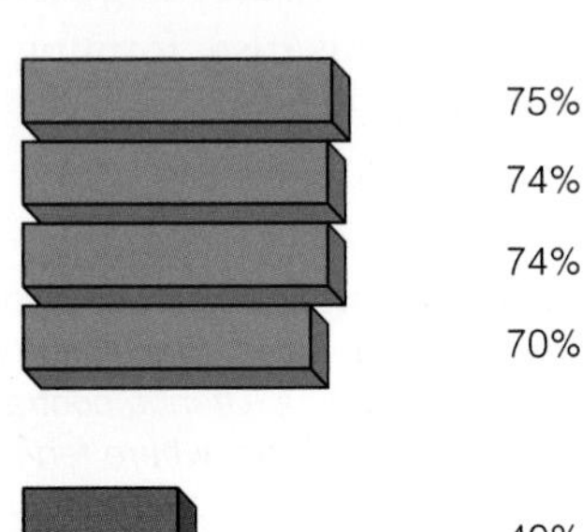

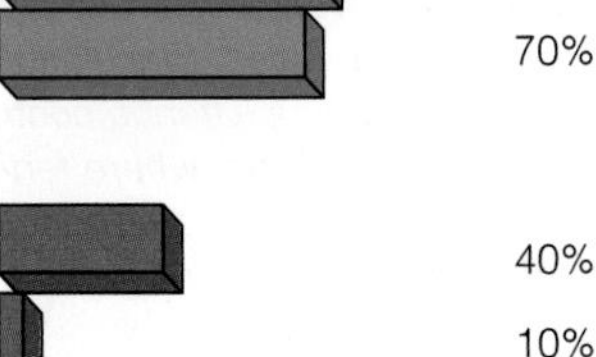

Figure 39.21 Comparison of the effectiveness of some methods of contraception. These percentages also indicate the number of unplanned pregnancies per 100 couples who use only that method of birth control for a year. For example, "94% effectiveness" for the oral contraceptives means that 6 of every 100 females will still become pregnant, on average.

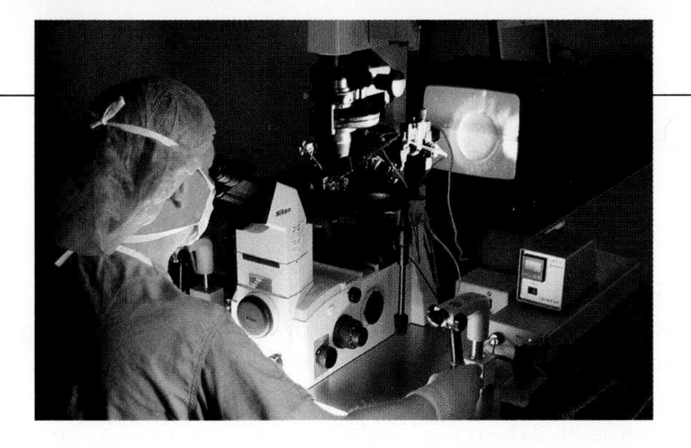

Figure 39.22 *In vitro* fertilization. A fertility doctor uses a micromanipulator to insert a human sperm into an oocyte. The video screen shows the view through the microscope,

condoms made of latex offer protection against sexually transmitted diseases (STDs). However, even the best ones can tear or leak, at which time they are useless.

A doctor must insert an intrauterine device, or *IUD*, in the uterus. Some IUDs release hormones that make cervical mucus thicken so sperm can't swim through it. Others shed copper, which interferes with implantation.

A *birth control pill*, with its synthetic estrogens and progesterone-like hormones, blocks maturation of oocytes and ovulation. "The Pill," the most common fertility control method in the West, reduces menstrual cramps. When used correctly, it is at least 94 percent effective, but it can cause nausea, headaches, and weight gain. Its use lowers the risk of ovarian and uterine cancer but raises the risk of breast, cervical, and liver cancer.

A *birth control patch* is a small, flat adhesive patch applied to skin. The patch delivers the same hormones as an oral contraceptive and blocks ovulation the same way. Like birth control pills, it is not for everyone. Some women, especially smokers, develop dangerous blood clots and other serious cardiovascular disorders.

Progestin injections or implants block ovulation. One Depo-Provera injection lasts for three months. Norplant works for five years. Both methods are quite effective, but they may cause sporadic, heavy bleeding.

Some women use *emergency contraception* after a condom tears, or after unprotected consensual sex or rape. These "morning-after pills" suppress ovulation and are available without a prescription to women eighteen or older. They work best if taken early but are somewhat effective up to five days after intercourse. They are not meant to be used on a regular basis. Nausea, vomiting, abdominal pain, headache, and dizziness are side effects.

Regarding Abortion About ten percent of women who learn that they are pregnant lose the embryo or fetus in a *spontaneous abortion*, or miscarriage. Many additional pregnancies end without ever having been detected. By some estimates, as many as 50 percent of all fertilized eggs are lost, most often because of genetic problems. Risk increases with maternal age.

Induced abortion is the deliberate dislodging and removal of an embryo or fetus from the uterus. In the United States, about half of all unplanned pregnancies end in induced abortion. Parents who find out through genetic tests that an embryo has a genetic abnormality may decide to terminate the pregnancy. About eighty percent of embryos diagnosed with Down syndrome are aborted (Section 11.6).

From a clinical standpoint, abortion usually is a brief, low-risk procedure, especially during the first trimester of pregnancy. *Mifepristone* (RU-486) and similar drugs can induce abortion during the first nine weeks. They interfere with how the body sustains the uterine lining, hence the pregnancy. Use of a suction device terminates pregnancies as late as fourteen weeks. Later abortions require more difficult surgical procedures. Many people find them more troubling because the fetus is closer to being fully developed.

Assisted Reproductive Technology About 15 percent of all couples in the United States experience infertility, in which a woman does not become pregnant or repeatedly miscarries. When a couple makes normal sperm and oocytes but cannot conceive naturally, they may turn to technology for assistance. Typically, the woman receives hormone injections that cause multiple oocytes to mature. For *in vitro fertilization*, a doctor withdraws oocytes and combines them with sperm in a dish or injects sperm into them (Figure 39.22). After fertilization, tiny clusters of cells form. Usually several are transferred to the uterus or an oviduct.

Assisted reproduction attempts are costly and usually fail. About a third of attempts in thirty-year-old women result in a birth. Only one in six attempts in forty-year-olds succeed. When assisted reproduction succeeds, it often leads to multiple births.

Table 39.3 Common Methods of Contraception

Method	Mechanism of Action
Abstinence	Avoid intercourse entirely
Rhythm method	Avoid intercourse in female's fertile period
Withdrawal	End intercourse before male ejaculates
Douche	Wash semen from vagina after intercourse
Vasectomy	Cut or close off male's vasa deferentia
Tubal ligation	Cut or close off female's oviducts
Condom	Enclose penis, block sperm entry to vagina
Diaphragm, cervical cap	Cover cervix, block sperm entry to uterus
Spermicides	Kill sperm
Intrauterine device	Prevent sperm entry to uterus or implantation
Oral contraceptives	Prevent ovulation
Hormone patches, implants, or injections	Prevent ovulation
Emergency contraception pill	Prevent ovulation

39.13 Sexually Transmitted Diseases

Unprotected sex exposes you to potential infection by any pathogens that your partner unknowingly may have picked up from a previous sexual partner.

CONSEQUENCES OF INFECTION

LINKS TO SECTIONS 19.2, 20.2, 35.10

Each year, pathogens that cause sexually transmitted diseases, or STDs, infect about 15 million people in the United States (Table 39.4). Two-thirds of those infected are under age twenty-five. One-fourth are teenagers. Over 65 million Americans now live with an incurable STD. Treating STDs and secondary complications costs a staggering $8.4 billion in an average year.

The social consequences are sobering. Females are more easily infected than males, and they develop more complications. For instance, pelvic inflammatory disease (PID) is a secondary outcome of some bacterial STDs. It affects about 1 million females annually. It scars the reproductive tract and can cause chronic pain, infertility, and tubal pregnancies (Figure 39.23*a*). Some fetuses acquire STDs before or during birth. They abort on their own or develop abnormally. Females commonly bestow chlamydia on their newborns (Figure 39.23*b*). Type II *Herpes* virus kills 50 percent of the fetuses it infects. It causes neural defects in 25 percent of the survivors.

Table 39.4 Estimated Annual STD Cases*

STD	U.S. Cases	Global Cases
HPV infection	5,500,000	20,000,000
Trichomoniasis	5,000,000	174,000,000
Chlamydia	3,000,000	92,000,000
Genital herpes	1,000,000	20,000,000
Gonorrhea	650,000	62,000,000
Syphilis	70,000	12,000,000
AIDS	40,000	4,900,000

* Global data on HPV and genital herpes were last compiled in 1997.

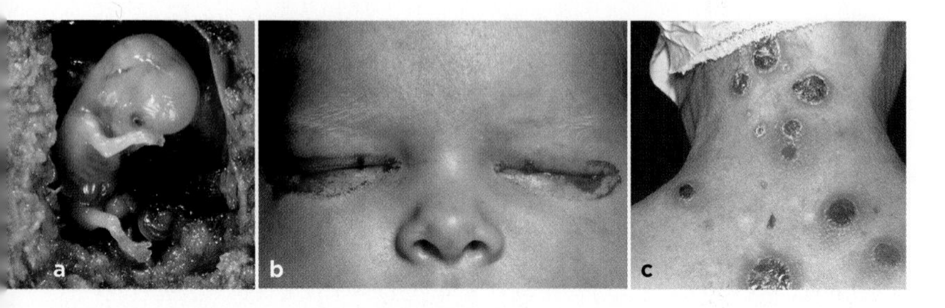

Figure 39.23 A few downsides of unsafe sex. (**a**) Tubal pregnancy. Scarring from STDs makes an embryo implant itself in an oviduct, not the uterus. Untreated tubal pregnancies can rupture an oviduct and cause bleeding, infection, and death. (**b**) An infant with chlamydia-inflamed eyes. Its mother transmitted the bacterial pathogen to it during birth. (**c**) Chancres typical of secondary syphilis.

MAJOR AGENTS OF STDS

HPV Human papillomavirus (HPV) infection is the most widespread and fastest growing STD in the United States. At least 20 million are already infected. A few of the 100 or so HPV strains cause genital warts: bumpy growths on external genitals and the area around the anus. Two strains, HPV16 and HPV18, cause cervical cancer. Sexually active females should have an annual pap smear to check for cervical changes. A vaccine can prevent infection if given before the first viral exposure.

Trichomoniasis *Trichomonas vaginalis*, a flagellated protist, causes the disease trichomoniasis (Section 20.2). Symptoms often include vaginal soreness, itching, and a yellowish discharge. Infected males usually show no symptoms. Untreated infections damage the urinary tract, cause infertility, and invite HIV infection. A single dose of an antiprotozoal drug quickly cures an infection. Both sexual partners should be treated.

Chlamydia Chlamydial infection is primarily a young person's disease. Forty percent of those infected are between ages fifteen and nineteen; 1 in 10 sexually active teenage girls is infected. *Chlamydia trachomatis* causes the disease (Figure 39.24*a*). Antibiotics quickly kill this bacterium. Most infected females are undiagnosed; they have no symptoms. Between 10 and 40 percent of those who are untreated will develop PID. Half of infected males have symptoms, such as abnormal discharges from the penis and painful urination. Untreated males risk an inflamed reproductive tract and infertility.

Genital Herpes About 45 million Americans have genital herpes, caused by type II *Herpes simplex* virus. Transmission to new hosts requires direct contact with active *Herpes* viruses or with sores that contain them. Mucous membranes of the mouth and genitals are vulnerable. Early symptoms are often mild or absent. Painful, small blisters may form on the vulva, cervix, urethra, or anal tissues of infected females. Blisters form on the penis and anal tissues of infected males. Within three weeks, the virus enters latency. Sores crust over and heal, but viral particles are hidden in the body.

The virus is reactivated sporadically, which causes painful sores at or near the original site of infection. Sexual intercourse, menstruation, emotional stress, or other infections trigger flare-ups. An antiviral drug can decrease healing time and often the pain.

Gonorrhea The STD gonorrhea is caused by *Neisseria gonorrhoeae* (Figure 39.24*b*). This bacterium commonly crosses mucous membranes of the urethra, cervix, or anal canal during sexual intercourse. An infected female may notice a vaginal discharge or burning sensation while urinating. If the bacterium enters her oviducts,

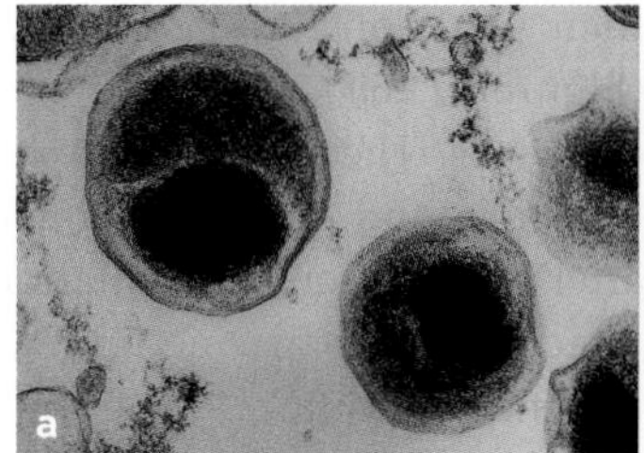
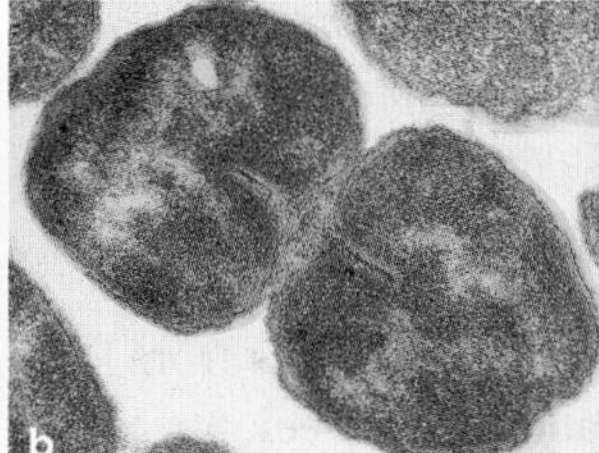
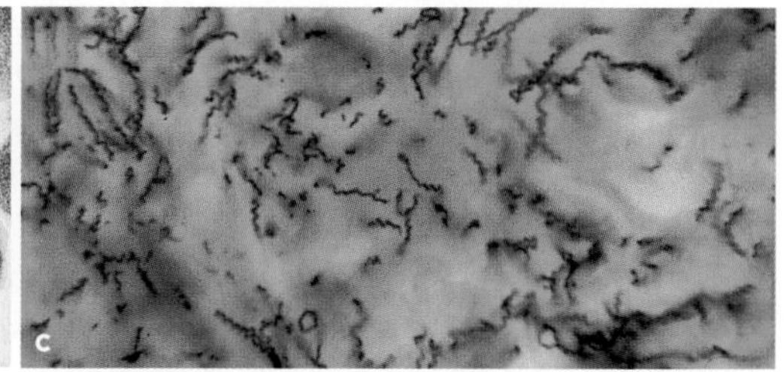

Figure 39.24 Light micrographs of three bacterial species that cause the common STDs called (**a**) chlamydia, (**b**) gonorrhea, and (**c**) syphilis.

it may cause cramps, fever, vomiting, and scarring that can lead to sterility. Less than one week after a male is infected, yellow pus oozes from his penis. Urination becomes more frequent, and it also may be painful.

Prompt treatment with antibiotics quickly cures this disease, yet it is rampant. Many ignore early symptoms or wrongly believe infection confers immunity. A person can contract gonorrhea repeatedly, probably because there are now at least sixteen strains of *N. gonorrhoeae*.

Syphilis A spirochete bacterium *Treponema pallidum* causes syphilis, a dangerous STD (Figure 39.24*c*). During sex with an infected partner these bacteria get onto the genitals or into the cervix, vagina, or oral cavity. They then slip inside the body through tiny cuts. One to eight weeks later, many *T. pallidum* cells are twisting about in a flattened, painless chancre (a localized ulcer).

The chancre is a sign of the primary stage of syphilis. It usually heals, but treponemes multiply inside the spinal cord, brain, eyes, bones, joints, and mucous membranes. In an infectious secondary stage, a skin rash develops and more chancres form (Figure 39.23*c*). In about 25 percent of the cases, immune responses succeed and symptoms subside. Another 25 percent are symptom-free. In the remainder, lesions and scars appear in the skin and liver, bones, and other organs. Few treponemes form during this tertiary stage, but the host's immune system is hypersensitive to them. Chronic immune reactions may damage the brain and spinal cord, and cause paralysis.

Possibly because the symptoms are so alarming, more people seek early treatment for syphilis than they do for gonorrhea. Later stages require prolonged treatment.

AIDS As you read earlier in Section 35.10, infection by HIV, the human immunodeficiency virus, leads to AIDS (acquired immune deficiency syndrome). The immune system almost always loses the battle with the virus. AIDS is presently incurable. There may be no symptoms at first. Five to ten years later, a set of chronic disorders develops. The immune system weakens, which opens the door to opportunistic infectious agents. Normally harmless bacteria already living in and on the body are the first to take advantage of the lowered resistance. Then dangerous pathogens take their toll. Over time, they overwhelm the compromised immune system.

Most often, HIV spreads by way of anal, vaginal, and oral intercourse, and intravenous drug use. Virus particles in blood, semen, urine, or vaginal secretions enter a new host through cuts and abrasions of the penis, vagina, rectum, or oral cavity. Oral sex is least likely to cause infection. Anal sex is five times more dangerous than vaginal sex and 50 times more dangerous than oral sex.

Free or low-cost, confidential testing for HIV exposure is available at public health facilities. It takes a few weeks to six months or more before the body forms detectable amounts of antibodies in response to the first exposure. Anyone who tests positive for HIV can spread the virus.

Public education may help slow the spread of HIV (Figure 39.25). Most health care workers advocate safe sex, although there is confusion over what "safe" means. The use of high-quality latex condoms together with a nonoxynol-9 spermicide helps prevent viral transmission. However, as mentioned earlier, the practice still carries a slight risk. Open-mouth kissing with an HIV-positive individual is risky. Caressing is not if there are no lesions or cuts where HIV-laden body fluids can enter the body. Skin lesions caused by any other sexually transmitted disease are vulnerable points of entry for the virus.

New, costly drug therapies are prolonging many lives. In the late 1990s, the rate of infection started to climb again, possibly because of the misperception that AIDS is no longer a lethal threat. But AIDS kills, and the viral agent keeps on mutating. How long today's drugs can keep a lid on deaths is anybody's guess.

Figure 39.25 NBA legend Magic Johnson, one of the torch bearers of the 2002 Winter Olympics. He was diagnosed as HIV positive in 1991. He contracted the virus through heterosexual sex, and credits his survival to AIDS drugs and informed medical care. He continues to campaign to educate others about AIDS.

39.14 Formation of the Early Embryo

Nine months or so after fertilization, a human female gives birth. Compared to other primates, humans take longer to develop inside the mother, and the young require more intense care for a longer period of time.

LINKS TO SECTION 24.6, CHAPTER 38 INTRODUCTION

Pregnancy lasts an average of thirty-eight weeks from the time of fertilization. It takes about one week for a blastocyst to form. A blastocyst is the type of blastula that forms in mammals. All major organs form in the *embryonic* period, which extends from the second week through the eighth week of pregnancy. At the end of this period, the developing individual is referred to as a fetus. In the *fetal* period, from the start of the ninth week until birth, organs grow and become specialized. Commonly, the first three months of a pregnancy is called the first trimester. The second trimester extends from the beginning of the fourth month to the end of the sixth, and the third trimester ends at birth.

CLEAVAGE AND IMPLANTATION

Cleavage begins within a day or two of fertilization (Figure 39.26*a*). A blastocyst forms by the fifth day. It consists of an outer layer of cells, a cavity filled with their secretions (a blastocoel), and an inner cell mass (Figure 39.26*d*). The embryo develops from the inner cell mass. The outer cells will help form membranes that surround the developing embryo. About six days after fertilization, implantation begins: The blastocyst attaches to the uterine lining, and then burrows into the endometrium. During implantation, the inner cell mass develops into two flattened layers of cells called the embryonic disk (Figure 39.26*e,f*).

EXTRAEMBRYONIC MEMBRANES

Membranes start forming outside the embryo during implantation (Table 39.5). A fluid-filled *amniotic* cavity opens up between the embryonic disk and part of the blastocyst surface (Figure 39.26*f*). Many cells migrate around the wall of the cavity and form the amnion, a membrane that will enclose the embryo. Fluid in the cavity will function as a buoyant cradle in which an embryo can grow, move freely, and be protected from abrupt temperature changes and hard impacts.

As the amnion forms, other cells migrate around the inner wall of the blastocyst and form a lining that becomes the yolk sac. In reptiles and birds, this sac

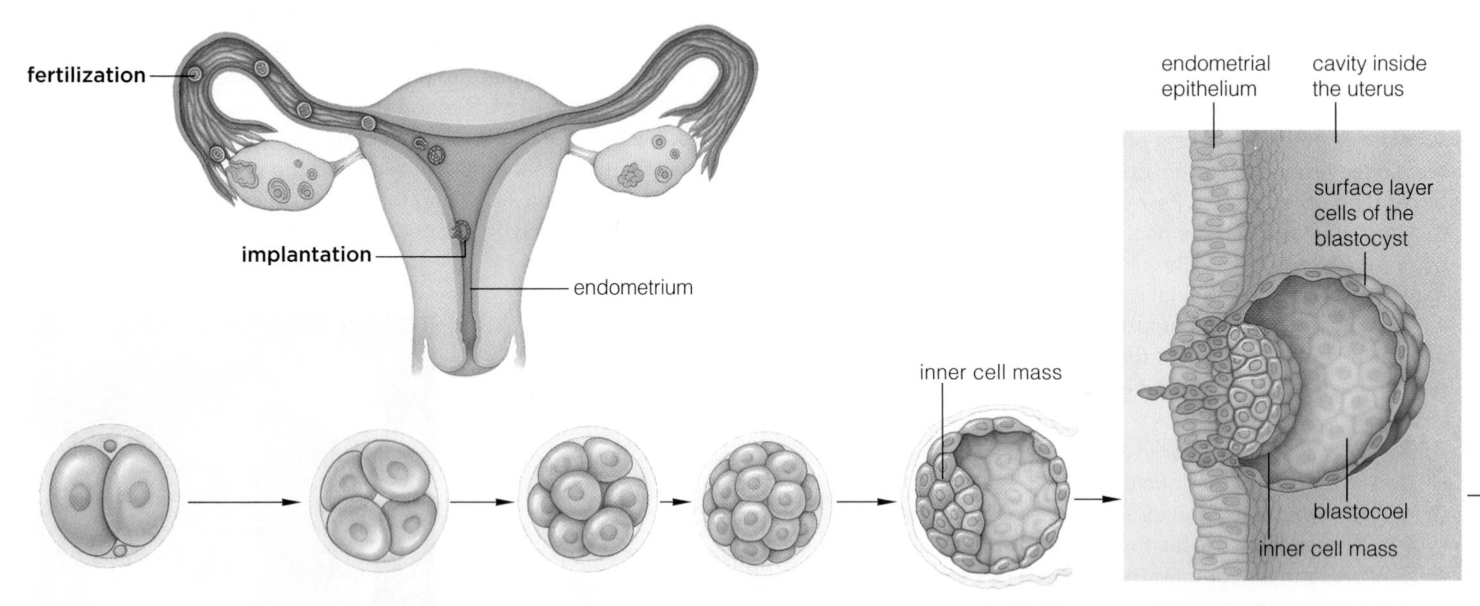

a **DAYS 1–2**. The first cleavage furrow extends between the two polar bodies. Later cuts are angled, so cells become asymmetrically arranged. Until the eight-cell stage forms, they are loosely organized, with space between them.

b **DAY 3**. After the third cleavage, cells abruptly huddle into a compacted ball, which tight junctions among the outer cells stabilize. Gap junctions formed along the interior cells enhance intercellular communication.

c **DAY 4**. By 96 hours there is a ball of sixteen to thirty-two cells shaped like a mulberry. It is a morula (after *morum*, Latin for mulberry). Cells of the surface layer will function in implantation and will give rise to a membrane, the chorion.

d **DAY 5**. A blastocoel (fluid-filled cavity) forms in the morula as a result of surface cell secretions. By the thirty-two-cell stage, differentiation is occurring in an inner cell mass that will give rise to the embryo proper. This embryonic stage is the blastocyst.

e **DAYS 6–7**. Some of the blastocyst's surface cells attach themselves to the endometrium and start to burrow into it. Implantation has started.

actual size 

holds nutritious yolk. In humans, cells of the yolk sac give rise to the embryo's blood cells and to germ cells, the forerunners of gametes.

Before a blastocyst is fully implanted, spaces open in maternal tissues and become filled with blood that seeps in from ruptured capillaries. In the blastocyst, a new cavity opens up around the amnion and yolk sac. The lining of the cavity becomes the chorion, a third membrane that balloons like fingers of many rubber gloves into maternal tissues. It will become part of the placenta. The placenta is an organ that functions in exchanges of materials between the bloodstreams of a mother and her developing child.

After the blastocyst is implanted, an outpouching of the yolk sac will become the fourth extraembryonic membrane—the allantois. It gives rise to the urinary bladder and the placenta's blood vessels.

EARLY HORMONE PRODUCTION

An implanted blastula prevents menstruation by the release of HCG (human chorionic gonadotropin). This hormone stimulates the corpus luteum to continue its secretion of progesterone and estrogens which, recall, maintain the uterine lining. After about three months, the placenta will take over the secretion of HCG.

Table 39.5 Human Extraembryonic Membranes

Amnion	Encloses, protects embryo in a fluid-filled, buoyant cavity
Yolk sac	Becomes site of red blood cell formation; germ cell source
Chorion	Lines amnion and yolk sac, becomes part of placenta
Allantois	Source of urinary bladder and blood vessels for placenta

HCG can be detected in a mother's urine as early as the third week of pregnancy. At-home pregnancy tests include a treated "dipstick" that changes color when exposed to urine containing HCG.

Cleavage of the human zygote produces a cluster of cells that develops into the blastocyst, which implants itself in the endometrium six or seven days after fertilization.

Projections from the blastocyst's surface invade maternal tissues, and connections start to form that in time will metabolically support the developing embryo.

Some parts of the blastocyst give rise to four external membranes: an amnion, yolk sac, chorion, and allantois. The extraembryonic membranes have different roles. Together, they contribute to the structural and functional development of the embryo.

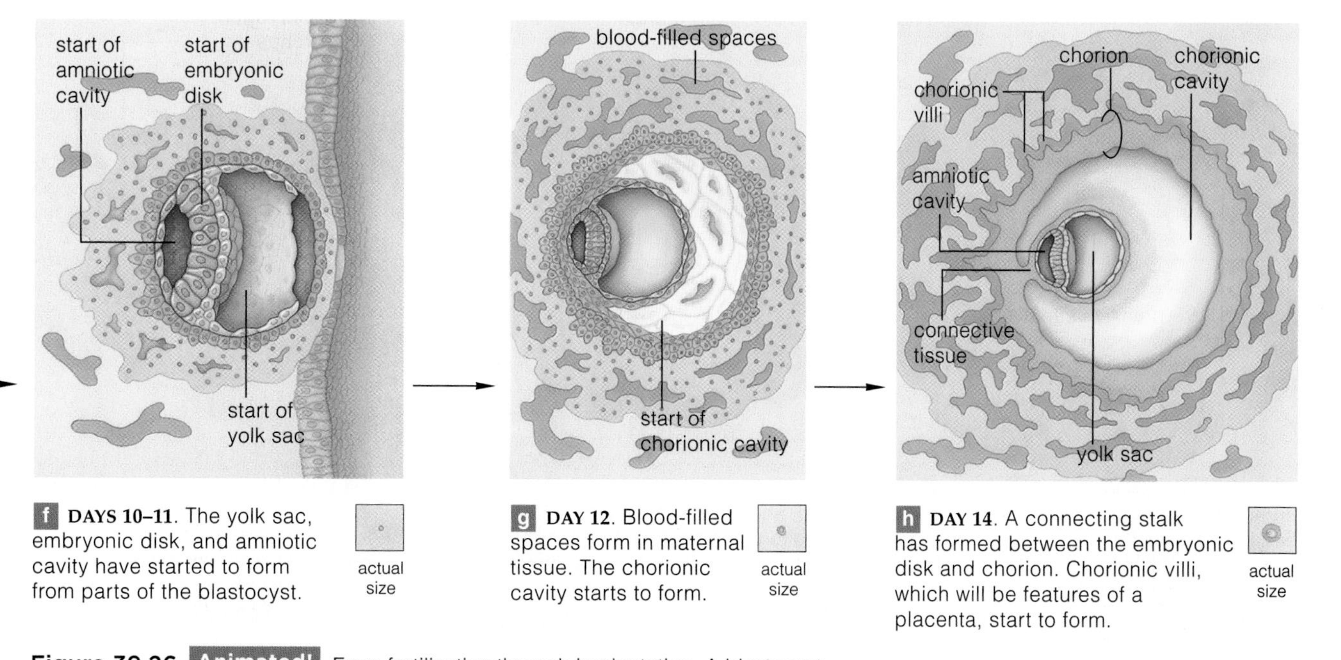

f **DAYS 10–11**. The yolk sac, embryonic disk, and amniotic cavity have started to form from parts of the blastocyst.

g **DAY 12**. Blood-filled spaces form in maternal tissue. The chorionic cavity starts to form.

h **DAY 14**. A connecting stalk has formed between the embryonic disk and chorion. Chorionic villi, which will be features of a placenta, start to form.

Figure 39.26 **Animated!** From fertilization through implantation. A blastocyst forms, and its inner cell mass becomes an embryonic disk two cells thick. It will later become the embryo. Three extraembryonic membranes (the amnion, chorion, and yolk sac) start forming. A fourth membrane (allantois) forms after implantation is over.

39.15 Emergence of the Vertebrate Body Plan

By the time a female misses a first menstrual period after fertilization, cleavage is over. Gastrulation is under way.

LINKS TO SECTIONS 14.1, 14.3, 16.8

By two weeks after fertilization, the inner cell mass of a blastocyst is a two-layered embryonic disk. During gastrulation, cells migrate inward along a depression, the primitive streak, that forms on the disk's surface (Figure 39.27*a*). The resulting three germ layers of the gastrula are the forerunners of all tissues (Table 39.6). The primitive streak's location establishes the body's front-to-back axis (Figure 39.27).

Many master genes are now being expressed and the tissues and organs are beginning to take shape. For example, by the eighteenth day after fertilization, the embryonic disk has two folds that will merge into a neural tube, the precursor of the spinal cord and brain (Figure 39.27*b*). Some mesoderm folds into another tube that develops into a notochord. This notochord acts as a structural model; bony segments form on it.

You may have heard of spina bifida. With this birth defect, the neural tube and one or more vertebrae do not form as they should. As a result, the spinal cord will protrude out of the vertebral column at birth.

Toward the end of the third week, multiple paired segments called somites arise from some mesoderm. They are embryonic sources of most bones as well as the head and trunk's skeletal muscles and overlying dermis. By now, pharyngeal arches start to form; they will contribute to the pharynx, larynx, and the face, neck, mouth, and nose (Figure 39.27*c*). Small spaces open up in certain parts of the mesoderm. Eventually, they will interconnect as a coelomic cavity.

Table 39.6 Derivatives of Human Germ Layers

Ectoderm (outer layer)	Outer layer (epidermis) of skin; nervous tissue
Mesoderm (middle layer)	Connective tissue of skin; skeletal, cardiac, smooth muscle; bone; cartilage; blood vessels; urinary system; gut organs; peritoneum (coelom lining); reproductive tract
Endoderm (inner layer)	Lining of gut and respiratory tract, and organs derived from these linings

The basic vertebrate body plan emerges early in the development of the new individual.

In all vertebrates, a primitive streak, neural tube, somites, and pharyngeal arches form during the embryonic period. Formation of the primitive streak establishes the body's anterior–posterior axis and its bilateral symmetry.

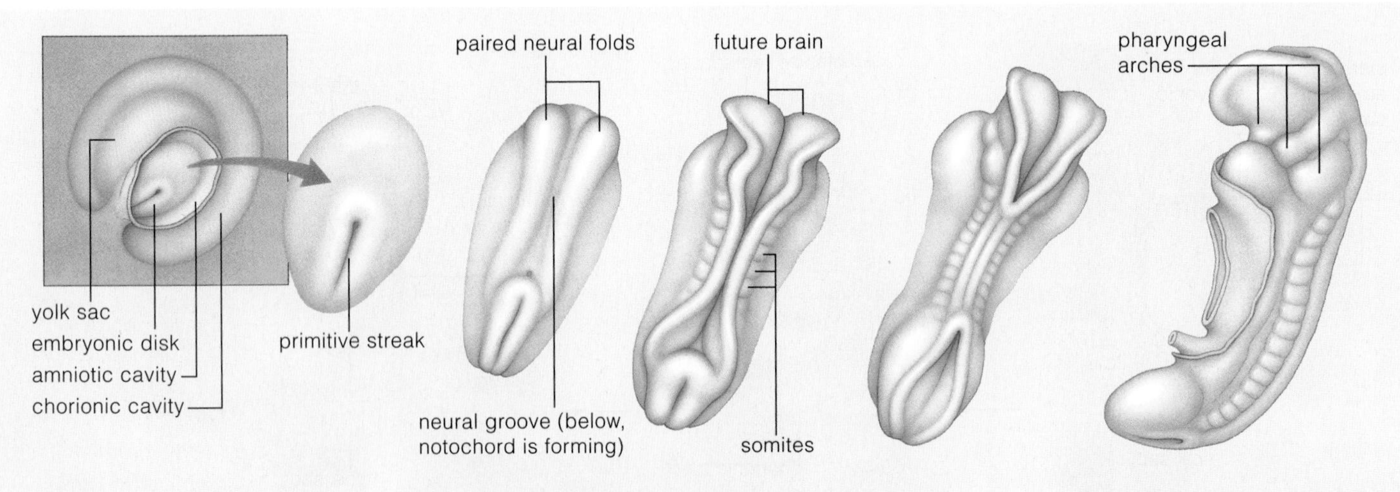

a DAY 15. A faint band appears around a depression along the axis of the embryonic disk. This band is the primitive streak, and it marks the onset of gastrulation in vertebrate embryos.

b DAYS 18–23. Organs start to form through cell divisions, cell migrations, tissue folding, and other events of morphogenesis. Neural folds will merge to form the neural tube. Somites (bumps of mesoderm) appear near the embryo's dorsal surface. They will give rise to most of the skeleton's axial portion, skeletal muscles, and much of the dermis.

c DAYS 24–25. By now, some embryonic cells have given rise to pharyngeal arches. These will contribute to the formation of the face, neck, mouth, nasal cavities, larynx, and pharynx.

Figure 39.27 Hallmarks of the embryonic period of humans and other vertebrates. A primitive streak and then a notochord form. Neural folds, somites, and pharyngeal arches form later. (**a**,**b**) Dorsal views of the embryo's back. (**c**) Side view.

39.16 The Function of the Placenta

How does an embryo floating in fluid inside its mother's uterus obtain essential nutrients and oxygen? By way of its umbilical cord and the placenta.

All exchange of materials between an embryo and its mother takes place by way of the placenta, a pancake-shaped, blood-engorged organ that consists of uterine lining and extraembryonic membranes. At full term, the placenta covers about one-fourth of the uterus's inner surface (Figure 39.28).

The placenta begins forming early in pregnancy. By the third week, maternal blood has begun to pool in spaces in the endometrial tissue. Chorionic villi, tiny fingerlike projections from the chorion, extend into the pools of maternal blood.

Embryonic blood vessels extend outward through the umbilical cord to the placenta, and then into the chorionic villi. Embryonic blood exchanges substances with the maternal blood, but the two bloodstreams do not mix. Oxygen and nutrients diffuse out of maternal blood and into embryonic blood vessels in the villi. Wastes such as urea and carbon dioxide diffuse the other way, and the mother's body disposes of them.

LINK TO SECTION 24.10

After the third month, the placenta produces large amounts of HCG, progesterone, and estrogens. These hormones encourage the ongoing maintenance of the uterine lining.

Vessels of the embryo's circulatory system extend through the umbilical cord to the placenta, where they run through pools of maternal blood. Maternal and embryonic blood do not mix; substances diffuse between the maternal and embryonic bloodstreams.

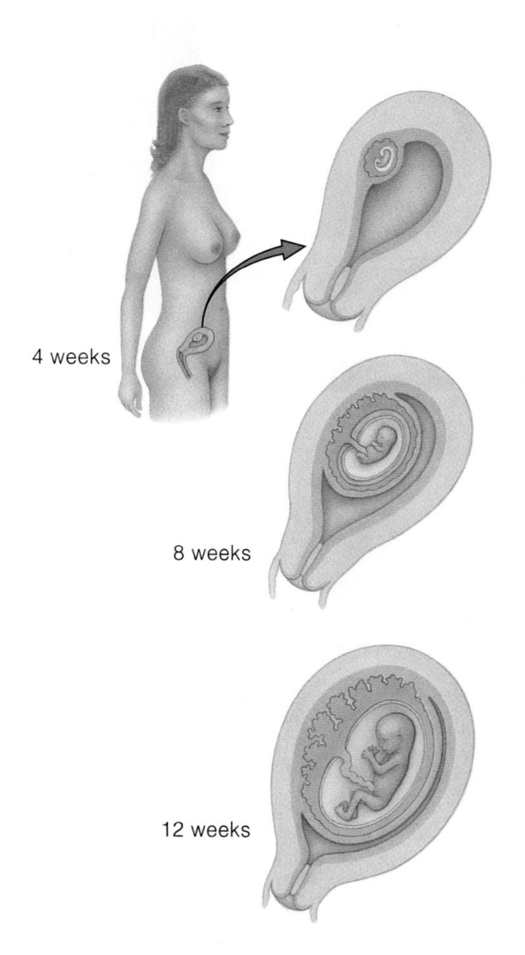

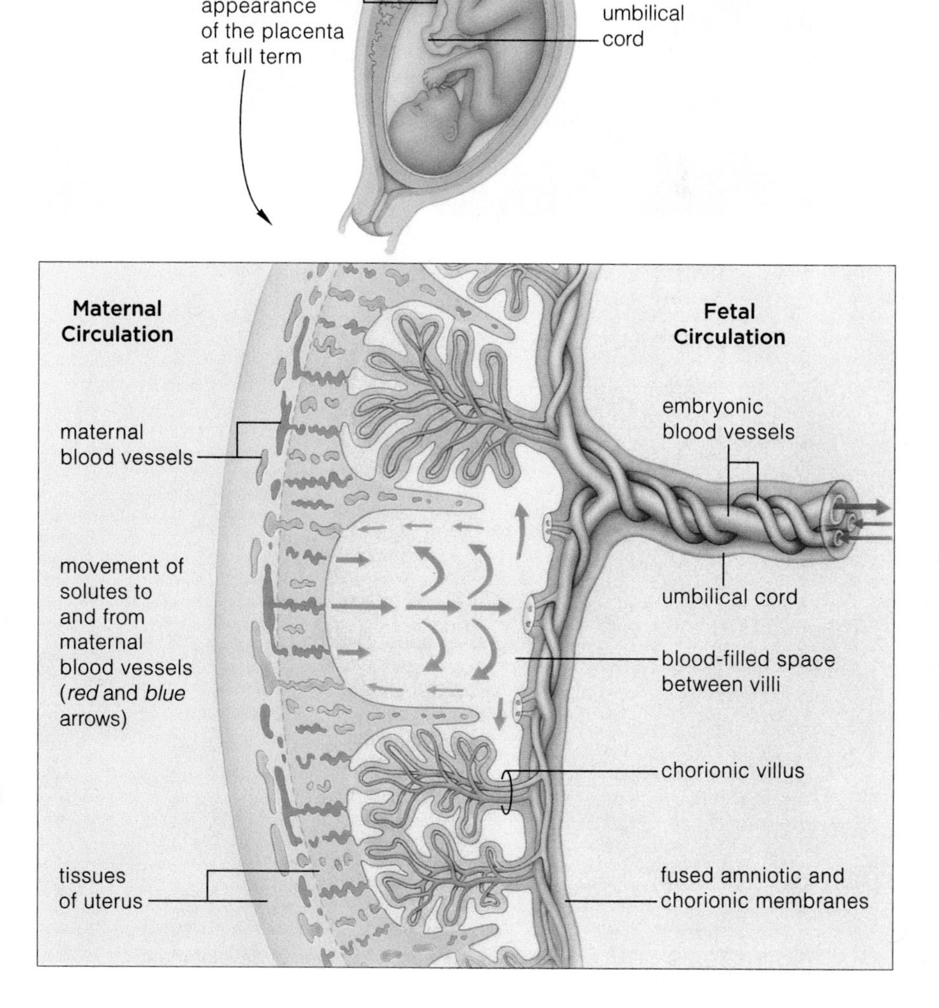

Figure 39.28 Relationship between fetal and maternal blood circulation in a full-term placenta. Blood vessels extend from the fetus, through the umbilical cord, and into chorionic villi. Maternal blood flows into spaces between villi. However, the two bloodstreams do not intermingle. Oxygen, carbon dioxide, and other small solutes diffuse across the placental membrane surface.

39.17 Emergence of Distinctly Human Features

Early on, a human embryo—with its phyaryngeal arches and tail—has a distinctly vertebrate appearance. The tail quickly disappears. By the beginning of the fetal period, the developing individual has distinctly human features.

LINKS TO SECTIONS 8.4, 11.1, 16.8, 25.5

When the fourth week ends, the embryo is 500 times its starting size, but still less than one centimeter in length. Growth slows as details of organs begin to fill in. Limbs form; paddles are sculpted into fingers and toes. The umbilical cord and the circulatory system develop. Growth of the head now surpasses that of all other regions (Figure 39.29). Reproductive organs start forming, as explained in Section 11.1. At the end of the eighth week, the individual is no longer just "a vertebrate." Its features define it as a human fetus.

In the second trimester, as developing nerves and muscles connect up, reflexive movements begin. Legs kick, arms wave about, and fingers grasp. The fetus

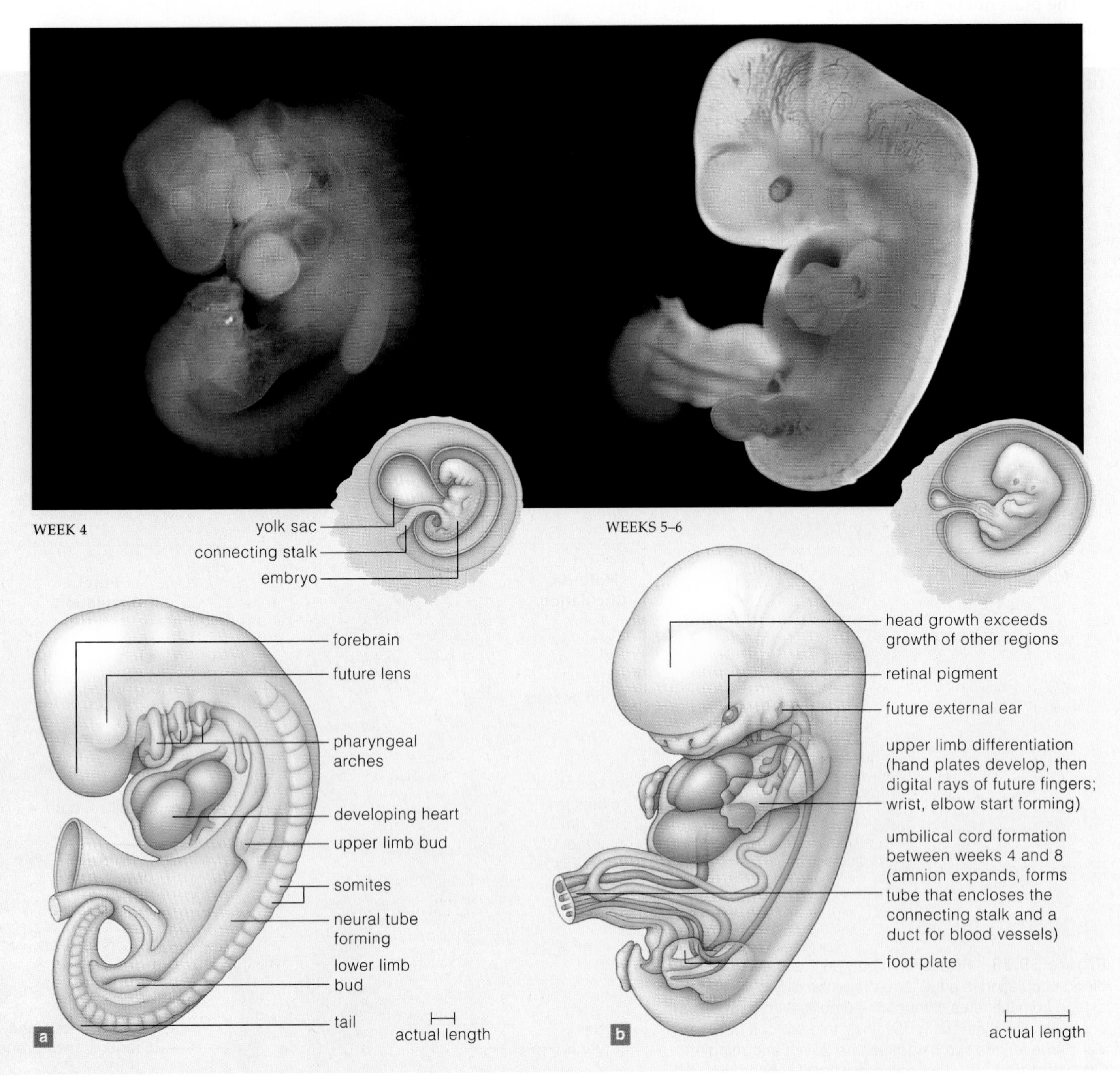

Figure 39.29 Human embryo at successive stages of development.

frowns, squints, puckers its lips, sucks, and hiccups. When the fetus is five months old, its heartbeat can be heard clearly through a stethoscope positioned on the mother's abdomen. The mother can sense movements of fetal arms and legs.

By now, soft, fetal hair (the lanugo) covers the skin; most will be shed before birth. A thick, cheesy coating protects the wrinkled, reddish skin from abrasion. In the sixth month, delicate eyelids and eyelashes form. Eyes open during the seventh month, the start of the final trimester. By this time, all portions of the brain have formed and have begun to function.

At one month, an embryo is less than a centimeter long, with a tail and tiny limb buds. A month later, at the onset of the fetal period, the tail is gone, limbs have formed, and the tiny individual appears distinctly human.

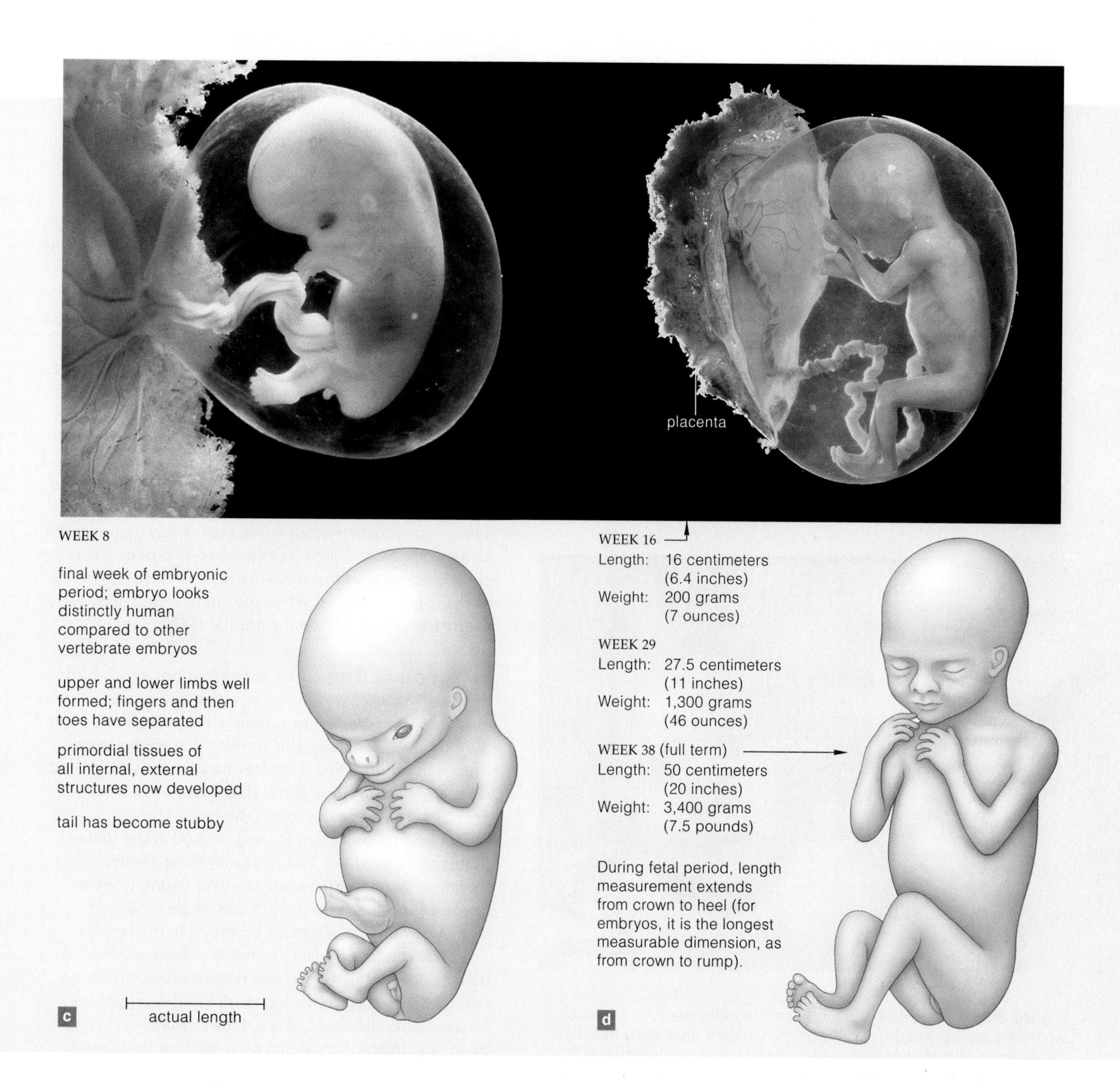

39.18 Mother as Provider and Protector

FOCUS ON HEALTH

Each mother-to-be is committing much of her body's resources to the growth and development of a new individual. From fertilization until birth, her future child is at the mercy of her diet, health habits, and life-style.

LINKS TO SECTIONS 30.6, 32.4, 35.5, 36.5

Nutritional Considerations A pregnant woman who eats a well-balanced diet supplies her future child with all of the proteins, carbohydrates, and lipids it needs for growth and development. Her own demands for vitamins and minerals increase, but both are absorbed preferentially across the placenta and taken up by the embryo. Taking B-complex vitamins in early pregnancy reduces the embryo's risk of neural tube defects. Folate (folic acid) is especially critical in this respect.

Dietary deficiencies affect many developing organs. For example, if a mother does not get enough iodine, her newborn may be affected by cretinism, a disorder that affects brain function and motor skills (Section 32.4).

Infectious Diseases As you know, IgG antibodies in a pregnant woman's blood cross the placenta and protect an embryo or fetus from bacterial infections (Section 35.5). But some viral diseases are dangerous in the early weeks after fertilization. Rubella, or German measles, is an example. A woman may sidestep the risk of passing on the rubella virus by getting vaccinated before she becomes pregnant.

A relative of the protist that causes malaria sometimes lurks in garden soil, cat feces, and undercooked meat. It causes toxoplasmosis. The disease often does not cause symptoms, so a pregnant woman may become infected and not realize it. If the parasite crosses the placenta, it can infect her child and lead to developmental problems, a miscarriage, or stillbirth. To minimize the risk, pregnant women should eat well-cooked meat and avoid cat feces.

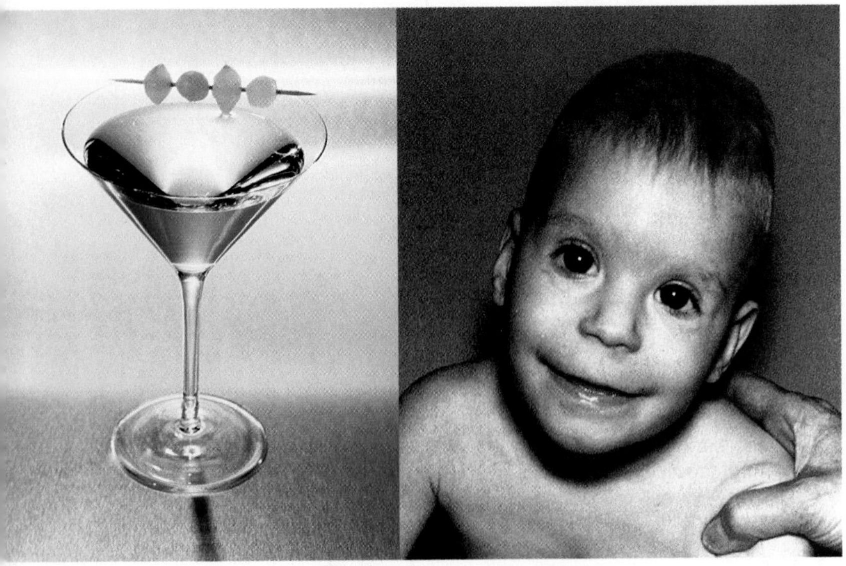

Figure 39.30 An infant with fetal alcohol syndrome—FAS. Obvious symptoms are low and prominently positioned ears, improperly formed cheekbones, and an abnormally wide, smooth upper lip. Growth-related complications and abnormalities of the nervous system can be expected.

Alcohol Use Remember the introduction to Chapter 5? Alcohol passes freely across cell membranes. It also passes across the placenta. If a pregnant female drinks alcohol, her embryo or fetus will quickly absorb it.

Alcohol intake invites fetal alcohol syndrome, or FAS. A small head and brain, facial abnormalities, slow growth, mental impairment, possible heart problems, and poor coordination characterize affected infants (Figure 39.30). The damage is permanent. Children affected by FAS never do catch up, physically or mentally.

Most doctors now advise women who are pregnant or attempting to become pregnant to avoid alcohol entirely. Even before a woman knows she is pregnant, tissues of the embryonic nervous system have begun forming, and alcohol can damage them. Even moderate drinking during pregnancy increases risk of miscarriage and stillbirth.

Smoking Smoking or exposure to secondhand smoke increases the risk of miscarriage and adversely affects fetal growth and development. Carbon monoxide in the smoke can outcompete oxygen for the binding sites on hemoglobin (Section 36.5), so the embryo or fetus of a smoker gets less oxygen than that of a nonsmoker. In addition, levels of the stimulant nicotine in amniotic fluid actually can be higher than those in the mother's blood.

Smoking's effects persist long after birth. For seven years, researchers tracked a group of children born in the same week. More children of smokers died of postdelivery complications, and the survivors were smaller, with twice as many heart defects. When the study ended, they were nearly half a year behind the normal reading age.

Prescription Drugs Some medications cause birth defects. Thalidomide administered as a tranquilizer is an infamous example. It was routinely prescribed in Europe during the 1960s. Infants of some of the women who used it during the first trimester had severely deformed arms and legs or none at all. Thalidomide never was prescribed as a tranquilizer in the United States.

Isotretinoin (Accutane) is widely used in the United States and elsewhere. This highly effective treatment for severe acne is often prescribed for young women. If taken early in a pregnancy, it can cause facial and cranial deformities and heart problems in the embryo.

Certain antidepressants increase the risk of birth defects. Paroxetine (Paxil) and related drugs inhibit the reuptake of serotonin. Taking them early in pregnancy increases the likelihood of heart malformations. Taking them later in pregnancy increases risk that infants will have fatal heart and lung disorders.

39.19 Birth and Lactation

As in other placental mammals, human fetuses are born live and are nourished with nutritious milk secreted from the mother's mammary glands. Shifts in the levels of hormones help control these processes.

GIVING BIRTH

A mother's body changes as her fetus nears full term, at about 38 weeks after fertilization. Until the last few weeks, her firm cervix helped prevent the fetus from slipping out of her uterus prematurely. Now cervical connective tissue becomes thinner, softer, and more flexible. These changes will allow the cervix to stretch enough to permit the fetus to pass out of the body.

The birth process is known as labor. Typically, the amnion ruptures right before birth, so amniotic fluid drains out from the vagina. The cervical canal dilates. Strong contractions propel the fetus through it, then out through the vagina (Figure 39.31).

A positive feedback mechanism operates during labor. When the fetus is nearing full term, it typically shifts position so that its head touches the mother's cervix. Receptors inside the cervix sense pressure and signal the hypothalamus, which signals the posterior lobe of the pituitary to secrete oxytocin. In a positive feedback loop, oxytocin binds to smooth muscle of the uterus, causing stronger uterine contractions that put additional pressure on the cervix. The added pressure triggers more oxytocin secretion, which causes more cervical stretching, and so on until a fetus is expelled. Synthetic oxytocin is often given to induce labor or to increase the strength of contractions.

Strong muscle contractions also detach and expel the placenta from the uterus as the "afterbirth." The umbilical cord that connects the newborn to this mass of expelled tissue is clamped, cut short, and tied. The short stump of cord left in place withers and falls off. The body's navel marks the former attachment site.

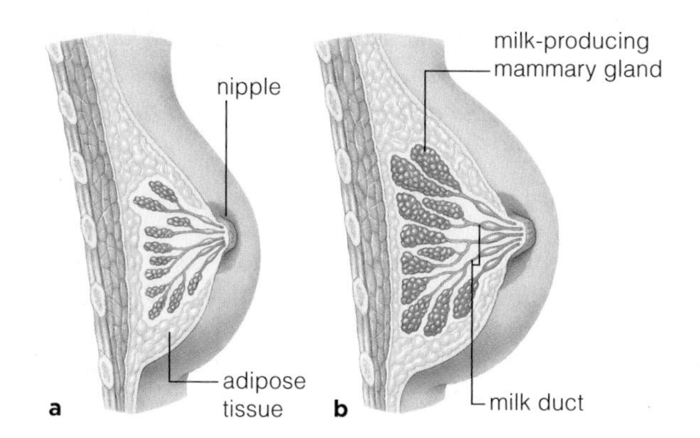

Figure 39.32 Cutaway views of (**a**) a breast of a human female who is not pregnant and (**b**) a breast of a lactating female.

NOURISHING THE NEWBORN

Before a pregnancy, a woman's breast tissue is largely adipose tissue. Milk ducts and mammary glands are small and inactive (Figure 39.32). During pregnancy, these structures enlarge in preparation for lactation, or milk production. Prolactin, a hormone secreted by the mother's anterior pituitary, triggers milk synthesis.

LINKS TO SECTIONS 24.10, 25.3, 32.3

After birth, a decline in progesterone and estrogens causes milk production to go into high gear. When a newborn suckles, neural signals cause the release of oxytocin. The hormone stimulates muscles around the milk glands to contract and force milk into the ducts.

Besides being nutrient-rich, human breast milk has antibodies that protect a newborn from some viruses and bacteria. Nursing mothers should know that drugs, alcohol, and toxic secretions can end up in milk.

During birth, hormonally stimulated muscle contractions force a fetus out of its mother's body. Hormones also stimulate milk production and secretion.

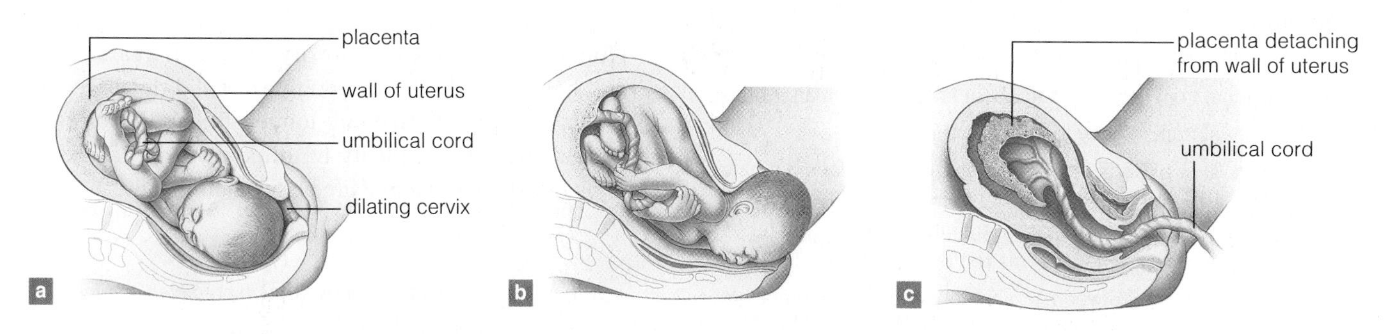

Figure 39.31 Expulsion of (**a**,**b**) a human fetus and (**c**) afterbirth during labor. The afterbirth consists of the placenta, tissue fluid, and blood.

39.20 Maturation, Aging, and Death

Human growth and development continue long after birth. Compared to other primates, humans have an extended period of dependency and a longer post-reproductive life.

STAGES OF THE LIFE CYCLE

After birth, humans continue to grow and change in proportions until they become sexually mature adults. Figure 39.33 shows a few of the proportional changes during the life cycle. Table 39.7 defines the prenatal (before birth) and postnatal (after birth) stages.

Births before 37 weeks are considered premature. A fetus born as early as 23 weeks can survive with the help of modern medicine but may have disabilities. A fetus born earlier than 22 weeks rarely survives.

Postnatal growth is most rapid between the ages of thirteen and nineteen years. Sexual maturation occurs at puberty, and bones stop growing shortly thereafter.

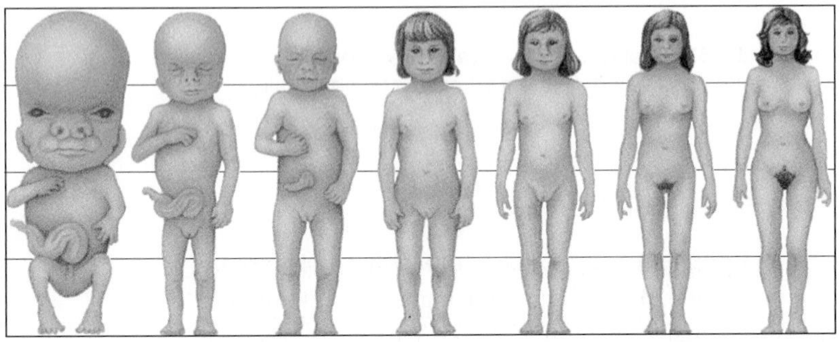

8-week embryo | 12-week embryo | newborn | 2 years | 5 years | 13 years (puberty) | 22 years

Figure 39.33 Observable, proportional changes in prenatal and postnatal periods of human development. Changes in overall physical appearance are slow but noticeable until the teens. For example, compared to an embryo, a teenager's legs are longer and the trunk shorter, so the head is proportionally smaller.

Table 39.7 Stages of Human Development

Prenatal period	
Zygote	Single cell resulting from fusion of sperm nucleus and egg nucleus at fertilization.
Morula	Solid ball of cells produced by cleavages.
Blastocyst (blastula)	Ball of cells with surface layer, fluid-filled cavity, and inner cell mass.
Embryo	All developmental stages from two weeks after fertilization until end of eighth week.
Fetus	All developmental stages from ninth week to birth (about 38 weeks after fertilization).
Postnatal period	
Newborn	Individual during the first two weeks after birth.
Infant	Individual from two weeks to about fifteen months after birth.
Child	Individual from infancy to about ten or twelve years.
Pubescent	Individual at puberty; secondary sexual traits develop; girls between 10 and 15 years, boys between 11 and 16 years.
Adolescent	Individual from puberty until about 3 or 4 years later; physical, mental, emotional maturation.
Adult	Early adulthood (between 18 and 25 years); bone formation and growth finished. Changes proceed slowly after this.
Old age	Aging processes result in expected tissue deterioration.

WHY WE AGE AND DIE

In developed countries, humans live about 80 years, on average. What limits our life span? There are two main hypotheses, and both may be partly right.

According to the programmed life span hypothesis, life span is genetically determined. Experiments with cultured cells support this view. Normal human cells do not divide more than fifty times. Why not? Cells, recall, duplicate all chromosomes before they divide. Chromosomes have telomeres: caps made from DNA and proteins. Each time a nucleus divides, enzymes remove a portion from each telomere. When only a nub of telomere remains, cells stop dividing and die.

Cancer cells and germ cells can divide indefinitely, but only because they make telomerase, an enzyme that lengthens telomeres. Expose other cultured cells to telomerase, and they too will keep dividing.

By the cumulative assaults hypothesis, aging is the outcome of damage at the molecular level. Recall that mutagens in the environment damage DNA, and free radicals degrade all biological molecules.

Studies of premature aging disorders can provide insight into normal aging. For example, people with Werner's syndrome have a mutated version of a gene that serves in DNA repair and accumulate mutations faster than normal. They also have short or missing telomeres in many cells. As a result of these factors, they typically die before age fifty of cancer.

In evolutionary terms, reproductive success means living long enough to produce and raise offspring. Humans reach sexual maturity in fifteen years or so and help their children through extended dependency. We do not *need* to live any longer. But we among all animals have the capacity to think about it, and most of us decide that we like life better than the alternative.

The human life span proceeds through predictable stages.

Aging may be an outcome of genetically determined factors, as well as damage at the molecular and cellular levels.

www.thomsonedu.com ThomsonNOW!

Summary

Section 39.1 Compared with asexual reproduction, reproducing sexually costs more in time and energy, and a parent does not have as many of its genes represented among offspring. However, producing variable offspring may be advantageous in environments where conditions fluctuate from one generation to the next.

Section 39.2 Most animal life cycles have six stages of development. Gametes form, fertilization takes place, cleavage produces a blastula, gastrulation results in an early embryo that has two or three primary tissue layers (ectoderm, mesoderm, and endoderm), organs form, and tissues and organs become specialized.

- *Use the animation on ThomsonNOW to track the development of a frog.*

Section 39.3 As an egg matures, substances become localized in its cytoplasm. Cleavage distributes different portions of the egg cytoplasm to different cells. It ends with formation of a blastula. A common blastula is a ball of cells with a fluid-filled cavity. Cleavage patterns vary among groups. The yolk volume influences the patterns.

- *Use the animation on ThomsonNOW to learn about cytoplasmic localization and control of gastrulation.*

Section 39.4 With cell differentiation, cells become specialized in their structure and function by activating subsets of the genome. Tissues and organs form by cell migrations, changes in cell shape, cell death, and other processes of morphogenesis.

Cytoplasmic localization and embryonic induction are triggers for cell differentiation. They give cells in certain tissues the means to signal cells in other tissues. Short-range and long-range signals selectively activate genes that cause tissues and organs to form in specific places. These signals are products of master genes, which are similar among all major animal groups.

- *Use the animation on ThomsonNOW to learn how the neural tube forms and a chick wing develops.*

Sections 39.5, 39.6 A human reproductive system consists of primary reproductive organs, or gonads, and accessory organs and ducts. The gonads produce gametes and sex hormones that influence reproduction, as well as development of gender-specific secondary sexual traits.

The male gonads are testes, which secrete testosterone and produce sperm. Feedback loops from the testes to the hypothalamus and pituitary gland control secretion of testosterone, which affects sperm formation. Sperm travel to the body surface in a series of ducts. Glands that empty into these ducts supply components of the semen.

- *Use the animation on ThomsonNOW to learn about the reproductive system of human males and how sperm form.*

Sections 39.7–39.10 The female gonads are ovaries, which produce oocytes and secrete the hormones estrogen and progesterone. A menstrual cycle is an approximately monthly recurring cycle of fertility. Feedback loops from ovaries to the hypothalamus and the anterior pituitary gland control it. In the cycle's follicular phase, FSH and LH stimulate maturation of an oocyte and the cells that surround it. FSH and LH also prompt ovaries to secrete hormones that cause thickening of the wall of the uterus. Ovulation, the release of a mature oocyte from an ovary, is triggered by a surge in LH. During the luteal phase, a corpus luteum forms from cells that surrounded the egg. Its hormonal secretions cause the uterine wall to thicken more. If fertilization does not occur, the corpus luteum degenerates and menstrual fluid flows out of the vagina as the cycle starts again. Menstrual cycles continue until a woman's fertility ends at menopause.

- *Use the animation on ThomsonNOW to observe cyclic changes in an ovary and learn about the effects of hormones on the menstrual cycle.*

Sections 39.11–39.13 Hormones and nerves govern the physiological changes that occur during arousal and intercourse. Millions of sperm are ejaculated, but usually only one penetrates the secondary oocyte. Fertilization ends when the sperm nucleus and egg nucleus fuse and form a zygote.

Humans prevent pregnancy by abstinence, surgery, physical or chemical barriers, and by affecting female sex hormones. Unsafe sex and other behaviors promote the spread of pathogens that cause STDs.

- *Use the animation on ThomsonNOW to see what happens during fertilization.*

Sections 39.14–39.18 Fertilization usually occurs in an oviduct. Cleavage starts and produces a blastocyst, a type of blastula with a fluid-filled cavity and an inner cell mass. The blastocyst implants itself in the uterine wall. Membranes that form outside the blastocyst support its development. A human gastrula has three germ layers (ectoderm, endoderm, and mesoderm), which give rise to all adult tissues. The first organ to form, the neural tube, gives rise to the brain and spinal cord. Somites are paired bumps of mesoderm. They develop into skeletal muscles, bones, and the dermis of the skin.

The placenta consists of extraembryonic membranes and endometrium. This organ allows embryonic blood to exchange substances with the maternal blood, although the two bloodstreams do not mix. Gases, nutrients, and wastes diffuse between them. Alcohol, drugs, and other harmful substances also cross the placenta, so a mother's health, nutrition, and life-style can affect the growth and development of her future child.

- *Use the animation on on ThomsonNOW to observe early human development.*

Section 39.19 Hormones induce labor at about 38 weeks. Positive feedback controls contractions that expel the fetus and then the afterbirth. Hormones also control maturation of the mammary glands and milk flow.

Section 39.20 Development and growth continue after birth and end at adulthood. Aging may be partly a result of a genetically limited number of cell divisions and partly an outcome of assaults on DNA.

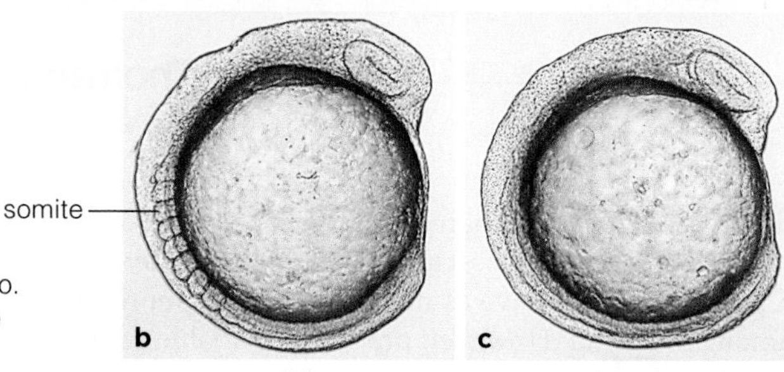

Figure 39.34 (**a**) An adult zebrafish. (**b**) Normal zebrafish embryo. Somites, a series of segments that give rise to bone and muscle, are clearly visible. (**c**) Mutant embryo that could not form somites.

Self-Quiz

Answers in Appendix III

1. Sexual reproduction _______ .
   a. requires internal fertilization
   b. produces offspring that vary in their traits
   c. is more efficient than asexual reproduction
   d. puts all of a parent's genes in each offspring

2. The typical end product of cleavage is a _______ .
   a. zygote   c. gastrula
   b. blastula   d. gamete

3. Is this statement true or false? Materials are randomly distributed in egg cytoplasm, so cleavage parcels out same kinds of cytoplasmic components to all cells.

4. Cells differentiate as a direct result of _______ .
   a. selective gene expression   c. gastrulation
   b. morphogenesis   d. all of the above

5. _______ help bring about morphogenesis.
   a. Cell migrations   c. Cell suicide
   b. Changes in cell shape   d. all of the above

6. Match each term with the most suitable description.
   ___ gamete formation   a. blastomeres form
   ___ fertilization   b. cellular rearrangements form primary tissues
   ___ cleavage   c. eggs and sperm form
   ___ gastrulation   d. sperm and egg nuclei fuse
   ___ pattern formation   e. cells migrate, change shape, commit suicide
   ___ morphogenesis processes   f. tissues, organs emerge in right order, in right places

7. Testosterone is secreted by the _______ .
   a. testes   c. prostate gland
   b. hypothalamus   d. all of the above

8. During a menstrual cycle, a midcycle surge of _______ triggers ovulation.
   a. estrogens   b. progesterone   c. LH   d. FSH

9. The corpus luteum forms from the _______ .
   a. first polar body   c. secondary oocyte
   b. follicle cells   d. zona pellucida

10. Sexually transmitted bacteria cause _______ .
   a. trichomoniasis   c. syphilis
   b. genital herpes   d. all of the above

11. A _______ implants in the lining of the human uterus.
   a. zygote   b. gastrula   c. blastocyst   d. fetus

12. The _______, a fluid-filled sac, surrounds and protects an embryo and keeps it from drying out.
   a. yolk sac   b. allantois   c. amnion   d. chorion

13. At full term, a placenta _______ .
   a. is composed of extraembryonic membranes alone
   b. directly connects maternal and fetal blood vessels
   c. keeps maternal and fetal blood vessels separated

14. During the second trimester of pregnancy, _______ .
   a. gastrulation begins   c. eyes open
   b. heartbeats start   d. all of the above

15. _______ stimulates milk synthesis in mammary glands.
   a. FSH   c. Testosterone
   b. Prolactin   d. Oxytocin

16. Match each term with the most suitable description.
   ___ testis   a. maternal and fetal tissues
   ___ anterior pituitary   b. stores mature sperm
   ___ cervix   c. secretes LH and FSH
   ___ placenta   d. produces testosterone
   ___ vagina   e. produces estrogens and progesterone
   ___ ovary   f. usual site of fertilization
   ___ oviduct   g. lining of uterus
   ___ epididymis   h. entrance to uterus
   ___ endometrium   i. birth canal

■ *Visit ThomsonNOW for additional questions.*

Critical Thinking

1. The zebrafish (*Danio rerio*) is special to developmental biologists. This small freshwater fish is easily maintained in tanks. A female produces hundreds of eggs, which can develop and hatch in three days. The transparent embryos let researchers directly observe developmental events (Figure 39.34). Single cells can be injected with dye to see how they change position, or they can be killed or injected with genes to observe the outcome. One mutant gene can prevent somites from forming. Explain why it is lethal.

2. Drugs that inhibit signals of sympathetic neurons may be prescribed for males who have high blood pressure. How might such drugs interfere with sexual performance?

3. By UNICEF estimates, each year 110,000 people are born with abnormalities as a result of rubella infections. Major symptoms of congenital rubella syndrome, or CRS, are deafness, blindness, mental impairment, and heart problems. A nonvaccinated woman who is infected in the first trimester of pregnancy is at risk, but not later. Review the developmental events that unfold during pregnancy and explain why this is the case.

4. The most common ovarian tumors in young women are ovarian teratomas. The name comes from the Greek word *teraton*, which means monster. What makes these tumors "monstrous" is the presence of well-differentiated tissues, most commonly bones, teeth, fat, and hair. Early physicians suggested that teratomas arose as a result of nightmares, witchcraft, or intercourse with the devil. Unlike all other tumors, which arise from somatic cells, teratomas arise from germ cells. Explain why a tumor derived from a germ cell is able to produce more differentiated cell types than one derived from a somatic cell.

VII Principles of Ecology

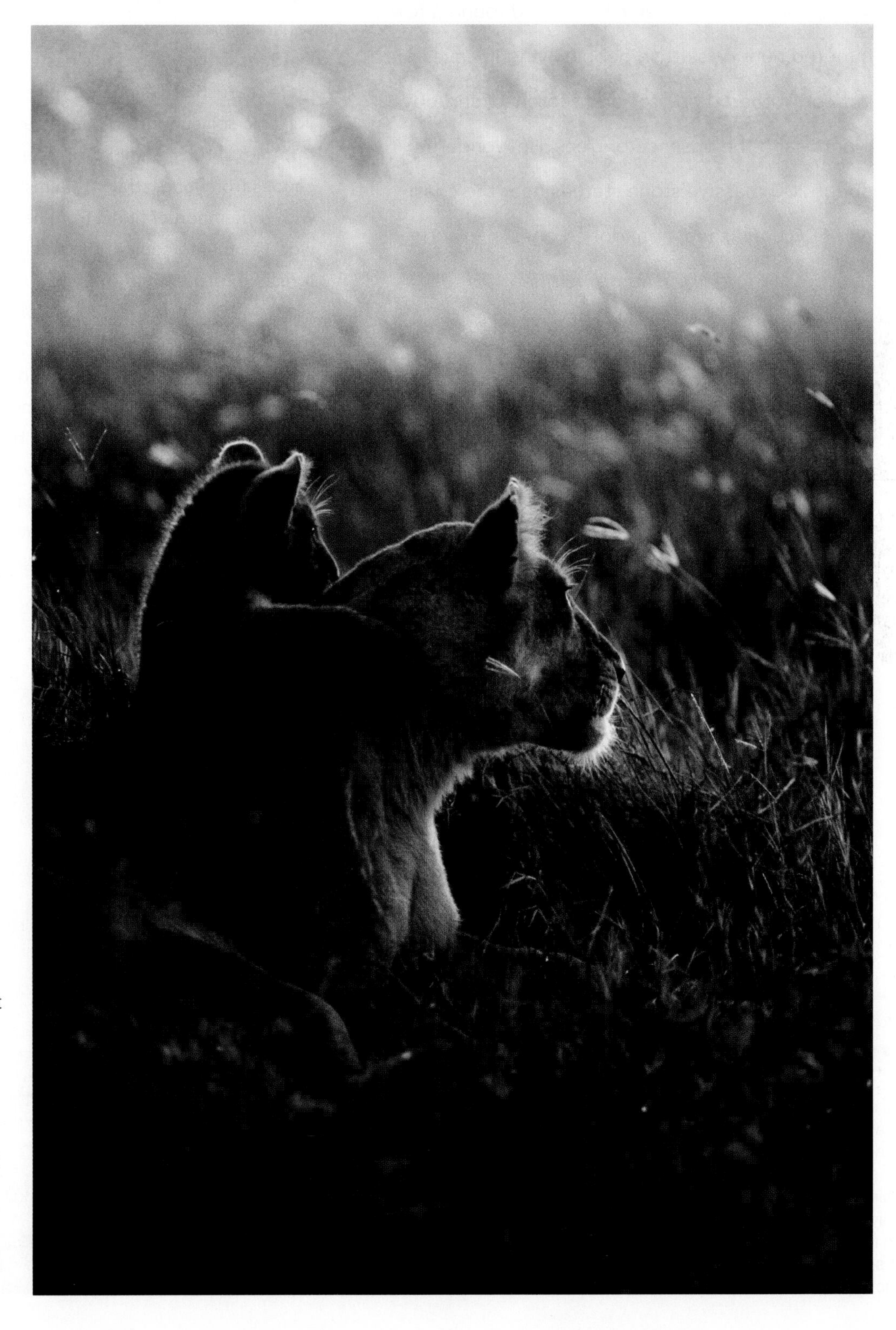

Lioness and her cub at sunset on the African savanna. What are the consequences of their interactions with each other, with other kinds of organisms, and with their environment? By the end of this last unit, you might find worlds within worlds in such photographs.

40 POPULATION ECOLOGY

IMPACTS, ISSUES The Numbers Game

In 1722, on Easter morning, a European explorer landed on a small volcanic island and found a few hundred hungry, skittish people living in caves. He noticed withered grasses and scorched, shrubby plants—and the absence of trees. He wondered about 200 massive stone statues near the coast and 500 unfinished, abandoned ones in inland quarries (Figure 40.1). Some weighed 100 tons and stood 10 meters (33 feet) high. There were no draft animals or wheeled carts on the island. How did the statues get from quarries to the coast? How were they erected?

Later, researchers solved the mystery of the statues. Easter Island, as it came to be called, is no larger than 165 square kilometers (64 square miles). Voyagers from the Marquesas discovered this eastern outpost of Polynesia more than 1,650 years ago. The place was a paradise. Its volcanic soil supported dense palm forests, hauhau trees, toromiro shrubs, and lush grasses. The new arrivals built canoes from long, straight palms that were strengthened with rope made of fibers from hauhau trees. They used wood as fuel to cook fishes and dolphins. They cleared forests to plant crops. They had many children.

By 1400, as many as 15,000 people were living on the island. Crop yields declined; ongoing harvests and erosion had depleted the soil of nutrients. Fish vanished from the waters close to the island, so fishermen had to sail farther and farther out on the open ocean.

Those in power built statues to appeal to the gods. They directed others to carve images of unprecedented size and move the new statues over miles of rough land to the coast. Like existing islanders, they probably lashed

See the video! Figure 40.1 Row of massive statues on Easter Island. Islanders set them up long ago, apparently as a plea for help after their once-large population wreaked havoc on their tropical paradise. Their plea had no effect whatsoever on reversing the loss in biodiversity on the island and in the surrounding sea. The human population did not recover, either.

How would you vote? Soaring numbers of white-tailed deer threaten forest plants and the animals that depend on them. Do you support encouraging deer hunting in regions where their overabundance is a threat? See ThomsonNOW for details, then vote online.

the statues to canoe-shaped rigs, which they then rolled along on a track assembled from greased logs.

Wars broke out over dwindling food and space. By 1550, no one ventured offshore to harpoon fishes or dolphins. They could not build canoes because there were no more trees.

Central authority crumbled. Dwindling numbers of islanders retreated to caves and launched raids against perceived enemies. Winners ate the losers and tipped over statues. Even if the survivors had wanted to, they had no way to get off the island. The once-flourishing population had collapsed.

Do you think Easter Island is just a quaint historical tale? Think of what is happening close to home in North America, where there are more white-tailed deer now than there were five centuries ago. Predators that held deer populations in check have been killed or scared off. Possibly 33 million deer are now munching leaves, bark, fruits, and seedlings all over the place. They threaten forest trees and forest communities. Year in, year out, deer accidentally cause 200 or so people to die on the highway and ruin $400 million worth of crops.

Fewer people hunt deer for sport. Hunting for the commercial market is banned. Injecting deer with birth control drugs is difficult and costly. Mountain lions and other predators could be reintroduced, but they would threaten people, livestock, and pets.

The point is, *certain principles govern the growth and sustainability of all populations over time*. The principles are the bedrock of ecology—the systematic study of how organisms interact with one another and with their environment. Those interactions start within and between populations and extend to communities, ecosystems, and the biosphere. They are the focus of this last unit. Later on in this chapter, we invite you to apply the same basic principles as we consider the past, present, and future of the human population.

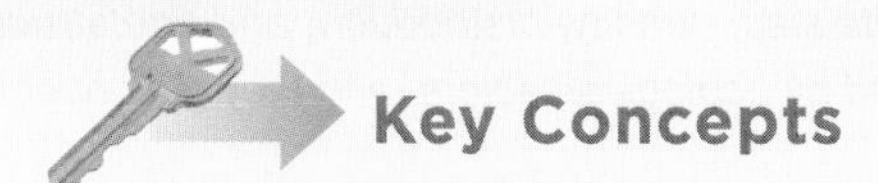

Key Concepts

THE VITAL STATISTICS

Ecologists explain population growth in terms of population size, density, distribution, and number of individuals in different age categories. They have methods of estimating population size and density in the field. **Sections 40.1, 40.2**

EXPONENTIAL RATES OF GROWTH

A population's size and reproductive base influence its rate of growth. When the population is increasing at a rate proportional to its size, it is undergoing exponential growth. **Section 40.3**

LIMITS ON INCREASES IN SIZE

Over time, exponential growth typically overshoots the carrying capacity—the maximum number of individuals of a population that environmental resources can sustain indefinitely. Some populations stabilize after a crash. Others never recover. **Section 40.4**

PATTERNS OF SURVIVAL AND REPRODUCTION

Resource availability, disease, and predation are major factors that can restrict population growth. These limiting factors differ among species and shape their life history patterns. **Sections 40.5, 40.6**

THE HUMAN POPULATION

Human populations have sidestepped limits to growth by way of global expansion into new habitats, cultural interventions, and technological innovations. No population can expand indefinitely. **Sections 40.7–40.10**

Links to Earlier Concepts

Earlier chapters defined and explored the evolutionary history and genetic nature of populations, including those of humans (Sections 1.4, 16.3, 17.1, 24.13). Now you will consider ecological factors that limit population growth. Again, science cannot address social issues—in this case, how exponential growth affects human lives (1.5). It can only help explain how ecological conditions and events sustain a population's growth, or put a stop to it.

40.1 Population Demographics

A population, again, is a group of individuals of the same species. Its size, density, distribution, and age structure are shaped by ecological factors, and they shift over time.

LINKS TO SECTIONS 10.7, 16.3, 17.1

So far, you have dipped into the gene pools and the evolutionary histories of populations. Turn now to their demographics: statistics that describe population size, age structure, density, distribution, and other factors.

Population size is the number of individuals that actually or potentially contribute to the gene pool. The age structure is the number of individuals in each of several age categories. For instance, individuals often are grouped by *pre-reproductive*, *reproductive*, and *post-reproductive* ages. Those in the first category have the capacity to produce offspring when mature. Together with individuals in the second category, they make up the population's reproductive base.

Population density is the number of individuals in some specified area or volume of a habitat. A habitat, remember, is the type of place where a species lives. We characterize a habitat by its physical and chemical features, and its particular array of species. Population distribution is the general pattern in which individuals are dispersed in a specified area.

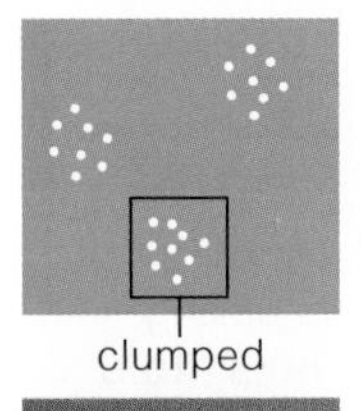

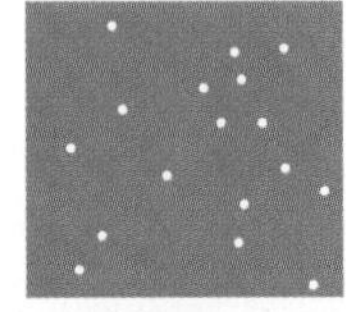

Crude density refers to how many individuals were counted in an area but not how they are dispersed through it. Even a habitat that looks uniform, such as a sandy shore, has variations in light, moisture, and many other variables. A given population may live in only a small part of the habitat, and it may do so all of the time or only some of the time.

Different species in a habitat typically compete for energy, nutrients, living space, and other resources. Such *interspecific* interactions influence the density and dispersal of neighboring populations.

Populations show clumped, nearly uniform, and random distribution patterns (Figure 40.2). Clumping occurs most frequently, for several reasons. First, the conditions and resources a species requires tend to be patchy. Animals cluster at a water hole, seeds sprout only in moist soil, and so on. Second, most seeds and some animal offspring cannot disperse far from their parents. Third, some animals gather in social groups that offer protection and other advantages.

With a nearly uniform distribution, individuals are more evenly spaced than we would expect on the basis of chance alone. Such distribution is relatively rare. It happens when competition for resources or territory is fierce, as in a nesting colony of seabirds.

We observe random distribution only when habitat conditions are nearly uniform, resource availability is fairly steady, and individuals of a population or pairs of them neither attract nor avoid one another. Each wolf spider does not hunt far from its burrow, which can be almost anywhere in forest soil (Figure 40.2).

The scale of the study area can affect the patterns we see. Seabirds often are spaced almost uniformly at a nesting site, but these sites may be clumped along a shoreline. Also, seabirds often crowd together when they breed, but disperse for the rest of the year.

Each population has characteristic demographics, such as size, density, distribution pattern, and age structure.

Environmental conditions and species interactions shape these characteristics, which may change over time.

Figure 40.2 Three patterns of population distribution: clumped, as in squirrelfish schools; more or less uniform, as in a royal penguin nesting colony; and random, as when wolf spiders dig their burrows almost anywhere in forest soil.

40.2 Elusive Heads to Count

FOCUS ON SCIENCE

Ecologists often do field studies to test assumptions about population dynamics and to monitor threatened or endangered populations.

Many, many white-tailed deer (*Odocoileus virginianus*) are dispersed through forests, fields, and suburbs of North America. How could you find out how many live near you?

A full count would be a careful measure of absolute population density. In the United States, census takers supposedly make such a count of human populations every ten years, although not everyone answers the door. Ecologists sometimes make counts of large species in small areas, such as birds in a forest, northern fur seals at their breeding grounds, and sea stars in a tidepool. More often, a full count would be impractical, so they sample part of a population and estimate its total density.

For instance, you could divide a map of your county into small plots, or quadrats. Quadrats are sampling areas of the same size and shape, such as rectangles, squares, and hexagons. You could count individual deer in several plots and, from that, extrapolate the average number for the county as a whole. Ecologists often do such estimates for plants and other species that stay put (Figure 40.3). They also make counts in small areas to help estimate the population sizes of migrating animals.

Ecologists use capture-recapture methods to estimate the population sizes of deer and other animals that do not stay put. First, they trap and mark individuals. Deer get collars, squirrels get tattoos, salmon get tags, birds get leg rings, butterflies get wing markers, and so forth (Figure 40.4). Marked animals are released at time 1. At time 2, traps are reset. The proportion of marked animals in the second sample is then taken to be representative of the proportion marked in the whole population:

$$\frac{\text{Marked individuals in sampling at time 2}}{\text{Total captured in sampling 2}} = \frac{\text{Marked individuals in sampling at time 1}}{\text{Total population size}}$$

Ideally, both marked and unmarked individuals of the population are captured at random, none of the marked animals are overlooked, and none die or otherwise depart during the study interval.

In the real world, recaptured individuals might not be a random sample; they might over- or underrepresent their population. Squirrels marked after being attracted to bait in boxes might now be trap-happy or trap-shy. Instead of mailing tags of marked fish to ecologists, a fisherman may keep them as souvenirs. Birds lose leg rings.

Estimates also may depend on the time of year that they are made. The distribution of natural populations varies with time, as when animals migrate in response to environmental rhythms. Few environments yield abundant resources all year long, so many populations must move between habitats as seasons change. Canada geese are like this. So are deer. In such cases, capture-recapture methods might be more revealing when used more than once a year, for several years.

Figure 40.3 Easy-to-count creosote bushes near the eastern base of the Sierra Nevada. They are an example of a relatively uniform distribution pattern. Individual plants compete for scarce water in this desert, which has extremely hot, dry summers and mild winters.

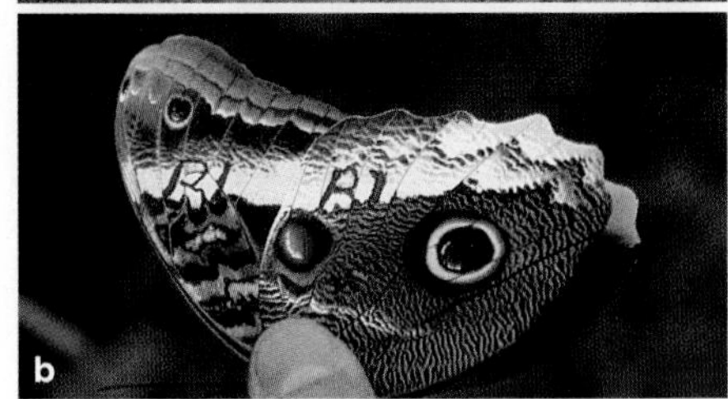

Figure 40.4 Two individuals marked for population studies. (**a**) Florida Key deer and (**b**) Costa Rican owl butterfly (*Caligo*).

40.3 Population Size and Exponential Growth

Populations are dynamic units of nature. Depending on the species, they may add or lose individuals every minute of every day, every season, or over years. New individuals may end up glutting the habitat. A habitat may be emptied of them when the population drives itself or is driven to extinction.

GAINS AND LOSSES IN POPULATION SIZE

LINKS TO SECTIONS 17.8, 19.1

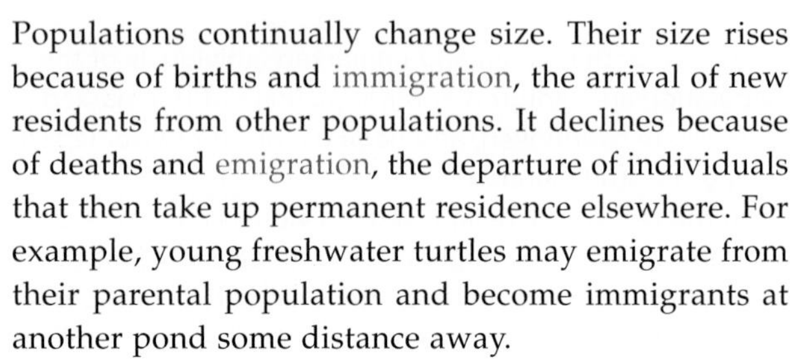

Populations continually change size. Their size rises because of births and immigration, the arrival of new residents from other populations. It declines because of deaths and emigration, the departure of individuals that then take up permanent residence elsewhere. For example, young freshwater turtles may emigrate from their parental population and become immigrants at another pond some distance away.

What about the individuals of species that migrate daily or seasonally? A migration is a recurring round-trip between regions, usually in response to expected shifts or gradients in environmental resources. Some or all members of a population leave an area, spend time in another area, then return. For our purposes, we may ignore these recurring gains and losses, because we can assume that they balance out over time.

FROM ZERO TO EXPONENTIAL GROWTH

Set aside the self-canceling effects of immigration and emigration, and we can define zero population growth as an interval during which the number of births is balanced by an equal number of deaths. Population size remains stable, with no net increase or decrease in the number of individuals.

Let's measure births and deaths in terms of rates per individual, or per capita. *Capita* means head, as in a head count. Subtract a population's per capita death rate (d) from its per capita birth rate (b) and you have the per capita growth rate, or r:

$$\underset{\text{(per capita growth rate)}}{r} = \underset{\text{(per capita birth rate)}}{b} - \underset{\text{(per capita death rate)}}{d}$$

Imagine 2,000 mice living in the same cornfield. If 1,000 mice are born each month, then the birth rate is 0.5 per mouse per month (1,000 births/2,000 mice). If 200 mice die one way or another each month, then the death rate is 200/2,000 = 0.1 per mouse per month. In this case, r is 0.5 – 0.1 = 0.4 per mouse per month.

As long as r remains constant, exponential growth will occur: Population size will increase by the exact

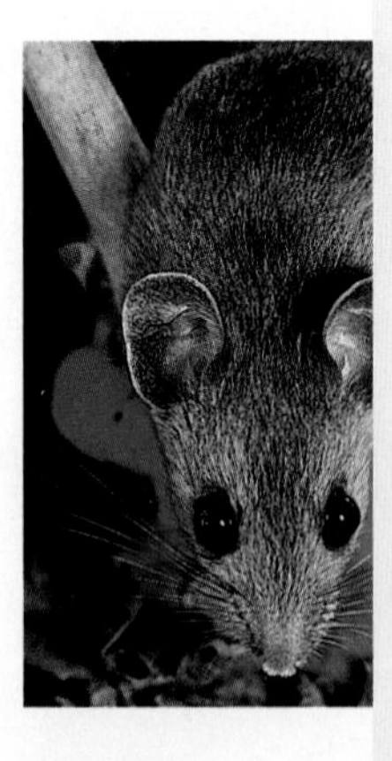

	Starting Population Size		Net Monthly Increase	New Population Size
$G = r \times$	2,000	=	800	2,800
$r \times$	2,800	=	1,120	3,920
$r \times$	3,920	=	1,568	5,488
$r \times$	5,488	=	2,195	7,683
$r \times$	7,683	=	3,073	10,756
$r \times$	10,756	=	4,302	15,058
$r \times$	15,058	=	6,023	21,081
$r \times$	21,081	=	8,432	29,513
$r \times$	29,513	=	11,805	41,318
$r \times$	41,318	=	16,527	57,845
$r \times$	57,845	=	23,138	80,983
$r \times$	80,983	=	32,393	113,376
$r \times$	113,376	=	45,350	158,726
$r \times$	158,726	=	63,490	222,216
$r \times$	222,216	=	88,887	311,103
$r \times$	311,103	=	124,441	435,544
$r \times$	435,544	=	174,218	609,762
$r \times$	609,762	=	243,905	853,667
$r \times$	853,667	=	341,467	1,195,134

a

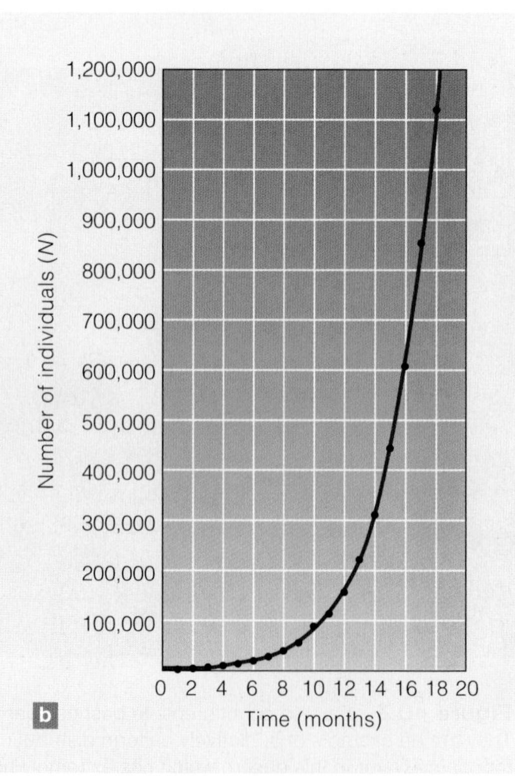

Figure 40.5 Animated! (**a**) Net monthly increases in a hypothetical population of mice when the per capita rate of growth (r) is 0.4 per mouse per month and the starting population size is 2,000.

(**b**) Graph these numerical data and you end up with a J-shaped growth curve.

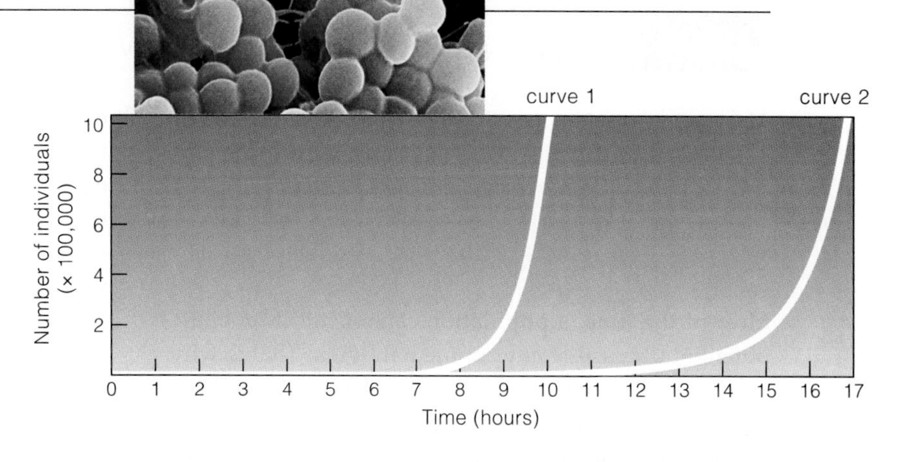

Figure 40.6 Effect of deaths on the rate of increase for two hypothetical populations of bacteria. Plot the population growth for bacterial cells that reproduce every half hour and you get growth curve 1. Next, plot the population growth of bacterial cells that divide every half hour, with 25 percent dying between divisions, and you get growth curve 2. Deaths slow the rate of increase, but as long as the birth rate exceeds the death rate, exponential growth will continue.

same proportion of its total in every successive time interval. We can calculate population growth (G) for each interval based on the per capita growth rate and the number of individuals (N):

$$\underset{\text{(population growth per unit time)}}{G} = \underset{\text{(per capita growth rate)}}{r} \times \underset{\text{(number of individuals)}}{N}$$

After one month, 2,800 mice are scurrying about in the field (Figure 40.5*a*). A net increase of 800 fertile mice has made the reproductive base larger. They all reproduce, so the population size expands, for a net increase of $0.4 \times 2{,}800 = 1{,}120$. Population size is now 3,920. At this growth rate, the number of mice would rise from 2,000 to more than 1 million in two years! Plot the increases against time and you end up with a curve, as shown in Figure 40.5*b*. Such a J-shaped curve is evidence of exponential growth.

With exponential growth, a population grows faster and faster, although the per capita growth rate stays the same. It is like the compounding of interest on a bank account. The annual interest *rate* stays fixed, yet every year the *amount* of interest paid increases. Why? The annual interest paid into the account adds to the size of the balance, and the next interest payment will be based on that balance.

In exponentially growing populations, r is like the interest rate. Although r remains constant, population growth accelerates as the population size increases. When 6,000 individuals reproduce, population growth is three times higher than it was when there were only 2,000 reproducers.

As another example, think of a single bacterium in a culture flask. After thirty minutes, the cell divides in two. Those two cells divide, and so on every thirty minutes. If no cells die between divisions, then the population size will double in every interval—from 1 to 2, then 4, 8, 16, 32, and so on. The time it takes for a population to double in size is its doubling time.

After 9–1/2 hours, or nineteen doublings, there are more than 500,000 cells. Ten hours (twenty doublings) later, there are more than a million. Curve 1 in Figure 40.6 is a plot of this outcome.

Will deaths put the brakes on exponential growth? Suppose 25 percent of the descendant cells die every thirty minutes. It now requires about seventeen hours, not ten, for that population to reach 1 million. *Deaths slow the rate of increase but do not stop exponential growth* (curve 2 in Figure 40.6). Exponential growth might be slow or fast, but as long as birth rates exceed death rates, it can continue.

WHAT IS THE BIOTIC POTENTIAL?

Now imagine a population living in an ideal habitat, free of all pollutants, predators, and pathogens. Every individual has plenty of shelter, food, and other vital resources. Under such conditions, a population would reach its biotic potential: the maximum possible per capita rate of increase for its species.

All species have a characteristic biotic potential. For many bacteria, it is 100 percent every half hour or so. For humans, it is about 2 to 5 percent per year.

Actual rates of growth depend on many factors. The population's age distribution, how often individuals reproduce, and how many offspring they produce in their lifetime are examples. The human population is not currently displaying its biotic potential, but it still is growing exponentially. The human population will be our focus later in the chapter.

The size of a population depends on its rates of births, deaths, immigration, and emigration.

Subtract the per capita death rate from the per capita birth rate to get r*, the per capita growth rate of a population. As long as* r *is constant and greater than one, a population will grow exponentially. With exponential growth, the number of individuals increases at an ever accelerating rate.*

The biotic potential of a species is its maximum possible population growth rate under optimal conditions.

40.4 Limits on Population Growth

Many complex interactions take place within and between populations in nature, and it is not always easy to identify all the factors that can restrict population growth.

ENVIRONMENTAL LIMITS ON GROWTH

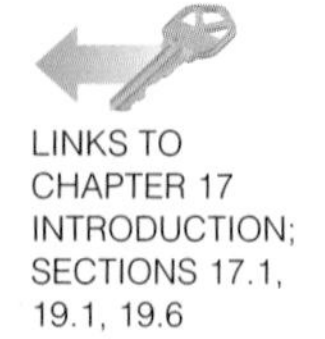

LINKS TO CHAPTER 17 INTRODUCTION; SECTIONS 17.1, 19.1, 19.6

Most of the time, a population cannot fulfill its biotic potential because of environmental limits. That is why sea stars—the females of which could make 2,500,000 eggs each year—do not fill the oceans with sea stars.

Any essential resource that is in short supply is a limiting factor on population growth. Food, mineral ions, refuge from predators, and safe nesting sites are examples (Figure 40.7). Many factors can potentially limit population growth. Even so, in any environment, one essential factor will kick in first, and it will act as the brake on population growth.

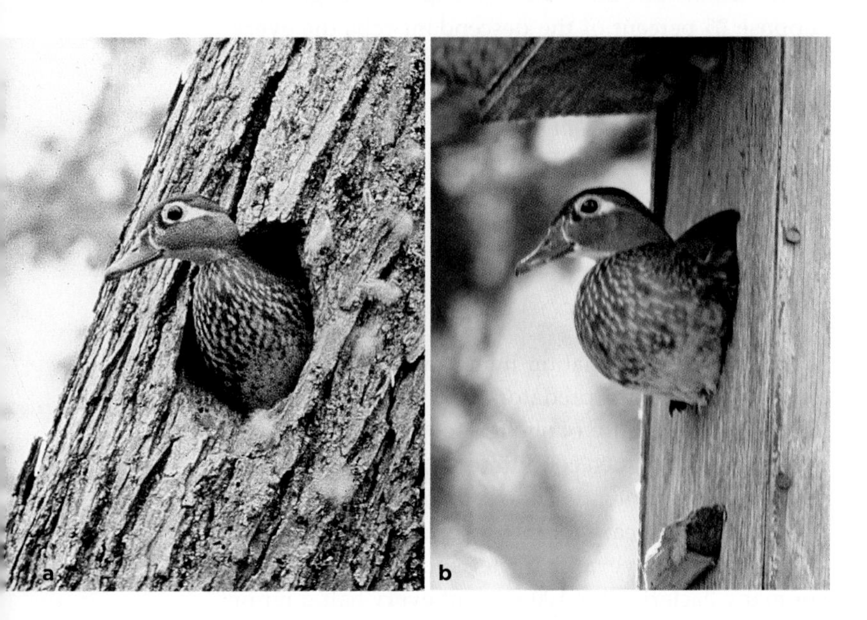

Figure 40.7 One example of a limiting factor. (**a**) Wood ducks build nests only inside cavities of specific dimensions. With the clearing of old growth forests, the access to natural cavities of the correct size and position is now a limiting factor on wood duck population size. (**b**) Artificial nesting boxes are being placed in preserves to help ensure the health of wood duck populations.

To get a sense of the limits on growth, start again with a bacterial cell in a culture flask, where you can control the variables. First, enrich the culture medium with glucose and other nutrients required for bacterial growth. Next, let many generations of cells reproduce.

Initially, growth may be exponential. Then it slows, and population size remains relatively stable. After a brief stable period, population size plummets until all the bacterial cells are dead. *What happened?* The larger population required more nutrients. In time, nutrient levels declined, and the cells could no longer divide. Even after cell division stopped, existing cells kept on taking up and using nutrients. Eventually, when the nutrient supply was exhausted, the last cells died out.

Suppose you kept freshening the nutrient supply. The population would still eventually collapse. Like other organisms, bacteria generate metabolic wastes. Over time, accumulation of this waste would pollute the habitat and halt growth. Adding nutrients simply substitutes one limiting factor for another. All natural populations run up against limits eventually.

CARRYING CAPACITY AND LOGISTIC GROWTH

Carrying capacity refers to the maximum number of individuals of a population that a given environment can sustain indefinitely. Ultimately, it means that the *sustainable* supply of resources determines population size. We can use the pattern of logistic growth, shown in Figure 40.8, to reinforce this point. By this pattern, a small population starts growing slowly in size, then it grows rapidly, then its size levels off as the carrying capacity is reached.

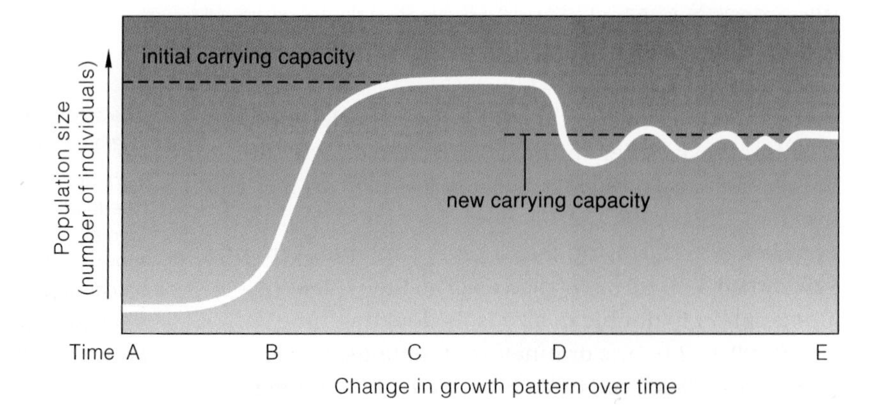

Figure 40.8 Animated! Idealized S-shaped curve characteristic of logistic growth. After a rapid growth phase (time B to C), growth slows and the curve flattens as carrying capacity is reached (time C to D). In the real world, growth curves vary more, as when a change in the environment lowers carrying capacity (time D to E). That happened to the human population of Ireland in the mid-1800s. Late blight, a disease caused by a water mold, destroyed the potato crop that was the mainstay of Irish diets (Section 20.6).

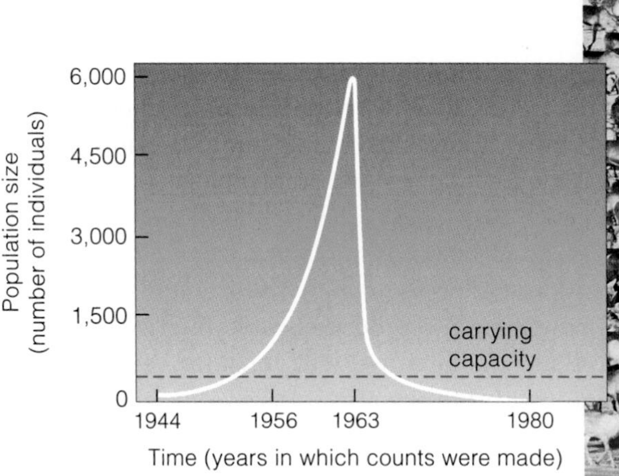

Logistic growth plots out as an S-shaped curve, as shown in Figure 40.8 (time A to C). In equation form,

population growth per unit time	=	maximum per capita population growth rate	×	number of individuals	×	proportion of resources not yet used

An S-shaped curve is simply an approximation of what goes on in nature. For instance, a population that grows too fast may drastically overshoot the carrying capacity. Figure 40.9 focuses on a small population of reindeer. As population size increased, more and more individuals competed for resources such as food and shelter, so each reindeer received a smaller share. More individuals died of starvation and fewer young were born. Deaths began to outnumber births. Finally, the death rate soared and the birth rate plummeted.

Figure 40.9 Graph of changes in a reindeer population that exceeded its habitat's carrying capacity (*blue* dashed line) and did not recover.

In 1944, during World War II, a United States Coast Guard crew established a station on St. Matthew, an island 320 kilometers (200 miles) west of Alaska in the Bering Sea. They brought in 29 reindeer as a back-up food source. Reindeer eat lichens. Thick mats of lichens cloaked the island, which is no more than 51 kilometers long and 6.4 kilometers (32 miles by 4 miles) across. World War II drew to a close before any reindeer were shot. The Coast Guard pulled out, leaving behind seabirds, arctic foxes, voles—and a herd of healthy reindeer with no predators big enough to hunt them down.

In 1957, biologist David Klein visited St. Matthew. On a hike from one end of the island to the other, he counted 1,350 well-fed reindeer and saw trampled and overgrazed lichens. In 1963, Klein and three other biologists returned to the island. They counted 6,000 reindeer. They could not help but notice the profusion of reindeer tracks and feces, and a lot of squashed lichens.

Klein returned to St. Matthew in 1966. Bleached-out reindeer bones littered the island. Forty-two reindeer were still alive. Only one was a male; it had abnormal antlers, which made it unlikely to reproduce. There were no fawns. Klein figured out that thousands of reindeer had starved to death during the unusually harsh winter of 1963–1964. By the 1980s, there were no reindeer on the island at all.

TWO CATEGORIES OF LIMITING FACTORS

Density-dependent factors lower reproductive success and appear or worsen with crowding. Competition for limited resources causes density-dependent effects, as does disease. Pathogens and parasites can spread more easily when hosts are crowded together. The chapter 17 introduction described how human populations in cities support huge numbers of rats that carry bubonic plague, typhus, and other deadly infectious diseases.

Density-dependent factors control population size through negative feedback. High density causes these factors to come into play, then their effects act to lower population density. A logistic growth pattern results from this feedback effect.

Density-independent factors decrease reproductive success too, but their likelihood of occurring and their magnitude of effect are unaffected by crowding. Fires, snow storms, earthquakes, and other natural disasters affect crowded and uncrowded populations alike. For example, in December of 2004, a powerful tsunami (a giant wave caused by an earthquake) hit Indonesia. It killed about 250,000 people. The degree of crowding did not make the tsunami any more or less likely to happen, or to strike any particular island. The logistic growth equation cannot be used to predict effects of density-independent factors.

Carrying capacity is the maximum number of individuals of a population that can be sustained indefinitely by the resources in a given environment.

With logistic growth, population growth is fast during times of low density, then it slows as the population approaches carrying capacity, where numbers level off.

Density-dependent factors such as disease promote a pattern of logistic growth. Density-independent factors such as natural disasters also affect population size.

40.5 Life History Patterns

By studying many kinds of species, researchers have identified age-specific adaptations that have impacts on individual survival, fertility, and reproduction.

LINKS TO SECTIONS 16.3, 17.1

So far, you have looked at populations as if all of their members are all identical with regard to age. For most species, however, individuals that make up a group are at many different stages of development. Often, those stages require different resources, as when caterpillars that eat leaves later develop into butterflies that sip nectar. In addition, individuals might be more or less vulnerable to danger at different stages.

In short, each species has a life history pattern. It has a set of adaptations that affect when an individual starts reproducing, how many offspring it has at one time, how often it reproduces, and other traits. In this section and the next, we will consider variables that underlie these age-specific patterns.

Table 40.1 Life Table for an Annual Plant Cohort*

Age Interval (days)	Survivorship (number surviving at start of interval)	Number Dying During Interval	Death Rate (number dying/ number surviving)	"Birth" Rate During Interval (number of seeds from each plant)
0–63	996	328	0.329	0
63–124	668	373	0.558	0
124–184	295	105	0.356	0
184–215	190	14	0.074	0
215–264	176	4	0.023	0
264–278	172	5	0.029	0
278–292	167	8	0.048	0
292–306	159	5	0.031	0.33
306–320	154	7	0.045	3.13
320–334	147	42	0.286	5.42
334–348	105	83	0.790	9.26
348–362	22	22	1.000	4.31
362–	0	0	0	0
		996		

* *Phlox drummondii*; data from W. J. Leverich and D. A. Levin, 1979.

Table 40.2 Life Table for Humans in the United States (based on 2003 conditions)

Age Interval	Number at Start of Interval	Number Dying During Age Interval	Life Expectancy at Start of Interval	Reported Live Births
0–1	100,000	687	77.5	
1–5	99,313	124	77.0	
5–10	99,189	73	73.1	
10–15	99,116	95	68.2	6,781
15–20	99,022	328	63.2	415,262
20–25	98,693	474	58.4	1,034,454
25–30	98,219	467	53.7	1,104,485
30–35	97,752	542	48.9	965,633
35–40	97,210	767	44.2	475,606
40–45	96,444	1,157	39.5	103,679
45–50	95,287	1,702	35.0	5,748
50–55	93,585	2,401	30.6	374
55–60	91,185	3,425	26.3	
60–65	87,760	5,092	22.2	
65–70	82,668	7,133	18.4	
70–75	75,535	9,825	14.9	
75–80	65,710	12,969	11.8	
80–85	52,741	15,753	9.0	
85–90	36,988	15,648	6.8	
90–95	21,340	12,363	5.0	
95–100	8,977	6,614	3.6	
100+	2,363	2,363	2.6	

LIFE TABLES

Each species has a characteristic life span, but only a few individuals survive to the maximum age possible. Death looms larger at particular ages. Individuals tend to reproduce during an expected age interval and, for most species, tend to die at particular times.

Age-specific patterns in populations intrigue life insurance and health insurance companies as well as ecologists. Such investigators often track a cohort—a group of individuals born during the same interval—from their time of birth until the last one dies (Table 40.1). They also record the number of offspring born to cohort members during each age interval.

A *periodic* life table is not based on an actual cohort. It uses information about current conditions to predict births and deaths for a hypothetical group. Table 40.2 is a periodic life table for humans based on conditions in the United States during 2003.

We often get useful information when we divide a population into age classes and establish age-specific birth rates and mortality risks. Unlike a crude head count, the data help us make decisions on issues such as pest management, habitats of endangered species, and social planning for human populations. Birth and death schedules for the northern spotted owl are one case in point. They were cited in federal court rulings that halted mechanized logging in the owl's habitat—old-growth forests of the Pacific Northwest.

PATTERNS OF SURVIVAL AND REPRODUCTION

We measure reproductive success of individuals in terms of number of surviving offspring (Section 16.3). Success turns on allocating energy and time to make gametes, secure mates, and often give parental care to offspring of one size or another. All organisms make trade-offs. Those with a massive body might be less vulnerable to predators, but they are slower to mature sexually and large offspring are expensive to produce. Small ones that produce many offspring increase the

Figure 40.10 Three generalized survivorship curves. (**a**) Elephants are Type I populations, with high survivorship until some age, then high mortality. (**b**) Snowy egrets are Type II populations, with a fairly constant death rate. (**c**) Sea star shows a Type III pattern, with survivorship lowest early on. Their tiny larvae, like this one, are highly vulnerable to predation.

odds some offspring will not get eaten. But they can allocate just a smidgen of resources, such as yolk, to each one.

Survivorship patterns reflect major differences. A survivorship curve is a graph line that emerges when you plot a cohort's age-specific survival in its habitat. Each species has a characteristic survivorship curve. Three types are common in nature.

Type I Curves A *type I* curve typifies a population with high survivorship until late in life, then a large increase in deaths. Large animals that bear one or, at most, a few offspring at a time and give them extended parental care show this pattern (Figure 40.10*a*). For example, a female elephant has four or five calves in her lifetime and cares for each one for several years.

Type I curves are typical of human populations when individuals have access to health care services. In regions where health care is poor, the survivorship curve plunges sharply, because of many infant deaths. After the plunge, the curve levels off from childhood to early adulthood.

Type II Curves A *type II* curve shows up when the death rate for the population is fairly constant at all ages. They are typical of organisms just as likely to be killed or die of disease at any age, such as lizards, small mammals, and large birds (Figure 40.10*b*).

Type III Curves A *type III* curve shows up when the death rate for a population is highest early in life. It is typical of species that produce many small offspring and provide little, or no, parental care. Figure 40.10*c* shows how the curve plummets for sea stars, which release great numbers of eggs. Sea star larvae are tiny and they drift in open ocean waters. Fish, snails, and sea slugs devour most of them before protective hard parts can develop. The type III curve is common for marine invertebrates, insects, fishes, fungi, as well as annual plants such as phlox (Table 40.1).

At one time, ecologists thought selection processes favored *either* early, rapid production of many small offspring *or* late production of a few large offspring. They now see that the two patterns are only extremes at opposite ends of a range of possible life histories.

Tracking a cohort (a group of individuals) from birth until the last one dies reveals patterns of reproduction, death, and migration that typify the populations of a species.

Survivorship curves can reveal differences in age-specific survival among species. In some cases, differences occur even between populations of the same species.

40.6 Natural Selection and Life Histories

Genetic differences underlie variations in life history traits. This means that, like other inherited traits, life history traits can be shaped by natural selection. Here we describe two studies that show how predation can serve as selective pressure that alters life histories.

LINKS TO SECTIONS 1.4, 16.3, 17.8

Predation on Guppies in Trinidad Several years ago, two evolutionary biologists drenched with sweat and clutching fishnets were wading through a stream. John Endler and David Reznick were in the mountains of Trinidad, an island in the southern Caribbean Sea. They wanted to capture guppies (*Poecilia reticulata*), small fishes that live in the shallow freshwater streams (Figure 40.11). The biologists were starting what would become a long-term study of guppy traits.

Male guppies are generally smaller and have more colorful scales than females of the same age. The colors function as visual signals during courtship rituals. The females are drab colored. Unlike males, they continue to grow after reaching sexual maturity.

Reznick and Endler were interested in how predators influence the guppy life history. For their study sites, they decided on streams with many small waterfalls. These waterfalls are barriers that prevent guppies in one part of a stream from moving easily to another. As a result, each stream holds several populations of guppies that have very little gene flow between them (Section 17.8).

a *Right*, guppy that shared a stream with killifishes (*below*).

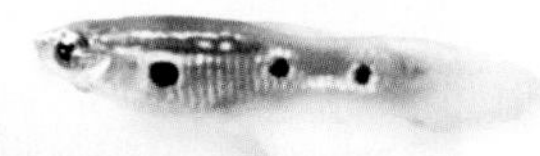

b *Right*, guppy that shared a stream with cichlids (*below*).

Figure 40.11 (**a**,**b**) Guppies and two guppy eaters, a killifish (**a**) and a cichlid (**b**). (**c**) Biologist David Reznick contemplating interactions among guppies and their predators in a freshwater stream in Trinidad.

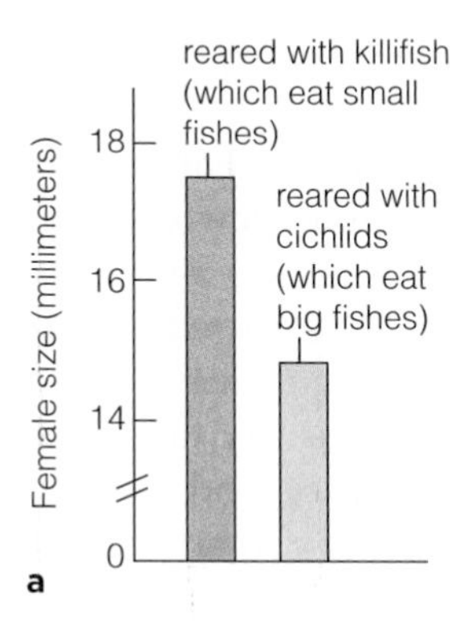

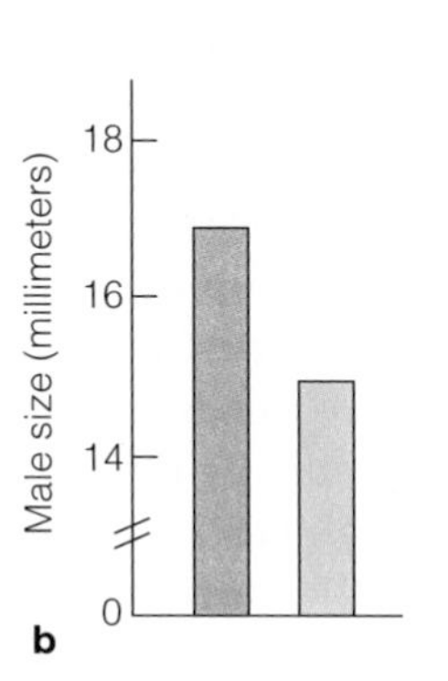

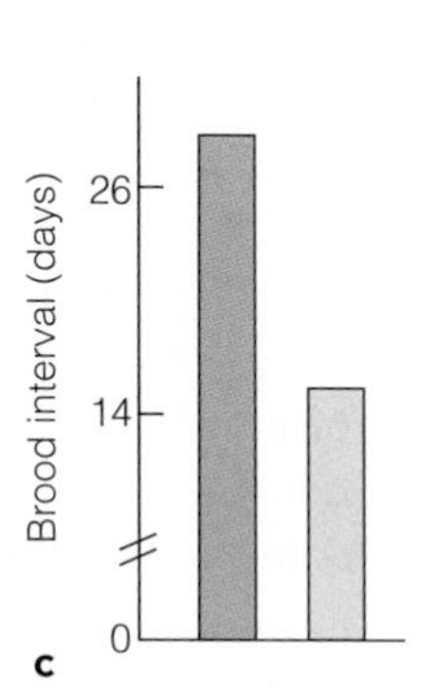

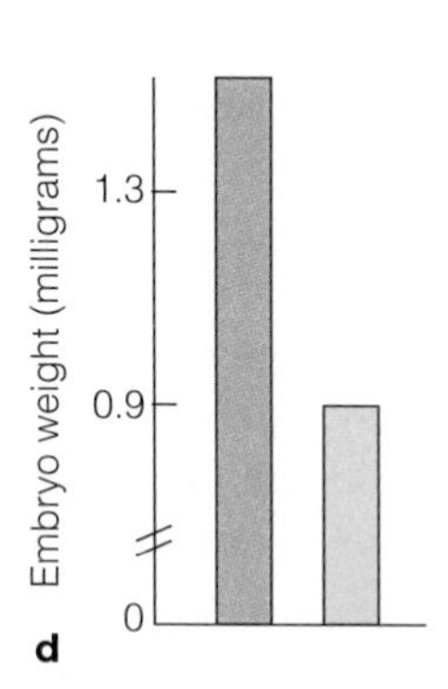

Figure 40.12 Experimental evidence of natural selection among guppy populations subject to different predation pressures. Compared to the guppies raised with killifish (*green* bars), guppies raised with cichlids (*tan* bars) differed in body size and in the length of time between broods.

The waterfalls also keep guppy predators from moving into different parts of the stream. In this habitat, the main guppy predators are a killifish species and cichlid species. The two differ in size and prey preferences. The killifish is relatively small and preys mostly on immature guppies. It ignores the larger adults. The cichlids are bigger fish. They tend to pursue mature guppies and pass on the small ones. Some parts of the streams hold one type of predator but not the other. Different guppy populations face different predation pressures.

As Reznick and Endler discovered, guppies in streams with cichlids grow faster and are smaller at maturity than those in streams with killifish (Figure 40.12). Also, guppies hunted by cichlids reproduce earlier, have more offspring at a time, and breed more frequently.

Were these differences in life history traits genetic, or did some environmental variation cause them? To find out, the biologists collected guppies from both cichlid- and killifish-dominated streams. They reared the groups in separate aquariums under identical predator-free conditions. Two generations later, the groups continued to show the differences observed in natural populations. The conclusion? *Differences between guppies preyed upon by different predators have a genetic basis.*

Reznick and Endler hypothesized that the predators are selective agents that have shaped guppy life history patterns. They made a prediction: If life history traits evolve in response to predation, then they will change when a population is exposed to a novel predator.

To test their prediction, they found a stream region above a waterfall that had killifish but no guppies or cichlids. They brought in some guppies from a region below the waterfall where there were cichlids but no killifish. At the experimental site, the guppies that had previously lived only with cichlids were now exposed to killifish. The control site was the downstream region below the waterfall, where relatives of the transplanted guppies still coexisted with cichlids.

Reznik and Endler revisited the stream over the course of eleven years and thirty-six generations of guppies. They monitored traits of guppies above and below the waterfall. The recorded data showed that guppies at the upstream experimental site were evolving. Exposure to a novel predator had caused big changes in their rate of growth, age at first reproduction, and other life history traits. By contrast, guppies at the control site showed no such changes. As Reznick and Endler concluded, *life history traits in guppies can evolve rapidly in response to the selective pressure exerted by predation.*

Overfishing and the Atlantic Cod The evolution of life history traits in response to predation pressure is not merely interesting. It has commercial importance. Just as guppies evolved in response to predators, so is the North Atlantic codfish (*Gadus morhua*) evolving—in this case, in response to humans. North Atlantic codfish are big (*below*). However, from the mid-1980s to early 1990s, fishing pressure on them increased. They became sexually mature at increasingly early ages, when they were much smaller in size. The fast-maturing, petite individuals were at a selective advantage. Commercial fisherman, not just sports fishermen, preferred to catch the big ones.

In 1992, the government of Canada banned cod fishing in some areas. That ban, and later restrictions, came too late to stop the Atlantic cod population from plummeting. The population still has not recovered.

Looking back, it seems as if life history changes were an early sign that the cod population was in trouble. Had biologists recognized the sign, they might have been able to save the fishery and protect the livelihood of more than 35,000 fishers and associated workers. Ongoing monitoring of the life history data for other economically important fishes may help prevent similar disastrous crashes in the future.

40.7 Human Population Growth

Human population size surpassed 6.6 billion in 2006. Take a look now at what the number means.

THE HUMAN POPULATION TODAY

LINKS TO SECTIONS 19.6, 24.13

In 2005, the estimated average rate of increase for the human population was 1.2 percent. As long as birth rates continue to exceed death rates, annual additions will drive a *larger* absolute increase each year into the foreseeable future.

Human population size continues to expand, even though limiting factors already confront more than 1 billion people. Many do not have adequate food, clean drinking water, shelter, access to health care systems, and waste treatment facilities. Overcrowded regions of the world are becoming ever more crowded. Figure 40.13 is a graphic clue to what the expansion means with respect to the carrying capacity.

Even if it were possible to double food supplies to keep pace with growth, living conditions would still be marginal for most people. At least 10 million would continue to die each year from starvation.

What happens when our population size doubles again? Can you brush the doubling aside as being too far in the future? *It is no farther removed from you than the sons and daughters of the next generation.*

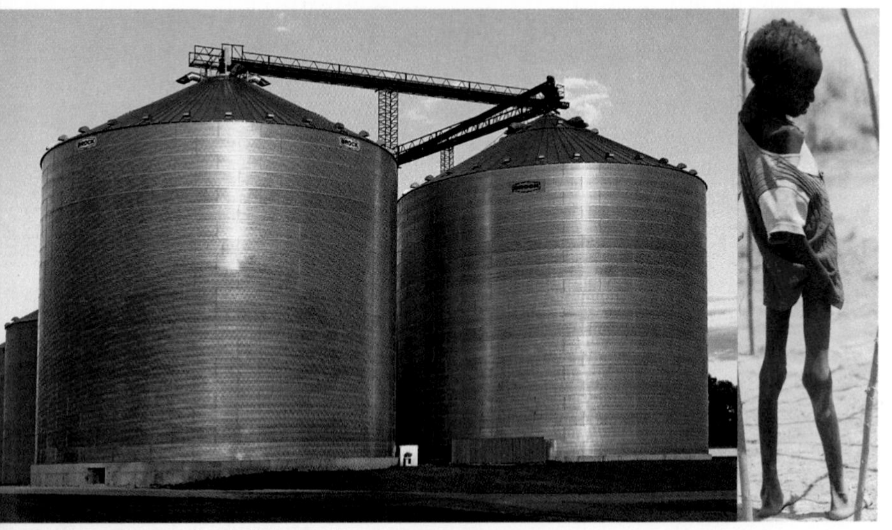

Banks of corn silos in Wisconsin

Figure 40.13 Far from well-fed humans in highly developed countries, an Ethiopian child shows the effects of starvation. Ethiopia is one of the poorest developing countries, with an annual per capita income of $120. Average caloric intake is more than 25 percent below the minimum necessary to maintain good health. Malnutrition stunts the growth, weakens the body, and impairs the brain development of about half of Ethiopia's children. Despite ongoing food shortages, Ethiopia's population has one of the highest annual rates of increase in the world. If growth continues at its current rate, the population of 75 million would be expected to double in less than 25 years.

EXTRAORDINARY FOUNDATIONS FOR GROWTH

How did we get into this predicament? For most of its history, the human population grew slowly. Things started to pick up about 10,000 years ago and in the past two centuries, growth rates have soared (Figure 40.14). Three trends promoted the big increases. First, humans gradually developed the capacity to expand into new habitats and climate zones. Second, humans increased the carrying capacity of existing habitats. Third, humans sidestepped some limiting factors that tend to restrain the growth of other species.

Geographic Expansion Reflect on the first trend. Early human populations evolved in dry woodlands, then in savannas. We assume they were vegetarians, mostly, but they also scavenged bits of meat. Bands of hunter–gatherers moved out of Africa about 2 million years ago. By 40,000 years ago, their descendants were established in much of the world (Section 24.13).

Few species can expand into such a broad range of habitats, but early humans had the necessary skills. They learned how to start fires, build shelters, make clothing, produce tools, and cooperate in hunts. With the advent of language, knowledge of such skills did not die with the individual. *Compared to most species, humans displayed a greater capacity to disperse fast over long distances and to become established in physically challenging new environments.*

Increased Carrying Capacity Starting about 11,000 years ago, bands of hunter–gatherers were shifting to agriculture. Instead of counting on the migratory game herds, they were settling in fertile valleys and other regions that favored seasonal harvesting of fruits and grains. They developed a more dependable basis for life. A pivotal factor was the domestication of wild grasses, including species ancestral to modern wheat and rice. Now people harvested, stored, and planted seeds all in one place. They domesticated animals as sources of food and to pull plows. They dug irrigation ditches and diverted water to croplands.

Agricultural productivity became a basis for increases in population growth rates. Towns and cities formed. Later, food supplies increased yet again. Farmers started to use chemical fertilizers, herbicides, and pesticides to protect their crops. Transportation and food distribution improved. *Even at its simplest, the management of food supplies through agricultural practices increased the carrying capacity for the human population.*

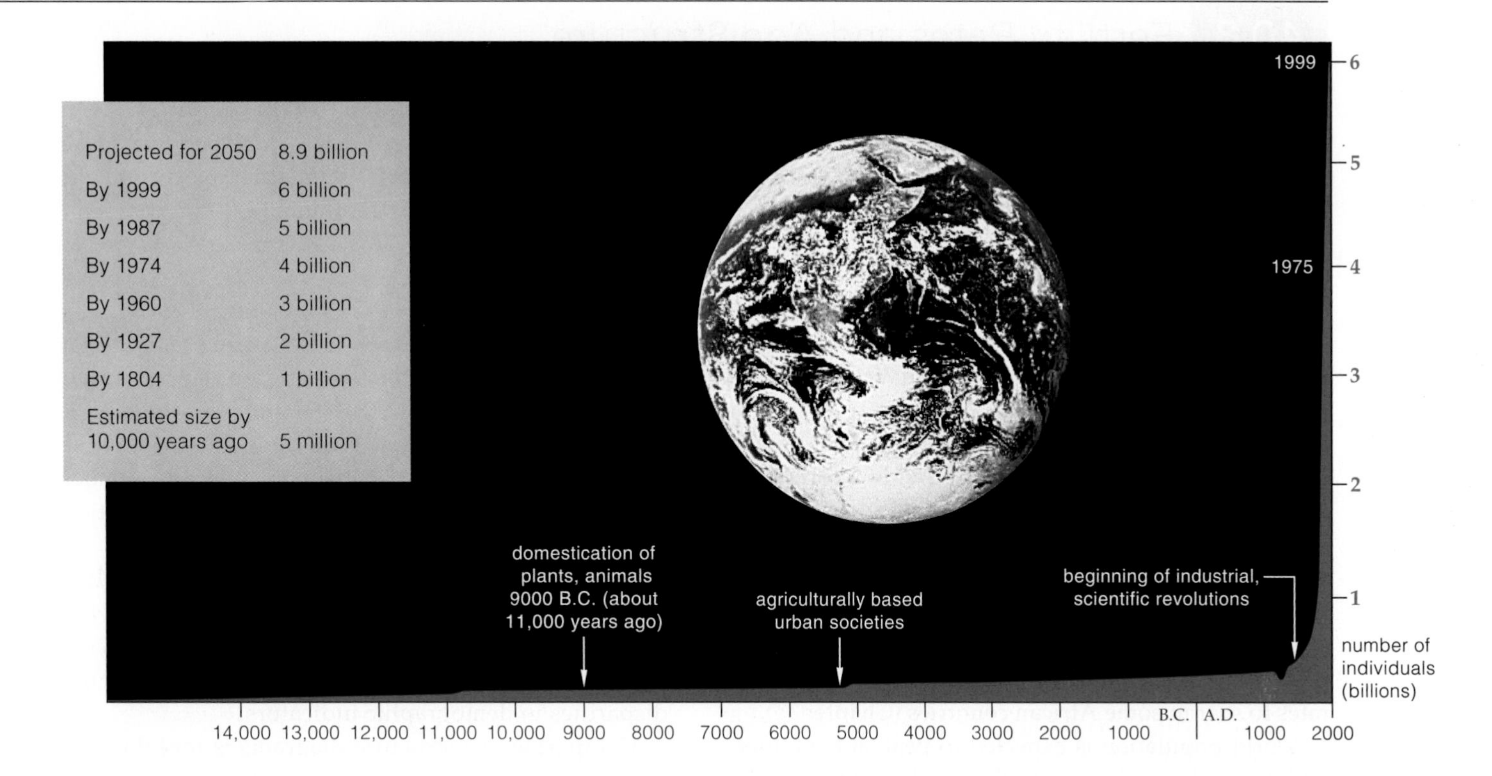

Figure 40.14 Growth curve (*red*) for the world human population. The *blue* box indicates how long it took for the human population to increase from 5 million to 6 billion. The dip between years 1347 and 1351 marks the time when 60 million people died during a pandemic that may have been a bubonic plague.

Sidestepped Limiting Factors Until about 300 years ago, malnutrition, poor hygiene, and infectious diseases kept death rates high enough to more or less balance birth rates. Infectious diseases were density-dependent controls. Epidemics swept through crowded settlements and cities infested with fleas and rodents. Then came plumbing and sewage treatment methods. Over time, vaccines, antibiotics, and other drugs were developed as weapons against many pathogens. Death rates dropped sharply. Births started to exceed them, and rapid population growth was under way.

The industrial revolution took off in the middle of the eighteenth century. People had discovered how to harness the energy of fossil fuels, starting with coal. Within decades, cities of western Europe and North America became industrialized. World War I sparked the development of more technologies. After the war, factories turned to mass-production of cars, tractors, and other affordable goods. Advances in agricultural practices meant that fewer farmers were required to support a larger population.

In sum, by controlling disease agents and tapping into fossil fuels—a concentrated source of energy—the human population sidestepped many factors that had previously limited its rate of increase.

Where have the far-flung dispersals and stunning advances in agriculture, industrialization, and health care taken us? Starting with *Homo habilis*, it took about 2.5 million years for human population size to reach 1 billion. As Figure 40.14 shows, it took just 123 years to reach 2 billion, 33 more to reach 3 billion, 14 more to reach 4 billion, and then 13 more to get to 5 billion. It took only 12 more years to arrive at 6 billion! Given the principles governing population growth, we may expect the rate of increase to decline as birth rates fall or death rates rise. Alternatively, the rates of increase may continue to rise if breakthroughs in technology expand the carrying capacity. *Even so, continued growth cannot be sustained indefinitely.*

Why not? Ongoing increases in population size are invitations to certain density-dependent controls. For instance, globe-hopping travelers introduce pathogens to dense urban areas all around the world in a matter of weeks (Section 19.6). Also, emigration away from economic hardship and civil strife have put 50 million individuals on the move within and between nations. Will relocations of so many individuals be peaceable? How much food, clean water, and other resources will become available to them, wherever they end up?

Through expansion into new habitats, cultural interventions, and technological innovations, the human population has temporarily skirted environmental resistance to growth.

Without technological breakthroughs, density-dependent controls will kick in and slow human population growth.

40.8 Fertility Rates and Age Structure

Acknowledgment of the risks posed by rising populations has resulted in increased family planning in almost every region. Putting the brakes on population growth is not easy, so expect ongoing increases in numbers.

SOME PROJECTIONS

LINK TO CHAPTER 35 INTRODUCTION

Most governments recognize that population growth, resource depletion, pollution, and quality of life are interconnected. Most are working to lower long-term birth rates, as with family planning programs. Details vary among countries, but most provide information about methods of fertility control.

For that reason and others, birth rates are slowing worldwide. Death rates are also falling in most regions. Improved diets and health care lower infant mortality rates: the number of infants per 1,000 who die in their first year. On the other hand, AIDS has caused death rates to soar in some African countries (Chapter 35).

World population is expected to peak at 8.9 billion by 2050 and possibly to decline as the century ends. Think of all the resources that will be required. We will have to boost food production, and find more energy and fresh water to meet basic needs, which is eluding half of the existing human population. Manipulations of resources on a larger scale will intensify pollution.

We expect to see the most growth in India, China, Pakistan, Nigeria, Bangladesh, and Indonesia, in that order. China (with 1.3 billion people) and India (with 1.09 billion) dwarf other countries; together, they make up 38 percent of the world population. Next in line is the United States, with 294 million.

SHIFTING FERTILITY RATES

The total fertility rate (TFR) is the average number of children born to the women of a population during their reproductive years. In 1950, the worldwide TFR averaged 6.5. Currently it is 2.7, which is still above the replacement level of 2.1—or the average number of children a couple must bear to keep the population at a constant level, given current death rates.

TFRs vary among countries. TFRs are at or below replacement levels in many developed countries; the developing countries in western Asia and Africa have the highest. Figure 40.15 has some examples of the disparities in demographic indicators.

Comparing age structure diagrams is revealing. In Figure 40.16, focus on the reproductive age category for the next fifteen years. The average range for childbearing years is fifteen to thirty-five. We can expect populations that have a broad base to grow faster. The United States population has a relatively narrow base and is growing slowly, which has implications for the 78 million baby-boomers (Figure 40.16*c*). This cohort started forming in 1946 when American soldiers came home after World War II and started to raise families.

Global increases seem certain. Even if every couple from this time forward has no more than two children, population growth cannot slow for sixty years. About 1.9 billion are about to enter their reproductive years. *More than a third of the world population is in the broad pre-reproductive base.*

China has the most wide-reaching family planning program. Its government discourages premarital sex. It urges people to delay marriage and limit families to one or two children. It offers abortions, contraceptives, and sterilization at no cost to married couples, which mobile units and paramedics provide even in remote areas. Couples who follow guidelines get more food, free medical care, better housing, and salary bonuses.

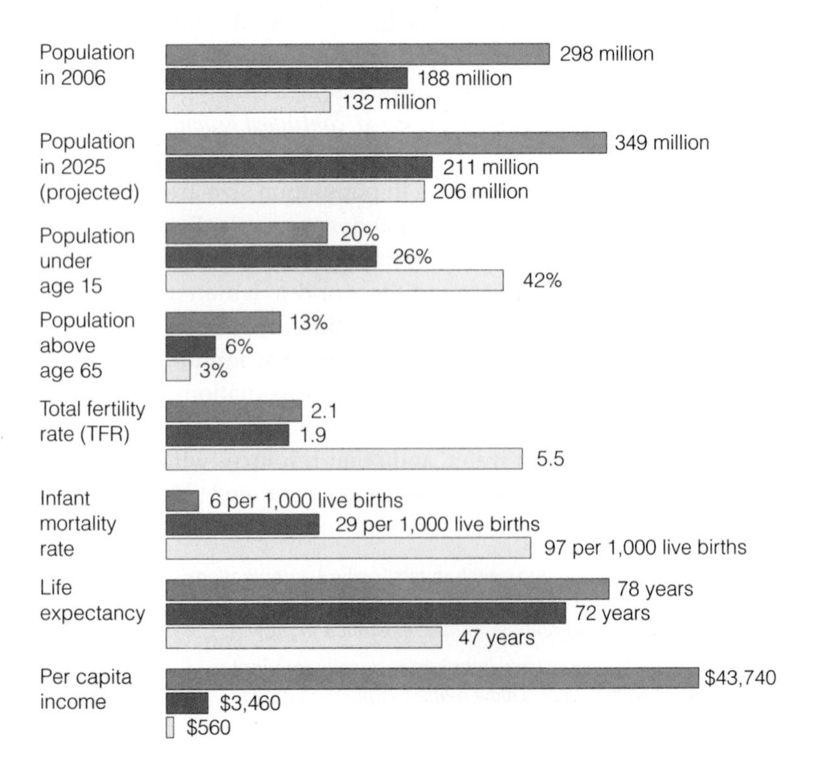

Figure 40.15 Key demographic indicators for three countries, mainly in 2006. The United States (*brown* bar) is highly developed, Brazil (*red* bar) is moderately developed, and Nigeria (*beige* bar) is less developed.

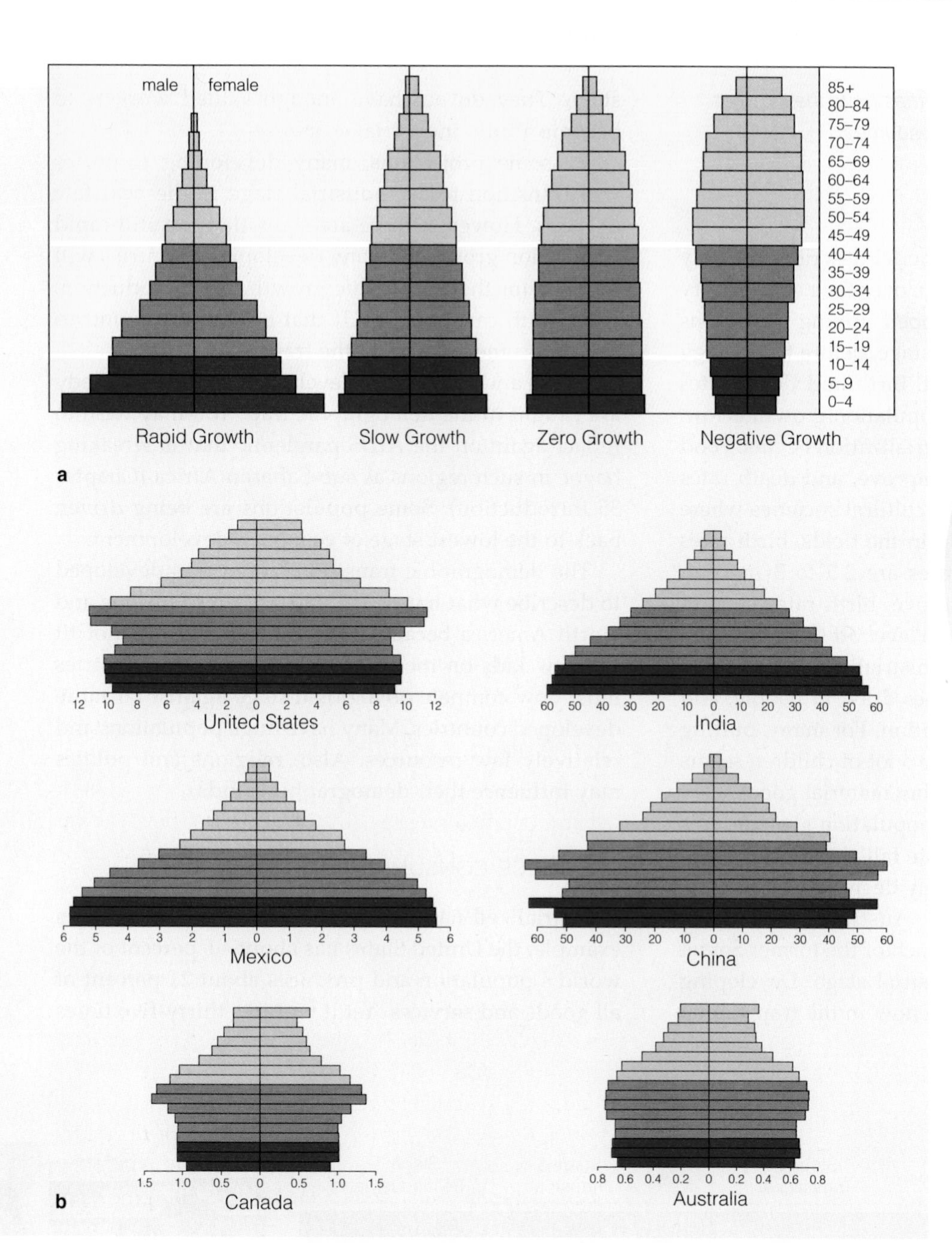

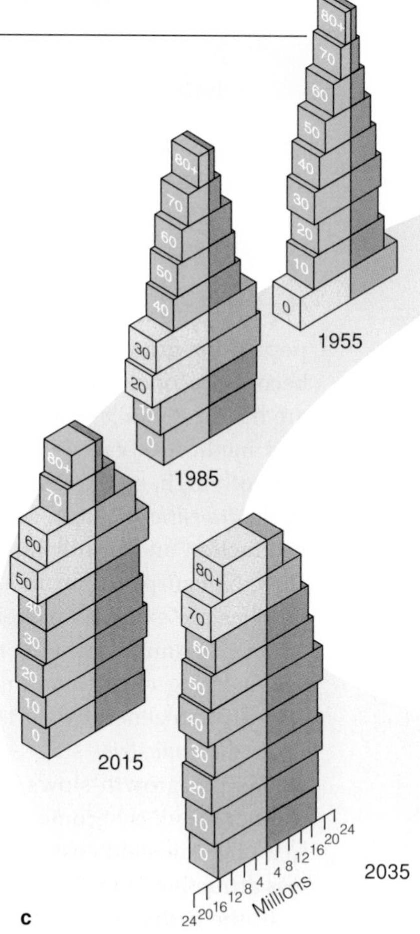

Figure 40.16 **Animated!** (**a**) General age structure diagrams for countries with rapid, slow, zero, and negative rates of population growth. The pre-reproductive years are *green* bars; reproductive years, *purple;* post-reproductive years, *light blue.* A vertical axis divides each graph into males (*left*) and females (*right*). The bar widths correspond to the proportions of individuals in each age group.

(**b**) 1997 age structure diagrams for six nations. Population sizes are measured in millions.

(**c**) Sequential age structure diagrams for the United States population. *Gold* bars track the baby-boomer generation.

Their offspring get free tuition and special treatment when they enter the job market. Parents having more than two children lose benefits and pay more taxes.

Since 1972, China's TFR has fallen sharply, from 5.7 to 1.6. An unintended consequence has been a shift in the country's sex ratio. Traditional cultural preference for sons, especially in rural areas, led some parents to abort developing females or even commit infanticide. Worldwide, 1.06 boys are born for every girl, but in China the latest census reports 1.19 boys for every girl. Also, more than 100,000 girls are abandoned each year. The government is offering additional cash and tax incentives to the parents of girls. In the meantime, the population time bomb keeps on ticking in China. About 150 million of its young females now make up the pre-reproductive age category.

The worldwide total fertility rate has been declining. It still is above the replacement level that would drive the growth rate of the human population toward zero.

Most countries support family planning programs of some sort. Even with the slowdowns, the human population will continue to increase; its pre-reproductive base is immense. At present, more than one-third of the human population is in a broad pre-reproductive base.

40.9 Population Growth and Economic Effects

The most highly developed countries have the slowest growth rates and use the most resources.

DEMOGRAPHIC TRANSITIONS

LINK TO CHAPTER 35 INTRODUCTION

The demographic transition model describes the way population growth rates change over time as a country becomes economically developed. Living conditions are harshest in a *preindustrial* stage, before technology and medical advances spread. Birth and death rates are both high, so the rate of population growth is low. In the *transitional* stage, industrialization begins. Food production and health care improve, and death rates slow. Not surprisingly, in agricultural societies where families are expected to help in the fields, birth rates are high. Annual growth rates are 2.5 to 3 percent. When living conditions improve, birth rates start to fall. Growth generally starts to level off (Figure 40.17).

In the *industrial* stage, industrialization is in full swing and growth slows. Cities draw in people, and smaller families become an option. For many, putting in all the time and cost to raise a lot of children seems less attractive than accumulating material goods.

In the *postindustrial* stage, population growth rates become negative. The birth rate falls below the death rate, and population size slowly decreases.

The United States, Canada, Australia, the bulk of western Europe, Japan, and much of the former Soviet Union have reached the industrial stage. Developing countries, such as Mexico, are now in the transitional stage. They do not have enough skilled workers to become a fully industrial economy.

By some projections, many developing countries will transition to an industrial stage in the next few decades. However, there are signs that the still-rapid population growth in many developing countries will overwhelm their economic growth, food production, and health care systems. If that occurs, the countries could become trapped in the transition stage.

Africa and some other developing countries already are caught in the demographic trap. You may wish to reflect again on the AIDS pandemic that is wreaking havoc in such regions as sub-Saharan Africa (Chapter 35 introduction). Some populations are being driven back to the lowest stage of economic development.

The demographic transition model was developed to describe what happened when western Europe and North America became industrialized. It may not fit the new kids on the block. Less developed countries must now compete on a global stage against far more developed countries. Many have huge populations and relatively few resources. Also, religions and politics may influence their demographic trends.

RESOURCE CONSUMPTION

Industrialized nations use the most resources. As an example, the United States has about 4.6 percent of the world's population and produces about 21 percent of all goods and services. Yet it requires thirty-five times

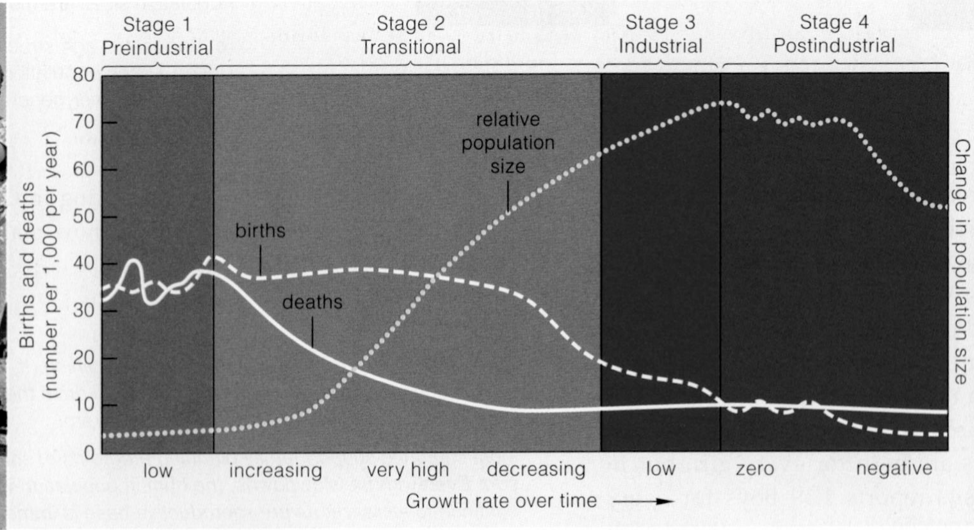

Figure 40.17 Animated! Demographic transition model for changes in population growth rates and sizes, correlated with long-term changes in the economy.

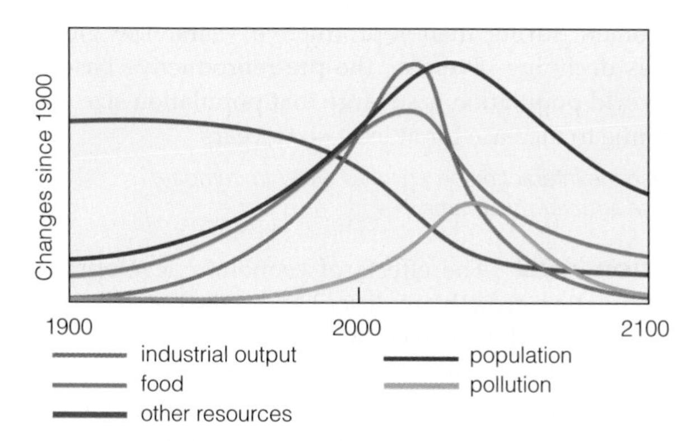

Figure 40.18 Computer-based projection of what might happen if human population size continues to skyrocket without dramatic policy changes and technological innovation. The assumptions were that the population has already overshot the carrying capacity and current trends will continue unchanged.

more goods and services than India. It uses about 25 percent of the world's minerals and energy supplies.

G. Tyler Miller once estimated that it would take 12.9 billion impoverished individuals living in India to have as much impact on the environment as 284 million people living at the time in the United States. Billions now living in India and other less developed nations dream of owning the same consumer goods as people in developed countries. The planet does not have enough resources to make that possible.

What will happen if the human population keeps on increasing as predicted? Will there be enough food, energy, water, and other basic resources to sustain so many individuals? Will regional governments be able to provide the necessary education, housing, medical care, and other social services? Some models suggest not (Figure 40.18). Other analysts claim we can adapt to a more crowded world if innovative technologies improve crop yields, if people rely less on meat for protein, and if those with an abundance of resources distribute more of them to others.

There are no easy answers. If you have not been doing so, start following the arguments in the media. It is a good idea to become an informed participant in a debate that will have impact on your future.

Differences in population growth and resource consumption among countries can be correlated with levels of economic development. Growth rates are typically greatest during the transition to industrialization.

Global conditions have so changed that the demographic transition model may no longer apply to many nations.

40.10 A No-Growth Society

For humans, as for all species, the biological implications of exponential population growth are sobering. So are the social implications of what would happen if human population growth were to approach zero.

In a growing population, most individuals tend to be in lower age brackets. When living conditions favor moderate growth, a future workforce is guaranteed because of the age distribution. Such distribution has social implications. Individuals in higher age brackets expect a younger workforce to support them. In the United States, most older people who retire receive social security payments and government-subsidized medical care. However, as one outcome of improved medicine and hygiene, people in upper age brackets are living far longer than seniors did when the social security program was established. The cash benefits being distributed to seniors exceed the contributions that these people paid in to the program when *they* were younger. When baby boomers begin to receive their benefits, the deficit will grow even more.

If the human population does reach and maintain zero growth, a larger proportion of individuals will end up in higher age brackets. Even slower growth poses problems. Will those people continue to receive goods and services as the workforce carries more and more of the economic burden? Put it to yourself. How much economic hardship are you willing to bear for the sake of your parents? For your grandparents? How much should the next generation bear for you?

We have arrived at a turning point, not only in our biological evolution but in our cultural evolution as well. The decisions awaiting us are among the most pressing and difficult we will ever have to make.

All species face limits to growth. We might think we are different from the rest, and in some respects we are. The uniquely human capacity to undergo rapid cultural evolution has helped us postpone the action of most factors that limit growth. However, the crucial word is *postpone*. On the basis of all the models that are available to us, we can be fairly sure of this: The resources humans need to live are in limited supply and we cannot escape the impacts of these limitations.

Not all countries show runaway population growth. When population growth slows, the percentage of individuals in the older age categories rises. This puts a financial burden on the younger members of the population.

Humans, like all species, face limits on population growth. We must decide how to best address these limits.

www.thomsonedu.com Thomson NOW!

Summary

Sections 40.1, 40.2 Each population is a group of individuals of the same species. Its growth is affected by its demographics—its size, density, distribution, and age structure, all of which are measurable. Most populations in nature have a clumped distribution pattern.

Counting the number of individuals in quadrats is a way to estimate the density of a population in a specified area. Capture–recapture methods can be used to estimate the population density for mobile animals.

- *Use the interaction on ThomsonNOW to learn how to estimate population size.*

Section 40.3 Many factors affect population growth rates, but births and deaths are easiest to track. The per capita birth rate minus the per capita death rate gives us r, the population's per capita growth rate. Here is the general growth equation: $G = rN$, where G is population growth and N is the number of individuals.

In cases of exponential growth, a population's growth is proportional to its size. Population size increases at a fixed rate in any given interval. Graphing population size against time produces a J-shaped growth curve. The rate of population growth may be slow or rapid. As long as per capita birth rate remains above its per capita death rate, the population shows exponential growth.

- *View the animation on ThomsonNOW to observe a pattern of exponential growth.*

Section 40.4 The carrying capacity is the maximum number of individuals of a given population that can be sustained indefinitely by resources in their environment. Density-dependent factors are conditions or events that can lower reproductive success and that get worse with crowding. Disease and competition for food are examples. Density-independent factors are conditions or events that can lower reproductive success, but their effect does not vary with crowding. Tsunamis are an example.

Unlike exponential growth, a logistic growth pattern plots out as an S-shaped curve. As one example, a small population starts growing slowly, then grows rapidly, then levels off once the carrying capacity is reached.

- *Watch the animation on ThomsonNOW to learn about logistic growth.*

Sections 40.5, 40.6 Each species has a life history pattern involving when an individual starts reproducing, how often it reproduces, the number of offspring each time, and other traits. Three types of survivorship curves are common: a high death rate late in life, a constant rate at all ages, or a high rate early in life. Life histories have a genetic basis and are subject to natural selection.

Section 40.7 The human population has surpassed 6.6 billion. Expansion into new habitats and agriculture allowed early increases. Later, medical and technological innovations helped raise the carrying capacity and got around limiting factors.

Section 40.8 Family planning programs are social efforts to slow population growth. A population's total fertility rate (TFR) is the average number of children born to women during their reproductive years. The global TFR is declining. Even so, the pre-reproductive base of the world population is so large that population size will continue to increase for at least sixty years.

- *Use the interaction on ThomsonNOW to compare age structure diagrams.*

Section 40.9 The effects of economic development on population growth are described by the demographic transition model. However, this model was developed on the basis of historical conditions that may not apply to currently developing nations. Per capita consumption of resources in developed nations is much higher than it is in developing nations.

- *Use the interaction on ThomsonNOW to learn about the demographic transition model.*

Section 40.10 Zero population growth has social repercussions. As one example, what happens when the number of older individuals in the population exceeds the number of young workers that supports them?

Self-Quiz

Answers in Appendix III

1. Most commonly, individuals of a population show a ______ distribution through their habitat.
   a. clumped c. nearly uniform
   b. random d. none of the above

2. The rate at which population size grows or declines depends on the rate of ______.
   a. births c. immigration e. a and b
   b. deaths d. emigration f. all of the above

3. Suppose 200 fish are marked and released in a pond. The following week, 200 fish are caught and 100 of them have marks. There are about ______ fish in this pond.
   a. 200 b. 300 c. 400 d. 2,000

4. A population of worms is growing exponentially in a compost heap. Thirty days ago there were 400 worms and now there are 800. How many worms will there be thirty days from now, assuming conditions remain constant?
   a. 1,200 b. 1,600 c. 3,200 d. 6,400

5. For a given species, the maximum rate of increase per individual under ideal conditions is its ______.
   a. biotic potential c. environmental resistance
   b. carrying capacity d. density control

6. ______ is a density-independent factor that influences population growth.
   a. Resource competition c. Predation
   b. Infectious disease d. Harsh weather

7. A life history pattern for a population is a set of adaptations that influence the individual's ______.
   a. longevity c. age at reproductive maturity
   b. fertility d. all of the above

8. The human population is now over 6.6 billion. It was about half that in ______.
   a. 2004 b. 1960 c. 1802 d. 1350

Figure 40.19 Saguaros (*Canegiea gigantea*) growing slowly in Arizona.

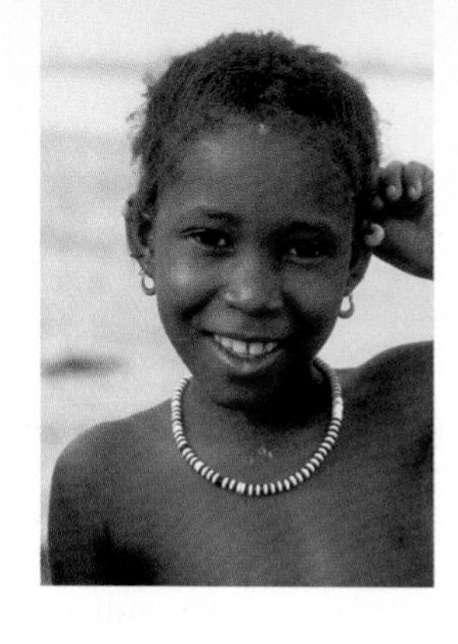

Figure 40.20 A young Malian, who has a 10 percent chance of becoming a mother by age fifteen, and a 50 percent chance before nineteen. At 7.1, Mali's total fertility rate is one of the world's highest.

9. Compared to the less developed countries, the highly developed ones have a higher ______ .
   a. death rate
   b. birth rate
   c. total fertility rate
   d. resource consumption rate

10. Match each term with its most suitable description.
   ___ carrying capacity
   ___ exponential growth
   ___ biotic potential
   ___ limiting factor
   ___ logistic growth

   a. maximum rate of increase per individual under ideal conditions
   b. population growth plots out as an S-shaped curve
   c. maximum number of individuals sustainable by the resources in a given environment
   d. population growth plots out as a J-shaped curve
   e. essential resource that restricts population growth when scarce

■ *Visit ThomsonNOW for additional questions.*

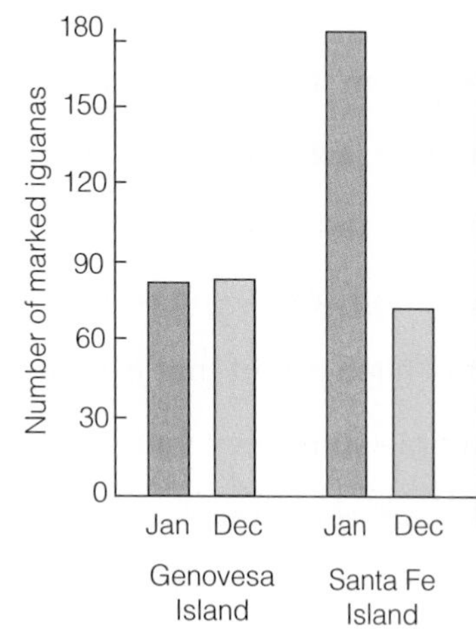

Figure 40.21 Shifting numbers of marked marine iguanas on two Galápagos islands. An oil spill occurred near Santa Fe island just before the January 2001 census (*green* bars). A second census was carried out in December 2001 (*tan*).

Critical Thinking

1. If house cats that are not neutered or spayed live up to their biotic potential, two can be the start of many kittens—12 the first year, 72 the second year, 429 the third, 2,574 the fourth, 15,416 the fifth, 92,332 the sixth, 553,019 the seventh, 3,312,280 the eighth, and 19,838,741 kittens the ninth year. Is this a case of logistic growth? Exponential growth? Irresponsible cat owners?

2. Think about Section 40.6. When researchers moved guppies from populations preyed on by cichlids to a habitat with killifish, the life histories of the transplanted guppies evolved. They came to resemble those of guppy populations preyed on by killifish. Males became gaudier; some scales formed larger, more colorful spots. How might a decrease in predation pressure on sexually mature fish cause this?

3. Each summer, a giant saguaro cactus produces tens of thousands of tiny black seeds. Most die, but a few land in a sheltered spot and sprout the following spring. The saguaro is a slow-growing CAM plant (Section 6.7). After fifteen years, it may be only knee high, and it will not flower for another fifteen years. It may live for 200 years. Saguaros share their habitat with annuals, such as poppies, that sprout, form seeds, and die in just a few weeks (Figure 40.19). Speculate on how these different life histories can both be adaptive in the same desert environment.

4. A third of the world population is younger than fifteen (Figure 40.20). Describe the effect this age distribution will have on the human population's growth rate. If you suspect it will have severe impact, what recommendations would you make to encourage individuals of this age group to limit their family size? What are some social, economic, and environmental factors that might prevent them from following the recommendations?

5. In 1989, Martin Wikelski started a long-term study of marine iguana populations in the Galápagos Islands (Section 16.2). He marked the iguanas on two islands—Genovesa and Santa Fe—and collected data on how their body size, survival, and reproductive rates varied over time. The iguanas eat algae and have no predators, so deaths are usually the result of food shortages, disease, or old age. His studies showed that numbers decline during El Niño events, when the surrounding waters heat up.

In January 2001, an oil tanker ran aground and leaked a small amount of oil into the waters near Santa Fe island. Figure 40.21 shows the number of marked iguanas that Wikelski and his team counted in their census of the study populations just before the spill and about a year later. How did populations of iguanas on each island change during this time? Wikelski concluded that changes on Santa Fe were the result of the oil spill, rather than sea temperature or other climate factors. Explain why his data support this interpretation.

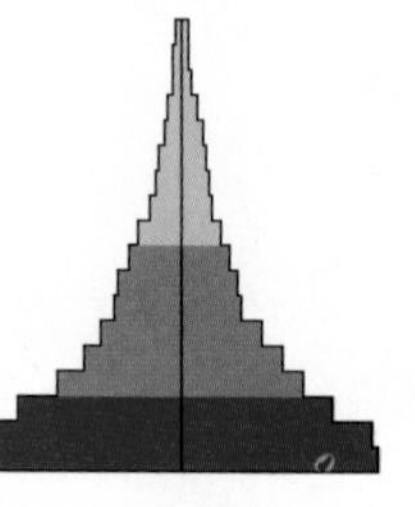

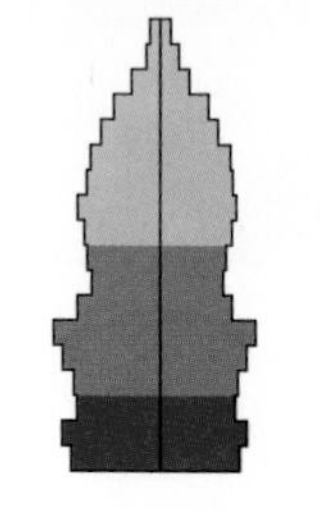

6. Two typical age structure diagrams for populations are shown at right. Select one of them, and write a short essay about the population it represents. Speculate on its current and future social and economic problems.

41 COMMUNITY STRUCTURE AND BIODIVERSITY

IMPACTS, ISSUES Fire Ants in the Pants

Accidentally step on a nest of red imported fire ants, *Solenopsis invicta*, and you will soon be sorry. The ants are quick to defend their nest against any perceived threat, including unwary walkers. Ants stream out from the ground, grab an intruder's flesh with their jaws, and inflict a series of stings. Venom injected by their stinger causes burning pain, kills body cells, and results in the formation of a pus-filled bump that is slow to heal (Figure 41.1). Multiple stings can cause nausea, dizziness, and—rarely—death.

Solenopsis invicta arrived in the United States from South America in the 1930s, probably as stowaways on a ship. The ants spread out from the Southeast and in time established colonies as far west as California and as far north as Kansas and Delaware.

These ants disrupt natural communities. They attack anything that disturbs them, including livestock, pets, and wildlife. They eat just about anything. They mow down seedlings, and prey on insects and other small animals that cannot escape them. They outcompete native ants and may also contribute to the decline of other native wildlife.

To give an example, the Texas horned lizard vanished from most of its home range when *S. invicta* moved in and displaced the native ants. Native ants are the Texas horned lizard's food of choice. It cannot tolerate eating the imported fire ants.

Invicta means "invincible" in Latin. So far, *S. invicta* is living up to its species name. Pesticides have not slowed

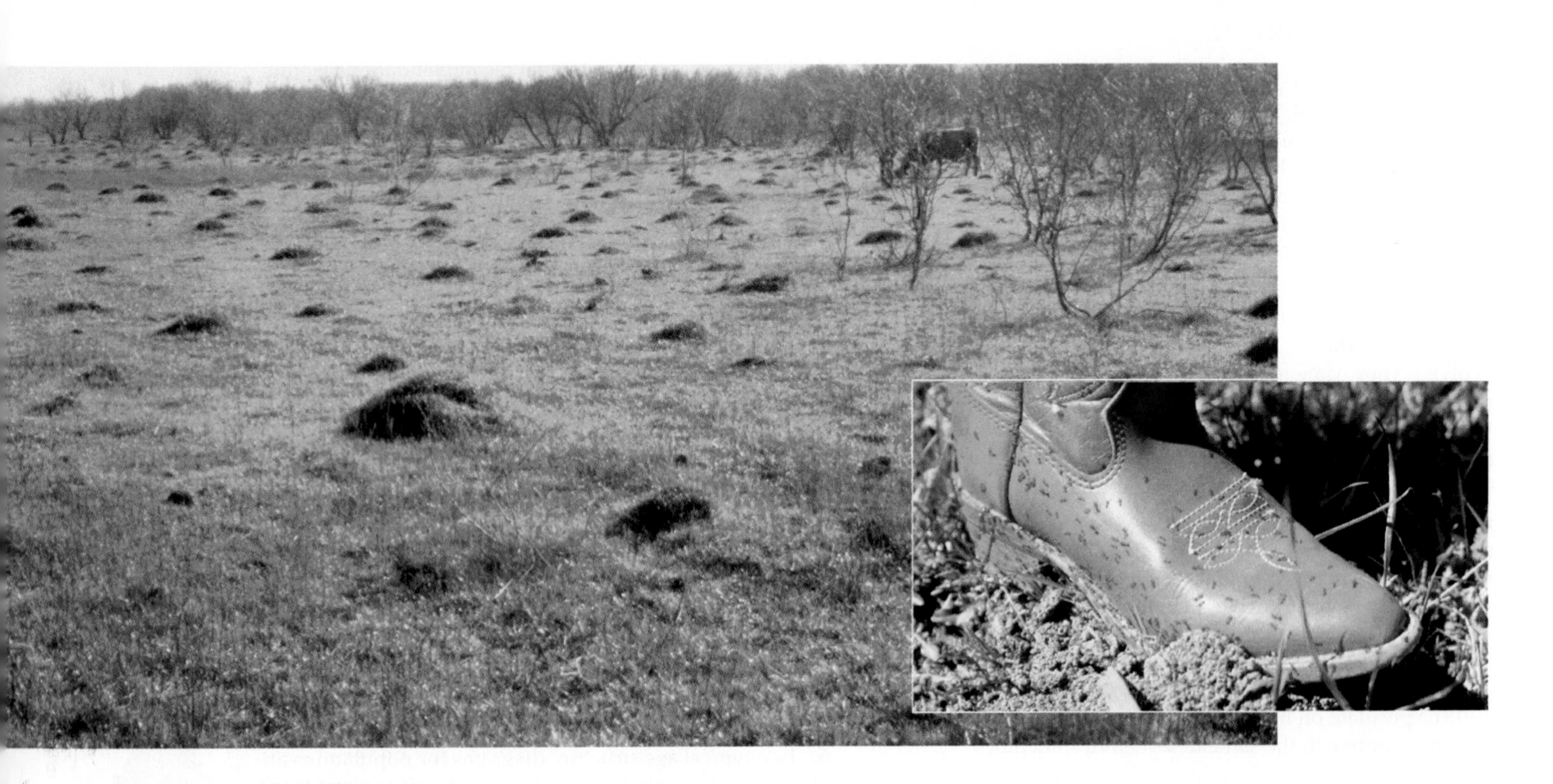

See the video! **Figure 41.1** Fire ant mounds in west Texas, and agitated fire ants swarming over a leather boot. *Facing page,* skin eruptions that typically follow a concerted attack by these exotic imports.

How would you vote? Currently, only a fraction of the crates imported into the United States are inspected for the inadvertent or deliberate presence of exotic species. Would the cost of added inspections be worth it? See ThomsonNOW for details, then vote online.

its dispersal. They might even be facilitating invasions by wiping out most of the native ant populations.

Ecologists are enlisting biological controls. Phorid fly species control *S. invicta* in its native habitat. The flies are parasitoids, a specialized type of parasite that kills its host in a rather gruesome way. A female fly pierces the cuticle of an adult ant, then lays an egg in the ant's soft tissues. The egg hatches into a larva, which grows and then eats its way through tissues to the ant's head. After the larva gets big enough, it secretes an enzyme that makes the ant's head fall off. The fly larva develops into an adult within the shelter of the detached ant head.

Several phorid fly species have now been introduced in various southern states. The flies seem to be surviving, reproducing, and increasing their range. They probably will never kill off all *S. invicta* in affected areas, but they are expected to reduce the density of colonies.

Ecologists are also exploring other options. They are testing effects of imported, pathogenic fungi or protists that infect *S. invicta* but not native ants. Another idea is to introduce a parasitic South American ant that invades *S. invicta* colonies and kills the egg-laying queens.

This example introduces the sometimes rough-and-tumble details of community structure, or patterns in the number of species and their relative abundances. As you will see, species interactions and disturbances to the habitat can shift community structure in small and large ways—some predictable, others unexpected.

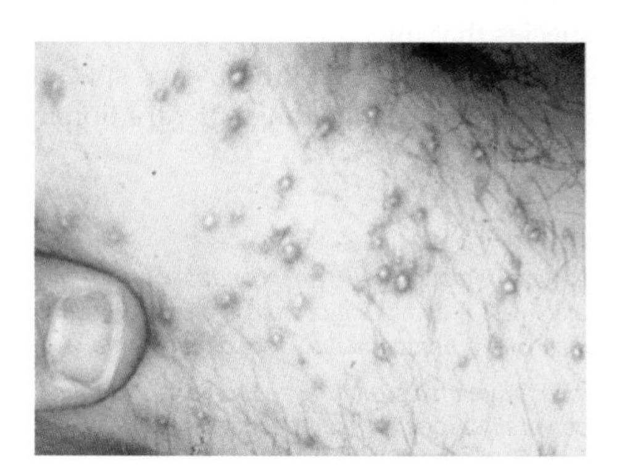

Key Concepts

COMMUNITY CHARACTERISTICS

A community consists of all species in a habitat. Each species has a niche—the sum of its activities and relationships. A habitat's history, its biological and physical characteristics, and interactions among species in the habitat affect community structure. **Section 41.1**

FORMS OF SPECIES INTERACTIONS

Commensalism, mutualism, competition, predation, and parasitism are interspecific interactions. They influence the population size of participating species, which in turn influences the community's structure. **Sections 41.2–41.7**

COMMUNITY STABILITY AND CHANGE

Communities show stability, as when certain species persist in a habitat. Communities also change, as when new species move into the habitat and others disappear. Physical characteristics of the habitat, species interactions, disturbances, and chance events affect how a community changes through time. **Sections 41.8–41.10**

GLOBAL PATTERNS IN COMMUNITY STRUCTURE

Biogeographers identify regional patterns in species distribution. They have shown that tropical regions hold the greatest number of species and that characteristics of islands can be used to predict how many species an island will hold. **Section 41.11**

ASSESSING AND PROTECTING SPECIES RICHNESS

Increases in habitat fragmentation and losses, species introductions, overharvesting, and illegal wildlife trading accelerate extinction rates. Conservation biologists are working to preserve species richness by resource management that is not in conflict with the human need for survival. **Sections 41.12, 41.13**

Links to Earlier Concepts

In this chapter, you will see how evidence from different fields of inquiry converges to explain patterns in nature. You will draw on the Unit IV survey of biodiversity. You will revisit biogeography and take a closer look at global patterns in species richness (Sections 16.1, 16.3). You will also see how species interactions lead to natural selection (17.4, 17.12, 21.7). You will see many examples of field experiments (1.7). You will also apply what you know about populations and factors affecting their growth (40.2–40.4).

41.1 Which Factors Shape Community Structure?

Community structure refers to the number and relative abundances of species in a habitat. It changes over time.

The type of place where a species normally lives is its habitat, and all species living in a habitat represent a community. Each community has a dynamic structure. It shows shifting patterns of species diversity: number and relative abundances of species in a habitat or any other specified region.

Many factors influence community structure. First, climate and topography affect a habitat's traits, such as temperature, soil, and moisture. Second, a habitat has only certain kinds and amounts of food and other resources. Third, the species themselves possess traits that are adaptive to habitat conditions, as in Figure 41.2. Fourth, the species interact in ways that cause shifts in their numbers and abundances. Finally, the timing and history of disturbances, both natural and human-induced, affect community structure.

THE NICHE

All species of a community share the same habitat—the same "address"—but each also has a "profession" or unique ecological role that sets it apart. This is the species' niche. We describe a species' niche in terms of the conditions, resources, and interactions necessary for its survival and reproduction. The temperatures it can tolerate, the kinds of foods it can eat, the types of places it can nest or hide are all aspects of an animal's niche. A description of a plant's niche would include its soil, water, light, and pollinator requirements.

Figure 41.2 Three of twelve fruit-eating pigeon species in Papua New Guinea's tropical rain forests: (**a**) pied imperial pigeon, (**b**) superb crowned fruit pigeon, and (**c**) the turkey-sized Victoria crowned pigeon. The forest's trees differ in the size of fruit and fruit-bearing branches. The big pigeons eat big fruit. Smaller ones, with smaller bills, cannot peck open big, thick-skinned fruit. They eat the small, soft fruit on branches too spindly to hold big pigeons.

Trees feed the birds, which help the trees. Seeds in fruit resist digestion in the bird gut. Flying pigeons disperse seed-rich droppings, often some distance from mature trees that would outcompete new seedlings for water, minerals, and sunlight. With dispersal, some seedlings have a better chance to take hold.

Table 41.1 Direct Two-Species Interactions

Type of Interaction	Effect on Species 1	Effect on Species 2
Commensalism	Helpful	None
Mutualism	Helpful	Helpful
Interspecific competition	Harmful	Harmful
Predation	Helpful	Harmful
Parasitism	Helpful	Harmful

CATEGORIES OF SPECIES INTERACTIONS

Dozens to hundreds of species interact even in simple communities. Table 41.1 lists the types of direct species interactions. Commensalism helps one species and has no effect on the other. Most bacteria in your gut are commensal. They benefit by living inside you, but do not help or harm you. Mutualism helps both species. Interspecific competition hurts both species. Predation and parasitism help one species at another's expense. Predators are free-living organisms that kill their prey. Parasites live on or in a host and usually do not kill it.

Parasitism, commensalism, and mutualism can all be types of symbiosis, which means "living together." Symbiotic species, or symbionts, spend most or all of their life cycle in close association with each other. An *endo*symbiont is a species that lives inside its partner.

Regardless of whether one species helps or hurts another, two species that interact closely for extended periods will coevolve. With coevolution, each species is a selective agent that shifts the range of variation in the other (Section 17.12).

A habitat is the type of place where an individual of a species normally lives. A community consists of all species in a habitat, each with a unique niche, or ecological role.

A community has a dynamic structure, reflected in shifting patterns of biodiversity. Species interactions, as well as abiotic factors such as climate, affect that structure.

Species interact in ways that can be beneficial, harmful, or have no effect on one another. Many are symbionts; they associate closely for most or all of their life cycles.

41.2 Mutualism

In a mutualistic interaction, two species take advantage of each other in ways that benefit both, as when one withdraws nutrients from the other while giving it shelter.

LINKS TO SECTIONS 18.4, 21.7, 22.5, 27.2, 28.1

Mutualists are common in nature. For example, birds, insects, bats, and other animals serve as pollinators of flowering plants (Sections 21.7 and 28.1). Pollinators feed on energy-rich nectar and pollen. In return, they transfer pollen between plants of the same species. Similarly, pigeons take food from rain forest trees but disperse their seeds to new sites (Figure 41.2).

In some mutualisms, neither species can complete its life cycle without the other. Yucca plants and the moths that pollinate them show such interdependence (Figure 41.3). In other cases, the mutualism is helpful but not a life-or-death requirement. Most plants, for example, use more than one pollinator.

Mutualists help most plants take up mineral ions (Section 27.2). Nitrogen-fixing bacteria living on roots of legumes such as peas provide the plant with extra nitrogen. Mycorrhizal fungi living in or on plant roots enhance the plant's mineral uptake.

Other fungi interact with photosynthetic algae or bacteria in lichens (Section 22.5). However, are they mutualists or parasites of a captive partner? There is some conflict between all mutualists. In a lichen, the fungus would do best by obtaining as much sugar as possible from its photosynthetic partner. That partner would do best by keeping as much sugar as possible for its own use.

For some mutualists, the main benefit is defense. A sea anemone has stinging cells (nematocysts), so most fishes avoid its tentacles. However, an anemone fish can nestle among tentacles in safety (Figure 41.4). A mucus layer shields the anemone fish from stings, and the tentacles keep it safe from predatory fish. The anemone fish repays its partner by chasing off the few fishes that are able to eat sea anemone tentacles.

Finally, reflect on a theory outlined in Section 18.4, whereby certain aerobic bacteria became mutualistic endosymbionts of early eukaryotic cells. The bacteria received nutrients and shelter. In time, they evolved into mitochondria and provided the "host" with ATP. Cyanobacteria living inside eukaryotic cells evolved into chloroplasts by a similar process.

Mutualism is a common form of symbiosis. Each species benefits as it exploits another in a way that helps ensure its own survival and reproductive success. In some cases, the mutualism is obligatory—one or both partners cannot complete the life cycle in the absence of the other.

Figure 41.3 Mutualism on a rocky slope of the high desert in Colorado.

Only one yucca moth species pollinates plants of each *Yucca* species; it cannot complete its life cycle with any other plant. The moth matures when yucca flowers blossom. The female has specialized mouthparts that collect and roll sticky pollen into a ball. She flies to another flower and pierces its ovary, where seeds will form and develop, and lays eggs inside. As she crawls out, she pushes a ball of pollen onto the flower's pollen-receiving platform.

After pollen grains germinate, they give rise to pollen tubes, which grow through the ovary tissues and deliver sperm to the plant's eggs. Seeds develop after fertilization.

Meanwhile, moth eggs develop into larvae that eat a few seeds, then gnaw their way out of the ovary. Seeds that larvae do not eat give rise to new yucca plants.

Figure 41.4 The sea anemone *Heteractis magnifica*, which shelters about a dozen fish species. It has a mutualistic association with the pink anemone fish (*Amphiprion perideraion*). This tiny but aggressive fish chases away predatory butterfly fishes that would bite off tips of anemone tentacles. The fish cannot survive and reproduce without the protection of an anemone. The anemone does not need a fish to protect it, but it does better with one.

41.3 Competitive Interactions

Where you observe limited supplies of energy, nutrients, living space, and other natural resources, you are likely to find organisms competing for a share of them.

LINKS TO SECTIONS 16.3, 40.4

As Charles Darwin clearly understood, competition for resources among individuals of the same species can be intense and leads to evolution by natural selection (Section 16.3). Focus now on competitive interactions between species. Such interspecific competition is not usually as intense. Why not? The requirements of two species might be similar, but they never can be as close as they are for individuals of the same species.

With *interference* competition, one species actively prevents another from accessing some resource. As an example, one species of chipmunk will chase another chipmunk species out of its habitat. It will not share seeds and shelter (Figure 41.5). As another example, some plants use chemical weapons against potential competition. Aromatic chemicals ooze from tissues of sagebrush plants, eucalyptus trees, and black walnut trees, and seep into the soil. These chemicals prevent other kinds of plants from germinating or growing.

In *exploitative* competition, species do not interact directly; each reduces the amount of resource available to the other by using that resource. For example, deer and blue jays both eat acorns in oak forests. The more acorns the birds eat, the fewer there are for the deer.

EFFECTS OF COMPETITION

Deer and blue jays share a fondness for acorns, but each also has other sources of food. Any two species differ in their resource requirements. Species compete most intently when the supply of a shared resource is the main limiting factor for both (Section 40.4).

In the 1930s, G. Gause conducted experiments with two species of ciliated protozoans (*Paramecium*) that compete for the same prey: bacteria. He cultured the *Paramecium* species separately and together (Figure 41.6). Within weeks, population growth of one species outpaced the other, which went extinct.

Experiments by Gause and others are the basis for the concept of competitive exclusion: Whenever two species require the same limited resource to survive or reproduce, the better competitor will drive the less competitive species to extinction in that habitat.

When resource needs of competitors are not exactly the same, species can coexist, but competition often reduces the population size of both competitors. For instance, Gause also studied two *Paramecium* species with differing food preferences. When grown together, one fed on bacteria suspended in culture tube liquid. The other ate yeast cells near the bottom of the tube. When grown together, population growth rates fell for both species, but they continued to coexist.

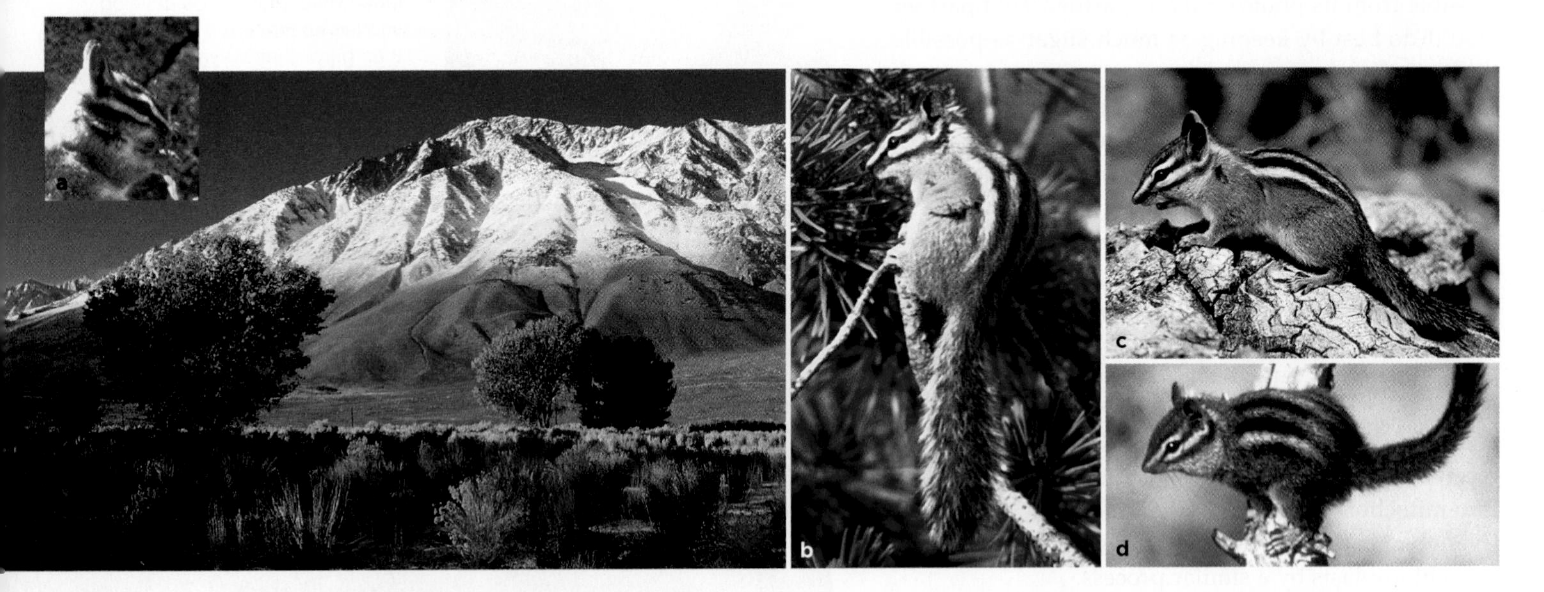

Figure 41.5 Example of interspecific competition in nature. On the slopes of the Sierra Nevada, competition helps keep nine species of chipmunks (*Tamias*) in different habitats.

The alpine chipmunk (**a**) lives in the alpine zone, the highest elevation. Below it are the lodgepole pine, piñon pine, and then sagebrush habitat zones. Lodgepole pine chipmunks (**b**), least chipmunks (**c**), and other species live in the forest zones. Merriam's chipmunk (**d**) lives at the base of the mountains, in sagebrush. Its traits would allow it to move up into the pines, but the aggressively competitive behavior of forest-dwelling chipmunks will not let it. Food preferences keep the pine forest chipmunks out of the sagebrush habitat.

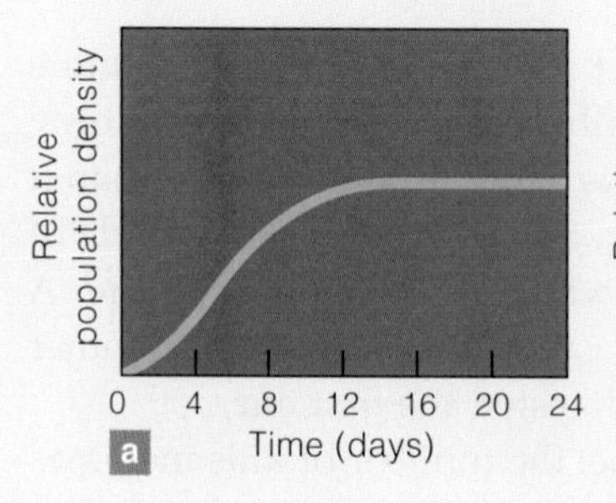

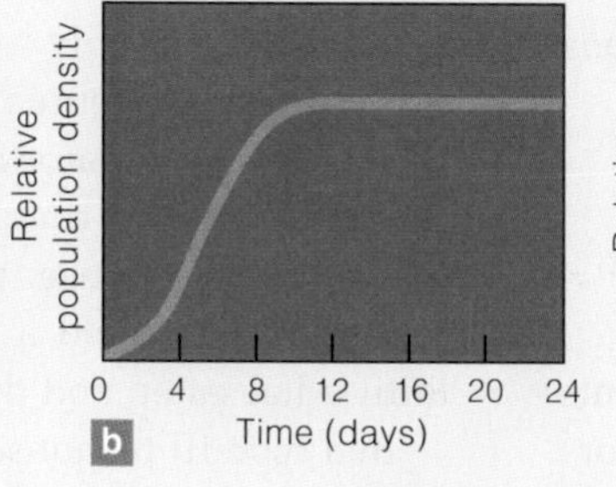

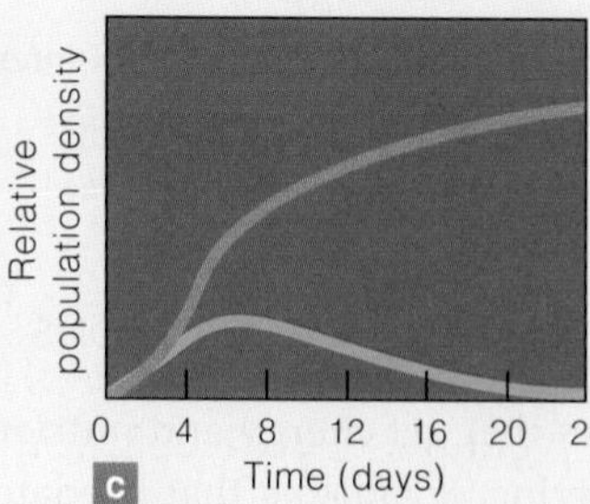

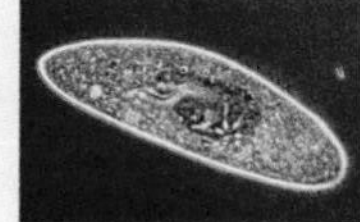

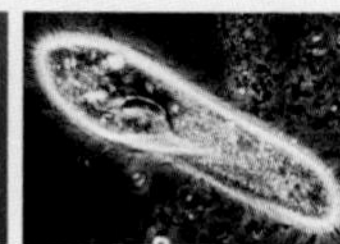

Figure 41.6 Animated! Results of competitive exclusion between two related species that compete for the same food. (**a**) *Paramecium caudatum* and (**b**) *P. aurelia* grown in separate culture flasks established stable populations. The S-shaped graph curves indicate logistic growth and stability. (**c**) For this experiment, the two species were grown together. *P. aurelia* (*brown* curve) drove *P. caudatum* toward extinction (*green* curve).

This experiment and others suggest that two species cannot coexist indefinitely in the same habitat *when they require identical resources.* If their requirements do not overlap much, one might influence the population growth rate of the other, but they may still coexist.

Experiments by Nelson Hairston showed the effects of competition between slimy salamanders (*Plethodon glutinosus*) and Jordan's salamanders (*P. jordani*). The salamanders coexist in wooded habitats (Figure 41.7). Hairston removed all slimy salamanders from certain test plots and Jordan's salamanders from others. He left a final group of plots alone, as unaltered controls.

After five years, the numbers and abundances of the two species had not changed in the control plots. In the plots with slimy salamanders alone, population density had soared. Numbers also increased in plots with Jordan's salamanders alone. Hairston concluded that whenever these salamanders coexist, competitive interactions suppress the population growth of both.

Figure 41.7 Two species of salamanders, *Plethodon glutinosus* (*above*) and *P. jordani* (*below*), that compete in areas where their habitats overlap.

RESOURCE PARTITIONING

Think back on those fruit-eating pigeon species. They all require fruit, but each eats fruits of a certain size. Their preferences are a case of resource partitioning: a *subdividing* of an essential resource, which reduces the competition among species that require it.

Similarly, three annual plant species live in the same field. They all require minerals and water, but their roots take them up at different depths (Figure 41.8).

When species with very similar requirements share a habitat, competition puts selective pressure on both. In each species, individuals who differ most from the competing species are favored. The outcome may be character displacement: Over the generations, a trait of one species diverges in a way that lowers the intensity of competition with the other species. Modification of the trait has promoted the partitioning of a resource.

In some competitive interactions, one species controls or blocks access to a resource, regardless of whether it is scarce or abundant. In other interactions, one is better than another at exploiting a shared resource.

When two species overlap too much in their requirements, they cannot coexist in the same habitat unless they share required resources in different ways or at different times.

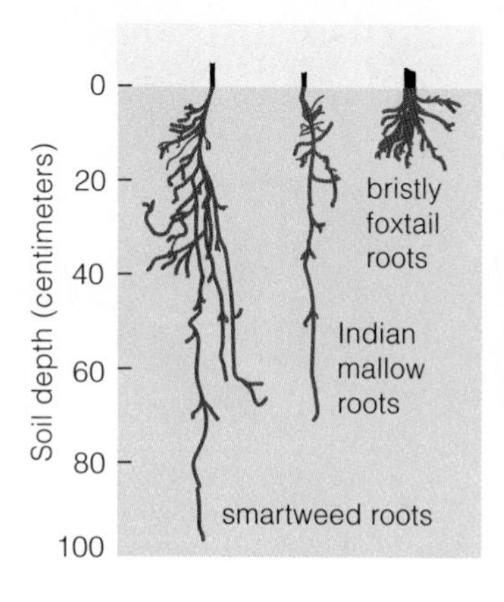

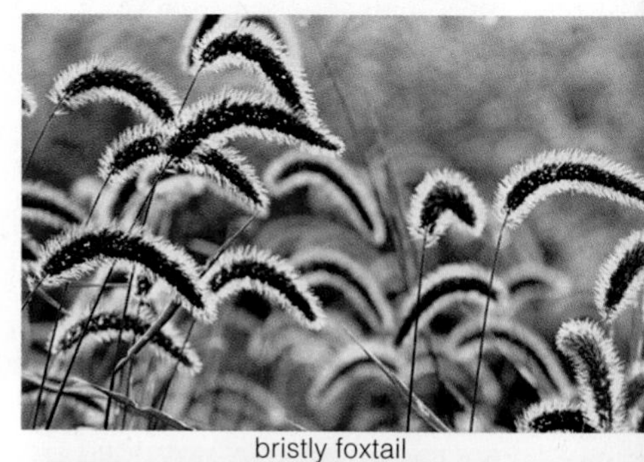

Figure 41.8 A case of resource partitioning among three annual plant species in a plowed but abandoned field. Roots of each species take up water and mineral ions from a different soil depth. This reduces competition among them and allows them to coexist.

41.4 Predator-Prey Interactions

The relative abundances of predator and prey populations of a community shift over time in response to species interactions and changing environmental conditions.

MODELS FOR PREDATOR-PREY INTERACTIONS

LINK TO SECTION 17.12

Predators are consumers that get energy and nutrients from prey, which are living organisms that predators capture, kill, and eat. The quantity and types of prey species affect predator diversity and abundance, and predator types and numbers do the same for prey.

The extent to which predators affect prey numbers depends partly on how individual predators respond to changes in prey density. Figure 41.9*a* compares models for three predator responses to increases in density.

In a type I response, the *proportion* of prey killed is constant, so the *number* killed in any given interval depends solely on prey density. Web-spinning spiders and other passive predators tend to show this type of response. As the number of flies in an area increases, more and more become caught in each spider's web. Filter-feeding predators also show a type I response.

In a type II response, the number of prey killed depends on the capacity of predators to capture, eat, and digest prey. When prey density increases, the rate of kills rises steeply at first because there are more prey to catch. Eventually, the rate of increase slows, because each predator is exposed to more prey than it can handle at one time. Figure 41.9*b* is an example of this type of response, which is common in nature. A wolf that just killed a caribou will not hunt another until it has eaten and digested the first one.

In a type III response, the number of kills increases slowly until prey density exceeds a certain level, then rises rapidly, and finally levels off. This response is common in nature in three situations. In some cases, a predator switches among prey, concentrating its efforts on the species that is most abundant. In other cases, the predators need to learn how to best capture each prey species; they get more lessons when more prey are around. In still other cases, the number of hiding places for prey is limited. Only after prey density rises and some individual prey have no place to hide, does the number of kills increase.

Knowing which type of response a predator makes to prey helps ecologists predict long-term effects of predation on a prey population.

THE CANADIAN LYNX AND SNOWSHOE HARE

In some cases, a time lag in the predator's response to prey density leads to cyclic changes in abundance of predators and prey. When prey density becomes low, the number of predators declines. As a result, prey are safer and their number increases. This allows an increase in predator numbers. Then predation causes another prey decline, and the cycle begins again.

Consider a ten-year oscillation in populations of a predator, the Canadian lynx, and the snowshoe hare

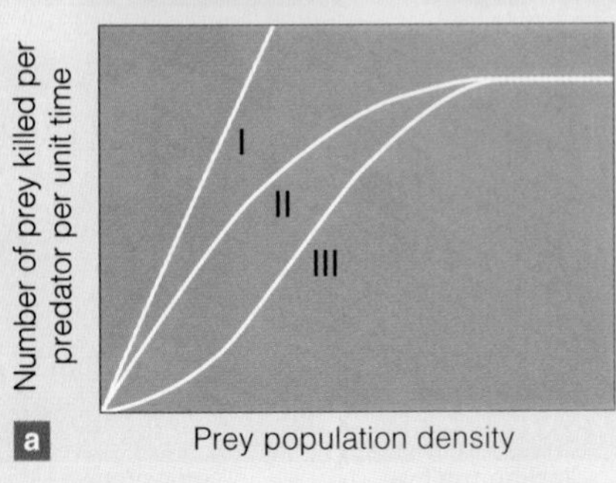

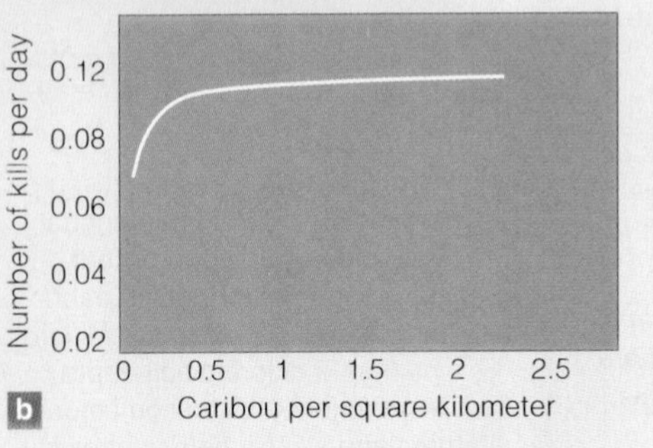

Figure 41.9 **Animated!** (**a**) Three models for responses of predators to prey density. Type I: Prey consumption rises linearly as prey density rises. Type II: Prey consumption is high at first, then levels off as predator bellies stay full. Type III: When prey density is low, it takes longer to hunt prey, so the predator response is low. (**b**) A type II response in nature. For one winter month in Alaska, B. W. Dale and his coworkers observed four wolf packs (*Canis lupus*) feeding on caribou (*Rangifer tarandus*). The interaction fit the type II model for the functional response of predators to the prey density.

Figure 41.10 Graph of the correspondence between abundances of Canadian lynx (*dashed* line) and snowshoe hares (*solid* line), based on counts of pelts sold by trappers to Hudson's Bay Company during a ninety-year period. Charles Krebs observed that predation causes heightened alertness among snowshoe hares, which continually look over their shoulders during the declining phase of each cycle. The photograph at *right* supports the Krebs hypothesis that there is a three-level interaction going on, one that involves plants.

The graph may be a good test of whether you tend to accept someone else's conclusions without questioning their basis in science. Remember those sections in Chapter 1 that introduced the nature of scientific methods?

What other factors may have had impact an on the cycle? Did the weather vary, with more severe winters imposing greater demand for hares (to keep lynx warmer) and higher death rates? Did the lynx compete with other predators, such as owls? Did the predators turn to alternative prey during low points of the hare cycle? When fur prices rose in Europe, did the trapping increase? When the pelt supply outstripped the demand, did trapping decline?

that is its main prey (Figure 41.10). To determine the causes of this pattern, Charles Krebs and coworkers tracked hare population densities for ten years in the Yukon River Valley of Alaska. They set up one-square-kilometer control plots and experimental plots. They used fences to keep predatory mammals out of some plots. Extra food or fertilizers that helped plants grow were used in other plots. The researchers captured and put radio collars on more than 1,000 snowshoe hares, lynx, and other animals, and then released them.

In predator-free plots, the hare density doubled. In plots with extra food, it tripled. In plots having extra food and fewer predators, it increased elevenfold.

The experimental manipulations delayed the cyclic declines in population density but did not stop them. Why not? Owls and other raptors flew over the fences. Only 9 percent of the collared hares starved to death; predators killed some of the others. Krebs concluded that a simple predator–prey or plant–herbivore model did not fully explain his results. Other variables were at work, in a multilevel interaction.

COEVOLUTION OF PREDATORS AND PREY

Interactions among predators and prey can influence characteristic species traits. If a certain genetic trait in a prey species helps it escape predation, that trait will increase in frequency. If some predator characteristic helps overcome a prey defense, it too will be favored. Each defensive improvement selects for a countering improvement in predators, which selects for another defensive improvement, and so on, in a never-ending arms race. The next section describes some outcomes.

Predator populations show three general patterns of response to changes in prey density. Population levels of prey may show recurring oscillations.

The numbers in predator and prey populations often vary in complex ways that reflect the multiple levels of interaction in a community.

Predator and prey populations tend to exert selective pressures on one another.

41.5 An Evolutionary Arms Race

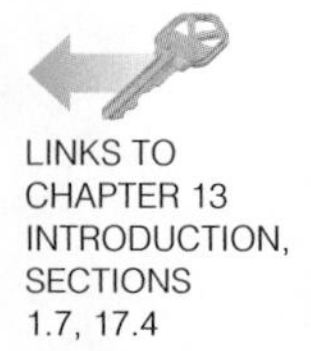

LINKS TO CHAPTER 13 INTRODUCTION, SECTIONS 1.7, 17.4

You just read that predators and prey exert selective pressure on one another. One must defend itself, and the other must overcome defenses. The traits of each evolve, by natural selection, in ways that help it survive long enough to reproduce.

Figure 41.11 Prey camouflage. (**a**) What bird? When a predator approaches its nest, the least bittern stretches its neck (which is colored like the surrounding withered reeds), points its bill upward, and sways like reeds in the wind. (**b**) An inedible bird dropping? No. This caterpillar's body coloration and its capacity to hold its body in a rigid position help camouflage it from predatory birds. (**c**) Find the plants (*Lithops*) hiding in the open from herbivores with the help of their stonelike form, pattern, and coloration.

PREY DEFENSES

Earlier chapters, including Chapter 23, introduced some examples of prey defenses. Many species have hard parts that make them difficult to eat. Spikes in a sponge body, clam and snail shells, lobster and crab exoskeletons, sea urchin spines—all of these traits help deter predators and thereby contribute to evolutionary success.

Also, many heritable traits function in camouflage: body shape, color pattern, behavior, or a combination of factors make an individual blend with its surroundings. Predators cannot eat prey they cannot find. Section 17.4 explains how selection favored alleles that improved the camouflage of a prey species, the desert pocket mouse.

Camouflage is widespread. Marsh birds called bitterns live among tall reeds. When threatened, a bittern points its beak skyward and blends with the reeds (Figure 41.11*a*). On a breezy day, the bird enhances the effect by swaying slightly. A caterpillar with mottled color patterns appears to be a bird dropping (Figure 41.11*b*). Desert plants of the genus *Lithops* usually look like rocks (Figure 41.11c). They flower only during a brief rainy season, when plenty of other plants tempt herbivores.

Many prey species contain chemicals that taste bad or sicken predators. Some produce toxins through metabolic processes. Others use chemical or physical weapons that they get from their prey. For instance, after sea slugs dine on a sea anemone or a jellyfish, they can store its stinging nematocysts in their own tissues (Figure 23.20*f*).

Leaves, stems, and seeds of many plants contain bitter, hard-to-digest, or toxic chemicals. Remember the Chapter 13 introduction? It explains how ricin acts to kill or sicken animals. Ricin evolved in castor bean seeds as a defense against herbivores. Caffeine in coffee beans and nicotine in tobacco leaves evolved as defenses against insects.

Many prey species advertise their bad-tasting or toxin-laden properties by warning coloration. They have flashy patterns and colors that predators learn to recognize and avoid. For instance, a toad might catch a yellow jacket once. But a painful sting from this wasp teaches the toad that black and yellow stripes mean *AVOID ME!*

Mimicry is an evolutionary convergence in body form; species come to resemble one another. In some cases, two or more well-defended organisms end up looking

a A dangerous model **b** One of its edible mimics **c** Another edible mimic **d** And another edible mimic

Figure 41.12 An example of mimicry. Edible insect species often resemble toxic or unpalatable species that are not at all closely related. (**a**) A yellow jacket can deliver a painful sting. It might be the model for nonstinging wasps (**b**), beetles (**c**), and flies (**d**) of strikingly similar appearance.

alike. In others, a tasty, harmless prey species comes to resemble an unpalatable or well-defended one (Figure 41.12). Predators may avoid the mimic after experiencing the disgusting taste, irritating secretion, or painful sting of the species it resembles.

When an animal is cornered or under attack, survival may turn on a last-chance trick. Many animals can startle predators. Section 1.7 describes an experiment that tested the peacock butterfly defenses—a show of eye-like spots and hissing. Other species puff up, bare sharp teeth, or flare neck ruffs (Figure 24.17*d*). Opossums "play dead." When cornered, many animals, including skunks, some snakes, many toads, and certain insects, secrete or squirt stinky or irritating repellents (Figure 41.13*a*).

ADAPTIVE RESPONSES OF PREDATORS

A predator's evolutionary success hinges on eating prey. Stealth, camouflage, and ways of avoiding repellents are countermeasures to prey defenses. For example, some edible beetles spray noxious chemicals at their attackers. A grasshopper mouse grabs the beetle and plunges the sprayer end into the ground, and then chews on the tasty, unprotected head (Figure 41.13*a,b*). Some evolved traits in herbivores are responses to plant defenses. The digestive tract of koalas can handle tough, aromatic eucalyptus leaves that would sicken other herbivorous mammals.

Also, a speedier predator catches more prey. Consider the cheetah, the world's fastest animal on land. One was clocked at 114 kilometers (70 miles) per hour. Compared with other big cats, a cheetah has longer legs relative to body size and nonretractable claws that act like cleats to increase traction. Thomson's gazelle, its main prey, can run longer but not as fast (80 kilometers per hour). Without a head start, it is toast, so to speak.

Camouflaging helps predators as well as prey. Think of white polar bears stalking seals on ice, striped tigers crouched in tall-stalked, golden grasses, and scorpionfish on the seafloor (Figure 41.13*c*). Camouflage can be quite stunning among predatory insects (Figure 41.13*d*). Even so, with each new, improved camouflaging trait, predators select for enhanced predator-detecting ability in prey.

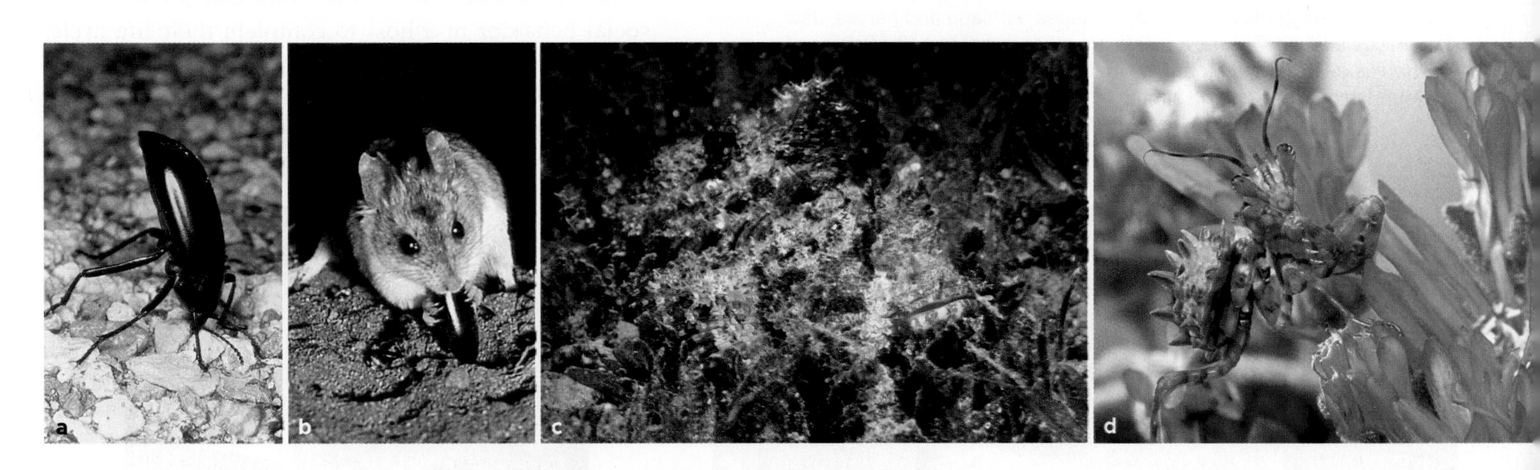

Figure 41.13 Predator responses to prey defenses. (**a**) Some beetles spray noxious chemicals at attackers, which deters them some of the time. (**b**) At other times, grasshopper mice plunge the chemical-spraying tail end of their beetle prey into the ground and feast on the head end. (**c**) Find the scorpionfish, a venomous predator with camouflaging fleshy flaps, multiple colors, and profuse spines. (**d**) Where do the pink flowers end and the pink praying mantis begin?

41.6 Parasite-Host Interactions

Return now to a point made in Section 23.5. A parasite completes its life cycle in one or more hosts, which it typically weakens but does not kill outright.

PARASITES AND PARASITOIDS

LINKS TO SECTIONS 19.6, 23.5, 23.8, 23.13

Parasites spend all or part of their life living in or on another organism or organisms, from which they steal nutrients. Most parasites are small, but they can have major impact on host populations. Many parasites are pathogens; they cause symptoms of disease in hosts. For example, *Myxobolus cerebralis* is a parasite of trout, salmon and related fishes. Following infection, a host develops deadly whirling disease (Figure 41.14).

Even when a parasite does not cause such dramatic symptoms, infection can weaken the host so it is more vulnerable to predation or less attractive to potential mates. Some parasitic infections cause sterility. Others shift the sex ratio of their host species. Parasites affect host numbers by altering birth and death rates. They also indirectly affect species that compete with their host. The decline in trout caused by whirling disease allows competing fish populations to increase.

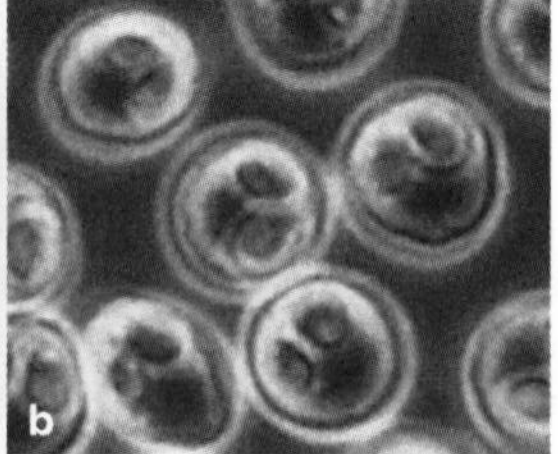

Figure 41.14 (**a**) A young trout with a twisted spine and darkened tail caused by whirling disease, which damages cartilage and nerves. Jaw deformities and whirling movements are other symptoms. (**b**) Spores of *Myxobolus cerebralis*, the parasite that causes the disease. They now are found in many lakes and streams in western and northeastern states.

Sometimes the gradual drain of nutrients during a parasitic infection indirectly leads to death. The host is so weak that it cannot fight off secondary infections. A rapid death is rare. Usually it happens only after a parasite attacks a novel host—one with no coevolved defenses—or after the body is overwhelmed by a huge population of parasites.

Nevertheless, in evolutionary terms, killing a host too quickly is bad for the parasite. Ideally, a host will live long enough to give the parasite time to produce some offspring. The longer the host survives, the more offspring can be produced. That is why we can predict that natural selection favors parasites with less-than-fatal effects on hosts (Section 19.6).

Unit IV describes many parasites. Some spend their entire life cycle in or on a single host species. Others have different hosts during different stages of the life cycle. Insects and other arthropods often act as vectors, or organisms that convey a parasite from host to host.

Even a few plants are parasitic. Nonphotosynthetic species, such as dodders, obtain energy and nutrients from a host plant (Figure 41.15). Other species carry out photosynthesis but steal nutrients and water from their host. Mistletoe is an example; its modified roots invade the vascular tissues of host trees.

Many tapeworms, flukes, and certain roundworms are parasitic invertebrates (Figure 41.16). So are ticks, many insects, and some crustaceans.

Parasitoids are insects that lay eggs in other insects. Larvae hatch, develop in the host's body, eat its tissue, and eventually kill it. The fire ant-killing phorid flies described in this chapter's introduction do this. As many as 15 percent of all insects may be parasitoids.

Social parasites are animals that take advantage of social behavior of a host to complete their life cycle. Cuckoos and North American cowbirds, as explained shortly, are social parasites.

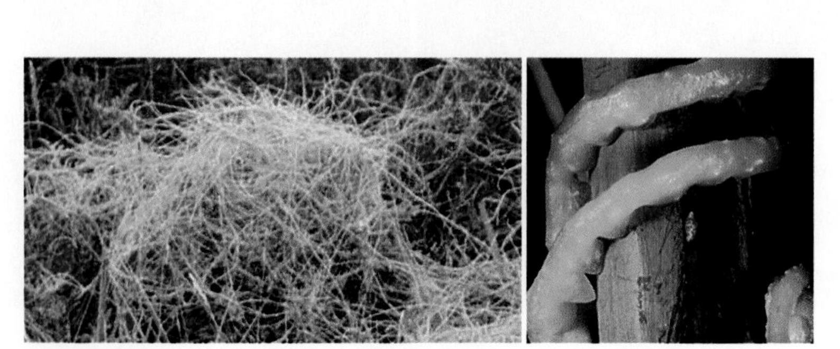

Figure 41.15 Dodder (*Cuscuta*), also known as strangleweed or devil's hair. This parasitic flowering plant has almost no chlorophyll. Leafless stems twine around a host plant during growth. Modified roots penetrate the host's vascular tissues and absorb water and nutrients from them.

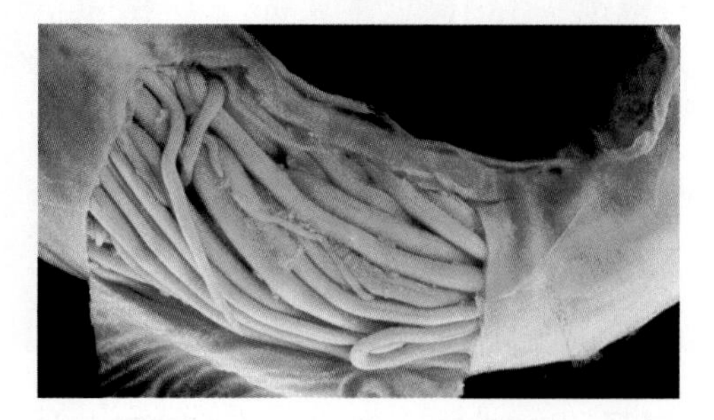

Figure 41.16 Adult roundworms (*Ascaris*), an endoparasite, inside the small intestine of a host pig. Sections 23.5 and 23.8 show more examples of parasitic worms.

Figure 41.17 Biological control agent: a commercially raised parasitoid wasp about to deposit an egg in an aphid. This wasp reduces aphid populations. After the egg it laid hatches, a wasp larva devours the aphid from the inside.

USES AS BIOLOGICAL CONTROLS

Parasites and parasitoids are commercially raised and released in target areas as biological controls. They are promoted as a workable alternative to pesticides. The chapter introduction and Figure 41.17 give examples.

Effective biological controls display five attributes. The agents are adapted to a specific host species and to its habitat. They are good at finding the hosts. Their population growth rate is high compared to the host's. Their offspring are good at dispersing. Also, they make a type III response to prey (Section 41.4), without much lag time after the prey or host population size shifts.

Biological control is not without risks of its own. Releasing more than one species of biological control agent in an area may invite competition among them, which can lower their effectiveness against an intended target. Also, introduced parasites sometimes go after nontargeted species in addition to, or instead of, those species they were introduced to control.

For example, parasitoids deliberately introduced to the Hawaiian Islands attacked the wrong target. They were brought in to control stink bugs that are pests of Hawaii's crops. Instead, the parasitoids decimated the population of koa bugs, Hawaii's largest native bug. Introduced parasitoids also have been implicated in ongoing declines of many native Hawaiian butterfly and moth populations.

Natural selection favors parasitic species that are not rapidly lethal, which favors an adequate supply of hosts.

Parasitic species occur in many groups, including bacteria, protists, invertebrates, and plants. Parasitoids are insects that eat other insects from inside out. Social parasites use the social behavior of another species to their own benefit.

41.7 Cowbird Chutzpah

FOCUS ON EVOLUTION

The brown-headed cowbird's genus name (Molothrus) means intruder in Latin. This bird intrudes, sneakily, into the life cycle of its hosts—other bird species.

Brown-headed cowbirds (*Molothrus ater*) evolved in the Great Plains of North America. They were commensal with bison. Great herds of these hefty ungulates stirred up plenty of tasty insects as they migrated through the grasslands, and, being insect-eaters, cowbirds wandered around with them (Figure 41.18*a*).

Cowbirds are social parasites that lay their eggs in the nests constructed by other birds, so young cowbirds are reared by foster parents. Many species became "hosts" to cowbirds; they did not have the capacity to recognize the differences between cowbird eggs and their own eggs. Concurrently, cowbird hatchlings became innately wired for hostile takeovers. Even before hatchlings open their eyes, they shove the owner's eggs out of the nest and demand to be fed by unwitting, and often smaller, foster parents (Figure 41.18*b*). For thousands of years, cowbirds have perpetuated their genes at the expense of hosts.

When American pioneers moved west, many cleared swaths of woodlands for pastures. Cowbirds now moved in the other direction. They adapted easily to a life with new ungulates—cattle—in the manmade grasslands; hence their name. They started to penetrate adjacent woodlands and exploit novel species. Today, brown-headed cowbirds parasitize at least fifteen kinds of native North American birds. Some of those birds are threatened or endangered.

Besides being successful opportunists, cowbirds are big-time reproducers. A female can lay an egg a day for ten days, give her ovaries a rest, do the same again, and then again in one season. As many as thirty eggs in thirty nests—that is a lot of cowbirds.

Figure 41.18 Oh give me a home, where the buffalo roam—brown-headed cowbirds (*Molothrus ater*) originally evolved as commensalists with bison, and as social parasites of other birds of the North American Great Plains. When conditions changed, they expanded their range. These imposters now nest in woodlands as well as grasslands in much of the United States.

41.8 Ecological Succession

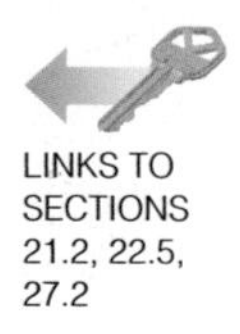

LINKS TO SECTIONS 21.2, 22.5, 27.2

In an older view, a community emerges as an outcome of competition and other interactions and stabilizes into a predictable array of species. However, chance events and abiotic forces, including fires, storms, and human-induced disturbances, also affect community structure.

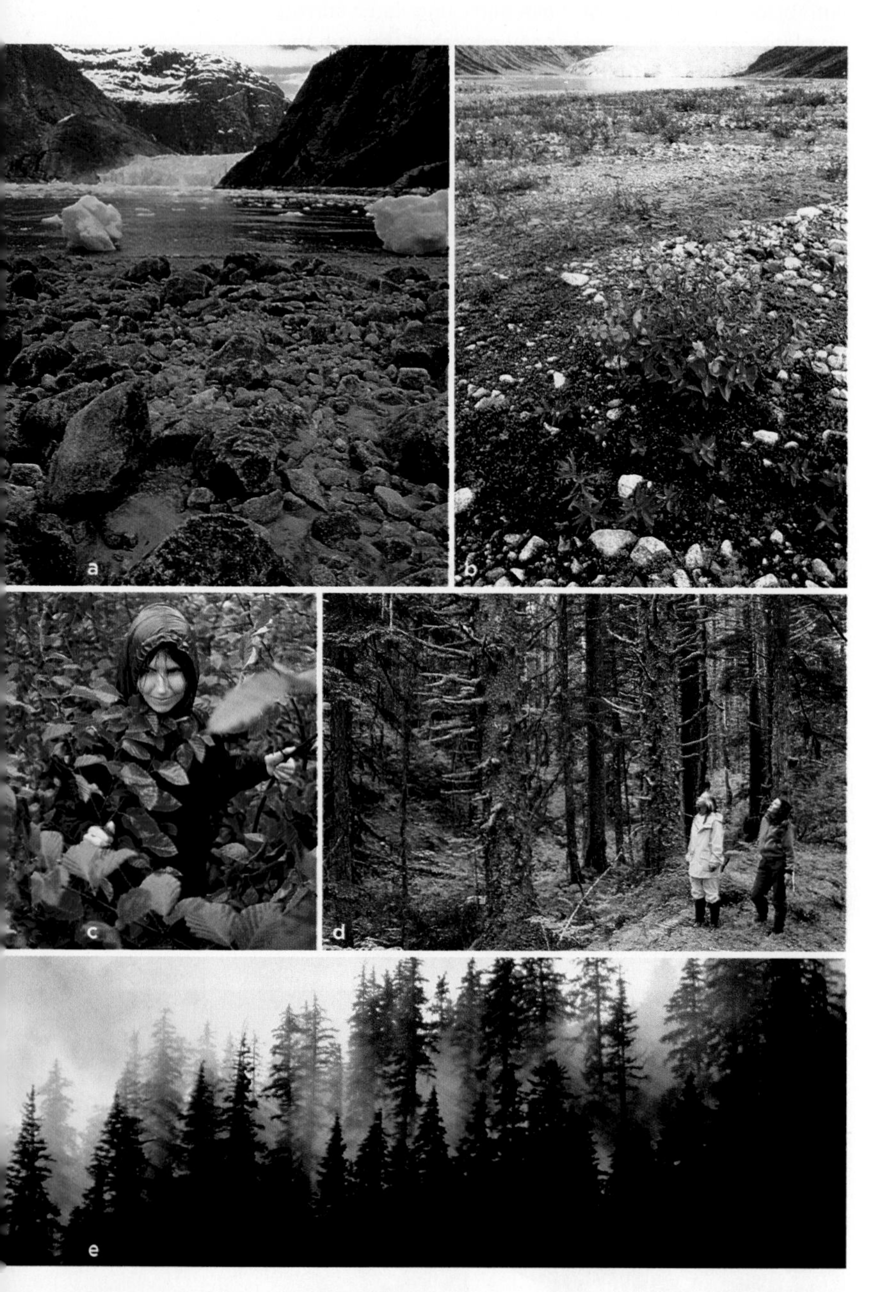

Figure 41.19 In Alaska's Glacier Bay region, one observed pathway of primary succession. (**a**) As a glacier retreats, meltwater leaches minerals from the rocks and gravel left behind. (**b**) Pioneer species include lichens, mosses, and some flowering plants such as mountain avens (*Dryas*), which associate with nitrogen-fixing bacteria. Within 20 years, alder, cottonwood, and willow seedlings take hold. Alders also have nitrogen-fixing symbionts. (**c**) Within 50 years, they form dense, mature thickets in which cottonwood, hemlock, and a few evergreen spruce grow. (**d**) After 80 years, western hemlock and spruce crowd out alders. (**e**) In areas deglaciated for more than a century, tall Sitka spruce dominate.

SUCCESSIONAL CHANGE

A concept of "nature in balance" once guided studies in community ecology. Researchers knew community structure starts with pioneer species, or opportunistic colonizers of new or newly vacated habitats. Pioneers have high dispersal rates, grow and mature quickly, and produce many offspring. More competitive species replace them. Then the replacements are replaced.

Primary succession is a process that begins when pioneer species colonize a barren habitat with no soil, such as a new volcanic island or land exposed by the retreat of a glacier (Figure 41.19). The earliest plants to colonize a new habitat are pioneer species, such as lichens and mosses (Sections 21.2 and 22.5). They are small, have a brief life cycle, and can tolerate intense sunlight, extreme temperature changes, and little or no soil. Flowering annual plants with wind-dispersed seeds are also among the pioneers.

Established pioneers help build and improve the soil. In doing so, they may set the stage for their own replacement. Many of the pioneers are mutualists with nitrogen-fixing bacteria, so they can grow in nitrogen-poor habitats. Seeds of later species find shelter inside mats of the pioneers, which do not grow high enough to shade out the new seedlings.

Organic wastes and remains accumulate over time. By adding volume and nutrients to soil, this material allows other species to take hold. Later successional species often crowd out early ones whose spores and seeds travel on winds and water—destined, perhaps, for another new but temporary habitat.

In secondary succession, a disturbed area within a community recovers. If improved soil is still present, secondary succession can be fast. It commonly occurs in abandoned fields, burned forests, and tracts of land where plants were killed by volcanic eruptions.

FACTORS THAT INFLUENCE SUCCESSION

In a traditional view, a predictable group of species in the habitat stabilizes as the *climax* community, after which not much changes. This community is adapted to many factors, such as topography, climate, soil, and species interactions, and it may show some variation along gradients of environmental conditions. In this view, even after a disturbance, the community reverts to the same climax species array.

Ecologists now realize that the species composition of a community changes frequently, in unpredictable ways. Communities do not journey along a well-worn path to some predetermined climax state.

Figure 41.20 A natural laboratory for succession after the 1980 Mount Saint Helens eruption (**a**). The community at the base of this Cascade volcano was destroyed. (**b**) In less than a decade, pioneer species took hold. (**c**) Twelve years later, seedlings of a dominant species, Douglas firs, took hold.

Random events can determine the order in which species arrive in a habitat and thus affect the course of succession. Arrival of a certain species may make it easier or more difficult for others to take hold. As an example, surf grass can only grow along a shoreline if algae has already colonized that area. The algae acts as an anchoring site for the grass. In contrast, the arrival of some species prevents others from moving in. After sagebrush gets established, chemicals it secretes into the soil keep most other plant species out.

Ecologists had an opportunity to investigate these factors after the 1980 eruption of Mount Saint Helens leveled about 600 square kilometers (235 square miles) of forest in Washington state (Figure 41.20). Ecologists recorded the natural pattern of colonization. They also carried out experiments in plots inside the blast zone. They added seeds of certain pioneer species to some plots and left other plots seedless. The results showed that some pioneers helped other later arriving plants become established. Different pioneers kept the same late arrivals out.

Disturbances also affect the species composition of many communities. By the intermediate disturbance hypothesis, species richness is the greatest in habitats where disturbances are moderate in their intensity or frequency. In such habitats, there is enough time for colonists to arrive and get established but not enough for competitive exclusion to cause extinctions:

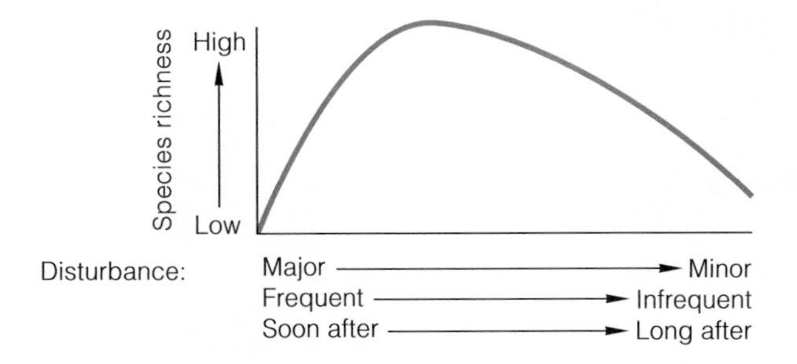

In short, the modern view of succession holds that the species composition of a community is affected by (1) physical factors such as soil and climate, (2) chance events such as the order in which species arrive, and (3) the extent of disturbances in a habitat. Because the second and third factors may vary even between two geographically close regions, it is generally difficult to predict what any given community will look like at any point in the future.

A community develops through a succession of stages, starting with pioneer species that may establish conditions for their own replacement. Biotic (biological) and abiotic (physical and chemical) factors affect community structure. Disturbances are unpredictable, and vary in frequency and size. By an intermediate disturbance hypothesis, species richness is greatest in between moderate disturbances.

41.9 Species Interactions and Community Instability

The loss or addition of even one species may destabilize the number and abundances of species in a community.

LINK TO SECTION 40.4

As you read earlier, short-term physical disturbances can knock a community out of equilibrium. Long-term changes in climate or another environmental variable also have destabilizing effects. Besides this, a shift in species interactions can tip the community out of its uneasy balance. Remember, resources are sustained as long as populations do not flirt dangerously with the carrying capacity (Section 40.4). Predators and prey coexist as long as neither wins. Competitors have no sense of fair play. Mutualists minimize their costs and maximize their benefits, as when plants make the least nectar necessary to attract pollinators and pollinators take as much nectar as they can.

Whether biotic or abiotic, a disturbance sometimes causes the number and relative abundances of species to shift irrevocably. For instance, if some occupants of the habitat happen to be rare or do not compete well with the others, they might be driven to extinction.

THE ROLE OF KEYSTONE SPECIES

The uneasy balancing of forces in a community comes into focus when we observe the effects of a keystone species. Such a species has a disproportionately large effect on a community relative to its abundance. Robert Paine was the first to describe the effect of a keystone species after his experiments on the rocky shores of California's coast. Species in the rocky intertidal zone withstand pounding surf by clinging to rocks. A rock to cling to is a big limiting factor. Paine set up control plots with the sea star *Pisaster ochraceus* and its main prey—chitons, limpets, barnacles, and mussels. Then he removed all sea stars from his experimental plots.

Mussels (*Mytilus*) happen to be the prey of choice for sea stars. In the absence of sea stars, they took over Paine's experimental plots; they became the strongest competitors and crowded out seven other species of invertebrates. In this intertidal zone, predation by sea stars normally keeps the number of prey species high because it restricts competitive exclusion by mussels.

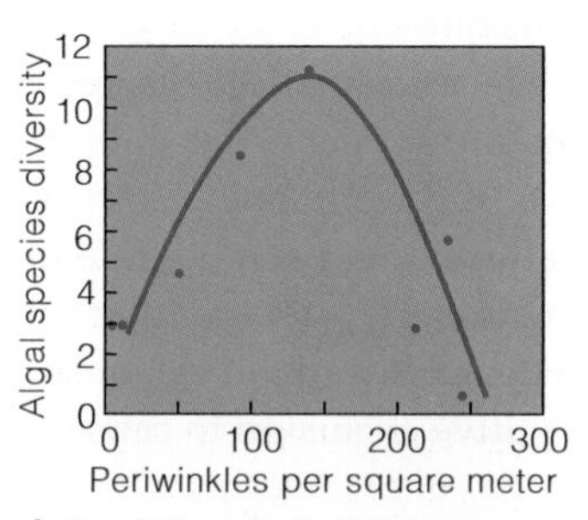

d Algal diversity in tidepools

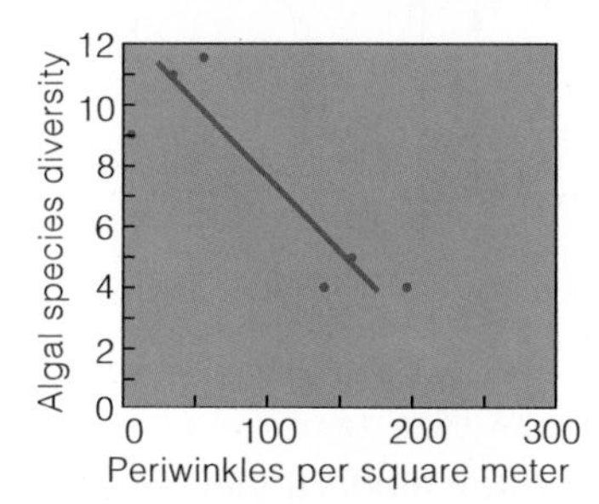

e Algal diversity on rocks that become exposed at high tide

Figure 41.21 Effect of competition and predation in an intertidal zone. (**a**) Grazing periwinkles (*Littorina littorea*) affect the number of algal species in different ways in different marine habitats. (**b**) *Chondrus* and (**c**) *Enteromorpha*, two kinds of algae in their natural habitats. (**d**) By grazing on the dominant alga in tidepools (*Enteromorpha*), the periwinkles promote the survival of less competitive algal species that would otherwise be overgrown. (**e**) *Enteromorpha* does not grow on rocks. Here, *Chondrus* is dominant. Periwinkles find *Chondrus* tough and dine instead on less competitive algal species. By doing so, periwinkles decrease the algal diversity on the rocks.

Table 41.2 Outcomes of Some Species Introductions Into the United States

Species Introduced	Origin	Mode of Introduction	Outcome
Water hyacinth	South America	Intentionally introduced (1884)	Clogged waterways; other plants shaded out
Dutch elm disease:			
Ophiostoma ulmi (fungus)	Asia (by way of Europe)	Accidental; on infected elm timber (1930)	Millions of mature elms destroyed
Bark beetle (vector)		Accidental; on unbarked elm timber (1909)	
Chestnut blight fungus	Asia	Accidental; on nursery plants (1900)	Nearly all eastern American chestnuts killed
Zebra mussel	Russia	Accidental; in ballast water of ship (1985)	Clog pipes and water intake valves of power plants; displacing native Great Lake bivalves
Japanese beetle	Japan	Accidental; on irises or azaleas (1911)	Close to 300 plant species (e.g., citrus) defoliated
Sea lamprey	North Atlantic	Accidental; on ship hulls (1860s)	Trout, other fish species destroyed in Great Lakes
European starling	Europe	Intentional release, New York City (1890)	Outcompete native cavity-nesting birds; crop damage; swine disease vector
Nutria	South America	Accidental release of captive animals being raised for fur (1930)	Crop damage, destruction of levees, overgrazing of marsh habitat

Remove all the sea stars, and the community shrinks from fifteen species to eight.

The impact of a keystone species can vary between habitats that differ in their species arrays. Periwinkles (*Littorina littorea*) are alga-eating snails that live in the intertidal zone. Jane Lubchenco found removing them can increase *or* decrease the diversity of algal species, depending on the habitat (Figure 41.21).

In tidepools, periwinkles prefer to eat a certain alga (*Enteromorpha*) which can outgrow other algal species. By keeping that alga in check, periwinkles help other, less competitive algal species survive. On rocks of the lower intertidal zone, *Chondrus* and other tough, red algae dominate. Here, periwinkles preferentially graze on competitively weaker algae. Periwinkles *promote* species richness in tidepools but *reduce* it on rocks.

HOW SPECIES INTRODUCTIONS TIP THE BALANCE

Instabilities are also set in motion when residents of an established community move out from their home range, then successfully take up residence elsewhere. This type of directional movement, called geographic dispersal, happens in three ways.

First, over a number of generations, a population might expand its home range by slowly moving into outlying regions that prove hospitable. Second, some individuals might be rapidly transported across great distances, an event called *jump* dispersal. This often takes individuals across regions where they could not survive on their own, as when insects travel from the mainland to Maui in a ship's cargo hold. Third, some population might be moved away from a home range by continental drift, at an almost imperceptibly slow pace over long spans of time.

Successful dispersal and colonization of a vacant adaptive zone can be remarkably rapid. Consider one of Amy Schoener's experiments in the Bahamas. She set out plastic sponges on barren sand at the bottom of Bimini Lagoon. How fast did aquatic species take up residence on or in the artificial habitats? Schoener recorded occupancy by 220 species within thirty days.

When you hear someone speaking enthusiastically about exotic species, you can safely bet the speaker is not an ecologist. An exotic species is a resident of an established community that dispersed from its home range and became established elsewhere. Unlike most imports, which never do take hold outside the home range, an exotic species permanently insinuates itself into a new community.

Following jump dispersal, more than 4,500 exotic species have become established in the United States. We put some of the new arrivals, including soybeans, rice, and wheat, to use as food crops. Others brighten gardens or provide textiles. An estimated 25 percent of Florida's plant and animal species are exotics. In Hawaii, 45 percent are exotic.

Both accidental and deliberate imports can change community structure. You already have learned about a few examples. Sudden oak death caused by an exotic protist threatens forests (Chapter 20 introduction). A parasite from Europe causes whirling disease in trout. Table 41.2 lists some additional imports, and the next section focuses on three notorious ones.

A keystone species is one that has a major effect on species richness and relative abundances in a habitat.

Species introductions and other biotic disturbances can permanently alter community structure.

41.10 Exotic Invaders

Nonnative species are on the loose in communities on every continent. They can alter habitats. Often they outcompete and displace native species.

THE ALGA TRIUMPHANT

LINK TO SECTION 20.8

They looked so perfect in saltwater aquariums, those long, green, feathery branches of *Caulerpa taxifolia*. So Stuttgart Aquarium researchers in Germany developed a hybrid, sterile strain of this green alga and generously shared it with other marine institutions. Was it from Monaco's Oceanographic Museum that the hybrid strain escaped into the wild? Some say yes, Monaco says no.

In any case, a small patch of the aquarium strain was found growing in the Mediterranean near Monaco in 1984. Boat propellers and fishing nets dispersed it. The alga now blankets tens of thousand of acres of seafloor in the Mediterranean and Adriatic (Figure 41.22*a*).

In 2000, scuba divers discovered *C. taxifolia* growing near the southern California coast. Someone might have drained water from a home aquarium into a storm drain or into the lagoon itself. The government and private groups quickly sprang into action. So far, eradication and surveillance programs have worked, but at a cost of more than $3.4 million.

Importing *C. taxifolia* or any closely related species of *Caulerpa* into the United States is now illegal. To protect native aquatic communities, aquarium water should never be dumped into storm drains or waterways. It should be discarded into a sink or toilet so wastewater treatment can kill any algal spores (Section 20.8).

Just how bad is *C. taxifolia*? The aquarium strain can thrive on sandy or rocky shores and in mud. It can live ten days after being discarded in meadows. Unlike its tropical parents, it can also survive in cool water and polluted water. It has the potential to displace endemic algae, overgrow reefs, and destroy marine food webs. Its toxin poisons invertebrates and fishes, including algae eaters that keep other algae in check. This algal strain has been nominated as one of the 100 worst exotic invaders.

THE PLANTS THAT ATE GEORGIA

In 1876, kudzu (*Pueraria montana*) was introduced to the United States from Japan. In its native habitat, this perennial vine is a well-behaved legume with a strong root system. It *seemed* like a good idea to use it for forage and to control erosion on slopes. But kudzu grew faster in the American Southeast. No native herbivores or pathogens were adapted to attack it. Competing plant species posed no serious threat to it.

With nothing to stop it, kudzu can grow sixty meters (200 feet) per year. Its vines now blanket streambanks, trees, telephone poles, houses, and almost anything else in their path (Figure 41.22*b*). Kudzu withstands burning, and grows back from its deep roots. Grazing goats and herbicides help. But goats eat most other plants along with it, and herbicides taint freshwater supplies. Kudzu invasions now stretch from Connecticut down to Florida and are reported in Arkansas. It crossed the Mississippi River into Texas. Thanks to jump dispersal, it is now an invasive species in Oregon.

Figure 41.22 (**a**) Aquarium strain of *Caulerpa taxifolia* suffocating yet another richly diverse marine ecosystem.

(**b**) Kudzu (*Pueraria montana*) taking over part of Lyman, South Carolina. This vine has become invasive in many states from coast to coast. Ruth Duncan of Alabama, who makes 200 kudzu vine baskets a year, just can't keep up.

Figure 41.23 Rabbit-proof fence? Not quite. This is part of a fence built to hold back the 200 million to 300 million rabbits that are wreaking havoc with the vegetation in Australia. It did not work.

On the bright side, Asians use a starch extracted from kudzu in drinks, herbal medicines, and candy. A kudzu processing plant in Alabama may export this starch to Asia, where the demand currently exceeds the supply. Also, kudzu may help save forests; it can be an alternative source for paper and other wood products. Today, about 90 percent of Asian wallpaper is kudzu-based.

THE RABBITS THAT ATE AUSTRALIA

During the 1800s, British settlers in Australia just couldn't bond with koalas and kangaroos, and so they imported familiar animals from home. In 1859, in what would be the start of a major ecological disaster, a landowner in northern Australia imported and released two dozen European rabbits (*Oryctolagus cuniculus*). Good food and great sport hunting—that was the idea. An ideal rabbit habitat with no natural predators—that was the reality.

Six years later, the landowner had killed 20,000 rabbits and was besieged by 20,000 more. The rabbits displaced livestock and caused the decline of native wildlife. Now 200 million to 300 million are hippity-hopping through the southern half of the country. They graze on grasses in good times and strip bark from shrubs and trees during droughts. Thumping hordes turn shrublands as well as grasslands into eroded deserts. Their burrows undermine the soil and set the stage for widespread erosion.

Rabbit warrens have been shot at, fumigated, plowed under, and dynamited. The first all-out assaults killed 70 percent of them, but the rabbits rebounded in less than a year. When a fence 2,000 miles long was built to protect western Australia, rabbits made it from one side to the other before workers could finish the job (Figure 41.23).

In 1951, the government introduced a myxoma virus that normally infects South American rabbits. The virus causes myxomatosis. This disease has mild effects on its coevolved host but nearly always kills *O. cuniculus*. Fleas and mosquitoes transmit the virus to new hosts. With no coevolved defenses against the import, European rabbits died in droves. But natural selection has since favored a rise in rabbit populations resistant to the imported virus.

In 1991, on an uninhabited island in Australia's Spencer Gulf, researchers released rabbits that were injected with a calicivirus. The rabbits died from blood clots in their lungs, heart, and kidneys. Then the test virus escaped from the island in 1995, perhaps on insect vectors.

By 2001, the rabbit population sizes stabilized at 80 to 85 percent below the peak values. Grasses, nonwoody shrubs, and woody shrubs are rebounding. Different kinds of herbivores are increasing in density.

The rabbit calicivirus was discovered in China in 1984 and is now found in Europe and other countries as well. To date, tests on more than forty animal species indicate that this pathogen replicates in rabbits alone. But other caliciviruses can and do cross species barriers. The jury is still out on the long-term impact of the viral releases.

As you might have deduced, *O. cuniculus* is another one of the 100 worst exotic invaders. Also on the list are two *Anopheles* species, mosquitoes that are vectors for malaria. The cane toad (*Bufo marinus*) made the list, too. It was introduced as a biological control of pests in fields of sugarcane and other crops all over the world, but it eats almost everything. The banana bunchy top virus is another one of the worst, despite its catchy name. House cats (*Felis catus*) gone wild are another problem. Finally, the house mouse (*Mus musculus*) probably has a greater distribution than any other mammal except humans. Populations of this prolific breeder destroy crops and consume or contaminate much of our food supplies. They are implicated in the extinction of many species. Interested in learning more? For some eye-openers, go to the Invasive Species Specialist Group at www.issg.org.

41.11 Biogeographic Patterns in Community Structure

The richness and relative abundances of species differ from one habitat or region of the world to another. Often the differences correspond to predictable patterns that have biogeographic and historical foundations.

LINKS TO SECTIONS 16.1, 24.13, 40.4

Biogeography is the study of how species richness is distributed in the natural world (Section 16.1). We see patterns that correspond with differences in sunlight, temperature, rainfall, and other factors that vary with latitude, elevation, or water depth. Still other patterns relate to the history of a habitat and the species in it. Each species has its own unique physiology, capacity for dispersal, resource requirements, and interactions with other species.

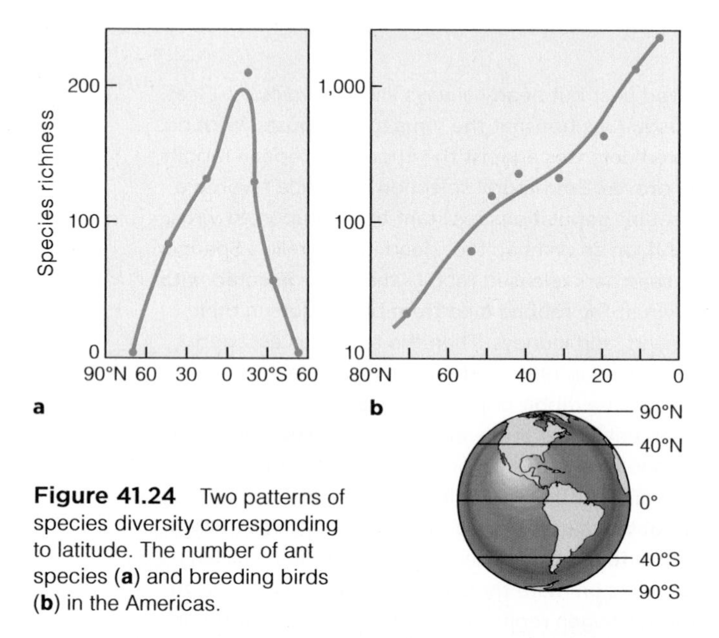

Figure 41.24 Two patterns of species diversity corresponding to latitude. The number of ant species (**a**) and breeding birds (**b**) in the Americas.

MAINLAND AND MARINE PATTERNS

Perhaps the most striking pattern of species richness corresponds with distance from the equator. *For most plants and animal groups, the number of coexisting species is at its maximum in the tropics, and that number declines from the equator to the poles.* Figure 41.24 illustrates two examples of this pattern. Consider just a few factors that help bring about such a pattern and maintain it.

First, for reasons explained in Section 43.1, tropical latitudes intercept more intense sunlight and receive more rainfall, and their growing season is longer. As one outcome, resource availability tends to be greater and more reliable in the tropics than elsewhere. One result is a degree of specialized interrelationships not possible where species are active for shorter periods.

Second, tropical communities have been evolving for a long time. Some temperate communities did not start forming until the end of the last ice age.

Third, species richness may be self-reinforcing. The number of species of trees in tropical forests is much greater than in comparable forests at higher latitudes. Where more plant species compete and coexist, more species of herbivores also coexist, partly because no single herbivore species can overcome all chemical defenses of all plants. In addition, more predators and parasites can evolve in response to more kinds of prey and hosts. The same principles apply to tropical reefs.

ISLAND PATTERNS

As you saw in Section 40.4, islands are laboratories for population studies. They have also been laboratories for community studies. For instance, in the mid-1960s volcanic eruptions formed a new island 33 kilometers

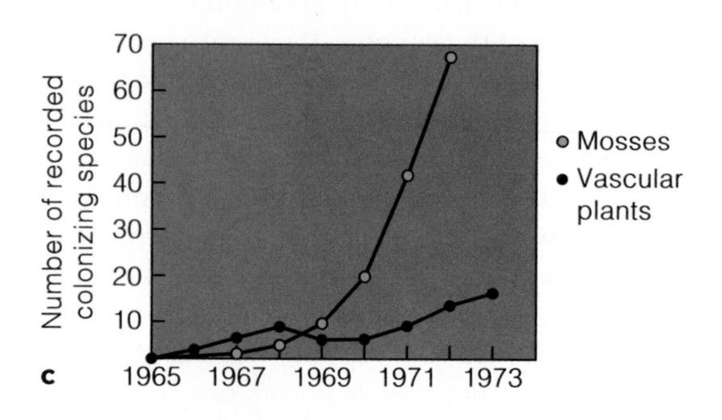

Figure 41.25 Surtsey, a volcanic island, at the time of its formation (**a**) and in 1983 (**b**). The graph (**c**) shows the number of colonizing plant species between 1965 and 1973.

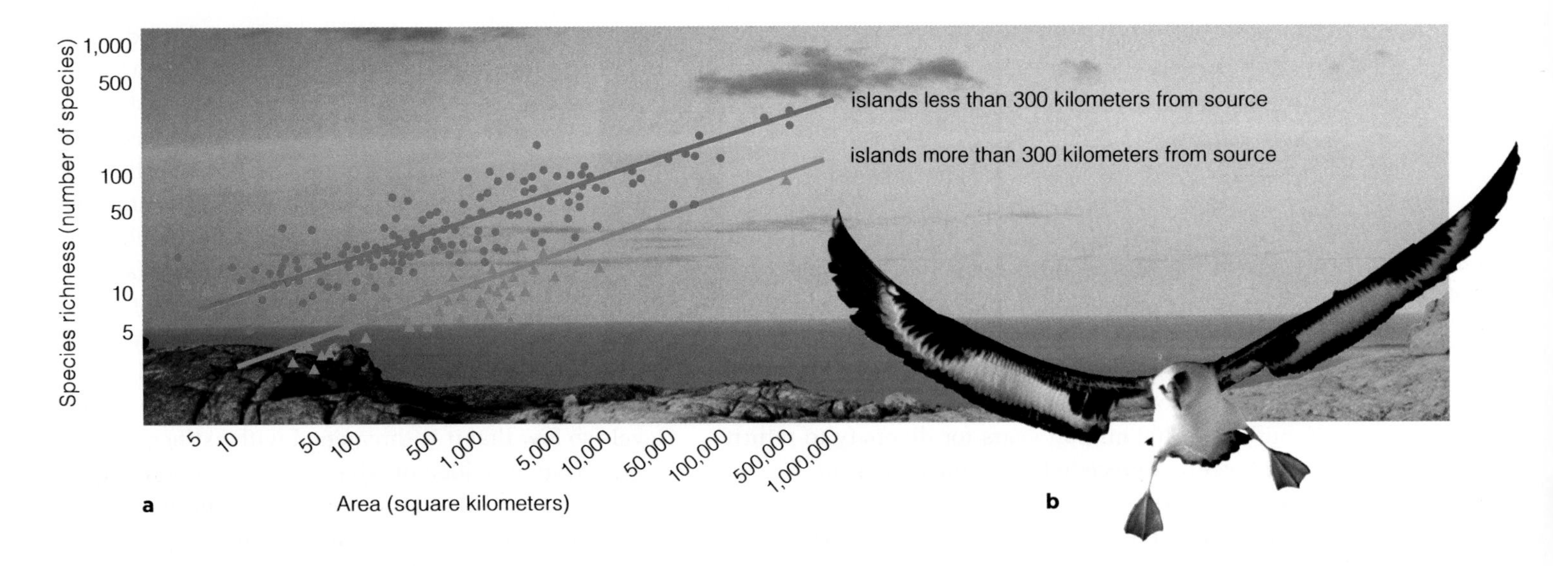

Figure 41.26 (**a**) Two island biodiversity patterns. Distance effect: Species richness on islands of a specified size declines with increasing distance from a source of colonizing species. *Green* circles signify islands less than 300 kilometers from the colonizing source. *Orange* triangles signify islands more than 300 kilometers (190 miles) from a source of colonists. Area effect: Among islands the same distance from a source of colonizing species, the larger ones support more species.

(**b**) Wandering albatross, one travel agent for jump dispersals. Seabirds that island-hop long distances often have seeds stuck to their feathers. Seeds that successfully germinate in a new island community may give rise to a population of new immigrants.

(21 miles) from the coast of Iceland. The island was named Surtsey (Figure 41.25). Bacteria and fungi were early colonists. Contrary to most expectations, the first plants to take root were vascular plants. Mosses came along two years later (Figure 41.25*c*). The first lichens were found five years later. As noted in the previous section, succession is highly unpredictable.

We can, however, predict that the number of species on Surtsey will not continue increasing forever. Can we estimate how many species there will be when the number levels off? Models based on studies of islands around the world suggest some answers.

The number of species living on any island reflects a balance between immigration rates for new species and extinction rates for established ones. The distance between an island and a mainland source of colonists affects immigration rates. An island's size affects both immigration rates and extinction rates.

Consider first the distance effect: Islands far from a source of colonists receive fewer immigrants than those closer to a source. Most species cannot disperse very far, so they will not turn up far from a mainland.

Species richness also is shaped by the area effect: Big islands tend to support more species than small ones. More colonists will happen upon a larger island simply by virtue of its size. Also, big islands are more likely to offer a variety of habitats, such as high and low elevations. These options make it more likely that a new arrival will find a suitable habitat. Finally, big islands can support larger populations of species than small islands. The larger a population, the less likely it is to become locally extinct because of random events.

Figure 41.26 illustrates how interactions between the distance effect and the area effect can influence the species richness on islands.

One more island pattern: Remember the miniature *Homo* species that was discovered on the Indonesian island Flores (Section 24.13)? It is not uncommon for species on an island to be bigger or smaller than their mainland ancestor. Islands hold different competitors, prey, and predators than the mainland, so becoming larger or smaller may be selectively advantageous. For a fascinating look at what biologists have learned by studying islands, pick up David Quammen's popular book *Song of the Dodo*.

Species richness reveals global patterns, as when it correlates with gradients in latitude, elevation, and water depth. Microenvironments along gradients often introduce variations in the overall patterns of species richness.

Generally, species richness is highest in the tropics and lowest at the poles. Tropical habitats have conditions that more species can tolerate and tropical communities have often been evolving for longer than temperate ones.

When a new island forms, species richness rises over time and then levels off. The size of an island and its distance from a colonizing source influence its species richness.

41.12 Threats to Biodiversity

We now recognize three levels of biodiversity—genetic diversity, species diversity, and ecosystem diversity. We are witnessing declines at all three levels. Expansion of the human population is causing a biodiversity crisis.

ON THE NEWLY ENDANGERED SPECIES

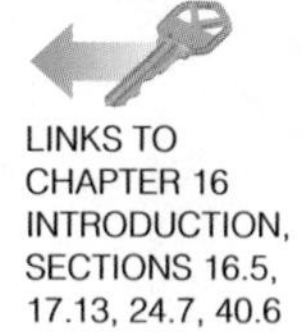

LINKS TO CHAPTER 16 INTRODUCTION, SECTIONS 16.5, 17.13, 24.7, 40.6

Five great mass extinctions mark the boundaries for geologic time periods (Section 16.5). After each event, biodiversity plummeted on land and in the seas. Then the surviving species underwent adaptive radiations. Each time, recovery was extremely slow. It took about 20 million to 100 million years for diversity to return to the level that preceded the extinction event.

No biodiversity-shattering asteroids have smashed into Earth for about 65 million years. Yet a sixth major extinction event is under way. Humans are driving many species over the edge. For example, mammals survived the asteroid impact that killed the last of the dinosaurs (Section 24.7). Many have not been as lucky in their encounters with humans. Shortly after the first humans migrated into the Americas, 75 percent of the large species of mammals vanished. Climate changes played a role, but hunting was a contributing factor.

About 10 percent of mammals, 5 percent of birds, and 20 percent of amphibians are now endangered. An endangered species is a species that originated in one region and is found nowhere else and has population levels so low that it is threatened with extinction.

We have little idea of what effect we are having on insects and the other less conspicuous species. Figure 41.27 shows approximate number of named species in some major groups. We know most about flowering plants, conifers, and vertebrates. We know more about life on land than in the seas. Astonishing numbers of invertebrates, fungi, protists, bacteria, and archaeans still await identification. We do not even know what is out there, so how can we assess what we are losing?

For now, start thinking about some practices that have had the most impact over the past forty years: species introductions, destruction and fragmentation of natural habitats, and overharvesting and poaching.

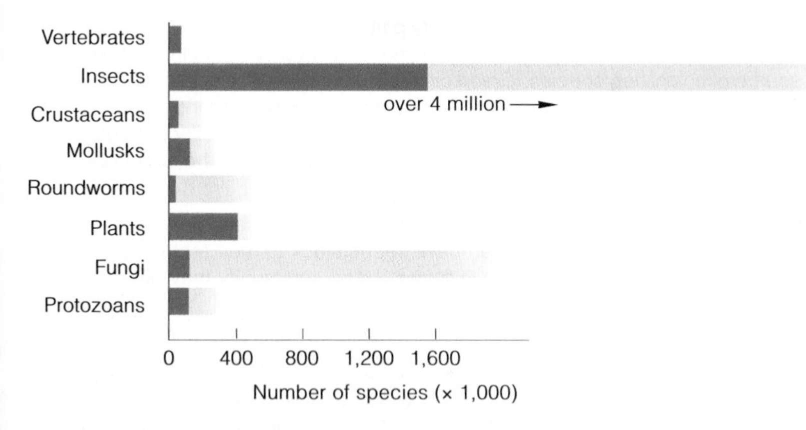

Figure 41.27 Current species diversity for a few major taxa. *Red* signifies the number of named species. *Gold* signifies the estimated number of species yet to be discovered and named.

Species Introductions Introduced species often outcompete natives, as when European rabbits overran Australian herbivores. As another example, gray squirrels from North America have largely replaced red squirrels in England, Scotland, Ireland, and Italy.

Introduced parasites and predators can wipe out an endemic species that did not evolve with any similar threats and that have no defenses against them. For example, Hawaiian birds evolved on remote islands that were free of mammalian predators. Later, ground-nesting species were driven to extinction by imported rats, cats, dogs, foxes, and mongooses (Section 17.13).

Habitat Loss and Fragmentation Naturalist Edward O. Wilson defines habitat loss as the physical reduction in suitable places to live, as well as the loss of suitable habitat because of pollution. Habitat loss is putting pressure on more than 90 percent of species now facing extinction.

Biodiversity is greatest in the tropics—the regions most threatened by deforestation. In the United States, 98 percent of the tallgrass prairies, 50 percent of the wetlands, and 95 percent of old-growth forests have disappeared. Even desert lands are not exempt from human encroachment (Figure 41.28).

Figure 41.28 Aerial view of naturally arid habitat in Nevada that has been modified to suit the now dominant species—humans.

Habitats often become fragmented, or chopped into patches, each with a separate periphery. Individuals at the habitat edges are more vulnerable to stresses, such as temperature shifts, wind, fire, and predation. Also, fragmentation may produce patches that are not large enough to support the population sizes required for successful breeding. Patches may be too small to have enough food or other resources to sustain a species.

Think back on the factors shaping species richness on islands. Although islands make up no more than a fraction of Earth's surface, about half of all plants and animals driven to extinction since 1600 were island natives that simply had nowhere else to go.

Models derived from studies of islands can help us make predictions about what might happen on habitat islands—in natural settings surrounded by a "sea" of degraded habitat. Many parks and wildlife preserves are habitat islands. The models can be used to estimate the size of an area that must be set aside as a protected reserve to ensure survival of a species.

Habitat damage and loss can affect different species in different ways. An indicator species is a species that alerts us to habitat degradation and impending loss of diversity when its health or numbers change. As one example, biologists often assess the health of a stream by monitoring its indicator species. Declines in a trout population can be an early sign of problems; trout do not tolerate pollutants or low oxygen levels. Lichens also function as indicators. Because all lichens absorb mineral ions from dust in the air, pollutants threaten them. The lichens absorb toxic metals, such as mercury and lead, and cannot get rid of them.

Overharvesting and Poaching The crash of the Atlantic codfish population, described in Section 40.6, is part of a trend. Biologist Boris Worm estimates that about 29 percent of commercially hunted populations of marine fishes and invertebrates have collapsed, by which he means the annual catch for these species has fallen below 10 percent of its recorded maximum. If this trend continues, populations of all commercially harvested marine species may collapse by 2050.

If that sounds overly pessimistic, consider that there were 3 billion to 5 billion passenger pigeons in North America when European settlers arrived. In the 1800s commercial hunting caused a steep decline in numbers. The last time anyone saw a wild passenger pigeon was 1900. He shot it. The last captive bird died in 1914.

In a sad commentary on human nature, the more rare a wild animal becomes, the more its value soars in the black market. Figure 41.29 gives an idea of the money that changes hands in illegal wildlife trading.

Figure 41.29 Confiscated tiger skins, leopard skins, rhinoceros horns, and elephant tusks. An illegal wildlife trade threatens the existence of more than 600 species, yet all but 10 percent of this illegal trade escapes detection. A few of the black market prices: Live, mature saguaro cactus ($5,000–15,000), bighorn sheep head ($10,000–60,000), polar bear ($6,000), grizzly ($5,000), bald eagle ($2,500), peregrine falcon ($10,000), live chimpanzee ($50,000), live mountain gorilla ($150,000), tiger hide ($100,000), and rhino horn ($28,600 per kilogram).

CONSERVATION BIOLOGY

Awareness of the current extinction rates gave rise to conservation biology. Goals of this field of research are to (1) survey systematically the range of biodiversity, (2) investigate its evolutionary and ecological origins, and (3) identify ways to maintain and use it in ways that benefit human populations. As you will see, the plan is to conserve as much biodiversity as possible and to use it in sustainable ways.

The fossil record tells us that biodiversity increases only slowly after a mass extinction.

We are in the midst of an accelerated decline in biodiversity caused by human activities. Habitat loss and fragmentation, species introductions, and overharvesting and poaching are major causes of the ongoing extinctions.

Studies of biodiversity may help us to sustain it.

41.13 Sustaining Biodiversity *and* Human Populations

Increasingly, efforts to save endangered species focus on identifying and protecting those communities that are both highly diverse and highly threatened.

IDENTIFYING AREAS AT RISK

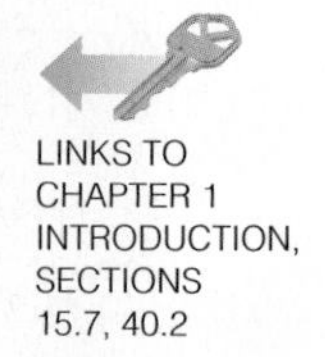

LINKS TO CHAPTER 1 INTRODUCTION, SECTIONS 15.7, 40.2

Conservation biologists now focus on identifying hot spots, habitats that hold many species that are found nowhere else and are in greatest danger of extinction.

Biologists first make an inventory of the organisms in an limited area, such as an isolated valley. They use sampling quadrats as well as mark–recapture studies to identify the species present and estimate the size of their populations (Section 40.2). The introduction to Chapter 1 highlights such an exploratory survey in New Guinea. More extensive study in the regions that contain multiple hot spots follows. For example, there are now research stations in many of Mexico's widely separated forests.

At the highest level, biologists work to define all ecoregions with multiple hot spots. An ecoregion is a land or ocean region defined by climate, geography, and producer species. Figure 41.30 shows ecoregions that are essential reservoirs of biodiversity. Of these, twenty-five are high-priority targets for conservation.

ECONOMICS AND SUSTAINABLE DEVELOPMENT

Every nation enjoys three forms of wealth—material, cultural, and biological wealth. Its biological wealth—biodiversity—can be a source of food, medicine, and other products. For instance, a particularly observant Mexican college student located *Zea diploperennis*, one wild maize species considered extinct. It had vanished from most of its former range. As it turned out, this species persists in 900 acres in the mountains. Unlike domesticated corn, it is perennial and highly resistant to viruses. The transfer of genes from *Z. diploperennis* into crop plants may boost corn production for hungry populations in Mexico and elsewhere.

Economic issues complicate the picture. The richest countries often have the least biological wealth and use more natural resources. Less developed countries with lots of biological wealth tend to have the fastest growing populations. As a result, people trapped in poverty often must choose between themselves and an endangered species. They hunt animals and plant crops in parks and reserves. Conservation biologists attempt to find ways for such people to earn a living from their biological wealth in ways that sustain them as well as other species.

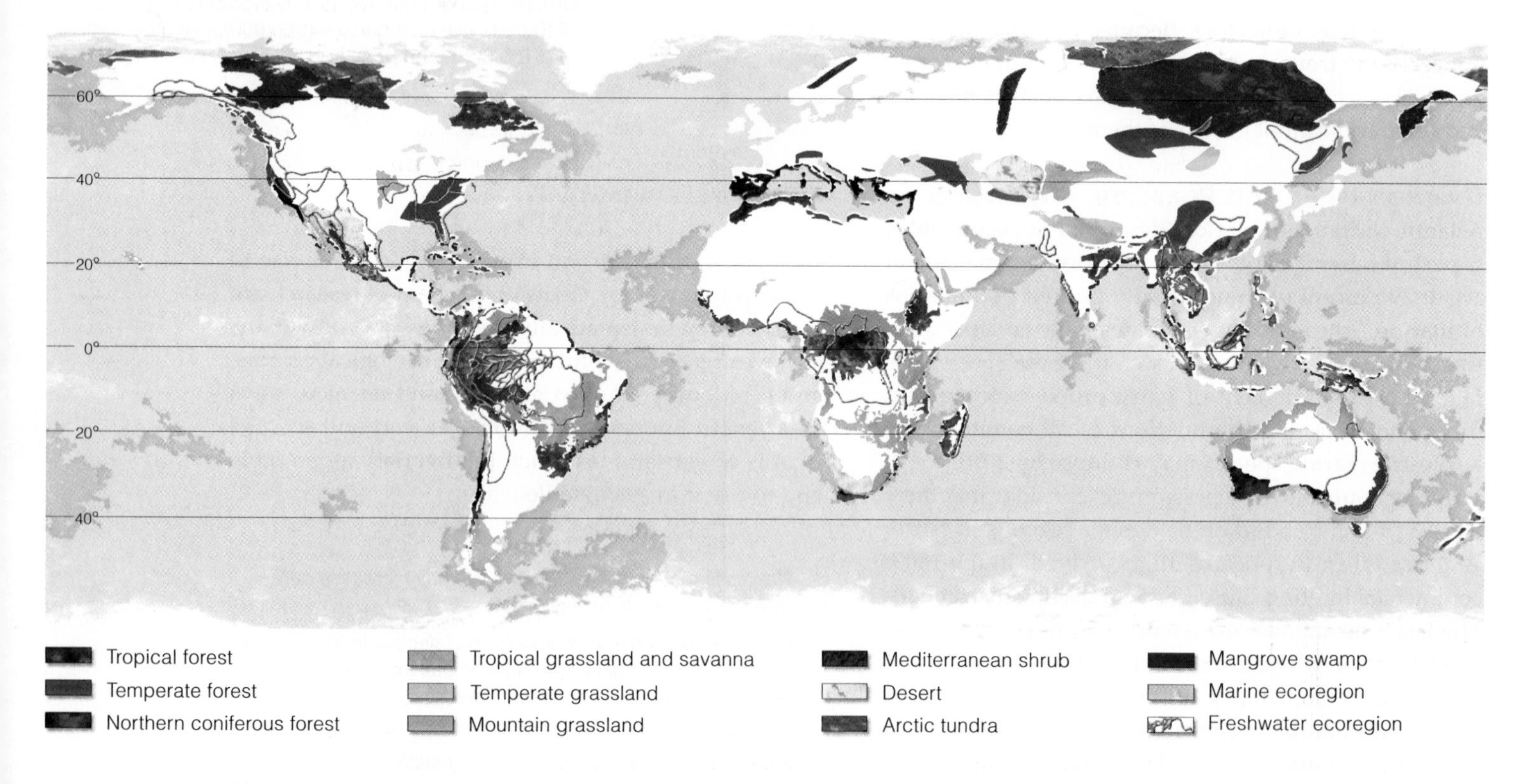

Figure 41.30 Map of the most vulnerable regions of land and seas, compiled by the World Wildlife Fund.

Figure 41.31 Strip logging. The practice may protect biodiversity as it permits logging on tropical slopes. A narrow corridor paralleling the land's contours is cleared. A roadbed is made at the top to haul away logs. After a few years, saplings grow in the cleared corridor. Another corridor is cleared above the roadbed. Nutrients leached from exposed soil trickle into the first corridor. There they are taken up by saplings, which benefit from all the nutrient input by growing faster. Later, a third corridor is cut above the second one—and so on in a profitable cycle of logging, which the habitat sustains over time.

Figure 41.32 Riparian zone restoration. The photos show Arizona's San Pedro River, before restoration (*above*) and after (*below*).

Ecotourism A nonprofit group owns Monteverde Cloud Forest Reserve in Costa Rica. During the 1970s, George Powell, a graduate student at the University of California, was studying birds in a forest that was rapidly being cleared. He got the idea to buy part of the forest as a nature sanctuary. His vision inspired individuals and conservation groups to donate funds. The reserve is now one of the few habitats remaining for endangered jaguars, ocelots, and pumas. It also is a highly popular ecotourism spot and a major income source for people of the region. Homes and schools have been built with tourism-based income.

Sustainable Logging A tropical forest yields wood for local needs and for export to developed countries. However, severe erosion often follows the logging of forested slopes. Gary Hartshorn devised a method to minimize erosion. As explained in Figure 41.31, strip logging allows cycles of logging. It yields sustainable economic benefits for local loggers, while minimizing effects of erosion and maximizing forest biodiversity.

Responsible Ranching Many cattle are raised on public lands and ranches in the American West. They need a lot of water, which they often get from streams and rivers. Cattle end up in riparian zones—narrow strips of land on either side of the waterway. Riparian zones provide wildlife with food, shelter, and shade, particularly in arid and semiarid regions. It takes only a few cattle to destroy a riparian zone. Ranchers help sustain biodiversity by fencing cattle out of riparian zones and providing water elsewhere. They rotate the herd from one area to another, so shore plants have a time to recover from grazing. Figure 41.32 shows how these practices can make a big difference.

Conserving biodiversity depends upon identifying and protecting regions that are especially rich in species. Biologists have already identified some of the most vulnerable and biologically valuable areas.

Managing biodiversity can sustain a nation's biological wealth and also provide monetary wealth to its people.

Summary

Section 41.1 Each species occupies a certain habitat characterized by physical and chemical features and by the array of other species living in it. All populations of all species in a habitat are a community. Each species in a community has its own niche, or way of living. Species interactions help shape community structure. Interacting species may coevolve.

Section 41.2 In a mutualism, two species interact and both benefit. Some mutualists cannot complete their life cycle without the interaction.

Section 41.3 Species compete by actively preventing others from accessing a resource or simply by using it. Species that need the exact same limited resource cannot coexist indefinitely. Species may coexist by partitioning a resource—using different portions of it or accessing it at different times.

■ *Use the animation on ThomsonNOW to learn about competitive interactions.*

Sections 41.4, 41.5 Predator and prey numbers can interact in complicated ways and often change in cycles. Carrying capacity, predator behavior, availability of other prey and other factors affect these cycles. Predators and their prey exert selective pressures on one another. The evolutionary results of such selection include chemical defenses, camouflage, and mimicry.

■ *Use the interaction on ThomsonNOW to learn about three alternative models for predator responses to prey density.*

Sections 41.6, 41.7 Parasites live in or on a host and withdraw nutrients from its tissues. Hosts may or may not die as a result. Parasitoids lay eggs on a host, then their larvae devour the host. Social parasites manipulate some aspect of a host's behavior.

Section 41.8 Ecological succession is the sequential replacement of one array of species by another over time. Primary succession happens in new habitats. Secondary succession occurs in disturbed ones. The first species of a community are pioneer species. Their presence may help, hinder, or have no effect on later colonists.

The older idea that all communities eventually reach a stable climax state has been replaced by models that emphasize continual change and the role of disturbances. Communities in which disturbances are of moderate intensity and frequency tend to have more species than those disturbed more or less often.

Sections 41.9, 41.10 Community structure reflects an uneasy balance of forces that operate over time. Major forces are competition and predation. Keystone species are especially important in maintaining the composition of a community. The removal of a keystone species or introduction of an exotic species—one that evolved in a different community—can alter community structure in ways that may be permanent.

Section 41.11 Species richness, the number of species in a given area, varies with latitude, elevation and other factors. Tropical regions tend to have more species than higher latitude regions. The species richness on islands varies with island area and distance from the mainland.

■ *Learn about the area effect and distance effect with the interaction on ThomsonNOW.*

Sections 41.12, 41.13 The range of biodiversity we see today is the result of historical mass extinctions and slow recoveries. Human activities are accelerating rates of extinction. Conservation biologists attempt to catalog biodiversity and devise methods that allow humans to use it in sustainable ways.

Self-Quiz

Answers in Appendix III

1. A habitat ______ .
   a. has distinguishing physical and chemical features
   b. is where individuals of a species normally live
   c. is occupied by various species
   d. all of the above

2. A species niche includes its ______ .
   a. habitat requirements
   b. food requirements
   c. reproductive requirements
   d. all of the above

3. Which cannot be a symbiosis?
   a. mutualism
   b. parasitism
   c. commensalism
   d. interspecific competition

4. Lizards and songbirds that share a habitat and both eat flies are an example of ______ competition.
   a. exploitative
   b. interference
   c. intraspecific
   d. interspecific
   e. both a and d

5. Two species may coexist indefinitely in some habitat when they ______ .
   a. differ in their use of resources
   b. share the same resource in different ways
   c. use the same resource at different times
   d. all of the above

6. Predator and prey populations ______ .
   a. always coexist at relatively stable levels
   b. may undergo cyclic or irregular changes in density
   c. cannot coexist indefinitely in the same habitat
   d. both b and c

7. Match the terms with the most suitable descriptions.
   ___ predation
   ___ mutualism
   ___ commensalism
   ___ parasitism
   ___ interspecific competition

   a. one free-living species feeds on another and usually kills it
   b. two species interact and both benefit by the interaction
   c. two species interact and one benefits while the other is neither helped nor harmed
   d. one species feeds on another but usually does not kill it
   e. two species attempt to utilize the same resource

Figure 41.33 Phasmids. (**a**) South African stick insect. (**b**) Leaf insect from Java. (**c**) Phasmid eggs often look like seeds.

Figure 41.34 One of the nominees for the worst 100 invaders: water hyacinths (*Eichhornia crassipes*) choking a Florida waterway.

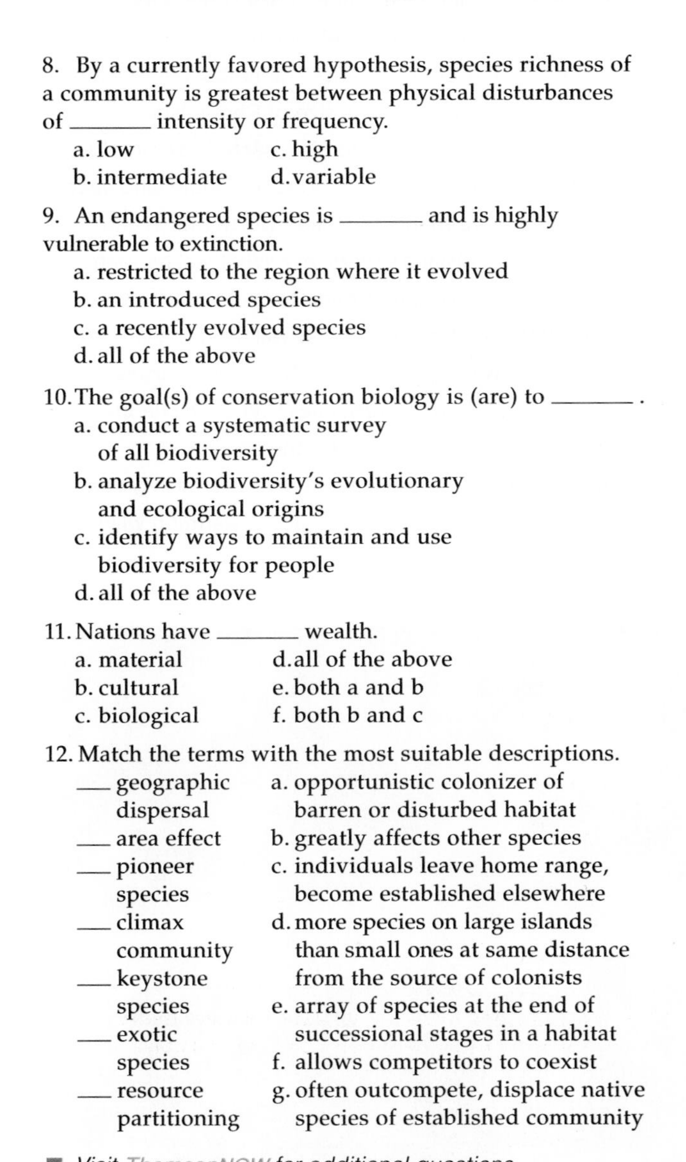

8. By a currently favored hypothesis, species richness of a community is greatest between physical disturbances of ______ intensity or frequency.
   a. low
   b. intermediate
   c. high
   d. variable

9. An endangered species is ______ and is highly vulnerable to extinction.
   a. restricted to the region where it evolved
   b. an introduced species
   c. a recently evolved species
   d. all of the above

10. The goal(s) of conservation biology is (are) to ______ .
   a. conduct a systematic survey of all biodiversity
   b. analyze biodiversity's evolutionary and ecological origins
   c. identify ways to maintain and use biodiversity for people
   d. all of the above

11. Nations have ______ wealth.
   a. material
   b. cultural
   c. biological
   d. all of the above
   e. both a and b
   f. both b and c

12. Match the terms with the most suitable descriptions.

___ geographic dispersal	a. opportunistic colonizer of barren or disturbed habitat
___ area effect	b. greatly affects other species
___ pioneer species	c. individuals leave home range, become established elsewhere
___ climax community	d. more species on large islands than small ones at same distance from the source of colonists
___ keystone species	e. array of species at the end of successional stages in a habitat
___ exotic species	f. allows competitors to coexist
___ resource partitioning	g. often outcompete, displace native species of established community

■ *Visit ThomsonNOW for additional questions.*

Critical Thinking

1. With antibiotic resistance rising, researchers are looking for ways to reduce use of antibiotics. Some cattle once fed antibiotic-laced food now get probiotic feed instead, with cultured bacteria that can establish or bolster populations of helpful bacteria in the animal's gut. The idea is that if a large population of beneficial bacteria is in place, then the harmful bacteria cannot become established or thrive. Which ecological theory is guiding this research?

2. Phasmids are a group of herbivorous insects that look like sticks or leaves (Figure 41.33). Most are motionless in the day, and feed at night. If disturbed, a phasmid will fall to the ground, as if dead. Speculate on selective pressures that may have shaped phasmid morphology and behavior. Suggest an experiment with one species to test whether its appearance and behavior may be adaptive.

3. The water hyacinth (*Eichhornia crassipes*) is an aquatic plant native to South America. Today, this plant lives in nutrient-rich waters from Florida to San Francisco. It has displaced many native species. It has also choked rivers and canals (Figure 41.34). Research and write a brief account of how it got from one continent to another.

4. Visit or study a riparian zone near where you live. Imagine visiting it ten years from now. Given its location, what kinds of changes do you predict for it?

5. Mentally transport yourself to a tropical rain forest of South America. Imagine you do not have a job. No jobs are available, even in overcrowded cities some distance away. You have no money or contacts that would allow you to relocate, and you are the sole supporter of a large family. Today a stranger approaches you. He tells you he will pay good money if you can capture alive a certain brilliantly feathered parrot in the forest. You know the parrot is seldom seen, but you have an idea of where it lives. What will you do?

6. As an alternative to formal burial on land, ashes of cremated people can be mixed with pH-neutral concrete and used to build artificial reef balls. Each perforated concrete ball weighs 400 to 4,000 pounds and is designed to last 500 years. So far, 100,000 of the artificial reefs are sustaining marine life in 1,500 locations. Proponents say this undersea burial is low cost and does not waste land. Do you think it's good for the environment?

7. Ecologist Robin Tyser once told us that students might find the current biodiversity crisis too overwhelming to contemplate. But people can make a difference. Many helped reestablish bald eagles in the continental United States. Others have reestablished wolves in northern Wisconsin. Daniel Janzen is working to re-create a dry forest ecosystem in Costa Rica. Do a library search or computer research and report on one or two success stories you find inspiring

42 ECOSYSTEMS

IMPACTS, ISSUES Bye-Bye, Blue Bayou

Each Labor Day, the coastal Louisiana town of Morgan City celebrates the Louisiana Shrimp and Petroleum Festival. The state is the nation's top shrimp harvester and the third-largest producer of petroleum, which is refined into gasoline and other fossil fuels. But the petroleum industry's very success may be contributing indirectly to the decline of the state's fisheries. Why? The lower atmosphere is warming up, thanks in part to fossil fuel burning (Section 6.8). As the climate heats up, the ocean's surface waters get warmer and expand, glaciers melt, and sea level rises.

If current trends continue, current coastal lowlands will be under water within fifty years. With more than 40 percent of the nation's saltwater wetlands, Louisiana has the most to lose. This state's marshes already have problems. Dams and levees keep back sediments that normally accumulate in coastal marshes and help hold back the sea (Figure 42.1). Since the 1940s, Louisiana has lost an area of marshland the size of Rhode Island.

Is an ecological and economic disaster now unfolding? What will happen to the livelihoods of people who each year harvest more than $3.5 billion worth of fish and shellfish from Louisiana's marshes? What will happen to the millions of birds that migrate to wetlands every winter? With marshes gone, what will keep the land from being swept away by the devastating storm surges that slam into coasts during hurricanes? What will happen to low-lying towns and cities along the coasts?

Researchers use models to study patterns in global climate. Some now predict that global warming will raise

See the video!
Figure 42.1 A fishing camp in Louisiana. It was built in a once-thriving marsh that has since given way to the open waters of Barataria Bay. *Facing page*, a marsh restoration project in Sabine National Wildlife Refuge. In marshland that has become open water, sediments are barged in and marsh grasses are planted on them.

How would you vote? Exhaust from motor vehicles contains greenhouse gases. The better mileage a vehicle gets, the fewer greenhouse gases it emits per mile. Should minimum fuel economy standards for cars and trucks be increased? See ThomsonNOW for details, then vote online.

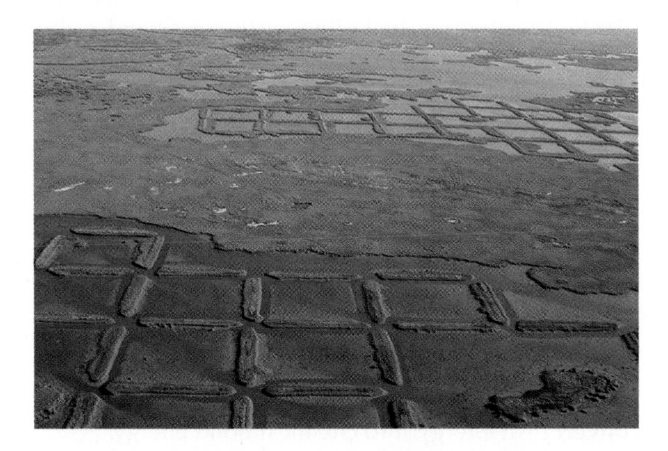

evaporation rates and alter weather patterns. Some of the consequences may already be in play. Over the past thirty years, the number of severe hurricanes in the Atlantic and cyclones in the Pacific has nearly doubled worldwide. In 2005, the category 5 hurricane Katrina slammed into the Gulf Coast. High winds and flooding ruined countless buildings. More than 1,700 people died.

Models also indicate that warming seas may promote algal blooms and huge fish kills. Warmer water fans the population growth of many pathogenic bacteria. More people get sick after swimming in contaminated water or eating contaminated shellfish.

Inland, heat waves and wildfires may become more intense. Deaths related to heat stroke will climb. Warmer temperatures will invite mosquitoes to extend their inland ranges. Some kinds of mosquitoes are vectors for agents of malaria, West Nile virus, and other diseases.

This chapter can get you thinking critically about the nature of ecosystems—the inputs and outputs of energy, and the inputs, cycling, and outputs of nutrient. It also returns to a related concept: We humans have become major players in the global flows of energy and nutrients even before we fully comprehend how ecosystems work. Decisions we make today about global warming and other environmental issues are likely to shape the quality of life and the environment far into the future.

Key Concepts

ORGANIZATION OF ECOSYSTEMS

An ecosystem consists of a community and its physical environment. A one-way flow of energy and a cycling of raw materials among its interacting participants maintain it. It is an open system, with inputs and outputs of energy and nutrients. **Section 42.1**

FOOD WEBS

Food chains are linear sequences of feeding relationships, from producers through consumers, decomposers, and detritivores. The chains cross-connect as food webs. Most of the energy that enters a food web returns to the environment, mainly as metabolic heat. Most of the nutrients are cycled; some reenter the environment.

Biological magnification is the increasing concentration of a substance in tissues of organisms as it moves up food chains. **Sections 42.2, 42.3**

PRIMARY PRODUCTIVITY

Primary productivity is the rate at which an ecosystem's producers capture and store energy in their tissues during a given interval. The number of producers and the balance between photosynthesis and aerobic respiration influence the amount stored. **Section 42.4**

CYCLING OF WATER AND NUTRIENTS

The availability of water, carbon, nitrogen, phosphorus, and other substances influences primary productivity. Ions or molecules of these substances move slowly in global cycles, from environmental reservoirs, into food webs, then back to reservoirs. Human activities can disrupt these cycles. **Sections 42.5–42.10**

Links to Earlier Concepts

This chapter takes a closer look at the participants in ecosystems, especially autotrophs (Section 5.5). You will be drawing on your understanding of the one-way flow of energy in nature (1.2, 5.1). You will consider nitrogen-fixing microbes, soil erosion, and leaching in the context of the global nitrogen cycle (19.2, 27.1, 27.2). You will draw on your knowledge of biochemistry, including chemical bonds (2.4) and pathways that secure and release energy (Chapters 6 and 7), You will see how slow movements of Earth's crust (16.6) influence the cycling of some nutrients.

42.1 The Nature of Ecosystems

The preceding chapter introduced the dynamic nature of community structure. Turn now to patterns of organization that arise as energy and nutrients from the environment flow among the species in a community. Ecologists work to identify these patterns and to make predictions about how they will shift over time.

OVERVIEW OF THE PARTICIPANTS

LINKS TO SECTIONS 1.2, 5.1, 27.1

Diverse natural systems abound on Earth's surface. In climate, soil type, array of species, and other features, prairies differ from forests, which differ from tundra and deserts. Reefs differ from the open ocean, which differs from streams and lakes. Yet, despite all these differences, all systems are alike in many aspects of their structure and function.

We define each ecosystem as an array of organisms and a physical environment, all interacting through a one-way flow of energy and a cycling of nutrients. It is an open system, because it requires ongoing inputs of energy and nutrients to endure (Figure 42.2).

All ecosystems run on energy captured by primary producers. These autotrophs, or "self-feeders," obtain energy from a nonliving source—usually sunlight—and use it to produce organic compounds from carbon dioxide and water. Plants and phytoplankton are the main producers. Chapter 6 explains how they obtain energy from the sun's rays and assemble sugars from carbon dioxide and water through photosynthesis.

Consumers are heterotrophs that get energy and carbon by feeding on tissues, products, and remains of producers and one another. We describe consumers by their diets. *Herbivores* eat plants. *Carnivores* eat the flesh of animals. *Parasites* live inside or on a living host and feed on its tissues. *Omnivores* eat both animals and plants. Detritivores, including earthworms and crabs, dine on small particles of organic matter, or detritus. Decomposers feed on organic wastes and remains and break them down into inorganic building blocks. The main decomposers are bacteria and fungi.

Energy transfers, remember, cannot be 100 percent efficient (Section 5.1). Energy flows one way—into an ecosystem, through its participants, then back to the physical environment. In time, all of the energy that the producers harnessed is lost to the surroundings, mainly in the form of heat generated by metabolism. Heat cannot be recycled because the producers cannot convert heat energy into chemical bond energy.

In contrast, many nutrients are cycled within an ecosystem. The cycle begins when producers take up hydrogen, oxygen, and carbon from inorganic sources, such as the air and water. They also take up dissolved nitrogen, phosphorus, and other minerals necessary for biosynthesis. Nutrients move from producers into the consumers who eat them. After an organism dies, decomposition returns nutrients to the environment, from which producers take them up again.

Not all nutrients remain in an ecosystem; typically there are gains and losses. Mineral ions are added to an ecosystem when weathering processes break down rocks, and when winds blow in mineral-rich dust from elsewhere. Leaching and soil erosion remove minerals (Section 27.1). Gains and losses of each mineral tend to balance out over time in a healthy ecosystem.

TROPHIC STRUCTURE OF ECOSYSTEMS

All organisms of an ecosystem have functional roles in a hierarchy of feeding relationships called trophic levels (*troph*, nourishment). Ecologists ask, "Who eats whom?" When organism B eats organism A, energy is transferred from A to B. All organisms at a particular trophic level are the same number of transfers away from the energy input into an ecosystem.

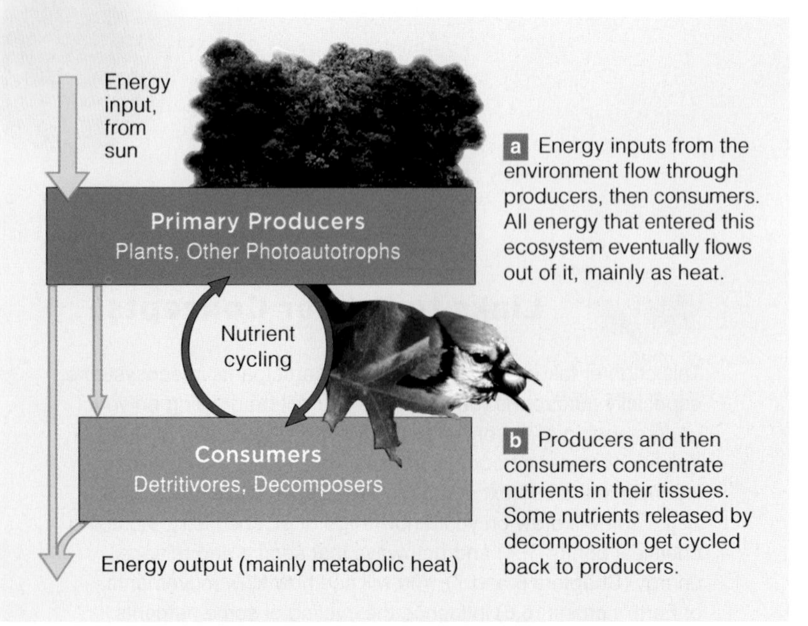

Figure 42.2 **Animated!** Model for ecosystems on land, in which energy flow starts with autotrophs that capture energy from the sun. Energy flows one way, into and out of the ecosystem. Nutrients get cycled among producers and heterotrophs.

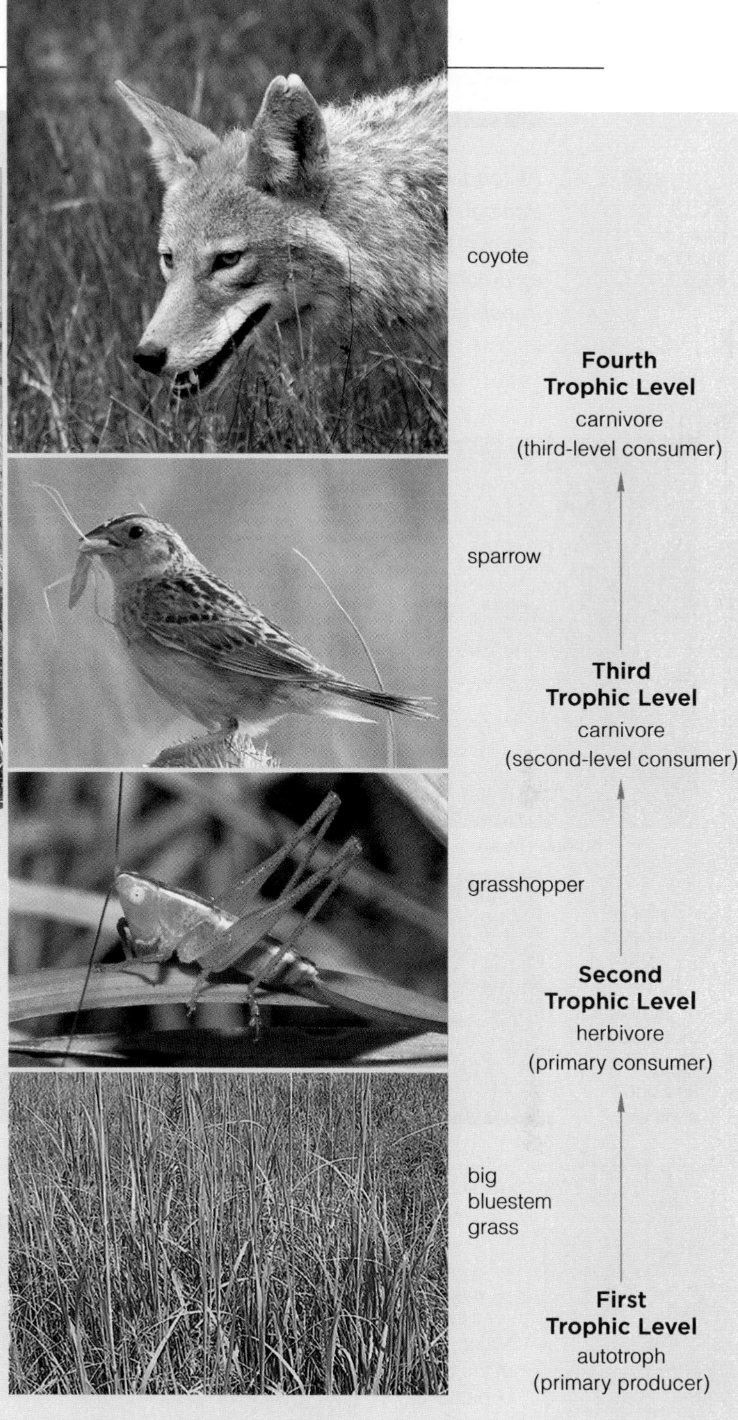

Figure 42.3 Example of a food chain and corresponding trophic levels in tallgrass prairie, Kansas.

A food chain is a sequence of steps by which some energy captured by primary producers is transferred to organisms at successively higher trophic levels. For example, big bluestem grass and other plants are the major primary producers in a tallgrass prairie (Figure 42.3). They are at this ecosystem's first trophic level. In one food chain, energy flows from bluestem grass to grasshoppers, to sparrows, and finally to coyotes. Grasshoppers are primary consumers; they are at the second trophic level. Sparrows that eat grasshoppers are second-level consumers and at the third trophic level. Coyotes are third-level consumers, and they are at the fourth trophic level.

At each trophic level, organisms interact with the same sets of predators, prey, or both. Omnivores feed at several levels, so we would partition them among different levels or assign them to a level of their own.

Identifying one food chain is a simple way to start thinking about who eats whom in ecosystems. Bear in mind, many different species usually are competing for food in complex ways. Tallgrass prairie producers (mainly flowering plants) feed grazing mammals and herbivorous insects. But many more species interact in the tallgrass prairie and in most other ecosystems, particularly at lower trophic levels. A number of food chains *cross-connect* with one another—as food webs —and that is the topic of the next section.

An ecosystem is a community of organisms that interact with one another and with their physical environment by a one-way energy flow and a cycling of materials.

Autotrophs tap into an environmental energy source and make their own organic compounds from inorganic raw materials. They are the ecosystem's primary producers.

Autotrophs are at the first trophic level of a food chain, a linear sequence of feeding relationships that proceeds through one or more levels of heterotrophs, or consumers.

In ecosystems, food chains cross-connect, as food webs.

42.2 The Nature of Food Webs

LINKS TO SECTIONS 5.1, 5.2

All food webs consist of cross-connecting food chains. Ecologists who untangled the chains of many food webs discovered patterns of organization. The patterns reflect environmental constraints and the inefficiency of energy transfers from one trophic level to the next.

HOW MANY TRANSFERS?

A food web diagram shows trophic interactions among many species in one particular ecosystem. The arctic food web shown in Figure 42.4 is a highly simplified example. Nearly all food webs are far more complex, as the diagram in Figure 42.5 makes clear.

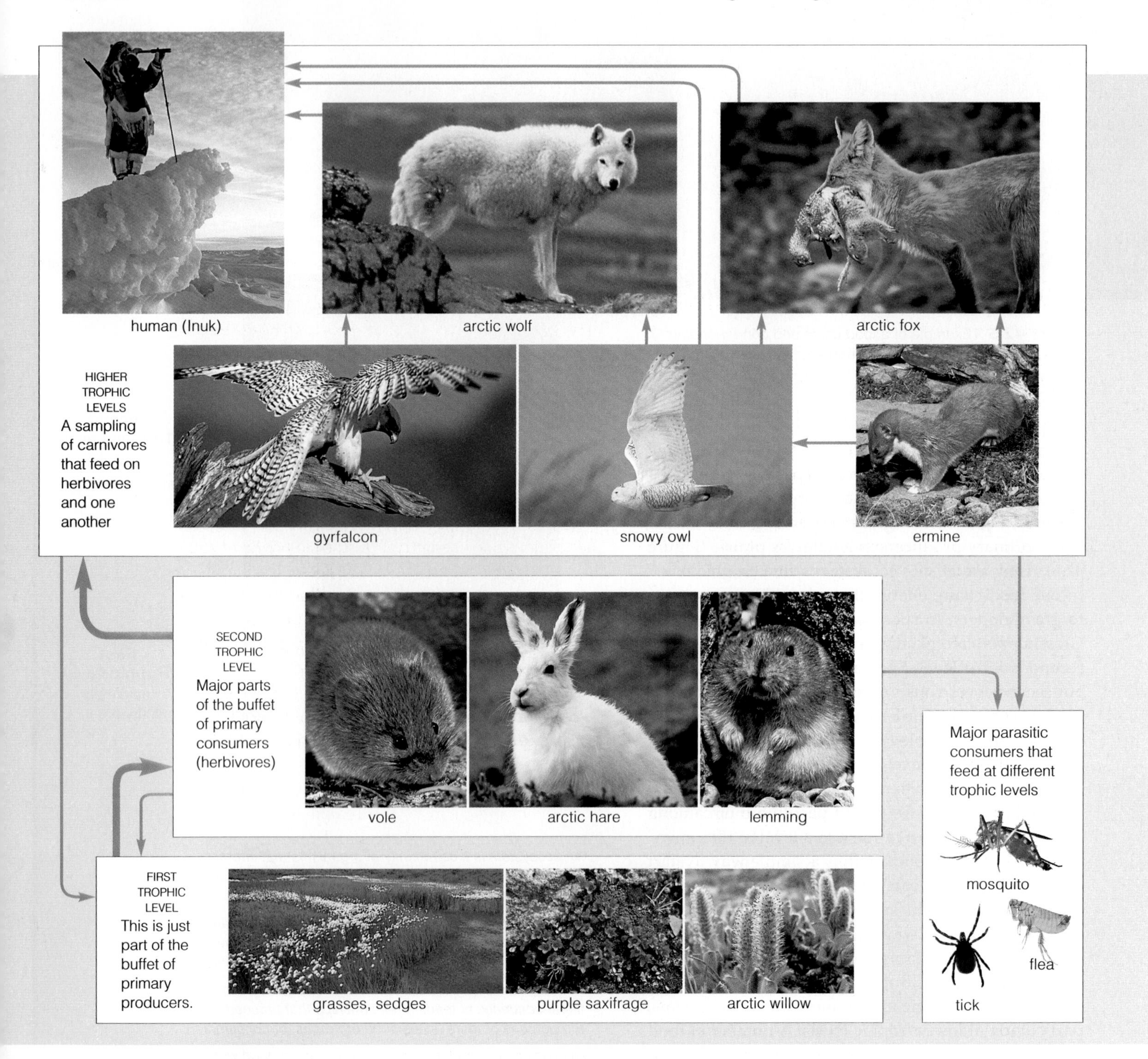

Figure 42.4 **Animated!** A very small sampling of organisms in an Arctic food web on land.

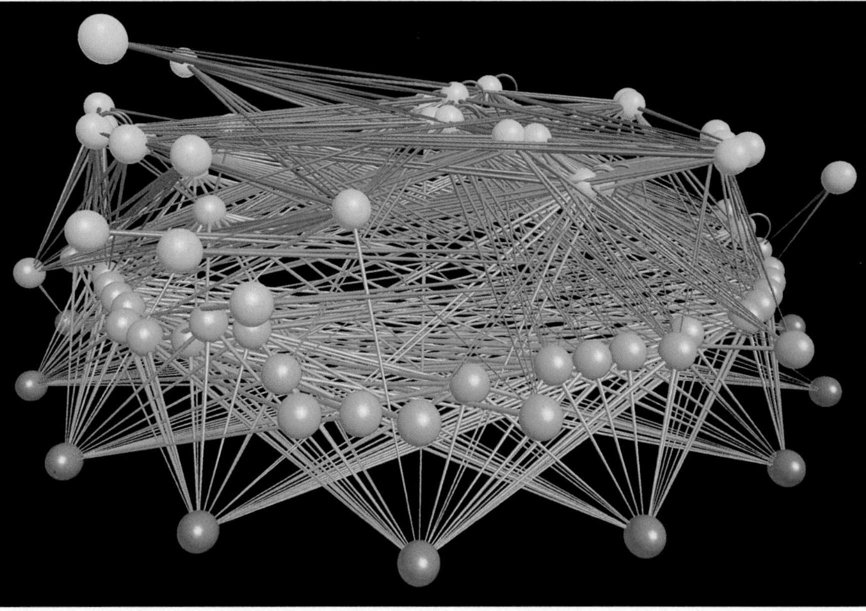

Figure 42.5 Computer model for a food web in East River Valley, Colorado. Balls signify species. Their colors identify trophic levels, with producers (coded *red*) at the bottom and top predators (*yellow*) at top. The connecting lines thicken, starting from an eaten species to the eater.

When ecologists compared food webs for a variety of ecosystems, patterns began to emerge. They found that the energy captured by producers usually passes through no more than four or five trophic levels. Even in ecosystems with many species, transfers are limited. Remember, energy transfers are not that efficient, and energy losses limit the length of a chain.

Field studies and computer simulations of aquatic and land food webs reveal more patterns. Food chains tend to be shortest in habitats where conditions vary widely over time. Chains tend to be longer in stable habitats, such as the ocean depths. The most complex webs tend to have a large variety of herbivores, as in grasslands. By comparison, the food webs with fewer connections tend to have more carnivores.

Diagrams of food webs help ecologists predict how ecosystems will respond to change. Neo Martinez and his colleagues constructed the one shown in Figure 42.5. By comparing different food webs, they realized that trophic interactions connect species more closely than people thought. On average, each species in any food web was two links away from all other species. Ninety-five percent of species were within three links of one another, even in large communities with many species. As Martinez concluded in his paper on these findings, "Everything is linked to everything else." He cautioned that extinction of any species in a food web has potential impact on many other species.

TWO CATEGORIES OF FOOD WEBS

Energy from organisms closest to a primary source—producers—flows into and out of two kinds of webs. In a grazing food web, energy flows mostly from the producers to herbivores, then carnivores, and finally to decomposers. In a detrital food web, energy flows mostly from the producers to detritivores and then to decomposers.

The amount of energy that moves through the two kinds of food webs differs among ecosystems, and it often varies with the seasons. In most cases, however, most of the energy stored in producer tissues moves through a detrital food web. Think of cattle that graze in a pasture. About half of the energy stored in grass enters cattle. But cattle cannot access all of the stored energy, because they cannot digest the bonds of some organic compounds in plant cell walls. Decomposers and detritivores go to work on the dead plants and on undigested plant parts. Similarly, in marshes, most of the energy stored in tissues of marsh grasses enters detrital food webs when the plants die.

In nearly all ecosystems, both kinds of webs cross-connect. For example, in some rocky intertidal zones, energy captured by algae flows into snails, which are eaten by herring gulls as part of a grazing food web. However, gulls also hunt crabs, which are among the primary consumers in a detrital food web.

The cumulative energy losses from energy transfers between trophic levels limits the length of food chains.

Even when an ecosystem has many species, trophic interactions link each species with many others.

Tissues of living plants and other autotrophs are the basis for grazing food webs. Remains and wastes of autotrophs are the main basis for detrital food webs.

Nearly all ecosystems have both a grazing food web and a detrital food web that interconnect.

42.3 Biological Magnification in Food Webs

We turn now to a premise that opened this chapter—that disturbances to one part of an ecosystem can have unexpected effects on other, seemingly unrelated parts.

LINKS TO SECTIONS 25.4, 41.5

Ecosystem Analysis Ecologists often use models to monitor and predict the outcome of disturbances to ecosystems. They work to identify all of the interacting biological, physical, chemical, and geologic factors that determine an ecosystem's processes and patterns. They might gather information by direct observations, satellite imaging and other remote sensing devices, and tests. Often they prepare mathematical models and computer programs to integrate pieces of available information on how the factors interact. Analysis of the results help them predict how the ecosystem will react to forces of change.

The analysis is most useful when *all* factors have been identified and accurately incorporated into a model for the ecosystem. The most crucial factor may be one that researchers do not yet know. A case in point follows.

Some Chemical Background Many plants repel herbivorous animals with natural toxins (Section 41.5). These organic compounds do not harm the plants, but they may repel or kill different species. We eat traces of natural plant toxins in hot peppers, potatoes, figs, celery, rhubarb, alfalfa sprouts, and many other foods. They do not sicken or kill us, because toxicity often depends on the concentration.

Just a few thousand years ago, farmers used sulfur, lead, arsenic, and mercury to help protect crop plants against insects. They freely dispensed these highly toxic metals until the late 1920s, when someone figured out they were poisoning people. Traces of toxic metals still turn up in contaminated croplands.

Farmers also used organic compounds extracted from leaves, flowers, and roots as natural pesticides. In 1945, scientists started to make synthetic toxins and identify how they work. *Herbicides*, such as synthetic auxins, kill weeds by disrupting metabolism and growth (Section 28.7). *Insecticides* clog the airways of a target insect, disrupt nerves and muscles, or prevent reproduction. *Fungicides* work against harmful fungi, including a mold that makes aflatoxin, one of the deadliest poisons. By 1995, people in the United States were spraying or spreading more than 1.25 billion pounds of synthetic chemicals each year through fields, gardens, homes, and industrial and commercial sites (Figure 42.6).

a *Malathion.* Like other organophosphates, this insecticide is cheap, breaks down faster than DDT, and is more toxic. Organophosphates represent half of all insecticides used in the United States. Some are now banned for crops; application of others must end at least three weeks before harvest. Farmers who are inconvenienced by this policy want the Environmental Protection Agency to consider economic and trade issues as well as human health.

b *2,4-D* (2,4-dichlorophenoxyacetic acid), a synthetic auxin widely used as a herbicide. Enzymes of weeds and microbes cannot easily degrade 2,4-D, compared with natural auxins.

c *Atrazine*, the best-selling herbicide, kills weeds within a few days, as do glyphosate (Roundup), alachlor, (Lasso), and daminozide (Alar). It now appears that atrazine causes abnormal sexual development in frogs, even in trace amounts below the level allowed in drinking water.

d Dichlorodiphenyltrichloroethane, or *DDT*. It takes two to fifteen years for this nerve cell poison to break down. Chlordane, another type of insecticide, also persists for a long time in the environment.

Figure 42.6 A few pesticides, some more toxic than others. The photograph shows one of the crop dusters that intervene in the competition for nutrients between crop plants and pests, including weeds.

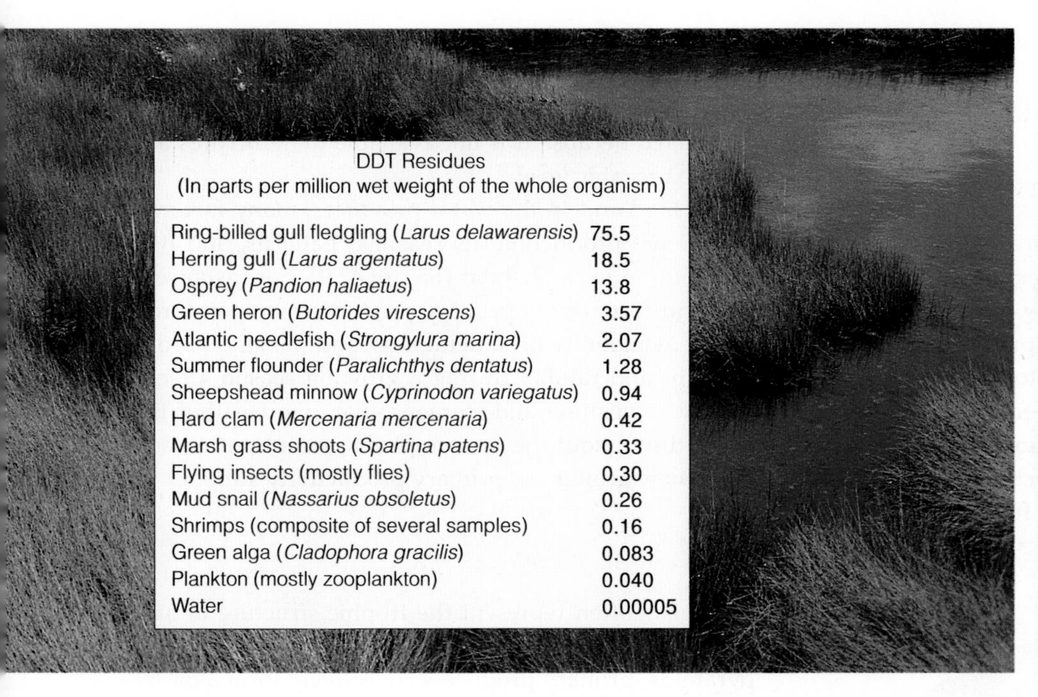

DDT Residues (In parts per million wet weight of the whole organism)	
Ring-billed gull fledgling (*Larus delawarensis*)	75.5
Herring gull (*Larus argentatus*)	18.5
Osprey (*Pandion haliaetus*)	13.8
Green heron (*Butorides virescens*)	3.57
Atlantic needlefish (*Strongylura marina*)	2.07
Summer flounder (*Paralichthys dentatus*)	1.28
Sheepshead minnow (*Cyprinodon variegatus*)	0.94
Hard clam (*Mercenaria mercenaria*)	0.42
Marsh grass shoots (*Spartina patens*)	0.33
Flying insects (mostly flies)	0.30
Mud snail (*Nassarius obsoletus*)	0.26
Shrimps (composite of several samples)	0.16
Green alga (*Cladophora gracilis*)	0.083
Plankton (mostly zooplankton)	0.040
Water	0.00005

Figure 42.7 Biological magnification in an estuary on the south shore of Long Island, New York, as reported in 1967 by George Woodwell, Charles Wurster, and Peter Isaacson. The researchers knew of broad correlations between the extent of DDT exposure and mortality. For instance, residues in birds known to have died from DDT poisoning were 30–295 ppm, and they were 1–26 ppm in several fish species. Some DDT concentrations measured during this study were below lethal thresholds but were still high enough to interfere with reproductive success.

Figure 42.8 Peregrine falcon, a top carnivore in some food webs. This raptor almost became extinct as a result of biological magnification of DDT. Its populations were restored by a successful wildlife management program. Peregrine falcons were reintroduced into wild habitats. They have adapted to cities. There, they hunt pigeons, large populations of which are a messy nuisance.

DDT in Food Webs The DDT molecule is a fairly stable hydrocarbon (Figure 42.6*d*). It is nearly insoluble in water. You might think—as many others did—that it would poison pests only where it was applied. However, wind easily disperses DDT molecules in vapor form, and water disperses them as fine particles. These molecules can disrupt metabolic activities and are toxic to many aquatic and terrestrial animals.

Given its molecular properties, DDT is highly soluble in fats, so it can accumulate in tissues of living organisms. That is why DDT can show biological magnification. By this process, a substance that degrades slowly or not at all becomes increasingly concentrated in the tissues of organisms at higher trophic levels. For example, if bugs eat DDT-coated plant parts, DDT ends up in their tissues. When a bird devours many of these bugs, it ends up getting multiple doses of DDT. A fox that eats many DDT-laden birds gets a still higher dose of DDT.

Several decades ago, DDT started to infiltrate food webs with unpredicted and worrisome effects. Where people sprayed DDT to halt Dutch elm disease, songbirds died. Where DDT was sprayed to kill budworm larvae, fishes in nearby streams died. In fields sprayed to control one species of pest, other pests increased. DDT killed predatory insects that kept pest populations in check.

Then side effects of biological magnification started to show up in habitats far removed from where DDT had been applied—*and much later in time*. Most vulnerable were brown pelicans, bald eagles, peregrine falcons, and other top carnivores of some food webs (Figures 42.7 and 42.8). Why? A product of DDT breakdown interferes with some physiological processes. As one outcome, bird eggs developed brittle shells; many chick embryos did not even hatch. Some species were facing extinction.

In the United States, DDT has been banned since the 1970s except where necessary to protect public health. Many species hit hardest have recovered. Some birds still lay thin-shelled eggs because they pick up DDT at their winter ranges in Latin America. As late as 1990, a fishery near Los Angeles was closed. DDT from industrial waste discharges that had stopped twenty years before was still contaminating that ecosystem.

Industrial pollutants and toxic metals, including lead and mercury, also accumulate in organisms. For example, water entering the San Francisco Bay contains mercury in runoff from abandoned gold mines further inland. The mercury accumulates as it moves through the bay's food chains. Fish such as striped bass that are top predators have the highest mercury levels. They are also the species most valued in sport fishing.

42.4 Studying Energy Flow Through Ecosystems

Ecologists measure the amount of energy and nutrients entering an ecosystem, how much is captured, and the proportion stored in each trophic level.

WHAT IS PRIMARY PRODUCTIVITY?

LINK TO SECTION 7.8

The rate at which producers capture and store energy in their tissues during a given interval is the primary productivity of an ecosystem. How much energy gets stored depends on (1) how many producers there are and (2) the balance between photosynthesis (energy trapped) and aerobic respiration (energy used). *Gross* primary production is all energy initially trapped by the producers. *Net* primary production is the fraction of trapped energy that producers funnel into growth and reproduction. Net ecosystem production is gross primary production *minus* the energy that producers, detritivores, and decomposers require. That energy is subtracted because it is not available to herbivores at the next trophic level.

On land and in the water provinces, many factors impact net production, its seasonal patterns, and its distribution in the habitat (Figure 42.9). For instance, size and form of the primary producers, temperature range, availability of mineral ions, and the amount of sunlight and rainfall during a growing season affect energy acquisition and storage. When the weather is harsh throughout the season, plants cannot put on as much new growth, so primary productivity suffers.

ECOLOGICAL PYRAMIDS

Ecologists often represent the trophic structure of an ecosystem in the form of ecological pyramids. In such pyramids, primary producers collectively form a base for successive tiers of consumers above them.

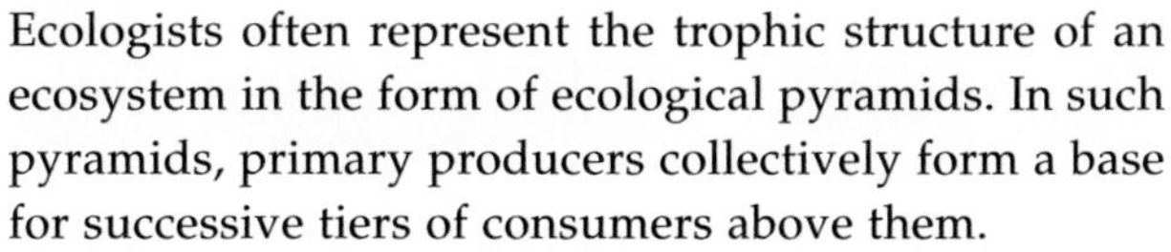

A biomass pyramid illustrates the dry weight of all organisms at each trophic level in a specific ecosystem. Figure 42.10*a* shows the biomass pyramid for a small aquatic ecosystem in Florida.

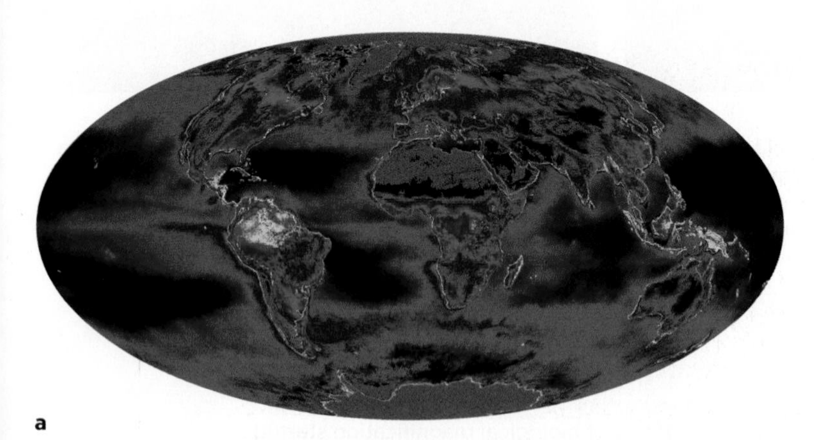

Figure 42.9 (**a**) Summary of satellite data on net primary productivity during 2002. Productivity is coded as *red* (highest) down through *orange*, *yellow*, *green*, *blue*, and *purple* (lowest). Although average productivity per unit of sea surface is lower than it is on land, total productivity on land and in the seas is roughly equal, because water covers most of Earth's surface.

(**b**) Examples of the seasonal shifts in net primary productivity for three categories of ecosystems in different regions of the ocean.

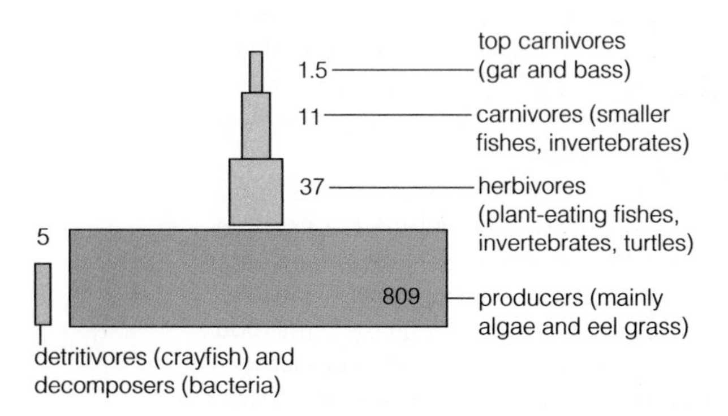

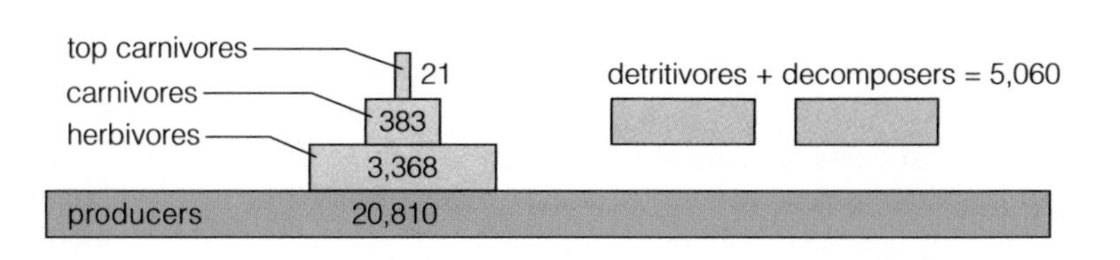

Figure 42.10 (**a**) Biomass (in grams/square meter) for Silver Springs, an aquatic ecosystem in Florida. In this ecosystem, producers make up the bulk of the biomass. In other ecosystems, producers are eaten almost as fast as they grow and reproduce. In such ecosystems, biomass accumulates faster in consumers, so the biomass pyramid appears "upside down." (**b**) Pyramid of annual energy flow through Silver Springs, in kilocalories/square meter.

Most commonly, primary producers make up most of the biomass in a pyramid, and top carnivores make up very little. However, some biomass pyramids have the smallest tier at the bottom. Such inverted pyramids occur when the primary producers get eaten as fast as they reproduce, so biomass cannot accumulate at this level. For example, in some aquatic habitats rapidly reproducing phytoplankton support a larger standing biomass of zooplankton.

An energy pyramid illustrates how the amount of usable energy diminishes as it is transferred through an ecosystem. Sunlight energy is captured at the base (the primary producers) and declines with successive levels to its tip (the top carnivores). Energy pyramids have a large energy base at the bottom, so they are always "rightside up." They depict the energy flow per unit of water (or land) per unit of time as shown in Figure 42.10*b*.

ECOLOGICAL EFFICIENCY

On average, only about 10 percent of energy in tissues of organisms at one trophic level ends up in tissues of those at the next level. Look once again at the biomass pyramid in Figure 42.10*a*. Only 4.6 percent ($37/809 \times 100$) of the biomass in primary producers ends up as biomass in first-level consumers.

Several factors influence the efficiency of transfers. For example, the proportion of digestible biomass can vary. Few herbivores are capable of digesting lignin and cellulose in plant parts. This limits the efficiency of any energy transfers from producers to herbivores. Warm-blooded animals lose more energy as heat than cold-blooded animals, and active animals lose more than less active ones. As a result, aquatic ecosystems tend to have more transfers than land systems. Algae are more easily digested than land plants, and aquatic ecosystems usually have more cold-blooded animals than land ecosystems. Higher efficiencies of transfer allow for longer food chains.

Primary productivity is the rate at which producers trap and store energy in tissues in a given interval. The net amount is the fraction they use in growth and reproduction.

Net ecosystem production is the amount of energy trapped by the producers minus the energy used by all producers, detritivores, and decomposers.

Ecological pyramids represent the trophic structure of ecosystems. Biomass pyramids may be top- or bottom-heavy, depending on the ecosystem. Energy pyramids always have the largest tier on the bottom.

42.5 Biogeochemical Cycles

Without water and the nutrients dissolved in it, there would be no primary productivity, and no life.

In a biogeochemical cycle, an essential element moves from one or more environmental reservoirs, through ecosystems, then back to the reservoirs (Figure 42.11).

LINKS TO SECTIONS 16.6, 27.1

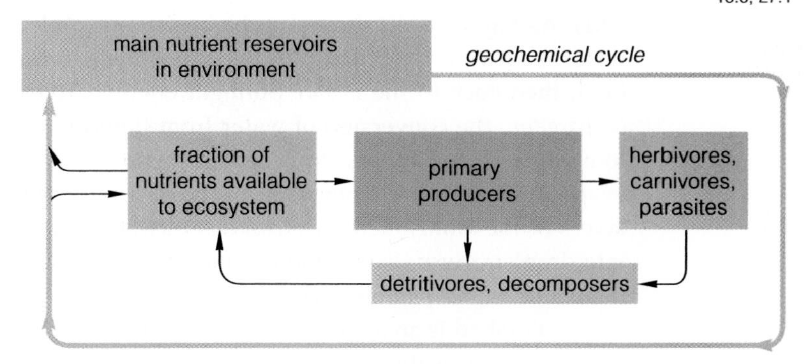

Figure 42.11 Overview of biogeochemical cycles.

As explained in the Chapter 2 introduction, oxygen, hydrogen, carbon, nitrogen, and phosphorus are some of the elements essential to all forms of life. We refer to these and other required elements as nutrients.

Transfers to and from reservoirs are far slower than rates of exchange among organisms of an ecosystem. Solid forms of elements are tied up in rocks. Nutrients enter and leave rocks by natural geologic processes, such as weathering. Erosion and runoff put nutrients into streams that carry them from their source. Plants take up nutrients dissolved in water.

Nutrient cycles depend on actions of decomposers. Various prokaryotic species help transform solids and ions into gases, then back again. Through their action, they convert some elements that function as nutrients to forms that primary producers can take up.

The next sections deal with biogeochemical cycles that are reservoirs for key elements. In the *hydrologic* cycle, oxygen and hydrogen move, on a global scale, in molecules of water. In *atmospheric* cycles, a gaseous form of a nutrient is available to ecosystems. Carbon and nitrogen cycles are examples. Nutrients that do not often occur as gases move in *sedimentary* cycles, such as the phosphorus cycle. They accumulate on the ocean floor and return to land by slow movements of Earth's crust (Section 16.6).

Primary productivity depends on water and nutrients that are dissolved in it. In a biogeochemical cycle, a nutrient moves slowly through the environment, then rapidly among organisms, and back to environmental reservoirs.

42.6 The Water Cycle

For autotrophs of land ecosystems, water and dissolved nutrients are not plentiful all of the time. Shifts in their availability influence primary productivity.

HOW AND WHERE WATER MOVES

LINK TO SECTION 27.3

The world ocean holds most of Earth's water (Table 42.1). As Figure 42.12 shows, in the hydrologic cycle, water moves from the ocean to the atmosphere, onto land, then back to the ocean. Sunlight energy drives *evaporation*, the conversion of water from liquid form to a vapor. *Transpiration*, explained in Section 27.3, is evaporation of water from plant parts. In cool upper layers of the atmosphere, *condensation* of water vapor into droplets gives rise to clouds. Later, clouds release the water as *precipitation*—as rain, snow, or hail.

A watershed is an area from which all precipitation drains into a specific waterway. It may be as small as a valley that feeds a stream, or as large as the Amazon River Basin, which encompasses 5.88 million square kilometers. The Mississippi River Basin covers about 41 percent of the continental United States.

Most precipitation falling in a watershed seeps into the ground. Some collects in aquifers, permeable rock layers that hold water. Groundwater is water in soil and aquifers. When the ground gets saturated, water becomes runoff; it flows over the ground into streams.

Flowing water moves dissolved nutrients into and out of a watershed. Experiments in New Hampshire's Hubbard Brook watershed illustrated that vegetation helps slow nutrient losses. Experimental deforestation caused a spike in loss of mineral ions (Figure 42.13).

Table 42.1 Environmental Water Reservoirs

Main Reservoirs	Volume (10^3 cubic kilometers)
Ocean	1,370,000
Polar ice, glaciers	29,000
Groundwater	4,000
Lakes, rivers	230
Soil moisture	67
Atmosphere (water vapor)	14

A GLOBAL WATER CRISIS

Our planet has plenty of water, but most of it is too salty to drink or use for irrigation. If all Earth's water filled a bathtub, the amount of fresh water that could be used sustainably in a year would fill a teaspoon.

Of the fresh water we use, about two-thirds goes to agriculture. Irrigation can make land less suitable for crops. Piped-in water often has high concentrations of mineral salts. When the water evaporates, these salts are left behind. Salinization, the build-up of mineral salts in soil, stunts crop plants and decreases yields.

Groundwater supplies drinking water to about half of the United States population. Pollution of this water now poses a threat. Chemicals leaching from landfills, hazardous waste facilities, and underground storage tanks often contaminate it. Unlike flowing rivers and streams, which can recover fast, polluted groundwater is difficult and expensive to clean up.

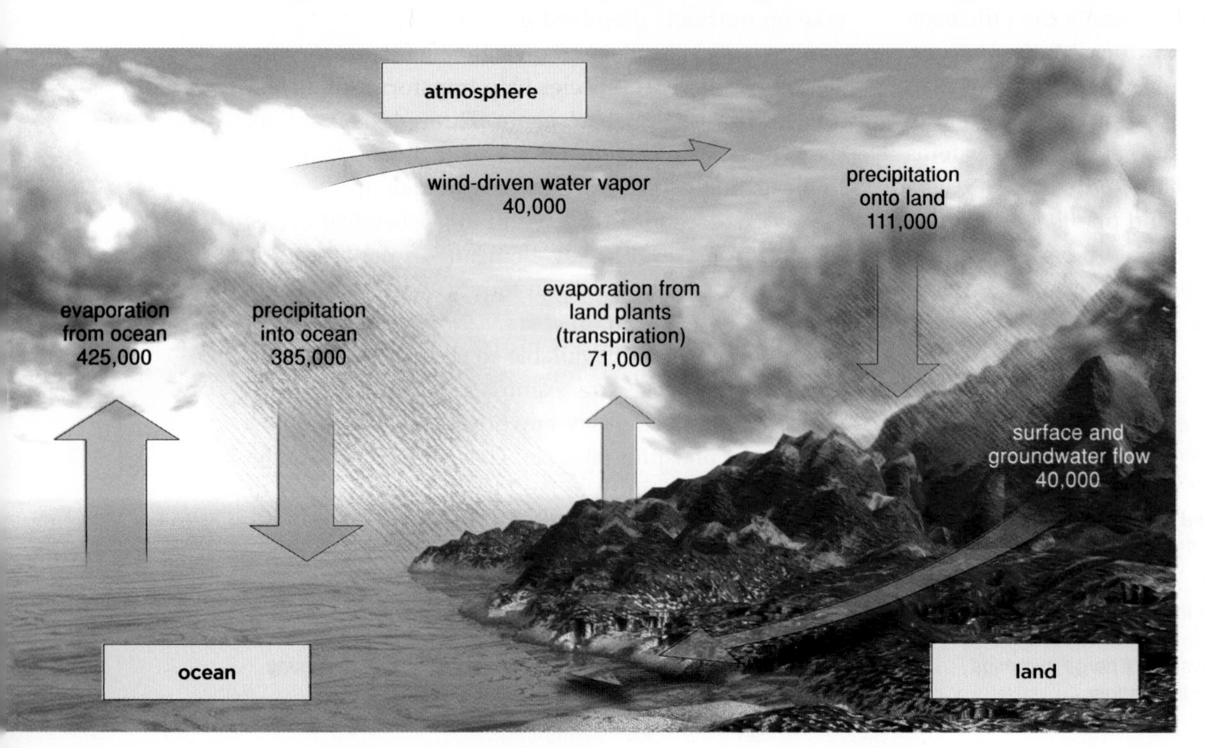

Figure 42.12 Animated! The hydrologic cycle. Water moves from the ocean to the atmosphere, land, and back. The arrows identify processes that move water. The numbers shown indicate the amounts moved, as measured in cubic kilometers per year.

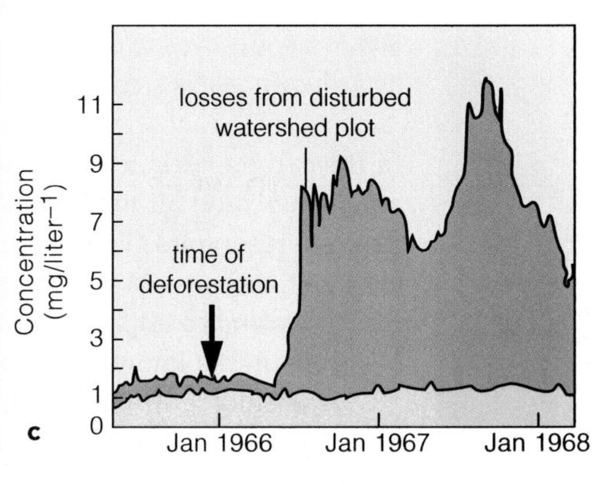

Figure 42.13 Hubbard Brook experimental watershed. (**a**) Runoff in this watershed is collected by concrete basins for easy monitoring. (**b**) This plot of land was stripped of all vegetation as an experiment. (**c**) After experimental deforestation, calcium levels in runoff increased sixfold (*purple*). A control plot in the same watershed showed no similar increase during this time (*light blue*).

Water overdrafts are now common; water is drawn from aquifers faster than natural processes replenish it. When too much fresh water is withdrawn from an aquifer near the coast, saltwater moves in and replaces it. Figure 42.14 highlights regions of aquifer depletion and saltwater intrusion in the United States.

Overdrafts have now depleted half of the Ogallala aquifer, which extends from South Dakota into Texas. This aquifer supplies the irrigation water for about 20 percent of the nation's crops. For the past thirty years, withdrawals have exceeded replenishment by a factor of ten. What will happen when water runs out?

Contaminants, such as sewage, animal wastes, and agricultural chemicals, make water in rivers and lakes unfit to drink. In addition, pollutants disrupt aquatic ecosystems, and in some cases they drive vulnerable species to local extinction.

Desalinization, the removal of salt from seawater, may help increase freshwater supplies. However, the process requires a lot of fossil fuel. Desalinization is feasible mainly in Saudi Arabia and other places that have small populations and very large fuel reserves. Most likely, desalinization will never be cost-effective for widespread use in the United States. In addition, the process produces mountains of waste salts that must be disposed of.

We may be in for upheavals and wars over water rights. Does that sound far-fetched? Turkey has built dams at the headwaters of the Tigris and Euphrates rivers. In the words of one of the dam site managers, Turkey can shut off water flow into Syria and Iraq "to regulate their political behavior." With such potential for conflict, regional, national, and global planning of water distribution seems long overdue.

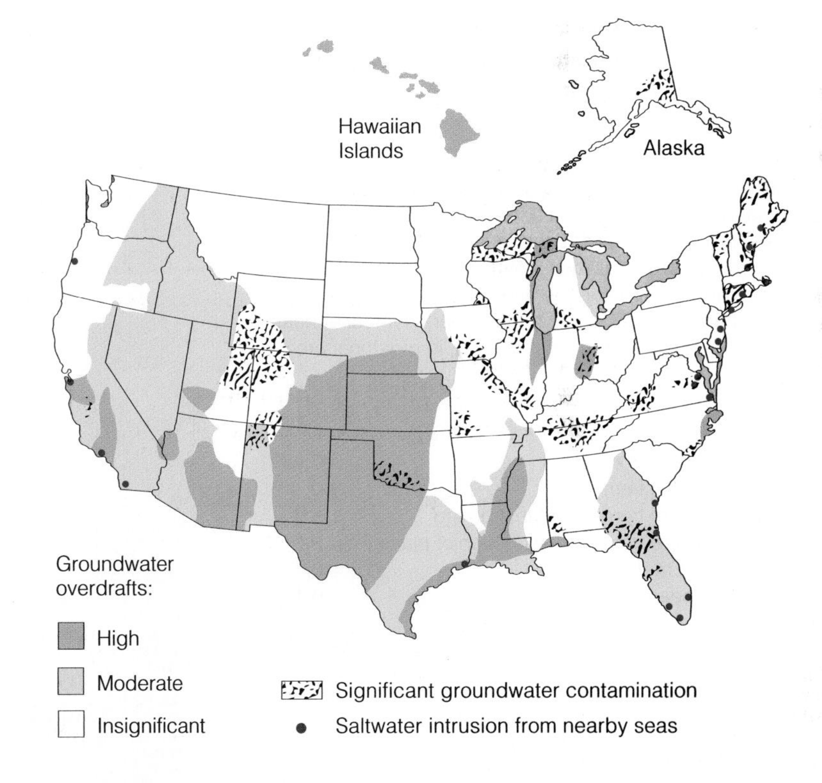

Figure 42.14 Groundwater troubles in the United States.

In the hydrologic cycle, water moves on a global scale. It moves slowly from the world ocean—the main reservoir—through the atmosphere, onto land, then back to the ocean.

Of the fresh water that human populations use, about two-thirds sustains agriculture.

Aquifers that supply much of the world's drinking water are becoming polluted and depleted. Regional conflicts over access to clean, drinkable water are likely to increase.

42.7 Carbon Cycle

Most of the world's carbon is locked in ocean sediments and rocks. It moves into and out of ecosystems as a gas, so its movement is said to be an atmospheric cycle.

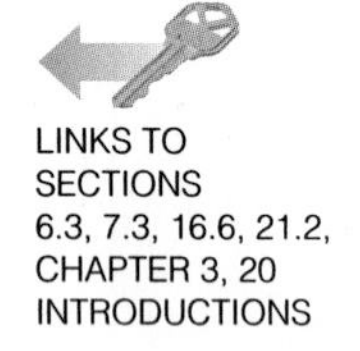

LINKS TO SECTIONS 6.3, 7.3, 16.6, 21.2, CHAPTER 3, 20 INTRODUCTIONS

In the carbon cycle, carbon moves through the lower atmosphere and all food webs on its way to and from its large reservoirs (Figure 42.15). Earth's crust holds the most carbon—66 million to 100 million gigatons. Each gigaton is a billion tons. There are 4,000 gigatons of carbon in the known fossil fuel reserves.

Organisms contribute to Earth's carbon deposits. Recall, from Chapter 20, that foraminiferans and many other single-celled protists form shells that are rich in calcium carbonate. Over many hundreds of millions of years, uncountable numbers of these cells died, sank, and were buried in seafloor sediments. Carbon in the remains has been cycled slowly. Movements of crustal plates uplifted portions of the seafloor, which drained off and became part of land ecosystems.

Most of the annual cycling takes place between the ocean and atmosphere. The ocean holds 38,000–40,000 gigatons of dissolved carbon, primarily in the form of bicarbonate and carbonate ions. The air holds about 766 gigatons of carbon, mainly combined with oxygen in the form of carbon dioxide (CO_2).

Detritus in soil holds another 1,500–1,600 gigatons of carbon atoms. Another 540–610 gigatons is present in biomass, or tissues of organisms. Methane hydrates, described in the introduction to Chapter 3, are another reservoir. Massive deposits of methane hydrates lie on the ocean floor near coasts of continents. Additional carbon is tied up beneath permafrost, the perpetually frozen soil of arctic regions.

Why doesn't all of the CO_2 dissolved in warm sea surface waters escape into the atmosphere? Driven by prevailing winds and regional differences in density, seawater flows in a gigantic loop from the surface of the Pacific and Atlantic oceans down to the Atlantic

Figure 42.15 Animated! *Right*, carbon cycling in (**a**) marine ecosystems and (**b**) land ecosystems. *Gold* boxes highlight the most important carbon reservoirs. The vast majority of carbon atoms are in sediments and rocks, followed by lesser amounts in seawater, soil, the atmosphere, and biomass (in that order). Typical annual fluxes in global distribution of carbon, in gigatons, are:

From atmosphere to plants by carbon fixation	120
From atmosphere to ocean	107
To atmosphere from ocean	105
To atmosphere from plants	60
To atmosphere from soil	60
To atmosphere from fossil fuel burning	5
To atmosphere from net destruction of plants	2
To ocean from runoff	0.4
Burial in ocean sediments	0.1

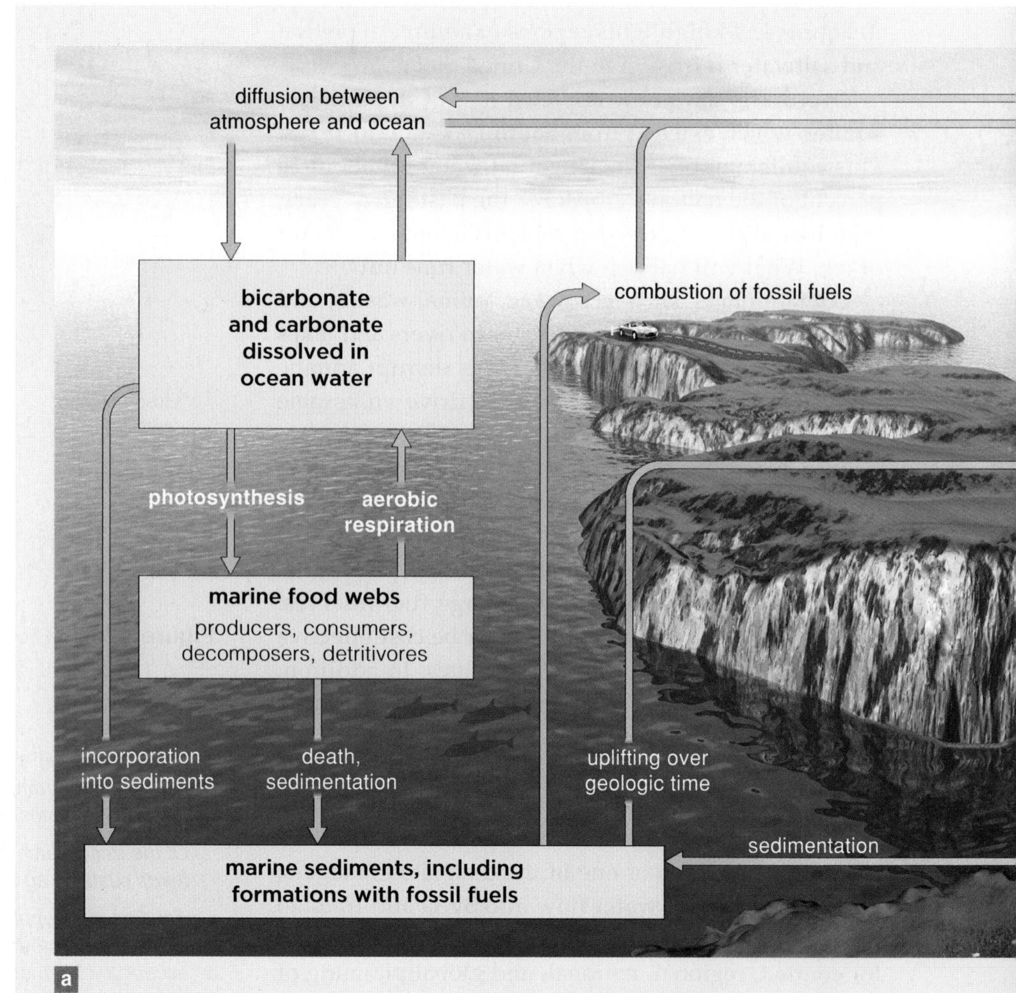

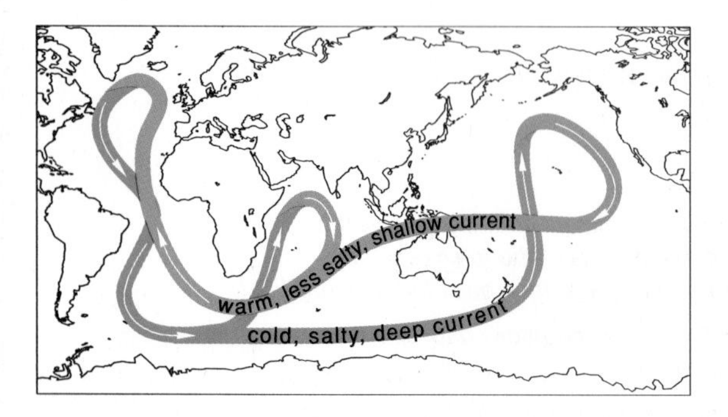

Figure 42.16 Loop that moves carbon dioxide to carbon's deep ocean reservoir. The loop sinks in the cold, salty North Atlantic. It rises in the warmer Pacific.

and Antarctic seafloors. There, CO_2 moves into deep storage reservoirs before water loops back up (Figure 42.16). The loop effectively mediates annual fluxes in the global distribution of carbon.

As you know, CO_2 is the key source of carbon for food webs in aquatic and land ecosystems. That is why biologists often refer to the global cycling of carbon, in the form of CO_2, as a carbon–oxygen cycle. Plants, phytoplankton, and some bacteria "fix" carbon when they engage in photosynthesis. Each year, they tie up billions of metric tons of carbon in sugars and other organic compounds. Breakdown of those compounds by aerobic respiration releases CO_2 into the air and water. More CO_2 escapes into the air when fossil fuels or forests burn and when volcanoes erupt.

The average time that an ecosystem holds a given carbon atom varies. As examples, organic wastes and remains decompose so fast in tropical rain forests that carbon does not build up at the soil surface. Bogs and other anaerobic habitats do not favor decomposition, so material accumulates, as in peat bogs (Section 21.2).

Each year, we withdraw 4 to 5 gigatons from fossil fuel reservoirs and our activities put about 6 gigatons more carbon in the air than can be cycled to the ocean reservoirs by natural processes. Only about 2 percent of the excess carbon entering the atmosphere becomes dissolved in ocean water. Carbon dioxide is one of the greenhouse gases, so increased outputs of it may be a factor in global warming. The next section looks at this possibility and some environmental implications.

Earth's crust is the largest carbon reservoir, followed by the world ocean. Most of the annual cycling of carbon occurs between the ocean and atmosphere.

In the carbon–oxygen cycle, carbon moves into and out of ecosystems mainly when combined with oxygen, as in carbon dioxide, bicarbonate, and carbonate.

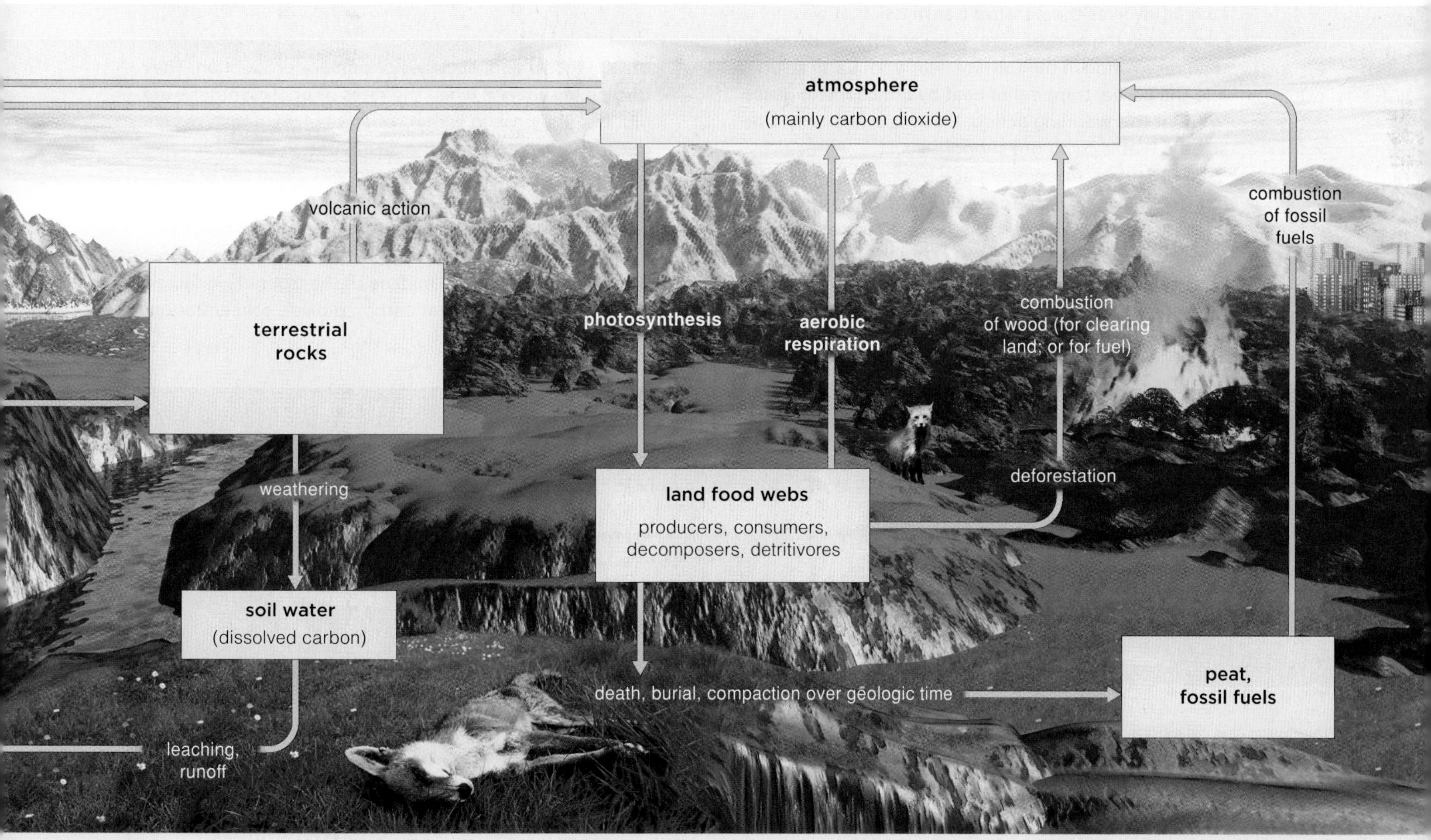

42.8 Greenhouse Gases, Global Warming

The atmospheric concentrations of gaseous molecules help determine the average temperature near Earth's surface. Human activities are contributing to increases that may cause dramatic climate change.

LINKS TO SECTION 6.8, CHAPTER 3 INTRODUCTION

Concentrations of various gaseous molecules profoundly influence the average temperature of the atmosphere near Earth's surface. That temperature, in turn, has far-reaching effects on global and regional climates.

Atmospheric molecules of carbon dioxide, water, nitrous oxide, methane, and chlorofluorocarbons (CFCs) are among the main players in interactions that can shift global temperatures. Collectively, the gases function like panes of glass in a greenhouse—hence the familiar name, "greenhouse gases." The gases absorb wavelengths of visible light and transmit them toward Earth's surface. The surface absorbs the wavelengths and then emits longer, infrared wavelengths—heat. Greenhouse gases impede the escape of heat energy from Earth into space. How? The gaseous molecules absorb the longer wavelengths, then radiate much of it back toward Earth (Figure 42.17).

By the greenhouse effect, Earth's lower atmosphere and surface warm up as sunlight energy gets converted to heat. It involves the nonstop transmission of wavelengths by gases in the atmosphere, wavelength absorption and then reradiation in the form of heat from Earth's surface, and the partial trapping of heat by atmospheric gases. Without this warming action, Earth's surface would be so cold that it could not support life.

In the 1950s, researchers at a laboratory on Hawaii's highest volcano began to measure the atmospheric concentrations of greenhouse gases. That remote site is almost free of local airborne contamination. It also is representative of atmospheric conditions for the Northern Hemisphere. What did they find? Briefly, concentrations of CO_2 follow annual cycles of primary production. They decline in summer, when the rates of photosynthesis are highest. They rise in winter, when photosynthesis declines but aerobic respiration and fermentation continue.

The alternating troughs and peaks along the graph line in Figure 42.18*a* are annual lows and highs of global CO_2 concentrations. For the first time, researchers saw the effects of carbon dioxide fluctuations for the entire hemisphere. Notice the midline of the troughs and peaks in the cycle. It shows that carbon dioxide concentration is

Figure 42.18 *Facing page*, graphs of recent increases in four categories of atmospheric greenhouse gases. A key factor is the sheer number of gasoline-burning vehicles in large cities. *Above*, Mexico City on a smoggy morning. With 10 million residents, it is the world's largest city.

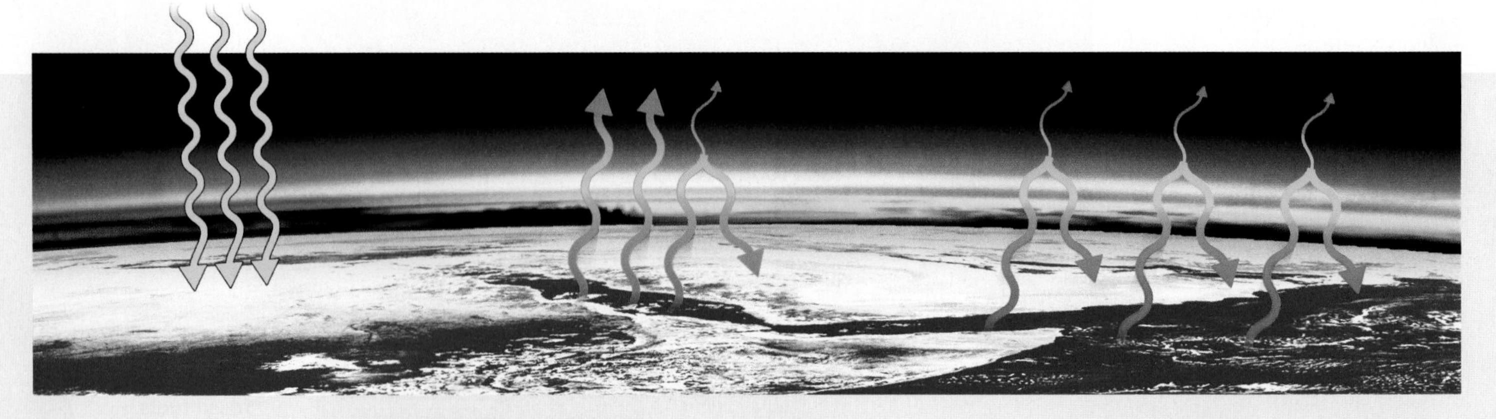

a Wavelengths in rays from the sun penetrate the lower atmosphere, and they warm Earth's surface.

b The surface radiates heat (infrared wavelengths) to the atmosphere. Some heat escapes into space. But greenhouse gases and water vapor absorb some infrared energy and radiate a portion of it back toward Earth.

c Increased concentrations of greenhouse gases trap more heat near Earth's surface. Sea surface temperatures rise, so more water evaporates into the atmosphere. Earth's surface temperature rises.

Figure 42.17 **Animated!** The greenhouse effect.

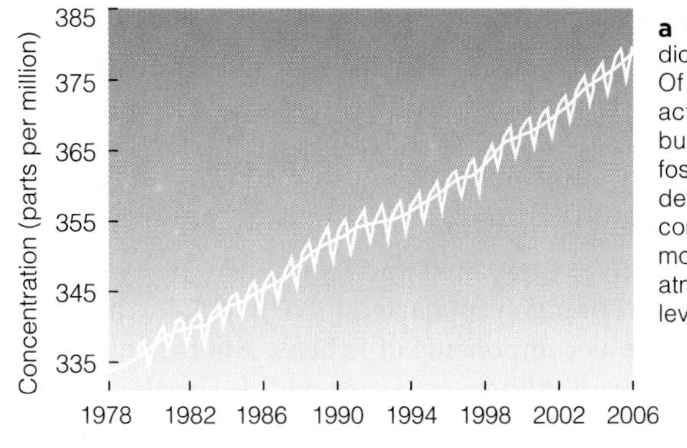

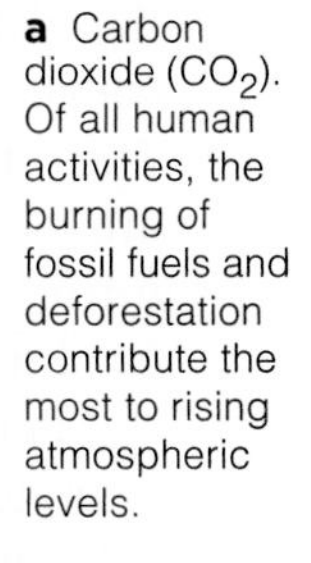
a Carbon dioxide (CO_2). Of all human activities, the burning of fossil fuels and deforestation contribute the most to rising atmospheric levels.

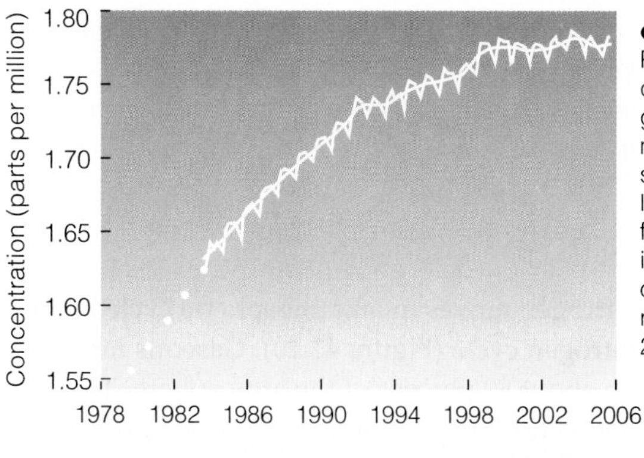

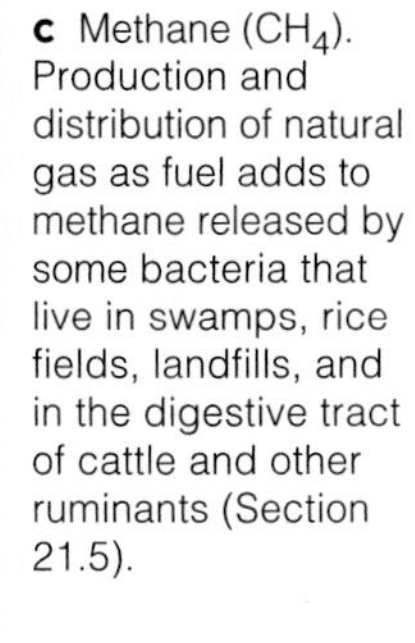
c Methane (CH_4). Production and distribution of natural gas as fuel adds to methane released by some bacteria that live in swamps, rice fields, landfills, and in the digestive tract of cattle and other ruminants (Section 21.5).

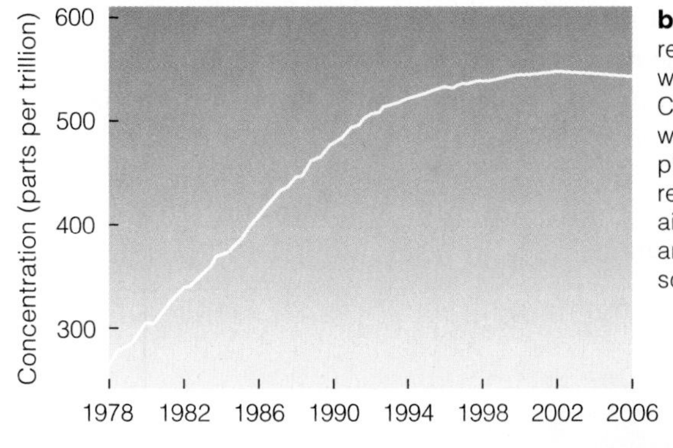

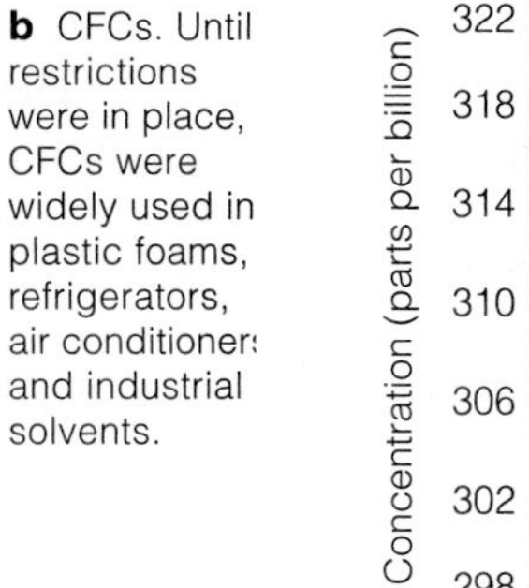
b CFCs. Until restrictions were in place, CFCs were widely used in plastic foams, refrigerators, air conditioners and industrial solvents.

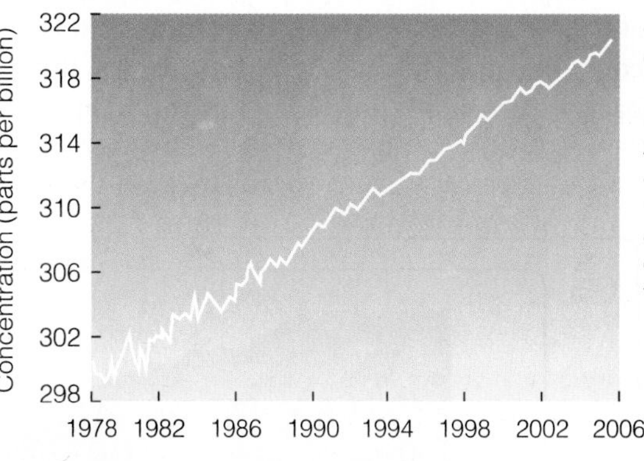

d Nitrous oxide (N_2O). Denitrifying bacteria produce N_2O in metabolism. Also, fertilizers and animal wastes release enormous amounts; this is especially so for large-scale livestock feedlots.

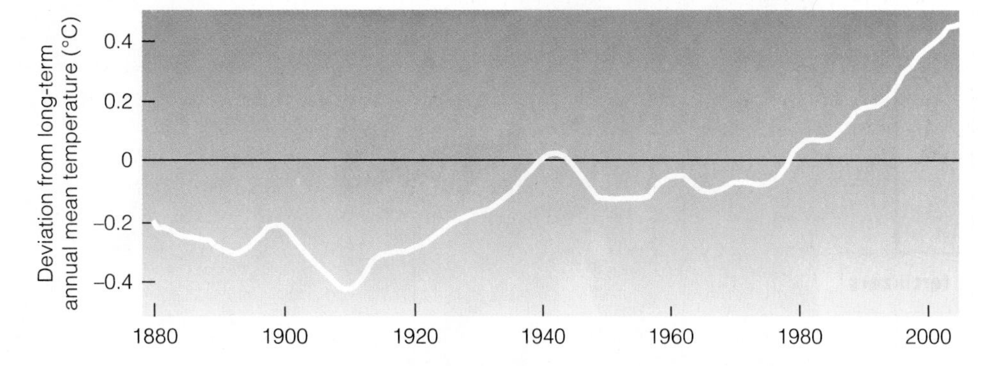

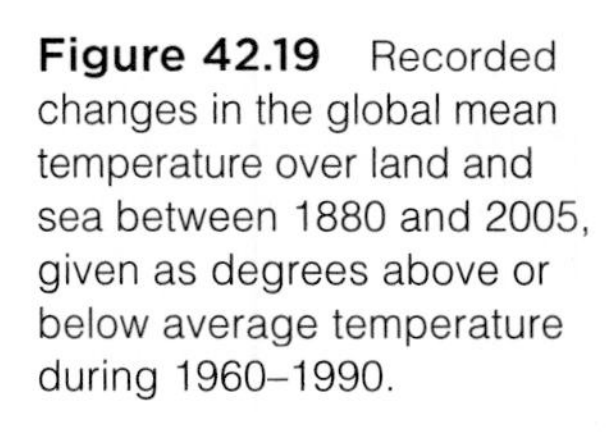
Figure 42.19 Recorded changes in the global mean temperature over land and sea between 1880 and 2005, given as degrees above or below average temperature during 1960–1990.

steadily increasing—as are concentrations of other major greenhouse gases.

Atmospheric levels of greenhouse gases are far higher than they were for most of the past. Carbon dioxide may be at its highest level since 420,000 years ago, possibly since 20 million years ago. There is scientific consensus that human activities—mainly the burning of fossil fuels—are contributing significantly to the current increases in greenhouse gases. The big worry is that the increase may have far-reaching environmental consequences.

The increase in greenhouse gases may be a factor in global warming, a long-term increase in temperature near Earth's surface (Figure 42.19). In the past thirty years, the global surface temperature increased at a faster rate, to 1.8°C (3.2°F) per century. Warming is most dramatic at the upper latitudes of the Northern Hemisphere.

Data from satellites, weather stations and balloons, research ships, and computer programs suggest that irreversible climate changes are already under way. Polar ice is melting fast, and ancient glaciers are retreating. This past century, the sea level may have risen as much as 20 centimeters (8 inches).

We can expect continued temperature increases to have drastic effects on climate. As evaporation increases, so will global precipitation. Intense rains and flooding probably will become more frequent in some regions, while droughts increase in others. Hurricanes probably will become more intense, and possibly more frequent.

It bears repeating: As investigations continue, a key research goal is to investigate all of the variables in play. With respect to consequences of global warming, the most crucial variable may be the one we do not know.

42.9 Nitrogen Cycle

Gaseous nitrogen makes up about 80 percent of the lower atmosphere. Successively smaller reservoirs are seafloor sediments, ocean water, soil, biomass on land, nitrous oxide in the atmosphere, and marine biomass.

INPUTS INTO ECOSYSTEMS

LINKS TO SECTIONS 2.4, 2.6, 19.2, 20.4, 27.1

Nitrogen moves in an atmospheric cycle known as the nitrogen cycle (Figure 42.20). Gaseous nitrogen makes up about 80 percent of the atmosphere. Triple covalent bonds hold its two atoms of nitrogen together (N_2, or N≡N). Volcanic eruptions and lightning can convert some N_2 into forms that enter food webs. Far more is converted through nitrogen fixation. By this process, bacteria break all three bonds in N_2, then incorporate the N atoms into ammonia (NH_3). Later, ammonia is converted into ammonium (NH_4^+) and nitrate (NO_3^-). These two nitrogen salts dissolve readily in water and are taken up by plant roots. Plants cannot use gaseous nitrogen, because they do not make the enzyme that can break its triple bond.

Many species of bacteria fix nitrogen (Section 19.2). Nitrogen-fixing cyanobacteria live in aquatic habitats, soil, and as components of lichens. Another nitrogen-fixing group, *Rhizobium*, forms nodules on the roots of peas and other legumes. Each year, the nitrogen-fixing bacteria collectively take up about 270 million metric tons of nitrogen from the atmosphere.

The nitrogen incorporated into plant tissues moves through the trophic levels of ecosystems. It ends up in

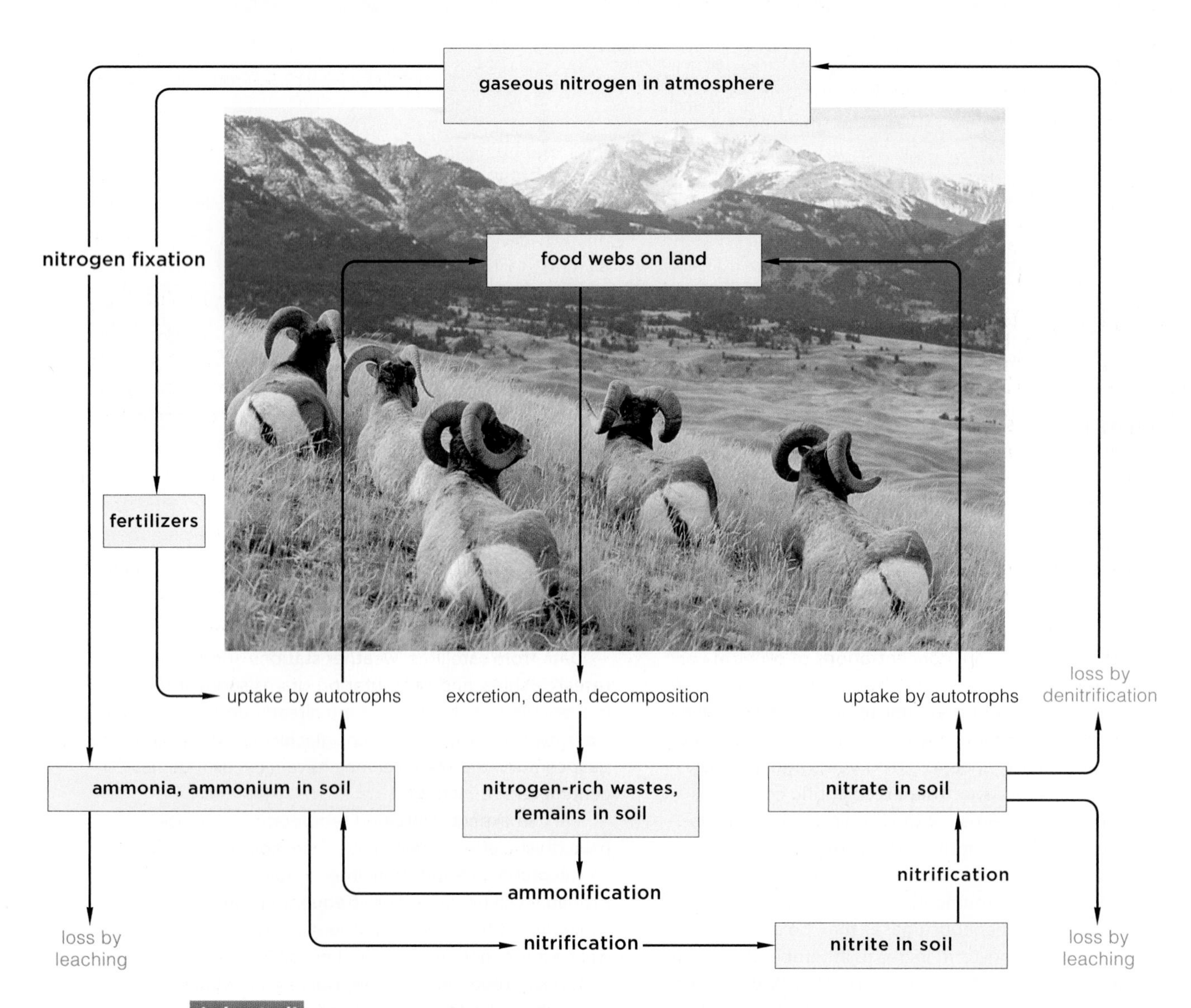

Figure 42.20 **Animated!** Nitrogen cycle in an ecosystem on land. Nitrogen becomes available to plants through the activities of nitrogen-fixing bacteria. Other bacterial species cycle nitrogen to plants. They break down organic wastes to ammonium and nitrates.

nitrogen-rich wastes and remains, which bacteria and fungi decompose (Sections 19.2 and 22.5). By a process called ammonification, ammonium forms as various organisms break down nitrogenous materials. These microbes use some of the ammonium and release the rest to soil. Plants and some nitrifying bacteria take it up. In the initial step of nitrification, certain bacteria strip electrons from ammonium, which leaves nitrite. Other nitrifying bacteria use the nitrite in reactions that end with the formation of nitrate (Figure 42.20).

NATURAL LOSSES FROM ECOSYSTEMS

Ecosystems lose nitrogen through denitrification. By this process, denitrifying bacteria convert nitrate or nitrite to gaseous nitrogen or to nitrogen oxide (NO_2). Most denitrifying bacteria are anaerobic; they live in waterlogged soils and aquatic sediments.

Ammonium, nitrite, and nitrate also are lost from a land ecosystem in runoff and by leaching, the removal of some nutrients as water trickles down through the soil (Section 27.1). Nitrogen-rich runoff enters streams and other aquatic ecosystems.

DISRUPTIONS BY HUMAN ACTIVITIES

Deforestation and grassland conversion for agriculture also cause big nitrogen losses. With each clearing and harvest, nitrogen in plant tissues is removed. Also, soil becomes more vulnerable to erosion and leaching.

Many farmers counter nitrogen losses by rotating crops. For example, they may plant corn and soybeans in the same field in alternating years. Nitrogen-fixing bacteria that associate with legumes such as soybeans add nitrogen to the soil (Section 27.2). In developed countries, farmers also spread synthetic nitrogen-rich fertilizers. As the result of these practices, crop yields have increased dramatically over the past forty years.

High temperature and pressure converts nitrogen and hydrogen gases to ammonia fertilizers. Although the manufactured fertilizers improve crop yields, they also modify soil chemistry. Adding ammonium to soil increases the concentration of hydrogen ions, as well as nitrogen. Rising acidity encourages ion exchanges; nutrient ions bound to particles of soil get replaced by hydrogen ions. As a result of ion exchanges, calcium and magnesium ions essential for plant growth trickle away in soil water.

Burning of fossil fuel in power plants and vehicles releases nitrogen oxides. These chemicals contribute to global warming and acid rain (Section 2.6). Winds frequently carry the pollutants far from their sources.

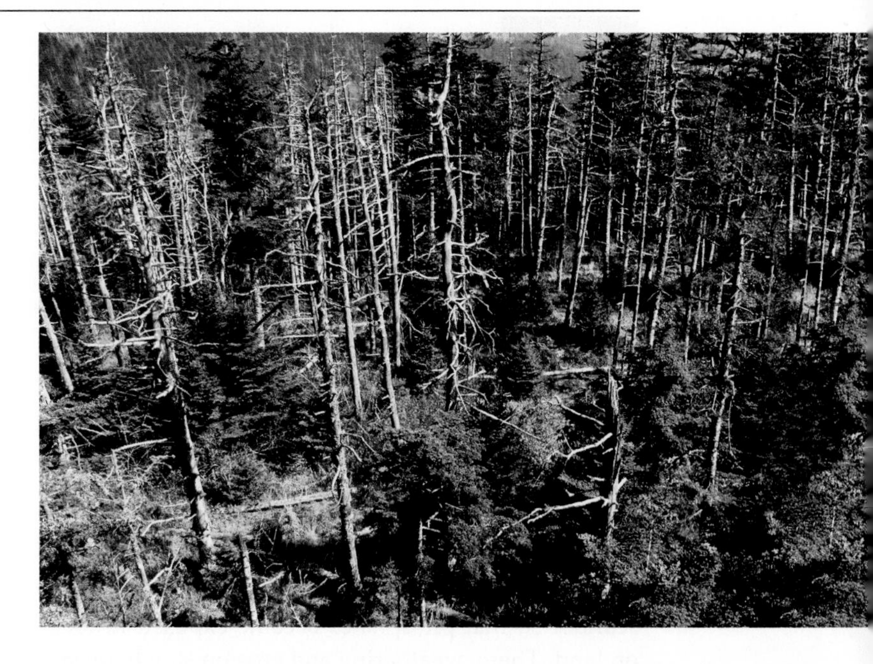

Figure 42.21 Dead and dying trees in Great Smoky Mountains National Park. Forests are among the casualties of nitrogen oxides and other forms of air pollution.

By some estimates, pollutants blowing into the Great Smoky Mountains National Park have increased the amount of nitrogen there sixfold (Figure 42.21).

Nitrogen in acid rain can have the same bad effects as overfertilization. Different plant species respond in different ways to changes in nitrogen levels. Increases in nitrogen can disrupt the balance among competing species in a community, causing diversity to decline. The impact can be especially pronounced in forests at high elevations or at high latitudes, where soils tend to be naturally nitrogen-poor.

Some human activities disrupt aquatic ecosystems through nitrogen enrichment. For instance, about half of the nitrogen in fertilizers applied to fields runs off into rivers, lakes, and estuaries. More nitrogen enters with sewage from cities and in animal wastes. As one result, nitrogen inputs promote algal blooms (Section 20.4). Phosphorus in fertilizers has the same negative effects, as explained in the next section.

The ecosystem phase of the nitrogen cycle starts with nitrogen fixation. Bacteria convert gaseous nitrogen in the air to ammonia and then to ammonium, which is a form that plants easily take up.

By ammonification, bacteria and fungi make additional ammonium available to plants when they break down nitrogen-rich organic wastes and remains.

By nitrification, bacteria convert nitrites in soil to nitrate, which also is a form that plants easily take up.

The ecosystem loses nitrogen when denitrifying bacteria convert nitrite and nitrate back to gaseous nitrogen, and when nitrogen is leached from soil.

42.10 The Phosphorus Cycle

Unlike carbon and nitrogen, phosphorus seldom occurs as a gas. Like nitrogen, phosphorus can be taken up by plants only in ionized form, and it, too, is often a limiting factor on plant growth.

LINKS TO SECTIONS 3.6, 16.6, 20.4

In the phosphorus cycle, phosphorus passes quickly through food webs as it moves from land to ocean sediments, then slowly back to dry land. Earth's crust is the largest reservoir of phosphorus.

The phosphorus in rocks is mainly in the form of phosphate (PO_4^{3-}). Weathering and erosion put these ions in streams and rivers, which deliver them to the sea (Figure 42.22). There, phosphates accumulate as deposits along continental margins. After millions of years, crustal movements may uplift a portion of the seafloor so that phosphates become exposed in rocks on land. There, weathering and erosion slowly release phosphates from the rocks and start the phosphorus cycle over again.

Phosphates are required building blocks for ATP, phospholipids, nucleic acids, and other compounds. Plants take up dissolved phosphates from soil water. Herbivores get them by eating plants; carnivores get them by eating herbivores. Animals lose phosphate in urine and in feces. Bacterial and fungal decomposers release phosphate from organic wastes and remains, then plants take them up again.

The hydrologic cycle helps move phosphorus and other minerals through ecosystems. Water evaporates from the ocean and falls on land. As it flows back to the ocean, it transports silt and dissolved phosphates that the primary producers require for growth.

Of all minerals, phosphorus most frequently acts as the limiting factor for plant growth. Only newly weathered, young soil is phosphorous-rich. In aquatic habitats, most phosphorus is locked up in sediments.

Many tropical and subtropical ecosystems already low in phosphorus are likely to be further depleted. In an undisturbed forest, decomposition releases the phosphorus stored in biomass. When forest is logged and converted to farmland, phosphorus stored in trees is lost to the ecosystem. Crop yields start out low and quickly become nonexistent. Later, after the fields are abandoned, regrowth remains sparse. Spreading finely ground, phosphate-rich rock can help restore fertility, but many developing countries lack this resource.

The developed countries have a different problem —too much phosphorus. It taints runoff from heavily fertilized fields. Sewage from cities and factory farms also contain phosphorus. Dissolved phosphorus that

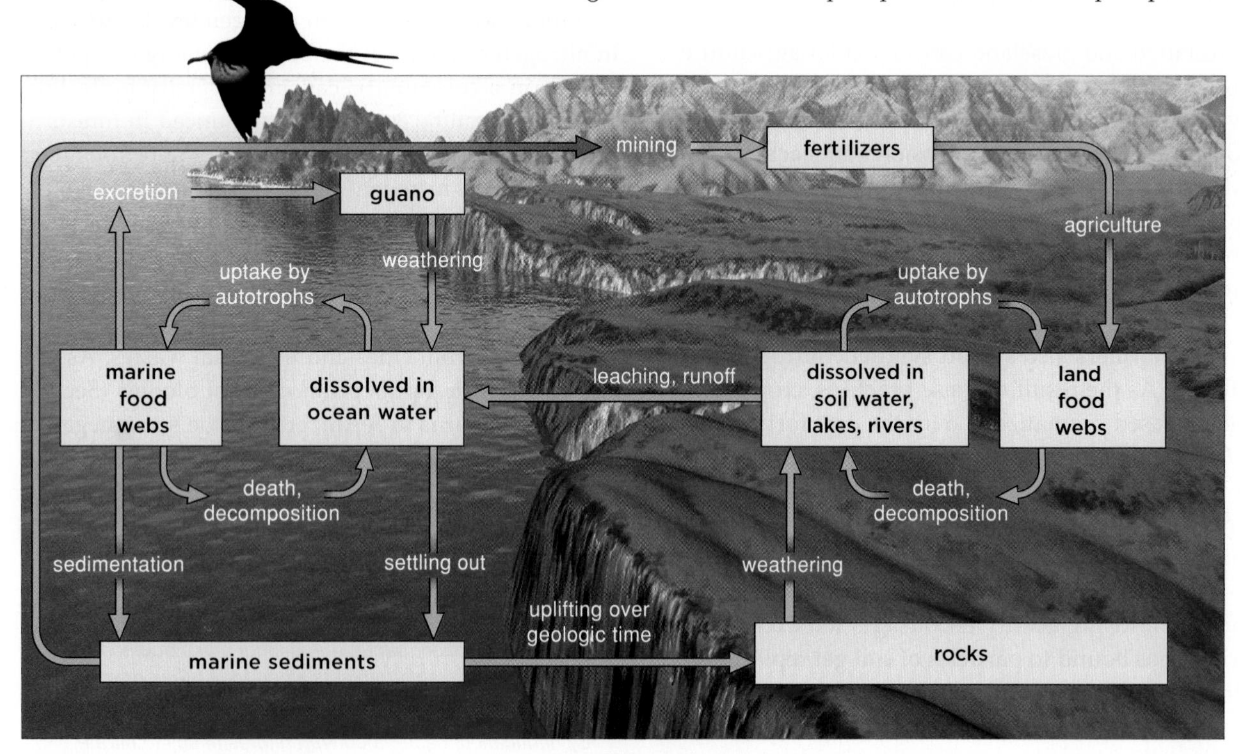

Figure 42.22 Animated! Phosphorus cycle. In this sedimentary cycle, phosphorus moves mainly in the form of phosphate ions (PO_4^{3-}) to the ocean. It moves through phytoplankton of marine food webs, then to fishes that eat plankton. Seabirds eat the fishes, and their droppings (guano) accumulate on islands. Humans collect and use guano as a phosphate-rich fertilizer.

gets into aquatic ecosystems can promote destructive algal blooms. Like the plants, algae require nitrogen, phosphorus, and other ions to keep growing. In many freshwater ecosystems, nitrogen-fixing bacteria keep the nitrogen levels high, so phosphorus becomes the limiting factor. When phosphate-rich pollutants pour in, algal populations soar and then crash. As aerobic decomposers break down remains of dead algae, the water becomes depleted of the oxygen that fishes and other organisms require.

Eutrophication refers to the nutrient enrichment of any ecosystem that is otherwise low in nutrients. It is a process of natural succession, but human activities can accelerate it, as the experiment shown in Figure 42.23 demonstrated. Eutrophication of a lake is hard to reverse. It can take many years for excess nutrients that encourage algal growth to be depleted.

Sedimentary cycles, in combination with the hydrologic cycle, move most mineral elements, such as phosphorus, through terrestrial and aquatic ecosystems.

Agriculture, deforestation, and other human activities upset the nutrient balances of ecosystems.

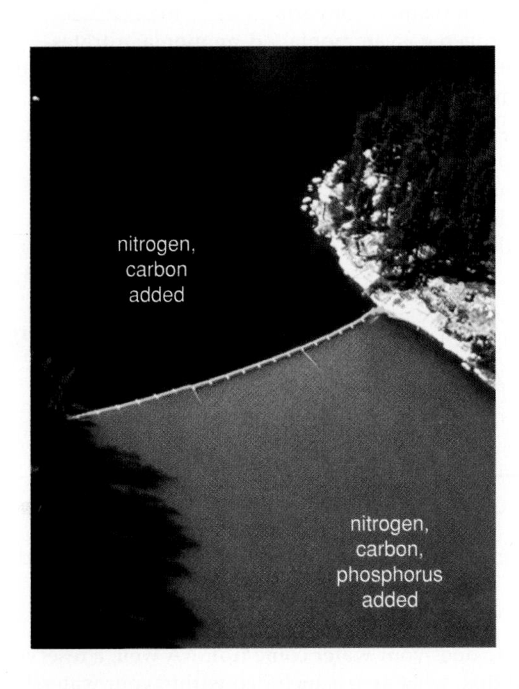

Figure 42.23 One of the eutrophication experiments. Researchers put a plastic curtain across a channel between two basins of a natural lake. They added nitrogen, carbon, and phosphorus on one side of the curtain (here, the *lower* part of the lake) and added nitrogen and carbon on the other side. Within months, the phosphorous-rich basin was eutrophic, with a dense algal bloom (*green*) covering its surface.

Summary

Section 42.1 An ecosystem consists of an array of organisms together with their physical and chemical environment. There is a one-way flow of energy into and out of an ecosystem, and a cycling of materials among resident species. All ecosystems have inputs and outputs of energy and nutrients.

Sunlight supplies energy to most ecosystems. Primary producers convert sunlight energy into chemical bond energy. They also take up nutrients that they, and all consumers, require. Herbivores, carnivores, omnivores, decomposers, and detritivores are consumers.

Organisms in an ecosystem are classified by trophic levels. Those at the same level are the same number of steps away from the energy input into the ecosystem. A food chain shows one path of energy and nutrient flow among organisms. It shows who eats whom.

- *Use the animation on ThomsonNOW to learn about energy flow and nutrient cycling.*

Section 42.2 Food chains interconnect as food webs. The efficiency of energy transfers is always low, so most ecosystems support no more than four or five trophic levels away from an original energy source. In a grazing food web, most energy captured by producers flows to herbivores. In detrital food webs, most energy flows from producers directly to detritivores and decomposers. Both types of food webs interconnect in nearly all ecosystems.

- *Use the animation on ThomsonNOW to explore a food web.*

Section 42.3 By biological magnification, a chemical substance is passed from organisms at each trophic level to those above and becomes increasingly concentrated in body tissues. DDT is an example.

Section 42.4 A system's primary productivity is the rate at which producers capture and store energy in their tissues. It varies with climate, seasonal changes, nutrient availability, and other factors.

Energy pyramids and biomass pyramids depict how energy and organic compounds are distributed among the organisms of an ecosystem. All energy pyramids are largest at their base. If producers get eaten as fast as they reproduce, the biomass of consumers can exceed that of producers, so the biomass pyramid is upside down.

- *Use the animation on ThomsonNOW to see how energy flows through one ecosystem.*

Section 42.5 In a biogeochemical cycle, water or a nutrient moves from an environmental reservoir, through organisms, then back to the environment.

Section 42.6 In the hydrologic cycle, evaporation, condensation, and precipitation move water from its main reservoir—the oceans—into the atmosphere, onto land, then back to oceans. Humans are disrupting the cycle.

- *Use the animation on ThomsonNOW to learn about the hydrologic cycle.*

Section 42.7 The carbon cycle moves carbon from reservoirs in rocks and seawater, through its gaseous form (CO_2) in the atmosphere, and through ecosystems. Deforestation and the burning of wood and fossil fuels are adding more carbon dioxide to the atmosphere than the oceans can absorb.

■ *Use the animation on ThomsonNOW to observe the flow of carbon through its global cycle.*

Section 42.8 Collectively, the greenhouse gases trap heat in the lower atmosphere, which helps make Earth's surface warm enough to support life. Natural processes and human activities are adding more greenhouse gases, including carbon dioxide, CFCs, methane, and nitrous oxide, to the atmosphere. The rise correlates with a rise in global temperatures and other climate changes.

■ *Use the animation on ThomsonNOW to explore the greenhouse effect and global warming.*

Section 42.9 The atmosphere is the main reservoir for N_2, a gaseous form of nitrogen that plants cannot use. In nitrogen fixation, some soil bacteria take up N_2 and form ammonia. Other reactions convert ammonia to ammonium and nitrate, which plants are able to take up. Some nitrogen is lost to the atmosphere by the action of denitrifying bacteria. Human activities add nitrogen to ecosystems; for example, through fertilizer applications and fossil fuel burning, which releases nitrogen oxides.

■ *Use the animation on ThomsonNOW to learn how nitrogen is cycled in an ecosystem.*

Section 42.10 Phosphorus moves in a sedimentary cycle. Earth's crust is the largest reservoir. Phosphorus is often the factor that limits population growth of plant and algal producers. Excessive inputs of phosphorus to an aquatic ecosystem can lead to eutrophication.

■ *Use the animation on ThomsonNOW to learn how phosphorus is cycled in an ecosystem.*

Self-Quiz

Answers in Appendix III

1. In most ecosystems, the primary producers use energy from ______ to build organic compounds.
   a. sunlight
   b. heat
   c. breakdown of wastes and remains
   d. breakdown of inorganic substances in the habitat

2. Organisms at the lowest trophic level in a tallgrass prairie are all ______.
   a. at the first step away from the original energy input
   b. autotrophs
   c. heterotrophs
   d. both a and b
   e. both a and c

3. Decomposers are commonly ______.
   a. fungi b. plants c. bacteria d. a and c

4. All organisms at the first trophic level ______.
   a. capture energy from a nonliving source
   b. are eaten by organisms at higher trophic levels
   c. would be at the bottom of an energy pyramid
   d. all of the above

5. Primary productivity on land is affected by ______.
   a. nutrient availability
   b. amount of sunlight
   c. temperature
   d. all of the above

6. If biological magnification occurs, the ______ will have the highest levels of toxins in their systems.
   a. producers
   b. herbivores
   c. primary carnivores
   d. top carnivores

7. Most of earth's freshwater is ______.
   a. in lakes and streams
   b. in aquifers and soil
   c. frozen as ice
   d. in bodies of organisms

8. Earth's largest carbon reservoir is ______.
   a. the atmosphere
   b. sediments and rocks
   c. seawater
   d. living organisms

9. Carbon is released into the atmosphere by ______.
   a. photosynthesis
   b. aerobic respiration
   c. burning fossil fuels
   d. b and c

10. Greenhouse gases ______.
    a. slow escape of heat energy from Earth into space
    b. are produced by natural and human activities
    c. are at higher levels than they were 100 years ago
    d. all of the above

11. The ______ cycle is a sedimentary cycle.
    a. hydrologic
    b. carbon
    c. nitrogen
    d. phosphorus

12. Earth's largest phosphorus reservoir is ______.
    a. the atmosphere
    b. guano
    c. sediments and rocks
    d. living organisms

13. Plant growth requires ______ uptake from the soil.
    a. nitrogen
    b. carbon
    c. phosphorus
    d. both a and c
    e. all of the above

14. Nitrogen fixation converts ______ to ______.
    a. nitrogen gas; ammonia
    b. nitrates; nitrites
    c. ammonia; nitrogen gas
    d. ammonia; nitrates
    e. nitrites; nitrogen oxides

15. Match each term with its most suitable description.
    ___ producers
    ___ herbivores
    ___ decomposers
    ___ detritivores
    a. feed on plants
    b. feed on small bits of organic matter
    c. degrade organic wastes and remains to inorganic forms
    d. capture sunlight energy

■ *Visit ThomsonNOW for additional questions.*

Critical Thinking

1. Marguerite has a vegetable garden in Maine. Eduardo has one in Florida. What are some of the variables that influence primary production in each place?

2. Where does your water come from? A well, a reservoir? Beyond that, what area is included within your watershed and what are the current flows like? Visit the *Science in Your Watershed* site at water.usgs.gov/wsc and research these questions.

3. Look around you and name all of the objects, natural or manufactured, that might be contributing to amplification of the greenhouse effect.

4. Polar ice shelves are vast, thickened sheets of ice that float on seawater. In March 2002, 3,200 square kilometers (1,250 square miles) of Antarctica's largest ice shelf broke free from the continent and shattered into thousands of icebergs (Figure 42.24). Scientists knew the ice shelf was shrinking and breaking up, but this was the single largest loss ever observed at one time. Why should this concern people who live in more temperate climates?

5. Fishes are a fine source of protein and of omega-3 fatty acids, which are necessary for the normal development of the nervous system. This would seem to make fish a good choice for pregnant women. But coal-burning power plants put mercury into the environment, and some of it ends up in fish. Eating mercury-tainted fish during pregnancy can adversely affect development of a fetal nervous system.

Tissues of predatory marine fishes, such as swordfishes, tunas, marlins, and sharks, have especially high levels of mercury. The Environmental Protection Agency has issued health advisories to pregnant women, suggesting that they limit their consumption of fish species most likely to be tainted with mercury. Although sardines are harvested from the same ocean, they have lower mercury levels and are not on the warning list. Explain why two species of fishes that live in the same place can have very different levels of mercury in their tissues.

6. In temperate areas, spring is a season of renewed plant growth. Trees leaf out and grassy fields turn green as days lengthen and temperatures rise. A similar seasonal change occurs in the seas, although we do not usually notice it.

As the introduction to Chapter 20 explained, the ocean's sunlit upper waters are home to countless photosynthetic protists, including foraminiferans, coccolithophores, dinoflagellates, and diatoms. Collectively they make up the phytoplankton, sometimes described as the "pastures of the seas."

Until NASA began gathering data from space satellites, we had no idea of the size and distribution of these marine pastures. Now we monitor them by recording the amount of chlorophyll. Figure 42.25*a* shows chlorophyll levels (one measure of photosynthetic activity) in the Atlantic Ocean during one winter. Figure 42.25*b* shows one spring bloom that stretched from North Carolina all the way past Spain!

Phytoplankton take up carbon dioxide from seawater and use it in photosynthesis. Carbon dioxide from the atmosphere dissolves in the water and replaces it. Thus, photosynthetic activity of phytoplankton decreases the level of carbon dioxide in the atmosphere. Each year, phytoplankton take up about a third of the excess carbon dioxide that humans produce by burning fossil fuels.

Unfortunately, phytoplankton populations are now in decline. During the past 20 years, numbers have fallen, especially in northern seas. In some places, they have declined as much as 30 percent. Some scientists fear that these declines are a result of global warming. Explain why—if this is the case—it could lead to a positive feedback cycle that would accelerate global warming. Diagram the potential feedback cycle.

7. Nitrogen-fixing bacteria live throughout the ocean, from its sunlit upper waters to 200 meters (650 feet) beneath its surface. Recall that nitrogen is a limiting factor in many habitats. What effect would an increase in populations of marine nitrogen-fixers have on primary productivity in the waters? What effect would that change have on carbon uptake in those waters?

Figure 42.24 Antarctica's Larsen B ice shelf in (**a**) January and (**b**) March 2002. About 720 billion tons of ice broke from the shelf, forming thousands of icebergs. (**c**) These are the just the tips of the icebergs, and they project 25 meters (82 feet) above the sea surface. About 90 percent of an iceberg's volume is hidden underwater.

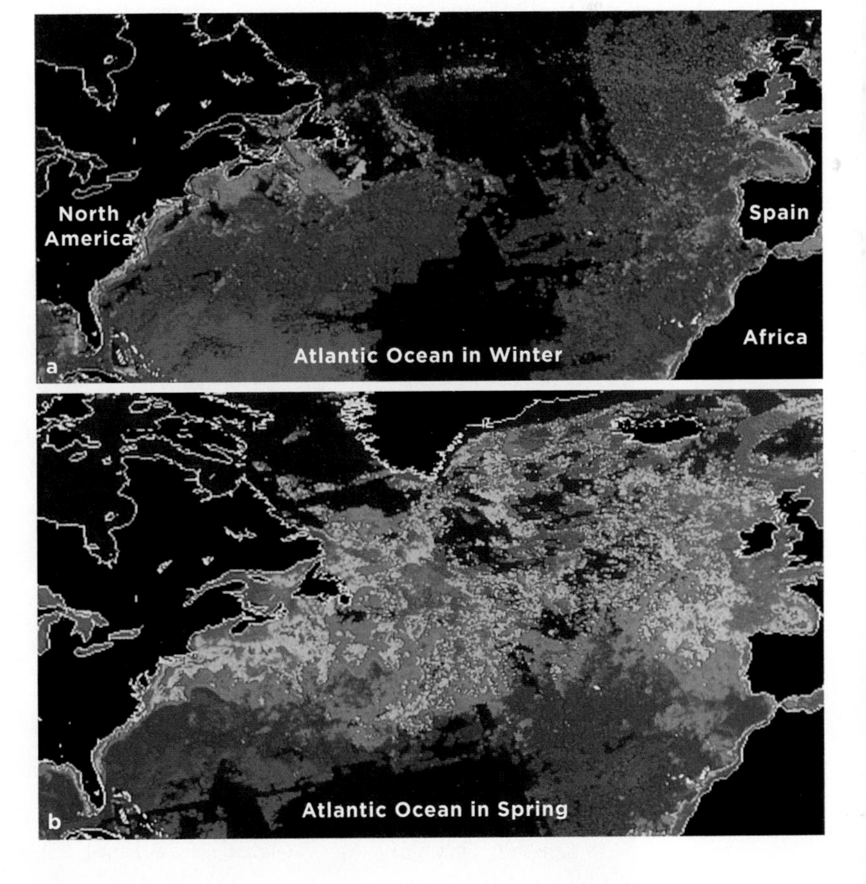

Figure 42.25 Two satellite images that convey the sheer magnitude of photosynthetic activity during springtime in the surface waters of the North Atlantic Ocean. Sensors in equipment launched with the satellite recorded concentrations of chlorophyll, which were greatest in regions coded *red*.

43 THE BIOSPHERE

IMPACTS, ISSUES Surfers, Seals, and the Sea

The winter of 1997–1998 was a fabulous time for surfers on the lookout for big waves and an awful time for seals and sea lions. Ken Bradshaw rode a monster wave (Figure 43.1), and storm surges wiped out half of the population of sea lions on the Galápagos Islands. In California, the number of liveborn northern fur seals plummeted. Many species in widely separate places had become connected by the fury of El Niño, a recurring event that ushers in an often spectacular seesaw in the world climate.

That winter, a massive volume of warm water from the southwestern Pacific moved east. As it piled into coasts from California down through Peru, it displaced currents that otherwise would have brought up nutrients from the deep. Without nutrients to sustain growth, primary producers for marine food webs declined in numbers. The dwindling food base and the warm water drove away consumers. Great populations of anchovies and some other small, cold-water fishes fled elsewhere. Fishes and squids that could not travel long distances starved to death. So did many seals and sea lions, because fishes and squids make up most of what they eat.

Marine mammals did not dominate the headlines, because the 1997–1998 El Niño gave humans plenty of other things to worry about. The El Niño started with

See the video! **Figure 43.1** Ken Bradshaw riding a wave more than 12 meters (29 feet) high, during the most powerful El Niño of the past century. In January 1998, a storm off the Siberian coast generated an ocean swell that, at the time, was the biggest wave known to have hit Hawaii.

How would you vote? We cannot stop an El Niño from happening, but we might be able to minimize its severity. Would you support the use of taxpayer dollars to fund research into the causes and effects of El Niño? See ThomsonNOW for details, then vote online.

changes in the equatorial Pacific ocean but had rippling effects on climates around the world. It battered Pacific coasts with high winds and torrential rains, which caused massive flooding and landslides. An ice storm stretching from New York and New England into central and eastern Canada shut down regional power grids. Three weeks later, 700,000 people still had no electricity. Meanwhile, the global seesaw brought drought-driven crop failures and raging wildfires to Australia and Indonesia.

All told, that one El Niño episode killed more than 2,000 people and drained tens of billions of dollars from economies around the world. Even so, an El Niño is not all bad. It fills reservoirs in the western United States, where water shortages are common. In Florida, Louisiana, and other Gulf states, hurricanes tend to be less severe.

So far in this unit, you have considered species at three levels of organization—populations, communities, and ecosystems. With El Niño, we invite you to move to the next level—to the abiotic factors that influence the distribution of species through the biosphere.

The biosphere includes all places where we find life on Earth. Many organisms live in the *hydrosphere*—the ocean, ice caps, and other bodies of water, liquid and frozen. Others live on and in sediments and soils of the *lithosphere*—Earth's outer, rocky layer. Many lift off into the lower region of the *atmosphere*—gases and airborne particles that envelop Earth. Biogeography, remember, is the study of the distribution and relative abundances of species through all three regions of the biosphere.

Key Concepts

AIR CIRCULATION PATTERNS

Air circulation patterns start with regional differences in energy inputs from the sun, Earth's rotation and orbit, and the distribution of land and seas. These factors give rise to the great weather systems and regional climates. **Sections 43.1, 43.2**

OCEAN CIRCULATION PATTERNS

Interactions among ocean currents, air circulation patterns, and landforms produce regional climates, which affect the distribution and dominant features of ecosystems. **Section 43.3**

LAND PROVINCES

Biogeographic realms are vast regions characterized by species that evolved nowhere else. They are subdivided into biomes, or regions characterized mainly by the dominant vegetation. Sunlight intensity, moisture, soil type, species interactions, and evolutionary history vary within and between biomes. **Sections 43.4–43.10**

WATER PROVINCES

Water provinces cover more than 71 percent of Earth's surface. All freshwater and marine ecosystems have gradients in light availability, temperature, and dissolved gases that vary daily and seasonally. The variations influence primary productivity. **Sections 43.11–43.15**

APPLYING THE CONCEPTS

Understanding interactions among the atmosphere, ocean, and land can lead to discoveries about specific events—in one case, recurring cholera epidemics—that impact human life. **Section 43.16**

Links to Earlier Concepts

With this chapter, you reach the highest level of organization in nature (Section 1.1). You will draw on earlier topics, from acid rain (2.6) to eutrophication (42.11), soils (27.1), the water cycle (42.7), global warming (42.9), primary productivity (42.4), and effects of deforestation (Chapter 21 introduction). You will deepen your knowledge of biogeography (16.1, 41.11). The chapter ends with an example of a scientific approach to problem solving (1.6, 1.7).

43.1 Global Air Circulation Patterns

The biosphere encompasses ecosystems that range from continent-straddling forests to rainwater pools in cup-shaped clusters of leaves. Except for a few ecosystems at hydrothermal vents, climate influences all of them.

CLIMATE AND TEMPERATURE ZONES

LINKS TO SECTIONS 21.4, 40.7

Climate refers to average weather conditions, such as cloud cover, temperature, humidity, and wind speed, over time. Climate is an outcome of global variations in sunlight intensity, the distribution of land masses and seas, and differences in elevation. These factors interact in ways that influence prevailing winds and ocean currents, even the composition of soils.

Different parts of Earth's surface intercept different amounts of sunlight (Figure 43.2). Equatorial regions get more annual sunlight than high latitudes for two reasons. First, fine particles of dust, water vapor, and greenhouse gases absorb specific wavelengths of light or reflect them back into space. They have more of an effect at high latitudes, where rays of light must pass through more atmosphere compared to the equator. Second, the angle of incoming light rays is greater at high latitudes than at the equator, so their energy is spread out over a greater surface area. As a result of these two factors, Earth's surface warms most at the equator and does not warm much at the poles.

The regional differences in surface warming are the start of global air circulation patterns. At the equator, air rises when it warms up. From there, the air flows toward the poles (Figure 43.3*a*). The prevailing winds have their origin in those masses of flowing air.

Earth's rotation and its curvature alter the initial air circulation pattern. Air masses moving north or south are not attached to Earth—which moves fastest at its equator during each spin around its axis. Air masses moving north or south will *seem* to be deflected east or west as Earth spins beneath them. This is the basis of prevailing east and west winds (Figure 43.3*b*).

Also, regional winds arise where land masses cause differences in air pressure. Land absorbs and releases heat faster than water does, so air rises and falls faster over land than it does over the ocean. Air pressure is lowest where air rises and greatest where air sinks.

Rainfall corresponds with air circulation patterns. Warm air holds more moisture than cooler air. As air

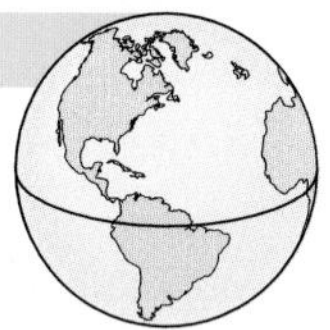

Figure 43.2 Variation in intensity of solar radiation with latitude. More solar energy arrives at the equator than at the poles, and that which does arrive is more concentrated; it falls on a smaller area of the surface.

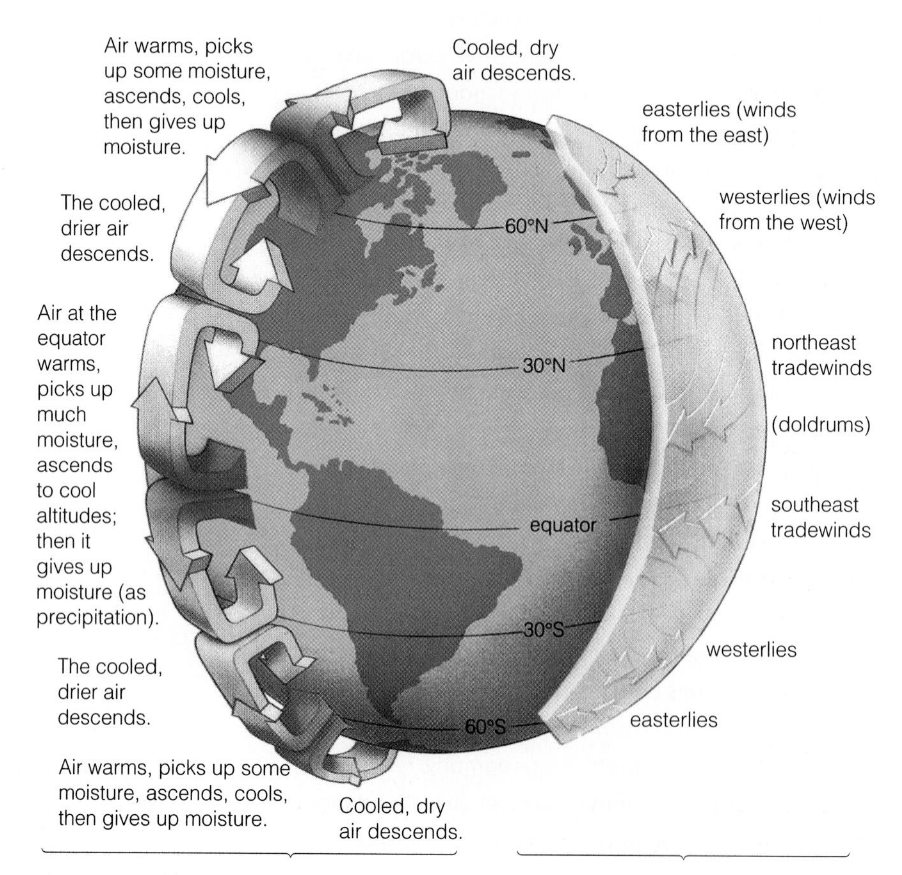

Figure 43.3 Animated! Global air circulation patterns and their effects on climate. (**a**) Latitudinal differences in the amount of solar heating sends air flowing to the north and south, away from the equator. (**b**) As Earth rotates beneath the moving air masses, its curvature deflects the north–south pattern, so prevailing east and west winds arise.

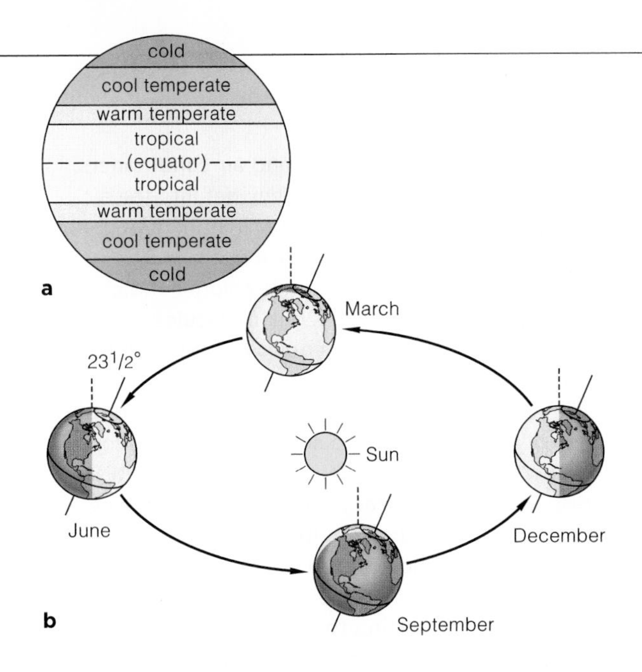

Figure 43.4 Animated! (**a**) World temperature zones. (**b**) Incoming solar radiation varies annually. The northern end of Earth's fixed axis tilts toward the sun in June and away from it in December, changing the equator's position relative to the day–night boundary. Variations in sunlight intensity and daylength cause seasonal temperature shifts.

Figure 43.5 (**a**) Arrays of electricity-producing photovoltaic cells in panels that collect solar energy. (**b**) A field of turbines harvesting wind energy.

warms near the equator, it picks up moisture from the seas, rises to cooler altitudes, and releases moisture as rain that supports lush tropical forests. The now-drier air flows from the equator and descends at latitudes of about 30°, where deserts often form. Farther north and south, air picks up moisture. It ascends, and then releases moisture at latitudes of about 60°. Frigid air descends in the polar regions. It holds little moisture, precipitation is sparse, and polar deserts form.

The effects of latitudinal variations in solar heating and modified patterns of air circulation form the basis for the world's major temperature zones. Figure 43.4*a* is a simple chart of these zones.

Seasonal changes in daylength and light intensity arise because Earth's axis tilts relative to the sun. The tilt mostly affects polar and temperate climate zones. In summer, the Northern Hemisphere tilts toward the sun, so rays arrive at less of an angle. In winter, it tilts away from the sun. Rays arrive at more of an angle, so the light is less intense (Figure 43.4*b*).

HARNESSING THE SUN AND WIND

Paralleling the human population's J-shaped growth curve is a steep rise in its energy consumption. Fossil fuels, including coal and gasoline, are nonrenewable (Section 21.4). Solar and wind energy are renewable. Annual incoming solar energy exceeds energy stored in all known fossil fuel reserves by about 10 times.

Photovoltaic cells convert light energy to electricity, which can be used to split water into gaseous oxygen and hydrogen (Figure 43.5*a*). The many proponents of solar–hydrogen energy argue that it could end smog, oil spills, and acid rain without the threats of nuclear power. Hydrogen gas can fuel cars and heat buildings. However, it is a small molecule that leaks easily from pipelines or containers. Increased hydrogen in the air might accelerate global warming.

We already harness the winds generated by solar energy (Figure 43.5*b*). Wind farms are practical where winds blow faster than 8 meters per second (17 miles per hour). Winds seldom blow constantly, but wind energy can charge batteries that supply power even on still days. It is estimated that the winds of North and South Dakota alone could supply 80 percent of all energy needed in the continental United States.

Wind farms have some drawbacks. Turbine blades chop up wayward birds and bats. Large facilities may alter local weather patterns. Also, to some people, the wind farms are a form of "visual pollution" that ruins otherwise scenic views and lowers property values.

Major temperature zones and climates start with global patterns of circulation, which arise through interacting factors: latitudinal variations in incoming solar radiation, Earth's rotation and annual path around the sun, and the distribution of land masses and seas.

43.2 Circulating Airborne Pollutants

By their air-polluting activities, human populations interact with global air circulation patterns in ways that have unintended consequences.

LINKS TO SECTIONS 2.6, 13.5, 18.3, 42.8

A pollutant is a natural or synthetic substance released into soil, air, or water in greater than normal amounts; it disrupts natural processes because organisms evolved in its absence, or are adapted to lower levels. Today, air pollution threatens biodiversity and human health.

A Fence of Wind and Ozone Thinning High in the atmosphere, ozone (O_3) and molecular oxygen (O_2) absorb most of the ultraviolet (UV) radiation streaming in with the sun's rays. Between 17 and 27 kilometers above sea level, the ozone concentration is so great that we call it the ozone layer (Figure 43.6*a*).

In the mid-1970s, scientists started to notice that the ozone layer was getting thinner. Its thickness had always varied a bit with the season, but now there was steady decline from year to year. By the mid-1980s, the spring ozone thinning over Antarctica was so pronounced that people were calling it an "ozone hole" (Figure 43.6*b*).

Declining ozone became an international concern. With a thinner ozone layer, people would be exposed to more UV radiation and get more skin cancers (Section 13.5). Higher UV levels also harm plants and wildlife, which do not have the option of rubbing on more sunscreen. Higher UV levels might even slow rates of photosynthesis and the release of oxygen into the atmosphere.

The chlorofluorocarbons, or CFCs, are the main ozone destroyers. These odorless gases were widely used as propellants in aerosol cans, as coolants in refrigerators and air conditioners, and in solvents and plastic foam. They get into the air, then interact with ice crystals and UV light high in the stratosphere. These reactions release free chlorine radicals that degrade ozone. A single chlorine radical can break apart thousands of ozone molecules.

Ozone thins the most at the poles because swirling winds concentrate CFCs over them during dark, cold polar winters. In the spring, increasing daylight and the presence of ice clouds allow a surge in the formation of chlorine radicals from the highly concentrated CFCs.

In response to the threat posed by ozone thinning, developed countries agreed in 1992 to phase out the production of CFCs and other ozone destroyers. As you learned from Figure 42.18*b*, the CFC concentrations in the atmosphere are starting to decline. However, they are expected to stay high enough to have significant impact on the ozone layer for the next twenty years.

No Wind, Lots of Pollutants, and Smog Often, weather conditions cause a thermal inversion: A layer of cool, dense air becomes trapped under a warm air layer. The stage is set for smog. In this atmospheric condition, air pollutants have accumulated to high concentrations beneath a thermal inversion layer; winds cannot disperse them (Figure 43.7). Thermal inversions have contributed to some of the worst air pollution disasters.

Industrial smog forms as a gray haze over cities that burn a lot of coal and other fossil fuels during cold, wet winters. Burning releases smoke, soot, ashes, asbestos, oil, particles of lead and other heavy metals, and sulfur oxides. Most industrial smog now forms in China, India, and eastern Europe.

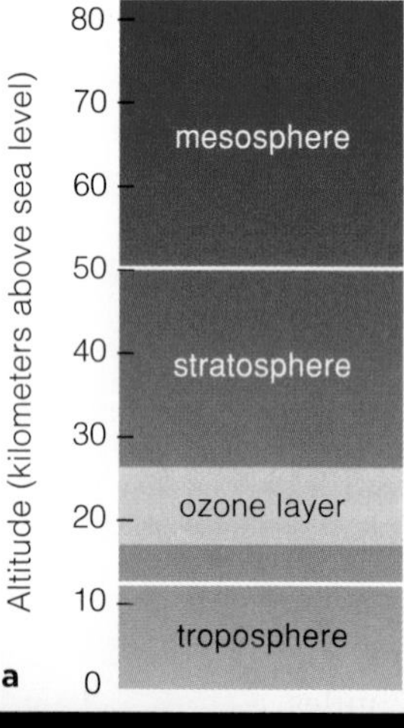

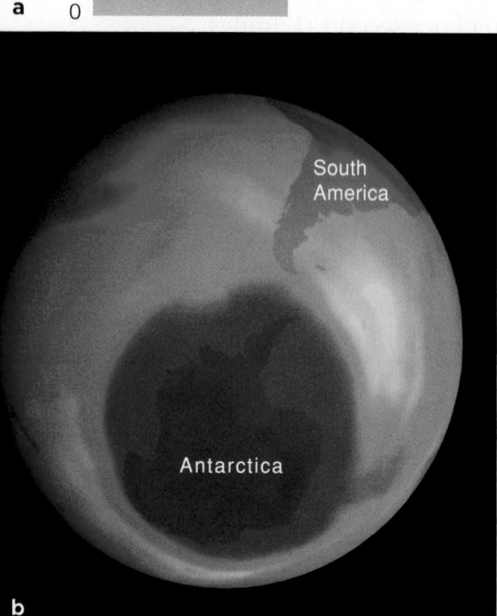

Figure 43.6 Animated! (**a**) The atmospheric layers. Ozone concentrated in the stratosphere helps shield life from UV radiation. (**b**) Seasonal ozone thinning above Antarctica in 2001. *Dark blue* represents the low ozone concentration, at the ozone hole's center.

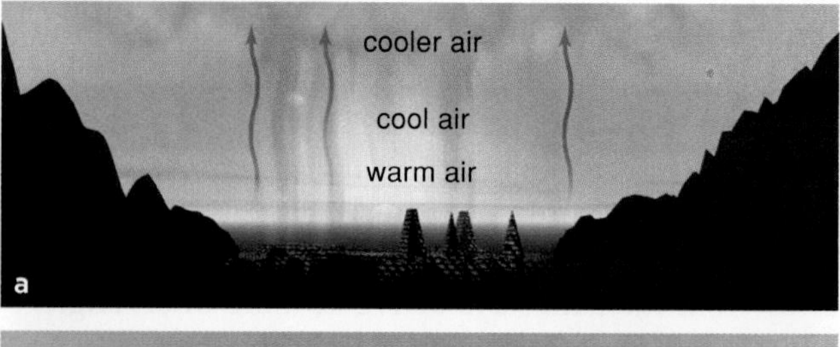

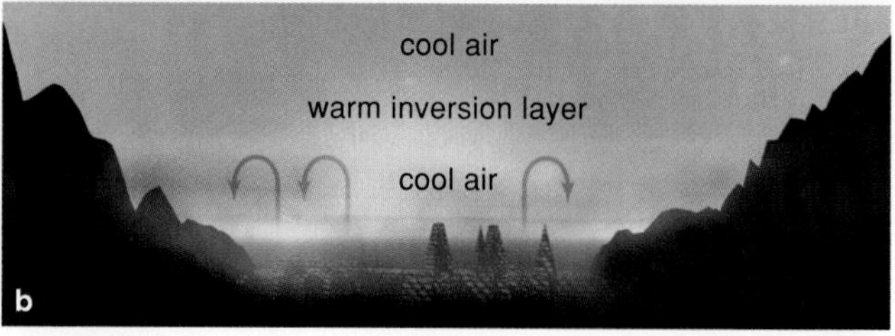

Figure 43.7 (**a**) Normal air circulation pattern in smog-forming regions. (**b**) Air pollutants trapped under a thermal inversion layer.

Photochemical smog is a brown haze that forms above big cities in warm climate zones. Photochemical smog is most dense when the city is in a natural topographic basin. Los Angeles and Mexico City are two classic cases. Exhaust fumes from vehicles contain nitric oxide, a major pollutant that forms nitrogen dioxide by combining with oxygen. When sunlight strikes it, nitrogen dioxide reacts with hydrocarbon gases in the air, forming photochemical oxidants. Most hydrocarbon gases are released from spilled or partly burned gasoline.

Winds and Acid Rain Coal-burning power plants, smelters, and factories emit sulfur dioxides. Vehicles, power plants that burn gas and oil, and nitrogen-rich fertilizers emit nitrogen oxides. In dry weather, airborne oxides fall as dry acid deposition. In moist air, they form nitric acid vapor, sulfuric acid droplets, and sulfate and nitrate salts. Winds typically disperse these pollutants far from their source. They fall to Earth in rain and snow. We call this wet acid deposition, or acid rain.

The pH of typical rainwater is about 5 (Section 2.6). Acid rain can be 10 to 100 times more acidic; it can be as potent as lemon juice! It corrodes metals, marble, rubber, plastics, nylon stockings, and other materials. It alters soil pH and can kill trees (Section 42.9) and other organisms.

Rain in much of eastern North America is thirty to forty times more acidic than it was even a few decades ago (Figure 43.8*a*). Heightened acidity has caused fish populations to vanish from more than 200 lakes in the Adirondack Mountains of New York (Figure 43.8*b*). It also is contributing to the decline of forests. Acid rain harms trees and symbiotic fungi that help them grow.

Windborne Particles and Health Pollen and other natural particles are carried aloft by the winds, along with pollutant particles of many sizes (Figure 43.9). Inhaling particles irritates nasal passages, the throat, and airways. It triggers asthma attacks and can increase their severity. The smallest particles are most likely to reach the lungs, where they can interfere with respiratory function. The resulting struggle for breath raises the risk of heart attacks and stroke, especially among the elderly.

Exhaust from vehicles is a major source of particulate pollution. Diesel-fueled engines are the worst offenders because they emit more of the smallest, most dangerous particles than their gasoline-fueled counterparts. Most trains, buses, boats, tractors, and construction vehicles use diesel fuel because it is less expensive than gasoline and provides better mileage per gallon.

Regardless of their source, air pollutants travel on the winds across continents and the open ocean. As you read this, pollutants are blowing into Oregon and California from diesel-powered trucks and coal-fired power plants in the rapidly expanding industrialized regions in China. *Winds do not stop at national boundaries.*

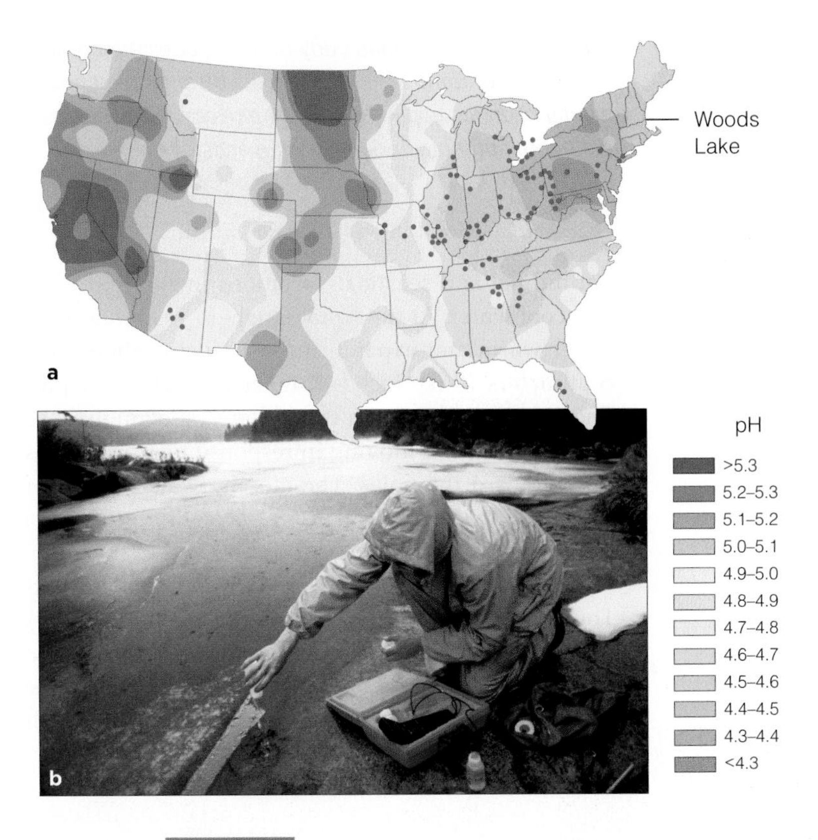

Figure 43.8 Animated! (**a**) The average 1998 precipitation acidities in the United States. *Red* dots mark large coal-burning power and industrial plants. (**b**) Biologist measuring the pH of New York's Woods Lake. In 1979, the lake water's pH was 4.8. The experimental addition of calcite to soil around the lake raised the pH to more than 6.

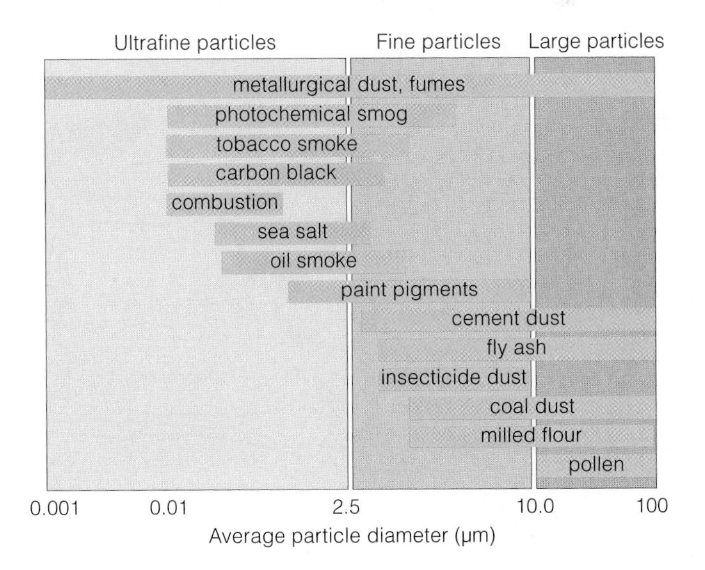

Figure 43.9 Suspended particulate matter. These solids and liquid droplets are small enough to stay aloft for variable intervals.

Ultrafine particles contribute to respiratory disorders. Carbon black is a powdered form of carbon used in paints, tires, and other goods. About 6 million tons are manufactured annually. Coal burning releases about 45 million tons of fly ash annually.

43.3 The Ocean, Landforms, and Climates

The ocean, a continuous body of water, covers more than 71 percent of Earth's surface. Driven by solar heat and wind friction, its upper 10 percent moves in currents that distribute nutrients through marine ecosystems.

OCEAN CURRENTS AND THEIR EFFECTS

LINKS TO SECTIONS 2.5, 16.6

Latitudinal and seasonal variations in sunlight warm and cool water. At the equator, where vast volumes of water warm and expand, the sea level is about eight centimeters (three inches) higher than at either pole. The volume of water in this "slope" is enough to get sea surface water moving in response to gravity, most often toward the poles. The moving water warms air parcels above it. At midlatitudes, it transfers *10 million billion* calories of heat energy per second to the air!

Enormous volumes of water flow as ocean currents. The directional movement of these currents is shaped by the pushing force of major winds, Earth's rotation, and topography. Surface currents circulate clockwise in the Northern Hemisphere and counterclockwise in the Southern Hemisphere (Figure 43.10).

Swift, deep, and narrow currents of nutrient-poor water flow away from the equator along the east coast of continents. Along the east coast of North America, warm water flows north, as the Gulf Stream. Slower, shallower, broader currents of cold water parallel the west coast of continents and flow toward the equator.

Ocean currents affect climate zones. Why are Pacific Northwest coasts cool and foggy in summer? The cold California Current is giving up moisture; warm winds at the coast are transferring heat to it. Why are Boston and Baltimore muggy in summer? Air masses pick up heat and moisture from the warm Gulf Stream, then southerly and easterly winds put them over the cities. Why are the winters milder in London and Edinburgh than they are in Ontario and central Canada—which are at the same latitude? Warm water from the Gulf Stream flows into the North Atlantic Current (Figure 43.10). That current gives up heat to prevailing winds, which warm northwestern Europe.

Ocean circulation patterns shift over geologic time as land masses move (Section 16.6). Some worry that global warming could also alter the patterns.

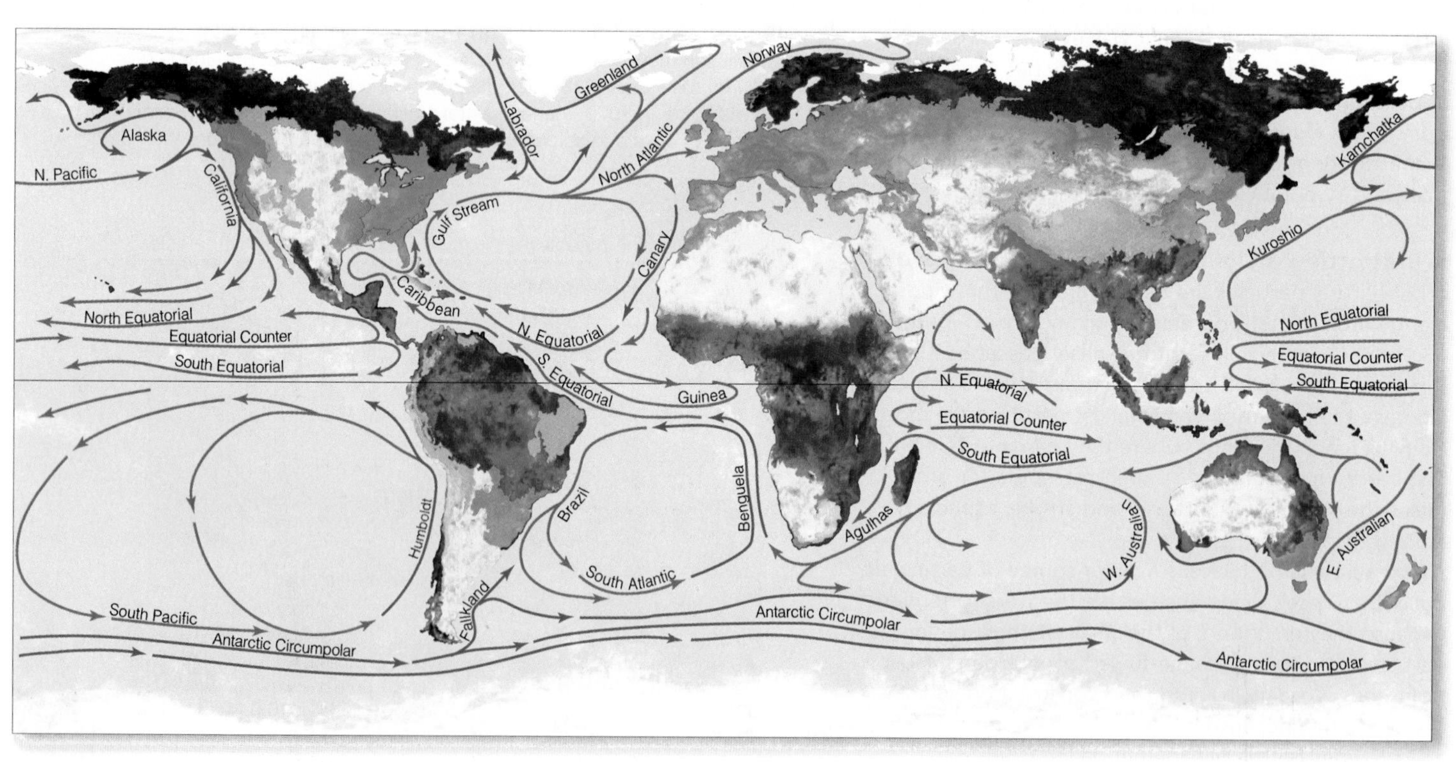

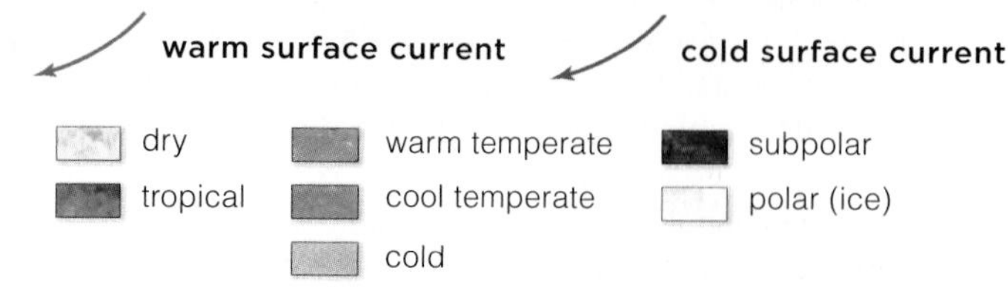

Figure 43.10 **Animated!** Major climate zones correlated with surface currents and surface drifts of the world ocean. Warm surface currents start moving from the equator toward the poles, but prevailing winds, Earth's rotation, gravity, the shape of ocean basins, and landforms influence the direction of flow. Water temperatures, which differ with latitude and depth, contribute to regional differences in air temperature and rainfall.

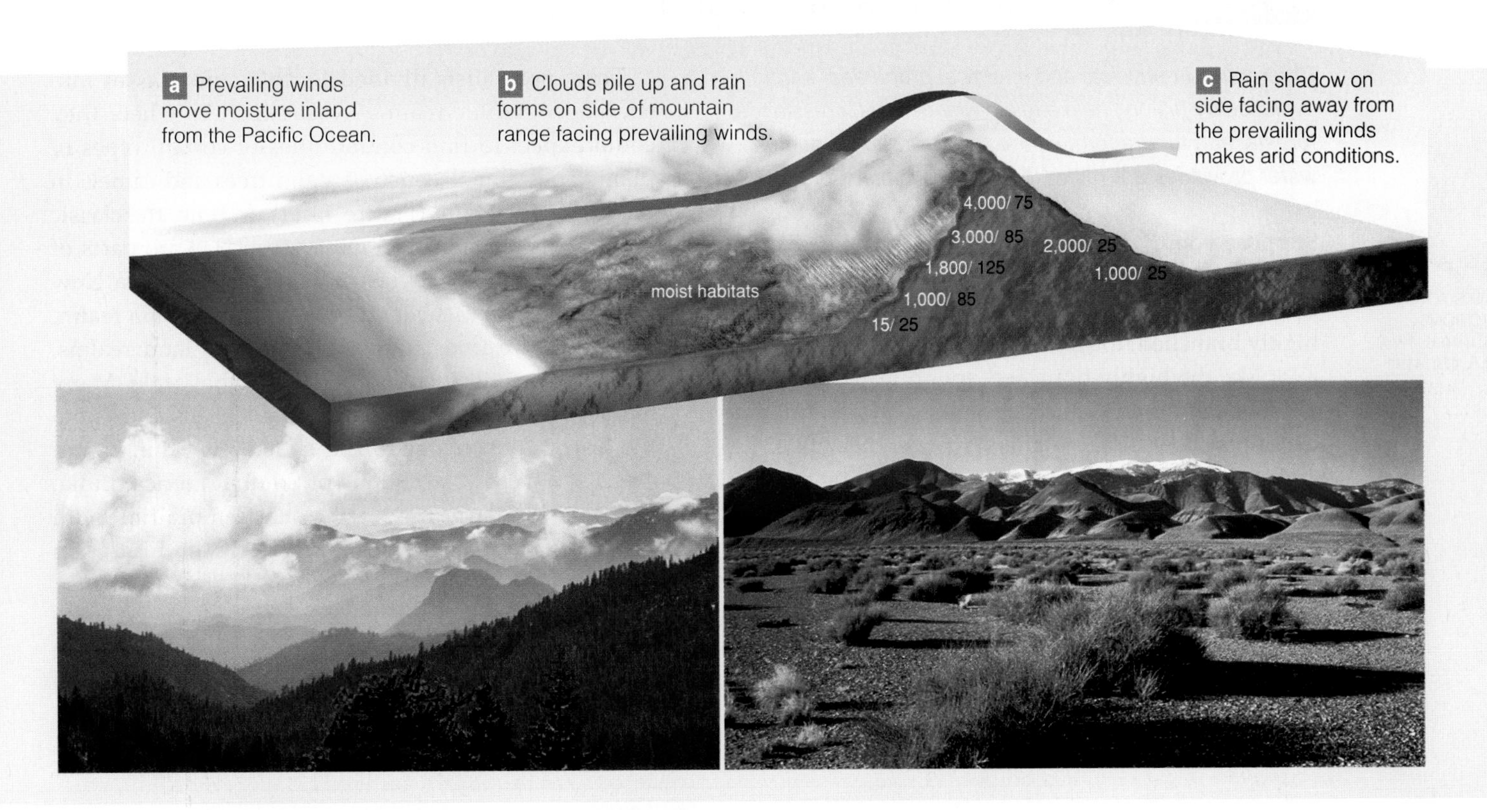

Figure 43.11 Animated! Rain shadow effect. On the side of mountains facing away from prevailing winds, rainfall is light. *Black* numbers signify annual precipitation, in centimeters, averaged on both sides of the Sierra Nevada, a mountain range. *White* numbers signify elevations, in meters.

RAIN SHADOWS AND MONSOONS

Mountains, valleys, and other surface features of the land affect the climate. Suppose you track a warm air mass after it picks up moisture off California's coast. It moves inland, as wind from the west, and piles up against the Sierra Nevada. This high mountain range parallels the distant coast. The air cools as it rises in altitude and loses moisture as rain (Figure 43.11). The result is a rain shadow—a semiarid or arid region of sparse rainfall on the leeward side of high mountains. *Leeward* is the side facing away from the wind. The Himalayas, Andes, Rockies, and other great mountain ranges cause vast rain shadows on continents.

The belts of vegetation at different elevations reflect differences in temperature and moisture. Grasslands form at the western base of the Sierra Nevada. Higher up, in the cooler air, deciduous and evergreen species dominate. Higher still are a few species of evergreens adapted to the cold habitat. Above the subalpine belt, only hardy, dwarfed plants withstand the even more extreme temperatures.

Air circulation patterns called monsoons affect the continents north or south of warm seas. For example, in Asia, the continental interior heats intensely during summer, so vast, low-pressure air parcels form above it. Low pressure draws in moisture from the ocean to the south. The resulting monsoon blows toward the north, bringing intense rains that often cause flooding. During winter, the continental interior is cooler than the ocean. As a result, dry winds blow from the north toward southern coasts, causing a seasonal drought.

Recurring coastal breezes are like mini-monsoons. Water and coastal land have different heat capacities. In the daytime, water does not warm as fast as land. Air warmed over the land rises, and cool offshore air moves in. After sundown, land loses heat faster than water, so breezes reverse and blow from land to sea.

Surface ocean currents influence regional climates and help distribute nutrients in marine ecosystems.

Air circulation patterns, ocean currents, and landforms interact in ways that influence regional temperatures and moisture levels. Thus they also influence the distribution and dominant features of ecosystems.

43.4 Biogeographic Realms and Biomes

Differences in physical and chemical properties, and in evolutionary history, help explain why deserts, grasslands, forests, and tundra form where they do, and why some water provinces are richer than others in biodiversity.

LINKS TO SECTIONS 16.1, 16.6, 17.1, 40.5, 41.1, 41.11

Suppose you live in the coastal hills of California and decide to tour the Mediterranean coast, the southern tip of Africa, and central Chile. In each region, you see highly branched, tough-leafed woody plants that look a lot like the highly branched, tough-leafed chaparral plants back home. Vast geographic and evolutionary distances separate the plants. Why are they alike?

You decide to compare their locations on a global map. You discover that American and African desert plants live about the same distance from the equator. Chaparral plants and their distant look-alikes all grow along the western and southern coasts of continents between latitudes 30° and 40°. You have noticed one of many patterns in the global distribution of species.

Early naturalists divided Earth's land masses into six biogeographic realms—vast expanses where they could expect to find communities of certain types of plants and animals, such as palm trees and camels in the Ethiopian realm (Figure 43.12). In time, the classic realms became subdivided, as when Hawaii, parts of Indonesia, Japan, Polynesia, Micronesia, Papua New Guinea, and New Zealand became the Oceania realm.

Biomes are finer subdivisions of the land realms, but they are still identifiable on a global scale. Many biomes occur on several continents. For instance, note the distribution of dry forest (coded *orange*) in Figure 43.12. It covers vast regions of South America, India, and Asia. Similarly, the North American prairie, South American pampa, southern Africa veld, and Eurasian steepe are all temperate grasslands (Figure 43.13).

The distribution of biomes is influenced by climate (especially temperature and patterns of rainfall), soil type, and interactions among the array of species that

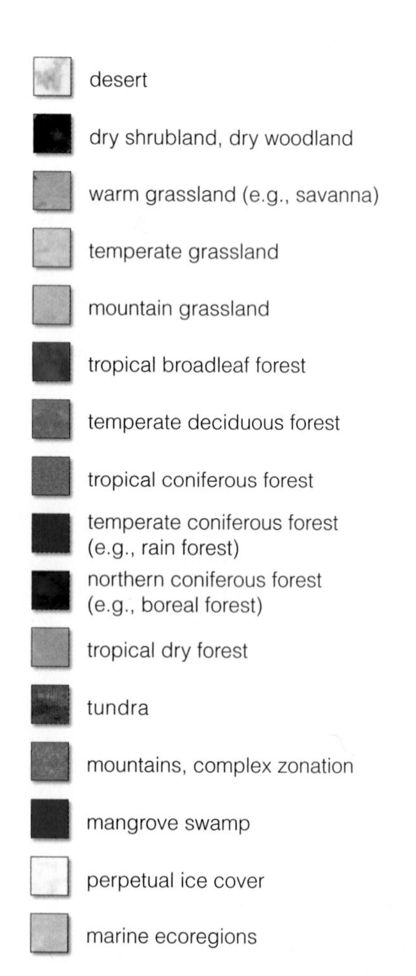

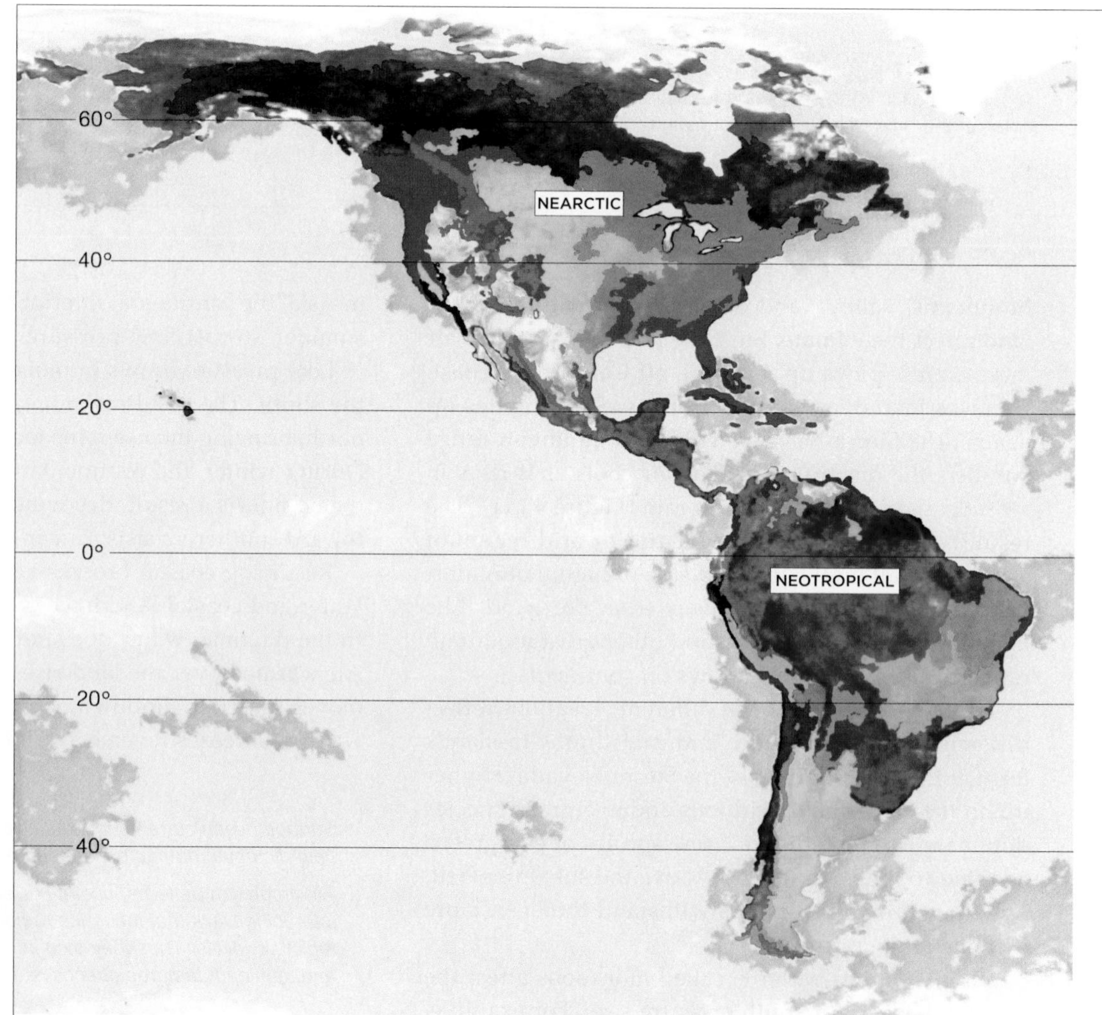

Figure 43.12 **Animated!** Global distribution of major categories of biomes and marine ecoregions.

make up their communities. Consumers are adapted to the dominant vegetation, as when pigeons that live in trees of a tropical rain forest partition the available food between the tree canopy and forest floor (Section 41.1). Each species, remember, shows adaptations in its form, function, behavior, and life history pattern.

The distribution of biomes has also been influenced by evolutionary history, as when species were rafted on different land masses after the breakup of Pangea (Section 16.6). Similarly, environmental conditions and evolutionary history helped shape the distribution of species in the seas. Figure 43.12 shows the key marine ecoregions as well as Earth's biomes.

Biomes are vast expanses of land dominated by distinct kinds of communities. Their global distribution is an outcome of topography, climate, species interactions, and evolutionary history. Marine ecoregions are realms of biodiversity in the seas.

Figure 43.13 Two examples of temperate grassland biome. *Above*, Argentine pampa. *Below*, Mongolian steppe. See also Figure 43.20.

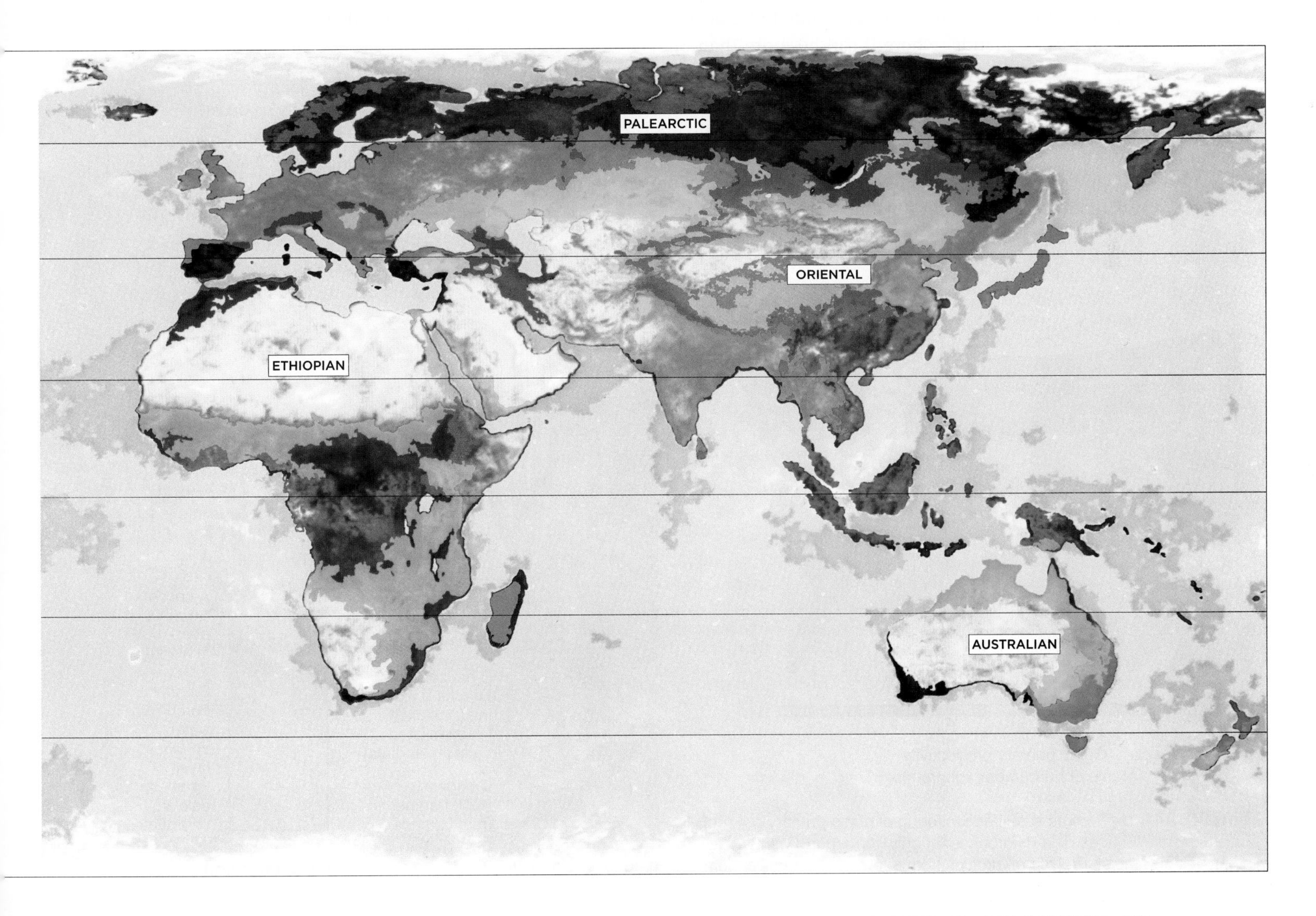

43.5 Availability of Sunlight, Soils, and Moisture

Biomes differ in their soils, in daily and seasonal supplies of water and other resources, and therefore in their primary productivity.

LINKS TO SECTIONS 27.1, 40.2

On land, plant growth depends on the availability of sunlight and on the water and dissolved mineral ions in soil. Soils, recall, are mixtures of mineral particles and varying amounts of decomposing organic matter called humus (Section 27.1). Weathering converts big rocks to coarse-grained gravel, then to sand, silt, and fine-grained clay. Water and air fill the spaces between soil particles. The types, proportions, and compaction of particles differ within and between biomes. Clay is richest in minerals, but its fine, close-packed particles drain poorly. It also does not have enough air spaces for roots to take up oxygen. In gravelly or sandy soils, leaching draws off water and mineral ions.

Each biome has a soil profile, a layered structure that develops over time (Figures 27.2 and 43.14). The top layers are leaf litter and topsoil, which is rich in humus but easily eroded. Topsoil tends to be deepest in natural grasslands, where it can be more than one meter thick. For this reason, grasslands are the prime choice for conversion to agriculture. In tropical forests, decomposition is fast and minimal topsoil accumulates above poorly draining lower layers. After such forests are cleared for agriculture, heavy rains quickly leach nutrients from their thin topsoil. Within a short time, crops cannot grow well unless nutrients are supplied by added fertilizers.

Reflect on Figure 43.14*a*. Where soil holds plenty of water, competition for sunlight is intense. That is why sun-drenched deciduous forests have a strong vertical structure. Dense stands of trees form a canopy high above the ground. Shorter, understory trees and shrubs compete for light filtering through the canopy.

Where desert biomes have formed, you seldom find competition for sunlight. Water is the major limiting factor. The wide spacing between plants is evidence of competition; the root systems of established plants do not leave much water for new ones (Section 40.2).

Some parts of desert biomes are monotonously low in biodiversity. Figure 43.15*a* shows part of hundreds of square kilometers of the Sonoran Desert in Arizona, where you see little more than widely spaced creosote bushes. Add some moisture and a higher elevation to the mix, and you see more diversity, as in this desert's wetter uplands (Figure 43.15*b*). Figure 43.16 is a brief glimpse into how desert soils form.

A biome's soil profile and its proportions of sand, silt, clay, gravel, and humus influence its primary productivity.

Even within the boundaries of the same biome, differences in water availability and temperature can bring about big differences in species richness.

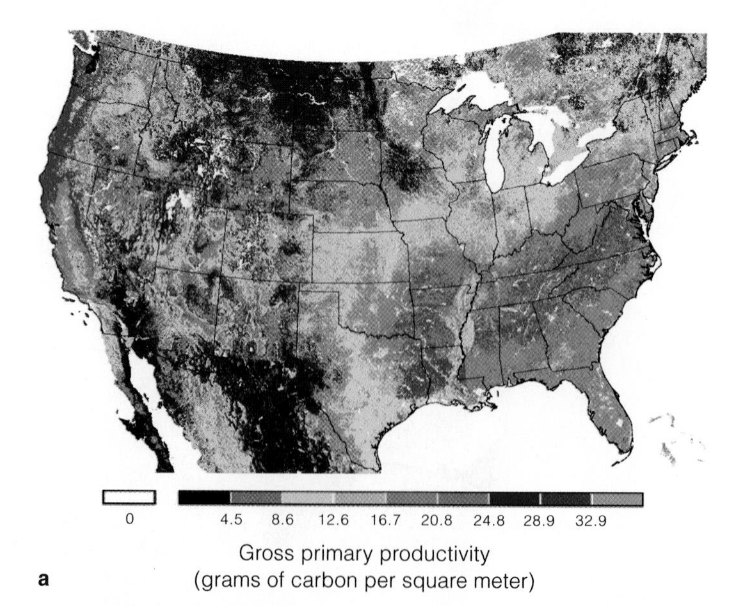

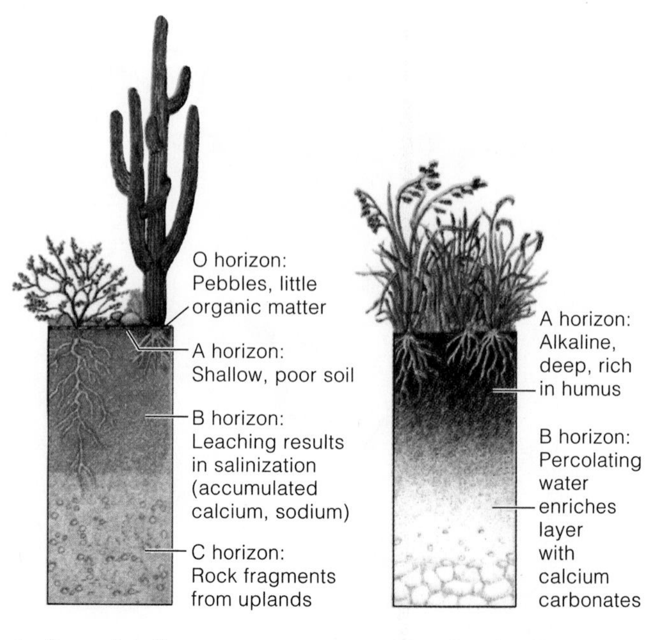

Figure 43.14 (**a**) Remote satellite monitoring of gross primary productivity across the United States. The differences roughly correspond with variations in soil types and moisture. (**b**) Soil profiles from a few representative biomes.

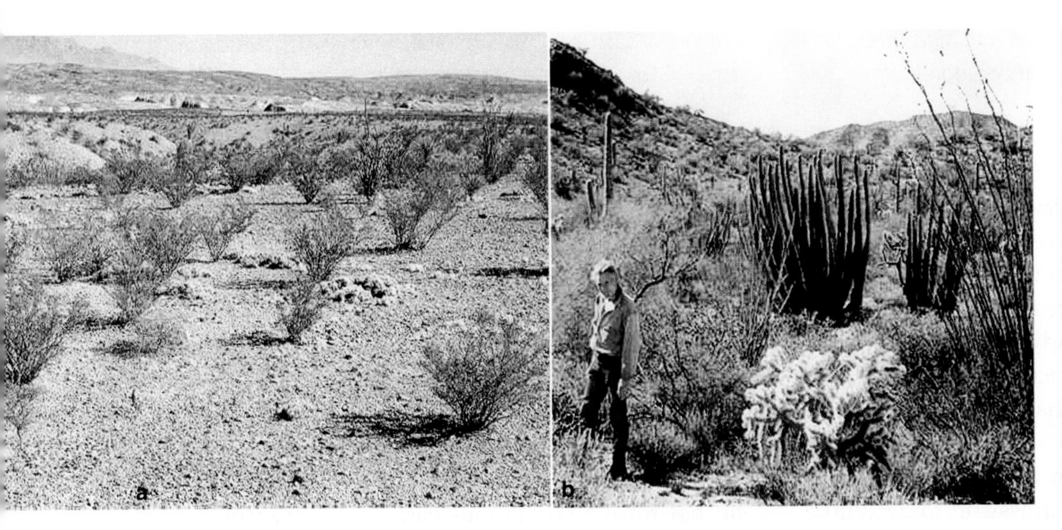

Figure 43.15 Two views of the same biome—the Sonoran Desert in Arizona. The sun's rays are just as intense in the desert lowlands (**a**) as in the uplands (**b**), but differences in water availability, temperature, and soil types influence plant growth.

Compared with wetter biomes, deserts are low in primary productivity. All desert soils are mineral-rich but are in an early stage of development. Particles in the upper horizons are loose and unstabilized. Little organic matter builds up on the surface, so decomposers cannot form much humus. Desert soils cannot hold onto water. Also, evaporative water loss is high, so dissolved salts rise to the soil surface. Extensive saltpans form and support little to no plant growth.

The uplands rarely frost over. They get heavy pulses of summer rain and prolonged, light rain in winter. Annuals and perennials that bloom in summer and winter are not as water-stressed as on hot, dry, flat lowlands. In lowlands, temperatures are high, rainfall is sparse, and widely spaced creosote bush (*Larrea*) and sometimes other woody plants dominate.

Figure 43.16 How desert soils start. Over millions of years, rain, wind, freezes and thaws, and chemical processes erode mountains. Runoff moves loose rocks and particles to the lowlands, where gravel, sand, and silt form. Lichens slowly enrich the soil (Section 22.5). Where a thin humus layer forms, soil stabilizes and holds water. Dune buggies racing across deserts destroy lichen crusts and allow nutrients to leach away.

The photograph shows debris in a gully at the base of mountains in the Chihuahuan Desert in Texas. The gully channels water draining from the slopes, and debris has stabilized enough to support woody shrubs and herbaceous plants. Plants are sparse on the gully's right slope. They are less water-stressed on the left side, which is shaded during the hottest time of day.

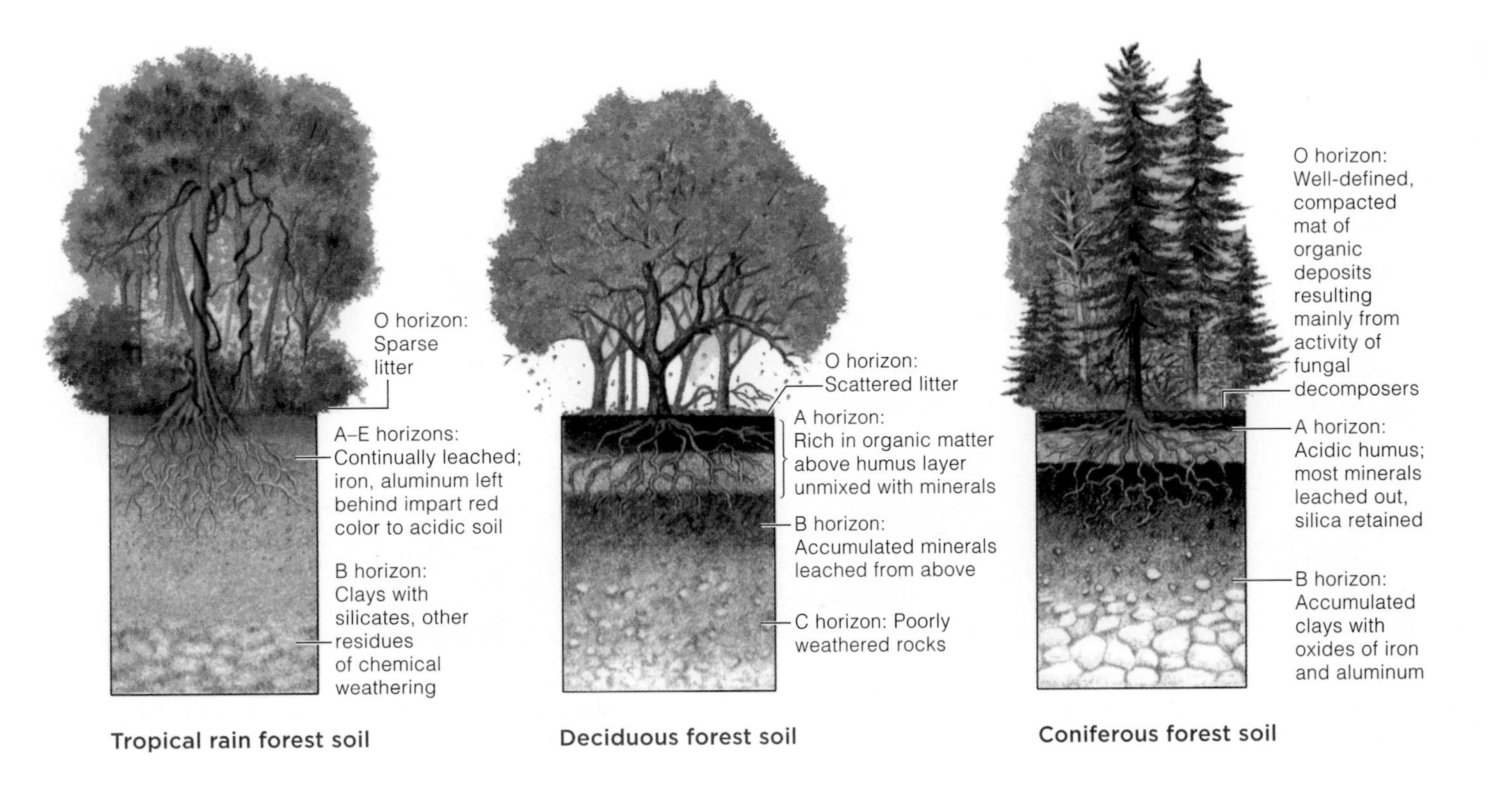

43.6 Moisture-Challenged Biomes

At certain latitudes with sparse rainfall and high rates of evaporation, drought-tolerant plants predominate.

DESERTS

LINKS TO SECTIONS 27.1, 42.1

Deserts tend to form near latitudes 30° north and south. Annual rainfall is less than 10 centimeters (4 inches). Humidity is low. Brief, infrequent rains erode topsoil. The ground heats fast during the day, then cools fast at night. The arid or semiarid conditions do not support big leafy plants, but species richness is high where moisture is available in more than one season. Figure 43.17 shows an example.

Climate changes and human activities can lead to desertification, a conversion of grassland to desertlike conditions. Shortsighted agricultural practices such as overgrazing encourage desertification. Drought speeds the process. Over the past fifty years, 9 million square kilometers (3.5 million square miles) of grassland have been converted to desert.

Desertification has far-reaching ecological impacts. Each year, winds blow hundreds of millions of tons of dust from northwest Africa across the Atlantic (Figure 43.18). Toxins and pathogens adhere to dust particles; they end up in Caribbean seas and may harm corals.

DRY SHRUBLANDS, WOODLANDS, AND GRASSLANDS

Dry shrublands receive less than 25 to 60 centimeters (10–24 inches) of rain annually. They dominate parts of the Mediterranean, South Africa, and California (Figure 43.19). Rains occur seasonally. Lightning starts fires during the dry season. Plants that resprout from their roots after a fire sweeps through have a selective advantage. Seeds of some chaparral species germinate only after a fire, when young seedlings will face little competition from other species.

Dry woodlands prevail where annual rainfall is 40 to 100 centimeters (16–40 inches). Trees are often tall, but do not form a continuous canopy. Eucalyptus forests of Australia and oak forests of California and Oregon are examples.

Grasslands form in the interior of continents in the regions between deserts and temperate forests (Figure 43.20). The summers are warm, and the winters cold. Annual rainfall of 25 to 100 centimeters (10–40 inches) keeps desert from forming, but is too little to support forests. The primary producers tolerate strong winds, sparse and infrequent rain, and intervals of drought. Dominant animals are grazers and burrowers species. Their activities, combined with infrequent fires, keep trees and shrubs from taking over.

Savannas form between the tropical forests and hot deserts of Africa, South America, and Australia. They are broad belts of grasslands with a few shrubs and

Figure 43.17 From the Mojave Desert of California, cactus (cholla), golden poppies, and—in the distance—tall saguaro cacti. The inset shows flowers of a claret cup cactus.

Figure 43.18 **Animated!** Dust blowing from the Sahara Desert across the Atlantic Ocean. Toxins and pathogens travel on the surface of the dust particles.

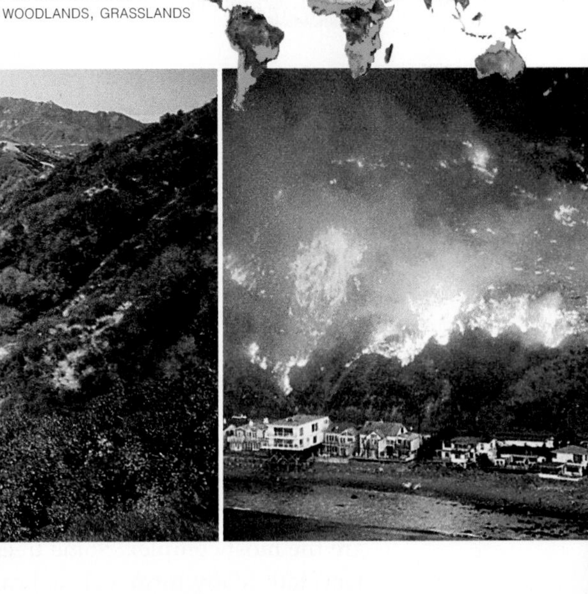

Figure 43.19 Two views of California chaparral, by far the state's most extensive natural community. Dominant plants are mostly branching, woody, and evergreen with leathery leaves. Most stand less than two meters tall and have adaptations that allow them to survive occasional fires. *Far right*, a firestorm racing through chaparral-covered hills above Malibu. Today, most fires that occur in this biome are caused by humans, not storms, and they are more frequent.

Figure 43.20 Three examples of grasslands. (**a**) African savanna. During dry seasons, water is scarce. Lakes and streams dry up, and the few remaining ponds beckon herbivores, including this big herd of wildebeest. The water and the herds are magnets for predators.

(**b**) Bison in a shortgrass prairie in South Dakota.

(**c**) Rare patch of natural tallgrass prairie in eastern Kansas. Refer also to Figure 42.3, which shows some of the organisms of this prairie.

trees (Figure 43.20*a*). Annual rainfall averages 90–150 centimeters (35–60 inches), and pronounced droughts are seasonal. Savannas grade into tropical woodlands.

Shortgrass and tallgrass prairie once covered much of the North American Great Plains (Figure 43.20*b*,*c*). Most was converted to cropland. In the 1930s, drought and poor agricultural practices turned the Great Plains into the Dust Bowl (Figure 27.3*b*). The about 2 percent of prairies that escaped the plow is now protected, and other patches are being restored.

Deserts, dry shrublands, dry woodlands, and grasslands form in regions of sparse rainfall, dry air, strong winds, and recurring episodes of drought and fire.

43.7 More Rain, Broadleaf Forests

In forest biomes, tall trees grow close together and form a fairly continuous canopy over a broad region. Rainfall and distance from the equator influence which trees dominate.

TROPICAL RAIN FORESTS

LINKS TO SECTIONS 6.1, 27.1

Evergreen broadleaf forests form between latitudes 10° north and south in equatorial Africa, the East Indies, Malaysia, Southeast Asia, South America, and Central America. Rainfall averages 130 to 200 centimeters (50 to 89 inches) each year. Regular rains, combined with an average temperature of 25°C, and high humidity support tropical rain forests of the sort shown on the facing page. In structure and diversity, these biomes are the most complex. Some trees are 30.5 meters (100 feet) tall. Many form a closed canopy that stops most sunlight from getting to the forest floor. Epiphytes, and vines abound. Epiphytes are plants that grow on another plant, but do not withdraw nutrients from it.

Decomposition and mineral cycling happen fast in these forests, so litter does not accumulate. Soils are highly weathered, heavily leached, and are very poor nutrient reservoirs. You read about the consequences of poor soils in Section 27.1.

SEMI-EVERGREEN AND DECIDUOUS BROADLEAF FORESTS

Away from the equatorial forests, between latitudes 10° and 25°, the dry season lasts longer, and broadleaf forests become less and less abundant. In the humid tropics of Southeast Asia and India, decomposition is slow, and *semi-evergreen forests* form.

Where less than 2.5 centimeters (1 inch) of rain falls during the dry season, *tropical deciduous* forests form. "Deciduous" refers to trees or shrubs that shed leaves once a year, prior to the season when very cold or dry conditions do not favor growth. In tropical deciduous forest, trees shed leaves at the onset of the dry season.

Temperate deciduous forests form in parts of eastern North America, western and central Europe, and parts of Asia, including Japan. About 50 to 150 centimeters (about 20–60 inches) of precipitation fall throughout the year. Leaves turn brilliant red, orange, and yellow before dropping in autumn (Figure 43.21 and Section 6.1). Having discarded their leaves, the trees become dormant during the cold winter, when water is locked in snow and ice. In the spring, when conditions again favor growth, deciduous trees flower and new leaves appear. Also during the spring, leaves shed the prior autumn decay and form a rich humus. Rich soil and a somewhat open canopy that allows sunlight through allows many understory plants to flourish.

Conditions in broadleaf forest biomes favor dense stands of trees that form a continuous canopy over broad regions. As in all other forest biomes, tree form and physiology are adapted to patterns in rainfall and temperature.

Figure 43.21 North American temperate deciduous forest. The series above shows changes in one deciduous tree's foliage from fall (*far left*) through winter, spring, and summer.

43.8 You and the Tropical Forests

FOCUS ON BIOETHICS

The Chapter 27 introduction mentioned the deforestation of northern conifer forests. We turn here to impacts of deforestation on the once-vast tropical forests.

Southeast Asia, Africa, and Latin America stretch across the tropical latitudes. There, developing nations have the fastest-growing populations and high demands for food, fuel, and lumber. Of necessity, they turn to their forests (Figure 43.22). Most of the tropical forests may vanish within our lifetime. That possibility concerns many people in highly developed nations—which, ironically, use most of the world's resources, including forest products.

On purely ethical grounds, the destruction of so much biodiversity *is* a concern. Tropical rain forests have the greatest variety and numbers of insects, and the world's largest ones. They are homes to the most species of birds and to plants with the largest flowers (*Rafflesia*). Forest canopies and understories support monkeys, tapirs, and jaguars in South America and apes, leopards, and okapis in Africa. Massive vines twist around tree trunks. Orchids, mosses, lichens, and other organisms grow on branches, absorbing minerals from rains. Communities of microbes, insects, spiders, and amphibians live, breed, and die in small pools of water that collect in furled leaves.

Also, products provided by rain forest species save and enhance human lives. Analysis of compounds in rain forest species can point the way toward new drugs. Quinine, an antimalarial drug, was first derived from an extract of *Cinchona* bark from a tree in the Amazonian rain forest. Two chemotherapy drugs, vincristine and vinblastine, were extracted from the Rosy periwinkle (*Catharanthus roseus*) a low-growing plant native to Madagascar's rain forests. Today, these drugs help fight leukemia, lymphoma, breast cancer, and testicular cancer. Many ornamental plants, spices, and foods, including cinnamon, cocoa, and coffee, originated in tropical forests. So did latex, gums, resins, dyes, waxes, and oils used in tires, shoes, toothpaste, ice cream, shampoo, compact discs, condoms, and perfumes.

Decline of rain forests could influence the atmosphere. The forests take up and store carbon and release oxygen. Burning enormous tracts of tropical forest to make way for agriculture releases carbon dioxide, which contributes to global warming (Section 42.8).

Conservation biologists decry the loss of forest species and their essential natural services. Yet tropical rain forest loss keeps accelerating. The amount of temperate forest is on the increase in North America, Europe, and China. But it cannot make up for staggering losses of tropical forests elsewhere. There are some encouraging signs. In Brazil, the country with the largest expanse of tropical rain forest, the rate of deforestation decreased in 2005 and 2006, after steady rises since the 1990s. A number of factors contributed to the decline, including better enforcement of anti-logging laws and decreases in the prices for crops typically planted on cleared forest land.

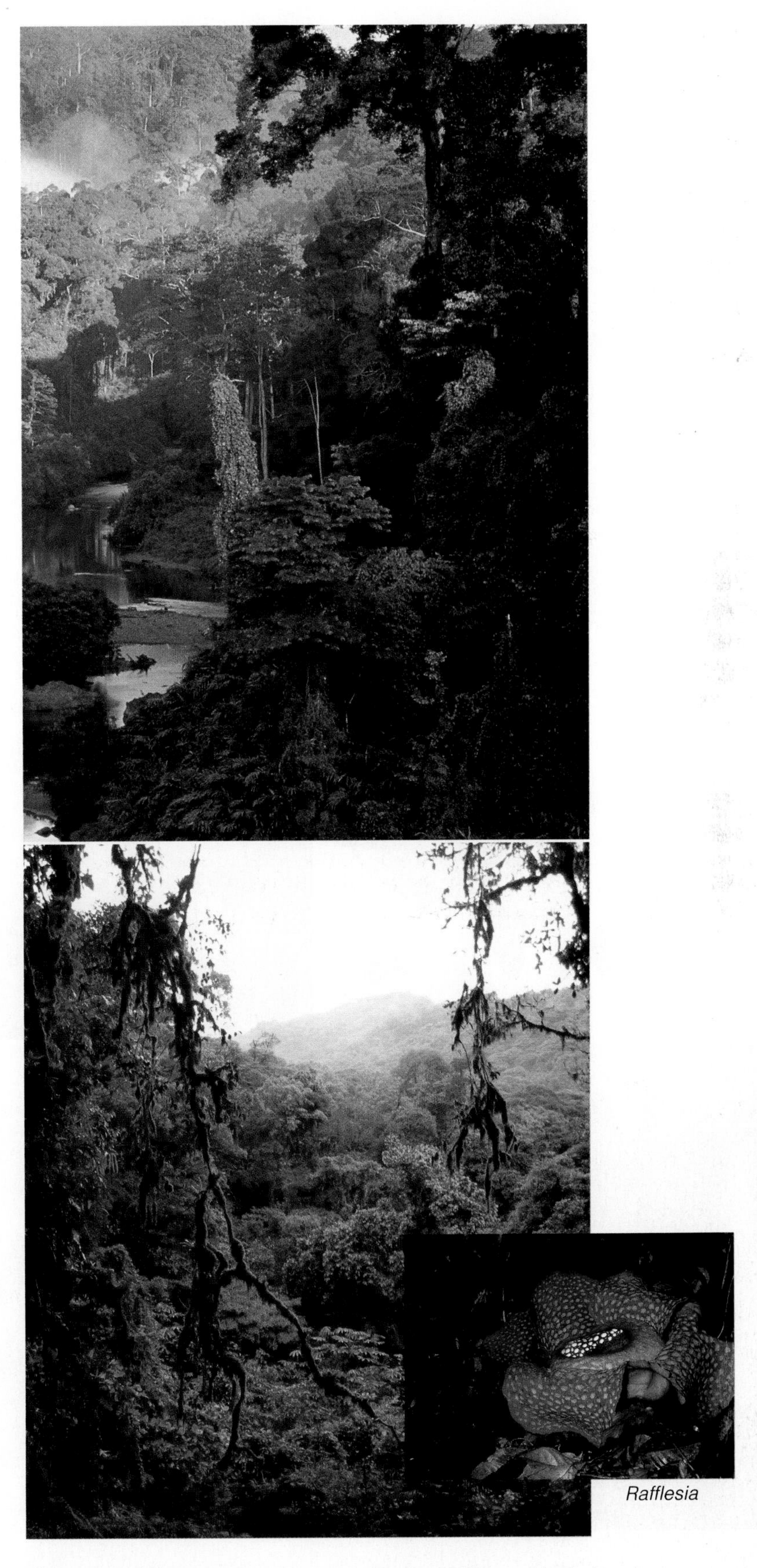

Figure 43.22 Tropical rain forests in Southeast Asia and Latin America.

43.9 Coniferous Forests

Conifers dominate boreal forests, montane coniferous forests, temperate rain forests, and pine barrens.

LINKS TO CHAPTER 21 INTRODUCTION, SECTION 21.6

Conifers—evergreen trees with seed-bearing cones—dominate *coniferous* forests. Their leaves are typically needle-shaped, with a thick cuticle. Stomata are sunk below the leaf surface. These adaptations help conifers conserve water during drought or times when the soil water is frozen. As a group, conifers tolerate poorer soils and drier habitats than deciduous trees.

In the boreal forests that stretch across northern Europe and Asia, and North America, the conifers are mostly pine, fir, and spruce. These forests are known as *taigas*, which means "swamp forests." They form in formerly glaciated regions where lakes and streams abound (Figure 43.23*a*). Most rain falls in the summer, and little water evaporates into the cool summer air. Winters are long, cold, and dry.

In the Northern Hemisphere, montane coniferous forests extend southward through the great mountain ranges (Figure 43.23*b*). Spruce and fir dominate at the highest elevations, with firs and pines taking over as you move farther down the slopes.

Conifers also dominate temperate lowlands along the Pacific coast from Alaska into northern California. These coniferous forests hold the world's tallest trees, Sitka spruce to the north and coast redwoods to the south. Large tracts have been logged (Chapter 21).

We find other conifer-dominated ecosystems in the eastern United States. About a quarter of New Jersey is *pine barrens*, a mixed forest of pitch pines and scrub oaks that grow in sandy, acidic soil. Pine forest covers about one-third of the Southeast. Fast-growing loblolly pines dominate these forests and are a major source of lumber and wood pulp. They can survive periodic fires that kill most hardwood species. When fires are suppressed, the hardwoods outcompete the pines.

Conifers prevail across the Northern Hemisphere's high-latitude forests, at high elevations, and in warm regions with nutrient-poor soils.

Figure 43.23 (**a**) Taiga in Alberta, Canada. (**b**) Montane coniferous forest near Mount Rainier, Washington.

ARCTIC TUNDRA

43.10 Brief Summers and Long, Icy Winters

Arctic tundra lies between the polar ice cap and belts of boreal forests in the Northern Hemisphere. It is the youngest biome. It appeared about 10,000 years ago, when glaciers retreated at the end of the last ice age.

The Northern Hemisphere's treeless plain, the *arctic tundra*, is dry and cold for most of the year. Annual snow and rain is usually is less than 25 centimeters (10 inches). During a brief summer, ground-hugging, shallow-rooted plants grow rapidly under the nearly continuous sunlight (Figure 43.24).

Only the surface soil thaws during summer. Below that lies permafrost, a frozen layer 500 meters (1,600 feet) thick in places. Permafrost prevents drainage, so the soil covering it remains perpetually waterlogged. The cool, anaerobic conditions slow nutrient cycling, so any plant remains decay slowly. Organic matter in the permafrost makes the arctic tundra one of Earth's greatest stores of carbon. However, this frozen layer is thawing. With global warming, much of the snow and ice that would otherwise reflect some sunlight is melting. Newly exposed dark soil absorbs heat from the sun's rays, which encourages more melting.

Alpine tundra is a similar biome, but it develops in high mountains throughout the world (Figure 43.25). At night, the below-freezing temperatures make it too difficult for trees to grow. Even in the summer, shaded patches of snow persist in this biome, but there is no permafrost. Alpine soil is thin and well drained, but it is nutrient-poor. As a result, primary productivity is low. Grasses, heaths, and small-leafed shrubs grow in patches where a bit of deeper soil has formed. These low-growing plants withstand strong winds.

Figure 43.25 Compact, low-growing, hardy plants typical of alpine tundra in the Washington Cascade range.

Arctic tundra prevails at high latitudes, where short, cold summers alternate with long, cold winters. Alpine tundra prevails in high, cold mountains regardless of seasonal differences in latitude.

Figure 43.24 (**a**) Arctic tundra in the summer. Hardy lichens and shallow-rooted, low-growing plants are a base for food webs that include voles, arctic hares, caribou, arctic foxes, wolves, and polar bears. Great numbers of migratory birds nest here in summer, when the air is thick with mosquitoes and other kinds of flying insects.

(**b**) Arctic tundra makes up about 4 percent of Earth's land mass. It is blanketed with snow for as long as nine months of the year. Most occurs in northern Russia and Canada, followed by Alaska and Scandinavia. Bands of humans have herded reindeer, hunted, and fished in these sparsely populated regions for hundreds of thousands of years. More people, and machines, are moving in to extract mineral and fossil fuels. If the operations alter the vegetation and soils, it might take decades for a region to recover. Why? Tundra plants grow very slowly, and their seasonal growth is limited to just a few months of each year.

43.11 Freshwater Ecosystems

Oceans, lakes, ponds, wetlands, reefs—such freshwater and saltwater provinces cover more of Earth's surface than all land biomes combined. No region is "typical." Ponds are shallow; Siberia's Lake Baikal is 1.7 kilometers (1 mile) deep. All aquatic ecosystems have gradients in light penetration, temperature, and dissolved gases, but values differ greatly. All we can do here is sample the diversity, starting with the freshwater ecosystems.

LAKES

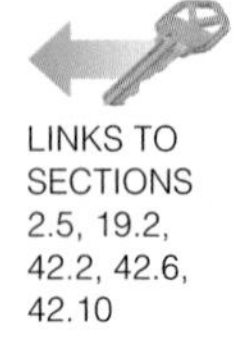

LINKS TO SECTIONS 2.5, 19.2, 42.2, 42.6, 42.10

A lake is a body of standing freshwater (Figure 43.26). It typically can be divided into zones that differ in their physical characteristics and species composition (Figure 43.27*a*). Near shore is the littoral zone. Here, sunlight penetrates all the way to the lake bottom and aquatic plants are primary producers. The lake's open waters include an upper, well-lit limnetic zone, and—if the lake gets deep enough—a dark profundal zone where light does not penetrate. Primary producers in the limnetic zone are members of the phytoplankton, a group of photosynthetic microorganisms, including green algae, diatoms, and cyanobacteria. They serve as food for rotifers and copepods, and other members of the zooplankton. In the profundal zone, there is not enough light for photosynthesis, so consumers here depend on food produced above. Debris drifts down from and feeds detritivores and decomposers.

Figure 43.26 A lake in Chile's spectacular Torres del Paine National Park.

Seasonal Changes Temperate zone lakes show seasonal variations in their density and temperature gradients, from surface to bottom. During the winter, a layer of ice forms at the lake surface. Unlike most substances, water is not most dense when it's solid. As water cools, its density increases, until it reaches 4°C (39°F). Below this temperature, any additional cooling decreases water's density—which is why ice floats on water (Section 2.5). In an ice-covered lake, water just under the ice is near its freezing point and its lowest density. The densest (4°C) water is at the lake bottom.

In spring, there are more daylight hours and the air warms. The ice melts, the temperature of the surface layer of water rises to 4°C, and this water sinks. Winds cause vertical currents that lead to a spring overturn, during which oxygen-rich water in the surface layers moves down and nutrient-rich water from the lake's depths moves up.

In summer, a lake has three layers (Figure 43.27*b*). The upper layer is warm and oxygen-rich. Below this is a thermocline, a thin layer where temperature falls rapidly. Beneath the thermocline is the coolest water. The thermocline acts as a barrier that keeps the upper and lower layers from combining. During the summer, decomposers deplete oxygen dissolved near the lake bottom and nutrients from the depths cannot escape into surface waters. In autumn, the upper layer cools and sinks, and the thermocline vanishes. During the fall overturn, oxygen-rich water moves down while nutrient-rich water moves up.

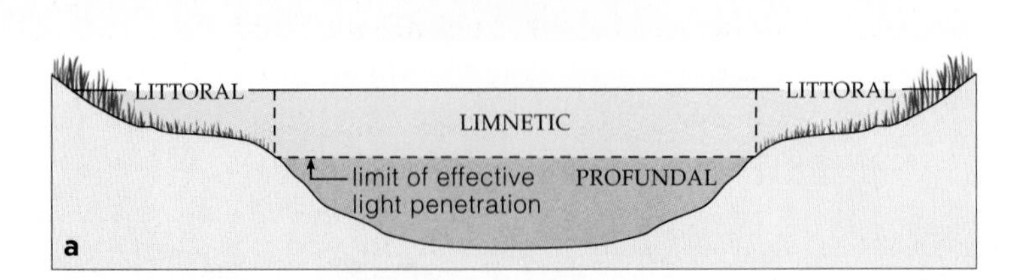

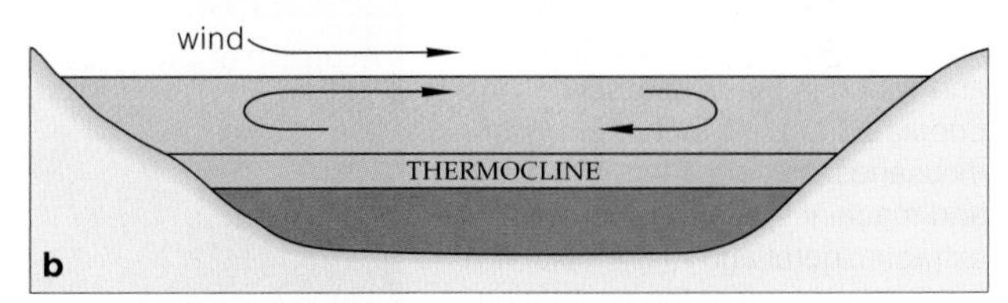

Figure 43.27 (**a**) Lake zonation. A lake's littoral zone extends all around the shore to a depth where rooted aquatic plants stop growing. Its limnetic zone is the open waters where light penetrates and photosynthesis occurs. Below that lie the cooler, dark waters of the profundal zone. (**b**) In temperate regions, thermal layering occurs in many lakes during the summer.

Figure 43.28 River habitats in North Carolina and Virginia. (**a**) Pool in autumn. Pools have a smooth surface; water is streaming slowly over a fine substrate. (**b**) A run, Sinking Creek. Runs are smooth surfaced but flow faster over rock and sand. (**c**) A riffle. Water flow is swift and turbulent over a rough stretch of the streambed. (**d**) Rapids, where rocks break up a swift current. Leaves add nutrients to the water.

Primary productivity is seasonal. After the spring overturn, longer daylengths and the cycled nutrients encourage primary productivity. During the growing season, vertical mixing ends. Nutrients do not move up, and photosynthesis slows. By late summer, these shortages limit growth. The overturn in the fall cycles nutrients to the surface and favors a burst of primary productivity. A sustained burst is not possible, given the fewer daylight hours. Primary productivity will not rise again until spring.

Nutrient Content and Succession Like land habitats, lakes undergo succession; they change over time (Section 41.8). A newly formed lake is naturally *oligotrophic*; deep, clear, and nutrient-poor, with low primary productivity. Later, as sediments accumulate, plants take root. The lake becomes more *eutrophic*; it becomes shallower, less clear, and nutrient-rich, with a higher primary productivity.

Eutrophication refers to processes, either natural or artificial, that enrich a body of water with nutrients. Section 42.10 described experimental eutrophication of one lake basin. Humans also turned Seattle's Lake Washington eutrophic. From 1941 to 1968, phosphate-rich sewage entered the lake, and nitrogen became the main limiting factor for phytoplankton. This favored cyanobacteria, which fix gaseous nitrogen (N_2) from the air (Section 19.2). Each summer until the sewage inputs ended, enormous slimy mats of cyanobacteria covered the lake. After phosphate levels declined, the cyanobacteria no longer had a competitive edge and nearly disappeared.

STREAMS AND RIVERS

All streams are flowing-water ecosystems that start as freshwater springs or seeps. As they flow downslope, they grow and merge. Rainfall, snowmelt, geography, altitude, and shade cast by plants affect flow volume and temperature. Streambed composition as well as human inputs affect a stream's solute concentrations.

A stream imports nutrients into many food webs. In forests, trees cast shade and hinder photosynthesis, but the litter sustains detrital food webs (Section 42.2). Aquatic species take up and release nutrients as water flows downstream. Nutrients move upstream only in the tissues of migratory fish and other animals. They cycle between aquatic organisms and the water as it flows on a one-way course to the sea.

A river is a flowing-water ecosystem that usually starts when streams converge. Along its length, a river has distinct areas that differ in water depth and speed of flow (Figure 43.28).

The land area from which water drains into a river is its watershed (Section 42.6). The Mississippi River watershed is the third largest in the world. It delivers water from thirty-one states and Canada to the Gulf of Mexico. Sediments, nutrients, and pollutants flow in the river to the sea. Each spring, nitrate fertilizer applied in the Midwest flows into the Mississippi and is carried into the Gulf of Mexico. The resulting algal blooms cause large offshore "dead zones."

Lakes show gradients in light, dissolved oxygen, and nutrients. Primary productivity varies with a lake's age and—in temperate zones—with the season.

Streams and rivers move sediments and nutrients, along with water, to the sea.

43.12 "Fresh" Water?

FOCUS ON HEALTH

Here are a few reasons why you might wish to investigate where your drinking water comes from, and what it has picked up from its surroundings.

LINKS TO SECTIONS 42.1, 42.2

Pollutants flow into rivers, lakes, and the groundwater from countless sources. They include sewage, animal wastes, industrial chemicals, fertilizers, and pesticides. Runoff from roads adds engine oil and antifreeze that dripped from vehicles, and rubber residues from tire wear. Leaky underground fuel tanks allow gasoline and other fuels to seep into groundwater.

How do we keep these poisons out of our drinking water? One safeguard is wastewater treatment. There are three stages of treatment. In primary treatment, screens and settling tanks remove large bits of organic material (sludge), which is dried, burned, dumped in landfills, or treated further. In secondary treatment, microbes break down any organic matter that remained after the primary treatment. Water is then treated with chlorine or exposed to ultraviolet light to kill disease-causing microorganisms. By now, most oxygen-demanding wastes are gone—but not all nitrogen, phosphorus, toxins, and heavy metals. Tertiary treatment uses chemical filters to remove these contaminants from water but it is expensive. Five percent of the nation's wastewater gets this level of treatment.

One variation on standard wastewater treatment is a solar-aquatic system such as the one constructed by biologist John Todd (Figure 43.29). Sewage enters tanks in which aquatic plants grow. Decomposers degrade wastes, releasing nutrients that promote plant growth. Heat from sunlight speeds the decomposition. Water next flows through an artificial marsh that filters out algae and organic wastes. Then it flows through other tanks filled with living organisms. After ten days, water flows into a second artificial marsh for final filtering and cleansing. Versions of this system are now being used to treat sewage and industrial wastes.

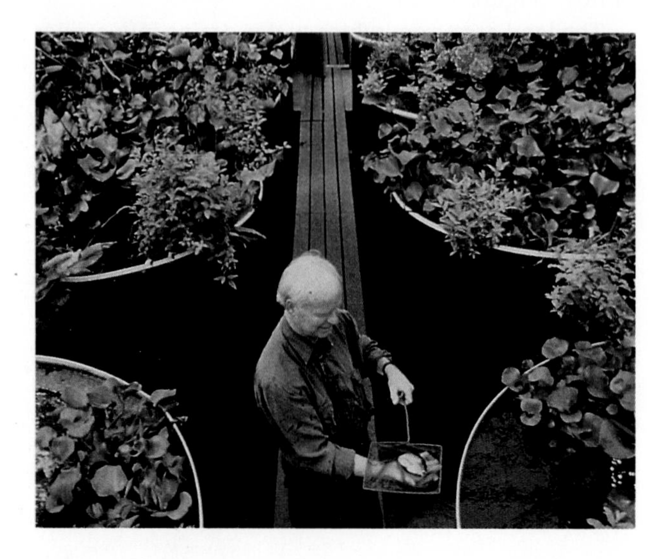

Figure 43.29 John Todd in the experimental solar-aquatic wastewater treatment facility he designed. Unlike traditional treatments, Todd's system does not require toxic chemicals or emit unpleasant odors. Bacteria, fungi, plants, invertebrates, and fish break down wastes. Solar-aquatic treatment systems are now in use in eight countries around the world.

43.13 Life at Land's End

Near the coasts of continents and islands, concentrations of nutrients support some of the world's most productive aquatic ecosystems.

WETLANDS AND THE INTERTIDAL ZONE

Like freshwater ecosystems, estuaries and mangrove wetlands have distinct physical and chemical features, including depth, water temperature, salinity, and light penetration. An estuary is an enclosed coastal region where seawater mixes with nutrient-rich fresh water from rivers and streams (Figure 43.30*a*). Water inflow continually replenishes nutrients, which is one reason estuaries can support highly productive ecosystems.

Primary producers include algae and other types of phytoplankton, and plants that tolerate submergence at high tide. Detrital food webs are common (Section 42.2). Estuaries are marine nurseries; many larval and juvenile stages of invertebrates and fishes develop in them. Migratory birds use estuaries as rest stops.

Estuaries range from broad, shallow Chesapeake Bay, Mobile Bay, and San Francisco Bay to the narrow, deep fjords of Norway and others like them in Alaska and British Columbia. Many face threats. Fresh water that should refresh them is diverted for human uses. Runoff from agriculture and pavement flows in.

In tidal flats at tropical latitudes, we find nutrient-rich *mangrove wetlands*. "Mangrove" refers to a forest of salt-tolerant woody plants in sheltered areas along tropical coasts. The plants have prop roots that extend out from the trunk (Figure 43.30*b*). Specialized cells at the surface of some roots allow gas exchange.

Increases in the human populations along tropical coastlines threaten mangroves. People cut these trees for firewood. Also, enormous stretches of mangrove wetlands have been converted to shrimp farms that supply the United States, Japan, and western Europe. The amount of mangrove wetlands has been halved and losses continue to accelerate. The disappearance of these wetlands threatens the migratory birds and fishes that depend upon them for shelter and food.

Figure 43.30 (**a**) South Carolina salt marsh. Marsh grass (*Spartina*) is the major producer. (**b**) In the Florida Everglades, a mangrove wetland lined with red mangroves (*Rhizophora*).

ROCKY AND SANDY COASTLINES

Rocky and sandy coastlines support ecosystems of the intertidal zone. Waves batter residents and the tides alternately submerge and expose them. The higher up they are, the more they dry out, freeze in winter, and bake in summer, and the less food comes their way. The lower they are, the more competition they have for limited living spaces.

Biologists can divide a shoreline into three vertical zones that differ in their physical characteristics and diversity. The *upper* littoral zone is submerged only at the highest tide of a lunar cycle. It holds the fewest species. The *mid*littoral zone is submerged during the highest regular tide and exposed at the lowest tide. The *lower* littoral zone, exposed by the lowest tide of the lunar cycle, has the most diversity.

You can easily see the zonation along a rocky shore (Figure 43.31). Here, strong wave action keeps detritus from building up, so grazing food webs prevail. Algae clinging to rocks are primary producers. The primary consumers they feed include a variety of snails.

Waves continually rearrange the loose sediments of sandy shores (Figure 43.32). Here, detrital food webs start with material washed ashore. Some crustaceans live on detritus in the upper littoral zone. Closer to the water, crabs, shrimps, bivalve mollusks, and worms burrow in the sediments. Sandy shores are essential to sea turtles. The females bury their eggs, which then develop beneath a layer of protective sand.

Wetlands and coral reefs show high primary productivity. Rocky and sandy shore communities have vertical zones, defined by how often they are submerged by high tides.

Intertidal zone's upper littoral; submerged only at highest tide of lunar cycle

midlittoral; submerged at each highest regular tide and exposed at lowest tide

lower littoral; exposed only at low tide of lunar cycle

a b

Figure 43.31 Rocky shores of the Pacific Northwest. (**a**) Vertical intertidal zonation. (**b**) Tidepool with algae and invertebrates, including red sea anemones.

Figure 43.32 Coral fragment washed up on the sandy shore of Heron Island, part of Australia's Great Barrier Reef system. Green sea turtles bury their eggs in this sand.

43.14 The Once and Future Reefs

What does it mean when someone says each species is part of a web of interactions? Here is a case in point.

LINKS TO SECTIONS 20.4, 23.4, 41.9

Coral reefs are wave-resistant formations that consist primarily of calcium carbonate secreted by generations of coral polyps (Section 23.4). Reef-forming corals live mainly in clear, warm waters between latitudes 25° north and 25° south (Figures 43.33*a* and 43.34). On many reefs, mineral-hardened cell walls of red algae contribute to the structural framework (43.34*d*). The resulting reef is home to living corals, algae, and other species.

Australia's Great Barrier Reef parallels Queensland for 2,500 kilometers (1,550 miles), and is the largest example of biological architecture. Actually it is a string of reefs, some 150 kilometers (95 miles) across. It supports 500 coral species, 3,000 fish species, 1,000 kinds of mollusks, and 40 kinds of sea snakes. Figure 43.34*e* only hints at the wealth of warning colors, tentacles, and stealthy behavior—all signs of fierce competition for resources among species jostling for the limited space.

Photosynthetic dinoflagellates live as symbionts inside the tissues of all reef-building corals (Section 23.4). The dinoflagellates find protection in the tissues. They provide the coral polyp with oxygen and sugars that it depends upon. When stressed, coral polyps expel the symbionts. When stressed for more than a few months, the corals die; only bleached hard parts remain (Figure 43.33*b*).

Abnormal, widespread bleaching in the Caribbean and the tropical Pacific began in the 1980s. So did increases in sea surface temperature, which might be a key stress factor. Is the damage one outcome of global warming? If so, as marine biologists Lucy Bunkley-Williams and Ernest Williams suggest, the future looks grim for reefs, which may be destroyed within three decades.

Also, people can directly destroy reefs, as by raw sewage discharges into nearshore waters of populated

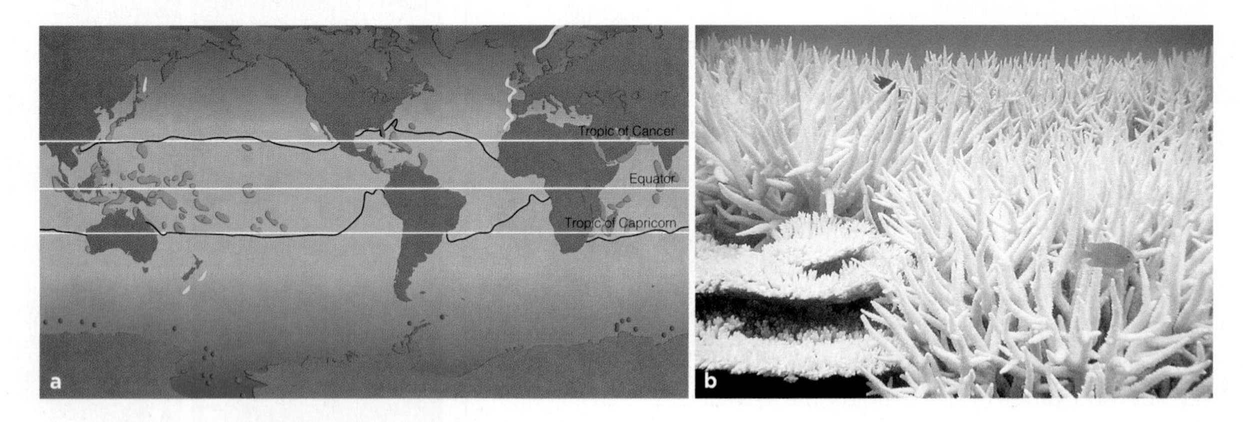

Figure 43.33 (**a**) Distribution map for coral reefs (*orange*) and coral banks (*yellow*). Nearly all reef-building corals live in warm seas, here enclosed in dark lines. Past latitudes 25° north and south, solitary and colonial corals (*red*) form coral banks in temperate seas and in cold seas above continental shelves. (**b**) Coral bleaching on Australia's Great Barrier Reef.

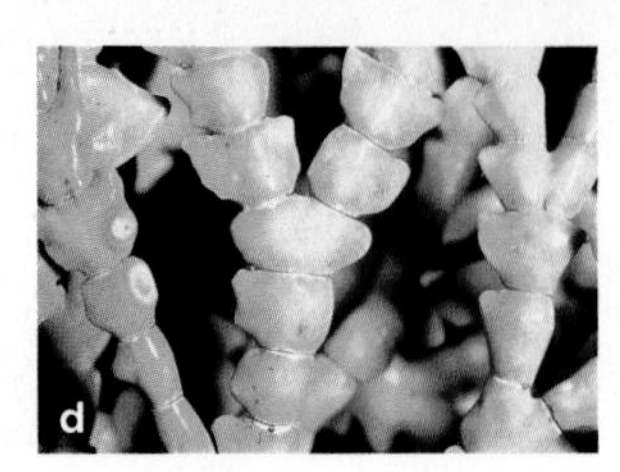

Figure 43.34 Three types of coral reef formations. (**a**) *Fringing* reefs form near land when rainfall and runoff are light, as on the downwind side of the youngest volcanic islands. Many reefs in the Hawaiian Islands and Moorea are like this. (**b**) *Barrier* reefs parallel the shore of continents and volcanic islands, as in Bora Bora. Behind them are calm lagoons. (**c**) Ring-shaped *atolls* consist of coral reefs and coral debris. They fully or partly enclose a shallow lagoon, often with a channel to open ocean. Biodiversity is not great in shallow water, which can get too hot for corals. (**d**) Coralline alga, one of the reef builders. (**e**) *Facing page*, a sampling of coral reef biodiversity.

islands. Massive oil spills, commercial dredging operations, and mining for coral rock have catastrophic impact.

LIONFISH

Fishing nets can break pieces off corals, but some fishermen prefer even more destructive practices. They drop dynamite in the water. Fish hiding in the coral are blasted out and float to the surface, some dead, but others only stunned. Sodium cyanide squirted into the water also stuns fish, which float to the surface. Most fish that survive being stunned with dynamite or cyanide are shipped off for sale in pet stores in the United States or Europe. Some large specimens are transported alive to restaurants in Asian cities, where they are served up as exorbitantly priced status symbols.

Invasive species also threaten reefs. In Hawaii, reefs are being overgrown by exotic algae, including several species imported for cultivation during the 1970s.

Reef biodiversity is in danger around the world, from Australia and Southeast Asia to the Hawaiian Islands, Galápagos Islands, Gulf of Panama, Florida, and Kenya. To give a final example of this, the biodiversity on the coral reef off Florida's Key Largo has been reduced by 33 percent since 1970.

PART OF A FIJIAN CORAL REEF

MORAY EEL

NUDIBRANCH

LONGNOSE HAWKFISH AND RED SEA FAN

e

BANDED CORAL SHRIMP

PURPLE TUBE SPONGE

GREEN CORAL POLYP

43.15 The Open Ocean

The world ocean has two vast provinces (Figure 43.35). Its benthic (bottom) province starts at continental shelves and extends to deep-sea trenches. Its pelagic province is the full volume of ocean water. The neritic zone is the volume above continental shelves; the oceanic zone is the volume above the ocean basins.

ASTOUNDING DIVERSITY

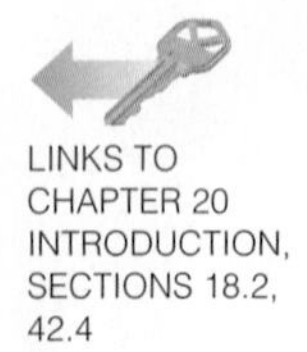

LINKS TO CHAPTER 20 INTRODUCTION, SECTIONS 18.2, 42.4

The ocean's upper, sunlit waters support most marine life. Members of the phytoplankton live in the upper sunlit waters. They include coccolithophores, diatoms, algae, and dinoflagellates. *Phyto*plankton also includes vast numbers of photosynthetic bacteria. As a group, members of the phytoplankton account for nearly all of the ocean's primary productivity. They feed marine primary consumers, such as copepod and krill, which are members of the *zoo*plankton.

Deeper ocean water is too dark for photosynthesis. There, food webs start with marine snow. These tiny bits of organic matter drift down from communities above. They are the base for staggering biodiversity; midoceanic water may be home to 10 million species!

In what may be the greatest circadian migrations, many species rise thousands of feet at night to feed in upper waters, then move down in the morning. At the top of food webs, carnivores range from the familiar sharks and squids to the bizarre deep-sea angler fishes and giant colonial cnidarians (Figure 43.36*a*,*b*).

The benthic province includes largely unexplored ecosystems on seamounts and at hydrothermal vents. Seamounts are undersea peaks standing 1,000 meters or more tall, but still far below the sea surface. They are areas of great species richness and often serve as habitat for species not seen elsewhere (Figure 43.36*c*).

At hydrothermal vents, superheated water rich in dissolved minerals spews out from an opening on the ocean floor. Where this mineral-rich water mixes with the cold deep sea water, minerals settle out and form extensive deposits. Prokaryotic chemoautotrophs get energy from these deposits. The prokaryotes serve as primary producers for food webs that include diverse invertebrates, including tube worms and brittle stars (Figure 43.36*d–f*). By one hypothesis, the first life could have originated in such heated, nutrient-rich places on the seafloor. Sections 18.2 and 18.3 discuss this idea.

UPWELLING AND DOWNWELLING

Where winds paralleling the west coasts of continents blow over an ocean, friction starts the surface waters moving. Upwelling occurs as Earth's rotation deflects the masses of slow-moving water away from a coast and cold, deep water moves up vertically in its place (Figure 43.37).

In the Southern Hemisphere, commercial fisheries depend on wind-induced upwelling along Peru and Chile. Prevailing winds from the south and southeast tug surface water away from the coast. Cold, deeper water of the Peru Current then moves up toward the continental shelf. This current dredges up nitrate and phosphate and carries them northward. The nutrients sustain phytoplankton that are the basis of one of the world's richest fisheries.

Every three to seven years, warm surface waters of the western equatorial Pacific Ocean move eastward. This massive displacement of warm water acts on the prevailing wind direction. The eastward flow speeds up so much that it hampers the vertical movement of water along the coasts of Central and South America.

Surface water piling into a coast is forced down and flows away from it. Near Peru's coast, prolonged downwelling of nutrient-poor water displaces cooler waters of the Peru Current and ends upwelling. The warmer current most often arrives around Christmas. Fishermen in Peru named it El Niño ("the little one,"

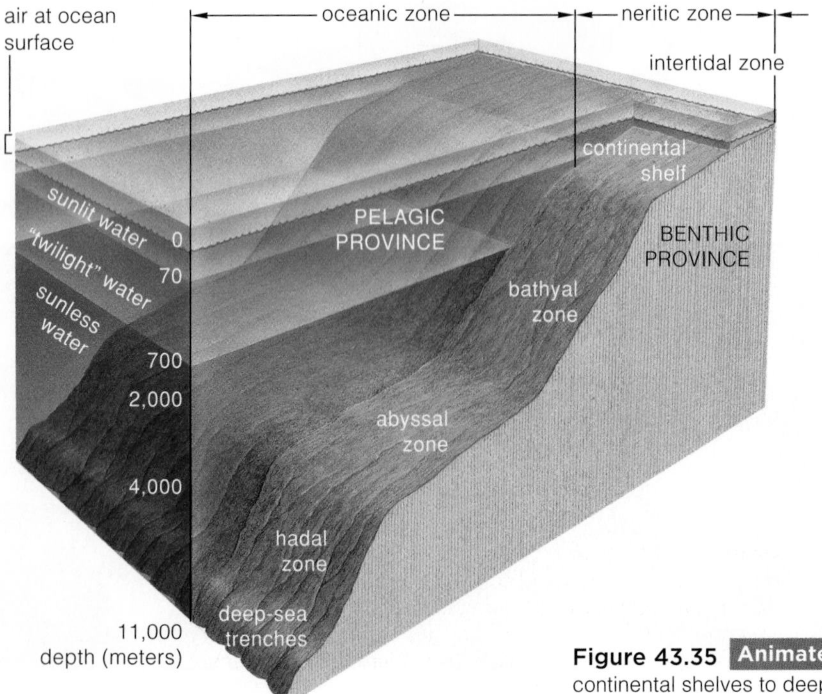

Figure 43.35 **Animated!** Oceanic zones. Seafloor extends out from the continental shelves to deep-sea trenches. Zone dimensions are not to scale.

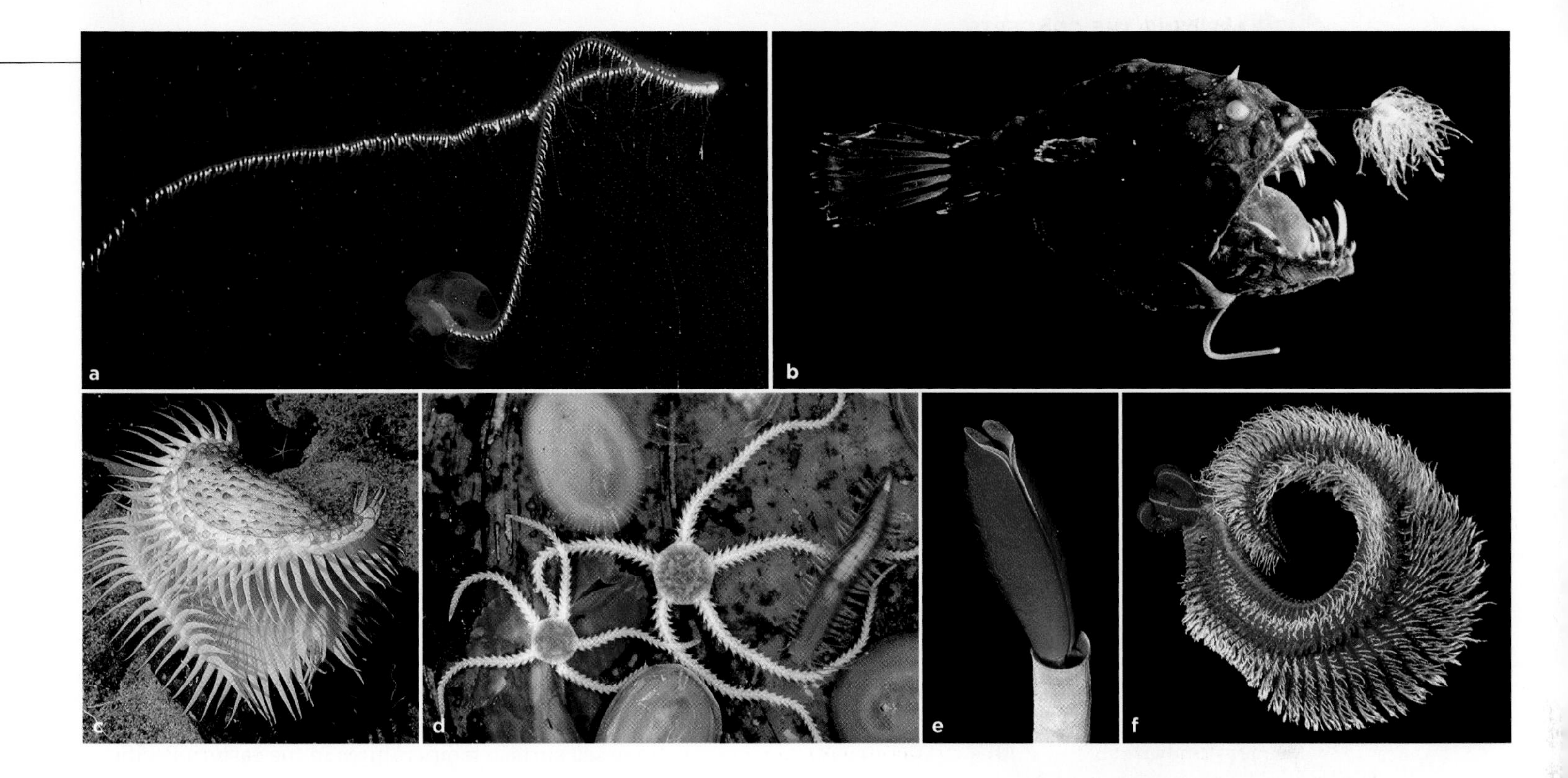

Figure 43.36 What lies beneath: a vast, largely unexplored world of marine life. (**a**) The siphonophore *Praya dubia*, one of the colonial relatives of corals and jellyfish. It can be 40 meters long. (**b**) Deep-sea angler fish with bioluminescent lures.

(**c**) A flytrap anemone, from Davidson seamount, just off the California coast. There are an estimated 30,000 seamounts. They support rich ecosystems and serve as rest stops for migrating fishes.

A few residents of thriving hydrothermal vent communities on the deep ocean floor: (**d**) brittle stars, limpets, and a many legged polychaete; (**e**) tube worm, a polychaete. (**f**) Pompei worm, another polychaete.

in reference to baby Jesus). The name became part of a more inclusive term: the El Niño Southern Oscillation, or ENSO. The next section takes a closer look at some of the consequences of this recurring event.

OCEAN AS GARBAGE DUMP

In the Northern Hemisphere, circular winds drag the ocean surface in an ever tightening spiral. This north central Pacific gyre, the most stable feature of Earth's weather systems, is now a floating garbage dump. Bits of debris cover an area of ocean the size of the state of Texas. Organic material decays, and sunlight degrades the plastic into sand-sized particles. There may be 2.7 kilograms (6 pounds) of debris for every .45 kilogram (1 pound) of plankton in this stretch of floating plastic sand. To field biologist Shawn Farry, the plastic, light bulbs, shoes, bottles, and fishing lines and fishing nets around remote Pacific islands was nothing new. What was shocking was the vast concentration he saw when he dove in the middle of the ocean. Even at depths of 30 meters (100 feet), researchers observed invertebrate filter-feeders eating or tangled in plastic debris. Trash washes ashore on even the most remote ocean islands. Think about that the next time you toss a plastic drink bottle into the trash, rather than the recycling bin.

From the ocean's coral reefs down to hydrothermal vents, throughout the pelagic province, we find astounding levels of primary productivity and biodiversity.

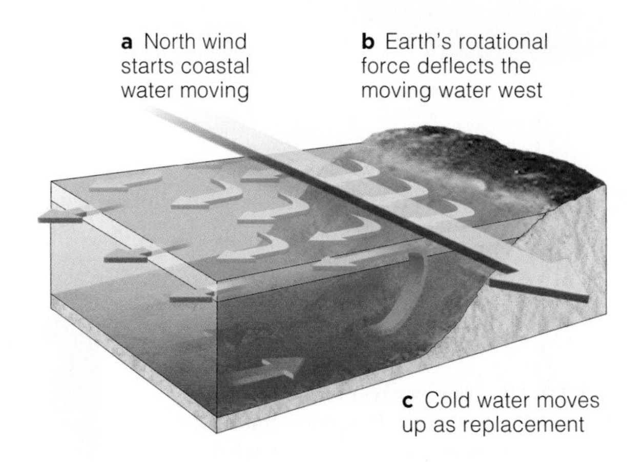

Figure 43.37 Coastal upwelling in the Northern Hemisphere. Prevailing winds blow from north to south (large arrow) and set the surface water moving. Movement of water away from the coast draws colder water in from adjacent depths to replace it. The cold upwelling water holds more nutrients than the warmer surface water that it displaces.

43.16 Climate, Copepods, and Cholera

We turn now to an application that reinforces a unifying ecological concept. Events in the atmosphere and ocean, and on land, interconnect in ways that can profoundly influence the world of life.

LINKS TO SECTIONS 19.6, 23.11

An El Niño Southern Oscillation, or ENSO, is defined by changes in sea surface temperatures and in the air circulation patterns. "Southern oscillation" refers to a seesawing of the atmospheric pressure in the western equatorial Pacific—Earth's greatest reservoir of warm water and warm air. It is the source of heavy rainfall, which releases enough heat energy to drive global air circulation patterns.

Between ENSOs, the warm waters and heavy rains move westward (Figure 43.38*a*). *During* an ENSO, the prevailing surface winds over the western equatorial Pacific pick up speed and "drag" surface waters east (Figure 43.38*b*). As they do, the westward transport of water slows down. Sea surface temperatures rise, evaporation accelerates, and air pressure falls. These changes have global repercussions.

El Niño episodes persist for 6 to 18 months. Often they are followed by a La Niña episode in which the Pacific waters become cooler than usual. Other years, waters are neither warmer nor colder than average.

As you read in the chapter opening, 1997 ushered in the most powerful ENSO event of the century. The average sea surface temperatures in the eastern Pacific rose by 5°C (9°F). This warmer water extended 9,660 kilometers (6,000 miles) west from the coast of Peru.

The 1997–1998 El Niño/La Niña roller-coaster had extraordinary effects on the primary productivity in the equatorial Pacific. With the massive eastward flow of nutrient-poor warm water, photoautotrophs were almost undetectable in satellite photos that measure primary productivity (Figure 43.39*a*).

During the La Niña rebound, cooler, nutrient-rich water welled up to the sea surface and was displaced westward all along the equator. As satellite images revealed, upwelling had sustained an algal bloom that stretched across the equatorial Pacific (Figure 43.39*b*).

During the 1997–1998 El Niño event, 30,000 cases of cholera were reported in Peru alone, compared with only 60 cases from January to August in 1997. People knew that water contaminated by *Vibrio cholerae* causes epidemics of cholera (Figure 43.40). The disease agent triggers severe diarrhea. Bacteria-contaminated feces enter the water supply and individuals who use the tainted water become infected.

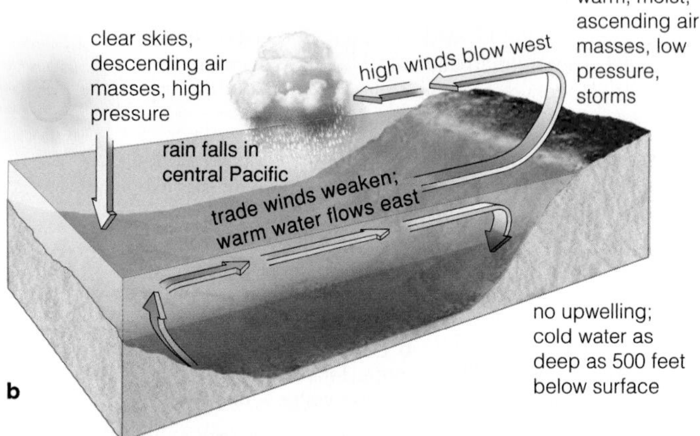

Figure 43.38 Animated! (**a**) Westward flow of cold, equatorial surface water between ENSOs. (**b**) Eastward dislocation of warm water during an ENSO.

Figure 43.39 Satellite data on primary productivity in the equatorial Pacific Ocean. The concentration of chlorophyll in the water was used as the measure. (**a**) During the 1997–1998 El Niño episode, a massive amount of nutrient-poor water moved to the east, and so photosynthetic activity was negligible. (**b**) During a subsequent La Niña episode, massive upwelling and westward displacement of nutrient-rich water led to a vast algal bloom that stretched all the way to the coast of Peru.

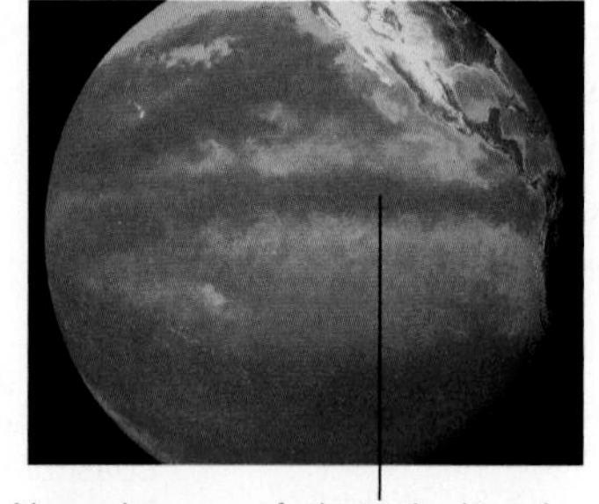

a Near-absence of phytoplankton in the equatorial Pacific during an El Niño.

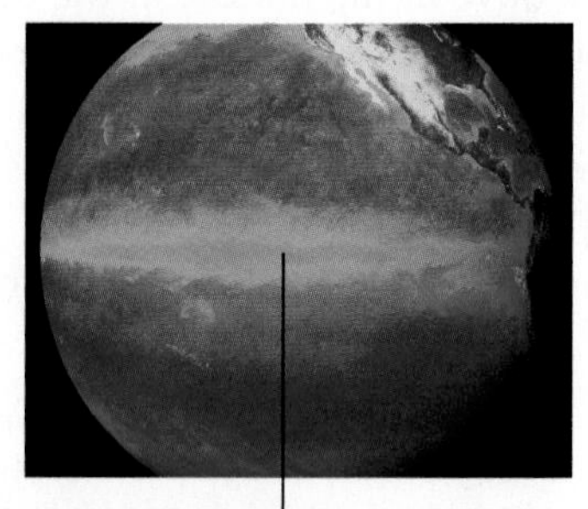

b Huge algal bloom in the equatorial Pacific in the La Niña rebound event.

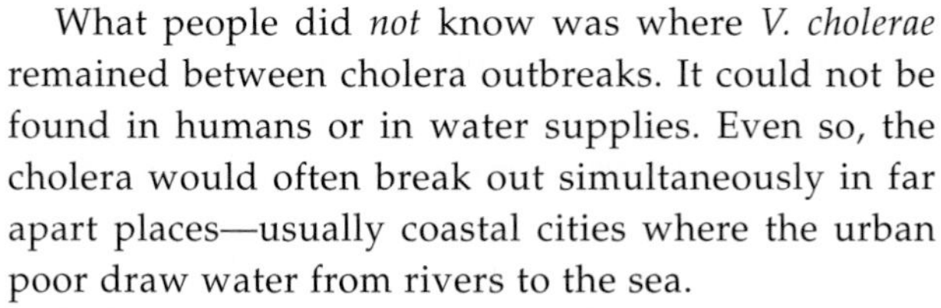

Figure 43.40 (**a**) Satellite data on rising sea surface temperatures in the Bay of Bengal correlated with cholera cases in the region's hospitals. *Red* signifies warmest summer temperatures. (**b**) *Vibrio cholerae*, agent of cholera. Copepods host a dormant stage of this bacterium that waits out adverse environmental conditions that do not favor its growth and reproduction. (**c**) A typical Bangladesh waterway from which water samples were drawn for analysis. (**d**) In Bangladesh, Rita Colwell comparing samples of unfiltered and filtered drinking water.

What people did *not* know was where *V. cholerae* remained between cholera outbreaks. It could not be found in humans or in water supplies. Even so, the cholera would often break out simultaneously in far apart places—usually coastal cities where the urban poor draw water from rivers to the sea.

Marine biologist Rita Colwell had been thinking about the fact that humans are not the host between outbreaks. Was there an environmental reservoir for the pathogen? Maybe. But nobody had detected it in water samples subjected to standard culturing.

Then Colwell had a flash of insight: What if no one could find the pathogen because it changes its form and enters a dormant stage between outbreaks?

During one cholera outbreak in Louisiana, Colwell realized that she could use an antibody-based test to detect a protein unique to *V. cholerae*'s surface. Later, tests in Bangladesh revealed bacteria in fifty-one of fifty-two samples of water. Standard culture methods had missed it in all but seven samples.

V. cholerae survives in rivers, estuaries, and seas. As Colwell knew, plankton also thrive in these aquatic environments. She decided to restrict her search for the unknown host to warm waters near Bangladesh, where outbreaks of cholera occur seasonally (Figure 43.40*c*). It was here that she discovered the dormant *V. cholerae* stage inside copepods. These tiny marine crustaceans graze on algae and other phytoplankton species. The number of copepods—and of *V. cholerae* cells inside them—increases and declines with shifts in phytoplankton abundance.

Colwell already knew about seasonal variations in sea surface temperatures. Remember the old saying, *Chance favors the prepared mind*? In one sense, she was prepared to recognize a connection between cholera cases and seasonal temperature peaks in the Bay of Bengal. She compared data from the 1990–1991 and 1997–1998 El Niño episodes and found the expected correlation. Reports of cholera cases rise four to six weeks after an increase in water temperature.

Today, Colwell and Anwarul Huq, a Bangladeshi scientist, are investigating salinity and other factors that may relate to outbreaks. Their goal is to design a model for predicting where cholera will occur next. They advised women in Bangladesh to use sari cloth as a filter to remove *V. cholerae* cells from the water (Figure 43.40*d*). The copepod hosts are too big to pass through the thin cloths, which can be rinsed in clean water, sun-dried, and used again. This inexpensive, simple method has cut cholera outbreaks by half.

Combine knowledge about life with knowledge about physical and chemical aspects of the biosphere, and who knows what you might discover.

www.thomsonedu.com ThomsonNOW!

Summary

Sections 43.1, 43.2 Global air circulation patterns affect climate and the distribution of communities. The patterns are set into motion by latitudinal variations in incoming solar radiation. The patterns are influenced by Earth's daily rotation and annual path around the sun, the distribution of landforms and seas, and elevations of landforms. Solar energy, and the winds that it causes, are renewable, clean sources of energy.

Human activities alter the atmosphere. Use of CFCs depletes ozone in the upper atmosphere and allows more UV radiation to reach Earth's surface.

Smog, a form of air pollution, occurs when fossil fuels are burned in warm, still air above cities. Coal-burning power plants also are big contributors to acid rain, which alters habitats and kills many organisms.

■ *Use animation on ThomsonNOW to learn how sunlight drives air circulation, how Earth's tilt affects seasons, how CFCs destroy ozone, and how acid rain forms.*

Section 43.3 Latitudinal and seasonal variations in sunlight warm sea surface water and start currents. The currents distribute heat energy worldwide and influence the weather patterns. Ocean currents, air currents, and landforms interact in shaping global temperature zones.

■ *Use the animation on ThomsonNOW to see the patterns of major ocean currents and learn what causes a rain shadow.*

Sections 43.4, 43.5 Biogeographic realms are areas that contain communities of plants and animals found nowhere else. Biomes are regions with a particular type of dominant vegetation. Regional variations in climate, elevation, soil types, and evolutionary history influence the distribution of biomes.

■ *Use the animation on ThomsonNOW to see the distribution of biomes and compare some of their soil profiles.*

Sections 43.6–43.10 Deserts form around latitudes 30° north and south. Desertification, conversion of lands to desert conditions, is a threat to biodiversity.

Slightly moister southern or western coastal regions support dry woodlands and shrublands. Vast grasslands form in the interior of midlatitude continents.

From the equator to latitudes 10° north and south, high rainfall, high humidity, and mild temperatures can support evergreen broadleaf tropical forests.

Semi-evergreen and deciduous broadleaf forests form between latitudes 10° and 25°, depending on how much of the annual rainfall occurs in a prolonged dry season.

Where a cold, dry season alternates with a cold, rainy season, coniferous forests dominate.

Low-growing, hardy plants of the tundra dominate at high latitudes and high altitudes.

■ *Use the animation on ThomsonNOW to see how desert dust from Africa is blown across the Atlantic ocean.*

Sections 43.11–43.14 Most lakes, streams, and other aquatic ecosystems show gradients in the penetration of sunlight, water temperature, and in dissolved gases and nutrients. These characteristics vary over time and affect primary productivity. Coastal zones and tropical reefs support diverse ecosystems. Coastal wetlands and coral reefs are especially productive.

Section 43.15 Life persists throughout the ocean. Diversity is highest in sunlit waters. Mineral-rich waters support communities at deep-sea hydrothermal vents. Upwelling is an upward movement of deep, cool, often nutrient-rich ocean water, typically along the coasts of continents. An El Niño event disrupts upwelling, and it triggers massive, reversible changes in rainfall as well as other weather patterns around the world.

■ *Use the interaction on ThomsonNOW to learn about the oceanic zones with the animation.*

Section 43.16 Drawing on knowledge of microbial ecology as well as biogeographic patterns, Rita Colwell found a crucial bit of information that led to effective countermeasures against cholera outbreaks.

■ *Use the interaction on ThomsonNOW to observe how an El Niño event affects ocean currents and upwelling.*

Self-Quiz

Answers in Appendix III

1. Solar radiation drives the distribution of weather systems and so influences ______ .
   a. temperature zones
   b. rainfall distribution
   c. seasonal variations
   d. all of the above

2. ______ shields living organisms against the sun's UV wavelengths.
   a. A thermal inversion
   b. Acid precipitation
   c. The ozone layer
   d. The greenhouse effect

3. Regional variations in the global patterns of rainfall and temperature depend on ______ .
   a. global air circulation
   b. ocean currents
   c. topography
   d. all of the above

4. A rain shadow is a reduction in rainfall ______ .
   a. on the inland side of a coastal mountain range
   b. during an El Niño event
   c. that occurs seasonally in the tropics

5. Acid rain is one outcome of ______ .
   a. coal burning
   b. gas and oil burning
   c. nitrogen-rich fertilizers
   d. all of the above

6. Biomes are ______ .
   a. water provinces
   b. water and land zones
   c. vast expanses of land
   d. partly characterized by dominant plants
   e. both c and d

7. Biome distribution depends on ______ .
   a. climate
   b. elevation
   c. soils
   d. all of the above

8. Grasslands most often predominate ______ .
   a. near the equator
   b. at high altitudes
   c. in interior of continents
   d. b and c

9. Permafrost underlies ______ , and is a vast store of carbon.
   a. arctic tundra
   b. alpine tundra
   c. coniferous forests
   d. all of the above

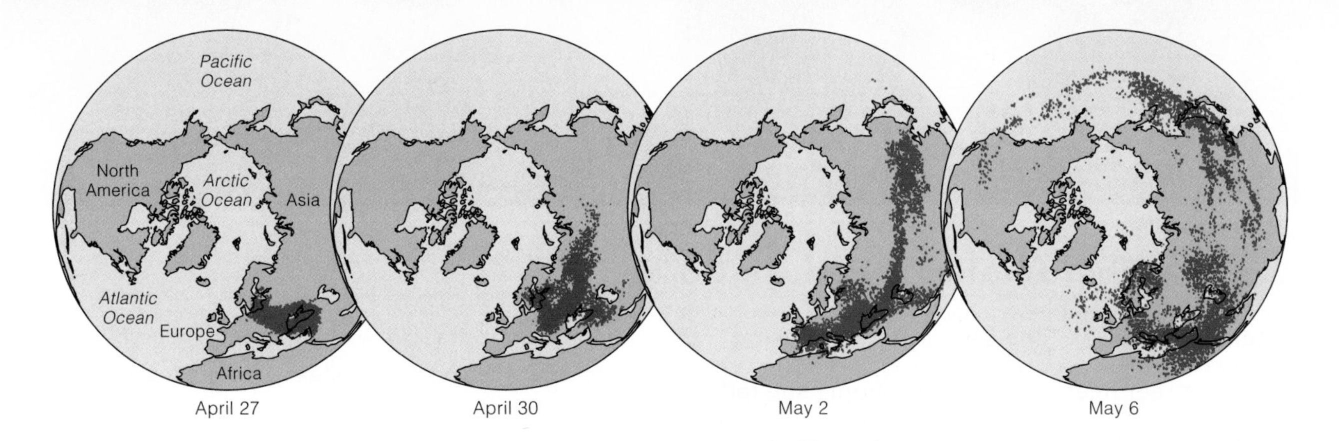

Figure 43.41 Global distribution of radioactive fallout after the 1986 meltdown of the Chernobyl nuclear power plant in Ukraine. The meltdown put 300 million to 400 million people at risk for leukemia and other radiation-induced disorders. By 1998, the rate of thyroid abnormalities in children living downwind from the site was nearly seven times as high as for those upwind; their thyroid gland concentrated the iodine radioisotopes.

10. During ______, deeper, often nutrient-rich water moves to the surface of a body of water.
    a. spring overturns c. upwellings
    b. fall overturns d. all of the above

11. Chemoautotrophic bacteria are the primary producers for food webs ______.
    a. in grasslands c. on coral reefs
    b. in deserts d. at hydrothermal vents

12. Match the terms with the most suitable description.

___ tundra	a. equatorial broadleaf forest
___ chaparral	b. partly enclosed by land where freshwater and seawater mix
___ desert	c. type of grassland with trees
___ savanna	d. has low-growing plants at high latitudes or elevations
___ estuary	e. at latitudes 30° north and south
___ boreal forest	f. mineral-rich, superheated water supports communities
___ tropical rain forest	g. conifers dominate
___ hydrothermal vents	h. dry shrubland

■ *Visit ThomsonNOW for additional questions.*

Critical Thinking

1. On April 26, 1986, in Ukraine, a meltdown occurred at the Chernobyl nuclear power plant. Nuclear fuel burned for nearly ten days and released 400 times more radioactive material than the atomic bomb that dropped on Hiroshima. Winds carried radioactive fallout around the globe (Figure 43.41). Thirty-one people died right after the meltdown. Thousands more are still likely to die from cancers and other harmful effects of radiation.

The Chernobyl accident stiffened opposition to nuclear power in the United States, but recent developments have some people reconsidering. Increasing the use of nuclear energy would diminish the country's dependence on oil from the politically unstable Middle East. Nuclear power does not contribute to global warming, acid rain, or smog. It does produce highly radioactive wastes. Investigate the pros and cons of nuclear power, and decide if you think the environmental benefits outweigh the risks. Would you feel differently if a nuclear power plant were about to be built 10 kilometers (6 miles) upwind from your home?

2. The use of off-road recreational vehicles may double in the next twenty years. Enthusiasts would like increased access to government-owned deserts. Some argue that it's the perfect place for off-roaders because "There's nothing there." Explain whether you agree, and why.

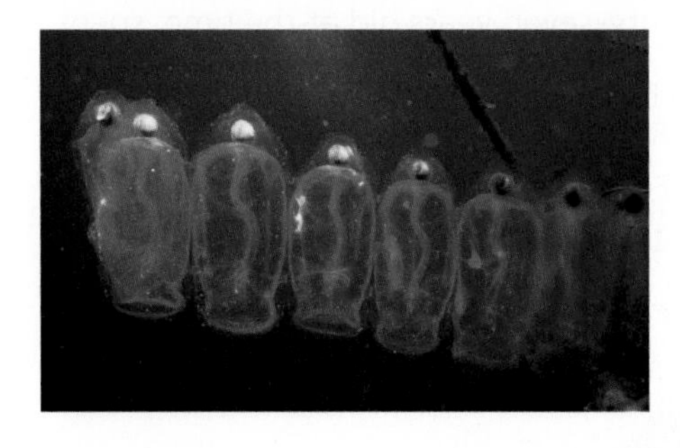

Figure 43.42 Chain of salps (*Thalia democratica*), a kind of tunicate (Section 24.1). Like you, salps have a nerve cord. The cord's anterior end develops into a rudimentary brain, with a light-sensitive eyespot.

3. Write a short description of how global warming may affect spring overturn and thermocline formation in a Minnesota lake. What would be some ecological effects?

4. *Thalia democratica*, a salp, is one of our remote chordate relatives (Figure 43.42). This urochordate is part of marine plankton, usually in warm and temperate seas but also in cold, deep water. Salps swim, separately or in loose chains, in staggering numbers. They range from 1.5 centimeters to a specimen (*Pyrostremma*) that reportedly was 20 meters long and big enough for a diver to swim through. Refer to Section 25.1 on the urochordate feeding mode. Then formulate a hypothesis on how all of the plastic sand floating in the ocean will affect the salp populations and, through them, marine food webs.

5. Southern pine forests dominate the coastal plains of the southern Atlantic and Gulf states. Many pine species are adapted to withstand periodic, lightning-sparked fires. But most fires are suppressed where human development encroaches on the forests. The result is an accumulation of dry undergrowth that can fuel uncontrollable wildfires, such as the one shown at *right*. However, using controlled burns is not a long-term solution. Understory plants grow back within three years or so. The effect of fire can be extended by following the burns with herbicide applications that stop natural postfire succession.

Do you support using controlled burns and herbicides to protect people who live near these forests? If not, can you suggest alternatives?

44 BEHAVIORAL ECOLOGY

IMPACTS, ISSUES My Pheromones Made Me Do It

One spring day, as Toha Bergerub was walking down a street near her Las Vegas home she felt a sharp pain above her right eye—then another, and another. Within a few seconds, hundreds of stinging bees covered the upper half of her body. Firefighters in protective gear rescued her, but she was stung more than 500 times. Bergerub, who was seventy-seven years old at the time, spent a week in the hospital, but recovered fully.

Bergerub's attackers were Africanized honeybees, a hybrid between gentle European honeybees and a more aggressive subspecies endemic to Africa (Figure 44.1). Bee breeders had imported African bees to Brazil in the 1950s. They thought cross-breeding might yield a mild-tempered but zippier pollinator for commercial orchards. However, some African imports escaped and mated with European honeybees that had become established in Brazil before them.

Then, in a grand example of geographic dispersal, some descendants of the hybrids buzzed all the way from Brazil to Mexico and on into the United States. So far, they have reached Texas, New Mexico, Nevada, Utah, California, Oklahoma, Louisiana, Alabama, and Florida.

Africanized honeybees became known as "killer bees," although they rarely kill humans. They have been in the United States since 1990, yet no more than fifteen people have died after being attacked.

All honeybees defend their hives by stinging. Each can sting only once, and all make the same kind of venom. Even so, compared with European honeybees, Africanized ones get riled up more easily, attack in greater numbers, and stay agitated longer. Some are known to have chased people for more than a quarter of a mile.

What makes Africanized bees so testy? Part of the answer is that they make a heightened response to alarm pheromone. A pheromone, recall, is a social cue, a type of signaling molecule emitted by one individual that can influence another individual of the same species. For instance, when a honeybee worker guarding the

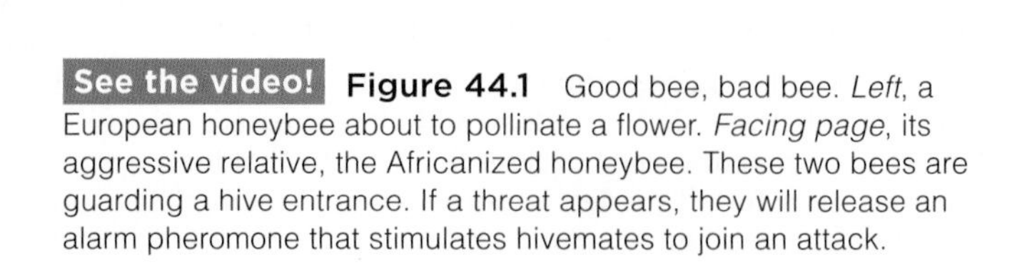

See the video! Figure 44.1 Good bee, bad bee. *Left*, a European honeybee about to pollinate a flower. *Facing page*, its aggressive relative, the Africanized honeybee. These two bees are guarding a hive entrance. If a threat appears, they will release an alarm pheromone that stimulates hivemates to join an attack.

How would you vote? Africanized bees are expanding their range in North America. Learning more about them may help us devise ways to protect ourselves. Should research into the genetic basis of their behavior be a high priority? See ThomsonNOW for details, then vote online.

entrance to a hive senses an intruder, it releases alarm pheromone. Pheromone molecules diffusing through the air excite other bees, which fly out and sting the intruder.

Researchers once studied hundreds of colonies of Africanized honeybees and European honeybees to quantify their responses to alarm pheromone. They positioned a seemingly threatening object, such as a scrap of black cloth, near the entrance of each hive. Then they released a small quantity of an artificial pheromone. The Africanized bees flew out of the hive and zeroed in on the perceived threat much faster. Those bees plunged six to eight times as many stingers into it.

The two kinds of honeybees show other behavioral differences. Africanized bees are less picky about where they establish a colony. They are more likely to abandon their hive after a disturbance. Of more pressing concern to beekeepers, the Africanized bees are less interested in storing large amounts of honey.

Such differences among honeybees lead us into the world of animal behavior—to coordinated responses that animal species make to stimuli. We invite you to reflect first on behavior's genetic basis, which is the foundation for its instinctive and learned mechanisms. Along the way, you will also come across examples of the adaptive value of behavior.

Key Concepts

FOUNDATIONS FOR BEHAVIOR

An individual's behavior starts with interactions among gene products, such as hormones and pheromones. Most forms of behavior have innate components, but they can be modified by environmental factors.

When behavioral traits have a heritable basis, they may evolve by way of natural selection. **Sections 44.1–44.3**

CUES FOR SOCIAL BEHAVIOR

Social behavior depends on evolved modes of communication. Communication signals hold clear meaning for both the sender and the receiver of signals. **Section 44.4**

COSTS AND BENEFITS OF BEHAVIOR

Life in social groups has reproductive benefits and costs. Not all environments favor the evolution of such groups. Self-sacrificing behavior has evolved among a few kinds of animals that live in large family groups. **Sections 44.5–44.7**

WHAT ABOUT HUMANS?

Human behavior was shaped by the same evolutionary forces that shape the behavior of other animals. Only humans consistently make moral choices about their behavior. **Section 44.8**

Links to Earlier Concepts

This chapter builds on your knowledge of sensory and endocrine systems (Sections 31.1, 32.1). You will revisit pheromones and the hormone oxytocin (32.3, 39.19). Be sure you understand the concepts of sexual selection (17.6) and adaptation (17.14). You will take a closer look at social parasites (41.7) and learn more about bird song (32.8). You may wish to review the sections on the evolution of primates and early hominids (24.11). You will be reminded once again of the limits of science (1.5).

44.1 Behavior's Heritable Basis

Like other traits, behavioral traits vary, and many have a heritable basis. Most often, the variation arises from genetic differences that affect the formation or activity of the nervous, endocrine, and muscular systems.

LINKS TO SECTIONS 31.1, 32.3, 32.8, 39.19, 41.7

GENETIC VARIATION IN BEHAVIOR

Animal behavior requires a capacity to detect stimuli. A stimulus, recall, is some type of information about the environment that a sensory receptor has detected (Section 31.1). Which types of stimuli an animal is able to detect and the types of responses it can make start with the structure of its nervous system. Indirectly, then, genes that deal with the structure and activity of the nervous system are the foundation for behavior.

As an example, Stevan Arnold discovered a genetic basis for differences in feeding behavior among garter snakes. Garter snakes that live in coastal forests of the Pacific Northwest prefer to eat banana slugs, a common mollusk on the forest floor (Figure 44.2*a*). There are no banana slugs inland. There, garter snakes eat fishes and tadpoles. Were the garter snakes born with their food preferences or did they learn it? To find out, Arnold offered newborn garter snakes a banana slug as a first meal. Most offspring of coastal snakes ate it. Offspring of inland snakes usually ignored it.

Newborn coastal snakes also flicked their tongue more often at a cotton swab soaked in slug juices, as in Figure 44.2*b*. (Tongue-flicking pulls molecules into the mouth.) Arnold hypothesized that inland snakes don't have a genetically determined mechanism to associate banana slug odor with "FOOD!" He predicted that if he crossed coastal snakes with inland snakes, hybrid offspring would make an *intermediate* response to slug odors. From such experimental crosses, he discovered that hybrid baby snakes tongue-flick at swabs with slug juices more than newborn inland snakes do—but not as often as newborn coastal snakes do.

Figure 44.2 (**a**) Banana slug, the food of choice for adult garter snakes of coastal California. (**b**) A newborn garter snake from a coastal population, tongue-flicking at a cotton swab that had been drenched with fluids from a banana slug.

HORMONES AND BEHAVIOR

Hormones are among the gene products that have big impacts on behavior. For instance, all mammals make and secrete oxytocin, or OT. This hormone has roles in labor and lactation (Section 32.3). In many species, OT also affects pair bonding, aggression, territoriality, and other forms of social behavior.

Among small rodents called prairie voles (*Microtus ochrogaster*), OT is the hormonal key that unlocks the female's heart. The female bonds with a male after a night of repeated matings, and she mates for life. In one experimental test of OT's influence, researchers injected pair-bonded female prairie voles with a drug that blocks the hormonal action. Females that got the injection immediately dumped their partners.

Vole species vary in the number and distribution of OT receptors. This variation may help explain some major differences in mating systems. As one example, prairie voles are monogamous; they mate for life. They have more OT receptors than mountain voles, which are highly promiscuous (Figure 44.3).

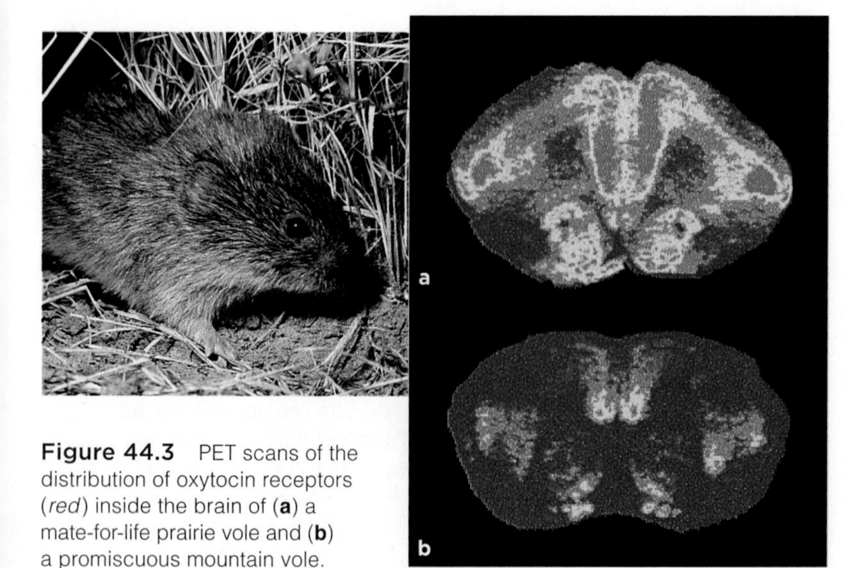

Figure 44.3 PET scans of the distribution of oxytocin receptors (*red*) inside the brain of (**a**) a mate-for-life prairie vole and (**b**) a promiscuous mountain vole.

Genes that affect the development and function of the nervous system influence an animal's ability to perceive stimuli and how it responds to those it does perceive.

Hormones are among the gene products that influence responses and thus affect behavior.

44.2 Instinct and Learning

Some behavioral responses are ready "out of the box"; they are in place at birth. Other behavior is shaped or modified by an animal's experiences, either during a short sensitive period, or throughout the lifetime.

INSTINCTIVE BEHAVIOR

All animals are born with the capacity for instinctive behavior—an innate response to a specific and usually simple stimulus. Such behavior is genetically based, not learned. A newborn coastal garter snake attacking a banana slug is an example. The cuckoo is another. Like cowbirds, this European bird is a social parasite (Section 41.7). Females lay eggs in nests of other birds. A newly hatched cuckoo is blind, but contact with an egg laid by its foster parent stimulates an instinctive response. That hatchling maneuvers the egg onto its back, then shoves it out of the nest (Figure 44.4*a*). The advantage is that the cuckoo gets all the attention.

A cuckoo's egg-dumping response is a fixed action pattern: a series of instinctive movements, triggered by a specific stimulus, that unfolds no matter what else is going on in the environment. Such fixed behavior has survival advantages; it permits a fast response to an important stimulus. It sometimes causes problems. The cuckoo's foster parents are not prewired to notice the color and size of offspring. A simple stimulus—a chick's gaping mouth—induces a fixed action pattern of parental feeding behavior (Figure 44.4*b*).

Figure 44.4 A case of instinctive behavior. (**a**) The cuckoo hatchling shoving an egg from a usurped nest. (**b**) A foster parent feeds the cuckoo chick in response to one simple cue—a gaping mouth.

LEARNED BEHAVIOR

Learned behavior develops after birth, and experience shapes it. Imprinting is a form of learning that occurs in a genetically determined time period. For example, a baby goose learns that the big object bending over it in response to its first peeps offers warmth and safety (Figure 44.5). Young geese trail after this object, which usually is their mother. When mature, they seek out a sexual partner that is similar to the imprinted object.

Figure 44.5 Nobel laureate Konrad Lorenz with geese that imprinted on him. The smaller photograph shows results of a more typical imprinting episode.

A genetic capacity to learn, combined with actual experiences in the environment, shapes most forms of behavior. Remember how the increased daylight hours in spring trigger hormonal changes that induce male songbirds to sing (Section 32.8)? Peter Marler showed that male songbirds often learn their species-specific song early in life by listening to songs of older males.

For one study, Marler raised white-crown sparrow nestlings in soundproof chambers so they could not hear adult males. The captives sang when mature, but their songs were oddly simple. Marler exposed other isolated white-crown sparrows to recordings of males of their species and another species. At maturity, the captives sang just the white-crown song. Instinctively, they paid attention to that song and ignored others.

Marler carried out many experiments. His results support the hypothesis that bird song starts out with a genetically based capacity to concentrate on specific acoustical information. Many birds seem to recognize the outlines of their species song. When they overhear a song in the environment, they use that outline as a guide to fill in details of their own song.

Instinctive behavior occurs without prior experience, is triggered by simple stimuli, and does not vary over time. Most vertebrate behavior involves the genetic capacity to learn based on experiences that vary with the environment.

FOUNDATIONS FOR BEHAVIOR

44.3 Adaptive Behavior

If a behavior varies and some of that variation has a genetic basis, then it will be subject to natural selection. Over time, the most adaptive versions of a behavioral trait will increase in frequency in the population.

LINKS TO SECTIONS 17.14, 32.1

To assess the adaptive value of a behavior, a biologist often considers how it might promote an individual's reproductive success. For example, Larry Clark and Russell Mason thought about the sprigs of aromatic plants that starlings (*Sturnus vulgaris*) tuck into nests. They hypothesized that the nest-decorating behavior suppresses populations of blood-sucking mites that could weaken nestlings. To test their hypothesis, they replaced nests in the wild with nests that were either decorated with wild carrot sprigs or sprig-free. They predicted that the decorated nests would have fewer mites than undecorated ones.

After the starling chicks left the nests, Clark and Mason recorded the number of mites left behind. The number was greatest in sprig-free nests (Figure 44.6). Why? As it turns out, one organic compound in the leaves of wild carrot prevents mites from maturing.

Mason and Clark concluded that decorating a nest with sprigs deters blood-sucking mites. They inferred that this nest-decorating behavior is adaptive because it promotes nestling survival, increasing reproductive success for the nest-decorating birds.

A behavior that increases the reproductive success of an individual tends to be favored by natural selection. Biologists often use controlled experiments to assess the adaptive value of behavioral traits.

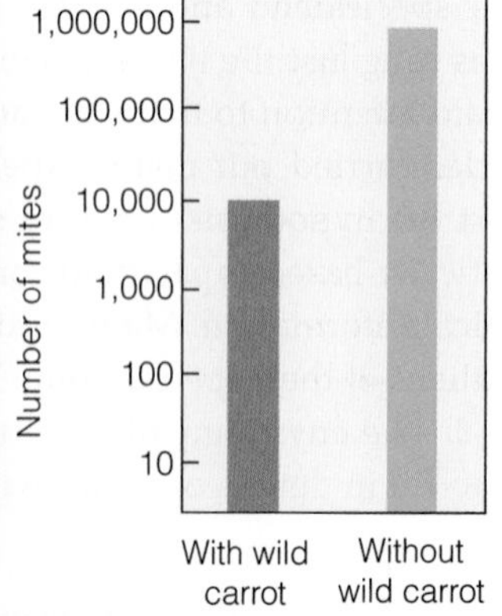

Figure 44.6 Results of an experiment to test the effect of wild carrot sprigs on the number of mites in starling nests. Nests with wild carrots had significantly fewer mites than those with no greenery. There may be a selective advantage to using wild carrot and other aromatic plants as nest-building materials.

CUES FOR SOCIAL BEHAVIOR

44.4 Communication Signals

Competing for food, defending territory, alerting others to danger, advertising sexual readiness, forming bonds with a mate, caring for the offspring—such intraspecific behaviors require unambiguous forms of communication.

Communication signals are cues for social behavior between members of a species. Chemical, acoustical, visual, and tactile signals transmit information from signalers to signal receivers.

Pheromones are *chemical* cues. Signal pheromones make a receiver alter its behavior fast. The honeybee alarm pheromone is an example. So are sex attractants that help males and females find each other. Priming pheromones cause longer term responses, as when a chemical dissolved in the urine of certain male mice triggers estrus in females of the same species.

Many *acoustical* signals, such as bird song, attract mates or define a territory. Others are alarm signals, such as a prairie dog's bark that warns of a predator.

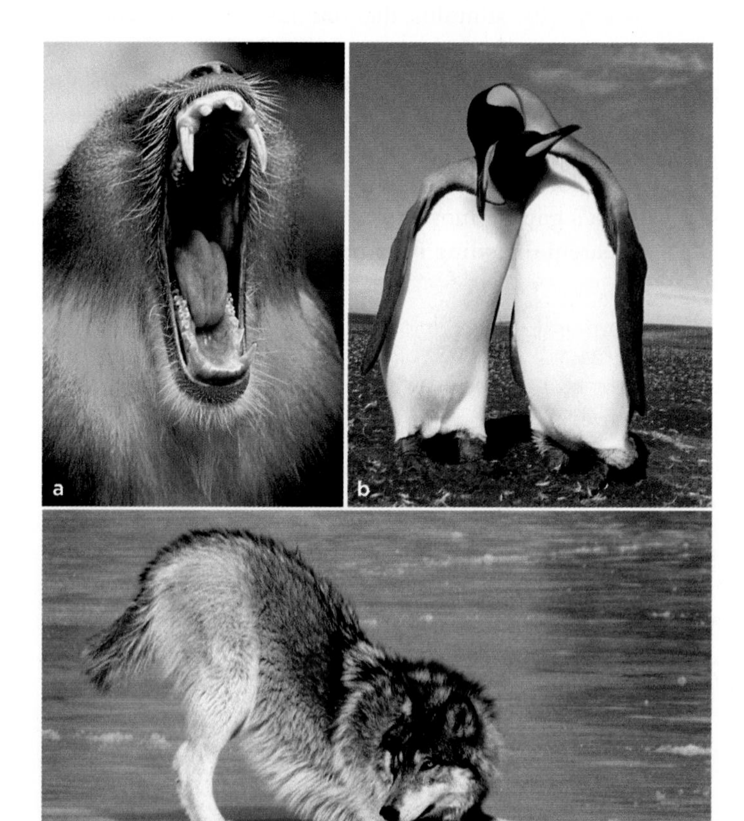

Figure 44.7 Visual signals. (**a**) A male baboon shows his teeth in a threat display. (**b**) Penguins engaged in a courtship display. (**c**) A wolf's play bow tells another wolf that behavior that follows is play, not aggression.

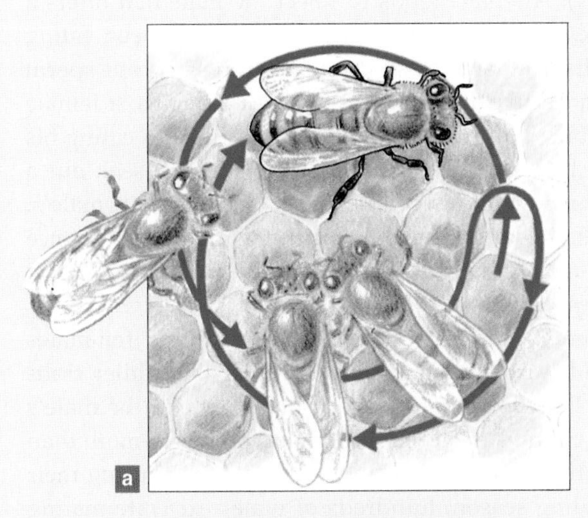

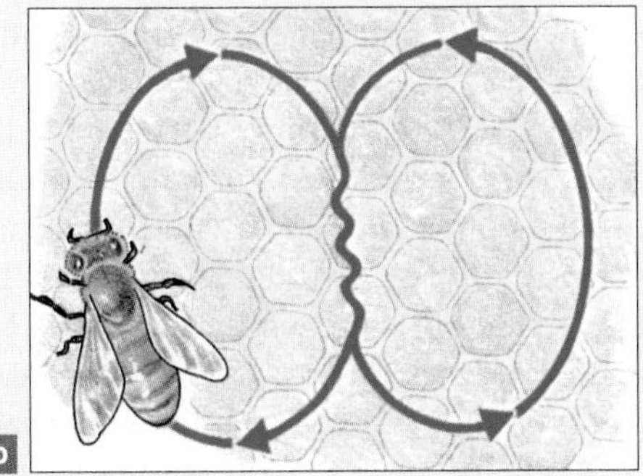

Figure 44.8 **Animated!** Honeybee dances, an example of a tactile display. (**a**) Bees that have visited a source of food close to their hive return and perform a *round* dance on the hive's vertically oriented honeycomb. Bees that maintain contact with the dancer fly out and search for food near the hive.

(**b**) A bee that visits a feeding source more than 100 meters (110 yards) from her hive performs a *waggle* dance. Orientation of an abdomen-waggling dancer in the straight run of her dance informs other bees about the direction of the food.

(**c**) If the food is in line with the sun, the dancer's waggling run proceeds straight up the honeycomb. (**d**) If food is in the opposite direction from the sun, the dancers waggle run is straight down. (**e**) If food is 90 degrees to the right of the direction of the sun, the waggle run is offset by 90 degrees to the right of vertical.

The speed of the dance and the number of waggles in the straight run provides information about distance to the food. When food is 200 meters away, a bee dances much faster, and with more waggles per straight run, compared with a dance inspired by a food source that is 500 meters away.

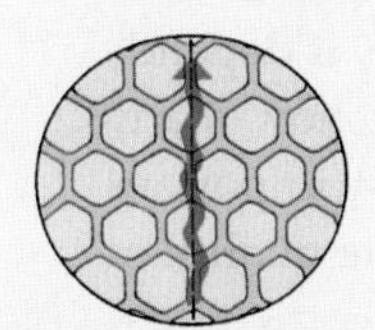

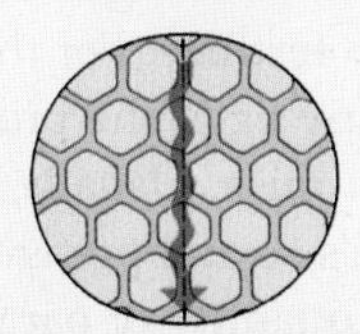

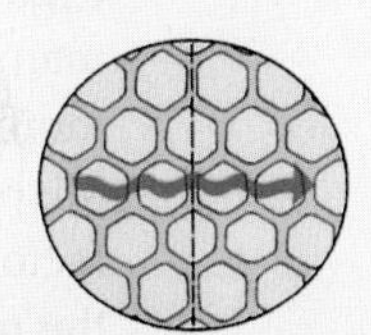

One *visual* signal is a male baboon threat display, which communicates readiness to fight a rival (Figure 44.7*a*). Visual signals are part of courtship displays that often precede mating in birds (Figure 44.7*b*). Selection favors unambiguous signals, so movements often get exaggerated and body form evolves in ways that draw attention to the movements. Figure 17.12 showed an example of feathers that enhance a courtship display.

With *tactile* displays, information is transmitted by touch. For example, after discovering food, a foraging honeybee worker returns to the hive and performs a complex dance. The bee moves in a defined pattern, jostling a crowd of other bees that surround her. The signals give other bees information about the distance and the direction of the food source (Figure 44.8).

The same signal sometimes functions in more than one context. For example, dogs and wolves solicit play behavior with a play bow (Figure 44.7*c*). Without the visual cue, a signal receiver may construe behaviors that follow as aggressive or sexual—but not playful.

A communication signal evolves and persists only if it benefits both sender and receiver. If the signal has disadvantages, then natural selection will tend to favor individuals that do not send or respond to it. Other factors can also select against signalers. For example, male tungara frogs attract females with complex calls, which also make it easier for frog-eating bats to zero in on the caller. When bats are near, male frogs call less, and usually with less flair. The subdued signal is a trade-off between locating a partner for mating and the need for immediate survival.

There are illegitimate signalers, too. For example, fireflies attract mates by producing flashes of light in a characteristic pattern. Some female fireflies prey on males of other species. When a predatory female sees the flash from a male of the prey species, she flashes back as if she were a female of his own species. If she lures him close enough, she captures and eats him.

A communication signal transfers information from one individual to another individual of the same species. Such signals benefit both the signaler and the receiver.

Some individuals of a different species act illegitimately as a communication signaler or receiver.

44.5 Mates, Offspring, and Reproductive Success

In studying behavior, it is often useful to consider the reproductive costs and benefits to each individual. We expect that each sex will evolve in ways that maximize benefits, and minimize costs, which can lead to conflicts.

SEXUAL SELECTION AND MATING BEHAVIOR

LINK TO SECTION 17.6

Males or females of a species often compete for access to mates, and many are choosy about their partners. Both situations lead to sexual selection. As explained in Section 17.6, this microevolutionary process favors characteristics that provide a competitive advantage in attracting and often holding on to mates.

But *whose* reproductive success is it—the male's or the female's? Male animals, remember, produce many small sperm, and females produce far larger but fewer eggs. For the male, success generally depends on how many eggs he can fertilize. For the female, it depends more on how many eggs she produces or how many offspring she can raise. Usually, the most important factor in a female's sexual preference is the quality of the mate, not the quantity of partners.

Female hangingflies (*Harpobittacus apicalis*) are an instructive example. The female prefers males that offer superior food. A male hunts and kills a moth or some other insect. Then he releases a sex pheromone, which attracts females to him and his "nuptial gift" (Figure 44.9*a*). A female tends to select the male that offers a large calorie-rich gift. Only after she has been eating for five minutes or so does she start to accept sperm from her partner. Even after mating begins, a female can break off from her suitor, if she finishes eating his gift. If she does end the mating, she will seek out a new male and his sperm will replace the first male's. Thus, the larger the male's gift, the greater the chance that mostly his sperm will actually end up fertilizing the eggs of his mate.

Females of certain species shop around for males who have appealing traits. Consider the fiddler crabs that live along many sandy shores. One of the male's two claws is enlarged; it often accounts for more than half his total body weight (Figure 44.9*b*). During their breeding season, hundreds of males excavate mating burrows near one another. Each male stands next to his burrow, waving his oversized claw. Female crabs stroll along, checking out males. If a female likes what she sees, she inspects her suitor's burrow. Only when a burrow has the right location and dimensions does she mate with its owner and lay eggs in his burrow.

Some female birds are similarly choosy. Male sage grouse (*Centrocercus urophasianus*) converge at a lek, a type of communal display ground, where each stakes out a few square meters. With tail feathers erect, the males emit booming calls by puffing and deflating big

Figure 44.9 (**a**) Male hangingfly dangling a moth as a nuptial gift for a potential mate. Females of some hangingfly species choose sexual partners that offer the largest gift to them. By waving his enlarged claw, a male fiddler crab (**b**) may attract the eye of a female fiddler crab (**c**). A male sage grouse (**d**) showing off as he competes for female attention at a communal display ground.

neck pouches (Figure 44.9*d*). As they do, they stamp about on their patch of prairie, looking like wind-up toys. Females tend to select and mate with one male sage grouse. Afterward, they go off to nest and raise any young by themselves. Often, many females favor the same male, so most of the males never do mate.

In another behavioral pattern, sexually receptive females of some species cluster in defendable groups. Where you come across such a group, you are likely to observe males competing for access to the clusters. Competition for ready-made harems has resulted in combative male lions, sheep, elk, elephant seals, and bison, to name a few types of animals (Figure 44.10).

Figure 44.10 Male bison locked in combat during the breeding season.

PARENTAL CARE

When females fight for males, we can predict that the males provide more than sperm delivery. Some, such as the male midwife toad, help with parenting. The male holds strings of fertilized eggs around his legs until the eggs hatch (Figure 44.11*a*). Once her eggs are being cared for, a female can mate with other males, if she can find some that are not already caring for eggs. Late in the breeding season, males without strings of eggs are rare, and females fight for access to them. The females even attempt to pry mating pairs apart.

Parental behavior uses up time and energy, which parents otherwise might spend on living long enough to reproduce again. However, for many animals, the benefit of immediate reproductive success outweighs the cost of parenting. Reproductive success might be more chancy later on.

Few reptiles provide care for young. Crocodilians, the reptiles most closely related to birds, are a notable exception. Crocodile parents bury their eggs in a nest. When young are ready to hatch, they call and parents dig them out and care for them for some time.

Most birds are monogamous, and both parents often care for the young (Figure 44.11*b*). In mammals, males typically leave after mating. Females raise the young alone, and males attempt to mate again or conserve energy for the next breeding season (Figure 44.11*c*). Mammalian species in which males help care for the young tend to be monogamous. About 5 percent of all mammals fall into this category.

Researchers use selection theory to explain some aspects of mating behavior.

Male or female preferences for certain behavioral traits can provide the individual's offspring with a competitive edge and promote its reproductive success.

Figure 44.11 (**a**) Male midwife toad with developing eggs wrapped around his legs. (**b**) A pair of Caspian terns cooperate in the care of their chick. (**c**) A female grizzly will care for her cub for as long as two years. The male takes no part in its upbringing.

44.6 Living in Groups

Survey the animal kingdom and you find evolutionary costs and benefits across a range of social groups.

DEFENSE AGAINST PREDATORS

LINKS TO SECTIONS 24.11, 40.4

In some groups, cooperative responses to predators reduce the net risk to all. Vulnerable individuals, too, can be on the alert for predators, join a counterattack, or engage in more effective defenses (Figure 44.12).

Birds, monkeys, meerkats, prairie dogs, and many other animals make alarm calls, as in Figure 44.12*a*. A prairie dog makes a particular bark when it sights an eagle and a different signal when it sights a coyote. Others dive into burrows to escape an eagle's attack or stand erect and observe the coyote's movements.

Sawfly caterpillars feed in clumps on branches and benefit by coordinated repulsion of predatory birds. When a potential predator approaches, they rear up and vomit partly digested eucalyptus leaves (Figure 44.12*b*). Birgitta Sillén-Tullberg showed that predatory birds prefer individuals to a wiggling mass of fluid-exuding caterpillars. When offered caterpillars one at a time, the birds ate an average of 5.6. Birds offered a cluster of twenty caterpillars ate an average of 4.1.

Whenever animals cluster, some individuals shield others from predators. Preference for the center of a group can create a selfish herd, in which individuals hide behind one another. Selfish-herd behavior occurs in bluegill sunfishes. A male sunfish builds a nest by scooping out a depression in mud on the bottom of a lake. Females lay eggs in these nests, and snails and fishes prey on eggs. Competition for the safest sites is greatest near the center of a group, with large males taking the innermost locations. Smaller males cluster around them and bear the brunt of the egg predation. Even so, small males are better off nesting at the edge of the group than out on their own.

IMPROVED FEEDING OPPORTUNITIES

Many mammals, including wolves, lions, wild dogs, and chimpanzees, live in social groups and cooperate in hunts (Figure 44.13). Are cooperative hunters more efficient than solitary ones? Often, no. In one study, researchers observed a solitary lion that caught prey about 15 percent of the time. Two lions cooperatively hunting caught prey twice as often but had to share it, so the amount of food per lion balanced out. When more lions joined a hunt, the success rate per lion fell. Wolves show a similar pattern. Among carnivores that hunt cooperatively, hunting success does not seem to be the major advantage of group living. Individuals hunt together, but they also may fend off scavengers, care for one another's young, and protect territory.

Group living also allows transmission of cultural traits, or behaviors learned by imitation. For example, chimpanzees make and use simple tools by stripping leaves from branches. They use thick sticks to make holes in a termite mound, then insert long, flexible "fishing sticks" into the holes (Figure 44.14). The long

Figure 44.12 Group defenses. (**a**) Black-tailed prairie dogs bark an alarm call that warns others of predators. Does this put the caller at risk? Not much. Prairie dogs usually act as sentries only after they finish feeding and happen to be standing next to their burrows. (**b**) Australian sawfly caterpillars form clumps and regurgitate a fluid (the yellow blobs) that predators find unappealing. (**c**) Musk oxen adults (*Ovibos moschatus*) form a ring of horns, often around their young.

Figure 44.13 Members of a wolf pack (*Canis lupus*). Wolves cooperate in hunting, caring for the young, and defending territory. Benefits are not distributed equally. Only the highest ranking individuals, alpha male and alpha female, breed.

stick agitates the termites, which attack and cling to it. Chimps withdraw the stick and lick off termites, as a high-protein snack. Different groups of chimpanzees use slightly different tool-shaping and termite-fishing methods. Youngsters of each group learn by imitating the adults.

DOMINANCE HIERARCHIES

In many social groups, subordinate individuals do not get an equal share of resources. Most wolf packs, for instance, have one dominant male that breeds with just one dominant female. The others are nonbreeding brothers and sisters, aunts and uncles. All hunt and carry food back to individuals that guard the young in their den.

Why would a subordinate give up resources and, often, breeding privileges? It might get injured or die if it challenges a strong individual. It might not be able to survive on its own. A subordinate might even get a chance to reproduce if it lives long enough or if its dominant peers are taken out by a predator or old age. As one example, some subordinate wolves move up the social ladder when the opportunity arises.

Figure 44.14 Chimpanzees (*Pan troglodytes*) using sticks as tools for extracting tasty termites from a nest. This behavior is learned by imitation.

REGARDING THE COSTS

If social behavior is advantageous, then why are there so few social species? In most habitats, costs outweigh benefits. For instance, packed-together individuals do compete more for resources (Section 40.4). Cormorants and other seabirds form dense breeding colonies, as in Figure 44.15. All compete for space and food.

Large social groups also attract more predators. If individuals are crowded together, they invite parasites and contagious diseases that jump from host to host. The individuals may also be at risk of being killed or exploited by others. Given the opportunity, a pair of breeding herring gulls will cannibalize the eggs and even the chicks of their neighbors.

Living in a social group can provide benefits, as through cooperative defenses or shielding against predators.

Group living has costs: increased competition, increased vulnerability to infections, and exploitation by others.

Figure 44.15 Nearly uniform spacing in a crowded cormorant colony.

44.7 Why Sacrifice Yourself?

Extreme cases of sterility and self-sacrifice have evolved in only two groups of insects and one group of mammals. How are genes of the nonreproducers perpetuated?

SOCIAL INSECTS

LINKS TO CHAPTER 41 INTRODUCTION, SECTION 1.5

True social insects include honeybees, termites, and fire ants (Chapter 41). All of these *eusocial* insects stay together for generations in a group that has a division of labor. Groups include permanently sterile workers that care cooperatively for the offspring of just a few breeding individuals. Such workers often are highly specialized in form and function (Figure 44.16).

A queen honeybee is the only fertile female in her hive. She is larger than her worker daughters, partly because of her enlarged ovaries (Figure 44.17*a*). She secretes a pheromone that makes all other female bees sterile. Female bees distribute it through the hive.

About 30,000 to 50,000 female workers feed larvae, clean and maintain the hive, and construct honeycomb from waxy secretions. Workers also gather the nectar and pollen that feeds the colony. They guard the hive and will sacrifice themselves to repel intruders.

Male stingless drones are produced seasonally and subsist on food gathered by their worker sisters. Each day, drones fly in search of a mate. If one is lucky, he will meet a virgin queen on her one flight away from a colony. He dies after mating. The queen mates with many males, then uses their stored sperm for years.

Like honeybees, termites live in enormous family groups with a queen specialized for producing eggs (Figure 44.17*b*). Unlike the honeybee hive, a termite mound holds sterile individuals of both sexes. A king supplies the female with sperm. Winged reproductive termites of both sexes develop seasonally.

SOCIAL MOLE-RATS

Sterility and extreme self-sacrifice are uncommon in vertebrates. The only eusocial mammals are African mole-rats. The best studied is *Heterocephalus glaber*, the naked mole-rat. Clans of this nearly hairless rodent build and occupy burrows in dry parts of East Africa.

A reproducing female dominates the clan and mates with one to three males (Figure 44.17*c*). Nonbreeding members live to protect and care for the "queen" and "king" (or kings) and their offspring. Sterile diggers excavate tunnels and chambers. When a digger finds an edible root, it hauls a bit back to the main chamber and chirps. Its chirps recruit others, which help carry food back to the chamber. In this way, the queen, her mates, and her young offspring get fed. Other sterile helpers guard the colony. When a predator appears, they chase and attack it at great risk to themselves.

EVOLUTION OF ALTRUISM

A sterile worker in a social insect colony or a naked mole-rat clan shows altruistic behavior: behavior that enhances another individual's reproductive success at the altruist's expense. How did this behavior evolve? According to William Hamilton's theory of inclusive fitness, genes associated with altruism are selected if they lead to behavior that promotes the reproductive success of an altruist's closest relatives.

A sexually reproducing, diploid parent caring for offspring is not helping exact genetic copies of itself. Each of its gametes, and each of its offspring, inherits one-half of its genes. Other individuals of the social group that have the same ancestors also share genes. Siblings (brothers or sisters) are as genetically similar

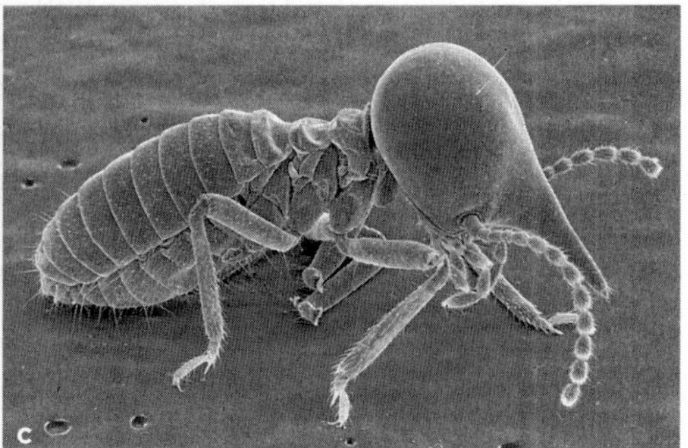

Figure 44.16 Specialized ways of serving and defending the colony. (**a**) An Australian honeypot ant worker. This sterile female is a living container for her colony's food reserves. (**b**) Army ant soldier (*Eciton burchelli*) with formidable mandibles. (**c**) Eyeless soldier termite (*Nasutitermes*). It bombards intruders with a stream of sticky goo from its nozzle-shaped head.

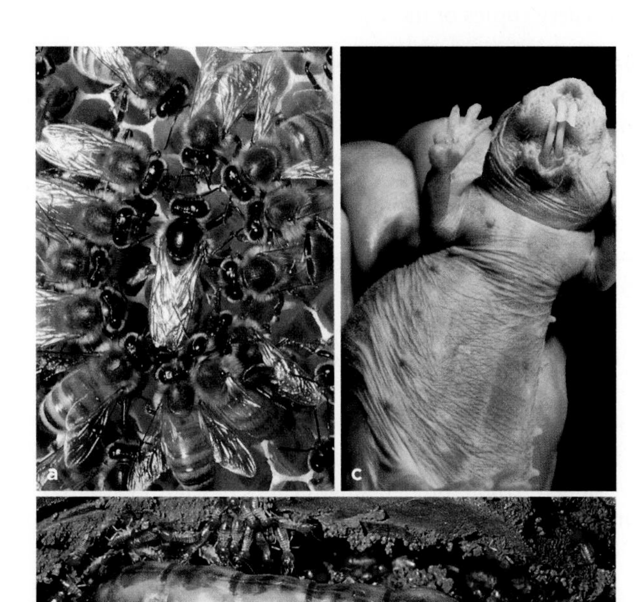

Figure 44.17 Three queens. (**a**) Queen honeybee with her sterile daughters. (**b**) A termite queen (*Macrotermes*) dwarfs her offspring and mate. Ovaries fill her enormous abdomen. (**c**) A naked mole-rat queen.

as a parent and offspring. Nephews and nieces share about one-fourth of their uncle's genes.

Sterile workers promote genes for "self-sacrifice" through behavior that can benefit very close relatives. In honeybee, termite, and ant colonies, sterile workers assist fertile relatives with whom they share genes. A guard bee will die after she stings, but her sacrifice preserves many copies of her genes in her hivemates.

Inbreeding increases the genetic similarity among relatives and may play a role in mole-rat sociality. A clan is highly inbred as a result of many generations of sibling, mother–son, and father–daughter matings. Researchers are now searching for other factors that select for eusocial behavior in mole-rats. According to one hypothesis, arid habitats and patchy food sources favor mole-rat genes that contribute to cooperation in digging burrows, searching for food, and fending off competitors of other species for resources.

Altruistic behavior may persist when individuals pass on genes indirectly, by helping relatives survive and reproduce.

By the theory of inclusive fitness, genes associated with altruistic behavior that is directed toward relatives may spread through a population in certain situations.

44.8 Human Behavior

FOCUS ON SCIENCE

Evolutionary forces shaped and continue to influence human behavior—but humans alone can make moral choices about their actions.

Hormones and Pheromones Are humans, too, influenced by hormones that contribute to bonding behavior in other mammals? Perhaps. Consider autism, a developmental disorder in which people have trouble making social contacts and often fixate on objects. The disorder is associated with low oxytocin levels. Oxytocin is known to affect bonding behavior in other mammals.

Also, pheromones may influence human behavior. When females live in proximity, as in college dormitories, their menstrual cycles typically become synchronized. Martha McClintock and Kathleen Stern demonstrated that one woman's menstrual cycle will lengthen or shorten after she has become exposed to sweat secreted by a woman who was in a different phase of the cycle.

Morality and Behavior If we are comfortable with studying the evolutionary basis of behavior of termites, naked mole-rats, and other animals, why do people often resist the idea of analyzing the evolutionary basis of human behavior? A common fear is that an objectionable behavior will be defined as "natural." To evolutionary biologists, however, "adaptive" does not mean "morally right." It simply means a behavior increases reproductive success. Scientific studies do not address moral issues (Section 1.5).

For example, infanticide is morally repugnant. Is it unnatural? No. It happens in many animal groups and all human cultures. Male lions often kill the offspring of other males when they take over a pride. Thus deprived of parenting tasks, the lionesses can now breed with the infanticidal male and increase its reproductive success.

Biologists would predict that unrelated human males are a threat to infants. Evidence supports the prediction. The absence of a biological father and the presence of an unrelated male increases risk of death for an American child under age two by more than sixty times.

What about parents who kill their own offspring? In her book on maternal behavior, primatologist Sarah Blaffer Hrdy cites a study of a village in Papua New Guinea in which parents killed about 40 percent of the newborns. As Hrdy argues, when resources or social support are hard to come by, a mother's fitness might increase if a newborn who is unlikely to survive is killed. The mother can allocate child-rearing energy to her other offspring or save it for children she may have in the future.

Do most of us find such behavior appalling? Yes. Can considering the possible evolutionary advantages of the behavior help us prevent it? Perhaps. An analysis of the conditions under which infanticide occurs tells us this: When mothers lack the resources they need to care for their children, they are more likely to harm them. We as a society can act upon such information.

www.thomsonedu.com ThomsonNOW!

Summary

Sections 44.1, 44.2 Behavior refers to coordinated responses that an animal makes to stimuli. It starts with genes that influence the development and activity of the nervous, endocrine, and muscular systems.

Instinctive behavior can occur without having been learned by experience. It is a fixed response and a simple environmental cue typically elicits it.

Animals learn when they process and integrate any information from experiences, then use that information to vary or change responses to stimuli. Imprinting is one form of learning that happens only during a sensitive period early in life.

Section 44.3 A behavior that has a genetic basis is subject to evolution by natural selection. Adaptive forms of behavior evolved as a result of individual differences in reproductive success in past generations.

Section 44.4 Communication signals are meant to change the behavior of individuals of the same species.

Chemical signals such as pheromones have roles in social communication. Acoustical signals are sounds that have precise, species-specific information. Visual signals are part of courtship displays and threat displays. Tactile signals are specific forms of physical contact between a signaler and a receiver.

■ *Use the animation on ThomsonNOW to explore the honeybee dance language.*

Section 44.5 Sexual selection favors traits that give an individual a competitive edge in attracting and often holding onto mates. Females of many species select for males that have traits or engage in behaviors they find attractive. When large numbers of females cluster in a defensible area, males may compete with one another to control the areas.

Parental care has reproductive costs in terms of future reproduction and survival. It is adaptive when benefits to a present set of offspring offset the costs.

■ *Read the InfoTrac article "Something Fishy in the Nest," Bryan Neff,* Natural History, *February 2004.*

Section 44.6 Animals that live in social groups may benefit by cooperating in predator detection, defense, and rearing the young. Benefits of group living are often distributed unequally. Species that live in large groups incur costs, including increased disease and parasitism, and increased competition for resources.

Section 44.7 Ants, termites, and some other insects as well as two species of mole-rats are eusocial. They live in colonies with overlapping generations and have a reproductive division of labor. Most colony members do not reproduce; they assist their relatives instead.

According to the theory of inclusive fitness, altruism is perpetuated because altruistic individuals share genes with their reproducing relatives. Altruistic individuals help perpetuate the genes that lead to their altruism by promoting reproductive success of close relatives that also carry copies of these genes.

Section 44.8 Hormones and possibly pheromones influence human behavior. A behavior that is adaptive in the evolutionary sense may still be judged by society to be morally wrong. Science does not address morality.

Self-Quiz

Answers in Appendix III

1. Genes affect the behavior of individuals by ______ .
   a. influencing the development of nervous systems
   b. affecting the kinds of hormones in individuals
   c. determining which stimuli can be detected
   d. all of the above

2. Stevan Arnold offered slug meat to newborn garter snakes from different populations to test his hypothesis that the snakes' response to slugs ______ .
   a. was shaped by indirect selection
   b. is an instinctive behavior
   c. is based on pheromones
   d. is adaptive

3. A behavior is defined as adaptive if it ______ .
   a. varies among individuals of a population
   b. occurs without prior learning
   c. increases an individual's reproductive success
   d. is widespread across a species

4. The honeybee dance language transmits information about distance to food by way of ______ signals.
   a. tactile
   b. chemical
   c. acoustical
   d. visual

5. A ______ is a chemical that conveys information between individuals of the same species.
   a. pheromone
   b. neurotransmitter
   c. hormone
   d. all of the above

6. In ______ , males and females typically cooperate in care of the young.
   a. mammals
   b. birds
   c. amphibians
   d. all of the above

7. Generally, living in a social group costs the individual, in terms of ______ .
   a. competition for food, other resources
   b. vulnerability to contagious diseases
   c. competition for mates
   d. all of the above

8. Social behavior evolves because ______ .
   a. social animals are more advanced than solitary ones
   b. under some conditions, the costs of social life to individuals are offset by benefits to the species
   c. under some conditions, the benefits of social life to an individual offset the costs to that individual
   d. under most conditions, social life has no costs to an individual

9. Eusocial insects ______ .
   a. live in extended family groups
   b. include termites, honeybees, and ants
   c. show a reproductive division of labor
   d. a and c
   e. all of the above

10. Helping other individuals at a reproductive cost to oneself might be adaptive if those helped are ______ .
  a. members of another species
  b. competitors for mates
  c. close relatives
  d. illegitimate signalers

11. Match the terms with their most suitable description.

___ fixed action pattern	a. time-dependent form of learning requiring exposure to key stimulus
___ altruism	b. genes plus actual experience
___ basis of instinctive and learned behavior	c. series of responses that runs to completion independently of feedback from environment
___ imprinting	d. assisting another individual at one's own expense
___ pheromone	e. one communication signal

■ *Visit ThomsonNOW for additional questions.*

Critical Thinking

1. Sexual imprinting is common in birds. During a short sensitive period in early life, the bird learns features that it will seek later, when ready to mate. Figure 44.18 shows an amorous rooster wading into the water after ducks. Speculate on what might have caused this behavior.

2. For billions of years, the only bright objects in the night sky were stars or the moon. Night-flying moths use them to navigate a straight line. Today, the instinct to fly toward bright objects causes moths to exhaust themselves fluttering around streetlights and banging against brightly lit windowpanes. This behavior clearly is not adaptive, so why does it persist?

3. Damaraland mole-rats are relatives of naked mole-rats (Figure 44.19). In their clans, too, nonbreeding individuals of both sexes cooperatively assist one breeding pair. Even so, breeding individuals in wild Damaraland mole-rat colonies usually are unrelated, and few subordinates move up in the social hierarchy to breeding status. Researchers suspect that ecological factors, not genetic ones, were the more important selective force in Damaraland mole-rat altruism. Explain why.

4. A cheetah scent-marks plants in its territory with certain exocrine gland secretions. What evidence would you require to demonstrate that the cheetah's action is an evolved communication signal?

5. A female chimpanzee engages in sex only during her fertile period, which is advertised by a swelling of her external genitals. Bonobos, the closest relatives of the chimpanzees, are more sexually active (Figure 44.20). Females mate even when they are not fertile. In both species, mating is promiscuous; each female mates with many males. By one hypothesis, female promiscuity may help prevent male infanticide. Chimpanzee males have been observed to kill infants in the wild, but this behavior has never been observed among bonobos.

Explain why it would be selectively disadvantageous for a male to kill the infant of a female with whom he had sex. How might the bonobo female's increased sexual activity lower likelihood of male infanticide still further?

Figure 44.18 Behaviorally confused rooster attempting to court a mallard duck.

Figure 44.19 Damaraland mole-rats (*Cryptomys damarensis*) in a burrow. Like the related naked mole-rats, they live in colonies with nonbreeding workers. Compared with naked mole-rats, these fuzzy burrowers are less inbred.

Figure 44.20 Female bonobo (*Pan paniscus*) with her offspring. Like humans, and unlike the chimpanzees, bonobo females have sexual organs that allow them to copulate facing their partner. Also like humans, bonobo females can be sexually receptive at any time, not only during their most fertile period.

Epilogue

BIOLOGICAL PRINCIPLES AND THE HUMAN IMPERATIVE

Molecules, single cells, tissues, organs, organ systems, multicelled organisms, populations, communities, ecosystems, and the biosphere. These are architectural systems of life, assembled in increasingly complex ways over the past 3.8 billion years. We are latecomers to this immense biological building program. And yet, within the relatively short span of 10,000 years, many of our activities have been changing the character of the land, ocean, and atmosphere, even the genetic character of species.

It would be presumptuous to think that we alone have had profound impact on the world of life. As long ago as the Proterozoic, photosynthetic organisms were irrevocably changing the course of biological evolution by enriching the atmosphere with oxygen. During the past as well as the present, competitive adaptations led to the rise of some groups whose dominance ensured the decline of others. Change is nothing new. What *is* new is the capacity of one species to comprehend what might be going on.

We now have the population size, technology, and cultural inclination to use up energy and modify the environment at rapid rates. Where will this end? Will feedback controls operate as they do, for instance, when population growth exceeds carrying capacity? In other words, will negative feedback controls come into play and keep things from getting too far out of hand?

Feedback control will not be enough, for it does not get under way until the deviation has reached a critical

threshold. Our patterns of resource consumption and our population growth are founded on an illusion of unlimited resources and a forgiving environment. A prolonged, global shortage of food or the passing of a critical threshold for the global climate can come too fast to be corrected; in which case the impact of the deviation may be too great to be reversed.

What about feedforward mechanisms, which might serve as early warning systems? For example, when sensory receptors near the surface of skin detect a drop in outside air temperature, each sends messages to the nervous system. That system responds by triggering mechanisms that raise the body's core temperature before the body itself becomes dangerously chilled. Extrapolating from this, if we develop feedforward control mechanisms, would it not be possible to start corrective measures before we do too much harm?

Feedforward controls alone will not work, for they operate after change is under way. Think of the DEW line—the Distant Early Warning system. It is like a vast sensory receptor for detecting missiles launched against North America. By the time it does what it is supposed to, it may be too late to stop widespread destruction.

It would be naive to assume we can ever reverse who we are at this point in evolutionary time, to de-evolve ourselves culturally and biologically, and become less complex in the hope of averting disaster. Yet there is reason to believe we can avert disaster by using a third kind of control mechanism—a capacity to anticipate events even before they happen. We are not locked into responding only after irreversible change has begun. We have the capacity to anticipate the future—it is the essence of our visions of utopia and hell. We all have the capacity to adapt to a future that we can partly shape.

For instance, we can stop trying to "beat nature" and learn to work with it. Individually and collectively, we can develop long-term policies that take into account biotic and abiotic limits on population growth. Far from being a surrender, this approach would be one of the most intelligent behaviors of which we are capable.

Having a capacity to adapt and using it are not the same thing. We have already put the world of life on dangerous ground because we have not yet mobilized ourselves as a species to work toward self-control.

Our survival depends on predicting possible futures. It depends on preserving, restoring, and constructing ecosystems that fit with our definition of basic human values and available biological models. Human values can change; our expectations can and must be adapted to biological reality. For the principles of energy flow and resource utilization, which govern the survival of all systems of life, do not change.

It is our biological and cultural imperative that we come to terms with these principles, and ask ourselves this: What will be our long-term contribution to the world of life?

Appendix I. Classification System

This revised classification scheme is a composite of several that microbiologists, botanists, and zoologists use. The major groupings are agreed upon, more or less. However, there is not always agreement on what to name a particular grouping or where it might fit within the overall hierarchy. There are several reasons why full consensus is not possible at this time.

First, the fossil record varies in its completeness and quality. Therefore, the phylogenetic relationship of one group to other groups is sometimes open to interpretation. Today, comparative studies at the molecular level are firming up the picture, but the work is still under way. Also, molecular comparisons do not always provide definitive answers to questions about phylogeny. Comparisons based on one set of genes may conflict with those comparing a different part of the genome. Or comparisons with one member of a group may conflict with comparisons based on other group members.

Second, ever since the time of Linnaeus, systems of classification have been based on the perceived morphological similarities and differences among organisms. Although some original interpretations are now open to question, we are so used to thinking about organisms in certain ways that reclassification often proceeds slowly.

A few examples: Traditionally, birds and reptiles were grouped in separate classes (Reptilia and Aves); yet there are compelling arguments for grouping the lizards and snakes in one group and the crocodilians, dinosaurs, and birds in another. Many biologists still favor a six-kingdom system of classification (archaea, bacteria, protists, plants, fungi, and animals). Others advocate a switch to the more recently proposed three-domain system (archaea, bacteria, and eukarya).

Third, researchers in microbiology, mycology, botany, zoology, and other fields of inquiry inherited a wealth of literature, based on classification systems that have been developed over time in each field of inquiry. Many are reluctant to give up established terminology that offers access to the past.

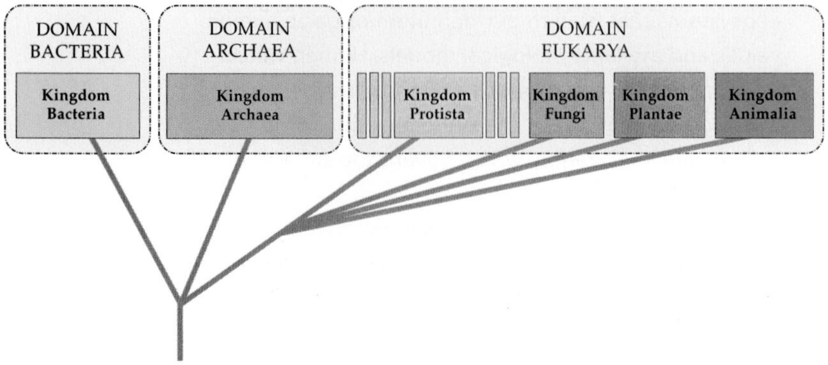

For example, botanists and microbiologists often use *division*, and zoologists *phylum*, for taxa that are equivalent in hierarchies of classification.

Why bother with classification frameworks if we know they only imperfectly reflect the evolutionary history of life? We do so for the same reasons that a writer might break up a history of civilization into several volumes, each with a number of chapters. Both are efforts to impart structure to an enormous body of knowledge and to facilitate retrieval of information from it. More importantly, to the extent that modern classification schemes accurately reflect evolutionary relationships, they provide the basis for comparative biological studies, which link all fields of biology.

Bear in mind that we include this appendix for your reference purposes only. Besides being open to revision, it is not meant to be complete. Names shown in "quotes" are polyphyletic or paraphyletic groups that are undergoing revision. For example, "reptiles" comprise at least three and possibly more lineages.

The most recently discovered species, as from the mid-ocean province, are not listed. Many existing and extinct species of the more obscure phyla are also not represented. Our strategy is to focus primarily on the organisms mentioned in the text or familiar to most students. We delve more deeply into flowering plants than into bryophytes, and into chordates than annelids.

PROKARYOTES AND EUKARYOTES COMPARED

As a general frame of reference, note that almost all bacteria and archaea are microscopic in size. Their DNA is concentrated in a nucleoid (a region of cytoplasm), not in a membrane-bound nucleus. All are single cells or simple associations of cells. They reproduce by prokaryotic fission or budding; they transfer genes by bacterial conjugation.

Table A lists representative types of autotrophic and heterotrophic prokaryotes. The authoritative reference, *Bergey's Manual of Systematic Bacteriology,* has called this a time of taxonomic transition. It references groups mainly by numerical taxonomy (Section 19.1) rather than by phylogeny. Our classification system does reflect evidence of evolutionary relationships for at least some bacterial groups.

The first life forms were prokaryotic. Similarities between Bacteria and Archaea have more ancient origins relative to the traits of eukaryotes.

Unlike the prokaryotes, all eukaryotic cells start out life with a DNA-enclosing nucleus and other membrane-bound organelles. Their chromosomes have many histones and other proteins attached. They include spectacularly diverse single-celled and multicelled species, which can reproduce by way of meiosis, mitosis, or both.

DOMAIN OF BACTERIA

KINGDOM BACTERIA

The largest, and most diverse group of prokaryotic cells. Includes photosynthetic autotrophs, chemosynthetic autotrophs, and heterotrophs. All prokaryotic pathogens of vertebrates are bacteria.

PHYLUM AQIFACAE Most ancient branch of the bacterial tree. Gram-negative, mostly aerobic chemoautotrophs, mainly of volcanic hot springs. *Aquifex.*

PHYLUM DEINOCOCCUS-THERMUS Gram-positive, heat-loving chemoautotrophs. *Deinococcus* is the most radiation resistant organism known. *Thermus* occurs in hot springs and near hydrothermal vents.

PHYLUM CHLOROFLEXI Green nonsulfur bacteria. Gram-negative bacteria of hot springs, freshwater lakes, and marine habitats. Act as nonoxygen-producing photoautotrophs or aerobic chemoheterotrophs. *Chloroflexus.*

PHYLUM ACTINOBACTERIA Gram-positive, mostly aerobic heterotrophs in soil, freshwater and marine habitats, and on mammalian skin. *Propionibacterium, Actinomyces, Streptomyces.*

PHYLUM CYANOBACTERIA Gram-negative, oxygen-releasing photoautotrophs mainly in aquatic habitats. They have chlorophyll *a* and photosystem I. Includes many nitrogen-fixing genera. *Anabaena, Nostoc, Oscillatoria.*

PHYLUM CHLOROBIUM Green sulfur bacteria. Gram-negative nonoxygen-producing photosynthesizers, mainly in freshwater sediments. *Chlorobium.*

PHYLUM FIRMICUTES Gram-positive walled cells and the cell wall-less mycoplasmas. All are heterotrophs. Some survive in soil, hot springs, lakes, or oceans. Others live on or in animals. *Bacillus, Clostridium, Heliobacterium, Lactobacillus, Listeria, Mycobacterium, Mycoplasma, Streptococcus.*

PHYLUM CHLAMYDIAE Gram-negative intracellular parasites of birds and mammals. *Chlamydia.*

PHYLUM SPIROCHETES Free-living, parasitic, and mutualistic gram-negative spring-shaped bacteria. *Borelia, Pillotina, Spirillum, Treponema.*

PHYLUM PROTEOBACTERIA The largest bacterial group. Includes photoautotrophs, chemoautotrophs, and heterotrophs; free-living, parasitic, and colonial groups. All are gram-negative.

Class Alphaproteobacteria. *Agrobacterium, Azospirillum, Nitrobacter, Rickettsia, Rhizobium.*

Class Betaproteobacteria. *Neisseria.*

Class Gammaproteobacteria. *Chromatium, Escherichia, Haemopilius, Pseudomonas, Salmonella, Shigella, Thiomargarita, Vibrio, Yersinia.*

Class Deltaproteobacteria. *Azotobacter, Myxococcus.*

Class Epsilonproteobacteria. *Campylobacter, Helicobacter.*

DOMAIN OF ARCHAEA

KINGDOM ARCHAEA

Prokaryotes that are evolutionarily between eukaryotic cells and the bacteria. Most are anaerobes. None are photosynthetic. Originally discovered in extreme habitats, they are now known to be widely dispersed. Compared with bacteria, the archaea have a distinctive cell wall structure and unique membrane lipids, ribosomes, and RNA sequences. Some are symbiotic with animals, but none are known to be animal pathogens.

PHYLUM EURYARCHAEOTA Largest archean group. Includes extreme thermophiles, halophiles, and methanogens. Others are abundant in the upper waters of the ocean and other more moderate habitats. *Methanocaldococcus, Nanoarchaeum.*

PHYLUM CRENARCHAEOTA Includes extreme theromophiles, as well as species that survive in Antarctic waters, and in more moderate habitats. *Sulfolobus, Ignicoccus.*

PHYLUM KORARCHAEOTA Known only from DNA isolated from hydrothermal pools. As of this writing, none have been cultured and no species have been named.

DOMAIN OF EUKARYOTES

KINGDOM "PROTISTA"

A collection of single-celled and multicelled lineages, which does not constitute a monophyletic group. Some biologists consider the groups listed below to be kingdoms in their own right.

PARABASALIA Parabasalids. Flagellated, single-celled anaerobic heterotrophs with a cytoskeletal "backbone" that runs the length of the cell. There are no mitochondria, but a hydrogenosome serves a similar function. *Trichomonas, Trichonympha.*

DIPLOMONADIDA Diplomonads. Flagellated, anaerobic single-celled heterotrophs that do not have mitochondria or Golgi bodies and do not form a bipolar spindle at mitosis. May be one of the most ancient lineages. *Giardia.*

EUGLENOZOA Euglenoids and kinetoplastids. Free-living and parasitic flagellates. All with one or more mitochondria. Some photosynthetic euglenoids with chloroplasts, others heterotrophic. *Euglena, Trypanosoma, Leishmania.*

RHIZARIA Formaminiferans and radiolarians. Free-living, heterotrophic amoeboid cells that are enclosed in shells. Most live in ocean waters or sediments. *Pterocorys, Stylosphaera.*

ALVEOLATA Single cells having a unique array of membrane-bound sacs (alveoli) just beneath the plasma membrane.

Ciliata. Ciliated protozoans. Heterotrophic protists with many cilia. *Paramecium, Didinium.*

Dinoflagellates. Diverse heterotrophic and photosynthetic flagellated cells that deposit cellulose in their alveoli. *Gonyaulax, Gymnodinium, Karenia, Noctiluca.*

Apicomplexans. Single-celled parasites of animals. A unique microtubular device is used to attach to and penetrate a host cell. *Plasmodium.*

STRAMENOPHILA Stramenophiles. Single-celled and multicelled forms; flagella with tinsel-like filaments.

Oomycotes. Water molds. Heterotrophs. Decomposers, some parasites. *Saprolegnia, Phytophthora, Plasmopara.*

Chrysophytes. Golden algae, yellow-green algae, diatoms, coccolithophores. Photosynthetic. *Emiliania, Mischococcus.*

Phaeophytes. Brown algae. Photosynthetic; nearly all live in temperate marine waters. All are multicellular. *Macrocystis, Laminaria, Sargassum, Postelsia.*

RHODOPHYTA Red algae. Mostly photosynthetic, some parasitic. Nearly all marine, some in freshwater habitats. Most multicellular. *Porphyra, Antithamion.*

CHLOROPHYTA Green algae. Mostly photosynthetic, some parasitic. Most freshwater, some marine or terrestrial. Single-celled, colonial, and multicellular forms. Some biologists place the chlorophytes and charophytes with the land plants in a kingdom called the Viridiplantae. *Acetabularia, Chlamydomonas, Chlorella, Codium, Udotea, Ulva, Volvox.*

CHAROPHYTA Photosynthetic. Closest living relatives of plants. Include both single-celled and multicelled forms. Desmids, stoneworts. *Micrasterias, Chara, Spirogyra.*

AMOEBOZOA True amoebas and slime molds. Heterotrophs that spend all or part of the life cycle as a single cell that uses pseudopods to capture food. *Amoeba, Entoamoeba* (amoebas), *Dictyostelium* (cellular slime mold), *Physarum* (plasmodial slime mold).

KINGDOM FUNGI Nearly all multicelled eukaryotic species with chitin-containing cell walls. Heterotrophs, mostly saprobic decomposers, some parasites. Nutrition based upon extracellular digestion of organic matter and absorption of nutrients by individual cells. Multicelled species form absorptive mycelia and reproductive structures that produce asexual spores (and sometimes sexual spores).

PHYLUM CHYTRIDIOMYCOTA Chytrids. Primarily aquatic; saprobic decomposers or parasites that produce flagellated spores. *Chytridium.*

PHYLUM ZYGOMYCOTA Zygomycetes. Producers of zygospores (zygotes inside thick wall) by way of sexual reproduction. Bread molds, related forms. *Rhizopus, Philobolus.*

PHYLUM ASCOMYCOTA Ascomycetes. Sac fungi. Sac-shaped cells form sexual spores (ascospores). Most yeasts and molds, morels, truffles. *Saccharomycetes, Morchella, Neurospora, Claviceps, Candida, Aspergillus, Penicillium.*

PHYLUM BASIDIOMYCOTA Basidiomycetes. Club fungi. Most diverse group. Produce basidiospores inside club-shaped structures. Mushrooms, shelf fungi, stinkhorns. *Agaricus, Amanita, Craterellus, Gymnophilus, Puccinia, Ustilago.*

"IMPERFECT FUNGI" Sexual spores absent or undetected. The group has no formal taxonomic status. If better understood, a given species might be grouped with sac fungi or club fungi. *Arthobotrys, Histoplasma, Microsporum, Verticillium.*

"LICHENS" Mutualistic interactions between fungal species and a cyanobacterium, green alga, or both. *Lobaria, Usnea.*

KINGDOM PLANTAE Most photosynthetic with chlorophylls *a* and *b*. Some parasitic. Nearly all live on land. Sexual reproduction predominates.

BRYOPHYTES (NONVASCULAR PLANTS)

Small flattened haploid gametophyte dominates the life cycle; sporophyte remains attached to it. Sperm are flagellated; require water to swim to eggs for fertilization.

PHYLUM HEPATOPHYTA Liverworts. *Marchantia.*

PHYLUM ANTHOCEROPHYTA Hornworts.

PHYLUM BRYOPHYTA Mosses. *Polytrichum, Sphagnum.*

SEEDLESS VASCULAR PLANTS

Diploid sporophyte dominates, free-living gametophytes, flagellated sperm require water for fertilization.

PHYLUM LYCOPHYTA Lycophytes, club mosses. Small single-veined leaves, branching rhizomes. *Lycopodium, Selaginella.*

PHYLUM MONILOPHYTA

Subphylum Psilophyta. Whisk ferns. No obvious roots or leaves on sporophyte, very reduced. *Psilotum.*

Subphylum Sphenophyta. Horsetails. Reduced scalelike leaves. Some stems photosynthetic, others spore-producing. *Calamites* (extinct), *Equisetum.*

Subphylum Pterophyta. Ferns. Large leaves, usually with sori. Largest group of seedless vascular plants (12,000 species), mainly tropical, temperate habitats. *Pteris, Trichomanes, Cyathea* (tree ferns), *Polystichum.*

SEED-BEARING VASCULAR PLANTS

PHYLUM CYCADOPHYTA Cycads. Group of gymnosperms (vascular, bear "naked" seeds). Tropical, subtropical. Compound leaves, simple cones on male and female plants. Plants usually palm-like. Motile sperm. *Zamia, Cycas.*

PHYLUM GINKGOPHYTA Ginkgo (maidenhair tree). Type of gymnosperm. Motile sperm. Seeds with fleshy layer. *Ginkgo.*

PHYLUM GNETOPHYTA Gnetophytes. Only gymnosperms with vessels in xylem and double fertilization (but endosperm does not form). *Ephedra, Welwitchia, Gnetum.*

PHYLUM CONIFEROPHYTA Conifers. Most common and familiar gymnosperms. Generally cone-bearing species with needle-like or scale-like leaves. Includes pines (*Pinus*), redwoods (*Sequoia*), yews (*Taxus*).

PHYLUM ANTHOPHYTA Angiosperms (the flowering plants). Largest, most diverse group of vascular seed-bearing plants. Only organisms that produce flowers, fruits. Some families from several representative orders are listed:

BASAL FAMILIES

Family Amborellaceae. *Amborella.*
Family Nymphaeaceae. Water lilies.
Family Illiciaceae. Star anise.

MAGNOLIIDS

Family Magnoliaceae. Magnolias.
Family Lauraceae. Cinnamon, sassafras, avocados.
Family Piperaceae. Black pepper, white pepper.

EUDICOTS

Family Papaveraceae. Poppies.
Family Cactaceae. Cacti.
Family Euphorbiaceae. Spurges, poinsettia.
Family Salicaceae. Willows, poplars.
Family Fabaceae. Peas, beans, lupines, mesquite.
Family Rosaceae. Roses, apples, almonds, strawberries.
Family Moraceae. Figs, mulberries.
Family Cucurbitaceae. Squashes, melons, cucumbers.
Family Fagaceae. Oaks, chestnuts, beeches.
Family Brassicaceae. Mustards, cabbages, radishes.
Family Malvaceae. Mallows, okra, cotton, hibiscus, cocoa.
Family Sapindaceae. Soapberry, litchi, maples.
Family Ericaceae. Heaths, blueberries, azaleas.
Family Rubiaceae. Coffee.
Family Lamiaceae. Mints.
Family Solanaceae. Potatoes, eggplant, petunias.
Family Apiaceae. Parsleys, carrots, poison hemlock.
Family Asteraceae. Composites. Chrysanthemums, sunflowers, lettuces, dandelions.

MONOCOTS

Family Araceae. Anthuriums, calla lily, philodendrons.
Family Liliaceae. Lilies, tulips.
Family Alliaceae. Onions, garlic.
Family Iridaceae. Irises, gladioli, crocuses.
Family Orchidaceae. Orchids.
Family Arecaceae. Date palms, coconut palms.
Family Bromeliaceae. Bromeliads, pineapples.
Family Cyperaceae. Sedges.
Family Poaceae. Grasses, bamboos, corn, wheat, sugarcane.
Family Zingiberaceae. Gingers.

KINGDOM ANIMALIA Multicelled heterotrophs, nearly all with tissues and organs, and organ systems, that are motile during part of the life cycle. Sexual reproduction occurs in most, but some also reproduce asexually. Embryos develop through a series of stages.

PHYLUM PORIFERA Sponges. No symmetry, tissues.

PHYLUM PLACOZOA Marine. Simplest known animal. Two cell layers, no mouth, no organs. *Trichoplax.*

PHYLUM CNIDARIA Radial symmetry, tissues, nematocysts.

Class Hydrozoa. Hydrozoans. *Hydra, Obelia, Physalia, Prya.*
Class Scyphozoa. Jellyfishes. *Aurelia.*
Class Anthozoa. Sea anemones, corals. *Telesto.*

PHYLUM PLATYHELMINTHES Flatworms. Bilateral, cephalized; simplest animals with organ systems. Saclike gut.

Class Turbellaria. Triclads (planarians), polyclads. *Dugesia.*
Class Trematoda. Flukes. *Clonorchis, Schistosoma.*
Class Cestoda. Tapeworms. *Diphyllobothrium, Taenia.*

PHYLUM ROTIFERA Rotifers. *Asplancha, Philodina.*

PHYLUM MOLLUSCA Mollusks.

Class Polyplacophora. Chitons. *Cryptochiton, Tonicella.*

Class Gastropoda. Snails, sea slugs, land slugs. *Aplysia, Ariolimax, Cypraea, Haliotis, Helix, Liguus, Limax, Littorina.*

Class Bivalvia. Clams, mussels, scallops, cockles, oysters, shipworms. *Ensis, Chlamys, Mytelus, Patinopectin.*

Class Cephalopoda. Squids, octopuses, cuttlefish, nautiluses. *Dosidiscus, Loligo, Nautilus, Octopus, Sepia.*

PHYLUM ANNELIDA Segmented worms.

Class Polychaeta. Mostly marine worms. *Eunice, Neanthes.*

Class Oligochaeta. Mostly freshwater and terrestrial worms, many marine. *Lumbricus* (earthworms), *Tubifex.*

Class Hirudinea. Leeches. *Hirudo, Placobdella.*

PHYLUM NEMATODA Roundworms. *Ascaris, Caenorhabditis elegans, Necator* (hookworms), *Trichinella.*

PHYLUM ARTHROPODA

Subphylum Chelicerata. Chelicerates. Horseshoe crabs, spiders, scorpions, ticks, mites.

Subphylum Crustacea. Shrimps, crayfishes, lobsters, crabs, barnacles, copepods, isopods (sowbugs).

Subphylum Myriapoda. Centipedes, millipedes.

Subphylum Hexapoda. Insects and sprintails.

PHYLUM ECHINODERMATA Echinoderms.

Class Asteroidea. Sea stars. *Asterias.*
Class Ophiuroidea. Brittle stars.
Class Echinoidea. Sea urchins, heart urchins, sand dollars.
Class Holothuroidea. Sea cucumbers.
Class Crinoidea. Feather stars, sea lilies.
Class Concentricycloidea. Sea daisies.

PHYLUM CHORDATA Chordates.

Subphylum Urochordata. Tunicates, related forms.

Subphylum Cephalochordata. Lancelets.

CRANIATES

Class Myxini. Hagfishes.

VERTEBRATES (SUBGROUP OF CRANIATES)

Class Cephalaspidomorphi. Lampreys.

Class Chondrichthyes. Cartilaginous fishes (sharks, rays, skates, chimaeras).

Class "Osteichthyes." Bony fishes. Not monophyletic (sturgeons, paddlefish, herrings, carps, cods, trout, seahorses, tunas, lungfishes, and coelocanths).

TETRAPODS (SUBGROUP OF VERTEBRATES)

Class Amphibia. Amphibians. Require water to reproduce.
Order Caudata. Salamanders and newts.
Order Anura. Frogs, toads.
Order Apoda. Apodans (caecilians).

AMNIOTES (SUBGROUP OF TETRAPODS)

Class "Reptilia." Skin with scales, embryo protected and nutritionally supported by extraembryonic membranes.

Subclass Anapsida. Turtles, tortoises.

Subclass Lepidosaura. *Sphenodon*, lizards, snakes.

Subclass Archosaura. Crocodiles, alligators.

Class Aves. Birds. In some classifications birds are grouped in the archosaurs.

Order Struthioniformes. Ostriches.
Order Sphenisciformes. Penguins.
Order Procellariiformes. Albatrosses, petrels.
Order Ciconiiformes. Herons, bitterns, storks, flamingoes.
Order Anseriformes. Swans, geese, ducks.
Order Falconiformes. Eagles, hawks, vultures, falcons.
Order Galliformes. Ptarmigan, turkeys, domestic fowl.
Order Columbiformes. Pigeons, doves.
Order Strigiformes. Owls.
Order Apodiformes. Swifts, hummingbirds.
Order Passeriformes. Sparrows, jays, finches, crows, robins, starlings, wrens.
Order Piciformes. Woodpeckers, toucans.
Order Psittaciformes. Parrots, cockatoos, macaws.

Class Mammalia. Skin with hair; young nourished by milk-secreting mammary glands of adult.

Subclass Prototheria. Egg-laying mammals (monotremes; duckbilled platypus, spiny anteaters).

Subclass Metatheria. Pouched mammals or marsupials (opossums, kangaroos, wombats, Tasmanian devils).

Subclass Eutheria. Placental mammals.

Order Edentata. Anteaters, tree sloths, armadillos.
Order Insectivora. Tree shrews, moles, hedgehogs.
Order Chiroptera. Bats.
Order Scandentia. Insectivorous tree shrews.
Order Primates.

Suborder Strepsirhini (prosimians). Lemurs, lorises.

Suborder Haplorhini (tarsioids and anthropoids).

Infraorder Tarsiiformes. Tarsiers.

Infraorder Platyrrhini (New World monkeys).

Family Cebidae. Spider monkeys, howler monkeys, capuchin.

Infraorder Catarrhini (Old World monkeys and hominoids).

Superfamily Cercopithecoidea. Baboons, macaques, langurs.

Superfamily Hominoidea. Apes and humans.

Family Hylobatidae. Gibbon.

Family "Pongidae." Chimpanzees, gorillas, orangutans.

Family Hominidae. Existing and extinct human species (*Homo*) and humanlike species, including the australopiths.

Order Lagomorpha. Rabbits, hares, pikas.

Order Rodentia. Most gnawing animals (squirrels, rats, mice, guinea pigs, porcupines, beavers, etc.).

Order Carnivora. Carnivores (wolves, cats, bears, etc.).

Order Pinnipedia. Seals, walruses, sea lions.

Order Proboscidea. Elephants, mammoths (extinct).

Order Sirenia. Sea cows (manatees, dugongs).

Order Perissodactyla. Odd-toed ungulates (horses, tapirs, rhinos).

Order Tubulidentata. African aardvarks.

Order Artiodactyla. Even-toed ungulates (camels, deer, bison, sheep, goats, antelopes, giraffes, etc.).

Order Cetacea. Whales, porpoises.

Appendix II. Annotations to A Journal Article

This journal article reports on the movements of a female wolf during the summer of 2002 in northwestern Canada. It also reports on a scientific process of inquiry, observation and interpretation to learn where, how and why the wolf traveled as she did. In some ways, this article reflects the story of "how to do science" told in section 1.5 of this textbook. These notes are intended to help you read and understand how scientists work and how they report on their work.

(1) ARCTIC

(2) VOL. 57, NO. 2 (JUNE 2004) P. 196–203

(3) Long Foraging Movement of a Denning Tundra Wolf

(4) Paul F. Frame,[1,2] David S. Hik,[1] H. Dean Cluff,[3] and Paul C. Paquet[4]

(5) *(Received 3 September 2003; accepted in revised form 16 January 2004)*

(6) **ABSTRACT** Wolves (*Canis lupus*) on the Canadian barrens are intimately linked to migrating herds of barren-ground caribou (*Rangifer tarandus*). We deployed a Global Positioning System (GPS) radio collar on an adult female wolf to record her movements in response to changing caribou densities near her den during summer. This wolf and two other females were observed nursing a group of 11 pups. She traveled a minimum of 341 km during a 14-day excursion. The straight-line distance from the den to the farthest location was 103 km, and the overall minimum rate of travel was 3.1 km/h. The distance between the wolf and the radio-collared caribou decreased from 242 km one week before the excursion to 8 km four days into the excursion. We discuss several possible explanations for the long foraging bout.

(7) *Key words:* wolf, GPS tracking, movements, *Canis lupus*, foraging, caribou, Northwest Territories

(8) **RÉSUMÉ** Les loups (*Canis lupus*) dans la toundra canadienne sont étroitement liés aux hardes de caribous des toundras (*Rangifer tarandus*). On a équipé une louve adulte d'un collier émetteur muni d'un système de positionnement mondial (GPS) afin d'enregistrer ses déplacements en réponse au changement de densité du caribou près de sa tanière durant l'été. On a observé cette louve ainsi que deux autres en train d'allaiter un groupe de 11 louveteaux. Elle a parcouru un minimum de 341 km durant une sortie de 14 jours. La distance en ligne droite de la tanière à l'endroit le plus éloigné était de 103 km, et la vitesse minimum durant tout le voyage était de 3,1 km/h. La distance entre la louve et le caribou muni du collier émetteur a diminué de 242 km une semaine avant la sortie à 8 km quatre jours après la sortie. On commente diverses explications possibles pour ce long épisode de recherche de nourriture.

Mots clés: loup, repérage GPS, déplacements, *Canis lupus*, recherche de nourriture, caribou, Territoires du Nord-Ouest

Traduit pour la revue *Arctic* par Nésida Loyer.

(9) Introduction

Wolves (*Canis lupus*) that den on the central barrens of mainland Canada follow the seasonal movements of their main prey, migratory barren-ground caribou (*Rangifer tarandus*) (Kuyt, 1962; Kelsall, 1968; Walton et al., 2001). However, most wolves do not den near caribou calving grounds, but select sites farther south, closer to the tree line (Heard and Williams, 1992). Most caribou migrate beyond primary wolf denning areas by mid-June and do not return until mid-to-late July (Heard et al., 1996; Gunn et al., 2001). Consequently, caribou density near dens is low for part of the summer.

(10) During this period of spatial separation from the main caribou herds, wolves must either search near the homesite for scarce caribou or alternative prey (or both), travel to where prey are abundant, or use a combination of these strategies.

(11) Walton et al. (2001) postulated that the travel of tundra wolves outside their normal summer ranges is a response to low caribou availability rather than a pre-dispersal exploration like that observed in territorial wolves (Fritts and Mech, 1981; Messier, 1985). The authors postulated this because most such travel was directed toward caribou calving grounds. We report details of such a long-distance excursion by a breeding female tundra wolf wearing a GPS radio collar. We discuss the relationship of the excursion to movements of satellite-collared caribou (Gunn et al., 2001), supporting the hypothesis that tundra wolves make directional, rapid, long-distance movements in response to seasonal prey availability.

[1] Department of Biological Sciences, University of Alberta, Edmonton, Alberta T6G 2E9, Canada

[2] Corresponding author: pframe@ualberta.ca

[3] Department of Resources, Wildlife, and Economic Development, North Slave Region, Government of the Northwest Territories, P.O. Box 2668, 3803 Bretzlaff Dr., Yellowknife, Northwest Territories X1A 2P9, Canada; Dean_Cluff@gov.nt.ca

[4] Faculty of Environmental Design, University of Calgary, Calgary, Alberta T2N 1N4, Canada; current address: P.O. Box 150, Meacham, Saskatchewan S0K 2V0, Canada

1 Title of the journal, which reports on science taking place in Arctic regions.

2 Volume number, issue number and date of the journal, and page numbers of the article.

3 Title of the article: a concise but specific description of the subject of study—one episode of long-range travel by a wolf hunting for food on the Arctic tundra.

4 Authors of the article: scientists working at the institutions listed in the footnotes below. Note #2 indicates that P. F. Frame is the *corresponding author*—the person to contact with questions or comments. His email address is provided.

5 Date on which a draft of the article was received by the journal editor, followed by date one which a revised draft was accepted for publication. Between these dates, the article was reviewed and critiqued by other scientists, a process called peer review. The authors revised the article to make it clearer, according to those reviews.

6 ABSTRACT: A brief description of the study containing all basic elements of this report. First sentence summarizes the *background* material. Second sentence encapsulates the *methods* used. The rest of the paragraph sums up the *results*. Authors introduce the main *subject* of the study—a female wolf (#388) with pups in a den—and refer to later *discussion* of possible explanations for her behavior.

7 Key words are listed to help researchers using computer databases. Searching the databases using these key words will yield a list of studies related to this one.

8 RÉSUMÉ: The French translation of the abstract and key words. Many researchers in this field are French Canadian. Some journals provide such translations in French or in other languages.

9 INTRODUCTION: Gives the background for this wolf study. This paragraph tells of known or suspected wolf behavior that is important for this study. Note that (a) major species mentioned are always accompanied by scientific names, and (b) statements of fact or *postulations* (claims or assumptions about what is likely to be true) are followed by references to studies that established those facts or supported the postulations.

10 This paragraph focuses directly on the wolf behaviors that were studied here.

11 This paragraph starts with a statement of the *hypothesis* being tested, one that originated in other studies and is supported by this one. The hypothesis is restated more succinctly in the last sentence of this paragraph. This is the *inquiry* part of the scientific process—asking questions and suggesting possible answers.

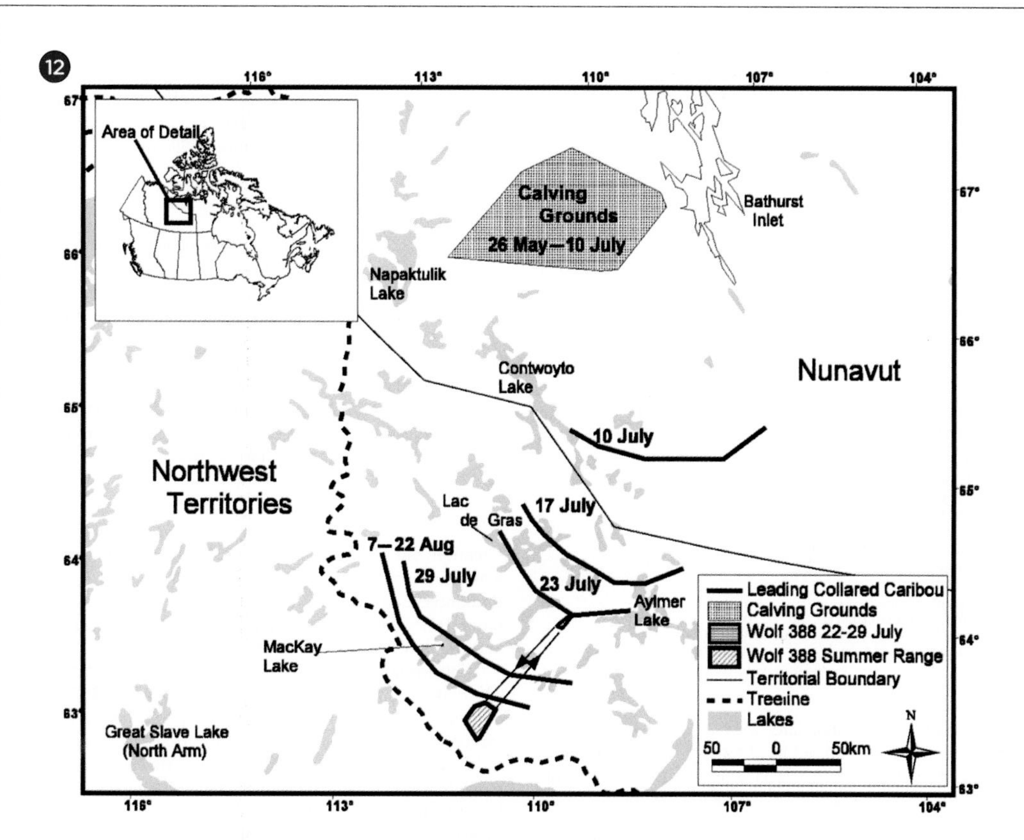

Figure 1. Map showing the movements of satellite radio-collared caribou with respect to female wolf 388's summer range and long foraging movement, in summer 2002.

13 Study Area

Our study took place in the northern boreal forest–low Arctic tundra transition zone (63° 30′ N, 110° 00′ W; Figure 1; Timoney et al., 1992). Permafrost in the area changes from discontinuous to continuous (Harris, 1986). Patches of spruce (*Picea mariana, P. glauca*) occur in the southern portion and give way to open tundra to the northeast. Eskers, kames, and other glacial deposits are scattered throughout the study area. Standing water and exposed bedrock are characteristic of the area.

 *Details of the Caribou-Wolf System*

The Bathurst caribou herd uses this study area. Most caribou cows have begun migrating by late April, reaching calving grounds by June (Gunn et al., 2001; Figure 1). Calving peaks by 15 June (Gunn et al., 2001), and calves begin to travel with the herd by one week of age (Kelsall, 1968). The movement patterns of bulls are less known, but bulls frequent areas near calving grounds by mid-June (Heard et al., 1996; Gunn et al., 2001). In summer, Bathurst caribou cows generally travel south from their calving grounds and then, parallel to the tree line, to the northwest. The rut usually takes place at the tree line in October (Gunn et al., 2001). The winter range of the Bathurst herd varies among years, ranging through the taiga and along the tree line from south of Great Bear Lake to southeast of Great Slave Lake. Some caribou spend the winter on the tundra (Gunn et al., 2001; Thorpe et al., 2001).

15 In winter, wolves that prey on Bathurst caribou do not behave territorially. Instead, they follow the herd throughout its winter range (Walton et al., 2001; Musiani, 2003). However, during denning (May–

12 This map shows the study area and depicts wolf and caribou locations and movements during one summer. Some of this information is explained below.

13 STUDY AREA: This section sets the stage for the study, locating it precisely with latitude and longitude coordinates and describing the area (illustrated by the map in Figure 1).

14 Here begins the story of how prey (caribou) and predators (wolves) interact on the tundra. Authors describe movements of these nomadic animals throughout the year.

15 We focus on the denning season (summer) and learn how wolves locate their dens and travel according to the movements of caribou herds.

16 Other variables are considered—prey other than caribou and their relative abundance in 2002.

17 METHODS: There is no one scientific method. Procedures for each and every study must be explained carefully.

18 Authors explain when and how they tracked caribou and wolves, including tools used and the exact procedures followed.

19 This important subsection explains what data were calculated (average distance ...) and how, including the software used and where it came from. (The calculations are listed in Table 1.) Note that the behavior measured (traveling) is carefully defined.

20 RESULTS: The heart of the report and the *observation* part of the scientific process. This section is organized parallel to the Methods section.

21 This subsection is broken down by periods of observation. Pre-excursion period covers the time between 388's capture and the start of her long-distance travel. The investigators used visual observations as well as telemetry (measurements taken using the global positioning system (GPS)) to gather data. They looked at how 388 cared for her pups, interacted with other adults, and moved about the den area.

Table 1. Daily distances from wolf 388 and the den to the nearest radio-collared caribou during a long excursion in summer 2002.

Date (2002)	Mean distance from caribou to wolf (km)	Daily distance from closest caribou to den
12 July	242	241
13 July	210	209
14 July	200	199
15 July	186	180
16 July	163	162
17 July	151	148
18 July	144	137
19 July[1]	126	124
20 July	103	130
21 July	73	130
22 July	40	110
23 July[2]	9	104
29 July[3]	16	43
30 July	32	43
31 July	28	44
1 August	29	46
2 August[4]	54	52
3 August	53	53
4 August	74	74
5 August	75	75
6 August	74	75
7 August	72	75
8 August	76	75
9 August	79	79

[1] Excursion starts.
[2] Wolf closest to collared caribou.
[3] Previous five days' caribou locations not available.
[4] Excursion ends.

August, parturition late May to mid-June), wolf movements are limited by the need to return food to the den. To maximize access to migrating caribou, many wolves select den sites closer to the tree line than to caribou calving grounds (Heard and Williams, 1992). Because of caribou movement patterns, tundra denning wolves are separated from the main caribou herds by several hundred kilometers at some time during summer (Williams, 1990:19; Figure 1; Table 1).

16 Muskoxen do not occur in the study area (Fournier and Gunn, 1998), and there are few moose there (H.D. Cluff, pers. obs.). Therefore, alternative prey for wolves includes waterfowl, other ground-nesting birds, their eggs, rodents, and hares (Kuyt, 1972; Williams, 1990:16; H.D. Cluff and P.F. Frame, unpubl. data). During 56 hours of den observations, we saw no ground squirrels or hares, only birds. It appears that the abundance of alternative prey was relatively low in 2002.

17 Methods

Wolf Monitoring

18 We captured female wolf 388 near her den on 22 June 2002, using a helicopter net-gun (Walton et al., 2001). She was fitted with a releasable GPS radio collar (Merrill et al., 1998) programmed to acquire locations at 30-minute intervals. The collar was electronically released (e.g., Mech and Gese, 1992) on 20 August 2002. From 27 June to 3 July 2002, we observed 388's den with a 78 mm spotting scope at a distance of 390 m.

Caribou Monitoring

In spring of 2002, ten female caribou were captured by helicopter net-gun and fitted with satellite radio collars, bringing the total number of collared Bathurst cows to 19. Eight of these spent the summer of 2002 south of Queen Maud Gulf, well east of normal Bathurst caribou range. Therefore, we used 11 caribou for this analysis. The collars provided one location per day during our study, except for five days from 24 to 28 July. Locations of satellite collars were obtained from Service Argos, Inc. (Landover, Maryland).

Data Analysis

Location data were analyzed by ArcView GIS software (Environmental Systems Research Institute Inc., Redlands, California). We calculated the average distance from the nearest collared caribou to the wolf and the den for each day of the study. **19**

Wolf foraging bouts were calculated from the time 388 exited a buffer zone (500 m radius around the den) until she re-entered it. We considered her to be traveling when two consecutive locations were spatially separated by more than 100 m. Minimum distance traveled was the sum of distances between each location and the next during the excursion.

We compared pre- and post-excursion data using Analysis of Variance (ANOVA; Zar, 1999). We first tested for homogeneity of variances with Levene's test (Brown and Forsythe, 1974). No transformations of these data were required.

Results 20

Wolf Monitoring

Pre-Excursion Period: Wolf 388 was lactating when captured on 22 June. We observed her and two other females nursing a group of 11 pups between 27 June and 3 July. During our observations, the pack consisted of at least four adults (3 females and 1 male) and 11 pups. On 30 June, three pups were moved to a location 310 m from the other eight and cared for by an uncollared female. The male was not seen at the den after the evening of 30 June. **21**

Before the excursion, telemetry indicated 18 foraging bouts. The mean distance traveled during these bouts was 25.29 km (± 4.5 SE, range 3.1–82.5 km). Mean greatest distance from the den on foraging

22

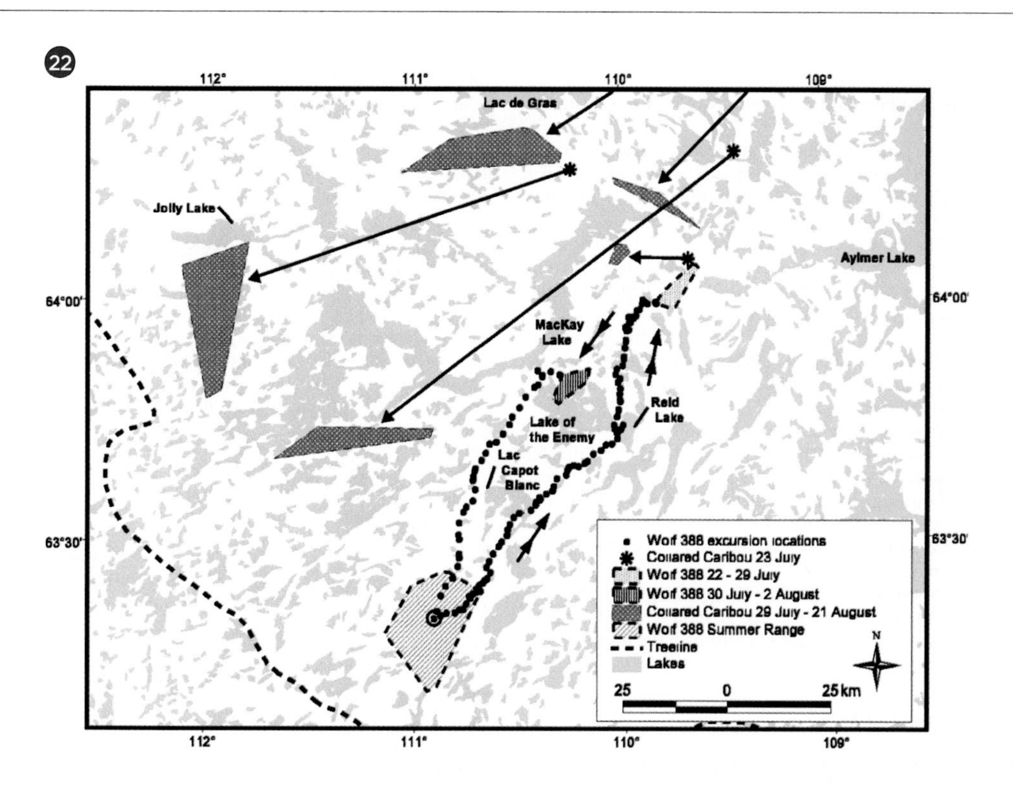

Figure 2. Details of a long foraging movement by female wolf 388 between 19 July and 2 August 2002. Also shown are locations and movements of three satellite radio-collared caribou from 23 July to 21 August 2002. On 23 July, the wolf was 8 km from a collared caribou. The farthest point from the den (103 km distant) was recorded on 27 July. Arrows indicate direction of travel.

22 The key in the lower right-hand corner of the map shows areas (shaded) within which the wolves and caribou moved, and the dotted trail of 388 during her excursion. From the results depicted on this map, the investigators tried to determine when and where 388 might have encountered caribou and how their locations affected her traveling behavior.

23 The wolf's excursion (her long trip away from the den area) is the focus of this study. These paragraphs present detailed measurements of daily movements during her two-week trip—how far she traveled, how far she was from collared caribou, her time spent traveling and resting, and her rate of speed. Authors use the phrase "minimum distance traveled" to acknowledge they couldn't track every step but were measuring samples of her movements. They knew that she went at least as far as they measured. This shows how scientists try to be exact when reporting results. Results of this study are depicted graphically in the map in Figure 2.

bouts was 7.1 km (± 0.9 SE, range 1.7–17.0 km). The average duration of foraging bouts for the period was 20.9 h (± 4.5 SE, range 1–71 h).

The average daily distance between the wolf and the nearest collared caribou decreased from 242 km on 12 July, one week before the excursion period, to 126 km on 19 July, the day the excursion began (Table 1).

23 **Excursion Period:** On 19 July at 2203, after spending 14 h at the den, 388 began moving to the northeast and did not return for 336 h (14 d; Figure 2). Whether she traveled alone or with other wolves is unknown. During the excursion, 476 (71%) of 672 possible locations were recorded. The wolf crossed the southeast end of Lac Capot Blanc on a small land bridge, where she paused for 4.5 h after traveling for 19.5 h (37.5 km). Following this rest, she traveled for 9 h (26.3 km) onto a peninsula in Reid Lake, where she spent 2 h before backtracking and stopping for 8 h just off the peninsula. Her next period of travel lasted 16.5 h (32.7 km), terminating in a pause of 9.5 h just 3.8 km from a concentration of locations at the far end of her excursion, where we presume she encountered caribou. The mean duration of these three movement periods was 15.7 h (± 2.5 SE), and that of the pauses, 7.3 h (± 1.5). The wolf required 72.5 h (3.0 d) to travel a minimum of 95 km from her den to this area near caribou (Figure 2). She remained there (35.5 km2) for 151.5 h (6.3 d) and then moved south to Lake of the Enemy, where she stayed (31.9 km^2) for 74 h (3.1 d) before returning to her den. Her greatest distance from the den, 103 km, was recorded 174.5 h (7.3 d) after the excursion

24 Post-excursion measurements of 388's movements were made to compare with those of the pre-excursion period. In order to compare, scientists often use *means*, or averages, of a series of measurements—mean distances, mean duration, etc.

25 In the comparison, authors used statistical calculations (F and df) to determine that the differences between pre- and post-excursion measurements were *statistically insignificant*, or close enough to be considered essentially the same or similar.

26 As with wolf 388, the investigators measured the movements of caribou during the study period. The areas within which the caribou moved are shown in Figure 2 by shaded polygons mentioned in the second paragraph of this subsection.

27 This subsection summarizes how distances separating predators and prey varied during the study period.

28 DISCUSSION: This section is the *interpretation* part of the scientific process.

29 This subsection reviews observations from other studies and suggests that this study fits with patterns of those observations.

30 Authors discuss a prevailing *theory* (CBFT) which might explain why a wolf would travel far to meet her own energy needs while taking food caught closer to the den back to her pups. The results of this study seem to fit that pattern.

began, at 0433 on 27 July. She was 8 km from a collared caribou on 23 July, four days after the excursion began (Table 1).

The return trip began at 0403 on 2 August, 318 h (13.2 d) after leaving the den. She followed a relatively direct path for 18 h back to the den, a distance of 75 km.

The minimum distance traveled during the excursionwas 339 km. The estimated overall minimum travel rate was 3.1 km/h, 2.6 km/h away from the den and 4.2 km/h on the return trip.

(24) **Post-Excursion Period:** We saw three pups when recovering the collar on 20 August, but others may have been hiding in vegetation.

Telemetry recorded 13 foraging bouts in the post-excursion period. The mean distance traveled during these bouts was 18.3 km (+ 2.7 SE, range 1.2–47.7 km), and mean greatest distance from the den was 7.1 km (+ 0.7 SE, range 1.1–11.0 km). The mean duration of these post-excursion foraging bouts was 10.9 h (+ 2.4 SE, range 1–33 h).

When 388 reached her den on 2 August, the distance to the nearest collared caribou was 54 km. On 9 August, one week after she returned, the distance was 79 km (Table 1).

Pre- and Post-Excursion Comparison

(25) We found no differences in the mean distance of foraging bouts before and after the excursion period ($F = 1.5$, $df = 1, 29$, $p = 0.24$). Likewise, the mean greatest distance from the den was similar pre- and post-excursion ($F = 0.004$, $df = 1, 29$, $p = 0.95$). However, the mean duration of 388's foraging bouts decreased by 10.0 h after her long excursion ($F = 3.1$, $df = 1, 29$, $p = 0.09$).

(26) *Caribou Monitoring*

Summer Movements: On 10 July, 5 of 11 collared caribou were dispersed over a distance of 10 km, 140 km south of their calving grounds (Figure 1). On the same day, three caribou were still on the calving grounds, two were between the calving grounds and the leaders, and one was missing. One week later (17 July), the leading radio-collared cows were 100 km farther south (Figure 1). Two were within 5 km of each other in front of the rest, who were more dispersed. All radio-collared cows had left the calving grounds by this time. On 23 July, the leading radio-collared caribou had moved 35 km farther south, and all of them were more widely dispersed. The two cows closest to the leader were 26 km and 33 km away, with 37 km between them. On the next location (29 July), the most southerly caribou were 60 km farther south. All of the caribou were now in the areas where they remained for the duration of the study (Figure 2).

A Minimum Convex Polygon (Mohr and Stumpf, 1966) around all caribou locations acquired during the study encompassed 85 119 km^2.

Relative to the Wolf Den: The distance from the nearest collared caribou to the den decreased from 241 km one week before the excursion to 124 km the day it began. The nearest a collared caribou came to the den was 43 km away, on 29 and 30 July. During the study, four collared caribou were located within 100 km of the den. Each of these four was closest to the wolf on at least one day during the period reported. (27)

(28) Discussion

Prey Abundance

Caribou are the single most important prey of tundra wolves (Clark, 1971; Kuyt, 1972; Stephenson and James, 1982; Williams, 1990). Caribou range over vast areas, and for part of the summer, they are scarce or absent in wolf home ranges (Heard et al., 1996). Both the long distance between radio-collared caribou and the den the week before the excursion and the increased time spent foraging by wolf 388 indicate that caribou availability near the den was low. Observations of the pups' being left alone for up to 18 h, presumably while adults were searching for food, provide additional support for low caribou availability locally. Mean foraging bout duration decreased by 10.0 h after the excursion, when collared caribou were closer to the den, suggesting an increase in caribou availability nearby. (29)

Foraging Excursion

One aspect of central place foraging theory (CPFT) deals with the optimality of returning different-sized food loads from varying distances to dependents at a central place (i.e., the den) (Orians and Pearson, 1979). Carlson (1985) tested CPFT and found that the predator usually consumed prey captured far from the central place, while feeding prey captured nearby to dependants. Wolf 388 spent 7.2 days in one area near caribou before moving to a location 23 km back towards the den, where she spent an additional 3.1 days, likely hunting caribou. She began her return trip from this closer location, traveling directly to the den. While away, she may have made one or more successful kills and spent time meeting her own energetic needs before returning to the den. Alternatively, it may have taken several attempts to make a kill, (30) which she then fed on before beginning her return trip. We do not know if she returned food to the pups, but such behavior would be supported by CPFT.

31 Other workers have reported wolves' making long round trips and referred to them as "extraterritorial" or "pre-dispersal" forays (Fritts and Mech, 1981; Messier, 1985; Ballard et al., 1997; Merrill and Mech, 2000). These movements are most often made by young wolves (1–3 years old), in areas where annual territories are maintained and prey are relatively sedentary (Fritts and Mech, 1981; Messier, 1985). The long excursion of 388 differs in that tundra wolves do not maintain annual territories (Walton et al., 2001), and the main prey migrate over vast areas (Gunn et al., 2001).

Another difference between 388's excursion and those reported earlier is that she is a mature, breeding female. No study of territorial wolves has reported reproductive adults making extraterritorial movements in summer (Fritts and Mech, 1981; Messier, 1985; Ballard et al., 1997; Merrill and Mech, 2001). However, Walton et al. (2001) also report that breeding female tundra wolves made excursions.

Direction of Movement

32 Possible explanations for the relatively direct route 388 took to the caribou include landscape influence and experience. Considering the timing of 388's trip and the locations of caribou, had the wolf moved northwest, she might have missed the caribou entirely, or the encounter might have been delayed.

A reasonable possibility is that the land directed 388's route. The barrens are crisscrossed with trails worn into the tundra over centuries by hundreds of thousands of caribou and other animals (Kelsall, 1968; Thorpe et al., 2001). At river crossings, lakes, or narrow peninsulas, trails converge and funnel towards and away from caribou calving grounds and summer range. Wolves use trails for travel (Paquet et al., 1996; Mech and Boitani, 2003; P. Frame, pers. observation). Thus, the landscape may direct an animal's movements and lead it to where cues, such as the odor of caribou on the wind or scent marks of other wolves, may lead it to caribou.

33 Another possibility is that 388 knew where to find caribou in summer. Sexually immature tundra wolves sometimes follow caribou to calving grounds (D. Heard, unpubl. data). Possibly, 388 had made such journeys in previous years and killed caribou. If this were the case, then in times of local prey scarcity she might travel to areas where she had hunted successfully before. Continued monitoring of tundra wolves may answer questions about how their food needs are met in times of low caribou abundance near dens.

34 Caribou often form large groups while moving south to the tree line (Kelsall, 1968). After a large aggregation of caribou moves through an area, its scent can linger for weeks (Thorpe et al., 2001:104). It is conceivable that 388 detected caribou scent on the wind, which was blowing from the northeast on 19–21 July (Environment Canada, 2003), at the same time her excursion began. Many factors, such as odor strength and wind direction and strength, make systematic study of scent detection in wolves difficult under field conditions (Harrington and Asa, 2003). However, humans are able to smell odors such as forest fires or oil refineries more than 100 km away. The olfactory capabilities of dogs, which are similar to wolves, are thought to be 100 to 1 million times that of humans (Harrington and Asa, 2003). Therefore, it is reasonable to think that under the right wind conditions, the scent of many caribou traveling together could be detected by wolves from great distances, thus triggering a long foraging bout.

Rate of Travel

35 Mech (1994) reported the rate of travel of Arctic wolves on barren ground was 8.7 km/h during regular travel and 10.0 km/h when returning to the den, a difference of 1.3 km/h. These rates are based on direct observation and exclude periods when wolves moved slowly or not at all. Our calculated travel rates are assumed to include periods of slow movement or no movement. However, the pattern we report is similar to that reported by Mech (1994), in that homeward travel was faster than regular travel by 1.6 km/h. The faster rate on return may be explained by the need to return food to the den. Pup survival can increase with the number of adults in a pack available to deliver food to pups (Harrington et al., 1983). Therefore, an increased rate of travel on homeward trips could improve a wolf's reproductive fitness by getting food to pups more quickly.

Fate of 388's Pups

36 Wolf 388 was caring for pups during den observations. The pups were estimated to be six weeks old, and were seen ranging as far as 800 m from the den. They received some regurgitated food from two of the females, but were unattended for long periods. The excursion started 16 days after our observations, and it is improbable that the pups could have traveled the distance that 388 moved. If the pups died, this would have removed parental responsibility, allowing the long movement.

Our observations and the locations of radio-collared caribou indicate that prey became scarce in

31 Here our authors note other possible explanations for wolves' excursions presented by other investigators, but this study does not seem to support those ideas.

32 Authors discuss possible reasons for why 388 traveled directly to where caribou were located. They take what they learned from earlier studies and apply it to this case, suggesting that the lay of the land played a role. Note that their description paints a clear picture of the landscape.

33 Authors suggest that 388 may have learned in traveling during previous summers where the caribou were. The last two sentences suggest ideas for future studies.

34 Or maybe 388 followed the scent of the caribou. Authors acknowledge difficulties of proving this, but they suggest another area where future studies might be done.

35 Authors suggest that results of this study support previous studies about how fast wolves travel to and from the den. In the last sentence, they speculate on how these observed patterns would fit into the theory of evolution.

36 Authors also speculate on the fate of 388's pups while she was traveling. This leads to . . .

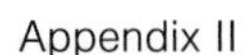

37 Discussion of cooperative rearing of pups and, in turn, to speculation on how this study and what is known about cooperative rearing might fit into the animal's strategies for survival of the species. Again, the authors approach the broader theory of evolution and how it might explain some of their results.

38 And again, they suggest that this study points to several areas where further study will shed some light.

39 In conclusion, the authors suggest that their study supports the hypothesis being tested here. And they touch on the implications of increased human activity on the tundra predicted by their results.

40 ACKNOWLEDGEMENTS: Authors note the support of institutions, companies and individuals. They thank their reviewers ad list permits under which their research was carried on.

41 REFERENCES: List of all studies cited in the report. This may seem tedious, but is a vitally important part of scientific reporting. It is a record of the sources of information on which this study is based. It provides readers with a wealth of resources for further reading on this topic. Much of it will form the foundation of future scientific studies like this one.

the area of the den as summer progressed. Wolf 388 may have abandoned her pups to seek food for herself. However, she returned to the den after the excursion, where she was seen near pups. In fact, she foraged in a similar pattern before and after the excursion, suggesting that she again was providing for pups after her return to the den.

(37) A more likely possibility is that one or both of the other lactating females cared for the pups during 388's absence. The three females at this den were not seen with the pups at the same time. However, two weeks earlier, at a different den, we observed three females cooperatively caring for a group of six pups. At that den, the three lactating females were observed providing food for each other and trading places while nursing pups. Such a situation at the den of 388 could have created conditions that allowed one or more of the lactating females to range far from the den for a period, returning to her parental duties afterwards. However, the pups would have been weaned by eight weeks of age (Packard et al., 1992), so nonlactating adults could also have cared for them, as often happens in wolf packs (Packard et al., 1992; Mech et al., 1999).

Cooperative rearing of multiple litters by a pack could create opportunities for long-distance foraging movements by some reproductive wolves during summer periods of local food scarcity. We have recorded multiple lactating females at one or more tundra wolf dens per year since 1997. This reproductive strategy may be an adaptation to temporally and (38) spatially unpredictable food resources. All of these possibilities require further study, but emphasize both the adaptability of wolves living on the barrens and their dependence on caribou.

Long-range wolf movement in response to caribou (39) availability has been suggested by other researchers (Kuyt, 1972; Walton et al., 2001) and traditional ecological knowledge (Thorpe et al., 2001). Our report demonstrates the rapid and extreme response of wolves to caribou distribution and movements in summer. Increased human activity on the tundra (mining, road building, pipelines, ecotourism) may influence caribou movement patterns and change the interactions between wolves and caribou in the region. Continued monitoring of both species will help us to assess whether the association is being affected adversely by anthropogenic change.

(40) Acknowledgements

This research was supported by the Department of Resources, Wildlife, and Economic Development, Government of the Northwest Territories; the Department of Biological Sciences at the University of Alberta; the Natural Sciences and Engineering Research Council of Canada; the Department of Indian and Northern Affairs Canada; the Canadian Circumpolar Institute; and DeBeers Canada, Ltd. Lorna Ruechel assisted with den observations. A. Gunn provided caribou location data. We thank Dave Mech for the use of GPS collars. M. Nelson, A. Gunn, and three anonymous reviewers made helpful comments on earlier drafts of the manuscript. This work was done under Wildlife Research Permit – WL002948 issued by the Government of the Northwest Territories, Department of Resources, Wildlife, and Economic Development.

(41) References

BALLARD, W.B., AYRES, L.A., KRAUSMAN, P.R., REED, D.J., and FANCY, S.G. 1997. Ecology of wolves in relation to a migratory caribou herd in northwest Alaska. Wildlife Monographs 135. 47 p.

BROWN, M.B., and FORSYTHE, A.B. 1974. Robust tests for the equality of variances. Journal of the American Statistical Association 69:364–367.

CARLSON, A. 1985. Central place foraging in the red-backed shrike (*Lanius collurio* L.): Allocation of prey between forager and sedentary consumer. Animal Behaviour 33:664–666.

CLARK, K.R.F. 1971. Food habits and behavior of the tundra wolf on central Baffin Island. Ph.D. Thesis, University of Toronto, Ontario, Canada.

ENVIRONMENT CANADA. 2003. National climate data information archive. Available online: http://www.climate.weatheroffice.ec.gc.ca/Welcome_e.html

FOURNIER, B., and GUNN, A. 1998. Musk ox numbers and distribution in the NWT, 1997. File Report No. 121. Yellowknife: Department of Resources, Wildlife, and Economic Development, Government of the Northwest Territories. 55 p.

FRITTS, S.H., and MECH, L.D. 1981. Dynamics, movements, and feeding ecology of a newly protected wolf population in northwestern Minnesota. Wildlife Monographs 80. 79 p.

GUNN, A., DRAGON, J., and BOULANGER, J. 2001. Seasonal movements of satellite-collared caribou from the Bathurst herd. Final Report to the West Kitikmeot Slave Study Society, Yellowknife, NWT. 80 p. Available online: http://www.wkss.nt.ca/HTML/08_ProjectsReports/PDF/Seasonal MovementsFinal.pdf

HARRINGTON, F.H., and ASA, C.S. 2003. Wolf communication. In: Mech, L.D., and Boitani, L., eds. Wolves: Behavior, ecology, and conservation. Chicago: University of Chicago Press. 66–103.

HARRINGTON, F.H., MECH, L.D., and FRITTS, S.H. 1983. Pack size and wolf pup survival: Their relationship under varying ecological conditions. Behavioral Ecology and Sociobiology 13:19–26.

HARRIS, S.A. 1986. Permafrost distribution, zonation and stability along the eastern ranges of the cordillera of North America. Arctic 39(1):29–38.

HEARD, D.C., and WILLIAMS, T.M. 1992. Distribution of wolf dens on migratory caribou ranges in the Northwest

Territories, Canada. Canadian Journal of Zoology 70:1504–1510.
HEARD, D.C., WILLIAMS, T.M., and MELTON, D.A. 1996. The relationship between food intake and predation risk in migratory caribou and implication to caribou and wolf population dynamics. Rangifer Special Issue No. 2:37–44.
KELSALL, J.P. 1968. The migratory barren-ground caribou of Canada. Canadian Wildlife Service Monograph Series 3. Ottawa: Queen's Printer. 340 p.
KUYT, E. 1962. Movements of young wolves in the Northwest Territories of Canada. Journal of Mammalogy 43:270–271.
———. 1972. Food habits and ecology of wolves on barren-ground caribou range in the Northwest Territories. Canadian Wildlife Service Report Series 21. Ottawa: Information Canada. 36 p.
MECH, L.D. 1994. Regular and homeward travel speeds of Arctic wolves. Journal of Mammalogy 75:741–742.
MECH, L.D., and BOITANI, L. 2003. Wolf social ecology. In: Mech, L.D., and Boitani, L., eds. Wolves: Behavior, ecology, and conservation. Chicago: University of Chicago Press. 1–34.
MECH, L.D., and GESE, E.M. 1992. Field testing the Wildlink capture collar on wolves. Wildlife Society Bulletin 20:249–256.
MECH, L.D., WOLFE, P., and PACKARD, J.M. 1999. Regurgitative food transfer among wild wolves. Canadian Journal of Zoology 77:1192–1195.
MERRILL, S.B., and MECH, L.D. 2000. Details of extensive movements by Minnesota wolves (*Canis lupus*). American Midland Naturalist 144:428–433.
MERRILL, S.B., ADAMS, L.G., NELSON, M.E., and MECH, L.D. 1998. Testing releasable GPS radiocollars on wolves and white-tailed deer. Wildlife Society Bulletin 26:830–835.
MESSIER, F. 1985. Solitary living and extraterritorial movements of wolves in relation to social status and prey abundance. Canadian Journal of Zoology 63:239–245.
MOHR, C.O., and STUMPF, W.A. 1966. Comparison of methods for calculating areas of animal activity. Journal of Wildlife Management 30:293–304.
MUSIANI, M. 2003. Conservation biology and management of wolves and wolf-human conflicts in western North America. Ph.D. Thesis, University of Calgary, Calgary, Alberta, Canada.
ORIANS, G.H., and PEARSON, N.E. 1979. On the theory of central place foraging. In: Mitchell, R.D., and Stairs, G.F., eds. Analysis of ecological systems. Columbus: Ohio State University Press. 154–177.
PACKARD, J.M., MECH, L.D., and REAM, R.R. 1992. Weaning in an arctic wolf pack: Behavioral mechanisms. Canadian Journal of Zoology 70:1269–1275.
PAQUET, P.C., WIERZCHOWSKI, J., and CALLAGHAN, C. 1996. Summary report on the effects of human activity on gray wolves in the Bow River Valley, Banff National Park, Alberta. In: Green, J., Pacas, C., Bayley, S., and Cornwell, L., eds. A cumulative effects assessment and futures outlook for the Banff Bow Valley. Prepared for the Banff Bow Valley Study. Ottawa: Department of Canadian Heritage.
STEPHENSON, R.O., and JAMES, D. 1982. Wolf movements and food habits in northwest Alaska. In: Harrington, F.H., and Paquet, P.C., eds. Wolves of the world. New Jersey: Noyes Publications. 223–237.
THORPE, N., EYEGETOK, S., HAKONGAK, N., and QITIRMIUT ELDERS. 2001. The Tuktu and Nogak Project: A caribou chronicle. Final Report to the West Kitikmeot/Slave Study Society, Ikaluktuuttiak, NWT. 160 p.
TIMONEY, K.P., LA ROI, G.H., ZOLTAI, S.C., and ROBINSON, A.L. 1992. The high subarctic forest-tundra of northwestern Canada: Position, width, and vegetation gradients in relation to climate. Arctic 45(1):1–9.
WALTON, L.R., CLUFF, H.D., PAQUET, P.C., and RAMSAY, M.A. 2001. Movement patterns of barren-ground wolves in the central Canadian Arctic. Journal of Mammalogy 82:867–876.
WILLIAMS, T.M. 1990. Summer diet and behavior of wolves denning on barren-ground caribou range in the Northwest Territories, Canada. M.Sc. Thesis, University of Alberta, Edmonton, Alberta, Canada.
ZAR, J.H. 1999. Biostatistical analysis. 4th ed. New Jersey: Prentice Hall. 663 p.

Appendix III. Answers to Self-Quizzes and Genetics Problems

Italicized numbers refer to relevant section numbers

CHAPTER 1

1. cell *1.1*
2. energy, nutrients *1.2*
3. Homeostasis *1.2*
4. Domains *1.3*
5. d *1.2, 1.4*
6. d *1.2*
7. Mutations *1.4*
8. adaptive *1.4*
9. a *1.6, 1.7*
10. c *1.1*
 e *1.4*
 d *1.6*
 b *1.6*
 a *1.6*

CHAPTER 2

1. False *2.1*
2. b *2.1*
3. d *2.2*
4. c *2.3*
5. a *2.4*
6. c *2.4*
7. b *2.4*
8. e *2.5*
9. d *2.6*
10. acid; base *2.6*
11. c *2.6*
12. e *Impacts, Issues*
 d *2.6*
 c *2.4*
 c *2.5*
 a *2.3*

CHAPTER 3

1. Carbohydrates *3.2*
 lipids *3.3*
 proteins/amino acids *3.4*
 nucleic acids/nucleotides *3.6*
2. d *3.1*
3. c *3.2*
4. f *3.2, 3.6*
5. b *3.3*
6. False *3.3*
7. b *3.3*
8. e *3.3*
9. d *3.4, 3.6*
10. d *3.5*
11. d *3.6*
12. b *3.6*
13. c *3.4*
 e *3.6*
 b *3.3*
 d *3.6*
 a *3.2*

CHAPTER 4

1. False *4.1*
2. c *4.1, 4.3*
3. c *4.1, 4.4, 4.6*
4. d *4.6*
5. b *4.7*
6. d *4.11*
7. False *4.11*
8. e *4.9*
 d *4.9*
 a *4.1*
 f *4.8*
 c *4.8*
 b *4.8*

CHAPTER 5

1. c *5.1*
2. a *5.1*
3. d *5.1*
4. d *5.4*
5. d *5.3*
6. d *5.6*
7. b *5.6, 5.7*
8. a *5.8*
9. a *5.9*
10. g *5.7*
 e *5.3*
 h *5.9*
 i *5.1*
 d *5.2*
 b *5.7*
 f *5.6*
 a *5.4*
 c *5.1*

CHAPTER 6

1. carbon dioxide; sunlight *Impacts, Issues*
2. b *6.1*
3. a *6.4*
4. b *6.3, 6.4*
5. c *6.4*
6. d *6.4*
7. c *6.6*
8. b *6.6*
9. e *6.6*
10. d *6.4, 6.5*
 a *6.6*
 b *6.4, 6.5*
 a *6.6*

CHAPTER 7

1. False *7.1*
2. d *7.1*
3. a *7.1*
4. c *7.2*
5. b *7.1*
6. d *7.3*
7. b *7.3*
8. c *7.4*
9. c *7.5*
10. b *7.5*
11. d *7.7*
12. see Figure 7.8 *7.4*
13. b *7.2*
 c *7.5*
 a *7.3*
 d *7.4*

CHAPTER 8

1. d *8.1*
2. b *8.1*
3. c *8.1*
4. d *8.2*
5. a *8.2*
6. c *8.2*
7. a *8.2*
8. see Figure 8.7 *8.3*
9. b *8.3*
10. d *8.3*
 b *8.3*
 c *8.3*
 a *8.3*

CHAPTER 9

1. c *9.2*
2. d *9.1*
3. d *9.2, 9.4*
4. b *9.2*
5. d *9.3*
6. d *9.3*
7. e *9.4*
8. Sister chromatids are still attached *9.3*
9. d *9.2*
 a *9.1*
 c *9.3*
 b *9.2*

CHAPTER 10

1. a *10.1*
2. b *10.1*
3. a *10.1*
4. b *10.1*
5. c *10.2*
6. a *10.2*
7. d *10.3*
8. c *10.5*
9. a *10.5*
10. b *10.3*
 d *10.2*
 a *10.1*
 c *10.1*

CHAPTER 11

1. d *11.1*
2. c *11.1*
3. b *11.2*
4. b *11.2*
5. b *11.2*
6. False *11.4*
7. d *11.4*
8. e *11.5*
9. d *11.6*
10. True *11.6*
11. c *11.6*
12. a *11.7*
13. c *11.6*
 e *11.5*
 f *11.6*
 b *11.5*
 a *11.1*
 d *11.6*

CHAPTER 12

1. c *12.2*
2. d *12.2*
3. c *12.2*
4. a *12.4*
5. 3′—CCAAAGAAGTTCTCT—5′ *12.2*
6. d *12.1*
 b *12.5*
 a *12.2*
 f *12.2*
 e *12.4*
 g *12.4*
 c *12.2*

CHAPTER 13

1. c *13.1*
2. b *13.1*
3. c *13.1*
4. b *13.1*
5. c *13.2*
6. a *13.2*
7. c *13.2*
8. a *13.3*
9. f *13.5*
10. e *13.5*
 c *13.4*
 a *13.1*
 f *13.2*
 d *13.3*
 g *13.1*
 b *13.2*

CHAPTER 14

1. d *14.1*
2. d *14.2*
3. b *14.1*
4. h *14.1*
5. c *14.1, 14.2, 14.3*
6. d *14.1*
7. b *14.2*
8. c *14.2*
9. c *14.3*
10. d *14.3*
11. b *14.4*
12. e *14.2*
 a *14.2*
 b *14.4*
 d *14.2*
 c *14.1*
 f *14.1*

CHAPTER 15

1. c *15.1*
2. plasmid *15.1*
3. b *15.1*
4. b *15.2*
5. b *15.2*
6. b *15.3, 15.4*
7. d *15.3*
8. b *15.10*
9. c *15.4*
 b *15.7*
 d *15.2*
 a *15.10*

CHAPTER 16

1. e *16.1*
2. d *16.2*
3. d *16.4*
4. a *16.5*
5. Gondwana *16.6*
6. d *16.7*
7. b *16.7*
8. d *16.8*
9. b *16.9*
10. c *16.10*
11. e *16.4*
 a *16.4*
 d *16.8*
 g *16.5*
 c *16.7*
 b *16.2*
 f *16.7*

CHAPTER 17

1. populations *17.1*
2. b *Impacts, Issues*
3. a *17.1*
4. a *17.3*
5. b *17.4*
6. c *17.5*
7. c *17.6*
8. b *17.7*
9. b *17.8*
10. c *17.8*
 d *17.3*
 a *17.1*
 b *17.7*
 e *17.12*

CHAPTER 18

1. c *18.1*
2. c *18.2*
3. a *18.2*
4. a *18.2*
5. c *18.3*
6. b *18.4*
7. d *18.4*
8. a *18.1*
 c *18.2*
 d *18.5*
 a *18.5*
 b *18.5*
 e *18.5*

CHAPTER 19

1. d *19.1*
2. c *19.1*
3. c *19.2*
4. d *19.2*
5. b *19.2*
6. c *19.2*
7. d *19.1*
8. c *19.4*
9. b *19.5*
10. d *19.4*
11. d *19.3*
 e *19.2*
 b *19.4*
 f *19.1*
 g *19.2*
 a *19.5*
 c *19.2*

CHAPTER 20

1. f *20.2*
2. a *20.3*
3. d *20.4*
4. b *20.6, 20.7*
5. f *20.2, 20.8*
6. d *20.10*
7. d *20.2*
 g *20.5*
 a *20.4*
 b *20.6*
 f *20.9*
 e *20.8*
 c *20.10*

CHAPTER 21

1. c *21.2*
2. a *21.3*
3. b *21.2*

Question	Answer	Section
4.	c	*21.3*
5.	a	*21.4*
6.	a	*21.2*
7.	b	*21.5*
8.	c	*21.5*
9.	c	*21.6*
	e	*21.1*
	g	*21.3*
	h	*21.7*
	f	*21.2*
	a	*21.1*
	b	*21.1*
	d	*21.7*

CHAPTER 22

Question	Answer	Section
1.	c	*22.1*
2.	a	*22.1*
3.	b	*22.2*
4.	c	*22.3*
5.	d	*22.4*
6.	c	*22.4*
7.	c	*22.4*
8.	b	*22.5*
9.	d	*22.6*
10.	d	*22.1*
	b	*22.1*
	d	*22.1*
	f	*22.3*
	g	*22.4*
	c	*22.5*
	e	*22.5*

CHAPTER 23

Question	Answer	Section
1.	b	*23.1*
2.	d	*23.1*
3.	a	*23.4*
4.	a	*23.5*
5.	b	*23.12*
6.	c	*23.5, 23.6*
7.	b	*23.10*
8.	c	*23.11*
9.	c	*23.12*
10.	b	*23.14*
11.	b	*23.2*
	j	*23.2*
	d	*23.3*
	i	*23.4*
	c	*23.5*
	a	*23.8*
	g	*23.6*
	e	*23.9*
	f	*23.7*
	h	*23.14*

CHAPTER 24

Question	Answer	Section
1.	c	*24.1*
2.	d	*24.1*
3.	a	*24.2*
4.	c	*24.1*
5.	d	*24.3*
6.	c	*24.6*
7.	f	*24.6*
8.	a	*24.6*
9.	c	*24.9*
10.	f	*24.12*
11.	c	*24.13*
12.	b	*24.1*
	i	*24.3*
	g	*24.4*
	f	*24.8*
	c	*24.9*
	d	*24.10*
	a	*24.10*
	h	*24.10*
	e	*24.11*

CHAPTER 25

Question	Answer	Section
1.	a	*25.1*
2.	c	*25.1*
3.	d	*25.2*
4.	d	*25.2*
5.	a	*25.2*
6.	c	*25.3*
7.	a	*25.3*
8.	d	*25.5*
9.	b	*25.4*
	d	*25.3*
	a	*25.5*
	c	*25.3*
	e	*25.3*
	f	*25.3*

CHAPTER 26

Question	Answer	Section
1.	eudicots, left; monocots, right	*26.1*
2.	a	*26.1, 26.3*
3.	d	*26.1, 26.6*
4.	eudicot; monocot	*26.1*
5.	c	*26.2*
6.	c	*26.4*
7.	b	*26.2*
8.	b	*26.2*
9.	d	*26.6*
10.	b	*26.1*
	d	*26.1*
	e	*26.2*
	c	*26.2*
	f	*26.5*
	a	*26.6*

CHAPTER 27

Question	Answer	Section
1.	e	*27.1*
2.	e	*27.2*
3.	b	*27.2*
4.	b	*27.2*
5.	c	*27.3*
6.	d	*27.3*
7.	a	*27.4*
8.	c	*27.4*
9.	c	*27.4*
	g	*27.1*
	e	*27.6*
	b	*27.2*
	d	*27.3*
	a	*27.3*
	f	*27.6*

CHAPTER 28

Question	Answer	Section
1.	b	*28.1*
2.	c	*28.1*
3.	b	*28.2*
4.	c	*28.2*
5.	a	*28.3*
6.	c	*28.3*
7.	c	*28.5*
8.	c	*28.7*
9.	e	*28.7*
10.	d	*28.8*
11.	a	*28.8*
12.	c	*28.8*
13.	c	*28.2*
	f	*28.1*
	g	*28.2*
	e	*28.1, 28.2*
	d	*28.1*
	b	*28.2*
	a	*28.2*

CHAPTER 29

Question	Answer	Section
1.	a. epithelium	*29.1*
	b. skeletal muscle	*29.3*
	c. loose connective tissue	*29.2*
	d. adipose tissue	*29.2*
2.	a	*29.1*
3.	c	*29.1*
4.	a	*29.1*
5.	b	*29.2*
6.	b	*29.2*
7.	c	*29.2*
8.	c	*29.3*
9.	d	*29.3*
10.	d	*29.4*
11.	b	*29.1*
	g	*29.1*
	a	*29.2*
	c	*29.5*
	d	*29.3*
	f	*29.2*
	e	*29.1*

CHAPTER 30

Question	Answer	Section
1.	a	*30.1*
2.	c	*30.2*
3.	d	*30.2*
4.	a	*30.4*
5.	c	*30.11*
6.	c	*30.8*
7.	b	*30.8*
8.	a	*30.5*
9.	a	*30.9*
10.	f	*30.7*
	d	*30.5*
	g	*30.10*
	b	*30.9*
	h	*30.10*
	a	*30.9*
	e	*30.11*
	i	*30.8*
	c	*30.9*

CHAPTER 31

Question	Answer	Section
1.	a	*31.1*
2.	c	*31.1*
3.	c	*31.2*
4.	e	*31.3*
5.	a	*31.2*
6.	b	*31.4*
7.	b	*31.5*
8.	See Figure 31.12	
9.	d	*31.7*
	g	*31.5*
	f	*31.6*
	a	*31.4*
	c	*31.7*
	e	*31.3*
	b	*31.4*
	h	*31.2*

CHAPTER 32

Question	Answer	Section
1.	f	*32.1*
2.	b	*32.3*
3.	a	*32.3*
4.	e	*32.1*
5.	b	*32.6*
6.	b	*32.4*
7.	b	*32.6*
8.	d	*32.4*
9.	b	*32.8*
10.	d	*32.5*
	f	*32.4*
	c	*32.4*
	e	*32.6*
	a	*32.8*
	b	*32.1*

CHAPTER 33

Question	Answer	Section
1.	a	*33.1*
2.	d	*33.2*
3.	b	*33.3*
4.	a	*33.2*
5.	a	*33.2*
6.	b	*33.4*
7.	b	*33.4*
8.	d	*33.4*
9.	d	*33.5*
10.	d	*33.5*
11.	e	*33.2*
	f	*33.5*
	g	*33.5*
	h	*33.2*
	j	*33.4*
	c	*33.2*
	b	*33.1*
	i	*33.4*
	d	*33.5*
	a	*33.5*

CHAPTER 34

Question	Answer	Section
1.	c	*34.1*
2.	b	*34.1*
3.	d	*34.2*
4.	b	*34.4*
5.	d	*34.2*
6.	b	*34.6*
7.	c	*34.6*
8.	a	*34.7*
9.	c	*34.7*
10.	d	*34.8*
11.	d	*34.10*
12.	f	*34.8*
	a	*34.10*
	e	*34.2*
	g	*34.6*
	c	*34.7*
	d	*34.5*
13.	See Figure 34.13	
14.	See Figure 34.17	

CHAPTER 35

Question	Answer	Section
1.	f	*35.2*
2.	d	*35.3*
3.	d	*35.1*
4.	e	*35.1*
5.	e	*35.4*
6.	d	*35.5*
7.	e	*35.5*
8.	b	*35.8*
9.	b	*35.6*
10.	b	*35.8*
11.	a, b, d, e, f	*35.9*
12.	c	*35.3*
	b	*35.5*
	a	*35.1*
	e	*35.5*
	d	*35.8*

CHAPTER 36

Question	Answer	Section
1.	a	*36.1*
2.	d	*36.1*
3.	a	*36.2*
4.	a	*36.3*
5.	c	*36.4*
6.	d	*36.6*
7.	a	*36.6*
8.	a	*36.5*
9.	c	*36.8*
10.	d	*36.4*
	h	*36.4*
	f	*36.5*
	e	*36.5*
	g	*36.4*
	c	*36.4*
	h	*36.4*
	a	*36.6*

CHAPTER 37

Question	Answer	Section
1.	d	*37.1*
2.	b	*37.4*
3.	c	*37.5*
4.	b	*37.4*
5.	a	*37.5*
6.	c	*37.6*
7.	a	*37.4*
8.	b	*37.5*
9.	b	*37.9*
10.	f	*37.4*
	b	*37.6*
	a	*37.4*
	d	*37.5*
	e	*37.4*
	c	*37.4*

CHAPTER 38

Question	Answer	Section
1.	See Figure 38.3, 38.9	
2.	c	*38.1*
3.	a	*38.3*
4.	b	*38.3*
5.	d	*38.4*
6.	b	*38.4*
7.	a	*38.4*
8.	a	*38.4*
9.	c	*38.3*
	a	*38.3*
	b	*38.3*
	e	*38.4*
	d	*38.4*
10.	d	*38.8*
11.	d	*38.8*
12.	True	*38.4*
13.	b	*38.7*
	a	*38.7*
	d	*38.7*
	c	*38.7*
	e	*38.7*

CHAPTER 39

Question	Answer	Section
1.	b	*39.1*
2.	b	*39.2*
3.	False	*39.3*
4.	a	*39.4*
5.	d	*39.4*
6.	c	*39.2*
	d	*39.2*
	a	*39.2*
	b	*39.2*
	f	*39.4*
	e	*39.4*
7.	a	*39.5*
8.	c	*39.9*
9.	b	*39.9*
10.	c	*39.13*
11.	c	*39.4*
12.	c	*39.4*
13.	c	*39.16*
14.	c	*39.17*
15.	b	*39.19*

16. d	*39.5*
c	*39.6*
h	*39.7*
a	*39.16*
i	*39.7*
e	*39.9*
f	*39.11*
b	*39.5*
g	*39.7*

CHAPTER 40

1. b	*40.1*
2. f	*40.1*
3. c	*40.2*
4. b	*40.3*
5. a	*40.3*
6. d	*40.4*
7. d	*40.5*
8. b	*40.7*
9. d	*40.9*
10. c	*40.4*
d	*40.3*
a	*40.3*
e	*40.4*
b	*40.4*

CHAPTER 41

1. d	*41.1*
2. d	*41.1*
3. d	*41.1*
4. e	*41.3*
5. d	*41.3*
6. b	*41.4*
7. a	*41.4*
b	*41.2*
c	*41.1*
d	*41.6*
e	*41.3*
8. b	*41.8*
9. a	*41.12*
10. d	*41.12*
11. d	*41.13*
12. c	*41.9*
d	*41.11*
a	*41.8*
e	*41.8*
b	*41.9*
g	*41.9*
f	*41.3*

CHAPTER 42

1. a	*42.1*
2. d	*42.1*
3. d	*42.1*
4. d	*42.1, 42.4*
5. d	*42.4*
6. d	*42.3*
7. c	*42.6*
8. b	*42.7*
9. d	*42.7*
10. d	*42.8*
11. d	*42.10*
12. c	*42.10*
13. d	*42.9, 42.10*
14. a	*42.9*
15. d	*42.1*
a	*42.1*
c	*42.1*
b	*42.1*

CHAPTER 43

1. d	*43.1*
2. c	*43.2*
3. d	*43.3*
4. a	*43.3*
5. d	*43.2*
6. e	*43.4*
7. d	*43.4*
8. c	*43.6*
9. a	*43.10*
10. d	*43.11, 43.15*
11. d	*43.15*
12. d	*43.10*
h	*43.6*
e	*43.6*
c	*43.6*
b	*43.13*
g	*43.9*
a	*43.7*
f	*43.15*

CHAPTER 44

1. d	*44.1*
2. b	*44.1*
3. c	*44.3*
4. a	*44.4*
5. a	*44.4*
6. b	*44.5*
7. d	*44.6*
8. c	*44.6*
9. e	*44.7*
10. c	*44.7*
11. c	*44.2*
d	*44.7*
b	*44.1*
a	*44.2*
e	*44.4*

CHAPTER 10: GENETIC PROBLEMS

1. a. Both parents are heterozygotes (*Aa*). Their children may be albino (*aa*) or unaffected (*AA* or *Aa*).

b. All are homozygous recessive (*aa*).

c. Homozygous recessive (*aa*) father, and heterozygous (*Aa*) mother. The albino child is *aa*, the unaffected children *Aa*.

2. Possible outcomes of an experimental cross between F_1 rose plants heterozygous for height (*Aa*):

	A	*a*
A	*AA* climber	*Aa* climber
a	*Aa* climber	*aa* shrubby

3:1 possible ratio of genotypes and phenotypes in F_2 generation

Possible outcomes of a testcross between an F_1 rose plant heterozygous for height and a shrubby rose plant:

Gametes shrubby plant: / Gametes F_1 hybrid:	*A*	*a*
a	*Aa* climber	*aa* shrubby
a	*Aa* climber	*aa* shrubby

1:1 possible ratio of genotypes and phenotypes in F_2 generation

3. a. *AB*

b. *AB, aB*

c. *Ab, ab*

d. *AB, Ab, aB, ab*

4. a. All offspring will be *AaBB*.

b. 1/4 *AABB* (25% each genotype)
1/4 *AABb*
1/4 *AaBB*
1/4 *AaBb*

c. 1/4 *AaBb* (25% each genotype)
1/4 *Aabb*
1/4 *aaBb*
1/4 *aabb*

d. 1/16 *AABB* (6.25% of genotype)
1/8 *AaBB* (12.5%)
1/16 *aaBB* (6.25%)
1/8 *AABb* (12.5%)
1/4 *AaBb* (25%)
1/8 *aaBb* (12.5%)
1/16 *AAbb* (6.25%)
1/8 *Aabb* (12.5%)
1/16 *aabb* (6.25%)

5. a. *ABC*

b. *ABC, aBC*

c. *ABC, aBC, ABc, aBc*

d. *ABC*
aBC
AbC
abC
ABc
aBc
Abc
abc

6. A mating of two M^L cats yields 1/4 *MM*, 1/2 M^LM, and 1/4 M^LM^L. Because M^LM^L is lethal, the probability that any one kitten among the survivors will be heterozygous is 2/3.

7. Yellow is recessive. Because F_1 plants have a green phenotype and must be heterozygous, green must be dominant over the recessive yellow.

8. A mating between a mouse from a true-breeding, white-furred strain and a mouse from a true-breeding, brown-furred strain would provide you with the most direct evidence. Because true-breeding strains of organisms typically are homozygous for a trait being studied, all F_1 offspring from this mating should be heterozygous. Record the phenotype of each F_1 mouse, then let them mate with one another. Assuming only one gene locus is involved, these are possible outcomes for the F_1 offspring:

a. All F_1 mice are brown, and their F_2 offspring segregate: 3 brown : 1 white. *Conclusion*: Brown is dominant to white.

b. All F_1 mice are white, and their F_2 offspring segregate: 3 white : 1 brown. *Conclusion*: White is dominant to brown.

c. All F_1 mice are tan, and the F_2 offspring segregate: 1 brown : 2 tan : 1 white. *Conclusion*: The alleles at this locus show incomplete dominance.

9. The data reveal that these genes do not assort independently because the observed ratio is very far from the 9:3:3:1 ratio expected with independent assortment. Instead, the results can be explained if the genes are located close to each other on the same chromosome, which is called linkage.

10. **a.** 1/2 red 1/2 pink ____ white
b. ____ red All pink ____ white
c. 1/4 red 1/2 pink 1/4 white
d. ____ red 1/2 pink 1/2 white

11. Because both parents are heterozygotes (Hb^AHb^S), the following are the probabilities for each child:

a. 1/4 Hb^SHb^S

b. 1/4 Hb^AHb^A

c. 1/2 Hb^AHb^S

12. 2/3

CHAPTER 12: GENETIC PROBLEMS

1. **a.** Human males (XY) inherit their X chromosome from their mother.

b. A male can produce two kinds of gametes. Half carry an X chromosome and half carry a Y chromosome. All the gametes that carry the X chromosome carry the same X-linked allele.

c. A female homozygous for an X-linked allele produces only one kind of gamete.

d. Fifty percent of the gametes of a female who is heterozygous for an X-linked allele carry one of the two alleles at that locus; the other fifty percent carry its partner allele for that locus.

2. Because Marfan syndrome is a case of autosomal dominant inheritance and because one parent bears the allele, the probability that any child of theirs will inherit the mutant allele is 50 percent.

3. **a.** Nondisjunction might occur during anaphase I or anaphase II of meiosis.

b. As a result of translocation, chromosome 21 may get attached to the end of chromosome 14. The new individual's chromosome number would still be 46, but its somatic cells would have the translocated chromosome 21 in addition to two normal chromosomes 21.

4. A daughter could develop this muscular dystrophy only if she inherited two X-linked recessive alleles—one from each parent. Males who carry the allele are unlikely to father children because they develop the disorder and die early in life.

5. In the mother, a crossover between the two genes at meiosis generates an X chromosome that carries neither mutant allele.

6. The phenotype appeared in every generation shown in the diagram, so this must be a pattern of autosomal dominant inheritance.

7. There is no scientific answer to this question, which simply invites you to reflect on the difference between a scientific and a subjective interpretation of this individual's condition.

Appendix IV. Periodic Table of the Elements

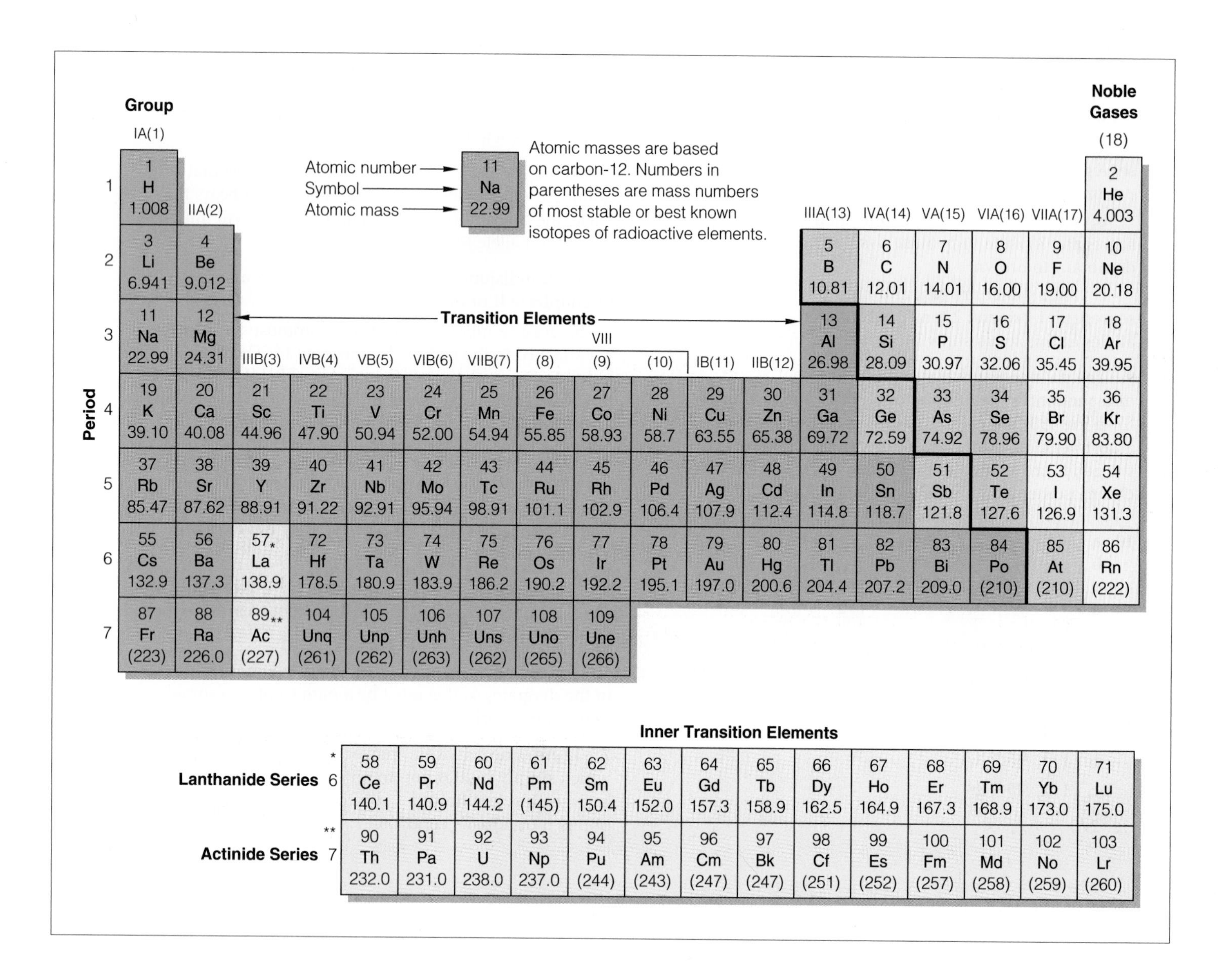

Atomic masses are based on carbon-12. Numbers in parentheses are mass numbers of most stable or best known isotopes of radioactive elements.

Key: Atomic number → 11; Symbol → Na; Atomic mass → 22.99

Transition Elements: groups IIIB(3) through IIB(12)

Period	IA(1)	IIA(2)	IIIB(3)	IVB(4)	VB(5)	VIB(6)	VIIB(7)	VIII (8)	VIII (9)	VIII (10)	IB(11)	IIB(12)	IIIA(13)	IVA(14)	VA(15)	VIA(16)	VIIA(17)	Noble Gases (18)
1	1 H 1.008																	2 He 4.003
2	3 Li 6.941	4 Be 9.012											5 B 10.81	6 C 12.01	7 N 14.01	8 O 16.00	9 F 19.00	10 Ne 20.18
3	11 Na 22.99	12 Mg 24.31											13 Al 26.98	14 Si 28.09	15 P 30.97	16 S 32.06	17 Cl 35.45	18 Ar 39.95
4	19 K 39.10	20 Ca 40.08	21 Sc 44.96	22 Ti 47.90	23 V 50.94	24 Cr 52.00	25 Mn 54.94	26 Fe 55.85	27 Co 58.93	28 Ni 58.7	29 Cu 63.55	30 Zn 65.38	31 Ga 69.72	32 Ge 72.59	33 As 74.92	34 Se 78.96	35 Br 79.90	36 Kr 83.80
5	37 Rb 85.47	38 Sr 87.62	39 Y 88.91	40 Zr 91.22	41 Nb 92.91	42 Mo 95.94	43 Tc 98.91	44 Ru 101.1	45 Rh 102.9	46 Pd 106.4	47 Ag 107.9	48 Cd 112.4	49 In 114.8	50 Sn 118.7	51 Sb 121.8	52 Te 127.6	53 I 126.9	54 Xe 131.3
6	55 Cs 132.9	56 Ba 137.3	57* La 138.9	72 Hf 178.5	73 Ta 180.9	74 W 183.9	75 Re 186.2	76 Os 190.2	77 Ir 192.2	78 Pt 195.1	79 Au 197.0	80 Hg 200.6	81 Tl 204.4	82 Pb 207.2	83 Bi 209.0	84 Po (210)	85 At (210)	86 Rn (222)
7	87 Fr (223)	88 Ra 226.0	89** Ac (227)	104 Unq (261)	105 Unp (262)	106 Unh (263)	107 Uns (262)	108 Uno (265)	109 Une (266)									

Inner Transition Elements

Series														
* Lanthanide Series 6	58 Ce 140.1	59 Pr 140.9	60 Nd 144.2	61 Pm (145)	62 Sm 150.4	63 Eu 152.0	64 Gd 157.3	65 Tb 158.9	66 Dy 162.5	67 Ho 164.9	68 Er 167.3	69 Tm 168.9	70 Yb 173.0	71 Lu 175.0
** Actinide Series 7	90 Th 232.0	91 Pa 231.0	92 U 238.0	93 Np 237.0	94 Pu (244)	95 Am (243)	96 Cm (247)	97 Bk (247)	98 Cf (251)	99 Es (252)	100 Fm (257)	101 Md (258)	102 No (259)	103 Lr (260)

Appendix V. Molecular Models

A molecule's structure can be depicted by different kinds of molecular models. Such models allow us to visualize different characteristics of the same structure.

Structural models show how atoms in a molecule connect to one another:

methane glucose

In such models, each line indicates one covalent bond: Double bonds are shown as two lines; triple bonds as three lines. Some atoms or bonds may be implied but not shown. For example, carbon ring structures such as those of glucose and other sugars are often represented as polygons. If no atom is shown at the corner of a polygon, a carbon atom is implied. Hydrogen atoms bonded to one of the atoms in the carbon backbone of a molecule may also be omitted:

glucose glucose

Ball-and-stick models show the relative sizes of the atoms and their positions in three dimensions:

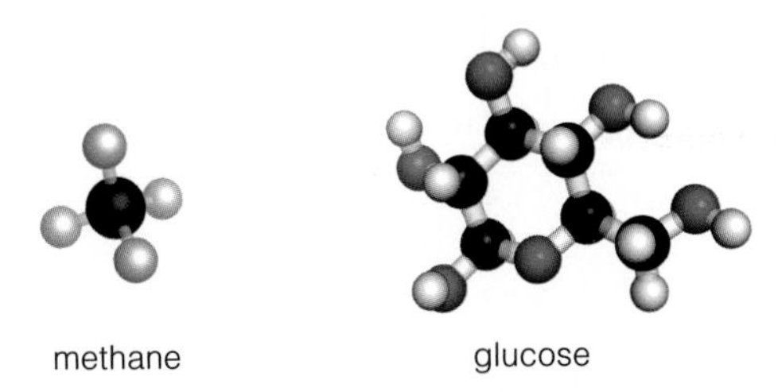

methane glucose

All types of covalent bonds (single, double, or triple) are shown as one stick. Typically, the elements in such models are coded in standardized colors:

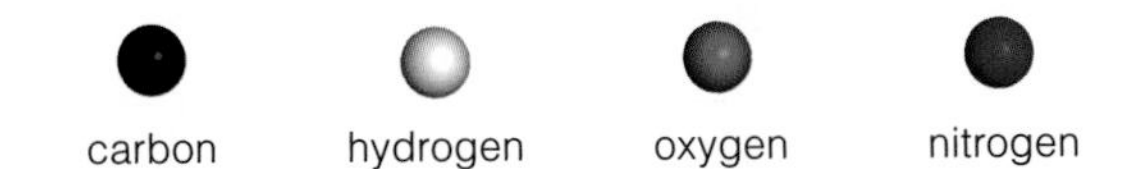

carbon hydrogen oxygen nitrogen

Space-filling models show the outer boundaries of the atoms in three dimensions:

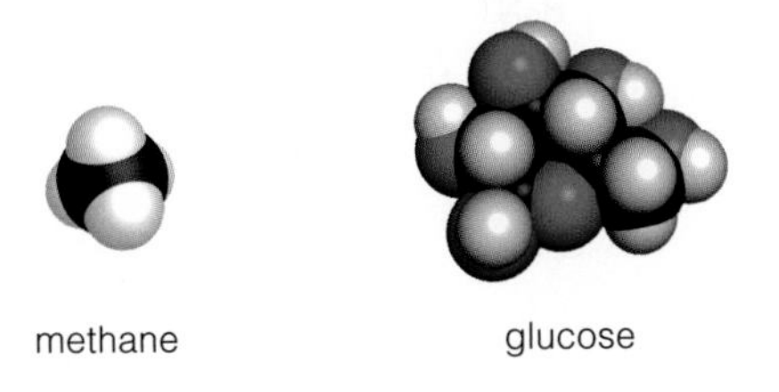

methane glucose

A model of a large molecule can be quite complex if all the atoms are shown. This space-filling model of hemoglobin is an example:

To reduce visual complexity, other types of models omit individual atoms. Surface models of large molecules can show features such as an active site crevice (Figure 5.7). In this surface model of hemoglobin, you can see two heme groups (red) nestled in pockets of the protein:

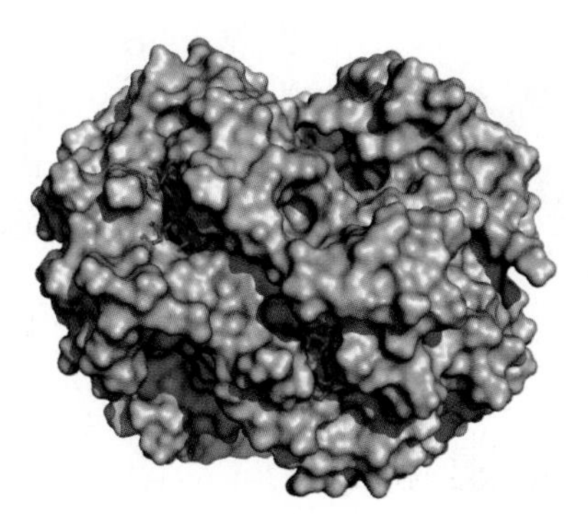

Large molecules such as proteins are often shown as ribbon models. Such models highlight secondary structure such as coils or sheets. In this ribbon model of hemoglobin, you can see the four coiled polypeptide chains, each of which folds around a heme group:

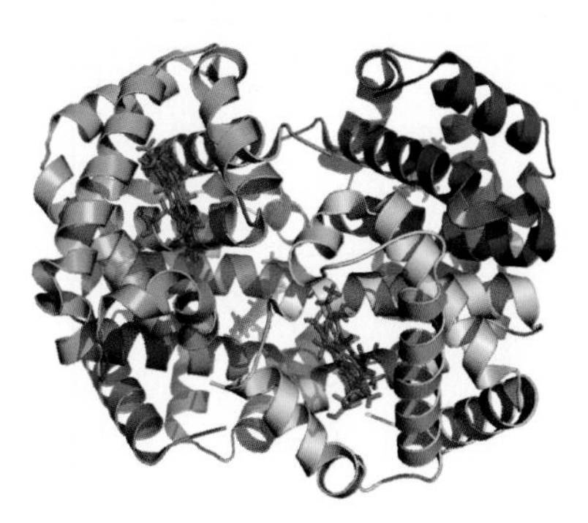

Such structural details are clues to how a molecule functions. Hemoglobin is the main oxygen carrier in vertebrate blood. Oxygen binds at the hemes, so one hemoglobin molecule can hold four molecules of oxygen.

The Amino Acids

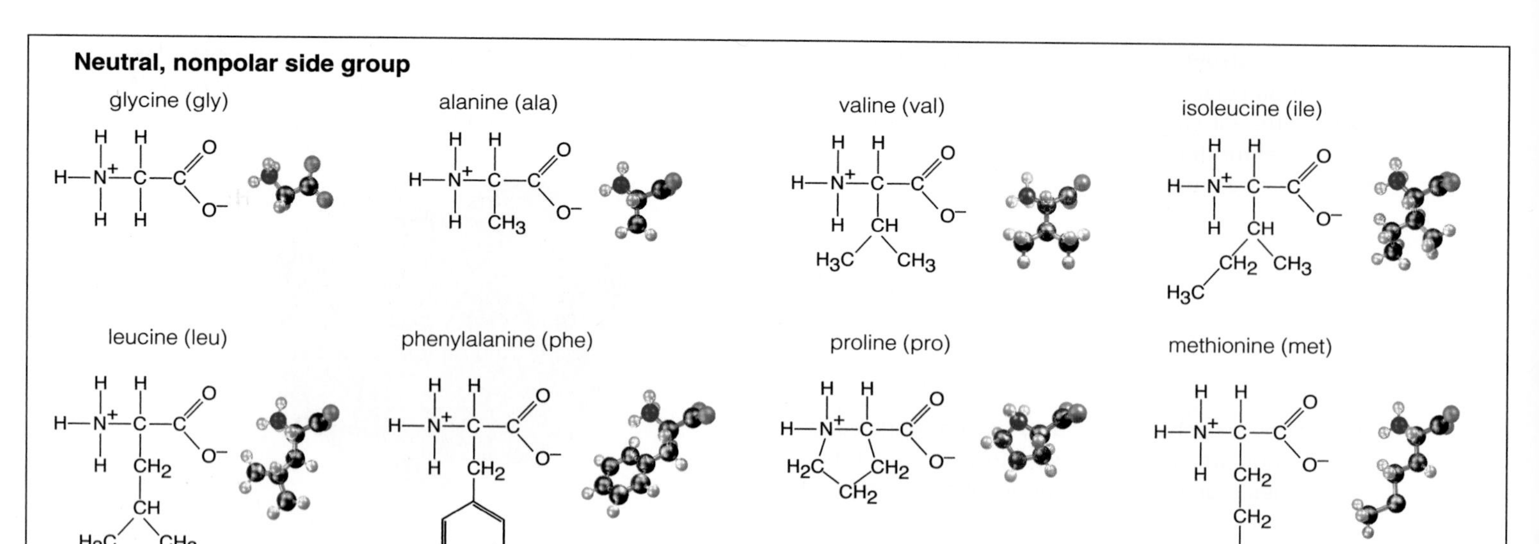

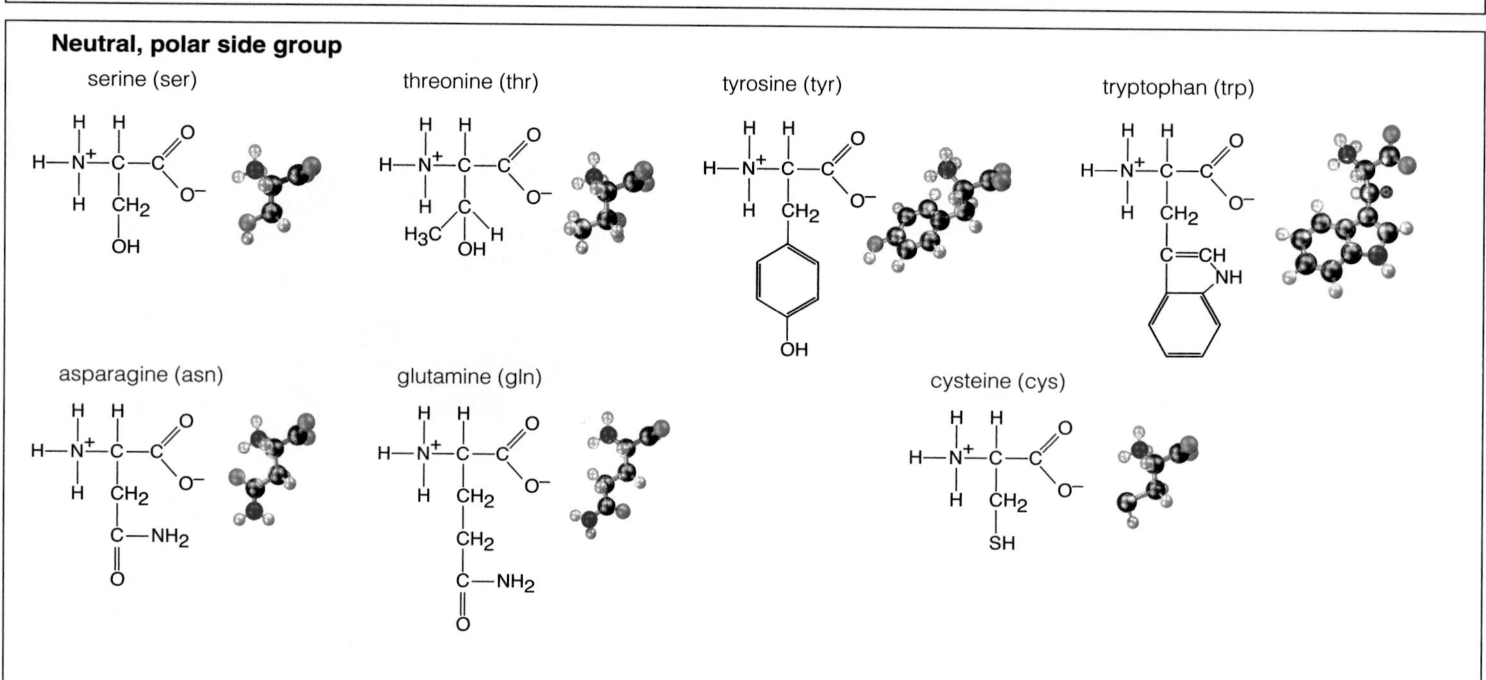

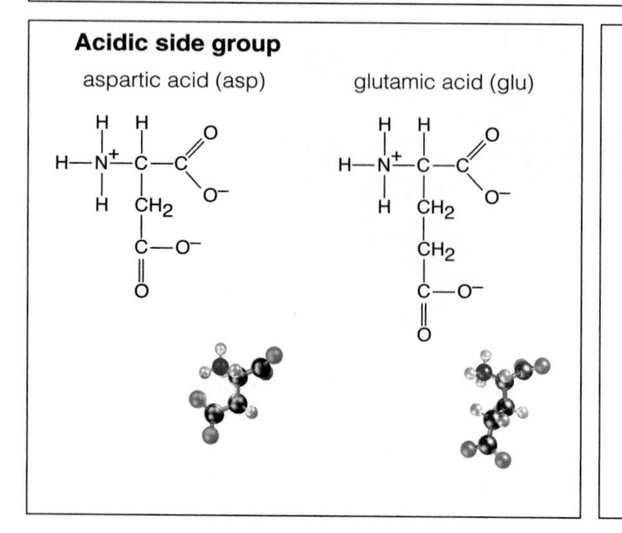

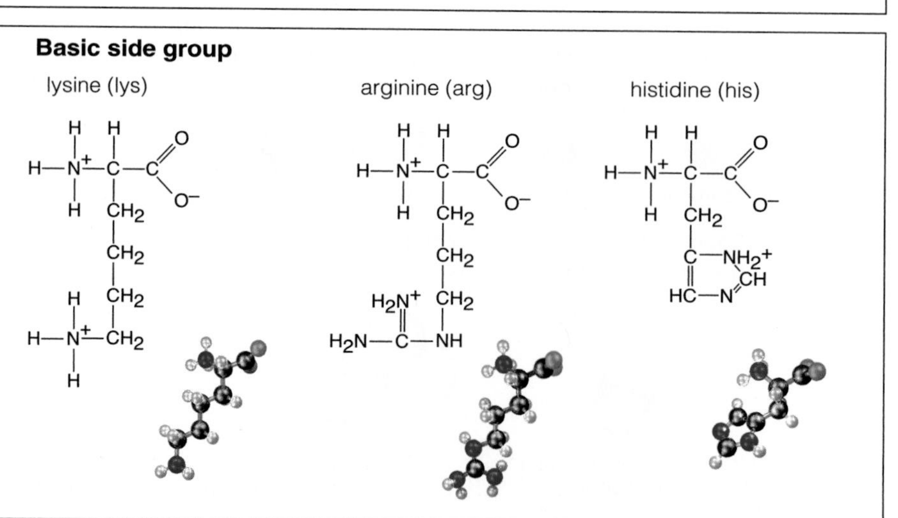

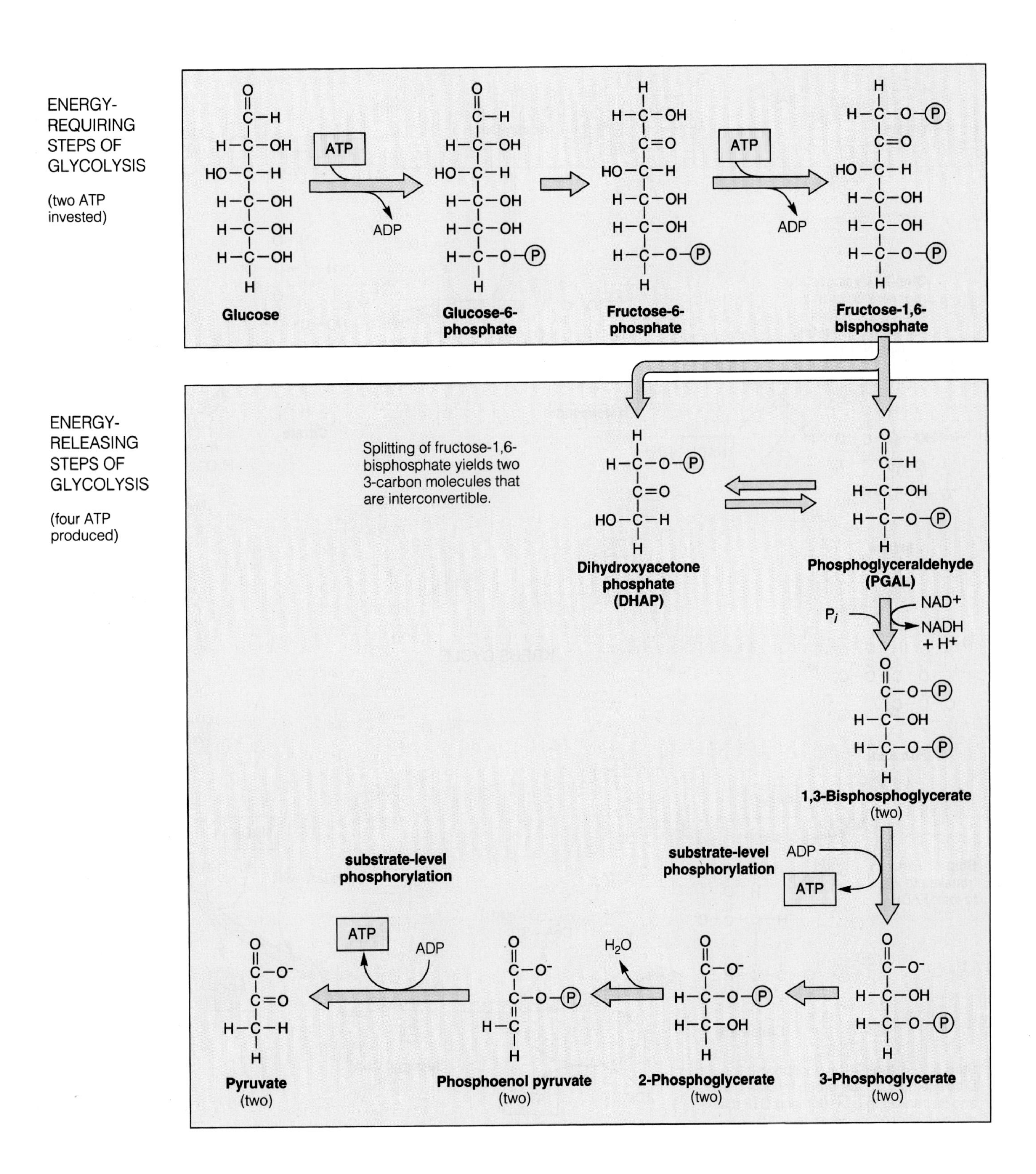

Figure A Glycolysis, ending with two 3-carbon pyruvate molecules for each 6-carbon glucose molecule entering the reactions. The *net* energy yield is two ATP molecules (two invested, four produced).

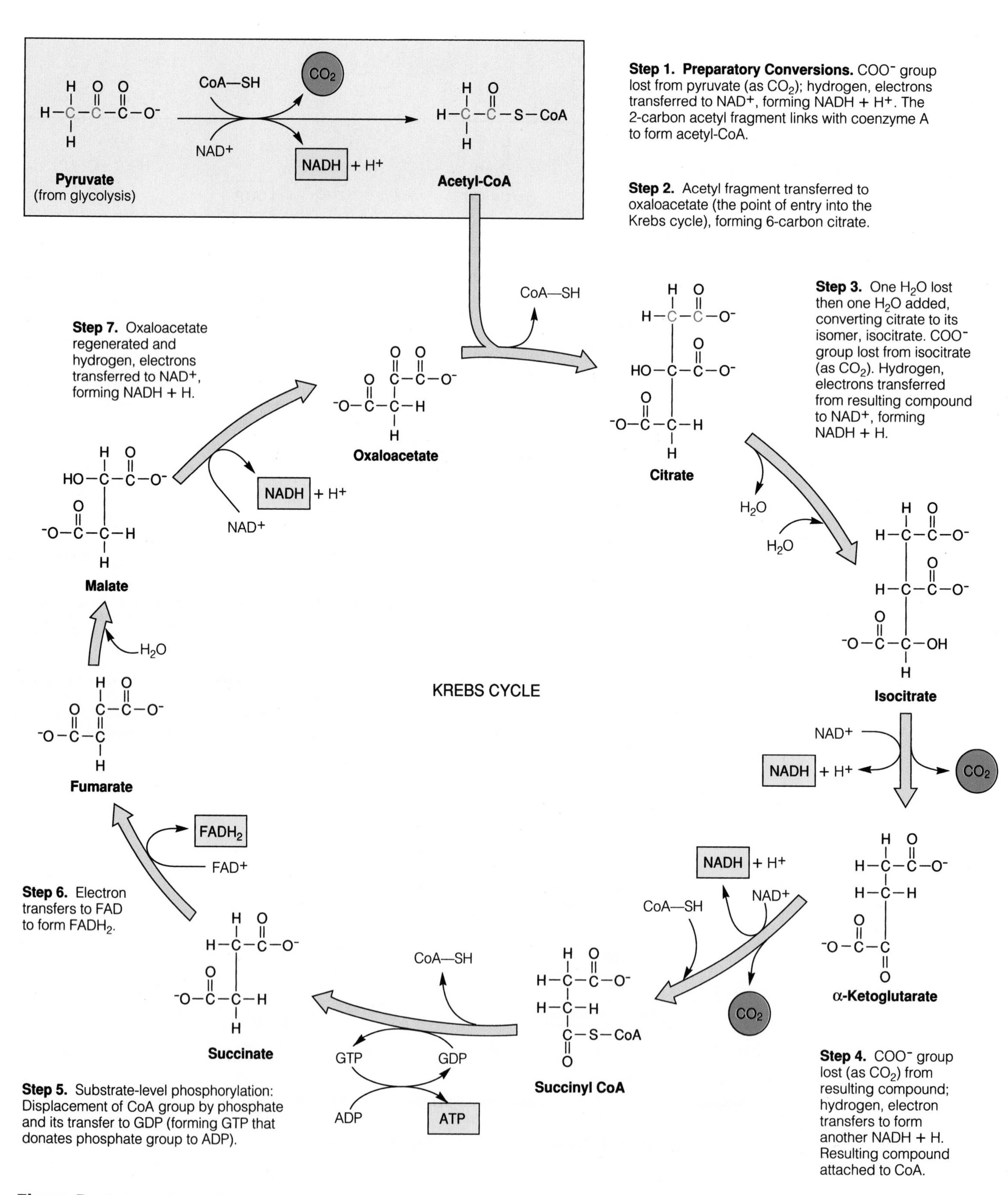

Figure B Krebs cycle, also known as the citric acid cycle. *Red* identifies carbon atoms entering the cyclic pathway (by way of acetyl-CoA) and leaving (by way of carbon dioxide). These cyclic reactions run twice for each glucose molecule that has been degraded to two pyruvate molecules.

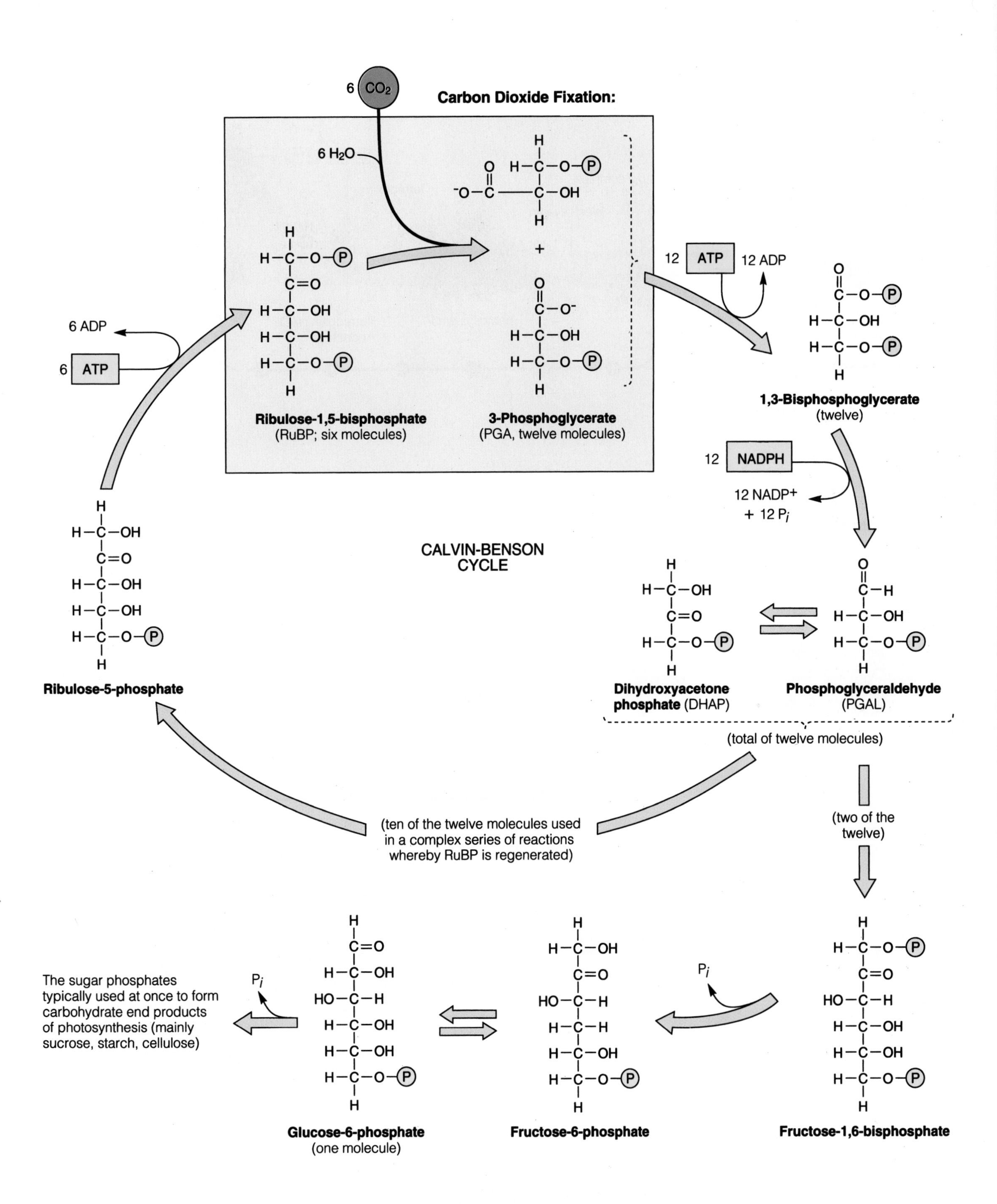

Figure C Calvin–Benson cycle of the light-independent reactions of photosynthesis.

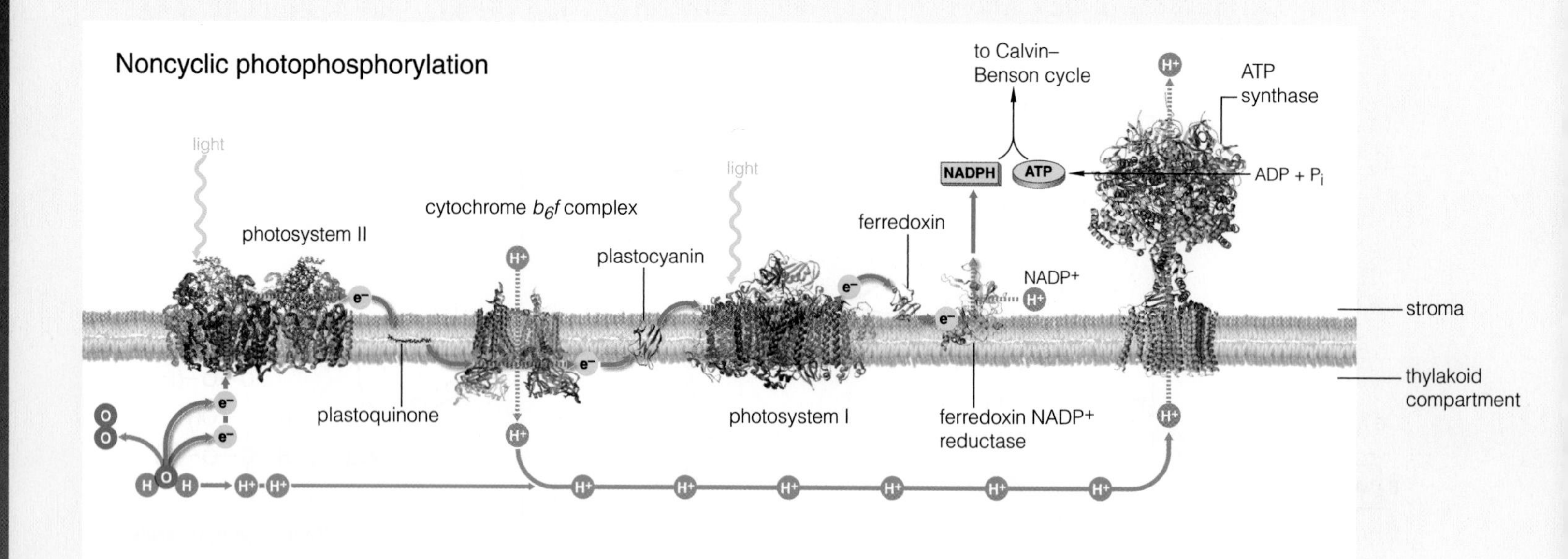

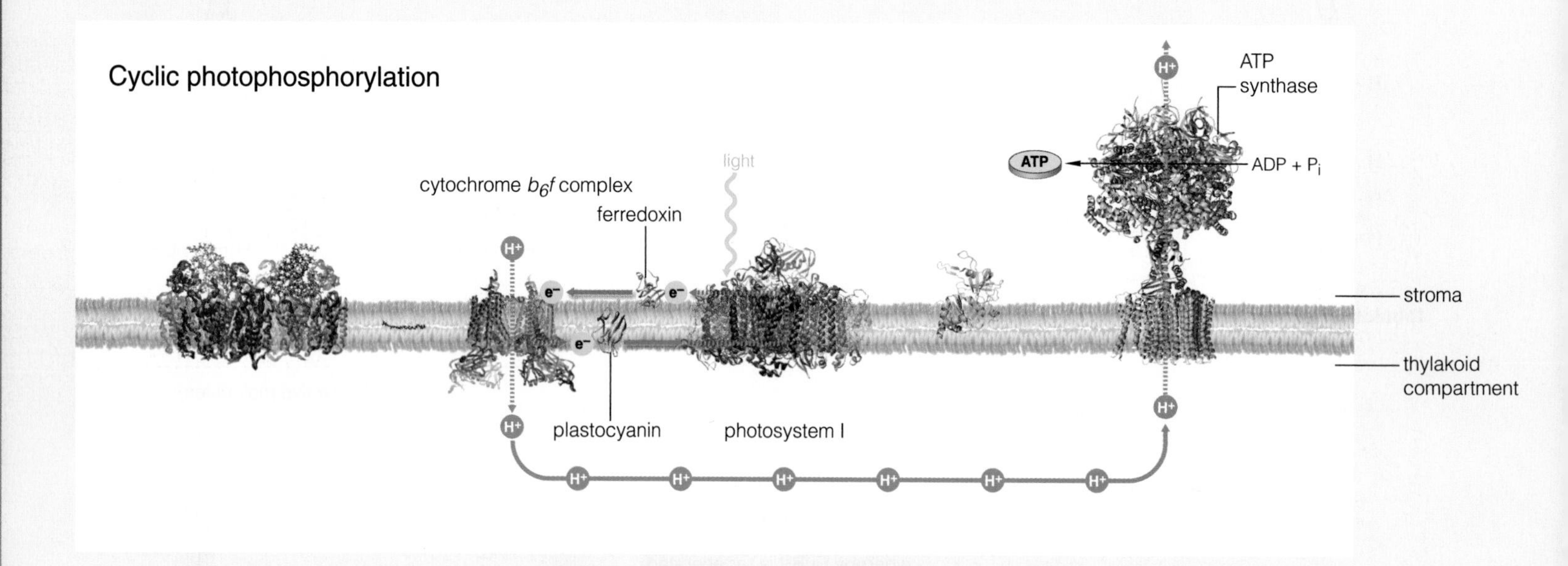

The arrangement of electron transfer chain components in highly folded thylakoid membranes maximizes the efficiency of ATP production. ATP synthases are positioned only on the outer surfaces of the thylakoid stacks, in contact with the stroma and its supply of NADP+ and ADP.

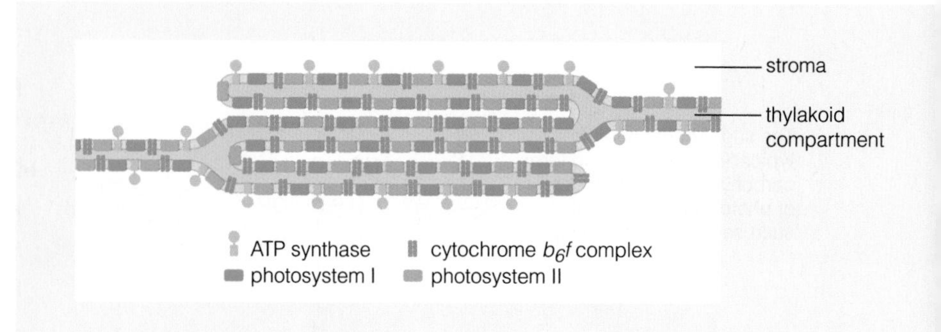

Figure D Electron transfer in the light-dependent reactions of photosynthesis. Members of the electron transfer chains are densely packed in thylakoid membranes; electrons are transferred directly from one molecule to the next. For clarity, we show the components of the chains widely spaced.

Appendix VII. A Plain English Map of the Human Chromosomes

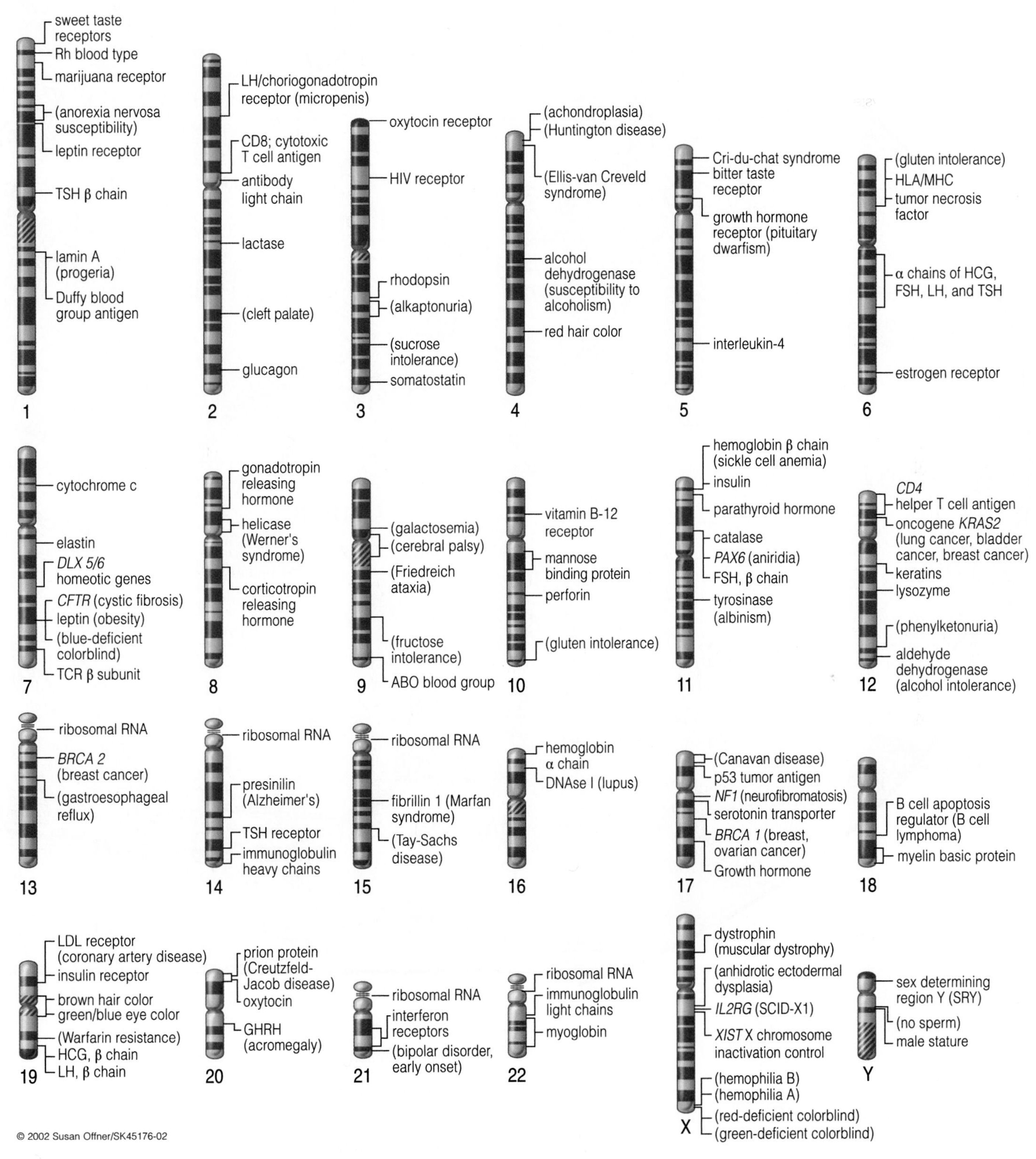

Haploid set of human chromosomes. The banding patterns characteristic of each type of chromosome appear after staining with a reagent called Giemsa. The locations of some of the 20,065 known genes (as of November, 2005) are indicated. Also shown are locations that, when mutated, cause some of the genetic diseases discussed in the text.

For more information go to http://www.thomsonedu.com.

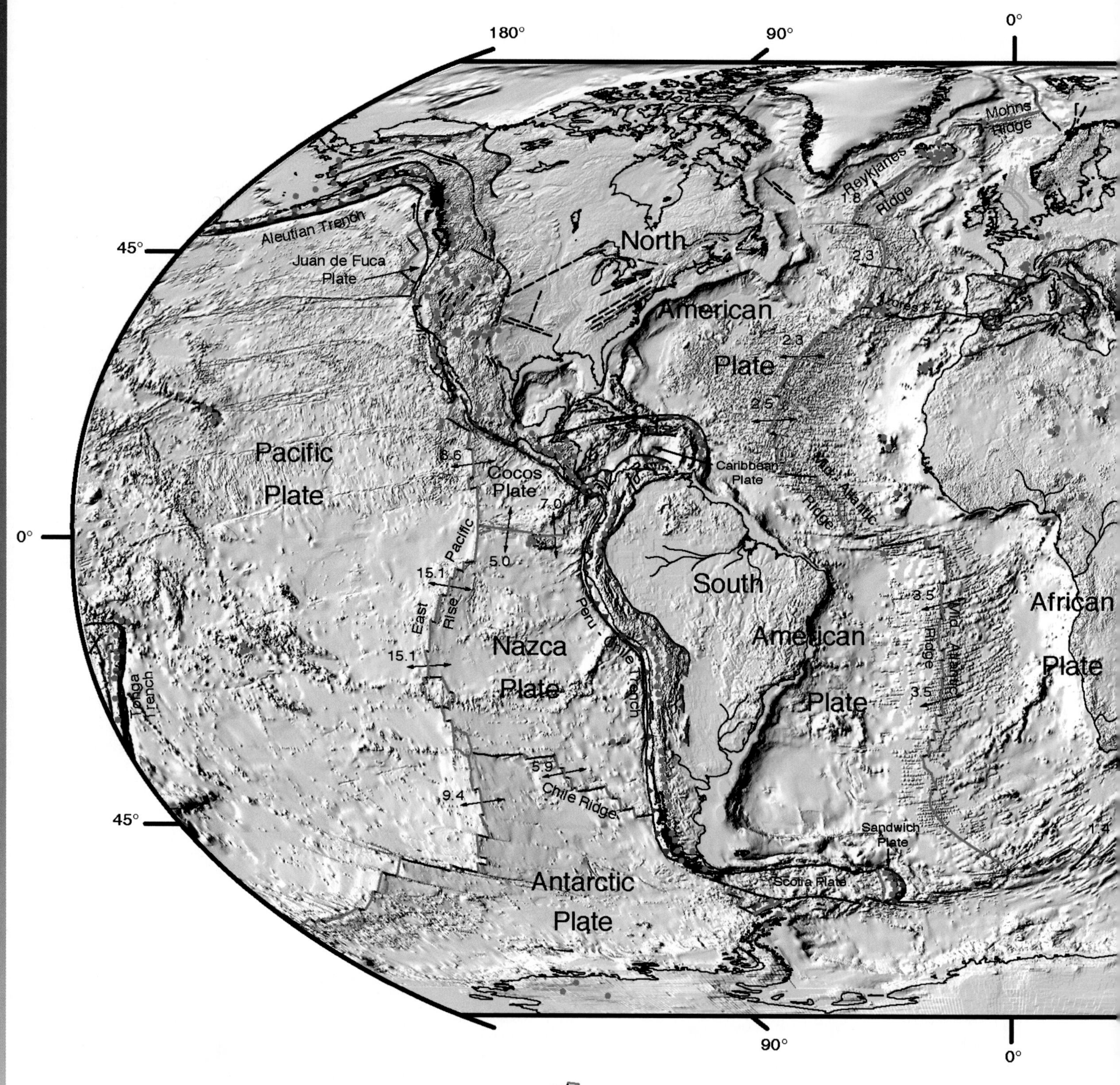

Actively-spreading ridges and transform faults

1.4 Total spreading rate, cm/year

Major active fault or fault zone; dashed where nature, location, or activity uncertain

Normal fault or rift; hachures on downthrown side

Reverse fault (overthrust, subduction zones); generalized; barbs on upthrown side

Volcanic centers active within the last one million years; generalized. Minor basaltic centers and seamounts omitted.

Appendix VIII. Restless Earth—Life's Changing Geologic Stage

This NASA map summarizes the tectonic and volcanic activity of Earth during the past 1 million years. The reconstructions at far right indicate positions of Earth's major land masses through time.

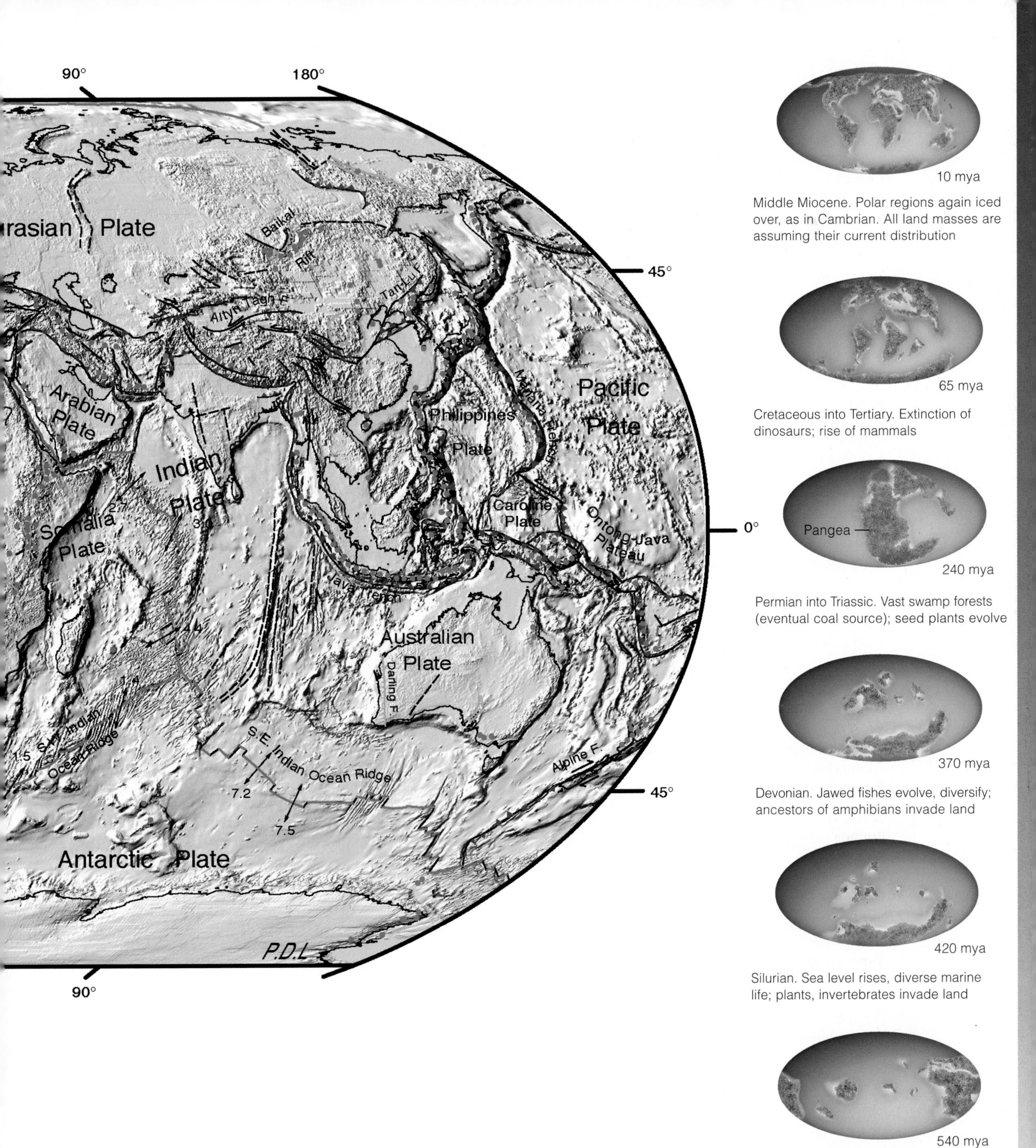

Middle Miocene. Polar regions again iced over, as in Cambrian. All land masses are assuming their current distribution

Cretaceous into Tertiary. Extinction of dinosaurs; rise of mammals

Permian into Triassic. Vast swamp forests (eventual coal source); seed plants evolve

Devonian. Jawed fishes evolve, diversify; ancestors of amphibians invade land

Silurian. Sea level rises, diverse marine life; plants, invertebrates invade land

Cambrian. Fragments of Rodinia, the first supercontinent. Major adaptive radiations in equatorial seas; icy polar regions

Appendix IX. Units of Measure

Length
1 kilometer (km) = 0.62 miles (mi)
1 meter (m) = 39.37 inches (in)
1 centimeter (cm) = 0.39 inches

To convert	*multiply by*	*to obtain*
inches	2.25	centimeters
feet	30.48	centimeters
centimeters	0.39	inches
millimeters	0.039	inches

Area
1 square kilometer = 0.386 square miles
1 square meter = 1.196 square yards
1 square centimeter = 0.155 square inches

Volume
1 cubic meter = 35.31 cubic feet
1 liter = 1.06 quarts
1 milliliter = 0.034 fluid ounces = 1/5 teaspoon

To convert	*multiply by*	*to obtain*
quarts	0.95	liters
fluid ounces	28.41	milliliters
liters	1.06	quarts
milliliters	0.03	fluid ounces

Weight
1 metric ton (mt) = 2,205 pounds (lb) = 1.1 tons (t)
1 kilogram (kg) = 2.205 pounds (lb)
1 gram (g) = 0.035 ounces (oz)

To convert	*multiply by*	*to obtain*
pounds	0.454	kilograms
pounds	454	grams
ounces	28.35	grams
kilograms	2.205	pounds
grams	0.035	ounces

Temperature
Celcius (°C) to Fahrenheit (°F) :

$$°F = 1.8\,(°C) + 32$$

Fahrenheit (°F) to Celsius:

$$°C = \frac{(°F - 32)}{1.8}$$

	°C	°F
Water boils	100	212
Human body temperature	37	98.6
Water freezes	0	32

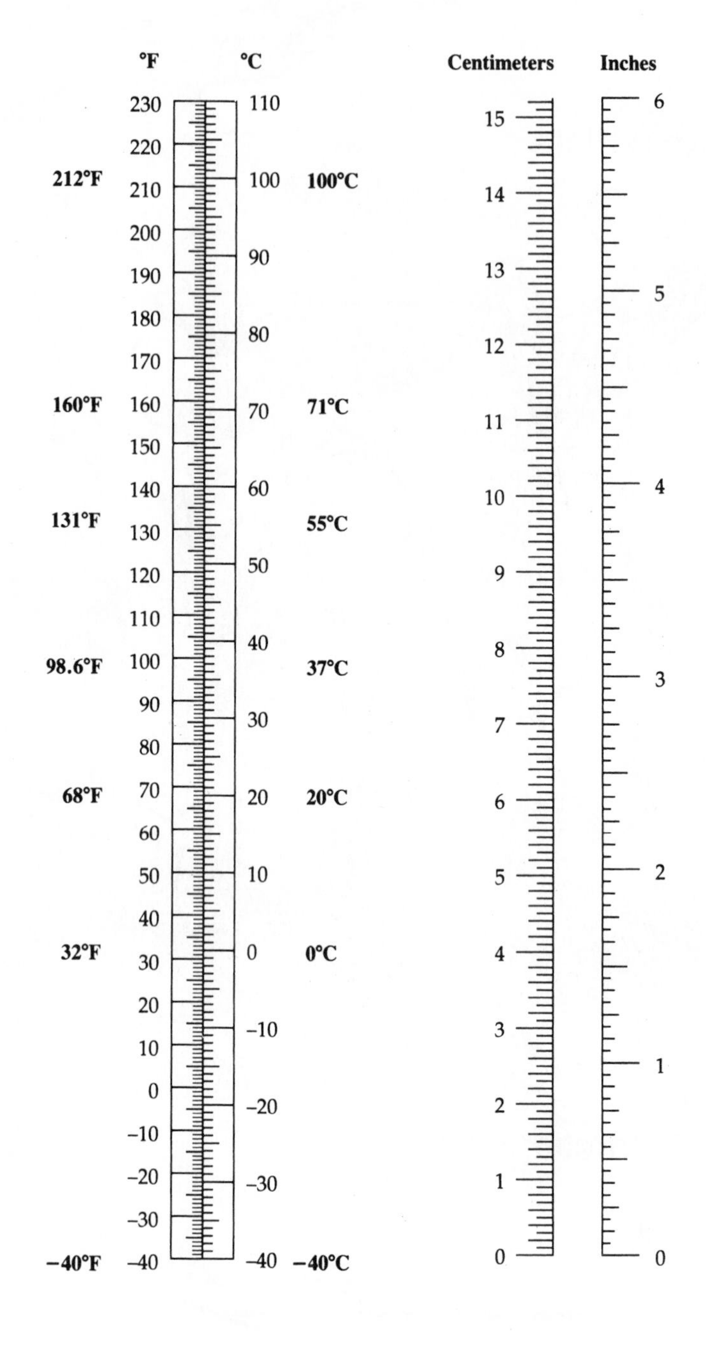

Appendix X. A Comparative View of Mitosis in Plant and Animal Cells

For step-by-step description of the stages of mitosis, refer to Figure 8.7.

Mitosis in a generalized animal cell. For simplicity, only two chromosomes are shown.

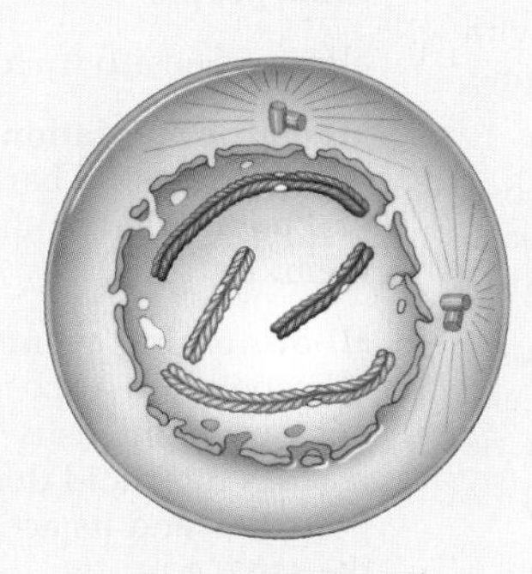

Prophase

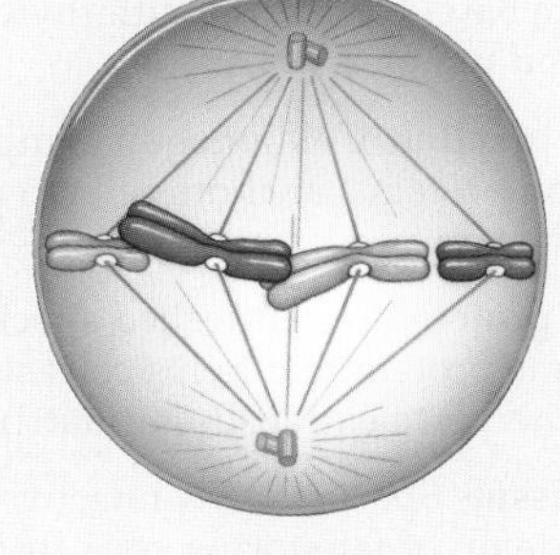

Metaphase

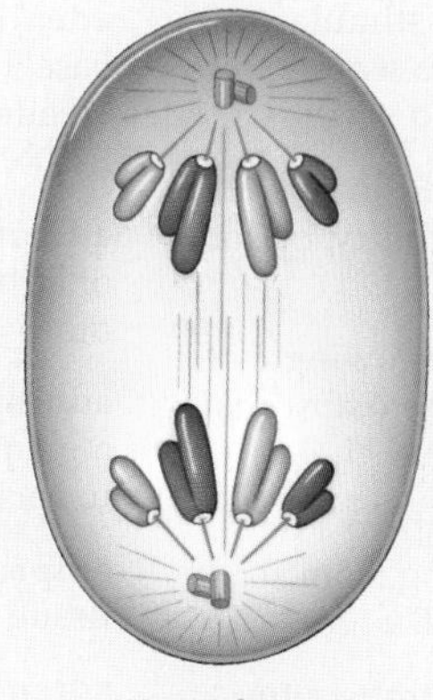

Anaphase

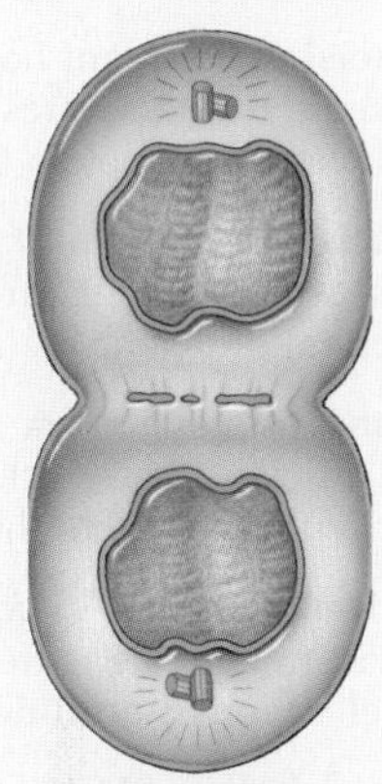

Telophase

Mitosis in a whitefish cell.

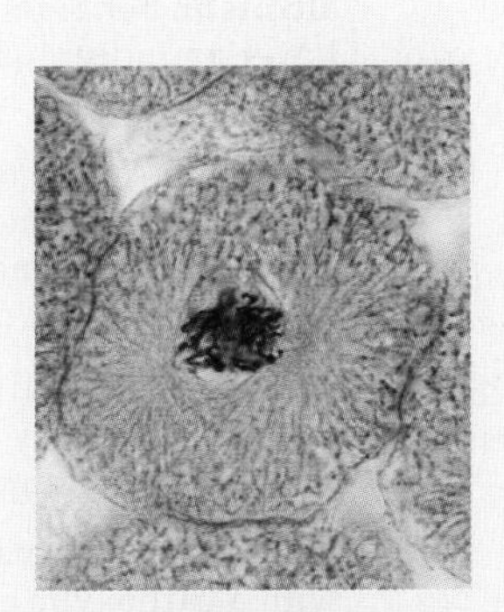

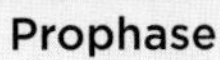

Prophase

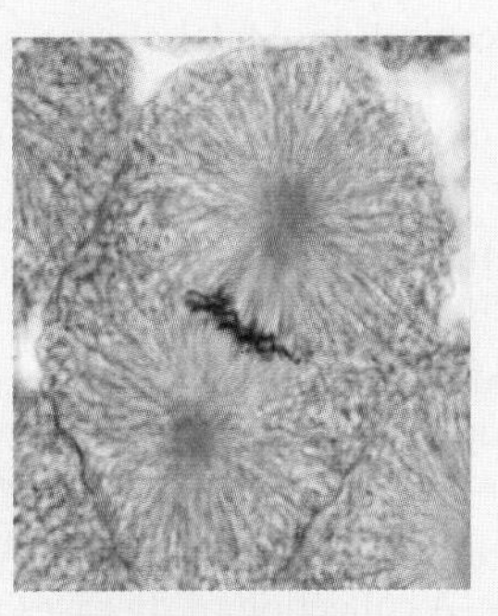

Metaphase

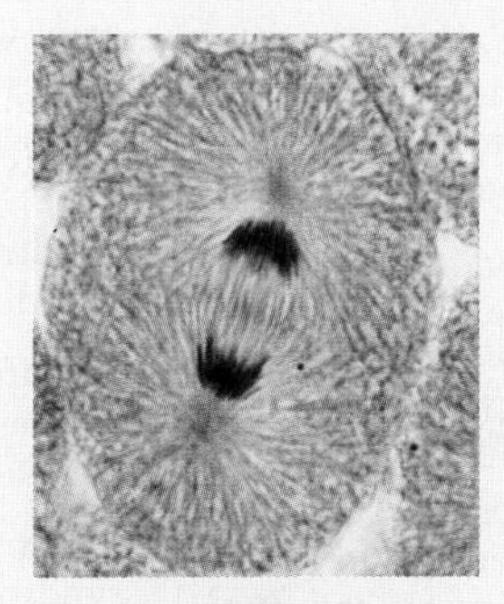

Anaphase

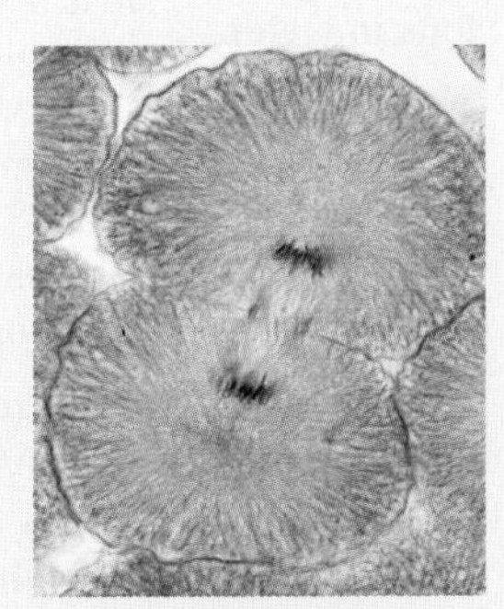

Telophase

Mitosis in a lily cell.

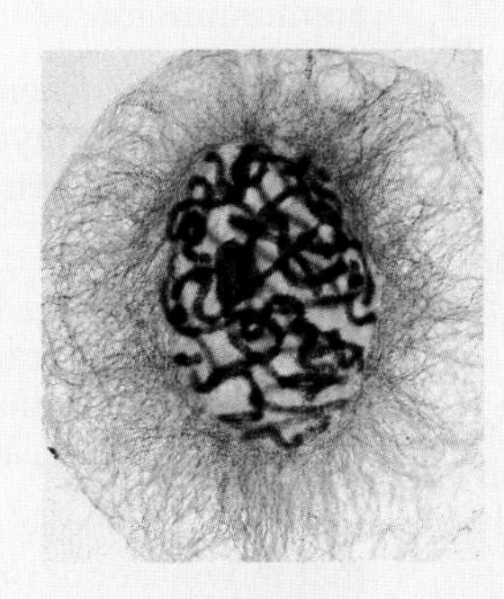

Prophase

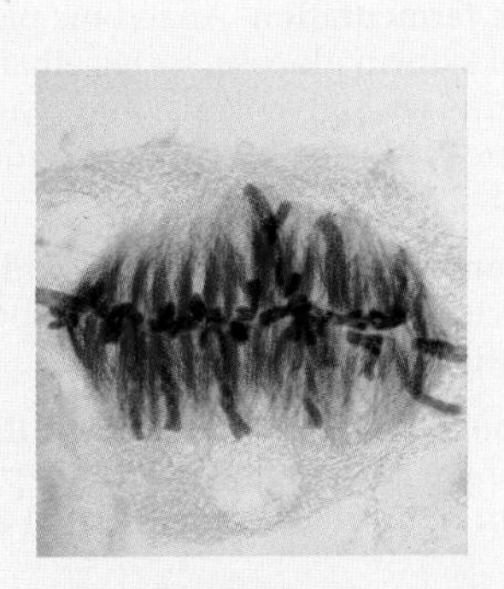

Metaphase

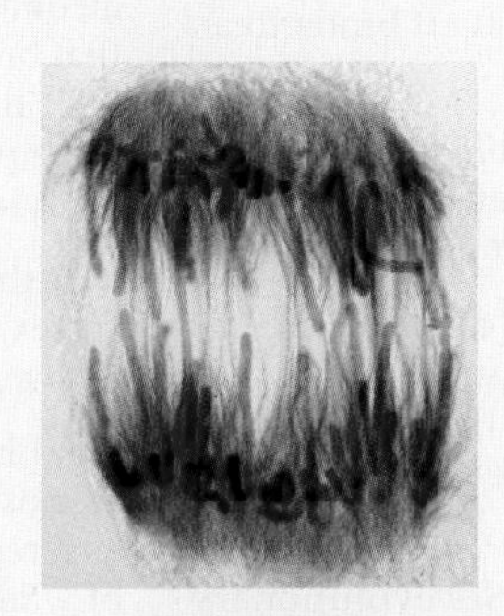

Anaphase

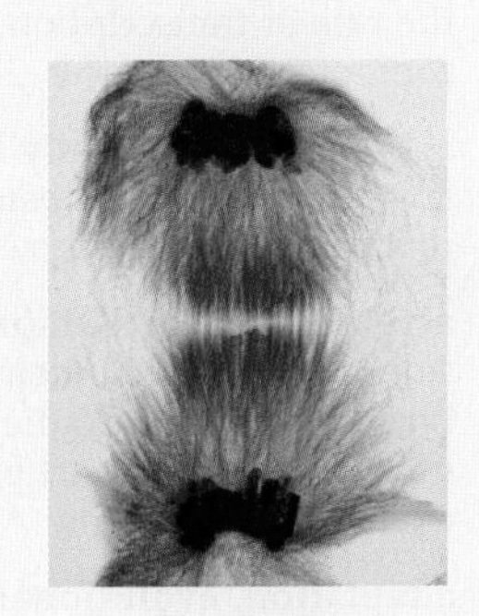

Telophase

Glossary of Biological Terms

ABC model Model for the genetic basis of flower formation; products of three master genes (*A*, *B*, *C*) control the development of sepals, petals, and stamens and carpels from meristematic tissue. *213*

ABO blood typing Method of identifying certain self-recognition proteins (A or B) on an individual's red blood cells; the absence of either type is designated O. *564*

abscisic acid Plant hormone; stimulates stomata to close in response to water stress; induces dormancy in buds and seeds. *467*

abscission Plant parts are shed in response to seasonal change, drought, injury, or some nutrient deficiency. *472*

absorption spectrum Graph showing the efficiency at which a pigment absorbs particular wavelengths of light. *96*

acclimatization A body adjusts to a new environment; e.g., after moving from sea level to a high-altitude habitat. *612*

acetylcholine (ACh) A neurotransmitter released by central and peripheral neurons of both the central and peripheral nervous systems; effects include skeletal muscle contraction. *498*

acid Any substance that releases hydrogen ions in water. *30*

acid–base balance Outcome of control over solute concentrations; extracellular fluid is neither too acidic nor too basic. *644*

acid rain Rain or snow made acidic by airborne oxides of sulfur or nitrogen. *759*

actin Globular protein; roles in cell shape, cell motility, and muscle contraction. *552*

action potential A brief, self-propagating reversal in the voltage difference across the membrane of a neuron or muscle cell. *494*

activation energy Minimum amount of energy required to start a reaction; enzymes lower it in metabolic reactions. *76*

activator A regulatory protein that increases the rate of transcription when it binds to a promoter or enhancer. *210*

active site Chemically stable crevice in an enzyme where substrates bind and a reaction can be catalyzed repeatedly. *76*

active transport Pumping of a specific solute across a cell membrane against its concentration gradient, through the interior of a transport protein. Requires energy input, as from ATP. *85, 414*

adaptation, evolutionary *See* evolutionary adaptation.

adaptation, sensory In a sensory neuron, a decline or cessation of action potentials in response to a continued stimulus. *514*

adaptive immunity Set of vertebrate immune responses characterized by self/nonself recognition, antigen specificity, antigen receptor diversity, and immune memory. Includes antibody-mediated and cell-mediated responses. *582*

adaptive radiation A burst of genetic divergences from a lineage gives rise to many new species. *284*

adenine (A) A type of nitrogen-containing base in nucleotides; also, a nucleotide with an adenine base. Base-pairs with thymine in DNA and uracil in RNA. *190*

adhering junction Cell junction composed of adhesion proteins; anchors cells to each other or to extracellular matrix. *479*

adhesion protein In multicelled species, a membrane protein that helps cells stick to each other or to extracellular matrix. *56*

adipose tissue Specialized connective tissue made up of fat-storing cells. *481*

adrenal cortex Outer zone of an adrenal gland; secretes steroid hormones, including aldosterol and cortisol. *535*

adrenal medulla Innermost zone of an adrenal gland; secretes epinephrine and norepinephrine. *535*

aerobic respiration Metabolic pathway that breaks down carbohydrates to produce ATP by using oxygen. Typical yield: 36 ATP per molecule of glucose. *108*

age structure Of a population, the number of individuals in each age category. *688*

agglutination Clumping of foreign cells, such as red blood cells, after antibodies bind to antigens on their surface. *564*

AIDS Acquired immune deficiency syndrome. A collection of diseases that develops after a virus (HIV) weakens the immune system. *594*

alcohol Organic compound having one or more hydroxyl groups; e.g., ethanol. *36*

alcoholic fermentation Anaerobic pathway that breaks down glucose, forms ethanol and ATP. Begins with glycolysis; end reactions regenerate NAD^+ so glycolysis continues. Net yield: 2 ATP per glucose. *116*

aldosterone Hormone secreted by adrenal cortex, acts in kidneys to promote sodium reabsorption. *643*

algal bloom Large increase in population size of single-celled photosynthetic protist as a result of nutrient enrichment of body of water. *323*

allantois An extraembryonic membrane of amniotes. In reptiles, birds, some mammals, it exchanges gases and stores wastes; in humans, it helps form a placenta. *675*

allele One of two or more molecular forms of a gene at a given locus; alleles arise by mutation and encode slightly different versions of the same trait. *140, 155*

allele frequency At a specific locus, the abundance of one allele relative to others among individuals of a population. *267*

allergen A normally harmless substance that provokes an immune response in some people. *592*

allergy Sensitivity to an allergen. *592*

allopatric speciation Speciation route in which a physical barrier separates members of a population, ending gene flow between them. *280*

all-or-nothing event An event that occurs in response to a specific trigger and does not vary in intensity; e.g., only if a neuron reaches threshold does it have an action potential, and its action potentials are all the same size. *496*

alternative splicing mRNA processing event in which some exons are removed or joined in various combinations. By this process, one gene can specify two or more slightly different proteins. *199*

altruistic behavior Social behavior that can lower an individual's reproductive success but improve that of others. *794*

alveolate A type of single-celled eukaryote with many tiny, membrane-bound sacs just beneath the plasma membrane; e.g., ciliate, apicomplexan, or dinoflagellate. *322*

alveolus, plural **alveoli** In a vertebrate lung, one of many tiny, thin-walled sacs where air exchanges gases with blood. *605*

amino acid A small organic compound with a carboxylic acid group, an amino group, and a characteristic side group (R); monomer of polypeptide chains. *42*

ammonification Process by which bacteria and fungi break down nitrogen-containing organic material and release ammonia and ammonium ions. *749*

amnion An extraembryonic membrane of amniotes; outer layer of a fluid-filled sac inside which the embryo develops. *674*

amniote Member of a vertebrate lineage that produces eggs having four extra-embryonic membranes (chorion, allantois, yolk sac, and amnion). Modern groups are reptiles, birds, and mammals. *394*

amoeba A solitary amoebozoan protist that moves about on pseudopods. All are predatory or parasitic. *329*

amphibian A thin-skinned vertebrate that spends time on land but lays eggs in water; e.g., a frog, a toad, a salamander. *392*

analogous structures Similar structures that evolved separately in different lineages after the lineages diverged; e.g., the flight surfaces of bat wings and fly wings. 253

anaphase Stage of mitosis in which sister chromatids separate and move to opposite spindle poles. 131

anemia Disorder resulting from having too few functional red blood cells. 564

aneuploidy A chromosome abnormality in which there are too many or too few copies of a particular chromosome; e.g., having three copies of chromosome 21, which causes Down syndrome. 178

angiosperm A flowering plant; it forms seeds inside a floral ovary, which develops into a fruit. 344

animal A multicelled heterotroph with unwalled cells. It develops through a series of embryonic stages and is motile during part or all of the life cycle. 9, 362

animal hormone *See* hormone.

annelid A bilateral invertebrate having a highly segmented body; major groups are polychaetes, oligochaetes, and leeches. 370

antennae In some arthropods, paired sensory appendages on the head that act in touch, smell, taste, and in detection of vibrations and temperature. 375

antibiotic Toxin that kills bacteria. 271

antibody Y- or T-shaped protein that can bind to a specific antigen; made only by B cells. 587

antibody-mediated immune response One of two arms of adaptive immunity in which antibodies are produced in response to a specific antigen; mediated by B cells. 587

anticodon Set of three nucleotide bases in a tRNA; base-pairs with mRNA codon. 201

antidiuretic hormone (ADH) Hormone released by posterior pituitary; induces water reabsorption by kidneys. 643

antigen A molecule or particle that the immune system recognizes as nonself; triggers an immune response. 582

antioxidant Substance that neutralizes free radicals and other strong oxidizers. 79

anus Waste-expelling, terminal opening of a complete digestive system. 621

aorta The main artery of human systemic circulation; receives blood from the left ventricle. 566

apical dominance Growth-inhibiting effect on lateral (axillary) buds, caused by auxin diffusing down a shoot from its tip. 466

apicomplexan A parasitic alveolate protist that penetrates the host cell using a unique microtubular structure; e.g., malaria-causing *Plasmodium* species. 323

apoptosis Programmed cell death. A cell commits suicide in response to molecular signals; part of a program of development and maintenance of an animal body. 420, 659

appendix Small, narrow outpouching from the cecum, vulnerable to infection. 626

aquifer Permeable rock layers that hold water. 742

archaean A member of the prokaryotic domain Archaea. Members have some unique features but also share some traits with bacteria and other traits with eukaryotic species. 8

archipelago A chain or cluster of islands in the ocean. 280

area effect Biogeographical pattern; larger islands support more species than smaller ones at equivalent distances from sources of colonizing species. 725

arteriole A blood vessel that carries blood from an artery to a capillary bed. 570

artery A thick-walled, muscular vessel that carries blood away from the heart. 570

arthropod Type of invertebrate having a hardened exoskeleton and specialized segments with jointed appendages; e.g., millipedes, spiders, lobsters, insects. 375

asexual reproduction Any reproductive mode by which offspring arise from one parent and inherit that parent's genes only; e.g., prokaryotic fission, transverse fission, budding, vegetative propagation. 140, 652

atom Particle that is a fundamental building block of matter; consists of varying numbers of electrons, protons, and neutrons. 4, 22

atomic number The number of protons in the nucleus of atoms of a given element. 22

ATP Adenosine triphosphate. Nucleotide that consists of an adenine base, the five-carbon sugar ribose, and three phosphate groups. The main energy carrier between reaction sites in cells. 46, 76

ATP/ADP cycle How a cell regenerates its ATP supply. ADP forms when ATP loses a phosphate group, then ATP forms as ADP gains a phosphate group. 76

australopith Member of one of many now-extinct species classified as hominids, but not as members of the genus *Homo*. 404

autoimmune response Immune response that targets one's own tissues. 593

autonomic nerve A peripheral nerve that carries signals regulating smooth muscle, cardiac muscle, and glands of viscera. 504

autosome Any chromosome other than a sex chromosome. 170

autotroph Organism that makes its own food using carbon from CO_2 and energy from light or inorganic substances. 80, 306

auxin A plant hormone; stimulates growth rate of cells in shoots and roots. Role in gravitropism and phototropism. 466

axon A neuron's signal-conducting zone; action potentials typically self-propagate away from the cell body along it. 494

B cell receptor Membrane-bound IgM or IgD antibodies on a new B cell. 588

B lymphocyte B cell. Type of white blood cell that makes antibodies. 582

bacteria Members of the prokaryotic domain Bacteria; the most diverse and most ancient prokaryotic lineage. 8

bacteriophage Type of virus that infects bacteria. 188

balanced polymorphism The maintenance of two or more alleles for a trait in some populations, as a result of natural selection against homozygotes. 275

bark In woody plants, secondary phloem and periderm. 436

Barr body A condensed X chromosome in cells of female mammals. 212

basal body An organelle that started out as a centriole, the source of a 9+2 array of microtubules in a cilium or flagellum. It remains below the finished array. 69

base A substance that accepts hydrogen ions as it dissolves in water. 30

base-pair substitution Type of mutation; a single base-pair change. 204

basophil White blood cell circulating in blood; role in inflammation. 582

bell curve Curve that typically results when range of variation for a continuous trait is plotted against frequency in the population. 164

big bang Model for the origin of universe, by a nearly instantaneous distribution of all matter and energy through space. 292

bilateral symmetry Body plan in which many appendages and organs are paired, one to each side of the main body axis. 362

bile Mix of salts, cholesterol, and pigments made in the liver, stored by the gallbladder, and used in fat digestion. 623

binary fission Asexual reproductive mode of some protists. 322

biofilm Community of different types of microorganisms living within a shared mass of slime. 59

biogeochemical cycle Slow movement of an element from environmental reservoirs, through food webs, then back. 741

biogeographic realm One of many vast expanses of land defined by the presence of certain types of plants and animals. 762

biogeography Scientific study of patterns in the geographic distribution of species and communities. *240, 724*

biological clock Internal time-measuring mechanism by which individuals adjust their activities seasonally, daily, or both in response to environmental cues. *470, 538*

biological magnification A pesticide or other chemical that becomes increasingly concentrated in the tissues of organisms at higher trophic levels. *739*

bioluminescence Light emitted as a result of reactions in a living organism. *89*

biomass pyramid Chart in which tiers of a pyramid depict the biomass (dry weight) in each of an ecosystem's trophic levels. *740*

biome A subdivision of a biogeographic realm; usually described in terms of the dominant plants; e.g., tropical broadleaf forest, grassland, tundra. *762*

biosphere All regions of Earth's waters, crust, and air where organisms live. *5*

biotic potential The maximum rate of increase per individual for a population growing under ideal conditions. *691*

bipedalism Habitually walking upright, as by ostriches and humans. *402*

bipolar spindle In eukaryotic cells, a dynamic array of microtubules that moves chromosomes with respect to its two poles during mitosis or meiosis. *129*

bird A warm-blooded, feathered amniote descended from certain dinosaurs. *398*

blastocyst A type of blastula with a surface layer of blastomeres, a cavity filled with their secretions, and an inner cell mass that develops into the embryo. *674*

blastula A ball of cells and a cavity filled with their own secretions; outcome of the cleavage stage of animal development. *654*

blood A fluid connective tissue that is the transport medium of circulatory systems. In vertebrates, it consists of plasma, blood cells, and platelets. *481, 560*

blood–brain barrier Specialized blood capillaries that protect the brain and spinal cord by exerting some control over which solutes enter cerebrospinal fluid. *506*

blood capillary *See* capillary, blood.

blood pressure Fluid pressure generated by heartbeats; causes blood circulation. *570*

bone tissue In vertebrates, a specialized connective tissue with a matrix hardened by calcium and other mineral ions. *481*

book lung The respiratory organ of some spiders; air and blood flow through spaces separated by thin sheets of tissue. *601*

bottleneck Severe reduction in population size; can reduce genetic diversity. *276*

Bowman's capsule Cup-shaped first part of a nephron; the water and solutes filtered out of glomerular capillaries enter it. *641*

brain stem The most evolutionarily ancient nerve tissue in a vertebrate brain. *506*

bronchiole One of many tiny airways that deliver air to the alveoli in a lung. *605*

bronchus, plural **bronchi** An airway that delivers air from the trachea to a lung. *605*

brown alga A stramenopile; a multicelled marine autotroph with an abundance of the pigment fucoxanthin; e.g., kelps. *326*

brush border cell Cell type specialized for absorption; found on the sides and tip of a villus in the small intestine. *624*

bryophyte Nonvascular land plant. The haploid stage dominates its life cycle, and its sperm require standing water to reach eggs. A moss, liverwort, or hornwort. *336*

buffer system Set of chemicals that can counter pH shifts in a solution by donating or accepting H^+ or OH^-. *31, 644*

C3 plant Type of plant that uses only the Calvin–Benson cycle to fix carbon. On dry days, photorespiration predominates. *102*

C4 plant Type of plant that minimizes photorespiration by fixing carbon twice, using a C4 pathway in addition to the Calvin–Benson cycle. *102*

calcium pump Active transport protein; pumps calcium ions across a cell membrane against their concentration gradient. *85*

Calvin–Benson cycle Light-independent reactions of photosynthesis; cyclic pathway that forms glucose from CO_2. *101*

CAM plant Type of C4 plant that conserves water by opening stomata only at night, when it fixes carbon by a C4 pathway. *102*

camouflage Body coloration, patterning, form, or behavior that helps predators or prey blend with the surroundings and possibly escape detection. *714*

cancer A malignant neoplasm; a mass of abnormally dividing cells that can leave their home tissue and invade and form new masses in other body regions. *135*

capillary, blood Smallest diameter blood vessel; blood exchanges substances with interstitial fluid across its wall, which is only one cell thick. *570*

capillary reabsorption The process by which water moves by osmosis from the interstitial fluid into protein-rich plasma at the venous end of a capillary bed. *573*

capture–recapture method Individuals of a mobile species are captured (or selected) at random, marked, then released so they can mix with unmarked individuals. One or more samples are taken. The ratio of marked to unmarked individuals is used to estimate total population size. *689*

carbohydrate Any molecule of carbon, hydrogen, and oxygen, typically in a 1:2:1 ratio. Main kinds are monosaccharides, oligosaccharides, and polysaccharides. *38*

carbon cycle Atmospheric cycle. Carbon moves from its environmental reservoirs (sediments, rocks, the ocean), through the atmosphere (mostly as CO_2), food webs, and back to the reservoirs. *744*

carbon fixation Process by which carbon from an inorganic source such as CO_2 is incorporated into an organic compound. Occurs in the light-independent reactions of photosynthesis. *101*

carbon–oxygen cycle The global cycling of carbon in the form of CO_2. *745*

carbonic anhydrase Enzyme in red blood cells that speeds the interconversion of CO_2 and water into bicarbonate. *607*

cardiac conduction system Specialized cardiac muscle cells that initiate and send signals that make other cardiac muscle cells contract. SA node, AV node, and junctional fibers that link them. *569*

cardiac cycle A recurring sequence of muscle contraction and relaxation that corresponds to one heartbeat. *568*

cardiac muscle tissue A contractile tissue present only in the heart wall. *482, 568*

cardiac pacemaker Sinoatrial (SA) node; a cluster of self-excitatory cardiac muscle cells that set the normal heart rate. *569*

carotenoid Type of accessory pigment in plants and some other photosynthetic organisms; e.g., carotene. *95*

carpel Female reproductive structure of a flower; a sticky or hairlike stigma, often stalked, above a chamber (ovary) in which one or more ovules mature into seeds. *456*

carrying capacity Maximum number of individuals of a species that a particular environment can sustain. *692*

cartilage Specialized connective tissue with fine collagen fibers in a rubbery matrix that resists compression. *480*

cartilaginous fish Jawed fish that has a cartilage skeleton; e.g., sharks. *390*

Casparian strip A waxy, impermeable band that seals abutting cell walls of endodermis (and exodermis) in roots; forces water and solutes to pass through cells, which helps control the type and amount of solutes that enter the vascular cylinder. *445*

catastrophism Now-abandoned hypothesis that geologic forces in the past were unlike those of the present day. *242*

cDNA DNA synthesized from RNA by the enzyme reverse transcriptase. *223*

cell Smallest unit with the properties of life—the capacity for metabolism, growth, homeostasis, and reproduction. *4, 52*

cell cortex Mesh of crosslinked cytoskeletal elements under the plasma membrane. *68*

cell count The number of cells of a given type present in one microliter of blood. *563*

cell cycle In eukaryotic cells, a series of events from the time a cell forms until it reproduces. A cycle consists of interphase, mitosis, and cytoplasmic division. *128*

cell differentiation The process by which unspecialized cells mature and become specialized in structure and function. *See also* selective gene expression. *210, 658*

cell junction Structure that connects a cell to another cell or to extracellular matrix; e.g., gap junction, adhering junction, tight junction. *67*

cell-mediated immune response Immune response involving cytotoxic T cells and NK cells that destroy infected or cancerous body cells. *587*

cell plate After nuclear division in a plant cell division, a disk-shaped structure that forms the cross-wall between the two new cells; will develop into a primary wall. *133*

cell theory All organisms consist of one or more cells; the cell is the smallest unit of organization with the properties of life; and each new cell arises from another cell. *54*

cell wall In many cells (not animal cells), a semirigid permeable structure around the plasma membrane; helps a cell retain its shape and resist rupturing. *58*

central nervous system Of vertebrates, the brain and spinal cord. *493*

central vacuole In many plant cells, a fluid-filled organelle that stores amino acids, sugars, and some wastes. *63*

centriole A barrel-shaped structure that organizes newly forming microtubules in a cilium or flagellum and in a spindle. *69*

centromere In a eukaryotic chromosome, a constricted region where microtubules of the spindle bind. *127*

centrosome Region in the cytoplasm of eukaryotic cells from which microtubules grow; usually includes centrioles. *130*

cephalization During the evolution of most kinds of animals, increasing concentration of sensory structures and nerve cells at the anterior end of the body. *362, 492*

cerebellum Hindbrain region with reflex centers that maintain posture and smooth out limb movements. *506*

cerebral cortex Surface layer of cerebrum; it receives, integrates, and stores sensory information and coordinates responses. *508*

cerebrospinal fluid Clear extracellular fluid that bathes and protects the brain and spinal cord; contained in a system of canals and chambers. *506*

cerebrum A forebrain region concerned with olfactory input and motor responses. In mammals, it evolved into a complex integrating center. *506*

character displacement Modifications of a trait of one species in a way that lowers intensity of competition with another species; occurs over generations. *711*

charophytes The lineage of green algae most closely related to land plants. *326*

chemical bond An attractive force that arises between two atoms when their electrons interact. *25*

chemical equilibrium State in which the concentrations of reactants and products do not change because the rate of reaction is about the same in both directions. *80*

chemical synapse Region where the axon endings of a neuron lie in close proximity to a cell that the neuron signals; a tiny gap separates the cells. *498*

chemoreceptor Sensory receptor; detects dissolved ions or molecules in fluid. *514*

chlamydias A group of bacteria that are intracellular parasites of animals; cannot make ATP, must get it from the host. *308*

chlorophyll *a* Main photosynthetic pigment in plants, algae, and cyanobacteria. *95*

chlorophyte Member of the most diverse lineage of green algae. *326*

chloroplast Organelle of photosynthesis in plants and some protists. Two outer membranes enclose a semifluid stroma. A third membrane forms a compartment that functions in ATP and NADPH formation; sugars form in the stroma. *64, 97*

choanoflagellates Protists that resemble sponge cells in having a microvilli collar around a single flagellum at their anterior end; closest living protistan relatives of the animals. *364*

chordate Animal with an embryo that has a notochord, a dorsal hollow nerve cord, gill slits in the pharynx wall, and a tail that extends past the anus. Some, none, or all of these traits persist in the adult. *386*

chorion An extraembryonic membrane of amniotes; in mammals it becomes part of the placenta. Villi form at its surface and facilitate the exchange of substances between the embryo and mother. *675*

chromatin All of the DNA molecules and associated proteins in a nucleus. *61*

chromosome A complete molecule of DNA and its attached proteins; carries part or all of an organism's genes. Linear in eukaryotic cells; circular in prokaryotes. *61*

chromosome number The sum of all chromosomes in a cell of a given type; e.g., it is 46 in body cells of humans. *129*

chyme Semidigested food in the gut. *622*

chytrid A type of fungus, the only fungal group with a flagellated stage. *352*

ciliate A heterotrophic alveolate protist with cilia at its surface; also known as a ciliated protozoan; e.g., *Paramecium*. *322*

cilium, plural **cilia** Short movable structure that projects from the plasma membrane of certain eukaryotic cells. *68*

circadian rhythm Any biological activity repeated about every 24 hours. *419, 470*

circulatory system Organ system that rapidly transports substances to and from cells; typically consists of a heart, blood vessels, and blood. Helps stabilize body temperature and pH in some animals. *560*

clade A group of species that share a set of derived traits. *258*

cladistics One method for determining evolutionary relationships. *258*

cleavage Early stage of development in animals. Mitotic cell divisions divide a fertilized egg into many smaller cells (blastomeres); the original volume of egg cytoplasm does not increase. *654*

climate Prevailing weather conditions of a region; e.g., temperature, cloud cover, wind speed, rainfall, and humidity. *756*

cloaca In fish, amphibians, reptiles, and birds, opening through which digestive and urinary wastes leave the body; may also function in reproduction. *396*

clone A genetically identical copy of DNA, a cell, or an organism. *140, 223*

cloning vector A DNA molecule that can accept foreign DNA, be transferred to a host cell, and get replicated in it. *222*

closed circulatory system Organ system in which blood flows continually inside blood vessels and does not come into direct contact with tissue fluids. *370*

club fungus Fungus that produces sexual spores in a club-shaped cell; most familiar mushrooms. *355*

cnidarian A type of radially symmetrical invertebrate that makes nematocysts; has two types of epithelial tissues and a saclike gastrovascular cavity; e.g., sea anemone, jellyfish, coral. *366*

coal A nonrenewable energy source that formed more than 280 million years ago from submerged, undecayed, and slowly compacted plant remains. *340*

coccolithophore A single-celled marine autotroph with calcium carbonate plates; one of the stramenopiles. *325*

cochlea A fluid-filled, coiled structure in the inner ear; transduces sound waves into action potentials. *518*

codominance Nonidentical alleles that are both fully expressed in heterozygotes; neither is dominant or recessive. *160*

codon In mRNA, a nucleotide base triplet that codes for an amino acid or stop signal during translation. *See* genetic code. *200*

coelom Of many animals, a tissue-lined cavity that lies between the gut and body wall. *363*

coenzyme An organic cofactor. *46, 79*

coevolution The joint evolution of two closely interacting species; each species is a selective agent that shifts the range of variation in the other. *284, 344, 708*

cofactor A metal ion or a coenzyme that associates with an enzyme and is necessary for its function; e.g., NAD^+. *79*

cohesion Tendency of molecules to stick together when a substance is under tension; a property of liquid water. *29*

cohesion–tension theory Explanation of how water moves from roots to leaves in plants; evaporation of water from leaves creates a continuous negative pressure (tension) that pulls water from roots upward in a cohesive column. *446*

cohort A group of individuals of the same age. *694*

collenchyma Simple plant tissue; alive at maturity. Lends flexible support to rapidly growing plant parts. *428*

colon Large intestine. *621*

commensalism An interspecific interaction in which one species benefits and the other is neither helped nor harmed. *708*

communication protein A membrane protein that helps form an open channel between cytoplasm of adjoining cells. *56*

communication signal A social cue that is encoded in stimuli, such as the body's surface coloration or patterning, odors, sounds, and postures. *788*

community All populations of all species in a habitat. *5, 708*

companion cell Specialized parenchyma cell that helps load sugars into conducting cells of phloem. *429*

comparative morphology Scientific study of the body plans and structures among groups of organisms. *241*

compartmentalization In some plants, a defense response in which an attacked region becomes walled off. *418*

competition, interspecific Interaction in which the individuals of different species compete for a limited resource; suppresses population size of both species. *708*

competitive exclusion When two species require the same limited resource to survive or reproduce, the better competitor will drive the less competitive one to extinction in the shared habitat. *710*

complement A set of proteins that circulate in inactive form in blood. When activated, they can destroy microbes or tag them for phagocytosis. *582*

complete digestive system A tubular digestive system; has a mouth at one end and an anus at the other. *618*

compound Type of molecule that has atoms of more than one element. *25*

concentration The amount of a substance in a given volume. *78*

concentration gradient Difference in the number per unit volume of molecules or ions between adjoining regions. *82*

condensation Chemical reaction in which two molecules become covalently bonded as a larger molecule; water often forms as a by-product. *37*

conduction Of heat: the transfer of heat between two objects in contact with one another. *645*

cone Reproductive structure of conifers; has clusters of scales. *342*

cone cell A vertebrate photoreceptor that responds to intense light and contributes to sharp vision and color perception. *522*

conifer A type of gymnosperm adapted to conserve water through droughts and cold winters. Cone-producing woody trees or shrubs with thickly cuticled needlelike or scalelike leaves. *342*

conjugation Among prokaryotes, transfer of a plasmid from one cell to another. *307*

connective tissue Most abundant type of animal tissue. Soft connective tissues differ in the amounts and arrangements of fibroblasts and extracellular matrix. Adipose tissue, cartilage, bone tissue, and blood are specialized types. *362, 480*

conservation biology Field of study that works to inventory and preserve species richness by resource management that is compatible with human needs. *727*

consumer Heterotroph that gets energy and carbon by feeding on tissues, wastes, or remains of other organisms. *6, 734*

continuous variation In a population, individuals show a range of small differences in a trait as the result of polygenic inheritance. *164*

contractile ring A thin band of actin and myosin filaments that wraps around the midsection of an animal cell undergoing cytoplasmic division. It contracts and pinches the cytoplasm in two. *132*

contractile vacuole In freshwater protists, an organelle that collects and then expels any excess water that moves into the cell by osmosis. *321*

convection Transfer of heat by moving molecules of air or water. *645*

coral reef A formation consisting mainly of calcium carbonate secreted by reef-building corals. *776*

cork Tissue component of bark with many suberized layers; waterproofs, insulates, and protects woody stem and root surfaces. *437*

cork cambium A lateral meristem, the descendants of which replace epidermis with periderm on woody plant parts. *436*

corpus luteum A glandular structure that forms from cells of a ruptured follicle after ovulation; its progesterone and estrogen secretions help thicken the endometrium in preparation for pregnancy. *666*

cortisol stress response Concentration of the hormone cortisol rises in times of injury, illness, or anxiety, when nervous signals override the normal feedback loop. *535*

cotyledon Seed leaf; part of a flowering plant embryo. *426*

countercurrent flow Any movement of two fluids in opposing directions. *602*

covalent bond Chemical bond in which two atoms share a pair of electrons. *26*

craniate A chordate that has its brain inside a cranium (brain case); any fish, amphibian, reptile, bird, or mammal. *387*

critical thinking Mental process of judging information. *11*

crossing over Process in which homologous chromosomes exchange corresponding segments during prophase I of meiosis. Puts nonparental combinations of alleles in gametes. *144*

culture Sum of behavior patterns of a social group, passed between generations by learning and symbolic behavior. *403*

cumulative assaults hypothesis The idea that aging is the outcome of damage at the molecular level. *682*

cuticle Of plants, a cover of waxes and cutin on the outer wall of epidermal cells. Of annelids, a thin, flexible secreted layer. Of arthropods, a lightweight exoskeleton hardened with chitin. *66, 334*

cyanobacterium A type of prokaryotic photoautotroph; carries out photosynthesis by the noncyclic pathway and so releases oxygen. *308*

cycad A gymnosperm of subtropical or tropical habitats; many resemble palms. *342*

cytokines Signaling molecules with major roles in vertebrate immunity. *582*

cytokinin A plant hormone; promotes cell division. *466*

cytoplasm The semifluid matrix between a cell's plasma membrane and its nucleus or nucleoid. *52*

cytoplasmic localization Accumulation of different materials in specific regions of a cell's cytoplasm. *656*

cytosine (C) A type of nitrogen-containing base in nucleotides; also, a nucleotide with a cytosine base. Base-pairs with guanine in DNA and RNA. *190*

cytoskeleton In eukaryotic cells, the dynamic framework of diverse protein filaments that structurally support, organize, and move the cell and internal structures. Prokaryotic cells have a few similar protein filaments. *68*

decomposer One of the prokaryotic or fungal heterotrophs that obtains carbon and energy by breaking down wastes or remains of organisms. *734*

deletion Loss of a part of a chromosome, ranging from a single nucleotide base to a larger segment. *176, 204*

demographics Statistics that describe a population; e.g., size, age structure. *688*

demographic transition model Model that correlates changes in population growth with stages of economic development. *702*

denature To unravel the three-dimensional shape of a protein or other macromolecule, as by high temperature or pH. *44*

dendrite In a neuron, one of the short, branching extensions that accept signals and conduct them to the cell body. *494*

dendritic cell Phagocytic white blood cell that patrols tissue fluids; presents antigen to T cells. *582*

denitrification Conversion of nitrate or nitrite to gaseous nitrogen (N_2) or nitrogen oxide (NO_2) by soil bacteria. *749*

dense, irregular connective tissue A type of animal tissue with fibroblasts and many fibers asymmetrically arrayed in a matrix. In skin and capsules around organs. *480*

dense, regular connective tissue A type of animal tissue with rows of fibroblasts between the parallel bundles of fibers. In tendons, elastic ligaments. *480*

density-dependent factor A factor that slows population growth, and either appears or worsens with crowding; e.g., disease, competition for food. *693*

density-independent factor A factor that slows population growth; its likelihood of occurring and magnitude of effect does not vary with population density. *693*

deoxyribonucleic acid *See* DNA.

derived trait A novel trait that appeared in a lineage but did not occur in the lineage's most recent ancestor. *258*

dermal tissue system Tissues that cover and protect all exposed plant surfaces. *426*

dermis Skin layer beneath the epidermis; mostly dense connective tissue. *486*

desalinization The removal of salt from saltwater. *743*

desertification Conversion of grassland or irrigated or rain-fed cropland to desertlike conditions. *766*

detrital food web A food web in which most energy in plants flows directly to detritivores and decomposers. *737*

detritivore Any animal that feeds on small particles of organic matter; e.g., a crab or earthworm. *734*

deuterostome A bilateral animal belonging to a lineage in which the second opening to appear on the embryo surface becomes the mouth; e.g., an echinoderm or chordate. *363*

development Of complex multicelled species, the series of stages that transforms a zygote into an adult. *7, 412*

diaphragm Broad sheet of smooth muscle beneath the lungs; partitions the coelom into a thoracic cavity and an abdominal cavity. Also, a birth control device inserted into the vagina to prevent sperm from entering uterus. *605*

diatom A photosynthetic stramenopile (protist) that lives as a single cell inside a two-part silica shell. *325*

diffusion Net movement of like ions or molecules from a region where they are most concentrated to an adjoining region where they are less concentrated. *82, 414*

digestive system Body sac or tube where food is digested and absorbed, and any undigested residues expelled. Incomplete systems have one opening; the complete systems have two (mouth and anus). *618*

dihybrid experiment An experiment that tests for dominant or recessive alleles at two gene loci; individuals with different alleles at two loci are crossed or self-fertilized; e.g., *AaBb* × *AaBb*. The ratio of phenotypes in the resulting offspring offers information about the alleles. *158*

dinoflagellate Alveolate protist typically having two flagella; deposits cellulose in alveoli. Heterotrophs and photoautotrophs; some cause red tides. *322*

dinosaur One of a group of reptiles that arose in the Triassic and were dominant land vertebrates for 125 million years. *395*

diploid Having two of each type of chromosome characteristic of the species ($2n$). *129, 155*

directional selection Mode of natural selection; forms at one end of a range of phenotypic variation are favored. *270*

disease Condition that arises when the body's defenses cannot overcome infection and activities of a pathogen interfere with normal body functions. *180, 308*

disruptive selection Mode of natural selection that favors extreme forms in the range of variation; intermediate forms are selected against. *273*

distal tubule In a kidney nephron, tube that conveys filtrate to collecting duct. *641*

distance effect A biogeographic pattern. Islands distant from a mainland have fewer species than those closer to the potential source of colonists. *725*

division of labor Specialization of cells, tissues, or organs on specific tasks that collectively support a body or colony. Also a splitting up of tasks among different stages of the life cycle, as in insects. *484*

DNA Deoxyribonucleic acid. Double-stranded nucleic acid twisted into a helix; hereditary material for all living organisms and many viruses. Information in its base sequence is the basis of an organism's form and function. *7, 46*

DNA chip Microscopic array of DNA fragments that collectively represent a genome; used to study gene expression. *229*

DNA cloning A set of procedures that uses living cells such as bacteria to make many identical copies of a DNA fragment. *222*

DNA fingerprint An individual's unique array of short tandem repeats. *227*

DNA ligase Enzyme that seals breaks in double-stranded DNA. *192*

DNA polymerase DNA replication enzyme; assembles a new strand of DNA based on the sequence of a DNA template. *192*

DNA repair mechanism One of several processes by which enzymes repair broken or mismatched DNA strands. *193*

DNA replication Process by which a cell duplicates its DNA molecules before it divides into daughter cells. *192*

DNA sequencing Method of determining the order of nucleotide bases in DNA. *226*

dominant With regard to an allele, having the ability to mask the effects of a recessive allele paired with it. *155*

dormancy Period of arrested growth. *472*

dosage compensation Theory that X chromosome inactivation equalizes gene expression between males and females. *213*

double fertilization Mode of fertilization in flowering plants. One sperm nucleus fuses with the egg, and a second sperm nucleus fuses with the endosperm mother cell. *458*

doubling time The time it takes for a population to double in size. *691*

downwelling Water moves down and away from a coast after winds make a surface current flow toward the coast. *778*

drug addiction Dependence on a drug, which takes on an "essential" role; follows habituation and tolerance. *501*

duplication Base sequence in DNA that is repeated two or more times. *176*

ecdysone Insect hormone with roles in metamorphosis, molting. *539*

echinoderm A radial invertebrate with some bilateral features and calcified spines or plates on the body wall; e.g., sea star. *381*

ecoregion Broad land or ocean province influenced by abiotic and biotic factors. *728*

ecosystem Community interacting with its environment through a one-way flow of energy and the cycling of materials. *5, 734*

ectoderm Outer primary tissue layer of animal embryos. *362, 484, 654*

ectotherm An animal that can stay warm mainly by absorbing environmental heat, as by basking in the sun. *645*

effector Muscle (or gland) that responds to neural or endocrine signals. *416*

egg Mature female gamete, or ovum. *146*

El Niño Eastward displacement of warm surface waters of the western equatorial Pacific. Recurs, alters global climates. *778*

electric gradient A difference in electric charge between adjoining regions. *82*

electron Negatively charged subatomic particle that occupies orbitals around the atomic nucleus. *22*

electronegativity A measure of an atom's ability to pull electrons away from other atoms. *25*

electron transfer chain Array of enzymes and other molecules in a cell membrane that accept and give up electrons in sequence, thus releasing the energy of the electrons in small, usable increments. *81*

electron transfer phosphorylation Third stage of aerobic respiration; electron flow through electron transfer chains in inner mitochondrial membrane sets up an H^+ gradient that drives ATP formation. *109*

element A substance that consists only of atoms with the same number of protons. *22*

embryonic induction Embryonic cells produce signals that alter the behavior of neighboring cells. *658*

embryophyte Member of the clade of land plants; its eggs and embryos develop in a multicelled reproductive structure. *337*

emergent property A property of a system that does not appear in its component parts; e.g., cells (which are alive) are composed of many molecules (which are not alive). *5*

emigration Permanent move of one or more individuals out of a population. *690*

emulsification In the small intestine, the coating of fat droplets with bile salts so that fats remain suspended in chyme. *623*

endangered species A species endemic (native) to a habitat, found nowhere else, and highly vulnerable to extinction. *726*

endocrine gland A ductless gland that secretes hormone molecules, which typically travel in blood to target cells. *479*

endocrine system Control system of cells, tissues, and organs that interacts intimately with the nervous system; secretes hormones and other signaling molecules. *528*

endocytosis Process by which a cell takes in a substance by engulfing it in a vesicle formed from a bit of plasma membrane. *88*

endoderm Innermost primary tissue layer of animal embryos. *362, 484, 654*

endodermis Cylindrical, sheetlike cell layer around root vascular cylinder; helps control water and solute uptake. *435*

endometrium Lining of the uterus. *664*

endophytic fungus One of the fungi that lives as a symbiont inside plant leaves and stems. *356*

endoplasmic reticulum (ER) Membranous organelle, a continuous system of sacs and tubes that is an extension of the nuclear envelope. Rough ER is studded with many ribosomes; smooth ER is not. *62*

endorphin A neuromodulator that acts a natural painkiller. *500*

endoskeleton In chordates, an internal framework consisting of cartilage, bone, or both; works with skeletal muscle to position, support, and move body. *546*

endosperm Nutritive tissue in the seeds of flowering plants only. *459*

endosperm mother cell In the ovule of a flowering plant, a cell that has two nuclei ($n + n$). At fertilization, a sperm nucleus will fuse with it, forming endosperm. *458*

endospore Of certain bacteria, a resting structure enclosing a bit of cytoplasm and the DNA; resists heat, irradiation, drying, acids, disinfectants, and boiling water. When conditions favor growth, it germinates and a bacterium emerges from it. *309*

endosymbiosis An intimate, permanent ecological interaction in which one species lives and reproduces in the other's body to the benefit of one or both. *298*

endotherm An animal warmed mainly by its own metabolically generated heat. *645*

energy A capacity to do work. *6, 74*

energy pyramid Diagram that depicts the energy stored in the tissues of organisms at each trophic level in an ecosystem. Lowest tier of the pyramid, representing primary producers, is always the largest. *741*

enhancer Binding site in DNA for proteins that enhance the rate of transcription. *210*

enkephalin A neuromodulator that is a natural painkiller. *500*

ENSO El Niño Southern Oscillation. A recurring seesaw of atmospheric pressure in the western equatorial Pacific that has global repercussions on climates. *780*

entropy Measure of energy dispersal. *74*

enzyme A type of protein that catalyzes (speeds) a reaction without being changed by it. Some RNAs are catalytic. *37,76*

eosinophil A white blood cell that combats parasitic infection. *582*

epidermis Outermost tissue layer of plants and nearly all animals. *429, 486*

epiglottis Flaplike structure between the pharynx and larynx; its positional changes direct air into the trachea or food into the esophagus. *605*

epiphyte A plant that grows on the trunk or branch of another plant but does not withdraw nutrients from it. *339*

epistasis Interacting products of two or more gene pairs influence a trait. *161*

epithelium Animal tissue that covers the external body surfaces and lines tubular organs and body cavities. *362, 478*

erythropoietin Kidney hormone; induces stem cells in bone marrow to give rise to red blood cells. *612*

esophagus A muscular tube between the pharynx (throat) and stomach. *621*

essential amino acid Any amino acid that an organism cannot synthesize for itself and so must obtain from food. *629*

essential fatty acid Any fatty acid that an organism cannot synthesize for itself and so must obtain from food. *628*

estrogen A sex hormone. It helps oocytes mature and primes the endometrium for pregnancy; affects growth, development, and female secondary sexual traits. *666*

estuary Partly enclosed coastal region where seawater mixes with fresh water and runoff from land, as in rivers. *774*

ethylene Gaseous plant hormone that promotes fruit ripening and leaf, flower, fruit abscission. *467*

eudicot Flowering plant having embryos with two cotyledons and floral parts in fours, fives, or multiples of these. *345*

euglenoid A flagellated protist with many mitochondria. Majority are heterotrophs, others photoautotrophs. *321*

eukaryotic cell Type of cell that starts life with a nucleus. *52*

eutherian Placental mammal. *400*

eutrophication Nutrient enrichment of a body of water; promotes population growth of phytoplankton. *751, 773*

evaporation Transition of a liquid to a gas; requires energy input. *29, 645*

evaporative heat loss A response to heat stress; panting or sweating increases loss of heat from moist respiratory surfaces and skin. *646*

evolution Change in a line of descent. *242*

evolutionary adaptation A heritable aspect of form, function, or behavior that improves an individual's likelihood of surviving and reproducing in its current environment. *286*

evolutionary tree A treelike diagram in which each branch point represents a divergence from a shared ancestor; each branch is a separate line of descent. *259*

exocrine gland Glandular structure that secretes a substance through a duct onto a free epithelial surface; e.g., sweat gland, mammary gland. *479*

exocytosis Fusion of a cytoplasmic vesicle with the plasma membrane; as it becomes part of the membrane, its contents are released to extracellular fluid. *88*

exodermis Cylindrical sheet of cells close to root epidermis of most flowering plants; helps control water and solute uptake. *445*

exon Segment of RNA that is not excised during transcript processing; interspersed with introns. *199*

exoskeleton An external skeleton; e.g., the hardened arthropod cuticle. *375, 546*

exotic species Species that has become established in a new community after dispersing from its home range. *721*

experiment, scientific A test carefully designed to support or falsify a prediction about the function, cause, or effect of a single variable in isolation. Involves experimental and control groups. *13*

exponential growth Population increases in size by the same proportion of its total in each successive interval. *690*

extinction Permanent loss of a species. *285*

extracellular fluid Body fluids not in cells; e.g., plasma and interstitial fluid. *413*

extracellular matrix Complex mixture of substances secreted by cells; it structurally and functionally supports tissues; e.g., bone, basement membrane. *66*

extreme halophile Organism adapted to a highly salty habitat; e.g., an archaean that lives in salt ponds. *310*

extreme thermophile Organism adapted to a hot habitat; e.g., an archaean that lives in a hot spring or at a hydrothermal vent. *310*

eye Sensory organ that incorporates a dense array of photoreceptors. *375, 520*

fall overturn During the fall, waters of a temperate zone mix. Upper, oxygenated water cools, gets dense, and sinks; nutrient-rich water from the bottom moves up. *772*

fat Lipid with one, two, or three fatty acid tails attached to a glycerol. *40*

fatty acid Organic compound that is a component of many lipids; carboxyl group joined to a backbone of four to thirty-six carbon atoms. Backbone of saturated types has single bonds only; that of unsaturated has double covalent bonds. *40*

feces Digestive waste that has been concentrated by action of the colon. *626*

feedback inhibition Mechanism by which a change that results from some activity decreases or stops the activity. *78*

fermentation pathway Metabolic pathway that breaks down carbohydrates to produce ATP without using oxygen. *See* alcoholic fermentation and lactate fermentation. *108*

fertilization Fusion of a sperm nucleus and an egg nucleus, the result being a single-celled zygote. *146, 654*

fetus In mammalian development, the stage after all major organ systems have formed (ninth week) until birth. *674*

fever An internally induced rise in core body temperature above the normal set point as a response to infection. *585, 646*

fibrous root system Adventitious and lateral roots arising from a stem. *434*

fight–flight response Response to danger or excitement. Parasympathetic input falls, sympathetic signals increase, and adrenal glands secrete epinephrine. This readies the body to fight or escape. *505*

filter feeder Animal that filters food from a current of water that flows through pores or slits of some body structure. *386*

fin An appendage that helps stabilize and propel most fishes in water. *389*

first law of thermodynamics Energy cannot be created or destroyed. *74*

fitness The degree of adaptation to the environment, measured by relative genetic contribution to future generations. *245*

fixation Of an allele. In a population, the loss of all alleles but one at a gene locus. *276*

fixed action pattern A series of instinctive movements, triggered by a simple stimulus, that continues no matter what else is going on in the environment. *787*

flagellated protozoan One of the single-celled heterotrophic protists having one or more flagella; e.g., a diplomonad. *321*

flagellum, plural **flagella** Long, slender cellular structure used for motility. In eukaryotes, it whips from side to side; in prokaryotes, it rotates like a propeller. *58*

flatworm Member of a group of bilaterally symmetrical, unsegmented invertebrates having organ systems derived from three primary tissue layers, but no coelom; e.g., a planarian, fluke, or tapeworm. *368*

flower Specialized reproductive shoot of an angiosperm. *344, 456*

fluid mosaic model A cell membrane has a mixed composition (mosaic) of lipids and proteins, the interactions and motions of which impart fluidity to it. *56*

food chain Linear sequence of steps by which energy stored in autotroph tissues enters higher trophic levels. *735*

food web Cross-connecting food chains consisting of producers, consumers, and decomposers, detritivores, or both. *735*

foramen magnum Opening in the skull where the vertebrate spinal cord and brain connect. Its position in bipedal species is different than that of species that walk on all fours. *402*

foraminiferan Heterotrophic single-celled protist that extends its pseudopods through a perforated calcium carbonate or silica shell. Most live on the ocean floor. *322*

fossil Physical evidence of an organism that lived in the past. *241*

fossilization How fossils form over time. An organism or evidence of it gets buried in sediments or volcanic ash; water slowly infiltrates the remains, and metal ions and other inorganic compounds dissolved in it replace the minerals in bones and other hardened tissues. *246*

founder effect A form of bottlenecking. Change in allele frequencies that occurs after a few individuals establish a new population. *276*

fruit Mature ovary, often with accessory parts, from a flowering plant. *346, 460*

FSH Follicle-stimulating hormone of the anterior lobe of pituitary gland; has reproductive roles in both sexes. *663*

functional group An atom or a group of atoms covalently bonded to carbon; imparts certain chemical properties to an organic compound. *36*

fungus, plural **fungi** Type of eukaryotic heterotroph; can be multicelled or single-celled; cell walls contain chitin; obtains nutrients by extracellular digestion and absorption. *8, 352*

gamete Mature, haploid reproductive cell; e.g., an egg or sperm. *140*

gametophyte A haploid, multicelled body in which gametes form during the life cycle of plants and some algae. *146, 334, 456*

ganglion, plural **ganglia** Group of neuron cell bodies; may function as an integrating center for signals. *368, 492*

gap junction Cell junction that forms an open channel across the plasma membrane of adjoining animal cells; permits rapid flow of ions and small molecules from the cytoplasm of one cell to another. *479*

gastric fluid Extremely acidic mixture of secretions from the stomach lining. *622*

gastrointestinal tract The gut. Starts at the stomach and extends through the intestines to the tube's terminal opening. *621*

gastrula Early animal embryo with two or three primary tissue layers. *654*

gastrulation Stage of animal development; embryonic cells formed by cleavage become arranged as two or three primary tissue layers in a gastrula. *654*

gel electrophoresis Method of separating DNA fragments according to length, or protein molecules according to size and charge. The molecules move apart while migrating through a semisolid gel in response to an electric current. *226*

gene Heritable unit of information in DNA; occupies a particular location (locus) on a chromosome. *140, 155*

gene expression Process by which the information contained in a gene becomes converted to a structural or functional part of a cell. *155, 210*

gene flow The movement of alleles into and out of a population, as by individuals that immigrate or emigrate. *277*

gene library Collection of host cells that contain different cloned DNA fragments representing all or most of a genome. *224*

gene pool All of the genes in a population; a pool of genetic resources. *266*

gene therapy The transfer of a normal or modified gene into an individual with the goal of treating a genetic disorder. *234*

genetic code Set of sixty-four mRNA codons, each of which specifies an amino acid or stop signal in translation. *200*

genetic drift Change in allele frequencies in a population due to chance alone. *276*

genetic engineering Process by which deliberate changes are introduced into an individual's chromosome(s). *230*

genetic equilibrium Theoretical state in which a population is not evolving with respect to a specified gene. *267*

genome The complete genetic material of a species. *224*

genomics The study of genes and their function. The structural branch determines the three-dimensional structure of proteins encoded by a genome; comparative branch compares genomes of different species. *229*

genotype The particular alleles carried by an individual. *155*

genus, plural **genera** A group of species that share a unique set of traits. *8*

geographic dispersal A movement of individuals out of the ancestral community and into a new one. *721*

geologic time scale Chronology of Earth history. *248*

germ cell Animal cell that can undergo meiosis and give rise to gametes. *140*

germ layer One of the primary tissue layers in an embryo (endoderm, ectoderm, or mesoderm). *484, 654*

germination The resumption of growth of a seed or a spore after dormancy, dispersal, or both. *464*

gibberellin Plant hormone; induces stem elongation, helps seeds break dormancy, has role in flowering in some species. *466*

gill A respiratory organ. In vertebrates, usually one of a pair of thin folds richly supplied with blood exchange gases with surrounding water. *388, 601*

ginkgo A gymnosperm; only surviving species is a deciduous tree that has fan-shaped leaves. *342*

gland A secretory organ derived from epithelium. Hormone-secreting endocrine glands have ducts; exocrine glands are ductless. *479*

global warming Long-term increase in temperature of Earth's lower atmosphere; rising levels of greenhouse gases contribute to the increase. *747*

glomerular filtration First step in urine formation; blood pressure forces water and solutes out of glomerular capillaries into Bowman's capsule. *642*

glomerulus In a kidney nephron, a cluster of capillaries from which fluid is filtered into Bowman's capsule. *641*

glottis Opening between vocal cords. *605*

glycemic index (GI) The ranking of foods by how they affect blood glucose level during the two hours after a meal. *628*

glycolysis First stage of carbohydrate breakdown pathways; glucose or other sugar is broken down to two pyruvates. Net yield: 2 ATP per glucose. *109*

gnetophyte A type of woody, vinelike, or shrubby gymnosperm. *342*

GnRH Gonadotropin-releasing hormone. Hypothalamic hormone that induces release of LH and FSH by the pituitary gland. *663*

golden alga Stramenopile protist; can be autotroph or heterotroph depending on the conditions. *325*

Golgi body Organelle of endomembrane system; enzymes inside its much-folded membrane modify new polypeptide chains and lipids; the products are sorted and packaged in vesicles for transport. *62*

gonad Primary reproductive organ in animals; produces gametes. *538, 660*

Gondwana Supercontinent that formed more than 500 million years ago. *251*

gradualism Idea that evolution occurs by slight changes over long time spans. *See also* punctuated equilibrium. *284*

Gram-positive bacteria Informal name for the mostly chemoheterotrophic bacteria that have a multilayered wall. *309*

gravitropism Plant growth in a direction influenced by gravity. *468*

gray matter Inner region of the brain and spinal cord. *505*

grazing food web Cross-connecting food chains in which most energy flows from plants to herbivores. *737*

greenhouse effect Some atmospheric gases absorb infrared wavelengths (heat) from the sun-warmed surface, and then radiate some back toward Earth, warming it. *746*

ground tissue system Tissue that makes up most of the plant body. *426*

groundwater Water contained in soil and in aquifers. *742*

growth Of multicelled species, increases in the number, size, and volume of cells. Of single-celled prokaryotes, increases in the number of cells. *412, 464*

growth factor Signaling molecules that cause target cells to divide, differentiate, or mature. *134, 510*

guanine (G) A type of nitrogen-containing base in nucleotides; also, a nucleotide with a guanine base. Base-pairs with cytosine in DNA and RNA. *190*

guard cell One of two cells that define a stoma across leaf or stem epidermis. *448*

gut A sac or tube in which food is digested. Also the gastrointestinal tract from the stomach onward. *362*

gymnosperm Nonflowering seed plant; forms its seeds on exposed surfaces of spore-producing structures; a gnetophyte, cycad, ginkgo, or conifer. *342*

habitat Place where an organism or species lives; described by physical and chemical features and array of species. *415, 708*

habitat loss Reduction in suitable living space; can cause species extinction. *726*

hair cell Hairlike mechanoreceptor; it fires when sufficiently bent or tilted. *517*

half-life The time it takes for half of a quantity of any radioisotope to decay. *248*

haploid Having one of each type of chromosome characteristic of the species (n); e.g., a human gamete is haploid. *141*

hardwood Strong, dense wood with many vessels, tracheids, and fibers in xylem. *437*

hearing Perception of sound. *518*

heart Muscular pump; its contractions circulate blood through an animal body. *560*

heartwood Dense, dark, aromatic tissue at the core of older tree stems and roots. *437*

hemoglobin Iron-containing respiratory protein. In humans, it occurs in red blood cells and carries the most oxygen. *600*

hemostasis Process that stops blood loss from a damaged vessel by coagulation, spasm, and other mechanisms. *574*

hermaphrodite An individual with male and female reproductive organs. *368*

heterotherm An animal that maintains its core temperature by controlling metabolic activity some of the time and allowing it to rise or fall at other times. *645*

heterotroph Organism that obtains carbon from organic compounds assembled by other organisms. *80, 306*

heterozygous Having two different alleles at a gene locus; e.g., *Aa*. *155*

histone Type of protein that structurally organizes eukaryotic chromosomes. Part of nucleosomes. *127*

homeostasis The collection of processes by which the conditions in a multicelled organism's internal environment are kept within tolerable ranges. *7, 413, 638*

homeotic gene Type of master gene; its expression controls formation of specific body parts during development. *214, 659*

hominid All humanlike and human species. *402*

homologous chromosome One of a pair of chromosomes in body cells of diploid organisms; except for the nonidentical sex chromosomes, members of a pair have the same length, shape, and genes. *141*

homologous structures Similar body parts among lineages; reflect shared ancestry. *252*

homozygous dominant Having a pair of dominant alleles at a locus on homologous chromosomes; e.g., *AA*. *155*

homozygous recessive Having a pair of recessive alleles at a locus on homologous chromosomes; e.g., *aa*. *155*

horizontal gene transfer Process by which a living cell acquires genes from another cell of the same or different species; e.g., by bacterial conjugation. *307*

hormone A signaling molecule that is secreted by one type of cell in a multicelled body and alters activities of other body cells that have receptors for it. *528*

hot spot A habitat that contains many species found nowhere else and at a high risk of extinction. *728*

human Member of the species *Homo*. *405*

humus Decomposing organic matter in soil. *442*

hybrid Heterozygote. Individual that has two different alleles at a gene locus. *155*

hydrogen bond Attraction that forms between a covalently bonded hydrogen atom and an electronegative atom taking part in another covalent bond. *27*

hydrologic cycle Biogeochemical cycle driven by solar energy; water moves through atmosphere, onto land, to the ocean, and back to the atmosphere. *742*

hydrolysis A cleavage reaction; an enzyme splits a molecule and attaches an —OH and an —H (both derived from water) to the exposed sites on the fragments. *37*

hydrophilic Dissolves easily in water. *28*

hydrophobic Resists dissolving in water. *28*

hydrostatic pressure *See* turgor.

hydrostatic skeleton A fluid-filled cavity on which muscle contractions act. *367, 546*

hydrothermal vent Underwater fissure where superheated, mineral-rich water is forced out under pressure. *294, 778*

hypertonic fluid Of two fluids, the one with the higher solute concentration. *86*

hypha, plural **hyphae** Of a multicelled fungus, a filament having chitin-reinforced walls; component of a mycelium. *352*

hypothalamus Forebrain region; a center of homeostatic control of internal environment (e.g., salt–water balance, core temperature); influences hunger, thirst, sex, other viscera-related behaviors, and emotions. *506, 532*

hypothesis, scientific Testable explanation of a natural phenomenon. *12*

hypotonic fluid Of two fluids, the one with the lower solute concentration. *86*

immigration One or more individuals move and take up residence in another population of its species. *690*

immunity The body's ability to resist and combat infections. *582*

immunization A process that is designed to promote immunity from disease; e.g., vaccination. *592*

implantation In mammalian pregnancy; a blastocyst burrows into uterine lining. *674*

imprinting A form of learning triggered by exposure to sign stimuli; time-dependent, usually occurs during a sensitive period while an animal is young. *787*

inbreeding Nonrandom mating among close relatives. *277*

incomplete digestive system Saclike gut; food enters and wastes leave through the same opening. *618*

incomplete dominance Condition in which one allele is not fully dominant over another, so the heterozygous phenotype is between the two homozygous phenotypes. *160*

independent assortment In meiosis, the genes on each pair of homologous chromosomes get sorted into gametes independently of how genes on other pairs of homologues are sorted. *158*

indicator species Any species which, by its abundance or scarcity, is a measure of the health of its habitat. *727*

induced-fit model Explanation of how some enzymes work; an active site bends or squeezes a substrate, and the tension destabilizes the substrate's bonds. *77*

inflammation Response to tissue invasion or injury. White blood cells release signals that attract phagocytes and increase local blood flow. Signs include redness, heat, swelling, pain. *584*

inheritance Transmission of DNA from parents to offspring. *7*

inhibitor A hormone that slows release of another hormone. *532*

innate immunity Set of vertebrate immune responses in which recognition of pathogen-associated molecular patterns triggers phagocytosis, causes inflammation, and activates complement. *582*

insertion A mutation by which one or more bases are introduced into DNA. *204*

instinctive behavior Behavior performed without having first been learned. *787*

integrator A control center that receives, processes, and stores sensory input, and coordinates the responses; e.g., a brain. *416*

integumentary exchange In some animals, gas exchange across thin, moistened skin or some other external body surface. *601*

intercostal muscles The skeletal muscles between the ribs; help change the volume of the thoracic cavity during breathing. *605*

intermediate disturbance hypothesis An explanation of community structure; holds that species richness is greatest in habitats where disturbances are moderate in intensity, frequency, or both. *719*

intermediate filament Cytoskeletal element that mechanically strengthens cells. *68*

internal environment The body fluid *not* in cells—extracellular fluid; in most animals, primarily blood and interstitial fluid. *560*

internal fertilization The union of sperm and egg within the female's body. *653*

interneuron Neuron that receives input from sensory neurons and sends signals to other interneurons or to motor neurons. *492*

interphase In a eukaryotic cell cycle, the interval between mitotic divisions when a cell grows in mass, roughly doubles the number of its cytoplasmic components, and replicates its DNA. *128*

interspecific competition *See* competition, interspecific.

interstitial fluid Fluid in between cells and tissues of a multicelled body. *416, 560*

intervertebral disk Cartilage disk that lies between adjacent vertebrae; acts as a flex point and shock absorber. *546*

intron Segment that is excised from RNA during transcript processing; intervenes between exons. *199*

inversion Structural rearrangement of a chromosome in which part of it becomes oriented in the reverse direction. *176*

ion Atom that carries a charge due to an unequal number of protons and electrons. *25*

ionic bond Type of chemical bond; strong mutual attraction between ions of opposite charge. *26*

ionizing radiation Radiation with enough energy to eject electrons from atoms. *204*

isotonic fluid Any fluid having the same solute concentration as another fluid to which it is being compared. *86*

isotope One of several forms of an element that differ in the number of neutrons. *22*

jaw Paired, hinged cartilaginous or bony feeding structures of most chordates. *388*

joint Area of contact between bones. *549*

karyotype Preparation of an individual's metaphase chromosomes arranged by size, length, shape, and centromere location. *171*

key innovation A change in some body structure or function that gives a species the opportunity to exploit the environment more efficiently or in a novel way. *284*

keystone species A species that influences community structure in disproportionately large ways relative to its abundance. *720*

kidney One of a pair of vertebrate organs that filter blood, remove wastes, and help maintain the internal environment. *638*

kilocalorie 1,000 calories of heat energy; amount needed to raise the temperature of 1 kilogram of water by 1°C at a standard pressure. Standard unit of measure for food's caloric content. *74*

kinase Enzyme that transfers a phosphate group to (phosphorylates) a substrate. *134*

knockout experiment An experiment in which an organism is genetically engineered so one of its genes does not function. *214*

Krebs cycle The second stage of aerobic respiration; as pyruvate from glycolysis is fully broken down to CO_2 and H_2O, two ATP and many coenzymes form. *109*

K–T asteroid impact hypothesis Idea that an asteroid impact was the cause of the mass extinction that marks the boundary between Cretaceous and Tertiary periods, 65 million years ago. *395*

La Niña Climatic event in which Pacific waters become cooler than average. *780*

labor The birth process. *681*

lactate fermentation Anaerobic pathway that breaks down glucose, forms ATP and lactate. Begins with glycolysis; end reactions regenerate NAD^+ so glycolysis continues. Net yield: 2 ATP per glucose. *117*

lactation Milk production and secretion by hormone-primed mammary glands. *681*

lake A body of standing fresh water. *772*

lancelet An invertebrate chordate, a small filter feeder with a fishlike shape. *386*

larva, plural **larvae** A free-living, immature stage between the embryo and adult in the life cycle of many animals. *365*

larynx Tubular airway leading to lungs; has vocal cords in some animals. *605*

lateral bud Axillary bud. A dormant shoot that forms in a leaf axil. *430*

leaching Removal of nutrients from soil as water percolates through it. *443*

learned behavior Enduring modification of a behavior as an outcome of experience in the environment. *787*

lens In camera eyes, a transparent body that bends light rays so they all converge suitably onto photoreceptors. *520*

lethal mutation Mutation that drastically alters phenotype; usually causes death. *266*

LH Luteinizing hormone. An anterior pituitary hormone; roles in reproductive function of males and females. *663*

lichen Symbiotic association between a fungus and a photoautotroph—an alga or cyanobacterium. *356*

life history pattern Of a species, pattern of when and how many offspring are produced during a typical lifetime. *694*

ligament A strap of dense connective tissue that bridges a skeletal joint. *549*

light-dependent reactions First stage of photosynthesis; one of two metabolic pathways (cyclic or noncyclic) in which light energy is converted to the chemical energy of ATP. NADPH and O_2 also form in the noncyclic pathway. *97*

light-independent reactions Second stage of photosynthesis; metabolic pathway in which the enzyme rubisco fixes carbon, and glucose forms. Runs on ATP and NADPH from the light-dependent reactions. *See also* Calvin–Benson cycle. *97*

lignin Organic compound that strengthens cell walls of vascular plants; reinforces stems and thus helps plant stand upright. *66, 334*

limbic system Centers in cerebrum that govern emotions; roles in memory. *509*

limiting factor Any essential resource that limits population growth when scarce. *692*

lineage Line of descent from an ancestor. *247*

linkage group All genes on a chromosome; tend to stay together during meiosis but may be separated by crossovers. *162*

lipid Fatty, oily, or waxy organic compound. *40*

lipid bilayer Structural foundation of cell membranes; mainly phospholipids arranged tail-to-tail in two layers. *52*

loam Soil best for plant growth; roughly equal amounts of sand, silt, and clay. *442*

lobe-finned fish Only bony fish having fleshy ventral fins supported by internal skeletal elements. *391*

local signaling molecule Chemical signal secreted into interstitial fluid. Has potent effects on nearby cells, but is inactivated fast; e.g., prostaglandins. *528*

logistic growth Population growth pattern. A population grows exponentially when small, then levels off in size once carrying capacity has been reached. *692*

loop of Henle Hairpin-shaped, tubular part of a nephron where water and solutes are reabsorbed from interstitial fluid. *641*

loose connective tissue Animal tissue with fibers and fibroblasts widely dispersed in the matrix; holds organs in place. *480*

lung Internal respiratory organ of all birds, reptiles, mammals, most amphibians, and some fish. *389, 602*

lungfish A type of bony fish that has a lung or lungs. *391*

lymph Interstitial fluid that has entered vessels of the lymphatic system. *576*

lymph node Lymphoid organ that is a key site for immune responses, as executed by its organized arrays of lymphocytes. *577*

lymph vascular system The portion of the lymphatic system that takes up and conducts excess tissue fluid, absorbed fats, and reclaimable solutes to blood. *576*

lysogenic pathway Viral replication mode in which viral genes get integrated into host chromosome and may be inactive through many host cell divisions before being replicated. *312*

lysosome Vesicle filled with enzymes that functions in intracellular digestion. *63*

lysozyme Antibacterial enzyme that occurs in body secretions; e.g., in tears. *583*

lytic pathway A rapid viral replication pathway. Viral genes direct the host cell to make new virus particles, which are released when the host cell dies. *312*

macroevolution Large-scale patterns, rates of change, and trends among lineages. *249*

macrophage Phagocytic white blood cell that patrols tissue fluids; presents antigen to T cells. *582*

magnoliid One of three major flowering plant groups; e.g., magnolias. *345*

Malpighian tubule One of many small tubes that help insects and spiders dispose of wastes without losing water. *376*

mammal Only amniote that makes hair and nourishes offspring with milk from the female's mammary glands. *400*

marine snow Organic matter that drifts down to ocean depths. *778*

marsupial Pouched mammal. *400*

mass extinction Loss of many lineages. *285*

mast cell White blood cell in connective tissue; factor in inflammation. *582*

master gene Gene encoding a product that affects the expression of other genes. Its product forms a gradient; cells express different genes depending on where they are within the gradient. *213*

mechanoreceptor Sensory cell that detects mechanical energy (a change in pressure, position, or acceleration). *514*

medulla oblongata Hindbrain region. Its reflex centers control respiration and other basic tasks; coordinate motor responses with complex reflexes, e.g., coughing. *506*

megaspore Haploid spore that forms in ovary of seed plants; gives rise to a female gametophyte with egg cell. *341, 458*

meiosis Nuclear division process that halves the chromosome number, to the haploid (n) number. Basis of sexual reproduction. *126, 140*

menopause Stage when a human female's fertility ends; menstruation ceases and secretion of sex hormones declines. *665*

menstrual cycle Approximately monthly cycle in human females of reproductive age. Hormonal changes lead to oocyte maturation and release, and prime the uterine lining for pregnancy. If pregnancy does not occur, this lining is shed and the cycle begins again. *665*

meristem Zone of undifferentiated, dividing plant cells; gives rise to differentiated cell lineages that form mature plant tissues. *426*

mesoderm Middle primary tissue layer (between endoderm and ectoderm) of most animal embryos. *362, 484, 654*

mesophyll Photosynthetic parenchyma; a plant tissue with many air spaces. *432*

messenger RNA (mRNA) RNA transcribed from genes that encode proteins. *198*

metabolic pathway A stepwise sequence of enzyme-mediated reactions by which cells construct, remodel, or break down organic molecules; e.g., photosynthesis. *80*

metabolism All the enzyme-mediated chemical reactions by which cells acquire and use energy as they construct, remodel, and break down organic molecules. *37*

metamorphosis Hormone-induced growth and tissue reorganization transforms larva into the adult form. *375*

metaphase Stage of mitosis; chromosomes are aligned at spindle equator. *131*

methanogen Any bacterium or archaean that produces methane gas. *310*

MHC marker Self-recognition protein on the surface of body cells. Adaptive immune response occurs when it becomes complexed with antigen fragments. *586*

micelle formation The combining of bile salts with fatty acids into tiny droplets. *625*

microevolution Of a population, small-scale changes in allele frequencies due to mutation, natural selection, genetic drift, and gene flow. *267*

microfilament Cytoskeletal element; a fiber of actin subunits. *68*

microspore Walled haploid spore of seed plants; gives rise to pollen grains. *458*

microsporidian Intracellular fungal parasite of aquatic habitats; only fungal group that forms flagellated spores. *352*

microtubule Cytoskeletal element; a hollow filament of tubulin subunits. *68*

microvillus, plural **microvilli** Slender extension from free surface of some cells such as brush border cells in the small intestine; increases surface area. *478, 624*

migration Of many animals, a recurring pattern of movement between two or more regions in response to seasonal change or other environmental rhythms. *398, 690*

mimicry Evolutionary convergence of body form; a close resemblance between species. A defenseless species may look like a well-defended one, or several well-defended species may all look alike. *714*

mineral In nutrition, an inorganic substance essential for survival and growth. *630*

mitochondrial DNA (mtDNA) DNA of mitochondria, distinct from nuclear DNA of a cell; has its own genetic code. *257*

mitochondrion Double-membraned organelle of ATP formation; site of second and third stages of aerobic respiration in eukaryotes. *64*

mitosis Nuclear division mechanism that maintains the chromosome number. Basis of body growth, tissue repair and replacement in multicelled eukaryotes, as well as asexual reproduction in some plants, animals, fungi, and protists. *126*

mixture Two or more types of molecules intermingled in proportions that vary. *25*

model Analogous system used to test an object or event that cannot itself be tested directly. *12*

molecular clock Method of estimating how long ago two lineages diverged; assumes that neutral mutations accumulate in DNA at a constant rate, measured as a series of predictable ticks back through time. The last tick stops close to the time the lineages diverged. *256*

molecule Atoms of the same or different elements joined by chemical bonds. *4, 25*

mollusk Only invertebrate with a mantle draped over a soft, fleshy visceral mass; most have an external or internal shell; e.g., gastropods, bivalves, cephalopods. *372*

molting Periodic shedding of worn-out or too-small body structures. *374, 539*

monocot Flowering plant characterized by embryo sporophytes having one cotyledon; floral parts in threes (or multiples of three); and often parallel-veined leaves. *345*

monohybrid experiment An experiment that tests for dominant or recessive alleles at one gene locus; individuals with different alleles at a single locus are crossed or self-fertilized; e.g., $Aa \times Aa$. The phenotype ratio of the resulting offspring offers information about the alleles. *156*

monomer A small molecule that is a repeating subunit in a polymer; e.g., glucose is a monomer of starch. *37*

monophyletic group An ancestor and all of its descendants. *258*

monotreme Egg-laying mammal. *400*

morphogen Master gene product; diffuses through embryonic tissues; the resulting gradient causes transcription of different genes in different parts of the embryo. *658*

morphogenesis Process by which tissues and organs form. *658*

morphological convergence Pattern of macroevolution in which similar body parts evolve separately in different lineages. *253*

morphological divergence Pattern of macroevolution in which a body part of an ancestor changes in its descendants. *252*

motor neuron Neuron that relays signals from the brain or spinal cord to muscle cells or gland cells. *492*

motor protein A type of protein that, when energized by ATP, interacts with elements of the cytoskeleton to move cell structures or the whole cell; e.g., myosin. *68*

motor unit A motor neuron and the muscle fibers that it controls. *554*

mouth The oral cavity. *621*

mucosa A mucus-secreting epithelium; e.g., the inner lining of the gut wall. *622*

multiple allele system Three or more alleles persist in a population. *160*

multiregional model Idea that modern humans evolved gradually from many different *Homo erectus* populations that lived in different parts of the world. *406*

muscle fatigue Decline in muscle tension when tetanic contraction is continuous. *554*

muscle fiber Long, multinucleated cell in a skeletal muscle. *550*

muscle spindle A sensory organ that detects muscle stretching. *503*

muscle tension Mechanical force exerted by a contracting muscle. *554*

muscle twitch A sequence of muscle contraction and relaxation in response to a brief stimulus. *554*

mutation Permanent, small-scale change in DNA. Primary source of new alleles and, thus, of life's diversity. *10, 155, 204*

mutualism An interspecific interaction that benefits both participants. *356, 708*

mycelium, plural **mycelia** Mesh of tiny, branching, food-absorbing filaments (hyphae) of a multicelled fungi. *352*

mycorrhiza "Fungus-root." A mutualism between a fungus and plant roots. *356, 444*

myelin sheath Lipid-rich wrappings around axons of some neurons; speeds propagation of action potential. *502*

myofibril In a muscle fiber, one of many long, thin structures that run parallel with the long axis of a muscle fiber; composed of sarcomeres arranged end-to-end. *552*

myoglobin Oxygen-storing protein, most abundant in muscle. *600*

myosin An ATP-driven motor protein that moves cell components on cytoskeletal tracks. Interacts with actin in muscle to bring about contraction. *553*

natural killer cell (NK cell) One of the lymphocytes; kills infected cells and tumor cells that evade other lymphocytes. *582*

natural selection Differential survival and reproduction of individuals of a population that differ in the details of heritable traits; a microevolutionary process. *10, 245, 269*

nature Everything in the universe except what humans have manufactured. *4*

nectar Sucrose-rich fluid of some flowers; attracts pollinators. *457*

negative feedback mechanism A major homeostatic mechanism by which some activity changes conditions in a cell or multicelled organism and thereby triggers a response that reverses the change. *416*

nematocyst Unique feature of cnidarians. Capsule that discharges a barbed or sticky thread when touched; has roles in feeding and defense. *367*

neoplasm Mass of cells (tumor) that lost control over the cell cycle. *134*

nephridium, plural **nephridia** Of annelids and some other i nvertebrates, one of many water-regulating units that help control the content and volume of tissue fluid. *370*

nephron Functional unit of the kidney; it filters water and solutes from blood, then reabsorbs adjusted amounts of both. *641*

nerve Bundles of axons enclosed in a sheath of connective tissue. *492*

nerve cord In bilateral animals, a line of communication that runs parallel with the anterior–posterior axis. In vertebrates, it develops as a hollow, neural tube that gives rise to the spinal cord and brain. *368*

nerve net Nervous system of cnidarians and some other invertebrates; asymmetrical mesh of neurons. *367, 492*

nervous system Organ system that detects internal and external stimuli, integrates information, and coordinates responses. *492*

nervous tissue Animal tissue consisting of neurons and often neuroglia. *483*

net ecosystem production In an ecosystem, the energy that producers capture in a specified time, *minus* the energy that producers and decomposers use. *740*

neuroglial cell, plural **neuroglia** Any of the nervous tissue cells that structurally and metabolically support neurons. *492*

neuromodulator Any signaling molecule that reduces or magnifies the influence of a neurotransmitter on target cells. *500*

neuromuscular junction A chemical synapse between a motor neuron and a skeletal muscle fiber. *498*

neuron A type of excitable cell; functional unit of the nervous system. *483, 492*

neurotransmitter An intercellular signaling molecule secreted by the axon endings of a neuron. *498, 528*

neutral mutation A mutation that has no effect on survival or reproduction. *266*

neutron Uncharged subatomic particle in the atomic nucleus. *22*

neutrophil Circulating phagocytic white blood cell. *582*

niche A species' unique ecological role; it is described in terms of the conditions, resources, and interactions necessary for survival and reproduction. *708*

nitrification One stage of the nitrogen cycle. Soil bacteria break down ammonia or ammonium to nitrite, then other bacteria break down nitrite to nitrate, which plants can absorb. *749*

nitrogen cycle An atmospheric cycle. Nitrogen moves from its largest reservoir (the atmosphere), then through the ocean, ocean sediments, soils, and food webs, then back to the atmosphere. *748*

nitrogen fixation Conversion of gaseous nitrogen to ammonia. *308, 444, 748*

nondisjunction Failure of sister chromatids or homologous chromosomes to separate during meiosis or mitosis. Resulting cells get too many or too few chromosomes. *178*

nonionizing radiation Radiation that carries enough energy to boost electrons to higher energy levels but not enough to eject them from an atom. *205*

nonshivering heat production In response to cold stress, mitochondria in cells of brown adipose tissue release energy as heat, rather than storing it in ATP. *647*

notochord A rod of stiffened tissue in chordate embryos; may or may not persist as a supporting structure in the adult. *386*

nuclear envelope A double membrane that encloses the nucleus. *61*

nucleic acid Single- or double-stranded chain of nucleotides joined by sugar–phosphate bonds; e.g., DNA, RNA. *46*

nucleic acid hybridization Base-pairing between DNA or RNA from different sources. *224*

nucleoid The region of a prokaryotic cell where the DNA is most concentrated. *52*

nucleolus In a nucleus, a roundish mass of material from which RNA and proteins are assembled into ribosomal subunits. *61*

nucleoplasm A semifluid portion of the nucleus; enclosed by nuclear envelope. *61*

nucleosome Smallest unit of structural organization in eukaryotic chromosomes; a length of DNA wound twice around a spool of histone proteins. *127*

nucleotide Organic compound with a five-carbon sugar, a nitrogen-containing base, and a phosphate group. Monomer of nucleic acids. *190*

nucleus In eukaryotic cells only, organelle with an outer envelope of two pore-ridden lipid bilayers; separates chromosomes from the cytoplasm. *52*

numerical taxonomy Method of classifying an unidentified microbe by comparing it with a known one; the more traits shared, the closer is the inferred relatedness. *307*

nutrient An atom or molecule that has an essential role in survival or growth. *6, 442*

obesity Having a health-threatening excess of fat in adipose tissue. *632*

olfaction The sense of smell. *516*

olfactory receptor Chemoreceptor for a water-soluble or volatile substance. *516*

oocyte An immature egg. *656*

oomycote Heterotrophic stramenopile; e.g., downy mildews, water molds. *325*

operator Part of an operon; a DNA binding site for a repressor. *217*

operon Group of genes together with a promoter–operator DNA sequence that controls their transcription. *217*

organ Body structure composed of tissues that interact in one or more tasks. *368, 412*

organ of equilibrium Organ that monitors a body's position and motion; e.g., human vestibular apparatus. *517*

organ system A set of organs that are interacting chemically, physically, or both in a common task. *368, 412*

organelle Structure that carries out a specialized metabolic function inside a cell; e.g., a nucleus in eukaryotes. *60*

organic compound Molecule with carbon and at least one hydrogen; most have functional groups. *36*

organic metabolism Reactions by which a body breaks down, uses, and stores the organic compounds it takes in. *627*

organism An individual consisting of one or more cells. *4*

osmoreceptor Sensory receptor that detects shifts in solute concentrations. *514*

osmosis Diffusion of water across a selectively permeable membrane from a region where the water concentration is higher to a region where it is lower. *86*

osmotic pressure Amount of hydrostatic pressure that will prevent osmosis from a hypotonic to a hypertonic fluid across a semipermeable membrane. *87*

osteoblast Bone-forming cell; it secretes matrix that gets mineralized. *548*

osteoclast Bone-digesting cell; it secretes enzymes that digest bone's matrix. *548*

osteocyte A mature bone cell, osteoblast that has become surrounded by its own secretions. *548*

ovary In animals, a female gonad. In flowering plants, the enlarged base of a carpel, inside which one or more ovules form and eggs are fertilized. *456, 664*

ovulation Release of a secondary oocyte from an ovary. *666*

ovule In a seed-bearing plant, structure in which a haploid, egg-producing female gametophyte forms; after fertilization, it matures into a seed. *341, 456*

ovum Mature egg. *669*

oxidation–reduction reaction Reaction in which one molecule accepts electrons (it becomes reduced) from another molecule (which becomes oxidized). *81*

oxyhemoglobin In red blood cells only, hemoglobin with bound oxygen. *606*

ozone layer An atmospheric layer of high ozone concentration. *758*

pain Perception of injury. *515*

pain receptor A sensory receptor that detects tissue damage. *514*

pancreas Glandular organ that secretes enzymes and bicarbonate into the small intestine; secretes the hormones insulin and glucagon into the blood. *536*

Pangea Supercontinent that formed about 237 million years ago and broke up 152 million years ago. *250*

parapatric speciation A speciation model in which different habitats spur divergences between parts of a single population. *283*

parasite An organism that obtains some or all the nutrients it needs from a living host, which it usually does not kill outright. *716*

parasitism Interaction in which a parasitic species benefits as it exploits and harms (but usually does not kill) the host. *708*

parasitoid A type of insect that, in a larval stage, grows inside a host (usually another insect), feeds on its tissues, and kills it. *716*

parasympathetic neuron A neuron of the autonomic nervous system. Its signals slow overall activities and divert energy to basic tasks; also works in opposition with sympathetic neurons to make small ongoing adjustments in the activities of internal organs that they both innervate. *504*

parathyroid gland One of four small glands in the back of the thyroid gland; regulates blood calcium level. *534*

parenchyma A simple plant tissue; makes up the bulk of the plant. Its living cells have roles in photosynthesis, storage, and other tasks. *428*

passive transport Diffusion of a solute across a cell membrane, through the interior of a transport protein. *84*

pathogen An agent that infects an organ and causes disease. *308*

pattern formation The process by which a complex body forms from local processes during embryonic development. *215, 659*

PCR Polymerase chain reaction. A method that rapidly generates many copies of a specific DNA fragment. *224*

peat bog Compressed, soggy, acidic mat of accumulated remains of peat mosses. *337*

pedigree Chart showing connections among individuals related by descent. *180*

pellicle A thin, flexible, protein-rich body covering of some single-celled eukaryotes such as euglenoids. *321*

peptide bond A bond between the amino group of one amino acid and the carboxyl group of another. *42*

per capita growth rate The rate obtained by subtracting a population's per capita death rate from per capita birth rate. *690*

periderm Layer that replaces epidermis on older stems and roots. *429*

periodic table of the elements Tabular arrangement of the elements by atomic number. *22*

peripheral nervous system All spinal and cranial nerves, branches of which extend through the body. *493*

peripheral vasoconstriction A narrowing of arterioles in the skin; blood flow to the body surface decreases. *647*

peripheral vasodilation A widening of the blood vessels in the skin; blood flow to the body surface increases. *646*

peritubular capillaries A set of blood capillaries that surround the tubular parts of a kidney nephron. *641*

permafrost A perpetually frozen layer that underlies arctic tundra. *771*

peroxisome Enzyme-filled vesicle that breaks down amino acids, fatty acids, and toxic substances such as ethanol. *63*

pH scale A measure of the acidity or alkalinity of a solution. pH 7 is neutral. *30*

pharynx Tube from oral cavity to the gut. In land vertebrates, it is the entrance to the esophagus and trachea. *368, 605, 621*

phenotype An individual's traits. *155*

pheromone Signaling molecule secreted by one individual that affects another of the same species; has roles in social behavior. *516, 528*

phloem Plant vascular tissue; distributes photosynthetic products through the plant body. *334, 429, 450*

phospholipid A lipid with a phosphate group in its hydrophilic head, and two nonpolar fatty acid tails; main constituent of cell membranes. *41*

phosphorus cycle A sedimentary cycle. Phosphorus (mainly phosphate) moves from land, through food webs, to ocean sediments, then back to land. *750*

phosphorylation Transfer of a phosphate group to a recipient molecule. *76*

photoautotroph Photosynthetic autotroph; e.g., nearly all plants, most algae, and a few bacteria. *94*

photolysis Reaction in which light energy breaks down a molecule. Photolysis of water molecules during noncyclic photosynthesis releases electrons and hydrogen ions used in the reactions, and molecular oxygen. *98*

photoperiodism Biological response to change in the relative lengths of daylight and darkness. *470*

photoreceptor A light-sensitive sensory receptor of invertebrates and vertebrates. *514*

photorespiration Rubisco-mediated reaction that occurs in the cells of a C4 plant when stomata close and oxygen levels rise; causes sugar production to be inefficient, and the cells lose carbon instead of fixing it. *102*

photosynthesis The metabolic pathway by which photoautotrophs capture light energy and use it to make sugars from CO_2 and water. *6*

photosystem In photosynthetic cells, a cluster of pigments and proteins that, as a unit, converts light energy to chemical energy in the first step of photosynthesis. *97*

phototropism Change in the direction of cell movement or growth in response to a light source. *468*

phytochrome A light-sensitive pigment that helps set plant circadian rhythms based on length of night. *470*

pigment An organic molecule that absorbs light of certain wavelengths. Reflected light imparts a characteristic color. *94*

pilomotor response Hairs or feathers become erect in response to cold. *647*

pilus, plural **pili** A protein filament that projects from the surface of some bacterial cells. *58*

pineal gland Light-sensitive, melatonin-secreting endocrine gland in the brain. *538*

pioneer species An opportunistic colonizer of barren or disturbed habitats. Adapted for rapid growth and dispersal. *718*

pituitary gland Vertebrate endocrine gland located inside the brain; interacts with the hypothalamus to control physiological functions, including activity of many other glands. Its posterior lobe stores and secretes hormones from the hypothalamus; anterior lobe makes and secretes its hormones. *532*

placenta In a placental mammal, organ that forms during pregnancy from maternal tissue and extraembryonic membranes. Allows a mother to exchange substances with a fetus but keeps their blood separate. *401, 677*

placozoan Simplest known animal, with no symmetry and just two cell layers. *364*

plankton Mostly microscopic autotrophs and heterotrophs of aquatic habitats. *322*

plant A multicelled photoautotroph, most with well-developed roots and shoots. The primary producers on land. *8*

plaque On teeth, a biofilm composed of bacteria, their extracellular products, and saliva glycoproteins. *591*

plasma Liquid portion of blood; mainly water with proteins, sugars, and other solutes, and dissolved gases. *562*

plasma membrane Outer cell membrane; encloses the cytoplasm. *52*

plasmid A small, circular DNA molecule in bacteria, replicated independently of the chromosome, having a few genes. *222, 307*

plate tectonics theory Plates of Earth's outer layer of rock float on the semi-molten mantle; they move slowly and have rafted continents to new positions over time. *250*

platelet Cell fragment that circulates in blood and functions in clotting. *563*

pleiotropy Effects of a single gene on two or more traits. *161*

polarity Any separation of charge into distinct positive and negative regions. *27*

pollen grain In seed-bearing plants, tiny structure that develops from a microspore; consists of a wall around a few cells that will develop into a mature, sperm-bearing, male gametophyte. *335, 456*

pollination The arrival of pollen on a receptive stigma of a flower. *341, 458*

pollination vector Any agent that moves pollen grains from one plant to another; e.g., wind, pollinators. *456*

pollinator An animal that delivers pollen grains from one plant to another; facilitates pollination; e.g., birds, bats, insects. *344*

pollutant A natural or synthetic substance released into soil, air, or water in greater than normal amounts; it disrupts natural processes because organisms evolved in its absence, or are adapted to lower levels. *758*

polygenic inheritance Inheritance pattern in which multiple genes affect the same heritable trait. *164*

polymer Large molecule of multiple linked monomers. *37*

polypeptide chain A chain of amino acids linked by peptide bonds. *42*

polyploid Having three or more of each type of chromosome characteristic of the species. *178, 282*

pons Hindbrain traffic center for signals between cerebellum and forebrain. *506*

population A group of individuals of the same species in a specified area. *5, 266*

population density Number of individuals of a population in a specified volume or area of a habitat. *688*

population distribution The pattern in which individuals of a population are dispersed through their habitat. *688*

population size The number of individuals that actually or potentially contribute to the gene pool of a population. *688*

positive feedback mechanism An activity changes some condition, which in turn triggers a response that intensifies the change. *417, 496*

predation Ecological interaction in which a predator kills and eats prey. *708*

predator A heterotroph that eats other living organisms (its prey). *712*

prediction A statement, based on some hypothesis, about a condition that should exist if the hypothesis is not wrong; often called the "if–then process." *12*

pressure flow theory Idea that organic molecules flow through phloem of vascular plants in response to pressure gradients between sources (where organic molecules are loaded into sieve tubes) and sinks (where the molecules are used or stored). *451*

pressure gradient Difference in pressure between two adjoining regions. *82*

prey An organism that a predator kills and eats. *712*

primary growth Plant growth from apical meristems in root and shoot tips. *427*

primary oocyte Of human females, an immature egg that is arrested in prophase I of meiosis until eight to ten hours before being released from an ovary. *666*

primary producer An autotroph at the first trophic level of an ecosystem. *734*

primary productivity The rate at which an ecosystem's primary producers secure and store energy in tissues. *740*

primary succession A community arises and species arrive and replace one another over time in an environment that was without soil, such as a newly formed island. *718*

primary wall The first thin, pliable wall of young plant cells. *66*

primate A type of mammal; a prosimian or an anthropoid. *402*

primer Short, single strand of DNA that is designed to hybridize with a template; DNA polymerases initiate synthesis at primers during PCR and DNA sequencing. *225*

prion A type of protein found in vertebrate nervous systems; becomes infectious when its shape changes. *314*

probability Measure of the chance that some outcome will occur; proportional to the total number of ways in which that outcome can be reached. *156*

probe Short fragment of DNA labeled with a tracer; designed to hybridize with a nucleotide sequence of interest. *224*

producer An autotroph. Most often a photosynthetic organism. *6*

product A molecule remaining at the end of a reaction. *75*

progesterone Sex hormone secreted by ovaries and the corpus luteum. *666*

proglottid One of many tapeworm body units that bud behind the scolex. *369*

programmed life span hypothesis The idea that life span is genetically set. *682*

prokaryotic cell Prokaryote. A single-celled organism in which the DNA is not contained in a nucleus; a bacterium or an archaean. *52*

prokaryotic chromosome Double-stranded, circular molecule of DNA together with a few attached proteins. *307*

prokaryotic fission Cell reproduction mechanism of prokaryotic cells. *307*

prolactin Hormone that induces synthesis of enzymes used in milk production. *681*

promoter Nucleotide sequence in DNA to which RNA polymerase binds. *198*

prophase Stage of mitosis and meiosis in which chromosomes condense and become attached to a newly forming spindle. *130*

protein Organic compound that consists of one or more polypeptide chains. *42*

proteobacteria A group of Gram-negative bacteria; the most diverse monophyletic group of prokaryotic cells. *308*

protists Informal name for eukaryotes that are not plants, fungi, or animals. *8, 320*

protocell A membrane-enclosed sac of molecules that captures energy, engages in metabolism, concentrates materials, and replicates itself. Presumed stage of chemical evolution that preceded living cells. *295*

proton Positively charged subatomic particle in the nucleus of all atoms. The number of protons defines an element. *22*

protostome A bilateral animal belonging to a lineage characterized in part by having the first indentation to form on the early embryo's surface become the mouth; e.g., mollusks, annelids, arthropods. *363*

proximal tubule Tubular portion of a nephron closest to Bowman's capsule. *641*

pseudopod A dynamic lobe of membrane-enclosed cytoplasm; functions in motility and phagocytosis by amoebas, amoeboid cells, and phagocytic white blood cells. *69*

puberty For humans, the post-embryonic stage when gametes start to mature and secondary sexual traits emerge. *538, 660*

pulmonary circuit Cardiovascular route in which oxygen-poor blood flows to lungs from the heart, gets oxygenated, then flows back to the heart. *561*

punctuated equilibrium Idea that most evolutionary change tends to occur over a brief time span, followed by long periods of little change. *See also* gradualism. *284*

Punnett square A diagram used to predict probable outcomes of a genetic cross. *157*

pyruvate Three-carbon compound that forms as an end product of glycolysis. *108*

quadrat One of a number of sampling areas of the same size and shape used to estimate population size. *689*

radial symmetry Animal body plan with parts arranged on a central axis like the spokes of a wheel. *362*

radioactive decay Process by which atoms of a radioisotope spontaneously emit energy and subatomic particles when their nucleus disintegrates. *23*

radioisotope Isotope with an unstable nucleus; decays into predictable daughter elements at a predictable rate. *23*

radiolarian Single-celled predatory protist with pseudopods that project through its perforated silica shell. Most drift in open upper ocean waters. *322*

radiometric dating Method of calculating the age of a rock or fossil by measuring the content and proportions of a radioisotope and its daughter elements. *248*

rain shadow Reduction in rainfall on the side of a high mountain range facing away from prevailing wind; results in arid or semiarid conditions. *761*

ray-finned fish A bony fish having fin supports derived from skin, a swim bladder, and thin, flexible scales. *391*

reactant Molecule that enters a reaction. *75*

receptor A molecule or structure that can respond to a form of stimulation such as light energy, or to binding of a signaling molecule such as a hormone. *6, 56*

recessive With regard to an allele, having effects that are masked by a dominant allele on the homologous chromosome. *155*

recognition protein Plasma membrane protein that identifies a cell as belonging to *self* (one's own body tissue). *56*

recombinant DNA A DNA molecule that contains genetic material from more than one organism. *222*

rectum Last part of the mammalian gut that stores feces before their expulsion. *621*

red alga An aquatic, usually multicelled autotrophic protist with an abundance of phycobilins. *328*

red blood cell Erythrocyte; hemoglobin-containing blood cell; transports oxygen. *562*

red marrow Site of blood cell formation in the spongy tissue of many bones. *548*

reflex Simple, stereotyped movement in response to a stimulus; sensory neurons synapse on motor neurons in the simplest reflex arcs. *503*

releaser Hypothalamic signaling molecule that enhances secretion of a hormone by the anterior pituitary. *532*

replacement model Idea that modern humans arose from a single *Homo erectus* population in sub-Saharan Africa within the past 200,000 years, then spread and replaced other hominids. *406*

repressor Transcription factor that blocks transcription by binding to a (eukaryotic) promoter or (prokaryotic) operator. *210*

reproduction An asexual or sexual process by which a parent cell or organism produces offspring. *7*

reproductive base The number of actually and potentially reproducing individuals of a population. *688*

reproductive isolating mechanism Any heritable feature of body form, function, or behavior that prevents interbreeding between species; aspect of speciation. *278*

reptile Not a formal taxon; amniotes that do not have features of birds or mammals; e.g., turtle, lizard. *394*

resource partitioning Use of different parts of a resource; permits two or more similar species to coexist in a habitat. *711*

respiration The sum of physiological processes that move O_2 from surroundings to metabolically active tissues in the body and CO_2 from tissues to the outside. *600*

respiratory cycle One inhalation and one exhalation. *608*

respiratory membrane Fused-together alveolar and blood capillary epithelia and the basement membrane in between; the respiratory surface in a human lung. *606*

respiratory protein A protein with one or more metal ions that binds O_2 in oxygen-rich animal tissues and gives it up where O_2 levels are lowest; e.g., hemoglobin. *600*

respiratory surface Any thin, moist body surface that functions in gas exchange. *600*

resting membrane potential The voltage difference across the plasma membrane of a neuron or other excitable cell that is not receiving outside stimulation. *494*

restriction enzyme Type of enzyme that cuts specific base sequences in DNA. *222*

retina In vertebrate and many invertebrate eyes, a tissue packed with photoreceptors and interwoven with sensory cells. *520*

reverse transcriptase A viral enzyme that catalyzes the assembly of nucleotides into DNA, using RNA as a template. *223*

Rh blood typing Method of determining whether Rh^+, a type of surface recognition protein, is present on an individual's red blood cells; if absent, the cell is Rh^-. *565*

rhizoid A rootlike absorptive structure of some bryophytes. *337*

rhizome A stem that grows horizontally underground. *338*

ribosomal RNA (rRNA) A type of RNA that becomes part of ribosomes; some catalyze formation of peptide bonds. *198*

ribosome Site of protein synthesis. An intact ribosome has two subunits, each composed of rRNA and proteins. *201*

ribozyme A catalytic RNA. *295*

riparian zone The narrow corridor of vegetation along a stream or river. *729*

river A flowing-water ecosystem that starts when streams converge. *773*

RNA Ribonucleic acid. Type of nucleic acid, typically single-stranded; important for transcription and translation; some are catalytic. *See also* ribosomal RNA, transfer RNA, messenger RNA, ribozyme. *47*

RNA interference Type of control over translation in which double-stranded RNA is chopped up, and mRNA complementary in sequence to the pieces is destroyed. *211*

RNA polymerase Enzyme that catalyzes transcription of DNA into RNA. *198*

RNA world Hypothetical time prior to the evolution of DNA in which RNA molecules stored genetic information and catalyzed protein synthesis. *295*

rod cell Vertebrate photoreceptor that detects very dim light; contributes to the coarse perception of movement. *522*

root Typically belowground plant part. It absorbs water and minerals, often anchors aboveground parts and stores food. *426*

root hair Hairlike, absorptive extension of a young cell of root epidermis. *435, 444*

root nodule Mutualistic association of nitrogen-fixing bacteria and roots of some legumes and other plants; infection leads to a localized tissue swelling. *444*

roundworm Bilateral invertebrate with a false coelom and complete digestive system in an unsegmented body. Most are decomposers; some are parasites. *374*

rubisco Ribulose biphosphate carboxylase, or RuBP. Carbon-fixing enzyme of light-independent photosynthesis reactions. *101*

ruminant Hoofed, herbivorous mammal that has multiple stomach chambers. *619*

runoff The water that flows into streams when the ground is saturated. *742*

sac fungus Fungus that produces sexual spores in sac-shaped cells. *354*

salinization Salt buildup in soil. *742*

salt Compound that dissolves easily in water and releases ions other than H^+ and OH^-. *31*

sampling error Difference between results derived from testing an entire group of events or individuals, and results derived from testing a subset of the group. *16*

sapwood Of an older stem or root, the moist secondary growth between the vascular cambium and heartwood. *437*

sarcomere One of many basic units of contraction along the length of a muscle fiber. It shortens by ATP-driven interactions between its parallel arrays of actin and myosin components. *552*

sarcoplasmic reticulum Specialized ER that forms flattened, membrane-bound chambers around muscle fibers; takes up, stores, and releases calcium ions. *554*

science Systematic study of nature. *11*

scientific theory Hypothesis that has not been disproven after many years of rigorous testing, and is useful for making predictions about other phenomena. *12*

sclerenchyma Simple plant tissue; dead at maturity, its lignin-reinforced cell walls structurally support plant parts. *428*

seamount Extinct seafloor volcano. *778*

second law of thermodynamics Energy tends to disperse spontaneously. *74*

second messenger Molecule produced by a cell in response to binding of a hormone at the plasma membrane; relays a signal into a cell signal; e.g., cyclic AMP. *530*

secondary growth A thickening of older stems and roots at lateral meristems. *427*

secondary oocyte A haploid cell produced by the first meiotic division of a primary oocyte; released at ovulation. *667*

secondary succession A community arises and changes over time in a habitat where another community existed previously. *718*

secondary wall Lignin-reinforced wall inside the primary wall of a plant cell. *66*

seed The mature ovule of a seed plant; contains embryo sporophyte. *335, 460*

segmentation In animal body plans, a series of repeated units. In tubular organs, oscillating movement produced by circular muscle in the wall. *363, 623*

segregation Mendelian theory that two genes on a pair of homologous chromosomes are separated from each other at meiosis, eventually to end up in different gametes. *157*

selective gene expression Different cell lineages of a multicelled individual express different subsets of genes; outcome of gene controls; basis of differentiation. *658*

selective permeability Property of cell membranes; some substances, but not others, can cross a cell membrane. *82*

selfish herd Animal group that forms when individuals each attempt to hide in the midst of others. *792*

senescence Of multicelled organisms, the phase in a life cycle from maturity until death; also applies to death of parts, such as plant leaves. *472*

sensation Conscious awareness of a stimulus. *514*

sensory adaptation Sensory neurons stop responding to a constant stimulus. *514*

sensory neuron Type of neuron that detects a stimulus and relays information about it toward an integrating center. *492*

sensory receptor Cells or cell parts that detect stimuli (specific forms of energy). *416*

sensory system Collectively, all sensory cells of a nervous system that detect and report information about external and internal stimuli to integrating centers. *514*

sex chromosome Member of a pair of chromosomes that differs among males and females. *170*

sexual dimorphism Distinct female and male phenotypes. *274*

sexual reproduction Production of genetically variable offspring by gamete formation and fertilization. *140, 652*

sexual selection Mode of natural selection in which some individuals outreproduce others of a population because they are better at securing mates. *274*

sexually transmitted disease (STD) Disease agent is transferred between individuals by sexual contact; e.g., syphilis, AIDS. *672*

shell model Model for how electrons are distributed in an atom; orbitals are shown as nested circles, electrons as dots. *24*

shivering response Rhythmic tremors in response to cold. *647*

shoot Aboveground plant parts; e.g., stems, leaves, flowers. *426*

short tandem repeat Stretch of DNA that consists of many copies of a short sequence; basis of DNA fingerprinting. *227*

sieve tube Conducting tube in phloem. *429*

sister chromatid One of two attached members of a duplicated eukaryotic chromosome. *126, 141*

six-kingdom classification system System of classification that groups all organisms into kingdoms Bacteria, Archaea, Protista, Fungi, Plantae, and Animalia. *258*

skeletal muscle tissue Contractile tissue that is the functional partner of bone. *482*

sliding-filament model Model for how the sarcomeres of muscle fibers contract. ATP-activated myosin heads repeatedly bind actin filaments (attached to Z lines) and tilt in short power strokes that slide actin toward the sarcomere's center. *553*

slime mold An amoebozoan; amoeba-like cells that cluster into a mass, differentiate, and form reproductive structures. *329*

small intestine Part of the vertebrate gut in which digestion is completed and from which most nutrients are absorbed. *621*

smog Atmospheric condition in which winds cannot disperse airborne pollutants trapped by a thermal inversion. *758*

smooth muscle tissue Contractile tissue in the wall of soft internal organs. *483*

social parasite An animal that takes advantage of its host's behavior, thus harming it; e.g., cuckoo. *716*

sodium–potassium pump Cotransporter that actively moves sodium out of a cell and passively moves potassium into it. *85*

softwood Wood with tracheids, no fibers or vessels; less dense than hardwood. *437*

soil Mixture of various mineral particles (sand, silt, clay) and decomposing organic matter (humus). *442, 764*

soil erosion A loss of soil under the force of wind and water. 443

soil profile Distinct soil layers that form over time in a biome. 764

solar tracking Circadian response; a plant part changes position in response to the sun's changing angle through the day. 470

solute A dissolved substance. 29

solvent Substance, typically a liquid, that dissolves other substances; e.g., water. 29

somatic nerve Nerve carrying information from skin, tendons, and skeletal muscles into the central nervous system. 504

somatic sensation Perception of stimuli detected by receptors located throughout the body; e.g., touch, warmth. 514

somite One of many paired segments in a vertebrate embryo that gives rise to most bones, skeletal muscles of the head and trunk, and the dermis. 676

sorus, plural **sori** Cluster of spore-forming chambers on underside of a fern frond. 339

speciation Formation of daughter species from a population or subpopulation of a parent species; the routes vary in their details and duration. 278

species A type of organism. Of sexually reproducing species, one or more groups of individuals that potentially can interbreed, produce fertile offspring, and are isolated reproductively from other groups. 8, 240

species diversity The number and relative abundances of species in a habitat or any other specified region. 708

sperm Mature male gamete. 146, 656

sphere of hydration A clustering of water molecules around a solute. 29

sphincter A ring of muscles that alternately contracts and relaxes, which closes and opens a passageway between two organs. 621

spinal cord The part of a central nervous system inside a vertebral canal. 505

spindle *See* bipolar spindle.

spirochete A motile, parasitic or symbiotic bacterium; looks like a stretched spring. 309

spleen The largest lymphoid organ, with phagocytic white blood cells and B cells; filters antigen and used-up platelets and worn-out or dead red blood cells. In embryos only, a site of red blood cell formation. 577

sponge Filter-feeding aquatic invertebrate with no symmetry, no tissues. 365

sporophyte Diploid, spore-producing body of a plant or multicelled alga. 146, 334, 456

spring overturn In temperate zone lakes, a downward movement of oxygenated surface water and an upward movement of nutrient-rich water in spring. 772

stabilizing selection Mode of natural selection; intermediate phenotypes are favored over extremes. 272

stamen The male reproductive part of a flowering plant; consists of a pollen-producing anther on a filament. 456

statolith Organelle that acts as a gravity-sensing mechanism. 468

stem cell Self-perpetuating, undifferentiated animal cell. A portion of its daughter cells become specialized. 562

sterol Lipid with a rigid backbone of four carbon rings and no fatty acid tails. 41

stimulus A specific form of energy that activates a sensory receptor able to detect it; e.g., pressure. 514, 786

stoma, plural **stomata** Gap that opens between two guard cells; lets water vapor and gases diffuse across the epidermis of a leaf or primary stem. 102, 334

stomach Muscular, stretchable sac; mixes and stores ingested food and helps break it apart mechanically and chemically. 621

stramenopile A single-celled or multicelled eukaryote with filaments on one of its two flagella; e.g., a water mold. 325

stratification Formation of sedimentary rock layers by deposition and compaction of silt, sand, and other particles. 247

stream A flowing-water ecosystem that starts out as a freshwater spring. 773

stroma The semifluid matrix between the thylakoid membrane and the two outer membranes of a chloroplast; site of light-independent photosynthesis reactions. 97

stromatolite Fossilized dome-shaped mats of aquatic photoautotrophic bacteria. 296

substance P A neuromodulator that enhances pain perception. 500

substrate A reactant molecule that is specifically acted upon by an enzyme. 76

substrate-level phosphorylation The direct transfer of a phosphate group from a substrate to ADP; forms ATP. 110

supernatural Not explainable by any known force or process of nature. 11

surface-to-volume ratio A relationship in which the volume of an object increases with the cube of the diameter, but the surface area increases with the square. 52

survivorship curve Plot of age-specific survival of a cohort, from the time of birth until the last individual dies. 695

swim bladder Adjustable flotation sac of some bony fish. 390

symbiosis Ecological interaction in which members of two species live together or otherwise interact closely; e.g., mutualism, parasitism, commensalism. 708

sympathetic neuron A neuron of the autonomic nervous system. Its signals cause increases in overall activities in times of stress or heightened awareness. Also works in opposition with parasympathetic neurons to make small ongoing adjustments in activities of internal organs they both innervate. 504

sympatric speciation A speciation model in which speciation occurs in the absence of a physical barrier; e.g., by polyploidy in flowering plants. 282

synapsid An amniote lineage of early mammal-like reptiles and mammals. 394

synaptic integration The summation of excitatory and inhibitory signals that arrive at an excitable cell's input zone at the same time. 499

syndrome The set of symptoms that characterize a medical condition. 180

system acquired resistance Of some plants, a mechanism that induces cells to produce and release compounds that will protect tissues from attack. 418

systemic circuit Cardiovascular route in which oxygenated blood flows from the heart through the rest of the body, where it gives up oxygen and takes up carbon dioxide, then flows back to the heart. 561

T lymphocyte T cell. White blood cell that regulates vertebrate immune responses by way of cytokines; cytotoxic T cells carry out cell-mediated immunity. 582

taproot system In eudicots, a primary root and all of its lateral branchings. 434

taste receptor A chemoreceptor that detects solutes in the fluid bathing it. 516

taxon, plural **taxa** An organism or set of organisms. 258

TCR (T cell receptor) Antigen-binding receptor on the surface of T cells; also recognizes MHC self-markers on body cells. 586

telomere A cap of repetitive DNA sequences on the end of a chromosome. Enzymes digest a bit of it during each nuclear division; cells stop dividing when it gets too short. 682

telophase Stage of mitosis during which chromosomes arrive at the spindle poles and decondense, and new nuclei form. 131

temperature Measure of molecular motion. 28

tendon A cord or strap of dense connective tissue that attaches a muscle to bone. 550

terminal bud Bud at a shoot tip; shoot's main zone of primary growth. 430

test Make systematic observations or conduct experiments. 12

testcross Method of determining unknown genotype; cross between an individual of unknown genotype and a homozygous recessive individual; the proportions of phenotypes in offspring are analyzed. 157

testis, plural **testes** One of a pair of male gonads where sperm form by meiosis; secretes the hormone testosterone. *660*

testosterone A sex hormone necessary for the development and functioning of the male reproductive system. *660*

tetanus Motor unit response to repeated stimulation; a strong prolonged contraction. Also, a disease caused by bacteria in which muscles stay contracted. *554*

thalamus Forebrain region; a coordinating center for sensory input and a relay station for signals to the cerebrum. *506*

theory of inclusive fitness The idea that genes associated with altruism are adaptive if they cause behavior that promotes the reproductive success of an altruist's closest relatives. *794*

theory of uniformity Idea that gradual repetitive processes occurring over long time spans shaped Earth's surface. *243*

thermal radiation Emission of heat from any object. *645*

thermocline Thermal stratification in a large body of water; a cool midlayer stops vertical mixing between warm surface water above it and cold water below it. *772*

thermoreceptor Type of sensory cell that detects a change in temperature. *514*

thigmotropism Redirected growth of a plant in response to contact with a solid object; e.g., vine curling around a post. *469*

three-domain system Classification system that groups all organisms into the domains Bacteria, Archaea, and Eukarya. *258*

threshold level Of a neuron, the voltage difference across the plasma membrane required for gated Na^+ channels to open and start an action potential. *496*

thylakoid membrane A chloroplast's inner membrane system, often folded as flattened sacs, that forms a continuous compartment in the stroma. In the first stage of photosynthesis, pigments and enzymes in the membrane function in the formation of ATP and NADPH. *97*

thymine (T) A type of nitrogen-containing base in nucleotides; also, a nucleotide with a thymine base. Base-pairs with adenine; does not occur in RNA. *190*

thymus gland Lymphoid organ located beneath the sternum; T cells formed in bone marrow move to it and mature under the influence of its hormonal secretions. *577*

thyroid gland Endocrine gland located at base of neck; its hormones influence growth, development, metabolic rate. *534*

tidal volume The volume of air that flows into and out of the lungs during a single inhalation and exhalation. *608*

tight junction Array of fibrous proteins that joins epithelial cells; collectively, these cell junctions prevent fluids from leaking between cells in epithelial tissues. *479*

tissue In multicelled organisms, a group of cells and matrixes interacting in the performance of one or more tasks. *412*

tissue culture propagation Laboratory method for cloning an entire plant from one cell. *463*

tooth In jawed vertebrates, a hardened appendage used to cut, shred, pierce, or pummel food. *621*

topsoil Uppermost soil layer; has the most nutrients for plant growth. *443*

total fertility rate (TFR) For humans, the average number of children born to a female during her lifetime. *700*

tracer A molecule with a detectable label attached; researchers can track it after delivering it into a cell or other system. *23*

trachea Airway that connects the pharynx with the bronchi leading to the lungs. *605*

tracheal system In insects and some other land arthropods, branching tubes that start at the body surface and end near cells; role in gas exchange. *601*

tracheid Type of tapered cell in xylem, dead at maturity; its perforated wall forms part of a water-conducting tube. *446*

transcription Process by which an RNA is assembled from nucleotides using a gene region in DNA as a template. First step in protein synthesis. *198*

transcription factor Regulatory protein that influences transcription; e.g., activator, repressor. *211*

transfer RNA (tRNA) Type of RNA that delivers amino acids to a ribosome during translation. Its anticodon pairs with an mRNA codon. *198*

transgenic organism Organism that has been genetically engineered to contain a gene from a different species. *230*

transition state The state in a chemical reaction when reactant bonds are at the breaking point. *77*

translation At ribosomes, information encoded in an mRNA guides synthesis of a polypeptide chain from amino acids. Second stage of protein synthesis. *202*

translocation Attachment of a piece of a broken chromosome to another chromosome. Also, a mechanism that moves organic compounds through phloem. *176, 450*

transpiration Evaporative water loss from a plant's aboveground parts. *446*

transporter Protein that spans a cell membrane and helps specific solutes move across the bilayer. *56*

transposable element Small DNA segment that can move spontaneously to a different, random location in the genome. *204*

triglyceride A lipid with three fatty acid tails attached to a glycerol backbone. *40*

trophic level All organisms that are the same number of transfer steps away from the energy input into an ecosystem. *734*

tubular reabsorption A process by which peritubular capillaries reclaim water and solutes that leak or are pumped out of a nephron's tubular regions. *642*

tubular secretion Transport of H^+, urea, other solutes out of peritubular capillaries and into nephrons for excretion. *642*

tumor Tissue mass of cells dividing at an abnormally high rate. Benign tumor cells stay in their home tissue; malignant ones metastasize, or slip away and invade other places in the body, where they may start new tumors. *See also* neoplasm. *134*

tunicate Filter-feeding, invertebrate chordate enclosed in a baglike secreted covering as an adult. *386*

turgor Hydrostatic pressure. Pressure that a fluid exerts against a wall, membrane, or some other structure that contains it. *86*

ultrafiltration In a capillary bed, pressure generated by the beating heart forces some protein-free plasma out of a blood capillary, into interstitial fluid. *573*

upwelling Upward movement of cool water from the depths, as when winds blow surface water away from a coast. *778*

uracil (u) A type of nitrogen-containing base in nucleotides; also, a nucleotide with a uracil base. Base-pairs with adenine; does not occur in DNA. *198*

urea Nitrogen-containing waste excreted in urine; forms in the liver when ammonia combines with CO_2. *638*

ureter A urine-conducting tube from each kidney to the urinary bladder. *640*

urethra Tube that drains the urinary bladder; opens at the body surface. *640*

urinary system Vertebrate organ system that adjusts the blood's volume and composition; rids the body of metabolic waste. *638*

urine Fluid consisting of excess water, wastes, and solutes; forms in kidneys by filtration, reabsorption, and secretion. *638*

uterus In a female placental mammal, a muscular, pear-shaped organ in which embryos are housed and nurtured during pregnancy. *664*

vaccine A type of antigen-containing preparation introduced into the body to prime the immune system to recognize the antigen before actual infection. *592*

vagina In female mammals, the organ that receives sperm, forms part of birth canal, and channels menstrual flow. *664*

variable In experiments, a characteristic or event that differs among individuals and that may change over time. *13*

vascular bundle Multistranded, sheathed cord of primary xylem and phloem in a stem or leaf. *431*

vascular cambium A lateral meristem that forms in older stems or roots. *436*

vascular cylinder Sheathed, cylindrical array of primary xylem and phloem in a root. *435, 445*

vascular tissue system All xylem and phloem in plants that are structurally more complex than bryophytes. *334, 426*

vector An insect or some other animal that carries a pathogen between hosts; e.g., a mosquito that transmits malaria. *See also* cloning vector. *321, 716*

vegetative growth Growth of a new plant from an extension or fragment. *462*

vein In plants, a vascular bundle in a stem or leaf. In animals, a large-diameter vessel that carries blood toward the heart. *433, 570*

venule A small blood vessel that connects several capillaries to a vein. *570*

vernalization Stimulation of flowering in spring by low temperature in winter. *471*

vertebra, plural **vertebrae** One of a series of hard bones that protects the spinal cord and forms the backbone. *388, 546*

vertebrate Animal with a backbone. *386*

vesicle Small, membrane-enclosed, saclike organelle in cytoplasm; different kinds store, transport, or degrade their contents. *62*

vessel member Type of cell in xylem, dead at maturity; its perforated wall forms part of a water-conducting tube. *446*

vestibular apparatus In vertebrates, an organ of equilibrium located in the inner ear. *517*

villus, plural **villi** One of many fingerlike multicellular projections that increases the surface area of some tissues in an animal body; e.g., small intestinal villi. *624*

viroid Infectious RNA molecule; most infect plants. *314*

virus Noncellular infectious particle that consists of DNA or RNA, a protein coat and, in some types, an outer lipid envelope; it can be replicated only after its genetic material enters a host cell and subverts the host's metabolic machinery. *312*

vision Perception of visual stimuli based on light focused on a retina and image formation in the brain. *520*

visual accommodation Adjustments in a lens position or shape that focus light rays on the retina. *521*

visual field The portion of the outside world that an animal sees. *520*

vital capacity Maximum volume of air that can be actively exhaled in a single breath after a maximal inhalation. *608*

vitamin Any organic substance that an organism requires in trace amounts but that it generally cannot produce. Many function as coenzymes. *630*

vomeronasal organ In many vertebrates, a cluster of sensory neurons in the nasal cavity that responds to pheromones. *516*

warning coloration In many toxic species and their mimics, bright colors, patterns, and other signals that predators learn to recognize and avoid. *714*

watershed A region of any specified size in which all precipitation drains into one stream or river. *742*

water–vascular system In echinoderms, a system of tube feet connected to canals; functions in movement, food handling. *381*

wavelength Distance between crests of two successive waves of radiant energy. *54, 94*

wax Water-repellent lipid with long fatty acid tails bonded to long-chain alcohols or carbon rings. *41*

white blood cell Leukocyte. Type of cell that functions in immune responses; e.g., macrophage, dendritic cell, eosinophil, neutrophil, basophil, T cell, B cell. *563*

white matter Outer portion of brain and spinal cord; includes myelinated axons. *505*

X chromosome inactivation Shutdown of one of the two X chromosomes in the cells of female mammals. *See also* Barr body; dosage compensation. *212*

xenotransplantation Transplantation of an organ from one species into another. *232*

x-ray diffraction image Film image of x-rays scattered by a crystalline sample; the resulting pattern of streaks and dots can be used to calculate the spacing between the atoms in the crystal lattice. *190*

xylem Complex tissue of vascular plants; conducts water and solutes through tubes that consist of the interconnected walls of dead cells. *334, 428*

yellow marrow In most mature bones, a fatty tissue that fills interior cavities; can convert to blood cell-producing red marrow if necessary. *548*

yolk Protein- and lipid-rich substance in many eggs; serves as the first food source for a developing embryo. *653*

zero population growth No net increase or decrease in population size during a specified interval. *690*

zygote Cell formed by fusion of gametes; first cell of a new individual. *140*

zygote fungi Fungal group; sexual spores are produced in a zygospore that forms after hyphae of opposite mating types meet and fuse; e.g., black bread mold. *353*

Art Credits and Acknowledgments

This page constitutes an extension of the book copyright page. We have made every effort to trace the ownership of all copyrighted material and secure permission from copyright holders. In the event of any question arising as to the use of any material, we will be pleased to make the necessary corrections in future printings. Thanks are due to the following authors, publishers, and agents for permission to use the material indicated.

Page i © Copyright 2003–2005 Minden Pictures.

TABLE OF CONTENTS **Page iv** left, upper, © Raymond Gehman/ Corbis; lower, © Bill Beatty/ Visuals Unlimited. **Page v** from left, © R. Calentine/ Visuals Unlimited; © Martin Barraud/ Stone/ Getty Images; © Larry West/ FPG/ Getty Images. **Page vi** from top, © Jim Cummins/ Corbis; Ed Reschke; © Ron Neumeyer, www.microimaging.ca. **Page vii** from left, © George Lepp/ Corbis; © Lauren Shear/ Photo Researchers, Inc.; © Dr. William Strauss; Photos by Victor Fisher, courtesy Genetic Savings & Clone. **Page viii** from top, © Courtesy of Golden Rice Humanitarian Board; Courtesy of Stan Celestian/ Glendale Community College Earth Science Image Archive; © Alan Solem. **Page ix** from left, © Chase Studios/ Photo Researchers, Inc.; © CAMR, Barry Dowsett/ Photo Researchers, Inc.; © Russell Knightly/ Photo Researchers, Inc.; © John Clegg/ Ardea, London. **Page x** from top, © R. J. Erwin/ Photo Researchers, Inc.; © Robert C. Simpson/ Nature Stock; © Eye of Science/ Photo Researchers, Inc. **Page xi** from left, © Karen Carr Studio/ www.karencarr.com; © Gary Bell/ Taxi/ Getty Images; © Cory Gray. **Page xii** from top, © David Cavagnaro/ Peter Arnold, Inc.; © Andrew Syred/ Photo Researchers, Inc.; © Robert Essel NYC/ Corbis. **Page xiii** from left, © Science Photo Library/ Photo Researchers, Inc.; From Neuro Via Clinicall Research Program, Minneapolis VA Medical Center; © Will & Deni McIntyre/ Photo Researchers, Inc. **Page xiv** from top, © Scott Camazine/ Photo Researchers, Inc.; © Ed Reschke. **Page xv** from left, © National Cancer Institute/ Photo Researchers, Inc.; © Juergen Berger/ Photo Researchers, Inc; © Francois Gohier/ Photo Researchers, Inc. **Page xvi** from top, © Ralph Pleasant/ FPG/ Getty Images; © Gary Head. **Page xvii** from top, © Ralph Pleasant/ FPG/ Getty Images; © Gary Head. **Page xviii** from top, © Bob Jensen Photography; © Jeff Vanuga/ Corbis. **Page xix** from left, Douglas Faulkner/ Sally Faulkner Collection; © Kevin Schafer/ Corbis; © Joseph Sohm, Visions of America/ Corbis.

INTRODUCTION NASA Space Flight Center

CHAPTER 1 **1.1** left, Courtesy of Conservation International; right, © Steve Richards. **1.2** Photo courtesy of Dr. Robert Zingg/ Zoo Zurich. **1.3** (a) Rendered with Atom In A Box, copyright Dauger Research, Inc.; (d) © Science Photo Library/ Photo Researchers, Inc.; (e) © Bill Varie/Corbis; (f–h) © Jeffrey L. Rotman/Corbis; (i) © Peter Scoones; (j–k) NASA. **1.4** above, Photodisc/ Getty Images; below, David Neal Parks. **1.5** © Y. Arthus-Bertrand/ Peter Arnold, Inc. **1.6** Photographs by Jack de Coningh. **1.7** © Jack de Coningh. **1.8** (a) clockwise from top left, © Dr. Richard Frankel; © David Scharf, 1999. All rights reserved; © Susan Barnes; © SciMAT/ Photo Researchers, Inc.; (b) left, © R. Robinson/ Visuals Unlimited, Inc.; right, © Dr. Harald Huber, Dr. Michael Hohn, Prof. Dr. K. O. Stetter, University of Regensburg, Germany; (c) above, left, clockwise from top, © Lewis Trusty/ Animals Animals; © Emiliania Huxleyi photograph, Vita Pariente, scanning electron micrograph taken on a Jeol T330A instrument at Texas A&M University Electron Microscopy center; © Carolina Biological Supply Company; © Oliver Meckes/ Photo Researchers, Inc.; Courtesy of James Evarts; right, © John Lotter Gurling/ Tom Stack & Associates; inset, © Edward S. Ross; below, left, from left, © Robert C. Simpson/ Nature Stock; © Edward S. Ross; right, © Stephen Dalton/ Photo Researchers, Inc. **1.9** left, Photographs courtesy Derrell Fowler, Tecumseh, Oklahoma; right, © Nick Brent. **1.10** (a) © Lester Lefkowitz/ Corbis; (b) Centers for Disease Control and Prevention; (c) © Raymond Gehman/ Corbis. **1.11** top, © Superstock. **1.12** (a) © Matt Rowlings, www.eurobutterflies.com; (b) © Adrian Vallin; (c) © Antje Schulte. **1.13** © Gary Head. **1.14** Scientific Paper; Adrian Vallin, Sven Jakobsson, Johan Lind and Christer Wiklund, Proc. R. Soc. B (2005 272, 1203, 1207). Used with permission of The Royal Society and the author.

Page 19 UNIT I © Wim van Egmond, Micropolitan Museum

CHAPTER 2 **2.1** © Owaki-Kulla/ CORBIS. **2.3** (c) Rendered with Atom In A Box, copyright Dauger Research, Inc. **2.5** (a) © CC Studio/ Photo Researchers, Inc.; (d) Harry T. Chugani, M.D., UCLA School of Medicine. **2.6** above, © Michael S. Yamashita/ CORBIS. **Page 25** © Hubert Stadler/ Corbis. **2.8** left, © Gary Head; center, © Bill Beatty/ Visuals Unlimited. **2.11** (a,b,c, left) PDB file from NYU Scientific Visualization Lab; (b, right) © Steve Lissau/ Rainbow; (c, right) © Dan Guravich/ Corbis. **2.13** (a) © Lester Lefkowitz/ CORBIS. **2.14** © JupiterImages Corporation, art by Lisa Starr. **2.15** left, © Michael Grecco/ Picture Group; right, © W. K. Fletcher/ Photo Researchers, Inc. **2.16** © National Gallery Collection; by kind permission of the Trustees of the National Gallery, London/ CORBIS. **2.17** © R. B. Suter, Vassar College. **Page 33** right, © JupiterImages Corporation.

CHAPTER 3 **3.1** left, © 2002 Charlie Wait/ Stone/ Getty Images; right, © Dr. W. Michaelis/ Universitat Hamburg. **3.2** (a,b), PDB file from NYU Scientific Visualization Lab; (c,d), PDB file from Klotho Biochemical Compounds Declarative Database. **3.4** left, Tim Davis/ Photo Researchers, Inc. **3.8** © Steve Chenn/ CORBIS. **3.9** © David Scharf/ Peter Arnold, Inc. **3.11** left, © Kevin Schafer/ CORBIS. **3.12** left, © ThinkStock/ SuperStock. **Page 41** Kenneth Lorenzen. **3.16** (a–d, bottom) PDB files from NYU Scientific Visualization Lab. **3.17** (b, right) After: *Introduction to Protein Structure*, 2nd ed., Branden & Tooze, Garland Publishing, Inc.; (c, left) PDB ID: 1BBB; Silva, M. M., Rogers, P. H., Arnone, A.; A third quaternary structure of human hemoglobin A at 1.7-Å resolution; J Biol Chem 267 pp. 17248 (1992); (c, right) After: *Introduction to Protein Structure*, 2nd ed., Branden & Tooze, Garland Publishing, Inc. **3.17** PDB ID: 1BBB; Silva, M. M., Rogers, P. H., Arnone, A.; A third quaternary structure of human hemoglobin A at 1.7-Å resolution; J Biol Chem 267 pp. 17248 (1992). **3.19** (a,b) PDB files from New York University Scientific Visualization Center; (c) © Dr. Gopal Murti/ SPL/ Photo Researchers, Inc.; (d) Courtesy of Melba Moore. **3.20** PDB files from Klotho Biochemical Compounds Declarative Database. **3.22** PDB ID:1BNA; H. R. Drew, R. M. Wing, T. Takano, C. Broka, S. Tanaka, K. Itakura, R. E. Dickerson; Structure of a B-DNA Dodecamer. Conformation and Dynamics; PNAS V. 78 2179, 1981. **3.23** right, © Professor P. Motta/ Department of Anatomy/ University La Sapienca, Rome/ SPL/ Photo Researchers, Inc. **3.24** left, PDB ID: 1AKJ; Gao, G. F., Tormo, J., Gerth, U. C., Wyer, J. R., McMichael, A. J., Stuart, D. I., Bell, J. I., Jones, E. Y., Jakobsen, B. K.; Crystal structure of the complex between human CD8alpha(alpha) and HLA-A2; Nature 387 pp. 630 (1997); right, Al Giddings/ Images Unlimited.

CHAPTER 4 **4.1** © Tony Brian and David Parker/ SPL/ Photo Researchers, Inc. **4.2** left, © Armed Forces Institute of Pathology; right, © The Royal Society. **4.6** Photographs: (hummingbird) © Robert A. Tyrrell; (human) © Pete Saloutos/ CORBIS; (redwood) © Sally A. Morgan, Ecoscene/ CORBIS. **4.7** (a) Leica Microsystems, Inc., Deerfield, IL; (b) © Geoff Tompkinson/ SPL/ Photo Researchers, Inc. **4.8** (a,b,d,e) Jeremy Pickett-Heaps, School of Botany, University of Melbourne; (c) © Prof. Franco Baldi. **4.11** (a) K. G. Murti/ Visuals Unlimited; (b) R. Calentine/ Visuals Unlimited; (c) Gary Gaard and Arthur Kelman. **4.12** (a,b); © University of California Museum of Paleontology; (c) © Courtesy Jack Jones, Archives of Microbiology, Vol. 136, 1983, pp. 254–261. Reprinted by permission of Springer-Verlag. **4.13** © Dr. David G. Davies and Peg Dirckx. **4.14** (a) © Micrograph, G. L. Decker; (b) M. C. Ledbetter, Brookhaven National Laboratory. **4.15** (a, top) © Micrograph, Gl L. Decker. **4.16** Micrographs: (a) Stephen L. Wolfe; (c,d) Don W. Fawcett/ Visuals Unlimited, computer enhanced; (e) Gary Grimes, computer enhanced. **4.17** right micrograph, Keith R. Porter. **4.18** © Dr. Jeremy Burgess/ SPL/ Photo Researchers, Inc. **4.20** (c) © Russell Kightley/ Photo Researchers, Inc. **4.21** (a) George S. Ellmore. **4.22** left, © Science Photo Library/ Photo Researchers, Inc.; right, Bone Clones®, www.boneclones.com. **4.23** © ADVANCELL (Advanced In Vitro Cell Technologies; S.L.) www.advancell.com. **4.24** (d) © Dylan T. Burnette and Paul Forscher. **4.26** (a) © Dow W. Fawcett/ Photo Researchers, Inc.; (b) © Mike Abbey/ Visuals Unlimited. **4.27** (a, left) After Stephen L. Wolfe, *Molecular and Cellular Biology*, Wadsworth, 1993; (a, right) © Don W. Fawcett/ Photo Researchers, Inc. **4.28** (a,b) From *Tissue and Cell*, Vol. 27, pp. 421–427, Courtesy of Bjorn Afzelius, Stockholm University.

CHAPTER 5 **5.1** left, © BananaStock/ SuperStock; right, Model by © Dr. David B. Goodin, The Scripps Research Institute; right, © Stockbyte/ SuperStock. **5.2** © Martin Barraud/ Stone/ Getty Images. **5.4** top, © Craig Aurness/

Corbis; bottom, © William Dow/ Corbis. **page 76** © JupiterImages Corporation. **5.7** Hemoglobin models: PDB ID: 1GZX; Paoli, M., Liddington, R., Tame, J., Wilkinson, A., Dodson, G.; Crystal structure of T state hemoglobin with oxygen bound at all four haems. *J.Mol.Biol.*, v256, pp. 775–792, 1996. **5.10** (b) © Scott McKiernan/ ZUMA Press. **5.11** (b) © Perennou Nuridsany/ Photo Researchers, Inc. **5.16** top, © Andrew Lambert Photography/ Science Photo Library/ Photo Researchers; Art, Raychel Ciemma. **5.19** PDB files from NYU Scientific Visualization Lab. **5.20** After: David H. MacLennan, William J. Rice and N. Michael Green, "The Mechanism of Ca2+ Transport by Sarco (Endo) plasmic Reticulum Ca2+-ATPases." *JBC* Volume 272, Number 46, Issue of November 14, 1997 pp. 28815–28818. **5.22** (a) Art, Raychel Ciemma; (b–d) © M. Sheetz, R. Painter, and S. Singer, *J of Cell Biol.*, 70:193 (1976) by permission, The Rockefeller University Press. **5.23** (a) © Gary Head; (b,c) © Perennou Nuridsany/ Photo Researchers, Inc. **5.25** © R. G. W. Anderson, M. S. Brown and J. L. Goldstein. *Cell* 10:351 (1977) **5.26** (a) © Biology Media/ Photo Researchers, Inc. **5.27** Sara Lewis, Tufts University; inset, Model by © Dr. David B. Goodin, The Scripps Research Institute. **5.29** © Frieder Sauer/Bruce Coleman Ltd. **5.30** © Prof. Marcel Bessis/ SPL/ Photo Researchers, Inc.

CHAPTER 6 **6.1** right, © Richard Uhlhorn Photography. **6.2** (a) © Photodisc/ Getty Images. **6.3** top, © Larry West/ FPG/ Getty Images. **6.5** Jason Sonneman. **6.6** (a) © left, Photodisc/ Getty Images. **6.11** left, (a) Courtesy of John S. Russell, Pioneer High School; (b) © Bill Boch/ FoodPix/ JupiterImages Corporation; (c) © Chris Hellier/ Corbis. **6.12** © JupiterImages Corporation. **6.14** (a) © Douglas Faulkner/ Sally Faulkner Collection; (b) © Herve Chaumeton/ Agence Nature. **Page 105** right, © E.R. Degginger; bottom, © JupiterImages Corporation.

CHAPTER 7 **7.1** left, © Professors P. Motta and T. Naguro/ SPL/ Photo Researchers, Inc.; right, © Louise Chalcraft-Frank and FARA. **7.2** left, clockwise from top, © Jim Cummins/ Corbis; © John Lotter Gurling/ Tom Stack & Associates; © Chase Swift/ Corbis. **7.10** (a,b) © Ben Fink/ Foodpix/ Jupiter Imges; (c) © Dr. Dennis Kunkel/ Visuals Unlimited. **7.11** © Randy Faris/ Corbis; inset, © Gladden Willis, MD/ Visuals Unlimited. **page 118** and **7.12** © Lois Ellen Frank/ Corbis.

Page 123 UNIT II © Francis Leroy, Biocosmos/ Science Photo Library/ Photo Researchers.

CHAPTER 8 **8.1** © Micrograph, Dr. Pascal Madaule, France. **8.2** Courtesy of the family of Henrietta Lacks. **8.4** (a) © Andrew Syred/ Photo Researchers, Inc.; (c) © B. Hamkalo; (d) © O. L. Miller, Jr., Steve L. McKnight. **8.6** (a) © L. Willatt, East Anglian Regional Genetics Service/ SPL/ Photo Researchers, Inc. **8.7** Micrographs, all, Ed Reschke. **8.8** (a) 3, © micrograph, D. M. Phillips/ Visuals Unlimited; (b) 3, © micrograph, R. Calentine/ Visuals Unlimited. **8.9** both, © Lennart Nilsson/ Bonnierforlagen AB. **8.10** © Phillip B. Carpenter, Department of Biochemistry and Molecular Biology, University of Texas—Houston Medical School. **8.11** © Science Photo Library/ Photo Researchers, Inc. **8.13** (a) © Ken Greer/ Visuals Unlimited; (b) © Biophoto Associates/ Science Source/ Photo Researchers, Inc.; (c) © James Stevenson/ SPL/ Photo Researchers, Inc. **8.14** A. S. Bajer, University of Oregon. **Page 137** left, David C. Martin, Ph.D.

CHAPTER 9 **9.1** (a) © Dan Kline/ Visuals Unlimited; (b) © George D. Lepp/ Corbis; (c) © Andrew Syred / Photo Researchers, Inc.; (d) AP/ Wide World Photos. **9.2** Image courtesy of Carl Zeiss MicroImaging, Thornwood, NY. **9.4** © Leonard Lessin/ Photo Researchers, Inc. **9.5** Photography, With thanks to the John Innes Foundation Trustees, computer enhanced by Gary Head; Art, Raychel Ciemma. **9.8** © Robert Potts, California Academy of Sciences **9.10** right © Francis Leroy, Biocosmos/ Science Photo Library/ Photo Researchers, Inc. **9.11** right, all © Jennifer W. Shuler/ Science Source/ Photo Researchers, Inc. **9.13** © Ron Neumeyer, www.microimaging.ca. **9.14** © Lisa O'Connor/ ZUMA/ Corbis.

CHAPTER 10 **10.1**, left, © Abraham Menashe; opposite, © Children's Hospital & Medical Center/ Corbis. **10.2** © The Moravian Museum, Brno. **10.3** © Jean M. Labat/ Ardea, London. **10.6, 10.7** White pea plant, © George Lepp/ Corbis. **10.10** © David Scharf/ Peter Arnold, Inc. **10.11** © JupiterImages Corporation. **10.12** © Ted Somes. **10.13** (a,c) © Michael Stuckey/ Comstock, Inc.; (b) Bosco Broyer, photograph by Gary Head. **10.14** © Bettmann/ Corbis. **10.15** © JupiterImages Corporation. **10.17** © Pamela Harper/ Harper Horticultural Slide Library. **10.18** (a) © Daan Kalmeijer; (b) © Dr. Christian Laforsch. **Page 164** from top, © Frank Cezus/ FPG/ Getty Images; © Frank Cezus/ FPG/ Getty Images; © Ted Beaudin/ Getty Images; © Michael Prince/ Corbis; © Lisa Starr. **10.19** (b,c) Courtesy of Ray Carson, University of Florida News and Public Affairs. **10.20** left, © Tom and Pat Leeson/ Photo Researchers, Inc.; right, © Rick Guidotti/ Positive Exposure. **10.21** (a) Courtesy of © www.waysidegardens.com; (b) © Gene Ahrens/ SuperStock; (c) © Karen Tweedy-Holmes/ Corbis; (d) © Clay Perry/ Corbis. **10.22** © Leslie Faltheisek. **Page 167** © Maximilian Stock Ltd./ Foodpix/ JupiterImages Corporation.

CHAPTER 11 **11.1** from left, © Reuters/ Corbis; George Griessman, www.president lincoln.com; © Hulton-Deutsch Collection/ Corbis. **11.2** (b) from M. Cummings, *Human Heredity: Principles and Issues*, 3rd Edition, p. 126. © 1994 by Brooks/Cole. All rights reserved; (c) after Patten, Carlson & others. **11.3** © University of Washington Department of Pathology. **11.4** above, © Frank Trapper/ Corbis Sygma. **11.5** © Lois Ellen Frank/ Corbis. **11.6** © Eddie Adams/ AP Wide World Photos. **11.8** © Bettmann/ Corbis. **11.9** left (both), Photo by Gary L. Friedman, www .FriedmanArchives.com. **Page 175** © Russ Schleipman/ Corbis. **11.10** Courtesy G. H. Valentine. **11.13** (a) © CNRI/ Photo Researchers, Inc. **11.14** right, © Lauren Shear/ Photo Researchers, Inc. **11.15** © UNC Medical Illustration and Photography. **11.16** © Stapleton Collection/ Corbis. **11.17** © Dr. Victor A. McKusick. **11.18** © Steve Uzzell. **11.19** © Saturn Stills/ SPL/ Photo Researchers, Inc. **11.20** © Lennart Nilsson/ Bonnierforlagen AB. **11.21** © Matthew Alan/ Corbis; inset, © Fran Heyl Associates/ Jacques Cohen, computer-enhanced by © Pix Elation. **11.22** Stefan Schwarz.

CHAPTER 12 **12.1** Photos by Victor Fisher, courtesy Genetic Savings & Clone. **12.2** C. Barrington Brown, 1968 J. D. Watson. **12.4** left, lower, © Eye of Science/ Photo Researchers, Inc. **12.6** PDB ID: 1BBB; Silva, M. M., Rogers, P. H., Arnone, A.: A third quaternary structure of human hemoglobin A at 1.7-Å resolution. *J Biol Chem* 267 pp. 17248 (1992). **12.10** (a–c) © James King-Holmes/ SPL/ Photo Researchers, Inc.; (d) © Mc Leod Murdo/Corbis Sygma. **12.11** Shahbaz A. Janjua, MD, Dermatlas; www .dermatlas.org.

CHAPTER 13 **13.1** right, © Vaughan Fleming/ SPL/ Photo Researchers, Inc. **13.3** (d) below, © Model by Dr. David B. Goodin, The Scripps Research Institute. **13.10** left, © Nik Kleinberg; right, P. J. Maughan. **13.11** © John W. Gofman and Arthur R. Tamplin. From *Poisoned Power: The Case Against Nuclear Power Plants Before and After Three Mile Island*, Rodale Press, PA, 1979. **13.13** © Dr. M.A. Ansary / SPL/ Photo Researchers, Inc.

CHAPTER 14 **14.1** Page 208, From the archives of www.breastpath.com, courtesy of J.B. Askew, Jr., M.D., P.A. Reprinted with permission, copyright 2004 Breastpath.com.; page 209, Courtesy of Robin Shoulla and Young Survival Coalition. **14.2** (b) From the collection of Jamos Werner and John T. Lis. **14.3** (b) From the collection of Jamos Werner and John T. Lis. **14.4** (a,b) © Dr. William Strauss; (c) © DermAtlas, www.dermatlas.org. **14.5** © Thinkstock Images/ JupiterImages Corporation. **14.6** (a) lower, © Juergen Berger, Max Planck Institute for Developmental Biology, Germany; (b) © Jose Luis Riechmann. **Page 214** © Lisa Starr. **14.7** (a) © Jürgen Berger, Max-Planck-Institut for Developmental Biology, Tübingen; (b) © Visuals Unlimited; (c) © Eye of Science/ Photo Researchers, Inc.; (d) far right, Courtesy of Edward B. Lewis, California Institute of Technology; all others, © Carolina Biological/ Visuals Unlimited. **14.8** (a) Palay/ Beaubois after Robert F. Weaver and Philip W. Hedrick, *Genetics*. © 1989 W. C. Brown Publishers; (b,c) © Jim Langeland, Jim Williams, Julie Gates, Kathy Vorwerk, Steve Paddock and Sean Carroll, HHMI, University of Wisconsin-Madison. **Page 217** © Lowe Worldwide, Inc. as Agent for National Fluid Milk Processor Promotion Board. **14.10** (a) © Jim Langeland, Jim Williams, Julie Gates, Kathy Vorwerk, Steve Paddock and Sean Carroll, HHMI, University of Wisconsin-Madison; (b) © Craig Brunetti and Sean Carroll, Howard Hughes Medical Institute, University of Wisconsin.

CHAPTER 15 **15.1** (a,b) © Courtesy of Golden Rice Humanitarian Board; page 221, © ScienceUV/ Visuals Unlimited. **15.3** (a) © Professor Stanley Cohen/SPL/Photo Researchers, Inc.; (b) with permission of © QIAGEN, Inc. **15.9** Courtesy of © Genelex Corp. **15.10** right, © Volker Steger/ SPL/ Photo Researchers, Inc. **15.11** © Ken Cavanagh/ Photo Researchers, Inc. **15.12** Courtesy of Joseph DeRisa. From *Science*, 1997 Oct. 24; 278 (5338) 680–686. **Page 230** Photo Courtesy of Systems Biodynamics Lab, P.I. Jeff Hasty, UCSD Department of Bioengineering, and Scott Cookson. **15.13** (d) © Lowell Georgis/ Corbis;

(e) © Keith V. Wood. **15.14** (a) The Bt and Non-Bt corn photos were taken as part of field trial conducted on the main campus of Tennessee State University at the Institute of Agricultural and Environmental Research. The work was supported by a competitive grant from the CSREES, USDA titled "Southern Agricultural Biotechnology Consortium for Underserved Communities," (2000–2005). Dr. Fisseha Tegegne and Dr. Ahmad Aziz served as Principal and Co-principal Investigators respectively to conduct the portion of the study in the State of Tennessee; (b) © Dr. Vincent Chaing, School of Forestry and Wood Projects, Michigan Technology University. **15.15** (a) © Adi Nes, Dvir Gallery Ltd.; (b) Transgenic goat produced using nuclear transfer at GTC Biotherapeutics. Photo used with permission; (c) Photo courtesy of MU Extension and Agricultural Infomation. **15.16** R. Brinster, R. E. Hammer, School of Veterinary Medicine, University of Pennsylvania. **Page 234** © Jeans for Gene Appeal. **15.17** (b) © Mike Stewart/ Corbis Sygma; (c) © Simon Kwong/ REUTERS/ Landov; (d) © Work of Atsushi Miyawaki, Qing Xiong, Varda Lev-Ram, Paul Steinbach, and Roger Y. Tsien at the University of California, San Diego.

Page 237, UNIT III © Wolfgang Kaehler/ Corbis.

CHAPTER 16 **16.1** (a) © Brad Snowder; (b) © David A. Kring, NASA/ Univ. Arizona Space Imagery Center. **16.2** (a,c) © Wolfgang Kaehler/ Corbis; (b) © Earl & Nazima Kowall/ Corbis; (d,e) © Edward S. Ross. **16.3** left, Gary Head; right above, © Bruce J. Mohn; inset, © Phillip Gingerich, Director, University of Michigan. Museum of Paleontology. **16.4** © Jonathan Blair/ Corbis. **16.5** (a) Courtesy George P. Darwin, Darwin Museum, Down House; (b) © Christopher Ralling; (e) © Dieter & Mary Plage/ Survival Anglia; above, page 243, © Heather Angel. **16.6** (a) © Joe McDonald/ Corbis; (b) © Karen Carr Studio/ www.karencarr.com. **16.7** (a) © Gerra and Sommazzi/ www.justbirds.org; (b) © Kevin Schafer/ Corbis; (c) © Alan Root/ Bruce Coleman Ltd. **16.8** © Down House and The Royal College of Surgeons of England. **16.9** (a) © H. P. Banks; (b) © Jonathan Blair; (c) Courtesy of Stan Celestian/ Glendale Comunity College Earth Science Image Archive. **16.10** © Jonathan Blair/ Corbis. **16.11** (a) Gary Head; (b) © Photodisc/ Getty Images; (c,d) Lisa Starr. **16.14** (a) NASA/ GSFC. **16.15** left, © Martin Land/ Photo Researchers, Inc.; right, © John Sibbick; (a–e) After A.M. Ziegler, C.R. Scotese, and S.F. Barrett, "Mesozoic and Cenozoic Paleogeographic Maps," and J. Krohn and J. Sundermann (Eds.), *Tidal Frictions and the Earth's Rotation II*, Springer-Verlag, 1983. **16.17** (a) © Taro Taylor, www.flickr.com/photos/tjt195; (b) © JupiterImages Corporation; (c) © Linda Bingham. **16.18** (a–b) Courtesy of Professor Richard Amasino, University of Wisconsin-Madison; (c) © Jose Luis Riechmann; (d) Courtesy of Professor Martin F. Yanofsky, UCSD. **16.19** (a) © Lennart Nilsson/ Bonnierforlagen AB; (b) Courtesy of Anna Bigas, IDIBELL-Institut de Recerca Oncologica, Spain; (c) From Embryonic staging system for the short-tailed fruit bat, *Carollia perspicillata, a model organism for the mammalian order Chiroptera, based upon timed pregnancies in captive-bred animals*, C.J. Cretekos et al., *Developmental Dynamics*, Vol. 233, Issue 3, July 2005, pp. 721–738. Reprinted with permission of Wiley-Liss, Inc., a subsidiary of John Wiley & Sons, Inc., (d) Courtesy of Prof. Dr. G. Elisabeth Pollerberg, Institut für Zoologie, Universität Heidelberg, Germany; (e) USGS. **16.20** (a) © Chip Clark; (b) above, Tait/ Sunnucks Peripatus Research; below and (c) below, © Jennifer Grenier, Grace Boekhoff-Falk and Sean Carroll, HMI, University of Wisconsin-Madison (c) above, © Herve Chaumeton/ Agence Nature; below; (d) above, © Peter Skinner/ Photo Researchers, Inc.; below, Courtesy of Dr. Giovanni Levi. **16.23** from left, © Kjell B. Sandved/ Visuals Unlimited; © Jeffrey Sylvester/ FPG/ Getty Images; © Thomas D. Mangelsen/ Images of Nature. **16.24** from left, © Science Photo Library/ Photo Researchers, Inc.; © Galen Rowell/ Corbis; © Kevin Schafer/ Corbis; Courtesy of Department of Entomology, University of Nebraska-Lincoln; Bruce Coleman, Ltd. **16.27** left from top, © Hans Reinhard/ Bruce Coleman, Inc.; © Phillip Colla, oceanlight .com; © Randy Wells/ Corbis; © Cousteau Society/ The Image Bank/ Getty Images; © Jack Jeffrey Photography. **16.29** Courtesy of Irving Buchbinder, DPM, DABPS, Community Health Services, Hartford CT **16.30** © John Klausmeyer, University of Michigan Exhibit of Natural History.

CHAPTER 17 **17.1** page 264, © Reuters NewMedia, Inc./ Corbis; page 265, © Rollin Verlinde/ Vilda. **17.2** (a) © Alan Solem; (b) third from left, © Roderick Hulsbergen/ http://www.photography.euweb.nl; all others, © JupiterImages Corporation. **17.3** top, © Photodisc/ Getty Images. **17.6** J. A. Bishop, L. M. Cook. **17.7** Courtesy of Hopi Hoekstra, University of California, San Diego. **17.10** © Peter Chadwick/ Science Photo Library/ Photo Researchers, Inc **17.11** © Thomas Bates Smith. **17.12** (a) Courtesy of Gerald Wilkinson; (b) © Bruce Beehler. **17.13** (a,b) After Ayala and others; (c) © Michael Freeman/ Corbis. **17.14** Adapted from S. S. Rich, A. E. Bell, and S. P. Wilson, "Genetic drift in small populations of Tribolium," *Evolution* 33:579–584, Fig. 1, p. 580, 1979. Used by permission of the publisher. **17.15** © Frans Lanting/ Minden Pictures (computer-modified by Lisa Starr). **17.16** left, © David Neal Parks; right, © W. Carter Johnson. **Page 278** © Alvin E. Staffan/ Photo Researchers, Inc. **17.19** left, Courtesy of Dr. James French; right, Courtesy of Joe Decruyenaere. **17.20** G. Ziesler/ ZEFA. **17.21** (a) © Graham Neden/ Corbis; (b) © Kevin Schafer/ Corbis; center, © Ron Blakey, Northern Arizona University (c) © Rick Rosen/ Corbis SABA. **17.22** Po'ouli, Bill Sparklin/ Ashley Dayer; All others, © Jack Jeffrey Photography. **17.24** (a) © Ian Hutton; (b) Courtesy of Peter Richardson; (c) © Jo Wilkins. **17.25** (a,b) Courtesy of Dr. Robert Mesibov. **17.26** Courtesy of Daniel C. Kelley, Anthony J. Arnold, and William C. Parker, Florida State University Department of Geological Science. **17.27** © Photo by Marcel Lecoufle. **Page 286** Image courtesy of the Image Analysis Laboratory, NASA Johnson Space Center. **17.29** from left, © Gary Head; © Dan Guravich/ Corbis; © Theo Allofs/Corbis. **17.30** from left, © Francois Gohier/ Photo Researchers, Inc.; © David Parker/ SPL/ Photo Researchers, Inc. **17.31** © Gulf News, Dubai, UAE. **Page 286** © JupiterImages Corporation.

CHAPTER 18 **18.1** © Peter Menzel/ Photo Researchers, Inc.; inset, Courtesy of Agriculture Canada. **18.2** © Philippa Uwins/ The University of Queensland. **18.3** © Jeff Hester and Paul Scowen, Arizona State University, and NASA. **18.4** (a) Painting by William K. Hartmann. **18.6** (a) © Eiichi Kurasawa/ Photo Researchers, Inc.; (b) © Dr. Ken MacDonald/ SPL/ Photo Researchers, Inc.; (c) © Micheal J. Russell, Scottish Universities Environmental Research Centre. **18.7** (a) © Sidney W. Fox; (b) From Hanczyc, Fujikawa, and Szostak, Experimental Models of Primitive Cellular Compartments: Encapsulation, Growth, and Division;www .sciencemag.org *Science* 24 October 2003; 302; 529, Figure 2, page 619. Reprinted with permission of the authors and AAAS. **18.8** (a) © Stanley M. Awramik; (b,c) © Bruce Runnegar, NASA Astrobiology Institute; (d) © N. J. Butterfield, University of Cambridge. **18.9** (a) © Christopher Scotese, PALEOMAP Project; (b) © Chase Studios/ Photo Researchers, Inc.; (c) © John Reader/ SPL/ Photo Researchers, Inc.; (d) © Sinclair Stammers/ SPL/ Photo Researchers, Inc.; (e) © Neville Pledge/ South Australian Museum. **18.10** (a) © CNRI/ Photo Researchers, Inc.; (b) © Robert Trench, Professor Emeritus, University of British Columbia. **18.11** (a) © CNRI/ Photo Researchers, Inc.; (b) © Robert Trench, Professor Emeritus, University of British Columbia.

Page 303 UNIT IV © Layne Kennedy/ Corbis.

CHAPTER 19 **19.1** © David Lees/Getty Images. **Page 305** © R. Sorensen/ J. Olsen/ Photo Researchers, Inc. **19.3** (a) © P. Hawtin, University of Southampton/ SPL/ Photo Researchers, Inc.; (b) © Dr. Dennis Kunkel/ Visuals Unlimited. **19.6** (a) © Dr. Jeremy Burgess/ SPL/ Photo Researchers, Inc.; (b) © P. W. Johnson and J. Mc N. Sieburth, Univ. Rhode Island/ BPS; (c) © Dr. Manfred Schloesser, Max Planck Institute for Marine Microbiology. **19.7** © Dr. Terry J. Beveridge, Department of Microbiology, University of Guelph, Ontario, Canada. **19.8** (a) © Stem Jems/ Photo Researchers, Inc.; (b) © California Department of Health Services; (c) © Bernard Cohen, M.D., DermAtlas; http://www.dermatlas.org. **19.10** (a) © Courtesy Jack Jones, *Archives of Microbiology*, Vol. 136, 1983, pp. 254–261. Reprinted by permission of Springer-Verlag; (b) © Dr. John Brackenbury/ Photo Researchers, Inc. **19.11** (a) © Martin Miller / Visuals Unlimited; (b) © Dr. Harald Huber, Dr. Michael Hohn, Prof. Dr. K. O. Stetter, University of Regensburg, Germany; (c) © Savannah River Ecology Laboratory; (d) © Alan L. Detrick, Science Source/ Photo Researchers, Inc. **19.12** (a) left, after Stephen L. Wolfe; right, © Dr. Harold Fisher/ Visuals Unlimited; (b) top © Dr. Hans Gelderblom/ Visuals Unlimited; bottom, after Stephen L. Wolfe. **19.14** © Russell Knightly/ Photo Researchers, Inc. **19.15** (a) Photo by Barry Fitzgerald, Courtesy of USDA; (b) Photo by Peggy Greb, Courtesy of USDA. **19.16** (a) © APHIS photo by Dr. Al Jenny; (b) © Lily Echeverria/ Miami Herald; (bottom) PDB ID: 1QLX; Zahn, R., Liu, A., Luhrs, T., Riek, R., Von Schroetter, C., Garcia, F. L., Billeter, M., Calzolai, L., Wider, G., Wuthrich, K.: NMR Solution Structure of the Human Prion Protein, Proc. Nat. Acad. Sci. USA 97 pp. 145 (2000). **Page 315** center, © Sercomi/ Photo Researchers, Inc.; others, © CAMR, Barry Dowsett/ Photo Researchers, Inc. **19.17** From left, Gary Head; E. A. Zottola, University of Minnesota. **19.18** Kenneth M. Corbett.

CHAPTER 20 **20.1** (a) © Wim van Egmond/ Visuals Unlimited; (b) © Adam Woolfitt/ Corbis; (c) © Ric Ergenbright/ Corbis. **Page 321** © Dr. Stan Erlandsen, University of Minnesota. **20.4** (a) © Dr. Dennis Kunkel/ Visuals Unlimited; (b) © Oliver Meckes/ Photo Researchers, Inc. **20.5** © Dr. David Phillips/ Visuals Unlimited. **20.6** (a) Courtesy of Allen W. H. Bé and David A. Caron; (b) © John Clegg/ Ardea, London. **20.7** (a) Redrawn from V. & M. Pearse and M. & R. Buchsbaum, *Living Invertebrates*, The Boxwood Press, 1987. Used by permission; (b) Courtesy James Evarts. **20.8** (a) left, © Wim van Egmond/ Micropolitan Museum; right, © Frank Borges Llosa/ www.frankley.com; (b) left, © Dr. David Phillips/ Visuals Unlimited; right, © Lexey Swall/ Staff from article, *Deep Trouble: Bad Blooms*, October 3, 2003 by Eric Staats. **20.9** (a) © Sinclair Stammers/ Photo Researchers, Inc.; (c) © London School of Hygiene & Tropical Medicine/ Photo Researchers, Inc.; (d) © Moredum Animal Health, Ltd./ Photo Researchers, Inc; (e) Micrograph Steven L'Hernaults. **Page 325** left, International Potato Center, Lima, Peru. **20.10** (a) © Susan Frankel, USDA-FS; (b) Heather Angel. **20.11** (a) Greta Fryxell, University of Texas, Austin; (b) © Wim van Egmond/ Visuals Unlimited; (c) © Emiliania Huxleyi. Photograph by Vita Pariente. Scanning electron micrograph taken on a Jeol T330A instrument at the Texas A & M University Electron Microscopy Center; (d) Ron Hoham, Dept. of Biology, Colgate University. **20.12** left, from T. Garrison, *Oceanography: An Invitation to Marine Science*, Brooks/Cole, 1993; right, © Lewis Trusty/ Animals Animals. **20.13** right, Courtesy of Professeur Michel Cavalla. **20.14** (a) Courtesy of Professor Astrid Saugestad; (b) © Lawson Wood/ Corbis; (c) Courtesy Microbial Culture Collection, National Institute for Environmental Studies, Japan; (d) © Wim van Egmond. **20.15** bottom, © PhotoDisc/ Getty Images. **20.16** © Wim van Egmond. **20.17** (a) © M I Walker/ Photo Researchers, Inc.; (b) © Edward S. Ross; (c) © Courtesy of www .hiddenforest.co.nz. **20.18** bottom, Courtesy Robert R. Kay from R. R. Kay, et al., *Development*, 1989 Supplement, pp. 81–90, © The Company of Biologists Ltd.; all others, © Carolina Biological Supply Company. **20.19** Gary W. Grimes and Steven L'Hernault. **20.20** © W. P. Armstrong; inset, Courtesy Brian Duval. **20.21** © Jeffrey Levinton, State University of New York, Stony Brook. **Page 330** Gary Head.

CHAPTER 21 **21.1** page 332 left, upper, © Jeri Hochman and Martin Hochman, Illustration by Zdenek Burian; lower, © Karen Carr Studio/ www.karencarr.com; right, © T. Kerasote/ Photo Researchers, Inc.; page 333, © Craig Allikas/ www.orchidworks.com. **Page 334** left, upper, © Reprinted with permission from Elsevier; lower, © Patricia G. Gensel. **21.2** (b) After E.O. Dodson and P. Dodson, *Evolution: Process and Product*, Third Ed., p. 401, PWS. **21.3** above, © Christopher Scotese, PALEOMAP Project. **21.4** © Craig Wood/ Visuals Unlimited. **21.5** top center, © Jane Burton/Bruce Coleman Ltd.; art, Raychel Ciemma. **21.6** (a) © Fred Bavendam/ Peter Arnold, Inc.; (b) © John D. Cunningham/ Visuals Unlimited. **21.7** (a) © University of Wisconsin-Madison, Department of Biology, Anthoceros CD; (b) left, © National Park Services, Paul Stehr-Green; right, © National Park Services, Martin Hutten; (c) both, © Wayne P. Armstrong, Professor of Biology and Botany, Palomar College, San Marcos, California. **21.8** (a) © Ed Reschke/ Peter Arnold, Inc. (b) © Gerald D. Carr; (c) © Colin Bates; (d) Photo by A. Murray, University of Florida, Center for Aquatic and Invasive Plants. Used with permission; (e) © Derrick Ditchburn/ Visuals Unlimited. **21.9** © A. & E. Bomford/ Ardea, London; art, Raychel Ciemma. **21.10** (a) © S. Navie (b) © David C. Clegg/ Photo Researchers, Inc.; (c) © Klein Hubert/ Peter Arnold, Inc. **21.11** right, © PaleoDirect.com. **21.12** © Field Museum of Natural History, Chicago (Neg. #7500C); inset, © Brian Parker/ Tom Stack & Associates. **Page 341** right, © George J. Wilder/ Visuals Unlimited, computer enhanced by Lachina Publishing Services, Inc. **21.13** (a) © Ralph Pleasant/ FPG / Getty Images; (b) © Earl Roberge/ Photo Researchers, Inc.; (c) © George Loun/ Visuals Unlimited; (d) Courtesy of Water Research Commission, South Africa. **21.14** (a) © Dave Cavagnaro/ Peter Arnold, Inc.; (b) © M. Fagg, Australian National Botanic Gardens; (c) © E. Webber/ Visuals Unlimited; (d) © Michael P. Gadomski/ Photo Researchers, Inc.; (e) © Sinclair Stammers/ Photo Researchers, Inc.; (f) Courtesy of © www.waysidegardens.com; (g) © Gerald & Buff Corsi/ Visuals Unlimited; (h) © Fletcher and Baylis/ Photo Researchers, Inc. **21.15** left, © Robert Potts, California Academy of Sciences (a) © Robert & Linda Mitchell Photography; (b) © R. J. Erwin/ Photo Researchers, Inc. **21.16** from top, © Ed Reschke; © Lee Casebere; © Robert & Linda Mitchell Photography; © Runk & Schoenberger / Grant Heilman, Inc. **21.18** (a) © Michelle Garrett/ Corbis (b) © Sanford/ Agliolo/ Corbis; (c) © Gregory G. Dimijian/ Photo Researchers, Inc.; (d) © Darrell Gulin/ Corbis; (e) © DLN/ Permission by Dr. Daniel L. Nickrent. **21.20** © Dan Fairbanks. **21.22** left, © Clinton Webb; right, Gary Head. **21.23** © Rod Planck/ Photo Researchers, Inc. **21.24** © William Campbell/ TimePix/ Getty Images.

CHAPTER 22 **22.1** page 350 both, © Charles Lewallen; page 351, © Jacques Langevin/ Corbis Sygma. **22.3** (a,d) © Robert C. Simpson/ Nature Stock. **22.4** (a,b) © Ed Reschke; below, © Dr. John D. Cunningham/ Visuals Unlimited. **22.5** (a) upper, © Michael Wood/ mykob.com; lower, © North Carolina State University, Department of Plant Pathology; (b) © Bill Beatty/ Visuals Unlimited; (c) © Dr. Dennis Kunkel/ Visuals Unlimited. **22.6** N. Allin and G. L. Barron. **22.7** left, Garry T. Cole, University of Texas, Austin/ BPS; right, © Eye of Science/ Photo Researchers, Inc.; art, After T. Rost, et al., *Botany*, Wiley 1979. **22.8** (a) Gary Head; (b) © Mark Mattock/ Planet Earth Pictures; (c) © Mark E. Gibson/ Visuals Unlimited; (d) After Raven, Evert, and Eichhorn, *Biology of Plants*, 4th ed., Worth Publishers, New York, 1986. **22.9** (a) © Gary Braasch; (b) © F. B. Reeves. **22.10** (a) © Dr. P. Marazzi/ SPL/ Photo Researchers, Inc.; (b) © Eric Crichton/ Bruce Coleman, Inc.; (c) © Harry Regin. **22.11** John Hodgin. **22.12** © Robert C. Simpson/ Nature Stock. **22.13** (a) © Jane Burton/ Bruce Coleman, Ltd.; (b) © Chris Worden.

CHAPTER 23 **23.1** (a) © K.S. Matz; (b) © Callum Roberts, University of York. **23.4** © The Natural History Museum (London). **22.5** (a,b) David Patterson, courtesy micro*scope/ http://microscope.mbl.edu; (c) © 2003 Ana Signorovitch. **23.7** (a) © David Sailors/ Corbis; (b) Marty Snyderman/ Planet Earth Pictures; (c) © Don W. Fawcett/ Visuals Unlimited; (d) © Bruce Hall. **23.9** (c) © Brandon D. Cole/ Corbis; (d) © Jeffrey L. Rotman/ Corbis. **23.10** (a) after Eugene Kozloff; (b) Courtesy of Dr. William H. Hamner. **23.11** (a) © Kim Taylor/ Bruce Coleman, Ltd.; (b) © A.N.T./ Photo Researchers, Inc.; inset, © Peter Parks/ Image Quest 3D. **23.12** After T. Storer, et al., *General Zoology*, Sixth Edition. **23.14** © James Marshall/ Corbis. **23.15** (c) © Andrew Syred/ SPL/ Photo Researchers, Inc. **23.16** (a,b) Adapted from Rasmussen, "Ophelia," Vol. 11, in Eugene Kozloff, *Invertebrates*, 1990; (c) © J. Solliday/ BPS; (d) © Jon Kenfield/ Bruce Coleman Ltd. **23.17** © J. A. L. Cooke/ Oxford Scientific Films. **23.18** above, © Cabisco/ Visuals Unlimited. **23.19** (a) © Science Photo Library/ Photo Researchers, Inc. **23.20** (a) Danielle C. Zacherl with John McNulty; (b) © B. Borrell Casals/ Frank Lane Picture Agency/ Corbis; (c) © Joe McDonald/ Corbis; (d) © Jeff Foott/ Tom Stack & Associates; (e) © Frank Park/ ANT Photo Library; (f) © Alex Kirstitch. **23.21** (a) Illustration by Zdenek Burian, © Jeri Hochman and Martin Hochman; (b) © Alex Kirstitch; (d) © Bob Cranston; (e) J. Grossauer/ ZEFA. **23.22** below, Micrograph, J. Sulston, MRC Laboratory of Molecular Biology. **23.23** (a) © L. Jensen/ Visuals Unlimited; (b) © Sinclair Stammers/ SPL/ Photo Researchers, Inc.; (c) Courtesy of © Emily Howard Staub and The Carter Center. **23.24** (a) © Dr. Chip Clark; (b) © Michael & Patricia Fogden/ Corbis; (c) © Jane Burton/ Bruce Coleman, Ltd. **23.25** (a) © Angelo Giampiccolo; (b) © Frans Lemmens/ The Image Bank/ Getty Images; (c) © Corbis; (d) © Andrew Syred/ Photo Researchers, Inc. **23.26** Redrawn from *Living Invertebrates*, V. & J. Pearse/M. & R. Buchsbaum, The Boxwood Press, 1987. Used by permission. **23.27** (a) © David Tipling/ Photographer's Choice/ Getty Images; (b) © Peter Parks/ Imagequestmarine.com; (c) © Science Photo Library/ Photo Researchers, Inc. **23.28** After D.H. Milne, *Marine Life and the Sea*, Wadsworth, 1995. **23.32** (a) © David Maitland/ Seaphot Limited/ Planet Earth Pictures; (b–g) Edward S. Ross; (h) © Mark Moffett/ Minden Pictures; (i) Marlin E. Rice, Iowa State University; (j) Courtesy of Karen Swain, North Carolina Museum of Natural Sciences; (k) © Chris Anderson/ Darklight Imagery; (l) © Joseph L. Spencer. **23.33** (a) © John H. Gerard; (b) © D. Suzio/ Photo Researchers, Inc.; (c) © Eye of Science/ Photo Researchers, Inc.; (d) Photo by James Gathany, Centers for Disease Control. **23.34** (a) © Fred Bavendam/ Minden Pictures; (b) © Jan Haaga, Kodiak Lab, AFSC/NMFS; (c) © Herve Chaumeton/ Agence Nature; (d) © George Perina, www.seapix.com; (e) right, © Herve Chaumeton/ Agence Nature. **23.35** © Walter Deas/ Seaphot Limited/ Planet Earth Pictures. **23.36** © Wim van Egmond/ Micropolitan Museum. **23.37** upper, © Dr. Dennis Kunkel/ Visuals Unlimited; lower, © Frank Romano, Jacksonville State University.

CHAPTER 24 **24.1** page 384, © Karen Carr Studio/ www.karencarr.com; page 385, © P. Morris/ Ardea London. **24.3** (a) © Gary Bell/ Taxi/ Getty Images; (b,c) Redrawn from *Living Invertebrates*, V. & J. Pearse and M. & R. Buchsbaum. The Boxwood Press, 1987. Used by permission. **24.4** above, © Patrick J. Lynch/ Photo Researchers, Inc. **24.5** below, © Brandon D. Cole/ Corbis. **24.6** (a) © John and Bridgette

Sibbick; (b,c) © Jenna Hellack, Department of Biology, University of Central Oklahoma. **24.7** (a–c) Adapted from A.S. Romer and T.S. Parsons, *The Vertebrate Body*, Sixth Edition, Saunders, 1986. **24.8** Photo of human by Lisa Starr; jawed fish courtesy of John McNamara, www.paleo direct.com. **24.9** (a) © Jonathan Bird/ Oceanic Research Group, Inc.; (b) © Gido Braase/ Deep Blue Productions; (c) from E. Solomon, L. Berg, and D.W. Martin, *Biology*, Seventh Edition, Thomson Brooks/Cole; (d) Robert & Linda Mitchell Photography; (e) © Ivor Fulcher/ Corbis; (f) Patrice Ceisel/ © 1986 John G. Shedd Aquarium. **24.10** (a) © Norbert Wu/ Peter Arnold, Inc.; (b) © Wernher Krutein/ photo vault.com; (c) © Alfred Kamajian; (d–f) © P. E. Ahlberg. **24.11** left, Adapted from A.S. Romer and T.S. Parsons, *The Vertebrate Body*, Sixth Edition, Saunders, 1986. (a) © Bill M. Campbell, MD; (b) © Stephen Dalton/ Photo Researchers, Inc.; (c) © John Serraro/ Visuals Unlimited. **24.12** © Juan M. Renjifo/ Animals Animals. **24.13** (a) © Pieter Johnson; (b) © Stanley Sessions/ Hartwick College. **24.14** (a) © D. Braginetz; (b) © Z. Leszczynski/ Animals Animals. **24.15** © Karen Carr Studio/ www.karencarr.com. **Page 395** right, © Julian Baum/ SPL/ Photo Researchers, Inc. **Page 396** left, © S. Blair Hedges, Pennsylvania State University. **24.17** (a) © Kevin Schafer/ Corbis (c) © Joe McDonald/ Corbis; (d) © David A. Northcott/ Corbis; (e) © Pete & Judy Morrin/ Ardea London; (f) © Stephen Dalton/ Photo Researchers, Inc.; (g) © Kevin Schafer/ Tom Stack & Associates. **24.18** (a) © Doug Wechsler/ VIREO; (b) With permission of the Australian Museum. **24.20** (a) © Gerard Lacz/ ANTPhoto.com.au; (b,c) Courtesy of Dr. M. Guinan, University of California-Davis, Anatomy, Physiology and Cell Biology, School of Veterinary Medicine; (d) © Kevin Schafer/ Corbis. **24.22** (a) © Sandy Roessler/ FPG/ Getty Images; (b) After M. Weiss and A. Mann, *Human Biology and Behavior*, 5th Edition, HarperCollins, 1990. **24.23** right above, Painting © Ely Kish. **24.24** (a) © D. & V. Blagden/ ANTPhoto.com.au; (b) © Nigel J. Dennis, Gallo Images/ Corbis; (c) © Tom Ulrich / Visuals Unlimited. **24.25** (a) © Alan and Sandy Carey; (b) © Merlin D. Tuttle/ Bat Conservation International; (c) © David Parker/ SPL/ Photo Researchers, Inc.; (d) © Mike Johnson. All rights reserved, www.earth window.com. **24.26** (a) © Larry Burrows/ Aspect Photolibrary; (c) © Dallas Zoo, Robert Cabello; (d) © Allen Gathman, Biology Department, Southeast Missouri State University; (e) © Bone Clones®, www.boneclones.com; (f) © Gary Head. **24.28** (a) © Rod Williams/ www.bciusa.com. **24.29** (a) © MPFT/ Corbis Sygma; (b–e) © Pascal Goetgheluck/ Photo Researchers, Inc. **24.30** (a) © Dr. Donald Johanson, Institute of Human Origins; (b,d) © Kenneth Garrett/ National Geographic Image Collection; (c) © Louise M. Robbins. **24.31** © Jean-Paul Tibbles, *Book of Life*, Ebury Press. **24.32** © John Reader/ Photo Researchers, Inc. **24.33** (a) © Pascal Goetgheluck/ Photo Researchers, Inc.; (b,c) © Peter Brown. **24.35** © Christopher Scotese, PALEOMAP Project. **24.37** © California Academy of Sciences. **24.38** © Jean Phillipe Varin/ Jacana/ Photo Researchers, Inc.

CHAPTER 25 **25.1** page 410, © Star Tribune/ Minneapolis-St. Paul; page 411, © Michael Davidson/Mortimer Abramowitz Gallery of Photomicrography/www.olympusmicro.com. **25.2** right, from top, Courtesy of Charles Lewallen; Dartmouth Electron Microscope Facility; Photo Courtesy of Prof. Alison Roberts, University of Rhode Island. **25.3** left, © PhotoDisc/ Getty Images, with art by Lisa Starr; right upper, © CNRI/ SPL/ Photo Researchers, Inc.; lower, © Dr. Robert Wagner/ University of Delaware, www.udel.edu/ Biology/ Wags. **25.4** left, © Montana Pritchard/ Getty Images Sport; right, © Darrell Gulin/ The Image Bank/ Getty Images. **24.5** (a) © Cory Gray; (b) © PhotoDisc/ Getty Images; (c) © Heather Angel; (d) © Biophoto Associates/ Photo Researchers, Inc. **25.6** (a) © Geoff Tompkinson/ SPL/ Photo Researchers, Inc.; (b) © Erwin & Peggy Bauer/ www.bciusa.com. **25.8** Right, © VVG/ Science Photo Library/ Photo Researchers, Inc. **25.9** © Galen Rowell/ Peter Arnold, Inc. **25.10** right, © Niall Benvie/ Corbis. **25.11** left, © Kennan Ward/ Corbis; right, © G. J. McKenzie (MGS). **25.12** © Frank B. Salisbury. **25.14** (a,b) Courtesy of Dr. Kathleen K. Sulik, Bowles Center for Alcohol Studies, the University of North Carolina at Chapel Hill. **25.15** © John DaSiai, MD/ Custom Medical Stock Photo. **25.16** (a) Courtesy of Dr. Consuelo M. De Moraes; (b–d) © Andrei Sourakov and Consuelo M. De Moraes.

Page 423 UNIT V © Jim Christensen, Fine Art Digital Photographic Images.

CHAPTER 26 **26.1** (a) © Michael Westmoreland/ Corbis; (b) © Charles O'Rear/ Corbis; (c) © Reuters/ Corbis. **26.3** (a) from left, © Bruce Iverson; © Ernest Manewal/ Index Stock Imagery; Courtesy of Dr. Thomas L. Rost; © Andrew Syred/ Photo Researchers, Inc.; (b) from left, © Mike Clayton/ University of Wisconsin Department of Botany; © Darrell Gulin/ Corbis; © Gary Head; © Andrew Syred/ Photo Researchers, Inc. **26.6** © Donald L. Rubbelke/ Lakeland Community College. **26.7** (a) © Dr. Dale M. Benham, Nebraska Wesleyan University; (b) © D. E. Akin and I. L. Risgby, Richard B. Russel Agricultural Research Center, Agricultural Research Service, U.S. Dept. Agriculture, Athens, GA; (c) © Kingsley R. Stern. **26.8** © Andrew Syred/ Photo Researchers, Inc. **26.9** © George S. Ellmore. **26.10** (d) above, © M. I. Walker/ Photo Researchers, Inc.; below, © Gary Head. **26.11** (a) center, © Mike Clayton/ University of Wisconsin Botany Department; right, © James W. Perry; (b) center, © Carolina Biological Supply Company; right, © James W. Perry. **26.13** © David Cavagnaro/ Peter Arnold, Inc. **Page 432** © JupiterImages Corporation. **26.14** (a) © N. Cattlin/ Photo Researchers, Inc.; (c) © C. E. Jeffree, et al., *Planta*, 172(1):20–37, 1987. Reprinted by permission of C. E. Jeffree and Springer-Verlag; (d) © Jeremy Burgess/ SPL/ Photo Researchers, Inc. **26.15** (a) Courtesy of Dr. Thomas L. Rost; (b) © Gary Head. **26.17** After Salisbury and Ross, *Plant Physiology*, Fourth Edition, Wadsworth. **26.18** (a) © Biodisc/ Visuals Unlimited; (b) © Brad Mogen/ Visuals Unlimited; (c) © Dr. John D. Cunningham/ Visuals Unlimited. **26.19** © Dr. John D. Cunningham/ Visuals Unlimited. **26.20** (a–c) © Omikron/ Photo Researchers, Inc. **26.22** (b) © Peter Gasson, Royal Botanic Gardens, Kew. **26.23** (a) © NOAA; (b) © David W. Stahle, Department of Geosciences, University of Arkansas. **26.24** © Edward S. Ross. **26.25** (a) © Peter Ryan/ SPL/ Photo Researchers, Inc.; (b) © Jon Pilcher; (c) © George Bernard/ SPL/ Photo Researchers, Inc.

CHAPTER 27 **27.1** (a) © OPSEC Control Number #4 077-A-4; (b) © Billy Wrobel, 2004; (c) Photo by Keith Weller, ARS, Courtesy of USDA. **Page 442** © Gary Head. **27.2** © William Ferguson. **27.3** (a) © Robert Frerck/ Stone/ Getty Images (b) Courtesy of NOAA. **27.4** (a) © Wally Eberhart/ Visuals Unlimited; (b) Mark E. Dudley and Sharon R. Long; (c) © NifTAL Project, Univ. of Hawaii, Maui. **27.5** © Andrew Syred/ Photo Researchers, Inc. **27.6** (b) © Dr. John D. Cunningham/ Visuals Unlimited; (c) © Francis Leroy, Biocosmos/ Photo Researchers, Inc. **27.7** (a) © Alison W. Roberts, University of Rhode Island; (b,c) © H. A. Core, W. A. Cote and A. C. Day, *Wood Structure and Identification*, 2nd Ed., Syracuse University Press, 1979. **27.8** left, © The Ohio Historical Society, Natural History Collections. **27.9** above, Courtesy of John S. Russell, Pioneer High School; below, micrograph, Bruce Iverson, computer-enhanced by Lisa Starr. **27.10, 27.11** Courtesy of E. Raveh. **27.12** © Don Hopey/ Pittsburgh Post-Gazette, 2002, all rights reserved. Reprinted with permission; inset, © Jeremy Burgess/ SPL/ Photo Researchers, Inc. **27.13** Photo by ARS, Courtesy of USDA. **27.14** (a) © James D. Mauseth, MCDB; (b) © J. C. Revy/ ISM/ Phototake. **27.15** © Martin Zimmerman, *Science*, 1961, 133:73–79, © AAAS. **27.18** (a,b) Robert & Linda Mitchell Photography; (c) John N. A. Lott, *Scanning Electron Microscope Study of Green Plants*, St. Louis: C. V. Mosby Company, 1976; (d) Robert C. Simpson/ Nature Stock.

CHAPTER 28 **28.1** page 454 upper, Courtesy of Caroline Ford, School of Plant Sciences, University of Reading, UK; lower, © James L. Amos/ Corbis; page 455, © Gary Head. **28.2** (a) upper left, © John McAnulty/ Corbis; right, © Robert Essel NYC/ Corbis. **28.3** (a) © David M. Phillips/ Visuals Unlimited; (b) © Dr. Jeremy Burgess/ SPL/ Photo Researchers, Inc.; (c) © David Scharf/ Peter Arnold, Inc. **28.4** (a) left, © John Alcock, Arizona State University; right, © Merlin D. Tuttle, Bat Conservation International; (b) © Thomas Eisner, Cornell University. **28.6** from left, © Michael Clayton, University of Wisconsin; Raychel Ciemma; © Michael Clayton, University of Wisconsin; © Dr. Charles Good, Ohio State University, Lima; © Michael Clayton, University of Wisconsin; © Michael Clayton, University of Wisconsin. **28.7** (a–c) Janet Jones; (e) © Dr. Dan Legard, University of Florida GCREC, 2000; (f) © Richard H. Gross; (g) © Andrew Syred/ SPL/ Photo Researchers, Inc.; (i) Mark Rieger. **28.8** (a) © Gregory K. Scott/ Photo Researchers, Inc.; (b) © Robert H. Mohlenbrock © USDA-NRCS PLANTS Database/ USDA SCS. 1989. *Midwest wetland flora; field office illustrated guide to plant species*. Midwest National Technical Center, Lincoln, NE; (c) © R. Carr. **28.9** © Darrell Gulin/ Corbis. **28.10** © Professor Dr. Hans Hanks-Ulrich Koop. **28.11** © Mike Clayton/ University of Wisconsin Department of Botany. **28.12** Right above, © Barry L. Runk/ Grant Heilman, Inc.; below, © James D. Mauseth, University of Texas. **28.13** right, © Herve Chaumeton/ Agence Nature. **28.14** © Sylvan H. Wittwer/ Visuals Unlimited. **28.16** left, © Robert Lyons/ Visuals Unlimited; right, @ mepr. **28.17** (a,b) © Michael Clayton, University of Wisconsin; (c,d) © Muday, GK and P. Haworth (1994) "Tomato root growth, gravitropism, and lateral development: Correlations with auxin transport." "Plant Physiology and Biochemistry" 32, 193–203 with permission from

Elsevier Science. **28.18** (a,b) Micrographs courtesy of Randy Moore from "How Roots Respond to Gravity," M. L. Evans, R. Moore, and K. Hasenstein, *Scientific American*, December 1986. **28.19** (c) © Cathlyn Melloan/ Stone/ Getty Images. **28.20** (c) © Gary Head. **28.21** Cary Mitchell. **28.22** Grant Heilman Photography, Inc. **28.24** (a) © Ray Evert, University of Wisconsin; (b) © Clay Perry/ Corbis; (c) © Eric Chrichton/ Corbis. **28.25** (a) © Clay Perry/ Corbis; (b) © Eric Chrichton/ Corbis. **28.26** Eric Welzel/ Fox Hill Nursery, Freeport, Maine. **28.27** left, © Roger Wilmshurst, Frank Lane Picture Agency/ Corbis; right, © Dr. Jeremy Burgess/ Photo Researchers, Inc. **28.28** Larry D. Nooden. **28.30** (a) © Edward S. Ross; (b,c) Gary Head.

Page 467 UNIT VI © Kevin Schafer.

CHAPTER 29 **29.1** © Dow W. Fawcett/ Photo Researchers, Inc.; inset, © Science Photo Library/ Photo Researchers, Inc. **29.2** left, © Ohlinger Jerry/ Corbis Sygma; right, © Sachs Ron/ Corbis Sygma. **29.3** (a) © Manfred Kage/ Bruce Coleman, Ltd.; (b) above, © Focus on Sports; (c) left, © Ray Simmons/ Photo Researchers, Inc.; center, © Ed Reschke/ Peter Arnold, Inc.; right, © Don W. Fawcett. **29.4** above, © Gregory Dimijian/ Photo Researchers, Inc; below, adapted from C.P. Hickman, Jr., L.S. Roberts, and A. Larson, *Integrated Principles of Zoology*, Ninth Edition, Wm. C. Brown, 1995. **29.6** above (a) © John Cunningham/ Visuals Unlimited; (b,c) © Ed Reschke; (d) © Science Photo Library/ Photo Researchers, Inc.; (e) © University of Cincinnati, Raymond Walters College, Biology; (f) © Michael Abbey/ Photo Researchers, Inc. **29.7** left, © Roger K. Burnard. **29.8** © Science Photo Library/ Photo Researchers, Inc. **29.9** above, © Tony McConnell/ Science Photo Library/ Photo Researchers, Inc.; (a,b) © Ed Reschke; (c) © Biophoto Associates/ Photo Researchers, Inc. **29.10** © Triarch/ Visuals Unlimited. **29.11** © Kim Taylor/ Bruce Coleman, Ltd. **29.14** (b) © John D. Cunningham/ Visuals Unlimited. **29.16** (b) © Frank Trapper/ Corbis Sygma; (c) © AFP/ Corbis. **29.17** © Pascal Goetgheluck/ Science Photo Library/ Photo Researchers, Inc. **Page 488** (a) © Ed Reschke/ Peter Arnold, Inc.; (b–d) © Ed Reschke. **29.18** © Keith Levit/ Alamy. **29.19** © David Macdonald. **29.20** © Dr. Preston Maxim and Dr. Stephen Bretz, Department of Emergency Services, San Francisco General Hospital.

CHAPTER 30 **30.1** page 490, © Jamie Baker/ Taxi/ Getty Images; page 491 left, © EMPICS; right, © Manni Mason's Pictures. **30.3** (a) Courtesy Dr. William J. Tietjen, Bellarmine University. **30.6** left, © Manfred Kage/ Peter Arnold, Inc. **30.9** (c) © Jeff Greenberg/ Index Stock Imagery. **30.10** (b) © Dr. Constantino Sotelo from *International Cell Biology*, p. 83, 1977. Used by copyright permission of the Rockefeller University Press. **30.11** left, Micrograph by Don Fawcett, Bloom and Fawcett, 11th edition, after J. Desaki and Y. Uehara/ Photo Researchers, Inc. **30.13** (a) AP/ Wide World Photos; (b,c) From Neuro Via Clinicall Research Program, Minneapolis VA Medical Center. **30.14** © E. D. London, et al., *Archives of General Psychiatry*, 47:567–574, 1990. **30.18** right, Washington University/ www.thalamus.wustl.edu. **30.20** (a) © Colin Chumbley/ Science Source/ Photo Researchers, Inc.; (b) © C. Yokochi and J. Rohen, *Photographic Anatomy of the Human Body*, 2nd Ed., Igaku-Shoin, Ltd., 197. **30.21** (a) after Penfield and Rasmussen, *The Cerebral Cortex of Man*, © 1950 Macmillan Library Reference. Renewed 1978 by Theodore Rasmussen; (b) © Colin Chumbley/ Science Source/ Photo Researchers, Inc. **30.22** (b) © Marcus Raichle, Washington Univ. School of Medicine. **30.25** © Nancy Kedersha/ UCLA/ Photo Researchers, Inc. **30.26** © Herve Chaumeton/ Agence Nature.

CHAPTER 31 **31.1** © AP/ Wide World Photos. **31.2** (a) © David Turnley/ Corbis. **31.3** left, after Penfield and Rasmussen, *The Cerebral Cortex of Man*, © 1950 Macmillan Library Reference. Renewed 1978 by Theodore Rasmussen; right, © Colin Chumbley/ Science Source/ Photo Researchers, Inc. **Page 516** left, © AFP Photo/ Timothy A. Clary/ Corbis. **31.9** (a) © Fabian/ Corbis Sygma; (d) Medtronic Xomed; (e) above, © Dr. Thomas R. Van De Water, University of Miami Ear Institute. **31.10** © Robert E. Preston, courtesy Joseph E. Hawkins, Kresge Hearing Research Institute, University of Michigan Medical School. **31.11** (a, below) After M. Gardiner, *The Biology of Vertebrates*, McGraw-Hill, 1972; (a) above, © E. R. Degginger; (b) G. A. Mazohkin-Porshnykov (1958). Reprinted with permission from *Insect Vision* © 1969 Plenum Press. **31.13** (b) © Bo Veisland/ Photo Researchers, Inc. **31.14** (a) above, © Lennart Nilsson/ Bonnierforlagen AB; (b) www.2.gasou.edu/psychology/courses/ muchinsky and www.occipita.cfa.cmu.edu. **31.17** above, © Will & Deni McIntyre/ Photo Researchers, Inc.; below, Courtesy of Dr. Bryan Jones, University of Utah School of Medicine. **31.18** © Eric A. Newman. **31.19** © Edward W. Bower/ The Image Bank/ Getty Images. **31.20** © Chase Swift.

CHAPTER 32 **32.1** left, © David Ryan/ SuperStock; right, © Catherine Ledner; page 527 © David Aubrey/ Corbis. **32.5** below, © Lisa Starr; right; Courtesy of Dr. Erica Eugster. **32.7** Left, © Gary Head. **32.8** (a) © Scott Camazine/ Photo Researchers, Inc.; (b) © Biophoto Associates/ SPL/ Photo Researchers, Inc. **32.11** left, © Ralph Pleasant/ FPG / Getty Images; right, © Yoav Levy/ Phototake. **32.12** © John S. Dunning/ Ardea, London. **32.13** © Frans Lanting/ Bruce Coleman, Ltd. **32.14** (a) Dr. Carlos J. Bourdony; (b) Courtesy of G. Baumann, MD, Northwestern University. **32.15** © Kevin Fleming/ Corbis.

CHAPTER 33 **33.1** left, © Michael Neveux; right, © Ed Reschke. **33.2** left, © Linda Pitkin/ Planet Earth Pictures; (a) above, © Stephen Dalton/ Photo Researchers, Inc. **33.4** left, © Yokochi and J. Rohen, *Photographic Anatomy of the Human Body*, 2nd Ed., Igaku-Shoin, Ltd., 1979. **33.5** (a) right, © Ed Reschke. **33.7** © Professor P. Motta/ Department of Anatomy/ La Sapienza, Rome/ SPL/ Photo Researchers, Inc. **33.8** © N.H.P.A./ ANT Photolibrary. **33.11** (a) below, © Dance Theatre of Harlem, by Frank Capri; (b,c) © Don Fawcett/ Visuals Unlimited, from D. W. Fawcett, *The Cell*, Philadelphia; W. B. Saunders Co., 1966. **33.16** Painting by Sir Charles Bell, 1809, courtesy of Royal College of Surgeons, Edinburgh. **33.17** (a) © Paul Sponseller, MD/ Johns Hopkins Medical Center; (b) Courtesy of the family of Tiffany Manning.

CHAPTER 34 **34.1** (a) From A. D. Waller, *Physiology: The Servant of Medicine*, Hitchcock Lectures, University of London Press, 1910; (b) Courtesy of The New York Academy of Medicine Library; (d) © Mark Thomas/ Science Photo Library/ Photo Researchers, Inc. **34.2** (a) left, © Darlyne A Murawski / Getty Images; (b) left, © Cabisco/ Visuals Unlimited. **34.3** (d) After Labarbera and S. Vogel, *American Scientist*, 1982, 70:54–60. **34.4** right, © National Cancer Institute/ Photo Researchers, Inc. **34.5** left, © EyeWire/ Getty Images; (Art) After *Bloodline Image Atlas*, University of Nebraska-Omaha, and Sherri Wicks, *Human Physiology and Anatomy*, University of Wisconsin Web Education System, and others. **34.6** From: Maslak P., Blast Crisis of Chronic Myelogenous Leukemia (posted online December 5, 2001). ASH Image Bank. Copyright American Society of Hematology, used with permission. **34.7** © Lester V. Bergman & Associates, Inc. **34.9** (a,b) After G. J. Tortora and N. Anagnostakos, *Principles of Anatomy and Physiology*, 6th ed. © 1990 by Biological Sciences Textbooks, Inc., A&P Textbooks, Inc., and Ellia-Sparta, Inc. Reprinted by permission of John Wiley & Sons, Inc. **34.13** (b) © C. Yokochi and J. Rohen, *Photographic Anatomy of the Human Body*, 2nd Ed., Igaku-Shoin, Ltd., 1979. **34.19** right, © Jose Pelaez, Inc./ Corbis. **34.20** left, © Sheila Terry/ SPL/ Photo Researchers, Inc.; right, Courtesy of Oregon Scientific, Inc. **34.21** left, © Biophoto Associates/ Photo Researchers, Inc. **34.22** (a) left, Lisa Starr, using © 2001 PhotoDisc, Inc./ Getty Images photograph; right, © Dr. John D. Cunningham/ Visuals Unlimited. **34.23** © Professor P. Motta/ Department of Anatomy/ University La Sapienza, Rome/ SPL/ Photo Researchers, Inc. **34.24** (a) © Ed Reschke; (b) © Biophoto Associates/ Photo Researchers, Inc. **34.25** left, © Lester V. Bergman/ Corbis. **34.28** left, © Lennart Nilsson/ Bonnierforlagen AB.

CHAPTER 35 **35.1** left, © NIBSC/ Photo Researchers, Inc.; right, © Lowell Tindell. **35.2** left, James Hicks, Centers for Disease Control and Prevention; right, © Eye of Science/ Photo Researchers, Inc. **35.3** After *Bloodline Image Atlas*, University of Nebraska-Omaha, and Sherri Wicks, *Human Physiology and Anatomy*, University of Wisconsin Web Education System, and others. **35.4** (a) © David Scharf, 1999. All rights reserved; (b) © Juergen Berger/ Photo Researchers, Inc. **35.5** (d) © Robert R. Dourmashkin, courtesy of Clinical Research Centre, Harrow, England. **35.6** below, © NSIBC/ SPL/ Photo Researchers, Inc. **35.7** (a) © Biology Media/ Photo Researchers, Inc. **35.14** © Dr. A. Liepins/ SPL/ Photo Researchers, Inc. **35.15** www.zahnarzt-stuttgart.com. **35.16** left, © David Scharf/ Peter Arnold, Inc.; right, © Kent Wood/ Photo Researchers, Inc. **35.17** © Greg Ruffing. **35.18** © Zeva Oelbaum/ Peter Arnold, Inc. **35.19** Left, © NIBSC/ Photo Researchers, Inc.; (a–e) After Stephen Wolfe, *Molecular Biology of the Cell*, Wadsworth. 1993 **35.20** © Kwangshin Kim/ Photo Researchers, Inc.

CHAPTER 36 **36.1** left, © Ariel Skelley/ Corbis; right, Courtesy of Dr. Joe Losos. **36.4** (a) © Peter Parks/ Oxford Scientific Films; (b) above, John Glowczwski/ University of Texas Medical Branch; below, Precisions Graphics; (c) left, © Ed Reschke; right, Redrawn from *Living Invertebrates*, V & J Pearse/ M & R Buchsbaum, The Boxwood Press, 1987; (d) left, © D. E. Hill;

right, redrawn from *Living Invertebrates*, V & J Pearse/ M & R Buchsbaum, The Boxwood Press, 1987. **36.8** left, © H. R. Duncker, Justus-Liebig University, Giessen, Germany. **36.10** Photographs, Courtesy of Kay Elemetrics Corporation. **36.11** (a) © R. Kessel/ Visuals Unlimited. **36.14** left, © PhotoDisc/ Getty Images (with art by Lisa Starr); (a,b) below, © Charles McRae, MD/ Visuals Unlimited. **36.15** right, © Joe McBride/ Getty Images. **36.16** © C. Yokochi and J. Rohen, *Photographic Anatomy of the Human Body*, 2nd Ed., Igaku-Shoin, Ltd., 1979. **36.17** (a) © Lennart Nilsson/ Bonnierforlagen AB; (b) © CNRI/ SPL/ Photo Researchers, Inc. **36.18** © O. Auerbach/ Visuals Unlimited. **36.19** © Francois Gohier/ Photo Researchers, Inc. **36.20** (a) Christian Zuber/ Bruce Coleman, Ltd.; (b) © Stuart Westmorland/ Stone/ Getty Images. **36.21** (a) © David Nardini/ Getty Images; (b) © John Lund/ Getty Images.

CHAPTER 37 **37.1** (a) © Jean-Paul Tibbles, *Book of Life*, Ebury Press; (b,c) Courtesy of Kevin Wickenheiser, University of Michigan. **37.2** Courtesy of Lisa Hyche. **37.5** (a) © W. Perry Conway/ Corbis; (a,b art) Adapted from A. Romer and T. Parsons, *The Vertebrate Body*, Sixth Edition, Saunders Publishing Company, 1986. **37.8** After A. Vander et al., *Human Physiology: Mechanisms of Body Function*, Fifth Edition, McGraw-Hill, 1990. Used by permission. **37.9** (a) © Microslide courtesy Mark Nielsen, University of Utah; (b) After A. Vander et al., *Human Physiology: Mechanisms of Body Function*, Fifth Edition, McGraw-Hill, 1990. Used by permission. **37.10** (a) right, © Microslide courtesy Mark Nielsen, University of Utah; (b) © D. W. Fawcett/ Photo Researchers, Inc.; Art, After Sherwood and others. **37.12** (b) National Cancer Institute. **37.14** page 628, © Ralph Pleasant/ FPG/ Getty Images; page 629 from left, © PhotoDisc/ Getty Images; © Paul Poplis Photography, Inc./ Stockfood America; © PhotoDisc/ Getty Images; © PhotoDisc/ Getty Images; © Gary Head. **37.15** © Gary Head. **37.16** © Dr. Douglas Coleman, The Jackson Laboratory. **37.17** © Reuters NewsMedia/ Corbis. **37.18** © Gunter Ziesler/ Bruce Coleman, Inc.

CHAPTER 38 **38.1** page 636, © Archivo Iconografico, S.A./ Corbis; page 637 left, © Ed Kashi/ Corbis; right, © Lawrence Lawry/ Science Photo Library/ Photo Researchers, Inc. **38.5** © Tom McHugh/ Photo Researchers, Inc. **38.9** © Air Force News/ Photo by Tech. Sgt. Timothy Hoffman. **38.10** (a) © Bob McKeever/ Tom Stack & Associates; (b) © S. J. Krasemann/ Photo Researchers, Inc. **38.11** © David Parker/ SPL/ Photo Researchers, Inc. **38.12** (a) © Dan Guravich/ Corbis; (b) © Corbis-Bettmann. **38.13** © Gary Head.

CHAPTER 39 **39.1** page 650, © 1999 Dana Fineman/ Corbis Sygma; page 651, © Lennart Nilsson/ Bonnierforlagen AB. **39.2** (a) © Fred SaintOurs/ University of Massachusetts-Boston; (b) © Martin Harvey/ Photo Researchers, Inc.; (c) © Marc Moritsch; (d) © Photodisc/ Getty Images. **39.3** (a) © Frieder Sauer/ Bruce Coleman, Ltd.; (b) © Matjaz Kuntner; (c) © Ron Austing, Frank Lane Picture Agency/ Corbis; (d) © Doug Perrine/ seapics.com; (e) © Carolina Biological Supply Company; (f) © Fred McKinney/ FPG/ Getty Images; (g) © Gary Head. **39.5** (b–i) © Carolina Biological Supply Company; (j–k) © David M. Dennis/ Tom Stack & Associates, Inc.; (l) © John Shaw/ Tom Stack & Associates. **39.7** right, © Carolina Biological Supply Company; all others, Dr. Maria Leptin, Institute of Genetics, University of Koln, Germany. **39.8** (a–b) After S. Gilbert, *Developmental Biology*, Fourth Edition; (c) © Professor Jonathon Slack. **39.9** (b) After B. Burnside, *Developmental Biology*, 1971, 26:416–441. Used by permission of Academic Press. **39.10** left, © Peter Parks/ Oxford Scientific Films/ Animals, Animals. **Table 39.1** page 660, © Laura Dwight/ Corbis. **39.13** (b) © Ed Reschke. **Page 665** © AJPhoto/ Photo Researchers, Inc. **39.17** (e) © Lennart Nilsson/ Bonnierforlagen AB. **39.19** © Marilyn Houlberg. **39.20** right, © David M. Phillips/ Photo Researchers, Inc. **39.22** Heidi Specht, West Virginia University. **39.23** (a) © Dr. E. Walker/ Photo Researchers, Inc.; (b) © Western Ophthalmic Hospital/ Photo Researchers, Inc.; (c) © CNRI/ Photo Researchers, Inc. **39.24** (a) © David M. Phillips/ Visuals Unlimited; (b) © CNRI/ SPL/ Photo Researchers, Inc.; (c) © John D. Cunningham/ Visuals Unlimited. **39.25** © Todd Warshaw/ Getty Images. **39.29** top, (all) © Lennart Nilsson/ Bonnierforlagen AB. **39.30** left, © Zeva Oelbaum/ Corbis; right, James W. Hanson, M.D. **39.33** Adapted from L.B. Arey, *Developmental Anatomy*, Philadelphia, W.B. Saunders Co., 1965. **39.34** (a) © David M. Parichy; (b,c) © Dr. Sharon Amacher.

Page 685 UNIT VII © Mitsuaki Iwago/ Minden Pictures.

CHAPTER 40 **40.1** © David Nunuk/ Photo Researchers, Inc. **40.2** from left, © Amos Nachoum/ Corbis; © A. E. Zuckerman/ Tom Stack & Associates; © Corbis. **40.3** from left, © E. R. Degginger; inset, © Jeff Foott Productions/ Bruce Coleman, Ltd. **40.4** (a) © Cynthia Bateman, Bateman Photography; (b) © Tom Davis. **40.5** left, © Jeff Lepore/ Photo Researchers, Inc. **40.6** above, © David Scharf, 1999. All rights reserved. **40.7** (a) © G. K. Peck; (b) © Rick Leche, www.flickr.com/photos/rick_leche. **40.9** right, © Peter Lija/ The Image Bank/ Getty Images. **40.10** (a) © Joe McDonald/ Corbis; (b) © Wayne Bennett/ Corbis; (c) © Douglas P. Wilson/ Corbis. **40.11** (a,b) © Hippocampus Bildarchiv; above, © David Reznick/ University of California-Riverside; computer enhanced by Lisa Starr; (c) © Helen Rodd. **Page 697** © Bruce Bornstein, www.captbluefin.com. **40.13** left, © Mark Harmel/ Photo Researchers, Inc.; right, © AP/ Wide World Photos. **40.14** NASA; Art by Precision Graphics. **40.16** (c) Data from Population Reference Bureau after G.T. Miller, Jr., *Living in the Environment*, Eighth Edition, Brooks/Cole, 1993. All rights reserved. **40.17** left, © Adrian Arbib/ Corbis; right, © Don Mason/ Corbis. **40.18** After G. T. Miller, Jr., *Living in the Environment*, Eighth Edition, Brooks/Cole, 1993. All rights reserved. **40.19** © John Alcock/ Arizona State University. **40.20** © Wolfgang Kaehler/ Corbis. **40.21** © Reinhard Dirscherl/ www.bciusa.com.

CHAPTER 41 **41.1** Page 706, Photography by B. M. Drees, Texas A&M University. http://fire ant.tamu; page 707, Daniel Wojak/ USDA. **41.2** left, © Donna Hutchins (a) © B. G. Thomson/ Photo Researchers, Inc.; (b) © Len Robinson, Frank Lane Picture Agency/ Corbis; (c) © Martin Harvey, Gallo Images/ Corbis. **41.3** upper, Harlo H. Hadow; lower, © Bob and Miriam Francis/ Tom Stack & Associates. **41.4** © Thomas W. Doeppner. **41.5** (a,d) © Don Roberson; (b) © Kennan Ward/ Corbis; (c) © D. Robert Franz/ Corbis; left, © Richard Cummins/ Corbis. **41.6** *Paramecium caudatum*, © Michael Abbey/ Photo Researchers, Inc.; *P. Aurelia*, © Eric V. Grave/ Photo Researchers, Inc. **41.7** © Stephen G. Tilley. **41.8** Art, After N. Weldan and F. Bazazz, *Ecology*, 56:681–688, © 1975 Ecological Society of America; upper, © Joe McDonald/ Corbis; lower, left, © Hal Horwitz/ Corbis; right, © Tony Wharton, Frank Lane Picture Agency/ Corbis. **41.9** (a,b) After Rickleffs & Miller, *Ecology*, Fourth Edition, page 459 (Fig. 23.13a) and page 461 (Fig. 23.14); photo, © W. Perry Conway/ Corbis. **41.10** left, © Ed Cesar/ Photo Researchers, Inc.; right, © Robert McCaw, www.robertmccaw.com. **41.11** (a) © JH Pete Carmichael; (b) © Edward S. Ross; (c) W. M. Laetsch. **41.12** (a,c) © Edward S. Ross; (d) © Nigel Jones. **41.13** (a,b) Thomas Eisner, Cornell University; (c) © Jeffrey Rotman Photography; (d) © Bob Jensen Photography. **41.14** (a) MSU News Service, photo by Montana Water Center; (b) © Karl Andree. **41.15** left, © The Samuel Roberts Noble Foundation, Inc.; right, Courtesy of Colin Purrington, Swarthmore College. **41.16** © C. James Webb/ Phototake USA. **41.17** © Peter J. Bryant/ Biological Photo Service. **41.18** (a) © Richard Price/ Getty Images; (b) © E.R. Degginger/ Photo Researchers, Inc. **41.19** (a) © Doug Peebles/ Corbis; (b) © Pat O'Hara/ Corbis; (c,d) © Tom Bean/ Corbis; (e) © Duncan Murrell/ Taxi/ Getty Images. **41.20** (a) R. Barrick/ USGS; (b) USGS; (c) P. Frenzen, USDA Forest Service. **41.21** (a,c) © Jane Burton/ Bruce Coleman, Ltd.; (b) © Heather Angel; (d,e) Based on Jane Lubchenco, *American Naturalist*, 112:23–19, © 1978 University of Chicago Press. Used with permission. **41.22** (a) © Pr. Alexandre Meinesz, University of Nice-Sophia Antipolis; (b) © Angelina Lax/ Photo Researchers, Inc.; right, © The University of Alabama Center for Public TV. **41.23** © John Carnemolla/ Australian Picture Library. **41.24** After W. Dansgaard et al., *Nature*, 364:218–220, July 15, 1993; D. Raymond et al., *Science*, 259:926–933, February 1993; W. Post, *American Scientist*, 78:310–326, July–August 1990. **41.25** (a) © Pierre Vauthey/ Corbis Sygma; (b) © Pierre Vauthey/ Corbis Sygma; (c) After S. Fridriksson, *Evolution of Life on a Volcanic Island*, Butterworth, London 1975. **41.26** (a) © Susan G. Drinker/ Corbis; (b) © Frans Lanting/ Minden Pictures (computer-modified by Lisa Starr). **41.28** © James Marshall/ Corbis. **41.29** left, © Bagla Pallava/ Corbis Sygma; right, © A. Bannister/ Photo Researchers, Inc. **41.31** © R. Bieregaard/ Photo Researchers, Inc.; © PhotoDisc/ Getty Images. **41.32**, © Bureau of Land Management. **41.33** (a) © Anthony Bannister, Gallo Images/ Corbis; (b) © Bob Jensen Photography; (c) © Cedric Vaucher. **41.34** © Heather Angel / Natural Visions.

CHAPTER 42 **42.1** page 732, © C. C. Lockwood/ Cactus Clyde Productions; page 733, Diane Borden-Bilot, U.S. Fish and Wildlife Service. **42.2** (a) © Photodisc/ Getty Images; (b) © David Neal Parks. **42.3** bottom right, © Van Vives; all others, © Dave Rintoul. **42.4** from left, top row, © Bryan & Cherry Alexander/ Photo Researchers, Inc.; © Dave Mech; © Tom & Pat Leeson, Ardea London Ltd.; 2nd row, © Tom Wakefield/ www.bciusa.com.; © Paul J. Fusco/ Photo Researchers, Inc.; © E. R. Degginger/ Photo Researchers, Inc.; 3rd row, © Tom J.

Ulrich/ Visuals Unlimited; © Dave Mech; © Tom McHugh/ Photo Researchers, Inc.; 4th row, © Jim Steinborn; © Jim Riley; © Matt Skalitzky; bottom right, mosquito, Photo by James Gathany, Centers for Disease Contro; flea, © Edward S. Ross; tick, © California Department of Health Services. **42.5** left, Courtesy of Dr. Chris Floyd; right, Graphic created by FoodWeb3D program written by Rich Williams courtesy of the Webs on the Web project (www.foodwebs.org). **42.6** left, © Inga Spence/ Tom Stack & Associates. **42.7** © Gary Head. **42.8** Craig Koppie, U.S. Fish and Wildlife Service. **42.9** (a) NASA/GSFC. **42.13** (a,b) USDA Forest Service, Northeastern Research Station; (c) After G. E. Likens and F. H. Bormann, "An Experimental Approach to New England Landscapes," in A. D. Hasler (ed.), *Coupling of Land and Water Systems*, Chapman & Hall, 1975. **42.14** Water Resources Council. **42.15** Lisa Starr after Paul Hertz; photograph © Photodisc/ Getty Images. **42.16, 42.17** Lisa Starr and Gary Head, based on NASA photographs from JSC Digital Image Collection. **42.18** left, © Yann Arthus-Bertrand/ Corbis; data, www.cmdl.noaa .gov. **42.19** Data, www.ncdc.noaa.gov. **42.20** © Jeff Vanuga/ Corbis. **42.21** © Frederica Georgia/ Photo Researchers, Inc. **42.22** Art, Gary Head and Lisa Starr; photograph, © Photodisc/ Getty Images. **42.23** Fisheries & Oceans Canada, Experimental Lakes Area. **42.24** (a,b) Courtesy of NASA's Terra satellite, supplied by Ted Scambos, National Snow and Ice Data Center, University of Colorado, Boulder; (c) Courtesy of Keith Nicholls, British Antarctic Survey. **42.25** NASA.

CHAPTER 43 **43.1** page 754, © Hank Fotos Photography; page 755, NASA. **43.5** © Alex MacLean/ Landslides, www.alexmaclean.com. **43.6** (b) NASA. **43.8** (a) Adapted from *Living in the Environment* by G. Tyler Miller, Jr., p. 428. © 2002 by Brooks/Cole, a division of Thomson Learning; (b) © Ted Spiegel/ Corbis. **43.10** NASA. **43.11** left, © Sally A. Morgan, Ecoscene/ Corbis; right, © Bob Rowan, Progressive Image/ Corbis. **43.12** NASA. **43.13** above, © Yves Bilat, Ardea London Ltd.; below, © Eagy Landau/ Photo Researchers, Inc. **43.14** (a) NASA's Earth Observatory; (b) After Whittaker, Bland, and Tilman. **43.15, 43.16** Courtesy of Jim Deacon, The University of Edinburgh. **43.17** © George H. Huey/ Corbis; inset, © John M. Roberts/ Corbis. **43.18** © Orbimage Imagery. Image provided by GeoEye and processing by NASA Goddard Space Flight Center. **43.19** left, © John C. Cunningham/ Visuals Unlimited.; right, AP/ Wide World Photos. **43.20** (a) © Jonathan Scott/ Planet Earth Pictures; (b) © Tom Bean Photography; (c) Ray Wagner/ Save the Tall Grass Prairie, Inc. **43.21** left, © James Randklev/ Corbis; all others, © Randy Wells/ Corbis. **43.22** upper, © Franz Lanting/ Minden Pictures; lower, Hans Renner; inset, Edward Ross. **43.23** (a) © Raymond Gehman/ Corbis; (b) © Thomas Wiewandt/ ChromoSohm Media, Inc./ Photo Researchers, Inc. **43.24** (a) © Darrell Gulin/ Corbis; (b) © Paul A. Souders/ Corbis. **43.25** © Pat O'Hara/ Corbis. **43.26** © Onne van der Wal/ Corbis. **43.27** After E. S. Deevy, Jr., *Scientific American*, October 1951. **43.28** (a–c) © E. F. Benfield, Virginia Tech; (d) © Bruce M. Herman/ Photo Researchers, Inc. **43.29** Ocean Arks International. **43.30** (a) © Annie Griffiths Belt/ Corbis; (b) © Douglas Peebles/ Corbis. **43.31** (a) Courtesy of J. L. Sumich, *Biology of Marine Life*, 7th ed., W. C. Brown, 1999; (b) © Nancy Sefton. **43.32** © Paul A. Souders/ Corbis. **43.33** (b) © Dr. Ray Berkelmans, Australian Institute of Marine Science. **43.34** (a) C. B. & D. W. Frith/ Bruce Coleman, Ltd.; (b) © Douglas Faulkner/ Photo Researchers, Inc.; (c) Douglas Faulkner/ Sally Faulkner Collection; (d) © Sea Studios/ Peter Arnold, Inc.; (e) lionfish, Douglas Faulkner/ Sally Faulkner Collection; all others, © John Easley, www.johneasley.com. **43.36** (a) Courtesy of © Montery Bay Aquarium Research Institute; (b) © Peter Herring/ imagequestmarine.com; (c) Image courtesy of NOAA and MBARI; (d,f) © Peter Batson/ imagequestmarine.com. **43.39** NASA, Goddard Space Flight Center Scientific Visualization Studio. **43.40** (a) CHAART, at NASA Ames Research Center; (b) © Eye of Science/ Photo Researchers, Inc.; (c) Courtesy of Dr. Anwar Huq and Dr. Rita Colwell, University of Maryland; (d) © Raghu Rai/ Magnum Photos. **43.41** After M. H. Dickerson, "ARAC: Modeling an Ill Wind," in *Energy and Technology Review*, August 1987. Used by permission of University of California Lawrence Livermore National Laboratory and U.S. Dept. of Energy. **43.42** © Lawson Wood/ Corbis. **Page 783** © Nigel Cook/ Dayton Beach News Journal/ Corbis Sygma.

CHAPTER 44 **44.1** page 784, © Stephen Dalton/ Photo Researchers, Inc.; page 785, © Scott Camazine. **44.2** (a) © Eugene Kozloff; (b) © Stevan Arnold. **44.3** left, © Robert M. Timm & Barbara L. Clauson, University of Kansas.; (a,b) Reprinted from *Trends in Neuroscience*, Vol. 21, Issue 2, 1998, L. J. Young, W. Zuoxin, T. R. Insel, *Neuroendocrine bases of monogamy*, pp. 71–75, © 1998, with permission from Elsevier Science. **44.4** (a) © Eric Hosking; (b) © Stephen Dalton/ Photo Researchers, Inc. **44.5** © Nina Leen/ TimePix/ Getty Images; inset, © Robert Semeniuk/ Corbis. **44.6** © Robert Maier/ Animals Animals. **44.7** (a) © Tom and Pat Leeson, leesonphoto.com; (b) © Kevin Schafer/ Corbis; (c) © Monty Sloan, www.wolfphotography.com. **44.8** © Stephen Dalton/Photo Researchers, Inc. **44.9** (a) © John Alcock, Arizona State University; (b,c) © Pam Gardner, Frank Lane Picture Agency/ Corbis; (d) © D. Robert Franz/ Corbis. **44.10** © Michael Francis/ The Wildlife Collection. **44.11** (a) © B. Borrell Casals, Frank Lane Picture Agency/ Corbis; (b) © Steve Kaufman/ Corbis; (c) © John Conrad/ Corbis. **44.12** (a) © Tom and Pat Leeson, leesonphoto.com; (b) © John Alcock, Arizona State University; (c) © Paul Nicklen/ National Geographic/ Getty Images. **44.13** © Jeff Vanuga/ Corbis. **44.14** © Steve Bloom/ stevebloom.com. **44.15** © Eric and David Hosking/ Corbis. **44.16** (a) © Australian Picture Library/ Corbis; (b) © Alexander Wild; (c) © Professor Louis De Vos. **44.17** (a) © Kenneth Lorenzen; (b) © Peter Johnson/ Corbis; (c) © Nicola Kountoupes/ Cornell University. **44.18** © F. Schutz. **44.19** © Dr.Tim Jackson, University of Pretoria. **44.20** © Gallo Images/ Corbis.

EPILOGUE Page 798–799 © Joseph Sohm, Visions of America/ Corbis.

Appendix V Hemoglobin models: PDB ID: 1GZX; Paoli, M., Liddington, R., Tame, J., Wilkinson, A., Dodson, G., Crystal structure of T state hemoglobin with oxygen bound at all four haems. *J.Mol.Biol.*, v256, pp. 775–792, 1996.

Appendix VI Electron transfer chains: PDB ID: 1A70; Binda, C., Coda, A., Aliverti, A., Zanetti, G., Mattevi, A., Structure of the mutant E92K of [2Fe-2S] ferredoxin I from Spinacia oleracea at 1.7 Å resolution. *Acta Crystallogr.*, Sect.D, v54, pp. 1353–1358, 1998. PDB ID: 1AG6; Xue, Y., Okvist, M., Hansson, O., Young, S., Crystal structure of spinach plastocyanin at 1.7 Å resolution. *Protein Sci.*, v7, pp. 2099–2105, 1998. PDB ID: 1ILX; Vasil`ev, S., Orth, P., Zouni, A., Owens, T.G., Bruce, D., Excited-state dynamics in photosystem II: insights from the x-ray crystal structure. *Proc.Natl.Acad.Sci.*, USA, v98, pp. 8602–8607, 2001. PDB ID: 1Q90; Stroebel, D., Choquet, Y., Popot, J.-L., Picot, D., An Atypical Haem in the Cytochrome B6F Complex, *Nature*, v426, pp. 413–418, 2003. PDB ID: 1QZV; Ben-Shem, A., Frolow, F., Nelson, N., Crystal structure of plant photosystem I, *Nature*, v426, pp. 630–635, 2003. PDB ID: 1IZL; Kamiya, N., Shen, J.-R., Crystal structure of oxygen-evolving photosystem II from Thermosynechococcus vulcanus at 3.7-Å resolution, *Proc.Natl.Acad.Sci.*, USA, v100, pp. 98–103, 2003. PDB ID: 1GJR; Hermoso, J.A., Mayoral, T., Faro, M., Gomez-Moreno, C., Sanz-Aparicio, J., Medina, M., Mechanism of coenzyme recognition and binding revealed by crystal structure analysis of ferredoxin-NADP+ reductase complexed with NADP+., *J.Mol.Biol.*, v319, pp. 1133–1142, 2002. pdb ID: 1C17; Rastogi, V.K., Girvin, M.E., Structural changes linked to proton translocation by subunit c of the ATP synthase., *Nature*, v402, pp. 263–268, 1999. PDB ID: 1E79; Gibbons, C., Montgomery, M.G., Leslie, A.G., Walker, J.E., The structure of the central stalk in bovine F(1)-ATPase at 2.4 Å resolution., *Nat.Struct.Biol.*, v7, pp. 1055–1061, 2000.

Index

Page numbers followed by an f *or* t *indicate figures and tables.* ■ *indicate applications. Bold terms indicate major topics.*

A

H

U